DEDICATED

TO THE ADVANCEMENT OF

THE PROFESSION

AND ITS ALLIED INDUSTRIES

Comments, criticisms, and suggestions regarding the subject matter are invited. Any errors or omissions in the data should be brought to the attention of the Editor. If required, an errata sheet will be issued at approximately the same time as the next Handbook. Notice of any significant errors found after that time will be published in the ASHRAE *Journal*.

ISBN 0-910110-80-8
ISSN 1041-2344

CONTENTS—1989 FUNDAMENTALS

CONTENTS—1988 EQUIPMENT

1991 ASHRAE HANDBO

Heating, Ventilating, and Air-Conditioning Applications

Inch-Pound Edition

American Society of Heating, Refrigerating and Air-Conditioning Engineers, Inc.
1791 Tullie Circle, N.E., Atlanta, GA 30329
404-636-8400

CONTENTS

ENERGY SOURCES

BUILDING OPERATION AND MAINTENANCE

GENERAL APPLICATIONS

ADDITIONS AND CORRECTIONS

INDEX

Composite index to the 1987 HVAC Systems and Applications, Chapters 1 through 16, 1988 EQUIPMENT, 1989 FUNDAMENTALS, 1990 REFRIGERATION Systems and Applications, and 1991 HVAC Applications volumes.

CONTRIBUTORS

In addition to the Technical Committees, the following individuals contributed significantly to this volume. The appropriate chapter numbers follow each contributor's name.

Van D. Baxter (1)
Oak Ridge National Laboratory

James G. Crawford (1)
Trane Company

Kerrin Kirkpatrick (1)
Amana Refrigeration, Inc.

David W. Treadwell (1)
Lennox Industries Inc.

John E. Wolfert (2, 38)
Melvin Simon & Associates

Chris P. Rousseau (3)
Newcomb & Boyd

George F. Evans (4)
Evans and Associates, Inc.

Reinhold Kittler (4)
Dectron Inc.

Lynn F. Werman (4)
Henningson, Durham & Richardson, Inc.

Edward D. Fitts (5, 6)
Sverdrup Corporation

Mark S. Lentz (6, 14)
PSJ Engineering, Inc.

John C. Mentzer (6)
Smith Hinchman and Grylls

Gary L. Merritt (6)
Odell Associates, Inc.

William A. Murray (6, 7)
Ellerbe Architects

Paul J. DuPont (7)
PAE Consulting Engineers, Inc.

John R. Lewis (7)
John R. Lewis & Associates

Mary Jane Phillips (7)
Naval Medical Command

David C. Allen (8)
Allen Associates

John Burgers (8, 9)
Long Manufacturing Ltd.

Robert Cummings (8)
Cummings Engineering Service

Marc-Jose Handfield (8)
Prevost Car Inc.

Andrew S. Lau (8)
Pennsylvania State University

Clifford A. Woodbury, III (8)
Louis T. Klauder and Associates

James J. Bushnell (8, 9)
General Dynamics, Convair

George Letton, Jr. (9)
Douglas Aircraft Company

Frederick H. Kohloss (12)
F.H. Kohloss & Associates

John F. Salsburg (12)

Alfred W. Woody (12)
Giffels Associates, Inc.

Alfred Brociner (13)
Port Authority of New York and New Jersey

Fred Horowitz (13)
Port Authority of New York and New Jersey

Norman I. Lesser (13)

A. York (13)
Port Authority of New York and New Jersey

Kenneth A. Brow (14)
National Institutes of Health

James Carty (14)
Eastman Kodak Company

Daniel A. Ghidoni (14)
The Baker Company

Dan Int-Hout, III (14)
Environmental Technologies

Victor A. Neuman (14)
Advance Information

George J. Sestak (14, 24)
ICI Americas, Inc.

Henry J. Vance (14)
E.I. DuPont de Nemours & Company

Norman Maxwell (16)
Gayle, King, Carr, & Lynch

Philip J. Naughton (16)
Motorola Inc.

William W. Dean (19)
Newcomb & Boyd

Carl B. Miller (19)

W.T. Colling (20)
Eastman Kodak Company

Donald L. Cooper (20)
Eastman Kodak Company

Robert C. Moyer (20, 28)
Eastman Kodak Company

Leslie L. Christianson (21)
University of Illinois

Harold Gray (21)
National Greenhouse Manufacturers Association

Gary L. Riskowski (21)
University of Illinois

Kenneth J. Hellevang (22)
North Dakota State University

Thomas Parker (23)
Weyerhaeuser Paper Co.

Burgess M. Allen (24)
Savannah River Site—Westinghouse

James A. Carlson (24, 49)
University of California

Richard A. Evans (24)
Westinghouse

John W. Jacox (24)
Jacox Associates

Robert E. Jensen (24)

William B. Paschal (24)
Sargent & Lundy

Frank H. Fuller (25)
Industrial Hygiene Associates, Inc.

Edgar L. Galson (25)
Galson Engineers/Architects

Howard D. Goodfellow (25)
Goodfellow Consultants Inc.

Phelps S. Eshelman (27)
Industrial Ventilation Consultants

Richard P. Hibbard (27)

Wayne M. Lawton (27)
Rockwell International

John R. Sosoka (27)
Sosoka & Associates

P. Gaston White (27)
Southern Company Services, Inc.

R.E. Fink (28)
Proctor & Schwartz

Robert C. Moyer (28)
Eastman Kodak Company

J.R. Thygeson (28)
Proctor & Schwartz

Kevin Rafferty (29)
Oregon Institute of Technology

Gordon M. Reistad (29)
Oregon State University

Charles J. Cromer (30, 37)
Florida Solar Energy Center

Larry D. Lister (30)
U.S. Army Corps of Engineers

Timothy J. Merrigan (30)
University of Central Florida

William J. Coad (31)
McClure Engineering Associates

David L. Grumman (31)
Grumman/Butkus Associates

Adam W. Hinge (32)
The Hinge Group

Lawrence G. Spielvogel (32)
L.G. Spielvogel, Inc.

Jeff S. Haberl (33, 36)
Texas A&M University

Ronald N. Jensen (33)
Jensen Engineering, Inc.

Nance C. Lovvorn (33)
Alabama Power Co.

Kenneth M. McGrath (33)
Edison Electric Institute

Dennis A. Smith (33)
Atlanta Gas & Light Co.

William J. Thomaston (33)
Alabama Gas Corporation

Richard Hegberg (34)
ITT Bell & Gossett

Rodney H. Lewis (34)
Rodney H. Lewis Associates, Inc.

Thomas A. Lutz (34)
Systems Analysis Incorporated

Gaylon Richardson (34)
Engineered Air Balance Co.

Dennis H. Tuttle (34)
Armstrong Pumps, Inc.

John D. Roach (35)
Draper and Kramer, Inc.

T. David Underwood (35)
Isotherm Engineering Ltd.

Frantisek Vaculik (35)
Public Works Canada

Albert W. Black (36)
MEDSI Computer Service

Hashem Akbari (37)
Lawrence Berkeley Laboratory

Jay D. Burch (37)
Solar Energy Research Institute

Ernest C. Freeman (37)
U.S. Department of Energy

J. Michael MacDonald (37)
Oak Ridge National Laboratory

Richard P. Mazzucchi (37)
Pacific Northwest Laboratory

Harry P. Misuriello (37)
The Fleming Group

William R. Mixon (37)
Oak Ridge National Laboratory

Mark P. Modera (37)
Lawrence Berkeley Laboratory

John Stoops (37)
Pacific Northwest Laboratory

Mashuri L. Warren (37)
ASI Controls

James E. Braun (38)
Johnson Controls

George E. Kelly (38)
National Institute of Standards and Technology

John W. Mitchell (38)
University of Wisconsin-Madison

William J. Wepfer (38)
Georgia Institute of Technology

Mark W. Fly (39)
Governair Corp.

Thomas R. Kroeschell (39)
Commonwealth Edison Co.

David R. Laybourn (39)
Reaction Thermal Systems, Inc.

Gemma Kerr (40)
Public Works Canada

Richard D. Rivers (40)
Environmental Quality Sciences, Inc.

Sam Silberstein (40)
National Institute of Science and Technology

Gaylen V. Atkinson (41)
Atkinson Electronics Inc.

Joe Dale Ball (41)
La Quinta Motor Inns, Inc.

Peter W. Brothers (41)
Johnson Controls Inc.

Steven T. Bushby (41)
National Institute of Science and Technology

Samuel F. Ciricillo (41)

Edward B. Gut (41)
Honeywell Incorporated

Robert M. Hadden (41)
Engineering Interface Ltd.

John I. Levenhagen (41)

David M. Schwenk (41)
USA CERL

Wilbert F. Stoecker (41)
University of Illinois

David M. Underwood (41)
USA CERL

Alfred L. Utesch (41)
CSM Corporation

Verle A. Williams (41)
V.A. Williams and Associates, Inc.

Warren E. Blazier, Jr. (42)
Warren Blazier Associates, Inc.

Robert M. Hoover (42)
Hoover & Keith Acoustical Consultants

Robert S. Jones (42)
Acentech Incorporated

Richard C. Kirchmeyer (43)

Carl B. Fliermans (43)
Environment America Incorporated

John A. Clark, Jr. (44)
Ace-Buehler Inc.

Wilbur L. Haag, Jr. (44)
Rheem Manufacturing Company

Allen J. Hanley (44, 45)

Robert M. Little, Sr. (44)
Tampa Electric Co.

Edwin A. Nordstrom (44)
Viessmann Manufacturing Co.

William H. Stephenson (44)
Insta-Help Inc.

Thomas I. Wetherington (44)
Florida Power Corp.

Lawrence H. Chenault (45)
Hume Snow Melting Systems, Inc.

John I. Woodworth (45)
The Hydronics Institute

Edward B. Hanf (46)
Munters Corporation

Theodore R. Kaly (46)
Mountain Air Sales, Inc.

Branislav Korenic (46)
Baltimore Aircoil Co Inc.

Gary W. McDonald (46)
A&C Enercom

Peter Morris (46)
Norsaire Corporation

John H. Klote (47)
National Institute of Science and Technology

John A. Clark (47)
Michaud, Cooley, Erickson & Associates

Richard Bourne (48)
Davis Energy Group

Norman A. Buckley (48)
Buckley Associates

J. Gordon Frye (48)
Ohio Power Company

Robert Genisol (48)
Gas-Fired Products, Inc.

Michael J. Hanson (48)
Airtex Corporation

Mark R. Imel (48)
Kansas State University

Luis H. Summers (48)
University of Colorado

Patrick J. Lama (49)
Mason Industries, Inc.

William Staehlin (49)
Office of Statewide Health, Planning and Development, California

Douglas G. Valerio (49)
Mason Industries, Inc.

ASHRAE TECHNICAL COMMITTEES AND TASK GROUPS

SECTION 1.0—FUNDAMENTALS AND GENERAL
1.1 Thermodynamics and Psychrometrics
1.2 Instruments and Measurements
1.3 Heat Transfer and Fluid Flow
1.4 Control Theory and Application
1.5 Computer Applications
1.6 Terminology
1.7 Operation and Maintenance Management
1.8 Owning and Operating Costs
1.9 Electrical Systems

SECTION 2.0—ENVIRONMENTAL QUALITY
2.1 Physiology and Human Environment
2.2 Plant and Animal Environment
2.3 Gaseous Air Contaminants
2.4 Particulate Air Contaminants
2.5 Air Flow Around Buildings
2.6 Sound and Vibration Control
TG Global Warming
TG Halocarbon Emission
TG Safety
TG Seismic Restraint Design

SECTION 3.0—MATERIALS AND PROCESSES
3.1 Refrigerants and Brines
3.2 Refrigerant System Chemistry
3.3 Contaminant Control in Refrigerating Systems
3.4 Lubrication
3.5 Sorption
3.6 Corrosion and Water Treatment
3.7 Fuels and Combustion

SECTION 4.0—LOAD CALCULATIONS AND ENERGY REQUIREMENTS
4.1 Load Calculation and Procedures
4.2 Weather Data Information
4.3 Ventilation Requirements and Infiltration
4.4 Thermal Insulation and Moisture Retarders
4.5 Fenestration
4.6 Building Operation Dynamics
4.7 Energy Calculations
4.8 Energy Resources
4.9 Building Envelope Systems
4.10 Indoor Environment Modeling
TG Cold Climate Design

SECTION 5.0—VENTILATION AND AIR DISTRIBUTION
5.1 Fans
5.2 Duct Design
5.3 Room Air Distribution
5.4 Industrial Process Air Cleaning
5.5 Air-to-Air Energy Recovery
5.6 Control of Fire and Smoke
5.7 Evaporative Cooling
5.8 Industrial Ventilation
5.9 Enclosed Vehicular Facilities

SECTION 6.0—HEATING EQUIPMENT, HEATING AND COOLING SYSTEMS AND APPLICATIONS
6.1 Hot Water and Steam Equipment and Systems
6.2 District Heating and Cooling
6.3 Central Forced Air Heating and Cooling
6.4 In Space Convection Heating
6.5 Radiant Space Heating and Cooling
6.6 Service Water Heating
6.7 Solar Energy Utilization
6.8 Geothermal Energy Utilization
6.9 Thermal Storage

SECTION 7.0—PACKAGED AIR-CONDITIONING AND REFRIGERATION EQUIPMENT
7.1 Refrigerators, Freezers, Water Coolers
7.2 Beverage Coolers
7.5 Room Air Conditioners and Dehumidifiers
7.6 Unitary Air Conditioners and Heat Pumps
TG Engine-Driven Heat Pumps

SECTION 8.0—AIR-CONDITIONING AND REFRIGERATION SYSTEM COMPONENTS
8.1 Positive Displacement Compressors
8.2 Centrifugal Machines
8.3 Absorption and Heat Operated Machines
8.4 Air-to-Refrigerant Heat Transfer Equipment
8.5 Liquid-to-Refrigerant Heat Exchangers
8.6 Cooling Towers and Evaporative Condensers
8.7 Humidifying Equipment
8.8 Refrigerant System Controls and Accessories
8.9 Valves
8.10 Pumps and Hydronic Piping
8.11 Electric Motors

SECTION 9.0—AIR-CONDITIONING SYSTEMS AND APPLICATIONS
9.1 Large Building Air-Conditioning Systems
9.2 Industrial Air Conditioning
9.3 Transportation Air Conditioning
9.4 Applied Heat Pump/Heat Recovery Systems
9.5 Cogeneration Systems
9.6 Systems Energy Utilization
9.7 Testing and Balancing
9.8 Large Building Air-Conditioning Applications
TG Building Commissioning
TG Clean Spaces
TG Laboratory Systems
TG Tall Buildings

SECTION 10.0—REFRIGERATION SYSTEMS
10.1 Custom Engineered Refrigeration Systems
10.2 Automatic Icemaking Plants
10.3 Refrigerant Piping
10.4 Ultra-Low Temperature Systems
10.5 Refrigerated Distribution and Storage Facilities
10.6 Transport Refrigeration
10.7 Food Display and Equipment
10.8 Refrigeration Load Calculations

SECTION 11.0—REFRIGERATED FOOD TECHNOLOGY AND PROCESSING
11.2 Foods and Beverages
11.5 Fruits, Vegetables and Other Products
11.9 Thermal Properties of Food

PREFACE

This handbook returns to a previous title to cover the heating, ventilating, and air-conditioning requirements for an extensive range of applications. Further, it describes the equipment needed to create desired conditions for particular building occupancies or to accomplish specific processes.

The information in this book reflects the work of hundreds of volunteers. Working through technical committees within the society, these volunteers reviewed all chapters from a previous volume and made major revisions to over one-third of them. In addition, they developed four new chapters, which include:

- Chapter 31, "Energy Resources," discusses present and projected sources of energy. It alerts the designer to the impact of the energy required by the HVAC and R equipment in a facility.
- Chapter 37, "Building Monitoring," presents guidelines for developing building monitoring protocols.
- Chapter 38, "Building Operating Dynamics," describes operating strategies that optimize energy consumption while maintaining comfortable conditions.
- Chapter 49, "Seismic Restraint Design," covers the design of restraints to limit the movement of HVAC and R equipment during an earthquake.

Chapters that have major revisions include:

Chapter 25, "Ventilation of the Industrial Environment," has an expanded section on dilution ventilation.

Chapter 29, "Geothermal Energy," includes more information on corrosion and the performance of materials in a geothermal environment.

Chapter 32, "Energy Management," has additional comprehensive summaries of commercial building and household energy use in the United States. Household energy consumption is summarized by region, climate zone, floor area, and age of construction.

Chapter 33, "Owning and Operating Costs," presents more information on estimating maintenance costs and has a revised section on economic analysis.

Chapter 35, "Operation and Maintenance Management," is completely revised to reflect today's sophisticated systems and equipment.

Chapter 36, "Computer Applications," is updated to include information on current computer technology for the industry.

Chapter 39, "Thermal Storage," describes current equipment and includes additional information on chilled water and ice storage.

Chapter 40, "Control of Gaseous Contaminants for Indoor Air," is completely rewritten and expanded.

Chapter 42, "Sound and Vibration Control," is revised to reflect current design criteria for sound control.

Chapter 46, "Evaporative Cooling," reintroduces information on effective temperature and comfort from a previous Handbook and describes several applications.

Chapter 48, "Radiant Heating and Cooling," is reorganized to include information from several previous chapters. Additional information on low, medium, and high intensity infrared applications is included.

Other chapters with significant revisions include Chapter 14 (Laboratories), Chapter 16 (Clean Spaces), Chapter 24 (Nuclear Facilities), Chapter 34 (Testing, Adjusting, and Balancing), Chapter 41 (Automatic Control), Chapter 43 (Corrosion Control and Water Treatment), Chapter 44 (Service Water Heating), and Chapter 45 (Snow Melting).

In addition to these changes, this Handbook has been converted from one edition with dual units of measurement to two editions. One edition contains Inch-Pound (I-P) units of measurement and the other, the International System of Units (SI).

Instead of being published in this volume, several systems-related chapters will be included in the 1992 ASHRAE *Handbook—HVAC Systems and Equipment.* These chapters will include information on (1) air-conditioning system selection, (2) all-air systems, (3) air-and-water systems, (4) all-water systems, (5) unitary refrigerant-based systems for air conditioning, (6) heat recovery systems, (7) panel heating and cooling systems, (8) cogeneration systems, (9) applied heat pump systems, (10) air distribution design for small heating and cooling systems, (11) steam systems, (12) central heating and cooling plant systems, (13) basic water system design, (14) chilled and dual-temperature water systems, (15) medium and high-temperature water heating systems, and (16) infrared radiant heating.

Robert A. Parsons
Handbook Editor

ASHRAE HANDBOOK COMMITTEE

Peter J. Hoey, Chairman

1991 HVAC Applications Volume Subcommittee: **Byron A. Hamrick, Jr.**, Chairman

Richard E. Batherman **Steven F. Bruning** **Oliver K. Lewis** **Giustino N. Mastro** **Charles J. Procell**

ASHRAE HANDBOOK STAFF

Robert A. Parsons, Editor **Claudia Forman**, Associate Editor

Andrea S. Andersen, Assistant Editor

Ron Baker, Production Manager

Gene A. Sweigart, **Nancy F. Thysell**, and **Clark L. Tomlin, Jr.**, Typography

Lawrence H. Darrow and **Susan M. Boughadou**, Graphics

W. Stephen Comstock, Publishing Director

CHAPTER 1

RESIDENCES

THE space conditioning systems selected for residential use vary with both local and application factors. Local factors include fuel availability (both present and projected), fuel prices, climate, socioeconomic circumstances, and the availability of people who have installation and maintenance skills. Application factors include housing type, construction characteristics, and building codes. As a result, many different systems are selected to provide combinations of heating, cooling, humidification, dehumidification, and air filtering. This chapter emphasizes the more common systems for space conditioning of both single (traditional site-built, modular or manufactured, and mobile homes) and multifamily residences. Generally, however, low-rise multifamily buildings follow single-family practice. Retrofit and remodeling construction also adopt the same systems as those for new construction, but site-specific circumstances may call for unique designs.

The common residential heating systems are listed in Table 1. Three groups generally recognized are central forced air, central hydronic, and zonal. System selection and design involves four key decisions: (1) the services provided, (2) distribution and delivery means, (3) source(s) of energy, and (4) conversion device(s).

Table 1 Residential Heating Systems

	Forced Air	Hydronic	Zonal
Source of Energy	Gas Oil Electricity Resistance Heat pump	Gas Oil Electricity Resistance Heat pump	Gas Electricity Resistance Heat pump
Heat Distribution Medium	Air	Water Steam	Air Water Refrigerant
Heat Distribution System	Ducting	Piping	Ducting, Piping, or None
Terminal Devices	Diffusers Registers Grilles	Radiators Fan-coil units	Included with product

Climate dominates the services provided. Heating and in many cases cooling are generally required. Air cleaning (by filtration or electrostatic devices) can be added to most systems. Humidification, which can be added to most systems, is generally provided only to heating when psychrometric conditions make it necessary for comfort. Cooling generally dehumidifies as well; the combination of cooling and dehumidification, commonly with air cleaning, is often called air conditioning. Typical residential installaions are shown in Figures 1 and 2.

Figure 1 includes a gas furnace, a split system cooling unit, a central system humidifier, and an air filter. The system (referring to the numbers in the figure) functions as follows: Air returns to the equipment through a return air duct (1). It passes initially through the air filter (2). The circulating blower (3) is an integral part of the gas furnace (4) that supplies heat during winter. An optional humidifier (10) adds moisture to the heated air, which is distributed throughout the home from the supply duct (9). When cooling is required, the circulating air passes across the evaporator coil (5) which removes heat from the air. Refrigerant lines (6) connect the evaporator coil to a remote condensing unit (7) located outdoors. Condensate from the evaporator drains away through the pipe (8).

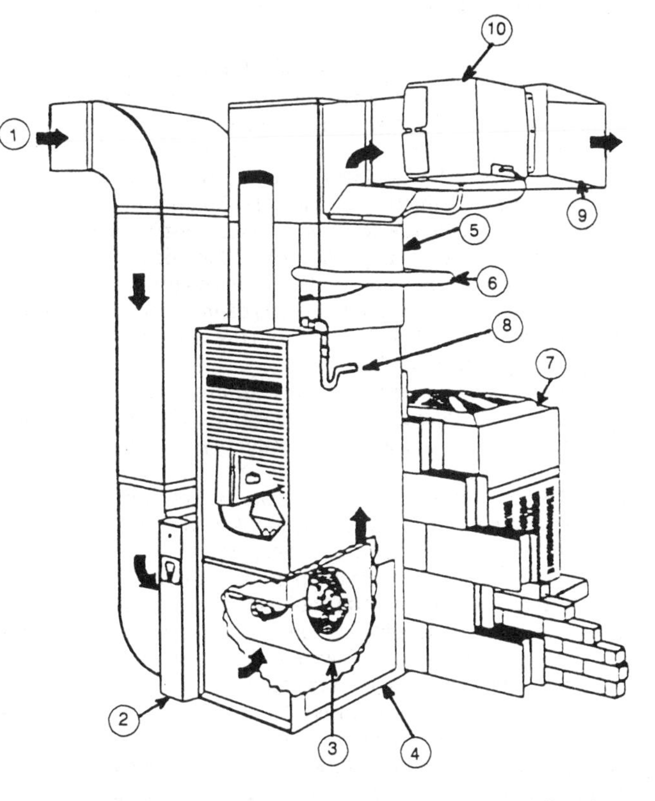

Fig. 1 Typical Residential Installation of Heating, Cooling, Humidifying, and Air Filtering System

Figure 2 includes a split system heat pump, an electric furnace, a central system humidifier, and an air filter. The system functions as follows: Air returns to the equipment through the return air

The preparation of this chapter is assigned to TC 7.6, Unitary Air Conditioners and Heat Pumps.

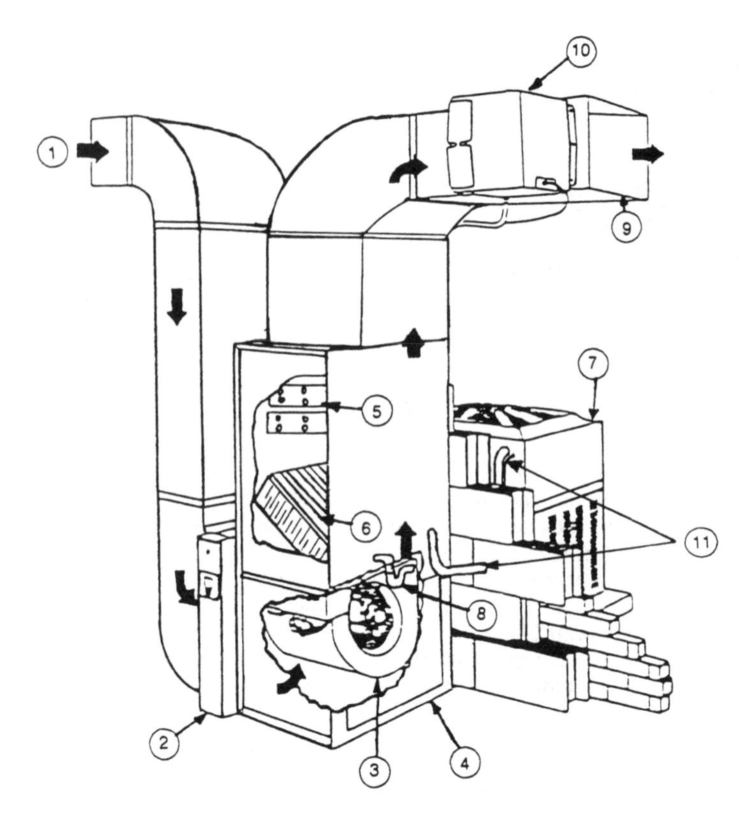

Fig. 2 Typical Residential Installation of Heat Pump Systems

duct (1) and passes through the air filter (2). The circulating blower (3) is an integral part of the heat pump indoor section (4) that supplies heat via the indoor coil (6) during the heating season. Optional electric heaters (5) supplement heat from the heat pump during periods of low ambient temperature and counteract airstream cooling during the defrost cycle. An optional humidifier (10) adds moisture to the heated air, which is distributed throughout the home from the supply duct (9). When cooling is required, the circulating air passes across the indoor coil (6), which removes heat from the air. Refrigerant lines (11) connect the indoor coil to the outdoor coil (7). Condensate from the indoor coil is drained away through the pipe (8).

Single-package systems (with all equipment contained in one cabinet) are also popular in the United States. They are used extensively in areas where residences have duct systems in crawlspaces beneath the main floor and in areas such as the Southwestern United States where they are typically rooftop-mounted. Crawlspace systems are connected to the supply and return ducts in the crawlspace. Rooftop units are connected to an attic duct system.

Central hydronic systems are popular both in Europe and in North America where central cooling is not normally provided; central cooling is generally not economical in hydronic systems for single-family houses with conventional equipment.

Zonal systems are designed to condition only part of a home at any one time. They may consist of individual room units or central systems with zoned distribution networks. Multiple central systems that serve individual floors or separately serve sleeping and communal portions of a home are also widely used in large single-family houses.

The source of energy is a major consideration in system selection. Gas and electricity are most widely used, followed by oil, wood, solar, coal, district thermal, and others. Local availability and assurance of future availability are the most limiting factors. Relative energy prices, safety, and environmental concerns (both indoor and outdoor) are further factors. Where various energy sources are available, economics strongly influence the selection.

CAPACITY SELECTION

The heat loss and gain of each conditioned room and crawlspace or basement in the structure must be accurately calculated to select equipment with the proper output and design. To determine heat loss and gain accurately, the floor plan and construction details must be known. The plan should include information on wall, ceiling, and floor construction and type and thickness of insulation. Window design and external door details are also needed. With this information, heat loss and gain can be calculated using the basic data in Chapters 22 through 28 of the 1989 ASHRAE *Handbook—Fundamentals*, and the Air-Conditioning Contractors of America *Manual* J or similar calculation procedures. To conserve energy, many jurisdictions require that the building be designed to meet or exceed the requirements of ASHRAE *Standard* 90A-1980, Energy Conservation in New Building Design, or similar requirements.

Proper matching of equipment capacity to the design heat loss and gain is essential. The heating capacity of air source heat pumps is usually supplemented by auxiliary heaters, most often of the electric resistance type; in some cases, however, fossil fuel furnaces or solar systems are used.

Undersized equipment results in an inability to maintain indoor design temperatures at outdoor design conditions. Grossly oversized equipment can cause discomfort due to short on-times, wide indoor temperature swings, and inadequate dehumidification when cooling. Gross oversizing may also contribute to higher energy use due to an increase in starting thermal transient losses, stopping thermal transient losses, and off-cycle losses. Variable capacity equipment (heat pumps, air conditioners, and furnaces) can match building loads over specific ambient temperature ranges, thus reducing these losses and improving comfort levels; in the case of heat pumps, it may reduce supplemental heat needs.

Recent trends toward heavily insulated, tightly constructed buildings with continuous vapor retarders and low infiltration can cause high indoor humidity conditions, and dehumidification may become necessary during the winter months. Air-to-air heat-recovery equipment may be used to provide tempered ventilation air to some tightly constructed houses. Outdoor air intakes connected to the return duct of central systems may also be used when lower installed costs are important.

SINGLE-FAMILY RESIDENCES

HEATING EQUIPMENT

Heat Pumps

Heat pumps may be classified by their respective thermal source and distribution medium in the heating mode. The most common equipment is air-to-air, followed, in order of declining application, by water-to-air, air-to-water, and water-to-water.

Heat pump systems, as contrasted to the actual heat pump equipment, are generally described as air-source or ground-source. The thermal sink for cooling is generally assumed to be the same as the thermal source for heating. Common sources and sinks for heat pumps are summarized in Table 2, Chapter 9 of the 1987 ASHRAE *Handbook—HVAC Systems and Applications*. Air-source systems using ambient air as the heat source/sink are generally less costly to install than other types and, thus, are the most commonly used. Ground-source systems may employ water-to-air or water-to-water heat pumps to extract heat from the ground via groundwater or a ground-coupled heat exchanger.

Groundwater (from individual wells or supplied as a utility from community wells) as a heat source/sink offers the following advantages compared to ambient air: (1) heat pump capacity is independent of ambient temperature, which reduces supplementary heating requirements; (2) no defrost cycle is required; and (3) for equal equipment efficiency, the seasonal efficiency is usually higher for heating and cooling. Ground-coupled systems offer the same advantages. However, they circulate brine or water to transfer heat from the ground via a buried heat exchanger. Direct ground-source systems, using direct expansion evaporators buried in the ground are occasionally used. The total number of ground-source systems is small but has been growing rapidly in recent years, particularly for the ground-coupled type. Water-source systems that extract heat from surface water (*e.g.*, lakes or rivers) or city (tap) water are also used where local conditions permit.

Water supply, quality, and disposal must be considered for ground-water systems. The ASHRAE publication *Design Manual on Closed Loop and Ground Coupled Heat Pumps* provides detailed information on these subjects. Secondary coolants for ground-coupled systems are discussed in this manual and in Chapter 18 of the 1989 ASHRAE *Handbook—Fundamentals*. Ground heat exchanger configurations may be horizontal or vertical, with the vertical including both multiple-shallow- and single-deep-well configurations. Ground-coupled systems avoid the water quality, quantity, and disposal concerns but are sometimes more expensive and usually slightly less efficient than groundwater systems (if water pumping power for the groundwater system is not excessive).

Heat pump systems using water as a distribution medium are identified as air-source hydronic, water-source hydronic, or ground-source hydronic systems.

Heat pumps for single-family houses are normally unitary systems; *i.e.*, they consist of one or more factory-built modules designed to be connected and used together. These differ from applied or built-up heat pumps, which require field engineering to select compatible components for complete systems.

Virtually all heat pumps commercially available (particularly in North America) are electrically powered. Supplemental heat is generally required during peak or defrost operation. In most cases, supplemental or backup heat is provided by electric resistance heaters, but add-on and unitary bivalent systems, which combine heat pumps with fuel-fired devices, are also used.

In add-on systems, a heat pump is added—often as a retrofit—to a furnace or boiler. The heat pump and combustion device are operated in one of two ways: (1) alternately, depending on which is most cost effective, or (2) in parallel. Unitary bivalent heat pumps combine the heat pump and combustion device into a common chassis and cabinets to provide similar benefits at lower installation costs.

Solar-assisted and solar-source heat pumps have been attempted in limited applications. Extensive research and development is being conducted to develop fuel-fired heat pumps. However, they are not currently actively marketed for residential use in North America.

Heat pumps may be equipped with desuperheaters (either integral or field added) to reclaim heat for domestic water heating. Integrated space-conditioning and water-heating heat pumps with an additional full-size condenser for water heating are also available.

Furnaces

Furnaces are fueled by gas (natural or propane), electricity, oil, wood, or other combustibles. Gas, oil, and wood furnaces may draw combustion air from the indoor space or be of the direct vent type (drawing combustion air from outdoors). The latter are used for mobile home applications and some high-efficiency equipment designs. Using outside air for combustion eliminates both infiltration losses associated with the use of indoor air for combustion and stack losses associated with atmospherically induced draft furnaces.

Two types of high-efficiency gas furnaces available are noncondensing and condensing. Both obtain increased efficiencies by added heat exchanger surface areas and by reduced heat loss during furnace off times. The higher efficiency condensing type also recovers more energy from the fuel by condensing water vapor from the flue products. The indoor air of many households contains chemicals from cleaning fluids and other materials that may cause early corrosion failures in certain furnace heat exchanger materials. Because of this, condensing models may use outdoor air for combustion. Such models may also require special materials for flue pipes and means to drain the condensate.

Wood-fueled furnaces are used in some areas. A recent advance in wood furnaces has been the addition of catalytic converters to enhance the combustion process, increasing furnace efficiency and producing cleaner exhaust.

Hydronic Heating Systems—Boilers

With the growth of demand for central cooling systems, hydronic systems have declined in popularity in new construction, but still account for a significant portion of existing systems in the northern climates. The fluid (hot water or steam) is heated in a central boiler and distributed by piping to individual fancoil units, radiators, or baseboard convectors located in each room. Most recent residential systems use a forced circulation, multiple zone, hot water system with a series loop piping arrangement. The equipment is described in Chapters 23 and 28 of the 1988 ASHRAE *Handbook—Equipment*. Hot water and steam heating systems are described in Chapters 11 and 13 of the 1987 ASHRAE *Handbook—HVAC Systems and Applications*.

Design water temperature is based on economic and comfort considerations. Generally, higher temperatures result in lower first costs because smaller distribution units are needed. However, losses tend to be greater, resulting in higher operating costs and reduced comfort because of the concentrated heat source. Typical design temperatures range from 180 to 200 °F. The preferred control system allows the water temperature to decrease as outdoor temperatures rise. Provisions for expansion and contraction of the piping and heat distributing units and elimination of air from the system are essential for quiet, leaktight operation.

Fossil fuel systems that condense water vapor from the flue gases must be designed for return water temperatures in the range of 120 to 130 °F for most of the heating season. Noncondensing systems must maintain high enough water temperatures in the boiler to prevent this condensation. If rapid heating is required, both the space heating surfaces and boiler size must be increased, but gross oversizing should be avoided.

Zonal Heating Systems

Zonal systems offer the potential for operating economies, since unoccupied areas can be kept at lower temperatures. For example, communal areas can be maintained at lower temperatures at night and sleeping areas kept at lower temperatures during the day.

One form of this system consists of individual heaters located in each room. These heaters are usually electric or gas-fired. Electric heaters are available in the following types: baseboard free-convection, wall insert free-convection, forced-fan, radiant panels for walls and ceilings, and radiant cables and inserts for walls, ceilings, and floors. Matching of equipment capacity to heating requirements is critical for individual room systems. Heating delivery cannot be adjusted by adjusting air or water flow, so greater precision on room-by-room sizing is needed.

Individual heat pumps for each room or group of rooms (zones) are another form of zonal electric heating. For example, two or more small unitary heat pumps are installed in two-story or large one-story homes. Air-to-air, water-to-air, or hydronic types are available.

The multisplit heat pump consists of a central compressor and outdoor heat exchanger to service up to five indoor zones. Each zone uses one or more fan coils with separate thermostatic control for each zone. Such systems are used in both new and retrofit construction.

Solar Heating

Both active and passive solar energy systems are used to heat residences. In typical active systems, flat plate collectors heat air or water. Air systems distribute heated air either to the living space for immediate use or to a thermal storage media (*i.e.*, a rock pile). Water systems pass heated water through a secondary heat exchanger and store extra heat in a water tank.

Trombe walls and sunspaces are two common passive systems. Glazing facing south (with overhanging eaves to reduce solar gains in the summer) and movable insulating panels reduce heating requirements. Like active systems, however, supplementary and backup heating ability is generally needed. Chapter 30 has information on sizing solar heating equipment.

AIR CONDITIONERS

Unitary Air Conditioners

In forced air systems, the same air distribution duct system can be used for both heating and cooling. Split central cooling is most widely used. It consists of an evaporator coil mounted in an indoor air-handling unit or furnace and a remote air-cooled condensing unit. Upflow, downflow, and horizontal airflow units are available. Condensing units are installed on a concrete pad outside and contain an electric motor-driven compressor, condenser, condenser fan and fan motor, and electrical controls. The condensing unit and evaporator coil are connected by refrigerant tubing that may be field-supplied. However, precharged, factory-supplied tubing with quick-connect couplings is common where the distance between components is not excessive. A distinct advantage of this system is that it can readily be added to existing forced air heating systems. Airflow rates may need to be increased above heating requirements to achieve good performance, but most existing heating duct systems are adaptable to cooling. Airflow rates of 350 to 450 cfm per nominal ton of refrigeration are normally recommended to achieve good cooling performance. As with heat pumps, these systems may be fitted with desuperheaters for domestic water heating.

Some forced-air heating equipment includes cooling as an integral part of the product. Year-round heating and cooling packages with a gas, oil, or electric furnace for heating and an electrically driven vapor-compression system for cooling are available. Also, air-to-air and water source heat pumps provide cooling and heating by reversing the flow of refrigerant.

Distribution Systems. Duct systems for cooling should be designed and installed in accordance with accepted practice. Useful information is found in ACCA (Air-Conditioning Contractors of America) *Manuals* D and G, and in Chapter 1 of the 1988 ASHRAE *Handbook—Equipment*. The duct sizes of year-round air-conditioning systems are usually larger than those of equivalent winter heating systems, since cooling air is supplied at only 15 to 25 °F below room temperature.

Reliable duct design methods that consider airflow and static pressure characteristics of the systems should be used. The type and size of supply and return inlets and their locations to effect satisfactory room air distribution must be considered. Return air ducts are not typically required for mobile home and many apartment applications because the air returns down the hallway to the closet containing the air circulation equipment. General practice is to calculate both cooling and heating airflow requirements and to design an air distribution system for the larger air quantity (usually the cooling airflow). Multispeed fans are sometimes included to reduce airflow when heat is delivered.

In one- or two-story homes, heating and cooling loads do not have the same proportional relationship for all rooms. Separate preliminary calculations should be made for the heating and cooling air distribution requirements, and then adjustments made to achieve a compromise solution. For example, it may be necessary to include additional cooling outlets in one or more of the upstairs rooms, in rooms that have large glass areas facing south or west, or in rooms with unusually high internal heat loads such as kitchens.

Both technical and economic aspects of residential cooling should be considered. An elaborately zoned air distribution system is rarely economical. However, because weather primarily influences the load, the cooling load in each room changes from hour to hour. Therefore, to obtain ultimate comfort, the owner or occupant should be able to make seasonal or more frequent adjustments to the air distribution system.

Such adjustments may involve the opening of additional outlets in second-floor rooms during the cooling cycle and throttling or closing heating outlets in some rooms during the winter. On deluxe applications, manually adjustable balancing dampers may be provided. Other refinements could be the installation of a heating and cooling system sized to meet heating requirements, with additional self-contained cooling units serving some or all of the second-floor rooms or rooms that have unusual load variations or high internal heat gains during the summer.

Operating characteristics of both heating and cooling equipment must be considered when zoning is used. For example, a reduction in the air quantity to one or more rooms may reduce the airflow across the evaporator to such a degree that frost forms on the fins. Reduced airflow on heat pumps during the heating season can cause overloading if airflow across the indoor coil is not maintained at 350 to 450 cfm per ton. A reduced air volume to a given room would reduce the air velocity from the supply outlet and could cause unsatisfactory air distribution in the room.

Special Considerations. In split-level houses, cooling and heating are complicated by air circulation that occurs through the large openings between various levels. In many such houses, the upper level tends to overheat in winter and undercool in summer. Multiple outlets, some near the floor and others near the ceiling on all levels, have been used with some success. To control airflow, the owner opens some outlets and closes others from season to season. Free circulation between floors can be reduced by locating returns high in each room and by keeping doors closed.

In existing homes, the cooling that can be added is limited by the air-handling capacity of the existing duct system. While it is usually satisfactory for normal occupancy, it may be inadequate when large-crowd entertainment occurs.

In all cases where cooling is added to existing homes, supply-air outlets must be checked for acceptable cooling air distribution. Upward airflow at an effective velocity is important to consider when converting existing heating systems with floor or baseboard outlets to both heat and cool. It is not necessary to change the deflection from summer to winter for registers located at the perimeter of a residence. Registers located at the inside walls of rooms also operate satisfactorily without changing of deflection from summer to winter.

Occupants of air-conditioned spaces usually prefer minimum perceptible air motion. Perimeter baseboard outlets with multiple slots or orifices effectively meet these requirements. Ceiling outlets with multidirectional vanes are also satisfactory.

A residence without a forced air heating system may be cooled by one or more central systems with separate duct systems or by individual room air conditioners (window-mounted or through-the-wall). Central systems may be single-package with all refrigeration components in one unit, or split systems with an outdoor condensing unit connected by refrigerant tubing to an evaporator coil in an indoor blower unit.

Cooling equipment must be located carefully. Because cooling systems require higher indoor airflow rates than most heating systems, the sound levels generated indoors are usually higher. Thus, indoor air-handling units located near sleeping areas may require sound attenuation. Outdoor noise levels should also be considered when locating the equipment. Many communities have ordinances regulating the sound level of mechanical devices, including cooling equipment. Many manufacturers of unitary air conditioners certify the sound level of their products in an ARI program (ARI *Standard* 270). ARI *Standard* 275 gives information on how to predict the dBA sound level when the ARI sound rating number, the equipment location relative to reflective surfaces, and the distance to the property line are known.

An effective and inexpensive way to reduce noise is to put distance and natural barriers, like walls or corners, between sound source and listener. However, airflow to and from air-cooled condensing units should not be obstructed. Outdoor units should be placed as far as is practical from porches and patios, which may be used when the house is being cooled. Locations near bedroom windows should also be avoided. In addition, neighbors should not be subjected to undesirable noise if they open their windows while the unit is on. Therefore, the rear of residences is usually the preferred location.

Evaporative Coolers

In dry climates, evaporative coolers can be used to cool residences. Further details on evaporative coolers can be found in Chapter 4 of the 1988 ASHRAE *Handbook—Equipment* and in Chapter 46 of this volume.

MULTIFAMILY RESIDENCES

Attached homes and low-rise multifamily apartments generally use heating and cooling equipment comparable to that used in single-family dwellings. Separate systems for each unit allow individual control to suit the occupant and facilitate individual metering of energy use.

Central Forced Air Systems

High-rise multifamily structures may also use unitary heating and cooling equipment comparable to that used in single-family dwellings. Equipment for this system may be installed in a separate mechanical equipment room in the apartment, or it may be placed in a soffit or above a drop ceiling over a hallway or closet.

Small residential warm air furnaces may also be used, but a method of providing combustion air and venting combustion products from gas- or oil-fired furnaces is required. It may be necessary to use a multiple-vent chimney or a manifold-type vent system. Local codes should be consulted. Direct vent furnaces that are placed near or on an outside wall are also available for apartments.

A popular concept for multifamily residences (also applicable to single-family dwellings) is a combined water heating/space heating system that utilizes water from the domestic hot water storage tank to provide space heating. Water circulates from the storage tank to a hydronic coil in the system air handler. Space heating is provided by circulating indoor air across the coil. A split system central air conditioner with the evaporator located in the system air handler can be included to provide space cooling.

Hydronic Central Systems

Individual heating and cooling units are not always possible or practical in high-rise structures. In this case, applied central systems are used. Hydronic central systems of the two- or four-pipe type are widely used in high-rise apartments. Each dwelling unit has individual room units located at the perimeter or interior, or ducted fan coil units.

The most flexible hydronic system with the most favorable operating costs is the four-pipe type, which provides heating or cooling for each apartment dweller. The two-pipe system is less flexible since it cannot provide heating and cooling simultaneously. This limitation causes problems during the spring and fall when some apartments in a complex require heating, while others require cooling due to solar or internal loads. In some systems, this problem is overcome by operating the two-pipe system in a cooling mode and providing the relatively low amount of heating that may be required by individual electric resistance heaters.

Through-the-Wall Units

Through-the-wall room air conditioners and packaged terminal air conditioner (PTAC) and packaged terminal heat pump (PTHP) versions of these products give the highest flexibility for conditioning each room in low- and high-rise apartments. In this application, each apartment room with an outside wall may have such a unit. These units are used extensively when renovating old apartment buildings because they are self-contained and do not require complex renovation for piping or ductwork.

Room air conditioners have integral controls and may include resistance or heat pump heating. PTACs and PTHPs have special indoor and outdoor appearance treatments, making them adaptable to a wider range of architectural needs. PTACs can include gas, electric resistance, hot water, or steam heat. Integral or remote wall-mounted controls are used for both PTACs and PTHPs. Further information may be found in Chapter 43 of the 1988 ASHRAE *Handbook—Equipment* and in ARI *Standards* 310 and 380.

Water Loop Heat Pump Systems

Any mid- and high-rise apartment structure having interior zones with high internal heat gains that require year-round cooling can efficiently use a water loop heat pump system. Such systems have the flexibility and control of a four-pipe system while using only two pipes. Water source heat pumps allow for individual metering of each apartment. The building owner pays only the utility cost (which can be prorated) for the circulating pump, cooling tower, and supplemental boiler heat. Economics permitting, solar or ground heat energy can provide the supplementary heat source in lieu of a boiler. The ground can also provide a heat sink which in some cases can eliminate the cooling tower.

Special Concerns for Apartment Buildings

Many ventilation systems are used in apartment buildings. Local building codes may govern air quantities. ASHRAE *Standard* 62-1989, Ventilation for Acceptable Indoor Air Quality, requires minimum outdoor air values of 50 cfm intermittent or 20 cfm continuous or operable windows for baths and toilets, and 100 cfm intermittent or 25 cfm continuous or operable windows for kitchens.

Some buildings with centrally controlled exhaust and supply systems operate the systems on time clocks for certain periods of the day. In other cases, the outside air is reduced or shut off during extremely cold periods. If known in advance, these factors should be considered when estimating heating load.

Buildings using exhaust and supply air systems 24 h a day may benefit from air-to-air heat recovery devices (see Chapter 34 of the 1988 ASHRAE *Handbook—Equipment*). Such recovery devices can reduce energy consumption by transferring 40 to 80% of the sensible and latent heat from the exhaust air to the supply air.

Infiltration loads in high-rise buildings without ventilation openings for perimeter units are not controllable on a year-round basis by general building pressurization. When outer walls are pierced to supply outdoor air to unitary or fan-coil equipment, combined wind and thermal stack effects create other infiltration problems.

Interior public corridors in apartment buildings need positive ventilation with at least two air exchanges per hour. Conditioned supply air is preferable. If necessary, some designs transfer air into the apartments through acoustically lined louvers to provide kitchen and toilet makeup air. Supplying air to, instead of exhausting air from, corridors minimizes odor migration from apartments into corridors.

Air-conditioning equipment must be isolated to reduce noise generation or transmission. The design and location of cooling towers must be chosen to avoid disturbing occupants within the building and neighbors in adjacent buildings. An important load, frequently overlooked, is heat gain from piping for hot water services.

In large apartment houses, a central panel allows individual apartment air-conditioning systems or units to be monitored for maintenance and operating purposes.

HUMIDIFIERS

For improved winter comfort, equipment that increases indoor relative humidity levels is often installed. In a ducted heating system, a central humidifier can be attached to or installed within a supply plenum or main supply duct, or installed between the supply and return duct systems. Exercise caution when applying supply-to-return duct humidifiers on heat pump systems to maintain proper airflow across the indoor coil. Self-contained humidifiers can be used in any residence. Even though this type introduces all the moisture to one area of the home, moisture will migrate and raise humidity levels in other rooms.

Central humidifiers are rated in accordance with ARI *Standard* 610. This rating is expressed in the number of gallons per day evaporated by 140°F entering air. Some manufacturers certify the performance of their product to the ARI standard, and these products are listed in the ARI *Directory of Certified Central System Humidifiers.*

Selecting the proper size humidifier is important and is outlined in ARI *Standard* 630, Selection, Installation and Servicing of Humidifiers. Since the maximum level of relative humidity that can be maintained without condensation forming is a function of the coldest surface in the living space, windows, double-pane glass, or storm windows must be installed. A humidifier should always be controlled by a humidistat set to keep the relative humidity below this maximum level. This also saves energy by preventing overhumidification. Caulking and weatherstripping significantly decrease outdoor air infiltration and, thus, humidification needs.

Since moisture migrates through all structural materials, insulated walls, ceilings, and floors should have a vapor retarder installed near the inside surface. Improper attention to this construction detail allows moisture to migrate from inside to outside, causing insulation to become damp, possible structural damage, and exterior paint blistering. Humidifier cleaning and maintenance schedules should be followed to maintain efficient operation and prevent bacteria buildup. Chapter 5 of the 1988 ASHRAE *Handbook—Equipment* has more information on residential humidifiers.

AIR FILTERS

Most comfort conditioning systems that circulate air incorporate some form of air filtration device. Usually, this consists of disposable or cleanable filters having relatively low air-cleaning efficiency. The comfort and cleanliness level of the residential system can be improved by adding higher efficiency air-filtering equipment such as a pleated media filter or an electronic air filter.

Air filters are mounted in the return air duct or plenum and operate whenever air circulates through the duct system. Air filters are rated in accordance with ARI *Standard* 680, Air Filter Equipment, based on ASHRAE *Standard* 52-1968(RA 76), Method of Testing Air-Cleaning Devices Used in General Ventilation for Removing Particulate Matter. Atmospheric dust spot efficiency levels are generally less than 20% for disposable filters and vary from 60 to 90% for electronic air filters.

To maintain optimum performance, the collector cells of electronic air filters must be cleaned periodically. Automatic indicators to signal the need for cleaning are often used. Electronic air filters have higher initial costs than disposable or pleated filters, but generally will last the life of the air-conditioning system. Chapter 10 of the 1988 ASHRAE *Handbook— Equipment* covers the design of residential air filters in more detail.

CONTROLS

Historically, residential heating and cooling equipment has been controlled by a wall thermostat. Today, simple wall thermostats with bimetallic strips are being replaced by microelectronic models that can control heating and cooling equipment at different temperature levels, depending on the time of day. This has led to night setback control to reduce energy demands and the cost of operation. For heat-pump equipment, electronic thermostats can incorporate night setback with an appropriate scheme to limit use of resistance heat during recovery. Chapter 41 contains more details about automatic control systems.

CHAPTER 2

RETAIL FACILITIES

THIS chapter covers the design and application of air-conditioning and heating systems for various retail merchandising facilities. Load calculations, systems, and equipment are covered elsewhere in the Handbook series.

GENERAL CRITERIA

To apply equipment properly, it is necessary to know the construction of the space to be conditioned, its use and occupancy, the time of day in which greatest occupancy occurs, the physical building characteristics, and lighting layouts.

The following must also be considered:

- Electric power—size of service
- Heating—availability of steam, hot water, gas, oil, or electricity
- Cooling—availability of chilled water, well water, city water, and water conservation equipment
- Rigging and delivery of equipment
- Structural considerations
- Obstructions
- Ventilation—opening through roof or wall for outside air duct, number of doors to sales area, and exposures
- Orientation of store

Specific design requirements, such as the increased outdoor air required for exhaust systems where lunch counters exist, must be considered. The requirements of ASHRAE ventilation standards must be followed. Heavy smoking and objectionable odors may require special filtering in conjunction with outdoor air intake and exhaust. Load calculations should be made with the procedure outlined in Chapters 25 and 26 of the 1989 ASHRAE *Handbook—Fundamentals*.

Almost all localities have some form of energy code in effect, which generally establishes strict requirements for insulation, equipment efficiencies, system designs, and so forth, and places strict limits on fenestration and lighting. The requirements of ASHRAE *Standard* 90 should be followed as a minimum guideline for retail facilities.

The selection and design of HVAC systems for retail facilities are normally determined by economics. HVAC system selections for small stores are usually determined by first cost; for large retail facilities, operating and maintenance costs are also considered. Generally, considerations for mechanical systems for retail facilities are based on a cash flow-type analysis rather than on a full life cycle analysis.

The preparation of this chapter is assigned to TC 9.8, Large Building Air-Conditioning Applications.

SMALL STORES

Small stores are often constructed with large glass areas in front, which may result in high peak solar heat gain, except for northern exposures. High heat loss may be experienced on cold, cloudy days. This portion of the small store should be designed to offset the greater cooling and heating requirements. Entrance heaters may be needed in cold climates.

Many new small stores are part of a shopping center. While exterior loads will differ in these stores, the internal loads will be similar, and the need for proper design is equally important.

DESIGN CONSIDERATIONS

System Design

Single-package rooftop equipment is common in store air conditioning. The use of multiple units to condition the store involves less ductwork and can maintain comfort in the event of equipment failure. Prefabricated and matching curbs simplify installations and ensure compatibility with roof materials.

The heat pump, offered as packaged equipment, readily adapts to small-store applications and has an economical first cost. Winter design conditions and utility rates should be compared against operating costs of conventional heating systems before deciding on this type of equipment.

Water-cooled unitary equipment is available in all capacities required for small-store air conditioning, but many communities in the United States have restrictions on the use of city water for condensing purposes and require installation of a cooling tower system. Water-cooled equipment generally operates efficiently and economically.

Air Distribution

External static pressures available in small-store air-conditioning units are limited, and duct systems should be designed to keep duct resistances low. Duct velocities should range between 800 to 1200 fpm and pressure drops between 0.07 to 0.10 in. of water per 100 ft of duct run. Average air quantities range from 350 to 450 cfm per ton refrigeration in accordance with the calculated internal sensible heat load.

Attention should be paid to suspended obstacles, such as lights and displays, which interfere with proper air distribution.

The duct system should contain enough dampers for air balancing. Dampers in the return duct and outdoor air duct should be installed for proper outdoor air-return air balance. Volume and splitter dampers should be installed in takeoffs from the main supply duct for air balance into branch ducts.

Control System

Controls for small-store systems should be kept simple while performing the required functions. Unitary equipment generally has factory-installed controls with terminal boxes for external installation of power and control wiring by an electrician.

Automatic dampers should be placed in the outdoor air intake to prevent outdoor air from entering the area when the fan is turned off.

Heating controls vary with the nature of the heating medium. Duct heaters are generally furnished with safety controls installed by the manufacturer. Space thermostats can control the heater with a summer-winter switch. All control circuits should be actuated through the load side of the fan circuit.

Maintenance

Air-conditioning units in small stores should be assigned to a reliable service company on a yearly contract to protect the initial investment and maintain maximum efficiency. The contract should clearly state the responsibility for filter replacements or cleaning, repair, and adjustment of controls, oil and grease, compressor maintenance, replacement of refrigerant, pump repairs, electrical maintenance, winterizing and startup of system, and replacement of materials and extra labor required for repairs.

Improving Operating Costs

Outdoor air economizers can improve the operating cost of cooling systems in most climates. These are generally available as factory options or accessories with roof-mounting unitary units.

Night setback should be evaluated for stores with more than 8 h of operation. This can be incorporated with economizer controls. Increased exterior insulation will generally reduce operating energy requirements and may, in some cases, allow for a reduction in the size of the equipment to be installed. Many types of insulation are available for roofs, ceilings, masonry walls, frame walls, slabs, and foundation walls. Many codes now have minimum requirements for insulation and fenestration materials.

DISCOUNT AND VARIETY STORES

Variety stores feature a wide range of merchandise, often including a large lunch counter or a separate restaurant, auto service area, and garden shop. Some stores sell pets, including fish and birds. This variety of merchandise must be considered in designing an air-conditioning system.

In addition to the sales area, such areas as stockrooms, rest rooms, offices, and special storage rooms for perishable merchandise may require air conditioning or refrigeration.

The design and application suggestions for small stores also apply to variety stores. The following design and application information should be considered in addition to that for small stores.

LOAD DETERMINATION

Operating economics and the spaces served often dictate the indoor design conditions for discount and variety store air conditioning. Some variety stores may base summer load calculations on higher inside temperatures, such as 80°F db, while the thermostats are set to control at 72 to 75°F db. This reduces the installed equipment size while providing the desired inside temperature most of the time.

If required, special rooms for candy storage are usually designed for a room temperature of 70°F, with a separate unitary air conditioner.

The heat gain from lighting will not be uniform throughout the entire area, with some areas such as jewelry and other specialty displays having as high as 6 to 8 W/ft^2 of floor area. For the entire sales area, an average value of 2 to 4 W/ft^2 may be used. For stockrooms and receiving, marking, toilet, and rest room areas, a value of 2 W/ft^2 may be used. When available, actual lighting layouts rather than average values should be used for load computation.

The store owner usually establishes the population density for a store based on the location, size, and past experience.

The food preparation and service areas in discount and variety stores range from small lunch counters with heat-producing equipment (ranges, griddles, ovens, coffee urns, toasters) in the conditioned space to large deluxe installations with separate kitchens beyond the conditioned space. For more specific information on HVAC systems for kitchen and eating spaces, see Chapter 3.

The heat released from special merchandising equipment, such as amusement rides for children or equipment used for preparing items such as popcorn, pizza, frankfurters, hamburgers, doughnuts, roasted chickens, and cooked nuts, should be obtained from the manufacturers.

Minimum outdoor air is generally based on 15 cfm per person in sales areas and 15 cfm per person for separate restaurant areas. A positive pressure should be maintained in the building.

Ventilation and outdoor air requirements must be provided as required in ASHRAE standards and local codes.

DESIGN CONSIDERATIONS

Discount and variety stores are generally constructed with a large open sales area and a partial glass storefront. The installed lighting is usually sufficient to offset the design roof heat loss. Therefore, the interior portion of these stores needs cooling during business hours throughout the year. The perimeter areas and especially the storefront and entrance areas may have a highly variable heating and cooling requirement.

Proper zone control and HVAC design are essential to meet the variable requirements of the storefront and entrance areas. Checkout lanes, entrances, and exits are generally located in this area, making proper environmental control even more important.

System Design

The important factors in selecting discount and variety store air-conditioning systems are (1) installation costs, (2) floor space required for equipment, (3) maintenance requirements and equipment reliability, and (4) simplicity of control. Roof-mounted unitary units are most commonly used for discount and variety store air conditioning.

Air Distribution

The air supply for large sales areas can generally be designed for the primary cooling requirement. Air distribution for the perimeter areas must consider the variable heating and cooling requirements.

Control System

The control system should be simple, dependable, and fully automatic, since it is usually operated by personnel who have little knowledge of air-conditioning systems. Many types of automatic control are used in chain stores. The system should be designed so that the store operator is only required to turn a switch to start or stop the system, or to change it to night operation.

Maintenance

Most variety stores do not employ trained maintenance personnel; instead they rely on service contracts with either the installer or a local service company. For improving operating costs, see information for small stores.

SUPERMARKETS

LOAD DETERMINATION

Heating and cooling loads may usually be calculated by the methods outlined in the 1989 ASHRAE *Handbook—Fundamentals*. Data for calculating the loads from people, lights, motors, and heat-producing equipment should be obtained from the store owner or manager or from the equipment manufacturer. Space conditioning in a supermarket is required for two reasons: human comfort and proper operation of refrigerated display cases. A minimum quantity of outdoor air should be introduced through the air-conditioning unit. The amount of outside air is the larger of the volume required for ventilation or to maintain slightly positive pressure in the space.

Many supermarkets are units of a large chain owned or operated by one company. Standardized construction, layout, and equipment used in designing many similar stores simplify load calculations.

It is important that the final air-conditioning load be correctly determined. Table 1 shows most general classifications of refrigerated display cases and tabulates the total heat extraction, sensible heat, latent heat, and the percentage of latent to total load. These data have been calculated from condensing unit data for actual operating conditions and from observed defrost water. Engineers report considerable fixture heat removal (case load) variation as the relative humidity and temperature vary in relatively small increments. Increases in relative humidity above 55% add substantial load, while reduced relative humidity substantially decreases the load, as shown in Figure 1.

The refrigerating effect, imposed by the display fixtures in the store, must be subtracted from the gross air-conditioning requirements for the building to produce a new total load and percentage of latent and sensible heat, which the air conditioning must handle.

Table 1 Refrigerating Effect Produced by Open Refrigerated Display Fixtures

	Refrigerating Effect (RE) on Building, Btu/h·ft of Fixture*			
Display Fixture Types	**Latent Heat, Btu/h·ft**	**%Latent to Total RE**	**Sensible Heat, Btu/h·ft**	**Total RE, Btu/h·ft**
Low temperature				
Frozen food				
Single deck	38	15	207	245
Single deck, double island	70	15	400	470
2 deck	144	20	576	720
3 deck	322	20	1288	1610
4 or 5 deck	400	20	1600	2000
Ice cream				
Single deck	64	15	366	430
Single deck, double island	70	15	400	470
Standard temperature				
Meats				
Single deck	52	15	298	350
Multideck	219	20	876	1095
Dairy				
Multideck	196	20	784	980
Produce				
Single deck	36	15	204	240
Multideck	192	20	768	960

*These figures are general magnitudes for fixtures adjusted for today's average desired product temperatures and apply to store ambients in front of the display cases of 72 to 74°F with 50 to 55% rh. Raising the dry bulb only 3 to 5°F and the humidity 5 to 10% can increase loads (heat removal) 25% or more. Equally lower temperatures and humidities, as found in stores in winter, have an equally marked effect on lowering loads and heat removal from the space. Consult display case manufacturer's data for the particular equipment to be used.

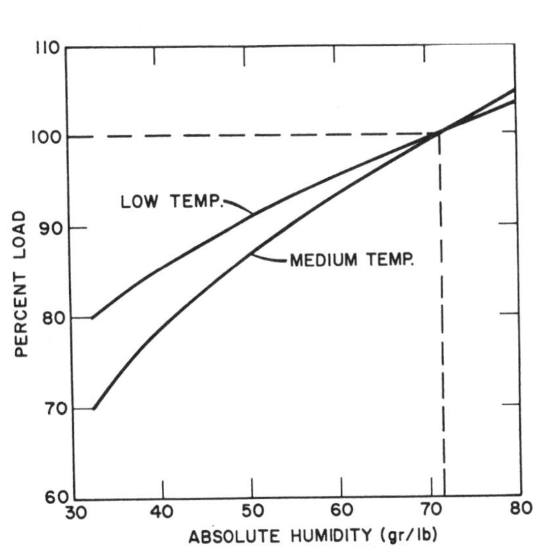

Fig. 1 Refrigerated Case Load Variation with Store Air Humidity

DESIGN CONSIDERATIONS

Store owners and operators frequently complain about cold aisles in stores, heating systems that operate even when the outdoor temperature is 78°F, and air-conditioning systems that operate infrequently. These problems are usually attributed to spillover of cold air from open refrigerated display equipment.

Although refrigerated display equipment may be the cause of cold stores, the problem is not that of excessive spillover or improperly operating equipment. To provide proper comfort and performance conditions, the design of heating and air-conditioning systems must compensate for the effects caused by open refrigerated display equipment. Design considerations include:

1. Increased heating requirements due to removal of large quantities of heat, even in summer, when only dehumidification is required.
2. Net air-conditioning load after deducting the latent and sensible refrigeration effect (Item 1). The load reduction and change in sensible-latent load ratio have a major effect on equipment selection.
3. Need for special air circulation and distribution to offset the heat removed by open refrigerating equipment.
4. Independent temperature and humidity control.

Each of these problems is present, to some degree, in every supermarket, although variations exist due to climate and store layout. The following sections discuss how to overcome these problems. While some of the instances may appear unusual, they may cause extremely high energy costs if the year-round air-conditioning system has not been designed to compensate for the effects of open equipment.

Heat Removed by Refrigerated Display Equipment

The display refrigerator not only cools a displayed product but envelops it in a blanket of cold air, which absorbs heat from the room air in contact with it. Approximately 80 to 90% of the heat removed from the room by vertical refrigerators is absorbed through the display opening. Thus, the open refrigerator acts as a large air cooler, absorbing heat from the room and rejecting it via the condensers outside the building. Occasionally, this conditioning effect can be more than the design air-conditioning capacity required by the store. The heat removed by the refrigerated equipment *must* be considered in the design of the air-conditioning and heating systems, because this heat is being removed constantly, day and night, summer and winter, without regard for the store temperature.

The display cases increase the heating requirement of the building and heat will often be required at times when not normally expected. The following example is an indication of the extent of this cooling effect. The desired store temperature is 75 °F. Store heat loss or gain is assumed to be 15,000 Btu/h per °F of temperature difference between outdoor and store temperature. (This value varies with store size, location, and exposure.) The heat removed by refrigerated equipment is 190,000 Btu/h. (This will vary with the number of refrigerators.) The latent heat removed is assumed to be 19% of the total (see Table 2), leaving 81% sensible or 154,000 Btu/h, which will cool the store 154,000/15,000 = 10 °F.

Therefore, by constantly removing sensible heat from its environment, the refrigerated equipment in this store will cool the store 10 °F below outdoor temperature in winter and summer. Thus, in mild climates, heat must be added to the store to maintain comfort conditions.

The designer has the choice of discarding the heat removed by refrigeration or reclaiming it. If economics and store heat data indicate that the heat should be discarded, heat extraction from the space must be added to the heating load calculations. If not, the heating system may not have sufficient capacity to maintain the design temperature under peak conditions.

The additional sensible heat removed by the cases may raise the air-conditioning latent load from 32% to as much as 50% of the net heat load. Removal of a 50% latent load by means of refrigeration alone is very difficult. Normally, it requires reheat or chemical adsorption.

Multishelf refrigerated display equipment requires 55% rh or less. In the dry-bulb ranges of average stores, humidity in excess of 55% can cause heavy coil frosting, product zone frosting with low-temperature cases, fixture sweating, and substantially increased power consumption for refrigeration.

Simple control systems can closely control humidity by using a humidistat during summer cooling, which transfers heat from the standard condenser to the heating coil. The store thermostat maintains proper summer temperature conditions, and the humidistat maintains proper humidity conditions. Override controls prevent a runaway situation between the humidstat and the thermostat.

The equivalent result can be accomplished with a conventional air-conditioning system by using three- or four-way valves and reheat condensers in the ducts. This system borrows heat from the standard condenser and is controlled by a humidistat. Desiccant dehumidifiers have also been used.

Humidity

Cooling from the refrigeration equipment does not preclude the need for air conditioning. On the contrary, it increases the need for humidity control.

With increases in humidity in the store, heavier loads are imposed on the refrigeration equipment, operating costs increase, more defrost periods are required, and the display life of products is decreased. The dew point rises with the relative humidity, and sweating can become profuse—to the extent that even non-refrigerated items such as shelving superstructures, canned products, mirrors, and walls may sweat.

There are two different methods of achieving humidity control. One is an electric vapor compression air conditioner, which cools air to a temperature below its dew point to remove moisture. This means overcooling the air to condense the moisture, and then reheating it to a temperature suitable to maintain comfort. Condenser waste heat is often used. Vapor compression holds humidity levels at 50 to 55% rh. The trends in store design, which include more food refrigeration and more efficient lighting, reduce the sensible component of the load even further.

The second method of dehumidification uses desiccant dehumidifiers. A desiccant absorbs or adsorbs moisture directly from air to its surface. The desiccant material is reactivated by passing hot air at 180 °F to 230 °F through the desiccant base. Condenser waste heat can provide as much as 40% of the heat required. Desiccant systems in supermarkets hold humidity levels between 30 and 40% rh, which results in significant savings in the operation of the refrigerated cases.

System Design

The same air-handling equipment and distribution system generally is used for cooling and heating. The entrance area of a supermarket is the most difficult section to heat. Many stores in the northern United States are built with vestibules provided with separate heating equipment to temper the cold air entering from outdoors. Auxiliary heat may also be provided at the checkout area, which is usually close to the front entrance.

Other methods used to heat entrance areas include (1) air curtains, (2) gas-fired or electric infrared radiant heaters, and (3) the use of waste heat from the refrigeration condensers. Air-cooled condensing units are most commonly used in supermarkets. Typically, a central air handler conditions the entire sales area. Specialty areas like bakeries, computer rooms, or warehouses are better served with a separate air handler. The loads in these areas vary and do not require the same control as the sales area. Most installations are made on the roof of the supermarket. If air-cooled condensers are located at ground level outside the store, they must be protected against vandalism as well as truck and customer traffic. If water-cooled condensers are used on the air-conditioning equipment and a cooling tower is required, provisions should be made to prevent freezing during winter operation.

Air Distribution

Designers overcome the concentrated load at the front of a supermarket by discharging at least 60% of the total air supply into the front third of the sales area.

The volume of air supply to the space with the vapor compression system has typically been 1 cfm/ft^2 of sales area. This should be calculated based on the sensible and latent internal loads. The desiccant system typically requires 0.5 cfm/ft^2 and is determined by the sensible load. This is due to the high moisture removal rate of a desiccant system and, in most cases, only about 40% of the circulation rate passes through the dehumidifier.

Being more dense, the air cooled by the refrigerators settles to the floor and becomes increasingly colder. The major effect of this is in the first 36 in. above the floor. If this cold air remains still, it will cause discomfort and serve no purpose, even though other areas of the store may need more cooling at the same time. To take advantage of the cooling effect of the refrigerators to provide an even temperature in the store, the cold air must be mixed with the entire store air. Cold floors or areas of the store cannot be eliminated by the addition of heat alone, and any reduction of air-conditioning capacity without circulation of the localized cold air would be analogous to installing an air conditioner without a fan.

To accomplish the necessary mixing, air returns should be located at the floor level; they should also be strategically placed to remove the cold air near concentrations of refrigerated fixtures. The returns should be designed and located to avoid creating drafts. Two general solutions to this problem are:

1. **Return Ducts in Floor.** This is the preferred method and can be accomplished in two ways. The floor area in front of the refrigerated display cases is the coolest area. All these cases have refrigerant lines run to them, usually in tubes or trenches. By enlarging the trenches or size of the tubes and having them open under the cases for air return, air can be drawn in from the cold area (see Figure 2). The air is returned to the air-handling unit through a tee connection to the trench before it enters the back room area. The opening where the refrigerant lines enter the back room should be sealed.

   If refrigerant line conduits are not used, the air can be returned through inexpensive underfloor ducts.

   Where fixtures do not have sufficient undercase air passage, check with the manufacturer. Often they can be raised off the floor approximately 1.5 in.

   Floor trenches can also be used as a duct for tubing, electrical supply, and so forth.

   Floor-level return relieves the problem of localized cold areas and cold aisles and uses the cooling effect for store cooling, or increases by the heating efficiency distributing the air in areas where most needed.

2. **Fans Behind Cases.** If ducts cannot be placed in the floor, circulating fans can draw air from the floor and discharge it into the upper levels (see Figure 3).

   However, while this approach will prevent objectionable cold aisles in front of the refrigerated display cases, it will often allow the area with a concentration of refrigerated fixtures to be colder than the rest of the store.

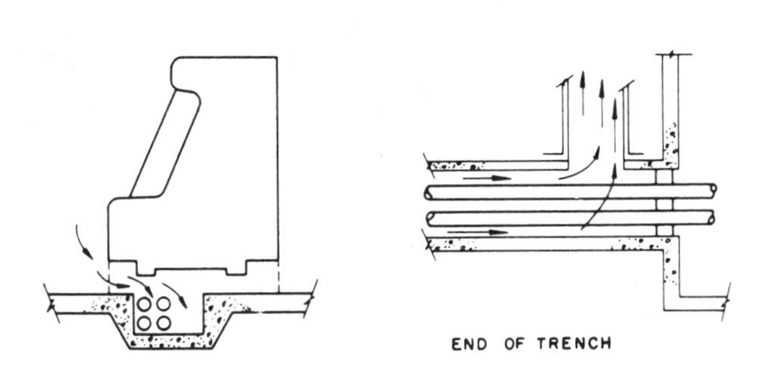

Fig. 2 Floor Return Ducts

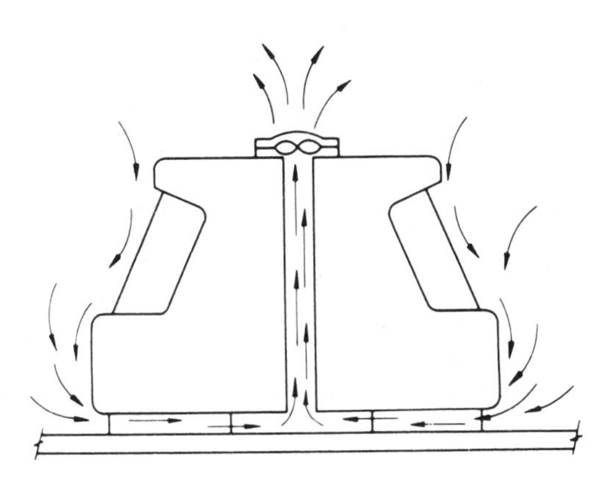

Fig. 3 Air Mixing Using Fans Behind Cases

Control Systems

The control system should be as simple as practicable so that store personnel are required only to change the position of a selector switch in order to start, stop, or change the system from heating to cooling or from cooling to heating. Control systems for heat recovery applications become more complex and should be coordinated with the equipment manufacturer.

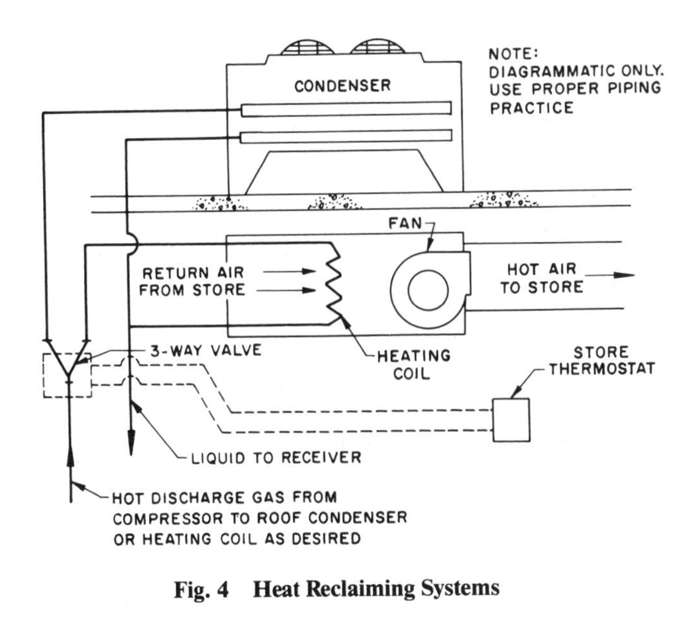

Fig. 4 Heat Reclaiming Systems

Maintenance and Heat Reclamation

Most supermarkets, except large chains, do not employ trained maintenance personnel but rely on service contracts with either the installer or a local service company. This relieves the store management of the personal responsibility of keeping the air-conditioning system in proper operating condition.

Heat extracted from the store plus the heat of compression may be reclaimed for heating cost savings. One method of reclaiming the rejected heat uses a separate condenser located in the air conditioner's air-handling system alternately or in conjunction with the main refrigeration condensers to provide heat, as required (see Figure 4). Another system uses water-cooled condensers and delivers its rejected heat to a water coil in the air handler.

The heat rejected by conventional machines using air-cooled condensers may be reclaimed by proper duct and damper design (see Figure 5). Automatic temperature controls can either reject this heat outdoors, recirculate it through the store, or store it in water tanks for future use.

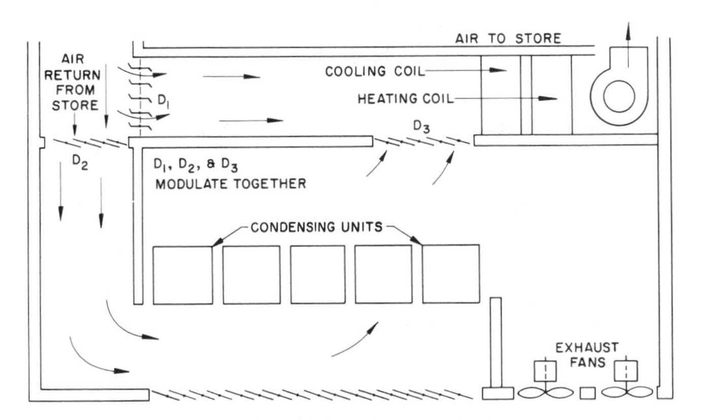

Fig. 5 Machine Room with Automatic Temperature Control Interlocked with Store Temperature Control

DEPARTMENT STORES

Department stores vary in size, type, and area of location so that the design of an air-conditioning system requires an individual solution for each store. An ample minimum quantity of outdoor air reduces or eliminates odor problems. Essential features of the quality system include (1) an automatic control system properly designed to compensate for load fluctuations, (2) zoned air distribution to maintain uniform conditions under shifting loads, and (3) use of outdoor air for cooling during intermediate seasons and peak sales periods. It is also desirable to adjust indoor temperatures for variations in outdoor temperatures. While the control of humidity to close tolerances is not necessary, a properly designed system should operate to maintain relative humidity not over 50% with a corresponding dry-bulb temperature of 78°F. This humidity limit eliminates musty odors and retards perspiration, particularly in fitting rooms.

LOAD DETERMINATION

Because the occupancy (except for store personnel) is transient, indoor conditions are commonly established not to exceed 78°F db and 50% rh at design outdoor summer design conditions, and 70°F db at design outdoor winter conditions. Winter humidification is seldom used in store air conditioning.

The number of customers and store personnel normally found on each conditioned floor must be ascertained, particularly in specialty departments or other areas having a greater-than-average concentration of occupants. Lights should be checked for wattage and types. Table 2 gives approximate values for lighting in various areas; Table 3 gives approximate occupancies.

Other sources of load, such as motors, beauty parlor and restaurant equipment, and any special display or merchandising equipment, should be determined.

The minimum outside air requirement should be as defined in the ASHRAE Ventilation Standards, which are generally acceptable and adequate for removing odors and keeping the building atmosphere fresh. However, local ventilation ordinances may require greater quantities of outdoor air.

Paint shops, alteration rooms, rest rooms, eating places, and locker rooms should be provided with positive exhaust ventilation, and their requirements must be checked against local codes.

Table 2 Approximate Lighting Loads for Department Stores

Area	Watts per ft^2
Basement	3-5
First floor	4-7
Upper floors, women's wear	3-5
Upper floors, house furnishings	2-3

Table 3 Approximate Occupancy for Department Stores

Area	ft^2 per person
Basement, metropolitan area	25-100
Basement, other with occasional peak	25-100
First floor, metropolitan area	25- 75
First floor, suburban	25- 75
Upper floors, women's wear	50-100
Upper floors, house furnishings	100 or more

DESIGN CONSIDERATIONS

Before making load calculations, the designer should examine the store arrangement to determine the various items that will affect the load as well as the system design. For existing buildings, a survey can be made of actual construction, floor arrangement, and load sources. For new buildings, examination of the drawings and discussion with the architect or owner will be required.

Larger stores may have beauty parlors, restaurants, lunch counters, or auditoriums. These special areas may operate during all store hours. Where the load from one of these areas is small in proportion to the total, this load may be included in the portion of the air-conditioning system serving the same floor. If, for any reason, present or future operation may be compromised by such a method, a separate air-conditioning system should serve this space. A separate air distribution system should be provided for the beauty parlor because of the concentrated load in this area.

The restaurant, because of the required service facilities, is generally centrally located. It is often used only during the noon hours. For control of odors, a separate air-handling system should be considered. Future plans for the store must be established, since they can have a great effect on the type of air-conditioning and refrigeration systems to be used.

System Design

Air-conditioning systems for department stores may be of the unitary or central-station type. The selection of the system should be based on owning and operating costs as well as any other special considerations for the particular store. The store hours, load variations, and size of load will affect the final selection of equipment and system.

Large department stores often use central-station systems consisting of air-handling units with chilled water cooling coils, hot water heating coils, fans, and filters. Air systems must have adequate zoning for varying loads, occupancy, and usage. Wide variations in people loads may permit consideration of variable volume air distribution systems. The water chilling and heating plants must distribute water to the various air-handling systems and zones and may take advantage of some diversity of loads in the total building.

The sales area should not be used for air-conditioning equipment; instead, ceiling, roof, and mechanical equipment room locations should be used wherever practicable. In selecting locations for equipment, maintenance and operation of the system after installation are important.

Air Distribution

All buildings must be studied for orientation, wind exposure, construction, and floor arrangement. These factors not only affect load calculations, but also zone arrangements and duct locations. In addition to planning for entrances, wall areas with significant glass, roof areas, and population densities, the expected locations of various departments (such as the lamp department) should be considered. Certain flexibility must be left in the duct design to allow for future movement of various departments. The preliminary duct layout should also be checked from the standpoint of winter heating to determine any special situations. In the case of entrances, it is usually necessary to design for separate air systems, particularly in northern areas. This is also true for storage areas where cooling is not contemplated.

Air curtain installations may be used at entrance doorways to limit or prevent infiltration of unconditioned air, while providing greater ease of entry.

Control Systems

The extent of automatic control depends on the type of installation and the extent to which it is zoned. The central station must be controlled so that air of the correct condition is delivered to the zones. Outdoor air should be automatically controlled to operate at minimum cost.

Partial or full automatic control should be placed on the refrigeration system to compensate for load fluctuations. Completely automatic refrigeration plants are now practical and should be considered carefully.

Maintenance

Most department stores employ operating personnel for routine operating and maintenance requirements. However, these stores normally rely on service and preventive maintenance contracts for the specialized requirements of refrigeration cycles, chemical treatment, central plant systems, and major repairs.

Improving Operating Costs

Outdoor air economizer systems can reduce the operating cost of the cooling system in many climates. These are generally available as factory options or accessories with the air-handling units or control systems. Heat recovery and thermal storage systems should also be analyzed.

CONVENIENCE CENTERS

Many small stores, variety stores, supermarkets, drugstores, theaters, and even department stores are located in convenience centers normally owned by a developer. The space for an individual store is usually leased. Arrangements for installing air-conditioning systems in leased space vary. In a typical arrangement, the developer has a shell structure built and provides the tenant with an allowance for a typical heating and cooling system and other minimum interior finish work. The tenant must then install an HVAC system under this leasing arrangement.

DESIGN CONSIDERATIONS

The developer or owner may establish standards for typical heating and cooling systems that may or may not be sufficient for the specified space requirements. Therefore, a tenant may have to install systems of different sizes and types than originally allowed for by the developer. The tenant must ascertain that power and other services will be available for its total intended requirements.

The use of party walls in convenience centers tends to reduce heating and cooling loads. However, the partition load while an adjacent space is unoccupied must be considered.

REGIONAL SHOPPING CENTERS

Regional shopping centers generally incorporate a heated and air-conditioned mall. These centers are normally owned by a developer, who may be an independent party or one of the major tenants in the center.

The major department stores are generally considered to be separate buildings, although they are attached to the mall. The space for individual stores is usually leased. Arrangements for installing air-conditioning systems in the individually leased spaces vary. Typically, a developer will have a completely finished mall and a shell strucure built for the individual stores. An allowance is provided to the tenant for a typical heating and cooling system and other minimum interior finish.

Table 4 presents typical factors that can be used as check figures and field estimates. However, this table should not be used for final determination of load, since the values are only averages.

DESIGN CONSIDERATIONS

The owner provides the air-conditioning system for the enclosed mall. The mall system may use a central plant or individual unitary units. The owner will generally require that the individual tenant stores connect to a central plant system and will include charges in the rent for heating and cooling. Where unitary systems are used, the owner generally requires that the individual tenant install a unitary system of similar design.

The owner may establish standards for typical heating and cooling systems that may or may not be sufficient for the specific space requirements. Therefore, a tenant may have to install systems of different sizes than originally allowed for by the developer.

Leasing arrangements may include language that has a detrimental effect on conservation (such as allowing excessive lighting and outdoor air or deleting requirements for economizer systems). The designer of HVAC systems for tenants in a shopping center must be fully aware of the lease requirements and work closely with leasing agents to guide these systems toward better energy efficiency.

Many regional shopping centers now have specialty food court areas. These areas require special considerations for odor control, outside air requirements, kitchen exhaust, heat removal, and refrigeration equipment.

Table 4 Typical Installed Capacity and Energy Usage in Enclosed Mall Centers[a] Based on 1979 Data—Midwestern United States

Type of Space	Installed Cooling Btu/h·ft^2	Annual Consumption, kWh/ft^2			
		Lighting[b]	Cooling[c]	Heating[d]	Miscellaneous
Candy store	44.8 to 78.0	23.9 to 29.6	7.8 to 23.5	9.6 to 5.7	2.4 to 70.1
Clothing store	37.3 to 45.5	14.1 to 24.9	8.0 to 9.6	8.9 to 5.1	1.2 to 6.7
Fast food	48.2 to 78.0	16.7 to 32.4	9.8 to 23.5	9.6 to 3.8	38.1 to 70.0
Game room	33.8 to 44.5	6.8 to 12.5	7.2 to 7.3	13.8 to 4.7	0.2 to 13.8
Gen. merchandise	32.9 to 43.6	13.2 to 23.3	6.3 to 9.7	6.3 to 5.6	1.4 to 9.3
Gen. service	39.3 to 50.3	15.7 to 17.4	7.8 to 9.2	12.0 to 5.9	7.6 to 8.2
Gift store	36.4 to 51.1	12.8 to 22.9	6.3 to 10.0	7.1 to 4.9	0.4 to 2.2
Grocery	69.5 to 86.2	10.9 to 19.3	5.3 to 8.9	6.0 to 7.0	18.6 to 21.0
Jewelry	53.5 to 66.1	37.2 to 44.9	10.6 to 12.3	5.6 to 3.6	7.6 to 9.3
Mall	30.0 to 48.0	5.4 to 12.0	8.3 to 13.7	11.5 to 5.6	0.2 to 1.5
Restaurant	40.0 to 53.0	5.0 to 22.0	7.8 to 12.3	19.0 to 16.7	17.2 to 21.0
Shoe store	36.9 to 50.6	20.4 to 32.1	7.4 to 11.4	7.2 to 4.6	1.0 to 2.3
Center average	30.0 to 48.0	18.0 to 28.0	8.0 to 15.0	6.0 to 3.0	1.0 to 3.0

[a] Operation of center assumed to be 12 h/day and 7 days/week.
[b] Lighting includes miscellaneous and receptacle loads.
[c] Heating includes blower and is for electric resistance heating.
[d] Cooling includes blower motor and is for unitary-type system.

System Design

Regional shopping centers vary widely in physical arrangements and architectural design. Single-level and smaller malls usually use unitary systems for mall and tenant air conditioning; larger and multilevel malls usually use central plant systems. The owner sets the design of the mall system and generally requires that similar systems be installed for tenant stores.

A typical central plant system may distribute chilled water or chilled air to the individual tenant stores and the mall air-conditioning system and use electric heating at the local use point. Some systems distribute both hot and chilled water. All-air systems have also been used; these systems distribute chilled or heated air to the individual tenant stores and to the mall air-conditioning system and use variable volume control at the local use point. The all-air systems provide improved efficiency and better overall economics; they also provide the basic components required for smoke control systems.

Air Distribution

Air distribution for individual stores should be designed for a particular space occupancy. Some tenant stores maintain a negative pressure relative to the mall for odor control.

The air distribution system should maintain a slight positive pressure relative to atmospheric pressure and a neutral pressure relative to most of the individual tenant stores. Smoke removal for fire protection should be considered. Exterior entrances should have vestibule entrances with independent heating systems.

Smoke removal and control systems are required by many building codes. Air distribution systems should be designed to easily accommodate smoke control requirements.

Maintenance

The center may employ operating personnel for routine operating and maintenance requirements of the mall. However, these centers normally rely on service and preventive maintenance contracts for the specialized requirements of refrigeration cycles, chemical treatment, central plant systems, and major repairs. Individual tenant stores may have to provide their own maintenance.

Improving Operating Costs

Outdoor air economizer systems can improve the operating cost of the cooling system in many climates. These are generally available as factory options or accessories with the air-handling units or control systems. Heat recovery and thermal storage systems should also be analyzed. Some shopping centers have successfully employed cooling tower heat exchanger economizers.

Central plant systems for regional shopping centers will typically provide much lower operating costs than unitary systems. The initial cost of the central plant system will typically be higher than that for the unitary installation.

MULTIPLE-USE COMPLEXES

Multiple-use complexes are being developed in most metropolitan areas. These complexes generally combine retail facilities with other functions such as office space, hotel space, residential space, or other commercial space into a single site. The combining of functions into a single site or structure provides benefits such as improved land use; structural savings; more efficient parking; utility savings; and opportunities for more efficient electrical, fire protection, and mechanical systems.

LOAD DETERMINATION

The various occupancies may have peak HVAC demands that occur at different times of the day and even at different times of the year. Therefore, the various occupancies should have HVAC loads determined independently. Where a combined central plant is to be considered, a block load should also be determined.

DESIGN CONSIDERATIONS

Retail facilities are generally located on the lower levels of multiple-use complexes, and other commercial facilities are on upper levels. Generally, the perimeter loads of the retail portion differ from those of the other commercial space. The greater lighting and population densities also make the HVAC demands different for the retail space than for the other commercial space.

The differences in HVAC characteristics for the various occupancies within a multiple-use complex indicate that separate air handling and distribution should be used for the separate functions. However, combining heating and cooling requirements into a central plant can achieve substantial savings.

A combined central heating and cooling plant for a multiple-use complex also provides good opportunities for heat recovery, thermal storage, and other similar functions that may not be economical in a single-use facility.

Many multiple-use complexes have atriums. The stack effect created by atriums requires specific design considerations for tenants and space on the main floor. Areas near entrances require special considerations to prevent drafts and accommodate extra heating requirements.

System Design

Individual air-handling and distribution systems should be designed for the various occupancies. The central heating and cooling plant may be sized for the block load requirements, which may be less than the total of each occupancy's demand.

Control Systems

Multiple-use complexes typically require a centralized control system. This may be dictated by requirements for fire and smoke control, security, remote monitoring, billing for central facilities use, maintenance control, building operations control, and energy management.

CHAPTER 3

COMMERCIAL AND PUBLIC BUILDINGS

THIS chapter is organized into eight sections. The first section, General Criteria, applies to all buildings and includes information on load characteristics, design concepts, and design criteria. Design criteria covers items such as comfort level; costs; local conditions and requirements; and fire, smoke, and odor control, which may apply to all building types.

The remaining sections present information applicable to specific buildings, including dining and entertainment centers, office buildings, libraries and museums, bowling centers, communication centers, transportation centers, and warehouses.

GENERAL CRITERIA

If properly applied, every system can be successful in any building. However, such factors as initial and operating costs, space allocation, architectural design, location, and the engineer's evaluation and experience limit the proper choices for a given building type.

Heating and air-conditioning systems that are simple in design and of the proper size for a given building generally have fairly low maintenance and operating costs. For optimum results, as much inherent thermal control as is economically possible should be built into the basic structure. Such control might include materials with high thermal properties, insulation, and multiple or special glazing and shading devices.

The relationship between the shape, orientation, and air-conditioning capacity of a building should also be considered. Since the exterior load may vary from 30 to 60% of the total air-conditioning load when the fenestration area ranges from 25 to 75% of the floor area, it may be desirable to minimize the perimeter area. For example, a rectangular building with a four-to-one aspect ratio requires substantially more refrigeration than a square building with the same floor area.

Proper design also considers controlling noise and minimizing pollution of the atmosphere and water into which the system will discharge. The quality of the indoor air is also a major factor to consider in design.

Retrofitting of existing buildings is also an important part of the construction industry because of increased costs of construction and the necessity of reducing energy consumption. Table 1 lists factors to consider before selecting a system for any building. The system choice, often made by the owner, may not be based on an engineering study. It is decided, to a great degree, by the engineer's ability to relate those factors involving higher first cost or lower life cycle cost and benefits that have no calculable monetary value.

Some buildings are constructed with only heating and ventilating systems. If such buildings provide for future cooling, humidification, or both, the design principles are the same as those for a fully air-conditioned building. For these buildings, greater design emphasis should be placed on natural or forced ventilation systems to minimize occupant discomfort during hot weather.

The general responsibility for this chapter is assigned to TC 9.8, Large Building Air-Conditioning Applications.

Load Characteristics

Any building analyzed for heat recovery or total energy systems requires sufficient load profile and load duration information on all forms of building input to (1) properly evaluate the instantaneous effect of one on the other when no energy storage is contemplated, and (2) evaluate short-term effects (up to 48 h) when energy storage is used.

Load profile curves consist of appropriate energy loads plotted against the time of day. Load duration curves indicate the accumulated number of hours at each load condition, from the highest to the lowest load for a day, a month, or a year. The area under load profile and load duration curves for corresponding periods is equivalent to load × time. These calculations must consider the type of air and water distribution systems in the building.

Load profiles for two or more energy forms during the same operating period may be compared to determine load-matching characteristics under diverse operating conditions. For example, when thermal energy is recovered from a diesel-electric generator at a rate equal to or less than the thermal energy demand, the energy can be used instantaneously, avoiding waste. But it may be worthwhile to store thermal energy when it is generated at a greater rate than demanded. A load profile study helps determine the economics of thermal storage.

Similarly, with internal source heat recovery systems, load matching must be integrated over the operating season with the aid of load duration curves for overall feasibility studies. These curves are useful in energy consumption analysis calculations as a basis for hourly input values in computer programs (see Chapter 28 of the 1989 ASHRAE *Handbook—Fundamentals*).

Economic feasibility of district heating and cooling systems (aside from environmental considerations) are influenced by load density and diversity factors for branch feeds to buildings along distribution mains. For example, the load density or energy per unit length of distribution main can be small enough in a complex of low-rise, lightly loaded buildings located at a considerable distance from one another, to make a central heating, cooling, or heating and cooling plant uneconomical.

Concentrations of internal loads peculiar to each application are covered later in this chapter and in Chapters 25 through 27 of the 1989 ASHRAE *Handbook—Fundamentals*.

Design Concepts

When a structure is characterized by several exposures and multipurpose use, especially with wide load swings and noncoincident energy use in certain areas, multiunit or unitary systems may be considered for such areas, but not necessarily for the entire building. The benefits of transferring heat absorbed by cooling from one area to other areas, processes, or services that require heat may enhance the selection of such systems. This is especially true if the systems have closed water loop heat pumps or other equipment.

When the cost of energy is included in rent with no means for permanent or check-metering, tenants tend to consume excess

Table 1 General Design Criteria[a]

General Category	Specific Category	Inside Design Conditions		Air Movement	Circulation, Air Changes per Hour
		Winter	Summer		
Dining and Entertainment Centers	Cafeterias and Luncheonettes	70 to 74 °F 20 to 30% rh	78 °F[e] 40% rh	50 fpm at 6 ft above floor	12 to 15
	Restaurants	70 to 74 °F 20 to 30% rh	74 to 78 °F 55 to 60% rh	25 to 30 fpm	8 to 12
	Bars	70 to 74 °F 20 to 30% rh	74 to 78 °F 50 to 60% rh	30 fpm at 6 ft above floor	15 to 20
	Nightclubs	70 to 74 °F 20 to 30% rh	74 to 78 °F 50 to 60% rh	below 25 fpm at 5 ft above floor	20 to 30
	Kitchens	70 to 74 °F	85 to 88 °F	30 to 50 fpm	12 to 15[h]
Office Buildings		70 to 74 °F 20 to 30% rh	74 to 78 °F 40 to 50% rh	25 to 45 fpm 0.75 to 2 cfm/ft^2	4 to 10
Libraries and Museums	Average	68 to 72 °F 40 to 55% rh		below 25 fpm	8 to 12
	Archival	See Special Considerations		below 25 fpm	8 to 12
Bowling Centers		70 to 74 °F 20 to 30% rh	75 to 78 °F 50 to 55% rh	50 fpm at 6 ft above floor	10 to 15
Communication Centers	Telephone Terminal Rooms	72 to 78 °F 40 to 50% rh	72 to 78 °F 40 to 50% rh	25 to 30 fpm	8 to 20
	Teletype Centers	70 to 74 °F 40 to 50% rh	74 to 78 °F 45 to 55% rh	25 to 30 fpm	8 to 20
	Radio and Television Studios	74 to 78 °F 30 to 40% rh	74 to 78 °F 45 to 55% rh	below 25 fpm at 12 ft above floor	15 to 40
Transportation Centers	Airport Terminals	70 to 74 °F 20 to 30% rh	74 to 78 °F 50 to 60% rh	25 to 330 fpm at 6 ft above floor	8 to 12
	Ship Docks	70 to 74 °F 20 to 30% rh	74 to 78 °F 50 to 60% rh	25 to 30 fpm at 6 ft above floor	8 to 12
	Bus Terminals	70 to 74 °F 20 to 30% rh	74 to 78 °F 50 to 60% rh	25 to 30 fpm at 6 ft above floor	8 to 12
	Garages	40 to 55 °F	80 to 100 °F	30 to 75 fpm	4 to 6 Refer to NFPA
Warehouses		Inside design temperatures for warehouses often depend on the materials stored inside.			1 to 4

Table 1 General Design Criteria (*Concluded*)

Minimum[b] Outdoor Air	Noise[c]	Filtering Efficiencies[d]	Energy Budget, 10^3 Btu/(ft^2 · year)	Load Profile	General
10 cfm per person	NC 40 to 50[f]	35% or better	50 to 400	Peak at 1 to 2 P.M. 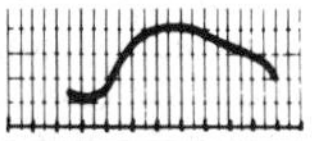	Prevent draft discomfort for patrons waiting in serving line
5 cfm per person	NC 35 to 40	35% or better	50 to 500	Peak at 1 to 2 P.M. 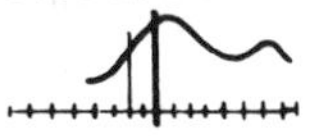	
100% or 10 cfm per person w/odor control	NC 35 to 50	Use charcoal for odor control with manual purge control for 100% outside air to exhaust +35% prefilters	50 to 400	Peak at 5 to 7 P.M. 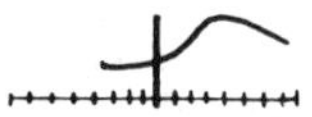	
25 cfm per person	NC 35 to 45[g]	Use charcoal for odor control with manual purge control for 100% outside air to exhaust ±35% prefilters	20 to 250	Peak after 8 P.M. off 2 A.M. to 4 P.M.	Provide good air movement but prevent cold draft discomfort for dancing patrons
100%	NC 40 to 50	10 to 15% or better	100 to 400	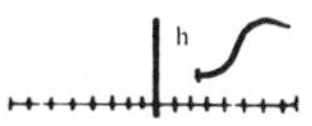 h	Negative air[j] pressure required for odor control
0.05 to 0.25 cfm/ft^2 or 5 cfm per person	NC 30 to 45	35 to 60% or better	25 to 300	Peak at 4 P.M. 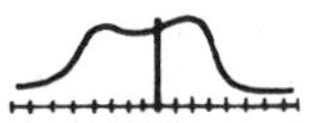	
5 cfm per person	NC 35 to 40	35 to 60% or better	150 to 250	Peak at 3 P.M. 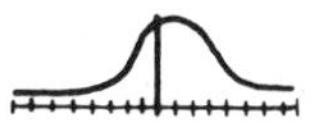	
5 cfm per person	NC 35	35% prefilters + charcoal filters 85 to 95% final[k]	25 to 100	Peak at 3 P.M. 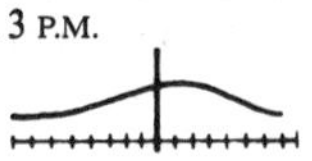	
20 to 30 cfm per person	NC 40 to 50	10 to 15%	100 to 200	Peak at 6 to 8 P.M. 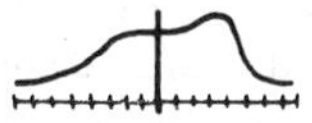	
5 cfm per person	to NC 60	85% or better	150 to 500	Varies with location and use	Constant temperature and humidity required
10 cfm per person	NC 40 to 50	85%	50 to 150	Varies with locations and use	
10 cfm per person	NC 15 to 25	35% or better	100 to 200	Varies widely from changes in lighting and people	
5 cfm per person	NC 35 to 50	35% or better and charcoal	100 to 150	Peak at 10 A.M. to 9 P.M. 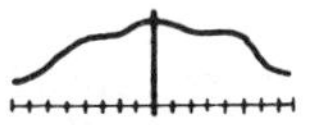	Positive air pressure required in terminal
5 cfm per person	NC 35 to 50	10 to 15%	25 to 100	Peak at 10 A.M. to 5 P.M. 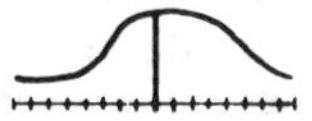	Positive air pressure required in waiting area
5 cfm per person	NC 35 to 50	35% with exfiltration	150 to 250	Peak at 10 A.M. to 5 P.M. 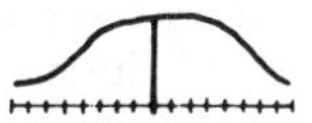	Positive air pressure required in terminal
1.5 cfm/ft^2 per person	NC 35 to 50	10 to 15%	20 to 200	Peak at 10 A.M. to 5 P.M. 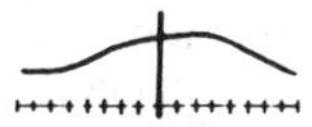	Negative air pressure required to remove fumes
0.01 cfm/ft^2 or 5 cfm per person	up to 75	10 to 35%	20 to 350[m]	Peak at 10 A.M. to 3 P.M. 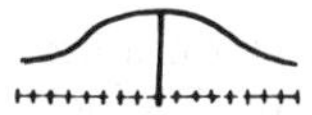	

Notes to Table 1, General Design Criteria

[a] This table shows design criteria differences between various commercial and public buildings. It should not be used as the sole source for design criteria. Each type of data contained here can be determined from the ASHRAE Handbooks and Standards.

[b] Governing codes should be consulted to determine minimum allowable requirements. Outdoor air requirements may be reduced if high efficiency adsorption equipment or other odor- or gas-removal equipment is used, but never below 5 cfm per person (see Chapter 34). Also, see Chapter 12 of the 1989 ASHRAE *Handbook—Fundamentals*, and ASHRAE *Standard* 62-1989.

[c] Refer to Chapter 42.

[d] Average Atmospheric Dust Spot Efficiency (see ASHRAE *Standard* 52 for method of testing).

[e] Food in these areas is often eaten more quickly than in a restaurant, so the turnover of diners is much faster. Since diners seldom remain for long, they do not require the degree of comfort necessary in restaurants. Thus, it may be possible to lower design criteria standards and still provide reasonably comfortable conditions. Although space conditions of 80°F and 50% rh may be satisfactory for patrons when it is 95°F and 50% rh outside, indoor conditions of 78°F and 40% rh are better.

[f] Cafeterias and luncheonettes usually have some or all of the food preparation equipment and food trays in the same room with the diners. These eating establishments are generally noisier than restaurants, so that noise transmission from the air-conditioning equipment is not as critical.

[g] In some nightclubs, the noise from the air-conditioning system must be kept low so that all patrons can hear the entertainment.

[h] Usually determined by kitchen hood requirements.

[i] Peak kitchen heat load does not generally occur at peak dining load, although in luncheonettes and some cafeterias where cooking is done in the dining areas, peaks may be simultaneous.

[j] NFPA 96-87 sets requirements for hood, fan, ducts, and fire protection.

[k] Methods for removal of chemical pollutants must also be considered.

[l] Also includes service stations.

energy. This energy abuse raises operating costs for the owner, decreases profitability, and has a detrimental effect on the environment. While design features can minimize excess energy penalties, they seldom eliminate abuse. For example, the U.S. Department of Housing and Urban Development field records in total-electric housing show that nationwide rent-included dwellings use approximately 20% more energy than those directly metered by a public utility company.

Diversity factor benefits for central heating and cooling in rent-included buildings may result in lower building demand and connected loads. However, energy abuse may easily result in load factors and annual energy consumption exceeding that of buildings where the individual has a direct economic incentive to reduce energy consumption.

Design Criteria

In many applications, design criteria are fairly evident, but in all cases, the engineer should understand the owner's and user's intent, since any one factor may influence system selection. The engineer's personal experience and judgment in the projection of future needs may be a better criterion for system design than any other single factor.

Comfort Level. Comfort, as measured by temperature, humidity, air motion, air quality, noise, and vibration, is not identical for all buildings, occupant activities, or use of space.

For spaces with a high population density, or with a sensible heat factor less than 0.75, lower dry-bulb temperature reduces the generation of latent heat. Reduced latent heat may further reduce the need for reheat and save energy. Therefore, an optimum temperature should be found.

Costs. Owning and operating costs can affect system selection and seriously conflict with other criteria. Therefore, the engineer must help the owner resolve such conflicts. The cost and availability of different fuels must also receive attention.

Local Conditions. Local, state, and national codes and regulations, and environmental concerns must be included in the design. Chapters 23 and 24 in the 1989 ASHRAE *Handbook—Fundamentals* give information on calculating the effects of weather in specific areas.

Automatic Temperature Control. Proper automatic temperature control maintains occupant comfort during the varying internal and external loads. Improper temperature control may mean a loss of customers in restaurants and other public buildings. An energy management control system can be combined with a building automation system to allow the owner to manage energy, lighting, security, fire protection, and other similar systems from one central control system. Chapters 32 and 41 include more details.

Fire, Smoke, and Odor Control. Fire and smoke can easily spread through elevator shafts, stairwells, and other means. Although an air-conditioning system can spread fire and smoke by fan operation, penetrations required in walls or floors, or by stack effect without fan circulation, a properly designed and installed system can be a positive means of fire and smoke control.

Knowledge of effective techniques for positive control after fire startup is limited, but increasing (see Chapter 47). Effective attention to fire and smoke control will also help prevent odor migration to unventilated areas (see Chapter 40).

DINING AND ENTERTAINMENT CENTERS

Load Characteristics

Air conditioning of restaurants, cafeterias, bars, and nightclubs presents common load problems encountered in comfort conditioning, with additional factors pertinent to dining and entertainment applications. Such factors include:

- Extremely variable loads with high peaks, in many cases occurring twice daily.
- High sensible and latent heat gains because of gas, steam, and electric appliances, people, and food.
- Localized high sensible and latent heat gains in dancing areas.
- Unbalanced conditions in restaurant areas adjacent to kitchens which, although not part of the conditioned space, still require special attention.
- Heavy infiltration of outdoor air through doors during rush hours.

Internal heat and moisture loads come from occupants, motors, lights, appliances, and infiltration. Separate calculations should be made for patrons and employees. The sensible and latent heat load must be proportioned in accordance with the design temperature selected for both sitting and working people, because latent to sensible heat ratio decreases as room temperature decreases for each category.

Hoods required to remove heat from appliances may also substantially reduce the space latent loads.

Infiltration is a considerable factor in many restaurant applications because of short occupancy and frequent door use. It is increased by the need for large quantities of air to replace air exhausted through hoods and for smoke removal. Wherever possible, vestibules or revolving doors should be installed to reduce such infiltration.

Design Concepts

The following factors influence system design and equipment selection:

- High concentration of food, body, and tobacco-smoke odors require adequate ventilation with proper exhaust facilities.

- Step control of refrigeration plants to give satisfactory and economical operation under reduced loads.
- Air exhausted at the ceiling removes smoke and odor.
- Building design and space limitations often favor one equipment type over another. For example, in a restaurant having a vestibule with available space above it, air conditioning with condensers and evaporators remotely located above the vestibule may be satisfactory. Such an arrangement saves valuable space, even though self-contained units located within the conditioned space may prove somewhat lower in initial cost. In general, small cafeterias, bars, and the like, with loads up to 35 kW, can be most economically conditioned with packaged units; larger and more elaborate establishments require central plants.
- If not required for kitchen exhaust, air required by an air-cooled or evaporative condenser may be drawn from the conditioned space. This eliminates the need for operating additional exhaust fans and also improves overall plant efficiency because of the lower temperature air entering the condenser. Proper water treatment is necessary if an evaporative condenser is used. It may also be necessary to bypass air when air is not required.
- The usual practice for the smaller restaurant with an isolated plant has been to use direct-expansion systems.
- Some air-to-air heat recovery equipment can reduce the energy required for heating and cooling ventilation air. Chapter 34 of the 1988 ASHRAE *Handbook—Equipment*, includes details.

Since eating and entertainment centers generally have low sensible heat factors and require high ventilation rates, fan-coil and induction systems are usually not applicable. All-air systems are more suitable. Space must be established for ducts, except for small systems with no ductwork. Large establishments are often served by central chilled water systems.

In cafeterias and luncheonettes, the air distribution system must keep food odors from the serving counters away from areas where patrons are eating. This usually means heavy exhaust air requirements at the serving counters, with air supplied into and induced from eating areas. Exhaust air must also remove the heat from hot trays, coffee urns, and ovens to minimize patron and employee discomfort and reduce air-conditioning loads. These factors often create greater air-conditioning loads for cafeterias and luncheonettes than for restaurants.

Odor Removal. Air transferred from dining areas into the kitchen keeps odors and heat out of dining areas and cools the kitchen. Outdoor air intake and kitchen exhaust louvers should be located so that exhaust air is neither drawn back into the system, nor causes discomfort to passersby.

Where odors may possibly be drawn back into dining areas, activated charcoal filters, air washers, or ozonators may be used to remove odors. No kitchen, locker room, toilet, or other odoriferous air should be recirculated unless air purifiers are used.

Hood Types. The air quantity for proper ventilation is a function of kitchen equipment heat release and kitchen hood size. While the heat release factor is more important, canopy-type hoods do not operate at maximum efficiency unless entrance air face velocity is at least 0.4 m/s. Face velocities of 0.4 to 0.5 m/s should be used for design, with 0.3 m/s as an absolute minimum.

Slot-type exhaust hoods are more efficient than overhead hoods, but they are more costly and may diminish valuable work area unless properly applied. Slot hoods require 230 to 310 L/(s·m) for proper operation but may substantially reduce kitchen exhaust air requirements (see Chapter 27). Slot hoods may also reduce overall kitchen ventilation system cost by obviating an additional makeup air system and related energy cost.

Makeup air exhaust hoods similar to slot-type hoods reduce the kitchen air-conditioning load. Outside air is supplied at the hood's perimeter or at the ceiling near the hood through slot diffusers. The exhaust hood then exhausts the supplied outside air plus approximately 15% additional air from the kitchen. Air colder than 16 °C entering the hood may coagulate grease on the grease filters, causing a high static pressure drop, unless they are regularly maintained. In addition, kitchen personnel may be uncomfortable if untempered outside air is directed from the hood toward their workstations. Hood makeup air should be tempered and filtered.

Exhaust hoods incorporating water-wash cleaning systems can recover heat from hot kitchen exhaust air (see Chapter 34 of the 1988 ASHRAE *Handbook—Equipment* for a discussion of heat recovery equipment).

Hood Fire Protection. In the past, carbon dioxide (CO_2), water sprinklers, and steam-smothering systems have been used. Most systems use water spray hoods or dry chemical extinguishing agents.

Provisions should be made to disconnect and isolate the fuel source from the kitchen cooking equipment under the exhaust hood and to shut off the hood exhaust fan if the hood's fire protection system is activated.

Hood Duct Construction. Some local codes require a refractory-type exhaust duct that allows grease fires to burn themselves out; others require mineral wool insulation (see NFPA 96-87).

Local authorities having jurisdiction should be consulted to determine the exact duct protection and clearance to combustible materials requirements. Many authorities consider plasterboard a combustible material due to the paper backing (even if it is a part of a fire-rated assembly).

In multistory structures, a significant problem is the space required for the range hood exhaust duct, which is considered a low-temperature chimney. The duct usually rises to the roof to prevent odor spread, grease streaking of walls, and to eliminate fire hazards. Chimney size and space should be determined early during building design to allow a direct path from the hood to the outside (see NFPA 96-87).

Hood Exhaust System Design. Ductwork should be designed for a velocity of 9 to 11 m/s to minimize the settling of grease particles. All turns should be made with elbows that have a minimum centerline radius of 1.5 times duct dimension in the turning direction. Where this is not possible, grease traps and cleanout panels may be provided in the ducts. Turning vanes should not be used.

The fan should be located at the discharge end of the duct run to minimize leaks, which could cause odor problems.

Ductwork design must allow for sufficient expansion caused by high temperatures during a fire (see Chapter 1 of the 1988 ASHRAE *Handbook—Equipment*.)

Kitchen Air Conditioning. Kitchens can often be air conditioned effectively, without excessive cost, if planned in the initial design phases. It is not necessary to meet the same design criteria as for dining areas, but kitchen temperatures can be reduced significantly. The relatively large number of people and food loads in dining and kitchen areas produce a high latent load. Additional cooling required to eliminate excess moisture increases refrigeration plant, cooling coils, and air-handling equipment size.

Self-contained units, advantageously located, with air distribution designed not to produce drafts off hoods and other equipment, can be used to spot cool intermittently. The costs are not excessive, and kitchen personnel efficiency can be improved greatly.

Evaporative cooling may be a good compromise (depending on the climate) between the expense of air conditioning and the lack of comfort in ventilated kitchens.

Special Considerations

In establishing design conditions, duration of individual patron occupancy should be considered, since patrons entering from outdoors are more comfortable in a room of high temperature than those who remain long enough to become acclimated. Nightclubs and deluxe restaurants should usually be operated at a lower effective temperature than cafeterias and luncheonettes.

Often, the ideal design condition must be rejected for an acceptable condition because of equipment cost or performance limitations. Restaurants are frequently affected in this way, since ratios of latent to sensible heat may result in uneconomical equipment selection, unless a combination of lower design dry-bulb temperature and higher relative humidity (giving equal effective temperature) is selected.

In severe climates, entrances and exits in any dining establishment should be completely shielded from diners to prevent drafts. Vestibules provide a measure of protection. However, both vestibule doors are often open simultaneously. Revolving doors or local means for heating or cooling infiltration air may be provided to offset drafts.

Employee comfort is difficult to maintain at a uniform level because of (1) temperature differences between kitchen and dining room, and (2) because employees are constantly in motion while patrons are seated. Since customer satisfaction is essential to a dining establishment's success, patron comfort is the primary consideration. However, maintenance of satisfactory temperature and atmospheric conditions for customers also helps alleviate employee discomfort.

One problem in dining establishments is the use of partitions to separate areas into modular units. Partitions create such varied load conditions that individual modular unit control is generally necessary.

Baseboard radiation or convectors, if required, should be located so as not to overheat patrons. This is difficult to achieve in some layouts because of movable chairs and tables. For these reasons, it is desirable to enclose all dining room and bar heating elements in insulated cabinets with top outlet grilles and baseboard inlets. With heating elements located under windows, this practice has the additional advantage of directing the heat stream to combat window downdraft and air infiltration. Separate smoking and nonsmoking areas should be considered. The smoking area should be exhausted or served by separate air-handling equipment.

Restaurants. In restaurants, people are seated and served at tables, while food is generally prepared in remote areas. This type of dining is usually enjoyed in a leisurely and quiet manner, so the ambient atmosphere should be such that the air conditioning is not noticed.

Bars. Bars are often part of a restaurant or nightclub. Where they are establishments on their own, they often serve food as well as drinks, and they should be classified as restaurants, with food preparation in remote areas. Alcoholic beverages produce pungent vapors, which must be drawn off. In addition, smoking at bars is generally considerably heavier than in restaurants. Therefore, outdoor air requirements are relatively high by comparison.

Nightclubs. Nightclubs may include a restaurant, bar, stage, and dancing area. The bar should be treated as a separately zoned area, with its own supply and exhaust system. People in the restaurant area who dine and dance may require twice the air changes and cooling than patrons who dine and then watch a show. The length of stay in nightclubs generally exceeds that encountered in most eating places. In addition, eating in nightclubs is usually secondary to drinking and smoking. Patron density usually exceeds that of conventional eating establishments.

Kitchens. The kitchen has the greatest concentration of noise, heat load, smoke, and odors; ventilation is the chief means of removing them and preventing these objectionable elements from entering dining areas. Kitchen air pressure should be kept negative relative to other areas, to ensure odor control. Maintenance of reasonably comfortable working conditions is important.

OFFICE BUILDINGS

Load Characteristics

Office buildings usually include both peripheral and interior zone spaces. The peripheral zone may be considered as extending from 12 to 18 ft inward from the outer wall toward the interior of the building and frequently has a large window area. These zones may be extensively subdivided. Peripheral zones have variable cooling loads in summer because of changing sun position and weather. These zone areas also require heating in winter. During intermediate seasons, one side of the building may require cooling, while another side simultaneously requires heating. However, the interior zone spaces whose thermal loads are derived almost entirely from lights, office equipment, and people require a fairly uniform cooling rate throughout the year. Often, interior space conditioning is done by an independent system, which has variable air volume control for low or no-load conditions.

Most office buildings are occupied from approximately 8:00 A.M. to 6:00 P.M.; many are occupied by some personnel from as early as 5:30 A.M. to as late as 7:00 P.M. Some tenants' operations may require night work schedules, usually not to extend beyond 10:00 P.M. Office buildings may contain printing plants, communications operations, broadcasting studios, and computing centers, which could operate 24 h a day. Therefore, for economical air-conditioning design, the intended uses of an office building must be very well established before design development.

Occupancy will vary considerably. In accounting or other sections where clerical work is done, the maximum density is approximately one person per 75 ft^2 of floor area. Where there are private offices, the density may be as little as one person per 200 ft^2. The most serious cases, however, are the occasional waiting rooms, conference rooms, or director's rooms where occupancy may be as high as one person per 20 ft^2.

The lighting load in an office building constitutes a significant part of the total heat load. Lighting and normal equipment electrical loads average from 2 to 5 W/ft^2 but may be considerably higher, depending on lighting type and the extent of equipment. Buildings with computer systems and other electronic equipment can have electrical loads as high as 5 to 10 W/ft^2. An accurate appraisal should be made of the amount, size, and type of computer equipment anticipated for the life of the building to size the air-handling equipment properly and provide for future installation of air-conditioning apparatus.

Where electrical loading is 6 W/ft^2 or more, heat should be withdrawn from the source by exhaust air or water tubing. About 30% of the total lighting heat output from recessed fixtures can be withdrawn by exhaust or return air and, therefore, will not enter into space conditioning supply air requirements. By connecting a duct to each fixture, the most balanced air system can be provided. However, this method is expensive, so the suspended ceiling is often used as a return air plenum with the air drawn from the space to above the suspended ceiling through the lights.

Miscellaneous allowances (for fan heat, duct heat pickup, duct leakage, and safety factors) should not exceed 12% of the total load.

Building shape and orientation are often determined by the building site, but variations in these factors can produce increases of 10 to 15% in refrigeration load. Shape and orientation should, therefore, be carefully analyzed in the early design stages.

Design Concepts

The variety of functions and range of design criteria applicable to office buildings have allowed the use of almost every available air-conditioning system. While multistory structures are discussed here, the principles and criteria are similar for all sizes and shapes of office buildings.

Attention to detail is extremely important, especially in modular buildings. Each piece of equipment, duct and pipe connections, and the like may be duplicated hundreds of times. Thus, seemingly minor design variations may substantially affect construction and operating costs. In initial design, each component must be analyzed not only as an entity, but as part of an integrated system. This systems design approach is essential to achieve optimum results.

There are several classes of office buildings, determined by the type of financing required and the tenants who will occupy the building. Design evaluation may vary considerably based on specific tenant requirements; it is not enough to consider typical floor patterns only. Included in many larger office buildings are stores, restaurants, recreation facilities, radio and television studios, and observation decks.

Built-in system flexibility is essential for office building design. Since business office procedures are constantly being revised, basic building services should be able to meet changing tenant needs.

The type of occupancy may have an important bearing on the selected air distribution system. For buildings with one owner or lessee, operations may be defined clearly enough so that a system can be designed without the degree of flexibility needed for a less well-defined operation. However, owner-occupied buildings may require considerable design flexibility, since the owner will pay for all alterations. The speculative builder can generally charge alterations to tenants. When different tenants occupy different floors, or even parts of the same floor, the degree of design and operation complexity increases to ensure proper environmental comfort conditions to any tenant, group of tenants, or all tenants at once. This problem is more acute where tenants have seasonal and variable overtime schedules.

Stores, banks, restaurants, and entertainment facilities may have hours of occupancy or design criteria that differ substantially from those of office buildings; therefore, they should have their own air distribution systems and, in some cases, their own refrigeration equipment.

Main entrances and lobbies are sometimes served by a separate system because they buffer the outside atmosphere and the building interior. Some engineers prefer to have a lobby summer temperature of 4 to 6°F above office temperature to reduce thermal shock to people entering or leaving the building. This also reduces operating costs.

The unique temperature and humidity requirements of data processing system installations, and the fact that they often run 24 h daily for extended periods, generally warrant separate refrigeration and air distribution systems. Separate backup systems may be required for data processing areas in case the main building HVAC system fails. Chapter 17 has further information.

The degree of air filtration required should be determined. The service cost and the effect air resistance has on energy costs should be analyzed for various types of filters. Initial filter cost and air pollution characteristics also need to be considered. Activated charcoal filters for odor control and reduction of outdoor air requirements is another option to consider.

There is seldom justification for 100% outdoor air systems for office buildings; therefore, most office buildings are designed to minimize outdoor air usage. However, recent attention to indoor air quality may dictate higher levels of ventilation air. In addition, a constant volume of ventilation air should be maintained in variable volume air-handling systems. Dry-bulb or enthalpy controlled economizer cycles should be considered for reducing energy costs.

When an economizer cycle is used, systems should be zoned so that energy abuse will not occur by having to heat outside air. This is often accomplished by a separate air distribution system for the interior and each major exterior zone.

Office buildings have traditionally used dual-duct, induction, or fan-coil systems. More recently, variable air volume systems and deluxe self-contained perimeter unit systems have also been used. Where fan-coil or induction systems have been installed at the perimeter, separate all-air systems have been generally used for the interior.

Many office buildings without an economizer cycle have a bypass multizone unit installed on each floor, with a heating coil in each exterior zone duct. Variable air volume variations of the bypass multizone and other floor-by-floor, all-air systems are also being used. These systems are popular because of low fan power, low initial cost, and energy savings, resulting from independent operating schedules, which are possible between floors occupied by tenants with different operating hours.

It may be more economical for smaller office buildings to have perimeter radiation systems with conventional, single-duct, low-velocity air-conditioning systems furnishing air from packaged air-conditioning units or multizone units. The need for a perimeter system should be carefully analyzed, since this system is a function of exterior glass percentage, external wall thermal value, and climate severity.

A perimeter heating system separate from the cooling system is preferable, since air distribution devices can then be selected for a specific duty, not a compromise between heating and cooling performance. The higher cost of additional air-handling or fan-coil units and ductwork may lead the designer to a less expensive option, such as fan-powered terminal units with heating coils serving perimeter zones in lieu of a separate heating system. Radiant ceiling panels for the perimeter zones are another option.

Interior space usage usually requires that interior air-conditioning systems allow modification to handle all load situations. Variable air volume systems have often been used. When using these systems, a careful evaluation of low load conditions should be made to determine if adequate air movement and fresh air can be provided without overcooling at the proposed supply air temperature. Increases in supply air temperature tend to nullify energy savings in fan power, which are characteristic of variable air volume systems.

In small to medium office buildings, air source heat pumps may be chosen. In larger buildings, internal source heat pump systems (water-to-water) are feasible with most types of air-conditioning systems. Heat removed from core areas is either rejected to a cooling tower or perimeter circuits. The internal source heat pump can be supplemented by a boiler on extremely cold days or over extended periods of limited occupancy. Removed excess heat may also be stored in hot water tanks.

Many heat recovery or internal source heat pump systems exhaust air from conditioned spaces through lighting fixtures. Approximately 30% of lighting heat can be removed in this manner. One design advantage is a reduction in required air quantities. In addition, lamp life is extended by operation in a much lower ambient temperature.

Suspended ceiling return air plenums eliminate sheet metal return air ductwork to reduce floor-to-floor height requirements. However, suspended ceiling plenums may increase the difficulty of proper air balancing throughout the building. Problems often connected with suspended-ceiling return plenums are as follows:

- Air leakage through cracks, with resulting smudges.
- Tendency of return air openings nearest shaft opening or collector duct to pull too much air, thus creating uneven air motion and possible noise.
- Air in suspended ceiling plenum may be blocked from return air shaft by beams or partitions.
- Effect on building structure fireproofing.
- Noise transmission between office spaces.

Air leakage can be minimized by proper workmanship. To overcome drawing too much air, return air ducts can be run in the suspended ceiling pathway from the shaft, often in a simple radial pattern. The ends of the ducts can be left open or dampered. Generous sizing of return air grilles and passages will lower the percentage of circuit resistance attributable to the return air path. This bolsters effectiveness of supply air-balancing devices and reduces the significance of air leakage and drawing too much air. Structural blockage can be solved by locating openings in beams or partitions with fire dampers, where required.

Total office building electromechanical space requirements are approximately 8 to 10% of the gross area. Clear height required for fan rooms varies from approximately 10 to 18 ft, depending on involved distribution system and equipment complexity. On typical office floors, perimeter units require approximately 1 to 3% of floor area, whereas interior shafts require 3 to 5%. Therefore, ducts, pipes, and equipment require approximately 3 to 5% of each floor's gross area. Electrical and plumbing space requirements per floor average an additional 1 to 3% of gross area.

Where large central units supply multiple floors, shaft space requirements depend on the number of fan rooms. In such cases, one mechanical equipment room usually furnishes air requirements for 8 to 20 floors (above and below for intermediate levels), with an average of 12 floors. The more floors served, the larger the duct shafts and equipment required. This results in higher fan room heights, greater equipment size and mass, and higher operating costs due to increased fan motor power.

The fewer floors served by an equipment room, the more equipment rooms will be required to serve the building. This axiom allows greater flexibility in serving changing floor or tenant requirements. Often, one mechanical equipment room per floor and complete elimination of vertical shafts requires no more total floor area than a few larger mechanical equipment rooms, especially when there are many small rooms and they are often the same height as typical floors. Equipment can also be smaller, although maintenance costs will be higher. Energy costs will be reduced, with more equipment rooms serving fewer areas, because the equipment can be shut off in unoccupied areas, and high-pressure ductwork will not be required.

Equipment rooms on upper levels generally cost more to install because of rigging and transportation logistics, but this cost must be balanced against lower level revenue and desirability of space by tenants.

In all cases, mechanical equipment rooms must be thermally and acoustically isolated from office areas.

Cooling towers are the largest single piece of equipment required for air-conditioning systems. Cooling towers require approximately 1 ft^2 of floor area per 400 ft^2 of total building area and are from 13 to 40 ft high. When towers are located on the roof, the building structure must be capable of supporting the cooling tower and dunnage, full water load (approximately 120 to 150 lb/ft^2), and wind load stresses.

Where cooling tower noise may affect neighboring buildings, towers should be designed to include sound traps or other suitable noise baffles. This may affect tower space, mass of the units, and motor power. Slightly oversizing cooling towers can also reduce noise and power consumption due to lower speeds, but this may increase initial cost.

Cooling towers are sometimes enclosed in a decorative screen for aesthetic reasons; therefore, calculations should ascertain that the screen has sufficient free area for the tower to obtain its required air quantity and to prevent recirculation.

If the tower is placed in a rooftop well, near a wall, or split into several towers at various locations, design becomes more complicated and initial and operating costs increase substantially. Also, towers should not be split and placed on different levels because hydraulic problems increase. Finally, the cooling tower should be built high enough above the roof so that the bottom of the tower and the roof can be maintained properly.

Special Considerations

Office building areas with special ventilation requirements include elevator machine rooms, electrical and telephone closets, electrical switchgear, plumbing rooms, refrigeration rooms, and mechanical equipment rooms. The high heat loads in some of these rooms may require air-conditioning units for spot cooling.

In larger buildings having intermediate elevator, mechanical, and electrical machine rooms, it is desirable to have these rooms on the same level or possibly on two levels. This may simplify the horizontal ductwork, piping, and conduit distribution systems and permit more effective ventilation and maintenance of these equipment rooms.

An air-conditioning system cannot prevent occupants at the perimeter from feeling direct sunlight. Venetian blinds and drapes are often provided but seldom used. External shading devices (screens, overhangs, etc.) or reflective glass are preferable.

Tall buildings in cold climates experience severe stack effect. No amount of additional heat provided by the air-conditioning system can overcome this problem. Features to help combat infiltration due to stack effect include:

- Revolving doors or vestibules at exterior entrances
- Pressurized lobbies
- Tight gaskets on stairwell doors leading to the roof
- Automatic dampers on elevator shaft vents
- Tight construction of exterior skin
- Tight closure and seals on all dampers opening to the exterior

LIBRARIES AND MUSEUMS

In general, libraries have stack areas, working and office areas, a main circulation desk, reading rooms, rare book vaults, and small study rooms. Many libraries also contain seminar and conference rooms, audiovisual rooms, record and tape listening rooms, special exhibit areas, computer rooms, and perhaps an auditorium. This wide diversity of functions requires careful analysis to provide proper environmental conditions.

Museums fall into several categories including:

- Art museums and galleries
- Natural and social history
- Scientific
- Specialized topics

In general, museums have exhibit areas, work areas, offices, and storage areas. Some of the larger museums may have shops, a restaurant, etc., but these areas are not basic to this type of building and are discussed in other sections.

Specialized topic museums, such as reconstructed or preserved residences, or industrial museums showing product development and growth, usually have less complex air-conditioning requirements. In some scientific museums, the necessity for reproducing the results of various exhibits or experiments may require close environmental control.

Most art museums and galleries, and some natural history museums, have their exhibits exposed within the viewing area. However, some exhibits are kept in enclosed cases, cubicles, or rooms. These exhibits may require special conditions that differ markedly from human comfort requirements. In this case, separate systems may be set up for maintaining the proper temperature and humidity.

Work areas in art museums consist of rooms for restoration and touch-up, picture framing, sculpture mounting, and repair. Paints, chemicals, plaster of paris, and other materials requiring special temperature, humidity, and air circulating conditions are used. Noise level is not critical but should not be objectionable to occupants.

A greater variety of functions, such as animal stuffing and reconstruction of fossils or cultural exhibits, may be performed in the work areas of natural and social history museums. Some museums have research facilities and laboratories, and odors and chemicals in these areas may require larger exhaust air quantities. Individual room or area zone control will generally be necessary.

Storage areas in most museums contain large numbers of articles for which exhibit space is not available, or articles that must be repaired. These areas may have to be kept within fairly close environmental conditions.

Load Characteristics

Many libraries, especially college libraries, operate up to 16 h a day and may run the air-conditioning equipment about 5000 h a year. Such constant usage requires the selection of heavy-duty, long life equipment, which requires little maintenance. Museums are generally open about 8 to 10 h a day, 5 to 7 days a week, and many people who visit museums do not remove their outer clothing.

The ambient conditions should not vary in temperature or relative humidity. The conditions should remain constant 24 h a day year-round. Cold or hot walls and windows, and hot steam or water pipes should be avoided. Object humidity may be destructive, even if the ambient relative humidity is under control. If the ambient dry-bulb temperature varies or if the collection is subjected to radiant effects, the temperature of objects will vary, always lagging behind the atmospheric changes.

Some of the specific factors of particular importance in determining the heating and cooling loads follow.

Sun Gain. Libraries and museums usually have windows, sometimes of stained glass, and skylights—more in traffic areas, than in book stacks or storage areas. Care must be taken to minimize the effects of sun; shortwave (actinic) rays are particularly injurious. Heat gain from skylights, often over artificially lighted frosted glass ceilings, can be reduced by a separate forced ventilation system.

Transmission. In winter, effects on objects located close to outside walls and possible condensation of moisture on the objects and the surface of outside walls must be evaluated. In summer, possible radiant effects from exposure should be considered.

People. Some areas may have concentrations as high as 10 ft^2 per person, while office space will have closer to 100 or 120 ft^2 per person, and book stack areas up to 1000 ft^2 per person.

When smoking is permitted, return air should be contained, and the recirculated part of the air should be deodorized with activated charcoal and similar odor-removal devices, or exhausted.

Lights. A detailed examination should be made of wattage provided in various rooms and the length of operation of the lights. In book stacks, various storage rooms, and vaults, lights may be discounted completely because of occasional use.

Stratification. Main reading rooms, large entrance halls, and large art galleries often have high ceilings that may allow the air temperature to stratify.

After individual room loads are evaluated at their optimum values to determine air requirements, the instantaneous refrigeration load should be calculated using proper diversity and storage factors.

Design Concepts

All-air systems are preferred in library areas where steam or water may ruin rare books, manuscripts, tapes, and so forth. This is also true for museums, because exhibit items are generally irreplaceable. However, there are many libraries that have used air-water systems with satisfactory results.

In some libraries with auditoriums, it is possible to use the spaces under the seats and behind the backs of the seats for handling supply and return air distribution. In picture galleries, the wall space below the rail height may be made available. Under some circumstances, the space above and below the wall cases may be used. In book stacks, each tier or deck should have individual air supply and return. Newly constructed book stacks would probably be interior spaces, with air ductwork and other services worked into the steel shelf-supporting structure. The most important consideration in designing air distribution should be to avoid stagnant spaces in all exhibits, especially in storage rooms and vaults. Steam or water piping should not run through exhibit or storage rooms, to avoid possible damage by accidental leaks and radiation.

In museums, patron traffic may follow a planned or random pattern, depending on the size of the museum, the number of exhibits and people, or the organization of the exhibits. The pattern may affect the type of air-conditioning system. People loads vary, depending on whether there is a new exhibit, the time of day, weather, and other factors. Thus, individually controlled zones are required to maintain optimal environmental conditions.

The most difficult problem encountered in designing an air-conditioning system for a museum is that partitioned areas may be radically changed from one exhibit to another. Attempts to establish a modular system for partitions have been only partially successful because of the wide range of sizes of items in the exhibits. Air distribution and lighting systems must be set up in the most flexible manner possible to minimize problems.

In art museums, particularly, partitions may create local pockets with hot air supply or exhaust; transfer grilles may be placed in the partitions to obtain some air movement.

Another problem is the location of room thermostats and humidistats. Sometimes it is not practicable to locate them either in the room or in the common return air duct because conditions may be typical of only a small area. One solution is to have the basic floor set up with small, individual zones. This, of course, is one of the most costly solutions. Other potential solutions are to locate the thermostats in return-air ducts and on aspirating diffusers.

Special Considerations

Many old manuscripts, books, museum exhibits, and works of art have been damaged or destroyed because they were not kept in a properly air-conditioned environment. The need for better preservation of such valuable materials together with a rising popular interest in using libraries and museums, requires that most of them, whether new or existing, be air conditioned.

Air-conditioning problems for museums and libraries are generally similar, but differ in design concept and application. The temperature and humidity ranges that are best for books, museum exhibits, and works of art do not usually fall within the human comfort range. Thus, compensations must be made to balance the value of preserving contents against human comfort, as well as initial and operating costs of air conditioning.

Design Criteria. In an average library or museum, less stringent design criteria are usually provided than for archives, because the value of the books and collections does not justify the higher initial and operating costs. Low-efficiency air filters are often provided. Relative humidity is held below 55%. Room temperatures are held within the 68 to 72 °F range.

Archival libraries and museums should have 85% or better air filtration, a relative humidity of 35% for books, and temperatures of 55 to 65 °F in book stacks and 68 °F in reading rooms. Canister-type filters or spray washers should be installed if chemical pollutants are present in the outdoor air.

Art storage areas are often maintained at 60 to 72 °F or lower, and 50% rh (±2%). Stuffed, fur-bearing animals should be stored at about 40 to 50 °F and 50% rh for maximum preservation; fossils and old bones are better preserved at higher humidities.

Museum authorities should be consulted to ensure optimal conditions for specific collections.

Building Contents. Because preservation of the collections housed within these buildings is so important, the reaction of each of the materials in the collections to room conditions should be carefully considered.

Paper used in books and manuscripts before the eighteenth century was very stable and was not significantly affected by room environment. Produced by a cottage industry, paper was made in small lots by breaking down the wood fibers by stamping, by using naturally alkaline water, and by applying a gelatin sizing. Industrialized production, in which wood fibers were cut with steel knives, ordinary water was used, and rosin sizing was substituted for gelatin, made a paper susceptible to deterioration because of its acid content. For archival preservation, this paper should be stored at very low temperatures. It is estimated that for each 10 °F dry bulb the room temperature is lowered, the life of the paper will double, and that any humidity reduction will also lengthen the life of paper.

Libraries, however, house more than books; they also store and use films and tapes. The dessication point for microfilm and magnetic tape is below 35% rh. This, then, is the lower limit for relative humidity in libraries, with the optimum humidity just above this point to minimize paper deterioration. The upper limit for humidity when the room dry-bulb temperature exceeds 65 °F is 67% rh, because mold forms above this point.

Many materials housed in museums are organically based and also benefit from lower room temperatures. Museums that display only part of their collection at one time and keep the rest in storage rooms should consider reducing storage temperature to a point below the comfort chart to lengthen the life of organic materials.

Chemical pollutants and dirt are other factors that affect the preservation of books and organic materials. Chemical pollutants causing the greatest concern include sulfur dioxide, the oxides of nitrogen, and ozone. Electrostatic filtration of the air is not recommended because it can generate ozone. Sulfur dioxide combines with water to form sulfuric acid. In the past, sulfur dioxide was removed from outdoor air by using a washer in which the spray water was kept at a pH value between 8.5 and 9.0. Some museum and library system designs use special canister-type filters to remove sulfur dioxide.

Effect of Ambient Atmosphere. The temperature and, particularly, the relative humidity (not humidity ratio) of the air have a marked influence on the appearance, behavior, and general quality of hygroscopic materials such as paper, textiles, wood, and leather, because the moisture content of these substances comes into equilibrium with the moisture content of the surrounding air.

This process is of particular importance because it can multiply the destructive effect of changes in the ambient atmosphere. If any object in a collection (such as a book, painting, tapestry, or other article on exhibit) is at a temperature higher or lower than that of the air in the museum or library, the relative humidity of the air close to the object will differ considerably from that of the ambient room air.

The *object humidity* is the relative humidity of the thin film of air in close contact with the surface of an object and at a temperature cooler or warmer than the ambient dry bulb (Banks 1974). Object humidity differs from that of the ambient air because the dry-bulb temperature of various layers of the air film approaches that of the object, while the dew-point temperature remains constant.

If objects in a museum are permitted to cool overnight, the next day they will be enveloped by layers of air having progressively higher relative humidities. These may range from the ambient of 45 to 60% to 97% immediately next to the object surface, thus effecting a change in material regain or even condensation. This, combined with the hygroscopic or salty dust often found on objects recovered from excavations, can be destructive. If the particular material is warmed, however, the object's humidity will be lower than the humidity of the surrounding space. This warming may be caused by spotlights or any hot, radiating surface.

Sound and Vibration. Air-conditioning equipment should be treated with sound and vibration isolation to ensure quiet comfort for visitors and staff. Acoustical isolation is also necessary to avoid transmitting or setting up resonant (sympathetic) vibration within objects on exhibit, which may be damaged by such motion. Sound level should be low, but not so low as to produce an environment where normal sounds will be objectionable. It should also be noted that exhibit spaces tend to be acoustically reverberant (see Chapter 42).

Case Breathing. Many objects and exhibits are housed in cases, and unless they are sealed tightly, the cases tend to breathe (Banks 1974). For example, the Declaration of Independence and the U.S. Constitution are inscribed on sheepskin parchment and enclosed within sealed receptacles filled with helium. The air in a case expands and contracts as air temperature is changed by variation in ambient temperature, lights, and atmospheric pressure. Consequently, the air within the case will change more or less frequently. The worst offender in this instance may be a spotlight thought to be installed far enough from the case as to have no effect.

Special Rooms. When a library or museum has seminar and conference rooms, audiovisual, record and tape listening rooms, and special exhibit areas, individual environmental room control is needed. Seminar and conference rooms may be exposed to heavy smoking, so auxiliary 100% exhaust should be provided. The other rooms listed may require a slightly quieter environment. Separate temperature and humidity controls should be provided for record and tape storage rooms.

Location of mechanical equipment rooms and air-handling equipment should be as remote as possible from the reading and exhibit areas to minimize the need for expensive sound and vibration isolation measures.

BOWLING CENTERS

Bowling centers may also contain a bar, restaurant, children's play area, offices, locker rooms, and other types of facilities. Such auxiliary areas are not discussed in this section, except as they may affect design for the bowling alley, which consists of alleys and a spectator area.

Load Characteristics

Bowling alleys usually have their greatest period of use in the evenings, but weekend daytime use may also be heavy. Thus, when designing for the peak air-conditioning load on the building, it is necessary to compare the day load and its high outside solar and off-peak people loads with the evening peak people load with no solar loads. Since bowling areas generally have little fenestration, the solar load may not be important.

If the building contains auxiliary areas, these areas may be included in the refrigeration and air distribution systems for the bowling alleys, with suitable provisions for zoning the different areas, as load analysis may dictate. Alternatively, separate systems may be established for each area having different load operation characteristics.

Heat buildup due to lights, external transmission load, and pinsetting machinery in front of the foul line can be reduced by exhausting some air above the alleys or from the area of the pinsetting machines; however, this gain should be compared against the cost of conditioning additional makeup air. In the calculation of the air-conditioning load, a portion of the unoccupied alley space load is included. Since this consists mainly of lights and some transmission load, about 15 to 30% of this heat load may have to be taken into account. The higher figure may apply when the roof is poorly insulated, no exhaust air is taken from this area, or no vertical baffle is used at the foul line. One estimate is 5 to 10 Btu/h · ft^2 of vertical surface at the foul line, depending mostly on the type and intensity of the lighting.

The heat load from bowlers and spectators may be found in Chapter 26 of the 1989 ASHRAE *Handbook—Fundamentals*. The proper heat gain should be applied for each person to avoid too large a design heat load.

Design Concepts

As with other building types having high occupancy loads, heavy smoke and odor concentration, and low sensible heat factors, all-air systems are generally the most suitable for bowling alley areas. Since most bowling alleys are almost windowless structures except for such areas as entrances, exterior restaurants, and bars, it is uneconomical to use terminal unit systems because of the small number required. Where required, radiation in the form of baseboard or radiant ceiling panels is generally placed at perimeter walls and entrances.

It is not necessary to maintain normal indoor temperatures down the length of the alleys; temperatures may be graded down to the pin area. Unit heaters are often used at this location.

Air Pressurization. Spectator and bowling areas must be well shielded from entrances so that no cold drafts are created in these areas. To minimize infiltration of outdoor air into the alleys, the exhaust and return air system should handle only 85 to 90% of the total supply air, thus maintaining a positive pressure within the space.

Air Distribution. Packaged units without ductwork produce uneven space temperatures, and unless they are carefully located and installed, they may cause objectionable drafts. Central ductwork systems are recommended for all but the smallest buildings, even where packaged refrigeration units are used. Since only the areas behind the foul line are air conditioned, the ductwork should provide comfortable conditions within this area.

The return and exhaust air systems should have a large number of small registers uniformly located at high points, or pockets, to draw off the hot, smoky, and odorous air. In some parts of the country and for larger bowling alleys, it may be desirable to use all outdoor air to cool during intermediate seasons.

Special Considerations

People in sports and amusement centers engage in a high degree of physical activity, which makes them feel warmer and increases their rate of evaporation. In these places, odor and smoke control are important environmental considerations.

Bowling alleys are characterized by the following:

- A large number of people concentrated in a relatively small area of a very large room. A major portion of the floor area is unoccupied.
- Heavy smoking, high physical activity, and high latent heat load.
- Greatest use from about 6:00 to 12:00 P.M.

The first two items make it mandatory that large amounts of outdoor air be furnished to minimize odors and smoke in the atmosphere.

The area between the foul line and the bowling pins need not be air conditioned or ventilated. Transparent or opaque vertical partitions are sometimes installed to separate the upper portions of the occupied and unoccupied areas so that air distribution is better contained within the occupied area.

COMMUNICATION CENTERS

Communication centers include telephone terminal buildings, teletype centers, radio stations, television studios, and transmitter and receiver stations.

Most telephone terminal rooms are air conditioned because constant temperature and relative humidity help prevent breakdowns and increase equipment life. In addition, air conditioning permits the use of a lower number of air changes, which, for a given filter efficiency, decreases the chances of damage to relay contacts and other delicate equipment.

Teletype centers are similar to telephone terminal rooms except that, since people operate the teletype machines, more care is required in the design of the air distribution systems.

Radio and television studios require critical analysis for the elimination of heat buildup and control of noise. Television studios have the added problem of air movement, lighting, and occupancy load variations. This section deals with television studios, since they encompass most of the problems found in radio studios.

Load Characteristics

The air-conditioning load for telephone terminal rooms is primarily equipment heat load, since human occupancy is limited. Teletype centers are similar, except for the load from people who operate the teletype machines.

Television studios have very high lighting capacities, which may fluctuate considerably in intensity over short periods. The operating hours may vary every day. In addition, there may be from one to several dozen people on stage for short times. The air-conditioning system must be extremely flexible and capable of handling wide load variations quickly, accurately, and efficiently, similar to the conditions of a theater stage. The studio may also have an assembly area with a large number of spectator seats. Generally, studios are located so that they are completely shielded from external noise and thermal environments.

Design Concepts

The critical areas of a television studio consist of the performance studio and control rooms. The audience area may be handled in a manner similar to that for a place of assembly. Each area should have its own air distribution system or at least its own zone control separate from the studio system. The heat generated in the studio area should not be allowed to permeate the audience atmosphere.

The air distribution system selected must have the capabilities of a dual-duct, single-duct with cooling and heating booster coils, variable air volume, or multizone system to satisfy design criteria. The air distribution system should be designed so that simultaneous heating and cooling cannot occur, unless such heating is achieved solely by heat recovery methods.

Studio loads seldom exceed 100 tons of refrigeration. Even if the studio is part of a large communications center or building, it is desirable for the studio to have its own refrigeration system in case of emergencies. In this size range, the refrigeration machine may be of the reciprocating type, which requires a remote location so that machine noise is isolated from the studio.

Special Considerations

On-Camera Studios. This is the stage of the television studio and requires the same general considerations as a concert hall stage. Air movement must be uniform and, since scenery, cameras, and equipment may be moved during the performance, ductwork must be planned carefully to avoid interference with proper studio functioning.

Control Rooms. Each studio may have one or more control rooms serving different functions. The video control room, which is occupied by the program and technical directors, contains monitoring sets and picture-effect controls. The room may require up to 30 air changes per hour to maintain proper conditions. The large number of air changes required and the low sound level that must be maintained require the air distribution system to be specially analyzed.

If a separate control room is furnished for the announcer, the heat load and air distribution problems will not be as critical as those for the program, technical, and audio directors.

Thermostatic control should be furnished in each control room, and provisions should be made to enable occupants to turn the air conditioning on and off.

Noise Control. Studio microphones are moved throughout the studio during a performance, so they may be moved past or set near air outlets or returns. These microphones are considerably more sensitive than the human ear; therefore, air outlets or returns should be located away from areas where microphones are likely to be used. Even a leaky pneumatic thermostat can be a problem.

Air Movement. It is essential that air movement within the stage area, which often contains scenery and people, be kept below 25 fpm within 12 ft of the floor. The scenery is often fragile and will move in air velocities above 25 fpm; also, actors' hair and clothing may be disturbed.

Air Distribution. Ductwork must be fabricated and installed so that there are no rough edges, poor turns, or improperly installed dampers to cause turbulence and eddy currents within the ducts. Ductwork should contain no holes or openings that might create whistles. Air outlet locations and the distribution pattern must be carefully analyzed to eliminate turbulence and eddy currents within the studio that might cause noise that could be picked up by studio microphones.

At least some portions of supply, return, and exhaust ductwork will require acoustical material to maintain noise criteria (NC) levels from 20 to 25. Any duct serving more than one room should acoustically separate each room by means of a sound trap. All ductwork should be suspended by means of neoprene or rubber in shear-type vibration mountings. Where ductwork goes through wall or floor slabs, the openings should be sealed with acoustically deadening material. The supply fan discharge and the return and exhaust fan inlets should have sound traps; all ductwork connections to fans should be made with nonmetallic, flexible material. Air outlet locations should be coordinated with ceiling-mounted tracks and equipment. Air distribution for control rooms may require a perforated ceiling outlet or return air plenum system.

Piping Distribution. All piping within the studio, as well as in adjacent areas that might transmit noise to the studio, should be supported by suitable vibration isolation hangers. To prevent transmission of vibration, piping should be supported from rigid structural elements to maximize absorption.

Mechanical Equipment Rooms. These rooms should be located as remotely from the studio as possible. All equipment should be selected for very quiet operation and should be mounted on suitable vibration-eliminating supports. Structural separation of these rooms from the studio is generally required.

Offices and Dressing Rooms. The functions of these rooms are quite different from each other and from the studio areas. It is recommended that such rooms be treated as separate zones, with their own controls.

Air Return. Whenever practicable, the largest portion of studio air should be returned over the banks of lights. This is similar to theater stage practice. Sufficient air should also be removed from studio high points to prevent heat buildup.

TRANSPORTATION CENTERS

The major transportation facilities are airports, ship docks, bus terminals, and passenger car garages. Airplane hangars and freight and mail buildings are also among the types of buildings to be considered. Freight and mail buildings are usually handled as standard warehouses.

Load Characteristics

Airports, ship docks, and bus terminals operate on a 24-h basis, although on a reduced schedule during late evening and early morning hours.

Airports. Terminal buildings consist of large, open circulating areas, one or more floors high, often with high ceilings, ticketing counters, and various types of stores, concessions, and convenience facilities. Lighting and equipment loads are generally average, but occupancy varies substantially. Exterior loads are, of course, a function of architectural design. The largest single problem often results from thermal drafts created by large entranceways, high ceilings, and long passageways, which have many openings to the outdoors.

Ship Docks. Freight and passenger docks consist of large, high-ceilinged structures with separate areas for administration, visitors, passengers, cargo storage, and work. The floor of the dock is usually exposed to the outdoors just above the water level. Portions of the side walls are often open while ships are in port. In addition, the large portion of ceiling (roof) area presents a large heating and cooling load. Load characteristics of passenger dock terminals generally require the roof and floors to be well insulated. Occasional heavy occupancy loads in visitor and passenger areas must be considered.

Bus Terminals. This building type consists of two general areas: the terminal building, which contains passenger circulation, ticket booths, and stores or concessions, and the bus loading area. Waiting rooms and passenger concourse areas are subject to a highly variable people load. Occupancy density could reach 10 ft^2 per person, and, at extreme periods, 3 to 5 ft^2 per person.

Design Concepts

Since heating and cooling plants may be centralized or provided for each building or group in a complex, these will not be discussed. In large, open circulation areas of transportation centers, any all-air system with zone control can be used. Where ceilings are high, air distribution will often be along the side wall to concentrate the air conditioning, where desired, and avoid disturbing stratified air. Perimeter areas may require heating by radiation, a fan-coil system, or hot air blown up from the sill or floor grilles, particularly in colder climates. Hydronic perimeter radiant ceiling panels may be especially suited to these high load areas.

Airports. Airports generally consist of one or more central terminal buildings connected by long passageways, or trains, to rotundas containing departure lounges for airplane loading. Most terminals have portable telescoping-type loading bridges connecting departure lounges to the airplanes. These passageways eliminate the heating and cooling problems associated with traditional permanent structure passenger loading.

Because of difficulties in controlling the air balance resulting from the many outside openings, high ceilings, and long, low passageways (which often are not air conditioned), the terminal building (usually air conditioned) should be designed to maintain a substantial positive pressure. Zoning will generally be required in passenger waiting areas, departure lounges, and at ticket counters to take care of the widely variable occupancy loads.

Main entrances may be designed with vestibules and windbreaker partitions to minimize undesirable air currents within the building.

Hangars must be heated in cold weather, and ventilation may be required to eliminate possible fumes (although fueling is seldom permitted in hangars). Gas-fired, electric, and low- and high-intensity heaters are used extensively in hangars because they provide comfort for employees at relatively low operating costs.

Hangars may also be heated by large air blast heaters or floor-buried heated liquid coils. Local exhaust air systems may be used to evacuate fumes and odors that result in smaller ducted systems. Under some conditions, exhaust systems may be portable and may possibly include odor-absorbing devices.

Ship Docks. In severe climates, occupied floor areas may contain heated floor panels. The roof should be well insulated and, in appropriate climates, evaporative spray cooling substantially reduces the summer load. Freight docks are usually heated and well ventilated but seldom cooled.

High ceilings and openings to the outdoors may present serious draft problems unless the systems are designed properly. Vestibule entrances or air curtains help minimize cross drafts. Air door blast heaters at cargo opening areas may be quite effective.

Ventilation of the dock terminal should prevent noxious fumes and odors from reaching occupied areas. Therefore, occupied areas should be under a positive pressure and the cargo and storage areas exhausted to maintain negative air pressure. Occupied areas should be enclosed to simplify the possibility of local air conditioning.

In many respects, these are among the most difficult buildings to heat and cool because of their large open areas. If each function is properly enclosed, any commonly used all-air or large fan-coil system could be suitable. If areas are left largely open, the best approach is to concentrate on proper building design and the heating and cooling of the openings. High-intensity infrared spot heating can often be advantageous (see Chapter 29 of the 1988 ASHRAE *Handbook—Equipment*). Exhaust ventilation from tow truck and cargo areas should be exhausted through the roof of the dock terminal.

Bus Terminals. Conditions are similar to those for airport terminals, except that all-air systems are more practical since ceiling heights are often lower, and perimeters are usually flanked by stores or office areas. The same types of systems are applicable as for airport terminals, but ceiling air distribution will generally be feasible.

Properly designed radiant hydronic or electric ceiling systems may be used if high occupancy latent loads are fully considered. This may result in smaller duct sizes than are required for all-air systems and may be advantageous where bus loading areas are above the terminal and require structural beams. This heating and cooling system reduces the volume of the building. In areas where latent load is a concern, heating-only panels may be used at the perimeter, with a cooling-only interior system.

The terminal area air-supply system should be under a high positive pressure to assure that no fumes and odors infiltrate from bus areas. Positive exhaust from bus loading areas is essential for a properly operating total system (see Chapter 13).

Special Considerations

Airports. Filtering of outdoor air with activated charcoal filters should be considered for areas subject to excessive noxious fumes from jet engine exhausts. However, locating outside air intakes as remotely as possible from airplanes is a less expensive and more positive approach.

Where ionization filtration enhancers are used, outdoor air quantities are sometimes reduced due to cleaner air. However, care must be taken to maintain sufficient amounts of outside air for space pressurization.

Ship Docks. Ventilation design must ensure that fumes and odors from forklifts and cargo in work areas do not penetrate occupied and administrative areas.

Bus Terminals. The primary concerns with enclosed bus loading areas are health and safety problems, which must be handled by proper ventilation (see Chapter 13).

Although diesel engine fumes are generally not as noxious as gasoline fumes, bus terminals often have many buses loading and unloading at the same time, and the total amount of fumes and odors may be quite disturbing.

Enclosed Garages. From a health and safety viewpoint, enclosed bus loading areas and car garages present the most serious problems in these buildings (see Chapter 13).

Three major problems are encountered. The first and most serious is carbon monoxide gas (CO) emission by cars and buses, which can cause serious illness and, possibly, death. The second problem is oil and gasoline fumes, which may cause nausea and headaches and can also create a fire hazard. The third involves lack of air movement and the resulting stale atmosphere that develops because of the increased carbon dioxide (CO_2) content in the air. This condition may cause headaches or grogginess.

Most codes require a minimum of four to six air changes per hour. This is predicated on maintenance of a maximum safe CO concentration in the air, assuming short periods of occupancy in the garage.

All underground garages should have facilities for testing the CO concentration or should have the garage checked periodically. Clogged duct systems, improperly operating fans, motors or dampers, clogged air intake or exhaust louvers, etc., may not allow proper air circulation. Proper maintenance is required to minimize any operational defects.

Carbon Monoxide Criteria. Minimum ventilation requirements, as set up by the National Institute of Standards and Technology (NIST) and ASHRAE, are primarily concerned with preventing buildup of noxious concentrations of carbon monoxide.

However, keeping the CO level within safe limits is no guarantee that patrons or employees in the garage will not experience discomfort. The air furnished will probably be sufficient to eliminate any atmospheric staleness, but the greatest discomfort can be caused by oil and gasoline fumes, which are particularly noticeable in areas with poor air circulation and at poorly ventilated ramps. Therefore, a properly designed air distribution system is essential to a comfortable and safe human environment.

WAREHOUSES

Warehouses are used to store merchandise and may be open to the public at times. They are also used to store equipment and material inventories as part of an industrial facility. The buildings are generally not air conditioned, but often have heat and ventilation sufficient to provide a tolerable working environment. Facilities such as shipping, receiving, and inventory control offices, associated with warehouses and occupied by office workers, are generally air conditioned.

Load Characteristics

Internal loads from lighting, people, and miscellaneous sources are generally low. Most of the load is thermal transmission and infiltration. An air-conditioning load profile would tend to flatten where materials stored are massive enough to cause the peak load to lag.

Design Concepts

Most warehouses are only heated and ventilated. Forced flow unit heaters may, in many instances, be located near heat entrances and work areas. Even though comfort for warehouse workers may not be desired, it may be necessary to keep the temperature above 40°F to protect sprinkler piping or stored materials from freezing.

Thermal qualities of the building, which would be included if the building might later be air conditioned, also minimize required heating and aid in comfort. For maximum summer comfort without air conditioning, excellent ventilation with noticeable air movement in work areas is necessary. Even greater comfort can be achieved in appropriate climates by adding roof spray cooling. This can reduce the roof's surface temperature by 40 to 60°F, thereby reducing ceiling radiation inside. Low- and high-intensity radiant heaters can be used to maintain minimum ambient temperature throughout a facility above freezing. Radiant heat may also be used for occupant comfort in areas permanently or frequently open to the outside.

Special Considerations

Powered forklifts and trucks using gasoline, propane, and other fuels are often used inside warehouses. Proper ventilation is necessary to alleviate the buildup of CO and other noxious fumes. Proper ventilation of battery-charging rooms for electrically powered forklifts and trucks is also required.

REFERENCES

Banks, P.N. 1974. Environmental standards for storage of books and manuscripts. *The Library Journal* 99 (3), February.

HUD *Bulletin* LR-11. Housing and Urban Development Agency, Washington, D.C.

Library of Congress. 1975. Environmental protection of books and related material. Library of Congress Preservation Leaflet No. 2 (February). Washington, D.C.

Smith, R. 1969. Paper impermanence as a consequence of pH and storage conditions. *Library Quarterly* 39 (2), April.

CHAPTER 4

PLACES OF ASSEMBLY

THIS chapter covers the design considerations associated with enclosed assembly buildings. (Chapter 3 covers general criteria that also apply to public assembly buildings.)

This chapter is organized into six sections. The first, "Common Characteristics," applies to all places of assembly. It includes information on general criteria and system considerations. Design criteria includes items such as ventilation lighting loads, indoor air conditions, filtration, and noise and vibration control.

The remaining sections apply to the following specific buildings: houses of worship, auditoriums, sports arenas, convention and exhibition centers, natatoriums, and fairs and other temporary exhibits. These sections include information on load characteristics, design concepts, criteria, and systems applicability, as appropriate.

COMMON CHARACTERISTICS

Assembly buildings are generally large, have relatively high ceilings, and are few in number for any given facility. They usually have a periodically high density of occupancy per unit floor area, as compared to other buildings, and thus have a relatively low design sensible heat ratio. Space volume per person is usually high, thus allowing fewer air changes than for many other building types.

GENERAL CRITERIA

Federal, state, and local guides relating to energy conservation have a major impact on system design as well as performance and require consideration.

Assembly buildings have relatively few actual hours of use per week. They may seldom be in full use when maximum outdoor temperatures or solar effects occur. The designer must obtain as much information as possible regarding the anticipated hours of use, particularly the times of full seating, so that simultaneous loads may be considered to obtain optimum air-conditioning loads and operating economy. These buildings often are fully occupied for as little as 1 to 2 h, and the load may be materially reduced by precooling. Latent cooling requirements should be considered before reducing equipment size. The intermittent or infrequent nature of the cooling loads may allow these buildings to benefit from thermal storage systems.

The occupants usually generate the major room cooling and ventilation load. The number of occupants is best determined from the seat count, but when this is not available, it can be estimated at 7.5 to 10 ft^2 per person for the entire seating area, including exit aisles but not the stage or performance areas or entrance lobbies.

The preparation of this chapter is assigned to TC 9.8, Large Building Air-Conditioning Applications.

Ventilation

Ventilation is a major contributor to total load. ASHRAE *Standard* 62-1989 lists outdoor air requirements for various occupancies. Typical minimum ventilation rates range from 15 to 60 cfm per person. These rates may be reduced somewhat if the facility is used for only short periods and where it can be flushed out between performances. Some building codes require higher ventilation rates.

Assembly buildings lend themselves to automatic recirculation and outdoor air control so that preoccupancy warm-up or cooling can be accomplished with low ventilation loads, and light occupancy use can be handled with reduced outdoor air loads.

The evaluation of ventilation loads is important in considering the effect of infiltration in any structure. Generally, sufficient outside air should be introduced into the air-handling system to offset the effect of infiltration and keep the structure under positive pressure, provided adjacent areas, such as ancillary facilities, are not adversely affected.

Lighting Loads

Lighting loads are one of the few major loads that vary from one type of assembly building to another. Lighting may be at the level of 150 footcandles in convention halls where color television cameras are expected to be used, but lighting is virtually absent during a presentation in a motion picture theater. In many assembly buildings, lights are controlled by dimmers or other means to present a suitably low level of light during performances, with much higher lighting levels during clean-up, when the house is near empty. The designer should ascertain what light levels will be associated with maximum occupancies, not only in the interest of economy but also to determine the proper room sensible heat ratio.

Indoor Air Conditions

Indoor air temperature and humidity should parallel the ASHRAE comfort recommendations (see Chapter 8 of the 1989 ASHRAE *Handbook—Fundamentals*); in addition, the following should be considered:

1. In sports arenas, gymnasiums, and some motion picture theaters, people generally dress informally in summer. The summer indoor conditions may favor the warmer end of the thermal comfort scale with no major ill effect. By the same reasoning, the winter indoor temperature may favor the cooler end of the scale.
2. In churches, concert halls, and legitimate theaters, people are usually fairly well dressed, with most men wearing jackets and ties and women often wearing suits. The temperature should favor the middle range of design, and there should be little summer-to-winter variation.

3. In convention and exhibition centers, the visiting public is continually walking. Here the indoor temperature should favor the lower range of comfort conditions both in summer and in winter.
4. For spaces of high population density or sensible heat factors of 0.75 or less, a lower dry-bulb temperature results in less latent heat from people, thus reducing the need for reheat and saving energy. Therefore, the optimum space dry-bulb temperatures should be the result of detailed design analysis.
5. Restrictions of energy-conserving codes must be considered in system design and operation.

Because of their low room sensible heat ratio, assembly areas generally require some form of reheat to maintain the relative humidity at a suitably low level during periods of maximum occupancy. Refrigerant hot gas or condenser water reject heat is frequently used for this purpose. Face and bypass control of low-temperature cooling coils is also effective. In colder climates, it may also be desirable to provide humidification. High rates of internal gain may make evaporative humidification attractive during economizer cooling.

Filtration

Most places of assembly are minimally filtered with filters rated at 30 to 35% efficiency, as tested in accordance with ASHRAE *Standard* 52-76. Where smoking is permitted, however, filters with a minimum rating of 80% are required before any effective amount of tobacco smoke is removed. Filters with 80% or higher efficiency are also recommended for those facilities having particularly expensive interior decor. Higher efficiency filters last longer, because of the few operating hours of these facilities, and can sometimes be economically justifiable for that reason alone. Low-efficiency prefilters are usually included with high-efficiency filters to extend their useful life. Ionization and chemically reactive filters should be considered where high concentrations of smoke or odors are present.

Noise and Vibration Control

The desired noise criteria (NC) vary with the type and quality of the facility. The need for noise control is minimal in a gymnasium or swimming amphitheater, but it is important in a quality concert hall. Some facilities are used for varied functions and require noise control evaluation over the entire spectrum of use.

In most cases, sound and vibration control is required for both equipment and duct systems, as well as in the selection of diffusers and grilles. When designing a project like a theater or concert hall, consultation with an experienced acoustics engineer is recommended. In these projects, the quantity and quality or characteristic of the noise is important.

Transmission of vibration and noise can be decreased by mounting pipes, ducts, and equipment on a separate structure independent of the music hall. If the mechanical equipment space is close to the music hall, it may be necessary to float the entire mechanical equipment room on isolators, including the floor slab, structural floor members, and other structural elements, supporting pipes, or similar materials that can carry vibrations. Properly designed inertia pads are often used under each piece of equipment, which is mounted on vibration isolators.

Manufacturers of vibration isolating equipment have devised methods to float large rooms and entire buildings on isolators. Where subway and street noise may be carried into the structure of the music hall, it is necessary to float the entire music hall on isolators. When the music hall is isolated from outside noise and vibration, it is necessary to isolate it from mechanical equipment and other internal noise and vibrations.

External mechanical equipment noise from equipment such as cooling towers should not be allowed to enter the building. Care should be taken to avoid air-conditioning designs that permit noises to enter the space through air intakes or reliefs and carelessly designed duct systems.

SYSTEM CONSIDERATIONS

Ancillary Facilities

Ancillary facilities are generally a part of any assembly building; almost all have some office space. Convention centers and many auditoriums and sports arenas have restaurants and cocktail lounges. Churches may have a religious school or apartments for the clergy. Many have parking structures. These varied facilities are mentioned individually in other chapters of this volume. However, for reasonable operating economy, these facilities should be served by separate systems when their hours of use are different from the main assembly areas.

Air-Conditioning Systems

Because of their characteristic large size and need for considerable ventilation air, assembly buildings are generally served by all-air systems, usually of the single-zone or variable-volume type. Separate air-handling units usually serve each zone, although multizone, dual-duct, or reheat types can also be applied with lower operating efficiency. In larger facilities, separate zones are generally provided for the entrance lobbies and arterial corridors that surround the seating space. In some assembly rooms, folding or rolling partitions divide the space for different functions, so a separate zone of control for each resultant space is best. In extremely large facilities, several air-handling systems may serve a single space, because of the limits of equipment size and also for energy and demand considerations.

Precooling

Cooling the building mass several degrees below the desired indoor temperature several hours before it is occupied allows it to absorb a portion of the peak heat load. This precooling reduces the equipment size needed to meet short-term loads. The effect can be used if precooling time (at least 1 h) is available prior to occupancy, and then only when the period of peak load is relatively short (2 h or less).

The designer must advise the owner that the space temperature is cold to most people as occupancy begins and then continues to climb as the performance progresses. This may be satisfactory, but it should be understood by all concerned before proceeding with a precooling design concept. Precooling is best applied when the space is used only occasionally during the hotter part of the day and when provision of full capacity for an occasional purpose is not economically justifiable.

Stratification

Because most applications involve relatively high ceilings, some heat may stratify above the occupied zone, thereby reducing the equipment load. Heat gain from lights can be stratified, except for the radiant effect (about 50% for fluorescent and 65% for incandescent or mercury-vapor fixtures). Similarly, only the radiant effect of upper wall and roof load (about 33%) reaches the occupied space. Stratification can be achieved only when air is admitted and returned at a sufficiently low elevation so that it does not mix with the upper air.

Conversely, stratification may increase heating loads during periods of minimal occupancy in winter months. In these cases, reduction of stratification by using ceiling fans, air-handling systems, or high/low air distribution may be desired. Balconies may also be affected by stratification and should therefore be well ventilated.

Air Distribution

In assembly buildings, people generally remain in one place throughout a performance and cannot avoid drafts. Good air distribution is essential to a successful installation.

Heating is seldom a major problem, except at entrances or during pre-occupancy warm-up. Generally, the seating area is isolated from the exterior by lobbies, corridors, and other ancillary spaces. It is practical to supply cooled air from the overhead space, where heat from lights can be directly absorbed and where much of the occupant heat can be aspirated and mixed with the supply air above the occupied zone. Return air openings can also aid air distribution. Air returns located below seating or at a low level around the seating can effectively distribute air with minimum drafts. Where returns are below the seats, register velocities in excess of 275 fpm may cause objectionable drafts and noise.

Because of the configuration of these spaces, it is sometimes necessary to install jet-type nozzles with long throw requirements of 50 to 150 ft for sidewall supplies. For ceiling distribution, downward throw is not critical provided returns are low. This approach has been successful in applications that are not particularly noise-sensitive, but the designer needs to carefully select air distribution nozzles. The application data should ensure proper performance for specific projects.

The air-conditioning systems must be quiet. This requirement is difficult to achieve if the air supply is expected to travel 30 ft or more from sidewall outlets to condition the center of the seating area. The large size of most houses of worship, theaters, and halls require high air discharge velocities from the wall outlets, which would create objectionable noise to those sitting near the outlets. Therefore, the concept of making the return air system do some of the work must be adopted, which means that the supply air is discharged from the air outlet (preferably at the ceiling) at the highest velocity consistent with an acceptable noise level. This velocity does not allow the conditioned air to reach the furthest portions of the audience. Therefore, return air registers, located in the vicinity of those not reached by the conditioned air, pull the air and cool or heat the audience, as required. In this way, the supply air blankets the seating area and is pulled down uniformly by the return air registers under or beside the seats.

A certain amount of exhaust air should be taken from the ceiling of the seating area, preferably over the balcony (when there is one) to prevent the formation of pockets of hot air, which can produce a radiant effect and cause discomfort, as well as increase the cost of air conditioning. Where the ceiling is close to the audience (*e.g.*, below balconies and mezzanines), specially designed plaques or air-distributing ceilings should be provided to absorb noise.

Regular ceiling diffusers placed more than 30 ft apart normally give acceptable results if careful engineering is applied in the selection of diffusers. Because large air quantities are generally involved and because the building is large, it is common to select fairly large capacity diffusers, which tend to be noisy due to the energy involved. Linear diffusers are more acceptable architecturally; they also perform well if selected properly. Integral dampers in diffusers should not be used as the sole means of balancing the system. These dampers, particularly in larger diffusers, generate intolerable amounts of noise.

Mechanical Equipment Rooms

The location of mechanical and electrical equipment rooms affects the degree of sound attenuation treatment required. Mechanical equipment rooms located near the seating area are more critical because of the normal attenuation of sound through space. Mechanical equipment rooms located near the stage area are critical because the stage is designed to project sound to the audience. If possible, mechanical equipment rooms should be located in an area separated from the main seating or stage area by buffers such as lobbies or service areas. The economies of the structure, attenuation, equipment logistics, and site must be considered in the selection.

At least one mechanical equipment room is placed near the roof to house the toilet exhaust, general exhaust, cooling tower, kitchen, and emergency stage exhaust fans, if any. Individual roof-mounted exhaust fans may be used, thus eliminating the need for a mechanical equipment room. However, to reduce sound problems, mechanical equipment should not be located on the roof over the music hall or stage but over offices, storerooms, or auxiliary areas.

HOUSES OF WORSHIP

Houses of worship seldom have full or near-full occupancy more than once a week, but they have considerable use for smaller functions (weddings, funerals, christenings, or daycare) throughout the balance of the week. It is important to determine how the building will be used, because this use varies considerably. When thermal storage systems are used, longer operation of equipment prior to occupancy may be required due to the high thermal mass of the structure. The seating capacity of houses of worship is usually well defined, except in those cases where a social function is separated from the main auditorium by a movable partition to form a single large auditorium for special holiday services. It is important to know when and how often this sort of maximum use is expected.

Because the design of many houses of worship is inspired by the classic Gothic cathedral, the high vaulted ceiling creates thermal stratification. Where stained glass is used, a shade coefficient is assumed to be approximately equal to solar glass (S.C. = 0.70).

Houses of worship test a designer's ingenuity to secure an architecturally acceptable solution to the problem of locating equipment and air-diffusion devices. When the occupants are seated, drafts and cold floors should be avoided.

Houses of worship may also have auxiliary rooms that should be air conditioned. The manner in which this is done depends on the relationship of the architectural layout and the systems selected to furnish the air conditioning. Privacy between adjacent areas is important in the air distribution scheme. Diversity in the total air-conditioning load requirements should be evaluated to take full advantage of the characteristics of each building.

In houses of worship and auditoriums, it is desirable to provide some degree of individual control for the platform, sacristy, and bema or choir area.

AUDITORIUMS

The types of auditoriums considered are the motion picture theater, the playhouse, and the concert hall. Other types of auditoriums in elementary schools and the large auditoriums found in some convention centers follow the same principles, with varying degrees of complexity.

Motion Picture Theaters

Motion picture theaters are the simplest of the auditorium structures mentioned here. They run continuously for periods of 4 to 8 h and, thus, are not a good choice for precooling techniques, except for the first matinee peak. They operate frequently at low occupancy levels, and low-load performance must be considered.

Motion picture sound systems make noise control less important than in other kinds of theaters. The lobby and exit passageways in a motion picture theater are seldom densely occupied, although some light to moderate congestion can be expected for short times in the lobby area. A reasonable design for the lobby space would be one person per 20 to 30 ft^2.

The lights are dimmed, since, at most times, the house is occupied; full lighting intensity is used only during cleaning. A reasonable judgment on lamp watts above the seating area during a performance would be 5 to 10% of the installed wattage. Designated smoking areas should be handled with separate exhaust or air-handling systems to avoid contamination of the entire facility.

Projection Booths

The projection booth represents the major problem in motion picture theater design. For large theaters using high intensity lamps, projection room design must follow applicable building codes. If no building code applies, the projection equipment manufacturer usually has specific requirements. The projection room may be air conditioned, but it is normally exhausted or operated at negative pressure. Exhaust is normally taken through the housing of the projectors. Additional exhaust is required for the projectionist's sanitary facilities. Other heat sources include the sound and dimming equipment, which requires a continuously controlled environment, necessitating a separate system.

Smaller theaters using 16-mm safety film have fewer requirements for projection booths. It is a good idea to condition the projection room with filtered supply air to avoid soiling lenses. Heat sources in the projection room include, in addition to the projector light, the sound equipment heat, as well as the dimming equipment heat—all of which are frequently located in the projection booth.

Legitimate Theaters

The legitimate theater differs from the motion picture theater in the following ways:

1. Performances are seldom continuous. Where more than one performance occurs in a day, the performances are separated by a period of 2 to 4 h. Accordingly, precooling techniques are applicable, particularly for afternoon performances.
2. Legitimate theaters seldom play to houses that are not full or near full.
3. Legitimate theaters usually have intermissions, and the lobby areas are used for drinking and socializing. Periods of occupancy are relatively short, however, seldom exceeding 15 to 20 min. During occupancy, the load may be as dense as one person per 5 ft^2.
4. Because sound amplification is less used than in a motion picture house, background noise control is more important.
5. Stage lighting contributes considerably to the total cooling load in the legitimate theater. Lighting loads can vary from performance to performance.

Stages

The stage presents the most complex problem. It consists of the following loads:

1. A heavy, mobile lighting load
2. Intricate or delicate stage scenery, which varies from scene to scene and presents difficult air distribution requirements
3. Actors performing tasks that require exertion

Approximately 40 to 60% of the lighting heat load can be negated by exhausting air around the lights. This procedure works for lights around the proscenium. However, for the light strips over the stage, it is more difficult to locate and arrange exhaust air ducts directly over the lights because of the scenery and light drops. Careful coordination is required to achieve an effective and flexible design layout.

Conditioned air should be introduced from the low side and back stages and returned or exhausted around the lights. Some exhaust air must be taken from the top of the tower directly over the stage containing lights and equipment (fly). The air distribution design is further complicated because pieces of scenery consist of light materials that flutter in the slightest air current. Even the vertical stack effect created by the heat from lights may cause this motion. Therefore, low air velocities are essential and air must be distributed over a wide area with numerous supply and return registers.

With multiple scenery changes, low supply or return registers from the floor of the stage are almost impossible to provide. However, some return air at the footlights and for the prompter should be considered. Air conditioning should also be provided for the stage manager and the control board areas.

One phenomenon encountered in many theaters with overhead flies is the billowing of the stage curtain when it is down. This situation is primarily due to the stack effect created by the height of the main stage tower, the heat from the lights, and the temperature difference between the stage and seating areas. Proper air distribution and balancing can minimize this phenomenon. Bypass damper arrangements with suitable fire protection devices may be feasible.

Loading docks adjacent to stages located in cold climates should be heated. The doors to this area may be open for long periods while scenery is being loaded or unloaded for a performance.

On the stage, local code requirements must be followed for emergency exhaust ductwork or skylight (or blow-out hatch) requirements. These openings are often sizable and should be incorporated in the early design concepts.

Concert Halls

Concert halls and music halls are similar to the legitimate theater in many ways. They normally have a full stage, complete with fly gallery, for the presentation of operas, ballet, and musical comedy shows. There are dressing areas for performers. Generally, the only differences between the two are size and decor, with the concert hall being larger and more plushly finished.

Air-conditioning design must consider that the concert hall is used frequently for special charity and civic events, which may be preceded by or followed by parties (which may include dancing) held in the lobby area. Concert halls often have cocktail lounge areas that become very crowded, with heavy smoking during intermissions. These areas should be equipped with flexible exhaust-recirculation systems. Concert halls may also have full restaurant facilities.

Noise control is important in concert halls as well as theaters. The design must avoid characterized or narrow-band noises in the level of audibility. Much of this noise is structure-borne resulting from inadequate equipment and piping vibration isolation. An experienced acoustical engineer is essential for help in the design of these applications.

SPORTS ARENAS

Functions performed in sports arenas may be quite varied, so the air-conditioning loads will vary. They are not only used for sporting events such as basketball, ice hockey, boxing, track meets, etc., but may also house circuses, rodeos, convocations, rock concerts, car, cycle, and truck events, and special exhibitions such as home, industrial, animal, or sports shows. For multipurpose operations, the designer must provide mechanical systems having a high degree of flexibility. High-volume ventilation systems may be satisfactory in many instances, depending on load characteristics and outside air conditions.

Load Characteristics

Depending on the degree of flexibility intended in the project's use, the load may vary from a very low sensible heat ratio for events such as boxing to a relatively high sensible heat ratio for industrial exhibitions. Often multispeed fans improve the performance at these two extremes and can aid in sound control for special events such as concerts or convocations. When using multispeed fans, the designer should consider the performance or air distribution devices and cooling coils when the fan is operating at lower speeds.

The designer must determine the degree of imperfection that can be tolerated in an all-purpose facility, or at least the kind of performances for which the facility is primarily intended.

As with other assembly buildings, seating and lighting combinations are the most important load considerations. Boxing, for example, may have the most seating, since the arena area is very small. For the same reason, however, the area that needs to be intensely illuminated is also small. Thus, boxing matches may represent the worst latent load situation. Other events that present latent load problems are rock concerts and large-scale testimonial dinner dances, although the audience at a rock concert is generally less concerned with thermal comfort. A good exhaust ventilation system is a must, however, since the audience at rock concerts may be eating and/or smoking during the performance to a much greater degree than is experienced at pop concerts. Good exhaust ventilation is also essential in the removal of fumes at car, cycle, and truck events. Circuses, basketball, and hockey have a much larger arena area and less seating. The sensible load from lighting the arena area does improve the sensible heat ratio. The large expanse of ice in hockey games represents a considerable reduction in both latent and sensible loads. High latent loads caused by occupancy or ventilation will create severe problems in ice arenas such as condensation on interior surfaces and fog. Special attention should be paid to the ventilation system, air distribution, and construction materials.

Enclosed Stadiums

An enclosed stadium may have either an operable or fixed roof. When the roof is open, mechanical ventilation is not required. However, when it is closed, ventilation is needed. Ductwork must be run in the permanent sections of the stadium. The large air volumes required and the long air throws make proper air distribution difficult to achieve; thus, the duct distribution systems must be capable of substantial flexibility and adjustment.

Some open stadiums have radiant heating coils in the floor slabs of the seating areas for use during cold weather. Another means of providing warmth is the use of gas-fired, electric, or infrared radiant heating panels located above occupants.

Open racetrack stadiums present a ventilation problem if the outside of the grandstand area is enclosed. The grandstand area may have multiple levels and be in the range of 1300 ft long and 200 ft deep. The interior (ancillary) areas must be ventilated because of toilet facilities, control of concession odors, and high population density. General practice is to provide about four air changes per hour for the stand seating area and exhaust the air through the rear of the service areas. More efficient ventilation systems may be selected if architectural considerations permit. Fogging of windows is a winter concern with glass-enclosed grandstands. This can be minimized by double glazing, humidity control, and dry air movement across the glass.

Air-supported structures require the continuous operation of a fan to maintain the structure in a properly inflated condition. The possibility of condensation on the underside of the air bubble should be considered. The U-value of the roof should be sufficient to prevent condensation at the lowest expected ambient temperature.

Heating and air-conditioning functions can either be incorporated into the inflating system or they can be separately furnished. Solar and radiation control is also possible through the structure's skin. Applications, though increasing rapidly, still require working closely with the enclosure manufacturer to achieve proper and integrated results.

Ancillary Spaces

The concourse areas of sports arenas contain concession stands that are heavily populated during entrance, exit, and intermission periods. Considerable odor is generated in these areas by food, drink, and smoke, requiring considerable ventilation. Where energy conservation is an important factor, carbon filters and controllable recirculation rates should be considered.

Ticket offices, restaurants, and similar facilities are often expected to be operative during hours that the main arena is closed, and, therefore, separate systems should be considered for these areas.

Locker rooms require little treatment other than excellent ventilation, not less than 2 or 3 cfm/ft^2. To reduce the outdoor air load, excess air from the main arena may be transferred into the locker room areas. However, reheat or recooling by water or primary air should be considered to maintain the locker room temperature. To maintain proper air balance under all conditions, locker room areas separate supply and exhaust systems.

Concourse area air systems should be considered for their flexibility of returning or exhausting air, since these areas are subject to heavy smoking between periods of sports events. Economics of this type of flexibility should be evaluated with regard to the associated problem of air balance and freeze-up in cold climates.

When an ice skating rink is designed into the facility, the concerns of groundwater conditions, site drainage, structural foundations, insulation, and waterproofing become even more important, with the potential of freezing of soil or fill under the floor and subsequent expansion. The rink floor may have to be strong enough to support heavy trucks. The floor insulation also must be strong enough to take this load. Ice-melting pits of sufficient size with steam pipes may have to be furnished. If the arena is to be air conditioned, the possibility of combining the air-conditioning system with the ice rink system may be analyzed. The designer should be aware that both systems operate at extremely different temperatures and may be operating at different capacity levels at any given time. The radiant effect of the ice on the people and the roof heat and light heat on the ice must be considered in the design and operation of the system. Also, low air velocity at the floor is related to minimizing refrigeration load. High air velocities will cause moisture to be drawn from the air by the ice sheet. Fog is caused by the uncontrolled introduction of airborne moisture through ventilation with warm, moist outside air and generally develops within the boarded area (playing area). Fog can be controlled by reducing outdoor air ventilation rates and appropriate air velocities bringing the air in contact with the ice and in combination with a dehumidification system. Air-conditioning systems have limited impact on reducing the dew-point temperature sufficiently to prevent fog.

The type of lighting used over ice rinks must be carefully considered when precooling is used prior to hockey games and between periods. Main lights should be capable of being turned off, if feasible. Incandescent lights require no warm-up time and are more applicable than types requiring warmup. Low emissivity ceilings with reflective characteristics successfully reduce condensation on roof structures; they also reduce lighting.

Gymnasiums

Smaller gymnasiums, such as those found in school buildings, are miniature versions of sports arenas and often incorporate multipurpose features.

Many school gymnasiums are not air conditioned. Most have a perimeter radiation system and either a central ventilation system with four to six air changes or unit heaters located in the ceiling area. Infrared heating is appearing more frequently in gymnasiums. Ventilation must still be addressed with the heavy activities and soiled equipment present.

Most gymnasiums are located in schools. However, public and private organizations and health centers may also have gymnasiums. During the day, most gymnasiums are used for physical culture activities, but in the evening and weekends, they may be used for sports events, social affairs, or meetings. Therefore, their activities fall within the scope of a civic center. More gymnasiums are being considered for air conditioning to make them more suitable for civic center activities.

The design criteria are similar to sports arenas and civic centers when used for nonstudent training activities. However, for schooltime use, space temperatures are often kept between 65 and 68 °F, when weather permits. Occupancy and degree of activity during daytime use does not usually require high quantities of outdoor air, but if used for other functions, system flexibility is required.

CONVENTION AND EXHIBITION CENTERS

Convention-exhibition centers perform more diverse functions than any other building type and present a unique challenge to the designer. The center generally is a high-bay, long-span space in which varied functions take place. With interior planning, and depending on the product type being exhibited, these centers can be changed weekly from an enormous computer room into a gigantic kitchen, a large machine shop, department store, automobile showroom, or miniature zoo. They can also be the site of gala banquets or major convention meeting rooms.

The income earned by these facilities is a direct function of the time it takes to change over from one activity to the next, so a highly flexible utility distribution system and air-conditioning system is needed.

Ancillary facilities include restaurants, bars, concession stands, parking garages, offices, television broadcasting rooms, and multiple meeting rooms varying in capacity from small (10 to 20 people) to large (hundreds or thousands of people). Often, an appropriately sized full-scale auditorium or sports arena will also be incorporated.

By their nature, the facilities are much too large and diverse in usage to be served by a single air-handling system. However, the situations previously noted and in other chapters are appropriate. Multiple air-handling systems with several chillers accommodate economical operation.

Load Characteristics

The main exhibition room undergoes a variety of loads, depending on the type of activity in progress. Industrial shows provide the highest sensible loads, which may have a connected capacity of 20 W/ft^2 along with one person per 40 to 50 ft^2. Loads of this magnitude are seldom considered because large power-consuming equipment is seldom in continuous operation at full load. An adequate design accommodates (in addition to lighting load) about 10 W/ft^2 and one person per 40 to 50 ft^2 as a maximum continuous load.

Alternative loads that are very different in character may be encountered. When the main hall is used as a meeting room, the load will be much more latent in character. Thus, multispeed fans or variable volume systems may provide a better balance of load during these high latent, low sensible periods of use. The determination of accurate occupancy and usage information is critical in any plan to design and operate such a facility efficiently and effectively.

System Applicability

The main exhibition hall would normally be handled by one or more all-air systems. These systems should be capable of operating on all outdoor air, because during set-up time, the hall may contain a number of highway-size trucks bringing in or removing exhibit material. There are also occasions when the space is used for equipment that produces an unusual amount of fumes or odors, such as restaurant or printing industry displays. It is helpful to build some flues into the structure to duct noxious fumes directly to the outside.

The groups of smaller meeting rooms are best handled with either separate individual room air-handling systems, or with variable-volume central systems, because these rooms have high individual peak loads but are in infrequent use. Constant volume systems of the dual- or single-duct reheat type waste considerable energy when serving empty rooms, unless special design features are incorporated.

The offices and restaurant spaces often operate for many more hours than the meeting areas or exhibition areas and should be served from separate systems. Storage areas can generally be conditioned by exhausting excess air from the main exhibit hall through these spaces.

NATATORIUMS

The materials used in the construction of the walls, floors, and roof, and their method of application, must be carefully analyzed and selected to ensure that the building will not be damaged by its humid, corrosive environment or by condensation. The careful selection of materials for the heating, ventilating, and/or air-conditioning systems and their controls is important for the same reason. The importance of energy conservation has emphasized the cost of ventilating the pool enclosure to control corrosion and condensation. In many cases, ventilation is the dominant source of heat loss. Thus heat recovery should be considered.

The design of the heating and air-conditioning or ventilation systems must be planned to provide comfort for spectators and swimmers, in or out of the pool. Excessive air motion or drafts in the pool area must be avoided.

Pools located in the interior of a building may provide less of a design problem than pools with wall and roof exposures. The use of glass on pool exposures complicates the design problem and adds cost, expecially in cold climates. Glass walls or exposures make cold draft conditions and condensation difficult and expensive to eliminate. The pool area should be isolated from adjacent building areas, if possible, by providing a negative pressure at the pool.

Load Characteristics

Natatoriums are characterized by high latent loads that should be controlled to minimize corrosion and condensation on the building construction. Outdoor air of proper moisture content may be used for this purpose. During periods when outdoor humidity is high, it may be necessary to provide some form of reheat for humidity control. Mechanical dehumidifiers with energy recycling features are increasingly used to control humidity as well as to heat pool water.

Pools used for open or free swimming have a large number of people engaged in fairly strenuous physical activity. Air must be introduced to the space without causing discomfort to occupants, in and out of the pool.

Design Concepts

The following three basic types of pools are discussed in this section:

1. Natatoriums with no spectator facilities
2. Natatoriums with spectator facilities
3. Therapeutic pools

Pools require humidity control to maintain comfort conditions. Pool air-handling systems are generally designed to use up to 100% outdoor air for cooling and/or dehumidification. On a winter cycle, when outdoor temperature and humidity are below pool design conditions, the amount of outdoor air can be controlled by a humidistat to maintain the desired humidity level. On a summer cycle, when outdoor temperature and humidity are above pool design conditions, minimum outdoor air is used. However, humidity and temperature condition may not be maintained when outside conditions are higher. Mechanical dehumidifier systems providing year-round humidity control make air conditioning possible for year-round temperature control and may be considered for year-round comfort control.

Spectator areas should have their own separate air supply, and pool air should not return or exhaust through the spectator area because of its high moisture content and chloramine odor.

System Applicability

All-air systems are required to remove large quantities of moisture in the air. Warm wall and floor surfaces increase occupant comfort and can be accomplished by application of insulation in walls, heating the perimeter tunnel where available, warm air curtains on perimeter walls and glass, and radiation and radiant panels.

Close attention should be paid to latent chloramine and moisture levels. High levels of humidity and corrosive elements are more likely to occur during periods of high activity. When systems have marginal moisture control and when pool water treatment is unbalanced, humidity levels and corrosive elements are again likely to increase. Building material selection is extremely critical. Selection criteria should include corrosion resistance and maintaining inside surface temperatures above dew point under design outside winter conditions.

Design Criteria

Design conditions for pools include:

Indoor air

Pleasure swimming	75 to 85 °F, 50 to 60% rh
Therapeutic	80 to 85 °F, 50 to 60% rh

Pool water

Pleasure swimming	75 to 85 °F
Therapeutic	85 to 95 °F
Competitive swimming	72 to 75 °F
Whirlpool/Spa	97 to 102 °F

Relative humidities should not be maintained below recommended levels because of the evaporative cooling effect on a person emerging from the pool. Humidities higher than recommended encourage corrosion and condensation problems as well as occupant discomfort. Lower-than-necessary relative humidity increases the rate of pool evaporation and pool heating requirements.

Air velocity at any point 8 ft above the walking deck of the pool should not exceed 25 fpm. In a diving area, air velocity around the divers should be below this level.

In spectator seating areas, air velocity may be increased to 40 to 50 fpm, unless seats are located in an area also occupied by swimmers.

Calculation for Minimum Air Requirements. The air supplied to the pool must be sufficient to remove the water that evaporates from the surface of the pool when using outside air for moisture removal. When using mechanical dehumidification only, the rate of evaporation needs to be determined in sizing dehumidification capacity. The rate of evaporation can be found from empirical Equation (1). This equation is for public pools at high to normal activity, allowing for splashing and a limited area of wetted deck. Other pool uses may have up to 50% less moisture evaporation.

$$w_p = A(95 + 0.425v)/Y\,[p_w - p_a] \qquad (1)$$

where

w_p = evaporation of water, lb/h
A = area of pool surface, ft^2
v = air velocity over water surface, fpm
Y = latent heat required to change water to vapor at surface water temperature, Btu/lb
p_a = saturation pressure at room air dew point, in. Hg
p_w = saturation vapor pressure taken at the surface water temperature, in. Hg

For values of Y about 1000 Btu/lb and values of v ranging from 10 to 30 fpm, Equation (1) can be reduced to:

$$w_p = 0.1\,A(p_w - p_a) \qquad (2)$$

The minimum air quantity required to remove this evaporated water can be found from Equation (3):

$$Q = w_p/60\rho(W_i - W_o) \qquad (3)$$

where

Q = quantity of air, cfm
ρ = standard air density, 0.075 lb/ft^3
W_i = humidity ratio of pool air at design criteria, lb/lb
W_o = humidity ratio of outdoor air at design criteria, lb/lb

The values of W can be obtained from the ASHRAE psychrometric charts.

When using makeup air for humidity control, design outdoor conditions should be reviewed carefully in cold climates, because this optimum condition for moisture removal occurs infrequently. Several outdoor conditions should be reviewed and calculated before establishing the minimum air volumes.

This calculated quantity of air often produces only one or two air changes per hour—a low rate of air movement. The following airflow rates are recommended, assuming the minimum air requirement from the previous calculation falls below the minimum recommendations. Most codes require six air changes per hour, except where air conditioning is furnished.

Pools with no spectator facilities	4 to 6 air changes
Spectator facilities	6 to 8 air changes
Therapeutic pools	4 to 6 air changes

Air volume above the minimum calculated value is recirculated, but the fan system incorporates return-exhaust fans and outdoor air, exhaust air, and return air controls to enable larger amounts (up to 100%) of outdoor air to be introduced during milder weather for temperature and humidity control. The return-exhaust fan facilitates the introduction of outdoor air and the maintenance of recommended pressure. Heating elements should be large enough to handle more than design amounts of outdoor air to control temperature during periods of heavy pool use. With mechanical dehumidification being used for moisture removal, greater control of both temperature and humidity can be achieved.

Air Distribution and Filtering

The type of air distribution system influences the amount of air introduced into the pool area. Air volume above the minimum calculated value may be recirculated, provided the recirculated air is dehumidified and filtered to reduce air contaminants to safe levels. Care should be taken in the choice of filtration used, especially where filter media may react with chloramines in the air. The air-handling system should have filters of 45 to 60% (based on ASHRAE Filter Test *Standard* 52-76) for occupant comfort and protection and to minimize streaking of walls and floors from dirt contacting moist surfaces.

Noise Level

Pool air system noise levels may be designed for NC45 to 50 without causing discomfort or annoyance. However, wall, floor, and ceiling surfaces should be evaluated for their contribution to increasing unwanted noise levels.

Special Considerations

Condensation and corrosion from the humid, corrosive atmosphere of the pool can cause damage and even failure of materials and equipment within or serving a natatorium. Ferrous metals should be eliminated from all areas of pool construction. Roofs and walls must be protected by a vapor barrier. Suspended ceilings should be discouraged because they provide a high humidity enclosure that requires separate ventilation. Despite such ventilation, the ceiling and appurtenances, such as lights and supports, are subject to hidden corrosion.

All components of the pool-heating, air-handling, and air distribution systems exposed directly or indirectly to the moist, corrosive pool atmosphere should be noncorrosive. Where it is economically unfeasible to do so, they should be protected with a high-quality corrosion-resistant coating.

All ductwork serving pool areas should be constructed of aluminum, galvanized coated steel, or other rust-resistant material. The high moisture content of the pool air being conveyed in ductwork requires extra design considerations. Return or exhaust ductwork located above ceilings should be insulated on the outside and sealed against moisture penetration.

The pool should be maintained at a slightly negative pressure of 0.05 to 0.15 in. of water to prevent moisture and chloramine odors from migrating to other areas of the building. Pool air may be used as make-up for showers and toilet rooms, but a separate system is more desirable. Locker rooms and offices should have separate supplies and a positive pressure relationship with respect to the pool. Openings from the pool to other areas should be minimized, and passageways should have a vestibule (air lock) or some other arrangement to discourage the passage of air and moisture.

A properly designed air distribution system is essential for uniform, draft-free conditions. Although it is more difficult to design, side wall distribution is generally less costly than overhead supply to provide draft-free conditions. Perimeter air distribution with air distribution upward against outside windows provides the best draft-free supply and is very effective in preventing condensation buildup. Any air outlet over the pool is difficult to adjust manually and may increase air motion over the water surface, thus increasing the rate of evaporation. It is not necessary for return and/or exhaust outlets to be located low to pick up the moisture, since the water vapor tends to rise.

In cold climates, the combination of high humidities and the use of glass requires careful design and engineering to avoid condensation problems. Double or triple glazing combined with radiation is relatively ineffective where ceiling air is directed toward high glass areas. This results in a forced downdraft condition, which will quickly overwhelm the natural convection caused by radiation. It is far more noticeable in pool areas than other occupancies due to the evaporative cooling that occurs on a body.

Where radiation from cold glass is not desirable, a radiant panel, perimeter floor radiant, or infrared system may be used. A combination of supplemental supply air to eliminate downdrafts may be used, since radiant floors alone may not be sufficiently effective against large glass areas. Locating the return slots at the bottom of the glass reduces draft effects away from the wall. The use of skylights should be discouraged in cold winter climates since it is very difficult to prevent condensation buildup.

Pool water-filtering equipment and chemical gases require special ventilation rates for occupational safety. This requirement is the decisive factor in the design of the pool equipment room air-handling system.

FAIRS AND OTHER TEMPORARY EXHIBITS

At frequent intervals, large-scale exhibits are constructed throughout various parts of the world to stimulate business, present new ideas, and provide cultural exchanges. Fairs of this type take years to construct, are open from several months to several years, and are sometimes designed with the thought of future use of some of the buildings, either at the fair site or elsewhere. Fairs, carnivals, or exhibits, which may consist of prefabricated shelters and tents that are moved from place to place and remain in a given location for only a few days or weeks, will not be covered here because they seldom require the involvement of architects and engineers.

DESIGN CONCEPTS

One consultant or agency should be responsible for setting uniform utility service regulations and practices to ensure proper organization and operation of all exhibits. Exhibits that remain open only during the intermediate spring or fall months require a much smaller heating or cooling plant than for peak summer or winter design conditions. This information is required in the earliest planning stages so proper analysis of system and space requirements can be established accurately.

Occupancy

Fair buildings have heavy occupancy during visiting hours, but patrons seldom stay in any one building for long periods. The approximate length of time that patrons stay in the building determines the design of the air-conditioning system. The shorter the anticipated stay, the greater the leeway in designing for less-than-optimum operating design conditions, equipment, and duct layout. If patrons wear outer garments while in the building, this will affect operating design conditions.

Equipment and Maintenance

Heating and cooling equipment used solely for maintaining proper comfort conditions and not for exhibit purposes may be secondhand or leased equipment, if available and of the proper capacities. Another possibility is to rent all air-conditioning equipment to reduce the exhibitors' capital investment and eliminate disposal problems when the fair is over.

Depending on the size of the fair, the length of time it will operate, the types of exhibitors, and the policies of the fair sponsors, it may be desirable to analyze the potential for a centralized heating and cooling plant versus individual plants for each exhibit. The proportionate cost of a central plant to each exhibitor, including utility and maintenance costs, may be considerably less than having to furnish space and plant utility and maintenance costs. The larger the fair, the more savings may result. It also makes it practical to consider making the plant a showcase, suitable for exhibit and possibly added revenue. Another feature of a central plant is that it may form the nucleus for the commercial or industrial development of the area after the fair is over.

If each exhibitor furnishes their own air-conditioning plant, it is advisable to analyze shortcuts that may be taken to reduce equipment space and maintenance aids. For a 6-month to 2-year maximum operating period, for example, tube pull or equipment removal space is not needed or may be drastically reduced. Higher fan and pump motor horsepowers and smaller equipment is permissible to save on initial costs. Ductwork and piping costs should be kept as low as possible because these are usually the most difficult items to salvage; cheaper materials may be substituted wherever possible. The job must be thoroughly analyzed to eliminate all unnecessary items and reduce all others to bare essentials.

The central plant may be designed for short-term use as well. However, if the plant is to be used after the fair closes, the central plant should be designed in accordance with the best practice for long-life plants. It is difficult to determine how much of the piping distribution system can be used effectively for permanent installations. For that reason, piping should be simply designed initially, preferably in a grid, loop, or modular layout, so that future additions can be made easily and economically.

Air Cleanliness

The efficiency of the filters for each exhibit is determined by the nature of the area served. Since the life of an exhibit is very short, it is desirable to furnish the least expensive filtering system. If possible, one set of filters should be selected to last for the life of the exhibit. In general, the filtering efficiencies do not have to exceed 30%. (See ASHRAE *Standard* 52-76 on atmospheric air.)

SYSTEM APPLICABILITY

If a central air-conditioning plant is not built, the systems installed in each building should be the least costly to install and operate for the life of the exhibit. These units and systems should be designed and installed to occupy the minimum usable space.

Whenever feasible, heating and cooling should be performed by one medium, preferably air, to avoid running a separate piping and radiation system for heating and a duct system for cooling. Air curtains used on an extensive scale may, on analysis, simplify the building structure and lower total costs.

Another possibility when both heating and cooling is required is a heat pump system, which may be less costly than separate heating and cooling plants. Economical operation may be possible, depending on the building characteristics, light, and people load. If well or other water is available, it may produce a more economical installation than an air source heat pump.

Since exhibits and buildings can serve many varied functions, when specific problems or applications arise, reference should be made to the building type most closely resembling the exhibit building.

CHAPTER 5

DOMICILIARY FACILITIES

GENERAL

ALL structures in this group are single-room or multiroom, long- or short-term dwelling (or residence) units that are stacked sideways and/or vertically. Ideally, each room of the unit should have equipment, distribution, and/or control that allows it to be cooled or heated and ventilated independently of any other room. If this ideal is not available, optimum air conditioning for each room will be compromised.

Domiciliary facilities include apartment houses (high- and low-rise rental, cooperative, and condominium, either as single-purpose structures or as a portion of a multipurpose building), dormitories, hotels, motels, nursing homes, and similar building types. Apartment houses are discussed in Chapter 1. Nursing homes are discussed in Chapter 7.

LOAD CHARACTERISTCS

1. Constantly operational, but not necessarily occupied at all times. Adequate flexibility designed into each unit's HVAC system allow cooling and ventilation to be shut off (except when humidity is required) and heating shut off or turned down.
2. Low lighting and occupancy concentrations; generally sedentary or light activity. Transient in nature with greater night use of bedrooms. Occasionally heavy occupancy; smoking and physical activity in dining rooms, living rooms, and other areas used for entertaining.
3. Potentially high appliance load, odor generation, and exhaust requirements in kitchens, whether integrated with or separated from residential quarters.
4. Generally, exterior rooms, except sometimes for kitchens, toilets, and dressing rooms; usually multiple exposure for the building as a whole and frequently for many individual dwelling units.
5. Toilet, washing, and bathing facilities almost always incorporated within dwelling units for hostelries and nursing homes; only occasionally in dormitories. Toilet areas requiring exhaust air usually are incorporated in each domicile unit.
6. Relatively high domestic hot water building demands, generally over short periods of an hour or two, several times a day. This can vary from a fairly moderate and level daily load profile for senior citizens' buildings, to sharp, unusually high peaks at about 6:00 p.m. in dormitories. Chapter 44 includes details on service hot water systems.
7. Load characteristics of rooms, dwelling units, and buildings can be well defined without the need for any substantial future design load changes other than the addition of a service such as cooling, which may not have been originally incorporated.

The preparation of this chapter is assigned to TC 9.8, Large Building Air-Conditioning Applications.

The predominance of exterior exposures with glass and shifting, transient, interior loads results in low diversity factors, while the long-hour usage results in fairly high load factors.

DESIGN CONCEPTS AND CRITERIA

Wide load swings and diversity in and between rooms require great care in designing a flexible system for 24-h comfort. Besides opening windows (available in typical dwelling units), the only positive technique for flexible temperature control is individual room components under individual room control, which can cool, heat, and ventilate independently of the equipment in any other room.

In some climates, summer humidity levels become objectionable because of the low internal sensible loads when cooling is on-off controlled. Modulated cooling and/or reheat may be required to achieve satisfactory comfort. Reheat should be avoided, unless some sort of heat recovery can be applied.

Odor control is impractical and uncontrollable within any single dwelling unit. Most designers strive simply to confine odors within the dwelling unit. Systems with pressurized public corridors from which makeup air is drawn into the dwelling unit are most successful. Because of the high operating cost of providing makeup air, high local odor concentrations, and codes that prohibit centralized air return from dwelling units to common supply air fans, centralized all-air systems are not in common use.

The noise threshold for some people is low enough that certain types of equipment disturbs their sleep; however, even unitary systems of poor quality (as long as they are not in need of repair) are acceptable to most people, especially if they must choose between a fairly steady noise (which effectively masks extraneous, intermittent noises) and no air conditioning at all. Poor noise levels may be acceptable in areas where there are low cooling degree days. Medium and better quality equipment is available with NC35 levels at 10 to 14 ft distances in medium to soft rooms, with little sound change when the compressor cycles.

Perimeter fan-coil systems are usually quieter than unitary systems. However, many occupants are more disturbed by equipment with wide changes in noise levels than in a continuous higher level. Fan-coil units often cycle the fans for thermostat control, so a high enough noise level occurring during *on* cycles can be more objectionable than the compressor cycling of a through-the-wall unit masked by its continuously operating fan.

SYSTEMS

ENERGY EFFICIENT SYSTEMS

The systems included in this type of occupancy should be energy efficient. Systems that are generally most efficient inlcude water source heat pumps and air source heat pumps. Water source heat pumps may be solar assisted in areas with ample solar radiation. Energy efficient equipment generally has the lowest operating cost and is relatively simple, an important factor in multiple-dwelling and domiciliary facilities, where skilled operating personnel are unlikely to be available. Most systems allow individual operation and thermostatic control. According to McClelland (1983), the typical system allows individual metering so that most, if not all, of the cooling and heating costs can be metered directly to the occupant. In existing buildings, individual metering can be done by BTU measuring meters and timers for fan motors.

The water loop heat pump system has favorable operating costs compared with air-cooled unitary equipment, especially where electric heat is used. Favorable installed cost encourages use of this system in mid- and high-rise buildings, where individual dwelling units have floor areas of 800 ft^2 or larger. Some systems incorporate sprinkler system piping as the water loop.

Except for a central circulating pump, heat rejector fans, and a supplementary heater, the system is predominantly decentralized; individual metering allows most of the HVAC operating costs to be paid by the occupants. System life should be longer than for other unitary systems because most of the mechanical equipment is protected within the building and not exposed to outdoor conditions; also, the refrigeration circuit is not subjected to as severe an operating duty because system water temperatures are controlled to operate under optimum conditions. Low operating costs are possible from inherent energy conservation as excess heat may be stored from daytime for the following night, and heat is transferred from one part of the building to another.

Simultaneous heating and cooling occurs during cool weather because heating is required in many areas, but cooling becomes necessary in some rooms because of solar or internal loads. Frequently, surplus heat throughout the building on a mild day is transferred into the hot water loop by air condensers on cooling cycle so that water temperature rises. The heat remains stored in water where it can be extracted at night without operating a water heater. This heat storage is improved by having a greater mass of water in the pipe loop; some systems include a storage tank for this purpose. Because the system is designed to operate during the heating season with water supplied at temperatures as low as 60°F, the water loop heat pump lends itself to solar assist with relatively high solar collector efficiencies resulting from the low water temperature.

Installed cost of the water loop heat pump system becomes higher in very small buildings. Where natural gas or fossil fuels are available at reasonable cost, the operating cost advantages of this system may diminish in severe cold climates with prolonged heating seasons, unless heat can be recovered from some other source such as solar collectors or internal heat loads from a commercial area served by the same system.

ENERGY NEUTRAL SYSTEMS

Energy neutral systems do not allow simultaneous cooling and heating. Systems of these types include Packaged Terminal Air Conditioners (PTAC) (through-the-wall units), window units for cooling (combined with finned or baseboard radiation for heating), unitary air conditioners with an integrated heating system, fan coils with remote condensing units, and variable air volume (VAV) systems with baseboard heating. To qualify as energy neutral systems, controls must prevent simultaneous operation of the cooling and heating cycles. In unitary equipment, this may be as simple as a heat-cool switch. In other types, dead-band thermostatic control may be required to separate cooling and heating.

PTAC are frequently applied in buildings where most of the individual units are relatively small, consisting of one or two rooms. A common arrangement for two rooms adds a supply plenum to the discharge of the PTAC unit so that some portion of the conditioned air serving one room can be diverted into the second, usually smaller, room. Multiple PTAC units are used in larger areas with more rooms, providing for additional zoning within the area. Additional radiation heat is sometimes specified in cold climates, where needed, to distribute heat around the perimeter.

The PTAC system can have electric resistance heat, for lowest initial cost, and a decentralized HVAC system, or have hot water or steam heating coils, where combustion fuels are available to provide lower operating costs. Despite relatively inefficient refrigeration circuits, operating costs of PTAC systems are quite reasonable because of the individual thermostatically controlled operation of each machine, which eliminates the use of reheat while still avoiding overheating or overcooling the space. Also, little power is devoted to circulating the room air, since the equipment is located in the space being served. Servicing is simple—a spare chassis replaces a defective machine, which can then be forwarded to a service organization for repair. Thus, building maintenance requires relatively unskilled personnel.

Noise levels are generally no higher than NC40, but some units are noisier than others. Installations near a seacoast should be specified with special construction (usually stainless steel or special coatings) to avoid accelerated corrosion to aluminum and steel components caused by salt. Installation in high-rise buildings above 12 stories requires special care, both in the design and construction of outside partitions and in the installation of air conditioners to avoid operating problems associated with leakage around and through the machines caused by stack effect.

This system offers reasonable operating costs and noise levels, which are tolerable but somewhat higher than those of the concealed water source heat pump system. Legislation may prescribe maximum acceptable outdoor equipment noise levels. Fan coil units with remote condensing units are used in smaller buildings. Units are located in closets, and ductwork distributes air to the rooms in the unit. Condensing units can be located on roofs, at ground level, or on balconies.

Frequently, the least expensive system to install is finned or baseboard radiation for heating and window-type room air conditioners for cooling. Often, the window units are purchased individually by the building occupants. This system offers reasonable operating costs and is relatively simple to maintain. Window units have the shortest equipment life, the highest operating noise level, and the poorest distribution of conditioned air of any of the systems compared in this section. This system may cause air and dirt leakage around window units.

Low capacity residential warm air furnaces may be used for heating only, but they require venting the products of combustion for gas or oil-fired units. In a one- or two-story structure, it is possible to use individual chimneys or flue pipes, but in a high-rise structure, a multiple-vent chimney or a manifold vent system must be used. Local codes should be consulted.

Sealed combustion furnaces have been developed for domiciliary use. These units draw all the combustion air from outside and discharge the flue products outside through a windproof vent. The unit must be located near an outside wall and exhaust gases must be directed away from windows and intakes. One- or two-story structures may also use outdoor units mounted on the roof or on a pad at ground level. All of these heating units can be obtained with cooling coils, either built-in or add-on. Evaporative-type cooling units are popular in motels, low-rise apartments, and residences of the southwestern United States.

INEFFICIENT ENERGY SYSTEMS

Inefficient energy systems allow simultaneous cooling and heating. Examples of these systems are terminal reheat, two-, three-, and four-pipe fan coil units, and induction systems. Some systems, such as the four-pipe fan coil, can be controlled so that they are energy neutral. However, their primary use is for humidity control.

The four-pipe system and two-pipe systems with electric heaters can be designed with complete temperature and humidity flexibility during summer and intermediate season weather, although none provides winter humidity control. Each of these systems provides full dehumidification and cooling with chilled water, reserving the other two pipes or electric coil for space heating or reheat. The systems and controls needed are expensive, and only the four-pipe system, with an internal-source heat-recovery design for the warm coil energy, can operate at low cost. Two-pipe systems with electric heat or four-pipe systems should be considered when year-round comfort is essential.

TOTAL ENERGY SYSTEMS

Any multiple or large housing facility with high year-round domestic hot water requirements is one of the more attractive applications for total energy systems. Total energy systems are a form of cogeneration in which all or most electrical and thermal energy needs are met by on-site systems as described in Chapter 8 of the 1987 ASHRAE *Handbook—HVAC Systems and Applications*. The thermal profiles for HVAC and service hot water can match the electrical generating profile better than many other applications, and the 24-h loads permit higher load factors and better equipment use. However, a detailed load profile must be analyzed before this application is recommended. Reliability and safety must also be considered with heat-recovery systems.

Any system described previously can provide the HVAC portion of total energy. The major considerations for choice of system as they apply to total energy are as follows:

1. Optimum use must be made of the thermal energy recoverable from the prime mover during all or most operating modes, not just during maximum HVAC design conditions.
2. Heat recoverable via the heat pump cycles may become less useful, since many of its potential operating hours will be satisfied with some or all of the heat recovered from the prime mover. It becomes more difficult, therefore, to justify the additional investment for heat pump or heat recovery cycles because operating savings are lower.
3. The best use for recovered waste heat is for those services using only heat (*i.e.*, domestic hot water, laundry facilities, and space heating).

SPECIAL CONSIDERATIONS

Local building codes govern ventilation air quantities for domiciliary buildings. Where they do not, average values are about 50 cfm for private bathrooms and a minimum of 75 cfm for private kitchens with nonexhausted hoods or 100 cfm for private kitchens with hoods exhausted to the outdoors. Outdoor air, usually slightly in excess of the exhaust quantities to pressurize the building, is introduced into the corridors.

Buildings using centrally controlled exhaust and supply systems operate the systems on a time clock or a central management system for certain periods of the day. In other cases, it is common practice to reduce or shut off the outside air during extremely cold periods, although this practice is not recommended and may be prohibited by local codes. If known in advance, these factors should be considered when estimating heating load.

Buildings using exhaust and supply air systems on a 24-h basis may merit consideration of air-to-air heat recovery devices (see Chapter 34 of the 1988 ASHRAE *Handbook—Equipment*). Such recovery devices can reduce the consumption of energy by capturing 60 to 80% of the sensible and latent heat extracted from the air source.

Infiltration loads in high-rise buildings without ventilation openings for perimeter units are not controllable on a year-round basis by general building pressurization. When outer walls are pierced for outdoor air to unitary or fan-coil equipment, combined wind and thermal stack-effect forces create problems. These factors must be considered for high-rise buildings (see Chapter 23 of the 1989 ASHRAE *Handbook—Fundamentals*).

Interior public corridors should have tempered supply air with transfer into individual area units, if necessary, through acoustically lined transfer louvers to provide kitchen and toilet makeup air requirements. Corridors, stairwells, and elevators should be pressurized for fire and smoke control (see Chapter 47).

Kitchens should not be exhausted but recirculated through hoods with activated charcoal filters whenever possible. Toilet exhaust can be VAV with installation of a damper operated by light switch. A controlled source of supplementary heat in each bathroom is recommended to ensure comfort while bathing.

Air-conditioning equipment must be isolated to reduce noise generated or transmitted. The cooling tower or condensing unit must be designed and located to avoid disturbing occupants within the building and neighbors in adjacent buildings.

An important load, frequently overlooked, is heat gain from piping for hot water services. At the least, minimum insulation thickness should follow the latest local energy codes and standards. In large, luxury-type buildings, installation of a central building or energy management system allows supervision of the individual air-conditioning unit(s) for operation and maintenance.

Some domiciliary facilities achieve energy conservation by reducing indoor temperatures. A reduction of indoor temperatures should be pursued with caution, however, if there are aged occupants because they are susceptible to hypothermia.

DORMITORIES

Dormitory buildings frequently have large commercial dining and kitchen facilities, laundering facilities, and common areas for interior recreation and bathing. Such ancillary loads improve the economic feasibility of heat pump or total energy systems—especially on campuses with year-round activity.

When dormitories are shut down during cold weather, the heating system must supply sufficient heat to prevent freezeup. If the dormitory contains administrative offices, eating facilities, etc., these facilities should be designed as a separate zone or system for optimum flexibility, economy, and odor control.

Subsidiary facilities should be controlled independently for flexibility and shutoff capability, but they may share common refrigeration and heating plants. With internal-source heat pumps, such interdependence of unitary systems permits reclamation of all possible internal heat that is usuable for building heating, domestic water preheating, and snow melting. It is easier and less expensive to use heat reclaim coils in air exhausted from the building than to use air-to-air heat-recovery devices. Also, heat reclaim can be easily sequence-controlled to add heat to the building's chilled water systems, when required.

HOTELS AND MOTELS

Both hotel and motel accommodations are usually single room with toilet and bath adjacent to a corridor, flanked on both sides by guest rooms. They may be single-story, low-, or high-rise. Multipurpose subsidiary facilities range from stores and offices to ballrooms, dining rooms, kitchens, lounges, auditoriums, and meeting halls. Luxury motels may be built with similar facilities. Occasional variations are seen, with kitchenettes, and, more frequently than for apartments, outside doors to patios and balconies, as well as multiroom suites.

LOAD CHARACTERISTICS

The diversity of use for all areas of hostelries makes load profile studies essential to avoid unnecessary oversizing and duplication, and low operating cost designs. Hostelries without guest room cooking facilities usually peak an hour or two earlier in the morning than do apartment houses and have lower noon and evening peaks. Electrical and cooling peaks for hostelries are highest between 6:00 and 8:00 p.m.

Because of the lower guest room peaks, substantial use of subsidiary facilities with noncoincident peaks, and more concentrated living areas, load factors for hostelries are higher than for apartment buildings.

DESIGN CONCEPTS AND CRITERIA

Air conditioning in hotel rooms should be quiet, easily adjustable, and draft-free. It must also provide ample outside air. The hotel business is competitive, and space is at a premium, so systems that require the least space and have low total owning and operating costs should be selected.

Imaginative use of energy profiles for each auxiliary area can lead to major investment savings due to common central plant and air distribution systems. When odor generation is common to the facilities on one air system, or when diverse odors are generated but controlled in a common apparatus, a VAV system with zone reheat sequencing can offer many advantages over the unitary system approach for each zone. Chapters 3 and 4 of this volume provide information on various public and assembly areas.

Systems must be trouble-free and allow easy maintenance because shutdown of systems will often cause loss of revenue. Multiple heating and refrigeration plant units ensures continuity of services. Emergency generators are occasionally required for protection against electrical power outages. Many utilities provide dual service and transfer devices to prevent long power outages.

The ballroom is the showplace of a hotel, and it requires special attention to include a satisfactory system within the interior design concept. Many ballrooms do not provide proper comfort for the functions they serve. Ballrooms generally have more odors and smoke than other places of assembly when used for banquets. Most ballrooms are designed as multipurpose rooms, and often the peak loads occur when they are used as meeting rooms. A ballroom should be served by its own system.

Economies in initial and operating costs are possible by proper transfer of air from air-conditioned areas to service areas such as kitchens, workshops, and laundry and storage rooms. Smaller air quantities may achieve satisfactory results and eliminate the need for separate supply systems.

In hotel guest wing towers, space is at a premium, so chases to the top of the building for kitchen and boiler flues and emergency generator exhaust stacks (which take away potential guest room space) should be analyzed and incorporated or alternative locations determined early in the design review. Such discharges should be located at least 100 ft away from the nearest outside air intake.

Hotels require good instrumentation for all major systems and services so that accurate operating data may be analyzed to ensure economy of operation. A good maintenance staff and well-instructed cleaning staff can regulate or shut off energy services in unoccupied areas. Many new hotels and motels incorporate some type of guest room energy management system, so units can be shut off in unoccupied rooms, temperature can be reset, or all functions can be controlled from the front desk and overridden by the occupant.

APPLICABILITY OF SYSTEMS

For guest rooms, most hotels use fan coil systems or deluxe self-contained (PTAC) unit systems. Depending on the architectural design and climate, the units may be located under the exterior windows, in the hung ceiling over the bathroom or vestibule, or vertically in the wall between the bathroom and bedroom.

Hotels lend themselves to VAV systems because they have simple room layouts and no guest cooking facilities. This is particularly true in tropical or high-humidity climates and when doors open to the outside, creating condensation problems on exposed casings of fan-coil units.

Each bathroom requires an average of a minimum of 2 cfm per square foot of design exhaust air. This air usually comes from the ventilation air delivered to each room or through the room's PTAC unit, if provided. Some systems supply outdoor air to the corridors, and this air is transferred into each room by the bathroom exhaust; however, this does not control odor and smoke. In tropical climates and where outdoor air supplied to the corridors has not been previously treated, this indirect method provides poor humidity control; in many instances, controlled air supply systems are added. In large systems, dessicant dehumidification used in conjunction with evaporative cooling can be very energy efficient in providing the required outdoor air for the facility.

Great latitude is available in system selection for other hotel areas, as long as the systems satisfy the space and hotel operating requirements. These systems are predominantly all-air because of high ventilation requirements.

With substantial public area cooling required during the heating season, as well as large heat sources from kitchens and laundries, internal source heat pumps with storage should be considered for hostelries. They can also be attractive for total energy systems because of their constant thermal demand.

SPECIAL CONSIDERATIONS

In some climates, guest bathrooms should have supplementary heat available to the guests. Guest rooms opening to the outdoors should have self-closing doors, so excess condensation does not form on the cooling coils within room units.

If the fan in fan coil units is shut off, the cooling coil control valve should also close to prevent possible damage from excess condensation.

Design information for administration, eating, entertainment, assembly, and other areas will be found in Chapters 3 and 4 of the 1987 ASHRAE *Handbook—HVAC Systems and Applications*. In tropical climates, special design techniques for proper insulation and ventilation are necessary to prevent excess moisture build-up in closets and storage areas and to avoid mildew and resultant damage to clothing and other materials. One effective device is to maintain a light in the closet and to louver the closet door.

MULTIPLE-USE COMPLEXES

These complexes combine retail facilities, office space, hotel space, residential space, or other commercial space into a single site. The various occupancies may have peak HVAC demands occuring at different times of the day and year. Loads should be determined independently for each occupancy. Where a combined central plant is considered, a block load should also be determined.

Separate air handling and distribution should be used for the separate functions. However, heating and cooling units can be combined economically into a central plant. A central plant provides good opportunities for heat recovery, thermal storage, and other similar sophistication that may not be economical in a single-use facility. The multiple-use complex is a good candidate for a central control system for fire and smoke control, security, remote monitoring, billing for central facility use, maintenance control, building operations control, and energy management.

REFERENCE

McClelland, L. 1983. Tennant Paid Energy Costs in Multi-Family Rental Housing. DOE, University of Colorado, Campus Box 108, Boulder, CO 80309.

CHAPTER 6

EDUCATIONAL FACILITIES

ALTHOUGH climate control systems for educational facilities may be similar to systems applied to other types of buildings, the operation of the various systems, as well as the school operations, must be understood to suit the mechanical systems to the special educational facilities. Year-round school programs, adult education, night classes, and community functions put ever-increasing demands on these facilities and require that the environmental control system be designed carefully to meet varying needs. Classrooms, lecture rooms, and assembly rooms have uniformly dense occupancies. Gymnasiums, cafeterias, laboratories, shops, and similar functional spaces have variable occupancies and also have different and varying ventilation and temperature requirements. Many schools include auditoriums serving as gymnasiums and community meeting rooms. System flexibility is more important because of this wide variation in use and occupancy.

In most cases, the installed cost of an effective heating and ventilating system approaches the cost of a complete air-conditioning system. Generally, a well-designed heating and ventilating system contains all the fans, dampers, controls, and other components necessary for air conditioning, except for the refrigeration cycle. Since school buildings almost always need cooling when occupied—not just during the summer months—the additional cost improves the usefulness and efficiency and prevents premature obsolescence.

Design for the indoor environment of educational facilities considers heat loss, heat gain, air movement, ventilation methods, and ventilation rates. Winter and summer design dry-bulb temperatures for the various spaces in schools are given in Table 1. These values should be evaluated for each application to avoid energy waste without sacrificing comfort. For example, reducing the winter design may cause discomfort near perimeter walls. Thus perimeter walls must be designed to reduce uncomfortable radiation transfer to the occupants.

Table 1 Recommended Winter and Summer Design Dry-Bulb Temperatures for Various Types of Spaces Common in Schools[a]

Space	Winter Design, °F	Summer Design, °F
Classrooms, Laboratories, Auditoriums, Libraries, Administration areas, etc.	72	78[b]
Shops	72	78
Locker, Shower rooms	75[d]	[c,d]
Toilets	72	[c]
Storage	65	[c]
Mechanical	60	[c]
Corridors	68	80[b]

[a] For spaces of high population density and where sensible heat factors are 0.75 or less, lower dry-bulb temperatures will result in generation of less latent heat, which may reduce the need for reheat and thus save energy. Therefore, optimum dry-bulb temperatures should be the subject of detailed design analysis.
[b] Frequently not air conditioned.
[c] Usually not air conditioned.
[d] Provide ventilation for odor control.

Automatic controls must quickly change from heating to cooling when a class is occupied after morning warmup because of the heat suddenly supplied by lights and students. Heating and cooling may cycle several times during the day if the school has an energy management system.

Odors should be controlled by proper ventilation (Table 2). Proper ventilation also reduces the spread of respiratory diseases.

Simple equipment that is reliable and requires minimal maintenance is particularly important for educational facilities. Possible damage to the building due to maintenance and service operations should be considered. For example, roofs should be protected from possible damage that may occur when roof-mounted equipment is serviced.

Table 2 Ventilation Requirements for Various School Spaces

Occupancy Type	Estimated Persons per 1000 ft² of Floor Area	Required Outdoor Air per Occupant (cfm/person)	(cfm/ft²)
Classrooms	50	15	
Laboratories	30	20	
Training shops[a]	30	20	
Music rooms	50	15	
Libraries	20	15	
Locker rooms[b]			0.50
Corridors			0.10
Auditoriums	150	15	
Smoking lounges	70	60	

Note: Air quantities are for outdoor air with adequate temperature control and filtration.
[a] Special contaminant control systems may be required.
[b] cfm/locker.

ENVIRONMENTAL AND REGULATORY CONSIDERATIONS

There are many different climate zones, and a system that is optimal for one may not be the best choice in another. Since climate affects energy consumption, the effect of the average temperature should be considered in selecting energy sources, equipment, and the design of the entire building for optimum life cycle cost; average temperature is a more important consideration for energy consumption than the extremes.

Many states have special regulations for school buildings. Fire, building, ventilation, and noise control codes all may have a great effect on design; ventilation requirements vary in different states and different localities. If the regulations are poor or nonexistent, good design practice generally dictates the minimum air change required and the outdoor air quantity. In more severe climates,

The preparation of this chapter is assigned to TC 9.8, Large Building Air-Conditioning Applications.

consideration may be given to treatment of the return air to reduce odors and fumes.

Energy studies are a legal requirement in many states. Computer programs are available to prepare such studies. (See Chapter 28 of the 1989 ASHRAE *Handbook—Fundamentals.*)

CONTROLS

The ideal climate-control system for educational facilities should serve diverse areas, regardless of the use of the various spaces involved. For example, when showing movies or video tapes with the lights turned off, heating may be required, while a nearby area may require cooling. During evenings and weekends, some areas may be used for special events or instructions while other rooms are unoccupied. These varying and often unpredictable conditions are best met by separate control of each space. Equipment should be capable of efficient operation at extremely low and variable loads. It may be desirable to design for future expansion with minimal investment.

Since most heat is required at night and weekends when buildings are at least partially unoccupied, *night set-back* is virtually mandatory.

A central control system is also desirable, especially one that ensures equipment is turned either to night set-back or off when not in use. Because of the large portion of both the heating and cooling load that can be attributed to the required ventilation, provision should be made in the control system to close the outside air dampers when the space is unoccupied.

FINANCIAL AND SPACE CONSIDERATIONS

Educational facility planners and owners generally agree that the highest quality construction that can be afforded should be built, since these facilities must serve 20 to 40 years or more. The taxpayers pay for all costs of public schools—initial, operating, and maintenance—simultaneously. Thus, schools pay no taxes, which make life cycle costs the only valid criteria as far as taxpayers are concerned, although many schools still use the most floor area per unit cost criteria.

Equipment requires building space, which may increase the cost of the usable floor space. Building costs are usually described in dollars per gross area. A better measure is the cost per 'usable' area per student or per classroom. The equipment selected should occupy minimum floor space in relatively inactive areas. This requirement has led to increasing use of roof-mounted equipment. Central equipment rooms are commonly large installations where the equipment is easily supervised, and a great deal of the maintenance is concentrated in one area. The engineer must evaluate these and other factors to arrive at an optimum choice for the particular project. Adequate all-year maintenance and service capability must be provided for all systems. Rooftop equipment designed to be installed in climates with severe cold weather must have adequate provisions for maintenance and service.

Protection of equipment from vandalism should be considered. All items accessible to students should be of heavy construction and resistant to tampering.

Flexibility of the mechanical system may be an important consideration. New classes and teaching methods may be adopted, requiring changes in ductwork, diffusers, and temperature control zones. The open classroom layout is an example; it may later be partitioned into conventional classroom spaces.

CLASSROOM LOAD PROFILE

Year-round cooling may be accomplished, at least partially, by use of an outdoor air economizer.

The heat flow characteristics of exterior rooms vary with internal heat gain, envelope construction, glass area, exposure, and outside temperature and insolation. It is common for different exterior rooms in the same facility to simultaneously experience the need for both heating and cooling during the heating season. An explanation of the dynamics of space heat flow characteristics can be found in Chapter 26 of the 1989 ASHRAE *Handbook—Fundamentals.*

The time of occupancy of educational facilities may be irregular. It is a widespread assumption that schools are fully occupied from 8:00 a.m. to 3:30 p.m. and then vacant. School custodians, adult education, and civic affairs such as band practice and scout meetings make the total occupancy time far longer than the normal school year of 1400 to 1500 h. Surveys show anywhere from 2000 to 3300 h of occupancy with *equivalent full occupancy* (EFO) of about 50%. Similar to equivalent full load, equivalent full occupancy is a mathematical integration of the total number of classroom hours per year divided by the number of rooms in the building. EFO is a measure of potential energy saving by turning off lights and using climate-control equipment in the unoccupied parts of the building.

ENERGY CONSIDERATIONS

Most new buildings have to meet ASHRAE *Standard* 90A-80 or ASHRAE *Standard* 90.1-1989 and the codes of appropriate jurisdiction. Usage patterns must be considered carefully in design selection of high efficiency equipment, proper insulation, and reduced fenestration, which are other ways to reduce energy consumption. In some cases, energy studies must be submitted. Simulated computer programs for energy studies are the most accurate for evaluating alternatives.

An *outdoor air economizer* cycle, which uses cool outdoor air during many hours of the year, can handle much of the high internal load. Proper sizing of the outdoor air dampers to balance the pressure loss between the indoor air return system and the outdoor air supply system is imperative for proper control of the economizer cycle. Low-leakage outdoor air dampers should be used.

Night set-back, activated by a central control system, conserves energy by locking out mechanical cooling and lowering the heating requirement during unoccupied periods. Since approximately 90% of the heat requirement in a school occurs when the building is unoccupied, this often represents a substantial saving. Intermittent use of the building at night may make individual area night set-back more economical.

Water-to-air heat pumps may conserve energy if the school has interior spaces. Excess heat generated during the day may be stored in water tanks to provide heating at night. Some large package water-to-air heat pumps now have built-in air economizer cycles.

Solar energy can supplement other energy sources. Air-to-air heat recovery devices should be considered for those areas that require substantial exhaust and outdoor air makeup.

Air-to-air heat pumps may make electrical heating costs competitive with fossil fuels. Life cycle costs of air-to-air heat pumps are generally comparable with other systems.

Proper cooling requires the distribution of relatively large amounts of air—substantially more air than is generally needed for heating—which may cause severe discomfort at temperatures around 55°F. Therefore, this air must be distributed evenly to eliminate drafts and excessive air noise.

In extremely cold areas, care also must be taken to heat window areas, although, with the trend to more insulation and smaller windows, the problem of downdrafts is less critical.

EQUIPMENT AND SYSTEM SELECTION

Architectural planning of educational facilities has considerable influence on the building's heating and cooling load and on the selection of equipment. The use of ASHRAE *Standard* 90.1-1989 can lead to sufficient reduction in loads to allow year-round air-conditioning systems to be installed at costs that compare favorably to systems used only for heating and ventilating.

No trends in educational facility design as related to heating, ventilating, and air-conditioning systems are evident. In smaller single-building facilities, centralized systems are often applied. These systems include unit ventilator, rooftop, and single and multizone-type units. Central station equipment, especially variable volume systems, continues to have wide application in larger facilities; water-to-air heat pumps have also been used.

Studies indicate that control of temperature is a more significant comfort factor than control of humidity, providing the latter is maintained within reasonable limits. Excessive variations in temperature may cause more discomfort than maintenance of slightly higher or lower temperatures with reasonable variations of ± 3 °F.

The *systems approach* is another method applied to educational facility construction. This method establishes a performance specification as the basis for bidding for a complete facility. Many of the building components are prefabricated in the factory, thereby reducing construction time on the job site. This allows a bidder greater latitude in selecting individual systems, as long as the overall performance requirement is met. However, close coordination of all systems proposed for the facility is required to ensure compatibility.

Single Duct with Reheat

For energy conservation, reheat systems have been restricted by ASHRAE *Standards* 90A-80 and 90.1-1989 and usually cannot be justified for schools, unless recovered energy is used for reheat.

Dual-conduit and dual-duct systems are recommended if they are designed to minimize air quantities to those required and incorporate adequate energy conservation features to make them as economical as other systems.

Multizone

Reset controllers should be used to receive signals from zone thermostats and maintain hot deck and cold deck at optimum temperatures, which minimizes mixing of airstreams. For energy conservation, it is important to select similar exposures for multizone units to minimize mixing of hot and cold airstreams. Multizone units with a neutral zone that mix either heated or cooled air with return air, but do not mix heated and cooled air, eliminate the energy-wasting practice of blending heated and cooled airstreams.

Factory-assembled units for rooftop installation are available in capacities from about 10 to 80 tons in multizone, dual-duct, and single-zone configurations. Options are offered for type of filtration, refrigerant reheat, heating source, return air fan, and control cycles. Routine servicing features may have to be added. Life expectancy may be less than for indoor equipment. However, some school districts have successfully operated rooftop units for over 15 years, with maintenance costs significantly less than schools with other systems.

Single Duct with Variable Volume

Supply air capacity is adjusted to space load by automatic volume control. Systems for exterior rooms are usually zoned by exposure.

Variable volume systems may have reheat, double-duct, or fan-powered terminals to serve each zone. Each unit is controlled in two stages. The first stage reduces the volume of air delivered to meet cooling loads until minimum ventilation rates are reached. The second stage introduces heat, either with hot air or heating at the terminal. Fan-powered terminals can continue to heat even when the primary air system is not operated during unoccupied periods.

Package Units

Unit ventilators or package units with either split or single, direct expansion air-conditioning units may serve one room or several rooms with a common exposure. Water-to-air (unitary or single package) heat pumps are also used. In this system, an uninsulated, constant flow unzoned water circuit serves each unit.

The closed loop incorporates a heat rejector (tower) and heat source (boiler). Certain operating conditions balance the loop; therefore, no heat should be added or rejected. Units simultaneously heat and cool according to room need. A water storage tank added to the circulating system can accumulate excess heat during the day for use at night.

Air-to-air heat pumps are also applicable to schools. Small rooftop and self-contained unit ventilators, featuring heat pump cycle, are also available.

SOUND LEVELS

Typical ranges of design levels are as follows:

	A-Sound Levels Decibels	Desired NC (Noise Criteria)
Libraries, Classrooms	35-45	30-40
Laboratories, Shops	40-50	35-45
Gyms, Multipurpose corridors	40-55	35-50
Kitchens	45-55	40-50

For further information, see Chapter 42 of this volume.

The lower end of the range of each room type is justifiable for buildings—either in quiet locations or where provisions have been made for reducing sound transmission through exterior walls. The upper end of the range is appropriate for buildings in relatively noisy locations, without adequate exterior wall sound transmission loss, or where the owner's requirements and budget indicate that lower costs are desirable.

REMODELING

To the consulting engineer, remodeling consists of two phases. First the engineer must ascertain what type of equipment is installed in the existing building, what should be reused or replaced, the overall facilities, and the possible space for future equipment. Existing drawings, if available, may not reflect the current building because of changes made after construction. The engineer must then design a system that is compatible with the contemplated changes to this building in terms of time, cost, energy consumption, and appearance. Remodeling is frequently attempted during normal vacation shutdowns and after school hours to keep the facilities available for service.

Central systems, rooftop equipment, window units, unitary systems, and self-contained unit ventilators have all been used successfully. Life cycle cost analyses have shown that lowest equipment cost does not result in the lowest total cost.

PRIMARY, MIDDLE, AND SECONDARY SCHOOLS

Load Characteristics

Primary and secondary schools are characterized by relatively small buildings, with the exception of some high schools, which may be large and approach the size and complexity of a small college. Most secondary school facilities, however, are low-rise structures of medium to small size.

Changes in teaching methods have affected educational facilities. Double-loaded corridor designs of individual classrooms have given way to open plan layouts and have reverted back again. Open plan schools often have numerous portable partitions that are frequently changed. The more flexible the HVAC system, the less chance for obsolescence and the easier it is to adapt to changing requirements.

Design Criteria

Design criteria are the same as those discussed in the section General Design Considerations, except that for kindergarten to fourth grade, special care must be taken to have warm floors.

Equipment

Equipment for small buildings should be simple to operate and require no skilled personnel. Specialized, trained operating personnel are often not available, and small schools usually find it necessary to hold down operating costs. Since the buildings are usually smaller and less elaborate in the lower grades, less elaborate HVAC systems are generally indicated.

Regional schools, on the other hand, may house large office areas, auditoriums, laboratories, shops, computer rooms, dispensaries, pools, gymnasiums, maintenance shops, and cafeterias. Usually more and better-skilled operating personnel are available, and systems can be more sophisticated and varied.

These larger buildings present many specialized problems that are covered individually in other sections of this volume.

COLLEGES AND UNIVERSITIES

The variety of buildings found on the campuses of colleges and universities resembles that of a small city. Consequently, the problems and requirements of colleges and universities are different from those found in secondary schools. Many colleges and universities have satellite campuses dispersed within the city or throughout the entire state. Therefore, the design requirements for campus buildings are influenced by many factors. Specific and special requirements for such buildings are established by the Planning Office or the Department of Physical Plant, also known as the Department of Buildings and Grounds.

The HVAC load for a given building on campus depends on the specific building use and the geographical location of the campus. The design criteria for a given campus building is established by the requirements of the users.

Following is a list of major building types found on college and university campuses:

1. Administrative buildings
2. Buildings housing animal colonies
3. Auditoriums and theaters
4. Computer sites
5. Classroom buildings
6. Dormitory buildings
7. Garages and auto repair shops
8. Gymnasiums
9. Freight handling and storage buildings
10. Ice rinks
11. Museums
12. Stores (book, gift, clothing, restaurants, and cafeterias)
13. Student and faculty apartment buildings
14. Natatoriums
15. Tennis pavilions
16. Laboratory buildings
17. Maintenance and repair shops
18. Central plants (boiler, refrigeration, power generation)
19. Barns
20. Poultry housing
21. Television studios
22. Radio stations
23. Hospitals
24. Chapels
25. Student unions

Although the college or university campus may consist of any combination of the building types listed, seldom is an entire campus constructed on a new location. This is unlike the secondary schools, where a new complex is erected to meet the needs of a new or expanding community. However, clusters of similar use buildings, such as research centers and dormitories, are generally erected on central locations. These clusters may have individual heating-cooling equipment or may be connected to the central plant designed for the particular center.

Central monitoring and control systems are coming into common use in colleges and school districts where monitoring of multiple systems within one or more buildings is required.

Central Heating and Refrigeration Plants

When planning a central heating and/or refrigeration plant for an existing college or university, various factors should be considered. In general, cost per given unit of energy is lower when it comes from the central plant. However, cost of distribution, piping (either direct, buried, or in utility tunnels), and pumping may offset the savings offered by a central plant. Detailed analyses of several central versus individual plant schemes, based on accurate long-term projections for the institution, are essential for the most desirable solution. These include such factors as:

1. Central maintenance versus individual building maintenance
2. Pollution and noise control
3. Topography and easement characteristics
4. Type of central distribution such as steam, high-temperature water, or bleed versus convertor takeoffs
5. Types of fuel and HVAC equipment available or projected
6. Schedules of operations for the various buildings including diversity factors
7. Future expansion

The uncertainty of fuel prices and potential unavailability of certain fuels mandate investigation of alternate energy sources for central plants and individual buildings, where feasible. One such source is solid waste, which could be used in central plants of larger campuses. Solar energy should also be considered.

Central Monitoring and Control System

The multibuilding campus requires constant monitoring of the various systems within these buildings. Manual operation of these systems requires large staffs and, consequently, becomes uneconomical. Central monitoring and control systems can often be justified on the basis of savings in labor (see Chapter 35). In addition, central monitoring and control systems offer some means of controlling energy costs using methods such as *scheduling* (which remotely shuts off equipment when it is not needed), *demand limiting* (which selectively switches off electric loads when demand limit is approached), and *duty cycling* (which cycles constant volume equipment during part-load operation).

The central monitoring and control systems can perform other functions, such as the following:

1. Monitoring and control of various accesses (building security)
2. Fire alarm system monitoring
3. Lighting control and maintenance scheduling
4. Optimization of operating equipment
5. Improved control system operation
6. More complete and accurate record-keeping of building, equipment, systems, and energy use information

REFERENCES

ASHRAE. 1980. Energy conservation in new building design. ASHRAE *Standard* 90A-1980.

ASHRAE. 1981. Thermal environmental conditions for human occupancy. ASHRAE *Standard* 55-1981.

ASHRAE. 1989. Ventilation for acceptable indoor air quality. ASHRAE *Standard* 62-1989.

ASHRAE. 1989. Energy efficient design of new buildings except low-rise residential buildings. ASHRAE *Standard* 90.1-1989.

CHAPTER 7

HEALTH FACILITIES

CONTINUOUS advances in medicine and technology require constant reevaluation of the air-conditioning needs of hospitals and medical facilities. While medical evidence has shown that proper air conditioning is helpful in the prevention and treatment of many conditions, the relatively high cost of air conditioning demands efficient design and operation to ensure economical energy management.

Health care occupancy classification, based on the latest occupancy guidelines from the National Fire Protection Association (NFPA), should be considered early in the design concept phase of the project. Health care occupancy is important for fire protection (smoke zones, smoke control), and for the future adaptability of the HVAC system for a more restrictive occupancy.

The specific environmental conditions required by a particular medical facility may vary from those in this chapter, depending on the agency responsible for the medical facility environmental standard. Among the agencies that may have standards are: state and local health agencies, Department of Health and Human Services, Indian Health Service, Public Health Service, Medicare/Medicaid, Department of Defense (Army, Navy, Air Force), the Veterans Administration, and the Joint Commission on Accreditation of Hospitals (JCAH). It is sometimes advisable to discuss infection control objectives with each hospital's infection control committee.

The general hospital was selected as the basis for the fundamentals outlined in the first section of this chapter because of the variety of services it provides. Environmental conditions and design criteria apply to comparable areas in other health facilities.

The general acute care hospital has a core of critical care spaces including operating rooms, labor rooms, delivery rooms, and a nursery. Usually the functions of radiology, laboratory, central sterile, and the pharmacy are located close by the critical care space. In-patient nursing, including intensive care nursing, is in the complex. The facility also incorporates emergency room, kitchen, dining and food service, a morgue, and central housekeeping support.

Because the growing trend in medical care is to outpatient services, criteria for outpatient facilities are given in the second section of this chapter. Outpatient surgery is performed with the anticipation that the patient will not stay overnight. An outpatient facility may be part of an acute care facility, a free-standing unit, or part of a medical facility such as a medical office building.

AIR CONDITIONING IN THE PREVENTION AND TREATMENT OF DISEASE

Dry conditions constitute a hazard to the ill and debilitated by contributing to secondary infection or infection totally unrelated to the clinical condition causing hospitalization. Clinical areas devoted to upper respiratory disease treatment, acute care, as well as the general clinical areas of the entire hospital, should be maintained at a relative humidity of 30 to 60%.

Patients with chronic pulmonary disease often have viscous respiratory tract secretions. As these secretions accumulate and increase in viscosity, the patient's exchange of heat and water dwindles. Under these circumstances, warm, humidified, inspired air is essential to prevent dehydration.

Patients needing oxygen therapy or those with a tracheotomy require special attention to ensure a warm humid supply of inspired air. Cold, dry oxygen or the bypassing of the nasopharyngeal mucosa presents an extreme situation. Rebreathing techniques for anesthesia and enclosure in an incubator are special means of treating the impaired heat loss that must be considered under therapeutic environments.

Hospital air conditioning assumes a more important role than just the promotion of comfort. In many cases, proper air conditioning is a factor in patient therapy, and, in some instances, it is the major treatment.

Studies show that patients in fully air-conditioned rooms generally have more rapid physical improvement than do those in hot and humid rooms. Patients with thyrotoxicosis do not tolerate hot, humid conditions or heat waves very well. A cool, dry environment favors the loss of heat by radiation and evaporation from the skin and may save the life of the patient.

Cardiac patients may be unable to maintain the circulation necessary to ensure normal heat loss. Therefore, air conditioning hospital wards and rooms of cardiac patients, particularly those with congestive heart failure, is necessary and is considered therapeutic. Individuals with head injuries, those subjected to brain operations, and those with barbiturate poisoning may have hyperthermia, especially in a hot environment, because of a disturbance in the heat regulatory center of the brain. Obviously, an important factor in recovery is an environment in which the patient can lose heat by radiation and evaporation—namely, a cool room with dehumidified air.

A hot, dry environment of 90°F dry bulb and 35% rh has been successfully used in treating patients with rheumatoid arthritis.

Burn patients need a hot environment and high relative humidity. A ward for severe burn victims should have temperature controls that permit adjusting the room temperature up to 90°F dry bulb and the relative humidity up to 95%.

HOSPITALS

Although proper air conditioning is helpful in the prevention and treatment of disease, the application of air conditioning to health facilities presents many problems not encountered in the usual comfort conditioning system.

The preparation of this chapter is assigned to TC 9.8, Large Building Air-Conditioning Applications.

The basic differences between air conditioning for hospitals (and related health facilities) and other building types stem from (1) the need to restrict air movement in and between the various departments; (2) the specific requirements for ventilation and filtration to dilute and remove contamination in the form of odor, airborne microorganisms and viruses, and hazardous chemical and radioactive substances; (3) the need for different temperature and humidity requirements for various areas; and (4) the need for sophistication in design to permit accurate control of environmental conditions.

Infection Sources and Control Measures

Bacterial Infection. Examples of bacteria that are highly infectious and transported within air or air and water mixtures are *Mycobacterium tuberculosis* and *Legionalla pneumophilia* (Legionnaire's disease). Wells (1934) showed that droplets (or infectious agents) that are 5 μm or less in size can remain airborne indefinitely. Isoard *et al.* (1980) and Luciano (1984) have shown that 99.9% of all bacteria present in a hospital are removed by 90 to 95% efficient filters (ASHRAE *Standard* 52-76). This is because bacteria are typically present in colony-forming units that are larger than 1 μm. Some authorities recommend that HEPA filters having DOP test filtering efficiencies of 99.97% be used in certain areas.

Viral Infection. Examples of viruses that are transported by and virulent within air are *Varicella* (chicken pox/shingles), *Rubella* (German measles), and *Rubeola* (regular measles). Epidemiological evidence and other studies indicate that the airborne viruses that transmit infection are so small that no known filtering technique is effective. Attempts to deactivate viruses with ultraviolet light and chemical sprays have not proven to be reliable or effective enough to be recommended by most codes as a primary infection control measure for viruses or bacteria. Therefore, isolation rooms and isolation anterooms with appropriate ventilation-pressure relationships are the primary infection control method used to prevent the spread of airborne viruses in the hospital environment.

Molds. Evidence indicates that some molds such as *Aspergillis* can be fatal to advanced leukemia, bone marrow transplant, and other immunocompromised patients.

Outdoor Air Ventilation. If outdoor air intakes are properly located and areas adjacent to outdoor air intakes are properly maintained, outdoor air, in comparison to room air, is virtually free of bacteria and viruses. Infection control problems frequently involve a bacterial or viral source within the hospital. Ventilation air is a dilutant of the viral and bacterial contamination within a hospital. Ventilation systems remove airborne infectious agents from the hospital environment if the systems are properly designed, constructed, and maintained to preserve the correct pressure relations between functional areas.

Temperature and Humidity. These conditions can inhibit or promote the growth of bacteria and activate or deactivate viruses. Some bacteria such as *Legionella pneumophilia* are basically waterborne and survive more readily in a humid environment. Codes and guidelines specify temperature and humidity range criteria in some hospital areas as an infection control measure, as well as a comfort criteria.

AIR QUALITY

Systems must also provide air virtually free of dust, dirt, odor, chemical, and radioactive pollutants. In some cases, outside air quality is hazardous to patients suffering from cardiopulmonary conditions or respiratory or pulmonary conditions. In such instances, systems that intermittently provide maximum allowable recirculated air should be considered.

Outdoor Intakes. These intakes should be located as far as is practical (on directionally different exposures whenever possible) but not less than 30 ft from exhaust outlets of combustion equipment stacks, ventilation exhaust outlets from the hospital or adjoining buildings, medical-surgical vacuum systems, plumbing vent stacks, or from areas that may collect vehicular exhaust and other noxious fumes. Outdoor intakes may be close to outlets that exhaust air suitable for recirculation; however exhaust air must not short-circuit into intakes of outdoor air units or fan systems that are used for smoke control. The bottom of outdoor air intakes serving central systems should be located as high as practical but not less than 6 ft above ground level, or if installed above the roof, 3 ft above the roof level (DHHS 1984a).

Exhaust Outlets. These exhausts should be located a minimum of 10 ft above ground, away from occupied areas or from doors and openable windows. Preferred location for exhaust outlets is at roof level projecting upward or horizontally away from outdoor intakes. Care must be taken in locating highly contaminated exhausts, *i.e.*, engines, fume hoods, biological safety cabinets, kitchen hoods, and paint booths. Prevailing winds, adjacent buildings, and discharge velocities must be taken into account (see Chapter 26 of the 1988 ASHRAE *Handbook—Equipment*). In critical applications, wind tunnel studies or computer modeling may be appropriate.

Air Filters. A number of methods are available for determining the efficiency of filters in removing particulates from an airstream (see Chapter 10 of the 1988 ASHRAE *Handbook—Equipment*). All central ventilation or air-conditioning systems should be equipped with filters having efficiencies no less than those indicated in Table 1. Where two filter beds are indicated, Filter Bed No. 1 should be located upstream of the air-conditioning equipment, and Filter Bed No. 2 should be downstream of the supply fan, any recirculating spray water systems, and water-reservoir type humidifiers. Appropriate precautions should be observed to prevent wetting of the filter media by free moisture from humidifiers. Where only one filter bed is indicated, it should be located upstream of the air-conditioning equipment. All filter efficiencies are based on ASHRAE *Standard* 52-76.

The following are guidelines for filter installations.

1. HEPA filters having DOP test efficiencies of 99.97% should be used on air supply systems serving rooms used for clinical treatment of patients with a high susceptibility to infection from leukemia, burns, bone marrow transplant, organ transplant, or acquired immunodeficiency syndrome (AIDS). They should also be used on the exhaust discharge air from fume hoods or safety cabinets in which infectious or highly radioactive materials are processed and should be designed and equipped to permit safe removal, disposal, and replacement of contaminated filters.
2. All filters should be installed to prevent leakage between the filter segments and between the filter bed and its supporting frame. A small leak that permits any contaminated air to escape through the filter can destroy the usefulness of the best air cleaner.
3. A manometer should be installed in the filter system to provide a reading of the pressure drop across each filter bank. This precaution furnishes a more accurate means of knowing when filters should be replaced than by relying on visual observation.
4. High efficiency filters should be installed in the system with adequate facilities provided for maintenance without introducing contamination into the delivery system or the area served.
5. Because high efficiency filters are expensive, the hospital should project the filter bed life and replacement costs, and incorporate these into their operating budget.
6. During construction, openings in ductwork and diffusers should be sealed to prevent intrusion of dust, dirt, and hazardous materials. Such contamination is often permanent and provides a medium for the growth of infectious agents. Existing or new filters may rapidly become contaminated by construction dust.

Table 1 Filter Efficiencies for Central Ventilation and Air-Conditioning Systems in General Hospitals

Minimum Number of Filter Beds	Area Designation	Filter Efficiencies, %		
		Filter Bed No. 1[a]	Filter Bed No. 2[a]	Filter Bed No. 3[b]
3	Orthopedic operating room Bone marrow transplant Organ transplant operating room	25	90	99.97[c]
2	General procedure operating rooms Delivery rooms Nurseries Intensive care units Patient care Treatment Diagnostic and related areas	25	90	
1	Laboratories Sterile storage	80		
1	Food preparation areas Laundries Administrative Bulk storage Soiled holding areas	25		

[a]Based on ASHRAE *Standard* 52-76.
[b]Based on DOP test.
[c]HEPA filters at air outlets.

Air Movement

The data given in Table 2 illustrate the degree of contamination that can be dispersed into the air of the hospital environment by one of the many routine activities for normal patient care. The bacterial counts in the hallway also clearly indicate the spread of this contamination.

Table 2 Influence of Bedmaking on Airborne Bacterial Count of Hospitals

Item	Count Per Cubic Foot	
	Inside Patient's Room	Hallway Near Patient's Room
Background	34	30
During bedmaking	140	64
10 min after	60	40
30 min after	36	27
Background	16	
Normal bedmaking	100	
Vigorous bedmaking	172	

Because of these necessary activities and the resultant dispersal of bacteria, air-handling systems should provide air movement patterns that minimize the spread of such contamination.

Controlling airflow in hospitals and other health facilities to minimize the spread of contaminants is desirable in design and operation. Undesirable airflow between rooms and floors is often difficult to control because of open doors, movement of staff and patients, temperature differentials, and stack effect accentuated by vertical openings such as chutes, elevator shafts, stairwells, and mechanical shafts common to hospitals. While some of these factors are beyond practical control, the effect of others may be minimized by terminating shaft openings in enclosed rooms and by designing and balancing air systems to create positive or negative air pressure within certain rooms and areas.

Systems serving highly contaminated areas such as contagious or immunocompromised isolation rooms and autopsy rooms should maintain a positive or negative air pressure within these rooms relative to adjoining rooms or the corridor (Murray 1988). The pressure is obtained by supplying more or less air to the area than is exhausted from it. This induces a flow of air into the area around the perimeters of doors and prevents an outward airflow. The operating room offers an example of an opposite condition. This room, which requires air that is free of contamination, must be pressurized relative to adjoining rooms or corridors to prevent any air movement into the operating room from these relatively highly contaminated areas.

Differentials in air pressure can be maintained only in an entirely closed room. Therefore it is important to obtain a reasonably close fit of all doors or closures of openings between pressurized areas. This is best accomplished by use of weather stripping and drop bottoms on doors. The opening of a door or closure between two such areas instantaneously reduces any existing pressure differential between them to such a degree as to nullify the effectiveness of the pressure. When such openings occur, a natural interchange of air takes place because of thermal currents resulting from temperature differences between the two areas.

For critical areas requiring the maintenance of pressure differentials to adjacent spaces while providing for personnel movement between the spaces and areas, the use of appropriate air locks or anterooms is indicated.

Figure 1 shows the bacterial count in a surgery room and its adjoining rooms during a normal surgical procedure. These bacterial counts were taken simultaneously in each of the rooms. The relatively low bacteria counts in the surgery room, compared with those of the adjoining rooms, are attributed to less activity within operating rooms and to higher air pressure.

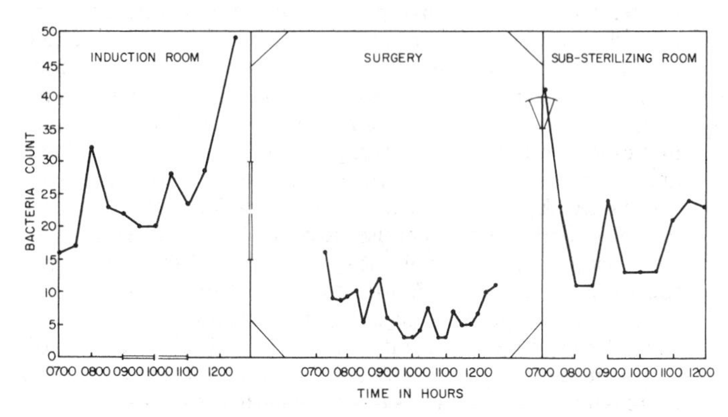

Fig. 1 Typical Airborne Contamination in Surgery and Adjacent Areas

In general, it is recommended that air supply outlets to sensitive ultraclean areas, as well as highly contaminated areas, be located on the ceiling with perimeter or several exhaust inlets near the floor. This provides a downward movement of clean air through the breathing and working zones to the contaminated floor area for exhaust. In this respect, the bottoms of return or exhaust openings should not be below 3 in. above the room floor.

The laminar airflow concept developed for industrial clean room use has attracted the interest of some medical authorities. There are advocates of both the vertical and horizontal laminar airflow systems with and without fixed or movable walls around the surgical team (Pfost 1981). Many medical authorities do not advocate laminar airflow for surgeries but encourage air systems similar to those in this chapter.

Laminar airflow in surgical operating rooms is defined as airflow that is predominantly unidirectional when not obstructed. The unidirectional laminar airflow pattern is commonly attained at a velocity of 90 $\pm$20 fpm.

Laminar airflow systems have shown promise for rooms used for the treatment of patients who are highly susceptible to infection. Among such patients would be the badly burned and those undergoing radiation therapy, concentrated chemotherapy, organ transplants, amputations, and joint replacement. Bench-type units may be used to a great extent in such areas as pharmacy, tissue banks, and laboratories.

Temperature and Humidity

Specific recommendations for design temperatures and humidities are given in the next section, Specific Design Criteria. Temperature and humidity requirements for other in-patient areas not covered should be 72 °F or less.

Pressure Relationships and Ventilation

Table 3 covers ventilation standards for comfort, as well as for asepsis and odor control, in areas of acute care hospitals that directly affect patient care. Table 3 does not necessarily reflect the criteria by DHHS or any other group. If specific organizational criteria must be met, refer to that organization's literature. Ventilation in accordance with ASHRAE *Standard* 62-1989, Ventilation for Acceptable Indoor Air Quality, shall be used for areas where specific standards are not given. In any case, where a higher outdoor air requirement is called for in ASHRAE *Standard* 62-1989 than in Table 3, the higher value shall be used. Specialized patient care areas including organ transplant units and burn units shall have additional ventilation provisions for air quality control as may be appropriate.

Design of the ventilation system must, as much as possible, provide air movement that is from clean to less clean areas. In critical care areas, constant volume systems should be employed to assure proper pressure relationships and ventilation except as noted below for unoccupied rooms. In noncritical patient care areas and staff rooms, variable air volume systems may be considered for energy conservation. When using variable air volume systems within the hospital, special care should be taken to assure that minimum ventilation rates (as required by codes) are maintained, and that pressure relationships between various departments are maintained. A method such as air volume tracking between supply, return, and exhaust could be used to control pressure relationships when using variable air volume systems (Lewis 1988). In Table 3, those areas that require continuous control are noted with "P", "N", or "E" for positive, negative, or equal pressure. Where ± is used, there is no requirement for continuous directional control.

The number of air changes may be reduced to 25% when the room is unoccupied if provisions are made to ensure that the number of air changes indicated is reestablished any time the space is occupied, and providing the pressure relationship to the surrounding rooms is maintained when the air changes are reduced.

Areas not indicated as requiring continuous directional control may have ventilation systems shut down when space is unoccupied and ventilation is not otherwise needed.

Because of the cleaning difficulty and potential for buildup of contamination, recirculating room units must not be used in areas marked "no". Note that the standard-type recirculating room unit may also be impractical for primary control where exhaust to the outside is required.

In rooms having hoods, extra air must be supplied for hood exhaust so that the designated pressure relationship is maintained. Refer to Chapter 14 for further discussion of laboratory ventilation.

For maximum energy conservation, use of a recirculated system is preferred. If an all-outdoor air system is used, an efficient heat recovery method should be considered.

Smoke Control

As the ventilation design is developed, proper smoke control strategy must be considered. Passive systems rely on fan shutdown, smoke and fire partitions, and openable windows. Proper treatment of duct penetrations must be observed.

Active smoke control systems use the ventilation system to create areas of positive and negative pressures which, along with fire and smoke partitions, limit the spread of smoke. The ventilation system may be used in a smoke removal mode where the products of combustion are exhausted by mechanical means. As design of active smoke control systems continues to evolve, the engineer and code authority should carefully determine system operation and configuration. Refer to Chapter 42, NFPA 90A, NFPA 92A, and NFPA 101.

SPECIFIC DESIGN CRITERIA

The environmental requirements of each of the departments/spaces within the six principal functions of an acute care general hospital, which include (1) surgery and critical care, (2) nursing, (3) ancillary, (4) diagnostic and treatment, (5) sterilizing and supply, and (6) service, differ to some degree according to their function and the procedures carried out in them. This section describes how these departments/spaces are actually used and covers details of design requirements. Close coordination with health care planners and medical equipment specialists is essential in the (mechanical) design and construction of health facilities to achieve the desired conditions.

Surgery and Critical Care

Surgery. No area of the hospital requires more careful control of the aseptic condition of the environment than does the surgical suite. The systems serving the operating rooms, including cystoscopic and fracture rooms, require careful design to reduce to a minimum the concentration of airborne organisms. The greatest amount of the bacteria found in the operating room comes from the surgical team and is a result of their activities during surgery.

During an operation, most members of the surgical team are in the vicinity of the operating table, which creates the undesirable situation of concentrating contamination in this highly sensitive area.

Studies of operating room air distribution systems and observation of installations in industrial clean rooms indicate that delivery of the air from the ceiling, with a downward movement to several exhaust inlets located on opposite walls, is probably the most effective air movement pattern for maintaining the concentration of contamination at an acceptable level. Completely perforated ceilings, partially perforated ceilings, and ceiling-mounted diffusers have been applied successfully (Pfost 1981).

Operating Rooms. In the average hospital, operating rooms are not in use more than 8 to 12 h per day (excepting emergencies). For this reason, and for energy conservation, the air-conditioning system should allow a reduction of air supplied to all or a portion of the operating rooms. However, positive space pressure should be maintained at reduced air volumes to ensure sterile conditions. Consultation with the hospital surgical staff will determine the feasibility of providing this feature.

A separate air-exhaust system or special vacuum system should be provided for removal of anesthetic trace gases (NIOSH 1975). Medical vacuum systems have been used for removal of nonflammable anesthetic gases (NFPA 99). One or more outlets in each operating room located to permit connection of the anesthetic machine scavenger hose may be used.

Although good results from air disinfection of operating rooms by irradiation have been reported, this method of disinfection is seldom used. The reluctance to use irradiation may be attributed to factors such as special designs required for installation, protective measures necessary for patients and personnel, constant vigilance relative to lamp efficiency, and the required maintenance.

Table 3 General Pressure Relationships and Ventilation of Certain Hospital Areas

Function Space		Pressure Relationships to Adjacent Areas	Minimum[a] Air Changes of Outdoor Air per Hour	Minimum Total Air Changes per Hour[b]	All Air Exhausted Directly to Outdoors	Recirculated Within Room Units
SURGERY AND CRITICAL CARE						
Operating room	(all-outdoor-air system)	P	15[c]	15	Yes	No
	(recirculating-air system)	P	5	25	Optional	No
Delivery room	(all-outdoor-air system)	P	15	15		
	(recirculating-air system)	P	5	25		
Recovery room		E	2	6	Optional	No
Nursery suite		P	5	12	Optional	No
Trauma room[d]		P	5	12	Optional	No
Anesthesia storage (see code requirements)		±	Optional	8	Yes	No
NURSING						
Patient room[e]		±	2	4	Optional	Optional
Toilet room[f]		N	Optional	10	Yes	No
Intensive care		P	2	6	Optional	No
Isolation[g]		±	2	6	Yes	No
Isolation alcove or anteroom[g]		±	2	10	Yes	No
Labor/delivery/recovery/postpartum (LDRP)		E	2	4	Optional	Optional
Patient corridor[e]		E	2	4	Optional	Optional
ANCILLARY						
Radiology	X-ray (surgery and critical care)	P	3	15	Optional	No
	X-ray (diagnostic and treatment)	±	2	6	Optional	Optional
	Darkroom	N	2	10	Yes[h]	No
Laboratory, general		N	2	6	Yes	No
Laboratory, bacteriology		N	2	6	Yes	No
Laboratory, biochemistry		P	2	6	Optional	No
Laboratory, cytology		N	2	6	Yes	No
Laboratory, glasswashing		N	Optional	10	Yes	Optional
Laboratory, histology		N	2	6	Yes	No
Laboratory, nuclear medicine		N	2	6	Yes	No
Laboratory, pathology		N	2	6	Yes	No
Laboratory, serology		P	2	6	Optional	No
Laboratory, sterilizing		N	Optional	10	Yes	No
Laboratory, media transfer		P	2	4	Optional	No
Autopsy		N	2	12	Yes	No
Nonrefrigerated body holding room[i]		N	Optional	10	Yes	No
Pharmacy		P	2	4	Optional	Optional
DIAGNOSTIC AND TREATMENT						
Examination room[e]		±	2	6	Optional	Optional
Medication room		P	2	4	Optional	Optional
Treatment room[e]		±	2	6	Optional	Optional
Physical therapy and hydrotherapy		N	2	6	Optional	Optional
Soiled workroom or soiled holding		N	2	10	Yes	No
Clean workroom or clean holding		P	2	4	Optional	Optional
STERILIZING AND SUPPLY						
Sterilizer equipment room		N	Optional	10	Yes	No
Central medical and surgical supply	Soiled or decontamination room	N	2	6	Yes	No
	Clean workroom and sterile storage	P	2	4	Optional	Optional
	Equipment storage	±	2 (Optional)	2	Optional	Optional
SERVICE						
Food preparation centers[j]		±	2	10	Yes	No
Warewashing		N	Optional	10	Yes	No
Dietary day storage		±	Optional	2	Optional	No
Laundry, general		N	2	10	Yes	No
Soiled linen sorting and storage		N	Optional	10	Yes	No
Clean linen storage		P	2 (Optional)	2	Optional	Optional
Linen and trash chute room		N	Optional	10	Yes	No
Bedpan room		N	Optional	10	Yes	No
Bathroom		N	Optional	10	Optional[f]	No
Janitor's closet		N	Optional	10	Optional	No

P = Positive N = Negative ± = Continuous directional control not required[e]

[a]Ventilation in accordance with ASHRAE *Standard* 62-1989, Ventilation for Acceptable Indoor Air Quality, shall be used for areas where specific standards are not given. In any case, where a higher outdoor air requirement is called for in ASHRAE *Standard* 62-1989 than in Table 3, the higher value shall be used.

[b]Total air changes indicated should be either supplied or, where required, exhausted.

[c]For operating rooms, use of 100% outside air should be limited to those cases where codes require it, and only if heat recovery devices are used.

[d]The term trauma room as used here is the first aid room and/or emergency room used for general initial treatment of accident victims. The operating room within the trauma center that is routinely used for emergency surgery should be treated as an operating room.

[e]Although continuous directional control is not required, variations should be minimized and in no case should a lack of directional control allow the spread of infection from one area to another. Boundaries between functional areas (wards or departments) should have directional control. Lewis (1988) describes some methods for maintaining directional control by applying air tracking controls.

[f]For a discussion of design considerations for central toilet exhaust systems, see text under Patient Rooms.

[g]The isolation rooms described in these standards are those that might be used for infectious patients in the average community hospital. The rooms are either positively or negatively pressurized depending on the patient. Some isolation rooms may have a separate anteroom. Refer to the discussion in the chapter for more detailed information.

[h]All air need not be exhausted if darkroom equipment has scavenging exhaust duct attached and meets ventilation standards of NIOSH, OSHA, and local employee exposure limits.

[i]The nonrefrigerated body-holding room would be applicable only for facilities that do not perform autopsies on-site and use the space for short periods while waiting for body transfer to be completed.

[j]Food preparation centers shall have ventilation systems that have an excess of air supply for positive pressure when hoods are not in operation. The number of air changes may be reduced or varied to any extent required for odor control when the space is not in use.

The following conditions are recommended for operating, catheterization, cystoscopic, and fracture rooms:

1. Variable temperature range capability of 68 to 76 °F.
2. Relative humidity of 50% minimum and 60% maximum.
3. Positive air pressure within the operating rooms relative to the pressure of any adjoining rooms by supplying 15% excess air.
4. Differential pressure indicating device installed to permit air pressure readings of the rooms. Thorough sealing of all wall, ceiling, and floor penetrations and tight-fitting doors are essential to maintain readable pressure.
5. Humidity indicator and thermometers located for easy observation.
6. Filter efficiencies in accordance with Table 1.
7. Entire installation should conform to the requirements of NFPA *Standard* 99-87, Health Care Facilities.
8. Supply all air at the ceiling and exhaust air or return from at least two locations near the floor (see Table 3 for minimum ventilating rates). Bottom of exhaust outlets should be not less than 3 in. above the floor. Supply diffusers should be unidirectional type. High induction ceiling or side wall diffusers should be avoided.
9. Acoustical materials should not be used as duct linings unless 90% efficient minimum terminal filters are installed downstream of the linings. Terminal units internal insulation may be encapsulated with approved materials. Duct-mounted sound traps should be packless type or have mylar linings over acoustical fill.
10. Any spray-applied insulation and fireproofing should be treated with fungi growth inhibitor.
11. Sufficient lengths of watertight stainless steel duct should be installed downstream of humidification equipment to assure complete evaporation of water vapor before air is discharged into the room.

Control centers that monitor and permit adjustment to temperature, humidity, and air pressure may be located at the surgical supervisor's desk.

Recovery Rooms. Postoperative recovery rooms used in conjunction with the operating rooms should be maintained at a temperature of 75 °F and a relative humidity of 50% minimum and 60% maximum. Because residual anesthesia odor sometimes creates an odor problem in recovery rooms, ventilation is important, and a balanced air pressure relative to the air pressure of adjoining areas should be provided.

Anesthesia Storage Room. The anesthesia storage room must be ventilated in conformance with NFPA *Standard* 99-87, Health Care Facilities. However, mechanical ventilation only is recommended.

Obstetrical. The pressure relationship of the obstetrical department should be positive or equal to other areas.

Delivery Rooms. The design for the delivery room should conform to the requirements of surgeries.

Nursery. Air conditioning is essential for nurseries to provide the constant temperature and humidity conditions essential to care of the newborn in a hospital environment. The air movement patterns in nurseries should be carefully designed to reduce the possibility of drafts.

All air supplied to nurseries should be at or near the ceiling and all air should be removed near the floor with bottom of openings located not less than 3 in. above the floor. Air system filter efficiencies should conform to Table 1. Finned tube radiation and other forms of convection heaters should not be used in nurseries.

Full-Term Nursery. A temperature of 75 °F with relative humidity from 30% minimum to 60% maximum is recommended for the full-term nursery, examination room, and work space. The maternity nursing section should be treated similarly to protect the infant during visits with the mother. The nursery should have a positive air pressure relative to the work space and examination room, and the rooms usually interposed between the nurseries and the corridor should be similarly pressurized relative to the corridor. This will prevent the infiltration of contaminated air from outside areas.

Special Care Nursery. The design conditions for this nursery require a variable range temperature capability from 75 to 80 °F and relative humidity from 30% minimum to 60% maximum. This nursery is usually equipped with individual incubators to regulate temperature and humidity. It is desirable to maintain these same conditions within the nursery proper to accommodate infants removed from the incubators, as well as those not placed in incubators. The pressurization of these nurseries should correspond to that of the regular nurseries.

Observation Nursery. Temperature and humidity requirements for this nursery are similar to those for the full-term nursery. Because infants in these nurseries have unusual clinical symptoms, the air from this area should not enter other nurseries. A negative air pressure relative to the air pressure of the workroom should be maintained in the nursery, and the workroom usually interposed between the nursery and the corridor should be pressurized relative to the corridor.

Emergency. This department, in most instances, is the most highly contaminated area in the hospital, as a result of the soiled condition of many patients on arrival and the relatively large number of persons accompanying them. The pressure relationship of the emergency department should be negative to other areas. Waiting rooms require a minimum ventilation rate of 10 air changes per hour. Temperatures and humidities of offices and waiting spaces should be within the comfort range. A negative air pressure should be maintained in the waiting room to contain tobacco smoke.

Trauma Rooms. Trauma rooms should be ventilated in accordance with requirements in Table 3. Emergency operating rooms located near the emergency department should have the same temperature, humidity, and ventilation requirements as those of a surgery.

Nursing

Patient Rooms. The recommendations given in Tables 1 and 3 for air filtration and air change rates should be followed when using central systems for air conditioning patients' rooms to reduce cross-infection and to control odor. Rooms used for isolation of infected patients should have all air exhausted directly outdoors. A winter design temperature of 75 °F and 30% rh is recommended; 75 °F and 50% rh is recommended for summer. Each patient room should have individual temperature control. The pressure relationship of patient suites should be neutral to other areas.

Most governmental agency design criteria and codes require all air from toilet rooms to be exhausted directly outdoors. The requirement appears to be based on odor control. Chaddock (1986) analyzed odor from central (patient) toilet exhaust systems of a hospital and found that large central exhaust systems generally have sufficient dilution to render the toilet exhaust practically odorless (Chaddock 1986). For this reason, and to conserve energy, consideration should be given (with agency approval) to recirculating some fraction of toilet room air where central systems with appropriate conditioning and filtering equipment are used.

Where room-unit-type systems are used, it is common practice to exhaust an amount of air equal to the amount of outdoor air brought into the room for ventilation through the adjoining toilet room. The ventilation of toilets, bedpan closets, bathrooms, and all interior rooms should conform to applicable codes.

Intensive Care Unit. This unit serves seriously ill patients with a variety of clinical conditions ranging from the postoperative to the coronary patient. A variable range temperature capability of 75 to 80 °F and a relative humidity of 30% minimum and 60% maximum with a positive air pressure is recommended.

Immunely Suppressed Patient Units. (This includes bone marrow or organ transplant, leukemia, burn, and acquired immunodeficiency syndrome patients.) Immunely suppressed patients are highly susceptible to diseases. Some physicians prefer an isolated laminar airflow unit to protect the patient. Others are of the opinion that the conditions of the laminar cell have psychologically harmful effects on the patient and prefer flushing out the room and reducing spores in the air. An air distribution of 15 air changes supplied through a nonaspirating diffuser is often recommended. The sterile air is drawn across the patient and returned near the floor, at or near the door to the room.

In cases where the patient is immunely suppressed but not contagious, a positive pressure should be maintained between the patient room and adjacent area. Some jurisdictions may require an anteroom, which maintains a negative pressure relationship with respect to the adjacent isolation room and an equal pressure relationship with respect to the corridor, nurse's station, or common area. Exam and treatment rooms should be treated in the same manner. A positive pressure should also be maintained between the entire unit and the adjacent areas to maintain a sterile condition.

When a patient is both immunely suppressed and contagious, isolation rooms within the unit may be permanently designed and balanced to provide an equal or negative pressure relationship with respect to the adjacent area or anteroom. Alternatively, when permitted by the jurisdictional authority, such isolation rooms may be equipped with controls to enable the room to be either positive, equal, or negative in relation to the adjacent area. However, in such instances, controls in the adjacent area or anteroom must maintain the correct pressure relationship with respect to the other adjacent room(s).

A separate, dedicated air-handling system to serve the immunely suppressed patient unit simplifies pressure control and quality (Murray 1988).

Isolation Unit. The isolation room, unless located in a strictly contagious disease ward, should either protect patients from an infectious disease patient or should protect a patient with very low resistance to infection from the normal bacterial flora of the remainder of the hospital. Rooms designed for infectious patients should be negatively pressurized while rooms designed for the low resistance patient should be positively pressurized.

Temperatures and humidities should correspond to those specified for patient rooms.

The designer should work closely with health care planners and the code authority to determine the appropriate isolation room design. When indicated by the program, it may be desirable to provide more complete control with a separate anteroom as an air lock to minimize the potential for airborne particulates from the patients' area reaching adjacent areas.

Some designers have provided isolation rooms that allow maximum space flexibility by using an approach that reverses the airflow direction by varying exhaust flow rate. This approach is useful only if appropriate adjustments can be ensured for different types of isolation procedures.

Treatment Rooms. Patients are brought to these rooms for special treatments that cannot be conveniently performed in the patient rooms. To accommodate the patient who may be brought from bed, the rooms should have individual temperature and humidity control. Temperatures and humidities should correspond to those specified for patients' rooms.

Clean Workrooms. The clean workroom serves as a storage and distribution center for clean supplies and should be maintained at a positive air pressure relative to the corridor.

The soiled workroom serves primarily as a collection point for soiled utensils and materials. It is considered a contaminated room and should have a negative air pressure relative to the air pressure of adjoining areas. Temperatures and humidities should be provided within the comfort range.

Floor Pantry. Ventilation requirements of this area depend on the type of food service adopted by the hospital. Where bulk food is dispensed and dishwashing facilities are provided in the pantry, the use of hoods over equipment, with exhaust to the outdoors, is recommended. Small pantries used for between-meal feedings require no special ventilation. The air pressure of the pantry should be in balance with that of adjoining areas to reduce the movement of air into or out of it.

Labor/Delivery/Recovery/Postpartum (LDRP). The procedures for normal childbirth are considered noninvasive and rooms are treated similarly to patient rooms. Some jurisdictions may require higher air change rates than a typical patient room. It is expected that invasive procedures such as cesarean section are performed in a nearby delivery room or surgery.

Ancillary

Radiology Department. The fluoroscopic, radiographic, therapy, and darkroom areas require special attention. Among conditions that make positive ventilation mandatory in these areas are the odorous characteristics of certain clinical conditions treated, special construction designed to prevent ray leakage, and lightproof window shades that reduce any natural light infiltration to the fluoroscopic area.

Fluoroscopic, Radiographic, and Deep Therapy Rooms. These rooms require temperatures of 75 to 80 °F and humidity of 40 to 50%. Depending on the location of air supply outlets and exhaust intakes, lead lining may be required in supply and return ducts at the points of entry to the various clinical areas to prevent ray leakage to other occupied areas.

The darkroom normally is in use for longer periods than the X-ray rooms and should have an independent system to exhaust the air to the outdoors. The exhaust from the film drier may be connected into the darkroom exhaust. To maintain the integrity of the film during processing, a 90% efficiency filter should be installed in the air supply system for this area.

Laboratories. Air conditioning for comfort and safety of the technicians is necessary in laboratories (Degenhardt and Pfost 1983). Chemical fumes, odors, vapors, heat from the equipment, and the undesirability of open windows all contribute to this need.

Particular attention should be given to the sizes and types of equipment used in the various laboratories as the heat gain from equipment usually constitutes the major portion of the cooling load.

The general air distribution and exhaust systems should be constructed of conventional materials following standard designs for the type of systems used. Exhaust systems serving hoods in which radioactive materials, volatile solvents, and strong oxidizing agents such as perchloric acid are used, should be fabricated of stainless steel. Washdown facilities should be provided for hoods and ducts handling perchloric acid. Perchloric acid hoods should have dedicated exhaust fans.

Hood use may dictate other duct materials. Hoods in which radioactive or infectious materials are to be used must be equipped with ultrahigh efficiency filters at the exhaust outlet of the hood and have equipment and a procedure for the safe removal and replacement of contaminated filters. Exhaust duct routing should be as short as possible with a minimum of horizontal offsets. This is especially true for perchloric acid hoods because of the extremely hazardous explosive nature of this material.

Determining the most effective, economical, and safest system of laboratory ventilation requires considerable study. Where the laboratory space ventilation air quantities approximate the air quantities required for ventilation of the hoods, the hood exhaust system may be used to exhaust all ventilation air from the laboratory areas. In situations where hood exhaust exceeds air supplied, a supplementary air supply for hood makeup may be used. The use of variable air volume supply/exhaust systems in the laboratory has gained acceptance but requires special care in design and installation.

This supplementary air supply, which need not be completely conditioned, should be provided by a system that is independent of the normal ventilating system. The individual hood exhaust system should be interlocked with the supplementary air system. However, should failure of the supplementary air system occur, the hood exhaust system should not shut off. Chemical storage rooms must have an individual exhaust air system with terminal fan.

Exhaust fans serving hoods should be located at the discharge end of the duct system to prevent any possibility of exhaust products entering the building. For further information on laboratory air-conditioning and hood exhaust systems, see Chapter 14 of this volume, NFPA *Standard* 99-87, Health Care Facilities, and Control of Hazardous Gases and Vapors in Selected Hospital Laboratories (Hagopian and Doyle 1984).

The exhaust air from the hoods in the unit for biochemistry, histology, cytology, pathology, glass washing/sterilizing, and serology-bacteriology should be discharged to the outdoors with no recirculation. Typically, exhaust fans discharge vertically a minimum of 7 ft above the roof at velocities up to 4000 fpm. The serology-bacteriology unit should be pressurized relative to the adjoining areas to reduce the possibility of infiltration of aerosols that might contaminate the specimens being processed. The entire laboratory area should be under slight negative pressure to reduce the spread of odors or contamination to other hospital areas. Temperatures and humidities should be within the comfort range.

Bacteriology Units. These units should not have undue air movement, and care should be exercised to limit air velocities to a minimum. The sterile transfer room, which may be within or adjoining the bacteriology laboratory, is a room where sterile media are distributed and where specimens are transferred to culture media. To maintain a sterile environment, an ultrahigh efficiency HEPA filter should be installed in the supply air duct near the point of entry to the room. The media room, essentially a kitchen, should be ventilated to remove odors and steam.

Infectious Disease and Virus Laboratories. These laboratories, found only in large hospitals, require special treatment. A minimum ventilation rate of six air changes per hour or makeup equal to hood exhaust volume is recommended for these laboratories, which should have a negative air pressure relative to any other area in the vicinity to prevent the exfiltration of any airborne contaminants from them. The exhaust air from fume hoods or safety cabinets in these laboratories requires sterilization before being exhausted to the outdoors. This may be accomplished by the use of electric or gas-fired heaters placed in series in the exhaust systems, and designed to heat the exhaust air to 600°F. A more common and less expensive method of sterilization of exhaust is to use ultrahigh efficiency filters in the system.

Nuclear Medicine Laboratories. Such laboratories administer radioisotopes to patients orally, intraveneously, or by inhalation to facilitate diagnosis and treatment of disease. There is little opportunity in most cases for airborne contamination of the internal environment, but exceptions warrant special consideration.

One important exception involves use of iodine-131 solution in vials or capsules to diagnose disorders of the thyroid gland. Another involves use of xenon-133 gas via inhalation to study patients with reduced lung function. Capsules of iodine-133 occasionally leak part of their contents prior to use. Vials emit airborne contaminants when opened for preparation of a dose.

It is common practice for vials to be opened and handled in a standard laboratory fume hood. A minimum face velocity of 100 fpm should be adequate for this purpose. This recommendation applies only where the stated quantities are handled in simple operations. Other circumstances may warrant provision of a glovebox or similar confinement.

Use of xenon-133 for patient study involves a special instrument that permits the patient to inhale the gas and to exhale back into the instrument. The exhaled gas is passed through a charcoal trap mounted in lead and then often (but not always) is vented outdoors. The process suggests some potential for escape of the gas into the internal environment.

Due to the uniqueness of this operation and the specialized equipment involved, it is recommended that system designers determine the specific instrument to be used and contact the manufacturer for guidance. Other guidance is available in the U.S. Nuclear Regulatory Commission (USNRC) Regulatory Guide 10.8 (1980). In particular, emergency procedures to be followed in case of accidental release of xenon-133 should include such considerations as temporary evacuation of the area and/or increasing the ventilation rate of the area.

Prior recommendations concerning pressure relationships, supply air filtration, supply air volume, no recirculation, and other attributes of supply and discharge systems for histology, pathology, and cytology laboratories are also relevant to nuclear medicine laboratories. There are, however, some special general ventilation system requirements imposed by the USNRC where radioactive materials are used. For example, USNRC Regulatory Guide 10.8 (1980) provides a computational procedure to estimate the airflow necessary to maintain xenon-133 gas concentration at or below specified levels. It also contains specific requirements as to the amount of radioactivity that may be vented to the atmosphere (with the disposal method of choice being adsorption onto charcoal traps).

Pathology Department. The areas of the pathology section that require special attention are the autopsy room and, in larger hospitals, the animal quarters.

Autopsy Rooms. These rooms are subject to heavy bacterial contamination and odor. Exhaust intakes should be located at both the ceiling and in the low sidewall. The exhaust system should discharge the air above the roof of the hospital. A negative air pressure relative to the air pressure of adjoining areas should be provided in the autopsy room to prevent the spread of this contamination. Where large quantities of formaldehyde are used, special exhaust hoods may be necessary to keep concentration levels below legal requirements. For smaller hospitals where the autopsy room is used infrequently, it may be desirable to provide local control of the ventilation system and odor control system with either activated charcoal or potassium permanganate impregnated activated alumina.

Animal Quarters. Principally because of odor, animal quarters require a mechanical exhaust system that discharges the contaminated air above the hospital roof. To prevent the spread of odor or other contaminants from the animal quarters to other areas, a negative air pressure of not less than 0.1 in. of water, relative to the air pressure of adjoining areas, must be maintained. Chapter 14 has further information on animal room air conditioning.

Pharmacy. The pharmacy should be treated for comfort and requires no special ventilation. Laminar airflow benches may be needed in pharmacies that prepare intravenous solutions.

Diagnostic and Treatment

Physical Therapy Department. The cooling load of the electrotherapy section is affected by the shortwave diathermy, infrared, and ultraviolet equipment used in this area.

Hydrotherapy Section. This section, with its various water treatment baths, is generally maintained at temperatures up to 80°F. The potential latent heat buildup in this area should not be overlooked. The exercise section requires no special treatment, and temperatures and humidities may be within the comfort zone. The air of these areas may be recirculated within the areas, and an odor control system is suggested.

Occupational Therapy Department. In this department, spaces for activities such as weaving, braiding, artwork, and sewing require no special ventilation treatment. Recirculation of the air in these areas using medium-grade filters in the system is permissible.

Larger hospitals or those specializing in rehabilitation have a greater diversification of skills and crafts such as carpentry, metalwork, plastics, photography, ceramics, and painting.

The air-conditioning and ventilation requirements of the various sections should conform to normal practice for such areas and to the code requirements relating to them. Temperatures and humidities should be maintained within the comfort zone.

Inhalation Therapy Department. Inhalation therapy is for treatment of pulmonary and other respiratory disorders. The air must be very clean and the area should have a positive air pressure relative to adjacent areas.

Central Sterilizing and Supply

Used and contaminated utensils, instruments, and equipment are brought to this unit for cleaning and sterilization prior to reuse. The unit usually consists of a cleaning area, a sterilizing area, and a storage area where supplies are kept until requisitioned. Where these areas are in one large room, air should flow from the clean storage and sterilizing areas toward the contaminated cleaning area. The air pressure relationships should conform to those indicated in Table 3. Temperature and humidity within the comfort range is recommended.

The following guidelines are important in the central sterilizing and supply unit:

1. Insulate sterilizers in these areas to reduce heat load.
2. Amply ventilate sterilizer equipment closets to remove excess heat.
3. Where ethylene oxide (ETO) gas sterilizers are used, provide a separate exhaust system with terminal fan (Samuals and Eastin 1980). Provide adequate exhaust capture velocity in the vicinity of sources of ethylene oxide leakage. Install an exhaust at sterilizer doors and over the sterilizer drain. Exhaust aerator and service rooms. Also, ethylene oxide concentration, exhaust flow sensors, and alarms should be provided.

   ETO sterilizers should be located in dedicated rooms that have a highly negative pressure relationship to adjacent spaces. Many jurisdictions require that ETO exhaust systems have equipment to remove ETO from exhaust air.
4. Maintain storage areas for sterile supplies at a relative humidity of not more than 50%.

Service Department

Service areas include dietary, housekeeping, mechanical, and employee facilities. Whether these areas are air conditioned or not, adequate ventilation is important to provide sanitation and a wholesome environment. Ventilation of these areas cannot be limited to exhaust systems only; provision for supply air must be incorporated in the design. Such air must be filtered and delivered at controlled temperatures. The best design exhaust system may prove ineffective without an adequate air supply. Experience has shown that reliance on open windows results only in dissatisfaction, particularly during the heating season. The incorporation of air-to-air heat exchangers in the general ventilation system offers possibilities for economical operation in these areas.

Dietary Facility. This area usually includes the main kitchen, bakery, dietitian's office, dishwashing room, and dining space. Because of the various conditions encountered (*i.e.,* high heat and moisture production and cooking odors), special attention in design is needed to provide an acceptable environment. Kitchen ventilation should conform to local codes. In many instances, hood exhaust quantities may dictate the quantity of supply air required.

Cooking equipment is usually grouped in one or more locations within the kitchen for efficient use. Hoods are provided over the equipment for removal of heat, odors, and vapors in accordance with code requirements. The entire system shall conform to the requirements of NFPA *Standard* 96-84, Installation of Equipment for the Removal of Smoke and Grease-Laden Vapors from Commercial Cooking Equipment.

The dietitian's office is often located within the main kitchen or immediately adjacent to it. It is usually completely enclosed to ensure privacy and noise reduction. Air conditioning is recommended for the maintenance of normal comfort conditions.

The dishwashing room should be enclosed and ventilated at a minimum rate to equal the dishwasher hood exhaust. It is not uncommon for the dishwashing area to be divided into a soiled area and a clean area. When this is done, the soiled area should be kept negative to the clean area.

Depending on the size of the area, all or the greatest part of exhaust air may be taken off through the dishwasher hood. Exhaust ductwork should be noncorrosive and watertight to handle condensation of steam and should be graded to drip into a convenient drain.

Kitchen Compressor/Condenser Space. Ventilation of this space should conform to all codes with the following additional considerations: (1) use 350 cfm of ventilating air per compressor power for units located within the kitchen; (2) condensing units should operate optimally at 90°F maximum ambient temperature; and (3) where air temperature or air circulation is marginal, combination air and water-cooled condensing units should be specified. It is often worthwhile to use condenser water coolers or remote condensers.

Dining Space. The ventilation of this space should conform to local codes. The reuse of dining space air for ventilation and cooling of food preparation areas in the hospital is suggested, providing the reused air is passed through 80% efficient filters. Where cafeteria service is provided, serving areas and steam tables are usually hooded. The air-handling capacities of these hoods should be at least 75 cfm/ft^2 of the perimeter area.

Laundry and Linen. Of these facilities, only the soiled linen storage room, soiled linen sorting room, soiled utility room, and laundry processing area require special attention.

The soiled linen storage room, provided for storage of soiled linen prior to pickup by commercial laundry, will be odorous and contaminated and should be well ventilated and maintained at a negative air pressure.

The soiled linen storage provides in-house laundry service. The soiled utility room is provided for in-patient services and is normally contaminated with noxious odors. This room should be exhausted directly outside by mechanical means.

In the laundry processing area, the washers, flatwork ironers, tumblers, and so forth, should have direct overhead exhaust to reduce humidity. Such equipment should be insulated or shielded whenever possible to reduce the high radiant heat effects. A canopy over the flatwork ironer and exhaust air outlets near other heat-producing equipment capture and remove heat best. The air supply inlets should be located so as to move air through the processing area toward the heat-producing equipment. The exhaust system from flatwork stoners and from tumblers should be independent of the general exhaust system and should be equipped with lint filters. Air should exhaust above the roof or where it will not be obnoxious to other areas. Heat reclamation of the laundry exhaust air may be desirable and practicable.

Where air conditioning is contemplated, a separate supplementary air supply, similar to that recommended for kitchen hoods, may be located in the vicinity of the exhaust canopy over the ironer, or spot cooling for the relief of personnel confined to specific areas should be considered.

Mechanical Facilities. The air supply to boiler rooms should provide both comfortable working conditions and the air quantities required for maximum combustion rates of the particular fuel used. Boiler and burner ratings establish maximum combustion rates, so the air quantities can be computed according to the type of fuel. Sufficient air must be supplied to the boiler room to supply the exhaust fans, as well as boilers.

At workstations, the ventilation system should limit temperatures to 90 °F effective temperature. When ambient outside air temperature is higher, maximum temperature may be that of outside air up to a maximum of 97 °F to protect motors from excessive heat.

Maintenance Shops. Carpentry, machine, electrical, and plumbing shops present no unusual ventilation requirements. Proper ventilation of paint shops and paint storage areas is important because of fire hazard and should conform to all applicable codes. Maintenance shops where welding occurs should have exhaust ventilation.

Administration. This department includes the main lobby, admitting and business offices, and medical records. This area requires no unusual treatment and should be conditioned for occupant comfort. A separate air-handling system is considered desirable to segregate this area from the hospital proper, since these areas are usually unoccupied at night.

CONTINUITY OF SERVICE AND ENERGY CONCEPTS

Zoning

Zoning of the air-handling systems may be desirable to compensate for exposures due to orientation or for other reasons imposed by a particular building configuration. Zoning—using separate air systems for different departments—may be indicated to minimize recirculation between departments, provide flexibility of operation, simplify provisions for operation on emergency power, and conserve energy.

By ducting the air supply from several air-handling units into a manifold, central systems can achieve a measure of standby capacity. To accommodate critical areas, which must operate continuously, air is diverted from noncritical or intermittently operated areas when one unit is shut down. Such provision or other means of standby protection is essential if the air supply is not to be interrupted by routine maintenance or component failure.

Separation of supply, return, and exhaust systems by department is often desirable; particularly for surgical, obstetrical, pathological, and laboratory departments. The desired relative balance within critical areas should be maintained by interlocking the supply and exhaust fans. For example, the surgical department exhaust should cease when the supply airflow is stopped.

Heating and Hot Water Standby Service

The number and arrangement of boilers should be such that when one boiler breaks down or routine maintenance requires that one boiler be temporarily taken out of service, the capacity of the remaining boilers is sufficient to provide hot water service for clinical, dietary, and patient use; steam for sterilization and dietary purposes; and heating for operating, delivery, birthing, labor, recovery, intensive care, nursery, and general patient rooms. However, reserve capacity is not required in areas where a design dry-bulb temperature of 25 °F represents no less than 99% of the total hours in any one heating period as noted in the Table of Climatic Conditions of the United States in Chapter 24 of the 1989 ASHRAE *Handbook—Fundamentals*.

Boiler feed pumps, heating circulating pumps, condensate return pumps, and fuel oil pumps should be connected and installed to provide normal and standby service. Supply and return mains and risers for cooling, heating, and process steam systems should be valved to isolate the various sections. Each piece of equipment should be valved at the supply and return ends.

Mechanical Cooling

The source of mechanical cooling for clinical and patient areas in a hospital should be carefully considered. The preferred method is an indirect refrigerating system using chilled water or antifreeze solutions. When using direct refrigerating systems, consult codes for specific limitations and prohibitions. Refer to ASHRAE *Standard* 15-1989, Safety Code for Mechanical Refrigeration.

Insulation

All hot piping, ducts, and equipment exposed to contact by the building occupants should be insulated to maintain the energy efficiency of all systems. To prevent condensation, ducts, casings, piping, and equipment with outside surface temperature below ambient dew point should be covered with insulation with an external vapor barrier. Insulation, including finishes and adhesives on the exterior surfaces of ducts, pipes, and equipment, should have a flame spread rating of 25 or less and a smoke developed rating of 50 or less, as determined by an independent testing laboratory in accordance with NFPA National Fire Code 255-84 as required by NFPA 90A-85. Smoke development rating for pipe insulation should not exceed 150 (DHHS 1984a).

Linings in air ducts and equipment should meet the Erosion Test Method described in Underwriters' Laboratories Inc., *Standard* 181-81. These linings, including coatings, adhesives, and insulation on exterior surfaces of pipes and ducts in building spaces used as air supply plenums, should have a flame spread rating of 25 or less and a smoke developed rating of 50 or less, as determined by an independent testing laboratory in accordance with ASTM *Standard* E 84.

Duct linings should not be used in systems supplying operating rooms, delivery rooms, recovery rooms, nurseries, burn care units, and intensive care units, unless terminal filters of at least 90% efficiency are installed downstream of linings.

When modifying existing systems, asbestos materials should be handled and disposed of in accordance with applicable regulations.

Energy

Health care is an energy-intensive, energy-dependent enterprise. Hospital facilities are different from other structures because they operate 24 h a day year-round, require sophisticated backup systems in case of utility shutdowns, use large quantities of outside air to combat odors and dilute microorganisms, and must deal with problems of infection and solid waste disposal. Similarly, large quantities of energy are required to power the diagnostic, therapeutic, and monitoring equipment, and the support services such as food storage, preparation, service, and laundry facilities.

Hospitals conserve energy in various ways such as using larger energy storage tanks, applying energy-saving measures, and employing energy conversion devices that transfer energy from hot or cold exhaust air from the building to heat or cool incoming air. Heat pipes, runaround loops, and other forms of heat recovery are receiving increased attention. Solid waste incinerators, which generate exhaust heat to develop steam for laundries and hot water for patient care, are becoming increasingly common.

The construction design of new facilities, including alterations of, and additions to, existing buildings, has a major influence on the amount of energy required to provide such services as heating, cooling, and lighting. The selection of building and system components for effective energy use requires careful planning and design. Integration of building waste heat into systems and with renewable energy sources such as solar under some climatic conditions will provide substantial savings. ASHRAE *Standard* 90A-1980 should also be considered for applicability.

OUTPATIENT SURGICAL FACILITIES

Outpatient surgical facilities may be a free-standing unit, part of an acute care facility, or part of a medical facility such as a medical office building (clinic). Surgery is performed without anticipation of overnight stay by patients (*i.e.*, the facility operates 8 to 10 h per day.)

Design Criteria

The system designer should refer to the following paragraphs from the section on hospitals:

1. Infection Sources and Control Measures
2. Air Quality
3. Air Movement
4. Zoning

The air-cleaning requirements are taken from Table 1 for operating rooms. The minimum ventilation rates, desired pressure relationships, desired relative humidity, and design temperature ranges

The design criteria for insulation applies equally to these facilities. Some owners may desire that the heating, air-conditioning, and service hot water systems have standby or emergency service capability, and that these systems be able to function after a natural disaster.

The following functional areas in an outpatient facility have similar design criteria as hospitals:

- Administration
- Surgical—operating rooms, recovery rooms, and anesthesia storage rooms
- Diagnostic and Treatment—generally a small radiology area
- Sterilizing and Supply
- Service—soiled workrooms, mechanical facilities, and locker rooms

To reduce utility costs, facilities should include energy conserving procedures such as recovery devices, variable air volume, load shedding, or systems to shut down or reduce ventilation of certain areas when unoccupied. Mechanical ventilation should take advantage of the outside air by using an economizer cycle, when appropriate, to reduce heating and cooling loads.

NURSING HOMES

Nursing homes may be classified as follows:

Extended care facilities for recuperation of hospital patients who no longer require hospital facilities but do require therapeutic and rehabilitation services by skilled nurses. This type of facility is either a direct hospital adjunct or a separate facility having close ties with the hospital. Clientele may be any age, usually stay from 35 to 40 days, and usually have only one diagnostic problem.

Skilled nursing homes for care of people who require assistance in daily activities, many of whom are incontinent and nonambulatory, and some of whom are disoriented. These homes may or may not offer skilled nursing care. Clientele come directly from home or residential care homes, generally are elderly (average age of 80), stay an average of 47 months, and frequently have multiple diagnostic problems.

Residential care homes are generally for the elderly who are unable to cope with regular housekeeping chores but are able to care for all their personal needs, lead normal lives, have no acute ailments, and move freely in and out of the home and the community. These homes may or may not offer skilled nursing care. The average length of stay is four years or more.

Functionally, these buildings have five types of areas that are of concern to the designer: (1) administrative and supportive areas, inhabited by the staff; (2) patient areas that provide direct normal daily services; (3) treatment areas that provide special medical-type services; (4) clean workrooms for storage and distribution of clean supplies; and (5) soiled workrooms for collection of soiled and contaminated supplies and for sanitization of nonlaundry items.

DESIGN CONCEPTS AND CRITERIA

Bacteria level in nursing homes does not command the same level of concern as it does in the acute care hospital. Nevertheless, the designer should be aware of the necessity for odor control, filtration, and airflow control between certain areas.

Table 4 lists recommended filter efficiencies for air systems serving specific nursing home areas. Table 5 lists recommended minimum ventilation rates and desired pressure relationships for certain areas in nursing homes.

Table 4 Filter Efficiencies for Central Ventilation and Air-Conditioning Systems in Nursing Homes[a]

Area Designation	Minimum Number of Filter Beds	Filter Efficiences (%) Main Filter Bed
Patient care, treatment, diagnostic, and related areas	1	80
Food preparation areas and laundries	1	80
Administrative, bulk storage, and soiled holding areas	1	30

[a]Ratings based on ASHRAE *Standard* 52-76.

Recommended interior winter design temperature for areas occupied by patients is 75 °F and 70 °F for nonpatient areas. Provisions for maintenance of minimum humidity levels for winter depend on the severity of the climate and is best left to the judgment of the designer. Where air conditioning is provided, the recommended interior summer design temperature and humidity is 75 °F, and 50% rh.

The general design criteria under the subheadings, "Heating and Hot Water Standby Service," "Insulation," and "Energy" for the acute care general hospital apply equally to nursing.

APPLICABILITY OF SYSTEMS

The occupants of nursing homes are usually frail and many are incontinent. They may be ambulatory, but some are bedridden, with illnesses in advanced stages. The selected system must dilute and control odors and should be free of drafts. Local climatic conditions, costs, and designer judgment determine the extent and degree of air conditioning and humidification. Odor control may indicate either relatively large volumes of outside air with heat recovery provisions or other air treatment conditioning, including activated carbon or potassium permanganate impregnated activated alumina, in the interest of energy conservation and reduced operating costs.

Temperature control should be on an individual room basis. Patients' rooms should have supplementary heat along exposed walls in geographical areas with severe climates. In moderate climates, *i.e.*, where outside winter design conditions are 30 °F or above, heating from overhead may be used.

REFERENCES

AIA. 1987. Guidelines for construction and equipment of hospital and medical facilities.

ASHRAE. 1986. *Terminology of heating, ventilation, air conditioning, and refrigeration.*

Chaddock, J.B. 1986. Ventilation and exhaust requirements for hospitals. ASHRAE *Transactions* 2(1).

DHHS. 1984a. Guidelines for construction and equipment of hospital and medical facilities. United States Department of Health and Human Services, Publication No. HRS-M-HF, 84-1.

DHHS. 1984b. Energy considerations for hospital construction and equipment. United States Department of Health and Human Services, Publication No. HRS-M-HF, 84-1A.

Table 5 Pressure Relationships and Ventilation of Certain Areas of Nursing Homes

Function Area	Pressure Relationship to Adjacent Areas	Minimum Air Changes of Outdoor Air per Hour Supplied to Room	Minimum Total Air Changes per Hour Supplied to Room	All Air Exhausted Directly to Outdoors	Recirculated within Room Units
PATIENT CARE					
Patient room	±	2	2	Optional	Optional
Patient area corridor	±	Optional	2	Optional	Optional
Toilet room	N	Optional	10	Yes	No
DIAGNOSTIC AND TREATMENT					
Examination room	±	2	6	Optional	Optional
Physical therapy	N	2	6	Optional	Optional
Occupational therapy	N	2	6	Optional	Optional
Soiled workroom or soiled holding	N	2	10	Yes	No
Clean workroom or clean holding	P	2	4	Optional	Optional
STERILIZING AND SUPPLY					
Sterilizer exhaust room	N	Optional	10	Yes	No
Linen and trash chute room	N	Optional	10	Yes	No
Laundry, general	±	2	10	Yes	No
Soiled linen sorting and storage	N	Optional	10	Yes	No
Clean linen storage	P	Optional	2	Yes	No
SERVICE					
Food preparation center	±	2	10	Yes	Yes
Warewashing room	N	Optional	10	Yes	Yes
Dietary day storage	±	Optional	2	Yes	No
Janitor closet	N	Optional	10	Yes	No
Bathroom	N	Optional	10	Yes	No

P = Positive N = Negative ± = Continuous Directional Control Not Required

Degenhardt, R.A. and J.F. Pfost. 1983. Fume hood design and application for medical facilities. ASHRAE *Transactions* 89 (2A and 2B).

Gustofson, T.L., *et al.* 1982. An outbreak of airborne Nosocomial Varicella. American Academy of Pediatrics 70(4).

Hagopian, J.H. and E.R. Hoyle. 1984. Control of hazardous gases and vapors in selected hospital laboratories. ASHRAE *Transactions* 90(2B).

Isoard, P., L. Giacomoni, and M. Payronnet. 1980. Proceedings of the 5th International Symposium on Contamination Control. Munich (September).

Lewis, J.R. 1988. Application of VAV, DDC, and smoke management to hospital nursing wards. ASHRAE *Transactions* 94(1).

Luciano, J.R. 1984. New concept in French hospital operating room HVAC systems. ASHRAE *Journal* (February).

Michaelson, G.S., D. Vesley, and M.M. Halbert. 1966. The laminar air flow concept for the care of low resistance hospital patients. Paper presented at the annual meeting of American Public Health Association, San Francisco (November).

Murray, W. A., *et al.* 1988. Ventilation protection of immune compromised patients. ASHRAE *Transactions* 94(1).

NIOSH. 1975. Elimination of waste anesthetic gases and vapors in hospitals. United States Department of Health, Education, and Welfare, Publication No. NIOSH 75-137 (May).

OSHA. Occupational exposure to ethylene oxide. United States Department of Labor. OSHA 29 CFR, Part 1910.

Pfost, J.F. 1981. A re-evaluation of laminar air flows in hospital operating rooms. ASHRAE *Transactions* 87(2).

Rhodes, W.W. 1988. Control of microbioaerosol contamination in critical areas in the hospital environment. ASHRAE *Transactions* 94(1).

Samuals, T.M. and M. Eastin. 1980. ETO exposure can be reduced by air systems. *Hospitals* (July).

Setty, B.V.G. 1976. Solar heat pump integrated heat recovery. *Heating, Piping and Air Conditioning* (July).

Wells, W.F. 1934. On airborne infection. Study II, Droplets and Droplet Nuclei. *American Journal of Hygiene*, 20:611.

CHAPTER 8

SURFACE TRANSPORTATION

AUTOMOBILE AIR CONDITIONING

WITH increased ease of operation, improved road isolation, lower noise levels, and increasing urban travel time, modern automobile users have become conscious of the in-vehicle environment. The use of truck cab air conditioning is increasing for long-distance haulage.

All passenger cars sold in the United States must meet federal defroster requirements, so ventilation systems and heaters are included in the basic vehicle design. Even though trucks are excluded from federal requirements, all manufacturers include heater/defrosters as standard equipment. Air conditioning remains an extra-cost option on nearly all vehicles.

ENVIRONMENTAL CONTROL

The environmental control system of modern automobiles consists of one or more of the following: (1) heater-defroster, (2) ventilation, and (3) cooling and dehumidifying (air-conditioning) systems. The integration of the heater-defroster and ventilation systems is common.

Heating

Outdoor air passes through a heater core, using engine coolant as a heat source. Interior air should not recirculate through the heater to avoid visibility-reducing condensation on the glass due to raised air dew point from occupant respiration and interior moisture gains.

Temperature control is achieved by either water flow regulation or heater air bypass and subsequent mixing. A combination of ram effect from forward movement of the car and the electrically driven blower provides the airflow.

Heater air is generally distributed into the lower forward compartment, along the floor under the front seat, and up into the rear compartment. Heater air exhausts through body leakage points. The increased heater air quantity at higher vehicle speeds (ram assist through the ventilation system) partly compensates for the subsequent infiltration increase. Air exhausters are sometimes installed to increase airflow and reduce the noise of air escaping from the car.

The heater air distribution system is usually adjustable between the diffusers along the floor and on the dash. Supplementary ducts are sometimes required when consoles, panel-mounted air conditioners, or rear seat heaters are installed. Supplementary heaters are frequently available for third-seat passengers in station wagons and for limousine and luxury sedan rear seats.

Defrosting

Some heated outside air is ducted from the heater core to defroster outlets situated at the bottom of the windshield. This air absorbs moisture from the windshield interior surface and raises the glass temperature above the interior dew point. Induced outdoor air has a lower dew point than the air inside the vehicle, which absorbs moisture from the occupants and car interior. Heated air also provides the heat necessary to melt ice or snow on the glass exterior. The defroster air distribution pattern on the windshield is developed by test for conformity with federal standards, satisfactory distribution, and fast defrost.

Some systems operate the air-conditioning compressor either to dry the induced outside air, to prevent the evaporator from increasing the dew point, or both. Some vehicles are equipped with side window demisters that direct a small amount of heated air and/or air with lowered dew point to the front side windows. Rear windows are primarily defrosted by heating wires embedded in the glass.

Ventilation

Fresh air ventilation is achieved by one of two systems: (1) ram air or (2) forced air. In both systems, air enters the vehicle through a screened opening in the cowl just forward of the base of the windshield. The cowl plenum is usually an integral part of the vehicle structure. Air entering this plenum can also supply the heater and evaporator cores.

In the ram air system, ventilation air flows aft and up toward the front seat occupants' laps, and then over the remainder of their bodies. Additional ventilation occurs by turbulence and air exchange through open windows. Directional control of ventilation air is frequently not available with ram systems. Airflow rate varies with relative wind-vehicle velocity but may be adjusted with windows or vents.

Forced air ventilation is available on many automobiles. The cowl inlet plenum and heater/air-conditioning blower are used together with instrument panel outlets for directional control. On air-conditioned vehicles, the forced air ventilation system uses the air-conditioning outlets. Vent windows and body air exhausts assist ventilation and exhaust air from the vehicle. With increased popularity of air conditioning and forced ventilation, most late model vehicles are not equipped with vent windows.

Air Conditioning

There are two basic types of systems: combination evaporator-heater and dealer-installed systems.

The combination evaporator-heater system in conjunction with the ventilation system dominates factory-installed air conditioning. This system is popular because (1) it permits dual use of components such as blower motors, outside air ducts, and structure; (2) it permits compromise standards where space considerations dictate (ventilation reduction on air-conditioned cars); (3) it generally reduces the number and complexity of driver controls; and (4) capacity control innovations such as automatic reheat are typical.

The preparation of this chapter is assigned to TC 9.3, Transportation Air Conditioning.

Outlets in the instrument panel distribute air to the car interior. These are individually adjustable, and some have individual shutoffs. The dash end outlets are for the driver and front seat passenger; center outlets are primarily for rear seat passengers.

The dealer-installed system is normally available only as a service or after-market installation. In recent designs, existing air outlets, blower, and controls are used. Evaporator cases are styled to look like factory-installed units. These units are integrated with the heater as much as possible to provide outside air and to take advantage of existing air-mixing dampers. Where it is not possible to use existing air ducts, custom ducts distribute the air in a manner similar to those in factory-installed units. On occasion, a slot is fabricated in the duct to accept an evaporator if cooling is ordered rather than having two ducts for new cars.

As most of the air for the rear seat occupants flows through the center outlets of the front evaporator unit, a passenger in the center of the front seat impairs rear seat cooling. Supplemental trunk and roof units, for luxury sedans and station wagons, respectively, improve the cooling of those passengers located behind the front seat.

The trunk unit consists of a blower-evaporator unit, complete with expansion valve, installed in the trunk of limousines and premium line vehicles. The unit uses the same high-side components as the front evaporator unit. It cools and recirculates air drawn from the base of the rear window (back light) of the car.

The roof-mounted unit is similar to the trunk unit, except that is it intended for station wagon use and is attached to the inside of the roof, toward the rear of the wagon. The advantages of the trunk unit also apply here.

GENERAL CONSIDERATIONS

General considerations include ambient temperatures and contaminants, vehicle and engine concessions, flexibility, physical parameters, durability, electrical power, refrigeration capacity, occupants, infiltration, insulation, solar effect, and noise.

Ambient and Vehicle Criteria

Ambient Temperature. Heaters are evaluated for performance from −40 to 70 °F. Air-conditioning systems with reheat are evaluated from 30 to 110 °F. Add-on units are evaluated from 50 to 100 °F, although ambient temperatures above 125 °F are occasionally encountered. Because the system is an integral part of the vehicle detail, factors resulting from vehicle heat and local heating must be considered.

Ambient Contaminants. Airborne bacteria, pollutants, and corrosion agents, as well as resistance to ambient temperature variations and extremes must be considered when selecting materials for seals and heat exchangers. Electronic air cleaners have seen application on premium automobiles.

Vehicle Concessions. Vehicle performance standards must be observed. Proper engine coolant temperature, freedom from gasoline vapor lock, adequate electrical charging system, acceptable vehicle ride, minimum surge due to compressor clutch cycling, and handling must be maintained.

Flexibility. Engine coolant pressure at the heater core inlet ranges up to 40 psig in cars and 55 psig on trucks. The engine coolant thermostat remains closed until 160 to 205 °F coolant temperature is reached. Coolant flow is a function of pressure differential and system restriction but ranges from 0.6 gpm at idle to 46 gpm at 60 mph (lower for water valve regulated systems because of the added restriction).

Present-day antifreeze coolant solutions have specific heats from 0.65 to 1.0 Btu/lb · °F and boiling points from 250 to 272 °F (depending on concentration) when a 15 psi radiator pressure cap is used.

Multiple-speed (usually four) blowers supplement the ram air effect through the ventilation system and produce the necessary velocities for distribution. Heater air quantities range from 125 to 190 cfm. Defroster air quantities range from 90 to 145 cfm. Other considerations are (1) compressor speed-engine speed, from 500 to 5500 rpm; (2) drive ratio, from 0.89:1 to 1.41:1; (3) condenser air from 50 to 125 °F and from 325 to 3000 cfm (corrected for restriction and distribution factors); and (4) evaporator air quantity from 100 to 300 cfm (limits established by design but selective at operator's discretion) and from 35 to 150 °F.

Physical Parameters

Parameters include engine rock, proximity to adverse environments, and durability.

Engine Rock. The engine moves relative to the rest of the car both fore and aft because of inertia and in rotation because of torque.

Rotational movements at the compressor may be as much as 0.75 in. due to acceleration and 0.5 in. because of deceleration; fore and aft movement may be as much as 0.25 in.

Proximity to Adverse Environments. Wiring, refrigerant lines, hoses, vacuum lines, and so forth, must be protected from exhaust manifold heat and sharp edges of sheet metal. Accessibility to the normal service items such as oil filler caps, power steering filler caps, and transmission dipsticks cannot be impaired. Removal of air-conditioning components should not be necessary for servicing other components.

Durability. Hours of operation are short compared to commercial systems (4000 h at 40 mph = 160,000 miles), but all the shock and vibration the vehicle receives or produces must not cause a malfunction or failure.

Systems are designed to meet the recommendations of SAE *Safe Working Practice* J639, which states that the burst strength of those components subjected to high side refrigerant pressure is at least 2.5 times the venting pressure or pressure equivalent to venting temperature of the relief device. Components for the low-pressure side frequently have burst strengths in excess of 300 psi. The relief device should be located as close to the discharge gas side of the compressor as possible, and, preferably, in the compressor itself.

Power and Capacity

Fan size is kept to a minimum, not only for space and weight effect, but because of power consumption. If a standard vehicle has a heater that draws 10 A and the air-conditioning system requires 20 A, an alternator and wiring system must be redesigned to supply this additional 10 A, with obvious cost and weight penalties.

The refrigeration capacity of a system must be adequate to reduce the vehicle interior temperature to the comfort temperature quickly, and then to maintain the selected temperature at reasonable humidity during all operating conditions and environments. A design is established by empirically evaluating all the known and predicted factors.

Occupancy per unit volume is high in automotive applications. The system (and auxiliary evaporators and systems) is matched to the intended vehicle occupancy.

Other Considerations

Infiltration varies with relative vehicle wind speed. It also varies with assembly quality. Body sealing is part of air-conditioning design for automobiles. Occasionally, seals beyond those required for dust, noise, and draft control are required.

Due to cost, insulation is seldom added to reduce thermal load. Insulation for sound control is generally considered adequate. Roof insulation is of questionable benefit as it retards heat loss during the nonoperating or soak periods. Additional dash and floor insulation is beneficial to reduce cooling load. Typical maximum ambient temperatures are 200 °F above mufflers and catalytic converters, 120 °F for other floor areas, 145 °F for dash and toe board, and 110 °F for sides and top. Solar effects must be added to the sides and top and must be considered in the following three phases:

Vertical. Maximum intensity occurs at or near the noon position of the sun. Windshield and back light vertical projections and angles are considered as substantial additions.

Horizontal and Reflected Radiation. The intensity is significantly less, but the glass area is large enough to merit consideration.

Surface Heating. The temperature of the surface is a function of the solar energy absorbed, the interior and ambient temperatures, and the velocity.

The temperature control system should not produce objectionable sounds. During maximum heating or cooling operation, a slightly higher noise level is acceptable. Thereafter, it should be possible to maintain comfort at a lower blower speed with acceptable noise level.

In air-conditioning systems, compressor-induced vibrations, gas pulsations, and noise must be kept to a minimum. Suction and discharge mufflers are often used to reduce noise. Belt-induced noises, engine torsional vibration, and compressor mounting all require particular attention.

COMPONENTS

The basic automobile air-conditioning system consists of the following key components:

Compressors

Much research and development has been conducted in the past several years to make compressors suitable for car cooling use.

Displacement. Today's compressors have displacements of 6.1 to 12.6 in.^{3}/rev. They are belt driven from the crankshaft at ratios ranging from 0.89:1 to 1.41:1.

Physical Size. Fuel economy, lower hood lines, and more engine accessories all contribute to less compressor installation space. These features along with smaller engines having less accessory power available all contribute to smaller compressors being used.

Speed Range. Since compressors are belt driven directly from the engine, they must withstand speeds over 6000 rpm and must be smooth and quiet down to 500 rpm. Unless some means of varying the compressor drive ratio is developed, the top speeds may go even higher to improve low-speed performance.

Torque Requirements. Because torque and vibration problems are so closely related, they impose difficult design problems. This does not preclude an economical single-cylinder compressor to reduce cost; however, any design must reduce peak torques and belt loads, which would normally be at a maximum in a single-cylinder design.

Refrigerant. Systems today use Refrigerant 12. Restrictions on R-12 use will require changing to an HCFC. A popular choice is R-134a for the close saturation pressure to temperature matching. Mineral oils are not miscible with HCFC-134a and poly-alkaline-glycol oils are recommended. HCFC-134a toxicity is yet unresolved. PAG oils should not be inserted into existing systems as the remaining free chloride breaks it down. Blends of HCFC-22, HCFC-152a, and HCFC-114 or HCFC-124, have better saturation matching. These should find favor in the service market. Refrigerant 22 was used in the past to gain additional refrigerating effect for a given displacement, but the high discharge pressures up to 650 psi at idle and slow car speeds were undesirable. Additionally, pressure pulses that generate vibration were more severe. Higher pressures and temperatures also accentuated compressor seal problems. With Refrigerant 12, the pressures at idle conditions seldom reach 400 psi.

Compressor Drives. A magnetic clutch, energized by power from the car engine electrical system, drives the compressor. The clutch is always disengaged when air conditioning is not required. The clutch can also be used to control evaporator temperature (see the section on Controls).

Variable Displacement Compressors. Wobble plate-type compressors are used for automobile air conditioning. The angle of the wobble plate is changed in response to the suction and discharge pressure to achieve a constant suction pressure just above freezing regardless of load. A bellows valve or electronic sensor-controlled valve controls the wobble plate.

Condensers

Condensers must be sized adequately. High discharge pressures reduce the compressor capacity and increase power requirements. Condenser air restriction must be compatible with the engine cooling fan and engine cooling requirements when the condenser is in series with the radiator. Generally, the most critical condition occurs at engine idle under high load conditions. An undersized condenser can raise head pressures sufficiently to stall small displacement engines. Condensers may be of the following designs: (1) aluminum plate fin and aluminum or copper tube; (2) skive fin; (3) brazed serpentine tube; and (4) brazed header tube. Aluminum is popular for lower cost and weight reduction.

An oversized condenser may produce condensing temperatures significantly below the engine compartment temperature. This can result in evaporation of refrigerant in the liquid line when the liquid line passes through the engine compartment (the condenser is ahead of the engine and the evaporator is behind it). Engine compartment air has not only been heated by the condenser but by the engine and radiator as well. Typically, this establishes a minimum condensing temperature of 30 °F above ambient.

Liquid flashing occurs more often at reduced load when the liquid line velocity decreases. This is more apparent on cycling systems than on systems that have a continuous liquid flow. It is audibly detected as gas entering the expansion valve. This problem can be reduced by adding a subcooler to the condenser.

Condensers generally cover the entire radiator surface to prevent air bypass. Accessory systems designed to fit several different cars are occasionally over-designed. Internal pressure drop should be minimized to reduce power requirements. Condenser to radiator clearances as low as 0.25 in. have been used, but 0.5 in. is preferable. Primary to secondary surface area varies from 8:1 to 16:1. Condensers are normally painted black so that they will not be visible through the vehicle's grille.

A condenser ahead of the engine-cooling radiator not only restricts air but also heats the air entering the radiator. The addition of air conditioning requires supplementing the engine cooling system. Radiator capacity is increased by adding fins, depth, or face area. It can also be raised by increasing coolant flow from higher pump speed. Pump cavitation at high speeds is the limiting factor. Increasing the speed of the water pump increases the engine cooling fan speed, which not only supplements the engine cooling system but also provides more air for the condenser. Fan size, number of blades, and blade width and pitch are frequently increased, and fan shrouds are often added when air conditioning is installed in an automobile. Increases in fan speed, diameter, and pitch raise the noise level and power consumption.

Temperature and torque-sensitive drives (viscous drives or couplings) or flex-fans reduce these increases in noise and power. They rely on ram air produced by the forward motion of the car to reduce the amount of air the radiator fan must move to maintain adequate coolant temperatures. As vehicle speed increases, fan requirements drop.

Front-wheel-drive vehicles typically have electric motor-driven cooling fans. Some vehicles also have side-by-side condenser and radiator designs, each with its own motor-driven fan.

Evaporator Systems

Current materials and construction include (1) copper or aluminum tube and aluminum fin, (2) brazed aluminum plate and fin, and (3) brazed serpentine tube and fin. Design parameters include air pressure drop, capacity, and condensate carryover. Fin spacing must permit adequate condensate drainage to the drain pan below the evaporator.

Condensate must drain outside the vehicle. The vehicle exterior is generally at a higher pressure than the interior at road speeds (1 to 2 in. of water). Drains are usually on the high-pressure side of the blower; they sometimes incorporate a trap and are as small as possible. Drains can become plugged not only by contaminants but also by road splash. Vehicle attitude (slope of the road and inclines), acceleration, and deceleration must be considered when designing condensate systems.

High refrigerant pressure loss requires externally equalized expansion valves. A bulbless expansion valve, which provides external pressure equalization without the added expense of an external equalizer, is available. The evaporator must provide stable refrigerant flow under all operating conditions and have sufficient capacity to ensure rapid cool-down of the vehicle after it has been standing in the sun.

The conditions affecting evaporator size and design are different from residential and commercial installations in that the average operating time, from a hot-soaked condition, is less than 20 min. Inlet air at the start of the operation can be as high as 150 °F, and it decreases as the duct system is ventilated. In a recirculating system, the temperature of inlet air decreases as the car interior temperature decreases; in a system using outdoor air, it decreases to a few degrees above ambient (perpetual heating by the duct system). During longer periods of operation, the system is expected to cool the entire vehicle interior rather than just produce a flow of cool air.

During sustained operation, vehicle occupants want less air noise and velocity, so the air quantity must be reduced but sufficient capacity preserved to maintain satisfactory interior temperatures. Ducts must be kept as short as possible and should preferably be insulated from engine compartment and solar-ambient heat loads. Thermal lag resulting from added heat sink of ducts and housings increases cool-down time.

Filters, Hoses, and Heater Cores

Air filters are not common. Coarse screening prevents such objects as facial tissues, insects, and leaves from entering fresh-air ducts. Studies show that wet evaporator surfaces reduce the pollen count appreciably. In one test an ambient of 23 to 96 mg/mm^3 showed 53 mg/mm^3 in a non-air-conditioned car and less than 3 mg/mm^3 in an air-conditioned car. Rubber hose assemblies are used where flexible refrigerant transmission connections are needed due to relative motion between components or because stiffer connections cause installation difficulties and noise transmission problems. Refrigerant effusion through the hose wall is a design consideration. Effusion occurs at a reasonably slow and predictable rate, which increases as pressure and temperature increase. Hose with a nylon core is less flexible (pulsation dampening), has a smaller O.D., is generally cleaner, and has practically no effusion.

The heat transfer surface in an automotive heater is usually cellular. It is a copper-brass assembly with lead-tin solder. The water course is brass (0.006 to 0.016 in.), and the air course is copper (0.003 to 0.008 in.). The tanks and connecting tubes are brass (0.026 to 0.034 in.). Both U-flow and straight-through flow are used. Capacity is adjusted by varying face area up to 120 in.^{2}, depth 2.5 in., and air-side surface geometry (for turbulence and air restriction) (Figure 1).

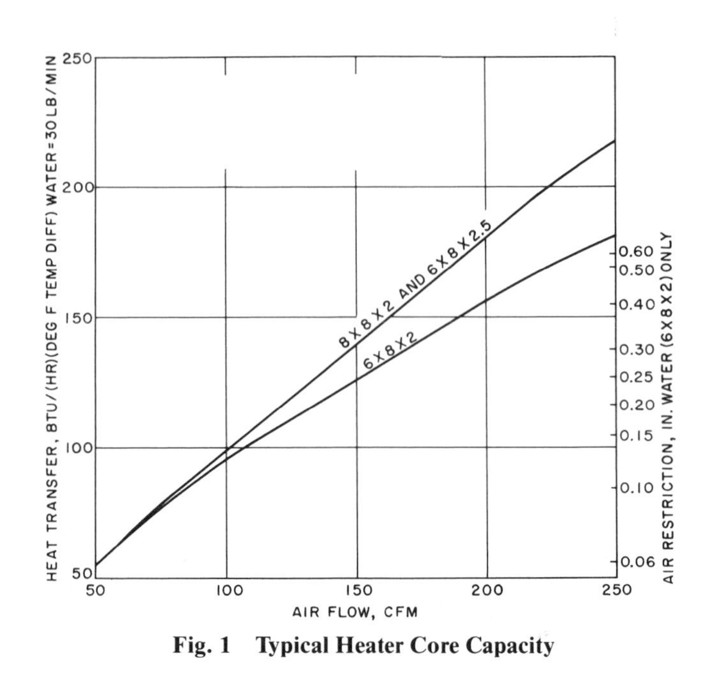

Fig. 1 Typical Heater Core Capacity

Receiver-Drier Assembly

The receiver-drier assembly accommodates charge fluctuations from changes in system load (refrigerant flow and density). It accommodates an overcharge of refrigerant (8 to 16 oz) to compensate for system leaks and hose effusion. The assembly houses the high-side filter and desiccant. Several types of desiccant are used, the most common of which are activated alumina, silica gel, and either granular or spherical molecular sieves. Mechanical integrity (freedom from powdering) is important because of the vibration to which the assembly is exposed. For this reason, molded desiccants have not obtained wide acceptance.

Moisture retention at elevated temperatures is also important. The rate of release with temperature increase and the reaction while accumulating high concentration should be considered. Design temperatures of at least 140 °F should be used.

The receiver-drier often houses a sight glass to provide a visual indication of the system charge level. It houses safety devices such as fusible plugs, rupture discs, or high-pressure relief valves. The last of these devices are gaining increasing acceptance, because they do not vent the entire charge. Location of the relief devices is important. Vented refrigerant should be directed in such a manner that it does not endanger personnel.

Receivers are usually (though not always) mounted on or near the condenser. They should be located so that they are ventilated by ambient air. Pressure drops should be minimal.

Expansion Valves

Thermostatic Expansion Valves (TXV) control the flow of refrigerant through the evaporator. These are applied as shown in Figures 4, 5, and 6. Both liquid and gas charged power elements are used. Also, internally and externally equalized valves are used as dictated by system design. Externally equalized valves are necessary where high evaporator pressure drops exist. A bulbless expansion valve, which senses evaporator outlet pressure without the need for an external equalizer, is now widely used. The trend toward a variable compressor pumping rate is also leading toward electrically controlled expansion valves.

Orifice Tubes

The use of an orifice tube instead of an expansion valve to control refrigerant flow through the evaporator has come into widespread use in original equipment installations. The components of the system must be carefully matched to obtain proper performance. Even so, there is some flood back of liquid refrigerant to the compressor under some conditions with this system.

Suction Line Accumulators

A suction line accumulator is required with an orifice tube system to ensure uniform return of refrigerant and oil to the compressor to prevent slugging and to cool the compressor. A typical suction line accumulator is shown in Figure 2. A bleed hole at the bottom of the standpipe meters oil and liquid refrigerant back to the compressor. The filter and desiccant are contained in the accumulator because no drier-receiver is used with this system. The amount of refrigerant charge is more critical when a suction line accumulator is used than it is with a drier-receiver.

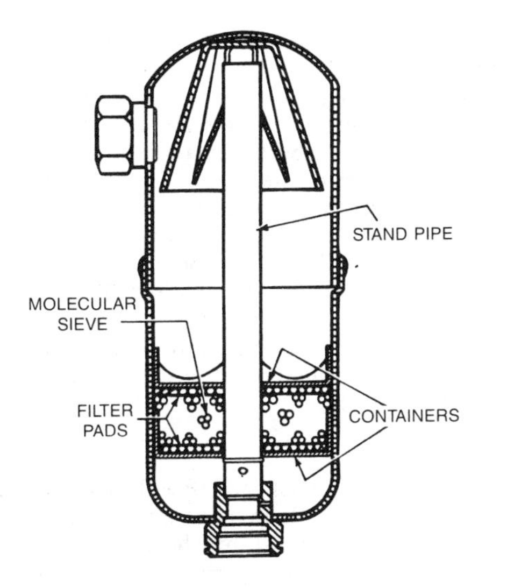

Fig. 2 Typical Suction Line Accumulator

Refrigerant Flow Control

The cycling clutch refrigerant systems shown in Figures 3 and 4 are common in late model cars for both factory- and dealer-installed units. The clutch is cycled by a thermostat sensing evaporator temperature or by a pressure switch sensing evaporator pressure. Some dealer-installed units use an adjustable thermostat, which controls car temperature by controlling evaporator temperature. The thermostat also prevents evaporator icing. Most units use a fixed thermostat or pressure switch set to prevent evaporator icing. Temperature is then controlled by blending air with warm air coming through the heater core.

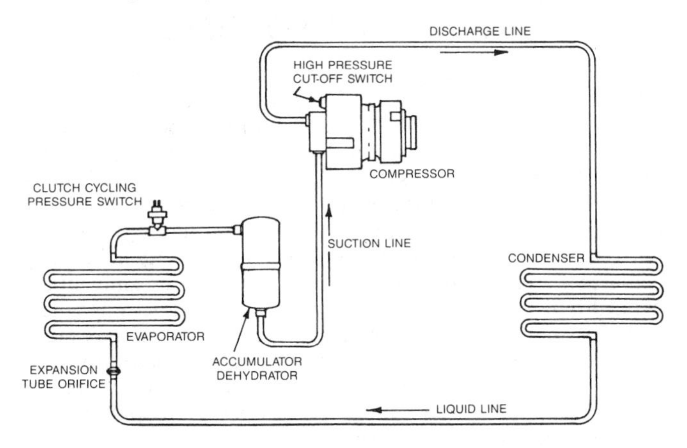

Fig. 3 Clutch Cycling Orifice Tube Air-Conditioning System Schematic

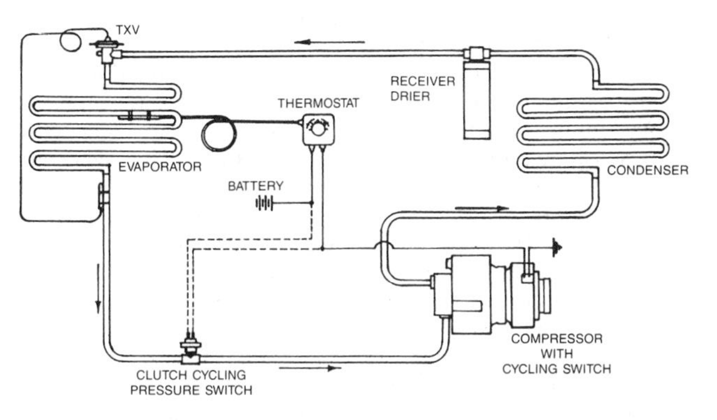

Fig. 4 Clutch Cycling System with Thermostatic Expansion Valve (TXV)

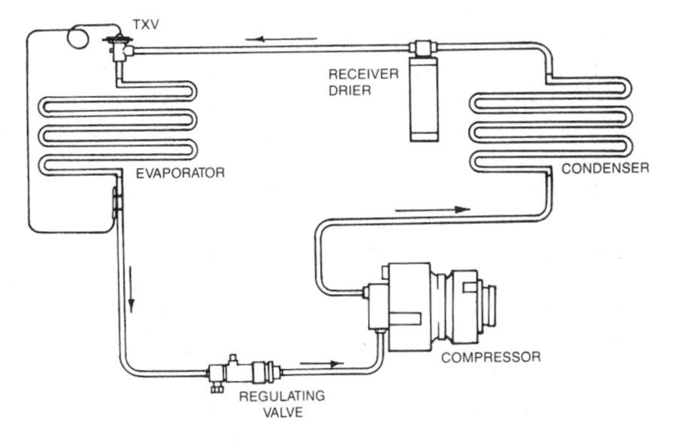

Fig. 5 Suction Line Regulation System

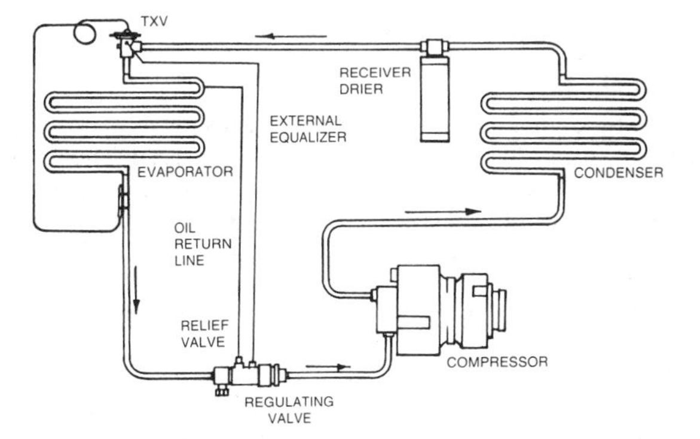

Fig. 6 Flooded Evaporator Suction Line Regulation System

Cycling the clutch sometimes causes noticeable surges as the engine is loaded and unloaded by the compressor. This is more evident in cars with smaller engines. Reevaporation of condensate from the evaporator during the off-cycle may cause objectionable temperature fluctuation or odor. This system cools faster and at lower cost than a continuous running system.

In orifice tube-accumulator systems, the clutch cycling switch disengages at about 25 psig and cuts in at about 45 psig. Thus, the evaporator defrosts on each off-cycle. The flooded evaporator has enough thermal inertia to prevent rapid clutch cycling. It is desirable to limit clutch cycling to a maximum of 4 cycles per minute because heat is generated by the clutch at engagement.

The pressure switch can be used with a TXV in a dry evaporator if the pressure switch is damped to prevent rapid cycling of the clutch.

Continuous running systems, as shown in Figures 5 and 6, have also been widely used. An evaporator pressure regulator (EPR) keeps the evaporator pressure above the condensate freezing level.

Temperature is controlled by reheat or air blending with warm air from the heater core. The valve, which is located on the downstream side of the evaporator, may be self-contained in its own housing and installed in the compressor suction line, or it may be placed in the suction cavity of the compressor. The device is pressure sensitive and operates to maintain a minimum evaporator pressure (saturation pressure). The setting depends on minimum airflow across the evaporator, the condensate draining ability of the evaporator coil, and the lowest ambient temperature at which it is anticipated that the system will be operated.

The continuously running system possesses neither of the disadvantages of the cycling clutch system mentioned previously, but it does increase the suction line pressure drop; hence, a slight reduction in system performance at maximum load. A solenoid version of this valve, which is controlled by an antifreeze switch that senses evaporator fin temperature, has also been used. The switch closes the valve electrically when the fin temperature drops to the control point.

Two basic refrigeration circuits use the evaporator pressure regulator (EPR). These are shown in Figures 5 and 6. Figure 5 shows the conventional dry-type system. Figure 6 shows a flooded evaporator system, which uses a plate-and-separator type of evaporator with a tank at the top and bottom. It is also a unique piping arrangement. The TXV external equalizer line is connected downstream of the evaporator pressure regulator valve. Also, a small oil return line containing an internal pressure relief valve is connected downstream of the EPR. These two lines may be connected to the housing of the EPR valve downstream of its valve mechanism. This piping arrangement causes the TXV to open wide when the EPR throttles, thus flooding the evaporator and causing the EPR to act as an expansion valve. The system allows the air conditioner to be run at ambients as low as 35°F to defog windows while maintaining adequate refrigerant flow and to ensure oil return to the compressor.

CONTROLS

Manual

The fundamental control mechanism is a flexible control cable, which transmits linear motion with reasonable efficiency and little hysteresis. Rotary motion is obtained by crank mechanisms. Common applications are (1) movement of damper doors, which control discharge air temperature in blend air systems; (2) regulation of the amount of defrost bleed; and (3) regulation of valves to control the flow of engine coolant through the heater core.

Vacuum

Vacuum provides a silent, powerful method of manual control, requiring only the movement of a lever or a switch by the operator. Vacuum is obtained from the engine intake manifold. A vacuum reservoir (25 to 250 $in.^3$) may be used to ensure an adequate source. Most systems are designed to function at minimum vacuum of 5 to 6 in. of Hg, even though as much as 26 in. of Hg may be available at times.

Linear or rotary slide valves select functions. Vacuum modulating, temperature compensating (bimetal) valves control thermostatic coolant valves. Occasionally, solenoid valves are used, but they are generally avoided because of their cost and associated wiring.

Electrical

Electrical controls regulate blower motors. Blowers have three or four speeds. The electrically operated compressor clutch frequently sees service as a secondary system control, operated by function (mode) selection integration or evaporator temperature sensing thermostats.

Temperature Control

Air-conditioning capacity control to regulate car temperature is achieved in one of two ways—the clutch can be cycled in response to an adjustable thermostat sensing evaporator discharge air, or the evaporator discharge air can be blended with or reheated by airflow through the heater core. The amount of reheat or blend is usually controlled by driver-adjusted damper doors.

Solid-state logic interprets system requirements and automatically adjusts to heating or air conditioning, depending on the operator's selection of temperature and on ambient temperature. Manual override enables the occupant to select the defrost function. The system regulates not only this function but also controls capacity and air quantity. An in-car thermistor measures the temperature of the air within the passenger compartment and compares it to the setting at the temperature selector. An ambient sensor, sometimes a thermistor, senses ambient temperature to prevent offset or droop. These elements, along with a vacuum supply line from an engine vacuum reservoir, are coupled to the control package, consisting of an electronic amplifier and a transducer. The output is a vacuum or electrical signal regulated by the temperature inputs.

This regulated signal is supplied to servo-units that control the quantity of coolant flowing in the heater core, the position of the heater blending door, or both; the air-conditioning evaporator pressure; and the speed of the blower. The same regulated signal controls sequencing of damper doors, resulting in the discharge of air on the occupant's feet or at waist level, or causing the system to work on the recirculation of interior air. A number of interlocking assurance devices are usually required. They prevent blower operation before engine coolant is up to temperature, prevent the air-conditioning compressor clutch from being energized at low ambient temperatures, and provide other features for passenger comfort.

Corrosion

R-12 is known to form HCl and HF with heat and the compressor lubricant. HCl and HF are noncorrosive as gases but form strong acids with system moisture. The more viscous 1.2×10^{-3} ft^2/s automotive oils are not required to pass the stringent sealed tube corrosion tests of refrigerator oils. Moisture may be present from inadequate evacuation or part storage on assembly or may effuse into the system at low ambients when the system is below ambient pressure. Acid may be detected by titration of an oil sample. A study of field failures indicates HCl concentrations above 0.05 mg/mL to be unacceptable. Also, the study observed that an activated alumina filter in the suction line will clear an acidic system. Some brazed plate fin evaporators are primarily vulnerable as their large end tanks reduce the flow velocity sufficiently to accumulate acid-laden oil. If an evaporator leaks, the system should be cleaned when it is replaced. U.S. Patent 4,599,185 states that a small amount of lecithin, a common food additive and preservative derived from soybeans, will retard acid formation.

BUS AIR CONDITIONING

Bus air-conditioning design must consider highly variable loads, from the standpoint of both passenger load and climate. It is usually not cost-effective to design for a specific climate. Therefore, the design should consider all likely climates, from the high ambient temperatures of the dry southwestern United States to the high humidity of the cooler east. Units should operate satisfactorily in ambient conditions up to 120°F. A unit designed for both extremes has a greater sensible cooling capacity in hot, dry climates than in humid climates. The quality of the ambient air must also be considered. Frequently, intakes are subjected to thermal contamination either from road surfaces, condenser air recirculation, or radiator air discharge. Vehicle motion also introduces pressure variables that affect condenser fan performance. In addition, engine speed will affect compressor capacity. Bus air-conditioning units are generally tested separately in climate-controlled test cells. With large vehicle test chambers now available, realistic as-installed testing is possible.

INTERURBAN BUSES

The following conditions have been adopted as a standard for interurban vehicles:

1. Capacity of 47 passengers
2. Insulation thickness of 1 to 1.5 in.
3. Double-pane tinted windows
4. Outdoor air intake of 400 cfm
5. Road speed of 60 mph
6. Inside design temperatures of 80°F dry-bulb and 67°F wet-bulb, or 20°F lower than ambient

Outside conditions typical of the United States give loads from 3.5 to 10 tons. Unless a specific area requires a custom design, the designer should consider extreme conditions. The design is sized for standby conditions. Using the engine as the compressor drive for standby results in excessive capacity over the road; appropriate unloading devices are necessary. Higher engine idle speed lessens the amount of compressor oversizing. Some interurban buses have a separate engine-driven air conditioner to give more constant performance; however, the space required, complexity, cost, and maintenance requirement are greater for these than for other units. Figure 7 shows a typical arrangement of equipment. A winter feature, including warm sidewalls and window diffusers, is important to offset downdrafts. The return air openings near the floor help reduce stratification. These features, while desirable, are not as important on urban buses, because passengers are exposed for only a short time and wear appropriate clothing. Some additional thermal losses occur with sidewall distribution and underside ducting; therefore, effective insulation and avoidance of thermal short circuits are necessary. Placing the condenser behind the front axle has been satisfactory. The condenser should not be placed low and near the rear axle on rear engine-driven buses because hot air (up to 140°F) from the radiator reduces the condenser capacity.

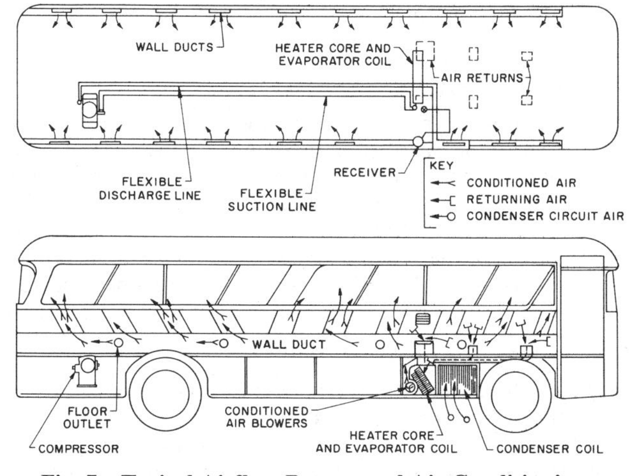

Fig. 7 Typical Airflow Pattern and Air-Conditioning Equipment Location in an Interurban Bus

Shock and Vibration

Most transport air-conditioning manufacturers design components for a shock loading between 3 and 5 *g* (acceleration of gravity). Vibration eliminators, flexible lines, and other shock-cushioning devices interconnect the various air-conditioning components. The vibration characteristics of each component are different. On a direct, engine-connected compressor, whether the connection is made by belts, flexible drives, gearing, or clutches, relative motion occurs between the main engine and the compressor. This motion may be taken care of by shock arms, vibration dampening devices, or automatic belt-tensioning devices. It is good practice to tie the compressor onto the engine by bracket extensions or other mounting devices so that the engine and compressor have no relative motion.

The relative motion between a compressor and other air-conditioning components is usually taken care of by flexible refrigerant hoses. The permeability of flexible lines have presented some difficulty with Refrigerant 22; thus Refrigerant 12, because it operates at lower pressures, is usually chosen. The use of R-22 then is limited to higher capacity systems such as double-decker and articulated interurban buses. The use of metallic bellow hoses is preferred.

Controls

Most interurban coaches have a simple driver control to select air conditioning or heating. In both modes, a thermal sensing element (thermistor, mercury tube, etc.) controls these systems with on/off circuitry and actuators. When dehumidification but not cooling is needed, the air-conditioning system is engaged and the air is reheated for comfort. Higher than needed engine power draw (18 HP) occurs with on/off control. More sophisticated controls and actuators to limit the need to reheat and lower the power are becoming popular. This improves comfort and fuel economy.

URBAN BUSES

Urban bus heating and cooling loads are greater than those of the interurban bus. A city bus may seat up to 50 people and carry a crush load of standees. The fresh air load is greater because of the number of door openings and the infiltration around doors. Table 1 shows the results of a test that recorded door openings. An urban bus stops frequently and may open both front and rear doors to take on or discharge passengers. Entering passengers bring with them the outside conditions, which could be considerably different from the stabilized conditions for which data is readily available.

Table 1 Door Operation of a City Bus

	Front Entrance Door	Center Exit Door
Door open		
Times per mile	7.5	5.5
Times per hour	70	44
% of operating time	35	15
Longest time open	55 s	34 s
Shortest time open	3 s	3 s
Average time open	13.5 s	12.5 s

Note: Observed during the rush period on a 35-passenger bus in San Antonio, Texas.

The cooling capacity required for the typical 50-seat urban bus is from 6 to 10 tons of refrigeration. The equipment must be flexible enough to range from low to maximum capacity promptly because the passenger load varies greatly. Buses with engine-driven compressors suffer a loss of compressor capacity at idle. This loss is compensated for by sizing other parts of the system somewhat larger than the load analysis indicates to meet a time average capacity equal to the comfort goal.

At maximum engine speed, the compressor may have a capacity in excess of the cooling load or the rating of the remainder of the system—a condition that must also be considered. In general, the amount of power available is a limiting factor and, more recently, fuel consumption has influenced specifications. Therefore, equipment must be designed for maximum efficiency and the vehicle, for minimum thermal losses.

System Types

Air-conditioning systems for urban buses generally fall into three categories. The newer, advanced design buses are usually equipped with either a roof- or rear-mounted package unit similar to that shown in Figures 8 and 9, which includes all system components except the compressors. The compressor is usually belt or shaft driven from the main traction engine.

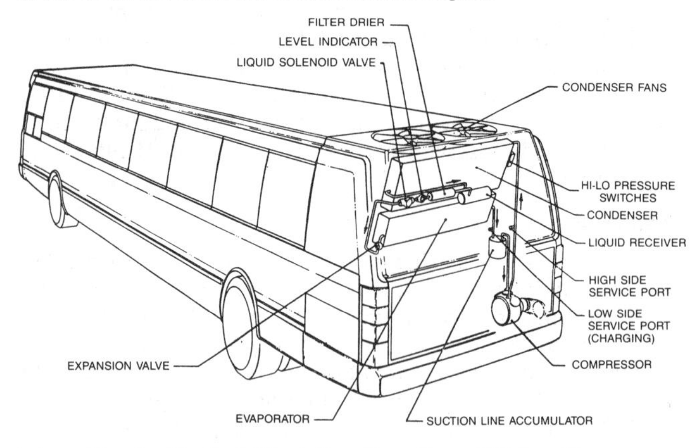

Fig. 8 Typical Mounting Location of Urban Bus Air-Conditioning Equipment

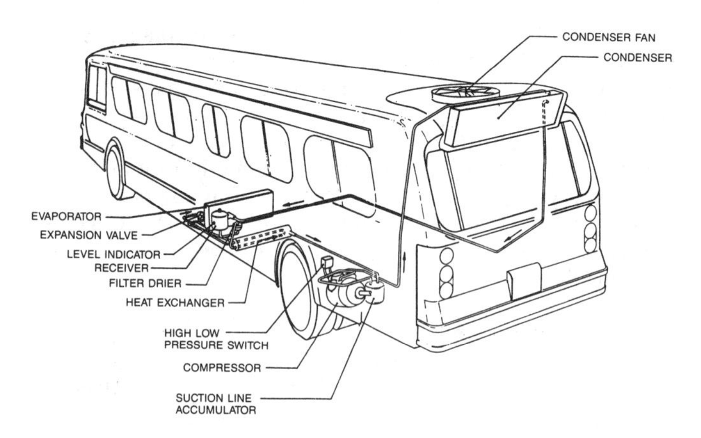

Fig. 9 Mounting Location of Air-Conditioning Equipment in Older Urban Buses

The heater is located just downstream of the evaporator. Hot coolant from the engine cooling system provides sufficient heat for most operations; however, additional sources may be required in colder climates for long durations of idle. Additional floor heaters may also be required to reduce the effects of stratification. Air distribution for these systems is either through linear diffuser(s) in the ceiling fed from a duct concealed in the space between the roof and ceiling panels, or by longitudinal ducts concealed in the sidewalls above the windows behind the lighting fixtures.

Many older urban buses, similar to the one shown in Figure 9, have an evaporator/blower unit mid-mounted below the bus floor, which feeds conditioned air through sidewall voids to slotted or perforated diffusers at the windowsills. These systems have either an upper, rear-mounted condenser, or the condenser is mounted in tandem with the engine radiator. The tandem installations are susceptible to recirculation of radiator discharge air, which can severely reduce capacity, especially at idle. Compressors are either shaft-driven from the transmission auxiliary power take-off, or belt-driven directly from the engine. In either case, a clutch connects or disconnects the drive for system control. The wide separation of components in this design requires considerable lengths of interconnecting piping and many fittings and joints that are susceptible to leakage from vibration and physical damage. In addition, the air distribution systems located in the lower sidewalls of the coach are easily damaged during a traffic accident and are most difficult to repair and keep airtight.

Retrofit Systems

As the public has demanded a more comfortable riding environment, there has been an upsurge in retrofitting older buses with air-conditioning systems and in replacement of older systems with newer, more efficient designs. Available systems follow the configuration of systems supplied for new buses in that roof- and rear-mount units can be fitted into old coaches. Some manufacturers also offer a simple system of interior evaporator/blower units, using two to four units per bus, depending on the bus size and the operating environment. These units are installed over the windows along one side of the bus. They sometimes replace the original lighting fixtures formerly at that location and provide the needed cooling without the need for duct installation. Single evaporator/blower units can be mounted in the rear of small buses, along with a skirt-mounted condenser (Figure 7), to supplement the factory-installed dash unit. Refrigerant piping in these systems is somewhat more extensive than in packaged units, but the overall installation can be made faster and at lower cost. In addition, renovations to the bus interior are kept to a minimum, and the structural concern of cutting an opening in the roof is eliminated.

Compressors and Drives

Several different types and styles of compressors and drives are seen on urban buses. The four- or six-cylinder reciprocating compressor, in which some cylinders are equipped with unloaders, is popular. The compressor is driven either from an auxiliary shaft extending from the transmission or mounted so that it can be directly belt-driven from the engine. In both cases, it is fitted with a clutch, either pneumatically or electrically actuated, by the temperature control system and refrigerant pressure. Several designs have one or more high-speed, axial piston or rotary compressors that are belt-driven and electric-clutch controlled. Helical screw compressors on urban transit buses have also been successfully demonstrated.

Articulated Coaches

Several urban transit authorities operate articulated buses in which each of the two sections is equipped with a separate roof- or rear-mount package unit. Each section has its own belt-driven compressor, and both units are usually located in the rear-section engine compartment. The package for the rear section has a dual condenser, while the package for the forward section has no condenser. Long runs of flexible refrigerant piping connect the front section to its compressor at the extreme rear of the rear section.

System Maintenance

Due to the extremely adverse operating environment of the urban transit bus, the cost of inspection and repairs is far greater than the initial cost of the system. For this reason, it is generally cost-effective to arrange the system for ease of access to the repairable parts and to provide convenient points for checking the critical pressures, fluid levels, and temperatures. Ease of access to the air filters for replacement or cleaning is essential, since this is the most often performed maintenance task. Refrigerant piping joints should be kept to a minimum, and all controls, safety devices, and accessory items should be heavy-duty and able to withstand the extreme conditions and environments to which the bus will be subjected.

Controls

The typical urban transit coach is relatively simply controlled. Cooling thermostats for full- and part-capacity and a heating thermostat to operate the pump or valve serving the heating core are usually included. Many current systems use solid-state control modules to interpret the bus interior and outside ambient temperatures and to generate signals to operate the full- or part-cooling or heating functions. In these systems, thermistor temperature sensors are used, which are usually more stable and reliable than electromechanical controls.

Control systems also include a ventilation of outside air cycle. This mode of operation maintains the coach interior comfort level using 100% outside air. The system is used during the temperature zone between heating and air conditioning. The ventilation cycle permits longer compressor off-cycle time, which benefits the coach operator in both fuel economy and compressor life.

RAILROAD AIR CONDITIONING

Railroad air-conditioning systems are electromechanical with direct expansion evaporators usually using R-12 or R-22. R-12 will be phased out with HCFC-134a or other refrigerants as CFCs are eliminated. Electronic solid-state automatic controls are common with a trend toward microprocessor control. Electric heating elements installed in the air-conditioning unit or supply duct temper outside air brought for ventilation and are often used as reheat for humidity control.

Passenger Car Construction

Passenger car design has emphasized lighter car construction to lower costs of building and to decrease operating and maintenance cost. The drive to reduce weight and cost has also reduced the size and weight of air-conditioning equipment and other auxiliaries.

Vehicle Types

Mainline railroads generally operate single and bilevel cars hauled by a locomotive. Locomotive-driven alternators or static inverters distribute power via an intercar cable power bus to the air-conditioning systems. A typical railcar has two evaporator-heater-fan units mounted above the ceiling with a common or two separate underfloor compressor-condenser units and a control package. Underfloor, interior, and roof-mounted package systems are less common.

Commuter cars operating around large cities are similar in size to mainline cars and carry similar types of air conditioning. These may be locomotive hauled but are often self-propelled by high voltage direct current (DC) or AC power supplied from an overhead catenary or from a DC-supplied third rail system. On such cars, the air conditioning may operate on AC or DC power. Diesel-driven vehicles still operate in a few areas.

Subways usually operate on a third rail DC power supply. The car air conditioning operates on the normal DC supply voltage or on three-phase AC supply from an alternator or inverter mounted under the car. Split air-conditioning systems are common with evaporators at roof level and underfloor-mounted condensing sections.

Streetcars, light rail vehicles, and downtown people movers usually run on the city AC or DC power supply and have air-conditioning equipment similar to subway cars. During rush hours, interior space is at a premium on these cars, so that roof-mounted packages are used more often than are split systems.

Equipment Selection

The source and type of power dictate the type of air-conditioning equipment installed on a railroad car; weight is also a major consideration. Thus, AC-powered semihermetic compressors, which are lighter than open machines with DC motor drives, are a common choice. Each car design must be carefully examined in this respect, since DC/AC inverters may increase the total system weight and power draw due to conversion losses.

Other aspects of equipment selection include space requirements, location, accessibility, reliability, and maintainability. Interior and exterior equipment noise levels must be considered during the early stages of design and later when coordinating the equipment with the car-builder's ductwork and grilles.

Design Limitations

Space underneath and inside a railroad car is at a premium, especially on self-propelled light rail vehicles, and subway and commuter cars, and generally rules out unitary interior or underfloor-mounted systems. Usually, the components of the system are built to fit the configuration of the available space. Overall car height, roof profile, ceiling cavity and under-car clearance restrictions determine the shape and size of equipment.

Since a mainline railroad car must operate in various parts of the country, air conditioning must handle the national seasonal extreme design days. Commuter cars and subway cars operate in a small geographical area, and only the local design temperatures and humidities need be considered.

Dirt and corrosion constitute an important design factor, especially if the equipment is beneath the car floor where it is subject to all types of weather, as well as severe dirt conditions. For this reason, corrosion-resistant materials and coatings must be selected wherever possible. Aluminum has not proved durable enough because the sandblasting effect destroys any surface treatment on aluminum installed beneath the car. Since dirt pickup cannot be eliminated, the equipment must be designed for quick and easy cleaning. Access doors are needed to get inside or behind coils to blow out the dirt. Evaporator and condenser fin spacing is usually limited to 8 to 10 fins per inch. Higher density causes more rapid dirt buildup and costs more to clean. Dirt, as well as other severe environmental conditions, must also be considered in selecting motors and controls.

Railroad equipment requires more maintenance and servicing than stationary units. A modern railroad car with sealed windows and a well-insulated structure becomes almost unusable if the air conditioning fails. For this reason, the equipment has many additional components to permit quick determination and correction of the failure. Motors, compressors, and valves, for example, must be easily accessible for inspection or repair. The liquid receiver should have sight glasses so that the amount of refrigerant can be quickly checked. Likewise, a readily accessible liquid charging valve should be available. Pressure gages and test switches allow a fast check of the system while the train is stopped at intermediate stations.

Safety must be considered, especially on equipment located beneath the car. Vibration isolators and supports, especially from bolt shears and loose nuts, should be fail-safe. A piece of equipment that hangs down or drops off could cause a train derailment. All belt drives or other revolving items must be safety guarded. High voltage controls and equipment must be labeled by approved warning signs. All pressure vessels and coils must meet ASME test specifications for protection of the passengers and maintenance personnel. Materials selection criteria should include low flammability, toxicity, and smoke emission.

Interior Comfort

Air-conditioning and heating comfort conditions may be selected in accordance with ASHRAE *Standard* 55-1981. However, the selected temperature and humidity levels must consider the passenger's metabolic rate on entry, clothing insulation, and journey time. Chapter 8 of the 1989 ASHRAE *Handbook—Fundamentals* has more details.

Vibration and dusty conditions preclude the use of commercial humidity controllers. Evaporator sectional staging and compressor cylinder unloading, coupled with electric reheat, is used to provide part-load humidity control. Future concepts under study are focusing on variable compressor and fan speed control to provide greater efficiency to inherently wasteful reheat control. A maximum relative humidity for optimum comfort is usually specified. In winter, humidity control is usually not provided.

The dominant summer cooling load is due to passengers. This is followed by ventilation, internal heat, car body transmission, and solar gain. Heating loads comprise car body losses and ventilation. The heating load calculation does not credit heat from passengers and internal sources. Comfortable internal conditions in ventilated non-air-conditioned cars can only be maintained when ambient conditions permit.

Due to the continuous variation in passenger and solar loads in mass-transit cars, the interior conditions are difficult to hold, and variation from the desired level of maximum acceptance occurs. Air-conditioning systems in North American cars are selected to maintain 73 to 76 °F, with a maximum relative humidity of 55 to 60%. In Europe and elsewhere, the conditions are usually set at a dry-bulb temperature from 0 to 10 °F below ambient, with a coincidental 50 to 66% rh. In the heating mode, the car interior is kept in the 65 to 70 °F. Outdoor conditions are chosen from Chapter 24 of the 1989 ASHRAE *Handbook—Fundamentals* and local climatic data.

System Requirements

Most cars are equipped with both overhead and floor heat. The overhead heat raises the temperature of the recirculated and ventilation air mixture to a temperature slightly above the car design. The floor heat offsets the heat loss through the car body and reduces temperature stratification.

Cooling and heating loads are calculated in accordance with Chapters 25 and 26 of the 1989 ASHRAE *Handbook—Fundamentals*. The times of maximum occupancy, outdoor ambient, and solar gain must be ascertained. For cooling loads on urban transit cars, this peak load usually coincides with the evening rush hour, and on intercity railroads, around midafternoon.

Heating capacity for the car depends on envelope construction, size, and the area averaged relative (to the car) design wind velocity. In some instances, minimum car warmup time may be the governing factor. In long-distance trains, the toilets, galley, and lounges often have exhaust fans. Ventilation airflow must be designed to sufficiently exceed forced exhaust air rates to maintain positive car pressure.

Minimum per occupant fresh air ventilation rates of 5.3 to 7.4 cfm all nonsmoking, and 10.6 to 14.8 cfm smoking permitted are desirable. Ventilation air pressurizes the car which reduces infiltration.

Air Distribution

The most common type of air distribution system is a centerline supply duct running the length of the car and located in the space between the ceiling and roof. The air outlets are usually ceiling-mounted linear slot air diffusers. Egg crate recirculation grilles are positioned in the ceiling beneath the evaporator units. The main supply duct must be insulated from the ceiling cavity to prevent summer thermal gain and winter condensation. The ventilation air should be taken from both sides of the roofline to overcome the effect of wind. Adequate snow and rain louvers must be installed on the outdoor air intakes. Separate outdoor air filters are usually paired with a return air filter. Disposable media or permanent, cleanable air filters are used, and they are usually serviced every month.

Piping Design

Standard refrigerant piping practice is followed for components. Pipe joints should be accessible for inspection and not concealed in car walls. When the complete system and piping has been installed, evacuation, leak testing, and dehydration must be completed successfully prior to charging. Piping should be supported adequately and installed without traps that could retard flow of oil back to the compressor. Pipe sizing and arrangement should be in accordance with Chapter 3 of the 1990 ASHRAE *Handbook—Refrigeration*. Evacuation, dehydration, charging, and testing should be performed as described in Chapter 22 of the 1988 ASHRAE *Handbook—Equipment*. Piping on packaged units should also conform to these standards.

Control Requirements

Car HVAC systems are automatically controlled for year-round comfort. In the cooling mode, load variations are handled by two-stage, direct expansion coils and compressors equipped with suction pressure unloaders or speed control. During low load conditions, cooling is provided by outdoor air supplied through the ventilation system. During part-load cooling, electric reheat activates to maintain humidity control. The system assumes the ventilation mode under low load when humidity control is not needed.

A pump-down cycle and low ambient lockout protect the compressor from damage caused by liquid slugging. In addition, the compressor may be fitted with a crankcase heater that is energized during the compressor off-cycle. During the heating mode, floor and overhead heaters are staged to maintain the car interior temperature.

Today's control systems use thermistors and solid-state auxiliary electronics instead of electromechanical devices. The control circuits are normally powered by low-voltage DC or, occasionally, by single-phase AC.

Future Trends

The demand for lighter, more efficient railway cars remains strong. The use of rooftop packaged air-conditioning units has reduced weight and improved reliability by eliminating long piping runs. Most manufacturers offer hermetic compressors, which are isolated to withstand the shock and vibration normally encountered. Some manufacturers market heat pump units. Neither hermetic compressors nor heat pumps have been fully field tested in North America. However, some operating authorities are considering them to reduce costs. Since other major costs include maintenance and replacement parts, recent specifications stipulate high reliability combined with low maintenance.

FIXED GUIDEWAY VEHICLE

Fixed guideway systems, commonly referred to as people movers, are increasing in popularity. They can be configured as monorails, or as rubber-tired cars running on an elevated or grade level guideway such as those seen at many airports and in downtown urban areas such as Miami. The guideway directs and steers the vehicle and provides the electrical power necessary to operate the car's traction motors, lighting, electronics, and air-conditioning and heating systems.

These systems are usually unmanned, and are computer controlled from some central point. The system's operations control determines the speed, headway, and length of time for door openings based on telemetry transmitted from the individual cars or trains. Therefore, providing a reliable and effective environmental control system is essential.

People movers are smaller than most other mass transit vehicles, generally having spaces for 20 to 40 seated passengers and generous floor space for standing passengers. Under some conditions of passenger loading, it has been found that a 40-ft car can accommodate 100 passengers. The wide range of passenger loads possible, and the moving of the car from full sunlight to deep shade in a short period of time, make it essential that the car's air-conditioning system be especially responsive to the amount of cooling required at a given moment.

System Types

The heating and air-conditioning systems provided for this type of vehicle are usually one of three types:

1. Conventional undercar condensing unit connected with refrigerant piping to an evaporator-blower unit mounted above the car's ceiling.
2. Packaged, roof-mounted unit having all components within one enclosure, and mating to an air distribution system built into the car's ceiling.
3. Packaged, undercar-mounted unit, mating to supply and return air ducts built into the car body.

Located at the extreme opposite ends of each car, two systems are usually provided; each provides one-half of the maximum cooling requirement. The systems, whether unitary or split, operate on the guideway's power supply, which is usually 60 Hz, 460 to 600 VAC.

Refrigeration Components

Due to the availability of commercial electrical power, standard semihermetic motor-compressors, as well as commercially available fan motors and other components, can be used. Compressors generally have one or two stages of unloaders and, in addition, hot gas bypass is used to maintain cooling at low levels of load requirement. Condenser and cooling coils are copper tube, copper, or aluminum fin units. Generally, flat fins are preferred for undercar systems to make it simpler to clean the coils. Evaporator-blower sections must often be configured to the shape of the car's ceiling and must be designed for the specific vehicle. The condensing units must also be arranged to fit within the limited space available and still ensure good airflow across the condenser coil. Almost universally, Refrigerant 22 is employed in these systems.

Heating

Where heating must be provided, electric resistance heaters are introduced into the system at the evaporator unit discharge. These heaters operate on the guideway power supply. One or two stages of heat control are used, depending on the size of the heaters.

Controls

A solid-state control system is usually used to maintain interior conditions. Temperature set points are usually desirable in the 74 to 76 °F range for cooling. For heating, set points will be at 68 °F or lower. Control systems are available to provide some humidity control by using the electric heat. Between cooling and heating set points, blowers continue to operate on a ventilation cycle. Often, two-speed blower motors are used, switching to low speed when on the heating cycle. Some control systems have been arranged with internal diagnostic capability; they are able to signal the system's operations center when a cooling or heating malfunction occurs.

Fresh Air

When using overhead air-handling equipment, fresh air is introduced into the return airstream at the evaporator entrance. Fresh air is usually taken from a grilled or louvered opening in the end or side of the car and, depending on the configuration of components, will be separately filtered or introduced so that the return air filter can handle both airstreams. For undercar systems, a similar procedure is used, except the air is introduced into the system through an intake in the undercar enclosure. In some cases, a separate ventilation fan is used to induce fresh air into the system. The amount of outside air ventilation to be provided is difficult to estimate due to the widely varying passenger loads to be provided for. Older vehicles provided an amount equivalent to 20 to 25% of the total air circulated. Newer designs attempt to relate this to the passenger load by providing 10 to 15 cfm per seated and comfortably standing passenger. In some cases, the outside air proportion will reach 30% or more of the total airflow.

Air Distribution

With overhead systems, air is distributed through linear ceiling diffusers often constructed as a part of the overhead lighting fixtures. Undercar systems usually make use of the void spaces in the car's sidewalls and below fixed seating. In all cases, the spaces used for air supply must be adequately insulated to prevent condensation on surfaces and, in the case of voids below seating, to avoid cold seating surfaces. Supply air discharge from undercar systems will typically be through a windowsill diffuser. Recirculation air from overhead systems use ceiling-mounted grilles. For undercar systems, return air grilles are usually found in the door wells or beneath seats.

Because of the small size of the vehicle and its low ceilings, extreme care must be taken in designing the air supply system to avoid direct impingement of supply air on passengers' heads or shoulders. High rates of diffusion are needed, and diffuser placement and arrangement should permit the discharge airstreams to hug the ceiling and wall surfaces of the car. A careful balance of total air quantity and discharge temperature must be made to avoid extreme cold drafts and air currents.

BIBLIOGRAPHY

Kuffe K.W. 1978. Air conditioning and heating systems for trucks. SP-425, SAE, 400 Commonwealth Drive, Warrendale, PA 15096.

Wojtkowski, E.F. 1964. System contamination and cleanup. ASHRAE *Journal* (June).

Towers, J.A. and R.H. Krueger. 1985. Refrigerant additive and method for reducing corrosion in refrigeration systems. U.S. Patent 4,599,185.

CHAPTER 9

AIRCRAFT

AIRCRAFT air-conditioning equipment must meet additional requirements beyond those for the air conditioning of buildings, although basic principles are applicable. This equipment must be compact, lightweight, accessible for quick inspection and servicing, highly reliable, and unaffected by airplane vibration and landing impact. The temperature control and pressure control systems must be capable of responding to rapid changes of ambient temperature and pressure as the airplane climbs and descends.

DESIGN CONDITIONS

The aircraft air-conditioning and pressurization system primarily maintains an aircraft environment that ensures the safety and comfort of passengers and crew and the proper operation of on-board electronic equipment. The system must maintain this environment through the wide range of aircraft ambient temperatures and pressures. The effects of system and component failures must be accounted for in maintaining passenger and crew safety and operation of flight-critical electronic equipment.

Ambient Temperature, Humidity, and Pressure

Figure 1 gives typical design ambient temperature profiles. The figure shows variations of ambient temperature for hot, standard, and cold days. The ambient temperatures used for the design of a particular aircraft may be higher or lower than those shown by

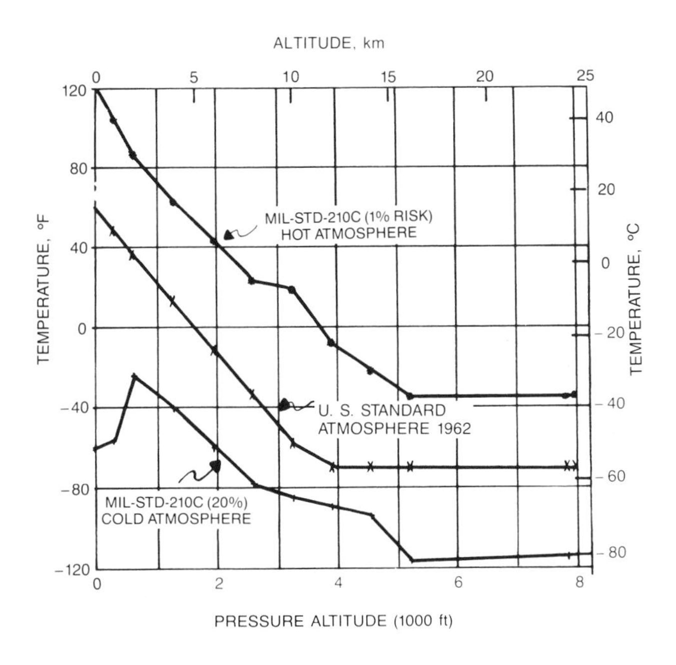

Fig. 1 Typical Ambient Temperature Profiles

Figure 1, depending on the regions in which the airplane may be operated. A recommended design ambient moisture variation with altitude for commercial aircraft is shown in Figure 2. Moisture content as high as 204 grains per pound of dry air at sea level are considered for military aircraft. The variation in ambient pressure with altitude is shown in Figure 3.

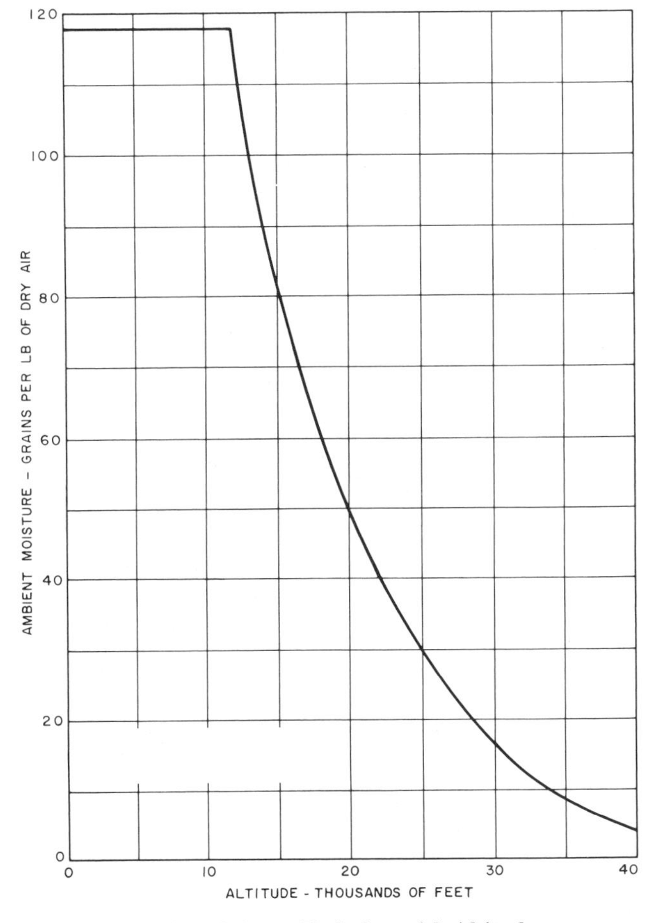

Fig. 2 Moisture Variation with Altitude

Air-Conditioning Performance

Air conditioning for passenger and crew compartments and cargo compartments should provide the following performance.

Cooling. During cruise, the system should maintain an average cabin temperature of 75 °F with full passenger load. During ground operations, the system should be capable of maintaining an average cabin no higher than 80 °F with full passenger load and all external doors closed. It should be capable of cooling the cabin on the ground to an average temperature of 80 °F within 30 min

The preparation of this chapter is assigned to TC 9.3, Transportation Air Conditioning.

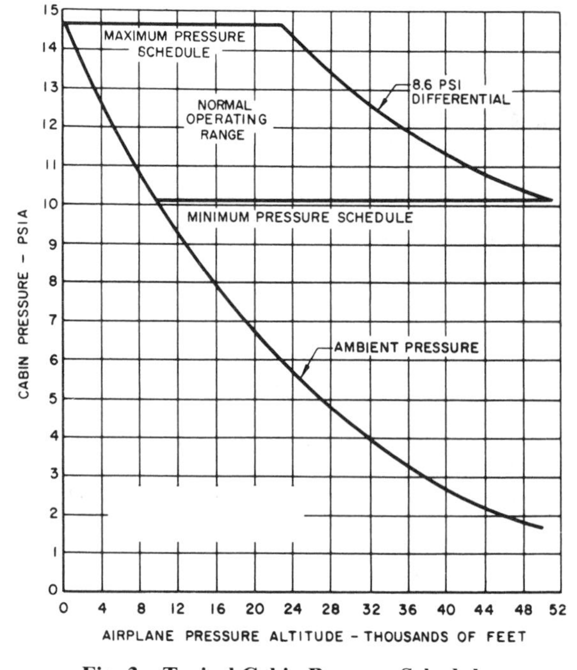

Fig. 3 Typical Cabin Pressure Schedule

starting with a cabin temperature of 115 °F, no passengers or other heat loads, and all external doors closed.

Heating. During cruise, the system should maintain an average cabin temperature of 75 °F, with a 20% passenger load. During ground operations, the system should be capable of heating the cabin on the ground to an average temperature of 70 °F within 30 min, starting with a cold soaked airplane at a temperature of −25 °F and an ambient temperature of −40 °F with no passengers or other internal heat loads and with all external doors closed. The system should maintain the cargo compartment bulk air temperature above 40 °F. Cargo floor temperatures should be above 32 °F to prevent freezing of cargo. These temperatures should be attainable during all cruise conditions.

Pressurization Performance

Figure 3 shows typical cabin pressure control operating range. Generally, aircraft designed for airline service are capable of maintaining a cabin altitude of 5000 to 7000 ft at typical operating altitudes and limiting cabin altitude to a maximum of 8000 ft at maximum cruise altitudes. Typically, cabin altitude rate of change is limited for passenger comfort to 500 ft/min for increasing altitude and 300 ft/min for decreasing altitude. Military fighter aircraft are normally designed to maintain a 5 psi differential at altitudes above approximately 23,000 ft.

Airplanes flying above 10,000 ft must be equipped with an oxygen system for use in case of loss of pressurization. To maintain cabin pressure control, air inflow to the pressurized cabin must exceed airflow leaking from the pressurized cabin. Leakage areas include controlled vents for the galley, toilets, and electronic equipment as well as uncontrolled leakage through door seals and structural joints.

Ventilation

Crew and passenger compartments should be ventilated whenever the aircraft is in operation. Actual ventilation rates in commercial aircraft range from 15 to 25 cfm/person based on a 100% passenger load. Generally, much higher ventilation rates are provided in the cockpit primarily to cool all the surrounding electrical and electronic equipment. A portion of the cabin ventilation air may be recirculated, but it should be filtered before reintroduction into the occupied areas. A ram air (outside air) or auxiliary ventilation source should be considered if reasonably probable failures could result in the loss of all ventilation. This auxiliary ventilation could consist of a ram air circuit supplying outside air through the normal distribution system, or provisions for manipulating cabin pressure control valves to draw ambient air through the cabin and flight deck, through door seals, negative relief valves, hatches, and so forth.

Air Quality

The quality of air in crew and passenger compartments depends on the quantity of fresh air supplied, contaminants that may be present in the air source, and the contamination generated within the aircraft compartments. Federal Aviation Regulations (FAR), Part 25, stipulate that crew and passenger compartment air be free of harmful or hazardous concentrations of gases or vapors. For military aircraft, air contaminant level within occupied compartments should not exceed the threshold limit values established by the American Conference of Governmental Industrial Hygienists and published in AFOSH *Standard* 161-8.

An important aspect of air quality in many commercial aircraft applications is the dilution of tobacco smoke to acceptable levels. A number of tests have been run to determine smoker and nonsmoker response to various dilution indices. The dilution index (DI) is defined as the liters of fresh air per mg of tobacco burned. Figure 4 shows how irritation varies with the DI, and Figure 5 shows the percentage of smokers and nonsmokers who reported the irritation levels as acceptable.

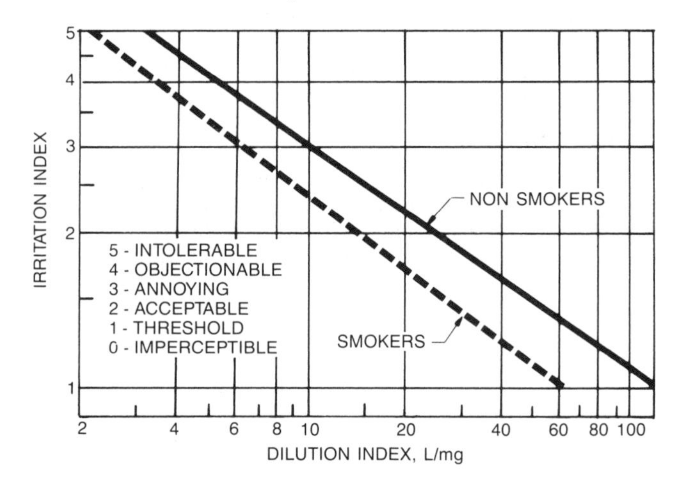

Fig. 4 Irritation versus Dilution of Smoke

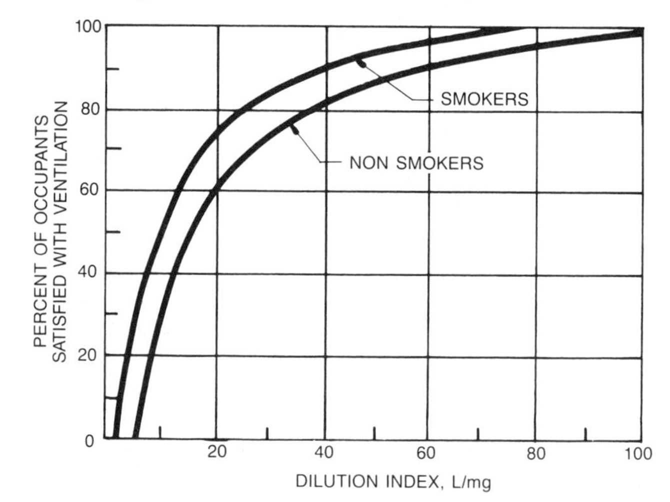

Fig. 5 Effect of Dilution Index on Occupant Satisfaction

Smoking Zones. The data in Figure 5 may be used, along with an estimated smoking rate of 17 mg/min per smoker to calculate the ventilation required to obtain any desired acceptance rate in a smoking zone.

Ventilation should be at a level so that at least 80% acceptance by smokers is achieved in smoking zones. For an 80% acceptance level, the DI is 24 L/mg for smokers in a smoking zone. At a smoking rate of 17 mg/min, a ventilation rate of 14.4 cfm per smoker is required.

Nonsmoking Zones. The minimum fresh air ventilation rate to limit CO_2 levels and to control odors is 15 cfm per occupant; however, temperature control and distribution requirements may establish higher ventilation rates. If air is recirculated from smoking zones into nonsmoking zones, it should be filtered to remove gaseous contaminants and aerosols, or it should be diluted with enough fresh air to achieve a DI of 60 L/mg, which would be acceptable to 90% of the nonsmokers. Equation (1) shows how the recirculation and fresh air requirements may be established in a nonsmoking zone.

$$\text{Fresh Air} = k(\text{Recirculated Air}) \tag{1}$$

where

$$k = [\text{DI Required/DI Smoking Zone} - 1]$$

Thus, if a 60 L/mg DI is required and the smoking zone DI is 24 L/mg, then 1.5 cfm of fresh air is required for every cfm of recirculated air.

Aircraft without Designated Smoking Zones. Smaller aircraft, such as business and executive aircraft, commuters, and small and large military transports, may not have designated smoking zones. The DI for these aircraft is based on the total smoke generated and total fresh air supplied. To achieve a 90% acceptance of ventilation by nonsmokers, a 60 L/mg DI is required for the entire cabin. The air distribution system for these types of aircraft should ensure good mixing to achieve a uniform DI.

Ozone. Aircraft operating at altitudes above approximately 30,000 ft may encounter atmospheric ozone concentrations of sufficient magnitude to affect cabin air quality adversely. Physiologically, ozone affects the soft tissues of the lung, causing pulmonary edema, dyspnea, and reduced lung capacity.

The effects of ozone are a function of ozone concentration and time of exposure. Currently, FAR Part 25 sets the following concentration and time of exposure limits for ozone in the cabins of transport aircraft:

1. 0.25 ppm by volume, sea level equivalent, at any time above 32,000 ft.
2. 0.1 ppm by volume, sea level equivalent, time-weighted average during any 3-h interval above 27,000 ft.

Ozone concentration in an airstream can be reduced by an absorption process, a chemical reaction with a filter surface, or a catalytic decomposition process.

Ozone decomposes at a rate determined by temperature. Thermal decomposition can be increased by allowing the air-ozone mixture to come in contact with metals or other materials that act as catalysts in decomposition reactions. Manganese dioxide, activated charcoal, stainless steel, and other common metals enhance the thermal decomposition of ozone.

Load Determination

The cooling and heating loads for a particular airplane must be determined from a heat transfer study and from an analysis of the solar and internal heat from occupants and electrical equipment. The study should consider all possible flow paths through the usually complex aircraft structure. Air film coefficients vary with altitude and should also be considered. For high-speed aircraft, the increase in air temperature and pressure due to ram effects is appreciable and may be calculated from Equations (2) and (3).

$$t_r = 0.2M^2 T_a F_r \tag{2}$$

$$p_r = (1 + 0.2M^2)^{3.5} p_a - p_a \tag{3}$$

where

F_r = recovery factor, dimensionless
t_r = increase in temperature due to ram effect, °F
M = Mach number, dimensionless
T_a = absolute static temperature of ambient air, °R
p_r = increase in pressure due to ram effect, psi
p_a = absolute static pressure of ambient air, psi

The average increase in airplane skin temperature for subsonic flight is generally based on a recovery factor F_r of 0.9.

Ground and flight requirements may be quite different. An aircraft sitting on the ground in bright sunlight will have local skin temperatures considerably higher than the ambient, where the surface is painted a dark color or unpainted and perpendicular to the sun's rays. Painting the upper portion of the fuselage with white paint will reduce this effect considerably, as will a breeze blowing across the airplane.

Other considerations for ground operations include cool-down or warmup requirements, time that doors are open for loading, and whether ground heating and cooling are provided using onboard equipment or equipment from an external source.

DESIGN APPROACHES

A trade study of all possible state-of-the-art approaches should be conducted when establishing the air-conditioning and pressurization system design for a particular aircraft. The design that meets performance requirements and is overall best from the standpoint of aircraft penalty, life cycle cost, and development risk should then be selected.

Air Distribution

The design should ensure air supply to, and exhaust from, the occupied compartments. The location of inlets is usually decided from previous experience and confirmed by quantified and subjective testing. Generally, passenger compartment air is introduced at a high level and exhausted at floor level. The design of air exhausts should preclude the possibility of blockage by luggage, clothing, or litter. The overall flow pattern should keep contaminated air generated by failures under the floor, behind furnishings, or in electronic equipment from entering occupied compartments. Air from toilets and galleys should not be exhausted into other occupied areas.

The flight crew should have means to direct and vary airflow. Such adjustment should not significantly affect the overall balance of air distribution; nor should it allow complete shutoff of air supply to the flight compartment.

With individual air supplies closed and the air-conditioning system operating normally, the air velocity in the vicinity of seated occupants should not exceed 60 fpm and should be between 20 and 40 fpm for optimum comfort. To avoid the sensation of no airflow, air velocities should not be less than 10 fpm at seated head height. Where individual air supplies are provided, the flow should be adjustable. The recommended jet velocity at seated head level should be at least 200 fpm. Cabin distribution duct and air inlets should be sized to limit air velocities so air noise levels are not objectionable to occupants. Longitudinal movement of air in the passenger cabin should be minimized.

Compartment air distribution should be such that, in stabilized temperature control system phases, the variation in temperature should not exceed 5°F, measured in a vertical plane from 2 in. above floor level to seated head height.

The temperature of air entering occupied compartments should, in normal operation, be no less than 35°F or more than

160°F. Where cabin air supplies combine, or where cold or hot air is added for temperature control, the various supplies should be effectively mixed to achieve uniform temperature distribution prior to distribution in occupied compartments.

Free water present in the air supply system should be removed to prevent it from being discharged onto passengers or equipment. Any drainage facilities should ensure that the water presents no hazards to equipment or the airplane structure.

Air Source

Engine compressor bleed air is the source of cabin pressurization and ventilation air on most current aircraft. Normally, bleed air contamination problems do not exist with current aircraft turbine engines. In most aircraft applications, the available compressor bleed air is more than adequate to meet the air-conditioning and pressurization requirement throughout the ground and flight operating envelope. Under many operating conditions, the bleed air temperature and pressure are higher than required and must be regulated to lower values. Use of engine compressor bleed air can be a significant penalty to the aircraft, and this penalty should be considered when conducting trade studies to select the system approach.

Auxiliary compressors can be used in lieu of engine bleed. These compressors can be shaft driven off the engine gearbox or a remote gear box. Auxiliary compressors may also be driven by pneumatic, hydraulic, or electrical power.

Use of engine bleed air is a convenient means of obtaining pressurized air. However, as a power source, bleed air use is generally not as efficient as other alternatives, partly because power often is dissipated by regulation to acceptable pressure levels. Electrical and hydraulic power can be provided at relatively high efficiencies, but it is limited by available sources and increases weight substantially. Direct mechanical power use is restricted by component location, but it is most efficient on high bypass ratio engines.

Refrigeration Systems

Both air cycle and vapor cycle refrigeration systems are used on aircraft. Air cycle systems are more commonly used because the lightweight, compact equipment and the readily available bleed air usually offsets its inherently low efficiency.

Air Cycle. The conventional open air cycle is characterized by the continuous extraction of engine compressor bleed air, which is processed by the air cycle system and dumped overboard after its use for cabin and avionics environmental control. Most current aircraft use ram air as the primary heat sink. In most cases, an air cycle machine turbine provides the required cooling capacity.

On most current aircraft, fuel is generally used as a heat sink for secondary power systems and engine oil, but it is not used as a heat sink for the air cycle refrigeration system. The operating penalties of the open air cycle result from the cost of bleeding a large amount of air from the engine, from use of ram air as a heat sink, and from the weight of the system hardware.

In the air cycle system, compressed air is cooled by expanding it through a turbine, which performs work. The turbine may drive a fan, which draws cooling air across an air-to-air heat exchanger (simple cycle), a compressor which raises the pressure of the air before it enters the turbine (bootstrap cycle), or both (simple/bootstrap or three-wheel bootstrap cycle). Figures 6, 7, and 8 show simplified schematics of each of these types of air cycle refrigeration systems.

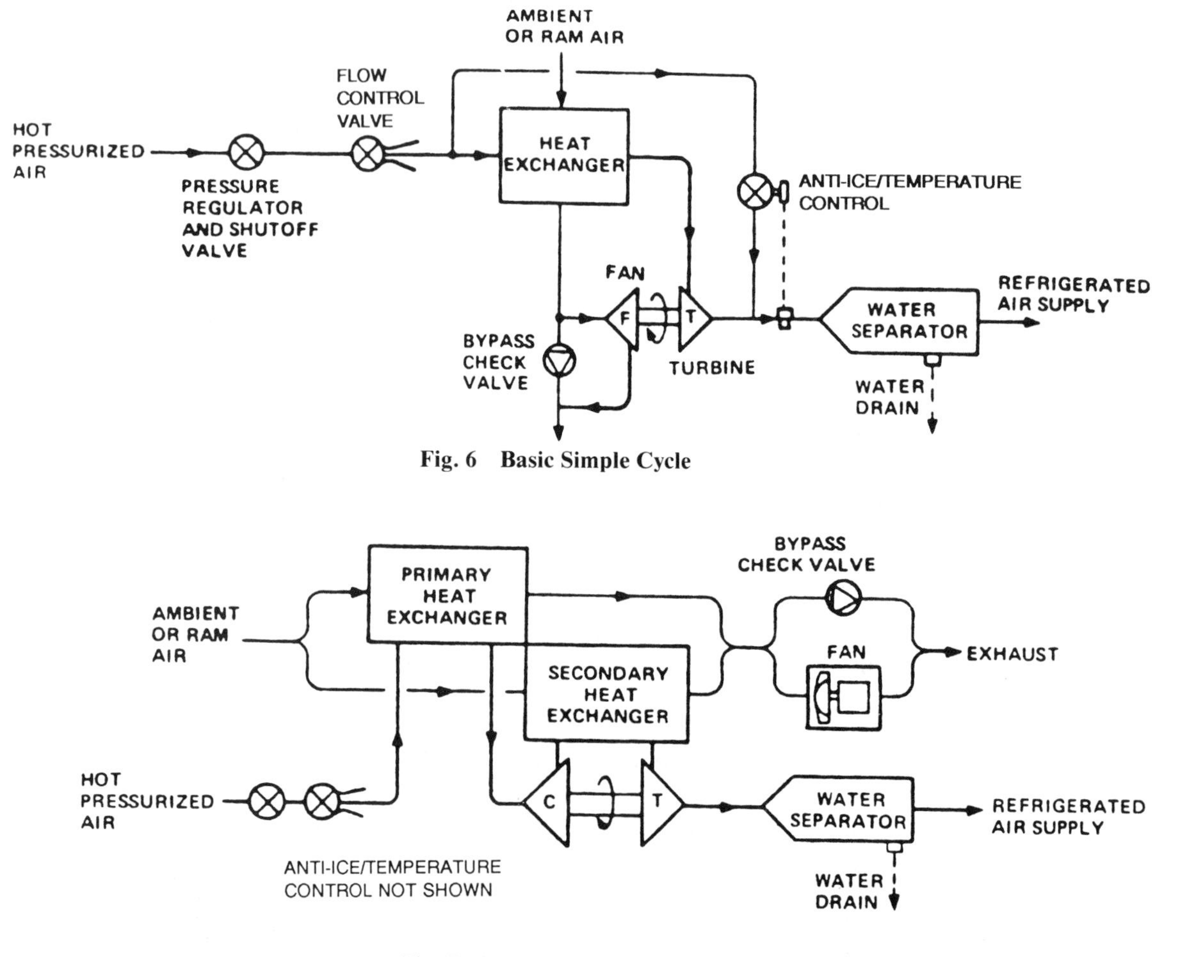

Fig. 6 Basic Simple Cycle

Fig. 7 Basic Bootstrap Cycle

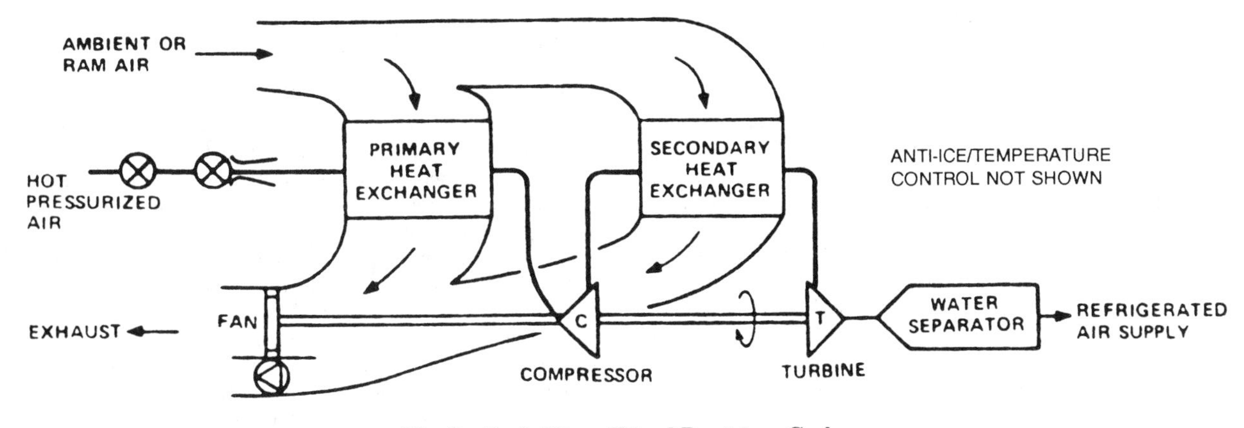

Fig. 8 Basic Three-Wheel Bootstrap Cycle

The advantages of an air cycle refrigeration system are its light weight, compact size, and high reliability, while the disadvantages are low efficiency and poor ground cooling. Ground cooling can be provided by an external air-conditioning cart. Cooling with the installed equipment can be obtained by an on-board auxiliary power unit, usually a gas turbine engine, to supply large quantities of high-pressure air. It is not practical to run the jet engines of an airplane on the ground to obtain cooling because of the higher power required, the associated noise, and the high rate of fuel consumption.

The newest open air cycle systems use high-pressure water removal, cabin air recirculation, and air-bearing cooling turbines.

Water Separation. The heart of a conventional low-pressure water separator in an air cycle system is a cloth bag interposed in the airstream at the turbine discharge. This bag coalesces the water particles formed in the turbine as the air is cooled. The resulting large droplets are separated from the airstream by centrifuge baffles and traps and is either drained overboard or reevaporated in the ram air heat sink supply to enhance the cycle efficiency.

The high-pressure water removal system condenses water at the turbine inlet. A large amount of the moisture from high humidity conditions may be condensed and removed at this point, primarily because of the higher dew-point temperature associated with the high pressures. Condensation is enhanced by cooling the turbine inlet temperature below the dew-point temperature in a small heat exchanger using turbine discharge air. The velocity is low, and the water may be induced to separate from the airstream and drain away. This scheme has the virtue of being maintenance free, whereas the coalescer bag of a low-pressure separator must be periodically cleaned. Figure 9 schematically shows a low-pressure water separator and a high-pressure separator.

Recirculation. Conventional air cycle systems are handicapped by their inability to use below-freezing turbine discharge temperatures effectively, since the frozen water particles would quickly block either a low-pressure water separator or the cabin supply duct system. If cabin air is returned to the turbine discharge and mixed with the turbine air, however, it has no influence on cooling capacity but raises the temperature of the air supply. By performing the anti-icing function with recirculated cabin air rather than hot bleed air, a greater amount of cooling can be achieved with a given quantity of bleed air.

Figure 10 shows this technique in conjunction with high-pressure water separation. The recirculation could be achieved with a jet pump or a motor-driven fan.

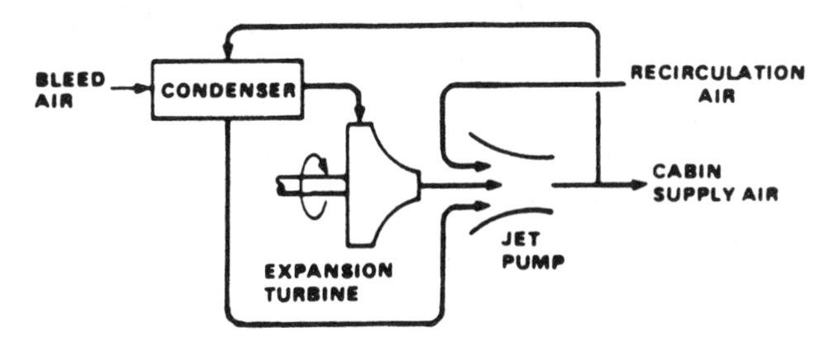

Fig. 10 Recirculation Technique as Implemented with High-Pressure Separation

Air-Bearing Cooling Turbine. The air-bearing cooling turbine is a recent innovation in the aircraft conditioning area. In this device, the oil-lubricated ball bearings of the cooling turbine are replaced by a set of journal and thrust bearings lubricated by air. Figure 11 shows the bearings for an air-bearing turbine. This advancement not only eliminates the maintenance associated with the oil lubrication system, but it also improves the reliability.

Air bearings and high-pressure water separation are responsible for totally eliminating the requirement for periodic maintenance of an air cycle environmental control system.

Vapor Cycle. The vapor cycle system has a higher efficiency than the air cycle system, but it is generally heavier. It has the additional

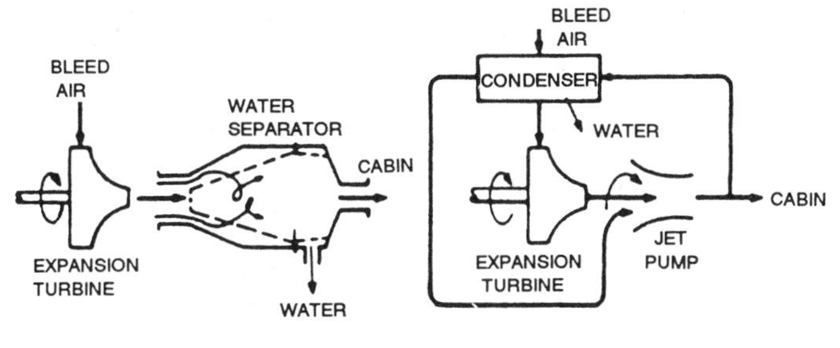

Fig. 9 Low-Pressure Water Separator and High-Pressure Water Separator

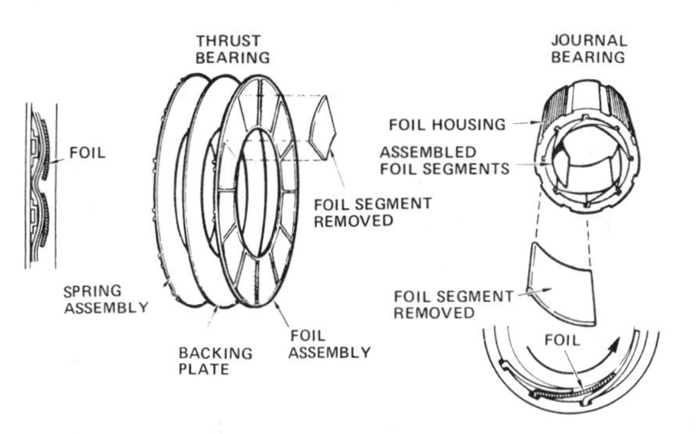

Fig. 11 Self-Acting Compliant Foil Air Bearing

advantage of providing ground cooling with just an electrical power source. The vapor compressor and fans also may be driven by air turbines, but lesser quantities of high-pressure air are required than for the air cycle system.

Several of the early jet transports had vapor cycle refrigeration systems with a cooling capacity of approximately 20 tons. Small capacity vapor cycle units are currently used on commercial air craft for galley refrigerators. Vapor cycle systems are widely used on general aviation aircraft and as supplementary systems for cooling electronic equipment on commercial and military aircraft.

Figure 12 shows a typical aircraft vapor cycle system. In this system, cabin air is normally recirculated by an electric fan through the evaporator when the airplane is on the ground. Capacity is modulated by an evaporator pressure regulator that raises or lowers the temperature level at which the refrigerant is evaporated. The expansion valve is thermostatic and must control to superheat sufficiently to ensure that no liquid enters the compressor. As flow is throttled at the evaporator pressure regulator or expansion valve, the surge control valve must open to bypass refrigerant and keep a minimum flow through the compressor.

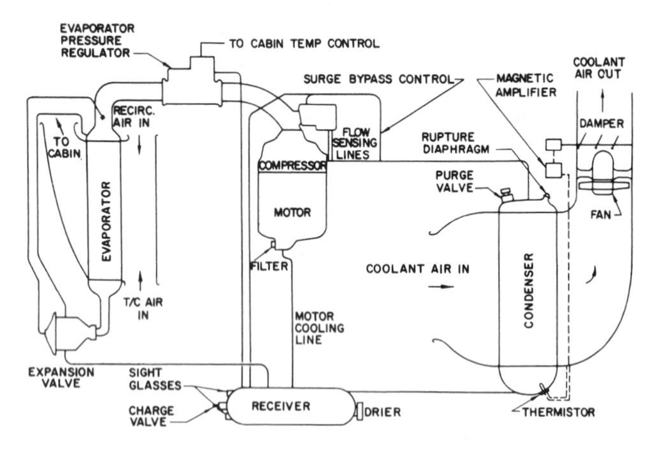

Fig. 12 Vapor Cycle Refrigeration Unit

Other methods of capacity modulation include bypassing air around the evaporator, throttling refrigerant at the evaporator inlet, varying compressor speed, or unloading the compressor. Compressor drive motor speed can be varied through use of a permanent magnet rotor in conjunction with a solid state variable frequency inverter power source. For evaporator inlet throttling with a motorized expansion valve and a centrifugal compressor, the function of the valve may be combined with the surge control on the same shaft. As the expansion valve closes, the bypass circuit opens.

Turbo-compressor air from outside or engine compressor bleed air may be manually selected after engines are started. It then switches on automatically at takeoff for cabin pressurization. The hot turbo-compressor or engine bleed air is first passed through an air-to-air heat exchanger, where it is partially cooled before entering the evaporator. Precooling is used to match the required in-flight capacity to that required on the ground and to improve system reliability.

The condenser fan provides cooling air on the ground and can also be turned on in flight to supplement ram air at low speeds and low altitude. In this system, the fan windmills in flight when turned off, but a separate circuit is provided for ram air on some airplanes. The cooling air is automatically modulated to maintain a minimum condensing pressure, both to reduce drag and to maintain pressure across the expansion valve at the low ambient encountered at high altitude.

The refrigerant must be nontoxic, odorless, and nonflammable. Refrigerants 11, 12, and 114 have been used in airplane systems, and other refrigerants or binary nonazeotropic refrigerant mixtures may have specific advantages.

Temperature Control System

Independent, automatic temperature controls should be provided for the flight crew and passenger compartments. In large airplanes, additional temperature controls should be provided, on a zonal basis, to cater for uneven thermal loading due, for example, to mixed cargo/passenger configuration or nonuniform occupancy. Flight crew compartment temperature control should not be affected significantly by these other compartment controls. Each compartment or zone air temperature should be selectable within the range 65 to 85 °F. The resultant temperature should be maintained within about 1 °F at the compartment sensor, when stabilized following initial switch-on or temperature reselection. Means should be provided for temperature sensors to sample compartment air at a rate compatible with control sensitivity requirements. The air sample should be taken from a location representative of average compartment temperature. Following reselection, compartment temperature should not overshoot the newly selected value by more than 3 °F. Compartment supply temperature should be controlled within limits appropriate to required heating and cooling rates and with consideration to occupant safety.

Figure 13 shows a basic temperature control system for a source of cold air from an air cycle refrigeration unit. Hot bleed air is added to the cold air source in response to the commands of a control system. A flow control valve will hold the total flow delivered to the cabin relatively constant no matter what the position of the modulating valve may be.

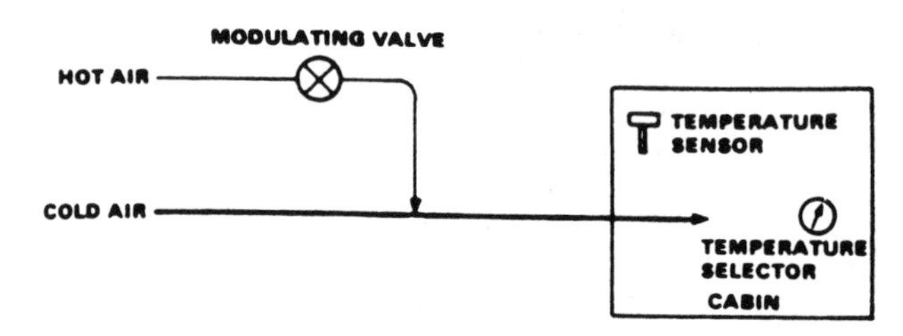

Fig. 13 Basic Temperature Control System

Basically, the crew selects a knob position that represents a desired cabin temperature. A sensor located at some point in the cabin measures the actual cabin temperature. The associated control equipment then compares these inputs and moves the valve to the position that will bring the cabin temperature to the selected value.

Figures 14 and 15 show variations of the basic concept and show a duct control system (which, instead of controlling the temperature in the cabin, controls the temperature of the air entering the cabin) and a two-compartment system, where the temperature of the cockpit and the cabin is controlled independently.

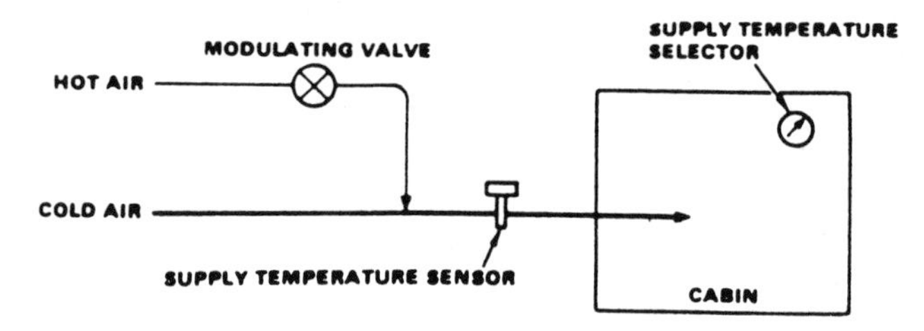

Fig. 14 Duct Control System Controlling Air Temperature before It Enters the Cabin

Cabin Pressurization Control System

The cabin pressurization control system must meter the exhaust and ventilating air to maintain the selected low-altitude cabin

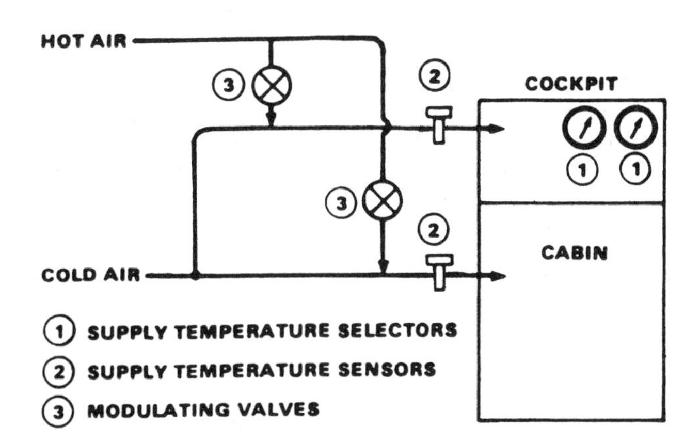

Fig. 15 Duct Control System Controlling Air Temperature to Cabin and Cockpit Separately

against the variables of airplane altitude and variable cabin inflow air. Both pneumatic and electronic systems have been used.

Cabin pressure control systems maintain the pressure in the cabin at levels acceptable to humans, change pressure levels during climb and descent at rates that are acceptable, and limit transient pressure changes, commonly called bumps, to magnitudes that are not annoying.

Reactions in the human ear cause the initial complaints about pressure changes. Stretching of the ear drum is the major factor causing unpleasant sensations that make pressure changes objectionable. Contraction of the eustachian tube dilator muscles, as from swallowing or yawning, aids in opening the tube and equalizing pressure. A positive pressure within the middle ear is easier to neutralize as it helps force the air through the tube. A negative pressure tends to collapse and seal the tube, hindering relief. For this reason, a lower rate of sustained pressure change is used for increasing ambient pressure than for decreasing it. Rates of 500 ft/min ascending and 300 ft/min descending have been the industry-accepted maximums for some time.

Positive pressure relief at some maximum pressure must be provided to protect the airplane in the event of a pressure control system failure. A negative (vacuum) pressure relief mechanism to let air in when outdoor pressure exceeds cabin pressure must also be provided.

Other desirable features are a barometric correction selector to help select the proper landing field altitude so that the pressure differential at landing may approach zero, and a limit control to maintain a maximum cabin altitude if other control components fail. An indicating system, including a rate of climb indicator, an altitude warning horn, an altimeter, and a differential pressure indicator, should be provided.

TYPICAL SYSTEM

Figure 16 shows the air cycle air-conditioning system for a current commercial transport aircraft. It operates from a source of preconditioned engine bleed air, creating a supply of conditioned air controlled to maintain the selected temperatures and ventilation rates within two passenger zones and the flight station. The various automatic temperature control functions are accomplished with electronic controllers and electric valve actuation. The two refrigeration packs are installed in the unpressurized area beneath the wing center section and are supplied with preconditioned bleed air from the two main engines or from the tail-mounted auxiliary power unit (APU). An underfloor distribution bay mixes conditioned air from the two packs with filtered recirculated cabin air and distributes it throughout the pressurized areas.

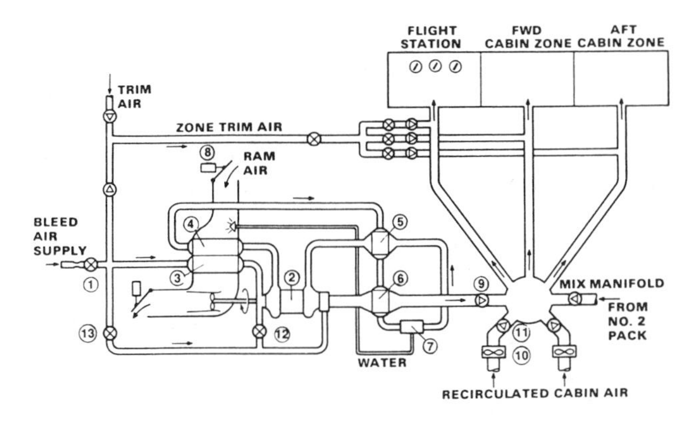

Fig. 16 Air-Conditioning Schematic for Commercial Airliner

Preconditioned bleed air enters each of the two parallel air-conditioning packs through a flow control valve (Item 1), which also functions as a pack shutoff valve. Most of the basic refrigeration equipment is assembled into a package ready for installation into the aircraft. Each of the two packs includes a three-wheel air cycle machine (ACM), four heat exchangers, and the pack temperature control valves, as well as protective devices and all necessary ducting and hardware.

The ACM (Item 2) has three rotating wheels—a compressor, a turbine, and a fan—which are mounted on a common shaft supported by air bearings. After the bleed air passes through the primary heat exchanger (Item 3), where it is cooled by ram air, it enters the compressor section of the ACM, where it is compressed to a higher pressure and temperature. The bleed air is cooled again by ram air in the secondary heat exchanger (Item 4) and, after passing through the reheater (Item 5), enters the condenser (Item 6). The air discharging from the condenser contains free moisture which is removed in the extractor (Item 7) before the air enters the other side of the reheater. The function of the reheater is to cool the air on the first pass, thereby reducing the amount of cooling required of the condenser, and to reheat the air before it enters the turbine. This reheating process evaporates small amounts of entrained moisture that may still be present and creates a higher temperature at the turbine inlet with an attendant increase in turbine power. The energy removed from the turbine airflow causes a substantial temperature reduction, permitting a turbine discharge temperature well below the ram temperature.

Ram cooling air for the heat exchangers enters through a variable area ram inlet scoop, passes through the secondary and primary heat exchangers in series, and exhausts through a variable-area exhaust door. Operation of the ram inlet and exhaust door actuators (both Item 8) is under control of the pack temperature controller.

Conditioned air from the two packs is delivered through bulkhead check valves (Item 9) to a manifold, where it mixes with recirculated cabin air for delivery to the flight deck zone and the two passenger zones. Both the pack and cabin zone temperature controls limit the total cabin supply flow to 35 °F minimum at all altitudes. Conditioned air for the flight deck is removed upstream of the mix manifold, and fresh air is used for flight deck cooling.

Cabin recirculation air is supplied to the manifold by electric fans (Item 10) through check valves (Item 11) that prevent reverse flow when the fans are not operating. Recirculation maintains the desired level of cabin ventilation while minimizing the use of bleed air.

Air-Conditioning Controls

Flow Control Operation. Maximum pack airflow is limited by flow control valves at the pack inlets. A dual flow schedule capability allows a pack flow to be increased 65% over a normal flow schedule. With the high flow schedule, one pack will maintain approximately 80% of normal cabin airflow. The high flow schedule is automatically selected through aircraft wiring circuits whenever a cooling pack or cabin recirculating air fan is shut off.

Pack Temperature Control. When the air-conditioning system is operating at maximum refrigeration capacity, the pack outlet temperature is determined by either the capabilities of the system or by the action of the low-limit controls. The maximum cooling capacity is necessary only for extremely hot conditions or to cool a heat-soaked aircraft. For most operating conditions, a warmer air supply is needed to satisfy the actual cabin cooling or heating demands, and the pack refrigeration capacity must be modulated.

The actual pack supply temperature requirement is determined by the zone controller and satisfies the cabin zone requiring the most cooling. Both packs are controlled to produce the required supply temperature. Each pack is operated independently by its own controller and temperature sensor, but both operate at the same condition and produce the same outlet temperature due to the single commanded temperature from the zone system. This results in the temperature control valve and ram doors operating at approximately the same relative position for both packs.

To adjust pack outlet temperature, the controller modulates the temperature control valve (Item 12) and the ram inlet and outlet actuators (both Item 8). Although each of these devices is independently actuated, the controller maintains a definite positional relationship between them. For maximum refrigeration, the ram doors are positioned fully open and the bypass valve is closed. To increase pack discharge temperature, the ram doors are partially closed, thus reducing cooling airflow while the bypass valve simultaneously begins opening to divert warm air around the air cycle machine. Proper scheduling of this pack control scheme minimizes ram air drag without exceeding equipment maximum temperature limitations.

The cabin air supply temperature leaving the mix manifold is limited to 35 °F minimum at all conditions. The temperature is measured by the duct air temperature sensor at the mix manifold outlet. The pack controller overrides the normal control schedule and modulates the temperature control valve and the ram air door actuators, as required.

An all-pneumatic backup control protects against icing that might occur during automatic control system failures. The pneumatic actuator of the low-limit valve (Item 13) is connected to sense the pressure difference across the condenser. The normal pressure drop across these items has no effect on the valve, but if an ice buildup starts, the increased pressure drop forces the low-limit valve open. Warm air then enters the turbine exit and stabilizes the ice accumulation at a low level. This feature also provides icing protection during manual mode of operation of the pack temperature control.

Cabin Zone Temperature Control. The cabin zone control subsystem incorporates a programmed duct air temperature control that accurately regulates during aircraft transient conditions. An inherent feature of the control is zone inlet duct temperature limitation, which results from electronic damping of the inlet temperature demand reference signal within predetermined limits.

Temperature is independently controlled in the flight station zone and in the fore and aft passenger zones by automatic feedback controls. The flight crew selects the desired temperature for each zone on selector units mounted in the flight station. Signals from each selector and corresponding zone temperature sensors are processed in the cabin zone controller, which produces a zone supply air temperature demand signal for each zone according to a predetermined schedule. The demand signal for each zone is appropriately supplied to each trim air control loop to provide the demanded temperature at each zone inlet.

The cabin zone controller also includes a discriminator function that selects the lowest of these demand signals and passes it to the pack controllers at the pack temperature command. Both packs are then modulated to provide this temperature. Thus, both packs are operated at the same supply duct temperature, which, when mixed with the cabin recirculation air, will satisfy the zone requiring the lowest inlet temperature. When the mix manifold outlet sensed temperature differs from the temperature demand, the temperature command signal to the packs is automatically changed until the sensed mix manifold outlet temperature equals the temperature demand from the zone requiring the greatest cooling. This guarantees that the demand of the zone requiring the greatest cooling is satisfied by the pack controls and that the trim air valve for that zone is closed.

BIBLIOGRAPHY

Crabtree, R.E, M.P. Saba, and J.E. Strang. 1980. The cabin air conditioning and temperature control system for the Boeing 767 and 757 airplanes. ASME-80-ENAS-55. American Society of Mechanical Engineers, New York.

Payne, G. 1980. Environmental control systems for executive jet aircraft. SAE 800607. Society of Automotive Engineers, Warrendale, PA.

SAE. 1976. Aircraft cabin pressurization control criteria. SAE *Standard* ARP 1270. Society of Automotive Engineers, Warrendale, PA.

Thayer, W.W. 1982. Tobacco smoke dilution recommendations for comfortable ventilation. Paper 7092. Douglas Aircraft Company, Long Beach, CA.

CHAPTER 10

SHIPS

THIS chapter covers air conditioning for oceangoing surface vessels including luxury liners, tramp steamers, and naval vessels. Although the general principles of air conditioning that apply to land installations also apply to marine installations, some systems are not suitable for ships due to excessive first cost or inability to meet shock and vibration requirements.

GENERAL CRITERIA

Air conditioning in ships provides an environment in which personnel can live and work without heat stress; increases crew efficiency; increases the reliability of electronic and similar critical equipment; and prevents rapid deterioration of special weapons equipment aboard naval ships.

Factors to consider in the design of an air-conditioning system for shipboard use are:

1. The system should function properly under conditions of roll and pitch.
2. The construction materials should withstand the corrosive effects of salt air and seawater.
3. The system should be designed for uninterrupted operation during the voyage and continuous year-round operation. Since ships en route cannot be easily serviced, some standby capacity, spare parts of all essential items, and extra refrigerant charges should be carried.
4. The system should have no objectionable noise or vibration, and must meet the noise criteria required by the shipbuilding specification.
5. The equipment should occupy a minimum of space commensurate with cost and reliability. Weight should be held to a minimum.
6. Since a ship may pass through one or more complete cycles of seasons on a single voyage and experience a change from winter operation to summer operation in a matter of hours, the system should be flexible enough to compensate for climatic changes with a minimum of attention by the ship's operating personnel.
7. Infiltration through weather doors is generally disregarded. However, specifications for merchant ships occasionally require an assumed infiltration load for heating steering gear rooms and the pilothouse.
8. Sun load must be considered on all exposed surfaces above the waterline. If a compartment has more than one exposed surface, the surface with the greatest sun load is used, and the other exposed boundary is calculated at outside ambient temperature.
9. Cooling load inside design conditions are given as a dry bulb with a maximum relative humidity. Cooling coil leaving air temperature for merchant ships is assumed to be 49 °F dry bulb. For naval ships, it is assumed to be 51.5 °F dry bulb. For both naval and merchant ships, the wet bulb is consistent with 95% rh. This off-coil air temperature is changed only in cases where humidity control is required in the cooling season.
10. When calculating winter heating loads, heat transmission through boundaries of machinery spaces in either direction is not considered.

The Society of Naval Architects and Marine Engineers bulletin, *Calculations for Merchant Ship Heating, Ventilation, and Air Conditioning Design*, gives sample calculation methods and estimated values.

MERCHANT SHIPS

DESIGN CRITERIA

Outdoor Ambient Temperatures

The service and type of vessel determines the proper outdoor design temperature. Some luxury liners make frequent off-season cruises where more severe heating and cooling loads may be encountered. The selection of the ambient design should be based on the temperatures prevalent during the voyage. In general, for the cooling cycle, outdoor design conditions for North Atlantic runs are 95 °F dry bulb and 78 °F wet bulb; for semitropical runs 95 °F dry bulb and 80 °F wet bulb; and for tropical runs, 95 °F dry bulb and 82 °F wet bulb. For the heating cycle, 0 °F is usually selected as the design temperature, unless the vessel will always operate in higher temperature climates. The design conditions for seawater is 85 °F summer and 28 °F winter.

Indoor Temperatures

Effective temperatures (ET) from 71 to 74 °F ET are generally selected as inside design conditions for commercial oceangoing surface ships.

Inside design temperatures range from 76 to 80 °F dry bulb and approximately 50% rh for summer and 65 to 75 °F dry bulb for winter.

Comfortable room conditions during intermediate outside ambient conditions should be considered in the design. Quality systems are designed to provide optimum comfort when outdoor ambient conditions of 65 to 75 °F dry bulb and 90 to 100% rh exist, and to ensure that proper humidity and temperature levels are maintained during periods when sensible loads are light.

Ventilation Requirements

Air-Conditioned Spaces. In public spaces, *i.e.*, mess rooms, dining rooms, lounges, and similar spaces, a minimum of 12 cfm of outdoor air per person or 0.33-h air change is required. In all other spaces, a minimum of 15 cfm or 0.5-h air change of outside air

The preparation of this chapter is assigned to TC 9.3, Transportation Air Conditioning.

Table 1 Heat Gain from Occupants

Degree of Activity at 80 °F	Heat Rate, Btu/h Sensible	Latent	Total
Dancing	245	605	850
Persons eating (mess rooms and dining rooms)	220	330	550
Waiters	300	700	1000
Moderate activity (lounge, ship's office, chart rooms)	200	250	450
Light activity (staterooms, crew's berthing)	195	205	400
Workshops	250	510	760

must be provided. However, the maximum outside air for any air-conditioned space is 50 cfm per person.

Ventilated Spaces. The fresh air to any space is determined by the required rate of change or the limiting temperature rise. The minimum quantity of air supplied to any space is limited to 30 cfm per occupant or 35 cfm minimum per terminal. In addition to these requirements, exhaust requirements must be balanced.

Load Determination. The cooling load estimate for air conditioning consists of those factors discussed in Chapter 26 of the 1989 ASHRAE *Handbook—Fundamentals*, including the following:

- Solar radiation
- Heat transmission through hull, decks, and bulkheads
- Heat (latent and sensible) dissipation of occupants
- Heat gain due to lights
- Heat (latent and sensible) gain due to ventilation air
- Heat gain due to motors or other electrical heat-producing equipment
- Heat gain from piping and other heat-generating equipment

The cooling effect of adjacent spaces is not considered unless temperatures are maintained with refrigeration or air-conditioning equipment. The latent heat from the scullery, galley, laundry, washrooms, and similar ventilated spaces is assumed to be fully exhausted overboard.

The heating load estimate for air conditioning should consist of the following:

- Heat losses through decks and bulkheads
- Ventilation air
- Infiltration (when specified)

No allowances are made for heat gain from warmer adjacent spaces.

Heat Transmission Coefficients

The overall heat transmission coefficient U between the conditioned space and the outside of the boundary in question, depends on the construction, material, and insulation. The composite structures common to shipboard construction do not lend themselves to theoretical derivation of such coefficients. They are most commonly obtained from full-scale panel tests. The Society of Naval Architects and Marine Engineers (SNAME) *Bulletin* 4-7, however, gives a method to determine coefficients, which may be used where tested data is unavailable.

Heat Dissipation from People

The rate at which heat and moisture dissipate from people depends on the ambient dry-bulb temperature and the people's state of activity. Values that can be used at 80 °F room dry bulb are listed in Table 1.

Heat Gain from Sources within a Space

The heat gain from motors, appliances, lights, and other equipment should be obtained from the manufacturer. Data found in Chapter 26 of the 1989 ASHRAE *Handbook—Fundamentals* may be used when manufacturer's data is unavailable.

EQUIPMENT SELECTION

General

The principal equipment required for an air-conditioning system can be divided into four categories:

1. Central station air-handling portion consisting of fans, filters, central heating and cooling coils, and sound treatment
2. Distribution network including air ductwork, and water and steam piping
3. Required terminal treatment consisting of heating and cooling coils, terminal mixing units, and diffusing outlets
4. Refrigeration equipment

Factors to be considered in the selection of equipment are:

- Installed initial cost
- Space available in fan rooms, passageways, machinery rooms, and staterooms
- Operation costs including system maintenance
- Noise levels
- Weight

High velocity air distribution offers many advantages. The use of unitary (factory assembled) central air-handling equipment and prefabricated piping, clamps, and fittings facilitates the installation for both new construction and conversions. Substantial space saving is possible as compared to the conventional low velocity sheet metal duct system. Maintenance is also reduced.

Fans must be selected for stable performance over their full range of system operation and should have adequate isolation to prevent transmission of vibration to the deck. Because fan rooms are often located adjacent to or near living quarters, effective sound treatment is essential.

In general, equipment used for ships is considerably more rugged than equipment for land applications, and it must withstand the corrosive environment of the salt air and sea. Materials such as stainless steel, nickel-copper, copper-nickel, bronze alloys, and hot-dipped galvanized steel are used extensively.

Fans

The Maritime Administration specifies a family of standard vaneaxial, tubeaxial, and centrifugal fans. Selection curves found on the standard drawings are used in selecting fans (Standard Plans S38-1-101, S38-1-102, S38-1-103). Belt-driven centrifugal fans must conform to these requirements, except for speed and drive.

Cooling Coils

Cooling coils must meet the following requirements:

- Maximum face velocity of 500 fpm
- Have at least six rows of tubes and 25% more rows than required by the manufacturer's published ratings
- Based on 42 °F inlet water temperature with a temperature rise of about 10 °F
- Construction and materials as specified by USMA (1965)

Heating Coils

Heat coils must meet the following requirements:

- Maximum face velocity of 1000 fpm
- Preheaters for supply systems have a final design temperature between 55 °F and 60 °F with outside air at 0 °F
- Preheaters for air-conditioning systems suit design requirements but with discharge temperature not less than 45 °F with 100% outside air at 0 °F
- Tempering heaters for supply systems have a final design temperature between 50 °F and 70 °F, with outside air at 0 °F

- Pressure drop for preheater and reheater in series, with full fan volume, does not exceed 0.5 in. of water
- Capacity of steam heaters is based on a steam pressure of 35 psi gage, less an allowed 5 psi gage line pressure drop and the design pressure drop through the control valve

Filters

Filters must meet the following requirements:

- All supply systems fitted with cooling and/or heating coils have manual roll, renewable marine-type air filters
- Maximum face velocity of 500 fpm
- Protected from the weather
- Located so that they are not bypassed when the fan room door is left open
- Clean, medium rolls are either fully enclosed or arranged so that the outside surface of the clean roll becomes the air entering side as the medium passes through the airstream
- Dirty medium wound with the dirty side inward

Medium rolls are 65 ft long and are readily available as standard factory-stocked items in nominal widths of 2, 3, 4, or 5 ft. A permanently installed, dry-type filter gage, graduated to read from 0 to 1 in. of water, is installed on each air filter unit.

Air-Mixing Boxes

If available, air-mixing boxes should fit between the deck beams. Other requirements include:

- Volume regulation over the complete mixing range must be within +5% of the design volume, with static pressure in the hot and cold ducts equal to at least the fan design static pressure
- Leakage rate of the hot and cold air valve(s) is less than 2% when closed
- Calculations for space air quantities allow for leakage through the hot air damper

Box sizes are based on manufacturer's data, as modified by the above requirements.

Air Diffusers

Air diffusers ventilate the space without creating drafts. In addition, diffusers must meet the following requirements:

- Diffusers serving air-conditioned spaces should be a high induction type and constructed so that moisture will not form on the cones when a temperature difference of 30 °F is used with the space dew point at least 9 °F above that corresponding to summer inside design conditions. Compliance with this requirement should be demonstrated in a mock-up test, unless previously tested and approved.
- The volume handled by each diffuser should not, in general, exceed 500 cfm.
- Diffusers used in supply ventilation systems are of the same design, except with adjustable blast skirts.
- Manufacturer's published data is used in selecting diffusers.

Air-Conditioning Compressors

Insofar as practicable, compressors should be the same as those used for ship's service and cargo refrigeration, except as to the number of cylinders or speed (see Chapter 30, "Marine Refrigeration," in the 1990 ASHRAE *Handbook—Refrigeration*).

TYPICAL SYSTEMS

General

Comfort air-conditioning systems installed on merchant ships are classified as (1) those serving passenger staterooms, (2) those serving crew's quarters and similar small spaces, and (3) those serving public spaces.

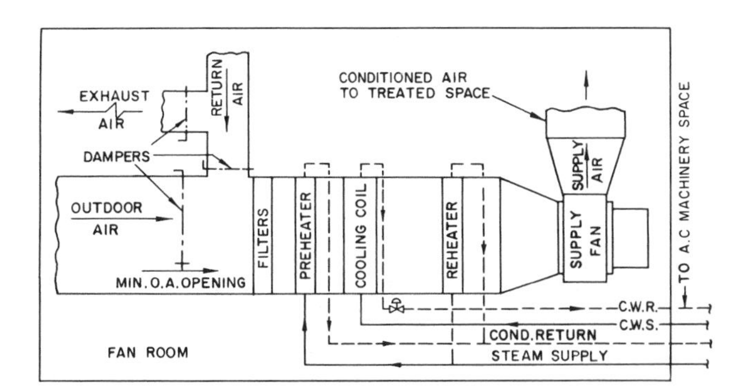

Fig. 1 Single-Zone Central (Type A) System

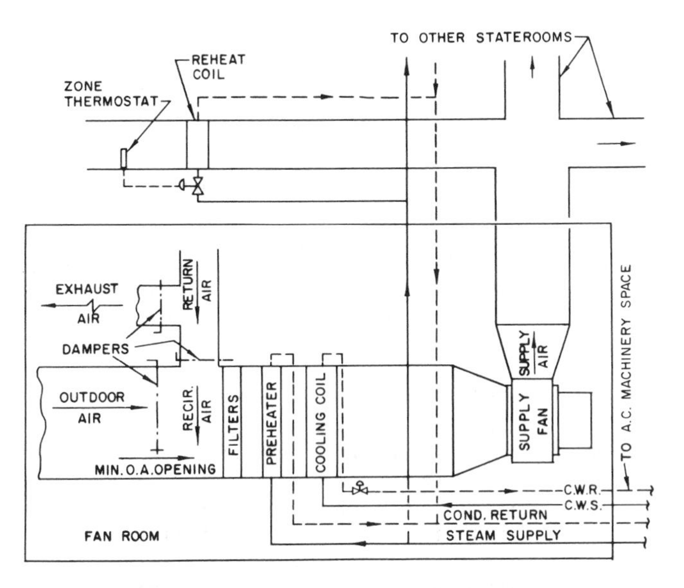

Fig. 2 Multizone Central (Type C) System

Single-Zone Central System

Public spaces are treated as one zone and are effectively handled by a single-zone central system. Exceptionally large spaces may require two systems. Generally, either built-up or factory-assembled fan coil central station systems are selected for large public spaces. A schematic of a typical system, which is also known as the *Type A* system, is shown in Figure 1. Under certain conditions, it is desirable to use all outdoor air, avoid the use of return ducts, and penalize the refrigeration plant accordingly. Outdoor air and return air are both filtered before being preheated, cooled, and reheated as required at the central station unit. A room thermostat maintains the desired temperature by modulating the valve regulating the flow of steam to the reheat coil. Humidity control is often provided. During mild weather, the dampers frequently are arranged to admit 100% outdoor air automatically.

Multizone Central System

A multizone central system, also known as the *Type C* system, usually is confined to crew's and officers' quarters (Figure 2). Spaces are divided into zones in accordance with similarity of loads and exposures. Each zone has a reheat coil to supply air at temperature adequate for all spaces served. These coils, usually steam, are controlled thermostatically. Manual control of air volume is the only means for occupant control of conditions. Filters, cooling coils, and dampers are essentially the same as *Type D* systems.

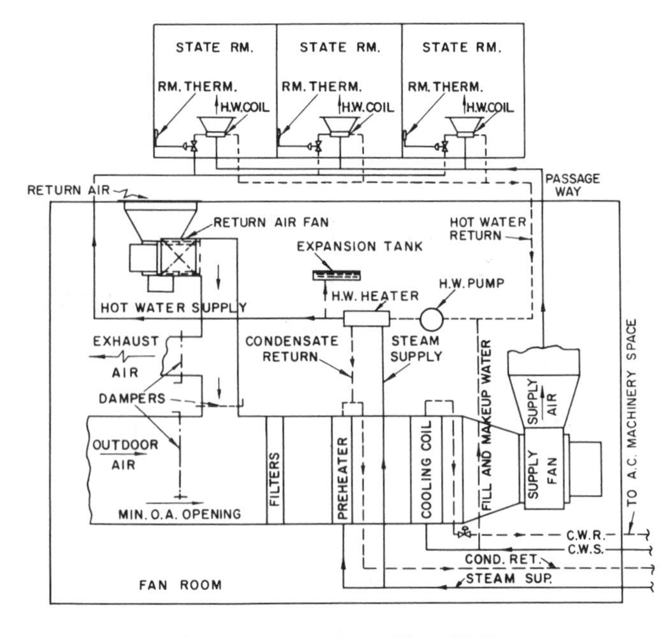

Fig. 3 Terminal Reheat (Type D) System

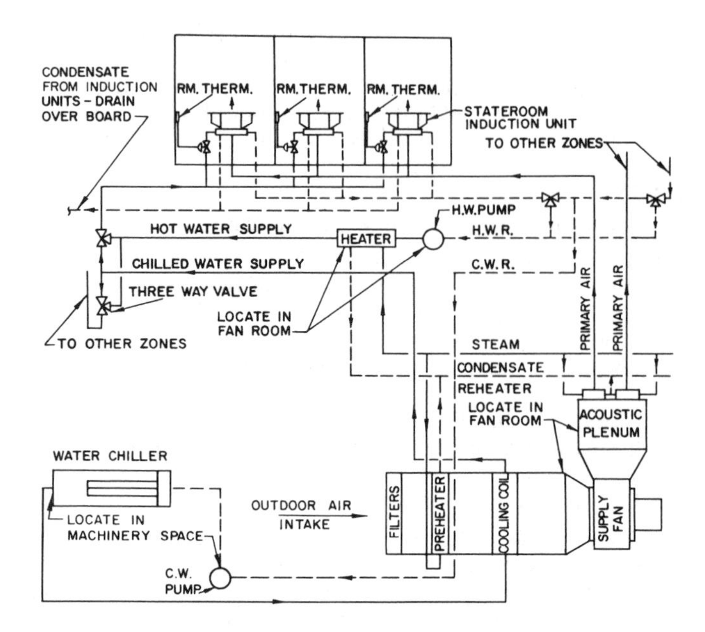

Fig. 4 Air-Water Induction (Type E) System

Each internal sensible heat load component, particularly solar and lights, varies greatly. The thermostatic control methods cannot compensate for these large variations; therefore, this system cannot always satisfy individual space requirements. Volume control is conducive to noise, drafts, and odors.

Terminal Reheat System

The terminal reheat system, or *Type D* system, is generally used for passengers' staterooms, officers' and crew's quarters, and miscellaneous small spaces (Figure 3). In this system, conditioned air is supplied to each space in accordance with its maximum design cooling load requirements. The room dry-bulb temperature is controlled by a reheater. A room thermostat automatically controls the volume of hot water passing through the reheat coil in each space. In this system, a mixture of outdoor and recirculated air circulates through the ductwork to the conditioned spaces. A minimum of outdoor air mixes with return or recirculated air in a central station system where it is filtered, dehumidified, and cooled by the chilled water cooling coils, and distributed by the supply fan through conventional ductwork to the spaces treated.

Dampers control the volume of outdoor air. No recirculated air is permitted for operating rooms and hospital spaces. When heating is required, the conditioned air is preheated at the control station to a predetermined temperature, and the reheat coils provide additional heating to maintain rooms at the desired temperature.

Air-Water Induction System

A second type of system for passenger staterooms and other small spaces is the air-water induction system, designated as the *Type E* system. This system is normally used for the same spaces as for the Type D, except where the sensible heat factor is low, such as in mess rooms. In this system, a central station dehumidifies and cools the primary outdoor air only (Figure 4).

The primary air is distributed to induction units located in each of the spaces to be conditioned. Nozzles in the induction units, through which the primary air passes, induce a fixed ratio of room (secondary) air to flow through a water coil and mix with the primary air. The mixture of treated air is then discharged to the room through the supply grille. The room air is either heated or cooled by the water coil. Water flow to the coil (chilled or hot) can be controlled either manually or automatically to maintain the desired room conditions.

This system requires no return or recirculated air ducts, because only a fixed amount of outdoor (primary) air needs to be conditioned at the central station equipment. This relatively small amount of conditioned air must be cooled to a sufficiently low dew point to take care of the entire latent load (outdoor air plus room air). It is distributed at high velocity and pressure and thus requires relatively little space for air distribution ducts. However, this space saving is offset by the additional space required for water piping, secondary water pumps, induction cabinets in staterooms, and drain piping.

Space design temperatures are maintained during intermediate conditions (outdoor temperature above changeover point, and chilled water at the induction units), with primary air heated at the central station unit according to a predetermined temperature schedule. Unit capacity in spaces requiring cooling must be sufficient to satisfy the room sensible heat load plus the load of the primary air.

When outdoor temperatures are below the changeover point (chiller secured), space design temperatures are maintained by circulating hot water to the induction units. Preheated primary air provides cooling for spaces that have a cooling load. Unit capacity in spaces requiring heating must be sufficient to satisfy the room heat load plus the load of the primary air. Detailed analysis is required to determine changeover point and temperature schedules for water and air.

High Velocity Dual-Duct System

The high velocity dual-duct system, also known as the *Type G* system, is normally used for the same kinds of spaces as Types D and E. In the Type G system, all air is filtered, cooled, and dehumidified in the central units (Figure 5). Blow-through coil arrangements are essential to ensure efficient design. A high pressure fan circulates air through two ducts or pipes at high velocities approaching 6000 fpm. One duct (or pipe) carries cold air while the other handles warm air. A steam reheater in the central unit heats the air according to requirements of the outdoor air temperature.

The supply of warm and cold air flows to an air-mixing unit in each space served. Each mixing unit has a control valve that proportions hot and cold air from the two ducts to satisfy the demand of the room thermostat. Any desired temperature, within the capacity limits of the equipment, can quickly be obtained

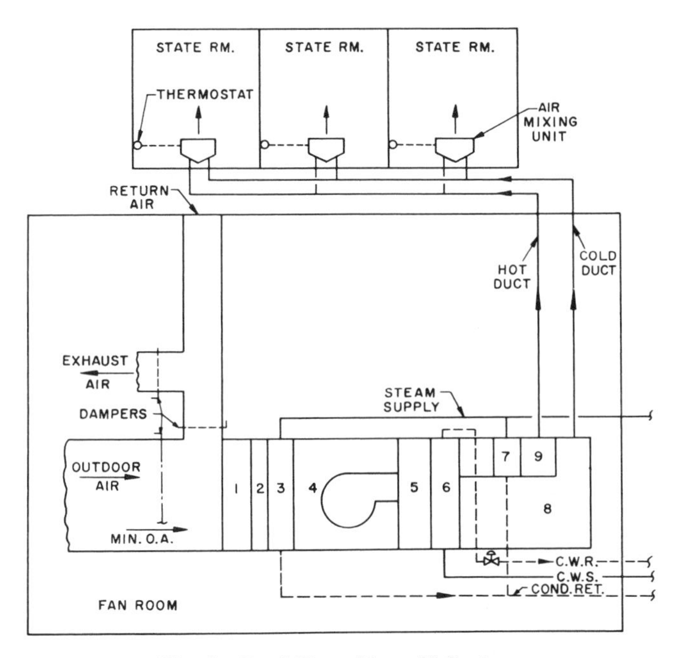

Fig. 5 Dual-Duct (Type G) System

and maintained with this arrangement, regardless of load variations in adjacent spaces. The air-mixing units incorporate self-contained regulators that maintain a constant volume of total air delivery to the various spaces, regardless of adjustments in the air supplied to rooms down the line.

Some of the advantages of this system include:

- All conditioning equipment is centrally located, simplifying maintenance and operation.
- The system can heat and cool adjacent spaces simultaneously without cycle changeover and with a minimum of automatic controls.
- Since only air is distributed from fan rooms, no water or steam piping, electrical equipment, or wiring appear in conditioned spaces.
- With all conditioning equipment centrally located, direct expansion cooling using halocarbon refrigerants is possible, eliminating all intermediary water-chilling equipment.

AIR DISTRIBUTION METHODS

Good air distribution in staterooms and public spaces is difficult because of low ceiling heights and compact space arrangements. The design should consider room dimensions, ceiling height, volume of air handled, air temperature difference between supply and room air, location of berths, and allowable noise level. On major installations, mock-up tests are often used to establish the exacting design criteria required for satisfactory performance.

Air usually returns from individual small spaces either by a sight-tight louver mounted in the door or by an undercut in the door leading to the passageway. An undercut door is confined to small air quantities of 75 cfm or less. Louvers are most commonly sized for a velocity of 400 fpm based on net area.

Ductwork

Ductwork on merchant ships is constructed of steel. Ducts, other than those requiring heavier construction because of susceptibility to damage or corrosion, are usually made with riveted seams sealed with hot solder or fire-resistant duct sealer, welded seams, or hooked seams and laps. They are fabricated of hot-dipped, galvanized, copper-bearing sheet steel, suitably stiffened

Table 2 Minimum Thickness of Steel Ducts

All vertical exposed ducts	16 USSG	0.0598 in.
Horizontal or concealed vertical ducts less than 6 in.	24 USSG	0.0239 in.
Horizontal or concealed vertical ducts 6.5 to 12 in.	22 USSG	0.0299 in.
Horizontal or concealed ducts ducts 12.5 to 18 in.	20 USSG	0.0359 in.
Horizontal or concealed vertical ducts 18.5 to 30 in.	18 USSG	0.0476 in.
Horizontal or concealed vertical ducts over 30 in.	16 USSG	0.0598 in.

externally. The minimum thickness of material is determined by the diameter for round ducts or the largest dimension of the rectangular ducts, as listed in Table 2.

The increased application of high velocity, high-pressure systems has resulted in a greater use of prefabricated round pipe and fittings, including spiral formed sheet metal ducts. It is important that the field fabrication of the ducts and fittings be such that all are airtight. Using factory fabricated fittings, clamps, and joints effectively minimizes air leakage for these high-pressure systems.

In addition to the space advantage, small ductwork saves weight, another important consideration for this application.

CONTROLS

The conditioning load, even on a single voyage, varies over a wide range in a short period. Not only must the refrigeration plant meet these variations in load, but the controls must readily adjust the system to sudden climatic changes. Accordingly, it is general practice to equip the plant with automatic controls. Since comfort is a matter of individual taste, adjustable room thermostats are placed in living spaces.

Manual volume control has also been used to regulate temperatures in cabins and staterooms. On the low velocity air distribution system, however, manual volume control tends to disturb the air balance of the remainder of the spaces served. Manual controls are also applied on high velocity single-duct systems where a constant volume regulator is installed in the terminal box.

Design conditions in staterooms and public spaces can be controlled by any one or a combination of the following:

Volume Control (Used In Type C Systems). This is the least expensive and simplest control. Basic disadvantages are an inability to meet simultaneous heating and cooling demands in adjacent spaces, unsatisfactory air distribution, objectionable noise, and inadequate ventilation because of reduction in air delivery.

Reheater Control. Zone reheaters of Type C systems are controlled by regulating the amount of steam to the zone coils by one of two methods. One method uses a room thermostat in a *representative* space with other spaces in the zone using manual dampers. Spaces served by this type of control do not always develop the desired comfort conditions. A second method uses a master-submaster control, which adjusts the reheater discharge temperature according to a predetermined schedule, so that full heat is applied at the design outside heating temperature and less heat as the temperature rises. This control does not adjust to meet variations in individual room loads. However, it is superior to the representative room thermostat method because it is more foolproof.

Preheater Control. A duct thermostat, which modulates steam through the valve, controls the preheaters. To prevent bucking, the set point usually is a few degrees below the design cooling off-coil setting.

Coil Control. Except for Type A systems with humidity control and Type E system primary air coils, cooling coils (water) are controlled by a dew-point thermostat to give a constant off-coil temperature during the entire cooling cycle. No control is provided for coils of Type E systems because they are in series with the flow-through induction units, and maximum dehumidification must be accomplished by the primary coil to maintain dry coil operation in the room units.

Damper Control. Outdoor, return, and exhaust dampers are either manually (as a group) or automatically controlled. If automatic, one of two methods is used. One controls the damper settings by thermostats exposed to weather air. A second method uses a duct thermostat that restricts the outdoor airflow only when the design temperature leaving the cooling coil cannot be achieved with full flow through the coil.

REGULATORY AGENCIES

Merchant vessels that operate under the United States flag come under the jurisdiction of the U.S. Coast Guard. Accordingly, the installation and components must conform to the Marine Engineering Rules and Marine Standards of the Coast Guard. Equipment design and installation must also comply with the requirements of the U.S. Public Health Service and Department of Agriculture. Principally, this involves ratproofing.

Comfort air-conditioning installations do not primarily come under the American Bureau of Shipping. However, equipment should be manufactured, wherever possible, to comply with the American Bureau of Shipping Rules and Regulations. This is important when the vessels are equipped for carrying cargo refrigeration, since the air-conditioning compressors may serve as standby units in the event of a cargo compressor failure. This compliance eliminates the necessity of a separate spare cargo compressor.

NAVAL SURFACE SHIPS

DESIGN CRITERIA

Outdoor Ambient Temperatures

Design conditions for naval vessels have been established as a compromise, considering the large cooling plants required for internal heat loads generated by machinery, weapons systems, electronics, and personnel. Temperatures of 90°F dry bulb and 81°F wet bulb are used as design requirements for worldwide applications, together with 85°F seawater temperatures. Heating season temperatures are assumed to be 10°F for outdoor air and 28°F for seawater.

Indoor Temperatures

Naval ships are generally designed for space temperatures of 80°F dry bulb with a maximum of 55% rh for most areas requiring air conditioning. The *Air Conditioning, Ventilation and Heating Design Criteria Manual for Surface Ships of the United States Navy* (USN 1969) gives design conditions established for specific areas. *Standard Specification for Cargo Ship Construction* (USMA 1965) gives temperatures for ventilated spaces.

Ventilation Requirements

Air-Conditioned Spaces. Naval ship design requires that air-conditioning systems serving living and berthing areas on surface ships replenish air in accordance with damage control classifications, as specified in USN (1969). These requirements are:

1. Class Z systems: 5 cfm per person.
2. Class W systems for troop berthing areas: 5 cfm per person.
3. All other Class W systems: 10 cfm per person. The flow rate is increased only to meet either a 75 cfm minimum branch requirement or to balance exhaust requirements. Outdoor air should be kept at a minimum to prevent the air-conditioning plant from becoming excessively large.

Load Determination

The cooling load estimate consists of coefficients from Design Data Sheet DDS511-2 or USN (1969) and includes allowances for the following:

- Solar radiation
- Heat transmission through hull, decks, and bulkheads
- Heat (latent and sensible) dissipation of occupants
- Heat gain due to lights
- Heat (latent and sensible) gain due to ventilation air
- Heat gain due to motors or other electrical heat-producing equipment
- Heat gain from piping, machinery, and other heat-generating equipment

Loads should be derived from requirements indicated in USN (1969). The heating load estimate for air conditioning should consist of the following:

- Heat losses through hull, decks, and bulkheads
- Ventilation air
- Infiltration (when specified)

Some electronic spaces listed in USN (1969) require 15% to be added to the calculated cooling load for future growth and that one-third of the cooling season equipment heat dissipation (less the 15% added for growth) be used as heat gain in the heating season.

Heat Transmission Coefficients. The overall heat transmission coefficient U, between the conditioned space and adjacent boundary, should be from Design Data Sheet DDS 511-2. Where new materials or constructions are used, new coefficients may be used from SNAME or calculated using methods found in DDS 511-2 and SNAME.

Heat Dissipation from People. USN (1969) gives heat dissipation values for people in various activities and room conditions.

Heat Gain from Sources within the Space. USN (1969) gives heat gain from lights and motors driving ventilation equipment. Heat gain and use factors for other motors and electrical and electronic equipment may be obtained from the manufacturer or Chapter 26 of the 1989 ASHRAE *Handbook—Fundamentals.*

EQUIPMENT SELECTION

The equipment described for merchant ships also applies for naval vessels, except as follows:

Fans

The navy has a family of standard vaneaxial, tubeaxial, and centrifugal fans. Selection curves used for system design are found on NAVSEA Standard Drawings 810-921984, 810-925368, and 803-5001058. Manufacturers are required to furnish fans dimensionally identical to the standard plan and within ±5% of the delivery. No belt-driven fans are included in the fan standards.

Cooling Coils

The navy uses eight standard sizes of direct expansion and chilled water cooling coils. All coils have 8 rows in the direction of airflow, with a range in face area of 0.6 to 10.0 ft^2.

The coils are selected for a face velocity of 500 fpm maximum, but sizes 54 DW to 58 DW may have a face velocity of up 620 fpm if the bottom of the duct on the discharge is sloped up at 15° for a distance equal to the height of the coil.

Chilled water coils are most commonly used and are selected using 45°F inlet water temperature with approximately 6.7°F rise in water temperature through the coil. This is equivalent to 3.6 gpm per ton of cooling.

Cfm quantity is based on lowest leaving air temperature from each size cooling coil at design entering air temperature.

Construction and materials are specified in MIL-C-2939.

Heating Coils

The navy has standard steam and electric duct heaters with specifications as follows:

Steam Duct Heaters

- Maximum face velocity is 1800 fpm.
- Preheaters leaving air temperature is 42 to 50°F.
- Steam heaters are served from the 50 psi gage steam system.

Electric Duct Heaters

- Maximum face velocity is 1400 fpm.
- Temperature rise through the heater is per MIL-H-22594A, but in no case more than 48°F.
- Power supply for the smallest heaters is 120 V, 3 phase, 60 Hz. All remaining power supplies are 440 V, 3 phase, 60 Hz.
- Pressure drop through the heater must not exceed 0.35 in. of water at 1000 fpm. Manufacturers tested data should be used in system design.

Filters

The navy uses seven standard filter sizes with the following characteristics:

- Filters are available in steel or aluminum.
- Filter face velocity is between 375 and 900 fpm.
- A filter cleaning station on board ship includes facilities to wash, oil, and drain filters.

Air Diffusers

The navy has standard diffusers for air-conditioning systems, but generally it uses a commercial type similar to those for merchant ships.

Air-Conditioning Compressors

The navy uses R-12 reciprocal compressors up to approximately 150 tons. For larger capacities, open, direct-drive centrifugal compressors using R-114 are used. Seawater is used for condenser cooling at the rate of 5 gpm per ton for reciprocal compressors and 4 gpm per ton for centrifugal compressors.

Typical Systems

On naval ships, zone reheat systems are used for most applications. Some ships with sufficient electric power have used low velocity terminal reheat systems with electric heaters in the space. Some newer ships have used a fan coil unit with fan, chilled water cooling coil, and electric heating coil in spaces with low to medium sensible heat per unit area of space requirements. The unit is supplemented by conventional systems serving spaces with high sensible or high latent loads.

Air Distribution Methods

Methods used on navy ships are similar to those discussed in the section for merchant ships. The minimum thickness of materials for ducts is listed in Table 3.

Table 3 Minimum Thickness of Materials for Ducts

Sheet for Fabricated Ductwork				
	Non-Watertight		Watertight	
Diameter or Longer Side	Galvanized Steel	Aluminum	Galvanized Steel	Aluminum
Up to 6	0.018	0.025	0.075	0.106
6.5 to 12	0.030	0.040	0.100	0.140
12.5 to 18	0.036	0.050	0.118	0.160
18.5 to 30	0.048	0.060	0.118	0.160
Above 30	0.060	0.088	0.118	0.160

Welded or Seamless Tubing		
Tubing Size	Non-Watertight Aluminum	Watertight Aluminum
2 to 6	0.035	0.106
6.5 to 12	0.050	0.140

Spirally Wound Duct (Non-Watertight)		
Diameter	Steel	Aluminum
Up to 8	0.018	0.025
Over 8	0.030	0.032

Note: All dimensions in inches.

Controls

The navy's principal air-conditioning control uses a two-position dual thermostat that controls a cooling coil and an electric or steam reheater. This thermostat can be set for summer operation and does not require resetting for winter operation.

Steam preheaters use a regulating valve with a weather bulb controlling approximately 25% of the valve's capacity to prevent freezeup, and a line bulb in the duct downstream of the heater to control the temperature between 42°F and 50°F.

Other controls are used to suit special system types, such as pneumatic/electric controls when close tolerance temperature and humidity control is required, *i.e.*, operating rooms. Thyristor controls are sometimes used on electric reheaters in ventilation systems.

REFERENCES

SNAME. Calculations for merchant ship heating, ventilation and air conditioning design. *Technical and Research Bulletin No. 4-16*. Society of Naval Architects and Marine Engineers, Jersey City, NJ.

SNAME. Thermal insulation report. *Technical and Research Bulletin No. 4-7*. Society of Naval Architects and Marine Engineers, Jersey City, NJ.

SNAME. 1971. *Marine engineering*. Chapter 19 by J. Markert. Society of Naval Architects and Marine Engineers, Jersey City, NJ.

USMA. 1965. Standard specification for cargo ship construction. U.S. Maritime Adminstration, Washington, D.C.

USMA. *Standard Plan* S38-1-101, *Standard Plan* S38-1-102, and *Standard Plan* S38-1-103. U.S. Maritime Administration, Washington, D.C.

USN. 1969. The air conditioning, ventilation and heating design criteria manual for surface ships of the United States Navy. Washington, D.C.

USN. NAVSEA Drawing No. 810-921984, NAVSEA Drawing No. 810-925368, and NAVSEA Drawing No. 803-5001058. Naval Sea Systems Command, Dept. of the Navy, Washington, D.C.

USN. *General specifications for building naval ships*. Naval Sea Systems Command, Dept. of the Navy, Washington, D.C.

Note: MIL specifications are available from Commanding Officer, Naval Publications and Forms Center, ATTN: NPFC 105, 5801 Tabor Ave., Philadelphia, PA 19120.

CHAPTER 11

ENVIRONMENTAL CONTROL FOR SURVIVAL

MOST heating, ventilating, and air-conditioning systems maintain comfort for occupants in enclosed spaces or create favorable environments for commercial processes. However, special applications exist in which survival, not comfort, is the main concern. This chapter gives background information, parameters, and requirements relating to control of the physical and chemical environment for survival of people in enclosed spaces, when adjacent areas may be neither safe nor habitable.

The primary source of information presented is based on theoretical and experimental data developed in the design and use of structures as protective shelters, both above and below ground. Evaluations of physiological stresses resulting from short- and long-term confinement of people in marginal environments were also derived. In this chapter, the term *shelter* is synonymous with any room, chamber, or enclosed space constructed and equipped to protect the health, safety, and lives of the occupants against certain hazards that may exist in the surroundings. Shelter requirements based on weapon effects remain, however, as a source and potential application for reliable design data.

FACTORS AFFECTING THE PHYSICAL ENVIRONMENT

With no planned control for the physical environment, the transient conditions in a shelter either approach a safe state of equilibrium or become intolerable. Intolerable conditions might be due to either a high carbon dioxide concentration, accompanied by a low oxygen content, or to an excessive effective temperature. Progressive deterioration of environmental conditions can be a physiological hazard, and remedial action would be needed before the individuals succumb to the extreme effects of vitiated air, elevated body temperature, organic strain, or dehydration. Some factors that influence or reflect the physical environment in an occupied shelter are as follows:

1. Number of occupants and duration of occupancy.
2. Metabolic traits of the people, *i.e.*, energy expenditure, sensible and latent heat transfer, oxygen consumed, and carbon dioxide produced.
3. Physiological and psychological reactions of people to the situation, *i.e.*, environmental stresses, deprivation, states of health, dehydration effects, and tolerance limits.
4. Clothing (insulating properties, absorptivity, porosity).
5. Diet (solid and liquid, including drinking water).
6. Air and mean radiant temperatures, humidity, and air motion in all spaces (effective temperature of environment).
7. Inside surface areas and temperatures and moisture condensation.
8. Interior heat and moisture sources other than people, such as lights, motors, engines, and appliances (continuous or intermittent use).
9. Heat exchange with adjacent structures or heat sources.
10. Thermal properties of the shelter and surrounding materials, *i.e.*, conductivity, density, specific heat, diffusivity, and moisture content.
11. Thickness and mass shielding properties of the shelter enclosure for protection from nuclear radiation.
12. Weather conditions regarding variations in temperature, humidity, solar radiation, wind, and precipitation.
13. Initial conditions of the shelter environment and its surroundings (temperature distribution and moisture).
14. Temperature, humidity, quality, quantity, and distribution of air supplied to shelter spaces from a safe source.
15. Versatility and reliability of environmental control systems (ventilation, cooling, heating, air conditioning, power supply).
16. Protection in adapted places of refuge from fire and smoke effects; exclusion of fumes by interspace pressure differentials.
17. Ability of the structure to resist applied loads.

The preparation of this chapter is assigned to TC 9.8, Large Building Air-Conditioning Applications.

PHYSIOLOGICAL ASPECTS

The physiological factors of concern in the design of environmental control systems for survival shelters arise from metabolic heat; moisture and carbon dioxide generated by the occupants; and the human tolerance limits for cold, heat, humidity, CO, CO_2, and O_2. Response to heat or cold is determined by the ability to dissipate metabolic energy by the combined mechanisms of evaporation, convection, and radiation (see Chapter 8 of the 1989 ASHRAE *Handbook—Fundamentals*). A rise or fall in normal body temperature may be caused by physical hazards associated with heat stress or cold stress. Removal of clothing and suppression of activity are guards against heat stress in shelters; increasing metabolic output by exercise and providing extra, dry clothing are guards against cold stress.

Metabolic heat production combined with warm climates and low airflow rates is a primary source of heat stress in the shelter environment. Since metabolic heat production is principally a function of activity, sedentary activity level is desirable. The value for a standard population sample is 275 Btu/h. The common value for sedentary metabolism in shelters is 400 Btu/h, the higher value providing an allowance for other than average groups and for increased metabolism for some working occupants (Pefley *et al.* 1969).

Heat stress causes increased blood flow through the skin, and thus a rise in skin temperature. Heat transfer by radiation, convection, and evaporation then proceeds at an increased rate until thermal balance is restored, if possible. The nude skin temperature rises until the onset of perspiration stabilizes the skin temperature at approximately 95 °F. Further rise in heat stress expands the role of evaporative cooling by increasing the perspiring area of the skin at about 95 °F until the entire body area is moist. Since the preferred mechanism for heat is sensible (convective and radiative) cooling, the body resorts to sweating only if sensible cooling is inadequate. Sensible cooling is the net effect of convection and radiation, which largely depends on the differences between the skin temperature and air dry-bulb temperature and the mean radiant temperature of the surroundings.

Ability to cope in a severe heat stress environment is a function of available drinking water, salt, acclimatization, age (the very young and old being weakest), and health. Deficiencies in any of these factors can trigger both failure of the thermoregulatory system and the onset of heat stroke. The values in Table 1 indicate limits beyond which the physiological strains resulting from heat stress or dehydration may be irreversible (Pefley *et al.* 1972, Blockley 1968).

Table 1 Physiological Limits for Healthy Male Subjects

Parameter	Conservative Values	Maximum Values
Sweat rate and heat of vaporization	1.06 quart/h	Slightly higher than conservative values during moderate to strenuous work
Increase in body temperature	2 to 3 °F	4 to 5 °F (rectal, esophageal, and tympanic membrane), within four hours during moderate work
Pulse rate	130 beats per min	150 to 190 beats per min within four hours
Skin temperature		102 to 103 °F within four hours while standing at ease
Cardiac output (estimated)		Slightly over 24 L/min within four hours while standing at ease
Dehydration limit		6% of body mass

Water Requirements

The amount of potable water needed daily to avoid dehydration in a sedentary man is shown in Figure 1 for conditions under which sweat can be freely and completely evaporated from body surfaces. If the ambient air is humid and hot, more sweat may be excreted than can be evaporated, and the requirement for replacement water is increased. These effects have been noted at effective temperatures greater than 85 °F. Women and children generally require smaller amounts of drinking water because their metabolic rates tend to be lower.

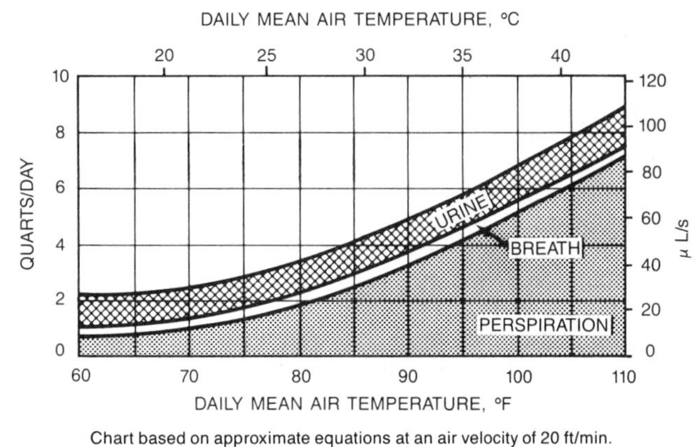

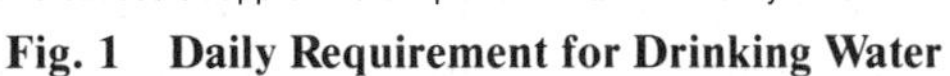
Chart based on approximate equations at an air velocity of 20 ft/min.

Fig. 1 Daily Requirement for Drinking Water to Avoid Dehydration in Men at Rest

Data analyses from some tests in which air motion around the subject was at a comparatively high and controlled level indicate a strong correlation of water intake with effective temperature (ET), as well as dry-bulb temperature. Tests conducted under typical shelter conditions with young, healthy men showed a freshwater intake of 2.7 quarts per day at an ET of 85 °F (Blockley 1968). Other tests under simulated conditions at the same effective temperature indicated the same per capita rate of freshwater intake, but an intake slightly over 3 quarts per day with stored water (McNall and Ryan 1968). At an ET of 82 °F, test results reported by both sources of data indicated a per capita range of 2.1 to 2.4 quarts per day for water usage.

The results of a long-term test indicate that healthy males under 50 years of age can survive for 10 days at an ET of 82 °F when the dry-bulb temperature is 86 °F, with a daily ration of 1.25 quarts per person (Gorton 1972). During this deprivation test, the daily dietary allowance was 1000 kilocalories of OCD rations per person. The loss of mass, attributed to both dehydration effects and deficient diet, was about 7% of initial body mass.

A prone resting position, well-ventilated body support, and minimum clothing aid in suppressing the onset of heat stroke. Fanning by others and the use of perspiration-wiping cloths effectively provide comfort to those experiencing extreme strain from the hot humid environment.

Cold stress causes the body to compensate by cutaneous vascular contraction, which reduces skin temperature and restores equilibrium between metabolic output and heat loss. This suppression of skin temperature also reduces to a minimum evaporative loss from the skin.

If vasoconstriction fails to maintain a heat balance through reduction in skin temperature, shivering occurs with a resulting increase in metabolism. If the environment is not too cold, the increased heat production arrests the lowering of body temperature. As the rectal temperature falls below 90 °F, the shivering mechanism begins to fail and may cease at rectal temperatures below 80 to 86 °F. Exercising, putting on additional dry clothing, eating, and bringing people into close proximity to each other and away from low-temperature zones are effective ways of combating cold stress.

Time-Temperature Tolerance

Predicted tolerance time limits are indicated in Table 2 for the exposure of properly clothed, healthy subjects at rest in a wide

Table 2 Tolerance Limits for Clothed, Healthy Subjects at Rest

Effective Temperature, ET

L3	L2	L1	A	H1	H2	H3

30 40 50 60 70 80 90 100 °F

Effective Temperature Ranges:

A = 68 to 72 °F
This is the desirable range for long-term comfort of normally clothed persons when air motion is minimal and the mean radiant temperature does not differ greatly from air temperature.

L1 = 68 to 50 °F
H1 = 72 to 82 °F
Most people tolerate environmental conditions within these ranges for periods of 14 days or more.

L2 = 50 to 35 °F
H2 = 82 to 90 °F
The physiological stresses associated with these ranges can be tolerated by most people for several hours and by motivated, hardy individuals for 24 h or longer.

L3 = Less than 35 °F
H3 = More than 90 °F
The severe physiological stresses associated with these ranges can be tolerated without injury for only a few hours at the threshold or a few minutes in extremely cold or hot humid environments. Procedures or a control system should be considered for health, safety, and survival in situations under which conditions in ranges L3 or H3 might develop.

range of environments (Pefley *et al.* 1972, Lind 1955). An effective temperature (ET) can be related to dry-bulb and wet-bulb temperatures by means of Figure 2. Conditions in range A, the comfort zone, do not impose any limits on the duration of occupancy.

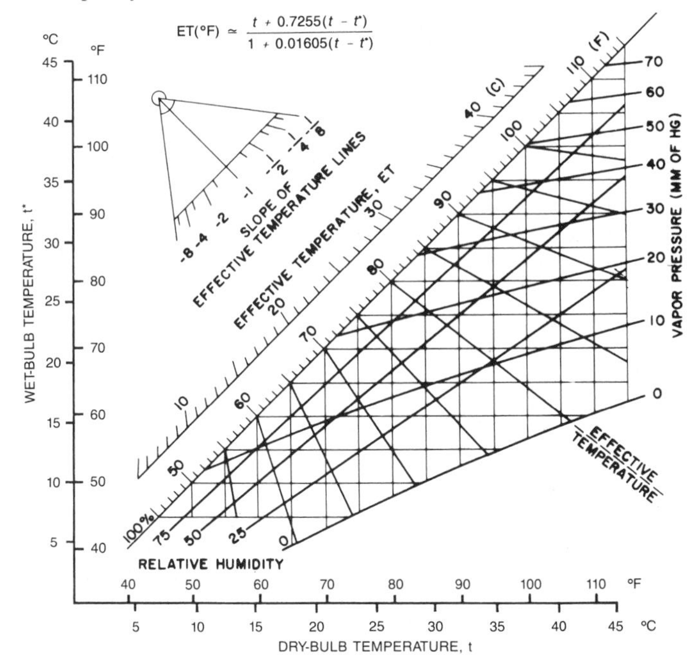

Fig. 2 Still-Air Effective Temperature

Near the outer bounds of ET ranges L1 and H1, susceptible individuals (including infants, the aged, and persons afflicted with arthritis, heart disease, or metabolic disorders) may experience difficulties (Lee and Henschel 1963, Henschel *et al.* 1968). At low temperatures in range L1, chilblains may appear. At higher temperatures in range H1, anxiety, sleeplessness, nausea, and heat rash will probably occur during prolonged exposure.

If exposure to the moderate cold stress in ET range L2 or heat stress in ET range H2 is likely to be extended beyond a few hours, special care should be taken to maintain safe body temperatures. An adequate diet and multilayered clothing adaptable to the prevailing temperature level are essential during prolonged exposure to conditions in ET range L2. Reliance on the warming effects of increased physical activity or shivering is not a desirable alternative.

Minimal clothing is appropriate in ET range H2, and, to avoid dehydration, potable water should be available for replacing metabolic losses. The ability to cope with heat stress varies among individuals; persons having subnormal sweat response are inherently susceptible. A posture that affords maximum opportunity for evaporation from skin surfaces is desirable in a hot environment; impermeable clothing and furnishings that impede evaporation should be avoided.

Exposure to the severe cold stress in ET range L3 and heat stress in ET range H3 cannot be safely prolonged beyond limits determined by body temperatures, dehydration, and individual sensitivities. With subfreezing temperatures, the fingers and toes of sedentary persons are most susceptible to frostbite, even with protective clothing. Exposure to relatively high dry-bulb temperatures can be endured for significant periods if the partial pressure of water vapor in the air is low. An environmental temperature of 176°F is tolerable for about an hour if the air is virtually dry. However, breathing is slightly painful in air having a dew point of 122°F, and such a condition is endurable for only a few minutes (Lind 1955).

Vitiation Factors

Carbon dioxide concentration should not exceed 3% by volume and preferably should be maintained below 0.5%. For a sedentary person, 3 cfm per person of fresh air will maintain a CO_2 concentration of 0.5%. At concentrations of 3% and above, performance deteriorates and basic physiological functions are affected. At 1.5%, basic performance and physiological functions are not affected, but slow adaptive processes have been observed that might induce pathophysiological states on long exposure. At 0.5 to 0.8%, no significant physiological or adaptive changes occur.

Oxygen level in a shelter is generally less critical than CO_2 levels in the absence of life-support systems. Oxygen concentration in normal air is about 21%; 17% is frequently taken as a limit for shelters.

An approximate relationship between physical activity, energy expenditure, oxygen consumption, carbon dioxide production, and rate of breathing is shown in Table 3. The respiratory quotient (RQ) is the volumetric ratio of carbon dioxide production to oxygen consumption. A value of 0.84 may be used for the conditions in Table 3. This value is representative; the actual RQ depends largely on diet and body chemistry. In the absence of metabolic disorders, values of RQ associated with the oxidation of carbohydrates, proteins, and fats are about 1.0, 0.8, and 0.7, respectively (Allen 1972, Harrow and Mazur 1958).

Table 3 Per Capita Rates of Energy Expenditure, Oxygen Consumption, Carbon Dioxide Production, and Pulmonary Ventilation for Man

Level of Physical Activity	Energy Expenditure; Metabolic Rate, Btu/h	Oxygen Consumption, ft^3/h	Carbon Dioxide Production, ft^3/h	Rate of Breathing, ft^3/h
Exhausting effort	3600	6.66	5.7	146
Strenuous work or sports	2400	4.44	3.8	97
Moderate exercise	1600	2.96	2.5	64
Mild exercise; light work	1000	1.84	1.55	40
Standing; desk work	600	1.10	0.93	24
Sedentary, at ease	400	0.74	0.62	16
Reclining, at rest	300	0.56	0.47	12

The effects of carbon monoxide must be considered, even though the amount of this gas produced by the body is negligibly small. In confined shelter spaces, the prime source of CO would be tobacco smoke, with pipes producing five times and cigars almost 20 times as much as cigarettes. Carbon monoxide can also come from fuel-burning devices in the shelter, from the exhaust gases of internal combustion engines, or through the ventilation intake from smoldering fires outside the shelter.

For industrial purposes, the allowable concentration of CO is 50 ppm by volume. This limit is based on an 8-h workday, five days per week. For exposure over longer sustained periods, lower limits are used. For submarines, the limit is 50 ppm or 0.005%, and for space cabins the design level is 10 ppm or 0.001%. Increased levels of carbon dioxide can increase the toxic effect of CO. The increased CO_2 results in deeper and more rapid breathing, which, in turn, increases the absorption of CO into the body (OCD 1969).

Odor arising from activities within the shelter, as well as toxic or explosive gases, should also be considered. In connection with austere shelters, minimal rates of air replacement are generally sufficient to dilute odors associated with human occupancy. Hydrocarbons from fuel leakage, hydrogen from batteries being discharged, and ingress of radioactive particulates, pathogenic organisms, or chemical agents are all possible hazards.

CLIMATE AND SOILS

Heat loss calculations for underground shelters require values of thermal conductivity and thermal diffusivity of earth. Information on earth temperature is also required, which is related to the thermal and physical properties of soil, as well as to the climatic conditions. Table 4 illustrates thermal conductivity and diffusivity of various types of soil with respect to moisture content. Earth temperature may be estimated from the soil temperature data at several selected stations throughout the United States, published in *Climatological Data* of the U.S. Weather Record Center, Asheville, NC.

Table 4 Thermal Properties of Soils, Rocks, and Concrete

Material	Thermal Conductivity, Btu/h·ft·°F	Thermal Diffusivity, ft^2/h	Density, lb/ft^3	Specific Heat, Btu/lb·°F
Dense rock	2.00	0.050	200	0.20
Average rock	1.40	0.040	175	0.20
Dense concrete	1.00	0.033	150	0.20
Solid masonry	0.75	0.025	143	0.21
Heavy soil, damp			131	0.23
Heavy soil, dry	0.50	0.020	125	0.20
Light soil, damp			100	0.25
Light soil, dry	0.20	0.011	90	0.20

Earth temperature beyond a depth of 3 ft is seldom affected by diurnal cycle of air, temperature, and solar radiation. However, the annual fluctuation of earth temperature extends to a depth of 30 to 40 ft. The integrated monthly average earth temperature from surface to a depth of 10 ft is insensitive to the thermal diffusivity of soil, as long as the diffusivity is larger than 0.02 ft^2/h. Table 5 presents annual maximum and minimum earth temperatures averaged over the surface to a depth of 10 ft for 47 stations throughout the United States, which may be used for an approximate calculation of underground heat transfer.

Table 5 Annual Maximums and Minimums for Integrated Average Earth Temperatures[a]

	Earth Temperatures, °F				
Location	Max.	Min.	Location	Max.	Min.
Auburn, AL	74	56	Bozeman, MT	56	32
Decatur, AL	71	48	Huntley, MT	64	36
Tempe, AR	81	59	Lincoln, NB	69	39
Tucson, AR	85	65	Norfolk, NB	66	40
Brawley, CA	90	68	New Brunswick, NJ	65	42
Davis, CA	76	56	Ithaca, NY	59	39
Ft. Collins, CO	64	36	Raleigh, NC	73	52
Gainesville, FL	80	69	Columbus, OH	65	41
Athens, GA	77	57	Coshocton, OH	64	40
Tifton, GA	80	62	Lake Hefner, OK	77	51
Moscow, ID	57	37	Pawhuska, OK	74	50
Lemont, IL	65	39	Ottawa, Ont., Canada	59	36
Urbana, IL	68	42	Corvallis, OR	66	46
West Lafayette, IN	66	38	Pendleton, OR	67	39
Burlington, IA	71	38	Calhoun, SC	76	52
Manhattan, KS	69	41	Union, SC	70	48
Lexington, KY	70	46	Madison, SD	61	33
Upper Marlboro, MD	70	42	Jackson, TN	71	49
East Lansing, MI	63	37	Temple, TX	83	59
St. Paul, MN	62	34	Salt Lake City, UT	63	40
State Univ., MS	79	55	Burlington, VT	63	35
Faucett, MO	65	43	Pullman, WA	60	36
Kansas City, MO	66	42	Seattle, WA	61	45
Sikeston, MO	71	43			

[a] Earth temperatures are integrated averages from surface to a depth of 10 ft derived to simulate observed phenomena, each for average amplitude and phase angle with earth thermal diffusivity, $\alpha = 0.025$ ft^2/h.

Other useful references on the subject of earth temperature and thermal and physical properties are listed in the bibliography of this chapter. Summer and winter design weather data from Chapter 24 of the 1989 ASHRAE *Handbook—Fundamentals* may be used as a guide in selecting design outdoor conditions for shelters.

CONTROL OF CHEMICAL ENVIRONMENT

Control of the physical environment may be regarded as two independent problems—control of the chemical environment and control of the thermal environment. Closed (buttoned-up) shelters without any air replacement from outside sources remain habitable for only a few hours, unless a life-support system is used. The permissible stay time (period of occupancy) is determined by the net volume of space per person, and is defined as the time required to raise the carbon dioxide concentration to 3% by volume. This is expressed as:

$$\theta_3 = 0.04 V_c \tag{1}$$

where

θt_3 = time to reach 3% carbon dioxide, h
V_c = unit volume of space, ft^3 per person

Thus, people can safely stay for 20 h in a closed shelter having a net volume of 500 ft^3 per person. The stay time can also be determined from Figure 3 for various conditions. In the figure, the relationship between the concentrations of carbon dioxide and oxygen in occupied spaces, the rate of ventilation per person, the net volume of space per person, and the time after entry is shown. The terminal values of carbon dioxide and oxygen concentration are tabulated for various rates of ventilation. Figure 2 is based on rates of oxygen consumption and carbon dioxide production representative of people in confined quarters (Allen 1960).

The example shown by dotted lines indicates that a carbon dioxide concentration of 3.5% by volume will develop in 10 h in an unventilated shelter having a net volume of 240 ft^3 per person, and that the oxygen content of the air will then be 16.2% by volume. Ventilation with pure outdoor air is the most economical method for maintaining the necessary chemical quality of air in a shelter. The recommended minimum ventilating rate of 3 cfm per person of fresh air will maintain a carbon dioxide concentration of about 0.5% and an oxygen content of approximately 20%, by volume, in a shelter occupied by sedentary people. However, an air replacement rate of 3 cfm per person is not sufficient to limit the resultant effective temperature to 85°F, under many conditions, unless the supply temperature is less than about 45°F. A ventilation capability for maintaining a low concentration of carbon dioxide and a correspondingly safe concentration of oxygen in a shelter has several advantages, including one or more of the following:

1. A longer stay time is gained for continued occupancy after shutdown of a ventilating system due to fire or for repair of disabled equipment.
2. Intermittent operation of a manual ventilating blower may become practicable.
3. Greater physical activity in the shelter becomes permissible.
4. Environmental conditions, such as temperature, humidity, moisture condensation, air distribution, air motion, and odors, as well as oxygen and carbon dioxide, may be improved without supplementary apparatus.

THERMAL ENVIRONMENT

Underground Shelters

As previously indicated, the thermal environment in a shelter depends on occupancy, construction features, climatic and soil conditions, and conditions of use. The internal environment represents a balance between heat generated inside the shelter, the

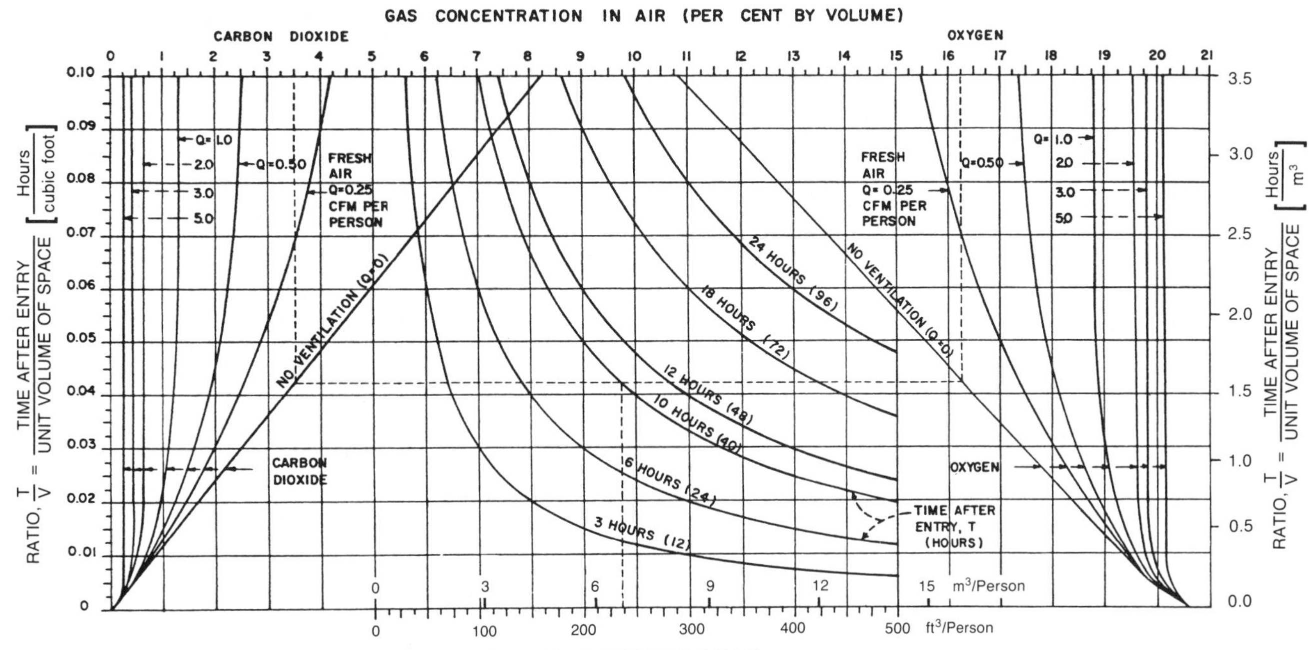

Terminal Values of Gas Concentration

Per Capita Ventilation Rate, cfm	Terminal Concentration (Percent by Volume) Carbon Dioxide	Oxygen
0.25	5.35	14.13
0.50	2.67	17.23
1.0	1.35	18.77
2.0	0.69	19.53
3.0	0.47	19.79
5.0	0.29	20.11

Physical Data (Per Capita Basis)

Oxygen consumption	0.9 ft³/h
Carbon dioxide production	0.75 ft³/h
Moisture loss from body	0.144 lb/h
Ventilation Air Properties	
Oxygen in fresh air	20.82%
Density	0.075 lb/ft³
Relative humidity	50%
Dry-bulb temperature	77°F
Atmospheric pressure	14.7 psi

Fig. 3 Carbon Dioxide and Oxygen in Occupied Spaces

heat conduction into the materials surrounding the shelter, and the heat exchange with the ventilating air. Because each of the heat exchanges include latent and sensible components that vary with time, computation of the temperature and humidity, for short periods of occupancy, is very complex.

The problem of keeping warm in shelters during winter conditions has not been considered acute since normal, healthy people can tolerate temperatures as low as 50°F for several days, if properly clothed. However, since people of all ages and in varying degrees of health will require shelter, and because it cannot be assumed that, on short warning, people will take adequate clothing into a shelter, the heating requirements of shelters should not be ignored. Generally, the need for heating will be greater in family-size shelters than in group or community shelters because of the greater surface area per occupant in the smaller shelters.

Because the steady-state rates of conducted heat are characteristically small for underground shelters, environmental temperatures less than 50°F can usually be avoided by arranging the ventilating system to use metabolic heat generated by the occupants. During cold weather, a mixture of fresh and recirculated air in varying proportions can be supplied to occupied spaces at a temperature of 50°F or more. This procedure is less effective if the shelter is only partially occupied (Allen 1972).

During the summer, the maintenance of suitable environmental conditions in a shelter is, in most instances, a question of survival rather than comfort. From considerations of survival, the environmental criterion for 14 days' duration has been selected at 83°F effective temperature, although higher or lower values may be used for different segments of the population.

Experimental measurements have been made in various shelters with real or simulated occupants to determine the temperature and humidity that would develop after one to two weeks in various climates, and with various ventilation rates and occupancies. Achenbach *et al.* (1962) and Kusuda and Achenbach (1963) compared computed and experimental results in a few shelters.

Figures 4 and 5 show the relative effects of ventilation rate, earth conductivity, and shelter size (Drucker and Cheng 1962, Drucker and Haines 1964, Baschiere *et al.* 1965). The effect of initial earth temperature is small in magnitude, averaging about 0.2°F shelter temperature per degree of earth temperature. The analog computer studies disclosed that after the tenth day, the temperature in the shelter would approximate 95% of the ultimate temperature rise.

Simplified Analytical Solutions

The thermal environment in a shelter is determined by the following energy balance relation:

$$q_m + q_i = q_v + q_w + q_r \qquad (2)$$

where

q_m = human metabolic heat
q_i = heat generated by lights, cooking appliances, motor-driven equipment, and auxiliary power apparatus

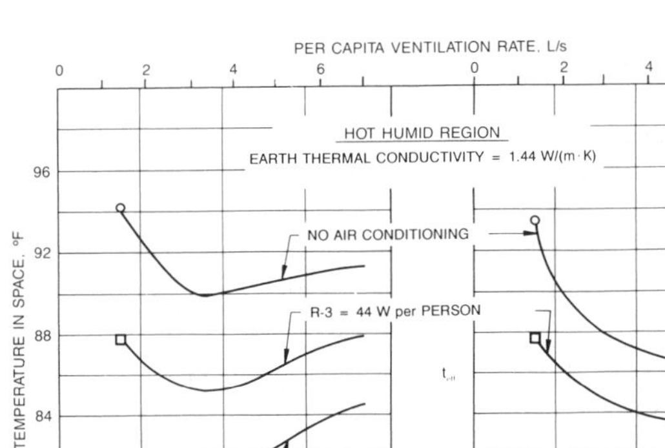
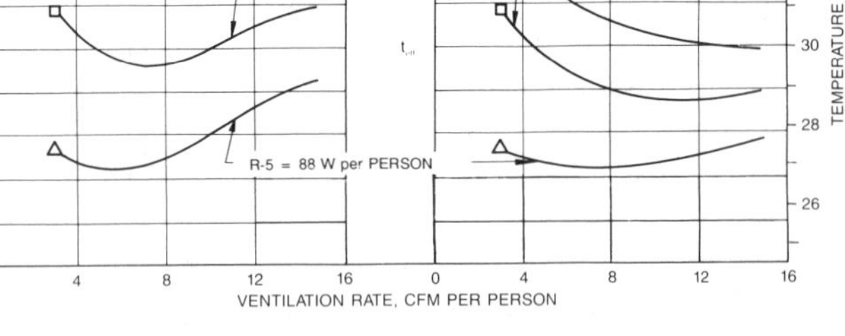

Fig. 4 Maximum Dry-Bulb and Effective Temperatures as Functions of Ventilation Rate

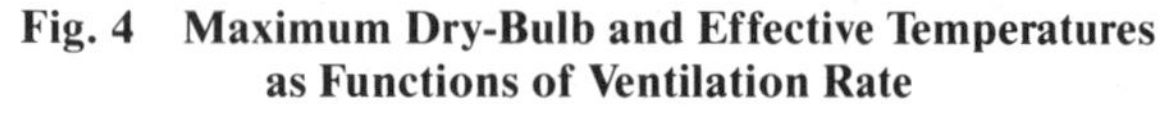
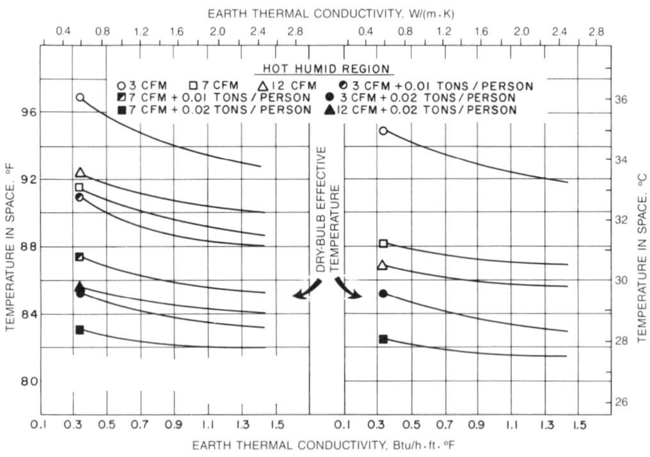

Fig. 5 Maximum Dry-Bulb and Effective Temperatures as Functions of Earth Thermal Conductivity

q_v = heat carried out by ventilation air
q_w = conduction heat loss to the surrounding media
q_r = heat absorbed by cooling equipment

Both the sensible and latent portions of the human metabolic heat q_m are functions of shelter temperature, although the sum of the two remains practically constant and varies in the range of from 300 to 600 Btu/h per person, depending on activity levels (see Chapter 8 of the 1989 ASHRAE *Handbook—Fundamentals*). The internal heat generated by lights and cooking q_i is usually equal to only a small fraction of the human metabolic heat. Under emergency conditions, illumination can, if necessary, be reduced to a low level, with a resultant heating effect of about 10 to 20 Btu/h per person (Allen 1962).

Three possible modes of heat dissipation for the shelter can be considered: (1) cooling by forced or natural ventilation with outdoor air (corresponding to q_v); (2) cooling by the effect of heat conduction into the surrounding media (corresponding to q_w); and (3) mechanical cooling and dehumidifying with refrigeration or well water (corresponding to q_r). Adsorption or absorption of humidity by desiccant or hygroscopic fluid has been considered, but it usually tends to increase, rather than decrease, the effective temperature.

In general, shelters do not have apparatus to maintain a preselected state point, and the environment depends on prevailing conditions for thermal equilibrium. Temperature and humidity of the environment, therefore, varies or tends to rise progressively as a result of diurnal variations in outdoor air conditions, attenuation of conductive heat transfer in surrounding masses of initially cool materials, and temperature-dependent changes in the sensible/latent ratio for metabolic heat.

Approximate techniques have been derived for three representative cases suitable for design and estimating purposes as follows:

Case 1 considers a ventilated shelter with insulated boundaries; no heat is transmitted through walls, ceiling, or floor, and heat-moisture loads are removed only by ventilating air.

Case 2 considers short-term occupancy of an unventilated underground shelter (sealed for a week or less) from which all heat is removed by conduction effects in surrounding masses of materials having an initially uniform temperature, and moisture is removed by condensation on relatively cool surfaces.

Case 3 considers a ventilated underground shelter from which heat and moisture are removed by the combined effects of ventilation and heat conduction.

The Case 1 solution, which requires a minimum of data, applies to any shelter in which solar radiation has little effect and is recommended for identified fallout shelters in normally heated basements and in aboveground core areas of buildings. Cases 2 and 3 include the cooling effects of earth conduction and inherently lead to a more economical system or a better environment in underground shelters, when earth properties, initial temperature, and shelter configuration are favorable. Case 2 or Case 3 shelters rely, to a minimum extent, on the ambient atmosphere for environmental control. Earth conduction alone may satisfy requirements for removal of metabolic heat in underground gallery-type shelters, which have a relatively low initial soil temperature, favorable thermal properties, and a large ratio of interior surface to floor area. Case 2 and Case 3 represent shelters having widely different shapes or sizes: the plane model for large chambers that are quite square in plan, the spherical model for small or cubical chambers, and the cylindrical model for elongated rectangular or gallery-type shelters.

Repeated trial solutions are necessary for determining the minimum ventilating rate for meeting stated effective temperature criteria. In most climatic areas, it may not be possible to prevent excessive effective temperatures, and ventilation system design should be based on coincident climatic data that may be exceeded during an acceptable portion of the year. If greater reliability is considered essential, apparatus for cooling may be indicated.

Case 1—Ventilated and Insulated Shelters

Case 1 applies to an occupied space with insulated (adiabatic) boundaries; the effects of heat transmitted by conduction through walls, floor, and overhead cover are virtually negligible relative to the thermal effects of internal loads associated with metabolic processes, lighting, and appliances that emit heat.

Ventilation with ambient air is the only means for removal of heat and moisture, and the system is presumed to create a uniform environment throughout the space. The method estimates capacities required for ventilation systems in shelters occupied during warm weather, when outside-inside temperature differentials may be minimal. The method is also useful for predicting the environmental conditions that might develop under various ambient conditions.

The Case 1 solution is based on the equations that follow. Ventilation rates referred to are standard dry air having a density of 0.75 lb/ft^3 and a specific heat of 0.24 Btu/lb·°F. Ventilation rates, metabolic parameters, and all internal heat exchangers are reduced to quantities per person.

The metabolic rate or total heat transferred to the environment from each occupant is assumed to be 400 Btu/h. The partition of metabolic heat q_m emitted by a normally clothed occupant into sensible q_s and latent q_l components can be adequately expressed as linear functions of dry-bulb temperature t_a of air in the space. When the ambient air t_a is 68 °F or more:

$$q_s = 10\,(100 - t_a) \quad (3)$$

$$q_l = 10\,(t_a - 60) \quad (4)$$

When t_a is less than 68 °F, q_s = 320 Btu/h and q_l = 80 Btu/h, both constant.

Equations (3) and (4) can be used in conjunction with Equation (2), with $q_w = q_r = 0$, to derive the following relationships for the temperature t_a and humidity ratio W_a of environmental air. Thus when t_a is 68 °F or more:

$$t_a = q_i + 1.08Gt_v + 1000 / (1.08G + 10) \quad (5)$$

$$W_a = W_v + (t_a - 60)/470G \quad (6)$$

When t_a is less than 68 °F:

$$t_a = t_v + (q_i + 320)/1.08G \quad (7)$$

$$W_a = W_v + 0.01702/G \quad (8)$$

where

G = ventilating rate, per person, cfm
q_i = lighting load, per person, Btu/h
t_v = dry-bulb temperature of supply air, °F
W_v = humidity ratio of supply air, pounds of moisture per pound of dry air

Example 1. Determine the environmental conditions in an occupied chamber ventilated with fresh air at the rate of 15 cfm per person. Air is supplied to the space at a dry-bulb temperature of 77 °F, wet-bulb temperature of 68 °F, humidity ratio of 0.01265 unit mass of moisture per unit mass of dry air, and a still-air effective temperature of 73 °F ET. The heat emitted by lights and equipment is 7.5 W (25.59 Btu/h) per person

Solution: The temperature of air in the space is higher than 68 °F. Then, from Equation (5),

$$t_a = \frac{25.59 + (1.08)(15)(77) + 1000}{(1.08)(15) + 10} = 86.76\,°\text{F}$$

and from Equation (6),

$$W_a = 0.01265 + \frac{86.76 - 60}{(470)(15)}$$

$$= 0.01645 \text{ lb of moisture per lb of dry air}$$

From ASHRAE Psychrometric Chart No. 1, at W_a = 0.01645 and t_a = 86.76 °F, the wet-bulb temperature is 75.4 °F. From Figure 2, the effective temperature is 80.4 °F with slow movement of air in the space.

Curves showing the results of equations similar to Case 1 calculations made to determine ventilation requirements to maintain various effective temperature rises are shown in Figure 6. Figure 7 shows the relationship of shelter and outdoor dry-bulb temperature with ventilation rate.

When earth temperature is below the shelter temperature, Case 1 solutions tend to overestimate the effective temperature as long as heat gain due to the solar heat load is negligible. Heat conduction loss to the surrounding media plays a significant role in dissipation of shelter heat during the sealed-up condition. Heat conduction to the surrounding media also helps decrease the shelter effective temperature or the ventilation air requirement during the normal shelter occupancy period.

The Case 1 solution is useful in estimating the ventilation air requirement for a given outdoor air condition to maintain the shelter effective temperature below the tolerance limit during the summer. Since the summer shelter thermal environment is affected as strongly by wet-bulb temperature as by the dry-bulb temperature of the ventilation air, a coincident design criterion of these two temperatures is needed for selecting the ventilation equipment

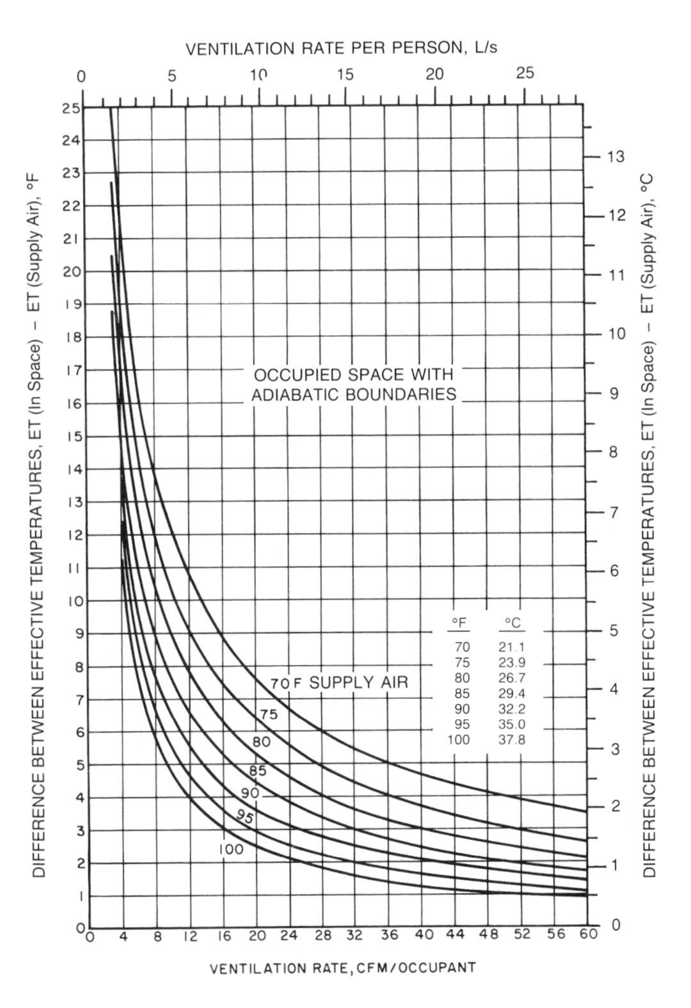

Fig. 6 Relation of Effective Temperature Difference to Unit Ventilation Rate

for a given climatic zone. Summer design weather data from Chapter 24 of the 1989 ASHRAE *Handbook—Fundamentals* do not meet this coincident requirement.

Ventilation adequacy factor accounts for the coincident occurrences of dry- and wet-bulb temperatures for a specified shelter effective temperature and a per capita ventilation air rate. The ventilation adequacy factor is defined as the percentage of annual hourly coincident occurrences of dry- and wet-bulb temperatures of outdoor air that could maintain the shelter effective temperature below a specified level for a specific per shelter occupant ventilation rate. Ventilation adequacy factors, which have been calculated using a modified Case 1 solution for several cities, are illustrated in Figures 8 and 9 for Houston, Texas and Minneapolis, Minnesota. In Houston, the outdoor air ventilation rate of 28 cfm per person is adequate for protecting the shelter occupant from effective temperatures exceeding 84 °F for 95% of the hours in the year.

The ventilation adequacy factor is a useful index in evaluating the cost effectiveness of the ventilating, as well as cooling, facilities from the standpoint of survival in protective shelters. Since such curves have not been developed for many areas, the shelter designer can use weather service data (Monthly Local Climatological Data Supplement), along with Figures 6 and 7, to develop local ventilation adequacy factors. The use of 24-h wet- and dry-bulb averages simplifies these compilations and introduces little error on actual daily average shelter conditions.

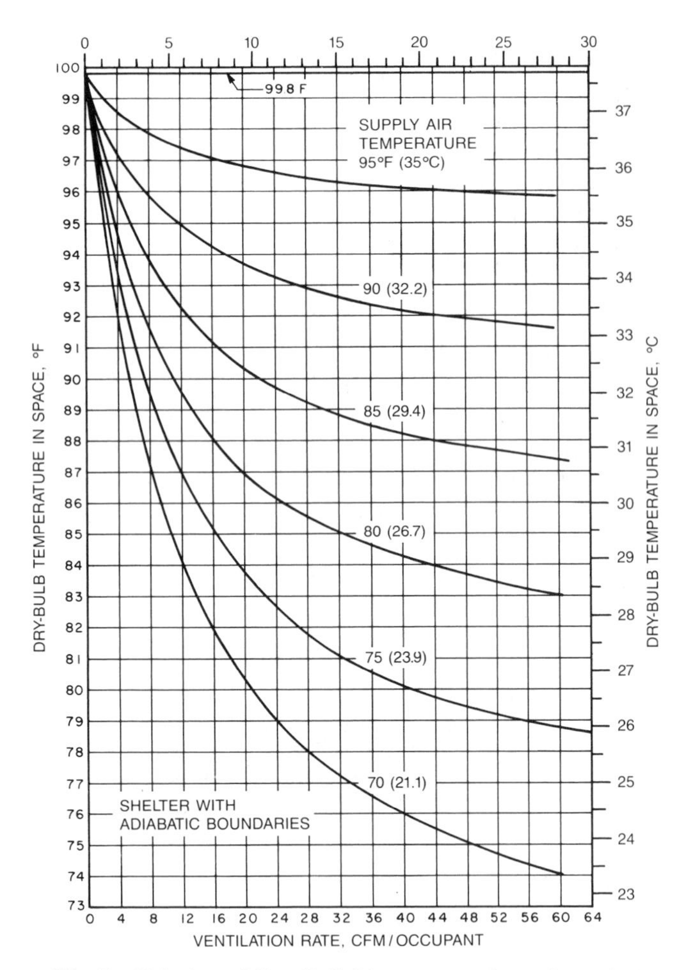

Fig. 7 Relation of Dry-Bulb Temperature in an Occupied Space to Unit Ventilation Rate

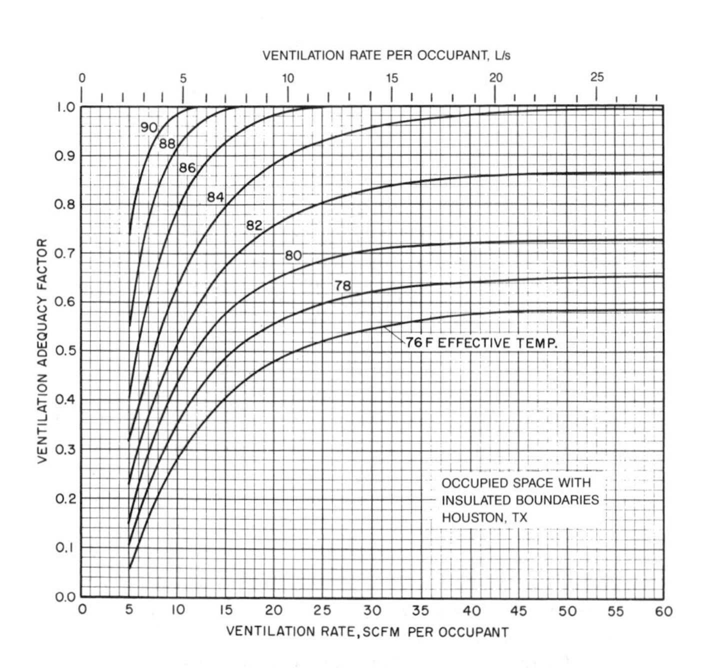

Fig. 8 Relation of Ventilation Rate to Annual Ventilation Adequacy Factors in Houston, Texas

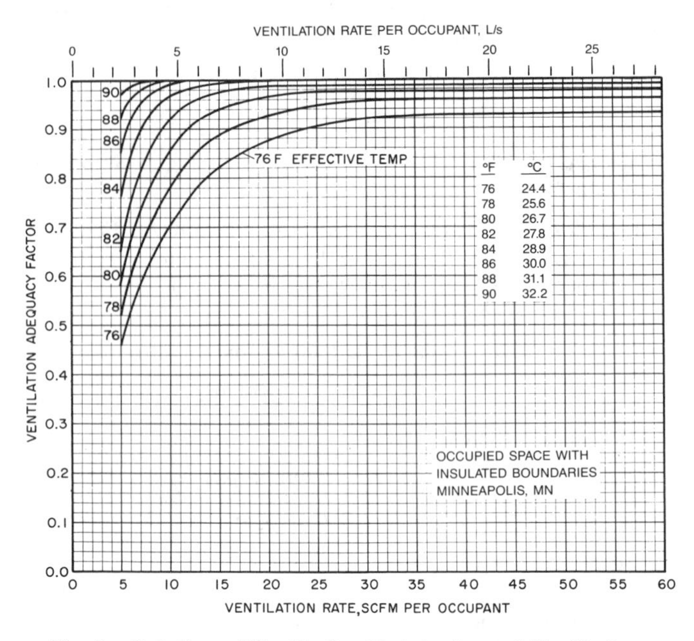

Fig. 9 Relation of Ventilation Rate to Annual Ventilation Adequacy Factors in Minneapolis, Minnesota

Figure 10 shows the map for 83 °F ET and 90% adequacy. At any point on an isoventilation line (for instance, the line for 15 cfm per person), an effective temperature of 83 °F ET would not be exceeded in an occupied shelter more than 10% of the time in a normal year. This map serves as the basis for design criteria recommended by DCPA for the ventilation of protective shelters. However, DCPA applies a standard adjustment of −1 °F ET, and refers to this set of data as the map for 82 °F ET (Figure 10). This downward adjustment is the result of observations made during simulated shelter occupancy tests. Figures 11 and 12 show maps for 80 °F ET, and for 90 and 95% adequacy, respectively. A comparison of data from these maps shows that a substantial increase in the ventilation rate is needed to obtain a moderate improvement in the probable environment.

Ventilating systems for shelters can be simulated by either of two analytical models. The first, the isostate model, is based on the assumption that the environment is spatially uniform. This model is the basis for the graphs and maps presented under the Case 1

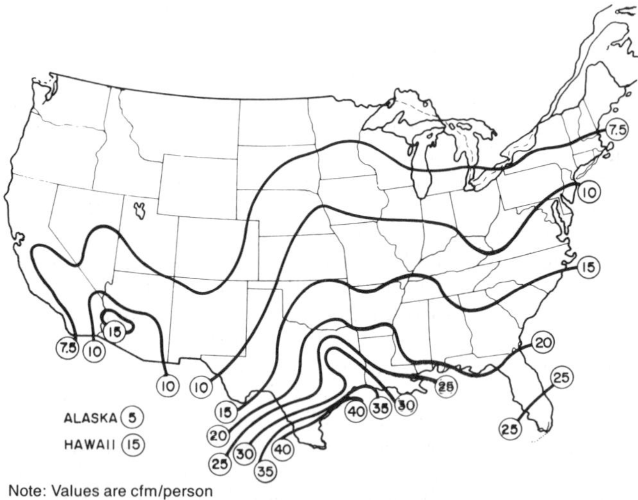

Fig. 10 Ventilation Required in Case 1 Shelter (Effective Temperature = 83 °F with 90% adequacy)

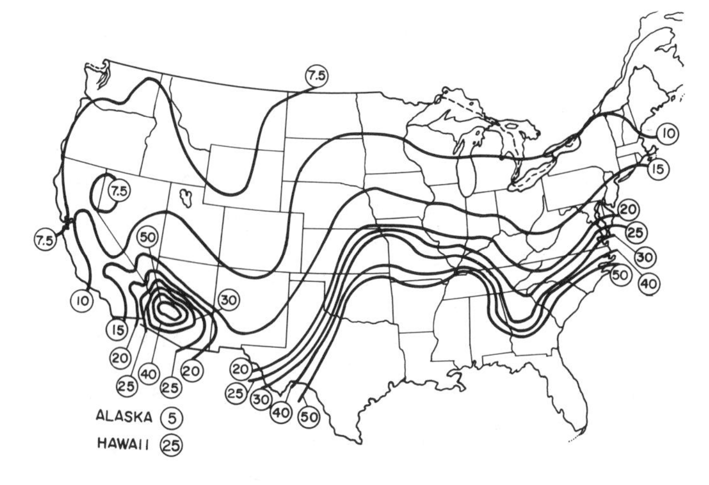

Note: Values are cfm/person

Fig. 11 Ventilation Required in Case 1 Shelter
(Effective Temperature = 80°F with 90% adequacy)

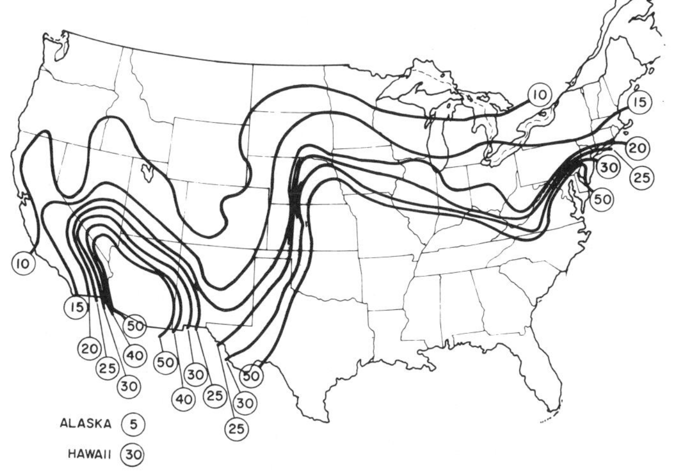

Fig. 12 Ventilation Required in Case 1 Shelter
(Effective Temperature = 80°F with 95% adequacy)

solutions. This state of uniformity can be approached only with a system that includes distribution ductwork and diffusion outlets. The second, the nonisostate or varistate model, is based on the incremental transfer of metabolic heat and moisture to the ventilating air as it passes in sequence through one or more occupied spaces. The varistate model is more realistic in the nonuniform environments with a minimum of air distribution ductwork and outlets. Relatively large changes in temperature and humidity are associated with low per capita rates of ventilation, and the predicted psychrometric states of air leaving such spaces may be quite different for the isostate and varistate models because the latent to total metabolic heat transfer is a function of environmental temperature (Allen 1970).

The degree of uniformity in the environment depends on the size and shape of the space, the arrangement of the air distribution system, the locations of heat loads and obstructions to airflow, and the unit rate of ventilation. Since the environment at one end of the space corresponds to the initial state of the air, it is beneficial to ventilate an unheated shelter with a mixture of fresh and recirculated air during cold weather (Allen 1970).

Case 2—Sealed Underground Shelters

Case 2 (sealed underground shelter with earth conduction effects) and Case 3 (ventilated shelter with earth conduction effects) have both been analyzed for deep underground shelter in which heat exchange near the surface does not materially influence the shelter thermal environment. Pratt and Davis (1958) and the Corps of Engineers (U.S. Army 1959) calculated the change of shelter dry-bulb temperature and the inner surface temperature from the initially uniform earth temperature by solving transient heat conduction equations for three simplified shelter models: (1) one-dimensional plane wall model, (2) cylindrical model, and (3) spherical model. The one-dimensional plane wall model can be applied to a large shelter where corner heat flow effect is small, whereas the spherical and cylindrical models are better for a small family-size shelter. The cylindrical model provides a good approximation for long underground tunnels.

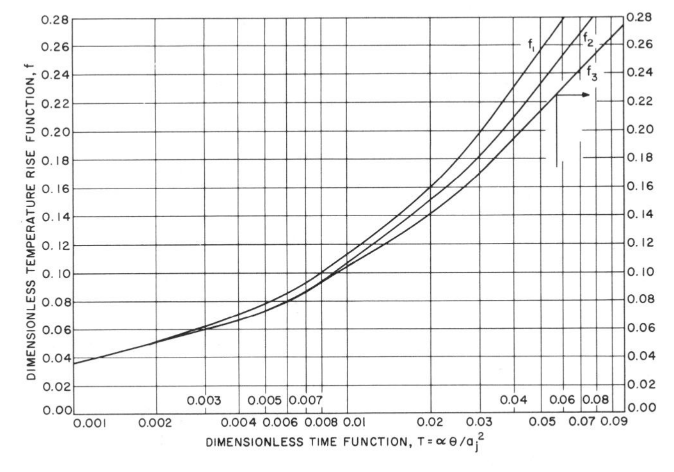

Fig. 13 Temperature Rise Function for Deep Underground Sealed Chambers

Functions for the approximate mathematical analysis for Case 2 are shown in Figure 13, where f_1, f_2, and f_3 represent temperature rise functions for a plane wall, cylindrical, and spherical shelter model, respectively. Dimensionless time function T is computed by:

$$T = \alpha\theta/a_j^2 \tag{9}$$

where

α = thermal diffusivity of soil, ft^2/h
θ = elapsed time, hours
a = equivalent radius of the shelter, ft

The subscript j refers to the model type, *e.g.*, 1, 2, or 3.

The equivalent radius of the shelter may be approximated by the following equations:

Plane Wall Model $$a_1 = S_i^{0.5} \tag{10}$$

Cylindrical Model $$a_2 = (S_c/\pi)^{0.5} \tag{11}$$

Spherical Model $$a_3 = (S_i/4\pi)^{0.5} \tag{12}$$

where

S_i = total inner surface area, ft^2
S_c = cross-sectional area, ft^2

The rise in temperature of the inner surface of the shelter is found from the following equation:

$$\Delta t_w = t_w - t_0 = (qa_j/S_i k)\, f_j \tag{13}$$

where

t_w = dry-bulb temperature of air in space, °F
t_0 = initial temperature of inner surface, °F
q = total heat generated in the sealed-up shelter, Btu/h
k = thermal conductivity of surrounding earth, Btu/h · ft · °F
f_j = temperature rise function (see Figure 13)

For $T < 0.001$, f_1, f_2, f_3 may be calculated by $f_i = 1.13\,(T)^{0.5}$.

For the sealed condition, the dew point of air in the space is approximately equal to the average inner surface temperature (Achenbach *et al.* 1962). The small difference Δt_x between average air and inner surface temperatures depends on the ratio of the per capita rate of heat transfer by convection q_c to the per capita area of inner surfaces S_p, as expressed in Equation (14):

$$\Delta t_x = t_a - t_w = q_c/h_c S_p \quad (14)$$

and, with the assumption that the convective part of the internal heat loads is one-half the entire sensible heat load, in Equation (15):

$$q_c = (q_s + q_i)/2 \quad (15)$$

where

t_a = dry-bulb temperature of air in space, °F
q_c = convective heat transfer at inner surfaces, Btu/h
h_c = coefficient for convective heat transfer, Btu/h · ft² · °F
S_p = unit area of inner surfaces, ft² per person
q_s = sensible part of metabolic heat, as defined by Equation (3) or (7)
q_i = lighting load, per person, Btu/h

The approximate temperature difference needed to maintain the rate of convective heat transfer to inner surfaces of the enclosure can be found by combining Equations (14) and (15) with Equation (3).

$$\Delta t_x = t_a - t_w = \frac{10(100 - t_w) + q_i}{2h_c S_p + 10} \quad (16)$$

In some cases, this temperature difference may be quite significant.

Example 2. Determine the effective temperature in a sealed underground chamber after two days of occupancy. The space is 12.5 ft long and 10 ft wide, with a ceiling height of 8 ft. There are 12 occupants, each having a metabolic heat rate of 400 Btu/h. Two 40-W lights in the space together contribute 22.7 Btu/h per person to the heat load. The earth has an initial temperature of 68 °F, a thermal diffusivity of 0.025 ft²/h, and a thermal conductivity of 0.75 Btu/h · ft · °F. The convection coefficient h_c is 0.70 Btu/h · ft² · °F.

Solution: Because the space is small and somewhat cubical, the spherical model is most applicable. The total inner surface area is,

$$S_i = 2[(10)(12.5) + (8)(10) + (8)(12.5)] = 610.0 \text{ ft}^2$$

and the unit area is $S_p = 50.83$ ft² per person.

From Equation (12),

$$a_3 = (610/4\pi)^{0.5} = 6.967 \text{ ft}$$

From Equation (9), when $\theta = 48$ h,

$$T = (0.025)(48)/(6.967)^2 = 0.0247$$

From Figure 13, $f_3 = 0.153$. From Equation (13),

$$\Delta t_w = [(12)(400 + 22.7)(6.967)/(610)(0.75)]\ (0.153) = 11.82\,°F$$

Then, the final temperature of inner surfaces is,

$$t_w = 68 + 11.82 = 79.82\,°F$$

This value, the inner surface temperature, is also the dew point of air in the room. The difference between air and boundary surface temperatures can be estimated by Equation (16),

$$t_a - t_w = \frac{(10)(100 - 79.82) + 22.7}{(2)(0.7)(50.83) + 10} = 2.77\,°F$$

and the temperature of air in the space is

$$t_a = 79.82 + 2.77 = 82.6\,°F$$

Using ASHRAE Psychrometric Chart No. 1, with a dry-bulb temperature of 82.6 °F and a dew point of 79.8 °F, the wet-bulb temperature is 80.4 °F. From Figure 2, the effective temperature is 81.3 °F ET.

Since the space is sealed (not ventilated), vitiation of the air may impose the most restrictive limit on stay time. The volume of space is 1000 ft³, and with 12 occupants, the unit volume is 83.3 ft³ per person. From Figure 3, for a carbon dioxide concentration of 3% by volume, the time/volume ratio is about 0.037 hours per ft³. For this criterion, the stay time is (0.037 × 83.3) = 3.08 h.

Therefore, the sealed chamber in Example 2 can be safely occupied by 12 people for two days, but only if appropriate means are provided to revitalize and replace polluted air in the space.

Case 3—Ventilated Underground Shelters

While the previous Case 2 solution example is useful in estimating shelter temperature rise during a relatively short period of sealed or buttoned-up condition, shelter thermal environment during the normal operation (where ventilation air is being introduced from outside) can be approximated as follows. Temperature rise of shelter air dry-bulb and shelter interior surface temperature can be calculated by ϕ_1 of Figure 14, ϕ_2 of Figure 15, and ϕ_3 of Figure 16, corresponding respectively to the plane wall, cylindrical, and spherical shelter models.

Equations (17) through (21) may be used with Figures 14, 15, and 16 to calculate shelter air temperature.

In Figures 14, 15, and 16, parameter N is evaluated by the following expression for a 400 Btu/h occupant:

$$N = h\ (a_j/k)\ (1.08G + 10)/(1.08G + 10 + hS_p) \quad (17)$$

where

h = surface heat transfer coefficient, Btu/h · ft³ · °F
S_p = inner surface area per person, ft²

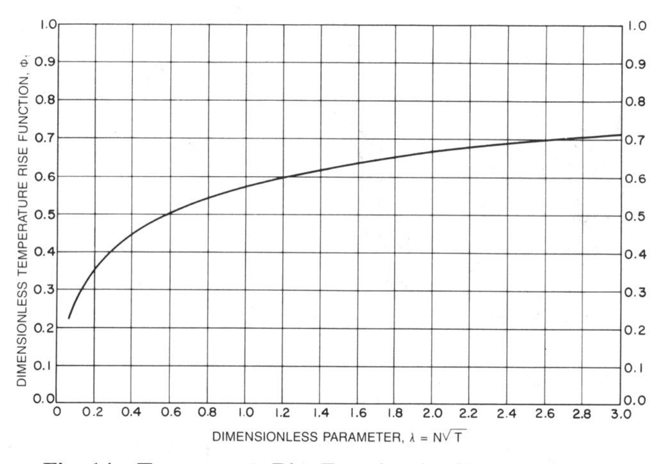

Fig. 14 Temperature Rise Function for Ventilated Deep Underground Chamber (Plane Semi-Infinite Model)

The temperature rise function ϕ_j (where $j = 1, 2,$ or 3) can be used to calculate the temperature rise of the inner surface.

$$\Delta t_w = t_w - t_o = u_o \phi_j \quad (18)$$

where

$$u_o = \frac{q_i + 10(100 - t_o) + 1.08G(t_v - t_o)}{1.08G + 10} \quad (19)$$

By denoting that

$$n = hS_p/1.08G + 10 + hS_p \quad (20)$$

the rise in dry-bulb temperature of air in the occupied space can be calculated with

$$\Delta t_a = t_a - t_o = [(1 - n) + n\phi_j]\ U_o \quad (21)$$

With the resultant value of t_a, the effective temperature in the space can then be determined if the humidity ratio of the air is known. By assuming that latent heat removed from the space by condensation of moisture on inner surfaces is a small part of the heat removed by conductive effects at inner surfaces, the humidity ratio can be estimated by means of Equation (7).

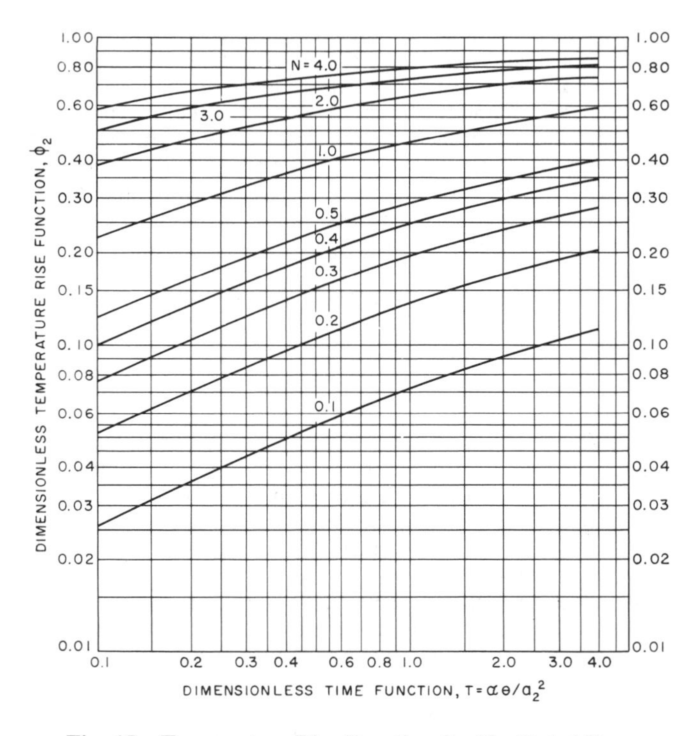

Fig. 15 Temperature Rise Function for Ventilated Deep Underground Chamber (Cylindrical Model)

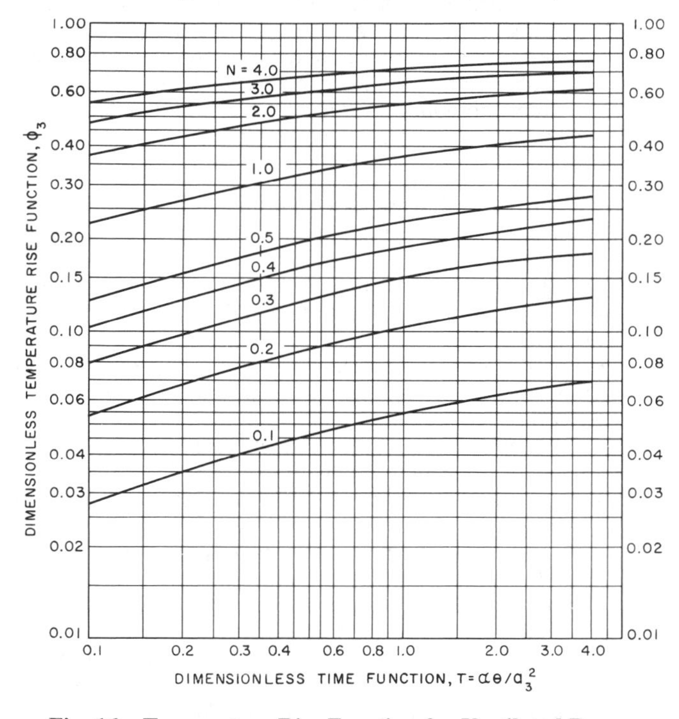

Fig. 16 Temperature Rise Function for Ventilated Deep Underground Chamber (Spherical Model)

Example 3. Determine the effective temperature at the end of a 14-day period of occupancy by 50 people in a ventilated underground chamber that is 25 ft long and 21 ft wide, with a ceiling height of 8 ft. Ventilation air having a dry-bulb temperature of 81.5 °F and a dew point of 68 °F is supplied to the space at a rate of 16 cfm per person. Per capita heat loads in the space are 400 Btu/h for metabolic effects, and 20.5 Btu/h for lighting. The adjacent earth has an initial temperature of 72.5 °F, a thermal diffusivity of 0.022 ft²/h, and thermal conductivity of 0.63 Btu/h · ft · °F. The surface coefficient of heat transfer is about 1.5 Btu/h · ft² · °F.

Solution: From an ASHRAE psychrometric chart, the humidity ratio of air supplied to the space is W_v = 0.01475 mass units of moisture per unit mass of dry air, and the wet-bulb temperature is 72 °F. From Figure 2, the apparent effective temperature of supply air is 76.7 °F ET.

The total inner surface area is,

$$S_i = 2[(21)(25) + (8)(21) + (8)(25)] = 1786 \text{ ft}^2$$
$$S_p = 1786/50 = 35.7 \text{ ft}^2 \text{ per person}$$

The plane semi-infinite model is most applicable. From Equation (10)

$$a_1 = (1786)^{0.5} = 42.3 \text{ ft}$$

From Equation (17)

$$N = \left[\frac{(1.5)(42.3)}{0.63}\right]\left[\frac{(1.08)(16) + 10}{(1.08)(16) + 10 + (1.5)(35.7)}\right]$$
$$= (100.62)(0.33737) = 33.94$$

a dimensionless number. From Equation (9)

$$T = (0.022)(14)(24)/1786 = 0.00414$$

Then, $N(T)^{0.5} = 33.94(0.00414)^{0.5} = 2.18$
and, from Figure 14, $\phi_1 = 0.68$.

From Equation (19),

$$u_o = \frac{20.5 + 10(100 - 72.5) + (1.08)(16)(81.5 - 72.5)}{(1.08)(16) + 10} = 16.5\,°\text{F}$$

From Equation (20)

$$n = \frac{(1.5)(35.7)}{(1.08)(16) + 10 + (1.5)(35.7)} = 0.663$$

From Equation (18)

$$\Delta t_w = (16.5)(0.68) = 11.2\,°\text{F}$$
$$\text{and } t_w = 72.5 + 11.2 = 83.7\,°\text{F},$$

From Equation (21)

$$\Delta t_a = [(1 - 0.663) + (0.663)(0.68)](16.5) = 13.0\,°\text{F}$$
$$\text{and } t_a = 72.5 + 13.0 = 85.5\,°\text{F}$$

From Equation (6),

$$W_a = 0.01475 + \frac{85.5 - 60}{(470)(16)}$$
$$= 0.01814 \text{ lb of moisture per lb of dry air}$$

From ASHRAE Psychrometric Chart No. 1, at the end of the 14-day period, the predicted dew point of air in the space is 73.8 °F and the wet-bulb temperature is 77 °F. From Figure 2, the effective temperature will then be 80.7 °F.

These examples show that the thermal properties of the soil, the initial earth temperature, and the stay time in the shelter have an important effect on the interior surface required for the adequate removal of heat from underground shelters by earth heat conduction. The conductive cooling effect in shelter roofs having shallow earth cover would be reduced in warm weather, and reduced to an even greater degree if the upper surface were exposed to solar radiation. This condition was not considered in the previous examples. In locating underground shelters, available shade and grass cover should be taken advantage of to minimize the heat gain from the earth surface.

Since the shelter interior surface area per person (s_p) tends to decrease as the size of a shelter increases, the cooling by earth heat conduction becomes proportionately less in large shelters. However, earth conduction in conjunction with adequate ventilation air may be sufficient to maintain a habitable thermal environment in an underground fallout shelter located in relatively cool earth temperature regions. Under favorable conditions, an expensive mechanical cooling system can be avoided.

When earth temperature and the outdoor ventilation air alone cannot maintain a habitable thermal environment, supplementary cooling must be used. The previous method of estimating the shelter thermal environment can still be applied if supply air temperature and humidity ratio are at the design exit conditions of the air-cooling systems.

The previous discussions for shelter thermal environment are all based on simplified models. More rigorous calculations, which are closer to the actual underground shelter than those used in this chapter, have been performed with reasonable success by Achenbach *et al.* (1962b), Kusuda and Achenbach (1963), Drucker and Cheng (1962), and Baschiere *et al.* (1965). These models can generally estimate transient shelter conditions to within 2 °F of observed temperatures. However, the method shown in this chapter can give approximate solutions that do not differ considerably from the actual thermal environment studied in experimental shelters containing simulated occupants.

Aboveground Shelters

The aboveground shelter is a more complex structure to analyze than an underground shelter because of the variety of surroundings with which the shelter can transfer energy. The shelter can be exposed to the ambient weather, other spaces within the structure, and to the soil, if the shelter is at grade level or partially below grade. In some respects, calculating the loads transmitted between the shelter and its surroundings is no more involved than determining the heating or cooling loads present in any conventional structure. The difference is that the shelter interior will not be maintained at constant dry- and wet-bulb temperatures as in conventional structures. Thus, the standard procedures of estimating transmission loads must be applied carefully.

Neglecting the transmission loads in calculating the ventilation requirements of aboveground shelters is generally a valid assumption because (1) as the ambient temperature rises, the inside to outside temperature differential becomes small and (2) as the shelter temperature rises, a smaller portion of the metabolic energy is given off as sensible heat. Therefore, in extremely hot weather, the transmission energy loss is a small percentage of the total energy to be eliminated from the shelter. In addition, the absorption of solar radiation will reduce the transmission of energy from the shelter, sometimes to the point of eliminating the transmission loss and creating a heat flow into the shelter. Heat storage within the shelter structure and its contents would also be expected to reduce the shelter environmental conditions to a lower level than calculated by Case 1. However, studies have shown that these effects are present only during the first week of shelter occupancy. In the second week, the heat storage effects are reduced considerably, if not eliminated.

NATURAL VENTILATION

Infiltration and natural ventilation are similar in that both move air by using air density differences and wind forces rather than fans driven by some source of power. The distinction is that infiltration makes use of random openings and cracks, whereas natural ventilation is planned to take advantage of building configuration, orientation, circulation paths, and openings. Natural ventilation is most applicable in aboveground fallout shelter spaces in existing buildings that have large openings or passageways necessary for moving large quantities of air by small pressure differentials. Since in most conceivable attacks almost all of the high-rise buildings will be exposed to a positive pressure sufficient to break out all windows, thermal forces will be quite weak compared with wind forces and generally will not provide adequate natural ventilation. Therefore, wind force is the only source of natural ventilation that can be considered.

An underground shelter having relatively cool walls may remain at a lower temperature than the outdoor air. This results in a stable condition that virtually eliminates air circulation or the effects of thermally induced natural ventilation (Ducar and Engholm 1965). However, under favorable conditions of orientation of shelter openings respective to adjacent structures, there may be a substantial natural ventilation effect due to kinetic wind forces.

An estimate of the magnitude of natural ventilation through an aboveground fallout shelter may be made by using the procedure in the section on natural ventilation in Chapter 23 of the 1989 ASHRAE *Handbook—Fundamentals*. However, because of the variable and complex environment, it is difficult to evaluate natural ventilation accurately.

VENTILATION SYSTEMS

Mechanical Ventilation

Mechanical ventilation is not subject to the uncertainties associated with natural ventilation. The major limitation is that a safe, effective temperature, such as 83 °F, cannot be maintained when the effective temperature of the outdoor air approaches 83 °F. Mechanical ventilation is most suitable for fallout shelters in which there is no impelling reason to minimize the size of openings or to provide extensive purification treatment for the fresh air supplied to the shelter.

A first step in designing a ventilating system for a shelter is to select a reliable power source. The possible choices include auxiliary electric power and muscle power. In general, commercial electric power should not be relied on as the sole source, since the supply may be disrupted when needed. The use of auxiliary electric power implies that the blowers would be driven by an electric motor with power supplied by an auxiliary engine-generator. Waste heat from an engine-generator can often be used in a shelter for tempering air in winter or heating domestic water. Fans or blowers with manual drives are limited in capacity by human factors and multiple units would be required in large shelters.

The system components for a shelter should be selected and arranged in accordance with the following objectives, wherever practicable:

1. Maintain a tolerable physical environment in the shelter.
2. Prevent or minimize the condensation of moisture on interior surfaces.
3. Facilitate operation, maintenance, and repair of all equipment.
4. Avoid awkward duct connections and the resultant noise and head losses.
5. Reuse waste air from the ceiling level of occupied spaces for scavenging service spaces such as equipment rooms, toilets, and entryways.
6. Protect people and equipment from effects of weapons and fire to a degree consistent with potential capabilities of the shelter.
    a) Maintain radiation barriers inviolate by shielding at points where ducts penetrate the structural shell.
    b) Provide a weatherproof air intake fixture or hood that tends to exclude particulates, and locate this fixture at a safe distance from combustible materials and above levels of ground turbulence and floodwater.
    c) Avoid contamination of interior spaces, equipment rooms, and entryways by radioactive particles and combustion gases from fires or fuel-burning equipment.
    d) Provide blast closures or attenuators for blast-resistant shelters.
    e) Provide filters for purifying fresh air to a degree consistent with the intended use of the shelter. Shield occupied spaces from air filters.
    f) Consider use of a life-support system for suitable closed shelters located in a potential fire area.
    g) Consider requirements for shock-mounting equipment in blast-resistant shelters.
7. Provide system flexibility to accommodate seasonal changes and variations in physical activity.

a) Reduce the quantity of fresh air in cold weather or temper the fresh air with waste heat to avoid overcooling.
b) Provide mixing dampers and plenum for partial recirculation of the air. This maintains air motion and tempers the air supplied to occupied spaces.
c) Provide means for adjustment of air distribution to correct objectionable drafts and to balance the system in accordance with space usage—that is, for sleeping or recreation.

8. Anticipate and facilitate probable future improvements or changes in shelter capabilities.
9. Achieve optimum cost-effectiveness—minimum cost consistent with adequate performance.

Environmental Control System

Figure 17 shows an environmental control system in an underground shelter. The various components are identified by numbers that correspond to the following itemized nomenclature, with brief comments. In general, blast-resistant shelter systems should have adequate provisions for shock mounting of components. Any equipment, such as heat-transfer apparatus, installed outside the protective structure must be blast-resistant.

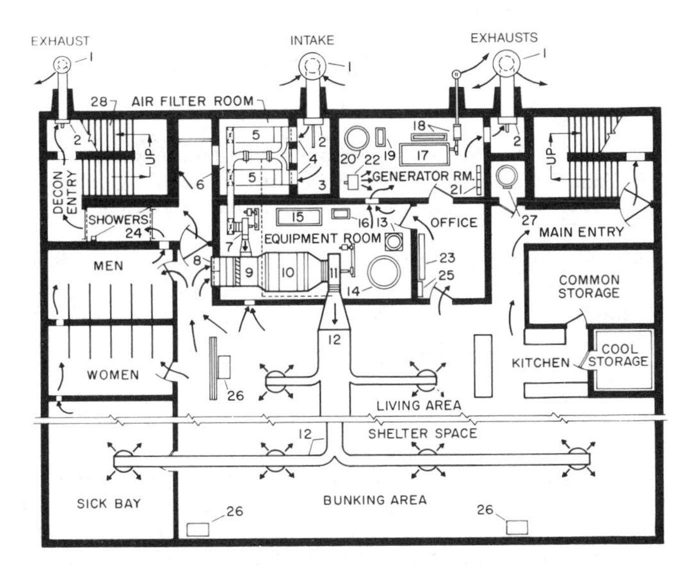

Fig. 17 Environmental Control System for Underground Community Shelter

Weatherproof hoods of the mushroom type for intake and exhaust air. Particle sizes having characteristic terminal velocities greater than the vertical component of the entering air velocity are not likely to be drawn into the ventilating system. Stormproof louvers in wall openings may also be used but are less effective in excluding particulates.

Automatic blast closures are consistent only in structures having appreciable blast resistance. Pressure-actuated blast valves close in a few milliseconds to prevent a destructive pressure rise within the shelter. Sensor-operated valves close before arrival of the shock wave and can be used when increased costs can be justified.

Fresh air intake plenums serve as a settling chamber for coarse particles that may be radioactive and as an expansion chamber to limit pressure rise due to leakage during closure of blast valves. Intervening concrete walls provide shielding from this secondary source of radiation.

Particulate filters or prefilters for fresh air. Panel-type filters with dry, pleated, glass-fiber media supported on both sides by pleated wire screens would be most suitable for service in blast-resistant structures.

Gas-particulate filter units. Each unit consists of an extreme-efficiency particulate filter and a special activated-charcoal filter mounted and sealed in a rectangular frame. These units can be used effectively only in tight shelters planned or adapted for protection against biological and chemical warfare agents.

Fresh air ductwork. For normal operation, a bypass duct with two open dampers may be provided for gas-particulate filter units. When these filter units are in service, the two bypass dampers are closed, and the length of duct between the two dampers can be pressurized by a pipe connection to the main-fan discharge transition to ensure that any damper leakage is uncontaminated air.

Fresh air blower with electric motor drive. If the shelter is cooled by means *other* than ventilation with outdoor air, the capacity of the blower can be 3 cfm per person. This amount also provides for some emergency overloading of the shelter. Since this blower pressurizes the shelter space and must operate against a rather high system resistance, if airflow is restricted by gas-particulate filters and blast valves, Class II construction may be indicated. Class I construction is adequate for fans in fallout shelters.

Grilled opening for recirculated air. Optional prefilters and activated-carbon filters may be provided, as shown behind the grille, for odor control. Alternatively, portable activated-carbon filter units may be used for this purpose.

Mixing plenum for fresh and recirculated air. Air quantities can be adjusted by changing damper positions or motor speeds.

Air-conditioning or heat-exchanger unit with extended-surface cooling and heating coils. Cooling coils can use well water or chilled water, and heating coils can use waste heat from the engine generator.

Main fan with electric motor drive. Required capacity of this fan depends on system design parameters. Adequate air distribution can be obtained with 10 to 15 cfm per person.

Air distribution ductwork with diffusion outlets. Although design of the duct system may be dictated by requirements for normal uses of the structure, a simple low-cost system that distributes all of the air along the remote end of the occupied space may be adequate for survival shelters.

Well and pump for potable and cooling water. A charging well for wastewater may be desirable. If well water is unavailable, other means must be substituted for removal of excess heat.

Hydropneumatic tank for pressure water system.

Optional *unitary water chiller* for alternative or supplementary use. This item may be needed for absolute control of the environment in hot humid climates where cool well water is not available. The system could be arranged for use of a unitary conditioner, which also avoids refrigerant lines between separated components.

Circulating pump for chilled water. This item may be an integral part of the package water chiller.

Emergency engine-generator set. The engine may be cooled by a heat exchanger using well water or, alternatively, by a remote radiator. The storage tank for fuel oil or gasoline may be buried adjacent to the shelter if provisions are made for differential movements of tank and structure.

Heat exchanger and muffler for engine cooling and waste heat recovery.

Hot water circulating pump.

Hot water storage tank. Hot water can be used to advantage for tempering fresh air or for showers.

Batteries for cranking engine or for emergency lighting.

Recirculating fan-coil unit for cooling the generator room with well water.

Control cabinet for functional control apparatus.

Decontamination facility. Showers are provided for personnel entering from contaminated areas and to promote welfare of occupants.

Cabinet for detection or test instruments and special tools.

Portable manually operated life-support packages. These optional units should provide for about 24 h of sealed operation during which the shelter may be subject to the effects of adjacent fires. Alternatively, a static life-support system using bottled oxygen and soda lime could be provided.

Incinerator for combustible waste materials.

Sewage sump and pump beneath the stairway.

The environmental control system (shown in Figure 17) has capabilities that are not likely to be essential for survival in most situations, and costs for a less comprehensive system are more acceptable. Minimum requirements for environmental control could be provided at much lower cost.

Air Intakes

During an emergency, the ambient atmosphere may be contaminated with noxious gases, fumes, or particulate matter that must be partially or completely removed from any ventilating air supplied to occupied spaces in a protective structure. Gas and/or particulate filters that have the necessary efficiencies and specific capabilities can be used to remove these contaminants, if the attendant costs are not prohibitive. These costs include not only the cost of the filter installation and maintenance, but also the additional costs for air-moving apparatus and power needed to accommodate the flow resistance of the filters.

When radioactive, irritant, toxic, or pathogenic particulates are removed, air intake facilities designed to use gravitational or inertial forces may be advantageous. If the hazardous particles are relatively coarse (mean diameter < 0.002 in.) and/or dense, a properly designed air intake facility may meet the requirement adequately without filters. If the installation includes particulate filters, the removal of coarse particles at the air intake facility would probably extend the service life of the filter elements.

Air intakes effective in removing particulates from ventilating air are grouped as (1) those that prevent the entry of particles into a fixture located outside the structure and (2) those that separate and collect the particles in a sump or chamber contained within the enclosed space.

External fixtures for intake air include gooseneck, mushroom, and sidewall types with projecting canopy. These share the characteristic that air from the atmosphere must enter the fixture in an upward direction with a vertical component of velocity low enough to capture only fine particles, while coarse particles fall to the ground.

Internal facilities for intake air include (1) vertical U-shaped vent shafts, in which the air changes direction abruptly and deposits coarse particles in a sump or basin from which they can occasionally be flushed with water down a drainpipe; and (2) gravity-separation chambers or passageways, through which the ventilating air flows horizontally at a low velocity so that coarse particles settle to the floor and accumulate until removed. Gravity separation techniques are discussed in Chapter 11 of the 1988 ASHRAE *Handbook—Equipment*. Underground passageways may also moderate the temperature of outside air supplied to occupied spaces. In any case, air intakes should be located above the level of ground turbulence and above the high water mark of any anticipated flood. Screens that serve to exclude birds and vermin are desirable.

EVAPORATIVE COOLING

Evaporative cooling either reduces the effective temperature or lowers the required ventilation rate in a shelter. The rate of change of effective temperature, with respect to dry-bulb temperatures, is positive over the ranges of dry- and wet-bulb temperatures that may be expected in shelters. Lowering the dry-bulb temperature through evaporative cooling lowers the effective temperature of the shelter for a fixed ventilation rate and inlet air condition. These trends have been verified by calculation of shelter conditions by a modified Case 1 solution.

Assuming that the evaporative cooler is at the inlet to the shelter, the adequacy of forced ventilation with evaporative cooling can be established for any evaporative cooler efficiency. Evaporative coolers with efficiencies of 80% can reduce the required ventilation rates 2 to 83% (Figure 18). The greatest reductions are in the regions that cannot be ventilated by natural means, or those in which extremely high ventilation rates are required (Baschiere *et al.* 1968).

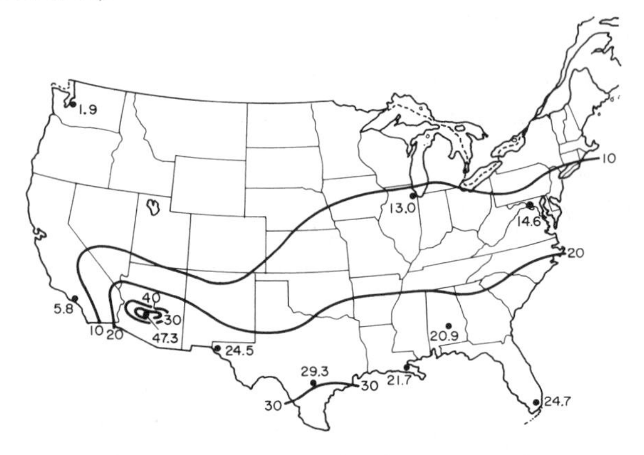

Fig. 18 Percent Reduction in Ventilation System Capacity Enabled by 80% Efficient Evaporative Coolers
(Effective Temperature limited to 83°F with 90% adequacy)

Increasing evaporation efficiency from 80 to 85% only reduces the required ventilation rate on additional 2 to 3%. Therefore, evaporative coolers with efficiencies as low as 80% are sufficient.

MOISTURE IN ENCLOSED SPACES

During standby or unoccupied periods, moisture may have to be controlled; if shelters are located in areas having a high water table, complete waterproofing is necessary.

As in all facilities unoccupied for long periods without lighting or other heat sources, mildew and other moisture effects are likely to occur. Conventional methods such as mechanical dehumidifiers, silica gel, and calcium chloride can control the standby environment. Conventional, residential-type mechanical package dehumidifiers are generally unsuited for operation at temperatures below 60°F due to frosting on the evaporator coils. Additional heat or alternate methods may have to be employed where colder temperatures are likely to occur.

Computer studies and actual shelter tests have shown that condensation on walls or ceilings is more likely to occur during summer, but, under certain conditions, may also occur in winter.

The interior surface temperatures of the shelter, as calculated in the Case 3 analysis, are helpful in predicting possible water vapor condensation along shelter walls.

For small shelters in moderate climatic regions, the condensation of water usually takes place during the initial phase of shelter occupancy while the soil temperature adjacent to the walls is relatively low. Drip from the ceiling, which is particularly objectionable, may be prevented by suspending metal pans under the ceiling to carry the condensed moisture into containers or a sump. Dehumidification by use of sorbent materials or mechanical dehumidifiers adds heat to the space as water is being removed, and the resultant effect is an increase in effective temperature.

FIRE EFFECTS

Fires ignited by thermal radiation may occur in all types of shelters. External fires could result where there is burning or smoldering material around or over a shelter or an intake opening. Basement shelters are vulnerable to fire and questionable in areas where direct or secondary fire effects are probable.

The probability of drawing in superheated air and toxic gases during a fire is reduced if the air intake opening is located at a distance from combustible buildings or other materials. Intakes at two different points with dampers reduce risk further. The most positive protection against fire effects can be provided by temporarily closing off the shelter openings, shutting down the ventilating system, and putting a life-support system into operation.

There is little evidence that oxygen is drawn from a shelter by a nearby fire, or that the oxygen in the area would be depleted to the point where life could not be sustained. If fire will burn, there is enough oxygen to breathe. The main danger comes from carbon monoxide and heated gases entering the air intake. Carbon monoxide and heated gases can also enter the shelter through porous walls and cracks in the shelter construction. The presence of hot air and toxic gases such as carbon monoxide would probably necessitate shutdown of the ventilation system. While there is sufficient air in a shelter to support life for a few hours without air replacement or revitalization, the attendant increase in carbon dioxide and decrease in oxygen concentration eventually requires that the shelter either be evacuated or have a life-support system placed in operation. For example, according to Figure 3, carbon dioxide concentration in an unventilated shelter having 100 ft^3 of free space per person would increase to 3% by volume in 3.7 h and to 5% in 6.1 h.

When the top of the slab or overburden above an underground shelter is blanketed with burning or smoldering material, the shelter may experience intense heat flux from the flame above, resulting in a peak surface temperature near 2000 °F. The pattern of the surface temperature and heat flux in the shelter roof region, however, depends on the characteristics of burning, heat content, and collapsing sequence of the burning structure. After the structure starts to cave in, the surface temperature may decrease, because the collapsed burn pile shields the shelter from the high-temperature flame, and the air that had been feeding the fire from beneath the suspended burn pile is restricted. This high surface temperature, however, will cause the shelter roof near the fire to rise in temperature during the fire and continue to rise for some time after the fire subsides. An analysis of heat conduction in the shelter roof is complex because the thermal properties depend on temperature and moisture content for large temperature change, and because some of the conducted heat is absorbed by vaporization of the water contained in the earth or slab.

An estimate of the temperature rise at the ceiling of the shelter can be obtained from Figure 19. The assumed fire model for these curves is based on sudden outside surface temperature rise to fire temperatures (t_F) at time = 0, followed by a linear return to the initial top slab surface temperature before the fire (t_o), as indicated in the insert of Figure 19. The elapsed time is designated as the fire period, θ_o. The temperature t of the lower surface of the slab continues to rise for a time during the postfire period and can be computed from the temperature rise function ϕ_F at any time θ after the start of the fire. The crest of each curve occurs at the time when the temperature of the lower surface is maximum.

Equations (22), (23), and (24) may be written to express these relationships.

$$T' = \alpha\phi_o / l^2 \quad (22)$$

$$\phi_F = (t - t_o)/(t_F - t_o) \quad (23)$$

$$T = \theta/\theta_o \quad (24)$$

where

α = thermal diffusivity, ft^2/h
θ_o = fire period, h
l = slab thickness, ft

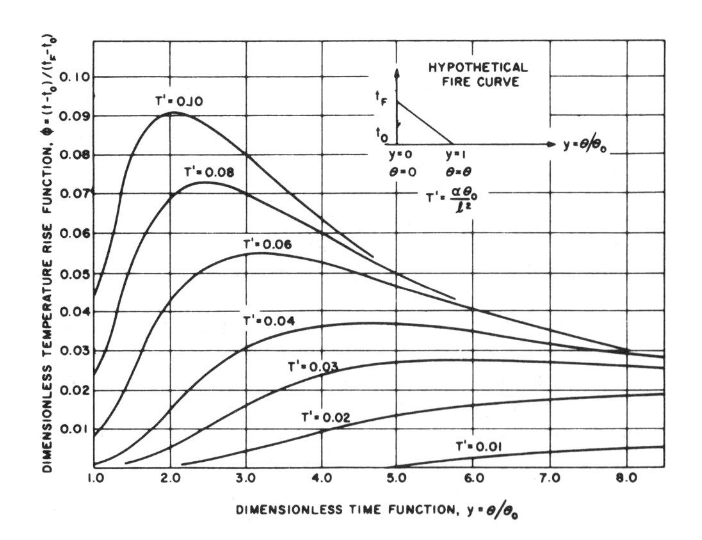

Fig. 19 Temperature Rise at Ceiling of Shelter Caused by Overhead Fire

t = temperature of the slab lower surface, °F
t_F = fire temperature, °F
t_o = initial top slab surface temperature, °F
θ = time after start of fire, h

Example 4. The structural roof slab over an underground chamber is covered with a layer of mineral fill to make a total thickness of l = 2 ft. The thermal diffusivity α of overhead materials is 0.030 ft^2/h. The upper surface is covered with burning debris from a collapsed building. During the fire, the temperature of the upper surface of cover materials increases suddenly from a normal of t_o = 80 °F to a maximum of t_F = 1500 °F, and then decreases linearly to normal over a fire period of θ_o = 8 h. Determine the maximum temperature of the ceiling surface in the chamber and the time after ignition at which the maximum temperature occurs.

Solution:
From Equation (22),

$$T' = (0.03)(8)/(2)^2 = 0.06$$

From Figure 19 at the peak of the 0.06 curve, ϕ_F is found to be 0.055. From Equation (23),

$$t - 80 = (0.055)(1500 - 80) = 78\,°F$$
$$t = 158\,°F = \text{maximum ceiling temperature}$$

Noting that on Figure 19 the peak of the curve occurs at Y = 3.25, the value of θ is determined from Equation (24).

$$\theta = (8)(3.25) = 26 \text{ h to maximum ceiling temperature}$$

Example 4 points out the importance of shelter location with regard to combustible structures or materials. A similar computation with increased overhead thickness would show the benefits to be derived from heavy earth cover.

In general, fire-safe refuges or sanctuaries can be established on each floor of noncombustible structures (Pefley and Allen 1982). Fire-safe refuges are conceived as a viable alternative that can be used in conjunction with, not as a replacement for, the sprinkler systems and escape routes that are commonly associated with fire protection in buildings. The ventilating system often provided for pressurizing a stairwell can be increased in capacity and extended to serve also as a source of pure air for pressurizing fire-safe refuges to exclude toxic gases and smoke. Potentially, the best spaces for secondary use as places of refuge are rest rooms, service corridors, and elevator lobbies in a building core that also includes a pressurized stairwell.

The design and use of environmental control systems for fire safety may be restricted by local building or fire codes, unless variances are approved.

SELECTION OF APPARATUS

The cost of virtually complete protection against all conceivable hazards is prohibitive. Therefore, the nature of the threat and the specific hazards against which protection is contemplated should be evaluated during preliminary planning. The system configuration and the components to be included can then be designed in accord with such guidance. Essential information on hazards and recommended or mandatory requirements for protective capabilities can be obtained from a cognizant government agency.

The need for air purification apparatus varies greatly among protective facilities, and the method selected can affect the system resistance to airflow, the power to drive air movers, and the capacity of any auxiliary power supply that may be needed. High efficiency air filters that remove fine particles would be needed if the threat included pathogenic organisms or irritant fumes. On the other hand, the acute hazard of radioactive fallout from nuclear explosions is associated with large particles that can be adequately separated from ventilating air by a suitable intake fixture. Many toxic gases can be removed from an airstream by activated charcoal impregnated with reactive or catalytic agents (Viessman 1954).

For survival under some unusual situations, a completely isolated environment with a self-sufficient life-support system may be necessary, including apparatus for limiting carbon dioxide concentration, for replacing the oxygen consumed, and for controlling the effective temperature in the space (Gates 1960).

REFERENCES

ACGIH. 1989. Threshold limit values and biological exposure indices for 1989-90. American Conference of Governmental Industrial Hygienists. Cincinnati, OH.

Achenbach, P.R., F.J.J. Drapeau, and C.W. Phillips. 1962a. Environmental characteristics of a small underground fallout shelter. ASHRAE *Journal* (January):21.

Achenbach, P.R., T. Kusuda, and F.J.J. Drapeau. 1962. Mathematical analysis of thermal environment underground shelters. ASHRAE Symposium on Survival Shelters, 9.

Allen, F.C. 1960. Control of shelter environment. *Proceedings*, Environmental Engineering in Protective Shelters, NAS-NRC (February):297.

Allen, F.C. 1962. Mechanical equipment requirements. ASHRAE Symposium on Survival Shelters, 131.

Allen, F.C. 1970. Ventilating and mixing processes in nonuniform shelter environments. Stanford Research Institute for OCD, AD 708-574.

Allen, F.C. 1972. Parametric study of thermal problems in densely populated shelters. Stanford Research Institute for Defense Civil Preparedness Agency AD 749-796. Menlo Park, CA.

Baschiere, R.J., M. Lokmanhekim, H.C. Moy, and G. Engholm. 1965. Analysis of aboveground fallout shelter ventilation requirements. ASHRAE *Transactions* 71(1):101.

Baschiere, R.J., C.E. Rathmann, and M. Lokmanhekim. 1968. Adequacy of evaporative cooling and shelter environmental prediction. General American Research Division, GATX, OCD, AD 679-874.

Blockley, W.V. 1968. Dehydration and survivability in warm shelters. Office of Civil Defense AD 673-857. Webb Associates. Malibu, CA.

Drucker, E.E. and H.S.Y. Cheng. 1962. Analog study of heating in survival shelters. ASHRAE Symposium on Survival Shelters, 35.

Drucker, E.E. and J.T. Haines. 1964. A study of thermal environment in underground survival shelters using an electronic analog computer. ASHRAE *Transactions* 70:70.

Ducar, G.J. and G. Engholm. 1965. Natural ventilation of underground fallout shelters. ASHRAE *Transactions* 71(1):88.

Glasstone, S. 1962. Effects of nuclear weapons. Prepared by Department of Defense and published by Atomic Energy Commission.

Gorton, R.L. 1972. Response of human subjects to reduced levels of water consumption under simulated civil defense shelter conditions. Civil Defense AD 739-562. Kansas State University, Manhattan, Kansas.

Harrow, B. and A. Mazur. 1958. *Textbook of biochemistry.* W.B. Saunders Co., Philadelphia, PA, 314.

Henschel, A. *et al.* 1968. An analysis of heat deaths in St. Louis during July 1966. National Center for Urban and Industrial Health for Office of Civil Defense *Report* TR-49.

Kusuda, T. and P.R. Achenbach. 1963. Numerical analysis of the thermal environment of occupied underground spaces with finite cover using a digital computer. ASHRAE *Transactions* 69:439.

Lee, D.H.K. and A. Henschel. 1963. Evaluation of thermal environment in shelters. Division of Occupational Health, Public Health Service, DHEW.

Lind, A.R. 1955. The influence of inspired air temperature on tolerance to work in the heat. *British Journal of Industrial Medicine* 12:126.

McNall, P.E. and P.W. Ryan. 1968. Water consumption during 24-hour exposure to hot humid environments. ASHRAE *Journal* 10(March):51.

Nevins, R.G. and P.E. McNall. 1967. Human physiological responses to shelter environment. Office of Civil Defense AD 659-403. Institute for Environmental Research, Kansas State University, KS.

OCD. 1962. Shelter design and analysis. Fallout Protection, Vol. 1. Office of Civil Defense, Washington, D.C.

OCD. 1969. Shelter design and analysis. Environmental Engineering for Shelters, TR-20, Vol. 3, Office of Civil Defense, Washington, D.C.

OCD. 1971. Design considerations for fallout shelter ventilating air intake systems. Technical Memorandum 71-1, Office of Civil Defense.

Pefley, R.K., J.F. Abel, and J.S. Dutton. 1972. Physiological effects of low ventilation rates, high temperatures and high humidities. Civil Defense Preparedness Agency AD 753-843. University of Santa Clara, Santa Clara, CA.

Pefley, R. and F.C. Allen. 1982. Fire-safe sanctuaries: A viable alternative for life safety in high-rise buildings. ASHRAE *Transactions* 88(2).

Pefley, R.K., E.T. Cull, and K.M. Sekins. 1969. Monoman calorimeter project. Office of Civil Defense AD 685-878. University of Santa Clara, Santa Clara, CA.

Pratt, A.W. and L.F. Davis. 1958. Heat transfer in deep underground tunnels. National Building Studies Research, London.

U.S. Army. 1959. Heating and air conditioning of underground installations. Corps of Engineers, Manual EM 1110-345-450.

BIBLIOGRAPHY

ASHRAE. 1983. Design of smoke control systems for buildings.

Buettner, K. 1957. Heat transfer and safe exposure time for man in extreme thermal environment. *Mechanical Engineering* 79(November):1031.

Everetts, J., D.R. Witt, and E.R. McLaughlin. 1970. The feasibility of augmenting below-grade shelter habitability with conditioned air. The Pennsylvania State University Report to Office of Civil Defense.

Glasstone, S. 1962. Effects of nuclear weapons. Prepared by Department of Defense and published by Atomic Energy Commission.

Hummell, J.D., D.E. Bearint, and L.J. Flanigan. 1964. Methods for disposing of excess shelter heat. Battelle Memorial Institute Report to Office of Civil Defense.

Kaiser, E.R. and J. Tolciss. 1962. A selective bibliography on environmental control and habitability of survival shelters. ASHRAE Symposium on Survival Shelters.

Kaiser, E.R. 1963. A selective bibliography on environmental control and habitability of survival shelters — Addendum No. 1. ASHRAE Symposium on Survival Shelters.

Peters, A. and N. Gentieu. Soil thermal properties: An annotated bibliography. Office of Civil Defense Research Report OCD-OS-6358 AD 431-604.

Speigal, W.F. 1968. Mechanical and electrical system design for protection shelters. Technical Manual Prepared for Office of Civil Defense.

Technical Standards for Fallout Shelters. Office of Civil Defense, Department of the Army, Technical Memorandum 69-1.

CHAPTER 12

INDUSTRIAL AIR CONDITIONING

INDUSTRIAL plants, warehouses, laboratories, nuclear power plants and facilities, or data processing rooms are designed for the specific processes that they enclose. Environmental conditions include proper temperature, humidity, air motion, and cleanliness. Airborne contaminants generated must be collected and treated before being discharged from the building or returned to the area.

Many industrial buildings require large quantities of energy, both in manufacturing and in the maintenance of building environmental conditions. Energy can be saved by proper insulation, ventilation, use of solar energy, and recovery of waste heat.

For worker efficiency, the environment should be comfortable, prevent fatigue, facilitate communication, and not be harmful to health. Equipment should control temperature and humidity or provide spot cooling to avoid heat stress; have low noise levels; provide adequate lighting; and control noxious and toxic fumes.

GENERAL REQUIREMENTS

Typical conditions of temperature, relative humidity, and specific filtration requirements for the storage, manufacture, and processing of various commodities are listed in Table 1. Requirements for a specific application may vary. Improvements in processes and increased knowledge may cause further variation; thus systems should be flexible to meet future requirements.

Inside temperature, humidity, filtration levels, and allowable variations should be established by agreement with the owner. Conditions may be limited either to heating and maximum humidity or to full air conditioning. A compromise between product or process conditions and comfort conditions may optimize quality and production costs.

A work environment that allows a worker to perform assigned duties without fatigue because of high or low temperatures and exposure to harmful airborne contaminants allows better, continued performance. Consequently, it may also improve worker morale and reduce absenteeism.

PROCESS AND PRODUCT REQUIREMENTS

Requirements for a process or product may be classified to control one or more factors such as (1) regain; (2) rate of chemical reactions; (3) rate of biochemical reactions; (4) rate of crystallization; (5) product accuracy and uniformity; (6) corrosion, rust, and abrasion of highly polished surfaces; (7) static electricity; and (8) air cleanliness and product formability.

Regain

In the manufacture or processing of hygroscopic materials such as textiles, paper, wood, leather, tobacco, and foodstuffs, air temperature and relative humidity have a marked influence on production rate, product mass, strength, appearance, and quality.

Moisture in vegetable or animal materials (and some minerals) reaches equilibrium content (by *regain*) with the moisture of the surrounding air. Regain is defined as the percentage of absorbed moisture in a material compared to its bone-dry mass. If a material sample with a mass of 110 g has a mass of 100 g after a thorough drying under standard conditions of 220 to 230°F, the mass of absorbed moisture is 10 g—10% of the sample's bone-dry mass. The regain, therefore, is 10%.

Table 2 lists typical values of regain for materials at 75°F in equilibrium at various relative humidities. Temperature change affects the rate of absorption or drying, which generally varies with the nature of the material, its thickness, and its density. Sudden temperature changes cause a slight regain change, even with fixed relative humidity; but the effect of temperature compared to relative humidity is comparatively small.

Hygroscopic Materials

In absorbing moisture from the air, hygroscopic materials deliver sensible heat to the air equal to the latent heat of the absorbed moisture. Moisture gains or losses by materials in processes are usually quite small, but if significant, the amount of liberated heat should be included in the load estimate. Actual values of regain should be obtained for a particular application. Manufacturing economy requires regain to be maintained at a level suitable for rapid and satisfactory manipulation. Uniformity allows high-speed machinery to operate with minimum loss.

Conditioning and Drying

Materials may be exposed to humidities desirable for treatment simultaneously with manufacture or processing, or they may be treated separately after conditioning and drying in special enclosures. Conditioning removes or adds hygroscopic moisture. Drying removes both hygroscopic moisture and free moisture in excess of that in equilibrium. Free moisture may be removed by evaporation, physically blowing it off, or by other means.

Drying and conditioning may be combined to remove moisture and accurately regulate final moisture content, for example in some textile products and tobacco. Conditioning or drying is frequently a continuous process of conveying the material through a tunnel and subjecting it to various controlled atmospheric conditions. Chapter 7 of the 1988 ASHRAE *Handbook—Equipment* describes dehumidification and pressure-drying equipment.

Rate of Chemical Reactions

Some processes require temperature and humidity control to regulate chemical reactions. In rayon manufacture, pulp sheets are conditioned, cut to size, and passed through a mercerizing process, where temperature controls the rate of reaction directly while relative humidity maintains a solution of constant strength and constant rate of surface evaporation.

The oxidizing process in drying varnish depends on temperature. Desirable temperatures vary with the type of varnish. High

The preparation of this chapter is assigned to TC 9.2, Industrial Air Conditioning.

Table 1 Temperatures and Humidities for Industrial Air Conditioning

Process	Dry Bulb (°F)	rh (%)
ABRASIVE		
Manufacture	79	50
CERAMICS		
Refractory	110 to 150	50 to 90
Molding room	80	60 to 70
Clay storage	60 to 80	35 to 65
Decalcomania production	75 to 80	48
Decorating room	75 to 80	48

Use high-efficiency filtration in decorating room. To minimize the danger of silicosis in other areas, a dust-collecting system or the proper level of mechanical filtration may be required.

Process	Dry Bulb (°F)	rh (%)
CEREAL		
Packaging	75 to 80	45 to 50
DISTILLING		
Storage		
Grain	6	35 to 40
Liquid yeast	32 to 33	—
General manufacturing	60 to 75	45 to 60
Aging	65 to 72	50 to 60

Low humidity and dust control are important where grains are ground. Use high-efficiency filtration for all areas to prevent mold spore and bacteria growth. Use ultrahigh efficiency filtration where bulk flash pasteurization is performed.

Process	Dry Bulb (°F)	rh (%)
ELECTRICAL PRODUCTS		
Electronics and X-ray		
Coil and transformer winding	72	15
Semiconductor assembly	68	40 to 50
Electrical instruments		
Manufacture and laboratory	70	50 to 55
Thermostat assembly and calibration	75	50 to 55
Humidistat assembly and calibration	75	50 to 55
Small mechanisms		
Close tolerance assembly	72*	40 to 45
Meter assembly and test	75	60 to 63
Switchgear		
Fuse and cutout assembly	73	50
Capacitor winding	73	50
Paper storage	73	50
Conductor wrapping with yarn	75	65 to 70
Lightning arrester assembly	68	20 to 40
Thermal circuit breakers assembly and test	75	30 to 60
High-voltage transformer repair	79	5
Water wheel generators		
Thrust runner lapping	70	30 to 50
Rectifiers		
Processing selenium and copper oxide plates	73	30 to 40

*Temperature to be held constant.

Dust control is essential in these processes. Minimum control requires medium-efficiency filters. Degree of filtration depends on the type of function in the area. Smaller tolerances and miniature components suggest high-efficiency filters.

Process	Dry Bulb (°F)	rh (%)
FLOOR COVERING		
Linoleum		
Mechanical oxidizing of linseed oil*	90 to 100	
Printing	80	
Stoving process	160 to 250	

*Precise temperature control required.

Air filtration is recommended for the stoving process.

Process	Dry Bulb (°F)	rh (%)
FOUNDRIES*		
Core making	60 to 70	
Mold making		
Bench work	60 to 70	
Floor work	55 to 65	
Pouring	40	
Shakeout	40 to 50	
Cleaning room	55 to 65	

*Winter dressing room temperatures. Spot coolers are sometimes used in larger installations.

In mold making, provide hoods at transfer points with wet-collector dust removal system. Use 600 to 800 cfm per hood.

In shakeout room, provide hoods with wet-collector dust removal system. Exhaust 400 to 500 cfm in grate area. Room ventilators are generally not effective.

In cleaning room, provide hoods for grinders and cleaning equipment with dry cyclones or bag-type collectors. In core making, oven and adjacent cooling areas require fume exhaust hoods. Pouring rooms require two-speed powered roof ventilators. Design for minimum of 2 cfm/ft^2 floor area at low speed. Shielding is required to control radiation from hot surfaces. Proper introduction of air minimizes preheat requirements.

Process	Dry Bulb (°F)	rh (%)
FUR		
Drying	110	—
Shock treatment	18 to 20	—
Storage	40 to 50	55 to 65

Shock treatment or eradication of any insect infestations requires lowering the temperature to 18 to 20°F for 3 to 4 days, then raising it to 60 to 70°F for 2 days, then lowering it again for 2 days and raising it to the storage temperature.

Furs remain pliable, oxidation is reduced, and color and luster are preserved when stored at 40 to 55°F.

Mold growth is prevalent with humidities above 80%, while hair splitting is common where humidity is lower than 55%.

Process	Dry Bulb (°F)	rh (%)
GUM		
Manufacturing	77	33
Rolling	68	63
Stripping	72	53
Breaking	73	47
Wrapping	73	58
LEATHER		
Drying	68 to 125	75
Storage, winter room temperature	50 to 60	40 to 60

After leather is moistened in preparation for rolling and stretching, it is placed in an atmosphere held at room temperature with a relative humidity of 95%.

Leather is usually stored in warehouses without temperature and humidity control. However, it is necessary to keep humidity sufficiently low to prevent mildew. Air filtration is recommended for fine finish.

Process	Dry Bulb (°F)	rh (%)
LENSES (OPTICAL)		
Fusing	75	45
Grinding	80	80
MATCHES		
Manufacture	72 to 73	50
Drying	70 to 75	60
Storage	60 to 63	50

Water evaporates with the setting of the glue. The amount of water evaporated is 18 to 20 lb per million matches. The match machine turns out about 750,000 matches per hour.

Table 1 Temperatures and Humidities for Industrial Air Conditioning (*Concluded*)

Process	Dry Bulb (°F)	rh (%)
MUSHROOMS		
Sweating-out period	110 to 140	—
Spawn added	60 to 72	nearly sat.
Growing period	50 to 60	80
Storage	32 to 35	80 to 85

As spawn starts to grow, it is necessary to cool the mushroom house abruptly by 18 °F in a 12-h period (approximately). Usually, this is the controlling factor in selecting refrigeration equipment, unless portable equipment is available.

Because of the deterioration of ferrous metals during the sweating-out period, ductwork is usually of wood.

In spawn rooms, high-efficiency (93 to 97% atmospheric dust spot) filters are required for controlling mold spores and bacteria.

Heat emission is 4 Btu/h · ft^2 of growing surface.

Ventilation is 10 ft^3/h per ft of growing surface.

Process	Dry Bulb (°F)	rh (%)
PAINT APPLICATION		
Lacquers: Baking	300 to 360	—
Oils paints: Paint spraying	60 to 90	80

The required air filtration efficiency depends on the painting process. On fine finishes, such as car bodies, high-efficiency filters are required for the outdoor air supply. Other products may require only low- or medium-efficiency filters.

Makeup air must be preheated. Spray booths must have 100 fpm face velocity if spraying is performed by humans; lower air quantities can be used if robots perform spraying. Ovens must have air exhausted to maintain fumes below explosive concentration. Equipment must be explosion-proof.

Process	Dry Bulb (°F)	rh (%)
PHOTO STUDIO		
Dressing room	72 to 74	40 to 50
Studio (camera room)	72 to 74	40 to 50
Film darkroom	70 to 72	45 to 55
Print darkroom	70 to 72	45 to 55
Drying room	90 to 100	35 to 45
Finishing room	72 to 75	40 to 55
Storage room (b/w film and paper)	72 to 75	40 to 60
Storage room (color film and paper)	40 to 50	40 to 50
Motion picture studio	72	40 to 55

The above data pertain to average conditions. In some color processes, elevated temperatures as high as 105 °F are used, and a higher room temperature is expected.

Conversely, ideal storage conditions for color materials necessitate refrigerated or deep-freeze temperature to ensure quality and color balance when long storage times are anticipated.

Heat liberated during printing, enlarging, and drying processes is removed through an independent exhaust system, which also serves the lamp houses and drier hoods. All areas except finished film storage require a minimum of medium-efficiency range filters.

Process	Dry Bulb (°F)	rh (%)
PLASTICS		
Manufacturing areas		
Thermosetting molding compounds	80	25 to 30
Cellophane wrapping	75 to 80	45 to 65

In manufacturing areas where plastic is exposed in the liquid state or molded, high-efficiency filters may be required. Dust collection and fume control are essential.

Process	Dry Bulb (°F)	rh (%)
PLYWOOD		
Hot pressing (resin)	90	60
Cold pressing	90	15 to 25
RUBBER-DIPPED GOODS		
Manufacture	90	—
Cementing	80	25 to 30*
Dipping surgical articles	75 to 80	25 to 30*
Storage prior to manufacture	60 to 75	40 to 50*
Laboratory (ASTM Standard)	73.4	50*

*Dew point of air must be below evaporation temperature of solvent.

Solvents used in manufacturing processes are often explosive and toxic, requiring positive ventilation. Volume manufacturers usually install a solvent-recovery system.

Process	Dry Bulb (°F)	rh (%)
TEA		
Packaging	65	65

Ideal moisture content is 5 to 6% for quality and mass. Low-limit moisture content for quality is 4%.

Process	Dry Bulb (°F)	rh (%)
TOBACCO		
Cigar and cigarette making	70 to 75	55 to 65*
Softening	90	85 to 88
Stemming and stripping	75 to 85	70 to 75
Packing and shipping	73 to 75	65
Filler tobacco casing and conditioning	75	75
Filter tobacco storage and preparation	77	70
Wrapper tobacco storage and conditioning	75	75

*Relative humidity fairly constant with range as set by cigarette machine.

Before stripping, tobacco undergoes a softening operation.

relative humidity retards surface oxidation and allows internal gases to escape as the chemical oxidizers cure the varnish from within. A bubble-free surface is maintained, with a homogeneous film throughout.

Rate of Biochemical Reactions

Fermentation requires temperature and humidity control to regulate the rate of biochemical reactions. Yeast develops best at 80 °F in the dough room. A minimum 75% rh holds the dough surface open to allow carbon dioxide from fermentation to pass through. This environment produces a bread loaf with an even, fine texture, free of large voids.

Rate of Crystallization

Cooling rate determines the size of crystals formed from a saturated solution. Both temperature and relative humidity control the cooling rate and change the solution density by evaporation.

In the coating pans for pills, gum, or nuts, a heavy sugar solution is added to the tumbling mass. As the water evaporates, sugar crystals cover each piece. Blowing the proper quantity of air at the correct dry- and wet-bulb temperatures forms a smooth opaque coating. If cooling and drying are too slow, the coating is rough, translucent, and unsatisfactory in appearance; if they are too fast, the coating chips through to the interior.

Product Accuracy and Uniformity

In the manufacture of precision instruments, tools, and lenses, air temperature and cleanliness affect the quality of work. Close temperature control prevents expansion and contraction of the material where tolerances of manufacture are within 0.0002 in. Constant temperature is more important than temperature level; thus conditions are usually selected for personnel comfort and to prevent a surface moisture film.

Corrosion, Rust, and Abrasion

In the manufacture of metal articles, temperature and relative humidity kept sufficiently low prevent sweating of hands and

Table 2 Regain of Hygroscopic Materials
Moisture Content Expressed in Percent of Dry Mass of the Substance of Various Relative Humidities—Temperature 75 °F

Classification	Material	Description	Relative Humidity								
			10	20	30	40	50	60	70	80	90
Natural Textile Fibers	Cotton	Sea Island—roving	2.5	3.7	4.6	5.5	6.6	7.9	9.5	11.5	14.1
	Cotton	American—cloth	2.6	3.7	4.4	5.2	5.9	6.8	8.1	10.0	14.3
	Cotton	Absorbent	4.8	9.0	12.5	15.7	18.5	20.8	22.8	24.3	25.8
	Wool	Australian Merino—skein	4.7	7.0	8.9	10.8	12.8	14.9	17.2	19.9	23.4
	Silk	Raw chevennes—skein	3.2	5.5	6.9	8.0	8.9	10.2	11.9	14.3	18.3
	Linen	Table cloth	1.9	2.9	3.6	4.3	5.1	6.1	7.0	8.4	10.2
	Linen	Dry spun—yarn	3.6	5.4	6.5	7.3	8.1	8.9	9.8	11.2	13.8
	Jute	Average of several grades	3.1	5.2	6.9	8.5	10.2	12.2	14.4	17.1	20.2
	Hemp	Manila and sisal rope	2.7	4.7	6.0	7.2	8.5	9.9	11.6	13.6	15.7
Rayons	Viscose nitro-cellulose	Average skein	4.0	5.7	6.8	7.9	9.2	10.8	12.4	14.2	16.0
	Cuprammonium cellulose acetate		0.8	1.1	1.4	1.9	2.4	3.0	3.6	4.3	5.3
Paper	M.F. newsprint	Wood pulp—24% ash	2.1	3.2	4.0	4.7	5.3	6.1	7.2	8.7	10.6
	H.M.F. writing	Wood pulp—3% ash	3.0	4.2	5.2	6.2	7.2	8.3	9.9	11.9	14.2
	White bond	Rag—1% ash	2.4	3.7	4.7	5.5	6.5	7.5	8.8	10.8	13.2
	Comm. ledger	75% rag—1% ash	3.2	4.2	5.0	5.6	6.2	6.9	8.1	10.3	13.9
	Kraft wrapping	Coniferous	3.2	4.6	5.7	6.6	7.6	8.9	10.5	12.6	14.9
Miscellaneous Organic Materials	Leather	Sole oak—tanned	5.0	8.5	11.2	13.6	16.0	18.3	20.6	24.0	29.2
	Catgut	Racquet strings	4.6	7.2	8.6	10.2	12.0	14.3	17.3	19.8	21.7
	Glue	Hide	3.4	4.8	5.8	6.6	7.6	9.0	10.7	11.8	12.5
	Rubber	Solid tires	0.11	0.21	0.32	0.44	0.54	0.66	0.76	0.88	0.99
	Wood	Timber (average)	3.0	4.4	5.9	7.6	9.3	11.3	14.0	17.5	22.0
	Soap	White	1.9	3.8	5.7	7.6	10.0	12.9	16.1	19.8	23.8
	Tobacco	Cigarette	5.4	8.6	11.0	13.3	16.0	19.5	25.0	33.5	50.0
Foodstuffs	White bread		0.5	1.7	3.1	4.5	6.2	8.5	11.1	14.5	19.0
	Crackers		2.1	2.8	3.3	3.9	5.0	6.5	8.3	10.9	14.9
	Macaroni		5.1	7.4	8.8	10.2	11.7	13.7	16.2	19.0	22.1
	Flour		2.6	4.1	5.3	6.5	8.0	9.9	12.4	15.4	19.1
	Starch		2.2	3.8	5.2	6.4	7.4	8.3	9.2	10.6	12.7
	Gelatin		0.7	1.6	2.8	3.8	4.9	6.1	7.6	9.3	11.4
Miscellaneous Inorganic Materials	Asbestos fiber	Finely divided	0.16	0.24	0.26	0.32	0.41	0.51	0.62	0.73	0.84
	Silica gel		5.7	9.8	12.7	15.2	17.2	18.8	20.2	21.5	22.6
	Domestic coke		0.20	0.40	0.61	0.81	1.03	1.24	1.46	1.67	1.89
	Activated charcoal	Steam activated	7.1	14.3	22.8	26.2	28.3	29.2	30.0	31.1	32.7
	Sulfuric acid	H_2SO_4	33.0	41.0	47.5	52.5	57.0	61.5	67.0	73.5	82.5

keeps fingerprints, tarnish, or etching from the finished article. The salt and acid in body perspiration can cause corrosion and rust within a few hours. Manufacture of polished surfaces usually requires better-than-average air filtering to prevent surface abrasion. This is also true of steel-belted radial tire manufacturing.

Static Electricity

In processing light materials such as textile fibers and paper, and where explosive atmospheres or materials are present, humidity reduces static electricity, which is often detrimental to processing and extremely dangerous in explosive atmospheres. Static electricity charges are minimized when the air in contact with the material processing is 35% rh or higher. The power driving the processing machines in conversion to heat raises the temperature in the machines above that of the adjacent air, where humidity is normally measured. Room relative humidity may need to be 65% or more to maintain sufficiently high humidity requirements in the machines.

Air Cleanliness

Each application must be evaluated to determine the filtration needed to counter the adverse effects of (1) minute dust particles on the product or process, (2) airborne bacteria, and (3) other air contaminants such as smoke, radioactive particles, spores, and pollen. These effects include chemically altering production material, spoiling perishable goods, or clogging small openings in precision machinery.

Product Formability

Pharmaceutical tablets manufacture requires close control of humidity for optimum tablet forming.

EMPLOYEE REQUIREMENTS

Space conditions required by health and safety standards to avoid excess exposure to high temperatures and airborne contaminants are often established by the American Conference of Governmental Industrial Hygienists (ACGIH). In the United States, the National Institute of Occupational Safety and Health (NIOSH) does research and recommends guidelines for workspace environmental control. The Occupational Safety and Health Administration (OSHA) sets standards from these environmental control guidelines and enforces them through compliance inspection at industrial facilities. Enforcement may be delegated to a corresponding state agency.

Standards for safe levels of contaminants in the work environment or in air exhausted from facilities do not cover all contaminants encountered. Minimum safety standards and facility design criteria are available, however, from various U.S. Department of Health, Education, and Welfare (DHEW) agencies such as the National Cancer Institute, the National Institute of Health, and the Public Health Service (Centers for Disease Control). For radioactive substances, standards established by the U.S. Nuclear Regulatory Commission (NRC) should be followed.

Thermal Control Levels

The common thermal range in industrial plants that needs no specific control for the product or process is from 68 to 100 °F and 25 to 60% rh. For a more detailed analysis, work rate, air velocity, quantity of rest, and effects of radiant heat must be considered (see Chapter 8 of the 1989 ASHRAE *Handbook—Fundamentals*). To avoid stress to workers exposed to high work rates and hot temperatures, the ACGIH established guidelines to evaluate high-temperature air velocity and humidity levels in terms of heat stress (Dukes-Dobos and Henschel 1971).

Where a comfortable environment is the concern rather than avoiding heat stress, the thermal control range becomes more specific (McNall, *et al.* 1967). In still air, nearly sedentary workers (120-W metabolism) prefer 72 to 75 °F dry-bulb temperature with 20 to 60% rh, and they can detect a 2 °F change per hour. Workers at a high rate of activity (300-W metabolism) prefer 63 to 66 °F dry bulb with 20 to 50% rh, but they are less sensitive to temperature change (ASHRAE *Standard* 55-1981). An increase in air velocity over a worker increases cooling, so a high metabolic rate activity can be handled in this manner also (ASHRAE *Standard* 55-1981).

Contaminant Control Levels

Toxic materials are present in many industrial plants and laboratories. In such plants, the air-conditioning and ventilation systems must minimize human exposure to toxic materials. When these materials become airborne, the body readily absorbs them and their range expands greatly, thus exposing more people. Chapter 11 of the 1989 ASHRAE *Handbook—Fundamentals*, current OSHA regulations, and *Threshold Limit Values of Airborne Contaminants*, published by ACGIH, give guides to evaluating the health impact of contaminants.

In addition to being a health concern, gaseous flammable substances must also be kept below explosive concentrations. Acceptable concentration limits are 20 to 25% of the lower explosive limit of the substance. Chapter 11 of the 1989 ASHRAE *Handbook—Fundamentals* includes information on flammable limits and means of control.

Instruments are available to measure concentrations of common gases and vapors. For less common gases or vapors, the air is sampled by drawing it through an impinger bottle or by inertial impaction-type air samplers, which supports biological growth on a nutrient gel, and which permits subsequent enumeration after incubation.

Gases and vapors are found near acid baths and tanks holding process chemicals. Machine processes, plating operations, spraying, mixing, and abrasive cleaning operations generate dusts, fumes, and mists. Many laboratory procedures, including grinding, blending, homogenizing, sonication, weighing, dumping of animal bedding, and animal inoculation or intubation, generate aerosols.

DESIGN CONSIDERATIONS

To apply equipment, the required environmental conditions, both for the product and personnel comfort, must be known. Consultation with the owner establishes design criteria such as temperature and humidity levels, energy availability and opportunities to recover it, cleanliness, process exhaust details, location and size of heat-producing equipment, lighting levels, frequency of equipment use, load factors, frequency of truck or car loadings, and sound levels. Consideration must be given to separating dirty processes from manufacturing areas that require relatively clean air. If not controlled by physical barriers, hard-to-control contaminants—such as mists from presses and machining, or fumes and gases from welding—can migrate to areas of final assembly, metal cleaning, or printing, and cause serious problems.

Because of changing mean radiant temperature, insulation should be evaluated for initial and operating cost, saving of heating and cooling, elimination of condensation (on roofs in particular), and comfort. When high levels of moisture are required within buildings, the structure and air-conditioning system must prevent condensation damage to the structure and ensure a quality product. Proper selection of insulation type and thickness, proper placing of vapor retarders, and proper selection and assembly of construction components to prevent thermal short-circuiting prevents condensation. Chapters 20 and 21 in the 1989 ASHRAE *Handbook—Fundamentals* have further details.

Personnel engaged in some industrial processes may be subject to a wide range of activity levels for which a broad range of indoor temperature and humidity levels are desirable. Chapter 8 of the 1989 ASHRAE *Handbook—Fundamentals* addresses recommended indoor conditions at various activity levels.

If layout and construction drawings are not available, a complete survey of existing premises and a checklist for proposed facilities is necessary (Table 3).

New industrial buildings are commonly single-story with flat roofs and with ample heights to distribute utilities without interfering with process operation. A common mounting height for fluorescent fixtures is up to 12 ft, for high-output fluorescent fixtures up to 20 ft, and for high-pressure sodium or metal halide fixtures above 20 ft.

Lighting design considers light quality, degree of diffusion and direction, room size, mounting height, and economics. Illumination levels should conform to recommended levels of the Illuminating Engineering Society of North America. Air-conditioning systems can be located in the top of the building. However, the designs require coordination because air-handling equipment and ductwork compete for space with sprinkler systems, piping, structural elements, cranes, material-handling systems, electric wiring, and lights.

Operations within the building must also be considered. Production materials may be moved through outside doors and allow large amounts of outdoor air to enter. Some operations require close control of temperature, humidity, and contaminants. A time schedule of operation helps in estimating the heating and cooling load.

LOAD CALCULATIONS

Table 1 and specific product chapters discuss product requirements. Chapters 25 and 26 of the 1989 ASHRAE *Handbook—Fundamentals* cover load calculations for heating and cooling.

Solar and Transmission

The solar load on the roof is generally the largest perimeter load and is usually a significant part of the overall load. Roofs should be light colored to minimize solar heat gain. Wall loads are often insignificant, and most new plants have no windows in the manufacturing area, so a solar load through glass is not present. Large windows in old plants may be closed in.

Internal Heat Generation

The process, product, facility utilities, and employees generate internal heat. People, power or process loads, and lights are internal sensible loads. Of these, production machinery often creates the largest sensible load. The design should consider anticipated brake power, rather than connected motor loads. The lighting load

Table 3 Facilities Checklist

Construction
1. Single or multistory
2. Type and location of doors, windows, crack lengths
3. Structural design live loads
4. Floor construction
5. Exposed wall materials
6. Roof materials and color
7. Insulation type and thicknesses
8. Location of existing exhaust equipment
9. Building orientation

Use of Building
1. Product needs
2. Surface cleanliness required; level of acceptable airborne contamination
3. Process equipment: type, location, and exhaust requirements
4. Personnel needs, temperature levels, required activity levels, and special workplace requirements
5. Floor area occupied by machines and materials
6. Clearance above floor required for material-handling equipment, piping, lights, or air distribution systems
7. Unusual occurrences and their frequency, such as large cold or hot masses of material moved inside
8. Loading frequency and length of time of doors open for loading or unloading
9. Lighting, location, type, and capacity
10. Acoustical levels
11. Machinery loads, such as electric motors (size, diversity), large latent loads, or radiant loads from furnaces and ovens
12. Potential for temperature stratification

Design Conditions
1. Design temperatures—indoor and outdoor dry and wet bulb
2. Altitude
3. Wind velocity
4. Makeup air required
5. Indoor temperature, allowable variance
6. Indoor relative humidity, allowable variance
7. Outdoor temperature occurrence frequencies
8. Operational periods, one, two, or three
9. Waste heat availability, energy conservation
10. Pressurization required
11. Mass loads from the energy release of productive materials

Code and Insurance Requirements
1. State and local code requirements for ventilation rates and other conditions
2. Occupational health and safety requirements
3. Insuring agency requirements

Utilities Available and Required
1. Gas, oil, compressed air (pressure), electricity (characteristics), steam (pressure), water (pressure), wastewater, interior and site drainage
2. Rate structures for each utility
3. Potable water

is generally significant. Heat gain from people is usually negligible in process areas.

Heat from operating equipment is difficult to estimate. Approximate values can be determined by studying the load readings of the electrical substations that serve the area.

In most industrial facilities, the latent heat load is minimal, with people and outdoor air being the major contributors. In these areas, the sensible heat factor approaches 1.0. Some processes release large amounts of moisture, such as in a paper-machine room. This moisture, including its condensation on cold surfaces, must be managed.

Stratification Effect

The cooling load may be dramatically reduced in a work space that takes advantage of temperature stratification; that is, it establishes a stagnant blanket of air directly under the roof by keeping air circulation to a minimum. The convective component of high energy sources such as the roof, upper walls, and high-level lights have little impact on the cooling load in the lower occupied zones. A portion of heat generated by high energy sources in the occupied zone such as process equipment is not part of the cooling load. This is due to radiation and a buoyancy effect which carries energy to the upper strata. The amount of energy lost varies from 20 to 60%. The magnitude of reduction depends on the temperature of the source, other surfaces, air, building construction, and rate of air movement.

Supply- and return-air ducts should be as low as possible to avoid mixing the warm boundary layers. The location of supply-air diffusers generally establishes the boundary of the warmer stratified air. For areas with supply air quantities greater than 2 cfm/ft^2, the return air temperature is approximately that of the air entrained by the supply airstream and only slightly higher than that at the end of the throw. With lower air quantities, the return air is much warmer than the supply air at the end of the throw. The amount depends on the placement of return inlets relative to internal heat sources. The average temperature of the space is higher, thus reducing the effect of outside conditions on heat gain. Spaces with a high area-to-employee ratio adapt well to low quantities of supply air and to spot cooling. For design specifics, refer to Chapter 25, Local Comfort Ventilation section.

Makeup Air

Makeup air, which has been filtered, heated, cooled, humidified, or dehumidified, is introduced to replace exhaust air, provide ventilation, and pressurize the building. For exhaust systems to function, air must enter the building by infiltration or the air-conditioning equipment. The space air-conditioning system must be large enough to heat or cool the outside air required to replace the exhaust. Cooling or heat recovery from exhaust air to makeup air can substantially reduce the outdoor air load.

Exhaust from air-conditioned buildings should be kept to a minimum by proper hooding or relocation of exhausted processes to areas not requiring air conditioning. Frequently, excess makeup air is provided to pressurize the building slightly, thus reducing infiltration, flue downdrafts, or ineffective exhaust under certain wind conditions. Makeup air and exhaust systems can be interlocked so that outdoor makeup air can be reduced as needed, or makeup air units may be changed from outdoor air heating to recirculated air heating when process exhaust is off.

In some facilities, outdoor air is required for ventilation because of the function of the space. Recirculation of air is avoided to reduce harmful gas concentrations, airborne bacteria, or air-carrying radioactive substances. Ventilation rates for human occupancy should be determined by ASHRAE *Standard* 62 and applicable codes.

Outdoor air dampers of industrial air-conditioning units should handle 100% supply air, so modulating the dampers in winter can satisfy room temperature. The infiltration of outdoor air often creates considerable load in industrial buildings because of poorly sealed walls and roofs. Infiltration should be minimized to conserve energy.

Fan Heat

The heat from air-conditioning return-air fans or supply fans goes into the refrigeration load. This energy is not part of the room sensible heat, except for supply fans downstream of the conditioning apparatus.

SYSTEM AND EQUIPMENT SELECTION

Industrial air-conditioning equipment includes heating and cooling sources, air-handling and air-conditioning apparatus, filters, and an air distribution system. To provide low life-cycle cost, components should be selected and the system designed for long life and low maintenance and operating costs.

Systems may be (1) for heating only in cool climates, where worker activity level is thermally satisfied with ventilation air; (2) air washer systems, where high humidities are desired and where the climate requires cooling; or (3) heating and mechanical cooling, where temperature and/or humidity control are required by the process and where the activity level is too great to be satisfied by other means of cooling. All systems include air filtration appropriate to the contaminant control required.

A careful evaluation will determine the zones that require control, especially in large, high-bay areas where the occupied zone is a small portion of the space volume. ASHRAE *Standard* 55-1981 defines the occupied zone as 3 to 72 in. high and more than 24 in. from the walls.

Air-Handling Units

Air-handling units heat, cool, humidify, and dehumidify air that is distributed to the workspace, and can supply all or any portion of outdoor air so that in-plant contaminants do not become too concentrated. Units may be factory assembled or field constructed.

HEATING SYSTEMS

Panel Heating

In industrial buildings, floor heating is often desirable, particularly in large high-bay buildings, garages, and assembly areas where workers must be near the floor, or where large or fluctuating outdoor air loads make maintenance of ambient temperature difficult. As an auxiliary to the main heating system, floors may be tempered to 65 or 70 °F by embedded hydronic systems, electrical resistance cables, or warm-air ducts.

The heating elements may be buried deep (6 to 18 in.) in the floor to permit slab warm-up at off-peak times, thus using the floor mass for heat storage to save energy during periods of high use. Floor heating may also be the primary or sole heating means, but floor temperatures above 85 °F are uncomfortable, so that such use is limited to small, well-insulated spaces.

Unit Heaters

Gas, oil, electricity, hot water, or steam unit heaters with propeller fans or blowers, are used for spot heating areas or are arranged in multiples for heating an entire building. Temperatures can be varied by individual thermostatic control. Unit heaters are located so the discharge (or throw) will reach the floor and flow adjacent to and parallel with the outside wall. They are spaced so that the discharge of one heater is just short of the next heater, thus producing a ring of warm air moving peripherally around the building. In industrial buildings with heat-producing processes, much heat stratifies in high-bay areas. In large buildings, additional heaters should be placed in the interior so that their discharge reaches the floor to reduce stratification. Downblow unit heaters in high bays and large areas may have a revolving discharge. Chapter 27 in the 1988 ASHRAE *Handbook—Equipment* includes more detail.

Gas- and oil-fired unit heaters should not be used where corrosive vapors are present. Unit heaters function well with regular maintenance and periodic cleaning in dusty or dirty conditions. Propeller fans generally require less maintenance than centrifugal fans. Centrifugal fans or blowers usually are required if heat is to be distributed to several areas. Gas- and oil-fired unit heaters require proper venting.

Ducted Heaters

Ducted heaters include large direct- or indirect-fired heaters, door heaters, and heating and ventilating units, and generally have centrifugal fans.

Code changes and improved burners have led to increased use of direct-fired gas heaters (the gas burns in the air supplied to the space) for makeup air heating. With correct interlock safety precautions of supply air and exhaust, no harmful effect from direct firing occurs. The high efficiency, high turndown ratio, and simplicity of maintenance make these units suitable for makeup air heating.

Common problems and solutions with ducted heaters in industrial applications are:

Steam Coil Freezeup. Steam-distributing type (sometimes called nonfreeze) coils, face-and-bypass control with steam valve wide open below 35 °F entering air; free condensate drainage; a thermostat in exit stops airflow when it falls below 40 °F.

Hot Water Coil Freezeup. Adequate circulation through drainable coil at all times; a thermostat in air and in water leaving the coil stop airflow under freezing conditions.

Temperature Override. Because of wiping action on coil with face-and-bypass control, zoning control is poor. Carefully locate face damper and use room thermostat to reset discharge air temperature controller.

Bearing Failure. Follow bearing manufacturers' recommendations in application and lubrication.

Insufficient Air Quantity. Require capacity data based on testing; keep forward-curved blade fans clean or use backward-inclined blade fans; keep filters clean.

Door Heating

Unit heaters and makeup air heaters commonly temper outdoor air that enters the building at open doors. These door heaters may have directional outlets to offset the incoming draft or may resemble a vestibule where air is recirculated.

Unit heaters successfully heat air at small doors open for short periods. They temper the incoming outdoor air through mixing and quickly bring the space to the desired temperature after the door is closed. The makeup air heater should be applied as a door heater in buildings where the doors are large (those that allow railroad cars or large trucks to enter) and open for long periods. They are also needed in facilities not tightly constructed or with a sizable negative pressure. These units help pressurize the door area, mix the incoming cold air and temper it, and bring the area quickly back to the normal temperature after the door is closed.

Often, door heater nozzles direct heated air at the top or down the sides of a door. Doors that create large, cold drafts when open can best be handled by introducing air in a trench at the bottom of the door. When not tempering cold drafts through open doors, some door heaters direct heated air to nearby spots. When the door is closed, it can switch to a lower output temperature.

The door heating units that resemble a vestibule operate with air flowing down across the opening and recirculating from the bottom, which helps reduce cold drafts across the floor. This type of unit is effective on doors routinely open and no higher than 10 ft.

Infrared

High-intensity infrared heaters (gas, oil, or electric) transmit heat energy directly to warm the occupants, floor, machines, or other building contents, without appreciably warming the air. Some air heating occurs by convection from objects warmed by the infrared. These units are classed as near- or far-infrared heaters, depending on the closeness of wavelength to visible light. Near infrared heaters emit a substantial amount of visible light.

Both vented and unvented gas-fired infrared heaters are available as individual radiant panels, or as a continuous radiant pipe with burners 15 to 30 ft apart and an exhaust vent fan at the end of the pipe. Unvented heaters require exhaust ventilation to remove flue products from the building, or moisture will condense on the roof and walls. Insulation reduces the exhaust required. Additional information on both electric and gas infrared is given in Chapter 29 of the 1988 ASHRAE *Handbook—Equipment*.

Infrared heaters are used in:

1. High-bay buildings where the heaters are usually mounted 10 to 30 ft above the floor, along outside walls, and tilted to direct maximum radiation to the floor. If the building is poorly insulated, the controlling thermostat should be shielded to avoid influence from the radiant effect of the cold walls.
2. Semiopen and outdoor areas, where people can comfortably be heated directly and objects can be heated to avoid condensation.
3. Loading docks for snow and ice control by strategic placement of near-infrared heaters.

COOLING SYSTEMS

Common cooling systems include refrigeration equipment, evaporative coolers, and high-velocity ventilating air.

For manufacturing operations, particularly heavy industry where mechanical cooling cannot be economically justified, evaporative cooling systems often provide good working conditions. If the operation requires heavy physical work, spot cooling by ventilation, evaporative cooling, or refrigerated air can be used. To minimize summer discomfort, high outdoor air ventilation rates may be adequate in some hot process areas. In all these operations, a mechanical air supply with good distribution is needed.

Refrigerated Cooling Systems

The refrigeration cooling source may be located at a central equipment area, or smaller packaged cooling systems may be placed near or combined with each air-handling unit. Central mechanical equipment uses positive displacement, centrifugal, or absorption refrigeration to chill water. Pumped through unit cooling coils, the chilled water absorbs heat, then returns to the cooling equipment.

Central system condenser water rejects heat through a cooling tower. The heat may be transferred to other sections of the building where heating or reheating is required, and the cooling unit becomes a heat recovery heat pump. Refrigerated heat recovery is particularly advantageous in buildings with a simultaneous need for heating exterior sections and cooling interior sections.

When interior spaces are cooled with a combination of outdoor air and chilled water obtained from reciprocating or centrifugal chillers with heat recovery condensers, hot water at temperatures up to 110°F is readily available. Large quantities of air at room temperature, which must be exhausted because of contaminants, can first be passed through a chilled water coil to recover heat. Heating or reheating obtained by refrigerated heat recovery occurs at a COP approaching 4, regardless of outdoor temperature, and can save considerable energy.

Mechanical cooling equipment should be selected in multiple units to match its reponse to load fluctuation and allow equipment maintenance during non-peak operation. Small packaged refrigeration equipment commonly uses positive displacement (reciprocating or screw) compressors with air-cooled condensers. These units usually provide up to 207 tons of cooling. Since equipment is often on the roof, the condensing temperature may be affected by warm ambient air, often 10 to 20°F higher than the design outdoor air temperature. In this type of system, the cooling coil may receive refrigerant directly.

The Safety Code for Mechanical Refrigeration, ASHRAE *Standard* 15-1989, limits the type and quantity of refrigerant in direct air-to-refrigerant exchangers. Fins on the air side improve heat transfer, but they increase the pressure drop through the coil, particularly as they get dirty.

Evaporative Cooling Systems

Evaporative cooling systems may be evaporative coolers or air washers. Evaporative coolers have water sprayed directly on a wetted surface through which air moves. An air washer recirculates water, and the air flows through a heavily misted area. Water atomized in the airstream evaporates and the water cools the air. Refrigerated water simultaneously cools and dehumidifies the air.

Evaporative cooling offers energy conservation opportunities, particularly in intermediate seasons. In many industrial facilities, evaporative cooling controls both temperature and humidity. In these systems, the sprayed water is normally refrigerated, and a reheat coil is often used. Temperature and humidity of the exit airstream may be controlled by varying the temperature of the chilled water and the reheat coil and by varying the quantity of air passing through the reheat coil with a dewpoint thermostat.

Care must be taken that accumulation of dust or lint does not clog the nozzles or evaporating pads of the evaporative cooling systems. It may be necessary to filter the air before the evaporative cooler. Fan heat, air leakage through closed dampers, and the entrainment of room air into the supply airstream all affect design.

AIR FILTRATION SYSTEMS

Air filtration systems remove contaminants from air supplied to or exhausted from building spaces. Supply air filtration (most frequently on the intake side of the air-conditioning apparatus) removes particulate contamination that may foul heat transfer surfaces, contaminate products, or present a health hazard to people, animals, or plants. Gaseous contaminants must sometimes be removed to prevent exposing personnel to harm or odor. The supply airstream may consist of air recirculated from building spaces and/or outdoor air for ventilation or exhaust air makeup. Return air with a significant potential for carrying contaminants should be recirculated only when filtered enough to minimize personnel exposure.

The supply air filtration system usually includes a collection medium or filter, media-retaining device or filter frame, and a filter house or plenum. The filter media are the most important components of the system; a mat of randomly distributed small diameter fibers is commonly used.

Depending on fiber material, size, density, and arrangement, fibrous filters have a wide range of performance. Low-density filter media with relatively large diameter fibers remove large particles, such as lint. These roughing filters collect a large percentage by mass of the particulates, but are ineffective in reducing the total particle concentration. The fibers are sometimes coated with an adhesive to reduce particle re-entrainment.

Small fiber, high-density filter media effectively collect essentially all particulates. Ultrahigh efficiency filters reduce total particle concentration by more than 99.9%. As filter efficiency increases, so does resistance to airflow, with typical pressure drop ranging from 0.05 to 1 in. of water. Conversely, dust-holding capacity decreases with increasing filter efficiency, and fibrous filters should not be used for dust loading greater than 100 $\mu g/ft^3$ of air. For more discussion of particulate filtration systems, refer to Chapter 10 of the 1988 ASHRAE *Handbook— Equipment*.

Exhaust Air Filtration Systems

Exhaust air systems are either (1) general systems that remove air from large spaces or (2) local systems that capture aerosols, heat, or gases at specific locations within a room and transport them to where they can be collected (filtered), inactivated, or safely discharged to the atmosphere. The air in a general system usually requires minimal treatment before discharging to the atmosphere. The air in local exhaust systems can sometimes be safely dispersed to the atmosphere, but sometimes contaminants must be removed so that the emitted air meets air quality standards.

Many types of contamination collection or inactivation systems are applied in exhaust air emission control. Fabric bag filters, glass-fiber filters, venturi scrubbers, and electrostatic precipitators all collect particulates. Packed bed or sieve towers can absorb toxic gases. Activated carbon columns or beds, often with oxidizing agents, are frequently used to absorb toxic or odorous organics and radioactive gases.

Outdoor air intakes should be carefully located to avoid recirculation of contaminated exhaust air. Because wind direction, building shape, and the location of the effluent source strongly influence concentration patterns, exact patterns are not predictable.

Air patterns resulting from wind flow over buildings are discussed in Chapter 14 of the 1989 ASHRAE *Handbook—Fundamentals*. The leading edge of a roof interrupts smooth airflow, resulting in reduced air pressure at the roof and on the lee side. To prevent fume damage to the roof and roof-mounted equipment, and to keep fumes from the building air intakes, fumes must be discharged either through (1) vertical stacks terminating above the turbulent air boundary or (2) short stacks with a velocity high enough to project the effluent through the boundary into the undisturbed air passing over the building. A high vertical stack is the safest and simplest solution to fume dispersal.

Contaminant Control

In addition to maintaining thermal conditions, air-conditioning systems should control contaminant levels to provide (1) a safe and healthy environment, (2) good housekeeping, and (3) quality control for the processes. Contaminants may be gases, fumes, mists, and airborne particulate matter. They may be created by a process within the building or found in the outside air.

Contamination can be controlled by (1) preventing the release of aerosols or gases into the room environment and (2) diluting room air contaminants. If the process cannot be enclosed, it is best to capture aerosols or gases near their source of generation with local exhaust systems that include an enclosure or hood, ductwork, fan, motor, and exhaust stack.

Dilution controls contamination in many applications but may not provide uniform safety for personnel within a space (West 1977). High local concentrations of contaminants can exist within a room, even though the overall dilution rate is quite high. Further, if tempering of outdoor air is required, high energy costs can result from the increased airflow required for dilution.

Exhaust Systems

The exhaust system draws the contaminant away from its source and removes it from the space. An exhaust hood that surrounds the point of generation contains the contaminant as much as is practical. The contaminants are transported through ductwork from the space, cleaned as required, and exhausted to the atmosphere.

The suction air quantity in the hood is established by the velocities required to contain the contaminant.

Design values for average and minimum face velocities are a function of the characteristics of the most dangerous material that the hood is expected to handle. Minimum values are prescribed in codes for exhaust systems. Contaminants with greater mass require higher face velocities for their control.

Properly sized ductwork keeps the contaminant flowing. This requires very high velocities for heavy materials. The selection of materials and the construction of exhaust ductwork and fans depend on the nature of the contaminant, the ambient temperature, the lengths and arrangement of duct runs, the method of hood fan operation, and the flame and smoke spread rating.

Exhaust systems remove chemical gases, vapors, or smokes from acids, alkalis, solvents, and oils. Care must be taken to minimize the following:

1. **Corrosion**, which denotes destruction of metal by chemical or electrochemical action; commonly used reagents in laboratories are hydrochloric, sulfuric, and nitric acid, singly or in combination, and ammonium hydroxide. Common organic chemicals include acetone, benzene, ether, petroleum, chloroform, carbon tetrachloride, and acetic acid.
2. **Dissolution**, which denotes a dissolving action. Coatings and plastics are subject to this action, particularly by solvent and oil fumes.
3. **Melting**, which can occur in certain plastics and coatings at elevated hood operating temperatures.

Low temperatures that cause condensation in ferrous metal ducts increase chemical destruction. Ductwork is less subject to attack when the runs are short and direct to the terminal discharge point. The longer the runs, the longer the period of exposure to fumes and the greater the degree of condensation. Horizontal runs allow moisture to remain longer than it can on vertical surfaces. Intermittent fan operation can contribute to longer periods of wetness (because of condensation) than continuous operation.

Maintenance of Components

All designs should allow ample room to clean, service, and replace any component quickly so that design conditions are affected as little as possible. Maintenance of refrigeration and heat-rejection equipment is essential for proper performance without energy waste.

For system dependability, water treatment is important. No air washer or cooling tower should be operated without water properly treated by specialists.

Maintenance of heating and cooling systems includes changing or cleaning system filters on a regular basis. Industrial applications are usually dirty, so frequent filter changing may be required. Dirt that lodges in ductwork and on forward-curved fan blades reduces air-handling capacity appreciably.

Fan and motor bearings require lubrication, and fan belts need periodic inspection. Infrared and panel systems usually require less maintenance than equipment with filters and fans, although gas-fired units with many burners require more attention than electric heaters.

The direct-fired makeup heater has a relatively simple burner requiring less maintenance than a comparable indirect-fired heater. The indirect oil-fired heater requires more maintenance than the comparable indirect gas-fired heater. With either type, the many safety devices and controls require periodic maintenance to ensure constant operation without nuisance cutout. Direct- and indirect-fired heaters should be inspected at least once a year.

Steam and hot water heaters have fewer maintenance requirements than comparable equipment having gas and oil burners. When used for makeup air in below-freezing conditions, however, the heaters must be correctly applied and controlled to prevent frozen coils.

REFERENCES

ACGIH. 1989. *Industrial ventilation, A manual of recommended practice*, 20th ed. American Conference of Governmental Industrial Hygienists, Cincinnati, OH.

AIHA. 1975. *Heating and cooling for men in industry*, 2nd ed. American Industrial Hygienist Association, Akron, Ohio.

ASHRAE. 1981. Thermal environmental conditions for human occupation. ANSI/ASHRAE *Standard* 55-1981.

ASHRAE. 1989. Ventilation for acceptable indoor air quality. ANSI/ASHRAE *Standard* 62-1989.

ASHRAE. 1989. Safety code for mechanical refrigeration. ASHRAE *Standard* 15-1989.

Azer, N.Z. 1982a. Design guidelines for spot cooling systems, Part 1—Assessing the acceptability of the environment. ASHRAE *Transactions* 88(2).

Azer, N.Z. 1982b. Design guidelines for spot cooling systems, Part 2—Cooling jet model and design procedure. ASHRAE *Transactions* 88(2).

Dukes-Dobos, F. and A. Henschel. 1971. *The modification of the WNGT Index for establishing permissible heat exposure limits in occupational work.* HEW, USPHS, ROSH joint publication TR-69.

Gorton, R.L. and Bagheri. 1987a. Verification of stratified air-conditioning design. ASHRAE *Transactions* 93(2):211-27.

Gorton, R.L. and Bagheri. 1987b. Performance characteristics of a system designed for stratified cooling operating during the heating season. ASHRAE *Transactions* 93(2):367-81.

Harstad, J., *et al.* 1967. Air filtration of submicron virus aerosols. *American Journal of Public Health* 57:2186-93.

IES. 1982. *Lighting handbook*, 6th ed. Illuminating Engineering Society of North America, New York.

McNall, P.E., J. Juax, F.H. Rohles, R.G. Nevins, and W. Springer. 1967. Thermal comfort (thermally neutral) conditions for three levels of activity. ASHRAE *Transactions* 73(1):I.3.1.

O'Connell, W.L. 1976. How to attack air-pollution control problems. *Chemical Engineering* (October), desktop issue.

West, D.L. 1977. Contaminant dispersion and dilution in a ventilated space. ASHRAE *Transactions* 83(1):125.

Whitby, K.T. and D.A. Lundgren. 1965. Mechanics of air cleaning. ASAE *Transactions* 8:3, 342. American Society of Agricultural Engineers, St. Joseph, MI.

Yamazaki, K. 1982. Factorial analysis on conditions affecting the sense of comfort of workers in the air-conditioned working environment. ASHRAE *Transactions* 88(1):241.

CHAPTER 13

ENCLOSED VEHICULAR FACILITIES

THIS chapter deals with the ventilation requirements for cooling, pollution control, and emergency smoke and temperature control for vehicular tunnels, rapid transit tunnels and stations, enclosed parking structures, and bus terminals. Also included is the design approach and type of equipment applied to these ventilation systems.

VEHICULAR TUNNELS

CONTROL BY DILUTION

All internal combustion engines produce exhaust gases that contain toxic compounds and smoke. To dilute the concentrations of obnoxious or dangerous contaminants to acceptable levels, vehicular tunnels require ventilation, which may be provided by natural means, traffic-induced piston effect, or mechanical equipment. The selected ventilation system should be the most economical, where construction and operating costs are concerned. Naturally ventilated and traffic-induced systems are considered adequate for tunnels of relatively short length and low traffic volume (or density). Long and heavily traveled tunnels should have mechanical ventilation systems.

The exhaust constituent of greatest concern is carbon monoxide (CO), because of its notorious asphyxiant nature. The ventilation system dilutes the CO content of the tunnel atmosphere to a safe and comfortable level. Tests and operating experience indicate that when CO has been properly diluted, the other dangerous and objectionable exhaust by-products are also diluted to acceptable levels. The section on bus terminals includes further information regarding diesel engine operation.

The ventilation discussed here will be incorporated as a permanent part of the finished tunnel and is intended primarily to serve the needs of the traveling public passing through the tunnel. Ventilation needed by workers during construction of the facility or while working in the finished tunnel is not covered. These ventilation requirements are specified in detail by state or local mining laws, industrial codes, or in the standards set by the U.S. Occupational Safety and Health Administration (OSHA).

ALLOWABLE CARBON MONOXIDE CONCENTRATIONS

In 1975, the U.S. Environmental Protection Agency (EPA) issued a supplement to their *Guidelines for Review of Environmental Statements for Highway Projects* which, until 1988, evolved into a design approach of a CO concentration of 125 ppm for a maximum of one hour exposure time for tunnels located at or below an altitude of 3280 ft (see also Minimum Ventilation Rate). In 1988, the EPA revised its recommendations for maximum CO levels in tunnels located at or below an altitude of 5000 ft to the following:

Max. 120 ppm for 15 min exposure time
Max. 65 ppm for 30 min exposure time
Max. 45 ppm for 45 min exposure time
Max. 35 ppm for 60 min exposure time

The new guidelines do not apply to existing tunnels.

Above an elevation of 3280 ft, the CO emission of vehicles is greatly increased, and human tolerance to CO exposure is reduced. For tunnels above 3280 ft, the designer should consult with medical authorities to establish a proper design value for CO concentrations. Unless specified otherwise, the material in this chapter refers to tunnels located at or below an altitude of 3280 ft.

CARBON MONOXIDE EMISSION

The CO content in exhaust gases of individual vehicles varies greatly, because of such factors as the age of the vehicle, carburetor adjustment, quality of fuel, engine power, level of vehicle maintenance, and the different driving habits of motorists. Despite these variables, several sources generally agree on average emission rates, although they apply only to vehicles without emission control devices. The bulk of the published data on CO emissions of controlled vehicles is not presently considered of sufficient accuracy for tunnel ventilation computations.

Due to the absence of suitable data, all suggested ventilation rates given in this chapter are derived using uncontrolled emission rates, but the results can be corrected, as indicated, to reflect the impact of emission control devices.

VENTILATING AIR REQUIRED

The ventilating system must have capacity enough to protect the traveling public during the most adverse and dangerous conditions, as well as during normal conditions. In addition to the problems of many uncontrollable variables, establishing air requirements is further complicated by an extensive number of possible vehicle combinations and traffic situations that could occur during the lifetime of the facility.

The preparation of this chapter is assigned to TC 5.9, Enclosed Vehicular Facilities.

Ventilating rates that meet the general criteria are computed in two parts. The first considers a lane of tunnel traffic composed entirely of passenger cars. The following assumptions are necessary:

1. A major traffic stoppage has occurred outside the tunnel, and traffic is blocked. This stoppage causes tunnel traffic to slow down from normal operating speeds and finally to stop with engines idling. When the stoppage is cleared, traffic is assumed to reverse the process, *i.e.*, idle first, then proceed at 6 mph, then up to 25 mph.
2. Although CO emission rates during acceleration and deceleration of vehicles are higher than at constant speed, the effect of speed changing is neglected. (The error introduced by this assumption will be offset by a 10% safety factor included in the computations.)
3. Traffic is assumed to move as a unit with spacing between vehicles remaining constant regardless of roadway grade.
4. Passenger cars are taken as 19 ft long (the usual length of a highway design vehicle), and spaces between cars at various speeds are shown in Table 1.

Table 1 Space Between Cars at Various Speeds

Speed, mph	Spacing, ft
25	72.8
20	54.4
15	36.0
10	17.6
5	9.0
Idle	4.0

From these computations, the governing ventilation rates for different roadway gradients from 6% downhill to 6% uphill are determined using the data plotted in Figure 1. In most cases, the critical traffic situation from a ventilation standpoint occurs when the lane is congested with vehicles moving at a speed of about 9.3 mph. On the steeper downgrades, the governing condition occurs when traffic is stopped and vehicles are bumper-to-bumper with engines idling. This condition calls for a constant ventilation rate regardless of roadway grade.

Figure 1 shows the results of a second set of computations made for a lane of tunnel traffic composed entirely of trucks and buses. About 40% of the vehicles are assumed to be diesel-powered, averaging 50 ft in length, and 60% are taken as the light- to medium-duty gasoline-powered trucks averaging 30 ft in length. The governing traffic conditions are similar to the passenger car lane except that on steeper upgrades, trucks traveling at a crawl speed of about 6 mph become the critical traffic condition.

In both cases, the ventilating rates given in Figure 1 are based on emission rates from vehicles not equipped with control devices and on an allowable concentration of CO not to exceed 125 ppm.

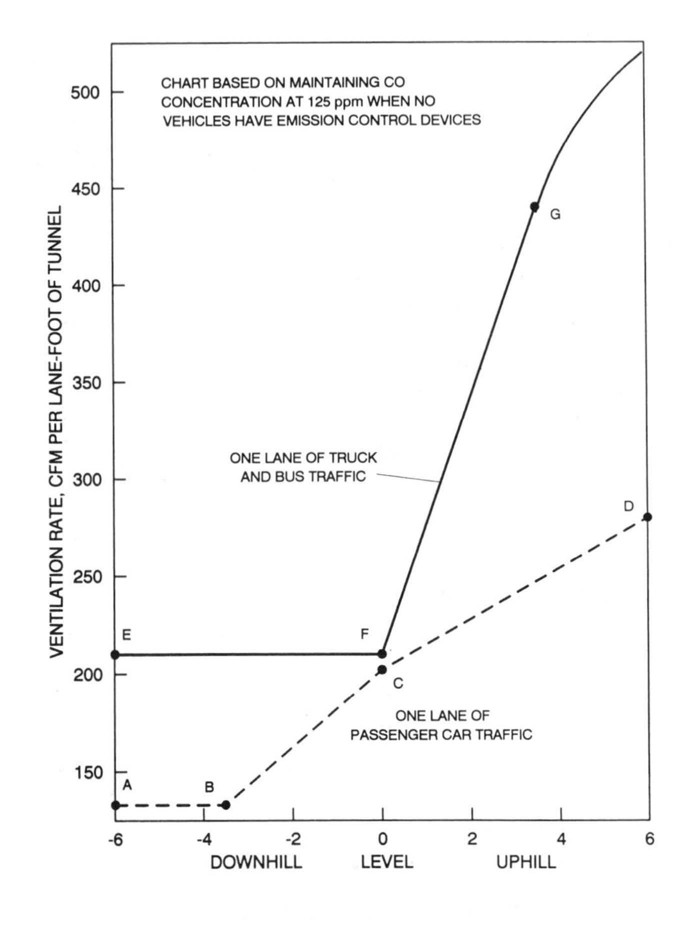

Key Values for Figure 1

Point	Grade	Ventilation Rate
A	−6.0	135
B	−3.5	135
C	zero	200
D	+6.0	280
E	−6.0	210
F	zero	210
G	+3.5	440

Fig. 1 Tunnel Ventilation Rates for Different Roadway Gradients

CONVERSION OF DIFFERENT TRAFFIC MIX

To correct the ventilation rate for emission control, the average emission factors for highway vehicles in Table 2 can be used; they are derived from EPA estimates for CO emission rate for 50,000 travel miles. These emission factors are for the vehicle population mix for the calendar year shown and not for vehicles of that model year only. The values are projected emission factors based (1) on actual test results of existing sources and control systems and (2) on projected values for future years based on required emission reductions, as stipulated in the present law (EPA emission standard for CO remains at 3.4 g/mi after the 1983 model year). The designer should be aware of any possible future revisions or corrections in the predicted factors and in emission control laws; actual emission rates versus predicted emission rates must also be evaluated.

Table 2 Emission Factors for Highway Vehicles

Calendar Year	Average Emission Factor (F)
1991	0.159
1992 and after	0.151

Converting to any desired calendar year may be accomplished with the following equation:

$$Q = VF \quad (1)$$

where

Q = converted ventilation rate
V = unconverted ventilation rate
F = emission factor for desired calendar year from Table 2

CONVERSION OF CARBON MONOXIDE CONCENTRATION

Converting ventilation rates to a CO level other than 125 ppm:

$$Q = V(125/C) \quad (2)$$

where

C = desired CO concentration, ppm

Adjustment of Ventilation Rate for Ambient CO Level

Ventilation rates shown in Figure 1 assume that the ventilating air contains little or no CO when it is introduced. If this is not the case, the ventilation rates may be adjusted by:

$$Q = 125V/(125 - E) \quad (3)$$

where

E = ambient CO level, ppm

Conversion of Truck Traffic Mix

The conversion constants given consider the size of the truck and the percentage of truck type in the traffic stream.

$$Q = V(1.22\,P_G + 0.85\,P_D) \quad (4)$$

where

P_G = ratio of gasoline-powered vehicles
P_D = ratio of diesel-powered vehicles

Minimum Ventilation Rate

In addition to the dilution of carbon monoxide, the ventilation system must provide sufficient ventilation for fire protection. Experience has shown that 100 cfm per lane foot of tunnel is sufficient for fire protection; however, an analysis of life safety and smoke control requirements could permit a reduction in this value.

Example 1. Calculate the volume flow rate of a single-bore tunnel with two lanes of unidirectional traffic. The tunnel, which will open in 1992, has the grades and dimensions shown below. It has a design level of 125 ppm and an ambient CO level of 5 ppm.

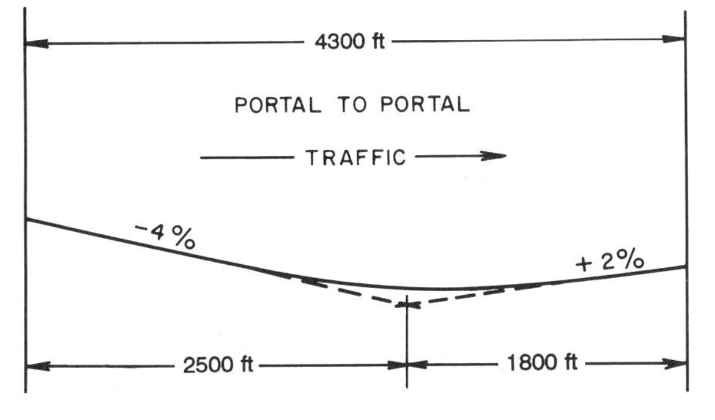

TUNNEL PROFILE

Solution (from Figure 1):

−4% grade	135 cfm per lane foot for cars	
	210 cfm per lane foot for trucks	
	345 cfm	
+2% grade	230 cfm per lane foot for cars	
	345 cfm per lane foot for trucks	
	575 cfm	

345 × 2500 = 862,000
575 × 1800 = 1,035,000
1,897,000 cfm

Adjusting for ambient CO level of 5.0 ppm [Equation (3)] and converting traffic mix for calendar year 1992 [Equation (1)]:

$$Q = VF[125/(125 - 5)]$$

From Table 2, F = 0.151

$$Q = 1{,}897{,}000 \times 0.151\,[125/(125-5)]$$
$$= 298{,}000 \text{ cfm}$$

Check for minimum ventilation rate:

For −4% grade: 345 × 0.151 (125/(125−5)] = 54 cfm
For +2% grade: 575 × 0.151 [125/(125−5)] = 90 cfm

Both grade requirements fall short of 100 cfm per lane foot; therefore, adequacy under fire conditions should be analyzed.

COMPUTER COMPUTATION PROGRAM

A computer computation program, TUNVEN, which solves the coupled one-dimensional steady-state tunnel aerodynamic and advection equations, is available. Using this program, predictions of the quasi-steady state longitudinal air velocities and the concentrations of carbon monoxide (CO), nitrogen oxides (NO_x) and total hydrocarbons (THC) may be obtained along a highway tunnel for a wide range of tunnel designs, traffic loads, and external ambient conditions. The program will model all common ventilation system types: natural, longitudinal, semitransverse and transverse. The program is available through the National Technical Information Service (NTIS).

VENTILATION SYSTEM TYPES

Natural Ventilation

Naturally ventilated tunnels rely chiefly on meteorological conditions to maintain a satisfactory environment within the tunnel. The piston effect of traffic provides additional airflow when the traffic is moving. The chief meteorological condition affecting environment is the pressure differential between two portals of a tunnel created by differences in elevation, ambient temperatures, or wind. Unfortunately, none of these factors can be relied on for continued, consistent results. A sudden change in wind direction or velocity can negate all these natural effects, including the piston effect. The sum total of all pressures must be of sufficient magnitude to overcome the tunnel resistance, which is influenced by tunnel length, coefficient of friction, hydraulic radius, and air density.

Airflow through a naturally ventilated tunnel can be portal-to-portal (Figure 2a) or portal-to-shaft (Figure 2b). Portal-to-portal flow functions best with unidirectional traffic, which produces a consistent, positive airflow. The air velocity within the roadway is uniform, and the contaminant concentration increases to a maximum at the exit portal. If adverse meteorological conditions occur, the velocity is reduced and the CO concentration is increased, as shown by the dashed line on Figure 2a. If bidirectional traffic is introduced into such a tunnel, further reductions in airflow result.

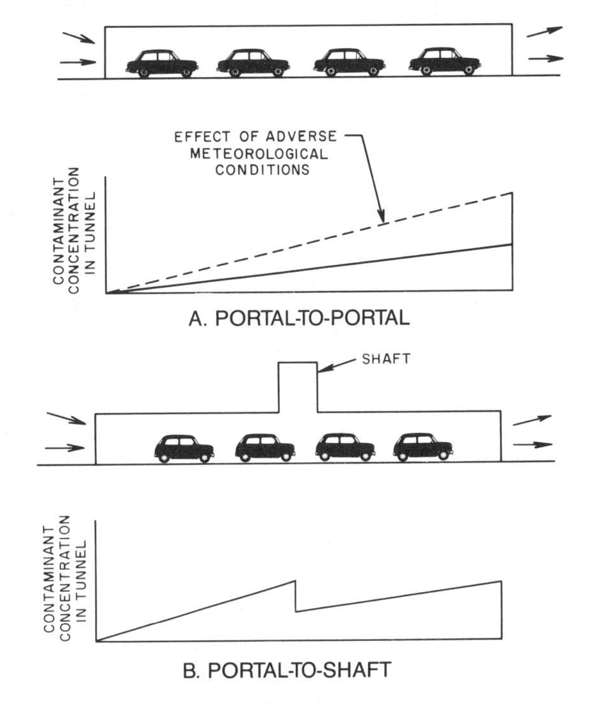

Fig. 2 Natural Ventilation versus Contaminant Concentration in Tunnel

The naturally ventilated tunnel with an intermediate shaft (Figure 2b) is best suited for bidirectional traffic. However, airflow through such a shafted tunnel is also at the mercy of the elements. The added benefit of the stack effect of the shaft depends on air and rock temperatures, wind, and shaft height. The addition of more than one shaft to a naturally ventilated tunnel is more of a disadvantage than an advantage, since a pocket of contaminated air can be trapped between the shafts and cause high contaminant levels.

Most naturally ventilated urban tunnels over 500 ft long require an emergency mechanical ventilation system to purge smoke and hot gases generated during an emergency and to remove stagnated, polluted gases during severe adverse meteorological conditions. Because of the aforementioned uncertainties, reliance on natural ventilation for all tunnels over 500 ft long should be thoroughly evaluated, especially the effect of adverse meteorological and operating conditions. This is particularly true for a tunnel with an anticipated heavy or congested traffic flow. If the natural ventilation is inadequate, consider installing a mechanical system with fans.

Mechanical Ventilation

The most appropriate mechanical ventilation systems for tunnels are longitudinal ventilation, semitransverse ventilation, and full transverse ventilation.

Longitudinal Ventilation. This applies to any system that introduces or removes air from the tunnel at a limited number of points, thus creating a longitudinal flow of air within the roadway. The injection-type longitudinal system has frequently been used in rail tunnels; however, it has also found application in vehicular tunnels. Air injected into the tunnel roadway at one end of the tunnel mixes with air brought in by the piston effect of the incoming traffic (see Figure 3a).

This system is most effective where traffic is unidirectional. The air velocity stays uniform throughout the tunnel, and the concentration of contaminants increases from zero at the entrance to a maximum at the exit. Adverse external atmospheric conditions can reduce the effectiveness of this system. The contaminant level increases at the exit portal as the airflow decreases or the tunnel length increases.

The longitudinal system with a fan shaft (see Figure 3b) is similar to the naturally ventilated system with a shaft, except that it provides a positive stack effect. Bidirectional traffic in a tunnel ventilated in this manner will cause a peak contaminant concentration at the shaft location. For unidirectional tunnels, however, the contaminant levels become unbalanced.

Another form of the longitudinal system has two shafts near the center of the tunnel: one for exhaust and one for supply (see Figure 3c). This arrangement reduces contaminant concentration in the second half of the tunnel. A portion of the air flowing in the roadway is replaced in the interaction at the shafts. Adverse wind conditions can reduce the airflow, which then causes the contaminant concentration to rise in the second half of the tunnel and short circuit flow from fan to exhaust.

In a growing number of tunnels, longitudinal ventilation is achieved with fans mounted at the tunnel ceiling (see Figure 3d). Such a system eliminates the space needed to house ventilation fans in the building; however, it may require a tunnel of greater height or width for the booster fans.

Standard longitudinal ventilation systems (excluding the booster fan system), with either supply or exhaust at a limited number of locations within the tunnel, are most economical because they require the least number of fans, place the least operating burden on these fans, and do not require distribution air ducts. As the length of the tunnel increases, however, the disadvantages of these systems, such as excessive air velocities in the roadway and smoke being drawn the entire length of the roadway

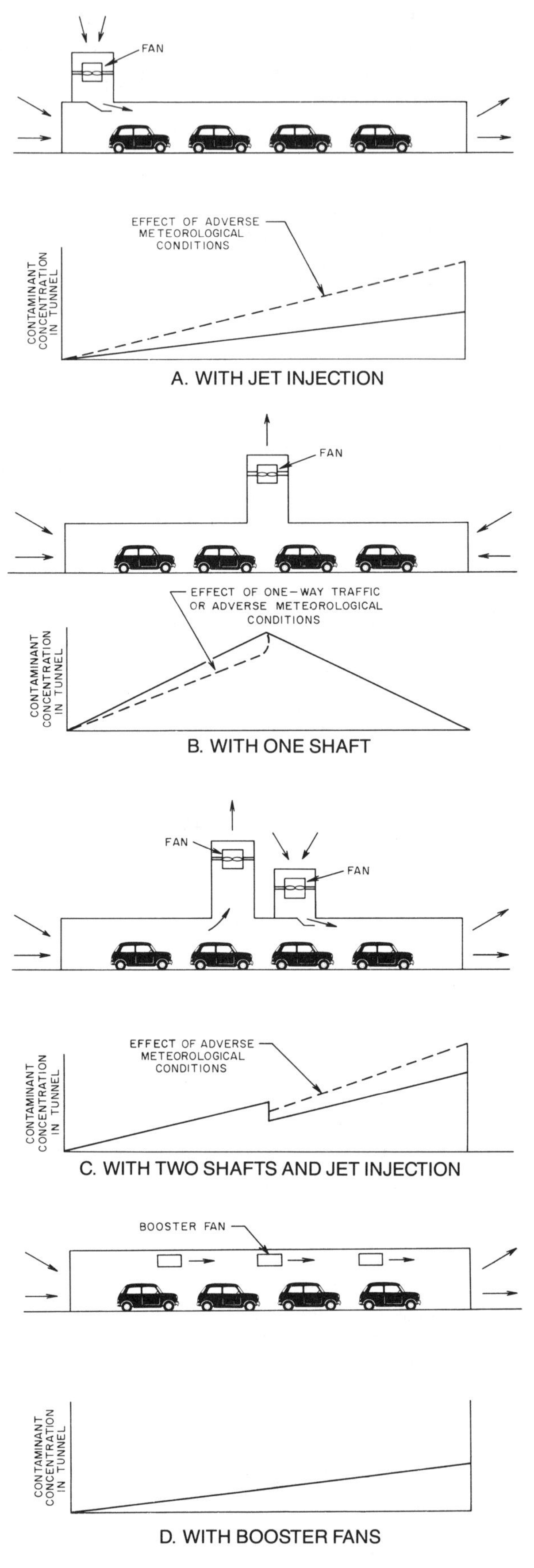

Fig. 3 Longitudinal Ventilation

during an emergency, become apparent. Uniform air distribution would alleviate these problems.

Semitransverse Ventilation. This system uniformly distributes or collects air throughout the length of a tunnel. The supply air version of the system (see Figure 4a) produces a uniform level of carbon monoxide throughout the tunnel because the air and the vehicle exhaust gases enter the roadway area at the same rate. In a tunnel with unidirectional traffic, additional airflow is generated within the roadway area.

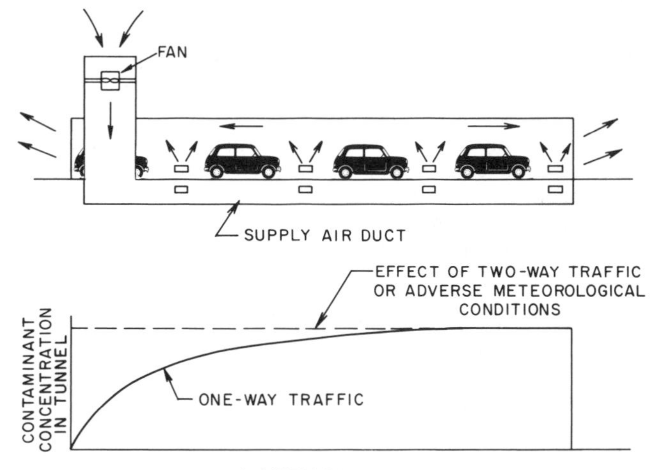

A. WITH SUPPLY DUCT

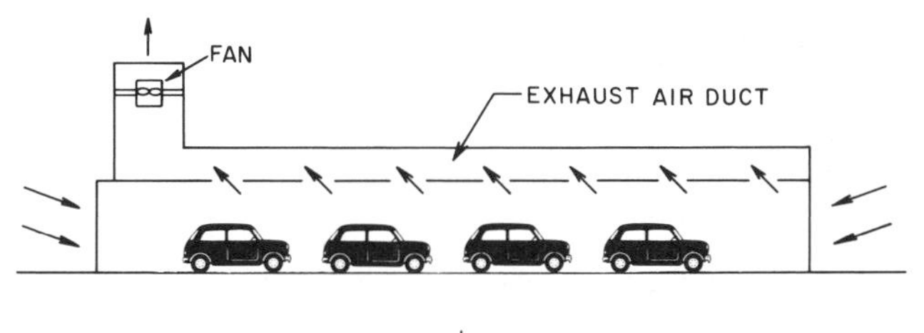

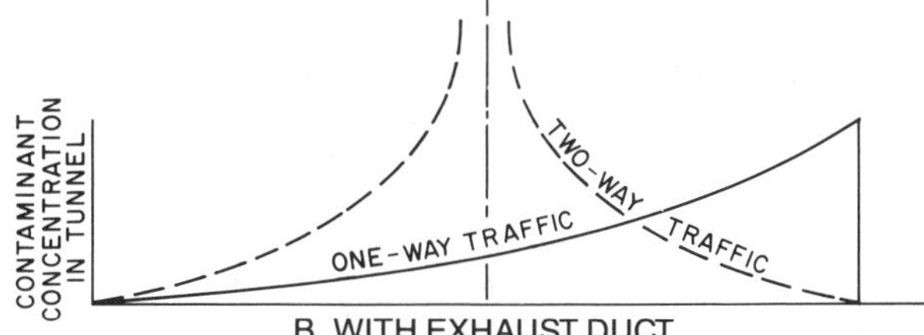

B. WITH EXHAUST DUCT

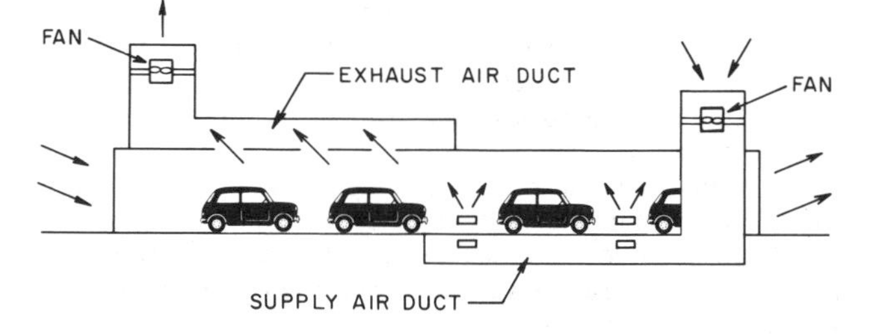

C. WITH SUPPLY AND EXHAUST DUCT

Fig. 4 Semitransverse Ventilation

Because of the fan-induced flow, this system is not adversely affected by atmospheric conditions. The air flows the length of the tunnel in a duct fitted with periodic supply outlets. Fresh air is best introduced at exhaust pipe level of the vehicles to dilute the exhaust gases immediately. An adequate pressure differential must be generated between the duct and the roadway to counteract piston effect and atmospheric winds.

If a fire occurs within the tunnel, the air supplied will dilute the smoke. To aid in fire-fighting efforts and in emergency egress, fresh air should enter the tunnel through the portals to create a respirable environment for these activities. Therefore, the fans in a supply semitransverse system should be reversible, and a ceiling supply should be considered. With a ceiling supply system and reversible fans, the smoke will be drawn upward.

The exhaust semitransverse system (Figure 4b) in a unidirectional tunnel produces a maximum contaminant concentration at the exit portal. In a bidirectional tunnel, the maximum level of contaminants occurs near the center of the tunnel. A combination supply and exhaust system (see Figure 4c) applies only in a unidirectional tunnel where the air entering the traffic stream is exhausted in the first half, and air supplied in the second half exhausts through the exit portal.

The supply semitransverse system is the only ventilation system not affected by adverse meteorological conditions or opposing traffic. Semitransverse systems are used in tunnels up to about 3,000 ft, at which point the tunnel air velocities near the portals become excessive.

Full Transverse Ventilation. This is used in large tunnels. A full exhaust duct added to a supply-type semitransverse system achieves uniform distribution of supply air and uniform collection of vitiated air (see Figure 5). With this arrangement, a uniform pressure will occur throughout the roadway, and no longitudinal airflow will occur except that generated by traffic piston effect, which tends to reduce contaminant levels. An adequate pressure differential between the ducts and the roadway is required to assure proper air distribution under all ventilation conditions.

Full-scale tests conducted by the U.S. Bureau of Mines showed that for rapid dilution of exhaust gases, supply air inlets should be at the level of vehicle emission and the exhaust outlets in the ceiling. The air distribution can be one- or two-sided.

Other Ventilation. Many variations and combinations of the systems described exist. Figure 6 shows a combined system for a unidirectional tunnel approximately 1,400 ft long. Section 3 uses a full transverse system because of the upgrade roadway; Section

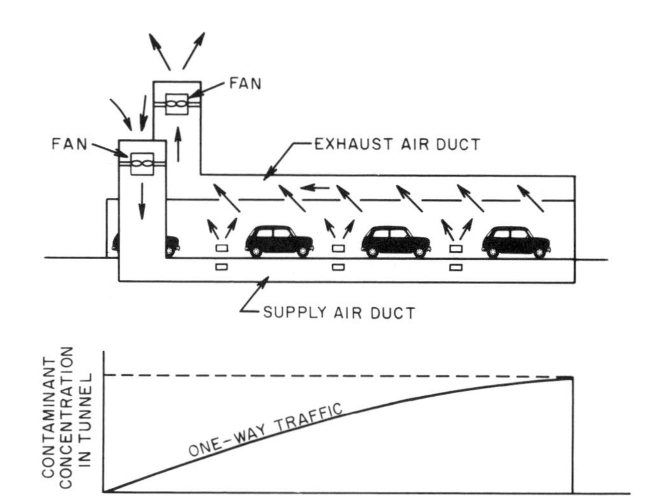

Fig. 5 Full Transverse Ventilation

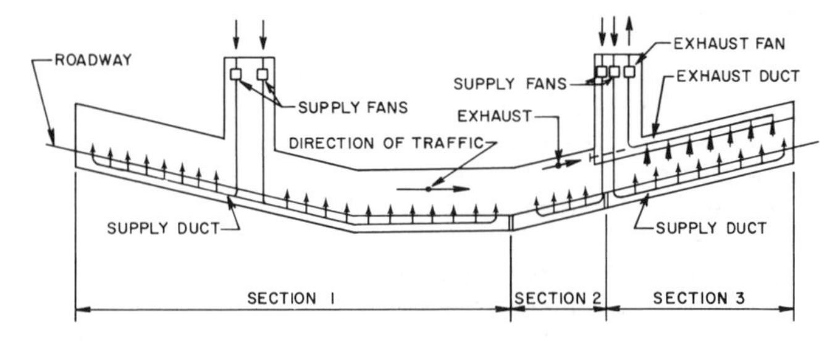

Fig. 6 Combined Ventilation System

2 uses a semitransverse supply with a longitudinal exhaust; and the remainder of the tunnel (Section 1) is a semitransverse supply system. This type of system is not recommended for long tunnels.

Emergency Conditions

An emergency condition in any vehicular tunnel, particularly one generating smoke and heat, such as during a fire, can lead to a disaster if the tunnel ventilation system is designed and operated improperly. The primary objective in such a situation is the rapid removal of smoke and heat from the tunnel to enable motorists to exit safely and fire fighters to reach the fire. Prior to the design of a tunnel, the criteria for emergency conditions, including the minimum level of ventilation required to create a safe environment, must be established.

PRESSURE EVALUATIONS

Air pressure losses in the tunnel duct system must be evaluated to compute fan pressure and drive requirements. Fan selection should be based on total pressure across the fans, not only on static pressure.

Fan total pressure (FTP) is defined by the Air Moving and Conditioning Association (AMCA) (ASHRAE *Standard* 51-1985) as the algebraic difference between the total pressures at the fan discharge (TP_2) and at the fan inlet (TP_1), as shown in Figure 7. The fan velocity pressure (FVP) is defined by AMCA as the pressure corresponding to the air velocity and air density at the fan discharge.

$$FVP = VP_2$$

The fan static pressure (FSP) equals the difference between the fan total pressure and the fan velocity pressure.

$$FSP = FTP - FVP$$

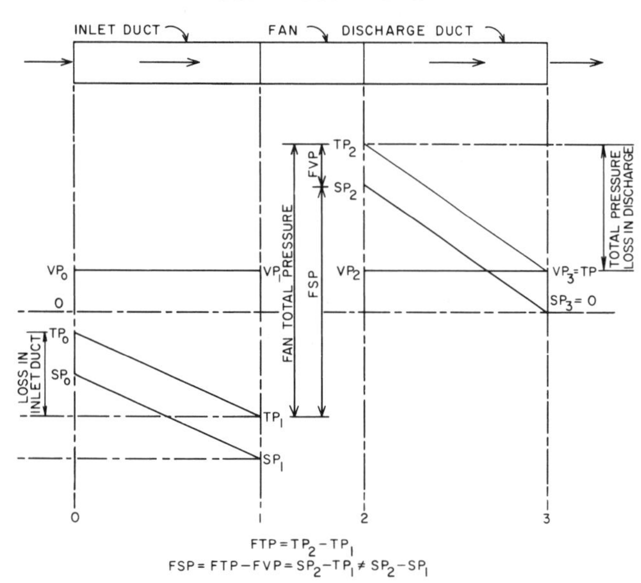

Fig. 7 Fan Total Pressure

The total pressure at the fan discharge (TP_2) must equal the total pressure losses ($\Delta TP_{2\text{-}3}$) in the discharge duct and the exit velocity pressure (VP_3).

$$TP_2 = \Delta TP_{2\text{-}3} - VP_3$$

Likewise, the total pressure at the fan inlet (TP_1) must equal the total pressure losses in the inlet duct system and the inlet pressure.

$$TP_1 = TP_0 + \Delta TP_{0\text{-}1}$$

Straight Ducts

Straight ducts in tunnel ventilation systems can be classified as: (1) those that transport air, thus having constant area and constant air velocity; and (2) those that uniformly distribute (supply) or uniformly collect (exhaust) air.

Several methods have been developed to predict pressure losses in a duct of constant cross-sectional area that uniformly distributes or collects air. The most widely used method was developed for the Holland Tunnel in New York (Singstad 1929). The following relationships give total pressure loss at any point in the duct.

For *supply duct*:

$$p = p_1 + \rho V_0^2/2[aLZ^3/3H - (1 - K)Z^2/2] \quad (5)$$

For *exhaust duct*:

$$p = p_1 + \rho V_0^2/2[aLZ^3/(3+c)H + 3Z^2/(2 + c)] \quad (6)$$

where

p = total pressure loss at any point in duct, in. of water
p_1 = pressure at last outlet, in. of water
ρ = density of air, lb/ft^3
V_0 = velocity of air entering duct, ft/s
a = constant related to coefficent of friction for concrete (0.0035)
L = total length of duct, ft
Z = $(L - X)/L$
X = distance from duct entrance to any location, ft
H = hydraulic radius, ft
K = constant that accounts for turbulence (0.615)
c = constant relating to turbulence of exhaust port = 0.25

For *transport duct*:

The pressure losses in a transport duct having constant cross-sectional area and constant velocity are due to friction alone and can be computed using the standard expressions for losses in ducts and fittings.

CARBON MONOXIDE ANALYZERS AND RECORDERS

The air quality in a tunnel should be constantly monitored at several key points. Carbon monoxide is the impurity usually selected as the prime indicator of tunnel air quality. Three types of CO analyzing instruments are applied: catalytic oxidation, infrared absorption, and electrochemical oxidation.

The *catalytic oxidation* (metal oxide) instrument, the most widely used in vehicular tunnels, offers reliability and stability at a moderate initial cost. Maintenance requirements are low, and the instruments can be calibrated and serviced by maintenance personnel after only brief instruction.

The *infrared analyzer* has the advantage of sensitivity and response but the disadvantage of high initial cost. Being a precise and complex instrument, it requires a highly trained technician for maintenance and servicing.

The *electrochemical analyzer* is precise, compact, and lightweight. The units are of moderate cost and are easily maintained.

No matter what type of CO analyzer is selected, each air sampling point must be located where significant readings can be obtained. For example, the area at or near the entrance of unidirectional tunnels usually has very low CO concentrations, and sampling points there yield little information for ventilation control. The length of piping between the sampling point and the CO analyzer should be as short as possible to maintain a reasonable air sample transport time.

In conditions where intermittent analyzers are suitable, provisions should be made to prevent the loss of more than one sampling point during periods of an air pump outage. It is advisable to provide each analyzer with a strip chart recorder to keep a permanent record of tunnel air conditions. Usually, recorders are mounted on the central control board.

Haze or smoke detectors have been used on a limited scale, but most of these instruments are optical devices and require frequent or constant cleaning with a compressed air jet. Should traffic be predominantly diesel powered, oxides of nitrogen and smoke haze require monitoring in addition to CO.

CONTROL SYSTEMS

To reduce the number of operating personnel at a tunnel, all ventilating equipment should be controlled at a central location. At many older tunnel facilities, fan operation is manual and controlled by an operator at the central control board. Many new tunnels, however, have the fan operation partially or totally automatically controlled.

CO Analyzer Control System

In this system, adjustable contacts within CO analyzers turn on additional fans or increase fan speeds as CO readings increase. The reverse occurs as CO levels decrease. The system requires a fairly complex wiring arrangement because fan operation must normally respond to the single highest level being recorded at several analyzers. In addition to a manual override, delay devices are required to prevent the ventilation system from responding to short-lived high or low CO levels.

Time Clock Control Systems

This type of automatic fan control is best suited for those installations that experience heavy rush-hour traffic. A time clock is set to increase the ventilation level in preset increments in advance of the anticipated traffic increase. The system is simple and is easily revised to suit changing traffic patterns. Because it anticipates an increased air requirement, the ventilation system can be made to respond slowly and, thus, to avoid expensive demand charges by the power company. As with the CO system, a manual override is needed to cope with unanticipated conditions.

Traffic Actuated Systems

Several automatic fan control systems based on recorded traffic flow information have been devised. Most of them require the installation of computers and other electronic equipment. These complex systems require concomitant maintenance expertise.

Local Fan Control

In addition to a central control board, each fan unit should have local control close to and within sight of the unit. It should be interlocked to permit positive isolation of the fan from remote or automatic operation during maintenance and servicing.

RAPID TRANSIT SYSTEMS

Most rapid transit systems run at least part of their lines underground, particularly in the central business districts of urban centers.

Older subway systems relied heavily on natural ventilation, with the primary air mover being the piston effect of the vehicles themselves. In recent years, the piston effect has been supplemented by forced mechanical ventilation. Most new systems use air-conditioned vehicles, but it is still important to provide a reasonable environment within subway tunnels and stations. A U.S. Department of Transportation Handbook (DOT 1976), based on unproven operating experience but validated by field and model tests, provides comprehensive and authoritative design aids.

DESIGN CONCEPTS

The factors to be considered, although interrelated, may be divided into three categories: natural ventilation, forced ventilation, and station air conditioning.

Natural Ventilation

Natural ventilation in subway systems (infiltration and exfiltration) is primarily the result of train operation in tightly fitting trainways, where air generally moves in the direction of train travel. The positive pressure in front of a train expels air from the system through portals and station entrances; the negative pressure in the wake of the train induces airflow into the system through these same openings.

Considerable short-circuiting occurs in subway structures where two trains traveling in opposite directions pass each other. It occurs especially in stations or in tunnels with perforated or no dividing walls. Such short-circuiting reduces the net ventilation rate and increases and possibly causes excess air velocities on station platforms and in station entrances. During the time of peak operation and peak ambient temperatures, it can cause an undesirable amount of heat to build up.

To help counter these negative effects, ventilation shafts are customarily placed near the interface between tunnels and stations. Shafts in the approach tunnel are often called *blast shafts*, through which part of the air pushed ahead of the train is expelled from the system. Shafts in the departure tunnel are called *relief shafts*, since they relieve the negative pressure created during the departure of the train and induce outside air through the shaft rather than through station entrances. Additional ventilation shafts may be provided between stations (or between portals for underwater crossings), as dictated by tunnel length. The high cost of such ventilation structures necessitates a design for optimum effectiveness. Internal resistance because of offsets and bends should be kept to a minimum, and shaft cross-sectional areas should be approximately equal to the cross-sectional area of a single-track tunnel (U.S. DOT 1976).

Mechanical Ventilation

Mechanical ventilation in subway systems (1) supplements the ventilation effect created by train piston action; (2) expels heated air from the system; (3) introduces cool outside air; (4) supplies makeup air for exhaust; (5) restores the cooling potential of the heat sink through extraction of heat stored during off-hours or system shutdown; (6) reduces the flow of air between the tunnel and the station; (7) provides outside air for passengers in stations or tunnels in an emergency or during other unscheduled interruptions of traffic; (8) purges smoke from the system in case of fire.

The most cost-effective design for mechanical ventilation is one that serves two or more purposes. For example, a vent shaft provided for natural (piston action) ventilation may also be used for emergencies if a fan is installed in a parallel with the bypass, or vice versa (Figure 8).

Several vent shafts may work together as a system capable of meeting many, if not all, of the aforementioned objectives. Depending on shaft location and the given train situation, the shaft may serve as a blast or relief shaft with the bypass damper open and the fan damper closed. With the fan in operation and the bypass damper closed, this arrangement can supply or exhaust air by mechanical ventilation, depending on direction of the fan rotation.

Except for emergencies, fan rotation is usually predetermined based on the overall ventilation concept. If subway stations are

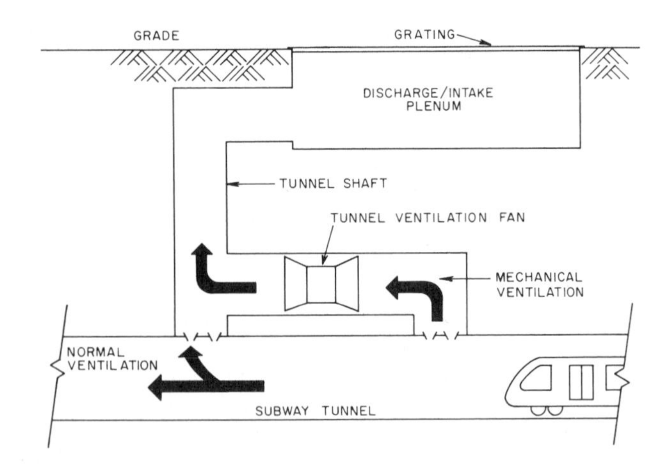

Fig. 8 Tunnel Ventilation Shaft

not air conditioned, the heated system air should be exchanged with cooler outside air at a maximum rate. If stations are air conditioned below ambient temperatures, the inflow of warmer outside air should be limited and controlled.

Figure 9 illustrates a typical tunnel ventilation system between two subway stations. Here the flow of heated tunnel air into a cooler station is kept to a minimum. The dividing wall separating Tracks No. 1 and 2 is discontinued in the vicinity of the emergency fans. Air pushed ahead of the train on Track No. 2 then partially diverts to the emergency fan bypasses and partially into the wake of a train on Track No. 1 as a result of pressure differences (Figure 9a). Figure 9b shows an alternative operation with the same ventilation system. When outdoor temperatures are favorable, the midtunnel fans operate as exhaust fans with makeup air introduced through the emergency fan bypasses. This concept can also provide or supplement station ventilation or both. To achieve this goal, emergency fan bypasses would be closed, and the makeup air for midtunnel exhaust fans would enter through station entrances.

A more direct ventilation concept removes station heat at its primary source, the underside of the train. Figure 10 illustrates a trackway ventilation system. U.S. DOT (1976) tests have shown that such systems not only reduce the upwelling of heated air onto platform areas, but also remove significant portions of the heat generated by dynamic braking resistor grids and air-conditioning condensers located underneath the train. Ideally, makeup air for the exhaust should be introduced at track level to provide a positive control over the direction of airflow (Figure 10a).

Underplatform exhaust systems without makeup supply air, as illustrated in Figure 10b, are least effective and, under certain conditions, may be detrimental, since such an arrangement could cause heated tunnel air to flow into the station. Figure 10c shows a cost-effective compromise where makeup air is introduced at the ceiling of the platform. Although the heat removal effectiveness may not be as good as that of the system illustrated in Figure 10a, it negates the inflow of hot tunnel air that might occur without supply air makeup.

Emergency Ventilation

Mechanical ventilation is a major control strategy in a subway tunnel fire. An increase in air supply over stoichiometric requirements will reduce the fire progression by lowering the flame temperature. Further, ventilation can control the direction of smoke

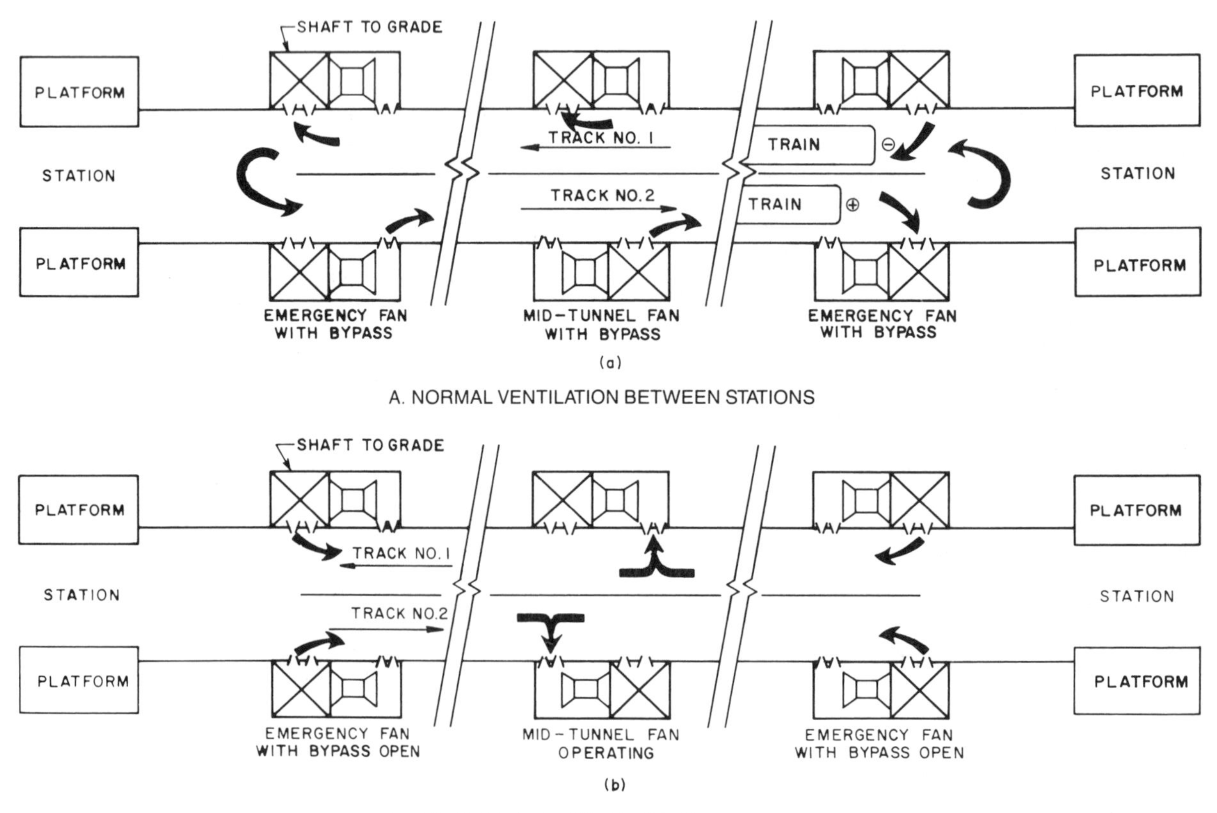

Fig. 9 Tunnel Ventilation Concept

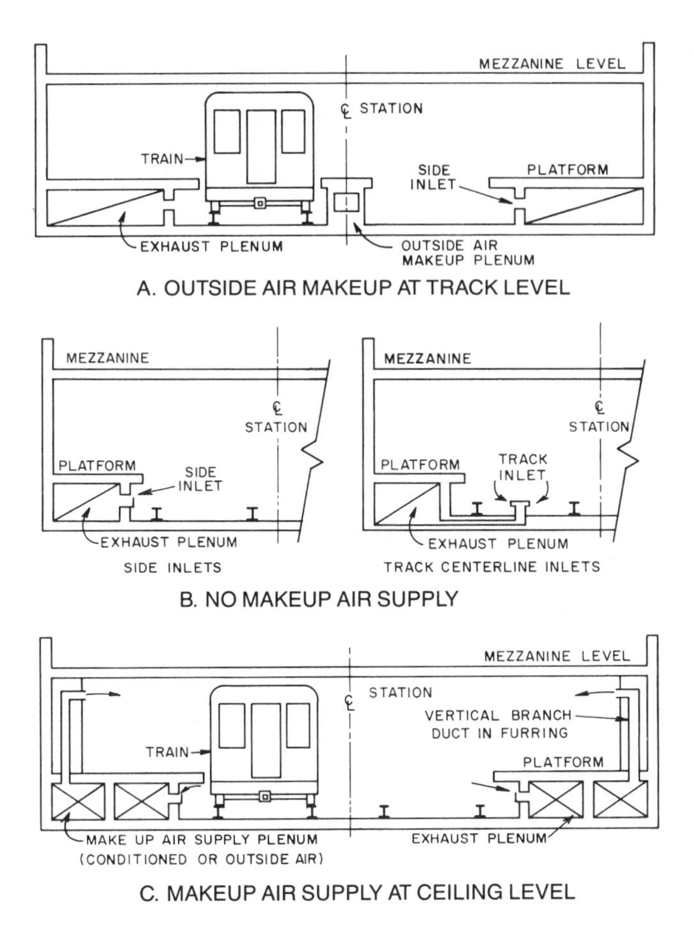

Fig. 10 Trackway Ventilation Concepts (Cross Sections)

emissions to permit safe evacuation of passengers and facilitate access by fire fighters.

Emergency ventilation must allow for the unpredictable location of a disabled train or the source of fire and smoke. Therefore, emergency ventilation fans should have nearly full reverse flow capability so that fans on either side of a stalled train operate together to control the direction of airflow and to counteract the migration of smoke. When a train is stalled between two stations and smoke is present, outside air is supplied from the nearest station and contaminated air is exhausted at the opposite end of the train (unless the location of the fire dictates otherwise). Then the passengers can be evacuated along walkways in the tunnel via the shortest route (Figure 11).

It is essential that provisions be made to (1) quickly assess any emergency situation; (2) communicate the situation to central control; (3) establish the location of the train; and (4) start, stop, and reverse emergency ventilation fans from the central console as quickly as possible to establish smoke control.

Midtunnel and station trackway ventilation fans may be used to enhance the emergency ventilation system; however, these fans must withstand elevated temperatures for a prolonged period and have reverse flow capacity.

Station Air Conditioning

Higher approach speeds and closer headways, made possible by computerized automatic train control systems, have increased the amount of heat gains. The net internal sensible heat gain in a typical double-track subway station, with 40 trains/h per track, traveling at top speeds of 50 mph may reach 5×10^6 Btu/h even after credit is taken for heat removal by the heat sink, by station underplatform exhaust systems, and by tunnel ventilation. To remove such a quantity of heat by station ventilation with outside air at a 3 °F temperature rise, for example, requires roughly 1.4×10^6 cfm.

Not only would such a system be costly, but the resulting air velocities on station platforms would be objectionable to passengers. The same amount of sensible heat gain, plus latent heat, plus outside air load with a 7 °F lower-than-ambient station design temperature could be handled by about 630 tons of refrigeration. Even if station air conditioning is more costly initially, the long-term benefits will result in (1) reduced design airflow rates; (2) improved environment for passengers; (3) increased equipment life; (4) reduced maintenance of equipment and structures; and (5) increased acceptance by subway passengers as a viable means of public transportation.

In addition to the station platform, other ancillary station areas such as concourses, concession areas, and transfer levels should be evaluated for air conditioning. However, unless these ancillary areas are designed to attract patronage to concessions, the air-conditioning cost for walk-through areas is not usually warranted.

The physical configuration of the platform level usually determines the cooling distribution pattern. High ceilings in domed stations, local hot spots as a result of train location, high-density passenger accumulation, or high-level lighting may need spot cooling. Conversely, where train length equals platform length and ceiling height above the platform is limited to 10 to 11.5 ft, isolation of heat sources and application of spot cooling is not usually feasible.

Use of available space in the station structure for air distribution systems should be of prime concern because of the high cost of underground construction. Overhead distribution ductwork, which adds to building height in commercial construction, could add to the depth of excavation in subway construction. The space beneath a subway platform normally offers an excellent area for low-cost distribution of supply, return, and/or exhaust air.

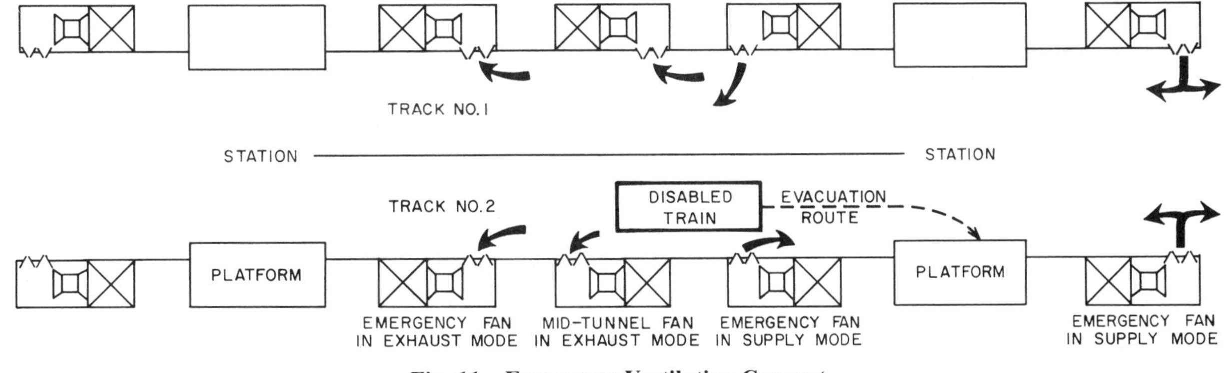

Fig. 11 Emergency Ventilation Concept

DESIGN APPROACH

A subway system may require two separate environmental criteria (1) for normal operations, and (2) for emergencies. Criteria for normal operations generally include limits on temperature and humidity for various times of the year, a minimum ventilation rate to dilute contaminants generated within the subway, and limits on air velocity and the rate of air pressure change to which commuters may be exposed. Some of these criteria are subjective and may vary on a demographic basis. Criteria for emergencies generally include a minimum purge time for sections of the subway in which smoke or fire may occur, and minimum and maximum fan-induced tunnel air velocities.

Given a set of criteria and a set of outdoor design conditions, coupled with appropriate tools for estimating interior heat loads, earth heat sink, ventilation, air velocity, and air pressure changes, design engineers then select the elements of the environmental control system. Selection should consider air temperature control, air velocity control, air quality control, and air pressure control. The system selected generally includes a combination of unpowered ventilation shafts, powered ventilation shafts, underplatform exhaust, and air-conditioning systems.

The train propulsion/braking systems and the configurations of tunnels and stations greatly affect the subway environment, which must often be considered during the early stages of design. The factors affecting a subway environmental control system are further discussed below. *The Subway Environmental Design Handbook* (U.S. DOT 1976) and NFPA *Standard* 130-88, Standard for Fixed Guideway Transit Systems, have additional information.

Comfort Criteria

Because of the transient nature of the environment experienced by a person during entry, passage through, and exit from a subway station, special considerations (as opposed to selecting comfort levels for spaces having continuous occupancy) are permissible.

As a general principle, the environment within a subway station should provide a smooth transition between conditions outside and those within the transit vehicles. People entering a subway station from the outside should not experience a substantial degradation of environmental conditions. Based on nuisance considerations, it is recommended that peak air velocities in public areas be limited to 1000 fpm.

Air Quality

Air quality within a subway system is influenced by many factors, some of which are not directly under the control of the HVAC engineer. Some particulates, gaseous contaminants, and odorants existing on the outside can be prevented from entering the subway by the judicious location of ventilation shafts. Particulate matter, including iron and graphite dust generated by train operations, is best controlled by a regular cleaning of the subway system. However, the only viable way to control gaseous contaminants such as ozone (from electrical equipment) and CO (generated by human respiration) is through adequate ventilation from outside. A minimum of 7.5 cfm of filtered outside air per person should be introduced into tunnels and stations to dilute gaseous contaminants.

Pressure Transients

The passage of trains through aerodynamic discontinuities in the subway system can cause time-varying changes in static pressure. These pressure transients can irritate passengers' ears and sinuses. Pressure transients may also cause additional load on various structures (*e.g.*, acoustical panels) and equipment (*e.g.*, fans). In consideration of potential nuisance to humans, it is recommended that if the total change in pressure is greater than 2.8 in. of water, the rate of static pressure change should be kept below 1.7 in. of water/s.

During emergencies, it is essential to provide ventilation to control smoke and reduce air temperatures to permit passenger evacuation and fire-fighting operations. The minimum air velocity within the tunnel section experiencing the fire emergency should be sufficient to prevent back layering of the smoke, *i.e.*, a flow of smoke in the upper cross section of the tunnel opposite in direction to the forced outside ventilation air. The method whereby this minimum velocity is ascertained is provided in DOT (1976). Further, the maximum air velocity experienced by evacuating passengers should be 2200 fpm.

Interior Heat Loads

Heat within a subway system is generated mostly by: (1) braking of trains; (2) acceleration of trains; (3) car air conditioning and miscellaneous accessories; (4) station lights, people, and equipment; and (5) ventilation (outside) air.

Deceleration. The majority (40 to 60%) of heat generated within a subway system arises from the braking of trains. Many rapid-transit vehicles use nonregenerative braking systems. For these systems, the kinetic energy of a train at the initiation of braking is dissipated as waste heat from dynamic and/or friction brakes, rolling resistance, and aerodynamic drag.

Acceleration. Heat is also generated as a result of train acceleration. Many operational trains use a cam-controlled variable resistance to regulate voltage across DC traction motors during acceleration. Electrical power dissipated by these resistors and by the third rail appears as heat within a subway system. Heat released during acceleration also includes that due to traction motor losses, rolling resistance, and aerodynamic drag. Heat released by train acceleration is generally from 10 to 20% of total heat released within a subway system.

For closely spaced stations, trains frequently undergo only acceleration or braking, with little operation at constant speed. In these cases, maximum train speed between station stops is controlled by the distance between stops.

Car Air-Conditioning Systems

Most new cars are fully climate-controlled. Air-conditioning equipment removes patron and lighting heat from the cars and, along with the condenser fan heat and compressor heat, deposits it into the subway system. Air-conditioning capacities generally range between 10 tons per car for the shorter cars (about 50 ft) up to about 20 tons for the longer cars (about 70 ft). Heat from car air conditioners and other accessories may range from 25 to 30% of total heat generated within a subway.

Other Sources

Within a subway system, heat is also released from people, lighting, induced outside air, and miscellaneous equipment (fare vending machines, escalators, and the like). These sources range from 10 to 30% of total heat generated within a subway system. For analysis of heat balance within a subway, it is convenient to define a control volume about each station, including the station and its approach and departure tunnels.

Heat Sink

The amount of heat flow from subway walls to subway air varies on a seasonal basis, as well as for morning and evening rush-hour operations. For portions of a subway not heated or air conditioned, short periods of abnormally high or low outside air temperature may temporarily cause a departure from the heat sink effect. Such outside air temperature phenomena (lasting up to several days) will, in turn, cause a related air temperature change within a subway. However, the departure of subway air temperature from normal is diminished by the thermal inertia of the subway structure. Thus, during abnormally hot periods, heat flow to the subway structure increases. Similarly, during abnormally cold periods, heat flow from the subway structure to the air increases.

For subways where daily station air temperatures are held constant by heating and cooling, heat flux from station walls is negligible. Depending on the infiltration of station air into the adjoining tunnels, heat flux from tunnel sections may also be reduced in magnitude. Other factors affecting the heat sink component are type of soil (dense rock or light dry soil), migrating groundwater, and the surface configuration of the tunnel walls (ribbed or flat).

Measures to Limit Heat Loads

Various measures have been proposed to limit the interior heat load within subway systems. Among these are regenerative braking, thyristor (chopper) motor controls, track profile optimization, and underplatform exhaust systems.

Electrical regenerative braking, which would otherwise appear as waste heat, is converted to electrical energy for use by other trains. Flywheel energy storage, an alternative form of regenerative braking, would store part of the braking energy of a train in high-speed flywheels for subsequent use in vehicle acceleration. Using these methods, the reduction in heat generated by braking is limited by present technology to approximately 25%.

Conventional cam-controlled propulsion systems apply a set of resistance elements to regulate traction motor current during acceleration. Electrical energy dissipated by these resistors appears as waste heat within a subway system. Thyristor motor controls replace the acceleration resistors by solid-state controls, which would reduce acceleration heat loss by from about 10% on high-speed subways to about 25% on low-speed subway systems.

Track profile optimization refers to a depressed trackway between stations which reduces vehicle heat emissions. In this way, less power is used for acceleration, since some of the train's potential energy, while in the station, is converted to kinetic energy as it accelerates toward the tunnel low point. Conversely, some of the kinetic energy of the train at maximum speed is converted to potential energy as it approaches the next station. Using this method, the maximum reduction in total vehicle heat loss from acceleration and braking is approximately 10%.

An underplatform exhaust system is essentially a *hooding* technique designed to prevent some of the heat generated by vehicle underfloor equipment (such as resistors and air-conditioning condensers) from entering the station environment. Exhaust ports beneath the station platform edge withdraw heated undercar air.

For preliminary calculations, it may be assumed that train heat release within the station box is about two-thirds of control volume heat load due to braking and the train air conditioning, and the underplatform exhaust system is about 50% effective.

A quantity of air equal to that withdrawn by the underplatform exhaust system enters the control volume from the outside. Thus, when station design temperature is below outside ambient, an underplatform exhaust system reduces the subway heat load by drawing off undercar heat but increases the heat load by drawing in outside air. To reduce the uncontrolled infiltration of outside air, a proposed technique is to provide a complementary supply of outside air on the opposite side of the trackway underplatform exhaust ports. While in principle the underplatform exhaust system, with complementary supply, tends to reduce the mixing of outside air with air in public areas of the subway, test results on such systems are unavailable.

PARKING GARAGES

Automobile parking garages are either fully enclosed or partially open. The fully enclosed parking areas are usually underground and require mechanical ventilation. The partially open parking levels are generally above-grade structural decks having open sides (except for barricades), with a complete deck above. Natural or mechanical ventilation, or a combination of both methods, can be used for partially open garages.

The operation of automobiles presents two concerns. The most serious is the emission of carbon monoxide, with its known risks. The second concern is the presence of oil and gasoline fumes, which may cause nausea and headaches, as well as present a fire hazard. Additional concerns regarding oxides of nitrogen and smoke haze from diesel engines may also require consideration. However, the ventilation required to dilute carbon monoxide to acceptable levels will also control the other contaminants satisfactorily.

Two factors are required to determine the ventilation quantity: the number of cars in operation and the emission quantities. Most codes simplify this determination by requiring four to six air changes per hour, or 0.75 to 1 cfm per square foot, for fully enclosed parking garages. For partially open parking garages, 2.5 to 5% of the floor area is required as a free opening to permit natural ventilation. Applicable codes and National Fire Protection Association (NFPA) standards should be consulted for specific requirements.

NUMBER OF CARS IN OPERATION

The number of cars in operation depends on the type of facility served by the parking garage. The variation is generally from 3% of the total vehicle capacity to 5% for a distributed, continuous use such as an apartment house or shopping area. It could reach 15 to 20% for peak use, such as in a sports stadium or short-haul airport.

The length of time that a car remains in operation within a parking garage is a function of the size and layout, as well as the number of cars attempting to enter or exit at a given time. This time could vary from 60 to 600 s, but on the average it ranges from 60 to 180 s.

A survey of existing parking garages serving a facility having similar use and physical characteristics as the proposed design should be conducted. Data can then be recorded on car entry and exiting rate by hour, as well as time in operation within the parking garage.

CONTAMINANT LEVEL CRITERIA

It is recommended that the ventilation rate be designed to maintain a CO level of 50 ppm, with peak levels not to exceed 125 ppm. The American Conference of Governmental Industrial Hygienists recommends a threshold limit of 50 ppm for an 8-h exposure, and the EPA has determined that, at or near sea level, a CO concentration of 125 ppm for exposure up to 1 h would be safe. For installations above 3500 ft, far more stringent limits would be required.

Design Approach

The operation of a car engine within a parking garage differs considerably from normal vehicle operation, including normal operation within a vehicular tunnel. On entry, the car travels slowly. On exiting, the engine is cold, and thus, at full choke. As it proceeds from the garage, the engine operates with a rich mixture and in low gear. Emissions for the cold start are considerably higher, and the distinction between hot and cold emission plays a critical role in determining the ventilation rate. Motor vehicle emission factors for hot and cold start operation are presented in Table 3. An accurate analysis requires correlation of CO readings with the survey data on car movements (Hama *et al.* 1974). Table 4 lists approximate data for vehicle movements. This data should be adjusted to suit the specific physical configuration of the facility.

Table 3 Predicted CO Emissions within Parking Garages

	Hot Emissions (Stabilized) g/min		Cold Emissions g/min	
Season	1991	1996	1991	1996
Summer (90°F)	2.54	1.89	4.27	3.66
Winter (32°F)	3.61	3.38	20.74	18.96

Results from EPA Mobil 3 version NYC-2.2; sea level location.
Note: Assumed vechicle speed is 5 mph.

Table 4 Average Entrance and Exit Times for Vehicles

Level	Average Entrance Time, s	Average Exit Time, s
1	35	45
3[a]	40	50
5	70	100

Source: Stankunas *et al.* (1980).
[a]Average pass-through time = 30 s.

While emission controls were instituted in 1968 with progressively more stringent controls programmed through 1990, the precise reduction for the modes of engine operation within parking garages has not been evaluated. Access tunnels or fully enclosed ramps should be designed in accordance with the recommendations for vehicular tunnels. When natural ventilation is used, the wall opening free area should be as large as possible. A portion of the free area should be placed at floor level. In parking levels with large interior floor areas, a central emergency smoke exhaust system should be considered. This measure would improve safety in the event that fume removal during calm weather or smoke removal is necessary.

The ventilation system, whether mechanical, natural, or both, should be designed to meet applicable codes and maintain an acceptable contaminant level. To conserve energy, fan systems should be controlled by carbon monoxide meters to vary the amount of air required, if permitted by local codes. For example, fan systems could consist of multiple fans with single- or variable-speed motors, or variable pitch blades on axial flow fans. In multilevel parking garages or single-level structures of extensive area, independent fan systems, each under individual control, are preferred.

Systems can be classified as supply-only, exhaust-only, or combined. Whichever system is chosen, consider: (1) the contaminant level of the outside air drawn in for ventilation; (2) avoiding short circuiting of supply air; (3) avoiding long flow fields that permit the contaminant levels to build up above an acceptable level at the end of the flow field; (4) providing short flow fields in areas of high pollutant emission, thereby limiting time to mix throughout the facility; (5) providing an efficient, adequate flow throughout the volume of the parking structure; and (6) stratification of engine exhaust gases.

Noise

The ventilation system in parking garages, in general, moves large quantities of air through large openings without extensive ductwork. These conditions, in addition to the highly reverberant nature of the space, contribute to high noise levels. For this reason, sound attenuation should be considered. This is a safety concern as well, since high noise levels may mask the sound of an approaching car.

Ambient Standards and Pollution Control

Some state and municipal authorities have developed ambient air quality standards. The exhaust system discharge should meet these requirements.

BUS TERMINALS

Bus terminals vary considerably in physical configuration. Most terminals consist of a fully enclosed space containing passenger waiting areas, ticket counters, and some food vending service. Buses load and unload outside the building, generally under a canopy to provide some weather protection. In larger cities, where space is at a premium and an extensive bus service exists, multiple levels may be required with the attendant busway tunnels and/or ramps, as well as extensive customer services.

Waiting rooms and consumer spaces should have a controlled environment in accordance with normal practice for public terminal occupancy. The space should be pressurized against intrusion of the busway environment. Waiting rooms and passenger concourse areas are subject to a highly variable people load. The occupant density may reach 10 ft^2 per person and, at extreme congestion periods, 3 to 5 ft^2 per person.

Basically, two types of bus service exist—urban-suburban and long distance. Urban-suburban service is characterized by frequent bus movements with rapid loading and unloading requirements. Therefore, the ideal passenger platforms are long, narrow, and the drive-through type, not requiring bus backup movements on departure. Long-distance operations, those with greater headways, generally use sawtooth gate configurations.

Ventilation systems to serve bus operating levels can be either natural or forced ventilation. When natural ventilation is selected, the bus levels should be open on all sides, and the slab-to-ceiling dimension should be sufficiently high, or contoured, to permit free air circulation. The installation of jet fans improves the natural airflow at a relatively low energy requirement. Mechanical systems that ventilate open platforms or gate positions should be configured to serve the bus operating areas, as shown in Figures 12 and 13.

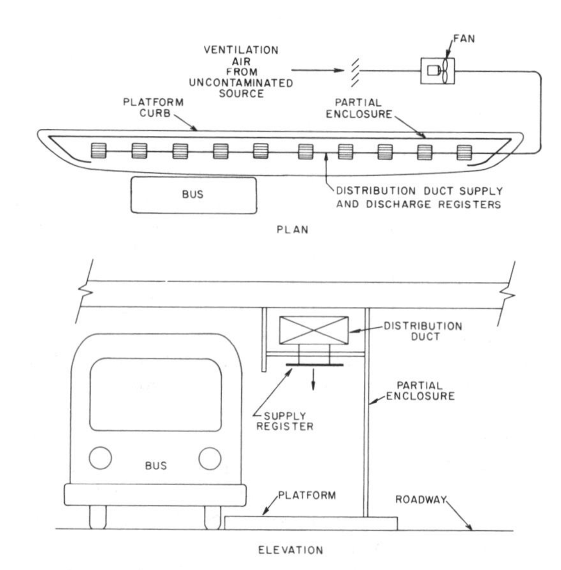

Fig. 12 Partially Enclosed Platform, Drive-Through Type

PLATFORMS

Where naturally ventilated and drive-through platforms may expose passengers to inclement weather and strong winds, enclosed platforms, except for an open front, should be considered and the appropriate mechanical systems provided. Even partially enclosed platforms, however, may trap contaminants and require mechanical ventilation.

Multilevel bus terminals have limited headroom, thus restricting natural ventilation. In this case, mechanical ventilation should be selected, and all platforms should be partially or fully enclosed. Ventilation supplied to partially enclosed platforms should be designed to minimize the induction of contaminated air from the busway. Figure 12 indicates a partially enclosed drive-through platform and an air distribution system. Supply air velocity should be limited to 250 fpm to avoid drafty conditions on the platform. Partially enclosed platforms require large amounts of outside air to provide an effective barrier against fume penetration. Present experience indicates that a minimum of 17 cfm/ft^2 of platform area is required during rush hours and about half of this quantity during the remaining time.

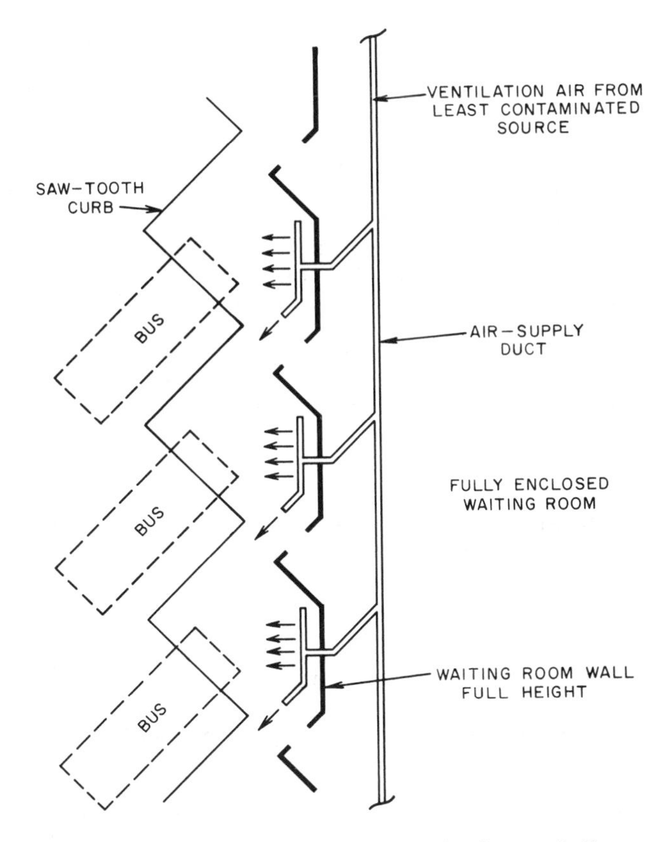

Fig. 13 Fully Enclosed Waiting Room with Sawtooth Gates

Platform air quality remains essentially the same as the ventilation air introduced. Because of the piston effect of buses, however, some momentary higher concentrations of pollutants will occur on the platform. Separate ventilation systems for each platform with two-speed fans permit operational flexibility. Fans should be controlled automatically to conform to bus operating schedules. In northern areas, mechanical ventilation could possibly be reduced during extreme winter weather.

Fully enclosed platforms are strongly recommended for large terminals with heavy bus traffic. They can be pressurized adequately and ventilated with the normal heating and cooling air quantities, depending on the tightness of construction, and the number of boarding doors and other openings. Air distribution can be of the conventional type used for air-conditioning systems. Air should not be recirculated. Openings around doors and in the enclosure walls are usually adequate to relieve air, without additional relief openings, unless platform construction is extraordinarily tight.

BUS OPERATION AREAS

Most buses are powered by diesel engines. Certain models have small auxiliary gasoline engines to drive the air-conditioning system. Tests performed on the volume and composition of exhaust gases emitted from diesel engines in various traffic conditions indicate large variations, which depend on engine type, temperature, humidity, manufacturer, size, adjustment, and fuel burned.

Contaminants

The components of diesel exhaust gases that affect the ventilation system design are oxides of nitrogen, hydrocarbons, formaldehyde, odor constituents, aldehydes, smoke particulates, and a relatively small amount of carbon monoxide. Operation of diesel engines in enclosed spaces, therefore, causes visibility obstruction and odors, as well as contaminants. Table 5 provides approximate

Table 5 Approximate Diesel Bus Engine Emissions (ppm)

Total Exhaust Gases	Idling 117 scfm	Accelerating 476 scfm	Cruising 345 scfm	Decelerating 302 scfm
Carbon monoxide	215	500	230	130
Hydrocarbons	390	210	90	330
Oxides of nitrogen (NO_x)	60	850	235	30
Formaldehydes (HCHO)	9	17	11	30

Note: For additional information for diesel bus and truck engine emissions, refer to Watson *et al.* (1988).

diesel engine exhaust gas data for the major contaminants that affect health. The nature of bus engines should be determined for each project, however.

The Federal Occupational Health and Environmental Control Regulation, Subpart G, sets the contaminant levels for an 8-h exposure as: carbon monoxide, 50 ppm, and nitric oxide, 25 ppm. Subpart G sets the ceiling (maximum) limit for nitrogen dioxide at 5 ppm. The American Conference of Governmental Industrial Hygienists also recommends these threshold limits and, in addition, lists a ceiling (maximum) limit for formaldehyde at 3 ppm.

Oxides of nitrogen occur in two basic forms: nitrogen dioxide (NO_2) and nitric oxide (NO). Nitrogen dioxide is the major contaminant to be considered in the design of a ventilation system. Exposure to concentrations of 10 ppm and higher will cause adverse health conditions. Furthermore, nitrogen dioxide affects light transmission, causing visibility reduction. It is intensely colored and absorbs light over the entire visible spectrum, especially in the shorter wavelengths. Odor perception is immediate at 0.42 ppm of NO_2 and can be perceived by some at levels as low as 0.12 ppm.

Terminal operation also affects the quality of surrounding ambient air. The dilution rate and the location and design of the intakes and discharges control the impact on ambient air quality. Also, the installation is subject to state and local regulations, which require consideration of local atmospheric conditions and ambient contaminant levels.

Calculation of Ventilation Rate

To calculate the ventilation rate, the total amount of engine exhaust gases should be determined. The bus operating schedule, the time that buses are in different modes of operation—that is, cruising, decelerating, idling, and accelerating—and the use of Table 5 permits computing the contaminant level. The design engineer must ascertain, from the configuration of the terminal, the grade (if any) within the terminal and whether the platforms are the drive-through type, drive-through with bypass lanes, or sawtooth type. The engineer must also evaluate bus headways and bus operating speeds and modes.

For instance, with sawtooth platforms, the bus on departure will have to accelerate backwards, brake, and then accelerate forward. The drive-through platform requires a different mode of operation. Certain codes prescribe a maximum idling time for engines, usually 180 to 300 s. However, it should be recognized that 60 to 120 s of engine operation are required to build up brake air pressure.

The discharged contaminant quantities should be diluted by natural and/or forced ventilation to acceptable, legally prescribed levels. To maintain visibility and odor control, the exhaust gas contaminants should be diluted in the proportion of 75 to 1 with outside air. Where urban-suburban operations are involved, the ventilation rate will vary considerably throughout the day and on weekdays compared to weekends. Fan speed or blade control should be used to conserve energy.

Source of Ventilation Air

Since dilution is the primary means of contaminant level control, the source of ventilation air is extremely important. The cleanest available ambient air, which is generally above the roof in an urban area, should be used. Surveys of ambient air contaminant levels should be conducted and the most favorable source located. Possible short-circuiting of exhaust air due to prevailing winds and building airflow patterns should be evaluated.

Control by Contaminant Level Monitoring

Time clocks or tapes coordinated with bus movement schedules and smoke monitors (obscurity meters) provide the most practical means of controlling the ventilation system.

Instrumentation is available for monitoring various contaminants. Also, control by instrumentation can be simplified by monitoring carbon dioxide (CO_2). Studies have shown a relationship between the levels of various diesel engine pollutants and carbon dioxide, thus permitting the use of CO_2 detection and control equipment. However, the mix and quantity of pollutants varies with the rate of operation and the maintained condition of the bus engines. Therefore, if CO_2 monitoring is employed, actual conditions should be obtained under specific bus traffic conditions (schedule) to verify the selected CO_2 settings.

Dispatcher's Booth

The dispatcher's booth should be kept under pressure with uncontaminated air to prevent the intrusion of engine fumes. Since it is occupied for sustained periods, normal interior comfort conditions as well as OSHA contaminant levels must be maintained.

EQUIPMENT

The ability of an enclosed vehicular facility to function depends mostly on the effectiveness and reliability of its ventilation system. The system must be completely effective under the most adverse environmental and traffic conditions and during periods when not all equipment is operational. For a tunnel, it is necessary to provide more than one dependable source of power to prevent an interruption of service.

FANS

Reserve Ventilation Capacity

The prime considerations in selecting the type and number of fans, include the total theoretical ventilating air capacity required and a reasonable factor of safety. This selection is also influenced by the manner in which reserve capacity is provided when a fan is inoperative, or during repair of equipment or of the power supply.

Selection of fans (number and size) to meet normal and reserve capacity requirements is based on the principles of parallel fan operation. Actual capacities can be determined by fan performance and system curves plotted on the same pressure-volume diagram.

It is important that fans selected for parallel operation operate in that region of their performance curves where transferral of capacity back and forth between fans will not occur. This is accomplished by selecting a fan size and speed where the duty point, no matter how many fans are operating, falls well below the unstable range of fan performance. Fans operating in parallel on the same system should be of equal size and, if multiple-speed, should always be operated at the same speed; *i.e.*, if one fan is operating at low speed, all other fans must be at low speed too.

Number and Size of Fans

At locations where no space limitations or other restrictions are placed on the structures that house the ventilation equipment, the number and size of fans should be selected by comparing several alternative fan arrangements. These comparisons should be based on the practicality and overall economy of the layouts, including an estimate of: (1) annual power costs to operate equipment; (2) annual capital cost of the ventilation equipment (usually capitalized over an assumed equipment life of 30 years); and (3) annual capital cost of the structure required to house equipment (usually capitalized over an arbitrary structure life of 50 years).

Two opposing views prevail on the proper number and size of fans. The first advocates a few large-capacity fans; the second prefers numerous small units. In most cases, a compromise arrangement produces a system with the greatest operating efficiency. Regardless of the design philosophy, the number and size of fans should be selected to build sufficient flexibility into the system to meet the varying air demands created by daily and seasonal traffic fluctuations.

Type of Fan

Normally, the ventilation system of a vehicular facility requires large air volumes working against relatively low pressures. Under these conditions, some fan designs have low efficiencies, thus, the choice of suitable fan type is often limited to either centrifugal or vane-axial.

Special Considerations. For rapid-transit systems, any fan installed in a duct leading from the train tube needs special consideration, since fans operate in the presence of the flow and pressure transients caused by train passage. If the transient tends to increase flow to the fan—*i.e.*, the positive flow in front of a train to an exhaust fan or negative flow behind the train to a supply fan—it is important that blade loading does not become so high as to produce long-term fatigue failures.

If the disturbance tends to decrease flow to the fan—*i.e.*, the negative flow behind the train to an exhaust fan or the positive flow in front of the train to a supply fan—the fan performance characteristic must have adequate margin to prevent aerodynamic stall of the fan.

In most cases, ability to reverse the rotation of tunnel supply, exhaust, and emergency fans rapidly is important in an emergency. The effects of this requirement must be considered in the selection and design of the fan and drive system.

DAMPERS

Shutoff dampers can be installed (1) to isolate any parallel nonoperating fan from those operating to prevent short-circuiting, with consequent pressure and flow loss through the inoperative fan; (2) to prevent serious windmilling of an inoperative fan; and (3) to provide a safe environment for maintenance and repair work on each fan.

Single-fan installations could have an isolating damper to prevent serious windmilling because of natural or piston effect drafts and to facilitate fan maintenance.

Two types of dampers have generally been used in ventilation systems: (1) the trapdoor type installed in a vertical duct so that the door lies flat when closed, and (2) the multiple-blade louver type, with parallel operating blades. Both types are usually driven by a gear motor, which is operated by the fan controller who closes the damper when the fan is off and reopens it when the fan is on.

The trapdoor damper is simple and works satisfactorily where a vertical duct enters a plenum-type fan room through an opening in the floor. The damper, usually constructed of steel plate with welded angle reinforcing hinged on one side, closes by gravity against an embedded angle frame of the opening. The opening mechanism is usually a shaft sprocket-and-chain device. The drive motor and gear drive must develop sufficient force to open the damper door against the maximum static pressure difference the fan can develop. This pressure can be obtained from fan performance curves. Limit switches start and stop the gear-motor drive at the proper position.

Dampers placed in ducts other than vertical should be the multiple-blade louver type. These dampers usually consist of a rugged channel frame whose flanges are bolted to the flanges of the fan, duct, or duct opening. Louver blades are mounted on shafts that turn in bearings mounted on the outside of the channel frame. This arrangement requires accessible space on the outside of the duct for bearing and shaft lubrication and maintenance, and space for operating linkages. Louver dampers should have edge and end seals to make them airtight.

The trapdoor-type damper, if properly fabricated, is inherently airtight due to its weight and overlap at its edges. However, louver dampers must be carefully constructed to ensure tightness on closing. A damper that leaks under pressure will cause the fan to rotate counter to its power rotation, thus making restarting dangerous and possibly damaging the drive motor.

SUPPLY AIR INTAKE

Supply air intakes require careful design to ensure that the quality of air drawn into the system is the best available. Such factors as recirculation of exhaust air or intake of contaminants from nearby sources should be considered.

Louvers or grillwork are usually installed over air intakes for aesthetic, security, or safety reasons. Bird screens are also important if openings between louver blades or grillwork are large enough to allow birds to enter.

Louvers with sufficiently low face velocities to be weatherproof may not be possible in certain ventilation systems due to the large quantities of intake air required. Therefore, intake plenums, shafts, fan rooms, and fan housings need water drains. Blowing snow can also fill the fan room or plenum with snowdrifts, but this usually will not stop the ventilation system from operating satisfactorily.

Situations may require installing noise elimination devices in fresh air intakes to keep fan and air noise from disturbing the outside environment. If sound reduction is required, it should be investigated as a total system—the fan, fan plenum, building, fan housing, and air intake (location and size). In designing the sound reduction system, fan selection should also be based on a total system, including the pressure drop that results from sound attenuation devices.

EXHAUST OUTLETS

The discharge of exhaust air should be remote from the street level or from areas with human occupancy. Contaminant concentrations in this exhaust air are not of concern if the system is working effectively. However, odors and entrained particulate matter make this air undesirable in occupied areas. Exhaust stack discharge velocity should be high enough to facilitate dispersion of contaminants into the atmosphere. A minimum of 2000 fpm is usually necessary.

In the past, evasé outlets were used to regain some static pressure and, thereby, to reduce the energy consumption by the exhaust fan. Unless the fan discharge velocity is in excess of 2000 fpm, however, the energy savings compared to the cost of the evase is questionable.

In a vertical or near-vertical exhaust fan discharge connection to an exhaust duct or shaft, rainwater will run down the inside of the stack into the fan. This water will dissolve material deposited from vehicle exhausts on the inner surface of the stack and become extremely corrosive. Therefore, fan housing should be fabricated of a corrosion-resistant material or be specially coated to protect the metal from corrosion.

REFERENCES

ACGIH. 1989. Industrial ventilation, A manual of recommended practice, 20th ed., Cincinnati, OH.

ASHRAE. 1985. Laboratory methods of testing fans for rating. ASHRAE *Standard* 51-1985 (AMCA *Standard* 210-85).

Ball, D. and J. Campbell. 1973. Lighting, heating and ventilation in multistory and underground car park. Paper presented at the Institution for Structural Engineers and The Institution of Highway Engineers' Joint Conference on Multi-story and Underground Car Parks (May).

DOT. 1976. *Subway environmental design handbook.* Urban Mass Transportation Administration, U.S. Government Printing Office.

EPA. Average emission factors for highway vehicles for selected calendar years. *Supplement No. 7 for compilation of air pollutant emission factors*, 3rd ed., Table I-3.

Federal Register. 1974. 39(125), June.

Hama, G.M., W.G. Frederick, and H.G. Monteith. 1974. How to design ventilation systems for underground garages. *Air engineering.* Study by the Detroit Bureau of Industrial Hygiene, Detroit (April).

NFPA. 1988. Standard for fixed guideway transit systems. NFPA *Standard* 130-88. National Fire Protection Association, Quincy, MA.

Ricker, E.R. *The traffic design of parking garages.* ENO Foundation for Highway Traffic Control.

Round, F.G. and H.W. Pearall. Diesel exhaust odor: Its evaluation and relation to exhaust gas composition. Research Laboratories, General Motors Corporation, *Society of Automotive Engineers Technical Progress Series* 6.

Singstad, O. 1929. *Ventilation of vehicular tunnels.* World Engineering Congress, Tokyo, Japan.

Stankunas, A.R., P.T. Bartlett, and K.C. Tower. 1980. Contaminant level control in parking garages. ASHRAE *Transactions* 86(2):584-605.

The Research Corporation of New England. 1979. Contaminant level control in parking garages. ASHRAE *Research Project*, Corporation of America (April).

Turk, A. 1963. Measurement of odorous vapors in test chambers: Theoretical. ASHRAE *Journal* (October).

Watson, A.Y., R.R. Bates, and D. Kennedy. 1988. Air pollution, the automobile, and public health. Sponsored by the Health Effects Institute. National Academy Press, Washington, D.C.

Wendell, R.E., J.E. Norco, and K.G. Croke. 1973. Emission prediction and control strategy: Evaluation and pollution from transportation systems. *Air Pollution Control Association Journal* (February).

CHAPTER 14

LABORATORIES

A research laboratory requires regulation of temperature, humidity, air pressure, air motion, air cleanliness, light, sound, and vibration. Information needed for design includes room conditions, research equipment heat loads, airflow patterns, contaminant control, sound levels, vibration limits, and any special user requirements. Local requirements about "energy use limits" are important in the predesign effort because laboratories require more energy for HVAC than do offices, stores, etc. The development of these parameters requires frequent communication between the designer, the researcher, and the researcher's safety office. The designer must obtain all HVAC performance information and explain the operational capabilities and limitations of the proposed design to the laboratory user before design can begin. The initial approval of the user's safety officer, maintenance/operations manager, and laboratory director are essential.

Because all research programs vary, the occupancy and arrangement of laboratory space are frequently altered. A successful laboratory can accommodate these changes (within the limits of the design) without major alteration and interference to adjacent areas. The HVAC system should have flexibility so that changes can be made with minimal alteration to HVAC equipment. The HVAC system must have the potential for, but not the initial capacity of, handling all of the ultimate needs.

Early mutual agreement can prevent redesign and improve the final result—a facility that meets the needs of current and future research programs. HVAC engineers should develop a 10-year concept of the laboratory use and then design a facility to meet this long-range plan.

Most aging laboratory facilities (over 15 years old) need extensive HVAC renovation. One survey indicated that 43% of the nation's research facilities are more than 20 years old and that they lack mechanical services to support current research needs (Kiil 1984). Due to rapidly changing research procedures and environmental requirements, many newer facilities may also need renovation.

This chapter focuses primarily on biomedical research laboratories. Laboratories in hospitals and schools are currently described in other chapters. Industrial process laboratories are not described; nor are clean rooms for microelectronic facilities.

RISK ASSESSMENT

Laboratory research always involves some risk. Relative risks for work with infectious agents, radioactive materials, and toxic substances have been documented. Pike (1976) summarizes laboratory-associated infections that occurred in the United States and in foreign countries through 1974. The data show that a high fatality rate from laboratory-acquired infections can occur; this points to the fact that risk assessment (involving HVAC design engineers) coupled with enforceable procedures for safe laboratory operation can make the difference between life and death for researchers. Currently, laboratory-associated infections have decreased from the peak decade of 1945 through 1954. This trend may be due to increased awareness of the hazards of working with infectious agents, as well as the increased use of safety devices (fume hoods and biological safety cabinets) and other safety features included in recent laboratory designs.

After the design purpose and scope have been determined, risk assessment must be completed before a laboratory can be designed. Table 1 presents guidelines to assess risk. A typical laboratory room may require several laboratory safety guidelines.

The design engineer must consult scientific staff and health and safety personnel to obtain a risk assessment of the materials and procedures to be used over the life of the proposed laboratories.

Table 1 Compilation of Laboratory Safety Guidelines

Laboratory safety	*CRC Handbook of Laboratory Safety*, 3rd ed. (CRC Press 1989)
	Prudent Practices for Handling Hazardous Chemicals in Laboratories (National Academy Press 1981)
	Guidelines for Laboratory Design (John Wiley & Sons 1987)
Microbiological and biomedical safety	*Biosafety in Microbiological and Biomedical Laboratories* (DHHS 1988)
Hospital laboratories	*Guidelines for Construction and Equipment of Hospital and Medical Facilities* (American Institute of Architects Press 1987)
	Medical Laboratory Planning and Design (College of American Pathologists 1985)
Chemical carcinogens	OSHA Safety and Health Standards (Current standards)
Fire safety	Fire Protection for Laboratories Using Chemicals, *Standard* 45 (National Fire Protection Association 1986)
	Health Care Facilities Handbook (National Fire Protection Association 1987)

The preparation of this chapter is assigned to TC 9.2, Industrial Air Conditioning.

The design team (architect, engineer, and manager) translates this risk assessment into a proper design approach. The design team should then visit similar laboratories to assess successful safe operating practices and the way that the design interacts with these practices. But since each research laboratory is unique, its design must be evaluated using current standards and practices rather than duplicating the designs of outmoded existing facilities.

The nature of the contaminant, the quantities present, the types of operations, and the degree of hazard dictate the type of containment and local exhaust devices. For personnel convenience, those operations posing less hazard are conducted in devices that use air currents for personnel protection, *e.g.*, laboratory fume hoods and biological safety cabinets; however, these devices do not provide absolute containment. Operations that have a significant hazard potential are conducted in devices that provide greater protection but are more restrictive, *e.g.*, sealed glove boxes. Glove boxes require far less exhaust air than do laboratory fume hoods and biological safety cabinets. Laboratories for low and moderately hazardous work may, therefore, have greater exhaust air requirements than do laboratories for highly hazardous work. Laboratory exhaust air requirements are determined by the type, number, size, and operating frequency of the containment devices. These requirements are critical in system design and in establishing supply air rates and flow patterns. Research laboratory design should include the following criteria:

1. Design is economical to construct and operate.
2. Design meets the requirements for safe management of the various hazards encountered.
3. Design facilitates research productivity.
4. Design is sufficiently flexible to accommodate changes in research programs.
5. Safety features designed into the laboratory closely match the assessed degree of risk of the research (West 1978).

DESIGN CONDITIONS AND THERMAL LOSS

Dry-bulb and wet-bulb temperature (or relative humidity), with specified tolerances for each value, define indoor conditions. These temperatures should be based on agreements with the research staff (users) and on a 10-year plan for facility use. Table 2 presents recommended design conditions for typical laboratories and offices. Indoor design conditions for laboratory animal rooms are presented later in this chapter. The designer should determine whether a stated set of conditions represents the limiting values or the levels to be maintained. For variable temperature rooms, it is necessary to establish humidity requirements for a specified dry-bulb range or selected dry-bulb temperatures.

Heat and vapor from laboratory equipment substantially add to the room's sensible and latent loads. Table 3 presents information about this major heat gain. Alereza and Breen (1984) provide additional information about other equipment. The designer needs to evaluate equipment nameplate ratings, applicable load and use factors, and the overall diversity factor. Heat released by equipment located in chemical fume hoods can be discounted. For equipment that is directly vented or water-cooled, appropriate reductions should be made in the heat released to the room. Any unconditioned auxiliary air that is not captured by the fume hood must be included in the room's load calculation.

Table 2 Comfortable Indoor Temperatures

Season	Temperature (Dry-bulb)
Winter	71.6 °F
Summer	72.7 °F

Source: Schiller and Arens (1988).

Laboratory (room) cooling and heating loads are always highly variable due to the operation of laboratory equipment. For this reason, individual laboratory rooms should always have separate thermostats.

Laboratories that contain harmful substances should be designed and field balanced so that air flows into the laboratory from adjacent (clean) spaces, offices, and corridors. This requirement for directional airflow into the laboratory is to contain odors and toxic chemicals, *i.e.*, negative pressurization. Air supplied to the corridor and adjacent clean spaces must be exhausted through the laboratory to achieve effective negative pressurization.

Exhausting contaminants from the laboratory to the atmosphere requires conditioning of large quantities of outdoor air. Building codes require most laboratories dealing with toxic or other hazardous materials to exhaust 100% of their supply air to the outdoors. These laboratories need large HVAC equipment and consume large amounts of energy. The selection of their outdoor and indoor conditions dictates the size and cost of refrigeration and heating facilities.

SUPPLY SYSTEMS

The minimum unit served by the air supply is the laboratory module, for example, a 10 by 20-ft room. The room size, physical arrangement, occupancy, type of exhaust system, and economics contribute to determining the type of supply air system.

Air may be supplied at high, medium, or low pressure and through single-duct, dual-duct, and terminal reheat systems. Air may be introduced into the laboratory through ceiling diffusers, sidewall grilles, perforated ceiling panels, or outlets under windows. The important factor is whether the air supply satisfies safe operating conditions.

Chapter 14 of the 1989 ASHRAE *Handbook—Fundamentals* has information on the locations for supply air system intakes and the airflow around buildings. It is important to consider airflow around laboratory buildings because exhaust air, which is often untreated, may reenter and contaminate the laboratory rooms.

Caution must be used when selecting a variable air volume (VAV) system. The room air pressure changes should not affect the performance of the chemical fume hoods or biological safety cabinets. The designer should determine that recirculated, filtered air is not contaminating the laboratory or its occupants. HVAC systems with 100% outside air are normally recommended for research laboratories to prevent toxic chemical vapors from entering the recirculated air.

Filtration

The filtration necessary for supply air depends on the activity in the laboratory. Conventional chemistry and physics laboratories commonly have 85% efficient filters (ASHRAE *Standard* 52-76 Test Method). Biomedical laboratories usually require 85 to 95% efficient filters. High-efficiency particulate air (HEPA) filters should be provided for special spaces where research materials or animals are particularly susceptible to contamination from external sources. HEPA filtration of the supply air is considered necessary in only the most critical applications such as environmental studies, specific pathogen-free (SPF) research animals, "nude mice," dust-sensitive work, and electronic assemblies. In many instances, biological safety cabinets (which are HEPA filtered), rather than HEPA filtration for the entire room, are satisfactory.

Air Distribution

Air supplied to a laboratory space must keep temperature gradients and air turbulence to a minimum, especially near the face of the laboratory fume hoods and biological safety cabinets. Air outlets must not discharge into the face of fume hoods. Also, cross-

Table 3 Recommended Rate of Heat Gain from Hospital Equipment Located in the Air-Conditioned Area

Appliance Type	Size	Maximum Input Rating, Btu/h	Recommended Rate of Heat Gain, Btu/h[a]
Autoclave (bench)	0.7 ft^3	4270	480
Bath, hot or cold circulating, small	1.0 to 9.7 gal, −22 to 212°F	2560 to 6140	440 to 1060 (sensible) 850 to 2010 (latent)
Blood analyzer	120 samples/h	2510	2510
Blood analyzer with CRT screen	115 samples/h	5120	5120
Centrifuge (large)	8 to 24 places	3750	3580
Centrifuge (small)	4 to 12 places	510	480
Chromatograph	—	6820	6820
Cytometer (cell sorter/analyzer)	1000 cells/s	73,230	73,230
Electrophoresis power supply	—	1360	850
Freezer, blood plasma, medium	13 ft^3, down to −40°F	340[b]	136[b]
Hot plate, concentric ring	4 holes, 212°F	3750	2970
Incubator, CO_2	5 to 10 ft^3, up to 130°F	9660	4810
Incubator, forced draft	10 ft^3, 80 to 140°F	2460	1230
Incubator, general application	1.4 to 11 ft^3, up to 160°F	160 to 220[b]	80 to 110[b]
Magnetic stirrer	—	2050	2050
Microcomputer	16 to 256 kbytes[c]	341 to 2047	300 to 1800
Minicomputer	—	7500 to 15,000	7500 to 15,000
Oven, general purpose, small	1.4 to 2.8 ft^3, 460°F	2120[b]	290[b]
Refrigerator, laboratory	22 to 106 ft^3, 39°F	80[d]	34[d]
Refrigerator, blood, small	7 to 20 ft^3, 39°F	260[b]	102[b]
Spectrophotometer	—	1710	1710
Sterilizer, freestanding	3.9 ft^3, 212 to 270°F	71,400	8100
Ultrasonic cleaner, small	1.4 ft^3	410	410
Washer, glassware	7.8 ft^3 load area	15,220	10,000
Water still	5 to 15 gal	14,500[e]	320[e]

[a] For hospital equipment installed under a hood, the heat gain is assumed to be zero.
[b] Heat gain per cubic foot of interior space.
[c] Input is not proportional to memory size.
[d] Heat gain per 10 ft^3 of interior space.
[e] Heat gain per gallon of capacity.

Source: Alereza and Breen (1984).

flows that impinge on the side of a hood more seriously alter airflow than do cross-flows in front of the hood (Schuyler and Waechter 1987). Large quantities of supply air can best be introduced through perforated plate air outlets or diffusers designed for large air volumes. The air supply should not discharge on a fire detector, as this slows its response.

Some general air distribution guidelines are (Caplan and Knutson 1978):

1. Terminal velocity of supply air jets (near hoods) is at least as important as hood face velocity in the range of 50 to 150 fpm face velocity.
2. The terminal throw velocity of supply air jets (near hoods) should be less than the hood face velocity, preferably no more than one-half to two-thirds the face velocity. Such terminal throw velocities are far less than those for conventional room air supply.
3. Perforated ceiling panels provide a better supply system than grilles or ceiling diffusers because the system design criteria are simpler and easier to apply, and precise adjustment of fixtures is not required. Ceiling panels also permit a greater concentration of hoods than do wall grilles or ceiling diffusers.
4. Wall grilles or registers should have double deflection louvers set for maximum deflection. The terminal velocity (near hoods) should be less than one-half of the face of the hood velocity.
5. If the wall grilles are located on the wall adjacent to the hood, the supply air jet should be above the top of the hood face opening. For equal terminal throw velocities, grilles on the adjacent wall cause less spillage than grilles located on the opposite wall.
6. The terminal throw velocity from ceiling diffusers at the hood face should be less than the hood face velocity.
7. Diffusers should be kept away from the front of the hood face. A larger number of smaller diffusers is advantageous, if the necessary low-terminal velocity can be maintained.
8. Blocking the quadrant of the ceiling diffuser blowing at the hood face results in less spillage.
9. Perforated ceiling panels should be sized so that the panel face velocity is less than the hood face velocity, preferably no more than two-thirds of the hood face velocity.
10. Perforated ceiling panels should be placed so that approximately one-third or more of the panel area is remote (more than 4 ft) from the hood.
11. Additional tests are needed to determine laboratory fume hood performance; Peterson *et al.* (1983) indicate that the only way to determine hood effectiveness is to test the specific hood under actual room conditions.
12. Room air distribution currents may be evaluated to determine if local air currents are high enough to cause discomfort or disturb experiments (ASHRAE 1990).

Variable Flow Supply Air Systems

A constant volume supply air system infers that a mass airflow balance exists and treats the specific pressure relationships between laboratory spaces and other occupied areas as a constant relationship. A variable airflow system is one in which the airflow may vary at any time. Variable flow systems have an additional degree of control freedom that requires building and space pressure control. The pressure control variable operates in parallel with thermal control requirements and causes competition between control variables for system control.

Pressure is controlled by introducing supply air into the controlled space to maintain specific pressure differentials across the space boundaries, or by the metered introduction of air to exactly replace a measured amount of air exhausted plus or minus a specific amount that is necessary to maintain the desired pressure relationship. The two approaches have distinctly different operating characteristics and different cost and construction implications.

Differential pressure (velocity) control requires tightly constructed and compartmentalized facilities. Engineering parameters are more difficult to predict. As differential pressure control works directly with the pressure relationship to be maintained, the control system is self-balancing but reacts to transient disturbances. Pressure controls recognize and compensate for unquantified disturbances such as stack effects, infiltration, and the influences of other systems within the same building. Expensive, complicated controls are not required, but the controls must be sensitive and reliable. Controls not required to be installed in corrosive environments can support any combination of exhaust applications, and are insensitive to minimum duct velocity conditions. Successful pressure control provides the desired directional airflow but cannot guarantee a specific mass balance.

Volumetric control requires that the flow of air at each supply and exhaust point from each space be controlled. Volumetric controls do not recognize or compensate for unquantified disturbances such as stack effects, infiltration, and influences of other systems within the same structure. Volumetric controls are insensitive to transient disturbances. Because engineering parameters are easy to predict, extremely tight construction is not necessary. Balancing is difficult and must be addressed across the full range of system operation for each space. Controls are often located in corrosive or contaminate environments; the controls may be subject to fouling, corrosive attack, and/or loss of calibration. Flow measurement controls are sensitive to minimum duct velocity conditions. Application of volumetric control does not guarantee directional airflow.

To meet space cooling needs of systems in which the primary makeup air also provides space cooling, a thermostatically-actuated exhaust may be required to induce the necessary amount of supply air. Systems in which heating and cooling systems are separate from the source of makeup air do not need a thermostatically actuated exhaust. Pressure controlled systems require careful sealing of the rooms served by these systems to ensure adequate control. The differential pressure necessary to induce a 100 fpm velocity airflow across a space boundary is aproximately 0.001 in. of water after accounting for orifice losses.

Unitary Systems

The unitary system form of air supply consists of a separate air-handling unit for each laboratory space. Each unit is made up of a fan and air treatment apparatus with a capacity equal to that required to maintain space temperature and to balance the exhaust air requirements. The unit typically contains a cooling coil, heating coil, humidifier, and filter. It is serviced with electricity, chilled water, steam, or hot water.

The unitary system shuts down when its operation is not needed. However, continuous, 100% outside air ventilation for laboratories containing hazardous chemicals is recommended. The system can be designed to match exhaust fan capacity and can be regulated to balance exhaust quantities, if they are variable. Each unit is capable of delivering sufficient heating and/or cooling to satisfy peak requirements of the space it serves. Each unit is also capable of separately reducing the heating and cooling capacity as the load in the laboratory space varies. Also, laboratories not requiring continuous service can be shut down without disrupting other laboratories.

The unitary system, when constructed to high quality standards, is initially expensive, takes considerable space, and is often costly to maintain. Its chief application is for isolated laboratory spaces and for buildings where operation hours are irregular. Unitary equipment with limited air treatment capabilities is used to supplement central apparatus in areas where the central system would be overloaded.

In biomedical laboratories, the installation of unitary systems is discouraged where the cooling coil with condensate drip pan and roughing filter would be located within the laboratory. The moisture associated with the coil and drip pan and the dust-collecting areas of the unit contribute to growth and dissemination of molds and other organisms commonly found in the environment. These organisms may be undesirable contaminants to experimental cultures, specimens, and biological products.

Central Systems

The simplest form of a central system that can be sucessfully applied to a laboratory subjected to variations in internal heat gain is a constant volume terminal reheat system in which the supply air is conditioned to (1) a dry-bulb temperature that satisfies the maximum sensible heat release in any space and (2) a dew point satisfactory for maintenance of room humidity within an acceptable range. Variations in heat gain in individual laboratory modules can be thermostatically controlled by reheat coils in the branch duct serving each space.

This system is economical if (1) close humidity control is not critical; (2) internal heat gains are moderate and fairly constant within a space and do not vary greatly between spaces; and (3) the exhaust air quantities are constant and in balance with the supply air necessary to maintain space conditions.

Since the entire central system must be in operation if any one space is being used, hours of occupancy or operation for each laboratory space should be approximately the same.

Central systems with supplementary conditioning or those with auxiliary air supply should be considered when (1) heat gains are high and subject to variation and (2) the exhaust air quantities are greater than supply air requirements for cooling. Auxiliary air should be tempered to within 5°F of room temperature and be filtered. Humidity variations in the laboratory due to the relatively unchanged auxiliary airstream should be calculated and presented to the research director for approval.

EXHAUST SYSTEMS

The total airflow rate required for some research laboratories is often dictated by the number and size of the fume hoods and biological safety cabinets; in other laboratories, the total airflow rate may be dictated by the laboratories' heat load. One fume hood may exhaust over 1000 cfm and thus determine the minimum amount of supply air required by the room. In other rooms, the sensible heat load from the research equipment may be 8000 Btu/h or higher for a 10 ft by 22 ft area. The HVAC engineer must discuss the fume hood requirements and the research equipment heat load requirements with the research staff. Often, a small hood or a hood with a reduced opening can meet the research needs while enabling a reduction in air volume requirements. Smaller hoods and those with reduced openings often provide better protection to the hood user. It is good practice to determine the ultimate needs of each laboratory so that possible increases in the hood exhaust volume can be handled by the installed HVAC systems. A prudent minimum recommendation is to allocate mechanical room and duct shaft space for future HVAC equipment and ductwork. One approach is to install the maximum size supply air ductwork and provide space for future exhaust ductwork. This planning minimizes expense and interruption of laboratory activities when future construction is required.

Economies in HVAC equipment and operation may be obtained by installing all the fume hoods and biological safety cabinets in separate support laboratories (common equipment rooms) that serve several researchers. The decision to remove any fume hood or safety cabinet from a research laboratory must be made by the research director and the safety officer.

Other possible economies with fume hood requirements are the selection of horizontal sliding sash fume hoods and/or reduced sash opening. The use of horizontal sliding sash (on dual tracks) fume hoods can reduce the effective face opening by 50%. This type of fume hood requires 50% less air. Another possible savings is to require that the maximum working opening of the standard hood (having a vertical sash) be half open and to design the air quantities based on this reduced opening size. This alternative should include instrumentation and alarms, and the operator should be safety trained.

Using the above experience with the normal density of fume hoods and research equipment, many research laboratories require about 20 to 30 air changes per hour of outside air. Often these laboratories require about 3 to 4.5 cfm per square foot of floor area.

Many institutions have established internal policies that require minimum air change rates for research laboratories. Generally, minimum rates must be within the range of 6 to 10 air changes per hour of 100% outside air. In most cases, the airflow requirements for the fume hoods and the biological safety cabinets exceed these minimum requirements. Also, the normal heat load from the research equipment usually exceeds the minimum requirements. (The minimum room air change rate policies do not limit the design airflow rate; the policies only establish a minimum rate that must be met.)

Laboratory exhaust systems can be classified as (1) constant volume or (2) variable volume, based on the laboratory fume hood and biological safety cabinet characteristics and the method of system operation and control. These classifications can be further divided into (1) individual, (2) central, or (3) combination systems, based on the arrangement of the major system components such as the fans, plenums, or duct mains and branches.

Degenhardt and Pfost (1983) describe a design of hospital laboratory fume hood exhaust system. Sessler and Hoover (1983) describe ways to avoid high noise levels sometimes generated by the laboratory fume hood exhaust system.

All laboratory fume hoods and biological safety cabinets must be equipped with visual and audible alarms to warn the laboratory workers of unsafe airflows. As alarms are among the most important safety features in a laboratory, they must be selected for proven reliability and accuracy.

Variable Volume Exhaust Systems

The decision to select a variable volume exhaust system should not be made without the understanding and approval of the research staff and local safety officials. The level of sophistication and ability of the maintenance staff to maintain such a complex system is also an important consideration.

In many laboratories, all hoods and safety cabinets are seldom needed at the same time. Thus, a laboratory that permits a usage diversity factor allows the exhaust system to have less capacity than that required for the full operation of all units. Even if the exhaust system is sized for full operation of all units, reducing the airflow during periods when some of the hoods and safety cabinets are not in use reduces operational (energy) costs.

Due to less than ideal conditions between the design and installation of systems, a generous safety factor should be added to the system capacity.

Exhaust air volume may be reduced by a face-velocity controlled hood. In addition, the hood face velocity remains constant when the hood face opening is partially closed. A sensing device responds to changes in hood sash geometry and operates a volume control device (duct damper control device, duct discharge damper, or a variable speed drive) to maintain the face velocity within the desired range. In large central systems, a volume control device gives satisfactory control for branch ducts, but system volume regulation must be supplemented by static pressure regulators (in the exhaust plenums) to control fan air volume. A hood served by an individual fan may obtain constant face velocity with either a duct control device or fan control device. Complete exhaust air system shutdown should be avoided as it may cause cold air and contaminants to enter the laboratory through the exhaust system. Variable volume exhaust systems allow more freedom in the installation of hoods and safety cabinets because the number of units connected is not entirely dependent on the capacity of the exhaust system.

Variable air volume systems are more difficult to balance and control and may be less stable in operation than constant volume systems. VAV systems require extensive instrumentation and controls that, in turn, cause high installation and maintenance costs. Control difficulties increase with system size and static pressure requirements. VAV systems are more dynamic than constant volume systems and may react to disturbances such as door openings. VAV systems require extensive instrumentation and controls, which cost more unless the system size is reduced to take advantage of the large diversities possible with these systems. Eventually, the payback in operation costs may outweigh these initial considerations. Balancing dampers in exhaust ducts are prohibited by codes for some applications. In corrosive atmospheres, variable air volume control equipment requires additional maintenance and must be installed where inspections and repairs can easily be made. The reliability of control equipment located in corrosive environments is questionable. If minimum duct velocities are to be maintained, a variable air volume system may not be suitable (see Table 4).

Diversified variable volume systems are potentially hazardous when the collective area of operating chemical fume hoods and biological safety cabinet face openings exceed design opening diversity values. If this condition occurs, design airflow cannot be maintained and laboratory personnel may be endangered. If credit is taken for diversity, it should only be assumed in central fan and primary distribution systems and not at individual terminals. If, on the other hand, total usage is less than design values, bypass devices may be required on hoods to maintain supply air rates, provide adequate thermal capacity, and ensure air balance and flow patterns. VAV fume hoods have additional advantages of providing accurate monitoring and alarm capabilities at the hood.

Accurate room pressurization controls to match supply and exhaust flows must be a part of the VAV hood system, however. To achieve better accuracy, more expensive, industrial-grade controls may be selected.

Individual Exhaust Systems

Individual exhaust systems include a separate exhaust connection, exhaust fan, and discharge duct for each laboratory fume hood or biological safety cabinet. This arrangement makes adding or moving hoods difficult in a large laboratory building without planning for the ultimate complement of hoods and restricting the locations of hoods. The system permits selective operation of

Table 4 Range of Design Duct Velocities for Exhaust System

Nature of Contaminant	Examples of Exhaust Materials	Design velocity, fpm
Vapors, gases, smoke	All vapors, gases, and smokes	1000 - 1200
Fumes	Zinc and aluminum oxide fumes	1400 - 2000
Very fine, light dust	Cotton lint, wood flour, litho powder	2000 - 2500
Dry dust and powders	Cotton dust, light shavings	2500 - 3500

Source: *Industrial Ventilation: A Manual of Recommended Practice.* American Conference of Governmental Industrial Hygienists, Cincinnati, OH, 4-7.

individual hoods and safety cabinets merely by starting or stopping the fan motor. However, stopping the fume hood exhaust fans may disrupt the negative pressurization maintained in the laboratory and allow the backflow of contaminants from the exhaust system into the laboratory. Shutdowns for repair or maintenance are localized. The unitary arrangement permits selective application of (1) special exhaust air filtration; (2) special duct and fan construction for corrosive effluents; (3) emergency power connections to selective fan motor; and (4) off-hour operation.

Individual exhaust fans are simple to balance and, when installed with a constant volume supply air system, provide a stable system. Maintaining negative pressurization requires periodic testing and adjustment of control dampers. The variation in exhaust flow between the fully open and fully closed hood sash must be considered. The recommended operation is to keep exhaust fans on at all times and to interlock them electrically with the supply fans so that if any critical exhaust fan is shut down, the supply fans shut down automatically. The individual exhaust system requires more fans than central systems and there are usually more overall duct shaft space requirements because of the many small ducts. The use of more fans also increases capital and maintenance costs.

Most research laboratories require directional airflow from the corridor into the laboratory to contain airborne contamination. The shutdown of individual exhaust systems upsets the proper directional airflow and causes hazardous contaminants and odors to flow out of the laboratory and into the corridor and adjacent rooms. If such a system is considered, appropriate precautions such as air locks or variable air volume controls should be installed to prevent reverse airflow.

Central Exhaust Systems

Central exhaust systems, which combine a number of fume hoods on the same exhaust duct, can increase overall safety by increased air dilution. These systems consist of one or more fans, a common exhaust (suction) plenum, and branch connections to multiple exhaust terminals. A backup fan or redundant fan capacity should be provided whenever possible. Central systems are difficult to balance initially and require periodic rebalancing to ensure proper airflow. The effects of mixing effluents from different research operations must also be considered with central systems.

The central exhaust system is best used to exhaust similar devices such as laboratory fume hoods. The exhausting of laboratory fume hoods, biological safety cabinets, and special filtered units (radioisotope fume hoods) with one central exhaust system can be difficult to control because pressure losses vary between the different devices. Solutions have been the use of constant volume regulators or boxes and the sizing of HEPA filters to provide a small pressure drop (or change) from their clean to their dirty conditions. The central exhaust system must be carefully controlled to maintain a flow within ± 5% to enable constant face velocity in hoods, biological safety cabinets, and so forth.

Research laboratories frequently change the demands made on the building exhaust system. The most common change is the adding of chemical fume hoods. After such changes, the entire system must be checked and rebalanced. Large central systems are difficult to rebalance.

Exhaust Fans and Ductwork

Exhaust fans that handle contaminants should always be located outside of occupied building areas and be close to the point of discharge. The ductwork on the discharge side of the fan should be airtight. Air leakage from an unsealed, pressurized exhaust duct can contaminate a mechanical equipment room or an entire building. The fan discharge should be connected directly to the vertical discharge stack without connections to other exhaust systems.

Ranges of duct velocities required for fume hood exhaust systems to carry different materials are presented in Table 4. However, many laboratory design engineers recommend a minimum velocity of 2000 fpm based on field observations of duct corrosion (Steere 1971, Koenigsberg and Seipp 1988).

Materials and Construction

The selection of materials and the construction of exhaust ductwork and fans depend on the following:

- Nature of the effluents
- Ambient temperature
- Lengths and arrangement of duct runs
- Method of hood fan operation
- Flame and smoke spread rating
- Duct velocities and pressures
- Effluent temperature

Effluents may be classified generically as organic or inorganic chemical gases, vapors, fumes, or smokes; and qualitatively as acids, alkalis, solvents, or oils. Exhaust system ducts, fans, dampers, flow sensors, and coatings are subject to (1) corrosion, which destroys metal by chemical or electrochemical action; (2) dissolution, which dissolves materials (coatings and plastics are subject to this action, particularly by solvent and oil effluents); and (3) melting, which can occur in certain plastics and coatings at elevated hood operating temperatures.

Commonly used reagents in laboratories include hydrochloric, sulfuric, and nitric acids (singularly or in combination), and ammonium hydroxide. Common organic chemicals include acetone, ether, petroleum ether, chloroform, and acetic acid.

Ambient temperature of the space housing ductwork and fans affects the condensation of vapors in the exhaust system. Condensation contributes to the corrosion of metals with or without the presence of chemicals.

Ductwork is less subject to corrosion when runs are short and direct and flow is maintained at reasonable velocities. The longer the duct, the longer the exposure to effluents and the greater the possibility of condensation. Horizontal runs provide surfaces where moisture can remain longer than on vertical surfaces. If condensation is probable, provide sloped ductwork and condensate drains. (The condensate drains may accumulate hazardous materials.)

Fan operation may be continuous or intermittent. Intermittent fan operation may allow longer periods of wetness due to condensation than would continuous fan operation.

Flame and smoke spread rating requirements established by codes and insurance underwriters must also be considered when selecting ductwork materials.

The procedures and recommendations for the selection of materials and construction are as follows:

1. Determine the types of effluents (and possible combinations of effluents) generated in the hood and handled by the exhaust system. Consider both present and future operations.
2. Classify the effluents as organic or inorganic and whether they occur in gaseous, vapor, or particulate form. Also, classify decontamination materials.
3. Determine the concentration of the reagents used and the temperature of the effluents at the hood exhaust port. In research laboratories, this determination is almost impossible.
4. Estimate the highest probable dew point of the effluents.
5. Determine the ambient temperature of the spaces where the ductwork is routed and in which the exhaust fans are located.
6. Consider the length and arrangement of duct runs and how they may affect the periods of exposure to fumes and the degree of condensation that may occur.
7. Consider the effects of intermittent versus continuous fan operation. If intermittent operation is desired, provide a time delay (about one hour) to dry wet surfaces before fan shutdown. Intermittent operation can easily unbalance airflows in the laboratory and cause unsafe conditions; continuous operation during working hours is better.
8. Determine whether insulation, watertight construction, slope, and drains are required.

9. Select materials and construction most suited for the application by considering the following:
   a) resistance to chemical attack
   b) weight
   c) flame and smoke spread rating
   d) installation and maintenance costs

Standard references and manufacturers have information on material properties. Materials for chemical fume exhaust duct systems and their characteristics include the following:

Glazed tile — Resistant to practically all corrosive agents except hydrofluoric acid; heavy in weight, limited to round sections, joint sealants subject to attack, considerable space required for directional changes; material costs low, installation costs high.

Cementatious material — Highly resistant, porous surface requires an internal impervious coating to prevent retention of potentially flammable materials; limited to round sections because of the difficulty in sealing joints, constructing directional changes and transitions, and bracing and supporting rectangular sections; joint sealants subject to attack; cost moderately high.

Galvanized iron — Subject to acid and alkali attack, particularly at cut edges and under wet conditions; easily formed; low in cost.

Stainless steel — Subject to acid and chloride compound attack varying with the chromium and nickel content of the alloy; relatively high in cost. (The higher the alloy content, the higher the resistance. Stainless steels range from the commercial 200 to the 400 alloy series, with ascending chromium and nickel content to proprietary alloys with custom chromium and nickel composition. Costs increase with chromium and nickel content.)

Asphaltum-coated steel — Resistant to acids; subject to solvent and oil attack; high flame and smoke spread rating; base metal vulnerable when exposed by coating imperfections and cut edges; moderate cost.

Epoxy-coated steel — Epoxy phenolic resin coatings on mild black steel can be selected for particular characteristics and applications; these have been successfully applied for both specific and general use, but no one compound is inert or resistive to all effluents; requires sand blast surface preparation for shop-applied coating and field touchup of coating imperfections and damage that occurs during shipment and installation; cost is moderate.

Fiberglass — Particularly good for acid applications including hydrofluoric, when additional glaze coats are provided.

Plastic materials — Have particular resistance properties to particular corrosive effluents; their limitations are in physical strength, flame spread rating, heat distortion, and high cost of fabrication.

10. Select fans constructed of the same materials as the ductwork or of mild steel with a suitable coating.
11. Provide outboard bearings, shaft seals, access doors, and multiple 200% rated belts for hood exhaust fans. Bearings should have a minimum rated life (L_{10}) of 100,000 h.
12. Design the ductwork and select the fan with consideration of potential fire and explosive hazard. Ductwork must meet flame spread and other requirements, as described in National Fire Protection Association standards.
13. For some systems, such as perchloric acid hoods, provide wash-down capability.
14. Develop layouts so that ducts may be easily inspected, decontaminated, and replaced, if necessary.

Exhaust Air Filtration

Depending on the hazard level associated with the laboratory operation and the degree of physical containment desired, filtration facilities for exhaust systems may be required. The hazardous or obnoxious pollutants to be removed from the exhaust air may be particulate and/or gaseous in nature.

Dry media filters, *e.g.*, 95% efficient by ASHRAE *Standard* 52-68 Test Method, or HEPA filters (99.97% efficient by DOP Test Method) may be required to meet specified design criteria. The filter assembly may include a prefilter for coarse particle separation and a filter enclosure arranged for ready access and easy transfer of the contaminated filter to a disposal enclosure. Manufactured filter enclosures that feature bag-in/bag-out filter changing should be considered for hazardous exhaust situations. A procedure for testing the filter system integrity and suitable test openings is also necessary. A damper is often added to balance airflow by compensating for the change in the resistance of the HEPA filter.

For convenient handling, replacement, and disposal, with minimum hazard to personnel, the filter should be: (1) located immediately outside the laboratory area, unless it is an integral part of a safety cabinet or hood; (2) located on the suction side of the exhaust fan; (3) installed in adequate space that provides free, unobstructed access; and (4) positioned at a convenient working height. Some installations require shutoff dampers and hardware for filter decontamination in the ductwork. The filter should be located on the suction side of the exhaust fan and as close as is practical to the source of contamination (laboratory) to minimize the length of contaminated ductwork. Parallel redundant filters and bypass arrangements should be considered for continuous operation of exhaust systems during filter change. Static pressure monitors should be placed across filter and prefilter banks to help determine time for filter change and damper adjustment.

Wet collectors or adsorption systems, such as activated charcoal, are often satisfactory for removing gas-phase toxic or odorous pollutants from exhaust air. When designing the installation of these devices, the safety of maintenance and service personnel, the potential concentrations of exhaust materials, and the frequency of filter replacement must be considered. When using charcoal filters, it is important to provide a gas monitoring system to determine time for changeout; otherwise, filters outgas when saturated.

Fire Safety for the Exhaust System

Most local authorities have laws that incorporate National Fire Protection Association (NFPA) *Standard* 45-1988, Fire Protection for Laboratories Using Chemicals. Laboratories located in patient care buildings require fire standards based on the NFPA *Health Care Facilities Handbook* (1987) in lieu of NFPA *Standard* 45.

Selected NFPA design criteria from NFPA *Standard* 45 include the following (references to specific items are noted):

Air Balance. "Laboratory. . . shall be maintained at an air pressure that is negative relative to the corridors or adjacent nonlaboratory areas." para. 6-4.2

Controls. "Controls and dampers. . . shall be of a type that, in the event of failure, will fail in an open position to assure continuous draft." para. 6-6.7

Diffuser Locations. "Care shall be exercised in the selection and placement of air supply diffusion devices to avoid air currents that would adversely affect the performance of laboratory hoods. . . " para. 6-4.3

Exhaust Stacks. "Exhaust stacks should extend at least 7 ft above the roof. . . " para. A-6-8.7

Fire Dampers. "Automatic fire dampers shall not be used in laboratory hood exhaust systems. Fire detection and alarm systems shall not be interlocked to automatically shut down laboratory hood exhaust fans. . ." para. 6-11.3

Hood Alarms. "Airflow indicators shall be installed on new laboratory hoods or on existing laboratory hoods, when modified." para. 6-9.7

Hood Placement. "A second means of access to an exit shall be provided from a laboratory work area if . . . a hood . . . is located adjacent to the primary means of exit access." para. 3-3.2 (d)

"For new installations, laboratory hoods shall not be located adjacent to a single means of access to an exit or high traffic areas." para. 6-10.2

Recirculation. "Air exhausted from laboratory hoods or other special local exhaust systems shall not be recirculated." para. 6-5.1.

"Air exhausted from laboratory work areas shall not pass unducted through other areas." para. 6-5.3

The designer should review the entire NFPA standard, determine if it is incorporated into the local code, and advise the other members of the design team of their responsibilities (such as fume hood placement).

The incorrect placement of fume hoods and biological safety cabinets is a frequent design error and a common cause of costly redesign work. The placement of the fume hood involves an analysis of the laboratory's work pattern and the location of casework, sinks, utilities, and so forth.

FUME STACK HEIGHT AND SAFETY

Peterson (1987) shows that short stacks on fume exhaust fans are a major hazard. A U.S. Environmental Protection Agency engineering practice recommends that the top of stacks that emit "large quantities" of pollutants be 2.5 times the building height. However, experience has shown no simple rule to determine safe stack heights for all situations. Further information is available in Chapter 14 of the 1989 ASHRAE *Handbook—Fundamentals*.

Fume exhausts must discharge straight up to transport the exhaust away from the building. Avoid rain caps, mushroom-shaped roof exhausts, and goosenecks that direct the exhaust down or sideways. An example of rain protection without a rain cap is shown on page 14.14 of the 1989 ASHRAE *Handbook—Fundamentals*. Failure to direct the stack discharge properly can cause fumes to recirculate into the building (Heider 1972).

Materials must discharge high enough above the building so that they do not (1) enter the outside air intake, (2) accumulate on the roof in dangerous concentrations, (3) enter the building doors or windows, or (4) present a danger to people on the roof or on the ground.

Sophisticated numerical models of the exhaust flow in conjunction with wind tunnel tests give the safest engineering design. These models show that discharge velocities usually should be greater than 2500 fpm, a velocity which was often recommended in the past. Instead, the discharge velocity should be 1.5 times greater than the wind speed that occurs 2% of the time or less. These studies also show that fewer stacks with greater volumes may have several times greater dilutions than smaller stacks with smaller volumes. In many cases, increasing stack height to about 20 to 30 ft (above the roof) gives good dilution.

AUXILIARY AIR SUPPLY

Exhaust air requirements for laboratory fume hoods and biological safety cabinets often exceed the supply air needed for air conditioning. Exhaust air requirements often dictate the supply air quantity for a laboratory. The supply air needed to maintain building air pressure can be obtained from primary and auxiliary systems. Primary supply air systems meet the air-conditioning needs of the laboratory. Auxiliary air supply systems augment the primary system to meet the air quantity requirements of hoods. Auxiliary air can be ducted directly to auxiliary air fume hoods or supplied directly to the room. Auxiliary air should be conditioned to meet the temperature and humidity needs of the laboratory. Operational costs may be reduced by keeping the heating, cooling, and humidifying energy requirements of the auxiliary air system to a minimum. First cost penalties for auxiliary air systems are usually higher than for other systems; operating costs are not necessarily reduced.

In some laboratory buildings, office and research areas are separated by a corridor. The offices are kept under pressure, so air exits from the offices into the corridor, and then into the laboratories. Transfer is simple, and air flows from less hazardous to more hazardous areas. Corridors may convey and distribute air as plenums within the limits established by NFPA 90A. Using this air from offices may eliminate or reduce the size of auxiliary air systems.

Constant Volume Auxiliary Air Supply

Airflow in a constant volume auxiliary airflow system remains unchanged during normal day-to-day operation. This system can be installed with excess capacity so that, as laboratory fume hoods or biological safety cabinets are added or relocated, appropriate changes in fan speed, inlet vane setting, or the distribution system can be made to maintain an air balance.

Constant airflow systems adapt to supplement cooling, particularly when high exhaust requirements and high internal heat gains occur simultaneously.

The change in exhaust volume due to changing the sash from fully open to fully closed must be considered. Some control of auxiliary air volume may be necessary because of changes in hood operation (on or off) and varying sash openings.

Variable Volume Auxiliary Air Supply

When exhaust air varies, the auxiliary supply air must vary to maintain an air balance. Central systems having many variable exhaust openings and a few central supply fans can be controlled by static pressure regulators cycling the fans, modulating the fan capacity, or by controlling dampers in a bypass duct section connecting the fan discharge and inlet. However, such systems are complex and difficult to balance.

Variable supplementary supply systems are not suitable for cooling unless the exhaust and supplementary supply air are directly proportional to the internal heat gain. This condition may exist in areas of extremely high and variable heat release, such as engine test cells, where the exhaust is primarily for heat removal.

The distribution of variable quantities of auxiliary air is complicated because it not only adds to the base system, but it is constantly changing. This problem can be solved if air can be safely supplied to an adjacent corridor and transferred into the laboratory through carefully sized and located louvers.

Unconditioned Auxiliary Air Supply

Some air-conditioned laboratories have unconditioned (outdoor) auxiliary supply air introduced directly adjacent to the exhaust hoods. This air must be used with great care. Air introduced within the fume hood can adversely affect safety and health by forcing fumes into the face of the researcher. Auxiliary air introduced outside the hood by a well-engineered terminal avoids this potential danger if adequate capture velocity is maintained at the face of the hood. Careful initial evaluation of the installed hood and subsequent monitoring ensures proper hood performance.

It is impossible to use an unconditioned auxiliary air system without changing the temperature and humidity in the laboratory. The added cost of this type of system may be greater than the value of the cooling capacity saved, since the air must be partially conditioned by heating to a reasonable temperature in the winter and, in some cases, by cooling in the summer.

RECIRCULATION

Air exhausted from chemical or biological laboratories should not be recirculated because it may expose both personnel and research materials to hazardous airborne contamination without warning. However, it is acceptable to recondition (*e.g.*, fan coil) air within individual laboratories.

In large facilities, the ventilation system serving chemistry and biological laboratories should discharge all exhaust air to the outdoors. Other laboratories in the same facility (*e.g.*, electronics) may be served by a recirculating system, providing that air quality is acceptable. Recirculation of a large percentage of air can cause odors or entrain contaminants. The HVAC design engineer should carefully review the laboratory safety guidelines listed in Table 1 and then obtain the final approval for this key design parameter from the laboratory's safety officer and research director before the project proceeds into the design phase. In the event of a chemical spill, all air should be exhausted to the outside.

Office areas in a chemical or biological research facility should be segregated from the laboratories and served by a conventional office recirculating ventilation system. These areas are not suitable for future use as laboratories without major upgrading of their HVAC systems.

AIR BALANCE AND FLOW PATTERNS

Control of airflow direction in research laboratories controls the spread of airborne contaminants, protects personnel from toxic and hazardous substances, and protects the integrity of experiments. In these facilities, the once-through principle of airflow is applied based on: (1) exhausting 100% of the supplied air; (2) maintaining constant volume airflow with all exhaust units operating at capacity; and (3) providing directional flow of air from areas of least contamination to those of greatest contamination. (In support facilities, such as office suites, where contamination is not likely, standard HVAC systems can be provided.) Determinants for air pattern control are: (1) type of research materials handled or generated in each space; (2) type, size, and number of laboratory fume hoods, biological safety cabinets, and auxiliary exhaust equipment in each space; and (3) permissibility of air transfer into or out of spaces.

Supply and exhaust airflow patterns may affect fire detector or suppression system operation. For example, a sidewall supply grille with a relatively high discharge velocity may retard the activation of a nearby detector or automatic sprinkler head. Coordination within the design team reduces this potential.

For critical air balance conditions, a personnel entry or exit air lock provides a positive means of air control. An air lock is an anteroom with airtight doors between a controlled and uncontrolled space. The air pattern in the air lock suits the foregoing laboratory space air balance requirements.

Supply air quantities are not fully established by the room cooling requirements and load characteristics. Additional supply air required to make up the differences between room exhaust requirements and primary supply may be designated: (1) infiltrated supply, if induced indirectly from the corridors and other spaces or (2) secondary supply, if conducted directly to the room.

Ventilation air quantities are often specified as air changes per hour. An air change occurs when a volume of outside air equal to the volume of the laboratory space passes through the space. For example, a 500 ft^2 laboratory with 10-ft ceilings has a volume of 5000 ft^3. If 500 cfm of outside air were supplied to (and exhausted from) that laboratory, an air change would occur every 10 min and there would be 6 air changes per hour. For some laboratories, a rate of 4 air changes per hour is a recommended minimum amount; for others, a rate of 12 air changes per hour is a guideline; for still other laboratories, a value of 20 air changes is required. Thus, the HVAC design engineer must carefully select the appropriate safety guideline from Table 1 and determine if the recommended *minimum* air change rate is sufficient to meet the needs of the safety officer, the chemical fume hoods, and the heat loads of the research equipment in the laboratory. Fume hoods in the space make the rate higher than the base rate. Certain special requirements, such as biohazard, radioactive hazard, fire hazard, and clean space, require higher rates. Refer to NFPA *Standards* 30 and 45 when quantities of flammable materials are handled in the laboratory. The possibilities of spills and broken glass containers should be considered in determining the ventilation airflow. It may be economical to have two-speed exhaust fans to give a higher flow possibility to prevent flammable mixture conditions in case of spills.

Risk assessment may indicate the need for explosion-proof electrical construction when the use or presence of flammable materials could provide a high potential for spills, resulting in flammable or explosive concentrations occurring in the space.

LABORATORY FUME HOODS

The Scientific Apparatus Makers Association (SAMA) defines a laboratory fume hood as "a ventilated enclosed work space intended to capture, contain, and exhaust fumes, vapors, and particulate matter generated inside the enclosure. It consists basically of side, back, and top enclosure panels, a work surface or counter top, an access opening called the face, a sash, and an exhaust plenum equipped with a baffle system for the regulation of airflow distribution" (SAMA 1980). The work opening has operable glass sash(es) for observation and shielding. A sash may be (1) vertically operable, (2) horizontally operable, or (3) vertically and horizontally operable. The latter provides maximum access for setting up apparatus, conserves energy due to the smaller face opening, and provides movable protective shielding. The horizontally operable sash-type fume hood has a similar energy-saving value. Laboratory fume hoods may be equipped with a variety of accessories such as internal lights, service outlets, sinks, air bypass openings, airfoil entry devices, flow alarms, special linings, a ventilated base unit (for storage of chemicals), and exhaust filters.

Figure 1 illustrates the basic elements of a general-purpose bench-type fume hood. Figures 2 and 3 show the basic features of a bypass-type fume hood and an auxiliary air fume hood, respectively.

Chemical fume hoods are manufactured to meet different research needs. Table 5 lists the different types and their typical application in laboratory facilities.

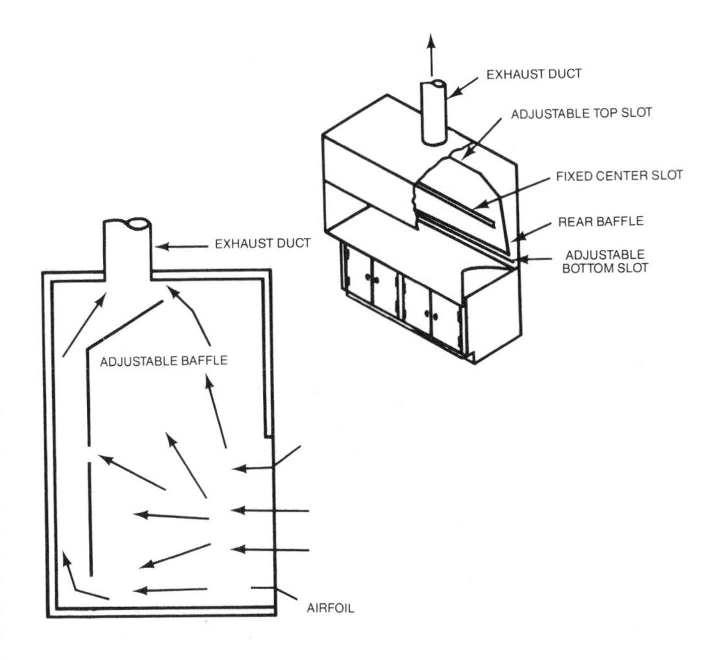

Fig. 1 Process Fume Hood

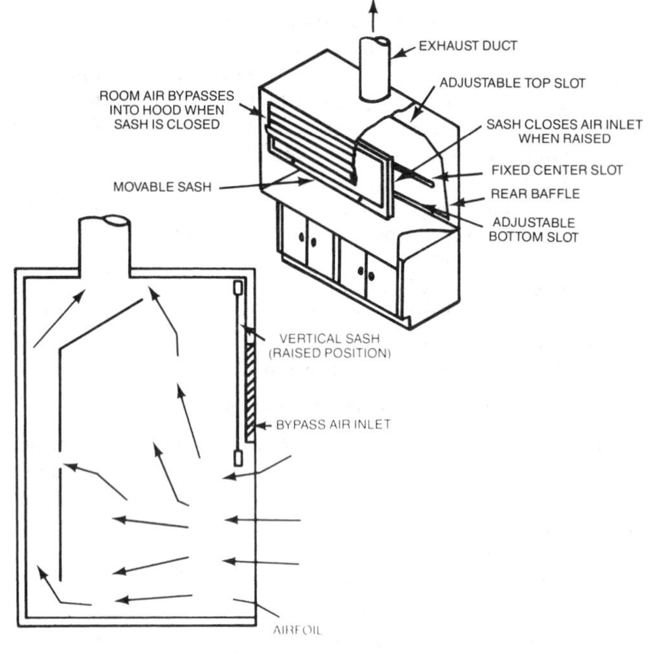

Fig. 2 Bypass Fume Hood with Vertical Sash and Bypass Air Inlet

Operating Principles

Containment of contaminants is based on the principle that a flow of air entering at the face, passing through the enclosure, and exiting at the exhaust port prevents the escape of airborne contaminants from the hood into the room. The degree to which this is accomplished depends on the design of the hood, its installation, and its operation. The critical design parameters of a hood are as follows:

Containment of contaminants:

- Face velocity
- Size of the face opening
- Shape of the opening surfaces
- Inside dimensions and location of work area relative to face edge
- Size and number of exhaust ports
- Back baffle and exhaust plenum arrangement
- Proportional bypass

Critical installation parameters:

- Distance from supply air outlets
- Type of air outlets
- Air velocity near hood
- Distance from doors
- Pedestrian traffic next to the hood face velocity
- Movements of researcher in hood face
- Location and type of research apparatus placed inside the hood

ASHRAE conducted tests to determine the interactions between room air motion and hood face velocity on the spillage of contaminants into the room (Caplan and Knutson 1977, 1978). Test conclusions included the observation that the effect of room air challenge is significant and of the same order of magnitude as the effect of hood face velocity. Consequently, improper design of replacement air supply can have a disastrous effect on the efficiency of a laboratory hood.

Hood Performance

Air currents external to a hood easily disturb its air pattern and cause contaminants to flow from the hood into the breathing zone of the researcher. Cross currents are generated by movements of the researcher, people walking past the hood, thermal convection, supply air movement, and rapid operation of room doors and windows. Terminal supply air velocity in the vicinity of the hood should be limited to one-half to two-thirds of the face velocity of the laboratory's hood (or biological safety cabinet). At 100 fpm face velocity, the room air velocity should be limited to about 50 fpm; however, if the hood velocity is 60 fpm, a room velocity of 50 fpm is much too high. It is very important to locate hoods away from doors and active aisles. This problem needs coordination between the designer and the research staff.

Performance Criteria

ASHRAE *Standard* 110-1985 covers performance testing of fume hoods. Performance criteria for fume hoods are: (1) flow control, (2) spillage, and (3) face velocity control. Flow is adjusted by horizontal slots in the back baffle. A slot at the bottom of the back baffle draws air across the working surface; another slot at the top exhausts the canopy; and a third slot is frequently midway on the baffle. These adjustable openings regulate exhaust distribution for specific operations; the openings should be set and locked by the certifying engineer.

All fume hoods should be tested annually by the engineer and their performance certified. Often, the vertical sash needs to be placed in a certain position for proper face velocity; the annual inspection should clearly mark this sash position.

Spillage of contaminates from hoods into the laboratory can be caused by: (1) drafts in the room; (2) eddy currents generated at hood opening edges, surface projections, or depressions; (3) thermal heads; and (4) high turbulence operations (blenders, mixers) within the hood. Corner and intermediate posts, deep deck lip depressions, sinks, and projecting service fittings near the face produce air turbulence and potential spillage conditions. Plain entrance edges produce a *vena contracta* within 1 in. of the surface and to a depth of 6 in. Fumes generated in this area are disturbed and may escape the hood enclosure. Air foil shapes at the entry edges correct this condition. Correcting this one feature on existing hoods has been the key to making satisfactory hoods from previously unacceptable units. Sinks and service fittings should be located at least 6 in. inside the hood face, and deck lips should have minimal projections.

Face velocity is affected by variations in the resistance of a hood exhaust system. Two common causes are: (1) variations in the face opening and (2) buildup of exhaust filter resistance. Increases in fume hood exhaust system pressures due to filter loading can range

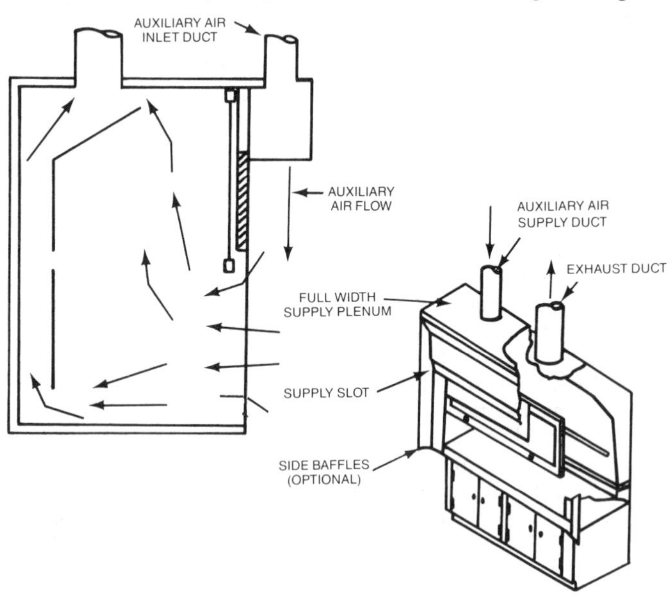

Fig. 3 Auxiliary Air Fume Hood

Table 5 Applications of Chemical Fume Hoods

Type of Hood	Research Applications
Standard	Research laboratories—continuous use. Moderate and high hazard processes; varying procedures. May have vertical, horizontal, or combination sash.
Process	Process laboratories—intermittent use. Low hazard processes; known procedures. No sash provided; use of specific operation.
Bypass	Research laboratories—continuous use. Moderate and high hazard processes; varying procedures. Has vertical sash.
Auxiliary air	Same application as *bypass hoods.*
Radioisotope	Process and research laboratories using radioactive isotopes. Special shielding and filters involved.
Perchloric acid	Process and research laboratories using perchloric acid. Mandatory use because of explosion hazard. Special washdown of hood and ductwork required.
California	For enclosing large and complex research apparatus. Size may be 6 ft wide by 8 ft high by 3 ft deep. Sash opening on both sides of hood.
Walk-in	Similar application as California fume hood. Only has sash on one side. Bottom of hood at floor level.
Canopy	Not a fume hood. Useful for heat removal over some work areas. Not to be substituted for chemical fume hood.

from 50 to 100% of the clean filter condition when high-efficiency filters are used. Pressure may be regulated by an automatic pressure controlled damper in the duct system, by a manually adjusted damper, or by exhaust fan vortex control. Most laboratory fume hoods do not require exhaust filters.

It is good practice to equip laboratory fume hoods with alarm devices to detect failure of exhaust airflow. Devices monitoring the rotation of the fan shaft are also recommended. An alarm, visual and/or audible, should be extended to all hoods served by the respective exhaust fan.

Performance Tests

Tests should be performed with the hood's sash open to three positions: (1) normal operating position, (2) 25% open, and (3) 50% open. Test by placing the usual amount of research equipment in the hood and using one of the common test methods. Some of the more common test methods include: (1) face velocity test, (2) outflow test, and (3) smoke test.

Face Velocity Test. The hood face velocity to be used as the test criteria is determined by each facility or company. The velocity is based on a balance between the safe operation of the fume hood, airflow requirements as dictated by the hood operation, and energy costs. To measure face velocity, form an imaginary grid pattern by dividing the vertical and horizontal dimensions to obtain one measurement per square foot. Take velocity readings with a calibrated heated wire anemometer, 4 in. behind the front face of the hood and/or the geometric face plane, at the intersections of the grid lines. The use of an instrument holder rather than a hand-held instrument is recommended to obtain an accurate measurement. The face velocity must be maintained at ±10% of the site-specified value. This average may be allowed to deteriorate to ±15% from the site-specified value before correction is required. The face velocity must be returned to the site-specified value. On standard fume hoods, individual velocity readings should not vary more than ±15% with the hood empty, or ±25% with research equipment in the hood. These ranges may be unattainable in fume hoods equipped with auxiliary air. In the case of auxiliary air hoods, additional tests are suggested to confirm acceptable airflow patterns.

Outflow Test. Swab a strip of titanium tetrachloride along both walls and the hood floor in a line parallel to the hood face and 6 in. back into the hood. (Titanium tetrachloride is corrosive to the skin and extremely irritating to the eyes and respiratory system.) Swab a large "A" on the back of the hood and on each side. Define air movement toward the face of the hood as reverse airflow and define the lack of movement as dead airspace. Swab the work surface of the hood, being sure to swab lines around all equipment in the hood. All smoke should be carried to the back of the hood and out. Test the operation of the bottom air bypass airfoil by running the cotton swab under the airfoil. Before going to the next test, move the cotton swab around the face of the hood; if there is any outfall, the exhaust capacity test should not be made.

The new ASHRAE outflow standard is a more quantitative test. It is documented in ASHRAE *Standard* ANSI/ASHRAE 110-1985, Method of Testing Performance of Laboratory Fume Hoods. This test determines the quantity of tracer gas that escapes from inside the hood and reaches the face of a mannequin standing at a working location at the hood. This test simulates the quantity of contaminants that escapes from inside the hood and reaches the face of the research worker standing at a working location.

Smoke Test. (To prevent false alarms, notify the laboratory director and the fire department before conducting this test.) Ignite and place a 30-s smoke bomb near the center of the work surface, making sure that the hole on the side of the bomb faces into the hood. After the smoke bomb begins to work, pick it up with tongs and move it around the hood. There should be no visual or odoriferous indications of the smoke outside the hood. Refer to ANSI/ASHRAE *Standard* 110-1985 for details regarding this test.

Airflow Indicator. Check the airflow indicator alarm to see if it is operating properly. Check to see that there is no leakage from the hood from the airflow indicator device.

Exhaust Fan. Check for proper performance.

Building Conditions. During the tests, the building air-conditioning or ventilating system should be operating in a normal fashion. During the smoke test, the room doors should be opened and closed to ensure that no leakage from the hood occurs.

Auxiliary Air Hoods

Auxiliary air hoods connect directly to an auxiliary air supply system. These hoods reduce the volume of conditioned room air exhausted, thereby reducing the overall cooling load. Auxiliary air should not be introduced within the hood because less air is drawn through the hood face and the face velocity is lowered correspondingly. When auxiliary air is introduced across or in front of the opening, the flow pattern of the auxiliary airstream is critical to hood performance. When auxiliary air is dispersed into the laboratory, it often causes undesirable changes to room temperature and humidity; additionally, condensation on cold surfaces may result. Air turbulence can occur if the airstream strikes personnel working at the hood or any hood surfaces. Make-up air to the auxiliary air hoods should be heated during the heating season so that cold, moist, dense outside air does not fall below the face opening. Cold air also makes the operator uncomfortable. Auxiliary air should not be used with horizontal sliding sash fume hoods.

The application of auxiliary air hoods should be based on the performance characteristics of the specific model, as determined by tests, and full consideration of the extreme level of toxicity that may occur within the hood. For proper hood performance, air balance must be carefully maintained. If the exhaust flow decreases while the auxiliary airflow remains constant, control may be lost. Also, laboratory airflow may reverse and flow from the laboratory into the corridor.

Special Laboratory Hoods and Exhausts

Perchloric acid fume hoods are required for research in which perchloric acid is volatized. These hoods have exhaust ducts of smooth, impervious, and cleanable materials resistant to acid attack. Stainless steels with high chromium and nickel content (not less than No. 316) or nonmetallic materials are recommended. Ductwork should be short, direct, and vertical to the terminal discharge point. Internal water spray systems for periodic washing of the duct surfaces are mandatory. Wash-down prevents the accumulation of perchloric acid deposits, which are a major explosion hazard. Since perchloric acid is an extremely active oxidizing agent, organic materials should not be used in the exhaust system in such places as joint gaskets. Joints should be welded and ground smooth. A perchloric acid hood should only be used for work involving the use of perchloric acid.

Radioactive hood exhaust ducts have flanged, neoprene-gasketed joints with quick disconnect fasteners that can be dismantled quickly for decontamination.

BIOLOGICAL SAFETY CABINETS

A biological safety cabinet (BSC) protects both the researcher and the research materials. BSCs are sometimes called safety cabinets, ventilated safety cabinets, laminar flow cabinets, and glove boxes. The BSCs are categorized into six groups:

Class I — Similar to chemical fume hood; no research material protection; 100% exhaust through a HEPA filter.

Class II, Type A — Frequent selection; can exhaust to the laboratory or to the outside, 70% recirculation within the cabinet; 30% exhaust through a HEPA filter; common plenum.

Class II, Type Bl — Exhausts to the outside; 70% exhaust through a HEPA filter, 30% recirculation within the cabinet; separate plenums.

Class II, Type B2 — Exhausts 100% to the outside through a HEPA filter.

Class II, Type B3 — Exhausts 30% to the outside through a HEPA filter; 70% recirculation within the cabinet; common plenum.

Class III — Selected for special applications; exhausts to the outside. This unit is also known as a glove box. There are physical barriers (gloves) between the researcher and the research material.

Several key decisions must be made by the researcher prior to selecting a biological safety cabinet (Eagleson 1984). An important difference in BSCs is their ability to properly handle chemical vapors (Stuart, *et. al.* 1983). Of special concern to the HVAC engineer is the proper placement of the BSC in the laboratory and the room's air distribution system. Rake (1978) concluded:

> "A general rule of thumb should be that, if the crossdraft or other disruptive room airflow exceeds the velocity of the air curtain at the unit face, then problems do exist. Unfortunately, in most laboratories such disruptive room airflows are present to various extents. Drafts from open windows and doors are the most hazardous sources because they can be far in excess of 200 fpm and accompanied by substantial turbulence. Heating and air-conditioning vents perhaps pose the greatest threat to the safety cabinet because they are much less obvious and therefore seldom considered. . . It is imperative then that all room airflow sources and patterns be considered before laboratory installation of a safety cabinet."

Biological safety cabinets should only be sited in the laboratory in compliance with the installation recommendations of the National Sanitation Foundation (NSF) *Standard* 49. Locate the units away from drafts, active walkways, and doorways. Block the section of diffuser opening that blows air at the face of the BSC.

A biological safety cabinet operates well when connected to its own exhaust system, which is designed and field balanced for the

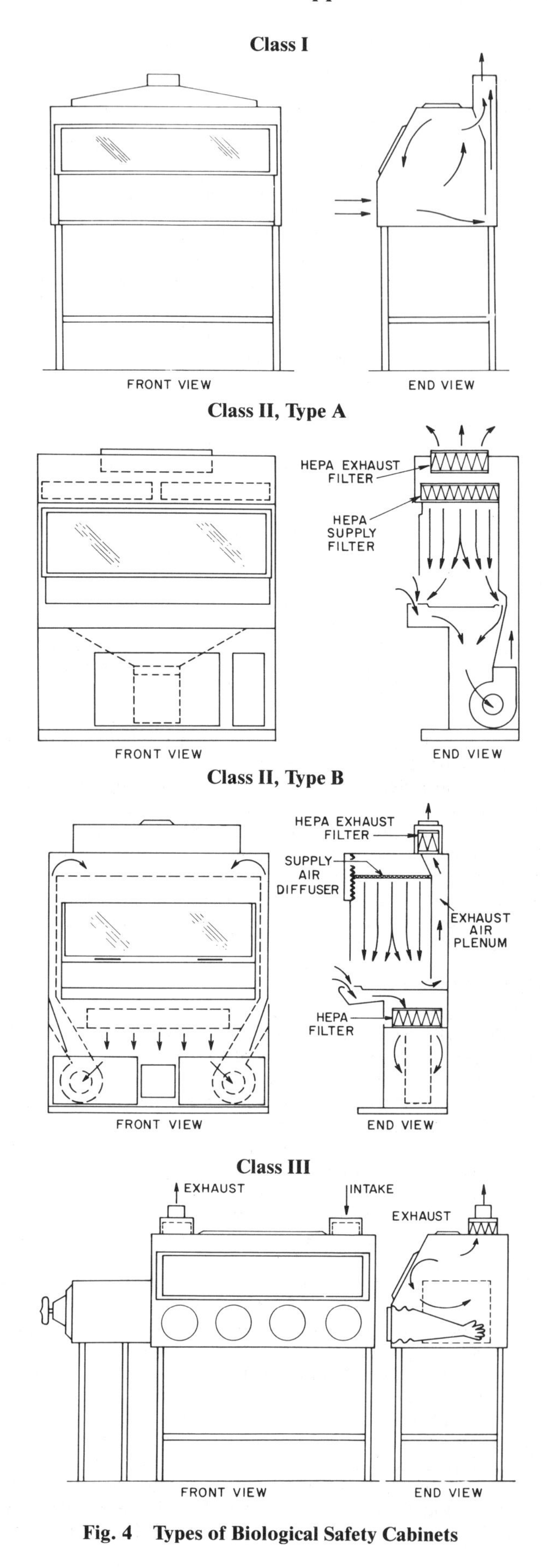

Fig. 4 Types of Biological Safety Cabinets

individual unit. Problems have occurred when several BSCs are connected to a common exhaust system serving more than one unit. The problems occur due to the differential filter loading and to intermittent operation of individual units by the researcher; these problems upset the exhaust system's balance and, therefore, the operation of the individual BSC units.

A BSC supplier's factory-fabricated exhaust transition duct piece (spool piece) can provide (1) access ports for testing and balancing and (2) an airtight damper for decontamination. The use of high quality, factory-fabricated duct pieces is recommended over field-fabricated sections of ductwork.

The National Sanitation Foundation *Standard* 49, Class II (Laminar Flow) Biohazard Cabinetry, is available to assist in procurement and testing of Class II cabinets. A listing of NSF-approved models may be obtained from the National Sanitation Foundation, P. O. Box 1468, Ann Arbor, Michigan 48106.

Class I Cabinets

The Class I cabinet is a partial containment cabinet designed for general research operations with low- and moderate-risk etiologic agents. It is useful for the containment of mixers, blenders, and other equipment. Room air flows through a fixed opening and prevents aerosols, which may be released within the cabinet enclosure, from escaping into the room. Entrained particles are removed by the exhaust air, which may be HEPA-filtered before being discharged from the cabinet to the exhaust system. The HEPA exhaust filter is optional and depends on the cabinet usage. For adequate personnel protection, the front opening of the cabinet, through which the worker operates, should be approximately 8 in. high. Air velocity through that opening must be a minimum of 75 fpm. This type of cabinet does not prevent contact exposure.

The Class I cabinet can be modified to contain chemical carcinogens by adding an appropriate exhaust-air treatment system and by increasing the velocity through the front opening to 100 fpm. Specific OSHA criteria should be consulted. Large pieces of research equipment such as centrifuges, blenders, and sonicators should be placed inside the cabinet and specially shielded. The cabinet, however, is not appropriate for containing experimental systems that are vulnerable to airborne contamination, since the inward flow of air can carry microbial contaminants into the cabinet. The Class I cabinet is also not recommended for use with highly infectious agents, since an interruption in the inward airflow, caused by drafts or fan failure, may allow aerosolized particles to escape.

Class II Cabinets

Class II cabinets provide protection to personnel, product, and environment. The cabinet features an open front with inward airflow and HEPA-filtered recirculated and exhaust air.

The Class II, Type A cabinet (formerly designated Type 1) has a fixed work opening with a minimum inflow velocity of 75 fpm. The average minimum downward air velocity is 75 fpm. Because this design recirculates approximately 70% of the total cabinet air, it should not be used with flammable solvents, toxic agents, or radioactive materials. Although the Class II, Type A cabinet can be installed to discharge exhaust air to the room, the cabinet should discharge the exhaust air to the laboratory exhaust system by a special canopy hood or "thimble unit" available from the manufacturer of the cabinet (NSF 1987). The Type A cabinet is suitable for work with agents meeting Biosafety Level 2 criteria in the absence of volatile or toxic chemicals and volatile radionuclides (National Safety Council 1979).

The Class II, Type B1 (formerly designated Type 2) cabinet has a vertical sliding sash and maintains an inward airflow of 100 fpm at a sash opening of 8 in. The average downward vertical air velocity is 50 fpm. The cabinet exhausts approximately 70% of the air flowing through the work area to the outdoors after passage through a HEPA filter. Type B1 cabinets are suitable for work with agents meeting Biosafety Level 3 criteria (DHHS 1984). The unit may also be used with biological agents treated with limited quantities of toxic chemicals and trace quantities of radionuclides, provided the work is performed in the direct exhausted area of the cabinet.

The Class II, Type B2 (referred to as total exhaust) cabinet is designed to maintain an inward airflow of 100 fpm through the work access opening. The downflow air is drawn from the laboratory or the outside and passes through a HEPA filter before entering the work space. There is no recirculation of air in these units, and the HEPA-filtered exhaust is discharged to the outdoors. The Type B2 cabinet may be used for the same level of work as the Type B1 with the added feature that the design permits use of toxic chemicals and radionuclides as adjuncts to microbiological studies.

The Class II, Type B3 cabinets are sometimes called convertible cabinets. They are designed to maintain an inward airflow of 100 fpm and are essentially the same as the Type A in performance.

Class II, Types A and B3 cabinets deliver air to the outlet where the exhaust system must begin to pull the air. Since the cabinet performance depends on a delicate balance of intake and exhaust, the exhaust system must not pull too much air nor restrict the flow of exhaust air from the cabinet.

With Class II, Types B1 and B2 cabinets, the exhaust system must pull the air from the cabinet's work space. The exhaust system must therefore overcome a certain amount of filter resistance (within the BSC) while pulling a specified flow. The amount of restriction introduced by the cabinet at rated flow can be obtained from the manufacturer. Types B1 and B2 units must be physically connected (hard-ducted) to the exhaust system; the thimble connection cannot be used.

With all Class II BSC units, the National Sanitation Foundation *Standard* 49 and/or the manufacturer must be consulted to obtain the acceptable airflow range for the units.

Class III Cabinets

The Class III biological safety cabinet is a gas-tight, negative-pressure containment system that physically separates the agent from the worker. These cabinets provide the highest degree of personnel protection. Work is performed through arm-length rubber gloves attached to a sealed front panel. Room air is drawn into the cabinet through HEPA filters. Particulate materials entrained in the exhaust air are removed by HEPA filtration or incineration before discharge to the atmosphere. A Class III system may be constructed to enclose and isolate incubators, refrigerators, freezers, centrifuges, and other research equipment. Double-door autoclaves, liquid disinfectant dunk tanks, and pass boxes are used to transfer materials into and out of the cabinet. Class III systems contain highly infectious materials and radioactive contaminants. Although there are operational inconveniences with these cabinets, they are the equipment of choice when a high degree of personnel protection is required. The use of Class III cabinets for research involving volatile substances has resulted in explosions.

LAMINAR FLOW HOODS (CLEAN BENCHES)

Horizontal (cross-flow) and vertical (down-flow) laminar flow clean benches, which discharge air out of the front opening and into the room, should not be used in a biomedical laboratory without critical assessment of risk. They provide product protection but expose the worker to potentially hazardous or allergenic substances because air is discharged across the research material directly into the face of the worker. Clean benches are not recommended for work involving any biological, chemical, or radionuclide materials.

LABORATORY ANIMAL ROOMS

Laboratory animals must be housed in comfortable, clean, and air-conditioned animal rooms, with animal welfare considered in the design.

It is also important that the air-conditioning system provide the desired microenvironment (animal cage environment) as specified by the facility's veterinarian (Woods 1980, Besch 1975, ILAR 1985). Early detailed discussions with the laboratory animal veterinarian concerning airflow patterns, cage layout, and risk assessment help ensure a successful animal room air-conditioning design. The elimination of research variables (fluctuating temperatures and humidities, drafts, and spreading of airborne diseases) is another reason for a high quality air-conditioning system.

Table 6 gives minimum ventilation rates recommended by the Institute of Laboratory Animal Resource (ILAR). The ability to control odor in animal facilities depends primarily on the number and species of animals housed, the amount of cleanable surfaces (including exposed ducts and pipes), the sanitation practices, and the proper design and operation of the air-conditioning system. The designer should not confuse these ventilation rates with the air-conditioning rate.

The air-conditioning flow rate for an animal room should be determined by the following factors: desired animal microenvironment (Besch 1975, 1980; ILAR 1985); species of animal(s); animal room population; recommended minimum ventilation rate (Table 6); recommended ambient temperature and humidity (Table 7); heat produced by motors on specialized animal housing units (such as laminar flow racks or HEPA-filtered air supply units for ventilated racks); and heat produced by the animals (Table 8). Additional design factors include: method of animal cage ventilation; operational use of a fume hood or a biological safety cabinet during procedures such as animal cage cleaning and animal examination; airborne contaminants (generated by animals, bedding, cage cleaning, and room cleaning); and institutional animal care standards (Besch 1980, ILAR 1985).

Due to the nature of research programs, air-conditioning design temperature and humidity set points are usually required. Research animal facilities require more precise control of the environment than farm animal or production facilities because variations affect the experimental results. An ideal system permits control of the temperature of individual rooms to within ±2°F for any temperature set point in a range of 64 to 85°F. Recommended ranges of room temperatures and relative humidities from which set points are selected are listed in Table 7. The relative humidity should be maintained between 30 to 70% throughout the year, according to the needs of the species.

Table 6 Recommended Ventilation for Laboratory Animal Rooms

Species	Minimum Room Air Changes per Hour (Using 100% Outside Air)	Reference
Mouse	15	ILAR (1977)
Hamster	15	ILAR (1977)
Rat	15	ILAR (1977)
Guinea pig	15	ILAR (1977)
Rabbit	10	—
Cat	10	ILAR (1978)
Dog	10	—
Nonhuman primate	10-15	ILAR (1980)

Table 7 Recommended Ambient Temperatures and Humidity Ranges for Animal Rooms

Species	Temperature, °F	Relative Humidity, %	Reference
Mouse	64-79	40-70	ILAR (1977)
Hamster	64-79	40-70	ILAR (1977)
Rat	64-79	40-70	ILAR (1977)
Guinea pig	64-79	40-70	ILAR (1977)
Rabbit	61-70	40-60	—
Cat	64-84	30-70	ILAR (1978)
Dog	64-84	30-70	—
Nonhuman primate	64-84	30-70	ILAR (1980)

Note: The above ranges permit the scientific personnel who will use the facility to select optimum conditions (set points). The ranges do not represent acceptable fluctuation ranges.

Table 8 Heat Generated by Laboratory Animals

Species	Weight, lb	Heat Generation, Normally Active Btu/h per Animal		
		Sensible	Latent	Total
Mouse	0.046	1.11	0.54	1.65
Hamster	0.260	4.02	1.98	6.00
Rat	0.62	7.77	3.83	11.6
Guinea pig	0.90	10.2	5.03	15.2
Rabbit	5.41	39.2	19.3	58.5
Cat	6.61	45.6	22.5	68.1
Nonhuman primate	12.0	71.3	35.1	106.
Dog	22.7	105.	56.4	161.
Dog	50.0	231.	124.	355.

If the entire animal facility or extensive portions of the facility are permanently planned for species with similar requirements, the range of individual adjustments should be reduced. Each animal room or group of rooms serving a common purpose should have temperature and humidity controls.

Control of air pressure in animal housing and service areas is important to ensure directional airflow. For example, quarantine, isolation, soiled equipment, and biohazard areas should be kept under negative pressure, whereas clean equipment and pathogen-free animal housing areas should be kept under positive pressure (ILAR 1985).

The animal facility and human occupancy areas should be conditioned separately. The human areas may use a return air HVAC system and may be shut down on weekends for energy conservation. Separation also prevents exposure of personnel to biological agents and odors present in animal rooms.

Supply air outlets should not cause drafts on research animals. A study by Neil and Larsen (1982) showed that the predesign evaluation of a full-size mock-up of the animal room and its HVAC system was a cost-effective way to select a system that distributes air to *all* areas of the animal-holding room effectively. Wier (1983) describes many typical design problems and their resolutions.

Room air distribution should be evaluated using ASHRAE *Standard* 113 procedures to evaluate drafts and temperature gradients.

Air-conditioning systems must remove the sensible and latent heat produced by laboratory animals. The literature concerning the metabolic heat production appears to be divergent, but new data is consistent. Current recommended values are given in Table 8. These values are based on experimental results and the following formula:

$$\text{ATHG} = 2.5\,M$$

where

ATHG = average total heat gain, Btu/h per animal
M = metabolic rate of animal, Btu/h per animal = $6.6\,W^{0.75}$
W = mass of animal, lb

Conditions in animal rooms must be maintained at constant values. This requires year-round availability of refrigeration and, in some cases, dual/standby chillers and emergency electrical power for motors and control instrumentation. The storage of critical spare parts is one alternative to installing standby refrigeration systems.

It is important that the HVAC ductwork and utility penetrations present few cracks in animal rooms so that all wall and ceiling surfaces can be easily cleaned. Exposed ductwork is not recommended. Joints around diffusers, grilles, and the like should be sealed. Return air grilles with 1-in. disposable filters are normally used.

An efficient air distribution system for animal rooms is essential; this may be accomplished effectively by supplying air through ceiling outlets and exhausting air at floor level (Hessler and Moreland 1984). Supply and exhaust systems should be sized to minimize noise.

The concept of multiple cubicles within animal rooms is being increasingly used to enhance the operational flexibility of the animal room (*i.e.*, housing multiple species in the same room, quarantine, and isolation). The same air exchange/balance, temperature, and humidity control considerations should be addressed for each cubicle as if it were a separate animal room.

ENERGY RECOVERY INTERFACE

Energy recovery is often justified in laboratory buildings with large quantities of exhaust air. For many research laboratory projects, either the coil energy-recovery loop or the twin-tower enthalpy recovery loop is selected because the fresh air intakes are located near ground level, and the exhaust discharges are located on the roof.

Leaks in coil loops causing contamination to migrate from the exhaust side to the supply side have been a source of field problems.

In the energy conservation analysis, most savings are obtained when a team (architect-engineer-owner) reevaluates the standard design parameters and makes reasonable changes. The following have an important effect on energy consumption in laboratory facilities:

1. Fume hood selection and operation. Consider fume hood diversity (Moyer 1983).
2. Fume hood sizing and airflow requirements. Consider horizontal sliding sash hoods; also, consider reducing the operating height of the vertical sash hoods.
3. Use of biological safety cabinets for some research procedures.
4. Thermal storage.
5. User experience with energy exchange units (Carnes 1984, Bridges 1980).
6. User experience with modification of existing terminal reheat systems (Cook 1980, Haines 1984).
7. Owner requirements for number of air changes per hour.
8. Monitoring of energy usage and increased maintenance (ASHRAE 1985).
9. Analysis of energy consumption (Cowan and Jarvis 1984).

SUPPLEMENTARY CONDITIONING

Supplementary conditioning equipment (fan coil units) handle cooling load when it: (1) varies sufficiently in intensity, frequency, or duration; or (2) occurs in a small percentage of the research spaces. High heat gains in limited areas may be handled more efficiently by adding cooling capacity in those specific areas than by cooling the air in a central system and then reheating it for all other spaces.

Supplementary conditioning equipment (fan coils) can be justified easily for spaces where the cooling requirements differ from those of the majority of spaces served by the central system.

SPECIAL REQUIREMENTS

Isolated rooms, such as cold rooms or warm rooms, that require the maintenance of special temperature, humidity, or other conditions for research are usually beyond the capability of the central systems and should have separate HVAC systems. These rooms should not upset the building air balance. If these rooms are personnel working areas, they must be ventilated.

MAINTAINABILITY

Careful consideration of HVAC equipment location must be made to ensure access and maintainability. Many indoor air quality problems are traced back to the lack of maintenance of equipment due to poor accessibility (Woods, *et al.* 1987).

Recent designs involving variable air volume (VAV) systems often include sophisticated control equipment that must be maintained by knowledgeable and specially trained personnel. Some VAV projects have failed due to improper maintenance of the systems. The design engineer should advise the owner that increased maintenance time and skills are required to successfully operate these systems.

CONTAINMENT LABORATORIES

With the initiation of biomedical research involving recombinant DNA technology, federal guidelines about laboratory safety were published for design teams, researchers, and others.

Containment describes safe methods for managing hazardous chemicals and infectious agents in laboratories. The three elements of containment are laboratory operational practices and procedures, safety equipment, and facility design. Thus, the HVAC design engineer helps decide two of the three containment elements during the design phase.

Biomedical research laboratory facilities usually meet one of the following three containment levels:

Biosafety Level One is suitable for work involving agents of no known or of minimal potential hazard to laboratory personnel and the environment. The laboratory is not required to be separated from the general traffic patterns in the building. Work may be conducted either on an open bench top or in a chemical fume hood. Special containment equipment is neither required nor generally used. The laboratory can be cleaned easily and contains a sink for washing hands. The federal guidelines for these laboratories contain no specific HVAC requirements, and typical college laboratories are usually acceptable. Many colleges and research institutions require directional airflow from the corridor into the laboratory, chemical fume hoods, and approximately 3 to 4 air changes per hour of outside air. These requirements protect the research materials from contamination, thereby improving research efficiency. Directional airflow from the corridor into the laboratory helps to control odors.

Biosafety Level Two is suitable for work involving agents of moderate potential hazard to personnel and the environment. Guideline DHHS (1988) contains lists that explain the levels of containment needed for various hazardous agents. Laboratory access is limited when certain work is in progress. The laboratory can be cleaned easily and contains a sink for washing hands. Biological safety cabinets (Class I or II) are used whenever:

1. Procedures with a high potential for creating infectious aerosols are conducted. The procedures include centrifuging, grinding, blending, vigorous shaking or mixing, sonic disruption, opening containers of infectious materials, inoculating animals intranasally, and harvesting infected tissues from animals or eggs.
2. High concentrations or large volumes of infectious agents are used. Federal guidelines for these laboratories contain minimum facility standards, as described above.

At this level of biohazard, most research institutions have a full-time safety officer (or safety committee) who establishes facility standards. The usual HVAC design criteria includes the following requirements:

1. 100% outside air systems
2. 6 to 15 air changes per hour
3. Directional airflow into the laboratory rooms

4. Site-specified hood face velocity at fume hoods (many institutions specify 100 fpm)
5. Research equipment heat load in a room of ±15 watts per assignable (net) square foot of laboratory space.

Biosafety Level Three applies to facilities in which work is done with indigenous or exotic agents that may cause serious or potentially lethal disease as a result of exposure by inhalation. All procedures involving the manipulation of infectious materials are conducted within biological safety cabinets. The laboratory has special engineering (HVAC) features such as airlocks and a separate ventilation system.

Most biomedical research laboratories are designed for Biosafety Level Two. However, the laboratory director must evaluate the risks and determine the correct containment level before design begins. Process laboratories may have hazardous materials and, therefore, the laboratories are designed to meet the criteria contained in DHHS guidelines (1988).

REFERENCES

Alereza, T. and J. Breen, III. 1984. Estimates of recommended heat gains due to commercial appliances and equipment. ASHRAE *Transactions* 90(2A):25-58.

AGCIH 1986. *Industrial ventilation: A manual of recommended practice.* American Conference of Governmental Industrial Hygienists, Cincinnati, OH.

American Institute of Architects, Committee on Architecture for Health. 1987. *Guidelines for construction and equipment of hospital and medical facilities.* American Institute of Architects Press, Washington, D.C.

ASHRAE. 1985. A biological evolution. ASHRAE *Journal* 27(3):62-64 (March).

ASHRAE. 1985. Method of testing performance of laboratory fume hoods. ANSI/ASHRAE *Standard* 110-1985.

ASHRAE. 1990. Method of testing for room air diffusion. ASHRAE *Standard* 113-1990.

Besch, E. 1975. Animal cage room dry bulb and dew point temperature differentials. ASHRAE *Transactions* 81(2):549-58.

Besch, E. 1980. Environmental quality within animal facilities. *Laboratory Animal Science* 30(2II):385-406.

Bridges, F. 1980. Efficiency study: Preheating outdoor air for industrial and institutional applications. ASHRAE *Journal* 22(2):29-31 (February).

Caplan, K. and G. Knutson. 1977. The effect of room air challenge on the efficiency of laboratory fume hoods. ASHRAE *Transactions* 83(1):141-56.

Caplan, K. and G. Knutson. 1978. Laboratory fume hoods: Influence of room air supply. ASHRAE *Transactions* 84(1):511-37.

Carnes, L. 1984. Air-to-air heat recovery systems for research laboratories. ASHRAE *Transactions* 90(2).

College of American Pathologists. 1985. *Medical laboratory planning and design.* College of American Pathologists, Skokie, IL.

Cook, E. 1980. An energy primer for terminal reheat. *Heating, Piping, and Air Conditioning* (July):83-85.

Cowan, J. and I. Jarvis. 1984. Component analysis of utility bills: A tool for the energy auditor. ASHRAE *Transactions* 90(1).

Degenhardt, R. and J. Pfost. 1983. Fume hood system design and application for medical facilities. ASHRAE *Transactions* 89(2B):558-70.

DHHS 1988. *Biosafety in microbiological and biomedical laboratories.* Pub. No. (CDC) 88-8395. U.S. Department of Health and Human Services, Public Health Service Centers for Disease Control and National Institutes of Health.

Di Beradinis, L., J. Baum, M. First, G. Gatwood, E. Groden, and A. Seth. 1987. *Guidelines for laboratory design: Health and safety considerations.* John Wiley & Sons, Boston.

Eagleson, J., Jr. 1984. Aerosol contamination at work. *The International Hospital Federation Yearbook*, 1984. Sabrecrown Publishing, London.

Haines, R. 1986. Retrofitting reheat-type systems. ASHRAE *Journal* 28(9):35-38 (September).

Hessler, J. and A. Moreland. 1984. Design and management of animal facilities. In *Laboratory Animal Medicine*, J. Fox, B. Cohen, F. Loew, eds. Academic Press.

ILAR. 1977. Laboratory animal management—Rodents. *ILAR News*, Vol. XX, No. 3. Institute of Laboratory Animal Resources, National Institutes of Health, Bethesda, MD.

ILAR. 1978. Laboratory animal management—Cats. *ILAR News*, Vol. XXI, No. 3. Institute of Laboratory Animal Resources.

ILAR. 1980. Laboratory animal management—Nonhuman primates. *ILAR News*, Vol. XXIII, No. 2-3. Institute of Laboratory Animal Resources.

ILAR. 1985. Guide for the care and use of laboratory animals. NIH *Publication No.* 85-23. Institute of Laboratory Animal Resources.

Kiil, L. 1984. Aging labs need extensive renovation. *Building Design and Construction* (November):75-78.

Koenigsberg, J. and H. Schaal. 1987. Upgrading existing fume hood installations. *Heating, Piping and Air Conditioning* (October):77-82.

Moyer, R. 1983. Fume hood diversity for reduced energy conservation. ASHRAE *Transactions* 89(2B).

NFPA. 1986. *Standard on fire protection for laboratories using chemicals. National Fire Code* 45-86. National Fire Protection Association, Quincy, MA.

NFPA. 1986. *Health care facilities.* In *Fire Protection*, 16th ed., *Section* 9-4. National Fire Protection Association, Quincy, MA.

NRC. 1981. *Prudent practices of handling hazardous chemicals in laboratories.* National Research Council. National Academy Press, Washington, D.C.

NSF. 1987. Class II (laminar flow) biohazard cabinetry. NSF *Standard* 49-87. National Sanitation Foundation, Ann Arbor, MI.

Neil, D. and R. Larsen. 1982. How to develop cost-effective animal room ventilation: Build a mockup. *Laboratory Animal Science* (Jan-Feb): 32-37.

Peterson, R. 1987. Designing building exhausts to achieve acceptable concentrations of toxic effluents. ASHRAE *Transactions* 93(2):2165-85.

Peterson, R., E. Schofer, and D. Martin. 1983. Laboratory air systems—Further testing. ASHRAE *Transactions* 89(2B).

Pike, R. 1976. Laboratory-associated infections: Summary and analysis of 3921 cases. *Health Laboratory Science* 13(2):105-14.

Rake, B. 1978. Influence of crossdrafts on the performance of a biological safety cabinet. *Applied and Environmental Microbiology* (August): 278-83.

Schyler, G. and W. Waechter. 1987. Peformance of fume hoods in simulated laboratory conditions. Report #487-1605 by Rowan Williams Davies & Irwin, Inc. under contract for Health and Welfare Canada.

SAMA. 1980. *SAMA standard for laboratory fume hoods.* Scientific Apparatus Makers Association, Washington, D.C.

Sessler, S. and R. Hoover. 1983. Laboratory fume hood noise. *Heating, Piping, and Air Conditioning* (September):124-37.

Schiller, G. and E. Arens. 1988. Thermal comfort in office buildings. ASHRAE *Journal* 30(10):26-32 (October).

Steere, N. (editor). 1971. *CRC Handbook for laboratory safety.* CRC Press, Boca Raton, FL.

Stuart, D., M. First, R. Rones, and J. Eagleston. 1983. Comparison of chemical vapor handling by three types of Class II biological safety cabinets. *Particulate & Microbial Control* (March/April).

West, D. 1978. Assessment of risk in the research laboratory: A basis for facility design. ASHRAE *Transactions* 84(1B).

Weir, R. 1983. Toxicology and animal facilities for research and development. ASHRAE *Transactions* 89(2B).

Woods, J. 1980. The animal enclosure—A microenvironment. *Laboratory Animal Science* 30(2II):407-13.

Woods, J., J. Janssen, P. Morey, and D. Rask. 1987. Resolution of the "sick" building syndrome. *Proceedings* of ASHRAE Conference: Practical Control of Indoor Air Problems, 338-48.

BIBLIOGRAPHY

Abramson, B. and T. Tucker. 1988. Recapturing lost energy. ASHRAE *Journal* 30(6):50-52 (June).

Albern, W., F. Darling, and L. Farmer. 1988. Laboratory fume hood operation. ASHRAE *Journal* 30(3):26-30 (March).

Anderson, C. 1988. HVAC controls in laboratories—A systems approach. ASHRAE *Transactions* 94(1B).

Anderson, C. and K. Cunningham. 1988. HVAC controls in laboratories —A systems approach. ASHRAE *Transactions* 94(1B).

Anderson, S. 1987. Control techniques for zoned pressurization. ASHRAE *Transactions* 93(2B):1123-39.

ASHRAE. 1987. ASHRAE energy award: Energy prescription, Emory clinic expansion and new eye clinic. ASHRAE *Journal* 29(3):30-32 (March).

Bertoni, M. 1987. Risk management considerations in design of laboratory exhaust stacks. ASHRAE *Transactions* 93(2B):2149-64.

Brow, K. 1989. AIDS research laboratories—HVAC criteria, ASHRAE *Symposium*: Building systems: Room air and air contaminant, 223-25.

Brown, A., P. Dobney, and J. Fricker. 1987. Calculation of energy targets—A practical method of setting energy targets and assessing project performance. ASHRAE *Journal* 29(10):24-27 (October).

Dahan, F. 1986. HVAC systems for chemical and biochemical laboratories. *Heating, Piping, and Air Conditioning* (May):125-30.

Davis, S. and R. Benjamin. 1987. VAV with fume hood exhaust systems. *Heating, Piping and Air Conditioning* (August):75-78.

DHHS. 1981. NIH guidelines for the laboratory use of chemical carcinogens. NIH *Publication No.* 81-2385. Department of Health and Human Services, National Institutes of Health, Bethesda, MD.

Gupta, V. 1987. A forum: Variable air volume. ASHRAE *Journal* 29(8):22-30 (August).

Habert, J., L. Smith, K. Cooney, and F. Stern. 1988. An expert system for building energy consumption analysis: Applications at a university campus. ASHRAE *Transactions* 94(1B).

Hayter, R. and R. Gorton. 1988. Radiant cooling in laboratory animal caging. ASHRAE *Transactions* 94(1B).

Knutson, G. 1984. Effect of slot position on laboratory fume hood performance. *Heating, Piping, and Air Conditioning* (February):93-96.

Knutson, G. 1987. Testing containment laboratory hoods: A field study. ASHRAE *Transactions* 93(2B):1801-12.

Koenigsberg, J. and E. Seipp. 1988. Laboratory fume hood—An analysis of this special exhaust system in the post "Knutson-Caplan" era. ASHRAE *Journal* 30(2):43-46 (February).

Lentz, M. and A. Seth. 1989. A procedure for modeling diversity in laboratory VAV systems. ASHRAE *Transactions* 95(1A).

McDiarmid, M. 1988. A quantitative evaluation of air distribution in full scale mock-up of animal holding rooms. ASHRAE *Transactions* 94(1B).

Marsh, C. 1988. DDC systems for pressurization, fume hood face velocity, and temperature control in variable air volume laboratories. ASHRAE *Transactions* 94(2B).

Maust, J. and R. Rundquist. 1987. Laboratory fume hood systems—Their use and energy conservation. ASHRAE *Transactions* 93(2B):1813-21.

Mikell, W. and L. Hobbs. 1981. Laboratory hood studies. *Journal of Chemical Education* 58(5):A165-A168 (May).

Moyer, R. and J. Dungan. Turning fume hood diversity into energy savings. ASHRAE *Transactions* 93(2B):1822-34.

Murray, W, A. Strifel, T. O'Dea, and F. Rhame. 1988. Ventilation for protection of immune compromised patients. ASHRAE *Transactions* 94(1B).

Neuman, V. 1989. Disadvantages of auxiliary air fume hoods. ASHRAE *Transactions* 95(1A).

Neuman, V. 1989. Design considerations for laboratory HVAC system dynamics. ASHRAE *Transactions* 95(1A).

Neuman, V. 1989. Health and safety in laboratory plumbing. *Plumbing Engineering* (March):21-24.

Neuman, V. and H. Guven. 1988. Laboratory building HVAC systems optimization. ASHRAE *Transactions* 94(2A).

Neuman, V., F. Sajed, H. Guven. 1988. A comparison of cooling thermal storage and gas air conditioning for lab building. ASHRAE *Transactions* 94(2A).

Neuman, V. and W. Rousseau. 1986. VAV for laboratory hoods—Design and costs. ASHRAE *Transactions* 92(1A).

Rhodes, W. 1988. Control of microbiological contamination in critical areas of the hospital environment. ASHRAE *Transactions* 94(1A).

Streets, R. and B. Setty. 1983. Energy conservation in institutional laboratory and fume hood systems. ASHRAE *Transactions* 89(2B).

Stuart, D., R. Greenier, R. Rumery, and J. Eagleson. 1982. Survey, use, and performance of biological safety cabinets. *American Industrial Hygiene Association Journal* 43:265-70.

Wilson, D. 1983. A design procedure for estimating air intake contamination from nearby exhaust vents. ASHRAE *Transactions* 89(2A).

CHAPTER 15

ENGINE TEST FACILITIES

INDUSTRIAL testing of internal combustion engines is done in test cells to test the engine itself and in chassis dynamometer rooms to test the engine in a complete vehicle. In both cases, the spaces are enclosed to control noise, heat, and fumes and to isolate the test for safety or security. When temperature, noise, and safety conditions can be resolved adequately, chassis dynamometers may be installed outdoors only with the control rooms enclosed. Large open areas in the plant are sometimes used for production testing and emissions measurements, but the principles of ventilation and safety for test cells generally apply.

Enclosed test cells are normally found in design facilities. Test cells may need instruments to measure cooling water flow and temperature, exhaust gas flow and temperature, fuel flow, power output, and combustion air volume and temperature.

VENTILATION SYSTEMS

Ventilation and air conditioning of test cells must (1) supply and exhaust proper quantities of air to remove heat and control temperature; (2) exhaust sufficient air at proper locations to prevent buildup of combustible vapors; (3) modulate large air quantities to meet changing conditions; (4) remove engine exhaust fumes; (5) supply combustion air; (6) prevent noise transmission through the system; and (7) provide for human comfort and safety during setup, testing, and tear-down procedures. Supply and exhaust systems for test cells are unitary, central, or a combination of both. Mechanical exhaust must be supplied in all cases, and ventilation is generally controlled on the exhaust side (Figure 1).

Constant-volume systems with variable supply temperatures can be used; however, variable-volume, variable-temperature systems are usually selected. Unitary variable-volume systems (Figure 1A) use an individual exhaust fan and makeup air supply for each cell. Supply and exhaust fans are interlocked, and operation is coordinated with the engine, usually by sensing room temperature. Some systems have only exhaust supply induced directly from outside (Figure 1B). Volume variation can be accomplished by changing fan speed or by various dampers.

Ventilation systems with central supply or exhaust fans, or both (Figure 1C) regulate air quantities by test cell temperature control of individual dampers or by two-position switches actuated by dynamometer operations; they also maintain air balance by static pressure regulation within the cell. Constant pressure in the supply duct is obtained by controlling inlet vanes, modulating dampers, or varying fan speed.

For individual exhaust fans combined with central supply, the exhaust system is regulated by cell temperature or a two-position switch actuated by dynamometer operation, while the central supply system is controlled by a static pressure device located within the cell to maintain room pressure (Figure 1D). Variable volume exhaust airflow should not drop below minimum requirements. The exhaust requirements should override cell temperature requirements; thus, reheat may be needed.

Ventilation should be interlocked with the fire protection system to shut down supply and exhaust to the cell in case of fire. Exhaust fans should be nonsparking, and makeup air should be heated by steam or hot water coils.

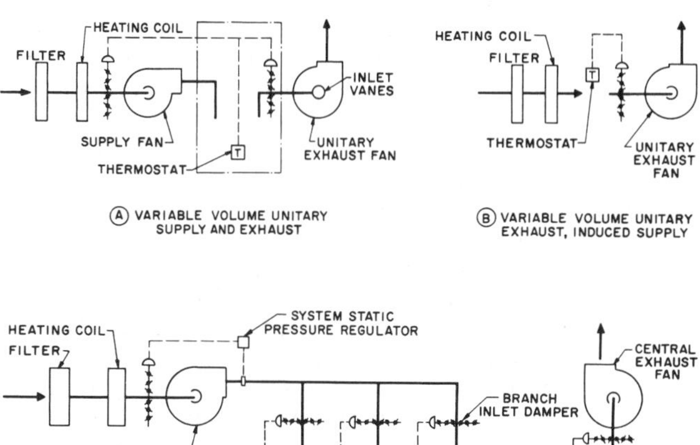

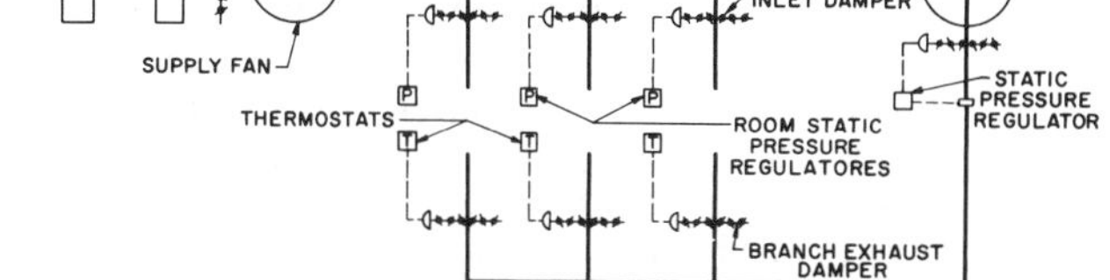

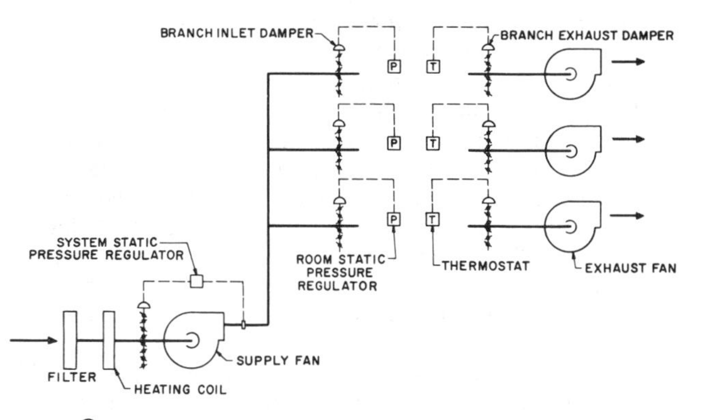

Fig. 1 Heat Removal Ventilation Systems

The preparation of this chapter is assigned to TC 9.2, Industrial Air Conditioning.

TEST CELL EXHAUST

Ventilation for test cells is based on exhaust requirements designed to satisfy (1) removal of heat generated by the engine when operating; (2) emergency purging (removal of fumes after fuel spills); and (3) cell scavenging, *i.e.*, a rate of continuous exhaust during nonoperating periods.

During engine operation, heat convects to the air in the cell and also radiates to surrounding surfaces. The flow of air through the cell to remove convected heat is determined by:

$$Q = 0.9H/(t_e - t_s) \quad (1)$$

where

Q = airflow, cfm
H = engine heat release, Btu/h
t_e = temperature of exhaust air, °F
t_s = temperature of supply air, °F

Heat radiated from the engine and exhaust piping must first warm the surrounding surfaces, which, in turn, release the heat to the air in the space by convection. The rate of release is governed by the temperature differences, film coefficients, and other factors discussed in Chapter 5 of the 1989 ASHRAE *Handbook—Fundamentals*. The value for $(t_e - t_s)$ in Equation (1) cannot be arbitrarily set when a portion of H is radiated heat. Determination of H is discussed in the section Engine Heat Release.

Vapor removal exhaust should be at the maximum rate to remove vapors as quickly as possible. The air required for engine heat removal is generally the quantity used for emergency purging; however, it should not be less than 10 cfm/ft^2 of floor area. Emergency purging should be by a manual overriding switch for each cell.

In the case of fire, the operation should (1) shut down all equipment, (2) close fire dampers at all openings, and (3) shut off the solenoid valves that control fuel flow. Fire and smoke control is discussed in Chapter 47.

Cell scavenging exhaust is the minimum required to keep combustible vapors of fuel leaks from accumulating, or the amount required for heating and cooling, whichever is greater. Generally 2 cfm per ft^2 of floor area is sufficient.

Exhaust grilles should be low (even when ceiling exhaust is used), since gasoline vapors are heavier than air. If exhaust is removed close to the engine, the convective heat that escapes into the cell is minimized. Some installations exhaust all air through a floor grating immediately surrounding the engine bedplate into a cubicle or duct below. In this case, supply slots in the ceiling over the bedplate are located to cover the engine with a curtain of air to remove the heat. This scheme has worked quite successfully in a number of installations and is particularly applicable for central exhaust systems (Figure 2). Water sprays in the cubicle or exhaust duct lessen the danger of fire or explosion in case of fuel spills.

Trenches or pits in test cells should be avoided. If they exist, they should be mechanically exhausted at all times, with a minimum rate of 10 cfm per ft^2 of horizontal area. Excessively long trenches require multiple takeoff exhaust points. The exhaust should sweep the entire area and leave no dead spaces. Test cells should have no suspended ceilings or basements located directly below where fuel spills and vapor accumulation can occur. If such spaces exist, they should be ventilated continuously, and fuel lines should never run through them.

Although the exhaust quantities should be determined individually for each test cell on the basis of heat to be removed, evaporation of possible fuel spills, and minimum downtime ventilation, a general idea of current practice can be obtained from Table 1.

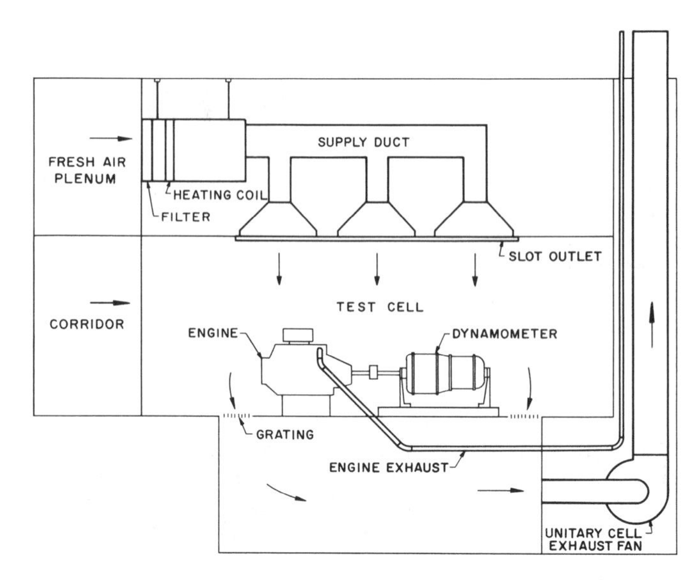

Fig. 2 Engine Test Cell Showing Direct Engine Exhaust—Unitary Ventilation System

TEST CELL SUPPLY

Air supplied to a test cell should be slightly less than that exhausted. Recirculation of test cell air during engine testing is not recommended. Return air from other nontest areas can be used when available, provided good ventilation practice is followed.

Air distribution methods for introducing large quantities of ventilation air are important, especially if the cell is occupied. It is desirable to project air to separate occupants from heat released from the engine. Slot-type outlets with automatic dampers to maintain constant discharge velocity have been used with variable-volume systems.

A variation of Systems C and D in Figure 1 includes a separate supply to the cell sized for the minimum downtime ventilation rate, with heated and chilled water coils regulated by a room thermostat to control the temperature within the cell. This is useful in installations where much time is devoted to setup and preparation of tests, or where constant temperature is required for complicated or sensitive instrumentation. Except for production and endurance testing, the actual engine operating time in test cells may be suprisingly low. Industry-wide test cells are used in about 15 to 20% of cases.

Filtration of all air to the cell is ordinarily required to remove atmospheric particulates and insects. The degree of filtration is a matter of choice or a requirement of the test. Facilities located in clean areas sometimes use outdoor air without filtration.

Table 1 Exhaust Quantities for Test Cells

	Minimum Range of Exhaust Rates per Unit Floor Area	
	cfm per ft^2	Air changes per h [a]
Engine testing—cell operating	10 to 20	60 to 120[b]
Cell idle	2 to 3	12 to 18
Trenches and pits	10	—
Accessory testing	4	24
Control rooms and corridors	1	6 to 10

[a] Based on a cell height of 10 ft.

[b] For chassis dynamometer rooms, this quantity is usually determined by test requirements.

Tempering supply air by heating coils is usually necessary for personnel if there is danger of freezing equipment or if low temperatures adversely affect the test performance.

ENGINE HEAT RELEASE

Engine heat released to the space is the total energy input to the engine less the energy transmitted to the dynamometer as work, heat removed by the jacket cooling water, and heat discharged in the exhaust gas. The relative quantities vary with the engine, operating speed and load, fuel used, type of dynamometer, and piping configurations external to the engine. The heat balance of an internal combustion engine at full load is approximately one-third useful work to the shaft, one-third to jacket cooling water, and one-third to exhaust; for a gas turbine at full load, it is roughly one-third to shaft and two-thirds to exhaust. In both cases, there are losses by convection and radiation from the engine itself and from the exhaust piping. Table 2 shows a representative breakdown of heat release by engine type and size.

The energy absorbed by the dynamometer as work is removed from the test cell, except for that portion transmitted to the space from the dynamometer. Absorbed energy is returned to the electrical system, dissipated in remote resistance grids, or removed by the cooling water when induction dynamometers are used.

Engine coolant system heat is normally removed by city or recirculated tower water through heat exchangers. Heat exchangers for engine oil are also required under many test conditions. Radiators within the cell that dissipate this heat require the exhaust air quantity to be increased accordingly.

For an L-head engine, heat rejected to the jacket water in Btu/h is 3.5 times RPM per cubic in. displacement, which is about 2520 Btu/h per brake horsepower output; for an overhead valve engine, it is 3.2 RPM = Btu/h per cubic in. displacement, which is 2280 Btu/h per brake horsepower output.

Heat lost by convection from engine block, head, and oil pan depends largely on engine coolant temperature, air temperature, and quantity passing over the engine. A surface temperature of 180 to 300 °F may be expected for internal combustion engines, and 400 to 500 °F for automotive gas turbines.

Engine exhaust pipes release considerable heat through radiation and convection, since exhaust gas temperatures are between 1200 and 1800 °F. Only a minimum amount of piping should be located within the cell. Commonly, engine exhaust pipes are placed in a vault below the cell. Catalytic converters and other emission devices are usually installed within the cell, adding to the heat load. Cooling the engine exhaust system by jacketing or injecting water directly into the pipe is common.

Methods for calculating convection and radiation losses from exhaust pipe surfaces are found in Chapter 3 of the 1989 ASHRAE *Handbook—Fundamentals*. Because exhaust piping, mufflers, and add-on devices are subject to change with the tests, it is usual to estimate this heat loss. A good estimate is to consider that the maximum heat loss will occur at wide-open throttle with the largest engine the installed dynamometer can handle. On this basis, the power going into the test cell ambient from the exhaust is considered to be 1% of the rating of the dynamometer per foot of exposed exhaust pipe (the exhaust manifold should be included in that calculation) for a diesel engine, and 1.5% per foot of pipe for a gasoline-powered engine. The actual diameter of the pipe has only a secondary effect because of several compensating factors.

In summary, the total engine heat release is:

$$BHP\,(L_d + L_e + L_c + C)$$

where

BHP = engine brake hp
L_d = dynamometer loss per bhp
L_e = exhaust pipe losses per bhp
L_c = engine connection losses per bhp
C = cooling system losses per bhp

Table 2 Heat Balance for Internal Combustion Engine at Wide-Open Throttle

Engine Type and Size	Engine Heat Balance, % Work Output	To Exhaust	To Coolant	Engine Convection, Radiation and Oil
Air-cooled I.C.				
To 15 hp	15	45	—	40
15 to 150 hp	20	43	—	37
Water-cooled I.C.				
150 to 500 hp	23.5	36	26	14.5
Diesel	32	36	22	10
G.T. regenerator				
To 400 hp	23	66	—	10

GAS TURBINE TEST CELLS

Aircraft gas turbine test cells must handle large quantities of air required by the turbine itself, attenuate the noise generated, and operate safely with the large flow of fuel required. Such cells are unitary in nature and use the gas turbine to draw in the untreated air and exhaust it through mufflers.

Small gas turbines for automotive and truck application can generally be tested in a conventional engine test cell with relatively minor modifications. Test cell ventilation supply and exhaust are sized for the heat generated by the turbine in the same manner as for conventional engines. Combustion air supply for the turbine is considerably more, but it may be drawn from the cell, from outdoors, or through separate conditioning units that handle air only for combustion.

Exhaust quantities are higher than from internal combustion engines and are usually ducted directly outdoors through suitable muffling devices which provide little restriction to flow. Water cooling of exhaust air may be used, as temperature may range from 300 to 700 °F, or for regenerative gas turbines, 500 to 600 °F.

CHASSIS DYNAMOMETER ROOMS

A chassis dynamometer (as shown in Figure 3) simulates road driving and acceleration conditions. The drive wheels of the vehicle rest on a large roll, which drives a dynamometer. To approximate the effect of operation at varying road speeds on the body of the vehicle (and for radiator cooling), air quantities calibrated to coincide with air velocities at that particular speed are blown across the front of the vehicle. Additional refinements may vary temperature and moisture conditions of the air within prescribed limits from ambient to 130 °F. Usually, air is introduced through an area approximating the frontal area of the vehicle. A return grille in a duct may be lowered to a position at the rear of the vehicle, causing the air to remain near the floor rather than short-cycle through a ceiling grille. The air is recirculated above the ceiling to the air-handling equipment.

Chassis dynamometers are also installed in cold rooms, where temperatures may be as low as −100 °F, and in full-sized wind tunnels with throat areas many times the cross-sectional area of the vehicle. Combustion air is drawn directly from the room, but a mechanical engine exhaust through the engine must be introduced into the facility in a way that will maintain the low temperatures and humidities.

ENGINE EXHAUST

Engine exhaust systems remove products of combustion, unburned fuel vapors, and water vapor resulting from injection

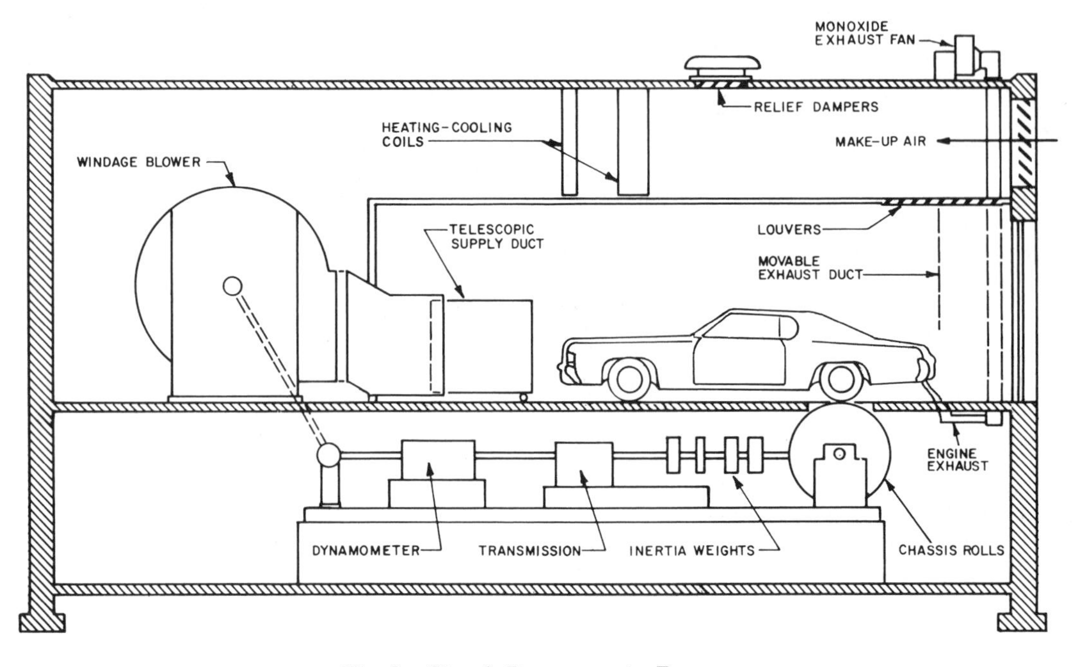

Fig. 3 Chassis Dynamometer Room

of water for gas cooling. Design criteria are flow loads and system operating pressure.

Flow loads are based on the number of engines, engine size, and load and use factors.

System operating pressure is governed by engine discharge pressures and the range of allowable suction or back pressure at the point of connection to the exhaust system. Systems may operate at positive pressure, using available engine tail pipe pressure to force flow of gas, or they may operate at negative pressure, with mechanically induced flow.

The simplest method for inducing engine exhaust from a test cell is to size the exhaust pipe to minimize pressure variations on the engine and to connect it directly to the atmosphere (Figure 4A).

Such direct systems are limited in length of run by the cost of adequate pipe size needed to keep back pressure to a minimum. Direct exhaust to outdoors is subject to wind currents and air pressures, may be hazardous because of positive pressures within the system, and cannot be as closely regulated as mechanically ventilated systems.

Mechanical engine exhaust systems are unitary or central. *Unitary systems* use a fan and conductors to serve an individual test cell and can be closely regulated to match operation to the engine's requirements (Figure 4B). *Central systems* use single or multiple fans, main ducts, and branch connections to individual cells (Figure 4D).

Pressures in all engine exhaust systems fluctuate with capacity of engine operation in relation to system design. The exhaust system should be designed so that load variations in individual cells exert a minimum effect on the system. Dampers and pressure regulators, if needed, keep the pressures within test tolerances. Engine characteristics and diversity of operation determine the maximum exhaust to be handled by the system, and allowable back pressure and tolerance for its variation are the basis for system size and regulation.

An indirect connection between the engine exhaust pipe and a mechanical removal system simplifies back pressure regulation (Figure 4C). The exhaust pipe terminates within an inlet to the

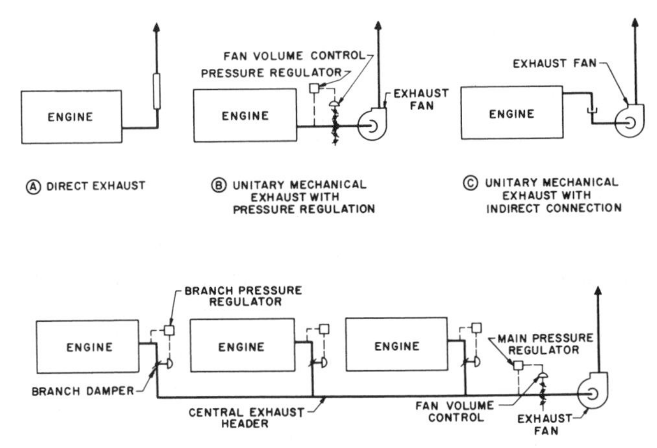

Fig. 4 Engine Exhaust Systems

collecting system with a clearance of 2 to 3 in. It then removes a mixture of exhaust fumes and room air from the test cell.

An indirect connection does require noise control if it is not within the test cell or in the open. Also, this connection can self-ignite when the engine is motored by the dynamometer, and a fuel rich mixture exhausts from the pipe at temperatures above 700°F.

Since the exhaust piping is open to the room, the system is essentially self-regulating. Room air mixes with the exhaust and cools the mixture. Materials for the indirect-connected system are not required to withstand the higher temperature conditions of a direct-connected system, although corrosion may be more severe.

To reduce exhaust gas temperatures (usually about 1500°F but sometimes as high as 1800°F), water may be injected directly into the exhaust pipe, the pipe may be water-jacketed, or a heat shield or shroud may be placed around the exhaust pipe, with air mechanically exhausted through the shrouded enclosure. Exhaust piping and fans must often use high strength, high-temperature alloys to withstand corrosion, thermal stresses, and pressure pulsations resulting from rapid flow fluctuations. The equipment must be adequately supported and anchored to relieve thermal expansion stresses.

Exhaust systems for chassis dynamometer installations must capture the high velocity exhaust from the tail pipe to prevent buildup of fumes in the test area.

Engine exhaust should discharge through a stack extending above the roof at an elevation and velocity sufficient to allow the fumes to clear the building windstream wake. If high stacks are not practical, a combination of shorter stacks and increased ejection velocity may propel the exhaust gases above the building wake. Chapter 14 of the 1989 ASHRAE *Handbook—Fundamentals* has further details. Ejection systems must keep the discharge velocity relatively constant, as the gas volume varies with load changes. Local or federal laws may require that exhaust gases be cleaned before they enter the atmosphere.

COOLING WATER SYSTEMS

Dynamometers absorb and measure the useful output of an engine or its components. Two basic classes of dynamometers are the water-cooled induction type and the electrical type. In the *water-cooled* dynamometer, engine work is converted to heat that is absorbed by a circulating water system. *Electrical* dynamometers convert engine work to electrical energy that can be used or dissipated as heat in resistance grids. Grids should be outdoors or adequately ventilated.

Heat loss from electric dynamometers is approximately 8% of the output measured, plus a constant load of about 5 kW for auxiliaries within the cell. Cooling water systems for absorbing heat from the engine jacket, oil coolers, and water-cooled dynamometers are usually designed for recirculation through a system of circulating pumps, cooling towers, or atmospheric coolers and hot and cold well-collecting tanks (Figure 5).

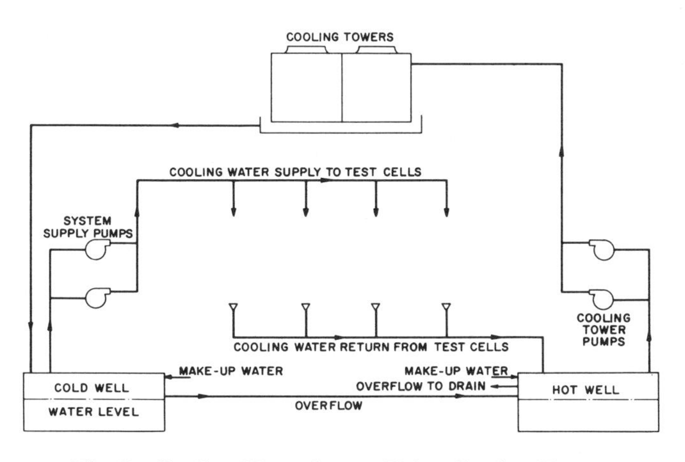

Fig. 5 Cooling Water System Using Cooling Towers

COMBUSTION AIR SUPPLY SYSTEMS

Combustion air is usually drawn from the test cell or introduced directly from outdoors. Where combustion air conditions must be closely regulated and conditioning of the entire test cell is impractical, separate units that condition only the air for combustion can be used. These units filter, heat, cool, and regulate supply air humidity and barometric pressure; they usually provide air directly to the carburetor. Combustion air systems may be central or portable package types of units.

NOISE

Characteristics of noise generated by internal combustion engines and gas turbines must be recognized in designing an air-handling system. Part of the engine noise is discharged through the tail pipe. If possible, internal mufflers should be installed at the engine to attenuate this noise at its source. Ventilation ducts or pipe trenches that penetrate the cells must be protected against sound transmission from the cells through the ducts to other building areas or to the outdoors. Attenuation should be applied to the duct penetrations equivalent to what is provided by the structure of the cell. Table 3 lists typical noise levels within test cells during engine operation.

Table 3 Typical Noise Levels in Test Cells

Type and Size of Engine	Decibel Reading 3 ft from Engine			
	Dba	124 Hz	500 Hz	2000 Hz
Diesel				
Full load	105	107	98	99
Part load	70	84	56	49
Gasoline engine 440 in^3 at 5000 rpm				
Full load	107	108	104	104
Part load	75	—	—	—
Rotary engine 100 hp				
Full load	90	90	83.5	86
Part load	79	78	75	72

BIBLIOGRAPHY

Associated Factory Mutual Fire Insurance Company. *Testing internal combustion engines and accessories.* Loss Prevention Bulletin 13.50.

Computer controls engine test cells. *Control Engineering* 16 (75): 69.

Factory Insurance Associates. 1952. *Recommended good practice for safeguarding combustion or jet engine test cells* (January).

Hazardous gases need ventilation for safety. 1968. *Power,* 112 (May): 92.

Heldt, P.M. 1956 *High speed combustion engines.* Chilton Co., Philadelphia.

Ricardo and Hempson. 1968. *The high speed internal combustion engine.* Blacke and Son Limited, London.

CHAPTER 16

CLEAN SPACES

CLEAN spaces or cleanrooms encompass much more than the traditional HVAC design of controlling temperature and humidity. In clean space design, particulate contamination control, airflow pattern control, sound and vibration control, as well as industrial engineering aspects and manufacturing equipment layout have to be considered.

The performance of a cleanroom depends on the quality of control concerning particulate concentration and dispersion, temperature, humidity, vibration, noise, airflow pattern, and construction. The objective of good cleanroom design is to control these parameters while maintaining reasonable installation and operating costs.

TERMINOLOGY

As-built cleanroom. A cleanroom that is complete and ready for operation, with all services connected and functional, but without production equipment or personnel within the room.

At-rest cleanroom. A cleanroom that is complete and has the production equipment installed and operating, but without personnel in the room.

Class 1. Particle count not to exceed 1 particle per cubic foot of a size 0.5 μm and larger, with no particle exceeding 5.0 μm (Figure 1). This criteria should be based on a large sampling of counts.

Class 10. Particle count not to exceed 10 particles per cubic foot of a size 0.5 μm and larger, with no particle exceeding 5.0 μm (Figure 1).

Class 100. Particle count not to exceed 100 particles per cubic foot of a size 0.5 μm and larger (Figure 1).

Class 1000. Particle count not to exceed 1000 particles per cubic foot of a size 0.5 μm and larger (Figure 1).

Class 10,000. Particle count not to exceed 10,000 particles per cubic foot of a size 0.5 μm and larger, or 65 particles per cubic foot of a size 5.0 μm and larger.

Class 100,000. Particle count not to exceed 100,000 particles per cubic foot of a size 0.5 μm and larger, or 700 particles per cubic foot of a size 5.0 μm and larger.

Cleanroom. A specially constructed enclosed area environmentally controlled with respect to airborne particulates, temperature, humidity, air pressure, airflow patterns, air motion, vibration, noise, viable organisms, and lighting.

Clean space. A defined area in which the concentration of airborne particles is controlled to specified limits.

Conventional flow cleanroom. A cleanroom with non-unidirectional or mixed airflow patterns and velocities.

Critical surface. The surface of the work or the part to be protected from particulate contamination.

Design conditions. The environmental conditions for which the clean space is designed.

DOP. Dioctyl phthalate, a chemical used in measuring filter efficiency, which creates a known particle size distribution when atomized. Conflicting comments about possible medical concerns using DOP have arisen.

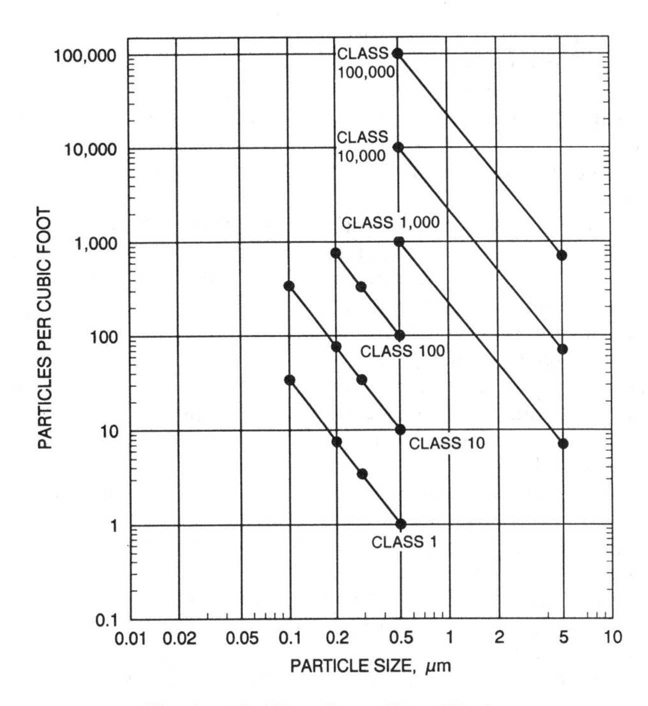

Fig. 1 Air Cleanliness Class Limits

First air. The air that issues directly from the HEPA filter before it passes over any work location.

High efficiency particulate air (HEPA) filter. A filter with an efficiency in excess of 99.97% of 0.3 micrometer particles, as determined by dioctyl phthalate (DOP) test.

Laminar airflow workstation. A workstation with airflow nominally in one direction through the work area.

Laminar flow cleanroom. A cleanroom with airflow in nominally one direction.

Makeup air. Air introduced to the secondary (recirculated) air system for ventilation, pressurization, and replacement of exhaust air.

Nonlaminar flow workstation. A workstation without uniform airflow patterns and velocities.

Operational cleanroom. A cleanroom in normal operation with all services functioning and with production equipment and personnel present and performing their normal work functions.

Particle concentration. The number of individual particles per unit volume of air.

Particle size. The apparent maximum linear dimension of a particle in the plane of observation.

Primary air. Air that recirculates through the workspace.

The preparation of this chapter is assigned to TC 9.2, Industrial Air Conditioning.

Secondary air. That portion of the primary air circulated through the air-conditioning equipment.

ULPA filter. A filter with an efficiency in excess of 99.9997% of 0.12 μm particles, as determined by dioctyl phthalate (DOP) test.

Workstation. An open or enclosed work surface with direct air supply.

CLEAN SPACES AND CLEANROOM APPLICATIONS

Applications of clean space environments for use in manufacturing, packaging, and research continue to grow as technological advances and the need for cleaner work environments increase. Demand for efficient cleanroom designs will continue as both existing users expand and new industries add clean spaces. A summary of the major industries employing clean spaces for their products follows.

Semiconductor Industry. The advances in semiconductor microelectronics continue to drive the state of the art in cleanroom design. Semiconductor facilities account for a significant percentage of all cleanrooms in operation in the United States, with most of the newer semiconductor cleanrooms being class 100 or cleaner.

Pharmaceutical and Biotechnology Industry. Preparation of pharmaceutical products, biological and medical products, and genetic engineering research are good examples of cleanroom application. Clean spaces are needed to control viable (living) particles that would produce undesirable bacterial growth.

Aerospace Industry. Cleanrooms developed for aerospace applications are used to manufacture and assemble satellites, missiles, and aerospace electronics. Most applications involve clean spaces of large volumes, with cleanliness levels of Class 10,000 or higher.

Miscellaneous Applications. Cleanrooms are also used in aseptic food processing and packaging, manufacture of artificial limbs and joints, automotive paint booths, laser/optic industries, and advanced materials research.

Hospital operating rooms may be classified as cleanrooms, although their primary function is to limit particular types of contamination rather than quantities of particles. Cleanrooms are used in patient isolation and surgery where risks of infection exist.

AIRBORNE PARTICLES AND PARTICLE CONTROL

Airborne particles occur in nature as pollen, bacteria, miscellaneous living and dead organisms, and from windblown dust and seaspray. Industry generates particles from combustion processes, chemical vapors, and friction in manufacturing equipment. People in a workspace are a prime source of particles in the form of skin flakes, lint, cosmetics, and respiratory emissions. These airborne particles vary in size from 0.001 μm to several hundred microns. Particles larger than 5 μm tend to settle quickly. With many manufacturing processes, these airborne particles are viewed as a source of contamination where contact between the particulate and the product will cause product failure.

Particle Sources in Clean Spaces

In general, particulate sources with respect to the clean space have been grouped into two categories—external and internal.

External Sources. External sources are those particles that enter the clean space from sources outside the clean space. The largest external source is outside makeup air entering through the air-conditioning system. Other sources are infiltration through doors, windows, wall penetrations for pipes, ducts, and so forth.

In an operational cleanroom, external particle sources will normally present little impact on overall cleanroom particle concentration. However, research has shown direct correlations between ambient particle concentrations and indoor particle concentrations for clean spaces "at rest." External particulate sources may be controlled primarily through makeup air filtration, room pressurization, and sealing all penetrations of the controlled space.

Internal Sources. Particulate contamination generated within the clean space is the result of people, cleanroom surface shedding, process equipment, material ingress, and the manufacturing process itself. Cleanroom personnel are potentially the largest source of internal particles, generating several thousand to several million particles per minute. These particles are usually skin flakes, moisture droplets, residually exhaled smoke, cosmetics, and hair. With airflow designed to continually "wash" the personnel with clean air, new cleanroom garments, and proper gowning procedures, personnel-generated particles are controlled. As personnel work in the cleanroom, their body movements may reentrain airborne particles from other sources. Other activities, such as writing, may also cause higher cleanroom particle concentrations.

Particle concentrations within the cleanroom may be used to define cleanroom class, but actual particle deposition on the product is of greater concern.

The science of aerosols, filter theory, and fluid motion aid in the understanding of contamination control. Although cleanroom designers may not be able to control or prevent internal particle generation completely, internal sources and design control mechanisms to limit their impact on the product can be anticipated.

Fibrous Air Filters. Most externally generated particles are prevented from entering the cleanroom with proper air filtration. Technology for high efficiency air filters are of two types—High Efficiency Particulate Air (HEPA) filters and Ultra Low Penetrating Air (ULPA) filters. HEPA filters have nominal efficiencies of 99.97 to 99.997% removal efficiency of 0.3 μm particles, and ULPA filters are 99.9997% efficient for 0.12 μm particles. Both HEPA and ULPA filters use glass fiber paper technology.

HEPA and ULPA filters are usually constructed in a deep pleated format, with aluminum, coated string, or filter paper as pleating separators. Filters may vary from 2 to 12 in. in depth; correspondingly higher media areas are available with deeper filters and higher pleat spacing.

Theories and models describing the different fibrous filter capture mechanisms have been developed and verified by empirical data. The consensus is that interception and diffusion are the dominant capture mechanisms for cleanroom HEPA filters. Fibrous filters have their lowest removal efficiency at the most penetrating particle size (MPPS), where filter fiber diameter, volume fraction or packing density, and air velocity determine the MPPS. For most HEPA filters, the MPPS is between 0.1 and 0.3 μm; thus HEPA and ULPA filters have rated efficiencies based on 0.3 and 0.12 μm particle sizes, respectively.

AIR PATTERN CONTROL

Air turbulence within the clean space is strongly influenced by air supply and return configurations, people traffic, and process equipment layout. Selection of the air pattern configurations is the first step for good cleanroom design. User requirements for cleanliness level, process equipment layout, available space for installation of air pattern control equipment (air handlers, clean workstations, environmental control components), and project financial considerations all influence the final air pattern design selection. Project financial aspects will govern air pattern control concepts where operating and capital costs may limit the type and size of air-handling equipment to be used. Even though numerous air pattern configurations are used today, they generally comprise two categories—unidirectional airflow (commonly, but incorrectly, referred to as laminar flow) and non-unidirectional airflow (turbulent airflow).

Although not truly laminar, *unidirectional airflow* is characterized by air flowing in a single pass in a single direction through a cleanroom or clean zone with generally parallel streamlines.

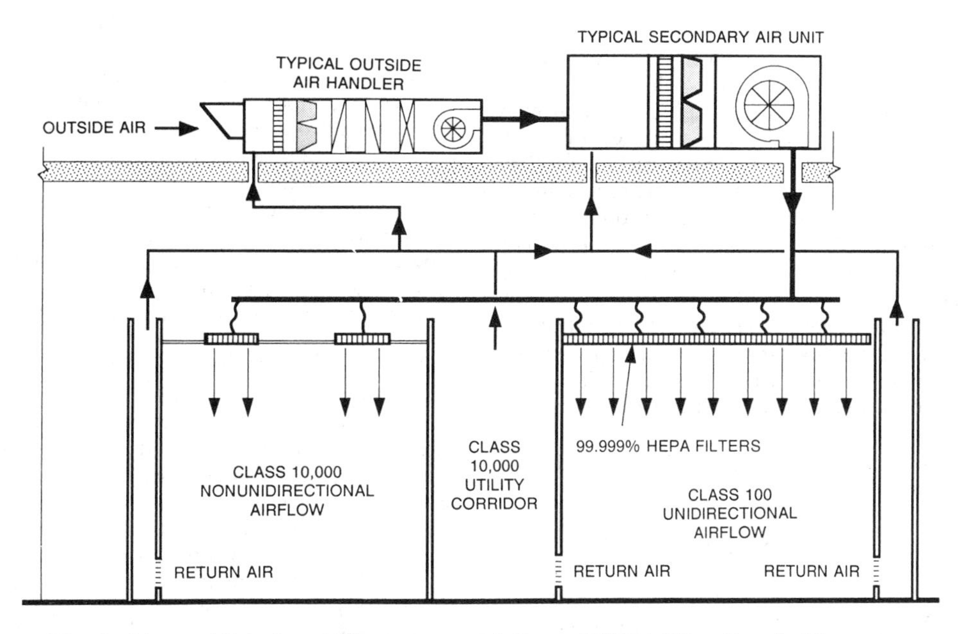

Fig. 2 Non-unidirectional Cleanrooms with Ducted HEPA Filter Supply Elements

Ideally, the flow streamlines are uninterrupted and, although personnel and equipment in the airstream do distort the streamlines, a state of constant velocity is approximated. Most particles that encounter an obstruction in a laminar airflow strike the obstruction and continue around it as the laminar airstream reestablishes itself downstream of the obstruction.

Non-unidirectional airflow does not meet the definition of unidirectional airflow by having either multiple-pass circulating characteristics or a nonparallel flow direction.

Non-unidirectional Airflow

Variations of non-unidirectional airflow are primarily based on the location of supply air inlets and outlets and air filter locations. Airflow is typically supplied to the space through supply diffusers with HEPA filters (Figure 2), through supply diffusers with HEPA filters in the ductwork or air handler (Figure 3), or air is prefiltered in the supply system components and HEPA filtered workstations are located in the clean space.

Non-unidirectional airflow may provide satisfactory contamination control for cleanliness levels of Class 1000 through Class 100,000. Attainment of desired cleanliness classes, with designs similar to those shown in Figures 2 and 3, presuppose that the major space contamination is from external sources (makeup air) and that contamination is removed in the air handler or ductwork filter housings or through HEPA filter supply devices. When internally generated particles are of primary concern, clean workstations are provided in the clean space.

Although air turbulence in flow control methods is harmful, it is needed to enhance the mixing of low and high particle concentrations to produce a homogeneous particle concentration level acceptable to the process.

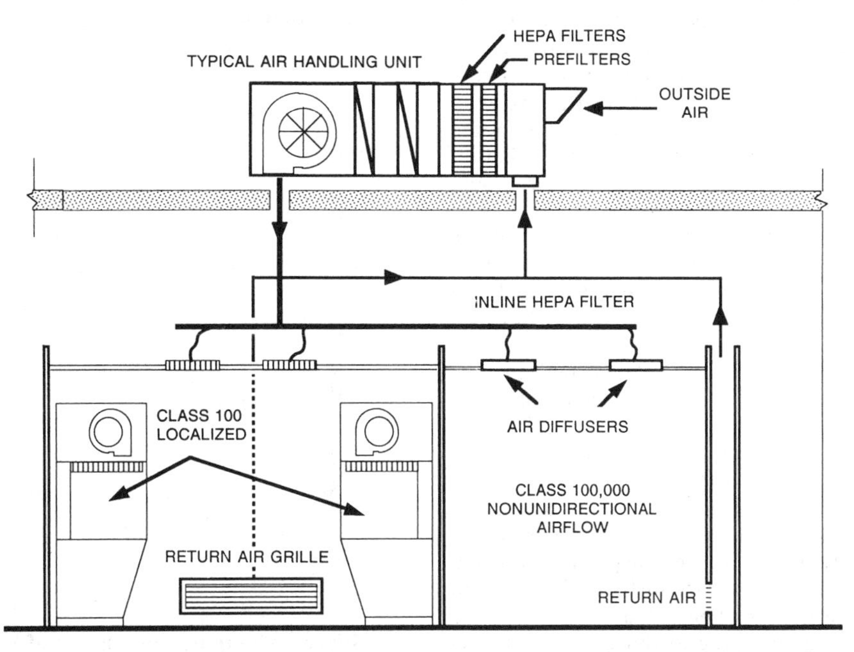

Fig. 3 Non-unidirectional Cleanroom with HEPA Filters Located at Air Handler

Unidirectional Airflow

Control of air patterns and air turbulence reduction is optimized in unidirectional airflow systems, where the cleanroom air is typically introduced through the ceiling and returned through a raised floor or at the base of sidewalls. This produces nominally parallel airflow, and the space is known as a vertical laminar flow room. A similar approach using a supply wall and return on the opposite wall is known as a horizontal flow cleanroom.

A downflow cleanroom (Figure 3) has a ceiling comprised of HEPA filters. As the class of the cleanroom gets lower, a greater percentage of the ceiling will require HEPA filters. For a Class 100 (or better) room, the entire ceiling will require HEPA filtration. Ideally, a grated or perforated floor serves as the air exhaust. Air in the downflow room moves uniformly from the ceiling to the floor. As it moves through the ceiling, it is filtered essentially free of all particles 0.3 μm and larger.

This type of airflow produces a uniform shower of air that bathes the entire room in a downward flow of ultraclean air. Contamination generated in the space tends not to move laterally against the downward flow of air (it is swept down and out through the return) and contributes only minutely to a contamination buildup in the room. Care must be taken in the design, selection, and installation of the system to seal the HEPA ceiling. Assuming that the HEPA filters installed in the ceiling have been properly sealed, this design can provide the cleanest working environment presently available.

A horizontal flow cleanroom uses the same filtration airflow technique as the downflow system, except that the air flows from one wall of the room to the opposite wall. The supply wall consists entirely of HEPA filters supplying air at approximately 90 fpm across the entire section of the room. The air then exits through the return wall at the opposite end of the room and recirculates within the system. As with the downflow room, this design removes contamination generated in the space at a rate equal to the air velocity and limits cross-contamination perpendicular to the airflow.

A major limitation to this horizontal flow design is that downstream contamination in the direction of airflow tends to increase. In this design, the air first coming out of the filter wall is as clean as the air in a downflow room. Process activities can be oriented to have the cleanest, most critical operations in the first air or at the clean end of the room, with progressively less critical operations located toward the return air, or dirtier, end of the room.

Federal *Standard* 209D, Clean Room and Work Station Requirements, Controlled Environment (1988), does not specify velocity standards. The actual velocity is expected to be as specified by the owner or agent. The previous standard of 90 fpm, as stated in Federal *Standard* 209B (1976), is still widely accepted in the cleanroom industry. Current research suggests that lower velocities may be possible. The benefit with lower flow rates is reduced air volume and reduced energy costs, with a possible decrease in cleanliness levels. Other reduced air volume designs use a mixture of high and low pressure drop HEPA filters, reduced coverage in traffic areas, or lower velocities in personnel corridor areas.

Unidirectional airflow systems result in a predictable airflow path followed by airborne particulates. Without good filtration practices, unidirectional airflow only ensures a predictable path for particulates. Superior cleanroom performance may be obtained as unidirectional airflow is better understood.

Unidirectional airflow produces air streamlines due to the HEPA filter pleating construction, uniform pressure drop, and low airflow velocities. These streamlines remain parallel (or within 18° of parallel) to below the normal work surface height of 30 to 36 in., but this flow deteriorates when the airflow encounters obstacles such as process equipment or workbenches. Personnel movement also brings about laminar flow degradation. The result is a cleanroom with areas of good laminar airflow and areas of turbulent airflow.

Turbulent zones are countercurrents of air with high velocities, reverse flow, or possibly no flow (stagnancy being present). Countercurrents produce stagnant zones where small particles may cluster and finally settle onto the product; they may also "lift" particles from contaminated surfaces. Once lifted, these particles may deposit on product surfaces.

Cleanroom mock-ups may help the designer avoid turbulent airflow zones and countercurrents. Smoke, DOP (dioctyl phthalate), neutral buoyancy helium-filled soap bubbles, and nitrogen vapor fogs can provide visible air streamlines within the cleanroom mock-up.

Computer-Aided Flow Modeling

Advances in computer simulation technology may make cleanroom mock-ups obsolete. Computer models of particle trajectories, transport mechanisms, and contamination propagation have been developed and are commercially available. Flow analysis with computer models may compare flow fields associated with different process equipment, workbenches, robots, and building structural design (Figures 4 and 5). Airflow design parameters may be modified to determine its effect on particle transport and flow streamlines, avoiding the cost of mock-ups. Airflow analysis of flow patterns and air streamlines are calculated by solving fundamental fluid mechanics equations of laminar and turbulent flow, where incompressibility and uniform thermophysical properties are assumed.

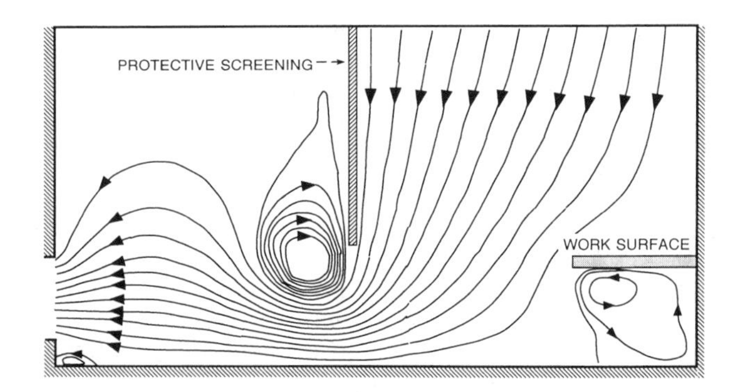

Fig. 4 Computer Modeling of Cleanroom Airflow Streamlines

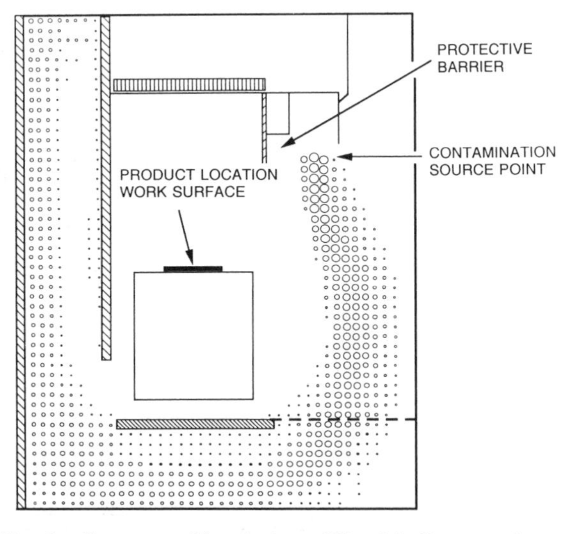

Fig. 5 Computer Simulation of Particle Propagation within Cleanroom

Major features and benefits associated with most computer flow models include:

- Two- or three-dimensional modeling of simple cleanroom configurations
- Modeling of both laminar and turbulent airflows
- Multiple air inlets and outlets of varying sizes and velocities
- Allowances for varying boundary conditions associated with walls, floors, and ceilings
- Aerodynamic effects of process equipment, workbenches, and people
- Prediction of specific airflow patterns of all or part of a cleanroom
- Reduced costs associated with new cleanroom design verification
- Graphical representation of flow streamlines and velocity vectors to assist in flow analysis
- Graphical representation of simulated particle trajectories and propagation

Research has shown excellent correlation between flow modeling by computer and that done in simple mock-ups. However, computer flow modeling software should be checked carefully for cleanroom design, as this is a new technique and inaccuracies may be discovered in the future.

TESTING CLEAN AIR AND CLEAN SPACES

Since early cleanrooms were designed and constructed largely for the U.S. government, the testing procedures have been dictated by government standards. Federal *Standard* 209 was widely accepted as it defined air cleanliness levels for clean spaces around the world. But aspects of cleanroom performance other than air cleanliness are no longer covered by Federal *Standard* 209D. Standardized testing methods and practices have since been developed and published by the Institute for Environmental Science (IES), ASTM, and other groups.

Three basic test modes for cleanroom systems are used to properly evaluate a facility—(1) as-built, (2) at rest, and (3) operational. A cleanroom facility cannot be fully evaluated until it has performed under full occupancy and the process to be performed is operational. Thus, the techniques for conducting initial performance tests and operational monitoring must be similar.

As previously described, sources of contamination are external and internal. For both unidirectional and non-unidirectional flow cleanrooms, the major source for external contamination is the primary air loop. Equipment that applies laser or light scattering principles may be used to detect minute particles. Some laser particle counters can detect particles as small as 0.1 μm. For particle sizes 5.0 μm and larger, microscopic counting can be used, with the particles collected on a membrane filter through which a sample of air has been drawn.

HEPA filters should be tested for pinhole leaks at the (1) filter media, (2) sealant between the media and the filter frame, (3) filter frame gasket, and (4) filter bank supporting frames. The area between the wall or ceiling and the frame should also be tested. A pinhole leak at the filter bank can be extremely critical, since the concentration of the leak varies inversely as the square of the pressure drop across the hole.

The Cleanroom Testing procedure of the Institute of Environmental Sciences (1984) describes 14 tests for cleanrooms. Which tests are applicable to a specific cleanroom project should be determined.

MICROBIOAEROSOL CONTROL IN CLEAN SPACES

Pharmaceutical cleanrooms are primarily interested in controlling bacteria and viruses rather than particles. In this regard, the materials of construction are critical. It is important to ensure that no materials will support biological growth. Actual particle counts are not as critical to the pharmaceutical cleanroom designer. Pharmaceutical cleanrooms tend to be Class 1000 and 10,000 rooms, with laminar flow hoods used to keep the product under Class 100 conditions.

Some biological cleanrooms deal with dangerous or infectious items which must be contained within that space. In such cases, the dangerous or infectious items are usually in hoods or safety cabinets, and the rooms are kept at a negative pressure to the surroundings to keep those items within the facility.

SEMICONDUCTOR CLEANROOMS

Most semiconductor facilities involved with microelectronic manufacturing are Class 100 and are able to provide Class 10 air; a few provide Class 1 air. As methods of contamination control have been developed, specialists have developed rules governing the dimensions of critical particles and how they relate to the manufacturing process.

Based on airborne particle measurements, today's semiconductor clean spaces are probably the cleanest manufacturing environments in the world; thus they tend to drive the optimized designs and engineering methods. As the sizes of microelectronic circuits reduce, the size of particles that may deposit on the circuit and produce a critical defect also become smaller. Contamination control specialists have coined the term *killer particle size* to define the minimum particle size that may cause circuit failure.

Configurations—Classes 100, 10, and 1

Semiconductor cleanrooms are designed using one of two major configurations—the clean tunnel or the open bay. The clean tunnel involves narrow modular cleanrooms, where each room may be completely isolated from another. Fully HEPA filtered pressurized plenums, ducted HEPA filters, or individual fan modules are used in clean tunnel installations. Production equipment may be located within the tunnel, or through-the-wall equipment installation may be used where a lower cleanliness level (nominally Class 10,000) service chase will be adjacent to the clean tunnel. The service chase is used in conjunction with either a sidewall or a raised floor. A basement return system is also used with a raised floor.

The primary advantage of the tunnel design is reduced HEPA filter coverage and ease of expanding additional tunnel modules into unfacilitated areas. The tunnel must be between 11 and 14 ft wide. If smaller, there is not enough room for production equipment on each side. If larger, the flow will become too turbulent and will tend to break toward the walls before it leaves the work plane. The tunnel approach has the drawback of restricting new equipment layouts and footprint sizes. Cleanroom flexibility is valuable to semiconductor manufacturing logistics. As processes change and new equipment is installed, the clean tunnel may restrict equipment location to the point that a new module must be added. The tunnel approach does not allow easy movement of product from one type of equipment to another.

The open bay design involves large (up to 50,000 ft^2) open construction cleanroom layouts. Interior walls may be placed wherever manufacturing logistics dictates, thus providing maximum equipment layout flexibility. Pressurized plenum or ducted filter modules are used, but pressurized plenum-type systems are becoming more common. When the pressurized plenum design is used, either one large plenum with multiple supply fans is used or small adjacent plenums may be configured. Small plenums provide the ability to shut down areas of the cleanroom without disturbing other clean areas. Small plenums may also employ one or more supply fans.

Major semiconductor facilities, with total manufacturing areas of 30,000 ft^2 and larger, may incorporate both open bays and tunnels. Flexibility for revising various equipment layouts warrants the open bay design. Process equipment suitable for through-the-wall installation, such as diffusion furnaces, may use the tunnel or open bay design. Process equipment, such as lithographic steppers and coaters, requires all equipment to be located under laminar flow conditions, which makes open bay designs more suitable. The cleanroom designer, production personnel, and contamination control specialist should decide which method to use.

History has shown that replacing process equipment with newer equipment is an ongoing process for most wafer fabrication facilities, where support services must be designed to handle different process equipment layouts and even changes in process function. Often manufacturers may completely redo their equipment layout if a new product is being made.

Many semiconductor facilities contain separate cleanrooms for process equipment ingress into the main factory. These ingress areas are staged levels of cleanliness. For instance, the equipment receiving area may be Class 100,000 for equipment uncrating, while the next stage may be Class 10,000 for preliminary equipment setup and inspection. The final stage prior to the entrance, where equipment is cleaned and final installation preparations made, will be Class 1000. In some cases, these staged cleanrooms must have adequate clear heights to allow for lifting of equipment subassemblies.

Vertical Unidirectional Airflow Systems

Current semiconductor industry cleanroom design is the vertical laminar airflow system (VLF). Vertical laminar flow produces a uniform shower of clean air throughout the entire cleanroom. With VLF systems, particles are swept from personnel and process equipment; contaminated air leaves at the floor level, thus producing clean air for all space above the work surface. In contrast, horizontal airflow systems produce decreasing cleanliness downstream from the air inlet.

In VLF systems with cleanliness classes of 1, 10, and 100, the cleanroom ceiling area consists of HEPA filters set in a nominal grid size of 2 by 4 ft. HEPA filters are set into a T-bar style grid, with gasket or caulked seals for many Class 100 systems; Class 1 and 10 systems use either low vapor pressure petrolatum fluid seals or silicone dielectric gel to seal the HEPA filters into a channel-shaped ceiling grid. Whether T- or channel-shaped grids are used, the HEPA filters will normally cover 92 to 95% of the ceiling area, with the remainder of the ceiling area composed of gridwork and fire protection sprinkler panels.

HEPA filters in VLF designs are installed with a pressurized plenum above the filters, through individually ducted filters, or with individually fan-powered filter modules. Plenum-type systems must provide even pressurization to maintain uniform airflow through each filter. Ducted systems typically have higher static pressure losses from the ducting and balance dampers. Higher maintenance costs may also be incurred due to the balance method involved with ducted systems. Individual fan-powered filter modules use fractional horsepower fans (usually forward curved fans), which provide airflow through one filter assembly. This method allows for variable airflows throughout the cleanroom and requires less space for mechanical components. The disadvantages of this method are the larger number of fans involved, lower fan and motor efficiencies due to the small sizes, potentially higher fan noises, and higher maintenance costs.

Vertically downward airflow is normally returned through the floor with perforated raised floor panels, or through floor grates in a waffle-type floor structure. When raised floor panels are used, vibration problems may occur due to the lack of rigidity seen in many floors of this type. Insufficiently raised floor height may also propagate air turbulence below the floor providing a means for particle uplift. Turbulence may also increase system static pressure requirements. When through-the-floor return grating is used, a basement return is normally provided, allowing for a more uniform return system and floor space for "dirty" production support equipment.

Sidewall returns are an alternate design to through-the-floor returns. With this approach, however, laminar airflow may not be uniform throughout the work area. As stated previously, this design is most applicable for cleanrooms with double-sidewall returns and widths of less than 14 ft, since this prevents air streamline trajectories from curving away from the work surface.

With any HEPA filtered air system, filter economics dictate adequate prefiltration to increase HEPA filter life. In VLF systems, prefilters are located in the recirculation airflow, either in the return basement or in the air handler. New cleanroom designs have incorporated the first-stage prefilters in the floor return grating. When the prefilters are below the exit plane of the cleanroom, elimination of potentially dirty turbulent subfloor airflow back into the cleanroom may be eliminated, but there may be disadvantages. If the prefilter system design requires the dirty prefilters to be removed through the cleanroom, a potential for contaminating the clean space exists. Other disadvantages include spill control procedures. In the event of a chemical spill, prefilters will become contaminated with hazardous chemicals, for example, complicating the cleanup procedure. With a basement return system, spill control may be handled more effectively.

Air Ionization Systems

In addition to cleanroom particle control with fibrous filters, air ionization technology is sometimes used to control particle deposition on product surfaces. These systems are expensive and may have some flaws, since they may deposit particulates on the product in question.

ENVIRONMENTAL SYSTEMS

Cooling Loads and Methods

Two major internal heat load components in semiconductor facilities are process equipment and fan energy. Since most cleanrooms are located entirely within the conditioned space, traditional heat sources of infiltration, fenestration, and heat conductance from adjoining spaces are typically less than 2 to 3% of the total load. Some cleanrooms have windows to the outside; this is usually for daylight awareness, with a corridor separating the cleanroom window from the exterior window.

The major types of cooling sources designed to remove cleanroom heat and maintain environmental conditions include makeup air units, primary and secondary air units, and the process equipment cooling water systems. Some process heat may be removed by the process exhaust system, but heat removed by the exhaust system is typically from electronic heat sources in process equipment computers and controllers.

Fan energy is a large heat source because of the large air volumes involved with VLF Class 100 or better cleanrooms. Recirculated airflow rates of 90 cfm per ft^2, which equates to 600 air changes per hour, are usual for Class 100 or better cleanrooms.

Latent loads are primarily associated with makeup unit dehumidification. The low dry-bulb leaving air temperature (35 to 45 °F) associated with makeup air-conditioning dehumidification supplements sensible cooling. Supplemental cooling by makeup air may account for as much as 300 Btu/h per square foot of cleanroom.

Process cooling water (PCW) is used in process equipment heat exchangers, performing simple heat transfer to cool internal heat sources, or in process-specific heat transfer, where the PCW contributes to the process reaction.

It is important to understand the diversity of manufacturing heat load sources, *i.e.*, how much of the total heat load is transferred to which cooling media. When bulkhead or ported production equipment is used, the equipment heat loss to support chases versus production area spaces will affect the cooling system design when the support chase is served by a different cooling system from the production area.

Makeup Air Systems

The control of makeup air and cleanroom exhaust affects cleanroom pressurization, humidity, and room cleanliness. The flow rate requirements of makeup air are dictated by replacement of process exhaust and air volumes for pressurization. Makeup air volumes are often much greater than the total process exhaust volume to provide adequate pressurization and safe ventilation standards.

Makeup air volumes are adjusted with zone dampers, makeup fan control using speed controllers, inlet vanes, and so forth. Opposed blade dampers should be specified with low leak characteristics and minimum hysteresis.

Makeup air is the primary source of external particulates in a cleanroom and should, therefore, be filtered prior to injection into the cleanroom. If the makeup air is injected upstream of the cleanroom HEPA filters, a minimum of 95% (ASHRAE Atmospheric Dust Test) efficient filters should be used to avoid high dust loading on the HEPA filters. In addition, prefilters of 30 and then 85% should be used for prolonging 95% filter life. When makeup air is injected downstream of the main HEPA filter system, filtering of this air should be of the same removal efficiency.

In addition to particle filtering, many makeup air-handling systems require filters for the removal of chemical contaminants in the outside air. These contaminants include salts and pollutants from industries and automobiles. Chemical filtration may be accomplished with chemical absorbers, such as activated carbon.

Makeup air is frequently introduced into the primary air path on the suction side of the primary fan(s). It may also be introduced into the pressurized HEPA plenum, but additional energy will be required to offset the HEPA plenum static pressure.

Process Exhaust Systems

Process exhaust systems used in semiconductor facilities handle acid, solvent, toxic, pyrophoric (self-igniting) fumes, and process heat exhaust. Process exhaust systems should be dedicated for each fume category by process area or based on the chemical nature of the fume and its compatibility with exhaust duct material. Typically, process exhausts are segregated into corrosive fumes using plastic or fiberglass reinforced plastic (FRP) ducts. Flammable gases (usually from solvents) and heat exhaust are ducted in metal systems. Care must be taken to ensure that gases are not able to combine into hazardous compounds that can ignite or explode within the ductwork. Segregated heat exhaust systems are sometimes installed to recover heat, or heat may be exhausted into the suction side of the primary air path.

Process exhaust volume requirements may vary from 1 cfm per ft^2 of cleanroom for photolithographic process areas to 10 cfm per ft^2 for wet etch, diffusion, and implant process areas. Areas using fume hoods could have even higher exhaust rates. When specific process layouts have not been designated prior to exhaust system design, an average of 5 cfm per ft^2 is normally acceptable for fan and abatement equipment sizing. Fume exhaust ductwork should be sized at low velocities (1000 fpm) to allow for future requirements.

For many airborne substances, the American Conference of Governmental Industrial Hygienists (ACGIH) has established requirements to avoid excessive worker exposure. Specific standards for allowable concentrations of airborne substances are set by the U.S. Occupational Safety and Health Administration (OSHA). These limits are based on working experience, laboratory research, and medical data, and are subject to constant revision. When evaluating a cleanroom exposure, refer to the latest standards. When limits are to be determined, refer to the ACGIH handbook, *Industrial Ventilation*, 20th ed.

Fire Safety. According to the Uniform Building Code (UBC), semiconductor facilities have been designated an occupancy class, Group H, Division 6, Semiconductor Fabrication Facilities. The H6 occupancy should be reviewed even if a municipality does not use the UBC. Currently, H-6 is the only major building code specially written for the semiconductor industry; hence, it can be considered usual practice, particularly if the municipality has few such facilities.

The H-6 requirements affect architectural, mechanical, and electrical designs significantly when compared to the previous B-2 class. The Uniform Fire Code (UFC), *Article* 51, Semiconductor Fabrication Facilities using Hazardous Production Materials, addresses the specific requirements for process exhaust systems relating to fire safety and minimum exhaust standards. UFC *Article* 80, Hazardous Materials, is relevant to many semiconductor cleanroom projects due to large quantities of stored hazardous materials. Ventilation and exhaust standards for production and storage areas cover control requirements, use of gas detectors, redundancy and emergency power, and duct fire protection.

Temperature and Humidity

Precise temperature control is required in most semiconductor cleanrooms. Specific chemical processes may change under different temperatures or masking of alignment errors may occur due to product dimensional changes as a result of the coefficient of expansion. Temperature tolerances of ±1°F are common, and precision of 0.1 to 0.5°F is likely in wafer or mask writing process areas. Wafer reticle writing by electron beam technology requires ±0.1°F, while photolithographic projection printers require a ±0.5°F tolerance. Specific process temperature control zones must be small enough to counteract the large air volume inertia in VLF cleanrooms.

Temperature control for many process areas is accomplished with internal environmental controls, which allows for room tolerances of ±1°F and larger temperature control zones.

Within the temperature zones of a typical semiconductor factory, latent heat loads are normally small enough to be offset by incoming makeup air. Sensible temperature control is accomplished with cooling coils in the primary airstream or with unitary sensible cooling units that bypass primary air through the sensible air handler and blend conditioned air with unconditioned primary air.

In most cleanrooms of cleanliness Class 1000 or better, production personnel wear full coverage protective smocks that require cleanroom temperatures of 68°F or less. Process temperature set points may be higher as long as tolerances are maintained. If full coverage smocks are not used, higher temperatures may not create employee comfort problems.

Semiconductor humidity levels vary from 30 to 50% rh. Humidity control and precision is a function of process requirements, prevention of condensation on cold surfaces within the cleanroom, and control of static electric forces. Humidity tolerances vary from 0.5 to 5% rh, which is primarily dictated by process requirements. Photolithographic areas have more precise standards and lower set points. Photoresists (chemicals used in photolithography) can be humidity sensitive, affecting their exposure time with varying relative humidities. Negative resists typically require low (35 to 45%) relative humidities, while positive resists tend to be more stable and allow the relative humidity to go up to 50%, creating fewer static electricity problems.

Independent makeup units should provide for dew-point control where direct expansion systems, chilled water/glycol cooling coils, or chemical dehumidification are used. Chemical dehumidification is rarely used in semiconductor facilities due to the high maintenance costs and potential chemical contamination in the cleanroom. Whether direct expansion systems or chilled water/glycol refrigeration are used depends on energy costs for each system. While an operating cleanroom does not generally require reheat, systems are typically designed with the ability to provide heating when new cleanrooms are being built and no production equipment has been installed.

Humidification of makeup air is accomplished by steam humidifiers, water spray nozzles, or evaporative coolers. Steam humidifiers are the most common. Care should be taken to avoid the potential for water treatment chemical release. High purity water, stainless steel piping, and stainless steel unitary packaged boilers have also been used. Located in the cleanroom return, water sprayers use air-operated water jet sprayers. Evaporative coolers have been used, taking advantage of the sensible cooling effect in dry climates. Evaporative coolers also eliminate the need to generate steam, as the energy to humidify can be accomplished with normal heating system hot water temperatures.

Pressurization

Pressurization of semiconductor cleanrooms is also used to control contamination, providing resistance to infiltration of external sources of contaminants. Particulate contaminations outside the cleanroom enter the space by infiltration through doors, cracks, pass-throughs, and other process-related penetrations for pipes, ducts, etc. Positive pressure is maintained within the cleanroom (as referenced to any less clean space) to ensure airflow from the cleanest space to less clean areas. With positive differential pressure in the cleanroom, unfiltered external sources of particulate contamination are inhibited from entering.

Differential pressure (DP) levels of 0.05 in. water is the widely used standard based on the recommendation of Federal *Standard* 209B. Differential pressure values between adjacent cleanroom cleanliness levels should have the cleanest cleanroom with the highest pressure and decreasing pressure levels corresponding to decreasing levels of cleanliness.

Pressure levels within the cleanroom are principally established by the balance between process exhaust and makeup air volumes, as well as supply and return volumes. Process exhaust requirements are dictated by process equipment vendors and industrial hygienists and cannot be changed without risking potential safety hazards. Cleanroom supply air volumes are also set by contamination control specialists. Control of makeup and return air volumes are the primary means available to cleanroom designers for pressure control. The pressure differences should be kept as low as possible while still creating the proper flow direction. Large pressure differences can create eddy currents at wall openings and cause vibration problems.

Static or active control methods based on pressure control tolerances are normally used in cleanrooms. Pressure level control precision is typically 0.01 to 0.03 in. of water, where the owner's contamination control specialist specifies the degree of precision required. Many semiconductor processes require a process chamber pressure precision of ± 0.0025 in. of water, where cleanroom pressure affects the process itself (*e.g.*, glass deposition with silane gas).

Static pressure control methods are suited for static or unchanging cleanroom environments, where the primary pressure control parameters, process exhaust, and supply air volumes do not change or change slowly over weeks or months. Static control systems provide initial room pressure levels; monthly or quarterly maintenance should adjust makeup and return volumes if the pressure level changes. Static systems may provide visual monitoring with differential pressure gages for maintenance personnel.

Active system designs are intended for cleanrooms where pressure control is critical. Opening of doors affects the differential pressure levels, and active system designs provide for closed loop control of the pressure system. Standard control systems do not normally have a quick enough response time to work with door openings. The requirements of this type of system should be carefully evaluated.

Air locks are also used to segregate pressure levels within a factory; typically, however, they are used only between uncontrolled personnel corridors and entrance foyers and the protective clothing (smocks) gowning area. Air locks may also be used between the gowning room and the main wafer fabrication area and for process equipment staging areas prior to ingress into the wafer fabrication facility. Air locks are rarely used within the main portion of the factory, because they restrict personnel access, evacuation routes, and traffic control.

System Sizing and Redundancy

The design of environmental systems must consider the future requirements of the factory. Semiconductor products can become obsolete in as few as two years, and process equipment may be replaced as new product designs dictate. As new processes are added or old ones deleted (*e.g.*, wet versus dry etch), the function of one cleanroom may change from high humidity requirements to low humidity, or the heat load may increase or decrease. Thus cleanrooms must be designed for flexibility and growth. Unless specific process equipment layouts are available, maximum cooling capability should be provided in all process areas at the time of installation, or space should be provided for future installations.

Since cleanroom space relative humidity must be held to close tolerances, and humidity excursions cannot be tolerated, the latent load removal air systems should be based on high ambient dew points and not on the high mean coincident dry- and wet-bulb data.

In addition to proper equipment sizing, system redundancy is also desirable when economics dictate it. Many semiconductor wafer facilities operate 24-h per day, seven days per week with shutdowns only during holidays and scheduled nonwork times. Mechanical and electrical system redundancy is required if the loss of such equipment would shut down critical manufacturing processes. For example, process exhaust fans must operate continuously for safety reasons. Highly hazardous exhaust should have two fans, both running. Most process equipment is computer controlled, with interlocks to provide safety to personnel and products. Electrical redundancy or uninterrupted power supplies may be necessary to prevent costly downtimes during power outages.

Energy Conservation

Electrical loads for environmental control, contamination control, and process equipment can be as much as 300 W/ft^2. Besides process equipment electrical loads, the major energy users are cooling systems, air movement, and process liquid transport [*i.e.*, deionized (DI) water and process cooling water (PCW) pumping].

The major operating costs associated with a cleanroom involve the conditioning of makeup air, air movement within the cleanroom, and process exhaust.

Fan Energy. With flow rates in typical semiconductor facilities being 90 to 100 times greater than those for conventional HVAC systems, the fan system should be closely examined for areas of energy conservation. System static pressures and total cfm requirements should be designed to reduce operating costs. Fan energy required to move cleanroom recirculation air may be reduced by reducing volumes or system static pressure.

Air volumes may be reduced with reduced HEPA filter coverage or by reducing the cleanroom average velocity. When air volumes are reduced, each square foot of reduced HEPA filter coverage saves 25 to 50 W/ft^2 in fan energy and an equal amount in cooling load. Reducing room average velocity from 90 to 80 fpm saves 5 W/ft^2 in fan energy and the identical amount in cooling energy. If air volume supplied to the cleanroom cannot be lowered, reductions in system static pressure can produce significant savings. With good fan selection and transport system design, up to 15 W/ft^2 can be saved per 1 in. of water reduction in static pressure. Static pressure reductions may be achieved with low pressure drop HEPA filters, pressurized plenums in lieu of ducted filters, and proper fan inlet and outlet conditions. Many cleanrooms have only one-shift operations. Reduced air volumes may be achieved during nonworking hours using two-speed motors, inverters, inlet vanes, variable pitch fans, or, in multifan systems, using only a portion of the fans.

Additional fan energy savings may be achieved by substituting standard efficiency motors with high efficiency motors. Good fan selection also influences system energy costs. The choice of forward curved centrifugal versus backward inclined, airfoil, or vaneaxial fans will affect system total efficiency. The number of fans used in a pressurized plenum design influences system redundancy as well as total energy usage. The size of the fan will also change the horsepower requirements. Different options for a prescribed cleanroom configuration should be investigated.

Makeup and Exhaust System Energy. As previously stated, process exhaust requirements in a typical semiconductor facility vary from 1 cfm per ft^2 to 10 cfm per ft^2. Correspondingly, the makeup air requirements also vary from a like amount plus an amount for leakage and pressurization. The amount of energy required to supply the conditioned makeup air can be large. The quantity of exhaust in a given facility is usually determined by the type of equipment installed by the user. It is important to provide a system design that efficiently uses low static duct system design and good fan selection. The use of heat recovery in process exhaust systems has been used effectively. When heat recovery is used, heat exchanger material selection is important due to the potentially corrosive atmosphere.

Reduction of makeup air quantities cannot normally be accomplished without reductions in process exhaust. Due to safety and contamination control requirements, reductions in process exhaust may be difficult to obtain. The costs of conditioning the makeup air should be investigated. Savings may also be accomplished using conventional HVAC methods of high efficiency chillers, good equipment selection, and precise control system design. One method employed in new facilities is the use of multiple temperature chillers to bring outdoor air temperatures to the desired dew point in steps.

NOISE AND VIBRATION

Noise is one of the most difficult variables to control. Particular attention must be given to the noise generated by contamination control equipment. For normal applications of laminar flow equipment, the noise level is designed to be below 65 decibels, as measured on the A scale of a calibrated noise-level meter. In applications where quiet is of utmost importance, noise levels may have to be reduced to about 55 NC level.

In normal applications of contamination control equipment, vibration displacement levels need not be dampened below 0.5 μm in the 1 to 50 Hz range. However, when electron microscopes and other ultrasensitive instrumentation are used, smaller deflections in different frequency ranges might become critical. Photolithography areas may prohibit floor deflections greater than 3 microinches. As a general rule, the displacement should not exceed one-tenth of the line width.

For highly critical areas, the use of vaneaxial fans should be considered, since they generate less noise in the lower frequencies; they may also be dynamically balanced to displacements of less than 0.00015 in., which decreases the likelihood of vibration transmittal to sensitive areas.

ROOM CONSTRUCTION AND OPERATION

Controlling particulate contamination from other than supply air depends on the classification of the space and the type of system and operation involved. Typical items, which may vary with the room class, include the following:

Construction Finishes

- **General.** Smooth, monolithic, cleanable, and chip resistant, with minimum seams, joints, and no crevices or moldings.
- **Floors.** Sheet vinyl, epoxy, or polyester coating with carried-up-wall base, or raised floor with and without perforations using the above materials.
- **Walls.** Plastic epoxy-coated drywall, baked enamel, polyester, or porcelain with minimum projections.
- **Ceilings.** Plaster covered with plastic, epoxy, or polyester coating or plastic-finished acoustical tiles, when entire ceiling is not fully HEPA filtered.
- **Lights.** Teardrop-shaped, single lamp fixtures mounted between filters or flush-mounted and sealed.
- **Service Penetrations.** All penetrations for pipes, ducts, conduit runs, etc., should be fully sealed or gasketed.
- **Appurtenances.** All doors, vision panels, switches, clocks, etc., should have either flush-mounted or sloped tops.

Personnel and Garments

- Hands and face are cleaned before entering area.
- Lotions and soap containing lanolin are used to lessen skin particles from being emitted.
- Wearing cosmetics and skin medications are not permitted.
- Smoking and eating are not permitted.
- Lint-free smocks, coveralls, gloves, and head and shoe covers are worn.

Materials and Equipment

- Equipment and materials are cleaned before entry.
- Nonshedding paper and ballpoint pens are used. Pencils and erasers are not permitted.
- Work parts are handled with gloved hands, finger cots, tweezers, vacuum wands, and other methods to avoid transfer of skin oils and particles.

Particulate-Producing Operations

- Grinding, welding, and soldering operations are shielded and exhausted.
- Containers are used for transfer and storage of materials.

Entries

- Air locks and pass-throughs are used to maintain pressure differentials and reduce contamination.

CHAPTER 17

DATA PROCESSING SYSTEM AREAS

DATA processing system areas contain computer equipment, as well as the ancillary equipment needed to meet a particular data processing function. Computers generate heat and contain components sensitive to extremes of temperature, humidity, and the presence of dust. Exposure to conditions outside prescribed limits can cause improper operation or complete shutdown of the equipment.

Ancillary spaces for activities directly related to the computer or for the storage of computer components and materials (including magnetic tape, disc packs and cartridges, data cells, paper, and punch cards) should have environmental conditions comparable to the areas housing the computers, although tolerances generally are wider and the degree of criticality is usually much less. If components and supplies are exposed to temperature and humidity levels outside the limits established by the manufacturer, they must be conditioned to the operating environment in accordance with the manufacturer's recommendations.

Other areas in the data processing complex may house such auxiliary equipment as engine generators, motor generators, uninterruptible power supplies (UPS), and transformers. The air-conditioning and ventilating quality requirements are less severe than those for computers, but the continuing satisfactory operation is vital to the proper functioning of the computer system.

DESIGN CRITERIA

The data processing system spaces that house the computers, computer personnel, and associated equipment require air conditioning to maintain proper environmental conditions for both the equipment and the comfort of personnel.

Computer room air conditioning does not require quick response to changes in set points for the environmental conditions, but *maintaining* conditions within established limits is essential to the operation. System reliability is so vital that the potential cost of system failure often justifies redundant capacity and/or components.

The environmental conditions required by computer equipment vary widely, depending on the manufacturer. Table 1 lists general recommendations on conditions for the computer room. Most manufacturers recommend that the computer equipment draw conditioned air from the room; some, however, permit or recommend direct cooling by supply air. Criteria for air introduced directly to computers differ from usual room air supply conditions; this air supply must remove computer equipment heat adequately and preclude the possibility of condensation within the equipment (see Table 2).

Because of the high energy needed to maintain proper environment in data processing areas, the design should hold *net* energy use as low as possible. For instance, heat energy recovery is quite feasible in many installations.

The preparation of this chapter is assigned to TC 9.2, Industrial Air Conditioning.

Computer Room Environment

Computer rooms should be kept at the lower end of the temperature tolerance of 72 ± 2°F for two reasons. First, the equipment distributes heat nonuniformly, which may not match the air distribution and/or response of the controls. Setting the controls no higher than 72°F generally assures that all equipment will be within the established range for satisfactory operation. Secondly, the lower control temperature provides a cushion for short-term peak load temperature rise without adversely affecting computer operation.

High relative humidity levels may cause improper feeding of cards and paper and, in extreme cases, condensation on machine surfaces. Low relative humidity in combination with other factors may result in static discharge, which can adversely affect the operation of data processing and other electronic equipment.

To maintain proper relative humidity in a computer room, vapor transmission retarders should be installed around the entire envelope sufficient to restrain moisture migration during the maximum expected vapor pressure differences between computer room and surrounding areas. Cable and pipe entrances should be sealed and caulked with a vaporproof material. Door jambs should fit tightly. Windows in colder climates should be double- or triple-glazed.

In localities where the outdoor air contains unusually high quantities of dust, dirt, salt, or corrosive gases, it may be necessary to pass it through higher efficiency filters or adsorption chemicals before it is introduced into the computer room.

Table 1 Typical Computer Room Design Conditions[a]

Temperature set point and offset	72 ± 2°F
Relative humidity set point and offset	50 ± 5%
Filtration quality (ASHRAE *Standard* 52-76 dust spot efficiency test)	45%, minimum 20%

[a] These conditions are typical and fall within the conditions recommended by most computer equipment manufacturers.

Table 2 Design Conditions for Air Supply Directly to Computer Equipment[a]

Conditions	Recommended Levels
Temperature	As required for heat removal but no lower than 60°F
Relative humidity	Maximum 65% (some manufacturers permit up to 80%)
Filtration quality (ASHRAE *Standard* 52-76, dust spot efficiency test)	45%

[a] These conditions are typical and fall within the conditions recommended by most computer equipment manufacturers, but under no circumstances should conditioned air be supplied directly to a computer unless it is designed and controlled within the manufacturer's limits for the particular piece of equipment.

The presence of dust can affect the operation of data processing equipment, so good quality filtration is important. Proper maintenance of filters is very important in the computer room. Dirty filters can reduce airflow, and thereby decrease the sensible heat ratio of computer room air-conditioning equipment. This, in turn, will load the computer room with energy intensive humidification, thereby needlessly increasing the operating cost.

Computer room systems should provide only enough outdoor air for personnel requirements and to maintain the room under a positive pressure relative to surrounding spaces. Since most computer rooms have few occupants and the quantity of conditioned air circulated is high compared with comfort applications, the need to maintain positive pressure is usually the controlling design criterion. In most computer rooms, an outdoor air quantity less than 5% of total supply air will satisfy ventilation requirements and assure against inward leakage. Outdoor air beyond the required minimum increases the cooling and heating loads, and makes control of atmospheric contaminants and winter humidity more difficult.

The air-conditioning system should operate within the noise levels of the computer equipment. This is ordinarily not a difficult criterion to satisfy. Vibration should be isolated to prevent structural transmission of the computer equipment. The computer manufacturer should be consulted regarding computer equipment sound levels and specific special requirements for vibration isolation.

Personnel comfort is also important. Temperature, humidity, and filtration requirements for the equipment are within the comfort range for the room occupants, but drafts and cold surfaces must be kept to a minimum in occupied areas.

Some manufacturers have established criteria for allowable rates of environmental change to prevent shock to the computer equipment. These can usually be satisfied by high quality commercially available controls responding to ±1°F and ±5% rh. The manufacturer's requirements should be reviewed to be sure that the system will function properly during normal operation and during periods of start-up and shutdown.

Computer equipment will tolerate a somewhat wider range of environmental conditions when not in operation; but to keep the room within those limits and to minimize thermal shock, it may be desirable to operate the air conditioning.

Computer technology is continually changing; during the life of a system, the computer equipment will almost certainly be changed and/or rearranged. The air-conditioning system must be sufficiently flexible to facilitate this rearrangement of components and permit expansion without requiring the system to be rebuilt. In the usual applications, it should be possible to make modifications without extensive air-conditioning shutdowns, and, in highly critical applications, with no shutdown at all.

Ancillary Spaces Environment

Spaces storing products such as paper, cards, and tape generally require the same environmental limits as the computer room itself.

Electrical power supply and conditioning equipment has more tolerance to variation in temperature and humidity than does computer equipment. Equipment in this category includes motor generators, uninterruptible power supplies (UPS), batteries, voltage regulators, and transformers. Normally, ventilation to remove heat from the equipment is sufficient. Manufacturers' data should be followed to determine the amount of heat release and design conditions for satisfactory operation.

Battery rooms require ventilation to remove hydrogen and control space temperature (optimum 77°F). Hydrogen accumulation must be no greater than 3% by volume, with the ventilation system designed to prevent pockets of concentration, particularly at ceilings (EXIDE 1972).

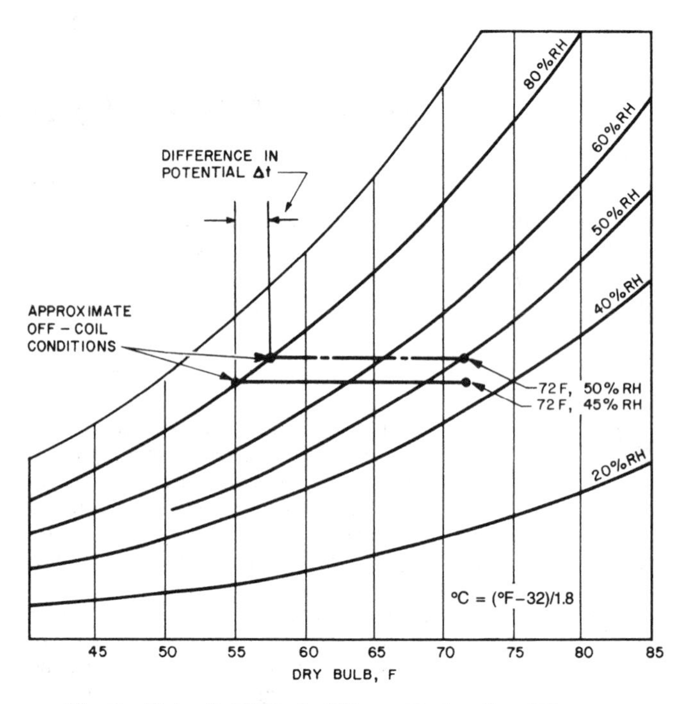

Fig. 1 Potential Effect of Room Design Conditions on Design Supply Air Quantity

Many computer installations include an engine generator for emergency power, which requires a large amount of ventilation when running. Starting is easier if low ambient temperatures are avoided.

COOLING LOADS

The equipment in a computer room is the major heat source, and it is highly concentrated and distributed nonuniformly. Heat gain from lights should be no greater than that for good quality office space; occupancy loads and outdoor air requirements will be low to moderate. Heat gains through the structure depend on the location and construction of the room. Transmission heat gain to supply spaces should be carefully evaluated and provided for in the design.

Information on computer equipment heat release should be obtained from the computer manufacturer. In general, the data is used without reduction for diversity, unless experience with a similar installation or the computer manufacturer recommends a reduction.

Because of the relatively low occupancy and the low proportion of outdoor air, computer room heat gains are almost entirely sensible. Because of this and the low room design temperatures, air supply quantity per unit of cooling load will be greater than for most comfort applications. Figure 1 shows how the choice of room design at 72°F, 45% rh can reduce air supply quantity by approximately 10% as compared to a slightly more humid room design of 72°F, 50% rh.

A sensible heat ratio, approximately 0.9 to 1.0, is common for computer room applications. This relatively small amount of latent cooling is adequate to handle the minimum moisture loads incurred and will not cause needless dehumidification.

AIR-CONDITIONING SYSTEMS

The air-handling apparatus for the computer air-conditioning system should be independent of other systems in the building, although it may be desired that systems be cross-connected, within or without the data processing area, to provide backup. Redun-

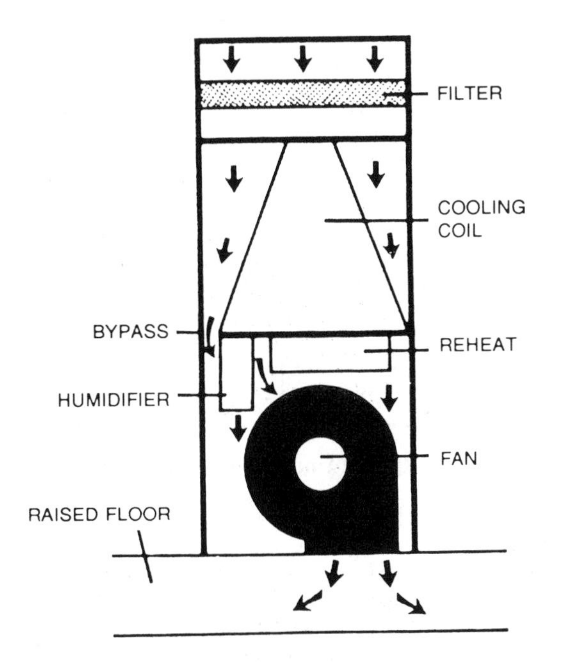

Fig. 2 Self-Contained Computer Room Air Conditioner

dant air-handling equipment is frequently used; normally automatic operation of such equipment is desired. The air-handling facilities should provide filtration, cooling and dehumidification, humidification, and heating of air.

The refrigeration systems should be independent of other systems and should be capable of year-round operation. It may be desirable to cross-connect refrigeration equipment, for backup, as suggested for air-handling equipment. Redundant refrigeration may be required; the extent of the redundancy will depend on the importance of the computer installation.

In many cases, standby power is justified for the computer room air-conditioning system. The system components to receive standby power require careful analysis to ensure that they provide 24 h a day, year-round operation.

Computer rooms are being successfully conditioned with a wide variety of systems including: (1) complete self-contained packaged units, with air-handling and refrigeration apparatus close-coupled within a common housing and installed within the computer room; (2) chilled water packaged units located within the computer room and served by remotely located refrigeration equipment; and (3) central station air-handling units with both air-handling and refrigeration equipment located outside the computer room.

Self-Contained Packaged Units

Self-contained units should be specifically designed for computer room applications. These units are built to higher overall standards of performance and reliability than conventional packaged air conditioners intended for comfort, although some major components are identical to those in standard unitary equipment.

Packaged units are available with (1) multiple reciprocating compressors and separate multiple refrigeration circuits through the cooling coil and condensers; (2) air filters to meet computer room criteria; (3) humidifiers; (4) a reheat coil; (5) corrosion-resistant construction for coils, humidifiers, and other components; (6) controls; (7) instrumentation such as indicator lights to show which equipment components are in operation, and alarms to signal dirty filters and component failure; and (8) filter gages to indicate the status of filter loading. Status and/or alarm devices may be connected to remote monitoring panels.

Although the placement of components varies with the manufacturer, a typical unit arrangement is shown in Figure 2. The units can supply air either downward to the computer room floor cavity or upward to overhead ducts and/or a ceiling plenum.

The refrigeration cycles for computer room units require a means for condensing, and this is provided by water-cooled condensers connected with a remote cooling tower, water- or glycol-cooled condensers connected with a remote radiator, or connections with remotely located air-cooled or evaporative condensers.

Self-contained air conditioners are usually located within the computer room, but may also be remotely located and ducted to the conditioned space. If they are remotely located, temperature and humidity controls should be located in the conditioned space. A major benefit of locating the air conditioner close to the load is the flexibility of such an arrangement, which accommodates the ever-changing load pattern within many computer rooms. The space occupied by these packaged units in the computer room may be expensive, but security considerations alone make it practical to place them there, since normally a central system serving a computer room has no security protection beyond normal building maintenance.

In systems with multiple unitary conditioners, it may be advantageous to introduce outdoor air through one conditioner that serves all of the spaces in the data processing area. Self-contained systems achieve redundancy by providing multiple units, so that the loss of one or more units will have a minimum effect on system performance. Additionally, expansion of the data processing facility is generally much easier to handle with self-contained units.

Chilled Water Packaged Units

These units are similar to complete, self-contained packaged units, except that they are served by remotely located refrigeration units, usually through chilled water connections, so they contain no refrigeration equipment. Since computer components in some systems require a source of chilled water for cooling, the use of chilled water packaged units may be more advantageous in this application. Chilled water supply temperatures suitable for water-cooled computer equipment can range from a low of 42°F, to a high of 60°F. These chilled water temperatures will normally be compatible with those required for computer room air-conditioning units.

When using chilled water packaged units, reliability of the remote refrigeration system must be considered. This system generally must be capable of operating 24 h a day, year-round. Low ambient operation must be provided in severe winter locations. Packaged units with chilled water coils, as well as direct expansion coils, are available and afford an alternate refrigeration source when the chilled water plant shuts down.

Chilled water packaged units occupy an area within the computer room, but since they do not contain refrigeration equipment, they require less servicing within the computer room than self-contained equipment.

Central Station Air-Handling Units

Central station supply systems permit using components with larger capacity than is available in self-contained equipment; also, since the equipment is not located within the computer room, it permits a greater variety of choices in air-conditioning system design and arrangement.

Central station air-handling equipment must satisfy computer room performance criteria; it should be arranged to facilitate servicing and maintenance. Redundancy can be achieved by cross-connecting systems, by providing standby equipment, or by a combination of these. No floor space in the computer room is required, and virtually all servicing and maintenance operations are performed in areas specifically devoted to air-conditioning equipment. However, as cooling equipment is outside the computer room, security is lessened.

Central station supply systems must be designed with expansion capability to accommodate additional loads in the computer areas. These systems must be complete with humidification, reheat, dehumidification, and controls. Ductwork penetration of the computer space must be sealed to prevent moisture migration to the computer areas. Central systems for computer rooms should not be tied together with building air-handling systems without special controls that provide for a year-round cooling operation and high sensible heat ratio design factors.

SUPPLY AIR DISTRIBUTION

Computer components that generate large quantities of heat are normally constructed with internal fans and passages to convey cooling air through the machine. The inlet usually draws the air from the computer room, but some manufacturers recommend that certain components of their systems take cooling air directly from the conditioned air supply before it reaches room conditions.

Computer room heat gains are often highly concentrated. For minimum room temperature gradients, supply air distribution should closely match load distribution, and the control thermostat must be located where it will sense average conditions in the area it serves. Return-air thermostats have generally best met computer requirements because of the constantly changing computer equipment arrangement. The distribution system should be sufficiently flexible to accommodate changes in the location and magnitude of the heat gains with a minimum amount of change in the basic distribution system.

Supply air systems usually require approximately 550 cfm per ton of cooling to satisfy computer room conditions. This airflow rate reduces the hot spots and maintains an even temperature distribution.

The construction materials and methods chosen for air distribution should recognize the need for a clean air supply. Duct or plenum material that may erode must be avoided. Access for cleaning is desirable.

Zoning

Computer rooms should be adequately zoned to maintain temperatures within the design criteria. The extent of zoning is reduced because of the relatively small number of rooms and the generally open character of the spaces involved. Except in the smallest computer rooms, some zoning is usually required to minimize temperature variations because of fluctuations in load. As a minimum, individual control for each major space is desirable and, in larger areas, potential for temperature variations may occur within a single room.

Packaged air-conditioning systems using the underfloor air supply plenum (Figure 3) adequately self-zone the large computer room area. These systems have return-air temperature and humidity contol, which automatically controls the air being circulated in its zone. The area of the zoning is controlled by the various floor registers and perforated floor panels. These systems give adequate flexibility for relocation of computer equipment and additional future heat loads.

Underfloor Plenum Supply

To facilitate interconnection of equipment components by electric cables, data processing equipment is usually set above a false floor, which affords a flat walking surface over the space where the connecting cables are installed. This space can be an air

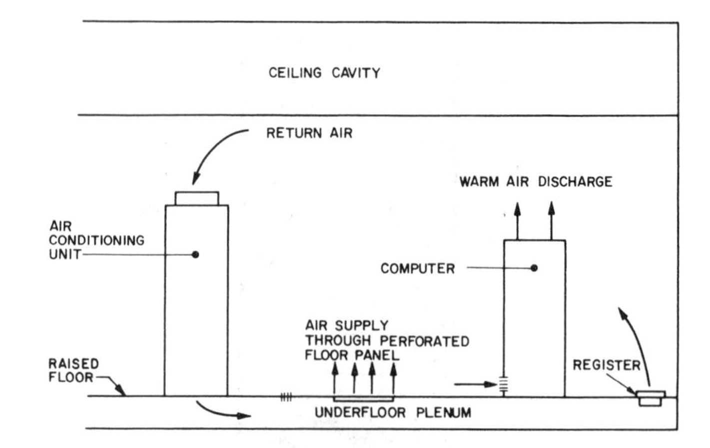

Fig. 3 Typical Underfloor Distribution

distribution channel for systems that use it either as a plenum or, less often, to accommodate ducts. Underfloor air is distributed to the room through perforated panels or registers set or built into the floor panels located around the room and in the vicinity of computer equipment with especially high heat release.

Some computer room floor manufacturers produce perforated floor panels similar in appearance to, and interchangeable with, conventional computer room floor panels. These panels allow the supply air to be easily changed to accommodate shifting equipment loads. The free area of various manufacturers' panels varies significantly. One configuration has slide-type dampers so airflow may be balanced in a manner comparable to a floor-mounted register (Figures 3 and 4).

Floor-mounted registers allow volume adjustment and are capable of longer throws and better directional control than perforated floor outlets. However, they are usually located outside traffic areas because some types are not completely flush-mounted and almost all tend to be drafty for nearby personnel. Perforated floor panels are especially suitable for installation in normal traffic aisles because they are completely level with the floors. In addition, when operated with moderate airflow, they produce a high degree of mixing near the point of discharge. For this reason, they may be located fairly close to equipment air inlets, with less danger of direct injection of unmixed conditioned air to the computer than with registers that have lower induction ratios.

Airflow is through all openings in the floor cavity; if direct flow to a computer unit is not desired, all openings between the unit

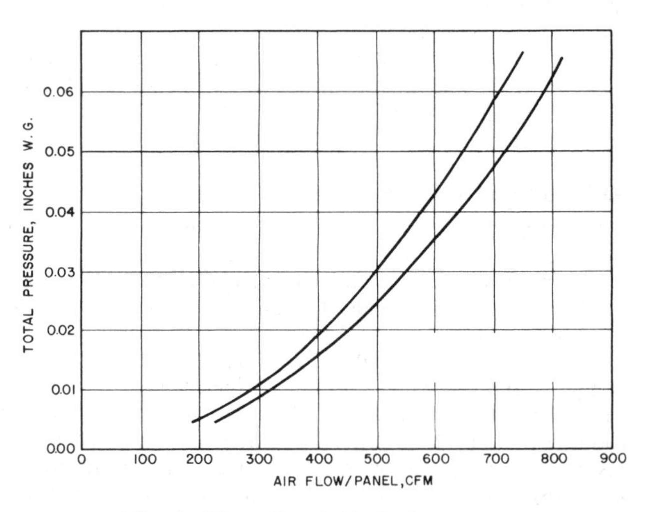

Fig. 4 Floor Panel Air Performance

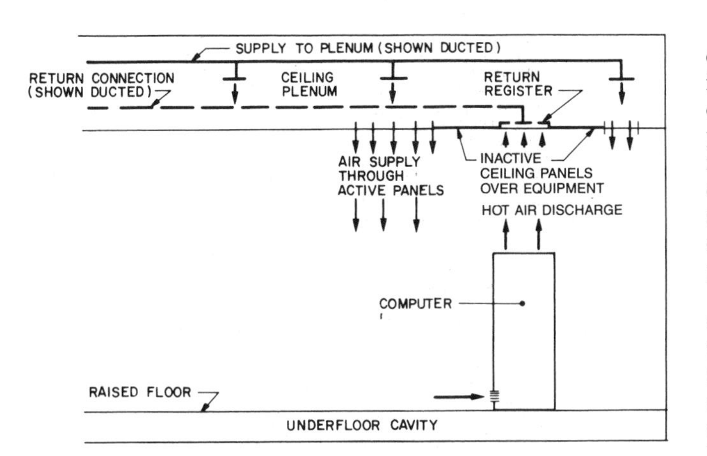

Fig. 5 Typical Ceiling Plenum Distribution

and the floor cavity should be sealed. Most openings between the underfloor space and the equipment accommodate cables; collars are available to fit the cable and seal the opening.

It is important that there be sufficient clearance area within the raised floor cavity to permit airflow. At least 12 in. clearance is desirable and 10 in. is the normal minimum; in applications where cabling is extensive and/or air quantities are especially high, additional clearance may be required.

The supply connection from unitary equipment to an underfloor cavity should allow minimum turbulence; turning vanes at the unit discharge sometimes helps accomplish this. The supply to the cavity should be central to the area served, where possible, and abrupt changes in direction should be avoided. Piping and conduit to unitary equipment should not interfere with airflow from the unit.

Where multiple zones are served from the underfloor plenum, dividing baffles may generally be avoided, as supply and return-airflow can be predicted to flow through the course of least resistance. If zone configurations require dividing baffles, or if codes require them to protect against spread of fire, they may inhibit the flexibility of the computer system as a whole, because computer units in different zones often require cabling interconnection, which requires penetration of the zone baffles whenever cabling changes are made.

Surfaces of underfloor plenums may require insulation or vapor barriers if (1) supply air temperatures drop low enough to cause condensation on either side of plenum surfaces or (2) temperatures on the floor above a ceiling plenum become so low as to cause discomfort to occupants. Insulation may also be required to reduce heat transmission through plenum surfaces. Plenums should be of airtight construction and be thoroughly cleaned and smoothly finished to prevent entrainment of foreign materials in the airstream. The use and method of construction of plenums may be limited by local codes or fire underwriter regulations.

The potential adverse effects of direct air supply on the computer equipment (condensation within the machine) are serious enough to discourage its use unless the manufacturer insists on it. If this occurs, sufficient controls and safety devices must be installed to ensure operation within specified limits.

Ceiling Plenum Supply

Overhead supply through perforated ceiling panels used as diffusers is occasionally suited to computer rooms; this arrangement may be used with either central station or packaged-unit unitary equipment. Such systems can satisfy equipment requirements, as well as personal comfort; however, they are generally not as flexible as underfloor plenum supply systems.

Distribution of air can be regulated either by selective placement of acoustical pads on the perforated panels of a metal panel ceiling or the location of *active* perforated sections of a lay-in acoustical tile ceiling. Where precise distribution is essential, active ceiling panels or air supply zones may be equipped with air valves. When required for zoning or to meet codes, overhead plenums may be divided by baffles without interference with underfloor computer system cabling. Ceiling plenums, if properly constructed and cleaned, are more likely to remain clean than are underfloor plenums.

Plenums must be sufficiently deep to permit airflow without turbulence; the depth will depend on air quantities. Best conditions may be achieved by the use of distribution ductwork, with the air discharged into the plenum through adjustable outlets above the ceiling (Figure 5). Sealing, surfaces, and insulation of ceiling plenums must be constructed as shown previously in the Underfloor Plenum Supply section.

Overhead Ducted Supply

Overhead ducted supply systems should be limited to applications where air supply concentrations are not high or the need for flexibility is not great. These factors impose a severe limitation on the use of overhead ducted supply systems. Where loads are high, overhead diffusers may cause drafts, especially where ceilings are low. Relocation of outlets can be difficult to accomplish in a facility that must remain in operation.

RETURN AIR

The more common return-air method has a minimum number of return openings. With packaged units, the use of a free return from supply outlets back to the return on the unit is typical (Figure 3). Inlets should be located near high heat loads, and the effect of future modifications of the computer installation should be considered.

Ceiling plenum returns have successfully captured a portion of the computer heat, as well as a portion of the heat from the lights, directly in the return airstream. This allows a reduction in the supply air circulation rate. A duct collar is usually provided to the top of the self-contained unit; the unit will draw the air from the return-plenum area, treat it, and discharge it into the underfloor supply plenum.

WATER-COOLED COMPUTER EQUIPMENT

Some computer equipment requires water cooling to maintain equipment environment within levels established by the manufacturer. Generally, a closed system circulates distilled water through passages in the computer to cool it. The manufacturer supplies this cooling system as part of the computer equipment, and a water-to-water heat exchanger attached to a chilled water connection to the air-conditioning system does the cooling.

In systems with water-cooled components, the majority of the total cooling is still accomplished by air. The overall heat release to the computer room from water-cooled computers is usually equal to (or greater than) that for most air-cooled computer systems.

Water-cooled equipment requires chilled water in the computer room. This can be provided either by a small separate chiller matched in capacity to the water-cooled computer equipment or by a branch of the chilled water serving central or decentralized air-handling units. Construction and insulation of chilled water piping and selection of the operating temperatures should be such that they minimize the possibility of leaks and condensation—especially within the computer room, while satisfying the requirements of the systems served.

Chilled water systems for water-cooled computer equipment must be designed for water temperature within the computer manufacturers' tolerances and for continuous availability. Refrigeration systems for chilled water must be capable of operating year-round, 24 h a day.

AIR-CONDITIONING COMPONENTS

Controls

A well-planned control system must coordinate the performance of the temperature and humidity equipment.

Controls of self-contained, packaged computer air-conditioning systems have been tested, coordinated, and most likely improved upon over several installations. Controls for systems not air conditioned by packaged equipment may be included in a project-specific control system, which may be computer based.

Space thermostats and humidistats should be carefully located to sample room conditions most accurately.

Where backup power is provided for the computer installation and some portion of the HVAC equipment, it is imperative that the necessary control equipment also have power backup. Liquid detectors, for underfloor plenum systems, are useful in warning of accidental water spillage, which otherwise could go undetected for some time.

Refrigeration

Separate refrigeration facilities for the data processing area are desirable because performance requirements often differ drastically from those for comfort systems. Refrigeration systems should generally satisfy the following criteria: (1) match the cooling load, be capable of expansion, and have a degree of flexibility comparable to other components; (2) be capable of year-round and continuous operation; (3) provide the degree of reliability and redundancy required by the particular operation; (4) be capable of being operated, serviced, and maintained without interfering with normal operation; and (5) be operable with emergency power if this is a system requirement. Fulfillment of these requirements results in the use of multiple units or cross-connections with other reliable systems.

Self-contained systems invariably use reciprocating compressors and direct-expansion cooling. Central systems and systems using decentralized air handling have a wider range of choice, because reciprocating, centrifugal, or absorption refrigeration equipment can be selected.

If the installation is especially critical, it may be necessary to install up to 100% standby refrigeration capacity or capacity adequate to provide the minimum requirements that can be tolerated until repair or replacement.

Condensing Methods

Heat rejection equipment must be designed and selected for *worst case* operating conditions to prevent computer shutdown during weather extremes.

In some areas, water-cooled equipment using cooling towers may be satisfactory; in other areas, the need for winterization and water treatment may preclude use of systems vulnerable to freezing or open to the atmosphere. Air-cooled condensers are widely used and are equipped with head pressure controls capable of operation through widely varying outdoor ambient temperatures. They ordinarily should not be used with separate condenser curcuits and separate hot gas and liquid lines for each refrigeration circuit in the system.

Glycol-cooled radiators, or water-cooled radiators in climates not subject to freezing, permit using a closed system on the condensing side and facilitate head pressure control by permitting control of the condensing medium. In systems where multiple refrigeration systems or units are remote from the point of heat rejection to the atmosphere, radiator-cooled systems may significantly simplify piping and capacity control (Figure 6).

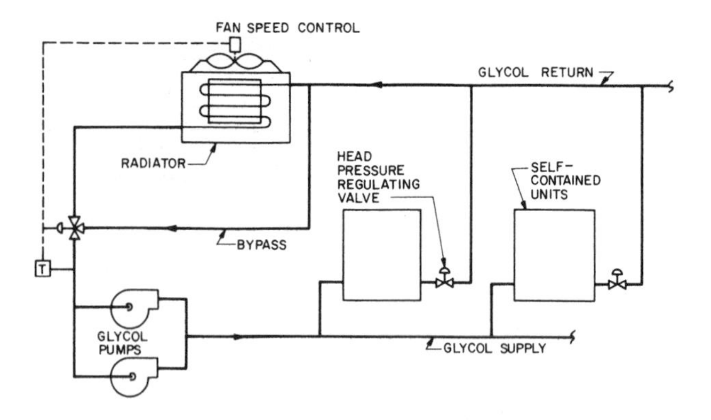

Fig. 6 Glycol System for Condensing

Glycol-cooled or water-cooled systems require water piping in the computer areas, and precautions must be taken to prevent and detect possible leaks.

Humidification

Many types of humidifiers are used to serve data processing system areas. Types of humidifiers include: steam with outside source of steam; steam-generating; pan with immersion element; pan with infrared (quartz lamp); and wetted-pad. Where a continuously available clean source is available, steam humidification should be considered. The humidification method chosen must be responsive to control, low in maintenance, and free of moisture carryover.

Chilled Water

Chilled water distribution systems should be designed to the same standards of quality, reliability, and flexibility as the rest of the system. Multiple units should be provided for those components that must be shut down for servicing or routine maintenance. The chilled water system should be designed for expansion or the addition of new conditioners without extensive shutdown where any likelihood of growth exists.

Figure 7 illustrates a looped chilled water system with sectional valves and multiple valved branch connections. The branches could serve air handlers or water-cooled computer equipment. The valves permit modifications or repairs without requiring complete shutdown.

Chilled water temperature should match the load and minimize the possibility of condensation, especially if portions of the system are installed within the computer room. Since computer room loads are primarily sensible, relatively high chilled water temper-

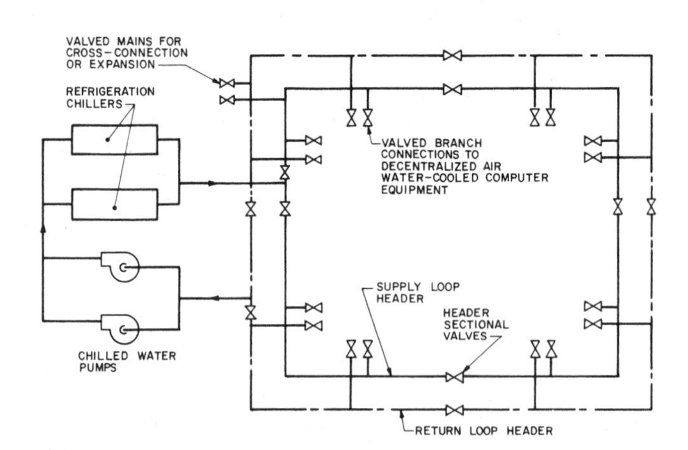

Fig. 7 Chilled Water Loop Distribution

atures should be circulated. Although this may result in some reduction in unit capacity or require a deeper cooling coil, the net effect on system operation and efficiency will be positive.

Water temperatures as high as 48 °F are still slightly below the dew point of a 72 °F, 45% rh room and several degrees below that of a 72 °F, 50% rh room. Because of this and because the operation of system controls may not regulate chilled water temperature precisely, the chilled water piping must be fully insulated.

The piping system should be pressure tested. The test pressure should be applied in increments in occupied areas if a leak could interfere with the operation of the computer system. Drip pans should be placed below any valves or other components that cannot be satisfactorily insulated within the computer room. A good quality strainer should be used to prevent clogging heat exchanger passages.

If cross-connections with other systems are made, the possible effects of the introduction of dirt, scale, or other impurities on the computer room system must be evaluated and handled.

INSTRUMENTATION

Because computer equipment malfunctions may be caused by, or attributed to, improper regulation of the computer room thermal environment, it may be desirable to keep permanent records of the space temperature and humidity. If direct air supply to the machines is used, these records can then be correlated with machine function. Alarms should be incorporated with the recorder to signal if temperature or humidity limitations are not satisfied. Where chilled water is supplied directly to computer units, records should be made of the chilled water temperature and pressure, both entering and leaving.

A sufficient number of indicating thermometers and pressure gages should be placed throughout the system so that operating personnel can tell, at a glance, when unusual conditions prevail. Properly maintained and accurate filter gages are simple devices that, when properly used, can help prevent loss of system capacity and maintain correct computer room conditions.

Sensing devices to indicate leaks or the presence of water in the computer room underfloor cavity are desirable, especially if glycol or chilled water distribution lines are installed under pressure within the computer room. Low points in drip pans installed below piping should also be monitored.

All monitoring and alarm devices should give local indication; if the system is in a building with a remote monitoring point, indications of system malfunctions should be transmitted to activate alarms in the remote location.

FIRE PROTECTION

Fire protection for the air-conditioning system should be fully integrated with fire protection for the computer room and the building as a whole. Applicable codes must be complied with, and the owner's insurers consulted. Automatic extinguishing systems afford the highest degree of protection. Fire underwriters often recommend an automatic sprinkler system (Jacobson 1967). Most computer system owners are reluctant to install such a system because of the use of water, so most computer rooms are not so protected.

Use of Halon fire-extinguishing systems in computer facilities have become more frequent (NFPA 1981). In facilities using Halon equipment, a prompt means of ventilation should be considered. Also, any openings, such as ventilation ducts, leading outside the Halon-protected space must have dampers that close immediately on discharge of the Halon.

At a minimum, sensing devices should be placed in both the room and the air-conditioned airstream to warn of fire. Sensing devices should be located in supply and return-air passages and in the underfloor cavity when electric cables are installed, whether or not this space is used for air supply. These devices should provide early warning to products of combustion, even if smoke is not visible and temperatures are at or near normal levels.

Any fire protection system should include a shutdown of computer power, either manual or automatic, depending on the criticality of the system and on the potential effect of shutdown because of false alarms (NFPA 1984).

HEAT RECOVERY AND ENERGY CONSERVATION

Computer room systems are attractive candidates for heat recovery systems because of their large, relatively steady year-round loads. If the heat removed by the conditioning system can be efficiently transferred and applied elsewhere in the building, some cost saving may be realized. However, reliability is a prime requirement in most computer installations, and any added complication to the basic conditioning system that might impair reliability of the system or otherwise adversely affect performance should be carefully considered before being incorporated into the design. Potential monetary loss from system malfunction or unscheduled shutdown will often outweigh savings because of increased operating efficiency.

Heat rejected for condensing can be used for space heating, domestic water heating, or other process heat. Packaged, self-contained units are available with such heat recovery components. When an air-side economizer is used in cooling computer areas, outdoor air humidification requirements should be considered. Use of a liquid heat exchanger between cool outdoor air and the interior air is economically feasible in some climates. Packaged, self-contained units are available with an additional cooling coil for chilled glycol solution.

Energy conservation can be achieved by effecting optimum operating conditions. For instance, coil surface temperatures should be maintained as high as possible to cool the space adequately, yet dehumidify no more than necessary. Coil temperatures lower than necessary waste energy, both through higher refrigeration energy consumption and over-dehumidifaction. Annual energy-use calculations or simulations to evaluate the economics of various designs are often desirable.

REFERENCES

Jacobson, D.W. 1967. Automatic sprinkler protection for essential electric and electronic equipment. *NFPA Fire Journal* (January):48.

NFPA. 1981. *Protection of electronic computer/data processing equipment.* National Fire Code 75-81. National Fire Protection Association, Quincy, MA.

NFPA. 1984. *National electrical code.* National Fire Protection Association, Quincy, MA.

NFPA. 1980. Halon 1301 fire extinguishing systems. National Fire Code 12A-80. National Fire Protection Association, Quincy, MA.

CHAPTER 18

PRINTING PLANTS

THIS chapter outlines air-conditioning requirements for key printing operations. Air conditioning of printing plants can provide controlled, uniform air moisture content and temperature in working spaces. Paper, the principal material used in printing, is hygroscopic and very sensitive to variations in the humidity of the surrounding air. Problems caused by expansion and contraction of paper in the printing process are solved by controlling moisture content and exposure of paper from the mill until printing is complete.

GENERAL DESIGN CRITERIA

The three basic methods of printing are as follows:

1. Letterpress (relief printing): ink is applied to a raised surface that does the printing.
2. Lithography: the inked surface that does the printing is neither in relief nor recessed.
3. Gravure (intaglio printing): the inked areas are recessed below the surface.

Figure 1 describes the general work flow through a printing plant. The operation begins at the publisher and ends with two products: (1) finished printing and (2) paper waste. Paper waste may be as high as 20% of the total paper used. The profitability of a printing operation requires efficient paper use. Without proper air conditioning, quality control is extremely difficult to achieve.

Sheetfed printing feeds individual sheets through the press from a stack or load of sheets, then collects the printed sheets. Webfed rotary printing uses a continuous web of paper, which is fed through the press from a roll. The printed material is cut, folded, and delivered from the press as signatures (sections of a book).

Sheetfed printing is a slow process in which the ink is essentially dry as the sheets are delivered from the press. *Offsetting,* the transference of an image from one sheet to another, is prevented by applying a powder or starch to each sheet as it is delivered from the press. The starch separates the sheets enough for sufficient drying to prevent offsetting. Starches present a housekeeping problem. The starch particles (30 to 40 μm in size) tend to fly and eventually settle on any horizontal surface.

Temperature and relative humidity have little to do with web breaks or runability of paper in a webfed press, if both are controlled within normal human comfort limits. At extremely low humidity, static electricity causes the paper to cling to the rollers, creating undue stress on the web, particularly with high-speed presses, and can be a hazard when flammable solvent inks are used.

The preparation of this chapter is assigned to TC 9.2, Industrial Air Conditioning.

Various areas in the printing plant require special attention to processing and heat loads. For example, an engraving department must have very clean air—not as clean as an industrial clean room, but cleaner than that required for an office.

Engraving and photographic areas require special ventilation standards because of the chemicals used. The nitric acid fumes used in powderless etching require stainless steel or aluminum ducts. Stereotype departments have very high heat loads. Composing room operations, which include computer equipment, should be given the same attention as similar office areas. The high loads created by lead pots on typecasting machines are no longer common. Excessive dust carried into the mailroom from the cutting operation in the press folders must be handled properly.

The pressroom exhaust air must be treated to eliminate pollutants (pigments, oils, resins, solvents, and dust) exhausted into the atmosphere. Ink mist suppression systems, applied as part of the presses, have reduced the ink mist carried into the exhaust filtration system.

Air-conditioning and air-handling equipment used in printing plants is conventional. Ventilation of storage areas should be about one-half air change per hour, and bindery ventilation should be about one air change per hour. Highly piled storage may need roof-mounted smoke and heat-venting devices. Air distribution in pressrooms is a compromise. The air supplied must not be close enough to the press equipment to cause flutter of the web and not too high to force contaminants or heat (which normally would be removed by roof vents) down to the occupied level.

In the bindery, loads of loose signatures are stacked near the equipment. Thus, it is difficult to supply air to occupants without scattering signatures. One approach is to run the main ducts at the ceiling with many supply branches dropped to within 8 to 10 ft of the floor. Conventional adjustable blow diffusers, often linear type, are used.

CONTROL OF PAPER

Control of the moisture content and temperature of paper is important in all printing, particularly in multicolor lithography. Paper should be received at the printing plant in moistureproof wrappers, which are not removed or broken until the paper is brought to pressroom temperature. Exposed paper at temperatures substantially below the room temperature rapidly absorbs moisture from the air, with resulting distortion. Figure 2 shows the time required to temperature-condition wrapped paper. Paper is usually ordered by the printer at a moisture content approximately in equilibrium with the relative humidity maintained in the pressroom. Papermakers find it difficult to supply paper in equilibrium with higher than 50% rh.

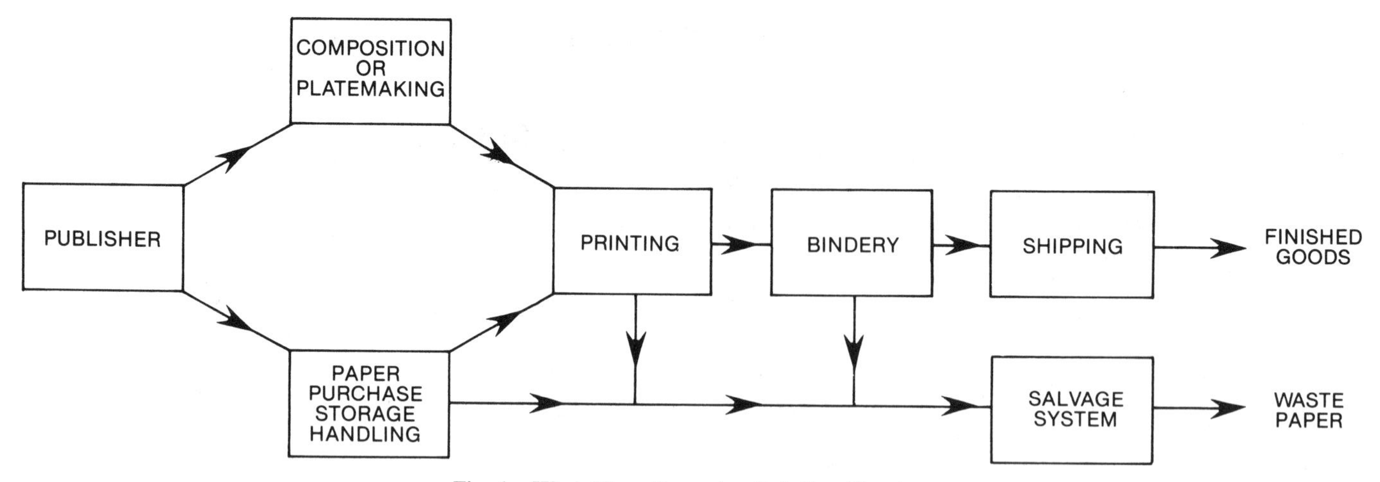

Fig. 1 Work Flow through a Printing Plant

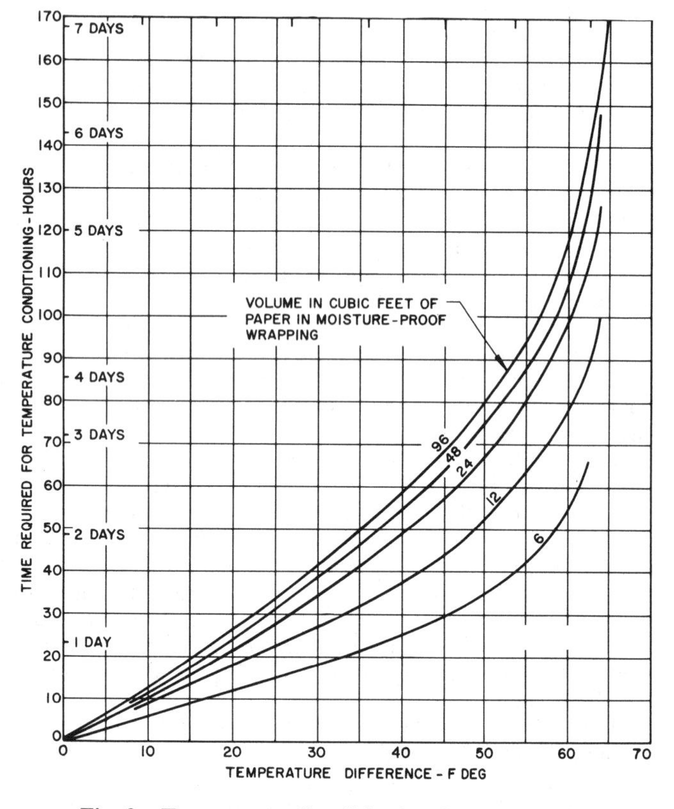

Fig. 2 Temperature Conditioning Chart for Paper

The sword hygrometer (or paper hygroscope) can check the hygroscopic condition of paper relative to the surrounding air. The blade contains a moisture-sensitive element, and its expansion or contraction actuates the pointer in a dial on the handle. The instrument is waved in the air until the pointer comes to rest, and the dial is set at zero. The sword is then inserted into the paper, and the pointer movement from zero indicates the paper's moisture relative to the surrounding air.

Figure 3 illustrates sword readings taken on three piles of paper in a pressroom. The first indicates that the paper is too dry to be exposed to room air without forming wavy edges. The second indicates that the paper is in equilibrium with the air, correct for all sheetfed printing, except multicolor lithography. The third was taken on paper wetter than the air in the pressroom. This paper

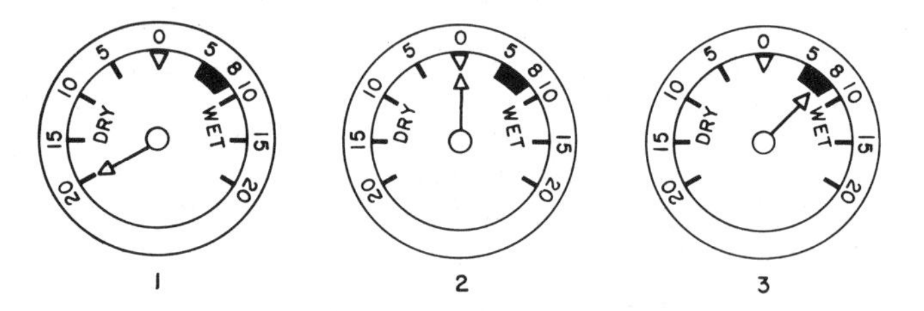

Fig. 3 Sword Hygroscope Readings Taken from Three Lots of Paper

was conditioned to balance with the air in the paper stockroom, where the relative humidity was 7% above that in the pressroom, optimal for multicolor lithographic printing.

Mill wrappings and roll tightness normally prevent detrimental effects on a paper roll for about six months. If the wrapper is damaged, moisture penetration of not more than 0.125 in. may be expected.

PLATEMAKING

Humidity and temperature control are important in making lithographic and collotype plates, photoengravings, and gravure plates and cylinders. The moisture content and temperature of the plates affect the sensitivity of the coating. The coatings increase in light sensitivity with increasing relative humidity and temperature, requiring adjustments in light intensity or length of exposure to give uniformity.

Maintaining constant dry-bulb temperature and relative humidity in platemaking rooms provides a plate at a known control point. A bichromated colloid coating starts to age and harden as soon as it is dry. The aging and hardening rate varies with the atmosphere, so exposures made a few hours apart may be quite different. This rate of reaction can be estimated more closely when the space is air conditioned. Exposure can then be reduced progressively to maintain uniformity. An optimum relative humidity of 45% or less substantially increases the useful life of bichromated colloid coatings; the relative humidity control should be within 2%. A dry-bulb temperature of 75 to 80°F maintained within 2°F is good practice. The ventilation air requirements of the plate room should be investigated. A plant with a large production of deep-etch plates should consider locating this operation outside the conditioned area.

Exhausts for platemaking operations consist primarily of lateral or downdraft systems at each operation. Because of their bulkiness or weight, plates or cylinders are generally conveyed by an overhead rail system to the work station, where they are lowered into the tank for plating, etching, or grinding. Exhaust ducts must be below or to one side of the working area, so lateral exhausts are generally used for open surface tanks.

Exhaust quantities vary, depending on the nature of the solution, but they should provide 100 cfm/ft^2 at standard conditions of surface of solution and/or a control velocity of 50 fpm at the side of the tank opposite the exhaust intake. To minimize air quantities and increase efficiency, tanks should be covered. Excessive supply air ventilation across open tanks should be avoided. Because of the nature of the exhaust, the duct construction in many systems must be acidproof and liquidtight to prevent moisture condensation.

Webfed offset operation and related departments are similar to webfed letterpress operation, without the heat loads created in the composing room and stereotype departments. Special attention should be given to air cleanliness and ventilation in platemaking to eliminate chemical fumes and dust errors in the plates.

A rotogravure plant can be hazardous because highly volatile solvents are used. Equipment must be explosionproof, and air-handling equipment must be sparkproof. Clean air must be supplied at controlled temperature and relative humidity, and the exhaust requires reclamation or destruction systems to keep photosensitive hydrocarbons from the atmosphere.

Currently available systems use activated carbon for continuous processing or eliminate the pollutants by rapid oxidation. The amount of solvents reclaimed may exceed that added to the ink.

LETTERPRESS

Letterpress printing relies on a raised surface to transfer ink by pressure directly to paper. Ink rollers apply ink only to the raised surface of the printing plate. Only the raised surface touches the paper to transfer the desired image.

Air conditioning in newspaper pressrooms and other web letterpress printing processes minimize problems caused by static electricity, ink mist, and expansion or contraction of the paper during printing. A wide range of operating conditions is satisfactory. The temperature should be selected for operator comfort.

At web speeds of 1000 to 2000 fpm, control of relative humidity is not essential because of the application of heat to dry inks. Moisture is applied to the web in some types of printing and passing the web over chill rolls further sets the ink.

The webfed letterpress ink is a heat-set ink, made with high boiling, slow evaporating synthetic resins and petroleum oils dissolved or dispersed in hydrocarbon solvents. The solvent has a narrow boiling range with a low volatility at room temperatures and a fast evaporating rate at elevated temperatures. The solvents are vaporized in the driers (part of the press) at temperatures of 400 to 500 °F, leaving the resins and oils on the paper. Webfed letterpress ink is dried after all colors of ink are applied to the web.

The paper, at speeds of 1000 to 2000 fpm, passes through driers of several types: open-flame gas cup, flame impingement, high-velocity hot air, or steam drum type.

Exhaust quantities through a press-drier system vary with the type of driers used and the speed of the press from about 7000 to 15,000 cfm at standard conditions. Exhaust temperatures range between 250 and 400 °F.

The solvent-containing exhaust is heated to temperatures of 1300 °F in an air pollution control device to incinerate the effluent. A catalyst can be used to reduce the temperature required for combustion from 1000 °F but requires periodic inspection and rejuvenation. Heat recovery is used to reduce the fuel required for incineration and to heat pressroom makeup air.

LITHOGRAPHY

Lithography prints with a grease-treated printing image receptive to ink, on a surface that is neither raised nor depressed. Both grease and ink repel water. This method applies water to all areas of the plate, except the printing image. Ink is then applied only to the printing image and transferred to the paper.

Offset printing transfers the image first to a rubber blanket and then to the paper. Sheetfed and web offset printing are similar to letterpress printing. The inks are similar but contain water-resistant vehicles and pigment. In web offset and gravure printing, relative humidity in the pressroom should be controlled at a low level, and the temperature should be selected for comfort or, at worst, to avoid heat stress. It is important to maintain steady conditions.

The pressroom for sheet multicolor offset printing has more exacting humidity requirements than other printing processes. The paper must remain flat with constant dimensions during multicolor printing in which the paper may make up to six or more passes through the press over a period of a week or more. If the paper does not have the right moisture content at the start, or if there are significant changes in atmospheric humidity during the process, the paper will not retain its dimensions and flatness, and misregister will result. In many cases of color printing, a register accuracy of 0.005 in. is required. Figure 4 shows the necessity of close control of the air relative humidity to achieve this result. The data shown in this figure are for composite lithographic paper.

Maintaining constant moisture content of the paper is complicated because paper picks up moisture from the moist offset blanket during printing—0.1 to 0.3% for each impression. When two or more printings are made in close register work, the paper at the start of the printing process should have a moisture content in equilibrium with air at 5 to 8% rh above the pressroom air. At this condition, the moisture evaporated from the paper into the air nearly balances the moisture added by the press. In obtaining register, it is important to keep the sheet flat and free from wavy or tight edges. To do this, the relative humidity balance of the paper should be slightly above that of the pressroom atmosphere. This balance is not as critical in four-color presses because the press moisture does not penetrate the paper fast enough between colors to affect sheet dimensions or sheet distortion.

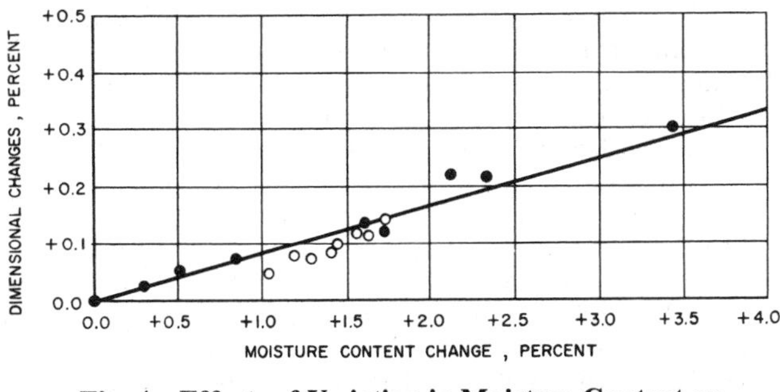

Fig. 4 Effects of Variation in Moisture Content on Dimensions of Printing Papers
(Weber and Snyder 1934)

Recommended Environment

The Graphic Arts Technical Foundation recommends 76 to 80 °F dry-bulb temperature and 43 to 47% rh as ideal conditions in a lithographic pressroom, with control ±2% rh and ±2 °F dry-bulb temperature (Reed 1970). The result of relative humidity variations on register may be estimated for offset paper from Figure 4. Closer relative humidity control of the pressroom air is required for multicolor printing of 76-in. sheets as compared with 22-in. sheets for the same register accuracy. Closer control is required for multicolor printing, where the sheet makes two or more trips through the press, than for one-color printing. Comfort and economy of operation influence the choice of temperature.

Drying of ink is affected by temperature and humidity, so uniform results and standardized procedures are difficult to obtain without control of the atmospheric conditions. Printing inks must dry rapidly to prevent offsetting and smearing. High relative humidity and high moisture content of paper tend to prevent ink penetration, and more ink remains on the surface than can be quickly oxidized. This affects drying time, intensity of color, and uniformity of ink on the surface. Relative humidity below 60% is favorable for drying at a comfortable temperature.

The air-conditioning system for the pressroom of a lithographic plant should control air temperature and relative humidity, filter the air, supply ventilation air, and distribute the air without pronounced drafts around the presses. Use of anti-offset sprays to set the ink creates an additional air-filtering load from the pressroom air. Drafts and high airflow over the press lead to excessive drying of the ink and water, which may cause operation troubles due to scumming.

The operating procedures of the pressroom should be analyzed to determine the heat removal load. The lighting load is high and constant throughout the day. The temperature of the paper brought into the pressroom and the length of time it is in the room should be considered to determine the sensible load from the paper. Figure 2 shows the hours required for wrapped paper to reach room temperature. The press motors usually constitute a large portion of the internal sensible heat gain.

Readings should be taken to obtain the running power load of the larger multicolor presses. The moisture content of the paper fed to the press and the relative humidity of the air must be considered when computing the internal latent heat gain. The printer usually specifies paper that is in equilibrium with air at a relative humidity somewhat higher than the pressroom condition. This means the paper will give up moisture to the space as it absorbs moisture from the press. If the moisture transfer is in balance, the water used in the printing process would be included in the internal moisture load. It is preferable to determine the water evaporation from the presses by test.

Air-Conditioning Systems

Precise multicolor offset lithography printing requires refrigeration with provision for separate humidity control, or sorption dehumidifying equipment for independent humidity control and provision for cooling. The need for humidity control in the pressroom may be determined by calculating the dimensional change of the paper for each percent change in relative humidity and checking this with the required register for the printing process.

The photographic department air conditioning is usually considered next in importance to the pressroom. Most of the work in offset lithography is done on film. Air conditioning controls cleanliness and comfort and holds the size of the film on register work. The quality of work and accuracy in this department carries over to other departments.

Air conditioning is important in the stripping department, both for comfort and for maintaining size and register. Curling of the film and flats, and shrinkage or stretch of materials can be minimized by maintaining constant relative humidity. This is particularly important on close register color work. The photographic area, stripping room, and platemaking area are usually maintained at the same conditions as the pressroom. Driers used for web offset printing are the same type as for webfed letterpress. Drying is less of a problem, partly because of the lesser amounts of ink applied and because of lower press speeds—800 to 1800 fpm.

ROTOGRAVURE

Rotogravure printing uses a cylinder with minute ink wells etched in the surface to form the printing image. Ink is applied to the cylinder, filling the wells. All excess ink is then removed from the cylinder surface by doctor blades, leaving only the ink in the wells, which make up the printing image. The image is then transferred to the paper as it passes between the printing cylinder and an impression cylinder.

Expansion, contraction, and distortion in sheetfed gravure printing, as in offset printing, should be prevented because of the importance of correct register. The paper need not be in equilibrium with air at a relative humidity higher than that of the pressroom, because no moisture is added to the paper in the printing process. The humidity and temperature control is exacting, as in offset printing. The relative humidity should be 45 to 50%, controlled within ±2%, and a comfort temperature within ±2°F.

Gravure printing ink dries principally by evaporating the solvent in the ink, leaving a solid film of pigment and resin. The solvent is low boiling hydrocarbon, and evaporation takes place rapidly, even without the use of heat. The solvents have closed-cup flash points from 22 to 80°F and are classified as Group I or special hazard liquids by local code and insurance company standards.

As a result, in areas adjacent to gravure press equipment and solvent and ink storage areas, electrical equipment must be Class I, Division 1 or 2, as described by the National Electrical Code, and ventilation requirements (both supply and exhaust) are stringent. Ventilating systems should be designed for high reliability, with sensors to detect unsafe pollutant concentrations and initiate alarm or safety shutdown.

Rotogravure printing units operate in tandem, each superimposing printing over that printed from a preceding unit. Press speeds range from 1200 to 2400 fpm. Each unit is equipped with its own drier to prevent subsequent smearing or smudging.

A typical drying system consists of four driers (or perhaps a total of 12 driers for 12 printing units) connected to an exhaust fan. Each dryer is equipped with recirculating fans and heating coils. An air quantity of 5000 to 8000 cfm at standard conditions is recirculated by a blower through a steam or hot water coil and then through jet nozzles at 130°F. The hot air impinges on the web and drives off the solvent-laden vapors from the ink. It is normal to exhaust half of this air. The system should be designed and adjusted to prevent solvent vapor concentration from exceeding 25% of its lower flammable limit (Marsailes 1970). Where this is not possible, constant lower-flammable-limit (LFL) monitoring, concentration control, and safety shutdown capability should be included.

In exhaust-system design for a particular process, solvent vapor should be captured from the printing unit where paper enters and leaves the drier, from the fountain and sump area, and from the printed paper, which continues to release solvent vapor as it passes from printing unit to unit. Details of the process, such as ink and paper characteristics and rate of use, are required to determine exhaust quantities.

When dilution-type ventilation is used, exhaust of 1000 to 1500 scfm at standard conditions at the floor is often provided between each unit. The makeup air units are adjusted to supply slightly less air to the pressroom than that exhausted to move air from surroundings into the pressroom.

OTHER PLANT FUNCTIONS

Flexography

Flexography is a type of printing that requires rubber-raised printing plates and functions much like a letterpress. Flexography is used principally in the packaging industry to print labels. Flexographic printing is also used to print on smooth surfaces, such as plastics and glass.

Collotype Printing

Collotype or photogelatin printing is a sheetfed printing process related to lithography. The printing surface is bichromated gelatin with varying affinity for ink and moisture, depending on the degree of light exposure received. There is no mechanical dampening as in lithography, and the necessary moisture in the gelatin printing surface is maintained by operating the press in an atmosphere of high relative humidity, usually about 85%. Since the tone values printed are very sensitive to changes in moisture content of the gelatin, relative humidity should be maintained within ±2%.

Temperature must also be closely maintained, since tone values are very sensitive to changes in ink viscosity; 80 ± 3 °F is recommended. Collotype presses are usually partitioned off from the main plant, which is kept at a lower relative humidity, and the paper is exposed to the high relative humidity only while it is being printed.

Salvage Systems

Salvage systems remove paper trim and shredded paper waste from production areas and carry airborne shavings to a cyclone collector, where they are baled for removal. Air quantities per pound of paper trim are 40 to 45 ft^3/lb and transport velocity in ductwork is 4500 to 5000 fpm (Marsailes 1970). Humidification may be provided to prevent the buildup of a static charge and consequent system blockage.

Air Filtration

Filters commonly used in ventilation and air-conditioning systems for printing plants are electronic automatic moving curtain filters with renewable media, having a weight arrestance of 80 to 90% (ASHRAE *Standard* 52-76).

In sheetfed pressrooms, a high performance bag-type after-filter is used to filter the starch particles, which require about 85% ASHRAE dust spot efficiency. In film processing areas where relatively dust-free conditions are needed, high efficiency air filters are installed, with 90 to 95% ASHRAE dust spot efficiency.

A different type of filtration problem in printing is ink mist or ink fly, common in newspaper pressrooms and not unusual in heatset letterpress or offset pressrooms. Minute droplets of ink are dispersed by ink rollers rotating in opposite directions. The cloud of ink droplets is electrostatically charged and of 5 to 10 μm size. Generally used are the ink mist suppressors, charged to repel the ink back to the ink roller. Additional control is provided by automatic moving curtain filters.

Binding and Shipping

After printing, some work must be bound. Two methods of binding are perfect binding and stitching. In perfect binding, sections of a book (signatures) are gathered, ruffed, glued, and trimmed. The product has a flat edge where the book was glued. Large books are more easily bound by this type of binding. A low-pressure compressed air system and a vacuum system are usually required to operate a perfect binder, and paper shavings must be removed by the trimmer. Using heated glue requires an exhaust system, since the glue fumes may be toxic.

In stitching, sections of a book are collected and stitched (stapled) together. Stitching requires that each signature be opened individually and laid over a moving chain. Careful handling of the paper is important. This system has the same basic air requirements as perfect binding.

Mailing areas of a printing plant wrap, label, and/or ship the manufactured goods. The wrapper machine can be affected by low humidity. In winter, humidification of the bindery and mailing area to about 40 to 50% rh may be necessary to prevent static buildup.

REFERENCES

ASHRAE. 1976. Methods of testing air cleaning devices used in general ventilation for removing particulate matter. ASHRAE *Standard* 52-76.

Marsailes, T.P. 1970. Ventilation, filtration and exhaust techniques applied to printing plant operation. ASHRAE *Journal* (December) 27.

Reed, R.F. 1970. What the printer should know about paper. Graphic Arts Technical Foundation, Pittsburgh, PA.

Weber, C.G. and L.W. Snyder, 1934. Reactions of lithographic papers to variations in humidity and temperature. *Journal of Research,* National Bureau of Standards 12 (January).

CHAPTER 19

TEXTILE PROCESSING

THIS chapter covers (1) the basic processes of making fiber, yarn, and fabric; (2) various types of air-conditioning systems used in textile manufacturing plants; (3) relevant health considerations; and (4) energy conservation procedures.

Most textile manufacturing processes may be put into one of three general classifications: (1) synthetic fiber manufacturing, (2) yarn making, and (3) fabric making. Synthetic fiber manufacturing is divided into staple processing, tow-to-top conversion, and continuous fiber processing; yarn making is divided into spinning and twisting; fabric making is divided into weaving and knitting. Although these processes vary, their descriptions reveal the principles on which air-conditioning design is based.

FIBER MAKING

Processes preceding fiber extrusion have diverse ventilating and air-conditioning requirements. However, principles that apply to air conditioning or ventilation of chemical plants apply to pre-extrusion areas as well.

Synthetic fibers are extruded from metallic spinnerets and solidified as continuous parallel filaments. This process, called *continuous spinning,* differs from mechanical spinning of fibers or tow into yarn, which is generally referred to as *spinning.*

Synthetic fibers may be formed by melt-spinning, dry-spinning, or wet-spinning. Melt-spun fibers are solidified by cooling the molten polymer; dry-spun fibers by evaporating a solvent, leaving the polymer in fiber form; and wet-spun fibers by hardening the extruded filaments in a liquid bath. Economic and chemical considerations determine which type of spinning is required. Generally, nylons, polyesters, and glass fibers are melt-spun, acetates dry-spun, rayons wet-spun, and acrylics dry- or wet-spun.

For melt-spun and dry-spun fibers, the filaments of each spinneret are usually drawn through a long vertical tube called a *chimney* or *quench stack,* within which solidification occurs. For wet-spun fibers, the spinneret is suspended in a chemical bath where coagulation of the fibers takes place. Wet-spinning is followed by washings, applying a finish, and drying.

Synthetic continuous fibers are extruded as a heavy denier tow if intended to be cut into short lengths called staple, or somewhat longer lengths for tow-to-top conversion; or they are extruded as light denier filaments if intended to be processed as continuous fibers. An oil is then applied to lubricate, give antistatic properties, and control fiber cohesion. The extruded filaments are usually drawn (stretched) both to align the molecules along the axis of the fiber and to improve the crystalline structure of the molecules, thereby increasing the fiber's strength and resistance to stretching.

Heat applied to the fiber when drawing heavy denier or high-strength synthetics releases a troublesome oil mist. In addition, the mechanical work of drawing generates a high localized heat load. If the draw is accompanied by twist, it is called *draw-twist.* If not, it is called *draw-wind.* After draw-twisting, continuous fibers may be given additional twist or may be sent directly to warping.

When tow is cut to make staple, the short fibers are allowed to assume random orientation. The staple, alone or in a blend, is then usually processed, as described for the cotton system. However, tow-to-top conversion, a more efficient process, has become more popular. The longer tow is broken or cut so as to maintain parallel orientation. Most of the steps of the cotton system are by-passed; the parallel fibers are ready for blending and mechanical spinning into yarn.

Glass fiber yarn is formed by attenuating molten glass through platinum bushings at high temperatures and speeds, as light denier multifilaments. These are drawn together while being cooled with a water spray. A chemical size is then applied to protect the fiber. This is all accomplished in a single process prior to winding the fiber for further processing.

YARN MAKING

The fiber length determines whether spinning or twisting must be used. Spun yarns are produced by loosely gathering synthetic staple, natural fibers, or blends into rope-like form; drawing them out to increase fiber parallelism, if required; and then applying twist. Twisted (continuous filament) yarns are made by twisting mile-long monofilaments or multifilaments. Ply yarns are made in a similar manner from spun or twisted yarns.

The principles of mechanical spinning are applied in three different systems: cotton, woolen, and worsted. The cotton system is used for all cotton, most synthetic staple, and many blends. Woolen and worsted systems are used to spin most wool yarns, some wool blends, and synthetic fibers such as acrylics.

Cotton System

The cotton system was originally developed for spinning cotton yarn, but now its basic machinery is used to spin all varieties of staple including wool, polyester, and blends. Most of the steps from raw materials to fabrics, along with the ranges of frequently used humidities, are outlined in Figure 1.

Opening, Blending, and Picking. The compressed tufts are partly opened, most foreign matter and some short fibers are removed, and the mass is put in an organized form. Some blending is desired to average the irregularities between bales or to mix different kinds of fiber. Synthetic staple, being cleaner and more uniform, usually requires less preparation. The product of the picker is pneumatically conveyed to the feed rolls of the card.

Carding. This process lengthens the lap into a thin web, which is gathered into a rope-like form called a *sliver.* Further opening and fiber separation is accomplished, as well as partial removal of short fiber and trash. The sliver is laid in an ascending spiral in cans of various diameters.

The preparation of this chapter is assigned to TC 9.2, Industrial Air Conditioning.

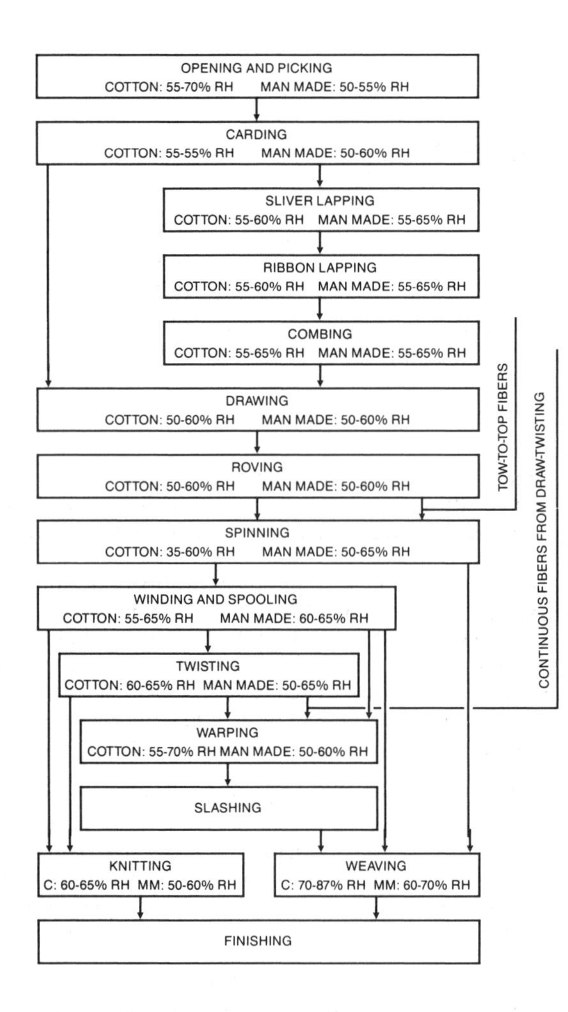

Fig. 1 Textile Process Flow Chart and Ranges of Humidity

For heavy, low-count (length per unit of mass) yarns of average or lower quality, the card sliver goes directly to drawing. For lighter, high-count yarns requiring fineness, smoothness, and strength, the card sliver must first be combed.

Lapping. In lapping, several slivers are placed side by side and drafted. In ribbon lapping, the resulting ribbons are laid one on another and drafted again. The doubling and redoubling averages out sliver irregularities, while drafting improves fiber parallelism. Some recent processes lap only once before combing.

Combing. After lapping, the fibers are combed with fine metal teeth. Combing substantially removes all fibers below a predetermined length, removes any remaining foreign matter, and improves fiber arrangement. The combed lap is then attenuated by drawing rolls and again condensed into a single sliver.

Drawing. Drawing follows either carding or combing and improves uniformity and fiber parallelism by doubling and drafting several individual slivers into a single composite strand. Doubling averages the thick and thin portions, while the drafting action further attenuates the mass and improves parallelism.

Roving. Roving continues the processes of drafting and paralleling until the strand is a size suitable for spinning. A slight twist is inserted and the strand is wound on large bobbins used for the next roving step or for spinning.

Spinning. Mechanical spinning simultaneously applies draft and twist. The packages (any form into or on which one or more ends can be wound) of roving are creeled at the top of the frame. The unwinding strand passes progressively through gear-driven drafting rolls, a yarn guide, the C-shaped traveler, and then to the bobbin. The vertical traverse of the ring causes the yarn to be placed in predetermined layers.

The difference in peripheral speed between the back and front rolls determines the draft. Twist is determined by the rate of front roll feed, spindle speed, and drag, which is related to the traveler weight.

The space between the nip or bite of the rolls is adjustable and must be slightly greater than the longest fiber. The speeds of front and back rolls are independently adjustable. Cotton spindles normally run at 8000 to 9000 rpm, but may exceed 14,000 rpm. In ring twisting, drawing rolls are omitted, and a few spindles run as high as 18,000 rpm.

Open-end, or turbine spinning, combines drawing, roving, lapping, and spinning. Staple fibers are fragmented as drawn from a sliver and fed into a fast-spinning, small centrifugal device. In this device, the fibers are oriented and discharged as yarn: twist is imparted by the rotating turbine. This system is faster, quieter, and less dusty than ring spinning.

Spinning is the final step in the cotton system, and the feature that distinguishes it from twisting is the application of draft. The amount and point of draft application accounts for many of the subtle differences that require different humidities for apparently identical processes.

Atmospheric Conditions. From carding to roving, the loosely bound fibers are vulnerable to static electricity. In most instances, static can be adequately suppressed with humidity, which, however, should not be so high as to cause other problems. In other instances, it is necessary to suppress electrostatic properties with antistatic agents. Wherever draft is applied, constant humidity is needed to maintain optimum frictional uniformity between adjacent fibers and, hence, cross-sectional uniformity.

Woolen and Worsted Systems

The woolen system generally makes coarser yarns, while the worsted system makes finer ones of a somewhat harder twist. Both may be used for lighter blends of wool, as well as for synthetic fibers with the characteristics of wool. The machinery used in both systems applies the same principles of draft and twist, but differs greatly in detail and is more complex than that used for cotton.

Wool fibers are dirtier, greasier, and more irregular. They must be scoured to remove grease and are then usually reimpregnated with controlled amounts of oil to make them less hydrophilic and to provide better interfiber behavior. Being scaly and curly, wool fibers are more cohesive and require different treatment. Wool differs from cotton and synthetic fibers in that it requires higher humidities in the processes up to and including spinning than it does in the processes that follow. Approximate humidities are presented in Table 1.

Table 1 Recommended Humidities for Wool Processing at 75 to 80 °F

Departments	Humidity, %
Raw wool storage	50 to 55
Mixing and blending	65 to 70
Carding—worsted	60 to 70
—woolen	60 to 75
Combing, worsted	65 to 75
Drawing, worsted—Bradford system	50 to 60
—French system	65 to 70
Spinning—Bradford worsted	50 to 55
—French (mule)	75 to 85
—woolen (mule)	65 to 75
Winding and spooling	55 to 60
Warping, worsted	50 to 55
Weaving, woolen, and worsted	50 to 60
Perching or clothroom	55 to 60

Twisting Filaments and Yarns

Twisting was originally applied to silk filaments; several filaments were doubled and then twisted to improve strength, uniformity, and elasticity. Essentially, the same process is used today, but it is now extended to spun yarns, as well as to single or multiple filaments of synthetic fibers. Twisting is widely used in the manufacture of sewing thread, twine, tire cord, tufting yarn, rug yarn, ply yarn, some knitting yarns, and others.

Twisting and doubling is done on a *down-* or *ring-twister,* which draws in two or more ends from packages on an elevated creel, twists them together, and winds them into a package. Except for the omission of drafting, down-twisters are similar to conventional ring-spinning frames.

Where yarns are to be twisted without doubling, an *up-twister* is used. Up-twisters are used primarily for throwing synthetic monofilaments and multifilaments to add to or vary elasticity, light reflection, and abrasion resistance. As with spinning, yarn characteristics are controlled by making the twist hard or soft, right (S) or left (Z). Quality is determined largely by the uniformity of twist, which, in turn, depends primarily on the tension and stability of atmospheric conditions (Figure 1 and Table 1). Since the frame may be double- or triple-decked, twisting requires concentrations of power. The frames are otherwise similar to those used in spinning and present the same problems in air distribution. In twisting, lint is not a serious problem.

FABRIC MAKING

Preparatory Processes

When spinning or twisting is complete, both types of yarn may be prepared for weaving or knitting. The principal processes include winding, spooling, creeling, beaming, slashing, sizing, dyeing, and so forth. The broad purpose is two-fold: (1) to transfer the yarn from the type of package dictated by the preceding process to a type suitable for the next, and (2) to impregnate some of the yarn with sizes, gums, or other chemicals that may not be left in the final product.

Filling Yarn. Filling yarn is wound on quills for use in a loom shuttle. It is sometimes predyed and must be put into a form suitable for package or skein dyeing before it is quilled. If the filling is of relatively hard twist, it may be put through a twist-setting or conditioning operation in which internal stresses are relieved by applying heat, moisture, or both.

Warp Yarn. Warp yarn is impregnated with a transient coating of size or starch, which strengthens the yarns' resistance to the chafing it will receive while in the loom. The yarn is first rewound onto a cone or other large package from which it will unwind speedily and smoothly. The second step is warping, which rewinds a multiplicity of ends in parallel arrangement on large spools, called warp or section beams. In the third step, slashing, the threads pass progressively through the sizing solution, squeeze rolls, and around cans or steam-heated drying cylinders—or, as an alternative, through an air-drying chamber. As much as several thousand pounds may be wound on a single loom beam.

Knitting Yarn. If hard spun, knitting yarn must be twist-set to minimize kinking. Filament yarns must be sized to reduce stripbacks and improve other running qualities. Both must be put in the form of cones or other suitable packages.

Uniform tension is of great importance in maintaining uniform package density. Yarns tend to hang up when unwound from a hard package or slough off from a soft one, and both tendencies are aggravated by spottiness. The processes that require air conditioning, along with recommended relative humidities, are presented in Figure 1 and Table 1.

Weaving

In the simplest form of weaving, harnesses raise or depress alternate warp threads to form an opening called a *shed.* A shuttle containing a quill is kicked through the opening, trailing a thread of filling behind it. The lay and the reed then beat the thread firmly into one apex of the shed and up to the fell of the previously woven cloth. Each shuttle passage forms a pick. These actions are repeated up to frequencies of five per second.

Each warp thread usually passes through a drop-wire, which is released by a thread break and automatically stops the loom. Another automatic mechanism inserts a new quill in the shuttle as the previous one is emptied, without stopping the loom. Other mechanisms are actuated by filling breaks, improper shuttle boxing, and the like, which stop the loom until manually restarted. Each cycle may leave a stopmark sufficient to cause an imperfection that may not be apparent until the fabric is dyed.

Beyond this basic machine and pattern are many complex variations in harness and shuttle control, which give intricate and novel weaving effects. The most complex is the jacquard, with which individual warp threads may be separately controlled. Other variations appear in looms for such products as narrow fabrics, carpets, and pile fabrics. In the Sulzer weaving machine, a special filling carrier replaces the conventional shuttle. In the rapier, a flat, spring-like tape uncoils from each side and meets in the middle to transfer the grasp on the filling. In the water jet loom, a tiny jet of high-pressure water carries the filling through the shed of the warp. Other looms transport the filling with compressed air.

High humidity increases the abrasion resistance of the warp. Many weave rooms require 80 to 85% humidity and even higher for cotton and up to 70% humidity for synthetic fibers. Many looms run faster when room humidity and temperature are precisely controlled.

In the weave room, power distribution is uniform, with an average concentration somewhat lower than in spinning. The rough treatment of fibers liberates many minute particles of both fiber and size, thereby creating considerable amounts of airborne dust. Due to high humidity, air changes average from four to eight per hour. It is necessary to make special provisions for maintaining conditions during production shutdown periods, usually at a lower relative humidity.

Knitting

Typical knitted products are seamless articles such as undershirts, socks, and hosiery made on circular machines, and those knitted flat, such as full-fashioned hosiery, tricot, milanese, and warp fabrics.

The basic process is one of generating fabric by forming millions of interlocking loops. In its simplest form, a single end is fed to needles, which are actuated in sequence. In more complex constructions, hundreds of ends may be fed to groups of elements that function more or less in parallel.

Knitting yarns may be either single or ply and must be of uniform high quality and free from neps or knots. These yarns, particularly the multifilament type, are usually treated with special sizes to keep broken filaments from stripping back and to provide lubrication.

Precise control of yarn tension, through controlled temperature and relative humidity, increases in necessity with the fineness of the product. For example, in finer gages of full-fashioned hosiery, a 2°F change in temperature is the limit, and a 10% change in humidity may add or subtract 3 in. in the length of a stocking. For knitting, desirable room conditions are approximately 76°F drybulb and 45 to 65% rh.

Dyeing and Finishing

Finishing is the final readying of a mill product for its particular market. Finishing ranges from cleaning to imparting special characteristics. The individual operations and the number involved vary considerably, depending on the type of fiber, yarn, or fabric, and the end product usage; operations are usually done in separate plants.

Inspection is the only finishing operation to which air conditioning is regularly applied, although most of the others require ventilation. Those that employ wet processes usually keep their solutions at high temperature and require special ventilation to prevent destructive condensation and fog. Where the release of sensible, latent, or radiant heat is great, spot cooling is necessary.

AIR-CONDITIONING DESIGN

Air washers are especially important in textile manufacturing. These may be either conventional low-velocity or high-velocity units in built-up systems. Unitary high-velocity equipment using rotating eliminators, although no longer in favor, may still be found in some plants.

Integrated Systems

Many mills use a refined air washer system that blends the air-conditioning system and the collector system (see later section) into an integrated unit. The air handled by the collector system fans is delivered back to the air-conditioning apparatus through a central duct, together with any air required to make up total return air. The air quantity returned by the individual yarn-processing machine cleaning systems must not exceed the air-conditioning supply air quantity. The air discharged by these individual suction systems is carried by return air ducts directly to the air-conditioning system. Before entering the duct, some of the cleaning system air usually passes over the yarn processing machine drive motor and through a special enclosure to capture the heat losses from the motor.

Integrated systems occasionally may exceed the supply air requirements of the area served. In such cases, the surplus air must be reintroduced after filtering.

Individual suction cleaning systems that can integrate with air conditioning are available for cards, drawing frames, lap winders, combers, roving frames, spinning frames, spoolers, and warpers. The following advantages result from this arrangement:

1. With a constant air supply, the best possible uniformity of air distribution can be maintained year-round.
2. Downward airflow can be controlled; crosscurrents in the room are minimized or eliminated; drift or fly from one process to another is minimized or eliminated. Room partitioning between systems serving different types of manufacturing processes further enhances the value of this arrangement controlling room air pattern year-round.

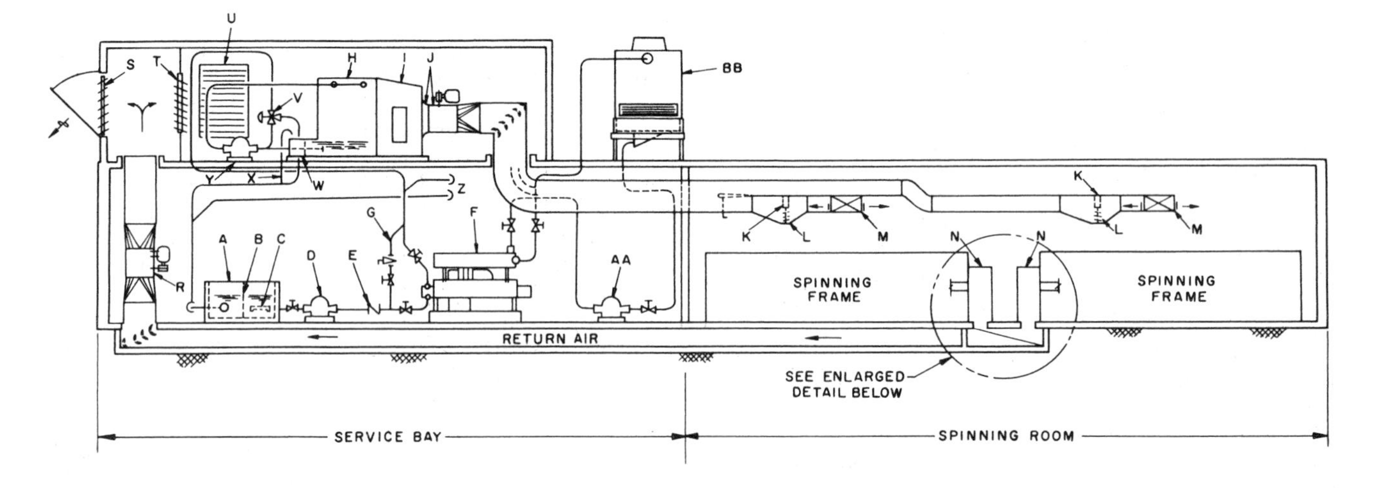

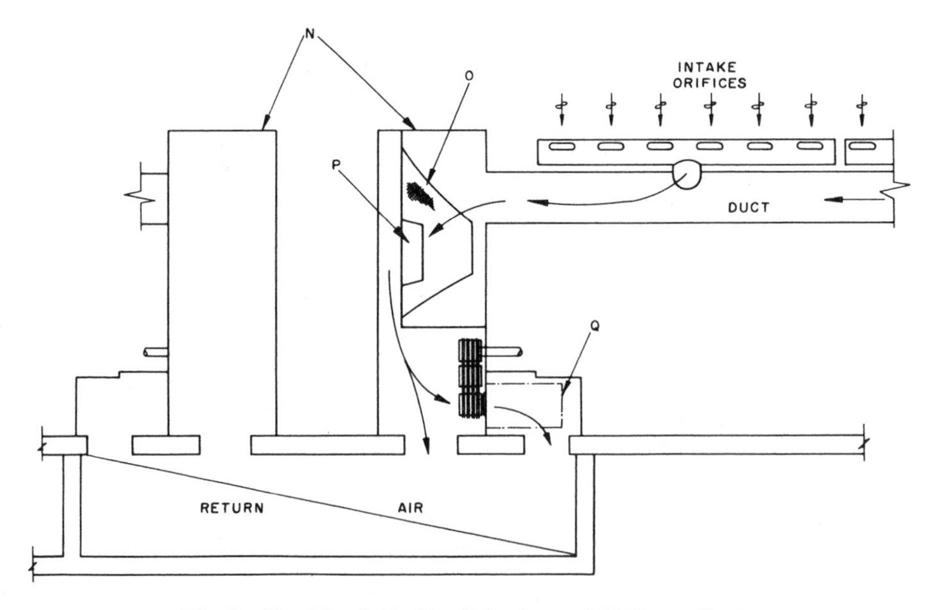

Fig. 2 Combined Air-Conditioning and Collector System

3. Heat losses of the yarn processing frame motor and any portion of the processing frame heat captured in the duct, as well as the heat of the collector system equipment, cannot influence room conditions; hot spots in motor alleys are eliminated, and although this heat goes into the refrigeration load, it does not enter the room, and, as a result, the supply air quantity is reduced.
4. Uniform conditions in the room improve production; conditioned air is drawn directly to the work areas on the machines, minimizing or eliminating wet or dry spots.
5. Maximum cleaning use is made of the air being moved.

A guide for cleaning air requirements is:

Pickers, per picker	2500 to 4000 cfm
Cards, per card	700 to 1500 cfm
Spinning, per spindle	4 to 8 cfm
Spooling, per spool	40 cfm

Collector Systems

A collector system is a waste-capturing device which uses many orifices operating at high suction pressures. Each production machine is equipped with suction orifices at all points of major lint generation. The captured waste is generally collected in a fan and filter unit either located on each machine, or centrally located to accept waste from a group of machines.

A collector in the production area may discharge waste-filtered air either back into the production area or into a return duct to the air-conditioning system. It then enters the air washer or is relieved through dampers to the outdoors.

Figure 2 shows a mechanical spinning room with air-conditioning and collector systems combined into an integrated unit. In this case, the collector system returns all of its air to the air-conditioning system. If supply air from the air-conditioning system exceeds the maximum that can be handled by the collector system, additional air should be returned by other means.

Figure 2 also shows return air entering the air-conditioning system through damper T, passing through air washer H, and being delivered by fan J to the supply duct, which distributes it to maintain conditions within the spinning room. At the other end of each spinning frame are unitary-type filter-collectors consisting of enclosure N, collector unit screen O, and collector unit fan P.

Collector fan P draws air through the intake orifices spaced along the spinning frame. This air passes through the duct running lengthwise to the spinning frame, through the screen O, and then is discharged into the enclosure base (beneath the fan and screen). The air quantity is not constant, but drops slightly as material builds up on the filter screen.

Since the return air quantity must remain constant and the air quantity discharged by fan P is slightly reduced at times, relief openings are necessary. Relief openings also may be required when the return air volume is greater than the amount of air the collector suction system can handle.

The discharge of fan P is split, so part of the air cools the spinning frame drive motor before rejoining the rest of the air in the return air tunnel. Regardless of whether the total return air quantity enters the return air tunnel through collector units, or through a combination of collector units and floor openings beneath spinning frames, return air fan R delivers it into the apparatus, ahead of return air damper T. Consideration should be given to filtering the return air prior to its delivery into the air-conditioning apparatus.

Mild-season operation causes more outdoor air to be introduced through damper U. This amount of air is relieved through motorized damper S, which opens gradually as outdoor damper U opens, while return damper T closes in proportion. All other components perform as typical central station air-washer systems.

A system of the general configuration of Figure 2 may also be used for carding; the collector system portion of this arrangement

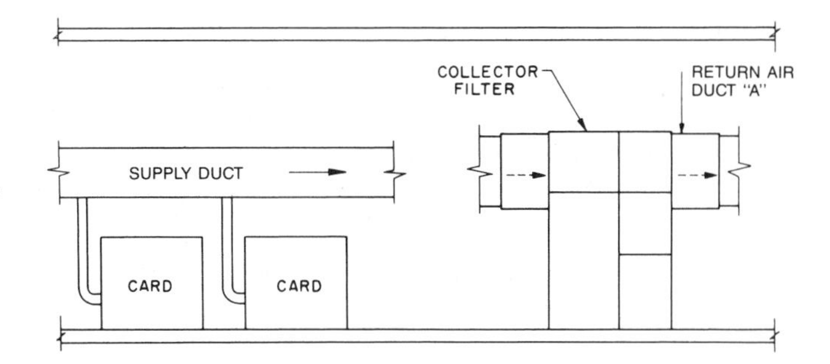

Fig. 3 Central Collector for Carding Machine

is shown in Figure 3. A central collector filters the lint-laden air taken from multiple points on each card. This air is discharged to return air duct A and then is returned to the air-conditioning system, exhausted outside, or returned directly to the room. A central collector filter may also be used with the spinning room system of Figure 2.

Air Distribution

Whenever *generally uniform distribution* is indicated, it is subject to the peculiarities of a particular situation, in which load concentration may require special handling.

Continuous Spinning Area. Methods of distribution here are diverse and generally not critical. However, spot cooling or localized heat removal may be required. This area may be cooled by air conditioning, evaporative cooling, or ventilation.

Chimney (Quench Stack). Carefully controlled and filtered air or other gas is delivered to the chimneys; it is returned for conditioning and recovery of valuable solvents, if present. Distribution of the air is of the utmost importance. Nonuniform temperature, humidity, or airflow disturbs the yarn, causing variations in fiber diameter, crystalline structure, and orientation. A fabric made of such fibers streaks when dyed.

For melt spinning, the concentration of solvent in chimney air must be maintained below the explosive limit of the solvent in the air. Even below the explosive limit, care must be exercised that the vapors are not ignited by a spark or flame. The air-conditioning system must be reliable, because an interruption of the spinning causes the solution to solidify in the spinnerets.

Wind-up or Take-up Areas of Continuous Spinning. These areas develop a heavy air-conditioning load. Air is often delivered through branch ducts alongside each spinning machine. Diffusers are of the low-velocity, low-aspiration type, sized so as not to agitate delicate fibers.

Draw-Twist or Draw-Wind Areas of Fiber Manufacture. These areas also have a heavy air-conditioning load. Distribution, diffusion, and return systems are similar to those for the continuous spinning take-up area.

Opening and Picking. Usually opening and picking require only a uniform distribution system. The area is subject to shutdown of machinery during portions of the day. Generally, an all-air system with independent zoning is installed.

Carding. Generally, a uniform distribution system is installed. Low air motion is required around the web. Older installations frequently used an overhead belt drive which sometimes made distribution difficult. Central lint collecting systems are available for this process, but must be considered in the system design. An all-air system is often selected for cotton carding.

In wool carding, air motion is frequently more important than in cotton carding, not only around the web but also to reduce cross-contamination between adjacent cards, since different colors of predyed wool may be run side by side on adjacent cards. A split system may be considered for wool carding to reduce air motion.

The method of returning air is also critical for achieving uniform conditions.

Drawing and Roving. Generally, a uniform distribution all-air system works well.

Mechanical Spinning Areas. These areas generate a heavy air-conditioning load, consisting of spinning frame power uniformly distributed along the frame length, and frame driver motor losses concentrated in the motor alley at one end of the frame.

Supply air ducts should run across the frames at right angles. Sidewall outlets between each of the two adjacent frames then direct the supply air down between the frames, where conditions must be maintained. Where concentrated heat loads occur, as in double motor alleys, placement of a supply air duct directly over the double motor alley should be considered. Sidewall outlets spaced along the bottom of the duct diffuse air into the motor alley.

The collecting system, whether unitary or central, with intake points distributed along the frame length at the working level, assists in pulling supply air down to the frame, where maintenance of conditions is most important. A small percentage of the air handled by a central collecting system may possibly be used to convey the collected lint and yarn to a central collecting point and the air would then be lost from the spinning room.

Winding and Spooling. Generally, a uniform distribution, all-air system is used in this area.

Twisting. This area has a heavy air-conditioning load. Distribution considerations are similar to those in spinning. Either all-air or split systems are installed.

Warping. This area has a very light load. Long lengths of yarn may be exposed unsupported in this area. Generally an all-air system with uniform distribution is installed. Diffusers may be low aspiration type. Return air is often near the floor.

Weaving. Generally, a uniform distribution system is necessary. Now that synthetic fibers are more commonly woven than natural fibers, lower humidity requirements allow the use of all-air systems rather than the previously common split system. When the lower humidity is coupled with the water jet loom, a high latent load results. In older weaving rooms, split systems are usually found.

Health Considerations

Control of Oil Mist. Whenever textiles coated with lubricating oils are heated above 200°F in drawing operations in ovens, tenterframes, or driers, an oil mist is liberated, which, if not collected at the source of emission and disposed of, produces a slightly odorous haze.

Various devices have been proposed to separate the oil mist from the exhaust air, such as process changes, fume incinerators, electrostatic precipitators, high energy scrubbers, absorption devices, high velocity filters, and condensers.

Control of Cotton Dust. Byssinosis, known as brown or white lung disease, is believed to be caused by a histamine-releasing substance in cotton, flax, and hemp dust. A cotton worker returning to work after a weekend experiences difficulty in breathing, that is not relieved until later in the week. After 10 to 20 years, the breathing difficulty becomes continuous, and even leaving the mill does not provide relief.

The U.S. Labor Department has continued to enforce an OSHA standard of lint-free dust. The most promising means of control for this requirement is improved exhaust procedures and filtration of recirculated air. Since the particles of concern are 1 to 15 μm in diameter, filtration equipment must be effective in this size range. Improvements in carding and picking, which leaves less trash in the raw cotton, also helps control dust.

Noise Control. The noise generated by HVAC equipment can be significant, especially if the textile equipment is modified to meet present safety criteria. For procedures to analyze and correct the noise due to ventilating equipment, see Chapter 42.

ENERGY CONSERVATION

Some of the steps taken to reduce energy consumption are as follows:

1. Heat recovery applied to water and air.
2. Automation of high-pressure driers to save heat and compressed air.
3. Lowering hot water and raising chilled water temperatures for rinsing and washing in the dyeing operations.
4. Replacing running washes with recirculating washes, where possible.
5. Revising double-bleaching procedures to single bleaching, where possible.
6. Eliminating rinses and final wash in dye operations, where possible.
7. Drying by means of the "bump and run" process.
8. Modifying the drying or curing oven air circulation systems to provide counterflow.
9. Using energy efficient textile machinery.

BIBLIOGRAPHY

Carrol-Porezynski. *Inorganic fibers.* Academic Press, New York.

Hearle and Peters. *Moisture in textiles.* Textile Book Publishers, Inc., New York.

Kirk and Othmer, eds. *Encyclopedia of chemical technology,* 2nd ed. Vol. 9. Interscience Publishers, New York.

Labarthe, J. *Textiles: Origins to usage.* Macmillan Co., New York.

Mark, J.F, S.M. Atlas, and E. Cernin, eds. *Man-made fibers: Science & technology,* Vol. 1. Interscience Publishers, New York.

Nissan. *Textile engineering processes.* Textile Book Publishers, Inc., New York.

Press, J.J., ed. *Man made textile encyclopedia.* Textile Book Publishers, Inc., New York.

Sachs, A. 1987. Role of process zone air conditioning. *Textile Month* (October):42.

Schicht, H.H. 1987. Trends in textile air engineering. *Textile Month* (May):41.

CHAPTER 20

PHOTOGRAPHIC MATERIALS

THE manufacture, processing, and storage of sensitized photographic products requires the temperature, humidity, and quality of air to be controlled.

MANUFACTURE

Light-sensitive photographic products generally consist of a flexible support, or *base*, coated with a gelatin emulsion containing salts of silver. For photographic film, a transparent base is cast from cellulose esters dissolved in solvents or made from synthetic polymers cast from a melt. Forming such a film base of required uniformity at safe and environmentally acceptable solvent concentration levels requires careful control of temperature and solvent vapor concentration in the coating machine. It also requires accurate temperature control of the coating wheel and systems to heat and stretch the base to the required thickness. Air in the heat-setting sections of the coating machine (also called casting machine) must be at the correct temperature for polymer crystallization and desired base characteristics.

Methods of solvent removal from the air systems during curing include condensation, activated carbon, molecular sieve adsorption, scrubbers, and incineration. Treatment selection depends on such factors as solvent recovery, safety, and environmental conservation.

Condensation systems can involve temperatures as low as −85 °F, using a circulating medium such as methylene chloride. At such low temperatures, alloy steels with a percentage of nickel are used. Required insulation thickness at −85 °F is 6-in. cork or its equivalent.

Other solvent treatment systems require materials resistant to the corrosive solvents and high temperatures. Safety considerations for handling flammable solvents must be observed. System performance under all possible operating conditions and potential problems due to malfunctions, new solvents, and changed quantities must be evaluated. This often requires fail-safe design, system or component redundancy, precise control sequence in the process, and fire detection and control systems.

Photographic paper is produced in a paper mill that controls stock consistency to about 1.1% and temperature at 130 °F, although the specific value is a function of the paper grade and speed of production. Dissolved and entrained air must be removed. Paper formation and subsequent operations of wet pressing, drying, calendaring, reeling, and winding all require precise control to produce paper free of photographic contamination, with high brightness, good permanence, high wet strength, stability under repeated wetting and drying, and high internal strength. Various aqueous or resin-extrusion coatings are applied to the paper to achieve desired surface characteristics before the light-sensitive emulsion is applied. These layers are applied by a series of application devices; then the paper is dried and sometimes further calendared. Many of these operations require accurately controlled air and water systems for temperature and moisture control.

The produced film or paper must be stored in conditioned areas to preserve the characteristics of the material and keep it satisfactory for subsequent operations (see Storing Unprocessed Photographic Materials).

In preparing light-sensitive photographic emulsions for film and paper, gelatin and one or more salts of the alkali halides form a colloidal suspension in water, and silver nitrate and other chemicals are added at a specific temperature. Some desired characteristics of finished photographic emulsions are obtained by the temperature and length of time the emulsion is allowed to ripen subsequent to precipitation.

Because of the nearly limitless variations of emulsion manufacturing processes, none is typical. The processes require close control of temperature, ingredient flow rate, agitation, precise addition of chemicals, and timing of sequences to fix the properties of the emulsion.

Held at the exact temperature in a liquid state, the emulsion is coated onto the film or paper base and passed directly into a chilling chamber, where it is gelled as quickly as possible at 50 °F or lower. After chilling, the emulsion-coated film or paper enters the drying section. Temperature, moisture content, air purity, and drying time are carefully controlled to provide a high quality product. When the film or paper has reached the correct moisture content (usually 1.5 to 3% for films and 5 to 8% for papers), it is wound in rolls and subsequently slit and cut into various sizes for packing.

All processes must be reviewed for energy conservation opportunities such as heat reclamation, enthalpy exchange, reduction in energy use, and reduction in peak energy demand.

Inert particles many times smaller than those visible to the human eye may become visible when photographic images are enlarged or projected. Photographically active contaminants (those that react chemically with light-sensitive products) may destroy an area of film or paper many times greater than the particle size. Active submicron particles can be more destructive than large inert particles. When minute particles agglomerate and impinge on sensitized products, they can cause the same imperfections as large particles. Radioactive dust is the largest source of active contaminants in the atmosphere. Most radioactive dust is less than 1 μm in size.

Where process air contacts photographic products, a high degree of cleaning is required to protect product quality. While intermediate efficiency filters may remove most harmful inert contaminants, the highest efficiency filters are required to reduce contamination effectively from radioactive dust.

Table 1 shows recommended filter efficiency requirements for general ventilation and process air systems. The filter requirements

The preparation of this chapter is assigned to TC 9.2, Industrial Air Conditioning.

Table 1 Filter Requirements in Manufacture of Sensitized Photographic Materials

Efficiency Requirement	Low	Intermediate	High
General Application	**Comfort, Cleanliness, Prefilters**	**Noncritical, but Cleanliness Required**	**Best Available Protection from All Contamination**
TYPES OF AREAS	Maintenance areas Office areas Storage of packaged products New-product laboratories	Processing and printing Storage of unpackaged products Finished product testing Process air	Manufacturing areas
Relative Efficiency (NBS Discoloration Test)	0 10 20 30 40	50 60 70 80	90 95 100
TYPES OF FILTERS	Panel throwaway Manual roll Automatic roll Pleated type with disposable cartridge	Deep bed with disposable media Electrostatic	AEC high efficiency (Complete unit disposable)

are classified as high, intermediate, and low efficiency. For general ventilation, the efficiency requirement is based on the use of the area, but for process air systems, the required efficiency depends on the particular process and product. In some operations, laminar flow systems with highest efficiency filters are needed to control contamination in the manufacturing process.

Dirt problems cannot be eliminated simply by efficient filtration for supply air systems. Atmospheric contaminants entering through supply air systems are often only a small portion of the overall particulate contamination in an area. Therefore, more efficient filters on the supply air would not significantly reduce the area dirt level. Dirt smudges on walls, ceilings, and around diffusers usually result from room air aspirating into low-pressure areas created by the velocity of the air supply. Keeping critical areas under positive pressure helps maintain cleanliness.

If dirt in the room is largely room-generated or carried in by occupants, efficient filtration of supply air will not significantly reduce these smudges, nor will it lower the possibility of indirect product damage by contaminants settling from the room air. If a large portion of room air is recirculated, efficient filters will reduce the built-up dirt levels in the area. In manufacturing areas, good control should be maintained over all sources of internal contaminant generation.

STORING UNPROCESSED PHOTOGRAPHIC MATERIALS

While virtually all photosensitive materials deteriorate with age, the rate of deterioration depends largely on the storage conditions. Deterioration increases with both high temperature and high relative humidity, and usually decreases with lower temperature and humidity.

High relative humidity alone is usually more harmful than high temperature alone. High humidity can accelerate loss of sensitivity and contrast, increase shrinkage, produce mottle, cause softening of the emulsion, and promote fungal growth. Low relative humidity can increase the susceptibility of the film or paper to static markings, abrasions, brittleness, and curl. Three-layer substractive-type color films and papers are more seriously affected than are black-and-white products, because heat and moisture usually affect the three emulsion layers to different degrees, causing a change in color balance and overall film and paper speed and contrast.

Film or paper for domestic consumption may be packaged in a container sufficiently moisture-resistant for protection in temperate zones only. Photographic film or paper intended for the tropics or other places of high humidity have extra protection against moisture and harmful gases. Packaging, often employing heat-sealed foil pouches, snap-cover plastic cans, and taped metal or plastic cans, is usually adequate to protect film or paper in tropical or areas of high humidity, as long as the original packaging remains intact. The packaging used for a particular product should be determined from the manufacturer, and the appropriate recommended storage conditions should then be maintained. Film products are packaged in equilibrium with air at a relative humidity between 40 to 60% $\pm 2\%$ at 75 °F.

Products packed in vaportight containers maintain their moisture content and do not require storage in an area with controlled humidity as long as the seal is unbroken. Products that are not in vaportight packages or that are in opened packages should not be stored in damp basements, iceboxes, refrigerators, or other high relative humidity locations. The relative humidity for storage should be between 40 and 60%, preferably near 40% and at 60 °F. A moderate temperature with low relative humidity, such as 60 °F and 40% rh, is better than a lower temperature with a high relative humidity, such as 40 °F with 80% rh.

When humid storage conditions cannot be avoided or when a refrigerator is used for cooling, products in opened or non-vaportight packages should be placed in a container that can be tightly sealed, or the original packaging material should be tightly resealed.

In tropical zones or in the summer in temperate zones, refrigerated storage is recommended to keep products cool if they are in vaportight packages or sealed in cans or jars. Normal film life is 18 months. Black-and-white papers should be stored at 70 °F or below, regardless of the storage time. Color films or papers kept for several months should be stored at 45 to 50 °F. Storage at temperatures above 70 °F for more than four weeks may change speed and color balance. To minimize a change in the latent image, process control strip films should be stored at 0 °F. Some special application products require stringent storage conditions, as recommended by labels or instruction sheets.

Films that must be stored for more than a year should be kept at 0 to -10 °F. This low temperature arrests changes in photographic characteristics almost completely if the product is adequately protected from background and stray radiation. The effects of adverse storage conditions between removal from refrigeration and exposure or between exposure and processing may cause unsatisfactory results in spite of previous low-temperature storage.

Except for motion picture films, temperatures below 32 °F are not harmful to photographic products. The water content of either film or paper is relatively small, and ice crystals do not form inside it at normal moisture levels, regardless of how low the temperature or how rapid the cooling. Motion picture film should be stored at 50 °F because, when it is stored at lower temperatures, it contracts and a loosely wound roll results (Carver *et al.* 1943).

Film or paper stored in a refrigerator or freezer should be removed some time before it is used, to allow it to warm to the outside temperature. Otherwise, moisture may condense on the cold film or paper when the sealed package is opened.

Table 2 shows the time needed to reach equilibrium for individual packages separated from each other. Cold packages stacked on top of each other would require much longer to warm up (proportional to total thickness).

Table 2 Suggested Warm-up Time for Film and Paper

Type of Film Package	Warm-up Time, h	
	20 °F Rise	75 °F Rise
Roll film, including 828	0.5	1
135 magazines, 110 and 126 cartridges	1	1.5
10 sheet box	1	1.5
50 sheet box	2	3
35 mm, any length	3	5
16 mm, any length	1	1.5

Paper Size	Warm-up Time, h		
	0 to 70 °F	35 to 70 °F	50 to 70 °F
16 × 20 in. (50 sheet box)	3	2	2
30 × 40 in. (50 sheet box)	3	2	2
16 in. × 250 ft roll	10	7	4
40 in. × 50 ft roll	12	8	5

Products not packaged in sealed foil envelopes or vaportight containers are vulnerable to contaminants. They must be kept away from formaldehyde vapor (emitted by particleboard, some insulation, some plastics, and some glues), industrial gases, motor exhausts, and vapors of solvents and cleansers. In hospitals, industrial plants, and laboratories, all photosensitive products, regardless of the type of packaging, must be protected from x-rays, radium, and other radioactive materials. For example, films stored 25 ft away from 100 mg of radium require the protection of 3.5 in. of lead around the radium.

Under extremely humid conditions, film or paper should be both exposed and processed as soon as possible after the package is opened. If exposed products cannot be processed the same day as they are exposed, they should be kept in a dehumidified cabinet or storeroom or desiccated and resealed in a moistureproof container. Color films or papers should also be stored at 50 °F or lower. Air conditioning with relative humidity control would provide the most desirable storage conditions for unprocessed products. An electric refrigerating dehumidifier controlled by a humidistat can be used for a storage cabinet or small storeroom. Also suitable is an electrically operated desiccating dehumidifier, which uses a desiccating agent such as silica gel or activated alumina with an automatic reactivation cycle.

PROCESSING AND PRINTING PHOTOGRAPHIC MATERIALS

Air Conditioning in Preparatory Operations

During receiving operations, exposed film is removed from protective packaging from presplicing and processing. Photographic emulsions become soft and can be mechanically damaged at high relative humidity. At excessively low relative humidity, the film base is prone to static, sparking, and curl deformation. Presplicing combines many individual rolls of film into a long roll to be processed. The presplice work area should be maintained between 50 to 55% rh and 70 to 75 °F dry bulb.

Air Conditioning and Processing Operation

Processing exposed films or paper involves a series of tempered chemical and wash tanks that emit heat, humidity, and fumes. A room exhaust system must be provided, together with local exhaust at noxious tanks. Air from the pressurized presplice rooms can be used as makeup for processing room exhaust to conserve energy. Further supply air should maintain the processing space at a maximum of 80 °F dry bulb and 50% rh.

The processed film or paper proceeds from the final wash to the drier to control the moisture remaining in the product. Too little drying will cause the film to stick when wound, while too much drying will cause undesirable curl. Drying can be regulated by controlling contact time, humidity, or, most often, temperature.

Air distribution to the drying area must provide a tolerable environment for operating personnel. The exposed sides of the drier should be insulated as much as is practical to reduce the large radiant and convected heat losses to the space. Return or exhaust grilles above the drier can directly remove much of its rejected heat and moisture. The supply air should be directed to offset the remaining radiant losses.

A solvent and wax mixture to provide lubrication is normally applied to motion picture film as it leaves the drier, with an exhaust to draw off the solvent vapor.

Air Conditioning for the Printing/Finishing Operation

In printing, where another sensitized product is exposed through the processed original, the amount of environmental control depends on the size and type of operation. For small-scale printing, close control of the environment is not necessary, except to minimize dust. In photo-finishing plants, printers for colored products emit substantial heat. The effect on the room can be reduced by removing the lamphouse heat directly. Computer-controlled electronic printers transport original film and raw film or paper at high speed. The proper temperature and humidity are especially important because, in some cases, two or three images from many separate films may be superimposed in register onto one film. For best results, the printing room should be maintained at between 70 and 75 °F and 50 to 60% rh to prevent curl, deformation, and static. Curl and film deformation affect the register and sharpness of the images produced. The elimination of the static charge prevents static marks and also helps to keep the final product clean.

Mounting of reversal film into slides is a critical operation of the finishing department, requiring a 70 to 75 °F dry-bulb temperature with 50 to 55% rh.

Particulates in Air

Air-conditioning systems for most photographic operations require 85% disposable bag-type filters, as indicated in Table 1, with lower efficiency prefilters to extend filter life. In critical applications (such as high-altitude serial films), and for microminiature images, filtering of foreign matter is extremely important. These products are handled in a laminar airflow room or workbench with HEPA filters plus prefilters.

Processing Temperature Control

The density of a developed image on photographic material depends on the emulsion characteristics, the exposure it has received, and the degree of development. With a particular emul-

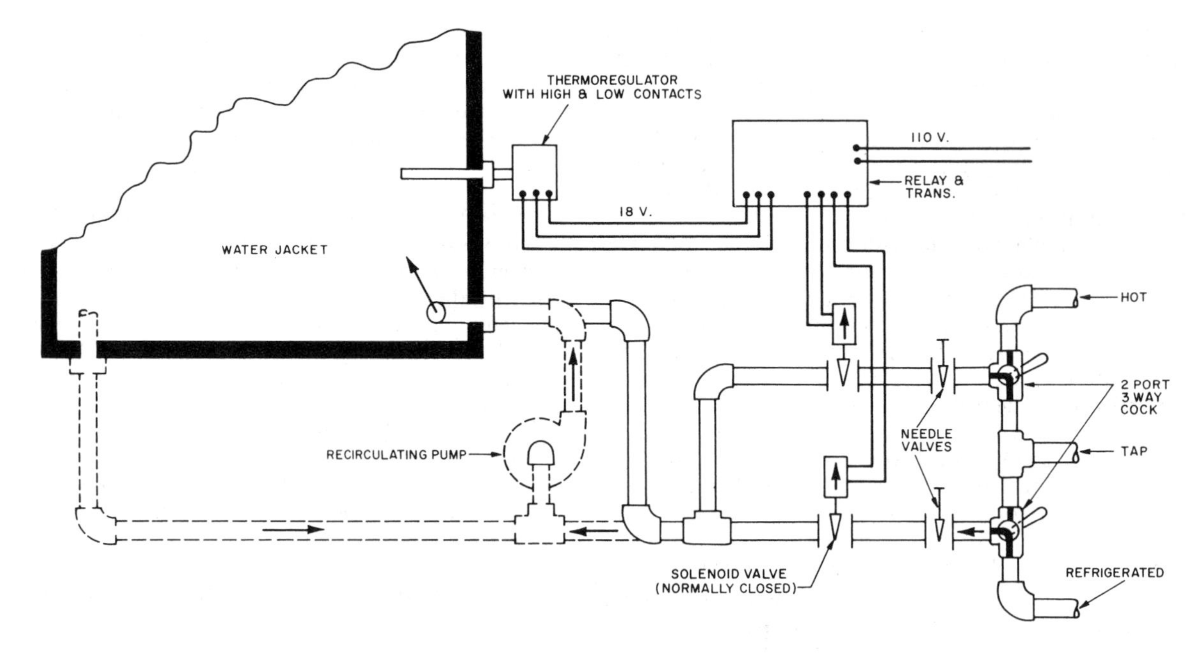

Fig. 1 Automatic Water Temperature Control in System with Recirculating Pump

sion, the degree of development depends on the time, temperature, degree of agitation, and developer activity.

With low developer temperature, the reaction is slow, and the development time recommended for the normal temperature would produce underdevelopment; at high temperature, the reaction is fast, and the same treatment time produces overdevelopment. Within limits, these changes in the rate of development can be compensated by increasing or decreasing the development time. Once a temperature for development is determined, it should be maintained within 1 °F for black-and-white products and within at least 0.5 °F for color materials. Other solutions in the processor can be maintained at a wider range of temperature tolerance.

Changes in the temperature of color film or paper developer cannot readily be compensated by changes in development time. Temperature change affects development characteristics of emulsion layers in different manners, upsetting the color balance. Contrast and speed are affected, and fog can occur.

Figure 1 shows a suggested method of controlling the temperature of a critical tank solution. A pump recirculates solution from the top of the tank to the bottom through a hot water heat exchanger. The heat input into the heat exchanger is regulated in response to the loop temperature. The recirculation eliminates temperature stratification, agitates the solution, allows continuous filtration, and brings the solution quickly to temperature at start-up.

Many processing machines have tanks with water jackets through which a continuous flow of controlled temperature water is circulated. To conserve water and energy, the heater water should be recirculated through an external heat exchanger or electric heater.

Wash tanks remove unwanted chemicals from the photographic image to prevent its degradation. Streams of continuously flowing water at a controlled temperature are supplied to the bottom of the final wash tank. The overflow of the final wash is connected to the bottom of the preceding wash tank, which saves water by countercurrent flow. A series of wash tanks can then be arranged.

The source of temperature-controlled wash water may be a central system for an entire laboratory or an individual self-contained thermostatic mixer for one processor. Thermostatic mixing valve performance depends on proper valve sizing and pressure equalization of the hot and cold supplies.

The mixing valve and recirculating systems may be combined for accuracy and economy. In one packaged unit, a thermostatic mixing valve supplies tempered water to all tanks, except the developer tank. After leaving the mixing valve, the water flows into the solution tank jackets and then to the wash tanks as rinse water, from which it drains to waste. The water in the developer tank jacket is in a completely closed system and recirculates through a temperature control unit comprising an air-cooled refrigeration unit, cooling coils, two immersion heaters, a pump, and a thermostat, which maintains a temperature differential of ± 0.2 °F.

STORING PROCESSED FILM AND PAPER

Storage of developed film and paper differs from storage of the raw stock, because the developed materials are no longer photosensitive, are seldom sealed against moisture, and are generally stored for much longer periods. Required storage conditions depend on (1) the value of the records, (2) the length of storage time, (3) whether the films are on nitrate or safety base, (4) whether the paper base is resin coated, and (5) the type of photographic image.

Photographic materials must be protected against fire, water, mold, chemical or physical damage, extreme relative humidity, and high temperature. Relative humidity is much more important than temperature. High relative humidity can cause films to stick together—particularly roll films, but also sheet films. High humidity also damages gelatin, encourages the growth of mold, increases dimensional changes, accelerates decomposition of nitrate support, and accelerates deterioration of both black-and-white and color images. Low relative humidity causes a temporary increase in curl and a decrease in flexibility, but this is usually reversed when the humidity rises again. An exception to this reversal occurs when motion picture film is stored for a long time in loosely wound rolls at very low humidities. The curl causes the film roll to take the shape of a polygon rather than a circle when viewed from the side. This *spokiness* occurs because a highly curled roll of film resists being bent in the length direction when it is already bent in the width direction. When a spoky roll is stored for a long time, the film permanently flows into the spoky condition, resulting in film distortion. Very low relative humidity in storage may also cause the film or paper to crack or break if handled carelessly.

Low temperature is desirable for film and paper storage provided that (1) the relative humidity of the cold air is controlled and (2) the material is warmed sufficiently before opening to avoid moisture condensation. High temperature can be harmful, accelerating the fading of dye images and accelerating film shrinkage, which may produce physical distortions. High temperature is also detrimental to the stability of nitrate film, which may still be in storage.

The storage life of processed photographic products is controlled by the composition, photographic processing, and storage conditions. Proper preservation of photographic materials is complicated, since each of these factors is controlled by different organizations. The composition of the film or paper is controlled by the manufacturer, the processing by the processing laboratory, and the storage conditions by the customer. The importance of all three must be recognized.

PHOTOGRAPHIC FILMS

ANSI (1984, 1985) specifies three levels of photographic film (medium-term, long-term, and archival) and two levels of storage conditions (medium-term and archival). *Medium-term* film is suitable for preservation for a minimum of 10 years when stored under medium-term conditions. *Long-term* film is suitable for the preservation of records for a minimum of 100 years when stored under archival conditions. *Archival* film, when stored under archival storage conditions, is suitable for the preservation of records having permanent value (Rhodes and Adelstein 1976).

Only films on safety base can qualify for these film types. No longer manufactured nitrate base films were not a suitable storage medium. Only silver-gelatin films can be classified as archival films, and diazo films can be specified as either medium- or long-term films (ANSI 1988).

Medium-term Storage

Medium-term storage rooms of safety base film should be protected from accidental water damage by rain, flood, or pipe leaks. Air conditioning with controlled relative humidity is desirable but not always essential in moderate climates. Extremes of relative humidity are detrimental to film.

The most satisfactory storage relative humidity for processed film is about 50%, although the range from 30 to 60% is satisfactory. Air conditioning is required where the relative humidity of the storage area exceeds 60% for any appreciable period. If air conditioning cannot be installed, a dehumidifier may be used for a small room. The walls should be coated with a vapor retarder and the controlling humidistat should be set at about 40% rh. If the prevailing relative humidity is under 25% for long periods, and problems are encountered from curl or brittleness, humidity should be controlled with a mechanical humidifier set with a controlling humidistat at 40%.

For medium-term storage, room temperature between 68 and 77 °F is recommended. Higher temperatures may cause dye fading, shrinkage, and distortion. Occasional peak temperatures of 95 °F should not be serious. Color films should be stored below 50 °F to reduce dye fading. Films stored below the ambient dew point should be allowed to warm up before opening to prevent moisture condensation (see Table 2).

An oxidizing or reducing atmosphere may deteriorate the film base and gradually fade the photographic image. Oxidizing agents may also cause microscopically small colored spots on fine grain film such as microfilm (Adelstein *et al.* 1970). Typical gaseous contaminants include hydrogen sulfide, sulfur dioxide, peroxides, ozone, nitrogen oxides, and paint fumes. When such fumes are present in the intended storage space, they must be eliminated, or the film must be protected from contact with the atmosphere. Chapters 27 and 40 have further information.

Archival Storage

For long-term films or for archival records that are to be preserved indefinitely, archival storage conditions should be used. The recommended space relative humidity varies between 15 and 50% rh, depending on the film type. When several film types are stored within the same area, 30% rh is a good compromise. The recommended storage temperature is below 70 °F. Low temperature aids preservation, but if storage temperature is below the dew point of the outdoor air, the records must be allowed to warm up in a closed container before they are used, to prevent condensation of moisture. Temperature and humidity conditions must be maintained year-round and should be continuously monitored.

Requirements of a particular storage application can be met by any one of several air-conditioning equipment/system combinations. Standby equipment should be considered. Sufficient outdoor air should be provided to keep the room under a slight positive pressure for ventilation and to retard the entrance of untreated air. The air-conditioning unit should be located outside the vault for ease of maintenance, with precautions taken to prevent water leakage into the vault. The conditioner casing and all ductwork must be well insulated. Room conditions should be controlled by a dry-bulb thermostat and either a wet-bulb thermostat, a hydrostat, or a dew-point controller (see Chapter 41).

Air-conditioning installations and fire dampers in ducts carrying air to or from the storage vault should be constructed and maintained to National Fire Protection Association recommendations for air conditioning (NFPA 1989) and for fire-resistant file rooms (NFPA 1986).

All supply air should be filtered with noncombustible HEPA filters to remove dust that may abrade the film or react with the photographic image. As with medium-term storage, gaseous contaminants such as paint fumes, hydrogen sulfide, sulfur dioxide, peroxides, ozone, and nitrogen oxides may cause slow deterioration of film base and gradual fading of the photographic image. When these substances cannot be avoided, an air scrubber, activated carbon adsorber, or other purification method is required.

Films should be stored in metal cabinets with adjustable shelves or drawers, and louvers or openings located to facilitate circulation of conditioned air through them. The cabinets should be spaced in the room to permit free circulation of air around them.

All films should be protected from water damage from leaks, fire sprinkler discharge, or flooding. Drains should have sufficient capacity to keep the water from sprinkler discharge from reaching a depth of 3 in. The lowest cabinet, shelf, or drawer should be at least 6 in. off the floor, constructed so that water cannot splash through the ventilating louvers onto the records.

When fire-protected storage is required, the film should be kept in either fire-resistant vaults or insulated record containers (Class 150). Fire-resistant vaults should be constructed in accordance with NFPA (1986). Although the NFPA advises against air conditioning in valuable-paper record rooms because of the possible fire hazard from outside, properly controlled air conditioning is essential for long-term preservation of archival films. The fire hazard introduced by the openings in the room for air-conditioning ducts may be reduced by fire and smoke dampers activated by ionization detectors in the supply and return ducts.

When the quantity of film is relatively small, insulated record containers (Class 150) may be used, as defined by Underwriters Laboratories, Inc. (UL 1983). Class 150 containers will not exceed an interior temperature of 140 °F and an interior relative humidity of 85% under a fire exposure test lasting from 1 to 4 h, depending on the classification. Insulated record containers should be on a ground-supported floor if the building is not fire resistant. For best fire protection, duplicate copies of film records should be placed in another storage area.

Storage of Nitrate Base Film

Although photographic film has not been manufactured on cellulose nitrate film base for several decades, many archives, libraries, and museums still have valuable records on this material. The preservation of the nitrate film is of considerable importance until printing on safety base has been accomplished.

Cellulose nitrate film base is chemically unstable and highly flammable. It decomposes slowly but continuously even under normal room conditions. The decomposition produces small amounts of nitric oxide, nitrogen dioxide, and other gases. Unless the nitrogen dioxide can escape readily, it reacts with the film base, accelerating the decomposition (Carrol and Calhoun 1955). The rate of decomposition is further accelerated by moisture and is approximately doubled with every 10°F increase in temperature.

All nitrate film must be stored in an approved vented cabinet or vault. Nitrate films should never be stored in the same vault with safety base films, because any decomposition of the nitrate film will cause decomposition of the safety film. Cans in which nitrate film is stored should never be sealed, since this traps the nitrogen oxide gas. Standards for the storage of nitrate film have been established (NFPA 1988). The National Archives and the National Institute of Standards and Technology have also investigated the effect of a number of factors on fires in nitrate film vaults (Ryan *et al.* 1956).

The storage temperature should be kept as low as economically possible. The film should also be kept at humidities below 50% rh. The temperature and humidity recommendations for the cold storage of color film in the following section also apply to nitrate film.

Storage of Color Film and Prints

All dyes fade in time. ANSI specifically excludes color film as an archival medium, and ANSI documents on photographic paper only include black-and-white images (1982, 1985). However, many valuable color films and prints exist, and it is important to preserve them for as long as possible.

Light, heat, moisture, and atmospheric pollution contribute to the fading of color photographic images. Storage temperature for the preservation of dyes should be as low as possible. For maximum permanence, the materials should be stored in light-tight sealed containers or in moistureproof wrapping materials at a temperature below freezing and at a relative humidity of 15 to 30%. The containers should be warmed to room temperature prior to opening to avoid moisture condensation on the surface. Photographic films can be conditioned to the recommended humidity by passing them through a conditioning cabinet with air circulating at about 15% rh for about 15 min.

An alternate procedure is the use of a storage room or cabinet controlled at a steady (noncycled) low temperature and maintained at the recommended relative humidity. This procedure eliminates the requirement for sealed containers but involves an expensive installation. The dye-fading rate decreases rapidly with decreasing storage temperature, as shown in Table 3.

Table 3 Effect of Temperature on Dye-Fading Rate (40% Relative Humidity)

Storage Temperature, °F	Approximate Relative Fading Rate	Approximate Relative Storage Time
86	2	0.5
75	1	1
66	0.5	2
54	0.2	5
45	0.1	10
14	0.01	100
−15	0.001	1000

PHOTOGRAPHIC PAPER PRINTS

Two main compositions of support are used for photographic papers. The fiber base generally has a baryta coating under the silver layer for improved smoothness. Resin-coated paper is also used, allowing shorter fixing, washing, and drying times.

The recommended storage conditions for processed paper prints are given by ANSI (1982). The optimum limits for relative humidity of the ambient air are 30 to 50%, but daily cycling between these limits should be avoided.

A variation in temperature can drive relative humidity beyond its acceptable range. A temperature between 59 and 77°F is acceptable, but daily variations of more than 7°F should be avoided. Prolonged temperature above 86°F should also be avoided. The degradative processes in black-and-white prints can be slowed considerably by low storage temperature.

As with color prints, moistureproof wrapping materials should be used. These packages must exceed the dew-point temperature before they are opened, to avoid moisture condensation.

Exposure to direct sunlight may lead to deterioration, especially in poorly processed prints. Avoid light sources containing high levels of ultraviolet radiation. Tungsten lights and ultraviolet-free fluorescent lamps are best for viewing or exhibiting.

Exposure to airborne particles and oxidizing or reducing atmospheres should also be avoided, as mentioned, for films. Protection from these atmospheres can be obtained by modifying the images or by converting them to a less reactive form. Treatment with toners that convert all or substantially all of the image to silver sulfide or silver selenide, or that modify the image by the deposition of gold, produce very stable images. Several commercially available toners can be used.

Display prints should be protected from accidental damage because of fire and water. Storage of collections should be undertaken with the same precautions as for archival storage of films.

REFERENCES

Adelstein, P.Z., C.L. Graham, and L.E. West. 1970. Preservation of motion picture color films having permanent value. *Journal of the Society of Motion Picture and Television Engineers* 79(November):1011.

ANSI. 1982. Photography (film and slides)—Practice for storage of black-and-white photographic paper prints. *Standard* PH 1.48-82. American National Standards Institute, New York.

ANSI. 1984. Photography (film)—Safety photographic film. *Standard* PH 1.25-84. American National Standards Institute, New York.

ANSI. 1985. Photography (film)—Processed safety film-storage. *Standard* PH 1.43-85. American National Standards Institute, New York.

ANSI. 1988. Specifications for stability of ammonia-processed diazo films. *Standard* IT 9.5-88. American National Standards Institute, New York.

Carrol, J.F. and J.M. Calhoun. 1955. Effect of nitrogen oxide gases on processed acetate film. *Journal of the Society of Motion Picture and Television Engineers* 64(September):601.

Carver, E.K., R.H. Talbot, and H.A. Loomis. 1943. Film distortions and their effect upon protection quality. *Journal of the Society of Motion Picture and Television Engineers* 41(July):88.

NFPA. 1986. Standard for the protection of records. *Standard* 232-86. National Fire Protection Association, Quincy, MA.

NFPA. 1988. Standard for the storage and handling of cellulose nitrate motion picture film. *Standard* 40-88. National Fire Protection Association, Quincy, MA.

NFPA. 1989. Standard for the installation of air conditioning and ventilating systems. *Standard* 90A-89. National Fire Protection Association.

Rhodes, J.B. and P.Z. Adelstein. 1976. Letters on archival permanence. *The Journal of Micrographics* 9(March):193.

Ryan, J.V., J.W. Cummings, and A.C. Hutton. 1956. Fire effects and fire control in nitro-cellulose photographic-film storage. Building Materials and Structures Report, No. 145. U.S. Department of Commerce, Washington, D.C. (April).

UL. 1983. Tests for fire resistance of record protection equipment, 12th ed. *Standard* 72-83. Underwriters Laboratories, Inc., Northbrook, IL.

CHAPTER 21

ENVIRONMENTAL CONTROL FOR ANIMALS AND PLANTS

THE design of plant and animal housing is complicated by the many environmental factors affecting the growth and production of living organisms and the financial constraint that equipment must repay costs through improved economic productivity. The engineer must balance the economic costs of modifying the environment against the economic losses of a plant or animal in a less-than-ideal environment.

However, economics are no longer the only concern relating to animals. Concerns for both workers and the care and welfare of animals influence design. State and federal laws on pollution, sanitation, and health assurance also affect the designs.

DESIGN FOR ANIMAL ENVIRONMENTS

Typical animal production systems modify the environment, to some degree, by housing or sheltering animals year-round or for parts of a year. The amount of modification is generally based on the expected increase in production. Animal sensible heat and moisture production data, combined with information on the effects of environment on growth, productivity, and reproduction, help designers select optimal equipment (Chapter 10 of the 1989 ASHRAE *Handbook—Fundamentals*). The Midwest Plan Service, *Structures and Environment Handbook* (1983) and ASAE *Monograph*-6, Ventilation of Agricultural Structures (1983), give more detailed information.

Design Approach

The environmental control system is typically designed to maintain thermal and air quality conditions within an acceptable range and as near the ideal for optimal animal performance as is practicable. Equipment is usually sized assuming steady-state energy and mass conservation equations. Experimental measurements confirm that heat and moisture production by animals is not constant and that there are important thermal capacitance effects in livestock buildings. Nevertheless, for most design situations, the steady-state equations are acceptable.

Achieving the appropriate fresh air exchange rate and distribution within the room are generally the two most important design considerations. The optimal ventilation rate is selected according to the ventilation rate logic curve (Figure 1).

During the coldest weather, the ideal ventilation rate is that required for maintaining indoor humidity at or below the maximum desired, and air contaminants within acceptable ranges. Supplemental heating is usually required to prevent the temperature from dropping below optimal levels.

The preparation of this chapter is assigned to TC 2.2, Plant and Animal Environment.

In milder weather, the ventilation rate required for maintaining optimal room air temperatures is greater than that required for moisture and air quality control (Figure 1). In hot weather, the ventilation rate is chosen to minimize the temperature rise above ambient and to provide optimal air movement over animals. Cooling is sometimes used in hot weather.

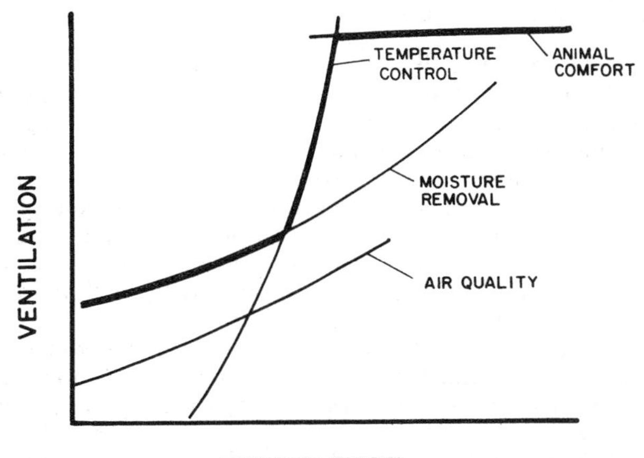

Fig. 1 Ventilation Rate Selection Curve as Influenced by Outside Temperature
(Christianson and Fehr 1983)

Temperature Control

The temperature within an animal structure is computed from the sensible heat balance of the system, usually disregarding transient effects. Nonstandard systems with low airflow rates and/or large thermal mass may require transient analysis. Steady-state heat transfer through walls, ceiling or roof, and ground is calculated as presented in Chapter 23 of the 1989 ASHRAE *Handbook—Fundamentals*.

Mature animals typically produce more heat per unit floor area than do young stock. Chapter 9 of the 1989 ASHRAE *Handbook—Fundamentals* presents estimates of animal heat loads. Lighting and equipment heat loads are estimated from power ratings and operating times. Typically, the designer selects indoor and outdoor design temperatures and calculates the ventilation rate to maintain the temperature difference. Outdoor design temperatures are given in Chapter 24 of the 1989 ASHRAE *Handbook—Fundamentals*. The section "Recommended Practices by Species" later in this chapter presents indoor design temperature values for various livestock.

Moisture Control

Moisture loads produced within an animal building may be calculated from data in the 1989 ASHRAE *Handbook—Fundamentals.* The mass of water vapor produced is then estimated by dividing the animal latent heat production by the latent heat of vaporization of water at the expected inside air temperature. Water spilled and evaporation of fecal water must be included in the estimates of animal latent heat production. Water vapor removed by ventilation from a totally slatted (manure storage beneath floor) swine facility may be up to 40% less than the amount removed from a conventional concrete floor (solid). If the floor is partially slatted, the 40% maximum reduction is reduced proportionally to the percentage of the floor that is slatted.

The ventilation system should remove enough moisture to prevent condensation but should not reduce the relative humidity so low (less than 50%) as to create dusty conditions. Design indoor relative humidities for winter ventilation are usually between 70 and 80%. The walls should have sufficient insulation to prevent surface condensation at 80% rh inside.

During cold weather, the ventilation needed for moisture control usually exceeds that needed to control temperature. Minimum ventilation must always be provided to remove animal moisture. Up to a full day of high humidity may be tolerated during extreme cold periods, when normal ventilation rates could cause an excessive heating demand. Humidity level is not normally controlled in mild and hot weather.

Air Quality Control

The amount of dust varies with animal density, size, and degree of activity, type of litter or bedding, type of feed, and relative humidity of the air. A high moisture content of 25 to 30%, wet basis, in the litter or bedding keeps dust to a minimum.

Gaseous contaminants within a building are often controlled as a result of ventilation for moisture or temperature control. Ammonia, which results from decomposition of manure, is the most important chronically present contaminant gas. Its production can be minimized by removing wastes from the room and keeping floor surfaces or bedding dry. Maintaining water over manure solids in gutters and pits also reduces ammonia by absorbing it. Ammonia should be maintained below 30 ppm and ideally, below 10 ppm.

Hydrogen sulfide, produced when stored manure is mechanically agitated, is the most important acute gas contaminant. During normal operation, hydrogen sulfide concentration is usually insignificant.

Hydrogen sulfide levels sometimes exceed 800 ppm during manure storage agitation. Since adverse effects on production begin to occur at 20 ppm, ventilation systems should be designed to maintain hydrogen sulfide levels below 20 ppm during agitation. When manure is agitated and removed from the storage, the building should be well ventilated and all animals and occupants evacuated because of potential deadly concentrations of gases.

Other gas contaminants can also be important. Carbon monoxide from improperly operating unvented space heaters sometimes reaches problem levels. Methane and carbon dioxide are other occasional concerns.

Disease Control

Airborne microbes can transfer disease-causing organisms among animals. For some situations, typically with young animals where there are low-level infections, it is important to minimize air mixing among pens. It is especially important to minimize air exchange between different animal rooms.

Air Distribution

Airspeeds should be maintained below 50 fpm for most animal species in cold and mild weather. Animal sensitivities to draft are comparable to human preferences, although some animals are more sensitive at different stages. Riskowski and Bundy (1988) document that air velocities for optimal gain rates and feed efficiencies can be below 25 fpm for young pigs at thermoneutral conditions.

Increased air movement during hot weather increases growth rates and improves heat tolerance. There are conflicting and limited data defining optimal air velocities in hot weather. Bond *et al.* (1965) and Riskowski and Bundy (1988) determined that both young and mature swine perform best when airspeeds are less than 200 fpm (Figure 2). Mount *et al.* (1980) did not observe performance penalties at airspeeds increased to a maximum of 150 fpm.

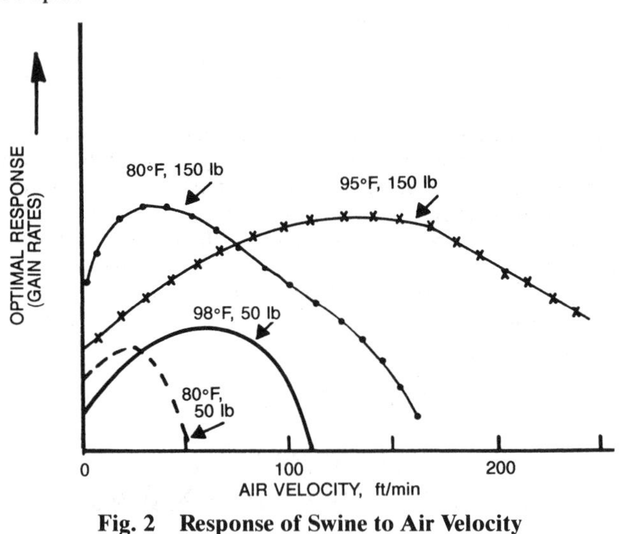

Fig. 2 Response of Swine to Air Velocity

Degree of Shelter

Livestock, especially young animals, need some protection from adverse climate. On the range, mature cattle and sheep need protection during severe winter conditions. In winter, dairy cattle and swine may be protected from precipitation, wind, and drafts with a three-sided, roofed shelter open on the leeward side in winter. The windward side should also have approximately 10% of the wall surface area open to prevent a negative pressure inside the shelter; this pressure could cause rain and snow to be drawn into the building on the leeward side. Such shelters do not protect against extreme temperatures or high humidity.

In warmer climates, shades often provide adequate shelter, especially for large, mature animals such as dairy cows. Shades are commonly used in Arizona; research in Florida has shown an approximate 10% increase in milk production and a 75% increase in conception efficiency for shaded versus unshaded cows. The benefit of shades has not been documented for areas with less severe summer temperatures. Although shades for beef are also common practice in the southwest, beef cattle are somewhat less susceptible to heat stress, and extensive comparisons of various shade types in Florida have detected little or no differences in daily weight gain or feed conversion.

The energy exchange between an animal and various areas of the environment is illustrated in Figure 3. A well-designed shade makes maximum use of radiant heat sinks, such as the cold sky, and gives maximum protection from direct solar radiation and high surface temperatures under the shade. Good design considers geometric orientation and material selection, including roof surface treatment and insulation materials on the lower surface.

An ideal shade has a top surface highly reflective to solar energy and a lower surface highly absorptive to solar radiation reflected from the ground. A white-painted upper surface reflects solar

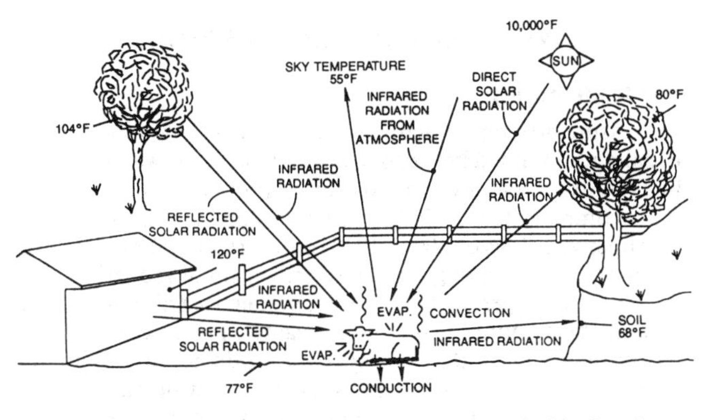

Fig. 3 Energy Exchange Between a Farm Animal and Its Surroundings in a Hot Environment

radiation, yet emits infrared energy better than aluminum. The undersurface should be painted a dark color to prevent multiple reflection of shortwave energy onto animals under the shade.

COOLING AND HEATING

Evaporative Cooling. Supplemental cooling of animals in intensive housing conditions may be necessary during heat waves to prevent heat prostration, mortality, or serious losses in production and reproduction. Evaporative cooling systems, which may reduce ventilation air to 80 °F or lower in most of the United States, are popular for poultry houses, both for breeders and layers. They are sometimes used for swine and dairy housing too.

Evaporative cooling is well-suited to animal housing because the high air exchange rates effectively remove odors and ammonia and increase air movement for convective heat relief. Initial cost, operating expense, and maintenance problems are all relatively low compared with other heating and ventilating equipment costs normally associated with controlled environment housing. Evaporative cooling works best in areas of low relative humidity, but significant benefits can be obtained even in the humid southeastern United States.

The pad area should be sized to maintain air velocities between 200 and 275 fpm through the pads. For most pad systems, these velocities produce evaporative efficiencies between 75 and 85%; they also increase the pressures against the ventilating fans from 0.04 to 0.12 in. of water.

Environment for Animals and Plants

The most serious problem encountered with evaporative pad systems for agricultural applications is clogging by dust and other airborne particles. Whenever possible, fans should exhaust away from pads on adjacent buildings. Regular preventive maintenance is essential. Water bleed-off and addition of algaecides to the water are recommended. When pads are not used in cool weather, they should be sealed to prevent dusty inside air from exhausting through them.

High-pressure fogging systems with water pressures of 500 psi are preferred to pad coolers for cooling the air in broiler houses with built-up litter. The high pressure creates a fine aerosol, causing minimal litter wetting. Timers and/or thermostats control the system. Evaporative efficiencies and installation costs are about one-half those of a well-designed evaporative pad system. Foggers can also be used with naturally ventilated, open-sided housing. Low-pressure systems are not recommended for poultry, but may be used during emergencies.

Nozzles that produce water mist or spray droplets to wet animals directly are used extensively during hot weather in swine confinement facilities with solid concrete or slatted floors. Currently, misting or sprinkling systems that directly wet the skin surface of the animals (not merely the outer portion of the hair coat) are preferred. Timers that operate a system periodically, *e.g.*, 30 s on a 5-min cycle, help to conserve water.

Mechanical Refrigeration. Mechanical refrigeration systems can be designed for effective animal cooling, but they are considered uneconomical for most production animals. Air-conditioning loads for dairy housing may require 2.5 kW or more per cow. Recirculation of refrigeration air is usually not feasible due to high contaminant loads in the air in the animal housing. Sometimes zone cooling of individual animals is used instead of whole-room cooling, particularly in swine farrowing houses where a lower air temperature is needed for sows than for unweaned piglets. Refrigerated air, 18 to 36 °F below ambient, is supplied through insulated ducts directly to the head and face of the animal. Air delivery rates are typically 10 to 30 cfm per sow for snout cooling, and 60 to 80 cfm per sow for zone cooling.

Earth Tubes

Some livestock facilities obtain cooling in summer and heating in winter by drawing ventilation air through tubing buried 6 to 13 ft below grade. These systems are most practical in the north central region for animals that benefit from both cooling in summer and heating in winter.

Goetsch and Muehling (1983) detail design procedures for these systems. A typical design uses 50 to 150 ft of 8-in. diameter pipe to provide 300 cfm of tempered air. Soil type and moisture, pipe depth, airflow, climate, and other factors affect the efficiency of buried pipe heat exchangers. The pipes must be sloped to drain condensation and must not have dips that could plug with condensation.

Heat Exchangers. Ventilation accounts for 70 to 90% of the heat losses in typical livestock facilities during winter. Heat exchangers can reclaim some of the heat lost with the exhaust ventilating air. However, predicting fuel savings based on savings obtained during the coldest periods will overestimate yearly savings from a heat exchanger. Estimates of energy savings based on air enthalpy can improve predictive accuracy.

Heat exchanger design must address the problems of condensate freezing and/or dust accumulation on the heat exchanging surfaces. These problems result in either reduced efficiency and/or the inconvenience of frequent cleaning. Finned heat exchangers should have a fin spacing of at least 4 per inch. Fouling problems are reduced, but not eliminated, in units without fins. Three basic heat exchanger designs have evolved for general use in animal confinement buildings: cross, diagonal, and counterflow.

Supplemental Heating. For poultry weighing 3.3 lb or more, for pigs heavier than 50 lb, and for other large animals such as dairy cows, the body heat of the animals at recommended space allocations is usually sufficient to maintain moderate temperatures, *e.g.*, above 50 °F, in a well-insulated structure. Combustion-type heaters are used to supplement heat for baby chicks and pigs. Supplemental heating also increases the moisture-holding capacity of the air, which reduces the quantity of air required for removal of moisture. Various types of heating equipment may be included in ventilation systems, but they need to perform well in dusty and corrosive atmospheres.

Insulation Requirements. The amount of building insulation required depends on climate, animal space allocations, and animal heat and moisture production. Usually, structures with an overall heat transmission coefficient of 0.08 to 0.12 Btu/h · ft^2 · °F are adequate for northern climates. In moderate climates, values ranging from 0.15 to 0.25 Btu/h · ft^2 · °F are adequate for adult animals, but may be decreased in cold areas to conserve fuel when heating for young animals. In warm weather, ventilation between the roof and insulation helps reduce the radiant heat load from the ceiling. Insulation in warm climates can be more important for reducing radiant heat loads in summer than reducing building heat loss in winter.

VENTILATION SYSTEMS

Mechanical Ventilation

Mechanical ventilation depends on fans to create a static pressure difference between the inside and outside of a building. Either positive pressure, with fans forcing air into a building, or negative pressure, with exhaust fans, is used in farm buildings. Some ventilation systems use a combination of positive pressure to introduce air into a building and separate fans to remove air. These zero-pressure systems are particularly appropriate for heat exchangers.

Positive Pressure Systems. Fans force humid air out through planned outlets, if any, and through leaks in walls and ceilings. If vapor barriers are not complete, moisture condensation will occur within the walls and ceiling during cold weather. Condensation causes deterioration of building materials and reduces insulation effectiveness. The energy used by fan motors and rejected as heat is added to the building—an advantage in winter but a disadvantage in summer.

Negative Pressure Systems. Air distribution in a negative pressure system is often less complex and costly. Simple openings and baffled slots in walls control and distribute air in the building. However, at low airflow rates, negative pressure systems may not distribute air uniformly because of air leaks and wind pressure effects. Supplemental air mixing may be necessary.

Allowances should be made for reduced fan efficiency because some dust accumulation on fan blades cannot be avoided (Bundy 1988). Totally enclosed fan motors are protected from exhaust air contaminants and humidity. Periodic cleaning helps prevent overheating. Negative pressure systems are more commonly used than positive pressure systems.

Ventilation systems should always be designed so that manure gases are not drawn into the building from manure storages connected to the building by underground pipes or channels.

Neutral Pressure Systems. Neutral pressure (push-pull) systems typically use fans to distribute air down the distribution duct to room inlets and exhaust fans to remove air from the room. Inlet and exhaust fan capacities should be matched.

Neutral pressure systems are often more expensive, but they achieve better control of the air. They are less susceptible to wind effects and to building leakage than are positive or negative pressure systems. Neutral pressure systems are most frequently used for young stock and for animals most sensitive to environmental conditions, primarily where cold weather is a concern.

Natural Ventilation

Either natural or mechanical ventilation systems are used to modify environments in livestock shelters. Natural ventilation is most common for mature animal housing, such as free-stall dairy and swine finishing houses. Natural ventilation depends on pressure differences caused by wind and temperature difference. A well-designed natural ventilation system keeps temperatures reasonably stable, if automatic controls regulate ventilation openings. Usually, a design includes an open ridge (with or without a rain cover) and openable sidewalls, which should cover at least 50% of the wall for summer operation. Ridge openings are about 2 in. wide for each 10 ft of house width, with a minimum ridge width of 6 in.

Openings can be adjusted automatically, with control based on air temperature. Some designs, referred to as flex housing, include a combination of mechanical and natural ventilation usually dictated by outside air temperature and/or the amount of ventilation required.

VENTILATION MANAGEMENT

Air Distribution

Pressure differences across walls and inlet or fan openings are usually maintained between 0.04 and 0.06 in. of water. (The exhaust fans are usually sized to provide proper ventilation at 0.12 in. to compensate for wind effects.) This pressure difference creates inlet velocities of 600 to 1000 fpm, sufficient for effective air mixing, but low enough to cause only a small reduction in air delivery. A properly planned inlet system distributes fresh air equally throughout the building. Negative pressure systems that rely on cracks around doors and windows do not distribute fresh air effectively. Inlets require adjustment, since winter airflow rates are typically less than 10% of summer rates. Automatic controllers are also available to regulate inlet area.

Positive pressure systems, with fans connected directly to perforated, polyethylene air distribution tubes, combine heating, circulation, and ventilation in one system. Air distribution tubes or ducts connected to circulating fans are sometimes used to promote air mixing in negative pressure systems.

Inlet Design

Inlet location and size most critically affect air distribution within a building. Continuous or intermittent inlets can be placed along the entire length of one or both outside walls. Building widths less than 20 ft may need only a single inlet along one wall. The total inlet area may be calculated by the system characteristic technique (described below). The distribution of the inlet area is based on the geometry and size of the building, which makes specific recommendations difficult.

System characteristic technique. This technique determines the operating points for the ventilation rate and pressure difference across inlets. Fan airflow rate as a function of pressure difference across the fan should be available from the manufacturer. Allowances should be made for additional pressure losses from fan shutters or other devices such as light restriction systems or cooling pads.

Inlet flow characteristics are available for hinged baffle and center-ceiling flat baffle slotted inlets (Figure 4). Airflow rates can be calculated for the baffles in Figure 4 by the following:

For Case A

$$Q = C_1 W p^{0.5} \tag{1}$$

For Case B

$$Q = C_2 W p^{0.5} \tag{2}$$

For Case C

$$Q = C_3 W p^{0.49} (D/T)^{0.08} e^{(-0.867\, W/T)} \tag{3}$$

where

Q = airflow rate, cfm per foot length of slot opening
W = slot width, in.
p = pressure difference across the inlet, in. of water gage
D = baffle width, in.
T = width of slot in ceiling, in.
C_1 = 285
C_2 = 183
C_3 = 320

Environment for Animals and Plants

In addition to airflow through the inlet, infiltration airflow should be included. Figure 5 illustrates infiltration rates for two types of dairy barn construction (appropriate for other animal buildings also). The infiltration rates can be described as:

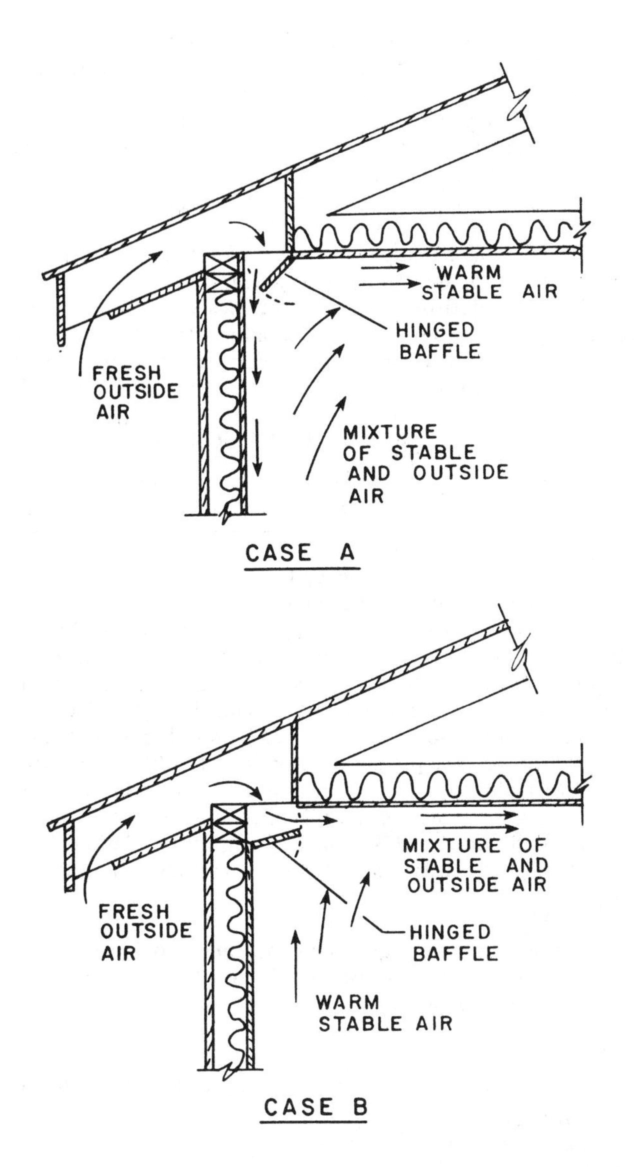

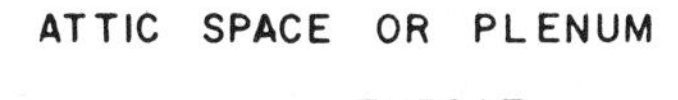

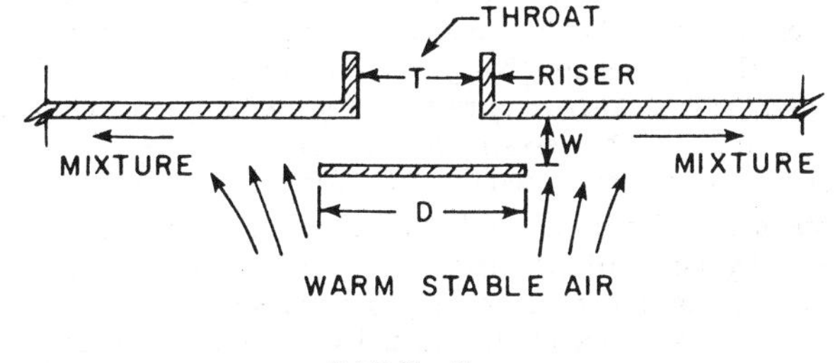

Fig. 4 Typical Livestock Building Inlet Configurations

Average construction:

$$I = C_4 p^{0.67} \quad (4)$$

Tight construction:

$$I = C_5 p^{0.67} \quad (5)$$

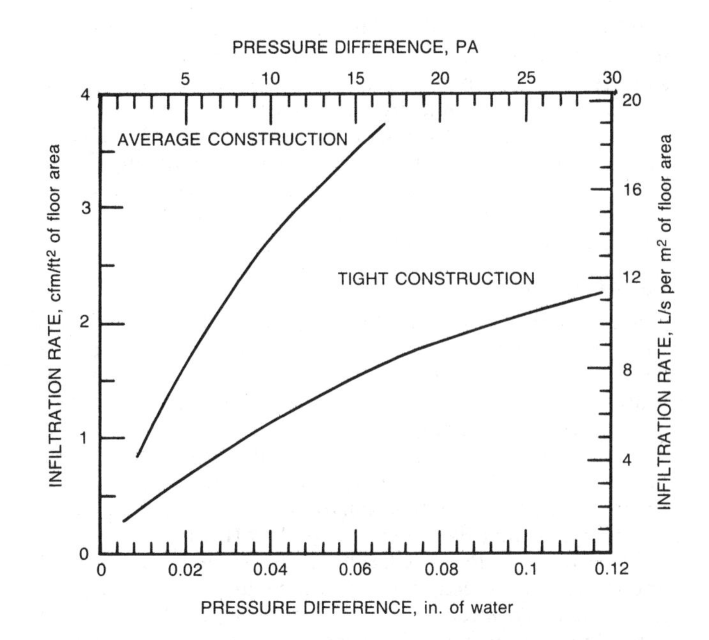

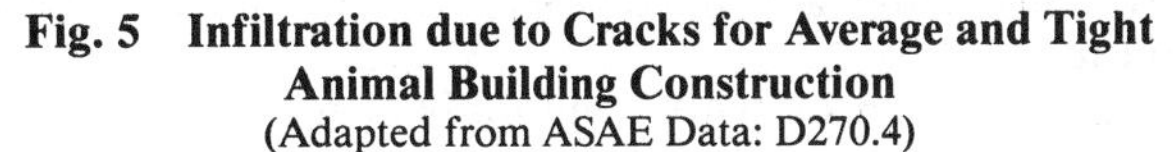

Fig. 5 Infiltration due to Cracks for Average and Tight Animal Building Construction
(Adapted from ASAE Data: D270.4)

where

I = infiltration rate, cfm per ft² of floor area
C_4 = 26.7
C_5 = 9.42

The above equation can be graphed as a system ventilation rate curve for various slot widths. Then, together with fan data, a ventilation schedule for expected weather and seasonal conditions can be developed.

Example. A dairy barn with a total floor area of 6000 ft² is to be ventilated with hinged baffle, wall flow, and slotted inlets. A total inlet length of 300 ft is available. The barn is considered to have tight construction. The total fresh air ventilation rate Q_{tot} is described by the following:

$$Q_{tot} = Q + I = 285\ W p^{0.5} \times 300 + 6000 \times 9.42\ p^{0.67}$$

$$Q_{tot} = 85{,}500\ W p^{0.5} + 56.500\ p^{0.67}$$

Room Air Velocity

The average air velocity inside a slot ventilated structure relates to the slot velocity, inlet slot width (or equivalent continuous length for boxed inlets), building width, and ceiling height. Estimates of air velocity within a barn, based on air exchange rates, may be very low due to the effects of jet velocity and recirculation. Conditions are usually partially turbulent, and there is no reliable way to predict room air velocity. General design guidelines keep throw distance less than 20 ft from slots and less than 10 ft from polyethylene tubes with holes.

Fans and Thermostats

Fans should not exhaust against prevailing winds. If structural or other factors require installing fans on the windward side, fans rated to deliver the required capacity against at least 0.12 in. of water static pressure and with a relatively flat power curve should be selected. The fan motor should withstand a wind velocity of 30 mph, equivalent to a static pressure of 0.4 in. of water, without overloading beyond its service factor. Wind hoods on the fans or windbreak fences reduce the effects of wind.

Control thermostats should be placed where they respond to a representative temperature as sensed by the animals. The thermostat needs protection and should be placed to prevent potential physical damage, *i.e.*, away from animals, ventilation inlets, water pipes, lights, heater exhausts, outside walls, or any other objects that will unduly affect performance. They are commonly placed near a fan exhaust. Control thermostats also require periodic adjustment based on accurate thermometer readings taken in the immediate area of the animal.

Flow Control

Since numbers and sizes of livestock and climatic conditions vary, means to modulate ventilation rates are often required beyond the conventional off/on thermostat switch. The minimum ventilation rate to remove moisture, reduce noxious gases, and keep water from freezing should always be provided. Methods include (1) intermittent fan operation—fans run on a percentage of time controlled by a percentage timer with a 10-min cycle; (2) staging of fans using multiple units or fans with high/low exhaust capability; (3) multispeed fans—larger fans (1/2 hp and up) with two flow rates, the lower being about 60% of the maximum rate; and (4) variable-speed fans—split capacitor motors designed to modulate fan speed smoothly from maximum down to 10 to 20% of maximum rate (the controller is usually thermostatically adjusted).

Generally, fans are spaced uniformly along the lee side of a building. Maximum distance between fans is between 115 and 165 ft. Fans may be grouped in a bank if this range is not exceeded. In housing with side curtains, exhaust fans that can be reversed or removed and placed inside the building in the summer are sometimes installed to increase air movement in combination with doors, walls, or windows being opened for natural ventilation.

Emergency Warning System

Animals housed in a high-density, mechanically controlled environment are subject to considerable risk of heat prostration if a power or ventilation equipment failure occurs. To reduce this danger, an alarm system and an automatic standby electric generator are highly recommended. Many alarm systems will detect failure of the ventilation system. These range from inexpensive "power off" alarms to systems that sense temperature extremes and certain gases. Automatic telephone dialing systems are effective as alarms and are relatively inexpensive. Building designs that allow some side wall panels (*e.g.*, 25% of wall area) to be removed for emergency situations are also recommended.

RECOMMENDED PRACTICES BY SPECIES

Mature animals readily adapt to a broad range of temperatures, but efficiency of production varies. Younger animals are more temperature sensitive. Figure 6 illustrates animal production response to temperature.

Relative humidity has not been shown to influence animal performance, except when accompanied by thermal stress. Relative humidity consistently below 50% may contribute to excessive dustiness; above 80%, it may increase building and equipment deterioration. Disease pathogens also appear to be more viable at either low or high humidities.

Dairy cattle

Dairy cattle shelters include confinement stall barns, freestalls, and loose housing. In the stall barn, cattle are usually confined to stalls approximately 4 ft wide, and all chores, including milking and feeding, are conducted there. Such a structure requires environmental modification, primarily through ventilation. Total space requirements are 50 to 75 ft^2 per cow. In free-stall housing, cattle are not confined to stalls but are free to move about. Here space requirements per cow are 75 to 100 ft^2. In loose housing, cattle are free to move within a fenced lot containing resting, feeding, and milking areas. Space required in sheltered loose housing is similar to that in free-stall housing. The shelters for resting and feeding areas are generally open-sided and require no air conditioning or mechanical ventilation, but supplemental air mixing is often beneficial during warm weather. The milking area is in a separate area or facility and may be fully or partially enclosed, thus requiring some ventilation.

For dairy cattle, climate requirements for minimal economic loss are broad and range from 35 to 75 °F, from 40 to 80% rh. Below 35 °F, production efficiency declines and management problems increase. However, the effect of low temperature on milk production is not as extreme as are high temperatures, where evaporative coolers or other cooling methods may be warranted.

Ventilation Rates for Each 1100-lb Cow

Winter	Spring/Fall	Summer
36 to 47 cfm	142 to 190 cfm	230 to 470 cfm

Required ventilation rates depend on specific thermal characteristics of individual buildings and internal heating load. The relative humidity should be maintained between 50 and 80%.

Both loose housing and stall barns require an additional milkroom to cool and hold the milk. Sanitation codes for milk production contain minimum ventilation requirements. The market being supplied should be consulted for all codes. Some state codes require positive pressure ventilation of milk rooms. Milk rooms are usually ventilated with fans at rates of 4 to 10 air changes per hour to satisfy requirements of local milk codes and to remove heat from milk coolers. Most milk codes require ventilation in the passageway (if any) between the milking area and the milk room.

Beef cattle

Beef cattle ventilation requirements are similar to those of dairy cattle on a unit weight basis. Beef production facilities often provide only shade and wind breaks.

Swine

Swine housing can be grouped into four general classifications:

1. Farrowing pigs from birth to 15 lbs
2. Nursery pigs from 15 to 50 lbs
3. Growing pigs from 50 lbs to market weight
4. Breeding and gestation

In farrowing barns, two environments must be provided: one for sows and one for piglets. Because each requires a different temperature, zone heating and/or cooling is used. The environment within the nursery is similar to that within the farrowing barn for piglets. Also, requirements for breeding stock housing will approach, but are less stringent than, those for growing barns.

Currently recommended practices for **farrowing houses** are:

Temperature: 50 to 68 °F, with small areas warmed for pigs to 82 to 90 °F by means of brooders, heat lamps, or floor heat. Avoid cold drafts and extreme temperatures. Hovers are sometimes used. Provide supplemental cooling (usually sprinklers or evaporative cooling systems) in extreme heat.

Relative humidity: up to 75% maximum

Ventilation rate: 40 to 530 cfm per sow and litter (about 400 lbs total weight). The low rate is for winter; the high rate is for summer temperature control.

Space: 34 ft^2 per sow and litter (stall); 65 ft^2 per sow and litter (pens).

Recommendations for **nursery barns** are:

Temperature: 79 °F for first week after weaning. Lower room temperature 3 °F per week to 72 °F. Provide warm, draft-free

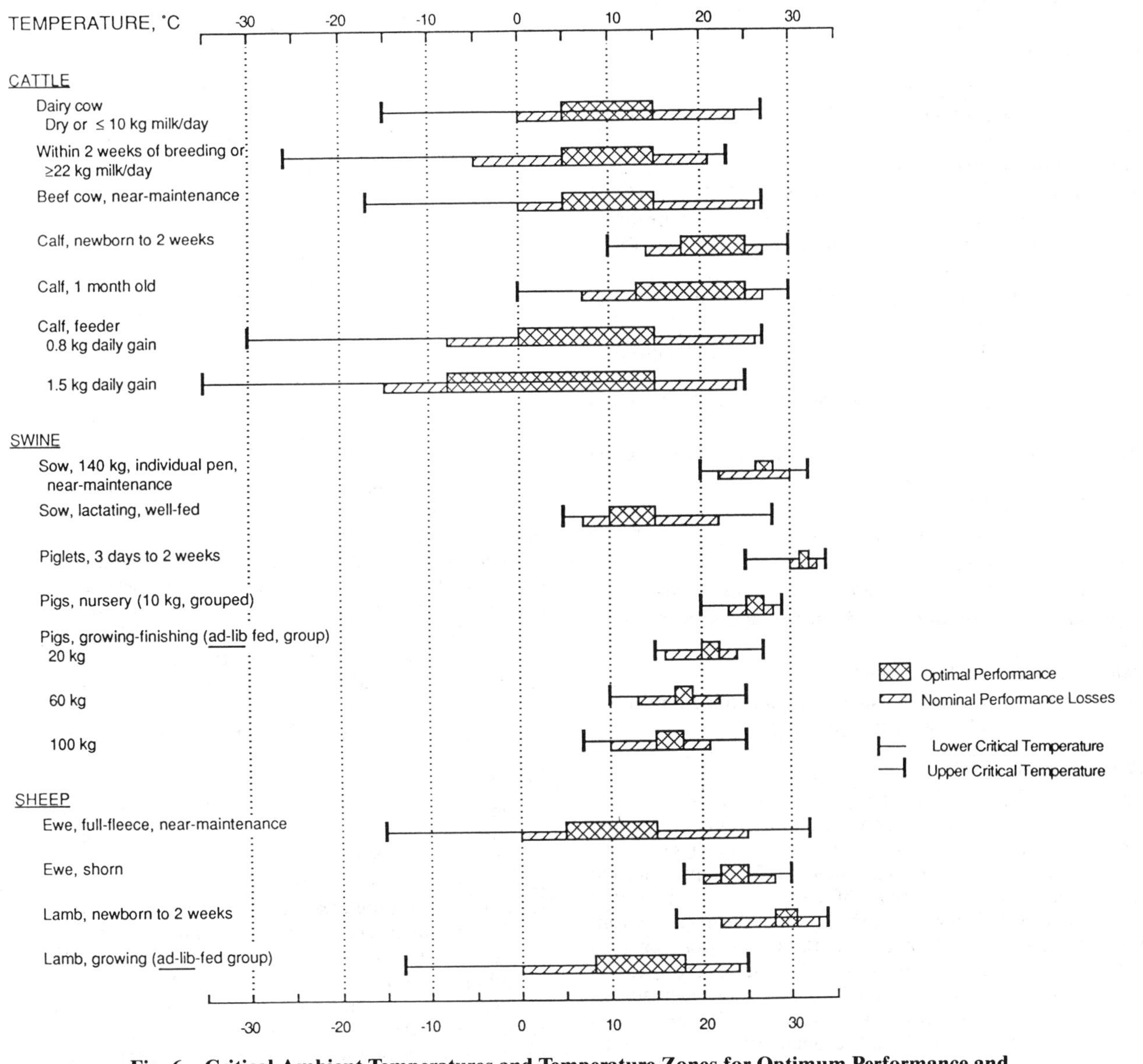

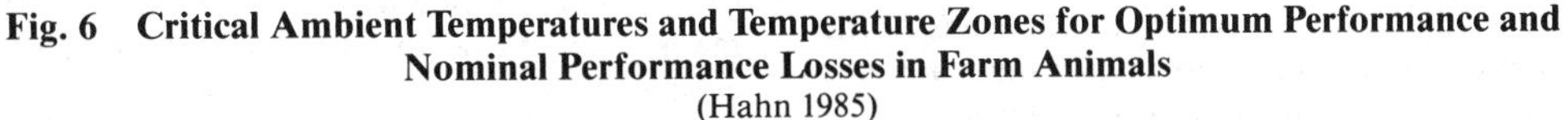

Fig. 6 Critical Ambient Temperatures and Temperature Zones for Optimum Performance and Nominal Performance Losses in Farm Animals
(Hahn 1985)

floors. Provide supplemental cooling for extreme heat (temperatures 80 °F and above).

Ventilation rate: 3 to 38 cfm per pig 13 to 79 lb each

Space: 1.7 to 3.9 ft^2 per pig 13 to 31 lb each

Recommendations for **growing/gestation** barns are:

Temperature: 55 to 72 °F preferred. Provide supplemental cooling (sprinklers or evaporative coolers) for extreme heat.

Relative humidity: 75% maximum in winter; no established limit in summer.

Ventilation rate:
Growing pig (75 to 150 lb)—6 to 85 cfm
Finishing pig (150 to 220 lb)—11 to 127 cfm
Gestating sow (240 to 500 lb)—0.5 cfm/lb

Space:
5.8 ft^2 per pig 75 to 150 lbs each
7.8 ft^2 per pig 150 to 220 lbs each
14 to 23.7 ft^2 per sow 240 to 500 lbs each

Poultry

In broiler and brooder houses, growing chicks require changing environmental conditions, and heat and moisture dissipation rates increase as the chicks grow older. Supplemental heat, usually from brooders, is used until sensible heat produced by the birds is adequate to maintain an acceptable air temperature. At early stages of growth, moisture dissipation per bird is low. Consequently, low ventilation rates are recommended to prevent excessive heat loss. Litter is allowed to accumulate over 3 to 5 flock placements. Lack of low-cost litter material may justify use of concrete floors. After each flock, caked litter is removed and fresh litter is added.

Housing for poultry may either be open, curtain-sided (dominant style in the South), or totally enclosed (dominant style in northern climates). Mechanical ventilation depends on the type of housing used. For open-sided housing, ventilation is generally natural airflow in warm weather, supplemented with stirring fans, and by fans with closed curtains in cold weather or during the brooding period. Totally enclosed housing depends on mechanical ventilation. Newer houses have smaller curtains and well-insulated construction to accommodate both natural and mechanical ventilation operation.

Recommendations for **broiler houses:**

Room temperature: 60 to 80 °F

Temperature under brooder hover: 86 to 91 °F, reducing 5 °F per week until room temperature is reached.

Relative humidity: 50 to 80%

Ventilation rate: Sufficient to maintain house within 2 to 4 °F of outside air conditions during summer. Generally, rates are about 0.1 cfm per lb live weight during winter and 1 to 2 cfm per lb for summer conditions.

Space: 0.6 to 1.0 ft^2 per bird (for the first 21 days of brooding, only 50% of floor space is used)

Light: Minimum of 10 lux to 28 days of age; 1 to 20 lux for growout (in enclosed housing)

Recommendations for **breeder houses** with birds on litter and slatted floors:

Temperature: 50 to 86 °F maximum; consider evaporative cooling if higher temperatures are expected

Relative humidity: 50 to 75%

Ventilation rate: Same as for broilers on live weight basis.

Space: 2 to 3 ft^2 per bird

Recommendations for **laying houses** with birds in cages:

Temperature, relative humidity, and ventilation rate: Same as for breeders.

Space: 50 to 65 in^2 per hen minimum

Light: Controlled day length using light-controlled housing is generally practiced (January through June).

Laboratory Animals

Proper management of laboratory animals includes any system of housing and care that permits animals to grow, mature, reproduce, behave normally, and be maintained in physical comfort and good health. Most recommendations for temperature and relative humidity are based on room temperature and humidity measurements, which may not be indicative of the microenvironment of the animal cage. Ventilation of the animal room (or cage) is necessary to regulate temperature and promote comfort.

Ideally, a system should permit individual compartment adjustments within ±2 °F for any temperature within a range of 64 to 86 °F. The relative humidity should be maintained throughout the year within a range of 30 to 75%, according to the needs of the species and local climatic conditions. Room air changes at a rate of 10 to 15 per hour are recommended for odor control.

DESIGN FOR PLANT FACILITIES

GENERAL

Greenhouses, plant growth chambers, and other facilities for indoor crop production overcome adverse outdoor environments and provide conditions conducive to economical crop production. The basic requirements of indoor crop production are (1) adequate light; (2) favorable temperatures; (3) favorable air or gas content; (4) protection from insects and disease; and (5) suitable growing media, substrate, and moisture. Because of their lower cost per unit of usable space, greenhouses are preferred over plant growth chambers for protected crop production. This section covers greenhouses and plant growth facilities and Chapter 9 of the 1989 ASHRAE *Handbook—Fundamentals* describes the environmental requirements in these facilities. Figure 7 shows the structural shapes of typical commercial greenhouses. Other greenhouses may have Gothic arches, curved glazing, or simple lean-to shapes. Glazing, in addition to traditional glass, now includes both film and rigid plastics. High light transmission by the glazing is usually

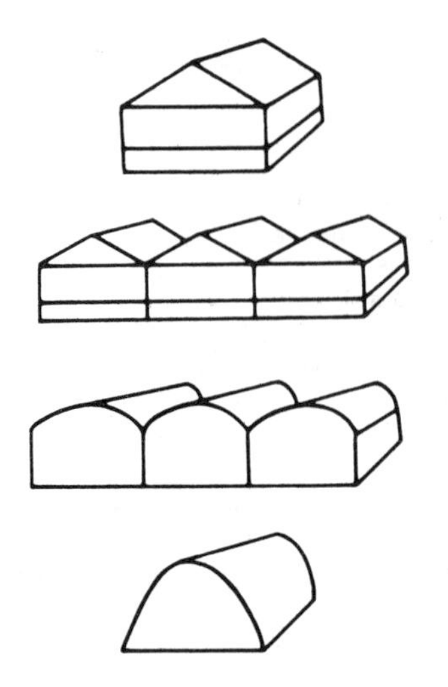

Fig. 7 Structural Shapes of Commercial Greenhouses

important; good location and orientation of the house are important in providing desired light conditions. Location also affects heating and labor costs, exposure to plant disease and air pollution, and material handling requirements. As a general rule in the northern hemisphere, a greenhouse should be placed at a distance of at least 2.5 times the height of the object closest to it in the eastern, western, and southern directions.

GREENHOUSES

Site Selection

Sunlight. Sunlight provides energy for plant growth and is often the limiting growth factor in greenhouses of the central and northern areas of North America during the winter. When planning greenhouses that are to be operated year-round, a designer should design for the greatest sunlight exposure during the short days of midwinter. The building site should have an open southern exposure, and if the land slopes, it should slope to the south.

Soil and Drainage. When plants are to be grown in the soil covered by the greenhouse, a growing site with deep, well-drained, fertile soil, preferably sandy loam or silt loam, should be chosen. Even though organic soil amendments can be added to poor soil, fewer problems occur with good natural soil. However, when a good soil is not available, growing in artificial media should be considered. The greenhouse should be level, but the site can and often should be sloped and well-drained to reduce salt buildup and insufficient soil aeration. A high water table or a hardpan may produce water-saturated soil, increase greenhouse humidity, promote diseases, and prevent effective use of the greenhouse. These problems can, if necessary, be alleviated by tile drain systems under and around the greenhouse. Ground beds should be level to prevent water from concentrating in low areas. Slopes within greenhouses also increase temperature and humidity stratification and create additional environmental problems.

Sheltered Areas. Provided they do not shade the greenhouse, surrounding trees act as wind barriers and help prevent winter heat loss. Deciduous trees are less effective than coniferous ones in midwinter, when the heat loss potential is greatest. In areas where snowdrifts occur, windbreaks and snowbreaks should be 100 ft or more away from the greenhouse to prevent damage.

Accessibility

Accessibility to utilities and transportation facilities are extremely important. Greenhouse operation is a year-round function requiring continuous service for all utilities and access to roads or other means of transportation for moving supplies into the facility and the crop to market.

Orientation

Generally in the northern hemisphere, for single-span greenhouses located north of 35 latitude, maximum transmission during winter is attained by an east-west orientation. South of 35 latitude, orientation is not important, provided headhouse structures do not shade the greenhouse. North-south orientation provides more light on an annual basis.

Gutter-connected or ridge and furrow greenhouses preferably are oriented with the ridge line north-south regardless of latitude. This orientation permits the shadow pattern caused by the gutter superstructure to move from the west to the east side of the gutter during the day. With an east-west orientation, the shadow pattern would remain north of the gutter, and the shadow would be widest and create the most shade during winter when light levels are already low. Also, the north-south orientation allows rows of tall crops, such as roses and staked tomatoes, to align with the long dimension of the house—an alignment that is generally more suitable to long rows and the plant support methods preferred by many growers.

The slope of the greenhouse roof is a critical part of greenhouse design. If the slope is too flat, a greater percentage of sunlight will reflect from the roof surface (Figure 8). A slope with a 1:2 rise-to-run ratio is the usual inclination for a gable roof.

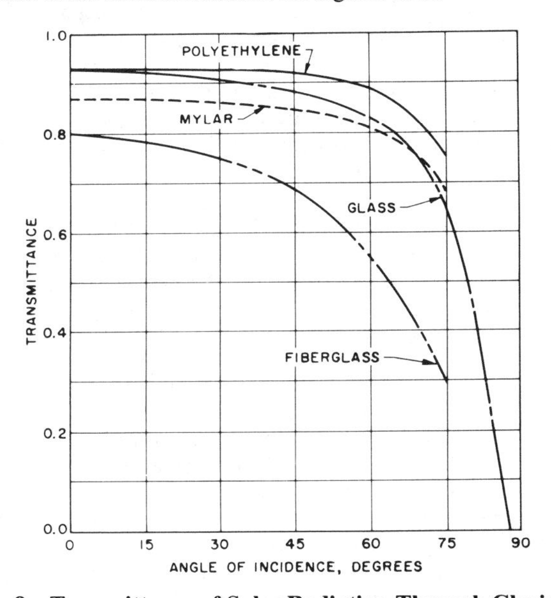

Fig. 8 Transmittance of Solar Radiation Through Glazing Materials for Various Angles of Incidence

Energy Balance

Structural Heat Loss. Estimates for heating and cooling a greenhouse consider conduction, infiltration, and ventilation energy exchange. In addition, the calculations must consider solar energy load and electrical input, such as light sources, which are usually much greater for greenhouses than for conventional buildings. Generally, conduction q_c plus infiltration q_i are used to determine the peak requirements q_t for heating.

$$q_t = q_c + q_i$$

$$q_c = UA\,(t_i - t_o)$$

$$q_i = 0.5\,VN\,(t_i - t_o)$$

where

U = overall heat loss coefficient, Btu/h · ft² · °F (Table 1)
A = exposed surface area, ft²
t_i = inside temperature, °F
t_o = outside temperature, °F
V = greenhouse internal volume, ft³
N = number of air exchanges per hour (Table 1)

Table 1 Suggested Heat Transmission Coefficients

		U, Btu/h · ft² · °F
Glass		
	Single-glazing	1.13
	Double-glazing	0.70
	Insulating	Manufacturers Data
Plastic film		
	Single film[a]	1.20
	Double film, inflated	0.70
	Single film over glass	0.85
	Double film over glass	0.60
Corrugated glass fiber		
	Reinforced panels	1.20
Plastic structured sheet[b]		
	16 mm thick	0.58
	8 mm thick	0.65
	6 mm thick	0.72

[a]Infrared barrier polyethylene films reduce heat loss; however, use this coefficient when designing heating systems because the structure could occasionally be covered with non-IR materials.
[b]Plastic structured sheets are double-walled, rigid plastic panels.

Type of Framing

The type of framing should be considered in determining overall heat loss. Aluminum framing and glazing systems may have the metal exposed to the exterior to a greater or lesser degree, and the heat transmission of this metal is higher than that of the glazing material. To allow for such a condition, the U-values of the glazing material should be multiplied by the factors shown in Table 2.

Table 2 Construction U-Value Multipliers

Metal frame and glazing system, 16 to 24 in. spacing	1.08
Metal frame and glazing system, 48 in. spacing	1.05
Fiberglass on metal frame	1.03
Film plastic on metal frame	1.02
Film or fiberglass on wood	1.00

Radiation Energy Exchange. Solar gain can be estimated using the procedures outlined in Chapter 27 of the 1989 ASHRAE *Handbook—Fundamentals.* As a guide, when a greenhouse is filled with a mature crop of plants, one-half the incoming solar energy is converted to latent heat, and one-quarter to one-third, to sensible heat. The rest is either reflected out of the greenhouse or absorbed by the plants and used in photosynthesis.

Radiation from a greenhouse to a cold sky is more complex. Glass admits a large portion of solar radiation but does not transmit long-wave thermal radiation in excess of approximately 5000 nm. Plastic films transmit more of the thermal radiation, but, in general, the total heat gains and losses are similar to those of glass. Newer plastic films containing infrared (IR) inhibitors reduce the thermal radiation loss. Plastic films and glass with improved radiation reflection are available at a somewhat higher cost.

Heating. Greenhouses may have a variety of heating systems. One is a convection system that circulates hot water or steam through plain or finned pipe. The pipe is most commonly placed along walls and occasionally beneath plant benches to create desirable convection currents. A typical temperature distribution pattern created by perimeter heating is shown in Figure 9. More uniform temperatures can be achieved when about one-third the total heat comes from pipes spaced uniformly across the house. These pipes can be placed above or below the crop, but temperature stratification and shading are avoided when they are placed below. Outdoor weather conditions affect temperature distribu-

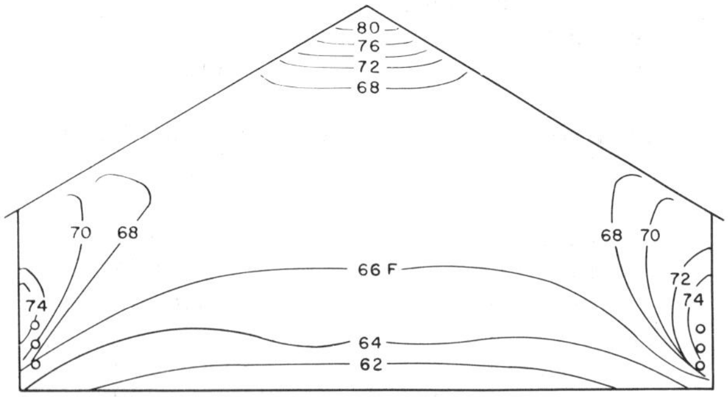

Fig. 9 Temperature Profiles in a Greenhouse Heated with Radiation Piping Along the Sidewalls

tion, especially on windy days in loosely constructed greenhouses. Manual or automatic overhead pipes are also used for supplemental heating to prevent snow buildup on the roof. In a gutter-connected greenhouse in a cold climate, a heat pipe should be placed under each gutter to prevent snow accumulation.

Overhead tube systems consist of a unit heater that discharges into 12- to 30-in. diameter plastic film tubing perforated to provide uniform air distribution. The tube is suspended at 6 to 10 ft intervals and extends the length of the greenhouse. Variations include a tube and fan receiving the discharge of several unit heaters. The fan and tube system is used without heat to recirculate the air and, during cold weather, to introduce ventilation air. However, tubes sized for heat distribution may not be large enough for effective ventilation during warm weather. A fan and tube system can also distribute fumigants and insecticides.

Perforated tubing, 6 to 10 in. in diameter, placed at ground level (underbench) can also improve heat distribution. Ideally, the ground-level tubing should draw air from the top of the greenhouse for recirculation or heating. Tubes on or near the floor have the disadvantage of being obstacles to workers and reducing usable floor space.

Underfloor heating can supply up to 25 or more of the peak heating requirements of northern greenhouses. A typical underfloor system uses 0.75-in. plastic pipe with nylon fittings in the floor, 4 in. below the surface, spaced 12 to 16 in. on centers, and covered 2 to 2.75 gpm/loop with regular gravel or porous concrete. Hot water, not exceeding 104 °F, circulates at a rate of 8 to 16 gpm. The pipe loops should generally not exceed 400 ft in length. This system can provide 16 to 20 $Btu/h \cdot ft^2$ from a bare floor, and about 75% as much when potted plants or seedling flats cover most of the floor.

Similar systems can heat soil directly, but root temperature must not exceed 77 °F. When used with water from solar collectors or other heat sources, the underfloor area can store heat. This storage consists of a vinyl swimming pool liner placed on top of insulation and a moisture barrier at a depth of 8 to 12 in. below grade, and filled with 50% void gravel. Hot water from solar collectors or other clean sources enters and is pumped out on demand. Some heat sources, such as cooling water from power plants, cannot be used directly but require a closed loop heat transfer system to avoid fouling the storage system and the power plant cooling water.

Greenhouses can also be bottom-heated with 0.25-in. diameter EPDM tubing (or variations of that method) in a closed loop system. The tubes can be placed directly in the growing medium of ground beds or under plant containers on raised benches. Best temperature uniformity is obtained by flow in alternate tubes in opposite directions. This method can supply all the greenhouse heat needed in mild climates.

Bottom heat, underfloor heating, and underbench heating are, because of the location of the heat source, more effective than overhead or peripheral heating and can reduce energy loss by 20 to 30%.

Unless properly located and aimed, overhead unit heaters, whether hydronic or direct fired, do not give uniform temperature at the plant level and throughout the greenhouse. Horizontal blow heaters positioned so that they establish a horizontal airflow around the outside of the greenhouse offer the best distribution. The airflow pattern can be supplemented with the use of horizontal blow fans or circulators.

When direct combustion heaters are used in the greenhouse, the combustion gases must be adequately vented to the outside to minimize the danger to plants and humans from products of combustion. One manufacturer recommends that combustion air must have access to the space through a minimum of two permanent openings in the enclosure, one near the bottom. A minimum of 1 in^2 of free area per 1000 Btu/h input rating of the unit, with a minimum of 100 in^2 for each opening, whichever is greater, is recommended. Unvented direct combustion units should not be used inside the greenhouse.

In many greenhouse heating systems, a combination of overhead and perimeter heating is used. Despite the type of heating, it is common practice first to calculate the overall heat loss, and then to calculate the individual elements such as the roof, sidewalls, and gables. It is then simple to allocate the overhead portion to the roof loss and the perimeter portions to the sides and gables, respectively.

The annual heat loss can be approximated by calculating the design heat loss and then, in combination with the annual degree-day tables using the 65 °F base, estimating an annual heat loss and computing fuel usage on the basis of the rating of the particular fuel used. If a 50 °F base is used, it can be prorated.

Heat curtains for energy conservation are becoming more important in greenhouse construction. Although this energy saving may be considered in the annual energy use, it should not be used when calculating design heat load; the practice is to open the heat curtains during snowstorms to facilitate the melting of snow, thereby nullifying its contribution to the design heat loss value.

Air-to-air and water-to-air heat pumps have been used experimentally on small-scale installations. Their usefulness is especially sensitive to the availability of a low-cost heat source.

Overhead steam, hot-water, or hot-air unit heaters can provide uniform temperature at plant level. Unit heaters must be carefully aimed to obtain uniform temperature within the greenhouse. Horizontal airflow fans or overhead circulators reduce the vertical temperature gradient while reducing temperatures in the peak of the greenhouse. Hot air, combustion unit heaters have been used successfully in small greenhouses as an alternative to a central boiler, but since the combustion device is located in the greenhouse, additional precautions must be taken. Combustion gases must be vented from the house and fresh air brought in. About 3 ft^2 of opening for make-up air must be added per therm of installed heating capacity.

Conventional, unvented gas and oil heaters should not be used for heating greenhouses because of adverse effects on the crops due to oxygen depletion and combustion fume toxicity. Specially designed, direct-combustion units are available (principally for CO_2 enrichment) that do not produce toxic gases. However, makeup air must still be provided for these units.

Ventilation. Greenhouse ventilation primarily prevents excessive temperature rise because of insolation. It also prevents CO_2 depletion and keeps relative humidity at a reasonable level. Ventilation (and cooling systems) should be designed to achieve uniform temperature distribution in the plant-growing zone. During cold weather, fresh air should normally be introduced through a variable-width slot inlet, keeping velocity between 700 and 1000 fpm and to retard the tendency of cool air to settle directly to the floor and produce large, vertical temperature gradients. As noted earlier, perforated polyethylene tubing can effectively distribute ventilation air in the winter. Perforations in the tubing at the 1000 and 1400 hour positions produce an upward airflow and prevent cold air from being directed onto the plants. Tubing of this kind can be used in either positive or negative pressure ventilation systems.

In summer, the cooler air should be directed across the greenhouse at or near plant level. Typical maximum ventilation rates provide for up to 1 air change per minute, and distances between fans and inlets should be limited to 200 ft to prevent excessive temperature rise. Such a system requires proper installation to ensure reasonably uniform airflow across all plants.

Most older greenhouses, and some newer ones, are ventilated by natural air exchange through ridge and side ventilators (Figure 10A). Automatic vent controls are used in some cases. With natural ventilation systems, the degree of temperature control depends on the house configuration, external wind speed, and outside air temperature. If natural ventilation is to occur, wind or thermal buoyancy must create a pressure difference. Vent openings on both sides of the greenhouse and the ridge take the best advantage of pressure differences created by the wind. Large vent openings supply sufficient ventilation when the wind is light or when thermal buoyancy forces are small. The total vent area should be from 15 to 25% of the floor area. For a single greenhouse, the combined sidewall vent area should equal the combined roof vent area for best ventilation by thermal buoyancy. Both sidewall and ridge vent openings should be continuous along the greenhouse. Sensors should be included in automatic control (or alarm) systems for times of rain and high wind, which could damage the crops.

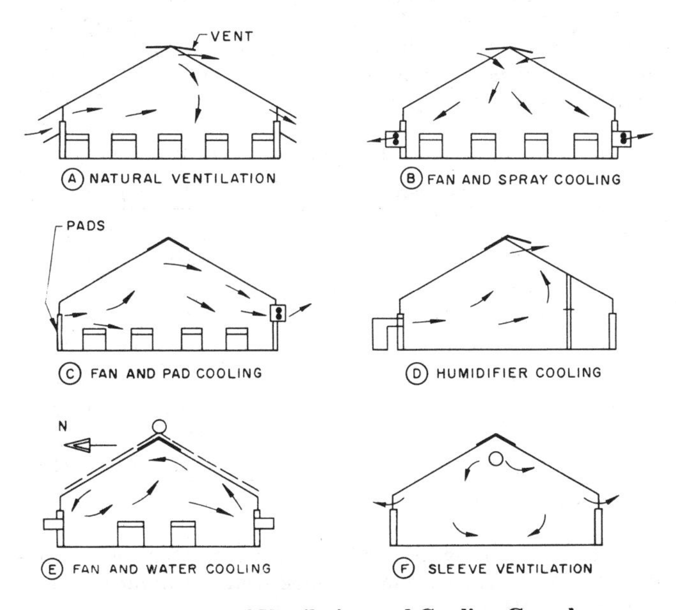

Fig. 10 Methods of Ventilating and Cooling Greenhouses

Fans give a more positive control over ventilation than do naturally vented systems (Figure 10). Generally, exhaust fans located in the side walls or ends of the greenhouse draw air through partially opened ridge vents, or end or side vents that may be equipped with evaporative pads. Inlet design is important, and inlets should be sized to maintain an airspeed between 700 and 1000 fpm at any ventilation stage. They should automatically adjust to maintain this speed as ventilation capacity changes. A common inlet is a top-hinged window vent located opposite the fans. Controls can be installed to vary the inlet opening automatically. Typically, the fans and inlets are controlled by (aspirated) thermostats located in the middle of the greenhouse at plant height.

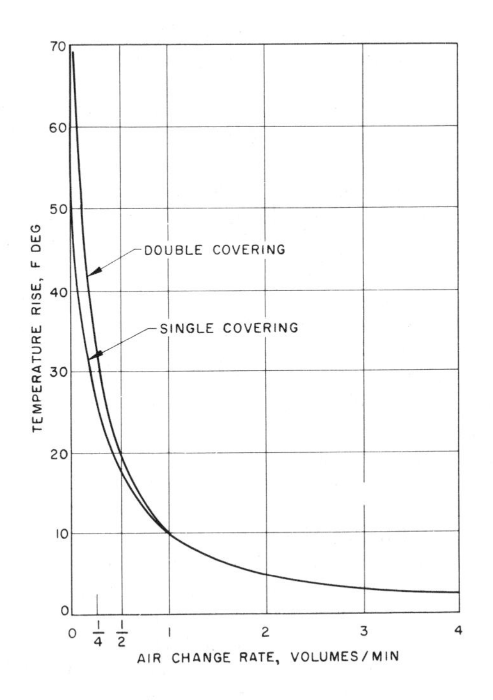

Fig. 11 Influence of Air-Exchange Rate on Temperature Rise in Single- and Double-Covered Greenhouses

The control over temperature rise achieved with fans relates directly to the air exchange rate. As shown in Figure 11, air exchange rates between 0.75 and 1 change per minute are effective. At lower airflow rates, the temperature differential increases rapidly. At higher airflow rates, the reduction of temperature rise is relatively small, fan power requirements increase, and plant damage can occur from increased airspeed. Suggested design air exchange rates are listed in Table 3.

Table 3 Suggested Design Air Changes (N)

New Construction	
Single glass lapped, unsealed	1.25
Single glass lapped, laps unsealed	1.0
Plastic film covered	0.6 to 1.0
Structured sheet	1.0
Film plastic over glass	0.9
Old Construction	
Good maintenance	1.5
Poor maintenance	2 to 4

Humidity Control. Maintaining sufficient relative humidity during the day and dissipating it at night is a problem in greenhouse crop production. Relative humidity should be kept high during the day to reduce water loss by transpiration, but 85 to 90% rh at night enhances growth of many disease organisms. Thus, a high relative humidity during the day condenses on plant leaf surfaces at night when the temperature drops. Opening ventilators or heating the greenhouse a few degrees at night are helpful but usually inadequate. Fan ventilation, especially when coupled to perforated polyethylene tubing, plus some greenhouse heating (even in mild weather) are reasonably effective in lowering the relative humidity.

Shading. Shading compounds are commonly applied to greenhouse glazings to reduce temperature during summer. Up to 50% shading can be achieved. Many shading compounds are forms of lime having varying durability. However, lime corrodes aluminum and damages some caulking. White latex paint diluted with water and sprayed rather than painted on the glass is more widely used with aluminum structures. For plastic glazings, the manufacturer can advise on the correct shading compound; some compounds cannot be removed from plastic. Ideally, the shading will wear away naturally during the summer and leave the glazing clean by fall when the natural light level declines to critical levels. In practice, some physical cleaning is usually needed.

Mechanically operated shade cloth and aluminum or wooden laths give the operator a great amount of control over summer weather conditions. The shade can be retracted on cloudy days in the summer and can insulate for energy conservation in winter.

Evaporative Cooling. Some form of cooling must be used during the day to maintain greenhouse temperatures equal to or less than those outside. Mechanical refrigeration is expensive and seldom used in commercial greenhouses because of the high solar heat loads. Evaporative cooling—either wet pads, unit coolers, or fog spray systems—can accomplish economical greenhouse cooling. However, evaporative cooling adds moisture to the air that, while reducing the potential for plant moisture stress, may also increase the incidence of disease.

Wetted pads of cellulose material in honeycomb format are typically placed in an end wall or side wall, and exhaust fans are placed in the opposite wall. Fans and pads should be no more than 160 ft apart. However, the designer should be aware that the performance of evaporative pads varies significantly with pad design. Table 4 shows data recommended for sizing the pad area. The water flow rate through the pad should be approximately 1.0 to 2.5 gpm per linear foot of pad (Table 5). The pads should be continuous along the length of the wall and high enough to obtain the required pad area. The pad material should operate at an evaporative efficiency of 85%. In practice, the outside air is cooled to within about 4°F of the wet-bulb temperature in a properly designed, well-managed system.

Table 4 Recommended Air Velocity Through Various Pad Materials

Type	Air Face Velocity Through Pad*, fpm
Corrugated cellulose, 4 in. thick	250
Corrugated cellulose, 6 in. thick	350

*Speed may be increased by 25% where construction is limiting.

Table 5 Recommended Water Flow and Sump Capacity for Vertically Mounted Cooling Pad Materials

Pad Type and Thickness	Minimum Water Rate per Lineal Foot of Pad, gpm	Minimum Sump Capacity per Unit Pad Area gal/ft²
Corrugated cullulose, 4 in. thick	0.5	0.8
Corrugated cellulose, 6 in. thick	0.8	1.0

The fog-evaporative system consists of fogging nozzles located in the upper portion of the house. A typical fogging arrangement locates fog nozzles on 3-ft centers and about 18 in. above the plants. Although the fog can be operated continuously, it is more common to operate on a timer of 5 to 10 s every 5 min. Air should flow from the fog nozzles to the plants. Water or air pressure of 200 to 500 psi develops a fine mist and keeps coarse droplets from falling onto the plants. Because some water, even with high-pressure nozzles, is likely to fall or condense on plant surfaces, all piping should be PVC rather than copper, which is toxic to some plants.

The amount of water required to saturate the air is highly variable. A high-pressure system does not have wide variations in water output. A system adequate for extreme conditions will generate excess fog at cooler times, thus wetting the crop. A fog system with nozzles in staged groups can be modulated to meet demand.

Other Environmental Factors

Carbon Dioxide Enrichment. Carbon dioxide enrichment is practiced in some greenhouse operations to increase growth and enhance yields. However, CO_2 enrichment is practical only when little or no ventilation is required for temperature control. Carbon dioxide can be generated from solid CO_2 (dry ice), bottled CO_2, and misting carbonated water. Bulk or bottled CO_2 gas is usually distributed through perforated tubing placed near the plant canopy. Carbon dioxide from dry ice is distributed by passing greenhouse air through an enclosure containing dry ice. Air movement around the plant leaf increases the efficiency with which the plant absorbs whatever CO_2 is available. One study found an airspeed of 100 fpm to be equivalent to a 50% enrichment in CO_2 without forced air movement.

Radiant Energy. Light is normally the limiting factor in greenhouse crop production during the winter. North of the 35th parallel, light levels are especially inadequate or marginal in fall, winter, and early spring. Artificial light sources, usually high-intensity discharge (HID) lamps, may be added to greenhouses to supplement low natural light levels. High-pressure sodium (HPS), metal halide (MH), low-pressure sodium (LPS), and occasionally mercury lamps coated with a color-improving phosphor are currently used. Since differing irradiance or illuminance ratios are emitted by the various lamp types, the incident radiation is best described as radiant flux density (W/ft^2) between 400 and 850 nm, or as photon flux density between 400 and 700 nm, rather than in photometric terms of lux or footcandles.

Table 6 shows conversions of HPS, MH, and LPS irradiance (W/ft^2) to illuminance (lux) and photon flux density to assist in relating irradiance to more familiar illuminance values. One footcandle is approximately 10 lux.

Table 6 Constants to Convert to W/m²

Light Source	klx	$\mu mol/s^2 \cdot m^2$
400-700 nm		
INC	3.99	0.20
FCW	2.93	0.22
FWW	2.81	0.21
HG	2.62	0.22
MH	3.05	0.22
HPS	2.45	0.20
LPS	1.92	0.20
Daylight	4.02	0.22

$1\ \mu mol/(s \cdot m^2) = 1\ einstein/(s \cdot m^2)$

Data in Table 7 suggest irradiance at the top of the plant canopy, duration, and time of day for supplementing natural light levels for specific plants.

Luminaires have been developed specifically for greenhouse use of HID lamps. The lamps in these special greenhouse luminaires are often placed in a horizontal position, which may decrease both light output and the life of the lamp. These drawbacks may be balanced by improved horizontal and vertical uniformity as compared to industrial parabolic reflectors.

Photoperiod Control. Artificial light sources are also used to lengthen the photoperiod during the short days of winter. Photoperiod control requires much lower light levels than those needed for photosynthesis and growth. Thus, photoperiod illuminance needs to be only 0.6 to 1.1 W/ft^2. The incandescent lamp is the most effective light source for this purpose due to its higher far red component. Lamps such as 150 W (PS-30) silverneck lamps

Table 7 Suggested Radiant Energy, Duration, and Time for Supplemental Lighting in Greenhouses

Plant and Stage of Growth	W/ft²	Duration	
		Hours	Time
African Violets early-flowering	1 to 2	12 to16	0600-1800 0600-2200
Ageratum early-flowering	1 to 4.5	24	
Begonias—fibrous rooted branching and early-flowering	1 to 2	24	
Carnation branching and early-flowering	1 to 2	16	0800-2400
Chrysanthemums	1 to 2	16	0800-2400
vegetable growth branching and multiflowering	1 to 2	8	0800-1600
Cineraria seedling growth (four weeks)	0.6 to 1	24	
Cucumber rapid growth and early-flowering	1 to 2	24	
Eggplant early-fruiting	1 to 4.5	24	
Foliage plants (Philodendron, Schefflera) rapid growth	0.6 to 1	24	
Geranium branching and early-flowering	1 to 4.5	24	
Gloxinia	1 to 4.5	16	0800-2400
early-flowering	0.6 to 1	24	
Lettuce rapid growth	1 to 4.5	24	
Marigold early-flowering	1 to 4.5	24	
Impatiens—New Guinea branching and early-flowering	1	16	0800-2400
Impatiens—Sultana branching and early-flowering	1 to 2	24	
Juniper vegetative growth	1 to 4.5	24	
Pepper early-fruiting, compact growth	1 to 2	24	
Petunia branching and early-flowering	1 to 4.5	24	
Poinsettia—vegetative growth	1	24	
branching and multiflowering	1 to 2	8	0800-1600
Rhododendron vegetative growth (shearing tips)	1	16	0800-2400
Roses (hybrid teas, miniatures) early-flowering and rapid regrowth	1 to 4.5	24	
Salvia early-flowering	1 to 4.5	24	
Snapdragon early-flowering	1 to 4.5	24	
Streptocarpus early-flowering	1	16	0800-2400
Tomato rapid growth and early-flowering	1 to 2	16	0800-2400
Trees (deciduous) vegetative growth	0.6	16	1600-0800
Zinnia early-flowering	1 to 4.5	24	

spaced 10 to 13 ft on centers and 13 ft above the plants provide a cost-effective system. Where a 13-ft height is not practical, 60-W extended service lamps on 6.5-ft centers are satisfactory. One method of photoperiod control is to interrupt the dark period by turning the lamps on at 2200 and off at 0200. The 4-h interruption, initially based on chrysanthemum response, induces a satisfactory long-day response in all photoperiodically sensitive species. Many species, however, respond to interruptions of 1 h or less. Demand charges can be reduced in large installations by operating some sections from 2000 to 2400 and others from 2400 to 0400. The biological response to these schedules, however, is much weaker than with the 2200 to 0200 schedule, so some varieties may flower prematurely. If the 4-h interruption period is used, it is not necessary to keep the light on throughout the interruption period. Photoperiod control of most plants can be accomplished by operating the lamps on light and dark cycles with 20% "on" times; for example, 12 s/min. The length of the dark period in the cycle is critical, and the system may fail if the dark period exceeds about 30 min. Demand charges can be reduced by alternate scheduling of the "on" times between houses or benches without reducing the biological effectiveness of the interruption.

Plant displays in showrooms or shopping malls, for example, require enough light for plant maintenance and a spectral distribution that will show the plants to best advantage. Metal halide lamps, with or without incandescent highlighting, are often used for this purpose. Fluorescent lamps, frequently of the special phosphor, plant-growth type, enhance color rendition but are more difficult to install in aesthetically pleasing designs.

Alternate Energy Sources and Energy Conservation

Limited progress has been achieved in heating commercial greenhouses with solar energy. Collecting and storing the heat requires a volume at least one-half the volume of the greenhouse. Passive solar units work at certain times of the year and, in a few localities, year-round.

If available, reject heat is a possible source of winter heat. Winter energy and solar (photovoltaic) sources are possible future energy sources for greenhouses, but the development of such systems is still in the research stage.

Energy Conservation. A number of energy-saving measures (*e.g.*, thermal curtains, double glazing, and perimeter insulation) have been retrofitted to existing greenhouses and incorporated into new construction. Sound maintenance is necessary to keep heating system efficiency at a maximum level.

Automatic controls, such as thermostats, should be calibrated and cleaned at regular intervals, and heating-ventilation controls should interlock to avoid simultaneous operation. Boilers that can burn more than one type of fuel permit using the lowest cost fuel available.

Modifications to Reduce Heat Loss

Film covers for heat loss reduction are widely used in commercial greenhouses, particularly for growing foliage plants and other species that grow under low light levels. Irradiance (intensity) is reduced 10 to 15% per layer of plastic film.

One or two layers of transparent 4 or 6 mil continuous-sheet plastic is stretched over the entire greenhouse (leaving some vents uncovered), or from the ridge to the sidewall ventilation opening. When two layers are used, (outdoor) air at a pressure of 0.2 to 0.25 in. of water is introduced continuously between the layers of film to maintain the airspace between them. When a single layer is used, an airspace can be established by stretching the plastic over the glazing bars and fastening it around the edges, or a length of polyethylene tubing can be placed between the glass and the plastic and inflated (using outside air) to stretch the plastic sheet.

Double-Glazing Rigid Plastic. Double wall panels are manufactured from acrylic and polycarbonate plastics, with walls separated by about 0.4 in. Panels are usually 48 in. wide and 96 in. or more long. Nearly all types of plastic panels have a high thermal expansion coefficient and require about 1% expansion space (0.12 in/ft). When a panel is new, light reduction is roughly 10 to 20%. Moisture accumulation between the walls of the panels must be avoided.

Double-Glazing Glass. The framing of most older greenhouses must be modified or replaced to accept double-glazing with glass.

Light reduction is 10% more than with single-glazing. Moisture and dust accumulation between glazings increases light loss. As with all types of double-glazing, snow on the roof melts slowly and increases light loss. Snow may even accumulate sufficiently to cause structural damage, especially in gutter-connected greenhouses.

Silicone Sealants. Transparent silicone sealant in the glass overlaps of conventional greenhouses reduces infiltration and may produce heat savings of 5 to 10% in older structures. There is little change in light transmission.

Precautions. The various methods described above reduce heat loss by reducing conduction and infiltration. They may also cause more condensation, higher relative humidity, lower carbon dioxide concentrations, and an increase in ethylene and other pollutants. Combined with the reduced light levels, these factors may cause delayed crop production, elongated plants, soft plants, and various deformities and diseases, all of which reduce the marketable crop.

Thermal blankets are any flexible material that is pulled from gutter to gutter and end to end in a greenhouse, or around and over each bench, at night. Materials ranging from plastic film to heavy cloth, or laminated combinations, have successfully reduced heat losses by 25 to 35% overall. Tightness of fit around edges and other obstructions is more important than the kind of material used. Some films are vaportight and retain moisture and gases. Others are porous and permit some gas exchange between the plants and the air outside the blanket. Opaque materials can control crop day length when short days are part of the requirement for that crop. Condensation may drip onto and collect on the upper sides of some blanket materials to such an extent that they collapse.

Multiple-layer blankets, with two or more layers separated by airspaces, have been developed. One such system combines a porous-material blanket and a transparent film blanket; the latter is used for summer shading. Another system has four layers of porous, aluminum foil-covered cloths, the layers separated by air.

Thermal blankets may be opened and closed manually as well as automatically. The opening and closing decision should be based on the irradiance level and whether it is snowing, rather than time of day. Two difficulties with thermal blankets are the physical problems of installation and use in greenhouses with interior supporting columns, and the loss of space due to shading by the blanket when it is not in use during the day.

Other Recommendations. While the foundation can be insulated, the insulating materials must be protected from moisture, and the foundation wall should be protected from freezing. All or most of the north wall can be insulated with opaque or reflective-surface materials. The insulation will reduce the amount of diffuse light entering the greenhouse, and, in cloudy climates, cause reduced crop growth near the north wall.

Ventilation fan cabinets should be insulated, and fans not needed during the winter should be sealed against air leaks. Efficient management and operation of existing facilities are the most cost-effective ways of reducing energy use. This includes continued contact with local Cooperative Extension personnel to obtain area-specific, current information on greenhouse operation techniques.

PLANT GROWTH ENVIRONMENTAL FACILITIES

Controlled-environment rooms (CER), also called plant growth chambers, include all controlled or partially controlled environmental facilities for growing plants, except greenhouses. CERs are indoor facilities. Units with floor areas less than 50 ft^2 may be moveable with self-contained or attached refrigeration units. CERs usually have artificial light sources, provide control of temperature and, in some cases, relative humidity and CO_2 level.

CERs are used to study all aspects of plant science. Some growers use growing rooms to increase seedling growth rate, produce more uniform seedlings, and grow specialized, high-value crops. The main components of the CER are (1) an insulated room or an insulated box with an access door; (2) a heating and cooling mechanism with associated air-moving devices and controls; and (3) a lamp module at the top of the insulated box or room. CERs are similar to walk-in cold storage rooms, except for the lighting and larger refrigeration system needed to handle heat produced by lighting.

Location

The location for a CER must have space for the outside dimensions of the chamber, refrigeration equipment, ballast rack, and control panels. Additional space around the unit is necessary for servicing the various components of the system, and, in some cases, for substrate, pots, nutrient solutions, and other paraphernalia associated with plant research. The location also requires an electrical supply (on the order of 28 W/ft^2 of controlled environment space), a water supply, and a compressed air supply.

Construction and Materials

Wall insulation should have a unit thermal conductance value of less than 0.026 Btu/h · ft^2 · °F. Materials should resist corrosion and moisture. The interior wall covering should be metal, with a high-reflectance white paint or specular aluminum with a reflectivity of not less than 80%. Reflective films or similar materials can be used but will require periodic replacement.

Floors and Drains

Floors that are part of the CER should be corrosion resistant. Tar or asphalt waterproofing materials and volatile caulking compounds should not be used because they will likely release phytotoxic gases into the chamber atmosphere. The floor must have a drain to remove spilled water and nutrient solutions. The drains should be trapped and equipped with screens to catch plant and substrate debris.

Plant Benches

Three bench styles for supporting the pots and other plant containers are normally encountered in plant growth chambers: (1) stationary benches; (2) benches or shelves built in sections that are adjustable in height; and (3) plant trucks, carts, or dollies on casters, which are used to move plants between chambers, greenhouses, and darkrooms. The bench supports containers filled with moist sand, soil, or other substrate and is usually designed for a minimum loading of 50 lb/ft^2. The bench or truck top should be constructed of nonferrous, perforated metal or metal mesh to allow free passage of air around the plants and to let excess water drain from the containers to the floor and subsequently to the floor drain.

Normally, benches, shelves, or truck tops are adjustable in height so that small plants can be placed close to the lamps and thus receive a greater amount of light. As the plants grow, the shelf or bench is lowered so that the tops of the plants continue to receive the original radiant flux density.

Controls

Environmental chambers require complex controls to provide the following:

1. Automatic transfer from heating to cooling with 2 °F or less dead zone and adjustable time delay.
2. Automatic daily switching of the temperature set point for different day and night temperatures (setback may be as much as 10 °F).
3. Protection of sensors from radiation. Ideally the sensors are located in a shielded, aspirated housing, but satisfactory performance can be attained by placing them in the return air duct.

4. Control of the daily duration of light and dark periods. Ideally, this control should be programmable to change the light period each day to simulate natural progression of day length. Photoperiod control, however, is normally accomplished with mechanical time clocks, which must have a control interval of 5 min or less for satisfactory timing.
5. Protective controls to prevent the chamber temperature from going more than a few degrees above or below the set point. Controls should also prevent short cycling of the refrigeration system, especially when the condensers are remotely located.
6. Audible and visual alarms to alert personnel of malfunctions.
7. Maintenance of relative humidity to prescribed limits.

Data loggers, recorders, or recording controllers are recommended to aid in monitoring daily operation of the system. Solid-state, microprocessor-based control is not yet widely used. However, programming flexibility and control performance are expected to improve as microprocessor control is developed for CER use.

Heating, Air Conditioning, and Airflow

When the lights are on, cooling will normally be required, and the heating system will rarely be called on to operate. When the lights are off, however, both heating and cooling may be needed. Conventional refrigeration systems are generally used with some modification. Direct expansion units usually operate with a hot-gas bypass to prevent numerous on-off cycles, and secondary coolant systems may use aqueous ethylene glycol rather than chilled water. Heat is usually provided by electric heaters, but other energy sources can be used, including hot gas from the refrigeration system.

The plant compartment is the heart of the growth chamber. The primary design objective, therefore, is to provide the most uniform, consistent, and regulated environmental conditions possible. Thus, airflow must be adequate to meet specified psychrometric conditions, but it is limited by the effects of high airspeeds on plant growth. As a rule, the average airspeed in CERs is restricted to about 100 fpm.

To meet the uniform conditions required by a CER, conditioned air is normally moved through the space from bottom to top, although an increasing number of CERs use top-to-bottom airflow. There is no apparent difference in plant growth between horizontal, upward, or downward airflow when the speed is less than 175 fpm. Regardless of the method, a temperature gradient is certain to exist, and the design should keep the gradient as small as possible. Uniform airflow is more important than the direction of flow; thus selection of properly designed diffusers or plenums with perforations is essential for achieving it.

The ducts or false side walls that direct air from the evaporator to the growing area should be small, but not so small that the noise level increases appreciably more than acceptable building air duct noise. CER design should include some provision for cleaning the interior of the air ducts.

Air-conditioning equipment for relatively standard chambers provides temperatures over 45 to 90 °F. Specialized CERs that require temperatures as low as −5 °F need low-temperature refrigeration equipment and devices to defrost the evaporator without increasing the growing area temperature. Other chambers that require temperatures as high as 115 °F need high-temperature components. The air temperature in the growing area must be controlled with the least possible variation about the set point. Temperature variation about the set point can be held to 0.5 °F using solid-state controls, but in most existing facilities, the variation is 1 to 2 °F.

The relative humidity in many CERs is simply an indicator of the existing psychrometric conditions and is usually between 50 and 80%, depending on temperature. Relative humidity in the chamber can be increased by steam injection, misting, hot water evaporators, and other conventional humidification methods. Steam injection causes the least temperature disturbance, and sprays or misting causes the greatest disturbance. Complete control of relative humidity requires dehumidification as well as humidification.

A typical humidity control system includes a cold evaporator or steam injection to adjust the chamber air dew point. The air is then conditioned to the desired dry-bulb temperature by electric heaters, a hot gas bypass evaporator, or a temperature-controlled evaporator. A dew point lower than about 40 °F cannot be obtained with a cold plate dehumidifier because of icing. Dew points lower than 40 °F require a chemical dehumidifier, usually in addition to the cold evaporator.

Lighting the Environmental Chambers

The type of light source and the number of lamps used in CERs are determined by the desired plant response. Traditionally, cool white fluorescent plus incandescent lamps that produce 10% of the fluorescent illuminance are used. Nearly all illumination data are based on either cool white or warm white fluorescent, plus incandescent. A number of fluorescent lamps have special phosphors hypothesized to be the spectral requirements of the plant. Some of these lamps are used in CERs, but there is little data to suggest that they are superior to cool white and warm white lamps. In recent years, high-intensity discharge lamps have been installed in CERs, either to obtain very high radiant flux densities, or to reduce the electrical load while maintaining a light level equal to that produced by the less efficient fluorescent-incandescent systems.

Table 8 Input Power Conversion of Light Sources

Lamp Identification		Total Input Power, W	Radiation (400-700 nm), %	Radiation (400-850 nm), %	Other Radiation, %	Conduction and Convection, %	Ballasts Loss, %
Incandescent							
(INC)	100A	100	7	15	75	10	0
Fluorescent							
Cool white	FCW	46	21	21	32	34	13
Cool white	FCW	225	19	19	34	35	12
Warm white	FWW	46	20	20	32	35	13
Plant growth A	PGA	46	13	13	35	39	13
Plant growth B	PGB	46	15	16	34	37	13
Infrared	FIR	46	2	9	39	39	13
Discharge							
Clear mercury	HG	440	12	13	61	17	9
Mercury deluxe	HG/DX	440	13	14	59	18	9
Metal halide	MH	460	27	30	42	15	13
High-pressure sodium	HPS	470	26	36	36	13	15
Low-pressure sodium	LPS	230	27	31	25	22	22

Conversion efficiency is for lamps without luminaire. Values compiled from manufacturers' data, published information, and unpublished test data by R.W. Thimijan.

One approach to design from the biological point of view is to base light source output recommendations on photon flux density μmol/(s·m^2) between 400 and 700 nm, or less frequently as radiant flux density between 400 and 700 nm, or 400 and 850 nm. Rather than basing illuminance measurements on human vision, this enables comparisons between light sources as a function of plant photosynthetic potential. Table 6 shows the conversion of various measurement units to W/m^2. However, instruments that measure the 400 to 850 nm spectral range are not generally available, and some controversy exists about the effectiveness of 400 to 850 nm as compared to the 400 to 700 nm range in photosynthesis. The power conversion of various light sources are listed in Table 8.

The design requirements for plant growth lighting differ greatly from that for vision lighting. Plant growth lighting requires a greater degree of horizontal uniformity and, usually, higher light levels than vision lighting. In addition, plant growth lighting should have as much vertical uniformity as possible—a factor rarely important in vision lighting. Horizontal and vertical uniformity are much easier to attain with linear or broad sources, such as fluorescent lamps, than with point sources, such as HID lamps. Tables 9 and 10 show the type and number of lamps, mounting height, and spacing required to obtain several levels of incident energy. Since the data were taken directly under lamps with no reflecting wall surfaces nearby, the incident energy is perhaps one-half of what the plants would receive if the lamps had been placed in a small chamber with highly reflective walls.

Table 9 Approximate Mounting Height and Spacing of Luminaires in Greenhouses

Lamp and Wattage	Irradiation, W/ft² 0.6	1.1	2.2	4.4
	Height and Spacing, in.			
HPS (400 W)	118	90	63	39
LPS (180 W)	94	67	47	31
MH (400 W)	106	79	55	35

Extended life incandescents or traffic signal lamps, with much longer life, will lower lamp replacement requirements. These lamps have lower lumen output, but they are nearly equivalent in the red portion of the spectrum. The short life of incandescent lamps is attributed to vibration in the chambers. Porcelain lamp holders and heat-resistant lamp wiring should be used. Lamps used for CER lighting include fluorescent lamps (usually 1500 mA), 250-, 400-, and occasionally 1000-W HPS and MH lamps, 180-W LPS, and various sizes of incandescent lamps. Extended life lamps may be used, but these sacrifice light output for longer life. In many installations, the abnormally short life of incandescent lamps is due to vibration from the lamp loft ventilation or from cooling fans. Increased incandescent lamp life under these conditions can be attained by using lamps constructed with a C9 filament.

Energy-saving lamps have approximately equal or slightly lower irradiance per input watt. Since the irradiance per lamp is lower, there is no advantage to using these lamps, except in tasks that can be accomplished with low light levels. Light output of all lamps declines with use, except perhaps for the low-pressure sodium lamps that appear to maintain approximately constant output but with an increase in input watts during use.

Fluorescent and metal halide designs should be based on 80% of the initial lumens. Most CER lighting systems have difficulty maintaining a relatively constant light level over considerable periods of time. Combinations of MH and HPS lamps compound the problem, because the lumen depreciation of the two light sources is significantly different. Thus, over time, the spectral energy distribution at plant level will shift toward the HPS. Lumen output can be maintained in two ways: (1) individual lamps, or a combination of lamps, can be switched off initially and activated as the lumen output decreases, and (2) by periodically replacing the oldest 25 to 33% of the lamps. Solid-state dimmer systems are commercially available only for low-wattage fluorescent lamps and for mercury lamps; in future, it may be possible to apply the technology to metal halide lamps.

Large rooms, especially those constructed as an integral part of the building and retrofitted as CERs, rarely separate the lamps from the growing area with a transparent barrier. Rooms designed as CERs at the time a building is constructed and freestanding rooms or chambers usually separate the lamp from the growing area with a barrier of glass or rigid plastic. Light output from fluorescent lamps is a function of the temperature of the lamp. Thus, the barrier serves a two-fold purpose: (1) to maintain optimum lamp temperature when the growing area temperature is higher or lower than optimum, and (2) to reduce the thermal radiation entering the growing area. Fluorescent lamps should operate in an ambient temperature and airflow environment that will maintain the tube wall temperature at 104°F. Under most conditions, the light output of HID lamps is not affected by ambient temperature. The heat must be removed, however, to prevent high thermal radiation from causing adverse biological effects (see Figure 12).

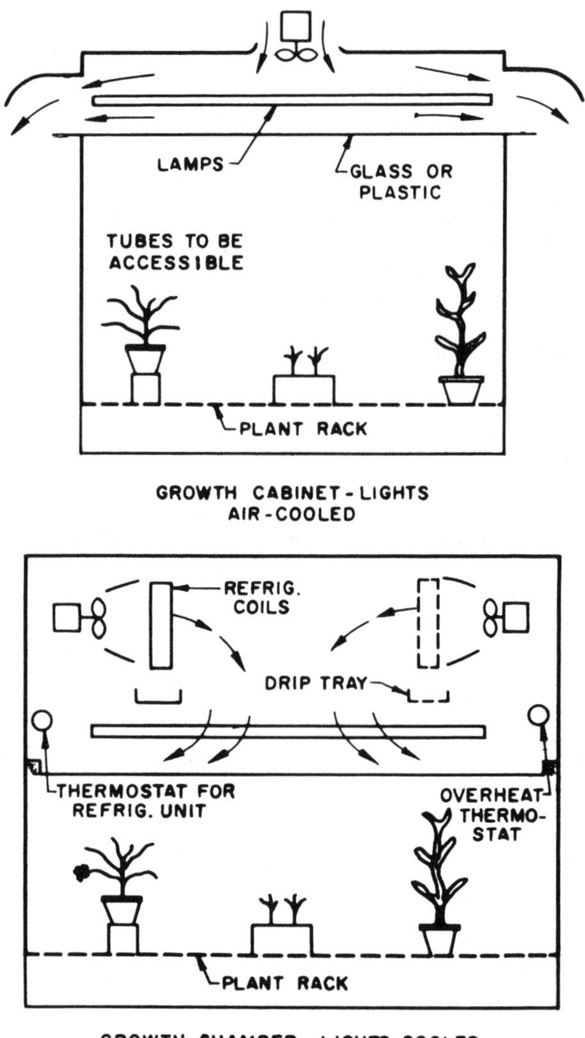

Fig. 12 Cooling Lamps in Growth Chambers

Table 10 Height and Spacing of Luminaires

Light Sources	Radiant Flux Density, W/ft²						
	0.03	0.08	0.28	0.84	1.67	2.5	4.6
Fluorescent—Cool White							
40 W single 4 ft lamp, 3.2 klm							
Radiant power, W/m², 400-700 nm	0.3	0.9	2.9	8.8			
Illumination, klx	0.10	0.30	1.0	3.0			
Lamps per 100 ft²	1.1	3.3	11	33	NA	NA	NA
Distance from plants, in.	114	67	36	21			
40 W 2-lamp fixtures (4 ft) 6.4 klm							
Radiant power, W/m², 400-700 nm	0.3	0.9	2.9	8.8			
Illumination, klx	0.10	0.30	1.0	3.0			
Fixtures per 100 ft²	0.6	1.7	5.5	16.7	NA	NA	NA
Distance from plants, in.	161	94	51	30			
215 Watt 2-8 ft lamps 31.4 klm							
Radiant power, W/m², 400-700 nm	0.3	0.9	2.9	8.8	17.6	23.5	49.0
Illumination, klx	0.10	0.30	1.0	3.0	6.0	8.0	16.7
Fixtures per 100 ft²	0.1+	0.4	1.2	3.6	7.1	9.3	20
Distance from plants, in.	346	201	110	63	43	39	28
High Intensity Discharge							
Mercury-1 400 W parabolic reflector							
Radiant power, W/m², 400-700 nm	0.28	0.84	2.80	8.39	16.8	22.4	46.6
Illumination, klx	0.1	0.32	1.1	3.2	6.4	8.6	18.0
Lamps per 100 ft²	0.2	0.5	1.6	4.8	9.3	13.0	27
Distance from plants, in.	299	173	94	55	39	31	24
Metal halide-1 400 W							
Radiant power, W/m², 400-700 nm	0.77	0.80	2.68	8.03	16.1	21.4	44.6
Illumination, klx	0.09	0.26	0.88	2.6	5.3	7.0	15.0
Lamps per 100 ft²	0.09	0.2	0.7	2.2	4.4	5.8	12.0
Distance from plants, in.	445	256	142	83	59	51	34
High-pressure sodium 400 W							
Radiant power, W/m², 400-700 nm	0.22	0.65	2.18	6.52	13.0	17.4	36.2
Illumination, klx	0.09	0.27	0.89	2.7	5.3	7.1	15.0
Lamps per 100 ft²	0.05	0.14	0.5	1.4	2.8	3.6	7.6
Distance from plants, in.	559	323	177	102	71	63	43
Low-pressure sodium 180 W							
Radiant power, W/m², 400-700 nm	0.26	0.79	2.64	7.93	15.9	21.1	44.0
Illumination, klx	0.14	0.41	1.4	4.1	8.3	11.0	23.0
Lamps per 100 ft²	0.08	0.24	0.8	2.4	4.9	6.5	13.6
Distance from plants, in.	421	244	134	9	55	47	33
Incandescent							
Incandescent 100 W							
Radiant power, W/m², 400-700 nm	0.14	0.41	1.38	4.14	8.28	11.0	23.0
Illumination, klx	0.033	0.10	0.33	1.0	2.0	2.7	5.6
Lamps per 100 ft²	0.5	1.6	5.2	15.8	32	42	87
Distance from plants, in.	165	94	51	30	21	18	13
Incandescent 150 W flood							
Radiant power, W/m², 400-700 nm	0.14	0.41	1.38	4.14	8.28	11.0	23.0
Illumination, klx	0.033	0.098	0.33	1.0	2.0	2.6	5.5
Lamps per 100 ft²	0.3	0.9	3.3	9.3	19.5	26	54
Distance from plants, in.	212	122	67	39	28	24	16
Incandescent-Hg 160 W							
Radiant power, W/m², 400-700 nm	0.14	0.41	1.38	4.14	8.28	11.0	23.0
Illumination, klx	0.050	0.15	0.50	1.5	3.0	4.0	8.3
Lamps per 100 ft²	0.7	2.0	6.9	20.4	42	56	111
Distance from plants, in.	146	83	47	26	18	16	11
Sunlight							
Radiant power, W per 100 ft²	2.0	6.2	20.5	61.7	124	164	714
Illumination, klx	0.054	0.16	0.54	1.6	3.2	4.3	8.9

Transparent glass barriers remove nearly all radiation from about 350 to 2500 nm. Rigid plastic is less effective than glass; however, the lighter weight and lower risk due to breakage makes it a popular barrier material. Ultraviolet is also screened by both glass and plastic (more by plastic). Special UV-transmitting plastic (which degrades rapidly) can be obtained if the biological process requires UV light. When irradiance is very high, especially from HID lamps or large numbers of incandescent lamps or both, rigid plastic can soften from the heat and fall from the supports. Further, very high irradiance and resulting high temperature can cause the plastic to darken, which increases absorptivity and temperature enough to destroy it. Under these conditions, heat-resistant glass may be necessary. The lamp compartment and barrier absolutely require positive ventilation regardless of the light source, and the lamp loft should have limit switches that will shut down the lamps if the temperature rises to a critical level.

OTHER PLANT ENVIRONMENTAL FACILITIES

Plants may be held or processed in warehouse-type structures prior to sale or use in interior landscaping. Required temperatures range from slightly above freezing for cold storage of root stock and cut flowers, to 68 to 77 °F for maintaining growing plants, usually in pots or containers. Provision must be made for venting fresh air to avoid CO_2 depletion.

Light duration must be controlled by a time clock. When they are in use, lamps and ballasts produce almost all heat required in an insulated building. Ventilation and cooling may be required. Illumination levels depend on plant requirements. Table 11 shows approximate mounting heights for two levels of illumination. Luminaires mounted on chains permit lamp height to be adjusted to compensate for varying plant heights.

Table 11 Mounting Height for Luminaires in Storage Areas

	Survival = 0.3 W/ft² Distance		Maintenance = 0.8 W/ft² Distance	
	ft	lux	ft	lux
Fluorescent (F)				
CWF 2-40 W	3.0	1000	2.5	3000
WWF	3.0	1000	2.5	3000
CWF 2-215 W	9.2	1000	5.2	3000
Discharge (HID)				
MH 400 W	10.8	800	6.6	2400
HPS 400 W	14.8	800	8.2	2400
LPS 180 W	11.2	1300	3.9	4000
Incandescent (INC)				
INC 160 W	4.3	350	1.0	1000
INC-HG 160 W	3.9	500	5.2	1500
DL	—	500	—	1500

The main concerns for interior landscape lighting are how it renders the color of plants, people, and furnishings, as well as the minimum irradiation requirements of plants. The temperature required for human occupancy is normally acceptable for plants. Light level and duration determine the types of plants that can be grown or maintained. Plants may be grouped into three levels based on level of irradiance. Plants grow when exposed to higher levels, but do not survive below the suggested minimum levels. Irradiance levels are as follows:

Low (Survival): A minimum light level of 0.07 W/ft² and a preferred level of 0.3 W/ft² irradiance for 8 to 12 h daily.

Medium (Maintenance): A minimum of 0.3 W/ft² and a preferred level of 0.8 W/ft² irradiance for 8 to 12 h daily.

High (Propagation): A minimum of 0.8 W/ft² and a preferred level of 2.2 W/ft² irradiance for 8 to 12 h daily.

Fluorescent (warm-white), metal halide, or incandescent lighting is usually chosen for public places. Table 10 lists irradiance levels with various light sources.

BIBLIOGRAPHY

ANIMALS

Handbooks and Proceedings

Albright, L.D. 1990. *Environment control for animals and plants, with computer applications.* American Society of Agricultural Engineers, St. Joseph, MI.

ASAE. 1982. *Dairy housing* II. ASAE Publication 4:83. Second National Dairy Housing Conference Proceedings, Madison, WI.

ASAE. 1982. *Livestock environment* II. ASAE Publication 3:82. Second International Livestock Environment Symposium, Ames, IA.

ASAE. 1988. *Livestock environment* III. Proceedings of the Third International Livestock Environment Symposium. American Society of Agricultural Engineers Publication No. 01-88.

ASAE. 1989. Design of ventilation systems for livestock and poultry shelters. American Society of Agricultural Engineers *Standard* ASAE D272.4.

Christianson, L.L., ed. 1989. *Building systems: Room air and air contaminant distribution.* American Society of Heating, Refrigerating and Air-Conditioning Engineers, Atlanta.

Christianson, L.L. and R.L. Fehr. 1983. Ventilation—Energy and economics. In "Ventilation of Agricultural Structures." ASAE, St. Joseph, MI, 335-49.

Curtis, S.E. 1983. *Environmental management in animal agriculture.* Iowa State University Press, Ames, IA.

Curtis, S.E., ed. 1988. Guide for the care and use of agricultural animals in agricultural research and teaching. Consortium for Developing a Guide for the Care and Use of Agricultural Animals in Agricultural Research and Teaching, 309 W. Clark Street, Champaign, IL 61820.

Hahn, G.L. 1985. "Management and housing of farm animals in hot environments." In *Stress physiology in livestock*, Vol. II. M.K. Yousef, ed. CRC Press, Boca Raton, FL, 151-76.

Hellickson, M.A. and J.N. Walker, eds. 1983. Ventilation of agricultural structures. ASAE *Monograph* No. 6. American Society of Agricultural Engineers, St. Joseph, MI.

HEW. 1978. Guide for the care and use of laboratory animals. U.S. Department of Health, Education and Welfare. Publication No. (NIH)78-23.

MWPS. 1983. *Structures and environment handbook.* Midwest Plan Service, MWPS-1, 11th ed. Ames, IA.

MWPS. 1990. *Ventilation handbook series.* Midwest Plan Service, Ames, IA.

Rechcigl, M., Jr., ed. 1982. *Handbook of agricultural productivity.* Vol. II, Animal productivity. CRC Press, Boca Raton, FL.

Air Cooling

Canton, G.H., D.E. Buffington, and R.J. Collier. 1982. Inspired-air cooling for dairy cows. *Transactions of* ASAE 25(3):730-34.

Hahn, G.L. and D.D. Osburn. 1969. Feasibility of summer environmental control for dairy cattle based on expected production losses. *Transactions of* ASAE 12(4):448-51.

Hahn, G.L. and D.D. Osburn. 1970. Feasibility of evaporative cooling for dairy cattle based on expected production losses. *Transactions of* ASAE 12(3):289-91.

Heard, L., D. Froelich, L. Christianson, R. Woerman, and R. Witmer. 1986. Snout cooling effects on sows and litters. *Transactions of* ASAE 29(4):1097-1101.

Kimball, B.A., D.S. Benham, and F. Wiersma. 1977. Heat and mass transfer coefficients for water and air in aspen excelsior pads. *Transactions of* ASAE 20(3):509.

Morrison, S.R., H. Heitman, Jr., and R.L. Givens. 1979. Effect of air movement and type of slotted floor on sprinkled pigs. *Tropical Agriculture* 56(3):257.

Morrison, S.R., M. Prokop, and G.P. Lofgreen. 1981. Sprinkling cattle for heat stress relief: Activation, temperature, duration of sprinkling, and pen area sprinkled. *Transactions of* ASAE 24(5):1299-1300.

Stewart, R.E., *et al.* 1966. Field tests of summer air conditioning for dairy cattle in Ohio. ASHRAE *Transactions* 72(1):271.

Timmons, M.B. and G.R. Baughman. 1983. Experimental evaluation of poultry mist-fog systems. *Transactions of* ASAE 26(1):207-10.

Timmons, M.B. and G.R. Baughman. 1984. A plenum concept applied to evaporative pad cooling for broiler housing. *Transactions of* ASAE 27:(6):1877-81.

Wilson, J.L., H.A. Hughes, and W.D. Weaver, Jr. 1983. Evaporative cooling with fogging nozzles in broiler houses. *Transactions of* ASAE 26(2):557-61.

Air Pollution in Buildings

Avery, G.L., G.E. Merva, and J.B. Gerrish. 1975. Hydrogen sulphide production in swine confinement units. *Transactions of* ASAE 18(1):149.

Bundy, D.S. and T.E. Hazen. 1975. Dust levels in swine confinement systems associated with different feeding methods. *Transactions of* ASAE 18(1):137.

Deboer, S. and W.D. Morrison. 1988. *The effects of the quality of the environment in livestock buildings on the productivity of swine and safety of humans—A literature review.* Department of Animal and Poultry Science, University of Guelph, Ontario, Canada.

Grub, W., C.A. Rollo, and J.R. Howes. 1965. Dust problems in poultry environment. *Transactions of* ASAE 8(3):338.

Logsdon, R.F. 1965. Methods of air filtration and protection of air-tempering equipment. *Transactions of* ASAE 8(3):345.

Van Wicklen, G. and L.D. Albright. 1982. An empirical model of respirable aerosol concentration in an enclosed calf barn. Proceedings of the Second International Livestock Environment Symposium. American Society of Agricultural Engineers, St. Joseph, MI, 534-39.

Effects of Environment on Production and Growth of Animals

Cattle

Anderson, J.F., D.W. Bates, and K.A. Jordan. 1978. Medical and engineering factors relating to calf health as influenced by the environment. *Transactions of* ASAE 21(6):1169.

Berry, I.L., M.C. Shanklin, and H.D. Johnson. 1964. Dairy shelter design based on milk production decline as affected by temperature and humidity. *Transactions of* ASAE 7:329-33.

Garrett, W.N. 1980. Factors influencing energetic efficiency of beef production. *Journal of Animal Science* 51(6):1434.

Gebremedhin, K.G., C.O. Cramer, and W.P. Porter. 1981. Predictions and measurements of heat production and food and water requirements of Holstein calves in different environments. *Transactions of* ASAE 24(3):715.

Gebremedhin, K.G., W.P. Porter, and C.O. Cramer. 1983. Quantitative analysis of heat exchange through the fur layer of Holstein calves. *Transactions of* ASAE 28(1):188-93.

Holmes, C.W. and N.A. McLean. 1975. Effects of air temperature and air movement on the heat produced by young Friesian and Jersey calves, with some measurements of the effects of artificial rain. *New Zealand Journal of Agricultural Research* 18(3):277.

Morrison, S.R., G.P. Lofgreen, and R.L. Givens. 1976. Effect of ventilation rate on beef cattle performance. *Transactions of* ASAE 19(3):530.

Morrison, S.R. and M. Prokop. 1983. Beef cattle performance on slotted floors: Effect of animal weight on space allotment. *Transactions of* ASAE 26(2):525-28.

Webster, A.J.F., J.G. Gordon, and J.S. Smith. 1976. Energy exchanges of veal calves in relation to body weight, food intake and air temperature. *Animal Production* 23(1):35.

Phillips, P.A. and F.V. MacHardy. 1983. Predicting heat production in young calves housed at low temperature. *Transactions of* ASAE 26(1):175-78.

General

Hahn, G.L. 1982. Compensatory performance in livestock: Influences on environmental criteria. Proceedings of the Second International Livestock Environment Symposium. American Society of Agricultural Engineers, St. Joseph, MI, 285-94.

Hahn, G.L. 1981. Housing and management to reduce climatic impacts on livestock. *Journal of Animal Science* 52(1):175-86.

Pigs

Boon, C.R. 1982. The effect of air speed changes on the group postural behaviour of pigs. *Journal of Agricultural Engineering Research* 27(1):71-79.

Bruce, J.M. and J.J. Clark. 1971. Models of heat production and critical temperature for growing pigs. *Animal Production* 13(2):285.

Christianson, L.L., D.P. Bane, S.E. Curtis, W.F. Hall, A.J. Muehling, and G.L. Riskowski. 1989. *Swine care guidelines for prot producers using environmentally controlled housing.* National Pork Producers Council, Des Moines, IA.

Close, W.H., L.E. Mount, and I.B. Start. 1971. The influence of environmental temperature and plane of nutrition on heat losses from groups of growing pigs. *Animal Production* 13(2):285.

Driggers, L.B., C.M. Stanislaw, and C.R. Weathers. 1976. Breeding facility design to eliminate effects of high environmental temperatures. *Transactions of* ASAE 19(5):903.

Holmes, C.W. and N.A. McLean. 1977. The heat production of groups of young pigs exposed to reflective and non-reflective surfaces on walls and ceilings. *Transactions of* ASAE 20(3):527.

McCracken, K.J., B.J. Caldwell, and N. Walker. 1979. A note on the performance of early-weaned pigs. *Animal Production* 29(3):423.

McCracken, K.J. and R. Gray. 1984. Further studies on the heat production and affective lower critical temperature of early-weaned pigs under commercial conditions of feeding and management. *Animal Production* 39:283-90.

Morrison, S.R., H. Heitman, Jr., and R.L. Givens. 1979. Effect of air movement and type of slotted floor on sprinkled pigs. *Tropical Agriculture* 56(3):257.

Mount, L.E. and I.B. Start. 1980. A note on the effects of forced air movement and environmental temperature on weight gain in the pig after weaning. *Animal Production* 30(2):295.

Nienaber, J.A. and G.L. Hahn. 1988. Environmental temperature influences on heat production of ad-lib-fed nursery and growing-finishing swine. Livestock Environment III. American Society of Agricultural Engineers, St. Joseph, MI, 73-78.

Phillips, P.A., B.A. Young, and J.B. McQuitty. 1982. Liveweight, protein deposition and digestibility responses in growing pigs exposed to low temperature. *Canadian Journal of Animal Science* 62:95-108.

Phillips, P.A. and F.V. MacHardy. 1982. Modelling protein and lipid gains in growing pigs exposed to low temperature. *Canadian Journal of Animal Science* 62:109-21.

Riskowski, G.L. and D.S. Bundy. 1988. Effects of air velocity and temperature on weanling pigs. *Livestock Environment* III. American Society of Agricultural Engineers, St. Joseph, MI, 117-24.

Poultry

Buffington, D.E., K.A. Jordan, W.A. Junnila, and L.L. Boyd. 1974. Heat production of active, growing turkeys. *Transactions of* ASAE 17(3):542.

Carr, L.E., T.A. Carter, and K.E. Felton. 1976. Low temperature brooding of broilers. *Transactions of* ASAE 19(3):553.

Riskowski, G.L., J.A. DeShazer, and F.B. Mather. 1977. Heat losses of white leghorn laying hens as affected by intermittent lighting schedules. *Transactions of* ASAE 20(4):727-31.

Siopes, T.D., M.B. Timmons, G.R. Baughman, and C.R. Parkhurst. 1983. The effect of light intensity on the growth performance of male turkeys. *Poultry Science* 62:2336-42.

Sheep

Schanbacher, B.D., G.L. Hahn, and J.A. Nienaber. 1982. Photoperiodic influences on performance of market lambs. Proceedings of the Second International Livestock Environment Symposium. American Society of Agricultural Engineers, St. Joseph, MI, 400-405.

Vesely, J.A. 1978. Application of light control to shorten the production cycle in two breeds of sheep. *Animal Production* 26(2):169.

Modeling and Analysis

Albright, L.D. and N.R. Scott. 1974a. An analysis of steady periodic building temperature variations in warm weather—Part I: A mathematical model. *Transactions of* ASAE 17(1):88-92, 98.

Albright, L.D. and N.R. Scott. 1974b. An analysis of steady periodic building temperature variations in warm weather—Part II: Experimental verification and simulation. *Transactions of* ASAE 17(1):93-98.

Albright, L.D. and N.R. Scott. 1977. Diurnal temperature fluctuations in multi-air spaced buildings. *Transactions of* ASAE 20(2):319-26.

Bruce, J.M. and J.J. Clark. 1979. Models of heat production and critical temperature for growing pigs. *Animal Production* 28:353-69.

Buffington, D.E. 1978. Simulation models of time-varying energy requirements for heating and cooling buildings. *Transactions of* ASAE 21(4):786.

Christianson, L.L. and H.A. Hellickson. 1977. Simulation and optimization of energy requirements for livestock housing. *Transactions of* ASAE 20(2):327-35.

Ewan, R.C. and J.A. DeShazer. 1988. Mathematical modeling the growth of swine. *Livestock Environment* III. American Society of Agricultural Engineers, St. Joseph, MI, 211-18.

Hellickson, M.L., K.A. Jordan, and R.D. Goodrich. 1978. Predicting beef animal performance with a mathematical model. *Transactions of* ASAE 21(5):938-43.

Teter, N.C., J.A. DeShazer, and T.L. Thompson. 1973. Operational characteristics of meat animals—Part I: Swine; Part II: Beef; Part III: Broilers. *Transactions of* ASAE 16:157-59; 740-42; 1165-67.

Timmons, M.B. 1984. Use of physical models to predict the fluid motion in slot-ventilated livestock structures. *Transactions of* ASAE 27(2):502-507.

Timmons, M.B., L.D. Albright, and R.B. Furry. 1978. Similitude aspects of predicting building thermal behavior. *Transactions of* ASAE 21(5):957.

Timmons, M.B., L.D. Albright, R.B. Furry, and K.E. Torrance. 1980. Experimental and numerical study of air movement in slot-ventilated enclosures. ASHRAE *Transactions* 86(1):221-40.

Shades for Livestock

Bedwell, R.L. and M.D. Shanklin. 1962. Influence of radiant heat sink on thermally-induced stress in dairy cattle. Missouri Agricultural Experiment Station Research Bulletin No. 808.

Bond, T.E., L.W. Neubauer, and R.L. Givens. 1976. The influence of slope and orientation of effectiveness of livestock shades. *Transactions of* ASAE 19(1):134-37.

Roman-Ponce, H., W.W. Thatcher, D.E. Buffington, C.J. Wilcox, and H.H. VanHorn. 1977. Physiological and production responses of dairy cattle to a shade structure in a subtropical environment. *Journal of Dairy Science* 60(3):424.

Transport of Animals

Ashby, B.H., D.G. Stevens, W.A. Bailey, K.E. Hoke, and W.G. Kindya. 1979. *Environmental conditions on air shipment of livestock.* USDA, SEA, Advances in Agricultural Technology, Northeastern Series No. 5.

Ashby, B.H., A.J. Sharp, T.H. Friend, W.A. Bailey, and M.R. Irwin. 1981. Experimental railcar for cattle transport. *Transactions of* ASAE 24(2):452.

Ashby, B.H., H. Ota, W.A. Bailey, J.A. Whitehead, and W.G. Kindya. 1980. Heat and weight loss of rabbits during simulated air transport. *Transactions of* ASAE 23(1):162.

Grandin, R. 1988. *Livestock trucking guide.* Livestock Conservation Institute, Madison, WI.

Jackson, W.T. 1974. Air transport of Hereford cattle to the People's Republic of China. *Veterinary Records* 9(1):209.

Scher, S. 1980. Lab animal transportation receiving and quarantine. *Lab Animal* 9(3):53.

Stermer, R.A., T.H. Camp, and D.G. Stevens. 1982. Feeder cattle stress during handling and transportation. *Transactions of* ASAE 25(1):246-48.

Stevens, D.G., G.L. Hahn, T.E. Bond, and J.H. Langridge. 1974. *Environmental considerations for shipment of livestock by air freight.* USDA, APHIS (May).

Stevens, D.G. and G.L. Hahn. 1981. Minimum ventilation requirement for the air transportation of sheep. *Transactions of* ASAE 24(1):180.

Laboratory Animals

NIH. 1978. Laboratory animal housing. Proceedings of a symposium held at Hunt Valley, MD, September 1976. National Academy of Sciences, Washington, D.C., 220.

McSheehy, T. 1976. *Laboratory animal handbook* 7—Control of the animal house environment. Laboratory Animals, Ltd.

Soave, O., W. Hoag, *et al.* 1980. The laboratory animal data bank. *Lab Animal* 9(5):46.

Ventilation Systems

Albright, L.D. 1976. Air flows through hinged-baffle, slotted inlets. *Transactions of* ASAE 19(4):728, 732, 735.

Albright, L.D. 1978. Air flow through baffled, center-ceiling, slotted inlets. *Transactions of* ASAE 21(5):944-47, 952.

Albright, L.D. 1979. Designing slotted inlet ventilation by the systems characteristic technique. *Transactions of* ASAE 22(1):158.

Goetsch, W.D., D.P. Stombaugh, and A.T. Muehling. 1984. *Earth-tube heat exchange systems.* AED-25, Midwest Plan Service, Ames, IA.

McGinnis, D.S., J.R. Ogilvie, D.R. Pattie, K.W. Blenkhorn, and J.E. Turnbull. 1983. Shell-and-tube heat exchanger for swine buildings. *Canadian Agricultural Engineer* 25(1):69-74.

Person, H.L., L.D. Jacobson, and K.A. Jordan. 1979. Effect of dirt, louvers and other attachments on fan performance. *Transactions of* ASAE 22(3):612-16.

Pohl, S.H. and M.A. Hellickson. 1978. Model study of five types of manure pit ventilation systems. *Transactions of* ASAE 21(3):542.

Randall, J.M. 1975. The prediction of air flow patterns in livestock buildings. *Journal of Agricultural Engineering Research* 20(2):199-215.

Randall, J.M. 1980. Selection of piggery ventilation systems and penning layouts based on the cooling effects of air speed and temperature. *Journal of Agricultural Engineering Research* 25(2):169-87.

Randall, J.M. and V.A. Battams. 1979. Stability criteria for air flow patterns in livestock buildings. *Journal of Agricultural Engineering Research* 24(4):361-74.

Sokhansanj, S., K.A. Jordan, L.A. Jacobson, and G.L. Messers. 1980. Economic feasibility of using heat exchangers in ventilation of animal buildings. *Transactions of* ASAE 23(6):1525-28.

Spengler, R.W. and D.P. Stombaugh. 1983. Optimization of earth-tube exchangers for winter ventilation of swine housing. *Transactions of* ASAE 26(4):1186-93.

Timmons, M.B. 1984. Internal air velocities as affected by the size and location of continuous inlet slots. *Transactions of* ASAE 27(5):1514-17.

Timmons, M.B. and G.R. Baughman. 1983. The FLEX house: A new concept in poultry housing. *Transactions of* ASAE 26(2):529-32.

Witz, R.L., G.L. Pratt, and M.L. Buchanan. 1976. Livestock ventilation with heat exchanger. *Transactions of* ASAE 19(6):1187.

Alarm Systems

Clark, W.D. and G.L. Hahn. 1971. Automatic telephone warning systems for animal and plant laboratories or production systems. *Journal of Dairy Science* 54(k6):932-35.

Natural Ventilation

Bruce, J.M. 1982. Ventilation of a model livestock building by thermal buoyancy. *Transactions of* ASAE 25(6):1724-26.

Jedele, D.G. 1979. Cold weather natural ventilation of buildings for swine finishing and gestation. *Transactions of* ASAE 22(3):598-601.

Timmons, M.B. and G.R. Baughman. 1981. Similitude analysis of ventilation by the stack effect from an open ridge livestock structure. *Transactions of* ASAE 24(4):1030-34.

Timmons, M.B., R.W. Bottcher, and G.R. Baughman. 1984. Nomographs for predicting ventilation by thermal buoyancy. *Transactions of* ASAE 27(6):1891-93.

PLANTS

Greenhouse and Plant Environment

Albright, L.D. 1990. Environment control for animals and plants, with computer applications. American Society of Agricultural Engineers, St. Joseph, MI.

Aldrich, R.A. and J.W. Bartok. 1984. *Greenhouse engineering.* Department of Agricultural Engineering, University of Connecticut, Storrs, CT.

ASAE. 1989. Engineering practice EP411: Guidelines for measuring and reporting environmental parameters for plant experiments in growth chambers. ASAE *Standards* 1986, American Society of Agricultural Engineers, St. Joseph, MI.

Clegg, P. and D. Watkins. 1978. *The complete greenhouse book.* Garden Way Publishing, Charlotte, VT.

Downs, R.J. 1975. *Controlled environments for plant research.* Columbia University Press, New York.

Hellmers, H. and R.J. Downs. 1967. Controlled environments for plant-life research. ASHRAE *Journal* (February):37.

Hellickson, M. and J. Walker. 1983. Ventilation of agricultural structures. *Monograph* 6, American Society of Agricultural Engineers, St. Joseph, MI.

Langhans, R.W. 1985. *Greenhouse management.* Halcyon Press, Ithaca, NY.

Mastalerz, J.W. 1977. *The greenhouse environment.* John Wiley & Sons, New York.

Nelson, P.V. 1978. *Greenhouse operation and management.* Reston Publishing Co., Reston, VA.

Pierce, J.H. 1977. *Greenhouse grow how.* Plants Alive Books, Seattle, WA.

Riekels, J.W. 1977. *Hydroponics.* Ontario Ministry of Agriculture and Food, Fact Sheet No. 200-24, Toronto, Ontario, Canada.

Riekels, J.W. 1975. *Nutrient solutions for hydroponics.* Ontario Ministry of Agriculture and Food, Fact Sheet No. 200-532, Toronto, Ontario, Canada.

Sheldrake, R., Jr. and J.W. Boodley. *Commercial production of vegetable and flower plants.* Research Park, 1B-82, Cornell University, Ithaca, NY.

Tibbitts, T.W. and T.T. Kozlowski, eds. 1979. *Controlled environment guidelines for plant research.* Academic Press, New York.

Light and Radiation

Armitage, A.M. and M.J. Tsugita. 1974. The effect of supplemental lights source, illumination, and quantum flux density on the flowering of seed propagated geraniums. *Journal of the American Society for Horticultural Science* 54:195.

Bickford, E.D. and S. Dunn. 1972. *Lighting for piant growth.* Kent State University Press, Kent, OH.

Boodley, J.W. 1970. Artificial light sources for gloxinia, African violet, and tuberous begonia. *Plants & Gardens* 26:38.

Carpenter, G.C. and L.J. Mousley. 1960. The artificial illumination of environmental control chambers for plant growth. *Journal of Agricultural Engineering Research* [England] 5:283.

Carpenter, G.A., L.J. Mousley, and P.A. Cottrell. 1964. Maintenance of constant light intensity in plant growth chambers by group replacement of lamps. *Journal of Agricultural Engineering Research* 9.

Campbell, L.E., R.W. Thimijan, and H.M. Cathey. 1975. Special radiant power of lamps used in horticulture. *Transactions of* ASAE 18(5):952.

Cathey, H.M. and L.E. Campbell. 1974. Lamps and lighting: A horticultural view. *Lighting Design & Application* 4:41.

Cathey, H.M. and L.E. Campbell. 1975. Effectiveness of five vision lighting sources on photo-regulation of 22 species of ornamental plants. *Journal of the American Society for Horticultural Science* 100(1):65.

Cathey, H.M. and L.E. Campbell. 1979. Relative efficiency of high- and low-pressure sodium and incandescent filament lamps used to supplement natural winter light in greenhouses. *Journal of the American Society for Horticultural Science* 104(6):812.

Cathey, H.M. and L.E. Campbell. 1980. Light and lighting systems for horticultural plants. *Horticultural Reviews* 11:491. AVI Publishing Co., Westport, CT.

Cathey, H.M., L.E. Campbell, and R.W. Thimijan. 1978. Comparative development of 11 plants grown under various fluorescent lamps and different duration of irradiation with and without additional incandescent lighting. *Journal of the American Society for Horticultural Science* 103:781.

Fonteno, W.C. and E.L. McWilliams. 1978. Light compensation points and acclimatization of four tropical foliage plants. *Journal of the American Society for Horticultural Science* 103:52.

Hughes, J., M.J. Tsujita, and D.P. Ormrod. 1979. *Commercial applications of supplementary lighting in greenhouses.* Ontario Ministry of Agriculture and Food, Fact Sheet No. 290-717, Toronto, Ontario, Canada.

Kaufman, J.E., ed. 1981. IES *Lighting handbook*, Application Volume. IES, New York.

Kaufman, J.E., ed. 1981. IES *Lighting handbook*, Reference Volume. IES, New York.

Poole, R.T. and C.A. Conover. Influence of shade and nutrition during production and dark storage stimulating shipment on subsequent quality and chlorophyll content of foliage plants. *HortScience* 14:617.

Robbins, F.V. and C.K. Spillman. 1980. Solar energy transmission through two transparent covers. *Transactions of* ASAE 23(5).

Sager, J.C., J.L. Edwards, and W.H. Klein. 1982. Light energy utilization efficiency for photosynthesis. *Transactions of* ASAE 25(6):1737-46.

Photoperiod

Cathey, H.M. and H.A. Borthwick. 1961. Cyclic lighting for controlling flowering of chrysanthemums. *Proceedings of* ASAE 78:545.

Heins, R.D., W.H. Healy, and H.F. Wilkens. 1980. Influence of night lighting with red, far red, and incandescent light on rooting of chrysanthemum cuttings. *HortScience* 15:84.

Piringer, A.A. and H.M. Cathey. 1960. Effect of photoperiod, kind of supplemental light, and temperature on the growth and flowering of petunia plants. *Proceedings of the American Society for Horticultural Science* 76:649.

Carbon Dioxide

Bailey, W.A., *et al.* 1970. CO_2 systems for growing plants. *Transactions of* ASAE 13(2):63.

Gates, D.M. 1968. Transpiration and leaf temperature. *Annual Review of Plant Physiology* 19:211.

Holley, W.D. 1970. CO_2 Enrichment for flower production. *Transactions of* ASAE 13(3):257.

Kretchman, J. and F.S. Howlett. 1970. Enrichment for vegetable production. *Transactions of* ASAE 13(2):252.

Pettibone, C.A., *et al.* 1970. The control and effects of supplemental carbon dioxide in air-supported plastic greenhouses. *Transactions of* ASAE 13(2):259.

Tibbitts, T.W., J.C. McFarlane, D.T. Krizek, W.L. Berry, P.A. Hammer, R.H. Hodgsen, and R.W. Langhans. 1977. Contaminants in plant growth chambers. *Horticulture Science* 12:310.

Watt, A.D. 1971. Placing atmospheric CO_2 in perspective. IEEE *Spectrum* 8(11):59.

Wittwer, S.H. 1970. Aspects of CO_2 enrichment for crop production. *Transactions of* ASAE 13(2):249.

Heating, Cooling, and Ventilation

ASAE. 1989. Engineering practice: EP406: Heating, ventilating and cooling greenhouses. ASAE *Standards*, American Society of Agricultural Engineers, St. Joseph, MI.

Albright, L.D., I. Seginer, L.S. Marsh, and A. Oko. 1985. InSitu thermal calibration of unventilated greenhouses. *Journal of Agricultural Engineering Research* 31(3):265-81.

Buffington, D.E. and T.C. Skinner. 1979. Maintenance guide for greenhouse ventilation, evaporative cooling, and heating systems. Publication No. AE-17, Department of Agricultural Engineering, University of Florida, Gainesville, FL.

Duncan, G.A. and J.N. Walker. 1979. Poly-tube heating ventilation systems and equipment. Publication No. AEN-7, Agricultural Engineering Department, University of Kentucky, Lexington, KY.

Elwell, D.L., M.Y. Hamdy, W.L. Roller, A.E. Ahmed, H.N. Shapiro, J.J. Parker, and S.E. Johnson. 1985. *Soil heating using subsurface pipes.* Department of Agricultural Engineering, Ohio State University, Columbus, OH.

Hanan, J.J. Ozone and ethylene effects on some ornamental plant species. *General Series Bulletin* No. 974. Colorado State University, Fort Collins, CO.

Heins, R. and A. Rotz. 1980. Plant growth and energy savings with infrared heating. *Florists' Review* (October):20.

Kimball, B.A. 1983. *A modular energy balance program including subroutines for greenhouses and other latent heat devices.* Agricultural Research Service, 4331 East Broadway, Phoenix, AZ 85040.

NGMA. *Standards for ventilating and cooling greenhouses.* National Greenhouse Manufacturers' Association, St. Paul, MN.

Roberts, W.J. and D. Mears. 1984. *Floor heating and bench heating extension bulletin for greenhouses.* Department of Agricultural and Biological Engineering, Cook College, Rutgers University, New Brunswick, NJ.

Roberts, W.J. and D. Mears. 1984. *Heating and ventilating greenhouses.* Department of Agricultural and Biological Engineering, Cook College, Rutgers University, New Brunswick, NJ.

Roberts, W.J. and D.R. Mears. 1979. *Floor heating of greenhouses.* Miscellaneous Publication, Rutgers University, New Brunswick, NJ.

Rogers, B.T. 1979. Wood and the curious case of the rock salt greenhouse. ASHRAE *Journal* (November):74 (see also Silverstein).

Rotz, C.A. and R.D. Heins. 1980. An economic comparison of greenhouse heating systems. *Florists' Review* (October):24.

Silverstein, S.D. 1976. Effect of infrared transparency on heat transfer through windows: A clarification of the greenhouse effect. *Science* 193:229 (see also Rogers).

Skinner, T.C. and D.E. Buffington. 1977. Evaporative cooling of greenhouses in Florida. Publication No. AE-14, Department of Agricultural Engineering, University of Florida, Gainesville, FL.

Walker, J.N. 1965. Predicting temperatures in ventilated greenhouses. *Transactions of* ASAE 8(3):445.

Walker, J.N. and G.A. Duncan. 1975. Greenhouse heating systems. Publication No. AEN-31, Agricultural Engineering Department, University of Kentucky, Lexington, KY.

Walker, J.N. and G.A. Duncan. 1979. Greenhouse ventilation systems. Publication No. AEN-30, Agricultural Engineering Department, University of Kentucky, Lexington, KY.

Walker, P.N. 1979. Greenhouse surface heating with power plant cooling water: Heat transfer characteristics. *Transactions of* ASAE 22(6):1370, 1380.

Energy Conservation

Badger, P.C. and H.A. Poole. 1979. *Conserving energy in Ohio greenhouses.* OARDC Special Circular No. 102, Ohio Agricultural Research and Development Center, Ohio State University, Wooster, OH.

Blom, T., J. Hughes, and F. Ingratta. 1978. Energy conservation in Ontario greenhouses. Publication No. 65, Ontario Ministry of Agriculture and Food, Toronto, Ontario, Canada.

Roberts, W.J., J.W. Bartok, Jr., E.E. Fabian, and J. Simpkins. 1985. *Energy conservation for commercial greenhouses.* NRAES-3. Department of Agricultural Engineering, Cornell University, Ithaca, NY.

Solar Energy Use

Albright, L.D., R.W. Langhans, G.B. White, and A.J. Donohoe. 1980. Enhancing passive solar heating of commercial greenhouses. Proceedings of the ASAE National Energy Symposium, St. Joseph, MI.

Albright, L.D., *et al.* 1980. Passive solar heating applied to commercial greenhouses. Acta Horticultura, Publication No. 115, Energy in Protected Civilization.

Cathey, H.M. 1980. Energy-efficient crop production in greenhouses. ASHRAE *Transactions* 86(2):455.

Duncan, G.A., J.N. Walker, and L.W. Turner. 1979. *Energy for greenhouses*, Part I: Energy Conservation. Publication No. AEES-16. College of Agriculture, University of Kentucky, Lexington, KY.

Duncan, G.A., J.N. Walker, and L.W. Turner. 1980. *Energy for greenhouses*, Part II: Alternative Sources of Energy. College of Agriculture, University of Kentucky, Lexington, KY.

Gray, H.E. 1980. Energy management and conservation in greenhouses: A manufacturer's view. ASHRAE *Transactions* 86(2):443.

Roberts, W.J. and D.R. Mears. 1980. Research conservation and solar energy utilization in greenhouses. ASHRAE *Transactions* 86(2):433.

Short, T.H., M.F. Brugger, and W.L. Bauerle. 1980. Energy conservation ideas for new and existing commercial greenhouses. ASHRAE *Transactions* 86(2):448.

Interior Plantscaping

Gaines, R.L. 1980. *Interior plantscaping*. Architectural Record, McGraw-Hill, New York.

Cathey, H.M. and L.E. Campbell. 1978. Indoor gardening artificial lighting, terrariums, hanging baskets and plant selection. USDA *Home and Garden Bulletin* No. 220.

CHAPTER 22

DRYING AND STORING FARM CROPS

PRESERVING the quality of farm crops as they move from the field to the market is critical. Control of the moisture content and temperature during storage is necessary to preserve product quality. Relative humidity and temperature affect mold growth, which is reduced to a minimum if the crop is kept cooler than 50°F and if the relative humidity of the air in equilibrium with the stored crop is less than 60% (Figure 1).

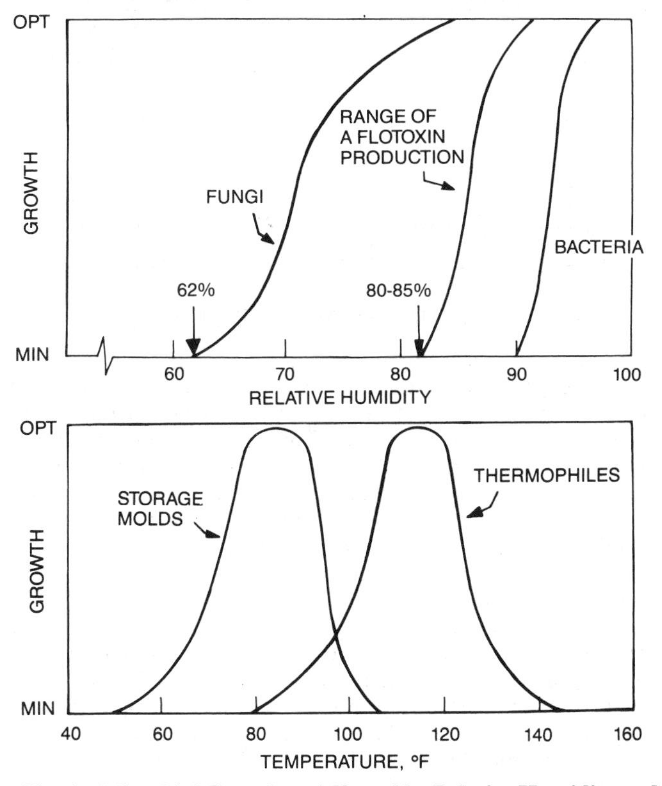

Fig. 1 Microbial Growth as Affected by Relative Humidity and Temperature (University of Kentucky 1973)

Mold growth and spoilage are a function of elapsed storage time, temperature, and moisture content above critical values. The approximate allowable storage life for cereal grains is shown in Table 1. For example, corn at 60°F and 20% moisture has a storage life of about 25 days. If the corn is dried to 18% after 12 days, one-half of its storage life has elapsed. Thus, the remaining storage life at 60°F and 18% moisture content is 25 days, not 50 days.

The preparation of this chapter is assigned to TC 2.2, Plant and Animal Environment.

Table 1 Approximate Allowable Storage Time (Days) for Cereal Grains

Moisture Content, %	Temperature, °F 30	40	50	60	70	80
14	*	*	*	*	200	140
15	*	*	*	240	125	70
16	*	*	230	120	70	40
17	*	280	130	75	45	20
18	*	200	90	50	30	15
19	*	140	70	35	20	10
20	*	90	50	25	14	7
22	190	60	30	15	8	3
24	130	40	15	10	6	2
26	90	35	12	8	5	2
28	70	30	10	7	4	2
30	60	25	5	5	3	1

Based on composite of 0.5% maximum dry matter loss calculated on the basis of USDA research at Iowa State University; *Transactions of* ASAE 333-337, 1972; and "Unheated Air Drying," Manitoba Agriculture Agdex 732-1, rev. 1986.
*Approximate allowable storage time exceeds 300 days.

Insects thrive in stored grain if the moisture content and temperature are too high. At low moisture contents and temperatures less than 50°F, insects remain dormant or die.

Most farm crops must be dried to, and maintained at, a moisture content of 12 to 13%, wet basis, depending on the specific crop, storage temperature, and length of storage. Oil seeds such as peanuts, sunflowers, and flaxseed must be dried to a moisture content of 8 to 9%. Grain stored for more than a year and seed stock should be dried to a lower moisture content. Moisture levels above these critical values lead to the growth of fungi, which may produce toxic compounds.

The maximum yield of dry matter can be obtained by starting harvest when corn has dried in the field to an average moisture content of 26%. Wheat can be harvested when it has dried to 20%. However, harvesting at these moisture contents requires expensive mechanical drying. Although field drying requires less expense than operating drying equipment, it may be more costly because field losses generally increase as the moisture content decreases.

The price of grain to be sold through commercial market channels is based on a specified moisture content, with price discounts for moisture levels above the specified amount. These discounts compensate for the weight of excess water, cover the cost of water removal, and control the supply of wet grain delivered to market. Grain dried below the market standard moisture content (15.5% for corn, 13.0% for soybeans, and 13.5% for wheat) generally does not receive a premium; thus, the seller loses the opportunity to sell water for the price of grain.

Grain Quantity

The bushel is the common measure used for marketing grain in the United States. Most driers are rated in bushels per hour for a specified moisture content reduction. Considerable confusion arises from the use of a bushel as a measure. Bushel is a volume measure equal to 1.244 ft^3. However, the bushel is used as a volume measure only to estimate holding capacity of bins, driers, and other containers.

For buying and selling grain, reporting production and consumption data, and for most other uses, the bushel weight is used. The legal weight of a bushel is set at 56 lb for corn and 60 lb for wheat. When grain is marketed, bushels are computed as the load weight divided by the bushel weight. For example, 56,000 lb of corn (regardless of moisture content) is 1000 bushels. Rice, grain sorghum, and sunflower are more commonly traded on the basis of hundredweight (100 lb), thus separating these measures from the volume connotation. The density of some crops is listed in Table 2.

Table 2 Calculated Densities of Grain and Seeds Based on Weights and Measures Used by the U.S. Department of Agriculture

	Bulk Density lb/ft³
Alfalfa	48.0
Barley	38.4
Beans, dry	48.0
Bluegrass	11.2 to 24.0
Clover	48.0
Corn[a]	
Ear, husked	28.0
Shelled	44.8
Cottonseed	25.6
Oats	25.6
Peanuts, unshelled	
Virginia type	13.6
Runner, Southeastern	16.8
Spanish	19.8
Rice, rough	36.0
Rye	44.8
Sorghum	40.0
Soybeans	48.0
Sudan grass	32.0
Sunflower	
Non-oil	19.2
Oil seed	25.6
Wheat	48.0

[a] 70 lb of husked ears of corn yields 1 bushel, or 56 lb of shelled corn. 70 lb of ears of corn occupies 2 bushels (2.5 ft^3).

Wet bushels and dry bushels sometimes refer to the volume of grain before and after drying. For example, 56,000 lb of 25% moisture corn may be referred to as 1000 wet bushels or simply 1000 bushels. When the corn is dried to 15.5% moisture content (m.c.), only 49,704 lb or 49,704/56 = 888 bushels remain. Thus, a drier rated on the basis of wet bushels (25% m.c.) shows a capacity 12.6% higher than if rated on the basis of dry bushels (15.5% m.c.).

The percent of initial weight lost due to water removed may be calculated by the following equation:

$$\text{Moisture shrink, \%} = \frac{M_o - M_f}{100 - M_f} \times 100$$

where

M_o = original or initial moisture content, wet basis
M_f = final moisture content, wet basis

Applying the formula to drying a crop from 25% to 15%:

$$\text{Moisture shrink} = \frac{25 - 15}{100 - 15} \times 100 = 11.76\%$$

In this case, the moisture shrink is 11.76%, or an average 1.176% weight reduction for each percentage point of moisture reduction. The moisture shrink varies depending on the final moisture content. For example, the average shrink per point of moisture drying from 20% to 10% is 1.111.

Economics

Producers generally have the choice of drying their grain on the farm before delivering it to market or delivering wet grain at a price discount for excess moisture. Costs of drying on the farm consist of fixed costs and variable costs. Once a given size of drier is purchased, depreciation, interest, taxes, and repairs are fixed and are minimally affected by volume. Costs of labor, fuel, and electricity are variable costs that vary directly with the volume dried. Total drying costs vary widely, depending on the volume dried, the drying equipment, and the fuel and equipment prices.

Energy consumption depends primarily on the type of drier. Generally, the faster the drying speed, the greater the energy consumption (Table 3).

Table 3 Estimated Corn Drying Energy Requirement

Drier Type	Btu/lb of Water Removed
Natural air	1000-1200
Low temperature	1200-1500
Batch-in-bin	1500-2000
High temperature	
Air recirculating	1800-2200
Without air recirculating	2000-3000

Note: Includes all energy requirements for fans and heat.

DRYING EQUIPMENT AND PRACTICES

Contemporary crop-drying equipment depends largely on mass and energy transfer between the drying air and the product to be dried. The drying rate depends on the initial temperature and moisture content of the crop, air-circulation rate, entering condition of the circulated air, length of flow path through the products, and elapsed time since the beginning of the drying operation. Outdoor air is frequently heated before it is circulated through the product. Heating increases the heat transfer rate to the product, raises its temperature, and increases the vapor pressure of the product moisture.

Most crop-drying equipment consists of: (1) a fan to move the air through the product; (2) a controlled heater to increase the ambient air temperature to the desired level; and (3) a container to distribute the drying air uniformly through the product. The exhaust air is vented to the atmosphere. Where climate and other factors are favorable, unheated air is used for drying and the heater is omitted.

Fans

The fan selected for a given drying application should meet the same requirements important in any air-moving application. It must deliver the desired amount of air against the static resistance of the product in the bin or column, the resistance of the delivery system, and the resistance of the air inlet and outlet.

Foreign material in the grain can significantly change the required air pressure in the following ways:

- Foreign particles larger than the grain (straw, plant parts, and larger seeds) reduce the airflow resistance.

- Foreign particles smaller than the grain (broken grain, dust, and small seeds) increase the airflow resistance. The effect in either case may be dramatic, even when little foreign material is present. The airflow rate may be increased by as much as 60% or more where large particles are present and by 100% or more where small particles are present.
- The type of filling method or agitation of the grain after it is placed in the drier can increase the pressure requirements up to 100%. High moisture in some grain causes less pressure drop than in dry grain.

Vane-axial fans are normally recommended when static pressures are less than 3 in. of water gage. Backward-curved centrifugal fans are commonly recommended when static pressures are higher than 4 in. of water. Low-speed centrifugal fans operating at 1750 rpm perform well up to about 7 in. of water, and high-speed centrifugal fans operating at about 3500 rpm have the ability to develop up to about 10 in. of water of static pressure. The in-line centrifugal fan has a centrifugal fan impeller mounted in the housing of an axial flow fan. A bell intake funnels the air into the impeller. The in-line centrifugal fan operates at about 3450 rpm and has the ability to develop pressures up to 10 in. of water on 7.5 hp or larger fans.

After functional considerations are made, initial cost of the drier fan should be considered. Drying equipment has a low percentage of annual use in many applications, so the cost of drier ownership per unit of material dried is sometimes greater than the energy cost of operation. Therefore, fan efficiency may be less important than initial cost. The same considerations apply to other components of the drier.

Heaters

Crop drier heaters are fueled by natural gas, liquified petroleum gas, or fuel oil; some electric heaters are also used. Driers using coal, biomass (such as corn cobs, stubble, or wood), and solar energy have also been built.

Fuel combustion in crop driers is similar to combustion in domestic and industrial furnaces. Heat is transferred to the drying air either indirectly by means of a heat exchanger or directly by combining the combustion gases with the drying air. Most grain driers use direct combustion. Indirect heating is used in drying products such as hay and sunflowers, because of their greater fire hazard.

Controls

In addition to the usual temperature controls for drying air, all heated air units must have safety controls similar to those found on space-heating equipment. These safety controls shut off the fuel in case of flame failure and stop the burner in case of overheating or excessive drying air temperatures. All controls should be arranged to operate safely in case of power failure.

SHALLOW LAYER DRYING

Batch Driers

The batch drier cycles through load, dry, cool, and unload of the grain. Fans force hot air through grain columns, which are generally about 12 in. thick. Drying time depends on the type of grain and the amount of moisture to be removed.

A general rule is 10 min per percentage point (percent) of moisture to be removed. Many driers circulate and mix the grain to prevent significant moisture content gradients from forming across the column. A circulation rate that is too fast or a poor selection of handling equipment may cause undue damage and loss of market quality. This type of drier is suitable for farm operations and is often portable.

Continuous Flow Driers

This type of self-contained drier passes a continuous stream of grain through the drying chamber. A second chamber cools the hot, dry grain prior to storage. Handling and storage equipment must be available at all times to move grain to and from the driers. The design of these driers can be classified as crossflow, concurrent flow, and counterflow.

Crossflow Driers. A crossflow drier is a column drier that moves air perpendicular to the grain movement. These driers commonly consist of two or more vertical columns surrounding the drying and cooling air plenums. The columns range in thickness from 8 to 16 in.

Airflow rates fall in the range of 40 to 160 cfm per ft^3 of grain. Thermal efficiency of the drying process increases as column width increases and decreases as airflow rate increases. However, moisture uniformity and drying capacity increase with an increase in airflow rate and a decrease in column width. Drier design obtains a desirable balance of the factors of airflow rate and column width for the expected moisture content levels and drying air temperatures. Performance is evaluated in terms of drying capacity, thermal efficiency, and dried product uniformity.

As with the batch drier, a moisture gradient forms across the column because the grain nearest the inside of the column is exposed to the driest air during the complete cycle. Several methods minimize the problem of uneven drying. One method includes turnflow devices in the columns, which split the grain stream and move the inside half of the column to the outside and the outside half to the inside. Although effective, turnflow devices tend to plug if the grain is trashy. Under these conditions, a scalper/cleaner should be used ahead of a drier to clean the grain. Another method divides the drying chamber into sections and ducts the hot air so that its direction through the grain is reversed in alternate sections. The effect is about the same as the turnflow devices. A third method divides the drying chamber into sections and reduces the drying air temperature in each consecutive section. This method is the least effective.

Rack-Type Driers. In this type of crossflow drier, grain flows over alternating rows of heated air supply ducts and air exhaust ducts (Figure 2). This action mixes the grain and alternates exposure to relatively hot drying air and air cooled by previous contact with the grain, promoting moisture uniformity and equal exposure of the product to the drying air.

Concurrent-Flow Driers. In the concurrent-flow drier, grain and drying air move in the same direction in the drying chamber. The drying chamber is then coupled to a counterflow cooling section. Thus, the hottest air is in contact with the wettest grain, allowing use of higher drying air temperatures (up to 450 °F). Rapid evaporative cooling in the wettest grain prevents the grain temperature

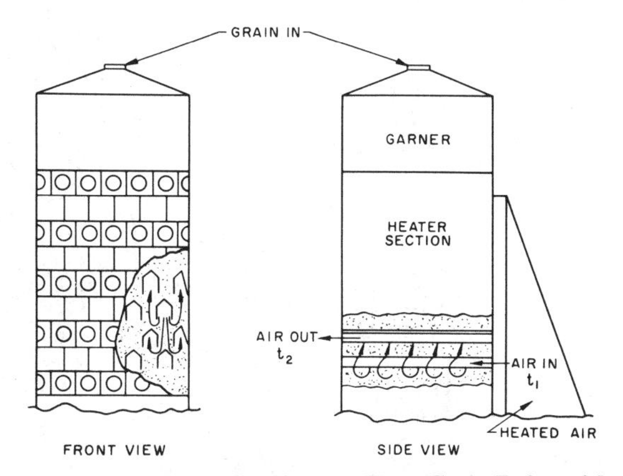

Fig. 2 Rack-Type Continuous-Flow Grain Drier with Alternate Rows of Air Inlet and Outlet Ducts

from reaching excessive levels. Because higher drying air temperatures are used, the energy efficiency is better than that obtained with a conventional crossflow drier. In the cooling section, the coolest air initially contacts the coolest grain. The combination of drying and cooling chambers results in lower thermal stresses in the grain kernels during drying and cooling. The result is a higher quality product.

Counterflow Driers. The grain and drying air move in opposite directions in the drying chamber of this drier. Counterflow is common for in-bin driers. Drying air enters from the bottom of the bin and exits from the top. The wet grain is loaded from overhead, and floor sweep augers can be used to bring the hot, dry grain to a center sump where it is removed by another auger. The travel of the sweep is normally controlled by temperature-sensing elements.

A drying zone exists only in the lower layers of the grain mass and is truncated at its lower edge so that the grain being removed is not overdried. As a part of the counterflow process, the warm, saturated, or near-saturated air leaving the drying zone passes through the cool incoming grain. Some energy is used to heat the cool grain, but some moisture may condense on the cool grain if the bed is deep and the initial grain temperature is low.

Reducing Energy Costs

Recirculation. Most commercially available driers have optional ducting systems to recycle some of the exhaust air from the drying and cooling chambers back to the inlet of the drying chamber (Figure 3). Variations in systems exist, but most make it possible to recirculate all the air from the cooling chamber and the air from the lower two-thirds of the drying chamber. Relative humidity of this air for most crossflow driers is less than 50%. Energy savings of up to 30% can be obtained in a well-designed system.

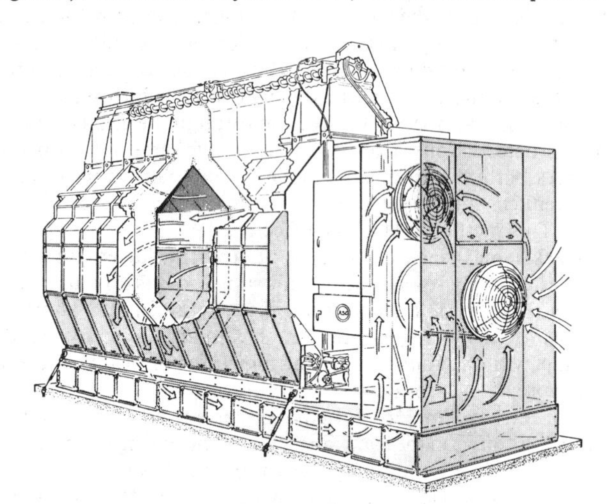

Fig. 3 Crop Drier Recirculation Unit
(Courtesy Farm Fans, Inc.)

Dryeration. This is another means of reducing energy consumption and improving grain quality. In this process, hot grain with a moisture content 1 or 2 percentage points above that desired for storage is removed from the drier (Figure 4). The hot grain is placed in a dryeration bin, where it tempers without airflow for at least 4 to 6 h. After the first grain delivered to the bin has tempered, the cooling fan is turned on as additional hot grain is delivered to the bin. The air cools the grain and removes 1 to 2% moisture before it is moved to final storage. If the cooling rate equals the filling rate, cooling is normally completed about 6 h after the last hot grain is added. The crop cooling rate should equal the filling rate of the dryeration bin. A faster cooling rate cools the grain before it has tempered. A slower rate may result in spoilage, since the allowable storage time of hot damp grain may be only a few days. The required airflow rate is based on drier capacity and crop density. An airflow rate of 12 cfm for each bushel per hour (bu/h) of drier capacity provides cooling capacity to keep up with the drier when drying corn that weighs 56 lb/bu. Recommended airflow rates for some crops are listed in Table 4.

Table 4 Recommended Airflow Rates for Dryeration

Crop	Weight, lb/bu	Recommended Dryeration Airflow Rate, cfm per bu/h
Barley	48	10
Corn	56	12
Durum	60	13
Edible beans	60	13
Flaxseed	56	12
Millet	50	11
Oats	32	7
Rye	56	12
Sorghum	56	12
Soybean	60	13
Non-oil Sunflower	24	5
Oil sunflower	32	7
HRS Wheat	60	13

Note: Basic air volume is 12.9 $ft^3/lb \cdot min$.

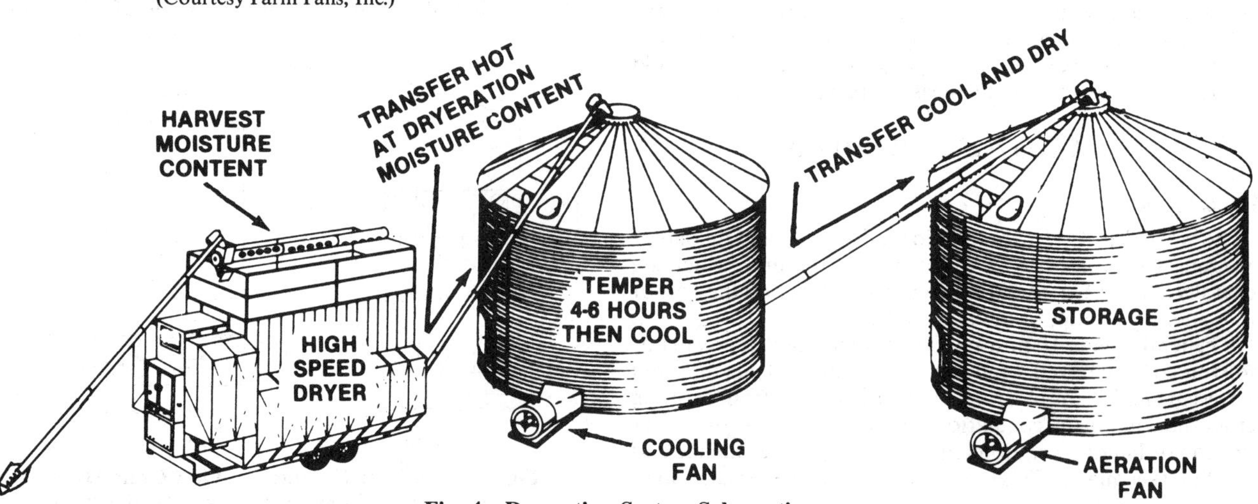

Fig. 4 Dryeration System Schematic

Combination Drying. This method was developed to improve drying thermal efficiency and corn quality. A continuous-flow drier dries corn to 18 to 20% moisture content, at which time it is transferred to a bin. The in-bin drying system then brings the moisture down to a safe storage level. For energy savings, operating temperatures of batch and continuous-flow driers are usually set at the highest level that will not damage the product for its particular end use. The energy requirements of a conventional crossflow drier as a function of drying air temperature and airflow rate are shown in Figure 5.

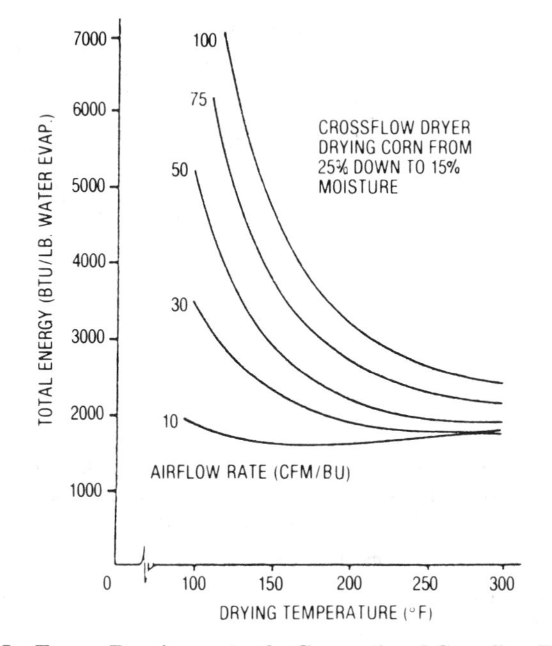

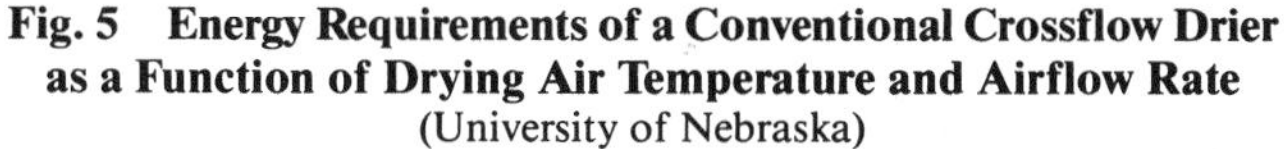

Fig. 5 Energy Requirements of a Conventional Crossflow Drier as a Function of Drying Air Temperature and Airflow Rate
(University of Nebraska)

DEEP BED DRYING

A deep bed drying system can be installed in any structure that holds grain. Most grain storage structures can be designed or adapted for drying by providing a means of distributing the drying air uniformly through the grain. This is most commonly done by either a perforated floor (Figure 6) or by duct systems placed on the floor of the bin (Figure 7).

Perforations in the floor should have a total area of at least 10% of the floor area; 15% is better. The perforated floor distributes air more uniformly and offers less resistance to airflow than do ducts, but the duct system is less expensive for larger floor area systems. Ducts can be removed after the grain is removed, and the structure can be cleaned and used for other purposes. Do not space ducts farther apart than one-half times the depth of the grain. The amount of perforated area and duct length will affect airflow distribution uniformity.

Air ducts and tunnels that disperse air into the grain should be large enough to prevent the air velocity from exceeding 2000 fpm; slower speeds are desirable. Sharp turns, obstructions, or abrupt changes in duct size should be eliminated, since they cause pressure loss. Operating methods for drying grain in storage bins may be classified as full-bin drying, layer drying, and batch drying.

Full-Bin Drying

Full-bin drying is generally done with unheated air or air heated up to 20 °F above ambient. A humidistat is frequently used to sense the humidity of the drying air and to turn off the heater if the

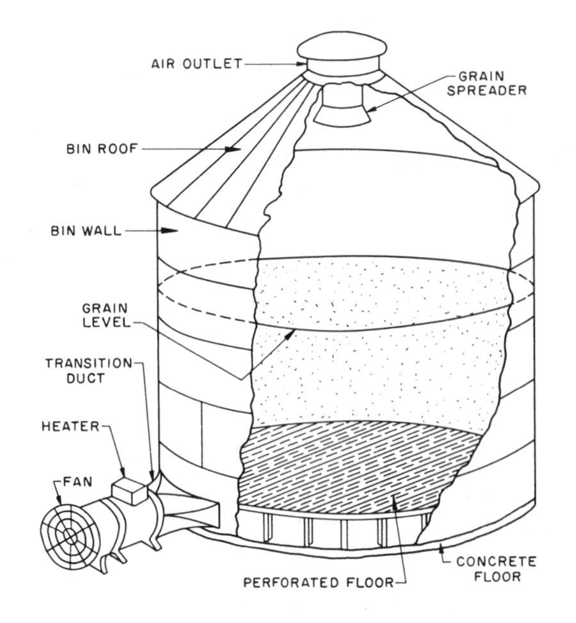

Fig. 6 Perforated Floor System for Bin Drying of Grain

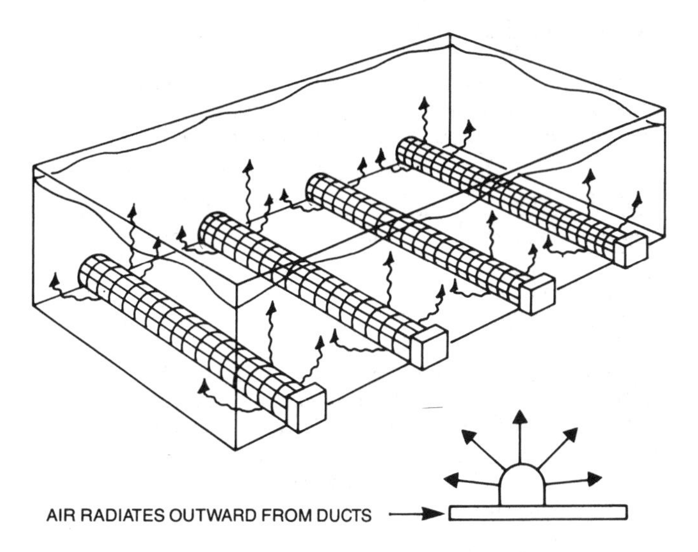

Fig. 7 Tunnel or Duct Air Distribution System

weather conditions are such that heated air would cause overdrying. A humidistat setting of 55% limits drying to approximately the 12% moisture level for most farm grains, assuming that the ambient relative humidity does not go below this point.

Airflow rate requirements for full-bin drying are generally made on the basis of cfm of air required per cubic foot or bushel of grain. The airflow rate recommendations depend on the weather conditions and on the type of grain and its moisture content.

Airflow rate is important for successful drying. Because faster drying results from higher airflow rates, the highest economical airflow rate should be used. The cost of drying should be calculated, since at high airflow rates the cost may exceed the cost of using column driers.

Recommendations for drying with unheated air are shown in Tables 5, 6, and 7. These recommendations apply to the principal production areas of the continental United States and are based on experience under average conditions; they may not be applicable under unusual weather conditions or even usual weather conditions in the case of late maturing crops. Much of the weather

Table 5 Maximum Corn Moisture Contents, Wet Mass Basis, for Single-Fill Natural Air Drying

Zone	Full-Bin Airflow Rate, cfm/bu	Harvest Date 9-1	9-15	10-1	10-15	11-1	11-15	12-1
		Initial Moisture Content, %						
A	1.0	18	19.5	21	22	24	20	18
	1.25	20	20.5	21.5	23	24.5	20.5	18
	1.5	20	20.5	22.5	23	25	21	18
	2.0	20.5	21	23	24	25.5	21.5	18
	3.0	22	22.5	24	25.5	27	22	18
B	1.0	19	20	20	21	23	20	18
	1.25	19	20	20.5	21.5	24	20.5	18
	1.5	19.5	20.5	21	22.5	24	21	18
	2.0	20	21	22.5	23.5	25	21.5	18
	3.0	21	22.5	23.5	24.5	26	22	18
C	1.0	19	19.5	20	21	22	20	18
	1.25	19	20	20.5	21.5	22.5	20.5	18
	1.5	19.5	20	21	22	23.5	21.5	18
	2.0	20	21	22	23	24.5	21.5	18
	3.0	21	22	23.5	24.5	25.5	22	18
D	1.0	19	19.5	20	21	22	20	18
	1.25	19	19.5	20.5	21	22.5	20.5	18
	1.5	19	19.5	21	22	23	21	18
	2.0	19.5	21	21.5	23	24	21.5	18
	3.0	20.5	21.5	23	24	25	22	18

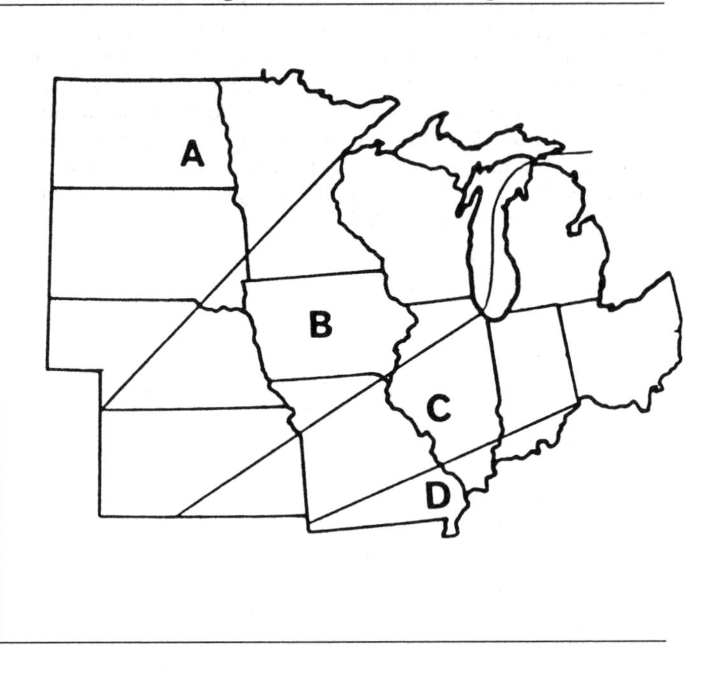

Developed by T.L. Thompson, University of Nebraska.

Table 6 Minimum Airflow Rates for Natural Air Low-Temperature Drying of Small Grains and Sunflower in the Northern Plains of the United States

Airflow Rate cfm/bu	cfm/ft³	Maximum Initial Moisture Content, % Wet Basis: Small Grains	Sunflower
0.5	0.4	16	15
1.0	0.8	18	17
2.0	1.6	20	21

hazard can be removed by the use of supplemental heat. Full-bin drying may not be applicable in some geographical areas.

The depth of grain (distance of air travel) is limited by the cost of the fan, motor, air distribution system, and power required. The maximum practical depth appears to be 20 ft for corn and beans, and about 15 ft for wheat when drying grain.

To ensure satisfactory drying, heated air may be used during periods of prolonged fog or rain. Burners should be sized to raise the temperature of the drying air up to only 20°F. In any case, the temperature should not exceed about 80°F after heating. Overheating the drying air causes the grain to overdry and dry nonuniformly; heat is only recommended when counteracting adverse weather conditions.

Drying takes place in a drying zone, which advances upward through the grain (Figure 8). Grain above this drying zone remains at or slightly above the initial moisture content, while grain below the drying zone is at a moisture content in equilibrium with the drying air. The equilibrium moisture content of a material is the moisture content that it approaches when exposed to air at a specified relative humidity.

As the direction of air movement does not affect the rate of drying, other factors must be considered in choosing the direction of air movement. A pressure system moves the moisture-laden air up through the grain, where it is discharged under the roof. If there are insufficient roof outlets, moisture may condense on the underside of metal roofs. During pressure system ventilation, the wettest grain is near the top surface and is easy to monitor. Fan and motor waste heat are put into the airstream and contribute to drying.

Table 7 Recommended Airflow Rates for Different Grains and Moisture Contents Using Unheated Air in the Southern United States

Type of Grain	Grain Moisture Content, %	Recommended Airflow Rate, cfm/ft³
Wheat	25	4.8
	22	4.0
	20	2.4
	18	1.6
	16	0.8
Oats	25	2.4
	20	1.6
	18	1.2
	16	0.8
Shelled Corn	25	4.0
	20	2.4
	18	1.6
	16	0.8
Ear Corn	25	6.4
	18	3.2
Grain Sorghum	25	4.8
	22	4.0
	18	2.4
	15	1.6
Soybeans	25	4.8
	22	4.0
	18	2.4
	15	1.6

Compiled from USDA Leaflet 332, 1952, and University of Georgia *Bulletin* NS 33, 1958.

A suction system moves the air down through the grain. The moisture-laden air discharges from the fan into the outside atmosphere; thus roof condensation is not a problem. However, the wettest grain is near the bottom of the mass and is difficult to

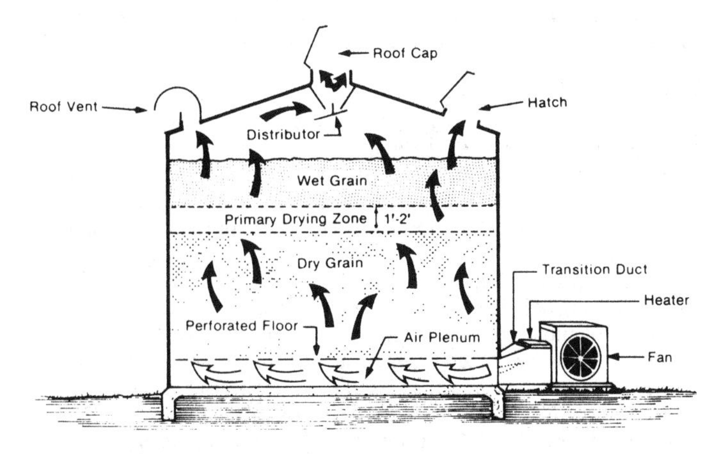

Fig. 8 Three Zones Within Grain During Natural Air Drying in a Typical Bin

sample. Of the two systems, the pressure system is recommended because it is easier to manage.

The following management practices must be observed to ensure the best performance of the drier:

1. Minimize foreign material. A scalper-cleaner is recommended for cleaning the grain to reduce wet pockets, and air pressure and energy requirements.
2. Distribute the remaining foreign material uniformly by installing a grain distributor on the end of the storage filling chute.
3. Place the grain in layers and keep it leveled.
4. Start the fan as soon as the floor or ducts are covered with grain.
5. Operate the fan continuously unless it is raining heavily or there is a dense ground fog. Once all the grain is within 1% of storage moisture content, run the fans only when the relative humidity is below 70%.

Layer Drying

Layer drying is done by placing successive layers of wet grain on top of dry grain. When the top 6 in. has dried to within 1% of the desired moisture content, another layer is added (Figure 9). In comparison to full-bin drying, layering reduces the time that the top layers of grain are left wet. Because effective airflow rate is greater for lower layers, allowable harvest moisture content can be greater than that in the upper levels. Either unheated air or air heated 10 to 20°F above the ambient is used, but the use of heated air controlled with a humidistat to prevent overdrying is most common. The first layer may be about 7 ft in depth, with successive layers of about 3 ft.

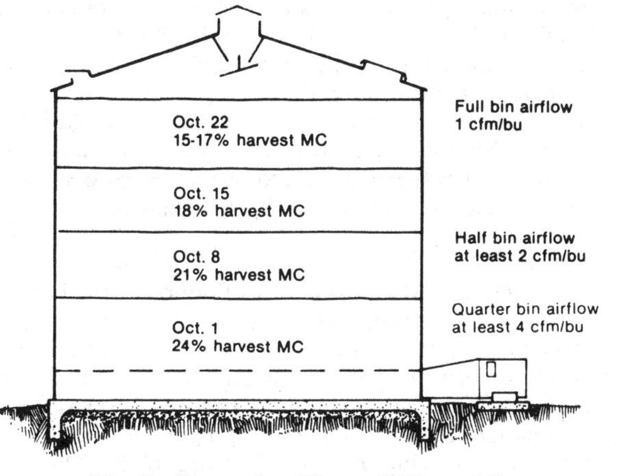

Fig. 9 Example of Layer Filling of Corn

Batch-in-Bin. A storage bin adapted for drying may be used to dry several batches of grain during a harvest season, if the grain is placed in shallow depths so that higher airflow rates and temperatures can be used. The bin is emptied, usually by a system of augers, and the cycle is repeated. The drying capacity of the batch system is greater than other in-storage drying systems. Typical operating conditions are those where corn in 3-ft depths is dried from an initial moisture content of 25% with 130°F air at the rate of about 20 cfm/ft^3. Considerable nonuniformity of moisture content may be present in the batch when drying is stopped; therefore, the grain should be well mixed as it is placed into storage. If it is mixed well, grain that is too wet equalizes in moisture with grain that is too dry before spoilage can occur.

Grain may be cooled in the drier to ambient temperature before it is placed in storage. Cooling is accomplished by operating the fan without the heater for about 1 h. Some additional drying occurs during the cooling process, particularly in the wetter portions of the batch.

Grain stirring devices are used with both full-bin and batch-in-bin drying systems. Typically, these devices consist of one or more open, 2-in. diameter, standard pitch augers suspended from the bin roof and side wall and extending to near the bin floor. The augers rotate and simultaneously travel horizontally around the bin. The device mixes the grain being dried to reduce moisture gradients and prevent overdrying the bottom grain. In addition, the grain is loosened, allowing a higher airflow rate for a given fan. Stirring equipment reduces bin capacity by about 10% and may cost as much as, and do no more than, a larger drier fan. Furthermore, commercial stirring devices are only available for round storage enclosures. Increased grain breakage can result from improper use of stirring devices.

Recirculating/Continuous Flow Bin Drier. This type of drier incorporates a tapered sweep auger that removes grain as it dries from the bottom of the bin (Figure 10). The dry grain is then redistributed on top of the bin or moved to a second bin for cooling. The sweep auger may be controlled by temperature or moisture sensors. When the desired condition is reached, the sensor starts the sweep auger, which removes a layer of grain. After a complete circuit of the bin, the sweep auger stops until the sensor again determines that another layer is dry. Some drying takes place in the cooling bin. Up to two percentage points of moisture may be removed, depending on the management of the cooling bin.

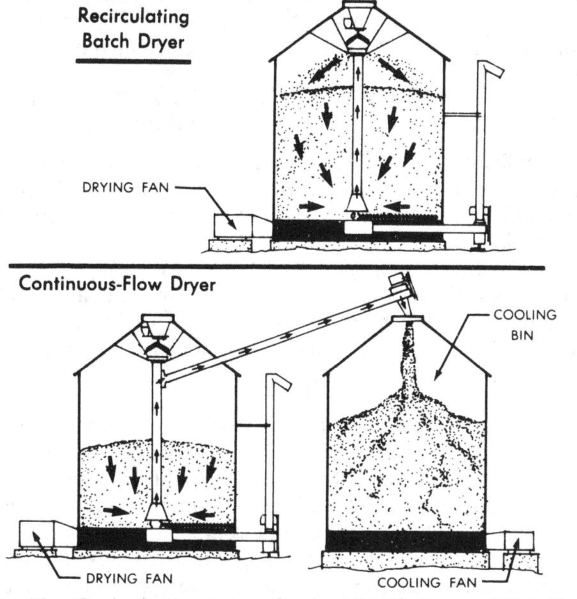

Fig. 10 Grain Recirculators Convert Bin Drier to a High-Speed Continuous Flow Drier

DRYING SPECIFIC CROPS

SOYBEANS

Soybeans usually need drying only when inclement weather occurs during the harvest season. Mature soybeans left exposed to rain or damp weather develop a dark brown color and a mealy or chalky texture. Seed quality deteriorates rapidly. Oil from weather-damaged beans is more costly to refine and is often not of edible grade. In addition to preventing deterioration, artificially drying soybeans offers the advantage of early harvest, which reduces chances of loss from bad weather. Early harvest also reduces natural and combine shatter loss. Soybeans harvested above 13.5% wet-basis moisture content exhibit less damage.

Drying Soybeans for Commercial Use

Conventional drying equipment for corn can be used for soybeans, with some limitations on heat input. Soybeans can be dried at 130 to 140°F for commercial use. Drying temperatures of 190°F reduce oil yield. Excessive seedcoat cracking occurs if the relative humidity of the drying air is below 40%. Cracks from drying cause many split beans in subsequent handling. Physically damaged beans can develop fungal growth and cause storage problems for processors, as well as a slight reduction in oil yield and quality. Flow-retarding devices should be used, and beans should not be dropped more than 20 ft onto concrete floors.

Drying Soybeans for Seed and Food

The relative humidity of drying air should be kept above 40%, regardless of the amount of heat used. Maximum drying temperature to avoid germination loss is 110°F. Natural air drying at a flow rate of 1.6 cfm/ft^3 is adequate for drying seed with starting moisture contents up to 16%.

Low-temperature drying in bins is another method used in the midwestern United States. Only enough heat to raise the ambient temperature not more than 5°F is used. This drying method is slow, but it results in excellent quality and avoids overdrying. However, drying must be completed before spoilage occurs. At an airflow rate of 1.6 cfm/ft^3, soybeans usually take about five days to dry from 16% to 13%. At higher moisture contents, good results have been obtained using an airflow rate of 3.2 cfm/ft^3 with humidity control. Data on allowable drying time for soybeans are unavailable. In the absence of better information, an estimate of storage life for oil crops might be made based on the values for corn, using an adjusted moisture content calculated by:

$$\text{Comparable moisture content} = \frac{\text{Oil seed moisture content}}{100 - \text{Seed oil content}} \times 100$$

Generally, use a corn moisture content 2% greater than soybeans to estimate allowable drying time, *i.e.*, 12% soybeans is comparable to 14% corn. Soybeans are not dried from as high an initial moisture content as corn.

Soybean seed technologists suggest drying high-moisture soybeans in a bin with air temperature controlled to maintain the relative humidity at 40% or higher. Airflow rates of 8.0 cfm/ft^3 are recommended, with the depth of beans not to exceed 4 ft.

HAY

Hay normally contains 65 to 80% wet-basis moisture at cutting. Field drying to 20% may result in a large loss of leaves. Alfalfa hay leaves average about 50% of the crop by weight, but they contain 70% of the protein and 90% of the carotene. The quality of hay can be increased and the risk of loss from weather reduced if the hay is put under shelter when partially sun dried (35% moisture content) and then artificially dried to a safe storage moisture content (about 20% moisture content). With good drying weather, hay conditioned by mechanical means can be dried sufficiently in one day and placed in the drier. Hay may be long, chopped, or baled for this operation; unheated or heated air can be used.

In-Storage Drying

Unheated air is normally used for in-storage or mow drying. The hay is dried in the field to 30 to 40% moisture content before being placed in the drier. For unheated air drying, the airflow should be at least 200 cfm per ton. The fan should be capable of delivering the required airflow against a static pressure of 1 to 2 in. of water.

Slotted floors, with at least 50% of the area open, are generally used for drying baled hay. For long or chopped hay, the center duct system is the most popular for mows less than 36 ft wide. A slotted floor should be placed on each side of the duct to within 5 ft of the ends and outside walls (Figure 11). If the width is greater than 36 ft, the mow should be divided crosswise into units of 28 ft or less. These should then be treated as individual driers. If storage depth exceeds about 13 ft, either vertical flues and/or additional levels of ducts may be used. If tiered ducts are used, a vertical air chamber, about 75% of the probable hay depth, should be used. The supply ducts are then connected at 7 to 10 ft vertical intervals as the mow is filled. With either of these methods, total hay depths up to 30 ft can be dried, considering the total depth of wet hay to be dried and the hay dried in previous loadings. The duct size should be such that the air velocity is less than 1000 fpm. The maximum depth of wet hay, which should be placed on a hay drying system at any time, depends on hay moisture content, weather conditions, the physical form of the hay, and the airflow rate.

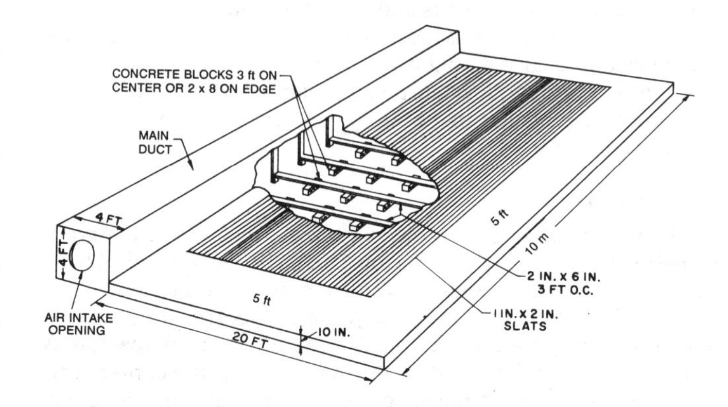

Fig. 11 Central Duct Hay-Drying System with a Lateral Slatted Floor for Wide Mows

The maximum safe depths are about 16 ft for long hay, 13 ft for chopped hay, and seven small rectangular bales deep for baled hay. Baled hay should have a density of about 8 lb/ft^3. For best results, bales should be stacked tightly together on edge to ensure that no openings exist between them.

For mow drying, the fan should run continuously during the first few days. Afterwards, it should be operated only during low relative humidity weather. During prolonged wet periods, the fan should be operated only enough to keep the hay cool.

Batch Wagon Drying

Batch drying can be done on a slotted floor platform; however, because this method has high labor requirements, wagon driers are more commonly used. With a wagon drier system, hay is baled at about 45% moisture content to a density of about 11 lb/ft^3. The hay is then stacked onto a wagon with tight, high sides and slotted or expanded metal floors. Drying is most efficiently accomplished by forcing the heated air (up to 158°F) down a canvas duct of a plenum chamber secured to the top of the wagon.

After 4 or 5 h of drying, the exhaust air is no longer saturated with moisture, and about 75% of it may be recirculated or passed through a second wagon of wet hay for greater drying efficiency.

With this method, the amount of hay harvested each day is limited by the capacity of the drying wagons. Since it is a 24-h process, the hay cut one day is stored the following day; only enough hay should be harvested each day to load the drying wagons.

The airflow rate is normally much higher with this method than when unheated air is used. About 40 cfm per ft^2 of wagon floor space is required. As with mow drying, the duct size should be such that the air velocity is less than 1000 fpm.

COTTON

Producers normally allow cotton to dry naturally in the field to a level of 12% or less before harvest. Cotton harvested under these conditions can be stored in trailers, baskets, or compacted stacks for extended periods with little loss in fiber or seed quality. Thus, cotton is not normally aerated or artificially dried prior to ginning. Cotton harvested during inclement weather, or stored cotton exposed to rain or snow, must be dried at the cotton gin within a few days to prevent self-heating and deterioration of the fiber and seed.

Even though cotton may be safely stored at moisture contents as high as 12%, moisture levels near the upper limit are too high for efficient ginning and for obtaining optimum fiber grade. The cleaning efficiency of cotton is inversely proportional to its moisture content, with the most efficient level being 5% fiber moisture content. However, fiber quality is best preserved when fiber seed separation is accomplished at moisture contents between 6.5 and 8%. Therefore, if cotton comes into the system below this level, it can be cleaned, but moisture should be added prior to fiber seed separation to improve the ginning quality. Driers in the cotton gins are capable of drying cotton to the desired moisture levels.

Although several types of driers are commercially available, the tower drier is most commonly used. This device operates on a parallel flow principle, where 14 to 24 cfm of drying air per pound of cotton also serves as the conveying medium. Cotton impacts on the drier walls as it moves through the drier's serpentine passages. This action agitates the cotton for improved drying and lengthens the exposure time. The drying time depends on many variables, but total exposure seldom exceeds 15 s. For extremely wet cotton, two stages of drying are necessary for adequate moisture control.

Wide variations in initial moisture content dictate different amounts of drying for each load of cotton. Rapid changes in drying requirements are accomplished by automatically controlling drying air temperature in response to moisture measurements taken before or after drying. These control systems prevent overdrying and reduce energy requirements. For safety and to preserve fiber quality, drying air temperature should not exceed 350°F in any portion of the drying system.

The germination of cottonseed is unimpaired by drying, if the internal cottonseed temperature does not exceed 140°F. This temperature is not exceeded in a tower drier; however, the moisture content of the seed after drying may be above the 12% level recommended for safe long-term storage. Wet cottonseed is normally processed immediately at a cottonseed oil mill. Cottonseed under 12% moisture is frequently stored for several months prior to milling, or prior to delinting and treatment at a planting seed processing plant. Aeration for cooling deep beds of stored cottonseed effectively maintains viability and prevents an increase in free fatty acid content. For aeration, ambient air is normally drawn downward through the bed at a minimum rate of 0.025 cfm/ft^3 of oil mill seed and 0.125 cfm/ft^3 of planting seed.

PEANUTS

Peanuts normally have a moisture content of about 50% at the time of digging. Allowing the peanuts to dry on the vines in the windrow for a few days removes much of this water. However, peanuts usually contain 20 to 30% moisture when removed from the vines, and some artificial drying is necessary. Drying should begin within 6 h after harvesting to keep the peanuts from self-heating. Both the maximum temperature and the rate of drying must be carefully controlled to maintain quality.

High temperatures result in off flavor or bitterness. Drying too rapidly without high temperatures results in blandness or nuts that do not develop flavor on roasting. High temperatures, rapid drying, or excessive drying cause the skin to slip easily and the kernels to become brittle. These conditions result in high damage rates in the shelling operation, but they can be avoided if the moisture removal rate does not exceed 0.5% per hour. Because of these limitations, continuous flow drying is not usually recommended for peanuts.

Peanuts can be dried in bulk bins using unheated air or air with supplemental heat. During poor drying conditions, unheated air may cause spoilage, so supplemental heat is preferred. Air should be heated no more than 13 to 14°F with a maximum temperature of 95°F. An airflow rate of 10 to 25 cfm/ft^3 of peanuts should be used, depending on the initial moisture content.

The most common method of drying peanuts is bulk wagon drying. Peanuts are dried in depths of 5 to 6 ft, using airflow rates of 10 to 15 cfm/ft^3 of peanuts and air heated 11 to 14°F above ambient. This method retains quality and usually dries the peanuts in three to four days. Wagon drying reduces handling labor but may require additional investment in equipment.

RICE

Of all grains, rice is probably the most difficult to process without quality loss. Rice containing more than 13.5% moisture cannot be safely stored for long periods, yet the recommended harvest moisture content for best milling and germination ranges from 20 to 26%. When harvested at this moisture content, drying must be started promptly to prevent the rice from souring. Heated air is normally used in continuous flow driers where large volumes of air are forced through 4- to 10-in. layers of rice. Temperatures as high as 130°F may be used, if the temperature drop across the rice does not exceed 20 to 30°F, the moisture reduction does not exceed two percentage points in a 0.5 h exposure, and the rice temperature does not exceed 100°F. Following drying, the rice should be aerated to ambient temperature before the next drying exposure. Aeration during the tempering period following each pass through the drier removes additional moisture and eliminates one to two drier passes that would be needed if the rice were tempered without cooling. It is estimated that full use of aeration following drier passes could increase maximum daily drying capacity by about 14%.

Unheated air or air with a small amount of added heat (13°F above ambient, but not exceeding 95°F) should be used for deep-bed rice drying. Too much heat overdries the bottom, resulting in checking and reduced milling qualities, and possible spoilage in the top. Because unheated air drying requires less investment and attention than supplemental heat drying, it is preferred when conditions permit. In the more humid rice-growing areas, supplemental heat is desirable to ensure that the rice can be dried. The time required for drying will vary with weather conditions, moisture content, and airflow rate. The recommended airflow rate is 0.2 to 2.4 cfm/ft^3 in California. Because of less favorable drying conditions in Arkansas, Louisiana, and Texas, greater airflow rates are recommended: *e.g.*, a minimum of 2.0 cfm/ft^3 is recommended in Texas. Whether unheated air or supplemental heat is used, the fan should be turned on as soon as rice uniformly covers the air distribution system. Then the fan should run continuously until the moisture content in the top 1 ft of rice is reduced to about 15%. When the moisture content has been reduced to this level, the supplemental heat should be turned off. The rice can then be dried to a safe storage level by operating the fan only when the relative humidity is below 75%.

STORAGE PROBLEMS AND PRACTICES

MOISTURE MIGRATION

Redistribution of moisture generally occurs in stored grain when grain temperature is not controlled (Figure 12). Localized spoilage can occur even though the grain is stored at a safe moisture level. Grain placed in storage in the fall at relatively high temperatures cools nonuniformly from the outer surfaces of the storage bin as winter approaches. Thus, the grain near the outside walls and top surface may be at cool outdoor temperatures while the grain nearer the center is still at nearly the same temperature it was at harvest. These temperature differentials induce air convection currents that flow downward along the outside boundaries of the porous grain mass and upward through the center. When the cool air from the outer regions contacts the warm grain in the interior, the air is heated and its relative humidity is lowered, increasing its capacity to absorb moisture from the grain as it moves up. Then, when the warm, humid air reaches the cool grain near the top of the storage, it again cools and transfers vapor to the grain. Under extreme conditions, water condenses on the grain. The moisture concentration near the center of the grain surface results in significant spoilage if moisture migration is uncontrolled. During spring and summer, the temperature gradients are reversed. The grain moisture content increases most at depths of 2 to 4 ft below the surface. Daily variations in temperature do not cause significant moisture migration. Aside from seasonal temperature variations, the size of the grain mass is the most important factor in fall and winter moisture migration. Storages containing less than 1200 ft^3 do not experience as much trouble with moisture migration. The problem becomes critical in large storages and is aggravated by incomplete cooling of artificially dried grain. Artificially dried grain should be cooled to near ambient temperature when dried.

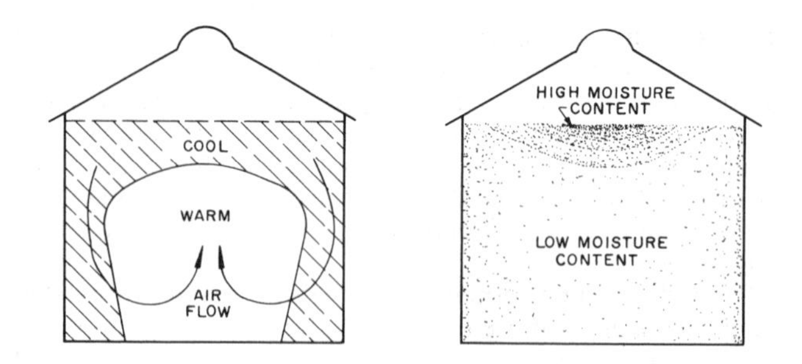

Fig. 12 Grain Storage Conditions Associated with Moisture Migration During Fall and Early Winter

GRAIN AERATION

Aeration by mechanically circulating ambient air through the grain mass is the best way to control moisture migration. Aeration systems are also used to cool grain after harvest, particularly in warmer climates where grain may be placed in storage at temperatures exceeding 100 °F. After harvest heat is removed, aeration may be continued in cooler weather to bring the grain to a temperature within 20 °F of the coldest average monthly temperature. The temperature must be maintained at less than 50 °F.

Aeration systems are not a means of drying because airflow rates are too low. However, in areas where the climate is favorable, carefully controlled aeration may be used to remove small amounts of moisture.

Commercial storages may have pockets of higher moisture grain because some batches of higher moisture wheat are delivered after a rain shower or early in the morning, for example. Aeration can control grain damage from heating in those higher moisture pockets.

Aeration Systems Design

Aeration systems include fans capable of delivering the required amount of air at the required static pressure, suitable ducts to distribute the air into the grain, and controls to regulate the operation of the fan.

The airflow rate determines how many hours are required to cool the crop (Table 8). Most aeration systems are designed with airflow rates between 0.05 and 0.2 cfm/bu.

Table 8 Airflow Rates Corresponding to Approximate Grain Cooling Time

Airflow Rate, cfm/bu	Cooling Time, h
0.05	240
0.1	120
0.2	60
0.3	40
0.4	30
0.5	24
0.6	20
0.8	15
1.0	12

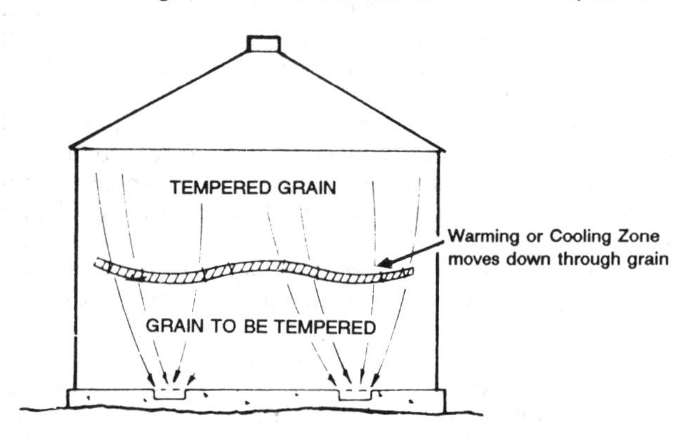

NEGATIVE PRESSURE AERATION

Downflow—cooling or warming zone moves down through the grain.

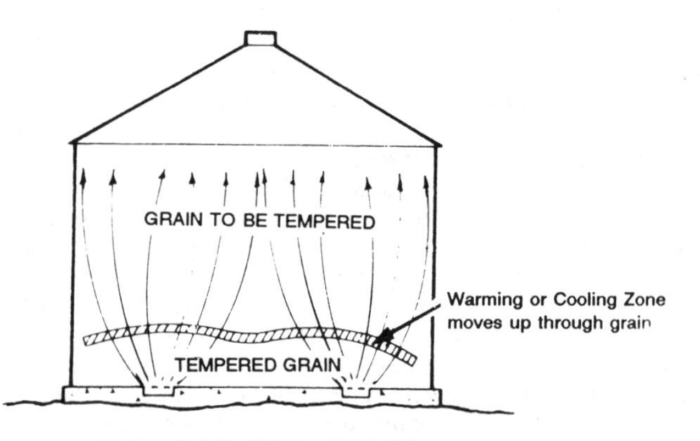

POSITIVE PRESSURE AERATION

Upflow—cooling or warming zone moves up through the grain.

Fig. 13 Aerating to Change Grain Temperature

Stored grain is aerated by forcing air up or down through the grain. Moving air up through the grain is more common because it is easier to observe that the cooling front has moved through the entire grain mass. Also, with upward airflow, there is little hazard of vent screens freezing over and fan suction collapsing the bin roof. Upward airflow results in more uniform air distribution than downdraft systems in large, flat storages with long ducts.

During aeration, a warming or cooling front moves through the crop (Figure 13); thus, it is important to run the fan long enough to move the front completely through the crop.

In tall tower storages, airflow rates should be limited or excessive power is required because of the long flow path through the grain. Crossflow aeration reduces the length of flow path, resulting in reduced system static pressure for a given aeration rate (Figure 14). Crossflow systems, however, are less easily adapted to aeration of partially filled bins than up- or downflow systems, and installation costs are higher.

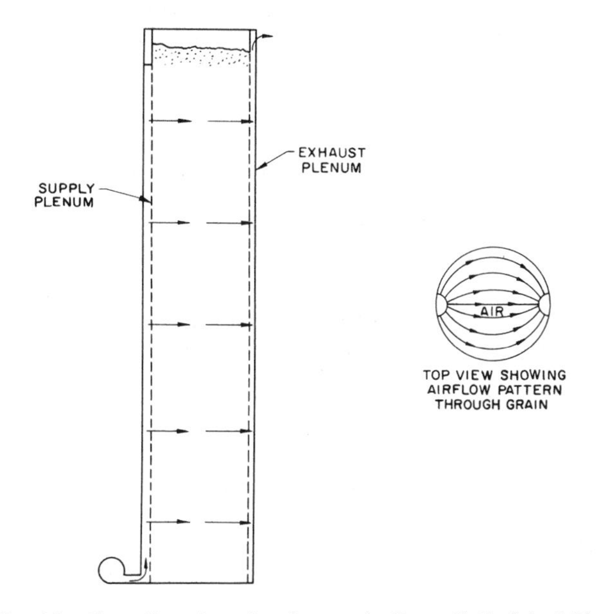

Fig. 14 Crossflow Aeration System in Deep Cylindrical Bin

Static pressure can be determined for an aeration system, using the airflow resistance information in Chapter 10 of the 1989 ASHRAE *Handbook—Fundamentals*. All common types of fans are used in aeration systems. Attention should be given to noise levels with fans that are operated near residential areas or where people work for extended periods. The supply ducts connecting the fan to the distribution ducts in the grain should be designed and constructed according to the standards of good practice for any air-moving application. A maximum air velocity of 2500 fpm may be used, but 1600 to 2000 fpm is preferred. In large systems, one large fan may be attached to a manifold duct that leads to several distribution ducts in one or more storages, or smaller individual fans may serve individual distribution ducts. Where a manifold is used, valves or dampers should be installed at each takeoff to allow adjustment or closure of airflow when part of the system is not needed.

Distribution ducts are usually perforated sheet metal with a circular or inverted U-shaped cross section, although many functional arrangements are possible. The area of the perforations should be at least 10% of the total duct surface. The holes should be uniformly spaced and small enough to prevent the passage of the grain into the duct. For example, 0.1-in. holes or 0.08-in. wide slots do not pass wheat.

Since most problems develop in the center of the storage and the crop cools naturally near the wall, the aeration system must provide good airflow in the center. If ducts placed directly on the floor are to be held in place by the crop, the crop flow should be directly on top of the ducts to prevent movement and damage. Flush floor systems work well in storages with sweep augers and unloading equipment. Ducts should be easily removable for cleaning. Duct spacing should not exceed the depth of the crop; the distance between the duct and storage structure wall should not exceed one-half the depth of the crop for bins and flat storages. Common duct patterns for round bins are shown in Figure 15. Duct spacing for flat storages is shown in Figure 16.

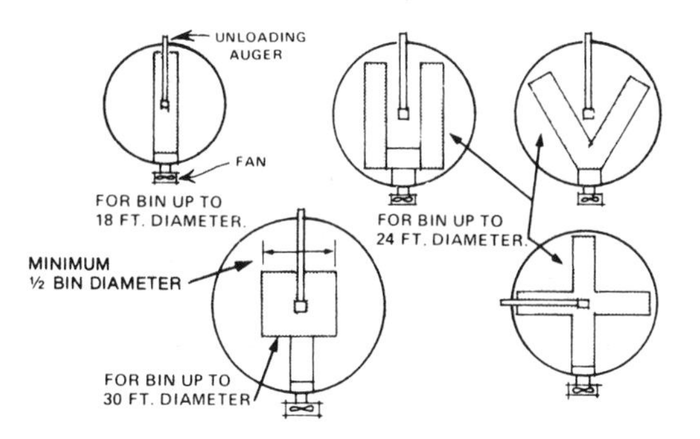

Fig. 15 Common Duct Patterns for Round Grain Bins

When designing the distribution duct system for any type of storage, consider: (1) the cross-sectional area and length of the duct, which influences the air velocity within the duct and the uniformity of air distribution; (2) the duct surface area, which affects the static pressure losses in the grain surrounding the duct; and (3) the distance between ducts, which influences the uniformity of airflow.

For upright storages where the distribution ducts are relatively short, distribution duct velocities up to 2000 fpm are permissible. Maximum recommended air velocities in ducts for flat storages are shown in Table 9. The air distribution uniformity can be improved by providing an effective outlet area equal to the duct cross-sectional area.

Table 9 Recommended Maximum Air Velocities within Ducts for Flat Storages

Grain	Airflow Rate, cfm/bu	Air Velocity within Ducts (fpm) for Grain Depths of: 10 ft	20 ft	30 ft	40 ft	50 ft
Corn, soybeans, and other large grains	0.05	—	750	1000	1250	1250
	0.1	750	1000	1250	1500	1750
	0.2	1000	1250	—	—	—
Wheat, grain sorghum, and other small grains	0.05	—	1000	1500	1750	2000
	0.1	750	1500	2000	—	—
	0.2	1000	2000	—	—	—

The duct surface area that is perforated or otherwise open for air distribution must be great enough so that the air velocity through the grain surrounding the duct is not high enough to cause excessive pressure loss. When a semicircular perforated duct is used, the entire surface area is effective, while only 80% of the area of a circular duct resting on the floor is effective. For upright storages, the air velocity through the grain near the duct (duct face velocity) should be limited to 30 fpm or less; in flat storages, to 20 fpm or less.

Duct strength and anchoring are important. Distribution ducts buried in the grain must be strong enough to withstand the pressure of the grain over them. In tall, upright storages, the static grain pressures may reach 10 psi. When ducts are located in the path of grain flow, as in a hopper, they may be subjected to many times this pressure during grain unloading.

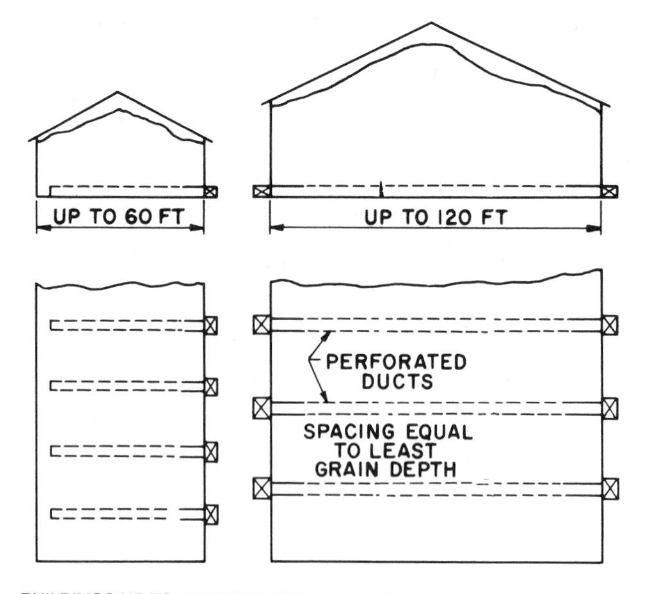

BUILDINGS UP TO 40 ft WIDE AND 100 ft LONG

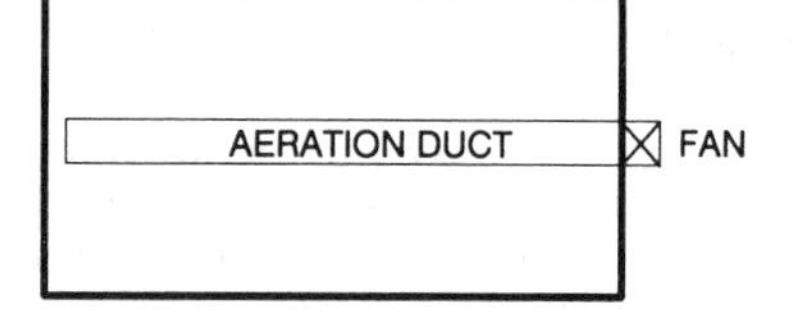

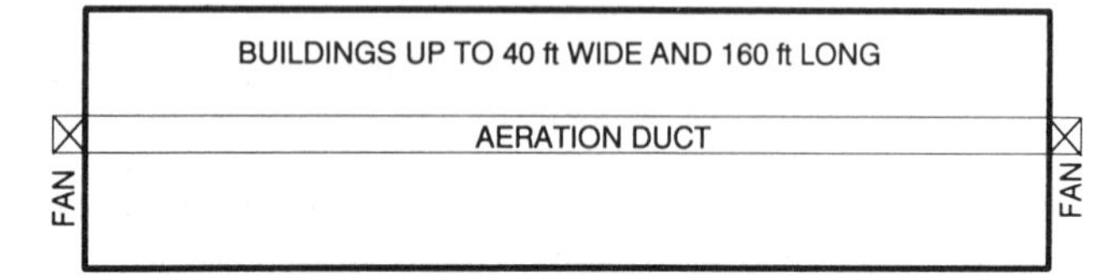

Grain Peaked with Shallow Depth Near Wall

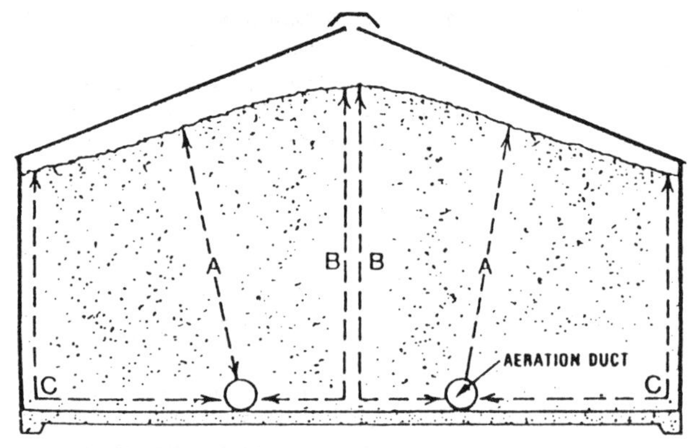

(A) IS THE SHORTEST AIR PATH
(B) AND (C) ARE LONGER AIR PATHS THAN (A)
(B) AND (C) SHOULD BE NO LONGER THAN 1.5 TIMES (A)

Fig. 16 Duct Arrangements for Large Flat Storages

Operating Aeration Systems

The operation of aeration systems depends largely on the objectives to be attained and the locality. In general, cooling should be carried out at any time the outdoor air temperature is about 15°F cooler than the grain. Do not aerate stored grain when the air humidity is much above the equilibrium humidity of the grain because moisture will be added. Running the fan just long enough to cool the crop limits the amount of grain that is rewetted. The fan should be operated long enough to cool the crop completely, but it should then be shut off and the fan covered.

Aeration fans should be started as soon as the storage is filled to cool the grain. Aeration to prevent moisture migration should be started whenever the air temperature is 10 to 15°F below the highest grain temperature. It is usually continued as weather permits until the grain is uniformly cooled to within 20°F of the average temperature of the coldest month or 30 to 40°F.

Grain temperatures of about 32 to 50°F are considered desirable. In the northern corn belt, aeration may be resumed in the spring to equalize and raise the grain temperatures to between 40 and 50°F. This reduces the risk of localized heating from moisture migration. The only reason to aerate when air temperatures are above 60°F is when there are problems in the storage. Aeration fans and ducts should be covered when they are not being used.

In storages where the fans are operated daily in the fall and winter months, automatic controls work well when the air is not too warm or humid. One thermostat usually prevents fan operation when the air temperature is too high, and another prevents operation when the air is too cold. A humidistat allows operation when the air is not too humid. Fan controllers that determine the equilibrium moisture content of the crop based on existing air conditions and regulate the fan based on entered information are available.

SEED STORAGE

Seed must be stored in a cool, dry environment to maintain viability. Most seed storages have refrigeration equipment to maintain a storage environment of 45 to 55°F. Seed conditions for storage must be achieved before mold and insect damage occur. Desired conditions can be met in 220 h at an airflow rate of 0.04 cfm/ft^3 and in 140 h at 0.08 cfm/ft^3.

BIBLIOGRAPHY

ASAE. 1989a. Density, specific gravity, and weight-moisture relationships of grain for storage. ASAE D241.2. American Society of Agricultural Engineers, St. Joseph, MI.

ASAE. 1989b. Moisture relationships of grain. ASAE D245.4. American Society of Agricultural Engineers, St. Joseph, MI.

ASAE. 1989c. Resistance of airflow through grains, seeds, and perforated metal sheets. ASAE D272.1. American Society of Agricultural Engineers, St. Joseph, MI.

Brooker, D.B., F. Bakker-Arkema, and C.A. Hall. 1974. *Drying cereal grains.* AVI Publishing, Westport, CT.

Cotton ginners handbook. 1977. Agricultural handbook No. 503, Agricultural Research Service, U.S. Department of Agriculture, Washington, D.C.

Midwest Plan Service. 1988. *Grain drying, handling and storage handbook.* MWPS-13. Iowa State University, Ames, IA.

Midwest Plan Service. 1980. *Low temperature and solar grain drying handbook.* MWPS-22. Iowa State University, Ames, IA.

Midwest Plan Service. 1980. *Managing dry grain in storage.* AED-20. Iowa State University, Ames, IA.

Foster, G.H. 1982. *Storage of cereal grains and their products,* 3rd ed. American Association of Cereal Chemists, St.Paul, MN.

Hall, C.A. 1980. *Drying and storage of agricultural crops.* AVI Publishing, Westport, CT.

Hellevang, K.J. 1989. *Crop storage management.* AE-791. NDSU Extension Service, North Dakota State University, Fargo, ND.

Hellevang, K.J. 1987. *Grain drying.* AE-701. NDSU Extension Service, North Dakota State University, Fargo, ND.

Hellevang, K.J. 1983. *Natural air/low temperature crop drying.* EB-35. NDSU Extension Service, North Dakota State University, Fargo, ND.

Henderson, S.M. and R.L. Perry. 1976. *Agricultural process engineering.* AVI Publishing, Westport, CT.

Schuler, Holmes, Straub, and Rohweder. 1986. *Hay drying.* A3380. University of Wisconsin-Extension, Madison, WI.

CHAPTER 23

AIR CONDITIONING OF WOOD AND PAPER PRODUCTS FACILITIES

THIS chapter covers some of the standard requirements for facilities involved in the manufacture of finished wood products, including pulp and paper manufacture.

GENERAL WOOD PRODUCT OPERATIONS

If a finished lumber product is to be used in a heated building, the stock storage areas should be heated 10 to 20°F above ambient. This provides sufficient protection for furniture stock, interior trim, cabinet material, and stock for the manufacture of such items as ax handles and glue-laminated beams. Air should be circulated within the storage areas. Lumber that is kiln dried to a moisture content of 12% or less can be controlled within a given moisture content range through storage in a heated shed. The moisture content can be regulated either manually or automatically by altering the dry-bulb temperature (Figure 1).

The preparation of this chapter is assigned to TC 9.2, Industrial Air Conditioning.

Some special materials require close control of moisture content. Musical instrument stock must be dried to a given moisture level and maintained there because the moisture content of the wood affects the harmonics of most stringed wooden instruments. This degree of control requires an air-conditioning system with reheat and a heating system with humidification.

Process Area Air Conditioning

Temperature and humidity requirements within wood product process areas vary according to product, manufacturer, and governing code. Only important design items are discussed in this section. For example, in match manufacturing, close temperature and humidity controls are required to dry or cure the match heat after dipping—primarily to avoid being close to the ignition point. Any process involving the application of flammable substances should follow the ventilation recommendations of the NFPA, the National Fire Code, and the U.S. Occupational Safety and Health Act.

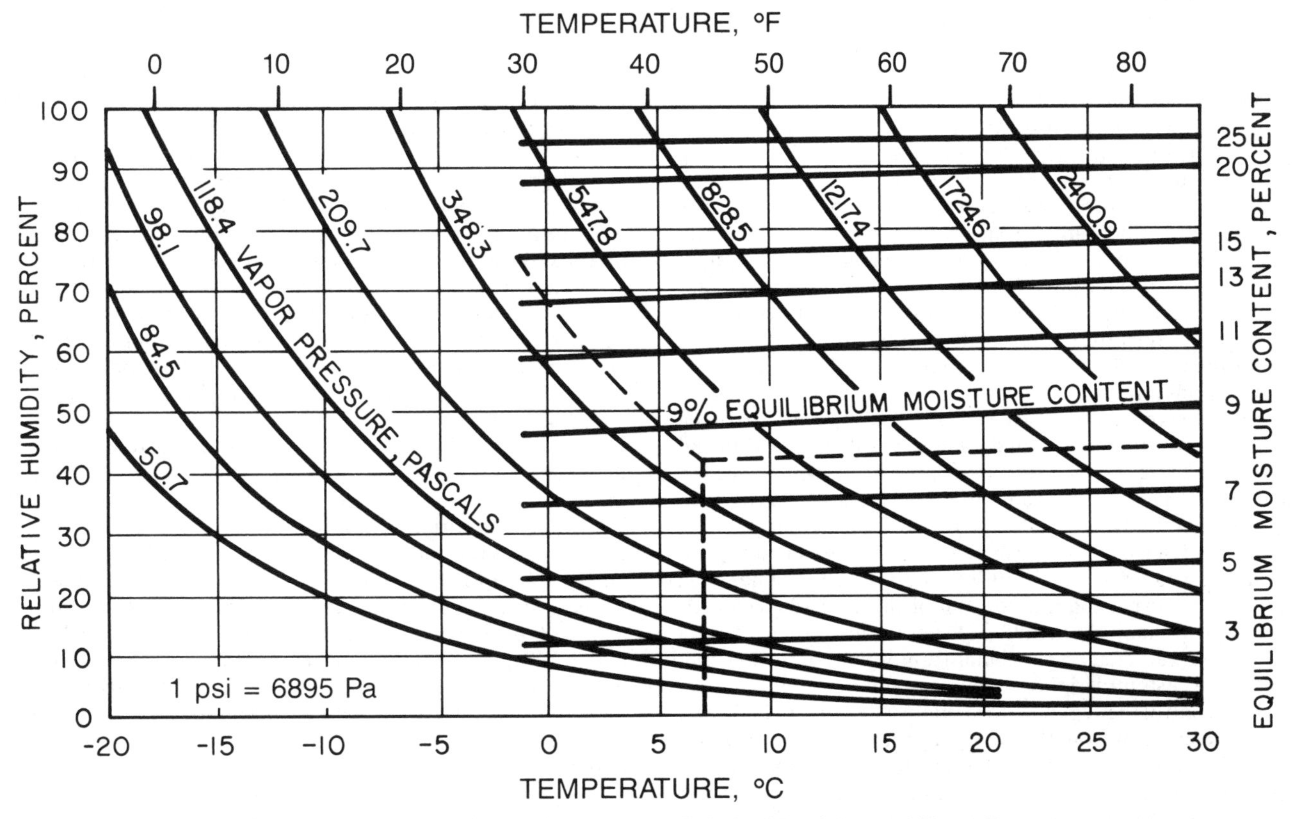

Fig. 1 Relationship between Temperature, Relative Humidity, and Vapor Pressure of Air and the Equilibrium Moisture Content of Wood

Finished Product Storage

Finished lumber products manufactured from predried stock (moisture content of 10% or less) regain moisture if exposed to high relative humidities for an extended period. All storage areas housing finished products, including furniture and musical instruments, should be conditioned as close as is practical to the environment in which the item is to be used.

Designers should be familiar with the client's entire operation. The designer should also remain aware of the potential problem of variable moisture regain in any wood product facility.

PULP AND PAPER OPERATIONS

The papermaking process comprises two basic steps: (1) wood is reduced to pulp, a water mass of wood fibers; and (2) the pulp is converted to paper. Wood may be reduced to pulp either by grinding (as in a groundwood-type pulp mill) or by chemical action.

Many different types of paper may be produced from the prepared pulp, ranging from the finest glossy finish to newsprint to bleached pulpboard to fluff pulp for disposable diapers. For example, to make newsprint, a mixture of the two pulps is fed into the paper machine. To make kraft paper (grocery bags, corrugated containers, etc.), however, only unbleached chemical pulp is used. Disposable diaper material and photographic film and paper require bleached chemical pulp with a very low moisture content of 6 to 9%.

Paper Machine Area

In the papermaking process, extensive air systems are required to support and enhance the process and provide reasonable comfort for machine personnel. Radiant heat from steam and hot water sources and mechanical energy dissipated as heat can result in summer temperatures in the machine room ranging as high as 104 to 120°F. In addition, high paper machine operating speeds from 2000 to 4500 ft/min and stock temperatures in the range of 122°F produce warm vapor in the machine room. A ventilation system should prevent condensation and maintain a safe, efficient environment for personnel.

Outside air makeup units absorb and remove water vapor released from the paper as it is dried (Figures 2 and 3). The makeup air is distributed to the working areas above and below the operating floor. Part of the air delivered to the basement will migrate to the operating floor through hatches and stairwells. Motor cooling systems should distribute cooler basement air to the paper machine DC drive motors. The intake to the personnel coolers should be taken in the basement below the warm stratified air that accumulates under the operating floor. A basement exhaust system for the wet end of the process should vent this stratum during the warmer months.

The most severe ventilation demand occurs in the area between the wet end forming section and water removal press section and the machine drier section. When the pulp fiber is introduced to the forming section, more than 70% water is deposited on a traveling screen. Water is removed sequentially by gravity, rolls, foils, vacuum, and steam boxes, and mechanically by three or more press roll nips at which point up to 50% of the water content has been uniformly removed. Evaporation of moisture at elevated temperatures and mechanical generation of vapor from turning rolls and cleaning showers create a very humid atmosphere at the wet end. Baffles and custom-designed exhaust systems in the forming section help control the vapor. A drive side exhaust system in the wet end area keeps heat from the DC drive motor vent air and vapor generated in the wet end to a minimum. To prevent condensation or accumulated fiber from falling on the traveling web, a false ceiling with duct connections to roof exhausters removes humid air, which has not been captured at a lower elevation. The wet end area

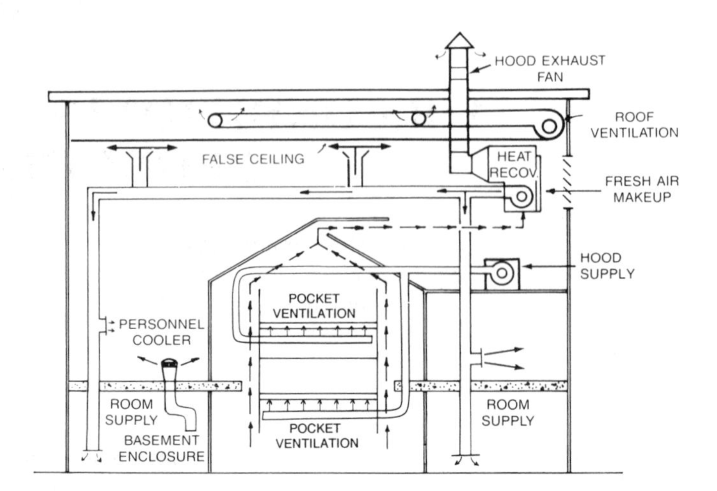

Fig. 2 Paper Machine Area

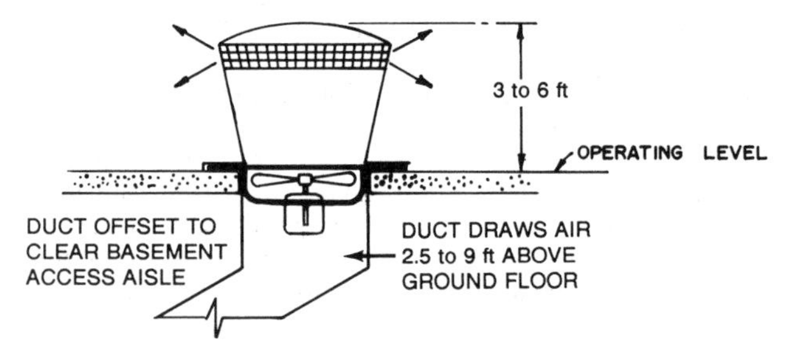

Fig. 3 Personnel Cooler

usually has a heated inside air circulation system that scrubs the underside of the roof to prevent condensation in cold weather. Additional roof exhaust may also remove accumulated heat from the drier section and the dry end area during warmer periods. Roof exhaust should be dominant in the wet end of the machine room.

In the drier section of a paper machine, steam-heated rotating drums dry the paper web traveling in a serpentine path in contact with the drier surface. Exhaust hoods control the heat from the driers and the moisture evaporated from the paper web. Most modern machines have enclosed hoods, which reduce the mass flow of air required to less than 50% of an open hood exhaust. Temperatures inside an enclosed hood range from 130 to 140°F at the operating floor level to 180 to 200°F in the hood exhaust plenum at 70 to 90% rh, with an exhaust rate generally ranging from 300,000 to 400,000 scfm.

Pocket ventilation air and hood supply air are drawn from the upper level of the machine room, where possible, to benefit from the preheating of the makeup air from process heat as it rises. The basement of the drier section is also enclosed to control infiltration of machine room air to the enclosed hood. The hood supply and pocket ventilation air typically operate at 200°F, while some systems run at temperatures as high as 250°F. Enclosed hood exhaust is typically in the range of 300 cfm per ton of machine capacity. The pocket ventilation and hood supply is designed for 75 to 80% of the exhaust, with the balance infiltrated from the basement and machine room. Large volumes of air are required to balance the paper machine's exhaust with building air balance (from 500,000 to 800,000 scfm).

Heat recovery potential from the hood exhaust air should be evaluated. Most of the energy in the steam supplied to the paper driers is converted to latent heat in the hood exhaust as water is evaporated from the paper web. Air-to-air heat exchangers are used where the air supply is located close to the exhaust. Air-to-liquid heat exchangers that recirculate water-glycol to heat remote makeup air units may also be used. Air-to-liquid systems provide more latent heat recovery, resulting in three to four times more total heat recovery than air-to-air units. Some machines use heat recovered from the exhaust air to heat process water. Ventilation rates in paper machine buildings range from 10 to 25 air changes per hour in northern mills to 20 to 50 in southern mills. Some plants use computer control of air systems that accounts for variable production rates and changes in the outside air temperature to optimize the operation of the total system and conserve energy.

After fine, bond, and cut papers have been bundled and/or packaged, they should be wrapped in a nonpermeable material. Most papers are produced with less than 10% moisture content by weight, the average being 7%. Dry paper and pulp are hygroscopic and will begin to noticeably swell and deform permanently when relative humidity exceeds 38%. Therefore, the finished products should be stored in an environmentally constant condition or encapsulated to maintain their uniform moisture content.

Finishing Area

To produce a precisely cut paper that will stabilize at a desirable equilibrium moisture content, the finishing areas require temperature and humidity controls. Further converting operations of printing and die cutting of the paper require optimum sheet moistures for efficient processing. Finishing room conditions will range from 70 to 75 °F db and from 40 to 45% rh. The system should maintain the selected conditions within reasonably close limits. Without precise environmental control, the paper equilibrium moisture content will vary, influencing dimensional stability, tendency to curl, and further processing.

PROCESS AND MOTOR CONTROL ROOMS

In most pulp and paper applications, process control, motor control, and switchgear rooms are located in separate rooms isolated from the process environment. Air conditioning removes heat generated by equipment, lights, etc., and reduces the air-cleaning requirements. (If the control room includes a computer, a computer terminal, or data processing equipment, refer to Chapter 17.) Ceiling grilles or diffusers should be located above access aisles to avoid the risk of condensation on control consoles or electrical equipment on start-up and on recovery after an air-conditioning shutdown. Electrical rooms are usually maintained in the range of 75 to 80 °F, with control rooms at 73 °F. Humidities are maintained in the 45 to 55% range in process control rooms and are not normally controlled in electrical equipment rooms.

More electronic equipment is currently being used in electrical control rooms for both distributed control and for control rooms with centralized computer process control. This equipment is susceptible to corrosion. The typical pulp and paper mill environment contains both particulate and vapor phase contaminants with sulfur- and chloride-based compounds. To protect the equipment, use multistage particulate and adsorbent filter systems that have treated activated charcoal and potassium permanganate impregnated alumina sections for vapor phase contaminants, as well as fiberglass and cloth media for particulate. A minimum amount of outside air should be used. Air conditioning and filtration of the outside air and a portion of the recirculated air will control the problem if filters are carefully maintained.

Switchgear and motor control centers are not as heat sensitive as control rooms, but the moisture-laden air carries chemical residue onto the contact surfaces. Arcing, corrosion, and general deterioration usually result. In general, a minimum of outside air is used, and air conditioning is provided to protect these areas.

Paper Test Laboratories

Design conditions within paper mill laboratories are critical and must be followed rigidly. The most recognized standard for testing environments for paper and paper products (paperboard, fiberboard, and containers made from them) is the Technical Association of Pulp and Paper Industry's *Standard* T402. Other standards include ASTM E171, Standard Atmospheres for Conditioning and Testing Materials, and ISO/TC 125, Enclosures and Conditions for Testing. The TAPPI T402 standard is discussed in this chapter.

Standard pulp and paper testing laboratories have three environments: a preconditioning atmosphere, a conditioning atmosphere, and a testing atmosphere. A sample tests differently physically if it is brought to a testing humidity from a high humidity than if it were brought to the same conditions from a lower humidity. Preconditioning at lower relative humidity tends to eliminate hysteresis. For a preconditioning atmosphere, TAPPI T402 recommends 10 to 35% rh and 72 to 104 °F db. This is usually accomplished in a controlled, conditioned cabinet.

Conditioning and testing atmospheres should also be maintained at 50 ± 2.0% rh and 73 ± 2 °F db. The designer should realize, however, that a change of 2 °F db at 73 °F makes the relative humidity fluctuate as much as 3%. A dry-bulb temperature tolerance of ±1 °F must be held to maintain a ± 2.0% rh. These low humidity variations suggest reheat or chemical driers of some type. Humidistatic and temperature instrumentation as well as graphical wet- and dry-bulb recorders should be provided for the laboratories.

Miscellaneous Areas

The pulp digester area contains many items contributing to high heat release and possibly dusty conditions. With batch digester use, the chip feeders are a source of dust requiring hooded exhaust and makeup air. The wash and screen areas have many items with hood exhausts that require considerable makeup air. Good ventilation is required to control fumes and limit humidity. The lime kiln feed-end is a room with extremely large heat releases requiring high ventilation rates or air conditioning.

Recovery-boiler and power-boiler buildings have conditions similar to those of many power plants, and the ventilation rates are similar. The control rooms are generally air conditioned. Used for making ground wood products, the grinding motor room contains many large motors that require ventilation to keep humidity low.

SYSTEM SELECTION

System and equipment selection for a pulp and paper project depends on variables unique to the particular application; *i.e.,* plant mediums available, plant layout, plant atmosphere, geographic location, roof and ceiling heights (which can exceed 100 ft in elevation), and degree of control desired.

Chilled water systems are economical and practical for most pulp and paper operations, because they require both large cooling capacity and precision of control to maintain the temperature and humidity requirements of the laboratories and the finishing areas. In the bleach plant, the manufacture of chlorine dioxide is enhanced by the use of 45 °F or lower temperature water, which is often supplied by the mill air-conditioning, chilled water system. If clean plant or process water is available, water-cooled chillers, supplemented by water-cooled direct expansion package units for remote, small areas, are satisfactory. However, if plant water is not clean enough for this application, a separate cooling tower and condenser water system must be installed for the air-conditioning systems.

Most manufacturers prefer water- over air-cooled systems because of gases and particulates present in most paper plant atmospheres. The most prevalent contaminants are chlorine gas, caustic soda, borax, phosphates, and sulfur compounds. With more efficient air cleaning, the air quality in and about most mills is adequate for properly placed air-cooled chillers or condensing units, which have well-analyzed and properly applied coil and housing coatings. Phosphor-free brazed coil joints are recommended in areas where sulfur compounds are present.

Heat is readily available from the processing operations and should be used whenever possible. Most plants have quality hot water and steam, which can be geared to unit heater, central station, or reheat use quite easily. Evaporative cooling should not be ignored. Newer plant air-conditioning methods, using energy conservation techniques, such as temperature destratification and stratified air conditioning, apply themselves well to this type of large structure. As in most industrial applications, absorption systems can be considered for pulp and paper plants, since they afford some degree of energy recovery from high-temperature steam processes.

BIBLIOGRAPHY

ACGIH. 1988. *Industrial ventilation. A manual of recommended practice,* 20th ed. American Conference of Governmental Industrial Hygienists, Cincinnati, OH.

ASTM. 1987. Standard specification for standard atmospheres for conditioning and testing materials. ASTM *Standard* E171-87. American Society for Testing and Materials, Philadelphia, PA.

Britt, K.W. 1970. *Handbook of pulp and paper technology,* 2nd ed. Van Nostrand Reinhold, New York.

Casey, J.T. 1960. *Pulp and paper chemistry and chemical technology,* Vol. 3. Interscience Publishers, Inc.

Houston Mill Guide. Southland Paper Company, Houston, TX.

ISO/TC 125. Enclosures and Conditions for Testing. International Organization for Standardization, Geneva, Switzerland.

Rasmussen, E.F. 1961. *Dry kiln operator's manual.* Agriculture Handbook No. 188, USDA.

Stephenson, J.N. 1950. *Preparation and treatment of wood pulp,* Vol. 1. McGraw-Hill Publishing Co., New York.

TAPPI. 1983. Standard conditioning and testing atmospheres for paper, board, pulp handsheets, and related products. Test Method T402 OM-83. Technical Association of the Pulp and Paper Industry, Atlanta, GA.

CHAPTER 24

NUCLEAR FACILITIES

THE heating, ventilating, and air-conditioning (HVAC) requirements for facilities using radioactive materials are discussed in this chapter. These facilities include nuclear power plants, fuel fabrication and processing plants, plutonium processing plants, and, to a lesser degree, hospitals, corporate and academic research facilities, and other facilities housing nuclear operations or materials. The information presented here serves as a guide to dealing with radioactive materials. However, no review can be all-inclusive; therefore, careful and individual analysis of each system is required.

BASIC TECHNOLOGY

Criticality, radiation, and regulation are three aspects of nuclear-related systems that differ in degree or technical detail from other special HVAC systems.

Criticality

Criticality considerations are unique to nuclear facilities. Criticality is the chain reaction of fissionable material that produces extreme radiation and heat as a minimum. Unexpected or uncontrolled criticalities must be prevented at all costs. Only a limited number and type of facilities—including fuel plants, weapons facilities, and some national laboratories—handle special nuclear material (SNM) subject to criticality concerns.

Radiation Fields

Radiation fields are found in all facilities using nuclear materials. The problems posed by radiation fields are personnel safety and material degradation. Although the problem of material degradation is usually covered under regulatory requirements, it must be considered in all designs. The amount of material degradation is affected by such factors as temperature, chemical compatibility, and pressure. The personnel safety hazard is more difficult to ascertain than is the amount of material degradation because a radiation field cannot be detected without special instruments. It is the responsibility of the designer and end-user to regulate radiation fields and personnel exposure.

Regulation

In the United States, the Department of Energy (DOE) regulates weapons-related facilities and national laboratories; the Nuclear Regulatory Commission (NRC) controls civil, industrial, and power facilities. Further complicating the issue of regulation is the addition of other government regulations at the local, state, and federal levels. For example, meeting an NRC requirement does not relieve the designer or operator of meeting Occupational Safety and Health Act (OSHA) requirements. All guidelines set by these agencies and the state, local, and federal government must be met when designing an HVAC system that will be used near radioactive materials.

The preparation of this chapter is assigned to TC 9.2, Industrial Air Conditioning.

As Low as Reasonably Achievable

As low as reasonably achievable (ALARA) means that all aspects of a nuclear facility are designed so that the workers are exposed to the minimum amount of radiation that is reasonably achievable. This does not refer to meeting legal minimums, but rather the lowest, below-legal minimum.

Design

HVAC requirements for a facility using radioactive materials depend on the type of facility and the specific service required. The following factors should be considered in the design:

1. Potential airborne radioactivity that could be encountered in the form of either particulate or gaseous matter.
2. Control of the HVAC system so that portions can be effectively shut down, as applicable, in the case of any event, accident, or natural catastrophe.

The design basis in nuclear facilities requires all systems and their components to have active control during and after any event, accident, or natural catastrophe.

Normal or Power Design Basis

The normal design basis, or *power design basis* for nuclear power plants covers normal plant operation, including normal operation mode and normal shutdown mode. This design requirement does not impose requirements on various systems or components above and beyond those imposed by the standard criteria specified for indoor conditions.

Safety Design Basis

A safety design basis must be defined if the facility cannot readily be shut down and isolated to an inactive state at any time. The safety design basis covers plant emergency operation. It establishes the special requirements necessary for a safe work environment and public protection from radiation exposure.

Any system designated as essential or safety related must mitigate the effect of an event, accident, or natural catastrophe that may cause the release of radioactivity to the surroundings or to the plant atmosphere where operating personnel are present. These safety systems must be available at all times. They must function during and after a design basis accident (DBA) or appropriate simultaneous events such as safe shutdown earthquake (SSE), tornado, loss of coolant accident (LOCA), and loss of off-site electrical power (LOEP). Nonsafety-related equipment should not adversely affect safety-related equipment. Therefore, the following additional requirements are imposed on safety-related systems and components. These requirements can be found in the safety analysis report (SAR) for an existing nuclear facility.

System Redundancy. Systems must be redundant so that the function can be performed even if a component of one system fails. Such failure should not cause a failure to the other subsystem. For additional redundancy requirements, refer to the Nuclear Regulatory Commission Facilities section.

Seismic Qualification. All safety-related components, including equipment, pipe, duct, and conduit, must be seismically qualified by testing or calculation to document that they are able to withstand and perform under the shock and vibration caused by a safe shutdown earthquake and operating basis earthquake (the largest earthquake postulated for the region in which the plant is located). This qualification includes any amplification by the building structure. In addition, any component whose failure could jeopardize the essential function of a safety-related component must be seismically qualified or restrained to prevent such failure.

Environmental Qualification. Components must often be environmentally qualified; *i.e.*, the useful life of the component in the environment in which it is operating must be determined through a program of accelerated aging. Environmental factors, such as temperature, humidity, pressure, and accumulated radioactivity, must be considered.

Quality Assurance. All designs and components of essential systems must comply with the requirements of a quality assurance (QA) program for design control, inspection, documentation, and traceability of material. Refer to 10 CFR 50, Appendix B, or ANSI/ASME NQA-1 for quality assurance program requirements.

Emergency Power. When required, all safety-related systems must be powered by a backup source such as an emergency diesel generator and provided with chilled or cooling water.

Outdoor Conditions

The 1989 ASHRAE *Handbook—Fundamentals*, the National Weather Service, or site meteorology provide information on outdoor conditions, temperature, humidity, solar load, altitude, and wind. DOE *Order* 6430.1A may also specify requirements.

Nuclear facilities generally consist of heavy structures with high thermal inertia. Time lag should be considered for solar loads. In some cases, 24-h averages suffice. Examples of these applications are diesel generator buildings or safety-related pump houses in nuclear power plants.

Indoor Conditions

Indoor temperatures are dictated by occupancy, equipment or process requirements, and personnel activities. HVAC system temperatures are dictated by the environmental qualification of the safety-related equipment located in the air-conditioned space and by ambient conditions during different operating modes of the equipment.

Indoor Pressures

A specific pressure in relation to the outside atmosphere or adjacent areas must be specified where control of air pattern is required. For process facilities where pressure zones are identified, the pressure relationships are specified in the Confinement System section.

In facilities where zoning is different from those of process facilities, and in cases where rooms are within the same zone but where potential airborne radioactivity must not spread, any airborne radioactivity must be controlled by airflow.

Airborne Radioactivity

The level of airborne radioactivity within a facility and the amount released to the surroundings must be controlled to meet the requirements of 10 CFR 20, 10 CFR 50, and 10 CFR 100 (see section on Codes and Standards).

Tornado Protection

Protection from tornados and the objects or missiles generated by tornados is normally required to prevent release of radioactive material to the atmosphere. When a tornado passes over a facility, a rapid, sharp drop in pressure occurs. If exposed to the transient pressure, ducts and filter housings could collapse because the pressure in the structure would be equal to that of the environment prior to the reduction caused by the tornado. Protection is usually provided by tornado dampers and missile barriers in all appropriate openings in the outside walls of the structure. Tornado dampers are heavy-duty, low leakage dampers designed for pressure differences in excess of 3 psi. Normally, tornado dampers are safety related and are environmentally and seismically qualified.

Fire Protection

Fire protection for the HVAC and filtration systems must be evaluated. Heat detectors and Halon systems should be considered for special operations, such as gloveboxes.

Specific Areas

The following areas must adhere to set safety requirements:

- Office areas
- Laboratories
- Change rooms and toilet facilities
- Mechanical and electrical equipment rooms
- Battery rooms
- Air locks
- Alarm stations

Control Room Habitability Zone

The HVAC system is a safety-related system that must fulfill the following requirements during all normal and postulated accident conditions to ensure that continuous occupancy can be maintained.

1. Maintain conditions comfortable to personnel and ensure the continuous functioning of control room equipment.
2. Protect personnel from exposure to potential airborne radioactivity present in the outside atmosphere or surrounding plant areas.
3. Protect personnel from exposure to potential toxic chemicals that are postulated to be released from the site or surrounding areas.
4. Protect personnel from the effects of high energy line breaks in the surrounding plant areas.
5. Protect personnel from combustion products emitted from fires on site.

Filtration

In nuclear power plants, HVAC filter systems remove radioactive particulate and radioactive iodine gases. They filter potentially contaminated exhaust prior to discharge to the environment and serve to protect personnel in plant areas. The filter train typically consists of the following components in series: demister, heater, prefilter, charcoal filter (adsorber), and high efficiency particulate air (HEPA) filter.

For nuclear processing facilities, the primary confinement zone, where contamination normally exists, is usually exhausted through multiple HEPA filters, a sand filter, or a combination of both. Secondary and tertiary confinement zones are exhausted through HEPA filters. The number of stages may differ depending on specific conditions and regulations.

Exhaust from other facilities using radioactive materials, such as a laboratory, are exhausted through HEPA filters.

In all facilities, the filter must contain the contaminant to be controlled. The filter systems must allow for reliable in-place testing and easy filter replacement.

The following types of filters are used in nuclear facilities:

Dust Filter/Prefilter. Dust filters are selected for the efficiency deemed necessary for the particular ventilation required. High efficiency dust filters are often used as prefilters for the special filters listed below to prevent them from being loaded with atmospheric dust and to minimize replacement costs.

HEPA Filters. These filters are generally used where there is a risk of particulate airborne radioactivity. They have a 99.97% minimum efficiency for removing 0.3 μm particles. They are tested by a standard procedure, using a cold-generated smoke of monodispersal dioctylphthalate (DOP).

Sand Filter. A sand filter consists of multiple beds of sand and gravel through which air is drawn. The air enters an inlet tunnel that runs the entire length of the filter. Smaller cross-sectional laterals running perpendicular to the inlet tunnel distribute the air across the base of the sand. The air then rises through the several layers of various sizes of sand and gravel, typically at the rate of 5 fpm. The air is then collected in the outlet tunnel for discharge to the atmosphere.

Charcoal Filters. Charcoal adsorbers use activated charcoal-based impregnated charcoal to remove particularly radioactive iodine, which is a vapor or gas. The charcoal bed depths are typically 2 or 4 in. These filters have an efficiency of 99.9% for elemental iodine and 95 to 99% for organic iodine. Charcoal filters rapidly lose efficiency as the relative humidity increases. They are preceded by a heating element to keep the relative humidity below 70% for entering air conditions up to saturation.

Argon is adsorbed on charcoal at extremely low (cryogenic) temperatures to control the argon content in the primary coolant. Helium is adsorbed in a high-temperature gas cooled reactor (HTGR) or in the off-gas system of a boiling water reactor (BWR).

Demisters (Mist Eliminators). If entrained moisture droplets are expected in the airstream, demisters are required to protect HEPA and charcoal filters. Demisters should be fire resistant. Heaters should be placed downstream of the demister to prevent saturated air from reaching the filter.

DEPARTMENT OF ENERGY FACILITIES

The HVAC system must be designed in accordance with DOE *Order* 6430.1A. Critical items and systems of plutonium processing facilities are designed to confine radioactive materials under normal operation and DBA conditions as required by 10 CFR 100. The degree of confinement must be sufficient to limit releases of radioactive materials to the environment to the ALARA level. In no case can the applicable exposure regulations be exceeded with respect to the operating personnel or the public at the site boundary or nearest point of public access.

The probability and effects of DBAs and SARs must be considered. Protection of personnel is the primary concern in all areas of the design. The nature of the material to be handled, including the isotopes of plutonium and/or other radioactive elements present, is taken into account in making these assignments. The engineer must be familiar with all regulations as applied to HVAC systems to avoid costly retrofit.

Confinement Systems (Zoning)

Typical process facility confinement systems are shown in Figure 1. Process facilities comprise several zones, discussed below.

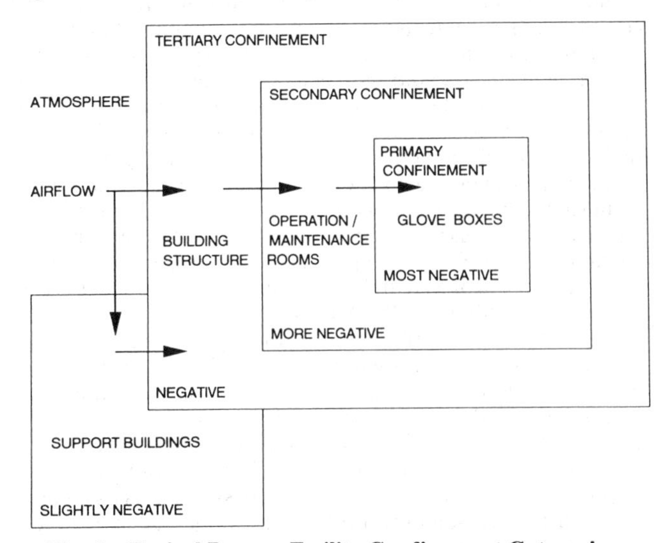

Fig. 1 Typical Process Facility Confinement Categories

Primary Confinement. This zone includes the interior of a hot cell, canyon, glovebox, or other containment for handling radioactive material. Containment features must prevent the spread of radioactivity within and release from the building under both normal and upset conditions up to and including a DBA for the facility. Complete isolation from neighboring facilities is necessary. Multistage HEPA and/or sand filtration of the exhaust is required.

Secondary Confinement. This zone is provided by walls, floors, roofs, and associated ventilation exhaust systems of the cell or enclosure surrounding the process material or equipment. Except in the case of glovebox operations, the area inside this barrier is usually unoccupied.

Tertiary Confinement. This zone is provided by the walls, floors, roofs, and associated ventilation exhaust systems of the facility. It provides a final barrier against release of hazardous material to the environment.

Uncontaminated Zone. This zone includes offices and cold shop areas. Radiation monitoring may be required at exit points.

Zone Pressure Control

Negative static pressure increases (becomes more negative) as zone classifications move from the uncontaminated zone to the primary confinement zone. Static pressure levels cause all air leakage to flow from uncontaminated to higher levels of potential contamination. All zones should be maintained at negative pressure with respect to the ambient atmospheric pressure.

Recirculating refers to the reuse of air in a particular zone or area. Room air recirculated from a space or zone can be returned to the primary air-handling unit for reconditioning and then, with the approval of health personnel, returned to the same space (zone). All secondary and tertiary recirculated air must be HEPA filtered prior to reintroducing the air to the same space or a zone of greater potential contamination. These systems must convert to 100% outside air as soon as contamination is detected downstream of the HEPA filtration. Recirculating air is not permitted in primary confinement areas.

Ventilation

Ventilation systems are designed to confine radioactive materials under normal and DBA conditions and to limit radioactive discharges to the required minimum. Where needed, inert atmospheres are used in enclosures. In such cases, recycle ventilation should be considered.

Suitable means of bypassing the recirculating system must be considered since it is required to discharge ventilation flow directly to the exhaust system. In general, recirculating air systems should be considered in nonradioactive or nontoxic environments. Exceptions may be considered only after a careful evaluation of comparative safety risks. Ventilation systems ensure that airflows are, under all normal conditions, toward areas (zones) of progressively higher potential radioactive contamination. Air-handling equipment should be sized conservatively so that upsets in the airflow balance do not cause air to reverse from higher to lower zone classification. Examples of an upset include improper use of an air lock, occurrence of a credible breach in the confinement barrier, or excessive loading of HEPA filters.

HEPA filters at ventilation inlets in all primary confinement zone barriers prevent movement of contamination from high to low level zones in the event of an airflow reversal. Ventilation system balancing should ensure that the building air pressure is always negative with respect to the outside atmosphere.

A safety analysis is necessary to establish minimum acceptable response requirements for the ventilation system and its components, instruments, and controls under normal, abnormal, and accident conditions.

The number of required exhaust filtration stages from any area of the facility is determined by analysis to limit quantities and concentrations of airborne radioactivity or toxic material released to the environment during normal and accident conditions in conformance with applicable standards, policies, and guidelines. Consult DOE *Order* 6430.1A and DOE *Order* 5480 for air-cleaning system criteria.

Ventilation Requirements

A partial recirculating ventilation system may be considered for economic reasons. However, such systems must be designed to prevent the entry of contaminated exhaust into the room air-recirculating systems. If the room air is recirculated, at least one stage of HEPA filtration is required.

Safety-related items of the ventilation system, usually the exhaust and the related fire suppression and detection system, are supplied with emergency power. Controls for these systems are generally supplied with an uninterruptible emergency power supply.

Sufficient redundancy and/or spare capacity ensures adequate ventilation during normal operations and DBA conditions. Failure of any single component or control function should not compromise minimum adequate ventilation.

The exhaust system is designed to clean radioactivity from the discharge air, to safely handle combustion products, and to maintain the building under negative pressure relative to the outside.

Provisions may be made for independent shutdown of ventilation systems or isolation of portions of the systems where it could be an advantage to operations, filter change, maintenance, or emergency procedures such as fire fighting. All possible effects of the shutdown on the airflows in other interfacing ventilation systems should be considered. Positive means must be provided to control backflow of air that might transport contamination. A HEPA filter installed at the interface between the enclosure and ventilation system minimizes contamination in the ductwork. A prefilter reduces HEPA filter loading. These filters are not to be considered the first stage of an airborne contamination cleaning system.

Ventilation Systems

Nuclear air-cleaning systems may include any or all of the following elements in the design of the overall air filtration and air-conditioning system.

- Air-sampling devices
- Carbon bed adsorbers
- Absorption, HEPA, sand, glass fiber filters, and prefilters
- Scrubbers
- Demisters
- Process vessel vent systems
- Condensers
- Distribution baffles
- Fire suppression systems
- Fire and smoke dampers
- Exhaust stacks
- Fans
- Coils
- Heat removal systems
- Pressure- and flow-measuring devices
- Radiation-measuring devices
- Critically safe drain systems
- Tornado dampers

The ventilation system and associated fire suppression system are designed for fail-safe operation. The ventilation system is appropriately instrumented and alarmed to report and record its behavior with readouts in control areas and in the utility services areas.

Both manual and automatic controls to alter system operation during unusual conditions should be considered. Provisions for convenient maintenance, decontamination, and/or replacement of components in the supply, exhaust, and filtration systems are necessary. Dampers/valves should be located so that a bank of filters can be completely isolated from the ventilation systems during filter element replacement operations. Filter redundancy may be required to maintain system operation.

Air and Gaseous Effluents Containing Radioactivity

All air and other gaseous effluents are exhausted through a ventilation system designed to remove radioactive particulates. Exhaust ducts or stacks downstream of final filtration that may contain radioactive contaminants must have two monitoring systems. One should be of the continuous type and the other a fixed sampler. These systems may be a combination unit. The probes are designed for isokinetic sampling and are located according to ANSI N13.1. Each monitoring system is connected to an emergency power supply.

NUCLEAR REGULATORY COMMISSION FACILITIES

NUCLEAR POWER PLANTS

The two basic commercial light-water power reactors used in the United States today are the pressurized water reactor (PWR) and the boiling water reactor (BWR). For both types of reactors, the main objective of the HVAC systems, in addition to personnel comfort and reliable equipment operation, is protecting operating personnel and the general public from airborne radioactive contamination during all normal and emergency operating modes of plant operation. Title 10 of the Code of Federal Regulations, Part 20 (10 CFR 20), sets forth the requirements to maintain radiation exposure as low as reasonably achievable (ALARA). The ALARA concept is used to accomplish the design objective of the HVAC system. In no case is the radiological dose allowed to exceed the limits as defined in 10 CFR 50 and 10 CFR 100.

The NRC has developed regulatory guides (RG) considered acceptable for meeting the design criteria that delineate techniques of evaluating specific problems and provide guidance to licensed applicants concerning information needed by the NRC to accomplish its review. Four regulatory guides that relate directly to HVAC system design are RG 1.52, RG 1.78, RG 1.95, and RG 1.140.

Deviations from RG criteria must be justified by the owner and approved by the NRC.

The design of the nuclear power generating station HVAC systems must ultimately be approved by the NRC staff in accordance with 10 CFR 50, Appendix A. The NRC developed standard review plans (SRP) as part of *Regulatory Report* NUREG-0800 to provide an orderly and thorough review. As such, the SRP provides a good basis or checklist for the preparation of a safety analysis report (SAR). The safety review plan is primarily based on the information provided by an applicant in an SAR as required by Section 50.34 of 10 CFR 50. Technical specifications for the nuclear power plant systems are developed by the owner and approved by the NRC as defined in Section 50.36 of 10 CFR 50. Technical specifications define safety limits, limiting conditions for operation, and surveillance requirements that must be met by all systems important to plant safety.

Minimum requirements for the performance, design, construction, acceptance testing, and quality assurance of equipment used in components in nuclear safety-related air and gas treatment systems in nuclear facilities are contained in the Code on Nuclear Air and Gas Treatment, ANSI/ASME AG-1.

PRESSURIZED WATER REACTOR HVAC SYSTEMS

Reactor Containment Building

The containment building houses the reactor in a nuclear power plant. The conditions for temperature and humidity are dictated by the nuclear steam supply system (NSSS). Generally, these are the criteria specified for three sets of operating conditions: normal operation, refueling operation, and postloss of coolant accident (LOCA) condition. General design requirements are contained in ANSI/ANS 56.6.

Normal Operating Conditions. Temperature and humidity requirements are specified by the NSSS supplier. Some power plants require recirculation filtration trains in the containment building to control the level of airborne radioactivity. Cooling is provided with a reactor containment fan cooling system.

Refueling Condition. The maximum temperature during refueling is influenced by the refueling personnel. Working in protective clothing, their activities are slowed by discomfort and, thus the refueling outage is prolonged. The required cooling can be accomplished by normal cooling units since the cooling load is low when the reactor is shut down. Ventilation with outside air is necessary.

Postloss of Coolant Accident Condition. A loss of coolant accident occurs if the primary cooling loop breaks. When this condition occurs, circulating water under high pressure and temperature flashes and fills the containment building with radioactive steam. The major source of radioactivity is radioactive iodine dissolved in the water.

Should such an accident occur, the primary actions taken are directed toward reducing the pressure in the containment building and lowering the amount of radioactive products in the containment atmosphere.

Pressure is reduced by the reactor containment fan coolers and/or the containment sprays to cool the atmosphere and condense the steam.

Containment Cooling. The following systems are typically provided for containment cooling:

Reactor containment fan coolers. These units remove most of the heat load. Distribution of the air supply depends on the containment layout and the location of the major heat sources.

Reactor cavity air-handling units or fans. Generally, these units are transfer fans without coils that provide cool air to the reactor cavity.

Control rod or element drive mechanism (CRDM or CEDM) air-handling units. The CRDM and CEDM are usually cooled by an induced draft system using exhaust fans. The flow rates, pressure drops, and heat loads are generally high; it is thus desirable to cool the air before it returns to the containment atmosphere.

Essential reactor containment fan-cooling units. Normally, the containment air-cooling system, or part of the containment air-cooling system, must provide cooling after a postulated accident. Therefore, the equipment must perform at high temperature, high pressure, high humidity, and a high level of radioactivity. Cooling coils are provided with essential service water.

System design must accommodate both normal and accident conditions. The ductwork must be analyzed to ensure that it will not collapse due to the rapid pressure buildup associated with accident conditions. Fan motors must be sized to handle the high density of air during accident conditions.

Radioactivity Control. Airborne radioactivity is controlled by the following means:

Essential containment air filtration units. For some older power plants, redundant filter units, powered from two Class IE buses, are relied on to reduce the amount of post-LOCA airborne radioactivity. The system consists of a demister, heater, HEPA filter, and a charcoal adsorber followed by a second HEPA filter. The electric heater is designed to reduce the relative humidity from 100% to less than 70% at the design inlet air temperature. All the components must be designed and manufactured to meet the requirements of a LOCA environment.

In the case of LOCA and the subsequent operation of the filter train, the charcoal becomes loaded with radioactive iodine such that the decay heat could produce self-ignition of the charcoal if the airflow stops. Therefore, a secondary fan maintains a minimum airflow through the charcoal bed to remove the heat generated by the radioactivity in case the primary fan stops. The decay heat fan is powered from a Class IE power supply. The filtration units are located inside the containment.

Containment power access purge or minipurge. It is necessary to ventilate during normal operation when the reactor is under pressure to control the level of airborne radioactivity within the containment. The maximum allowed opening size in the containment boundary during normal operation is 8 in.

The system consists of a supply fan, double containment isolation valves in each of the containment wall penetrations (supply and exhaust), and an exhaust filtration unit with a fan. The filtration unit has a HEPA filter and a charcoal adsorber followed by a second HEPA filter.

This system should not be connected to any duct system inside the containment. It should have a debris screen over the inlet and outlet ducts within the containment so that the containment isolation valves can close, even when blocked by debris or collapsed ducts.

Containment refueling purge. Ventilation is required to control the level of airborne radioactivity during refueling. Since the reactor is not under pressure during refueling, there are no restrictions on the allowed size of the penetrations through the containment boundary. Large openings, 42 to 48 in., each protected by double containment isolation valves, may be provided. The required ventilation rate is typically based on one air change per hour.

The system consists of a supply air-handling unit, double containment isolation valves at each supply and exhaust containment penetration, and an exhaust fan. Filters are recommended.

Containment combustible gas control. In case of LOCA, when water with a strong solution of sodium hydroxide or boric acid is sprayed into the containment, various metals react and produce hydrogen. Also, the fuel rod cladding can react with steam at elevated temperatures if some of the fuel rods are not covered with water. The reaction could also release hydrogen into the containment. Therefore, redundant hydrogen recombiners are needed to remove the air from the containment atmosphere, recombine the hydrogen with the oxygen, and return the air to the containment. The recombiners could be backed up by special exhaust filtration trains.

BOILING WATER REACTOR HVAC SYSTEMS

Primary Containment

The boiling water reactor (BWR) primary containment is a low leakage, pressure-retaining structure that surrounds the reactor pressure vessel and related piping. Sometimes referred to as the *drywell*, it is designed to withstand, with minimum leakage, the postulated effect of a major reactor coolant line break that would cause high temperature and pressure conditions. General design requirements are contained in ANSI/ANS 56.7.

The primary containment HVAC system consists of recirculating fan/cooler units. It normally recirculates and cools the primary containment atmosphere to maintain the environmental conditions specified by the NSSS supplier. In an accident, the system has a safety-related function of recirculating the air to prevent stratification of hydrogen that may be generated. The cooling function may or may not be safety-related, depending on the specific plant design.

Primary containment temperature problems have been experienced at many BWRs due to temperature stratification effects and underestimation of heat loads. The ductwork should adequately mix the air to prevent stratification. Heat load calculations should include a sufficient safety factor to account for deficiencies in insulation installation. In addition, a temperature monitoring system should be installed in the primary containment to ensure that bulk average temperature limits are not exceeded.

REACTOR BUILDING

The reactor building completely encloses the primary containment, auxiliary equipment, and refueling area. During normal conditions, the reactor building HVAC maintains the design space conditions and minimizes the release of radioactivity to the environment. The HVAC system consists of a 100% outside air-cooling system. Outside air is filtered, heated, or cooled as required, prior to being distributed throughout the various building areas. The exhaust air flows from areas with the potential of least contamination to the areas of most potential contamination. Prior to exhausting to the environment, potentially contaminated air is filtered with HEPA filters and charcoal adsorbers, and all exhaust air is monitored for radioactivity. To ensure that no unmonitored exfiltration occurs during normal operations, the ventilation systems maintain the reactor building at a negative pressure relative to the atmosphere.

On detecting abnormal plant conditions such as a line break, high radiation in the ventilation exhaust, or loss of negative pressure, the HVAC system's safety-related function is to isolate the reactor building to limit radioactive releases to the environment. Once isolated via fast-closing, gas-tight isolation valves, the reactor building serves as a secondary containment boundary.

The secondary containment boundary is designed to low leakage criteria and to contain any leakage from the primary containment or refueling area following an accident.

On isolation, the secondary containment pressure rises due to the loss of the normal ventilation system and the thermal expansion of confined air. A safety-related exhaust system, the standby gas treatment system (SGTS), is started to reduce and maintain the negative pressure in the building.

The SGTS exhausts air from the secondary containment and filters it through HEPA filters and charcoal adsorbers prior to release to the environment. The capacity of the SGTS is determined from an analysis of how much exhaust air is required to reduce and maintain the secondary containment at the design negative pressure, depending on the secondary containment leakage rates and required drawdown times.

In addition to the SGTS, some designs include safety-related recirculating air systems within the secondary containment to mix, cool, and/or treat the air during accident conditions. These recirculating systems use portions of the normal ventilation system ductwork; therefore, the ductwork must be classified as safety related.

If the isolated secondary containment area is not cooled during accident conditions, it is necessary to determine the maximum temperatures attained during the accident. All safety-related components in the secondary containment must be environmentally qualified to operate at these temperatures. In most plant designs, safety-related unit coolers handle the high heat release with the operation of the emergency core cooling system (ECCS) pumps.

Turbine Building

Only the BWR supplies radioactive steam directly to the turbine, which could cause a direct release of airborne radioactivity to the surroundings. Therefore, a part of the turbine building for a BWR should be enclosed, at least in the areas where release of airborne radioactivity is a possibility. These areas must be ventilated and the exhaust filtered to ensure that no radioactivity is released to the surrounding atmosphere. Filters consist of trains of a prefilter, a HEPA filter, and a charcoal adsorber followed by a second HEPA filter. The extent of filtration necessary to meet the requirements is based on the plant and site configuration.

Areas Common to Both PWRs and BWRs

All areas located outside the primary containment are designed to the general requirements contained in ANSI/ANS 59.2.

Auxiliary Building

The auxiliary building contains a large amount of support equipment, much of which handles potentially radioactive material. The building is air conditioned for equipment protection, and the exhaust is filtered to prevent the release of potential airborne radioactivity. The filtration trains consist of prefilters, HEPA filters, and charcoal adsorbers followed by a second HEPA filter.

The HVAC system is a once-through system, as needed for general cooling. Ventilation is augmented by local recirculation air-handling units located in the individual equipment rooms where additional cooling is required due to localized heat loads from the equipment. The building is maintained under negative pressure relative to the outside.

If the equipment in these rooms is nonsafety related, it is cooled by normal air-conditioning units. If the equipment is safety related, the area is cooled by safety-related or essential air-handling units powered from the same Class IE (per ANSI/IEEE *Standard* 323-1983) power as the equipment in the room.

The normal and essential unit may be combined into one unit, with both a normal and essential cooling coil and a safety-related fan served from a Class IE bus. The normal coil is served with chilled water from a normal chilled water system and the essential coil operates with chilled water from a safety-related chilled water system.

Control Room

The control room HVAC system serves the control room habitability zone—those spaces required to be habitable following a postulated accident in order to allow an orderly shutdown of the reactor—and functions to:

- Control indoor environmental conditions
- Provide pressurization to prevent infiltration
- Reduce radioactivity of the influent

- Protect the zone from hazardous chemical fume intrusion
- Protect the zone from fire
- Remove noxious fumes, such as smoke

The design requirements are described in detail in SRP 6.4, SRP 9.4.1, and a paper by Murphy and Campe (1974). Regulatory guides that directly affect control room design are RG 1.52, RG 1.78, and RG 1.95. NUREG-CR-3786 provides a summary of the documents affecting control room system design. ANSI/ASME N509 also provides guidance on the design of control room habitability systems and methods of analyzing pressure boundary leakage effects.

Control Cable Spreading Rooms

These rooms, which contain many cables, are directly above and below the control room. They usually are served by the same air-handling units that serve the electric switchgear room or the control room.

Diesel Generator Building

Nuclear power plants have auxiliary power plants to generate electric power for all essential and safety-related equipment in the event of loss of off-site electrical power. The auxiliary power plant consists of at least two independent diesel generators, each sized to meet the emergency power load. The heat released by the diesel generator and associated auxiliary systems is normally removed with outside air ventilation.

Emergency Electrical Switchgear Rooms

These rooms house the electrical switchgear that controls essential or safety-related equipment. Switchgear located in these rooms must be protected from excessive temperatures to prevent loss of power circuits required for proper operation of the plant, especially its safety-related equipment, and to retain a reasonable equipment life in accordance with the environmental qualification of the equipment.

Fuel-Handling Building

New and spent fuel is stored in the fuel-handling building. The building is air conditioned for equipment protection. The facility is ventilated with a once-through air system to control potential airborne radioactivity. Normally, the level of airborne radioactivity is so low that the exhaust need not be filtered. However, exhaust is monitored. If significant airborne radioactivity is detected, the building is sealed and kept under negative pressure by exhaust through filtration trains powered from Class IE buses.

Personnel Facilities

For nuclear power plants, this area usually includes decontamination facilities, laboratories, and medical treatment rooms.

Pump Houses

Cooling water pumps are protected by houses that are often ventilated by fans to remove the heat from the pump motors. If the pumps are essential or safety related, the ventilation equipment must also be safety related.

Radwaste Building

Except for spent fuel, radioactive waste is stored, shredded, baled, or packaged for disposal in this building. The building is air conditioned for equipment protection and ventilated for the control of potential airborne radioactivity. The ventilation may require filtration through HEPA filters and/or charcoal adsorbers prior to release to the atmosphere.

Technical Support Center

The technical support center (TSC) is an outside facility located close to the control room and is used by plant management and technical support personnel to provide assistance to control room operators during accident conditions.

The TSC HVAC system must maintain comfort conditions as well as provide the same radiological habitability conditions as the control room under accident conditions. The system is generally designed to commercial HVAC standards. An outside air filtration system (HEPA-charcoal-HEPA) pressurizes the facility with filtered outside air during emergency conditions. The TSC HVAC system does *not* have to be designed to safety-related standards.

NUCLEAR NONPOWER MEDICAL AND RESEARCH REACTORS

The requirements for HVAC and filtration systems for nuclear nonpower medical and research reactors are set by the NRC. The criteria for the HVAC and filtration systems depend on the type of reactor, which could range from a nonpressurized swimming pool type to a 10 MW or more pressurized reactor, the type of fuel, the degree of enrichment, and the type of facility and environment. To a certain degree, many of the same requirements discussed in the various nuclear power plant sections apply to these reactors. Therefore, it is imperative for the designer to become familiar with the NRC requirements for the reactor under design.

LABORATORIES

The requirements for HVAC and filtration systems for laboratories involved with the use of radioactive materials are determined by the DOE and/or the NRC. Laboratories located on DOE facilities are governed by DOE regulations. All other laboratories using radioactive materials are regulated by the NRC. Other agencies responsible for the regulation of other toxic and carcinogenic materials present in the facility may be required.

Laboratory containment equipment for nuclear-processing facilities are treated as primary, secondary, or tertiary containment zones depending on the level of radioactivity anticipated for the area and the materials to be handled.

Gloveboxes

Gloveboxes are windowed enclosures equipped with one or more flexible gloves for manual handling of material inside the enclosure from the outside. The gloves are attached to a porthole in the enclosure, thereby totally sealing the enclosure from the surrounding environment. Hence, gloveboxes permit hazardous materials to be manipulated by hand while preventing the release of material to the environment.

Since the glovebox is usually used to handle hazardous materials, the exhaust is HEPA filtered locally before leaving the box and prior to entering the main exhaust duct. In nuclear processing facilities, a glovebox is considered Zone I, or primary confinement (Figure 1), and is, therefore, subject to the criteria governing those areas. For nonnuclear processing facilities, the designer should be aware of the designated use of the glovebox and design the system according to the regulations governing that particular application.

Radiobenches

Radiobenches have the same geometric shape as a glovebox except that in lieu of the panel for the gloves, the glove area is open. Air velocity across the opening is generally the same as for laboratory hoods. The level of radioactive contamination handled in a radiobench is much lower than that handled in a glovebox.

Laboratory Fume Hoods

Laboratory fume hoods are similar to those used in nonnuclear applications. Air velocity across the hood opening must be sufficient to capture and contain all contaminants in the hood. Excessive air velocities should be avoided. High hood face velocities cause contaminants to escape when an obstruction such as an operator is positioned at the hood face. For fume hood testing, refer to ASHRAE *Standard* 110-1985, Method of Testing Performance of Laboratory Fume Hoods.

CODES AND STANDARDS

ANSI N13.1	Guide for Sampling Airborne Radioactive Materials
ANSI/ANS 56.6	Pressurized Water Reactor Containment Ventilation Systems
ANSI/ANS 56.7	Boiling Water Reactor Containment Ventilation Systems
ANSI/ANS 59.2	Safety Criteria for HVAC Systems Located Outside Primary Containment
ANSI/ASME AG-1	Code on Nuclear Air and Gas Treatment
ANSI/ASME NQA-1	Quality Assurance Program Requirements for Nuclear Facilities
ANSI/ASME N509	Nuclear Power Plant Air Cleaning Units and Components
10 CFR	Title 10 of the Code of Federal Regulations
Part 20	Standards for Protection Against Radiation (10 CFR 20)
Part 50	Domestic Licensing of Production and Utilization Facilities (10 CFR 50) Appendix A General Design Criteria for Nuclear Power Plants
Part 100	Reactor Site Criteria (10 CFR 100)
DOE Order 5480 Series	
DOE Order 6430.1A	Department of Energy General Design Criteria
ERDA 76-21	Nuclear Air Cleaning Handbook
NFPA 801	Recommended Fire Protection Practice for Facilities Handling Radioactive Materials
Murphy and Campe	Murphy, K.G. and K.M. Campe. 1974. Nuclear Power Plant Control Room Ventilation System Design for Meeting General Design Criterion 19. Paper 6-6, 13th AEC Air Cleaning Conference. (August).
NUREG-0696	Functional Criteria for Emergency Response Facilities
NUREG-0800	Standard Review Plan
SRP 6.2.3	Secondary Containment Functional Design
SRP 6.2.5	Combustible Gas Control in Containment
SRP 6.4	Control Room Habitability Systems
SRP 6.5.1	ESF Atmosphere Cleanup Systems
SRP 6.5.3	Fission Product Control Systems and Structures
SRP 9.4.1	Control Room Area Ventilation System
SRP 9.4.2	Spent Fuel Pool Area Ventilation System
SRP 9.4.3	Auxiliary and Radwaste Building Ventilation Systems
SRP 9.4.4	Turbine Area Ventilation System
SRP 9.4.5	Engineered Safety Feature Ventilation System
NUREG-CR-3786	A Review of Regulatory Requirements Governing Control Room Habitability
Regulatory Guides	Nuclear Regulatory Commission
RG 1.52	Design, Testing, and Maintenance Criteria for Engineered Safety Feature Atmospheric Cleanup System Air Filtration and Adsorption Units of LWR Nuclear Power Plants
RG 1.78	Assumptions for Evaluating the Habitability of Nuclear Power Plant Control Room During a Postulated Hazardous Chemical Release
RG 1.95	Protection of Nuclear Power Plant Control Room Operators Against Accidental Chlorine Release
RG 1.140	Design, Testing, and Maintenance Criteria for Normal Ventilation Exhaust System Air Filtration and Adsorption Units of LWR Nuclear Power Plants

CHAPTER 25

VENTILATION OF THE INDUSTRIAL ENVIRONMENT

GENERAL ventilation controls heat, odors, and hazardous chemical contaminants, which could affect the health and safety of industrial workers. In most cases, heat and contaminants generated in a space should be exhausted at their sources because better control is attained at lower airflows than those required by general (dilution) ventilation (Goldfield 1980). Chapter 27 supplements this chapter.

General ventilation may be provided by natural draft, by a combination of general supply and exhaust air fan and duct systems, by exhaust fans only (with makeup [replacement] air through inlet louvres and doors), or by supply fans only (exhaust through relief louvers and doors).

Many industrial ventilation systems must handle simultaneous exposures to hazardous substances and heat. In such cases, ventilation may consist of a combination of local, general supply, and exhaust air systems. Air supply makeup for hood exhaust may be insufficient for control of heat exposure. Therefore, the ventilation engineer must carefully analyze supply and exhaust air requirements to determine the optimum solution to achieve a balance between supply and exhaust airflows. An important design parameter is to account for seasonal conditions (*i.e.*, summer/winter) on the performance of ventilation systems.

Specification of acceptable toxic chemical and heat exposure levels for design is usually the responsibility of the industrial hygienist or industrial hygiene engineer. Such specifications may be made by using appropriate government standards along with guidelines given either in this chapter or in reference material. For most chemicals and heat exposures, standards are time-weighted averages that allow excursions above the limit so long as they are compensated for by equivalent excursions below the limit during the workday. However, exposure level standards for heat and contaminants are not fine lines of demarcation between safe and unsafe exposures. Rather, they represent conditions under which it is believed that nearly all workers may be repeatedly exposed day after day without adverse effects (ACGIH 1989). A small percentage of workers may be overly stressed at exposure values below the standards. Thus, it is prudent for the ventilation engineer to design for exposure levels below the limits.

Unfortunately, in the case of exposures to toxic chemicals, it is rare that the number of contaminant sources, their generation rates, and effectiveness of exhaust hoods are known. Consequently, the ventilation engineer must rely on common ventilation/industrial hygiene practice when designing toxic chemical controls. Close cooperation among the industrial hygienist, the process engineer, and the ventilation engineer is required (Schroy 1986).

The preparation of this chapter is assigned to TC 5.8, Industrial Ventilation.

This chapter describes principles of good ventilation practice and includes other information to help the engineer appreciate the hygiene concerns involved in the industrial environment. Various publications from NIOSH (1983), British Occupational Hygiene Society (1987), National Safety Council (1988), and the U.S. Department of Health and Human Services (1986) provide in-depth coverage of industrial hygiene principles and their application. Ventilation control measures alone, for example, are frequently inadequate for meeting heat stress standards. Optimum solutions may involve additional controls, such as spot air conditioning, changes in work-rest patterns, radiation shielding, etc. Goodfellow and Smith (1982) summarized the technical progress being made in the industrial ventilation field by different investigators throughout the world. Proceedings from international symposiums (Ventilation '85 and Ventilation '88) are also valuable sources of information on ventilation technology.

Supplemental information is included in Chapters 1 through 6, 9, 10, and 11 of the 1988 ASHRAE *Handbook—Equipment*. Chapters 13 through 24 of this volume include ventilation requirements for specific applications, and Chapter 40 covers control of gaseous contaminants. Fundamentals of heating, cooling, and ventilation are covered in the 1989 ASHRAE *Handbook—Fundamentals*.

HEAT CONTROL IN INDUSTRIAL WORK AREAS

Ventilation for Heat Relief

Many industrial work situations involve processes that release high amounts of heat and moisture to the environment. In such cases, maintaining comfort conditions (ASHRAE 1981), particularly during the hot summer months, may be economically infeasible. But comfortable conditions are not physiologically necessary. They are necessary only for the body to be in thermal balance with the environment, and this can occur at temperature and humidity conditions well above the comfort zone. Even in cases where heat and moisture gains from the process are low to moderate, comfort conditions may not be provided simply because personnel exposures are infrequent and of short duration. In such cases, ventilation is one of many controls that may be necessary to prevent excessive physiological strains resulting from heat stress conditions.

The engineer must distinguish between the control needs for *hot-dry* industrial areas and *warm-moist* conditions. In the first case, the process gives off only sensible and radiant heat without adding moisture to the air. The heat load on exposed workers is thereby increased, but the rate of cooling by evaporation of sweat is not reduced. Heat balance may be maintained, although possibly at the expense of excessive sweating. In the warm-moist situation, the wet process gives off mainly latent heat. The rise in the heat load on the worker may be insignificant, but the increase in moisture content of the air seriously reduces heat loss by evaporation of sweat. The warm-moist condition is potentially more hazardous than the hot-dry condition.

Hot-dry work situations occur around hot furnaces, forges, metal-extruding and rolling mills, glass-forming machines, and so forth. Typical warm-moist operations are found in many textile mills, laundries, dye houses, and deep mines where water is used extensively for dust control.

The industrial heat problem varies with local climatic conditions. Solar heat gain and an elevated outdoor temperature increase the heat load at the workplace, but these contributions may be insignificant when compared with the locally generated process heat. The moisture content of the outdoor air is an important climatic factor affecting hot-dry work situations, and, on a moist summer day, it seriously restricts an individual's evaporative cooling. For the warm-moist job, solar heat gain and elevated outdoor temperature are more important because, compared with the moisture release on the job, the release contributed by the outdoor air is of little significance.

Heat Stress—Thermal Standards

Heat stress is the thermal condition of the environment which, in combination with the metabolic heat generation of the body, causes the deep body temperature to exceed 100.4°F. The recommended heat stress index for evaluating an environment's heat stress potential is the wet-bulb globe temperature (WBGT), defined as follows—

Outdoors with solar load:

$$WBGT = 0.7\, t_{nw} + 0.2\, t_g + 0.1\, t_{db} \qquad (1)$$

Indoors or outdoors with no solar load:

$$WBGT = 0.7\, t_{nw} + 0.3\, t_g \qquad (2)$$

where

t_{nw} = natural wet-bulb temperature (no defined range of air velocity; not the same as adiabatic saturation temperature or psychrometric wet bulb)
t_{db} = dry-bulb temperature (shielded thermometer)
t_g = globe temperature (Vernon bulb thermometer 6-in. diameter)

The threshold limit value (TLV) for heat stress is set for different levels of physical stress, as shown in Figure 1 (ACGIH 1989). This chart depicts the allowable work regime, in terms of rest periods and work periods each hour, for different levels of work, over a range of WBGT. For applying Figure 1, it is assumed that the rest area has the same WBGT as the work area. If the rest area is at or below 75°F WBGT, the resting time is reduced by 25%. The curves are valid for people acclimatized to heat. Refer to criteria of the National Institute for Occupational Safety and Health (NIOSH 1986) for recommended WBGT ceiling value and time weighted average exposure limits for both acclimatized and unacclimatized workers.

The WBGT index is an international standard (ISO 7243) for evaluation of hot environments. It is recommended that the WBGT index and activity values be evaluated on 1-h mean values, *i.e.*, WBGT and activity are measured and estimated as time-weighted averages on a 1-h basis for continuous work, or on a 2-h basis when the exposure is intermittent. Although recommended by NIOSH, the WBGT has not been accepted as a legal standard by the Occupational Safety and Health Act (OSHA). It is generally used in conjunction with other methods to determine heat stress.

While Figure 1 is useful for evaluating heat stress, it is of limited use for control purposes or for evaluation of comfort. Air velocity and psychrometric wet-bulb measurements are usually needed in order to specify proper controls, and neither is measured in

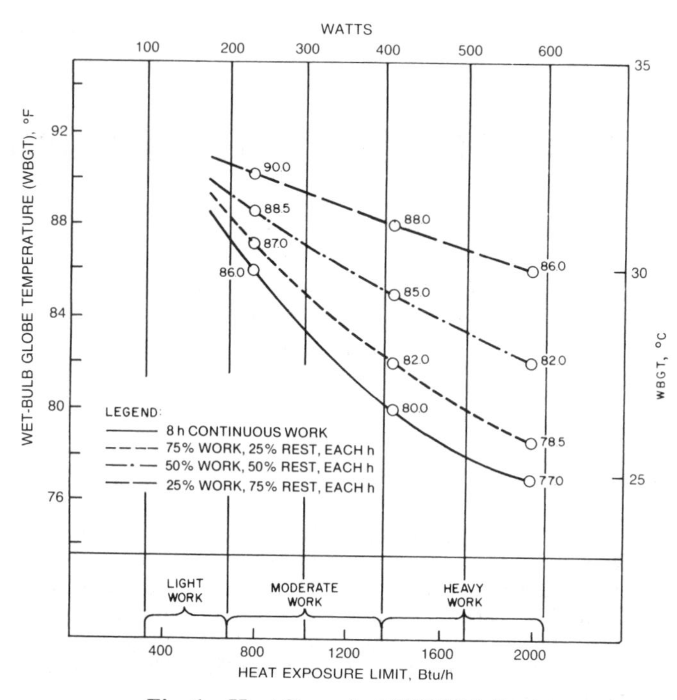

Fig. 1 Heat Stress by WBGT Method

WBGT determinations. However, Harris (1988) uses the adiabatic wet-bulb temperature line on the psychrometric chart to represent natural wet bulb as a conservative substitution in heat stress situations. More useful tools, including the Heat Stress Index (HSI), may be found in Chapter 8 of the 1989 ASHRAE *Handbook—Fundamentals* and in ISO *Standards* 7730 and 7933.

The thermal relationship between humans and their environment depends on four independent variables: air temperature, radiant temperature, moisture content of the air, and air velocity. These may combine in various ways, together with the rate of internal heat production (metabolic rate), to create different degrees of heat stress. Heat stress index formulas include calculations of the relative contributions to stress resulting from rate metabolism; radiant heat gain (or loss); convective heat gain (or loss); and evaporative (sweat) heat gain (or loss) (AIHA 1975). For supplemental information on evaluation and control of heat stress covering such methods as (1) reduction of radiation, (2) changes in work-rest pattern, (3) spot cooling, and (4) cooling vests and suits, refer to NIOSH (1986), ACGIH (1988), Constance (1983), Caplan (1980), AIHA (1975), and Brief (1971).

HEAT EXPOSURE CONTROL

Control at Source

The magnitude of heat exposure can be reduced by insulating hot equipment, locating such equipment in zones with good general ventilation within buildings or outdoors, covering steaming water tanks, providing covered drains for direct removal of hot water, and maintaining tight joints and valves where steam may escape.

Local Exhaust Ventilation

Where appropriate, local exhaust ventilation can remove the natural convection column of heated air rising from a hot process with a minimum of air from the surrounding space. Chapter 47 covers design of exhaust hoods and duct systems for local exhaust.

Radiation Shielding

In some industries, the major environmental heat load is radiant heat. Sources are hot objects and surfaces, *e.g.*, furnaces, ovens, furnace flues and stacks, boilers, molten metal, hot ingots, castings or forgings, and other hot surfaces. Because air temperature has no significant influence on radiant heat flow, ventilation is of little help in controlling such exposure. The only effective control is reducing the amount of radiant heat impinging on the workers. Radiant heat exposures can be reduced by insulating or placing radiation shields around the emitting source (Chapter 3, 1989 ASHRAE *Handbook—Fundamentals*).

Radiation shields are effective in the following forms (AIHA 1975):

Reflective Shielding. Sheets of reflective material or insulating board, semipermanently attached to the hot equipment or arranged in a semiportable floor stand.

Absorptive Shielding (Water-Cooled). The shields absorb and remove the heat from the hot equipment.

Transparent Shields. Heat reflective tempered plate glass, reflective metal chain curtains, and close mesh wire screens moderate radiation without obstructing visual contact with the hot equipment.

Flexible Shielding. Aluminum-treated fabrics give a high degree of radiation shielding.

Protective Clothing. Reflective garments such as aprons, gauntlet gloves, and face shields provide moderate radiation shielding. For extreme radiation exposures, complete suits with vortex tube cooling may be required.

If the shield is a good reflector, it will remain relatively cool in severe radiant heat. Bright or highly polished tinplate, stainless steel, and ordinary flat or corrugated aluminum sheets are efficient and durable. Foil-faced plasterboard, though less durable, gives good reflectivity on one side. To be efficient, however, the reflective shield must remain bright.

The best radiation shields are infrared reflectors, which must be installed properly to avoid transferring a radiant heat load to the wrong place. The shield should reflect the radiant heat back to the primary source, where it is removed by local exhaust. However, unless the shield completely surrounds the primary source, some of the infrared energy will be reflected into the cooler surroundings and possibly into an occupied area. Therefore, the direction of the reflected heat should be studied before shielding is installed.

VENTILATION DESIGN PRINCIPLES

General Ventilation

General ventilation supplies and/or exhausts air to provide heat relief, to dilute contaminants to an acceptable level, and to replace (makeup) exhaust air. Outdoor air is unacceptable for ventilation if it is known to contain any contaminant at a concentration above that shown in ASHRAE *Standard* 62-1989. If air is thought to contain any contaminant not listed, guidance on acceptable exposure levels should be obtained by referring to OSHA standards. General ventilation rates exceed the air quality required to dilute CO_2 produced and expired from the lungs.

General ventilation may be provided with either natural or mechanical supply and/or exhaust systems. Some factors to consider in selection and design are as follows:

1. Local exhaust systems provide general ventilation for the work area.
2. A balance of the supply and exhaust systems is required for either system to function as designed.
3. Natural ventilation systems are most applicable when internal heat loads are high and the building is tall enough to produce a significant stack effect (*e.g.*, oxygen steel-making plants, glass-melting furnances). Goodfellow (1985) outlined the design methodology for industrial ventilation systems.
4. To provide effective general ventilation for heat relief by either natural or mechanical supply, the air must be delivered in the work zones (below 10 ft) with an appreciable air velocity. A sufficient exhaust volume is necessary to remove the heat liberated in the space. Local relief systems may require supplementary supply air for heat removal.
5. Supply and exhaust air cannot be used interchangeably. Supply air can be delivered where it is wanted at controlled velocities, temperature, and humidity. Exhaust systems should be used to capture heat and fumes at the source.
6. General building exhaust may be required in addition to local exhaust systems.
7. The exhaust discharge, whether local or general, should be located where it will not be recirculated (see Chapter 14 of the 1989 ASHRAE *Handbook—Fundamentals*).

Need for Makeup Air

For safe, effective operation, most industrial plants require makeup air to replace the large volumes of air exhausted to provide conditions for personnel comfort, safety, and process operations. If makeup air is provided consistently with good air distribution, more effective cooling can be provided in the summer, and more efficient and effective heating will result in the winter. Using windows or other inlets that cannot be used in stormy weather should be discouraged. The most important needs for makeup air can be summarized as follows:

1. To replace air being exhausted through combustion processes and local and general exhaust systems (see Chapter 27).
2. To eliminate uncomfortable cross drafts by proper arrangement of supply air and prevent infiltration (through doors, windows, and similar openings) that may make hoods unsafe or ineffective, defeat environmental control, bring in or stir up dust, or adversely affect processes by cooling or disturbance.
3. To obtain air from the cleanest source. Supply air can be filtered; infiltration air cannot.
4. To control building pressure and airflow from space to space. Such control is necessary for three reasons:
   a) To avoid positive or negative pressures that will make it difficult or unsafe to open doors and to avoid the conditions mentioned in Items 1 and 2 (see Table 1).
   b) To confine contaminants and reduce their concentration and to control temperature, humidity, and air movement positively.
   c) To recover heat and conserve energy.

Table 1 Negative Pressures which May Cause Unsatisfactory Conditions within Buildings

Negative Pressure, in. of water	Adverse Conditions which May Result
0.01 to 0.02	Worker Draft Complaints—High-velocity drafts through doors and windows.
0.01 to 0.05	Natural Draft Stacks Ineffective—Ventilation through roof exhaust ventilators, flow through stacks with natural draft greatly reduced.
0.02 to 0.05	Carbon Monoxide Hazard—Back drafting will take place in hot water heaters, unit heaters, and other combustion equipment not provided with induced draft.
0.03 to 0.10	General Mechanical Ventilation Reduced—Airflows reduced in propeller fans and low pressure supply and exhaust systems.
0.05 to 0.10	Doors Difficult to Open—Serious injury may result from nonchecked, slamming doors.
0.10 to 0.25	Local Exhaust Ventilation Impaired—Centrifugal fan fume exhaust flow reduced.

Source: ACGIH. 1988. *Industrial ventilation—A manual of recommended practice*

Wherever possible, building pressure should be controlled by flow from one space to another rather than by an appreciable pressure differential. Otherwise, very large volumes of makeup air will be required. Chapter 27 has more information.

Ventilation Airflow for Temperature Control

To determine air volumes required, both sensible and latent heat loads from all sources and an appropriate air temperature rise above design outdoor temperature conditions must be estimated. In many industrial areas where people work, a 5 to 10°F temperature rise is acceptable and practical even when outdoor conditions are above 90°F. For some applications in the metals industry, a design temperature rise as high as 30 to 35°F can be used if all heat sources are at a low level and there are no workers at higher elevations in the plant. The outdoor design conditions may already be above acceptable thermal standards for even sedentary work, so additional control measures may be necessary. Generally, ventilation design of an area is established by experience with successful existing systems similar to the proposed installation. Estimates based on experience for ventilation (*e.g.*, air changes per hour and ventilation air volume per unit of floor area), properly interpreted, are useful guides to check ventilation design.

For sensible heat loads, the volume of ventilation air required may be estimated from the following equations:

$$Q = q_s/1.08\ \Delta t \quad (3)$$

where

Q = required ventilation rate, cfm
q_s = sensible heat to be removed, Btu/h
Δt = temperature rise of the air, °F

The quantity q_s must include all internal heat from equipment, process, lights, occupancy, and solar and roof transmission gains. Airflow requirements can be reduced somewhat in buildings with high roofs, if air is supplied at a low level and exhausted at the roof.

The ventilation rate for latent loads is as follows:

$$Q = q_l/0.67\ \Delta W \quad (4)$$

where

q_l = latent heat to be removed, Btu/h
ΔW = moisture content difference, indoor design and supply air conditions, grains per lb of dry air

The air quantity required is the larger of the two calculations. In cases where occupancy is limited, the air change method is often used to determine ventilation airflow (Constance 1983).

Ventilation is often expressed as volume per unit of floor area. This is a more rational approach than air changes per hour, since appropriately designed supply ventilation will provide the desired heat relief independent of the ceiling height of the space. A ventilation rate of 2 cfm/ft^2 gives reasonably good results for many plants having an internal load of 100 to 125 Btu/h · ft^2.

When local exhaust ventilation is required for contaminant control, the exhaust rates are frequently greater than the required general ventilation for comfort control. Under these conditions, the exhaust rate determines makeup air rates. A supply system with air distribution arranged so that the makeup air is provided without disturbance at the hoods and process is mandatory.

Caplan and Knutson (1978) and Peterson *et al.* (1983) established that the manner of air supply to a room with a hood has a major impact on the hood performance. The hood performance factor (or hood index) is defined as the logarithm of the ratio of contamination concentration within the hood to the contamination concentration just outside the hood. Adequate hood performance factors range from 4 to 6, with the higher number indicating a more effective hood installation (Fuller and Etchells 1979).

Computer models, fluid dynamic scale modeling, and tracer gas studies are now used frequently to predict air velocity, contaminant concentration, and temperature at arbitrary points within a work space containing equipment, furniture, and people. For example, ventilation computer models can be used to study the effects of different climatic conditions (*i.e.*, summer/winter, wind speed, wind direction, etc.) on the ventilation characteristics for a plant (Goodfellow 1987). Problems such as contamination due to cross drafts or high temperatures can be identified quickly and corrective measures taken. However, further testing and development are needed before computer models become widely used.

Fluid dynamic scale models are used to quantify flow patterns for a wide variety of industrial ventilation problems (Goodfellow 1987). Air and water are the common fluids used. A detailed analysis of the scaling parameters for the possible fluid systems is required to select the best modeling technique for the specific ventilation problem.

Tracer gases such as sulfur hexafluoride are also used as a design tool for ventilation systems (general ventilation and local exhaust ventilation). Both the exponential decay method and the constant injection rate method are used.

Roof Ventilators

Roof ventilators are basically heat escape ports located high in a building and properly enclosed for weathertightness. Stack effect plus some wind induction are the motive forces for gravity operation of continuous and round ventilators. The latter can be equipped with fan barrel and motor, thus permitting gravity operation or motorized operation.

Many ventilator designs are available; two designs are the low ventilator that consists of a stack fan with a rainhood, and the ventilator with a split butterfly closure that floats open to discharge air and self-closes. Both use minimum enclosures and have little or no gravity capacity. Split butterfly dampers tend to make the fans noisy and are subject to damage because of slamming during strong wind conditions. Because noise is frequently a problem in many powered roof ventilators, the manufacturer's sound rating should be reviewed.

Roof ventilators can be listed in diminishing order of heat removal capacity. The continuous ventilation monitor most effectively removes substantial concentrated heat loads. An efficient type is the streamlined continuous ventilator. It is designed to prevent backdraft, is weathertight, and usually has dampers that may be readily closed in winter to conserve building heat. Its capacity is limited only by the available roof area and the proper location and sizing of low level air inlets. Gravity ventilators have the advantage of low operating costs, do not generate noise, and are self-regulating (*i.e.*, a higher heat release results in higher airflow through the ventilators). Care must be taken to ensure that a positive pressure exists at the ventilators, otherwise outside air will enter the ventilators. This is of particular importance during the heating season.

Next in capacity are: (1) round gravity or windband ventilator, (2) round gravity with fan and motor added, (3) low hood powered ventilator, and (4) vertical upblast powered ventilator. The shroud for the vertical upblast design has a peripheral baffle to deflect the air up instead of down. Vertical discharge is highly desirable to reduce roof damage caused by the hot air, if it contains condensable oil or solvent vapor. Ventilators with direct-connected motors are desirable, because of the locations of the units and the belt maintenance required for units having short shaft centerline distances. Round gravity ventilators have a low capacity and are applicable to warehouses with light heat loads and to manufacturing areas having high roofs and light loads.

Streamlined continuous ventilators must operate effectively without mechanical power. Efficient ventilator operation is generally obtained when the difference in elevation between the average air inlet level and the roof ventilation is at least 30 ft and the exit temperature is 25°F above the prevailing outdoor temperature. Chapter 23 of the 1989 ASHRAE *Handbook—Fundamentals* has further details. Under these conditions and with a wind velocity of 5 mph, the ventilator throat velocity will be about 375 fpm and it will remove $1.08 \times 25\,°F \times 375 = 10{,}000$ Btu/h·ft².

To ensure this level of performance, sufficient low level openings must be provided for the incoming air. Manufacturers recommend 250 to 450 fpm inlet velocity. Insufficient inlet area and significant air currents are the most common reasons gravity roof ventilators malfunction. A positive supply of air around the hot equipment may be necessary within large buildings where the external wall inlets are remote from the equipment.

The cost of electrical power for mechanical ventilation is offset by the advantage of constant airflow. Mechanical ventilation can also create the pressure differential necessary for good airflow, even with small inlets. Inlets should be sized correctly to avoid infiltration and other problems caused by high negative pressure in the building. Often, a mechanical system is justified to supply enough makeup air to maintain the work area under positive pressure.

Careful study of airflow around buildings is necessary to avoid reintroducing contaminants from the exhaust into the ventilation system. Even discharging the exhaust at the roof level or from an area opposite the outside intake may still allow it to reenter the building even when intakes are inside walls opposite the exhaust discharge point. Chapter 14 of the 1989 ASHRAE *Handbook—Fundamentals* describes the nature of airflow around buildings.

Fusible link dampers, which close in case of fire, may be required by building codes.

Heat Conservation and Recovery

Because of the large volumes of ventilation required for industrial plants, heat conservation and recovery should be used, and will provide substantial savings (see Chapter 40). For example, assume a plant required 106,000 cfm makeup air in a heating season of 4300 h. If No. 2 fuel oil is used, with a heat content of 131,500 Btu/gal and an overall efficiency of 75%, then 1100 barrels of oil can be saved for each 9 °F reduction in heating temperature. The savings each hour are:

$$H = (1.08)(9)(106{,}000) = 1{,}030{,}000 \text{ Btu/h}$$

The fuel savings, with 4300 h of operation are:

$$\frac{(1{,}030{,}000 \text{ Btu/h})(4300 \text{ h})}{(131{,}500 \text{ Btu/gal})(42 \text{ gal/bbl})0.75} = 1100 \text{ bbl}$$

In some cases, it is possible to provide unheated or partially heated makeup air to the building. Rotary, regenerative heat exchangers recover up to 80% of the heat represented by the difference between the exhaust and the outdoor air temperatures. Reductions of 9 to 18 °F in the heating requirement for most industrial systems should be routine. While most of the heat conservation and recovery methods outlined in this section apply to heating, the saving possibility of air-conditioning systems is equally impressive. Heat conservation and recovery should be incorporated in preliminary planning for an industrial plant. Some methods are:

1. In the original design of the building, process, and equipment, provide insulation and heat shields to minimize heat loads. Vaporproofing and reduction of glass area may be required. Changes in process design may be required to keep the building heat loads within reasonable bounds. Review the exhaust needs for hoods and process and keep those to a practical, safe minimum.
2. Design the supply air systems for efficient distribution by delivering the air directly to the work zones; by mixing the supply with hot building air in the winter; by using recirculated air within the requirements of winter makeup; and by bringing unheated or partially heated air to hoods or process whenever possible (ACGIH 1988a, Holcomb and Radia 1986).
3. Design the system to achieve highest efficiency and lowest residence time for contaminants. In the design, consider that it is psychologically sound and good practice to permit workers to adjust and modify the air patterns to which they are exposed; people desire direct personal control over their working environment.
4. Conserve exhaust air by using it; *e.g.*, office exhaust can be directed first to work areas, then to locker rooms or process areas, and finally, to the outside. Clean, heated air can be used from motor or generator rooms after it has been used for cooling the equipment. Similarly, the cooling systems for many large motors and generators have been arranged to discharge into the building in the winter to provide heat and to the outside in summer to avoid heat loads.
5. Supply air can be passed through air-to-air, liquid-to-air, or hot-gas-to-air heat exchangers to recover building or process heat. Rotary, regenerative, and air-to-air heat exchangers are discussed in Chapter 34 of the 1988 ASHRAE *Handbook—Equipment*.
6. Operate the system for economy. Shut the systems down at night or weekends whenever possible, and operate the makeup air in balance with the needs of operating process equipment and hoods. Keep supply air temperatures at the minimum for heating and the maximum for cooling, consistent with the needs of process and employee comfort. Keep the building in balance so that uncomfortable drafts do not require excessive heating.

COMFORT VENTILATION

Effective comfort ventilation is based on the principle that air must be delivered directly to the work zone with sufficient air motion and at a low enough temperature to cool the worker by convection and evaporation. In most cases, the objective is to provide tolerable working conditions rather than complete comfort. In each case, the methods described for comfort ventilation are based on the assumption that exhaust ventilation, radiation shielding, equipment insulation, and possible changes in process design have been fully used to minimize the heat loads. In addition, supply air must not blow on or at hot equipment or hoods nor through the layers of hot ceiling air before reaching the work zone. In the former case, the ventilation disturbs the capture velocity of the hood or thermal rise from the hot equipment and distributes hot and/or contaminated air into the workroom. In the latter case, huge volumes of hot and possibly contaminated air are entrained and brought down to the work zone.

Ventilation Methods

The following two types of ventilation for heat relief are used:

General Ventilation. This might be termed *low-level* or *displacement ventilation* because the supply air is delivered at the 8 to 12 ft level and displaces the warm air rising from equipment, lights, and occupants. The internal heat is picked up as the supply air leaves the work zone, and the work zone may be maintained within a few degrees of the supply air temperature with relatively low supply air volumes. Such low level systems should be applied to large work areas having a high, uniform worker population. These systems are local in the sense that no attempt is made to maintain conditions in the upper levels of the building (see Figures 2 and 3).

Local-Area or Spot-Cooling Ventilation. In buildings having only a few work areas, it would be impractical and wasteful to treat the entire building. In such cases, relief may be provided by the following methods:

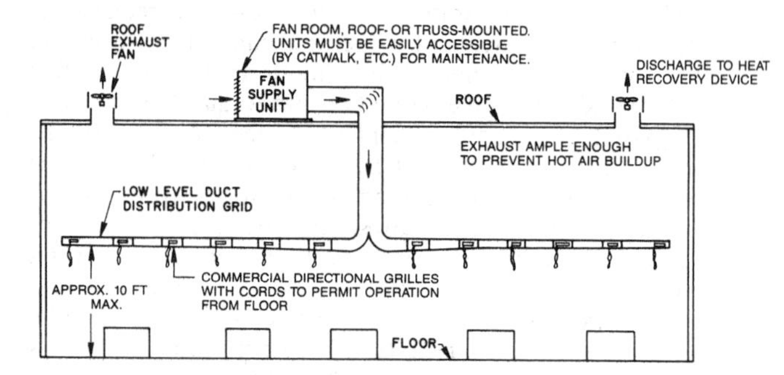

Fig. 2 Low Level Air Distribution System for Plant Heating, Relief Ventilation, and Heat Recovery

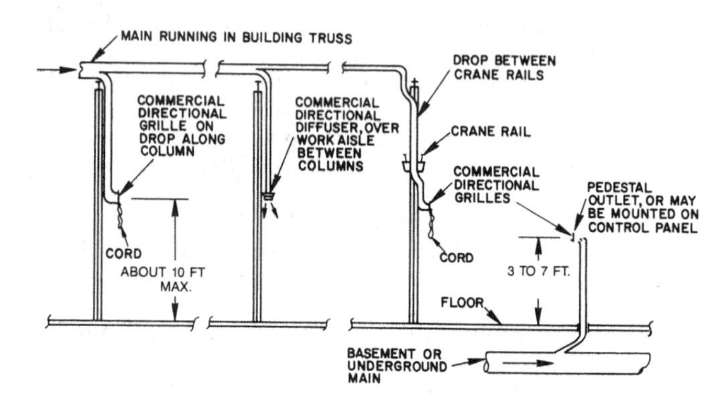

Fig. 3 Additional Means for Providing Low Level Heating and Summer Relief Ventilation

1. Provide a complete enclosure around the worker with separate ventilation to maintain cooler working conditions; *e.g.*, a control room, small shelter booth, or ventilated crane cab. In effect, this is *localized* general ventilation, differing only in the conditions of air temperature, humidity, and motion required.
2. Surround the worker with a relatively cool atmosphere by a direct supply of air introduced at a low level over a small area of the plant. In such cases, the temperature at higher levels in the space is of little or no concern (see Figures 3, 4, and 5). The cool supply air at the face of a hood will also aid in the control of dusts and fumes by bathing the worker in clean air and by minimizing the effects of back eddy diffusion (Volkavein *et al.* 1988, Chamberlin 1988).
3. Direct a high-velocity airstream at the worker to increase the convective and evaporative cooling effect. This method, called *spot cooling*, will incorporate varying degrees of Method 2, depending on the number of employees and the distribution of the workstations.

A workstation enclosure of Method 1 is the most desirable because it completely controls the environment. Method 2 is effective in large areas with many workstations, such as machine shops and assembly lines. Method 3 cools large spaces with scattered workstations and localized heat sources.

Physiological Aspects

Two different heat load situations that affect design are when (1) radiant heat sources are unimportant and (2) radiant heat sources are important.

Radiant Heat is Unimportant. Where no important sources of radiant heat are located within or close to the work area, the relief air needs only be introduced into the work space to displace the hot air, thus surrounding the worker with acceptable temperature and air motion.

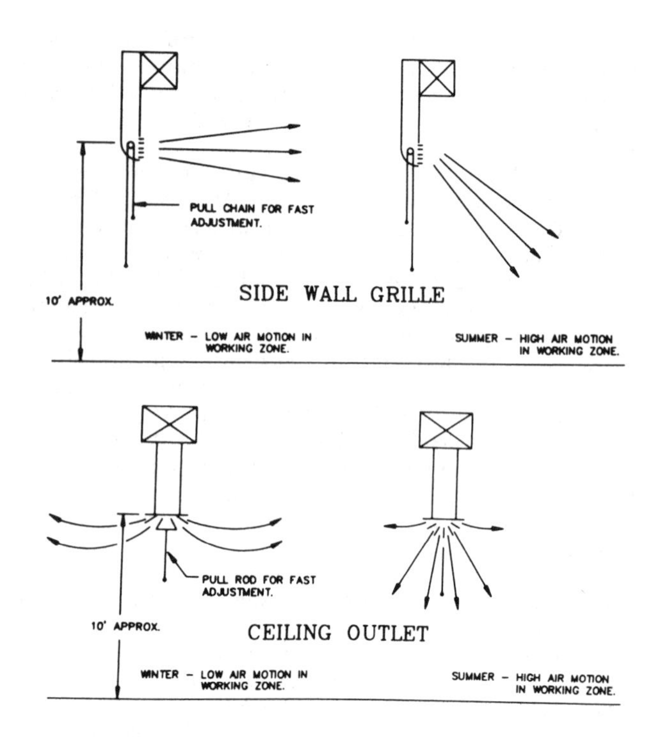

Fig. 4 Seasonal Air Control for Comfort
(ACGIH. 1988. *Industrial ventilation—A manual of recommended practice.*)

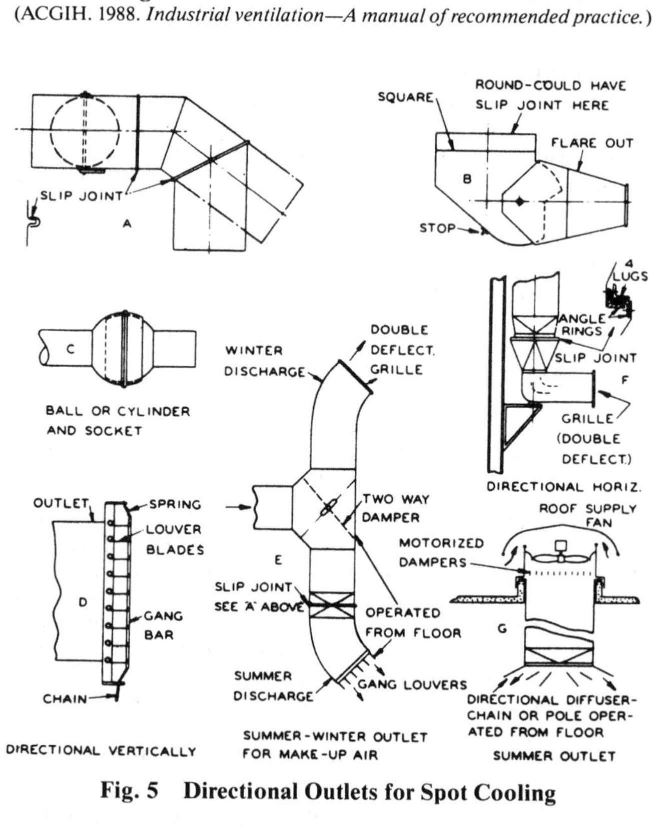

Fig. 5 Directional Outlets for Spot Cooling

Radiant Heat is Important. Where important sources of radiant heat are present and cannot be entirely controlled by radiation shielding, the relief air must greatly increase heat removal by convection and evaporation from the worker to offset the radiation load. Where the radiant load cannot be offset, intermittent work periods may also be necessary, which further emphasizes the importance of shielding to reduce the load. Radiant heat shields or other barriers are the only reasonable protection from radiant heat. In other cases, the supply air temperature may be reduced by refrigeration or evaporative cooling to a level at which the worker can maintain body heat balance without strain.

Regardless of the heat source or load, if the temperature of the relief air exceeds skin temperature, a convective load is then added to any existing radiative load, and these, together with metabolic heat, must be removed by evaporation of sweat. It is important to direct the airflow to the parts of the body exposed to the radiation. Additional velocity over the worker simply increases the convective load, making the working condition less tolerable.

While increased velocity will also increase evaporation, a critical point is reached when the convective addition exceeds heat removal by evaporation. At low velocities, below the critical point, maximum cooling may not occur because of incomplete evaporation. Again, cool air is very important for high heat level conditions and is the reason that evaporatively cooled supply air systems are used so frequently in hot industrial areas.

The level of air motion at the worker is important. At fixed work positions with light activity, particularly when the individual is seated at a desk or bench, the impingement velocity should not exceed 200 fpm for continuous exposure. With high work levels and intermittent exposures (relief stations), velocities of 400 to 800 fpm may be used. In hot industries such as foundries and steel mills, velocities as high as 3000 to 4000 fpm are common (see Table 2). When high-velocity air is used, it is important to avoid the undesirable effects or hot air convection, dust entrainment with its eye hazard potential, and the disturbance of local exhaust ventilation systems.

Table 2 Acceptable Air Motion at the Worker

Exposure	Air Velocity, fpm
Continuous	
Air-conditioned space	50 to 75
Fixed workstation, general ventilation, or spot cooling	
Sitting	75 to 125
Standing	100 to 200
Intermittent, spot cooling, or relief stations	
Light heat loads and activity	1000 to 2000
Moderate heat loads and activity	2000 to 3000
High heat loads and activity	3000 to 4000

The temperature at the workplace and not at the outlet should be in the range of 68 to 86 °F, depending on the heat load. Evaporative cooling ventilation systems usually provide this. Preferably, the air should be directed to the front of the torso and not on the back of the head, neck, and shoulders. The air may need warming in winter, except in areas of heat load where air may need to be cooled.

People vary considerably in their tolerance to air motion, temperature, and humidity, and this tolerance varies with the season. An air motion level that feels comfortable and refreshing in hot weather may feel disagreeable and drafty in the winter. Therefore, the air supply outlets for most local ventilation systems should be adjustable in direction and permit reduction in outlet velocity. This will increase the acceptance of spot cooling (Olesen and Nielsen 1980, 1983).

System Types and Design Recommendations

Supply air can be provided by outdoor air introduced directly or after dehumidification or cooling (evaporative or mechanical). It can also be provided by combinations of outdoor and recirculated air.

Local cooling fans should be used with caution other than in light heat load areas where ambient temperature is below skin temperature. Where an elevated ambient temperature exists, the high velocity may add considerably to the convective heat load and, thus, seriously increase the demand for evaporative cooling, with little or no relief for the worker.

A supply system using outdoor air will provide excellent relief in most industrial areas. However, where the outdoor air temperature exceeds the skin temperature, the direct supply of outdoor air is obviously reduced in effectiveness. Such a system is best used in climates where hot weather periods are brief.

Evaporative cooling systems offer greater relief for workers in that the discharge temperature can be lowered within 5 °F of the wet-bulb temperature to obtain adequate convective body cooling. Because of the increase in wet-bulb temperature of evaporative cooling, it is usually unwise to recirculate evaporatively cooled air.

For economic reasons, evaporative cooling systems generally are used to provide heat relief. Separate supply systems or roof ventilators provide additional makeup air and heat removal. However, evaporatively cooled systems supplying 100% of the ventilation air have been used with excellent results in laundries and similar applications. Chapter 4 of the 1988 ASHRAE *Handbook—Equipment*, Chapter 46 of this volume, and Phillips (1955) have further information.

An air supply with mechanical refrigeration offers the greatest relief. Such systems have high initial cost, but they are used increasingly in precision work and testing areas, in areas requiring constant ambient conditions for product uniformity or control, and where increased worker efficiency is reflected in a reasonable return on the investment.

Outdoor air should be brought as directly as feasible to the proximity of the workstation. It should not impinge on hot equipment or mix with hot ambient air. Supply ducts passing through hot areas should be insulated. Aluminum ducts keep heat gain from radiant sources to a minimum. For large work areas with a high rate of ventilation, local relief can best be obtained by locating the outlets as close to the floor as possible (Azer 1982a, 1982b, 1984; Tillman *et al.* 1984).

The outside air intake should be located with great care to avoid cross contamination from external sources. Frequently, this is a problem in the industrial environment (Wilson 1982).

Outlet Design

When designing outlets for local relief, consider the following:

Location. For general low level ventilation, the outlets should be at about the 10 ft level, although 8 to 12 ft is acceptable. For spot cooling, the outlets should be kept close to the worker to minimize mixing with the warmer air in the space. In most spot cooling installations, the outlets should be brought down to the 7 ft level.

Discharge Velocity. Discharge velocity may be as high as necessary (consistent with good practice) to obtain the desired velocity at the workstation. For the throw of the outlets, consult the manufacturer's data for the particular outlet. Velocities of 1000 to 2000 fpm are most frequently used for outlets at the 10 ft level. When the supply air is cooled, the velocity through an outlet directly at or over a worker should be kept low—about 100 fpm or lower. These recommended velocities are for conditions of maximum heat load. For more moderate weather and ambient conditions, the worker will want to reduce the velocities. Outlet dampers for volume and directional control should always be provided.

Discharge Volume. The outlet volume required will vary widely, depending on whether the system is designed to provide spot cooling or general ventilation. Generally, 1500 to 2500 cfm per station will be adequate for moderate loads, and 3000 to 5000 cfm for areas having high heat loads. With remote outlets, large air volumes are required to ensure adequate relief because of the mixing of the supply air with the warmer surrounding air through which it is projected.

The airstream from a large outlet will maintain an appreciable core of air at the original supply air temperature for a considerable distance from the outlet. Small outlets and slot outlets have small cores that rapidly dissipate through induction. In small enclosures or semi-enclosures, the low induction characteristics of perforated panel supply outlets make them very effective for mechanically cooled installations.

Direction Control. Directional outlets can be directed down in the summer when cooling is needed and up in winter for ventilation, heating, and makeup air. From the production standpoint, control may be necessary to direct the relief ventilation so as not to disturb the product or upset the performance of local exhaust hoods (see Figure 4).

Types of Outlets

Chapters 31 and 32 of the 1989 ASHRAE *Handbook—Fundamentals* and Chapter 2 of the 1988 ASHRAE *Handbook—Equipment* should be reviewed before selecting outlets for a specific application.

Commercial grilles and diffusers are used for many industrial applications, for fully air-conditioned spaces where directional adjustment and vigorous air motion are not needed or desired. Directional grilles, diffusers, and nozzles designed specifically for industrial relief systems are also available. To permit control of the discharge direction and the air motion at the worker, these outlets should be readily adjustable from the floor. Frequent directional adjustment is required for most systems to provide desirable working conditions. Because of frequent adjustment, a high quality damper should be used, and it must be maintained in working order.

Directional grilles are used for long horizontal throws in large open areas. Circular diffusers may be used for similar applications and also provide good spot cooling. Nozzle outlets such as the ordinary ball and socket, the punkah louver (a small ball and socket outlet frequently seen in aircraft), and drum outlets are often used for spot cooling. The drum outlet also provides good general ventilation and is frequently used for this purpose. This outlet is similar to the punkah louver but is rectangular, has only horizontal adjustment, and is fitted with louvers for manual vertical adjustment.

The variety of custom outlets is limited only by the designer's imagination and ingenuity. The important thing is that such outlets should do the required job. Good velocity distribution through the outlet face area is necessary, or predicted performance must be modified to suit the nonuniform flow. Outlets in the sides and bottoms of ducts should have turning vanes. Diffuser outlets should have turning vanes at the drop connections.

Figure 5 shows some of the directional outlets that have been used for low level general ventilation and spot cooling. Outlet *A* (the Navy Type E) in various forms has been used for many years in ship machinery spaces. Outlets *B* and *C* are excellent for local area or spot cooling. The adjustable louvers of *D* applied to directional outlets such as *E* or *F* provide excellent control. The commercial directional grilles serve the same purpose. The two-way damper arrangement *E* is used in local area or aisleway ventilation, to direct the supply air upward in winter to mix the discharge air with warm or hot air rising from internal sources.

Outlet *G* indicates the use of a commercial directional diffuser that can be adjusted to provide a variable downward airflow pattern from flat to vertical. The outlet is shown on a roof supply fan, which is a common application, but the outlet drop must extend through the layer of hot ceiling air to provide effective relief ventilation.

Where overhead installations are not possible because of interferences, outlets near the floor may be used. These may be grilles located in pilasters, control panels, tables, equipment, or in the floor itself. Such outlets have been successful in welding and foundry areas and at work locations in kitchens.

DILUTION VENTILATION

Dilution ventilation is general ventilation used to dilute contaminated air with uncontaminated air within the building or workroom. In many cases, dilution ventilation is less satisfactory than local exhaust ventilation (see Chapter 27), from either a health hazard control or energy conservation point of view. Dilution ventilation may be preferred when (1) moving sources are involved, (2) many small sources are scattered throughout the workroom and effectively prohibit local exhaust ventilation, and (3) when general ventilation is required for comfort control.

Dilution ventilation may be inappropriate if (1) large volumes of contaminant are released or the contaminant has a high degree of toxicity, (2) employees must work close to the source so that insufficient dilution occurs before the contaminated air passes through their breathing zone, and (3) operator exposure is highly variable and not predictable.

Dilution ventilation required by ventilation systems that address the problems of fire, smoke, and explosion needs careful study.

Design of Dilution Systems

The first step in the design of a dilution ventilation system is to determine an acceptable level of exposure. These levels could be permissible exposure limits (PELs), threshold limit values (TLVs), or in-house TLVs. It is desirable to set design objectives below statutory PELs and TLVs because of the variability of individual sensitivity to contaminants and the knowledge that acceptable limits do change—usually in the direction of lower values.

The second step is to determine the generation rate of the contaminant. Production records, material balance, existing ventilation rate, contaminant exposure, similar operations, and experienced engineering judgment can assist in determining the generation rate. However obtained, the acceptable exposure level and the contaminant generation rate are necessary to design a dilution ventilation system properly. Design based on the number of air changes per hour, or other estimates, are inadequate and could lead to unacceptably high exposure levels or to unnecessarily high installation costs and/or energy consumption.

The third step in calculating the dilution rate is to determine the mixing factor *K*. Typically, *K* ranges between 1 and 10 and is required to ensure that employee exposure is maintained below the acceptable level, not just that the average concentration in the room or the exhaust duct is sufficiently low. The value of *K* depends on several factors: (1) toxicity of the contaminant, (2) physiological effects of overexposure, (3) uniformity of contaminant generation, (4) effectiveness of the ventilation system, (5) geometry of the work area, and (6) contaminant concentration in the uncontaminated replacement air.

However, because ventilation costs vary almost linearly with size, selection of a *K* factor that is too high will make costs prohibitive. Caplan (1986) indicates *K* factors of 2 to 4 from a limited number of tests. Constance (1983) provides some guidelines on selection of *K* values as a function of air distribution and toxicity, but the basis for these data is unknown (see Tables 3 and 4). Fluid dynamic scale modeling can be used to predict building ventilation rates (Davis 1980, Anderson and Mehos 1988, Hortsman 1988, Skaret 1986, and Curd 1987). Also refer to ACGIH's *Industrial ventilation—A manual of recommended practice*, 20th ed. (1988).

Table 3 *K* Ranges versus Distribution

K Values	Distribution System
1.2 to 1.5	Perforated ceiling
1.5 to 2.0	Air diffusers
2.0 to 3.0	Duct headers along ceiling with branch jets pointing downward
3.0 and above	Window, wall fans, etc.

Source: *Power*, February 1976.

Table 4 K Factors for Four Distribution Conditions versus Toxicity

Distribution Condition	Toxicity		
	Slight	Moderate	High
Poor	7	8	11
Average	4	5	8
Good	3	4	7
Excellent	2	3	6

Source: *Power*, February 1976.

Dilution ventilation is applied most often to the dilution of solvent vapors. In this case, the required dilution rate, at normal temperature and pressure, is given by:

$$Q_v = 4.03 \times 10^8 \text{ SG} \times K/(M \times \text{TLV}) \quad (5)$$

$$Q_m = 3.87 \times 10^8 \text{ SG} \times K/(M \times \text{TLV}) \quad (6)$$

where

Q_v = dilution rate required for solvents, ft^3/pt solvent evaporated
Q_m = dilution rate required for solvents, ft^3/lb solvent evaporated
SG = specific gravity of liquid
M = molecular weight of solvent
TLV = threshold limit value, ppm
K = mixing factor

Dilution ventilation for dust and fumes is frequently unsuccessful because (1) the lower TLV (higher toxicity of the material) requires excessive exhaust rate, (2) the rate of evolution and violence of dispersion frequently makes dilution ventilation ineffective, and (3) quantity or release rate of contaminant is extremely difficult to obtain.

Effective Specific Gravity and Location of Exhaust Grilles

In general ventilation, the contaminant concentration in the exhaust air is not significantly different than the general room air concentration. In most situations with health hazards, hazardous dusts, fumes, vapors, and gases are airborne and follow air currents in spite of the fact that the contaminant may be more dense than air. For example, the effective specific gravity of perclene (C_2Cl_4) at 100 ppm by volume is calculated as follows:

Molecular weight of perclene = 153.84
Molecular weight of air = 28.94
100 ppm = 0.01% by volume

The molecular weight of the mixture is:

$$\begin{aligned} 0.0001 \times 153.84 &= 0.0154 \\ 0.9999 \times 28.94 &= \underline{28.9371} \\ & 28.9525 \end{aligned}$$

Therefore, the effective specific gravity of the mixture is:

$$\text{SG} = 28.9525/28.94 = 1.0004$$

Thus, for general dilution ventilation, exhaust grilles are usually located above floor level since room air generally rises. If, however, general ventilation is provided to handle only large liquid spills onto the floor, exhaust grilles should be located near the floor. In many industrial situations, both locations are used.

Mixtures

When more than one hazardous material is present in the workplace, the combined effect of the contaminants must be considered. In the absence of contrary data, their effect should be considered as additive. When designing a dilution ventilation system, this requires calculating the dilution rate of each component and using the sum of the rates as the dilution rate for the mixtures.

Air Curtains and Air Jets

An air curtain is a two-dimensional jet that can be used to contain and convey conditioned air and particulates. Air curtains work by entraining surrounding air. The volume of airflow constantly increases and the velocity decreases with increasing distance from the air curtain nozzle. The important design parameters for air curtains are (1) the width of the jet, (2) the jet velocity, and (3) the angle of the jet to opposing forces. The design equations are given by:

$$M_o = M_x$$

$$\rho_o V_o^2 A_o = \rho_x V_x^2 A_x$$

where

M_o = momentum at jet nozzle, lb/s
M_x = momentum at distance x from the jet nozzle, lb/s
ρ = density of fluid, lb/ft^3
V = velocity of jet, lb/s
A = area of jet, ft^2

Powlesland (1971) discusses the use of air curtains as airflow regulators, thermal barriers, and for dust and fume control. For all these applications, no hoods or ducts were used. Bintzen (1976) described the use of air curtains on the slag door of an electric steelmaking furnace.

Locker Room, Toilet, and Shower Space Ventilation

The ventilation of locker rooms, toilets, and shower spaces is important in modern industrial facilities to remove odor and reduce humidity. In some industries, adequate control of workroom contamination requires prevention of ingestion, as well as inhalation, so adequate hygienic facilities, including appropriate ventilation, may be required in locker rooms, change rooms, showers, lunchrooms, and break rooms. State and local regulations should be consulted at early stages of design.

Supply air may be introduced through door or wall grilles. In some cases, plant air may be so contaminated that filtration or, preferably, mechanical ventilation, may be required. When control of workroom contaminants is inadequate or not feasible, the total exposure to employees can be reduced by ensuring that the level of contamination in the locker rooms, lunchrooms, and break rooms is minimized by pressurizing these areas with excess supply air.

When mechanical ventilation is used, the supply system should have supply fixtures such as wall grilles, ceiling diffusers, or supply plenums to distribute the air adequately throughout the area.

In the locker rooms, the exhaust should be taken primarily from the toilet and shower spaces, as needed, and the remainder from the lockers and the room ceiling. In the absence of specific codes, Table 5 provides a guide for ventilation of these spaces.

Table 5 Ventilation for Locker Rooms, Toilets, and Shower Spaces

Description	Units
Locker Rooms	
Coat hanging or clean change room for non-laboring shift employees with clean work clothes	1 cfm/ft^2
Change room for laboring employees with wet or sweaty clothes	2 cfm/ft^2; 7 cfm exhausted from each locker
Change room for laborers or workers assigned to heavy work and where clothes will be wet or pick up odors	3 cfm/ft^2; 10 cfm exhausted from each locker
Toilet Spaces	2 cfm/ft^2; at least 25 cfm per toilet facility; 200 cfm minimum
Shower Spaces	2 cfm/ft^2; at least 50 cfm per shower head; 200 cfm minimum

Note: The source of this table is unknown. Nevertheless, the information has been used with apparent success for many years (ASHRAE *Guide and Data Book* 1970). Also refer to Article 16 of the BOCA *National Building Code* (1990) and ASHRAE *Standard* 62-1989.

REFERENCES

ACGIH. 1988. *Industrial ventilation—A manual of recommended practice*, 20th ed. American Conference of Governmental Industrial Hygienists, Cincinnati, OH.

ACGIH. 1989. Threshold limit values and biological exposure indices for 1989-90. American Conference of Governmental Industrial Hygienists, Cincinnati, OH.

AIHA. 1975. *Heating and cooling for man in industry*, 2nd ed. American Industrial Hygiene Association, Akron, OH.

Anderson, R. and M. Mehos. 1988. Evaluation of indoor air pollutant control techniques using scale experiments. ASHRAE Indoor Air Quality Conference.

ASHRAE. 1981. Thermal environmental conditions for human occupancy. ASHRAE *Standard* 55-1981.

Azer, N.Z. 1982a. Design guidelines for spot cooling systems: Part 1—Assessing the acceptability of the environment. ASHRAE *Transactions* 88(2).

Azer, N.Z. 1982b. Design guidelines for spot cooling systems: Part 2—Cooling jet model and design procedure. ASHRAE *Transactions* 88(1).

Azer, N.Z. 1984. Design of spot cooling systems for hot industrial environments. ASHRAE *Transactions* 90(1).

BOCA. 1990. BOCA *National building code*, 11th ed. Building Officials and Code Administrators International, Country Club Hills, IL.

Cameron, D.B. and A.B. Konzen. 1982. A unique jet augmented ventilation design. American Industrial Hygiene Association, American Industrial Hygiene Conference, Cincinnati, OH.

Caplan, K.J. and G.W. Knutson. 1978. Laboratory fume hoods: Part 2—Influence of room air supply. ASHRAE *Transactions* 84(2):522.

Caplan, K.J. 1986. Research and development trends and needs. Ventilation '85. Elsevier Science Publishers B.V., Amsterdam.

Chamberlin, L.A. 1988. Use of controlled low velocity air patterns to improve operator environment at industrial work stations. Masters thesis, University of Massachusetts, September.

Constance, J.D. 1983. Controlling in-plant airborne contaminants. Marcel Dekker, Inc., New York.

Curd, E.F. 1984. Industrial ventilation design by hydraulic modeling. International Environmental and Safety Conference, London (March).

Davis, J.A. and W.D. Baines. 1980. Hydraulic modelling—A powerful ventilation tool. 19th Annual Conference of Metallurgists, Nova Scotia.

Goldfield, J. 1980. Contaminant concentration reduction general ventilation versus local exhaust ventilation. *American Industrial Hygienists Association Journal* 41(November).

Goodfellow, H.D. 1985. Advanced design of ventilation systems for containment control. Elsevier Science Publishers B.V., Amsterdam.

Goodfellow, H.D. 1987. Ventilation, industrial. *Encyclopedia of Physical Science and Technology* (14). Academic Press, Inc., San Diego, CA.

Harris, R.L. 1988. Design of dilution ventilation for sensible and latent heat. *Applied Industrial Hygiene* 3(1).

Heinsohn, R.J. and M.S. Choi. 1986. Advanced design methods in industrial ventilation. Ventilation '85. Elsevier Science Publishers, B.V., Amsterdam.

Holcomb, M.L. and J.T. Radia. 1986. An engineering approach to feasibility assessment and design of recirculating exhaust systems. Ventilation '85. Elsevier Science Publishers B.V., Amsterdam.

Horstman, R.H. 1988. Predicting velocity and contamination distribution in ventilated volumes using Navier-Stokes equations. ASHRAE Indoor Air Quality Conference.

ISO. 1982. Hot environments—estimation of the heat stress on a working man, based on the WBGT—Index (wet bulb globe temperature). Geneva.

ISO. 1984. Moderate thermal environments—Determination of the PMV and PPD indices and specifications of the conditions for thermal comfort. ISO *Standard* 7730, Geneva.

ISO. 1989. Hot environments—An analytical determination and interpretation of thermal stress using calculation of required sweat rate. ISO *Standard* 7933, Geneva.

Kato, S. and S. Murakamii. 1988. New ventilation efficiency scales based on spatial distribution of contaminant concentration aided by numerical simulation. ASHRAE *Transactions* 94.

McDermott, H.J. 1976. *Handbook of ventilation for contamination control*. Ann Arbor Science Publishers, Inc.

Mehta, M.P., H.E. Ayer, B.E. Saltzman, and R. Ronk. 1988. Predicting concentration for indoor chemical spills. ASHRAE Indoor Air Quality Conference.

NIOSH. 1986. Occupational exposure to hot environments. Revised criteria. National Institute for Occupational Safety and Health, Washington, D.C.

Olesen, B.W. and R. Nielsen. 1980. Spot cooling of workplaces in hot industries. Final Report to the European Coal and Steel Union. Research Programme Ergonomics—Rehabilitation III, No. 7245-35-004 (October):88.

Olesen, B.W. and R. Nielsen. 1981. Radiant spot cooling of hot places of work. ASHRAE *Transactions* 87(1).

Olesen, B.W. and R. Nielsen. 1983. Convective spot-cooling of hot working environments. Proceedings of the XVIth International Congress of Refrigeration Paris, (September).

Peterson, R.L., E.L. Schofer, and D.W. Martin. 1983. Laboratory air systems—Further testing. ASHRAE *Transactions* 89(2B):571.

Phillips, R.E., Jr., *et al.* 1955. Evaporative cooling—A symposium on heating, piping and air conditioning. ASHRAE *Journal* (August):141.

Powlesland, J.W. 1971. Air curtains. *Canadian Mining Journal* (October):84-93.

Sanberg, M. Ventilation efficiency. ASHRAE *Transactions* 89(2B)455.

Schroy, J.M. 1986. A philosophy on engineering controls for workplace protection. *Annals of Occupational Hygiene* 30(2):231-36.

Skaret, E. Ventilation efficiency. ASHRAE *Transactions* 89(2B):480.

Skaret, E. 1986. Industrial ventilation—Model tests and general development in Norway and Scandinavia. Ventilation '85. Elsevier Science Publishers B.V., Amsterdam.

Stephanov, S.P. 1986. Investigation and optimization of air exchange in industrial halls ventilation. Ventilation '85. Elsevier Science Publishers B.V., Amsterdam.

Tillman, F.A., C.L. Hwang, and M.J. Lin. 1984. Optimal design of an air jet for spot cooling. ASHRAE *Transactions* 90(1B):476.

Volkavein, J.C., M.R. Engle, and T.D. Raether. 1988. Dust control with clean air from an overhead air supply island (oasis). *Applied Industrial Hygiene* 3(August):8.

Wilson, D.J. 1982. A design procedure for estimating air intake contamination from nearby exhaust vents. ASHRAE *Transactions* 89(2A):136.

BIBLIOGRAPHY

Bintzer, W. and F.A. Malehom. 1976. Air curtains on electric arc furnaces at Lukens Steel Co. *Iron and Steel Engineer* (July):53-55.

BOCA *National Mechanical Code*, 6th ed. 1987.

Brief, R.S., S. Lipton, S. Amarmani, and R.W. Powell. 1983. Development of exposure control strategy for process equipment. *Annals of the American Conference of Governmental Industrial Hygienists* (5).

Caplan, K.J. 1980. Heat stress measurements. *Heating, Piping and Air Conditioning* (February):55-62.

Constance, J.D. 1983. *Controlling in-plant airborne contaminants*. Marcel Dekker, Inc., New York.

Curd, E.F. 1981. Possible applications of wall jets in controlling air contaminants. *Annals of Occupational Hygiene* 24(1):133-46.

Eto, J.H. and C. Meyer. 1988. The HVAC costs of fresh air ventilation. ASHRAE *Journal* (September):31-35.

Fuller, F.H. and A.W. Etchells. 1979. The rating of laboratory hood performance. ASHRAE *Journal* (October):49-53.

Goodfellow, H.P. and J. W. Smith. 1982. Industrial ventilation—A review and update. *American Industrial Hygiene Association Journal* 43(March):175-84.

National Safety Council. 1988. *Fundamentals of industrial hygiene*, 3rd ed. Chicago, IL.

NIOSH. 1973. The industrial environment—its evaluation and control. National Institute for Occupational Safety and Health, Washington, D.C.

U.S. Department of Health and Human Services. 1986. Advanced industrial hygiene engineering PB87-229621. Cincinnati, OH.

Ventilation '85, Proceedings. Elsevier Science Publishers B.V., Amsterdam.

Ventilation '88, Proceedings. Elsevier Science Publishers B.V., Amsterdam.

CHAPTER 26

MINE AIR CONDITIONING AND VENTILATION

EXCESS humidity, high temperature, and the need for adequate oxygen have always been points of concern in underground mines. A combination of heat and humidity lowers efficiency and productivity and can cause illness and death. Air cooling and ventilation are needed in deep underground mines to minimize heat stress. As mines have become deeper, heat removal and ventilation problems have become more difficult to solve.

WORKER HEAT STRESS

Mine cooling must maintain air temperature and humidity at levels that maintain the health and comfort of the miners so that they may work safely and efficiently. Chapter 8 of the 1989 ASHRAE *Handbook—Fundamentals* addresses human response to heat and humidity. The upper temperature limit for humans at rest in still, saturated air is about 90 °F. If air is moving at 200 fpm, the upper limit shifts to 95 °F. A relative humidity of less than 80% in a hot, humid mine environment is desirable.

Hot, humid environments are improved to a limited extent by providing air movement of 150 to 500 fpm. A greater air volume may lower the mine temperature, but air velocity has limited value in increasing the individual's comfort.

Indices for defining acceptable temperature limits include:

Effective Temperature Scale. An 80 °F effective temperature is the upper limit for worker comfort and high efficiency.

Wet-Bulb Globe Temperature (WBGT) Index. An 80 °F WBGT is the permissible temperature exposure limit during moderate continuous work, and 77 °F WBGT for heavy continuous work.

Figure 1 shows acceptable heat exposure limits for different levels of work efforts.

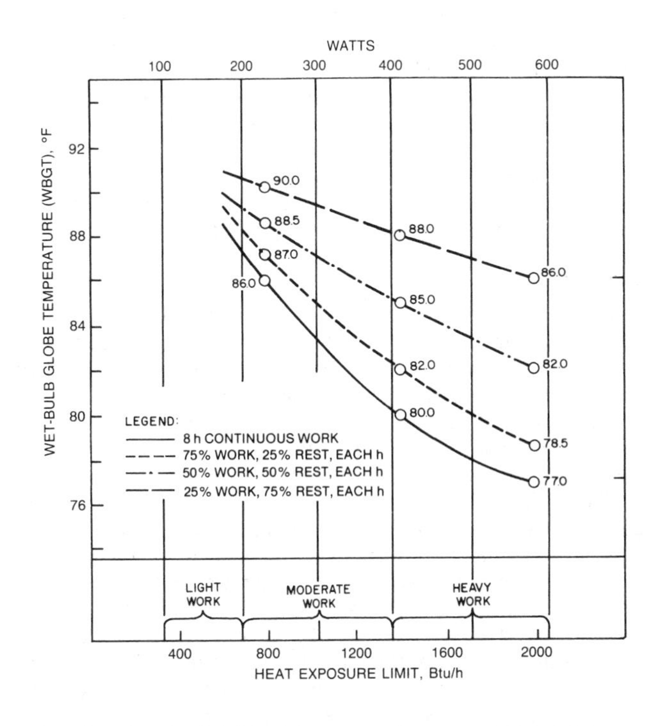

Fig. 1 Permissible Heat Exposure Threshold Limit Value

DESIGN CONSIDERATIONS: SOURCES OF HEAT ENTERING MINE AIR

Adiabatic Compression

Air descending a shaft increases in pressure (due to the mass above it) and temperature. As air flows down a shaft, with no heat interchange between the shaft and air and no evaporation of moisture, it is heated as if compressed in a compressor.

One Btu is added to each pound of air for every 778 ft elevation or is subtracted for the same elevation increase. For dry air, the dry-bulb temperature change is $1/(0.24 \times 778) = 778 = 0.00535$ °F/ft or 1 °F/187 ft elevation. For constant air-vapor mixtures, the change in dry-bulb temperatures equals $(1 + W)/(0.24 + .45W)$ per 778 ft of elevation, depending on the humidity ratio W (pounds of water per pound of air).

Theoretically, when 100,000 cfm of standard air is delivered underground via an inlet airway, the heat of autocompression for every 1000 ft of depth is:

$$100{,}000 \times 0.075 \times \frac{1 \text{ Btu/lb}}{778 \text{ ft}} \times 1000 \text{ ft} = 9640 \text{ Btu/min}$$

$$= 48.2 \text{ tons refrigeration}$$

Autocompression of air may be masked by the presence of other heating or cooling sources, such as shaft wall rock, groundwater, air and water lines, or electrical facilities. The actual temperature of air descending a shaft does not usually match the theoretical adiabatic temperature increases, due to the following:

- The night cool air temperature effect on the rock or lining of the shaft
- Temperature gradient of ground rock related to depth
- Evaporation of moisture within the shaft, which decreases the temperature while increasing the moisture content of the air

The seasonal variation in surface air temperature has a major effect. If surface air temperature is high, much heat is liberated to the shaft walls so that the temperature rise may not reach the adiabatic rate. When the surface temperature is low, heat is taken from the shaft walls and temperature increases more than the adiabatic rate. Similar diurnal variations may occur. As air flows

The preparation of this chapter is assigned to TC 9.2, Industrial Air Conditioning.

down a shaft and increases in temperature and density, its volume and cooling ability decrease. Additionally, the mine ventilation requirements increase with depth. Underground fan pressures up to 10 in. water gage static pressure are common in mine ventilation and also raise the temperature of the air about 0.45 °F per inch of water fan static pressure.

Electromechanical Equipment

Underground power-operated equipment transfers the heat to air. Common underground mine power systems are electricity, diesel, and compressed air.

Heat produced by underground diesel equipment equals about 90% of the heat value of the fuel consumed. Therefore, a value of 125,000 Btu/gal for fuel consumed by diesel equipment is dissipated to the ventilating airstream as heat. If the exhaust gas is bubbled through a wet scrubber, thus cooling the gases by adiabatic saturation, both the sensible heat and the moisture content of the air are increased.

Vehicles with electric drive or an electric-hydraulic system may release from one-third to one-half the heat of diesel-driven equipment.

All energy used in a horizontal plane appears as heat added to the mine air. Energy required to elevate a load gives potential energy to the material and will not appear as heat.

Groundwater

Transport of heat by groundwater is the largest variable in mine ventilation. Groundwater is ordinarily the same temperature as the virgin rock temperature. If a ditch containing hot water is not covered, there may be more heat picked up by the ventilation cooling air from the ditch water than from the heat conducted through the hot wall rock. It is very important to contain warm drainage water in pipelines or in a covered ditch.

The significance of heat from open ditches increases as airways get older and the flow of heat from the surrounding rock decreases. In one Montana mine, water in open ditches was 40 °F cooler than when it issued from the wall rock; the heat was transferred to the air. Evaporation of water from wall rock surfaces lowers the surface temperature of the rock, which increases the temperature gradient of the rock, depresses the dry-bulb temperature of the air, and allows more heat to flow from the rock. Most of this extra heat is usually expended in evaporation.

WALL ROCK HEAT FLOW

Heat flow from the wall rock into an airway is complex. The heat flow from wall rock with constant thermal conductivity is considerably higher after a mine opening is first excavated (transient heat flow) than several years later, when steady-state conditions have developed. An empirical equation can be used for the calculations:

$$Q = UA\,\Delta t$$

where

Q = heat flow in Btu/h
U = overall coefficient of heat transfer, Btu/h·ft²·°F (from Figure 2 or 3)
A = area
Δt = temperature difference between the virgin rock and the dry-bulb air temperature

The above equation in conjunction with mine ventilation rates, virgin rock temperature, and the age of airways will permit wall rock heat flow calculations to be made throughout a mine.

In young mining areas, heat inflow must be computed by transient heat flow techniques. A heat flow graph from wall rock to air versus time can be plotted. Figures 2 and 3 show examples of heat flow with time from two types of mines. The temperature difference is not constant along the entire length of the airway; it is thus advisable to treat the long airway as a series of consecutive lengths. Figures 2 and 3 are based on the following:

- Calculations of mine openings with cross-sectional areas range from 225 to 450 ft².
- The heat transfer coefficient decreases from the time the heading is excavated and ventilated.
- Steady-state heat transfer coefficients are approximately 0.10 to 0.15 Btu/h·ft²·°F for salt or granite envelopes.

The Starfield and Goch-Patterson methods of calculating heat rate from underground exposed rock surfaces are often used. The Starfield method is often difficult to apply because of its dependence on air velocity criteria. The errors in obtaining or projecting velocity, length, and cross-sectional dimensions are often worse than this theoretically more accurate method.

The Goch-Patterson method, which assumes that the rock face temperature is the same as that of the air, is widely used. Heat load calculated for airways is overestimated by 10 to 20%, so a contingency is normally not added to the heat load. When using the Goch-Patterson tables, care should be taken that the correct value of instantaneous versus average heat load data is used for shafts and older connecting airways. Time average heat load data should be used for excavations that are continuous and active.

Since heat from the wall rock is one of the major sources of heat into a mine, its rate may be accurately estimated horizontally and vertically. It is desirable that virgin rock temperatures (VRT) be taken of the wall rock at various elevations throughout the mine.

Before the actual gradient can be constructed to estimate the temperatures at various elevations, a base point for measurement must be determined. It is generally agreed that a virgin rock temperature taken 50 ft below the earth's surface represents a good datum point. This value is normally equal to the mean surface air temperature (dry-bulb) taken over a number of years.

Figure 4 graphically shows approximate geothermal gradient of various mining districts of the world. Table 1 lists maximum virgin rock temperatures at the bottom of various mines.

Table 1 Maximum Virgin Rock Temperatures
(Fenton 1972)

Mining District	Depth, ft	Temperature, °F
Kolar Gold Field, India	11,000	152
South Africa	10,000	125 to 130
Morro Velho, Brazil	8,000	130
N. Broken Hill, Australia	3,530	112
Great Britain	4,000	114
Braloroe BC Canada	4,100	112.5
Kirkland Lake, Ontario	4,000 to 6,000	66 to 81
Falconbridge Mine, Ontario	4,000 to 6,000	70 to 84
Lockerby Mine, Ontario	3,000 to 4,000	67 to 96
Levac Borehold (Inco) Ontario	7,000 to 10,000	99 to 128
Garson Mine, Ontario	2,000 to 5,000	54 to 78
Lake Shore Mine, Ontario	6,000	73
Holiinger Mine, Ontario	4,000	58
Creighton Mine, Ontario	2,000 to 10,000	60 to 138
Superior, AZ	4,000	140
San Manuel, AZ	4,500	118
Butte, MT	5,200	145 to 150
Ambrosia Lake, NM	4,000	140
Brunswick, No. 12, New Brunswick, CA	3,700	73
Belle Island Salt Mine, LA	1,400	88

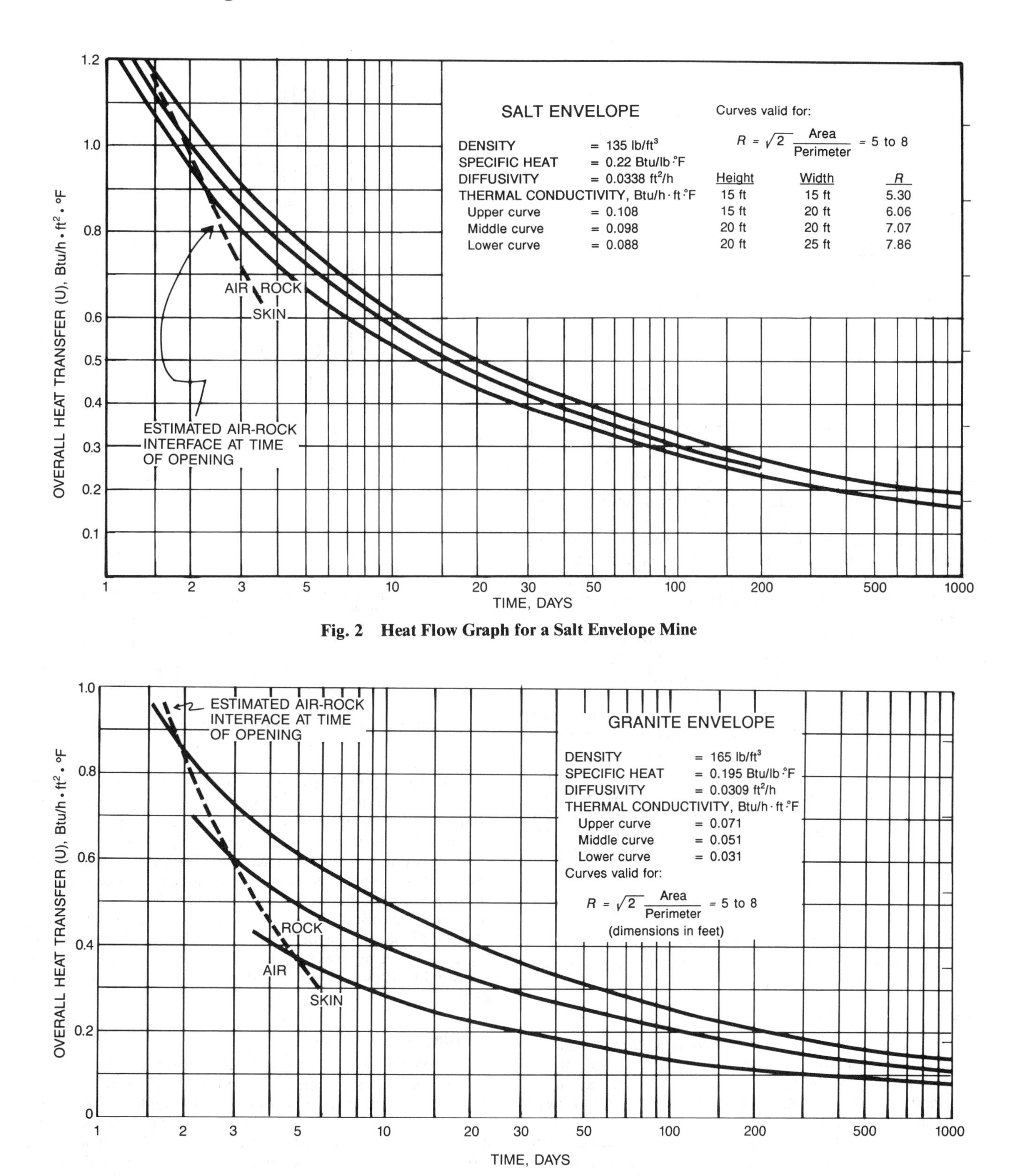

Fig. 2 Heat Flow Graph for a Salt Envelope Mine

Fig. 3 Heat Flow Graph for a Granite Envelope Mine

Heat from Blasting

Short-term heat produced by blasting can be appreciable. Virtually all the explosive energy is converted to heat during detonation. The typical heat potential in various types of explosives is similar to that of 60% dynamite, about 1800 Btu/lb.

MINE AIR COOLING AND DEHUMIDIFICATION

Sources of underground mine heat are hot wall rock, hot water issuing from underground sources, adiabatic compression of the inlet air down deep shafts, operation of electromechanical equipment, heat exchange between hot-compressed air lines or hot

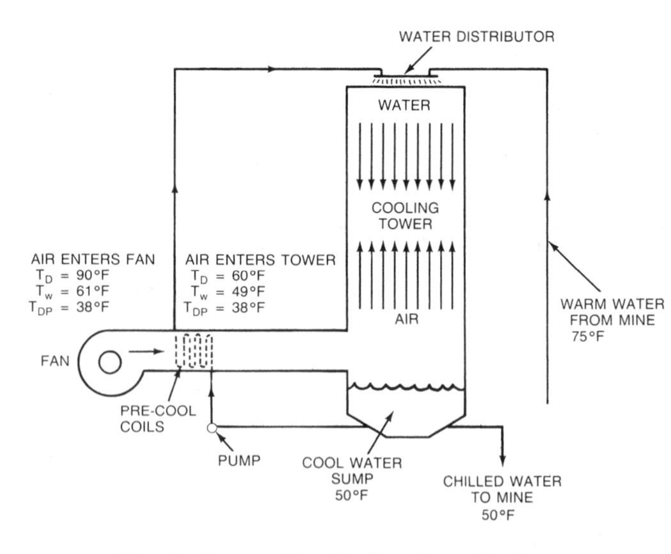

Fig. 4 Evaporative Cooling Tower System
(Richardson 1950)

mine-water draining lines and the cool ventilation air in the entrance airway or shaft, oxidation of timber and sulfide minerals, heat liberated by blasting, friction and shock losses of moving air, body metabolism, friction heat from rock movement, and heat generated by internal combustion engines (Fenton 1972).

Geothermic gradients of the world's major mining districts vary from approximately 0.5 °F to over 4 °F per 100 ft of elevation decrease.

When air is cooled underground, the heat taken from this air must be moved in some manner. If enough cool, uncontaminated air is available, the warmed air can be mixed with it and the resulting mixture can then be removed from the mine. In a water heat exchanger, the water is warmed while the air is cooled. The warm water is either returned to the cooling plant (cooling tower, underground mechanical refrigeration units, or heat exchangers) in a closed circuit where it is again cooled, discharged to the mine drainage system, or sprayed into the mine exhaust air system.

If available, water is the best heat transfer medium in a mine. A relatively small pipeline can remove as much heat as would be removed by a large volume of air that requires a large shaft and inlet airways.

EQUIPMENT AND APPLICATIONS

Air-Conditioning Components and Practices

Figures 4 through 7 show components normally used in underground air-conditioning systems.

Cooling Surface Air

If cold air is available on the surface, as in far northern and far southern hemisphere winters, it can be forced into the mine. It may even be necessary to heat the intake air in winter. In deep mines with small cross-sectional airways, this supply of cold air may be insufficient for adequate cooling.

Cooling of entering mine air on the surface is the least expensive method of cooling. Factors favoring surface cooling are low adiabatic compression (not too deep shaft) and low heat gain in the shaft and intake airways. The main energy-saving advantage of cooling inlet air on the surface rather than an underground cooling plant is that no electric power is needed to return cooling water from the plant to the surface.

In a surface plant installation, air is chilled by cold water or brine produced from mechanical refrigeration units. It is then delivered to the intake shaft to be taken underground. Condenser

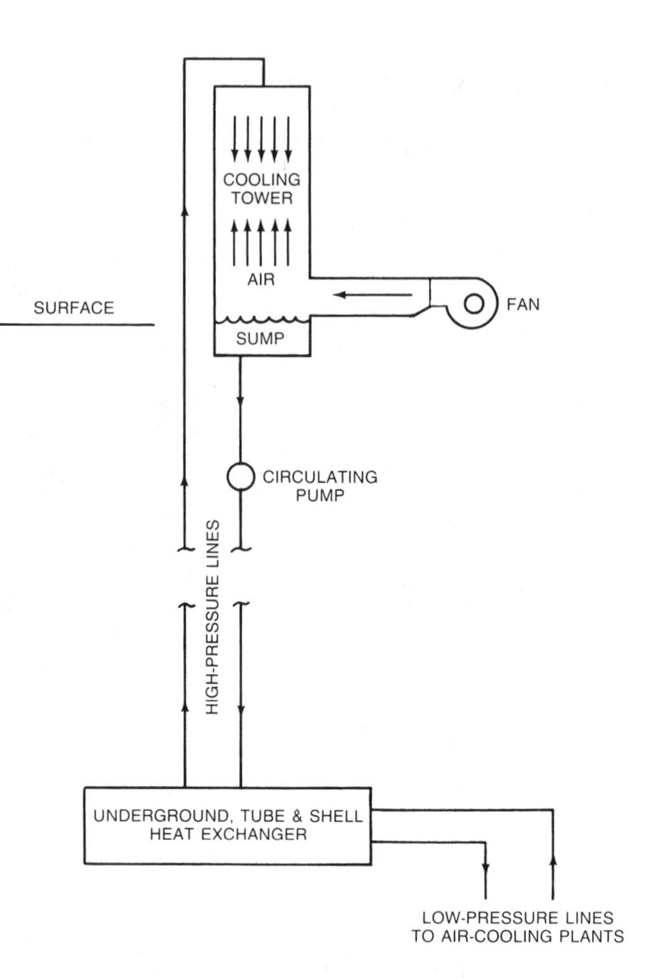

Fig. 5 Underground Heat Exchanger, Pressure Reduction System

water is usually cooled in towers or spray ponds. Such installations have been used in Morro Velho, Brazil; Robinson Deep Mine, Rand, South Africa; Kolar, India; and Rieu du Coeur, Belgium.

When air reaches the working headings, it has warmed considerably by autocompression and heat from the wall rock. Further, a plant of sufficient size for peak load has an annual load factor as low as 60%. For these reasons, no new surface plant systems have been installed recently. The trend is to install air-cooling plants underground, with the cooling coils as close to the working areas of the mine as possible. If a large supply of cold water is available, it can be piped underground to air-conditioning plants.

Cooling Surface Water

The concept of delivering surface-made ice or an ice slurry in 32 °F water down a pipeline is intriguing. Chilled water could be used for drilling, wetting down muckpiles, and other activities where men are working. Chilled water at 32 °F available underground reduces the circulated flow rate to 50 to 70%, which reduces pumping capital and power costs of the usual chilled water air-conditioning plants.

Ice-making machines, however, have a lower coefficient of performance and higher capital costs compared to normal mechanical refrigeration units used to chill mine water.

In one Canadian mine, large, cold, mined-out areas are sprayed with water during the winter months, allowing ice to form in the openings. In summer, intake air is cooled by drawing it over this ice.

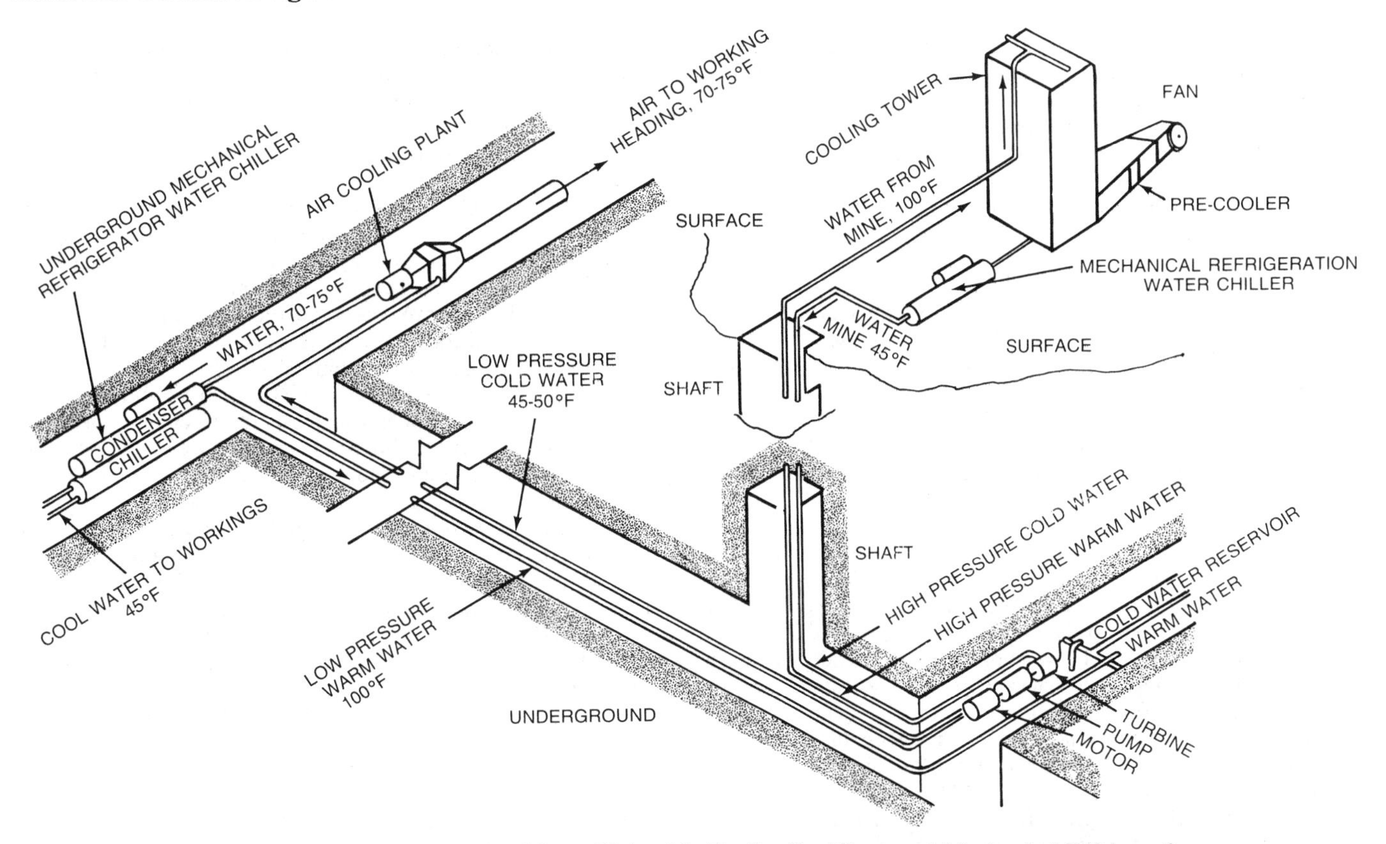

Fig. 6 Layout for a Turbine-Pump-Motor Unit with Air-Cooling Plants and Mechanical Refrigeration

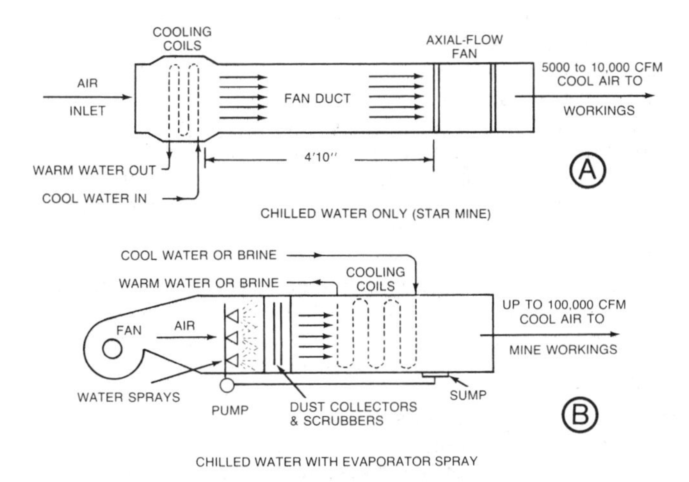

Fig. 7 Underground Air-Cooling Plants

Evaporative Cooling of Mine Chilled Water

Geographic areas with both low average winter temperatures and low relative humidity during the summer have natural cooling capacity adaptable to mine air-cooling methods. A system in such an area cools water or brine on the surface, carries it through closed-circuit piping to an underground air-cooling plant close to the working zone, and returns it to the cooling tower through a second pipeline. At the mine underground air-cooling plant, heat is adsorbed by the water circulating through the system and is dissipated to the surface atmosphere in the surface cooling tower. The closed-circuit piping balances the hydrostatic head, so pumping power must only overcome frictional resistance.

An evaporative cooling tower was installed at a mine in the northwest United States. This "dew-point" cooling system reduces the temperature of the cooling medium to below the wet-bulb temperature of the surface atmosphere (Figure 4).

Precooling coils were installed between the fan and the cooling tower. Some cool water from the sump at the base of the cooling tower is pumped through the coils to the top of the cooling tower, where this heated flow joins the warm return water from the airflow; moisture content in the airstream passing over the coils is unchanged, so that the dew-point temperature remains constant. The heat content of the air is reduced and the equivalent heat is added to the water circulating in coil adsorbers. The dry- and wet-bulb temperature of the air entering the bottom of the cooling tower is less than the temperature of the air entering the fan. The water leaving the tower approaches the wet-bulb temperature of the surface atmosphere as an economic limit.

Evaporative Cooling Plus Mechanical Refrigeration

On humid summer days, the wet-bulb temperature may increase over extended periods, severely hampering the effectiveness of evaporative cooling. This factor, plus the warming of the air entering the mine, may necessitate the series installation of a mechanical refrigeration unit to chill the water delivered underground.

Performance characteristics at one northern United States mine on a spring day were as follows:

Evaporative cooling tower	1224 tons
Mechanical refrigeration unit (in series)	816 tons
Water volume circulated	1750 gpm
Water temperature entering mine	40 °F
Water temperature leaving mine	68 °F

Combination Systems

Components may be arranged in various ways for the greatest efficiency. For example, air-cooling towers may be used to cool water during the cool months of the year and to supplement a mechanical refrigerator during the warm months of the year.

Surface-installed mechanical refrigeration units provide the bulk of cooling in summer. In winter, much of the cooling comes from the precooling tower when the ambient wet-bulb temperature is usually lower than the temperature of water entering the tower. The precooling tower is normally located above the return water storage reservoir. An evaporative cooling tower is more cost-effective (capital and operating costs) than mechanical refrigeration with comparable capacity.

Reducing High-Pressure Liquids

Use of underground refrigerated water chillers is increasing, due to system efficiency because of their location close to the work. Transfer of heat from condensers is the major problem with these systems. If hot mine water is used to cool the condenser, efficiency is lost due to high condensing temperature, probable corrosion, and fouling. If surface water is used, it must be piped both in and out of the mine after use. If water is noncorrosive and nonfouling, fairly good chiller efficiencies can be obtained up to 125 °F entering condenser water.

Surface water is usually allowed to flow into tanks at intervals down the shaft to break the high water head that develops in a vertical pipe. Energy is wasted and the temperature will rise about 1 °F for every 1000 ft of drop. The water must be pumped out of the mine, which adds cost. The water pressure can be reduced for use at the mine level and then discharged to the mine drainage system, however, the costs for pumping the water to the surface are high, thus offsetting the convenience of low-pressure mine cooling.

In a pipe 4000 ft high filled with water (density = 62.4 lb/ft^3), the pressure would be $62.4 \times 4000/144 = 1730$ psi. In an open piping system, the pressure at the bottom is further increased by the head necessary to raise the water up the pipe and out of the mine. Water pipes and coils in deep mines must resist this high pressure. Fittings and pipe specialties are costly, and safety precautions and care must be taken with such high-pressure equipment. Closed-circuit piping has the same static pressures, but pumps must only overcome pipe friction.

Frequent movement of surface cooling towers, a desired feature for shifting mining operations, results in high construction costs. Closed-circuit systems have been used in various mines in the United States to overcome the cost of pumping brine or cooling water out of the mine.

To achieve the advantage of both low-pressure and a closed circuit, the Magma mine in Arizona has installed heat exchangers underground at the mining horizon. The shell-and-tube heat exchanger converts surface chilled water in a high-pressure closed circuit to a low-pressure chilled water system on mine production levels. Air-cooling plants and chilled water lines can be constructed of standard materials, permitting frequent relocation (Figure 5). Although desirable, this system has not been widely used.

Energy-Recovery Systems

Pumping costs can be reduced by combining a water turbine with the pump. The energy of high-pressure water flowing to a lower pressure drives the pump needed for the low-pressure water circuit. Rotary-type water pumps have been developed to pump against a 5000-ft head, and water turbines are also available to operate under head. Figure 6 shows a turbine pump-motor combination. Only the shaft and the pipe and fittings to the unit on the working level need strong pipes. The system connects to underground refrigeration water-chilling units, whose return chilled water is used for condensing before being pumped out.

Two types of turbines are suitable for mine use—the Pelton wheel and a pump in reverse. The Pelton wheel has a high-duty efficiency of about 80%, is simply constructed, and is readily controlled. A pump in reverse is only 10 to 15% efficient, but mine maintenance and operating personnel are familiar with this equipment. Turbine energy recovery may have a problem when operating on chilled mine service water. Mine demands fluctuate widely, often outside the operating range of the turbine. Operating experience shows that coupling a Pelton wheel to an electric generator is the best approach.

South African mines have mechanical refrigeration units, surface heat-recovery systems, and turbine pumps incorporated into their air-cooling plants in a closed circuit.

A surface-sited plant has two disadvantages. Chilled water delivered underground at low operating pressure heats up at a rate of 1 °F per 1000 ft of shaft depth. The pumping of this water back to the surface is expensive. The installation of energy-recovery turbines underground and a heat-recovery system on the surface helps to justify the installation economically.

The descending chilled water is fed through a turbine mechanically linked to pumps operating in the return chilled water line. The energy recovery turbine reduces the rate of temperature increase in the descending chilled water column to about 0.3 °F/1000 ft of shaft depth.

Precooling towers on the surface reduce the water temperature a few degrees before it enters the refrigeration plant. Operating cost of a surface refrigeration plant is about one-half that of a comparable underground plant because of the unlimited supply of relatively cool ambient air for heat rejection.

In a uranium mine in South Africa, waste heat comes from surface refrigeration units and a high condensing heat pump. Condensing water from the refrigeration units is the source for the heat pump, which discharges 130 °F water. This water can be used as service hot water or as preheated feedwater for steam generation in the uranium plant. The total additional costs of a heat pump over a conventional refrigeration plant have a simple payback of about 18 months.

Pelton turbines are used by the South African gold-mining industry to recover energy from chilled water flowing down the shafts. About 1000 kW might be recovered at a typical installation, thus partially offsetting the power requirements for pumping return water to surface.

One mining company in the United States has installed an energy-recovery system of two separate units at the 2500-ft level and at the 5000-ft level. Each installation consists of turbines directly connected to 200 hp, 3600 rpm induction motors operating as generators. Rated output is 144 kW at 550 gpm water flow at approximately 2500-ft head.

A surface level installation chills the service water used throughout the mining operation. A 6-in. chilled water line feeds chilled water to the turbine through a pneumatically controlled valve. The water level in an adjacent discharge reservoir is sensed to modulate the position of this valve proportional to the water demand in the reservoir. When the system is shut down, the inlet valve to the turbine slowly closes by spring pressure. Simultaneously, the bypass valve opens slowly and water discharges through an orifice into the sump. Typical operating data are as follows:

Bypass discharge temperature	43.7 °F
Turbine inlet temperature	40.2 °F
Turbine discharge temperature	40.8 °F
Turbine output	144 kW
Volume through turbine	550 gpm

MECHANICAL REFRIGERATION PLANTS

In most underground applications, mechanical refrigeration plants (Figure 6) provide chilled water for delivery underground where some or all may be supplied condensed to direct-expansion air-to-air cooling systems. Air that is passed over the direct-expansion evaporator coils is cooled and delivered to working headings.

Underground mechanical refrigeration plants avoid heat added by autocompression of air coming down the shafts. The main disadvantage of these units is disposal of heat from the condensers. The condenser discharge water can either be run into the mine drainage system, returned to the surface for cooling, or sprayed into the mine discharge air system.

Hot water discharged from condensers of underground refrigeration units may be ejected into the sumps of the mine water-pumping system. This procedure requires a constant resupply of condenser water from freshwater lines or mine drainage sources and increases mine pumping costs. Underground cooling towers are seldom used, since they suffer from air shortage, which results in higher condensing temperatures underground than on the surface. It is difficult to predict the performance of underground cooling towers and condensing temperatures of equipment over the life of a mine. High condenser temperatures cause excessive power consumption. In addition, the water is often contaminated with dust and fumes, which cause fouling, scaling, and corrosion of the piping system and condenser tubes. These problems make placing mechanical refrigeration units on the surface more favorable.

Spot Cooling

Spot cooling permits the driving of long headings for exploration or extended development prior to establishment of primary ventilation equipment. Using mechanical refrigeration systems with direct-expansion air coolers, they service a single heading or localized mining area. Spot coolers allow the mine to advance development headings more rapidly and under more desirable conditions. Spot cooling of selected development headings may also be required in many mines where rock temperatures exceed 100 °F.

UNDERGROUND HEAT EXCHANGERS

Two types of heat exchangers are used underground—air-to-water and water-to-water. Air-to-water exchanges include those with water sprays in the airstream and finned tubes.

Air Cooling versus Working Place Cooling

Figure 7 shows a typical underground central plant to serve a large area of the mine with air cooled by chilled water. Air flows through ventilation ducts or headings from the chiller to the working headings. When working headings are distant, the air warms and, in some cases, picks up moisture in transit. This limits available cooled air in the production area. The performance characteristics of one plant are as follows:

Air volume cooled	100,000 cfm
Entering air temperature	
Dry bulb	80.5 °F
Wet bulb	74.3 °F
Discharge air temperature	56.4 °F
Entering water temperature	42 °F
Discharge water temperature	67 °F
Heat extracted from air	508 tons

If one of the air-conditioning plants from the workings becomes a problem, it is preferable to pump chilled water in closed circuit to cooling plants close to or serving individual productive headings.

Another underground cooling plant arrangement (Figure 7A) at a mine in Wallace, Idaho, is a light, portable unit which attaches to the normal auxiliary ventilation system and requires no extra excavation. This system uses no separate spray and dust-collection system, and the condensed water drips from the duct. The fan gives good service downstream from the cooler. A mechanical refrigeration unit located on the same or a nearby level chills water to cool the air, using potable water in the condenser and discharging the same to the mine discharge water sumping system.

This system avoids the central plant problem in which air is cooled and then exposed to wall rock as it is delivered many hundreds or several thousand feet down an airway for pickup by auxiliary ventilation equipment. The cooling coils are normally located within 300 ft of the working place and reduce overall air-cooling plant requirements. Table 2 shows typical performance characteristics of the units.

Table 2 Typical Performance of Portable, Underground Cooling Units

Size Rating	30 by 48 in. 40 ton		24 by 36 in. 20 ton	
Location	Drift	Shaft	Stope	Stope
Entering air temperature	80 °F sat.	80 °F sat.	80 °F sat.	80 °F sat.
Discharge air temperature	70 °F sat.	64 °F sat.	68 °F sat.	80 °F sat.
Volume, cfm	12,000	12,000	6,000	6,000
Calculated tons	45	60	22.5	26

Cooling Coils and Fan Position

The fan may be upstream or downstream of the cooling coils in underground cooling plants. An upstream fan provides a few degrees lower discharge temperature, but a downstream fan distributes air more efficiently over the coils.

WATER SPRAYS AND EVAPORATIVE COOLING SYSTEMS

Finned tube heat exchangers need periodic cleaning, especially when they are upstream from the fan. The downstream position has spray water, which helps to wash some of the dirt away.

Another type of bulk-spraying cooling plant in South Africa consists of a spray chamber serving a section of isolated drift up to several hundred feet long. Chilled water is introduced through a manifold of spray nozzles. Warm mild air flows countercurrent to the various stages of water sprays. Air cooled by this direct air-to-water cooling system is delivered to the active mine workings by the primary and secondary ventilation system. These bulk-spray coolers are efficient and economical.

At one uranium mine, a portable bulk-spray cooling plant was developed that could be advanced with the working faces to overcome the high heat load between a stationary spray chamber and the production heading.

The 300-ton capacity plant has a 81-in. long, 87-in. wide, and 102-in. high stainless steel chamber containing two stages of spray nozzles and a demister baffle. The skid-mounted unit weighs about 2200 lb and is divided into four components (spray chamber, demister assembly, and two sump halves).

Portable bulk-spray coolers have a wide range of cooling capacity and are cost-effective. A new, smaller portable spray cooling plant has been developed to cool mine air adjacent to the workplace. It air cools and cleans through direct air-to-water contact. The cooler is tube-shaped and is normally mounted in a remote location. The mine inlet and discharge air ventilation ducts are connected to duct transitions from the unit. Chilled water is piped to an exposed manifold, and warm water is discharged from the unit into a sump drain.

Hot, humid air enters the cooler at the bottom; it then slows down and flows through egg crate flow straighteners. Initial heat exchange occurs as the air passes through plastic mesh and contacts suspended water droplets.

Vertically sprayed water in a spray chamber then directly contacts the ascending warm air. The air passes through the mist eliminator, which removes suspended water droplets. Cool, dehumidified air exits from the cooler through the top outlet transition. The warmed spray water drops to the sump and discharges through a drainpipe.

BIBLIOGRAPHY

Anonymous. 1980. Surface refrigeration proves energy efficient at Anglo mine. *Mine Engineering* (May).

Bell, A.R. 1970. Ventilation and refrigeration as practiced at Rhokana Corporation Ltd., Zambia. *Journal of the Mine Ventilation Society of South Africa* 23(3):29-35.

Beskine, J.M. 1949. Priorities in deep mine cooling. *Mine and Quarry Engineering* (December):379-84.

Bossard, F.C. 1983. *A manual of mine ventilation design practices.*

Bossard, F.C. and K.S. Stout. Underground mine air-cooling practices. USBM Sponsored Research Contract G0122137.

Bromilow, J.G. 1955. Ventilation of deep coal mines. *Iron and Coal Trades Review.* Part I, February 11:303-08; Part II, February 18:376; Part III, February 25:427-34.

Brown, U.E. 1945. Spot coolers increase comfort of mine workers. *Engineering and Mining Journal* 146(1):49-58.

Caw, J.M. 1953. Some problems raised by underground air cooling on the Kolar Gold Field. *Journal of the Mine Ventilation Society of South Africa* 2(2):83-137.

Caw, J.M. 1957. Air refrigeration. *Mine and Quarry Engineering* (March):111-17; (April):148-56.

Caw, J.M. 1958. Current ventilation practice in hot deep mines in India. *Journal of the Mine Ventilation Society of South Africa,* 11(8):145-61.

Caw, J.M. 1959. Observations at an underground air conditioning plant. *Journal of the Mine Ventilation Society of South Africa,* 12(11):270-74.

Cleland, R. 1933. Rock temperatures and some ventilation conditions in mines of Northern Ontario. C.I.M.M. *Bulletin Transactions Section* (August):370-407.

Fenton, J.L. 1972. Survey of underground mine heat sources. Masters Thesis, Montana College of Mineral Science and Technology.

Field, W.E. 1963. Combatting excessive heat underground at Bralorne. *Mining Engineering* (December):76-77.

Goch, D.C. and H.S. Patterson. 1940. The heat flow into tunnels. *Journal of the Chemical Metallurgical and Mining Society of South Africa* 41(3):117-28.

Hartman, H.L. 1961. *Mine ventilation and air conditioning.* The Ronald Press Company, New York.

Hill, M. 1961. Refrigeration applied to longwall stopes and longwall stope ventilation. *Journal of the Mine Ventilation Society of South Africa* 14(5):65-73.

Kock, H. 1967. Refrigeration in industry. *The South African Mechanical Engineer* (November):188-96.

Le Roux, W.L. 1959. Heat exchange between water and air at underground cooling plants. *Journal of the Mine Ventilation Society of South Africa* 12(5):106-19.

Marks, John. 1969. Design of air cooler—Star Mine. Hecla Mining Company, Wallace, ID.

Minich, G.S. 1962. The pressure recuperator and its application to mine cooling. *The South African Mechanical Engineer* (October):57-78.

Muller, F.T. and M. Hill. 1966. Ventilation and cooling as practiced on E.R.P.M. Ltd., South Africa. *Journal of the South African Institute of Mining and Metallurgy.*

Richardson, A.S. 1950. A review of progress in the ventilation of the mines of the Butte, Montana District. Quarterly of the Colorado School of Mines (April), Golden, CO.

Sandys, M.P.J. 1961. The use of underground refrigeration in stope ventilation. *Journal of the Mine Ventilation Society of South Africa* 14(6):93-95.

Schlosser, R.B. 1967. The Crescent Mine cooling system. Northwest Mining Association Convention (December).

Short, B. 1957. Ventilation and air conditioning at the Magma Mine. *Mining Engineering* (March):344-48.

Starfield, A.M. 1966. Tables for the flow of heat into a rock tunnel with different surface heat transfer coefficients. *Journal of the South African Institute of Mining and Metallurgy* 66(12):692-94.

Thimons, E., R. Vinson, and F. Kissel. 1980. Water spray vent tube cooler for hot stopes. USBM TPR 107.

Thompson, J.J. 1967. Recent developments at the Bralorne Mine. *Canadian Mining and Metallurgy Bulletin* (November):1301-1305.

Torrance, B. and G.S. Minish. 1962. Heat exchanger data. *Journal of the Mine Ventilation Society of South Africa* 15(7):129-38.

Van der Walt, J., E. de Kock, and L. Smith. Analyzing ventilation and cooling requirements for mines. Engineering Management Services, Ltd., Johannesburg, Republic of South Africa.

Warren, J.W. 1958. The science of mine ventilation. Presented at the American Mining Congress, San Francisco (September).

Warren, J.W. 1965. Supplemental cooling for deep-level ventilation. *Mining Congress Journal* (April):34-37.

Whillier, A. 1972. Heat—A challenge in deep-level mining. *Journal of the Mine Ventilation Society of South Africa* 25(11):205-13.

CHAPTER 27

INDUSTRIAL EXHAUST SYSTEMS

INDUSTRIAL exhaust ventilation systems collect and remove airborne contaminants consisting of particulates (dust, fumes, mists, fibers), vapors, and gases that can create an unsafe, unhealthy, or undesirable atmosphere. They also salvage usable material and improve plant housekeeping. Chapter 11 of the 1989 ASHRAE *Handbook—Fundamentals* covers definitions, particle sizes, allowable concentrations from a health point of view, and upper and lower explosive limits of various air contaminants.

There are two types of exhaust systems: (1) *general exhaust*, in which an entire work space is exhausted without considering specific operations; and (2) *local exhaust*, in which the contaminant is controlled at its source. Local exhaust is preferable because it offers better contaminant control with minimum air volumes, thereby lowering the cost of air cleaning and replacement air equipment. Chapter 25 of this volume and Chapter 2 of *Industrial Ventilation—A Manual of Recommended Practice* (ACGIH 1989) detail steps to determine the air volumes necessary to dilute the contaminant concentration for general exhaust.

Replacement air, which is usually conditioned, provides air to the work space to replace exhausted air. Therefore, the systems are not isolated from each other. Refer to Chapter 25 for further information on replacement and makeup air. A complete industrial ventilation program includes replacement air systems that provide a total volumetric flow rate equal to the total exhaust rate. If insufficient replacement air is provided, the pressure of the building will be negative relative to local atmospheric pressure. Negative pressure allows air to infiltrate through open doors, window cracks, and combustion equipment vents. As little as 0.05 in. of water negative pressure can cause drafts and might cause downdrafting of combustion vents, thereby creating a potential health hazard. Negative plant pressures can also cause excessive energy use. If workers near the plant perimeter complain about cold drafts, unit heaters are often installed. Heat from these units is usually drawn into the plant interior because of the velocity of the infiltration air, leading to overheating. Too often, the solution is to exhaust more air from the interior, causing increased negative pressure and more infiltration. Negative plant pressures reduce the exhaust volumetric flow rate because of increased system resistance. Balanced plants that have equal exhaust and replacement air rates use less energy. Wind effects on building balance are discussed in Chapter 14 of the 1989 ASHRAE *Handbook—Fundamentals.*

FLUID MECHANICS

Chapters 2 and 32 of the 1989 ASHRAE *Handbook—Fundamentals* describe basic fluid mechanics and its applications to duct systems. A thorough understanding of this subject is essential to an economical local exhaust system design.

The equation for *volumetric flow rate* (volume) is

$$Q = VA \tag{1}$$

The preparation of this chapter is assigned to TC 5.8, Industrial Ventilation.

where

Q = volumetric flow rate, cfm
V = average flow velocity, fpm
A = flow cross-sectional area, ft^2

Another equation relates velocity to velocity pressure:

$$V = 1097\,(p_v/\rho)^{0.5} \tag{2}$$

where

p_v = velocity pressure, in. of water
ρ = density, lb/ft^3

If the air temperature is 68 °F ±30 °F, the ambient pressure is standard 14.7 psia, the duct pressure is no more than 20 in. of water different from the ambient pressure, the dust loading is low (0.1 to 1.0 grains/ft^3), and moisture is not a consideration, then the density in Equation (2) is standard without significant error. For a standard air density of 0.075 lb/ft^3, Equation (2) simplifies to:

$$V = 4005\sqrt{p_v} \tag{3}$$

LOCAL EXHAUST SYSTEM COMPONENTS

Local exhaust systems have five basic components: (1) the *hood*, or entry point, of the system; (2) the *duct system*, which transports air; (3) the *air-cleaning device*, which removes contaminants from the airstream; (4) the *air-moving device*, which provides motive power for overcoming system resistance; and (5) the *exhaust stack*, which discharges system air to the atmosphere.

Hoods

The most effective hood uses the minimum exhaust volumetric flow rate to provide maximum contaminant control. Knowledge of the process or operation is essential before a hood can be designed.

Hoods are either *enclosing* or *nonenclosing* (Figure 1). Enclosing hoods provide better and more economical contaminant control because the exhaust rate and the effects of room air currents are minimal compared to those with a nonenclosing hood. There should be access for inspection and maintenance where required. Hood access openings should be as small as possible and placed out of the natural path of the contaminant. Hood performance (*i.e.*, how well it controls the contaminant) should be checked by an industrial hygienist.

Where access requirements make it necessary to leave all or part of the process open, a nonenclosing hood may have to be used. Careful attention to airflow patterns around the process and hood and to the characteristics of the process is required to make nonenclosing hoods functional.

Another consideration in the design of nonenclosing hoods is the operator. Hoods should be located so that the contaminant is drawn away from the operator's breathing zone. Canopy hoods should not be used where the operator must bend over the tank or process.

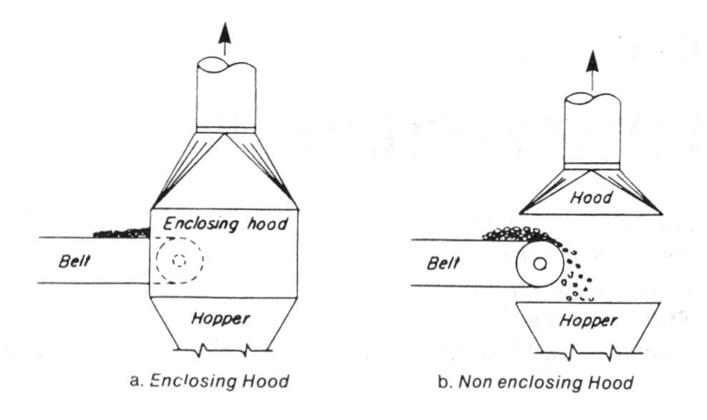

Fig. 1 Enclosing and Nonenclosing Hoods

Capture Velocities

To select an adequate volumetric flow rate (Q) to withdraw air through a hood, designers use the concept of capture velocities. Capture velocities are air velocities at the point of contaminant generation upstream of a hood. The contaminant enters the moving airstream at the point of generation and is conducted along with the air into the hood. Table 1 shows ranges of capture velocities for several industrial operations. These capture velocities are based on successful experience under ideal conditions. If velocities anywhere upstream of a hood are known [$v = f(Q, x, y, z)$], such as for flanged circular openings, the capture velocity is set equal to v at points (x, y, z) where contaminants are to be captured and Q is found. To ensure that contaminants enter an inlet, the transport equations have to be solved between the source and hood.

Hood Volumetric Flow Rate

After the hood configuration and capture velocity are determined, the exhaust volumetric flow rate can be calculated.

For *enclosing hoods*, the exhaust volumetric flow rate is the product of the hood face area and the velocity required to prevent outflow of the contaminant [Equation (1)]. The inflow velocity is typically 100 fpm. However, research with laboratory hoods indicates that lower velocities can reduce the vortex downstream of the human body and, therefore, lessen the reentrainment of contaminant into the operator's breathing zone (Caplan and Knutson 1977, Fuller and Etchells 1979). These lower face velocities require the replacement air supply to be distributed to minimize the effects of room air currents. This is one reason why replacement air systems must be designed with exhaust systems in mind. Because air must enter the hood uniformly, interior baffles are sometimes necessary (Figure 2).

Table 1 Range of Capture Velocities

Condition of Contaminant Dispersion	Examples	Capture (Control) Velocity, fpm
Released with essentially no velocity into still air	Evaporation from tanks, degreasing, plating	50 to 100
Released at low velocity into moderately still air	Container filling, low-speed conveyor transfers, welding	100 to 200
Active generation into zone of rapid air motion	Barrel filling, chute loading of conveyors, crushing, cool shakeout	200 to 500
Released at high velocity into zone of very rapid air motion	Grinding, abrasive blasting, tumbling, hot shakeout	500 to 2000

In each category above, a range of capture velocities is shown. The proper choice of values depends on several factors (Alden and Kane 1982):

Lower End of Range	Upper End of Range
1. Room air currents are favorable to capture.	1. Distributing room air currents.
2. Contaminants of low toxicity or of nuisance value only.	2. Contaminants of high toxicity.
3. Intermittent, low production.	3. High production, heavy use.
4. Large hood; large air mass in motion.	4. Small hood; local control only.

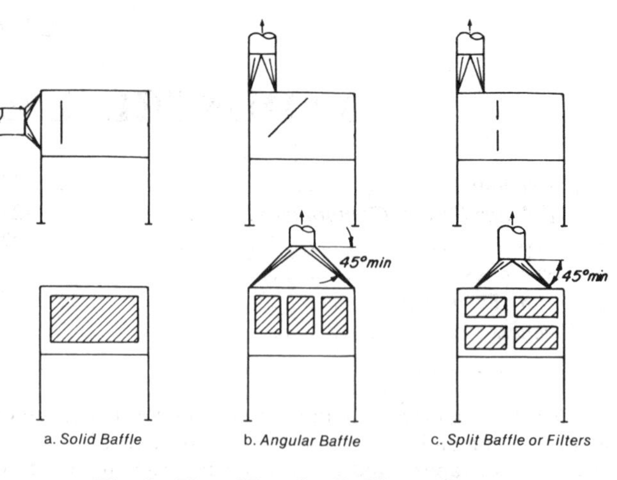

Fig. 2 Use of Interior Baffles to Ensure Good Air Distribution

For *nonenclosing hoods*, the air velocity at the point of contaminant release must equal the capture velocity and be directed so that the contaminant enters the hood. The simplest form of nonenclosing hood is the *plain opening* shown in Figure 3 (Alden and Kane 1982). For unflanged round and rectangular openings, the required flow rate can be approximated by:

$$Q = V(10X^2 + A) \tag{4}$$

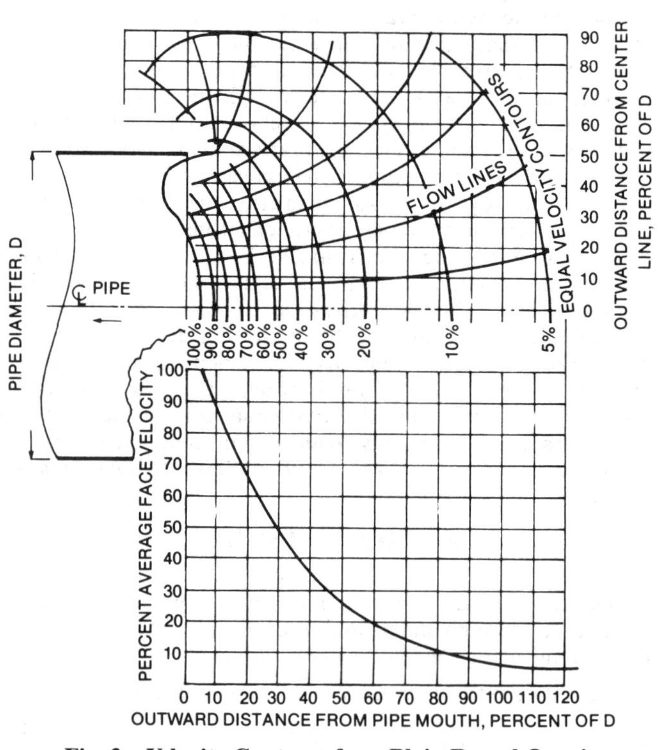

Fig. 3 Velocity Contours for a Plain Round Opening

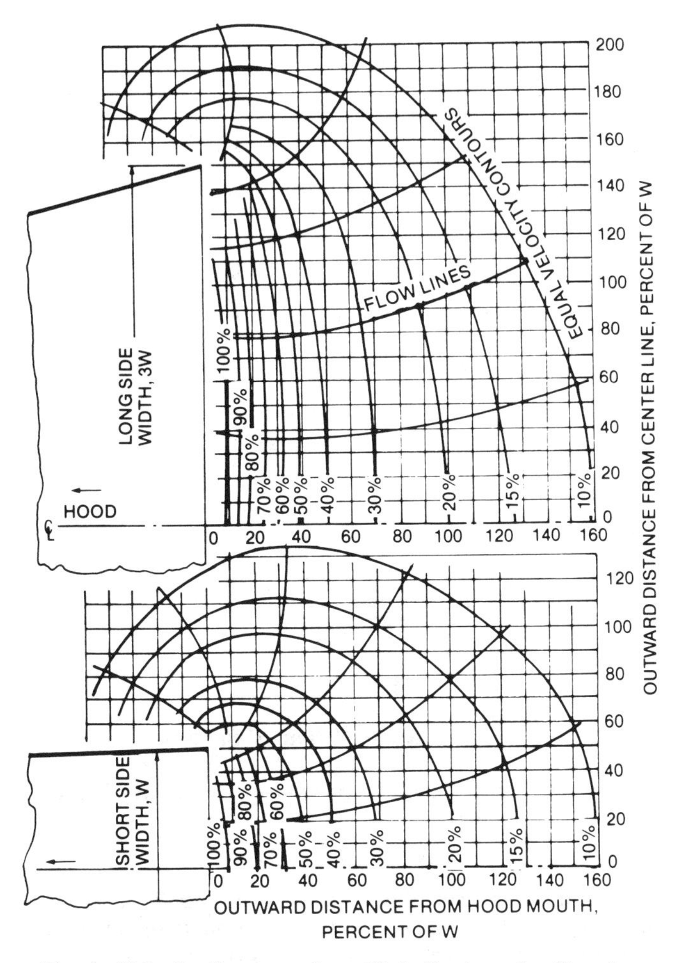

Fig. 4 Velocity Contours for a Plain Rectangular Opening with Sides in a 1:3 Ratio

where

Q = flow rate, cfm
V = capture velocity, fpm
X = centerline distance from the hood face to the point of contaminant generation, ft
A = hood face area, ft^2

Figure 3 shows lines of equal velocities (velocity contours) for a plain round opening. The velocities are expressed as percentages of the hood face velocity. Studies have established the *principle of similarity of contours*, which states that the positions of the velocity contours, when expressed as a percentage of the hood face velocity, are functions only of the hood shape (DallaValle 1952).

Figure 4 (Alden and Kane 1972) shows velocity contours for a rectangular hood with an aspect ratio of 0.333 (sides in the ratio of 1 to 3). The profiles are similar to those for the round hood but are more elongated.

If the aspect ratio is lower than about 0.2 (0.15 if flanged), the shape of the flow pattern in front of the hood changes from approximately spherical to approximately cylindrical, and Equation (4) is no longer valid. For this type of nonenclosing hood, commonly called a *slot hood*, the required flow rate is predicted by an equation for openings of 0.5 to 2 in. in width (Silverman 1942):

$$Q = 3.7\, LVX \tag{5}$$

where

L = long dimension of the slot, ft

The boundary between the plain opening and the slot hood is not precisely defined.

In Figures 3 and 4, air migrates from behind the hood. If a *flange* (defined as a solid barrier to reduce flow from behind the hood face) is installed, the required flow rate for plain openings is about 75% of that for the corresponding unflanged plain opening (Hemeon 1963). Therefore, a flange can reduce the required exhaust volumetric flow rate for a given capture velocity. Alden and Kane (1982) state that the flange size should be approximately equal to the hydraulic diameter (four times the area divided by the perimeter) of the hood face. Brandt (1947) states that the flange should equal the capture distance.

For flanged slots with aspect ratios less than 0.15 and flanges greater than 3 times the slot width, Silverman (1942) reported that:

$$Q = 2.6LVX \tag{6}$$

Multiple slots are often used on a hood face. The slot hood equations [(5) or (6)] are *incorrect* in this situation. The volumetric flow rate should be determined by Equation (4), where A is the overall face area, because the slots on such a hood are for air distribution over the hood face only.

Another device that helps increase the effectiveness of a nonenclosing hood is a *baffle*, a solid barrier that prevents airflow from unwanted areas in front of the hood. For example, if a rectangular hood is placed on a bench surface, flow cannot come from below the hood and the hood is more effective. Equation (4) can be modified for this situation by assuming that a mirror image of the flow exists on the underside of the bench. This will yield:

$$Q = V(5X^2 + A) \tag{7}$$

This is known as the DallaValle half-hood equation (DallaValle 1952). For better collection, the hood should be flanged. Flanging and baffling a nonenclosing hood create a situation more similar to an enclosure.

Nonenclosing hoods should be placed as near as possible to the source of contamination to prevent excessive exhaust rates. Figures 3 and 4 show that the velocity decreases extremely quickly as the distance in front of the hood increases, reaching 10% of the face velocity at approximately one hydraulic diameter in front of the hood. Because the volumetric flow rate is a direct function of distance squared (except for single slots), the required flow rate increases rapidly with capture distance. Typically, nonenclosing hoods are ineffective if the capture distance is greater than about 3 ft. Large capture distances can decrease contaminant control. Room air currents caused by thermal currents, supply air grilles, mechanical action of the process, or personnel-cooling fans can disturb the flow toward the hood enough to deflect the contaminant-laden airstream away from the hood. If capture distances greater than 3 ft are required, the designer should consider a push-pull ventilation system, as discussed below.

Example 1. A nonenclosing hood is to be designed to capture a contaminant that is liberated with a very low velocity 2 ft in front of the face of the hood. The nature of the operation requires hood face dimensions of 1.5 by 4 ft. The hood rests on a bench, and a flange is placed on the sides and top of the face (Figure 5). Determine the volumetric flow rate required to capture the contaminant. Assume that (1) the room air currents are variable in direction but less than 50 fpm, (2) the contaminant has low toxicity, and (3) the hood is used continuously.

Solution: From Table 1, a capture velocity of 50 to 100 fpm is required. The selected capture velocity must be higher than the room air currents; assume that 80 fpm is sufficient. The flow rate required can be predicted by modifying Equation (7) to account for the flanges. Thus,

$$Q = 0.75V(5x^2 + A)$$

adequately predicts the flow pattern in front of the hood. For this example, the required flow rate is:

$$Q = (0.75)(80)[(5)(2)^2 + (1.5)(4)](1) = 1560 \text{ cfm}$$

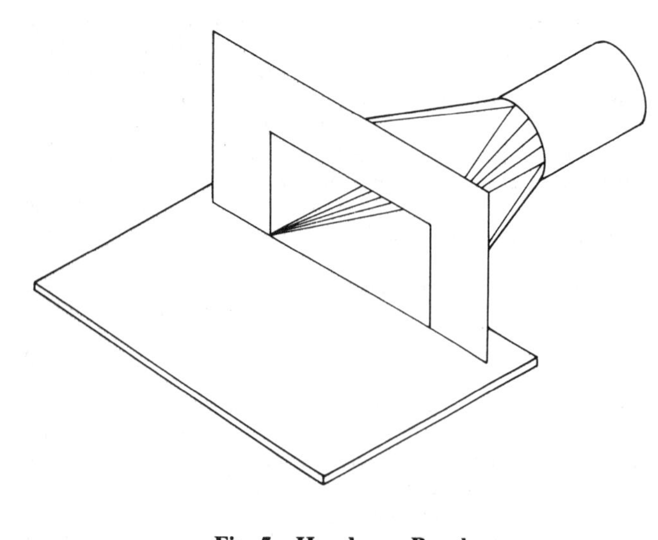

Fig. 5 Hood on a Bench

Hot Process Hoods. The exhaust from hot processes requires special consideration because of the buoyant effect of heated air near the hot process. The minimum exhaust rate is obtained by completely enclosing the process and placing the exhaust connection at the top of the enclosure. Determining the exhaust rate for hot processes requires knowing the convectional heat transfer rate (see Chapter 3 of the 1989 ASHRAE *Handbook—Fundamentals*) and the physical size of the process.

If the process cannot be completely enclosed, place the canopy hood above the process so that the natural path of the contaminant is directed toward the hood. Canopy hoods should be designed with caution; contaminants may be drawn across the operator's breathing zone. The height of the hood above the process should be kept to a minimum to reduce the total exhaust air rate.

The *low canopy hood*, which is within 3 ft of the process, is the nonenclosing hood that requires the least volumetric flow rate. The *high canopy hood*, which is more than 10 ft above the process, requires more volumetric flow rate because room air is entrained in the column of hot, contaminated air rising from the process. This situation should be avoided. Where lateral exhaust must be used, the airflow toward the hood must overcome the tendency of the air to rise and, therefore, generally requires the highest exhaust volumetric flow rate.

Hemeon (1963) lists Equations (8) through (12) for determining the volumetric flow rate of hot gases for low canopy hoods. Note: Canopy hoods located 3 to 10 ft above the process cannot be analyzed by these equations. Equation (8) can be used to estimate the flow of heated air rising from a hot body, if the actual temperature of the hot air leaving the body is approximately the same as the average temperature of the air as it rises past the body (this is normally acceptable):

$$Q_o = (3600 \times 2gR/pc_p)^{1/3}(qLA_p^2)^{1/3} \quad (8)$$

where

Q_o = volumetric flow rate, cfm
g = gravitational acceleration, 32.2 ft/s²
R = air gas constant, 53.352 ft • lb/lb • °R
p = local atmospheric pressure, psf abs
c_p = constant pressure specific heat, 0.24 Btu/lb • °R
q = convection heat transfer rate, Btu/min
L = vertical height of hot object, ft
A_p = cross-sectional area of airstream at the upper limit of the hot body, ft

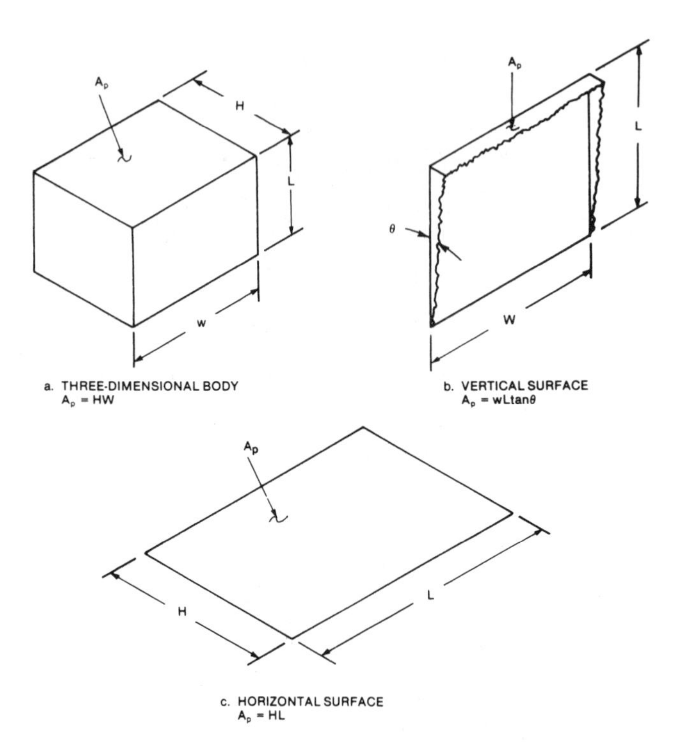

Fig. 6 A_p for Various Situations

For a standard atmospheric pressure of 2117 psf abs, Equation (8) can be written as:

$$Q_o = 29(qLA_p^2)^{1/3} \quad (9)$$

For three-dimensional bodies, the area A_p in Equation (8) is approximated by the plan view area of the hot body (Figure 6a). For horizontal cylinders, A_p is the product of the length times the diameter of the rod.

For vertical surfaces, the area A_p in Equation (8) is the area of the airstream (viewed from above) as the flow leaves the vertical surface (Figure 6b). As the airstream moves upward on a vertical surface, it appears to expand at an angle of approximately 4 to 5°. Thus, A_p is given by:

$$A_p = wL \tan \theta \quad (10)$$

where

w = width of vertical surface, ft
L = height of vertical surface, ft
θ = angle at which the airstream expands

For horizontal heated surfaces, A_p is the surface area of the heated surface and L is the longest length (conservative) of the horizontal surface, or its diameter if round (Figure 6c).

Where the heat transfer is caused by steam from a hot water tank,

$$q = h_{fg}GA_p \quad (11)$$

where

q = heat transferred, Btu/h
h_{fg} = latent heat of vaporization, Btu/lb
G = steam generation rate, lb/(min • ft²)
A_p = surface area of the tank, ft²

At 212°F, the latent heat of vaporization is 970.3 Btu/lb. Using this value and Equation (11), Equation (8) simplifies to

$$Q_o = 287A_p(GL)^{1/3} \quad (12)$$

The exhaust volumetric flow rate determined by Equation (8) or (12) is the required exhaust flow rate when: (1) a low canopy

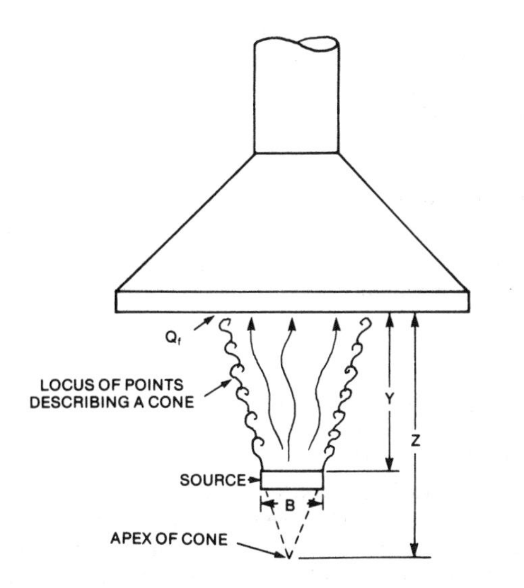

Fig. 7 High Canopy Hood Parameters for Y>10 ft

hood of the same dimensions as the hot object or surface is used and (2) side and back baffles are used to prevent room air currents from disturbing the rising air column. If side and back baffles cannot be used, increase the canopy hood size and the exhaust flow rate to reduce the possibility of spillage from the hood. A good design provides a low canopy hood overhang equal to 40% of the distance from the hot process to the hood face on all sides (ACGIH 1989). The hood flow rate can be increased using:

$$Q_T = Q_o + V_f (A_f - A_p) \tag{13}$$

where

Q_T = total flow rate entering hood, cfm
Q_o = the flow rate determined by Equations (8) or (12)
V_f = the desired indraft velocity through the perimeter area, fpm
A_f = hood face area, ft^2
A_p = plan view area of Equations (8) or (12), as discussed above

A minimum indraft velocity of 100 fpm should be used for most design conditions. However, when room air currents are appreciable or if the contaminant discharge rate is high and the design exposure limit is low, higher values of V_f might be required.

Sutton (1950) reported that the volumetric flow rate for a high canopy hood (more than 10 ft above the process) can be predicted empirically from round, square, or nearly square sources by:

$$Q_z = 7.4\, Z^{3/2} q^{1/3} \tag{14}$$

where

Q_z = volumetric flow rate at any elevation Z, cfm
Z = vertical distance (Figure 7), ft
q = convection heat transfer rate, Btu/min

Z can be obtained from:

$$Z = Y + 2B \tag{15}$$

where

Y = distance from the hot process to the hood face (Figure 7), ft
B = maximum dimension of the hot process plan view (Figure 7), ft

At Z, the rising airstream will be nearly round, with a diameter determined by:

$$D_z = 0.5 Z^{0.88} \tag{16}$$

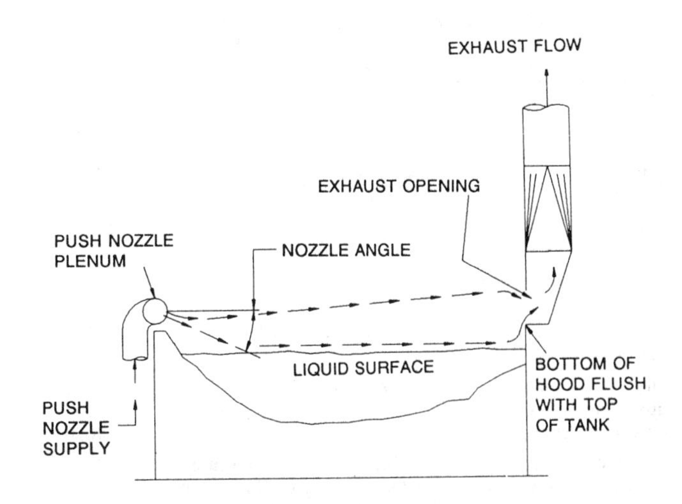

Fig. 8 Push-Pull Tank Ventilation

(Courtesy American Conference of Governmental Industrial Hygienists, Inc.)

where

D_z = flow diameter at any elevation Z above the apparent point source, ft

High canopy hoods are extremely susceptible to room air currents. Therefore, they are typically much larger (often twice as large) than indicated by Equation (16) and are used only if the low canopy hood cannot be used. The total flow rate exhausted from the hood should be evaluated using Equation (13), if Q_o is replaced by Q_z.

Lateral Ventilation Systems

For open vessels, contaminants can be controlled by a lateral exhaust hood, which exhausts air through slots on the periphery of the vessel. Alternatively, air can be blown through one slot and exhausted from another opposite slot. This concept is known as push-pull (Figure 8). The inlet is large to accommodate the jet and the room air entrained by the jet. Push-pull systems accommodate wider vessels than do lateral exhaust hoods. Chapter 10 of the *Industrial Ventilation Manual* (ACGIH 1989) provides a design procedure for these systems.

Special Situations

Some operations may require exhaust flow rates different from those developed by the aforementioned equations. Typical reasons for different flow rates include the following:

1. Induced air currents are created whenever anything is projected into an airspace. For example, high-speed rotating machines such as pulverizers, high-speed belt material transfer systems, falling granular materials, and escaping compressed air from pneumatic tools all produce air currents. The size and direction of the airflow should be considered in hood design.
2. Exhaust flow rates that are insufficient to dilute combustible vapor-air mixtures to less than about 25% of the lower explosive limit of the vapor (NFPA 1981).
3. Room air currents caused by cross drafts, compensating air, spot cooling, or motion of the machinery of operators. This is especially significant when designing high canopy hoods for hot process exhaust.

Duct Considerations

The second component of a local exhaust ventilation system is the duct through which contaminated air is transported from the hood(s). Round ducts are preferred because they (1) offer a more

Table 2 Contaminant Transport Velocities
Adapted from *Industrial Ventilation—A Manual of Recommended Practices* (*ACGIH 1988*)

Nature of Contam-inant	Examples	Minimum Transport Velocity, fpm
Vapors, gases, smoke	All vapors, gases, smokes	Usually 1000 to 2000
Fumes	Welding	2000 to 2500
Very fine light dust	Cotton lint, wood flour, litho powder	2500 to 3000
Dry dusts and powders	Fine rubber dust, Bakelite molding powder dust, jute lint, cotton dust, shavings (light), soap dust, leather shavings	3000 to 4000
Average industrial dust	Grinding dust, buffing lint (dry), wool jute dust (shaker waste), coffee beans, shoe dust, granite dust, silica flour, general material handling, brick cutting, clay dust, foundry (general), limestone dust, packaging and weighing asbestos dust in textile industries	3500 to 4000
Heavy dusts	Sawdust (heavy and wet), metal turnings, foundry tumbling barrels and shakeout, sandblast dust, wood blocks, hog waste, brass turnings, cast-iron boring dust, lead dust	4000 to 4500
Heavy or moist dusts	Lead dust with small chips, moist cement dust, asbestos chunks from transite pipe cutting machines, buffing lint (sticky), quicklime dust	4500 and up

uniform air velocity to resist settling of material and (2) can withstand the higher static pressures normally found in exhaust systems. When design limitations require rectangular ducts, the aspect ratio (height to width ratio) should be as close to unity as possible.

Minimum transport velocity is that velocity required to transport particulates without settling. Table 2 lists some generally accepted transport velocities as a function of the nature of the contaminants [*Industrial Ventilation Manual* (ACGIH 1989)]. The values listed are typically higher than theoretical and experimental values to account for: (1) damages to ducts, which would increase system resistance and reduce volume and duct velocity; (2) duct leakage, which tends to decrease velocity in the duct system upstream of the leak; (3) fan wheel corrosion or erosion and/or belt slippage, which could reduce fan volume; and (4) reentrainment of settled particulate caused by improper operation of the exhaust system. Design velocities can be higher than the minimum transport velocities but should never be significantly lower. When particulate concentrations are low, the effect on fan power is negligible.

Standard duct sizes and fittings should be used for economic and delivery time reasons. Information on available sizes and the cost impact of nonstandard sizes should be obtained from the contractor(s).

Duct Size Determination

The size of the round duct attached to the hood can be calculated by using Equation (1), the volumetric flow rate, and the minimum transport velocity.

Example 2. Suppose the contaminant captured by the hood in Example 1 requires a minimum transport velocity of 3000 fpm. What diameter round duct should be specified?

Solution: Using Equation (1), the duct area required is:

$$A = 1560/3000 = 0.52 \text{ ft}^2$$

The area calculated generally will not correspond to a standard duct size. The area of the standard size chosen should be less than that calculated above. For this example, a 9-in. diameter with an area of 0.4418 ft² should be chosen. The actual duct velocity is therefore:

$$V = 1560/0.4418 = 3530 \text{ fpm}$$

Hood Entry Loss

When air enters a hood, a loss of total pressure occurs because of dynamic losses. This is called the *hood entry loss* and might have several components. Each component is given by:

$$h_e = C_o p_v \tag{17}$$

where

h_e = hood entry loss, in. of water
C_o = loss factor, dimensionless
p_v = appropriate velocity pressure, in. of water

Loss factors for various hood shapes are given in Figure 9. This graph shows an optimum hood entry angle to minimize the entry losses. However, this total included angle of 45° is impractical in many situations because of the required transition length. A 90° angle, with a corresponding loss factor of 0.25 (for rectangular openings), is standard for most tapered hoods.

Total pressure is difficult to measure in a duct system, since it varies from point to point across a duct, depending on the local velocity. On the other hand, static pressure remains constant across a straight duct. Therefore, a single measurement of static pressure in a straight duct downstream of the hood can monitor the volumetric flow rate. The absolute value of this static pressure, *hood suction*, is given by:

$$p_{hs} = p_v + h_e \tag{18}$$

where

p_{hs} = hood suction, in. of water

Simple Hoods

A simple hood has only one dynamic loss. In this situation, the hood suction becomes:

$$p_{hs} = (1 + C_o)p_v \tag{19}$$

where p_v is the duct velocity pressure.

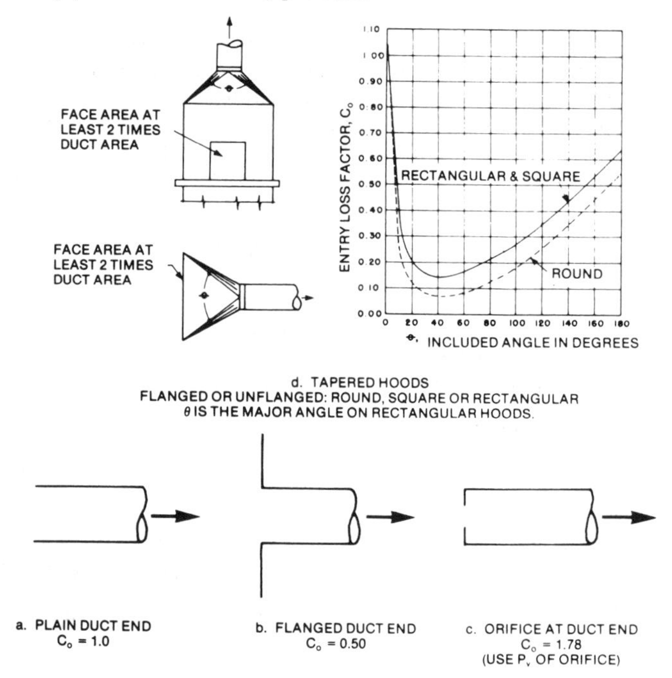

Fig. 9 Entry Losses for Typical Hoods

Example 3. Suppose that the hood in Example 1 was designed with the largest angle of transition between the hood face and the duct equal to 90°. What is the suction for this hood? Assume standard air density.

Solution: The two transition angles cannot be equal. Whenever this is true, the largest angle is used to determine the loss factor from Figure 9. Because the transition piece originates from a rectangular opening, the curve marked "rectangular" must be used. This corresponds to a loss factor of 0.25. The duct diameter and the velocity required were determined in Example 2. Equation (3) can be used to determine the duct velocity pressure because the air is standard air, or:

$$p_v = (3531/4005)^2 = 0.78 \text{ in. of water}$$

Equation (19) then gives:

$$p_{hs} = (1 + 0.25)(0.78) = 0.98 \text{ in. of water}$$

Compound Hoods

The losses for multislotted hoods (see Figure 10) or single-slot hoods with a plenum (called *compound hoods*) must be analyzed somewhat differently. The slots distribute air over the hood face and do not influence capture efficiency. The slot velocity should be approximately 2000 fpm to provide the required distribution at the minimum energy cost. Higher velocities dissipate more energy.

Losses occur when air passes through the slot and when air enters the duct. Because the velocities, and, therefore, the velocity pressures, can be different at the slot and at the duct entry locations, the hood suction must reflect both losses and is given by:

$$p_{hs} = p_v + (C_o p_v)_s + (C_o p_v)d \qquad (20)$$

where the first p_v is generally the higher of the two velocity pressures, s refers to the slot, and d refers to the duct entry location.

Example 4. A multislotted hood has 3 slots, each 1 by 40 in. At the top of the plenum is a 90° transition into the 10-in. duct. The volumetric flow rate required for this hood is 1650 cfm. Determine the hood suction. Assume standard air.

Solution: The slot velocity (V_s) from Equation (1) is:

$$V_s = (1650)(144)/[(3)(1)(40)] = 1980 \text{ fpm}$$

Substituting this velocity in Equation (3) gives:

$$p_v = (1980/4005)^2 = 0.24 \text{ in. of water}$$

The duct area is 0.5454 ft^2. Therefore, the duct velocity is given by Equation (1) as:

$$V_d = 1650/0.5454 = 3025 \text{ fpm}$$

Substituting the velocity in Equation (3) gives:

$$p_v = (3028/4005)^2 = 0.57 \text{ in. of water}$$

For a 90° transition into the duct, the loss factor is 0.25. For the slots, the loss factor is 1.78 (Figure 9). Therefore, using Equation (20):

$$p_{hs} = 0.57 + (1.78)(0.24) + (0.25)(0.57) = 1.14 \text{ in. of water}$$

Hood suction is the negative static pressure measured about 3 duct diameters downstream of the hood, the larger distance required for included angles of 180° or larger. Note that the duct velocity pressure is added to the sum of the two losses because it is the larger.

Exhaust volume requirements, minimum duct velocities, and entry loss factors for many specific operations are given in Chapter 10 of the *Industrial Ventilation Manual* (ACGIH 1989).

DUCT LOSSES

Chapter 32 of the 1989 ASHRAE *Handbook—Fundamentals* describes friction losses and how they are calculated, and presents loss coefficients for many fitting types. Many of these fittings do not apply to exhaust systems and should not be used. Table 3 lists the fitting shown in Chapter 32, of the 1989 ASHRAE *Handbook—Fundamentals*, that apply to industrial exhaust systems.

Elbows with a large center line radius-to-diameter (r/D) ratio (greater than 1.5) are the most suitable. If the r/D is 1.5 or less, abrasion in dust-handling systems can reduce the life of the fitting. The data in the 1989 ASHRAE *Handbook—Fundamentals* are limited to an r/D of 2 for gored elbows. A more complete compilation of the elbow loss data, especially for larger r/D elbows, is given in Table 4. Elbows are often made of 7 or more gores, especially in larger diameters. The loss data for 5-gore elbows applies to these elbows.

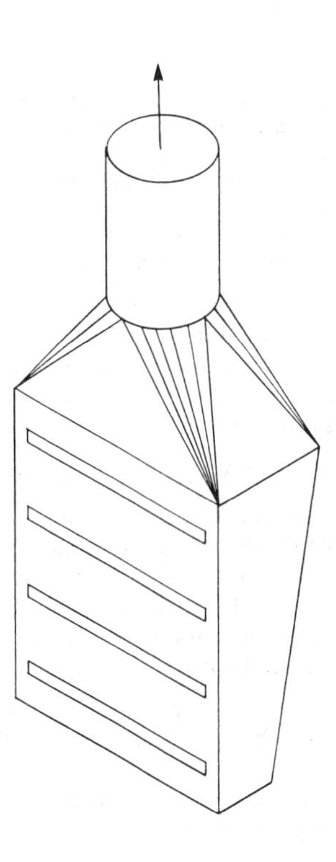

Fig. 10 Multislotted Nonenclosing Hood

Sepsy and Pies (1973) recommend an entry angle of 30° because it has less loss than a 45° entry, and there is less abrasion in dust-handling systems. Fittings 5-1 and 5-4 in Chapter 32 of the 1989 ASHRAE *Handbook—Fundamentals* include loss data on converging fittings with 30° angles.

Expansion or contraction losses depend on the conditions both upstream and downstream of the fitting. Chapter 32 also presents data for contractions and expansions both within a duct segment and at the end of a system (an evasé).

A system can be designed by calculating either the static pressure loss or the total pressure loss through the system. Chapter 32 calculates total pressure loss, while the *Industrial Ventilation Manual* (ACGIH 1989) calculates static pressure loss. Both use the same technique (but with different numerical values) for calculating fitting losses.

Where exhaust systems handling particulates must allow for a substantial increase in future capacity, required transport velocities can be maintained by providing open-end stub branches in the main through which air is admitted into the systems at the proper pressure and volumetric flow rate until the future connection is installed. Figure 11 shows such an air bleed-in. The use of outside air minimizes replacement air requirements. The size of the opening can be calculated by first determining the pressure drop required across the orifice from the duct calculations. Then the orifice velocity pressure can be determined from:

$$p_{v,o} = \Delta p_{t,o}/C_o \qquad (21a)$$

or

$$p_{v,o} = \Delta p_{s,o}/(C_o + 1) \qquad (21b)$$

where

$P_{v,o}$ = orifice velocity pressure, in. of water
$\Delta P_{t,o}$ = total pressure to be dissipated across the orifice, in. of water
$\Delta P_{s,o}$ = static pressure to be dissipated across the orifice, in. of water
C_o = orifice loss coefficient referenced to the velocity at the orifice cross-sectional area, dimensionless (Figure 9)

Equation (21a) must be used if the total pressure is calculated through the system; use Equation (21b) if the static pressure is calculated through the system. Once the velocity pressure is known, Equation (2) or (3) can be used to determine the orifice velocity. Equation (1) can then be used to determine the orifice size.

EXHAUST STACKS

The exhaust stack must be designed and located to prevent the reentrainment of discharged air into supply system inlets. The building's shape and surroundings determine the atmospheric airflow over it. Chapter 14 of the 1989 ASHRAE *Handbook—Fundamentals*, has more details on exhaust stack design. The following guidelines can be used:

1. For many one- or two-story industrial or laboratory buildings, 16-ft stacks may be adequate for discharging above the roof cavity.
2. The minimum stack height should be at least 8 ft above the roof for safety purposes.

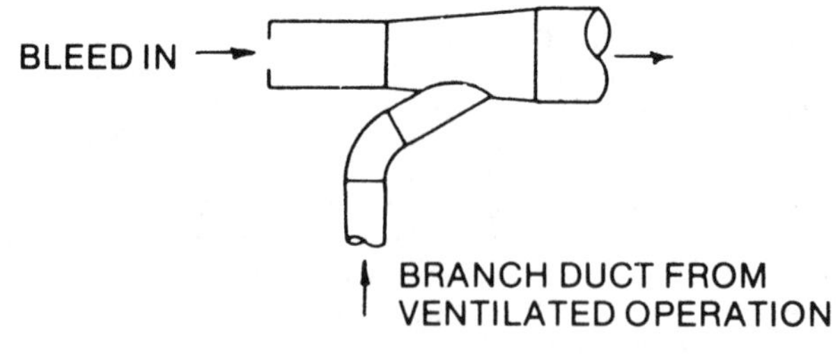

Fig. 11 Air Bleed-In

3. When the effluents are highly toxic or their odors are perceived at very low concentrations, a pollution problem will probably occur. The discharge should be above the point where wind flow is unaffected by the building.

If shorter stacks must be used, the stack discharge velocity should be high enough to project the contaminant-laden air above the roof recirculation zone. This usually requires wastefully high velocities (up to 8000 fpm), however. Wind velocities of 10 to 15 mph reduce exhaust plume rise by 85 to 90% compared to no wind condition.

If rain protection is important, stackhead design is preferable to weathercaps. Weathercaps, which are not recommended, have three disadvantages:

1. They deflect air downward, increasing the chance that contaminants will recirculate into air inlets.

Table 3 Classification of Fittings for Industrial Exhaust Systems

This table lists some of the fittings in Chapter 32 of the 1989 ASHRAE *Handbook—Fundamentals*. They are classified here by their applicability to industrial exhaust systems. Note that "Figure 1-1" Refers to the figure in the Fittings Loss Coefficients section of Chapter 32.
Codes are as follows: F = Frequently used; O = Occasionally used; N/A = Not applicable; A = Avoid

Figure	Code	Notes
1-1	F	1
1-2	O	2
1-3	O	2
1-4	N/A	
1-5	N/A	
1-6	N/A	
1-7	F	
1-8	O	
2-1	F	18
2-2	F	18
2-3	A	5
2-4	A	5
2-5	A	5
2-6	A	5
2-7	A	5
2-8	A	5
2-9	A	5
2-10	A	3
2-11	A	4,5
2-17	A	
3-1	F	(See Table 4)
3-2	F	(See Table 4)
3-3	A	4,6
3-4	O	
3-5	O	9,21
3-6	A	5,7,8
3-7	A	9
3-8	A	9
3-9	A	9
3-10	N/A	
3-11	A	4,5,6,8
3-12	A	4,5,6,8
3-13	O	
3-14	O	
3-15	A	8
4-1	O	10,11
4-2	A	8
4-3	A	8
4-4	A	8
4-5	O	8,12
4-6	O	
4-7	N/A	19
5-1	A	7, 15
5-2	A	7,14,15
5-3	A	4,6,7,14
5-4	F	
5-5	O	14
5-6	A	8,14
5-7	N/A	
5-8	N/A	
5-9	N/A	
5-10	N/A	
5-11	N/A	
5-12	N/A	
5-13	N/A	
5-14	N/A	
5-15	N/A	
5-16	N/A	
5-17	N/A	
5-18	N/A	
5-19	N/A	
5-20	N/A	
5-21	N/A	
5-22	N/A	
5-23	N/A	
5-24	N/A	
5-25	N/A	
5-26	N/A	
5-27	N/A	
5-28	N/A	
5-29	N/A	
5-30	N/A	
5-31	N/A	
5-32	O	8
5-33	F	8
5-34	O	8
5-35	N/A	
5-36	N/A	
6-1	A	6,16,17
6-2	A	6,8,16,17
6-3	O	16,17
6-4	A	8,16,17
6-5	N/A	
6-6	N/A	
6-7	N/A	
6-8	N/A	
6-9	O	Try to avoid
6-10	A	7
6-11	A	7,8
7-8	A	
7-9	A	3
7-10	A	3
7-11	O	12
7-12	A	7,12
7-13	A	7,12
7-14	A	7,12
7-15	0	12

Notes:

1. Hood data if wall is considered as a flange
2. Expensive to build
3. Can be used if discharge is vertical
4. High loss
5. Reentrainment of contaminants probable with its use
6. High wear for systems carrying particulate
7. Poor design
8. Round duct and fitting preferred
9. Elbows with vanes are not recommended due to plugging in dust-handling systems and corrosion in some vapor/mist systems
10. Use $\theta \leqslant 20°$
11. Do not use sudden change (see also notes 3 and 5)
12. Keep $\theta \leqslant 20°$
13. Keep $\theta \leqslant 20°$
14. Recommended entry angle is 30°
15. $A_s = A_c$ recommended only for A_b very much less than A_s
16. Use only for off-on applications
17. Dampers must have no leakage at design pressure
18. If vertical with no wall and total pressure is used to design system, this is the loss for a stackhead. If static pressure is used, ignore loss coefficient
19. With $\theta = 180°$, fitting is sometimes used as exhaust from a plenum behind multislotted hood
20. Used when multiple inlets to a collector are required
21. Avoid $r/w < 1.5$

Table 4 Elbow Losses

r/D	Smooth	5-Gore
0.5	0.71	0.98
0.75	0.33	0.46
1.0	0.22	0.33
1.25	0.17	0.27
1.5	0.15	0.24
1.75	0.14	0.22
2.00	0.13	0.19
2.25	0.13	0.17
2.50	0.12	0.16
2.75	0.12	0.15

Notes:
1. Expanded data for Fitting 3-1 and 3-2 in Chapter 32 of the 1989 ASHRAE *Handbook—Fundamentals*.
2. Data for 5-gore elbows also applies to elbows with more than 5 gores.
3. *r* is the center line radius of the elbow.

2. They have high friction losses.
3. They provide less rain protection than a properly designed stackhead.

Figure 12 contrasts the flow patterns of weathercaps and stackheads. Loss data for weathercaps are presented in Chapter 32 of the 1989 ASHRAE *Handbook—Fundamentals*. Losses in the straight duct form of stackheads (Figure 17 F or G, Chapter 14 of the 1989 ASHRAE *Handbook—Fundamentals*) are balanced by the pressure regain at the expansion to the larger diameter stackhead.

INTEGRATING DUCT SEGMENTS

Most systems have more than one hood. If the pressures are not designed to be the same for merging parallel airstreams, the system will adjust by itself to achieve pressure equality at the common point; however, the flow rates of the two merging airstreams will not necessarily be the same as designed. As a result, the hoods can fail to control the contaminant adequately and expose workers to potentially hazardous contaminant concentrations.

Two design methods ensure that the two pressures will be equal. The preferred method is a design that will self-balance without external aids. The second method uses adjustable balance devices, such as blast gates or dampers. This method is not recommended, especially when conveying abrasive material.

AIR CLEANERS

Air-cleaning equipment is usually selected to (1) conform to federal, state, or local emission standards and regulations; (2) prevent reentrainment of contaminants to work areas, where they may become a health or safety hazard; (3) reclaim usable materials; (4) permit cleaned air to recirculate to work spaces and/or processes; and (5) prevent annoying neighbors and/or physically damaging adjacent property.

Factors to consider when selecting air-cleaning equipment include type of contaminant (number of components, particulate versus gaseous, and concentration), contaminant removal efficiency required, disposal method, and air or gas stream characteristics. Chapter 11 of the 1988 ASHRAE *Handbook—Equipment*, covers industrial gas cleaning and air pollution control equipment such as dry centrifugal collectors, fabric collectors, electrostatic precipitators, and wet collectors. Consult a qualified applications engineer when selecting equipment.

The collector's pressure loss must be added to overall system pressure calculations. In some collectors, specifically some fabric filters, the loss varies as operation time increases. The system should be designed with the maximum pressure drop of the collector, or hood flow rates will be lower than designed during most of the duty cycle. Also, fabric collector losses are usually given only to the clean air plenum. A reacceleration to the duct velocity, with the associated entry losses, must be calculated in the design phase.

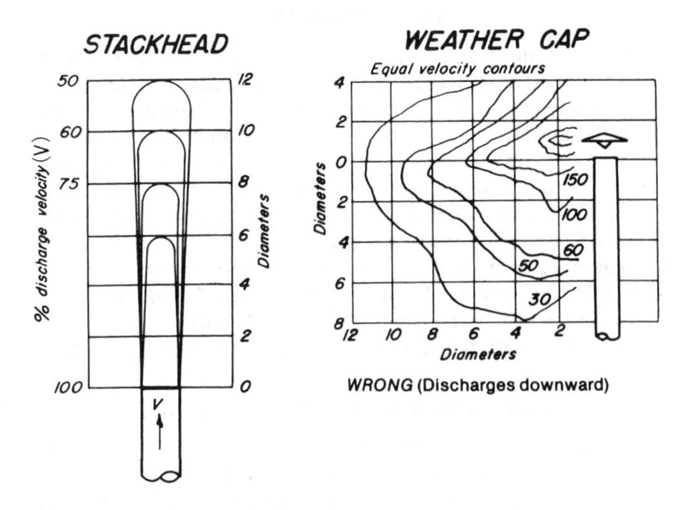

Fig. 12 Comparison of Flow Patterns for Stackheads and Weathercaps

Most other collectors are rated flange-to-flange and include reacceleration in the loss.

AIR-MOVING DEVICES

The type of air-moving device used depends on the type and concentration of contaminant, the pressure rise required, and allowable noise levels. Fans are usually selected. Chapter 3 of the 1988 ASHRAE *Handbook—Equipment* describes available fans and refers the reader to the Air Movement and Control Association (AMCA) *Publication* 201, Fans and Systems, for proper connection of the fan(s) to the system. The fan should be located downstream of the air cleaner whenever possible (1) to reduce possible abrasion of the fan wheel blades and (2) to create a negative pressure in the air cleaner so that air will leak into it and maintain positive control of the contaminant. In some instances, however, the fan is located upstream from the cleaner to help remove dust. This is especially true when using cyclone collectors, such as in the woodworking industry.

If explosive, corrosive, flammable, or sticky materials are handled, an *injector* can transport the material to the air-cleaning equipment. Injectors create a shear layer that induces airflow into the duct. Injectors are a last resort because their efficiences seldom exceed 10%.

ENERGY RECOVERY

Transferring energy from exhausted air to replacement air may be economically feasible, depending on (1) the location of the exhaust and replacement air ducts, (2) the temperature of the exhausted gas, and (3) the nature of the contaminants being exhausted.

DUCT CONSTRUCTION

Elbows and converging flow fittings should be made of thicker material than the straight duct, especially if abrasives are conveyed. Some cases require elbows constructed with a special wear strip in the heel of the elbow.

When corrosive material is present, alternatives such as special coatings or different duct materials (fiberglass, stainless steel, or special coatings) can be used. Industrial duct construction is described in Chapter 1 of the 1988 ASHRAE *Handbook—Equipment*. For further construction details, refer to SMACNA industrial duct construction standards (1977 and 1980).

SYSTEM TESTING

After installation, every exhaust system should be tested to ensure that it operates properly with the required flow rates through each hood. If the actual installed flow rates are different from the design values, correct the errant volumetric flow rates before using the system. Testing is also necessary to obtain baseline data to determine (1) compliance with federal, state, and local codes; (2) by periodic inspections, whether maintenance on the system is needed to ensure design operation; (3) if a system has sufficient capacity for additional airflow; and (4) whether or not system leakage is acceptable. Chapter 34, AMCA *Bulletin* 203, and Chapter 9 of *Industrial Ventilation* (ACGIH 1989) contain detailed information on the preferred methods of testing systems.

OPERATION AND MAINTENANCE

Periodic inspection and maintenance is required for proper operation. Systems are often changed or damaged after installation, resulting in low duct velocities and/or incorrect volumetric flow rates. Low duct velocities can cause the contaminant to settle and plug the duct, which reduces volumes at the affected hoods. Installing additional hoods in an existing system can change volumetric flow at the original hoods. In both cases, changed hood volumes can increase worker exposure and health risks.

The maintenance program should include: (1) ductwork inspection for particulate accumulation and damage to the ductwork by erosion or physical abuse, (2) checking exhaust hoods for proper volumetric flow rates and physical condition, (3) checking fan drives, and (4) maintaining air-cleaning equipment according to manufacturers' guidelines.

REFERENCES

Alden, J.L. and J.M. Kane. 1982. *Design of industrial ventilation systems*, 5th ed. Industrial Press, Inc., New York, 21.

ACGIH. 1989. *Industrial ventilation—A manual of recommended practice*, 20th ed. Committee on Industrial Ventilation, American Conference of Governmental Industrial Hygienists, Chapters 2, 3, and 10.

Brandt, A.D., R.J. Steffy, and R.G. Huebscher. 1947. Nature of air flow at suction openings. ASHRAE *Transactions* 53:55.

Caplan, K.J. and G.W. Knutson. 1977. The effect of room air challenge on the efficiency of laboratory fume hoods. ASHRAE *Transactions* 83(1):141.

DallaValle, J.M. 1952. *Exhaust hoods*. Industrial Press, Inc., New York, 22.

Flynn, M.R. and M.J. Ellenbecker. 1985. The potential flow solution for air flow into a flanged circular hood. *American Industrial Hygiene Journal* 46(6):318-22.

Fuller, F.H. and A.W. Etchells. 1979. The rating of laboratory hood performance. ASHRAE *Journal* 21(10):49-53.

Heinsohn, R.J., K.C. Hsieh, and C.L. Merkle. 1985. Lateral ventilation systems for open vessels. ASHRAE *Transactions* 91(1B):361-82.

Hemeon, W.C.L. 1963. *Plant and process ventilation*. Industrial Press, Inc., New York, 77.

Huebener, D.J. and R.T. Hughes. 1985. Development of push-pull ventilation. *American Industrial Hygiene Association Journal* 46(5): 262-67.

NFPA. 1981. Standard for ovens and furnaces—Design, location and equipment. National Fire Protection Association, *Standard* 86A, Item 4-2.1.

Sepsy, C.F. and D.B. Pies. An experimental study of the pressure losses in converging flow fittings used in exhaust systems. Prepared by Ohio State University for National Institute for Occupational Health, Document PB 221 130.

Shibata, M., R.H. Howell, and T. Hayashi. 1982. Characteristics and design method for push-pull hoods: Part 1—Cooperation theory of air flow; Part 2—Streamline analysis of push-pull flow. ASHRAE *Transactions* 88.

Silverman, L. 1942. Velocity characteristics of narrow exhaust slots. *Journal of Industrial Hygiene and Toxicology* 24 (November):267.

SMACNA. 1977. *Round industrial duct construction standards*. Sheet Metal and Air Conditioning Contractors' National Association, Inc., Vienna, VA.

SMACNA. 1980. *Rectangular industrial duct construction standards*. Sheet Metal and Air Conditioning Contractors' National Association, Inc., Vienna, VA.

Sutton, O.G. 1950. The dispersion of hot gases in the atmosphere. *Journal of Meteorology* 7(5):307.

Zarouri, M.D., R.J. Heinsohn, and C.L. Merkle. 1983. Computer-aided design of a grinding booth for large castings. ASHRAE *Transactions* 89(2A):95-118.

Zarouri, M.D., R.J. Heinsohn, and C.L. Merkle. 1983. Numerical computation of trajectories and concentrations of particles in a grinding booth. ASHRAE *Transactions* 89(2A):119-35.

CHAPTER 28

INDUSTRIAL DRYING SYSTEMS

DRYING removes water and other liquids from gases, liquids, or solids. Drying is most commonly used, however, to describe the removal of water or solvent from solids by thermal means. *Dehumidification* describes the drying of gases, usually by condensation or absorption by drying agents (see Chapter 19 of the 1989 ASHRAE *Handbook—Fundamentals*). *Distillation*, particularly *fractional distillation*, refers to the drying of liquids.

It is more economical to separate as much water as possible from solid materials before drying. Mechanical methods such as filtration, screening, pressing, centrifuging, or settling require less power and less capital outlay per unit mass of water removed.

This chapter describes systems used for industrial drying and their advantages, disadvantages, relative energy consumption, and applications.

MECHANISM OF DRYING

When a solid dries, two simultaneous processes occur: (1) the transfer of heat to evaporate the liquid and (2) the transfer of mass as vapor and internal liquid. Factors governing the rate of each process determine drying rate.

The principal objective in commercial drying is to supply the required heat most efficiently. Heat transfer can occur by convection, conduction, radiation, or by a combination of these. Types of industrial driers differ in the method used for transferring heat to the solid. In general, heat must flow first to the outer surface of the solid and then into the interior. An exception is drying with high-frequency electrical currents, where heat is generated within the solid, producing a higher temperature at the interior than at the surface and causing heat to flow from inside the solid to the outer surfaces.

APPLYING HYGROMETRY TO DRYING

In many drying applications, recirculating the medium improves thermal efficiency. Determining the optimum recycled air proportion requires balancing the lower heat loss with more recirculation and the higher drying rate with less recirculation.

Since the recycle ratio affects the humidity of drying air, air humidity throughout the drier must be analyzed to determine whether the predicted moisture pickup of the air is physically attainable. The maximum ability of air to absorb moisture corresponds to the difference between saturation moisture content at wet-bulb (or adiabatic cooling) temperature and moisture content at supply air dew point. Actual moisture pickup of air is determined by heat and mass transfer rate and is always less than the maximum attainable.

ASHRAE psychrometric charts for normal and high temperatures (No. 1 and No. 3) can be used for most drying calculations. The process will not exactly follow the adiabatic cooling lines, since some heat is transferred to the material by direct radiation or by conduction from the metal tray or conveyor.

Example 1. Assume a drier with a capacity of 90.5 lb of bone-dry gelatin per hour. Initial moisture content is 228% bone-dry basis, and final moisture content is 32% bone-dry basis. For optimum drying, supply air is at 120°F dry bulb and 85°F wet bulb in sufficient quantity so that the condition of exhaust air is 100°F dry bulb and 84.5°F wet bulb. Makeup air is available at 80°F dry bulb and 65°F wet bulb.

Find (1) the required amount of makeup and exhaust air, and (2) the percentage of recirculated air.

Solution: In this example, the humidity in each of the three airstreams is fixed; hence, the recycle ratio is also determined. Refer to ASHRAE Psychrometric Chart No. 1, and obtain the humidity ratio of makeup air and exhaust air. To maintain a steady-state condition in the drier, water evaporated from the material must be carried away by exhaust air. Therefore, the difference between the humidity ratio of exhaust air and that of makeup air (known as pickup) is equal to water evaporated from the material divided by the pounds per hour of dry air in the exhaust.

Step 1. From ASHRAE Psychrometric Chart No. 1, the humidity ratios can be found as follows:

	Dry bulb, °F	Wet bulb, °F	Humidity ratio, lb/lb dry air
Supply air	120	85	0.018
Exhaust air	100	84.3	0.022
Makeup air	80	65.2	0.010

Moisture pickup is 0.022 − 0.010 = 0.012 lb per pound of dry air. The amount of water evaporated in the drier is:

$$90.5\,(228 - 32)/100 = 177 \text{ lb/h}$$

The mass of dry air (makeup and exhaust) required to remove the evaporated water = 177/0.012 = 14,750 lb/h.

Step 2. Assume x percentage of recirculated air and $(100 - x)$ = percentage of makeup air. Then:

(Humidity ratio of exhaust and recirculated air) $x/100$
+ (Humidity ratio of makeup air)$(100 - x)/100$
= Humidity ratio of supply air.

Hence:

$$0.022(x/100) + 0.010\,(100 - x)/100 = 0.018$$

$x = 66.7\%$ recirculated air
$100 - x = 33.3\%$ makeup air

DETERMINING DRYING TIME

Three methods of finding drying time are listed below, in order of preference.

1. Conduct tests in a laboratory drier simulating conditions in the commercial machine, or obtain performance data directly from the commercial machine.
2. If the specific material is not available, obtain drying data on similar material by either of the above methods. This is subject to the investigator's experience and judgment.

The preparation of this chapter is assigned to TC 9.2, Industrial Air Conditioning.

3. Estimate drying time from theoretical equations (see Bibliography).

When designing commercial equipment, tests are conducted in a laboratory drier that simulates commercial operating conditions. Sample materials used in the laboratory tests should be identical to the material found in the commercial operation. Results from several tested samples should be compared for consistency. Otherwise, the test results may not reflect the drying characteristics of the commercial material accurately.

When laboratory testing is impractical, commercial drying data can be based on the equipment manufacturer's experience—an important source of data.

Since estimating drying time from theoretical equations yields only approximate values, care should be taken in using this method.

Commercial Drying Time

When selecting a commercial drier, the estimated drying time determines what size machine is needed for a given capacity. If the drying time has been derived from laboratory tests, the following should be considered:

- In a laboratory drier, considerable drying may be the result of radiation and heat conduction. In a commercial drier, these factors are usually negligible.
- In a commercial drier, humidity conditions may be higher than in a laboratory drier. In drying operations with controlled humidity, this factor can be eliminated by duplicating the commercial humidity condition in the laboratory drier.
- Operating conditions are not as uniform in a commercial drier as in a laboratory drier.
- Because of the small sample used, the test material may not be representative of the commercial material.

Thus, the designer must use experience and judgment to correct the test drying time to suit commercial conditions.

Drier Calculations

For preliminary cost estimates for a commercial drier, determine the following:

Circulating Air. The required circulating or supply airflow rate is established by the optimum air velocity relative to the material. This can be obtained from laboratory tests or previous experience, keeping in mind that the air also has an optimum moisture pickup. (See the section Applying Hygrometry to Drying.)

Makeup and Exhaust. The makeup and exhaust airflow rate required for steady-state conditions within the drier is also discussed under Applying Hygrometry to Drying. In a *continuously operating* drier, the relation between the moisture content of the material and the quantity of makeup air is given by:

$$G_T(W_2 - W_1) = M(w_1 - w_2), \text{ in which } W_2 \text{ is constant} \quad (1)$$

where

G_T = dry air supplied as makeup air to the drier, lb/h
M = stock dried in a continuous drier, lb/h
W_1 = humidity ratio of entering air, pounds of water vapor per pound of dry air
W_2 = humidity ratio of leaving air, pounds of water vapor per pound of dry air
w_1 = moisture content of entering material dry basis, pounds of water per pound
w_2 = moisture content of leaving material dry basis, pounds of water per pound

In *batch* driers, the drying operation is given as:

$$G_T(W_2 - W_1) = M_1\,(dw/d\theta) \quad (2)$$

where

M_1 = mass of material charged in a discontinuous drier, pounds per batch.
$dw/d\theta$ = instantaneous rate of evaporation corresponding to w. W_2 varies during a portion of the cycle.

The makeup air quantity is constant and is based on the average evaporation rate. Equation (2) then becomes identical to Equation (1), where $M = M_1/\theta$. Under this condition, the humidity in the *batch drier* varies from a maximum to a minimum during the drying cycle, whereas in the *continuous drier*, the humidity is constant with constant load.

Heat Balance. To estimate the fuel requirements of a drier, a heat balance consisting of the following is needed:

- Radiation and convection losses from the drier.
- Heating of the commercial dry material to the leaving temperature (usually estimated).
- Vaporization of the water being removed from the material (usually considered to take place at the wet-bulb temperature).
- Heating of the vapor from the wet-bulb temperature in the drier to the exhaust temperature.
- Heating of the total water in the material from the entering temperature to the wet-bulb temperature in the drier.
- Heating of the makeup air from its initial temperature to the exhaust temperature.

The energy absorbed must be supplied by the fuel. The selection and design of the heating equipment is an essential part of the overall design of the drier.

Example 2. Magnesium hydroxide is to dried from 82% to 4% moisture content (dry basis) in a continuous conveyor drier witha fin-drum feed, as shown in Figure 7. The desired production rate is 3000 lb/h. The drier is heated with steam at 50 psig. The optimum circulating air temperature for drying is 160°F, which is not limited by the existing steam pressure.

Step 1. Laboratory tests indicate the following:

Specific heats	
air (c_a)	= 0.24 Btu/lb·°F
material (c_m)	= 0.3 Btu/lb·°F
water (c_w)	= 1.0 Btu/lb·°F
water vapor (c_v)	= 0.45 Btu/lb·°F
Temperature of material entering drier	= 60°F
Temperature of makeup air	
dry bulb	= 70°F
wet bulb	= 60°F
Temperature of circulating air	
dry bulb	= 160°F
wet bulb	= 100°F
Air velocity through drying bed	= 250 ft/min
Drier bed loading	= 6.82 lb/ft²
Test drying time	= 25 min

Step 2. Previous experience indicates that the commercial drying time is 70% greater than the time obtained in the laboratory test. Therefore, the commercial drying time is 1.7 × 25 = 42.5 min.

Step 3. Calculate the holding capacity of the drier bed as follows:

$$3000\,(42.5/60) = 2125 \text{ lb at } 4\% \text{ (dry basis)}$$

Calculate required conveyor area as 2125/6.82 = 312 ft^2.

Assuming the conveyor is 8 ft wide, the length of the drying zone is 312/8 = 39 ft.

Step 4. The amount of water in the material entering the drier is:

$$3000\,[82/(100 + 4)] = 2370 \text{ lb/h}$$

The amount of water in the material leaving is:

$$3000\,[4/(100 + 4)] = 115 \text{ lb/h}$$

Thus, the moisture removal rate is 2370 − 115 = 2255 lb/h.

Step 5. The air circulates perpendicular to the perforated plate conveyor, so the air volume is the face velocity times the conveyor area, or:

$$\text{Air volume} = 250 \times 312 = 78{,}000 \text{ cfm}$$

ASHRAE Pyschrometric Charts 1 and 3 show the following air properties:

Supply air (160 °F db, 100 °F wb)
Humidity ratio = 0.0285 lb/lb of dry air
Specific volume = 16.33 ft^3/lb of dry air

Makeup air (70 °F db, 60 °F wb)
Humidity ratio W_1 = 0.0086 lb/lb of dry air

The mass flow rate of dry air is:

$$(78{,}000 \times 60)/16.33 = 286{,}500 \text{ lb/h}$$

Step 6. The amount of moisture pickup is:

$$2255/286{,}500 = 0.0079 \text{ lb/lb of dry air}$$

Therefore, the humidity ratio of the exhaust air is:

$$W_2 = 0.0285 + 0.0079 = 0.0364 \text{ lb/lb of dry air}$$

Substitute in Equation (1) and to calculate G_T as follows:

$$G_T(0.0364 - 0.0086) = (3000/1.04)(82 - 4)/100$$

$$G_T = 81{,}000 \text{ lb dry air per hour}$$

Therefore,
Makeup air = 100 × 81,000/286,500 = 28.2%
Recirculated air = 71.8%

Step 7. Heat Balance

Sensible heat of material $= M(t_{m2} - t_{m1})c_m$
= (3000/1.04)(100 − 60) 0.3
= 34,600 Btu/h

Sensible heat of water $= M_{w1}(t_w - t_{m1})\, c_w$
= 2370 (100 − 60) 1.0
= 94,800 Btu/h

Latent heat of evaporation $= M(w_1 - w_2)\, H$
= 2255 × 1037
= 2,338,400 Btu/h

Sensible heat of vapor $= M(w_1 - w_1)(t_2 - t_w)\, c_v$
= 2255 (160 − 100) 0.45
= 60,900 Btu/h

Required heat for material = 2,528,700 Btu/h

The temperature drop $(t_2 - t_3)$ through the bed is:

$$\frac{\text{Required heat}}{\text{Supplied air, lb/h} \times c_a} = \frac{2{,}528{,}700}{286{,}500 \times 0.24} = 37\,°\text{F}$$

Therefore, the exhaust air temperature is 160 − 37 = 123 °F

Required heat for makeup air $= G_T(t_3 - t_1)\, c_a$
= 81,000 (123 − 70) 0.24
= 1,030,000 Btu/h

The total heat required for material and makeup air is:

$$2{,}528{,}700 + 1{,}030{,}000 = 3{,}559{,}000 \text{ Btu/h}$$

Additional heat must be provided for radiation and convection losses, which can be calculated from the known construction of the drier surfaces.

DRYING SYSTEM SELECTION

A general procedure for selecting a drying system consists of the following:

1. Survey of suitable driers.
2. Preliminary cost estimates of various types.
   a) Initial investment
   b) Operating cost
3. Drying tests conducted in prototype or laboratory units of the most promising equipment available. Sometimes a pilot plant is justified.
4. Summary of tests to evaluate quality and samples of the dried products.

Some items can overshadow the operating or investment cost, such as the following:

- Product quality, which should not be sacrificed.
- Dusting, solvent, or other product losses.
- Space limitation.
- Product's bulk density, which can affect packaging cost.

Friedman (1951) and Parker (1963) discuss additional aids to drier selection.

TYPES OF DRYING SYSTEMS

Radiant Infrared Drying

Thermal radiation may be applied by infrared lamps, gas-heated incandescent refractories, steam-heated sources, and, most often, by electrically heated surfaces. Since infrared heats only near the receiver's surface, it is best used to dry material in thin sheets.

Using infrared heating to dry webs of material, such as uncoated material, has been relatively unsuccessful because of process control problems. Thermal efficiency can be low, since heat transfer depends on the emitter's characteristics and configuration, and on the properties of the material to be dried.

Radiant heating is used for drying ink or other coatings on paper, textile fabrics, paint films, and lacquers. The development of inks specifically formulated to react with tuned or narrow wavelength infrared radiation has revived infrared drying. Flammable material must not get too close to the heat source.

Ultraviolet Radiation Drying

Ultraviolet (UV) drying uses electromagnetic radiation. Inks and other coatings based on monomers are cure dried when exposed to UV radiation. Ultraviolet drying of inks has been justified because of superior final printing properties (Chatterjee and Ramaswamy 1975). The print resists scuff, scratch, acid, alkali, and some solvents. Web printing can be done at higher speeds without damage to the web.

A major barrier to wider acceptance of UV drying is the high capital installation cost and increased ink cost. The cost and frequency of replacing UV lamps are greater than for infrared oven maintenance.

Overexposure to radiation and ozone, formed by UV radiation's effect on atmospheric oxygen, can cause severe sunburn and possibly blood and eye damage. Safety measures include fitting the lamp housings with screens, shutters, and exhausts.

Conduction Drying

Drying rolls or drums (Figure 1), flat surfaces, open kettles, and immersion heaters are examples of direct-contact drying. The heating surface *must* have close contact with the material, and agitation may increase uniform heating or prevent overheating.

Conduction drying is used to manufacture and dry paper products. It (1) does not provide a high drying rate, (2) does not furnish uniform heat and mass transfer conditions, (3) usually results in a poor moisture profile across the web, (4) lacks proper control, (5) is costly to operate and install, and (6) usually creates undesira-

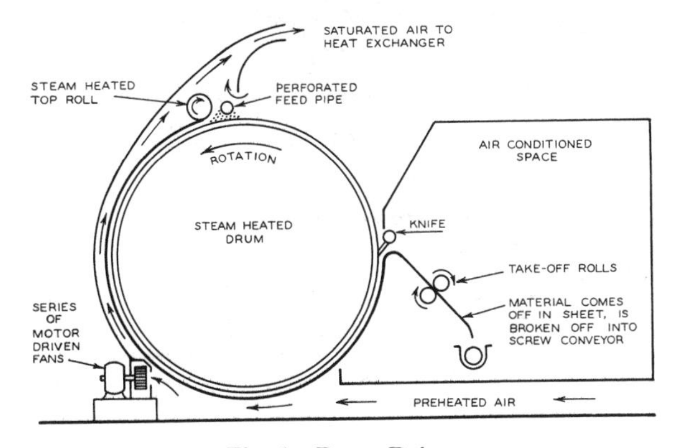

Fig. 1 Drum Drier

ble working conditions in areas surrounding the machine. Despite these disadvantages, replacing existing systems with other forms of drying is expensive. For example, Joas and Chance (1975) report that RF (dielectric) drying of paper compared to steam cylinder conduction drying is approximately four times the capital cost, six times the operating (heat) cost, and five times the maintenance cost. However, augmenting conduction drying with dielectric drying sections offsets the high cost of RF drying and may produce savings and increased profits from greater production and higher final moisture content.

Further use of large conduction drying systems depends on reducing heat losses from the drier, improving heat recovery, and incorporating other drying techniques to improve final product quality.

Dielectric Drying

When wet material is placed in a strong, high-frequency (2 to 100 MHz) electrostatic field, heat is generated within the material. More heat is developed in the wetter areas than in the drier areas, resulting in automatic moisture profile correction. Water is evaporated without unduly heating the substrate. Therefore, in addition to its leveling properties, dielectric drying provides uniform heating throughout the web thickness.

Dielectric drying is controlled by varying field or frequency strength; varying field strength is easier and more effective. Response to this variation is quick, with no time nor thermal lag in heating. The dielectric heater is a sensitive moisture meter.

There are several electrode configurations. The platen type (Figure 2) is used for drying and baking foundry cores, heating plastic preforms, and drying glue lines in furniture. The rod or stray field types (Figure 3) are used for thin web material such as paper and textile products. The double-rod types (over and under material) are used for thicker webs or flat stock, such as stereotype matrix board and plywood.

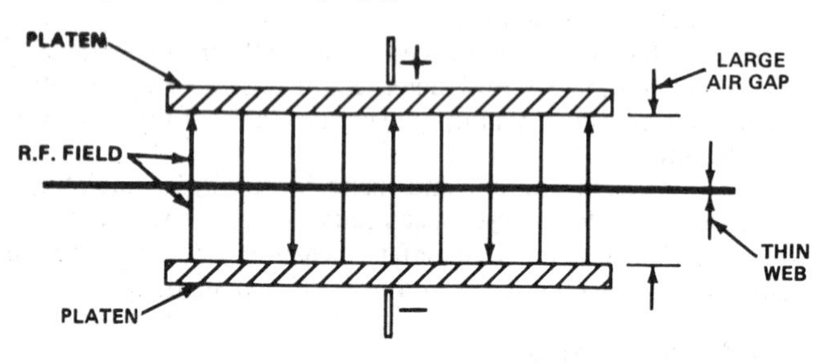

Fig. 2 Platen Type

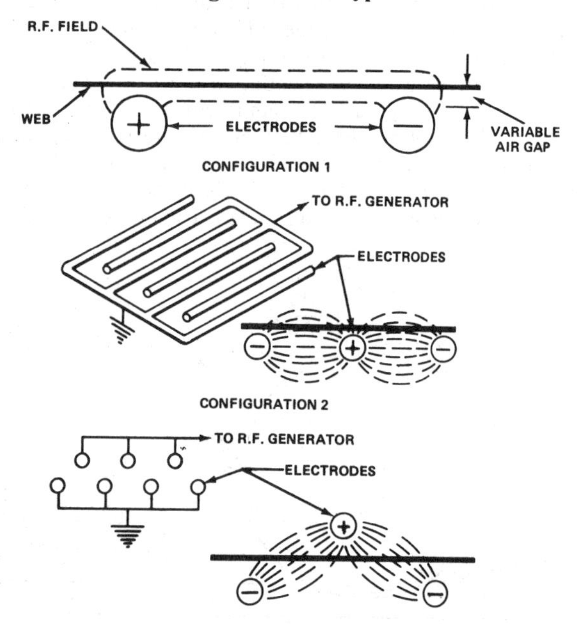

Fig. 3 Rod Type

Dielectric drying is popular in the textile industry. Because air is entrained between fibers, convection drying is slow and uneven. This can be overcome by dielectric drying after yarn drying. Because the yarn is usually transferred to large packages immediately after drying, even and correct moisture content can be obtained by dielectric drying. Knitting wool seems to benefit from internal steaming in hanks.

Warping caused by nonuniform drying is a serious problem for plywood and linerboard. Dielectric drying yields warp-free products.

Dielectric drying is uneconomical for overall paper drying, but has advantages when used at the dry end of a conventional steam drum drier. It corrects moisture profile problems in the web without overdrying. This conventional/dielectric combination is synergistic; the effect of adding the two is greater than the sum of the two effects achieved independently. This is more pronounced in thicker web materials, accounting for as much as a 16% line speed increase and a corresponding 2% energy input increase.

Microwave Drying

Microwave drying or heating uses ultrahigh frequency (900 to 5000 MHz) power. It is a form of dielectric heating, since it is applied to heating nonconductors. Because of its higher frequency, microwave equipment is capable of generating extreme power densities.

Microwave drying is applied to thin materials in strip form by passing the strip through the gap of a split waveguide. Entry and exit shielding requirements make continuous process applications difficult. Its many safety concerns make microwave drying more expensive than dielectric drying. Control is also difficult because microwave drying lacks the self-compensating properties of dielectrics.

Convection Drying (Direct Driers)

Some convection drying occurs in almost all driers. True convection driers, however, use circulated hot air or other gases as the principal heat source. Each means of mechanically circulating air or gases has some virtue.

Rotary Driers. These cylindrical drums cascade the material being dried through the airstream (Figure 4). The driers are heated directly or indirectly, and air circulation is parallel or counterflow. A variation is the rotating-louver drier, which introduces air beneath the flights, achieving close contact.

Cabinet and Compartment Driers. These batch driers range from the heated loft (with only natural convection and usually poor and nonuniform drying) to self-contained units with forced draft and properly designed baffles. Several systems may be evacuated to dry delicate or hygroscopic materials at low temperatures. These driers are usually loaded with material spread in trays to increase the exposed surface. Figure 5 shows a drier that can dry water-saturated products.

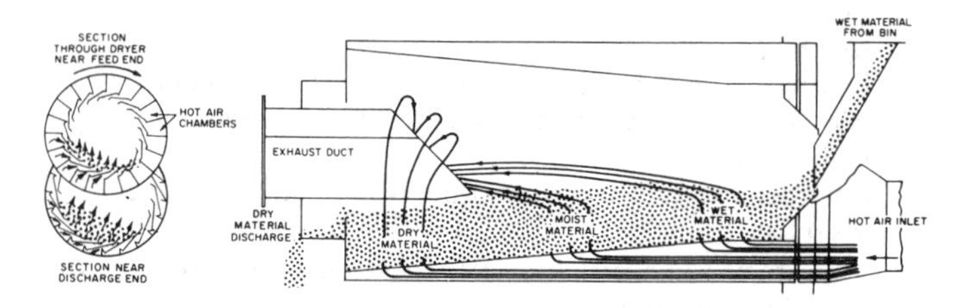

Fig. 4 Cross Section and Longitudinal Section of Rotary Drier

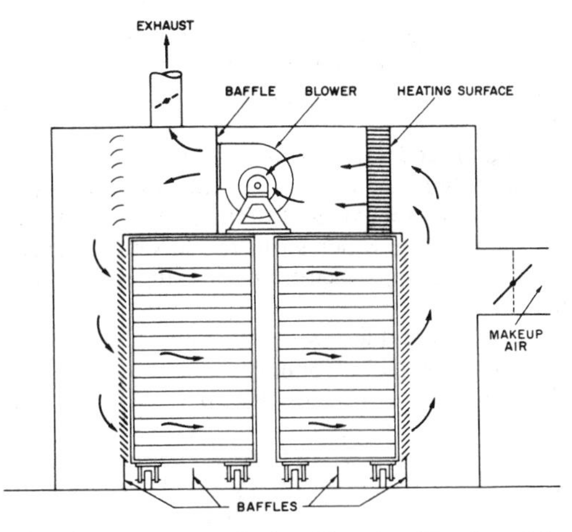

Fig. 5 Compartment Drier, Showing Trucks with Air Circulation

When designing driers to process products saturated with solvents, special features must be included to prevent explosive gases from forming. Safe operation requires exhausting 100% of the air circulated during the initial drying period or during any part of the drying cycle when the solvent is evaporating at a high rate. At the end of the purge cycle, the air is recirculated and heat is gradually applied. The amount of air circulated, the cycle lengths, and the rate that heat is applied to prevent circulated air from becoming explosive is determined in the laboratory drier for each product. Recirculating air as soon as it is feasible in the drying cycle is economical when using costly dehumidified air. The air *must not* recirculate when cross-contamination of products is prohibited.

Driers must have special safety features in case any part of the drying cycle fails. *Industrial Ovens and Driers* (FMEA 1990) is one source of information listing safety features to use when designing driers. For example:

1. Each compartment must have separate supply and exhaust fans and an explosion-relief panel.
2. The exhaust fan blade tip speed should be 5000 fpm for a forward-inclined blade, 6800 fpm for a radial-tip blade, and 7500 fpm for a backward-inclined blade. These speeds produce high static pressures at the fan, ensuring constant air exhaust volumes under conditions such as negative pressures in the building or downdrafts in the exhaust stacks.
3. An airflow failure switch in the exhaust duct must shut off both fans and the heating coil, and sound an alarm.
4. An airflow failure switch in the air supply system must shut off both fans and the heating coil, and sound an alarm.
5. A high-temperature limit controller in the supply duct must shut off the heat to the heating coil and sound an alarm.
6. An electric interlock on the drier door must cause the drying cycle to repeat if the door is opened beyond a set point, such as wide enough for a person to enter for product inspection.

Tunnel Driers. Tunnel driers are modified compartment driers that operate continuously or semicontinuously. Heated air or combustion gas is circulated by fans. The material is handled on trays or racks on trucks and moves through the drier either intermittently or continuously. The airflow may be parallel, counterflow, or a combination obtained by center exhaust (Figure 6). Air may also flow across the tray surface, vertically through the bed, or in any combination of directions. By reheating the air in the drier or recirculating it, a high degree of saturation is reached before air is exhausted, reducing sensible heat loss.

A variation is the strictly continuous drier having one or more mesh belts that carry the product through it, as shown in Figure 7. Many combinations of temperature, humidity, air direction, and velocity are possible. Hot air leaks at the entrance and exit can be minimized by baffles or inclined ends, where the material enters and leaves from the bottom.

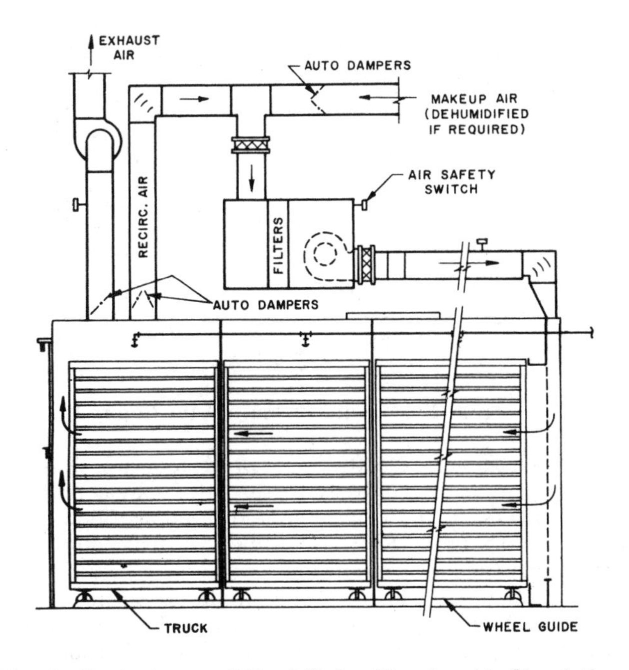

Fig. 6 Explosionproof Truck Drier, Showing Air Circulation and Safety Features

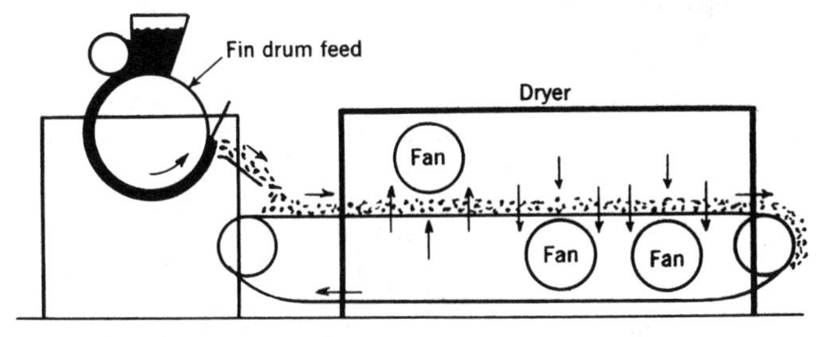

Fig. 7 Section of Continuous Drier, Blow-Through Type

High-Velocity Driers. High-velocity hoods or driers have been tried as supplements to conventional cylinder driers for drying paper. When used with conventional cylinder driers, web instability and lack of process control result. Applications such as thin permeable webs, where internal diffusion is not the controlling factor in the drying rate, offer more promise.

Spray Driers. Spray driers have been used in the production of dried milk, coffee, soaps, and detergents. Because the dried product (in the form of small beads) is uniform and the drying time is short (5 to 15 s), this drying method has become more important. When a liquid or slurry is dried, the spray drier has high production rates.

Spray drying involves the atomization of a liquid feed in a hot-gas drying medium. The spray can be produced by a two-fluid nozzle, a high-pressure nozzle or a rotating disc. Inlet gas temperatures range from 200 to 1400°F, with the high temperatures requiring special construction materials. Since thermal efficiency increases with the inlet gas temperature, high inlet temperatures are desirable. Even heat-sensitive products can be dried at higher temperatures because of the short drying time. Hot-gas flow may be either concurrent or countercurrent to the falling droplets. Dried particles settle out by gravity. Fine material in the exhaust air is collected in cyclone separators or bag filters. Figure 8 shows a typical spray drying system.

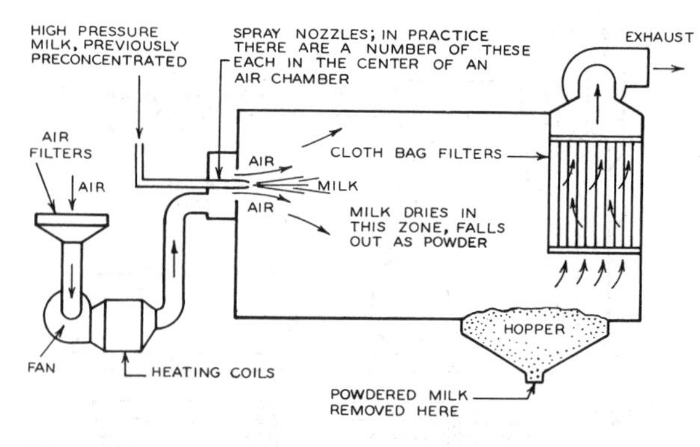

Fig. 8 Spray Drier of the Pressure-Spray Rotary Type

The bulk physical properties of the dried product, such as particle size, bulk density, and dustiness, are affected by atomization characteristics and the temperature and direction of flow of the drying gas. The product's final moisture content is controlled by the humidity and temperature of the exhaust gas stream.

Currently, pilot-plant or full-scale production operating data are required for design purposes. The drying chamber design is determined by the nozzle's spray characteristics and heat and mass transfer rates. There are empirical expressions that approximate mean particle diameter, drying time, chamber volume, and inlet and outlet gas temperatures.

Freeze Drying

Freeze drying has been applied to pharmaceuticals, serums, bacterial and viral cultures, vaccines, fruit juices, vegetables, coffee and tea extracts, seafoods, meats, and milk.

The material is frozen, then placed in a high vacuum chamber connected to a low-temperature condenser or chemical desiccant. Heat is slowly applied to the frozen material by conduction or infrared radiation, allowing the volatile constituent, usually water, to sublime and condense or be absorbed by the desiccant. Most freeze-drying operations occur between 14 and −40°F under minimal pressure. While this process is expensive and slow, it has advantages for heat-sensitive materials (see Chapter 9 of the 1990 ASHRAE *Handbook—Refrigeration* and Perry and Chilton 1978).

Vacuum Drying

Vacuum drying takes advantage of the lowered boiling point of water as the pressure is lowered. Vacuum drying of paper has been partially investigated. Serious complications arise if the paper breaks and massive sections must be removed to gain access. Vacuum drying is used successfully for pulp drying, where lower speeds and higher weights make breakage relatively infrequent.

Fluidized-Bed Drying

A fluidized-bed system contains solid particles through which a gas with a velocity higher than the incipient fluidizing velocity but lower than the entrainment velocity flows. Heat transfer between the individual particles and the drying air is efficient, since there is close contact between powdery or granular material and the fluidizing gas. This contact makes it possible to dry sensitive materials without danger of large temperature differences.

The dry material is free-flowing and unlike convection-type driers, it is not encrusted on trays or other heat-exchanging surfaces. Automatic charging and discharging are possible, but the greatest advantage is reduced process time. Only simple controls are important, *i.e.*, control over fluidizing air or gas temperatures and the drying time of the material.

All fluid-bed driers should have explosion-relief flaps. Both the pressure and flames of an explosion are dangerous. Also, when toxic materials are used, uncontrolled venting to the atmosphere is prohibited. Explosion suppression systems, such as pressure-actuated ammonium-phosphate extinguishers, have been used instead of relief venting. An inert drier atmosphere is preferable to suppression systems because it prevents explosive mixtures from forming.

When organic and inflammable solvents are used in the fluid-bed system, the closed system offers advantages other than explosion protection. A portion of the fluidizing gas is continuously run through a condenser, which strips the solvent vapors and greatly reduces air pollution problems, thus making solvent recovery convenient.

Materials dried in fluidized-bed installations include coal, limestone, cement rock, shales, foundry sand, phosphate rock, plastics, medicinal tablets, and foodstuffs. Leva (1959) and Othmer (1956) discuss the theory and methods of fluidization of solids. Clark (1967) and Vanecek *et al.* (1966) develop design equations and cost estimates.

Agitated-Bed Drying

Uniform drying is ensured by periodically or continually agitating a bed of preformed solids with a vibrating tray or conveyor, a mechanically operated rake, or, in some cases, by partial fluidization of the bed on a perforated tray or conveyor through which recycled drying air is directed. Drying and toasting cereals is an important application.

Drying in Superheated Vapor Atmospheres

When drying solids with air or another gas, the vaporized solvent (water or organic liquid) must diffuse through a stagnant gas film to reach the bulk gas stream. Since this film is the main resistance to mass transfer, the drying rate depends on the solvent vapor diffusion rate. If the gas is replaced by solvent vapor, resistance to mass transfer in the vapor phase is eliminated, and the drying rate depends only on the heat transfer rate. Drying rates in solvent vapor, such as superheated steam, are greater than in air for equal temperatures and mass flow rates of the drying media (Chu *et al.* 1953).

This method also has higher thermal efficiency, easier solvent recovery, a lower tendency to overdry, and eliminates oxidation or other chemical reactions that occur when air is present. In drying cloth, superheated steam reduces the migration tendency of resins and dyes. Superheated vapor drying cannot be applied to heat-sensitive materials because it requires high material temperatures.

Commercial drying equipment having recycled solvent vapor as the drying medium is available. Installations have been built to dry textile sheeting and organic chemicals.

Flash Drying

Finely divided solid particles that are dispersed in a hot-gas stream can be dried by flash drying, which is rapid and uniform. Commercial applications include drying pigments, synthetic resins, food products, hydrated compounds, gypsum, clays, and wood pulp.

REFERENCES

Brown, G.G. and Associates. 1950. *Unit operations.* John Wiley & Sons, New York, 564.

Chatterjee, P.C. and R. Ramaswamy. 1975. Ultraviolet radiation drying of inks. *British Ink Maker* 17(2):76.

Chu, J.C., A.M. Lane, and D. Conklin. 1953. Evaporation of liquids into their superheated vapors. *Industrial and Engineering Chemistry* 45:1586.

Clark, W.E. 1967. Fluid bed drying. *Chemical Engineering* 74, March 13:177.

Friedman, S.J. 1951. Steps in the selection of drying equipment. *Heating and Ventilating* (February):95.

Friedman, S.J., R.A. Gluckert, and W.R. Marshall, Jr. 1952. Centrifugal disk atomization. *Chemical Engineering Progress* 48:181.

Haley, N.A. 1976. High frequency heating on a linerboard machine. Tappi Papermakers Conference (Atlanta) Preprint, April 26 to 29, 217.

Heating and Ventilating. 1942. What the air conditioning engineer should know about drying (December).

Joas, J.G. and J.L. Chance. 1975. Moisture leveling with dielectric, air impingement and steam drying—A comparison. *Tappi* 58(3):112.

Leva, M. 1959. *Fluidization*. McGraw-Hill Book Co., New York.

Othmer, D.F. 1956. *Fluidization*. Reinhold Publishing Corp., New York.

Parker, N.H. 1963. Aids to drier selection. *Chemical Engineering* 70 June 24:115

Perry, R.H. and C.H. Chilton, ed. 1978. *Chemcial engineers' handbook*, 5th ed. Section 17, Sublimination, and Section 20, Gas-Solid Systems. McGraw-Hill Book Co., New York.

Simon, E. Containment of hazards in fluid bed technology. *Manufacturers of Chemcial Aerosol News* 49(1):23.

Vanecek, Markvart, and Drbohlav. 1966. *Fluidized bed drying*. Chemical Rubber Company, Cleveland, OH.

BIBLIOGRAPHY

[*NOTE: In the following items, ABIPC stands for Abstract Bulletin of the Institute of Paper Chemistry, Appleton, WI.*]

Alt, C. 1964. A comparison between infrared and other ink drying methods. *Polygraph* 17(4):200; ABIPC 34:1455.

Appel, D.W. and S.H. Hong. 1969. Condensate distribution and its effect on heat transfer in steam heated driers. *Pulp and Paper Canada* 70(4):66, T51; ABIPC 39:943.

Balls, B.W. 1970. The control of drying cylinders. *Paper Technology* 1(5):483. ABIPC 31:1119.

Bell, J.R. and P. Grosberg. 1962. The movement of vapor and moisture during the falling rate period of drying of thick textile materials. *Journal of the Textile Institute*, Transactions 53(5):T250; ABIPC 33: 72.

Booth, G.L. 1970. Factors in selecting an air heating system for drying coatings. *Paper Trade Journal* 154(23); Graphic Arts, 47; Abstracts 24, no. 7:71.

Booth, G.L. 1970. General principles in the drying of paper coatings. *Paper Trade Journal* 154(17):48; Graphic Arts Abstract 24, no. 7:72.

Chu, J.C., S. Finelt, W. Hoerrner, and M.S. Lin. 1959. Drying with superheated steam-air mixtures. *Industrial and Engineering Chemistry* 51:275.

Church, F. 1968. How dielectric heating helps to control moisture content. *Pulp and Paper International* 10(2):50; ABIPC 39:202.

Daane, R.A. and S.T. Han. 1961. An analysis of air-impingement drying. *Tappi* 44(1):73; C.A. 55:8855, ABIPC 31:1120.

Dooley, J.A. and R.D. Vieth. 1965. High velocity drying gives new impetus to solution coatings. *Paper, Film, Foil Converter* 39(4):53; ABIPC 36.

Dyck, A.W.J. 1969. Focus on paper drying. *American Paper Industry* 51(6):49; ABIPC 40:458.

FMEA. 1990. Industrial ovens and driers. Data Sheet No. 6-9. Factory Mutual Engineering Association, Norcross, GA.

Foust, A.S. *et al.* 1962. *Simultaneous heat and mass transfer II: Drying. Principles of unit operations*. John Wiley & Sons, New York.

Gardner, T.A. 1964. Air systems and Yankee drying. *Tappi* 47(4):210; ABIPC 34:1787.

Gardner, T.A. 1968. Pocket ventilation and applied fundamentals spell uniform drying. *Paper Trade Journal* 152(4) January 22, J. 152, No. 4:46, 48, 51-2, January 22; ABIPC 39:115.

Gavelin, G. 1970. New heat recovery system for paper machine hood exhaust. *Paper Trade Journal* 154(8):38; ABIPC 41:1225.

Gavelin, G. 1964. Paper and paperboard drying—Theory and practice (monograph). Lockwood Trade Journal Co., Inc., New York, 85 pp.; ABIPC 35:480.

Hoyle, R. 1963. Thermal conditions in a steam drying cylinder. *Paper Technology* 4(3):259; ABIPC 34:341.

Janett, L.G., A.J. Schregenberger, and J.C. Urbas. 1965. Forced convection drying of paper coatings. *Pulp and Paper Canada* 66(January): T20; ABIPC 35:1433.

Larsson, T. 1962. Comparing high velocity driers—Aspects of theory and design. *Paper Trade Journal* 146(38), September 17, p. 36; ABIPC 33:547.

Marshall, W.R., Jr. *Drying section in Encyclopedia of Chemical Technology*, 2nd ed., Vol. 7. Interscience Publishers, New York, 326.

Metcalf, W.K. 1970. What pocket ventilation systems can do and what they cannot. *Pulp and Paper* 44(2):95; No. 2:95-7, Feb.; ABIPC 41:3231.

Mill, D.N. 1961. High velocity air drying. Paper Technology 2(4), August; ABIPC 32:588.

Nissan, A.H. 1968. Drying of sheet materials. *Textile Research Journal* 38:447.

Nissan, A.H. and D. Hansen. 1962. Fundamentals of drying of porous materials. Errata. *Tappi* 45(7):608; ABIPC 33:395.

Olmedo, E.B. 1966. *Steam control in paper machine driers.* ATCP 6(2):142.

Priestly, R.J. 1962. Where fluidized solids stand today. *Chemical Engineering* July 9, 125.

Scheuter, K.R. 1968. Drier theory and dryer systems. *Druckprint* 105(12):939; ABIPC 40:24.

Sloan, C.E., T.D. Wheelock, and G.T. Tsao. 1967. Drying. *Chemical Engineering* 74:167.

Spraker, W.A., G.B. Wallis, and B.R. Yaros. 1969. Analysis of heat and mass transfer in the Yankee drier. *Pulp and Paper Canada* 70(1):55; T1-5, January 3; ABIPC 39:947.

Stangl, K. 1966. Progress in flash drying. *Pulp and Paper National* 8(6):65; ABIPC 37:305.

Streaker, W.A. 1968. Drying pigmented coatings with infrared heat. *Tappi* 51(10):105; ABIPC 39: 659.

Tarnawski, Z. 1962. Drying of paper and calculation of the drying surface of the paper machine. *Przeglad Papier* 18(7):218; ABIPC 34:639.

Wen, C.Y. and W.E. Loos. 1969. Rate of veneer drying an a fluidized bed. *Wood Science and Technology* 3.

Wilhoit, D.L. 1968. Theory and practice of drying aqueous coatings with a high velocity air drier. *Tappi* 51(1), ABIPC 38:803, C.A. 68:4969, Pkg. Abstr. 35:360.

Yoshida, T., and T. Hyodo. 1963. Superheated vapor as a drying agent in spinning fiber. *Industrial and Engineering Chemistry, Process Design and Development* January:52.

CHAPTER 29

GEOTHERMAL ENERGY

EMPHASIS on the use of geothermal energy has been directed toward production of electricity, although efforts have also been made to use geothermal energy for space and domestic water heating, industrial processing, and cooling.

An overall geothermal system includes the resources, user system, and disposal of the effluent. The resource location and its characteristics (temperature, allowable fluid flow rate, fluid quality) are important because (1) geothermal energy is not available at all localities, and geothermal fluids cannot be economically transmitted over more than a few miles, and (2) the characteristics of the available resource may or may not be appropriate for the particular application for which an energy supply is being sought. For the most economical and satisfactory operation, the system should be designed specifically for use of geothermal fluids. While the equipment is off-the-shelf equipment, it is different from that which has traditionally been used in the same application operated with conventional energy supplies.

Thus, equipment designed for geothermal energy use is substantially different from the design for conventionally fueled systems. For geothermal energy systems, the design must consider (1) the available resource temperature and flow rate; (2) an appropriate temperature drop of the fluid, which is normally much greater than that specified for fluid loops in conventional systems; and (3) the fluid composition. The transmission and distribution system and the peaking/backup system are designed by conventional techniques to provide economical and reliable operation. The main concerns in disposal are that the fluid be disposed of in an environmentally acceptable manner or injected into the reservoir, as necessary, to maintain production.

This chapter illustrates how geothermal energy can be used for space heating and cooling, domestic water heating, and industrial processing by (1) describing the types of geothermal resources and their general extent in the United States; (2) considering the potential market that may be served with geothermal energy; and (3) illustrating the evaluation considerations, special design aspects, and approaches for geothermal energy use in each of the applications.

RESOURCE

Geothermal energy is the thermal energy within the earth's crust—the thermal energy in rock and the fluid (water, steam, or water containing large amounts of dissolved solids) that fills the pores and fractures within the rock. Calculations show that the earth, originating from a completely molten state, would have cooled and become completely solid many thousands of years ago, had there not been an additional energy input other than from the sun. It is believed that the ultimate source of geothermal energy is radioactive decay within the earth (Bullard 1973).

Through plate motion and vulcanism, some of this energy is concentrated at high temperature near the surface of the earth. In addition, energy transfer from the deeper parts of the crust to the earth's surface by conduction (and also by convection in regions where geological conditions and the presence of water permit) results in the general condition of thermal energy of elevated temperature at depth.

Because of variations in volcanic activity, radioactive decay, rock conductivities, and fluid circulation, different regions have different heat flows (through the crust to the surface), as well as different temperatures at a particular depth. The *normal* increase of temperature with depth (*i.e.*, the normal geothermal gradient) is about 13.7 °F/1000 ft of depth, with gradients of about 5 to 27 °F/1000 ft being common. The areas with the higher temperature gradients and/or higher-than-average heat flow rates constitute the most viable economic resources. However, with the presence of certain geological features, areas with normal gradients may also be valuable resources.

Geothermal resources in the United States are categorized into the following types:

- Igneous point sources
- Deep convective circulation in areas of high regional heat flow
- Geopressured
- Concentrated radiogenic heat sources
- Deep regional aquifers in areas of near normal gradient

Igneous point resources are associated with magma bodies, which result from volcanic activity. These bodies heat the surrounding and overlying rock by conduction and convection, as permitted by the rock permeability and fluid content in the rock pores.

Deep circulation of water in areas of high regional heat flow can result in hot fluids near the surface of the earth. Known as *hydrothermal convection systems*, this type of geothermal resource is widely used. The fluids near the surface have risen from natural convection circulation between the hotter, deeper formation and the cooler formations near the surface. The passageway that allows this deep circulation must consist of fractures and faults of adequate permeability.

The geopressured resource, present over a wide region in the Gulf Coast area, consists of regional occurrences of confined hot

The preparation of this chapter is assigned to TC 6.8, Geothermal Energy Utilization.

water in deep sedimentary strata, where pressures of 11,000 psi are common. This resource also contains methane dissolved in the geothermal fluid.

Radiogenic heat sources exist in various regions as granitic plutonic rocks that are relatively rich in uranium and thorium. These plutons have a higher heat flow than the surrounding rock, and if the plutons are blanketed by sediments of low thermal conductivity, elevated temperatures can result at the base of the sedimentary section. This resource has been identified in the eastern United States. Such systems have also been identified in the western United States, but there they are of secondary importance to both igneous point sources and regions of high heat flow.

Deep regional aquifers of commercial value can occur in deep sedimentary basins, even in areas of normal temperature gradient. To be of commercial value, (1) the basins must be deep enough to allow usable temperature levels at the prevailing gradient, and (2) the permeabilities within the aquifer should be adequate for flow in the aquifer.

The thermal energy in geothermal resource systems exists primarily in the rocks and only secondarily in the fluids that fill the pores and fractures within them. Thermal energy is usually extracted by bringing to the surface the hot water or steam that occurs naturally in the open spaces in the rock. Where rock permeability is low, the energy extraction rate is low. To extract thermal energy from the rock itself, a recharge of water into the system must occur as the initial water is extracted. In permeable aquifers, or where natural fluid conductors occur, the produced fluid may be injected back into the aquifer some distance from the production hole to pass through the aquifer again and recover some of the energy in the rock. Such a system is termed a *stimulated* or *forced geoheat recovery system* (Bodvarsson and Reistad 1976). This type of system is presently in operation in France (BRGM 1978) and at a few sites in the United States. For recovering energy from impermeable rock, research is now being conducted to evaluate the feasibility of creating artificial permeability by fracturing (hydraulic and thermal stress) and then extracting the thermal energy by injecting cold water into the fractured system through one well and removing the heated fluid through a second well. Because it is directed at hot rock bodies containing little or no water, this technology is referred to as hot dry rock (Brown *et al.* 1979).

TEMPERATURES

The temperature of fluids in the earth's crust and used for their thermal energy content varies from about 60 to 680 °F. The lower value represents the fluids used as the low temperature energy source for heat pumps, and the higher temperature represents an approximate value for the HGP-A well at Hilo, Hawaii.

Figure 1 shows examples of the temperature as a function of depth for the cases of (A) near normal gradient of 13.7 °F/1000 ft; (B) high gradient of 37.3 °F/1000 ft; and (C) and (D), convective systems (Combs *et al.* 1980). Temperature reversals, as in (C), are not unusual and indicate lateral flow in permeable strata at the higher temperature nodes. In conductive systems, temperature is relatively constant with depth throughout the permeable horizon but increases with depth above and below this. To achieve high temperatures, either deep drilling or convective systems that originate at depth and provide circulation to shallower regions are necessary.

The following classification system of resources by temperature level is used in the geothermal industry:

High temperature	$T > 300\,°F$
Intermediate temperature	$194\,°F < T < 300\,°F$
Low temperature	$60\,°F < T < 194\,°F$

Electricity generation is generally not economically feasible for resources with temperatures below about 300 °F, which is the reason for the division between high- and intermediate temperature

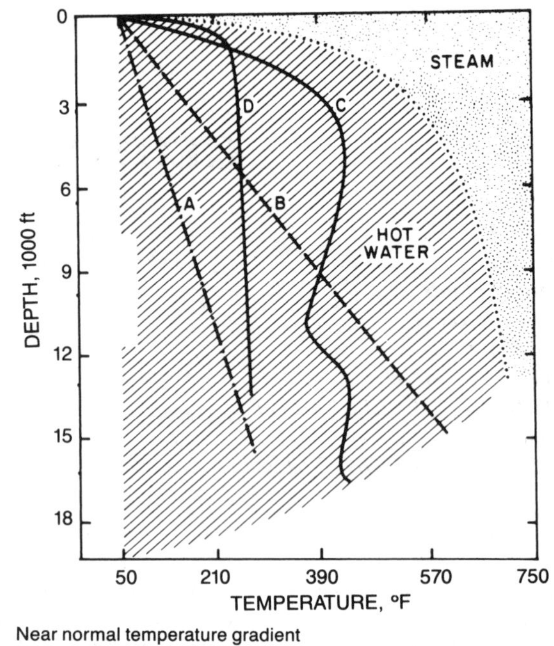

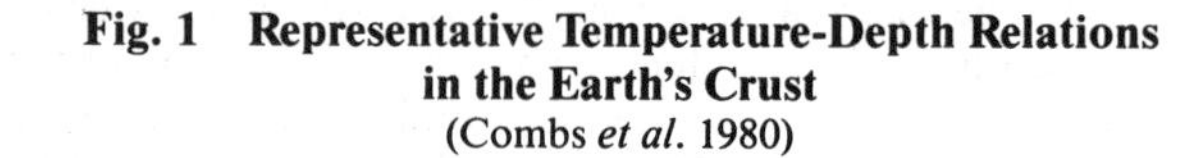

Fig. 1 Representative Temperature-Depth Relations in the Earth's Crust
(Combs *et al.* 1980)

systems. However, binary power plants, with the proper set of circumstances, have demonstrated that it is possible to generate electricity economically above 230 °F. In 1988, there were 86 binary plants worldwide, generating a total of 126.3 MW (Di Pippo 1988).

The 194 °F division between intermediate and low temperatures is common in resource inventories but, it is somewhat arbitrary. At 194 °F and above, applications such as district heating can be readily implemented with equipment used in conventional applications of the same type, while at lower temperatures, these applications require redesign to take the greatest advantage of the geothermal resource.

Geothermal systems at lower temperature levels are more common. The frequency of identified convective systems by reservoir temperature for temperatures above 194 °F is shown in Figure 2. A quantitative estimation of thermal energy recoverable from low-temperature (< 194 °F) geothermal systems within the United States is shown in Figure 3 (Reed 1982).

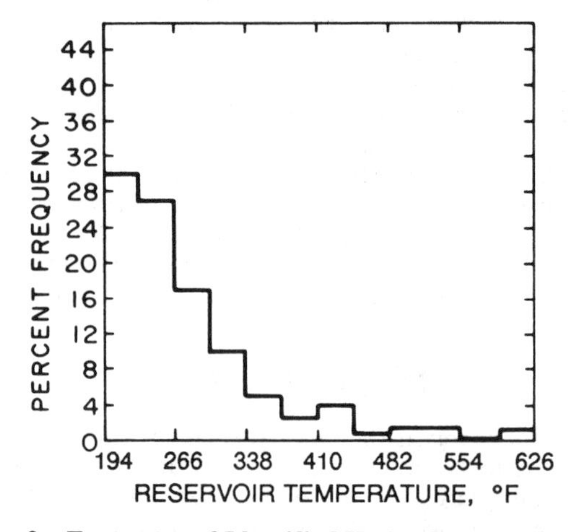

Fig. 2 Frequency of Identified Hydrothermal Convection Systems by Reservoir Temperature
(Muffler *et al.* 1980)

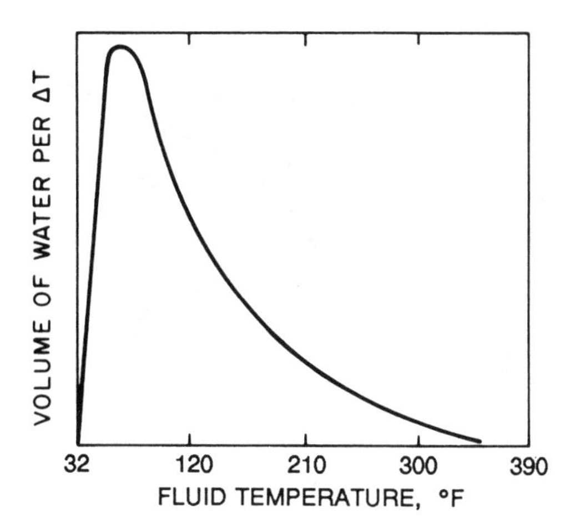

Fig. 3 Probable Distribution for Geothermal Fluids
(Reed 1982)

GEOTHERMAL FLUIDS

Geothermal energy is extracted from the earth through some fluid medium. This medium is the naturally occurring fluids in rock pores and fractures; in future, however, it may also be an additional fluid introduced into the geothermal system and circulated through it to recover energy. The fluids being produced are either steam, hot liquid water, or a two-phase mixture of both. These may contain various amounts of impurities, notably dissolved gases and dissolved solids.

Geothermal systems that produce essentially dry steam are referred to as *vapor dominated.* While these systems are valuable resources, they are rare. Hot water systems (liquid dominated) are much more common than vapor-dominated systems and can be produced either as hot water or as a two-phase mixture of steam and hot water, depending on the pressure maintained on the production system. If the pressure in the production casing or in the formation around the casing is reduced below the saturation pressure at that temperature, some of the fluid will flash, and a two-phase fluid will result. If the pressure is maintained above the saturation pressure, the fluid will remain as a single phase. In water-dominated systems, both dissolved gases and dissolved solids are significant. For such fluids, the quality varies from site to site and varies from water of a few hundred ppm to fluids that have over 300,000 ppm dissolved solids. The U.S. Geological Survey classifies the degree of salinity of mineralized waters as follows:

Dissolved Solids, ppm	Classification
1,000 to 3,000	Slightly saline
3,000 to 10,000	Moderately saline
10,000 to 35,000	Very saline
More than 35,000	Brine

Thus, geothermal fluids range from nonsaline to brine, depending on the particular resource.

Table 1 presents the composition of fluids from a number of geothermal wells in the United States. The list illustrates the types of substances and the range of concentrations that can be expected in the fluids. Although there is great site dependency, the harshness of the fluid increases with increasing temperature. This chapter concentrates on systems produced as a single-phase hot liquid.

LIFE OF THE RESOURCE

Although the radioactive decay that appears to be the ultimate source of geothermal energy is continuous, geothermal energy in a specific locality is limited. The limiting factor is usually thermal water, the medium used to transfer the energy from the rock to the surface. If production rates of thermal water exceed natural recharge rates, water levels can decline, and the resource should be developed with a reservoir management plan that includes injection wells to maintain reservoir pressure. Reservoir life is difficult to determine and involves expensive reservoir engineering techniques. The usual procedure is to expand the area to be developed in stages, monitoring the water levels in wells, and then to apply proper reservoir management methods as additional capacity is required and/or initial energy production rates start to decline.

Table 1 Representative Fluid Compositions from Geothermal Wells in Various Resource Areas of the United States

Material	Boise, ID[a]	Klamath Falls, OR[b]	Salton Sea, CA
	Concentrations are in parts per million		
Temperature, °F	176	192	482
Total dissolved solids	290	795	220,000
SiO_2	160	48	350
Na	90	205	5,100
K	1.6	4.3	12,500
Ca	1.7	26	23,000
Mg	0.05	—	150
Cl	10	51	133,000
F	14	1.5	13
SO_4	23	330	5
NO_3	—	4.9	—
NH4	—	1.3	—
H_2S	trace	1.5	—
HCO_3	70	20	7,025
CO_3	4	15	—
CO_2	0.2	—	—
B	0.14	—	350
Fe	0.13	0.3	1,300
O_2	0.0029	0.2	—

Source: Boise and Salton Sea data from Cosner and Apps (1978); Klamath Falls data from Ellis and Conover (1981).
[a]Well name unknown, but near old penitentiary.
[b]Wendling Well (Lund *et al.* 1976).

ENVIRONMENTAL ASPECTS

Geothermal resources directly coupled to production must consider the overall environmental aspects of the use and production of geothermal energy. The primary environmental issues and a brief discussion of each are presented in Table 2. Bloomquist (1988) and Lunis (1989) consider the concerns regarding direct applications.

PRESENT USE AND POTENTIAL DEVELOPMENT

Recent discoveries of concentrated radiogenic heat sources and deep regional aquifers in areas of near normal temperature gradient indicate that 37 states in the United States have economically exploitable geothermal resources (Interagency Geothermal Coordinating Council 1980).

The geysers resource area in northern California is the largest single geothermal development in the world. The total electricity generated by geothermal development in the world was 5175 MW in 1988 (Di Pippo 1988). The direct application of geothermal energy for space heating and cooling, water heating, agricultural growth- related heating, and industrial processing represents about 8600 MW worldwide in 1988 (Lienau *et al.*, 1988 and Gudmundsson, 1985). In the United States in 1988, direct-use installed capacity amounted to 5.7×10^9 Btu/h providing 17×10^{12} Btu/year (Lienau 1988).

The major uses of geothermal energy in agricultural growth applications are for heating greenhouse and aquaculture facilities.

Table 2 Primary Environmental Issues in Specific Applications of Geothermal Resources

Issue	Potential Environmental Impact	Comments
Air quality	Emission of various gases into the atmosphere	Certain resources have some H_2S, radon, or other noncondensable gases, which require proper design.
Noise	Noise pollution	Primarily a problem during drilling or testing. Proper noise abatement procedures should be followed.
Surface water quality	Degradation of water quality from thermal, chemical, or natural radioactive properties of disposed fluids.	Proper disposal system design and planning for accidental releases are required.
Land use	Conflict of geothermal use of land with uses such as agriculture, recreation, etc.	Since surface area required for geothermal development is relatively small, this issue can usually be resolved.
Geological alteration	Subsidence and/or induced seismic activity.	Subsidence can occur in sedimentary resource areas when fluid injection is not used. Induced seismicity is not a major concern, except for cases of deep high-pressure fluid injection.
Water supply and hot springs alteration	Alteration of existing and potential water supplies or hot springs activities because of withdrawal of geothermal fluid and/or energy or injection of geothermal fluids.	Geothermal development near water supplies and hot springs may be restricted or prohibited because hydrological information is inadequate to predict the impact of the development.

The principal industrial use of geothermal energy in the United States is for food processing; it is presently used in vegetable dehydration. Worldwide, the main applications include space and water heating, space cooling, agricultural growth, and food processing. The exceptions are diatomaceous earth processing in Iceland and pulp and paper processing in New Zealand.

POTENTIAL IMPACT

In terms of the fluids produced, geothermal energy is restricted to temperature levels substantially lower than those from fossil fuels. The maximum temperature of a producing field is 680°F, and the usual resource is expected at much lower temperatures. Space heating and cooling, as well as sanitary water heating, represent uses that are both significant in scale and readily accommodated by the low- and intermediate temperature geothermal resources. Many process heat requirements occur at these temperature levels. Several estimates of the amount of energy consumed at various temperature levels in the process industries are summarized in Figure 4. The figure is a cumulative plot of energy use at or below a particular temperature.

DIRECT APPLICATION SYSTEMS

Because many processes require thermal energy at a temperature level compatible with geothermal energy and, furthermore, because geothermal energy can be exploited in many areas, it is a good choice of energy for many applications. However, an evaluation of using geothermal energy, as well as the proper design

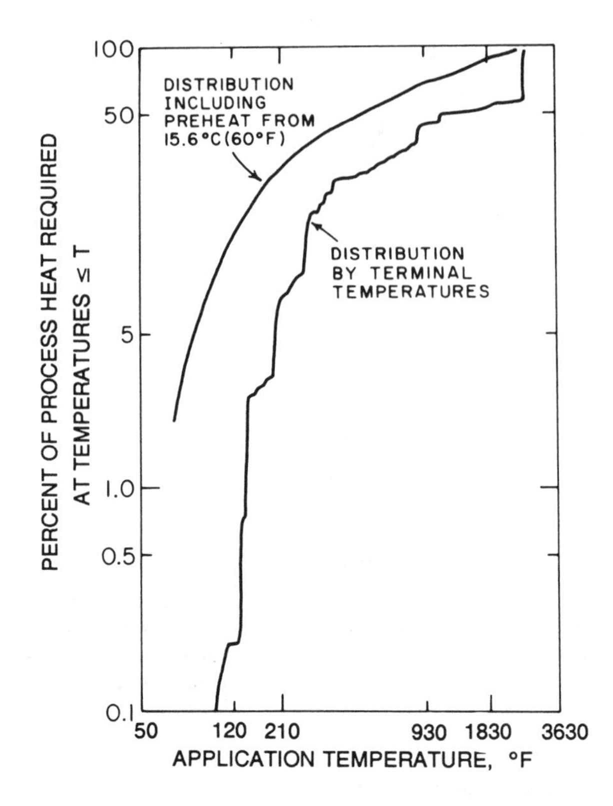

Fig. 4 Cumulative Distribution of Process Heat Requirements (Intertechnology Corporation 1977)

for its use in specific applications, requires consideration of the characteristics of geothermal systems and the interaction with specific equipment. This section presents geothermal system characteristics and the use of geothermal energy in residential, commercial, and industrial applications.

GENERAL

Figure 5 schematically illustrates a typical direct-use system. Such a system may consist of five subsystems: (1) the production system, including the producing wellbore and associated wellhead equipment; (2) the transmission and distribution system, which transports the geothermal energy from the resource site to the user site and then distributes it to the individual user loads; (3) the user system; (4) the disposal system, which can be either surface disposal or injection back into a formation, which may or may not be the same as that from which it was originally produced; and (5) a peaking/backup system. None of the 14 major geothermal district systems in the United States includes a peaking/backup component as part of the main distribution system. Backup is most commonly included in end user systems, and peaking is rarely employed.

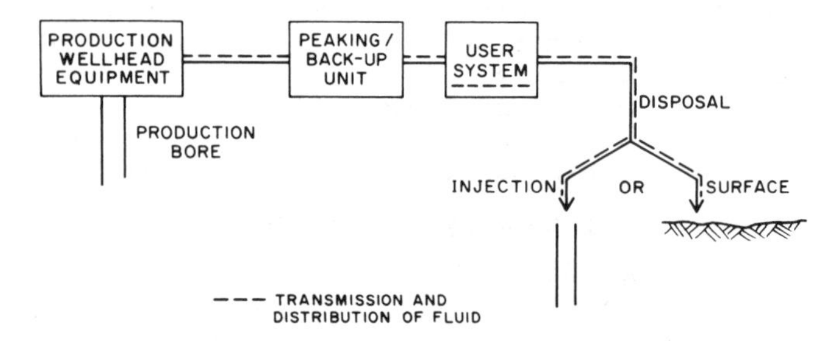

Fig. 5 Basic Geothermal Direct-Use System

In a typical system of this type, the geothermal fluid is produced from the production borehole by using a lineshaft multistage centrifugal pump. (Some wells may free-flow adequate quantities of fluid so that a pump is not required. However, the more common commercial-sized operations are expected to require pumping to provide the required flow rate.) When the geothermal fluid reaches the surface, it is delivered to the application site through transportation and distribution systems.

In the system shown in Figure 6, the geothermal production and disposal system are closely coupled, and they are both separated from the remainder of the system by a heat exchanger. This limits the contact of the geothermal fluid with system equipment, thereby reducing corrosion and scaling. A secondary loop fluid is heated by geothermal fluid in the heat exchanger. This secondary fluid, usually treated water, is the medium for transferring the energy to the application. The rest of the system shown in Figure 6 remains the same as that in Figure 5, except that the equipment designs can be based on the properties of the secondary loop fluid for equipment in Figure 6, but they must be based on the geothermal fluid properties for application as illustrated in Figure 5. The secondary loop is desirable, especially when the geothermal fluid is particularly harsh in terms of corrosion and/or scaling. Such a system arrangement is also advantageous when the geothermal fluid is clean, and the application requires direct use of the heated fluid in a process where water quality is of utmost concern, such as in the food processing industry. Environmentally, this type of system is advantageous because the geothermal fluid is pumped directly back into the ground without loss to the surrounding surface environment.

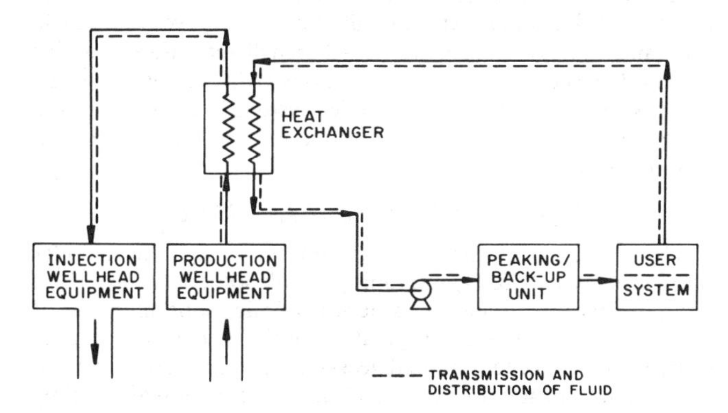

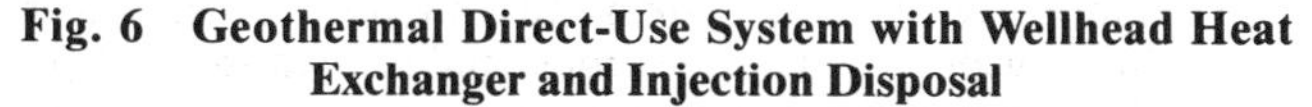

Fig. 6 Geothermal Direct-Use System with Wellhead Heat Exchanger and Injection Disposal

CHARACTERISTICS

Geothermal energy systems have several characteristics that greatly influence their applicability and the design for their use. These characteristics arise from (1) the resource, (2) the application, and (3) the interaction between the resource and application. The characteristics take the form of either constraints or design variables. The constraints are fixed for a particular resource and application, and although they cannot be varied for the application under construction, they influence the feasibility of using the geothermal resource in that particular application. Design variables are those parameters varied to improve the feasibility of geothermal energy use.

The characteristics influencing the cost of energy delivered from geothermal systems are as follows:

- Depth of resource
- Distance between resource location and application site
- Cost of capital
- Well flow rate
- Temperature of resource
- Temperature drop
- Load size
- Load factor
- Composition of fluid
- Ease of disposal
- Resource life

Many of these characteristics have a major influence because the costs of geothermal systems are primarily front-end capital costs, and the annual operating costs are relatively low.

Depth of the Resource

The well cost is usually one of the larger items in the overall cost of a geothermal system, and as the depth of the resource increases, so does the cost of the overall system. Compared to many other geothermal areas worldwide, the western United States is fortunate in that well depth requirements are relatively shallow; most larger geothermal systems in that area operate with production wells of less than 2000 ft, and many at less than 1000 ft.

Distance Between Resource Location and Application Site

The direct use of geothermal energy must occur near the resource location. The reason is primarily economic; although the geothermal fluid (or a secondary fluid) could be transmitted over moderately long distances (greater than 60 mi.) without a great temperature loss, such transmission would generally not be economically feasible. Most existing geothermal projects are characterized by transmission distances of less than 1 mile.

Cost of Capital

In a system where the costs are primarily capital costs, an increase in the interest rate has nearly the same influence as an equal percentage increase in the capital costs. Consequently, as the interest rate increases, the economic position of a particular geothermal system decreases.

Well Flow Rate

The energy output from a production well varies directly with the flow rate of fluid. Thus, the energy cost at the wellhead varies inversely with the well flow rate. Typical good resources have production rates of about 400 to 800 gpm per production well, although individual geothermal direct-use wells have been designed to produce up to 2000 gpm.

Temperature

In geothermal systems, the available temperature is an approximately fixed value for a given resource. Although the temperature could increase with deeper drilling, natural convection in the fluid-dominated systems keeps the temperature relatively uniform throughout the depth of the resource (Figure 1). If drilling is continued through the producing region into a region that is not permeable enough to permit natural convection, the temperature may increase or decrease, but usually inadequate flow occurs to yield an economic resource. Deeper drilling at the same area can, however, result in recovery of energy at a higher temperature if deeper separate aquifers (producing zones) occur. Such an increase in temperature is also theoretically possible in the yet unproved hot dry-rock type systems.

The temperature limitation can present a severe restriction on potential applications. It often requires a reevaluation of accepted application temperatures, since these were developed in systems served by conventional fuels where the application temperature could be selected at any value within a relatively broad range (without a major change in the overall system design or energy efficiency). When geothermal energy is used, the application temperature must be lower than the produced fluid temperature; however, when heat pumps are used, the application temperature may be somewhat higher than the produced fluid temperature.

Temperature Drop

The power output from a geothermal well is directly proportional to the temperature drop of the geothermal fluid effected by the user system, since the well flow rate is limited. Consequently, a larger temperature drop results in lower energy cost at the wellhead. If there is a loop fluid, such as in Figure 6, and the maximum loop fluid temperature approaches the geothermal supply fluid temperature, the loop fluid must also have a reasonably large temperature drop across the user system. This is in great contrast to many conventional and solar systems which circulate a heating fluid with a small temperature drop.

Consequently, a different design philosophy and different equipment are required.

Although it is important to have a goal of a large temperature difference Δt, the maximum Δt in a single application is not always the most desirable because of the expense of heat exchanger equipment at low approach temperatures. Thus, cascading the geothermal fluid to uses with lower temperature requirements can be advantageous in achieving a large Δt. Most geothermal systems have been designed for a Δt between 30 and 50°F, although one system was designed for 100°F Δt with a 190°F resource temperature.

Load Size

Large-scale applications gain from economy of scale, particularly in regard to reduced resource development and transmission system costs. However, the applications are often not extremely large, and the more usual application has one to several production wells. For these smaller developments, matching the size of the application with the production rate from the geothermal system is important because the output comes in increments of one well's output.

Load Factor

Defined as the ratio of the average load to the design capacity of the system, the load factor effectively reflects the fraction of time that the initial investment in the system is working. Again, because the geothermal system costs are primarily initial costs rather than operating costs, this factor significantly affects the viability of a geothermal system. As this factor increases, so does the economic position of using geothermal energy. The two main ways of increasing the load factor are (1) to select applications where it is naturally high and (2) to use peaking equipment so that the design load is not the application peak load, but rather a reduced load that occurs over a longer period.

Composition of Fluid

The quality of the produced fluid is site specific and may vary from less than 1000 ppm total dissolved solids (TDS) to heavily brined. The quality of the fluid influences two aspects of the system design—(1) material selection to avoid corrosion and scaling effects, and (2) disposal or ultimate end use of the fluid.

Many direct-use geothermal systems operate with fluids containing less than 1000 ppm TDS. In spite of this, such fluids can create substantial corrosion and scaling problems. It is thus advisable to isolate the geothermal fluid from the balance of the system. While it is more expensive than using the fluid directly in the process or heating equipment, isolation is the preferred design for minimizing long-term system maintenance requirements.

Ease of Disposal

Most systems dispose of the geothermal effluent on the surface, which includes discharge to irrigation systems, rivers, and lakes. This method of disposal is considerably less expensive than the construction of injection wells. However, the magnitude of geothermal development in certain areas (*e.g.*, Klamath Falls, Oregon and Boise, Idaho) has caused water levels to decline in aquifers where surface disposal has historically been employed. As a result, regulatory authorities in these and many other areas are beginning to favor injection to maintain reservoir fluid levels.

In addition to hydraulic considerations, geothermal fluids sometimes contain chemicals that cause surface disposal to become a problem. Some of these are listed in Table 3.

Table 3 Selected Chemical Species Affecting Fluid Disposal (Lunis 1989)

Hydrogen sulfide (H_2S)	Odor
Boron (B)	Damage to agricultural crops
Fluoride (F)	Level limited in drinking water sources
Radioactive species	Levels limited in air, water, and soil

If injection is required, the depth at which the fluid can be injected affects well costs substantially. Some jurisdictions allow considerable latitude in terms of injection level; others require the fluid to be returned to the same or similar aquifers. If the latter applies, it may be necessary to construct the injection well to the same depth as the production well.

The costs associated with disposal, particularly when injection is involved, can substantially affect development costs.

Resource Life

The life of the resource has a direct bearing on the economic viability of a particular geothermal application. There is little experience on which to base projections of resource life for heavily developed geothermal resources. However, resources can readily be developed in a manner that will allow useful lives of 30 to 50 years and more. In some heavily developed direct-use areas, major systems have been in operation for many years. For example, the Boise Warm Springs Water District system (a district heating system serving about 240 residential users) has been in continuous operation since 1892.

EQUIPMENT AND MATERIALS

The primary equipment used in geothermal systems includes pumps, heat exchangers, and piping. While some aspects of these components are unique to geothermal applications, many are of routine design. However, the great variability and general aggressiveness of the geothermal fluid necessitates limiting corrosion and scale buildup rather than system cleanup. Corrosion and scaling can be limited by (1) proper system and equipment design, which is routine in geothermal applications, and (2) treatment of the geothermal fluid, which is generally precluded by environmental regulations relating to disposal.

Key Corrosive Species

Geothermal fluids commonly contain seven key chemical species that produce a significant corrosive effect (Ellis 1989). The key species include:

- Oxygen (generally from aeration)
- Hydrogen ion (pH)
- Chloride ion
- Sulfide species
- Carbon dioxide species
- Ammonia species
- Sulfate ion

The principal effects of these species are summarized in Table 4. Except as noted, the described effects are for carbon steel.

Two of these species are not reliably detected by standard water chemistry tests and deserve special mention. Dissolved oxygen does not naturally occur in low-temperature (120 to 220°F) geothermal fluids that contain traces of hydrogen sulfide. However, because of slow reaction kinetics, oxygen from air in-

Table 4 Principal Effects of the Key Corrosive Species
(Ellis 1989)

Key Corrosive Species	Principle Effects
Oxygen	• Extremely corrosive to carbon and low alloy steels; 30 parts per billion (ppb) shown to cause four-fold increase in carbon steel corrosion rate. • Concentrations above 50 ppb cause serious pitting. • In conjunction with chloride and high temperature, <100 ppb dissolved oxygen can cause chloride-stress corrosion cracking (chloride-SCC) of some austenitic stainless steels.
Hydrogen ion (pH)	• Primary cathodic reaction of steel corrosion in air-free brine is hydrogen ion reduction. Corrosion rate decreases sharply above pH 8. • Low pH 5 promotes sulfide stress cracking (SSC) of high strength low alloy (HSLA) steels and some other alloys coupled to steel. • Acid attack on cements.
Carbon dioxide species	• Dissolved carbon dioxide lowers pH, increasing carbon and HSLA steel corrosion.
(dissolved carbon dioxide, bicarbonate ion, carbonate ion)	• Dissolved carbon dioxide provides alternative proton reduction pathway, further exacerbating carbon and HSLA steel corrosion. • May exacerbate SSC. • Strong link between total alkalinity and corrosion of steel in low-temperature geothermal wells.
Hydrogen sulfide species	• Potent cathodic poison, promoting SSC of HSLA steels and some other alloys coupled to steel.
(hydrogen sulfide, bisulfide ion, sulfide ion)	• Highly corrosive to alloys containing both copper and nickel in any proportions.
Ammonia species (ammonia, ammonium ion)	• Causes SCC of some copper-based alloys.
Chloride ion	• Strong promoter of localized corrosion of carbon, HSLA, stainless steel, and other alloys. • Chloride dependent threshold temperature for pitting and SCC. Different for each alloy. • Little if any effect on SSC. • Steel passivates at high temperature in pH 5, 6070 ppm chloride solution with carbon dioxide. 133,500 ppm chloride destroys passivity above 300 °F.
Sulfate ion	• Primary effect is corrosion of cements.

leakage may persist for some minutes. Once the geothermal fluid is produced, it is extremely difficult to prevent oxygen contamination, especially if pumps other than downhole submersible or lineshaft turbine pumps are used to move the geothermal fluid. Even though the fluid systems may be maintained at positive pressure, air in-leakage at the pump seals is likely, particularly at the low level of maintenance in many installations.

Hydrogen sulfide is ubiquitous at low parts-per-million (ppm) or parts-per-billion (ppb) levels in geothermal fluids above 120 °F. This corrosive species also occurs naturally in many cooler groundwaters. For alloys such as cupronickels, which are strongly affected by it, hydrogen sulfide concentrations in the low ppb range may have a serious detrimental effect, especially if oxygen is also present. At these levels, the characteristic rotten egg odor of hydrogen sulfide may be absent, and field methods are required for detection. Hydrogen sulfide levels down to 50 ppb can be detected using a simple field kit; however, absence of hydrogen sulfide at this low level may not preclude damage by this species. Field spectrophotometry, which requires a spectrometer and different kit, has a detection limit of less than 10 ppb.

Two other key species should also be measured in the field—pH and carbon dioxide. This is necessary because most geothermal fluids will rapidly off-gas carbon dioxide, causing a rise in pH.

Kindle and Woodruff (1981) present recommended procedures for complete chemical analysis of geothermal well waters.

Performance of Materials

Carbon Steel. The Ryznar Index has traditionally been used to estimate the corrosivity and scaling tendencies of potable water supplies. However, one study found no significant correlation (at the 95% confidence level) between carbon steel corrosion and the Ryznar Index. Therefore, the Ryznar and other indices based on calcium carbonate saturation should not be used to predict corrosion in geothermal systems. The Ryznar Index may be predictive of calcium carbonate scaling tendencies.

In Class Va geothermal fluids, as described by Ellis (1989), [< 5000 ppm Total Key Species (TKS), total alkalinity 207 to 1329 ppm as $CaCO_3$, ph 6.7 to 7.6], corrosion rates of about 5 to 20 mils per year (mpy) can be expected, often with severe pitting.

In Class Vb geothermal fluids, as described by Ellis (1989), (< 5000 ppm TKS, total alkalinity < 210 ppm as $CaCO_3$, ph 7.8 to 9.85), carbon steel piping has given good service in a number of systems, provided the system design rigorously excludes oxygen. However, introduction of 30 ppb oxygen under turbulent flow conditions causes a four-fold increase in uniform corrosion. Saturation with air often increases the corrosion rate at least 15 times. Oxygen contamination at the 50 ppb level often causes severe pitting. Chronic oxygen contamination causes rapid failure. These fluids are characteristic of those used by Icelandic district heating systems. In those systems, steel piping has been generally successful, but carbon steel shell-and-tube heat exchangers have been satisfactory only in systems where 10 ppm excess sodium sulfite is added continuously as an oxygen scavenger.

In the case of buried steel pipe, the external surfaces must be protected from contact with groundwater. Groundwater is aerated, and has caused pipe failures by external corrosion. Required external protection can be obtained by using coatings, pipe-wrap, or preinsulated piping, provided the selected material will resist the system operating temperatures and thermal stresses.

At temperatures above 135 °F, galvanizing (zinc coating) will not reliably protect steel from geothermal fluid or groundwater.

Hydrogen sulfide inhibits the reaction of atomic hydrogen on the steel surface. Atomic hydrogen is a product of the proton reduction step of corrosion processes, during which molecular hydrogen forms. The atomic hydrogen enters the steel lattice and, in the case of carbon (mild) steels, accumulates as molecular hydrogen in microvoids, causing hydrogen blistering. Hydrogen blistering can be prevented by using void-free ("killed") steels.

Low alloy steels (steels containing no more than 4% alloying elements) have corrosion resistance similar, in most respects, to carbon steels. As in the case of carbon steels, sulfide promotes entry of atomic hydrogen into the metal lattice. If the steel exceeds a hardness of Rockwell C22, sulfide stress cracking may occur.

Copper and Copper Alloys. Copper fan coil units and copper-tubed heat exchangers have a consistently poor performance due to traces of sulfide species found in geothermal fluids in the United States. Copper tubing rapidly becomes fouled with cuprous sulfide films more than 1 mm thick. Serious crevice corrosion occurs at cracks in the film, and uniform corrosion rates of two to six may appear typical, based on failure analyses. These corrosion rates were measured by coupons in low-velocity (60 fpm) fluid at the same or similar geothermal resources.

Experience in Iceland also indicates that copper is unsatisfactory for heat exchange service and that most brasses (Cu-Zn) and bronzes (Cu-Sn) are even less suitable. Cupronickels often perform more poorly than does copper in low-temperature geothermal service because of trace sulfide.

Much less information is available regarding copper and copper alloys in nonheat transfer service. Copper pipe shows corrosion behavior similar to copper heat exchange tubes under conditions of moderate turbulence (Reynolds numbers of 40,000 to 70,000). The internals of the few yellow brass valves analyzed showed no significant corrosion. However, silicon bronze CA 875 (12-16-Cr, 3-5-Si, < 0.05-Pb, < 0.05-P), an alloy normally resistant to dealloying, failed in less than three years when used as a pump impeller. Leaded red brass (CA 836 or 838) and leaded red bronze (SAE 67) appear to be viable as pump internals. Based on tests at a few Class Va sites, aluminum bronzes have shown potential for corrosion in heavy-walled components.

Solder is yet another problem area associated with copper equipment. Lead-tin solder (50-Pb, 50-Sn) was observed to fail by dealloying after a few years exposure. Silver solder (1Ag-7P-Cu) was completely removed from joints in less than two years. However, if a designer elects to accept this risk, solders containing at least 70% tin should be used.

Stainless Steel. Unlike copper and cupronickels, stainless steels are unaffected by traces of hydrogen sulfide. Their most likely application should be as heat exchange surfaces. For economic reasons, most heat exchangers are probably of the plate-and-frame type, most of which are fabricated with Type 304 or Type 316 austenitic stainless steel, since these are the two standard alloys. In addition, some pump and valve trim is fabricated from these or other stainless steels.

These alloys are subject to pitting and crevice corrosion above a threshold chloride level which depends on the chromium and molybdenum content of the alloy and on the temperature of the geothermal fluid. Above this temperature, the passivation film, which gives the stainless steel its corrosion resistance, is ruptured in local areas and/or in crevices. These ruptured areas corrode in the form of pitting and crevice corrosion.

Figure 7 shows the relationship between temperature, chloride, and occurrence of localized corrosion of Type 304 and Type 316 stainless steel. For example, this figure indicates that localized corrosion of Type 304 may occur in 80 °F geothermal fluid if the chloride level exceeds approximately 210 ppm, while Type 316 is resistant at that temperature until the chloride level reaches approximately 510 ppm. Type 316 is always more resistant to chlorides than is Type 304, due to its 2 to 3% molybdenum (Mo) content. The fact that localized corrosion can occur does not predict the rate, but more severe attack should be expected as the chloride temperature conditions intrude further and further into the localized corrosion region. These alloys can be used in this region provided that oxygen is rigorously excluded, but there would be a risk of rapid failure should even traces of oxygen intrude.

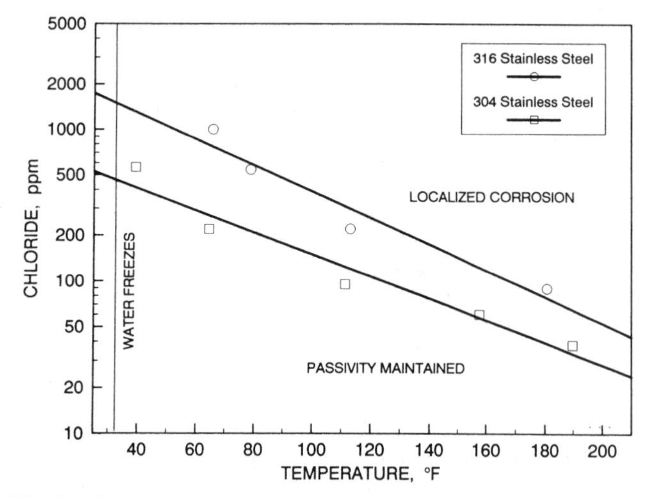

Fig. 7 Chloride Concentration Required to Produce Localized Corrosion of Stainless Steel as a Function of Temperature
(Efrid and Moeller 1978)

These alloys can also fail by stress corrosion cracking above approximately 140 °F. In practice, however, no such failures have occurred in low-temperature (120 to 220 °F) geothermal applications. As a precaution, heat exchanger plates should be stress relieved after forming.

Other austenitic stainless steels with increased chromium and molybdenum contents, compared to Type 316, can be expected to be resistant to pitting, crevice corrosion, and stress corrosion cracking under virtually any conditions encountered with resources of this fluid class, but they are not routinely available as heat exchanger plates.

Nonaustenitic stainless steels are generally resistant to chloride stress corrosion cracking, but many, especially grades containing about 12% chromium and less than 2% molybdenum, will probably pit severely, especially in aerated environments. More highly alloyed "super ferritics" offer considerable promise but have not been tested in these environments. Type 444 (18Cr-2Mo) has pitting resistance similar to Type 316. SeaCure (26Cr-3Mo), Allegheny-Ludlum 29-4 (29Cr-4Mo), and Allegheny-Ludlum 29-4-2 (29Cr-4Mo-2Ni) should resist any environment resulting from use of fluids of this class. ASTM XM27 (26Cr-1Mo) is of intermediate resistance. All these alloys resist chloride stress corrosion cracking.

Aluminum. Generally, aluminum alloys are not acceptable in most cases because of catastrophic pitting.

Titanium. This material has extremely good corrosion resistance and can be used for heat exchanger plates in any low-temperature geothermal fluid, regardless of dissolved oxygen content. Great care is required if acid cleaning is to be performed. The vendor's instructions must be followed. Care must be taken to avoid scratching the titanium with iron or steel tools, since this can cause pitting.

CPVC (Chlorinated Polyvinyl Chloride) and FRP (Fiber Reinforced Plastic). These materials offer ease of fabrication and are not adversely affected by oxygen intrusion. External protection against groundwater is not required. Their mechanical properties at higher temperatures may vary greatly from ambient temperature properties, and the mechanical limits of the materials should not be exceeded. The usual mode of failure is creep rupture, the strength of which decays with time. Design data are available from manufacturers, based on extrapolation of 10,000-h test results to 100,000 h. The effect on mechanical properties of exposures longer than 100,000 h is not known.

The manufacturer's directions for joining should be followed to avoid premature failure of joints.

Elastomeric Seals. Tests on O-ring materials in a low-temperature system in Texas indicate that fluroelastomer is the best material for piping of this nature; Buna-N is also acceptable. Neoprene, which developed extreme compression set, failed the test. Natural rubber and Buna-S should also be avoided. EPDM has been used successfully in gasket, O-ring, and valve seat applications in many systems.

Corrosion Engineering and Design. The design of the geothermal system is as critical to controlling corrosion as the selection of suitable materials (Ellis 1989). Furthermore, the design/material selection process is interactive, in that certain design decisions force the use of certain materials, while selection of materials may dictate design. In all cases, the objective should be the same—to produce an adequately reliable system with the lowest possible lifetime cost.

Three basic corrosion engineering design philosophies for geothermal systems are apparent:

1. Use corrosion-resistant materials throughout the system.
2. Exclude or remove oxygen, and use carbon steel throughout the system.
3. Transfer the heat via an isolation heat exchanger to a noncorrosive working medium so that the kind and number of components contacting the geothermal fluid are minimized, and make those components of corrosion-resistant materials.

The first philosophy produces a reliable, low-maintenance system. However, the cost would be high, and many of the desired components may not be available in the required alloys.

The second philosophy may be considered for district-sized heating projects with attendant surface storage and potential for oxygen intrusion, because it may be economical to inhibit the geothermal fluid by continuous addition of excess sulfite as an oxygen scavenger. When this is done, carbon steel can be used for heat exchange equipment, provided fluid pH is greater than eight. This approach is widely and successfully used in municipal heating systems in Iceland. If it is used, system design should minimize the introduction of oxygen to reduce sulfite costs. Use of vented tanks should also be minimized.

The second philosophy has three major drawbacks:

1. The sulfite addition plant is relatively complex and requires careful maintenance and operation.
2. Failure of the sulfite addition plant, or insufficient treatment, is likely to cause rapid failure of the carbon steel heat exchangers.
3. Sulfite addition will probably not be economical for smaller systems because of the complexity and excessive maintenance and operation requirements of the plants. Without oxygen scavenging, and even with careful design, some oxygen contamination will occur, and carbon steel heat exchangers will probably not be satisfactory. In addition, environmental considerations regarding disposal of the treated fluid may be a problem.

The third philosophy—transferring the heat via isolation heat exchangers to a noncorrosive secondary heat transfer medium and minimizing the kind and number of components exposed to the geothermal fluids—has several advantages for systems of all sizes. The system is much simpler to operate, and therefore, more reliable than those mentioned above. Only a small number of geothermal-resistant components are required, and it is feasible to design systems in which oxygen exclusion is not critical. Retrofitting of existing fossil-fired hot water systems is also simplified, since the isolation heat exchanger may entirely replace the existing heater or be located in-line with the water supply to the existing heater. Even with careful material selection and design, geothermal heating systems require more maintenance than conventional systems. Minimizing the number and kind of components in contact with geothermal fluid will further reduce maintenance costs.

All geothermal components should be easily disassembled for maintenance. Because scale deposits from geothermal fluid may make threaded joints almost impossible to disassemble, this type of joint should be avoided. Similarly, plate and frame heat exchangers may be desirable, because they have much higher heat transfer effectiveness than tube-and-shell units, and because they are easy to clean and inspect.

Table 5 illustrates materials choice/design constraint interactions for piping, valves, and heat exchangers. Corrosion and design cautions should be followed to obtain favorable results. Where specific reference is not made to geothermal corrosivity Class Va or Class Vb, the data presented are equally applicable to either class of resource.

Pumps. Pumps are used for production, circulation, and disposal. For circulation and disposal, whether surface disposal or injection, standard hot water circulating pumps, almost exclusively of the centrifugal design, are used. These are routine engineering design selections, the only special consideration being the selection of appropriate materials. In many applications, small circulating pumps (of the in-line variety) have employed all-iron construction with acceptable performance. For larger pumps (base mounted, vertical in-line, double suction), cast-iron impellers and volute with stainless steel shaft, screws, keys, and washers are typically employed (Rafferty 1989). In addition, mechanical seals instead of packing are preferred.

Production well pumps are among the most critical components in a geothermal system and have, in the past, been the source of much system downtime. Therefore, proper selection and design of the production well pump for a geothermal system is extremely important.

Well pumps are available for larger systems in two general configurations—lineshaft and submersible. The lineshaft type is most often used for direct-use systems (Rafferty 1989).

Lineshaft pumps. Lineshaft pumps are similar to those typically used in irrigation applications. An aboveground driver, typically an electric motor, rotates a vertical shaft extending down the well to the pump. The shaft rotates the pump impellers within the pump bowl assembly, which is positioned at such a depth in the wellbore that adequate net positive suction head (NPSH) is available when the unit is operating.

Two designs for the shaft/bearing portion of the system are available—open and enclosed.

In the open lineshaft design, the shaft bearings are supported in "spiders," which are anchored to the pump column pipe at 5- to 10-ft intervals. The shaft and bearings are exposed to and lubricated by the fluid flowing up the pump column. In geothermal applications, bearing materials for open lineshaft designs have consisted of both bronze and various elastomer compounds. Shaft material is typically stainless steel. Experience with this type of design in geothermal applications has been mixed. Two large district heating systems have successfully operated such pumps for approximately six years; many more systems, however, initiated operation with open lineshaft pumps and have subsequently changed to enclosed lineshaft design after numerous failures.

Open lineshaft pumps are generally less expensive than enclosed lineshaft pumps for the same application. In addition, the open lineshaft type is easier to apply to artesian wells.

In an enclosed lineshaft pump, an enclosing tube protects the shaft and bearings from exposure to the pumped fluid. As a result, a second fluid must be supplied to lubricate the bearings. For geothermal application, oil and water have been used for this purpose. The lubricating fluid is admitted to the enclosing tube at the wellhead. It flows down the tube lubricating the bearings and exits at the base of the pump column where the column attaches to the bowl assembly. In both open and enclosed lineshaft pumps, the bowl shaft and bearings are lubricated by the pumped fluid. Oil-lubricated enclosed lineshaft pumps have the longest service life in low-temperature direct-use applications.

Materials typically employed in these pumps include carbon or stainless steel shafts and bronze bearings in the lineshaft assembly, and stainless steel shafts and leaded red bronze bearings in the bowl assembly. Keyed-type impeller connections (to the pump shaft) are superior to collet-type connections (Rafferty 1989).

The reliability of lineshaft pumps decreases as the pump-setting depth increases because of the lineshaft bearings. Nichols (1978) indicates that at depths greater than about 800 ft, reliability is questionable, even under good pumping conditions.

Submersible pumps. The electric submersible pump system consists of three primary components located downhole—the pump, the drive motor, and the motor protector. The pump is a

Table 5 Material/Design Interaction for Various Components (Ellis 1989)

Material	Corrosion Comments	Design Comments
PIPING		
Carbon steel	*Grade I* In class Vb resources	• Oxygen contamination may cause very serious localized corrosion.
	Caution: Oxygen must be rigorously excluded. Aeration will cause tenfold or greater increase in corrosion rate	• Minimize oxygen into the system. • Do not use vented tanks, especially recirculation tanks, unless oxygen scavengers are also used.
	Grade III In class Va resources, but use Schedule 80.	• Protect exterior surfaces from groundwater.
Galvanized steel	*Grade IV* Zinc not protective at operating temperatures and may cause rapid pitting.	
Copper	*Grade III* May be acceptable in thick-walled applications. Caution: Crevice corrosion at cracks in the cuprous sulfide corrosion product scale has been observed. Caution: No suitable solders have been verified.	• If used, design should provide for easy replacement. • Limit fluid velocity to < 3 ft/s.
CPVC and fiberglass	*Grade V* Oxygen intrusion should have no effects. Failure by degradation of creep rupture strength with time. No data are available for estimating allowable stress after 100,000 h (11.4 years). Caution: Properties deteriorate significantly at elevated temperature.	• Observe manufacturer's temperature, pressure, and stress limits. • Fabricate joints exactly as prescribed by manufacturer.
VALVES		
Carbon steel body with carbon steel trim	*Grade IV* Trim life not adequate in many cases.	• Probable failure mode trim related. Valves should be easy to remove and maintain. Flange or wafer design favored. Minimize threaded parts.
	Localized corrosion at stem/seal/air interface. Caution: Aeration will cause failure.	• Plug (globe) valves not recommended for frequent cycle duty because of plug/stem motion which causes seal failure or seizing.
Carbon steel body with nonaustenitic stainless steel trim	*Grade III* Pitting of trim significant risk.	• Gate valves not recommended for frequent cycle duty because of reciprocating stem motion, which causes seal failure or seizing.
Carbon steel body with austenitic stainless steel	*Grade I* Caution: Aeration will cause rapid corrosion of body.	• Ball or butterfly valves recommended for frequent cycle duty because rotation of stem minimized stem/seal problems. austenitic stainless steel or elastomeric (Buna N, Viton, TFE) seat satisfactory.
Brass body with brass trim and brass or Austenitic stainless steel	*Grade III* Caution: Cathodic to steel, but effect probably not severe in most cases. Caution: Dezincification may be significant risk. Use only red brass or red bronze.	
CPVC body and trim	*Grade V*	
Fiber reinforced plastic body and trim	Failure of materials by degradation of creep rupture strength with time. Data not available for estimating allowable stress after 100,000 h (11.4 years)	• Observe comments above. • Observe manufacturer's limitations on temperature, pressure, and stress.
	Oxygen intrusion should have no effect. Caution: Properties deteriorate greatly at elevated temperatures.	
HEAT EXCHANGERS		
Carbon steel	*Grade IV* if oxygen scavenger is not used.	• Continuous use of oxygen scavenger required.
	Grade II if the resource is Class Vb and the water is continuously treated with excess sulfite.	
	Grade IV if the resource is Class Va.	
Copper	*Grade IV* Caution: Heat exchange corrosion rates typically order of magnitude greater than coupon rates. Serious degradation by ppb H_2S. Suitable solders not identified.	
Aluminum	*Grade IV* Extremely severe corrosion rates with catastrophic pitting.	
Austenitic stainless steel	*Grade II* Caution: SSC of T304 possible above 140 °F. Resistance to SSC increases with molybdenum additions.	• Plate-type heat exchangers recommended for ease of cleaning and economic reasons.
	Caution: Pitting and crevice corrosion of T304 possible at high temperature, high chloride conditions, with oxygen. Pitting and crevice corrosion resistance increases with molybdenum additions (see Figure 8). Can be acid cleaned.	• Stress relief after forming is recommended. • Fibrous gaskets should be avoided. Viton gaskets recommended. • Minimum aeration recommended.
Titanium	*Grade I* in absence of fluoride ion. Caution: Care required when acid cleaned. Avoid scratching with steel tools.	• Plate-type heat exchangers recommended over tube-and-shell for economic reasons and ease of cleaning.

Grade I: Can be reliably used in most cases, with little or no testing, provided corrosion and design cautions are observed.
Grade II: Acceptable in many cases, but confirmatory tests are advisable. Corrosion and design caution must be observed.
Grade III: Acceptable in a limited number of cases. Confirmatory tests strongly advisable. Corrosion and design caution must be observed.
Grade IV: Probably not acceptable.
Grade V: Long-term suitability has not been verified.

vertical multistage centrifugal type. The motor is usually a three-phase induction type that is oil filled for cooling and lubrication. The motor is cooled by heat transfer to the pumped fluid moving up the well. The motor protector is located between the pump and the motor and isolates the motor from the well fluid while allowing for pressure equalization between the pump intake and the motor cavity.

The electrical submersible pump has several advantages over lineshaft pumps, particularly for wells requiring greater pump bowl setting depths. As the well gets deeper, the cost of the submersible decreases and it becomes easier to install. Moreover, it is less sensitive to vertical well deviation. The breakover point is at a pump depth of about 800 ft, with the submersible desirable at pump depths greater than this and the lineshaft preferred at shallower pump settings.

Submersible pumps have not demonstrated acceptable lifetimes in most geothermal applications. Although they are commonly used in high-temperature, downhole applications in the oil and gas industry, the acceptable overhaul interval in that industry is much shorter than it is in a geothermal application. In addition, most submersibles are 3600 rpm machines, which results in greater susceptibility to erosion in aquifers that produce moderate amounts of sand. None of the 14 largest geothermal district heating systems in the United States employs submersibles.

They have, however, been applied in geothermal projects where it is necessary to use an existing well of relatively small diameter. Since the submersibles operate at 3600 rpm, they are capable of providing greater flow capacity for a given bowl size than an equivalent 1750 rpm lineshaft pump.

Variable-speed drives. In many geothermal applications, well flow requirements vary over a substantial range, particularly in space heating applications. To avoid inefficient throttling of the pump to meet those requirements, variable speed drives are often applied. In most cases, one of two varieties of controls has been employed—variable-frequency drive or fluid coupling. One major geothermal system employs a two-speed induction motor on the production pump.

The fluid coupling was most often used prior to the general availability of electronic frequency controls. These units are typically installed between the electric motor and the pump shaft on lineshaft pumps. They cannot be installed on submersible installations. Advantages of the fluid coupling include relatively low cost and mechanical simplicity, so that this equipment can generally be serviced by local pump contractors.

The efficiency of the fluid coupling is a function of the ratio of the output rpm to the input rpm. As a result, in applications in which a large turndown is required (less than about 70% of full speed), operating costs suffer. In most geothermal applications, the minimum pump speed is controlled by the necessity to generate sufficient head to raise the fluid from the pumping level to the wellhead, resulting in only moderate turndown requirements.

Variable-frequency drives are commonly used on geothermal production well pumps. The variable frequency unit is used in conjunction with an inductive motor drive. This provides a more efficient overall drive than the fluid coupling approach described above.

The savings to be achieved by using a variable-speed drive in geothermal applications are subject to several unique pumping considerations. Most well pumping applications involve a considerable static head requirement (lifting the fluid out of the well). This static requirement is not subject to the usual cubic relationship between flow and power requirement experienced in a 100% friction head situation. As a result, for an equal reduction in flow rate, the well pump system pump energy savings are less than those for a circulating pump (friction head) system.

Heat exchangers. The systems shown in Figures 5 and 6 use heat exchangers in contact with the geothermal fluid. For the system in Figure 6, one or more large heat exchangers are located near the wellhead; in the system in Figure 5, air and/or a process fluid is heated by the geothermal fluid in a heat exchanger. The trend is to isolate the geothermal fluid coming in contact with either complicated systems or systems that cannot readily be designed to be compatible with the geothermal fluids. For instance, the geothermal fluid is used directly to heat processing water in many industries; however, it would not usually be used directly in the evaporator of a heat pump (because of the complicated and expensive system) or in the extended surface coils of a building heat system (because the extended surface coils are made of copper-based materials that are incompatible with most geothermal fluids and also because the building system is complex and expensive).

The principal types of heat exchangers used in transferring energy from the geothermal fluid are (1) plate, (2) shell-and-tube, and (3) downhole.

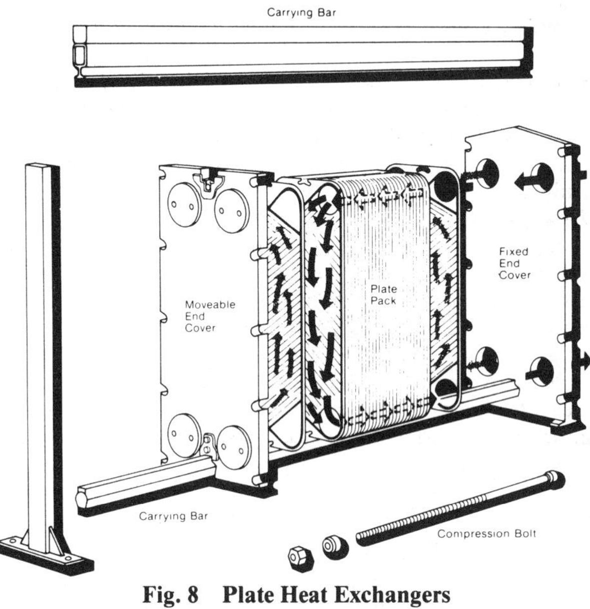

Fig. 8 Plate Heat Exchangers
(Courtesy Alfa-Laval)

Plate heat exchangers. In a plate heat exchanger (Figure 8), the geothermal fluid is routed along one side of each plate, and the heated fluid is routed along the other. The plates can be readily manifolded for various combinations of series and parallel flow. Plate heat exchangers have been widely used for many years in the food-processing industry and in marine applications. They have two main characteristics that make them desirable for many geothermal applications:

- They are readily cleaned. By loosening the main bolts, the header plate and the individual heat exchanger plates can be removed and cleaned.
- The stamped plates are thin and may be made of a wide variety of materials. When expensive materials are required, the thinness of the plates keeps the cost of this type of heat exchanger lower than for other types.

Plate heat exchangers have additional characteristics that influence their selection in specific applications:

- Approach temperature differences are usually smaller than those for shell-and-tube heat exchangers; this is particularly important in low-temperature geothermal applications. Many geothermal applications involve approach temperatures of 10°F and some as low as 5°F.
- Pressure drops can be smaller than those for shell-and-tube heat exchangers since, for a given duty, fewer passes are required through a plate exchanger and fewer entrance and exit losses result.

- Because overall heat transfer coefficients approach 1000 Btu/h · ft² · °F in water-to-water applications, heat transfer per unit volume is usually larger than for shell-and-tube heat exchangers.
- Applications are restricted to temperatures of less than 500 °F because of limitations on elastomeric gaskets.
- Increased capacity can be accommodated easily; only the addition of plates is required.

In most low-temperature direct-use applications, plate heat exchangers are constructed of 316 stainless steel plates and Buna-N gaskets (Rafferty 1989). Stainless steel plates have failed in at least two applications in which geothermal fluid was used to heat swimming pool water. This probably resulted from the high chlorine content in the pool water, which promoted pitting of the plates.

Shell-and-tube heat exchangers. This type of heat exchanger is used in a limited number of geothermal applications. Its application is limited because plate heat exchangers are more economical when specialized materials are required to minimize corrosion. However, when mild steel shells and copper or silicon bronze tubes can be used, shell-and-tube heat exchangers are more economical. With the geothermal fluid passing through the tube side of a shell-and-tube heat exchanger, the tubes should be in a straight configuration to facilitate mechanical cleaning.

Downhole heat exchangers (Culver 1989). The downhole heat exchanger (DHE) consists of an arrangement of pipes or tubes suspended in a wellbore. A secondary fluid circulates from the users system through the exchanger and back to the system in a closed loop. The primary advantage of a DHE is that only heat is extracted from (and with heat pumps, heat is rejected to) the earth eliminating the need for disposal of spent fluids. Other advantages are the elimination of pumps with their initial, operating, and maintenance costs; the potential for depletion of groundwater; and elimination of environmental and institutional restrictions of surface disposal.

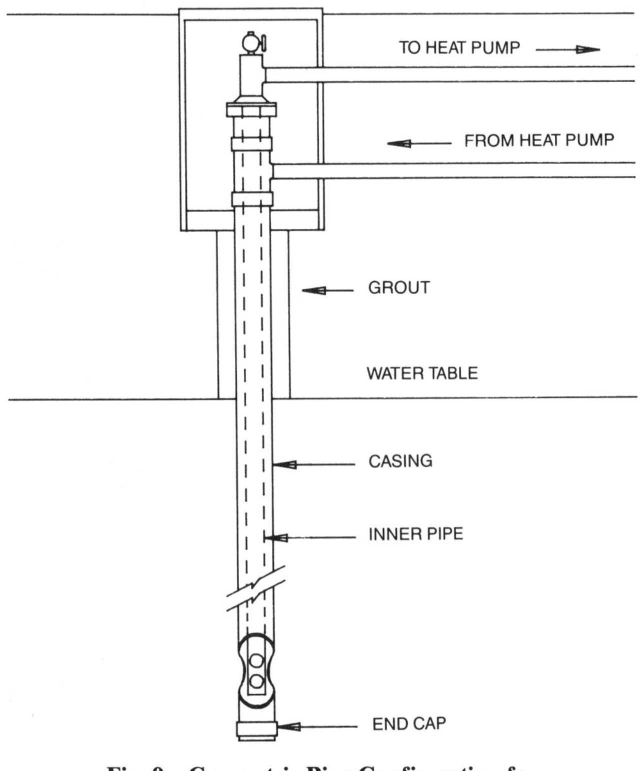

Fig. 9 Concentric Pipe Configuration for Vertical Heat Exchanger
(Braud 1982)

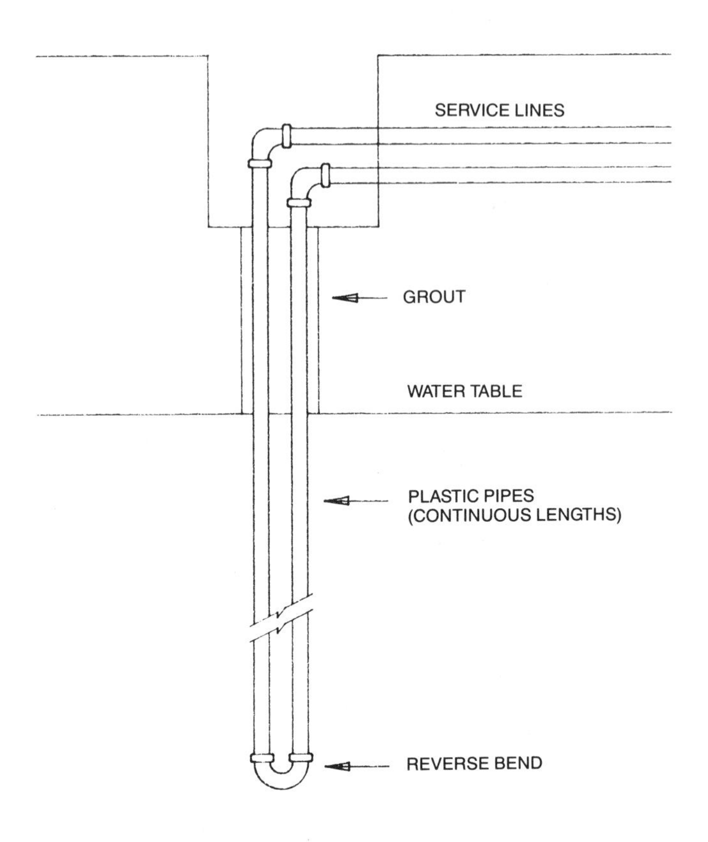

Fig. 10 Two-Pipe, Side-by-Side Configuration for Vertical Heat Exchanger
(Braud 1982)

One disadvantage of DHEs is the limited amount of heat that can be extracted from or rejected to the well. This largely depends on the hydraulic conductivity of the aquifer and well design.

Downhole heat exchangers are used in at least 13 countries worldwide. Most are used for heat pump applications; there are an estimated 20,000 in the United States (Lienau *et al.* 1988) and well over 2000 in Switzerland, where there is a growing concern about legal and environmental factors, since a typical installation removes heat faster than the geothermal heat flux and solar radiation replaces it in the small building plots (Rybach 1985).

The largest capacity DHE reported is 20×10^6 Btu/h in Turkey. The well naturally discharges a two-phase mixture of steam and water and is highly scaling, which prohibits use of the water in typical aboveground exchangers. Steam and a small amount of water are vented to maintain output. A 68,000 Btu/h, 16-ft prototype heat pipe DHE was successfully tested in the Agnano field in Italy, and 340, 100-ft deep DHEs thermosyphoning ammonia provide deicing to an interstate highway ramp in Cheyenne, Wyoming.

Two configurations of DHEs are used in heat pump installations—concentric pipe (Figure 9) and U-bend (Figure 10). The pipes are usually PVC since they are more economical and corrosion resistant than steel. In some instances, steel well casing is used for the outer pipe in the concentric tube arrangement. Bose (1988) presents methods and data for calculating DHE length required for given heat pump operating conditions.

Low-moderate temperature DHEs. DHEs in low- to moderate temperature geothermal wells are similar to heat pump DHEs except that they are in a casing (Figure 11). The U-bend design is most common, but the concentric tube design is also used. Pipes are usually black iron or steel; however, epoxy-fiberglass DHEs have also been installed.

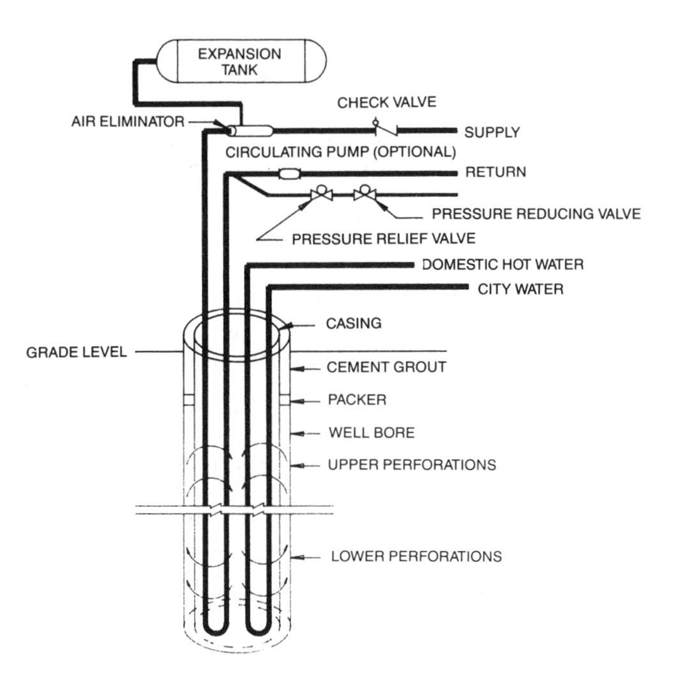

Fig. 11 Typical Connection of a Downhole Heat Exchanger for Space and Domestic Hot Water Heating
(Reistad *et al.* 1979)

DHEs with higher outputs rely on water circulation within the well, whereas heat pump and lower output DHEs rely on earth conduction. Circulation within the well can be accomplished by one of two methods, both of which rely on the density difference between the water surrounding the DHE and that in the aquifer.

Figure 11 shows well construction in competent formations, *i.e.*, where the wellbore stands open without casing. An undersized casing with perforations at the lowest producing zone (usually near the bottom) and just below the static water level is installed. A packer near the top of the competent formation permits the installation of cement between it and the surface. When the DHE is installed and heat is extracted, thermosyphoning causes cooler water inside the casing to move to the bottom, and hotter water moves up the annulus outside the casing.

Circulation provides the following advantages:

1. Water circulates around the DHE at velocities that, at optimum conditions, can approach those in the shell of a shell-and-tube exchanger.
2. Hot water moving up the annulus heats the upper rocks, and the well becomes nearly isothermal.
3. Some of the cool water, being more dense than the water in the aquifer, sinks into the aquifer and is replaced by hotter water which flows up the annulus.

Since most DHEs are used for space heating (an intermittent operation), the heated rocks in the upper portion of the well provide heat storage for the next cycle.

In areas where the well will not stand open without casing, the convection tube can be used. This is a pipe one-half the diameter of the casing, either hung with its lower end above the well bottom and its upper end below the surface or set on the bottom with perforations at the bottom and below the static water level. If a U-bend DHE is used, it can be outside the convection tube or have one leg in the convection tube. A concentric tube DHE should be outside the convection tube.

DHEs operate best in high hydraulic conductivity aquifers, which provide the water movement for heat and mass transfer.

Valves. Resilient lined butterfly valves have long been the design of choice for geothermal applications. The lining material protects the valve body from exposure to the geothermal fluid. The rotary, rather than the reciprocating motion of the stem, is less susceptible to leakage and the buildup of scale deposits. In addition, such valves are generally less expensive to purchase and install than gate valves, and are capable of both throttling and shutoff duty.

For many direct-use applications, materials employed in these valves include Buna-N or EPDM seats, stainless steel shafts, and bronze or stainless steel discs.

Gate valves have also been employed in some larger geothermal systems, but have been subject to stem leakage and seizure. After several years of use, they are no longer capable of 100% shutoff. This can cause problems when isolation of system branches is necessary for maintenance.

Piping. Due to the aggressive nature of most geothermal fluids, nonmetallic piping has been widely applied to direct-use projects. Although steel distribution and transmission piping have been used occasionally, such materials as asbestos cement, fiberglass, and polybutylene are much more common. Asbestos cement piping has been widely applied; however, it is now being phased out.

Fiberglass reinforced piping (FRP)—available in both epoxy resin and polyester resin formulations and in a wide variety of joining methods—has been used in a number of projects. The epoxy adhesive bell and spigot joint has seen the widest use, along with limited mechanical joining. One recent project employed a threaded product.

Fiberglass piping offers lighter weight and smoother interior surfaces for lower pressure drop compared to metallic piping. Although it can be formulated for service as high as 300°F, most cost-competitive products are limited to approximately 200 to 250°F. The adhesive-type joining procedures are frequently unfamiliar to contractors, and the specification of factory representatives for training purposes may be useful. Mechanical joining methods require less labor time and skill, but increase material costs.

The rate of expansion for this type of piping is about twice that of steel. Due to the low modulus, however, the forces exerted are only about 3 to 5% that of steel. As a result, in buried application, expansion loops and joints are generally not required.

Fitting costs, particularly in large sizes, can constitute a significant portion of the overall material costs for a system. Fiberglass piping is generally available in sizes larger than 2 in.

Polybutylene (PB) piping has been widely used in small diameter (<2 in.) sizes for service lines on geothermal district heating systems. A large diameter size (2 to 6 in.) was used successfully in one geothermal distribution system.

Polybutylene is a member of the polyolefin family, and, as such, cannot be joined using solvent welding. The preferred method of joining for small sizes and the only method for larger sizes is that of thermal fusion. Small diameter piping can be socket fused; large piping, butt fused.

Polybutylene is a flexible material and, in small sizes, is available in rolls. As with many other plastics, its pressure rating depends on temperature and is influenced by wall thickness. The pipe is available in various wall thicknesses as designated by the

Standard Dimension Ratio (SDR), a ratio of the nominal diameter to the wall thickness. For example, an SDR 13.5 pipe carries a pressure rating of 104 psi at 150°F, while an SDR 11 pipe's rating is 130 psi at 150°F.

Polybutylene piping can be used to a temperature of approximately 200°F.

Polyethylene, another polyolefin, has seen only limited use in geothermal applications, although it has been used in some non-geothermal district heating systems in Europe. It shares many of the same considerations discussed for polybutylene, except for a lower maximum service temperature (140 to 150°F). It is much lower in cost than polybutylene.

Both these polyolefin products can be employed in direct buried applications without the use of expansion joints or loops.

Steel piping is susceptible to corrosion from both soil moisture (external) and geothermal fluid (internal). As such, it has not been widely used in direct burial applications for transporting geothermal fluid. In two large projects in which steel piping was used for geothermal transmission lines, both cases involved tunnel or in-building type installation. As a result, the piping was not exposed to soil moisture.

Steel piping is employed in many cases for mechanical room and in-building piping. In these applications, Schedule 40 materials with welded or grooved end connections have been most common.

It is critical that oxygen be carefully excluded from the system when using steel pipe to transport geothermal fluid.

Preinsulated pipe. Most large transmission and distribution piping systems for transporting geothermal fluids employ preinsulated piping products. These products consist of a carrier pipe through which the geothermal fluid flows, a layer (nominal 1 to 2 in.) of polyurethane insulation and a jacket for protecting the insulation from physical damage and moisture penetration. The most common material combinations are AC carrier/AC jacket, FRP carrier/FRP jacket, FRP carrier/PVC jacket, steel carrier/FRP jacket, steel carrier/PE jacket.

In most existing direct buried applications, the connections between lengths of pipe are left uninsulated unless steel piping is used.

Installation. Transportation and distribution lines for geothermal fluids can be installed both above and below grade. Below-grade installations are generally preferred for reasons of piping protection and aesthetics.

Two types of buried installations have been employed—concrete tunnel and, most commonly, direct burial of the line. The direct burial method is considerably less expensive, but it lacks the accessibility offered by the tunnel system.

RESIDENTIAL AND COMMERCIAL APPLICATIONS

The primary applications for the direct use of geothermal energy in the residential and commercial area are space heating, sanitary water heating, and space cooling. While space and sanitary water heating are widespread, space cooling is rare. Sanitary water heating, or at least preheating, is accomplished almost universally when space heating is accomplished.

SPACE AND SANITARY WATER HEATING

Figure 12 illustrates the use of geothermal fluid at 170°F (Austin 1978). This geothermal fluid is used in two main equipment components for heating of the structures: (1) a plate heat exchanger that supplies energy to a closed heating loop previously heated by a natural gas boiler (the natural gas boiler remains as a standby unit), and (2) a water-to-air coil used for preheating ventilation air.

In this system, proper control is crucial for economical operation. One of the major goals of control is to extract a large amount of energy from each unit of geothermal fluid by discharging at the lowest feasible temperature at part-load conditions as well as at the design point (Phillips *et al.* 1977). After it leaves the heat transfer equipment, the geothermal fluid goes to a storm drain that leads to a spray cooling pond and fluid discharge to the river.

The average temperature of the discharged fluid is 120 to 130°F. The geothermal fluid is used directly in the preheat terminal equipment within the buildings; this would probably not be applied if the system were being designed today. Several corrosion problems have arisen in the direct use, mainly because of the action of hydro-

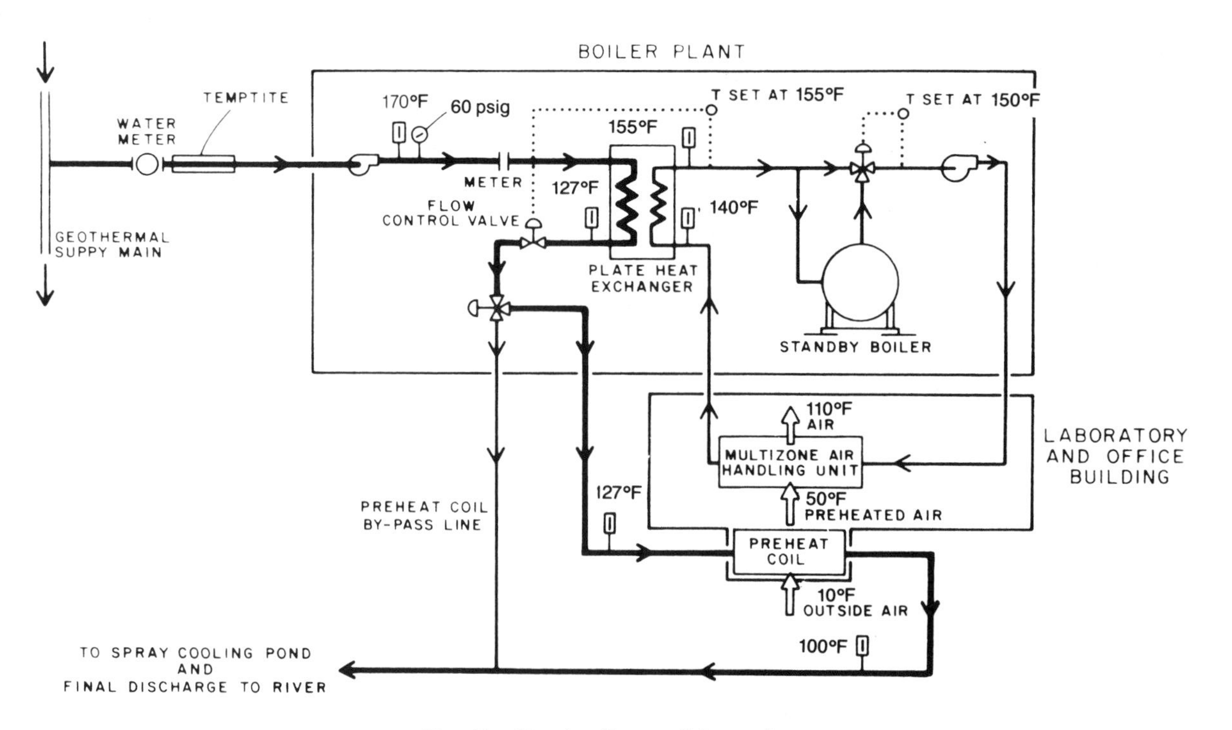

Fig. 12 Heating System Schematic

gen sulfide on copper-based equipment parts (Mitchell 1980). Even with these difficulties, the geothermal system appears to be highly cost-effective (Lienau 1979).

Individual residence systems, such as those in the Klamath Falls, Oregon, area, use downhole heat exchangers (Figure 11). Water from the community supply system circulates in a closed loop for space heating, with makeup through a pressure-reducing valve. The sanitary hot water is supplied by connecting the city water supply to one leg of a downhole heat exchanger; the other leg is connected to the hot water supply piping within the residence. These systems have the advantage of being simple, requiring a minimal amount of wellhead equipment, and avoiding any disposal problems. However, such systems are limited in overall applicability; their best application is for shallow resources and applications that have a thermal power requirement less than the output of a one-well installation (Reistad *et al.* 1979).

A geothermal district heating system is illustrated in Figure 13. The geothermal fluid is produced from three wells. Depending on the load, one, two, or three wells may be in operation. Each well is equipped with a single-speed electric motor and is capable of producing a constant flow rate. The main system flowmeter controls which pumps are in operation. System flow rate is also controlled by a large throttling valve located in the main production line. This valve responds to a signal from a pressure transducer located in the downtown area on the supply side of the distribution system.

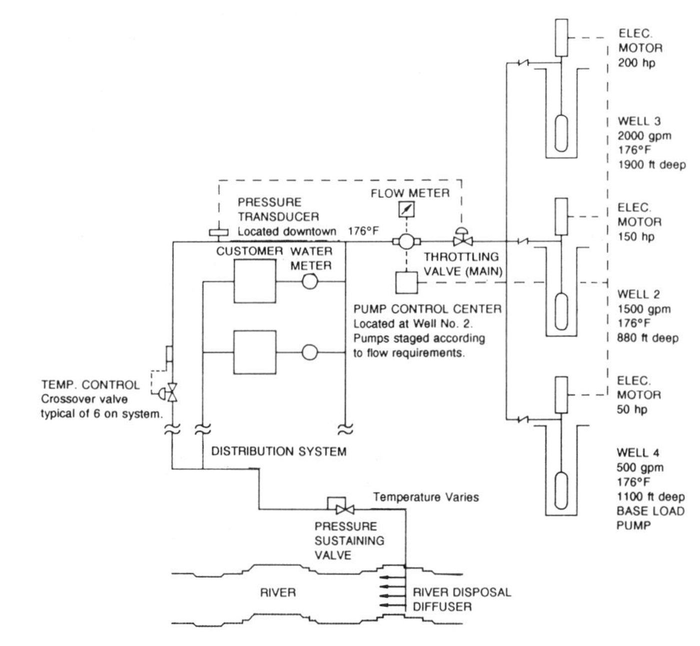

Fig. 13 Open-Type Geothermal District Heating System
(Rafferty 1989)

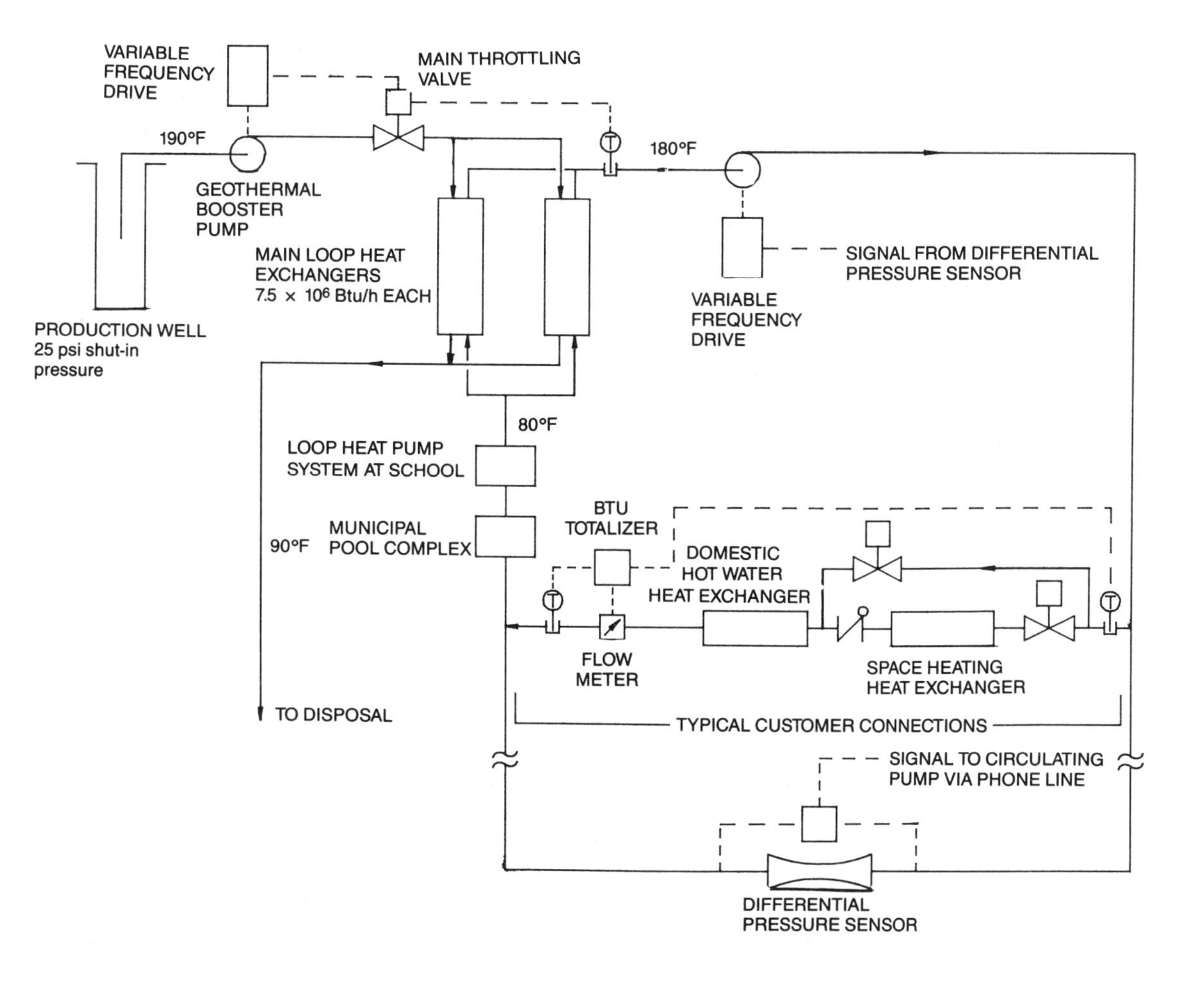

Fig. 14 Closed-Type Geothermal District Heating System
(Rafferty 1989)

The open-type distribution system is composed of preinsulated pipe for supply fluid and uninsulated pipe for disposal fluid. Temperature-sensitive, self-actuated valves are located at six strategic points around the distribution system. Where necessary, these valves open to maintain acceptable supply water temperatures in the system. Water meters are located at each valve to allow system operators to monitor the amount of fluid required for temperature maintenance.

Customers are equipped with hot water meters for billing purposes. Rates are based on water volume at an assumed 35 °F Δt, which encourages customers to use a heating system that is capable of achieving at least a 35 °F Δt.

At present, this system disposes of spent fluid into a nearby river. A pressure sustaining valve maintains a back pressure of 35 psi in the disposal system. From this point, the cooled geothermal fluid is delivered to a diffuser submerged in the river. This diffuser serves to evenly distribute the geothermal fluid for adequate mixing to ensure the required chemical and thermal dilution.

Figure 14 shows a geothermal district heating system that is unique in terms of its design based on a peak load Δt of 100 °F using a 190 °F resource. It is a closed loop design with central heat exchangers.

The production well has an artesian shut-in pressure of 25 psi. As a result, the system operates with no production pump for most of the year. During colder weather, a surface centrifugal pump located at the wellhead boosts the pressure.

Geothermal flow from the production well is initially controlled by a throttling valve on the supply line to the main heat exchanger, which responds to a temperature signal from the supply water on the closed loop side of the heat exchanger. When the throttling valve has reached the full open position, the production booster pump is enabled. The pump is controlled through a variable frequency drive which responds to the same supply water signal as the throttling valve. The booster pump is designed for a peak flow rate of 300 gpm of 190 °F water.

The two main heat exchangers serve to isolate the "clean" distribution loop from the geothermal fluid. These units are staged according to system load. Each is designed for an approach of 10 °F and a peak load of 7.5×10^6 Btu/h.

The main loop circulating pump is also controlled by a variable frequency drive. This drive responds to a signal generated by a differential pressure unit located at the end of the distribution system. The signal is relayed via telephone line to the main mechanical room.

The distribution system consists of direct buried, preinsulated steel piping (Schedule 40 steel, polyurethane insulation, fiberglass jacket).

A number of techniques were used to accomplish the 100 °F system Δt. Individual customers are equipped for both space and domestic hot water heating and these uses are placed in series to obtain a larger Δt. Two-way control valves are used exclusively for control to complement the variable-speed circulating pump.

The system was designed for a limited and predetermined customer base. This resulted in close coordination between the building retrofit design and the distribution system design. As a result, the system includes two users, which are connected in series with the distribution return line. These users—a municipal pool complex and a loop heat pump system—are capable of operating with very low input temperature, which substantially increases the overall system Δt.

Disposal for the system is to the surface at one of two points. Under normal conditions, the fluid is delivered to a wastewater treatment plant effluent line for delivery to an irrigation system and percolation ponds. When this is not possible, a pressure relief valve discharges the fluid to a storm sewer.

Types of Terminal Heating Equipment

The terminal equipment used in geothermal systems is the same as that used in nongeothermal heating systems. However, certain types of equipment are better able to accommodate general considerations for geothermal design than others.

In many cases, buildings that derive their heating needs from geothermal fluids operate a heating system at less than conventional supply water temperatures because of low resource temperatures and the use of heat exchangers to isolate the fluids from the building loop. In addition, it is frequently advisable to design for a larger than normal Δt in geothermal systems.

The selection of equipment to accommodate these considerations can enhance the feasibility of using a geothermal source.

Finned coil, forced air systems are generally the most capable of functioning under the low-temperature/high Δt situation described above. Lower supply water temperatures are compensated for through one or two additional rows of coil depth. While increased Δt affects coil circuiting, it improves controllability. This type of system should be capable of using supply water temperatures as low as 100 to 110 °F.

Radiant floor panel systems are able to use very low water temperatures, particularly in industrial applications with little or no floor covering. The availability of new nonmetallic piping has renewed the popularity of this type of system. In industrial settings, with a bare floor and relatively low space temperature requirements, average water temperatures could conceivably be as low as 95 °F. For higher space temperatures and/or thick floor coverings, much higher water temperatures may be required.

Baseboard convectors and similar equipment are the least capable of operating at low supply water temperature. Derating factors for this type of equipment at 150 °F average water temperatures are on the order of 0.43 to 0.45. As a result, the quantity of equipment needed to meet the design load at low temperatures is generally uneconomical. This type of equipment can be operated at low temperatures to provide baseload heating capacity from the geothermal source. Peak load can be supplied by a conventional boiler.

Heat pump systems take advantage of the lowest temperature geothermal resources. Loop heat pump systems operate with water temperatures in the 60 to 90 °F range, and central station heat pump plants (supplying four-pipe systems) in the 45 °F range.

Feasibility of Space Heating

In addition to the general characteristics of geothermal applications, several factors have a major influence on the feasibility of space heating from geothermal resources. These factors include the following:

- Type of terminal unit
- Density of units
- Alternative energy costs
- Total number of units
- Climate
- Financing for distribution system
- Type of residential unit

The type of terminal unit and its interaction with temperature, was discussed earlier. The alternative energy cost factor refers to the fact that as competing energy costs rise, the feasibility of the geothermal application increases, if all other factors remain constant. The remaining factors affect the cost of (1) the distribution system and (2) the load characteristics. Factors affecting the distribution system costs are important because in district heating systems (whether based on geothermal, solar, or conventional energy supplies), it represents a significant part of the heating cost. Like the geothermal production system costs, these distribution costs are, to a large extent, due to initial capital expenses, and several factors that influence the production system capital costs also apply to the distribution system capital costs. Load characteristics influence the feasibility of geothermal space heating. The feasibility generally increases as both the peak requirement and the load duration (*i.e.*, load factor) increase. It increases as the peak requirement increases, because the average has increased and the system can benefit from economy of scale, which is particularly important in the user lines of the distribution system. The feasibility also increases as the load factor increases, because an increase in load factor improves the economics of such primarily capital cost systems as the geothermal production and distribution systems.

An outdoor temperature versus duration curve for temperatures at which space heating is required for a particular location is shown in Figure 15. The residence heating system is designed not for the minimum temperature, but for the ASHRAE 97.5 percentile design point; the resulting load represents the peak design load. Unless auxiliary heating within the space is provided (not usual), the district heating system must also meet this peak load. Since the energy requirement is approximately proportional to the difference between the inside and outside temperatures, a power scale can be superimposed on Figure 15, as illustrated. The power curve also includes a contribution for the yearly average water heating rate. The geothermal system may be designed to meet only about 60% of the peak design load, which, as illustrated in Figure 15 (crosshatched area), satisfies about 85 to 90% of the yearly energy requirements. Designed in such a manner, the distribution system operates with a load factor dictated by the climate and system served, with representative values being 0.25 to 0.35 for locations like New York; Portland, Oregon; and Boise, Idaho. The geothermal production system operates with a load factor of 1/0.6 = 1.67 times that of the distribution system.

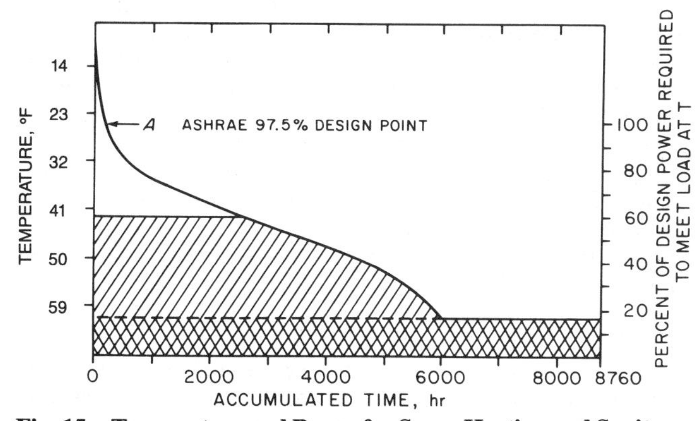

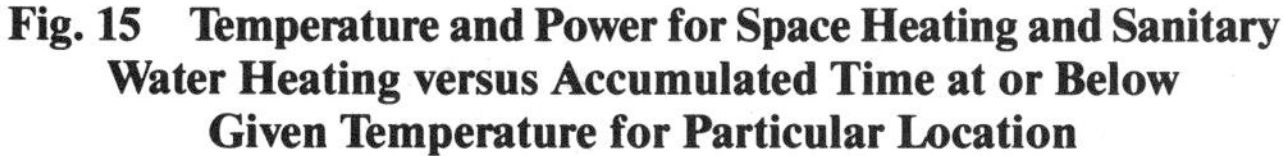

Fig. 15 Temperature and Power for Space Heating and Sanitary Water Heating versus Accumulated Time at or Below Given Temperature for Particular Location

The type of residential unit served influences the feasibility of the geothermal application because it determines the amount of energy delivered to each site. As the load at a point increases, the cost to deliver the energy, per unit of energy, decreases. For developments of a fixed housing density (number of housing units per unit area), developments with high energy requirements per housing unit have greater feasibility for district heating than those with small energy requirements. Also, as the density of heating units increases, so does the feasibility of district heating. Therefore, the feasibility of district heating is much better for apartments than for suburban residence heating.

Peaking and Energy Storage. District heating systems have substantial load variations, which have to be met by the overall system design. In addition to the basic geothermal district heating system, peaking and energy storage systems can be incorporated to meet the load variations. These load variations arise from (1) annual cycles of temperature, (2) daily cycles of temperature, and (3) personal habits such as the lowering of thermostats at night, different working hours, and shower times.

Annual and daily temperature variations result in the temperature duration curve previously considered in the discussion on climate. Because significant daily temperature variation cannot be relied on during periods when the design condition is approached, and the design condition may occur for an extended period, peaking methods, rather than storage, have been used to meet the primary load variations resulting from temperature change. Peaking is achieved by (1) a fossil-fired peaking station, and (2) variable pumping of the geothermal resource with consequent large drawdown of the geothermal reservoir (Bodvarsson and Reistad 1979).

On the other hand, storage has been used mainly to meet short-term load increases that occur on a daily basis, primarily because of personal habits. The storage is in large tanks, which, using the geothermal experience in Iceland, are designed to hold about 20% of the peak flow over a 24-h interval (Olson, *et al.* 1979). No geothermal district heating systems in the United States employ such peaking methods. Due to the relatively shallow, high flow capabilities of U.S. resources in comparison to European sources, the cost of additional wells is competitive with peaking stations.

Domestic Water Heating

Domestic water heating in a district space heating system is beneficial because it increases the overall size of the energy load, the energy demand density, and the load factor (Figure 15). For those resources where domestic water heating to the required temperature is not feasible, preheating is usually desirable. Whenever possible, the domestic hot water load should be placed in series with the space heating load to reduce system flow rates.

Space Cooling

Geothermal energy has seldom been used for cooling, although emphasis on solar energy and waste heat has created interest in systems that cool with thermal energy. Although a number of methods for producing cooling effect using a heat source are available, the absorption cycle is most often used. Lithium-bromide/water absorption machines are commercially available in a wide range of capacities. At least two geothermal/absorption space cooling applications are in successful operation in the United States.

Temperature and flow requirements for absorption chillers run counter to the general design philosophy for geothermal systems. They require high supply water temperatures and small Δt on the hot water side. Figure 16 illustrates the effect of reduced supply

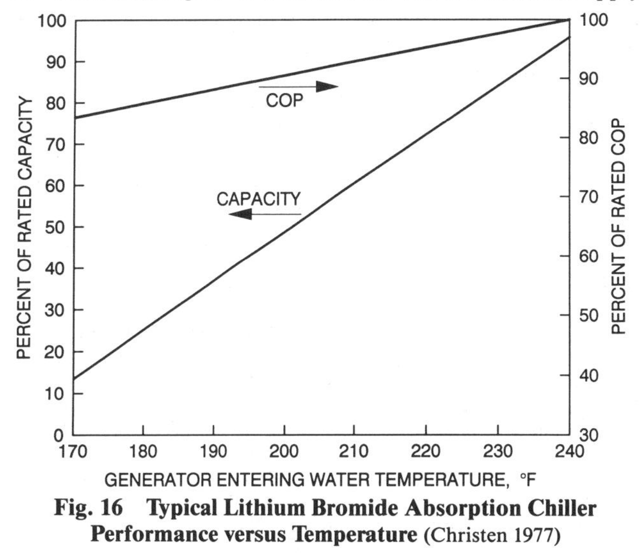

Fig. 16 Typical Lithium Bromide Absorption Chiller Performance versus Temperature (Christen 1977)

water temperature on machine performance. The machines are rated at a 240°F input temperature and derating factors must be applied if the machine is operated below this temperature. For example, operation at a 200°F supply water temperature would result in a 50% decrease in capacity, which seriously affects the economics of absorption cooling at low resource temperatures.

Coefficient of performance (COP) is less seriously affected by reduction in supply water temperatures. The nominal COP of a single-stage machine at 240°F is 0.65 to 0.70; *i.e.*, for each ton of cooling output, a heat input of 12000 Btu/h divided by 0.65 or 18,460 Btu/h is required.

Most absorption equipment has been designed for steam input (an isothermal process) to the generator section. As a result, when operated from a hot water resource, a relatively small Δt must be employed. This creates a mismatch between the flow requirements of a building for space heating and cooling. For example, assume a 200,000 ft^2 building is to use a geothermal resource for heating and cooling. At 25 Btu/h · ft^2 and a design Δt of 40°F, the flow requirement for heating would amount to 250 gpm. At 30 Btu/h · ft^2, a Δt of 15°F, and a COP of 0.65, the flow requirement for cooling would be 1230 gpm.

Some small capacity (3 to 25 ton) absorption equipment has been optimized for low-temperature operation in conjunction with solar systems. While this equipment could also be applied to geothermal resources, the prospects for this are questionable. Small absorption equipment would generally compete with package direct expansion units in this range. In comparison, the absorption equipment requires a great deal more mechanical auxiliary equipment for a given capacity. The cost of the chilled water piping, pump and coil, cooling water piping, pump and tower, and the hot water piping substantially raises the capital cost of the absorption system. As a result, only in large sizes (> 10 tons) and in areas with high electric rates (> $ 0.12/kW) and high cooling requirements (> 2000 full load hours) would this type of equipment offer an attractive investment to the owner (Rafferty 1989).

INDUSTRIAL APPLICATIONS

Geothermal energy for industrial applications, including agricultural facilities, requires design philosophies similar to those for space conditioning. However, these applications have the potential for more economical use of the geothermal resource, primarily because (1) they operate year-round, which gives them the potential for greater load factors than possible with most space conditioning applications; (2) they do not require an extensive (and expensive) distribution to dispersed energy consumers, as is common in district heating; and (3) they often require various temperature levels and, consequently, may be able to make greater use of a particular resource than space conditioning restricted to a specific temperature level. Lienau (1988) summarizes the industrial application of geothermal energy in the United States (Table 6).

Table 6 Geothermal Industrial Uses in the United States
(Lienau 1988)

Application	Number of Sites	Resource Temperature, °F
Sewage digester heating	2	130; 170
Laundry	4	104 to 181 (Varies with site)
Vegetable dehydration	1	270
Mushroom growing	1	235
Greenhouse heating	37	95 to 210 (Varies with site)
Aquaculture	8	61 to 205 (Varies with site)
Grain drying	1	200
Enhanced oil recovery	3	170 to 200 (Varies with site)
Highway deicing	3	47; 190
Gold mine heap leaching	2	238; 186

POTENTIAL APPLICATIONS

From an engineering viewpoint, a primary prerequisite is that the temperature of the geothermal resource matches the application. Thus, for a direct use, the temperature of the geothermal resource must be greater than the temperature requirement of the process. However, with the use of heat pumps, the geothermal resource temperature may be below the required temperature of the application. Because the temperature requirement is so important, it provides an initial screening parameter for considering potential processes.

Applications with major requirements below about 250°F are most suited to use geothermal energy because (1) above this temperature, the geothermal resource has potential for generation of electrical power and any direct use will face stiff competition from such use, and (2) higher temperature applications are more efficiently met with conventional fuels than are the lower temperature applications, while the reverse is true for geothermal resources over a wide temperature range. Industrial processes below 250°F can be categorized as occurring in (1) washing; (2) cooking, blanching, and peeling; (3) sterilization; (4) evaporation and distillation; (5) drying; (6) preheating; (7) miscellaneous heating; (8) refrigeration; (9) greenhouse heating; and (10) aquaculture.

Washing

Considerable low-temperature thermal energy in the range of about 100 to 200°F is used for washing, principally in the food processing, textile, and metal-fabricating industries. The plastics and leather industries represent smaller consumers of such energy. Washing may or may not consume the washing fluid. In consumption use (usual in the food processing and textile industries), fresh wash water is heated from the temperature of the available water supply to the required wash temperature. This allows efficient use of the geothermal resource, since a large temperature drop of the geothermal resource can be realized. For nonconsumption use of the wash fluid (usual in the metal-fabricating industry), the fluid is recirculated for heating, with a 10 to 20°F temperature change being representative of the reheating process. Such a restricted heating temperature change requires (1) that the resource temperature be greater than the required temperature, or (2) that additional measures be taken to achieve a wide temperature change for the geothermal fluid.

Cooking, Blanching, and Peeling

The food processing industry uses thermal energy to cook, blanch, and aid in the peeling of many foods. These processes are done in either a batch or continuous-flow mode. In the blanching or peeling operation, the produce comes in direct contact with a hot fluid. The hot fluid must have closely controlled properties and has to be heated through a heat exchanger if geothermal energy is to be used. These processes occur at temperatures in the range of 170 to 220°F. Cooking is accomplished both where the product comes in direct contact with the hot fluid and where the product is in containers, which are in turn heated by the hot fluid. Cooking occurs over the temperature range of 170 to 220°F, with most of the cooking occurring at about 212°F.

Sterilization

Thermal energy is required at temperatures ranging from 220 to 250°F for sterilization in a wide range of processes. These processes can use geothermal energy to heat the sterilizing water. While much of the sterilization occurs continuously, equipment washdown and sterilization often occur periodically.

Evaporation and Distillation

Many industries use evaporators and distillers for concentrating solutions or separating various products. The temperature requirements may vary over a wide range, depending on the products involved and the specific designs chosen. In many applications, water is being evaporated, and the typical operating temperatures are in the range of 180 to 250°F.

For geothermal application, optional economical operation requires that more stages of evaporation at lower temperatures be designed as compared to designs based on conventional fuels. Geothermal energy can potentially be used for evaporation and distillation in sugar and organic liquor processing.

Drying

The drying of products occurs in many industries and is a large consumer of thermal energy in the temperature range appropriate for direct applications of geothermal resources. In most drying applications, heated air is passed around or through the product to achieve the desired drying rate. Products that require large amounts of drying energy include pulp and paper, textiles, farm products (grain, beet pulp, beverage malt, alfalfa, tobacco, and soybean meal), lumber and plywood, and food products (sugar and dehydrated foods).

Preheating

Many industries consume large quantities of steam or high-temperature water, heated from the local water supply temperature to the required use temperature. Geothermal energy can preheat this water, thereby decreasing the load on conventionally fueled heating equipment.

Miscellaneous Heating

Thermal energy is used to maintain a space or product at a temperature elevated relative to its surroundings. Examples include space heating in industries, warming of plant beds (open field soil warming and mushroom growing), and maintaining sewage digester tanks at operating temperatures.

Refrigeration

Industrial cooling can be accomplished with geothermal energy by several methods. The required refrigeration temperature strongly influences the type of system that will be selected. For refrigeration temperatures above the freezing point of water, the situation is analogous to space cooling, and the lithium bromide water system appears to be most applicable. For refrigeration temperatures below about 32°F, either the Rankine engine/vapor compression system or an absorption system with a refrigerant other than water is the logical choice. Figure 17 presents estimated temperature requirements for industrial refrigeration. As indicated, temperature requirements are in the same range as those employed for binary-type power generation plants.

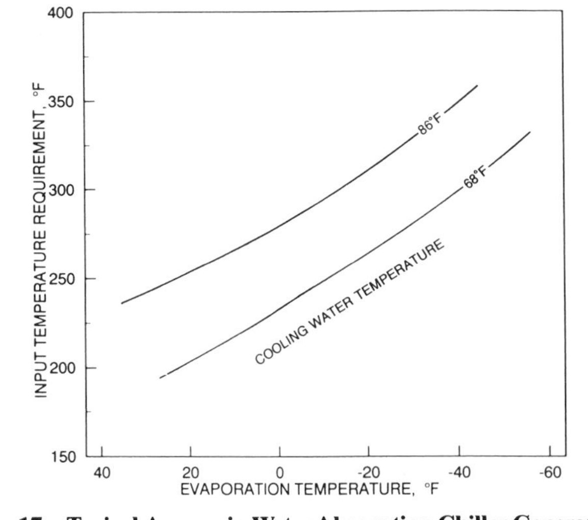

Fig. 17 Typical Ammonia-Water Absorption Chiller Generator Input Temperature Requirements
(Hirai 1982)

Greenhouse Heating

Greenhouse space heating is the most common direct application of geothermal energy next to building space heating. It is an attractive use of the resource due to the high heating requirements of greenhouses and their ability to take advantage of very low resource temperatures.

In an attempt to reduce costs, many greenhouse systems have used geothermal fluid directly in the terminal equipment. This has been successful in only a few cases; in the remainder, serious corrosion problems have occurred. As with other geothermal systems, the heating loop should be isolated from the geothermal fluid via a plate heat exchanger. When selecting equipment, a 5 to 10°F temperature loss across the exchanger must be included. A variety of terminal equipment, including fan coil units, unit heaters (horizontal and vertical), radiant panel, finned pipe convectors, and bare tubing, has been applied to greenhouses.

Fan coil units selected for large temperature drops keep airflow to a minimum. These units are generally installed at the end of the house and deliver air through a collapsible polyethylene tube distribution system. In the absence of other considerations, the fan coil system is the preferred design for a geothermal greenhouse.

Unit heaters are installed in a manner similar to fan coil units. In standard configurations, however, they are considerably less capable of dealing with low supply water temperatures and low water flow rates. If a high-temperature resource is available, unit heaters are more economical to install than fan coil units.

Radiant panels are seldom used for greenhouse heating. In most cases, the heat loss of the greenhouse requires floor temperatures (of the radiant panel) in excess of maximum recommended values. As a result, these systems should be used for baseload operation, with a separate system for peak load periods.

Finned pipe systems are not suitable for temperatures below approximately 150°F because they must be too long to provide enough heat.

Overhead, bare tube systems are common in geothermal greenhouses. In most cases, small diameter (1/2 to 1 in.) polybutylene or polyethylene (depending on the temperature) tubes can provide 100% of the structure's heating requirements. The tubes should be suspended separately in the air, however. Some installations have bundled the tubing together on the floor; this reduces the amount of effective surface area available from each tube.

Most greenhouse operators have a preference for a specific type of heating system. In addition, the type of crop grown can influence their selection.

Aquaculture

Next to greenhouses, aquaculture is the most popular industrial use of geothermal energy. A variety of species, from catfish to prawns, are currently grown commercially in geothermal fluids. In most applications, aquaculture operators prefer to use hot springs or artesian wells to avoid the potential of pump failure causing loss of the aquaculture crop.

The heating system simply involves a control system, which admits the flow of geothermal fluid to the ponds or raceways in such a manner as to regulate the temperature at a given value. Table 7 presents a summary of temperature requirements for various species.

REFERENCES

Allen, E. 1980. Preliminary inventory of western U.S. cities with proximate hydrothermal potential. Report, Eliot Allen and Associates, Inc., Salem, OR.

Table 7 Temperature Requirements and Growth Periods for Selected Aquaculture Species
(Behrends 1973)

Species	Tolerable Extremes, °F	Optimum Growth, °F	Growth Period to Market Size, Months
Oysters	32 to 97	76 to 78	24
Lobsters	32 to 88	72 to 75	24
Penaeid shrimp			
Kuruma	40 to ?	77 to 87	6 to 8 typical
Pink	52 to 140	75 to 85	6 to 8
Salmon (Pacific)	40 to 77	59	6 to 12
Freshwater prawns	75 to 90	83 to 87	6 to 12
Catfish	35 to 95	82 to 87	6
Eels	32 to 97	73 to 86	12 to 24
Tilapia	47 to 106	72 to 86	—
Carp	40 to 100	68 to 90	—
Trout	32 to 89	63	6 to 10
Yellow perch	32 to 86	72 to 82	10
Striped bass	? to 86	61 to 66	6 to 8

Anderson, D.A. and J. Lund, eds. 1980. *Direct utilization of geothermal energy.* Technical Handbook, Geothermal Resource Council Special Report No. 7.

Austin, J.C. 1978. A low temperature geothermal space heating demonstration project. Geothermal Resources Council *Transactions* 2(2). Geothermal Resources Council, Davis, CA.

Behrends, L.L. 1978. Waste heat utilization for agriculture and aquaculture. Tennessee Valley Authority, Muscle Shoals, AL.

Bloomquist, G. 1989. "Regulatory and commercial aspects." In *Geothermal direct use engineering and design guidebook*, Chapter 19. Oregon Institute of Technology, Geo Heat Center, Klamath Falls, OR.

Bodvarsson, G. 1974. Geothermal resource energetics. *Geothermics* 3.

Bodvarsson, G. and G.M. Reistad. 1976. Econometric analysis of forced geoheat recovery for low-temperature uses in the Pacific Northwest. Proceedings of the Second United Nations Symposium on the Development and Use of Geothermal Resources, San Francisco, Vol. 3. Lawrence Berkeley Laboratory, Berkeley, CA.

Bose, J. 1988. Closed-loop/ground source heat pump systems, installation guide. Oklahoma State University, Stillwater, OK.

Braud, H.J. 1982. Earth coupled heat exchange for heat pumps. Paper presented at 1982 Summer Meeting, American Society of Agriculture Engineers, University of Wisconsin, Madison, WI.

Breindel, B., R.L. Harris, and G.K. Olson. 1979. Geothermal absorption refrigeration for food processing industries. ASHRAE *Transactions* 85(1).

BRGM. 1978. *La geothermie en France.* Orleans, France.

Brown, M.C., R.B. Duffield, C.L.B. Siciliana, and M.C. Smith. 1979. Hot dry rock geothermal energy development program. Annual Report, Fiscal Year 1978. DOE Report LA-7807-HDR. Los Alamos Scientific Laboratory, University of California.

Bullard, E. 1973. *Basic theories* (Geothermal Energy; Review of Research and Development). UNESCO, Paris, France.

Christen, J.E. 1977. Central cooling—Absorption chillers. Oak Ridge National Laboratories, Oak Ridge, TN.

Combs, J., J.K. Applegate, R.O. Fournier, C.A. Swanberg, and D. Nielson. 1980. Exploration, confirmation and evaluation of the resource. Geothermal Resources Council, Special Report No. 7, Direct utilization of geothermal energy: Technical handbook.

Cosner, S.R. and J.A. Apps. 1978. A compilation of data on fluids from geothermal resources in the United States. DOE Report LBL-5936. Lawrence Berkeley Laboratory, Berkeley, CA.

Costain, J.K. 1979. Geothermal exploration methods and results—Atlantic Coastal Plain. Geothermal Resources Council, Special Report No. 5, A Symposium of Geothermal Energy and Its Direct Uses in the Eastern United States.

Costain, J.K., L. Glover, III, and A.K. Sinha. 1977. Evaluation and targeting of geothermal energy resources in the Southeastern United States. DOE Report VPI-SU-5103-5. Virginia Polytechnic Institute and State University, Blacksburg, VA.

Costain, J.K., L. Glover, III, and A.K. Sinha. 1977. Evaluation and targeting of geothermal energy resources in the United States. DOE Report VPI-SU-5648, 1 through 5. Virginia Polytechnic Institute and State University, Blacksburg, VA.

Costain, J.K., G.V. Keller, and R.A. Crewdson. 1979. Geological and geophysical study of the origin of the warm springs in Bath County, Virginia. DOE Report TID-28271. Virginia Polytechnic Institute and State University, Blacksburg, VA.

Coulbois, P. and J. Herault. 1976. Conditions for the competitive use of geothermal energy in home heating. Proceedings of the Second United Nations Symposium on the Development and Use of Geothermal Resources, San Francisco (May), Vol. 3. Lawrence Berkeley Laboratory, Berkeley, CA.

Culver, G.G. 1976. Optimization of geothermal home heating systems. Geoheat Utilization Center, Oregon Institute of Technology, Klamath Falls, OR.

Culver, G.G. and G.M. Reistad. 1978. Evaluation and design of downhole heat exchangers for direct applications. USDOE Report No. RLO-2429-7.

Dan, F.J., D.E. Hersam, S.K. Khoa, and L.R. Krumland. 1975. Development of a typical generating unit at The Geysers Geothermal Project—A case study. Proceedings of the Second United Nations Symposium on the Development and Use of Geothermal Resources, San Francisco.

Di Pippo, R. 1988. Industrial developments in geothermal power production. *Geothermal Resources Council Bulletin* 17(5).

Efrid, K.D. and G.E. Moeller. 1978. Electrochemical characteristics of 304 and 316 stainless steels in fresh water as functions of chloride concentration and temperature. Paper 87, Corrosion/78, Houston, TX (March).

Einarsson, S.S. 1973. Geothermal district heating. Geothermal Energy: Review of Research and Development, UNESCO, Paris, France.

Ellis, P. 1989. Materials selection guidelines. *Geothermal Direct Use Engineering and Design Guidebook,* Ch. 8. Oregon Institute of Technology, Geo-Heat Center, Klamath Falls, OR.

Ellis, P.F. and M.F. Conover. 1981. Material selection guidelines for geothermal energy utilization systems. DOE Report RA/27026-1. Radian Corporation, Austin, TX.

Grim, P.J., C.R. Nichols, P.N. Wright, G.W. Berry, and J. Swanson. 1978. State maps of low-temperature geothermal resources. Geothermal Resources Council Transactions 2.

Gudmundsson, J.S. 1985. Direct uses of geothermal energy in 1984. Geothermal Resources Council Proceedings, 1985 International Symposium on Geothermal Energy, International Volume, Davis, CA.

Hirai, W.A. 1982. Feasibility study for an ice making and cold storage facility using geothermal waste heat. Oregon Institute of Technology, Geo-Heat Center, Klamath Falls, OR.

Interagency Geothermal Coordinating Council. Geothermal energy, research, development and demonstration program. DOE Report RA-0050, IGCC-5. U.S. Department of Energy, Washington, D.C.

Intertechnology Corporation. 1977. Analysis of the economic potential of solar thermal energy to provide industrial process heat. Final Report, Vols. 1, 2 and 3. Col. 28-29-1 NTIS, Springfield, VA.

Kindle, C.H. and E.M. Woodruff. 1981. Techniques for geothermal liquid sampling and analysis. Batelle Pacific Northwest Laboratory, Richland, WA.

Kunze, J.F., A.S. Richardson, K.M. Hollenbaugh, C.R. Nichols, and L.L. Mind. 1976. Nonelectric utilization project, Boise, Idaho. Proceedings of the Second United Nations Symposium on the Development and Use of Geothermal Resources, San Francisco, May 1975, Vol 3. Lawrence Berkeley Laboratory, Berkeley, CA.

Lienau, P.J. 1979. Materials performance study of the OIT geothermal heating system. Geo-Heat Utilization Center Quarterly Bulletin, Oregon Institute of Technology, Klamath Falls, OR.

Lienau, P.J. 1984. Geothermal district heating projects. District Heating 70(1 and 2). The International District Heating Association, Washington, D.C.

Lienau, P., G. Culver, and J. Lund. 1988. Geothermal direct use developments in the United States. Oregon Institute of Technology, Geo-Heat Center, Klamath Falls, OR.

Lund, J.W., P.J. Lienau, G.G. Culver, and C.V. Higbee. 1979. Klamath Falls geothermal heating district. Geothermal Resources Council Transactions 3.

Lunis, B. 1986. Geothermal direct use program opportunity notice projects—Lessons learned. EG&G Idaho, Inc., Idaho Falls, ID.

Lunis, B. 1989. Environmental considerations. Geothermal direct use engineering and design guidebook, Ch. 20. Oregon Institute of Technology, Geo-Heat Center, Klamath Falls, OR.

McDonald, C.L. 1977. An evaluation of the potential for district heating in the United States. Proceedings of the Miami International Conference on Alternative Energy Sources, Clean Energy Research Institute, University of Miami, Coral Gables, FL.

Mitchell, D.A. 1980. Performance of typical HVAC materials in two geothermal heating systems. ASHRAE *Transactions* 86(1).

Muffler, L.J.P., ed. 1979. Assessment of geothermal resources of the United States—1978. U.S. Geological Survey Circular No. 790.

Nichols, C.R. 1978. Direct utilization of geothermal energy: DOEs resource assessment program. Direct Utilization of Geothermal Energy: A Symposium Geothermal Resources Council.

Nichols, K.E. and A.J. Malgieri. 1978. Technology assessment of geothermal pumping equipment. DOE Report ALO-4162-2. Barber-Nichols Engineering Company, Arvada, CO.

Olson, G.K., D.L. Benner-Drury, and G.R. Cunnington. 1979. Multi-use geothermal energy system with augmentation for enhanced utilization: A non-electric application of geothermal energy in Susanville, CA. Final Report DOE-ET-248447-1. Aerojet Energy Conversion Company, Sacramento, CA.

Peterson, E.A. 1979. Possibilities for direct use of geothermal energy. ASHRAE *Transactions* 85(1).

Phillips, S.L., A.K. Mathur, and R.E. Doebler. 1977. A study of brine treatment. EPRI Report ER-476. Lawrence Berkeley Laboratory, Berkeley, CA. (Contains literature search of 348 citations.)

Rafferty, K. 1988. Geothermal heating plants. IEA Handbook on advanced heat source technologies for district heating. Draft chapter.

Rafferty, K. 1989. A materials and equipment review of selected U.S. geothermal district heating systems. Oregon Institute of Technology, Geo-Heat Center, Klamath Falls, OR.

Rafferty, K. 1989. Absorption refrigeration. In Geothermal direct use engineering and design guidebook, Ch. 14. Oregon Institute of Technology, Geo-Heat Center, Klamath Falls, OR.

Reed, M.J., ed. 1983. Assessment of low-temperature geothermal resources of the United States—1982. Geological Survey Circular 892, United States Department of Interior, Alexandria, VA.

Reistad, G.M., G.G. Culver, and M. Fukuda. 1979. Downhole heat exchangers for geothermal systems: Performance, economics and applicability. ASHRAE *Transactions* 85(1).

Rybach, L. 1979. Geothermal resources: An introduction with emphasis on low-temperature reservoirs. Geothermal Resources Council, Special Report No. 5, A Symposium of Geothermal Energy and Its Direct Uses in the Eastern United States.

Rybach, L. 1985. Overview of geothermal activities in Switzerland. Geothermal Resources Council *Transactions* 19:5. Geothermal Resources Council, Davis, CA.

Smith, C.S. and P.F. Ellis. 1983. Addendum to materials selection guidelines for geothermal energy utilization systems. Radian Corporation, Austin, TX.

USDOE. 1979. Direct heat application summary. Geothermal Resources Council Annual Meeting.

Zoega, J. 1974. The Reykjavik municipal heating system. Proceedings of the International Conference on Geothermal Energy for Industrial, Agricultural and Commercial-Residential Uses. Oregon Institute of Technology, Klamath Falls, OR.

BIBLIOGRAPHY

Anderson, D.A. and J. Lund, eds. 1980. Direct utilization of geothermal energy: Technical handbook. Geothermal Resources Council, Special Report No. 7.

Armstead, H.C.H., ed. 1973. Geothermal energy: Review of research and development. UNESCO Press, Paris, France.

Bloomster, C.H., L.L. Fassbender, and C.L. McDonald. 1977. Geothermal energy potential for district and process heating applications in the U.S.: An economic analysis. BNWL-2311. Battelle Pacific Northwest Laboratories, Richland, WA.

Geothermal Resources Council. 1977. Geothermal: State of the art. Geothermal Resources Council Transactions 1. San Diego, CA.

Geothermal Resources Council. 1978. Geothermal energy: A novelty becomes a resource. Geothermal Resources Council Transactions 2(1 and 2). Hilo, Hawaii.

Geothermal Resources Council. 1979. Expanding the geothermal frontier 3. Reno, NV.

Howard, J.H., ed. 1975. Present status and future prospects for non-electrical uses of geothermal resources. ERDA Report UCRL-51926. Lawrence Livermore Laboratory, University of California, Livermore, CA.

Kruger, P. and C. Ott, eds. 1972. Geothermal energy. Stanford University Press.

Lienau, P.J. and J.W. Lund, eds. 1974. Multipurpose use of geothermal energy. Proceedings of the International Conference on Geothermal Energy for Industrial, Agricultural and Commercial-Residential Uses. Geo-Heat Utilization Center, Klamath Falls, OR.

Reistad, G.M. 1980. Direct application of geothermal energy. ASHRAE SP No. 26.

United Nations. 1961. Proceedings of the United Nations Conference on New Sources of Energy, Vols. 2 and 3. Rome, Italy.

United Nations. 1970. Proceedings of the Second United Nations Symposium on the Development and Utilization of Geothermal Resources. United Nations, Pisa, Italy. Published in Special Issue of Geothermics, Vols. 1 and 2.

United Nations. 1975. Proceedings of the Second United Nations Symposium on the Development and Use of Geothermal Resources, Vols 1, 2, and 3. Lawrence Berkeley Laboratory, Berkeley, CA.

CHAPTER 30

SOLAR ENERGY UTILIZATION

SOLAR water heaters first appeared in the United States in the early 1900s. However, the introduction of low-cost water heaters using fossil fuels and electricity ended the use of solar water heaters except for a few applications where conventional fuels were unavailable or cost prohibitive. In the 1920s and 1930s, interest in the use of solar radiation for space heating increased. Early solar houses (located north of the equator) used large expanses of south-facing glass to admit winter sunshine. While the extra glass admitted large amounts of solar radiation, it also lost so much heat at night and on cloudy days that actual energy savings was minimal.

Recent passive solar homes use movable insulation and efficient glazing systems to minimize the heat loss when the sun is not shining. Installing solar energy collectors on roofs or walls, and using water or air as a heat transfer medium with rock beds or heat-of-fusion materials for energy storage has received equal interest.

The major obstacles encountered in solar heating and cooling are economic, resulting from the high cost of the equipment needed to collect and store solar energy. In some cases, the cost of owning solar equipment is greater than the resulting saving in fuel cost. Some problems that must be overcome are inherent in the nature of solar radiation. For example:

- It is relatively low in intensity, rarely exceeding 300 Btu/h·ft². Consequently, when large amounts of energy are needed, large collectors must be used.
- It is intermittent because of the inevitable variation in solar radiation intensity from zero at sunrise to a maximum at noon and back to zero at sunset. Some means of energy storage must be provided at night and during periods of low solar irradiation.
- It is subject to unpredictable interruptions because of clouds, rain, snow, hail, or dust.
- Systems should be chosen that make maximum use of the solar energy input by effectively using the energy at the lowest temperatures possible.

QUALITY AND QUANTITY OF SOLAR ENERGY

Solar Constant, I_{sc}

Solar energy approaches the earth as electromagnetic radiation extending from X rays 0.1 μm in wavelength to 100-m long radio waves. The earth maintains a thermal equilibrium between the annual input of shortwave (0.3 to 2.0 μm) radiation from the sun and the outward flux of longwave radiation (3.0 to 30 μm). This equilibrium is global rather than local, because at any given time, some places are too cold for human existence while others are too hot.

Only a limited band need be considered in terrestrial applications, because 99% of the sun's radiant energy is contained between 0.28 and 4.96 μm. The most probable value of the solar constant (defined as the intensity of solar radiation on a surface normal to the sun's rays, beyond the earth's atmosphere at the average earth-sun distance) is 429.5 Btu/h·ft², ±1.5%. The radiation scale in use since 1956 is considered to be 2% too low and varies slightly because of changes in the sun's output of ultraviolet radiation.

The major variations in solar radiation intensity and air temperature are the results of the slightly elliptical nature of the earth's orbit around the sun and, with respect to the orbital plane, the tilt of the axis about which the earth rotates (see Figure 1). The sun is located at one focus of the earth's orbit and is 91.5 × 10⁶ miles away in late December and early January, while the earth-sun distance on July 1 is about 94.4 × 10⁶ miles. Because radiation follows the inverse square law, the normal incidence intensity on an extraterrestrial surface varies from 444.1 Btu/h·ft² on January 1 to 415.6 Btu/h·ft² on July 5.

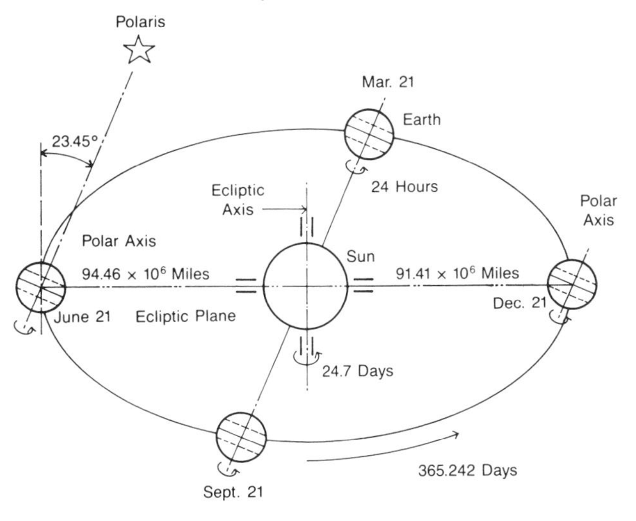

Fig. 1 Annual Motion of the Earth about the Sun

Solar Angles

The axis about which the earth rotates is tilted at an angle of 23°27.14′ to the plane of the ecliptic that contains the earth's orbital plane and the sun's equator. The earth's tilted axis results in a day-by-day variation of the angle between the earth-sun line and the earth's equatorial plane, called the *solar declination* δ. This angle varies with the date, as shown in Table 1, for the year 1964 and in Table 2 for 1977. For other dates, the declination may be estimated by:

$$\delta = 23.45 \sin [360 (284 + N)/365] \quad (1)$$

The preparation of this chapter is assigned to TC 6.7, Solar Energy Utilization.

Table 1 Date, Declination, and Equation of Time for the 21st day of Each Month; with Data* (A, B, C) Used to Calculate Direct Normal Radiation Intensity at the Earth's Surface

Month	Jan	Feb	Mar	Apr	May	June	July	Aug	Sept	Oct	Nov	Dec
Day of the year	21	52	80	111	141	172	202	233	264	294	325	355
Declination δ, degrees	−19.9	−10.6	0.0	+11.9	+20.3	+23.45	+20.5	+12.1	0.0	−10.7	−19.9	−23.45
Equation of time, minutes	−11.2	−13.9	−7.5	+1.1	+3.3	−1.4	−6.2	−2.4	+7.5	+15.4	+13.8	+1.6
Solar noon		late			early		late				early	
A, Btu·h/ft^2	390	385	376	360	350	345	344	351	365	378	387	391
B, dimensionless	0.142	0.144	0.156	0.180	0.196	0.205	0.207	0.201	0.177	0.160	0.149	0.142
C, dimensionless	0.058	0.060	0.071	0.097	0.121	0.134	0.136	0.122	0.092	0.073	0.063	0.057

*Apparent solar irradiation at air mass zero for each month. Atmospheric extinction coefficient. Ratio of the diffuse radiation on a horizontal surface to the direct normal irradiation.

where

N = year day, with January 1 = 1 (For values of N, see Tables 1 and 2.)

The relationship between δ and the date varies to an insignificant degree. The daily change in the declination is the primary reason for the changing seasons, with their variation in the distribution of solar radiation over the earth's surface and the varying number of hours of daylight and darkness.

The earth's rotation causes the sun's apparent motion (Figure 2). The position of the sun can be defined in terms of its altitude β above the horizon (angle HOQ) and its azimuth ϕ, measured as angle HOS in the horizontal plane.

At solar noon, the sun is, by definition, exactly on the meridian, which contains the south-north line and, consequently, the solar azimuth ϕ is 0.0°. The *noon altitude* (β_N) is:

$$\beta_N = 90° - \text{LAT} + \delta \qquad (2)$$

where

LAT = latitude
δ = solar declination

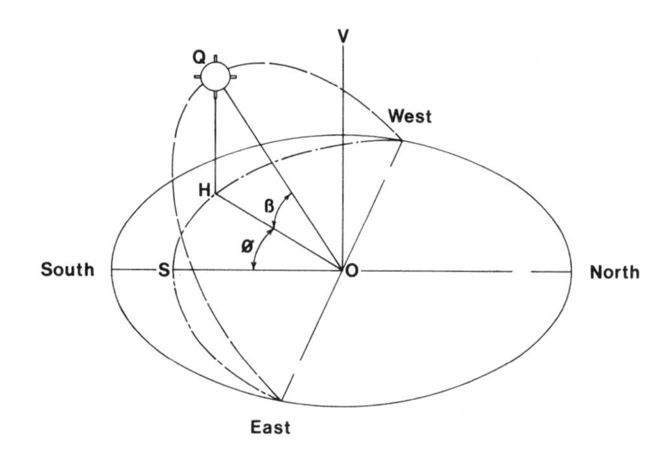

Fig. 2 Apparent Daily Path of the Sun Showing Solar Altitude (β) and Solar Azimuth (ϕ)

Because the earth's daily rotation and its annual orbit around the sun are regular and predictable, the solar altitude and azimuth may be readily calculated for any desired time of day as soon as the latitude, longitude, and date (declination) are specified. Apparent solar time (AST) must be used, expressed in terms of the hour angle H, where:

Table 2 Solar Position Data for 1977

Date	Month	Jan	Feb	Mar	Apr	May	June	July	Aug	Sept	Oct	Nov	Dec
1	Year Day	1	32	60	91	121	152	182	213	244	274	305	335
	Declination δ	−23.0	−17.0	−7.4	+4.7	+15.2	+22.1	+23.1	+17.9	+8.2	−3.3	−14.6	−21.9
	Eq of Time	−3.6	−13.7	−12.5	−4.0	+2.9	+2.4	−3.6	−6.2	+0.0	+10.2	+16.3	+11.0
6	Year Day	6	37	65	96	126	157	187	218	249	279	310	340
	Declination δ	−22.4	−15.5	−5.5	+6.6	+16.6	+22.7	+22.7	+16.6	+6.7	−5.3	−16.1	−22.5
	Eq of Time	−5.9	−14.2	−11.4	−2.5	+3.5	+1.6	−4.5	−5.8	+1.6	+11.8	+16.3	+9.0
11	Year Day	11	42	70	101	131	162	192	223	254	284	315	345
	Declination δ	−21.7	−13.9	−3.5	−8.5	+17.9	+23.1	+22.1	+15.2	+4.4	−7.2	−17.5	−23.0
	Eq of Time	−8.0	−14.4	−10.2	−1.1	+3.7	+0.6	−5.3	−5.1	+3.3	+13.1	+15.9	+6.8
16	Year Day	16	47	75	106	136	167	197	228	259	289	320	350
	Declination δ	−20.8	−12.2	−1.6	+10.3	+19.2	+23.3	+21.3	+13.6	+2.5	−8.7	−18.8	−23.3
	Eq of Time	−9.8	−14.2	−8.8	+0.1	+3.8	−0.4	−5.9	−4.3	+5.0	+14.3	+15.2	+4.4
21	Year Day	21	52	80	111	141	172	202	233	264	294	325	355
	Declination δ	−19.6	−10.4	+0.4	+12.0	+20.3	+23.4	+20.6	+12.0	+0.5	−10.8	−20.0	−23.4
	Eq of Time	−11.4	−13.8	−7.4	+1.2	+3.6	−1.5	−6.2	−3.1	+6.8	+15.3	+14.1	+2.0
26	Year Day	26	57	85	116	146	177	207	238	269	299	330	360
	Declination δ	−18.6	−8.6	+2.4	+13.6	+21.2	+23.3	+19.3	+10.3	−1.4	−12.6	−21.0	−23.4
	Eq of Time	−12.6	−13.1	−5.8	+2.2	+3.2	−2.6	−6.4	−1.8	+8.6	+15.9	+12.7	−0.5

Source: ASHRAE *Standard* 93-1986. Units for declination are angular degrees; units for equation of time are minutes of time. Values of declination and equation of time will vary slightly for specific dates in other years.

$$H = \text{(number of hours from solar noon) } 15° = \text{(number of minutes from solar noon)/4} \quad (3)$$

Solar Time

Apparent solar time generally differs from local standard time (LST) or daylight saving time (DST), and the difference can be significant, particularly when DST is in effect. Because the sun appears to move at the rate of 360° in 24 h, its apparent rate of motion is 4 min per degree of longitude. The procedure for finding AST is shown in Equation (4) as follows:

$$\text{AST} = \text{LST} + \text{Equation of Time} + (4\text{ min})(\text{LST Meridian} - \text{Local Longitude}) \quad (4)$$

The longitudes of the seven standard time meridians that affect North America are Atlantic ST, 60°; Eastern ST, 75°; Central ST, 90°; Mountain ST, 105°; Pacific ST, 120°; Yukon ST, 135°; and Alaska-Hawaii ST, 150°.

Equation of time is the measure, in minutes, of the extent by which solar time, as told by a sundial, runs faster or slower than local standard time (LST), as determined by a clock that runs at a uniform rate. Table 1 gives values of the declination of the sun and the equation of time for the 21st day of each month for the year 1964 (when the ASHRAE solar radiation tables were first calculated), while Table 2 gives values of δ and the equation of time for six days each month for the year 1977.

Example 1. Find AST at noon DST on July 21 for Washington, D.C., longitude = 77°, and for Chicago, longitude = 87.6°.

Solution: Noon DST is actually 11:00 A.M., LST. For Washington, in the eastern time zone, the LST meridian is 75°; for July 21, the equation of time = −6.2 min. Thus noon, Washington DST, is actually:

$$11{:}00 - 6.2 + 4\,(75 - 77) = 10{:}45.8 \text{ AST} = 10.76 \text{ h}$$

For Chicago, in the central time zone, the LST meridian is 90°, so noon, central daylight saving time, is:

$$11{:}00 - 6.2 + 4\,(90 - 87.6) = 11.03.4 \text{ AST} = 11.06 \text{ h}$$

The hour angles H, for these two examples (see Figure 2) are:

for Washington, $H = (12.00 - 10.76)\,15 = 18.6°$ east

for Chicago, $H = (12.00 - 11.06)\,15 = 14.10°$ east

To find the solar altitude β and the azimuth ϕ when the hour angle H, the latitude LAT, and the declination δ are known, the following equations may be used:

$$\sin\beta = \cos(\text{LAT})\cos\delta\cos H + \sin(\text{LAT})\sin\delta \quad (5)$$

$$\sin\phi = \cos\delta\sin H/\cos\beta \quad (6)$$

or

$$\cos\phi = [\sin\beta\sin(\text{LAT}) - \sin\delta]/\cos\beta\cos(\text{LAT}) \quad (7)$$

Tables 3 through 11 in Chapter 27 of the 1989 ASHRAE *Handbook—Fundamentals* give values for latitudes from 0 to 64° north. For any other date or latitude, interpolation between the tabulated values will give sufficiently accurate results. More precise values, with azimuths measured from the north, are given in the U.S. Hydrographic Office Bulletin No. 214 (1958).

Incident Angle

The incident angle θ between the line normal to the irradiated surface (OP′ in Figure 3) and the earth-sun line OQ is important in solar technology because it affects the intensity of the direct component of the solar radiation striking the surface and the ability of the surface to absorb, transmit, or reflect the sun's rays.

To determine θ, the surface azimuth ψ and the surface solar azimuth γ must be known. The surface azimuth (angle POS in Figure 3) is the angle between the south-north line SO and the normal PO to the intersection of the irradiated surface with the horizontal plane, shown as line OM. The surface-solar azimuth,

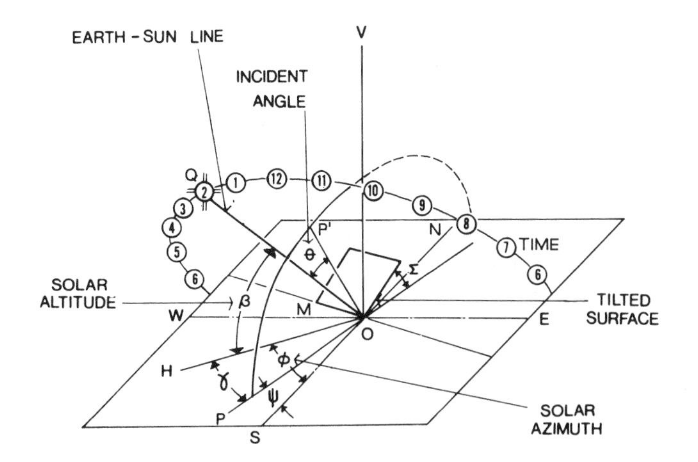

Fig. 3 Solar Angles with Respect to a Tilted Surface

angle HOP, is designated by γ and is the angular difference between the solar azimuth ϕ and the surface azimuth ψ. For surfaces facing *east* of south, $\gamma = \phi - \psi$ in the morning and $\gamma = \phi + \psi$ in the afternoon. For surfaces facing *west* of south, $\gamma = \phi + \psi$ in the morning and $\gamma = \phi - \psi$ in the afternoon. For south-facing surfaces, $\psi = 0°$, so $\gamma = \phi$ for all conditions. The angles δ, β, and ϕ are always positive.

For a surface with a tilt angle Σ (measured from the horizontal), the angle of incidence θ between the direct solar beam and the normal to the surface (angle QOP′ in Figure 3) is given by:

$$\cos\theta = \cos\beta\cos\gamma\sin\Sigma + \sin\beta\cos\Sigma \quad (8)$$

For vertical surfaces, $\Sigma = 90°$, $\cos\Sigma = 0$, $\sin\Sigma = 1.0$, thus:

$$\cos\theta = \cos\beta\cos\gamma \quad (9)$$

For horizontal surfaces, $\Sigma = 0°$, $\sin\Sigma = 0$, and $\cos\Sigma = 1.0$, thus:

$$\theta_H = 90 - \beta \text{ deg} \quad (10)$$

Example 2. Find θ for a south-facing surface tilted upward at 30° to the horizontal at 40° north latitude at 4:00 P.M., AST, on August 21.

Solution: From Equation (3) at 4 P.M. on August 21:

$$H = 4 \times 15 = 60°$$

From Table 1, $\delta = 12.1°$.

From Equation (5):

$$\sin\beta = \cos 40\cos 12.1\cos 60 + \sin 40\sin 12.1$$
$$\beta = 30.6°$$

From Equation (6):

$$\sin\phi = \cos 12.1\sin 60/\cos 30.6$$
$$\phi = 79.7°$$

Since the surface faces south, $\phi = \gamma$. Then using Equation (8):

$$\cos\theta = \cos 30.6\cos 79.7\sin 30 + \sin 30.6\cos 30$$
$$\theta = 58.8°$$

Tabulated values of θ are given in Tables A-7 through A-12 of ASHRAE *Standard* 93-86, Methods of Testing to Determine the Performance of Solar Collectors, for horizontal surfaces and for south-facing surfaces tilted upward at angles equal to the latitude − 10°, the latitude, the latitude + 10°, and the latitude + 20°. These tables cover the latitudes from 24° to 64° north, by 8° intervals.

Solar Spectrum

Beyond the earth's atmosphere, the effective black body temperature of the sun is 10,370°R. The maximum spectral intensity occurs at 0.48 μm in the green portion of the visible spectrum

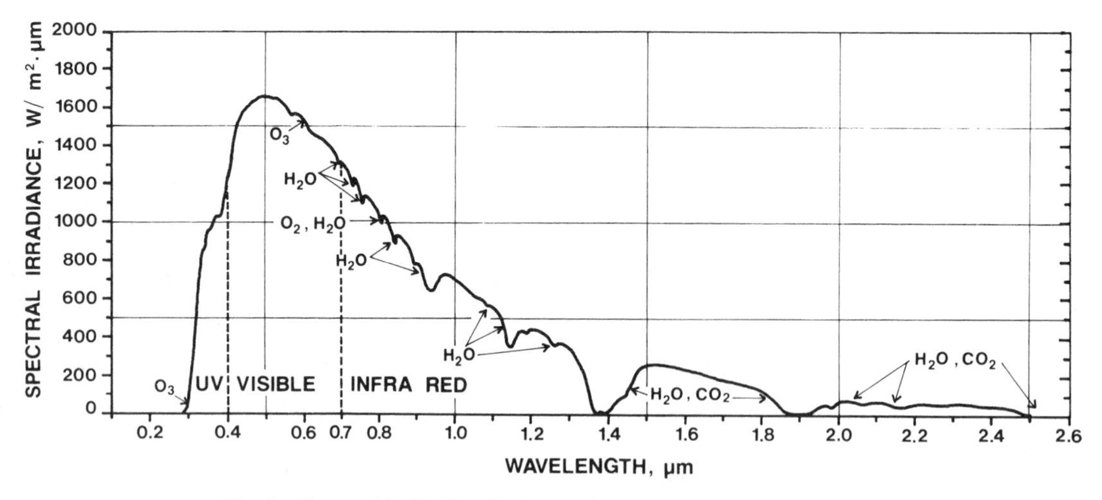

Fig. 4 Spectral Solar Irradiation at Sea Level for Air Mass = 1.0

(Figure 4). Thekaekara (1973) presents tables and charts of the sun's extraterrestrial spectral irradiance from 0.120 to 100 μm, the range where, for practical purposes, all of the sun's radiant energy is contained. The ultraviolet portion of the spectrum below 0.40 μm contains 8.73% of the total, another 38.15% is contained in the visible region between 0.40 and 0.70 μm, and the infrared region contains the remaining 53.12%.

Solar Radiation at the Earth's Surface

In passing through the earth's atmosphere, some of the sun's direct radiation I_D is scattered by nitrogen, oxygen, and other molecules, which are small compared to the wavelength of the radiation; and by aerosols, water droplets, dust, and other particles with diameters comparable to the wavelength (Gates 1966). This scattered radiation causes the sky to be blue on clear days, and some of it reaches the earth as diffuse radiation I_d.

Attenuation of the solar rays is also caused by absorption, first by the ozone in the outer atmosphere, which causes a sharp cutoff at 0.29 μm of the ultraviolet radiation reaching the earth's surface. In the longer wavelengths, there are a series of absorption bands, caused by water vapor, carbon dioxide, and ozone. The total amount of attenuation at any given location is determined by the length of the atmospheric path, which the rays traverse, and by the composition of the atmosphere. The path length is expressed in terms of the air mass m, which is the ratio of the mass of atmosphere in the actual earth-sun path to the mass that would exist if the sun were directly overhead at sea level (m = 1.0). For all practical purposes, at sea level, $m = 1.0/\sin\beta$. Beyond the earth's atmosphere, $m = 0$.

Prior to 1967, solar radiation data was based on an assumed solar constant of 419.7 Btu/h·ft² and on a standard sea level atmosphere containing the equivalent depth of 2.8 mm of ozone, 20 mm of precipitable moisture, and 300 dust particles per cm³. Threlkeld and Jordan (1958) considered the wide variation of water vapor in the atmosphere above the United States at any given time, and particularly the seasonal variation, which finds three times as much moisture in the atmosphere in midsummer as in December, January, and February. The basic atmosphere was assumed to be at sea level barometric pressure, with 2.5 mm of ozone, 200 dust particles per cubic centimetre, and an actual precipitable moisture content that varied throughout the year from 8 mm in midwinter to 28 mm in mid-July. Figure 5 shows the variation of the direct normal irradiation with solar altitude, as estimated for clear atmospheres and for an atmosphere with variable moisture content.

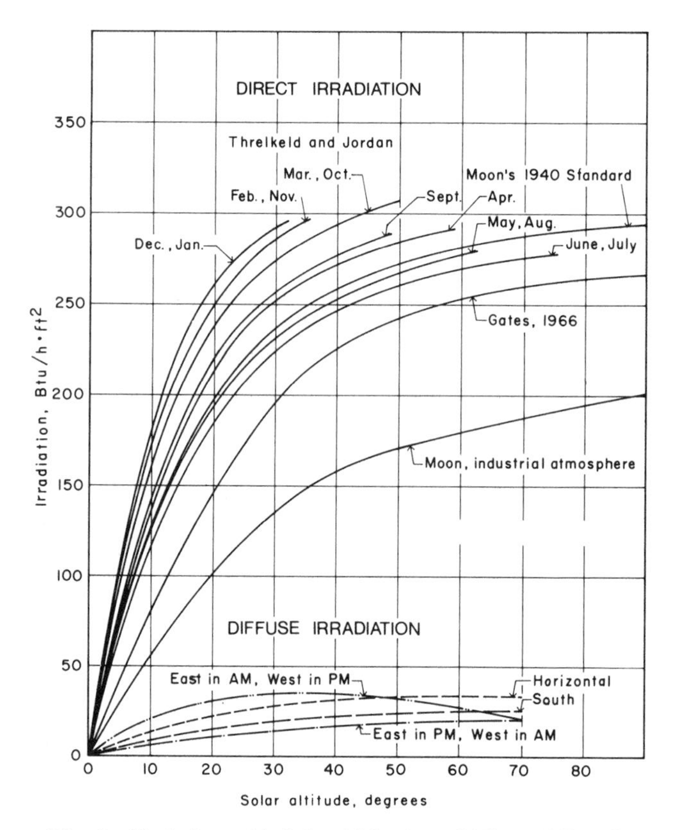

Fig. 5 Variation with Solar Altitude and Time of Year for Direct Normal Irradiation

Stephenson (1967) shows that the intensity of the direct normal irradiation at the earth's surface on a clear day can be estimated by:

$$I_{DN} = A/\exp(B/\sin\beta) \qquad (11)$$

where A, the apparent extraterrestrial irradiation at $m = 0$, and B, the atmospheric extinction coefficient, are functions of the date which take into account the seasonal variation of the earth-sun distance and the air's water vapor content.

The values of the parameters A and B given in Table 1 were selected so that the resulting value of I_{DN} would be in close agreement with the Threlkeld and Jordan values on average cloudless days. The values of I_{DN}, given in Tables 3 through 11 in Chapter 27 of the 1989 ASHRAE *Handbook—Fundamentals*, were obtained by using Equation (11) and data from Table 1. The values of the solar altitude β and the solar azimuth ϕ may be obtained by using Equations (5) and (6).

Because local values of atmospheric water content and elevation can vary markedly from the sea level average, the concept of *clearness number* was introduced to express the ratio between the actual clear-day direct radiation intensity at a specific location and the intensity calculated for the standard atmosphere for the same location and date.

Figure 6 shows the Threlkeld-Jordan map of winter and summer clearness numbers for the continental United States. Irradiation values should be adjusted by the clearness numbers applicable to each particular location.

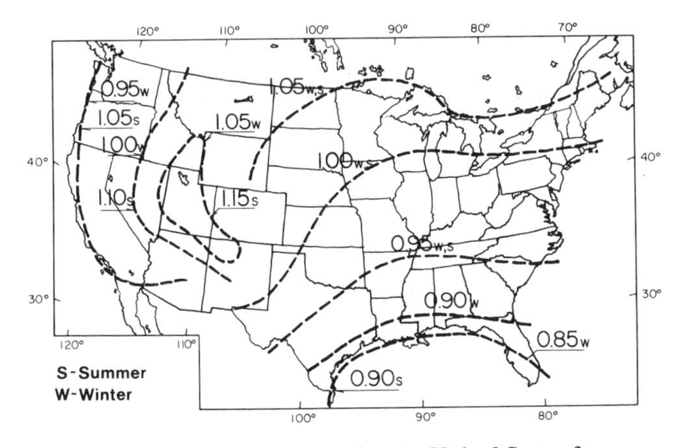

Fig. 6 Clearness Numbers for the United States for Summer (S) and Winter (W)

Design Values of Total Solar Irradiation

The total solar irradiation of a terrestrial surface of any orientation and tilt, with an incident angle θ designated as $I_{t\theta}$, is the sum of the direct component $I_{DN}\cos\theta$ plus the diffuse component coming from the sky $I_{d\theta}$ plus whatever amount of reflected shortwave radiation I_r may reach the surface from the earth or from adjacent surfaces:

$$I_{t\theta} = I_{DN}\cos\theta + I_{d\theta} + I_r \quad (12)$$

The diffuse component is difficult to estimate because of its nondirectional nature and its wide variations. Figure 5 shows typical values of diffuse irradiation of horizontal and vertical surfaces. For clear days, Threlkeld (1963) has derived a dimensionless parameter (designated as C in Table 1), which depends on the dust and moisture content of the atmosphere and thus varies throughout the year.

$$C = I_{dH}/I_{DN} \quad (13)$$

where I_{dH} is the diffuse radiation falling on a horizontal surface under a cloudless sky.

To estimate the amount of diffuse radiation $I_{d\theta}$ that reaches a tilted or vertical surface, Equation (14) may be used:

$$I_{d\theta} = C\,I_{DN}\,F_{ss} \quad (14)$$

where F_{ss} is the angle factor between the surface and the sky and:

$$F_{ss} = (1 + \cos\Sigma)/2 \quad (15)$$

$$F_{sg} = (1 - \cos\Sigma)/2 \quad (16)$$

where F_{sg} is the angle factor between the surface and the earth. The reflected radiation I_r from the foreground is:

$$I_r = I_{tH}\,\rho_g\,F_{sg} \quad (17)$$

where ρ_g is the reflectance of the foreground.

The intensity of the reflected radiation that reaches any surface depends on the nature of the reflecting surface and on the incident angle between the sun's direct beam and the reflecting surface. Many measurements made of the reflection (albedo) of the earth under varying conditions show clean, fresh snow has the highest reflectance (0.87) of any natural surface.

Threlkeld (1963) gives values of reflectance for commonly encountered surfaces at solar incident angles from 0 to 70°. Bituminous paving generally reflects less than 10% of the total incident solar irradiation; bituminous and gravel roofs reflect from 12 to 15%; concrete, depending on its age, reflects from 21 to 33%. Bright green grass reflects 20% at θ = 30° and 30% at θ = 65°.

The maximum daily amount of solar irradiation that can be received at any given location is that which falls on a flat plate with its surface kept normal to the sun's rays so it receives both direct and diffuse radiation. For fixed flat-plate collectors, the total amount of clear day irradiation depends on the orientation and slope. As shown by Figure 7 for 40° north latitude, the total irradiation of horizontal surfaces reaches its maximum in midsummer, while vertical south-facing surfaces experience their maximum irradiation during the winter. These curves show the combined effects of the varying length of days and changing solar altitudes.

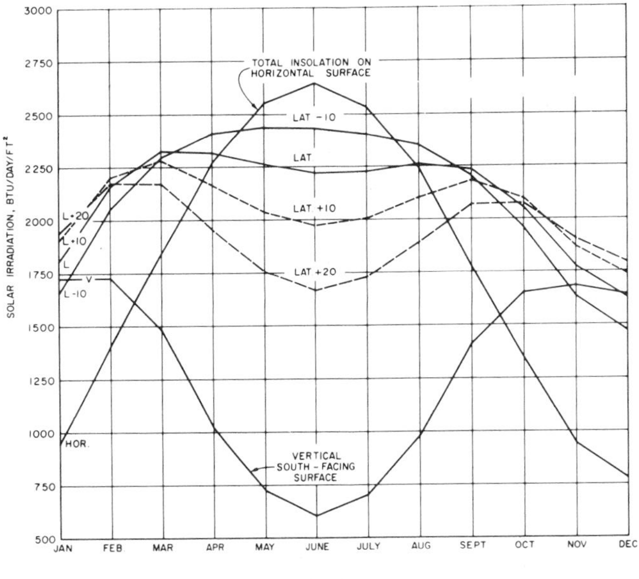

Fig. 7 Total Daily Irradiation for Horizontal, Tilted, and Vertical Surfaces at 40° North Latitude

In general, flat-plate collectors are mounted at a fixed tilt angle Σ (above the horizontal) to give the optimum amount of irradiation for each purpose. Collectors intended for winter heating benefit from higher tilt angles than those used to operate cooling systems in summer. Solar water heaters, which should operate satisfactorily throughout the year, require an angle that is a compromise between the optimal values for summer and winter. Figure 7 shows the monthly variation of total day-long irradiation on the 21st day of each month at 40° north latitude for flat surfaces with varying tilt angles.

Tables E.2 through E.7 of ASHRAE *Standard* 93-1986 give the total solar irradiation for the 21st day of each month at latitudes 24 to 64° north on surfaces with the following orientations: normal to the sun's rays (direct normal, DN, data *do not* include diffuse irradiation); horizontal; south-facing, tilted at (LAT−10), LAT, (LAT+10), (LAT+ 20), and 90° from the horizontal. The day-long total irradiation for fixed surfaces is highest for those that face south, but a deviation in azimuth of 15 to 20° causes only a small reduction.

Solar Energy for Flat-Plate Collectors

The preceeding data apply to clear days. The irradiation for average days may be estimated for any specific location by referring to publications of the U.S. Weather Service. The *Climatic Atlas of the United States* gives maps of monthly and annual values of percentage of possible sunshine, total hours of sunshine, mean solar radiation, mean sky cover, wind speed, and wind direction.

The total daily horizontal irradiation data reported by the U.S. Weather Bureau for approximately 100 stations prior to 1964 show that the percentage of total clear-day irradiation is approximately a linear function of the percentage of possible sunshine. The irradiation is not zero for days when the percentage of possible sunshine is reported as zero, because substantial amounts of energy reach the earth in the form of diffuse radiation. Instead, the following relationship exists:

$$\frac{\text{Daylong actual } I_{tH}}{\text{Clear day } I_{tH}} 100 = a + b \text{ (possible sunshine percent)} \quad (18)$$

where a and b are constants for any specified month at any given location. [See also Jordan and Liu (1977) and Duffie and Beckman (1974).]

Longwave Atmospheric Radiation

In addition to the shortwave (0.3 to 2.0 μm) radiation that it receives from the sun, the earth also receives longwave radiation (4 to 100 μm, with maximum intensity near 10 μm) from the atmosphere. In turn, a surface on the earth emits longwave radiation in accordance with the Stefan-Boltzmann Law:

$$q_{Rs} = e_s \sigma (T_s/100)^4 \quad (19)$$

where

e_s = surface emittance
σ = constant, 0.1713
T_s = absolute temperature of the surface, °R

For most nonmetallic surfaces, the longwave hemispheric emittance is high, ranging from 0.84 for glass and dry sand to 0.95 for black built-up roofing. For highly polished metals and certain selective surfaces, e_s may be as low as 0.05 to 0.20.

Atmospheric radiation comes primarily from water vapor, carbon dioxide, and ozone (Bliss 1961); very little comes from oxygen and nitrogen, although they make up 99% of the air.

Approximately 90% of the incoming atmospheric radiation comes from the lowest 300 ft. Thus, the air conditions at ground level largely determine the magnitude of the incoming radiation. The downward radiation from the atmosphere may be expressed as:

$$q_{Rat} = e_{at} \sigma (T_{at}/100)^4 \quad (20)$$

The emittance of the atmosphere is a complex function of air temperature and moisture content. The dew point of the atmosphere near the ground determines the total amount of moisture in the atmosphere above the place where the dry-bulb and dew-point temperatures of the atmosphere are determined (Reitan 1963). Bliss (1961) found that the emittance of the atmosphere is related to the dew-point temperature, as shown by Table 3.

Table 3 Sky Emittance and Amount of Precipitable Moisture versus Dew-Point Temperature

Dew Point, °F	Sky Emittance, e_a	Precipitable Water, in.
−20	0.68	0.12
−10	0.71	0.16
0	0.73	0.18
10	0.76	0.22
20	0.77	0.29
30	0.79	0.41
40	0.82	0.57
50	0.84	0.81
60	0.86	1.14
70	0.88	1.61

If the apparent sky temperature is defined as that temperature which, radiating as a blackbody, emits radiation at the rate actually emitted by the atmosphere at ground level temperature with its actual emittance e_{at}, then:

$$\sigma (T_{sky}/100)^4 = e_{at} \sigma (T_{at}/100)^4 \quad (21)$$

or

$$T_{sky}^4 = e_{at} T_{at}^4 \quad (22)$$

Example 3. Consider a summer night condition when the ground level temperatures are 65 °F dew point and 85 °F dry bulb. By interpolation from Table 3, e_{at} at 65 °F dew point is 0.87, and the apparent sky temperature becomes:

$$T_{sky} = 0.87^{0.25} (85 + 459.6) = 526.0\,°\text{R}$$

Thus, T_{sky} = 526.0 − 459.6 = 66.4 °F, which is 18.6 °F below the ground level dry-bulb temperature.

For a winter night in Arizona, when the temperatures at ground level are 60 °F dry bulb and 25 °F dew point, the emittance of the atmosphere would be 0.78 by interpolation from Table 3, and the apparent sky temperature would be 488.3 °R or 28.7 °F.

A simple relationship, which ignores vapor pressure of the atmosphere, may also be used to estimate the apparent sky temperature, as:

$$T_{sky} = 0.0552 (T_{at})^{1.5} \quad (23)$$

where T is in kelvins.

If the temperature of the radiating surface is assumed to equal the atmospheric temperature, the loss of heat from a black surface (e_s = 1.00) may be found from Figure 8.

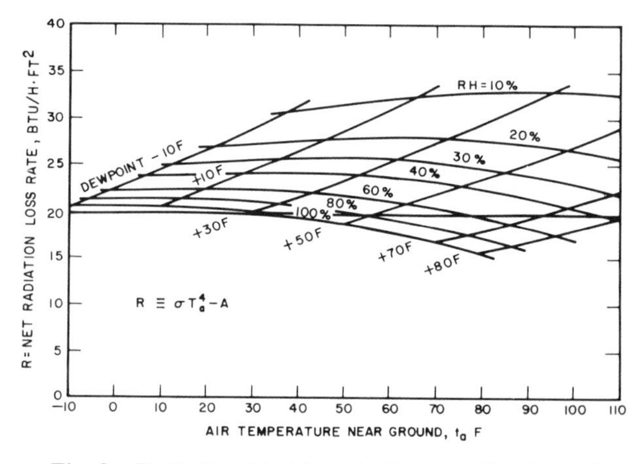

Fig. 8 Radiation Heat Loss to Sky from Horizontal Blackbody

Example 4. For the conditions used in the foregoing summer example, 85 °F dry bulb and 65 °F dew point, the rate of radiative heat loss would be about 23 Btu/h · ft². For the winter example, with 60 °F dry bulb and 25 °F dew point, the heat loss would be about 27 Btu/h · ft².

Where a bare, blackened roof is used as a heat dissipater, the rate of heat loss rises rapidly as the surface temperature goes up. For the summer example, a black-painted metallic roof, e_s = 0.96, at 100°F will have a heat loss rate of:

$$q_{RAD} = 0.96 \times 0.1713\,[(559.6/100)^4 - (526.0/100)^4]$$
$$= 35.4\ \text{Btu/h}\cdot\text{ft}^2$$

This analysis shows that radiation alone is not a potent means of dissipating heat under summer conditions of high dew-point and high ambient temperature. In spring and fall, when both dew-point and dry-bulb temperatures are relatively low, radiation becomes much more effective.

On overcast nights, when the cloud cover is low, the clouds act much like blackbodies at ground level temperature, and virtually no heat can be lost by radiation. The exchange of longwave radiation between the sky and terrestrial surfaces occurs in the day-time as well as at night, but the much greater magnitude of the solar irradiation masks the longwave effects.

SOLAR ENERGY COLLECTION

Solar energy can be used by (1) heliochemical, (2) helioelectrical, and (3) heliothermal processes. The first process, through photosynthesis, produces food and converts CO_2 to O_2. The second process, using photovoltaic converters, powers spacecraft and is useful for many terrestrial applications. The third process, the primary subject of this chapter, provides thermal energy for space heating and cooling, domestic water heating, power generation, distillation, and process heating.

Solar Heat Collection by Flat-Plate Collectors

The solar irradiation data presented in the foregoing sections may be used to estimate how much energy is likely to be available for collection at any specific location, date, and time of day by either a concentrating device, which uses only the direct rays of the sun, or by a flat-plate collector, which can use both direct and diffuse irradiation. Because the temperatures needed for space heating and cooling do not exceed 200°F, even for absorption refrigeration, they can be attained by carefully designed flat-plate collectors. Single-effect absorption systems can, depending on the load and ambient temperatures, use energizing temperatures of 110 to 230°F.

A flat-plate collector generally consists of the following components shown in Figure 9:

Glazing. One or more sheets of glass or other diathermanous (radiation-transmitting) material.

Tubes, fins, or *passages*. Conduct or direct the heat transfer fluid from the inlet to the outlet.

Absorber plates. Flat, corrugated, or grooved plates, to which the tubes, fins, or passages are attached. The plate may be integral with the tubes.

Headers or *manifolds*. Admit and discharge the fluid.

Insulation. Minimizes heat loss from the back and sides of the collector.

Container or *casing*. Surrounds the aforementioned components and keeps them free from dust, moisture, etc.

Flat-plate collectors have been built in a wide variety of designs from many different materials (Figure 10). They have been used to heat fluids such as water, water plus an antifreeze additive, or gases such as air. Their major objective has been to collect as much solar energy as possible at the lowest possible total cost. The collector should also have a long effective life, despite the adverse effects of the sun's ultraviolet radiation; corrosion or clogging because of acidity, alkalinity, or hardness of the heat-transfer fluid; freezing or air-binding in the case of water, or deposition of dust or moisture in the case of air; and breakage of the glazing because of thermal expansion, hail, vandalism, or other causes. These problems can be minimized by the use of tempered glass.

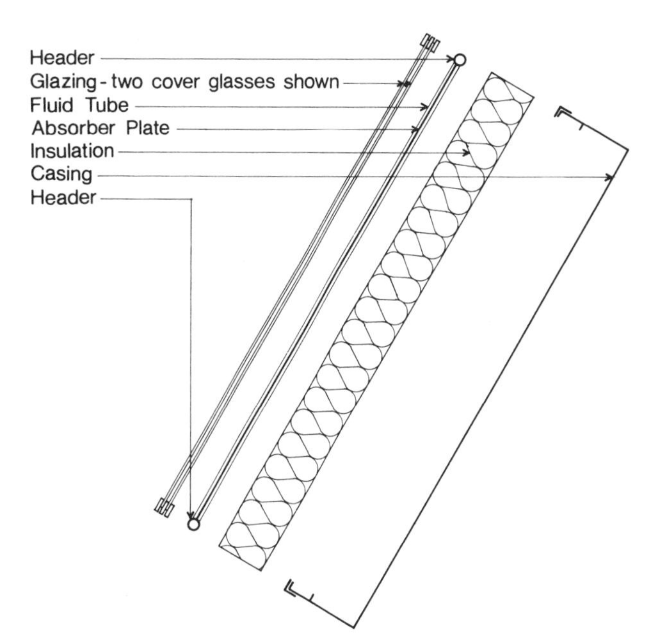

Fig. 9 Exploded Cross Section through Double-Glazed Solar Water Heater

Glazing Materials

Glass has been widely used to glaze solar collectors because it can transmit as much as 90% of the incoming shortwave solar irradiation while transmitting virtually none of the longwave radiation emitted outward by the absorber plate. Glass with low iron content has a relatively high transmittance for solar radiation (approximately 0.85 to 0.90 at normal incidence), but its transmittance is essentially zero for longwave thermal radiation (5.0 to 50 μm) emitted by sun-heated surfaces.

Plastic films and sheets also possess high shortwave transmittance, but because most usable varieties also have transmission bands in the middle of the thermal radiation spectrum, they may have longwave transmittances as high as 0.40.

Plastics are also generally limited in the temperatures they can sustain without deteriorating or undergoing dimensional changes. Only a few can withstand the sun's ultraviolet radiation for long periods. They can withstand breakage by hail and other stones, and in the form of thin films, they are completely flexible and have a low mass.

The glass generally used in solar collectors may be either single-strength (0.085 to 0.100 in. thick) or double-strength (0.115 to 0.133 in. thick). The commercially available grades of window and greenhouse glass have normal incidence transmittances of about 0.87 and 0.85, respectively. For direct radiation, the transmittance varies markedly with the angle of incidence, as shown by Table 4, which gives transmittances for single- and double-glazing using double-strength clear window glass.

The 4% reflectance from each glass-air interface is the most important factor in reducing transmission, although a gain of about 3% in transmittance can be obtained by using water-white glass. Antireflection coatings and surface texture can also improve transmission significantly. The effect of dirt and dust on collector glazing may be quite small, and the cleansing effect of occasional rain is adequate to maintain the transmittance with 2 to 4% of its maximum value.

The function of the glazing is to admit as much solar irradiation as possible and to reduce the upward loss of heat to the lowest attainable value. Although glass is virtually opaque to the longwave radiation emitted by collector plates, the absorption of that

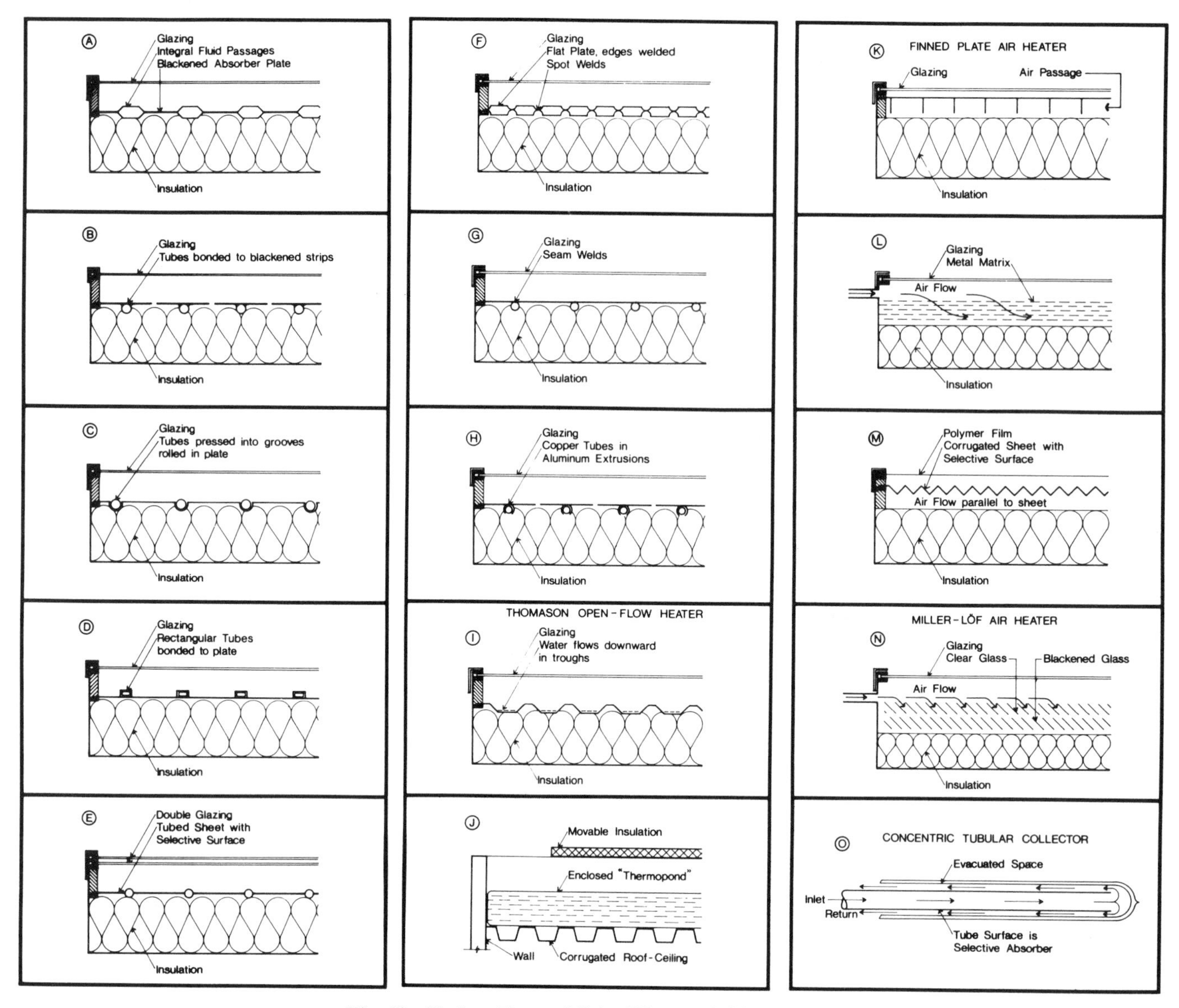

Fig. 10 Various Types of Solar Water and Air Heaters

Table 4 Variation with Incident Angle of Transmittance for Single and Double Glazing and Absorptance for Flat Black Paint

Incident Angle, Deg	Transmittance		Absorptance for Flat Black Paint
	Single Glazing	Double Glazing	
0	0.87	0.77	0.96
10	0.87	0.77	0.96
20	0.87	0.77	0.96
30	0.87	0.76	0.95
40	0.86	0.75	0.94
50	0.84	0.73	0.92
60	0.79	0.67	0.88
70	0.68	0.53	0.82
80	0.42	0.25	0.67
90	0.00	0.00	0.00

radiation causes the glass temperature to rise and thus to lose heat to the surrounding atmosphere by radiation and convection. This type of heat loss can be reduced by using an infrared reflective coating on the underside of the glass; however, such coatings are expensive, and they also reduce the effective solar transmittance of the glass by as much as 10%.

In addition to serving as a heat trap by admitting shortwave solar radiation and retaining longwave thermal radiation, the glazing also reduces heat loss by convection. The insulating effect of the glazing is enhanced by the use of several sheets of glass, or glass plus plastic. The upward heat loss may be expressed by:

$$Q_{up} = A_p U_L (t_p - t_{at}) \tag{24}$$

where

Q_{up} = heat loss upward from the absorber, Btu/h
A_p = absorber plate area, ft^2
U_L = upward heat loss coefficient, Btu/h·ft^2·°F
t_p = absorber plate temperature, °F
t_{at} = ambient air temperature, °F

The loss from the back of the plate rarely exceeds 10% of the upward loss.

Collector Plates

The collector plate absorbs as much as possible of the irradiation, reaching it through the glazing while losing as little heat as possible upward to the atmosphere and downward through the back of the casing, and transfers the retained heat to the transport fluid. The absorptance of the collector surface for shortwave solar radiation depends on the nature and color of the coating and on the incident angle, as shown in Table 4 for a typical flat black paint.

By suitable electrolytic or chemical treatments, surfaces can be produced with high values of solar radiation absorptance α and low values of longwave emittance e_s. Essentially, typical selective surfaces consist of a thin upper layer, which is highly absorbent to shortwave solar radiation but relatively transparent to longwave thermal radiation, deposited on a substrate that has a high reflectance and a low emittance for longwave radiation. Selective surfaces are particularly important when the collector surface temperature is much higher than the ambient air temperature.

For fluid-heating collectors, passages must be integral with or firmly bonded to the absorber plate. A major problem is obtaining a good thermal bond between tubes and absorber plates without incurring excessive costs for labor or materials. Materials most frequently used for collector plates are copper, aluminum, and steel. UV-resistant plastic extrusions are used for low-temperature application. If the entire collector area is in contact with the heat transfer fluid, the thermal conductance of the material is not important.

Whillier (1964) concluded that steel tubes are as effective as copper if the bond conductance between tube and plate is good. Potential corrosion problems should be considered for any metals. Bond conductance can range from a high of 1000 Btu/h·ft^2·°F for a securely soldered or brazed tube to a low of 3 for a poorly clamped or badly soldered tube. Plates of copper, aluminum, or stainless steel with integral tubes are among the most effective types available.

Figure 10, adapted from Van Straaten (1961) and other sources, shows a few of the solar water and air heaters that have been used with varying degrees of success. Figure 10A shows a bonded sheet design, in which the fluid passages are integral with the plate, thus ensuring good thermal contact between the metal and the fluid. Figures 10B and 10C show fluid heaters with tubes soldered, brazed, or otherwise fastened to upper or lower surfaces of sheets or strips of copper, steel, or aluminum. Copper tubes are used most often because of their superior resistance to corrosion.

Thermal cement, clips, clamps, or twisted wires have been tried in the search for low-cost bonding methods. Figure 10D shows the use of extruded rectangular tubing to obtain a larger heat transfer area between tube and plate. Mechanical pressure, thermal cement, or brazing may be used to make the assembly. Soft solder must be avoided because of the high plate temperatures encountered at stagnation conditions.

Figure 10E shows a double-glazed collector with a tubed copper sheet, while 10F shows the use of thin parallel sheets of malleable metal (copper, aluminum, or stainless steel) that are seam-welded along their edges and spot-welded at intervals to provide fluid passages that are developed by expansion. Nontubular types are limited in the internal pressure they can sustain; they are more adapted to space heating than to domestic hot water heating because of the relatively high city water pressures used in the United States.

Figure 10G shows a seam-welded stainless steel absorber with integral tubes. Figure 10H shows copper tubes pressed into appropriately shaped aluminum extrusions. Differential thermal expansion may break the thermal bond.

Figure 10I shows a proprietary open flow collector, which uses black-painted corrugated aluminum sheets, generally mounted on steeply pitched south-facing roofs. Water is distributed to the channels by perforated copper or plastic piping running along the peak of the roof. The sun-warmed water is collected at the bottom of the channels by a galvanized or plastic trough and conducted through plastic pipe to the basement storage tank.

Figure 10J shows another proprietary collector that consists of horizontally arranged transparent plastic bags filled with treated water. A corrugated metal roof ceiling with a black waterproof liner supports these water-filled bags. The water is thus in thermal contact with the ceiling. Movable horizontal insulating panels cover the ponds on winter nights; during winter days, they are opened to admit solar radiation. In summer, the operation is reversed: the panels are opened at night to dissipate heat but are closed during the day to keep out unwanted solar radiation.

Air or other gases can be heated with flat-plate collectors, particularly if some type of extended surface (Figure 10K) is used to counteract the low heat-transfer coefficients between metal and air. Metal or fabric matrices (Figure 10L), or thin corrugated metal sheets (Figure 10M) may be used, with selective surfaces applied to the latter when a high level of performance is required. The principal requirement is a large contact area between the absorbing surface and the air. The Miller-Löf air heater, shown in Figure 10N, provides this area by overlapping glass plates, the upper being clear and the lower blackened to absorb the incoming solar radiation.

Reduction of heat loss from the absorber can be accomplished either by a selective surface to reduce radiative heat transfer or by suppressing convection. Francia (1961) showed that a honeycomb made of transparent material, placed in the airspace between the glazing and the absorber, was beneficial.

Tubular collectors (Figure 10O) with evacuated jackets have demonstrated that the combination of a selective surface and an effective convection suppressor can result in good performance at temperatures higher than a flat-plate collector can attain.

Concentrating Collectors

Temperatures far above those attainable by flat-plate collectors can be reached if a large amount of solar radiation is concentrated on a relatively small collection area. Simple flat reflectors can markedly increase the amount of direct radiation reaching a flat-plate collector, as shown in Figure 11A.

Because of the apparent movement of the sun across the sky, conventional concentrating collectors must follow the sun during its daily motion. There are two methods by which the sun's motion can be readily tracked. The altazimuth method requires the tracking device to turn in both altitude and azimuth; when performed properly, this method enables the concentrator to follow the sun exactly. Paraboloidal solar furnaces, Figure 11B, generally use this system. The polar, or equatorial, mounting points the axis of rotation at the North Star, tilted upward at the angle of the local latitude. By rotating the collector 15°/h, it follows the sun perfectly (on March 21 and September 21). If the collector surface or aperture must be kept normal to the solar rays, a second motion is needed to correct for the change in the solar declination. This motion is not essential for most solar collectors.

The maximum variation in the angle of incidence for a collector on a polar mount will be ±23.5° on June 21 and December 21; the incident angle correction would then be cos 23.45° = 0.917.

Horizontal reflective parabolic troughs, oriented east and west, as shown in Figure 11C, require continuous adjustment to compensate for the changes in the sun's declination. There is inevitably some morning and afternoon shading of the reflecting surface if the concentrator has opaque end panels. The necessity of moving the concentrator to accommodate the changing solar declination can be reduced by moving the absorber or by using a trough with two sections of a parabola facing each other, as shown in Figure 11D. Known as a *compound parabolic concentrator* (CPC), this design can accept incoming radiation over a relatively

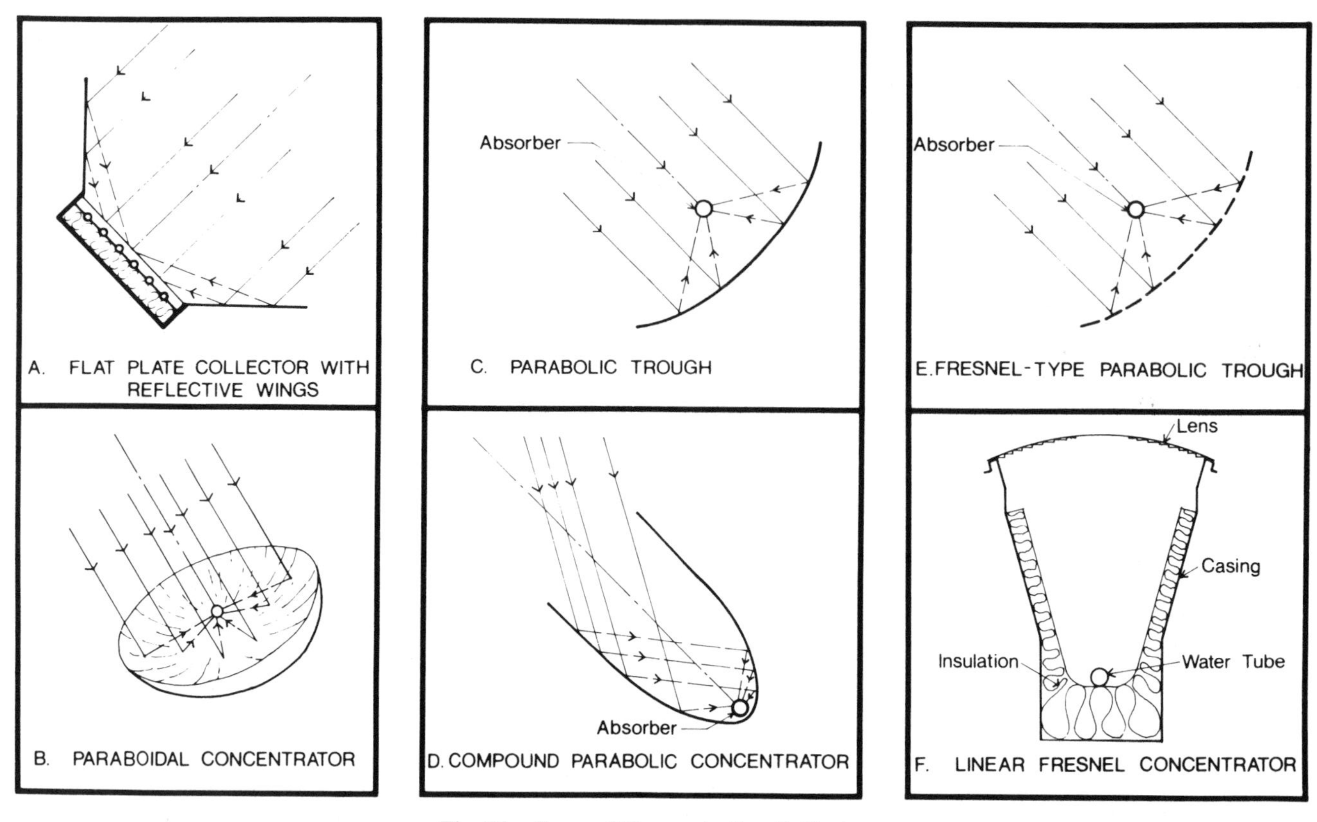

Fig. 11 Types of Concentrating Collectors

wide range of angles. By using multiple internal reflections, any radiation that is accepted finds its way to the absorber surface located at the bottom of the apparatus. By filling the collector shape with a highly transparent material having an index of refraction greater than 1.4, the acceptance angle can be increased. By shaping the surfaces of the array properly, total internal reflection is made to occur at the medium-air interfaces, which results in a high concentration efficiency. Known as a *dielectric compound parabolic concentrator* (DCPC), this device has been applied to photovoltaic generation of electricity (Cole *et al.* 1977).

The parabolic trough of Figure 11C can be simulated by many flat strips, each adjusted at the proper angle so that all reflect onto a common target. By supporting the strips on ribs with parabolic contours, a relatively efficient concentrator can be produced with less tooling than the complete reflective trough.

Another concept applied this segmental idea to flat and cylindrical lenses. A modification is shown in Figure 11F, in which a linear Fresnel lens, curved to shorten its focal distance, can concentrate a relatively large area of radiation onto an elongated receiver. Using the equatorial sun-following mounting, this type of concentrator has been used as a means of attaining temperatures well above those that can be reached with flat-plate collectors.

Concentrating collectors have both bad and good features: (1) except at low concentration ratios, they can use only the direct component of solar radiation, since the diffuse component cannot be concentrated by most types; and (2) in summer, when the sun rises and sets well to the north of the east-west line, the sun-follower, with its axis oriented north-south, can begin to accept radiation directly from the sun long before a fixed, south-facing flat plate can receive anything other than diffuse radiation from the portion of the sky that it faces. Thus, at 40° north latitude, for example, the cumulative *direct* radiation available to a sun-follower on a clear day is 3180 Btu/ft^2, while the *total* radiation falling on the flat plate tilted upward at an angle equal to the latitude is only 2220 Btu/ft^2 each day. Thus, in relatively cloudless areas, the concentrating collector may capture more radiation per unit of aperture area than a flat-plate collector.

For extremely high inputs of radiant energy, a multiplicity of flat mirrors, or *heliostats,* using altazimuth mounts, can be used to reflect their incident direct solar radiation onto a common target. Using slightly concave mirror segments on the heliostats, large amounts of thermal energy can be directed into the cavity of a steam generator to produce steam at high temperature and pressure.

Collector Performance

The performance of collectors may be analyzed by a procedure originated by Hottel and Woertz (1942) and extended by Whillier (ASHRAE 1977). The basic equation is:

$$q_u = I_{t\theta}\,(\tau\alpha)_\theta - U_L(t_p - t_{at}) = \dot{m}\,c_p\,(t_{fe} - t_{fi})/A_{ap} \quad (25)$$

Equation (25) also may be used with concentrating collectors:

$$q_u/A_{ap} = I_{DN}\,(\tau\alpha)_\theta\,(\rho\Gamma) - U_L\,(A_{abs}/A_{ap})\,(t_{abs} - t_a) \quad (26)$$

where

q_u = heat usefully gained by collector, Btu/h·ft^2
$I_{t\theta}$ = total irradiation of collector, Btu/h·ft^2
I_{DN} = direct normal irradiation, Btu/h·ft^2
$(\tau\alpha)_\theta$ = transmittance of cover times absorptance of plate at prevailing incident angle, θ
U_L = loss factor, Btu/h·ft^2·°F
t_p, t_{at} = temperatures of the absorber plate and the atmosphere, °F
$\dot{m}$ = fluid flow rate, lb/h

t_{fe}, t_{fi} = temperatures of the fluid leaving and entering the collector, °F

$\rho\Gamma$ = reflectance of the concentrator surface times fraction of reflected or refracted radiation that reaches the absorber

A_{abs}, A_{ap} = areas of absorber surface and of aperture that admit or receive radiation, ft²

The total irradiation and the direct normal irradiation for clear days may be taken from ASHRAE *Standard* 93-1986. The transmittance for single and double glazing and the absorptance for flat black paint may be found in Table 4 for incident angles from 0 to 90°. These values, and the products of τ and α, are also shown in Figure 12. Little change occurs in the solar-optical properties of the glazing and absorber plate until θ exceeds 30°, but, since all values reach zero when $\theta = 90°$, they drop off rapidly for values of θ beyond 40°.

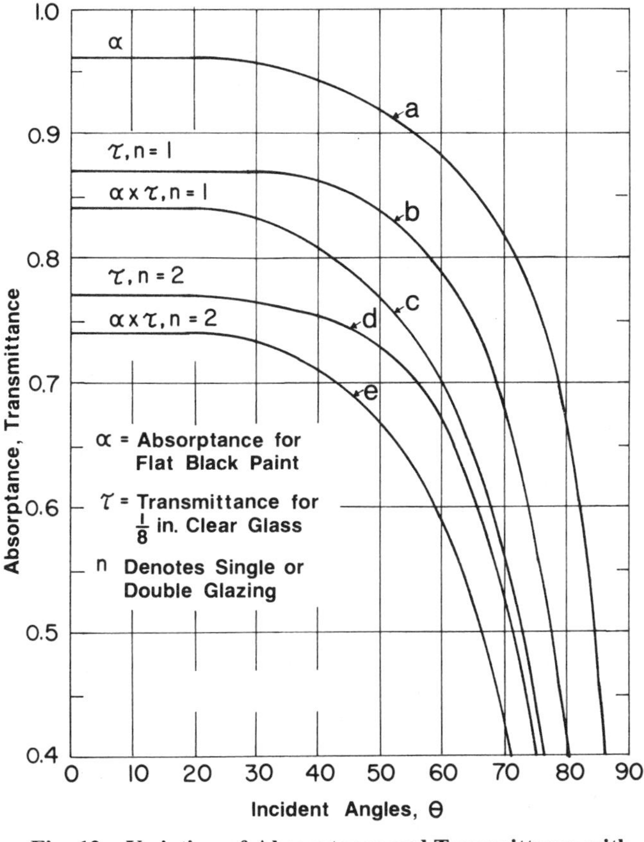

Fig. 12 Variation of Absorptance and Transmittance with Incident Angle

For nonselective absorber plates, U_L varies with the temperature of the plate and the ambient air, as shown in Figure 13. For selective surfaces, which effect major reductions in the emittance of the absorber plate, U_L will be much lower than the values shown there. Manufacturers of such surfaces should be asked for values applicable to their products, or test results should be consulted that give the necessary information.

Example 5. A flat-plate collector is operating in Denver, latitude = 40° north, on July 21 at noon solar time. The atmospheric temperature is assumed to be 85°F, and the average temperature of the absorber plate is 140°F. The collector is single-glazed with flat black paint on the absorber. The tilt angle is 30° from the horizontal, and the collector faces south. Find the rate of heat collection and the collector efficiency. Neglect the back and side losses from the collector.

Solution: From Table 2, $\delta = 20.6°$

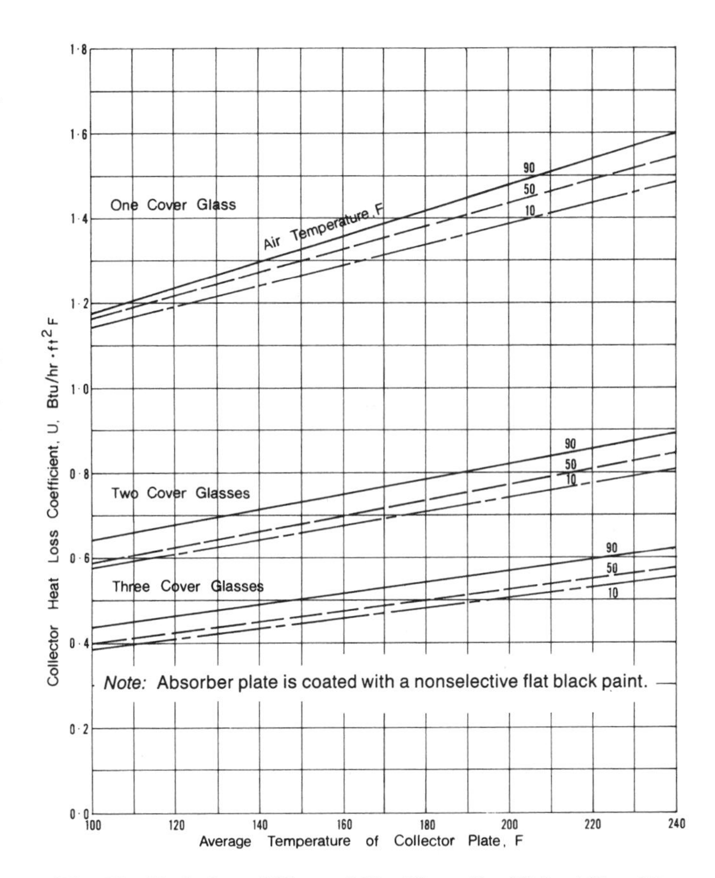

Fig. 13 Variation of Upward Heat Loss Coefficient U_L with Collector Plate Temperature and Ambient Air Temperatures for Single-, Double-, and Triple-Glazed Collectors

From Equation (2)

$$\beta_N = 90 - 40 + 20.6 = 70.6°$$

By geometry, $\theta = 10.6°$

From Table 1, A = 344 Btu/h·ft², B = 0.207, and C = 0.136. Then from Equation (11):

$$I_{DN} = 344/\exp(0.207/\sin 70.6) = 276 \text{ Btu/h·ft}^2$$

Combining Equations (14) and (15) gives

$$I_{d\theta} = 0.136 \times 276 (1 + \cos 30)/2 = 35 \text{ Btu/h·ft}^2$$

Using Equation (12) and assuming $I_r = 0$ gives a total solar irradiation on the collector of:

$$I_{t\theta} = 276 \cos/0.6 + 35 = 306 \text{ Btu/h·ft}^2$$

From Figure 12, $\tau = 0.87$ and $\alpha = 0.96$

From Figure 13, $U_L = 1.3$ Btu/h·ft²·°F

Then from Equation (25):

$$q_u = 306 (0.87 \times 0.96) - 1.3 (140 - 85) = 184 \text{ Btu/h·ft}^2$$

The collector efficiency η is:

$$184/306 = 0.60$$

The general expression for collector efficiency is:

$$\eta = (\tau\alpha)_\theta - U_L (t_P - t_{at})/I_{t\theta} \quad (27)$$

For incident angles below about 35°, the product τ times α is essentially constant and Equation (27) is linear with respect to the parameter $(t_p - t_{at})/I_{t\theta}$, as long as U_L remains constant.

Whillier (ASHRAE 1977) suggested that an additional term F_R be introduced to permit the use of the fluid inlet temperature in Equations (25) and (27) to give:

$$q_u = F_R [I_{t\theta} (\tau\alpha)_\theta - U_L (t_{fi} - t_{at})] \quad (28)$$

$$\eta = F_R (\tau\alpha)_\theta - F_R U_L (t_{fi} - t_{at})/I_{t\theta} \quad (29)$$

where F_R = heat actually delivered by collector/heat that would be delivered if the absorber were actually at t_{fi}. F_R is the *collector heat removal factor* and its value is found from the results of a test performed in accordance with ASHRAE *Standard* 93-1986.

The results of such a test are plotted in Figure 14. When the parameter is zero, because there is no temperature difference between the fluid entering the collector and the atmosphere, the value of the Y-intercept equals $F_R (\tau\alpha)$. The slope of the efficiency line equals the heat loss factor U_L multiplied by F_R. For the single-glazed, nonselective collector with the test results shown in Figure 14, the Y-intercept is 0.82, and the X-intercept is 0.69 °F·ft²·h/Btu. This collector used high transmittance single glazing, τ = 0.91, and the black paint has an absorptance of 0.97, so F_R = 0.82/(0.91 × 0.97) = 0.93.

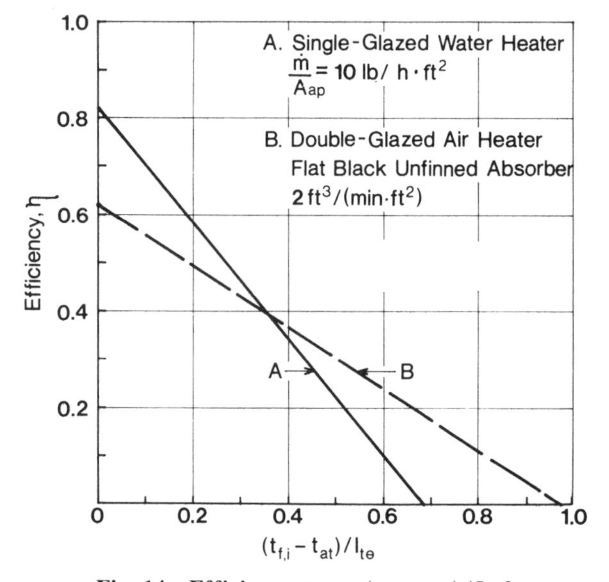

Fig. 14 Efficiency versus $(t_{f,i} - t_{at})/I_{t\theta}$ for Single-Glazed Solar Water Heater and Double-Glazed Solar Air Heater

Assuming that the relationship between η and the parameter is actually linear, as shown, the slope is −0.82/0.69 = −1.19; thus U_L = 1.19/F_R = 1.19/0.93 = 1.28 Btu/h·ft²·°F. The tests for which the results are shown in Figure 14 were run indoors. Factors that affect the measured efficiency are wind speed and fluid velocity.

Figure 14 also shows the efficiency for a double-glazed air heater with an unfinned absorber coated with flat black paint. The Y-intercept for the air heater *B* is considerably lower than it is for the water heater *A* because (1) transmittance of the double glazing used in *B* is lower than the transmittance of the single glazing used in *A*; and (2) F_R is lower for *B* than for *A* because of the lower heat-transfer coefficient between the air and the unfinned metal absorber.

The X-intercept for air heater *B* is higher than it is for the water heater *A* because the upward loss coefficient U_L is much lower for the double-glazed air heater than for the single-glazed water heater. The data for both *A* and *B* were taken at near-normal incidence with high values of $I_{t\theta}$. For Example 5, using a single-glazed water heater, the value of the parameter would be close to (140 − 85)/307 = 0.18 °F·ft²·h/Btu, and the expected efficiency, 0.60, agrees closely with the test results.

As ASHRAE *Standard* 93-1986 shows, the incident angles encountered with south-facing tilted collectors vary widely throughout the year. Considering a surface located at 40° north latitude with a tilt angle Σ = 40°, the incident angle θ will depend on the time of day and the declination δ. On December 21, δ = 23.456°; at 4 h before and after solar noon, the incident angle is 62.7°, and it remains close to this value for the same solar time throughout the year. The total irradiation at these conditions varies from a low of 45 Btu/h·ft² on December 21 to approximately 140 Btu/h·ft² throughout most of the other months.

When the irradiation is below about 100 Btu/h·ft², the losses from the collector may exceed the heat that can be absorbed. This situation varies with the temperature difference between the collector inlet temperature and the ambient air, as suggested by Equation (28).

When the incident angle rises above 30°, the product of the transmittance of the glazing and the absorptance of the collector plate begins to diminish; thus, the heat absorbed also drops. The losses from the collector are generally higher as the time moves farther from solar noon, and consequently the efficiency also drops. Thus, the daylong efficiency is lower than the near-noon performance. During the early afternoon hours, the efficiency is slightly higher than at the comparable morning time, because ambient air temperatures are lower in the morning than in the afternoon.

ASHRAE *Standard* 93-1986 describes the *incident angle modifier,* which may be found by tests run when the incident angle is set at 30, 45, and 60°. Simon (1976) showed that for many flat-plate collectors, the incident angle modifier is a linear function of the quantity (1/cos θ − 1). For evacuated tubular collectors, the incident angle modifier may grow with rising values of θ.

ASHRAE *Standard* 93-1986 specifies that the efficiency be reported in terms of the gross collector area A_g rather than the aperture area A_{ap}. The reported efficiency will be lower than the efficiency given by Equation (29), but the total energy collected is not changed by this simplification:

$$\eta A_g = \eta_{ap} A_{ap}/A_g \quad (30)$$

HEAT STORAGE SYSTEMS

Storage may be part of solar heating, cooling, and power generating systems. For approximately one-half of the 8760 h per year any location is in darkness, heat storage is necessary if the system must operate continuously. For some applications, such as swimming pool heating, daytime air heating, and irrigation pumping, intermittent operation is acceptable, but most other uses of solar energy require operating at night and when the sun is obscured by clouds. Chapter 39 provides further information on general thermal storage technologies.

Packed Rock Beds

A packed bed, pebble bed, or rock pile storage uses the heat capacity of a bed of loosely packed particles through which a fluid, usually air, is circulated to add heat to, or remove heat from, the bed. While a variety of solids may be used, rock of 0.75 to 2 in. in size is most common. Well-designed packed rock beds have several desirable characteristics for energy storage. The heat transfer coefficient between the air and the solid is high, the cost of the storage material is low, the conductivity of the bed is low when airflow is not present, and a large heat transfer area is achieved at low cost by using small storage particles. Figure 15 is a schematic of a packed bed storage unit. Essential features include a container, a porous structure (such as a wirescreen) to support the bed, and air plenums at both inlet and outlet. The container can be made of wood, concrete blocks, or poured concrete. Insulation requirements outside the bed are small for short-term storage, because the lateral thermal conductivity of the bed is low. For a rock bed used for daily cycling, the 24-h heat loss should not exceed 5%; it should preferably be only 2%.

Fluid flows through the bed in one direction during the addition of heat and in the opposite direction during the removal of heat. Therefore, heat cannot be added to and removed from these storage devices simultaneously. Packed bed storage devices operated in this manner thermally stratify naturally. For efficiency, the bed temperature should rise during downward heat flow and lower during upward heat flow. This flow (hot top, cool bottom) inhibits convection currents during standby periods, which decreases storage efficiency. To prevent channeling (in which a narrow jet of heat transfer fluid travels from inlet to outlet without sweeping through the entire bed), both the inlet and outlet should have plenum chambers. To further equalize pressure drops along all flow paths, it is best to arrange inlet and outlet on opposite sides of the storage bed (see Figure 15).

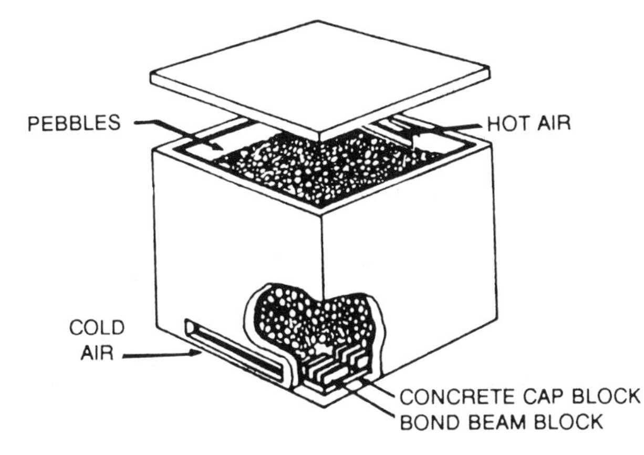

Fig. 15 Vertical Flow Packed Rock Bed

The plenum chambers distribute the flow over the entire cross section of the bed. Tests by Jones and Loss (1982) on a packed bed like the one in Figure 15 showed that the inlet jet shoots across the rock top, hits the far wall, and forms two counter-rotating flow cells along the side walls. The flow distribution is improved by placing a block slightly larger than the inlet opening inside the top plenum at a distance of 1.5 times the vertical inlet dimension.

Because of the large surface areas of the pebbles exposed to the air passing through the bed, pebble beds exhibit good heat transfer between the air and the storage medium. An empirical relation for the volumetric heat transfer coefficient h_v is given by the following equation (Lof and Hawley 1948):

$$h_v = 0.79\,(G/D)^{0.7} \tag{31}$$

$$D = [(6/\pi)\,(\text{net volume of particles})/(\text{number of particles})]^{1/3} \tag{32}$$

where

h_v = volumetric heat transfer coefficient, Btu/h · ft³ · °F
G = superficial mass velocity, lb/ft² · h
D = equivalent spherical diameter of the particles, ft

The superficial mass velocity is the mass flow per unit time through the rock bed per unit area of bed face.

The particle size in a packed bed should be uniform to obtain a large void fraction and thus minimize the pressure drop through the bed. Figure 16 (Cole *et al.* 1980) shows the variation of that pressure drop as a function of face velocity and particle size diameter. Analytical or numerical methods can be used to evaluate the performance of packed bed storage devices, but the calculations for an arbitrary time-dependent inlet air temperature variation is laborious. Hughes *et al.* (1976) and Mumma and Marvin (1976) give methods of solution.

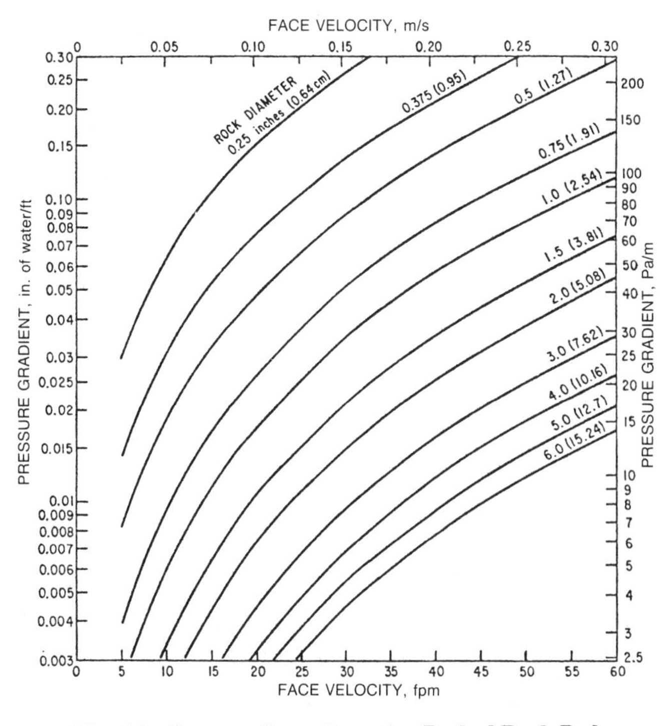

Fig. 16 Pressure Drop through a Packed Rock Bed
(Cole *et al.* 1980)

SOLAR WATER HEATING SYSTEMS

A solar water heater includes a solar collector array that absorbs solar radiation and converts it to heat, which is then absorbed by a heat transfer fluid (water, a nonfreezing liquid, or air) that passes through the collector. The heat transfer fluid's heat can be stored or used directly.

Since portions of the solar energy system are exposed to the weather, they must be protected from freezing. The systems must also be protected from overheating caused by high insolation levels during periods of low energy demand.

In solar water heating systems, potable water heated directly in the collector, or indirectly by a heat transfer fluid that is heated in the collector, passes through a heat exchanger, and transfers its heat to the domestic or service water. The heat transfer fluid is transported by either natural or forced circulation. Natural circulation occurs by natural convection (thermosiphoning), whereas forced circulation uses pumps or fans. Except for thermosiphon systems, which need no control, solar domestic and service hot water systems are controlled using differential thermostats.

Six types of solar energy systems that are used to heat domestic and service hot water are: thermosiphon, direct circulation, drain-back, indirect, integral collector storage, and air. The term recirculation has been used to describe a freeze protection method for direct solar water heating systems. Similarly, the term drain-down has also been used to describe a freeze protection method for direct circulation systems.

Thermosiphon Systems

Thermosiphon systems (Figure 17) heat potable water or a heat transfer fluid and use natural convection to transport it from the collector to storage. For direct systems, pressure-reducing valves

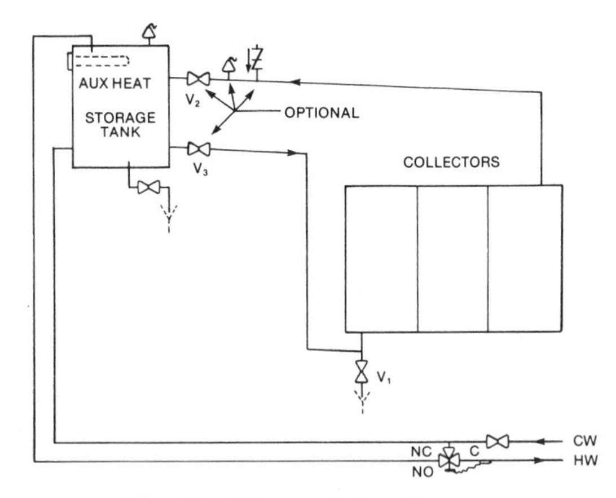

Fig. 17 Thermosiphon System

are required when city water pressure is greater than the working pressure of the collectors. In a thermosiphon system, the storage tank must be elevated above the collectors, which sometimes requires designing the upper level floor and ceiling joists to bear this additional load. Also, extremely hard or acidic water can cause scale deposits that clog or corrode the absorber fluid passages.

Since thermosiphon flow is induced whenever there is sufficient sunshine, these systems do not need pumps.

Direct Circulation Systems

These systems (Figure 18) are direct water heating systems that pump potable water from storage to the collectors when there is enough solar energy available to warm it, and then return it to the storage tank until needed. Since a pump circulates the water, the collectors can be mounted either above or below the storage tank. Direct circulation systems are feasible only in areas where freezing is infrequent. Freeze protection for extreme weather conditions is provided either by recirculating warm water from the storage tank or by flushing the collectors with cold water. Direct water heating systems should not be used in areas where water is extremely hard or acidic. Scale deposits may clog or corrode the absorber fluid passages, rendering the system inoperable.

This type of system is exposed to city waterline pressures and must be assembled to withstand test pressures, as required by local codes. Pressure-reducing valves and pressure relief valves are required when the city water pressure is greater than the working pressure of the collectors. Direct circulation systems often use a single storage tank for both solar energy storage and the auxiliary water heater, but two-tank storage systems can be used.

Drain-down systems (Figure 19) are direct circulation, water heating systems in which potable water is pumped from storage to the collector array where it is heated. Circulation continues until usable solar heat is no longer available. When a freezing condition is anticipated or a power outage occurs, the system drains automatically by isolating the collector array and exterior piping from city water pressure and draining it using one or more valves. The solar collectors and associated piping *must* be carefully sloped to drain the collector's exterior piping.

This type of system is exposed to city water pressures and must be assembled to withstand test pressures, as required by local codes. Pressure-reducing valves and pressure relief valves are required when city water pressure is greater than the working pressure of the collectors. One- or two-tank storage systems can be used. Scale deposits and corrosion can occur in the collectors with hard or acidic water.

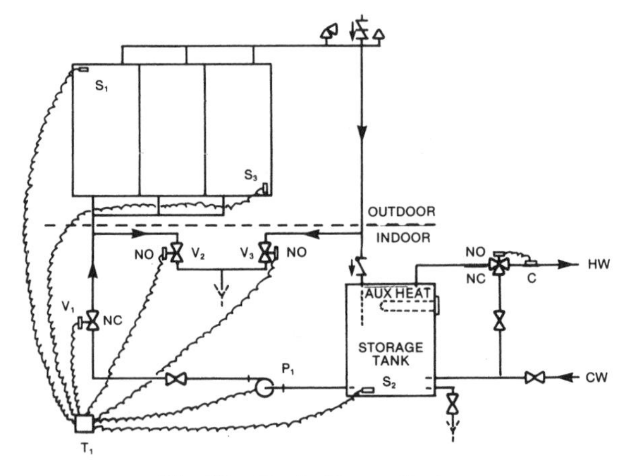

Fig. 19 Drain-Down System

Drain-Back Systems

Drain-back systems are generally indirect water heating systems that circulate treated or untreated water through the closed collector loop to a heat exchanger, where its heat is transferred to the potable water. Circulation continues until usable energy is no longer available. When the pump stops, the collector fluid drains by gravity to a storage tank or drain-back tank. In a pressurized system, the tank also serves as an expansion tank when the system is operating and must be protected from excessive pressure with a temperature and pressure relief valve. In an unpressurized system (Figure 20), the tank is open and vented to the atmosphere.

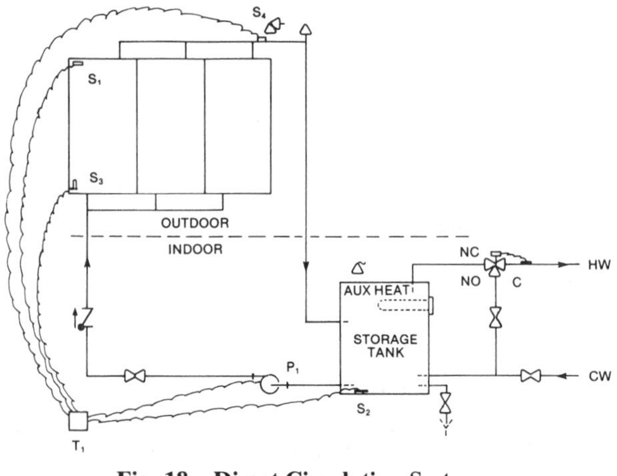

Fig. 18 Direct Circulation System

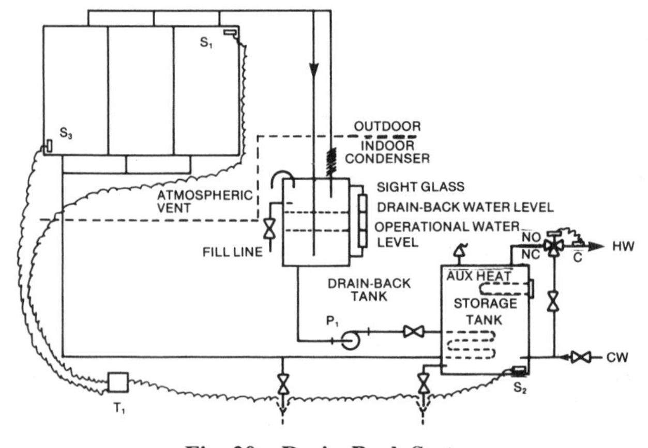

Fig. 20 Drain-Back System

Since the collector loop is isolated from the potable water, valves are not needed to actuate draining, and scaling is not a problem. The collector array and exterior piping must be sloped to drain completely and pumping pressure must be sufficient to lift water to the top of the collector array.

Indirect Water Heating Systems

Indirect water heating systems (Figure 21) circulate a freeze-protected heat transfer fluid through the closed collector loop to a heat exchanger, where its heat is transferred to the potable water. The most commonly used heat transfer fluids are water/ethylene glycol and water/propylene glycol solutions, although other heat transfer fluids such as silicone oils, hydrocarbons, or refrigerants can also be used (ASHRAE 1983). These fluids are nonpotable, sometimes toxic, and normally require double wall heat exchangers. The double wall heat exchanger can be located inside the storage tank, or an external heat exchanger can be used. The collector loop is closed and therefore requires an expansion tank and a pressure relief valve. A one- or two-tank storage system can be used. Additional over-temperature protection may also be needed to prevent the collector fluid from decomposing or becoming corrosive.

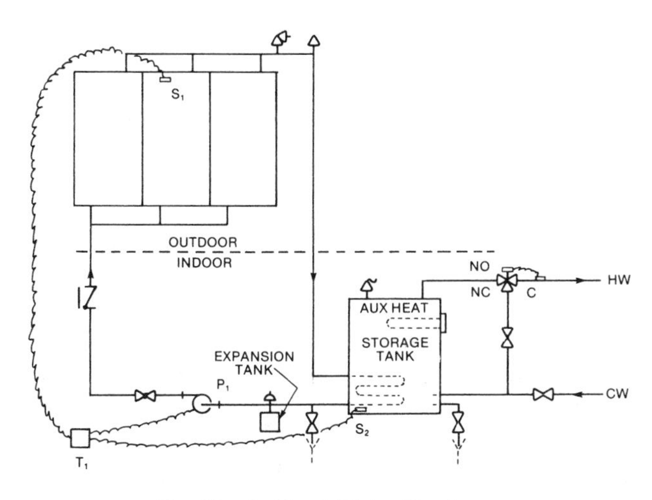

Fig. 21 Indirect Water Heating

Designers should avoid automatic water makeup in systems using water/antifreeze solutions because a significant leak may induce enough water into the system to raise the freezing temperature of the solution above the ambient temperature, causing the collector array and exterior piping to freeze. Also, antifreeze systems with large collector arrays and long pipe runs may need a time-delayed bypass loop around the heat exchanger to avoid freezing the heat exchanger on startup.

Integral Collector Storage Systems

Integral collector storage (ICS) systems use hot water storage as part of the collector. Some types use the surface of a single tank as the absorber, and others use multiple, long, thin tanks placed side-by-side horizontally to form the absorber surface. In this type of ICS, hot water is drawn from the top tank and cold replacement water enters the bottom tank. Because of the greater nighttime heat loss from ICS systems, they are typically less efficient than pumped systems, and selective surfaces are recommended. ICS systems are normally installed as a solar preheater without pumps or controllers. Flow through the ICS system occurs on demand, as hot water flows from the collector to a standard hot water auxiliary tank within the structure.

Air Systems

Air systems (Figure 22) are indirect water heating systems that circulate air through the collectors via ductwork to an air-to-liquid heat exchanger. There, heat is transferred to the potable water, which is pumped through the tubes of the exchanger and returned to the storage tank. Circulation continues as long as usable heat is available. Air systems can use single or double storage tank configurations. The two-tank storage system is used most often, since air systems are generally used for preheating domestic hot water and may not be capable of reaching 120 to 160 °F delivery temperatures.

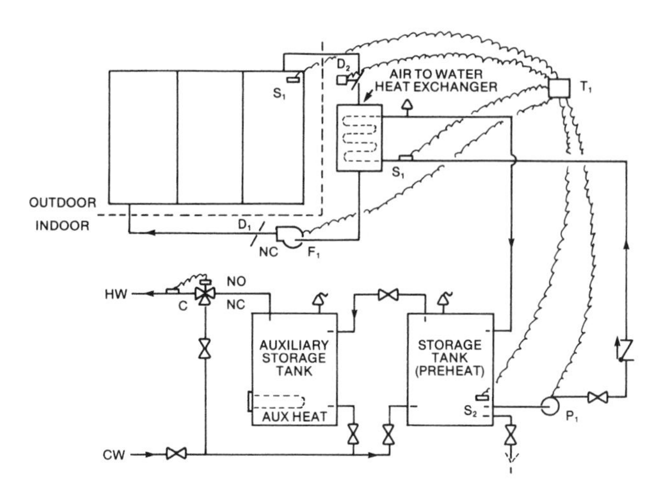

Fig. 22 Air System

Air does not need to be protected from freezing or boiling, is noncorrosive, and is free. However, air ducts and air-handling equipment need more space than piping and pumps. Ductwork is laborious to seal, and air leaks are difficult to detect. Power consumption is generally higher than that of a liquid system because of high collector and heat exchanger static pressure loss. All dampers installed in air systems must fit tightly to prevent leakage and heat loss.

In areas with freezing temperatures, tight dampers are needed in the collector ducts to prevent reverse thermosiphoning at night, which could freeze the water in the heat exchanger coil. No special precautions are needed to control overheating conditions in air systems.

Pool Heating Systems

Solar pool heating systems require no separate storage tank, since the pool itself serves as storage. In most cases, the pool's filtration pump forces the water through the solar panels or plastic pipes. In some retrofit applications, a larger pump may be required to handle the needs of the solar system, or a small pump is added to boost the pool water to the solar collectors.

Automatic controls may be used to direct the flow of filtered water to the collectors when solar heat is available; this may also be accomplished manually. Normally, solar systems are designed to drain down into the pool when the pump is turned off, which provides the collectors with freeze protection.

Four primary types of collector designs used for swimming pool heat are: (1) rigid black plastic panels (polypropylene), usually 4 ft by 10 ft or 4 ft by 8 ft; (2) tube-on-sheet panels, which usually have a metal deck (copper or aluminum) with copper water tubes; (3) EPDM rubber mat, extruded with the water passages running its length; and (4) arrays of black plastic pipe, usually 1.5-in. diameter ABS plastic (Root *et al.* 1985).

DOMESTIC HOT WATER RECIRCULATION SYSTEMS

Domestic hot water (DHW) recirculation systems (Figures 23 and 24), which continuously circulate domestic hot water throughout a building, are common in motels, hotels, hospitals, dormitories, office buildings, and other commercial buildings. The recirculation heat losses in these systems are usually a significant part of the total water heating load. A properly integrated solar energy system can make up much of this heat loss.

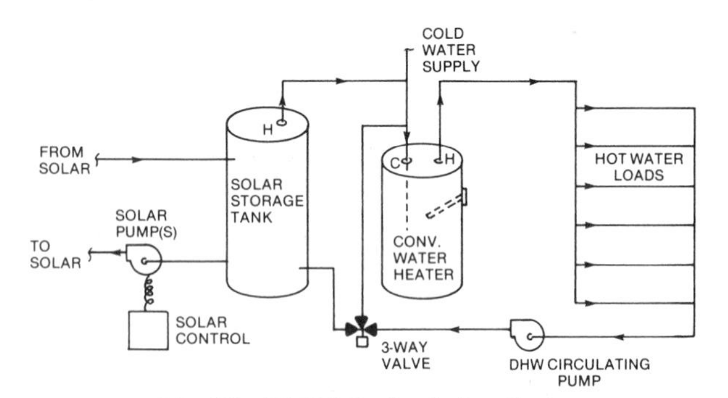

Fig. 23 DHW Recirculation System

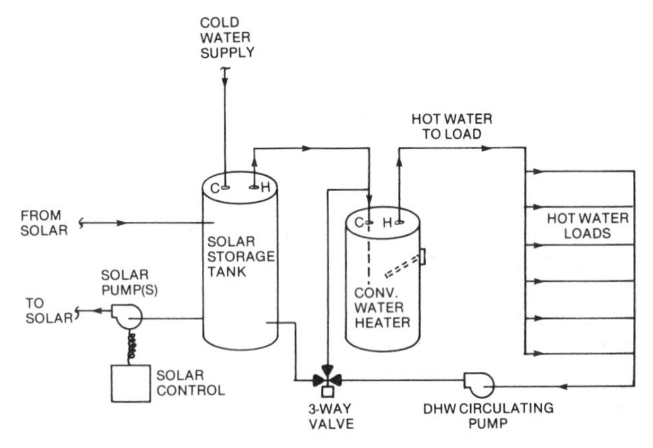

Fig. 24 DHW Recirculation System with Makeup Preheat

SYSTEM COMPONENTS

A solar domestic hot water system has a variety of components, which are combined, depending on function, component compatibility, climate conditions, required performance, site characteristics, and architectural requirements. For optimum system performance, the components of a solar energy system must fit and function together satisfactorily. The system must also be properly integrated into the conventional domestic or service hot water system. The major components involved in the collection, storage, transportation, control, and distribution of solar energy are discussed in this section.

Collectors

Flat-plate collectors are most commonly used in water heating applications because of the year-round load requiring temperatures of 80 to 180 °F. For discussions of other collectors and applications, see ASHRAE *Standard* 93-1986 and previous sections of this chapter. Collectors must withstand extreme weather conditions (such as freezing, stagnation, and wind) as well as system pressures.

Heat Transfer Fluid

Heat transfer fluids transport heat from the solar collectors to the domestic water. Potential safety problems exist in the transfer of heat energy from solar collectors to potable hot water supplies. The problems are both chemical and mechanical in nature and apply primarily to liquid transfer and storage systems in which a heat exchanger interface exists with the potable water supply. Both the chemical compositions of the heat transfer fluids (pH, toxicity, and chemical durability), as well as their mechanical properties (specific heat and viscosity) must be considered.

Except in unusual cases or when potable water is being circulated, the energy transport fluid is nonpotable and has the potential for contaminating potable water during the heat transfer process. Even potable or nontoxic fluids in closed systems are likely to become nonpotable because of contamination from metal piping, solder joints, and packing, or by the inadvertent installation of a toxic fluid at a later date.

Trade-offs between thermal efficiency, cost effectiveness, and risk for both heat transfer fluids and heat exchangers may be necessary to provide acceptable safety. Existing codes regulate the need and design of heat exchangers.

Thermal Energy Storage

Thermal energy (heat) storage in hydronic solar domestic and service heating systems is virtually always liquid stored in tanks. All storage tanks and bins should be well insulated so that the collected heat is not lost to the surroundings, which are normally unoccupied.

In domestic hot water systems, thermal energy is usually stored in one or two tanks. The hot water outlet is at the top of the tank and cold water enters the tank through a dip tube that extends down to within 4 to 6 in. of the tank bottom. The outlet on the tank to the collector loop should be approximately 4 in. above the tank bottom to prevent scale deposits from being drawn into the collectors. Water from the collector array returns to the upper portion of the storage tank. This plumbing arrangement may take advantage of thermal stratification, depending on the delivery temperature from the collectors and the flow rate through the storage tank.

Single-tank electric auxiliary systems often incorporate storage and auxiliary heating within the same vessel. Conventional electric water heaters commonly have two heating elements: one near the top and one near the bottom. If a dual element tank is used in a solar energy system, the bottom element should be disconnected and the top left functional to take advantage of fluid stratification. Standard gas- and oil-fired water heaters should not be used in single-tank arrangements. In gas and oil water heaters, heat is added to the bottom of the tanks, which reduces both stratification and collection efficiency in single-tank systems.

Dual-tank systems often use the solar domestic hot water storage tank as a preheat tank. The second tank is normally a conventional domestic hot water system tank and contains the auxiliary heat source. Multiple tank systems are sometimes used in large institutions, where they operate similarly to dual-tank systems. Using a two-tank system may increase collector efficiency and the fraction of the total heating load supplied by the solar energy system (solar fraction). However, it increases tank heat losses. The water inlet to these tanks is usually a dip tube that extends close to the bottom of the tank.

Estimates for sizing storage tanks usually range from 1 to 2.5 gal/ft^2 of the solar collector area. The most often used figure is 1.8 gal/ft^2 of collector area, which will usually provide enough heat for a sunless period of a day or so. Storage volume should be analyzed and sized according to the project water requirements and draw schedule; however, solar applications typically require larger tanks than would normally be used.

Some electric utilities offer reduced electric rates to customers willing to charge their tanks during off-peak hours only. The charge time is controlled by a time clock or by a signal sent by the utility either over its power lines (ripple control) or by radio.

Depending on the nature of the hot water draw by the individual user, a tank larger than the one normally used is recommended. Thus solar heating is an ideal adjunct to this off-peak water heating application.

Heat Exchangers

Heat exchangers transfer thermal energy (heat) between two fluids. All solar energy systems using indirect water heating require one or more heat exchangers. Potential contamination problems exist in the transfer of heat energy from solar collectors to potable hot water supplies.

Heat exchangers influence the effectiveness with which collected energy is made available to heat domestic water. They also separate and protect the potable water supply from contamination when nonpotable heat transfer fluids are used. Like transport fluid selection, heat exchanger selection considers thermal performance, cost effectiveness, reliability and safety, and the following characteristics:

- Heat exchange effectiveness
- Pressure drop, operating power, and flow rate
- Physical design, design pressure, configuration, size, materials, and location in the system
- Cost and availability
- Reliable protection of the potable water supply from contamination by the heat transfer fluids
- Leak detection, inspection, and maintainability
- Material compatibility with other system elements such as metals and fluids
- Thermal compatibility with system design parameters such as operating temperatures, flow rate, and fluid thermal properties

Heat exchanger selection depends on the characteristics of the fluids that pass through the heat exchanger and the properties of the exchanger itself. Fluid characteristics to consider are fluid type, temperature, specific heat, and mass flow rate. Physical properties of the heat exchanger are the overall heat transfer coefficient of the heat exchanger and the heat transfer surface area. When these variables are known, the heat transfer rate can be determined. The heat transfer fluid temperatures of interest are the hot and cold fluid inlet temperatures and the hot and cold fluid outlet temperatures.

For most solar domestic hot water system designs, only the hot and cold inlet temperature are known; the other temperatures must be calculated, which requires knowing the physical properties of the heat exchanger.

This information can be used to evaluate two quantities that are useful in determining the heat transfer in a heat exchanger and the performance characteristics of the collector when combined with a given heat exchanger. These quantities are (1) the fluid capacitance rate, which for a given fluid is the product of the mass flow rate and the specific heat of the fluid passing through the heat exchanger, and (2) heat exchanger effectiveness, which relates the capacitance rate of the two fluids to the fluid inlet and outlet temperatures. The effectiveness is equal to the ratio of the actual heat transfer rate to the maximum heat transfer rate theoretically possible. Generally, a heat exchanger effectiveness of 0.4 or greater is recommended.

Expansion Tanks. Indirect solar water heating systems operating in a closed collector loop require an expansion tank to control excessive pressure increases. Fluid in solar collectors under stagnation conditions can easily boil, resulting in excessive pressure increases in the collector loop. Solar system expansion tanks must be sized to account for this design condition. The ASHRAE expansion tank sizing formulas for closed loop hydronic systems found in Chapter 13 of the 1987 ASHRAE *Handbook—HVAC Systems and Applications*, may be used for solar system expansion tank sizing, but the expression for volume change due to temperature increase should be replaced with the total volume of fluid in the solar collectors and any piping located above the collectors, if significant. This sizing method provides a passive means for eliminating fluid loss due to over temperature or stagnation, a common problem in closed loop solar systems. This results in expansion tanks larger than those typically found in hydronic systems, although the increase in cost is small compared to the savings in fluid replacement and maintenance costs (Lister and Newell 1989).

Pumps. Pumps circulate heat transfer liquid through collectors and heat exchangers. The two most commonly used fluid circulators are centrifugal pumps and positive displacement pumps. In solar domestic hot water systems, the pump is usually a centrifugal circulator driven by a small fractional horsepower motor. The flow rate for collectors generally ranges from 0.015 to 0.04 gpm/ft^2.

Pumps used in drain-back systems must provide the lift head up to the collectors in addition to the friction head. These heads must be added for systems with open return piping, and the pump should be able to satisfy both head requirements.

System Piping. Piping can be plastic, copper, galvanized steel, or stainless steel. The most widely used is sweat-soldered L-type copper tubing. M-type copper is also acceptable if permitted by local building codes. If water/glycol is the heat transfer fluid, galvanized pipes or tanks must not be used because unfavorable chemical reactions will occur; copper piping is recommended instead. Also, if glycol solutions or silicone fluids are used, they may leak through joints where water would not. System piping should be compatible with the collector fluid passage material, *e.g.*, copper or plastic piping should be used with collectors having copper fluid passages.

Piping that carries potable water can be plastic, copper, galvanized steel, or stainless steel. In indirect systems, corrosion inhibitors must be checked and adjusted routinely, preferably every three months. Inhibitors should also be checked if the system overheats during stagnation conditions. If dissimilar metals are joined, use dielectric or nonmetallic couplings. The best protection is sacrificial anodes or getters in the fluid stream. Their location depends on the material to be protected, the anode material, and the electrical conductivity of the heat transfer fluid. Sacrificial anodes consisting of magnesium, zinc, or aluminum are often used to reduce corrosion in storage tanks. Because there are so many possibilities, each combination must be evaluated on its own merits. A copper-aluminum or copper-galvanized steel joint is unacceptable because of severe galvanic corrosion. Systems containing aluminum, copper, and iron metals have a greatly increased potential for corrosion.

Insufficient consideration of air elimination requirements, pipe expansion control, and piping slope can cause serious system failures. Collector pipes (particularly manifolds) should be designed to allow expansion from stagnation temperatures to extreme cold weather temperatures. Expansion control can be achieved with offset elbows in piping, hoses, or expansion couplings. Expansion loops should be avoided unless they are installed horizontally, particularly in systems that must drain for freeze protection. The collector array piping should slope 0.06 in. per foot for drainage (DOE 1978).

Air can be eliminated by placing air vents at all piping high points and by air purging during filling. Flow control, isolation, and other valves in the collector piping must be chosen carefully so that these components do not restrict drainage significantly or back up water behind them. The collectors must drain completely.

Valves and Gages. Valves in solar domestic hot water systems must be located to ensure system efficiency, satisfactory performance, and the safety of equipment and personnel. Drain valves must be ball-type; gate valves may be used if the stem is installed horizontally.

Check valves or other valves used for freeze protection or for reverse thermosiphoning must be reliable to avoid significant damage.

Auxiliary Heat Sources. On sunny days, a typical solar energy system should supply water at a predetermined temperature, and the solar storage tank should be large enough to hold sufficient water for a day or two. Because of the intermittent nature of solar radiation, an auxiliary heater must be installed to handle hot water requirements. Operation of the auxiliary heater can be timed to take advantage of off-peak utility rates, if a utility is the source of auxiliary energy. The auxiliary heater should be carefully integrated with the solar energy system to obtain maximum solar energy use. For example, the auxiliary heater should not destroy any stratification that may exist in the solar-heated storage tank, which would reduce collector efficiency.

Fans. Fans circulate air in air systems by forcing air that has been heated in the solar collectors through ductwork to an air-to-water heat exchanger. The heat from the air is then transferred to the water that is pumped through the coil section of the heat exchanger. For detailed information on fan performance and selection, refer to Chapter 3 of the 1988 ASHRAE *Handbook— Equipment*.

Ductwork, particularly in systems with air-type collectors, must be sealed carefully to avoid leakage in duct seams, damper shafts, collectors, and heat exchangers. Ducts should be sized using conventional air duct design methods.

Control Systems. Control systems regulate solar energy collection by controlling fluid circulation, activate system protection against freezing and overheating, and initiate auxiliary heating when it is required.

The three major control components are sensors, controllers, and actuated devices. Sensors detect conditions or measure quantities, such as temperatures. Controllers receive output from the sensors, select a course of action, and signal a system component to adjust the condition. Actuated devices are components, such as pumps, valves, dampers, and fans, that execute controller commands and regulate the system.

Temperature sensors measure the temperature of the absorber plate near the collector outlet and near the bottom of the storage tank. The sensors send signals to a controller, such as a differential temperature thermostat, for interpretation.

The differential thermostat compares the signals from the sensors with adjustable set points for high and low temperature differentials. The controller performs different functions, depending on which set points are met. In liquid systems, when the temperature difference between the collector and storage reaches a high set point, usually 20 °F, the pump starts, automatic valves are activated, and circulation begins. When the temperature difference reaches a low set point, usually 4 °F, the pump is shut off and the valves are deenergized and returned to their normal positions. To restart the system, the high-temperature set point must again be met. If the system has either freeze or overtemperature protection, the controller opens or closes valves or dampers and starts or stops pumps or fans to protect the system when its sensors detect that either a freezing or an overheating condition is about to occur.

Collector loop sensors can be located on the absorber plate, in a pipe above the collector array, on a pipe near the collector, or in the collector outlet passage. Sensors must be selected to withstand high temperatures, such as those that may occur during collector stagnation. Although any of these locations may be acceptable, attaching the sensor on the collector absorber plate is recommended. When attached properly, it gives accurate readings, can be installed easily, and is basically unaffected by ambient temperature, as are sensors mounted on exterior piping.

A sensor installed on an absorber plate reads temperatures 3 to 4 °F higher than the temperature of the fluid leaving the collector. However, such temperature discrepancies can be compensated for in the differential thermostat settings.

The sensor must be attached to the absorber plate with good thermal contact. If a sensor access cover is provided on the enclosure, it must be gasketed for a watertight fit. Adhesives and adhesive tapes should not be used to attach the sensor to the absorber plate.

The storage temperature sensor should be near the bottom of the storage tank to detect the temperature of the fluid before it is pumped to the collector or the heat exchanger. The storage fluid is usually coldest at that location because of thermal stratification and the location of the makeup water supply. The sensor should be either securely attached to the tank and well insulated, or immersed inside the tank near the collector supply.

The freeze protection sensor, if required, should be located so that it will detect the coldest liquid temperature when the collector is shut down. Common locations are the back of the absorber plate at the bottom of the collector, the collector intake or return manifolds, or attached to the center of the absorber plate. The center absorber plate location is recommended because reradiation to the night sky will freeze the collector heat transfer fluid, even though the ambient temperature is above freezing. Some systems, such as the recirculation system, have two sensors for freeze protection, while others, such as the drain-down, use only one.

Control on-off temperature differentials affect system efficiency. The turn-on temperature differential must be selected properly because, if the differential is too high, the system starts later than it should, and if it is too low, the system starts too soon. The turn-on differential for liquid systems usually ranges from 15 to 30 °F and is most commonly 20 °F. For air systems, the range is usually 25 to 45 °F.

The turn-off temperature differential is more difficult to estimate. Selection depends on a comparison between the value of the energy collected and the cost of collecting it. It varies with individual systems, but a value of 4 °F is typical.

Water temperature in the collector loop depends on ambient temperature, solar radiation, radiation from the collector to the night sky, and collector loop insulation. Freeze protection sensors should be set to detect temperatures of 40 °F.

Sensors are an important but often overlooked control system component. They must be selected and installed properly because no control system can produce accurate outputs from unreliable sensor inputs. Sensors are used in conjunction with a differential temperature controller and are usually supplied by the controller manufacturer. Sensors must survive the anticipated operating conditions without physical damage or loss of accuracy. Low-voltage sensor circuits must be located away from 120/240 VAC lines to avoid electromagnetic interference. Any sensors attached to the collector should be able to withstand stagnation temperatures.

Sensor calibration, which is often overlooked by installers and maintenance personnel, is critical to system performance; a routine calibration maintenance schedule is essential.

Another control system is the photovoltaic (PV) pumped system. A PV panel provides direct current (DC) electricity converted from sunlight directly to a small DC circulating pump. No additional sensing is required as the PV panel and pump output increase with sunlight intensity and stop when no sunlight (collector energy) is available. Cromer (1984) has shown that with proper matching of pump and PV electrical characteristics, PV panel sizes as low as 5 W per 40 ft^2 of thermal panel may be used successfully. Difficulty with late starting and running too long in the afternoon can be alleviated by tilting the PV panel slightly to the east.

System Performance Evaluation Methods

The performance of any solar energy system is directly related to (1) heating load requirements, (2) the amount of solar radiation available, and (3) the solar energy system characteristics. Various calculation methods use different procedures and data when considering available solar radiation. Some simplified methods consider only average annual incident solar radiation, while complex methods use hourly data.

Solar energy system characteristics, as well as individual component characteristics, are required to evaluate performance. The degree of complexity with which these systems and components are described varies from system to system.

The cost effectiveness of a solar domestic and service hot water heating system depends on initial cost and energy cost savings. A major task is to determine how much energy is saved. The *annual solar fraction*—the annual solar contribution to the water heating load divided by the total water heating load—can be used to estimate these savings. It is expressed as a decimal fraction or percentage and generally ranges from 0.3 to 0.8 (30 to 80%), although more extreme values are possible.

Analysis and Economic Evaluation Methods

System performance and economic evaluation methods are generally categorized according to complexity and include (1) hand calculation methods, (2) simplified computer methods, and (3) complex hour-by-hour computer methods.

Water Heating Load Requirements

The amount of water required must be estimated as accurately as possible because it affects system component selection. Chapter 44 gives methods to determine the load.

Oversized storage may result in low-temperature water that requires auxiliary heating to reach a desired supply temperature. Undersizing can prevent the collection and use of available solar energy.

Integration of the solar energy system into a recirculating hot water system, if one is used, must also be considered carefully to maximize solar energy use (see Figures 23 and 24).

COOLING BY SOLAR ENERGY

A review of solar-powered refrigeration by Swartman (1974) emphasizes various absorption systems. Newton (Jordan and Liu 1977) discusses commercially available water vapor/lithium bromide absorption refrigeration systems. Standard absorption chillers are generally designed to give rated capacity for activating fluid temperatures well above 200 °F at full load and design condenser water temperature. Few flat-plate collectors can operate efficiently in this range; therefore, lower hot fluid temperatures are used when solar energy provides the heat. Both the temperature of the condenser water and the percentage of design load are determinants of the optimum energizing temperature, which can be quite low, sometimes below 120 °F. Proper control can raise the coefficient of performance (COP) at these part-load conditions.

Many large commercial or institutional cooling installations must operate year-round, and Newton (Jordon and Liu 1977) showed that the low-temperature cooling water available in winter enables the LiBr − H_2O to function well with hot fluid inlet temperatures below 190 °F. Residential chillers in sizes as low as 1.5 tons, with inlet temperatures in the range of 175 °F, have been developed.

COOLING BY NOCTURNAL RADIATION AND EVAPORATION

Bliss (1961) used unglazed tube-in-strip collectors mounted on south-facing roofs to collect solar heat in winter and to reject heat in summer by radiation to the night sky. He used radiant ceilings with water tanks for storage and heat pumps to raise or lower the temperature of the circulating water as needed. Nocturnal radiation can produce cooling at the rate of 20 to 30 $Btu/h \cdot ft^2$ during nights when the dew-point temperature is low. Radiation cooling effectiveness is greatly reduced when the dew-point temperature is high; little radiant cooling can be accomplished under overcast skies. Evaporation must be used when high rates of heat dissipation are needed.

In Australia, school cooling systems that evaporate water in the discharge air to chill the rocks in a switched bed rock-filled recuperator have operated successfully. The incoming fresh air required for ventilation in Australian schools is generally 20 to 25 cfm per person, the latter for infant and primary schools. With the rock bed regenerator (RBR) system, 100% makeup air is used during the summer months, and the ventilation rate per pupil ranges from 28 to 85 cfm.

In the evaporative RBR, the incoming air cools as it passes through rocks chilled by the evaporatively cooled exhaust air that flows through the RBR. The airflow is switched every 10 min. Water sprays for only 10 s at the beginning of each cycle, and this is adequate to provide the necessary cooling. Power for the two fans used in the 2000 cfm system is only 600 W, because the pressure drop through the 5-in. deep pebble bed (0.25-in. mesh screen) is only 0.13 in. of water.

The conventional evaporative cooler is used to cool industrial buildings and low-cost housing, but the high indoor humidities caused by direct evaporative cooling produce nearly as much discomfort as the high temperatures with which it contends.

SOLAR HEATING AND COOLING SYSTEMS

The components and subsystems discussed earlier may be combined to create a wide variety of solar heating and cooling systems. There are two principal categories of such systems: passive and active. *Passive* systems require little, if any, nonrenewable energy to make them function (Yellott 1977, Yellott *et al.* 1976). Every building is passive in the sense that the sun tends to warm it by day and cool it at night. *Active* systems must have a continuous availability of nonrenewable energy, generally in the form of electricity, to operate pumps and fans. *Hybrid* systems require some nonrenewable energy, but the amount is so small that they can maintain a coefficient of performance of about 50.

Passive Systems

Passive systems may be divided into several categories. The first residence to which the name *solar house* was applied used a large expanse of south-facing glass to admit solar radiation; this was known as a *direct gain* passive system.

Indirect gain solar houses use the south-facing wall surface or the roof of the structure to absorb solar radiation, which causes a rise in temperature, which, in turn, conveys heat into the building in several ways. This principle was applied to the pueblos and cliff dwellings of the southwestern United States. Glass has led to modern adaptations of the indirect gain principle (Trombe 1977, Balcomb *et al.* 1977).

By glazing a large south-facing, massive masonry wall, solar energy can be absorbed during the day, and conduction of heat to the inner surface provides radiant heating at night. The mass of the wall and its relatively low thermal diffusivity delays the arrival of the heat at the indoor surface until it is needed. The glazing reduces the loss of heat from the wall back to the atmosphere and increases the collection efficiency of the system.

Openings in the wall, near the floor, and near the ceiling allow convection to transfer heat to the room. The air in the space between the glass and the wall warms as soon as the sun heats the outer surface of the wall. The heated air rises and enters the building through the upper openings. Cool air flows through the lower openings, and convective heat gain can be established as long as the sun is shining.

In another indirect gain passive system, a metal roof-ceiling supports transparent plastic bags filled with water (Hay and Yellott 1969). Movable insulation above these water-filled bags is rolled away during the winter day to allow the sun to warm the stored water. The water then transmits heat indoors by convection and radiation. The insulation remains over the water bags at night or during overcast days.

During the summer, the water bags are exposed at night to cool them by (1) convection to the cool night air, (2) radiation cooling to the night sky, and (3) evaporative cooling from exposed water added to the water bags. The insulation covers the water bags during the day to protect them from unwanted irradiation.

Pittenger *et al.* (1978) tested a building where water rather than insulation was moved to provide summer cooling and winter heating.

Add-on greenhouses can be used as solar attachments when the orientation and other local conditions are suitable. The greenhouse can provide a buffer between the exterior wall of the building and the outdoor conditions. During daylight hours, warm air from the greenhouse can be introduced into the house by natural convection or a small fan.

In most passive systems, control is accomplished by moving a component that regulates the amount of solar radiation admitted into the structure. Manually operated window shades or venetian blinds are the most widely used and simplest controls.

Active Systems

Active systems absorb solar radiation with collectors and convey it to storage by a suitable fluid. As heat is needed, it is obtained from storage via heated air or water. Control is exercised by several types of thermostats, the first being a differential device that starts the flow of fluid through the collectors when they have been sufficiently warmed by the sun. It also stops the fluid flow when the collectors no longer gain heat. In locations where freezing conditions occur only rarely, a low-temperature sensor on the collector controls a circulating pump when freezing impends. This process wastes some stored heat, but it prevents costly damage to the collector panels. This system is not suitable for regions where subfreezing temperatures persist for long periods.

The space heating thermostat is generally the conventional double-contact type that calls for heat when the temperature in the controlled space falls to a predetermined level. If the temperature in storage is adequate to meet the heating requirement, a pump or fan is started to circulate the warm fluid. If the temperature in the storage subsystem is inadequate, the thermostat calls on the auxiliary or standby heat source.

Space Heat and Service Hot Water

Figure 25 shows one of the many systems for service hot water and space heating. In this case, a large, atmospheric pressure storage tank is used, from which water is pumped to the collectors by pump P_1 in response to the differential thermostat T_1. The drain-back system is used to prevent freezing, since the amount of antifreeze required in such a system would be prohibitively expensive. Service hot water is obtained by placing a heat exchanger coil in the tank near the top, where, if stratification is encouraged, the hottest water will be found.

An auxiliary water heater boosts the temperature of the sun-heated water when required. Thermostat T_2 senses the indoor temperature and starts P_2 when heat is needed. If the water in the storage tank becomes too cool to provide enough heat, the second contact on the thermostat calls for heat from the auxiliary heater.

Standby heat becomes increasingly important as heating requirements increase. The heating load, winter availability of solar radiation, and cost and availability of the auxiliary energy must be determined. It is rarely cost effective to do the entire heating job for either space or service hot water by using the solar heat collection and storage system alone.

Electric resistance heaters have the lowest first cost, but may have high operating costs. Water-to-air heat pumps, which use the sun-heated water from the storage tank as the evaporator energy source, are an alternative auxiliary heat source. The heat pump's COP is high enough to yield 10,000 to 14,000 Btu of heat for each kilowatt-hour of energy supplied to the compressor. When summer cooling as well as winter heating is needed, the heat pump becomes a logical solution, particularly in large systems where a cooling tower is used to dissipate the heat withdrawn from the system.

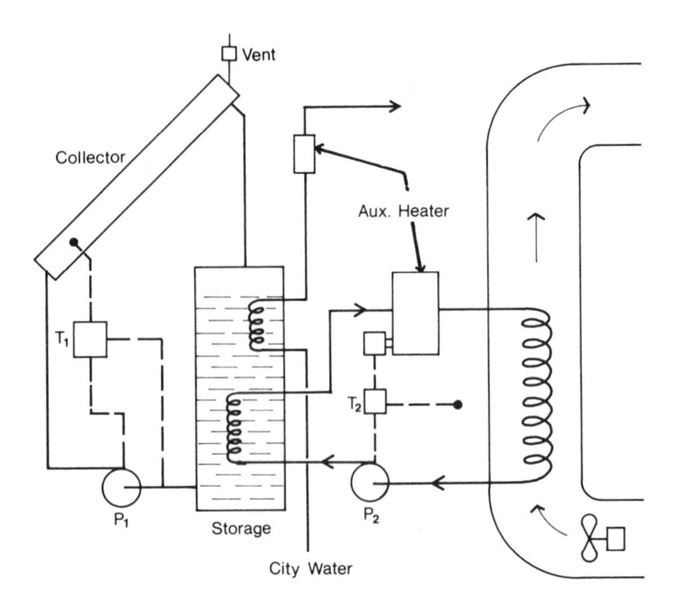

Fig. 25 Solar Collection, Storage, and Distribution System for Domestic Hot Water and Space Heating

The system shown in Figure 25 may be retrofitted into a warm air furnace. In such systems, the primary heater is deleted from the space heating circuit and the coil is located in the return duct of the existing furnace. Full backup is thus obtained and the auxiliary heater provides only the heat not available at the storage temperature of the solar system.

Solar Cooling with Absorption Refrigeration

When solar energy is used for cooling as well as for heating, the absorption system shown in Figure 26, or one of its many modifications, may be used. The collector and storage subsystems must operate at temperatures approaching 200 °F on hot summer days when the water from the cooling tower exceeds 80 °F, but considerably lower operating water temperatures may be used when cooler water is available from the tower. The controls for the collection, cooling, and distribution subsystems are generally separated, with the circulating pump P_1 operating in response to the collector thermostat T_1, which is located within the air-conditioned space. When T_2 calls for heating, valves V_1 and V_2 direct the water flow from the storage tank through the unactivated auxiliary heater to the fan coil in the air distribution system. The fan F_1 in this unit may respond to the thermostat also, or it may have its own control circuit so that it can bring in outdoor air when suitable temperature conditions are present.

When thermostat T_2 calls for cooling, the valves direct the hot water into the absorption unit's generator, and pumps P_3 and P_4 are activated to pump the cooling tower water through the absorber and condenser circuits and the chilled water through the cooling coil in the air distribution system. A relatively large hot water storage tank allows the unit to operate when no sunshine is available. A chilled water storage tank (not shown) may be added so that the absorption unit can operate during the day whenever water is available at a sufficiently high temperature to make the unit function properly. The COP of a typical lithium-bromide-water absorption unit may be as high as 0.75 under favorable conditions, but frequent on-off cycling of the unit to meet a high

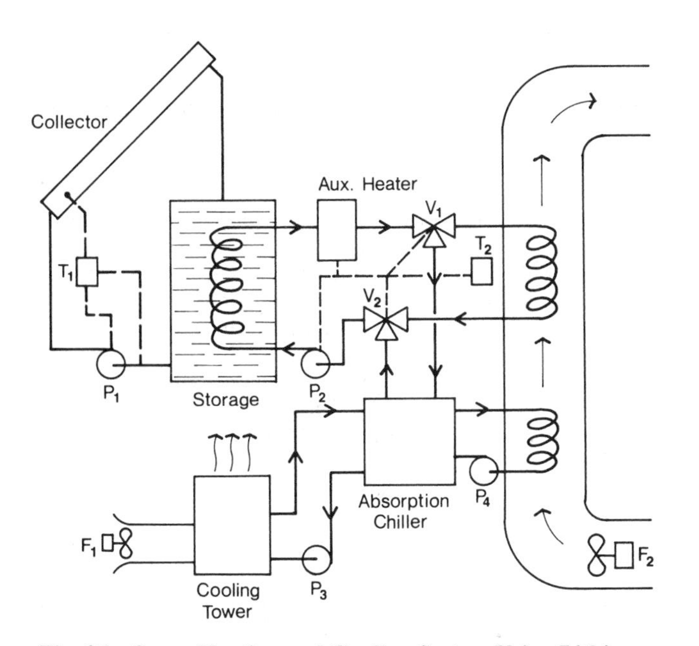

Fig. 26 Space Heating and Cooling System Using Lithium-Bromide-Water Absorption Chiller

variable cooling load may cause significant loss in performance because the unit must be heated to operating temperature after each shutdown. Modulating systems are analyzed differently than are on-off systems.

Water-cooled condensers are required with the absorption cycles, since the lithium-bromide-water cycle operates with a relatively delicate balance among the temperatures of the three fluid circuits—cooling tower water, chilled water, and activating water. The steam-operated absorption systems, from which solar cooling systems are derived, customarily operate at energizing temperatures of 230 to 240 °F, but these are above the capability of most flat-plate collectors. The solar cooling units are designed to operate at considerably lower temperatures, but unit ratings are also lowered.

Smaller domestic units may operate with natural circulation, or *percolation*, used to carry the lithium-bromide-water solution from the generator (to which the activating heat is supplied) to the separator and condenser; there, the reconcentrated Li-Br is returned to the absorber while the water vapor goes to the condenser before being returned to the evaporator where cooling takes place. Larger units use a centrifugal pump to transfer the fluid.

SIZING COLLECTOR AND STORAGE SUBSYSTEMS

The basic approach to sizing solar water heating and space heating and cooling systems is to find the monthly heat requirements by using Weather Service temperature data and the ASHRAE procedure for estimating heating loads. For service water heating, the daily and monthly requirements and the delivery temperatures are first established, and the probable temperature of the supply water is found from local records for each month.

The availability of solar radiation for any specific location must next be determined, and the clear day data (given in ASHRAE *Standard* 93-1986) are a good starting point. Multiplication of the clear day irradiation for the selected surface tilt angle by the number of days in the month and the probable percentage of possible sunshine will give an estimated amount of solar radiation that reaches the collector on average days. Collector performance can be estimated for each month by using an average daylight ambient air temperature, insolation data, and the characteristics of the proposed collector. The slope-intercept method is outlined earlier under Collector Performance.

If only a hand calculator is available, one day per month is about the limit of the calculating capability, but when higher powered computation facilities are available, hour-by-hour estimates can be made, using weather data available from the National Weather Service at Asheville, North Carolina. (ASHRAE-developed bin and degree hour summaries are available.) A *standard year* is sometimes employed, for which the average horizontal irradiation and the average ambient air temperatures come close to matching the Weather Service averages for that location. This method must be approached with care, however, because of changing weather patterns and the absence of long-term solar irradiation data for most locations.

Methods of estimating solar system performance have been proposed that greatly simplify the task of selecting the most economical combinations of collector orientation tilt and area with storage subsystems of varying capacity. Beckman *et al.* (1977) describe the *f*-chart method validated by computer simulations based on data established by the Weather Service for locations around the United States.

Installation Guidelines

Most solar system components are the same as those in present HVAC and hot water systems (pumps, piping, valves, and controls), and their installation is not much different from conventional system installation.

Solar collectors are the most unfamiliar component used in a solar energy system. They are located outdoors, which necessitates penetration of the building envelope. They also require a structural element to support them at the proper tilt and orientation toward the sun.

Site considerations must be taken into account. The collectors should be located so that shading is minimized and installed so that they are attractive on and off site. They should also be located to minimize vandalism and avoid a safety hazard.

Collectors should be placed as near to the storage tank as possible to reduce piping costs and heat losses. The collector and piping must be installed so that they can be drained without trapping fluid within the system.

For best annual performance, collectors should be installed at a tilt angle above the horizontal appropriate for the local latitude. They should be oriented toward true south, not magnetic south. Small variations in tilt (± 10 °) and orientation (± 20 °) are acceptable without significant performance degradation.

Collector Mounting

Solar collectors are usually mounted on the ground or on flat or pitched roofs. A roof location necessitates the penetration of the building envelope by mounting hardware, piping, and control wiring. Ground or flat roof-mounted collectors are generally rack-mounted.

Pitched roof mounting can be done several ways. Collectors can be mounted on structural *standoffs,* which support them at an angle other than that of the roof to optimize solar tilt. In another pitched roof-mounting technique known as *direct mounting,* collectors are placed on a waterproof membrane on top of the roof sheeting. The finished roof surface together with the necessary collector structural attachments and flashing are then built up around the collector. A weatherproof seal between the collector and the roof must be maintained to prevent leakage, mildew, and rotting.

Integral mounting can be done for new pitched roof construction. The collector is attached to and supported by the structural framing members. The top of the collector then serves as the finished roof surface. Weather tightness is crucial to avoid damage and mildew.

Collectors should support the snow loads that occur on the roof area they cover. The collector tilt usually expedites snow sliding with only a small loss in efficiency. The roof structure should be free of objects that could impede snow sliding, and the collectors should be raised high enough to prevent snow buildup over them.

The mounting structure should be built to withstand winds of at least 100 mph, which impose a wind load of 40 lb/ft^2, on a vertical surface or an average of 25 lb/ft^2 on a tilted roof (HUD 1977). Wind load requirements may be higher, depending on local building codes. Flat-plate collectors mounted flush with the roof surface should be constructed to withstand the same wind loads.

When mounted on racks, the collector array becomes more vulnerable to wind gusts as the angle of the mount increases. Collectors can be uplifted by wind striking the undersides. This wind load, in addition to the equivalent roof area wind loads, should be determined according to accepted engineering procedures.

Expansion and contraction of system components, material compatibility, and the use of dissimilar metals must be considered. Collector arrays and mounting hardware (bolts, screws, washers, and angles) must be well protected from corrosion. Steel-mounting hardware in contact with aluminum, and copper piping in contact with aluminum hardware are both examples of high corrosion potential combinations.

Dissimilar metals can be separated by washers made of fluorocarbon polymer, phenolic, or neoprene rubber.

Freeze Protection

Freeze protection is important and is often the determining factor when selecting a system. Freezing can occur at ambient temperatures as high as 42°F because of night sky radiation.

One simple way of protecting against freezing is to drain the fluid from the collector array and interior piping when a potential freezing condition exists. The drainage may be automatic, as in drain-down and drain-back systems, or manual, as in direct thermosiphon systems. Automatic systems should be capable of fail-safe drainage operation—even in the event of pump failure or power outage. Special pump considerations may also be required to permit water to drain back through the pump and to permit system refilling without causing cavitation.

In areas where freezing is infrequent, recirculating water from storage to the collector array can be used as freeze protection.

Freeze protection can be provided by using fluids that resist freezing. Fluids such as water/glycol solutions, silicone oils, and hydrocarbon oils are circulated by pumps through the collector array and double wall heat exchanger. Draining the collector fluid is not required, since these fluids have a freezing point well below the coldest anticipated outdoor temperature.

In mild climates where recirculation freeze protection is used, a second level of freeze protection can be provided by flushing the collector with cold supply water when the collector approaches near-freezing temperatures. This can be accomplished with a temperature-controlled valve that is set to open a small port at a near-freezing temperature of about 40°F and then close at a slightly higher temperature.

Overtemperature Protection

During periods of high insolation and low hot water demand, overheating can occur in the collectors or storage tanks. Protection against overheating must be considered for all portions of the solar hot water system. Liquid expansion or excessive pressure can burst piping or storage tanks. Steam or other gases within a system can restrict liquid flow, making the system inoperable.

The most common methods of overheat protection stop circulation in the collection loop until the storage temperature decreases, discharge the overheated water from the system and replace it with cold makeup water, or use a heat exchanger as a means of heat rejection. Some freeze protection methods can also provide overheat protection.

For nonfreezing fluids such as glycol antifreezes, overtemperature protection is needed to limit fluid degradation at high temperatures during collector stagnation.

Safety

Safety precautions required for installing, operating, and servicing a solar domestic hot water system are essentially the same as those required for a conventional domestic hot water system. One major exception is that some solar systems use nonpotable heat transfer fluids. Local codes may require a double wall heat exchanger for potable water installations.

Pressure relief must be provided in all parts of the collector array that can be isolated by valves. The outlet of these relief valves should be piped to a container or drain and not where workers could be affected.

Startup Procedure

After completing the installation, certain tests must be performed before charging or filling the system. The system must be checked for leakage, and pumps, fans, valves, and sensors must be checked to see that they are functional. Testing procedures vary with system type.

Closed-loop systems should be hydrostatically tested. The system is filled and pressurized to 1.5 times the operating pressure for one hour and inspected for leaks and any appreciable pressure drop.

Drain-down systems should be tested to be sure that all water drains from the collectors and piping located outdoors. All lines should be checked for proper pitch so that gravity drains them completely. All valves should be verified to be in working order.

Drain-back systems should be tested to ensure that the collector fluid is draining back to the reservoir tank when circulation stops and that the system refills properly.

Air systems should be tested for leaks before insulation is applied by starting the fans and checking the ductwork for leaks.

Pumps and sensors should be inspected to verify that they are in proper working order. Proper cycling of the system pumps can be checked by a running time meter. A sensor that is suspected of being faulty can be dipped alternately in hot and cold water to see if the pump starts or stops.

Following system testing and before filling or charging it with heat transfer fluid, the system should be flushed to remove debris.

System Maintenance

All systems should be checked at least once a year in addition to any periodic maintenance that may be required for specific components. A log of all maintenance performed should be kept, along with an owner's manual that describes system operational characteristics and maintenance requirements.

The collectors' outer glazing should be cleaned periodically by hosing. Leaves, seeds, construction dirt, and other debris should be carefully swept from the collectors. Care should be taken not to damage plastic covers.

Without opening up a sealed collector panel, the absorber plate should be checked for surface coating damage caused by peeling, crazing, or scratching. Also, the collector tubing should be inspected to ensure that it contacts the absorber. If the tubing is loose, the manufacturer should be consulted for repair instructions.

Heat transfer fluids should be tested and replaced at intervals suggested by the manufacturer. Also, the solar energy storage tank should be drained about every six months to remove sediment.

Performance Monitoring/Minimum Instrumentation

Temperature sensors and temperature differential controllers are required to operate most solar systems. However, additional instruments should be installed for system monitoring, checking, and troubleshooting.

Thermometers should be located on the collector supply and return lines so that the temperature difference in the lines can be determined visually.

A pressure gage should be inserted on the discharge side of the pump. The gage is used to monitor the pressure that the pump must work against and to indicate if the flow passages are blocked.

Running time meters on pumps and fans may be installed to determine if the system is cycling properly.

DESIGN, INSTALLATION, AND OPERATION GUIDE

The following checklist is for designers of solar heating and cooling systems. Specific values have not been included because these vary for each application. The designer must decide whether design figures are within acceptable limits for any particular project (see DOE 1978 for further information). The review order listed does not reflect their precedence or importance during design.

Collectors

- Check flow rate for compliance with manufacturer's recommendations.
- Check that collector area matches application and claimed solar participation.
- Review collector instantaneous efficiency curve and check match between collector and system requirements.
- Relate collector construction to end use; two cover plates are not required for low-temperature collection in warm climates and may, in fact, be detrimental. Two cover plates are more efficient when the temperature difference between the absorber plate and outdoor air is high, such as in severe winter climates or when collecting at high temperatures for cooling. Radiation losses only become significant at relatively high absorber plate temperatures. Selective surfaces should be used in these cases. Flat black surfaces are acceptable and sometimes more desirable for low collection temperatures.
- Check match between collector tilt angle, latitude, and collector end use.
- Check collector azimuth.
- Check collector location for potential shading and exposure to vandalism or accidental damage.
- Review provisions made for high stagnation temperatures. If not used, are liquid collectors drained or left filled in the summer?
- Check for snow hang-up and ice formation. Will casing vents become blocked?
- Review precautions, if any, against outgassing.
- Check access for cleaning covers.
- Check mounting for stability in high winds.
- Check for architectural integration. Do collectors on roof present rainwater drainage or condensation problems? Do roof penetrations form potential leak problems?
- Check collector construction for structural integrity and durability. Will materials deteriorate under operating conditions? Will any pieces fall off?
- Are liquid collector passages organized in such a way as to allow natural fill and drain? Does mounting configuration affect this?
- Does air collector duct connection configuration promote a balanced airflow and an even heat transfer? Are connections potentially leaky?

Hydraulics

- Check that the flow rate through the collector array matches system parameters.
- If antifreeze is used, check that flow rate has been modified to allow for viscosity and specific heat.
- Review properties of proposed antifreeze. Some fluids are highly flammable. Check toxicity, vapor pressure, flash point, and boiling and freezing temperatures at atmospheric pressure.
- Check means of makeup into antifreeze system. An automatic water makeup system can result in freezing.
- Check that provisions are made for draining and filling the system. (Air vents at high points, drains at low points, pipes correctly graded in-between, drain-back systems vented to storage or expansion tank.)
- If system uses drain-back freeze protection, check that:
  1. Provision is made for drain-back volume and back venting.
  2. Pipes are graded for drain back.
  3. Solar primary pump is sized for lift head.
  4. Pump should be self-priming if tank is below pump.
- Check that collector pressure drop for drain-back system is slightly higher than static pressure between the supply and return headers.
- Optimum pipe arrangement is reverse return with collectors in parallel. Series collectors reduce flow rate and increase head. A combination of parallel/series can sometimes be beneficial, but check that equipment has been sized and selected properly.
- Cross-connections under different operating modes sometimes result in pumps operating in opposition or tandem, causing severe hydraulic problems.
- If heat exchangers are used, check that approach temperature differential has been recognized in the calculations.
- Check that adequate provisions are made for water expansion and contraction. Use specific volume/temperature tables for calculation. Each unique circuit must have its own provision for expansion and contraction.
- Three-port valves tend to leak through the closed port. This, together with reversed flows in some modes, can cause potential hydraulic problems. As a general rule, simple circuits and controls are better.

Airflow

- Check that the flow rate through the collector array matches the system design parameters.
- Check temperature rise across collectors using air mass flow and specific heat.
- Check that duct velocities are within the system parameters.
- Check that cold air or water cannot flow from collectors by gravity under "no-sun" conditions.
- Verify duct material and construction methods. Ductwork must be sealed to reduce losses.
- Check duct configuration for balanced flow through collector array.
- More than two collectors in series can reduce collection efficiency.

Thermal Storage

- Check that thermal storage capacity matches parameters of collector area, collection temperature, utilization temperature, and system load.
- Verify that thermal inertia does not impede effective operation.
- Check provisions made to promote temperature stratification during both collection and use.
- Check that pipe and duct connections to storage are compatible with the control philosophy.
- If liquid storage is used for high temperatures (above 200 °F), check that tank material and construction can withstand the temperature and pressure.
- Check that storage location does not promote unwanted heat loss or gain and that adequate insulation is provided.
- Verify that liquid storage tanks are treated to resist corrosion. This is particularly important in tanks that are partially filled.

- Check that provision is made to protect liquid tanks from exposure to either an overpressure or vacuum.

Uses

Domestic Hot Water

- Characteristics of domestic hot water loads include short periods of high draw interspersed with long dormant periods. Check that domestic hot water storage matches solar heat input.
- Check that provisions have been made to prevent reverse heating of the solar thermal storage by the domestic hot water backup heater.
- Check that system allows cold makeup water preheating on days of low solar input.
- Verify that tempering valve limits domestic hot water supply to a safe temperature during periods of high solar input.
- Depending on total dissolved solids, city water heated above 150 °F may precipitate a calcium carbonate scale. If collectors are used to heat water directly, check provisions made to prevent scale formation in absorber plate waterways.
- Check if the system is required to have a double wall heat exchanger and that it conforms to appropriate codes, if the collector uses nonpotable fluids.

Heating

- Warm air heating systems have the potential of using solar energy directly at moderate temperatures. Check that air volume is sufficient to meet the heating load at low supply temperatures and that the limit thermostat has been reset.
- At times of low solar input, solar heat can still be used to meet part of the load by preheating return air. Check location of solar heating coil in system.
- Baseboard heaters require relatively high supply temperatures for satisfactory operation. Their output varies as the 1.5 power of the log mean temperature difference and falls off drastically at low temperatures. If solar is combined with baseboard heating, check that supply temperature is compatible with heating load.
- Heat exchangers imply an approach temperature difference that must be added to the system operating temperature to derive the minimum collection temperature. Verify calculations.
- Water-to-air heat pumps rely on a constant solar water heat source for operation. When the heat source is depleted, the backup system must be used. Check that storage is adequate.

Cooling

- Solar activated absorption cooling with fossil fuel backup is currently the only commercially available active cooling. Be assured of all design criteria and a large amount of solar participation. Verify calculations.
- Storing both hot water and chilled water may make better use of available storage capacity.

Controls

- Check that control philosophy matches the desired modes of operation.
- Verify that collector loop controls recognize solar input, collector temperature, and storage temperature.
- Verify that controls allow both the collector loop and the utilization loop to operate independently.
- Check that control sequences are reversible and will always revert to the most economical mode.
- Check that controls are as simple as possible within the system requirements. Complex controls increase the frequency and possibility of breakdowns.
- Check that all controls are fail-safe.

Performance

- Check building heating, cooling, and domestic hot water loads as applicable. Verify that building thermal characteristics are acceptable.
- Check solar energy collected on a monthly basis. Compare with loads and verify solar participation.

REFERENCES

ASHRAE. 1986. Methods of testing to determine the thermal performance of solar collectors. ANSI/ASHRAE *Standard* 93-1986.

Angstrom, A. 1915. A study of the radiation of the atmosphere. *Smithsonian Miscellaneous Collection* 65(3).

Balcomb, D. *et al.* 1977. Thermal storage walls in New Mexico. *Solar Age* 2(8):20.

Beckman, W.A., S.A. Klein, and J.A. Duffie. 1977. *Solar heating design by the F-chart method.* John Wiley, New York.

Bennett, I. 1965. Monthly maps of daily insolation in the U.S. *Solar Energy* 9(3):145.

Bennett, I. 1967. Frequency of daily insolation in Anglo North America during June and December. *Solar Energy* 11(1):41.

Bliss, R.W. 1961. Atmospheric radiation near the surface of the earth. *Solar Energy* 59(3):103.

Butler, C.P. *et al.* 1964. Surfaces for solar spacecraft power. *Solar Energy* 8(1):2.

Climatic atlas of the U.S. 1968. U.S. Government Printing Office, Washington, D.C.

Cole, R.L. *et al.* 1977. Applications of compound parabolic concentrators to solar energy conversion. Report No. AMLw42. Argonne National Laboratory, Chicago.

Cromer, C.J. 1984. Design of a DC-pump, photovoltaic-powered circulation system for a solar domestic hot water system. Florida Solar Energy Center (June).

Duffie, J.A. and W.A. Beckman. 1974. *Solar energy thermal processes.* John Wiley, New York.

Edwards, D.K. *et al.* 1962. Spectral and directional thermal radiation characteristics of selective surfaces. *Solar Energy* 6(1):1.

Francia, G. 1961. A new collector of solar radiant energy. U.N. Conference on New Sources of Energy, Rome, 4:572.

Gates, D.M. 1966. Spectral distribution of solar radiation at the earth's surface. *Science* 151(3710):523.

Gier, J.T. and R.V. Dunkle. 1958. Selective spectral characteristics as an important factor in the efficiency of solar energy collectors. Transactions of the Conference on Scientific Uses of Solar Energy, 1955, 2(1-A):41.

Hall, I.J., R.R. Prairie, H.E. Anderson, and E.C. Boes. 1979. Generation of typical meteorological years: 426 Solmet stations. ASHRAE *Transactions* 85(2):507.

Hay, H.R. and J.I. Yellott. 1969. Natural air conditioning with roof ponds and movable insulation. ASHRAE *Transactions* 75(1):165.

Hottel, H.C. and B.B. Woertz. 1942. The performance of flat-plate solar collectors. *Transactions of* ASME 64:91.

HUD. 1977. Intermediate minimum property standards supplement for solar heating and domestic hot water systems. U.S. Department of Housing and Urban Development, SD Cat. No. 0-236-648.

Jordan, R.C. and B.Y.H. Liu, eds. 1977. Applications of solar energy for heating and cooling of buildings. ASHRAE Publication GRP 170.

Lister, L. and T. Newell. 1989. Expansion tank characteristics of closed loop, active solar energy collection systems; Solar engineering—1989. ASME, New York.

Morrison, C.A. and E.A. Farber. 1974. Development and use of solar insolation data for south facing surfaces in northern latitudes. ASHRAE *Transactions* 80(2):350.

Mumma, S.A. 1985. Solar collector tilt and azimuth charts for rotated collectors on sloping roofs. Proceedings Joint ASME-ASES Solar Energy Conference, Knoxville, TN.

Newton, A.B. and S.F. Gilman. 1982. Solar collector performance manual. ASHRAE Publication SP 32.

Newton, A.B. U.S. Patents 2,343,211 and 2,396,338.

Parmalee, G.V. and W.W. Aubele. 1952. Radiant energy transmission of the atmosphere. ASHVE *Transactions* 58:85.

Pittenger, A.L., W.R. White, and J.I. Yellott. 1978. A new method of passive solar heating and cooling. Proceedings of the Second National Passive Systems Conference, Philadelphia, ISES and DOE.

Reitan, C.H. 1963. Surface dew point and water vapor aloft. *Journal of Applied Meteorology* 2(6):776.

Root, D.E., S. Chandra, C. Cromer, J. Harrison, D. LaHart, T. Merrigan, and J.G. Ventre. 1985. *Solar water and pool heating course manual*, 2 vols. Florida Solar Energy Center, Cape Canaveral, FL.

Simon, F.F. 1976. Flat-plate solar collector performance evaluation. *Solar Energy* 18(5):451.

Stephenson, D.G. 1967. Tables of solar altitude and azimuth; Intensity and solar heat gain tables. Technical Paper No. 243, Division of Building Research, National Research Council of Canada, Ottawa.

Tables of computed altitude and azimuth. 1958. Hydrographic Office Bulletin No. 214, Vols. 2 and 3. U.S. Superintendent of Documents, Washington, D.C.

Tables of radiation powers. Paper No. 105, Division of Building Research, National Research Council of Canada, Ottawa.

Tabor, H. 1958. Selective radiation. I. Wavelength discrimination. Transactions of the Conference on Scientific Uses of Solar Energy (1955) 2(1-A):1, University of Arizona Press.

Telkes, M. 1949. A review of solar house heating. *Heating and Ventilating*, 68.

Thekaekara, M.P. 1973. Solar energy outside the earth's atmosphere. *Solar Energy* 14(2):109 (January).

Threlkeld, J.L. 1963. Solar irradiation of surfaces on clear days. ASHRAE *Transactions* 69:24.

Threlkeld, J.L. and R.C. Jordan. 1958. Direct radiation available on clear days. ASHRAE *Transactions* 64:45.

Trombe, F. *et al.* 1977. Concrete walls for heat. *Solar Age* 2(8):13.

Van Straaten, J.F. 1961. Hot water from the sun. Ref. No. D-9, National Building Research Institute of South Africa, Council for Industrial and Scientific Research, Pretoria, South Africa.

Whillier, A. 1964. Thermal resistance of the tube-plate bond in solar heat collectors. *Solar Energy* 8(3):95.

Yellott, J.I. 1977. Passive solar heating and cooling systems. ASHRAE *Transactions* 83(2):429.

Yellott, J.I., D. Aiello, G. Rand, and M.Y. Kung. 1976. Solar-oriented architecture. Arizona State University Architecture Foundation, Tempe, AZ.

Zarem, A.M. and D.D. Erway. 1963. *Introduction to the utilization of solar energy.* McGraw-Hill, New York.

BIBLIOGRAPHY

ASHRAE. 1977. *Applications of solar energy for heating and cooling of buildings.*

ASHRAE. 1983. *Solar domestic and service hot water manual.*

ASHRAE. 1986. Methods of testing to determine the thermal performance of flat-plate solar collectors containing a boiling liquid. ASHRAE *Standard* 109-1986.

ASHRAE. 1987. Methods of testing to determine the thermal performance of solar domestic water heating systems. ASHRAE *Standard* 95-1981. (Reaffirmed 1987).

ASHRAE. 1988. *Active solar heating systems design manual.*

ASHRAE. 1989. Methods of testing to determine the thermal performance of unglazed flat-plate liquid-type solar collectors. ASHRAE *Standard* 96-1980. (Reaffirmed 1989).

Colorado State University. 1980. Solar heating and cooling of residential buildings: Design of systems. Superintendent of Documents, U.S. Government Printing Office, Washington, D.C.

Colorado State University. 1980. Solar heating and cooling of residential buildings: Sizing, insulation and operation of systems. Superintendent of Documents, U.S. Government Printing Office, Washington, D.C.

Cook, J., ed. 1989. *Passive cooling.* MIT Press, Cambridge, MA.

Daniels, F. 1964. *Direct use of the sun's energy.* Yale University Press, New Haven, CT.

Diamond, S.C. and J.G. Avery. 1986. Active solar energy system design, installation & maintenance: Technical applications manual. LA-UR-86-4175.

Durlak, E.R. 1986. Evaluation of installed solar systems at navy, army and air force bases. U.S. Naval Civil Engineering Laboratory. NCEL TN-1750.

Foresti, F.G. 1981. Corrosion and scaling in solar heating systems. U.S. Department of Energy, Vitro Laboratories Division, Automation Industries, Inc. Solar/0909-81/70; DE82006139.

HUD. 1979. Solar domestic hot water. U.S. Department of Housing and Urban Development, HUD 000-1230.

HUD. 1980. Installation guidelines for solar DHW systems in one- and two-family dwellings. U.S. Department of Housing and Urban Development, 2nd ed. (May).

HUD. 1980. Solar terminology. U.S. Department of Housing and Urban Development, HUD-PDR-465(2) (March).

Hunn, B.D., N. Carlisle, G. Franta, and W. Kolar. 1987. Engineering principles and concepts for active solar systems. Solar Energy Research Institute, Golden, CO. SERI/SP-271-2892.

ITT. 1976. *Solar heating systems design manual.* International Telephone and Telegraph Corporation.

Knapp, C.L., T.L. Stoffel, and S.D. Whitaker. 1980. Insolation data manual. Solar Energy Research Institute, Golden, CO. SERI/SP-755-789.

Kreider, J.F. 1989. Solar design: Components, systems, economics. Hemisphere Publication Corporation, New York.

Lameiro, G.F. and P. Bendt. 1978. The GFL method for designing solar energy space heating and domestic hot water systems, 2.1. Proceedings at the 1978 Annual Meeting of the American Section of the International Solar Energy Society, Inc., Denver, CO.

Lane, G.A. 1986. *Solar heat storage: Latent heat materials*, 2 vols. CRC Press, Inc., Boca Raton, FL.

Lof, G.O., J.A. Duffie, and C.D. Smith. 1966. World distribution of solar radiation. Solar Energy Laboratory, University of Wisconsin, Madison, WI. Report No. 21.

Mueller Associates, Inc. 1980. Economic analysis of commercial solar combined space-heating and hot-water systems. Argonne National Laboratory. MAI Report No. 204.

Mueller Associates, Inc. 1980. Economic analysis of residential and commercial solar heating and hot water systems; Summary report. Argonne National Laboratory. MAI Report No. 206.

Mueller Associates, Inc. 1985. Active solar thermal design manual. U.S. Department of Energy, Solar Energy Research Institute and ASHRAE.

Mumma, S., L. Milnarist, and J. Rodriquez-Anza. 1973. Innovative double walled heat exchanger for use in solar water heating. EWxG-03.

Ruegg, R.T., G.T. Sav, J.W. Powell, and E.T. Pierce. 1982. Economic evaluation of solar energy systems in commercial buildings; Methodology and case studies. U.S. National Bureau of Standards. NBSIR 82-2540.

Solar Energy Research Institute. 1981. Solar design workbook—Solar federal buildings program. U.S. Department of Energy and Los Alamos Scientific Laboratory. SERI/SP-62-308.

Solar Energy Research Institute. 1981. Solar radiation energy resource atlas of the United States. Golden, CO. SERI/SP642-1037.

Solar Environmental Engineering Co., Inc. 1981. Solar domestic hot water system inspection and performance evaluation handbook. Solar Energy Research Institute. SERI/SP-98189-1B.

U.S. Department of Energy. 1978. DOE Facilities solar design handbook. DOE/AD-0006/1.

U.S. Department of Energy. 1978. SOLCOST—Solar hot water handbook; A simplified design method for sizing & costing residential & commercial solar service hot water systems, 3rd ed. DOE/CS-0042/2.

U.S. National Bureau of Standards. 1982. Performance criteria for solar heating and cooling systems in residential buildings. NBS Building Science Series 147.

U.S. National Bureau of Standards. 1984. Performance criteria for solar heating and cooling systems in commercial buildings. NBS Technical Note 1187.

Ward, D.S. and H.S. Oberoi. 1980. Handbook of experience in the design and installation of solar heating and cooling systems. ASHRAE (July).

CHAPTER 31

ENERGY RESOURCES

BUILDINGS and facilities of various types may be heated, ventilated, air conditioned, and refrigerated—using systems and equipment designed for that purpose and using the site energy forms commonly available—without concern for the original energy resources from whence those energy forms came. Since the energy used in buildings and facilities comprises a significant amount of the total energy used for all purposes, and since the use of this energy has an impact on energy resources, ASHRAE recognizes the "effect of its technology on the environment and natural resources to protect the welfare of posterity" (ASHRAE 1984).

Many governmental agencies regulate energy conservation legislation for obtaining building permits (Conover 1984). The application of specific values to building energy use situations has a considerable effect on the selection of HVAC and R systems and equipment and how they are applied.

CHARACTERISTICS OF ENERGY AND ENERGY RESOURCE FORMS

The HVAC and R industry deals with energy forms as they occur on or arrive at a building site. Generally, these energy forms are fossil fuels (natural gas, oil, and coal) and electricity. Solar energy and wind energy are also available at most sites, and geothermal energy (earth heat) is available at some. These are the prime forms of energy used to power or heat the improvements on a site.

Forms of On-Site Energy

Fossil fuels and electricity are commodities that are usually metered or measured for payment by the facility owner or operator. On the other hand, solar or wind energy which might be considered a distributed energy form in its natural state (*i.e.*, not requiring central processing or a distribution network) cost nothing for the commodity itself, but do incur cost for the means to make use of it. Geothermal energy, which is not universally available, may or may not be a sold commodity, depending on the particular locale and local regulations (Chapter 29).

Some prime on-site energy forms require further processing or conversion into other forms more directly suited for the particular systems and equipment needed in a building or facility. For instance, natural gas or oil is burned in a boiler to produce steam or hot water, a form of thermal energy which is then distributed to various use points (such as heating coils in air-handling systems, unit heaters, convectors, fin-tube elements, steam-powered cooling units, humidifiers, and kitchen equipment) throughout the building. Although electricity is not converted in form on-site, it is nevertheless used in a variety of ways, including lighting, running motors for fans and pumps, powering electronic equipment and office machinery, and space heating. While the methods and efficiency with which these processes take place fall within the scope of the HVAC and R designer, the process by which a prime energy source arrives at a given facility site is not under direct control of the professional. On-site energy choices, if available, may be controlled by the designer based, in part, on the present and future availability of the associated resource commodities.

The basic energy source for heating may be natural gas, oil, coal, or electricity. Cooling may be produced by electricity, thermal energy, or natural gas. If electricity is generated on-site, the generator may be turned by an engine using natural gas or oil, or by a turbine using steam or gas directly.

The term *energy source* refers to on-site energy in the form in which it occurs as it arrives at or occurs on a site (*e.g.*, electricity, gas, oil, or coal). *Energy resource* refers to the raw energy, which (1) is extracted from the earth (wellhead or mine-mouth), (2) is used in the generation of the energy source delivered to a building site (coal used to generate electricity), or (3) occurs naturally and is available at a site (solar, wind, or geothermal energy).

Depletable and Nondepletable Energy Resources

From the standpoint of energy conservation, energy resources may be classified in two broad categories: (1) depletable (or discontinuous) resources of energy, which have definite, although sometimes unknown, limitations; and (2) nondepletable (or continuous) resources of energy, which can generally be freely used without depletion or have the potential to renew in a reasonable period. Resources used most in industrialized countries, both now and in the past, are depletable (Gleeson 1951).

Depletable resources of energy include:

- Coal
- Petroleum
- Natural gas
- Uranium 235 (atomic energy)
- Wood (Note: Formerly thought to be nondepletable, it is now apparent that extensive use would deplete the supply faster than it could be renewed.)

Nondepletable sources of energy include:

- Hydropower
- Solar
- Wind
- Earth heat (geothermal)
- Tidal power
- Ocean thermal
- Atmosphere or large body of water (as used by the heat pump)

The above list might also include waste products and biomass, which are becoming more plentiful as waste products increase. Such resources, however, are still limited.

Characteristics of Fossil Fuels and Electricity

Most on-site energy for buildings in developed countries involves electricity and fossil fuels as the prime on-site energy sources. Both fossil fuels and electricity can be described in terms of their energy

The preparation of this chapter is assigned to TC 4.8, Energy Resources.

content (quadrillions of Btus). However, this implies that the two energy forms are comparable and that an equivalence can be established. In reality, however, fossil fuels and electricity are only comparable in energy terms when they are used to generate heat. Fossil fuels, for example, cannot directly drive motors or energize light bulbs. Conversely, electricity gives off heat as a by-product regardless of whether it is used for running a motor or lighting a light bulb, and regardless of whether that heat is needed or not. Thus, electricity and fossil fuels have different characteristics, uses, and capabilities besides any differences relating to their derivation.

Beyond the building site, further differences between these energy forms may be observed, such as methods of extraction, transformation, transportation, delivery, and characteristics of the resource itself. Natural gas arrives at the site in virtually the same form in which it was extracted from the earth. Oil is processed (distilled) before arriving at the site, having been extracted as crude oil, and arrives on a given site as, for example, No. 2 oil or diesel fuel. Electricity is created (converted) from a different energy form, often a fossil fuel, which itself is first converted to a thermal form. The total electricity conversion, or generation, process includes energy losses governed largely by the laws of thermodynamics.

Fuel cells, although used only on a small scale, convert a fossil fuel to electricity by chemical means.

Fossil fuels undergo a conversion process by combustion (oxidation) and heat transfer to thermal energy in the form of steam or hot water. The equipment used is a boiler or a furnace in lieu of a generator, and it usually occurs on a project site rather than off-site. (District heating is an exception.) Inefficiencies of the fossil fuel conversion occur on-site, while the inefficiencies of most electricity generation occur off-site, *i.e.*, before the electricity arrives at the building site. (Cogeneration is an exception.)

ENERGY RESOURCE USE

Global energy consumption in 1985 was estimated at 220 quads (1 quad = 10^{15} Btu), of which about 80 quads were consumed by the United States. Globally, 50% of energy use is derived from oil, and 20% from coal and natural gas. In addition to predicted substantial increases in oil consumption in both the United States and countries that do not produce their own oil, oil consumption is expected to grow even more rapidly for developing countries than for industrialized nations. The United States is the world's largest user and is second in projected use; it is, however, lower in reserves than many other countries.

Nuclear energy provides electricity on a worldwide basis, with more than 370 nuclear power plants in operation in 1985 and approximately 432 operating in 1990. France, Belgium, Taiwan, Sweden, and Switzerland derive a significant portion of their total electricity from nuclear energy, and Japan and eastern block countries will increase their nuclear capacity significantly by the year 2000. In the United States, nuclear energy is second only to coal in generating electricity. However, a new nuclear plant has not been ordered in the United States since 1978.

Coal supplies are abundant in many countries. United States coal reserves are said to be 500 billion tons, which would constitute several hundred years of domestic supply at current consumption rates. Historically, however, energy consumption has grown at an exponential rate. If that growth trend continues, reserves that would last for decades or centuries will be rapidly depleted (Bartlett 1983).

Since renewable energy is a nondepletable energy resource, it is projected to play a larger role in meeting the world's total energy needs. Although hydroelectric facilities fall into this category, the major growth in renewable resources is expected from new technologies. The emphasis on energy conservation of the 1970s and early 1980s encouraged many research and demonstration projects dealing with geothermal, solar, and wind energy. The types of renewable energy technologies expected to contribute to U.S. energy supplies in future include geothermal, solar, wind, biofuels, small-scale hydropower, and ocean energy.

The basic sources for United States energy data are EIA (1985) and OPPA (1988). Another data source with worldwide energy data is *Energy Security* (USDOE 1987), which is a report to the president of the United States by the U.S. Department of Energy. Figures 1, 2, and 3 (ISPE 1989) summarize data from these sources. Note: The trend/projection lines of Figures 1 and 2 are cumulative; that is, the differences between lines represent the energy consumed, not the distance from each line to the baseline. Also, data for 1985 was the most recent available when EIA (1985) and OPPA (1988) were prepared.

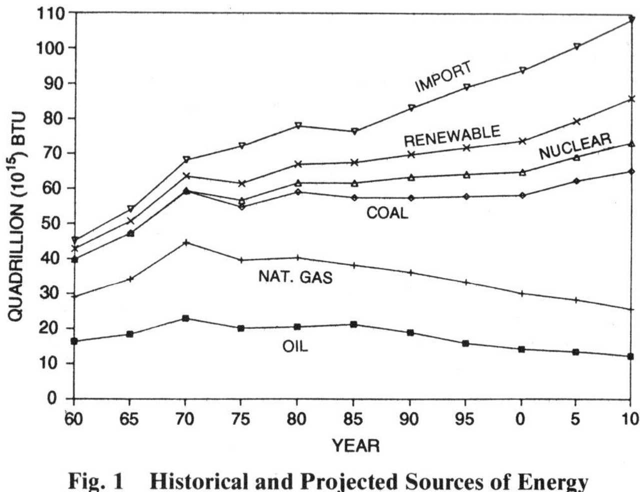

Fig. 1 Historical and Projected Sources of Energy in the United States

With respect to Figure 1, several observations are significant.

- Fossil fuels (oil, natural gas, and coal) comprise most of the total energy consumption.
- Domestic oil contributions to U.S. energy use have remained about constant and are projected to decline in the future; as a percentage of the total, they have declined.
- Use of coal as an energy resource has been increasing slowly and, relative to other resources, is expected to increase markedly in the future.
- The contribution of nuclear power has increased slightly since 1960, but it is not projected to increase significantly in the future.
- Use of imported energy (primarily oil) ballooned in the early 1970s, was held in check and even declined somewhat in the mid-1980s, but is projected to play an expanded and significant role in the future.
- Use of renewable energy has increased slightly in absolute terms, but its proportion of the total share is not projected to increase much in the future.
- Emphasis on energy conservation in the mid-1970s and early 1980s had some impact. Total energy resource use declined between 1980 and 1985, but it is projected to resume growth at about the same rate that prevailed during 1970 to 1980. This consumption growth rate is lower than prevailed in the latter part of the 1960s.

Figure 2 illustrates energy use by sector, the major use sectors being identified as residential, commercial, industrial, and transportation. HVAC and R engineers are primarily concerned with the residential, commercial, and industrial sectors. Significant observations include:

- Transportation is a major user of energy compared to other sectors.

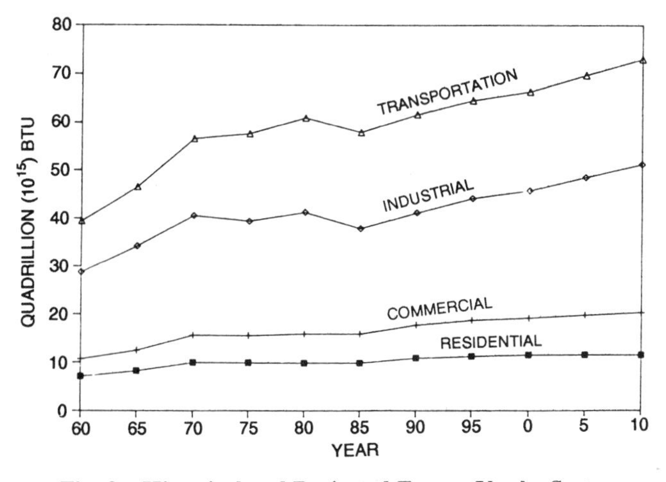

Fig. 2 Historical and Projected Energy Use by Sector in the United States

- Residential and commercial sectors have experienced slight growth historically and are projected to continue to grow at a similar rate.
- The industrial sector, also a major use sector, experienced a leveling off in the 1970s and early 1980s; it is expected to resume a steady growth in energy use at a reduced rate of increase compared to that of the 1960s.

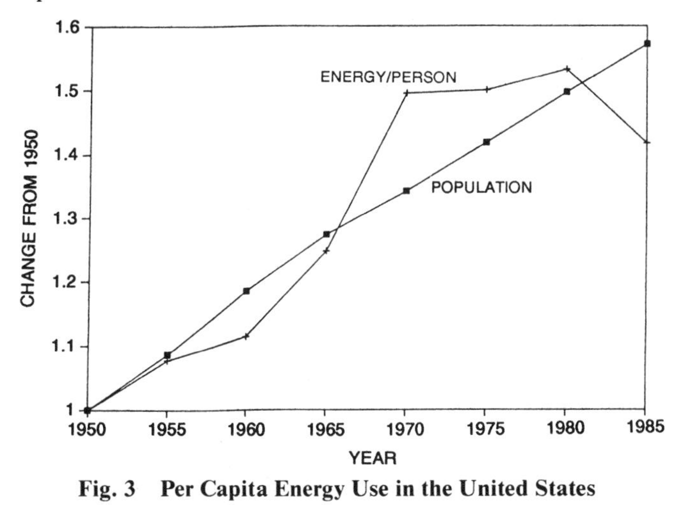

Fig. 3 Per Capita Energy Use in the United States

Figure 3 superimposes per capita energy use over a comparable graph of population growth to give it a frame of reference. In contrast to population growth, per capita energy use has varied significantly. The 1960s experienced a sharp increase in per capita energy use, which leveled off during the 1970s, due to higher energy prices and the emphasis on energy conservation. In the early 1980s, however, a significant drop in per capita energy use occurred as industrial sluggishness increased, efficiency of use improved, and global economic pressures took root.

RELATIONSHIPS

In designing the systems required for a facility, an HVAC and R designer sooner or later must consider the use of one or more forms of prime energy. Most likely, these would be depletable energy sources (fossil fuels and electricity), although installations are sometimes designed using a single energy source (*e.g.*, only a fossil fuel or only electricity).

Solar energy normally impinges on the site (and on the facilities to be put there), so it will have an impact on the energy consumption of the facility. The designer must account for this impact as well as decide whether to make active use of solar energy. When solar energy is used beneficially, it can reduce the requirements for depletable energy forms. Other naturally occurring and distributed nondepletable forms such as wind power and earth heat (if available) might also be considered.

Designers must be aware of the relationship between on-site energy sources and raw energy resources, if they are to understand and be concerned with the earth's energy resources—including how these resources are used and what they are used for. The relationship between energy sources and energy resources involves two parts: (1) quantifying the energy resource units expended and (2) considering the societal impact of the depletion of one energy resource (caused by on-site energy use) with respect to others. The following two sections describe those parts in more specific terms.

Quantifiable Relationships

As stated previously, as on-site energy sources are consumed, a corresponding amount of resources are consumed to produce that on-site energy. For instance, for every 1000 gal of No. 2 oil consumed by a boiler at a building site, some greater number of gallons of crude oil was extracted from the earth to yield that No. 2 oil. On leaving the well, the oil is transported and processed into its final form, perhaps stored, and then transported to the site where it will be used.

Even though gas requires no significant processing, it is transported, often over long distances, to reach its final destination, which causes some energy loss. Electricity may have as its raw energy resource a fossil fuel, uranium, or an elevated body of water (hydroelectric generating plant).

Data to assist in determining the amount of resource use per delivered on-site energy source unit is available. In the United States, data is available from entities within the U.S. Department of Energy and from the agencies and associations listed at the end of this chapter.

The amount of energy resources consumed as a result of consuming one or more energy sources on a site may be determined by a Resource Utilization Factor (RUF). The RUF is the ratio of resources consumed to energy delivered (for each form of energy) to a building site. Specific RUFs may be determined for various depletable energy sources normally consumed on-site. These resources include coal, gas, oil, electricity, and other nondepletable sources (such as solar, geothermal, waste, and wood energy). With electricity, which may derive from several sources depending on the particular fuel mix of the generating stations in the region served, the overall RUF is the weighted combination of individual factors applicable to electricity and a particular energy resource. ASHRAE (1989) and Grumman (1984) give specific formulas for calculating RUFs.

While a designer does not need to determine the amount of energy resources attributable to a given building or building site for its design or operation, this information may be helpful when planning the long-range availability of energy for a building or the building's impact on energy resources. Presently, factors or fuel-quantity-to-energy resource ratios are used, which suggests that energy resources are of concern to the HVAC and R industry.

Intangible Relationships

Energy resources should not simply be converted into common energy units [*e.g.*, quadrillion (10^{15}) Btu] because the commonality thus established gives a misleading picture of the "equivalence" of these resources. Other differences and limitations of each of the resources defy easy quantification but are nonetheless real.

For instance, consider electricity that arrives and is used on a site and the resources from which it is derived. Electricity can be generated from coal, oil, natural gas, uranium, or elevated bodies of water (hydropower). The end result is the same: electricity at

X kilovolts, Y hertz. However, is a megajoule of electricity generated by hydropower equal in societal impact to that same megajoule generated by coal? by uranium? by domestic oil? or by imported oil? In other words, electricity generated by hydropower, though identical with the same quantity of electricity generated with imported oil, might be considered more desirable from a societal impact standpoint.

Intangible factors such as safety, environmental acceptability, availability, and national interest also are affected in different ways by the consumption of each resource. Heiman (1984) lists some intangible factors as follows:

National/Global Considerations

- Balance of trade
- Environmental impacts
- International policy
- Employment
- Minority employment
- Availability of supply
- Alternative uses
- National defense
- Domestic policy
- Effect on capital markets

Local Considerations

- Exterior environmental impact—air
- Exterior environmental impact—solid waste
- Exterior environmental impact—water resources
- Local employment
- Local balance of trade
- Use of distribution infrastructure
- Local energy independence
- Land use
- Exterior safety

Site Considerations

- Reliability of supply
- Indoor air quality
- Aesthetics
- Interior safety
- Anticipated changes in energy resource prices

The value of these factors is subjective, but a procedure for weighting and evaluating them has been proposed by ASHRAE (1989).

Summary

In designing HVAC and R systems, the need to address immediate issues such as economics, performance, and space constraints often prevents designers from fully considering the energy resources affected. Today's energy resources are less certain because of issues such as availability, safety, national interest, environmental concerns, and the world political situation. As a result, the reliability, economics, and continuity of many common energy resources over the potential life of a building being designed are unclear. For this reason, the designer of building energy systems must consider the energy resources on which the long-term operation of the building will depend. If the continued viability of those resources is reason for concern, the design should provide for, account for, or address such an eventuality.

AGENCIES AND ASSOCIATIONS

American Gas Association
Arlington, VA

Bureau of Mines
Department of Interior
Washington, D.C.

Council on Environmental Quality
Washington, D.C.

Edison Electric Institute
Washington, D.C.

Electric Power Research Institute
Palo Alto, CA

National Coal Association
Washington, D.C.

North American Electric Reliability Council
Princeton, NJ

Petroleum Institute
Washington, D.C.

REFERENCES

ASHRAE. 1984. Energy policy (February).

ASHRAE. 1989. Standard procedures for estimating the impact of the choice of fuels and energy sources required for building systems. BSR/ASHRAE (Proposed) *Standard* 112P.

Bartlett, A.A. 1983. *Forgotten fundamentals of the energy crisis.* Boulder, CO.

Conover, D.R. 1984. Accounting for energy resource use in building regulations. ASHRAE *Transactions* 90(1B):547-63.

EIA. 1985. *Annual energy review.* DOE/EIA—0384(85). Energy Information Administration, U.S. Department of Energy, Washington, D.C.

Gleeson, G.W. 1951. Energy—Choose it wisely today for safety tomorrow. ASHVE *Transactions* 57:523-40.

Grumman, D.L. 1984. Energy resource accounting: ASHRAE *Standard* 90C-1977R. ASHRAE *Transactions* 90(1B):531-46.

Heiman, J.L. 1984. Proposal for a simple method for determining resource impact factors. ASHRAE *Transactions* 90(1B):564-70.

ISPE. 1989. A summary review by the energy committee of the Illinois Society of Professional Engineers of USDOE 1987. Illinois Society of Professional Engineers, Springfield, IL (June/July).

OPPA. 1988. *Long range energy projections to 2010.* Office of Policy, Planning and Analysis, U.S. Department of Energy, Washington, D.C. (July).

USDOE. 1987. *Energy security, A report to the President of the United States.* U.S. Department of Energy, Washington, D.C. (March).

BIBLIOGRAPHY

Anderson, R.J. 1984. The energy resource picture in 1984 in the U.S. and abroad. ASHRAE *Transactions* 90(1B):521-30.

Pacific Northwest Laboratory. 1987. *Development of whole-building energy design targets for commercial buildings phase 1 planning.* PNL-5854, Vol. 2. U.S. Department of Energy, Washington, D.C.

USDOE. 1979. *Impact assessment of a mandatory source-energy approach to energy conservation in new construction.* U.S. Department of Energy, Washington, D.C.

CHAPTER 32

ENERGY MANAGEMENT

ENERGY conservation can be defined as more efficient or effective use of energy. As fuel costs rise and environmental concerns grow, more efficient energy conversion and utilization technologies become cost-effective. However, technology alone cannot produce sufficient results without a continuing management effort. Energy management begins with the commitment and support of an organization's top management. A suggested flow chart for developing an energy management program is shown in Figure 1.

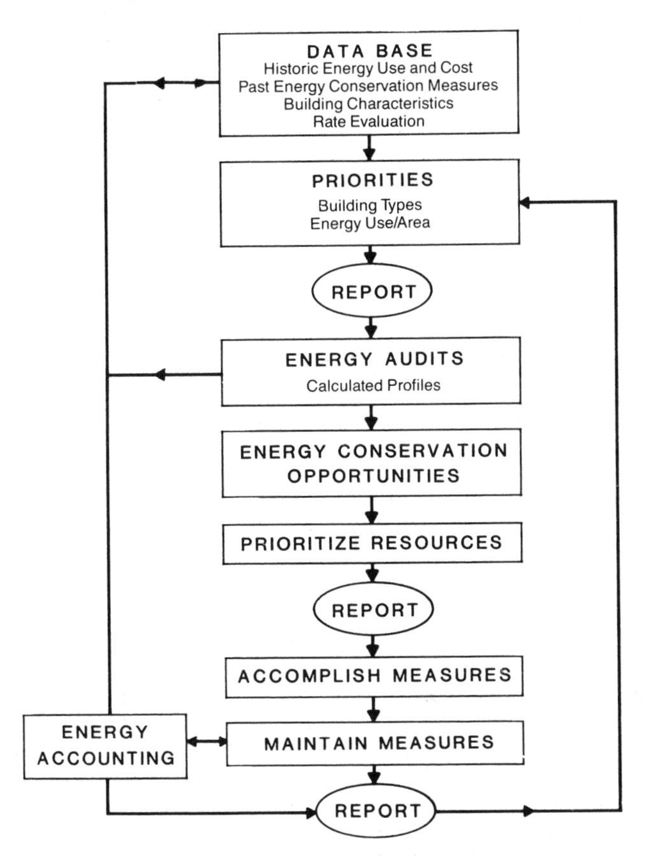

Fig. 1 Energy Management Program

ORGANIZATION

Since energy management is performed in existing facilities, most of this chapter is devoted to these facilities. Information on energy conservation in new design can be found in all volumes of the ASHRAE *Handbook*. However, a true energy conserving design allows the principles of energy management to be applied within the facility without modification. The area most likely to be overlooked in new design is the ability to measure and monitor energy consumption and trends for each energy use category.

To be effective, energy management must be given the same emphasis as management of any other cost/profit center. In this regard, the functions of top management are as follows:

- To establish the energy cost/profit center
- To assign direct responsibility for the program
- To hire or assign an energy manager
- To allocate resources
- To ensure that the energy management program is promulgated to all departments to provide necessary support to achieve effective results
- To monitor the cost-effectiveness of the program
- To set up an ongoing reporting and analysis function to monitor the energy management program

An effective energy management program requires that the manager (supported by a suitable budget) act and be held accountable for those actions. It is common for a facility to allocate 3 to 10% of the annual energy cost for the administration of an energy management program. In addition to salaries and other administrative expenses, the budget should include continuing education for the energy manager and staff. However, other resources are needed. In fact, money is often the least constraining factor, since properly managed energy conservation activities soon pay for themselves. The necessary resources are management attention, skills (both inside and outside the organization), manpower, and money.

If it is not possible to add a full-time, first-line manager to the staff, an existing employee should be considered for either a full- or part-time position. This person must be trained to organize an energy management program. Energy management should not be an alternate or collateral duty of a person who is already fully occupied.

Figure 2 shows the organizational functions for effective energy management. The solid lines indicate normal reporting relationships within the organization; the dotted lines indicate new relationships created by energy management. The arrows indicate the primary directions in which initiatives related to energy conser-

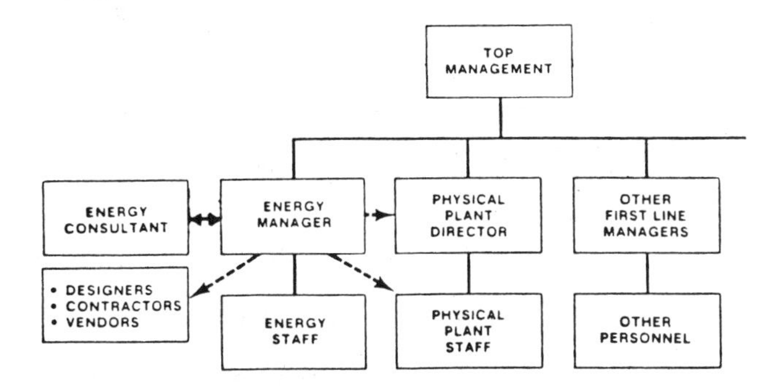

Fig. 2 Organizational Relationships

The preparation of this chapter is assigned to TC 9.6, Systems Energy Utilization.

vation normally occur. This chart shows the ideal organization. In actuality, however, the functions shown may overlap, especially in smaller organizations. For example, the assistant principal of a high school may also serve as the energy manager and physical plant director. However, it is important that all functions be covered.

Energy Manager

The functions of an energy manager fall into the following broad categories:

1. Technical
2. Policy-related
3. Planning and purchasing
4. Public relations

Technical functions include:

1. Conducting energy audits and identifying energy conservation opportunities (ECOs)
2. Establishing a baseline from which energy-saving improvements can be measured
3. Acting as an in-house technical consultant on new energy technologies, alternative fuel sources, and energy-efficient practices
4. Evaluating the energy efficiency of proposed new construction, building expansion, remodeling, and new equipment purchases
5. Setting performance standards for efficient operation and maintenance of machinery and facilities
6. Reviewing state-of-the-art energy management hardware
7. Selecting the most appropriate technology
8. Reviewing operation and maintenance
9. Implementing energy conservation measures (ECMs)
10. Establishing an energy accounting program for continuing analysis of energy use and the results of energy conservation measures
11. Maintaining the effectiveness of energy conservation measures

Policy-related functions include:

1. Fulfilling energy policy established by top management
2. Monitoring federal and state legislation and regulatory activities, and recommending policy/response on such issues
3. Representing the organization in energy associations
4. Administering government-mandated reporting programs

Planning and purchasing functions include:

1. Monitoring energy supplies and costs to take advantage of fuel-switching and load management opportunities
2. Ensuring that energy-using systems and equipment are purchased based on economics and ability to perform the required functions, not simply on the lowest initial cost
3. Negotiating or advising on major utility contracts
4. Developing contingency plans for supply interruptions or shortages
5. Forecasting the organization's short- and long-range energy requirements and costs
6. Developing short- and long-range energy conservation plans and budgets
7. Periodic reporting to top management

Public relations functions include:

1. Making fellow employees aware of the benefits of efficient energy use
2. Establishing a mechanism to elicit and evaluate energy conservation suggestions from employees
3. Recognizing successful energy conservation projects through awards to plants or employees
4. Setting up a formal reporting mechanism to top management
5. Establishing an energy communications network within the organization, including bulletins, manuals, and conferences
6. Making the community aware of the energy conservation achievements of the organization with press releases and appearances at civic groups

General qualifications of the staff energy manager include:

1. A technical background, preferably in engineering, with experience in energy-efficient design of building systems and processes
2. Practical, hands-on experience with systems and equipment
3. Goal orientation
4. Ability to work with people at all levels—from operations and maintenance personnel to top management

Desirable educational and professional qualifications of the staff energy manager include:

1. Bachelor of Science degree from an accredited four-year college, preferably in mechanical, electrical, industrial, or chemical engineering
2. Thorough knowledge of the principles and practices of energy resource planning and conservation
3. Familiarity with the administrative governing organization
4. Ability to analyze and compile technical and statistical information and reports with particular regard to energy usage
5. Knowledge of resources and information relating to energy conservation and planning
6. Ability to develop and establish effective working relationships with other employees and to motivate people to act without direct control
7. Ability to function as a goal-oriented manager
8. Ability to interpret plans and specifications for buildings facilities
9. Knowledge of the basic types of automatic controls and systems instrumentation
10. Knowledge of energy-related metering equipment and practices
11. Knowledge of the organization's manufacturing processes
12. Knowledge of building systems design and operation and/or maintenance of building systems
13. Personal qualities of interest and enthusiasm for efficient energy use and the ability to present ideas to all organization levels

Energy Consultant

In many instances, an energy manager needs outside assistance to conduct the entire energy management program. An energy consultant may be called on to assist in any of the energy manager's functions and, in addition, may be responsible for training the energy management, operations, and maintenance staffs. In some cases, the energy consultant may be responsible for implementation and monitoring of the energy management program.

The basic qualifications of an energy consultant should be similar to those of the energy manager. The consultant should be an objective party or engineer with no connections to the sale of product equipment or systems.

Specialists with narrow areas of expertise may not contribute effectively to the system interaction of a comprehensive and integrated energy management program. For example, a lighting engineer could develop methods to reduce lighting energy while simultaneously causing other systems that interact with the lighting system to increase their energy consumption. Optimum energy conservation can only be accomplished if system interaction is thoroughly understood and accounted for by the energy consultant and manager.

Motivation

The success of an energy management program depends on the interests and motivation of the people implementing it (Turner 1982). Participation and communication are key ingredients. Employees can also be stimulated to support an energy management program through awareness, by informing them of:

1. The amount of energy they use
2. Costs
3. The critical part energy has in the continued viability of their jobs
4. What energy saving means in their operations
5. The relationship between the production rate and energy consumption
6. Benefits, such as greater comfort, if they participate

To the extent that it is practical, energy management activities can be made a part of each supervisor's performance or job standards. If the supervisor knows that top management is solidly behind the energy management program and the overall performance rating depends, to some extent, on the energy savings the department or group achieves, the supervisor will motivate employee interest and cooperation.

IMPLEMENTING A PROGRAM

There are five basic stages in implementing an energy management program:

1. Develop a thorough understanding of how energy is used.
2. Conduct a planned, comprehensive search to identify all potential opportunities for energy conservation activities.
3. Identify, acquire, allocate, and prioritize the resources necessary to implement and maintain energy conservation opportunities.
4. Accomplish the energy conservation measures in rational order. This is usually a series of independent activities that take place over a period of years.
5. Monitor and maintain the energy conservation measures that have been taken. Reevaluate them as building functions change over time.

Energy efficiency and/or energy conservation efforts should not be equated with discomfort, nor should they interfere with the primary function of the organization or facility. Energy conservation activities that disrupt or impede normal functions of workers and/or processes and adversely affect productivity constitute false economies.

Database

In developing an energy management program, a base of past energy usage and cost should be developed. Any reliable utility data that is applicable should be examined. Generally, this is monthly data; it should be analyzed over several years. A base year should be established to be used as a reference point for future energy conservation and energy cost avoidance activities. In tabulating such data, the actual dates of meter readings should be recorded; any periods during which consumption was estimated rather than measured should be noted.

If energy is available for more than one building and/or department within the authority of the energy manager, each of these should be tabulated separately. Initial tabulations should include both energy (Btu) and cost per unit area. (In an industrial facility, this may be energy and cost per unit of goods produced.) Available information on variables that may have affected past energy use should also be tabulated. These might include heating or cooling degree-days, percent occupancy for a hotel, or quantity of goods produced in a production facility. Since such variables may not be directly proportional to energy use, it is best to plot information separately or to superimpose one plot over another, rather than developing values as Btu per square foot per degree day, for example. As such data are tabulated, energy accounting procedures for regular collection and use of future data should be developed.

Comparing a building's energy use with many different buildings is a valuable way to check its relative efficiency. The Energy Information Administration (EIA) within the U.S. Department of Energy collects data on buildings in all sectors. This information is summarized in DOE/EIA-0246(86) and DOE/EIA-0318 (86) for nonresidential buildings and in DOE/EIA-0321/1(87) for households.

Tables 1 and 2 list physical characteristics of the buildings surveyed. Table 3 lists measured energy consumption by building type. Table 4 lists energy sources for various end uses. The EIA also collects data on household energy consumption, which is summarized in Table 5.

All the data presented in these tables are derived from detailed reports of consumption patterns in buildings. Before using the data, however, it is important to understand how it was derived. For example, all the household energy consumption data presented in Table 5 are average data, and they may not reflect the variations in appliances or fuel situations for different buildings. Therefore, when using the data, verify the correct use of it with the original EIA documents. In addition, because these surveys are performed regularly, newer data may be available.

Table 1 Commercial Building Characteristics in the United States (1986)

Building Type	Size, ft²				% Buildings Less than 50%	
	Per Building[a]		Per Worker[b]			
	Mean	Median	Aggregate	Median	Heated[c]	Cooled[d]
Assembly	12,800	5,600	1,706	2,902	12	44
Educational	30,300	10,000	1,072	933	4	49
Food sales	7,000	2,500	471	480	10	23
Food service	6,400	4,400	360	375	12	26
Health care	40,500	5,000	417	423	8	15
Lodging	20,400	8,000	1,223	2,001	7	36
Mercantile/Service	9,900	4,500	807	901	23	61
Office	15,500	4,700	382	438	10	21
Public order/Safety	12,400	5,400	498	875		
Warehouse	16,400	5,100	1,684	3,002	72	94
Other	16,800	4,900	1,102	1,601		
Vacant	12,300	4,800	3,934		54	71
All buildings	14,000	5,000	793	1,001	26	53

Mean Area for Specific Functions[e]	
Inpatient health	1040 ft²/bed
Skilled nursing	452 ft²/bed
Education	92 ft²/seat
Food service	53 ft²/seat
Lodging	518 ft²/room

Source: DOE/EIA-0246 (86). Based on 6,072 buildings.
[a]Listed in Table 17 of the source document.
[b]Listed in Table 20.
[c]Listed in Table 43.
[d]Listed in Table 44.
[e]Listed in Table 9.

ANSI/ASHRAE *Standard* 105-1984, Standard Methods of Measuring and Expressing Building Energy Performance, contains information that allows uniform, consistent expressions of energy consumption, both in proposed and in existing buildings. Its use is recommended. However, the data presented here are not in accordance with this standard.

The quality of published energy consumption data for buildings varies because they are collected for different purposes by people with different levels of technical knowledge of buildings. The data presented here are primarily national data. In some cases, local energy consumption data may be available from local utility companies or state or provincial energy offices.

Table 2 1986 Commercial Building Characteristics in the United States

		Percent
Predominant Lighting		
T13	Standard flourescent	64
	Flourescent	27
	Standard incandescent	41
	Efficient incandescent	10
	High intensity discharge	6
	Other	1
District Heat Cool		
T6	Any source	2
	Steam	2
	Hot water	0
	Chilled water	0
Weekly Schedules		
T10	Closed	8
	Less than 24 h	
	Mon to Fri	28
	Mon to Sat	27
	Mon to Sun	17
	Open 24 h	8
	Other	11
Weekly Operating Hours		
T11	39 or fewer	19
	40 to 48	27
	49 to 60	23
	61 to 84	14
	85 to 167	9
	Continuous	8
Building Size, ft^2		
T11	1001 to 5000	53
	5001 to 10,000	23
	10,001 to 25,000	14
	25,001 to 50,000	6
	50,001 to 100,000	3
	100,001 to 200,000	1
	200,001 to 500,000	1
	Over 500,000	0
Occupancy Nongovernment Building		
T13	Owner occupied	
	Single	88
	Multiple	12
	Nonowner occupied	
	Single	72
	Multiple	24
	Vacant	4
Year Constructed		
T11	1900 or before	4
	1901 to 1920	6
	1921 to 1945	15
	1946 to 1960	21
	1961 to 1970	18
	1971 to 1973	6
	1974 to 1979	14
	1980 to 1983	9
	1984 to 1986	7
Number of Floors		
T13	One	64
	Two	24
	Three	8
	More than three	4

		Percent
Energy Sources Used		
T11	Electricity	99
	Natural gas	55
	Fuel oil	13
	District heat	2
	District cool	0
	Propane	9
	Other	4
Heating Sources Used		
T11	Electricity	29
	Natural gas	51
	Fuel oil	13
	District heat	2
	Propane	6
	Other	4
	Not heated	9
Cooling Sources Used		
T11	Electricity	68
	Natural gas	3
	District cool	0
	Other	4
	Not cooled	28
Water Heating Sources		
T11	Electricity	35
	Natural gas	33
	Fuel oil	3
	District heat	1
	Propane	3
	None	28
Cooking Energy Sources		
T11	Electricity	6
	Natural gas	8
	Propane	2
	None	28
Percent Heated		
T11	Not heated	8
	1 to 50% ft^2	15
	51 to 99% ft^2	11
	100% ft^2	65
Percent Cooled		
T11	Not cooled	28
	1 to 50% ft^2	24
	51 to 99% ft^2	13
	100% ft^2	36
Percent Lit Open Hours		
T13	Not lit	2
	1 to 50% ft^2	16
	51 to 99% ft^2	16
	100% ft^2	66
Heating Equipment		
T13	Furnaces	45
	Boilers	16
	In space	27
	Packaged	13
	Air heat pump	8
	District heat	2

		Percent
Cooling Equipment		
T13	Central	28
	Individual	23
	Packaged	18
	Air heat pumps	8
	District cooling	0
HVAC Distribution		
T13	Forced air	
	Heating only	15
	Cooling only	4
	Heat and cool	44
	Baseboard/Radiators	
	Steam	6
	Hot water	7
	Fan coil units	
	Heating only	5
	Cooling only	1
	Heat and cool	4
	Heating panels	5
Shell Conservation		
T13	Any	82
	Roof/Ceiling insulation	68
	Wall insulation	50
	Multiglazing	31
	Glass film/Shade	22
	Awnings/Shading	32
	Weatherstripping	63
	Other	3
HVAC Conservation		
T13	Any	54
	Preventive maintenance	52
	Heat recovery	4
	Energy management system	5
	Time clock/Thermostat	2
	Economizer cycle	0
	Other	2
Lighting Conservation		
T13	Any	36
	High efficiency ballasts	25
	Delamping	8
	Natural light sensors	4
	Other controls	10
	Other	2
In Response to Audit		
T59	Any	5
	HVAC	2
	Building shell	2
	Lighting	3
Occupant HVAC Control		
T13	Heating only	16
	Cooling only	2
	Both	50
Reduction in Off Hours		
T13	Heating only	19
	Cooling only	3
	Both	58

Source: T# designates the number of the table from DOE/EIA-0318(86), except Tables 6, 10, and 59, which are from DOE/EIA-0246(86).

Table 3 Commercial Building Energy Use in the United States (1986)

Type	Percent by Bldg	Percent by Area	All Buildings Using Any Fuel[a] Building Elec	Gas	Total	Electric[b] Peak W/ft²	Electric[b] Total kWh/ft²	Intensity By Region, 1000 Btu/ft² Major Fuel Energy NE	NW	SO	WE	Electric Energy NE	NW	SO	WE	Natural Gas NE	NW	SO	WE	Total Electric Use[a] Heat	HVAC	AC	Total Gas Heat[a]
Assembly	14	13	23	31	44	4.2	4.7	76	58	41	59	28	12	22	33	33	44	24	23	29	33	22	35
Education	6	13	25	46	59	3.4	6.0	103	103	67	74	23	23	26	28	33	63	35	47	32	35	23	52
Food sales	3	1	139	92	202	11.2	49.9		229				129	156		104				163	163	107	82
Food service	5	2	94	124	184	11.3	30.6	186	209	157	286	64	104	96	103	144	120	73	189	132	138	83	120
Health care	1	4	63	127	160	4.3	9.7	233	314	158	188	62	73	65	45	97	161	103	134	41	39	71	135
Lodging	3	4	43	56	81	3.9	10.1	99	143	105	102	38	28	54	46	46	67	46	69	56	64	32	68
Mercantile/ Service	31	22	42	38	68	3.8	7.7	79	85	80	62	34	40	49	42	29	45	42	26	51	52	39	40
Office	15	16	68	45	94	5.0	10.6	109	118	92	107	68	56	69	78	27	70	37	33	72	74	67	48
Public order/ Safety	1	1	45	64	81	4.4	9.5					31										42	71
Warehouse	13	15	30	30	46	2.1	3.6	52	64	58	35	23	22	40	26	24	45	26	13	37	40	35	31
Other	2	3	49	52	76	2.6	7.3	85	94	119	104	33	47	63	54		41		61	30	37	91	70
Vacant	6	5	17	31	35	2.8	5.0	38	55	30	52	15	15	14	27		41	23	31	24	27	20	35
All buildings	100	100	42	46	72	4.2	8.2	91	102	79	85	38	37	46	48	35	60	41	41	51	56	43	49
Source table	1	1	3	3	11	43	43	19	19	19	19	20	20	20	20	21	21	21	21	30	31	32	33

Special Occupancy	Elec	Gas	Total
Inpatient health	67	123	190
Skilled nursing	49	66	115

By Special Measures, Millions of Btu	Elec	Gas	Total
Inpatient health per bed	69	127	196
Skilled nursing per bed	22	29	51
Education per seat	2	4	6
Food service per seat	5	7	12
Lodging per room	22	30	52

Source: DOE/EIA-0318(86). Data based on 6072 buildings. The standard error for the nationwide total electricity or natural gas consumption is 6%. Missing values represent inadequate data.

[a]Values are in thousands of Btu per square foot of building.

[b]Total = (Total electricity + Total energy for gas)/Total floorspace using energy or gas with other fuels.

[c]Values are the median for demand metered buildings only.

At this point in the development of an energy management program, it is useful to compile a list of previously accomplished energy conservation measures and the actual energy and/or cost savings of such measures. These items should be studied during subsequent energy audits to determine their present effectiveness and the effort(s) necessary to maintain and/or improve them.

Since most energy management activities are dictated by economics, the energy manager must understand the utility rates that apply to each facility. Special rates are commonly applied for such variables as interruptible service, on peak/off peak, summer/winter, and peak demand, to name a few. There are more than 1000 electric rate variations in the United States. The energy manager should work with local utilities to develop the most cost-effective methods of metering and billing and to enable energy cost avoidance to be calculated effectively.

For example, it is common for electric utilities to meter both electric consumption (kWh) and demand (kWd). Demand is the peak rate of consumption, typically integrated over a 15- or 30-min. period. Electric utilities may also establish a ratchet billing procedure for demand. A simplified example of ratchet demand billing would state that the billing demand is established as either the *actual demand* for the month in question or a percentage of the *highest demand* during the previous 11 months, whichever is greatest. Figure 3 illustrates a gas heated, electrically cooled building with the highest electric demands occurring in the summer, and shows actual demand versus billing demand under an 85% ratchet. The winter demand is approximately 700 kWd each month, with chiller operation causing the August demand to peak at 1000 kWd. Since an 85% ratchet applies, all months following August with actual demand lower than 850 kWd are billed at 850 kWd. Therefore, the following can be concluded:

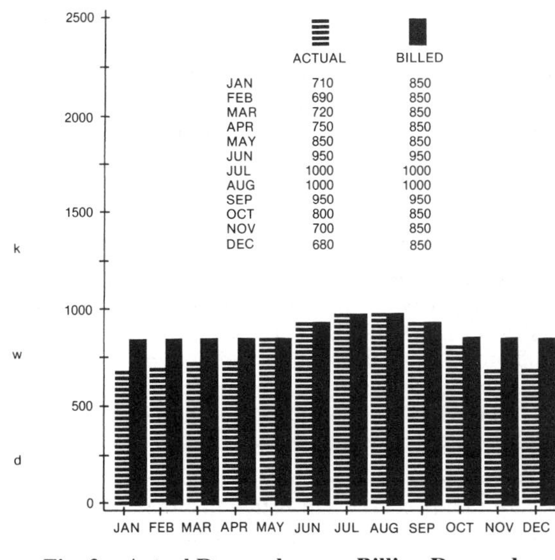

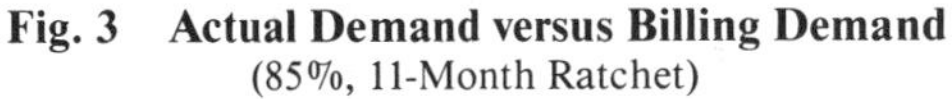
Fig. 3 Actual Demand versus Billing Demand
(85%, 11-Month Ratchet)

1. Energy management that reduces winter demand would not reduce the billed cost of demand, unless it also reduced sum-

Table 4 Energy Sources for End Uses—Percent of Buildings

Sources	Buildings Using Energy	Space Heating	Water Heating	Cooling	Cooling	Manufacturing
Electricity	99	29	35	68	6	3
Natural gas	55	51	33	3	8	1
Fuel oil	13	13	3	0	0	0
Propane	9	96	3	0	2	0
District heat	2	2	1	0	0	0
Chilled water	0	0	0	0	0	0
Minor fuels	4	4	0	4	0	0

Source: DOE/EIA-0318 (86), Table 12.

mer demand; and if such an opportunity was implemented in September, it would not produce savings in demand billings for the first 11 months.

2. Energy management that reduces peak demand in each of the summer months (for example, 50 kWd reduction in chiller peaks) results in savings in demand billings throughout the year.

There are many variations of the above billing methods, and it is important to understand applicable rates.

Priorities

Having established a database, the energy manager should assign priorities to future work efforts. If there is more than one building or department under the energy manager's care, the database for energy use and cost for each should be compared on an overall basis and on the basis of energy use and cost per unit area, cost per unit of production, or some other index that demonstrates an acceptable level of accuracy. Comparisons should also be made with realistic energy targets, if they are known. From such comparisons, it is often possible to set priorities that use the available resources most effectively.

At this point, a report should be prepared for top management outlining the data collected, the priorities assigned, and plans for continued development of the energy management program and projected budget needs. This should be the beginning of a regular monthly, quarterly, or semiannual reporting procedure.

Energy Audits

Three levels of energy audits or analysis have been defined (ASHRAE 1989). Depending on the physical and energy-use characteristics of a building and the needs and resources of the owner, these steps require different levels of effort. Depending on the level of effort expended in performing each of these steps, energy analysis can generally be classified into the following three categories, *following* a preliminary energy use evaluation.

Level I—Walk-Through Assessment. Assess a building's energy cost and efficiency by analyzing energy bills and conducting a brief survey of the building. A Level I energy analysis will identify and provide a savings and cost analysis of low-cost/no-cost measures. It will also provide a listing of potential capital improvements that merit further consideration, along with an initial judgment of potential costs and savings.

Level II—Energy Survey and Analysis. This includes a more detailed building survey and energy analysis. A breakdown of energy use within the building is provided. A Level II energy analysis identifies and provides the savings and cost analysis of all practical measures that meet the owners constraints and economic criteria, along with a discussion of any effect on operation and maintenance procedures. It also provides a listing of potential capital-intensive improvements that require more thorough data collections and analysis, along with an initial judgment of potential costs and savings. This level of analysis will be adequate for most buildings and measures.

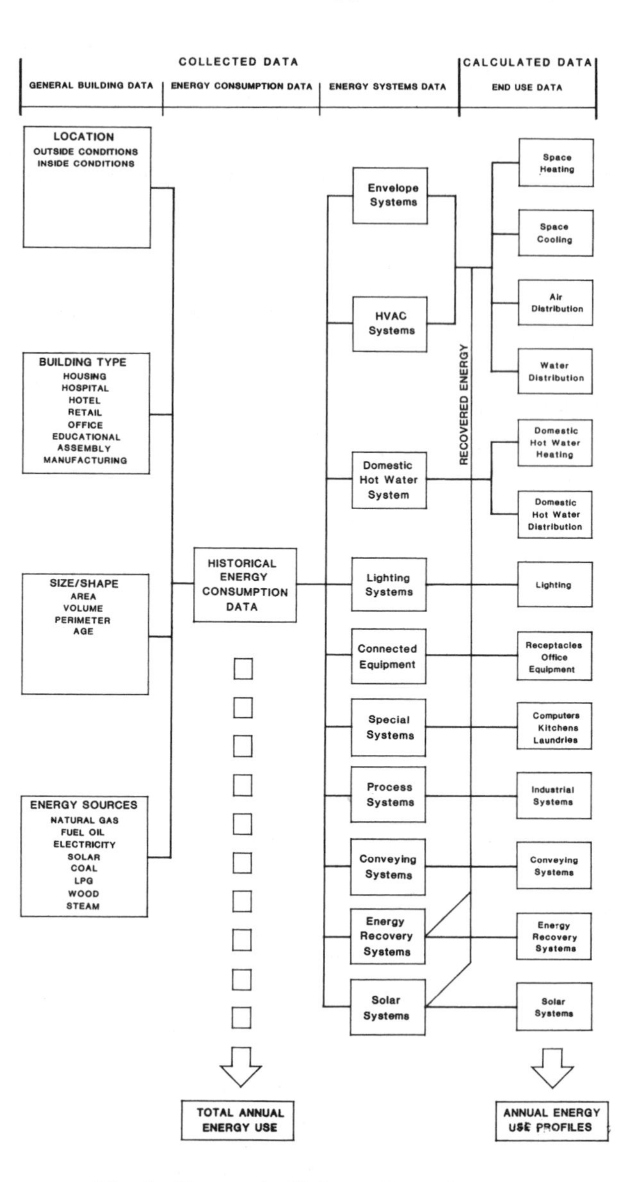

Fig. 4 Energy Audit Input Procedures

Level III—Detailed Analysis of Capital-Intensive Modifications. This level of analysis focuses on potential capital-intensive projects identified during Level II and involves more detailed field data gathering and engineering analysis. It provides detailed project cost and savings information with a high level of confidence sufficient for major capital investment decisions.

These levels do not have sharp boundaries between them. They are general categories for identifying the type of information that can be expected and an indication of the level of confidence in the results; that is, various measures may be subjected to different levels of analysis during an energy analysis of a particular building.

In the complete development of an energy management program, Level II audits should be performed on all facilities, while Level I audits are useful in establishing the program. Figure 4 illustrates Level II energy audit input procedures in which the following data are collected:

1. General building data
2. Historic energy consumption data
3. Energy systems data

The collected data is used to calculate an energy use profile that includes all end use categories. From the energy use profiles, it is possible to develop and evaluate energy conservation opportunities.

In conducting an energy audit, a thorough *systems* approach produces the best results. This approach has been described as beginning at the end rather than at the beginning. As an example of this approach, consider a factory with steam boilers in constant operation. An expedient (and often cost-effective) approach would be to measure the combustion efficiency of each boiler and to improve boiler efficiency. However, beginning at the end would require observing all or most of the end uses of steam in the plant. It is possible that this would result in the discovery of considerable quantities of steam being wasted by venting to the atmosphere, venting through defective steam traps, uninsulated lines, and passing through unused heat exchangers. Elimination of such end use waste could produce greater savings than those easily and quickly developed by improving boiler efficiency. When using this approach, care must be taken to make cost-effective use of the energy auditor's time. It may not be cost-effective to track down every end use.

When conducting an energy audit, it is important to become familiar with operating and maintenance procedures and personnel; the energy manager can then recommend, through the appropriate departmental channels, energy-saving operating and maintenance procedures. The energy manager should determine, through continued personal observation, the effectiveness of the recommendations.

Stewart *et al.* (1984) tabulated 139 different energy audit input procedures and forms for 10 different building types, in each of which 62 factors are used. They discuss features of selected audit forms and can help in developing or obtaining an audit procedure.

To calculate the energy cost avoidance of various energy conservation opportunities, it is helpful to develop an energy cost distribution chart similar to that shown for a hospital in Figure 5. Preliminary information of this nature can be developed from monthly utility data by calculating end use energy profiles (Shehadi *et al.* 1984).

Much of the information needed to develop this energy cost (or use) distribution can be estimated from utility bills.

Analysis of electrical operating costs starts with the recording of data from the bills on a form similar to that in Table 6. By dividing the consumption by the days between readings, the average daily consumption can be calculated. This consumption should

Fig. 5 Energy Cost Distribution

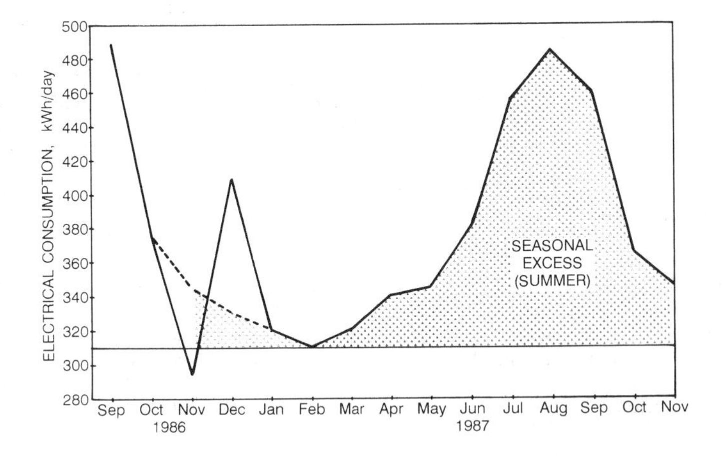

Fig. 6 Average Daily Electrical Consumption

be plotted to detect errors in meter readings or reading dates and to detect consumption variances (see Figure 6). For this example, 312 kWh/day is chosen as the "base electrical consumption" to cover year-round electrical needs such as lighting, business machines, domestic hot water, terminal reheat, security, and safety lighting. At this point, consumptions or spans that appear to be in error should be reexamined and corrected, as necessary. If the reading date for the 10Nov86 bill in Table 6 was 05Nov86, the curve in Figure 6 would be more continuous. On the basis of a 05Nov86 reading, the minimum daily consumption of 312 kWh on the continuous curve (Figure 6) occurred in the February billing.

To start the analysis, the monthly base consumption is calculated (base daily consumption times billing days) and is subtracted from each monthly total to obtain the difference. These differences fall under either summer excess or winter excess, depending on the season. Excess consumption in the summer is primarily due to the air-conditioning load.

A similar analysis is made of the actual monthly demand. In Table 6, the base demand is 33.0 kW, and it is usually found in the same or adjacent months as the month with the base consumption.

Substantial errors might arise if missing bills are not accounted for.

The base consumption can be further analyzed by calculating the electrical load factor (ELF) associated with this consumption. If the base demand had operated 24 h a day, then base consumption would be:

$$33.0 \text{ kW} \times 24 \text{ h/day} = 792 \text{ kWh/day} \tag{1}$$

But if the daily base consumption is 312 kWh/day, then the electrical load factor is:

$$\text{ELF} = \frac{\text{Base consumption}}{\text{Base demand} \times 24} = \frac{312}{792} = 0.394 \text{ or } 39.4\% \tag{2}$$

The example indicates that this electrical load factor (39.4%) is higher than the occupancy factor (29.8%).

$$\text{Occupancy factor} = \frac{\text{Occupied hours}}{24 \text{ h} \times 7 \text{ days}} = \frac{50}{168} = 0.298 = 29.8\% \tag{3}$$

One reason for this difference may be that the lights are left illuminated beyond the occupied hours.

Because of the extra air-conditioning demand, summer demand is 46.8 kW. If this additional demand of 13.8 kW had operated 24 h each day, the summer extra would equal 13.8 × 24 or 331.2

Table 5 Household Energy Consumption (1987)

Characteristics	Table 11 Millions of Btu per Household								By Energy End Use, Million Btu							
	Use of All Fuels			Use of Electricity					Table 24 Natural Gas				Table 25 Electricity			
	All	Main Heat Gas*	Main Heat Oil	All	With Gas* Heat	With Elec HVAC	With Elec Heat	With Oil Heat	All	Space Heat	Water Heat	Appli-ances	Space Heat	Water Heat	Appli-ances	Air Cond.
Total U.S. Households	100.8	117.9	126.4	30.5	24.3	52.5	40.2	25.7	84.3	59.0	19.2	5.9	3.1	3.5	19.0	4.8
Census Region and Division																
Northeast	124.4	138.8	134.9	23.3	18.8	49.1	51.2	20.4	88.5	62.8	18.4	7.3	2.5	2.2	16.6	2.0
New England	121.0	130.2	135.9	22.9	16.9	47.6	42.2	21.9	86.8	60.0	20.3	6.6	2.3	2.6	16.9	1.2
Middle Atlantic	125.4	140.3	134.5	23.4	19.2	49.4	57.2	19.7	88.8	63.3	18.1	7.4	2.6	2.1	16.5	2.3
Midwest	122.5	132.1	122.6	27.4	22.5	64.9	63.5	36.0	106.8	82.4	19.2	5.1	2.0	2.7	18.8	3.9
East North Central	125.3	134.5	126.4	26.4	20.9	65.2	65.9	37.2	110.5	85.2	19.8	5.4	2.1	2.9	18.4	3.1
West North Central	115.8	126.1	108.0	29.8	26.5	64.3	Q	31.4	97.5	75.2	17.8	4.5	1.8	2.3	19.9	5.8
South	84.3	111.0	103.9	39.5	32.6	53.4	35.8	35.4	72.3	49.1	17.5	5.1	4.4	5.3	20.9	8.9
South Atlantic	81.0	113.9	103.2	38.9	29.7	53.8	29.7	33.7	77.0	56.4	16.4	4.2	4.4	6.3	20.4	7.9
East South Central	84.1	112.0	106.0	43.7	33.7	60.6	46.1	45.2	76.1	55.8	15.6	4.6	6.8	6.9	21.4	8.7
West South Central	90.1	108.0	Q	37.5	34.7	46.7	Q	Q	66.9	40.4	19.2	6.2	2.7	2.6	21.4	10.8
West	77.8	91.5	92.3	26.4	21.1	43.6	35.3	39.6	65.5	37.0	21.7	6.4	3.0	2.7	18.7	2.0
Mountain	103.5	116.1	Q	27.8	22.5	50.4	42.7	Q	90.2	63.9	21.0	4.4	2.1	2.1	20.6	2.9
Pacific	69.6	82.2	89.4	26.0	20.6	40.4	34.8	40.6	56.8	27.6	22.0	7.1	3.3	2.8	18.1	1.7
Climate Zone																
<2000 CDD, >7000 HDD	110.4	126.8	121.5	27.2	22.1	61.8	46.6	27.9	100.1	80.7	16.2	3.2	2.2	4.7	18.9	1.5
5500 to 7000 HDD	126.0	136.6	145.7	26.5	20.7	63.5	51.9	26.9	109.5	82.8	20.7	5.8	2.9	2.5	18.5	2.7
4000 to 5499 HDD	106.7	127.6	123.3	30.6	24.9	57.3	42.4	22.0	83.1	58.7	17.9	6.5	4.0	3.5	18.5	4.5
Under 4000 HDD	77.4	87.7	93.6	29.4	22.9	48.3	23.3	33.6	59.7	32.9	20.5	6.3	3.1	3.6	18.4	4.3
>2000 CDD, <4000 HDD	73.7	98.6	68.9	39.5	35.2	49.2	22.3	37.6	58.1	33.9	17.5	5.8	2.9	4.2	21.4	11.0
By Status of Unit																
Single family detached	114.5	132.4	139.0	35.7	29.3	66.7	56.2	32.5	98.2	70.1	21.4	6.6	3.3	3.9	22.8	5.7
Owned	116.6	134.4	142.5	36.8	30.2	67.4	56.5	32.4	99.2	71.3	21.1	6.6	3.4	3.9	23.4	6.1
Rented	101.3	120.6	108.6	28.5	23.3	57.4	54.6	33.1	92.2	62.5	23.2	6.4	2.4	4.1	18.5	3.5
Building 5 or more units	64.2	77.6	96.8	18.5	12.7	31.4	23.9	8.9	48.5	30.2	14.2	3.5	2.4	2.0	10.4	3.6
Owned	67.1	97.8	83.7	23.6	19.5	32.9	Q	9.9	48.6	30.3	14.0	4.3	3.0	1.6	14.9	4.0
Rented	64.0	76.6	97.9	18.1	12.4	31.2	24.3	8.8	48.5	30.1	14.2	3.4	2.4	2.0	10.1	3.5
Heated Area of Residence																
Fewer than 600	63.1	74.3	99.0	16.5	11.9	27.5	20.1	9.7	48.8	29.1	15.4	4.2	1.9	2.5	10.2	1.9
600 to 999	77.2	91.3	96.0	22.9	17.1	39.5	35.9	19.7	65.4	43.5	16.7	4.9	2.5	3.2	13.8	3.4
1000 to 1599	92.8	108.7	119.5	32.3	24.0	53.7	48.1	26.6	78.5	53.8	18.9	5.6	3.6	4.2	19.5	5.1
1600 to 1999	113.9	131.4	127.9	35.5	28.2	63.7	60.2	27.8	95.7	69.0	20.0	6.6	3.8	3.6	22.0	6.0
2000 to 2399	129.7	144.1	143.8	36.7	31.6	66.1	Q	28.9	106.0	76.0	22.9	7.1	3.3	3.1	24.4	6.0
2400 to 2999	145.0	160.7	152.1	37.6	32.4	79.6	Q	33.1	121.2	90.0	22.7	7.8	3.2	3.2	25.3	5.8
3000 or more	169.0	184.1	207.4	46.8	38.6	87.9	Q	42.7	132.3	100.7	24.0	7.4	4.0	3.7	30.5	8.7
Year of Construction																
1939 or Before	120.4	131.9	137.3	22.5	19.4	47.9	33.2	23.0	97.4	72.6	18.7	6.1	1.2	2.5	16.5	2.3
1940 to 1949	103.9	113.3	122.1	26.2	22.8	52.9	44.5	23.6	80.1	56.1	17.4	6.6	1.5	3.1	17.8	3.8
1950 to 1959	109.6	117.8	129.1	29.0	26.1	53.0	39.7	27.9	86.1	59.0	20.3	6.3	1.7	2.7	19.9	4.8
1960 to 1969	99.7	110.4	114.1	30.2	26.8	54.6	32.7	24.7	77.7	52.5	18.9	5.7	2.8	2.7	18.8	5.9
1970 to 1974	95.3	116.9	106.1	32.6	25.2	53.4	33.0	29.6	83.7	57.0	20.8	5.8	3.9	3.8	19.4	5.4
1975 to 1979	86.1	116.5	126.3	41.7	30.8	53.9	46.9	35.5	72.2	49.1	18.7	4.5	7.1	5.1	22.2	7.4
1980 to 1984	71.2	97.0	89.1	37.9	25.3	50.1	48.0	31.1	65.5	42.4	18.9	4.3	5.4	5.4	21.1	6.0
1985 or after	71.1	108.6	Q	39.9	22.7	51.7	Q	Q	82.0	50.7	23.9	7.5	6.6	6.4	20.8	6.1

Source: DOE/EIA-0321/1(87), October 1989.

Notes: 1. Gas* does not include LPG; site value of electricity calculated as 3412 Btu per kWh.
2. Estimates subject to sampling error (see source Report).
3. Q = inadequate data.

kWh/day. The ratio of each summer month's excess as a percent of 331.2 kWh yields the summer electrical load factor. Summer ELFs higher than the occupancy factor indicate that air conditioning is not shut off as early as possible in the evening. Winter excess demand and consumption are analyzed in the same way to yield winter monthly electrical load factors.

Base load energy use is the amount of energy consumed independent of weather. When a building has electric cooling and no electric heating, the base load energy use is normally the energy consumed during the winter months. The opposite is true for heating. The annual estimated base load energy consumption can be obtained by establishing the average monthly consumption during the nonheating or cooling months and multiplying by 12. Then, by subtracting the base load energy consumption from the total annual energy consumption, an accurate estimate of the heating or cooling energy consumption can be obtained for many buildings. This approach is not valid when building use differs from summer to winter; when there is cooling in operation 12 months of the year; or when there is space heating use during the summer months, as for reheat. In many cases, the baseload energy use analysis can be improved by using hourly load data that may be available from the utility company.

Although it is difficult to relate heating and cooling energy used in commercial buildings directly to severity of weather, Shehadi *et al.* (1984) suggest that this is possible, using a curve fitting method to calculate the balance point of a building (the balance point of

Table 5 Household Energy Consumption (1987) (*Concluded*)

	By Energy End Use										
	Table 26 Oil		Table 28	Table 29	Table 30	Table 31			Table 32 Air Conditioning		
			Btu/(HDD·ft²)			Water Heating Fuel, Million Btu				Type, Btu/(CDD·ft²)	
Characteristics	All, 10^6 Btu	Space Heat, 10^6 Btu	Main Gas*	Heating Electric	Fuel Oil	Gas*	Electric	LPG	Consump, 10^6 Btu	Central	Room
Total U.S. Households	70.3	60.5	10.0	3.4	9.1	22.2	9.8	18.5	7.6	3.1	3.1
Census Region and Division											
Northeast	96.0	78.0	9.6	2.5	9.4	23.7	9.3	16.6	3.8	3.0	3.1
New England	102.1	83.6	10.2	3.1	7.4	24.3	9.1	18.4	2.7	3.7	3.8
Middle Atlantic	93.6	75.8	9.6	2.4	10.4	23.6	9.4	Q	4.0	3.0	2.9
Midwest	51.8	51.4	8.7	3.3	6.2	21.5	10.3	19.5	5.7	3.1	3.1
East North Central	54.2	53.7	8.8	3.5	6.3	22.1	10.8	20.3	5.0	3.1	3.3
West North Central	41.6	41.2	8.3	2.8	5.7	20.2	9.0	18.2	7.2	3.3	2.8
South	37.0	36.2	12.6	3.9	12.6	21.1	9.8	17.9	10.9	3.4	3.3
South Atlantic	41.1	40.0	11.4	3.9	12.6	21.0	9.6	16.7	9.9	3.1	2.9
East South Central	25.8	25.7	13.2	3.9	12.6	21.3	10.4	17.1	10.3	3.8	3.5
West South Central	Q	Q	13.9	3.8	Q	21.0	9.3	19.7	12.7	3.6	3.7
West	31.1	29.9	10.1	3.4	5.4	23.2	9.8	18.9	5.3	2.4	2.7
Mountain	Q	Q	9.7	3.8	Q	22.6	9.2	20.1	7.4	2.1	1.8
Pacific	32.9	31.2	10.3	3.3	5.0	23.4	10.0	18.4	4.7	2.7	3.0
Climate Zone											
<2000 CDD, >7000 HDD	74.5	68.5	7.2	3.2	6.1	21.1	9.8	19.4	3.7	3.0	3.5
5500 to 7000 HDD	88.2	74.8	9.2	3.3	8.1	22.8	10.5	18.2	4.5	3.1	3.1
4000 to 5499 HDD	72.5	60.6	9.8	2.9	11.1	22.3	9.7	17.7	6.7	3.4	3.1
Under 4000 HDD	29.4	28.9	12.4	3.7	15.5	23.5	10.2	21.2	7.2	3.3	3.6
>2000 CDD <4000 HDD	12.6	Q	14.3	4.4	11.2	19.6	9.1	15.3	13.3	3.3	2.9
By Status of Unit											
Single family detached	71.8	63.5	9.7	3.5	8.0	24.5	10.6	19.0	9.0	2.8	2.9
Owned	74.9	65.6	9.2	3.4	7.9	24.2	10.5	19.9	9.2	2.8	2.8
Rented	50.5	48.5	13.5	5.0	9.5	26.8	11.4	15.8	7.4	3.2	3.6
Building 5 or more units	70.1	50.2	12.1	2.6	15.4	17.0	7.1	Q	5.0	4.5	3.8
Owned	64.5	47.1	7.8	1.8	14.7	15.6	6.4	Q	4.8	3.1	3.5
Rented	70.5	50.4	12.5	2.8	15.5	17.1	7.2	Q	5.0	4.8	3.9
Heated Area of Residence											
Fewer than 600	63.6	48.1	20.6	4.6	22.8	18.0	8.2	12.2	4.5	5.8	4.7
600 to 999	53.3	45.4	14.8	4.2	13.9	19.6	8.9	16.1	5.4	4.8	4.1
1000 to 1599	59.7	51.5	12.1	3.9	11.4	22.5	9.9	21.3	7.9	3.5	3.3
1600 to 1999	69.2	62.5	9.4	3.2	8.9	22.9	9.9	17.9	8.7	3.2	2.7
2000 to 2399	82.8	73.0	8.2	2.6	7.6	25.3	11.0	22.8	8.6	2.8	2.7
2400 to 2999	85.1	73.3	7.6	2.5	6.0	25.7	11.2	18.6	9.2	2.8	2.6
3000 or more	127.7	112.8	5.5	2.1	5.9	25.9	12.6	23.5	11.6	2.6	2.0
Year of Construction											
1939 or Before	84.3	74.9	11.3	3.9	9.9	22.1	9.7	16.5	4.9	3.3	2.8
1940 to 1949	63.0	54.5	11.0	4.6	9.9	21.9	10.0	22.0	6.8	3.6	3.1
1950 to 1959	81.1	68.6	10.9	4.6	8.9	23.3	9.8	19.3	7.6	3.3	3.0
1960 to 1969	67.1	54.1	9.7	3.9	9.4	21.9	9.7	14.5	8.4	3.0	3.5
1970 to 1974	50.0	43.1	8.7	3.9	9.1	22.6	10.3	18.3	7.6	3.4	3.7
1975 to 1979	49.4	42.5	7.9	3.3	5.7	20.7	10.0	23.1	9.9	3.0	3.8
1980 to 1984	35.2	29.8	7.1	2.9	4.8	21.6	9.6	17.1	8.1	2.9	3.0
1985 or after	Q	Q	7.1	2.8	Q	26.1	9.4	Q	7.9	2.5	2.6

Source: DOE/EIA-0321/1(87) October 1989.
Notes: 1. Gas* does not include LPG; site value of electricity calculated as 3412 Btu per kWh.
2. Estimates subject to sampling error (see source Report).
3. Q = inadequate data.
4. Heating (HDD) and cooling (CDD) degree days calculated from 65 °F base.

a building is discussed in Chapter 28 of the 1989 ASHRAE *Handbook—Fundamentals*). The pitfalls of such an analysis are: (1) using estimated rather than actual utility usage data, (2) the necessity for using the actual dates of the metered information, and (3) nonregularity of building use and/or operation.

Having used only monthly data, a more detailed breakdown of energy usage requires some metered data to be collected on a daily basis (winter days versus summer days, weekdays versus weekends) and some hourly information to be collected to develop profiles for night (unoccupied), morning warmup, day (occupied), and the shutdown period. For large facilities, some submetering may also be desirable. For relatively constant loads, routine indications may be acceptable. For variable loads, kilowatt-hour meters are necessary. For more information on building energy use monitoring, see Chapter 37.

Energy Conservation Opportunities

It is possible to quantitatively evaluate various energy conservation opportunities from the end use energy profiles. Important considerations in this process are as follows:

- System interaction
- Utility rate structure
- Comprehensiveness
- Quality or usability of energy
- Life of the measure

Table 6 Example of Analytical Method for Analyzing Electrical Operating Costs

		Consumption, kWh			Air Conditioning			Demand, kW		
Billing Date	Billing Days	Total Actual	Actual per Day	Base[b]	Difference	ELF	Excess	Actual	Winter Excess	Summer Excess
12Aug86								46.2		0.6
11Sep86	30	14,700	490	9,360	5,340	53.7	1,859	46.8		1.2
10Oct86	29	10,860	374.5	9,048	1,812	18.9		46.8		1.2
10Nov86	31	9,120	294.2	8,112	1,008[a]	11.7		45.6		0.0
06Dec86	26	10,680	410.8	9,672	1,008[a]			33.0[c]	0.0	
09Jan87	34	10,860	319.4	10,608	252			33.0	0.0	
13Feb87	35	10,920	312[b]	10,920	000			33.6	0.6	
11Mar87	27	8,700	322.2	8,424	276			33.0	0.0	
10Apr87	30	10,140	338	9,360	780			33.6	0.6	
12May87	32	11,020	344.4	9,984	1,036	9.8		45.6[d]		0.0
12Jun87	31	11,760	379.4	9,672	2,088	20.3		46.8		1.2
13Jul87	31	14,160	456.8	9,672	4,488	43.7	893	46.8		1.2
11Aug87	29	14,340	494.5	9,048	5,292	55.2	1,937	46.8		1.2
10Sep87	30	13,740	458	9,360	4,380	44.1	904	46.2		0.6
14Oct87	34	12,120	356.5	10,608	1,512	13.4		45.6		0.0
10Nov87	27	9,360	346.7	8,424	936			33.6	0.6	

[a]Estimated from corrected monthly consumption from Figure 6
[b]Base electrical consumption = 312 kWh/day
[c]Base winter demand = 33 kW
[d]Base summer demand = 45.6 kW. Base summer excess demand = (45.6 − 33) = 12.6

- Maintainability
- Impact on building operation and appearance
- Relative payback

Accurate energy savings calculations can be made only if system interaction is fully understood and proper allowances are made for such interaction. Annual simulation models may be necessary to accurately estimate the interactions between various energy conservation opportunities. The resultant remaining energy use calculated should be verified against a separately calculated zero-based energy target.

Further, the actual energy cost avoidance may not be proportional to the energy saved, depending on the method of billing for energy used. Using average costs per unit of energy in calculating the energy cost avoidance of a particular measure is likely to result in incorrect values.

Figure 7 is one example of a comprehensive list of potential energy conservation opportunities; the references also contain such lists. An example of the quality of energy is the analysis of equipment requiring steam at various pressures; that is, steam at higher pressures is intrinsically more valuable.

In addition, previously accomplished energy conservation measures should be evaluated—first to ensure that they have remained effective, and second to consider revising them to reflect changing technology and/or building use.

Prioritize Resources

After establishing a list of energy conservation opportunities, the necessary resources should be evaluated, prioritized, and implemented.

In establishing priorities, the capital cost, cost-effectiveness, and resources available must be considered. Factors involved in evaluating the desirability of a particular energy conservation retrofit measure are as follows:

- Rate of return (simple payback, life cycle cost)
- Total savings (energy, cost avoidance)
- Initial cost (required investment)
- Other benefits (safety, comfort, improved system reliability, and improved productivity)
- Liabilities (increased maintenance costs and potential obsolescence)
- Risk of failure (confidence in predicted savings, rate of increase in energy costs, maintenance complications, and success of others with the same measures)

The resources available to accomplish an energy conservation retrofit measure should include the following:

- Management attention, commitment, and follow-through
- Skills
- Manpower
- Investment capital

Energy conservation measures may be financed with the following:

- Profit/investment
- Borrowing
- Rearranging budget priorities
- Energy savings
- Shared savings plans with outside firms and investors
- Utility incentives
- Grant programs
- Tax avoidance

When all these considerations have been weighed and a prioritized list of recommendations is developed, a report should be prepared for management. Each recommendation should include the following:

- Present condition of the system or equipment to be modified
- Recommended action
- Who should accomplish the action
- Necessary documentation or follow-up required
- Potential interferences to successful completion of the recommendation
- Staff effort required
- Risk of failure
- Interactions
- Economic analysis (including payback, investment cost, and estimated savings figures) using corporate economic evaluation criteria
- Schedule for implementation

The energy manager must be prepared to sell the plans. Every organization has limited funds available and must use these funds in the most effective way. Energy conservation measures generally must be financially justified if they are to be adopted. In this regard, the energy manager is competing with others in the organization for the same funds. A successful plan is presented in a form easily understood by the decision makers. Finally, the energy

BOILERS	OUTSIDE AIR VENTILATION
BOILER AUXILIARIES	VENTILATION LAYOUT
CONDENSATE SYSTEMS	ENVELOPE INFILTRATION
WATER TREATMENT	WEATHERSTRIPPING
FUEL ACQUISITION	CAULKING
FUEL SYSTEMS	VESTIBULES
CHILLERS	ELEVATOR SHAFTS
CHILLER AUXILIARIES	SPACE INSULATION
STEAM DISTRIBUTION	VAPOR BARRIER
HYDRONIC SYSTEMS	GLAZING
PUMPS	INFRARED REFLECTION
PIPING INSULATION	WINDOWS
STEAM TRAPS	WINDOW TREATMENT
DOMESTIC WATER HEATING	SHADING
LAVATORY FIXTURES	VEGETATION
WATER COOLERS	TROMBE WALLS
FIRE PROTECTION SYSTEMS	THERMAL SHUTTERS
SWIMMING POOLS	SURFACE COLOR
COOLING TOWERS	ROOF COVERING
CONDENSING UNITS	LAMPS
CITY WATER COOLING	FIXTURES
AIR HANDLING UNITS	BALLASTS
COILS	SWITCH DESIGN
OUTSIDE AIR CONTROL	PHOTO CONTROLS
BALANCING	INTERIOR COLOR
AIR VOLUME CONTROL	DEMAND LIMITING
SHUTDOWN	CURRENT LEAKAGE
AIR PURGING	POWER FACTOR
MINIMIZING REHEAT	TRANSFORMERS
AIR HEAT RECOVERY	POWER DISTRIBUTION
FILTERS	COOKING PRACTICES
DAMPERS	HOODS
HUMIDIFICATION	REFRIGERATION
DUCT RESISTANCE	DISHWASHING
SYSTEM AIR LEAKAGE	LAUNDRY
DIFFUSERS	VENDING MACHINES
SYSTEM INTERACTION	CHILLER HEAT RECOVERY
SYSTEM RECONFIGURATION	HEAT STORAGE
SPACE SEGREGATION	TIME-OF-DAY RATES
EQUIPMENT RELOCATION	COMPUTER CONTROLS
FAN-COIL UNITS	COGENERATION
HEAT PUMPS	ACTIVE SOLAR SYSTEMS
RADIATORS	STAFF TRAINING
SYSTEM INFILTRATION	OCCUPANT INDOCTRINATION
RELIEF AIR	DOCUMENTATION
SPACE HEATERS	MANAGEMENT STRUCTURE
CONTROLS	FINANCIAL PRACTICES
THERMOSTATS	BUILDING GEOMETRY
SETBACK	SPACE PLANNING
INSTRUMENTATION	

Fig. 7 Comprehensiveness

manager must present benefits other than the financial ones, such as improved quality of product or the possibility of postponing other expenditures by implementing a particular measure.

Accomplish Measures

From the above prioritized list, and following approval of management, the energy manager directs the completion of selected energy conservation retrofit measures. Certain measures require that an engineer prepare plans and specifications for the retrofit work. As such, the package of services required from the engineer usually includes drawings, specifications, assistance in obtaining competitive bids, evaluation of bids, selection of the best bid, construction observation, final checkout, and assistance in training of personnel in the proper use and application of the revisions.

Maintain Measures

Once energy conservation measures have begun, it is necessary to establish procedures to record, on a frequent and regular basis, energy consumption and costs for each building and/or end use category, in a manner consistent with functional cost accountability. Additional metering may be needed to accomplish this work accurately. Metering can be in the form of devices that automatically read and transmit data to a central location or in the form of less expensive metering devices that require routines for building maintenance and/or security personnel to assist in a regular meter-reading program. Many energy managers find it beneficial to collect energy consumption information as often as daily or twice daily (*i.e.*, shift changes). However, if the energy manager is not able to evaluate data as frequently as it is collected, it may be more practical to collect data less frequently. The energy manager must review data while it is current and take immediate action if profiles indicate that a trend is in the wrong direction. Such trends could be caused by control systems requiring recalibration, changes in operating practices, or failures of mechanical systems that should be isolated and corrected as soon as possible.

Energy Accounting

The energy manager continues the meter reading, monitoring, and tabulation of facility energy use and profiles. These tabulations indicate the cost of energy management efforts and the resulting energy cost avoidance. In conjunction with this effort, the energy manager periodically reviews pertinent utility rates, rate structures, and their trends as they affect the facility. The energy manager provides periodic reports of the energy management efforts to top management, summarizing the work accomplished, the cost-effectiveness of such work, the plans and suggested budget for future work, and projections of future utility costs. If energy conservation measures are to retain their cost-effectiveness, continued monitoring and periodic reauditing is necessary, since many energy conservation measures become less effective if they are not carefully monitored and maintained.

BUILDING ENERGY USE REDUCTION

The need for occasional reductions in energy use during specific periods has become more common due to rising energy costs and sometimes due to supply reductions or equipment failures. Emergency periods include a short-term shortage of a particular energy source or sources brought about by factors such as natural disasters, extreme weather conditions, utility system equipment disruptions, labor strikes, world political activities, or other forces beyond the control of the building owner and operator; by failures in building systems or equipment; or because of self-imposed cutbacks in energy use. This section provides information for building owners and operators to help maintain near normal operation of facilities during energy emergencies.

The following terms are applicable to such programs:

Energy Emergency—A period where energy supply reductions and/or climatic and natural forces or equipment failures preclude the normal operation of a building and necessitate a reduction in building energy use.

Level of Energy Emergency—A measure of the emergency severity that calls for the implementation of various energy use reduction measures. Typical levels include the following:

1. Green (Normal)—All occupant or building functions maintained and systems operating under the normal operating circumstances for the building
2. Blue—All or most of the occupant or building functions maintained with reductions in building system output that may result in borderline occupancy comfort
3. Yellow—A minor reduction of occupant or building functions with reductions in output
4. Red—A major reduction of occupant or building functions and systems that barely maintains building occupancy capability

5. Black—The orderly shutdown of systems and occupancy that maintains the minimum building conditions needed to protect the building and its systems

Implementation

Each building owner, lessor, and operator should use the energy team approach and identify an individual with the necessary authority who will review and fit recommendations into a plan for the particular building. For each class of emergency, the responsible party recommends a specific plan to reduce building energy use that still maintains the best building environment under the given circumstances. Implementation of the particular recommendations should then be coordinated through the building operator with assistance from the responsible party and the building occupants, as necessary. The plan should be tested occasionally.

Depending on the type of building, its use, the form of the energy source(s) for each function, and local conditions such as climate and availability of other similar buildings, the following steps should be taken in developing a building energy plan:

1. Develop a list of measures similar to the list below that are applicable to the building.
2. Estimate the amount and type of energy savings for each of the measures and appropriate combination of measures (*e.g.*, account for air-conditioning savings that result from reducing lighting and other internal loads). Tabulate demand and usage savings separately for response to different types of emergencies.
3. For the various levels of energy emergency, develop a plan that would maintain the best building environment under the circumstances. Include both short- and long-term measures in the plan. Operational changes may be implemented quickly and prove adequate for short-term emergencies.
4. Experiment with the plan developed above, record energy consumption and demand reduction data, and revise the plan, as necessary. Much of the experimentation may be done on weekends to minimize disruptive effects.
5. Meet with the local utility company(s) to review the plan.

Depending on the level of energy emergency and the building priority, the following actions may be considered in developing the plan for emergency energy reduction in the building:

- Change operating hours
- Move personnel into other building areas (consolidation)
- Shut off nonessential equipment

Thermal Envelope
- Use all existing blinds, draperies, etc., during summer
- Install interior window insulation
- Caulk and seal around unused exterior doors and windows
- Install solar shading devices in summer
- Seal all unused vents and ducts to outside

Heating, Ventilating, and Air-Conditioning Systems and Equipment
- Modify controls or control set points to raise and lower temperature and humidity, as necessary
- Shut off or isolate all nonessential equipment
- Tune up equipment
- Lower thermostat set points in winter
- Reduce the level of reheat or eliminate it in winter
- Reduce or eliminate ventilation and exhaust airflow
- Raise thermostat set points in summer
- Reduce the amount of recooling in summer

Lighting Systems
- Remove lamps or reduce lamp wattage
- Use task lighting, where appropriate
- Move building functions to exterior or daylight areas
- Turn off electric lights in areas with adequate natural light
- Lower luminaire height, where appropriate
- Wash all lamps and luminaires
- Replace fluorescent ballasts with high efficiency or multilevel ballasts
- Revise building cleaning and security procedures to minimize lighting periods
- Consolidate parking and turn off unused parking security lighting

Special Equipment
- Take transformers off line during periods of nonuse
- Shut off or regulate the use of vertical transportation systems
- Shut off unused or unnecessary equipment, such as photocopying equipment, music, typewriters, and computers
- Reduce or turn off hot water supply

Building Operation Demand Reduction
- Sequence heating or air-conditioning systems
- Disconnect or turn off all nonessential loads
- Turn off some lights
- Preheat or precool prior to the emergency period

REFERENCES

ASHRAE. 1989. Development of a guide for analyzing and reporting building characteristics and energy use in commercial buildings. *Special Project Report* 56.

DOE/EIA. 1989. Household energy consumption and expenditures 1987. Part 1: National Data. DOE/EIA-0321/1(87).

DOE/EIA. 1989. Nonresidential buildings energy consumption survey: Characteristics of commercial buildings 1986. DOE/EIA-0246(86).

DOE/EIA. 1989. Nonresidential buildings energy consumption survey: Commercial buildings consumption and expenditures 1986. DOE/EIA-0318(86).

Shehadi, Cowan, Spielvogel, and Wulfinghoff. 1984. Energy use evaluation. Symposium AT84-8. ASHRAE *Transactions* 90(1B):401-50.

Stewart, R., S. Stewart, and R. Joy. 1984. Energy audit input procedures and forms. ASHRAE *Transactions* 90(1A):350-62.

Turner, W.C. 1982. *Energy management handbook*. John C. Wiley & Sons, Inc., New York, 11.

BIBLIOGRAPHY

Freund, J.K. 1980. Selling energy management to an owner's engineer's management. *Heating, Piping and Air-Conditioning* (September).

Guide to a Successful Project Energy Conservation & Management, Model Competitive Procurement Procedure and *Professional Selection of Professional Engineers.* NSPE-PEPP, 2029 K Street, N.W., Washington, D.C. 20006.

Landsberg, D. and R. Stewart. 1980. *Improving energy efficiency in buildings.* State University of New York Press, Albany, NY.

Total Energy Management. 1979. National Electrical Contractors Association, Washington, D.C.

CHAPTER 33

OWNING AND OPERATING COSTS

SINCE HVAC systems are a normal part of a facility, the HVAC owning and operating cost information should be part of the total owning and operating cost of the facility, *i.e.*, a required part of the investment plan of the facility. HVAC owning and operating cost information can also be used for preparing annual budgets, managing assets, and selecting design options. A representative form that summarizes these costs is shown as Table 1.

A properly engineered system must also be economical. Economics are difficult to assess because of the complexities surrounding the effective management of money and the inherent difficulty of predicting future operating and maintenance expenses. Complex tax structures and the time value of money can affect the final engineering decision. This does not imply the use of either the cheapest or the most expensive system; instead, it demands an intelligent analysis of financial objectives and requirements of the owner. Therefore, the engineer is responsible for evaluating the proper use of money, as dictated by specific circumstances.

Certain tangible and intangible costs or benefits must also be considered when assessing owning and operating costs. Local codes may require highly skilled or certified operators for specific types of equipment. This could be a significant cost over the life of the system. Similarly, such intangible items as aesthetics, acoustics, comfort, security, flexibility, and environmental impact may be important to a particular building or facility.

OWNING COSTS

Three elements must be established to calculate annual owning costs: (1) initial cost, (2) service life, and (3) insurance. Once established, these elements are coupled with operating costs to develop an economic analysis, which may be a simple payback evaluation or an in-depth analysis such as outlined in the Economic Analysis Techniques section.

INITIAL COST OF SYSTEM

Major decisions affecting annual owning and operating costs for the life of the building must generally be made prior to the completion of contract drawings and specifications. To achieve the best performance and economics, comparisons between alternate methods of solving the engineering problems peculiar to each project must be made in the early stages of architectural design. Oversimplified estimates can lead to substantial errors in evaluating the system.

A thorough understanding of the installation costs and accessory requirements must be established. Detailed lists of materials, controls, space and structural requirements, services, installation labor, and so forth can be prepared to increase the accuracy in preliminary cost estimates. A reasonable estimate of the cost of components may be derived from cost records of recent installations of comparable design or from quotations submitted by manufacturers and contractors. A representative checklist for initial cost items is shown as Table 2.

SERVICE LIFE

Service life is the median time during which a particular system or component remains in its original service application and then is replaced. Replacement may occur for any reason, including, but not limited to, failure, general obsolescence, reduced reliability, excessive maintenance cost, and changed system requirements due to such influences as building characteristics or energy prices.

Table 3 lists representative estimates of the service lives of various system components. The service life may not be the same as the depreciation period or the study period.

INSURANCE

Insurance reimburses a property owner for a financial loss so that equipment can be repaired or replaced. Financial recovery may also include replacing the loss of income, rents, or profits resulting from property damage.

Some government authorities regulate the activities of insurance companies doing business within their jurisdiction and determine the premium rates that may be charged for various forms of insurable property. Some of the principal factors that influence the total annual premium are building size, construction material, amount and size of mechanical equipment, and policy deductibles. A property owner should consult an insurance specialist to select an appropriate insurance program.

OPERATING COSTS

Operating costs result from the actual operation of the system. They include fuel and electrical costs, wages, supplies, water, material, and maintenance parts and services. Chapter 28 of the 1989 ASHRAE *Handbook—Fundamentals*, outlines how fuel and electrical requirements are estimated. Note that total energy consumption cannot generally be multiplied by a per unit energy cost to arrive at annual utility cost.

ELECTRICAL ENERGY

The total cost of electrical energy is usually a combination of several components: energy consumption charges, fuel adjustment charges, special allowances or other adjustments, and demand charges.

The preparation of this chapter is assigned to TC 1.8, Owning and Operating Costs.

Table 1 Owning and Operating Cost Data and Summary

OWNING COSTS

I. Initial Cost of System
- A. Equipment (see Table 2 for items included) __________
- B. Control systems—Complete __________
- C. Wiring and piping costs attributable to system __________
- D. Any increase in building construction cost attributable to system __________
- E. Any decrease in building construction cost attributable to system __________
- F. Installation costs __________

TOTAL INITIAL COST __________

II. Annual Fixed Charges
- A. Equivalent uniform annual cost __________
- B. Income taxes __________
- C. Property taxes __________
- D. Insurance __________
- E. Rent __________

TOTAL ANNUAL FIXED CHARGES __________

OPERATING COSTS

III. Annual Maintenance Allowances
- A. Replacement or servicing oil, air, or water filters __________
- B. Contracted maintenance service __________
- C. Lubricating oil and grease __________
- D. General housekeeping cost __________
- E. Replacement of worn parts (labor and material) __________
- F. Refrigerant __________

TOTAL ANNUAL MAINTENANCE ALLOWANCE __________

IV. Annual Energy, Fuel, and Water Costs
- A. Electric Energy Costs
  1. Chiller or compressor __________
  2. Pumps
     - a) Chilled water __________
     - b) Heating water __________
     - c) Condenser or tower water __________
     - d) Well water __________
     - e) Boiler auxiliaries (including fuel oil heaters) __________
  3. Fans
     - a) Condenser or tower __________
     - b) Inside air handling __________
     - c) Exhaust __________
     - d) Makeup air __________
     - e) Boiler auxiliaries and equipment room ventilation __________

IV. Annual Energy, Fuel and Water Costs (*continued*)
  4. Resistance heaters (primary or supplementary) __________
  5. Heat pump __________
  6. Domestic water heating __________
  7. Lighting __________
  8. Cooking and food service equipment __________
  9. Miscellaneous (*e.g.*, elevators, escalators, and computers) __________
- B. Gas, Oil, Coal, or Purchased Steam Costs
  1. On-site generation of the electrical power requirements under A of this section __________
  2. Heating __________
     - a) Direct heating __________
     - b) Ventilation
       - (1) Preheaters __________
       - (2) Reheaters __________
     - c) Supplementary heating (*i.e.*, oil preheating) __________
     - d) Other __________
  3. Domestic water heating __________
  4. Cooking and food service equipment __________
  5. Air conditioning
     - a) Absorption __________
     - b) Chiller or compressor
       - (1) Gas/Diesel engine driven __________
       - (2) Gas turbine driven __________
       - (3) Steam turbine driven __________
  6. Miscellaneous __________
- C. Water
  1. Condenser makeup water __________
  2. Sewer charges __________
  3. Chemicals __________
  4. Miscellaneous __________

TOTAL ANNUAL FUEL, ENERGY, AND WATER COSTS __________

V. Wages of engineers and operators __________

SUMMARY

II. Total Annual Fixed Charges __________

III. Total Annual Maintenance Costs __________

IV. Total Annual Energy, Fuel, and Water Costs __________

V. Annual Wages for Engineers and Operators __________

TOTAL ANNUAL OWNING AND OPERATING COSTS __________

Energy Consumption Charges

Most utility rates have step rate schedules for consumption, and the cost of the last unit of energy consumed may be substantially different from the first. The last unit may be cheaper than the first because the fixed costs to the utility may already have been recovered from earlier consumption costs. Alternatively, the last unit of energy may be sold at a higher rate to encourage conservation.

To reflect time-varying operating costs, some utilities charge different rates for consumption according to the time of use and season; typically, costs rise toward the peak period of use. This may justify the cost of shifting the load to off-peak periods.

Fuel Adjustment Charge

Due to substantial variations in fuel prices, electric utilities may apply a fuel adjustment charge to recover costs. This adjustment may not be reflected in the rate schedule. The fuel adjustment is usually a charge per unit of energy, and may be positive or negative depending on how much of the actual fuel cost is recovered in the energy consumption rate.

Power plants with multiple generating units that use different fuels typically have the greatest effect on this charge (especially during peak periods, when more expensive units must be brought on-line). Although this fuel adjustment charge can vary monthly, the utility should be able to estimate an accurate annual or seasonal fuel adjustment average for calculations.

Allowances or Adjustments

Special allowances may be available for customers who can receive power at higher voltages or for those who own transformers or similar equipment. Special rates may be available for specific interruptible loads such as domestic water heaters.

Certain facility electrical systems may produce a low power factor which means that the utility must supply more current on an intermittent basis, thus increasing their costs. These costs may be passed on as an adjustment to the utility bill if the power factor is below a level established by the utility. The power factor is the ratio of active (real) kilowatt power to apparent (reactive) kVA power.

When calculating power bills, utilities should be asked to provide detailed cost estimates for various consumption levels. The final

Table 2 Initial Cost Checklist

Energy and Fuel Service Costs
Fuel service, storage, handling, piping, and distribution costs
Electrical service entrance and distribution equipment costs
Total energy plant
Heat-Producing Equipment
Boilers and furnaces
Steam-water converters
Heat pumps or resistance heaters
Make-up air heaters
Heat-producing equipment auxiliaries
Refrigeration Equipment
Compressors, chillers, or absorption units
Cooling towers, condensers, well water supplies
Refrigeration equipment auxiliaries
Heat Distribution Equipment
Pumps, reducing valves, piping, piping insulation, etc.
Terminal units or devices
Cooling Distribution Equipment
Pumps, piping, piping insulation, condensate drains, etc.
Terminal units, mixing boxes, diffusers, grilles, etc.
Air Treatment and Distribution Equipment
Air heaters, humidifiers, dehumidifiers, filters, etc.
Fans, ducts, duct insulation, dampers, etc.
Exhaust and return systems
System and Controls Automation
Terminal or zone controls
System program control
Alarms and indicator system
Building Construction and Alteration
Mechanical and electric space
Chimneys and flues
Building insulation
Solar radiation controls
Acoustical and vibration treatment
Distribution shafts, machinery foundations, furring

calculation should include any applicable special rates, allowances, taxes, and fuel adjustment charges.

Demand

Electric rates may also have demand charges based on the customer's peak kilowatt demand. While consumption charges typically cover the operating costs of the utility, demand charges typically cover the owning costs.

Where demand charges are applied, they may be formulated in a variety of ways:

1. Straight charge—$/kW per month, charged for the peak demand of the month.
2. Excess charge—$/kW above a base demand (*i.e.*, 50 kW), which may be established each month.
3. Maximum demand (ratchet)—$/kW for the maximum annual demand, which may be reset only once a year. This established demand may either benefit or penalize the owner.
4. Combination demand—In addition to a basic demand charge, utilities may include further demand charges as demand-related consumption charges, in cents per hour of operation of the demand.

The actual level of demand represents the peak energy use averaged over a specific period, usually 15, 30, or 60 min. Accordingly, high electrical loads of only a few minutes' duration may never be recorded at the full instantaneous value. Alternatively, demands may be established during a short period (*i.e.*, 5 min out of each hour), where peak demand is recorded as the average of several consecutive short periods.

The particular method of demand metering and billing is important when load shedding or shifting devices are considered. The portion of the total bill attributed to demand may vary widely, from 0% to as high as 70%, depending on the duration of the demand (*i.e.*, the electric load factor).

Table 3 Estimates of Service Lives of Various System Components

Equipment Item	Median Years	Equipment Item	Median Years	Equipment Item	Median Years
Air conditioners		Air terminals		Air-cooled condensers	20
Window unit	10	Diffusers, grilles, and registers	27	Evaporative condensers	20
Residential single or split package	15	Induction and fan-coil units	20	Insulation	
Commercial through-the-wall	15	VAV and double-duct boxes	20	Molded	20
Water-cooled package	15	Air washers	17	Blanket	24
Heat pumps		Ductwork	30	Pumps	
Residential air-to-air	15[a]	Dampers	20	Base-mounted	20
Commercial air-to-air	15	Fans		Pipe-mounted	10
Commercial water-to-air	19	Centrifugal	25	Sump and well	10
Roof-top air conditioners		Axial	20	Condensate	15
Single-zone	15	Propeller	15	Reciprocating engines	20
Multizone	15	Ventilating roof-mounted	20	Steam turbines	30
Boilers, hot water (steam)		Coils		Electric motors	18
Steel water-tube	24 (30)	DX, water, or steam	20	Motor starters	17
Steel fire-tube	25 (25)	Electric	15	Electric transformers	30
Cast iron	35 (30)	Heat Exchangers		Controls	
Electric	15	Shell-and-tube	24	Pneumatic	20
Burners	21	Reciprocating compressors	20	Electric	16
Furnaces		Package chillers		Electronic	15
Gas- or oil-fired	18	Reciprocating	20	Valve actuators	
Unit heaters		Centrifugal	23	Hydraulic	15
Gas or electric	13	Absorption	23	Pneumatic	20
Hot water or steam	20	Cooling towers		Self-contained	10
Radiant heaters		Galvanized metal	20		
Electric	10	Wood	20		
Hot water or steam	25	Ceramic	34		

Source: Data obtained from a national survey of the United States by ASHRAE Technical Committee TC 1.8 (Akalin 1978). Data updated by TC 1.8 in 1986.

[a]See Lovvorn and Hiller (1985) and Easton Consultants (1986) for further information.

NATURAL GAS

Rates

Conventional natural gas rates are usually a combination of two main components: (1) utility rate for gas consumption, and (2) purchased gas adjustment (PGA) charges.

Although gas is usually metered by volume, it is often sold by energy content (therm). The utility rate is the amount the local distribution company charges per unit of energy to deliver the gas to a particular location. This rate may be graduated in steps, so that the first 100 therms of gas consumed may not be the same price as the last 100 therms. The PGA is an adjustment for the cost of the gas per unit of energy to the local utility. It is similar to the electric fuel adjustment charge. The total cost per therm is then the sum of the appropriate utility rate and the PGA, plus taxes and other adjustments.

Interruptible Gas Rates and Contract/Transport Gas

Large industrial plants usually have the ability to burn alternate fuels at their facility and can qualify for special interruptible gas rates. During peak periods of severe cold weather, these customers may be curtailed by the gas utility and may have to switch to propane, fuel oil, or some other backup fuel. The utility rate and PGA are usually considerably cheaper for these interruptible customers than they are for firm rate (noninterruptible) customers.

Deregulation of the natural gas industry now allows end users to negotiate for gas supplies on the open market. The customer actually contracts with a producer or gas broker and pays for the gas at the source. Then, transport fees must be negotiated with the pipeline companies carrying the gas to the customer's local gas utility. This can be a very complicated administrative process and is usually only economically feasible for large gas users. Some local utilities have special rates for delivering contract gas volumes through their system; others simply charge a standard utility fee (PGA is not applied since the customer has already negotiated with the supplier for the cost of the fuel itself).

When calculating natural gas bills, be sure to determine which utility rate and PGA and/or contract gas price is appropriate for the particular interruptible or firm customer. As with electric bills, the final calculation should include any taxes, prompt payment discounts, or other adjustments that are applicable.

OTHER FUELS

Propane, fuel oil, and diesel are examples of other fuels in widespread use today. Calculating the cost of these fuels is usually much easier than calculating typical utility rates. When these fuels are used, other items that can affect owning or operating costs must be considered.

The cost of the fuel itself is usually a simple per unit volume or per unit mass charge. The customer is free to negotiate for the best price. However, trucking or delivery fees must also be included in final calculations. Some customers may have their own transport trucks, while most shop around for the best "delivered" price. Rental fees for storage tanks must be considered if they are not customer-owned. Periodic replacement of diesel-type fuels may be necessary due to storage or shelf-life limitations and must also be considered. The final fuel cost calculation should include any of these costs, where applicable, as well as the appropriates taxes.

MAINTENANCE COSTS

The quality of maintenance and maintenance supervision can be a major factor in the energy cost of a building. Chapter 35 covers the maintenance, maintainability, and reliability of systems. Dohrmann and Alereza (1986) obtained maintenance costs and HVAC system information from 342 buildings located in 35 states in the United States. In 1983 U.S. dollars, data collected showed a mean HVAC system maintenance cost of \$0.32 per square foot, with a median cost of \$0.24/ft^2. The age of the building has a statistically significant but minor effect on HVAC maintenance costs. When analyzed by geographic location, the data revealed that location does not significantly affect maintenance costs. Analysis also indicated that building size is not statistically significant in explaining cost variation.

The type of maintenance program or service agency that the building management contracts for can also have a significant effect on total HVAC maintenance costs. While extensive or thorough routine and preventive maintenance programs cost more to administer, they usually produce benefits such as extended equipment life, improved reliability, and less system downtime.

ESTIMATING MAINTENANCE COSTS

Using data from Table 4, the following method may be used for estimating or comparing the total building HVAC maintenance costs for various equipment combinations. (Manufacturers and other industry sources should be consulted for current information on new types of equipment). Several important limitations of the data presented here should be noted:

- Only data selected from Table 12 of the report by Dohrmann and Alereza (1986) are presented here.
- Data were collected for office buildings only.
- Data measure total HVAC building maintenance costs and do not measure the costs associated with maintaining particular items or individual pieces of equipment.
- The cost data are not intended for use in selecting new HVAC equipment or systems. This information is for equipment and systems already in place and may not be representative of the maintenance costs expected with newer equipment.

The premise of this method assumes that the base HVAC system in the building consists of fire-tube boilers for heating equipment, centrifugal chillers for cooling equipment, and VAV distribution systems. The total building HVAC maintenance cost for this system is 33.38 ¢/ft^2. Adjustment factors from Table 4 are then applied to this base cost to account for building age and variations on type of HVAC equipment as follows:

C = Total building HVAC maintenance cost (¢/ft^2) =
Base system maintenance costs
+ (Age adjustment factor) × (age in years n)
+ Heating system adjustment factor h
+ Cooling system adjustment factor c
+ Distribution system adjustment factor d

or: $C = 33.38 + 0.18n + h + c + d$

Table 4 HVAC Maintenance Cost Adjustment Factors (in cents per square foot, 1983 U.S. dollars)

Age + 0.18 (per year n)	
Heating Equipment h	
Water tube boiler	+0.77
Cast iron boiler	+0.94
Electric boiler	−2.67
Heat pump	−9.69
Electric resistance	−13.3
Cooling Equipment c	
Reciprocating chiller	−4.0
Absorption chiller (single stage)[a]	+19.25
Water source heat pump	−4.72
Distribution System d	
Single zone	+8.29
Multizone	−4.66
Dual duct	−0.29
Constant volume	+8.81
Two-pipe fan coil	−2.77
Four-pipe fan coil	+5.80
Induction	+6.82

[a]These results pertain to buildings with older, single-stage absorption chillers. The data from the survey are not sufficient to draw inferences about the costs of HVAC maintenance in buildings equipped with new, double-stage absorption chillers.

Example 1: Estimate the total building HVAC maintenance costs per square foot for a building that is 10 years old, has an electric boiler, a reciprocating chiller, and a constant volume distribution system.

$C = 33.38 + 0.18(10) - 2.67 - 4.0 + 8.81$
$C = 37.32$ ¢/ft² in 1983 dollars

This estimate can be adjusted to current dollars by multiplying the maintenance cost estimate by the current Consumer Price Index (CPI) divided by the CPI in July 1983. In July 1983, the CPI was 100.1. Monthly CPI statistics are recorded in *Survey of Current Business*, a U.S. Department of Commerce publication. This estimating method is limited to one equipment variable per situation. That is, the method can estimate maintenance costs for a building having either a centrifugal chiller or a reciprocating chiller, but not both. Assessing the effects of combining two or more types of equipment within a single category requires a more complicated statistical analysis.

ECONOMIC ANALYSIS TECHNIQUES

Definition of terms:

C_e = cost of energy to operate the system for one period
$C_{s,assess}$ = initial assessed system value
$C_{s,salv}$ = system salvage value at the end of its useful life in constant dollars
$C_{s,init}$ = initial system cost
C_y = annualized system cost in constant dollars
$D_{k,SL}$ or $D_{k,SD}$ = amount of depreciation at the end of period k depending on the type of depreciation schedule used, where $D_{k,SL}$ is the straight line depreciation method and $D_{k,SD}$ represents the sum-of-digits depreciation method in constant dollars
F = future value of a sum of money
i_d = discount rate
i_m = market mortgage rate (real rate + general inflation rate)
$i_m P_k$ = interest charge at the end of period k
i' = $(i_d - j)/(1+j)$ = effective discount rate adjusted for general inflation j, sometimes called the real discount rate
i'' = $(i_d - j_e)/(1+j_e)$ = effective discount rate adjusted for energy inflation j_e
I = annual insurance costs
ITC = investment tax credit for energy efficiency improvements, if applicable
j = general inflation rate per period
j_e = energy inflation rate per period
k = end of period(s) in which replacement(s), repair(s), depreciation, or interest is calculated
M = periodic maintenance cost
n = number of period(s) under consideration
P = a sum of money at the present time, *i.e.*, its present value
P_k = outstanding principle of the loan for $C_{s,init}$ at the end of period k in current dollars
R_k = net replacement(s) or repair cost(s) at the end of period k in constant dollars
T_{inc} = (state tax rate + federal tax rate) − (state tax rate × federal tax rate) where rates are based on the last dollar earned, *i.e.*, the marginal rates
T_{prop} = property tax rate
T_{salv} = tax rate applicable to salvage value of the system

For any proposed capital investment, the capital and interest costs, salvage costs, replacement costs, energy costs, taxes, maintenance costs, insurance costs, interest deductions, depreciation allowances, and other factors must be weighed against the value of the services provided by the system.

Single Payment

A common method for analyzing the impact of a future payment is to reduce it to its present value or present worth. The primary underlying principle is that all monies (those paid now and in the future) should be evaluated according to their present purchasing power. This approach is known as discounting.

The future value F of a present sum of money P over n periods with compound interest rate i is:

$$F = P(1+i)^n \qquad (1)$$

The present value or present worth P of a future sum of money F is given by:

$$P = F/(1+i)^n = F \times \mathrm{PWF}(i,n) \qquad (2)$$

where PWF(i,n), the present worth factor, is defined by:

$$\mathrm{PWF}(i,n) = 1/(1+i)^n \qquad (3)$$

Example 2: Calculate the future value of a system presently valued at $10,000, in 10 years, at 10% interest.

$$F = P(1 + i)^n = \$10{,}000 \times (1 + 0.10)^{10}$$
$$= \$10{,}000 \times 2.593742 = \$25{,}937.42$$

Example 3: Using the present worth factor PWF(i = 0.10, n = 10), calculate the present value of a future sum of money valued at $10,000.

$$P = F \times \mathrm{PWF}(i,n) = 10{,}000 \times 1/(1 + 0.1)^{10} = \$3{,}855.43$$

Accounting for Varying Inflation Rates

Inflation, which accounts for the rise in cost of a commodity over time, is a separate issue from the time value of money—the basis for discounting. Inflation must often be accounted for in an economic evaluation. Further complexities are added when one considers that different economic goods inflate at different rates. One way to account for this is to use effective interest rates that account for varying rates of inflation.

The effective interest rate i', sometimes called the real rate, accounts for the general inflation rate j and the discount rate i_d, and can be expressed as follows (Kreith and Kreider 1978, Kreider and Kreith 1982):

$$i' = \frac{1 + i_d}{1 + j} - 1 = \frac{(i_d - j)}{(1+j)} \qquad (4)$$

Such an expression can be adapted to account for energy inflation by considering the general discount rate i_d and the energy inflation rate j_e, thus:

$$i'' = \frac{1 + i_d}{1 + j_e} - 1 = \frac{(i_d - j_e)}{(1+j_e)} \qquad (5)$$

The discount equations (1) through (3) can be revised to consider the effects of varying inflation rates. The future value F, using constant currency of an invested sum P with a discount rate i_d under inflation j during n periods now becomes:

$$F = P[1 + i_d/1 + j]^n = P(1 + i')^n \qquad (6)$$

The present worth P, in constant dollars, of a future sum of money F with discount rate i_d under inflation j during n periods is then expressed as:

$$P = F/[(1 + i_d)/(1 + j)]^n \qquad (7)$$

In constant currency, the present worth P of a sum of money F can be expressed with an effective interest rate i' which is adjusted for inflation by:

$$P = F/(1 + i')^n = F \times \mathrm{PWF}(i',n) \qquad (8)$$

where the effective present worth factor is given by:

$$\mathrm{PWF}(i',n) = 1/(1 + i')^n \qquad (9)$$

Example 4: Calculate the effective interest rate taking into consideration a discount rate i_d of 10% and a general inflation rate j of 5%.

$$i' = (i_d - j)/(1 + j) = (0.1 - 0.05)/(1 + 0.05) = 0.04762 = 4.62\%$$

Example 5: Using the effective interest rate in **Example 4**, calculate the future value of $10,000 10 years from now.

$$F = P(1 + i')^n = \$10{,}000\,(1 + 0.04762)^{10} = \$15{,}923.47 \text{ (constant dollars)}$$

Recovering Capital as a Series of Payments

Another important economic concept is the recovery or acquisition of capital as a series of uniform payments—the capital recovery factor. The capital recovery factor is commonly used to describe periodic uniform mortgage or loan payments. It is the ratio of the periodic payment to the total sum being repaid. The discounted sum S of such an annual series of payments P_{ann} invested over n periods with interest rate i is given by:

$$S = P_{ann}\,[1 - (1 + i)^{-n}]/i \qquad (10)$$

This series of uniform annual payments P_{ann} over n periods is equivalent to the present value sum S divided by the capital recovery factor, with interest rate i expressed as:

$$P_{ann} = S\,i/\,[1 - (1 + i)^{-n}] \qquad (11)$$

This multiplier is the capital recovery factor CRF(i,n), which can be expressed as:

$$\text{CRF}(i,n) = i/\,[1 - (1 + i)^{-n}] = i(1 + i)^n/[(1 + i)^n - 1] \qquad (12)$$

Example 6: Calculate the discounted sum of ten periodic payments of $1,000 with a discount rate of 10%.

$$S = P_{ann}\,[1 - (1 + i)^{-n}]/i = 1000\,[1 - (1 + 0.10)^{-10}]/0.10 = \$6144.57$$

Example 7: Calculate the capital recovery factor for **Example 6.**

$$\text{CRF}(i,n) = i/1 - (1 + i)^{-n} = 0.10/1 - (1 + 0.10)^{-10} = 0.162\,745$$

Annualized Costs

The economic analysis methods presented to this point have included only single cash flows or a simple series of payments that were adjusted for inflation. A detailed cash flow analysis of a mechanical system should consider all positive and negative cash flows throughout the life of the system, including single payments, series of payments, and increasing or decreasing payments, and must also consider varying inflation rates. Such analysis should take into account taxes, tax credits, mortgage payments, and all other costs associated with a particular system. One convenient way of accomplishing this is to use annualized system costs as presented by Kreith and Kreider (1978, 1982) and Kreider (1979).

Total annualized mechanical system costs depend on initial investments, salvage values, replacement costs, energy costs, property taxes, property tax deductions, interest tax deductions, maintenance costs, replacement costs, and insurance costs. Such a representation lends itself to optimization of system costs and can be applied to residential and commercial systems.

Annualized mechanical system owning, operating, and maintenance costs (for a profit-making firm) can be expressed in constant currency as:

C_y = − capital and interest + salvage value − replacements
− operating energy − property tax
− maintenance − insurance
+ interest tax deduction + depreciation (for commercial systems) (13)

where

$(C_{s,init} - \text{ITC})\,\text{CRF}(i',n)$ = capital and interest

$C_{s,salv}\,\text{PWF}(i',n)\,\text{CRF}(i',n)\,(1 - T_{salv})$ = salvage value

$\sum_{k=1}^{n} [R_k\,\text{PWF}(i',k)]\,\text{CRF}(i',n)\,(1 - T_{inc})$ = replacements

$C_e\,[\text{CRF}(i',n)\,/\,\text{CRF}(i'',n)]\,(1 - T_{inc})$ = operating energy

$C_{s,assess}\,T_{prop}\,(1 - T_{inc})$ = property tax

$M(1 - T_{inc})$ = maintenance

$I\,(1 - T_{inc})$ = insurance

$T_{inc} \sum_{k=1}^{n} [i_m P_k\,\text{PWF}(i_d,k)]\,\text{CRF}(i,n)$ = interest tax deduction

$T_{inc} \sum_{k=1}^{n} [D_k\,\text{PWF}(i',k)]\,\text{CRF}(i'n)$ = depreciation (for commercial systems)

The outstanding principle P_k during year k at market mortgage rate i_m is given by:

$$P_k = (C_{s,init} - \text{ITC})\left[(1 + i_m)^{k-1} + \frac{(1 + i_m)^{k-1} - 1}{(1 + i_m)^{-n} - 1}\right] \qquad (14)$$

Note: P_k is in current dollars and must, therefore, be discounted by the discount rate i_d not i'.

Likewise, the summation term for interest deduction can be expressed as:

$$\sum_{k=1}^{n}[i_m P_k/(1 + i_d)^k] = \left[\frac{\text{CRF}(i_m,n)}{\text{CRF}(i_d,n)} + \frac{1}{(1 + i_m)}\,\frac{i_m - \text{CRF}(i_m,n)}{\text{CRF}[(i_d - i_m)/(1 + i_m),n]}\right] \times (C_{s,init} - \text{ITC}) \qquad (15)$$

and if $i_d = i_m$,

$$\sum_{k=1}^{n}[i_m P_k/(1 + i_d)^k] = \left[1 + \frac{n}{1 + i_m}\,[i_m - \text{CRF}(i_m,n)]\right](C_{s,init} - \text{ITC}) \qquad (16)$$

Depreciation terms commonly used include depreciation calculated by the straight line depreciation method, which is:

$$D_{k,SL} = (C_{s,init} - C_{s,salv})/n \qquad (17)$$

and the sum-of-digits depreciation method:

$$D_{k,SD} = (C_{s,init} - C_{s,salv})\,[2(n - k + 1)]/n(n + 1) \qquad (18)$$

Riggs (1977) and Grant *et al.* (1982) present further information on advanced depreciation methods. Certified accountants may also be consulted for information regarding accelerated methods allowed by the IRS.

Example 8: Calculate the annualized system cost using constant dollars for a $10,000 system considering the following factors: a 5-year life, a salvage value of $1,000 at the end of the 5 years, ignore investment tax credits, a $500 replacement in year 3, a discount rate i_d of 10%, a general inflation rate j of 5%, a fuel inflation rate j_e of 8%, a market mortgage rate i_m of 10%, an annual operating cost for energy of $500, a $100 annual maintenance cost, a $50 insurance cost, straight line depreciation, an income tax rate of 50%, a property tax rate of 1% of assessed value, an assessed system value equal to 40% of the initial system value, and a salvage tax rate of 50%.

Effective interest rate i'

$$i' = (i_d - j)/(1 + j) = (0.10 - 0.05)/(1 + 0.05) = 0.047619$$

Effective interest rate i''

$i' = (i_d - j_e)/(1 + j_e) = (0.10 - 0.08)/(1 + 0.08) = 0.018519$

Capital recovery factor CRF(i',n)

$\text{CRF}(i',n) = i'/[1 - (1 + i')^{-n}]$
$= 0.047619/[1 - (1.047619)^{-5}] = 0.229457$

Capital recovery factor CRF(i_m,n)

$\text{CRF}(i_m,n) = i_m/[1 - (1 + i_m)^{-n}] = 0.10/1 - (1.10)^{-5} = 0.263797$

Capital recovery factor CRF(i'',n)

$\text{CRF}(i'',n) = i''/[1 - (1 + i'')^{-n}]$
$= 0.018519/1 - (1.018519)^{-5} = 0.211247$

Present worth factor PWF(i_d, years 1 to 5)

$\text{PWF}(i_d,1) = 1/(1.10)^1 = 0.909091$
$\text{PWF}(i_d,2) = 1/(1.10)^2 = 0.826446$
$\text{PWF}(i_d,3) = 1/(1.10)^3 = 0.751315$
$\text{PWF}(i_d,4) = 1/(1.10)^4 = 0.683013$
$\text{PWF}(i_d,5) = 1/(1.10)^5 = 0.620921$

Present worth factor PWF(i', years 1 to 5)

$\text{PWF}(i',1) = 1/(1.047619)^1 = 0.954545$
$\text{PWF}(i',2) = 1/(1.047619)^2 = 0.911157$
$\text{PWF}(i',3) = 1/(1.047619)^3 = 0.869741$
$\text{PWF}(i',4) = 1/(1.047619)^4 = 0.830207$
$\text{PWF}(i',5) = 1/(1.047619)^5 = 0.792471$

Capital and interest

$(C_{s,init} - \text{ITC})\,\text{CRF}(i',n) = (\$10{,}000 - \$0)0.229457 = \$2{,}294.57$

Salvage value

$C_{s,salv}\,\text{PWF}(i',n)\,\text{CRF}(i',n)\,(1 - T_{salv})$
$= \$1{,}000 \times 0.792471 \times 0.229457 \times 0.5 = \$90.92$

Replacements

$\sum_{k=1}^{n} (R_k\,\text{PWF}(i',k)]\,\text{CRF}(i',n)\,(1 - T_{inc})$

$= \$500 \times 0.869741 \times 0.229457 \times 0.5 = \$49.89$

Operating energy

$C_e\,[\text{CRF}(i',n)/\text{CRF}(i'',n)]\,(1 - T_{inc})$
$= 500\,[0.229457/0.211247]\,0.5 = \$271.55$

Property tax

$C_{s,assess}\,T_{prop}\,(1 - T_{inc}) = \$10{,}000 \times 0.40 \times 0.01 \times 0.5 = \$20.00$

Maintenance

Given: $100.00

Insurance

Given: $50.00

Interest deduction

$$T_{inc} \sum_{k=1}^{n} [i_m P_k\,\text{PWF}(i_d,k)]\,\text{CRF}(i',n) = \ldots$$

To begin with, annual payments, interest payments, outstanding principal, discounted interest, and discounted principal are calculated. Annual payments are the sum of the initial system cost $C_{s,init}$ and the capital recovery factor CRF(i_m,5). These values are shown in Table 5.

Note: Equation (15) can be used to calculate the total discounted interest deduction directly.

Next, apply the capital recovery factor CRF(i',5) and tax rate T_{inc} to the total of the discounted interest sum.

$\$2554.66\,\text{CRF}(i',5)\,T_{inc} = \$2554.66 \times 0.229457 \times 0.5 = \$293.09$

Depreciation

$$T_{inc} \sum_{k=1}^{n} [D_{k,SL}\,\text{PWF}(i',k)]\,\text{CRF}(i',n) \ldots$$

Use the straight line depreciation method to calculate depreciation.

$D_{k,SL} = (C_{s,init} - C_{s,salv})/n = (\$10{,}000 - \$1000)/5 = \$1800.00$

Next, discount the depreciation.

Year	$D_{k,SL}$	PWF(i_d,k)	Discounted Depreciation
1	$1800.00	0.909091	$1636.36
2	$1800.00	0.826446	$1487.60
3	$1800.00	0.751315	$1352.37
4	$1800.00	0.683013	$1229.42
5	$1800.00	0.620921	$1117.66
Total	—	—	$6823.42

Finally, the capital recovery factor and tax are applied.

$\$6823.42\,\text{CRF}(i',n)\,T_{inc} = \$6823.42 \times 0.229457 \times 0.5 = \$782.84$

Summary of terms

Capital and interest	−$ 2294.57
Salvage value	+$ 90.92
Replacements	−$ 49.89
Operating costs	−$ 271.55
Property tax	−$ 20.00
Maintenance	−$ 100.00
Insurance	−$ 50.00
Interest deduction	+$ 293.09
Depreciation deduction	+$ 782.84
Total annualized cost	−$ 1619.16

Table 5 Interest Deduction Summary

Year	Payment Amount (current $)	Interest Payment (current $)	Principal Payment (current $)	Outstanding Principal (current $)	PWF(i_d,k)	Discounted Interest (discounted $)	Discounted Payment (discounted $)
0	—	—	—	10,000.00	—	—	—
1	2637.97	1000.00	1637.97	8362.02	0.909091	909.09	2398.17
2	2637.97	836.20	1801.77	6560.26	0.826446	691.07	2180.14
3	2637.97	656.03	1981.95	4578.31	0.751315	492.89	1981.95
4	2637.97	457.83	2180.14	2398.17	0.683013	312.70	1801.77
5	2637.97	239.82	2398.17	0	0.620921	148.91	1637.97
Total	—	3189.88	10,000.00	—	—	2554.66	10,000.00

REFERENCES

Akalin, M.T. 1978. Equipment life and maintenance cost survey. ASHRAE *Transactions* 84(2):94-106.

Dohrmann, D.R. and T. Alereza. 1986. Analysis of survey data on HVAC maintenance Costs. ASHRAE *Transactions* 92(2A):550-565.

Easton Consultants. 1986. Survey of residential heat pump service life and maintenance issues. Available from American Gas Association, Arlington, VA (Catalog No. S-77126).

Grant, E., W. Ireson, and R. Leavenworth. 1982. *Principles of engineering economy.* John Wiley and Sons, New York.

Kreider, J. and F. Kreith. 1982. *Solar heating and cooling.* Hemisphere Publishing Corporation, Washington, D.C.

Kreider, J. 1979. *Medium and high temperature solar processes.* Academic Press, New York.

Kreith, F. and J. Kreider. 1978. *Principles of solar engineering.* Hemisphere Publishing Corporation, Washington, D.C.

Lovvorn, N.C. and C.C. and Hiller. 1985. A study of heat pump service life. ASHRAE *Transactions* 91(2B):573-88.

Riggs, J.L. 1977. *Engineering economics.* McGraw-Hill, New York.

U.S. Department of Commerce, Bureau of Economic Analysis. Survey of current business. U.S. Government Printing Office, Washington, D.C.

BIBLIOGRAPHY

ASTM. 1985. Definition of terms relating to building economics. ASTM *Standard* E833-85. ASTM, Philadelphia.

Kurtz, M. 1984. *Handbook of engineering economics: A guide for engineers, technicians, scientists, and managers.* McGraw-Hill, New York.

Quirin, D.G. 1967. *The capital expenditure decision.* Richard D. Win, Inc., Homewood, IL.

Van Horne, J.C. 1980. *Financial management and policy.* Prentice Hall, Englewood Cliffs, NJ.

CHAPTER 34

TESTING, ADJUSTING, AND BALANCING

THE environment within a building is a dynamic entity that changes with time and must be rebalanced accordingly. The designer must consider initial and supplementary testing and balancing requirements for commissioning. Complete and accurate operating and maintenance instructions and manuals that include intent of design and how to test, adjust, and balance the building systems are essential. Building operating personnel must be well trained or qualified operating service organizations must be employed to ensure optimum comfort, proper process operations, and economy of operation.

This chapter does not suggest which groups or individuals should perform the functions of a complete testing, adjusting, and balancing procedure. However, the procedure must produce repeatable results that meet the intent of the designer and accurately reflect the requirements of the owner. Overall, one source must be responsible for testing, adjusting, and balancing all systems. As part of this responsibility, the testing organization should check the performance data of all equipment under field conditions to ensure compliance.

During the life of a building's systems, the testing and balancing should be updated as the systems are renovated and changed. The testing of boilers and other pressure vessels for compliance with safety codes is not the primary function of the testing and balancing firm; rather it is to verify and adjust operating conditions in relation to design conditions for flow, temperature, pressure drop, noise, and vibration (ASHRAE *Standard* 111-1988).

DEFINITIONS

System testing, adjusting, and balancing is the process of checking and adjusting all the environmental systems in a building to produce the design objectives. This process includes: (1) balancing air and water distribution, (2) adjusting the total system to provide design quantities, (3) electrical measurement, (4) establishing quantitative performance of all equipment, (5) verifying automatic controls, and (6) sound and vibration measurement. These are accomplished by (1) checking installations for conformity to design, (2) measuring and establishing the fluid quantities of the system, as required to meet design specifications, and (3) recording and reporting the results.

The following definitions are used in this chapter:

Test. To determine quantitative performance of equipment.

Balance. To proportion flows within the distribution system (submains, branches, and terminals) according to specified design quantities.

Adjust. To regulate the specified fluid flow rate and air patterns at the terminal equipment (*e.g.*, reduce fan speed, adjust a damper).

Procedure. An approach and execution of a sequence of work operations to yield repeatable results.

Report forms. Test data sheets arranged for collecting test data in logical order for submission and review. The data sheets should also form the permanent record to be used as the basis for any future testing, adjusting, and balancing.

Terminal. A point where the controlled medium (fluid or energy) enters or leaves the distribution system. In air systems, these may be variable air or constant volume boxes, registers, grilles, diffusers, louvers, and hoods. In water systems, these may be heat transfer coils, fan coil units, convectors, or finned-tube radiation or radiant panels.

GENERAL CRITERIA

Effective and efficient testing, adjusting, and balancing require a systematic, thoroughly planned procedure implemented by experienced and qualified staff. All activities, including organization, calibrated instrumentation, and execution of the actual work, should be scheduled. Because many systems function differently on a seasonal basis, and because temperature performance is significant, it is important to coordinate air-side with water-side work. Preparatory work includes planning and scheduling all procedures, collecting necessary data (including all change orders), reviewing data, studying the system to be worked on, preparing forms, and making preliminary field inspections.

Duct systems must be designed, constructed, and installed to minimize and control air leakage. All duct systems should be sealed and tested for air leakage and water piping should be tested during construction. Any leakage can have a marked effect on system testing, adjusting, and balancing.

Design Considerations

Testing, adjusting, and balancing begin as design functions, with most of the devices required for adjustments being integral parts of the design and installation. To ensure that proper balance can be achieved, the engineer should show and specify a sufficient number of dampers, valves, flow measuring locations, and flow balancing devices; these must be properly located in required straight lengths of pipe or duct for accurate measurement. The testing procedure depends on the system's characteristics and layout. The interaction between individual terminals varies with the system pressures, flow requirements, and control devices.

The preparation of this chapter is assigned to TC 9.7, Testing and Balancing.

AIR VOLUMETRIC MEASUREMENT METHODS

General

The pitot-tube traverse is the generally accepted method of measuring airflow in duct systems. Other methods of measuring airflow at individual terminals are described by the various terminal manufacturers. The primary objective is to establish repeatable measurement procedures that correlate with the pitot-tube traverse.

Laboratory tests, data, and techniques prescribed by equipment and air terminal manufacturers must be reviewed and corroborated for accuracy, applicability, and repeatability of the results. Conversion factors that correlate field data with laboratory results must be developed to predict the equipment's actual field performance.

Air Devices

Generally, K factors of air diffuser manufacturers should be checked for accuracy by field measurement, comparing actual flow measured by pitot-tube traverse to actual measured velocity. Air diffuser manufacturers usually base their volumetric test measurements on a deflection vane anemometer. The velocity is multiplied by an empirical effective area to obtain the air diffuser's delivery. Accurate results are obtained by measuring at the vena contracta with the probe of the deflection vane anemometer.

The methods advocated for measuring the airflow of troffer-type terminals is similar to the methods described for air diffusers. The capture hood is frequently used to measure device airflows, primarily of diffusers and slots. K factors should be established for hood measurements with varying flow rates and deflection settings. If the air does not fill the measurement grid, the readings will require establishing a correction factor (similar to the K factor).

Rotating vane anemometers are commonly used to measure airflow from sidewall grilles. Effective areas (K factors) should be established with the face dampers fully open and deflection set uniformly on all grilles. Correction factors are required when measuring airflow in open ducts, *i.e.,* damper openings and fume hoods (Sauer and Howell 1990).

Duct Flow

Most procedures for testing, adjusting, and balancing air-handling systems rely on measuring volumes in the ducts rather than at the terminals. These measurements are more reliable than those obtained at the terminals, which are based on manufacturer's data. In such procedures, terminal measurements are relied on only for proportionally balancing the distribution within a space or zone.

The preferred method of duct volumetric flow measurement is the pitot-tube traverse average. Care should be taken to obtain the maximum straight run to the traverse station. Test holes should be located as shown in Chapter 13 of the 1989 ASHRAE *Handbook—Fundamentals* and ANSI/ASHRAE *Standard* 111 to obtain the best duct velocity profile. Where factory-fabricated volume measuring stations are used, the measurements should be checked against a pitot-tube traverse for field calibration.

The power input to a fan's driver should be used only as a guide to indicate its delivery. It may be used to verify performance determined by a reliable method (*e.g.*, pitot-tube traverse of system's main) considering system effects that may be present. The flow rate from some fans is not proportional to the power needed to drive them. In some cases, as with forward-curved blade fans, the same power is required for two or more flow rates. The backward-curved blade centrifugal fan is the only type with suitable characteristics *i.e.*, flow rate that varies directly with the power input. If an installation has an inadequate straight length of ductwork or no ductwork to allow a pitot-tube traverse, multiple face velocities across the coil using the vane anemometer and determining the K factor may be read. The velocity readings must be taken exactly as prescribed by Sauer and Howell (1990), using procedures from the airflow measurements at coil faces.

Mixture Plenums

Approach conditions are often so unfavorable that the air quantities comprising a mixture (*e.g.*, outdoor air and return air) cannot be determined accurately by volumetric measurements. In such cases, the temperature of the mixture indicates the balance (proportions) between the component airstreams. The temperature of the mixture can be calculated from Equation (1) as follows:

$$Q_t t_m = Q_o t_o + Q_r t_r \tag{1}$$

where

Q_t = total air quantity, %
Q_o = outside air quantity, %
Q_r = return air quantity, %
t_m = temperature of outside and return mixture, °F
t_o = outdoor temperature, °F
t_r = return temperature °F

Pressure Measurements

The pressures involved with air measurements are barometric pressure, static pressure, velocity pressure, total pressure, and differential pressure. Pressure measurement for field evaluation of air-handling system performance should be taken as recommended in ANSI/ASHRAE *Standard* 111 and analyzed together with the manufacturers' fan curves and system effect as predicted from application of methods in AMCA *Standard* 210. When taken in the field, pressure readings, air quantity, and power input often do not correlate with the manufacturers' certified performance curves unless proper correction is made.

Pressure drops through equipment such as coils, dampers, or filters should not be used to measure airflow. Pressure is an acceptable means of establishing flow volumes only where it is required by, and performed in accordance with, the manufacturer certifying the equipment.

Stratification

Normal design minimizes conditions causing air turbulence to produce the least friction, resistance, and consequent pressure losses in the system. However, under certain conditions, air turbulence is desirable and necessary. For example, two airstreams of different temperatures can stratify in smooth, uninterrupted flow conditions. In this situation, mixing should be promoted in the design. The return and outside airstreams at the inlet side of the air-handling unit tend to stratify where enlargement of the inlet plenum or casing size decreases the air velocity. Without a deliberate effort to mix the two airstreams (*i.e.*, in cold climates, placing the outdoor air entry at the top of the plenum and the return air at the bottom of the plenum to allow natural mixing), stratification can exist and be carried throughout the system (*e.g.*, filter, coils, eliminators, fans, and ducts). Stratification can cause damage by freezing coils and rupturing tubes. It can also affect the temperature control in plenums, spaces, or both.

Stratification can also be reduced by adding vanes to break up and mix the two airstreams. No solution to stratification problems is guaranteed; each condition must be evaluated by field measurements and experimentation.

BALANCING PROCEDURES FOR AIR DISTRIBUTION SYSTEMS

General procedures for testing and balancing are described here, although no one established procedure is applicable to all systems. The bibliography lists sources of additional information.

Instrumentation for Testing and Balancing

The minimum instruments necessary for air balance are:

- Micromanometer calibrated in 0.005 in. of water divisions
- Combination inclined and vertical manometer (0 to 10 in. of water)
- Pitot tubes in various lengths, as required
- Tachometer, direct contact, self-timing type, or strobe light
- Clamp-on ampere meter with voltage scales (rms type)
- Deflecting vane anemometer
- Rotating vane anemometer
- Flow hood
- Dial thermometers (2 in. diameter minimum and 1°F graduations mininimum) and glass stem thermometers (1°F graduations minimum)

The instrumentation must be evaluated periodically to verify its accuracy and repeatability prior to use in the field.

Preliminary Procedure for Air Balancing

Before operating the system, the following steps should be performed:

1. Obtain as-built design drawings and specifications and become thoroughly acquainted with the design intent.
2. Obtain copies of approved shop drawings of all air-handling equipment, outlets (supply, return, and exhaust), and temperature control diagrams including performance curves.
3. Compare design to installed equipment and field installation.
4. Walk the system from the air-handling equipment to terminal units to determine variations of installation from design.
5. Check filters and dampers (both volume and fire) for correct and locked position, and temperature control for completeness of installation before starting fans.
6. Prepare report test sheets for both fans and outlets. Obtain manufacturer's outlet factors and recommended testing procedure. A summation of required outlet volumes permits a cross-checking with required fan volumes.
7. Determine best locations in main and branch ductwork for most accurate duct traverses.
8. Place all outlet dampers in the open position.
9. Prepare schematic diagrams of system as-built ductwork and piping layouts to facilitate reporting.
10. Prepare estimate of system effect of installed ductwork on fan performance.
11. For variable volume air systems, develop a plan to simulate diversity.

Equipment and System Check

1. Place all fans in operation (supply, return, and exhaust) and immediately check the following items:
   a) Motor amperage and voltage to guard against overload.
   b) Fan rotation.
   c) Operability of static pressure limit switch.
   d) Automatic dampers for proper position.
   e) Air and water resets operating to deliver required temperatures.
   f) Air leaks in the casing and in the scarfing around the coils and filter frames should be checked by moving a bright light along the outside of the duct joints while observing the darkened interior of the casing. Any leaks should be caulked. Note points where piping enters the casing to ensure that escutcheons are right. Do not rely on pipe insulation to seal these openings because the insulation may shrink. In prefabricated units, check that all panel-fastening holes are filled to prevent whistling.
2. Traverse the main supply ductwork whenever possible. All main branches should also be traversed where duct arrangement permits. Selection of traverse points and method of traverse should be as follows:
   a) Traverse each main or branch after the longest possible straight run for the duct involved.
   b) For test hole spacing, refer to Chapter 13 of the 1989 ASHRAE *Handbook—Fundamentals.*
   c) Traverse using a pitot tube and manometer where velocities are over 600 fpm. Below this velocity, use either a micromanometer and pitot tube or a recently calibrated thermal anemometer.
   d) Note temperature and barometric pressure to determine if they need to be corrected for standard air quantity. Corrections are normally insignificant below 2000-ft elevation; however, where accurate results are desirable, corrections would justify them.
   e) After establishing the total air being delivered, adjust the fan speed to obtain the design airflow, if necessary. Check power and speed to see that motor power, critical fan speed, or both have not been exceeded.
   f) Proportionally adjust branch dampers until each has the proper air volume.
   g) With all the dampers and registers in the system open and with the supply, return, and exhaust blowers operating at or near design speed, set the minimum outdoor and return air ratio. If duct traverse locations are not available, this can be done by measuring the mixture temperature with thermometers in the return air, outdoor air louver, and filter section. As an approximation, the temperature of the mixture may be calculated from Equation (1).

      The greater the temperature difference between hot and cold air, the easier it is to get accurate damper settings. Take the temperature at many points in a uniform traverse to be sure there is no stratification. When washable filters are installed, take a reading at the center of each filter for a simple, but effective, traverse.

      After the minimum outdoor air damper has been set for the proper percentage of outdoor air, take another traverse of mixture temperatures and install baffling if the variation from the average is more than ±5%. Remember that stratified mixed air temperatures vary greatly with the outdoor temperature in cold weather, while return air temperature has only a minor effect.
3. Carefully set the system for balance using the prescribed procedures as follows:
   a) Adjust the system with mixing dampers positioned under the minimum outdoor air conditions.
   b) When adjusting multizone or double-duct constant volume systems, establish the ratio of the design volume through the cooling coil to total fan volume to achieve the desired diversity factor. Keep the proportion of cold to total air constant during the balance. However, check each zone or branch with this component on full cooling. If the design calls for full flow through the cooling coil, the entire system should be set to full flow through the cooling side while making tests. Perform the same procedure for the hot air side.
4. Balance the terminal outlets in each control zone in proportion to each other. The following steps may be followed to balance the terminals.
   a) Once the preliminary fan quantity is set, proportion the terminal outlet balance from the outlets into the branches to the fan. Concentrate on proportioning the flow rather than the absolute quantity. As changes are made to the fan

settings and branch dampers, the outlet terminal quantities remain proportional. Branch dampers should be used for major adjusting and terminal dampers for trim, or minor, adjustment only. It may be necessary to install additional subbranch dampers to decrease the use of terminal dampers that create objectionable noise.

b) Normally, several passes through the entire system are necessary to obtain proper outlet values.

c) Totaling the tested outlet air quantity acts as a possible indicator of duct leakage when compared to duct traverse air quantities.

d) With total design air established in the branches and at the outlets, perform the following: (1) take new fan motor amperage readings, (2) find static pressure across the fan, (3) read and record static pressure across each component (intake, filters, coils, and mixing dampers), and (4) take a final duct traverse.

Dual-Duct Systems

Most constant-volume dual-duct systems are designed to handle a portion of the total system's supply through the cold duct and smaller air quantities through the hot duct. Balancing should be accomplished as follows:

1. Check the leaving air temperature at the nearest terminal to verify that the hot and cold damper inlet leakage is not greater than the maximum allowable leakage established.
2. Check apparatus and main trunks, as outlined in the Equipment and System Check section in this chapter.
3. Determine if the static pressure at the end of the system (the longest duct run) is at or above the minimum required for mixing box operation. Proceed to the extreme end of the system and check the static pressure drop across the last three boxes with an inclined manometer. The drop across the box should exceed the minimum static pressure recommended by the manufacturer. Additional static pressure is required for the low-pressure distribution system downstream of the box.
4. Proportionately balance the diffusers or grilles on the low-pressure side of the box, as described for low-pressure systems.
5. Change the control settings to full heating, and make certain that the controls and dual-duct boxes function properly. Spot-check the airflow at several diffusers. Check for stratification.
6. If the engineer has included a diversity factor in selecting the main apparatus, it will not be possible to get full flow from all boxes simultaneously, as outlined above.

VARIABLE VOLUME SYSTEMS

Many types of variable volume systems have been developed to conserve energy. These systems can be categorized as pressure dependent or pressure independent.

Pressure-dependent systems incorporate air terminal boxes that have a thermostat signal controlling a damper actuator. The air volume to the space varies to maintain the space temperature, while the air temperature supplied to the terminal boxes remains constant. The balance of this system constantly changes with loading changes; therefore, any balancing procedure will not produce repeatable data unless changes in system load are simulated by using the same configuration of thermostat settings each time the system is tested—*i.e.*, the same terminal boxes are fixed in the minimum and maximum positions for the test.

Pressure-independent systems incorporate air terminal boxes that have a thermostat signal used as a master control to open or close the damper actuator and a velocity controller used as a submaster control to maintain the maximum and minimum amounts of air to be supplied to the space. The air volume to the space varies to maintain the space temperature, while the air temperature supplied to the terminal remains constant. Care should be taken to verify the operating range of the damper actuator as it responds to the velocity controller to prevent dead bands or overlap of control in response to other system components (*e.g.*, double-duct VAV, fan-powered boxes, and retrofit system). Care should also be taken to verify the action of the thermostat with regard to the damper position, as the velocity controller can change the control signal ratio or reverse the control signal.

In a pressure-dependent system, the setting of minimum airflows to the space, other than at no flow, is not suggested, unless the terminal box has a normally closed damper and the manufacturer of the damper actuator provides adjustable mechanical stops. The pressure-independent system requires verification that the velocity controller is operating properly. Inlet duct configuration can adversely affect the operation of the velocity controller (Griggs *et al.* 1990). The primary difference between the two systems is that the pressure-dependent system supplies a different amount of air to the space as the pressure upstream of the terminal box changes. If the thermostats are not calibrated properly to meet the space load, several zones may overcool or overheat. When the zones overcool and receive greater amounts of supply air than required, they decrease the amount of air that can be supplied to overheated zones. The pressure-independent system is not affected by improper thermostat calibration in the same way that a pressure-dependent system is, because the minimum and maximum airflow limits may be set for each zone.

System Static Control

System static control is important to achieve fan energy savings and to prevent overpressurizing the duct system. The system pressure in either the pressure-dependent system or the pressure-independent system can be controlled in the following ways:

No Fan Volumetric Control. This is sometimes referred to as "riding the fan curve." This type of system should be limited to systems with minimum airflows of 50% of peak design and flat forward-curved fans. Pressure and noise are potential problems.

System Bypass Control. As the system pressure increases due to terminal boxes closing, a relief damper bypasses the system air back to the fan inlet. With this type of control, the economy of varied fan output is nonexistent, and the relief damper is usually a major source of duct leakage and noise. The relief damper should be modulated to maintain a minimum duct static pressure.

Discharge Damper. System losses and noise should be considered with this system.

Vortex Damper. System losses due to inlet air conditions are a problem, and the vortex damper does not completely close. The minimum expected airflow should be evaluated.

Varying Fan Speed Mechanically. Slippage loss of belts, cost of belt replacement, and the initial cost of the components are of concern.

Variable Pitch-in-Motion Fans. This system is comparable to varying fan speed. Maintenance and the means to prevent the fan running in the stall condition must be evaluated.

Varying Fan Speed Electrically. This system is usually the most efficient and is accomplished by varying the voltage or the frequency to the fan motor. Some versions of motor drives may cause electrical noise and affect other devices.

In controlling these fan systems, the location of the static pressure sensors is critical and should be field verified to give the most representative point of operation. After the terminal boxes have been proportioned, the static pressure control can be verified by observing static pressure changes at the fan discharge and the static pressure sensor as the load is simulated from maximum airflow to minimum airflow (*i.e.*, set all terminal boxes to balanced airflow conditions and verify if any changes in static pressure occur by placing one terminal box at a time to minimum airflow, until all terminals are placed at the minimal airflow setting). Care should be taken to verify that the maximum to minimum air volume changes are within the fan curve performance (rpm or total pressure).

Diversity

Diversity may be used on a VAV system where it is assumed that the total system airflow volume, by design, is lower, assuming that all of the system terminal boxes will never open fully at the same time. Care should be taken to avoid any duct leakage. All ductwork upstream of the terminal box should be considered as medium pressure ductwork, whether in a low- or medium-pressure system. (See Chapter 43 in the 1988 ASHRAE *Handbook—Equipment*.)

A procedure to test the total air on the system should be established by setting terminal boxes to the zero or minimum position nearest to the fan on either type of system. During peak load conditions, care should be taken to verify that an adequate pressure is available upstream of the terminal boxes to achieve design airflow to the spaces.

Outside Air Requirements

Maintaining the space under a slight positive or neutral pressure to atmosphere is difficult with all variable volume systems. In most systems, the exhaust requirement for the space is constant; hence, the outside air used to equal the exhaust air and meet the minimum outside air requirements for the building codes must also remain constant. Due to the location of the outside air intake and the changes in pressure, this does not usually happen. The outside air should enter the fan at a point of constant pressure (*i.e.*, supply fan volume can be controlled by proportional static pressure control, which can control the volume of the return air fan).

Return Air Fans

If return air fans are required in series with a supply fan, the type of control and sizing of the fans is most important, as serious over- and underpressurization can occur, especially during the economizer cycle.

Types of VAV Systems

Single-Duct VAV. This system incorporates a pressure-dependent or independent terminal and usually has reheat at some predetermined minimal setting on the terminal unit or separate heating system.

Bypass System. This system incorporates a pressure-dependent damper, which, on demand for heating, closes the damper to the space and opens to the return air plenum.

This system sometimes incorporates a constant bypass airflow or a reduced amount of airflow bypassed to the return plenum in relation to the amount supplied to the space. No economical value can be obtained by varying the fan speed with this system. A control problem can exist if any return air sensing is done to control a warm-up or cool-down cycle.

VAV System Using Single-Duct VAV and Fan-Powered, Pressure Dependent Terminals. This system has a primary source of air from the fan to the terminal and a secondary powered fan source, which pulls air from the return to the air plenum before the additional heat source. This system places additional maintenance of terminal filters, motors, and capacitors on the building owner. In certain fan-powered boxes, back-draft dampers are a source of system duct leakage when the system calls for the damper to be fully closed. Typical applications include geographic areas where the ratio of heating hours to cooling hours is low.

Double-Duct VAV. This type of terminal incorporates two single-duct variable terminals and can be controlled by velocity controllers in sequence with one another so that both hot and cold ducts are closed or can be controlled by a velocity controller. This controller is used as a downstream sensor in the terminal unit to maintain either heating or cooling airflow. Often, low system pressure, either in the hot or cold duct, causes mixing of air in the two ducts, which results in excess energy use or discomfort in the space.

Balancing the VAV System

The general procedure for balancing a VAV system is:

1. Determine the required maximum air volume to be delivered by the supply and return air fans. Diversity of load usually means that the volume will be somewhat less than the outlet total.
2. Obtain fan curves on these units and request information on surge characteristics from the fan manufacturer.
3. If an inlet vortex damper control is to be used, obtain the fan manufacturer's data pertaining to the deaeration of the fan when used with the damper. If speed control is used, find the maximum and minimum speed that can be used on the project.
4. Determine the minimum and maximum operating pressures for terminal or variable volume boxes to be used on the project from the manufacturer.
5. Construct a theoretical system curve, including an approximate surge area. The system curve starts at the minimum inlet static pressure of the boxes, plus system loss at minimum flow, and terminates at the design maximum flow. The operating range using an inlet vane damper is between the surge line intersection with the system curve and the maximum design flow. When variable speed control is used, the operating range is between (a) the minimum speed that can produce the necessary minimum box static at minimum flow still in the fan's stable range and (b) the maximum speed necessary to obtain maximum design flow.
6. Position the terminal boxes to the proportion of maximum fan air volume to total installed terminal maximum volume.
7. Set the fan to operate at approximate design speed (increase about 5% for a full open inlet vane damper).
8. Check a representative number of terminal boxes. If a wide variation in static pressure is encountered, or if the airflow at a number of boxes is below minimum at maximum flow, check every box.
9. Run a total air traverse.
10. Increase the speed if either or both static pressure and volume are low. If the volume is correct, but the static is high, reduce the speed. If the static is high or correct, but the volume is low, check for system effect at the fan. If there is no system effect, go over all terminals and adjust them to the proper volume.
11. Run steps (7) through (10) with the return or exhaust fan set and traversed, and with the system set on minimum outdoor air.
12. Proportion the outlets and verify the design volume with the VAV box on the maximum flow setting. Verify the minimum flow setting.
13. Set the terminals to minimum, and adjust the inlet vane or speed controller until minimum static pressure and airflow are obtained.
14. The temperature control and the balancing personnel should agree on the final placement of the sensor for the static pressure controller. This sensor must be placed in the supply duct far enough from the fan discharge to represent the average static pressure in the system.
15. Check the return air fan speed or its inlet vane damper that tracks or adjusts to the supply fan airflow to ensure proper outside air volume.
16. Operate the system on 100% outside air (weather permitting) and check supply and return fans for proper power and static pressure.

Induction Systems

Most induction systems use high-velocity air distribution. Balancing should be accomplished as follows:

1. Perform steps outlined under the basic procedures common to all systems for apparatus and main trunk capacities.
2. Determine the primary airflow at each terminal unit by reading the unit plenum pressure with a manometer and locating the point on the charts (or curves) of air quantity versus static pressure supplied by the unit manufacturer.
3. Normally, about three complete passes around the entire system are required for proper adjustment. Make a final pass without adjustments to record the end result.
4. To provide the quietest possible operation, adjust the fan to run at the slowest speed that provides sufficient nozzle pressure to all units with minimum throttling of all unit and riser dampers.
5. After balancing each induction system with minimum outdoor air, reposition to allow maximum outdoor air and check power and static pressure readings.

Report Information

To be of value to the consulting engineer and owner's maintenance department, the air-handling report should consist of at least the following items:

1. Design
- a) Air quantity to be delivered
- b) Fan static pressure
- c) Motor power
- d) Percent of outside air under minimum conditions
- e) Rpm of the fan
- f) Power required to obtain this air quantity at design static pressure

2. Installation
- a) Equipment manufacturer
- b) Size of unit installed
- c) Arrangement of the air-handling unit
- d) Class fan
- e) Nameplate power, nameplate voltage, phase, cycles, and full-load amperes of the motor installed

3. Field tests
- a) Fan rpm
- b) Power readings (voltage, amperes of all *legs* at motor terminals)
- c) Total pressure differential across unit components
- d) Fan suction and fan discharge static pressure (equals fan total pressure)
- e) Plot of actual readings on manufacturer's fan performance curve to show the installed fan operating point

It is important to establish the initial static pressures accurately for the air treatment equipment and the duct system so that the variation in air quantity due to filter loading can be calculated. It enables the designer to ensure that the total fan quantity will never be less than the minimum requirements. It also serves as a check of dirt loading in coils, since the design air quantity for peak loading of the filters has already been calculated.

4. Terminal Outlets
- a) Outlet by room designation and position
- b) Outlet manufacture and type
- c) Outlet size (using manufacturer's designation to ensure proper factor)
- d) Manufacturer's outlet factor (Where no factors are available, or field tests indicate the listed factors are incorrect, a factor must be determined in the field by traverse of a duct leading to a single outlet. A traverse also ensures that no installation air leakage exists.)
- e) Design air quantity and the required velocity in fpm to obtain this cfm
- f) Test velocities and resulting air quantity
- g) Adjustment pattern for every air terminal

5. Additional Information

The following information is desirable under applicable circumstances:
- a) Air-handling units
  - (1) Belt number and size
  - (2) Drive and driven sheave size
  - (3) Belt position on adjusted drive sheaves (bottom, middle, and top)
  - (4) Motor speed under full load
  - (5) Motor heater size
  - (6) Filter type and static pressure at initial use and full load; time to replace
  - (7) Variations of velocity at various points across the face of the coil
  - (8) Existence of vortex or discharge dampers, or both
- b) Distribution system
  - (1) Unusual duct arrangements
  - (2) Branch duct static readings in double-duct and induction system
  - (3) Ceiling pressure readings where plenum ceiling distribution is being used; tightness of ceiling
  - (4) Relationship of building to outdoor pressure under both minimum and maximum outdoor air
  - (5) Induction unit manufacturer and size (including required air quantity and plenum pressures for each unit) and a test plenum pressure and resulting primary air delivery from the manufacturer's listed curves
- c) All equipment nameplates visible and easily readable

Many independent firms have developed detailed procedures suitable to their own operations and the area in which they function. These procedures are often available for information and evaluation on request (see Bibliography).

PRINCIPLES AND PROCEDURES FOR BALANCING HYDRONIC SYSTEMS

Both air- and water-side balance techniques must be performed with sufficient accuracy to ensure that the system operates economically, with minimum energy, and with proper distribution. Air-side balance requires a precise flow measuring technique, because air, which is usually the prime heating or cooling transport medium, is more difficult to measure in the field. Reducing the airflow to less than the design requirement directly reduces the heat transfer. In contrast, the rate of heat transfer for the water side of the terminal does not vary linearly with the water flow rate. Therefore, the proper balancing valve with the correct control characteristics must be provided at each terminal to proportionally balance the water flow rate according to the design conditions.

Heat Transfer at Reduced Flow Rate

The typical heating-only hydronic terminal gradually reduces its heat output as flow is reduced (Figure 1). Decreasing water flow to 50% of design reduces the heat transfer to 90% of that at full design flow. The control valve must reduce the water flow to 10% to reduce the heat output to 50%. The reason for the relative insensitivity to changing flow rates is that the governing coefficient for heat transfer is the air-side coefficient. A change in internal or water-side coefficient with flow rate does not materially affect the overall heat transfer coefficient. This means that heat transfer for water-to-air terminals is established by the mean air-to-water temperature difference, the heat transfer is measurably changed,

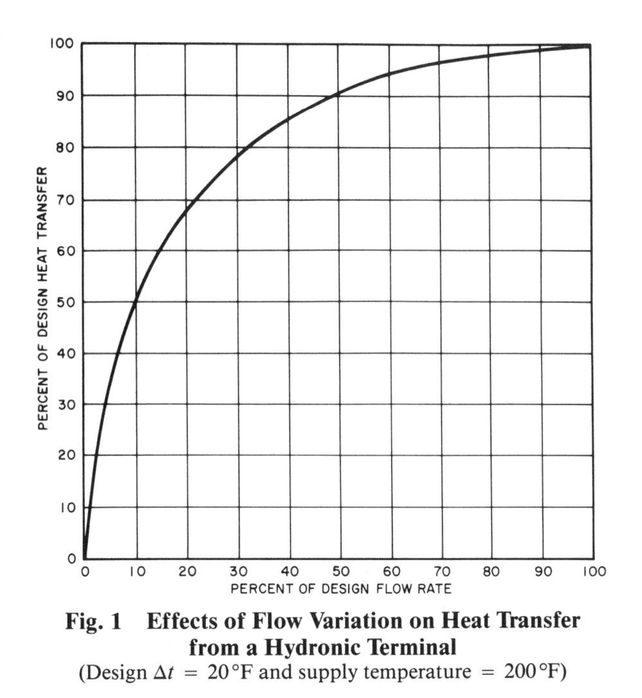

Fig. 1 Effects of Flow Variation on Heat Transfer from a Hydronic Terminal
(Design Δt = 20°F and supply temperature = 200°F)

and a change in the mean water temperature requires a greater change in the water flow rate.

A secondary safety factor also applies to heating terminals. Unlike chilled water, hot water can be supplied at a wide range of temperatures. So, in some cases, an inadequate terminal heating capacity caused by insufficient flow can be overcome by raising the system supply water temperature. Design within the temperature limits of 250°F of the low pressure code, for example, must be considered, however.

The previous comments apply to heating terminals selected for a 20°F temperature drop (Δt) and with a supply water temperature of about 200°F. Figure 2 shows the flow variation when 90% terminal capacity is acceptable. Note that heating system tolerance decreases with temperature and flow rates and that chilled water terminals are much less tolerant to flow variation than hot water terminals.

Dual-temperature heating/cooling hydronic systems are sometimes completed and started during the heating season. Adequate heating ability in the terminals may suggest that the system is balanced. Figure 2 shows that 40% of design flow through the terminal provides 90% of design heating with 140°F supply water and a 10°F temperature drop. Increased supply water temperature establishes the same heat transfer at terminal flow rates of less than 40% design.

In some cases, dual-temperature water systems may experience a decreased flow during the cooling season because of the chiller pressure drop; this could cause a flow reduction of 25%. For example, during the cooling season, a terminal that originally heated satisfactorily would only receive 30% of the design flow rate.

While the example of reduced flow rate at Δt = 20°F only affects the heat transfer by 10%, this reduced heat transfer rate may have the following negative affects:

1. The object of the system is to deliver (or remove) heat where required. When the flow is reduced from the design rate, the system must supply heating or cooling for a longer period to maintain room temperature.
2. As the load reaches design conditions, the reduced flow rate is unable to maintain room design conditions.

Terminals with lower water temperature drops have a greater tolerance to unbalanced conditions. However, larger water flows

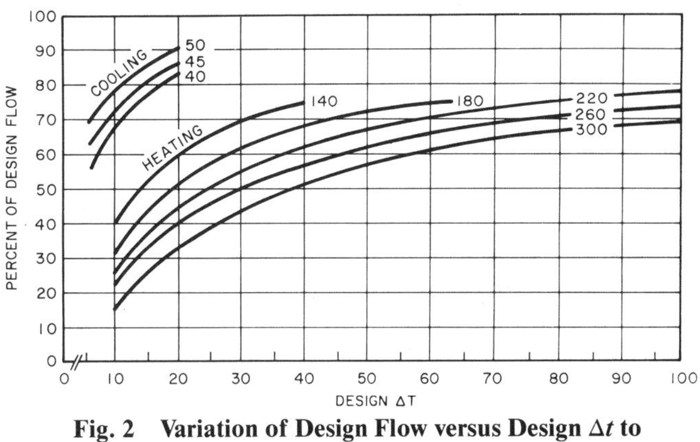

Fig. 2 Variation of Design Flow versus Design Δt to Maintain 90% Terminal Heat Transfer

are necessary, requiring larger pipes, pumps, and pumping costs. Also, automatic valve control is more difficult.

System balance becomes more important in terminals with a large Δt. Less water flow is required, which reduces the size of pipe, valves, and pumps as well as pumping cost. A more linear emission curve gives better system control.

Generalized Chilled Water Terminal—Flow versus Heat Transfer

The heat transfer for a typical chilled water coil in an air duct versus water flow rate is shown in Figure 3. The curves shown are based on ARI rating points: 45°F inlet water at a 10°F rise with entering air at 80°F dry bulb and 67°F wet bulb.

The basic curve applies to catalog ratings for lower dry-bulb temperatures, providing a consistent entering air moisture content *e.g.*, 75°F dry bulb, 65°F wet bulb. Changes in inlet water temperature, temperature rise, air velocity, and dry-bulb and wet-bulb temperatures will cause terminal performance to deviate from the curves. Figure 3 is only a general representation of the total heat transfer change versus flow for a hydronic cooling coil and does

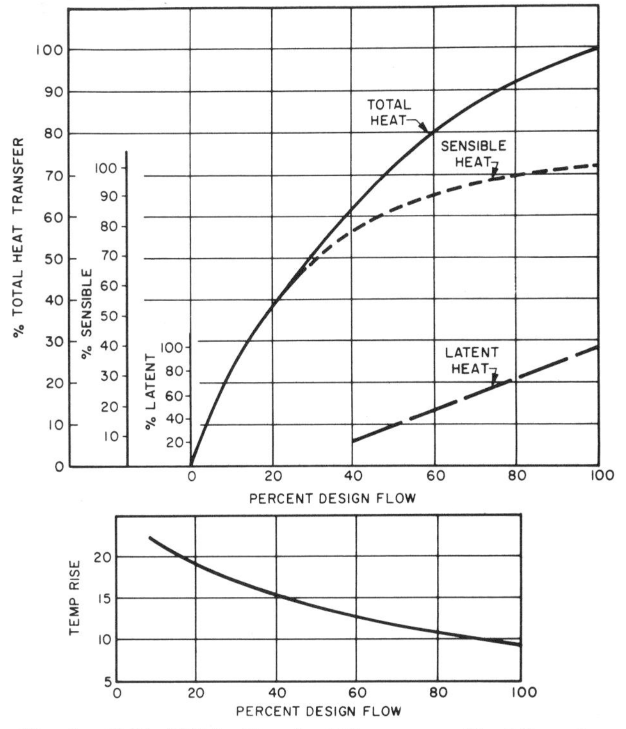

Fig. 3 Chilled Water Terminal Flow versus Heat Transfer

Table 1 Load Flow Variations

Load Type	% Design Flow at 90% Load	Other Load, Order of %		
		Sensible	Total	Latent
Sensible	65	90	84	58
Total	75	95	90	65
Latent	90	98	95	90

Note: Dual-temperature systems are designed to chilled flow requirements and often operate on a 10 °F temperature drop at full-load heating.

not apply to all chilled water terminals. Comparing Figure 3 with Figure 1 indicates the similarity of the nonlinear heat transfer in comparison with flow for both the heating and the cooling terminal.

Table 1 shows that if the coil is selected for the load and the flow is reduced to 90% of the load, three flow variations can be interpreted to satisfy the reduced load at various sensible and latent combinations.

Flow Tolerance and Balance Procedure

The design procedure rests on a design flow rate and an allowable flow tolerance. The designer must define both the terminal's flow rates and feasible flow tolerance, bearing in mind that the cost of balancing rises with tightened flow tolerance.

Figure 4 illustrates flow tolerance guidelines as a function of supply water temperature and water Δt and is based on achieving from 97 to 101.5% of expected terminal heat transfer. A maximum flow tolerance band of ±10% is suggested, even though some hot water systems can tolerate a higher flow deviation. Some systems require a tighter flow tolerance; for example, terminals using low-temperature water from heat recovery may require a ±5% at a supply of 110 °F with a Δt of 20 °F.

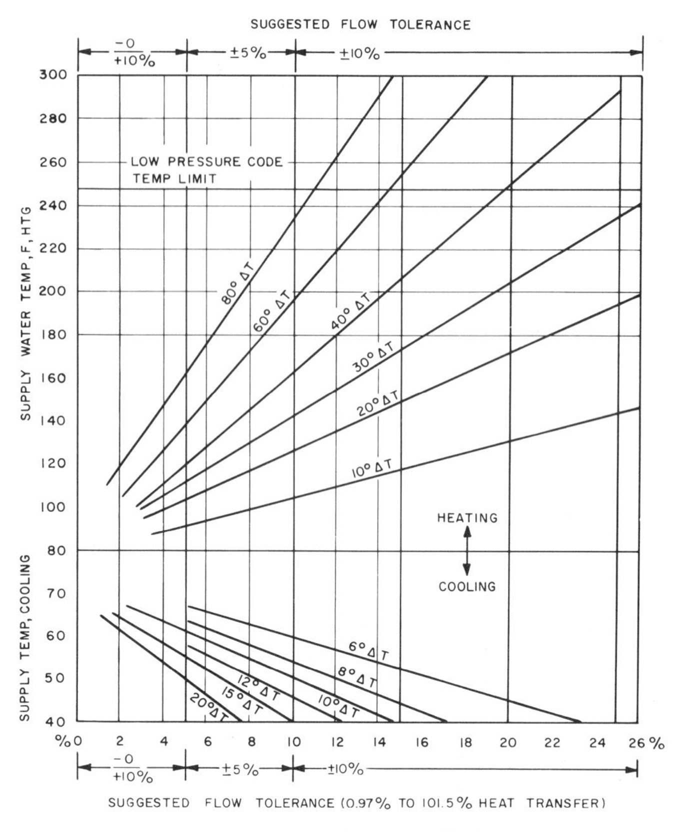

Fig. 4 Flow Tolerance Plot

WATER-SIDE BALANCING

The water side should be balanced by direct flow measurement. This approach is accurate because it avoids the compounding errors introduced by temperature difference procedures. Measuring the flow at each terminal enables proportional balancing and ultimate matching of the pump to the actual system requirements (by trimming the pump impeller or reducing the pump motor power, for example). In many cases, reduction in pump operating cost will pay for the cost of water-side balancing.

Equipment

Proper equipment selection and preplanning is needed to successfully balance hydronic systems. Circumstances sometimes dictate that flow, temperature, and pressure be measured. The designer should specify the water flow balancing devices for installation during construction and testing during the balancing of the hydronic system. The devices may consist of all or a combination of the following:

- Flow meters (venturi, orifice plate, multiported pitot tubes, and flow indicators)
- Balancing valves for measuring and setting flow
- Manometers, differential pressure gages (either analog or digital)
- Portable digital meter to measure flow and pressure drop
- Test pressure taps, pressure gages, thermometers, and wells
- System components used as flow meters (terminal coils, chillers, heat exchangers, or control valves if using the manufacturer's factory-certified flow versus drop curves)
- Flow-limiting or regulator devices (to add a variable load to the pump head)
- Pumps with factory-certified pump curves
- Control valve with factory flow coefficient C_v or flow versus pressure drop tables
- Balancing valve with a factory-rated flow coefficient C_v, a flow versus handle position and pressure drop table, or a slide rule flow calculator.

Record Keeping

Balancing requires keeping accurate records while making field measurements. The actual dated and signed field test report helps the designer or customer in approval of the work and the owner will have a valuable reference when documenting future changes.

Sizing Balancing Valves

A balancing valve is placed in the system to adjust water flow to a terminal, branch, zone, riser, or main. A common valve-sizing approach is to select for the line size; this may be unwise, however, as control may be lost at low flows or too much pressure drop may have to be added at high flow settings, thus requiring larger valves. Many balancing valves and measuring meters can give an accuracy of ±5% down to a pressure drop of 12 in. of water with the balancing valve in the wide open position. Some designer's select a balancing valve such that the pressure drop is between 5 to 10% of the total loop pressure drop at design flow when the valve is wide open. Too large a balancing valve pressure drop will affect the authority and flow characteristic of the control valve. Equation (2) may be used to determine the flow coefficient C_v for a balancing valve. The equation may also be used to size a control valve.

The flow coefficient C_v is defined as the number of gallons per minute (U.S.) of water that flows through a wide-open valve with a pressure drop of 1 psi at 60 °F. This is shown as:

$$C_v = Q(s_f/\Delta p)^{0.5} \quad (2)$$

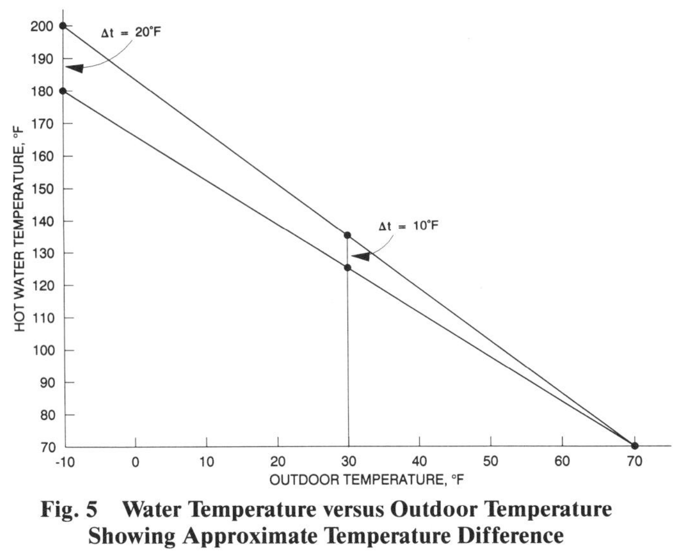

Fig. 5 Water Temperature versus Outdoor Temperature Showing Approximate Temperature Difference

If pressure drop is determined in feet of water, Equation (2) can be shown as:

$$C_v = 1.5Q/(s_f/\Delta h)^{0.5} \quad (3)$$

where

C_v = flow coefficient at 1 psi drop
Q = design flow for terminal or valve, gpm
Δp = pressure drop, psi
Δh = pressure drop, ft of water
s_f = specific gravity of fluid

Hydronic Balancing Methods

A variety of techniques for balancing hydronic systems is used. The following are the most common:

- Balance by temperature difference
- Balance by read-and-set method
- Water balance by preset pressure differential
- Water balance by proportional method

Balance by Temperature Difference

This often-used balancing procedure is based on water temperature difference measurement between supply and return at the terminal. The designer selects the cooling and/or heating terminal for a calculated design load at full-load conditions. At less than full load, which is true for most operating hours, the temperature drop is proportionately less. Figure 5 demonstrates this for a heating system at a design Δt = 20°F for outside design of −10°F and room design of 70°F.

At any outside temperature, other than design, the balancing technician should construct a similar chart and read off the Δt for balancing. For example, at 50% load, or 30°F outdoor air, the Δt required is 10°F or 50% of the design drop.

The temperature balance method is a rough approximation and should not be used where great accuracy is required. It is not accurate enough for a cooling system or for high-temperature drop heating systems.

Balance by Read-and-Set Method

In a simple application, the pressure difference is read across the balancing valve taps. By noting the valve size and the handle degrees or position markings and referring to the manufacturer's valve data sheet, nomogram, or slide rule, the flow can be read directly.

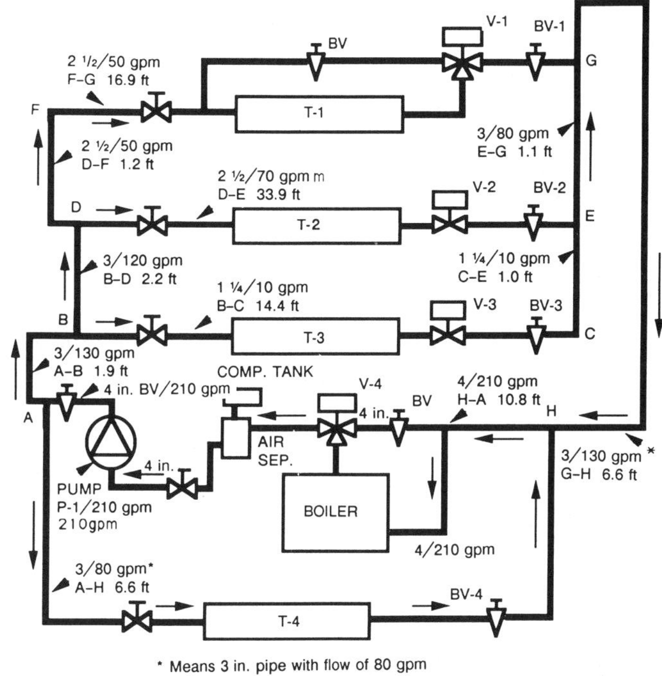

Fig. 6 Preset Example Flow Diagram

Table 2 Preset Example—Pipe Section Flow and Estimated Head Loss

Pipe Section	Flow Rate, gpm	Estimated Pressure Drop, ft of water
A–B	130	1.9
B–D	120	2.2
D–F	50	1.2
C–E	10	1.0
E–G	80	1.1
G–H	130	6.6
F–G (T_1)	50	16.9
D–E (T_2)	70	33.9
B–C (T_3)	10	14.4
A–H (T_4)	80	6.6
A–H (Equipment Room)	210	10.8

Water Balance by Preset Pressure Differential

Using the project's shop drawings and flow diagrams and starting from the pump, list the pressure drop of each component including flow stations, heat exchangers, coils, pipe, and fittings. Assumed control valve C_v and temperature drops should also be listed. See Figure 6 and Tables 2 through 5 for an example. Once the required pressure drop through each circuit has been determined, settings of the balancing valves can be calculated and preset to establish an equal pressure drop in all circuits.

If variable orifice flow regulators are used for balancing, size them in a similar manner. These regulators should not be used in series or in systems with diversity or variable flow.

Balancing by the preset method is done from a design point of view; as-built variations need to be considered when testing at the job-site.

Water Balance by Proportional Method

The proportional method of water-side balance uses as-built conditions and adapts well to design diversity factors. Circuits in a system are proportionately balanced to each other by a flow

Table 3 Preset Example—Head Loss Total Per Circuit

A-H through T-1		A-H through T-2		A-H through T-3		A-H through T-4	
Pipe Section	Head Loss, ft	Pipe Section	Head Loss, ft	Pipe Section	Head Loss, ft	Pipe Section	Head Loss, ft
A-B	1.9	A-B	1.9	A-B	1.9	A-H (T_4)	6.6
B-D	2.2	B-D	2.2	B-C (T_3)	14.4		
D-F	1.2	D-E (T_2)	33.9	C-E	1.0		
F-G (T_1)	16.9	E-G	1.1	E-G	1.1		
G-H	6.6	G-H	6.6	G-H	6.6		
Total required circuit head loss	28.8		45.7[a]		25.0		6.6

[a]Highest head loss

Table 4 Preset Example—Determination of Balancing Valve Head Loss

CIRCUIT	A→H through T_1	A→H through T_2	A→H through T_3	A→H through T_4
Balance governing head loss, ft	47.4	47.4	47.4	47.4
Total required head loss/circuit, ft	−28.8	−45.7	−25.0	−6.6
Circuit head loss difference, ft	18.6	1.7	22.4	40.8
Balancing valve flow, gpm	50	70	10	80
Balancing valve no.	BV-1	BV-2	BV-3	BV-4
Balancing valve preset	18.6 ft at 50 gpm	1.7 ft at 70 gpm	22.4 ft at 10 gpm	40.8 ft at 80 gpm

Table 5 Preset Example—Balancing Valve Preset Setting

Balancing Valve No.	BV-1	BV-2	BV-3	BV-4
Valve size, in.	2-1/2	2-1/2	1-1/4	3
Setting, degrees	36	Open	39	45

quotient, which is the actual flow rate divided by the design flow rate through a circuit.

$$\text{Flow quotient} = \frac{\text{Actual flow rate}}{\text{Design flow rate}} \tag{4}$$

To balance a branch system proportionally, first open the balancing valves and the control valves in that branch to their wide-open position and calculate each balancing valve's quotient based on actual measurements at each valve. Record these values on the test form, and note the circuit with the lowest flow quotient.

The circuit with the lowest quotient is the circuit with the greatest head loss and is the reference circuit. Adjust all the other balancing valves in that branch until they have the same quotient as the reference circuit.

Note that when a second valve is adjusted, the flow quotient in the reference valve will also change, and continued adjustment is required to make their flow quotients equal. Once they are equal, they will remain equal or in proportional balance to each other while other valves in the branch are adjusted or there is a change in pressure or flow.

When all the balancing valves are adjusted to their branch's respective flow quotient, the total system water flow is adjusted to the design by setting the balancing valve at the pump discharge to a flow quotient of one.

The pressure drop across the balancing valve at the pump discharge is pressure produced by the pump that is not required to provide the design flow rate to the system. This excess pressure can be removed by trimming the pump impeller or reducing the pump speed. Once the trimming is done, the pump discharge balancing valve must be reopened to its wide-open position to provide the design flow.

As in variable speed pumping, diversity and flow changes are well accommodated by a system that has been proportionately balanced. Since the balancing valves have been balanced to each other at a particular flow (design), any changes in that flow are proportionately distributed.

Balancing of the water side in a system that uses diversity must be done at full flow. The system components are selected based on heat transfer at full flow, so they must be balanced to this point. To accomplish full-flow proportional balance, shut off part of the system while balancing the remaining sections. When a section has been balanced, shut it off and open the section that was open originally to complete full balance of the system. (Proper care should be taken when balancing, if the building is occupied or if near full load conditions exist).

Variable Speed Pumping. To achieve hydronic balance, full flow through the system is required during the balancing procedure. Once balance is achieved, the system can be placed on automatic control and the pump speed allowed to change. Note that flow in a balancing device or control valve can become so low that it does not generate a measurable pressure drop. When this occurs, the flow rate cannot be determined or controlled (see section on Sizing Balancing Valves). To check flows in a variable speed system, the system may need to be returned to its full flow condition.

Other Balancing Techniques

Flow Balancing by Rated Differential Procedure. This procedure depends on deriving a performance curve for the test coil that compares water temperature difference (Δt_w) to entering water temperature t_{ew} minus entering air temperature t_{ea}. One point of the desired curve can be determined from the manufacturer's ratings since these are published as ($t_{ew} - t_{ea}$). A second point is established by observing the heat transfer from air to water is zero (and consequently $\Delta t_w = 0$) when ($t_{ew} - t_{ea}$) is zero. With these two points, an approximate performance curve can be drawn (see Figure 5). Then, for any other ($t_{ew} - t_{ea}$), this curve is used to determine the appropriate Δt_w.

Example 1. From manufacturer's data:

Capacity = 100 Btu/h
t_{ew} = 200°F
t_{ea} = 60°F
Water flow = 1.5 gpm

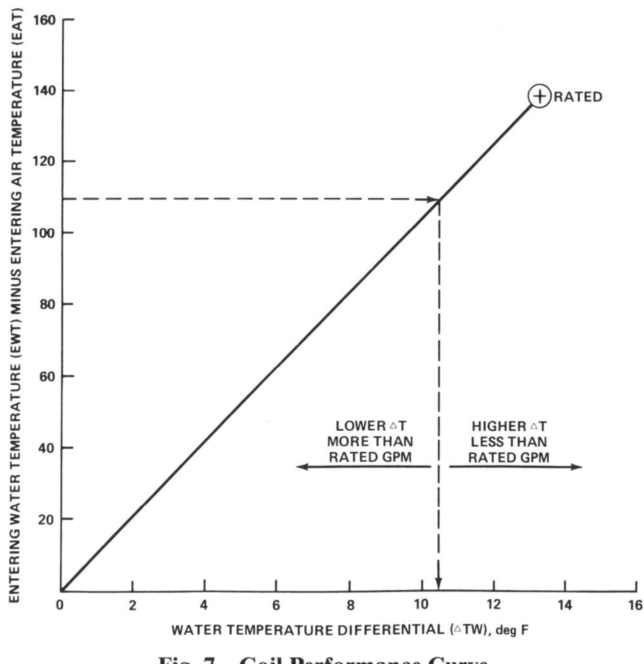

Fig. 7 Coil Performance Curve
(Derived from manufacturer's data)

Solution:

1. Calculate rated Δt_w

$$\Delta t_w = \frac{10{,}000 \text{ Btu/h}}{1 \text{ Btu/lb} \cdot {}^\circ\text{F} \times 8.33 \text{ lb/gal} \times 60 \text{ min/h} \times 1.5 \text{ gpm}}$$

$$= 13.33\,^\circ\text{F}$$

2. Construct a performance curve as illustrated in Figure 7.
3. From test data:

$$t_{ew} = 180\,^\circ\text{F}$$
$$t_{ea} = 70\,^\circ\text{F}$$
$$t_{ew} - t_{ea} = 110\,^\circ\text{F}$$

4. From Figure 7 read $\Delta t_w = 10.5\,^\circ$F, which is required to balance water flow at 1.5 gpm. The water temperature difference may also be calculated as proportion of the rate value as follows:

$$\frac{(t_{ew} - t_{ea})_{test}}{(t_{ew} - t_{ea})_{rated}} (\Delta t_w)_{rated} = (\Delta t_w)_{required}$$

$$\frac{(180 - 70)}{(200 - 60)} 13.3 = 10.5\,^\circ\text{F}$$

This procedure is useful for balancing terminal devices such as finned tube convectors, where flow measuring devices do not exist and where airflow measurements cannot be made. It may also be used for cooling coils for sensible transfer (dry coil).

Flow Balancing by Total Heat Transfer. This procedure determines water flow by running an energy balance around the coil. From field measurements of airflow, wet- and dry-bulb temperatures both upstream and downstream of the coil, and entering and leaving water temperatures Δt_w, water flow can be determined by the following equations:

$$Q_w = Q/500\,\Delta t_w \tag{5}$$

$$q_{cooling} = 4.5\,Q_a\,(h_1 - h_2) \tag{6}$$

$$q_{heating} = 1.08\,Q_a\,(t_1 - t_2) \tag{7}$$

where

Q_w = water flow rate, gpm
q = load, Btu/h
$q_{cooling}$ = cooling load, Btu/h
$q_{heating}$ = heating load, Btu/h
Q_a = airflow, cfm
h = enthalpy, Btu/lb
t = temperature, °F

For example:

Test data
$t_{ewb} = 68.5\,^\circ$F
$t_{lwb} = 53.5\,^\circ$F
$Q_a = 22{,}000$ cfm
$t_{lw} = 59.0\,^\circ$F
$t_{ew} = 47.5\,^\circ$F

From psychrometric chart
$h_1 = 32.84$ Btu/lb
$h_2 = 22.32$ Btu/lb

$$Q_w = \frac{4.5 \times 22{,}000\,(32.84 - 22.32)}{500\,(59.0 - 47.5)} = 181 \text{ gpm}$$

The desired water flow is achieved by successive manual adjustments and recalculations.

Balance by Flow Measurement

All the variations of balancing hydronic systems cannot be listed; however, the general approach should balance the system while minimizing operating cost. Excess pump pressure (excess operating power) can be eliminated by trimming the pump impeller, rather than by allowing the excess pressure to be absorbed by throttle valves, which adds a lifelong operating cost penalty to system operation.

Balance for lowest cost operation can be achieved either by (1) presetting the calibrated balance valves followed by final adjustment and impeller trim, or by (2) setting the system balance valves while simultaneously controlling the pumped flow with the pump throttle valve. When the balance valves are finally set, excess pump head is eliminated by trimming the pump impeller and then reopening the pump throttle valve.

The following is a general procedure based on setting the balance valves on the site:

1. Develop a flow diagram if one is not included in the design drawings. Illustrate all balance instrumentation, and include any additional instrumentation requirements.
2. Compare pumps, primary heat exchangers, and terminal units specified and determine whether a design diversity factor can be achieved.
3. Examine the control diagram and determine the control adjustments needed to obtain design flow conditions.

Balance Procedure—Primary and Secondary Circuits

1. Inspect the system completely to ensure that (a) it has been flushed out and is clean, (b) all manual valves are open or in operating position, (c) all automatic valves are in their proper position and operative, and (d) the expansion tank is properly charged.
2. Place the controls in position for design flow.
3. Examine the flow diagram and piping for obvious short circuits; check flow and adjust the balance valve.
4. Take suction, discharge, and differential pressure readings at both full flow and no flow.
5. Read amperage and voltage and determine approximate power.

6. Establish a pump curve and determine the approximate flow rate.
7. If a total flow station exists, check the pressure differential, determine the flow, and compare this flow with the pump curve flow.
8. If possible, set the total flow about 10% high; maintain pumped flow to a constant value as balance proceeds by adjusting the pump throttle valve.
9. If branch main flow stations exist, these should be tested and set, starting by setting the shortest runs low as balancing proceeds to the longer branch runs.
10. If the system incorporates primary and secondary pumping circuits, a reasonable balance must be obtained in the primary loop before the secondary loop can be considered. The secondary pumps must be running and terminal units must be open to flow when the primary loop is being balanced, unless the secondary loop is decoupled.

SYSTEM COMPONENTS AS FLOW METERS

Flow Measurement Based on Manufacturer's Data

Any system component (terminal, control valve, or chiller) that has an accurate, factory-certified, flow-pressure drop relationship can be used as a flow-indicating device. The flow and pressure drop may be used to establish an equivalent flow coefficient as shown in Equation (3).

Based on the Bernoulli equation, pressure drop varies as the square of the velocity or flow rate (assuming density is constant), or:

$$Q_1^2 / Q_2^2 = \Delta h_1 / \Delta h_2 \quad (8)$$

For example, a chiller has a certified flow pressure drop of 25 ft of water at 100 gpm. The calculated flow with a field measured pressure drop of 30 ft is:

$$Q_2 = Q_1(\Delta h_2 / \Delta h_1)^{0.5} = 100(30/25)^{0.5} = 109.5 \text{ gpm}$$

However, flow results calculated in this manner have limited accuracy. Accuracy of system components used as flow indicators depends on (1) the accuracy of cataloged information concerning flow-pressure drop relationships, and (2) the accuracy of the pressure differential readings. As a rule, the component should be factory-certified flow tested if it is to be used as a flow indicator.

Pressure Differential Readout by Gage

Either gages or manometers can be used to read differential pressures. Gages are usually used for high differential pressures and mercury manometers for lower differentials. Accurate gage readout is diminished when two gages are used, especially when the gages are permanently mounted and, as such, subject to malfunction.

A single high quality gage should be used for differential readout (Figure 8). This gage should be alternately valved to the high- and low-pressure side to establish the differential. A single gage needs no static height correction and errors caused by gage calibration are eliminated.

Differential pressure can also be read from differential gages, thus eliminating the need to subtract outlet from inlet pressures to establish differential pressure. Differential pressure gages are usually dual gages mechanically linked to read differential pressure. The differential pressure gage readout can be stated in terms psi or feet of head of 60 °F water.

Conversion of Differential Pressure to Head

Pressure gage readings can be restated to fluid head, which is a function of fluid density. The common hydronic system conversion factor is related to water density at about 60 °F, 1 psi equal to 2.31 ft. Pressure gages can be calibrated to feet of water head

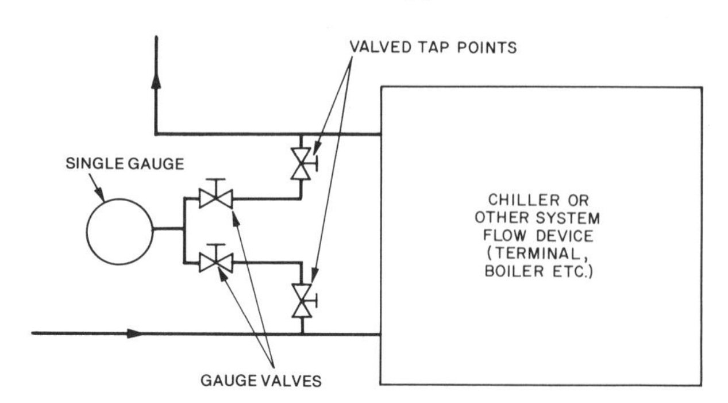

Fig. 8 Single Gage for Reading Differential Pressure

Table 6 Differential Pressure Conversion to Feet of Head

Fluid Specific Gravity	Corresponding Water Temp., °F	Foot Fluid Head Equal to 1 psi[a]	Correction Factor when Gage is Stated to Feet of Water (60 °F)[b]
1.5		1.54	
1.4		1.65	
1.3		1.78	
1.2		1.93	
1.1		2.10	
1.0	60	2.31	1.00
0.98	150	2.36	1.02
0.96	200	2.41	1.04
0.94	250	2.46	1.065
0.92	300	2.51	1.09
0.90	340	2.57	1.11
0.80		2.89	
0.70		3.30	
0.60		3.85	
0.50		4.63	

[a]Differential psi readout is multiplied by this number to obtain feet fluid head when gage is calibrated in psi.

[b]Differential feet water head readout is multiplied by this number to obtain feet fluid head when gage calibration is stated to feet head of 60 °F water.

using this conversion. Because the calibration only applies to water at 60 °F, the readout may require correction when the gage is applied to water at a significantly higher temperature.

Conversion factors and correction factors for pressure gages for various fluid densities are shown in Table 6. The differential gage readout should only be defined in terms of fluid head of the fluid actually causing the flow pressure differential. When this is done, the resultant fluid head can be applied to the C_v to determine actual flow rate through any flow device, if the manufacturer has correctly stated the flow to fluid head relationship.

For example, a manufacturer may test a boiler or control valve with 100 °F water. If the test differential pressure is converted to head at 100 °F, a C_v independent of test temperature and density may be calculated. Differential pressures from another test made in the field at 250 °F may be converted to head at 250 °F. The C_v calculated with this head is also independent of temperature. The manufacturer's data can then be directly correlated with the field test to establish flow rate at 250 °F.

A density correction must be made to the gage reading when differential heads are to be used to estimate pump flows as in Figure 9. This is because of the shape of the pump curve. An incorrect head difference entry into the curve, as caused by an uncorrected gage reading, can cause a major error in the estimated pumped flow. In this case, gage readings for a pumped liquid that has a specific gravity of 0.9 (2.57 ft liquid/psi) were not corrected; the gage conversion assumed at 2.31 ft liquid/psi. A 50% error in flow estimation is shown.

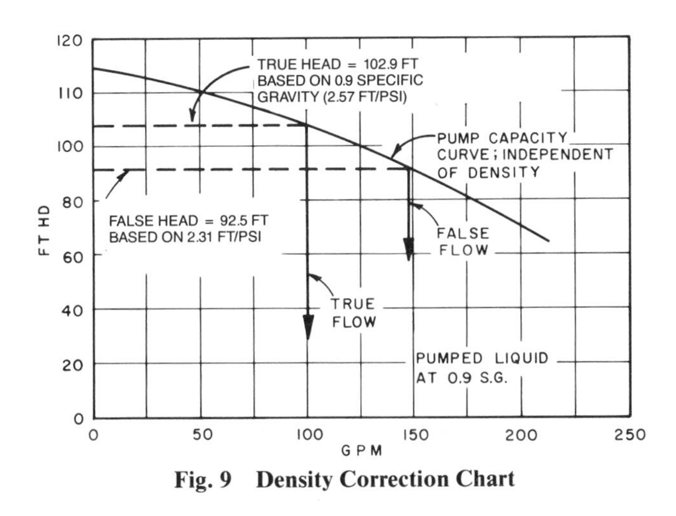

Fig. 9 Density Correction Chart

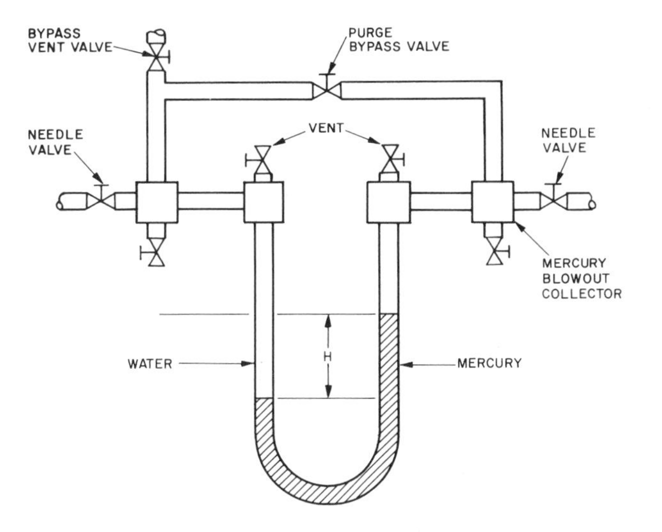

Fig. 10 Mercury Manometer Arrangement for Accurate Reading and Blowout Protection

Differential Head Readout with Mercury Manometers

Mercury manometers are also used for differential pressure readout, especially when very low differentials, great precision, or both, are required. Mercury manometers should not be used for field testing because mercury could blow out into the water system, which would rapidly deteriorate the components. Mercury manometers must be handled with care. A proposed manometer arrangement is shown in Figure 10.

Reference to Figure 10 and the following instructions provide accurate manometer readings with minimum risk of mercury blowout.

1. Make sure that both legs of the manometer are filled with water.
2. Open the purge bypass valve.
3. Open valved connections to high and low pressure.
4. Open the bypass vent valve slowly and purge air here.
5. Open manometer block vents and purge air at each point.
6. Close the needle valves. The mercury columns should zero in if the manometer is free of air. If not, vent again.
7. Open the needle valves and begin throttling the purge bypass valve slowly, watching the mercury columns. If the manometer has an adequate available mercury column, the valve can be closed and the differential reading taken. However, if the mercury column reaches the top of the manometer before the valve is completely closed, this indicates insufficient manometer height, and further throttling will blow mercury into the blowout collector. A longer manometer or the single gage readout method should then be used.

Given that an accurate height of mercury differential is established, an error is often introduced when converting inches of mercury to feet of water. Conversion tables almost always state that 1 in. Hg equals 1.13 ft of 60°F water and this conversion is usually applied. While correct for a mercury column exposed to air, the 1.13 conversion factor is incorrect for liquid differential readout because it disregards the fact that the mercury differential column (*H* in Figure 10) is partially counterbalanced by an equal height of flow liquid. When a mercury manometer is used to determine differential water head in a water flow test, the conversion factor is 1 in. of mercury for 1.046 ft of water when the water is on the order of 60°F.

The conversion factor changes with fluid test temperature, density, or both. Conversion factors, as shown, are to a water base; counterbalance water height *H* is considered to be at room temperature. Fluid flow density changes, with temperature only, are shown in Table 7.

Table 7 Conversion Table

Water Temperature, °F	Ft Head Differential per in. Hg Differential
60	1.046
150	1.07
200	1.09
250	1.11
300	1.15
340	1.165

Other Flow Information Devices, Orifice Plates, and Flow Indicators

Manufacturers provide flow information for several devices used in hydronic system balance. In general, the devices can be classified as (1) orifice flow meters, (2) venturi flow meters, (3) impact or velocity head flow meters, (4) pitot-tube flow meters, (5) bypass spring impact flow meters, and (6) calibrated balance valves.

The *orifice flow meter* is widely used and is extremely accurate. The meter is calibrated and shows differential pressure versus flow. Accuracy generally increases as the pressure differential across the meter increases. The differential pressure readout instrument may be a manometer, differential gage, or a single gage (Figure 8).

The *venturi flow meter* has lower pressure loss than the orifice plate meter, since a carefully formed flow path increases velocity head recovery. The venturi flow meter is placed in a main flow line where it can be read continuously.

Velocity impact meters have precise construction and calibration. The meters are generally of specially contoured glass or plastic, which permits observation of a flow float. As flow increases, the flow float rises in the calibrated tube to indicate flow rate. Velocity impact meters generally have high accuracy.

A special version of the velocity impact meter is applied to hydronic systems. This version operates on the velocity head difference between the pipe side wall and the pipe center, which causes fluid to flow through a small flow meter. Accuracy depends on the location of the impact tube and on a velocity profile that corresponds to theory and the laboratory test calibration base. Generally, the accuracy of this *bypass flow impact* or differential velocity head flow meter is less than a flow-through meter, which can operate without creating a pressure loss in the hydronic system.

The *pitot-tube flow meter* is also used for pipe flow measurement. The pitot is either traversed or a calibrated averaging tube

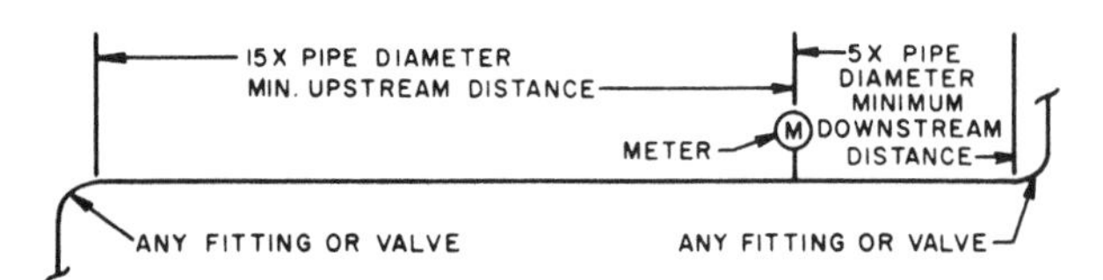

Fig. 11 Minimum Installation Dimensions for Flow Meter

is used. Manometers are generally used to measure velocity head differences because these differences are low.

The *bypass spring impact flow meter* uses a defined piping pressure drop to cause a correlated bypass side branch flow. The side branch flow pushes against a spring that increases in length with increased side branch flow. Each individual flow meter is calibrated to relate extended spring length position to main flow. The bypass spring impact flow meter has, as its principal merit, a direct readout. However, dirt on the spring reduces accuracy. The bypass is opened only when a reading is made. Flow readings can be taken at any time.

The *calibrated balance valve* is an adjustable orifice flow meter. Balance valves can be calibrated so that a flow pressure drop relationship can be obtained for each incremental setting of the valve. A ball, rotating plug, or butterfly valve may have its setting expressed in percent open or degree open; a globe valve, to percent open or number of turns. The calibrated balance valve must be manufactured with precision and care to ensure that each valve of a particular size has the same calibration characteristics.

If any of the above meters are to be useful, the minimum distance of straight pipe upstream and downstream as recommended by the meter manufacturer and flow measurement handbooks must be adhered to. Figure 11 presents minimum installation suggestions.

Using a Pump as an Indicator

Although the pump is not a meter, it can be used as an indicator of flow together with the other system components. Differential pressure readings across a pump can be correlated with the pump curve to establish the pump flow rate. Accuracy depends on (1) accuracy of readout, (2) pump curve shape, (3) actual conformance of the pump to its published curve, (4) pump operation without cavitation, (5) air-free operation, and (6) velocity head correction.

When a differential pressure reading must be taken, a single gage with manifold provides the greatest accuracy (Figure 12). The pump suction to discharge differential can be used to establish pump differential pressure and, consequently, pump flow rate. The single gage and manifold may also be used to check for strainer clogging by measuring the pressure differential across the strainer.

Pressure differential, as obtained from the gage reading, is converted to feet of head, which is pressure divided by the fluid

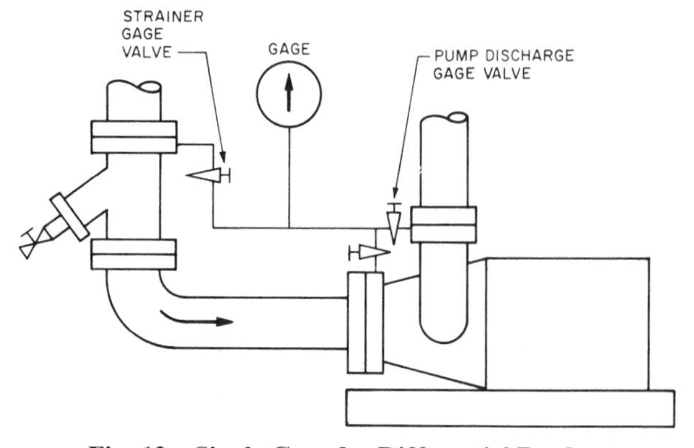

Fig. 12 Single Gage for Differential Readout across Pump and Strainer

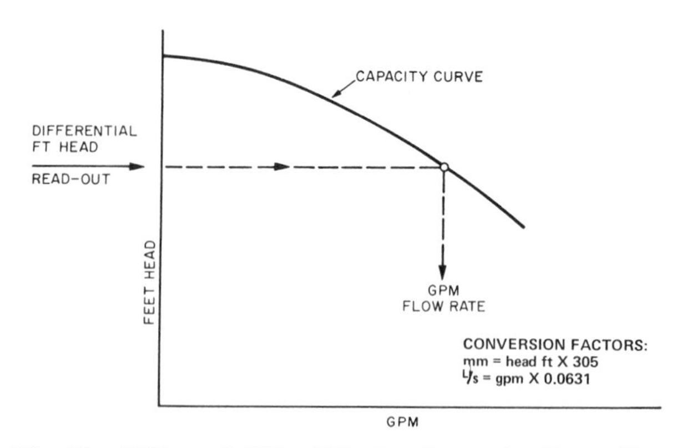

Fig. 13 Differential Head Used to Determine Pump Flow

density (use the conversion Table 7). The pump differential head is then used to determine pump flow rate (Figure 13). As long as the differential head used to enter the pump curve is expressed as head of the fluid being pumped, the pump curve shown by the manufacturer should be used as described. The pump curve may state that the curve was defined by test with 85°F water. This is unimportant, since the same curve applies unchanged from 60 to 250°F water, or to any fluid within a broad viscosity range.

Generally, pump-derived flow information, as established by the performance curve, is questionable, unless the following precautions are observed:

1. The installed pump should be factory calibrated by a test to establish the actual flow-pressure relationship for that particular pump. Production pumps can vary from the cataloged curve because of minor changes in impeller diameter, interior casting tolerances, and machine fits.
2. When a calibration curve is not available for a centrifugal pump being tested, the discharge valve can be closed briefly to establish the no-flow shutoff pressure and then compared to the published curve. If the shutoff pressure differs from that published, draw a new curve parallel to the published curve. While not exact, the new curve will usually fit the actual pumping circumstance more accurately. Clearance between the impeller and casing minimize the danger of damage to the pump during a no-flow test, but verify with the manufacturer to be sure.
3. Differential head should be determined as accurately as possible, especially for pumps with flat flow curves.
4. The pump should be operating air-free and without cavitation. A cavitating pump will not operate to its curve, and differential readings will provide false results.
5. Ensure that the pump is operating above the minimum net positive suction head.
6. Power readings can be used as a check for the operating point when the pump curve is flat or as a reference check when the pump is suspected to be cavitating or providing false readings because of air.

The power draw should be measured in watts. Ampere readings cannot be trusted because of voltage and power factor problems. If motor efficiency is known, the wattage drawn can be related to pump brake power, as described on the pump curve, and the operating point determined.

Central Plant Chilled Water Systems

In existing installations, the need for establishing thermal load profiles accurately is of prime importance in establishing proper primary chilled water supply temperature and flow. In new installations, actual load profiles can be compared with design load profiles to obtain valid operating data.

Table 8 Instruments for Monitoring a Water System

Point of Information	Manifold Gage	Single Gage	Thermometer	Test Well	Pressure Tap
Pump—Suction, discharge	x				
Strainer—In, out					x
Cooler—In, out		x	x		
Condensers—In, out		x	x		
Concentrator—In, out		x	x		
Absorber—In, out		x	x		
Tower cell—In, out				x	x
Heat exchanger—In, out	x		x		
Coil—In, out				x	x
Coil bank—In, out		x	x		
Booster coil—In, out					x
Cool panel—In, out					x
Heat panel—In, out				x	x
Unit heater—In, out					x
Induction—In, out					x
Fan coil—In, out					x
Water boiler—In, out			x		
Three-way valve—All ports					x
Zone return main			x		
Bridge—In, out			x		
Water makeup		x			
Expansion tank		x			
Strainer pump					x
Strainer main	x				
Zone three-way—All ports				x	x

To perform proper testing and balancing, all interconnecting points between the primary and secondary system must be designed with sufficient temperature, pressure, and flow connections so that adequate data may be indicated and/or recorded.

Water Flow Instrumentation

As indicated previously, the proper location and use of system instrumentation is vital to the accuracy of the system balance. A suggested approach to temperature and pressure test points is listed in Table 8. With a table of this nature, instrumentation can be tailored to a specific design with ease and accuracy. Flow-indicating devices should be placed in water systems as follows:

- At each major heating coil bank (10 gpm)
- At each major cooling coil bank (10 gpm or more)
- At each bridge in primary-secondary systems
- At each main pumping station
- At each water chiller evaporator
- At each water chiller condenser
- At each water boiler outlet
- At each floor takeoff to booster reheat coils, fan coil units, induction units, ceiling panels, and radiation (Do not exceed 25 terminals off any one zone meter probe.)
- At each vertical riser to fan coil units, induction units, and radiation
- At the point of tie-in to existing systems

STEAM DISTRIBUTION SYSTEMS

Procedures for Steam Balancing Variable Flow Systems

A steam distribution system cannot be balanced by adjustable flow-regulating devices. Flow regulation is accomplished by fixed restrictions built into the piping system in accordance with carefully designed pipe and orifice sizes.

It is important to have a balanced distribution of steam to all portions of the steam piping system at all load levels. This is best accomplished by the proper design of the steam distribution piping system by carefully considering steam pressure, steam quantities required by each branch circuit, pressure drops, steam velocities, and pipe sizes. Just as other flow systems are balanced, steam distribution systems are balanced by ensuring that the pressure drops are equalized at design flow rates for all portions of the piping system. Only marginal balancing can be done by pipe sizing. Therefore, additional steps must be taken to achieve a balanced performance.

Steam flow balance can be improved by using spring-type packless supply valves equipped with precalibrated orifices. The valves should have a tight shutoff between 25 in. of Hg to 60 psig. These valves have a nonrising stem, are available with a lockshield, and have a replaceable disc. Orifice flanges can also be used to regulate and measure steam flow at appropriate locations throughout the system. The orifice sizes are determined by the pressure drop required to provide a given flow rate at a given location in the system. A schedule should be prepared showing (1) orifice sizes, (2) valve or pipe sizes, (3) required flow rates, and (4) corresponding pressure differentials for each flow rate. It may be useful to calculate pressure differentials for several flow rates for each orifice size. Such a schedule should be maintained for future reference.

After the appropriate regulating orifices are installed in the proper locations, the system should be tested for tightness by sealing all openings in the system and applying a vacuum of 20 in. of Hg, which should be held for 2 h. Next, the system should be readied for warm-up and pressurizing with steam following the procedures outlined in Section VI of the ASME *Boiler and Pressure Vessel Code*. After the initial warm-up and system pressurization, evaluate system steam flow and compare it to system requirements. The orifice schedule calculated earlier will now be of value should any of the orifices need to be changed.

Steam Flow Measuring Devices

Many devices are available for measuring flow in steam piping systems: (1) steam meters, (2) condensate meters, (3) orifice plates, (4) venturi fittings, (5) steam recorders, and (6) manometers for reading differential pressures across orifice plates and venturi fittings. Some of these devices are permanently affixed to the piping system to facilitate taking instantaneous readings that may be necessary for proper system operation and control. A surface pyrometer used in conjunction with a pressure gage is a convenient way to determine steam saturation temperature and the degree of superheat at various locations in the system. Such information can be used to evaluate performance characteristics of the system.

COOLING TOWERS

Field testing cooling towers is a demanding and difficult task. ASME *Standard* PTC 23 (1986) and CTI *Standard Specification* ATC-105 (1982) establish procedures for these tests. Certain general guidelines for testing cooling towers are as follows:

Conditions at Time of Test

- Water flow within 15% of design flow
- Heat load within 30% of design heat load and stabilized
- Entering wet bulb within 12°F of entering design wet bulb

Using the above limitations and as accurate field readings as possible, a projection to design conditions produces an accuracy of ±5% of tower performance.

Conditions for Performing Test

- Water-circulating system serving the tower should be thoroughly cleaned of all dirt and foreign matter. Samples of water should be clear and indicate clear passage of water through pumps, piping, screens, and strainers.
- Fans serving the cooling tower should operate in proper rotation. Foreign obstructions should be removed. Permanent obstruction should be noted.
- Interior filling of cooling tower should be clean and free of foreign materials such as scale, algae, or tar.

- Water level in the tower basin should be maintained at the proper level. Visually check the basin sump during full flow to determine that the centrifugal action of the water is not causing entrainment of air, which could cause pump cavitation.
- Water-circulating pumps should be tested with full flow through the tower. If flow exceeds design, it should be valved down until design is reached. The flow finally set should be maintained throughout the test period. All valves, except necessary balancing valves, should be in the full-open position.
- If makeup and blowdown have facilities to determine flow, set them to design flow at full flow through the tower. If flow cannot be determined, shut off both.

Instruments

Testing and balancing agencies provide instruments to perform the required tests. Mechanical contractors provide and install all components such as orifice plates, venturis, thermometer wells, gage cocks, and corporation cocks. Designers specify measuring point locations.

The instruments used should be recently calibrated from the following:

Temperature

- Mercury with divisions of 0.2 °F in a proper well for water should be used.
- Mercury thermometer with solar shield and 0.2 °F division or thermocouple with 0.2 °F reading having mechanical aspiration for wet-bulb readings should be used.
- Sling psychrometer may be used for rough checks.
- Mercury thermometers with 0.2 °F should be used for taking dry-bulb readings.

Water Flow

- Orifice or venturi drops can be read using a water-over-mercury manometer or a recently calibrated differential pressure gage.
- Where corporation cocks are installed, a pitot tube and manometer traverse can be made by trained technicians.

Test Method

The actual test consists of the following steps:

1. Conduct water flow tests to determine volume of water on the tower, volume of makeup, and blowdown water.
2. Conduct water temperature tests, if possible, in suitable wells as close to the tower as possible. Temperature readings at pumps or the condensing element are not acceptable in tower evaluation. If there are no wells, surface pyrometer readings are acceptable.
3. Take makeup water volume and temperature readings at the point of entry to the system.
4. Take blowdown volume and temperature readings at the point of discharge from system.
5. Take inlet and outlet dry- and wet-bulb temperature readings using the prescribed instruments.
   a) Use wet bulb entering and leaving to determine tower actual performance as against design.
   b) Use wet bulb and dry bulb entering and leaving to determine evaporation involved.
6. If the tower has a ducted inlet or outlet, where a reasonable duct traverse can be made, use this air volume as a cross-check of tower performance.
7. Take wet- and dry-bulb temperature readings between 3 and 5 ft from the tower on all inlet sides. These readings shall be taken halfway between the base and the top of the inlet louvers at no more than 5-ft spacing horizontally; then they should be averaged. Note any unusual inlet conditions.
8. Note wind velocity and direction at the time of test.
9. Take test readings continually with a minimum of time lapse between readings.
10. If the first test indicates a tower deficiency, perform two additional tests to verify the original readings.

TEMPERATURE CONTROL VERIFICATION

The test and balance technician should work closely with the temperature control installer to ensure the project is complete. The balancing technician need only verify proper operation, not adjust, relocate, or recalibrate any controls.

On completing the testing, adjusting, and balancing of all HVAC systems, the automatic control system(s) should be staged to prove its capability of matching system capacity to varying load conditions. In the event all flow balancing is completed in a particular season of operation, such as the cooling cycle, arrangements for the opposite season (heating cycle) control verification should be implemented and completed prior to project acceptance.

Suggested Procedure

1. Obtain design drawings and specifications and become thoroughly acquainted with the design intent.
2. Obtain copies of approved shop drawings of control diagrams.
3. Compare design to installed equipment and field installation.
4. Obtain manufacturer's recommended operating and testing procedure.
5. Verify that all controllers are calibrated and in control.
6. Check for proper location of transmitters and controllers. Note any adverse conditions that would affect control. Suggest relocation, if necessary.

Pneumatic Systems

1. Verify main control supply air for proper pressure and observe compressor and drier operation.
2. Verify calibration of all controllers and sensitivity of each controller, and note any overlap in controlled devices.
3. Compare all control terminations with design drawings.
4. Verify operation of all limiting controllers (*i.e.*, firestats, freezestats, preheat thermostats, and high and low thermostats).
5. Activate controlled devices, checking for free travel and proper operation of dampers. Verify proper application of normally open (N.O.) and normally closed (N.C.) positions.
6. Verify operation of pilot positioners, sequence of damper operators, and operation of control valves to ensure proper relationship.
7. Check adjustment of all pressure/electric end switches and mercury switches for proper setting and operation for the seasonal cycle of operation in effect. Simulate conditions to activate sequences used in the opposite season.
8. Check level and zero of inclined gage or U-tube manometers. Verify proper location of sensors.
9. Verify operation of lockout or interlock system.
10. Verify the span of control from a normally closed position to a normally open position, observing any dead bands, excessive pressures, etc.
11. Verify sequence of operations (*i.e.*, night setback, switchovers, resets, cooling tower control, etc.).

Electric Systems

For high- and low-voltage systems, complete all of the Suggested Procedure steps.

1. With voltmeter, verify control voltage.
2. Set thermostat in cool position and turn to lowest setting. Verify proper operation of contactor, damper motor, etc.

3. Set thermostat to highest setting. Verify proper action of damper motors, end switches, and resistance heat sequences.
4. Activate solenoid valves, low-limit thermostats, and lockout devices to verify proper action.

Direct Digital Control (DDC)

1. Check software algorithms for each control loop for accuracy and correct application.
2. Check all control loops and their individual field points for correct response.
3. Check calibration of all field sensors.
4. Check calibration and response time on transducers.
5. Check fail-safe modes [normally open (N.O.), normally closed (N.C.), etc.] of all control devices.
6. Manually stroke each damper and control valve.
7. Check lightning protection and system battery backup.
8. Check phone modem.
9. With system in full operation, test each control loop at both ends of control range.

Electronic Digital

1. Check all control loops and their individual field points for correct response.
2. Check calibration of all field sensors.
3. Check calibration and response time of all transducers.
4. Check fail-safe modes [normally closed (N.C.), normally open (N.O.), etc.] of all control devices.
5. Manually stroke each damper and control valve.
6. Check lightning protection.
7. With system in full operation, test each control loop at both ends of control range.

Energy Management Systems (EMS)

The calibration and verification of sequences usually does not include verification of sensors used on the energy management system. After the total system control has been checked and made fully operational, the energy management system contractor should verify that readouts of all sensors and transmitters are within the range of the control. The energy management system contractor should contact the control representative and the test balance technician for help in tracing any sensor problem found after a thorough check.

FIELD SURVEY FOR ENERGY AUDIT

An energy audit is an organized survey of a specific building to identify and measure all energy uses, determine probable sources of energy losses, and list energy conservation opportunities. This is usually performed as a team effort under the direction of a qualified energy engineer. Gathering the field data can be done by firms employing technicians trained in testing, adjusting, and balancing.

Instruments

To determine a building's energy use characteristics, an accurate measurement of existing conditions must be made with proper instruments. Accurate measurements not only point out opportunities to reduce waste, but also provide a record of the actual conditions in the building before energy conservation measures were taken. They provide a compilation of data of installed equipment and a record of equipment performance prior to changes.

Remember that judgments will be made based on the information gathered during the field survey; that which is not accurately measured cannot be properly evaluated.

Generally, the instruments required for performing testing, adjusting, and balancing are sufficient for energy conservation surveying. Possible additional instruments include a power factor meter, a light meter, combustion testing equipment, refrigeration gages, and equipment for recording temperatures, fluid flow rates, and energy use over time. Only high quality instruments should be used.

Observation of system operation and any information the technician can obtain from the operating personnel pertaining to the operation should be included in the report.

Data Recording

Organized record keeping is extremely important. A camera is also helpful. Photographs of building components and mechanical and electrical equipment can be reviewed later when the data is analyzed.

Data sheets needed for energy conservation field surveys contain different and, in some cases, more comprehensive information than those used for testing, adjusting, and balancing. Generally, the energy engineer determines the degree of fieldwork to be performed; data sheets should be compatible with the instructions received.

Building Systems

The most effective way to reduce building energy waste is to identify and define the energy load by building system. This provides an orderly procedure for tabulating the load. Also, the most effective energy conservation opportunities can be achieved more quickly because high priorities can be assigned to systems that consume the most energy.

For this purpose, load is defined as the quantity of energy used in a building, or by one of its subsystems, for a given period.

A building can be divided into nonenergized systems and energized systems. Nonenergized systems do not require outside energy sources such as electricity and fuel. Energized systems require outside energy sources. Examples might be mechanical systems and electrical systems. Energized and nonenergized systems can be divided into subsystems defined by function. Nonenergized subsystems are (1) building site, envelope, interior, and subsystem; (2) building utilization subsystem; and (3) building operation subsystem.

Building Site, Envelope, and Interior

The site, envelope, and interior are surveyed to determine how they can be changed to reduce the building load that the mechanical and electrical systems must meet without adversely affecting the building's appearance. This requires uncovering energy-wasting items and recording any existing conditions affecting the practicability of making changes to eliminate waste.

It is important to compare actual conditions with conditions assumed by the designer, so that the mechanical and electrical systems can be adjusted to balance their capacities to satisfy the real needs.

Building Use

The functioning of people within the building envelope subsystem is most important when considering the building load; it must be observed because the action of people affects the energy usage of all other building subsystems.

Building-use loads can be classified as (1) people occupancy loads and (2) people operation loads. People occupancy loads are related to schedule, density, and mixing of occupancy types (*e.g.*, process and office). People operation loads are varied, such as (1) operation of manual window shading devices; (2) setting of room thermostats; and (3) conservation-related habits such as turning off lights, closing doors and windows, turning off energized equipment when not in use, and not wasting domestic hot or chilled water.

Building Operation Subsystem

This subsystem consists of the operation and maintenance of all the building subsystems. The load on the building operation subsystem is affected by factors such as (1) the time at which janitorial services are performed, (2) janitorial crew size and time required to do the cleaning, (3) amount of lighting used to perform janitorial functions, (4) quality of the equipment maintenance program, (5) system operational practices, and (6) equipment efficiencies.

Building Energized Systems

The energized subsystems of the building are generally plumbing, heating, ventilating, cooling, space conditioning, control, electrical, and food service. Although these systems are interrelated and often use common components, it is important to evaluate the energy use each subsystem as independently as possible, for a logical organization of data. In this way, proper energy conservation measures for each subsystem can be developed.

Process Loads

In addition to building subsystem loads, the process load in most buildings must be evaluated. The energy field auditor must be able to determine its impact.

Most tasks not only require energy for performing a service, but they also affect the energy consumption of other building subsystems. For example, if a process releases large amounts of heat to the space, the process consumes energy and also imposes a large load on the cooling system.

Guidelines for Developing a Field Study Form

A brief checklist that outlines requirements for a field study form needed to conduct an energy audit follows.

Inspection and observation of all systems. Record the following physical and mechanical conditions:

- Fan blades, fan scroll, drives, belt tightness, and alignment
- Filters, coils, and housing tightness
- Ductwork (equipment room and space, where possible)
- Strainers
- Insulation ducts and piping
- Makeup water treatment and cooling tower

Interview physical plant supervisor. Record conditions to the following survey:

- Is the system operating as designed? If not, what changes have been made to ensure its performance?
- Have there been changes, modifications, or additions to the system?
- If the system has been a problem, list problems by frequency of occurrence.
- Are any systems cycled? If so, which systems and when, and would building load permit it?

Recording system information. Record the following system/equipment identification:

- Type of system—single-zone, multizone, double-duct, low or high-velocity, reheat, variable volume, or other
- System arrangement—fixed minimum outside air, no relief, gravity or power relief, economizer gravity relief, exhaust return, or other
- Air-handling equipment—fans (supply, return, or exhaust): manufacturer, model and size, type, class; dampers (vortex, scroll, or discharge); motors: manufacturer, power requirement, full load amperes, voltage, phase, and service factor
- Chilled and hot water coils—area, tubes on face, fin spacing, and number of rows (coil data necessary when shop drawings are not available)
- Terminals—high-pressure mixing box: manufacturer, model, type (reheat, constant volume, variable volume, induction); grilles, registers, and diffusers: manufacturer, model, style, and AK factor to convert field-measured velocity to flow rate
- Main heating and cooling pumps, over 5 hp—manufacturer, pump service and identification, model and size, impeller diameter, rpm, flow rate, head at full flow, head at no flow; motor data: power, rpm, voltage, amperes, and service factor
- Refrigeration equipment—chiller manufacturer, type, model, serial number, nominal tons, brake horsepower, total heat rejection, motor (horsepower, amperes, volts), chiller pressure drop, entering and leaving chilled water temperatures, condenser pressure drop, condenser entering and leaving water temperatures, running amperes and volts, no load running amperes and volts
- Cooling tower—manufacturer, size, type, nominal tons, range, flow rate, and entering wet-bulb temperature
- Heating equipment—boiler (small through medium) manufacturer, fuel, energy input (rated), and heat output (rated)

Recording test data. Record the following test data:

- Systems in normal mode of operation (if possible)—fan motor: running amperes and volts and power factor (over 5 hp); fan: rpm, total air (pitot-tube traverse where possible), and static pressure (discharge static minus inlet total); static profile drawing (static pressure across filters, heating coil, cooling coil, and dampers); static pressures at ends of runs of the system (identifying locations)
- Cooling coils—entering dry- and wet-bulb temperatures, leaving dry and wet bulb, entering and leaving water temperature, coil pressure drop (where pressure taps permit and manufacturer's ratings can be obtained), flow rate of coil (when other than fan), outdoor wet and dry bulb, time of day, and conditions (sunny or cloudy)
- Heating coils—entering and leaving dry-bulb temperature, entering and leaving water temperatures, coil pressure drop (where pressure taps permit and manufacturer's ratings can be obtained), and flow rate through the coil (when other than fan)
- Pumps—no flow head, full-flow discharge pressure, full-flow suction pressure, full-flow differential pressure, motor running amperes and volts, and power factor (over 5 hp)
- Chiller (under cooling load conditions)—chiller pressure drop, entering and leaving chilled water temperatures, condenser pressure drop, entering and leaving condenser water temperature, running amperes and volts, no load running amperes and volts, chilled water on and off, and condenser water on and off
- Cooling tower—water flow rate on tower, entering and leaving water temperature, entering and leaving wet bulb, fan motor [amperes, volts, power factor (over 5 hp) ambient wet bulb]
- Boiler (full fire)—input energy (if possible), percent CO_2, stack temperature, efficiency, and complete Orsat test on large boilers
- Boiler controls—description of the operation
- Temperature controls—operating and set point temperatures for mixed air controller, leaving air controller, hot deck controller, cold deck controller, outdoor reset, interlock controls, and damper controls; description of complete control system and any malfunctions
- Outside air intake versus exhaust air—total airflow measured by pitot-tube traverses of both outside air intake and exhaust air systems obtained where possible. Determine if there is an imbalance in the exhaust system that causes infiltration. Observe exterior walls to determine if outside air can infiltrate into the return air system (record outside air temperature, dry and wet bulb; return air temperature, dry and wet bulb; and return air plenum temperature, dry and wet bulb). The greater the differential between outside and return air, the more evident the problem will appear.

TESTING FOR SOUND AND VIBRATION

Testing for sound and vibration ensures that equipment is operating satisfactorily and that no objectionable noise and vibration are transmitted to the building structure and occupied space. Although sound and vibration are specialized fields that require expertise not normally developed by the HVAC engineer, the procedures to test HVAC systems are relatively simple and can be performed with a minimum of equipment by following the steps outlined in this section. Although useful information is provided for resolving common noise and vibration problems, this section does not provide information on problem solving or the design of HVAC systems (see Chapter 42).

TESTING FOR SOUND

The present state of the art does not permit tests to determine if equipment is operating with desired sound levels. Field tests can determine only sound pressure levels, and equipment ratings are almost always in terms of sound power levels. Until new techniques are developed, the testing engineer can determine only if sound pressure levels are within desired limits and, if not, determine which equipment, systems, or components are the source of excessive or disturbing transmission.

Sound-Measuring Instruments

Although an experienced listener can often determine whether or not systems are operating in an acceptably quiet manner, sound-measuring instruments are necessary to determine whether system noise levels are in compliance with specified criteria and, if not, to obtain and report detailed information to evaluate the cause of noncompliance. Instruments normally used in field testing are as follows:

The *precision sound level meter* is used to measure sound pressure level. The most basic sound level meters measure overall sound pressure level and have up to three weighted scales that provide limited filtering capability. The instrument is useful in assessing outdoor noise levels in certain situations and can provide limited information on the low-frequency content of overall noise levels, but it provides insufficient information for problem diagnosis and solution. Its usefulness in evaluating indoor HVAC sound sources is thus limited.

Proper evaluation of HVAC sound sources requires a sound level meter capable of filtering overall sound levels into frequency increments of one octave or less.

Sound analyzers provide detailed information about the sound pressure levels at various frequencies through filtering networks. The most popular sound analyzers are the octave band and center frequency types, which break the sound into the eight octave bands of audible sound. Instruments are also available for 0.33, 0.1, and narrower spectrum analysis; however, these are primarily for laboratory and research applications. Sound analyzers (octave band or center frequency) are required where specifications are based on noise criterion (NC) curves or similar criteria based on frequency, and for problem jobs where a knowledge of frequency is necessary to determine proper corrective action.

Personal computers are a versatile sound-measuring tool. Software used in conjunction with portable computers has all the functional capabilities described above, plus many that previously required a fully equipped acoustical laboratory. This type of sound-measuring system is many times faster and much more versatile than conventional sound level meters. With suitable accessories, it can also be used to evaluate vibration levels.

A *stethoscope* is an invaluable instrument in measuring sound levels and tracking down problems, as it enables the listener to determine the direction of the sound source.

Regardless of the sound-measuring system used, it should be calibrated prior to each use. Some systems have built-in calibration; others use external calibrators. Much information is available on the proper application and use of sound-measuring instruments.

Air noise caused by air flowing at a velocity of over 1000 fpm or winds over 12 mph can cause substantial error in sound measurements due to wind effect on the microphone. For outdoor measurements or in places where air movement is prevalent, a wind screen for the microphone or a special microphone is required.

Sound Level Criteria

In the absence of specified values, the testing engineer must determine if sound levels are within acceptable limits (Chapter 42). Note that complete absence of noise is seldom a design criterion, except for certain critical locations such as sound and recording studios. In most locations, a certain amount of noise is desirable to mask other noises and provide speech privacy; it also provides an acoustically pleasing environment, since few people can function effectively in extreme quiet. Chapter 7, Table 2, of the 1989 ASHRAE *Handbook—Fundamentals* lists typical sound pressure levels. In determining allowable HVAC equipment noise, it is as inappropriate to demand 30 dB for a factory where the normal noise level is 75 dB as it is to specify 60 dB for a private office where normal noise level might be 35 dB.

Most field sound-measuring instruments and techniques yield an accuracy of ±3 dB, the smallest difference in sound pressure level that the average person can discern. A reasonable tolerance for sound criteria is 5 dB, and, if 35 dBA was considered the maximum allowable noise, the design engineer should specify 30 dBA.

The measured sound level of any location is a combination of all sound sources present, including sound generated by HVAC equipment as well as sound from other sources such as plumbing systems and fixtures, elevators, light ballasts, and outside noises. In testing for sound, all sources from other than HVAC equipment are considered background or ambient noise.

Background sound measurements generally have to be made when (1) the specification requires that the sound levels from HVAC equipment only, as opposed to the sound level in a space, not exceed a certain specified level; (2) the sound level in the space exceeds a desirable level, in which case the noise contributed by the HVAC system must be determined; and (3) in residential locations where little significant background noise is generated during the evening hours and where generally low allowable noise levels are specified or desired. Because background noise from outside sources such as vehicular traffic can fluctuate widely, sound measurements for residential locations are best made in the normally quiet evening hours.

Sound-Testing Procedures

Ideally, a building should be completed and ready for occupancy before sound level tests are taken. All spaces in which readings will be taken should be furnished with drapes, carpeting, and furniture, as these affect the room absorption and the subjective quality of the sound. In actual practice, since most tests have to be conducted before the space is completely finished and furnished for final occupancy, the testing engineer must make some allowances. Since furnishings increase the absorption coefficient and reduce the sound pressure level that can be expected between most live and dead spaces to 4 dB, the following guidelines should suffice for measurements made in unfurnished spaces. If the sound pressure level is 5 dB or more over specified or desired criterion, it can be assumed that the criterion will not be met, even with the increased absorption provided by furnishings. If the sound pressure level is 0 to 4 dB greater than specified or desired criterion, recheck when the room is furnished to determine compliance.

Follow this general procedure:

1. Obtain a complete set of accurate, as-built drawings and specifications, including duct and piping details. Review specifications to determine sound and vibration criteria and any special instructions for testing.
2. Visually check systems for noncompliance with plans and specifications, obvious errors, and poor workmanship. Turn system on for audible check. Listen for noise and vibration, especially duct leaks and loose fittings.
3. Adjust and balance equipment, as described in other sections, so that final acoustical tests are made with the system as it will be operating. It is desirable to perform acoustical tests for both summer and winter operation, but where this is not practical, make tests for the summer operating mode, as it usually has the potential for higher sound levels. Tests must be made for all mechanical equipment and systems, including standby.
4. Check calibration of instruments.
5. Measure sound levels in all areas as required, combining measurements as indicated in Item 3 if equipment or systems must be operated separately. Before final measurements are made in any particular area, survey the area using an A-weighted scale reading (dBA) to determine the location of the highest sound pressure level. Indicate this location on a testing form and use it for test measurements. Restrict the preliminary survey to determine location of test measurements to areas that can be occupied by standing or sitting personnel. For example, measurements would not be made directly in front of a diffuser located in the ceiling, but they would be made as close to the diffuser as standing or sitting personnel might be situated. In the absence of specified sound criteria, the testing engineer should measure sound pressure levels in all occupied spaces to determine compliance with critieria, as indicated in Chapter 42, and to determine any sources of excessive or disturbing noise.
6. Determine if background noise measurements must be made.
   a) If specification requires determination of sound level from HVAC equipment only, it will be necessary to take background noise readings by turning HVAC equipment off.
   b) If specification requires compliance with a specific noise level or criteria (for example, sound levels in office areas shall not exceed 35 dBA), ambient noise measurements need be made only if the noise level in any area exceeds the specified value.
   c) For residential locations and areas requiring very low noise levels, such as sound recording studios and locations that are used during the normally quieter evening hours, it is usually desirable to take sound measurements in the evening and/or take ambient noise measurements.
7. For outdoor noise measurements to determine noise as radiated by outdoor or roof-mounted equipment such as cooling towers and condensing units, the Sound Control for Outdoor Equipment Installations section of Chapter 42, which presents proper procedure and necessary calculations.

Noise Transmission Problems

Regardless of the precautions taken by the specifying engineer and the installing contractors, situations can occur where the sound level exceeds specified or desired levels, and there will be occasional complaints of noise in completed installations. A thorough understanding of Chapter 42 and the Vibration Testing section of this chapter is desirable before attempting to resolve any noise and vibration transmission problems. The following is intended as an overall guide rather than a detailed problem-solving procedure.

All noise transmission problems can be evaluated in terms of the source-path-receiver concept. Objectionable transmission can be resolved by (1) reducing the noise at the source by replacing defective equipment, repairing improper operation, proper balancing and adjusting, and replacing with quieter equipment; (2) attenuating the paths of transmission with silencers, vibration isolators, and wall treatment to increase transmission loss; and (3) reducing or masking objectionable noise at the receiver by increasing room absorption or introducing a nonobjectionable masking sound. The following discussion includes (1) ways to identify actual noise sources, using simple instruments or no instruments and (2) possible corrections.

When the engineer is troubleshooting in the field, it is important to listen to the offending sound. The best instruments are no substitute for careful listening, as the human ear has the remarkable ability to identify certain familiar sounds such as bearing squeak or duct leaks and is able to discern small changes in frequency or sound character that might not be apparent from meter readings only. The ear is also a good direction and range finder, since noise generally gets louder as one approaches the source, and direction can often be determined by turning the head. Hands can also identify noise sources. Air jets from duct leaks can often be felt, and the sound of rattling or vibrating panels or parts often changes or stops when these parts are touched.

In trying to locate noise sources and transmission paths, the engineer should consider the location of the affected area. In areas remote from equipment rooms containing significant noise producers but adjacent to shafts, noise is usually the result of structure-borne transmission through pipe and duct supports and anchors. In areas adjoining, above, or below equipment rooms, noise is usually caused by openings (acoustical leaks) in the separating floor or wall or by improper, ineffective, or maladjusted vibration isolation systems.

Unless the noise source or path of transmission is quite obvious, the best way to identify it is by eliminating all sources systematically as follows:

1. Make sure that the objectionable noise is caused by the HVAC system by turning off all equipment. If the noise stops, the HVAC system components (compressors, fans, and pumps) must be operated separately to determine which are contributing to the objectionable noise. Where one source of disturbing noise predominates, the test can be performed starting with all equipment in operation and turning off components or systems until the disturbing noise is eliminated. Tests can also be performed starting with all equipment turned off and operating various component equipment singularly, which permits evaluation of noise from each individual component.

   When testing with a sound level meter, any system component can be termed a predominant noise source if, when the equipment is shut off, the sound level drops 3 dBA or, if measurements are taken with equipment operating individually, the sound level is within 3 dBA of the overall objectionable measurement.

   When a sound level meter is not used, it is best to start with all equipment operating and shut off equipment one at a time, since the ear can reliably detect differences and changes in noise but not absolute levels.
2. When some part of the HVAC system is established as the source of the objectionable noise, try to further isolate the source. By walking around the room, determine whether the noise is coming from the air outlets or returns, the hung ceiling, or through the floors or walls.
3. If the noise is coming through the hung ceiling, check that ducts and pipes are isolated properly and not touching the hung ceiling supports or electrical fixtures, which would provide large noise radiating surfaces. If ducts and pipes are

the source of noise and are isolated properly, the possible remedies are to reduce the noise by changing flow conditions, installing silencers, and/or wrapping the duct or pipe with an acoustical barrier such as a lead blanket.

4. If noise is coming through the walls, ceiling, or floor, check for any openings to adjoining shafts or equipment rooms, and check vibration isolation systems to ensure that there is no structure-borne transmission from nearby equipment rooms or shafts.
5. Noise traced to air outlets or returns usually requires careful evaluation by an engineer or acoustical consultant to determine the source and proper corrective action (see Chapter 42). In general, air outlets can be selected to meet any acoustical design goal by keeping the velocity sufficiently low. For any given outlet, the sound level increases about 2 dB for each 10% increase in airflow velocity over the vanes, and doubling the velocity increases the sound level by about 16 dB. Also, the sound approach conditions caused by improperly located control dampers or improperly sized diffuser necks can easily increase sound levels by 10 to 20 dB.

   A simple, yet effective, instrument that aids in locating noise sources is a microphone mounted on a pole. It can be used to localize noises in hard-to-reach places, such as hung ceilings and behind heavy furniture.
6. If the noise is traced to an air outlet, measure the A-sound level close to it but with no air blowing against the microphone. Then, remove the inner assembly or core of the air outlet and repeat the reading with the meter and the observer in exactly the same position as before. If the second reading is more than 3 dB below the first, a significant amount of noise is caused by airflow over the vanes of the diffuser or grille. In this case, check whether the system is balanced properly. As little as 10% too much air will increase the sound generated by an air outlet by 2.5 dB. As a last resort, a larger air outlet could be substituted to obtain lower air velocities, and hence less turbulence for the same air quality. Before considering this, however, the air approach to the outlet should be checked.

   Noise far exceeding the normal rating of a diffuser or grille is generated when a throttled damper is installed close to it. Air jets impinge on the vanes or cones of the outlet and produce *edge tones* similar to the hiss heard when blowing against the edge of a ruler. The material of the vanes has no effect on this noise, although loose vanes may cause additional noise from vibration.

   When balancing air outlets with integral volume dampers, consider the static pressure drop across the damper, as well as the air quantity. Separate volume dampers should be installed sufficiently upstream from the outlet so that there is no jet impingement. Plenum inlets should be brought in from the side, so that the jets do not impinge on the outlet vanes.
7. If the air outlets are eliminated as sources of excessive noise, inspect the fan room. If possible, change the fan speed by about ±10%. If resonance is involved, this small change can make a significant difference.
8. Sometimes fans are poorly matched to the system. If a belt-driven fan delivers air at a higher static pressure than is needed to move the design air quantity through the system, reduce the fan speed by changing sheaves. If the fan does not deliver enough air, consider increasing the fan speed only after checking the duct system for unavoidable losses. Turbulence in the air approach to the fan inlet not only increases the fan sound generation, but decreases its air capacity. Other parts that may cause excessive turbulence are dampers, duct bends, and sudden enlargements or contractions of the duct.

   When investigating fan noise, seek assistance from the supplier or manufacturer of the fan.
9. If additional acoustical treatment is to be installed in the ductwork, obtain a frequency analysis. This involves the use of an octave-band analyzer and should generally be left to a trained acoustician.

TESTING FOR VIBRATION

Vibration testing is necessary to ensure that (1) equipment is operating with satisfactory vibration levels and (2) objectionable vibration and noise are not transmitted to the building structure. Although these two factors are interrelated, they are not necessarily interdependent. A different solution is required for each, and it is essential to test both the isolation system and the vibration levels of the equipment.

General Procedure

The general order of steps in vibration testing are as follows:

1. Make a visual check of all equipment for obvious errors that must be corrected immediately.
2. Make sure all isolation systems are *free floating* and not short-circuited by any obstruction between equipment or equipment base and building structure.
3. Turn on the system for an audible check of any obviously rough operation. Check bearings with a stethoscope. Bearing check is especially important because bearings can become defective in transit and/or if equipment was not properly stored and maintained. Defective bearings should be replaced immediately to avoid damage to the shaft and other components.
4. Equipment and systems should be adjusted and balanced, so that final vibration tests are made on equipment as it will actually be operating.
5. Test equipment vibration.

Instrumentation

Although instruments are not required to test vibration isolation systems, they are essential to test equipment vibration properly.

Sound level meters and *computer-driven sound-measuring systems* are the most useful instruments for measuring and evaluating vibration. Usually, they are fitted with accelerometers or vibration pickups for a full range of vibration measurement and analysis. Other instruments used for testing vibration in the field are described below.

Reed vibrometers are relatively inexpensive instruments often used for testing vibration, but relative inaccuracy limits their usefulness.

Vibrometers are moderately priced instruments that measure vibration amplitude by means of a light beam projected on a graduated scale.

Vibration meters are moderately priced electronic instruments that measure vibration amplitude on a meter scale and are very simple to use.

Vibrographs are moderately priced mechanical instruments that measure both amplitude and frequency. They are useful for analysis and testing because they provide a chart recording showing amplitude, frequency, and actual wave form of vibration. They can be used for simple yet accurate determination of the natural frequency of shafts, components, and systems by a *bump test.*

Vibration analyzers are relatively expensive electronic instruments that measure amplitude and frequency, usually incorporating a variable filter.

Strobe lights are often used with many of the aforementioned instruments for analyzing and balancing rotating equipment.

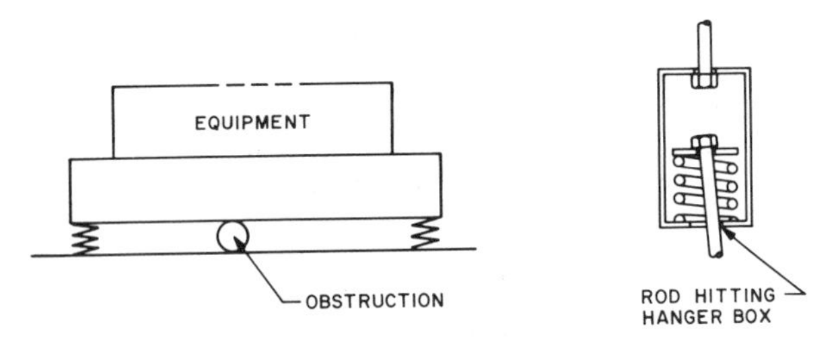

Fig. 14 Obstructed Isolation Systems

Stethoscopes that amplify sound are available as inexpensive mechanic's type (basically, a standard stethoscope with a probe attachment); relatively inexpensive types incorporating a tuneable filter; and moderately priced powered types, which electronically amplify sound and provide some type of meter and/or chart recording. Also, stethoscopes are often used to determine whether bearings are bad.

The choice of instrumentation depends on the test. A stethoscope should be part of every tester's kit as it is one of the most practical, yet least expensive, instruments and one of the best means of checking bearings. The vibrometers and vibration meters can be used to measure vibration amplitude as an acceptance check. Since they cannot measure frequency, they cannot be used for analysis and primarily function as a go/no go instrument. The best acceptance criteria consider both amplitude and frequency. However, since vibrometers and vibration meters are moderately priced and easy to use, they are widely used. Anyone seriously concerned with vibration testing should use an instrument that can determine frequency, as well as amplitude, such as a vibrograph or vibration analyzer.

Testing Vibration Isolation Systems

The following steps should be taken to ensure that vibration isolation systems are functioning properly:

1. Ensure that the system is *free floating* by applying an unbalanced load, which should cause the system to move freely and easily. On floor-mounted equipment, check that there are no obstructions between the base or foundation and the building structure that would cause transmission while still permitting equipment to rock relatively free because of the application of an unbalanced force (Figure 14). On suspended equipment, check that hanger rods are not touching the hanger. Rigid connections such as pipes and ducts can prohibit mounts from functioning properly and from providing a transmission path. Note that because the system is free floating, does not mean that the isolators are functioning properly. For example, a 500 rpm fan installed on isolators

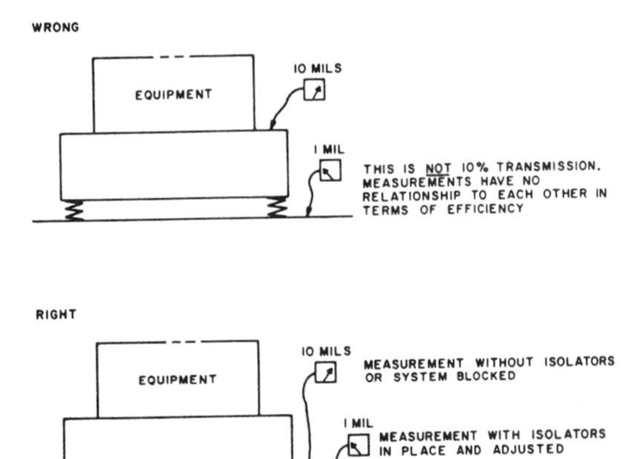

Fig. 15 Testing Isolation Efficiency

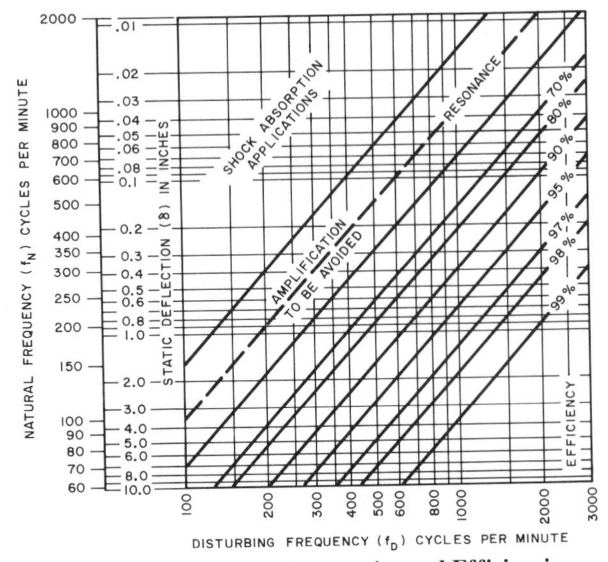

Fig. 16 Isolator Natural Frequencies and Efficiencies

having a natural frequency of 500 cpm could be free floating but would actually be in resonance, resulting in transmission to the building and excessive movement.

2. Determine if isolators are adjusted properly and providing desired isolation efficiency. All isolators supporting a piece of equipment should have approximately the same deflection, *i.e.*, compressed the same under the equipment. If not, they have been improperly adjusted, installed, or selected; this should be corrected immediately. Note that isolation efficiency cannot be checked by comparing vibration amplitude on equipment to amplitude on the structure as shown in Figure 15.

   The only accurate check of isolation efficiencies is to compare vibration measurements of equipment operating with isolators to measurements of equipment operating without isolators. Since these tests are usually impractical, it is best to check isolator deflection to determine if deflection is as specified, and if specified or desired isolation efficiency is being provided. Figure 16 shows natural frequency of isolators as a function of deflection and indicates the theoretical isolation efficiencies for various frequencies at which the equipment operates.

While it is easy to determine the deflection of spring mounts by measuring the difference between the free heights with a ruler (information as shown on submittal drawings or available from a manufacturer), such measurements are difficult with most pad or rubber mounts. Further, most pad and rubber mounts do not lend themselves to accurate determination of natural frequency as a function of deflection. For such mounts, the most practical approach is to check that there is no excessive vibration of the base and no noticeable or objectionable vibration transmission to the building structure.

If isolators are in the 90% efficiency range and there is transmission to the building structure, the equipment is operating roughly or there is a flanking path of transmission, such as connecting piping or obstruction, under the base.

Testing Equipment Vibration

Testing equipment vibration is necessary as an acceptance check to determine if equipment is functioning properly and to ensure that objectionable vibration and noise are not transmitted. Although a person familiar with equipment can determine when it is operating roughly, instrumentation is usually required to determine accurately if vibration levels are satisfactory.

Vibration Tolerances

Vibration tolerance criteria are provided in Table 33, Chapter 42. These criteria are based on equipment installed on vibration isolators and can be met by any reasonably smoothly running equipment.

Procedure for Testing Equipment Vibration

The following steps should be taken to ensure that equipment vibration is properly tested:

1. Determine operating speeds of equipment from nameplates, drawings, or a speed-measuring device such as a tachometer or strobe, and indicate them on the test form. For any equipment where the driving speed (motor) is different from the driven speed (fan wheel, rotor, impeller) because of belt drive or gear reducers, indicate both driving and driven speeds.
2. Determine acceptance criteria from specifications and indicate them on the test form. If specifications do not provide criteria, use those shown in Chapter 42.
3. Ensure that the vibration isolation system is functioning properly (see Testing Vibration Isolation Systems section).
4. Operate equipment and make visual and aural checks for any apparent rough operation. Check all bearings with a stethoscope. Any defective bearings, misalignment, or obvious rough operation should be corrected before proceeding further. If not corrected, equipment should be considered unacceptable.
5. Measure and record on the test form vibration at bearings of driving and driven components in horizontal, vertical, and, if possible, axial directions. There should be at least one axial measurement for each rotating component (fan motor, pump motor).
6. Evaluate measurements as described below.

Evaluating Vibration Measurements

Vibration measurements are evaluated in accordance with the type of measurements made as follows:

Amplitude Measurement. When specification for acceptable equipment vibration is based on amplitude measurements only and measurements are made with an instrument that measures only amplitude, such as a vibration meter or vibrometer:

1. No measurement shall exceed specified values or values shown in Table 34 of Chapter 42, taking into consideration reduced values for equipment installed on inertia blocks
2. No measurement shall exceed values shown in Table 34, Chapter 42, for driving and driven speeds, taking into consideration reduced values for equipment installed on inertia blocks. For example, with a belt-driven fan operating at 800 rpm and having an 1800 rpm driving motor, amplitude measurements at fan bearings must be in accordance with values shown for 800 cpm, and measurements at motor bearings must be in accordance with values shown for 1800 cpm. If measurements at motor bearings exceed specified values, take measurements of the motor only with belts removed to determine if there is feedback vibration from the fan.
3. No axial vibration measurement should exceed maximum radial (vertical or horizontal) vibration at the same location.

Amplitude and Frequency Measurements. When specification for acceptable equipment vibration is based on both amplitude and frequency measurements and measurements are made with instruments that measure both amplitude and frequency, such as a vibrograph or vibration analyzer:

1. No amplitude measurements at driving and driven speeds shall exceed specified values or values shown in Table 34, Chapter 42, taking into consideration reduced values for equipment installed on inertia blocks. Measurements that exceed acceptable amounts may be evaluated as explained in the Vibration Analysis section.
2. No axial vibration measurement shall exceed maximum radial (vertical or horizontal) vibration at the same location.
3. The presence of any vibration at frequencies other than driving or driven speeds is generally reason to rate operation unacceptable and such vibration should be analyzed as explained in the Vibration Analysis section that follows.

Vibration Analysis

The following guide covers most vibration problems that may be encountered.

Axial Vibration Exceeds Radial Vibration. When amplitude of axial vibration (parallel with shaft) at any bearing exceeds radial vibration (perpendicular to shaft—vertical or horizontal), it usually indicates misalignment; this should be checked carefully. This is most common on direct-driven equipment because, although flexible couplings that accommodate parallel and angular misalignment of shafts are generally used, such misalignment can generate forces that cause axial vibration. As axial vibration can cause premature bearing failure, misalignment should be corrected promptly. Other possible causes of large amplitude axial vibration are resonance, defective bearings, insufficient rigidity of bearing supports or equipment, and loose hold-down bolts.

Vibration Amplitude Exceeds Allowable Tolerance at Rotational Speed. The allowable vibration limits established by Table 33, Chapter 42 are based on vibration caused by rotor imbalance, which results in vibration at rotational frequency. While vibration caused by imbalance must be at the frequency at which the part is rotating, a vibration at rotational frequency can be, but does not have to be, caused by imbalance. An unbalanced rotating part develops centrifugal force, which causes it to vibrate at rotational frequency, but vibration at rotational frequency can also result from other conditions such as bent shaft, eccentric sheave, misalignment, and resonance. If vibration amplitude exceeds allowable tolerance at rotational frequency, the following steps should be taken before performing field balancing of rotating parts.

1. Check vibration amplitude as equipment goes up to operating speed and as it coasts to a stop. Any significant peaks at or near operating speed, as shown in Figure 17, indicate a probable condition of resonance, *i.e.*, some part having a natural frequency close to the operating speed, resulting in greatly amplified levels of vibration.

   A bent shaft or eccentricity will usually cause imbalance, and will also result in significantly higher vibration amplitude at lower speeds, as shown in Figure 18, whereas vibration caused by imbalance generally increases as speed increases.

   If bent shaft or eccentricity is suspected, check the dial indicator. A bent shaft or eccentricity between bearings as shown in Figure 19A, can usually be compensated for by field balancing, although some axial vibration might remain. Field balancing cannot correct vibration caused by bent shaft on direct-connected equipment, on belt-driven equipment where the shaft is bent at the location of sheave, or if the sheave is eccentric (Figure 19B). This is because the center-to-center distance of the sheaves will fluctuate, each revolution resulting in vibration.

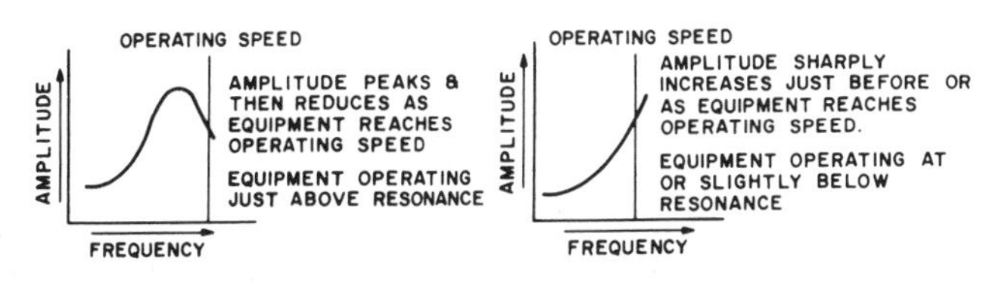

Fig. 17 Vibration from Resonant Condition

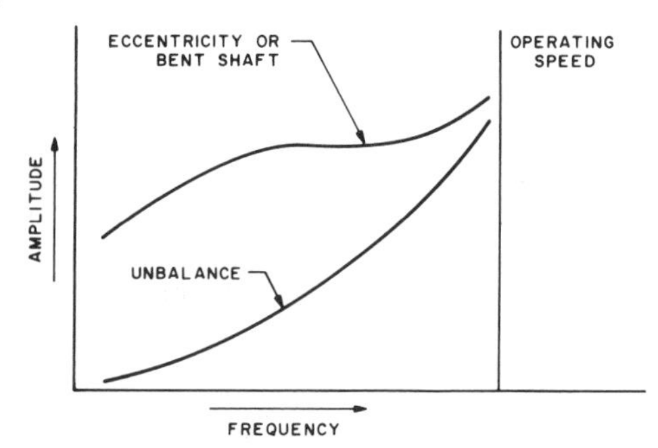

Fig. 18 Vibration Caused by Eccentricity

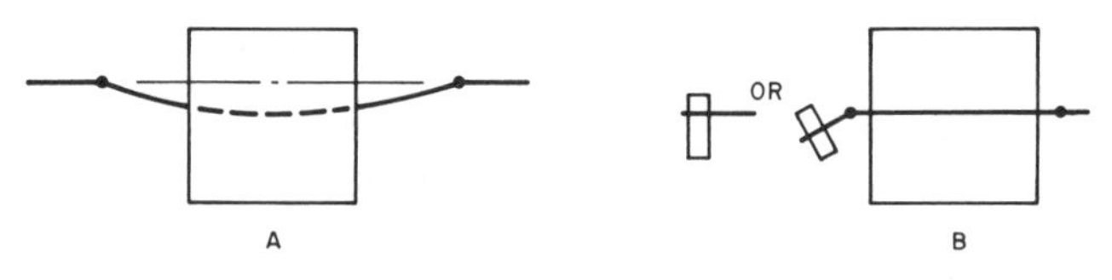

Fig. 19 Bent Shafts

2. For belt- or gear-driven equipment where vibration is at motor driving frequency rather than driven speed, it is best to disconnect the drive to perform tests. If the vibration amplitude of the motor operating by itself does not exceed specified or allowable values, excessive vibration (when connected) is probably a function of bent shaft, misalignment, eccentricity, resonance, or loose hold-down bolts.
3. Vibration caused by imbalance can be corrected in the field by firms specializing in this service or by testing personnel if they have appropriate equipment and experience.

Vibration at Other than Rotational Frequency. Vibration at frequencies other than driving and driven speeds is generally considered unacceptable. Table 9 shows some common conditions that can cause vibration at other than rotational frequency.

Resonance. If resonance is suspected, determine which part of the system is in resonance.

Isolation Mounts. The natural frequency of the most commonly used spring mounts is a function of spring deflection, as shown in Figure 12, Chapter 7 of the 1989 ASHRAE *Handbook—Fundamentals*, and it is relatively easy to calculate by determining the difference between the free and operating height of the mount, as explained in the Testing Vibration Isolation Systems section. This technique cannot be applied to rubber, pad, or fiberglass mounts, which have a natural frequency in the 300 to 3000 cpm range. Natural frequency for such mounts is determined by

Table 9 Common Causes of Vibration other than Unbalance at Rotation Frequency

Frequency	Source
0.5 × rpm	Vibration at approximately 0.5 rpm can result from improperly loaded sleeve bearings. This vibration will usually suddenly disappear as equipment coasts down from operating speed.
2 × rpm	Equipment not tightly secured or bolted down.
2 × rpm	Misalignment of couplings or shafts usually results in vibration at twice rotational frequency and generally a relatively high axial vibration.
Many × rpm	Defective antifriction (ball, roller) bearings usually result in low amplitude, high frequency, erratic vibration. Since defective bearings usually produce noise rather than any significantly measurable vibration, it is best to check all bearings with a stethoscope or similar listening device.

a bump test. Any resonance with isolators should be immediately corrected as it results in excessive movement of equipment and more transmission to the building structure than if equipment were attached solidly to the building (installed without isolators).

System Component. Resonance can occur with any system component shaft, structural base, casing, and connected piping. The easiest way to determine natural frequency is to perform a bump test with a vibrograph. This test consists of bumping the part and measuring with an instrument; the part will vibrate at its natural frequency, which is recorded on instrument chart paper. Similar tests, though not as convenient or accurate, can be made with a reed vibrometer or a vibration analyzer. However, most of these instruments are restricted to frequencies above 500 cpm and, therefore, cannot be used to determine natural frequencies of most isolation systems, which usually have natural frequencies lower than 500 cpm.

Checking for Vibration Transmission. The source of vibration transmission can be checked by determining frequency with a vibration analyzer and tracing back to equipment operating at this speed. However, the easiest and usually the best method (even if test equipment is being used) is to shut off equipment one at a time until the source of transmission is located. Most transmission problems cause disturbing noise; listening is the most practical approach to determine a noise source, since the ear is usually better than sound-measuring instruments at distinguishing small differences and changes in character and amount of noise. Where disturbing transmission consists solely of vibration, a measuring instrument will probably be helpful, unless vibration is significantly above the sensory level of perception. Vibration below the sensory level of perception will generally not be objectionable.

If equipment is located near the affected area, check isolation mounts and equipment vibration. If vibration is not being transmitted through the base or if the affected area is remote from equipment, the probable cause is transmission through connected piping and/or ducts. Ducts can usually be isolated by isolation hangers. However, transmission through connected piping, which is very common, presents numerous problems that should be understood before attempting to correct them.

Vibration and Noise Transmission in Piping Systems

Vibration and noise in connected piping can be caused by mechanical vibration generated by (1) equipment (*e.g.*, pump, compressor) and transmitted through the walls of pipes; (2) mechanical vibration and transmitted by a water column; or (3) flow (velocity) noise and vibration. Flexible pipe connectors, which provide system flexibility to permit isolators to function properly and protect equipment from stress caused by misalignment and thermal expansion, can be useful in attentuating mechanical vibration transmitted through a pipe wall. However, they rarely suppress flow vibration and noise and only slightly attenuate mechanical vibration as transmitted through a water column.

Tie rods are often used with flexible rubber hose and rubber expansion joints (Figure 20). While they accommodate thermal

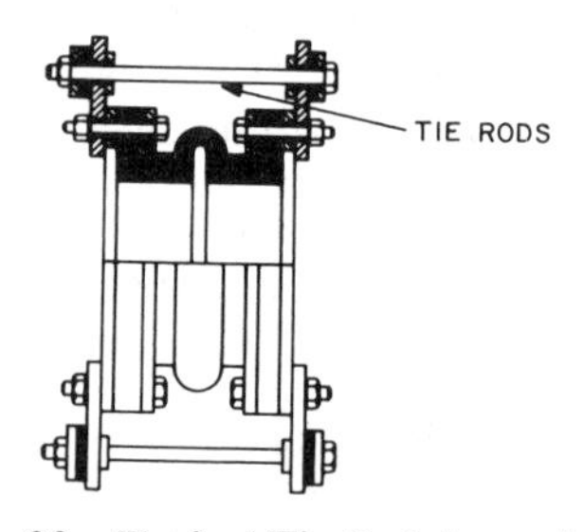

Fig. 20 Typical Tie Rod Assembly

movements, they hinder the isolation of vibration and noise. This is because pressure in the system causes the hose or joint to expand until resilient washers under tie rods are virtually rigid. To isolate noise adequately with a flexible rubber connector, tie rods and anchor piping should not be used. However, this technique generally cannot be used with pumps that are on spring mounts because they would still permit the hose to elongate. Flexible metal hose can be used with spring isolated pumps, since wire braid serves as tie rods; metal hose controls vibration but not noise.

Problems of transmission through connected piping are best resolved by changes in the system to reduce noise (improve flow characteristics, turn down impeller) or by completely isolating piping from the building structure. Note, however, that it is almost impossible to isolate piping completely from the structure, as required resiliency is inconsistent with rigidity requirements of pipe anchors and guides. Chapter 42 contains information on flexible pipe connectors and resilient pipe supports, anchors, and guides, which should help resolve any piping noise transmission problems.

REFERENCES

AMCA. 1985. Laboratory methods of testing fans for rating ANSI/AMCA *Standard* 210-85. Also ANSI/ASHRAE *Standard* 51-1985. Air Movement and Control Association, Arlington Heights, IL.

ASHRAE. 1989. Guideline for commissioning of HVAC systems. ANSI/ASHRAE *Guideline* 1-1989.

ASHRAE. 1988. Practices for measurement, testing, adjusting and balancing of building heating, ventilation, air conditioning and refrigeration systems. ANSI/ASHRAE *Standard* 111-1988.

ASME. 1986. Atmospheric water cooling equipment. ASME *Standard* PTC 23-86. American Society of Mechanical Engineers, New York.

CTI. 1982. Standard specifications for thermal testing of wet/dry cooling towers. CTI *Standard Specification* ATC-105-82. Cooling Tower Institute, P.O.Box 73383, Houston, TX.

Griggs, E.I., W.B. Swim, and H.G. Yoon. 1990. Placement of air control sensors. ASHRAE *Transactions* 96(1).

Sauer, H.J. and R.H. Howell. 1990. Airflow measurements at coil faces with vane anemometers: Statistical correction and recommended field measurement procedure. ASHRAE *Transactions* 96(1):502-11.

BIBLIOGRAPHY

AABC. 1967. Field balancing depends on dampers. Associated Air Balance Council, Washington, D.C. (January).

AABC. 1984. National standards for total system balance, 4th ed. Associated Air Balance Council, Washington, D.C.

AMCA. 1987. Fan application manual. Air Movement and Control Association, Arlington Heights, IL.

Armstrong Pump. 1986. Technology of balancing hydronic heating & cooling systems. Armstrong Pump, North Tonawanda, NY.

ASA. 1983. Specification for sound level meters. ANSI *Standard* S1, 4-83, or ASA *Standard* 47-83. Acoustical Society of America, New York.

ASHRAE. 1974. Energy conservation pumping systems. ASHRAE *Journal* 16(6).

Carlson, G.F. 1974. Liquid viscosity effects on pumping systems, Part I and II. ASHRAE *Journal* 16(7, 9).

Carlson, G.F. 1968-1969. Hydronic systems analysis and evaluation. ASHRAE *Journal* (October through March).

Carlson, G.F. 1972. Central plants chilled water systems. ASHRAE *Journal* (February through April).

Carrier. *Carrier air conditioning manual.* Syracuse, NY.

Choat, E.E. 1976. An evaluation of the temperature difference method for balancing hydronic coils. ASHRAE *Transactions* 82(1).

Coad, W.J. 1985. Variable flow in hydronic systems for improved stability, simplicity and energy economics. ASHRAE *Transactions* 91(1B):224-37.

Crane Co. Flow of fluids through valves, fittings and pipe. *Technical Paper* 410. Crane Co, King of Prussia, PA.

Eads, W.G. 1983. Testing, balancing and adjusting of environmental systems. In *Fan engineering*, 8th ed. Buffalo Forge Company, Buffalo, NY.

Gladstone, J. 1981. *Air conditioning—Testing and balancing A field practice manual.* Van Nostrand and Reinhold Company, New York.

Gupton, G. 1987. HVAC controls, operation and maintenance.

Haines, R.W. 1987. *Control systems for heating, ventilating, and air conditioning*, 4th ed. Van Nostrand Reinhold Company, New York.

Hansen, E.G. 1985. *Hydronic system design and operation.* McGraw-Hill, New York.

Hayes, F.C. and W.F. Stoecker. 1966. The effect of inlet conditions on flow measurements at ceiling diffusers. ASHRAE *Transactions* 72(2).

Hayes, F.C. and W.F. Stoecker. 1966. Tables of application factors for flow measurement at return intakes. ASHRAE *Transactions* 72(2).

Hightower, G.B. 1971. Testing, balancing and adjusting of HVAC induction systems. ASHRAE *Journal* 13(6).

ITT Bell & Gossett. Balance procedure manual. TEB-985. ITT Bell & Gossett, Morton Grove, IL.

McQuiston, F.C. and J.D. Parker. 1988. *Heating, ventilating and air conditioning, analysis & design*, 3rd ed. John Wiley & Sons, New York.

Miller, R.W. 1983. *Flow measurement engineering handbook.* McGraw-Hill, New York.

National Joint Steamfitter-Pipefitter Apprenticeship Committee. 1976. *Start, test & balance manual.* Washington, DC.

NEBB. 1983. Procedural standards for testing, balancing and adjusting of environmental systems, 4th ed. National Environmental Balancing Bureau, Vienna, VA.

NEBB. 1986. Testing, adjusting, balancing manual for technicians, 1st ed. (April).

Nevins, R.G. and E.D. Ward. 1968. Room air distribution with an air distribution ceiling. ASHRAE *Transactions* 74(1).

SMACNA. 1977. Round industrial duct construction standards. Sheet Metal and Air Conditioning Contractors' National Association, Inc., Merrifield, VA.

SMACNA. 1980. Rectangular industrial duct construction standards.

SMACNA. 1983. HVAC Systems—Testing, adjusting and balancing.

SMACNA. 1985. HVAC Duct construction standards—Metal and flexible.

SMACNA. 1985. HVAC Air duct leakage test manual.

Tamura, G.T. and A.G. Wilson. 1966. Pressure differences for a nine-story building as a result of chimney effect and ventilating system operation. ASHRAE *Transactions* 72(1).

Tamura, G.T. and A.G. Wilson. 1968. Pressure differences caused by wind on two tall buildings. ASHRAE *Transactions* 74(2).

Tamura, G.T. and A.G. Wilson. 1967. Pressure differences caused by chimney action and mechanical ventilation. ASHRAE *Transactions* 73(2).

Tamura, G.T. and A.G. Wilson. 1967. Building pressures caused by chimney action and mechanical ventilation. ASHRAE *Transactions* 73(2).

The Trane Company. 1988. *Trane air conditioning manual.* The Trane Company, LaCrosse, WI.

Yerges, L.F. 1978. *Sound, noise and vibration control*, 2nd ed. Van Nostrand Reinhold, New York.

CHAPTER 35

OPERATION AND MAINTENANCE MANAGEMENT

THE success of operation and maintenance programs depends on how the resources of these systems are managed. Maintenance management planning must include a proper life cycle cost analysis and a process to ensure that occupant comfort, energy planning, and safety and security systems are optimal for all facilities. Appropriate technical expertise—in-house or contracted—is also important. The following issues are addressed in this chapter:

- Life cycle costs
- Variations in approaches according to the number and size of buildings or systems within a program
- Organizational frameworks according to the number and size of buildings or systems in a program
- Documentation and record keeping
- Responsibilities of designers, contractors, manufacturers/suppliers, and owners related to operation and maintenance

Operation and maintenance of all HVAC&R systems should be considered when the original design of a building is being formulated. Any successful operation and maintenance program must include proper documentation of the design criteria. Newly installed systems must be commissioned to ensure that they are functioning as designed; it is then the responsibility of operation and maintenance staff to retain that function throughout the life of the building. Existing systems may need to be recommissioned to accommodate changes.

While mechanical maintenance was once the responsibility of trained technical personnel, increasingly sophisticated systems and equipment require overall management programs to handle organization, staffing, planning, and control. These programs should meet present and future energy management requirements, upgrade management skills, and increase communication among those who benefit from cost-effective operation and maintenance.

TERMS

System operation defines the parameters under which the building or systems operator can adjust components of the system to satisfy the tenant comfort or process requirements and the strategy for optimum energy use and minimum maintenance.

The *maintenance program* defines maintenance in terms of time and resource allocation. It documents the objectives, establishes the criteria for evaluation, and commits the maintenance department to basic areas of performance, such as prompt response to mechanical failure and attention to planned functions that protect a capital investment and minimize downtime or failure response.

The preparation of this chapter is assigned to TC 1.7, Operation and Maintenance Management.

Failure response classifies maintenance department resources expended or reserved to handle interruptions in the operation or function of a system or equipment under the maintenance program. This classification has two types of response—repair and service.

Repair is to make good, or to restore to good or sound condition with the following constraints: (1) operation must be fully restored without embellishment, and (2) failure must trigger the response.

Service provides what is necessary to effect a maintenance program short of repair. It is usually based on procedures recommended by manufacturers.

Planned maintenance classifies maintenance department resources that are invested in prudently selected functions at specified intervals. All functions and resources attributed to this classification must be planned, budgeted, and scheduled. It embodies two concepts—preventive and corrective maintenance.

Preventive maintenance classifies resources allotted to ensure proper operation of a system or equipment under the maintenance program. Durability, reliability, efficiency, and safety are the principal objectives.

Corrective maintenance classifies resources, expended or reserved, for predicting and correcting conditions of impending failure. Corrective action is strictly remedial and always performed before failure occurs. An identical procedure performed in response to failure is classified as a repair. Corrective action may be taken during a shutdown caused by failure, providing it is optional and unrelated.

Predictive maintenance is a function of corrective maintenance. Statistically supported objective judgment is implied. Nondestructive testing, chemical analysis, vibration and noise monitoring, as well as routine visual inspection and logging are classified under this function, providing that the item tested or inspected is part of the planned maintenance program.

Durability is the average expected service life of a system or facility. Table 3 in Chapter 33 lists median years of service life of various equipment. More specifically, individual manufacturers quantify it as design life, which is the average number of hours of operation before failure, extrapolated from accelerated life tests and stressing critical components to economic destruction.

Reliability implies that a system or facility will perform its intended function for a specified period of time without failure.

QUANTITATIVE MANAGEMENT CONCEPT

The following life cycle management concept is recommended for operation and maintenance programs. The concept can be used for planning during the construction cycle as well as for day-to-day operation and maintenance management. Derived from

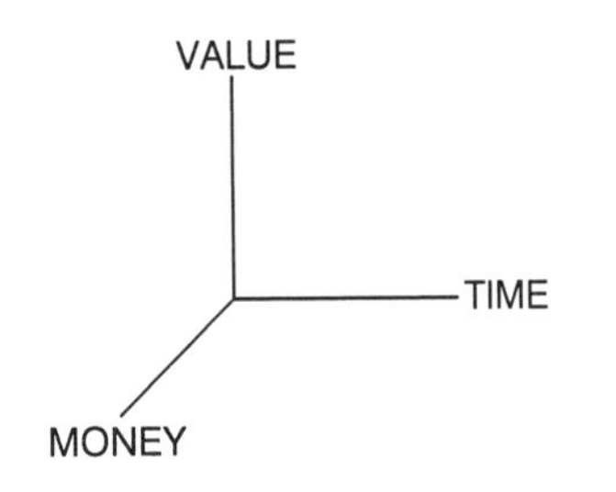

Fig. 1 Life Cycle Concept of Management

value engineering, this life cycle concept of management involves three dimensions—effectiveness (value), durability (time), and life cycle cost (money)—which are interdependent (Figure 1). Their numerical values are affected by the type and extent of the operation and maintenance programs.

The effectiveness of a system is the probability of successfully providing required services over a given period when operated under specified conditions. To be perfectly effective, a facility must provide the required services satisfactorily and must operate dependably without failure for the duration of the operation. Mathematically, effectiveness E is the product of capability C and dependability D.

$$E = CD \tag{1}$$

Effectiveness may be assigned a numerical value according to the type of service required to meet a tenant's functional need. Table 1 is a guide for assigning effectiveness values. Note: The values given in this table are examples only, and values must be developed for each project to create proper relationships between the different systems in each facility.

Table 1 Examples of Effectiveness Values

Range of Effectiveness	Effectiveness Category	Example of Service
0.50 to 0.75	Regular	Environmental control
0.76 to 0.85	Essential	Freeze protection
0.86 to 1.00	Critical	Smoke control

Capability is the measure of a system's ability to satisfactorily provide required service. It is the probability of meeting functional requirements, providing the system is operated under previously designated operating conditions. An example of capability is the ability of a heating system to cope with heating load at the design winter temperature. The capability must be verified when the system is first commissioned and then, whenever the functional requirements change.

Capability can be calculated from the time the system cannot meet the requirements N and the time it is expected to operate in one year Y:

$$C = 1 - (N/Y) \tag{2}$$

Dependability is the measure of a system's condition. Assuming the system was operative at the beginning of its service life, dependability is the probability of it being operative at any other time until the end of its life.

For those systems that cannot be repaired during use, dependability is the probability that there will be no failure during use. For systems that can be repaired, dependability is governed by the ease and rapidity with which repairs can be made. This ease is the system maintainability, defined below.

Dependability D can be expressed as a product of reliability R and maintainability M.

$$D = RM \tag{3}$$

Reliability implies that a system will perform its function, without a failure, for a desired portion of a specified period. Reliability of two or more systems is calculated as follows:

1. If the systems are arranged in series, where the output of one is the input of the next (Figure 2), their combined reliability is:

$$R = R_1 R_2 \ldots R_n \tag{4}$$

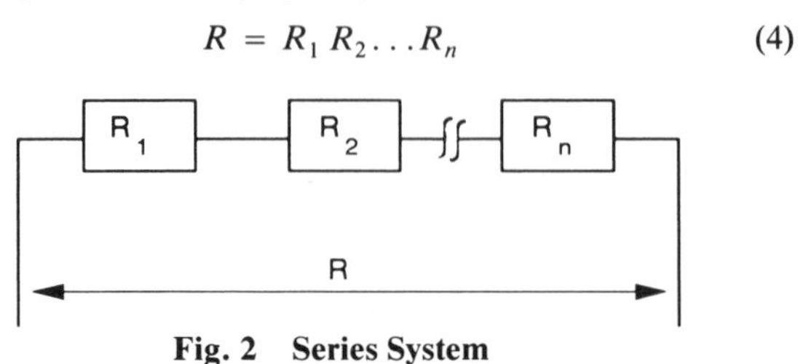

Fig. 2 Series System

2. If the systems are arranged in parallel, such as dual pumping systems with automatic changeover on failure of one of the two pumps (Figure 3), their combined reliability is calculated from the following equation:

$$R = 1 - (1 - R_1)(1 - R_2)\ldots(1 - R_n) \tag{5}$$

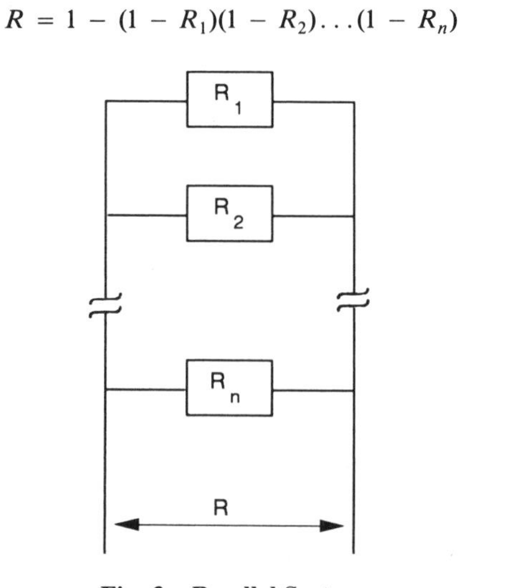

Fig. 3 Parallel System

Maintainability is the ease and rapidity with which a system can be repaired and complements reliability by defining the specific time that a system can operate in a fully restored condition.

The value of maintainability is calculated from the time the system is being repaired in regular occupancy B and the time the system is expected to operate in one year Y.

$$M = 1 - (B/Y) \tag{6}$$

The above terms can be interrelated by an algorithm. Shown in Figure 4, this algorithm can be manipulated either manually or by computer.

DOCUMENTATION

Operation and maintenance documentation should be prepared throughout the delivery cycle of any project. Information should be documented as soon as it becomes available to support design, construction activities, and training of operation and maintenance staff in preparation for system commissioning.

A complete operation and maintenance documentation package consists of the following documents:

- Complete set of design criteria and summary results
- Complete set of specifications, including all addenda and all approved and applied changes
- Complete set of approved shop drawings, including all subsequent modifications

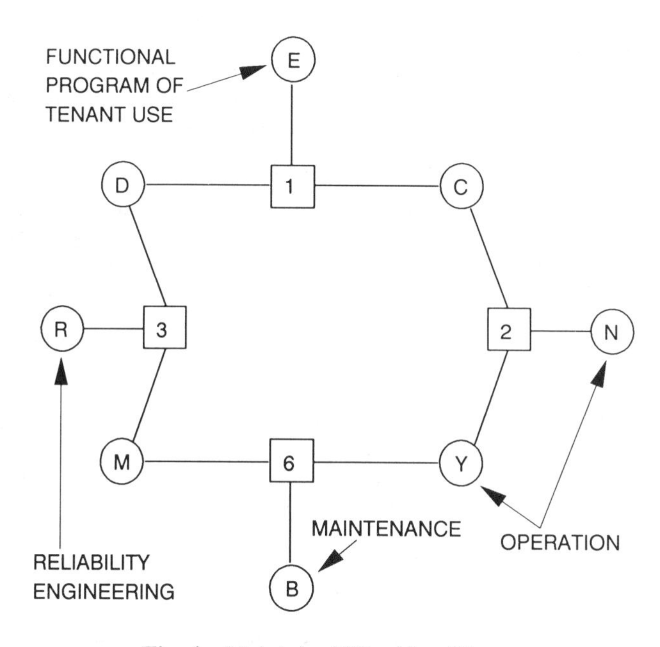

Fig. 4 Maintainability Algorithm

- Complete set of as-built drawings
- Operating manual, which includes control sequences of operation and, if direct digital control, a flowchart and hard copy of software code/data base
- Maintenance manual, including proposed equipment maintenance programs to facilitate staffing
- Electrical power coordination report
- Air and water balancing report
- Performance verification report
- Copies of all certificates by the inspectors representing authorities having jurisdiction
- Copy of the commissioning report

All the above documents should be available for large, complex facilities. However, small projects require only documentation appropriate for their size and complexity.

The most important documents are operation and maintenance manuals. The following suggested organization of these manuals reflects the needs of their various users.

Operation Manual

This manual should consist of two parts. Part I, Operation Instruction, contains information a qualified operator needs (1) to start and stop equipment, (2) to control and monitor the performance of the equipment in normal modes of operation, (3) to change from one mode of operation to another, and (4) to operate equipment in emergency situations. Operation procedures for all integrated systems with proper flowcharts are also required. The system function should be represented pictorially and in writing.

Part II, Performance Verification Procedures, should contain all the information a qualified operator needs to verify equipment and overall system performance at the design load as well as at part loads, where applicable. The design calculations are required to allow the performance verification and should be included in this manual.

Maintenance Manual

This manual should also consist of two parts. Part I, Inventory, should contain a listing of all systems and pieces of equipment to be maintained as well as all the technical information needed to order spare parts. Manufacturers' catalogs should be considered useful adjuncts only. Part II, Maintenance Program, should contain information necessary to perform maintenance under the selected type(s) of program—breakdown, preventive, or predictive. These programs require written information regarding when or how often to perform maintenance in the most efficient and economical fashion to satisfy tenant needs.

KNOWLEDGE AND SKILLS

To be effective and economical, operation and maintenance requires the right combination of technical and managerial skills. Technical skills include not only the skills of operation and maintenance mechanics, but also the engineering skills of the physical plant engineer. Managerial skills involve managing the facility in life cycle terms as well as on a day-to-day basis. This level of management administers both the accommodation contract with the tenant and the administration of service contracts, including maintenance service for all equipment. Specialized contractual maintenance companies require yet another level of management.

Physical plant engineers require a variety of skills, since they need to coordinate the equipment selected by the designer, the operating and maintenance personnel, and the requirements of the investment plan. Good physical plant engineering solutions are developed through an iterative process that should start when the investment plan is being formulated and continue throughout the life of the facility.

TYPES OF BUILDINGS

The size and complexity of the system to be operated and maintained need to be considered, since buildings range in size and complexity from houses through commercial offices, institutional buildings and complexes, processing plants, refrigerated storage facilities, and large central plants. All installed systems, no matter how simple, should have a commissioning plan; this gives the building or system owner(s) and operator(s) a means of operating the system in the most economical manner—to minimize energy consumption and maintenance costs while meeting user requirements. Since these costs may not be mutually exclusive, the commissioning plan should identify the tradeoffs in operating a system that provides comfort while affecting the frequency of various types of maintenance.

Small Buildings

Small systems have relatively simple operating procedures such as adjusting a single zone thermostat from heating to cooling if it is not automatic and resetting temperatures as desired by the occupants. Maintenance procedures for this type of system are limited to those recommended by most manufacturers.

Most building managers call on-site maintenance staff to change filters, belts, and motors. However, cleaning of condenser and evaporator coils and assessing refrigeration and control systems require a more qualified technician. In most small facilities, maintenance contractors provide this service on a maintenance program basis. The frequency of maintenance depends on hours of system use, proximity to major transportation routes (dirt accumulation), and type of operation in the facility.

Smaller facilities may also have complex control systems, especially when zoning is critical for a variety of load conditions. Whenever the operator does not have the knowledge to service and repair the systems or components installed, the owner should ensure that qualified contractors are used.

Furthermore, dirty evaporator and condenser coils and filters require additional energy to operate due to higher head pressures and inefficient heat transfer. This condition also requires longer operating time to satisfy space conditions. While these items are variable and difficult to quantify, proper unit and system maintenance improves both the system operation and equipment life cycle, regardless of system size.

Medium-sized Buildings

The systems in medium-sized buildings are generally more complex, with the interaction of several pieces of mechanical equipment acting together through a control system to provide a variety of comfort zones.

Commissioning is increasingly important with this type of system to ensure optimum comfort at minimal energy cost. Without detailed documentation, the operation staff cannot consider energy budgets while they are addressing building occupants' comfort complaints.

To manage maintenance programs in medium-sized facilities, the programs must be detailed for the maintenance personnel as well. Maintenance programs should be implemented on a sequenced basis to reduce the failure response required. Over time, the maintenance programs can be adjusted for the particular characteristics of the system. Computerized maintenance programs can assist management in overseeing the effectiveness of the program.

For medium-sized systems, some operation and maintenance personnel may be employed by the building owner or leaseholder. They should have the technical knowledge to operate these systems if they observed a full system commissioning process.

The operating personnel may also be responsible for system maintenance. The maintenance procedures for the system should be detailed in the commissioning documentation (with individual equipment maintenance frequency detailed in manufacturers' literature). The maintenance program should consider the unique nature of each building. Again, it may be necessary to contract out certain maintenance programs for highly technical pieces of equipment should the staff not have the technical knowledge.

Large Buildings

Large buildings, including central plants, require a management structure to staff, direct, and control two staffs—one for operation personnel, and the other for maintenance personnel.

The operations budget should be large enough to support computerized maintenance programs, which detail proper timing of system maintenance procedures so that all systems operate at maximum efficiency while maintaining occupant comfort. Annual and life cycle cost planning are essential to ensure the most cost-effective operation and maintenance of these systems.

To facilitate proper commissioning, management should be involved with the design process and construction of the facility. Because facilities are long-term investments, any first-cost compromises must consider both life cycle cost and the ability to satisfy occupant comfort while maintaining reasonable energy budgets.

Logged information can be used with proper data-base management systems as predictive maintenance to reduce failure response requirements. These systems may be used to indicate the weaknesses in the systems so that management can make appropriate decisions or system changes. As with small and medium-sized systems, large systems may require specific outside contractor or manufacturer's support for specific equipment.

RESPONSIBILITIES

The building owner should be apprised of the design intent of the system to allow self-management. The system design must include proper operation and maintenance information. Flowcharts and instrumentation requirements must be indicated. The installing contractor must provide the director of operations and maintenance with an organized and comprehensive turnover of the systems that have been installed.

An effective director should be able to organize, staff, train, plan, and control the operation and maintenance of a facility with the cooperation of senior management and all departments. A manager's responsibilities include administering the operation and maintenance budget and protecting the life cycle objectives. Before selecting a least-cost alternative, a manager should determine the consequential effect on durability and loss prevention (Loveley 1973).

NEW TECHNOLOGY

Operation and maintenance programs are based on the technology available at the time of their preparation. The programs should be adhered to throughout the required service life of the facility or the system. In the course of the service life, a new technology may become available that would affect the operation and/or maintenance program. When this occurs, the switch from the existing to new technology must be assessed in life cycle terms. The existing technology must be assessed for the degree of loss due to shorter return on investment; the new technology must be assessed (1) for all initial and operation and maintenance cost, (2) for the correlation between its service life and the remaining service life of the facility, and (3) for the cost of conversion, including the revenue losses due to the associated downtime.

REFERENCE

Loveley, J.D. 1973. Durability, reliability, and serviceability. ASHRAE *Journal* 15(1):67.

BIBLIOGRAPHY

Fuchs, S.J. 1982. *Complete building equipment maintenance desk book.* Prentice Hall, Englewood Cliffs, NJ.

Lawson, C.N. 1989. Commissioning—The construction phase. ASHRAE *Transactions* 95(1):887-94.

Trueman, C.J. 1989. Commissioning: An owner's approach for effective operation. ASHRAE *Transactions* 95(1):895-99.

Petrocelly, K.L. 1989. *Physical plant operations handbook.* Fairmont Press, Englewood Cliffs, NJ.

Criswell, J.W. 1987. *Planned maintenance for productivity and energy conservation*, 2nd ed. Fairmont Press, Englewood Cliffs, NJ.

CHAPTER 36

COMPUTER APPLICATIONS

THE use of digital computers in the heating, refrigeration, and air-conditioning industry has come about because of the wide variety of easily used engineering analysis programs for the HVAC industry, an even larger number and range of programs for business use, and the low cost of powerful computers on which to run them. The ordinary calculations required in the HVAC industry such as heating and cooling loads can be accomplished easily and inexpensively on a computer. In addition, computers sometimes allow the solution of more complicated problems, which otherwise would be impractical to solve. The operation and maintenance of buildings has also benefited from more powerful computers that monitor, control, and, in certain cases, diagnose problems in HVAC equipment. This chapter discusses (1) alternatives for obtaining computing capability, (2) available computer hardware, and (3) computer programs or software for HVAC applications. ASHRAE (1990) lists many commercially available programs.

Throughout any discussion of computers, the terms hardware and software are used. They refer to the distinction between the physical equipment (hardware) and the programs or sets of instructions that direct the equipment to perform the desired tasks (software).

HARDWARE OPTIONS

Hardware options in computing may be examined by considering the classes of equipment that can be used: large (mainframes), intermediate (minicomputers and workstations), small (microcomputers and personal computers), and programmable calculators. These categories roughly indicate relative rankings in computing speed, number of simultaneous users, and cost.

Large computers require a substantial commitment to personnel who direct the use of the computers. Therefore, any organization with plans to use such equipment needs more expertise and guidance than is given here. On the other hand, a computer that someone else owns and maintains may be leased. The following two approaches may be considered:

1. Contract the work to a computer service bureau.
2. Work with a time-sharing service company.

The first method, contracting the work out-of-house, may be the easiest approach and the fastest to implement. The responsibilities and problems are placed on the computer bureau, there is no need to purchase expensive hardware and software packages, and no need to establish training programs for in-house personnel or to acquire experienced computer staff. In most cases, costs are high, but they can be included in the total project cost.

This method also has some disadvantages. In-house personnel gain no experience for future projects. The money spent is not recoverable for the next job, and the availability and time span of job turnaround is not readily controllable. A firm that does little work requiring computers and plans to perform only short-term projects finds this approach reasonable, but a firm performing medium to large projects and requiring skilled personnel to handle a large variety of tasks in a relatively short time is limited by this approach.

The second method, working with a time-sharing service, requires the purchase or lease of a minimum amount of hardware (a terminal and telephone coupler) and the use of a telephone-linked time-sharing network. As with the first approach, there is little or no first-cost expense involved. However, the responsibilities and problems are not transferred as with the computer bureau; the firm using the service does the greater part of the work. Most computer service firms supply some technical support in varying degrees. A total service provides consulting, programming, training, and documentation. In addition, interactive and batch processing are available, providing quick response and versatility, as well as economy and computer power. Probably the greater advantages to this approach are the accessibility of a large software library and the enormous computing power of a large mainframe computer at relatively low cost. There is no need to acquire computer specialists, update equipment and software, or contract the work to another firm.

However, there are some disadvantages to using this method. Training for in-house personnel to use the service efficiently is essential. It is necessary to allow enough time to become familiar with the services as well as the specific program(s) before they are actually used. In addition, keeping track of operating and storage charges is required to reduce hidden costs.

The costs involved in using a time-sharing service are related to the computer time used, plus any royalty fees for third-party programs and access fees to enhanced programs. The computer time is often taken as a function of time spent on the computer, type of computing done, size of memory space used, and operating expense of personnel and equipment. Other surcharges encountered are telephone connect time, charges for printed output, and storage charges for retaining information on discs or tapes. Because most time-sharing services charge separately for each resource, careful oversight can reduce costs substantially. Unnecessary connect time, on-line storage of unneeded files, interactive processing of large programs, and output to slow terminals can all escalate costs dramatically.

The preparation of this chapter is assigned to TC 1.5, Computer Applications.

Support services, such as consulting, training, and programming, are often considered part of the total service included in the computer time that is used (usually referred to as time and maintenance) as long as it is not too extensive. However, support services can result in additional cost, normally depending on the type or level of support. Such examples would include in-depth training sessions, specific modifications to software, and technical consulting on a specific program application or project. Normally, before those additional charges are assessed, an overall cost or stated rate is presented to the user.

A personal computer is often the most promising for a first-time user. The experience and knowledge gained from using these computers can later be applied to a larger computer system. A personal computer can also interface with the large mainframe computer of the time-sharing service. The personal computer system can operate at lower costs to provide interactive usage and storage for small tasks that a first-time user generally takes on. Data can be transferred for processing by a specific program available on a time-shared computer. This combination can increase the capabilities of a firm to handle a larger number and variety of projects than would be possible with time-sharing or in-house computing alone.

One use of time-sharing for which there is no substitute is the accessing of specialty data bases. The amount and diversity of information available on these systems is vast: data bases that contain information on corporate, financial, legal, medical, scientific, cultural, bibliographic, and many other types of data are readily available, although the cost of using the more specialized data bases is significant.

The user connects to the data base with a terminal or a microcomputer with a modem and terminal program, initiates a search for the information desired, and receives the result of the query. In many instances, the considerable cost of a specialized data-base system (to search for legal precedents, for example) is offset by the hours of expensive professional effort that is otherwise required.

In-House Mini and Microcomputers

The availability of inexpensive workstations and personal computers at prices within the range of any company has made in-house computing an attractive option for even the smallest firm. The obvious advantages of rapid turnaround and unlimited computing for a fixed investment are compelling. Time-sharing and service bureau costs appear to be sufficiently high so that an in-house system could be paid for out of savings.

What cannot be overlooked are additional costs incurred by having the computer in-house, such as maintenance, software, and personnel. Software must either be bought or developed, and as expensive as it may seem to buy, it is almost always more expensive to develop. Also, a firm should have at least one person knowledgeable about the system and the software.

The cost of a large computer system can be a burden on a small firm and is thus impossible to write off on the first few projects. In addition to hardware and maintenance costs, software purchases and maintenance add substantially to the total operating cost of a computer system. Replacing and updating hardware and software must also be accounted for in the overall costs. To justify an in-house computer system, a firm must use it frequently. However, with the increasing power and networking capabilities of personal computers, many of the functions previously performed only on a large mainframe can now be performed in-house.

Microcomputers or personal computers offer the small office enormous power at low cost. These machines should be considered in the class of business machines, in that they offer possibilities far beyond technical analysis (see sections Administrative Uses of Computers and Productivity Tools).

Table 1 summarizes benefits of the four computing approaches.

Programmable Calculators

Calculators exist that can be programmed to perform programs of hundreds of steps and can be outfitted with printers and readers. A range of commercial programs is available for these machines. Program listings may be found in engineering magazines and several manufacturers are also making programs available. Top-of-the-line programmable calculators have magnetic card readers; these machines can store user-created programs on cards and can read-in programs created by the user or by a vendor in machine readable form. Long programs can be chained; that is, the program may be on several cards, which are read in succession as each program

Table 1 Ways of Using Computers

	Computer Bureau	In-House Computer	Micro- or Personal Computer	Time-Sharing Service
Equipment required	—None—	Computer, terminal, printer, hard copier, storage devices	Computer, monitor, printer, storage (floppy or hard disc)	Terminal, modem
Personnel	—None—	Computer specialists, systems analysts, trained users	Trained users	Trained users
Program types	Usually specializing in a few programs	Dependent on availability of public domain programs or leased arrangements from private companies or locally developed programs	Wide variety for technical and business applications from software vendors, manufacturers, and user's groups	Public domain programs, third-party programs w/royalty fee, company program w/access fees, locally developed programs
Charges and costs	Large costs for entire work, personnel costs, computer charges, etc.	Low operating costs, high first cost, cost of maintenance and update to avoid obsolescence	Low operating cost, modest first cost and maintenance cost, cost of initial training, obsolescence occurs only when applications required cannot be accomplished on computer	Computer time costs, printing charges, telephone connect costs, storage charges
Type of user	Small firms seldom requiring projects with large amounts of data to be analyzed	Frequent users, large to medium size firms with several uses for computer system besides outside projects	Small to large firms	Frequent to nonfrequent users requiring access to large computing power of a mainframe or to extensive library of software programs
Advantages	Easiest approach for not being involved with computer technology	Variety of in-house uses, lowest operating costs	Low cost, fast turnaround time, can be used for many functions besides technical	Accessibility, support services

segment is completed. Some manufacturers also have program modules that plug into the calculator; others have optical bar code readers that can read printed programs.

Primary advantages of the programmable calculator are low cost and portability, although with a printer attached, the calculators are not nearly as portable. For field calculations, they are unexcelled, although they are being challenged by battery-powered, notebook-size laptop computers and even hand-held computers.

Disadvantages of some include the awkwardness of creating programs (they must be programmed in their own machine language) and the fact that they simply cannot handle the size of programs a computer does; nor do they work as fast as a computer. Thus, they represent an entirely new tool that encroaches little on the areas in which a computer excels.

DESIGN CALCULATIONS

Although computers are now widely used in the design process, most programs do not perform design, but simulation. That is, the engineer proposes a design and the computer program calculates the consequences of that design. When alternatives are easily cataloged, a program may design by simulating a range of alternatives and then selecting the best according to predetermined criteria. Thus, a program to calculate annual energy usage of a building requires a definition of the building and its systems; it then simulates the performance of that building under certain conditions of weather, occupancy, and scheduling. A duct-design program may actually size ductwork, but an engineer must still decide air quantities, duct routing, and so forth.

Because computers do repetitive calculations rapidly, accurately, and tirelessly, it is possible for the designer to explore a wider range of alternatives and to use selection criteria based on annual energy costs or life cycle costs, procedures much too tedious for wide usage without a computer.

Heating and Cooling Loads

The calculation of design thermal loads in a building is a necessary step in the selection of HVAC equipment for virtually any building project. Peak heating loads are usually calculated at steady-state conditions without solar or internal heat gains to ensure that heating equipment can maintain satisfactory building temperature under all conditions. This relatively simple calculation can also be calculated manually without a computer.

Peak cooling loads, however, are more transient than heating loads. Radiative heat transfer within a space and thermal storage cause the thermal loads to lag behind instantaneous heat gains and losses. This lag can be important, especially with cooling loads in that the peak is both reduced in magnitude and delayed in time compared to the heat gains that cause it. Early methods of calculating peak cooling loads tended to overestimate loads; this resulted in oversized cooling equipment with penalties of both first cost and part-load operating inefficiencies. Today various calculation methods are used to account for the transient nature of cooling loads. Some widely used methods of performing loads analysis for building elements include the following:

- Transfer Function Method (TFM)
- Cooling Load Temperature Difference/Cooling Load Factor (CLTD/CLF) Method
- Total Equivalent Temperature Differential/Time Averaging (TETD/TA) Method
- Steady-State Heat Transfer Method
- Response Factor Method
- Finite Difference Method

Chapter 26 of the 1989 ASHRAE *Handbook—Fundamentals* describes some of these methods in detail.

Both the TETD/TA and the TFM methods require a history of gains and loads. Since histories are not initially known, they are assumed zero and the building under analysis is taken through a number of daily cycles of weather and occupancy to establish a proper 24-h load profile. Thus, procedures required for the TFM and TETD/TA methods involve so many individual calculations that noncomputerized calculations are almost ruled out. The CLTD/CLF method, on the other hand, is meant to be a manual calculation method. The CLTD and CLF tables presented in the 1989 ASHRAE *Handbook—Fundamentals* are, in fact, based on application of the TFM to certain geometries and building constructions.

An automated TFM cooling load calculation is the preferred method of calculation. An automated TETD/TA method, however, can provide a good approximation with significantly less computational effort than the TFM. As a tabular form of the TFM, the CLTD/CLF method is intended to be applied as a manual method; a spreadsheet program can be used in its implementation.

Characteristics of a Loads Program. In general, a loads program requires user input for most or all of the following:

- Full building description, including the construction of the walls, roof, windows, etc., and the geometry of the rooms, zones, and building. Shading geometries may also be included.
- Sensible and latent internal loads due to people, lights, and equipment and their corresponding operating schedules.
- Indoor and outdoor design conditions.
- Geographic data such as latitude and elevation.
- Ventilation requirements and amount of infiltration.
- Number of zones per system and number of systems.

With this input, loads programs will calculate both the heating and cooling loads as well as perform a psychrometric analysis. Output typically includes peak room and zone loads, supply air quantities, and total system (coil) loads.

Selecting a Loads Program. In addition to general considerations such as hardware and software requirements, ease of use (including menu-driven versus command-driven), availability of manuals and support, and cost, some loads program-specific characteristics should be considered when selecting a loads program. Among the items to be considered are:

- Type of building to be analyzed—residential versus commercial. Residential-only loads programs tend to be simpler to use than the more general programs meant for commercial and industrial use. However, the residential-only programs are limited in their abilities.
- Method of calculation for the cooling load, as discussed previously in this section.
- Program limits on such items as number of rooms, zones, and systems and number of surfaces per room, etc.
- Sophistication of modeling techniques such as the capability of handling exterior or interior shading devices, tilted walls, and skylights, for example.
- Units of input and output: inch-pounds and/or SI.
- Complexity of program. In general, the more sophisticated and flexible programs require more input and are somewhat more difficult to use than the simpler programs.

Energy and System Simulation

Building energy simulation programs differ from peak loads programs in that in the former, loads are integrated over time (usually a year), the systems serving the loads are considered, and the energy required by the equipment to support the system is usually the end result. Most energy programs simulate the performance of systems already designed, although programs are now available that make selections formerly left to the designer, such as equipment sizes, system air volume, and fan power. Energy programs are fundamental in making decisions regarding building energy use, and along with life cycle costing routines, quantify the impact of proposed energy conservation measures in existing buildings. In new building design, energy programs aid in determining the type and size of building systems and components; they also explore the effects of design tradeoffs and can be used to evaluate the benefits of innovative control strategies and new equipment.

Energy programs that track building energy use accurately can help in determining whether a building is operating efficiently or wastefully. They have also been used to allocate costs from a central heating/cooling plant among customers of the plant.

Characteristics of Building Energy Simulation Programs. Because energy, as a time-integrated quantity, is much less dependent on instantaneous occurrences than are design loads, many techniques for calculating loads are employed. Most programs simulate a wide range of building, mechanical equipment, and control options.

While the importance of approximating solar loads on cloudy days is widely recognized, computational results differ substantially from program to program. The shading effect from overhangs, side projections, and adjacent buildings is frequently a factor in energy consumption of buildings; however, the diversity of approaches to the problem results in a wide range of answers.

Depending on the requirements of each program, various weather data are used. These can be broken down into four groups:

1. Hourly data for one year only
2. Hourly data for one year, as well as design conditions for typical design days
3. Reduced weather, commonly a typical day or days per month for the year
4. Reduced weather, nonserial or bin format

Space temperature variations can also be taken into account. When the space temperature is not constant, the rates of heat extraction and space temperatures, rather than cooling and heating loads, are calculated. Temperature variations occur due to temperature control systems' response requirements when building systems are shut off and when loads are greater than system capacity.

Heat extraction is the rate at which the HVAC system removes heat from a conditioned space. This rate equals the cooling load when the space temperature is kept constant; since this rarely happens, heat extraction is generally either smaller or larger than the cooling load. This concept is important for the analysis of intermittently operated HVAC systems; it provides information on the relationship between load and space temperature and leads to the calculation of preheat/cool load for various combinations of equipment capacity and preheat/cool periods.

Both air-side and energy conversion simulations are required to handle the wide variations among central heating, ventilation, and air-conditioning systems. System simulations are performed independently for each combination of system design, operating scheme, and control sequence for proper estimation of energy use.

System Simulation Techniques. Two basic approaches are currently used in computer simulation of energy systems—the fixed schematic technique and the component relation technique.

The fixed-schematic-with-options technique is the most prevalent program organization. Used in the development of the first generally available energy analysis programs, this technique involves writing a calculation procedure that defines a given set of systems. The system schematic is then fixed, with the user having some options that are usually limited to equipment performance characteristics, fuel types, and the choice of certain components.

Component Relation Technique. Advances in system simulation and increased interest in special and innovative systems have prompted interest in the component approach to system simulation. The component relation technique differs from the fixed schematic in that it is organized around components rather than systems. Each of the components is described mathematically and placed in a library for use in constructing the system. The user input includes the definition of the system schematic in addition to equipment characteristics and capacities. Once all the components of a system have been identified and a mathematical model for each has been formulated, the components may be connected together in the desired manner and information transferred between them. Although certain inefficiencies are built into this approach because of its more general organization, this technique does offer versatility in defining system configurations.

Selecting an Energy Program. In selecting an energy analysis program, factors such as cost, availability, ease of use, technical support, and accuracy are important. There is, however, another fundamental consideration—whether the program will do what is required of it. It should be sensitive to the parameters of concern, and its output should include the necessary data. For other considerations, see Chapter 28 of the 1989 ASHRAE *Handbook—Fundamentals*.

In the past, computer simulation technology concentrated on thermal load calculation techniques. System simulation was limited to a number of common systems. Simulation development in the future will likely focus on system assimilation techniques, *i.e.*, the means by which any system can be configured, given the mathematical models of its components.

Comparisons of Energy Programs. Since most energy analysis computer programs take different approaches in their method of calculation, significant differences in the results can be found. Many comparisons, verifications, and validations of simulation programs have been made and reported (see Bibliography). Conclusions from these reports can be summarized as follows:

- The results obtained by using several computer programs on the same building range from good agreement to no agreement at all. The degree of agreement depends on the interpretations of the program user and the ability of the computer programs to model the building.
- Several people using several programs on the same building will probably not agree on the results of an energy analysis.
- The same person using different programs on the same building may or may not have good agreement, depending on the complexity of the building and its systems and on the ability of the computer programs to handle specific conditions in that building.
- "Forward" computer simulation programs that calculate the performance of a building given a set of descriptive inputs, weather conditions, and occupancy conditions are best used for design purposes.
- Calibration of hourly forward computer simulation programs is possible but can require considerable effort for a moderately complex commercial office building. Detailed information concerning the scheduled use, equipment set points, and even certain on-site measurements may be necessary for a closely calibrated model. Special-purpose graphic plots are useful in calibrating a simulation program with data from monthly, daily, or hourly measurements.
- "Inverse," empirical, or system parameter identification models may also be useful in determining characteristics of building energy usage. Such models determine the relevant building parameters from a given set of actual performance data—the inverse approach from that of traditional building modeling, hence the name.

Energy Programs for Modeling Existing Buildings. Computer energy analysis for existing buildings can handle complex situations, evaluate many alternatives, evaluate concepts that could have positive and/or negative energy impacts, and predict the relative magnitude of energy use.

Once simulation of a building has been performed, several alternatives may be evaluated, such as changes in control settings, occupancy, and equipment performance. Many programs are available with widely varying costs, degrees of complexity, and ease of use (Degelman 1987).

A general procedure should be followed in using computer programs for existing buildings. First, energy consumption data must be obtained for a one- to two-year period. These data usually consist of metered electrical energy consumption and demand on a month-by-month basis, as a minimum. For natural gas, the data are in a form similar to those for electricity and are almost always on a monthly basis. For other types of fuel, such as oil and coal, data are available by the delivery.

Unless fuel use is metered or measured daily or monthly, consumption for any specific period less than one season or year is difficult to determine. The data should be converted to a per-day usage or adjusted to account for differences in the length of metering periods. The data tell how much energy went into the building on a gross basis. Unless extensive submetering is used, it is nearly impossible to determine when and how that energy was used and what it was used for. Such meters may need to be installed to determine energy use.

Thermal and electrical characteristics of a building and its energy-consuming systems as a function of ambient conditions, time, and occupancy must also be determined. Most computer programs can use as much detailed information as is available on the building and its mechanical and electrical system. Where the energy implications of these details are significant, it is worth the effort of obtaining them. If time and resources permit, testing fan systems for air quantities, pressures, control set points, and actions can provide valuable information on deviations from design conditions. Test information on pumps can also be useful.

Building occupancy is among the most difficult information to obtain. Since most energy analysis computer programs simulate the building on an hourly basis for a one-year period, it is necessary to know how the building is used for each of those hours. Frequent observations of the building during days, nights, and weekends show what energy-consuming systems are being used and to what degree.

Weather data, usually one year of hour-by-hour weather data, is necessary for simulation (see Chapter 24 of the ASHRAE *Handbook—Fundamentals*). Also, if available, the actual weather data for the year for which energy consumption data are available may be chosen. Where the energy-consuming nature of the building is related more to internal than to external loads, the selection of weather data is relatively unimportant; however, with residential buildings or buildings with large outside air loads, the selection of weather data can affect results significantly. The purpose of making the simulation should also be considered when choosing weather data; *i.e.*, either specific year data, data representative of long-term averages, or data of temperature extremes may be needed, depending on the end use of the simulation.

It is likely that the results of the first computer runs will not agree with the actual metered energy consumption data. Possible reasons for this are:

- Insufficient detail for the energy-consuming systems that create the greatest use.
- Information on the occupancy and time of building use is not accurate.
- Design information on air quantities, set points, and control sequences is not appropriate.

The input data must be adjusted and trial runs continued to match the actual energy use. It is usually difficult to match the metered energy consumption precisely. Results within 10 to 20% in any month are considered adequate using the aforementioned procedures. It is more important to match the character of the energy consumption on an hourly and monthly basis than it is to match the annual total precisely.

Having a simulation of the building as it is being used permits subsequent computer runs to evaluate the energy impact on various alternatives or modifications. The evaluation may be accomplished simply by changing the input parameters and rerunning the program. An evaluation may then be made of the various alternatives, so that an appropriate one may be selected.

Duct Design

Two major needs exist in duct design: sizing and flow distribution. Duct sizing and equipment selection are required for a new duct system design. Flow distribution is a problem of calculating flows through the duct sections and terminals for an existing system with known cross sections and fan characteristics.

Duct Sizing. Two major approaches are followed for computerized duct sizing: (1) application of manual procedures, which, although computerized, are still limited in capability, and (2) optimization. No available commercial programs use optimization techniques.

Selecting and Using a Program. Duct design involves laying out ductwork, selecting fittings, and sizing the ducts. Computer programs comply with many duct system constraints that require recomputation of the duct size. Computer printouts provide detailed documentation. Any calculation requires preparing accurate estimates of pressure losses in duct sections and defining these interrelations of velocity heads, static pressures, total pressures, and fitting losses.

The general computer procedure is to designate nodes (the beginning and end of duct sections) by number. Details about each node (*e.g.*, divided flow fitting and terminal) and each section of duct between nodes (maximum velocity, flow rate, length, fitting codes, size limitation, insulation, and acoustic liner) are used as input data (Figure 1).

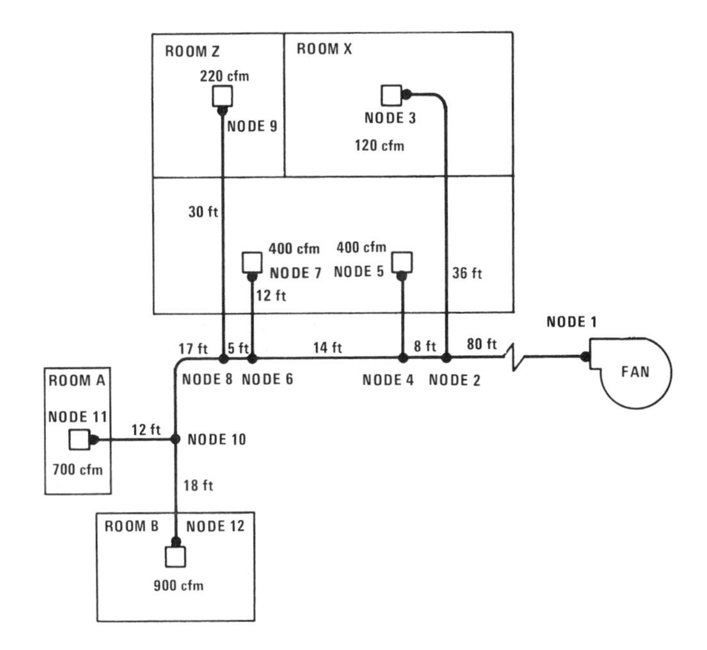

Fig. 1 Example of Duct System Node Designation

Some characteristics of a duct design program include the following:

- Calculations for supply, return, and exhaust systems
- Sizing by constant friction, velocity reduction, static regain, and constant velocity methods
- Analysis of existing duct systems
- Inclusion of fitting codes for a variety of common fittings
- Selection of duct run with the highest pressure loss, and tabulation of all individual losses in each run
- Printout of all input data for verification
- Provision for error messages
- Calculation and printout of airflow for each duct section
- Printout of velocity, fitting pressure loss, duct pressure loss, and total static pressure change for each duct section
- Printout of a schematic or line diagram indicating duct size, shape, flow rate, and temperature in the duct system
- Calculation of heat gain/loss in the system and correction of temperatures and flow rates, including possible resizing of the system
- Specification of maximum velocities, size constraints, and insulation thicknesses
- Consideration of insulated or acoustically lined duct
- Bill of materials for sheet metal, insulation, and acoustic liner
- Acoustic calculations for each section of the system

Since many duct design programs are available, the following factors should be considered in program selection:

- Maximum number of branches that can be calculated
- Maximum number of terminals that can be calculated
- Types of fittings that can be selected
- Number of types of fittings that can be accommodated in each branch
- Ability of the program to balance pressure losses in branches
- Ability to size a double-duct system
- Ability to handle draw-through and blow-through systems
- Ability to prepare cost estimates
- Ability to calculate fan motor power
- Provision for determining acoustical requirements at each terminal
- Ability to update the fitting library

Optimization Techniques for Duct Sizing. Optimized duct design selects fan pressure and duct cross sections by minimizing an objective function. This function is the life cycle cost of the system, which includes the initial cost and energy cost. A large number of constraints, including constant pressure balancing, acoustic restrictions, and size limitations, must be satisfied. Duct optimization is a mathematical programming problem with a nonlinear objective function and many nonlinear constraints. The solution must be taken from a set of standard diameters and standard equipment. Several numerical methods for duct optimization exist, such as the T-Method (Tsal *et al.* 1988), Coordinate Descent (Tsal and Chechick 1968), Lagrange Multipliers (Stoecker *et al.* 1971, Kovarik 1971), Dynamic Programming (Tsal and Chechick 1968), and Reduced Gradient (Arklin and Shitzer 1979). Unfortunately, these methods have not been incorporated into commercial programs.

Flow Distribution. Another problem is the prediction of airflows in each section of a presized system with known fan characteristics. This is called the flow distribution or air duct simulation problem. The need to calculate flow distribution in a duct system occurs whenever a retrofit to an existing duct system is considered. An HVAC engineer may then ask the following questions:

- How will this influence the flow at existing terminals?
- Is it possible to change only the motor and leave the same fan?
- What is the new working point on the fan characteristic?
- Which ducts should change in size?
- What are the new duct sizes and what are the flows in the system with fully opened dampers?
- What is the best way to connect additional diffusers to an existing system?

A simulation program can help answer these questions. In addition, a simulation program can analyze the efficiency of a control system effectively, check the effect of the performance of a number of parallel fans if one is not running, and predict the flows during field air balancing. The T-Method (Tsal *et al.* 1988) and the Gradient Steepest Descent Method (Tsal and Chechik 1968) have been used for simulating a duct system.

Piping Design

A large number of computer programs exist to size piping systems or to calculate the flexibility of piping systems. The sizing programs normally size piping and estimate pump head for systems based on velocity and pressure drop limits, and some consider heat gain or loss from piping sections. Several programs produce bill of materials or cost estimates for the piping system. Piping flexibility programs assist in a stress and deflection analysis of piping systems. Many of the piping design programs can handle thermal effects in pipe sizing, as well as deflections, stresses, and moments.

The general technique for computerizing piping design problems is to represent the three-dimensional system as a set of nodes and links. Each takeoff tee or terminal is a node; each set of nodes is linked by pipe and fittings. A typical piping problem in its nodal representation is shown in Figure 2.

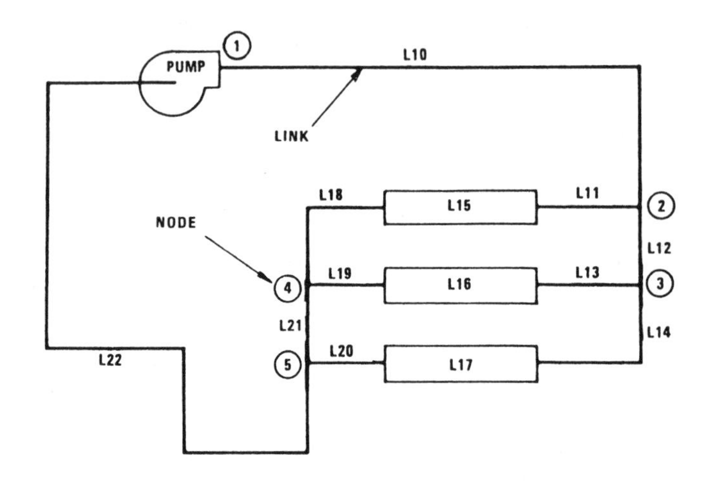

Fig. 2 Nodal System for Piping System

Useful piping programs provide sufficient design information. They also:

- Perform calculations for both open and closed systems
- Calculate the flow, pipe size, and pressure drop in each section of the system
- Handle three-dimensional piping systems
- Provide a wide selection of commonly used valves and fittings, including such specialized types as solenoid and pressure-regulating valves
- Consider different piping materials such as steel, copper, and plastic by including generalized friction factor routines
- Accommodate liquids, gases, and steam by providing property information for a multiplicity of fluids
- Calculate pump capacity and head required for liquids
- Calculate the available terminal pressure for nonreturn pipe systems

- Calculate the required expansion tank size
- Estimate heat gain/loss for each portion of the system
- Prepare a system cost estimate, including costs of pipe and insulation materials and associated labor
- Print out a bill of material for the complete system
- Calculate balance valve requirements
- Perform a pipe flexibility analysis
- Perform a stress analysis for the pipe system
- Print out a graphic display of the system
- Allow customization of specific design parameters such as maximum and minimum velocities, maximum pressure drops, and design conditions such as condensing temperature, superheat temperature, and subcooling temperature
- Allow evaluation of piping systems for off-design conditions
- Provide links for calling by other programs, such as equipment simulation programs

Some limiting factors to consider in piping program selection include the following:

- Maximum number of terminals the program can accommodate
- Maximum number of circuits the program can handle
- Maximum number of nodes each circuit can have
- Maximum number of nodes the program can handle
- Compressibility effects for gases and steam
- Provision for two-phase fluids

Acoustic Calculations

Chapter 42 presents a concise summary of sound generation and attenuation in HVAC systems. Application of these data and methodology to the analysis of noise in HVAC systems is straightforward, but it is complex in the amount of computation involved. All sound generation mechanisms and sound transmission paths are potential candidates for analysis. Compounding the computational work load is the necessity of extending the analysis over, at a minimum, octave bands 1 through 8 (63 Hz through 8 kHz). Thus a computer can save a great amount of work in the analysis of any noise situation, not only in HVAC systems. However, the HVAC system designer should be wary of using unfamiliar software.

Caution and critical acceptance of analytical results is mandatory at all frequencies, but particularly at low frequencies. Many manufacturers of equipment and sound control devices do not provide data below 125 Hz. Thus, the HVAC system designer conducting the analysis and the programmer developing computer software must make assumptions based on experience in these critical low-frequency ranges.

The designer/analyst should be well satisfied if predictions are within 5 dB of field-measured results. Results within 10 dB in the low-frequency "rumble" regions are often as accurate as can be expected, particularly in areas of fan discharge. Thus, conservative analysis and application of results is necessary, especially if the acoustic environment of the space being served is critical.

Both manual and computerized computations relating to noise in HVAC duct systems should not concentrate only on the obvious noise generators such as fans; they should also consider the potential for flow generation in duct elements (*e.g.*, tees and branches) and downstream elements (VAV boxes, etc.). Whenever flow velocities exceed 1500 fpm, flow-generated noise is a significant possibility.

Sound in ducts propagates upstream as well as downstream and sound from the fan of a single system travels down at least four paths (Figure 3):

- Fan discharge sound travels down the supply duct through the diffusers into the space (path A).
- Fan discharge sound can break out of the supply duct. This path is significant for occupants within 50 ft of the fan discharge (path B).

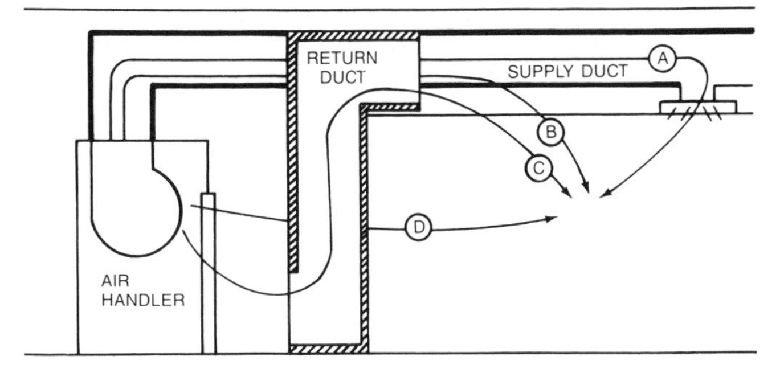

Ⓐ Discharge Airborne Sound
Ⓑ Discharge Breakout
Ⓒ Inlet or Return
Ⓓ Radiated

Fig. 3 Significant Indoor Air Handler Sound Paths

- Fan inlet sound travels down the return duct, through the return grilles or ceiling and into the occupied space (path C).
- Finally, radiated fan sound can penetrate the unit casing and mechanical room wall and enter the occupied space (path D).

The resulting sound pressure level in a space is the composite of all sound paths to the space. For example, one diffuser in a room may meet NC-30, but the combined sound pressure level of ten in a room may be NC-40 or greater.

Several currently available acoustics programs are generally easy to use but are often less detailed than the custom programs developed by acoustic consultants for their own use. Acoustics programs are designed for comparative sound studies and allow the system designer to design a comparatively quiet system. Acoustic analysis by hand or computer should address the following key areas of the HVAC system:

- Sound generation by HVAC equipment
- Sound attenuation and regeneration in duct elements
- Wall and floor sound attenuation
- Ceiling sound attenuation
- Sound break-out or break-in of ducts or casings
- Room absorption effect (relation of sound power criteria to sound pressure experienced)

Algorithm-based programs are preferred because they cover more situations (see Chapter 42). Conversely, assumptions are an essential ingredient of algorithms. These basic algorithms plus sound data from the acoustics laboratories of equipment manufacturers are incorporated in acoustics programs to various degrees. HVAC equipment sound levels used in acoustics programs should come from the manufacturer and be based on measured data, since similar equipment has a wide variation in sound. Some generic equipment sound generation data, which may be used as a last resort in the absence of specific measured data, are found in Chapter 42. Whenever possible, equipment sound power by octave band (including 32 Hz and 63 Hz) should be obtained for the path under study. A good sound prediction program relates all performance data so that it creates reasonably accurate predictions.

Many other nongeneric acoustics programs are available. Various manufacturers provide equipment selection programs that not only make optimum equipment selection for a specific application, but also provide associated sound power data by octave bands. These programs relate to a specific application or item of equipment and can help in the design of a specific aspect of a job. Data from these programs should be incorporated in the analysis. For example, duct design programs may contain sound predictions for discharge airborne sound based on the discharge sound power of the fans, noise generation/attenuation of duct fittings, attenuation and end reflections of VAV terminals, attenuation of ceiling tile, and room effect. VAV terminal selection programs generally contain subprograms that estimate space NC level near the VAV unit in the occupied space. However, projected space NC levels alone are not acceptable substitutes for octave-band data. The designer/analyst should be aware of assumptions (such as room effect) made by the manufacturer in the presentation of acoustical data.

Acoustic predictive software allows system designers to look at HVAC-generated sound in a realistic, affordable time frame, thus making it part of the services offered to building owners. HVAC-oriented acoustic consultants generally assist designers by providing cost-effective sound control ideas in sound-critical applications. A well-executed analysis of the various components and sound paths enables the designer to assess the relative importance of each and to direct corrective measures, where indicated, to the most critical items. Computer-generated results should supplement the designer's skills, not replace them.

Equipment Selection and/or Simulation

Three types of equipment-related computer programs include equipment selection, equipment optimization, and equipment simulation.

Equipment selection programs are basically computerized catalogs. The program locates an existing equipment model that satisfies entered criteria. The output is a model number, performance data, and may include alternative selections.

Equipment optimization programs select all possible equipment alternatives and then let the user establish ranges of performance data or first cost to narrow the selection. The user keeps narrowing the performance ranges until the best selection is found. Performance data used for optimizing selections varies by product family.

Equipment simulation programs select specific equipment and calculate full and part-load performance over time, generally one year. The equipment performance is matched against an equipment load profile to determine energy requirements per hour. Utility rate structures and related economic data are then used to project equipment operating cost, life cycle cost, and comparative payback.

Some advantages of equipment programs include the following:

1. High speed and accuracy of the selection procedure
2. Pertinent data presented in an orderly fashion
3. More consistent selections than with manual procedures
4. More extensive selection capability
5. Multiple or alternate solutions
6. Small changes in specifications or operating parameters easily and quickly evaluated

Simulation programs have the advantage of (1) projecting part-load performance quickly and accurately, (2) establishing minimum part-load performance, and (3) projecting operating costs and payback-associated higher performance product options.

Programs have been written for nearly every type of HVAC equipment. The more common programs and their optimization parameters include the following:

Air distribution units	Pressure drop, first cost, sound, throw
Air-handling units, rooftop units	Power, first cost, sound, filtration, heating and cooling capacity
Boilers	First cost, efficiency, stack losses
Cooling towers	First cost, design capacity, power, flow rate, air temperatures
Chillers	Condenser head, evaporator head, capacity, power input, first cost, compressor size, evaporator size, condenser size
Coils	Capacity, first cost, water pressure drop, air pressure drop, rows, fin spacing
Fan coils	Capacity, first cost, sound, power
Fans	Volume flow, power, sound, first cost, minimum volume flow
Heat recovery equipment	Capacity, first cost, air pressure drop, water pressure drop (if used), effectiveness
Pumps	Capacity, head, impeller size, first cost, power
Air terminal units (variable and constant volume flow)	Volume flow rate, air pressure drop, sound, first cost

Some extensive selection programs have evolved. For example, coil selection programs can select steam, hot water, chilled water, and refrigerant (direct expansion) coils. Generally, they select coils according to procedures in ARI *Standards* 410 and 430 (ARI 1981, 1985).

Chiller and refrigeration equipment selection programs can choose optimal equipment based on such factors as lowest first cost, highest efficiency, best load factor, and best life cycle performance. In addition, some manufacturers have modular equipment that is arranged for customization of their product. This type of equipment is ideal for computer selection.

However, equipment selection programs have limitations. The logic of most of the programs is proprietary and not available to the user. All programs have built-in approximations or assumptions, some of which may not be known to the user. Qualify equipment selection programs before using them.

ADMINISTRATIVE USES OF COMPUTERS

Word Processing

The first application of microcomputers to gain almost universal usage was word processing, which creates documents electronically rather than physically, as in typing. As such, editing text is easier and less expensive, since the entire document does not have to be retyped with each draft. The documents are usually stored on magnetic disc, although optical disc storage systems with vast capacities are now becoming available.

Text can be entered continuously without the need to type carriage returns (word wrap); it can be corrected or reformatted; and it can be moved, copied, and deleted in blocks. Heading and footing text can be specified for all pages in a report. Global search and replace functions locate and replace words or phrases wherever desired in a document, sometimes even reformatting paragraphs where the changes took place.

Advanced word processors can do arithmetic with columns of figures, renumber paragraphs to match an outline, and organize notes, either as footnotes on the page of the reference or as endnotes at the end of the document. It is also possible to incorporate images from graph or drawing programs into printed output.

Word processing software is available for almost all computers. In addition, many companies offer dedicated word processors—computer-like machines that are optimized for word processing functions. The distinctions between general purpose computers and dedicated word processors are becoming less clear as word processing programs are written as look-alikes to dedicated word processors and the dedicated word processors are supplied with arithmetic and accounting capabilities.

Commonly found additional functions are as follows:

Mail-merging, the ability to insert text, such as names and addresses, from one file into a form letter; thus, everyone on the list receives an individualized copy of the letter.

Spelling checking, in which words in the text are compared with a master dictionary and errors are flagged; this may be done either after completing a document or as the text is entered, depending on the program. Similarly, on-line thesaurus programs allow the user to ask for a list of equivalent words. Most spelling checkers allow the addition of specialty words to the main standard dictionary or to an auxiliary dictionary for future reference.

Grammar and style checking, in which long sentences, excessive use of passive voice, incomplete sentences, incorrect choice of homonyms (*e.g.*, there, their, they're), improper capitalization, unpaired punctuation (*i.e.*, lack of closing quotation marks or parentheses), lack of subject-verb agreement, and many other potential flaws are flagged. If these programs have a fault, it is in flagging too many things; they can often bog the user down with excessive citations. Nonetheless, they can give valuable guidance.

Indexing, in which programs can create indexes or tables of contents from specially marked section and paragraph headings in the text.

Specification Writing

The efficiency and flexibility of a computerized master specification in conjunction with a word processor make it a valuable tool for preparing project specifications. Those who use the master specification should standardize language to avoid repetitive editing of text. The need to minimize extensive rewriting and the insertion of new material makes a master specification most beneficial to firms that do not specialize in unique, one-of-a-kind designs. If the master can be used with only minor revisions for the majority of specifications, the cost and time spent in editing may be justifiable. Generally, the larger the size and the greater the number of documents produced, the greater the savings in time and money compared to traditional preparation and typing.

Even without a master specification, however, it is possible for the writer and typist to save considerable time by "cutting and pasting" from previous specifications with minor revisions.

Desktop Publishing

The term desktop publishing usually refers to the creation of documents that integrate text with graphics. Many high-end word processors have the ability to import graphic images and to print in decorative typefaces, but the full range of abilities to arrange page layouts on a graphics screen are only found with desktop publishing programs.

Most desktop publishing programs are limited in their ability to create text (they normally import text created by a word processing program) and are also limited in the ability to create drawings (they normally import images from separate drawing programs).

The advent of high-resolution graphic displays for personal computers and the laser printer have allowed the creation and printing of high-quality graphic images and has been the impetus for desktop publishing.

Management Planning and Decision Making

Computers are ideal for such repetitive operations as inventory control, accounting, purchasing, and project control, providing managers with up-to-date information. The ability to store and manipulate large amounts of data provides a powerful technique for forecasting. With sufficient historical data, such concerns as sales, inventory needs, production capability, manpower needs, and capital requirements may be forecast using statistical methods. Econometric modeling may be included to help forecast the effects of varying inflation rates, for example.

Much of the data required to implement some of the aforementioned techniques may not be in the company's files and must be found elsewhere. Technical papers, market trends, industry reports, financial forecasts, codes, standards, regulatory data, and demographic or geographic information are all examples of the information needed for engineering management that can be searched for electronically by computer. In most cases, only a local phone call must be made, and the company is charged a time-sharing fee for time spent on the system. Management information systems should also have capabilities that provide special or supporting information should it become necessary to delve deeper into an operation.

Employee Records and Accounting

All corporate accounting functions from payroll, aging accounts receivable, to taxes can be automated on even the smallest microcomputers with a wide choice of engineering-specific accounting software. In addition, a firm that has unique accounting requirements may often meet them by developing specific applications called templates, or overlays, for common spreadsheet programs. Continuously updated financial reports needed for project evaluation and government use can be handled more readily by computer than by manual methods or off-site accounting services. Labor expense for various projects or departments can easily be separated and monitored on a regular basis. Personnel records for each employee can also be maintained on a computer.

Project Scheduling and Job Costing

The success of any project, no matter what its size, depends on good scheduling and control, so that manpower and materials will be where they are supposed to be at the proper time. Whenever portions of a project can be done simultaneously, are dependent on the completion of one or more prior parts, or must share a limited amount of manpower and equipment, a project scheduling technique such as CPM (Critical Path Method) or PERT (Program Evaluation and Review Technique) can determine which portions are critical. These techniques are well implemented on microcomputers. The current microcomputer programs can also perform manpower leveling to make the best use of available resources, and some may provide graphics suitable for client presentations as well.

Computerized scheduling techniques commercially available quickly provide updated schedules and critical items, even if the critical path does change due to a delay in one part of the job. This capability is valuable for determining if scheduled completion dates and costs can be met and for determining possible alternatives. Once a computerized system of personnel, payroll, and accounting functions has been established, a history of past project costs, manpower requirements, and material requirements can be obtained. A computer system can categorize and tabulate them for future reference, permitting more accurate estimates of future projects.

Security and Integrity

While computers are generally reliable, provision must be made for potential loss of important data in the event of equipment failure, theft, sabotage, natural catastrophe, or a previously undiscovered problem in the software. The most common method of protection is periodic duplication of master tape or disc files. For the most important data files, such as the firm's accounts, this duplication should take place with every posting. These duplicate files should be kept in a secure area, such as a fireproof vault on the premises or a totally off-site storage.

Unauthorized access to particular data is another security concern. Implementing a formal, written policy of data security, no matter how minimal it is initially, is a critical step in the computerization of any firm. Making certain that only authorized personnel have access to confidential files requires careful planning. Most major accounting packages, as well as many popular spreadsheet and data-base management programs, contain built-in safeguards such as multilevel password protection to avoid this problem.

PRODUCTIVITY TOOLS

Spreadsheet Software

One important type of microcomputer program routinely used by managers and engineers is the spreadsheet. Spreadsheets are two-dimensional or three-dimensional electronic tables containing thousands of cells in which the user may enter labeling text, numbers, formulas, or even macro-commands that can transfer control to another program. Automatic recalculation of all cells with formulas following data entry keeps all areas of the spreadsheet current. Some advantages of spreadsheets include:

- Rapid assembly and testing of a chain of equations
- Preformatted output/input and "template" structures
- Prepackaged or "canned" graphics
- Easy file transfer between spreadsheets or between different proprietary spreadsheet programs
- The ability to sort lists of information
- Matrix and statistical computations
- Additional programs with special features that "add-on" to the existing features of a spreadsheet

Spreadsheets also allow for the numeric and alphanumeric data to be imported or exported from other nonspreadsheet programs. Graphs from spreadsheets can also be imported into sophisticated graphics presentation programs for additional editing and preparation.

Graphics Software

Microcomputer software graphics packages now readily produce X-Y graphs, bar and pie charts, and CAD drawings and often have a library of graphic elements that allow for rapid custom presentations. Graphics files can be imported and exported and incorporated into desktop publishing documents. A host of specialty programs now allow for graphics files to be transferred between different proprietary application programs and even allow for digitized images (from photographs) to be imported into an existing graphics drawing.

Communications Software

Electronic communication is also an evolving technology. Communications between computers can take place through telephone lines or through networks of various configurations. Such networks can use coaxial cables, twisted pair wiring, superimposed A/C carrier frequencies, fiber optics, radio transmissions, infrared light, or even the ionized trails of meteors (*i.e.*, meteor bursts). Features available on such networks include bulletin boards, on-line retrieval services, shared services, electronic mail, and file transfer.

Electronic communication using existing telephone lines is rapidly becoming possible anywhere there is telephone service, even through overseas exchanges or with cellular phones. Although telephone systems are being upgraded to provide for digital capabilities, which will vastly expand capabilities, it will take many years before this is completed throughout the United States (Bellamy 1982).

Electronic communication continues to require both appropriate software and a modem, which performs the translation of the digital computer signals into analog signals capable of being transferred by current telephone lines. The modems most commonly used at present transfer data at 30, 120, or 240 characters per second using one of several standards for transmission.

Communications programs, at the least, allow messages typed at the keyboard to be sent to a remote computer, allow the user to specify a file in a local storage to be sent to the remote computer, or allow the user to receive a file from the remote computer and save it locally. Other possible capabilities include executing error-checking protocols with the remote computer to eliminate transmission errors and allowing the remote computer to take control of the local one, sometimes referred to as terminal emulation.

Data-Base Software

Microcomputer data-base systems allow the user to record, organize, and store information. Most of these systems originated in the 1970s relational data bases, which allow users to store and retrieve data from a data base that is a collection of relations or tables (Date and White 1989). Such systems can be viewed as computerized record-keeping systems (Date 1988). Most languages include a query-based command language, as well as a programming language, which can be used to create applications. Some advanced programs allow for query-by-example, windowing, or even an icon-driven user interface.

One example of an application could be a client list, containing the following for each client: firm name, street address, city, state, mail code, telephone number, contact person, dates of contact, and links to others in the list; each "record" in the data-base file would contain the above information for one client. Data-base programs allow for very large lists of records to be alphabetized, selected by specific index (*i.e.*, city, state, etc.), and printed onto mailing labels or incorporated into a report.

Special Purpose Software

Almost daily, new types of special purpose software are created, some specific to the HVAC industry. Some programs, which coexist with other programs in the computer memory, provide the following functions: notepads; appointment calendars; battery charge (in laptop PCs); name and address files; an alarm clock; binary, octal, and hexadecimal calculators; telephone dialers and on-line dictionaries; spell checkers; thesaurus; and "last command" recall. Specialty programs also include file organizers, routine backup programs, file restoration programs, and hardware diagnostics. Many software packages integrate word publishing programs, spreadsheets, communication programs, and graphic programs.

A memory resident (RAM) electronic data manager that contains over 300 HVAC engineering tables taken directly from the 1989 ASHRAE *Handbook—Fundamentals* has been developed (ASHRAE *Access* 1989). Many other special purpose HVAC application programs exist, including proprietary equipment selection packages, microcomputer versions of the DOE-2 program, BLAST program, TRNSYS program, and other powerful simulation programs previously available only on large mainframe computers.

Many non-HVAC advanced analysis programs are also available, including statistical and mathematical packages, specialty artificial intelligence or knowledge-based system development shells, and neural network development programs.

Advanced Input and Output Options

With the rapid expansion of input and output options has come advanced input capabilities, including digitizing of images and text, scanners and bar code input, downloading of data directly from portable field instruments, and downloading of information from facsimile (fax) machines.

Traditionally, only text and numbers have been stored and manipulated in computers. Internally, each character of text can be represented by a single byte of storage. Graphic images are more difficult to store and manipulate. Representation can be in bitmap or in vector format. In the bitmap or raster representation, the image is decomposed into points in a rectangular array (pixels). Storage requires a minimum of one bit per pixel for monochrome images; color requires several bits per pixel. In vector representation, only end points of line segments are stored (as numbers). While output of bitmap images is simple on raster devices such as graphic printers, scaling of images does not always produce satisfactory output. Vector representations plot easily on pen plotters and scale to a high degree of accuracy, but must be converted to raster format for output on graphic printers.

CAD and other drawing systems use a vector representation, which allows easy manipulation. Paint programs use bitmap images. Scanners that scan an image on paper and produce a raster representation (text and images are both in bitmap format) are now available. Paint programs can take scanned images directly and manipulate them. It is possible to scan drawings, representing them in a raster format, and then use conversion software to change the raster representation into vector form for use in CAD systems. Similarly, Optical Character Representation (OCR) software exists to take a raster image of text and convert it to individual text characters in the computer for input to word processing programs.

Facsimile equipment scans input images into a raster representation and sends the images bit-by-bit over telephone lines to a receiver, which prints them out as raster images. Fax boards on personal computers allow internal text to be converted to raster format and sent without first converting it to an image on paper. Similarly, computers can capture incoming fax images, allowing either conversion to internal text form or manipulation of the images.

Just as input capabilities have improved, so have output options, including the porting of graphic routines to VCR tapes for assembly into training tapes, direct 35-mm slide production, and direct transmission of output via fax machine. Newly available hardware and specialty software for laptops allow the user to preassemble and display an animated electronic slide show, digitized images, or even CAD drawings using a special translucent output screen that is placed on top of a traditional overhead projector. Certain add-in cards even allow the viewing of a television image (in a separate window) while working on a document or spreadsheet. Such an application might be the precursor of a combined video-energy management and control system (EMCS).

COMPUTER-AIDED DESIGN AND DRAFTING

Capabilities

Computer-aided design and drafting (CAD) systems give designers computerized tools for the basic drafting of HVAC contract and shop drawings, and also simplify HVAC analytical and design tasks. Further, computer graphics of buildings and their mechanical and other systems help coordinate interdisciplinary design and simplify modifications. The distinction is sometimes drawn between CAD (Computer-Aided Drafting) and CADD (Computer-Aided Design and Drafting) in that the former is a subset of the latter: CADD encompasses both the creation of graphic images and the storing and use of attributes of the graphic elements, such as automated area takeoff or the storage of design characteristics in a data base. This section designates CAD for all aspects of design and drafting systems.

Scale drawings express not only physical size information, but also how different elements are connected and where they are located within an area. Computerized drawings of buildings make it easy to extract measurement of lengths, areas, and volumes, as well as show the orientation and connectedness (topology) of the building parts. Drawings can be easily changed and do not require redrawing, as in manual drafting. Building walls or ductwork, for example, can be added, removed, enlarged, reduced, and moved to a new location on a drawing. A whole building can be rotated on the site plan without redrawing the building or destroying the previous drawing.

Some computer-aided design systems also aid in (1) material and cost estimating, (2) design and analysis of HVAC systems, (3) visualization and interference checking, and even (4) manufacturing of HVAC components such as ductwork. The systems are now customized to HVAC applications; for instance, a two-line duct layout can be generated automatically from a one-line drawing and section size information.

The computer-generated building drawing parts can be linked to nongraphic characteristics to automatically extract data about the building for reports, schedules, and specifications. For example, an architectural designer can draw wall partitions and windows, and then have the computer automatically extract the number and size of windows, the areas and lengths of walls, and the areas and volumes of rooms and zones. Similarly, an operator can store information about a drawing for later use in reports, schedules, design procedures, or drawing notes. For example, a fan displayed on a drawing can have its airflow, voltage, weight, manufacturer, model number, cost, and other data stored in a file or data base. The linkage between the graphics and attributes makes it possible to review the characteristics of any item or to generate schedules of items located in a certain area on a drawing. Conversely, the graphics-to-data links allow the designer to enhance the graphic items that have a particular characteristic by searching for those characteristics in the data files and then making the associated graphics stand out as a brighter, flashing, bolder, or different color display.

Computer-generated drawings, especially those created in layers, can also help the designer visualize the building and its systems and check for interferences, such as the architecture and ductwork occupying the same space. This feature also helps an HVAC designer or drafter coordinate designs with other disciplines.

CAD HVAC drawings and their associated data can be used by building owners and maintenance personnel for ongoing facilities management, strategic planning, and maintenance. They are useful for computer-aided manufacturing, such as duct construction. Two-dimensional CAD drawings can be used to create a three-dimensional model of the building and its components, which can then be used to visualize all or part of the building. The three-dimensional building model can also aid in developing sections and details for building drawings. Further, CAD can automate cross-referencing of drawings, drawing notation, and building documents.

Hardware Considerations

Computer-aided design can be implemented on a wide range of computers, terminals, and display screens using many output and storage devices. The nature of the user's work, such as the size and type of buildings, may influence the CAD hardware selection. Small microcomputer-based systems have basic CAD capabilities but are somewhat limited in the size, number, and coordination of drawings that they can work with. Microcomputer systems are also inherently single-user systems, which means that other designers generally cannot share those drawings while the single user is working on them. On the other hand, minicomputer and mainframe computer-based CAD systems are inherently multiuser, multitasking systems that are designed for sharing drawings and data among several users as well as managing a large inventory of drawings of almost unlimited size and complexity. These differences are important for both small single-discipline design firms and larger, integrated, multidisciplinary design firms that may need CAD drawings available to all in-house designers working on a common project. Other hardware considerations include a system's speed, cost, memory and storage limitation, and ease of use. Turnkey CAD systems are available where the hardware is developed, installed, maintained, and upgraded by the CAD supplier, thus allowing the operator to use the system with little startup and support time.

Software Considerations

Selecting a CAD system that lets a user perform design and drawing tasks with the least effort is perhaps the most critical consideration. CAD software tailored to HVAC design and drafting functions is preferred to software that has only basic graphics and data management capabilities. Further, CAD software with flexible user-definable features is advantageous to users who may need to customize or enhance the CAD HVAC task. Built-in error checking and error reporting features of CAD software are important to keep the design process moving ahead when problems occur. Software support, check-out, training, and enhancement are also important and can make the difference between needing an in-house computer programmer or not. Turnkey CAD systems are generally customized to particular applications such as HVAC design and offer more substantial software check-out, support, enhancements, and training.

Another software consideration is transferability of drawings and data among different CAD systems, because the building owner, architect, and other engineers may have different CAD systems. Several graphic exchange standards make this exchange possible. The Initial Graphics Exchange Specification (IGES) and Standard Interchange Format (SIF) are the two prominent formats. Software interfaces to these formats are important.

COMPUTER GRAPHICS AND MODELING

Computer graphics programs may be used to create and manipulate pictorial information between the computer and the user (Rankin 1989).

Information can be assimilated more easily when presented in the form of graphical displays, diagrams, and models. Historically, the building of scale models has improved the design engineer's understanding and analysis of early prototype designs. The psychrometric chart has also been invaluable to HVAC engineers for many years (Li *et al.* 1986).

Combining alphanumerics with computer graphics enables the design engineer to quickly evaluate design alternatives. Computers can pictorially simulate design problems with three-dimensional color displays that can predict the performance of mechanical systems before they are constructed. Simulation graphic software is also used to test design conditions normally infeasible with scale models due to high test costs or time constraints. Therefore, product manufacturability and economic feasibility may be determined without a working prototype ever being built. Where only dozens of scale model tests may be performed prior to full scale manufacturing, combined computer and scale model testing allows hundreds or thousands of tests to be performed, thus improving reliability in product design.

Integrating computer graphics software with expert system software can enhance design time even more. Using integrated CAD and expert systems can help the building designer and planner with construction design drawings and with construction simulation for planning and scheduling complex building construction scenarios (Potter 1987).

A simple yet helpful tool available to the HVAC engineer is the graphical representation of thermodynamic properties and thermodynamic cycle analysis. These programs may be as simple as computer-generated psychrometric charts with cross-hair cursor retrieval of properties and computer display zooming or magnification of chart areas for easier data retrieval. More complex software is also available. Using graphic-assisted numeric analysis, these programs provide graphic representation of thermodynamic cyclic paths overlaid on two-or three-dimensional thermodynamic property graphs such as *P-v*, *P-h-T*, or *T-h-s* diagrams. With simultaneous display of the calculated results of Carnot efficiency, COP, and first and second law of thermodynamics compliance, the user can quickly understand the cycle fundamentals and practicality of the cycle synthesis (Abtahi *et al.* 1986).

HVAC engineers use complex simulation software for assisting the design of large district heating or cooling systems. Typical piping and duct system software produces large amounts of data to be analyzed. But, by introducing computer graphics into the simulation process, the task is simplified. Pictorially enhanced simulation output with graphic displays of piping system curves, pump curves, and load curves speeds the sizing and selection process for the engineer. In addition, using what-if scenarios with numeric/graphic output increases the designer's understanding and helps avoid system design problems that may occur after installation (Chen 1988). Graphic-assisted fan and duct system design and analysis is similarly available to system designers (Mills 1989).

As the numerical analysis of system design becomes more complex, *i.e.*, where analyses involve three-dimensional simulations with highly complex coupling between the various system variables, massive amounts of data are generated. When the volume of generated data becomes so large that analysis and understanding are inhibited, the use of graphic simulation becomes ideal or even required to improve the visualization effectiveness of the process and allow for the study of complex problems such as fluid flow field analysis. Using graphic simulation and analysis may help avoid (or at least complement) the time and cost associated with scale model testing in wind and water tunnels normally associated with these problems.

Graphic post-processing of numeric data provides the user with a graphical form suitable for easy understanding of the flow analysis versus the prototype experiments currently used (Sturgess *et al.* 1986). For example, graphic modeling and simulation of complex fluid flow fields is used in the simulation of airflow fields present in unidirectional clean room design. This simulation technology may make obsolete the need and expense of clean room mock-ups (Busnaina *et al.* 1988). Computer models of particle trajectories, transport mechanisms, and contamination propagation are also commercially available (Busnaina 1987).

Airflow analysis of flow patterns and air streamlines are calculated by solving the fundamental equations of fluid mechanics for laminar and turbulent flows where incompressibility and uniform thermophysical properties are assumed (Kuehn 1988). For example, the Navier-Stokes equations for conservation of momentum and the continuity equation for conservation of mass can be used. Finite element modeling techniques are used to produce two- or three-dimensional pictorial-assisted displays, where the velocity vectors and velocity pressures at individual nodes are solved and numerically displayed.

An example of pictorial output is seen in Figure 4 of Chapter 16, where calculated airflow streamlines have been displayed overlaid against a graphic input of the two-dimensional clean room. Using this display, the clean room designer can foresee potential problems with this configuration, such as the circular flow pattern in the lower left and right.

Major features and benefits associated with most computer flow models are:

- Two- or three-dimensional modeling of simple clean room configurations
- Modeling of both laminar and turbulent airflows
- CAD of air inlets and outlets of varying sizes and room construction features
- Allowances for varying boundary conditions associated with walls, floors, and ceilings
- Pictorial display of aerodynamic effects of process equipment, workbenches, and people
- Prediction of specific airflow patterns of all or part of a clean room

- Reduced costs associated with new clean room design verification
- Providing graphical representation of flow streamlines and velocity vectors to assist in flow analysis

A graphical representation of simulated particle trajectories and propagation is shown in Figure 5 of Chapter 16. With this type of graphic data, the user inputs a simulated concentrated particle contamination and the program simulates the propagation of particle populations. In this example, the circles represent particle sources, where the diameter is a function of the concentration present. Simulations such as this assist the clean room designer in placing protective barriers.

While research has shown excellent correlation between flow modeling by computer and flow modeling done in simple mock-ups, the modeling software should not be considered a panacea for clean room design; flow simulation around complex shapes is still being developed.

The acceptance and success of computer graphic-assisted programs depends on graphic output that is correct and on the method used to convey the display. Whether the output is two-dimensional black and white, or fully animated, color, three-dimensional simulations, the choice of display type will influence user acceptance. Many three-dimensional color outputs are no longer used only for attractive presentations, but are necessary design tools that increase the productivity of the user (Mills 1989).

KNOWLEDGE-BASED SYSTEMS

A knowledge-based, or expert, system (KBS) has the ability to replicate human decision-making processes. Two main approaches are followed—rule-based, where human knowledge is encoded as a set of "if-then" rules, and model-based, where a model of the system is used to examine potential actions. Hybrid systems contain both models and rules. Because decisions are made based on information, KBS are usually integrated with data bases, spreadsheets, or data acquisition systems. Harmon and King (1985) give a good introduction to the technology and Feigenbaum *et al.* (1988) describe successful uses of KBS in a variety of business and technical applications.

The rapid advancement of the field has been made possible in part by the use of KBS "shells," *i.e.*, user-friendly environments in which a KBS can be created. Shells fill the same role for KBS that spreadsheets and data bases do for algorithmic programming (Harmon *et al.* 1988).

KBS techniques are generally used in two areas—design and diagnosis. In terms of building applications, the idea of intelligent design tools has attracted much attention (Brambley *et al.* 1988). Work has also been done in the diagnostic area, for both comfort and energy performance problems (Brothers and Cooney 1989, Culp 1989, Haberl *et al.* 1989).

A survey of activity in KBS applications for HVAC is given by Hall and Deringer (1989). KBS applications are in regular use in the HVAC field; however, these are restricted to company in-house programs and are not released for general use.

COMMUNICATIONS

Computer communications usually involves the passing of digital information. In the traditional computer sense, communications includes the connection of computer terminals to time-sharing systems, word processing workstations to a central time-shared processor, one computer to another, and the connection of a computer to various peripherals such as plotters, printers, or mass storage units. In the building services industry, this includes connecting controllers to supervisors or other controllers, as well as connecting the HVAC systems to computers, or to fire or security alarm systems. The types of information passed are sensor and status information needed for control or coordination, data logging, system tuning or maintenance, and set point or schedule information for performance modification or remote programming.

Typical applications requiring computer or digital communications can be summarized as follows:

- Several computers may share peripheral resources that would be underused by one computer; this would include printers, plotters, and mass storage devices.
- Small computers may access a central time-sharing system when heavy computing is needed.
- Computers may access a central data base containing software, product, customer, inventory, or vendor information.
- Computers may be networked so that all computers on the network may share storage and peripherals, as well as memory and computational capabilities.
- Electronic mail through networks and bulletin boards through telephone links allow efficient distribution of information among individuals.
- Building controls share occupancy, performance, and set point information for control coordination, energy-saving strategies, remote monitoring, and programming capabilities.

Eight digits, or bits, which are the fundamental units of information, are grouped together as bytes. Most computers represent information with single characters of which each character is represented by a byte. The two most common character sets are ASCII and EBCDIC. These codes assign digits, alphabetic characters, and punctuation to specific bit patterns in a byte.

Generally, computers and peripherals communicate to each other through ports. A so-called parallel port uses eight signal lines plus control lines for sending a complete character at one time. While this is an advantage for speed, it becomes expensive over long distances and cannot take advantage of telephone lines.

A serial port sends characters one bit at a time or serially; this requires as few as two or three wires. Serial communication requires a scheme for the receiver and transmitter to coordinate the beginning and end of each character, as well as knowledge of the bit rate (speed) at which the transmitter is sending information. To this end, communication can be done synchronously or asynchronously. With asynchronous communication, each character has a specific length that includes a start bit and one or two stop bits. The receiver resynchronizes on every character sent. Synchronous communications generally begin with a string of bits in a predetermined pattern. This pattern allows the receiver to synchronize with the transmitter, so that a long string of bytes may be sent one after the other without the overhead of start and stop bits for each character. Although there is a time penalty of about 25% for sending start and stop bits for each character with asynchronous serial communication, most smaller computers do it because of the simplicity of both the hardware and the software.

When communication errors are a concern and/or multiple devices are using a network to exchange information, a protocol is needed to check the data for errors, to correct the data if errors are found, to provide equitable access for all devices on the network, and to assure that only one device transmits at a time. A communication network limited to an office, a building, or to several buildings close together is called a Local Area Network (LAN). LANs are typically used to share data (such as sharing drawings among CAD users or sharing a common data base with an order entry system) or peripherals (such as sharing printers, plotters, or large disc drives); workstations on a LAN can even share processing power. LAN options include medium (twisted pair, coaxial, or fiber optic cabling), interconnection scheme or topology (bus, ring, or star configurations), access method (contention, token passing, or time slice), and operating system. The LAN operating system can be added to the normal computer operating system, which handles the communication functions on the network.

Gateways are devices that interconnect similar or dissimilar LANs, dissimilar large computers, LANs to large computers, and so forth. A Wide-Area Network (WAN) connects disparate elements, especially over a large geographic area.

When using switched telephone lines or fiber optic cables, the voltage levels used for short-distance communications cannot be used. A modem (modulator/demodulator) is used to interface the two media by converting bits between voltage levels and tones or light. The quality of the medium determines the communication speed. Current technology is providing up to 9600 bits per second over normal telephone lines. As more information is passed around buildings, voice, computer data, building services data, and CATV networks may be called on to integrate these forms of information interchange to reduce installation and reconfiguration costs (Tanenbaum 1981, Bellamy 1982).

DATA ACQUISITION

Data Acquisition System Hardware

Data from the field is rapidly replacing estimates derived from simulations, nomograms, and tables. Most microcomputer and data logger combinations can be classified as (1) analog data loggers (strip charts); (2) digital data loggers (analog and digital data are captured in electronic format, stored internally, and transferred to a microcomputer via serial communications lines); or (3) internal add-on data acquisition cards for a microcomputer. Data Acquisition Systems (DASs) offer a large number of options, including the number of channels, types of channels (*e.g.*, digital, analog, and pulse counting), scanning rate, and error checking. Data acquisition wiring commonly delivers the sensor signal to the data acquisition system and can be accomplished with (1) direct wiring, (2) RF carrier systems, (3) power-line carrier systems, and, in special instances, (4) infrared communications.

Data Acquisition System Software

Software for a data acquisition system can be purchased from DAS hardware vendors or specialty software vendors; it can also be adapted from public domain software (Ryan 1986, Feuerman and Kempton 1987) or developed in-house.

Software developed in-house requires establishing an experiment plan, developing an interface to the data-gathering hardware, and selecting and calibrating analog sensors and digital inputs. Coding the DAS software includes: (1) selecting the programming language, (2) defining the data channels, (3) selecting the time intervals, (4) designing the archive data base, (5) developing error detection and correction algorithms, (6) analyzing the data, (7) graphic considerations, (8) transfer of the data (*i.e.*, modem, RF, floppy), and (9) final storage of the data (*e.g.*, tape, optical disc, floppy).

MONITORING AND CONTROL

Direct Digital Control (DDC) of HVAC components gives more accurate control and greater flexibility than the proportional control commonly used in pneumatics. Such flexibility permits controllers to be tuned or allows control algorithms to be replaced or extended (Nesler 1986). For example, DDC can control temperature in a variable air volume terminal box, with control based only on a dry-bulb temperature sensor. The same microprocessor can be modified to monitor relative humidity and mean radiant temperature and to change the dry-bulb set point to better maintain comfort conditions (Int-Hout 1986). It can adjust airflows and temperatures based on occupancy indicators in an individual office. By transferring data back to a computer controlling the central fans, the microprocessor also facilitates minimizing the fan power required to deliver a given flow of air (Englander and Norford 1988). Finally, it can control flow on the basis of CO_2 measurements to provide a supply of outdoor air matched to the number of occupants in the space.

Microprocessor-based devices are currently used to turn on chillers and boilers at an optimal time to recover from a period when a building is unconditioned. The required programs measure the building's thermal behavior to adjust the equipment start time; such programs are examples of parameter estimation routines and give the controller an adaptive capability. This kind of optimization has been performed off-line (Hackner *et al.* 1985) and, more effectively, as part of an on-line control system (Cumali 1988). The control could be extended to include weather forecasting algorithms, which could improve control of thermal storage or schedule precooling by night ventilation (Shapiro *et al.* 1988).

Linking local microprocessor controllers with a central, supervisory computer makes it possible to integrate HVAC control with such services as security, life safety monitoring, and lighting. The network can be extended to the electric utility, with the utility providing spot pricing information as input to load-shedding programs and even dropping specified equipment in the event of power shortages; this technology has been demonstrated in pilot projects (Peddie and Bulleit 1985).

Central computers are already used to collect data from individual controllers or meters, but perform a minimal amount of data analysis. Analysis performed off-site by consultants identifies long-term trends in energy consumption, isolates beneficial or harmful changes in equipment operation, and normalizes energy consumption for changes in weather. These analysis programs can be incorporated into on-site computers, providing operators and management with up-to-date, readily available information (Anderson *et al.* 1989).

Hardware is available in a wide range of sizes and capabilities from specialized packaged units, such as individual air handler controls, through intelligent field panels or programmable controllers that gather information from several inputs and control multiple outputs, to stand-alone computers, which not only monitor and control but are also suitable for program development.

SOFTWARE

A computer must have both hardware and software to perform useful tasks. Osborne (1980), Norton (1986), and Goodman (1988) provide further information. Software can be segmented into four major categories: systems software, languages, utilities, and application programs.

System software, otherwise known as the operating system, is the environment in which other programs run. It handles the input and output (keyboard, video display, and printer), and the file transfer between discs and memory; it also supports the operation of other programs. Operating system software can be supplied by the computer manufacturer or from software companies. The operating system is specific to a particular type of computer.

Languages are used to write computer programs. These range from assembly language (which allows coding to occur at the machine instruction level) to high-level languages such as FORTRAN, BASIC, Pascal, or C. Many high-level languages exist to satisfy various programming requirements. FORTRAN is useful for scientific or mathematical applications. BASIC established microcomputers (personal computers) as viable business machines. New dialects of BASIC are overcoming many of its previous limitations. Pascal is a structured language originally designed for teaching programming. C is emerging as a preferred standard for professional programming in many industries, including the HVAC industry. It compiles to a very efficient and fast code, and the C source code can be recompiled with other C compilers to run on many types of machines. The major disadvantage of C is that a fairly high level of programming skill is required to create programs.

Utility software programs do standard organizing and data-handling tasks for a specific computer, such as copying files from one disc to another, printing directories of files on a disc, printing files, and merging files (putting two or more files together in some specific order). Utility software generally performs one or two specific functions, while applications software has a particular application which can include many functions and utilities.

Application programs are designed to use the computer's power to calculate items such as loads, energy, piping design, accounting, and word processing. General-purpose applications software is discussed in the Productivity Tools section. Another specialized area of applications software is that of expert or knowledge-based systems, discussed in Knowledge-Based Systems.

Purchased Software

While manufacturers of computer equipment also offer software, other distribution channels also exist as well. For personal computers (PCs), software can be purchased from the manufacturers, software companies, distributors, and discount houses. The decision on where to purchase software should be influenced by price, the level of support needed, return policies, and how the software is distributed. Most software companies support their product in some way. Some companies have a poor reputation for supporting customers, so potential purchasers should determine if the level of support from a particular vendor is acceptable. Some vendors allow customers to try software and then return it if it is not acceptable; others make demonstration or limited function versions available for a minimal charge.

Another area of concern is how the software is distributed. The media need to be compatible with the computer that they will be loaded into. Also, some vendors still use copy protection schemes, which prevent the user from running the software on more computers than the one licensed. In reality, this copy protection causes inconvenience to the user, and the trend is to move away from these protection methods.

When PC software is purchased, the user is given a license with specific restrictions on how that software may be used, typically that it is to be used only on one computer. Often the user implicitly agrees to the terms of the license by opening the package. A separate signed license is not as common but is occasionally used. Because the distribution of unauthorized copies has been so costly to software companies, many are actively prosecuting users for using the software illegally.

Before purchasing general-purpose software such as data base, spreadsheet, or word processor programs, the user should look at how the software runs and how easy it is to use. A full-service computer store may offer advice and demonstrations not available from mail order or discount software sources. The user interface is important. Some programs are very abstruse and difficult to operate and understand, while others are very easy to use. Computer magazines often publish comparisons of most general-purpose software which can be used for a first appraisal. The final determination needs to be made by the user. Once a package is chosen, the user may have a difficult time changing to another package because of all of the data that is entered into the chosen program.

Occasionally suppliers offer the program's source code, which is a set of human readable instructions in a computer language. Skilled programmers can modify a source code to change the operation of the program. This is generally not recommended, because once changed, support must be supplied internally.

Public Domain Software

Public domain software is software that is available to the public either without charge or for a minimal support charge (usually for maintenance and support). Government-supported developments, universities, and individuals are the source for these programs. The source code for most public domain programs is available, but it is poorly supported. However, some are well-documented and supported.

Public domain software can be obtained from computer bulletin boards, other individuals, and companies that distribute this software for a small duplication charge. Care should be used when obtaining public domain software, since its origin is difficult to trace. Some public domain programs have computer viruses. Programs on bulletin boards are particularly suspect. A virus is a small program inserted into another program that can destroy stored data, lock up the computer, and so forth (McAfee and Haynes 1989).

Royalty Software in Time-Sharing

Unused computers cost almost as much to own as heavily used computers. This fact created a large industry in the 1960s—computer time-sharing, which involves more than one user using a computer or family of computers simultaneously. Most major time-sharing service organizations have networks of mainframe computers and communications equipment capable of serving thousands of customers on an international basis. Frequently, telephone communications are used between customers and the computer centers, allowing a customer to access a computer across the country simply by dialing a telephone number.

Time-share organizations sell computer resources. Purchased software can be installed on the networks for the private use of individual customers. Under this arrangement, computer time is used and the cost varies with the activity level. The software is either developed by the time-share company or by third party authors who sublicense to the time-share company. The costs are billed to the user from the time-share company.

The advent of the PC has reduced the number of time-share users since the cost of a PC is low compared to the cost of a mainframe computer. More and more of the programs that used to run only on mainframe computers are now being run on PCs.

Custom Programming

Three major strategies can be used to obtain custom programs. These are contracting to an outside firm, using an outside firm to provide internal consultation and help, or developing the programs internally.

Contracting to an outside firm for the entire programming effort should be considered if the host organization does not have the people or desire to support the software on an on-going basis. Funds should be budgeted to have the outside organization support any modifications or enhancements, which almost always occur. Contracting outside is a good approach for an organization that does not want to get involved with programming. The main drawbacks are the expense and lack of control of the program. Licensing issues and ownership should also be carefully spelled out in the contract.

Using an outside firm to provide internal consulting is viable if internal skill is not available and long-term support and maintenance of the software is to be done internally. Outside firms can provide the expertise to get a project going quickly. The design specification is critical to the success of the development.

Developing the program internally is viable if the skill and resources are available. Internal projects are easier to control because the people on the project are usually under one roof. Most of the major vendors of HVAC systems, which are software based, develop software internally with occasional consultation from outside firms.

No matter which approach is chosen, the user needs to provide a detailed functional design specification. The calculations, human interface, reports, user documents, and testing procedures need to be carefully agreed on by all parties before the development begins. To create a useful software program, a thorough understanding of the subject matter and a solid knowledge of computer programming are required. Software testing also needs to be specified at the beginning of the project to avoid a common problem of low quality because of hasty and inadequate testing. Design testing needs to address the human interface, a wide range of input values (including improper inputs), the algorithm, and any outputs (to paper, disc, or other media). Field testing needs to be done with final users of the software under field conditions.

For any of the development approaches, good understandable documentation is required. If the software cannot be supported adequately for any reason, the program will either stop being used, be replaced, or be redone. Any of these results can cause a substantial drain on an organization.

Custom programming should be used only to create programs or features not available with existing software: no matter what the cost of existing software, custom software will cost more in the long run to produce comparable results.

Sources for HVAC Programs

ASHRAE *Journal's* HVAC&R *Software Directory* (1990) is a comprehensive listing of HVAC-related software. A brief description of each program is given, along with information on availability.

BIBLIOGRAPHY

Abtahi, H., T.L. Wong, and J. Villanueva, III. 1986. Computer aided analysis in thermodynamic cycles. *Proceedings* of the 1986 ASME International Computers in Engineering Conference 2.

Anderson, D., L. Graves, W. Reinert, J.F. Kreider, J. Dow, and H. Wubbena. 1989. A quasi-realtime expert system for commercial building HVAC diagnostics. ASHRAE *Transactions* 95(2).

ARI. 1987. Forced-circulation air-cooling and air-heating coils. ARI *Standard* 410-87. Air-Conditioning and Refrigeration Institute, Arlington, VA.

ARI. 1989. Central station air-handling units. ARI *Standard* 430-89. Air-Conditioning and Refrigeration Institute, Arlington, VA.

Arklin, H. and A. Shitzer. 1979. Computer aided optimal life-cycle design of rectangular air supply duct systems. ASHRAE *Transactions* 85(1).

ASHRAE. 1979. Cooling and heating load calculation manual. ASHRAE GRP 158.

ASHRAE. 1989. ASHRAE *Access*.

ASHRAE. 1990. ASHRAE *Journal's* HVAC&R *Software Directory*.

Bellamy, J. 1982. *Digital telephony*. John Wiley & Sons, New York.

Brambley, M.R., D.B. Crawley, D.D. Hostetler, R.C. Stratton, M.S. Addison, J.J. Deringer, J.D. Hall, and S.E. Selkowitz. 1988. Advanced energy design and operation technologies research. Pacific Northwest Laboratory *Report* PNL-6255.

Brothers, P.W. 1988. Knowledge engineering for HVAC expert systems. ASHRAE *Transactions* 94(1).

Brothers, P.W. and K.P. Cooney. 1989. A knowledge-based system for comfort diagnostics. ASHRAE *Journal* 31(9).

Busnaina, A.A. 1987. Modeling of clean rooms on the IBM personal computer. 1987 *Proceedings of the Institute of Environmental Sciences*, 292-97.

Busnaina, A.A., S. Abuzeid, and M.A.R. Sharif. 1988. Three-dimensional numerical simulation of fluid flow and particle transport in a clean room. 1988 *Proceedings of the Institute of Environmental Sciences*, 326-30.

Chen, T.Y.W. 1988. Optimization of pumping system design. *Proceedings* of the 1988 ASME International Computers in Engineering Conference.

Christensen, C. 1984. Digital and color energy maps for graphic display of hourly data. *Proceedings* of the 9th Annual Passive Solar Conference, Columbus, OH (September).

Christensen, C. and K. Ketner. 1986. Computer graphic analysis of class B hourly data. *Proceedings* of the 11th Annual Passive Solar Conference. Boulder, CO (June).

Culp, C.H. 1989. Expert systems in preventive maintenance and diagnosis. ASHRAE *Journal* 31(8).

Cumali, Z. 1988. Global optimization of HVAC system operations in real time. ASHRAE *Transactions* 94(1).

Date, C. 1988. *Database—A primer.* Addison Wesley Publishing Company, Reading, MA.

Date, C. and C. White. 1987. SQL/DS—A user's guide to the IBM product structured query language/data system. Addison Wesley Publishing Company, Reading, MA.

Diamond, S.C., C.C. Cappiello, and B.D. Hunn. 1985. User-effect validation tests of the DOE-2 building energy analysis computer program. ASHRAE *Transactions* 91(2).

Diamond, S.C. and B.D. Hunn. 1981. Comparison of DOE-2 computer program simulations to metered data for seven commercial buildings. ASHRAE *Transactions* 87(1).

Englander, S.L. and L.K. Norford. 1988. Fan energy savings: Analysis of a variable speed drive retrofit. *Proceedings* of the American Council for an Energy-Efficient Economy, 1988 summer study on energy efficiency in buildings, Asilomar, CA.

Feigenbaum, E.A., P. McCorduck, and H.P. Nii. 1988. *The rise of the expert company*. The York Times Books, New York.

Feuermann, D. and W. Kempton. 1987. ARCHIVE: Software for management of field data. Center for Energy and Environmental Studies, *Report* No. 216. Princeton University, Princeton, NJ.

Goodman, D. 1988. *The complete hypercard handbook*. Bantam Books, New York.

Haberl, J.S. and D. Claridge. 1985. Retrofit energy studies of a recreation center. ASHRAE *Transactions* 91(2).

Haberl, J.S., M. MacDonald, and A. Eden. 1988. An overview of 3-D graphical analysis using DOE-2 hourly simulation data. ASHRAE *Transactions* 94(1).

Haberl, J.S., L.K. Norford, and J.S. Spadero. 1989. Expert systems for diagnosing operation problems in HVAC systems. ASHRAE *Journal* 31(6).

Hackner, R.J., J.W. Mitchell, and W.A. Beckman. 1985. System dynamics and energy use. ASHRAE *Journal* (June).

Hall, J.D. and J.J. Deringer. 1989. Overview of knowledge-based systems research and development activities in the HVAC industry. ASHRAE *Journal* 31(7).

Harmon, P. and D. King. 1985. *Expert systems: Artificial intelligence in business*. John Wiley and Sons, New York.

Harmon, P., R. Maus, and W. Morrisey. 1988. *Expert systems: Tools and applications*. John Wiley and Sons, New York.

Hsieh, E. 1988. Calibrated computer models of commercial buildings and their role in building design and operation. *Report* No. 230, Center for Energy and Environmental Studies, Princeton University, Princeton, NJ.

Int-Hout, D., III. 1986. Microprocessor control of zone comfort. ASHRAE *Transactions* 92(1).

Jones, L. 1979. The analyst as a factor in the prediction of energy consumption. Proceedings of the Second International CIB Symposium on Energy Conservation in the Building Environment, Session 4, Copenhagen, Denmark.

Judkoff, R. 1988. International energy agency design tool evaluation procedure. Solar Energy Research Institute *Report* No. SERI/TP-254-3371 (July).

Judkoff, R., D. Wortman, and B. O'Doherty. 1981. A comparative study of four building energy simulations, phase II: DOE-2.1, BLAST-3.0, SUNCAT-2.4 and DEROB, Solar Energy Research Institute *Report* No. SERI/TP-721-1326 (July).

Kaplan, M., J. McFerran, J. Jansen, and R. Pratt. 1990. Reconciliation of a DOE-2.1C model with monitored end-use data for a small office building. ASHRAE *Transactions* 96(1).

Kovarik, M. 1971. Automatic design of optimal duct systems. Use of computers for environmental engineering related to buildings. National Bureau of Standards, Building Science Series 39 (October).

Kuehn, T.H. 1988. Computer simulation of airflow and particle transport in cleanrooms. *The Journal of Environmental Sciences*, 31.

Kusuda, T. and J. Bean. 1981. Comparison of calculated hourly cooling load and indoor temperature with measured data for a high mass building tested in an environmental chamber. ASHRAE *Transactions* 87(1).

Kusuda, T., T. Pierce, and J. Bean. 1981. Comparison of calculated hourly cooling load and attic temperature with measured data for a Houston test house. ASHRAE *Transactions* 87(1).

Li, K.W., W.K. Lee, and J. Stanislo. 1986. Three-dimensional graphical representation of thermodynamic properties. *Proceedings* of the 1986 ASME International Computers in Engineering Conference 1.

May, W.B., Jr. and L.G. Spielvogel. 1981. Analysis of computer simulated thermal-performance of the Norris Cotton Federal Building. ASHRAE *Transactions* 87(1).

McAfee, J. and C. Haynes. 1989. Computer viruses, worms, data diddlers, killer programs and other threats to your system. St. Martin's Press, New York.

Mills, R.B. 1989. Why 3D graphics? *Computer Aided Engineering*. Penton.

Milne, M. and S. Yoshikawa. 1978. Solar-5 an interactive computer-aided passive solar building design system. *Proceedings* of the Third National Passive Solar Conference, Newark, DE.

Nesler, C.G. 1986. Automated controller tuning for HVAC applications. ASHRAE *Transactions* 92(2).

Norton, P. 1986. *Inside the IBM PC*. Brady Books, Prentice Hall, New York.

Olgyay, V. 1963. *Design with climate: Bioclimatic approach to architectural regionalism*. Princeton University Press, Princeton, NJ.

Osborne, A. 1980. *An introduction to microcomputers*, Vols. 1, 2, and 3. Osborne-McGraw Hill, Berkeley, CA.

Peddie, R.A. and D.A. Butteit. 1985. "Managing electricity demand through dynamic pricing." In Energy sources: Conservation and renewables. Conference Proceedings 135. American Institute of Physics, New York.

Potter, C.D. 1987. CAD in construction. *Computer Aided Engineering*. Penton.

Rabl, A. 1988. Parameter estimation in buildings. Methods for dynamic analysis of measured energy use. ASME *Journal of Solar Energy Engineering*, 110.

Rankin, J.R. 1989. *Computer graphics software construction*. Prentice Hall, New York, 1-4.

Reddy, A. 1989. Application of dynamic building inverse models to three occupied residences monitored non-intrusively. *Proceedings* of the Thermal Performance of the Exterior Envelopes of Buildings IV. ASHRAE/DOE/BTECC/CIBSE (December).

Reiter, P. 1986. Early results from commercial ELCAP buildings: Schedules as a primary determinant of load shapes in the commercial sector. ASHRAE *Transactions* 92(2).

Robertson, D.K. and J. Christian. 1985. Comparison of four computer models with experimental data from test buildings in New Mexico. ASHRAE *Transactions* 91(2).

Ryan, L. 1986. Documentation of DAS: A data acquisition software package. Center for Energy and Environmental Studies—*Working Paper* No. 87. Princeton University, Princeton, NJ.

Schley, M. 1984. CAD Buyers checklist. Architectural Technology, Summer, Washington, D.C.

Shanmugavelu, I., T.H. Kuehn, and B.Y.H. Liu. 1987. Numerical simulation of flow fields in clean rooms. 1987 *Proceedings* of the Institute of Environmental Sciences, 298-303.

Shapiro, M., A. Yager, and T. Ngan. 1988. Test hut validation of a microcomputer predictive HVAC control. ASHRAE *Transactions* 94.

Sorrell, F., T. Luckenback, and T. Phelps. 1985. Validation of hourly building energy models for residential buildings. ASHRAE *Transactions* 91(2).

Spielvogel, L.G. 1975. Computer energy analysis for existing buildings. ASHRAE *Journal* (August):40.

Spielvogel, L.G. 1977. Comparisons of energy analysis computer programs. ASHRAE *Transactions* 83(2).

Stoecker, W.F., *et al.* 1971. Optimization of an air-supply duct system. Use of computers for environmental engineering related to buildings. National Bureau of Standards, Building Science Series 39 (October).

Sturgess, G.J., W.P.C. Inko-Tariah, and R.H. James. 1986. Postprocessing computational fluid dynamic simulations of gas turbine combustor. *Proceedings* of the 1986 ASME International Computers in Engineering Conference 3.

Subbarao, K. 1988. PSTAR Primary and secondary terms analysis and renormalization, a unified approach to building energy simulations and short-term monitoring. Solar Energy Research Institute *Report* No. SERI/TR-254-3175.

Tanenbaum, A.S. 1981. *Computer networks*. Prentice-Hall, New York.

Tsal, R.J. and E.I. Chechik. 1968. Use of computers in HVAC systems. Budivelnick Publishing House, Kiev. Available from the Library of Congress, Service -TD153.T77 (Russian).

Tsal, R.J., H.F. Behls, and R. Mangel. 1988. T-method duct design: Part I, optimization theory; Part II, calculation procedure and economic analysis. ASHRAE *Technical Data Bulletin* (June).

Wortman, D. and B. O'Doherty. 1981. The implementation of an analytical verification technique on three building energy analysis codes: SUNCAT-2.4, DOE-2.1 and DEROB III. Solar Energy Research Institute *Report* No. SERI/TP-721-1008 (January).

Yuill, G. 1985. Verification of the BLAST computer program for two houses. ASHRAE *Transactions* 91(2).

Yuill, G. and E. Phillips. 1981. Comparison of BLAST program predictions with the energy consumption of two buildings. ASHRAE *Transactions* 87(1).

CHAPTER 37

BUILDING ENERGY MONITORING

COLLECTING empirical data to determine building energy performance is an important but complex and costly activity. Careful, thorough planning facilitates timely and cost-effective data collection and analysis. Despite varying project objectives and scopes, several issues are common to building energy monitoring projects, which make possible standardized methodologies and procedures (monitoring protocols).

This chapter provides general guidelines for developing building monitoring protocols, which, if applied, yield quality empirical data. The protocols are designed to ensure that data products are planned to meet project objectives. The intended audience comprises building energy monitoring practitioners and data end users, such as energy suppliers, energy end users, building system designers, public and private research organizations, utility program managers and evaluators, equipment manufacturers, and officials who regulate residential and commercial building energy systems.

TYPES OF PROJECTS

Building energy monitoring projects develop empirical information using field data to better understand building energy performance. Monitoring projects can be noninstrumented (performance tracking using utility billing data) or instrumented (billing data supplemented by additional sources of data, such as an installed instrumentation package or building energy management system). Noninstrumented approaches are generally simpler and less costly, than instrumented approaches, but they are subject to more uncertainty in interpretation. The analyst must decide whether the extra cost of an instrumented approach, which can provide greater specificity and accuracy, is necessary.

Instrumented field monitoring projects generally involve data acquisition systems, which are typically comprised of various sensors, and data recording devices, such as data loggers. Field monitoring projects may involve a single building or hundreds of buildings, and may be carried out over periods ranging from weeks to years. Most monitoring projects involve the following activities:

- Project planning
- Site installation of data acquisition equipment
- Ongoing data collection
- Data analysis and reporting

These activities often require support by various professional disciplines (such as engineering, data analysis, and management) and construction trades (electrical installers or plumbers).

Useful building energy performance data extend over a wide range of time and space. Space dimensions range from individual end use (lighting) or components (HVAC equipment, walls) in a building, to the building itself (meter readings) and to the utility network (utility load factors, excess capacity). Time dimensions of the data range from seconds (controller actuation) to many years (building/component lifetimes). This wide range of building data corresponds to the wide range of data uses and analysis methods.

Current monitoring practices vary considerably. For example, a utility load research project tends to characterize the average performance of buildings with relatively few data points per building while an elaborate and complex monitoring project involves a test of new technology performance. Practitioners should develop accepted standards of monitoring practices to communicate results. A key element in this process is to classify the types of project monitoring and to build consensus on the purposes, approaches, and problems associated with each type (Misuriello 1987).

Monitoring projects can be broadly categorized by goals, objectives, experimental approach, and level of monitoring detail (Table 1). Other factors such as data analysis procedures, duration and frequency of data, and instrumentation are common to most, if not all, projects.

Aggregate Energy Use

Aggregate energy use projects monitor the average energy use of a representative sample of buildings or building systems. Typically, these projects have a small number of data points in each building. A large number of buildings is statistically analyzed to provide results with the desired level of accuracy. Utility load research projects, using special demand meters, and evaluations of state or utility energy conservation programs, using billing data analysis, are examples of this type of project.

Energy End Use

Energy end use projects focus on energy systems in particular buildings, typically for large samples. Monitoring usually requires a separate meter or collection channel for each end use, and analysts must account for all factors that may affect energy use. Examples of this approach include detailed utility load research efforts and projects to verify design or to simulate building energy systems. Depending on project objectives, the frequency of data collection may range from one-time measurements of full-load operation, to 15-min data for electric load analysis or to monthly submeter readings (Mazzucchi 1987).

Technology Assessment

Technology assessment projects monitor the field performance of specific equipment or technologies that affect building energy use, including envelope retrofit measures, major end uses (such as lighting), or mechanical equipment.

The typical goal of retrofit performance monitoring projects is to measure actual savings resulting from the retrofit, excluding the effects of weather, indoor conditions, or occupant behavior. The frequency and complexity of data collection depend on

The preparation of this chapter is assigned to TC 9.6, Systems Energy Utilization.

Table 1 Characteristics of Major Monitoring Project Types

Characteristic	Aggregate Energy Use	—Project Type— Energy End Use	Technology Assessment	Diagnostics
Goals and objectives	Infer average performance of large sample of buildings or of technology applications.	Determine characteristics of specific energy end uses within building.	Measure field performance of building system technology, or retrofit measure in individual buildings.	Solve problems. Measure physical or operating parameters that affect energy use, or that are needed to model building or system performance.
General approach	Large, statistically designed sample with small number of data points per building. Often uses whole-building utility meters.	Often uses large statistically designed sample. Monitor energy demand or use profile of each end use of interest within building.	Characterize individual building or technology, occupant, and operation. Account and correct for variations.	Typically uses one-time and/or short-term measurement with special methods, such as infrared imaging, flue gas analysis, blower door, and coheating.
Level of detail	Average performance of sample, not individual building. Cannot explain variations between buildings. Often uses seasonal or annual data.	Detailed data on end uses metered. Collect building and operating data that affect end use.	Uses detailed audit, submetering, indoor temperature, on-site weather, and occupant surveys. May use weekly, hourly, or short-term data.	Focused on specific building component or system. Amount and frequency of data varies widely between projects.
Examples	Load forecasting. Conservation or demand-side management program evaluation. Rate design.	Load forecasting. Identify energy conservation opportunities. Rate design.	Technology evaluation. Retrofit performance. Validate models and predictions.	Energy audit. Identify and solve operation and maintenance, indoor air qualities, or system problems. Provide input for models. Building commissioning.

project objectives. Projects in this category typically assess performance variations between individual buildings.

Field tests of end use equipment are characterized by detailed monitoring of all critical performance parameters and operational modes. Data are often needed to develop or verify algorithms predicting technology performance. The project scope may include reliability, maintenance, design, energy efficiency, sizing, and environmental effects (Hughes and Clark 1986).

Building Diagnostics

Diagnostic projects are designed to measure physical and operating parameters that determine the energy use of buildings and systems. The project goal is usually to determine the cause of problems, provide parameters to model energy performance, or isolate the effects of components. Diagnostic tests frequently involve one-time measurements or short-term monitoring (*i.e.*, one hour to one week); however, they sometimes require intermittent or long-term continuous data collection.

The residential sector, particularly single-family dwellings, probably has the greatest number of diagnostic measurement procedures. Typical measurements for single-family residences include (1) flue gas analysis to determine steady-state furnace efficiency, (2) fan pressurization tests to measure and locate building envelope air leakage, and (3) infrared thermography to locate thermal defects in the building envelope.

Some evolving diagnostic measurements for single-family residences include techniques to measure or determine (1) the efficiency of major end use appliances, such as furnaces, air conditioners, water heaters, and refrigerators; (2) the airtightness of air distribution systems (Modera 1989); (3) overall energy loss *UA*, *i.e.*, electric coheating or the dynamic signature of a building envelope (Subbarao 1986, 1988); and (4) the thermal signature of building envelope components (Duffy *et al.* 1988).

Diagnostic measurements in multifamily residential buildings are mostly in the developmental stage (Porterfield 1988). Some techniques are designed to determine the operating efficiency of steam and hot water boilers and to measure the air leakage between apartments (Modera *et al.* 1986).

Several levels of diagnostic measurements are used in commercial buildings. At the research level, techniques have been designed to measure the overall airtightness of building envelopes and the thermal performance of walls (Persily and Grot 1988). Practicing engineers also employ a host of diagnostic techniques to analyze energy performance. For example, short-term chart-recording measurements can be used to diagnose poor commercial building chiller performance (Misuriello 1988).

Diagnostic measurements are also well suited to support the development and implementation of building energy management programs (Chapter 32). In this role, diagnostic measurements can be used in conjunction with energy audits to identify existing energy usage patterns, to determine energy performance parameters of building systems and equipment, and to identify and quantify areas of inefficient energy usage (Misuriello 1988). Diagnostic measurement projects can generally be designed using procedures adapted to specific project requirements as appropriate (see Design Methodology).

Diagnostic measurement equipment may be installed temporarily or permanently for energy management data collection. Designers should consider providing permanent or portable check metering of major electrical loads in new building designs (ASHRAE 1989). The same concept can be extended to fuel and thermal energy use.

COMMON MONITORING ISSUES

Field monitoring projects are costly and involve unique challenges. A strong, lasting project commitment is necessary, as failing to anticipate or deal effectively with problems at any stage of implementation may endanger the usability of data.

Field monitoring projects also require effective management of various professional skills. Project staff must understand the end use equipment, data acquisition, and sensor technology. The logistics of field monitoring projects pose additional challenges, such as coordinating equipment procurement, delivery, and installation, as well as data collection, processing, and analysis.

Effective Planning

Many common problems can be avoided in monitoring projects by effective and comprehensive planning.

Project goals. Project goals and data needs should be established before selecting hardware. Unfortunately, projects are often hardware-oriented, rather than driven by project objectives, because monitoring hardware must be ordered several months before data collection or because project initiation procedures are lacking. As a result, the hardware may be inappropriate for the particular monitoring task or critical data points may be overlooked.

Data products. It is important to establish what data products are needed before selecting data points. Failure to plan these products first may lead to failure to answer critical research questions.

Commitment. A long-term commitment regarding personnel and resources is necessary. Project success depends on daily, long-term attention to detail, and staff continuity is essential.

Accuracy requirements. The accuracy requirements of the final data and the experimental design needed to meet these requirements should be determined early on. After specifying the required accuracy, the iterative process of choosing the sample size (number of buildings, control buildings, or pieces of equipment), and of determining the required measurement precision (including error propagation into the final data products), must be performed. This process is often complicated by a large number of independent variables (occupants, operating modes) and the stochastic nature of many variables (weather).

Implementation and Verification

The following steps can facilitate smooth project implementation and data verification:

- Track sensor performance on a regular basis. Detecting sensor failure or calibration problems quickly is essential. Ideally, this should be a daily task because dispatching troubleshooters to the site can be time-consuming. Moreover, the value of lost data is high, as reconstructing it takes time.
- Generate data on a timely, periodic basis. Problems often occur in developing final data products. These include missing data from failed sensors, data points not installed due to planning oversights, and anomalous data for which there are no explanatory records. By specifying data products as part of general project planning and by producing these products periodically, production problems can be identified and resolved as they occur.

Follow Through on Data Analysis and Reporting

Merely obtaining data is insufficient. Diligent analysis and reporting facilitate success. Make sure that the data collected and analyzed are the data needed, particularly by focusing on analytical selection of data points.

Pay close attention to resource allocation, ensuring that adequate resources are dedicated to data analysis tasks. As a quality control procedure and to make data analysis more manageable, these activities should be distributed evenly over time.

DESIGN METHODOLOGY

A methodology for designing effective field monitoring projects is described in this section. The task components and relationships among the nine activities constituting this methodology are identified in Figure 1. These activities fall into four categories: project management; project development; resolution and feedback; and production quality and data transfer. Field monitoring projects (resources, goals and objectives, data product requirements, or others) vary, thereby affecting how the methodology should be applied. Nonetheless, the methodology provides a proper framework, helping to avoid or minimize project implementation problems by up-front planning.

An iterative approach to planning activities is best. Accuracy and techniques can be adjusted based on cost estimates and resource assessments. The initial design should therefore be performed simply and quickly to estimate cost and evaluate resources. Formulation and reassessment are also suggested in specifying data products and analytical methods—a crucial step in obtaining high-quality results. After a reasonable design is developed, pretesting of the overall monitoring process (from data collection to producing final data products) is beneficial to identify and mitigate potential problems before full-scale monitoring begins.

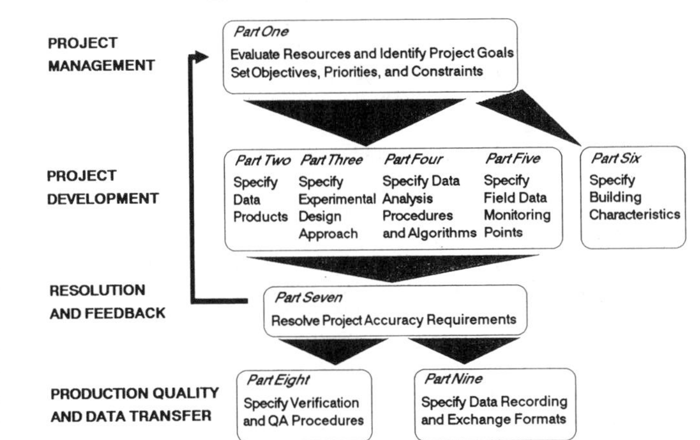

Fig. 1 Methodology for Designing Field Monitoring Projects

An iterative approach to project planning is also necessary when desired levels of instrumentation exceed resources available for the project. The planning process should identify and resolve various trade-offs necessary to execute the project within a given budget. In many instances, it is preferable to reduce the scope of the project rather than to relax instrumentation specifications or accuracy requirements.

Part One: Identify Project Goals, Objectives, and Research Questions

First formulate a statement of project outcome(s). The goal and objectives statement determines the overall direction and scope of the data collection and analysis effort. The statement should also list the research questions to be answered by the empirical data, noting the error or uncertainty associated with the desired result. A realistic assessment of error is needed, since too small an uncertainty will lead to an overly complex and expensive project.

Research questions can have varying scope and level of detail addressing entire systems or specific components as follows.

- Classes of buildings
  Example: To an accuracy of 20%, how much energy has been saved by using a mandated building construction/performance standard in this state?
- Particular buildings
  Example: Has a lapse in building maintenance caused energy performance to degrade?
- Particular components
  Example: What is the average reduction in demand charges due to installation of an ice storage system in this building?

Research questions also vary widely in technical complexity, generally in two levels:

1. How does the building/component perform?
2. Why does the building/component perform as it does?

The former question can sometimes be answered without detailed monitoring and analysis, whereas the latter usually requires more detailed monitoring and analysis, and thus more detailed planning.

Part Two: Specify Data Products and Project Output

This activity identifies the specific data products needed to answer critical research questions and provides a clear specification of the data to be delivered by the project.

Evaluation results can be presented in many forms, often as interim and final reports (by heating and/or cooling season), technical notes, or technical papers. These documents must convey specific results of the field monitoring clearly and concisely.

Data presentations and analysis summaries should be determined early to ensure that no critical parameters are overlooked (Hough *et al.* 1987). For instance, mock-ups of data tables, charts, and graphs should be prepared. Data products should also be prioritized to accommodate possible cost trade-offs or revisions resulting from error analysis (see Part 7). A worksheet can be used to determine information needs, facilitate organization of specific research questions and technology issues by project objectives, and identify corresponding information needs. Next, specific data presentation formats should be determined.

It is important to identify the type and amount of comparative and evaluation work to be performed. For example, any need for data normalization for weather or use profiles, along with the method to be employed, should be determined up front.

In specifying project data products, future information needs should be anticipated to provide project personnel with a more comprehensive understanding of testing requirements. For example, a project may have a short- and long-term data need (*e.g.*, demonstrating peak electrical demand reductions versus demonstrating cost-effectiveness or reliability to a target audience). The initial results on demand reductions may not be the ultimate goal, but rather a step towards later presentations on energy consumption and initial cost. Thus, it is prudent to consider long-term and potential future data needs so that additional qualitative data needs, such as arranging for photographs or testimonials, may be identified.

Part Three: Specify Experimental Design Approach

A general experiment design that defines two interacting factors—the number of buildings admitted to the study and the experimental approach—must be developed. For example, a less accurate approach can be considered if the number of buildings is increased, and vice versa. Note that if the goal is related to a specific product, the experiment design must isolate the effects of that product.

This step is particularly important because the total building performance is a complex function of several variables, all of which are subject to changes that are difficult to monitor and to translate into performance. When using a simpler approach, uncertainties and errors (noise sources) can make it difficult to detect performance changes of less than 20 to 30% (Fracastoro and Lyberg 1983).

In some cases, judgment should be used in selecting the number of buildings involved in the project. If an owner seeks information about a particular building, the choice is simple—the number of buildings in the experiment is fixed at one. However, for other monitoring applications, such as drawing conclusions regarding effects in a sample population of buildings, some degree of choice is involved. Generally, the error in the derived conclusions decreases as the square root of the number of buildings (Box 1978).

Accuracy requirements depend on the effect to be measured and on the specified experimental design and analysis methods. Without special measurements, energy savings of less than 20% of total consumption are difficult to detect from an analysis of utility bills. A specific project may be directed at:

1. Fewer buildings or systems with more detailed measurements
2. Many buildings or systems with less detailed measurements
3. Many buildings or systems with more detailed measurements

With projects of the first type, project accuracy requirements are usually resolved first by determining the expected excursions of measured quantities from the average response of the dependent variables to the driving signal of independent data quantities. For buildings, a typical concern is the response of heating and cooling loads (dependent variables) to temperature or solar input (independent variables). The response of building lighting energy to daylighting systems using natural light is also an example of the relation and effect of dependent and independent variables. The fluctuations in response are caused by outside influences not quantified by the measured data and by limitations and uncertainties associated with measurement equipment and procedures. Thus, accuracy must often be determined using statistical methods to describe mean tendencies of dependent data variables.

For projects of the second and third type, the increased number of buildings facilitates improved confidence in the mean tendencies of the dependent response(s) of interest. Larger sample sizes are also needed to use experimental designs with control groups, which are used to adjust for some outside influences. For further information, see Box (1978), Fracastoro (1983), and Palmer (1989).

Most monitoring procedures can be characterized as using one or several of the following general experimental approaches:

On-Off. If the retrofit or product can be activated or deactivated at will, energy consumption can be measured in a number of repeated on-off cycles. The on-period consumption is then compared to the off-period consumption (Woller 1989, Cohen *et al.* 1987).

Before-After. Building energy consumption is monitored before and after a new component or retrofit is installed. Changes in the weather and building operation during the two periods must be accounted for, requiring a model-based analysis (Hirst *et al.* 1983, Fels 1986, Robison and Lambert 1989, Sharp and MacDonald 1990).

Test-Reference. The building energy consumption data of two identical buildings, one with the product or retrofit, are compared. Because buildings cannot be absolutely identical (*e.g.*, different air leakage distributions, insulation effectiveness, temperature settings, and solar exposure), measurements should be taken prior to installation as well, allowing calibration. Once installed, the deviation from the calibration relationship can be attributed to the product (Levins and Karnitz 1986, Fracastoro and Lyberg 1983).

Simulated Occupancy. In some cases, the desire to reduce noise can lead the experimenter to postulate certain standard profiles for temperature set points, internal gains, moisture release, or window manipulation and to introduce this profile into the building by computer-controlled devices. The reference is often given by the test-reference design. In this case, both occupant and weather variations are nearly eliminated in the comparison (Levins 1986).

Nonexperimental Reference. A reference for assessing the performance of a building can be derived nonexperimentally by (1) using a normalized, stratified performance data base—energy use per unit area by some building-type classification (MacDonald and Wasserman 1989); or (2) by using a reasonable standard building, simulated by a calculated building energy performance model subject to the same weather and occupancy as the monitored building.

Engineering Field Test. When the experiment is focused on the performance of a test on a particular piece of equipment, the total building performance is not of primary interest. The building provides a realistic environment for testing the equipment, where reliability, maintenance, comfort, and noise may be considered in addition to energy. The whole-building energy advantages of the

Table 2 Advantages and Disadvantages of Common Experimental Approaches

Mode	Advantages	Disadvantages
Before/ After	No reference building required. Same occupants implies smaller occupant variations. Modeling processes will be mostly identical before/after.	Weather different before/after. More than one heating/cooling season may be needed. Model is required to account for weather and other changes.
Test reference	One season of data may be adequate. Small climate difference between buildings.	Reference building required. Calibration phase required (may extend testing to two seasons). Occupants in either or both buildings can change behavior.
On/Off	No reference building required. One season may be adequate. Modeling processes will be mostly identical before/after. Most occupancy changes will be small.	Requires reversible product. Cycle may be too long if time constants are large. Model is required to account for weather differences in cycles. Dynamic model accounting for transients may be needed.
Simulated occupancy	Noise from occupancy effects is eliminated. A variety of standard schedules can be studied.	Not "real" occupants. Expensive apparatus required. Building must be unoccupied with attendant extra cost.
Nonexperimental reference	Cost of actual reference building eliminated. With simulation, weather variation is eliminated.	Data base may be lacking in strata entries. Simulation errors and definition of reference problematic. With data base, weather changes usually not possible.
Engineering field test	Information focused on the product of interest. Minimal number of buildings required. Same occupants during the test.	Extensive instrumentation of product processes required. Models required to extrapolate to other buildings and climates. Occupancy effects not determined.

product are then derived using one of the above approaches. The equipment may be instrumented extensively.

Some of the general advantages and disadvantages of these approaches are listed in Table 2 (Fracastero 1983). Note that experimental design choices are not exclusive; for example, before-after and test-reference approaches have been successfully combined.

Some of the questions to be considered in the choice of experimental approach are:

- Can the product be turned on/off at will? If so, the on-off design offers considerable advantages.
- Are occupancy and occupant behavior critical? Changing building tenants, use schedules, internal gains, temperature set points, and natural or forced ventilation practices should be considered, since any one of these can ruin an experiment if not constant or accounted for.
- Are actual baseline energy performance data critical? In before-after designs, time must be allotted to characterize the before case to the same precision as the after case.
- Is it a test of an individual technology, or are multiple technologies, installed as a package, being tested? If individual impacts are desired, detailed component data and careful model-based analyses are required.
- Does the technology have a single or multiple mode(s) of operation? Can the modes be controlled to suit the experiment? If many modes are involved, testing over a variety of conditions and conducting model-based analysis will be necessary.

Part Four: Specify Data Analysis Procedures and Algorithms

Data are useless unless they are distilled into meaningful products allowing conclusions to be drawn. Too often, data are collected and never analyzed. This planning step focuses on specifying the data analysis procedures and algorithms and detailing how collected data will be processed to produce desired data products. This step determines which parameters will be assumed as constants and which will be calculated by actual continuous data or be monitored continuously as a field data point. Based on this information, monitoring practitioners should:

- Prepare a list of sensors and analysis constants to be determined in the field (*e.g.*, fan power, steady-state efficiency, and performance).
- Develop engineering calculations and equations (algorithms) necessary to convert field data to end products, including statistical methods and simulation modeling.
- Specify detailed items, such as the frequency of data collection and whether certain data must be obtained at different intervals, *i.e.*, 15-min demand data intervals within hourly data strings.

Data analysis procedures and algorithms should take into account established testing and rating standards. However, it is usually not practical to implement standards in the field. For example, maintaining the length of straight ductwork required for an airflow sensor is usually difficult in the field, necessitating compromise.

Algorithm inputs can be assumed values (such as the energy value of a unit volume of natural gas), one-time measurements (the leakage area of a house), or time-series measurements (fuel consumption and outside and inside temperatures at the site). The spatial level of the algorithms may pertain to (1) utility level aggregates of buildings, (2) particular whole-building performance, or (3) performance of instrumented components.

Analysis methods can generally be classified as empirical or model-based.

Empirical Methods. While empirical methods are the simplest, they have large uncertainty and generate little or no cause-effect information except for large sample sizes. The simplest empirical methods are based on annual consumption values, tracking the annual numbers and looking for degradation. Questions about building performance relative to other buildings are based on comparing certain performance indices between the building and an appropriate reference (for example, ANSI/ASHRAE/IES *Standard* 105-1984).

For commercial buildings, the most common index is the energy use intensity (EUI)—the annual consumption summed over all fuel types, divided by the floor area. Comparison is often made only on the basis of general building type, implying large variations in climate, building occupancy, and HVAC systems. Such variations can be accommodated to some degree by stratifying the data base from which the reference EUI is chosen. Although no empirical data base is currently large enough, generating such a data base is one of the motivations for standardizing monitoring practices. A common alternative is to use computer simulations to set reasonably well-stratified standards. Generally, however, simple empirical methods are not suited to retrofit applications because variations between the building and the reference cannot be accounted for.

Monthly EUI or billing data can also be used to detect billing errors, improper equipment operation during unoccupied hours, and seasonal space condition problems. Daily data are often used in these analyses, and raw hourly total building consumption data, when available, provides information on occupied versus unoccupied performance. Hourly, daily, monthly, and annual EUI across buildings can be directly compared when reduced to average power per unit area (power density).

Other normalizations besides floor area have been suggested. If the analyst is dealing mainly with heating concerns, it is useful to divide the annual EUI by heating degree-days, producing the building performance index (BPI). The BPI allows comparison across different climates or weather years. Other normalizations include dividing consumption by occupancy hours or by units of some basic product or service, such as meals served or rooms occupied.

Model-Based Methods. These techniques allow a wide range of additional data normalization to improve the accuracy of comparisons and provide cause-effect relations. To generate a model-based analysis, the analyst must carefully define the system and postulate a useful form of the governing energy balance equation. Explicit terms are retained for equipment or processes of particular interest. As part of the data analysis, whole building data (driving forces and thermal or energy response) are used to determine the model's thermally descriptive parameters. The parameters themselves can provide insight for diagnosing the building, although parameter interpretation can be difficult, particularly with time-integrated billing data methods. The model can then be used for a number of normalization processes as well as future diagnostic and control applications. Two general classes of models are used in analysis methods—time-integrated methods and dynamic techniques (Burch 1986).

Time-integrated methods. Based on algebraic calculation of the building energy balance, time-integrated methods are often used prior to data comparison to correct annual consumption for variations in outdoor temperature, internal gains, and internal temperature (Fels 1986, Busch *et al.* 1984, Haberl and Claridge 1987). This type of correction is essential for most retrofit applications.

Time-integrated methods can be used with whole-building energy consumption data (billing data) or with submetered end use data. For example, standard time-integration methods are often used to separately integrate end use consumption data (heating, cooling, domestic water heating) for comparison and analysis. Time-integrated methods are generally reliable, as long as three conditions are accounted for:

- *Appropriate time step.* Generally, the time step should be as long, or longer, than the response time of the building or building system for which energy use is being integrated using a model-based method. For example, the response of daylighting controls to natural illumination levels (which result in changes to lighting energy use) can be rapid, allowing short time steps for data integration. By contrast, the response of cooling systems energy use to changes in cooling load (from internal gains, ambient temperature, and solar gains) can be comparatively slow. Thus, the choice of time step depends on building and system characteristics. In this instance, a sufficiently long time step should be used that will average over these slow variations, or a dynamic model should be used instead. In general, selecting an appropriate time step should account for the physical behavior of the energy system(s) and how this behavior is expressed in model parameters.
- *Linearity of model results.* Generally, time-integrated models should not be applied to model parameters used to estimate nonlinear effects. Air infiltration, for example, is nonlinear when estimated using wind speed and indoor/outdoor temperature difference parameters in certain models. Estimation errors would result if these parameters were independently time-integrated and then used to calculate air infiltration. This problem can be avoided by modeling such nonlinear effects at each time step (1 h).
- *End use uniformity within data set.* Analysts who apply time-integrated models to end use data sets should ensure that the particular data set is uniform, *i.e.*, that it does not inadvertently contain observations with measurements of end uses other than those intended. During mild weather, for example, HVAC systems may provide both heating and cooling over the course of a day, creating data observations of both heating and cooling measurements. In a time-integrated model of heating energy usage, these cooling energy observations will lead to error. In such cases, observations should be identified or otherwise flagged as to their true end use.

For whole-building energy consumption data (billing data), reasonable results can be expected from heating analysis models when the building is dominantly responsive to inside-outside temperature difference. Note that the billing data analysis method yields little of interest when internal gains are large compared to skin loads, as in large commercial buildings and industrial applications. Daily, weekly, and monthly whole-building heating season consumption integration steps have been employed (Fels 1986, Ternes 1986, Sharp 1990). As might be expected, cooling analysis results have been less reliable, since cooling load is not strictly proportional to cooling degree-days (Fels 1986). Problems will also arise when solar gains are dominant and vary by season.

Dynamic Techniques. Dynamic models—macrodynamic (whole-building model) or microdynamic (component specific)—offer great promise for reducing monitoring duration and increasing conclusion accuracy. Further, individual effects from multiple measures and system interactions can be examined explicitly. Dynamic whole-building analysis will generally accompany detailed instrumentation of specific technologies.

Dynamic techniques pose a dynamic physical model for the building, adjusting the parameters of the model to fit the experimental data (Subbarao 1988, Duffy 1988). In residential applications, computer-controlled electric heaters can be used to maintain a steady interior temperature overnight, extracting from this data an experimental value for the building steady-state load coefficient. These methods can also use a cool down period to extract information on building internal thermal mass storage. Daytime data can be used to renormalize the response (computed from a microdynamic model) of the building to solar radiation, which is particularly appropriate for buildings with glazing areas over 10% of the building floor area (Subbarao 1986). Once the data with electric heaters has been taken, the building can be used as a dynamic calorimeter to assess the performance of auxiliary heating and cooling systems.

Similar techniques have been applied to commercial buildings (Norford *et al.* 1985, Burch *et al.* 1990). In these cases, delivered energy from the HVAC must be directly monitored in lieu of using electric heaters. Because ventilation is a major, variable term in the building energy balance, the outside airflow rate should also be directly monitored. Simultaneous heating and cooling, common in large buildings, requires a multizone treatment which has not been adequately tested in any of the dynamic techniques.

Table 3 provides a guide to analysis method selection. The error quotations are intended as rough estimates for a single building scenario.

Table 3 Whole Building Analysis Guidelines

	Class of Method		
Project Goal	**Empirical (Billing Data)**	**Time-Integrated Model**[a]	**Dynamic Model**
Building evaluation	Yes, but expect fluctuations in 20 to 30% range.	Yes, extra care needed beyond 15% uncertainty.	Yes, extra care needed beyond 10% uncertainty.
Building retrofit evaluation	Not generally applicable. Requires various normalization techniques for reasonable accuracy.	Yes, but difficult beyond 15% uncertainty. Method cannot distinguish multiple retrofit effects.	Yes, can resolve 5% change with short-term tests. Can estimate multiple retrofit effects.
Component evaluation	Not applicable.	Not applicable unless submetering is done to supplement.	Yes, about 5% accuracy, but best with submetering.

Note: Error figures are approximate for total energy use in a single building. All methods improve with selection of more buildings.

[a]Accuracy can be improved by decreasing time step to weekly or daily. These methods are of little use when $T_{outdoor}$ approaches $T_{balance}$.

Part Five: Specify Field Data Monitoring Points

Careful specification of field monitoring points is critical in planning and organizing a monitoring project. Its purpose is to identify the variables that need to be monitored or measured in the field to produce the required data.

Because metering projects are dynamic, special consideration must be given to identifying and controlling significant changes in climate, systems, and operation during the monitoring period. Additional monitoring points may be required to measure variables that are assumed to be constant, insignificant, or related to other measured variables in order to draw sound conclusions from the measurements. Since the necessary data may be obtained in several ways, it is recommended that the data analysts, equipment installers, and data acquisition system engineers work together to develop the tactics that best suit the project requirements.

The costs of data collection are nonlinear functions of the number, accuracy, and duration of measurements which must be considered while planning within budget constraints. If the extent of data applications, such as in research projects, is unknown, consideration should be given to the value of other concurrent measurements, since the incremental cost of alternative analyses may be small.

For any project involving large amounts of data, data quality should be verified in an automated fashion. While this approach may require adding monitoring points to facilitate energy balances or redundancy checks, the added costs are likely to be offset by savings in data verification for large projects.

If multiple sites are to be monitored, common protocols for the selection and description of all field monitoring points should be established so that data can be more readily verified, normalized, compared, and averaged. The protocol also adds consistency in selecting monitoring points. Pilot installations should be conducted to provide data for a test of the system and to ensure that the necessary data points have been properly specified and described.

Certain general considerations include:

- Evaluating the equipment thoroughly under actual test conditions before committing to large-scale procurement. Pay particular attention to any sensitivity to power outages and lightning.
- Avoiding state-of-the-art and untried systems. Concern regarding the ability or reliability of data acquisition equipment should be avoided.
- Considering the costs and benefits of remote data interrogation and programming.
- Verifying vendor claims by calling references or obtaining performance guarantees.

An examination of the analysis method will determine the data to be measured in the field. The simplest methods require no on-site instrumentation; as the methods become more complex, data channels increase. For engineering field tests conducted with dynamic building methods, up to 100 data channels may be required.

Particular attention must be paid to sensor location. For example, if the method requires an average indoor temperature, examine the potential for internal temperature variations, which often require data from several temperature sensors to be averaged.

After the channel list is compiled, sensor accuracy and scan rates should be assigned. Some sensors, such as indoor and outdoor temperature sensors, may require low scan rates (once every 5 min); others, such as total electric or solar radiation sensors, may contain high-frequency transients that require rapid sampling (every 10 s). The maximum sampling rate is usually programmed into the logger, and averages are stored at a specified time step (hourly). Some loggers can scan different channels at different rates. The logger's interrupt capability can also be used for rapid, infrequent transients. Interrupt channels will signal the data logger to start monitoring an event only once it begins. In some cases, the on-line computation of derived quantities must be considered. For example, if heat flow in an air duct is required, it might be computed from a differential temperature measurement multiplied by an air mass flow rate that was determined from a one-time measurement. However, it should only be computed and totaled when the fan is operating.

Once the required field data monitoring points are specified, these requirements should be clearly communicated to all members of the project team to ensure that the actual monitoring points are accurately described. This can be accomplished by publishing handbooks for measurement plan development and equipment installation and by outlining procedures for diagnostic tests and technology assessments.

Scanning and recording intervals. The frequency of data measurements and data storage can have an impact on the accuracy of project output. Scanning differs from storage in that data channels may be read (scanned) once per second, for example, while data may be recorded and stored every 15 min. Most data loggers maintain temporary storage registers, accumulating an integrated average of channel readings from each scan. The average is then recorded at the specified storage interval.

The data scan rate must be sufficiently fast to ensure that all significant effects are monitored; the frequency of data recording varies by data channel. As mentioned previously, transient and/or random events are often best monitored using the interrupt capabilities of modern data loggers.

Sensors and data acquisition systems. Working from the field data list, select sensors to obtain each measurement. Next, specify conversion and proportion constants for each sensor type on the field data list, and note the accuracy, resolution, and repeatability of each sensor.

Because hardware needs vary considerably by project, specific selection guidelines are not provided here. However, general

Table 4 General Characteristics of Data Acquisition Systems

Types of Data Acquisition Systems (DAS)	Typical Use	Typical Data Retrieval	Comments
Manual readings	Total energy use	Monthly or daily written logs	Human factors may affect accuracy and reading period. Data must be manually entered for computer analysis.
Pulse counter, cassette tapes (1 to 4 channels)	Total energy use (some end use)	Monthly pickup of cassette tapes	Data loss due to cassette is a common problem. Pulse data must be read and converted before it can be analyzed.
Pulse counter, solid state (1,4, or 8 channels)	Total energy use (some end use)	Telephone protocols to mainframe or minicomputer	Computer hardware and software for transfer and conversion of pulse data is needed and can be expensive. Can handle large numbers of sites. User friendly.
Plug-in A/D boards for personal computers (PC)	Diagnostics, technology assessment, and control	On-site real-time collection and storage	Usually small quantity, unique applications. PC programming capability needed to set up data software and configure boards for application.
Simple field DAS (usually 16 to 32 channels)	Technology assessment, residential end use (some diagnostics)	Phone retrieval to host computer for primary storage (usually daily to weekly)	Can use PCs as hosts for data retrieval. Good A/D conversion available. Low cost per channel. Requires programming skills to set up field unit and configure communications for data transfer.
Advanced field DAS (usually >40 channels/units)	Diagnostics, energy control systems commercial end use	On site real-time collection and data storage, or phone retrieval	Usually designed for single buildings. Can be PC-based or stand-alone unit. Can run applications/diagnostic programs. User friendly.

characteristics of generic data acquisition hardware components are shown in Table 4. Some typical concerns for selecting data acquisition hardware are outlined in Table 5.

Part Six: Specify Building Characteristics

The relevance of measured energy data cannot be fully understood unless the characteristics of the building(s) being monitored and its use (activities) are known. As part of planning the monitoring project, develop a data structure to describe the buildings involved.

Building characteristics can be collected at many levels of detail, but it is important to provide at least enough detail to understand:

- General building configuration and envelope (particularly energy-related aspects)
- Building occupant information (number, occupancy schedule, activities)
- Internal loads
- Type and quantity of energy-using systems
- Any building changes over the course of the monitoring project

The minimum level of detail is known as summary characteristics data (Ternes 1986). Simulation level characteristics—detailed information collected for hourly simulation model input—may be desirable for some buildings. Regardless of the level of detail, however, the data should provide a context for analysts, who may not be familiar with the project, to understand the building(s) and how energy is used.

Building characteristics information should be collected in four areas:

1. Building descriptive information, summarizing key building envelope and internal heat gain parameters to be collected
2. HVAC system descriptive information, characterizing key parameters affecting HVAC system performance
3. Entrance interview information, focusing on the energy-related behavior of building occupants prior to monitoring
4. Exit interview information, documenting physical or lifestyle changes at the test site that may affect data analysis

Part Seven: Resolve Data Product Accuracies

System accuracy requirements and equipment selection can be determined by addressing the following factors, typically in an iterative manner.

Determining Measurement System Uncertainties. To determine the overall accuracy and precision of the final data products, all sources of bias and uncertainty must be included. The precision of the final data products can usually be directly determined from the precision of the sensor/data acquisition systems; the relationship between the variables being measured due to stochastic variables such as weather, occupant behavior, operational variations, and the length of the measurement period; and the number of cases studied.

On the other hand, the accuracy (or potential bias) of the final data cannot be improved by monitoring more buildings or including more heating seasons and is usually not dominated by the accuracies associated with the sensors. Rather, the accuracy of the data is usually controlled by limitations on the number and type of measurements made (single-point temperature measurements, one-time flow measurements), and on the simplifications required to analyze and interpret the data. Many of these simplifications are associated with analyses or physical models

Table 5 Practical Concerns for Selecting Data Acquisition Hardware

FDAS Component	Field Application Concerns
Data logger unit and peripherals	• Select equipment for field application. • Equipment should store data in electronic form such as on floppy disc or magnetic tape. • Remote programming capability should be available to minimize on-site software modifications. • Avoid equipment with cooling fans. • Use high quality, proven modems.
Cabling and interconnection hardware	• Use only signal grade cable-shielded, twisted-pair with drain wire. • Mitigate sources of common mode and normal mode signal noise.
Sensors	• Use rugged, reliable sensors that are rated for field application. • Use signal splitter if sharing existing sensors or signals with other recorders or energy management control system (EMCS). • Operate sensors at 50 to 75% of full scale. • Choose sensors that do not require special signal conditioning.

used (see Part 4) to account for the stochastic or random nature of the parameters being measured and for the indirect nature of most of the measurements.

Once a specific procedure (algorithm or equation) for obtaining final data from physical measurements has been established, standard techniques that incorporate measurement uncertainties into the final data product are available. The general procedure for determining the uncertainty in a data product Y as a function of the uncertainties in the measured parameters x_i is to assume that the uncertainties of the measured parameters are small compared to the mean parameter values (this is not always true), in which case the equation for the final data products can be expanded in a Taylor series, neglecting the higher order terms. Thus, the overall uncertainty (or standard deviation) in the data product Y can be expressed in terms of the uncertainties in the measured parameters σx_i as:

$$\sigma_Y = \sqrt{\left(\frac{\partial Y}{\partial x_1}\right)^2 \sigma_{x_1}^2 + \left(\frac{\partial Y}{\partial x_2}\right)^2 \sigma_{x_2}^2 + \left(\frac{\partial Y}{\partial x_1}\right)\left(\frac{\partial Y}{\partial x_2}\right)\sigma_{x_1 x_2} + \ldots} \quad (1)$$

As the errors in a given measurement do not correlate with (are independent of) the errors in another measurement, the covariance terms, $\sigma_{x_i x_j}$ are usually zero. For example, if the errors in x_1 and x_2 are uncorrelated, the percentage uncertainty in $Y = x_1 x_2$ reduces to:

$$\frac{\sigma_Y}{\bar{Y}} = \sqrt{\left(\frac{\sigma_{x_1}}{\bar{x}_1}\right)^2 + \left(\frac{\sigma_{x_2}}{\bar{x}_2}\right)^2} \quad (2)$$

On the other hand, if the errors are still uncorrelated, but $Y = x_1 + x_2$, the percentage uncertainty in Y becomes:

$$\frac{\sigma_Y}{\bar{Y}} = \sqrt{\left(\frac{\sigma_{x_1}}{\bar{Y}_1}\right)^2 + \left(\frac{\sigma_{x_2}}{\bar{Y}_2}\right)^2} \quad (3)$$

Estimating the accuracy (or potential bias) of the data products based on a given experimental plan is more complicated. The usual technique for quantifying the accuracy (or bias) in final data products is to make a statistically significant set of Monte Carlo simulations of building/system operation and performance. By simultaneouly enacting the measurement plan, the bias of the final data products is determined by comparing the mean values of the experimentally determined data products with those based on the known (assumed) performance of the building and system. Such a procedure also provides a direct measure of the overall uncertainty in the final data products based on the observed scatter of the measurement-based results.

This step represents one part of the iterative procedure associated with proper experimental design. If the final data product uncertainty based on the above evaluation procedure is unacceptable, the uncertainty can be reduced by (1) reducing overall measurement uncertainty (improving sensor precision), (2) increasing the duration of the monitoring period (to average out stochastic variations), and/or (3) increasing the number of buildings tested. On the other hand, if simulations indicate that the expected bias in the final data products is unacceptable, the bias may be reduced by (1) employing additional sensors to get an unbiased measurement of the desired quantity, (2) using more detailed models and analysis procedures, and/or (3) increasing the data acquisition frequency (in combination with a more detailed model) to eliminate biases due to sensor or system nonlinearities.

Review System Accuracy and Cost Requirements. The process of specifying system accuracies is iterative. After the first round of analysis and decision making, cost constraints and accuracy levels should be reviewed. Adjustments should be made based on accuracy needs and costs. Several iterations of determining the overall system accuracy and specific equipment options may be necessary.

Part Eight: Specify Verification and Quality Control Procedures

Establishing and using data quality assurance (QA) procedures is essential to the successful implementation of a field test. The entire data path, from sensor installation to procedures that generate results for the final report, must be subjected to rigorous verification tests. Moreover, the data flow path should be checked routinely for failure of sensors or test equipment, as well as unexpected or unauthorized modifications to equipment.

Quality assurance requires often complex data handling of the monitored projects. Any building energy monitoring project collects data from sensor(s) and manipulates that data into results. The process entails several steps—the data must be read, collected, stored, verified, processed (into a suitable format), archived, applied in analysis, incorporated in results, and published. In a project with only a few sensors and required readings, this process can be handled with a relatively simple data flow on paper. Computers, which are generally used in one or more stages of the process, require a different level of process documentation since much of what occurs has no direct paper trail.

Computers also facilitate the collection of large data sets and increase project complexity. To achieve the maximum automation for the entire process, computers require the development of specific software. Often, separate computers are involved in each step, so passing information from one computer system to another must be automated in large projects. To move the data as smoothly as possible, an automated data pipeline should be developed. This pipeline minimizes the time delay from data collection to the production of results and maximizes the cost-effectiveness of the entire project.

An automated data quality or data verification procedure should be incorporated in the system. If possible, this procedure should operate in close to real-time mode and automatically notify system operators of data quality failures. This method of operation minimizes data loss due to equipment failures and/or changes at the instrumented site. It also allows the processed information to be applied quickly.

The following QA actions should take place:

1. Calibrate hardware to establish a good control procedure to facilite the collection of good data.
2. Verify data, check for reasonableness, and prepare a summary report to assure the quality of the data after it is collected.
3. Diagnose the data, since significant findings may lead to changes in data quality checking procedures. The diagnostic procedure itself must be disciplined and well documented.
4. Thoroughly document and control procedures applied to remedy problems. These procedures may entail changes in hardware or collected data (such as data reconstruction), which can have a fundamental impact on the collected data.

Table 6 presents the three aspects of a monitoring project that require quality assurance (in developing and applying QA procedures)—hardware, engineering data, and characteristics data. Three QA reviews are necessary for each of these aspects. Initial QA confirms that the project starts correctly; ongoing QA is necessary to confirm that the information collected by the project continues to satisfy quality requirements; and periodic QA are additional checks, established at the beginning of the project, which ensure continued performance at acceptable levels of quality.

Information about the data quality and the quality assurance process should be readily available to data users. Otherwise,

Table 6 Table of Quality Assurance Elements

Timeframe	Hardware	Engineering Data	Characteristics Data
Initial start-up	Bench calibration (1) Field calibration (1) Installation verification (1)	Installation verification (1) Collection verification (1,2) Processing verification (1,2) Result production (1,2)	Field verification (1) Completeness check (1) Reasonableness check (1,2) Result production (1,2)
Ongoing	Functional testing (1) Failure mode diagnosis (3) Repair/maintenance (4) Change control (1)	Quality checking (2) Reasonableness checking (2) Failure mode diagnosis (3) Data reconstruction (4) Change control (1)	Problem diagnosis (3) Data reconstruction (4) Change control (1)
Periodic	Preventative maintenance (1) Calibration (1)	Summary report preparation and review (2)	Scheduled updates/resurveys (1) Summary report preparation and review (2)

(1) Actions to ensure good data. (2) Actions to check data quality. (3) Actions to diagnose problems. (4) Actions to repair problems.

significant analytical resources may be expended to determine data quality, or the analyses may never be performed due to uncertainties.

Part Nine: Specify Recording and Data Exchange Formats

This step involves specifying the formats in which the data will be supplied to the end user or other data analysts. Both raw and processed (adjusted for missing data or anomalous readings) data formats should be specified. In addition, if any supplemental analyses are planned, the media and format to be used (magnetic tape, disc, spreadsheet, ASCII) should be specified. These requirements can be determined by analyzing the software data format specifications. A common format for raw data is comma-delineated ASCII, which does not require data conversion.

Data documentation is essential on all monitoring projects, especially when several organizations are involved. Data usability is facilitated by specifying and adhering to data recording and exchange formats. Most data transfer problems are related to inadequate documentation. Other problems include hardware or software incompatibilities, errors in the tape or disc, errors or inconsistencies in the data, and transmittal of the wrong data set. Some of these problems can be mitigated by the following:

- Provide documentation to accompany the data exchange (Table 7). Because these guidelines apply to general data, models, programs, and other types of information, all items listed in Table 7 may not apply to every case.
- Provide documentation of transfer media, including computer operating system; software used to create files; media format (ASCII, EBCDIC, binary); tape or disc characteristics (tracks, density, record length, block, size).
- Provide procedures to check the accuracy and completeness of the data transfer, including statistics or frequency counts for variables, hard copy versions of the file, and test input data and corresponding output results for models on other programs.

PROTOCOLS FOR RETROFIT PERFORMANCE MONITORING

Residential Retrofit Monitoring

Protocols for residential building retrofit performance can answer specific research questions associated with the actual measured performance (Ternes 1986). Discrepancies between predicted and actual performance, as measured by the energy bill, are common. Protocols improve on previous methods in two ways: (1) internal temperature is monitored, which eliminates a major source of noise or unknown variable in data interpretation; and (2) data are taken more frequently than monthly, which potentially shortens the monitoring duration. (Bill analysis generally requires a full season of pre- and postaction data; this approach may require only a single season.)

For example, Ternes (1986) developed a single-family retrofit monitoring protocol, which consists of a data specification guideline that identifies important parameters to be measured. Both one-time and time-sequential data parameters are covered, and the parameters are defined sufficiently to ensure consistency and comparability between experiments. The guideline identifies minimum data; this must be collected in all field studies that use the guideline. It also describes optional extensions to the minimum data set that can be used to study additional issues, such as those related to occupant behavior, effects of microclimate, and effects of the distribution system on energy performance. Szydlowski and Diamond (1989) have developed a similar method for multifamily buildings.

The single-family retrofit monitoring protocol recommends a before-and-after experimental design approach, and the minimum data set allows performance to be measured on a normalized basis with weekly time-series data. The minimum time-series data for heating and/or cooling consumption are total consumption by fuel type and indoor temperature. The protocol also allows hourly recording intervals for time-integrated parameters—an extension of the basic data requirements in the minimum data set. The

Table 7 Documentation Included with Computer Data to be Transferred

1. Title and/or acronym
2. Contact person (name, address, phone number)
3. Description of file (number of records, geographic coverage, spatial resolution, time period covered, temporal resolution, sampling methods, uncertainty/reliability)
4. Definition of data values (variable names, units, codes, missing value representation, location, method of measurement, variable derivation)
5. Original uses of file
6. Size of file (number of records, bytes)
7. Original source (person, agency, citation)
8. Pertinent references (complete citation) on materials providing additional information
9. Appropriate reference citation for this file
10. Credit line (for use in acknowledgments)
11. Restrictions on use of data/program
12. Disclaimer (examples listed below)
    (a) Unverified/unvalidated data; use at your own risk
    (b) Draft data; use with caution
    (c) Clean data to the best of our knowledge. Please let us know of any possible errors or questionable values.
    (d) Program under development
    (e) Program tested under the following conditions...(conditions specified by author)

Table 8 Data Parameters for Residential Retrofit Monitoring

	Recording Period	
	Option 1	Option 2
Basic Parameters		
House description	a	a
Space conditioning system description	a	a
Entrance interview information	a	a
Exit interview information	a	a
Pre- and post-retrofit infiltration rates	a	a
Metered space conditioning system performance	a	a
Retrofit installation quality verification	a	a
Heating and cooling equipment energy consumption	weekly	hourly
Weather station climatic information	weekly	hourly
Indoor temperature	weekly	hourly
Indoor humidity	—	hourly
House gas or oil consumption	weekly	hourly
House electricity consumption	weekly	hourly
Wood heating use	—	hourly
Domestic hot water energy consumption	weekly	hourly
Optional Parameters		
Occupant behavior		
Additional indoor temperatures	weekly	hourly
Heating thermostat set point	—	hourly
Cooling thermostat set point	—	hourly
Indoor humidity	weekly	—
Microclimate		
Outdoor temperature	weekly	hourly
Solar radiation	weekly	hourly
Outdoor humidity	weekly	hourly
Wind speed	weekly	hourly
Wind direction	weekly	hourly
Shading	a	a
Shielding	a	a
Distribution system		
Evaluation of ductwork infiltration	a	a

[a]One-time measurements

minimum one-time data measurements cover descriptive information on house and space conditioning systems, occupant interviews at the beginning and end of the field study period, infiltration measurements at the beginning and end of the field study period, performance measurement of space conditioning systems, and an assessment of retrofit installation quality.

The minimum data set may also be extended through optional data parameter sets for users seeking more analytical information. This protocol has standardized the experimental design and data collection specifications, enabling independent researchers to compare project results more readily. Moreover, the approach of both minimum and optional data sets and two recording intervals accommodates research projects of varying financial resources.

The data parameters in this protocol have been grouped into four data sets: basic, occupant behavior, microclimate, and distribution system (Table 8). The minimum data set consists of a weekly option of the basic data parameter set. Time-sequential measurements are monitored continuously throughout the field study period. These are all time-integrated parameters, *i.e.*, the appropriate average value, rather than the instantaneous value, of a parameter over the recording period.

This protocol also addresses instrumentation installation, accuracy, and measurement frequency and expected ranges for all time-sequential parameters (Table 9). The minimum data set (weekly option of the basic data) must always be collected. At the user's discretion, hourly data may be collected, which allows two optional parameters to be monitored. Parameters from the optional data sets may be chosen, or other data not described in the protocol added, to arrive at the final data set.

Commercial Retrofit Monitoring

A protocol has been developed for use in field monitoring studies of energy improvements (retrofits) for commercial buildings (MacDonald *et al.* 1989). Similar to the residential protocol, it addresses data requirements for monitoring studies. Commercial buildings are, however, more complex, with a diverse array of potential efficiency improvements. Consequently, the approach to specifying measurement procedures, describing buildings, and determining the range of analysis must differ.

The strategy used for this protocol is to specify data requirements for projects, analysis, performance data with optional extensions, and a building core data set that describes the field performance of efficiency improvements. The key advance made in developing this protocol is the specification of those building characteristics deemed important in determining the building energy consumption. Since data are often collected by different investigators, this protocol helps to ensure that data from different projects are reasonably consistent and comparable.

When using this protocol, data should be reported in the following categories:

1. Project or program description—general information including identification of the project or program, why it was conducted, and what improvements were made to the buildings or systems studied.
2. Analysis methods and results—summary of analysis (evaluation) methods, experimental design, and project results.
3. Performance data—summary of monthly (billing) data, submetered or detailed energy consumptions, inclusion of demand data (if any), and temperature and weather data.
4. Building description data—survey data describing each building and associated building systems, functional use areas, tenants, schedules, base energy data, and energy improvements. Core data must be provided to describe each building, before and after any energy improvements are made (two distinct time periods).

This protocol requires that the approach used for analyzing building energy performance be described. The description must cover the data used in the performance calculations and the methods used to account for any performance variations caused by changes in building characteristics. The model or equations used to describe building or system performance must also be specified.

In addition, the general experimental design must be indicated. Common designs such as side-by-side (test-reference) testing, before-and-after testing, on/off testing, or control-treatment group testing, can simply be listed on the form. Other designs should be fully described.

The analysis of retrofit performance must be provided. Results must include both cost and energy savings (including demand savings for electricity, if applicable). Savings may be reported as a percentage of some cost or energy quantity, but the absolute savings ($/year and energy use per year) must be reported. Cost savings must indicate the breakdown between energy cost savings, electric kilowatt demand cost savings, and operation and maintenance (O&M) cost savings. The results should indicate any useful data extractions, aggregations, combinations, and normalization methods.

The cost for implementing energy improvement, and any additional costs for increased demand or changes in O&M should also be reported. In addition, the cost of implementation and O&M changes should be used to report measures of cost-effectiveness.

The level of confidence in the results should also be reported. Confidence can be either a statistical determinant or a qualitative description and should also include a discussion of the expected longevity of the results. For example, if an initial study on building performance included some short-term testing (assume 4

Table 9 Time-Sequential Parameters for Residential Retrofit Monitoring

Data Parameter	Accuracy[a]	Range	Stored Value per Recording Period	Scan Rate[b]	
				Option 1	Option 2
		Basic Parameters			
Heating and cooling equipment energy consumption	3%		Total consumption	15 s	15 s
Indoor temperature	1.0°F	50 to 95°F	Average temperature	1 h	1 min
Indoor humidity	5% rh	10 to 95% rh	Average humidity		1 min
House gas or oil consumption	3%		Total consumption	15 s	15 s
House electricity consumption	3%		Total consumption	15 s	15 s
Wood heating use	1.0°F	50 to 800°F	Average surface temperature or total use time		1 min
Domestic hot water	3%		Total consumption	15 s	15 s
		Optional data parameter sets			
Occupant behavior					
Additional indoor temperatures	1.0°F	50 to 95°F	Average temperature	1 h	1 min
Heating thermostat set point	1.0°F	50 to 95°F	Average set point		1 min
Cooling thermostat set point	1.0°F	50 to 95°F	Average set point		1 min
Indoor humidity	5% rh	10 to 95% rh	Average humidity	1 h	
Microclimate					
Outdoor temperature	1.0°F	−40 to 120°F	Average temperature	1 h	1 min
Solar radiation	10 Btu/h·ft²	0 to 350 Btu/h·ft²	Total horizontal radiation	1 min	1 min
Outdoor humidity	5% rh	10 to 95% rh	Average humidity	1 h	1 min
Wind speed	0.5 mph	0 to 20 mph	Average speed	1 min	1 min
Wind direction	5°	0 to 360°	Average direction	1 min	1 min

[a]All accuracies are ± stated values.
[b]Applicable scan rates if nonintegrating instrumentation is employed.

months after the improvement was made), the results should indicate their validity for the approximately 4 months after the improvement was made. If a later report covered a 2-year period after the change was made, this length of time would be indicated in a follow-up report. Availability of results data and contact information should also be reported.

The objectives and resources of different projects usually influence the amount and detail of monitored data at two levels: (1) monthly (billing), and (2) at more detailed field recording intervals. Monthly data are required as a minimum, but submetered data or more detailed whole-building data are often necessary to provide more meaningful results.

Since monitoring at short time intervals can produce a large volume of data, the maximum amount of submetered or detailed monitored data that can be accepted varies, depending on the recording interval. The necessary performance data, including identification of a minimum data set, are outlined in Table 10.

The minimum data set was selected to obtain as much information about a building and its performance as possible, while maximizing the number of projects that could participate. By selecting readily available billing and outdoor temperature data as minimum requirements, the buildings where energy improvements have been made without submetering will also provide data.

Pre- and post-retrofit measuring periods should be consecutive, ensuring that outside operational conditions remain as constant as possible (*e.g.*, minimize changes in occupancy, operating schedules, operating modes, building use, or other variables). If these variables could all be held constant, a change in energy use could be attributed to the effect of the retrofit.

The protocol is intended to promote more consistent results to better understand the field performance of energy improvements in these buildings, to facilitate comparisons of energy use and savings between buildings, to support continued advances in reporting of energy improvement results, and to provide data for use in estimating regional benefits.

EXISTING TEST PROCEDURES AND POTENTIAL PROTOCOL APPLICATIONS

In addition to the specialized protocols for particular monitoring applications, a number of specific laboratory and field measurement standards exist (see Chapter 50) and many monitoring sourcebooks are also in circulation (*e.g.*, Fracastoro and Lyberg 1983).

REFERENCES

ANSI/ASHRAE. 1984. Standard methods of measuring and expressing building energy performance. ANSI/ASHRAE *Standard* 105-1984.

ASHRAE. 1989. Energy efficient design of new buildings except low-rise residential buildings. ASHRAE/IES *Standard* 90.1.

Box, G.E.P. 1978. *Statistics for experimenters: An introduction to design, data analysis and model-building*. John Wiley, New York.

Burch, J.D. 1986. Thermal performance monitoring methods. *Passive Solar Journal* 3.

Burch, J.D., K. Subbarao, A. Lekov, M. Warren, L. Norford, and M. Krarti. 1990. Short-term energy monitoring in a large commercial building. ASHRAE *Transactions* 96(1):1459-77.

Busch, J.F., A.K. Meier, and T.S. Nagpal. 1984. Measured heating performance of new, low-energy homes: Updated results from the BECA-A database. LBL-17883, Lawrence Berkeley Laboratory, Berkeley, CA.

Cohen, R.R., P.W. O'Callaghan, S.D. Probert, N.M. Gibson, D.J. Nevrala, and G.F. Wright. 1987. Energy storage in a central heating system: Spa school field trial. *Building Service Engineering Research Technology* (Great Britain) 8:79-84.

Duffy, J.J., D. Saunders, and J. Spears. 1988. Low-cost method for evaluation of space heating efficiency of existing homes. Proceedings of the 12th Passive Solar Conference, ASES, Boulder, CO.

Fels, M.F., ed. 1986. Measuring energy savings: The scorekeeping approach. *Energy and Buildings* 9.

Haberl, J.S. and D.E. Claridge. 1987. An expert system for building energy consumption analysis: Prototype results. ASHRAE *Transactions* 93(1):979-98.

Hirst, E., D. White, and R. Goeltz. 1983. Comparison of actual electricity savings with audit predictions in the BPA residential weatherization pilot program. ORNL/CON-142. Oak Ridge National Laboratory, Oak Ridge, TN. (See also ORNL/CON-174.)

Table 10 Performance Data Requirements of the Commercial Retrofit Protocol

Projects with Submetering			
	Before Retrofit	**After Retrofit**	
Utility billing data (for each fuel)	12 month minimum	3 month minimum (optional update to 12 months)	
Submetered data (for all recording intervals)	All data for each major end use up to 12 months	All data for each major end use up to 12 months	
	Type	**Recording Interval**	**Period Length**
Temperature data (daily maximum and minimum must be provided for any periods without integrated averages)	Maximum and minimum —or— Integrated averages	Daily —or— Same as for submetered data but not longer than daily	Same as billing data length —or— Length of submetering
Projects without Submetering			
	Before Retrofit	**After Retrofit**	
Utility billing data (for each fuel)	12 month minimum	12 month minimum	
	Type	**Recording Interval**	**Period Length**
Temperature data	Maximum and minimum —or— Integrated averages	Daily	Same as billing data length

Hough, R.E., P.J. Hughes, R.J. Hackner, and W.E. Clark. 1987. Results-oriented methodology for monitoring HVAC equipment in the field. ASHRAE *Transactions* 93(1):1569-79.

Hughes, P. and W. Clark. 1986. Planning and design of field data acquisition and analysis projects: A case study. Proceedings of the National Workshop on Field Data Acquisition for Building and Equipment Energy Use Monitoring, Oak Ridge National Laboratory, Oak Ridge, TN (March).

Levins, W.P. and M.A. Karnitz. 1986. Cooling-energy measurements of unoccupied single-family houses with attics containing radiant barriers. ORNL/CON-200. Oak Ridge National Laboratory, Oak Ridge, TN. (See also ORNL/CON-2136.)

MacDonald, J.M. and D.M. Wasserman. 1989. Investigation of metered data analysis methods for commercial and related buildings. ORNL/CON-279. Oak Ridge National Laboratory, Oak Ridge, TN.

MacDonald, J.M., T.R. Sharp, and M.B. Gettings. 1989. A protocol for monitoring energy efficiency improvements in commercial and related buildings. ORNL/CON-291. Oak Ridge National Laboratory, Oak Ridge, TN (September).

Mazzucchi, R.P. 1987. Commercial building energy use monitoring for utility load research. ASHRAE *Transactions* 93(1).

Misuriello, H. 1987. A uniform procedure for the development and dissemination of monitoring protocols. ASHRAE *Transactions* 93(1).

Misuriello, H. 1988. Instrumentation applications for commercial building energy audits. ASHRAE *Transactions* 95(2).

Modera, M.P. 1989. Residential duct system leakage: Magnitude, impacts, and potential for reduction. ASHRAE *Transactions* 95(2).

Modera, M.P., R.C. Diamond, and J.T. Brunsell. 1986. Improving diagnostics and energy analysis for multifamily buildings: A case study. LBL-20247, Lawrence Berkeley Laboratory, Berkeley, CA.

Norford, L.K, A. Rabl, R.H. Socolow. 1985. Measurement of thermal characteristics of office buildings. ASHRAE Conference on the Thermal Performance of Building Envelopes. Clearwater Beach, FL.

Palmer, J. 1989. Energy performance assessment: A guide to procedures, Vols. I and II. Energy Technology Support Unit, Harwell, United Kingdom.

Persily, A.K. and R. Grot. 1988. Diagnostic techniques for evaluating office building envelopes. ASHRAE *Transactions* 94(1).

Porterfield, J.M. 1988. Alternatives for monitoring multidwelling energy measures. ASHRAE *Transactions* 94(1).

Robison, D.H. and L.A. Lambert. 1989. Field investigation of residential infiltration and heating duct leakage. ASHRAE *Transactions* 95(2):542-50.

Sharp, T.R. and J.M. MacDonald. 1990. Effective, low-cost HVAC controls upgrade in a small bank building. ASHRAE *Transactions* 96.

Subbarao, K. 1988. PSTAR—A unified approach to building energy simulations and short-term monitoring. SERI/TR-254-3175. Solar Energy Research Institute, Golden, CO.

Subbarao, K., J.D. Burch, and H. Jeon. 1986. Building as a dynamic calorimeter: Determination of heating system efficiency. SERI/TR-254-2947. Solar Energy Research Institute, Golden, CO.

Szydlowski, R.F. and R.C. Diamond. 1989. Data specification protocol for multifamily buildings. LBL-27206, Lawrence Berkeley Laboratory, Berkeley, CA.

Ternes, M.P. 1986. Single-family building retrofit performance monitoring protocol: Data specification guideline. ORNL/CON-196. Oak Ridge National Laboratory, Oak Ridge, TN.

Ternes, M.P. 1987. A data specification guideline for DOE's single-family building energy retrofit research program. ASHRAE *Transactions* 93(1):1607-18.

Woller, B.E. 1989. Data acquisition and analysis of residential HVAC alternatives. ASHRAE *Transactions* 95(1):679-86.

BIBLIOGRAPHY

Alereza, T., D.R. Dohrmann, M. Martinez, and D. Mort. End-use metered data for commercial buildings. ASHRAE *Transactions* 96(1):1004-10.

ASHRAE. 1986. Energy use in commercial buildings: Measurements and models. *Technical Data Bulletin* 2(4).

ASHRAE. 1987. Energy measurement applications. *Technical Data Bulletin* 3(3).

Baird, G., M.R. Donn, F. Pool, W.D.S. Brander, and C.S. Aun. 1984. Energy performance of buildings. CRC Press, Boca Raton, FL.

Burch, J., K. Subbarao, A. Lekov, M. Warren, and L. Norford. 1990. Short-term energy monitoring in a large commercial building. ASHRAE *Transactions* 96(1):1459-77.

Cowan, J.D. and I.A. Jarvis. 1984. Component analysis of utility bills: A tool for the energy auditor. ASHRAE *Transactions* 90(1B):411-23.

Douglass, J.C. 1989. Assessing in situ performance of advanced residential heat recovery ventilator systems: A case study of monitoring protocol. ASHRAE *Transactions* 95(1):697-705.

EPRI. 1983. Monitoring methodology handbook for residential HVAC systems. EPRI Em-3003, Electric Power Research Institute (May).

EPRI. 1985. Survey of utility commercial sector activities. EPRI EM-4142, Electric Power Research Institute (July).

EPRI. 1985. Survey of residential end-use projects. EPRI Em-4578, Electric Power Research Institute (May).

Fracastoro, G.V. and M.D. Lyberg. 1983. Guiding principles concerning design of experiments, instruments, instrumentation, and measuring techniques. Swedish Council for Building Research, Stockholm, Sweden (December).

Heidel, J. 1985. Commercial building end-use metering inventory. PNL-5027. Pacific Northwest Laboratory, Richland, WA (March).

Jones, J.R. and S. Boonyatikarn. 1990. Factors influencing overall building efficiency. ASHRAE *Transactions* 96(1):1449-58.

Kaplan, M.B., J.B. McFerran, J. Jansen, and R.G. Pratt. 1990. Reconciliation of a DOE2.1C model with monitored end-use data for a small office building. ASHRAE *Transactions* 96(1):98193.

Piette, M.A. 1986. A comparison of measured end-use consumption for 12 energy-efficient new commercial buildings. Proceedings of ACEEE 3:176.

Pratt, R.G. 1990. Errors in audit predictions of commercial lighting and equipment loads and their impacts on heating and cooling estimates. ASHRAE *Transactions* 96(1):994-1003.

Sharp, T.R. and J.M. MacDonald. 1990. Effective, low-cost HVAC controls upgrade in a small bank building. ASHRAE *Transactions* 96(1):1011-17.

Shehadi, M.T. 1984. Automated utility management system. ASHRAE *Transactions* 90(1B):401-10.

Spielvogel, L.G. 1984. One approach to energy use evaluation. ASHRAE *Transactions* 90(1B):424-36.

Subbarao, K., J.D. Balcomb, J.D. Burch, C.E. Hancock, and A. Lekov. 1990. Short-term energy monitoring: Summary of results from four houses. ASHRAE *Transactions* 96(1)1478-83.

Van Hove, J. and L. van Loon. 1990. Long-term cold energy storage in aquifer for air conditioning in buildings. ASHRAE *Transactions* 96(1):1484-88.

Wulfinghoff, D.R. 1984. Common sense about building energy consumption analysis. ASHRAE *Transactions* 90(1B):437-47.

CHAPTER 38

BUILDING OPERATING DYNAMICS

THIS chapter describes HVAC system operating strategies that optimize energy consumption without penalizing comfort within the controlled environment. Strategies should provide (1) energy conservation, (2) effective environmental control, (3) optimal equipment loading, (4) predictive load requirements, and (5) optimal equipment performance. Dynamic control, which is the continuous adjustment of operating conditions in response to load and weather changes, is an important element of any operating strategy.

Examples of some tradeoffs in designing and implementing a control strategy include:

- Increasing chilled water supply temperature may slightly improve chiller efficiency and greatly increase air-handling unit energy consumption
- Decreasing condenser entering water temperature may result in improved chiller efficiency and increased tower fan energy consumption
- Reducing chilled water distribution quantities may reduce fluid-handling energy requirements, reduce heat transfer rates, and increase heat transfer surface area requirements
- Decreasing cooling supply air temperatures may reduce air-handling energy requirements, improve humidity removal, reduce ventilation rates and cold drafts, while increasing the potential for water condensation on air distribution systems and components.

The optimal HVAC system operating strategy requires the best combination of temperatures and flow quantities so that the overall cost is minimized.

The efficiency of the building's HVAC system and comfort are improved by using building mass for thermal storage. Resetting temperature set points and flows by using dynamic control and predictive load methods is an emerging art. However, control methods are being devised to automatically take advantage of the dynamic characteristics of buildings.

The methods and results presented in this chapter are based on tests and simulations performed on several buildings and facilities. The strategies for achieving optimal control were derived from field data and simulations, so that designers and building operators may apply the strategies to other facilities.

AIR-HANDLING SYSTEMS

Air-handling systems in commercial buildings provide conditioned air to building spaces to maintain human comfort or the proper environment for special-purpose equipment, such as computers. The performance of these systems has a significant impact on the overall energy consumption of a building. In discussing the operation of these systems, this chapter separates local loop (or process control) from supervisory control (or control strategy selection) and from the techniques employed for airflow control.

The preparation of this chapter is assigned to TC 4.6, Building Operating Dynamics.

HVAC Process Control

The manner in which valves and dampers on air-handling equipment are controlled can directly affect comfort conditions within a building, the amount of energy consumption, and wear and tear on mechanical equipment. Factors with the most impact include: (1) choice of set points, (2) ability of individual loop controllers to maintain these set points, (3) response of the system being controlled, and (4) quality of control achieved.

Proper setting and resetting of zone set points by a central energy management system or by local controllers is critical to maintaining comfort while minimizing the associated energy costs. In either case, the objectives should be the same: (1) to maintain the highest space temperature during the cooling cycle and the lowest space temperature during the heating cycle compatible with occupant comfort and equipment requirements, (2) to minimize simultaneous heating and cooling to reduce the cost of operation, and (3) to provide minimum or no conditioning to unoccupied spaces, where possible. When changing from heating to cooling, the space temperature should be allowed to float from one limit to the other—hence, the concept of floating space temperature or deadband control.

Hittle (1979) used the building simulation program BLAST to study a hypothetical, 100 ft long by 100 ft wide, 10-story office building in the Washington, D.C. area. The building was of light metal wall construction, with double-glazed windows representing 34% of the wall area. The HVAC system was designed to maintain a floating space temperature between 67 and 78 °F on workdays between 7 a.m. and 6 p.m., and an unoccupied heating setback temperature of 57 °F. The two HVAC systems investigated were a constant air volume (CAV) terminal reheat system with proportional controls and a variable air volume (VAV) system with additional heating coils in the exterior zones that operated in sequence with the zone dampers in the VAV terminal units.

Schematics for these two systems are shown in Figure 1. The results of varying the mixed air temperature set point and the cooling coil discharge air temperature set point from their desired values are shown in Figures 2 and 3, respectively (Kao 1985). For the mixed air temperature set point, lower set point temperatures for both HVAC systems had no effect on the cooling energy consumed but used more heating energy at approximately 6.4% per 1 °F. Higher set point temperatures result in a slight energy savings for both systems. The main difference between the reheat system and the VAV system shows up in the effect of higher mixed air set point temperatures on the building cooling energy consumption.

Approximately 11% more cooling energy is used per 1 °F increase in the mixed air set point temperature for the reheat system (Figure 3), while the VAV system shows a 2.2% increase per 1 °F increase in the mixed air set point temperature. The effect of changing the cooling coil discharge air temperature was equally

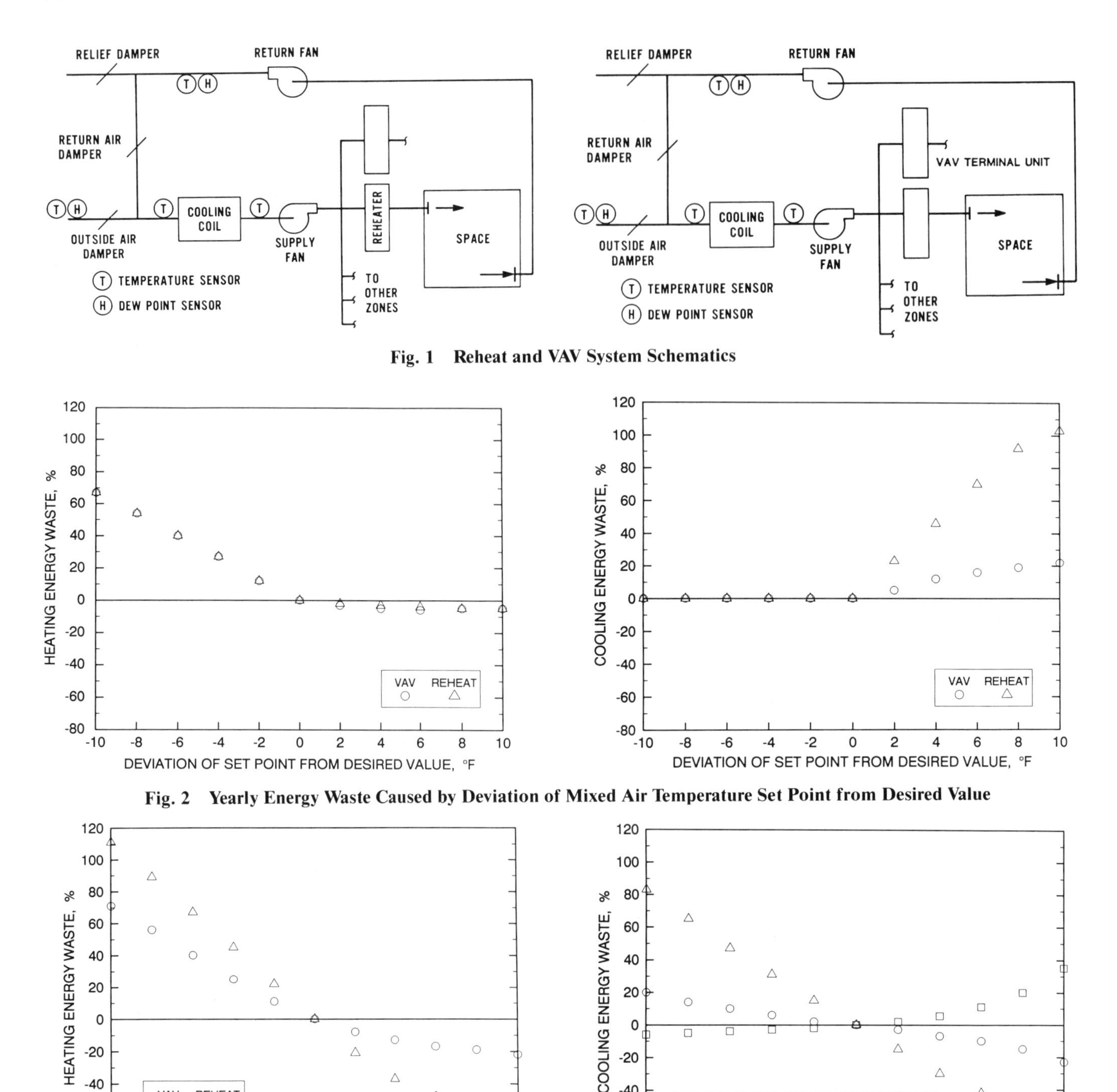

Fig. 1 Reheat and VAV System Schematics

Fig. 2 Yearly Energy Waste Caused by Deviation of Mixed Air Temperature Set Point from Desired Value

Fig. 3 Yearly Energy Waste Caused by Deviation of Coil Discharge Air Temperature Set Point from Desired Value

as significant. Lower set point temperatures increased energy use about 7.2% per 1°F of the heating energy and 2.1% per 1°F of the cooling energy for the VAV system. This is considerably less than the 11.2% per 1°F heating energy use and 8.5% per 1°F cooling energy use for the reheat system. The smaller energy use of the VAV system results from the assumption that the terminal units on the VAV system could cut down the extra cooling capacity by maintaining the desired space temperature. When the cooling coil discharge set point is above the desired value, some heating and cooling energy is saved for both systems. However, the prescribed space conditions in some thermal zones were not satisfied. In addition, the fan energy consumption for the VAV system increased at elevated cooling temperatures (Figure 3) due to higher airflow rate to some zones to compensate for higher supply air temperatures.

Control accuracy or the ability of a controller to maintain a given set point is a function of the control method (or algorithm) used, the characteristics of the process being controlled, and the

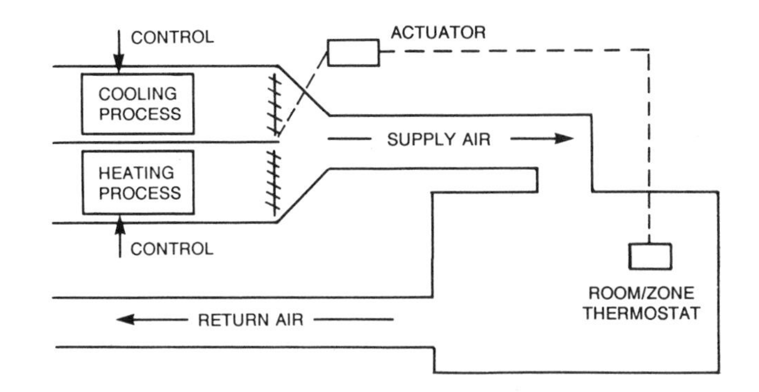

Fig. 4 Single Zone System

disturbances affecting the process. The room or zone control schematic for a multizone HVAC system is shown in Figure 4 (Miller 1980). For a constant supply air temperature, control remains closer to the set point as the gain on a proportional control algorithm is increased—at the expense of an increasingly oscillatory room temperature. The energy used over the time considered is shown in Figure 5 for a typical conference room or classroom. Increasing the gain decreases the energy consumption of the system by 23 and 37% for gain changes from 0.05 to 0.1 units/°F and from 0.05 to 0.2 units/°F, respectively. The gains considered can be compared to 1 psi/°F, 2 psi/°F, and 4 psi/°F, which are the range of gains common to pneumatic room controllers.

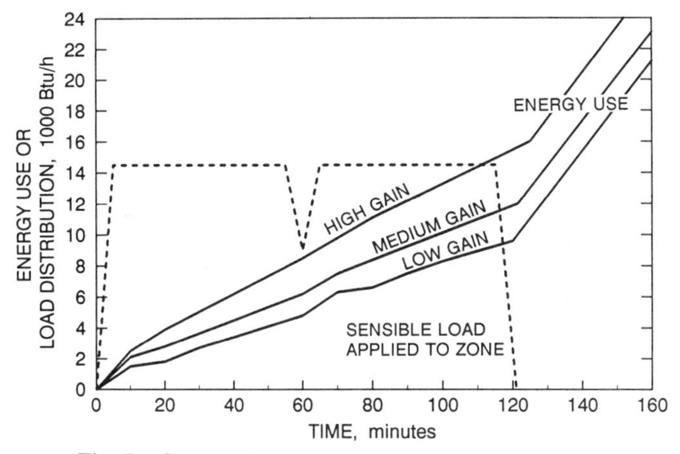

Fig. 5 Control Offset Relationship to Energy Use

Additional increases in the gain above 0.2 units/°F further reduces energy consumption, but at the expense of continuously oscillatory temperature conditions, which are unacceptable for comfort as well as mechanical wear on the HVAC system. Adding proportional plus integral control, in which the integral term tends to bring the control variable to the set point, reduced energy consumption 16% over the high gain case shown in Figure 5 (Miller 1980). This response is similar to the mid-gain proportional controller.

Shavit and Brandt (1982) conducted a study on a similar HVAC system which showed the effect of various integral gains KI on the performance of a system employing an equal percent valve with a positioner (no hysteresis). A lack of integral gain causes droop (offset from the set point), while the optimum integral gain forces the response of the system to operate uniformly around the set point (Figure 6). As the integral gain increases slightly above the optimum (KI = 0.001 units/°F), the response initially deviates from the set point but still matches the set point at steady state. As the integral gain further increases, the system becomes unstable.

The response of the system without a positioner is even more unstable. Figure 7 shows the valve displacement and actuator output signal of a proportional controller and a proportional plus integral controller for an equal percent valve without a positioner (10% hysteresis). The temperature of the air leaving the coil is steady, but the actuator movement is affected by hysteresis. For proportional control, once the system reaches equilibrium, all variables reach a steady-state condition, and the system freezes in this position unless it is disturbed. In the proportional plus integral case, as the valve reaches equilibrium, deviation from the set point (error signal) may occur if the change in controller signal is not significant enough to alter the position of the valve due to the presence of hysteresis.

The characteristics of the valve can also affect the stability of the control system. For many heat exchangers with an equal per-

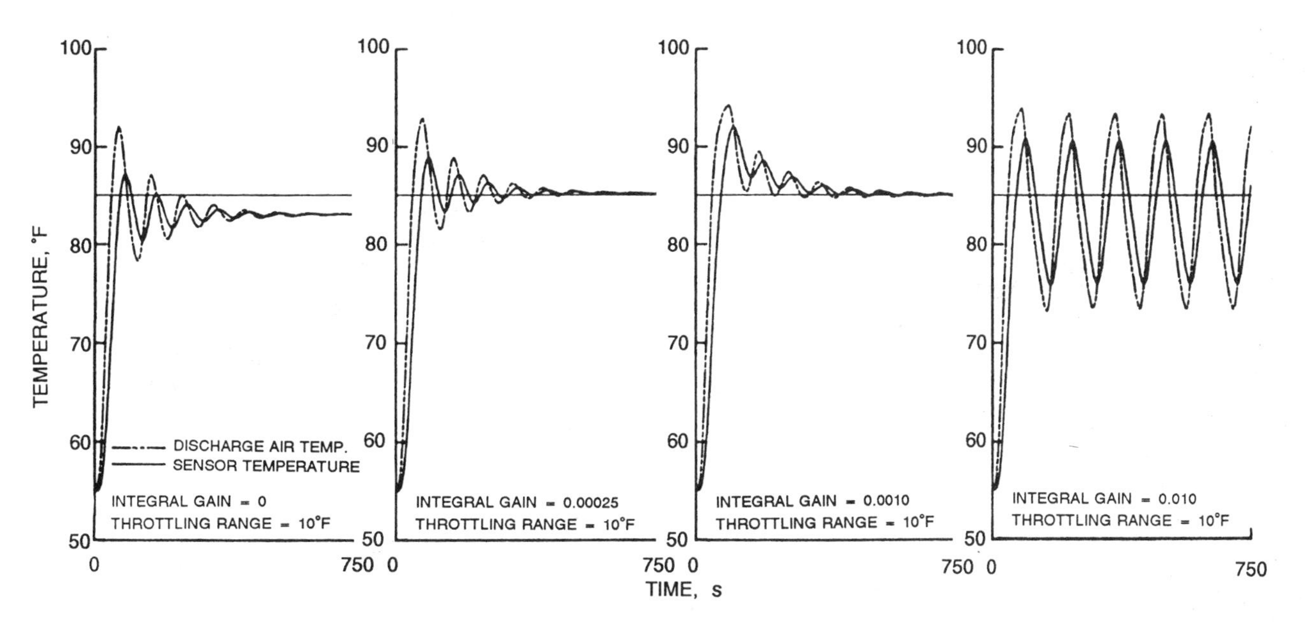

Fig. 6 Effect of Integral Gain on PI Control with Hysteresis
(Using an Equal Percent Valve)

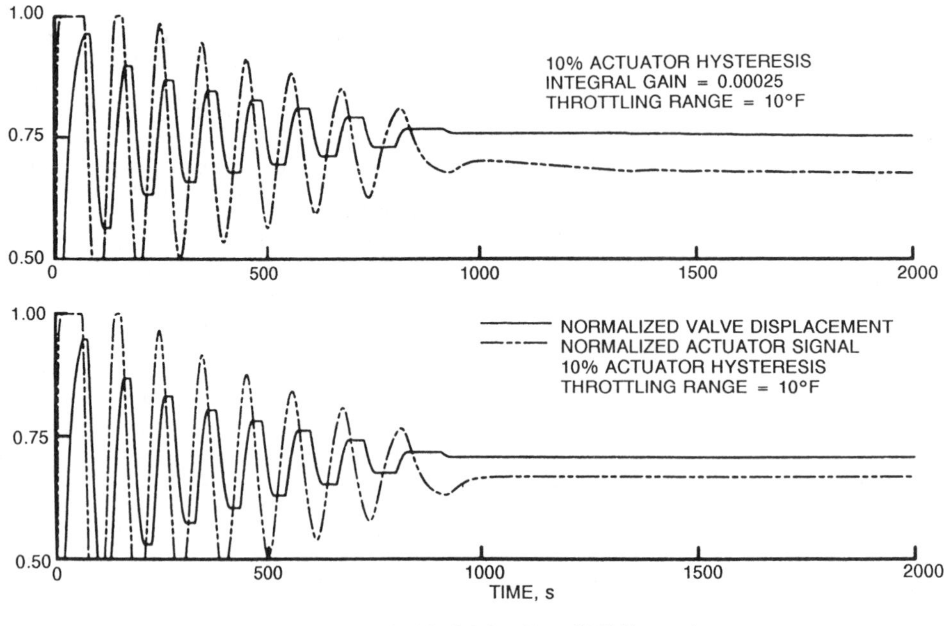

Fig. 7 Hysteresis Model for P and PI Control
(Using an Equal Percent Valve)

centage valve, the energy output as a function of valve position is linear. This results in a valve-coil system with a (approximate) fixed gain. However, the static characteristics of the valve-coil system with a linear valve often cause the gain to vary with valve position. The gain is likely to be larger than the gain with the equal percentage valve when the valve is less than 50% open, and smaller when the valve is more than 50% open. When the valve is less than 50% open, the system with the linear valve is unstable. When the valve is more than 50% open, the system stabilizes faster than the system with an equal percentage valve.

Maintaining a set point reset schedule and the proper sequencing of valves and dampers are important functions of process control systems on building air-handling units. Examples of reasonably good and poor set point control are shown in Figure 8 for two pneumatic control systems on two air handlers serving an actual office building (Bushby and Kelly 1988). The control system in Figure 8A performs best, maintaining the supply air temperature in a tight temperature band of about 2 °F and following the prescribed reset schedule closely for outdoor temperatures ranging from 35 to 55 °F. At temperatures below 35 °F, the actual maintained set point continues its linear trend instead of maintaining the constant set point supply temperature normally implemented on this particular system.

The response of the air handler shown in Figure 8B indicates a similar pattern, except that the data has a much greater spread due to sequencing trouble in this air-handling unit. An analysis of data from this unit shows that outdoor air dampers tend to oppose the steam preheat valve in the temperature range between 23 and 50 °F. In fact, the outdoor air dampers and preheat valve have a tendency to swing open and closed in alternating fashion in a very short time. Figure 9 shows the position of the steam preheat valve and outdoor dampers for about 20 min (Bushby and Kelly 1988). The solid line indicates the position of the outdoor air dampers, and the dashed line represents the position of the steam preheat valve.

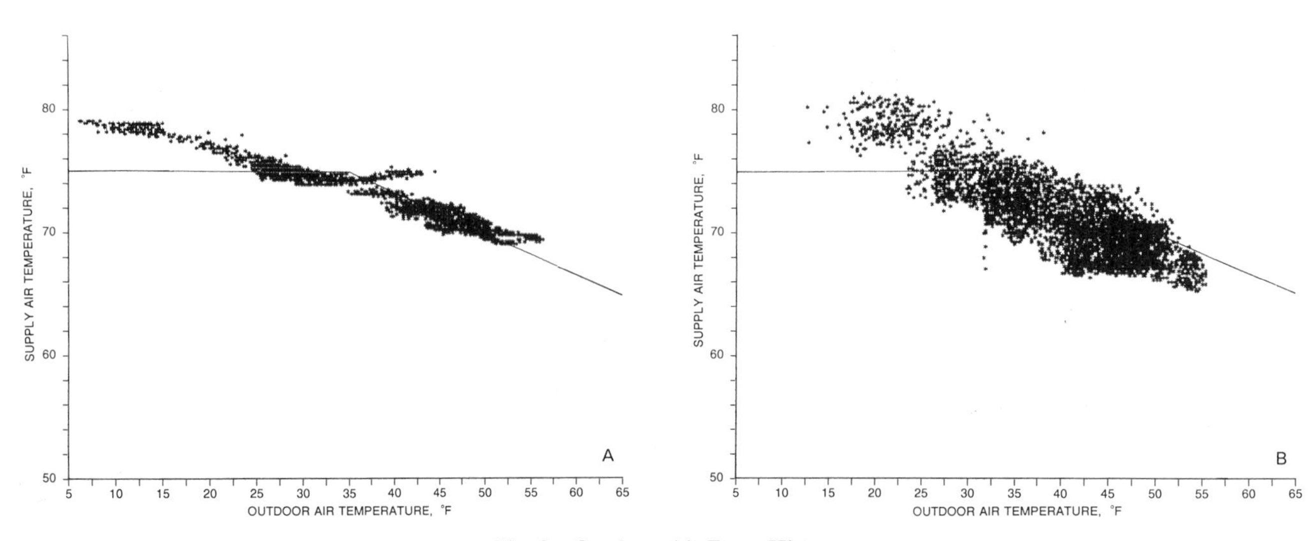

Fig. 8 Outdoor Air Reset History

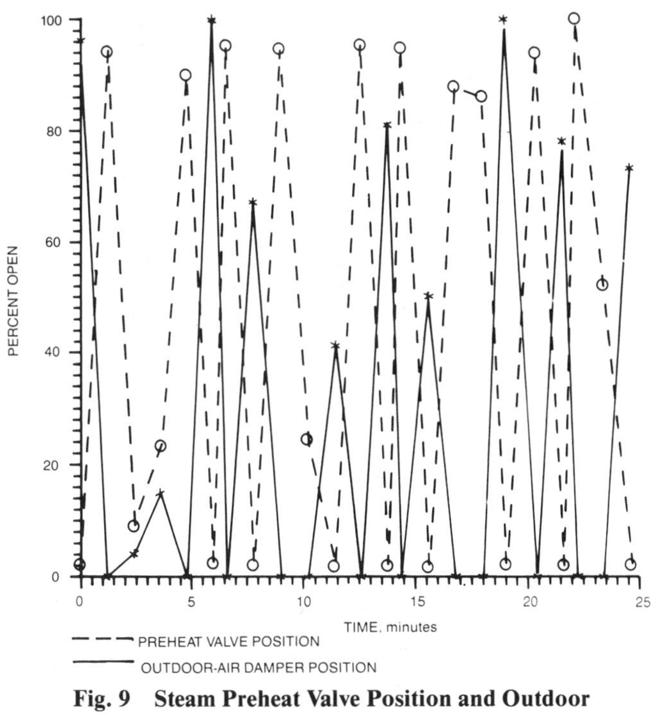

Fig. 9 Steam Preheat Valve Position and Outdoor Air Damper Position

Both the steam preheat valve and the outdoor air dampers swing open and closed in a cyclic fashion in just less than 2 min. The two cycles are roughly 180° out of phase. The effect of swinging on the mixed air temperature and the supply air temperature is shown in Figure 10. Both these temperatures oscillate in conjunction with the steam valve and the dampers. The worst peak-to-peak mixed air temperature swing approaches 27°F; the worst supply air temperature swings are smaller, approaching 4.5°F. Clearly, such control system performance wastes energy by unnecessary use of preheating and has a negative impact on comfort conditions; in addition, it causes needless wear and tear on valves, dampers, and actuators.

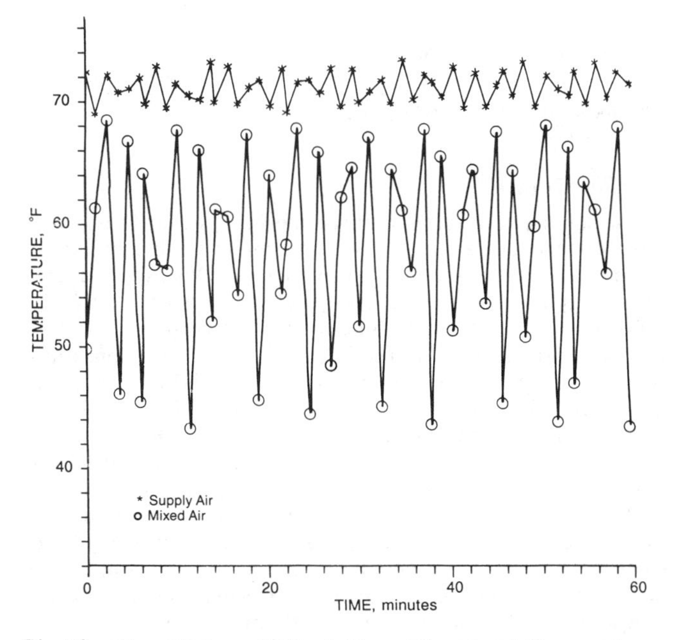

Fig. 10 Time History of Mixed Air and Supply Air Temperatures

Control Strategies

The energy use of a building is influenced by its shell construction, how it is used, heating and cooling equipment efficiencies, the type of HVAC system employed, and the way the building equipment and systems are operated. Control strategies applied to air-handling systems play a key role in determining the effectiveness of these systems. Kao (1985) investigated the energy effect of the most commonly used control strategies on different air-handling systems in four different building types. The results were generated using the building energy program BLAST (Hittle 1979) and are useful as general guidelines for the design or operation of HVAC systems; however, a more detailed analysis is required to incorporate the unique aspects of a particular building and its intended use.

The control strategies studied include two different techniques: (1) resetting of supply air temperature by either outdoor air temperature or space demand, and (2) a combination of these approaches. The heating, cooling, and fan energy consumption was computed at the air-handling system level and thus does not include plant and distribution efficiencies. Typical Meteorological Year (TMY) hourly weather data were used in simulations for six cities in the United States—Lake Charles, LA; Madison, WI; Nashville, TN; Santa Maria, CA; Seattle, WA; and Washington, D.C. In all cases, the air-handling system was operational only during occupied periods in the cooling season, and used space temperature set back at night during the heating season.

Figures 11 and 12 (Kao 1985) show the yearly cooling and the yearly heating energy per square foot of floor area in a large office building. Cases 1 through 6 simulate a terminal reheat system for the entire building. Adding a dry-bulb economy cycle (case 2) allows supply air temperature to be maintained by using cooler outside air, consequently saving a large amount of cooling energy while using a substantial amount of heating. Changing to an enthalpy economizer (case 3) results in more cooling savings and greater heat use. The increase in heating energy for both economizer cycles results mainly from lowering the supply air temperature by the proportional controls during the period that the economy cycles were in operation. Both the absolute amount and the percentage of cooling savings are greater in low cooling degree-day areas than in high degree-day areas.

Resetting the supply air temperature benefits both heating and cooling, since reheat is not needed if the extra cooling capacity is reduced. Resetting by sensing zone temperature (case 5) reduces cooling energy approximately twice as much as resetting by using outside air temperature (case 4). Heating energy reduction is even more dramatic when resetting is done by sensing zone air temper-

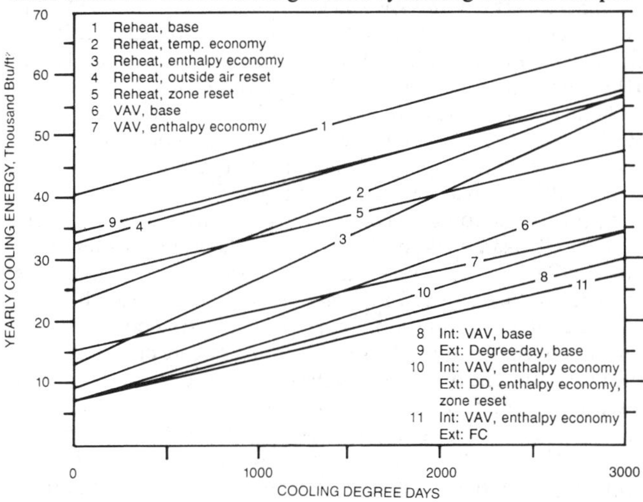

Fig. 11 Cooling Energy Consumption of Large Office Building

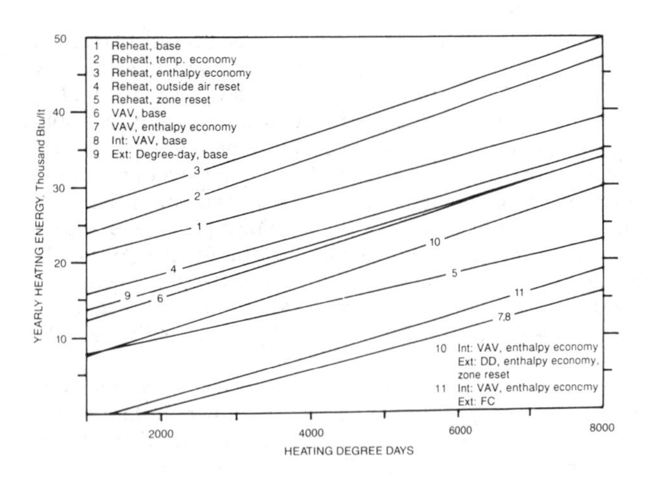

Fig. 12 Heating Energy Consumption of Large Office Building

ature. By adding an enthalpy "economy" cycle to the supply air temperature reset strategy (case 6), cooling energy is further reduced substantially, while heating energy increases for the reason cited previously.

Cases 7 and 8 use variable air volume (VAV) systems for the entire building. The perimeter zones have reheat coils which operate in sequence with the zone dampers; these are allowed to reduce the supply air to 20%. The interior zones have only damper controls. The base VAV system (case 7), which has no special strategy, has comparable cooling energy consumption relative to the best reheat case—case 6, which uses an enthalpy economy cycle and supply air temperature reset by zone demand. When an enthalpy economy cycle is added to the VAV system, 15 (Lake Charles) to 59% (Seattle) of cooling energy is saved. Contrary to results obtained with the reheat system, adding enthalpy cycles to the VAV system (case 8) does not significantly increase heating energy in most cities. The exceptions were Santa Maria (20% increase) and Lake Charles (7%).

Case 9 simulates a base dual-duct system for the perimeter zones and a simple VAV system for interior zones. Case 10 is the result of adding an enthalpy economy cycle to the entire building, with both the hot and the cold air of the dual-duct system reset by sensing the space temperature. It is difficult to pinpoint the effects of the individual strategies and systems for these two cases. The overall results show that much cooling and heating energy is saved as compared to the base VAV and dual-duct systems (case 9). The perimeter zones are also simulated with a four-pipe fan-coil system having both the cold and hot water available year-round (case 11). A fixed amount of outside air is introduced directly to the fan-coil units during operating hours. The interior system remains a VAV system with an economy cycle. The energy consumption for both heating and cooling is similar to that in case 8.

Kao (1985) also analyzed the cooling and heating energy consumption of the combined classroom and office areas of a school building. Reheat, dual-duct, variable air volume, and cooling-type unit ventilator systems were simulated for these areas. Roughly the same relative pattern as for the large office building was observed for the cooling consumption of the first three systems. The cooling lines of the various VAV system strategies were close together, the best being the one applying the enthalpy economy cycle and resetting the supply air temperature by zone sensing. The unit ventilator system performs slightly poorer on cooling than do the VAV systems. Unlike the large office building discussed previously, applying economy cycles to the reheat systems of this building does not increase the heating energy much.

Kao (1985) indicates that most of the discussion on the previous two buildings also applies to retail and small office buildings. For the retail store, among the best strategies and systems are a VAV system with the supply air temperature reset by the outside air temperature and the packaged DX system, both with enthalpy economy cycles. A VAV system with enthalpy economy cycle and supply air temperature reset by zone demand sensing may be the best choice for the small office building.

Table 1 (Kao 1985) illustrates the energy consumption ratios of some selected cases for Washington, D.C. (1415 cooling degree-days and 4211 heating degree-days). The ratios are defined as the building energy consumption obtained using a particular control strategy, divided by the energy consumption of the base reheat case. These data indicate that drastic improvements in a building's energy use may be obtained using different control strategies and air-handling systems. For example, when the enthalpy economy cycle is employed on the VAV system of the large office building, it consumes only 25% of the cooling energy and less than one-third of the heating energy consumed by a reheat system using a fixed amount of outside air year-round.

Table 1 Comparisons of Building Energy Use in Washington, D.C.

	Small Office		Large Office		School[a]		Retail Store	
Strategies	**Cool**	**Heat**	**Cool**	**Heat**	**Cool**	**Heat**	**Cool**	**Heat**
	Ratios Relative to Base Reheat Cases							
Reheat with enthalpy economy	0.57	1.08	0.58	1.24	0.61	1.03	n.a.	
Reheat with enthalpy economy and zone reset	n.a.		0.43	0.76	0.43	0.73	n.a.	
Reheat with enthalpy economy and OA reset	n.a.		n.a.		n.a.		0.54	0.67
VAV with enthalpy economy	0.36	0.26	0.25	0.31	0.37	0.45	n.a.	
VAV with enthalpy economy and OA reset	0.33	0.20	n.a.		n.a.		0.43	0.17
VAV with enthalpy economy and zone reset	0.33	0.19	n.a.		0.33	0.43	n.a.	
Fan-coil for perimeter	n.a.		0.32	0.29	n.a.		n.a.	
Unit ventilator	n.a.		n.a.		0.42	0.10	n.a.	

[a]Classroom, office, and library area only
n.a.—not available

Airflow Control

The popularity of variable air volume (VAV) systems has grown rapidly due to their ability to save large amounts of heating, cooling, and fan energy in comparison with other HVAC systems. The fan energy consumed by these systems is, however, strongly influenced by the control technique used to vary the airflow rate. Brothers and Warren (1986) compared the fan energy consumption for three different ways in which airflow is typically regulated—by system dampers on the outlet side of the fan, by inlet vanes on the fan, and by variable speed control of the fan motor. The investigation consisted of using the building simulation program DOE-2.1 to simulate the annual energy consumption of a 10,000 ft^2 commercial building employing a VAV system with these three different flow control methods. Both centrifugal and vaneaxial fans were considered, even though system dampers are not usually used with vaneaxial fans as they may overload the fan motor. The VAV system used an enthalpy economy cycle and provided cooling only during working hours—7 a.m. to 5 p.m., five days a week. The analysis was performed for five different cities—Fresno, CA; Forth Worth, TX; Miami, FL; Phoenix, AZ; and Washington, D.C.

Values for the annual energy consumption are presented in Table 2. In all locations, the centrifugal fan uses less energy than the vaneaxial fan. The vaneaxial fan has a higher efficiency at the full load design point, but the centrifugal fan has better off-design characteristics that lead to lower annual energy consumption.

Table 2 Annual Energy Use with Different Flow Throttling Techniques

Type of Fan	Flow Control	Fan Energy Use, 10^3 kWh				
		Fresno	Ft. Worth	Miami	Phoenix	Washington, D.C.
Backward—Inclined Centrifugal	System damper	12.6	12.9	13.9	17.9	11.5
	Inlet vane	10.2	10.4	10.1	14.4	9.5
	Variable speed	4.1	4.5	5.9	5.8	3.3
Vaneaxial	System damper[a]	16.0	16.2	15.5	22.8	15.0
	Inlet vane	10.8	11.0	10.9	15.4	10.0
	Variable speed	6.8	7.0	7.7	9.6	5.9

[a]Not normally used with vaneaxial fans due to possible overload of fan motor.

System damper control uses an average (over the five locations) of 25% more energy than inlet vane control with a centrifugal fan. With variable speed control on a centrifugal fan system, the fan energy savings range from 42% for Miami to 65% for Washington, D.C., with an average of 57% for the five locations.

OPTIMIZING COOLING PLANT DISTRIBUTION SYSTEMS

Many building complexes are heated and cooled by centrally located facilities. Because the total annual fuel costs for large plants are extremely high, even a small relative savings in the energy requirements of such plants translates into a significant reduction in operating costs.

A centralized cooling plant consists of one or more chillers, cooling towers, pumps, and, in some cases, air handlers controlled to satisfy the cooling requirements of one or more buildings. Figure 13 shows a typical centralized chilled water system. Return air from the zones is mixed with an outside ventilation airstream and is cooled and dehumidified through cooling coils. In a variable air volume system, the air handler flow rate is typically adjusted based on static pressure changes in the terminal units (VAV boxes). Supply air temperatures are controlled by modulating the water flows through each cooling coil with control valves. Heat extracted from the air flowing across the cooling coils warms the water returning to the chiller. The chilled water supply temperature is maintained by controlling the chiller refrigerant flow through modulation of the compressor and expansion device. Cooling towers with multiple cells sharing common sumps are typically used to reject heat from the condenser of a chiller to the environment.

Design retrofits to existing systems may provide significant savings in operating costs. Many older systems were designed with little concern for energy costs. For instance, most large cooling systems have pumps and cooling tower fans that operate at fixed speeds. Modulation of flow rates in response to changing load conditions is limited to the on/off cycling of individual equipment operated in parallel or series. A more efficient method of operation continuously varies flow rates using variable-speed equipment. Treichler (1985) concluded that variable-speed pumping is

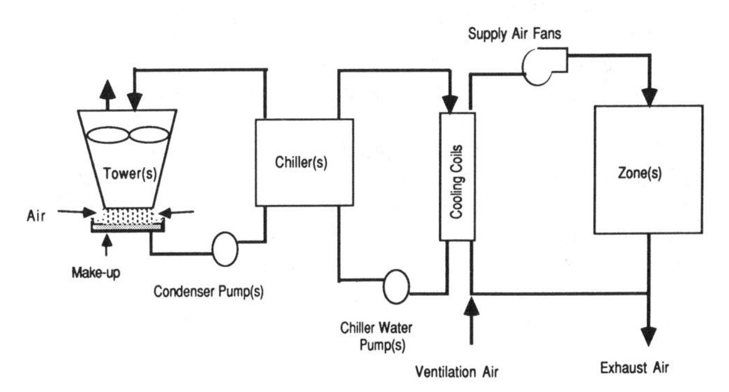

Fig. 13 Schematic of a Typical Chilled Water System

economically attractive for both chilled water distribution and condenser water systems.

Centrifugal chillers are the largest consumers of energy in central cooling systems in which the economizer cycle operation provides only a small fraction of the cooling. Many chillers are operated at fixed speeds and the chilled water set point temperature is maintained by varying the position of prerotation inlet vanes to the centrifugal compressor. Significantly higher efficiencies and consequent operational savings can be realized at part-load conditions by retrofitting the compressor with variable-speed control.

In some central plants, overall chiller performance may be improved by changing the refrigerant. As an existing plant improves its energy management practices and applies conservation methods to reduce the building load, the load requirement of the chiller is reduced. As a result, the optimal refrigerant choice may change from that of the original design.

If properly sized and controlled, adding chilled water storage can significantly reduce operating costs (ASHRAE 1985). For systems with electrically driven chillers, storage can be used to take advantage of time-of-day charges and limit demand charges by shifting the load requirements. Storage may also reduce costs by shifting the load to times when environmental conditions result in improved chiller performance. For further information, refer to Chapter 39.

Plant operating costs can also be reduced through better control. Optimal control of a central chilled water system may be thought of as having a two-level hierarchical structure. The first involves local (loop) control in response to prescribed set points. For example, a first-level control variable is the compressor speed for a variable-speed chiller; the second-level control involves supervisory management of those independent variables that may be adjusted to minimize operating costs, while still satisfying load requirements. For example, the set point for the chilled water supply temperature is a second-level variable.

The dynamics of the first-level (local loop) control must be considered in order to maintain prescribed set points in an efficient manner. However, the dynamics of equipment such as chillers, fans, and pumps may be neglected when determining the optimal second-level control set points (Hackner 1985). With supervisory (second-level) control, the local loop (first-level) may be considered to meet the set points exactly.

Optimal control of a central HVAC system minimizes total power consumption of the chiller, cooling tower fans, condenser water pumps, chilled water pumps, and the air-handling fans at each instant of time with respect to the independent, continuous and discrete control, while maintaining the desired zone conditions and ensuring that the control variables are within acceptable bounds. Discrete control variables are not continuously adjustable, but have discrete settings, such as the number of operating chillers, cooling tower cells, condenser water pumps, and chilled water pumps and the number of relative speeds for multispeed fans or pumps. Independent, continuous control variables include chilled water and supply air set temperatures, relative water flow rates to the chillers (evaporators and condensers), cooling tower cells, and cooling coils, and the speeds for variable-speed fans or pumps.

The optimal operating point changes over time as cooling requirements and ambient conditions change. Currently, operators of central cooling plants determine control practices that yield reasonable operating costs by experience gained through trial and error. Implementation of on-line optimal control is in its beginning stages.

Most control studies have been concerned with the local-loop control of an individual component or subsystem needed to maintain a prescribed set point. Global optimum plant control has been studied by Marcev *et al.* (1980), Sud (1984), Lau *et al.* (1985), Hackner *et al.* (1984, 1985), Johnson (1985), and Nugent *et al.* (1988).

Lau *et al.* (1985) studied the effect of control strategies on overall energy costs of a large facility located in Charlotte, NC through the use of annual simulations. The operating costs associated with the existing control strategy were compared with those resulting from optimal control of condenser flow rates, tower fan flows, the number of chillers operating, and the use of storage. The reduction in the utility bill through optimal control was about 5.2% of the total. Hackner *et al.* (1985) investigated optimal control strategies for a cooling system without storage at a large office building in Atlanta, GA. Optimal control of the chiller plant resulted in a reduction of about 8.7% in the utility bill when compared with the existing control strategy. If the plant had been operated with fixed set points, similar to conventional practice, the optimal control would have resulted in a cost reduction of about 19%. Similar conclusions were reached by Marcev *et al.* (1984). These studies demonstrate potential savings by using optimal control in plants without storage.

General methodologies for optimal control have been attempted. Nizet (1984) applied a state-space model to a building zone to minimize the energy consumption with respect to the supply airflow rate to the zone. The authors concluded that it was not practical to apply this methodology to more complicated systems in an on-line application. Braun (1988) and Braun *et al.* (1989) investigated optimal control as applied to typical chilled water systems and developed control guidelines useful to plant engineers. They developed a simple methodology for near-optimal control, which is described later in this chapter.

BUILDING DYNAMIC CONTROL

Optimal start algorithms determine the times for turning equipment on so that the building zones reach the desired conditions when the building becomes occupied. The goal of these algorithms is to minimize the precool (or preheat) time. During occupied hours, zone conditions are typically maintained at specified set points (Figure 14). For conventional strategies, the building mass actually increases operating costs. A massless building would require no time for precooling (or preheating) and would have lower overall cooling (or heating) loads than would actual buildings.

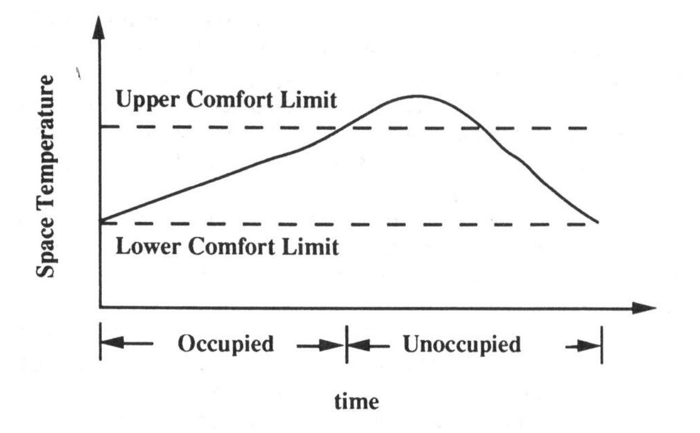

Fig. 14 Typical Space Temperature Variation for Building Precooling

In some situations, the thermal mass of a building represents a storage medium that may be used for reducing operating costs to some extent. Dynamic building control strategies attempt to manage the thermal mass of the building within acceptable comfort limits to (1) limit peak electrical demands and (2) minimize the daily operating costs in response to favorable utility rates and outside air cooling possibilities. Although Hartmann (1988) and Shapiro *et al.* (1988) have presented some simple strategies for using building thermal mass, the potential cost savings and comfort effects associated with the use of dynamic building control strategies have not been well documented.

Recovery from Night Setback

For buildings that are not continuously occupied, a significant savings in operating costs may be realized by raising the building set point temperature for cooling and by lowering the set point for heating during unoccupied times. Bloomfield and Fish (1977) have shown a potential energy savings of 12% for heavyweight buildings and 34% for lightweight buildings.

An optimal controller returns zone temperatures to the comfort range when the building becomes occupied. Seem *et al.* (1989) compared seven different algorithms for determining the optimal time for return from night setback. Each of these methods requires the estimation of parameters from measurements of the actual time for return from night setback.

Seem *et al.* (1989) showed that the optimal return time for cooling was not strongly influenced by the outdoor temperature. The following quadratic function of the initial zone temperature was found to be adequate for estimating the return time:

$$\tau = a_o + a_1 t_{z,i} + a_2 t_{z,i}^2 \quad (1)$$

where τ is the estimate of optimal return time, $t_{z,i}$ is initial zone temperature at the beginning of the return period, and a_o, a_1, and a_2 are empirical parameters. The parameters of Equation (1) may be estimated by applying linear least squares techniques to the difference between the actual return time and the estimates. These parameters may be continuously corrected using recursive updating schemes as outlined by Ljung and Söderstom (1982).

For heating, ambient temperature has a significant effect on the return time. One relationship for estimating the return time for heating is:

$$\tau = a_o + (1 - w)(a_1 t_{z,i} + a_2 + a_2 t_{z,i}^2) + w a_3 t_a \quad (2)$$

where t_a is the ambient temperature, a_o, a_1, a_2, and a_3 are empirical parameters, and w is a weighting function given by:

$$w = 1000^{-(t_{z,i} - t_{occupied})/(t_{occupied} - t_{unoccupied})} \quad (3)$$

where $t_{unoccupied}$ and $t_{occupied}$ are the zone set points for unoccupied and occupied periods. Within the context of Equation (2), this function weights the outdoor temperature more heavily when the initial zone temperature is close to the set point temperature during the unoccupied time. Again, the parameters of Equation (2) may be estimated by applying linear least squares techniques to the difference between the actual return time and the estimates.

Ideally, separate equations should be used for zones that have significantly different return times. Equipment operation is initiated for the zone with the earliest return time. In a building with a central cooling system, the equipment should be operated above some minimum load limit. With this constraint, some zones need to be returned to their set point earlier than the optimum time.

Optimal start algorithms often use a measure of the building mass temperature rather than the space temperature for determining return time. Although use of space temperature results in lower energy costs (*i.e.*, shorter return time), the mass temperature may result in better comfort conditions at the time of occupancy.

Building Precooling

One method for reducing building operating costs uses "free" cool night air to reduce cooling requirements for the next day. In reality, the use of outside air is not free, since energy is required to operate the air-handling fans. Precooling with outside air should only be considered if (1) heating is not required during the occupancy period, (2) the humidity of the ambient air is lower than an acceptable comfort limit, and (3) the cost of operating air-handling fans is less than the reduction in operating costs associated with mechanical cooling during the occupied period. The task of evaluating the cost savings of precooling with outside air is not straightforward. For example, mechanical precooling may reduce both demand and on-peak charges. Depending on the cooling requirements and equipment design, precooling with outside air is usually advantageous when the temperature difference between the zones and the ambient air is greater than about 5 to 10 °F. However, no other general guidelines for using building mass are available.

Figure 14 shows an example of the space temperature variation for a precooled building. During the unoccupied time, the building space temperature is allowed to float above an upper comfort limit until the ambient temperature is low enough to economically use 100% outside air to cool the building. The goal is to cool both the space temperature and building mass to the lower acceptable comfort limit at the time when the building becomes occupied and to do so in an optimal manner.

For a constant volume system, the best strategy reduces the operating time of the fans to a minimum. Since the ambient temperature is generally lowest during the hours leading up to the occupied time, it is best to delay precooling of the building as long as possible, while still reaching the lower limit at occupancy. An optimal start algorithm that attempts to reduce the building's temperature to the minimum set point at occupancy could be used for this purpose. However, a simpler and more conservative strategy is to control fans to achieve and maintain the zone temperature at the lower limit whenever ambient cooling is available.

For a variable-air-volume system, the best precool control strategy is difficult to determine. Since the fan power varies in a nonlinear fashion (*e.g.*, power as the cube of speed), it is better to operate the fans at a lower flow over a longer precool period as compared with the constant volume system, except under extreme conditions. Again, a simpler and more conservative strategy is to control the fan flow to achieve and maintain the lower zone set point whenever ambient "free" cooling is attractive.

After precooling a building, the space temperature set point may be adjusted upward (within the comfort zone) to take advantage of the thermal storage of the building to reduce both the peak and total power consumption required to operate the mechanical cooling equipment. The best strategy for discharging energy stored in the building mass depends on the load requirements and ambient conditions for that day. For a day where the peak power consumption could affect the demand charges for that month, the set point should be adjusted to minimize the peak. As with a partial thermal storage system, a good strategy is to operate the chillers at a constant load during the on-peak portion of the occupied cycle. However, with this operating strategy, comfort conditions cannot be maintained for the entire occupancy period. Typically, both the maximum ambient temperature and the peak building load occur during the mid to late afternoon. A reasonable strategy is to maintain the zone temperature at the minimum set point until the on-peak period begins and then to adjust the set point incrementally upward to reach the upper limit at the end of the occupancy. The set point may also be overridden by demand limiting of the mechanical equipment. For those days where demand charges are not an issue, the set point should be immediately adjusted to the upper comfort limit after the onset of the on-peak period, if mechanical cooling is performed.

CONTROL GUIDELINES FOR SYSTEMS WITHOUT STORAGE

Figure 13 illustrates a variable air volume (VAV) system. The central cooling facility, which consists of multiple centrifugal chillers, cooling towers, and pumps, provides chilled water to a number of air-handling units in order to cool air that is supplied to building zones. At any given time, it is possible to meet the cooling needs with many different modes of operation and set points.

For some of the independent control variables, guidelines may be developed that, when implemented, yield near-optimal performance. These guidelines simplify the selection of the control process required to achieve optimum performance and may readily be implemented by supervisory controllers.

Multiple Chillers

Multiple chillers normally operate in parallel and are controlled to supply chilled water at identical temperatures. For parallel chiller combinations, this control is either optimal or near-optimal (when not constrained by equipment). Besides the chilled water set point, additional control variables are the relative chilled and condenser water flow rates.

In general, the condenser water flow to each chiller should be controlled to give identical leaving condenser water temperatures for all chillers. This condition approximately corresponds to relative condenser flow rates equal to the relative loads on the chillers, even if the chillers are loaded unevenly. Figure 15 shows results for four sets of two chillers operated in parallel from data at the Dallas-Fort Worth (D/FW) airport (Braun 1988, Braun *et al.* 1989, and Hackner 1984). The curves in Figures 15 and 16 represent data from four chillers at three different installations: (1) the Dallas-Fort Worth Airport, in Dallas, Texas, which consists of one 5500-ton variable-speed chiller; (2) the IBM Building in Atlanta, Georgia (Hackner *et al.* 1984, 1985), which has a 550-ton fixed-speed chiller (the performance is scaled up for comparison with the D/FW chillers); and (3) the IBM Building in Charlotte, North

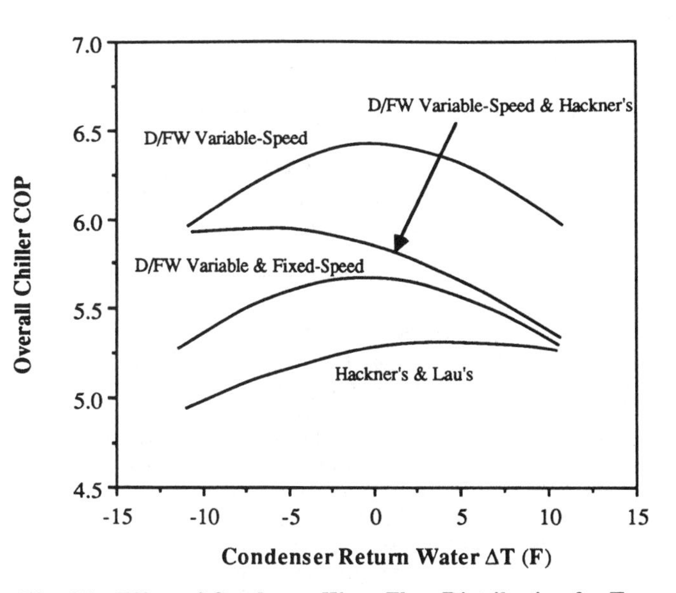

Fig. 15 Effect of Condenser Water Flow Distribution for Two Chillers in Parallel

Carolina (Lau *et al.* 1985), which has a 1250-ton fixed-speed chiller (the performance is scaled up for comparison to the D/FW chillers).

The overall chiller coefficient of performance (COP) is plotted versus the difference between the condenser water return temperatures for equal loadings on the chillers. For identical chillers, either variable-speed or fixed-speed, the optimal temperature difference is almost zero. For situations in which chillers do not have identical performance, equal leaving condenser water temperatures result in chiller performance that is close to the optimum. Even for the variable and fixed-speed chiller combinations, which have very different performance characteristics, the penalty associated with the use of identical condenser leaving water temperatures is small.

For chillers with different cooling capacities, but identical part-load characteristics, each chiller should be loaded according to the ratio of its capacity to the sum total capacity of all operating chillers. For a given chiller *i*, the part load should be:

$$f_{L,i} = q_{cap,i}/\Sigma q_{cap,i} \text{ for } i = 1 \text{ to } n \tag{4}$$

where $q_{cap,i}$ is the cooling capacity of the *i*th chiller, and *n* is the total number of chillers.

The relative loadings determined with Equation (4) could result in either minimum or maximum power consumptions. However, this solution gives a minimum when the chillers are operating at loads greater than the point at which the maximum COP occurs. Typically, but not necessarily, the maximum COP occurs at loads that are about 40 to 60% of a chiller's cooling capacity. Generally, with loads greater than about 50% of cooling capacity, the control results in a minimum power consumption.

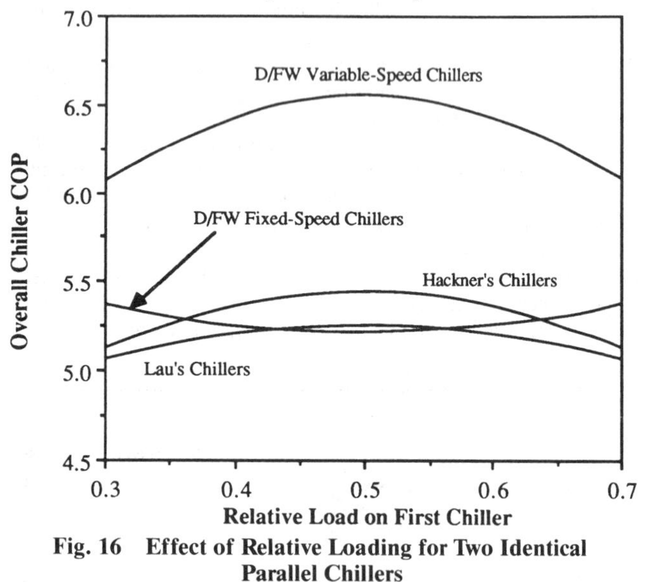

Fig. 16 Effect of Relative Loading for Two Identical Parallel Chillers

Figure 16 shows the effect of the relative loading on chiller COP for different sets of identical chillers loaded at approximately 70% of their overall capacities. Three of the chillers have maximum COPs when evenly loaded, while the fourth (D/FW fixed-speed) obtains a minimum at that point. The part-load characteristic of the D/FW fixed-speed chiller is unusual in that the maximum COP occurs at its maximum capacity. This chiller was retrofit with a different refrigerant and drive motor, which derated its capacity from 8700 to 5500 tons. As a result, the evaporators and condensers are oversized for its current capacity.

One of the important issues concerning control of multiple chillers is chiller sequencing, *i.e.*, determining the conditions at which specific chillers are brought on-line or off-line. The optimal sequencing of chillers depends primarily on their part-load characteristics. Chillers should be brought on-line at conditions where the total power of operating with the additional chiller would be less than without it. For dedicated chilled water and condenser pumps, it is optimal to operate chillers to their full capacity. For nondedicated pumps, the optimal sequencing of chillers may not be decoupled from the optimization of the rest of the system. The characteristics of the system change when a chiller is brought on-line or off- line due to changes in the system pressure drops and overall part-load performance. The optimal point for switching chiller operation may differ significantly from switch points determined if only chiller performance were considered at the conditions before the switch takes place.

Multiple Air Handlers

A large central chilled water facility may provide cooling to several buildings, each of which may have a number of air-handling units in parallel. The penalty of identical supply air temperatures for all air handlers is relatively small as compared with the optimal temperatures, even when the loading on the various cooling coils differs significantly.

Figure 17 shows a comparison between individual and identical set point control values for a system with two identical air handlers. The system coefficient of performance (COP) associated with optimal control is plotted versus the relative loading on one of the air handlers. The difference between individual and identical set point control is about 10% over the practical range of relative loadings. The use of identical set points applies to air handlers in parallel and to nonidentical designs. However, it is not appropriate if any of the air handlers are operating at a minimum level of flow.

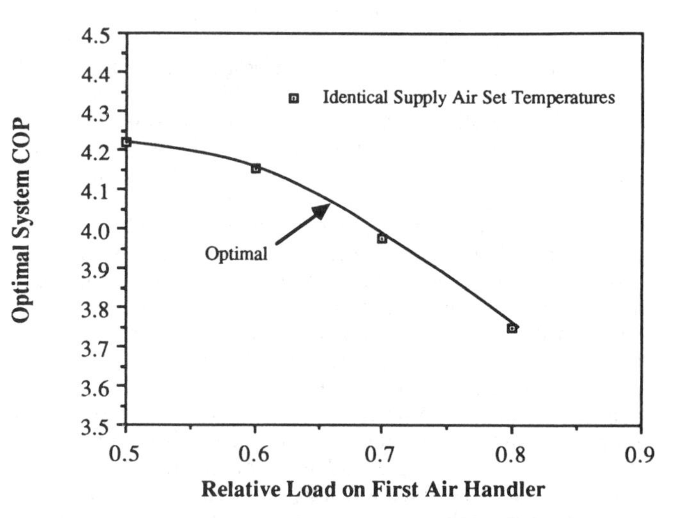

Fig. 17 Comparison of Optimal Performance for Individual Supply Air Set Points with that for Identical Values

For the purpose of determining optimal control, air handlers may be combined into a single effective air handler under the conditions that all zones are maintained at the same air temperature and the heat transfer characteristics of the coils are similar.

Multiple Cooling Tower Cells

The power consumption of a chiller is sensitive to the condensing water temperature, which is, in turn, affected by both the condenser water and cooling tower airflow rates. Increasing either of these flows reduces the chiller power requirement, but at the expense of an increase in the pump or fan power consumption.

Braun *et al.* (1987) and Nugent *et al.* (1988) showed that for variable-speed fans, operating all cooling tower cells under most conditions consumes the minimum amount of power. The power consumption of the fans depends on the cube of the fan speed. Thus, for the same total airflow, operating all cells in parallel rather than some at high flow and some off, allows for lower individual fan speeds and reduced overall fan power consumption. An additional benefit associated with full-cell operation is a lower pressure drop across the water spray nozzles, which results in lower pumping power requirements. However, at very low pressure drops, inadequate spray distribution may adversely affect the thermal performance of the cooling tower. Another economic consideration is the greater water loss associated with full-cell operation.

Most current cooling towers use multiple-speed fans rather than continuously adjustable variable-speed fans. In this case, it is not optimal to operate all tower cells under all conditions. The optimal number of cells operating and individual fan speeds will depend on the system characteristics and ambient conditions. However, simple relationships exist for the best sequencing of cooling tower fans as capacity is added or removed. In almost all cases, when additional tower capacity is required, the tower fan operating at the lowest speed (including fans that are off) should be increased in speed first. Similarly, for removing tower capacity, the highest fan speeds are the first to be reduced.

These guidelines are derived from evaluating the tradeoff between the incremental power increase associated with increasing the fan currently on or switching on another fan. For two-speed fans, the incremental power increase associated with adding a low-speed fan is less than that for increasing one to high speed, if the low speed is less than 79% of the high fan speed. In addition, if the low speed is greater than 50% of the high speed, the incremental increase in airflow is greater (and therefore thermal performance is better) for adding the low-speed fan. Most commonly, the low speed of a two-speed cooling tower fan is between one-half and three-quarters of full speed. In this case, all tower cells should be brought on-line at low speed before any operating cells are set to high speed. Similarly, all the fan speeds should be reduced to low speed before any cells are brought off-line.

For three-speed fans, where the low speed is greater than or equal to one-third of full speed and the difference between the high and intermediate speeds is equal to the difference between the intermediate and low speeds, then the best strategy is to increment the lowest fan speeds first when adding tower capacity and to decrement the highest fan speeds when removing capacity. Typical three-speed combinations that satisfy this criterion are (1) one-third, two-thirds, and full speed or (2) one-half, three-quarters, and full speed.

Another issue related to control of multiple cooling tower cells having multiple-speed fans concerns the distribution of water flow to the individual cells. Typically, the water flow is divided equally among the operating cells. However, the best overall thermal performance of the cooling tower occurs when the flow is divided so that the ratio of water flow rate to airflow rates is identical for all cooling tower cells. In comparing a strategy of equal flow rates to one with equal flow rate ratios, a maximum 5% difference between the heat transfer effectiveness for a combination of one cell operating at one-half speed and the other cell at full speed was found. Depending on the conditions, these differences generally result in a less than 1% change in the chiller power. These differences should also be contrasted with the lower water pressure drop across the spray nozzles (lower pumping power) associated with equally divided flow. In addition, the performance differences are smaller for greater than two-cell operation, when a majority of cells are operating at the same speed. Overall, equal water flow distribution between cooling tower cells is near-optimal.

Multiple Pumps

A common control strategy for sequencing both condenser and chilled water pumps is to bring pumps on-line or off-line with chillers. In this case, both a condenser and a chilled water pump are associated with each chiller. For fixed-speed pumps, this strategy is not optimal. When a chiller is brought on-line in parallel, the pressure drops and the subsequent flow increases through both the condenser and chilled water loops if the pump control is not changed. The increased flow rates tend to improve the overall chiller performance. However, if the pumps are operated near their peak efficiency before the additional chiller is brought on-line, the pump efficiency drops when the additional chiller is added while holding the pump control constant. Most often, the improvements in chiller performance offset the degradation in pump performance, so that no additional pump is needed at the chiller switch point.

Figure 18 shows the optimal system performance for different combinations of chillers and fixed-speed pumps in parallel as a function of load for a given wet bulb. The optimal switch point for a second pump occurs at a much higher relative chilled load (0.62) than the switch point for adding or removing a chiller (0.38). If the second pump were sequenced with the second chiller, the optimal switch point would occur at the maximum chiller capacity and an approximate 10% penalty in performance would result. The optimal control for sequencing fixed-speed pumps depends on the load and the ambient wet-bulb temperature and should not be directly coupled to the chiller sequencing.

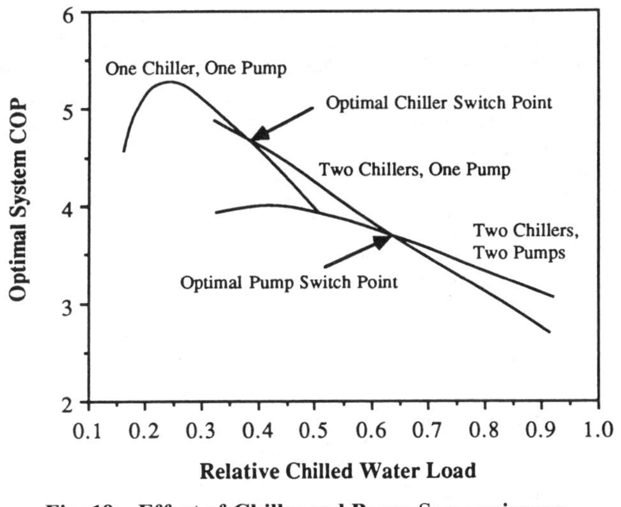

Fig. 18 Effect of Chiller and Pump Sequencing on Optimal System Performance

The sequencing of variable-speed pumps is more straightforward than that for fixed-speed pumps. For a given set of operating chillers and tower cells (*i.e.*, given system pressure drop characteristics), variable-speed pumps have relatively constant efficiencies over their range of operation. As a result, the best sequencing strategy is to select pumps that operate near their peak efficiencies for each possible combination of chillers. Since the system pressure drop characteristics change when chillers are added or removed, the sequencing of variable-speed pumps should be directly coupled to the sequencing of chillers. For identical variable-speed pumps oriented in parallel, the best overall efficiency is obtained by operating them at identical speeds. For nonidentical pumps, near-optimal efficiency is realized by operating them at equal fractions of their maximum speed.

EFFECTS OF LOAD AND AMBIENT CONDITIONS ON OPTIMAL CONTROL SETTINGS

For a given system in which the relative load in each zone is relatively constant with time, the optimal control variables are primarily a function of the total sensible and latent gains to the zones and of the ambient dry- and wet-bulb temperatures. The effect of the ambient dry-bulb temperature alone is insignificant since air enthalpies depend primarily on wet-bulb temperatures and the performance of wet surface heat exchangers are driven primarily by enthalpy differences. Typically, the zone latent gains are on the order of 15 to 25% of the total zone gains. In this range, the effect of changes in latent gains have a relatively small effect on system performance for a given total load. Consequently, results for overall system performance and optimal control may be correlated in terms of only the ambient wet-bulb temperature and total chilled water load. If the load distribution between zones changes significantly over time, the load distribution must also be included as a correlating variable.

Chilled Water Loop

Both the optimal chilled water and supply air temperatures decrease with increasing load for a fixed wet-bulb temperature, because the rate of change in air handler fan power with respect to load changes is larger than that for the chiller at optimal control points. As the wet-bulb temperature increases for a fixed load, the optimal set point temperatures also increase.

Figure 19 shows the sensitivity of system power consumption to chilled water and supply air set point temperatures for a given load and wet-bulb temperature. Within about 3°F of the optimum, the power consumption is within 1% of the minimum. Outside of this range, sensitivity to the set points increases significantly. The penalty associated with operation away from the optimum is greater in the direction of smaller differences between the supply air and chilled water set points. As this temperature difference is reduced, the required flow of chilled water to this coil increases and the chilled water pumping power is greater. For a given chilled water or supply air temperature, the temperature difference is limited by the heat transfer characteristics of the coil. Below this limit, the required water flow and pumping power would approach infinity if the pump were not constrained. It is generally better to have too large rather than too small a temperature difference between the supply air and chilled water set points.

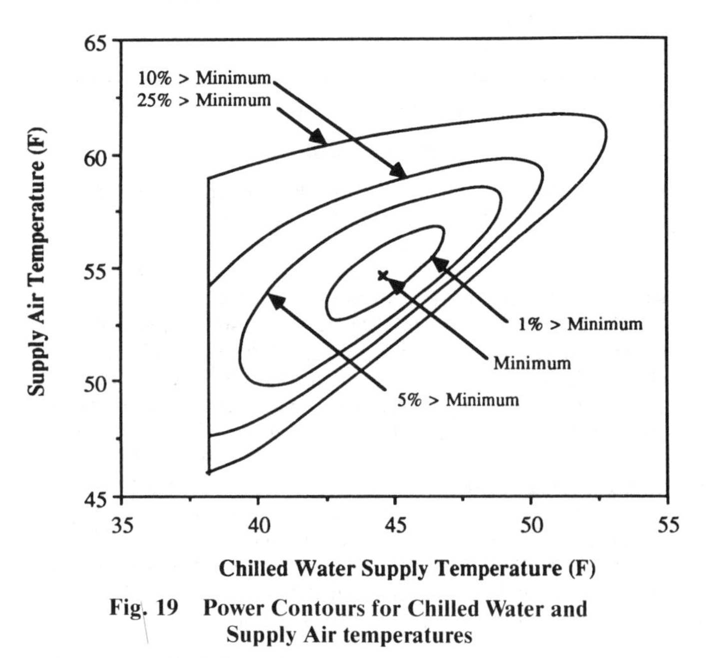

Fig. 19 Power Contours for Chilled Water and Supply Air temperatures

Condenser Water Loop

The primary controllable variables associated with heat rejection to the environment are the condenser water and tower airflow rates. Both optimal air and water flows increase with load and wet-bulb temperature. Higher condensing temperatures and reduced chiller performance result from either increasing loads or wet-bulb temperatures for a given control. Increasing the air and water flow under these circumstances reduces the chiller power consumption at a faster rate than the increases in fan and pump power.

Figure 20 shows the sensitivity of the total power consumption to the tower fan and condenser pump speed. Contours of constant power consumption are plotted versus fan and pump speeds. Near the optimum, power consumption is not sensitive to either of these control variables, but increases significantly away from the optimum. The rate of increase in power consumption is particularly large at low condenser pump speeds. A minimum pump speed is necessary to overcome the static pressure associated with the height of the water discharge in the cooling tower above the takeup from the sump. As the pump speed approaches this value, the condenser flow approaches zero and the chiller power increases dramatically. A pump speed that is too high is generally better than a pump speed that is too low. The broad area near the optimum indicates that it is not necessary to accurately determine the optimal setting.

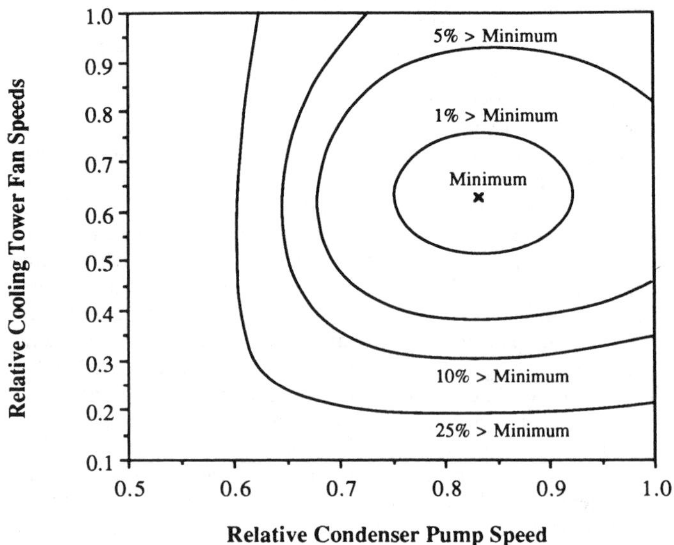

Fig. 20 Power Contours for Condenser Loop Control Variables

OPTIMAL VERSUS ALTERNATIVE CONTROL STRATEGIES

No general strategy has been established for controlling chilled water systems. Most often, the chilled water and supply air set point temperatures are constant with time, while in some applications, these set points vary according to the ambient dry-bulb temperature. Generally, no attempt is made to control the cooling tower and condenser water flow in response to changes in the load and ambient wet bulb. One strategy for controlling these flow rates is to maintain constant temperature differences between the cooling tower outlet and the ambient wet bulb (approach) and between the cooling tower inlet and outlet (range), regardless of the load and wet bulb. In some applications, humidity along with temperature is controlled within the zones.

No fixed values of chilled water temperatures and supply air set point temperatures and tower approach and range are optimal or near-optimal over a wide range of conditions. In addition, it is not obvious how to choose fixed values that work best overall. One simple, yet reasonable, approach is to determine fixed values that result in near-optimal performance at design conditions. Figure 21 shows a comparison between the performance with optimal control and three alternative strategies:

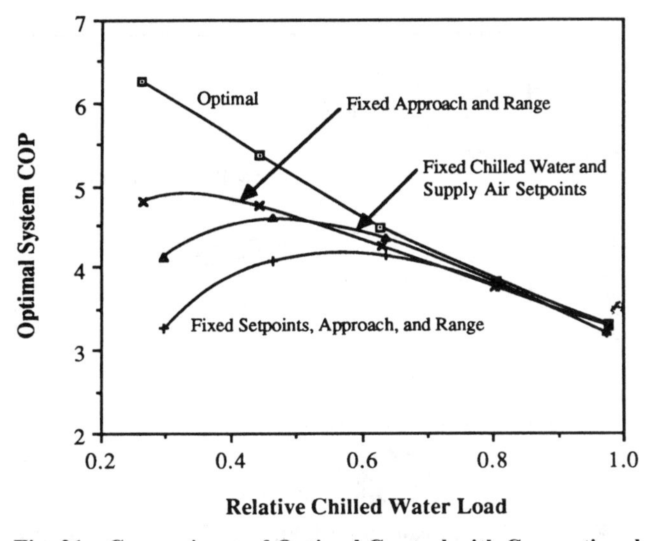

Fig. 21 Comparisons of Optimal Control with Conventional Control Strategies

1. Fixed chilled water and supply air temperature set points (40 and 52°F), respectively, with optimal condenser loop control
2. Fixed tower approach and range (5 and 12°F), respectively, with optimal chilled water loop control
3. Fixed set points, approach, and range

The results are given as a function of load for a fixed wet-bulb temperature. Since the fixed values are chosen to be appropriate at design conditions, the differences in performance as compared with optimal control are minimal at high loads. However, at part-load conditions, Figure 21 shows that the savings associated with the use of optimal control become significant. Optimal control of the chilled water loop results in greater savings than that for the condenser loop for part-load ratios less than about 50%.

The overall savings over a cooling season for optimal control depends on the time variation of the load. If the cooling load were relatively constant, fixed values of temperature set points, approach, and range could be chosen to give near-optimal performance. For typical building loads with daily and seasonal variations, the penalty for using a fixed set point control strategy is in the range of 5 to 20% of the HVAC system energy.

Humidity Control

In a variable air volume system, it is generally possible to adjust the chilled water temperature, supply air temperature, and airflow in order to maintain both temperature and humidity. For a given chilled water temperature, only one combination of the supply air temperature, airflow rate, and water flow rate will maintain both the room temperature and humidity set point.

Chapter 8 in the 1989 ASHRAE *Handbook—Fundamentals* defines acceptable bounds on the room temperature and humidity for human comfort. For a zone being cooled, the equipment operating costs are minimized when the zone temperature is at the upper bound of the comfort region. However, operation also at the humidity upper limit does not minimize costs. Figure 22 shows a comparison of costs and room humidity associated with fixed and free-floating zone humidity as a function of the load. Over the range of loads for this system, the free-floating humidity operates within the comfort zone at both lower cost and humidity, with the largest differences occurring at the high loads. Operation at the upper humidity bound results in lower latent loads, but the addition of this humidity control constraint requires higher supply air temperatures than that associated with free-floating humidities. In turn, the higher supply air temperature results in greater

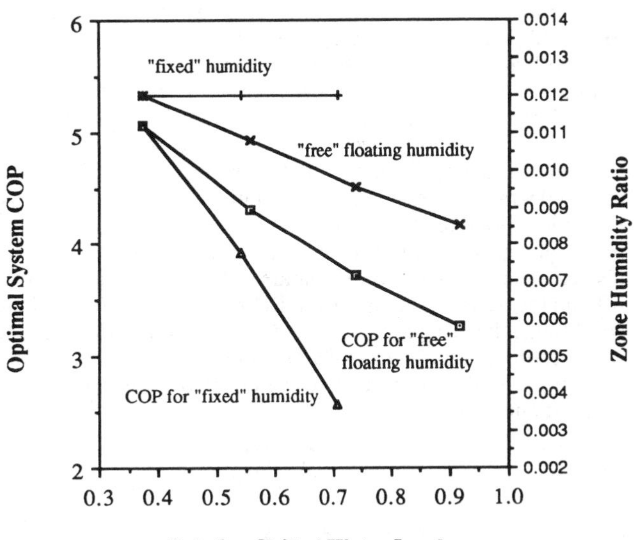

Fig. 22 Comparison of Free-Floating and Fixed Humidity Control

air handler power consumption. When determining optimal control points, the humidity should be allowed to float freely unless it falls outside the bounds of human comfort.

USING VARIABLE-SPEED EQUIPMENT

Variable-speed equipment such as chillers, fans, and pumps facilitates control and reduces operating costs. The overall savings associated with the use of variable-speed equipment over a cooling season depends on the time variation of the load. Overall, the use of fixed-speed equipment results in operating costs that are 20 to 50% higher than for equipment with variable-speed drives.

The part-load performance of a centrifugal chiller is generally better for variable-speed than it is with vane control. Figure 23 gives a comparison between the overall optimal system performance for variable speed and variable vane control of a large chilled water facility (Braun 1988, Braun *et al.* 1989). At part-load conditions, the performance associated with the use of variable-speed control is up to 25% better. However, the power requirements are similar at conditions associated with peak loads. This is expected, since at this condition, the vanes are wide open and the speed under variable-speed control approaches that of fixed-speed operation.

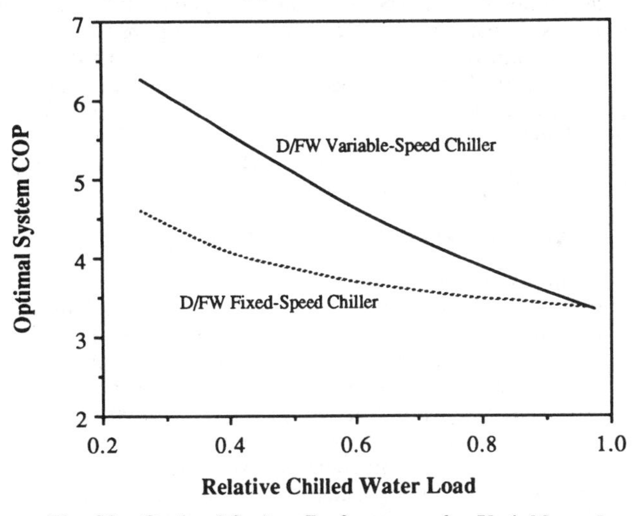

Fig. 23 Optimal System Performance for Variable and Fixed-Speed Chillers

The most common design for cooling towers places multiple tower cells in parallel with a common sump. Each tower cell has a fan with one, two, or possibly three operating speeds. Although multiple cells with multiple fan settings offer wide flexibility in control, the use of variable-speed tower fans can provide additional improvements in overall system performance.

Figure 24 shows a comparison of optimal system performance for single-speed, two-speed, and variable-speed tower fans as a function of load for a given wet bulb. The variable-speed option results in higher system COP under all conditions. In contrast, for discrete fan control, the tower cells are isolated when their fans are off and the performance is poorer. Below about 70% of full-load conditions, the difference between one-speed and variable-speed fans becomes significant (15% on total energy consumption). With two-speed fans, the differences are on the order of 3 to 5% over the entire range.

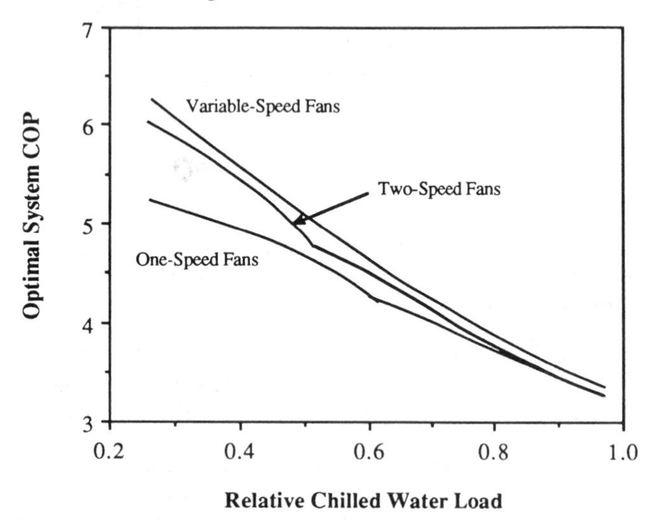

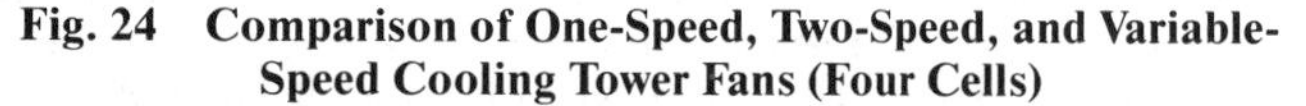

Fig. 24 Comparison of One-Speed, Two-Speed, and Variable-Speed Cooling Tower Fans (Four Cells)

Fixed-speed pumps, which are sized to give proper flow to a chiller at design conditions, are oversized for part-load conditions. As a result, the system will have higher operating costs than with a variable-speed pump of the same design capacity. Using a smaller fixed-speed pump for low load conditions improves the flexibility in control and can reduce overall power consumption. The optimal system performance for variable-speed and fixed-speed pumps applied to both the condenser and chilled water flow loops is shown in Figure 25. Large fixed-speed pumps are sized for design conditions, while small pumps have one-half the flow capacity of the large pumps. Below about 60% of full-load conditions, variable-speed pumps show a significant improvement over single fixed-speed pumps. With the addition of small fixed-speed pumps, the improvements with variable-speed pumps become significant at about 40% of the maximum load.

SIMPLIFIED OPTIMAL CONTROL METHOD

Optimal supervisory control of central cooling systems (without thermal storage) determines the control variables that minimize total power plant consumption at each time (or over each decision interval). In practice, optimization techniques are applied to a steady-state model of the power consumption. The plant model may take any number of different forms. Cumali (1988) described a global optimization procedure that involves a detailed mechanistic simulation of the plant components.

Braun (1988) and Braun *et al.* (1989) present a component-based method for determining plant optimal control. This optimization tool was used to develop a simpler method for determining the

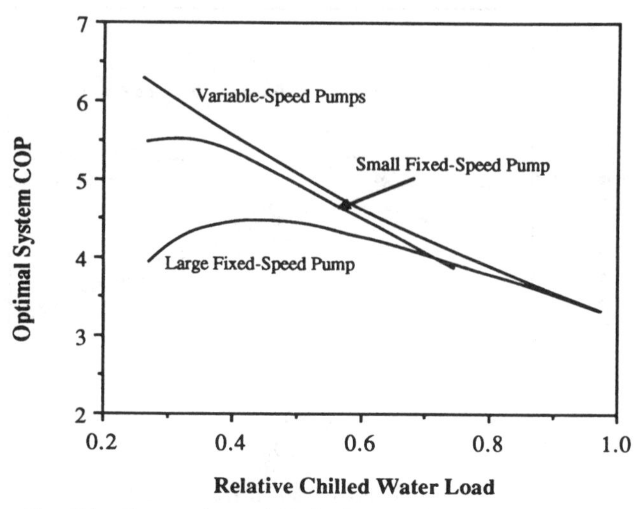

Fig. 25 Comparison of Variable and Fixed-Speed Pumps

optimal control that requires only one empirical correlating function for the plant power consumption.

In the vicinity of any optimal control point, the plant power consumption may be estimated with a quadratic function of the continuous control variables for each of the operating modes (*i.e.*, discrete control mode). A quadratic function also correlates power consumption in terms of the uncontrolled variables (*i.e.*, load, ambient temperature) over a wide range of conditions. The following cost function may be used to determine optimal control points:

$$J(f,M,u) = u^T A u + b^T u + f^T C f + d^T f + f^T E u + g \quad (5)$$

where J is the total operating cost, u is a vector of continuous control variables, f is a vector of uncontrolled variables, M is a vector of discrete control variables, and the superscript T designates the transpose vector. The A, C, and E are coefficient matrices, $\boldsymbol{b}$ and $\boldsymbol{d}$ are coefficient vectors, and g is a scalar. The empirical coefficients of the above cost function depend on the operating modes so that these constants must be determined for each feasible combination of discrete control modes.

A solution for the optimal control vector that minimizes the cost may be determined analytically by applying the first-order condition for a minimum. Equating the Jacobian of Equation (5) with respect to the control vector to zero and solving for the optimal control gives:

$$u^* = k + Kf \quad (6)$$

where

$k = (1/2) A^{-1} b$
$K = -(1/2) A^{-1} E$

The cost associated with the unconstrained control of Equation (6) is:

$$J^* = f^T \theta f + \sigma^T f + \tau \quad (7)$$

where

$\theta = K^T A k + EK + C$
$\sigma = 2KAk + Kb + Ek + d$
$\tau = K^T A k + b^T k + g$

The control defined by Equation (6) results in a minimum power consumption if A is positive definite. If this condition holds and if the system power consumption is adequately correlated with Equation (5), then Equation (6) dictates that the optimal continuous control variables vary as a nearly linear function of the uncontrolled variables. However, a different linear relationship applies to each feasible combination of discrete control modes. The minimum cost associated with each mode combination must be computed from Equation (7). The costs for each combination are compared in order to identify the minimum.

Many coefficients must be determined empirically. One approach for determining these constants would be to apply regression techniques directly to measured total power consumption. Since the cost function is linear with respect to the empirical coefficients, linear regression techniques may be used. A set of experiments could be performed on the system over the expected range of operating conditions. Possibly, the regression could be performed on-line using least-squares recursive parameter updating (Ljung and Söderstron 1983).

Rather than fitting empirical coefficients of the system cost function of Equation (5), the coefficients of the optimal control Equation (6) and the minimum cost function of Equation (7) could be estimated directly. At a limited set of conditions, optimal values of the continuous control variables could be estimated through trial-and-error variations in the system. Only three independent conditions would be necessary to determine coefficients of the linear control law given by Equation (6) if the load and wet bulb are the only uncontrolled variables. The coefficients of minimum cost function could then be determined from system measurements with the linear control law in effect. The disadvantage of this approach is that there is no direct way to handle constraints on the controls.

Using the guidelines for optimal control described earlier, the important independent control variables are (1) supply air set temperature, (2) chilled water set temperature, (3) relative tower airflow, (4) relative condenser water flow, and (5) the number of operating chillers.

The supply air and chilled water set points are continuously adjustable control variables. However, since the chilled water flow requirements depend on these controls, discrete changes in power consumption may be associated with these controls if the pump operation undergoes discrete control changes. For the same total flow rate, the overall pumping efficiency changes with the number of operating pumps. However, this has a relatively small effect on the overall power consumption and may be neglected in fitting the overall cost function to changes in the control variables.

For variable-speed cooling tower fans and condenser water pumps, the relative tower air and condenser water flows are continuous control variables. Analogous to the chilled water flow, the overall condenser pumping efficiency changes with the number of operating pumps, so that the power consumption associated with continuous changes in the overall relative condenser water flow may have a discontinuity. This discontinuity may also be neglected in fitting the overall cost function to changes in this control variable.

With variable-speed pumps and fans, the only significant discrete control variable is the number of operating chillers. A chiller mode defines which of the available chillers are to be on-line. The optimization involves determining optimal values of only four continuous control variables for each of the feasible chiller modes. The chiller mode giving the minimum overall power consumption represents the optimum. For a chiller mode to be feasible, the specified chillers must operate safely within their capacity and surge limits. In practice, abrupt changes in the chiller modes should also be avoided. Large chillers should not be cycled on or off, except when the savings associated with the change is significant.

For fixed-speed cooling tower fans and condenser water pumps, the relative flows are discrete values. One method of handling these variables is to consider each of the discrete combinations as separate modes. However, for multiple cooling tower cells with multiple fan speeds, the number of possible combinations may be large. A simpler approach that works well is to treat the relative flows as continuous control variables and to select the discrete relative flow that is closest to that determined with the continuous optimization. At least three relative flows (discrete flow modes) are necessary for each chiller mode in order to fit the quadratic cost function. The number of possible sequencing modes for fixed-speed pumps is generally much more limited than that for cooling tower fans, with two or three possibilities (at most) for each chiller mode. In fact, with many current designs, individual pumps are physically coupled with chillers, and it is impossible to operate more or fewer pumps than the number of operating chillers. Thus, it is generally best to treat the control of fixed-speed condenser water pumps with a set of discrete control possibilities rather than using a continuous control approximation.

The methodology for near-optimal control of a chilled water system may be summarized as follows:

1. Change the chiller operating mode if system operation is at the limits of chiller operation (near surge or maximum capacity).
2. For the current set of conditions (load and wet bulb), estimate the feasible modes of operation that would avoid operating the chiller and condenser pump at their limits.
3. For the current operating mode, determine optimal values of the continuous controls using Equation (6).
4. Determine a constrained optimum if controls exceed their bounds.
5. Repeat steps 3 and 4 for each feasible operating mode.
6. Change the operating mode if the optimal cost associated with the new mode is significantly less than that associated with the current mode.
7. Change the values of the continuous control variables. When treating multiple-speed fan control with a continuous variable, use the discrete control closest to the optimal continuous value.

If the linear optimal control Equation (6) is directly determined from optimal control results, then the constraints may be handled directly. A simple solution is to constrain the individual control variables as necessary and neglect the effects of the constraints on the optimal values of the other controls and the minimum cost function. The variables of primary concern with regard to constraints are the chilled water and supply air set temperatures. These controls must be bounded for proper comfort and safe operation of the equipment. On the other hand, the cooling tower fans and condenser water pumps should be sized so that the system performs efficiently at design loads and constraints on control of this equipment should only occur under extreme conditions.

The optimal value of the chilled water supply temperature is coupled to the optimal value of the supply air temperature, so that decoupling these variables in evaluating constraints is generally not justified. However, optimization studies indicate that when either control is operated at a bound, the optimal value of the other "free" control is approximately bounded at a value that depends only on the ambient wet-bulb temperature. The optimal value of this free control (either chilled water or supply air set point) may be estimated at the load at which the other control reaches its limit. Coupling between optimal values of the chilled water and condenser water loop controls is not as strong, so that interactions between constraints on these variables may be neglected.

Braun *et al.* (1987) correlated the power consumption of the Dallas/Fort Worth airport chiller, condenser pumps, and cooling tower fans with the quadratic cost function given by Equation (5) and showed good agreement. Since the chilled water loop control was not considered, the chilled water set point was treated as a known uncontrolled variable. The discrete control variables associated with the four tower cells with two-speed fans and the three condenser pumps were treated as continuous control variables. The optimal control determined by the near optimal Equation (6) also agreed well with that determined using a nonlinear optimization applied to a detailed simulation of the system. Figures 26 and 27 show values of the chilled water and supply air temperatures obtained from the quadratic approach and the true optimal values. The chilled water temperature was constrained

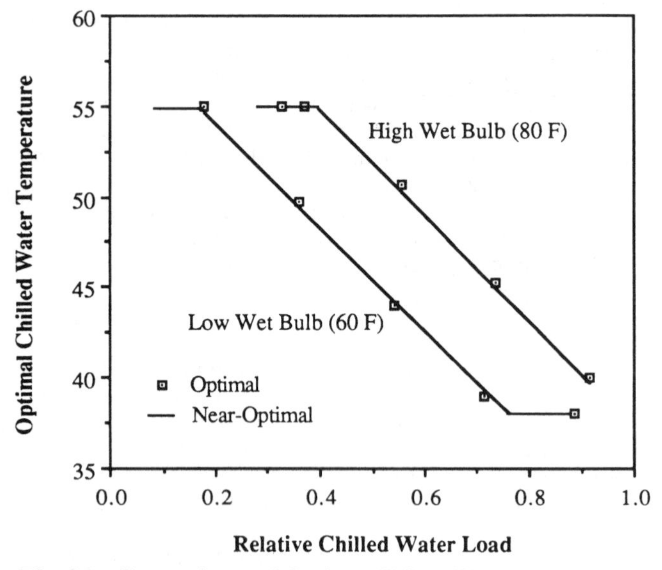

Fig. 26 Comparisons of Optimal Chilled Water Temperature

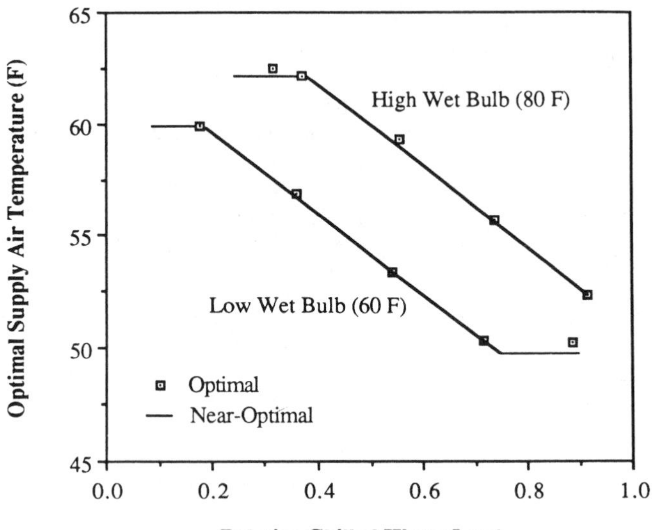

Fig. 27 Comparisons of Optimal Supply Air Temperature

between 38 and 55 °F, while the supply air set point was allowed to float freely. Also, for conditions where the chilled water temperature is constrained, the optimal supply air temperature is also bounded at a value that depends on the ambient wet bulb.

REFERENCES

ASHRAE. 1985. Thermal Storage. ASHRAE *Technical Data Bulletin* (January).

Bloomfield, D.P. and D.J. Fisk. 1977. The optimization of intermittent heating. *Buildings and Environment* 12:43-55.

Braun, J.E. 1988. Methodologies for the design and control of central cooling plants. Ph.D. Dissertation, University of Wisconsin-Madison.

Braun, J.E., J.W. Mitchell, S.A. Klein, and W.A. Beckman. 1987. Performance and control characteristics of a large central cooling system. ASHRAE *Transactions* 93(1):1830-52.

Braun, J.E., S.A. Klein, J.W. Mitchell, and W.A. Beckman. 1989. Methodologies for optimal control to chilled water systems without storage. ASHRAE *Transactions* 95(1).

Braun, J.E., S.A. Klein, J.W. Mitchell, and W.A. Beckman. 1989. Applications for optimal control to chilled water systems without storage. ASHRAE *Transactions* 95(1).

Brothers, P.W. and M.L. Warren. 1986. Fan energy use in variable air volume systems. ASHRAE *Transactions* 92(2).

Bushby, T.B. and G.E. Kelly. 1988. Comparison of digital control and pneumatic control systems in a large office building. National Bureau of Standards, NBSIR 88-3739.

Cumali, Z. 1988. Global optimization of HVAC system operations in real time. ASHRAE *Transactions* 95(1).

Hackner, R.J., J.W. Mitchell, and W.A. Beckman. 1984. HVAC system dynamics and energy use in buildings—Part I. ASHRAE *Transactions* 90(2B):523-35.

Hackner, R.J., J.W. Mitchell, and W.A. Beckman. 1985. HVAC system dynamics and energy use in buildings—Part II. ASHRAE *Transactions* 91(1B):781.

Hartman, P.E. 1988. Dynamic control: A new approach. *Heating/Piping/Air Conditioning.*

Hittle, D.C. 1979. The building loads analysis and systems thermodynamics (BLAST) program, Version 2.0. U.S. Army Construction Engineering Research Laboratory, Users Manual, Vol. 1. (Available from NTIS, Springfield, VA 22151.)

Johnson, G.A. 1985. Optimization techniques for a centrifugal chiller plant using a programmable controller. ASHRAE *Transactions* 91(2).

Kao J.Y. 1985. Sensor errors. ASHRAE *Journal* (January); also J.Y. Kao. 1985. Control strategies and building energy consumption. ASHRAE *Transactions* 91(2).

Lau, A.S., W.A. Beckman, and J.W. Mitchell. 1985. Development of computer control—Routines for a large chilled water plant. ASHRAE *Transactions* 91(1).

Ljung, L. and T. Söderstron. 1983. *Theory and practice of recursive identification*. MIT Press, Cambridge, MA.

Marcev, C.L., C.R. Smith, and D.D. Bruns. 1984. Supervisory control and system optimization of chiller-cooling tower combinations via simulation using primary equipment component models—Part II, Simulation and optimization. Proceedings of the Workshop on HVAC Controls Modeling and Simulation, Georgia Institute of Technology (February).

Marcev, C.L. 1980. Steady-state modeling and simulation as applied to energy optimization of a large office building integrated HVAC system. MS thesis, University of Tennessee.

Miller, D.E. 1980. The impact of HVAC process dynamics on energy use. ASHRAE *Transactions* 86(2).

Nizet, J.L., J. Lecomte, and F.X. Litt. 1984. Optimal control applied to air conditioning in buildings. ASHRAE *Transactions* 90(1B):587-600.

Nugent, D.R., S.A. Klein, and W.A. Beckman. 1988. Investigation of control alternatives for a steam turbine driven chiller. ASHRAE Winter Annual Meeting, Dallas.

Seem, J.E., P.R. Armstrong, and C.E. Hancock. 1989. Comparison of seven methods for forecasting the time to return from night setback. ASHRAE *Transactions* 95(2).

Shapiro, M.M., A.J. Yager, and T.H. Ngan. 1988. Test hut validation of a microcomputer predictive HVAC control. ASHRAE *Transactions* 94(1).

Shavit, G. and S.G. Brandt. 1982. The dynamic performance of a discharge air-temperature system with a P-I controller. ASHRAE *Transactions* 88(2):826-38.

Sud, I. 1984. Control strategies for minimum energy usage. ASHRAE *Transactions* 90(2).

Treichler, W.W. 1985. Variable speed pumps for water chillers, water coils, and other heat transfer equipment. ASHRAE *Transactions* 91(1).

CHAPTER 39

THERMAL STORAGE

THERMAL storage is the temporary storage of high- or low-temperature energy for later use. Examples of thermal storage are the storage of solar energy for night heating, the storage of summer heat for winter use, the storage of winter ice for space cooling in the summer, and the storage of heat or coolness generated electrically during off-peak hours for use during subsequent peak hours. Most thermal storage applications involve a 24-h storage cycle, although weekly and seasonal cycles are also used.

When the period of energy availability is longer than the period of energy use, thermal storage permits the installation of smaller heating or cooling equipment than would otherwise be required. In many cases, storage can provide benefits for both heating and cooling, either simultaneously or at different times of the year. Conditions favoring thermal storage include high loads of relatively short duration; high electric power demand charges; low-cost electrical energy during off-peak hours; need for cooling backup in case of power outage or a refrigeration plant failure (only chilled water pumps must be powered by the emergency generator); need to provide cooling for small after-hour loads such as cooling of restaurants, midheight elevator equipment rooms, individual offices, and computer rooms; building expansion (installation of storage may eliminate the need for heating/cooling plant expansion); need to provide a fire fighting reservoir (water stored in a water or ice system is available for fire fighting in an emergency); need to supplement a limited capacity cogeneration plant; and others.

Definitions

In **sensible heat storage**, storage is accomplished by raising or lowering the temperature of the storage medium, *i.e.*, water, rock beds, bricks, sand, or soil.

In **latent heat storage**, storage is accomplished by a change in the physical state of the storage medium, usually from liquid to solid (heat of fusion) or vice versa. Typical materials are water/ice, salt hydrates, and certain polymers. Energy densities for latent heat storage are greater than those for sensible heat storage, which results in smaller and lighter storage devices and lower storage losses.

Storage efficiency is the ratio of energy that can be withdrawn from storage, divided by the amount put into storage. Storage efficiencies up to 90% can be achieved in well-stratified water tanks that are fully charged and discharged on a daily cycle. All storage devices are subject to standby losses, which are generally proportional to the exposed surface area of the storage vessel.

Storage Media

For HVAC and refrigeration purposes, water and phase change materials (PCMs)—particularly ice—constitute the principal storage media. Soil, rock, and other solids are also used. Water has the advantage of universal availability, low cost, and transportability through other system components. Phase change materials have the advantage of approximately 80% less volume than that of water storage for a *temperature swing* of 18°F, which is the difference in temperature between a fully charged and a fully discharged storage vessel. Some phase change materials are viscous and corrosive and must be segregated within the container for the heat transfer medium. If heating and cooling storage is required, two phase change materials must be provided, unless heat pumping is used.

Operating Modes

The five modes of operation of a thermal energy storage system for cooling which store and release thermal energy are illustrated in Figure 1 as follows:

1. *Charging storage.* Involves operating the refrigeration system to prepare the storage vessel for its cooling function.
2. *Simultaneous recharging storage and live-load chilling.* Some of the cooling capacity is used by the load at the same time that storage is being recharged.
3. *Live-load chilling.* Identical to the normal operation of a conventional system, as it provides cooling capacity as needed.
4. *Discharging and live-load chilling.* The refrigeration plant operates and the storage vessel discharges at the same time to satisfy cooling demand.
5. *Discharging.* Cooling needs are met only from storage, with no operation of the refrigeration plant.

Open versus Closed Systems

An open system is one where the storage is not made to be pressurized; therefore, the heat transfer fluid returning to the tank must be at atmospheric pressure. The heat transfer fluid (usually water) is also normally the storage medium (water or ice). These tanks are site-built concrete or prefabricated steel. In low-rise buildings, the extra energy used for pumping is minimal. As the height of the building increases, other options must be studied. Heat exchangers are often used to separate the system's heat transfer fluid from the storage medium.

In a closed system, a separation is built between the storage medium and the heat transfer fluid. This separation allows the system's heat transfer fluid to be pressurized and keeps the system closed to the atmosphere. Thus, the static head of the buildings is not added to the pumping requirements.

Storage Containers

Storage for one- or two-story buildings can often be "in-line" with the circuits served, because the cost premium for the storage container to withstand 7 psi is negligible. Higher pressures may

The preparation of this chapter is assigned to TC 6.9, Thermal Storage.

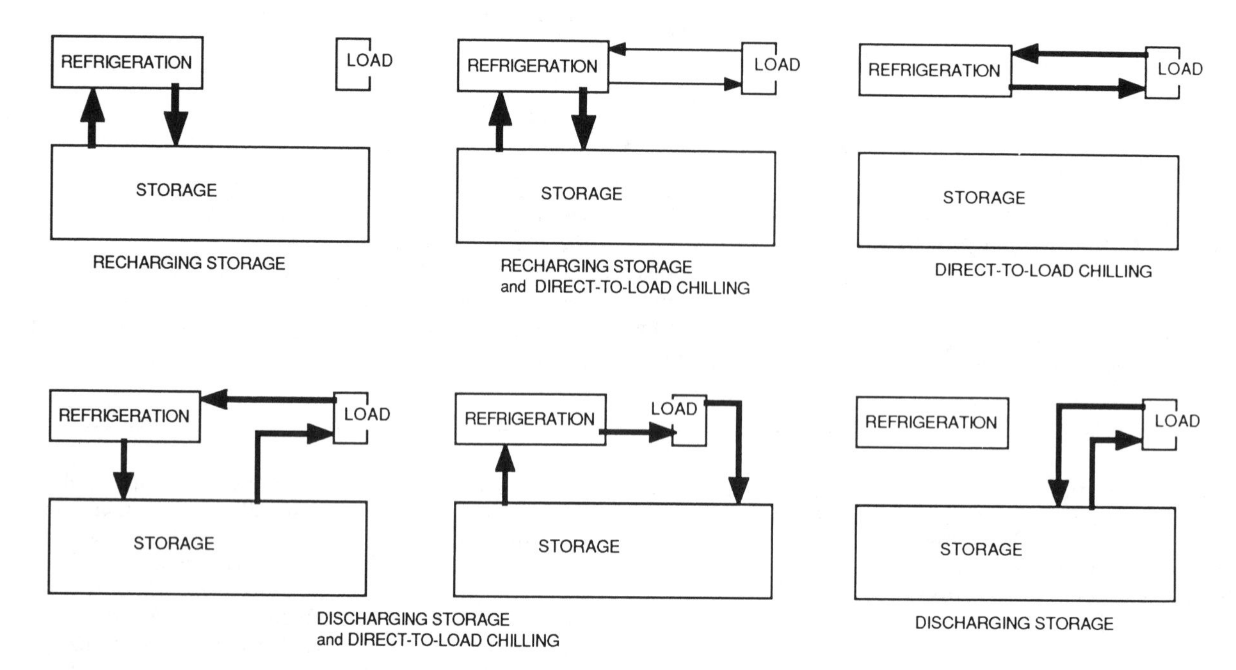

Fig. 1 Operating Strategies of Cool Storage

require the use of steel tanks with reinforced dished ends. Many storage installations are vented to the atmosphere. Concrete has been the preferred construction material for vented chilled water containers because of its universal availability and adaptability to underground conditions. Prefabricated steel tanks are available in virtually all sizes. Plastic tanks are often used for ice and PCM storage media, or the PCM material may be encased in plastic containers which, in turn, are immersed in chilled water or antifreeze tanks.

Storage Location

In high-rise buildings, water tanks are usually located at ground level or below and are used as a secondary water supply for fire protection. If possible, the storage should be near the machine room. Installation outside the building above grade may be subject to aesthetic considerations. If storage is located below grade, the units must be anchored to secure them against uplift from groundwater, if it is anticipated that they may be drained for maintenance, repair, or inspection.

Direct Pumping from Open Storage

Open steel or concrete storage tanks are common because of the high cost of pressure vessels. For this reason, some storages are located on the roof to reduce transfer pumping energy. In others, a heat exchanger is placed between the storage and the piping system. But the initial cost and temperature and pressure losses of heat exchangers have limited their application to buildings more than 30 stories high.

The height of the building must be added to the pressure drop of systems that pump from an open storage at the base of the building. Also, a pressure sustaining valve (PSV) must be installed to drop the system pressure as water returns to storage (Figure 2). Glove and plug valves are best suited for this duty because they can handle large pressure drops with little cavitation. The valve manufacturer can help in selecting the proper valve.

An automatic, positive shutoff valve in series with the PSV should be placed in systems subject to regular shutdown. In addition to the positive shutoff valve, a small pump could be installed to restore pressure at the top of the circuit in case the PSV or pump check valve leaks during shutdown.

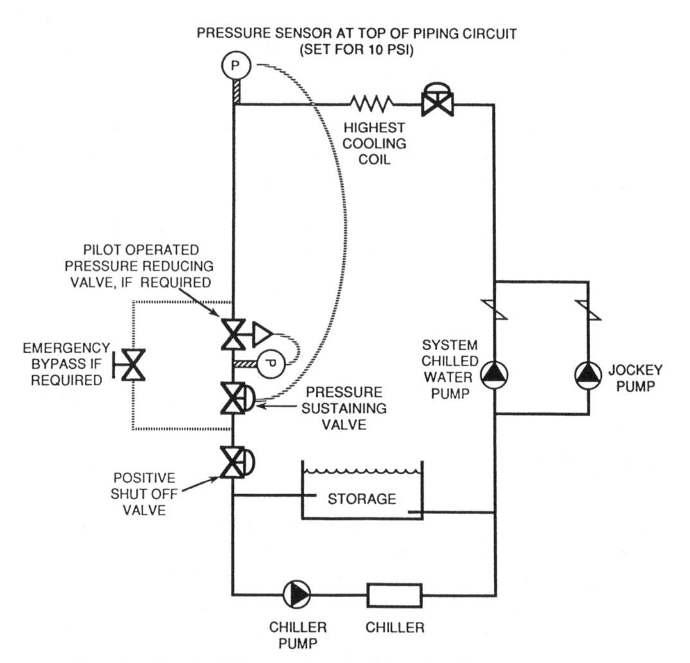

Fig. 2 Typical Circuit with Pressure Sustaining Valve

When upstream pressure exceeds about 50 psi, the PSV should have a pilot-operated pressure-reducing valve in front of it. The downstream PSV should then respond to a sensor at the top of the piping circuit set for 10 psi.

With large flows, consider installing two PSVs in parallel for better flow modulation. One valve might be sized for two-thirds flow and the other for one-third of the flow. For very tall buildings, consider installing several automatic PSVs in series, an upstream energy recovery turbine, or a reverse running pump. The cost of these extra devices for an open system should be checked against the cost of installing a pressure system instead. However, do not overlook the cost of temperature and pressure losses in the heat exchanger for the pressure system.

Controls

In full-storage systems, the chilling plant is either on or off and always operates at full power. In partial-storage plants, the chilling plant and the storage unit share the load, and the fraction of the load supplied by either requires more control than a simple on-off device.

Time clocks are frequently used to shut down the heating/cooling plant during on-peak periods. Malfunction of mechanical clocks has caused unwanted plant operation during on-peak periods, so electronic time clocks, which are more reliable, should be installed. System operators should not be required to switch compressors on and off manually because, if the operator forgets to shut off the compressor before the beginning of the on-peak period even once, it could result, under certain utility tariffs, in a billing penalty for an entire year.

Sensors

Monitoring sensors require regular maintenance and frequent recalibration. Proper sensor performance is more important in thermal storage installations than in other HVAC systems, as the advantages of storage can only be realized through exact timing and accurate sensing of storage water temperature or ice inventory.

Water Treatment

Water treatment is essential in both open and closed water systems. Treatment is especially necessary for open systems subject to atmospheric pollutants and poor quality makeup water. Ice storage tanks and evaporator coils are subject to corrosion and require proper cleaning and water treatment, especially if the tank is emptied during the off-season. Brine systems require that the antifreeze material used be inhibited from causing corrosion. Chapter 43 covers this subject in detail.

Operator Training

Operator training is more critical with thermal storage systems than with conventional HVAC systems, because the systems are more complicated and unfamiliar to most operators. Some storage systems have not saved money because either the original operators had left and were replaced by new, unskilled operators, or the original operators had not been properly trained how to manage these systems.

Operators should be carefully trained by personnel who understand the system fully (designer, contractor, factory representative), and their performance should be closely monitored for at least one full season. Some ice storage installations may require longer operating training periods than chilled water systems, because they may contain elements unfamiliar to most HVAC system operators.

ECONOMICS

Thermal storage is installed for two major reasons: (1) to lower initial costs and (2) to lower operating costs. Lower initial costs are usually obtained when the load to meet is of short duration and there is a long time before the load returns. A church or sports facility are examples of such a load. These buildings have a relatively large space conditioning load for less than 6 h per day, and only on a few days per week. The relatively small refrigeration plant for these applications would operate continuously for up to 100 h or more to recharge a thermal storage. The initial cost of such a design could be less than a standard cooling plant designed to meet the highest instantaneous cooling load.

Secondary capital costs may also be lower for thermal storage. For example, the electrical service entrance size can sometimes be reduced because energy demand is lower. If the low temperature from an ice bank thermal storage can be circulated through the air distribution system, smaller fans and ducts can sometimes be used. Also, the floor-to-floor height of the building can sometimes be reduced while providing more usable space.

An example of this reduction in equipment size is shown in Figure 3. The load profiles shown are for a 100,000 ft^2 commercial building. If the 2040 ton-hour load is met by a conventional air-conditioning system, as shown in Figure 3A, a 220-ton unit is required to meet the peak cooling demand.

If a load-leveling partial storage is used, as shown in Figure 3B, an 85-ton chiller meets the demand. The design-day cooling load in excess of the chiller output (1020 ton-hours) is supplied by the storage.

If a full-storage system is installed, as shown in Figure 3C, the entire peak load is shifted to the storage and a 120 ton chiller would be required. Although larger than the load-leveling system (Figure 3B), a full-storage system offers substantial savings over the conventional system (Figure 3A).

The other reason for installing a thermal storage system is to reduce operating costs. Most electric utilities charge less during night or weekend off-peak hours, and more during the times of highest electrical demand. For many utilities, peak demands occur on hot, summer afternoons because of high air-conditioning use. Electric rates are normally divided into demand charges and consumption charges. Demand charges are based on the highest recorded demand for electricity, measured over a brief period (1 to 15 min). A demand charge based on the highest demand during the year is called ratchet billing. It is assessed even if the maximum demand reached in any particular month is less than the annual peak.

Consumption charges are based on measured use of electricity in kilowatt-hours (kWh), and are tied to the utility's costs of fuel to operate its generation facilities. In some cases, consumption charges are lower during off-peak hours, reflecting the lower production costs to the utility because a higher proportion of the electricity is generated by baseload plants that are less expensive to operate. These rates are known as time-of-day billing structures.

To evaluate the savings, annual operating costs of each system being considered must be estimated. This estimate is made by adding all energy to be used during the year and multiplying by the electric rate per kilowatt-hour. If time-of-day billing is used, the kilowatt-hours must be grouped by the time-of-day period to make an accurate estimate. Next, the peak demand for the system is estimated for each month. Monthly peak demand is then multiplied by the demand charge, totaled for the year, and adjustments are made for any ratchet billing.

Using the load profiles for the building described in Figure 3 as an example, the first step is to determine the annual utility cost for the conventional air-conditioning system. The energy used during the off-peak period (corresponding to 700 ton · h of cooling) would be multiplied by the off-peak electrical rate. The energy used during the on-hour on-peak period (1340 ton · h) is multiplied by the on-peak electrical rate. The peak demand, which occurred at 3 p.m., is multiplied by the demand charge and adjusted for the ratchet billing.

This procedure is repeated (1) for the average day in each month to determine the energy charges and (2) for the peak in each month to determine the demand charges.

Annual costs for a load-leveling partial thermal storage system are made the same way. Using the load profile for the load-leveling partial storage system as an example, the energy used during the off-peak period (935 + 510 ton · h) is multiplied by the off-peak electrical rate. The energy consumed during the on-peak period (595 ton · h) is multiplied by the on-peak electrical rate. The peak demand, which is 85 ton × 3.52 kW/ton = 300 kW in this example, is multiplied by the demand charge and adjusted for any ratchet billing. This process is repeated for each month. The demand charges for the load-leveling system are lower than those

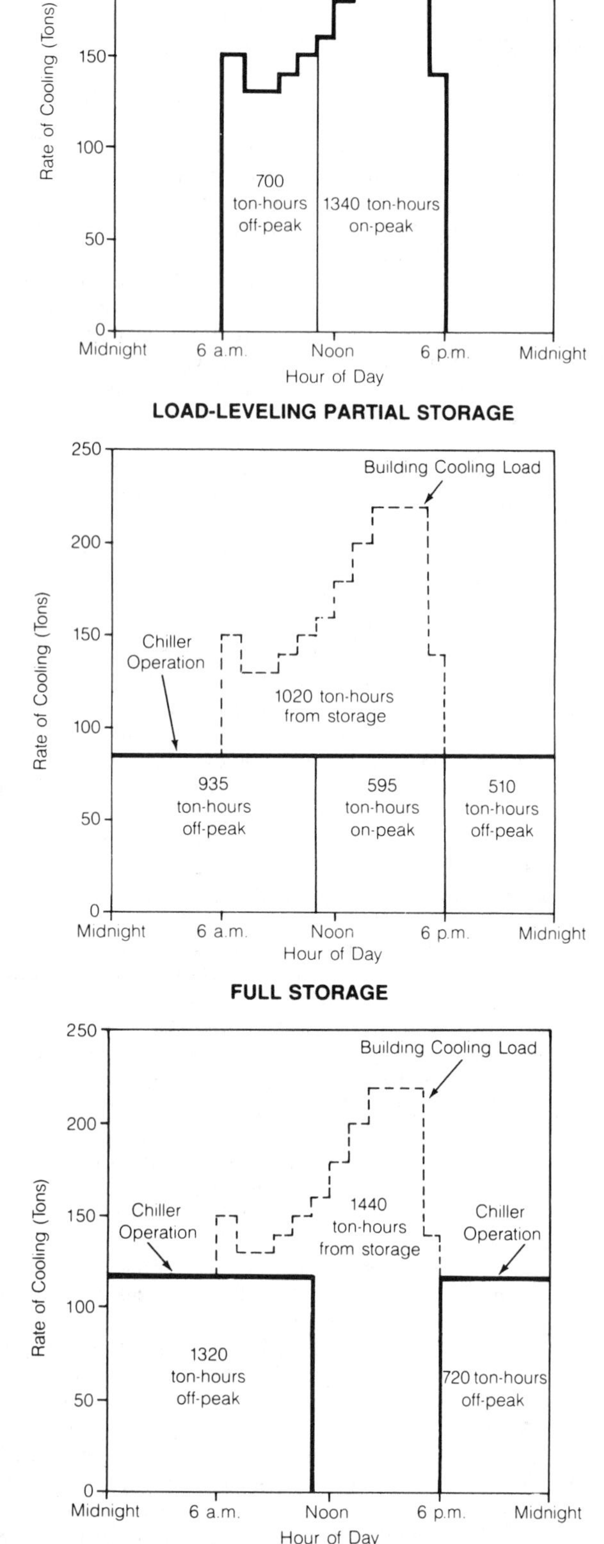

Fig. 3 Hourly Cooling Load Profiles

for a conventional system. If time-of-day billing is used, the total energy charges are lower because less energy was used during the on-peak hours.

The full-storage system eliminates all energy use during the on-peak hours except energy used by the auxiliary equipment needed to transfer stored energy into the system. The energy costs for the off-peak periods (1320 + 720 ton·h) are multiplied by the off-peak rate. This procedure should be repeated for the average day in each month to determine the annual energy costs. Some utilities assess demand charges only for demand occurring during the on-peak hours. If this is the case, this example system design would avoid any demand charges for the refrigeration equipment.

In order to complete the analysis, the initial costs must be determined. Equipment costs should be obtained from each of the manufacturers under consideration and an estimate of installation cost made.

The cost savings along with the net capital costs should be analyzed using the life cycle cost method or other suitable method to determine which system is best for the project. Other items to be considered are space requirements for the system, reliability of the system, and ease of interface to the delivery system planned for the project. An optimal thermal energy storage application will maximize the savings in utility charges and minimize the initial cost of the installation needed to achieve the savings.

THERMAL STORAGE TECHNOLOGIES

WATER TANKS

Conceptually, the simplest thermal storage is a water tank. During off-peak periods, hot or cold water is generated and stored in the tank, and then withdrawn during peak periods as needed. Water has the highest specific heat of all common materials—1 Btu/lb—making it well-suited for thermal storage. Introducing solid materials, such as rocks, into a water tank decreases its thermal storage capacity. ASHRAE *Standard* 94.3-1986 gives procedures for measuring the thermal performance of water tanks.

Tanks may be vertical or horizontal cylinders, or they may be rectangular in shape. Steel is the most common material for aboveground tanks, and concrete for buried tanks. Metal tanks, especially cylinders with dished ends, can withstand hydraulic pressure exerted by the building piping circuits. The cost and size limitations of dished ends, however, restrict most tank installations to unpressurized water storage.

Water stored in tanks may be stratified thermally with the lighter, warmer water on top. The buoyancy differential becomes smaller, however, as water reaches its densest condition at 39°F (Table 1). In chilled water storage, which commonly ranges from 41 to 59°F, the density differential is small enough that some care is required to prevent mixing of stored chilled water with warmer return water. Mixing is less critical in hot water storage where the density differential is much larger. Tamblyn (1980) describes several methods for achieving temperature separation in water storage.

Table 1 Chilled Water Density

°F	lb/ft³	°F	lb/ft³	°F	lb/ft³
32	62.419	44	62.424	58	62.378
34	62.424	46	62.421	60	62.368
36	62.426	48	62.417	62	62.357
38	62.427	50	62.411	64	62.344
39	62.428	52	62.404	66	62.331
40	62.427	54	62.396	68	62.316
42	62.426	56	62.387		

Performance of Chilled Water Storage

A perfect or 100% effective storage would deliver water at the same temperature at which it was stored. It would also require that the water returning to storage neither mix nor exchange heat with the stored water. In practice, both situations occur, and some expense is incurred to reduce blending to acceptable limits.

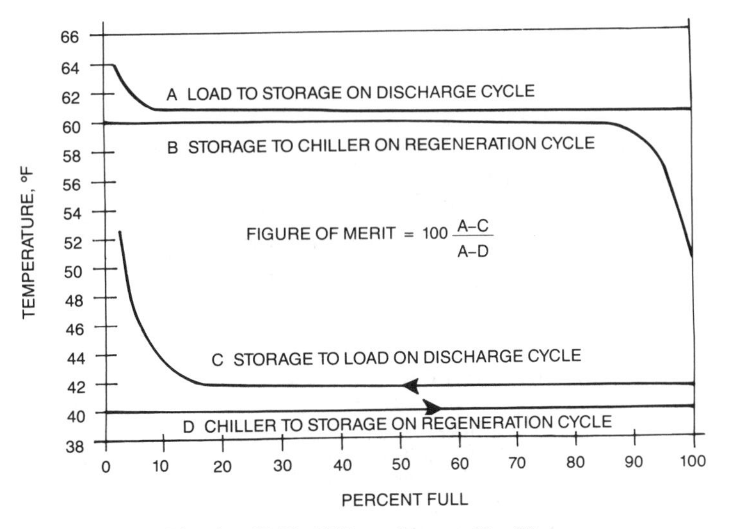

Fig. 4 Chilled Water Storage Profiles

Typical temperature profiles of water entering and leaving storage are shown in Figure 4. Tran *et al.* (1989) tested several large chilled water storage systems and developed the following figure of merit to compare systems and qualify usefulness.

$$\text{Figure of Merit (\%)} = \frac{\text{Area between A and C}}{\text{Area between A and D}} \times 100$$

The following techniques have been developed to minimize temperature blending to maintain a figure of merit above 90%.

1. A labyrinth with both horizontal and vertical traverse. This design takes the form of successive cubicles about 6 ft^2 with high and low ports. Tanks with successive vertical weirs and strings of cylindrical tanks are also used.
2. A horizontal diaphragm that floats in the tank to separate stored and returning water.
3. A series of compartments in the tank. Pumping is scheduled so that one compartment is always empty to receive water returning from the system during the occupied period and from the chiller during storage regeneration. In this way, separate compartments store water at the different temperatures, and blending is kept to a minimum.
4. Stratified storage, which floats the warmer, less dense returning water on top of the stored chilled water. This cost-effective technique supplies and withdraws the water for storage very gently so that buoyancy forces overcome any inertia effect. Water become less dense as its temperature is lowered from 39 °F to freezing (Table 1). Therefore, stratification would not work in this range, although stratified storages have been charged with water as low as 37.8 °F with few problems (Tran *et al.* 1989).

   When the stratified storage is charged, water between 39 and 44 °F enters through a double pipe or grid-type diffuser (Figures 5 and 6) at the bottom of the tank and returns to the chiller through the diffuser at the top of the tank (Wildin and Truman 1989). Typically, the incoming water forms a 1- to 3-ft thick blended zone. This zone then rises to the top as recharging continues. The blended zone or thermocline deepens somewhat, due to conduction and heat recycling, to the walls of the tank.

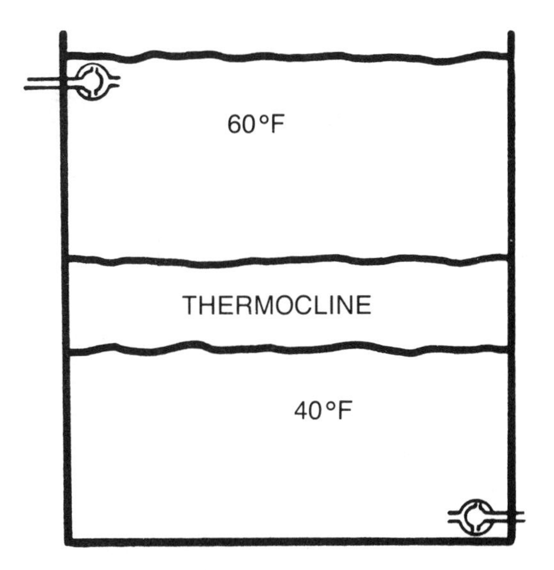

Fig. 5 Thermally Stratified Storage With Concentric Pipe Diffusers

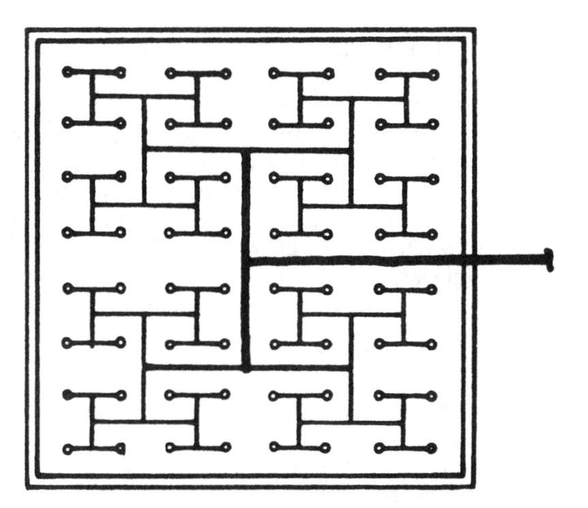

Fig. 6 Plan View of Distributed Nozzle Inlet/Outlet

Stratified storages become more efficient as the tank becomes deeper because the blended zone becomes a smaller fraction of the height of the tank. The storage may have any cross section, as long as the walls are nearly vertical. Horizontal, cylindrical tanks are not good candidates for stratified storage, although they have been used in series with internal vertical baffles to reduce blending.

Temperature Range and Storage Size

The volume of chilled water storage relates directly to the temperature range. A storage with a 10 °F range contains half the cooling capacity of one with a 20 °F range. For cost-effective storage, the cooling coils should provide a 20 °F or greater range. The cost for the extra coil surface required to provide this range can be offset by savings in pipe size, insulation, and pumping energy. Further, fan energy costs are reduced if the extra coil surface is added to the face area rather than as extra rows of coils. Chillers should cool water at least to 40 °F, and coils should, at maximum load, be able to return water to storage at least as high as 60 °F.

Large chilled water tanks have less area per unit volume of water stored than smaller ice storages. Thus the initial cost for water storage compares favorably with ice storage on large installations.

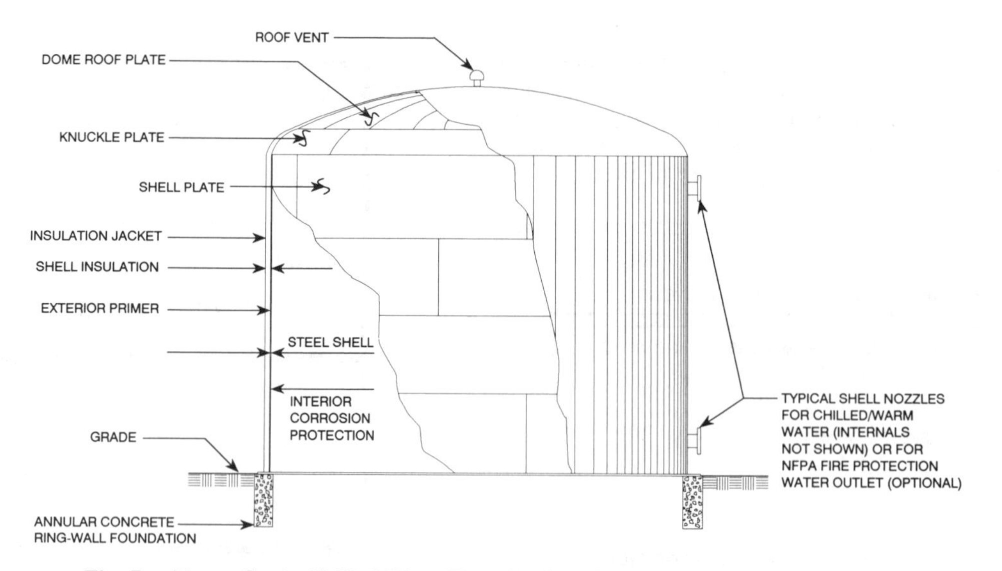

Fig. 7 Above-Grade Chilled Water Tank for Stratified Thermal Energy Storage

Specialized contractors use posttensioning techniques to warrant concrete tanks against leakage. Both steel and concrete tanks are available with manufacturer's warranties against leakage. A surface-mounted steel tank is shown in Figure 7.

Stratified Diffuser Design

Inlet and outlet streams must be kept at sufficiently low velocity so that buoyancy forces predominate over inertia forces to produce a gravity current (density current) across the bottom or the top of the tank. For this pupose, the diffuser dimensions should be selected to create an inlet Froude number between 1.0 and 2.0 (Wildin and Truman 1985a, 1985b; Yoo and Wildin 1986). The inlet Froude number F is defined in Equation (2).

$$F = Q/\sqrt{gh^3 (\Delta\rho/\rho)} \quad (2)$$

where

Q = volume flow rate per unit length of diffuser, ft^3/s
g = gravitational acceleration, ft/s^2
h = inlet opening height, ft
ρ = inlet water density, lb/ft^3
$\Delta\rho$ = difference in density between stored water and incoming or outflowing water, lb/ft^3

Since the density difference $\Delta\rho$ is always very small, its value can be obtained from Table 1. Yoo and Wildin (1986) give more information concerning the design of stratified storage hardware.

Example 1. Find the slot height for a 20-ft long diffuser supplying 40°F water at 80 gpm to a tank with water stored at 60°F.

Solution: From Equation (2)

$$h = (Q/F)^{2/3}/(g\Delta\rho/\rho)^{1/3}$$

$$h = \frac{[80 \text{ gpm} / (60 \text{ s/min} \times 7.48 \text{ gal/ft}^3 \times 20 \text{ ft} \times 1)]^{2/3}}{[32.2 \text{ ft/s}^2 (62.427 - 62.368)/62.368]^{1/3}}$$

$$h = 0.138 \text{ ft (1.65 in)}$$

Experiments by Wildin and Truman (1989) have shown that a low inlet Reynolds number can minimize mixing of the inlet side of the thermocline. Observations suggest that optimal stratification can be achieved with an inlet Reynolds number in the range of 240 to 800.

$$\text{Re} = Q/\nu$$

where

Q = volume flow per unit length, ft^2/s
ν = kinematic viscosity, ft^2/s

$$\nu = 1.67 \times 10^{-5} \text{ at } 40°F$$

$$\nu = 1.22 \times 10^{-5} \text{ at } 60°F$$

Then $\text{Re} = 80/60 \times 7.48 \times 20 \times 1.22 \times 10^{-5} = 730$

Conclusion: A diffuser with an opening height of 1.7 in. and a length of 20 ft would suffice to stratify the tank initially and to maintain stratification throughout charging and discharging.

Storage Tank Insulation

Well-stratified storage tanks have storage efficiencies of 90% and higher under daily complete charge/discharge cycles, and between 80 and 90% under partial charge/discharge cycles. To achieve high efficiency, the full design temperature swing must be maintained. Heat gains or losses through the tank enclosure decrease the effective temperature swing. For this reason, surface-mounted tanks have always been insulated. It may also be cost-effective to insulate buried chilled water tanks where the ground is warm and wet, as in some parts of Florida. To date, however, it has not been viable to insulate buried tanks, except on exposed walls and within partitions separating 104°F and 41°F compartments.

Heat transfer between the stored water and the tank walls is the primary source of thermal losses. To counter it, water storage should not be charged sooner than necessary. Not only does the stored fluid lose (or gain) heat to the ambient by conduction through the walls, but there is also a vertical heat flow along the tank walls from the warmer to the cooler region. The higher the conductivity of the tank wall material and the thicker the walls, the larger the heat flow. Thus, walls should not be made thicker than necessary for structural strength. Exterior insulation of the tank walls does not inhibit this heat transfer. Therefore, when the tank walls are made of a highly conductive material like steel, they should be insulated on the inside.

Thus far, the only materials suitable for interior insulation with completely closed cell construction, such as foamed glass, have been too expensive for the results achieved. Heat transfer to the ambient is inhibited by either exterior or interior insulation; interior insulation is more effective. Heat loss or gain through the container walls is especially important in storage compartments smaller than 50,000 gal because the ratio of enclosure area to stored volume is high. It is also a greater problem with empty tank and labyrinth concepts, because they contain more partitions, and with steel tank walls because of their high thermal conductivity. It becomes less important with large compartments of 100,000 gal or more. Heat is also transferred across the thermocline by conduction.

Other Factors

The cost of chemicals for water treatment is a significant budget item, especially if the tank is filled more than once. A filter system to provide complete filtration of the storage system at least once every 48 h is a useful accessory to keep the stored water clean. Atmospheric exposure of the storage tank may require occasional addition of biocides.

Many large concrete installations leak initially, and because some localities are short of water, it has been good practice to include at least one partition in the storage tank. This feature gives extra security against the need for complete drainage in all cases and the opportunity to store chilled and hot water simultaneously for greater energy savings in colder climates. Leaks in water tanks have been fixed with pressure grouting, sealants, or liners. The storage pumps should be below the lowest water level of storage for flooded suction. It is equally important not to place equipment between the tank and the suction side of the pumps to avoid stalling of the water flow.

PACKED ROCK BEDS

A packed bed, pebble bed, or rock pile storage uses the heat capacity of a bed of loosely packed particles through which a fluid, usually air, is circulated to add heat to, or remove heat from, the bed. A variety of solids may be used; rock of 0.75 to 2 in. in size is most prevalent. Well-designed packed rock beds have several desirable characteristics for energy storage. The heat transfer coefficient between the air and the solid is high, the cost of the storage material is low, the conductivity of the bed is low when airflow is not present, and a large heat transfer area is achieved at low cost by using small storage particles.

While rock beds store either heat or coolness, the overwhelming number of applications in North America is for heat storage in solar heating systems (Cole *et al.* 1980). Additional information is provided in Chapter 30. When rock beds are used for cool storage in a humid climate, fungus growth can occur in the bed, which can cause major indoor environmental problems. While they have been used successfully to store coolness in Australia (Close *et al.* 1968), this application is not recommended for North America, except in dry regions like the Southwest.

ICE STORAGE

Thermal energy can be stored in the latent heat of fusion of ice. Water has the highest latent heat of fusion among all common materials on a weight basis—144 Btu/lb. This considerably reduces the volume requirement compared with sensible heat storage, which contributes to lower capital cost.

Water/Ice Storage on Refrigerant Coils

Several types of ice storage units are available. The type in longest use is the direct expansion (DX) ice builder, which consists of refrigerant coils inside a storage tank filled with water. A compressor and evaporative condenser freeze the tank water on the outside of the coils to a thickness of up to 2.5 in. Ice is melted from the outside of the formation by circulating return water through the tank whereby it again becomes chilled. Stirrers or air bubblers agitate the water to promote uniform ice buildup.

Ice-on-Coil Systems

Major considerations for control of ice-on-coil systems unique to this storage method are: (a) to limit ice thickness and thus excess compressor energy during the build cycle, and (b) to minimize the bridging of ice between individual tubes in the ice bank. Bridging restricts the free circulation of water during the discharge cycle. While not physically damaging to the tank, the blockage reduces the performance, which is manifested as higher leaving water temperature, due to the reduced heat transfer surface.

Whichever refrigeration method is used—direct expansion, pumped liquid overfeed, or indirect brine cooling—the compressor is controlled by a time clock, which restricts operation to the specific periods dictated by the utility rate structure, and an ice thickness override control, which stops the compressor(s) at a predetermined thickness. A minimum of one ice thickness device per ice bank should be installed and connected in series where multiple ice banks are used. Where multiple refrigeration circuits per ice bank are used, one ice thickness control per circuit should be installed. To minimize bridging, placement of the ice thickness device(s) should be dictated by the ice bank manufacturer based on circuit geometry and flow pressure drop.

Ice thickness controls are either mechanically or electrically operated. Mechanical controls typically consist of a fluid-filled probe that is positioned at the desired ice thickness above the coil. As ice builds, it encapsulates the probe, causing the fluid to freeze and apply pressure. The pressure signal controls the refrigeration system via a P-E switch. Electric controls sense the three times greater electrical conductivity of ice as compared to water. Multiple probes are installed at the desired thickness, and the change in current flow between probes provides a control signal. Consistent water treatment maintains a constant conductivity and thus accurate control.

Ice-on-Coil Inventory Measurement

Since energy use is related to ice thickness on the coil, a partial-load ice inventory management system should be considered. This feature maintains the ice inventory at a minimum level to supply immediate future cooling needs, rather than topping up the ice inventory after each discharge cycle. This method also helps prevent bridging by ensuring that the tank is completely discharged at regular intervals, thereby allowing ice to build evenly.

The most common form of inventory measurement is based on the principle that ice has a greater volume than water; thus a water level change can be sensed to indicate the ice stored in the tank. Devices for measuring water level can be either an electrical probe or a pressure-sensing transducer.

Water/Ice Storage on Brine Coils

Instead of passing refrigerant through the coils inside the storage tank, brine can be pumped through them. Brine has the advantage of greatly decreasing the refrigerant inventory. However, a refrigerant-to-brine heat exchanger is required between the chiller and the storage tank.

Solid Ice Brine Coil

In this type of storage device, plastic mats containing brine coils are tightly rolled and placed inside a cylindrical water tank (Figure 8). The mats occupy approximately 10% of the tank volume; another 10% of the volume is left empty to allow for the expansion of the water upon freezing, and the rest is filled with water. A brine solution, *e.g.*, 25% ethylene glycol and 75% water, cooled by a liquid chiller circulates through the coils and freezes the water in the tank. Ice is built up to a thickness of 0.4 to 0.5 in. on the coils. During discharge, the cool brine solution circulates to fan coils and returns to the storage to be cooled again.

A standard brine chiller provides the refrigeration for these systems. During the charging cycle, the chilled brine exits at a constant 25 to 26 °F and brine returns at 31 °F. When the tanks are 90% charged, the chiller inlet and outlet temperatures fall rapidly because little water is left to freeze. When the chiller exit temperature reaches approximately 22 °F, the chiller is shut down and locked off for the night so that it does not short cycle through the pipes. The chiller thermostat is set at 22 °F exit temperature; it thus stays fully loaded through the entire cycle and keeps the system running at its most efficient condition.

Freezing and melting is done by the same heat transfer surface; therefore, there is no penalty for freezing completely every night.

A temperature modulating valve at the outlet of the tanks keeps a constant flow of liquid to the load (see Figure 20). When the storage system is full, the chiller is kept off, and the modulating

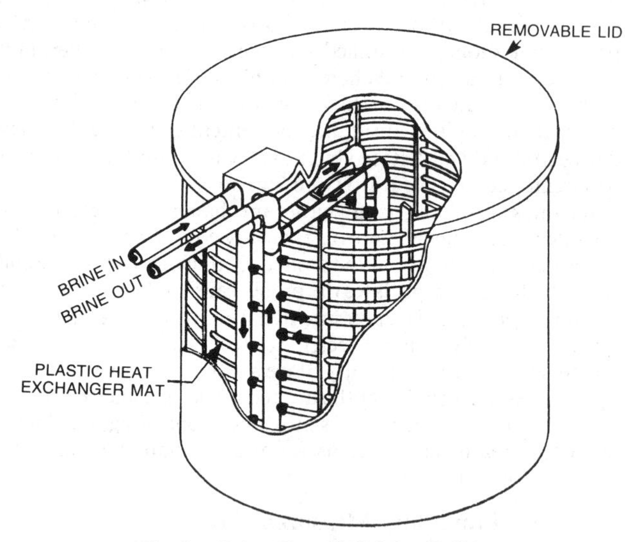

Fig. 8 Brine-Type Solid Ice Builder
(Courtesy: Calmac Manufacturing Corporation)

valve allows some fluid to bypass the tanks to supply the load as needed. If a partial storage system is required, the chiller thermostat is reset from the 22 °F setting up to the design temperature to the cooling coils (*e.g.*, 44 °F). If the load on the building is low, the chiller meets the 44 °F setting by itself. If the load is greater than the chiller's capacity, the exiting brine temperature rises, and the temperature modulating valve automatically opens to maintain the design temperature to the coils.

Because water increases 9% in volume when it turns to ice, the water level in the tank varies in direct proportion to the amount of ice in the tank, as long as the ice is kept submerged. This displaced water must not be frozen or it will trap ice above the original water level. Therefore, no heat exchange surface area can be above the original water level of the tank.

The change in water level is typically 6 in. in each modular tank. This change can be measured with either a pressure gage for visual output, or a standard transducer for an electrical signal.

Most projects have more than one tank; a reverse-return piping system ensures uniform flow through all the tanks. Thus, measuring the level on only one tank is sufficient.

Ice in Containers

This type of thermal storage system relies on primary plastic containers filled with deionized water and an ice nucleating agent.

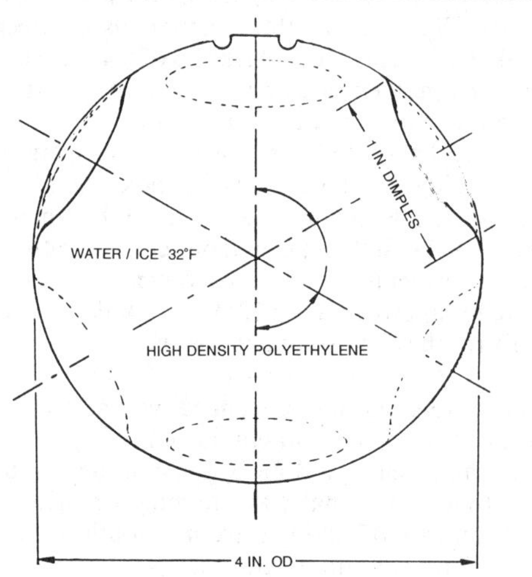

Fig. 9 Spherical Container for Ice Thermal Storage

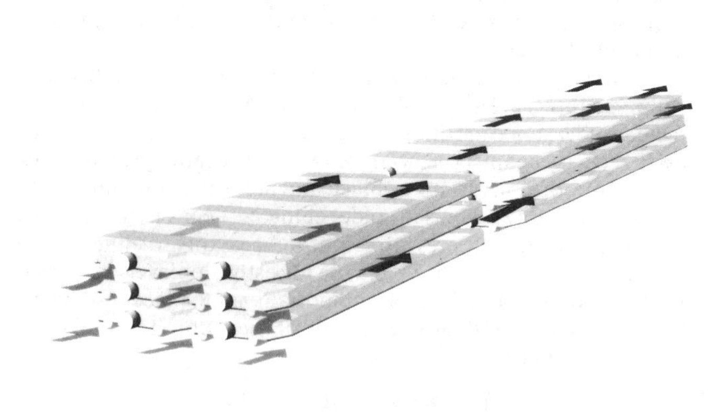

Fig. 10 Ice in Rectangular Container
(Courtesy: Reaction Thermal Systems, Inc.)

The commerically available systems include either spherical containers approximately 4 in. in diameter (Figure 9) or rectangular containers with dimensions of 1.375 in. by 12 in. by 30 in. (Figure 10). The primary containers are placed in storage tanks, which may be either steel pressure vessels, open concrete tanks, or suitable fiberglass or polyethylene tanks. This approach to ice storage allows any size, shape, or type of storage tank to be used, and the system may be packaged with any properly rated liquid chiller.

A brine solution, cooled to 24 to 26 °F by a liquid chiller, circulates through the tank and over the outside surface of the plastic containers, forming ice inside the containers. The primary plastic containers are designed to be flexible, either through preformed dimples in the surface of the spherical product or through direct flexure of walls of the rectangular container, to allow for the change of shape during ice formation. During discharge, cool brine solution flows directly to fan coils or a load side heat exchanger, thereby removing heat from the load and melting the ice encapsulated within the plastic containers. As ice melts, the plastic containers return to their original shape.

Ice inventory is measured and controlled by an inventory/expansion tank piped directly to the main storage tank. As ice forms, the plastic containers flex to allow for expansion and force the surrounding brine solution into the inventory tank. The liquid level within the inventory tank may be monitored to control the system to account for the ice available at any point during the charge or discharge cycle.

Ice Harvesting Systems

Ice harvesting systems consist of an ice producing section and an ice/cold water storage section. Each section has its own control. The ice producing section consists of multiple flat or cylindrical evaporators, a compressor, and a recirculating water pump. Ice is formed on the outside of the evaporators at an economical thickness between 0.25 and 0.40 in. The ice is harvested by introducing hot gas into an evaporator long enough to break the bond and cause the ice to drop into storage. This control prevents operation during periods specified by the utility rate structure and prevents the ice from growing too thick before dropping into storage.

When the temperature in the water distribution line reaches a set point of about 32 °F, the hot gas defrost cycle is activated. A time clock harvests the ice at preset intervals; normally every 29 min. The evaporators are grouped in sections and defrosted singly so that heat or rejection from active sections provides energy for defrost.

Thin sheets of ice continue to drop into storage until it is full, as determined by a high level controller. When the controller determines that storage is full, the ice producing section is shut down. The high ice level sensor can be mechanical, optical, or electrical. A timer also prevents operation during periods specified by the utility rate structure.

Ice inventory measurement requires indirect methods, because when ice is floating in the tank, the water level is constant. Two methods of measurement may be used: the water conductivity method or the heat balance method.

Because ice buildup is controlled to a constant optimum thickness, the compressor always operates at the most efficient point. The ice storage tank can then be filled to the maximum during each utility off-peak period, regardless of the amount of cooling used in the previous period. The inventory control procedure simply uses as much cooling as required in one peak period and replaces that cooling with ice during the next off-peak period.

Refrigeration systems may be reciprocating, screw, or centrifugal. Refrigerant suction temperatures are relatively constant between 20 and 22 °F. Condensing temperatures vary with the type of heat rejection used and ambient conditions.

Ice Slurry System

In this system, antifreeze solution (typically water and ethylene glycol) is pumped from storage to the top of an ice generator where it flows as a thin film down the inside surface of tubes whose outer surfaces are cooled by evaporating refrigerant. When the aqueous ethylene glycol solution is cooled to its freezing point, discrete ice crystals form in the fluid film. The resultant slurry is pumped to storage, where the ice crystals form a floating porous ice pack. The remaining concentrated solution is recirculated to the ice generator. As the process continues, the concentration of the brine solution in storage increases. The tank is fully charged when the predetermined amount of ice is in storage (Figure 11).

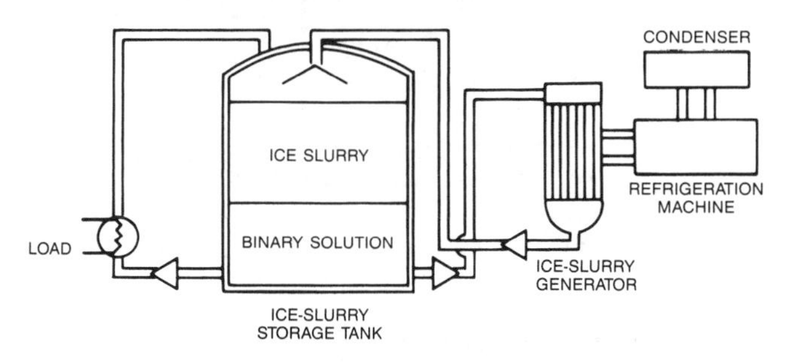

Fig. 11 Ice Slurry Storage
(Courtesy: CBI Industries, Inc.)

As in all ice storage systems, compressor operation can be controlled to operate only during utility off-peak periods. During the ice making mode of operation, the binary solution is continually circulated through the evaporator, continuously forming ice crystals, until the ice storage is full to a predetermined amount of ice. This is determined by measuring the temperature of the solution in storage at equilibrium conditions.

Because ice production is continuous, no defrost cycle is required. Compressors are allowed to run at maximum capacity as long as the maximum heat flux across the tubes is not exceeded. The maximum heat flux is the rate of heat transfer at which ice would begin to adhere to the tubes. Throughout the ice building cycle, the compressor suction temperature will gradually decrease as the storage solution temperature decreases. High seasonal efficiencies are achieved by allowing the condensing pressure to float with ambient temperature. This can be done because the system has no expansion valves that require higher condensing pressures to operate properly. The ice storage may be completely filled during each off-peak period, and only sufficient chilled water may be withdrawn from the system for the required cooling during the peak period.

Low-Temperature Carbon Dioxide

Compact latent heat thermal energy storage can be provided with carbon dioxide (CO_2) material at its triple point of −70 °F and 60 psig. Operation at the triple point (the unique combination of temperature and pressure at which all three phases—solid, liquid, and vapor—are in thermodynamic equilibrium) permits the use of a single substance as: (1) latent heat of fusion storage medium; (2) liquid overfeed refrigerant (when discharging storage); and (3) vapor compression refrigerant (when charging storage).

Triple point properties of CO_2 permit compact low-temperature storage: liquid density is 73.5 lbs/ft^3; solid density is 94.4 lbs/ft^3; and latent heat of fusion is 84.1 Btu/lb. With solid fraction typically varying 70 or 80% by mass between full charge and full discharge conditions, thermal storage capacity is over 60 Btu/lb of fluid stored, requiring only 2.7 ft^3/ton·h.

A typical closed-loop installation is shown schematically in Figure 12. Recharging storage is accomplished with a cascade refrigeration cycle, using CO_2 for the low stage refrigerant and NH_3 (or a common halocarbon) for the high stage refrigerant. In some applications, it may be advantageous to expend part of the stored CO_2 liquid; atmospheric venting of the CO_2 vapor, as is common in food freezing applications, can provide refrigeration at temperatures as low as −109 °F.

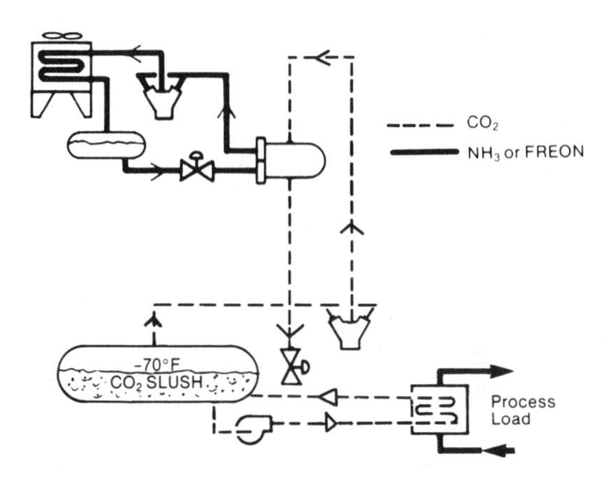

Fig. 12 Typical Carbon Dioxide System
(Courtesy: Chicago Bridge and Iron Company)

Other Phase Change Materials (PCMs)

Like ice and chilled water storage systems, eutectic salts have been in use for many decades. Phase change materials (PCMs) have been developed with various phase change points, such as eutectics, which melt and freeze at about 0 °F for low-temperature applications. To date, the most commonly used eutectic for cool storage applications melts and freezes at 47 °F. This material is a mixture of inorganic salts (the primary salt being sodium sulfate), water, and nucleating and stabilizing agents. It has a latent heat of fusion of 41 Btu/lb and a density of 93 lb/ft^3.

The 47 °F PCM requires conventional chilled water temperatures (42 °F) to charge the storage system. These temperatures allow any new or existing centrifugal, screw, or reciprocating chiller to be used to charge the storage system at refrigeration conditions comparable with standard air conditioning, and allow eutectics to be particularly appropriate for retrofit applications. Like ice storage systems, eutectic systems rely on a phase change. Use of the PCM's latent heat of fusion requires about 5.5 ft^3/ton·h for the entire tank assembly, which includes piping headers and water in the tank.

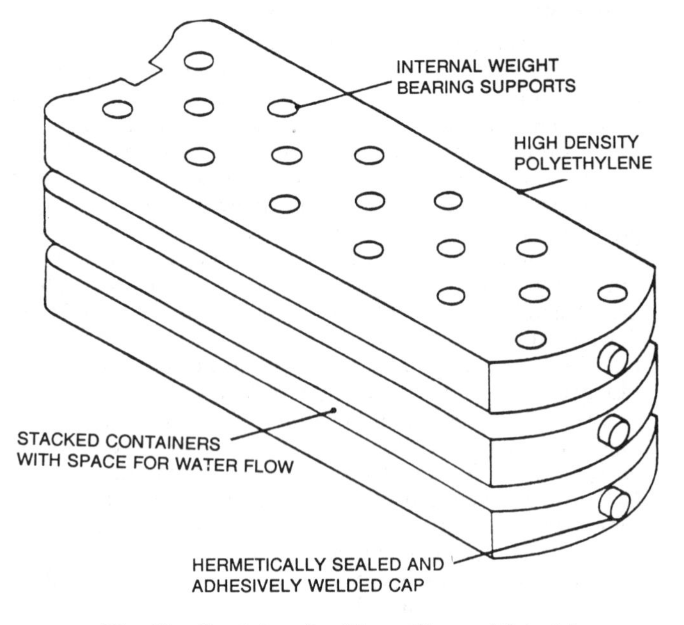

Fig. 13 Container for Phase Change Material
(Courtesy: Transphase Systems)

As shown in Figure 13, the PCM is filled into rugged, self-stacking, water impermeable containers made of high density polyethylene. The containers measure 24 in. by 8 in. by 1-¾ in., and are hermetically and redundantly sealed. The containers are designed with a surface-to-volume ratio of 24 for maximum heat transfer, and provide ¼ in. of space for water to pass between the container in a meandering flow pattern. The eutectic salt does not expand or contract when it freezes or melts.

The containers are placed into a tank, typically below-grade concrete or gunite. The containers displace about two-thirds of the tank's volume, so that one-third of the tank is occupied by the water used as the heat-transfer medium. The eutectic salt is about 1.5 times the density of water, so when the PCM freezes, the containers do not float or expand, and the heat transfer/container spacing arrangement is maintained throughout the melt/freeze cycle. The top of the tank is usually designed for heavy truck traffic, and either relandscaped or restored for use as a parking lot. Thus, the tank does not use or alter above-grade space and may be placed remote from the chiller plant.

Typically, 40 to 42 °F chilled water is used to charge the storage tank. The leaving tank temperature remains a relatively flat 47 °F, until the eutectic salts are almost completely frozen, at which point the leaving tank temperature is the same as the tank entering charging temperature, *e.g.*, 42 °F. The 6 °F temperature difference across the tank might correspond to the chiller being 60% loaded, which is in the most efficient part of the centrifugal chillers' performance curve. It is usually preferable to charge the last 10 to 20% of the tank's capacity when there is a cooling load because of the low 3 to 4 °F temperature difference across the tank.

During the tank discharge, water temperatures leaving the tank begin at 42 °F and rise to a 46 to 48 °F plateau. A 49 °F leaving water temperature occurs at the end of the discharge period, when, in many cases, the building load has already begun to drop off and the coils can accept slightly higher water temperatures.

Part-storage, precooler designs allow for the benefits of part-storage along with colder water temperatures to the load. The 47 °F PCM can be used with the chiller downstream of storage while still supplying 44 °F water to the load.

PCM storage devices can be tested according to procedures of ASHRAE *Standard* 94.1-1985, Methods of Testing Latent Heat Storage Devices Based on Thermal Performance.

ELECTRICALLY CHARGED HEAT STORAGE DEVICES

Thermal energy can also be stored in electrically charged, thermally discharged storage devices. The size of this equipment is usually specified by the nominal power rating (to the nearest kW) of the internal heating elements, and the nominal storage capacity is taken as the amount of energy supplied (to the nearest kWh) during an 8-h full-power charge period. For example, a 5-kW heater would have a nominal storage capacity of 40 kWh. ASHRAE *Standard* 94.2-1981 gives methods of testing these devices.

Room Storage Heaters

Room storage heaters have olivine or magnesite brick cores encased in shallow metal cabinets that can fit under windows (Figure 14). The core is heated to 1400 °F during off-peak hours by resistance heating elements located throughout the cabinet. Storage heaters are discharged by natural convection, radiation, and conduction (static heaters); or by a fan. The air flowing through the core is mixed with room air to limit the outlet air temperature to a comfortable range. Room storage heaters tend to overheat on mild winter days unless equipped with a controller that adjusts the thermal charge dependent on the outdoor temperature. These room heaters are used extensively in European homes, and many units of European manufacture have been installed in the United States. These heaters have storage capacities of 14 to 50 kWh with a power draw of 2 to 6 kW, respectively.

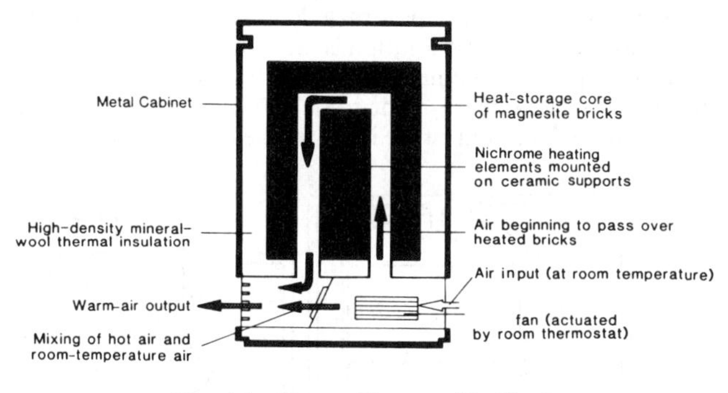

Fig. 14 Room Storage Air Heater
(Courtesy: Control Electric Corporation)

Central Storage Air Heaters

Central storage heaters operate on the same principle as room storage heaters; they contain a brick core that is heated to 1400 °F during off-peak hours. They are suitable for buildings with forced air systems since they are discharged by a fan through the building's warm air ducts. They also contain a night heating section, which provides thermal comfort during the off-peak period when the core is being charged. A damper section mixes return air with the heated air circulating through the core section to maintain a comfortable outlet air temperature. A charge controller senses outdoor air temperature as well as core temperature to determine the required charge and to prevent overheating. The units store from 112 to 240 kWh and have a charging draw of 14 to 30 kW, respectively, in addition to the power required for night heating. The core heating elements are activated in steps to prevent current surges. A manual override permits activating the night section during the on-peak period in case the storage is exhausted.

Pressurized Water Storage Heaters

This storage device consists of a cylindrical steel tank containing submersed electrical resistance elements near its bottom and a water-to-water heat exchanger near its top (Figure 15). The tank is insulated. During off-peak periods, the resistance elements are sequentially energized until the storage water reaches a maximum temperature of 280 °F, corresponding to a gage pressure of 35 psi. The ASME Boiler Code considers such vessels "unpressurized," and they are not required to meet the provisions for pressurized vessels. The heaters are controlled by a pressure sensor, which eliminates problems that could be caused by unequal temperature distributions. A thermal controller gives high-limit temperature control protection. Heat is withdrawn from storage by running service water through the heat exchangers and a tempering device that controls the output temperature to a predetermined level. The storage capacity of the device is the sensible heat of water between 280 and 20 °F above the desired output water temperature. The output water can be used for space heating or as service hot water. The water in the storage tank is permanently treated and sealed, requires no makeup water, and does not interact with the service water. The units are custom-made in sizes from 240 to 18,000 gal, with input power from 20 to 5200 kW.

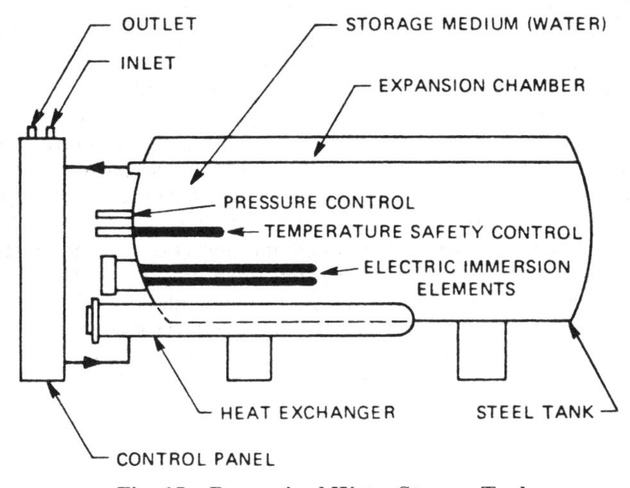

Fig. 15 Pressurized Water Storage Tank
(Courtesy: Megatherm Corporation)

Underfloor Storage Heaters

This storage consists of electric resistance cables buried in a bed of sand 1 to 3 ft below the floor of a building. It is suitable for single-story commercial buildings, such as factories and warehouses. This type of storage acts as a flywheel; while it is charged only during the nightly off-peak, it always maintains the top of the floor slab at a constant temperature slightly higher than the desired space temperature. Since the heat from the mats spreads in all directions, they do not have to cover the entire slab area. For most buildings, a mat location of 18 in. below the floor elevation is optimum. The sand bed should be insulated along its perimeter with 2 in. of rigid closed-cell foam insulation to a depth of 4 ft (see Figure 16). Even with a well-designed storage system of this type, 10% or more of the input heat may be lost to the ground.

GROUND-COUPLED STORAGE

Ground coupling uses the earth as a storage medium or heat source/sink, usually for space conditioning. There are two types of ground-coupled systems: direct heating/cooling systems and heat pump systems. In direct heating/cooling systems, heat or coolness is stored in a buried vessel or in a localized volume of earth and removed when needed to heat or cool the load directly. In heat pump systems, a heat pump removes heat from, or rejects heat to, the ground for space heating and cooling, respectively (Bose *et al.* 1985).

Ground coupling can also increase the capacity of underground thermal storage tanks. In such cases, the tanks are not insulated, and the surrounding soil acts as an extension of the storage tank (Metz 1982).

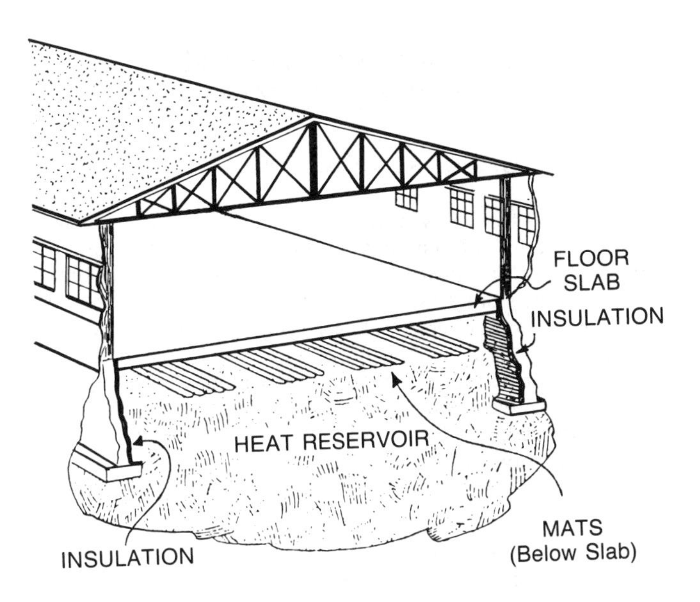

Fig. 16 Underfloor Heat Storage
(Courtesy: Smith-Gates Corporation)

Seasonal Storage

Seasonal thermal storage over prolonged periods, ideally from summer to winter and vice versa, requires large quantities of energy; therefore, the energy sources must be inexpensive. Recovered heat, waste incineration, industrial reject heat, nuclear cooling water, solar heat, and winter chill are examples of suitable energy sources. Examples of media used for seasonal storage include mined caverns, open ponds, rock beds, and large volumes of soil.

Seasonal Ice Storage. Ice can store winter chill for summer cooling. Thin sheets of water are sprayed on a growing ice block indoors, or commercial snow-making equipment forms a pile of snow/ice during the winter. The ice can then be used for summer air conditioning.

STORAGE IN AQUIFERS

Aquifers have the potential to store large quantities of thermal energy. Aquifers are water-bearing rocks found near or at nominal depths below the earth's surface. To be an aquifer, the rock must be saturated with water. Many aquifers, especially near the surface, consist of a mixture of sand and gravel, saturated with water. Approximately 60% of the surface of the continental United States is underlain by aquifers adequate for thermal energy storage. Aquifer storage is best for large systems, mainly communities and major buildings. A large volume is required to ensure reasonable energy recovery and, to be economical, the well cost should be distributed over a large system. The primary application of aquifers has been for annual storage, but projects using daily and weekly storage are being developed.

Aquifer storage systems are charged and discharged by transferring warm or cold water to and from the aquifer. For example, water is chilled during cold weather in cooling towers or ponds, injected into storage, and retrieved for air conditioning, when required. During charging, cold water is injected in one well, and warm water is withdrawn from another well. During discharge, the direction of water flow is reversed. Depending on withdrawal and injection rates, a spacing of a few hundred feet between wells is sufficient. Water loss or gain on an annual basis is negligible because injected and recovered water quantities are equal.

Aquifers require several seasons before achieving their full potential. Storage efficiency can be greater than 80% for aquifers 50 ft thick with zero natural flow. However, most aquifer storage systems can be expected to be 40 to 60% efficient.

THERMAL STORAGE APPLICATIONS

GENERAL CONSIDERATIONS

Figure 17 from Engineering Interface (1986) shows a water storage and building circuit in its simplest form. The "chiller" in the figure may produce hot or cold water, depending on the side to which the storage device is connected. During recharge, only the lower loop is operative. During on-peak operation with full storage, only the upper loop is operative. With partial storage, both loops may operate in parallel, with part of the flow passing through storage and part passing through the chiller. A temperature modulating valve adjusts the flow to the load to achieve the desired supply temperature.

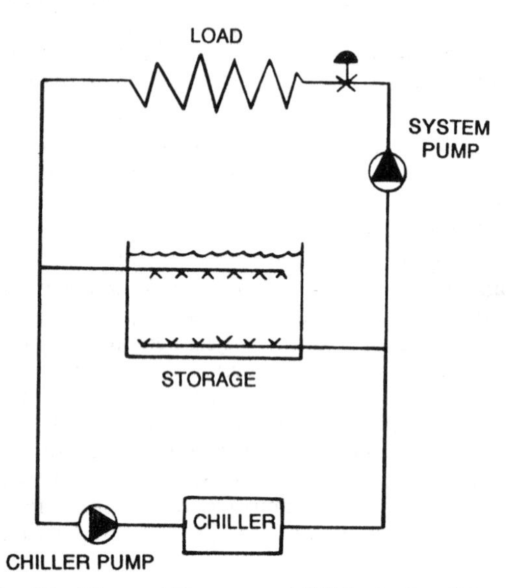

Fig. 17 Thermal Storage and Chiller in Parallel

Storage of Heat in Cool Storage Units

Cool storage installations can also store heat. Since most commercial buildings require cooling some time during the day even in the winter, the refrigeration plant may be in operation all year. When cool storage is charged during the off-peak period, the compressor heat of water-cooled condensers can be stored for use during morning warm-up on the following day.

If more than one storage unit is installed, one or more of these units can be switched from summer cool storage to winter heat storage in the fall and switched back in the spring. In such an installation, both heat and cool storage can be accomplished simultaneously by operating the compressor during nighttime off-peak periods. The stored heat can be used for morning warm-up (and late afternoon and evening heating, if required), and the stored coolness can provide midday cooling. If both chiller and storage are large enough, all compressor operation can be avoided during the on-peak period, and the required heating or cooling can be satisfied from storage. This method of operation replaces the air-side "economizer cycle," which consists of using outside air to cool the building when its enthalpy is lower than that of the return air. Heat and cool storage is usually lower in first cost than the equipment required for an economizer cycle.

Cool storage with heat pumps or heat-recovery equipment may also be used during the heating season. Cooling is withdrawn from storage, as needed. When heat is demanded, the cool storage is recharged by using it as the heat source for the heat pump. If needed, additional heat can be obtained from off-peak heating, solar collectors, exhaust air, or other waste heat sources. Excess energy can be rejected through a cooling tower if more cooling than heating is required. The cooling portion of this cycle is usually referred to as "free cooling."

The extra cost of equipping a chilled water storage facility for heat storage is small. The necessary plant additions are a partition in the storage tank to convert some portion to storing warm water, some additional controls, and a chiller equipped for heat reclaim duty. The additional cost may be readily offset by savings in heating energy. In fact, adding heat storage capability may increase the economic attractiveness of cool storage systems.

Circuitry for Ice Storage Systems

Figures 18 and 19 (Tamblyn 1977) show charge and discharge operation for ice storage devices suitable for partial storage. During off-peak periods, the building loop containing the cooling coils is bypassed, and the chiller charges the ice storage unit. Since the outlet water from an ice storage unit is usually 34°F and most building loops are designed for 42 to 48°F chilled water supply, the return water from the building loop is only partially cooled in the chiller. A part of that flow is then passed through the ice storage unit where it is cooled to 34°F. In this case, the remainder of the flow bypasses storage, and a downstream modulating valve maintains the desired chilled water supply temperature to the building loop at 44°F. If demand limiting is desired, the chiller electrical demand must also be controlled. It is set for the maximum permissible electrical demand and provides chilled water at constant temperature but variable volume. The rest of the chilled water flow is provided from storage. Individual cooling coil demand is controlled by thermostats, and the total pump flow is controlled either through a variable speed pump or through a bypass.

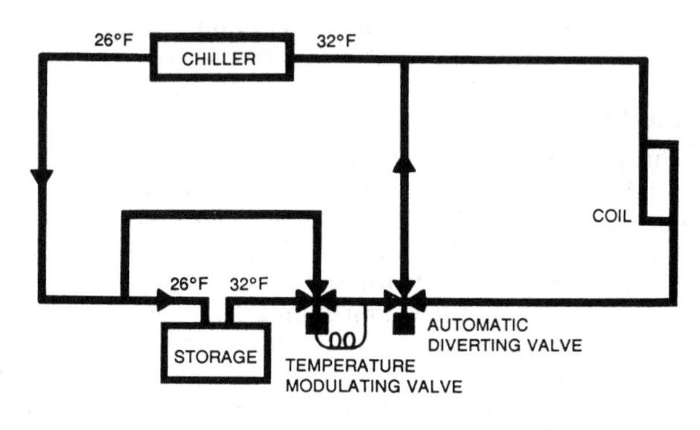

Fig. 18 Partial Storage—Ice-Making Mode

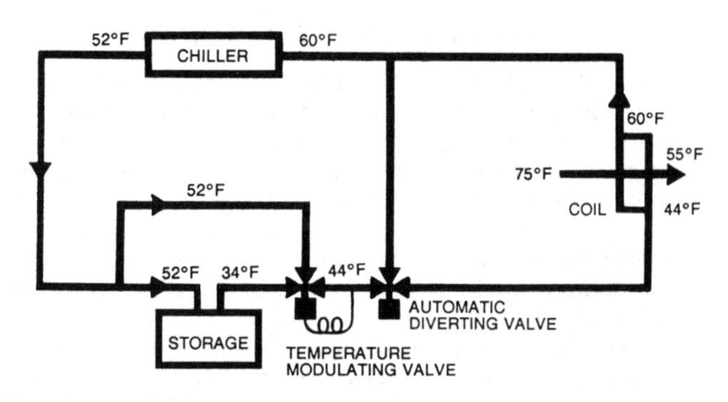

Fig. 19 Partial Storage—Cooling Mode with Water Tempered to Standard Coil Temperature

Cold Air Distribution

Distribution air at a lower temperature is attractive because smaller fans, ducts, pumps, and piping can be used, resulting in lower intial costs. In addition, the reduced ceiling space required for ductwork can significantly reduce the building height, particularly in high-rise construction. These cost reductions can make thermal storage systems competitive with conventional systems on the basis of initial cost.

The optimum supply air temperature should be determined by an economic analysis that considers changes in initial costs and operating costs for various design options. In many cases, operating costs compared to conventional systems will be lower, even when net energy consumption increases, because relatively expensive on-peak fan energy is traded for relatively inexpensive off-peak cooling energy.

The minimum achievable supply air temperature is determined by the chilled water temperature and the temperature rise between the cooling plant and the terminal units. The stored fluid rises in temperature with some ice storage equipment as the storage is discharged; therefore, the minimum supply temperatures normally achievable with various types of ice storage plants must be carefully investigated with the equipment supplier.

Temperature rises between the cooling plant and terminal units are the result of heat gains and imperfect heat transfer. A heat exchanger, sometimes required with storage tanks at atmospheric pressure, will add 2 to 4 °F to the final chilled water supply temperature. The rise in chilled water temperature between the cooling plant discharge and the chilled water coil depends on the length of piping and amount of insulation, and could be up to 0.5 °F on long exposed runs.

The difference between chilled water temperature entering the cooling coil and supply air temperature leaving the coil can generally be estimated at 6 °F. Closer approach can be achieved with more rows or a larger face area on the cooling coil, but this extra heat transfer surface is often uneconomical. A 3 °F temperature rise due to duct heat gain between the cooling coil and the terminal units can be assumed for preliminary design analysis. This rise can be reduced to as low as 1 °F with careful design and adequate insulation.

A blow-through configuration provides the lowest supply temperatures and the minimum supply air volume. The minimum supply air temperature achievable with a draw-through configuration is 2 to 3 °F higher, because the fan heat goes into the supply air rather than directly from the mixed air to the cooling coil. Use a draw-through configuration if space between the fan and coil for flow straightening is limited, or if packaged air handlers are used.

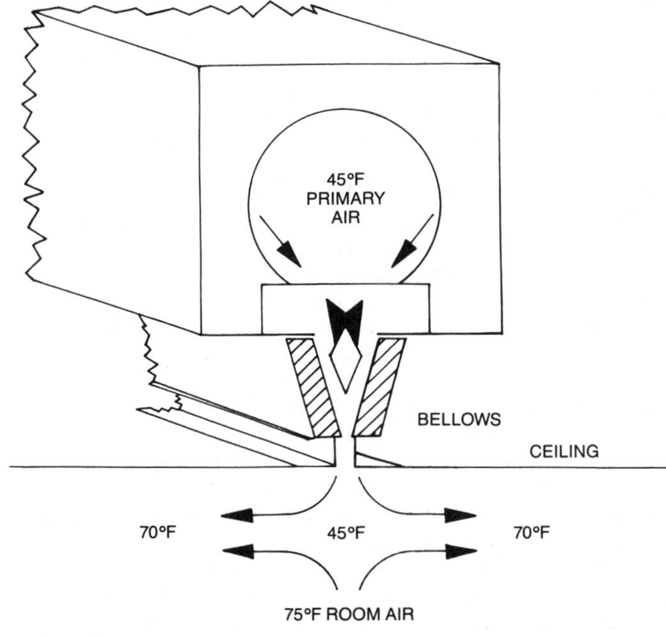

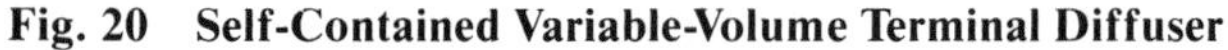

Fig. 20 Self-Contained Variable-Volume Terminal Diffuser

Cold primary air can be tempered with room or plenum return air using fan-powered mixing boxes or induction boxes before supplying it to the space. The energy use of continuously operating fan-powered mixing boxes is significant, and in some cases, it can negate the savings due to downsized central supply fans. Some diffusers are capable of supplying air as cold as 45 °F directly to the space without dumping or cold drafts, eliminating the need for fan-powered boxes. Figure 20 shows one type of self-contained variable-volume terminal diffuser that induces large amounts of room air within a short distance from the diffuser.

If supply airflow to occupied spaces is expected to be below 0.4 cfm/ft^2, fan-powered or induction boxes should be used to boost the air circulation rate. At supply air rates of 0.4 to 0.6 cfm/ft^2, a diffuser with a high ratio of induced room air to supply air should be used to ensure adequate dispersion of ventilation air throughout the space. A diffuser that relies on turbulent mixing rather than induction to temper the primary air may not achieve adequate ventilation effectiveness at these flow rates.

Cold air distribution systems normally maintain space humidity at 30 to 45% rh, as opposed to the 50 to 60% levels typical for conventional systems. With the lower humidity, equivalent comfort conditions are provided at higher dry-bulb temperatures. An increased dry-bulb set point generally results in decreased energy consumption, although an equivalent comfort level is maintained.

All surfaces that may be cooled below the ambient dew point should be insulated, including air-handling units, ducts, and terminal boxes. All vapor-barrier penetrations should be sealed to prevent migration of moisture into the insulation. Prefabricated, insulated round ducts should be externally insulated at joints where internal insulation is not continuous. When ducts are lined internally, access doors should also be lined.

Duct leakage is undesirable because it represents cooling capacity that is not delivered to the conditioned space. In cold air distribution systems, leaking air can cool nearby surfaces to the point where condensation forms. Designers should specify acceptable methods of sealing ducts, as well as allowable leakage rates and test procedures. These specifications must be followed up with on-site supervision and inspection during construction.

The required face velocity through the cooling coil determines the size of the coil for a given supply air volume. This, in turn, determines the size of the air-handling unit. A lower face velocity allows a lower supply air temperature. A higher face velocity results in smaller equipment and lower costs, but the velocity is limited by moisture carryover from the coil. The recommended face velocity for cold air distribution systems is 350 to 450 fpm, with a maximum limit of 550 fpm.

A thorough commissioning process is important for optimum operation of any large space-conditioning system, particularly with cold air distribution. Reductions in first costs and operating costs are major selling points for cold air distribution, but the commitment to provide a successful system must not be compromised by the desire to reduce costs. While a commissioning procedure may appear to involve additional expense, it actually saves future costs of system malfunctions and troubleshooting, and provides increased value in the form of optimal system operation.

Refrigeration Design

Packaged compression or absorption equipment is available for most of the cool storage technologies discussed in this chapter. The maximum cooling load that can be satisfied with a single package, and the number of packages that can be installed in parallel to yield greater capacity, depend on the specific technology.

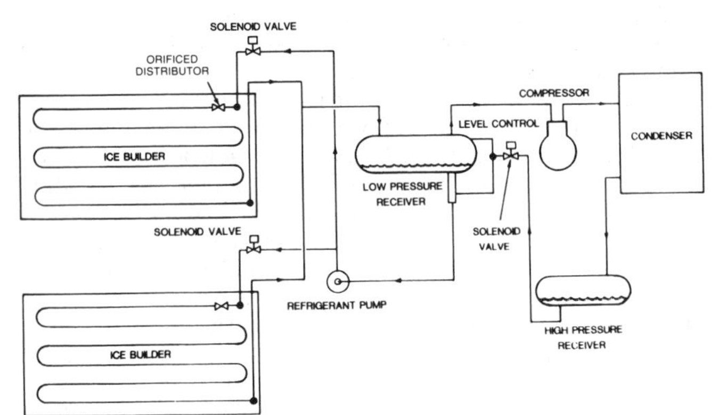

Fig. 21 Typical Liquid Refrigerant and Ice Storage System (Donovan)

Packaged equipment is often uneconomical due to the large size of a project, site limitations, or special application requirements. In these cases, individual components such as chiller bundles, falling film chillers, ice makers, compressors, and condensers must be interconnected on site (Figure 21). The design and installation of such field-erected systems is straightforward and well understood; nevertheless, care must be taken by all parties involved. Particular emphasis should be placed on the following considerations:

Design for part-load operation. Various components in a cool storage system do not always operate at full load. Refrigerant flow rates, pressure drops, and velocities are reduced during part-load operation. Components and piping must be designed so that, at all load conditions, control of the system can be maintained and oil can be returned to the compressors.

Design for pull-down load. Chilling or ice making equipment designed to operate at water temperatures near 32 °F imposes much higher loads on the refrigeration system during the initial startup when water temperatures may be 75 °F. The components must be sized to handle this high load if fast pull-down is required, or perhaps more economically, provision must be made to run at reduced capacity until pull-down is complete.

Plan for chilling versus ice making. Some equipment, particularly harvesting ice makers, have much higher instantaneous chilling capacities than ice making capacities. This higher chilling capacity can be used advantageously by the system designer and owner, but only if the compression or absorption equipment and interconnecting piping is properly sized to handle the higher chilling load.

Protect compressor from liquid slugging. Ice builders and ice harvesters tend to contain more refrigerant than do conventional chillers of similar capacity, because ice builders contain more circuits and plate ice harvesters contain more valves. These factors combine to provide more opportunity for compressor liquid slugging; thus, care should be taken to oversize suction accumulators and equip them with high level compressor cutouts and suction heat exchangers to evaporate any remaining liquid.

Size the high-pressure receiver for full-system refrigerant charge. Since the economics of many cool storage systems depend on shifting part or all of the load, every opportunity should be taken to make the system easy to maintain and service.

Prevent oil trappings. Refrigerant lines should be arranged to prevent the trapping of large amounts of oil anywhere in the system and to ensure its return to the compressor under all operating conditions, especially during periods of light compressor loads. All suction lines should slope toward the suction line accumulators, and all discharge lines should pitch toward the oil separators. Oil tends to accumulate in the evaporator coils of the storage units, as that is the location of lowest temperature and pressure. Since refrigerant accumulators trap oil in addition to liquid refrigerant, the larger accumulators required for ice storage systems necessitate special provisions to ensure adequate oil return to the compressor.

Fully analyze design tradeoffs. When comparing cool storage to other available cost and energy-saving opportunities, make equal comparisons. If energy-efficient evaporative condensing is assumed for a cool storage design, compare it to a conventional system that also uses evaporative condensing. Use current design applications, whether they are based on HCFCs or ammonia, or use any of the proven refrigeration techniques including direct expansion, flooded, or liquid recirculation.

OFF-PEAK REFRIGERATION AND PROCESS COOLING

Refrigeration and process cooling typically require lower temperatures than air conditioning. A storage medium with a relatively low supply temperature, such as water ice, low-temperature eutectic ice, or carbon dioxide, must be used.

Applications such as vegetable hydrocooling, milk cooling, carcass spray cooling, or storage room dehumidification can use cold water from an ice storage system. Lower temperatures can be obtained for freezing food, storing frozen food, and so forth, by using either a low-temperature eutectic charged and discharged by brine or a nonaqueous PCM such as CO_2.

These applications have traditionally used heavy-duty industrial equipment. The size of the loads has often required custom-engineered, field-erected systems, while the type of occupancy has often allowed the use of ammonia. Cool storage systems for these applications can be optimized by choice of refrigerant (halocarbon or ammonia) and choice of refrigerant feed control (direct expansion, flooded, mechanically pumped, gas pumped).

Some of the advantages of cool (thermal) storage may be particularly important for refrigeration and process cooling applications. The energy losses often associated with ice storage may be negligible when 34 °F water is required by the load. Cool storage systems can provide steady supply temperatures regardless of large and fast changes in cooling load or return temperature. Charging equipment and discharging pumps can be placed on a separate electrical service to economically provide cooling during a power blackout, or to take advantage of low, interruptible electrical rates. Storage tanks can be oversized to provide water for emergency cooling or fire-fighting water as well as load-leveling or load-shifting. Cooling can be extracted very quickly from harvested ice storage or chilled water storage to satisfy a very large load, and then be recharged slowly using a small charging system. System design may be simplified when the load profile is determined by production scheduling rather than occupancy and outdoor temperatures.

Suitability of Storage Devices to Retrofits

The smaller volume required for latent heat storage (ice and PCM) compared to water storage favors them for retrofit installations where space is limited. Latent heat storage is also preferred for building sites where rock or a high water table make deep excavation for a tank expensive. The small module size of brine/ice units is an advantage in instances where access to the equipment room is limited or where the storage units can be distributed throughout the building and located near the cooling coils. Chilled water tanks incur external thermal losses of 2 to 5% per day, while latent heat storage tanks, because of their smaller size, lose less. Existing installations that already contain chilled water piping and conventional water chillers are more easily and cheaply retrofitted to accept chilled water storage than latent heat storage. The requirement for brine in some ice storage systems may complicate pumping and heat exchange for existing equipment. Brine temperatures colder than chilled water temperatures in the original design offset the reduction in heat transfer.

Ice storage has generally been used with positive displacement compressors because they are more easily modified for ice making. Centrifugal chillers, with their low cost for large sizes and their reputation for reliability, are more suited to chilled water generation than to ice making. To equip a centrifugal compressor for ice making requires a change of compressor wheel because a higher pressure ratio is needed to make ice than to chill water. This greater pressure also increases the required compressor power, making it necessary either to derate the compressor or to replace the motor with a higher kilowatt motor. Valves and other components of the cooling plant may also require replacement because of the higher power. In many cases, it may be more economical to replace the entire cooling plant with one designed for ice storage than to attempt to retrofit a centrifugal chiller plant for that purpose. On the other hand, PCMs having freezing temperatures near 47 °F can be retrofitted into existing centrifugal chiller systems without modification or derating. Thus, these devices are well-suited to retrofit applications.

OFF-PEAK HEATING

Service Water Heating

The tank-equipped service water heater, which is the standard water heater in North America, is a thermal storage device. Some electric utilities provide incentives for off-peak water heating. The heater is then equipped with a control system that is activated by the electric utility to curtail power use during periods of peak demand. It is advisable to increase the tank size of off-peak water heaters over the size chosen for conventional heaters, although the need for that depends on the habits of the users.

An alternate method of off-peak water heating connects two tanks in series so that the hot water outlet of the first tank supplies water to the second tank (Oak Ridge 1985). This arrangement minimizes the mixing of hot and cold water in the second tank, because that tank benefits from stratification in the first tank. Tests performed on this configuration show that it can supply 80 to 85% of its rated capacity at temperatures suitable for domestic needs, compared to 70% for single-tank configurations. The wiring of the four heating elements in the two water heaters must be modified to accommodate the dual-tank configuration.

Solar Space and Water Heating

Either rock beds, water tanks, or PCMs can be used, depending on the type of solar system. Various applications are described in Chapter 30.

Space Heating

Thermal storage can be carried out by (1) radiant floor heating, (2) brick storage heaters, or (3) water storage heaters. The first is generally applicable to single-story buildings. The choice of the other two depends on the type of heating system, with air systems using brick and water systems using water or a PCM. Water systems can be charged either electrically or thermally.

Popular locations for storage in residential applications include basements, crawl spaces, garages, and outdoors. Installation can be difficult in basements and crawl spaces if the storage unit must be assembled in place in an area with limited access. Garages impose insulation problems, but complete tanks can easily pass through their doors. Outdoor storage above grade requires additional insulation, which must also be protected from the weather. Groundwater is the main problem with outdoor storage below grade. Porous insulation should not be used below grade, and even closed-cell foams may have their insulating value reduced by one-half in the presence of groundwater. Drainage tiles and a sump pump may be required where the soil drains poorly.

Insulation is more important for hot storage than for cool storage because the difference in temperatures between storage and occupied space is much greater. Methods of determining the economic thickness of insulation are given in Chapter 20 of the 1989 ASHRAE *Handbook--Fundamentals*.

Cole (1980) graphed the relation between nominal rock size, face velocity (volumetric flow rate divided by cross-sectional area of the bed), and pressure gradient through rock bed storage units. The rock bed cover is a potentially serious source of leakage in an air system, since the force on a 6 by 8 ft cover pressurized to 1 in. of water is approximately 250 lb. Covers should be sealed with silicone sealants or a soft rubber gasket and should be screwed and bolted to the container sides.

Figure 22 shows the operating characteristics of electrically charged room storage heaters. Curve 1 represents their theoretical performance. In reality, heat is continually transferred to the room during charging by radiation and convection from the exterior surface of the device. This *static discharge* is shown in Curve 2. When the thermostat calls for heat, the internal fan is started, and the resulting faster *dynamic discharge* corresponds to Curves 3 and 4.

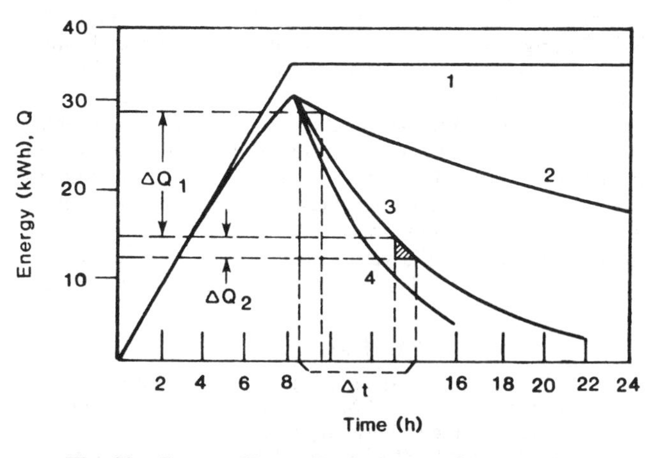

Fig. 22 Storage Heater Performance Characteristics
(Hersh 1981)

The operating characteristics of these devices differ from conventional electric heaters in two important respects. First, because the heating elements are energized only during the off-peak period, they must store the total daily heating requirement during that period, typically 8 or 10 h. Second, the rate at which heat can be delivered to the heated space decreases as the charge level decreases. Thus, the heaters are less able to meet the building heat load toward the end of the discharge period. This applies primarily to room heaters and less to central heaters that contain tempering dampers.

A rational design procedure requires an hourly simulation of design heat load and discharge capacity of the storage device. The first criterion for satisfactory performance is that the discharge rate is no less than the design heating load at any hour. During the off-peak period, energy is added to storage at the rate determined by the connected load of the resistance elements, while the design heating load is subtracted from storage. At the end of a 24-h period, the second criterion of satisfactory performance, the daily energy balance, is checked. Assuming that the design day is preceded and followed by similar days, the storage energy level at the end of the simulated day should equal the starting level.

A simpler procedure is recommended by Hersh *et al.* (1982) for typical residential heating system designs. For each zone of the building, the design heat loss is calculated in the usual manner and multiplied by the selected sizing factor. The resulting value is rounded to the next higher kilowatt and designated as the required storage heater capacity. Sizing factors in the United States range from 2.0 to 2.5 for an 8-h charge period and from 1.6 to

2.0 for a 10-h charge period. The lower end of the range is marginal for the northeastern United States.

Central ceramic brick storage heaters usually have a night heating section that supplies heat to the occupied space during the off-peak period while storage is being charged. Thus, the amount of heat that must be put into storage is equal to the space heating needs only during the peak period. For best efficiency, the unit should not be charged to a higher temperature than required to satisfy the heating needs of the following day. Therefore an outdoor anticipatory temperature sensor is recommended. Electrically charged pressurized water storage tanks must recharge the full daily heating and hot water requirement during the off-peak period, since they do not have a night heating section. Their heat exchanger design allows them to discharge at a constant rate, which does not decrease near the end of the cycle.

Thermally charged hot water storage tanks are similar in design to that of cool storage tanks described under the section on off-peak cooling; most of these tanks are also used for cooling during part of the year. Contrary to off-peak cooling, off-peak heating seldom permits a reduction in the size of the heating plant. The size of storage used for both heating and cooling depends more on cooling than on heating for lowest life cycle cost, except in some residential applications.

Care should be taken to avoid thermal shock of concrete tanks. Tamblyn (1985) showed that the seasonal change from hot to cool storage caused sizable leaks to develop if the cool-down period was accelerated below five days. Raising a concrete tank's temperature does not cause problems, because this generates compressive stresses, which concrete is well able to sustain. Cool-down, on the other hand, causes tensile stresses, under which concrete has low strength.

REFERENCES

Abhat, A. 1978. Performance studies of a finned heat pipe latent heat thermal energy storage system. *SUN, Mankind's Future Source of Energy*, Vol. 1. Pergamon Press, Elmsford, NY, 541-46.

ASHRAE. 1981. Methods of testing thermal storage devices with electrical input and thermal output based on thermal performance. ASHRAE *Standard* 94.2-1981.

ASHRAE. 1985. Methods of testing active latent heat storage devices based on thermal performance. ASHRAE *Standard* 94.1-1985.

ASHRAE. 1986. Methods of testing active sensible heat thermal energy storage devices based on thermal performance. ASHRAE *Standard* 94.3-1986.

Ayres Associates. 1980. *A guide for off-peak cooling in buildings*. Southern California Edison Company, Rosemead, CA.

Ayres Associates. 1985. *Performance of commercial cool storage systems*. Vol. 1, Early case histories; Vol. 2, Cipher Data Products ice storage system improvements. EPRI EM 4044 (June).*

Ayres, J.M., H. Lau, and J.R. Scott. 1984. Sizing of thermal storage systems for cooling buildings with time-of-use electric rates. ASHRAE *Transactions* 90(1).

Ayres Sowell Associates. 1982. *Field performance of an ice storage off-peak cooling installation*. Southern California Edison Company, Rosemead, CA.

Bose, J.E., J.D. Parker, and F.C. McQuiston. 1985. Design/data manual for ground-coupled heat pumps. ASHRAE *Publication* GCHP.

Calmac Manufacturing Corporation. 1982. *Bulk storage of PCM*. Report to Argonne National Laboratory, NTIS PB82-805862, National Technical Information Service, Springfield, VA.

Carbado, J.J. 1985. A direct-contact charged, direct-contact discharged cool storage system using gas hydrate. ASHRAE *Transactions* 91(2).

Chubb, T.A., *et al*. 1980. DOE Conference 801055 (October):68.

Close, D.J., R.V. Dunkle, and K.A. Robeson. 1968. Design and performance of a thermal storage air conditioning system, MC4:45. Mechanical and Chemical Engineering Transactions, Institute of Civil Engineers, Australia.

Cole, R.T., *et al*. 1980. Design and installation manual for thermal energy storage, 2nd ed. ANL-79-15. Argonne National Laboratory.

Cole, R.L. and F.O. Bellinger. 1982. Natural thermal stratification in tanks. Phase I, Final Report. ANL-82-5. Argonne National Laboratory.

Comstock and Wescott, Inc. 1962. Promising heat storage material found through research. EEI *Bulletin* (February):49-50.

Di Lauro, G.F. and R.E. Rice. 1981. Conceptual design of a latent heat thermal energy storage subsystem for a saturated steam solar receiver and load. Proceedings of the 6th Annual Thermal and Chemical Storage Contractors' Review Meeting, DOE Conf-810940 (September).

Donovan, J.F. *The ice builder, An off-peak approach*. Chester-Jensen Company, Chester, PA.

Engineering Interface, Ltd. 1986. *Commercial cool storage: Presentation material*. Vol. 1, Seminar Handbook; Vol. 2, Slide Package. EPRI EM 4405 (February).*

EPRI. 1987. Field evaluation of cold air distribution systems. EPRI EM 5447. Electric Power Institute, Palo Alto, CA, 4.17, 4.21.

EPRI. 1988a. Cold air distribution design guide. EPRI EM 5730. Electric Power Research Institute, Palo Alto, CA.

EPRI. 1988b. Proceedings of workshop on cold air/water utilization with cool storage systems. Electric Power Research Institute, Palo Alto, CA.

EPRI. 1988c. Comfort criteria in a low-humidity environment. Research Project RP2732-10. Electric Power Research Institute, Palo Alto, CA.

Gatley, D.P. and J.J. Riticher. 1985. Successful thermal storage. ASHRAE Technical Data Bulletin, Thermal Storage, TDB TH1 (January):37-49.

GPU Service Corporation and Enviro-Management & Research Co. 1985. Commercial cool storage design guide. EPRI EM 3981, Electric Power Research Institute, Palo Alto, CA.

Hersh, H.N. 1981. Optimal sizing of heating systems that store and use thermal energy. ANL/SPG-18, Argonne National Laboratory (June).

Hersh, H., G. Mirchandani, and R. Rowe. 1982. Evaluation and assessment of thermal energy storage for residential heating. ANL SPG-23, Argonne National Laboratory (April).

Hughes, P.J., S.A. Klein, and D.A. Close. 1976. Packed bed thermal storage models for solar air heating and cooling systems. ASME *Journal of Heat Transfer* 98:336.

Jones, D.E. and W. Loss. 1982. Flow distribution improvements in vertical rock beds for solar energy thermal storage. ASHRAE *Transactions* 88(2).

Knebel, D.E. 1986. A showcase on cost savings. ASHRAE *Journal* (May):28.

Kohler, J. and D. Lewis. 1983. Phase change products for passive homes. *Solar Age* 65.

Lane, G.A. 1982. Congruent-melting phase-change heat storage materials. ASHRAE *Transactions* 88(2).

Lane, G.A. 1985. PCM Science and technology: The essential connection. ASHRAE *Transactions* 1(2).

Lof, G.O.G. and R.W. Hawley. 1948. Unsteady heat transfer between air and loose solids. *Industrial and Engineering Chemistry* 40:1061.

Metz, P.D. 1982. The use of gound-coupled tanks in solar-assisted heat pump systems. ASME *Transactions, Journal of Solar Energy Engineering* 104:366.

Metz, P.D. 1983. A simple computer program to model three-dimensional underground heat flow with realistic boundary conditions. ASME *Transactions, Journal of Solar Energy Engineering* 105:42.

Mumma, S.M. 1985. Field testing of systems using controls to enhance thermal stratification during solar collection. ASHRAE *Transactions* 91(2).

Mumma, S.A. and W.C. Marvin. 1976. A method of simulating the performance of a pebble bed thermal storage and recovery system. ASME *Paper* 76-HT-13.

Oak Ridge National Laboratory. 1985. Field performance of residential thermal storage systems. EPRI EM 4041 (May).*

Page, J.K.R., *et al*. 1981. *Thermal storage materials and components for solar heating*. International Solar Energy Society, Solar World Forum, Brighton, U.K. (August).

Schaetzle, W.J., *et al*. 1980. *Thermal energy storage in aquifers: The design and applications*. Pergamon Press, Elmsford, NY.

Tamblyn, R.T. 1977. Thermal storage, A sleeping giant. ASHRAE *Journal* (June).

Tamblyn, R.T. 1985. College Park thermal storage experience. ASHRAE *Transactions* 91(1).

Tamblyn, R.T. 1980. Thermal storage, Resisting temperature blending. ASHRAE *Journal* (January):65.

Telkes, M. 1947. Solar house heating—A problem of heat storage. *Heating and Ventilating* 44:68-75.

Todd, D.K. 1963. *Groundwater hydrology*. John Wiley & Sons, New York.

Tomlinson, J.J. 1985. Clathrates and conjugating binaries: New materials for thermal storage. ASHRAE *Transactions* 91(2).
Tran, N., J.F. Kreider, and P. Brothers. 1989. Filed measurement of chilled water thermal performance. ASHRAE *Transactions* 95(1):1106-12.
Wildin, M.W. and C.R. Truman. 1985. Evaluation of stratified chilled water storage techniques. EPRI EM 4352, Electric Power Research Institute (December).*
Wildin, M.W. and C.R. Truman. 1985. A summary of experience with stratified chilled water tanks. ASHRAE *Transactions* 91(1).
Wildin, M.W. and C.R. Truman. 1989. Performance of stratified vertical cylindrical thermal storage tanks, Part I: Scale model tank. ASHRAE *Transactions* 95(1):1086-95.
Yoo, J. and M.W. Wildin. 1986. Initial formation of a thermocline in stratified thermal storage tanks. ASHRAE *Transactions* 92(2).

BIBLIOGRAPHY

Phase Change Materials

Abhat, A. 1983. Low temperature latent heat thermal energy storage: Heat storage materials. *Solar Energy* 30(4):313-32.
Altman, M., H. Yeh, and H.G. Lorsch. 1973. Conservation and better utilization of electric power by means of thermal energy storage and solar heating. University of Pennsylvania Report, NTIS PB 210359, National Technical Information Service, Springfield, VA.
Boer, K.W., J.H. Higgins, and J.H. O'Connor. 1975. Solar One, Two years experience. Proceedings of the 10th IECEC, 7.
Hale, D.V., M.J. Hoover, M.J. O'Neill. 1971. *Phase change materials handbook*. NASA CR-61363.
Herrick, C.S. 1982. Melt-freeze cycle life testing of Glauber's salt in a rolling cylinder heat store. *Solar Energy* 28:99-104.
Jones, D.E. and J.E. Hill. 1979. An evaluation of ASHRAE *Standard* 94-77 for testing pebble bed and phase change thermal energy storage devices. ASHRAE *Transactions* 85(2):607.
Janz, G.J., *et al.* 1976. Eutectic data. ERDA TID-27163. Molten Salts Data Center, Rensselaer Polytechnic Institute, Troy, NY.
Kauffman, K.W. and I.J. Gruntfest. 1973. Congruently melting materials for thermal energy storage. Report NCEMP-20, National Center for Energy Management and Power, University of Pennsylvania.
Lane, G.A. and H.E. Rossow. 1976. Encapsulation of heat of fusion storage materials. Procedures of the 2nd Southeastern Conference on Applications of Solar Energy, O.A. Arnas, ed. Baton Rouge, LA, 442.
Lane, G.A. 1983. *Solar heat storage: Latent heat materials*. Vol. I, Background and Scientific Principles. CRC Press, Boca Raton, FL.
Lane, G.A. 1986. *Solar heat storage: Latent heat materials*. Vol. II, Technology. CRC Press, Boca Raton, FL.
Laybourn, D.R. 1988. Thermal energy storage with encapsulated ice. ASHRAE *Transactions* 94(1):1971-88.
MacCracken, C.D. 1984. Design considerations for modular glycol-ice storage systems. ASHRAE *Transactions* 90(1).
MacCracken, C.D. 1984. Solid state phase change for passive solar applications. 9th National Passive Solar Conference, American Solar Energy Society, Columbus, OH (September).
MacCracken, C.D. 1985. Control of brine-type ice storage systems. ASHRAE *Transactions* 91(1).
Marks, S.B. 1983. The effect of crystal size on the thermal energy storage capacity of thickened Glauber's salt. *Solar Energy* 30:45-49.
Rueffel, P.G. 1980. U.S. Patent 4,211,885.
Schroder, J. *Some materials and measures to store latent heat.* Philips GmbH Forschungslaboratorium, Aachen, West Germany (in English).
Telkes, M. 1974. Solar heat storage. ASHRAE *Journal* 16:38.
Telkes, M. 1980. U.S. Patent 4,187,189.

Ground-Coupled Storage

ASTM. 1985. Classification of soils for engineering purposes. ASTM 2487. American Society for Testing and Materials, Philadelphia.
Biehl, R.A. 1977. The annual cycle energy system: A hybrid heat pump cycle. ASHRAE *Journal* (July):20.
Bose, J.E. 1982. Earth coil/heat pump research at Oklahoma State University. 6th Heat Pump Technology Conference, Oklahoma State University, Stillwater, OK.
Buies, S. 1984. The "Fabrikaglace" process to make and store natural ice. Third International Workshop on Ice Storage for Cooling Applications, Argonne National Laboratory, ANL/CNSV-TM-77, 4.
Foster, L.J. 1985. A review of the design and system performance. ASHRAE *Transactions* 91(1).
Francis, C.E. 1985. The production of ice with long-term storage. ASHRAE *Transactions* 91(1).
Hansen, K.K., P.N. Hansen, and V. Ussing. 1985. Stratified operation of a 500 m^3 test pit. Proceedings, Enerstock '85, 3rd International Conference on Energy Storage for Building Heating and Cooling, Public Works Canada, Ottawa K1A 0M2, 157.
Kusuda, T. and P.R. Achenbach. 1965. Earth temperatures and thermal diffusivity at selected stations in the United States, Part I. ASHRAE *Transactions* 71:61-75.
McGarity, A.E., D.L. Kirkpatrick, and L.K. Norford. 1987. Design and operation of an ice pond for cooling a large commercial office building. ASHRAE *Transactions* 92(1).
Metz, P.D. 1983. Ground coupled heat pump system experimental results. ASHRAE *Transactions* 89(2B).
Nordell, B., *et al.* 1985. The borehole heat store at Lulea. Proceedings, Enerstock '85, 3rd International Conference on Energy Storage for Building Heating and Cooling, Public Works Canada, Ottawa, Canada K1A 0M2, 70, 71, 131.
Ostensson, B.A. 1985. HVAC system based on seasonal storage. Proceedings, Enerstock '85, 3rd International Conference on Energy Storage for Building Heating and Cooling, Public Works Canada, Ottawa, K1A 0M2, 132.
Proceedings of the Nordic Symposium on Earth Heat Pump Systems, Chalmers University of Technology, Goteborg, Sweden, October, 1979.
Wijsman, A.J. and J. Havings. 1986. The Groningen Project: 96 solar houses with seasonal heat storage in the soil. Proceedings of the annual meeting, American Solar Energy Society, Boulder, CO, 530.

Aquifers

Brett, C.E. and W.J. Schaetzle. Experience with chilled water storage in a water table aquifer. Proceedings, Enerstock '85, 3rd International Conference on Energy Storage for Building Heating and Cooling, Public Works Canada, Ottawa, Canada K1A 0M2.
DOE. 1981-83. Proceedings DOE Physical and Chemical Energy Storage Annual Contractors' Review Meetings, Conf-810940, 1981; Conf-820827, 1982; Conf-830974, 1983. Available from NTIS, Springfield, VA.
Ebeling, L., *et al.* 1979. The effect of system size on the practicality of aquifer storage. Proceedings of the Silver Jubilee Congress, International. Solar Energy Society. Available from American Solar Energy Society, Boulder, CO.
LBL. 1978. Thermal energy storage in aquifers workshop. LBL Report No. 8431. Lawrence Berkeley Laboratory, Berkeley, CA.
Trinity University. 1975. Proceedings, Solar Energy Storage Options.
Vail, L.W. 1983. Numerical model for analysis of multiple well aquifer thermal energy storage systems. DOE Physical and Chemical Energy Storage Annual Contractors Review Meeting, Conf-830974.

Applications

ASHRAE. 1985. Thermal storage. ASHRAE TDB-TH-1.
Electric Power Research Institute. 1983. Issues in residential load management. EM 2991 (April).*
Hersh, H.N. 1985. Current trends in commercial cool storage. EPRI, EM 4125 (July).*
Lorsch, H.G. and M.A. Baker. 1984a. Survey of thermal energy storage installations in the United States and Canada. ASHRAE.
Lorsch, H.G. and M.A. Baker. 1984b. A description of six representative thermal storage installations, Oak Ridge National Laboratory, ORNL/Sub/83-28915/1 (May).
PEPCO/DOE/EPRI. 1984. Thermal energy storage: Cooling commercial buildings using off-peak energy. Seminar Proceedings, EPRI EM 2244 (February).*
San Diego Gas and Electric Company. 1984. Thermal energy storage-inducement program for commercial space cooling. San Diego, CA (April).

---

*Available from Research Reports Center, P.O. Box 50490, Palo Alto, CA 94303; Phone (415) 965-4081. Additional bibliographic references on thermal storage are listed in Chapter 6.

CHAPTER 40

CONTROL OF GASEOUS CONTAMINANTS FOR INDOOR AIR

AIR throughout the world contains nearly constant amounts of nitrogen (78% by volume), oxygen (21%), and argon (0.9%), with varying amounts of carbon dioxide (0.03%) and water vapor (up to 3.5%). In addition, trace quantities of inert gases (neon, xenon, krypton, helium, etc.) are always present. Beyond this list, other gases present are usually considered contaminants or pollutants. Their concentrations are almost always small, but they may have serious effects on building occupants, construction materials, or contents. Removal of these gaseous pollutants is often desirable or necessary.

Traditionally, indoor gaseous contaminants are controlled with ventilation air drawn from outdoors; but minimizing outdoor airflow by using a high recirculation rate and filtration is an attractive means of energy conservation. However, recirculated air cannot be made equivalent to fresh outdoor air by removing only particulate contaminants. Noxious, odorous, and toxic gaseous contaminants must also be removed by gaseous contaminant control equipment, which is different from particulate filtration equipment. In addition, available outdoor air may also contain undesirable gaseous contaminants at unacceptable concentrations. If so, it too will require treatment by gaseous contaminant removal equipment.

This chapter deals with design procedures for gaseous pollutant control systems for occupied spaces only. The control of gaseous pollutants from industrial processes and stack gases is covered in Chapter 11 of the 1988 ASHRAE *Handbook—Equipment*.

GASEOUS CONTAMINANT CHARACTERISTICS

Before a gaseous contaminant control system can be designed, the contaminants present in significant concentrations must be identified, and at least the following items of information must be determined:

1. Exact chemical identity of the contaminants
2. Rates at which the contaminants are generated in the space, and the rates at which they are brought into the space with outdoor air

This information is often difficult to obtain. Designers must often make do with a chemical family name (aldehydes) and a qualitative description of generation rates (from plating tank) or perceived concentration (at odorous levels). System design under these conditions is uncertain. When the exact chemical identity of a contaminant is known, the chemical and physical properties influencing its collection by control devices can usually be obtained from handbooks and technical publications. Factors of special importance are:

- Molecular weight
- Normal boiling point (*i.e.*, at 1 standard atmosphere)
- Heat of vaporization
- Polarity
- Chemical reactivity and chemisorption velocity

With this information, the performance of control devices on contaminants for which no specific tests have been made may be estimated. The following sections on control methods discuss the usefulness of such estimations.

Some gaseous contaminants have unique properties that must be considered in system design. Ozone, for example, will reach an equilibrium concentration in a ventilated space without a filtration device. It does so partly because an ozone molecule will join with other ozone molecules to form normal oxygen, but also because it reacts with people, plants, and materials in the space. This oxidation is harmful to all three, and therefore natural ozone decay is not a satisfactory way to control ozone except at low concentrations (< 0.2 mg/m^3). Fortunately, activated carbon adsorbs ozone readily, both reacting with it and catalyzing its conversion to oxygen.

Radon is a radioactive gas that decays by alpha-particle emission, eventually yielding individual atoms of polonium, bismuth, and lead. These atoms form extremely fine aerosol particles called radon daughters or radon progeny, which are also radioactive; they are especially toxic, in that they lodge deep in the human lung, where they emit cancer-producing alpha particles. Radon progeny, both attached to larger aerosol particles and by themselves (unattached), can be captured by particulate air filters. Radon gas itself may be removed with activated carbon (Thomas 1974), but this method costs too much for the benefit derived in HVAC systems. Control of radon emission at the source and ventilation are the preferred methods of radon control.

Another gaseous contaminant that often appears in particulate form is sulfur trioxide (SO_3), which should not to be confused with sulfur dioxide (SO_2). At ambient temperatures, it reacts rapidly with water vapor to form a fine mist of sulfuric acid. If this mist collects on a particulate filter and no means is provided to remove it, the acid will vaporize and reenter the protected space. Because many such conversions and desorptions are possible, the designer must understand the problems they cause.

Major chemical families of gaseous pollutants and examples of specific compounds are given in Table 1. The *Merck Index* (Windholtz 1983), the *Toxic Substances Control Act Chemical Substance Inventory* (EPA 1979), and *Dangerous Properties of Industrial Materials* (Sax and Lewis 1988) are all useful in identifying contaminants, including some known by trade names only.

The preparation of this chapter is assigned to TC 2.3, Air Contaminants: Gaseous and Particulate.

Table 1 Major Chemical Families of Gaseous Air Pollutants (with Examples)

Inorganic—Pollutants

1. Single-Element Molecules
    - chlorine
    - radon
    - mercury
2. Oxidants
    - ozone
    - nitrogen dioxide
    - nitrous oxide
    - nitric oxide
3. Reducing Agents
    - carbon monoxide
4. Acid Gases
    - sulfuric acid
    - hydrocloric acid
    - nitric acid
5. Nitrogen Compounds
    - ammonia
6. Sulfur Compounds
    - hydrogen sulfide
7. Miscellaneous
    - arsine

Organic Pollutants

8. n-Alkanes
    - methane
    - n-butane
    - n-hexane
    - n-octane
    - n-hexadecane
9. Branched Alkanes
    - 2-methyl pentane
    - 2-methyl hexane
10. Alkenes and Cyclohexanes
    - 1-octene
    - 1-decene
    - cyclohexane
11. Chlorofluorocarbons
    - R-11 (trichlorofluoromethane)
    - 1,1,1 trichloroethane
    - R-114 (dichlorotetrafluoroethane)
12. Halide Compounds
    - carbon tetrachloride
    - chloroform
    - methyl bromide
    - methyl iodide
    - phosgene
    - carbonyl sulfide
13. Alcohols
    - methanol
    - ethanol
    - 2-propanol
    - isopropanol
    - phenol
    - cresol
    - diethylene glycol
14. Ethers
    - vinyl ether
    - methoxyvinyl ether
    - n-butoxyethanol
15. Aldehydes
    - formaldehyde
    - acetaldehyde
    - acrolein
    - benzaldehyde
16. Ketones
    - 2-butanone (MEK)
    - 2-propanone
    - acetone
    - methyl isobutyl ketone
    - chloroacetophenone
17. Esters
    - ethyl acetate
    - n-butyl acetate
    - di-ethylhexyl phthalate (DOP)
    - di-n-butyl phthalate
    - butyl formate
    - methyl formate
18. Nitrogen Compounds, and Other Than Amines
    - nitromethane
    - acetonitrile
    - acrylonitrile
    - pyrrole
    - pyridine
    - hydrogen cyanide
    - peroxyacetal nitrate
19. Aromatic Hydrocarbons
    - benzene
    - toluene
    - ethyl benzene
    - naphthalene
    - p-xylene
    - benz-alpha-pyrene
20. Terpenes
    - 2-pinene
    - limonene
21. Heterocylics
    - furan
    - tetrahydrofuran
    - methyl furfural
    - nicotine
    - 1,4 dioxane
    - caffeine
22. Organophosphates
    - malathion
    - tabun
    - sarin
    - soman
23. Amines
    - methylamine
    - diethylamine
    - n-nitroso-dimethyamine
24. Monomers
    - vinyl chloride
    - methyl formate
    - ethylene
25. Mercaptans and Other Sulfur Compounds
    - bis-2-chloroethyl sulfide (mustard gas)
    - ethyl mercaptan
    - methyl mercaptan
    - carbon disulfide
26. Miscellaneous
    - ethylene oxide

Note that the same chemical compound, especially organic compounds, may have several scientific names.

HARMFUL EFFECTS OF GASEOUS CONTAMINANTS

The only reason to remove a gaseous contaminant from an airstream is that it has harmful or annoying effects on the ventilated space or its occupants. These effects are noticeable at different concentration levels. In most cases, contaminants become annoying by their odors before they reach levels toxic to humans, but this is not always true. For example, the highly toxic (even deadly) contaminant carbon monoxide has no odor.

Harmful effects may be divided into four categories: toxicity, odor, irritation, and material damage. These effects depend on the concentrations of the contaminants and are usually expressed in the following units:

ppm = parts of contaminant by volume per million parts of air by volume
ppb = parts of contaminant by volume per billion parts of air by volume
mg/m^3 = milligrams of contaminant per cubic metre of air
$\mu g/m^3$ = micrograms of contaminant per cubic metre of air

The conversions between ppm and mg/m^3 are:

$$ppm = 62.32(mg/m^3)(273.15 + t)/(Mp) \quad (1)$$

$$mg/m^3 = 0.01605(ppm)(Mp)/(273.15 + t) \quad (2)$$

where

M = molecular weight of the contaminant
p = mixture pressure, mm Hg
t = mixture temperature, °C

Concentration data is often reduced to standard temperature and pressure (*i.e.*, 25 °C and 760 mm Hg), in which case:

$$ppm = 24.45(mg/m^3)/M \quad (3)$$

Toxicity

The harmful effects of gaseous pollutants on people depend on both short-term peak concentrations and the time-integrated exposure received by a person. The allowable concentration for short exposures is higher than for long exposures. The Occupational Safety and Health Administration (OSHA) has defined three periods for concentration averaging and has assigned allowable levels that may exist in these categories in workplaces for over 490 compounds, mostly gaseous contaminants. The abbreviations for concentrations for the three averaging periods are:

AMP = acceptable maximum peak for a short exposure
ACC = acceptable ceiling concentration, not to be exceeded during an 8-h shift, except for periods where AMP applies
TWA8 = time-weighted average, not to be exceeded in any 8-h shift of a 40-h week.

In non-OSHA literature, ACC is sometimes called STEL (short-term exposure limit), and TWA8 is sometimes called TLV (threshold limit value). The medical community disagrees on what values should be assigned to AMP, ACC, and TWA8. OSHA values, which change periodically, are published yearly in the *Code of Federal Regulations* (29 CFR 1900, Part 1900.1000 ff) and intermittently in the *Federal Register.* A similar list is available from the American Conference of Governmental Industrial Hygienists (ACGIH 1989b). The National Institute for Occupational Safety and Health (NIOSH) is charged with researching toxicity problems, and it greatly influences the legally required levels. NIOSH annually publishes the *Registry of Toxic Effects of Chemical Substances* as well as numerous *Criteria for Recommended Standard for Occupational Exposure to (compound).* Some compounds not in the OSHA list are covered by NIOSH literature, and their recommended levels are sometimes lower than the legal requirements set by OSHA. The NIOSH *Pocket Guide to Chemical Hazards* (Mackison *et al.* 1978) is a condensation of these references and is convenient for engineering purposes. This publication also lists values for the following toxic limit:

IDLH = immediately dangerous to life and health

Although this toxicity limit is rarely a factor in HVAC design, HVAC engineers should consider it when deciding how much recirculation is safe in a given system. Ventilation airflow must never be so low that the concentration of any gaseous contaminant could rise to the IDLH level. The levels set by OSHA and NIOSH define acceptable occupational exposures, and cannot be used by themselves as acceptable standards for residential or commercial spaces. They do, however, suggest some upper limits for contaminant concentrations for design purposes.

Personnel exposure to radioactive gases is not addressed by OSHA regulations, but by rules promulgated by the Nuclear Regulatory Commission. Allowable concentrations, in terms of radioactivity, are listed annually in the *Code of Federal Regulations*, Section 10, Part 50 (10 CFR 50, 1990). Radioactive gases are controlled in much the same way as nonradioactive gases, but their high toxicity demands more careful system design. Chapter 24 introduces the procedures unique to nuclear systems.

Another toxic effect that may influence design is the loss of sensory acuity due to gaseous contaminant exposure. Carbon monoxide, for example, affects psychomotor responses, and could be a problem in areas such as air traffic control towers. Clearly, waste anesthetic gases should not be allowed to reach levels in operating suites where the alertness of any of the personnel is affected. NIOSH recommendations are frequently based on such subtle effects.

Odors

Concentrations as high as those set by OSHA regulations are rarely encountered in commercial or residential spaces. In such spaces, gaseous contaminant problems usually appear as complaints about odors or stuffiness and are the result of concentrations considerably below TWA8 values. Each individual has different sensitivities to odors, and this sensitivity decreases even during a relatively brief exposure. One pollutant may enhance or mask the odor of another. Odors are usually stronger when relative humidity is high. In addition, the human olfactory response S is nonlinear; perceived intensities are approximate power functions of the contaminant concentration C:

$$S = kC^n \quad (4)$$

where n is typically about 0.6.

This nonlinear response means that the concentration of some odorants must be reduced substantially before the odor level is perceived to change. For these reasons, determining acceptable concentrations of pollutants on the basis of their odors is as imprecise as for their toxicity. Far fewer data are available on odors, since odors are more of an annoyance than a hazard. In fact, a toxic or explosive material that has an odor threshold well below toxic levels is desirable, for this can warn building occupants that the pollutant is present.

At some low concentration of an odorant, an individual ceases to be aware of its presence. Odor studies seek to establish the level at which a percentage of the general population—usually 50%—is no longer aware of the odor of the compound studied. This concentration is the *odor threshold* for that compound. Fazzalari (1978) compiled odor thresholds for a large number of gaseous compounds in commercial and industrial situations. Additional values, reflecting newer measurement techniques but largely limited to compounds found only in industrial workplaces, are listed in AIHA (1989), Moore and Houtala (1983), and Van Gemert and Mettenbreijer (1977). Table 2 includes odor threshold values taken from these lists to show the wide range of threshold values. Chapter 12 of the 1989 ASHRAE *Handbook—Fundamentals* discusses odor perception.

Table 2 Characteristics of Selected Gaseous Air Pollutants

Pollutant	Allowable Concentration, mg/m³ IDLH	AMP	ACC	TWA8	Odor Threshold, mg/m³	Chemical and Physical Properties Family	BP,°C	M	Retentivity, %
Acetaldehyde	18000			360	1.2	15	21	44	8*
Acetone	4800		3200	2400	47	16	56	58	16
Acetonitrile	7000	105		70	>0	18	82	41	1
Acrolein	13		0.75	0.25	0.35	15	52	56	
Acrylonitrile	10			45	50	18	77	53	3
Allyl chloride	810		9	3	1.4	12	44	77	
Ammonia	350		35	38	33	5	−33	17	0
Benzene	10000		25	5	15	19	80	78	12
Benzyl chloride	50			5	0.2	12	179	127	0.3
2-Butanone (MEK)	8850			590	30	16	79	72	12
Carbon dioxide	90000		54000	9000	00	4	−78	44	
Carbon monoxide	1650		220	55	00	3	−192	28	0
Carbon disulfide	1500	300	90	60	0.6	6	46	76	4*
Carbon tetrachloride	1800	1200	150	60	130	12	77	154	8
Chlorine	75		1.5	3	0.007	1	−34	71	2*
Chloroform	4800		9.6	240	1.5	12	124	119	11
Chloroprene	1440		3.6	90		12	120	89	
p-Cresol	1100			22	0.056	13	305	108	5*
Dichlorodifluromethane	250000			4950	5400	11	−30	121	
Dioxane	720			360	304	21	100	68	13
Ethylene dibromide	3110	271	233	155		12	131	188	11
Ethylene dichloride	4100	818	410	205	25	12	84	99	12
Ethylene oxide	1400		135	90	196	21	10	44	0
Formaldehyde	124	12	6	4	1.2	15	97	30	0.4*
n-Heptane	17000			2000	2.4	8	98	100	7*
Hydrogen chloride	140		7	7	12	4	−121	37	0.7*
Hydrogen cyanide	55			11	1	18	26	27	0.4
Hydrogen fluoride	13		5	2	2.7	4	19	20	1*
Hydrogen sulfide	420	70	28	30	0.007	6	−60	34	0.8*
Mercury	28			0.1	00	1	357	201	13
Methane	ASPHY					8	−164	16	0
Methanol	32500			260	130	13	64	32	6*
Methyl chloride	59500		1783	1189	595	12	74	133	
Methylene chloride	7500		3480	1740	750	12	40	85	9
Nitric acid	250			5		4	84	63	3*
Nitric oxide	120	45		30	>0	2	−152	30	
Nitrogen dioxide	90		1.8	9	51	2	21 46	2*	
Ozone	20			2	0.2	2	−112	48	(note)
Phenol	380		60	19	0.18	13	182	94	5*
Phosgene	8		0.8	0.4	4	7	8	90	1
Propane	36000				1800	8	−42	44	5*
Sulfur dioxide	260			13	1.2	6	−10	64	1.7*
Sulfuric acid	80			1	1	4	270	98	1.9*
Tetrachloroethane	1050			35	24	12	146	108	
Tetrachloroethylene	3430	2060	1372	686	140	12	121	166	
o-Toluidene	440			22	24	23	199	107	
Toluene	7600	1900	1140	760	8	17	111	92	17*
Toluene diisocyanate	70		0.14	0.14	15	18	251	174	
1,1,1 Trichloroethane	2250			45	1.1	11	113	133	5
Trichloroethylene	5410	1620	1080	541	120	12	87	131	
Vinyl chloride monomer			0.014	0.003	1400	24	−14	63	0.13
Xylene	43500		870	435	2	19	137	106	16*

Notes: IDLH, AMP, ACC, and TWA8 are defined in the text.

BP = boiling point at 1 atmosphere pressure

M = molecular mass

Codes for chemical families are as given in Table 1.

ASPHY = Simple asphyxiant, causes breathing problems when concentration reaches about 1/3 atmospheric pressure.

Retentivities are for typical commercial-grade activated carbon, either measured at TWA8 levels or corrected to TWA8 inlet concentrations using the expression for breakthrough time given in Nelson and Correia (1976):

$$t_{b2} = t_{b1} (C_2/C_1)^{-2/3}$$

Multiplying breakthrough time by inlet concentration gives retentivity, so:

$$R_2 = R_1 (C_2/C_1)^{1/3}$$

Both concentrations must be in the same units, here mg/m³.

Retentivities marked (*) were calculated from values given in ASTM *Standard* D 1605-60 (1988) and Turk (1954), assuming that the listed retentivities were measured at 1000 ppm.

Ozone life is extremely long; activated carbon assists the essentially complete conversion of ozone to normal oxygen by both chemisorption and catalysis.

References:

ASTM *Standard* D 1605-60 (1988)
Balieu *et al.* (1977)
Dole and Klotz (1946)
Freedman *et al.* (1973)
Gully *et al.* (1969)
Miller and Reist (1977)
Revoir and Jones (1972)
Turk (1954)

Irritation

Although gaseous pollutants may have no discernible continuing health effects, exposure to such elements may irritate building occupants. Coughing; sneezing; eye, throat, and skin irritation; nausea; breathlessness; drowsiness; headaches; and depression have all been attributed to air pollutants. Rask (1988) suggests that when 20% of a single building's occupants suffer such irritations, the structure is said to suffer from the "sick building syndrome" or the "tight building syndrome." For the most part, case studies of such occurrences have consisted of analyses of questionnaires submitted to building occupants. Some attempts to relate irritations to gaseous contaminant concentrations are reported (Lamm 1986, Cain *et al.* 1986, Berglund *et al.* 1986, Molhave *et al.* 1982). The correlation of reported complaints with gaseous pollutant concentrations is not strong; many factors affect these less serious responses to pollution. In general, physical irritation does not occur at odor threshold concentrations.

Damage to Materials

Material damage from gaseous pollutants may take such forms as corrosion, embrittlement, and discoloration. Since such effects usually involve chemical reactions that need water, material damage from air pollutants is less in the relatively dry indoor environment than outdoors, even with similar gaseous contaminant concentrations. Indoor condensation should be avoided to maintain this advantage. However, damage to some materials can be significant, especially from ozone, hydrogen peroxide and other oxidants, sulfur dioxide, and hydrogen sulfide.

These effects are most serious in museums, as any loss of color or texture changes the nature of the object. Libraries and archives are also vulnerable, as are pipe organs and textiles. Ventilation is often a poor solution to protecting collections of rare objects; facilities are usually located in the centers of cities that have relatively polluted ambient outdoor air.

Filtration systems for rare-object protection must be applied with great care. Pollution-control methods that claim to modify, but not remove, damaging contaminants, offer no protection to fragile materials. Likewise, delaying the passage of a contaminant—capturing it at a relatively high concentration and releasing it slowly at low concentration—may reduce its odor or health effects, but has almost no effect on material damage. A control involves a chemical reaction that converts a pollutant into another compound, which may reduce or increase material damage, depending on the reaction product. Finally, the performance of the pollution-control device must be demonstrated by actual tests at concentrations and environmental conditions expected for the ventilation air.

Caution must be taken with systems incorporating electrostatic air cleaners, which can generate ozone. In rare-object protection systems, such equipment must be followed by an effective ozone adsorber; the entire system must operate and be maintained so that ozone levels in the ventilation air downstream of the purification equipment is always below an acceptable level. Various concerns on material damage by indoor air pollutants are discussed by Walsh *et al.* (1977), Chiarenzelli and Joba (1966), NTIS (1982, 1984), Braun and Wilson (1970), Jaffe (1967), Grosjean *et al.* (1987), American Guild of Organists (1966), Mathey *et al.* (1983), Haynie (1978), Graminski *et al.* (1978), and Thomson 1986.

CONTAMINANT SOURCES AND GENERATION RATES

Tobacco Smoke

Tobacco smoke is a prevalent and potent source of indoor air pollutants. Cigar and pipe smoking produce somewhat different compounds than cigarette smoking, but almost all tobacco pollution arises from cigarette smoking. Table 3 lists some of the important compounds found in the gas phase of cigarette smoke and gives average concentrations for the two types of smoke produced—mainstream and sidestream. Mainstream smoke is that which goes directly from the cigarette to the respiratory tract of the smoker. The part that is not trapped returns to the room air when exhaled by the smoker and joins the sidestream smoke that has not been directly inhaled by the smoker (including the smoke produced between puffs). The type of tobacco and cigarette configurations vary widely; the data represent averages of values from several references. The effect of cigarette tip filters on these values is not striking.

Table 3 Major Gaseous Compounds in Typical Cigarette Smoke

Pollutant	Mainstream Smoke Generation Rate[a], mg/cigarette	Sidestream Smoke Generation Rate[b], mg/cigarette	References
Acetaldehyde	0.7	4.4*	2,3
Acetone	0.4	—	1
Acetonitrile	0.1	0.4	1
Acrolein	0.1	0.6*	2,3
Ammonia	0.1	5.9	1
2-Butanone	0.1	0.3	1
Carbon dioxide	45	360	1
Carbon monoxide	17	43	1,4
Ethane	0.4	1.2*	2
Ethene	—	0.8*	2
Formaldehyde	0.05	2.3	2,3
Hydrogen cyanide	0.4	0.1	1
Isoprene	0.3	3.1	2
Methane	1.0	0.4	1
Methyl chloride	0.4	0.8	1
Nitrogen dioxide	—	0.2	1
Nitric oxide	0.2	1.8	1
Propane	0.2	0.7*	2
Propene	0.2	0.8	1

[a]Averages of values reported in listed references.
[b]Entries without asterisks were obtained by multiplying mainstream values by SS/MS ratios given in Surgeon General (1979), p. 11-6, 14-3, and 14-37. Entries with asterisks were obtained by using average chamber concentrations in Loefroth *et al.* (1989), using a chamber volume of 24 m^3.

References:
1. Surgeon General (1979)
2. Loefroth *et al.* (1989)
3. Newsome *et al.* (1965)
4. Cohen *et al.* (1971)

Cleaning Agents and Other Consumer Products

Commonly used liquid detergents, waxes, polishes, spot removers, and cosmetics contain organic solvents that volatilize slowly or quickly. Mothballs and other pest control agents emit organic volatiles. Knoeppel and Schauenburg (1989), Black and Bayer (1986), and Tichenor (1989) report data on the release of these volatile organic compounds (VOCs). Field studies have shown that such products contribute significantly to indoor pollution; however, a large variety of compounds are in use, and few studies have been made that allow calculation of typical emission rates. Pesticides, both those applied indoors and those applied outdoors to control termites, also pollute building interiors.

Building Materials and Furnishings

Particleboard, which is usually made from wood chips bonded with a phenol-formaldehyde or other resin, is widely used in current construction, especially for mobile homes, carpet underlay, and case goods. These materials, along with ceiling tiles, carpeting, wall coverings, office partitions, adhesives, and paint finishes emit formaldehyde and other VOCs. Latex paints containing mercury have been shown to emit mercury vapor. While the emission rates for these materials decline steadily with age, the half-life of emissions is surprisingly long. Black and Bayer (1989), Nelms (1985), and Molhave (1982) report on these sources. Measured emission rates are listed in Table 4.

Equipment

Commercial and residential spaces have internal sources of gaseous contaminants, though generation rates are substantially lower than in the industrial environment. As equipment sources are rarely hooded, emissions go directly to the occupants. In commercial spaces, the chief sources of gaseous contaminants are office equipment, including electrostatic copiers (ozone), diazo printers (ammonia and related compounds), carbonless copy paper (formaldehyde), correction fluids, inks, and adhesives (various VOCs). Medical and dental activities generate pollutants from the escape of anesthetic gases (nitrous oxide and halomethanes) and sterilizers (ethylene oxide). The potential for asphyxiation is always a concern when compressed gases are present, even if that gas is nitrogen.

In residences, the main sources of equipment-derived pollutants are gas ranges, wood stoves, and kerosene heaters. Venting is helpful, but some pollutants escape into the occupied area. The pollutant contribution by gas ranges is somewhat mitigated by the fact that they operate for shorter periods than heaters. The same is true of showers, which can contribute to radon and halocarbon concentrations indoors.

Traynor *et al.* (1985a, 1985b) report on emission rates from indoor combustion devices (Table 5). Emission rates for equipment depend greatly on the type of equipment in use and the way it is used. Typical values are difficult to obtain (Traynor 1987). If significant sources of equipment-generated pollutants are suspected, hoods should be installed. Filtration is not likely to be an economical or safe solution to such problems.

Occupants

Humans and animals emit a wide array of pollutants by breath, sweat, and flatus. Some of these emissions are conversions from solids or liquids within the body. Many volatile organics emitted are, however, reemissions of pollutants inhaled earlier, with the tracheobronchial system acting like a gas chromatograph or a saturated physical adsorber. For such pollutants, the occupant may be considered a filter, and, in that sense, a pollutant-control device. Some data on lung filter efficiency are listed in Table 6. Several studies (mostly in relation to spacecraft habitability) have measured pollutant generation by humans (Table 7).

Table 4 Generation of Gaseous Pollutants by Building Materials

Contaminant	Average Generation Rate, μg/(h · m²) Caulk	Adhesive	Linoleum	Carpet	Paint	Varnish	Lacquer
c-10 Alkane	1200						
n-Butanol	7300						760
n-Decane	6800						
Formaldehyde			44	150			
Limonene		190					
Nonane	250						
Toluene	20	750	110	160	150		310
Ethyl benzene	7300						
Trimethyl benzene		120					
Undecane						280	
Xylene	28						310

Contaminant	Average Generation Rate, μg/(h · m²) GF Insulation	GF Duct Liner	GF Duct Board	UF Insulation	Particleboard	Underlay	Printed Plywood
Acetone					40		
Benzene					6		
Benzaldehyde					14		
2-butanone					2.5		
Formaldehyde	7	2	4	340	250	600	300
Hexanal					21		
2-propanol					6		

GF = glass fiber UF = ureaformaldehyde foam
Sources: Matthews *et al.* (1983, 1985), Nelms *et al.* (1986), and White *et al.* (1988).

Table 5 Generation of Gaseous Pollutants by Indoor Combustion Equipment

	Generation Rates, μg/kJ CO_2	CO	NO_2	NO	HCHO	Typical Heating Rate, 1000 Btu/h	Typical Use, hour/day	Vented or Unvented	Fuel
Convective heater	51 000	83	12	17	1.4	31	4	U	Natural gas
Controlled combustion wood stove		13	0.04	0.07		13	10	V	Oak, pine
Range oven		20	10	22		32	1.0	U	Natural gas
Range-top burner		65	10	17	1.0	9.5/burner	1.7	U	Natural gas

Notes: Sterling and Kobayashi (1981) found that gas ranges are used for supplemental heating by about 25% of users in older apartments. This increases the time of use per day to that of unvented convective heaters.

Sources: Wade *et al.* (1975), Sterling and Kobayashi (1981), Cole (1983), Knight *et al.* (1986), Traynor *et al.* (1985b), Leaderer *et al.* (1987), and Moschandreas and Relwani (1989).

Table 6 Efficiency of Lung/Tracheobronchial System in Sorption of Some Volatile Organic Pollutants

Pollutant	Average Concentrations, mg/m^3 Ambient	Breath	Lung and Tracheobronchial Efficiency, %	References
Benzene	5	1.5	70	1
Chloroform	4	1.8	80	1
Vinylidene chloride	7	1.5	79	1
1,1,1 Trichloroethylene	60	5	92	1
Trichloroethylene	5.5	1	82	1
	5.5	1.7	70	7
n-dichlorobenzene	3	2	33	1
Formaldehyde	< 0.005		98	3
Acetaldehyde	< 0.0004		60	2
Acrolein	< 0.002		82	3
Carbon monoxide			55	4,5
Sulfur dioxide	14.3	4.3	70	6
Mercury	0.8		74	8
Furan	310		80	9

Sources:
1. Wallace *et al.* (1983)
2. Egle (1971)
3. Egle (1972)
4. Cohen *et al.* (1971)
5. Surgeon General (1979)
6. Wolff *et al.* (1975)
7. Stewart *et al.* (1974)
8. Hursh *et al.* (1976)
9. Egle (1979)

Table 7 Total-Body Emission of Some Gaseous Pollutants by Humans

Contaminant	Typical Emission, μg/h	Contaminant	Typical Emission, μg/h
Acetaldehyde	35	Methane	1710
Acetone	475	Methanol	6
Ammonia	15 600	Methylene chloride	88
Benzene	16	Propane	1.3
2-Butanone (MEK)	9700	Tetrachloroethane	1.4
Carbon dioxide	32×10^6	Tetrachloroethylene	1
Carbon monoxide	10 000	Toluene	23
Chloroform	3	1,1,1 Trichloroethane	42
Dioxane	0.4	Vinyl chloride monomer	0.4
Hydrogen sulfide	15	Xylene	0.003

Sources: Anthony and Thibodeau (1980), Brugnone *et al.* (1989); Cohen *et al.* (1971), Conkle *et al.* (1975), Gorban *et al.* (1964), Hunt and Williams (1977), and Nefedov *et al.* (1972).

Table 8 Primary Ambient Air Quality Standards for the United States

Contaminant	Long Term Concentration, ug/m^3	Long Term Averaging Period	Short Term Concentration, μg/m^3	Short Term Averaging Period, h
Sulfur dioxide	80	1 year	365	24
Carbon monoxide			10000	1
			40000	8
Nitrogen dioxide	100	1 year		
Ozone[a]			235	1
Hydrocarbons				
Total particulate (PM10)	75	1 year	260	24
Lead particulate	1.5	3 months		

[a]Standard is met when the number of days per year with maximum hour-period concentration above 235μg/m^3 is less than one.

Table 9 Typical Outdoor Concentrations of Selected Gaseous Air Pollutants

Pollutant	Typical Concentration, μg/m^3	Pollutant	Typical Concentration, μg/m^3
Acetaldehyde	20	Methylene chloride	2.4
Acetone	3	Nitric acid	6
Ammonia	1.2	Nitric oxide	10
Benzene	8	Nitrogen dioxide	51
2-Butanone (MEK)	0.3	Ozone	40
Carbon dioxide	612 000[a]	Phenol	20
Carbon monoxide	3000	Propane	18
Carbon disulfide	310	Sulfur dioxide	240
Carbon tetrachloride	2	Sulfuric acid	6
Chloroform	1	Tetrachloroethylene	2.5
Ethylene dichloride	10	Toluene	20
Formaldehyde	20	1,1,1 Trichloroethane	4
n-Heptane	29	Trichloroethylene	15
Mercury (vapor)	0.005	Vinyl chloride monomer	0.8
Methane	1100	Xylene	10
Methyl chloride	9		

[a]Normal concentration of carbon dioxide in air. The concentration in occupied spaces needs to be maintained at no greater than three times this level (1000 ppm).

Sources: Braman and Shelley (1980), Casserly *et al.* (1987), Chan *et al.* (1990), Cohen *et al.* (1989), Coy (1987), Fung *et al.* (1987), Hakov *et al.* (1987), Hartwell *et al.* (1985), Hollowell *et al.* (1982), Lonnemann *et al.* (1974), McClenny *et al.* (1987), McGrath and Stele (1987), Nelson *et al.* (1987), Sandalls and Penkett (1977), Shah and Singh (1988), Singh *et al.* (1981), Wallace *et al.* (1983), and Weschler and Shields (1989).

Outdoor Air

The following ventilation model developed by Janssen (1988) allows calculation of the effect of outdoor pollution on indoor air quality. Outdoor concentrations of gaseous contaminants must be included as inputs to this model. For a few pollutants, these concentrations are available for cities in the United States from EPA summaries of the National Air Monitoring Stations network data and in similar reports for other countries (EPA 1989). A condensed summary of this data is available (EPA 1990). In many cases, the concentrations listed in the National Ambient Air Quality Standards, which are the levels the EPA considers acceptable for six pollutants (Table 8), give satisfactory results. This is the procedure recommended by ASHRAE *Standard* 62-1989.

Table 9 gives outdoor concentrations for gaseous pollutants at urban sites. These values are typical; however, they may be exceeded if the building under consideration is located near a fossil-fueled power plant, refinery, chemical production facility, sewage treatment plant, municipal refuse dump or incinerator, animal feed lot, or any major source of gaseous contaminants. If such sources will have a significant influence on the intake air, a field survey or dispersion model must be run. Many computer programs have been developed to expedite such calculations.

INFLUENCE OF VENTILATION PATTERNS

A recirculating air-handling system is shown schematically in Figure 1. In this case, mixing is not perfect; the horizontal dashed line represents the boundary of the region close to the ceiling through which air passes directly from the inlet diffuser to the return-air intake. Ventilation effectiveness E_v is the fraction of the total air supplied to the space that mixes with the room air and does not bypass the room along the ceiling. Meckler and Janssen (1988) suggest a value of 0.8 for E_v. Any people in the space are additional sources and sinks for gaseous contaminants. In the ventilated space, the steady-state concentration for a single gaseous contaminant is:

$$C_{ss} = a/b \quad (5)$$

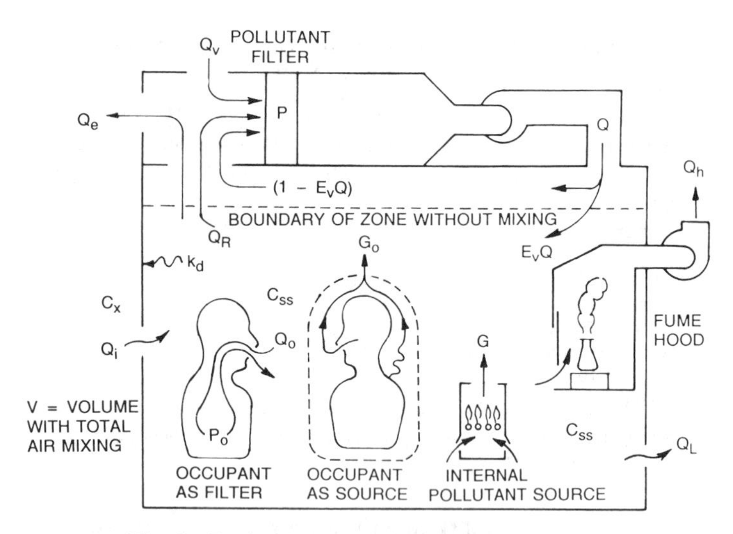

Fig. 1 Recirculatory Air-Handling and Gaseous Pollutant System

where

$$a = C_x(Q_i + 0.01P\,E_vQ_v/f) + 2119\,(G_i + N\,G_o) \quad (6)$$

$$b = Q_e + Q_h + Q_L + k_dA + NQ_o(1 - 0.01P_o) + (E_vQ - Q_v)(1 - 0.01P)/f \quad (7)$$

and

Q = system total flow, cfm
Q_e = exhaust airflow, cfm
Q_v = ventilation (makeup) airflow, cfm
Q_h = hood flow, cfm
Q_i = infiltration flow, cfm
Q_L = leakage (exfiltration) flow, cfm
Q_o = average respiratory flow for a single occupant, cfm
P = filter penetration for this pollutant, %
P_o = penetration of pollutant through human lung, %
A = surface area on which pollutant can be adsorbed inside the ventilated space, ft^2
k_d = deposition velocity on A for pollutant, ft/min
C_x = outdoor concentration of pollutant, mg/m^3
C_{ss} = steady-state indoor concentration of pollutant, mg/m^3
G_o = generation rate for pollutant by an occupant, mg/s
G_i = generation rate for pollutant, nonoccupant sources, mg/s
N = number of occupants
E_v = ventilation effectiveness, fraction
f = $1 - 0.01P(1 - E_v)$

Flow continuity allows the expression for b to be simplified to the following alternate form, which may make it easier to determine flows:

$$b = Q_i + Q_v + k_dA + NQ_o(1 - 0.01P_o) \quad (7a)$$

The parameters for this model must be determined carefully so that nothing significant is ignored. The leakage Q_L, for example, may include flow up chimneys or toilet vents.

The steady-state concentration is of interest in the design of a system. It may also be interesting to know how rapidly the concentration changes when conditions change suddenly. The dynamic equation for Figure 1 is:

$$C\theta = C_{ss} + (C_o - C_{ss})e^{-bt/V} \quad (8)$$

where

V = volume of the ventilated space, ft^3
C_o = concentration in the space at time $t = 0$
$C\theta$ = concentration in the space θ minutes after a change of conditions

C_{ss} is given by Equation (5), and b by Equations (7) or (7a), with the parameters for the new condition inserted.

Since the reduction in air infiltration, leakage, and ventilation air needed to reduce energy consumption raises concerns about indoor contaminant buildup, a low-leakage structure may be simulated by letting $Q_i = Q_L = Q_h = 0$. Then:

$$C_{ss} = \frac{0.01PE_vQ_vC_x/f + 2119(G_i + NG_o)}{Q_v + k_dA + NQ_o(1 - 0.01P_o) + (E_vQ - Q_v)(1 - 0.01P)/f} \quad (9)$$

Even if ventilation air Q_V is reduced to zero, a low penetration (high efficiency) gaseous contaminant filter and high recirculation rate help lower the internal contaminant concentration. In commercial structures, infiltration and exfiltration are never zero. The only inhabited spaces operating on 100% recirculated air are space capsules, undersea structures, and the like, where life-support systems eliminate carbon dioxide, carbon monoxide, and supply oxygen.

Real systems may have many rooms, with multiple sources of gaseous contaminants and complex room-to-room air changes. In addition, there may be mechanisms (other than adsorption) that eliminate gaseous contaminants on building interior surfaces. Nazaroff and Cass (1986) provide estimates for k_d in Equations (5) through (9) that range from 0.0006 to 0.12 ft/min for surface adsorption only. A worst-case analysis, giving the highest estimate of indoor concentration, is obtained by setting $k_d = 0$. Sparks (1988) and Nazaroff and Cass (1986) describe computer programs to handle these calculations.

The assumption of bypass and mixing used in the models presented here approximates the multiple-room case, since gaseous contaminants are readily dispersed by airflows. In addition, a contaminant diffuses from a zone of high concentration to a zone of low concentration, thus averaging concentrations, even with low rates of turbulent mixing.

Quantities appropriate for the various flows in Equations (5) through (9) are discussed under the sections Hooding and Local Exhaust and General Ventilation. Infiltration flow can be determined approximately by the techniques described in Chapter 22 of the 1989 ASHRAE *Handbook—Fundamentals*, or, for existing buildings, by tracer or blower-door measurements. ASTM *Standard* E 741-83 defines procedures for tracer-decay measurements. ASTM STP 719 (1980) covers tracer and blower-door techniques; DeFrees and Amberger (1987) describe a useful variation on the blower-door technique applicable to large structures.

CONTROL TECHNIQUES

Elimination of Sources

This approach to gaseous contaminant control is one of the most effective and least expensive. Control of radon gas begins by installing traps in sewage drains and sealing and venting leaky foundations and crawl spaces to prevent entry of the gas into the structure (EPA 1986, 1987). Prohibition of smoking in a building, or its isolation to limited areas, greatly reduces indoor pollution, even when these rules are poorly enforced (Elliott and Rowe 1975, Lee *et al.* 1986). The use of waterborne materials instead of those requiring organic solvents may reduce volatile organic contaminants, although Girman *et al.* (1984) show that the reverse is sometimes true. The substitution of compressed carbon dioxide for halocarbons in spray-can propellants is an example of the use of a relatively innocuous pollutant instead of a more troublesome one. The growth of mildew and other organisms that emit odorous pollutants can be restrained by controlling condensation and applying fungicides and bactericides.

Hooding and Local Exhaust

When substantial sources of gaseous pollutants are generated within a building, direct exhaust through hoods is more effective than control by general ventilation. When these pollutants are toxic, irritating, or strongly odorous, hooding is essential. The minimum transport velocity that exists in the capture of large particles differs from that required to capture gaseous contaminants; otherwise, the problems of capture are the same for both.

Hoods are normally provided with exhaust fans and stacks that vent to the outdoors. Hoods are great consumers of ventilation air, requiring large fan power for exhaust and makeup, and wasting heating and cooling energy. Makeup for air exhausted by a hood should be supplied forcibly, so that the general ventilation system balance is not upset when a hood exhaust fan is turned on. Back diffusion from an open hood to the general work space can be eliminated by surrounding the work space near the hood with an isolation enclosure (Figure 2). Such a space not only isolates the pollutants, but also tends to keep unnecessary personnel out of the area. Glass walls for the enclosure decrease the claustrophobic effect of working in a small space.

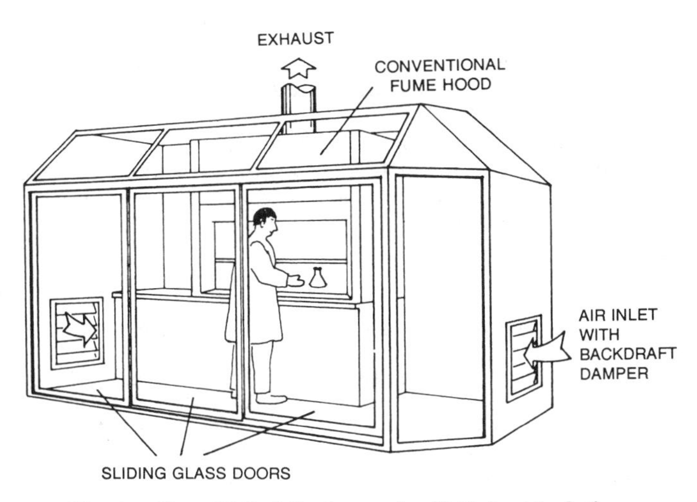

Fig. 2 Glass-Walled Enclosure for Pollutant Isolation

Increasingly, filtration of hood exhausts is required to prevent toxic releases to the outdoors. Hoods should be equipped with controls that decrease their flows when maximum protection is not needed. Hoods are sometimes arranged to exhaust air back into the occupied space, thus saving heating and cooling of that air. This practice must be limited to sources of the most innocuous pollutants, because of the risk of filtration system failure. The design and operation of effective hoods are described in Chapters 14, 25, and 27 and in *Industrial Ventilation—A Manual of Recommended Practice* (ACGIH 1989a).

General Ventilation

In residential and commercial buildings, the chief use of hoods and local exhaust occurs in kitchens, bathrooms, and occasionally around distinctive point sources, such as diazo printers. Where there is no local exhaust of contaminants, the general ventilation distribution system provides control. Such systems must meet both thermal load requirements and pollutant control standards. Complete mixing and a relatively uniform air supply per occupant is desirable for both purposes. The guidelines for air distribution system design given in Chapters 31 and 32 of the 1989 ASHRAE *Handbook—Fundamentals* are appropriate for pollution control by general ventilation. In addition, the standards set by ASHRAE *Standard* 62-1989 must be met.

When local exhaust is combined with general ventilation, a proper supply of makeup air must equal the exhaust flow for any hoods present. Supply fans may be needed to provide enough pressure to maintain flow balance. Systems (such as clean rooms) are sometimes designed so that static pressure forces the air to flow from clean to less clean spaces. With these systems, the effects of door openings, wind pressures, etc. may require back-draft dampers. Chapter 16 covers clean rooms in detail.

Absorbers

Air washers, whose prime purpose is to control the temperature and humidity of the air, also remove some water-soluble gaseous pollutants by absorption (Swanton *et al.* 1971). Absorbers are used in life-support systems in spacecraft and in undersea craft. There, both solid and liquid reactive absorbers reduce carbon dioxide and carbon monoxide to carbon and return the oxygen to the system. Liquid absorption devices (scrubbers) are also used to remove gaseous pollutants from stack gases.

In an absorption process, the gaseous pollutant is dissolved in or reacts with the body of an absorbing medium, which can be either a porous solid or a liquid. Pollutant gases are absorbed in liquids when the partial pressure of the pollutant in the bulk gas flow is greater than the solution vapor pressure for the pollutant. The rate of transfer of pollutant to liquid is a complex function of the physical properties of the pollutant, carrier gas, and liquid; the geometry of the gas/liquid contactor device; and the temperature and velocity of the gas flow through the contactor. The effectiveness of such absorbers can be improved by adding reagents to the scrubbing liquid. Chapter 11 of the 1988 ASHRAE *Handbook—Equipment* introduces liquid absorber design, and many chemical engineering texts also cover the subject.

The use of liquid absorbers in occupied spaces has several disadvantages. The airstream tends to be saturated with the vapor of the scrubbing liquid, which then becomes a pollutant itself; or, if the liquid is water, relative humidity may be raised to undesirable levels. To avoid excess humidification, some systems have been installed that use chilled water and subsequent thermal recovery (Pedersen and Fisk 1984). Systems with dessicants, which are regenerated, and heat exchanger wheels, which recover the energy expended, have also been tried (Novosel *et al.* 1987). Unfortunately, dessicants that have enough affinity for the pollutants present are not readily available.

Physical Adsorbers

Adsorption, in contrast to absorption, is a surface phenomenon, similar in many ways to condensation. Pollutant gas molecules that strike a surface and remain bound to it for an appreciable time are said to be adsorbed by that surface. This effect takes place to some degree on all surfaces, but gaseous contaminant control adsorption media have their available surfaces expanded in two ways. First, these adsorbers are provided in granular or fibrous forms to increase the gross surface exposed to an airstream. Second, the surface of the adsorber is treated to develop pores of microscopic dimensions, which greatly increases the surface available for molecular contact. Typical treated (activated) aluminas have surface areas of 200 m^2 per gram of adsorber; typical activated carbons have areas from 1000 to 1500 m^2 per gram of carbon. The various sizes and shapes of pores also provide minute traps that can fill with multiple layers of immobile pollutant molecules. These pores are filled with condensed vapors of the adsorbed pollutants.

In physical adsorption, the forces binding the pollutant molecules to the adsorber surface are the same forces that give the pollutant its thermophysical properties important to heat and mass transfer. Several steps must occur before a molecule can be adsorbed (Figure 3):

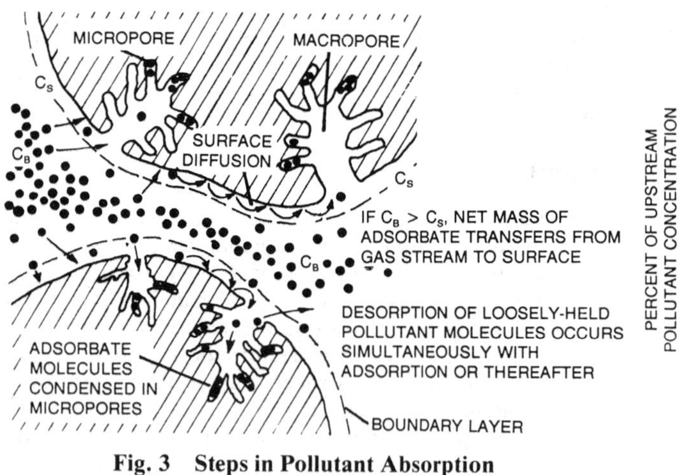

Fig. 3 Steps in Pollutant Absorption

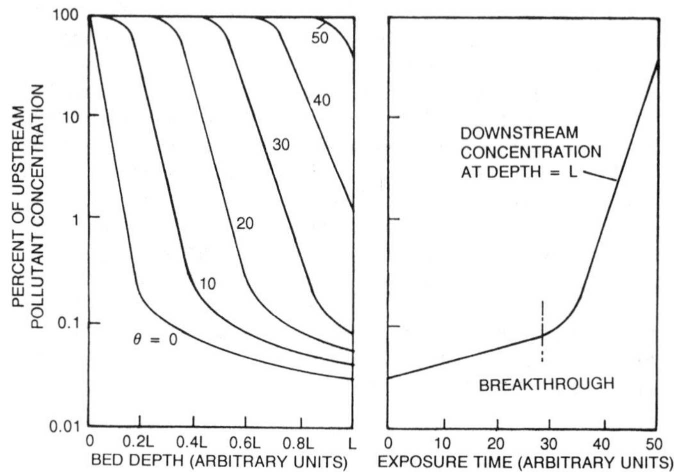

Fig. 4 Dependence of Adsorbate Concentrations on Bed Depth and Exposure Time

1. The molecule must be transported from the carrier gas stream across the boundary layer surrounding the adsorber granule. This process is random, with molecular movement both to and from the surface; the net flow of molecules cannot be toward the surface unless the concentration of the pollutant in the gas flow is greater than its concentration at the granule surface. For this reason, adsorption action decreases as pollutant load on the adsorber surface increases. Also, very low concentrations in the gas flow result in low adsorption rates.
2. The molecules of the pollutant must diffuse into the pores if they are to occupy that portion of the surface.
3. The pollutant molecules must be bound to the surface.

Any one of these steps may determine the rate at which adsorption occurs. In general, step (3) is very fast for physical adsorption, but reversible. This means that an adsorbed molecule can be desorbed at a later time, either when relatively cleaner air passes through the adsorber bed, or by the arrival of another pollutant that binds more tightly to the adsorber surface. Coupled with the above effects are thermal effects, because a physically adsorbed molecule gives up energy when it is bound to the adsorber surface. This energy raises the temperature of the adsorber and the carrier gas stream.

When a pollutant is fed to an adsorber at constant concentration and constant gas flow rate, the resulting gas stream concentration varies with time and bed depth as shown in Figure 4A. When bed loading begins ($\theta = 0$), the pollutant concentration decreases logarithmically with bed depth. Deeper into the bed, the pollutant concentration is lower, and the slope of the concentration-versus-bed depth curve decreases. At later times, the entrance portion of the adsorber bed becomes loaded with pollutant, and higher concentrations appear at each bed depth.

If the bed is only of depth L, the concentration-versus-time pattern downstream of the bed is as shown in Figure 4B. Usually the downstream concentration is very low until time θ_{b1}, then the concentration rises rapidly until the downstream concentration is the same as the upstream. This penetration is called *breakthrough*, because it tends to occur suddenly. Not all adsorber-pollutant combinations show as sharp a breakthrough as Figure 4B indicates; therefore, when given a breakthrough time, the entering pollutant concentration, downstream concentration at breakthrough, bed depth, and the velocity, temperature, and relative humidity of the airstream used in the test must also be specified.

In spite of the complexity of physical adsorption, comparatively simple expressions have been developed to describe its behavior. Expressions fitting available test data over a wide range of operating conditions for many pollutants are given in articles by Nelson and Correia (1976) and by Yoon and Nelson (1984a, 1984b, 1988). The expressions apply to the case of constant flow and constant pollutant inlet concentration. Nelson and Correia developed a semi-empirical expression:

$$t_{10} = K_a C^{-2/3} \quad (10)$$

where

t_{10} = breakthrough time for a downstream concentration equal to 10% of upstream concentration
K_{al} = constant for a given adsorber/pollutant combination; factors within this constant include adsorption media mass per unit airflow and pollutant molecular mass and boiling point
C = pollutant concentration entering bed, ppm

Yoon and Nelson's articles present a more rigorous expression, giving the breakthrough time at any desired downstream/upstream concentration ratio:

$$t = t_{50} + K_{a2} \ln[C_b/(C - C_b)] \quad (11)$$

where

t_{50} = breakthrough time for a downstream concentration equal to 50% of upstream concentration
C_b = downstream concentration, ppm
K_{a2} = a constant determined experimentally for the adsorber/pollutant combination by tests run at two inlet concentrations

Most studies were run at concentrations of interest for short-term exposures (~ 1000 ppm), which may be misleading for the low concentrations met in HVAC applications. Low concentration studies are reported by Ensor *et al.* (1988), Miller and Reist (1977), Ostajic (1985), Nelson and Correia (1976), Nunez *et al.* (1989), Stampfer (1982), Stankavich (1969), and Jonas and Rehrmann (1972). All showed that breakthrough time depends on inlet concentration. Yoon and Nelson (1984a) showed the equivalence between their expressions and those of Wheeler (1969) and Mecklenberg (1925) for low bed loadings. These theories, assisted by the Dubinin-Raduskevich isotherm theory (Dubinin 1975), relate dynamic adsorption breakthrough times to isotherm data obtained under static equilibrium conditions. (An isotherm is a plot of equilibrium vapor mass adsorption as a function of vapor pressure.) Isotherms are, however, normally plotted over the complete pressure range from 0 to 1 atmosphere. The range of interest for indoor pollutants is from about 10^{-9} to 10^{-2} atmospheres (1 ppb to 10,000 ppm), which is barely visible on such plots. The behavior at very low pressures cannot be reliably extrapolated to zero from higher pressures (Figure 5).

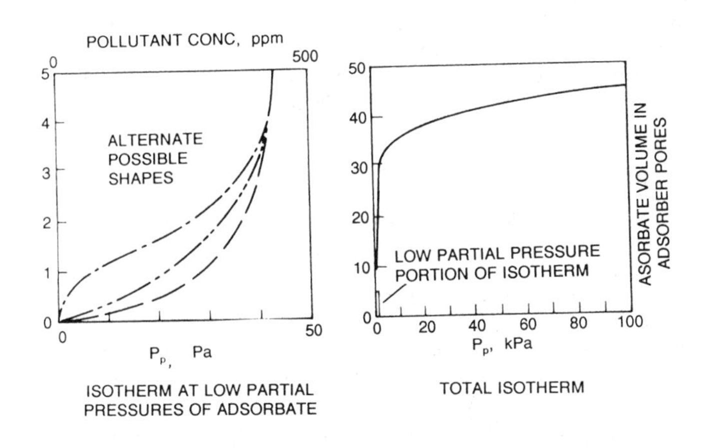

Fig. 5 Uncertainty in Isotherm Shapes at Low Partial Pressures of Adsorbate

Yoon and Nelson (1988) and Underhill *et al.* (1988) discuss the effect of relative humidity on physical adsorption. In essence, water vapor is a second pollutant, which alters the adsorption parameters, reducing the amount of the first pollutant that can be held by the adsorber and shortening the breakthrough times. The simultaneous adsorption of several pollutants shortens breakthrough times in similar fashion. Jonas *et al.* (1983) calculate mixture breakthrough times at low bed loadings by determining the volumes of the individual components the adsorber will hold separately, then summing the actual partial volumes, depending on the partial pressures of each, to obtain an effective volume held for the mixture.

Several sources list maximum possible retentivity (mass of adsorbate held per mass of adsorber) for various pollutants on typical activated charcoals. These values were normally obtained by tests with concentrations on the order of 1000 ppm, and therefore grossly overestimate the amount of pollutants trapped at low indoor concentrations. Calculations based on such loadings should be viewed skeptically in determining the economic advantages of using adsorbers versus using ventilation. The maximum breakthrough time for an adsorber, assuming no concentration effect and 100% collection of pollutant, is:

$$\theta_{max} = 35{,}300 f_a W_a / FC \qquad (12)$$

where

θ_{max} = maximum breakthrough time, min
W_a = mass of adsorber, g
f_a = ratio of maximum pollutant mass adsorbed to adsorber mass, g/g
F = flow through the adsorber, cfm
C = pollutant concentration, mg/m^3

ASTM *Standards* D 1605-60 and D 3686-89 list maximum adsorptivities for about 130 vapors in high-quality activated carbons. In D 3686-89, the column headed "Recommended Maximum Tube Loading, mg" represents the maximum percentage retentivity for the listed pollutants at TWA8 levels. The percentage retentivities in ASTM D 1605-60 should be corrected to TWA8 levels by Equation (13), derived from Equation (10):

$$R_{\mathrm{TWA8}} = R_{1000}(\mathrm{TWA8}/1000)^{1/3} \qquad (13)$$

where

R_{TWA8} = percent retentivity at pollutant concentration = TWA8 in ppm
R_{1000} = percent retentivity at pollutant concentration = 1000 ppm (test level assumed used in ASTM D 1605-60)

Table 10 lists some available physical adsorbers used for indoor air applications. Activated carbon is by far the most popular, because of the wide range of pollutants it can adsorb.

Table 10 Low-Temperature Absorbers, Chemisorbers, and Catalysts

Material	Impregnant	Typical Vapors or Gases Captured
Physical Adsorbers		
Activated carbon	none	organic vapors, ozone, acid gases
Activated alumina	none	polar organic compounds[a]
Activated bauxite	none	polar organic compounds
Silica gel	none	water, polar organic compounds
Molecular sieves (Zeolites)	none	carbon dioxide, iodine
Porous polymers	none	various organic vapors
Chemisorbers		
Activated alumina	$KMnO_4$	hydrogen sulfide, sulfur dioxide
Activated carbon	I_2, Ag, S	mercury vapor
Activated carbon[b]	I_2, KI_3, amines	radioactive iodine and organic iodides
Activated carbon	$NaHCO_3$	nitrogen dioxide
LiO_3, NaO_3, KO_3	none	carbon dioxide
LiO_2, NaO_2, KO_2, $Ca(O_2)_2$	none	carbon dioxide
Li_2O_2, Na_2O_2	none	carbon dioxide
LiOH	none	carbon dioxide
$NaOH + C^a(OH)_2$	none	acid gases
Activated carbon	KI, I_2	mercury vapor
Catalysts		
Activated carbon	none	ozone
Activated carbon[c]	Cu+Cr+Ag+NH_4	acid gases, chemical warfare agents
Activated alumina	$CuCl_2 + PdCl_2$	formaldehyde
Activated alumina	Pt,Rh oxides	carbon monoxide

[a] polar organics = alcohols, phenols, aliphatic and aromatic amines, etc.
[b] Mechanism may be isotopic exchange as well as chemisorption.
[c] "ASC Whetlerite"

Chemisorbers

Chemisorption is similar in many ways to physical adsorption. The three steps shown in Figure 3 must also take place in chemisorption. However, the third step in chemisorption is different from physical adsorption, for surface binding in chemisorption is by chemical reaction with electron exchange between the pollutant molecule and the chemisorber. This action differs in the following ways from physical adsorption:

1. It is highly specific; only certain pollutant compounds will react with a given chemisorber.
2. Chemisorption improves as temperature increases; physical adsorption improves as temperature decreases.
3. Chemisorption does not generate heat, but instead may require heat input.
4. Chemisorption is not generally reversible. Once the adsorbed pollutant has reacted, the original pollutant cannot be desorbed. However, one or more reaction products, different from the original pollutant, may be formed in the process. These products may themselves have undesirable effects.
5. Water vapor often helps chemisorption, or is necessary for it; it usually hurts physical adsorption.
6. Chemisorption *per se* is a monomolecular layer phenomenon, and the pore-filling effect that takes place in physical adsorption does not occur, except where adsorbed water condensed in the pores forms a reactive liquid.

Although the overall pattern of the concentration-time-bed depth relations in chemisorption is the same as described under physical adsorption, and the same equations may be used to describe chemisorption, the rate constant k depends on the kinetics of the chemical reaction between the pollutant and the chemisorber. Test data on individual chemisorber/pollutant combinations at operating conditions provide the only way to determine chemisorber performance in typical applications. However, the upper limit of capacity for a chemisorber/pollutant combination may be determined if the chemical reaction at the surface and the amount of reactant available are both known. A chemical reaction cannot involve more than stoichiometric amounts of the reactants, unless catalytic effects are present.

Equation (12) may be used for chemisorbers with the following substitutions:

W_a = mass of reactant available for reaction, g
F_a = number of grams of pollutant that combine stoichiometrically with each gram of reactant

Solid chemisorbers may be porous, chemically homogeneous materials, but this is usually wasteful and ineffective, because only the exposed, generally small, surface reacts with the pollutant. For this reason, most chemisorptive media are formed from a highly porous support (such as activated alumina or activated carbon) coated or impregnated with a chemical reactant. The mass W_a is the mass of the reactant, not the entire support. The support may have physical adsorption ability that comes into play when chemisorptive action ceases. Table 10 lists a few chemisorbers used in indoor air pollution control.

Catalysis

A catalytic air purifying medium is one that operates at ambient temperatures, or at least generates its own activation energy by stimulating an exothermic chemical reaction. Catalytic combustion, which enables a pollutant to burn (oxidize) at lower temperatures than is possible with unassisted combustion, is described later; it requires substantial heat input and elevation of the temperature of the whole gas stream. Only a few catalysts are known to be effective at ambient temperatures for gaseous pollutant removal; some of these are listed in Table 10.

Catalytic air purification is closely related to chemisorption, for chemical reactions occur at the surface of a catalyst. A significant difference exists, however, in that the gaseous pollutant does not react stoichiometrically with the catalyst itself. The catalyst assists whatever reaction takes place, but is not used up in the reaction. This feature offers the potential for much longer lives from catalysts than from adsorbers or chemisorbers, if the reaction that occurs converts the gaseous pollutant to something innocuous. This conversion may divide the gaseous pollutant into smaller molecules, or it may join the pollutant to a supplied chemical. The most satisfactory supplied reactant is oxygen, which is present in unlimited quantity in the carrier airstream.

An attractive feature of catalysts is that some are available to convert carbon monoxide to carbon dioxide, which is not controlled by physical adsorbers such as activated carbon (Collins 1986). Thus, sequential beds can be used to remove carbon monoxide by catalysis, carbon dioxide by absorption, and odors by adsorption from truly 100% recirculatory systems. This type of filtration is found in spacecraft, but the systems are complex and are not competitive for more conventional ventilation applications.

Combustion

Combustion is widely used to remove toxic gaseous pollutants from stack gases. In many cases, volatile organics can be converted to innocuous concentrations of steam and carbon dioxide. The entire carrier gas stream must, however, be raised to a temperature above which substantial oxidation of the pollutant occurs (from 1200 to 1500 °F for most organic pollutants). In most applications, the needed temperatures are produced by direct firing of natural gas. Even in stack gas combustors, regenerators are frequently installed to reduce the amount of heating gas required. Still more cooling and heat recovery would be needed to apply direct combustion to indoor environmental control; and the unavoidable combustion products would defeat the purpose of a gaseous pollution control system.

Catalytic Combustion

Gaseous pollutants can sometimes be oxidized at temperatures somewhere between ambient and the temperature of ignition of the pollutant by passing the gas stream through a heated bed of catalyst. Nuclear submarines incorporate a heat recovery unit and return the airstream temperature to ambient following a catalytic combustor (Piatt and White 1962). These systems oxidize volatile organic contaminants at the same time that they oxidize the carbon monoxide and hydrogen in the submarine atmosphere. In general, such complexity is unlikely to be economical in commercial HVAC systems. Millions of catalytic combusters are used in automobile exhaust pipes and in wood stoves to oxidize unburnt hydrocarbons that would otherwise reduce the quality of urban ventilation air.

Nwanko and Turk (1975) describe a system in which styrene, toluene, and methyl ethyl ketone are adsorbed at 77 °F on activated carbon impregnated with palladium. The carrier stream is then heated slowly to 590 to 790 °F, which almost completely oxidizes the organics without destroying the activated carbon. Such a cyclic system can possibly be adapted to HVAC applications.

Cryogenic Condensation

Physical adsorption is far more effective at low temperatures than at ambient temperatures. For this reason, some adsorbers in nuclear safety systems are operated at cryogenic temperatures for capture of noble gases. It is also possible to condense a gas or vapor out of an airstream by cooling the airstream below the boiling point of the gas. However, the gas partial pressures are so low at normal odor thresholds and TWA8 concentration levels that temperatures far below boiling points are necessary to obtain sufficient mass transfer to the cooling surface. The resulting air dessication, surface icing, and cost make this scheme impractical for HVAC systems.

Plants

Plants can accomplish some of the same pollution-control activity indoors that they perform outdoors. Wolverton *et al.* (1989) describe a series of experiments in which benzene, formaldehyde, and trichloroethylene were removed from a sealed chamber atmosphere by various decorative plants (see results in Table 11). The specific combination of plant and pollutant is important. These authors also describe a scheme in which air is drawn through the potting soil of several plants, and pollutants are adsorbed by activated carbon in the soil. They postulate that plant root bacteria are able to regenerate the activated carbon and maintain its effectiveness far longer than is normal.

Occupants

The model shown in Figure 1 includes the inhalation and exhalation of gas by a single occupant. A typical person engaged in office work has an average breathing flow of about 0.49 cfm. About 70% of this flow reaches the lung alveoli, where volatile organics as well as water-soluble inorganic pollutant gases are absorbed to a considerable degree. Wallace *et al.* (1983) show the relation between ambient concentrations and exhaled air concentrations for seven organic pollutants. The high removal efficiency (>70%) for these subjects shows distinct adsorption in the tracheobronchial system as well as in the alveoli (see Table 6).

Table 11 Adsorption of Pollutants from a Sealed Chamber by Houseplants

Plant	Formaldehyde				Benzene				Trichloroethylene			
	Total Leaf Area, m^2	Concentration, mg/m^3 $\theta=0$	Concentration, mg/m^3 $\theta=24$ h	Deposition Velocity, mm/s	Total Leaf Area, m^2	Concentration, mg/m^3 $\theta=0$	Concentration, mg/m^3 $\theta=24$ h	Deposition Velocity, mm/s	Total Leaf Area, m^2	Concentration, mg/m^3 $\theta=0$	Concentration, mg/m^3 $\theta=24$ h	Deposition Velocity, mm/s
None (control)	0.0	23	22	0.039	0.0	65	61	0.056	0.0	109	98	0.083
Mass cane (Dracenaena massangeana)	?	25	7.5	?		45	36		?	87	76	
Pot chrysanthemum (Chrysanthemum morifolium)	?	22	8.7	0.42		188	88	5.30	0.72	109	71	5.14
Gerbera daisy (Gerbera jamesonii)	?	20	10		?	210	68		0.46	109	71	3.00
Wareckei (Dracenaena deremensis W.)	?	10	5.0		0.72	87	42	8.72	0.72	109	98	1.28
Ficus (Ficus benjamina)	?	24	12		?	65	45		?	103	92	

Source: Wolverton *et al.* (1989). Deposition rates are calculated by use of the equation $Dc/dt = -(KA/V)C$, which gives K (deposition velocity, mm/s) = $-1000(V_\theta/A)\ln(C_\theta/C_0)$, where V is the chamber volume, A the area of the deposition surface, θ the time for the final concentration measurement, and C_0 and C_θ the concentrations at time 0 and time θ, respectively. The volume V of the chamber used in calculation was 0.44 m^3, and area A was 3.47 m^2. Only the leaf areas were used in the calculation of deposition velocities when plants were in the chamber.

The body is not a perfect trap for such organics, but behaves somewhat like a physical adsorber and desorbs some of the captured, but nonmetabolized, pollutants when cleaner air is subsequently breathed. The time span for this process can be several days. Inorganics (acid gases and so forth) are captured more efficiently than volatile organics by the breathing system and are more readily taken up by the body.

Odor Counteractants and Odor Masking

An odor counteractant is a vapor supplied to a space to reduce the sensitivity of the nose to a known odor present in the space. An odor mask is a vapor supplied in sufficient quantity to overshadow the existing odor. Both of these solutions apply only to specific odors and may be effective for a limited percentage of the occupants of a given space. They offer no improvement in toxic effects or material damage and are pollutants themselves.

CONTROL SYSTEM CONFIGURATIONS

Adsorbers, chemisorbers, and catalysts are supplied as powders and granular or pelletized media. They must be held in a structure that allows the treated air to pass through the media with an acceptable pressure drop at the operating airflow. Typical configurations of units where granular or pelletized media are held between perforated metal sheets or wire screens are shown in Figure 6. The perforated metal or screens must have holes smaller than the smallest diameter particle of the active media. A margin without perforations must be left around the edges of retaining sheets or screens to minimize the amount of air that can slip around the active media. Media must be tightly packed within the structure so that open passages do not develop through the beds. Aluminum, stainless steel, painted, plated, or conversion-coated steel, and plastics are all used for retainers. Adsorptive media are also retained in fibrous filter media, which can then be pleated into large filter structures as shown in Figure 6B. The active media must be bound to the fibers in such a way that media micropores are not damaged and adequate overall adsorptive capacity is maintained. These factors must be evaluated by tests on the complete media structure, not the base adsorptive media alone.

Figure 6A is used for such applications as collection of noble gases under cryogenic conditions in nuclear installations. Velocities are low and contact times long (typically >1 s). Similar configurations are used for portable, temporary filter systems during cleanup of chemical spills. Figures 6B, 6C, and 6D are typical fixed-bed units used in nuclear safety systems and highly toxic situations, such as the disposal of chemical warfare agents. Contact times here are in the order of 0.25 s, and the active media particles are finer than those used in commercial HVAC systems to maximize the effective mass transfer. In addition, great attention is paid to keeping the active media tightly packed and to maintaining highly reliable perimeter gaskets.

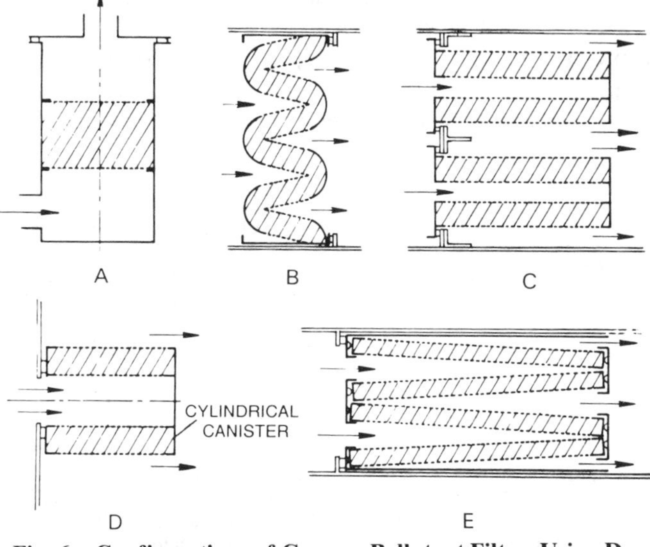

Fig. 6 Configurations of Gaseous Pollutant Filters Using Dry Granular Adsorbers, Chemisorbers, and Catalysts

Figure 6E is typical of commercial HVAC equipment. The retaining elements are flat panels or trays, which slide into a holding rack that has gaskets to seal all joints. This construction allows the trays to be removed easily for replacement or regeneration of the active media. Contact times are usually from 0.05 to 0.25 s; efficiency is higher for longer contact times.

Large systems can be supplied with banks of cells of the types shown in Figures 6B through 6E, in both face-access and side-access arrangements. Clamping and gasketing between cells prevents the bypass of untreated air.

ENVIRONMENTAL INFLUENCES ON CONTROL SYSTEMS

High relative humidity in the treated airstream lowers the gaseous-pollutant removal efficiency of physical adsorbers. Very low relative humidities may make some chemisorption activity impossible. Therefore, the performance of these media must be tested over the expected range of operation, and the relative humidity of an operating system must be held within limits acceptable for the sorption media. Temperatures beyond acceptable limits must also be prevented.

All adsorption and catalytic media have a modest ability to capture dust particles and lint, which eventually plug the openings between the media granules and cause a rapid rise in the pressure drop across the media or a decrease in system airflow. All granular gaseous adsorption beds need to be protected against particulate buildup by installing appropriate particulate filters upstream. Vibration breaks up the granules to some degree, depending on the granule hardness. ASTM *Standard* D 3802-79 is a test procedure for measuring the resistance of activated carbon to abrasion, and ASTM *Standard* D 4058-87 specifies the procedure for measuring the same property of somewhat harder catalysts and catalyst carriers. Critical systems using activated carbon require hardness numbers above 92% as described by ASTM D 3802-79.

Adsorption media, chemisorbers, and catalysts sometimes accelerate the corrosion of metals that they touch. For this reason, media holding cells should be made of organically coated or stainless steel components. Liquid-scrubbing absorption devices must be made of corrosion-resistant materials, especially when reagents are present in water solutions.

CONTAMINANT MEASUREMENT AND MONITORING

To determine whether the gaseous contaminant quality of air is acceptable, the concentration of various pollutants in the air must be measured. Such testing may show that air quality is acceptable, even when occupants complain. It may also show that gaseous contaminant control through hood exhaust, filtration, or added ventilation is needed to maintain safe or pleasant conditions.

Conditions acceptable from the standpoint of carbon dioxide content and odor are defined by ASHRAE *Standard* 62-1989. The alternative Indoor Air Quality Procedure specified in that standard, which applies where outdoor air is reduced to low levels by filtering recirculated air, sets limits for indoor concentrations for several contaminants (see the section on energy consumption). The standard also cautions that if unusual contaminants are present, they must be controlled to one-tenth the TLV levels specified by ACGIH (1989b). Therefore, the actual gaseous contaminant concentrations must be measured to be sure that ASHRAE *Standard* 62-1989 is met when outdoor air levels are reduced. The measurement of gaseous contaminant concentrations at levels acceptable for indoor air is not as straightforward as measurement of temperature or humidity. Relatively costly analytical equipment is needed, and this equipment must be calibrated and operated carefully by experienced personnel.

Measurement of gaseous contaminant concentrations has two aspects: sampling and analysis. Currently available sampling techniques are listed in Table 12. Some analytical or detection techniques are specific to a single pollutant compound; others are

Table 12 Gaseous Contaminant Sampling Techniques

Technique	Advantages	Disadvantages
1. Direct flow to detectors	•Real-time readout, continuous monitoring possible •Several pollutants possible with one sample (when coupled with chromatograph, spectroscope, or multiple detectors)	•Average concentration must be determined by integration •No preconcentration possible before detector; sensitivity may be inadequate •On-site equipment is complicated, expensive, intrusive
2. Capture by pumped flow through solid adsorbent; subsequent desorption for concentration measurement	•On-site sampling equipment relatively simple and inexpensive •Preconcentration and integration over time inherent in method •Several pollutants possible with one sample	•Sampling media and desorption techniques are compound specific •Interaction between captured compounds and between compounds and sampling media, bias may result •Gives only average values over sampling period, no peaks
3. Colorimetric detector tubes	•Very simple, relatively inexpensive equipment and materials •Immediate readout •Integration over time •Several pollutants possible with one sample	•Rather long sampling period normally required •One pollutant per sample •Relatively high detection limit •Poor precision
4. Collection in evacuated containers	•Very simple on-site equipment •No pump (silent) •Several pollutants possible with one sample	
5. Collection in nonrigid containers (plastic bags) held in an evacuated box	•Simple, inexpensive on-site equipment (pumps required) •Several pollutants possible with one sample	•Cannot hold some pollutants
6. Cryogenic condensation	•Wide variety of organic pollutants can be captured •Minimal problems with interferences and media interaction •Several pollutants possible with one sample	
7. Passive diffusional samplers	•Simple, unobtrusive, inexpensive •No pumps; mobile; may be worn by occupants to determine average exposure	•Gives only average over sampling period, no peaks
8. Liquid impingers (bubblers)	•Integration over time •Several pollutants possible with one sample	•May be noisy

All techniques except 1 and 3 require laboratory work after completion of field sampling. Only procedure 1 is adaptable to continuous monitoring and able to detect short-term excursions.

Sources: NIOSH (1977, 1984), Lodge (1988), Taylor *et al.* (1977), and ATC (1990).

capable of presenting a concentration spectrum for many compounds simultaneously (Table 13). Pollutant concentration measurement instruments should be able to detect contaminants of interest at about one-tenth TWA8 levels; and if odors are of concern, detection sensitivity must be at odor threshold levels. As indicated in Table 12, pollutants may be accumulated or concentrated over time so that very low average concentrations can be measured. Procedures for evaluating odor levels are given in Chapter 12 of the 1989 ASHRAE *Handbook—Fundamentals*.

When sampling and analytical procedures appropriate to the application have been selected, a pattern of sampling locations and times must be carefully planned. The building and air-handling

Table 13 Gaseous Contaminant Concentration Measurement Methods

Method	Description	Typical Application
1. Gas chromatography	• Separation of gas mixtures by time of passage down an adsorption column	
(using the following detectors) Flame ionization	• Change in flame electrical resistance due to ions of pollutant gas	• Volatile, nonpolar organics
Flame photometry	• Measures light produced when pollutant is ionized by a flame	• Sulfur, phosphorous compounds
Photoionization	• Measures ion current for ions created by ultraviolet light	• Most organics (except methane)
Electron capture	• Radioactively generated electrons attach to pollutant atoms; current measured	• Halogenated organics • Nitrogenated organics
Mass spectroscopy	• Pollutant atoms are charged, passed through electrostatic, magnetic fields in a vacuum; path curvature depends on mass of atom, allowing separation and measure of number of each type	• Volatile organics
2. Infrared spectroscopy and Fourier-Transform Infrared Spectroscopy (FTIR)	• Adsorption of infrared light by pollutant gas in a transmission cell; a range of wavelengths used, allowing identification and measurement of individual pollutants	• Acid gases • Many organics
3. High-Performance Liquid Chromatography (HPLC)	• Pollutant is captured in a liquid, which is then passed through a liquid chromatograph, analogous to a gas chromatograph	• Aldehydes, ketones • Phosgene • Nitrosamines • Cresol, phenol
4. Colorimetry	• Chemical reaction with pollutant in solution yields a colored product whose light adsorption is measured	• Ozone • Nitrogen oxides • Formaldehyde
5. Fluorescence and Pulsed fluorescence	• Pollutant atoms are stimulated by a monochromatic light beam, often ultraviolet; they emit characteristic fluorescence wavelengths, whose intensity is measured	• Sulfur dioxide • Carbon monoxide
6. Chemiluminescence	• Reaction (usually with a specific injected gas) results in photon emission proportional to concentration	• Ozone • Nitrogen compounds • Several organics
7. Electrochemical	• Pollutant is bubbled through reagent/water solution changing its conductivity or generating a voltage	• Ozone • Hydrogen sulfide • Acid gases
8. Titration	• Pollutant is absorbed into water	• Acid gases
9. Ultraviolet adsorption	• Adsorption of UV light by a cell through which the polluted air passes	• Ozone • Aromatics • Sulfur dioxide • Oxides of nitrogen • Carbon monoxide
10. Atomic adsorption	• Contaminant is burned in a hydrogen flame. A light beam with a spectral line specific to the pollutant is passed through the flame; optical adsorption of the beam is measured	• Mercury vapor

Sources: NIOSH (1977, 1984), Lodge (1988), Taylor *et al.* (1977), and ATC (1990).

system layout and the space occupancy and use patterns must be considered so that representative concentrations will be measured. Traynor (1987) and Nagda and Rector (1983) offer guidance in planning such surveys.

DEVICE TESTING

No standard procedure has been developed to test gaseous contaminant filters for general ventilation applications. Many articles have, however, described procedures where a steady concentration of a single contaminant is fed to a filter and the downstream concentration is determined as a function of the total contaminant captured by the filter (Nelson and Correia 1976; Mahajan 1987, 1989; and ASTM 1989). Rivers (1988) summarizes the desirable elements of a standardized test and some problems involved in formulating one. One problem with gaseous contaminant filter testing is that tests need to be run at the low concentrations comparable to those met in occupied spaces (TWA8 or odor threshold).

Testing at high concentrations does not automatically evaluate performance at low concentrations; the corrections described earlier under physical adsorption must be applied to the test data. If attempts are made to speed the test process by interrupting high-concentration loading with short, low-concentration feeds, pollutant desorption from the filter may confuse the results (Ostajic 1985). Also, an adsorber cannot be tested for every pollutant, nor is there general agreement on what contaminants should be considered typical. Nevertheless, tests run according to the previously mentioned references do give useful measures of filter performance on single contaminants, and they do give a basis for estimates of filter penetrations and filter lives. Filter penetration data thus obtained can be used in Equation (9) to estimate steady-state indoor concentrations.

Since physical adsorbers, chemisorbers, and catalysts are affected by the temperature and relative humidity of the carrier gas and the moisture content of the filter bed, they should be tested over the range of conditions expected in the application. In addition, the extraction of captured contaminants and the generation of reaction products needs to be evaluated by sweeping the test filter with unpolluted air and measuring downstream concentrations. Reaction products may be as toxic, odorous, or corrosive as captured pollutants.

Gaseous contaminant control devices have been tested in sealed chambers by recirculating contaminated air through them and measuring the decay of an initial contaminant concentration in time. This method introduces factors extraneous to the device itself, such as possible errors introduced by adsorption on the chamber surfaces, leaks in the facility, and the drawing of test samples. Daisey and Hodgson (1988) compared the pollutant decay rate with and without the control device operating to overcome these uncertainties.

In critical applications, such as chemical warfare protective devices and nuclear safety applications, a sorption media is evaluated in a small canister, using the same media carrier-gas velocity as in the full-scale unit. The full-scale unit is then checked for leakage through gaskets, structural member joints, and thin spots or gross open passages in the sorption media by feeding a readily adsorbed contaminant to the filter and probing for its presence downstream (see ANSI *Standard* N 510).

The test used to evaluate fundamental properties of sorption media is a static test, the measurement of the adsorption isotherm. In this test, a small sample of the adsorber is exposed to the pollutant vapor at sucessively increasing pressures and the mass of pollutant adsorbed at each pressure is measured. The low-pressure section of an isotherm can be used to predict kinetic behavior, although the calculation is not simple. One test widely used in specifying activated carbons provides a single point on the isotherm for a single pollutant, carbon tetrachloride (ASTM *Standard* D 3467-88). It is a qualitative measure of performance at other conditions, and a useful quality control procedure. Another qualitative measure of performance is the surface area determined by measuring the mass of an adsorbed monolayer of nitrogen, the BET method (ASTM *Standard* D 4567-86). The results of this test are reported in square metres per gram of sorbent or catalyst. This number is often used as an index of adsorber quality, with high numbers indicating high quality.

ENERGY CONSUMPTION

The choice between using outside air only and outside air plus filtration may be made on the basis of technical or maintenance factors, convenience, economics, or a combination of these. An energy consumption calculation must be made before an economic analysis is possible. Presumably, a building load calculation will be available, which will include the various flows in Figure 1 and the total heating and cooling energy requirements for various seasons of the year.

Enough ventilation air must be supplied to every building space to satisfy the metabolic requirements of the occupants, including the dilution of carbon dioxide to nontoxic levels. This is only 0.11 cfm per person. ASHRAE *Standard* 62-1989, however, requires a minimum of 15 cfm of ventilation air per person; higher flows are required for certain spaces, such as where smoking is expected. The minimum quality of outdoor air to be used is specified in Table 8. If the amount of outdoor air used is less than that specified in *Standard* 62, a different set of criteria must be met. The air in the occupied space must be no worse than the quality given in Table 8, and it must also have concentrations less than those stated for four additional pollutants, as given in Table 14. Systems may use outdoor air or filtered recirculated air in any ratio, provided the air quality level is maintained. The logic of these requirements is discussed by Janssen (1989).

Table 14 Additional Indoor Air Quality Criteria Specified by ASHRAE *Standard* 62-89 when Ventilation Air is Recirculated

Pollutant	Maximum Allowed Concentration	Sampling Period
Carbon dioxide	1.8 g/m^3	Continuous
Chlordane	5 μg/m^3	Continuous
Ozone	100 μg/m^3	Continuous
Radon	0.027 WL	Annual average

WL = Working Level, a unit of measure of human exposure to radon gas and radon progeny.

Where building habitability can be maintained with ventilation alone, economizer cycles are recommended in appropriate outdoor conditions. When a system is in the economizer mode, some energy is required to move air through the system. For fixed-flow systems, this energy requirement is almost constant. For VAV systems, power consumption varies, depending on the flow rate and on the design of the control, fan, and motor system.

The resistance of the filtration device as a function of airflow must be provided by the manufacturer, as must the resistance of the heating/cooling coils. In addition, the resistance of any air filters required to protect the gaseous contaminant filter must be included in the energy analysis.

VAV systems involve considerably more calculation than fixed-flow systems; fortunately, computer programs can help in the calculations. Treating a VAV system as operating at two or three conditions representative of heating, cooling, and economizer modes should, however, show which systems are economically sound.

ECONOMIC CONSIDERATIONS

In general, most meaningful comparisons are made with a discounted cash flow analysis. In this case, the capital and operating costs of the two competing systems are estimated separately. If the system with lower capital cost also costs less to operate, it is the obvious choice. But if the more expensive system costs less to operate, some time will be required to recoup the added investment. The difference in cash flows generated by each system must be estimated for each year of operation. Usually the cash flow is not constant because inflation increases the costs for energy, material, and labor. The analysis is run for 20 years, for example, which is a conservative estimate of the life of the equipment. Then a constant compound interest rate is calculated for this period to discount each year's cash flow difference from the last year back to the first. The sum of the discounted cash flow differences equals the investment difference. If this rate of return on the added investment is sufficient, the building owner should buy the more expensive system.

All capital and operating costs for each competing system need to be identified. Table 15 is a checklist of these items; the actual values to be used change by location and by year. Also, the life of any filtration media used is an important factor in this economic analysis, as is the cost of utilities for heating and cooling. Often the cost of replacing the filtration media can be reduced by sending spent media to a reactivation service. However, some fresh media must always be added to maintain performance, and reactivation is not permissible for highly toxic pollutants.

Table 15 Checklist of Items Included in Economic Comparisons Between Competing Gaseous Contaminant Control Systems

Capital	Operating
Ventilation Costs	
Fans	Utilities
Motors	Water treatment
Coils	Maintenance labor
Plenum	
Chiller	
Boiler	
Piping	
Pumps	
Controls	
Installation	
Floor space	
Filtration Costs	
Added filtration equipment	Replacement or reactivation of gaseous pollutant filter media
Fan?	Disposal of spent gaseous pollutant filter media
Motor?	Added electric power
Controls?	Maintenance labor
Plenum?	
Space?	
Spare media holding units	
Floor space	

(?) Indicates item that may or may not be present, depending on design.

REFERENCES

ACGIH. 1989a. *Industrial ventilation—A manual of recommended practice*, 20th ed. American Conference of Governmental Industrial Hygienists, Cincinnati, OH.

ACGIH. 1989b. *Threshold limit values and biological exposure indices for* 1989-90. American Conference of Governmental Industrial Hygienists, Cincinnati, OH.

AIHA. 1989. *Odor thresholds for chemicals with established occupational health standards.* American Industrial Hygiene Association, Akron, OH.

American Guild of Organists. 1966. Air pollution and organ leathers: Panel discussion. AGO *Quarterly* 11(2):62-73.

ANSI. 1989. Testing of nuclear air treatment systems. ANSI *Standard* N 510-89. American Society of Mechanical Engineers, New York.

Anthony, C.P. and G.A. Thibodeau. 1980. *Textbook of anatomy and physiology.* C.V. Mosby Co., St. Louis, MO.

ASHRAE. 1989. Ventilation for acceptable indoor air quality. *Standard* 62-1989.

ASTM. 1979. Recommended practices for sampling atmospheres for analysis of gases and vapors. *Standard* D 1605-60. ASTM, Philadelphia.

ASTM. 1980. Building air change rate and infiltration measurements. STP 719. ASTM, Philadelphia.

ASTM. 1983. Standard test method for determining air leakage rate by tracer dilution. *Standard* E 741-83. ASTM, Philadelphia.

ASTM 1986. Test method for ball-pan hardness of activated carbon. *Standard* D 3802-79. ASTM, Philadelphia.

ASTM. 1986. Test method for single point determination of specific surface area of catalysts using nitrogen adsorption by continuous flow method. *Standard* D 4567-86. ASTM, Philadelphia.

ASTM. 1987. Standard test method for attrition and abrasion of catalysts and catalyst supports. *Standard* D 4058-87. ASTM, Philadelphia.

ASTM. 1988. Standard test method for carbon tetrachloride activity of activated carbon. *Standard* D 3467-88. ASTM, Philadelphia.

ASTM. 1989. Standard practice for sampling atmospheres to collect organic compound vapors (activated charcoal tube adsorption method). *Standard* D 3686-89. ASTM, Philadelphia.

ASTM. 1989. Standard test method for nuclear-grade activated carbon. *Standard* D 3803-89. ASTM, Philadelphia.

ATC. 1990. Technical assistance document for sampling and analysis of toxic organic compounds in ambient air. EPA/600/8-90-005. Environmental Protection Agency, Research Triangle Park, NC.

Balieu, E., T.R. Christiansen, and L. Spindler. 1977. Efficiency of b-filters against hydrogen cyanide. Staub-Reinhalt. Luft 37/10:387-90.

Berglund, B., U. Berglund, and T. Lindvall. 1986. Assessment of discomfort and irritation from the indoor air. In IAQ '86: *Managing Indoor Air for Health and Energy Conservation*, 138-49. ASHRAE, Atlanta.

Black, M.S. and C.W. Bayer. 1986. Formaldehyde and other VOC exposures from consumer products. In IAQ '86: *Managing Indoor Air for Health and Energy Conservation.* ASHRAE, Atlanta.

Braman, R.S. and T.J. Shelley. 1980. Gaseous and particulate ammonia and nitric acid concentrations: Columbus, Ohio area—summer 1980. PB 81-125007. NTIS, Springfield, VA.

Braun, R.C. and M.J.G. Wilson. 1970. The removal of atmospheric sulphur by building stones. *Atmospheric Environment* 4:371-78.

Brugnone, F., L. Perbellini, F.B. Faccini, G. Pasini, G. Maranelli, L. Romeo, M. Gobbi, and A. Zedde. 1989. Breath and blood levels of benzene, toluene, cumene and styrene in non-occupational exposure. International Archives of Environmental Health 61:303-11.

Cain, W.S., L.C. See, and T. Tosun. 1986. Irritation and odor from formaldehyde chamber studies, 1986. In IAQ '86: *Managing Indoor Air for Health and Energy Conservation.* ASHRAE, Atlanta.

Casserly, D.M. and K.K. O'Hara. 1987. Ambient exposures to benzene and toluene in southwest Louisiana. *Paper* 87-98.1. Air and Waste Management Association, Pittsburgh, PA.

Chan, C.C., L. Vanier, J.W. Martin, and D.T. Williams. 1990. Determination of organic contaminants in residential indoor air using an adsorption-thermal desorption technique. *Journal of Air and Waste Management Association* 40(1):62-67.

Chiarenzelli, R.V. and E.L. Joba. 1966. The effects of air pollution on electrical contact materials: a field study. *Journal of the Air Pollution Control Association* 16(3):123-27.

Code of Federal Regulations. 1990. 29 CFR 1900. U.S. Government Printing Office, Washington, D.C.

Code of Federal Regulations. 1990. National primary and secondary ambient air quality standards. 40 CFR 50. U.S. Government Printing Office, Washington, D.C.

Code of Federal Regulations. 1990. 10 CFR 50. U.S. Government Printing Office, Washington, D.C.

Cohen, M.A., P.B. Ryan, Y. Yanagisawa, J.D. Spengler, H. Ozkaynak, and P.S. Epstein. 1989. Indoor/outdoor measurements of volatile organic compounds in the Kanawha Valley of West Virginia. *Journal of the Air Pollution Control Association* 39(8):1086-93.

Cohen, S.I., N.M. Perkins, H.K. Ury, and J.R. Goldsmith. 1971. Carbon monoxide uptake in cigarette smoking. *Archives of Environmental Health* 23(6):427-33.

Cole, J.T. 1983. Constituent source emission rate characterization of three gas-fired domestic ranges. APCA *Paper* 83-64.3. Air & Waste Management Association, Pittsburgh.

Collins, M.F. 1986. Room temperature catalyst for improved indoor air quality. In *Indoor Air Quality in Cold Climates*, D.S. Walkinshaw, ed., 448-60. Air & Waste Management Association, Pittsburgh.

Conkle, J.P., B.J. Camp, and B.E. Welch. 1975. Trace composition of human respiratory gas. *Archives of Environmental Health* 30(6):290-95.

Coy, C.A. 1987. Regulation and control of air contaminants during hazardous waste site remediation. APCA *Paper* 87-18.1. Air and Waste Management Association, Pittsburgh.

Daisey, J.M. and A.P. Hodgson. 1988. Efficiencies of portable air cleaners for removal of nitrogen dioxide and volatile organic compounds. *Report* LBL-24964, Lawrence Berkeley Laboratory, Berkeley, CA.

DeFrees, J.A. and R.F. Amberger. 1987. Natural infiltration analysis of a residential high-rise building. In IAQ 87: *Practical Control of Indoor Air Problems*, 195-210. ASHRAE, Atlanta.

Dole, M. and I.M. Klotz. 1946. Sorption of chloropicrin and phosgene on charcoal from a flowing gas stream. *Industrial Engineering Chemistry* 38(12):1289-97.

Dubinin, M.M. 1975. Physical adsorption of gases and vapors in micropores. *Progress in Surface and Membrane Science* 9:1-70.

Egle, J.L. 1971. Single-breath retention of acetaldehyde in man. *Archives of Environmental Health* 23(6):427-33.

Egle, J.L. 1972. Retention of inhaled formaldehyde, proprionaldehyde, and acrolein in the dog. *Archives of Environmental Health* 23(6):427-33.

Egle, J.L. 1979. Respiratory retention and acute toxicity of furan. *Archives of Environmental Health* 40(4):310-14.

Elliott, L.P. and D.R. Rowe. 1975. Air quality during public gatherings. *Journal of the Air Pollution Control Association* 25(6):635-36.

Ensor, D.S., A.S. Viner, J.T. Hanley, P.A. Lawless, K. Ramanathan, M.K. Owen, T. Yamamoto, and L.E. Sparks. 1988. Air cleaner technologies for indoor air pollution. In IAQ '88: *Engineering Solutions to Indoor Air Problems*, 111-29. ASHRAE, Atlanta.

EPA. 1979. Toxic substances control act chemical substance inventory, Volumes I-IV. Environmental Protection Agency, Office of Toxic Substances, Washington, D.C.

EPA. 1986. Radon reduction techniques for detatched houses: Technical guidance. EPA/625/5-86/O19. US Environmental Protection Agency, Center for Environmental Research Information, Cincinnati.

EPA. 1987. Radon reduction methods: A homeowner's guide, 2nd ed. OPA-87-010. US Environmental Protection Agency, Center for Environmental Research Information, Cincinnati.

EPA. 1990. National Air Quality and Emission Trends Report, 1988. U.S. Environmental Protection Agency, Research Triangle Park, NC.

Fazzalari, F.A., ed. 1978. Compilation of odor and taste threshold values data. DS 48A. ASTM, Philadelphia.

Foresti, R., Jr. and G. Dennison. 1986. Formaldehyde originating from foam insulation. In IAQ '86: *Managing Indoor Air for Health and Energy Conservation*, 523-37. ASHRAE, Atlanta.

Freedman, R.W., B.I. Ferber, and A.M. Hartstein. 1973. Service lives of respirator cartridges versus several classes of organic vapors. *American Industrial Hygiene Association Journal* (2):55-60.

Fung, K. and B. Wright. 1990. Measurement of formaldehyde and acetaldehyde using 2,4-dinitrophenylhydrazine-impregnated cartridges. *Aerosol Science and Technology* 12(1):44-48.

Girman, J.R., A.P. Hodgson, A.F. Newton, and A.P. Winks. 1984. Emissions of volatile organic compounds from adhesives for indoor application. *Report* LBL-17594, Lawrence Berkeley Laboratory, Berkeley, CA.

Gorban, C.M., I.I. Kondrat'yeva, and L.Z. Poddubnaya. 1964. Gaseous activity products excreted by man in an airtight chamber. In *Problems of Space Biology.* JPRS/NASA. NTIS, Springfield, VA.

Graminski, E.L., E.J. Parks, and E.E. Toth. 1978. The effects of temperature and moisture on the accelerated aging of paper. NBSIR 78-1443. National Institute of Standards and Technology, Gaithersburg, MD.

Grosjean, D., P.M. Whitmore, C. Pamela de Moor, G.R. Cass, and J.R. Druzik. 1987. Fading of alizarinad-related artist's pigments by atmospheric ozone. *Environmental Science and Technology* 21:635-43.

Gully, A.J., R.M. Bethea, R.R. Graham, and M.C. Meador. 1969. Removal of acid gases and oxides of nitrogen from spacecraft cabin atmospheres. NASA-CR-1388. National Aeronautics and Space Administration, NTIS, Springfield, VA.

Hakov, R., J. Jenks, and C. Ruggeri. 1987. Volatile organic compounds in the air near a regional sewage treatment plant in New Jersey. *Paper* 87-95.1. Air & Waste Management Association, Pittsburgh.

Hartwell, T.D., J.H. Crowder, L.S. Sheldon, and E.D. Pellizzari. 1985. Levels of volatile organics in indoor air. *Paper* 85-30B.3. Air & Waste Management Association, Pittsburgh.

Haynie, F.H. 1978. Theoretical air pollution and climate effects on materials confirmed by zinc corrosion data. In *Durability of Building Materials and Components* (ASTM STP 691). ASTM, Philadelphia.

Hollowell, C.D., R.A. Young, J.V. Berk, and S.R. Brown. 1982. Energy-conserving retrofits and indoor air quality in residential housing. ASHRAE *Transactions* 88(1).

Hunt, R.D. and D.T. Williams. 1977. Spectrometric measurement of ammonia in normal human breath. *American Laboratory* (June): 10-23.

Hursh, J.B., T.W. Clarkson, M.G. Cherian, J.J. Vostal, and R. Vandermallie. 1976. Clearance of mercury (Hg-197,Hg-203) vapor inhaled by human subjects. *Archives of Environmental Health* 31(6):302-309.

Jaffe, L.S. 1967. The effects of photochemical oxidants on materials. *Journal of the Air Pollution Control Association* 17(6):375-78.

Janssen, J.H. 1989. Ventilation for acceptable indoor air quality. ASHRAE *Journal* 31(10):40-54.

Jonas, L.A. and J.A. Rehrmann. 1972. The kinetics of adsorption of organo-phosphorous vapors from air mixtures by activated carbons. *Carbon* 10:657-63.

Jonas, L.A., E.B. Sansone, and T.S. Farris. 1983. Prediction of activated carbon performance for binary vapor mixtures. *American Industrial Hygiene Association Journal* 44:716-19.

Knoeppel, H. and H. Schauenburg. 1989. Screening of household products for the emission of volatile organic compounds. *Environment International* 15:413-18.

Lamm, S.H. 1986. Irritancy levels and formaldehyde exposures in U.S. mobile homes. In *Indoor Air Quality in Cold Climates*, 137-47. Air and Waste Management Association, Pittsburgh.

Leaderer, B.P., R.T. Zagranski, M. Berwick, and J.A.J. Stolwijk. 1987. Predicting NO_2 levels in residences based upon sources and source uses: A multi-variate model. *Atmospheric Environment* 21(2):361-68.

Lee, H.K., T.A. McKenna, L.N. Renton, and J. Kirkbride. 1986. Impact of a new smoking policy on office air quality. In *Indoor Air Quality in Cold Climates*, 307-22. Air & Waste Management Association, Pittsburgh.

Lodge, J.E., ed. 1988. *Methods of air sampling and analysis*, 3rd ed. Lewis Publishers, Chelsea, MI.

Loefroth, G., R.M. Burton, L. Forehand, S.K. Hammond, R.L. Seila, R.B. Zweidlinger, and J. Lewtas. 1989. Characterization of environmental tobacco smoke. *Environmental Science and Technology* 23(5):610-14.

Lonnemann, W.A., S.L. Kopczynski, P.E. Darley, and F.D. Sutterfield. 1974. Hydrocarbon composition of urban air pollution. *Environmental Science and Technology* 8(3):229-35.

Mackinson, F.W., ed. 1978. *NIOSH/OSHA pocket guide to chemical hazards*. U.S. Department of Labor, OSHA, Washington, D.C. Users should read Cohen (1983) before using this reference.

Mahajan, B.M. 1987. A method of measuring the effectiveness of gaseous contaminant removal devices. NBSIR 87-3666. National Institute of Standards and Technology, Gaithersburg, MD.

Mahajan, B.M. 1989. A method of measuring the effectiveness of gaseous contaminant removal filters. NBSIR 89-4119. National Institute of Standards and Technology, Gaithersburg, MD.

Mathey, R.G., T.K. Faison, and S. Silberstein. 1983. Air quality criteria for storage of paper-based archival records. NBSIR 83-2795. National Institute of Standards and Technology, Gaithersburg, MD.

Matthews, T.G., T.J. Reed, B.J. Tromberg, C.R. Daffron, and A.R. Hawthorne. 1983. Formaldehyde emissions from combustion sources and solid formaldehyde resin-containing products. In *Proceedings of Engineering Foundation Conference on Management of Atmospheres in Tightly Enclosed Spaces*. ASHRAE, Atlanta.

Matthews, T.G., W.G. Dreibelbis, C.V. Thompson, and A.R. Hawthorne. 1985. Preliminary evaluation of formaldehyde mitigation studies in unoccupied research homes. In *Indoor Air Quality in Cold Climates*, 137-47. Air & Waste Management Association, Pittsburgh.

McGrath, T.R. and D.B. Stele. 1987. Characterization of phenolic odors in a residential neighborhood. *Paper* 87-75A.5. Air & Waste Management Association, Pittsburgh.

Mecklenberg, W. 1925. *Z. Elektrochem.* 31:488.

Meckler, M. and J.E. Janssen. 1988. Use of air cleaners to reduce outdoor air requirements. In IAQ '88: *Engineering Solutions to Indoor Air Problems*, 130-47. ASHRAE, Atlanta.

Miller, G.C. and P.C. Reist. 1977. Respirator cartridge service lives for exposure to vinyl chloride. *American Industrial Hygiene Association Journal* 38:498-502.

Molhave, L., I. Anderson, G.R. Lundquist, and O. Nielson. 1982. Gas emission from building materials. *Report* 137, Danish Building Research Institute, Copenhagen.

Moore, J.E. and E. Houtala. 1983. Odor as an aid to chemical safety: odor thresholds compared with threshold limit values and volatility for 214 industrial chemicals in air and water dilution. *Journal of Applied Toxicology* 3:272.

Moschandreas, D.J. and S.M. Relwani. 1989. Field measurement of NO_2 gas-top burner emission rates. *Environment International* 15:489-92.

Nagda, N.L. and H.E. Rector. 1983. Guidelines for monitoring indoor-air quality. EPA 600/4-83-046. Environmental Protection Agency, Research Triangle Park, NC.

Nazaroff, W.W. and G.R. Cass. 1986. Mathematical modeling of chemically reactive pollutants in indoor air. *Environmental Science and Technology* 20:924-34.

Nefedov, I.G., V.P. Savina, and N.L. Sokolov. 1972. Expired air as a source of spacecraft carbon monoxide. *Paper*, 23rd International Aeronautical Congress, International Astronautical Federation, Paris.

Nelms, L.H., M.A. Mason, and B.A. Tichenor. 1986. The effects of ventilation rates and product loading on organic emission rates from particleboard. In IAQ '86: *Managing Indoor Air for Health and Energy Conservation*, 469-85. ASHRAE, Atlanta.

Nelson, G.O. and A.N. Correia. 1976. Respirator cartridge efficiency studies: VIII. Summary and conclusions. *American Industrial Hygiene Association Journal* 37:514-25.

Nelson, E.D.P., D. Shikiya, and C.S. Liu. 1987. Multiple air toxics exposure and risk assessment in the south coast air basin. *Paper* 87-97.4. Air & Waste Management Association, Pittsburgh, PA.

Newsome, J.R., V. Norman, and C.H. Keith. 1965. Vapor phase analysis of tobacco smoke. *Tobacco Science* 9:102-10.

NIOSH. 1977. NIOSH *Manual of sampling data sheets*. U.S. Department of Health and Human Services, National Institute for Occupational Safety and Health, Washington, D.C.

NIOSH. Annual registry of toxic effects of chemical substances. U.S. Department of Health and Human Services, National Institute for Occupational Safety and Health, Washington, D.C.

NIOSH (intermittent) criteria for recommended standard for occupational exposure to (compound). U.S. Department of Health and Human Services, National Institute for Occupational Safety and Health, Washington, D.C.

Novosel, D., S.M. Relwani, and D.J. Moschandreas. 1987. Development of a dessicant-based environmental control unit. In IAQ '87: *Practical Control of Indoor Air Problems*, 195-210. ASHRAE, Atlanta.

NTIS. 1982. Air pollution effects on materials. *Published Search* PB82-809427. National Technical Information Service, Springfield, VA.

NTIS. 1984. Air pollution effects on materials. *Published Search* PB84-800267. National Technical Information Service, Springfield, VA.

Nunez, C., M. Kosusko, and B. Daniel. 1989. Effect of humidity on carbon adsorption performance in removing organics from contaminated air streams. *Paper* 89-29.2. Air & Waste Management Association, Pittsburgh.

Nwanko, J.N. and A. Turk. 1975. Catalytic oxidation of vapors adsorbed on activated carbon. *Environmental Science & Technology* 9(9):846-49.

Ostajic, N. 1985. Test method for gaseous contaminant removal devices. ASHRAE *Transactions* 91(2):11-31.

Pedersen, B.S. and W.J. Fisk. 1984. Air washing for the control of formaldehyde in indoor air. In *Proceedings* of the 3rd International Conference on Indoor Air and Climate. Swedish Council for Building Research, Stockholm.

Piatt, V.R. and J.C. White. 1962. The present status of chemical research in atmosphere purification and control on nuclear-powered submarines. NRL *Report* 5814, U.S. Naval Research Laboratory, Washington, D.C.

Rask, D. 1988. Indoor air quality and the bottom line. *Heating, Piping and Air Conditioning* 60(10).

Revoir, W.H. and J.A. Jones. 1972. Superior adsorbents for removal of mercury vapor from air. AIHA *Conference Paper.* American Industrial Hygiene Association, Akron, OH.

Rivers, R.D. 1988. Practical test method for gaseous gontaminant removal devices. In *Proceedings of the Symposium on Gaseous and Vaporous Removal Equipment Test Methods* (NBSIR 88-3716). NIST, Gaithersburg, MD.

Sandalls, F.J. and S.A. Penkett. 1977. Measurements of carbonyl sulfidae and carbon disulphide in the atmosphere. *Atmospheric Environment* 11:197-99.

Sax, N.I. and R.J. Lewis, Sr. 1988. *Dangerous properties of industrial materials*, 6th ed. 3 vols. Van Nostrand Reinhold, New York.

Shah, J.J. and H.B. Singh. 1988. Distribution of volatile organic chemicals in outdoor and indoor air. *Environmental Science and Technology* 22(12):1381-88.

Singh, H.B., L.J. Salas, A. Smith, R. Stiles, and H. Shigeishi. 1981. Atmospheric measurements of selected hazardous organic chemicals. PB 81-200628, NTIS, Gaithersburg, MD.

Sparks, L.E. 1988. Indoor air quality model, version 1.0. EPA-600/8-88-097a, U.S. Environmental Protection Agency, Research Triangle Park, NC.

Stampfer, J.F. 1982. Respirator canister evaluation for nine selected organic vapors. *American Industrial Hygiene Association Journal* 43(5):319-28.

Stankavich, A.J. 1969. The capacity of activated charcoal under dynamic conditions for selected atmospheric contaminants in the low parts-per-million range. In ASHRAE Symposium *Odors and Odorants: The Engineering View*. ASHRAE, Atlanta.

Sterling, T.D. and D. Kobayashi. 1981. Use of gas ranges for cooking and heating in urban dwellings. *Journal of the Air Pollution Control Association* 31(2):162-65.

Stewart, R.D., C.L. Hake, and J.E. Peterson. 1974. Use of breath analysis to monitor trichloroethylene exposures. *Archives of Environmental Health* 29(1):6-13.

Surgeon General. 1979. *Smoking and health*. U.S. Department of Health and Human Services—NIOSH, Washington, D.C.

Taylor, D.G., R.E. Kupel, and J.M. Bryant. 1977. *Documentation of the NIOSH validation tests*. U.S. Department of Health and Human Services-NIOSH, Washington, D.C.

Thomas, J.W. 1974. Evaluation of activated carbon canisters for radon protection in uranium mines. HASL-280, NTIS, Springfield, VA.

Thomson, G. 1986. *The museum environment*, 2nd ed. Butterworths, London.

Tichenor, B.A. 1989. Measurement of organic compound emissions using small test chambers. *Environment International* 15:389-96.

Traynor, G.W. 1987. Field monitoring design considerations for assessing indoor exposures to combustion pollutants. *Atmospheric Environment* 21(2):377-83.

Traynor, G.W., J.R. Girman, M.G. Apte, J.F. Dillworth, and P.D. White. 1985a. Indoor air pollution due to emissions from unvented gas-fired space heaters. *Journal Air Pollution Control Association* 35:231-37.

Traynor, G.W., I.A. Nitschke, W.A. Clarke, G.P. Adams, and J.E. Rizzuto. 1985b. A detailed study of thirty houses with indoor combustion sources. *Paper* 85-30A.3, Air & Waste Management Association, Pittsburgh.

Turk, A. 1954. Odorous atmospheric gases and vapors: Properties, collection, and analysis. *Annals of the New York Academy of Sciences* 58:193-214.

Underhill, D.W., G. Micarelli, and M. Javorsky. 1988. Effects of relative humidity on adsorption of contaminants on activated carbon. In *Proceedings of the Symposium on Gaseous and Vaporous Removal Equipment Test Methods* (NBSIR 88-3716), NIST, Gaithersburg, MD.

Van Gemert, L.J. and A.H. Mettenbreijer. 1977. *Compilation of odor threshold values in air and water.* Central Institute for Nutrition and Food Research, TNO Delft (Nederland) or National Institute for Water Supply, Zeist/Voorburg (Nederland).

Wade, W.A., W.A. Cote, and J.E. Yocum. 1975. A study of indoor air quality. *Journal Air Pollution Control Association* 25(9):933-39.

Wallace, L.A., E.D. Pellizzari, T.D. Hartwell, C. Sparacino, and H. Zelon. 1983. Personal exposure to volatile organics and other compounds indoors and outdoors—The TEAM-study, 1983. APCA *Paper*, Air & Waste Management Association, Pittsburgh.

Walsh, M., A. Black, A. Morgan, and G.H. Crawshaw. 1977. Sorption of SO2 by typical indoor surfaces including wool carpets, wallpaper and paint. *Atmospheric Environment* 11:1107-11.

Weschler, C.J. and H.C. Shields. 1989. The effects of ventilation, filtration, and outdoor air on the composition of indoor air at a telephone office building. *Environment International* 15:593-604.

White, J.B., J.C. Reaves, P.C. Reist, and L.S. Mann. 1988. A data base on the sources of indoor air pollution emissions. In IAQ '88: *Engineering Solutions to Indoor Air Problems.* ASHRAE, Atlanta.

Windholz, M. ed. 1983. *The Merck Index*, 10th ed. Merck & Company, Rahway, NJ.

Wolverton, B.C., A. Johnson, and K. Bounds. 1989. Interior landscape plants for indoor air pollution abatement. NASA, Stennis Space Center, MS.

Yoon, Y.H. and J.H. Nelson. 1984a. Application of gas adsorption kinetics: I. A theoretical model for respirator cartridge service life. *American Industrial Hygiene Association Journal* 45(8):509-16.

Yoon, Y.H. and J.H. Nelson. 1984b. Application of gas adsorption kinetics: II. A theoretical model for respirator cartridge service life and its practical applications. *American Industrial Hygiene Association Journal* 45(8):517-24.

Yoon, Y.H. and J.H. Nelson. 1988. A theoretical study of the effect of humidity on respirator cartridge service life. *American Industrial Hygiene Association Journal* 49(7):325-32.

BIBLIOGRAPHY

Cohen, H.J., ed. 1983. AIHA's respiratory protection committee's critique of NIOSH/OSHA pocket guide to chemical hazards. *American Industrial Hygiene Association Journal* 44(9):A6-A7.

Mehta, M.P., H.E. Ayer, B.E. Saltzman, and R. Romk. 1988. Predicting concentrations for indoor chemical spills. In IAQ '88: *Engineering Solutions to Indoor-Air Problems*, 231-50. ASHRAE, Atlanta.

NBS. 1959. Maximum permissible body burdens and maximum permissible concentrations of radionuclides in air and water for occupational exposure. NBS *Handbook* 69. U.S. Government Printing Office, Washington, D.C.

Riggen, R.M., W.T. Wimberly, and N.T. Murphy. 1990. Compendium of methods for determination of toxic organic compounds in ambient air. EPA/600/D-89/186, U.S. Environmental Protection Agency, Research Triangle Park, NC. Also second supplement, 1990, EPA/600/4-89/017.

Spicer, C.W., R.W. Coutant, G.F. Ward, A.J. Gaynor, and I.H. Billick. 1986. Removal of nitrogen dioxide from air by residential materials. In IAQ '86: *Managing Indoor Air for Health and Energy Conservation*, 584-90. ASHRAE, Atlanta.

Sterling, T.D., E. Sterling, and H.D. Dimich-Ward. 1983. Air quality in public buildings with health related complaints. ASHRAE *Transactions* 89(2A):198-212.

Webb, P. 1964. *Bioastronautics data book.* NASA SP-3006. NTIS, Springfield, VA.

Woods, J.E., J.E. Janssen, and B.C. Krafthefer. 1986. Rationalization of equivalence between the ventilation rate and air quality procedures in ASHRAE standard 62. In IAQ '86: *Managing Indoor Air for Health and Energy Conservation.* ASHRAE, Atlanta.

AUTOMATIC CONTROL

AUTOMATIC control of HVAC systems and equipment usually includes temperature, humidity, pressure, and flow rate control. Automatic control primarily modulates stages or sequences of the equipment capacity to meet load requirements and provides safe operation of the equipment. It uses pneumatic, mechanical, electrical, and electronic control devices, and implies that human intervention is limited to starting and stopping equipment and adjusting control set points.

This chapter focuses on controls that are normally custom-designed by a control system designer. It covers (1) control fundamentals, including terminology; (2) the different types of control components; (3) the methods of connecting these components together to form various types of individual control loops or subsystems; (4) the synthesis of these subsystems into a complete control system; and (5) commissioning, operation, and maintenance.

CONTROL FUNDAMENTALS

CONTROL

A *closed loop* control, or feedback control system, measures actual changes in the controlled variable and actuates the control device to bring about a change. The corrective action continues until the variable is brought to a desired value within the design limitations of the controller. This system of transmitting the value of the controlled variable back to the controller is known as *feedback*.

An *open loop* or *feedforward* control system does not have a direct link between the value of the controlled variable and the controller. *Feedforward* control anticipates how an external variable will affect the system. An outdoor thermostat arranged to control heat flow to a building in proportion to the calculated load caused by changes in outdoor temperature is one example. In essence, the designer of this system presumes a fixed relationship between outside air temperature and the heat requirement of the building and takes control action based on the outdoor air temperature. The actual space temperature has no effect on this controller. Because there is no feedback of the controlled variable (space temperature) in this case, the control is known as open loop.

Feedforward control can also be used effectively on a feedback controller to change the characteristics of the control based on both the feedback variable and an external variable affecting the control loop. The controller is then a feedback controller with feedforward compensation.

TERMINOLOGY

The *control loop*, shown in Figure 1 illustrates the components of a typical control loop.

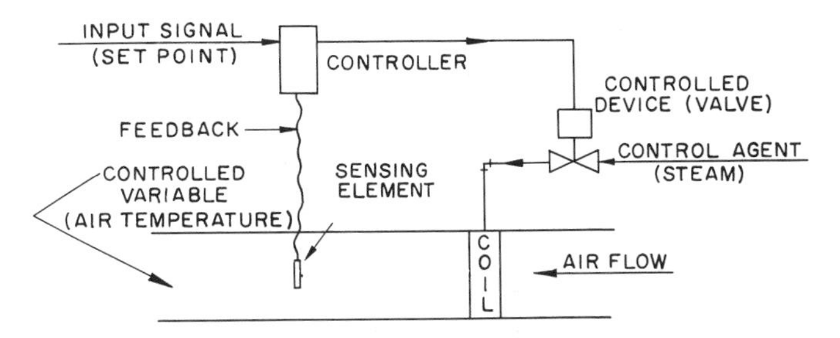

Fig. 1 Discharge Air Temperature Control—Example of Feedback Control System

The *sensor* measures the controlled variable and conveys values to the controller. The *controller* compares the value of the controlled variable with set point and generates a signal to the controlled device for corrective action. Thermostats, humidistats, and pressure controls are examples of controllers.

The *set point* is the desired value of the controlled variable. The controller seeks to maintain this set point.

The *controlled device* reacts to signals received from the controller to vary the flow of the control agent. It may be a valve, damper, electric relay, or a motor driving a pump or a fan.

The *control agent* is the medium manipulated by the controlled device. It may be air or gas flowing through a damper; gas, steam, or water flowing through a valve; or an electric current.

The preparation of this chapter was assigned to TC 1.4, Control Theory and Application.

The *process plant* is the air-conditioning apparatus being controlled. It reacts to the output of the control agent and affects the change in the controlled variable. It may include a coil, air duct, fan, or occupied space of the building.

The *controlled variable* is that variable temperature, humidity, or pressure being controlled.

A control loop can be represented in the form of a block diagram, in which each component of the control loop is modeled and represented in its own block (Figure 2). The flow of information from one component to the next is shown by lines between the blocks. The figure shows the set point being compared to the feedback of the controlled variable. This difference, or error, is fed into the controller, which sends a control signal to the controlled device. In this case, the controlled device is a valve. The valve can change the amount of steam flow through the coil of Figure 1. The amount of steam flow is the input to the next block, which represents the process plant. From the process plant block comes the controlled variable, which is temperature. The controlled variable is sensed by the sensing element and fed to the controller as feedback, completing the loop.

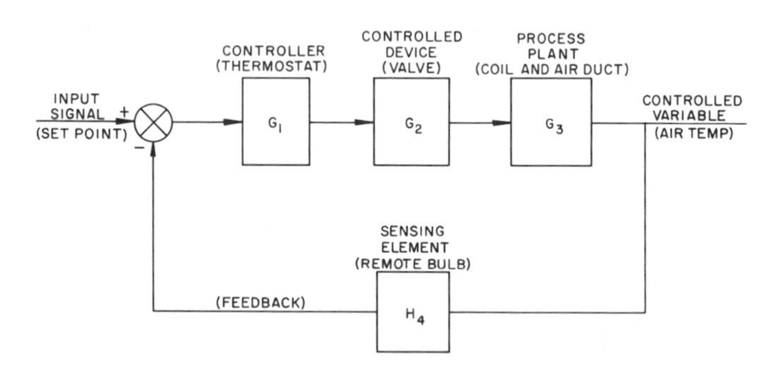

Fig. 2 Block Diagram of Discharge Air Temperature Control System

Each component of Figure 2 can be represented by a transfer function, which is an idealized mathematical representation of the relationship between the input and output variables of the component. The transfer function must be sufficiently detailed to cover both the dynamic and static characteristics of the device. The dynamics of the component are represented in the time domain by a differential equation. In environmental control systems, the transfer function of many of the components can be adequately described by a first order differential equation, implying that the dynamic behavior is dominated by a single capacitance factor. The differential equation is converted to its *LaPlace transform* or *z-transform*.

The time constant is defined as the time it takes for the output to reach 63.2% of its final value when a step change in the input is effected. When the time constant of the component is small, its output will react rapidly to reflect changes in the input; conversely, components with a larger time constant will be sluggish in responding to changes in the input.

Deadtime is a nonlinearity, which can cause control and modeling problems. Deadtime is the time between a change in the process input and when that change affects the output of the process. Deadtime can occur in the control loop of Figure 1 due to the transportation time of the air. The temperature the sensor sees represents the conditions at the coil some time in the past, because the air it is sensing went through the coil some time in the past. Deadtime can also occur due to a slow sensor, a time lag in the signal from the controller, or the transportation time of the control agent. If the deadtime is small, it can be ignored in the model of the control system; if it is significant, it has to be considered.

The *gain* of a transfer function is the amount by which the output of the component will change for a given change of input under steady-state conditions. If the element is linear, its gain remains constant. However, many control system components are nonlinear and have varying gains, depending on the operating conditions.

Figure 3 shows the response of a first order plus deadtime process to a step change of the input signal. Notice that the process shows no reaction during the deadtime, followed by a response that resembles a first order exponential.

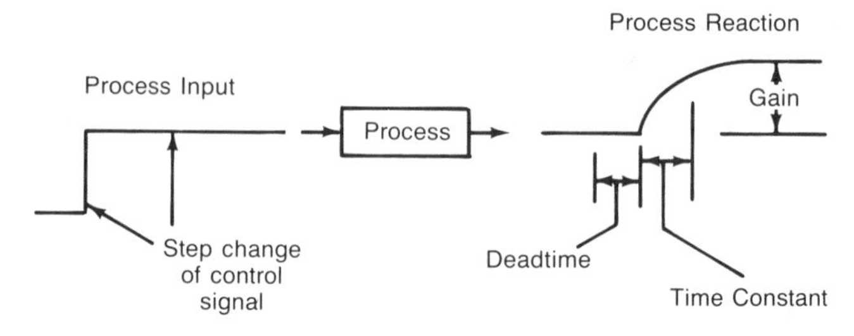

Fig. 3 Process Subjected to a Step Input

The principle behind controllers is that the sensor sends a signal (pneumatic, electric, or electronic) whose pressure, voltage, or current, is proportional to the value of the variable being measured. The controller contains circuitry that compares this signal from the sensor to the desired value, and outputs a control signal based on this comparison. The controller is an analog device that receives and acts on data continuously.

TYPES OF CONTROL ACTION

Closed loop control systems are commonly classified by the type of corrective action the controller is programmed to take when it senses a deviation of the controlled variable from the set point. Following are the most common types of control action.

Two-Position Action. The controlled device shown in Figure 4 can be positioned only to a maximum or minimum state, or can be either on or off. A typical home thermostat that starts and stops a furnace is a good example of two-position action.

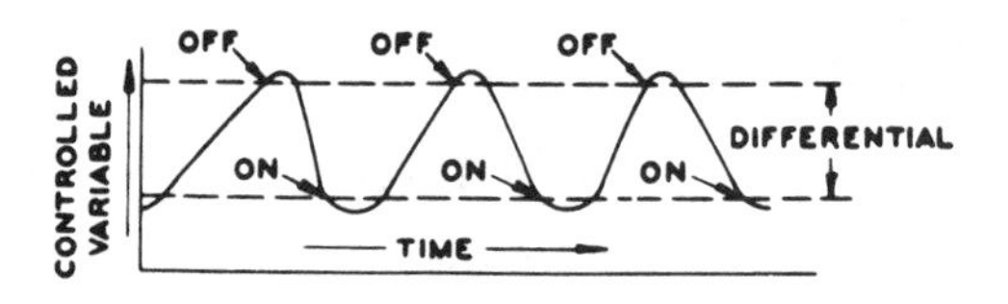

Fig. 4 Two-Position Control

Controller differential, as it applies to two-position control action, is the difference between a setting at which the controller operates to one position and a setting at which it operates to the other. Thermostat ratings usually refer to the differential in degrees that become apparent by raising and lowering the dial setting. This differential is known as the *manual differential* of the thermostat. When the same thermostat is applied to an operating system, the total change in temperature that occurs between a call for more heat and a call for less heat is usually greater than the manual differential caused by thermostat lag. The differential encountered on the job under control is the *operating differential.* For example, a thermostat with a 2 °F differential when raising and lowering the dial setting may actually produce temperature variations of 3 °F between the "system on" and "system off" states.

Timed Two-Position Action. This is a common variation of straight two-position action that is often used on room thermostats to reduce operating differential. In heating thermostats, a heater element is energized during "on" periods, prematurely shortening the on-time as the heater falsely warms the thermostat. This is known as heat anticipation. The same anticipating action can be obtained in cooling thermostats by energizing a heater during thermostat "off" periods. In either case, the percentage of on-time is varied in proportion to the system load, while the total cycle time remains relatively constant.

Floating Action. In floating action, the controller can perform only two operations—moving the controlled device toward either its open or closed position, usually at a constant rate (see Figure 5). Generally, a neutral zone between the two positions allows the controlled device to stop at any position whenever the controlled variable is within the differential of the controller. When the controlled variable gets outside the differential of the controller, the controller moves the controlled device in the proper direction.

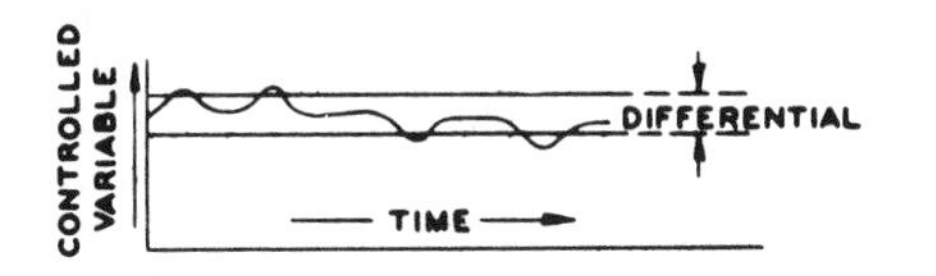

Fig. 5 Floating Control Showing Variations in Controlled Variable as Load Changes

Proportional Action. In proportional action, the controlled device is positioned proportionally in response to changes in the controlled variable (Figure 6). A proportional controller can be described by:

$$V_p = K_p e + V_o$$

where

V_p = output of the proportional controller
K_p = proportional gain (proportional to 1/throttling range)
e = error signal, or offset
V_o = offset adjustment parameter

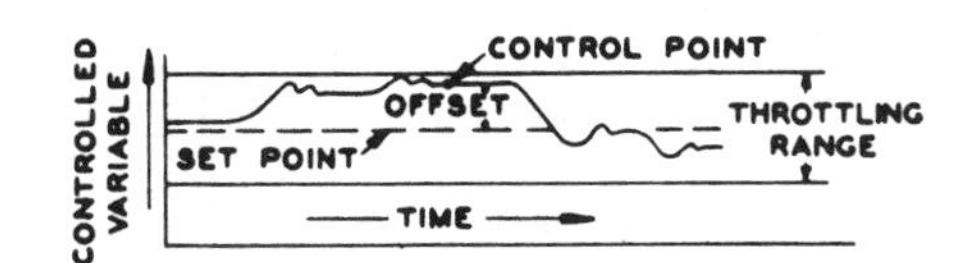

Fig. 6 Proportional Control Showing Variations in Controlled Variable as Load Changes

The output of the controller is proportional to the difference between the sensed valve and its set point. The controlled device is normally adjusted to be in the middle of its control range at set point, by using an offset adjustment. This control is similar to that shown in Figure 1.

Throttling range is the amount of change in the controlled variable required for the controller to move the controlled device from one extreme to the other. It can be adjusted to meet job requirements. Throttling range is inversely proportional to proportional gain.

Control point is the actual value of the controlled variable at which the instrument is controlling. It varies within the throttling range of the controller and changes with changing load on the system and other variables.

Offset or *error signal*, is the difference between the set point and the actual control point under stable conditions. This is sometimes called drift, deviation, droop, or steady-state error.

Proportional Plus Integral. Proportional plus integral (PI) control improves on simple proportional control by adding another component to the control action that eliminates the offset typical of proportional control (Figure 7). The reset action is most easily shown by the equation:

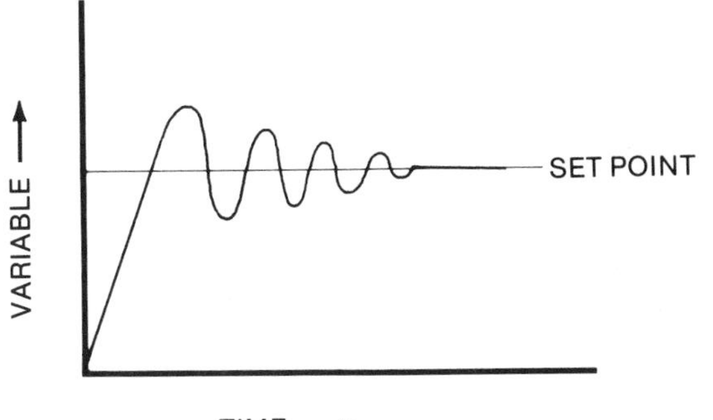

Fig. 7 Proportional Plus Integral (PI) Control

$$V_p = K_p e + K_i \int e\,dt + V_o$$

where

V_p = output of the controller
K_i = integral gain
t = time
e = error
V_o = offset adjustment parameter

The second term in the equation implies that the longer the period during which the error (e) exists, the more the controller output will change in attempting to eliminate the error.

Selecting proportional and integral gain constants is critical to system stability. Proper selection eliminates offset, obtaining greater control accuracy. Also, energy efficiency is improved by PI control in applications such as VAV fan control, chiller control, and hot and cold deck control.

Proportional-Integral-Derivative (PID). This type of control is PI control with a term added to the controller. It varies with the value of the derivative of the error. The equation for PID control is:

$$V_p = K_p e + K_i \int e\,dt + K_d\,de/dt + V_o$$

where

K_d = derivative gain of controller
de/dt = derivative of error
V_o = offset adjustment parameter

The other terms are defined above. Adding the derivative term gives some anticipatory action to the controller, which will result in a faster response and greater stability. However, the addition of the derivative term also makes the controller more sensitive to noisy signals and harder to tune than a PI controller. Most HVAC control loops perform satisfactorily with PI control, without the need for adding the derivative term. Adaptive control, or self-tuning, is a form of PID control, where the gain factors (K_p, K_i, and K_d) are continuously or periodically modified automatically to compensate for the control loop offset.

PERFORMANCE REQUIREMENTS OF CONTROL SYSTEMS

The control system performance in an air-conditioning application is evaluated in terms of speed of response and stability. A stable control loop will keep the controlled variable near set point while avoiding long-term oscillations. A control loop with a fast speed of response quickly responds to process disturbances. The requirements of accuracy, speed, and stability are often contradictory, *i.e.*, a change made in one of the component parameters could adversely affect others. Hence, the performance level of such systems must be selected to suit the application and must be evaluated in terms of control, comfort, and energy conservation.

CLASSIFICATIONS OF CONTROL SYSTEMS

Control components can be classified into the following three categories according to the primary source of energy:

- *Pneumatic* components use compressed air, usually at a pressure of 15 to 35 psig, as an energy source. The air is generally supplied to the controller, which regulates the pressure supplied to the controlled device.
- *Electric* components use electrical energy, either low or line voltage, as the energy source. The electric energy supplied to the controlled device is regulated by the controller. Controlled devices in this category include relays; contactors; electromechanical, electromagnetic, and hydraulic actuators, and solid-state regulating devices. The components that include signal conditioning, modulation, and amplification in their operation are classified as electronic.

  A direct digital controller receives electronic signals from the sensors, converts the electronic signals to numbers, and performs mathematical operations on these numbers inside the computer. The output from the computer takes the form of a number, which can be converted to an electric or pneumatic signal to operate the actuator. The digital controller must sample its data because the computer must have time for other operations besides reading data. If the sampling interval for the digital controller is properly chosen, no significant degradation in control performance will be seen due to sampling.
- *Self-powered* components apply the power of the measured system to induce the necessary corrective action. The measuring system derives its energy from the process under control, without any auxiliary source of energy. Temperature changes at the sensor result in pressure or volume changes of the enclosed media, which are transmitted directly to the operating device of the valve or damper. A component using a thermopile in a pilot flame to generate electrical energy is also self-powered.

This method of classification can be extended to individual control loops and to complete control systems. For example, the room temperature control for a particular room that includes a pneumatic room thermostat and a pneumatically actuated reheat coil would be referred to as a pneumatic control loop. Most complete control systems use a combination of some or all of the above components and are more accurately called *hybrid systems*. As an example, the control system for an air handler could include electric components for on/off control of the fan, pneumatic components for control of the heating and cooling coils, and self-powered safety controls (*e.g.*, a freezestat).

CONTROL COMPONENTS

While control components may be classified in several ways, this section groups components by their function within a complete control system. The first subsection considers the controlled device or final control element, examples of which are relays, valves, dampers, and VAV boxes (which contain a damper or damper-like mechanism). Actuators, which are used to drive the valve or damper assembly, are also covered.

The next subsection considers the sensing element that measures changes in the controlled variable. Specific examples of sensor types included are temperature, humidity, water and air pressure, water and air differential pressure, and water and airflow rate. While many other kinds of special sensors are available, these types represent the majority of those found in the HVAC control systems and subsystems described in Section IV.

In the third subsection, various types of controllers are reviewed. Controllers are classified according to the control action they cause to maintain the desired condition (set point)—whether they are two-position, floating control, proportional control, proportional plus integral (PI) control, or proportional plus integral plus derivative (PID) control. In addition, this section describes the various techniques available for making the control decision in a modulating control system, such as pneumatic, electronic, and digital controllers. Thermostats (devices that combine a temperature sensor and controller into a single unit) are also described.

Many control systems can be constructed using only the types of components described in the first three sections. In practice, however, a fourth group is sometimes necessary. The members of this group are neither sensing elements nor controlled devices or controllers, but are referred to as auxiliary control components, including transducers, relays, switches, power supplies, and air compressors.

CONTROLLED DEVICES

The controlled device is most frequently used to regulate or vary the flow of steam, water, or air within an HVAC system. Water and steam flow regulators are known as *valves*, and airflow control devices are called *dampers*; both types perform essentially the same function and must be properly sized and selected for the particular application. The control system's link to the valve or damper is a component referred to as an *operator*, or *actuator*. This device uses electricity, compressed air, or hydraulic fluid to power the motion of the valve stem or damper linkage through its operating range.

Valves

An *automatic valve* is designed to control the flow of steam, water, gas, and other fluids and may be considered as a variable orifice positioned by an electric or pneumatic operator in response to impulses, or signals, from the controller. It may be equipped with a throttling plug or V-port specially designed to provide desired flow characteristics.

Renewable composition discs are common. They are made of materials best suited to the media handled by the valve, the operating temperature, and the pressure. For high pressures or for superheated steam, metal discs are often used. Internal parts of valves, such as the seat ring, throttling plug or V-port skirt, disc holder, and stem, are sometimes made of stainless steel or other hard and corrosion-resistant metals for use in severe service.

Various types of automatic valves include the following:

A *single-seated valve* (Figure 8A) is designed for tight shutoff. Appropriate disc materials for various pressures and media are used.

A *double-seated* or *balanced valve* (Figure 8B) is designed so that the media pressure acting against the valve disc is essentially balanced, reducing the operator force required. It is widely used where fluid pressure is too high to permit a single-seated valve to close. It cannot be used where tight shutoff is required.

A *three-way mixing valve* (Figure 9A) has two inlet and one outlet connection and a double-faced disc operating between two seats. It is used to mix two fluids entering through the inlet con-

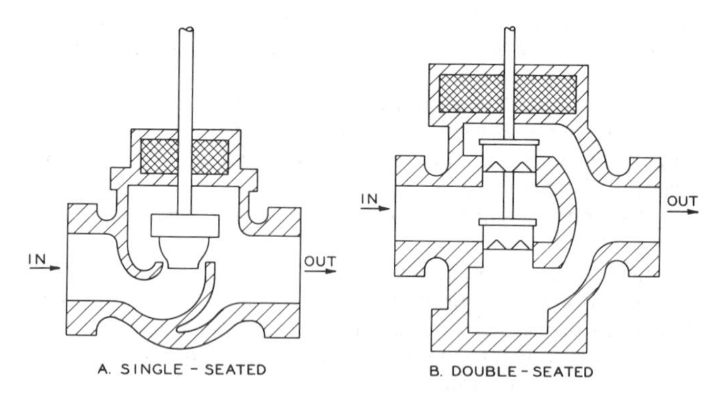

Fig. 8 Typical Single- and Double-Seated Two-Way Valves

nections and leaving through the common outlet, according to the position of the valve stem (and disc).

A *three-way diverting valve* (Figure 9B) has one inlet and two outlet connections, and two separate discs and seats. It is used to divert the flow to either of the outlets or to proportion the flow to both outlets.

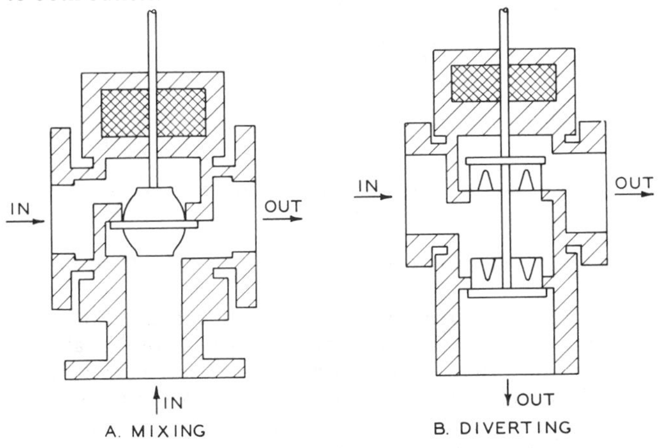

Fig. 9 Typical Three-Way Mixing and Diverting Valves

A *butterfly valve* consists of a heavy ring enclosing a disc that rotates on an axis at or near its center and is similar to a round single-blade damper in principle. The disc seats against a ring machined within the body or a resilient liner in the body. Two butterfly valves can be used together to act like a 3-way valve for mixing or diverting.

Characteristics

The performance of a valve is expressed in terms of its flow characteristics as it operates through its stroke, based on a *constant pressure drop.* Three common characteristics are shown in Figure 10 and are defined as follows:

Quick opening. Maximum flow is approached rapidly as the device begins to open.
Linear. Opening and flow are related in direct proportion.
Equal percentage. Each equal increment of opening increases the flow by an equal percentage over the previous value.

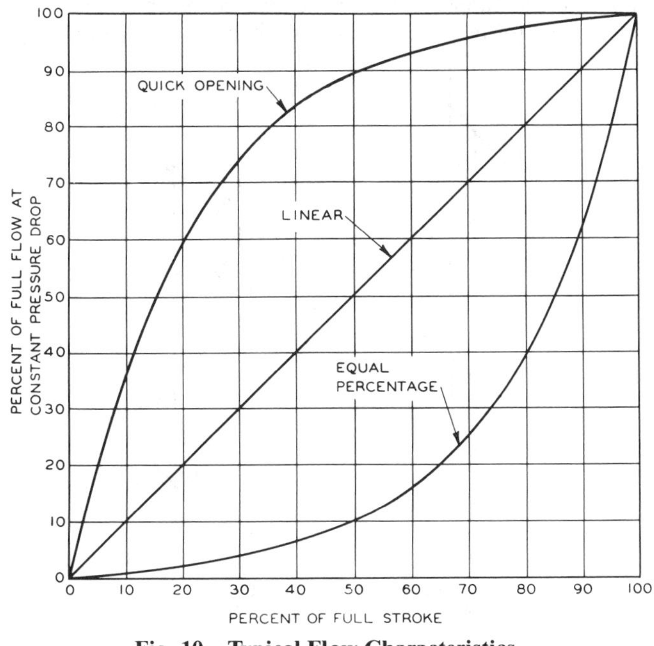

Fig. 10 Typical Flow Characteristics

Since the pressure drop across a valve seldom remains constant as its opening changes, actual performance usually deviates from the published characteristic curve. The magnitude of the deviation is determined by the overall system design. For example, in a system arranged so that control valves or dampers can shut off all flow, the pressure drop across a controlled device increases from a minimum at design conditions to the total system pressure drop at no flow. Figure 11 shows the extent of the resulting deviations for a valve or damper designed with a linear characteristic, when selection is based on various percentages of total system pressure drop. To approximate the designed characteristic of the valve or damper, the design pressure drop should be a reasonably large percentage of the total system pressure drop, or the system should be designed and controlled so that this pressure drop remains relatively constant.

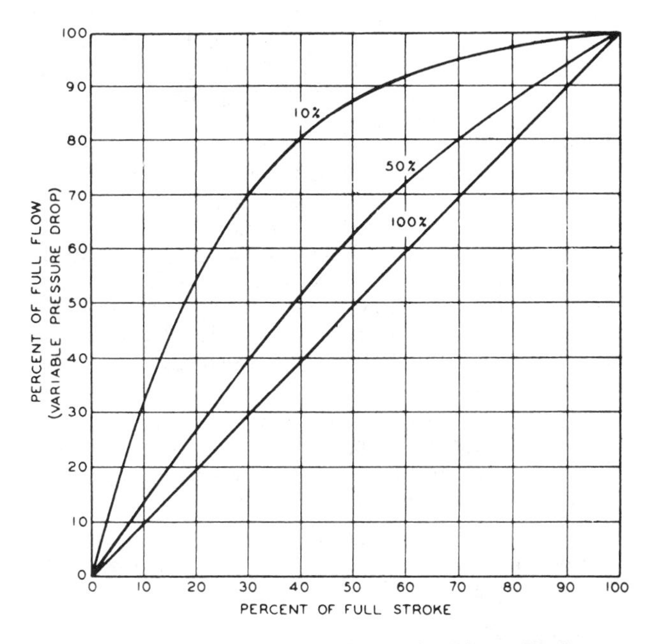

Fig. 11 Typical Performance Curves for Linear Devices at Various Percentages of Total System Pressure Drop

Higher pressure drops for controlled devices are obtained by using smaller sizes with a possible increase in size of other equipment in the system. Since sizing techniques are different for steam, water, and air, each is discussed separately.

OPERATORS

Valve operators are of the following four general types:

1. A *solenoid* consists of a magnetic coil operating a movable plunger. Most are for two-position operation, but modulating solenoid valves are available with a pressure equalization bellows or piston to achieve modulation. Solenoid valves are generally limited to relatively small sizes (up to 4 in.).
2. An *electric motor* operates the valve stem through a gear train and linkage. Electric motor operators are classified in the following three types:

   a) *Unidirectional*—for two-position operation. The valve opens during one-half revolution of the output shaft and closes during the other one-half revolution. Once started, it continues until the half revolution is completed, regardless of subsequent action by the controller. Limit switches built into the operator stop the motor at the end of each stroke. If the controller has been satisfied during this interval, the operator will continue to the other position).

b) *Spring-return*—for two-position operation. Electric energy drives the valve to one position and a spring returns the valve to its normal position.
c) *Reversible*—for floating and proportional operation. The motor can run in either direction and can stop in any position. It is sometimes equipped with a return spring. In proportional control applications, a feedback potentiometer for rebalancing the control circuit is also driven by the motor.

3. A *pneumatic operator* consists of a spring-opposed, flexible diaphragm or bellows attached to the valve stem. An increase in air pressure, above the minimum point of the spring range, compresses the spring and simultaneously moves the valve stem. Springs of various ranges, in terms of air pressure required to compress the spring completely, can sequence the operation of two or more devices by proper selection or adjustment of the springs. For example, a chilled water valve operator may modulate the valve from fully closed to fully open over a spring range of 3 to 8 psig, while a sequenced steam valve may operate from 8 to 13 psig.
4. *Springless pneumatic operators*, using two opposed diaphragms or two sides of a single diaphragm, are also used, but they are generally limited to special applications involving large valves or high pressures. Pneumatic operators are used primarily for proportional control. Two-position control is accomplished using a two-position controller or a two-position pneumatic relay to apply either full air pressure or no pressure to the valve operator. Pneumatic valves and valves with spring-return electric operators can be classified as normally open or normally closed.

   a) A *normally open* valve will assume an open position, providing full flow, when all operating force is removed.
   b) A *normally closed* valve will assume a closed position, stopping flow, when all operating force is removed.

5. An *electric-hydraulic* actuator is similar to a pneumatic one, except that it uses an incompressible fluid, which is circulated by an internal electric pump.

Selection and Sizing

Steam Valves. Steam-to-water and steam-to-air heat exchangers are typically controlled through regulation of steam flow rate using a two-way throttling valve. One-pipe steam systems require a line-size two-position valve for proper condensate drainage and steam flow, while two-pipe steam systems can be controlled by two-position or modulating (throttling) valves.

Water Valves. Valves used for water service may be two- or three-way and two-position or proportional. Proportional valves are used most often, but two-position valves are not unusual and are sometimes essential (*e.g.*, on steam preheat coils). Two-position valves are normally the quick opening type, while proportioning valves are normally the linear or equal percentage type.

The *flow coefficient* C_v is generally used to compare valve capacities. C_v is defined as the volume of water flow in gpm at 60 °F through a control valve in the fully open position with a 1 psi differential across the valve.

While it is possible to design a water system in which the pressure differential from supply to return is kept constant, it is seldom done. It is safer to assume that the pressure drop across the valve will increase as it modulates from fully open to fully closed. Figure 12 shows the effect in a simple system with one pump and one two-way control valve with its heat exchanger. The *system curve* represents the pressure or head loss in the piping system and heat exchanger at various flow rates. The *pump curve* is the typical curve for a centrifugal pump. At design flow rates, the valve is selected for a specific pressure drop, A-A.′ At part load, the valve must partially close to provide a higher pressure drop, B-B.′ The

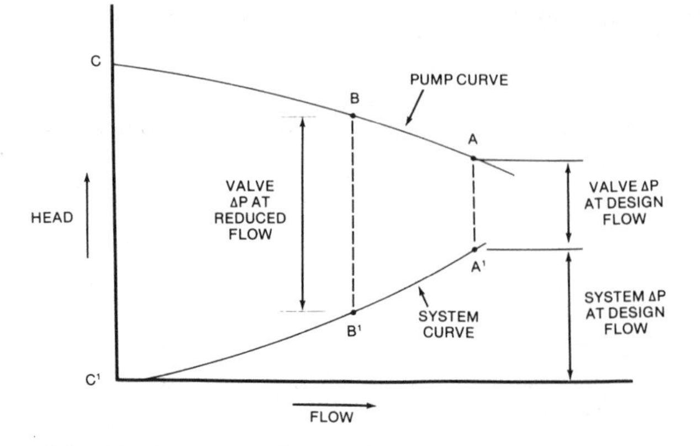

Fig. 12 Pump and System Curves with Valve Control

ratio between the design pressure drop A-A′ and the zero flow pressure drop C-C′ influences the control capability of the valve.

Better control at part load is obtained by using equal percentage valves, particularly in hot water coils where the heat output of the coil is not linearly related to flow. As flow is reduced, a greater amount of heat is transferred from each unit volume of water, counteracting the reduction in flow. Reset of the supply water temperature with the outdoor temperature can partially correct this.

Two-way control valves should be sized to provide from 20 to 60% of the total system pressure drop. The valve operator should be sized to close the valve against the full pump head pressure to ensure complete shut-off during no-flow condition.

DAMPERS

Types and Characteristics

Automatic dampers are used in air-conditioning and ventilation systems to control airflow. They may be used for modulating control to maintain a controlled variable such as mixed air temperature or supply air duct static pressure, or a two-position controller to initiate system operation (such as opening minimum outside air dampers when a fan system is started).

Two damper arrangements are used for air-handling system flow control—parallel-blade and opposed-blade (see Figure 13). Parallel-blade dampers are adequate for two-position control and can be used for modulating control when they are the primary

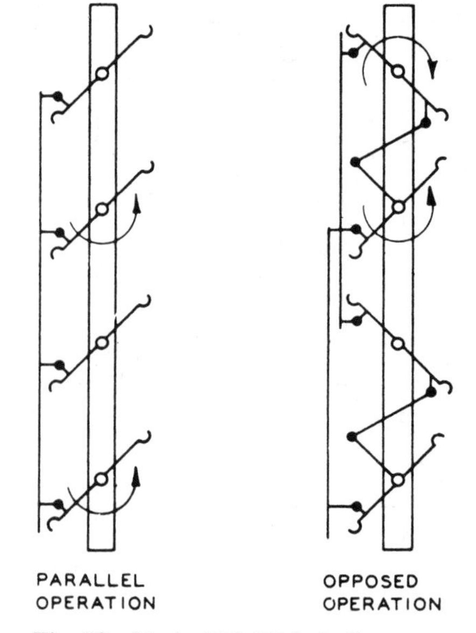

Fig. 13 Typical Multiblade Dampers

source of system pressure drop. However, opposed-blade dampers are preferable, since they normally provide better control (Figures 14 and 15). In these figures, the parameter α is the ratio of the system pressure drop to the drop across the damper at maximum (fully open) flow.

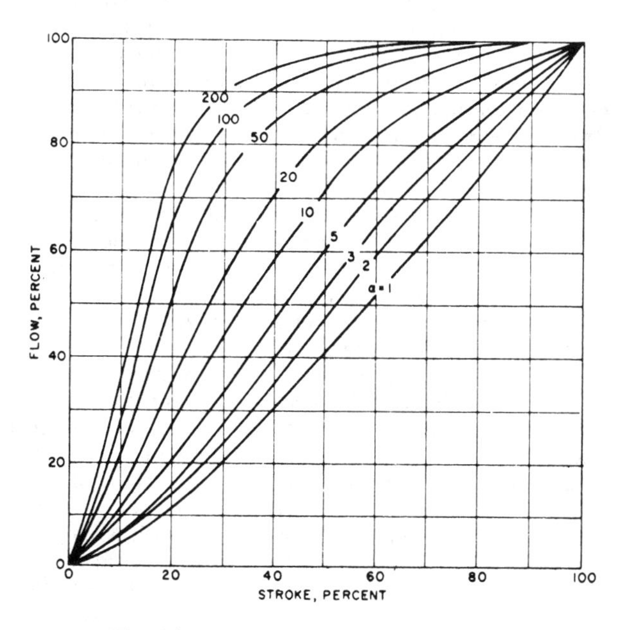

Fig. 14 Installed Characteristic Curves of Parallel Blade Dampers

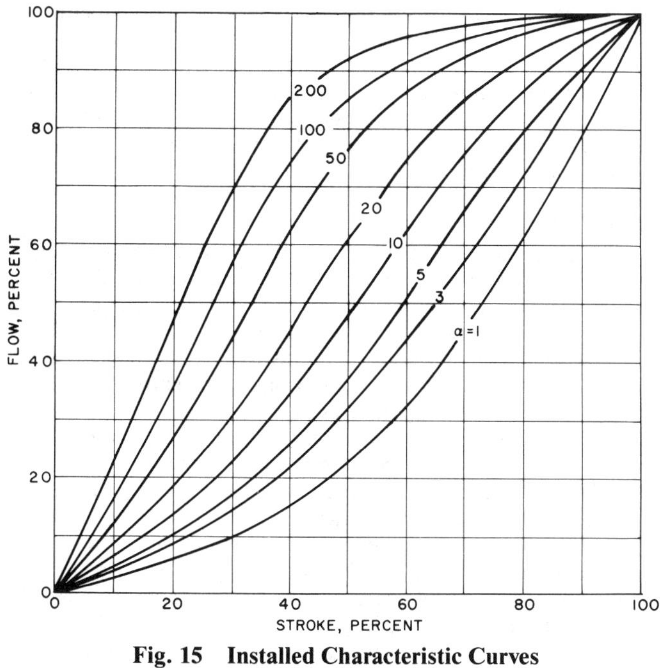

Fig. 15 Installed Characteristic Curves of Opposed Blade Dampers

Damper leakage is important, particularly where tight shutoff is required. For example, an outdoor air damper must close tightly to prevent coils and pipes from freezing. Low leakage dampers are more costly and require larger operators because of the friction of the seals in the closed position; therefore, they should be used only when necessary, including any location where the tight closing damper will reduce energy consumption significantly.

Operators

Like valve operators, damper operators are available using either electricity or compressed air as a power source.

Electric damper operators, or *actuators*, can be either unidirectional, spring return, or reversible. The reversible type is frequently used for accurate control in modulating damper applications. A reversible electric actuator has two sets of motor windings within the unit's housing. Energizing one set of windings moves the actuator output shaft in a clockwise direction, and energizing the other causes the shaft to turn in a counterclockwise direction. When neither set of windings is energized, the shaft remains in its last position. The simplest form of control for this actuator is a floating point controller, which causes a contact closure to drive the motor in a clockwise and counterclockwise direction. This type of actuator is available with a wide range of options for rotational shaft travel (expressed in degrees of rotation) and timing (expressed in the number of seconds to move through the rotational range). In addition, a variety of standard electronic signals from electronic controllers (such as 4-20 milliamps dc or 0-10 volts dc) can be used to drive this type of modulating actuator. A spring return actuator will move in one direction whenever voltage is applied to its internal windings, and, when no power is present, the actuator is returned via spring force to its normal position. Depending on how the actuator is connected to the dampers, this will open or close the dampers.

Pneumatic damper operators are similar to pneumatic valve operators, except that they have a longer stroke, or the stroke is increased by a multiplying lever. Increasing the air pressure produces a linear motion of the shaft, which, through a linkage, moves the crank arm to open or close the dampers. Normally open or normally closed operation refers to the position of the dampers when no air pressure is applied at the operator. This position depends on how the operator is mounted and how the linkage is connected.

Mounting. Damper operators are mounted in several different ways, depending on the damper size, accessibility, and power required to move the dampers. They can be mounted in the airflow on the damper frame and connected by a linkage directly to a damper blade, or they can be mounted outside the duct and connected to a crank arm attached to a shaft extension of one of the blades. On large dampers, two or more operators may be needed. In this case, they are usually mounted at separate points on the damper. An alternative is to install the damper in two or more sections, each section being controlled by a single damper operator; however, proper airflow control is easier with a single modulating damper. Positive positioners may be required for proper sequencing. A small damper with a two-position operator may be used for minimum outside flow, with a large damper being independently controlled for economy cycle cooling.

POSITIVE POSITIONERS

A pneumatic operator may not respond quickly or accurately enough to small changes in control pressure due to friction in the actuator or load, or to changing load conditions such as wind acting on a damper blade. Where accurate positioning of a modulating damper or valve in response to load is required, positive positioners should be used.

A positive positioner provides up to full main control air pressure to the actuator for any change in position required by the controller. This is achieved by the sample arrangement shown schematically in Figure 16. An increase in branch pressure from the controller (A) moves the relay lever (B), opening the supply valve (C). This allows main air to flow to the relay chamber and the actuator cylinder, moving the piston (not shown). The piston movement is transmitted through a linkage and spring (D) to the other end of the lever (B), and when the force due to movement

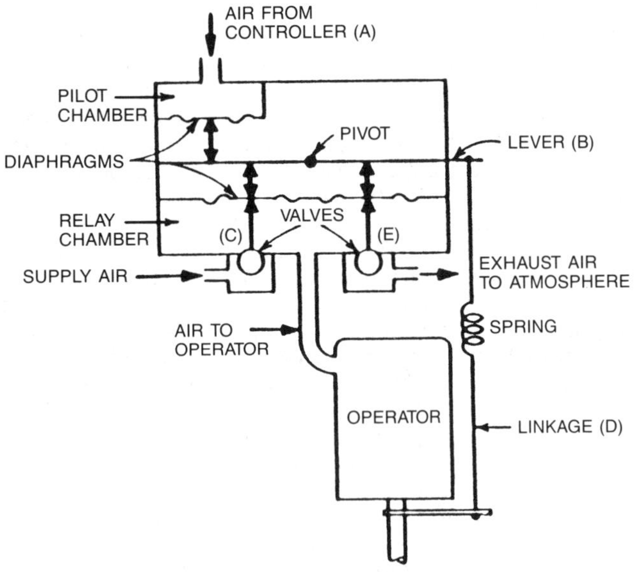

Fig. 16 Positive Positioner

balances out the control force, the supply valve closes, leaving the actuator in the new position. A decrease in control pressure allows the exhaust valve (E) to open until a new balance is obtained. Thus, full main air pressure is available, if needed, even though the control pressure may have changed only a fraction of that amount. The movement feedback linkage may be mounted internally or externally. Positioners may be connected for direct or reverse action.

A positive positioner provides finite and repeatable positioning change and permits adjustment of the control range (spring range of the actuator) to provide a proper sequencing control of two or more controlled devices.

SENSORS

A sensor is the component in the control system that measures the value of the controlled variable. A change in the controlled variable (such as the temperature of water flowing in a pipe) produces a change in some physical or electrical property of the primary sensing element, which is then available for translation or amplification by mechanical or electrical signal. When the sensor uses a conversion from one form of energy (mechanical or thermal) to another (electrical), the device is known as a transducer. In some cases, the sensing element is a transducer, such as a thermistor, in which a change in electrical resistance occurs as a direct result of a change in temperature.

With the trend toward electronic miniaturization and solid-state sensing elements, sensor selection has become a specialty. New measurement technologies and manufacturers are emerging regularly, expanding the options available to the control system designer and outdating older texts on sensor application. This section does not describe all the technologies for measurement and transmission of even the common variables of temperature, humidity, pressure, and flow rate. Chapter 13 of the 1989 ASHRAE *Handbook—Fundamentals*, manufacturers' catalogs and tutorials, and the bibliography give specific applications information.

In selecting a particular sensor product for a specific application, the following elements should be considered:

Operating Range of the Controlled Variable. The sensor must be capable of providing a detectable and significant change in its output signal over the expected operating range. In the case of an office room temperature measurement, for example, the output of a pneumatic transmitter may change from 3 to 15 psig over a temperature range of 55 to 85 °F.

Compatibility of the Controller Input. Pneumatic receiver controllers for HVAC systems will typically accept input signals of 3 to 15 psig. Electronic and digital controllers accept various ranges and types of electronic signals. The selection of an electronic sensor must consider the specific controller to be used; if this is not known, an industry "standard" signal, such as 4-20 milliamps dc or 0-10 volts dc, should be used.

Set Point Accuracy and Consistency. Some control applications require the controlled variable to be maintained within a narrow band around a desired set point. Both the accuracy and sensitivity of the sensor selected must reflect this requirement, although an accurate sensor alone cannot maintain the set point if the controller is unable to resolve the input signal, the controlled device cannot be positioned accurately, or the controlled device exhibits excessive hysteresis.

System Response Time (or Process Dynamics). Associated with a sensor/transducer arrangement is a response curve, which describes the response of the sensor output to a change in the controlled variable. If the time constant of the process being controlled is short and stable accurate control is important, the sensor selected must have a fast response time.

Control Agent Properties and Characteristics. The control agent is the medium to which the sensor is exposed, or with which it comes in contact, for measuring a variable such as temperature or pressure. If the agent acts on the sensor so as to corrode or otherwise degrade its performance, a different sensor should be selected, or the sensor must be isolated or protected from direct contact with the agent.

Ambient Environment Characteristics. When isolated from direct contact with the control agent, the ambient environment in which the sensor's components are located must be considered. The temperature and humidity range of the ambient environment must not adversely affect the sensor or its accuracy. Likewise, the presence of certain gases, chemicals, and electromagnetic interference (EMI) can cause component degradation. In such cases, a special sensor or transducer housing can be used to protect the element, while ensuring a true indication of the control variable.

Temperature Sensors

Temperature-sensing elements fall into three general categories—(1) those that use a change in relative dimension due to differences in thermal expansion, (2) those that use a change in state of a vapor or liquid-filled bellows, and (3) those that use a change in some electrical property. Within each category, there are a variety of sensing element configurations to measure room, duct, water, and surface temperatures.

The specific types of temperature-sensing technologies commonly used in HVAC applications are as follows:

- A *bimetal* element is composed of two thin strips of dissimilar metals fused together. Because the two metals have different coefficients of thermal expansion, the element bends as the temperature varies and produces a change in position. Depending on the space available and the movement required, it may be a straight strip, U-shaped, or wound into a spiral. This element is commonly applied in room, insertion, and immersion thermostats.
- A *rod-and-tube* element consists of a high expansion metal tube containing a low expansion rod with one end attached to the rear of the tube. The tube changes length with changes in temperature, causing the free end of the rod to move. This element is commonly used on certain types of insertion and immersion thermostats.
- A *sealed bellows* element is either vapor-, gas-, or liquid-filled after being evacuated of air. Temperature changes cause changes

in pressure or volume of the gas or liquid, resulting in a change in force or movement. This element is often used in room thermostats.

- A *remote bulb* element is a sealed bellows or diaphragm to which a bulb or capsule is attached by means of a capillary tube; the entire system is filled with vapor, gas, or liquid. Temperature changes at the bulb result in volume or pressure changes that are communicated to the bellows or diaphragm through the capillary tube. The remote bulb element is useful where the temperature measuring point is remote from the desired thermostat location. It usually is provided with fittings suitable for insertion into a duct, pipe, or tank. *Averaging bulbs* are used in large ducts where a single sensing point may not be representative.
- A *thermistor* makes use of the change of electrical resistance of a semiconductor material for a representative change in temperature. The characteristic curve of a thermistor is nonlinear over a wide range. It has a negative temperature coefficient, *i.e.*, the resistance decreases as the temperature increases. For electronic control systems, a variety of techniques are available to provide a linear change over a particular temperature range. In a digital control system, one technique to linearize the curve over certain ranges is to store a computer "look-up table" that maps the temperatures corresponding to the measured resistance. Thermistors are used because of their relatively low cost and the large change in resistance possible for a small change in temperature.
- A *resistance temperature detector* (RTD) is another sensor that changes electrical resistance with temperature. Most metallic materials increase in resistance with increasing temperature. Over limited ranges, this variation is linear for certain metals such as platinum, copper, tungsten, and some nickel/iron alloys. Platinum, for example, is linear within ±0.3% from 0 to 300°F. The RTD sensing element is available in several different forms for surface or immersion mounting. For direct measurement of fluid temperatures, the winding of resistance wire is encased in a stainless-steel bulb to protect it from corrosion. Flat grid windings are used for measurements of surface temperatures. The RTD measurement circuit typically consists of three wires to correct for line resistance. In many cases, the three-wire RTD is mated to an electronic circuit to produce a 4 to 20 milliamp current signal over a finite temperature range.
- A *thermocouple* is formed by the junction of two wires of dissimilar metals. An electromotive force dependent on the wire materials and junction temperature exists between the wires. When the wires are joined at two points, a thermocouple circuit is formed. As long as one junction is kept at a constant temperature, different from the other junction, an electric current flows through the circuit as a result of the difference in voltage potential developed by the two junctions. The constant temperature junction is called the cold junction. Various systems are used to provide cold junction compensation. Advances in solid-state circuitry have produced thermocouple transmitters with built-in cold junction and linearization circuits. Attaching the proper thermocouple to this circuit is all that is required to provide a linearized signal (such as 4 to 20 milliamp dc) to a controller or an indicating digital meter.

Humidity Sensors

Humidity sensors or hygrometers can be used to measure the relative humidity or dew point of ambient or moving air. Materials that respond directly to atmospheric moisture detect relative humidity directly. Two basic types are available for use in central systems—mechanical hygrometers and electronic hygrometers. A mechanical hygrometer operates on the principle that a hygroscopic material, when exposed to water vapor, retains moisture and expands. The change in size or form is detected by a mechanical linkage and converted to a pneumatic or electronic signal. The most direct indicator of relative humidity is the change in length of an untampered human hair. Human hair will exhibit a total change of about 2.5% of original length as the relative humidity changes from 0 to 100%. A human hair hygrometer can measure relative humidity from 5 to 100% to within 5% at temperatures above 32°F. Other materials used in humidity sensors include organic materials (wood fibers, paper, cotton) and manufactured materials (nylon).

Electronic hygrometers can be of the resistance or capacitance type. The resistance type uses a conductive grid coated with a hygroscopic (or water-absorbent) substance. The conductivity of the grid varies with the water retained; thus, the resistance varies according to the relative humidity. The conductive element is arranged in an AC-excited wheatstone bridge and responds to humidity changes quickly. The capacitance type is a stretched membrane of nonconductive film coated on both sides with metal electrodes and mounted within a perforated plastic capsule. The change of the sensor's capacity versus relative humidity is nonlinear with progressive characteristics, *i.e.*, with rising relative humidity. The signal is linearized and temperature is compensated in the amplifier circuit to provide an output signal as the relative humidity changes from 0 to 100%.

Pressure Transducers, Transmitters, and Controllers

A pneumatic pressure transmitter converts a change in absolute, gage, or differential pressure to a mechanical motion using a bellows, diaphragm, or Bourdon tube mechanism. When corrected through appropriate linkage, this mechanical motion produces a change in the air pressure to a controller. In some instances, the sensing and control functions are combined in a single component, a *pressure controller*.

Electronic pressure transducers may use the mechanical actuation of a diaphragm or Bourdon tube device to produce a displacement detected by a potentiometer or differential transformer. The change in resistance, in the form of a potentiometer, is then converted using electronic circuitry to a voltage or direct current compatible with the input of the controller. Another type of transducer uses a strain gage bonded to a diaphragm. The strain gage detects the displacement resulting from the force applied to the diaphragm. Electronic circuitry for temperature compensation and amplification produces a standard output signal.

Flow Rate Sensing

The following basic sensing principles and devices are used to sense water or fluid flow—orifice plate, pitot tube, venturi, turbine meter, magnetic flow meter, vortex shedding meter, and Doppler effect meter. Each of these has characteristics of rangeability, accuracy, cost, and suitability for use with clean or dirty fluids that make it appropriate for a particular application. In general, the pressure differential devices (orifice plates, venturi, and pitot tubes) are less expensive and simpler to use but have limited range; thus, their accuracy depends on how they are applied and where they are located in a system.

More sophisticated flow devices, such as turbine, magnetic, and vortex shedding meters, usually have better range and are more accurate over a wide range. In the case of a flowmeter being retrofit into an existing piping system, the expense of shutting down a system and cutting into a pipe must sometimes be considered. In this case, a noninvasive meter, such as a Doppler effect meter, can be cost-effective.

CONTROLLERS

Controllers take the sensor effect, compare it with the desired control condition (set point), and regulate an output signal to cause a control action on the controlled device. The controller and sensor can be combined in a single instrument, such as a room thermostat, or they may be two separate devices. When separate pneumatic units are used, the pneumatic controller is usually referred to as a *receiver-controller.*

Electric/Electronic Controllers

For two-position control, the controller output may be a simple electrical contact that starts a burner or pump, or one that actuates a spring-return valve or damper operator. Single-pole, double-throw (SPDT) switching circuits are used to control a three-wire unidirectional motor operator. SPDT circuits are also used for heating-cooling applications. Either single-pole, single-throw (SPST), or SPDT circuits can be modified for timed two-position action.

For floating control, the controller output is an SPDT switching circuit with a neutral zone where neither contact is made. This control is used with reversible motor operators.

Proportional control gives continuous or incremental changes in output signal to position an electrical actuator or controlled device.

Indicating or Recording Controllers

Controllers can be of the indicating or recording type. The nonindicating controller is most common in HVAC work and includes all types in which the sensing element does not provide a visual indication of the value of the controlled variable. For an indication, a separate thermometer, relative humidity indicator, or pressure gage is required. With room thermostats, for example, a separate thermometer is often attached to the cover.

An *indicating* controller has a pointer added to the sensing element or attached to it by a linkage, so that the value of the controlled variable is indicated on a suitable scale.

A *recording* controller is similar to an indicating controller, except that the indicating pointer is replaced by a recording pen that provides a permanent record on a special recorder chart paper.

Pneumatic Receiver Controllers

Pneumatic receiver controllers are normally combined with sensing elements with a force or position output to obtain a variable air pressure output. The control mode is usually proportional, but other modes such as proportional-integral can be used. These controllers are generally classified as nonrelay, relay direct, or reverse-acting.

The *nonrelay* pneumatic controller uses a restrictor in its air supply and a bleed nozzle. The sensing element positions an air exhaust flapper that varies the nozzle opening, causing a variable air pressure output applied to the controlled device, usually a pneumatic operator. The response time is relatively long, since all air must flow through the small orifice to position the actuator.

A *relay-type* pneumatic controller, directly or indirectly through a restrictor, nozzle, and flapper, actuates a relay device that amplifies the air volume available for control. This arrangement provides quicker response to a measured variable change.

Controllers are further classified by construction as direct- or reverse-acting.

Direct-acting controllers increase the output signal as the controlled variable increases. For example, a direct-acting pneumatic thermostat increases output pressure when the sensing element detects a temperature rise.

Reverse-acting controllers increase the output signal as the controlled variable decreases. A reverse-acting pneumatic thermostat increases output pressure when the temperature drops.

Direct Digital Controllers

A direct digital controller uses a digital computer (such as a microprocessor or minicomputer) to implement control algorithms on one or multiple control loops. It is fundamentally different from pneumatic or electronic controllers in that the control algorithm is stored as a set of program instructions in memory (software or firmware). The controller itself calculates the proper control signals digitally rather than using an analog circuit or mechanical change.

A digital controller can be either a single- or multiloop controller. Interface hardware allows the digital computer to process signals from various input devices such as electronic temperature, humidity, or pressure sensors described in the section on sensors. Based on the digitized equivalents of the voltage or current signals produced by the inputs, the control software calculates the required state of the output devices, such as valve and damper actuators and fan starters. The output devices are then positioned to the calculated state via interface hardware which converts the digital signal from the computer to an analog voltage or current required to position the actuator or energize a relay.

The operator enters parameters such as set points, proportional or integral gains, minimum on and off times, or high and low limits. The control algorithms stored in the computer's memory in conjunction with actual input values make the control decisions. The computer scans the input devices, executes the control algorithms, and then positions the output device(s) in a time-multiplex scheme. Digital controllers can be classified with regard to the way control algorithms are stored in memory.

Preprogrammed control routines are typically stored in permanent memory, such as PROM (programmable read only memory). To prevent unauthorized alteration, the operator can modify parameters such as set points, limits, and minimum off times within the control routines, but the program logic cannot be changed without replacing the memory chips.

User-programmable controllers allow the algorithms to be changed by the user. A programming language, provided with the controller, can vary from a derivation of a standard language (such as Pascal or Basic) to a custom language developed by the controller's manufacturer. Preprogrammed routines for proportional, proportional plus integral, Boolean logic, timers, and so forth, are typically included in the language. Standard energy management routines may also be preprogrammed and may interact with other control loops where appropriate.

A terminal allows the user to communicate with and, where applicable, modify the program in the controller. These terminals can range from hand-held units with an LCD display and several buttons or a full-size console with a CRT (cathode-ray tube) and typewriter-style keyboard. The terminal can be limited in function to allow only the display of sensor and parameter values, or powerful enough to allow changing or reprogramming the control strategies. In some instances, an operator's terminal can communicate remotely with one or more controllers, thus allowing centralized system displays, alarms, and commands.

Thermostats

Thermostats combine the control and sensing function into a single device. Because they are so prevalent, this section describes the various types and operating characteristics.

- The *occupied-unoccupied* or *dual-temperature room* thermostat controls at a reduced temperature at night. It may be indexed (changed from occupied to unoccupied operation) individually or in groups by a manual or time switch from a remote point. Some electric types have an individual clock and switch built into the thermostat.

  The *pneumatic day-night* thermostat uses a two-pressure air supply system—the two pressures often being 13 and 17 psig, or 15 and 20 psig. Changing the pressure at a central point from one value to the other actuates switching devices in the thermostat and indexes it from occupied to unoccupied or vice versa. Supply air mains are often divided into two or more circuits, so that switching can be done in various areas of the building at different times. For example, a school building may have separate circuits for classrooms, offices and administrative areas, the auditorium, the gymnasium, and locker rooms.

- The *heating-cooling* or *summer-winter* thermostat can have its action reversed and its set point changed by indexing. It is used to actuate controlled devices, such as valves or dampers, that regulate a heating source at one time and a cooling source at another. Often it is indexed manually in groups by a switch, or automatically by a thermostat sensing the temperature of the control agent (outdoor temperature or another suitable variable).

  The *pneumatic heating-cooling* thermostat uses a two-pressure air supply, as described for occupied-unoccupied thermostats.
- *Multistage* thermostats are arranged to operate two or more successive steps in sequence.
- A *submaster* thermostat has its set point raised or lowered over a predetermined range, in accordance with variations in output from a master controller. The master controller can be a thermostat, manual switch, pressure controller, or similar device. For example, a master thermostat measuring outdoor air temperature can be used to readjust a submaster thermostat set point controlling the water temperature in a heating system. Master-submaster combinations are sometimes designated as having single-cascade action. When this action is accomplished by a single thermostat with more than one measuring element, it is known as *compensated control.*
- A *wet-bulb* thermostat is often used for humidity control with proper control of the dry-bulb temperature. A wick or other means for keeping the bulb wet with pure (distilled) water, and rapid air motion, to ensure a true wet-bulb measurement, are essential. Because of maintenance problems, wet-bulb thermostats are seldom used.
- A *dew-point* thermostat is designed to control dew-point temperatures. Dew point is measured in several ways—the most accurate is to measure the temperature of an electrically heated chemical layer.
- A *dead band* thermostat has a wide differential over which the thermostat remains neutral, requiring neither heating nor cooling. This differential may be adjustable up to 10°F. The thermostat then controls to maximum or minimum output over a small differential at each end of the dead bank, as shown in Figure 17.

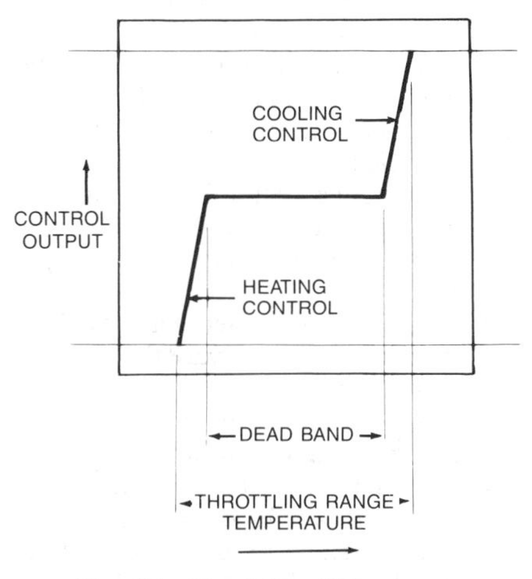

Fig. 17 Dead Band Thermostat

AUXILIARY CONTROL DEVICES

In addition to conventional controllers and controlled devices, many control systems require auxiliary devices including:

- *Transformers* to provide current at the required voltage.
- *Electric relays* to control electric heaters or to start and stop oil burners, refrigeration compressors, fans, pumps, or other apparatus for which the electrical load is too large to be handled directly by the controller. Other uses include time-delay and circuit-interlocking safety applications.
- *Potentiometers* for manual positioning of proportional control devices, for remote set point adjustment of electronic controllers, and for feedback.
- *Manual switches* for performing several operations. These can be two-position or multiple-position, with single or multiple poles.
- *Auxiliary switches* on valve and damper operators for providing a selected sequence of operation.

Auxiliary control equipment for pneumatic systems includes the following:

- *Air compressors* and accessories, including driers and filters, to provide a source of clean, dry air at the required pressure.
- *Electropneumatic relays*, electrically actuated air valves for operating pneumatic equipment in accordance with variations in electrical input to the relay.
- *Pneumatic-electric switches*, actuated by air pressure to make or break an electrical circuit.
- *Pneumatic relays*, actuated by the pressure from a controller to perform several functions. They may be divided into two groups:
  1. *Two-position relays*, which permit a controller actuating a proportional device to also actuate one or more two-position devices.
  2. *Proportional relays*, which are used to reverse the action of a proportional controller, select the higher or lower of two or more pressures, average two or more pressures, respond to the difference between two pressures, add or subtract pressures, and amplify or retard pressure changes.
- *Positive positioning relays* are devices for ensuring accurate positioning of a valve or damper operator in response to changes in pressure from a controller. They are affected by the position of the operator and the pressure from the controller. Whenever the two are out of balance, the relays will use full control air pressure to change the pressure applied to the operator until balance is restored.
- *Switching relays* are pneumatically operated air valves for diverting air from one circuit to another, or for opening and closing air circuits.
- *Pneumatic switches* are manually operated devices for diverting air from one circuit to another or for opening and closing air circuits. They can be two-position or multiple-position.
- *Gradual switches* are proportional devices for manually varying the air pressure in a circuit.
- *Logic networks* and *square root extractors.*

Auxiliary control devices common to both electric and pneumatic systems include the following:

- *Step controllers* for operating a number of electric switches in sequence by means of a proportional electric or pneumatic operator. They are commonly used for controlling several steps of refrigeration capacity and may be arranged to prevent simultaneous starting of compressors and to alternate the sequence to equalize wear. They may also be used for sequence operation of electric heating elements and other equipment.
- *Power controllers* for controlling electrical power input to resistance-type electric heating elements. The final controlled device may be a variable autotransformer, a saturable-core reactor, or a solid-state power controller. They are available with various ratings for single- or three-phase heater loads and are usually arranged to regulate power input to the heater in response to the demands of proportional electronic or pneumatic controllers. However, solid-state controllers may also be used in two-position control modes.
- *Clocks* or *timers* for turning apparatus on and off at predetermined times, for switching control systems from day to night operation, and for other time sequence functions.

- *Transducers*, which consist of combinations of electric or pneumatic control devices, may be required. For these applications, transducers are used to convert electric signals to pneumatic output or vice versa. Transducers may convert proportional input to either proportional or two-position output.

The electronic-to-pneumatic (E-to-P) transducer has come into widespread use (Figure 18). It converts a proportional electronic output signal into a proportional pneumatic signal and can be used to combine electronic and pneumatic control components to form a control loop, as illustrated in Figure 19. Electronic components are used for sensing and signal conditioning, while pneumatics are used for actuation. The electronic controller can either be analog or digital.

The E-to-P transducer presents a special option for retrofit applications. An existing HVAC system with pneumatic controls can be retrofit with electronic sensors and controllers while retaining the existing pneumatic actuators (Figure 20). One

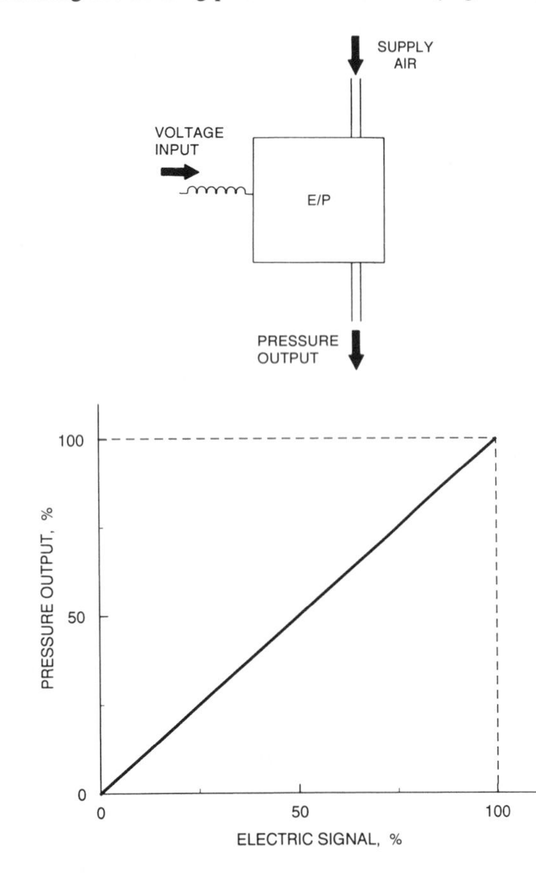

Fig. 18 Electronic-to-Pneumatic Transducer (E/P)

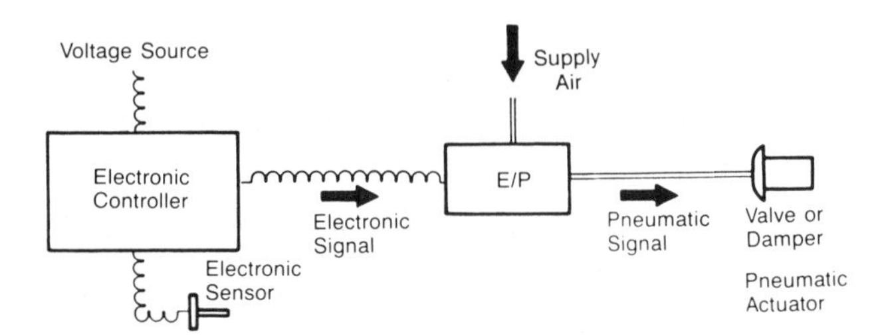

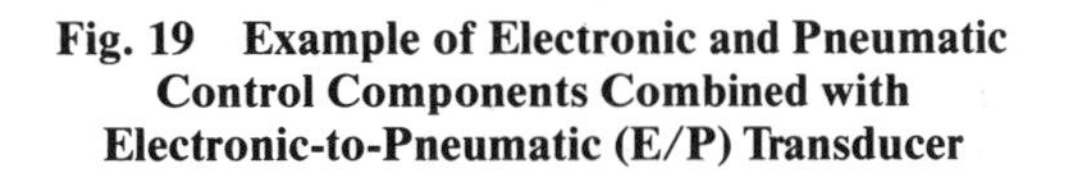

Fig. 19 Example of Electronic and Pneumatic Control Components Combined with Electronic-to-Pneumatic (E/P) Transducer

advantage of this approach is that the retrofit can be accomplished with only a minor interruption to the operation of the control system.

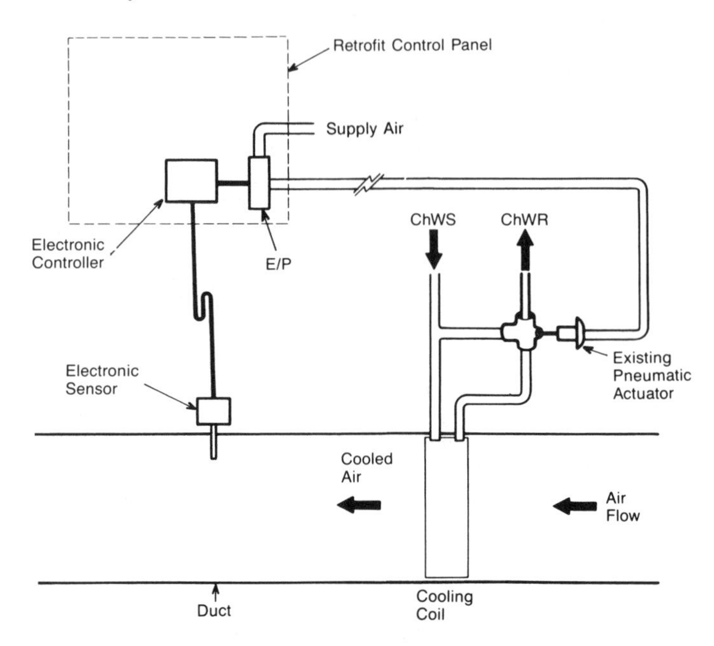

Fig. 20 Retrofit of Existing Pneumatic System with Electronic Sensors and Controllers

CONTROL OF CENTRAL SUBSYSTEMS

CONTROL OF OUTDOOR AIR QUANTITY

Fixed minimum outdoor air control provides ventilation air, space pressurization (exfiltration), and makeup for exhaust fans.

For systems *without* return fans, the outdoor air damper (Figure 21A) is interlocked to open only when the supply fan operates. The outdoor air damper should open quickly when the fan turns on to prevent excessive negative duct pressurization. In some systems, the fan's on-off switch opens the outdoor air damper before the fan is started. Rate of outdoor airflow is determined by damper opening and the pressure difference between the mixed air plenum and the outdoor air plenums.

Fixed minimum outdoor air control for systems *with* return fans has two variations. Minimum outdoor airflow is determined by the pressure (airflow) difference between the supply and return fans (Figure 21B).

If the outdoor air supplied is greater than the difference between the supply and return fan airflows, a variation of economizer cycle control is used (Figure 21C).

With systems using 100% outdoor air, all air goes to the fan and no air is returned. The outdoor air damper (Figure 22) is interlocked and usually opens before the fan starts.

Economizer cycle control reduces cooling costs when outside conditions are suitable; *i.e.*, when the outdoor air temperature is low enough to be used as a cooling medium. If outdoor air is below a high-temperature limit, typically 65°F, the return, exhaust, and outdoor air dampers modulate to maintain a ventilation cooling set point, typically 55 to 60°F (Figure 23). The relief dampers are interlocked to close and the return air dampers to open when the supply fan is not operating. When the outdoor air temperature exceeds the high-temperature limit set point, the outdoor air damper is closed to a fixed minimum and the exhaust and return air dampers close and open, respectively.

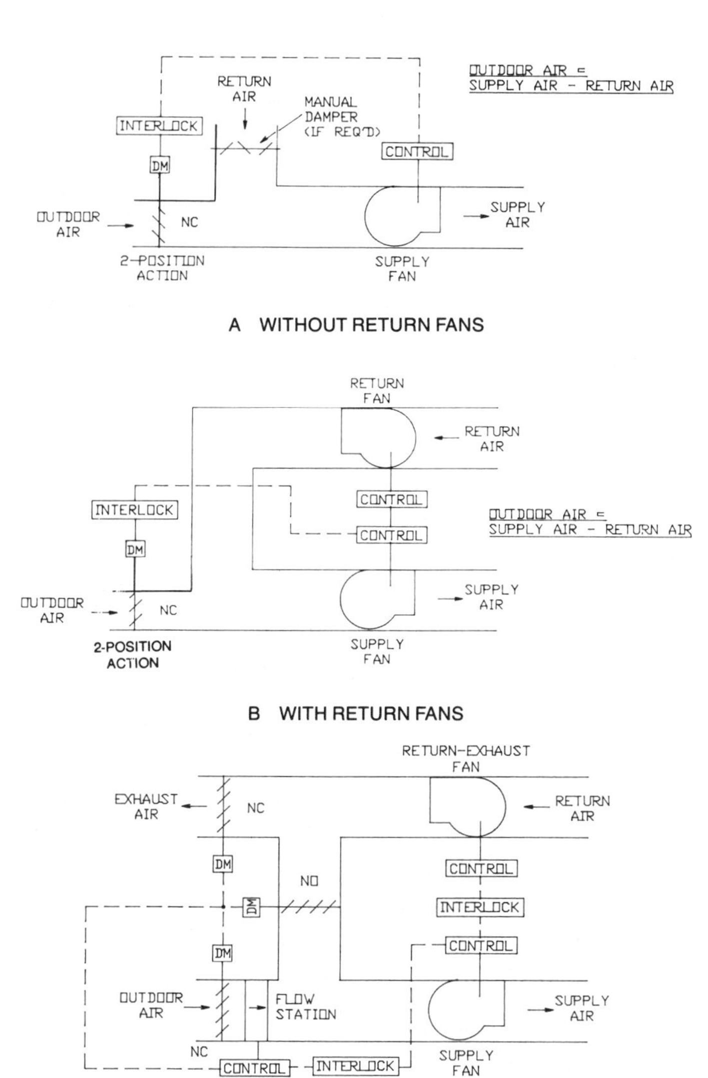

Note: OUTDOOR AIR = SUPPLY AIR − RETURN AIR + EXHAUST AIR

C WITH RETURN-EXHAUST FANS—OUTDOOR AIR IS GREATER THAN DIFFERENCE BETWEEN SUPPLY AND RETURN AIRFLOWS

Fig. 21 Fixed Minimim Outdoor Air for Various Systems

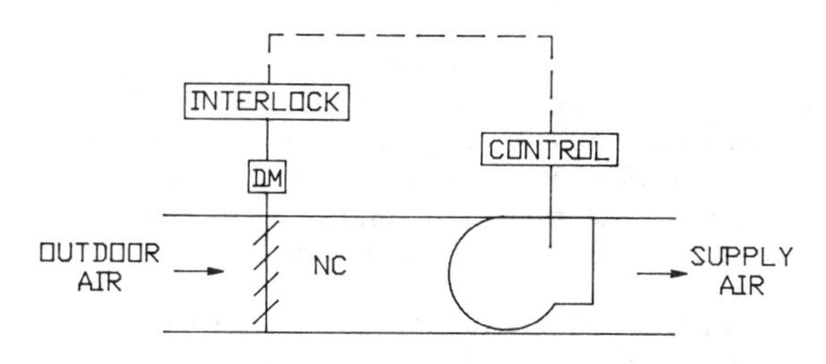

Fig. 22 100% Outdoor Air Control

Enthalpy economizer control replaces the high-temperature limit of the economizer cycle to further reduce energy costs when latent loads are significant. The interlock function (Figure 23) can be based on (1) a fixed enthalpy high limit, (2) a comparison with return air so as not to exceed return air enthalpy, or (3) a combination of enthalpy and high-temperature limits.

VAV warm-up control during unoccupied periods requires no outdoor air. Typically, outdoor and exhaust dampers remain

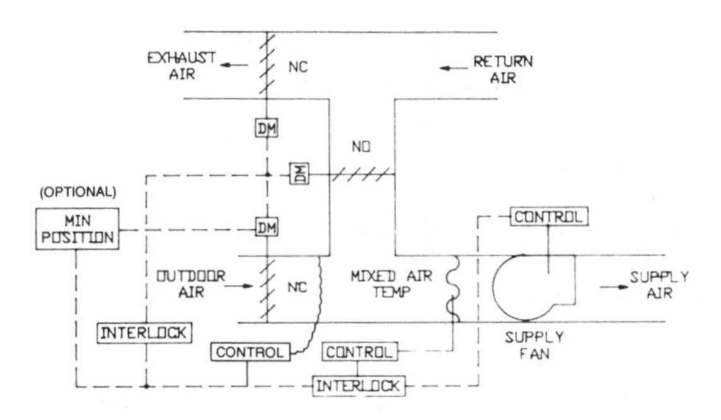

Fig. 23 Economizer Cycle Control

closed. However, systems with a return fan (Figure 24) should position the outdoor air damper at its minimum position and limit supply airflow (volume) to return air airflow (volume) to prevent excessive positive or negative duct pressurization. The Control System Design and Application section has information on fan control during warm-up.

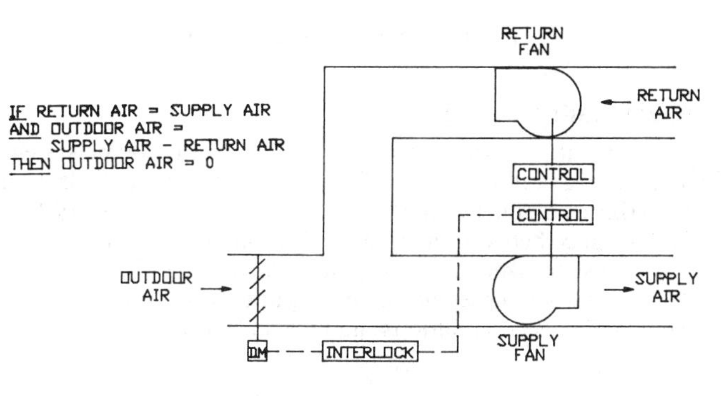

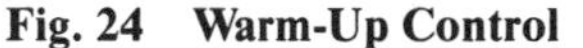

Fig. 24 Warm-Up Control

Night cool-down control (night purge) provides 100% outdoor air for cooling during unoccupied periods (Figure 25). The space is cooled to a space set point, typically 9 °F above outdoor air temperature. Limit controls prevent operation if outdoor air is above space dry-bulb, if outdoor air dew-point temperature is excessive, or if outdoor air dry-bulb temperature is too cold, typically 50 °F. The night cool-down cycle is initiated before sunrise, when overnight outside temperatures are usually the coolest, prior to the beginning of the optimum start program—typically 4 h before occupancy.

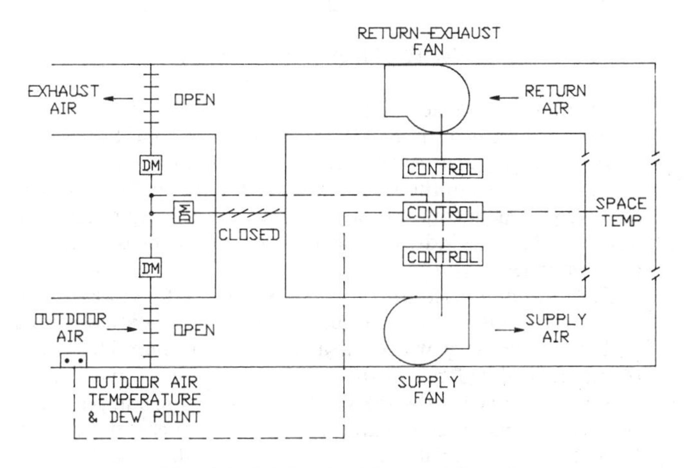

Fig. 25 Night Cool Down Control

FAN CONTROL

Fan modulation is provided by the following methods:

- Fan inlet or discharge dampers
- Fan scroll dampers
- Variable speed
- Fan runaround or scroll bypass
- Controllable pitch
- Fan inlet vanes

Constant Volume Control

Constant volume control fixes the airflow rate when duct resistances vary (Figure 26).

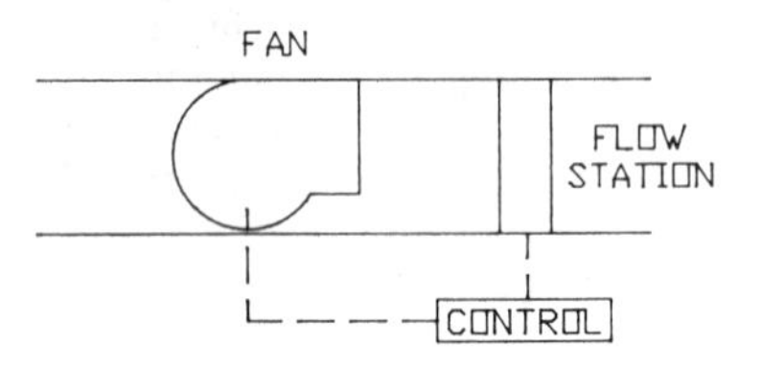

Fig. 26 Constant Volume Control

Duct Static Control

Duct static control for VAV and other terminal-type systems maintains a static pressure at a location of measurement. The location (based on duct layout) is typically between 75 and 100% of the distance between the first and last air terminal (Figure 27). Care must be taken in selecting the "reference static" sensor location. Controller upset due to opening and closing doors, elevator shafts and other conditions should also be prevented.

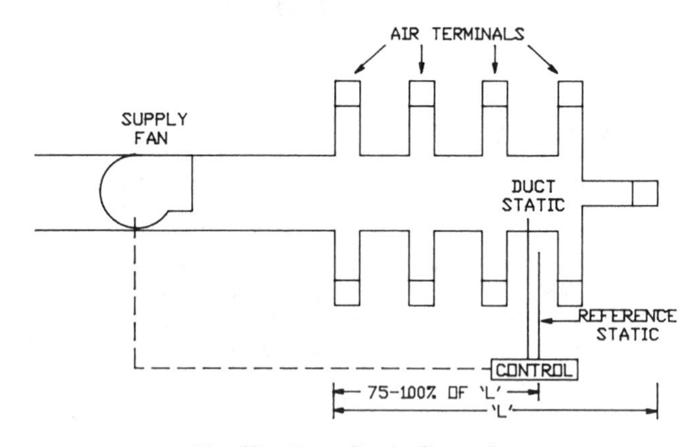

Fig. 27 Duct Static Control

The pressure selected provides minimum static pressure to all air terminal units during all supply fan design conditions.

Multiple static sensors (Figure 28) are required when more than one duct runs from the supply fan. The sensor with the lowest static requirement controls the fan. Since duct run-outs may vary, a control that uses individual set points for each measurement is recommended.

Duct static limit control prevents excessive duct pressures—usually at the discharge of the supply fan. Two variations are used —(1) a *fan shutdown*, which is a safety high limit control that turns the fans off, and (2) a *controlling high limit*, which is used in systems having zone fire dampers. When the zone fire damper closes (Figure 29), duct pressure drops, causing the duct static control to increase fan modulation; however, the controlling high limit will override.

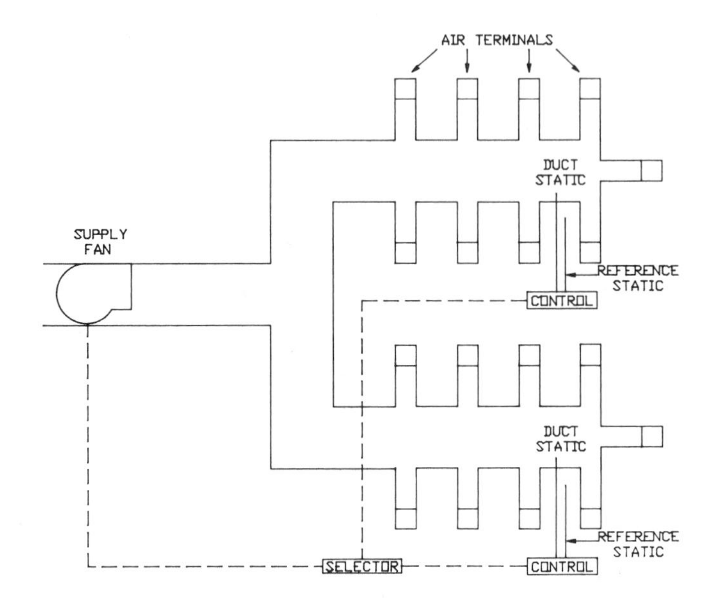

Fig. 28 Multiple Static Sensors

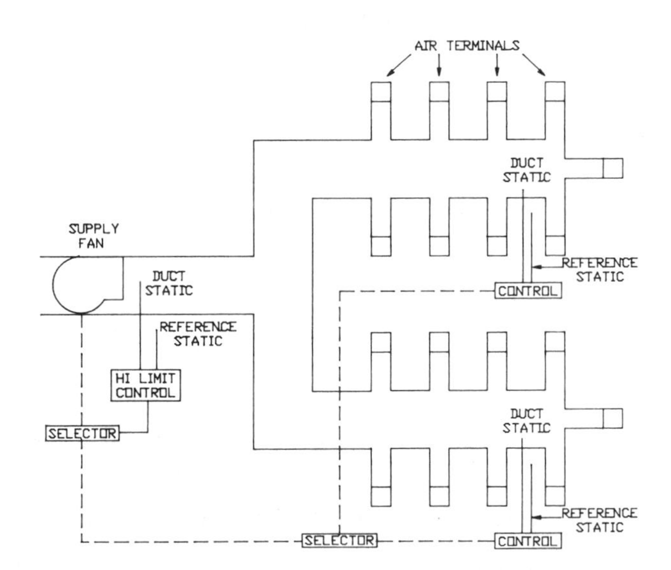

Fig. 29 Duct Static Limit Control: Controlling High Limit Type

Supply fan warm-up control for systems having a return fan must prevent the supply fan from delivering more airflow than the return fan maximum capacity during warm-up mode (Figure 30). If supply fan airflow exceeds return fan airflow maximum capacity, excessive duct pressurization can occur. Limiting pressure controllers can assist in preventing damage from over- or under-pressurizing the ductwork.

Return fan static control from returns having local (zoned) flow control is identical to the supply fan static control concept (Figure 27).

Return Fan Control

Return fan control for VAV systems provides proper building pressurization and minimum outdoor air.

Duct static control to the supply-fan-modulation means is sent to the return-fan-modulation means (Figure 31). This open loop (without feedback) control requires similar supply and return fan airflow modulation characteristics. The return fan airflow is adjusted at minimum and maximum airflow conditions. System

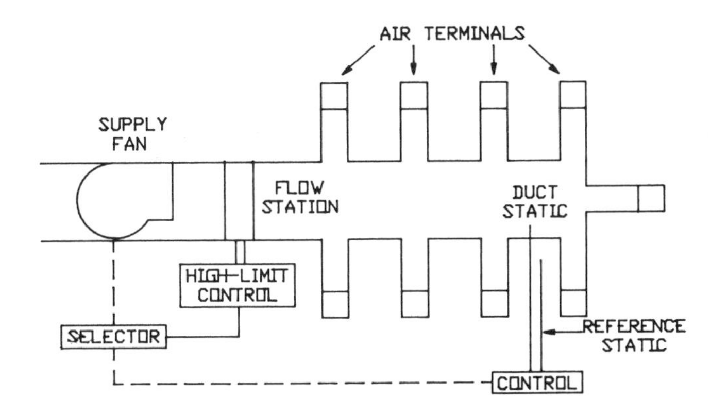

Fig. 30 Supply Fan Warm-Up Control

airflow turndown should not be excessive, typically a maximum of 50%. Provisions for warm-up and exhaust fan switching are impractical.

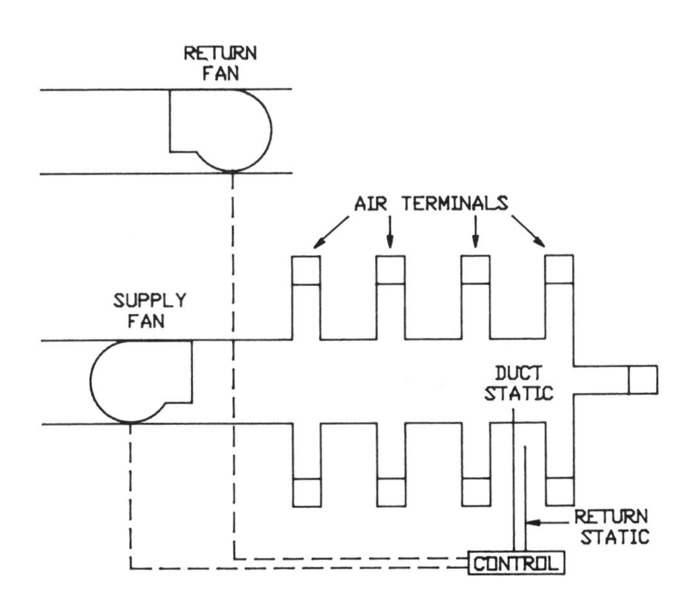

Fig. 31 Duct Static Control of Return Fan

Direct building pressurization control of the return fan is accomplished by measuring the space and outdoor static pressures (Figure 32). The indoor static location must be selected carefully—away from doors and openings to outside, away from elevator lobbies, and, in a large representative area using a sensor, shielded from air velocity effects. Likewise, the outdoor static location must be selected carefully, typically 10 to 15 ft above the building and oriented to minimize wind effects from all directions. The amount of minimum outdoor air varies with building permeability and exhaust fan switching. During the warm-up mode, the building static pressure is reset to zero differential pressure, and all exhaust fans are turned off.

Airflow tracking uses duct airflow measurements to control the return air fans (Figure 33). Sensors called flow stations are typically multiple-point, pitot tube, and averaging. To maintain pressurization of the building, provisions must be made for exhaust fan switching. Warm-up is accomplished by setting the return airflow equal to the supply fan airflow—usually with exhaust fans turned off—and limiting supply fan volume to return fan capability.

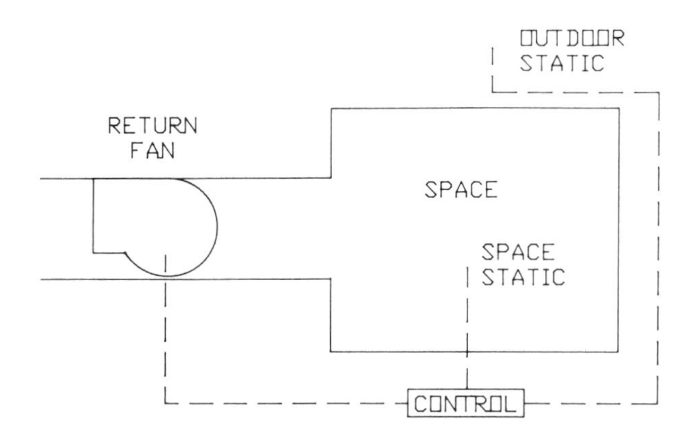

Fig. 32 Direct Building Pressurization Control

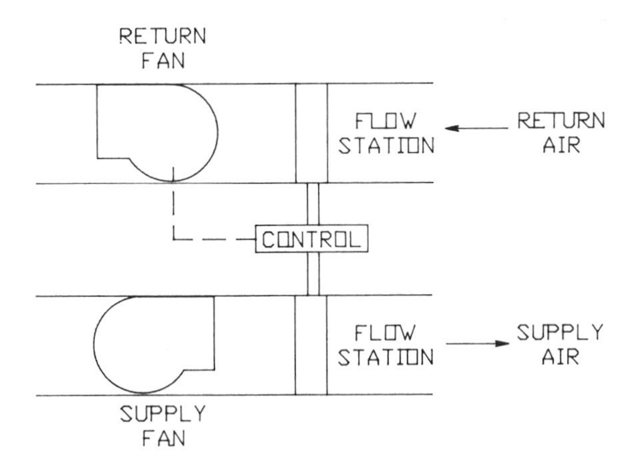

Fig. 33 Airflow Tracking Control

During night cool-down, the return fan operates in the normal mode.

Sequencing Fans

Sequencing of fans for VAV systems gives greater airflow reduction, resulting in greater operating economy and stable fan operation if airflow reductions are significant. Alternation of fans usually provides greater system reliability.

Centrifugal fans are controlled to keep system disturbances to a minimum when additional fans are started. The added fan is started and slowly brought to capacity while the capacity of operating fans is *simultaneously* reduced. The output of all fans combined then equals the output before fan addition.

Vane axial fans usually cannot be sequenced in the same manner as centrifugal fans. To avoid stall, operating fans must be reduced to some minimum level. Hence, additional fans may be started and both fans modulated to achieve equilibrium.

Unstable Fan Operation

Fan instability for VAV systems can usually be avoided by proper fan sizing. However, if airflow reduction is large (typically over 60%) a technique for maintaining airflow within the fan's stable range, such as sequencing of fans, is usually required.

Coil reset avoids fan instability by resetting the cooling coil discharge temperature higher (Figure 34) so that the building cooling loads require greater airflow. Since a time lag occurs between temperature reset and demand for more airflow, the value at which reset starts should be selected on the safe side of the fan instability point. When this technique is used, dehumidification requirements should be checked to be sure they can be maintained.

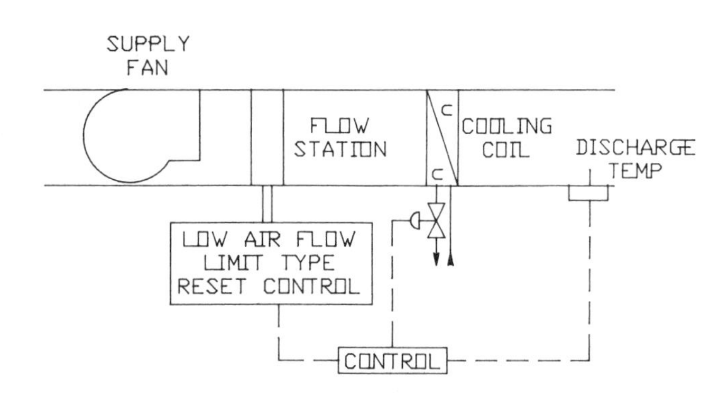

Fig. 34 Coil Reset Control to Prevent Supply Fan Instability

Fan bypass allows airflow to "short circuit" around the fan so that a minimum airflow can be maintained (Figure 35). This technique uses constant volume control to limit the low airflow.

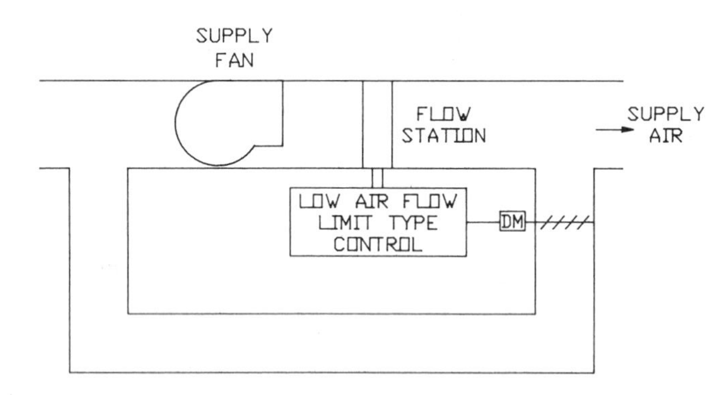

Fig. 35 Fan Bypass Control to Prevent Supply Fan Instability

CONTROL OF HEATING COILS

Heating coils in central air-handling units preheat, reheat, or heat, depending on the climate and the amount of minimum outdoor air needed.

Preheating Coils

Control of steam or hot water preheating coils must include protection against freezing, unless the minimum outdoor air quantity is small enough to keep the mixed air temperature above freezing and there is enough mixing to prevent stratification. Even though the average mixed air temperature is above freezing, inadequate mixing may allow a freezing air stratum to impinge on the coil.

Steam preheat coils should have two-position valves and vacuum breakers to prevent a buildup of condensate in the coil. The valve should be fully open when outdoor air (or mixed air) temperature is below freezing. This causes unacceptably high coil discharge temperatures at times, so that face and bypass dampers are required for final temperature control (Figure 36). The bypass damper should be sized to provide the same pressure drop at full bypass airflow as the combination of face damper and coil at full airflow.

Hot water preheat coils must maintain a minimum water velocity of 3 fps in the tubes to prevent freezing. A two-position valve combined with face and bypass dampers can usually be used to control the water velocity. More commonly, a secondary pump control in one of two configurations (Figures 37 and 38) is used. The control valve modulates to maintain the desired coil air discharge temperature, while the pump operates to maintain the minimum tube water velocity when outdoor air is below freezing. The system in Figure 38 uses less pump power, allows variable flow in the hot water supply main, and is preferred for energy conservation. The system in Figure 37 may be required on small systems with only one or two air handlers, or where constant main water flow is required.

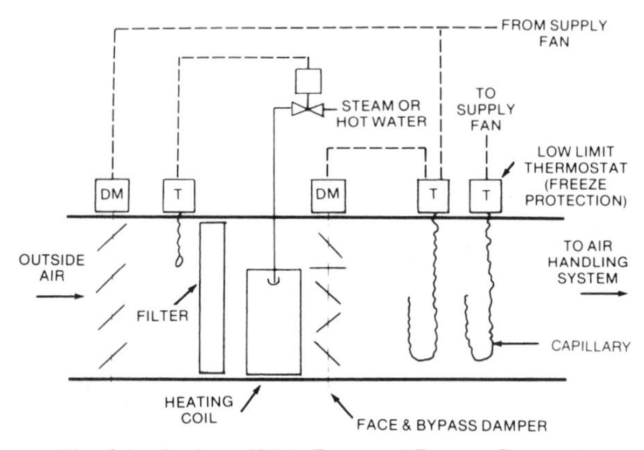

Fig. 36 Preheat With Face and Bypass Dampers

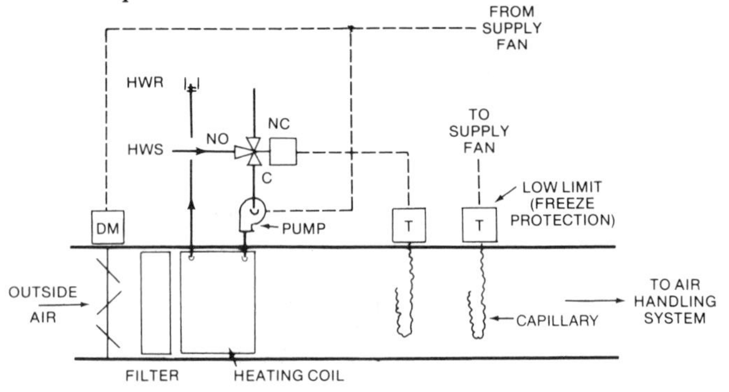

Fig. 37 Preheat-Secondary Pump and Three-Way Valve

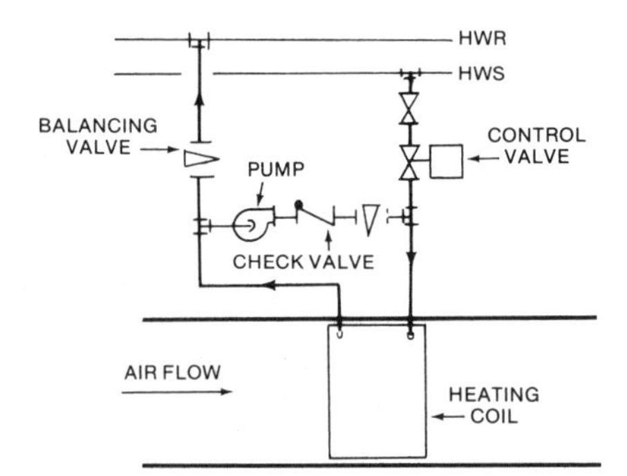

Fig. 38 Preheat-Secondary Pump and Two-Way Valve

Some systems may use a glycol solution in combination with any of these methods.

Reheat and Heating Coils

Steam or hot water reheat and heating coils not subject to freezing can be controlled by simple two- or three-way modulating valves (Figure 39). Steam distributing coils are required to ensure proper steam coil control. The valve is controlled by coil discharge air temperature or by space temperature, depending on the HVAC system. Valves are usually open to allow heating if control power fails. On many systems, outdoor air temperature resets the heating discharge controller.

Electric heating coils are controlled in either a two-position or modulating mode. Two-position operation uses power relays

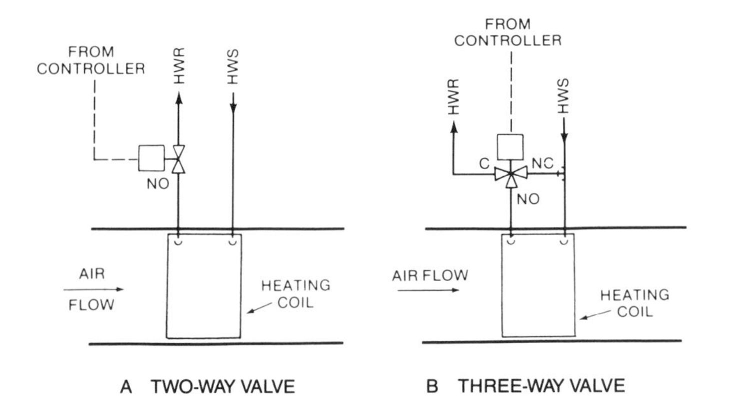

Fig. 39 Heating Control

with contacts sized to handle the power required by the heating coil. Timed two-position control requires a timer and contactors. The timer can be electromechanical, but it is usually electronic and provides a time base of 1 to 5 min. Step controllers provide cam-operated sequencing control of up to ten stages of electric heat. Each stage may require a contactor, depending on the step controller contact rating. Thermostat demand determines the percentage of "on" time. Since rapid cycling of mechanical or mercury contactors can cause maintenance problems, solid-state controllers like SCRs (silicon control rectifiers) or triacs are preferred. These devices make cycling so rapid that the control is proportional. For safety (and code requirements), an electric heater must have a minimum airflow switch and high-temperature limit sensors—one with manual reset and one with automatic reset. Therefore, face and bypass dampers are not used. A control system with a solid-state controller and safety controls is shown in Figure 40.

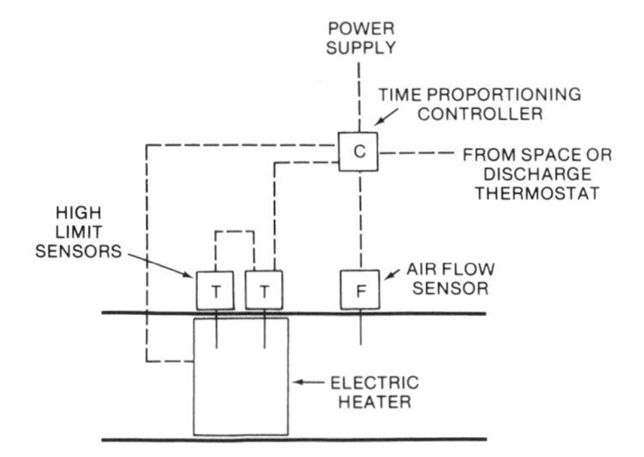

Fig. 40 Electric Heat: Solid-State Controller

CONTROL OF COOLING COILS

Cooling coils in central air-handling units use chilled water, brine, glycol, or refrigerant (direct expansion) as the cooling medium. Most comfort cooling processes involve some dehumidification. The amount of dehumidification that occurs is a function of the effective coil surface temperature and is limited by the freezing point of water or other coolant. If water condensing out of the airstream freezes on the coil surface, airflow is restricted and, in severe cases, may be shut off. The practical limit is about 40 °F dew point on the coil surface. As indicated in Figure 41, this results in a relative humidity of about 30% at a space temperature of 75 °F, which is adequate for most industrial processes. When lower humidities are required, chemical dehumidifiers are necessary (see Humidity Control).

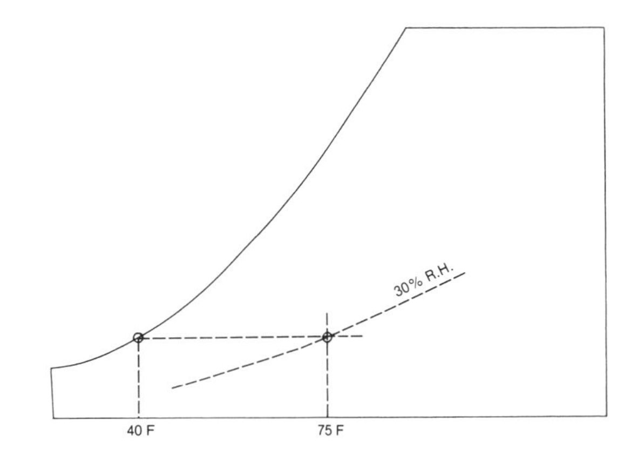

Fig. 41 Cooling and Dehumidifying—Practical Low Limit

Chilled water or brine cooling coils are controlled by two- or three-way valves (Figures 42). Valves are similar to those used for heating control but are usually closed to prevent cooling when the fan is off. The valve typically modulates in response to coil air discharge temperature or space temperature. Outdoor air reset is sometimes used.

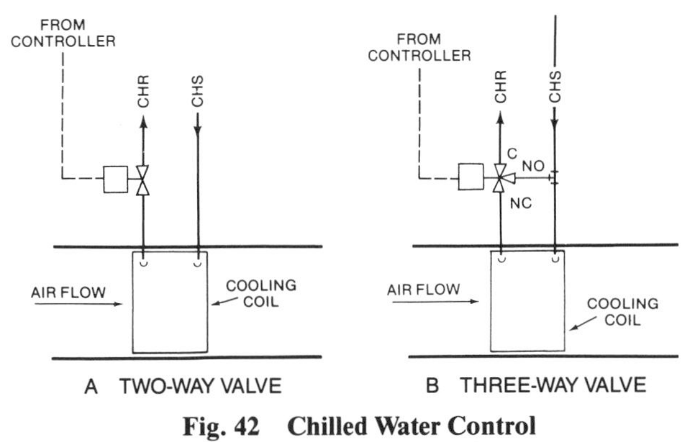

Fig. 42 Chilled Water Control

When maximum relative humidity control is required, a space or return air humidistat is provided with the space thermostat. To limit maximum humidity, a control function selects the higher of the two output signals and controls the cooling coil valve accordingly. A reheat coil must maintain space temperature (Figure 43). If humidification is also provided, this cycle is sometimes referred to as a "constant temperature, constant humidity" cycle.

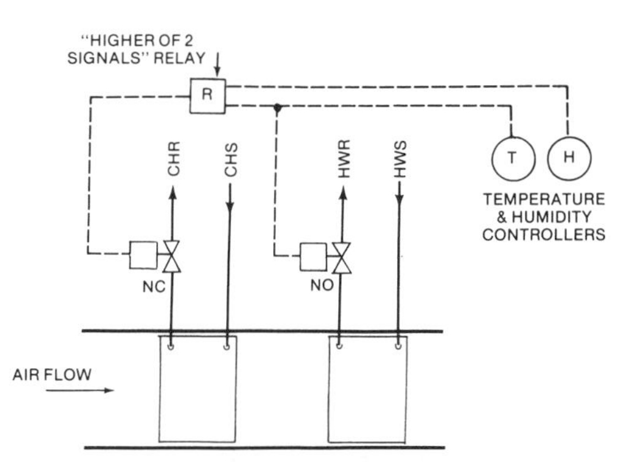

Fig. 43 Cooling and Dehumidifying with Reheat

Direct expansion (DX) cooling coils are usually controlled by solenoid valves in the refrigerant liquid line (Figure 44). Face and bypass dampers are not recommended because ice tends to form on the coil when airflow is reduced. Control can be improved by using two or more stages, with the solenoid valves controlled in sequence and with a differential of 1 or 2°F between stages (Figure 45). The first stage should be the first coil row on the entering air side, with the following rows forming the second and succeeding stages. Side-by-side stages tend to generate icing on the stage in use, with reduction of airflow and loss of control. Modulating control is achieved using a variable suction pressure controller (Figure 46). This type of control is uncommon but necessary if accurate control of discharge or space temperature is needed. If this type control is required, a refrigerant compressor capacity reduction control should be supplied.

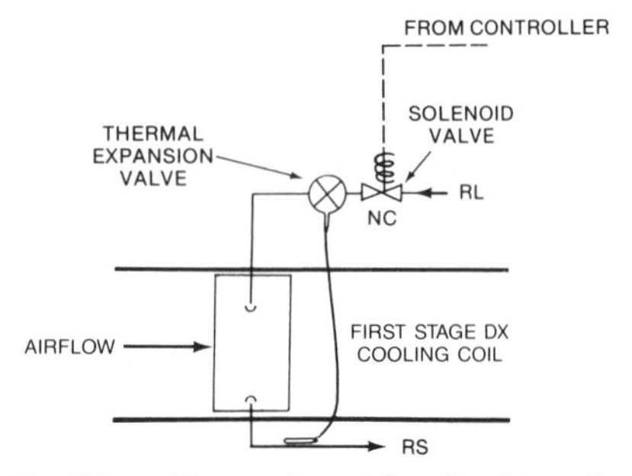

Fig. 44 Direct Expansion—Two-Position Control

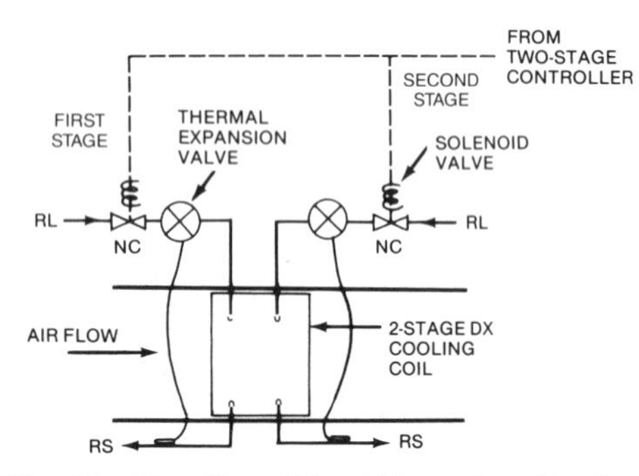

Fig. 45 Two-Stage Direct Expansion Cooling

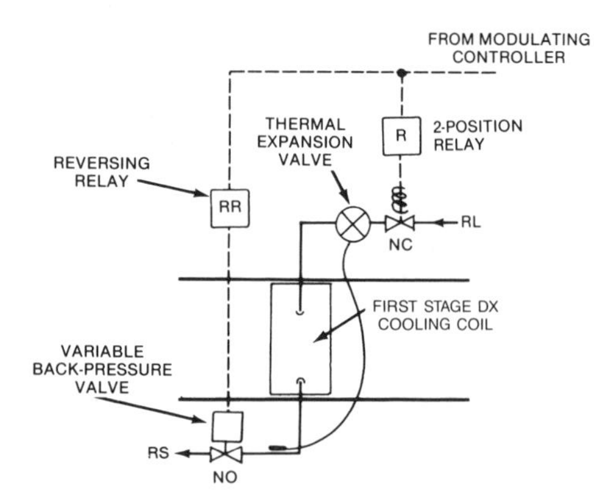

Fig. 46 Modulating Direct Expansion Cooling

Evaporative cooling can also provide sensible cooling using standard evaporative coolers or air washers (Figure 47). Process efficiency is described by the ratio of dry-bulb temperature difference between inlet and outlet divided by the difference between inlet dry-bulb and the inlet wet-bulb temperature. Air washers are usually 90 to 95% efficient, while evaporative coolers vary from 50 to 90%. The space temperature controls the spray pump (Figure 48), which causes high relative humidity in the space; humidity cannot be controlled, since it is primarily a function of outdoor wet-bulb temperature.

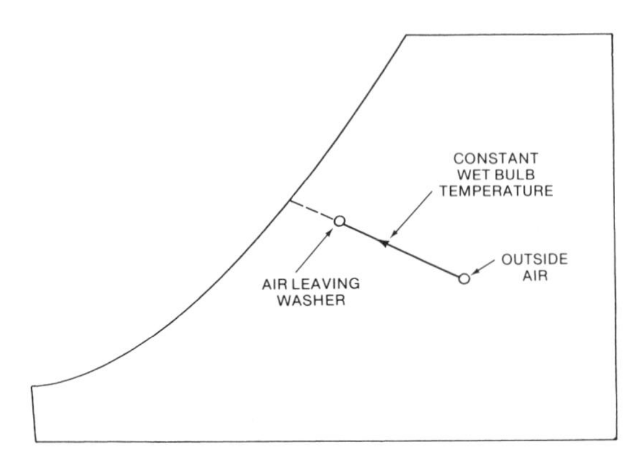

Fig. 47 Psychrometric Chart: Evaporative Cooling

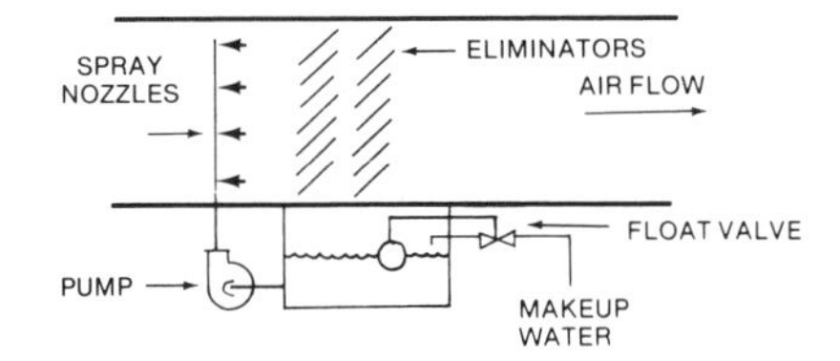

Fig. 48 Evaporative Cooling: Air Washer

HUMIDITY CONTROL

While simple cooling by refrigeration maintains an upper limit to space humidity, it does not control humidity without additional equipment.

Dehumidification

The *sprayed coil dehumidifier* (Figure 49) was formerly used for dehumidification. The systems in Figures 48 and 49 have essentially the same effect (Figure 50). Space relative humidities ranging from 35 to 55% at 75°F can be obtained with these systems; however, maintenance and operating costs caused by reheat and solid deposition on the coil make this system undesirable.

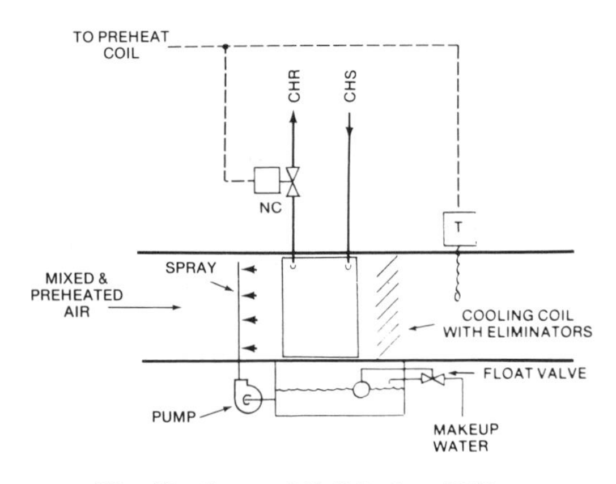

Fig. 49 Sprayed Coil Dehumidifier

Chemical dehumidifiers can develop space humidities below those possible with cooling/dehumidifying coils. These devices adsorb moisture using silica gel or a similar material. For continuous operation, heat is added to regenerate the material. The adsorption process also generates heat (Figure 51). Figure 52 shows a typical control system.

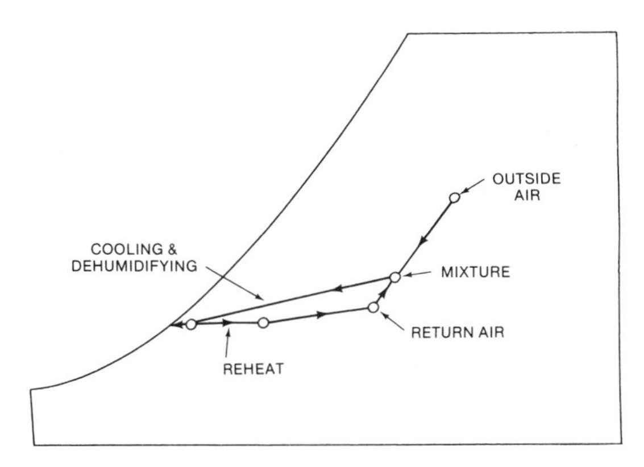

Fig. 50 Psychrometric Chart for Air Washer Evaporative Cooling and Sprayed Coil Dehumidification

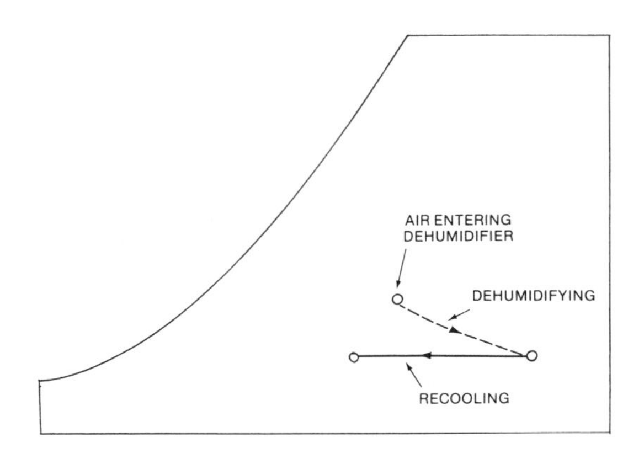

Fig. 51 Psychrometric Chart: Chemical Dehumidification

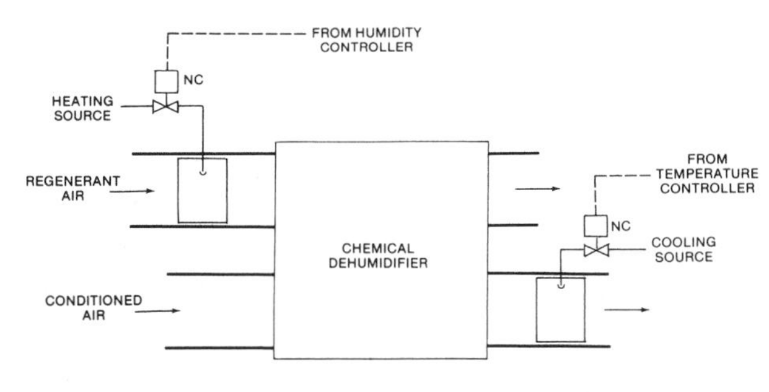

Fig. 52 Chemical Dehumidifier

Humidification

Evaporative pans (usually heated), *steam jet*, and *atomizing spray tubes* are all used to humidify a space. A space or return air humidistat is used for control. A high limit duct humidistat should also be used to minimize moisture carryover or condensation in the duct (Figure 53). Proper use and control of humidifiers can achieve high space humidity, although humidifiers more often maintain design minimum humidity during the heating season.

CONTROL OF SPACE CONDITIONS

A space thermostat controls *single-zone* heating and cooling directly. If used, a humidifier is controlled by the space humidistat.

Multizone and *dual-duct* units have mixing dampers controlled by the zone space thermostat for each zone (Figure 54). If used,

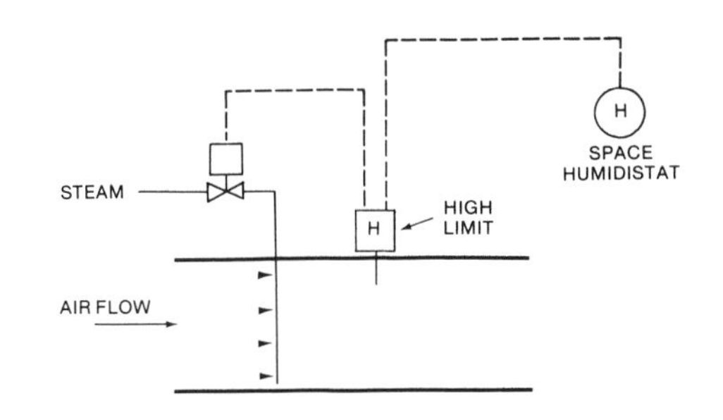

Fig. 53 Steam Jet Humidifier

humidifiers are usually controlled by a return air humidistat. Some states no longer permit mixing hot and cold air to provide simultaneous heating and cooling. A three-deck unit may be used as an alternative in a multizone system (Figure 55). Zone dampers in this unit operate with sequenced damper motors (DM) either (1) to mix hot supply air with bypass air when the cold deck damper is closed or (2) to mix cold supply air with bypass air when the hot deck damper is closed.

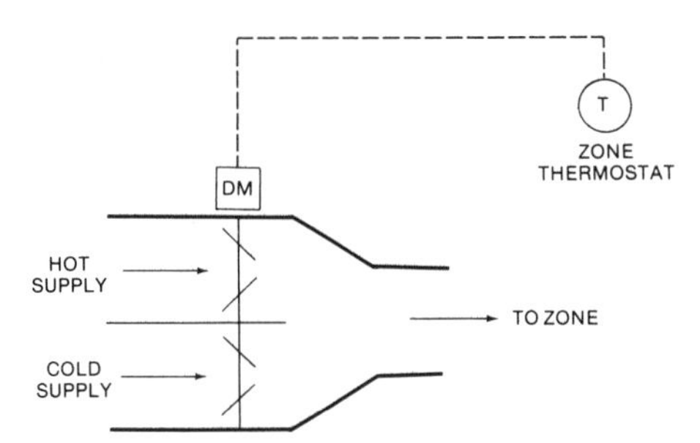

Fig. 54 Zone Mixing Dampers—Multizone System

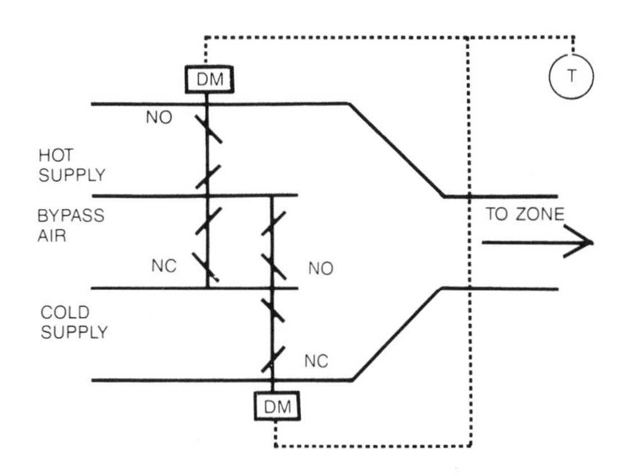

Fig. 55 Zone Mixing Dampers—Three-Deck Multizone System

Variable air volume units have motorized dampers in each zone supply duct. A related zone space thermostat controls each damper by a flow sensor/controller that is reset by the thermostat (Figure 56). If used, humidifiers are controlled by a return air humidistat or a humidistat in a representative critical zone.

The system layout can make the control of economizers and static pressure difficult. The interaction of the return exhaust fan and the static pressure and volume control is a particular concern in the control system layout.

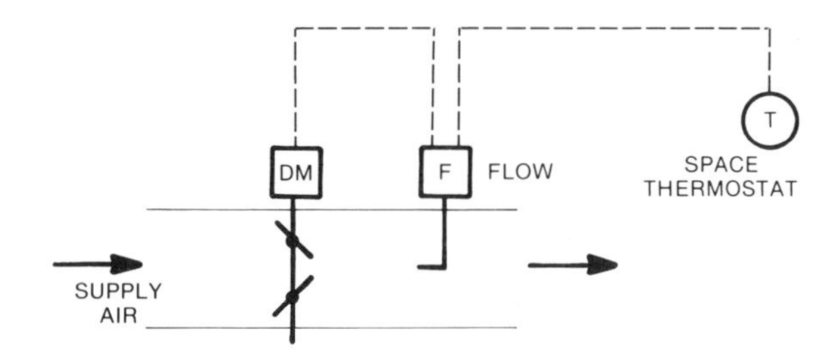

Fig. 56 Zone Variable Air Volume Damper

Pressure Control

The most common application for static pressure controls is fan capacity control in VAV systems. Static pressure controls can also be used to pressurize a building or space relative to adjacent spaces or outdoors. Typical applications include clean rooms (positive to prevent infiltration), laboratories (positive or negative, depending on use), and various manufacturing processes, such as spray-painting rooms. The pressure controller usually modulates dampers in the supply ducts to maintain desired pressures as exhaust volumes change.

CONTROL SYSTEM DESIGN AND APPLICATION

CENTRAL AIR-HANDLING SYSTEMS

Variable Air Volume

Variable air volume systems vary the amount of air supplied by terminal units in individual zones as the load varies in those zones. Hybrid systems that use bypass terminal units to vary air volume to the space while handling a constant air volume from the central fan are treated as constant volume (CV) systems.

From a control standpoint, VAV systems can further be classified as:

1. Single-duct cooling only (Figure 57)
2. Single-duct heating/cooling
3. Dual-duct (see Dual-Duct Systems in this section)
   a) Dual-duct single supply fan
   b) Dual-duct dual supply fan

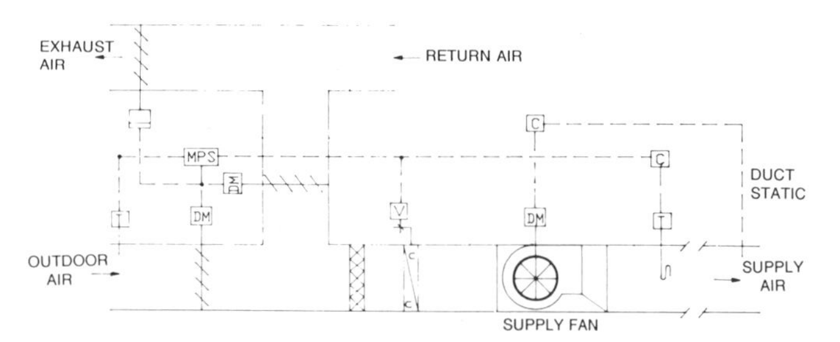

Fig. 57 VAV, Single-Duct Cooling Only

Single-Duct Cooling Only

Fan Control. In a VAV system having a supply fan with no means of modulation, as terminal units reduce total airflow, the duct static pressure increases as the fan moves up its operating curve. If uncontrolled, this pressure can damage ductwork. Even in strong ducts, terminal unit dampers must work against a higher pressure, which results in poor control, increased noise, and increased fan energy when compared to reducing the pressure. Therefore, supply fan volume controls should be used.

Fan volume control is based on supply duct static pressure. To conserve fan energy, the static pressure controller should be set at the lowest point that permits proper air distribution at design conditions. The controller requires proportional-integral (PI) control because the low proportional gain required to stabilize fan control loops in proportional-only (P) control allows static pressure to offset upward as the load decreases, thus imposing more terminal unit pressure and consuming more fan energy. PI control will eliminate offset while maintaining stability. The pressure sensor must be properly placed to maintain optimum pressure throughout the supply duct. Experience indicates that most systems perform satisfactorily with the sensor located at 75 to 100% of the distance from the first to the most remote terminal. If located less than 100% of the distance to the most remote terminal, the control set point should be set to account for the pressure loss between the sensor and the remote terminal.

In addition to the remote static pressure controllers, a high limit static pressure controller should be placed at the fan discharge to turn the fan off or limit discharge static pressure in the event of excessive duct pressure, as would be the case if a fire or smoke damper closes between the fan and the remote sensor. Supply fan static pressure control devices such as inlet guide vanes and variable-speed drives, should be interlocked to go to the minimum flow or closed position when the fan is not running; this precaution prevents fan overload or damage to ductwork on start-up.

Temperature and Ventilation Control. VAV systems are typically designed to supply constant temperature air at all times. To conserve central plant energy, supply temperature can be raised in response to demand from the zone with the greatest load (load analyzer control). However, since more cool air must then be supplied to match a given load, mechanical cooling energy saved by raising the supply temperature may be offset by an increase in fan energy. Equipment operating efficiency should be studied closely before using temperature reset on cooling-only VAV systems.

Ventilation dampers (outside air, return air, and exhaust air) are controlled for free cooling as a first-stage cooling in sequence with a cooling coil from the discharge temperature controller. When outdoor air temperature rises to the point that it can no longer be used for cooling, an outdoor air limit (economizer) control overrides the discharge controller and moves ventilation dampers to the minimum ventilation position. An enthalpy control system can replace outdoor air limit economizer control for climatic areas, where applicable.

Single-Duct Heating/Cooling. Single-duct VAV systems, which supply warm air to all zones when heating is required and cool air to all zones when cooling is required, have limited application and are used where heating is required only for morning warm-up. They are not recommended if some zones require heating and others require cooling simultaneously. These systems are generally controlled during occupancy, like single-duct cooling-only systems.

During warm-up periods, as determined by a time clock or manual switch, a constant heating supply air temperature is maintained. Since the terminal unit may be fully open, uncontrolled overheating can result. It is preferable to allow unit thermostats to maintain complete control of their terminal units by reversing their action to the unit. During warm-up and unoccupied cycles, outdoor air dampers should be closed.

Constant Volume

Constant volume (CV) systems supply a constant amount of variable temperature air to individual zones. In all single-duct CV systems, fans and ductwork must be sized for design conditions in all zones simultaneously.

Humidity control is commonly from a representative zone humidistat or a return air humidity sensor-controller and is interlocked with the humidifying system to operate only during the heating season. A high limit humidity sensor-controller in the supply air duct prevents excessive duct moisture.

Ventilation dampers (OA, RA, and EA) are controlled for free cooling as a first-stage cooling in sequence with the cooling coil from the discharge temperature controller. When outdoor air temperature rises to the point that it can no longer be used for cooling, an outdoor air limit control overrides the discharge controller and moves ventilation dampers to the minimum ventilation position, as determined by the minimum positioning switch. An enthalpy control system can replace outdoor air limit control for applicable areas.

Constant volume systems can be classified as follows:

Single-duct single zone (see Single-Zone Systems in this section)
Single-duct with zone reheat
Single-duct bypass
Dual-duct multiple zone (see Dual-Duct Systems in this section)

Single-duct with zone reheat systems (Figure 58) use a single central CV fan system to serve multiple zones. All air delivered to zones is cooled to satisfy the greatest cooling load. Air delivered to other zones is then reheated with heating coils in individual zone ducts. Because these systems consume more energy than VAV systems, they are generally limited to those applications with larger ventilation needs, such as hospitals and special process or laboratory applications. Some states no longer permit simultaneous heating and cooling.

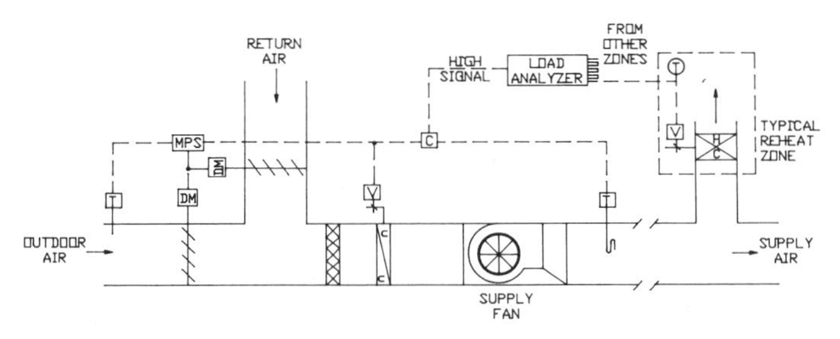

Fig. 58 Single Duct with Zone Reheat

No *fan control* is required, since the design, selection, and adjustment of fan system components determine system air volume and duct static pressure.

These systems generally supply *constant temperature* air at all times. To conserve energy, supply temperature can rise in response to demand from the greatest cooling load (load analyzer control) for less than design loads.

Single-duct with zone bypass systems (Figure 59) are a compromise between single-duct VAV systems and CV reheat systems. The primary systems and all ducts supply a constant volume of air. During partial load conditions in the space, however, the terminal unit diverts some air directly back to the return system instead of reheating it, thus bypassing the space. These terminals are often added to single-zone CV systems to provide zoning to packaged air-handling systems without the energy penalty for reheat.

Control is similar to that for single-duct multiple-reheat zone systems. Supply temperature should be reset from load analyzer control that monitors individual space temperatures or from a return air temperature sensor to prevent a high percentage of the air bypassing the space, reducing air circulation and comfort.

To provide unoccupied heating or preoccupancy warm-up, the central fan system can include a heating coil. During warm-up or unoccupied periods, a constant supply-duct heating temperature is maintained with all bypass terminal units closed to the bypass and the cooling coil valve closed. An unoccupied mode zone thermostat can cycle the fan, or the terminal unit thermostat action can be reversed to prevent overheating.

Dual-Duct

Dual-duct systems may be either CV or VAV, depending on whether terminal mixing box units operate their dampers together or individually, thus allowing zone air volume to vary with variations in load conditions.

VAV dual-duct single supply fan systems (Figure 60) use a single fan to supply separate heating and cooling ducts. Terminal mixing box units in which the heating and cooling dampers operate in sequence are used to satisfy space load requirements. Frequently, the space thermostat provides for zero energy band operation for greater energy savings.

Fan Control. Static control is similar to VAV single-duct systems, except that it is necessary to use static pressure sensors in each supply duct. Through a comparator control, the sensor sensing the lowest pressure controls the fan volume control system, thus ensuring adequate static pressure under all load conditions to supply the necessary air for all zones, whether they are supplied from the hot or cold duct.

If the system includes a return air fan, its volume control considerations are similar to those described under Return Fan Control. To sense total supply airflow, flow stations are usually located in each supply duct, and a signal corresponding to the sum of the two airflows is transmitted to the RA fan volume controller to establish the set point of the return fan controller.

Temperature Control. The hot deck has its own heating coil, and the cold deck has its own cooling coil. Each coil is controlled

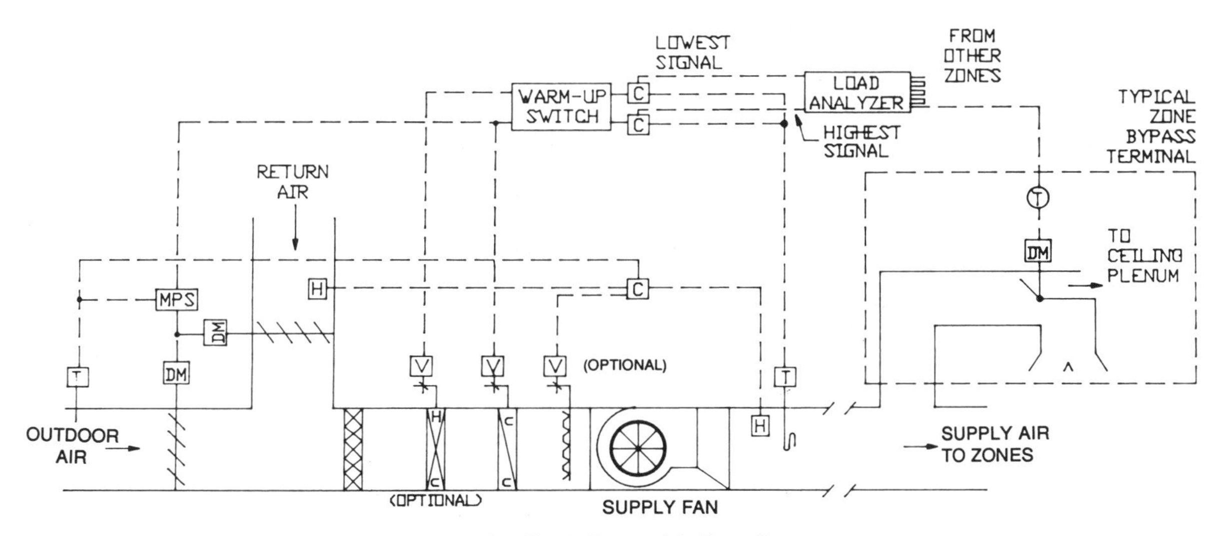

Fig. 59 Single Duct with Zone Bypass

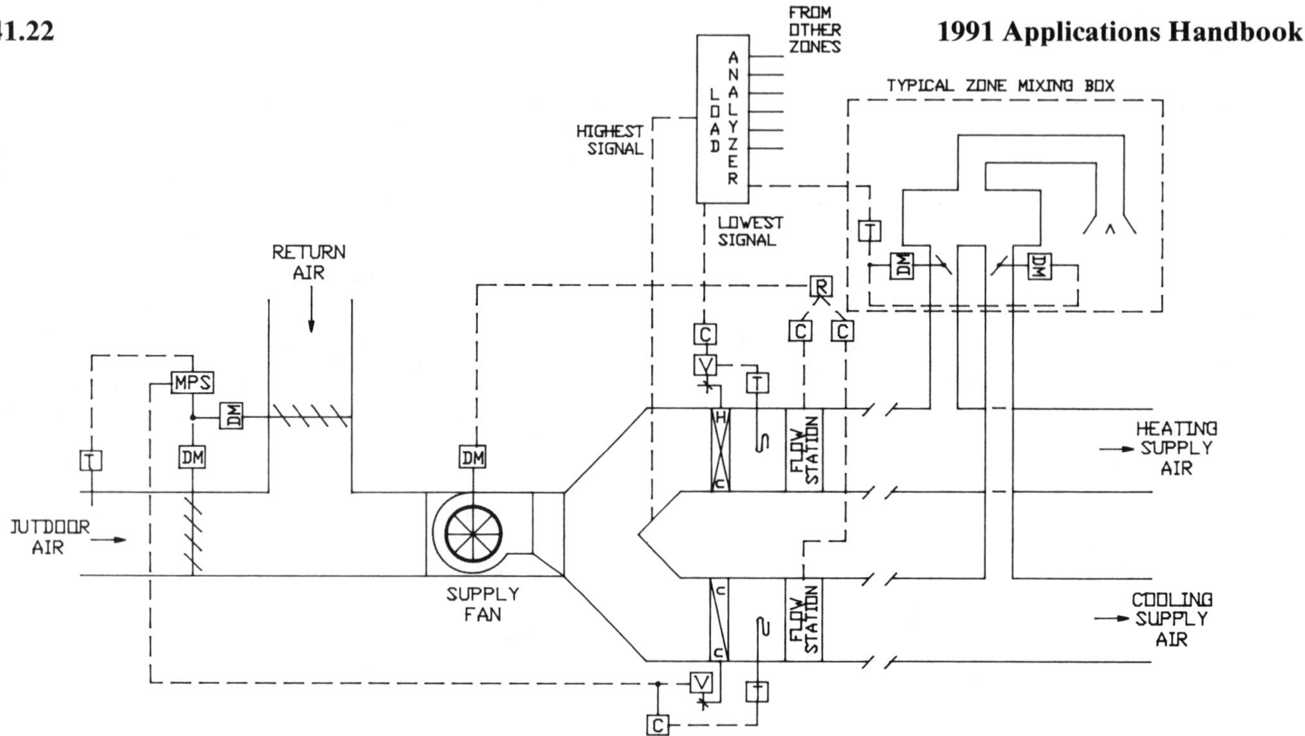

Fig. 60 VAV, Dual-Duct, Single Fan System

from its own discharge air temperature controller. The controller set point may be reset based on the greatest representative demand zone as dictated by the zone temperature—the hot deck may be reset from the zone with the greatest heating demand and the cold deck, from the zone with greatest cooling demand. An alternative to greatest-demand reset is reset from a parameter that predicts load, such as OA temperature, duct air volume, or RA temperature. In any case, positive control is required to prevent simultaneous heating and cooling of the air (a code mandate in many states).

Control based on the zone requiring the most heating or cooling increases operation economy because it reduces the energy level delivered at less-than-maximum load conditions. Figure 60 shows this as reset control of duct temperatures. However, the expected economy is lost if (1) air quantity to a zone is undersized, (2) thermostats in some spaces are reset to an extreme value by occupants, (3) the thermostat is placed so that it senses spot loads (coffee pots, solar, copier), or (4) if the thermostat malfunctions. In these cases, a weighted average of zone signals can regain the benefit at the expense of some comfort in specific zones.

In specific applications, the highest demand heating thermostat should not dictate supply air temperature, because the highest demand may come from a temporarily unoccupied zone.

Ventilation Control. Ventilation dampers (OA, RA, and EA) are controlled for cooling, with outside air as the first stage of cooling in sequence with the cooling coil from the cold deck discharge temperature controller. When outdoor air temperature rises to the point that it can no longer be used for cooling, an outdoor air limit control overrides a discharge controller and moves the ventilation dampers to the minimum ventilation position, as determined by the minimum positioning switch. An enthalpy control system can replace OA limit control for applicable areas. A more accurate OA flow measuring system can replace the minimum positioning switch.

Humidity is commonly controlled based on the return air humidity. It is coordinated with the OA temperature sensor so the humidifying system only operates during the heating season. In addition, a high limit humidistat in the supply air duct prevents excessive duct moisture.

VAV dual-duct dual supply fan systems (Figure 61) use separate supply fans for the heating and cooling ducts. Static pressure control is similar to that for VAV dual-duct single supply fan systems, except that each supply fan has its own separate static pressure sensor and control system. If the system has a return air fan, its volume control considerations are similar to those described under Return Fan Control in the section Control of Central Subsystems. Temperature, ventilation, and humidity control considerations are similar to those for VAV dual-duct single supply fan systems.

Constant volume dual-duct multiple-zone systems (Figure 62) use a single CV supply fan and multiple zone mixing dampers in terminal units that supply a constant volume of air to individually controlled zones. They require neither automatic control of duct static pressure nor fan volume as these operating characteristics are set by system design and component selection and adjustment. Considerations for temperature, ventilation, and humidity control are similar to those for VAV dual-duct single supply fan systems.

Multizone Units

Multizone units are the same as CV dual-duct multiple-zone systems, except for the location of zone mixing dampers. These dampers are at the central air-handling unit and feed individual zone supply ducts that extend to the conditioned spaces. Fan, temperature, ventilation, and humidity control considerations are the same as those for CV dual-duct multiple-zone systems.

Single Zone

Single-zone systems (Figure 63) use a CV air-handling unit (usually factory packaged) and are controlled from a single space thermostat. No fan control is required because fan volume and static pressure in the ducts are set by system design and component selection.

Temperature and Ventilation Control. A single space controller or thermostat controls the heating coil, ventilation dampers, and cooling coil in sequence, as thermal load varies in the conditioned space. Ventilation dampers (OA, RA, and DA) are controlled for outside air cooling as a first stage cooling. When outdoor air temperature rises to the point that it can no longer be used for cooling, an outdoor air limit control overrides the signal to the ventilation dampers and moves them to the minimum ventilation

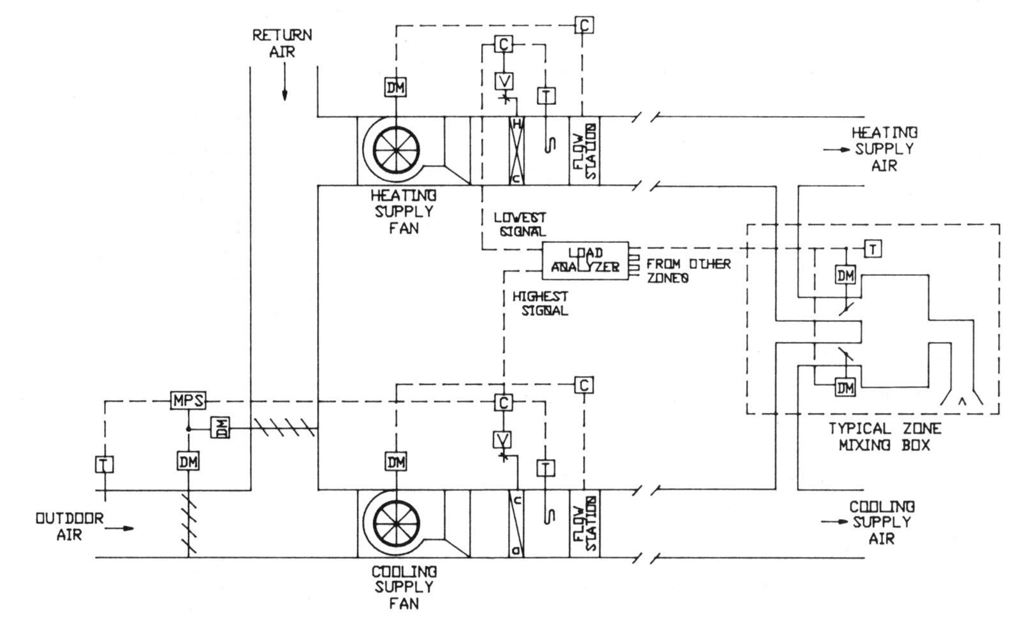

Fig. 61 VAV, Dual-Duct, Dual-Supply Fan System

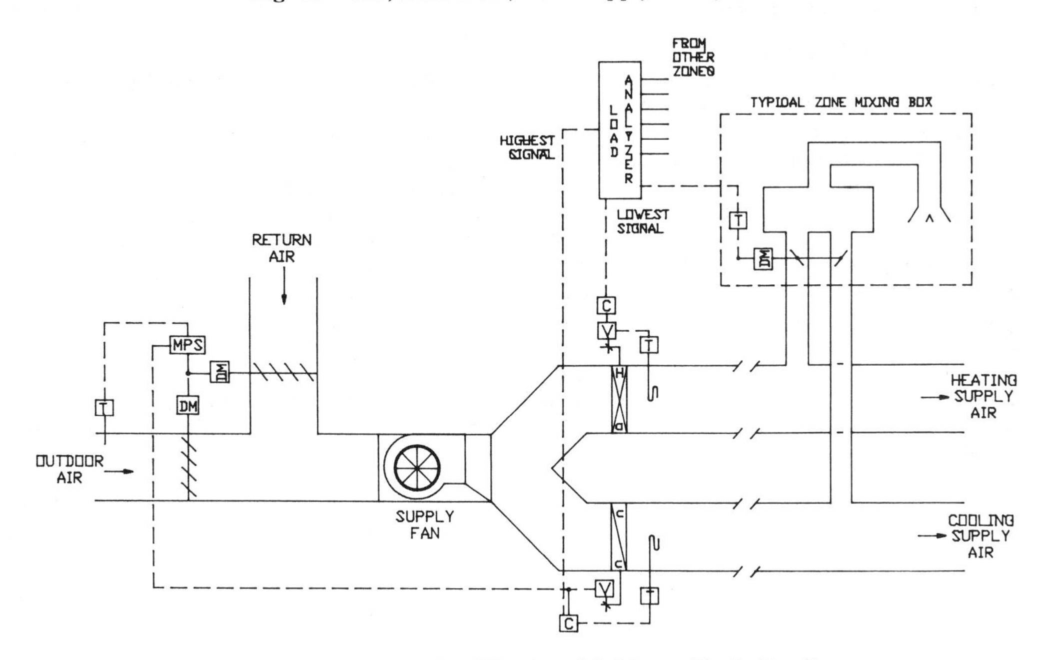

Fig. 62 Constant Volume, Dual Duct, or Multizone Single-Fan System

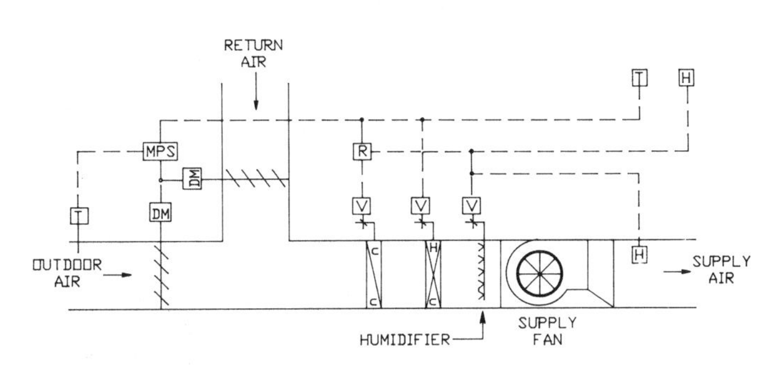

Fig. 63 Single-Zone Fan System

position, as determined by the minimum positioning switch. Where applicable, an enthalpy control system can replace the OA limit control for appropriate climatic areas. A zero energy band thermostat can separate the heating and cooling control ranges, thus saving energy.

Humidity control is from either a space or return air humidistat. It controls a moisture-adding device (steam jet or pan-type) for winter operation and can also override space thermostat control of the cooling coil for dehumidification during summer operation. In the latter case, the space thermostat brings on the heating coil for reheat. The high limit humidistat in the supply air duct prevents excessive duct moisture.

Makeup Air Systems

Makeup air systems (Figure 64) replace air exhausted from the building through exfiltration or by laboratory or industrial processes. Air must enter the space at or near space conditions to minimize uncomfortable air currents. The fan is usually turned on, either manually or automatically, as exhaust fans are turned on.

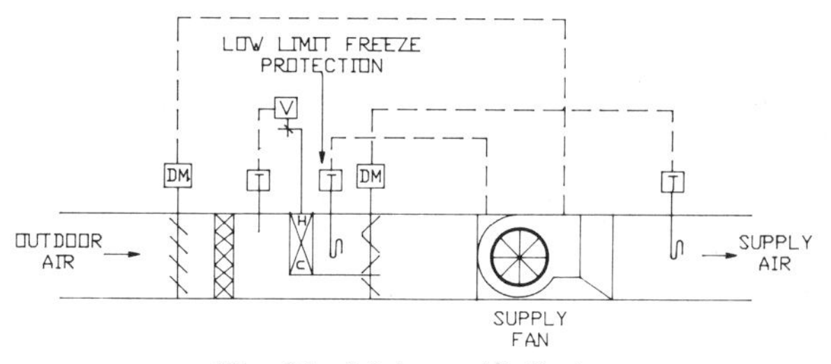

Fig. 64 Makeup Air System

The two-position outdoor air damper is closed, except when the makeup fan is in operation. The outdoor air limit control opens the preheat coil valve when outdoor air temperature drops to the point where the air requires heating to raise it to the desired supply air temperature. The discharge temperature controller then positions the face and bypass dampers to maintain the desired supply air temperature. A capillary element thermostat located adjacent to the coil shuts the fan down for freeze protection should air temperature approach freezing at any spot along the sensing element.

HYDRONIC SYSTEMS

Proper control of hot and chilled water flow rates in hydronic systems must consider both the hydraulic and thermal requirements of the distribution system. Failing to consider hydraulic principles in piping network design and pump and control valve selection can make the hydronic system uncontrollable. Proper location and use of the compression tank, dynamic separation of primary and secondary pump circuits, design allowance for adequate pressure drop through the automatic control valves, and stable system pressures are important for stable system control.

Of the load control methods used in larger systems, valve control for variable flow in the load circuit is the most popular. Heating or cooling coil capacity in air systems can be controlled using a three- or two-way throttling valve. The three-way valve is usually a mixing valve located in the coil outlet piping. As the valve mixes two flow streams, total flow rate through the branch piping is essentially constant. Figure 65 shows an example of three-way valve control for a central chilled water plant system.

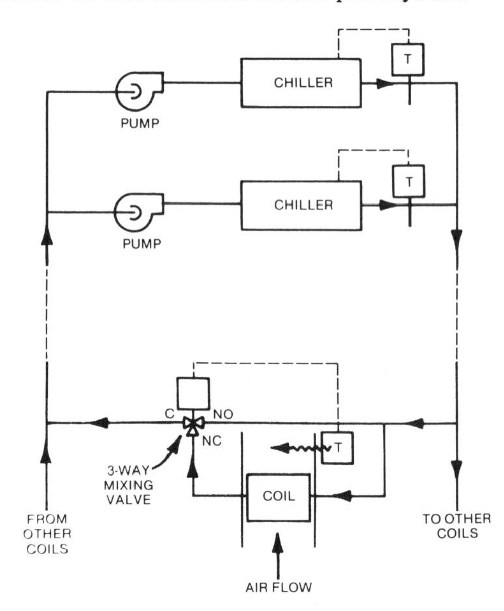

Fig. 65 Three-Way Valve Control of a Coil

A two-way throttling valve reduces the flow rate and increases the pressure differentials across the pump and load circuits. Throttling requires more attention to valve and pump selection because the pump operating point moves as the valves reduce system flow rate (Figure 66). Because of the more exacting design requirements of two-way valve control at terminal units, the three-way valve has been widely used until recently. Three-way valve control requires more energy because the flow rate is constant; thus, pump horsepower is constant, regardless of cooling load. With constant flow, there are many operating hours when temperature difference across the chiller is 2 to 3 °F relative to design chilled water ranges of 10 to 12 °F.

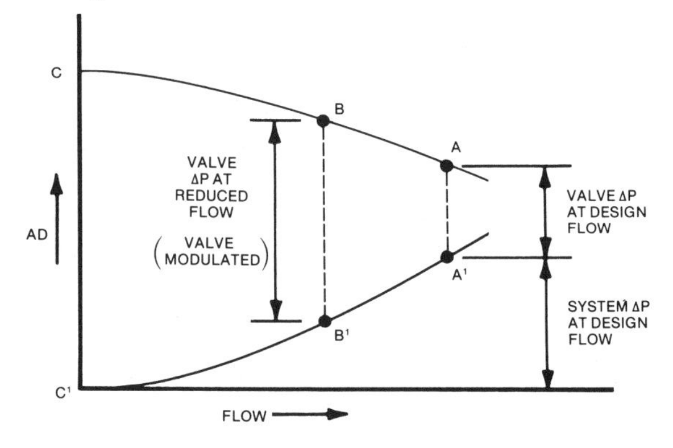

Fig. 66 Pump and System Curves with Valve Control

The schematic in Figure 67 solves the problem of variable water flow to terminal units and constant flow through the chiller while operating the chiller plant efficiently. Although only two are shown, any number of chillers can be used. At full load, both chillers are on-line, and full flow goes to the terminal units. As terminal unit valves modulate from decreased load, flow decreases and the pressure drop from supply to return mains increases. The pressure differential controller senses this change and partially opens the bypass valve to compensate. The bypass valve is sized to match the flow through one chiller, enabling one chiller and pump to shut down when the bypass is fully open. As load increases to a point where the bypass is closed completely, the second chiller and pump can restart.

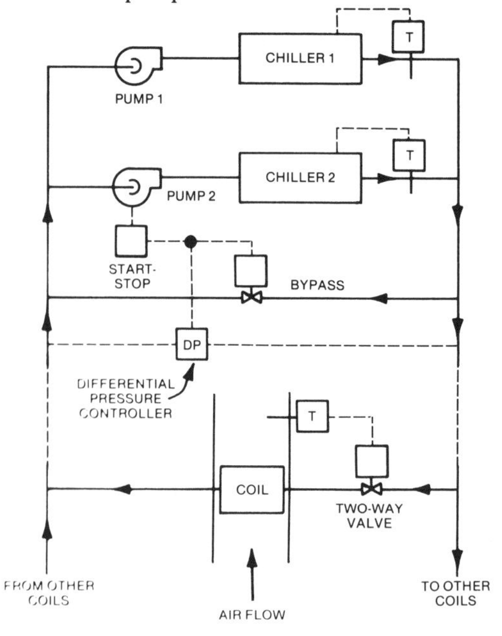

Fig. 67 Two-Way Valve with Pump Bypass

ZONE CONTROL SYSTEMS

VAV Terminal Units

VAV terminal units are used in conjunction with VAV central air-handling fan systems to vary the volume of air into individual zones as required by thermal load on the area. They are available in several configurations; therefore, control considerations vary. VAV terminal unit controls are discussed in the following categories:

Single-Duct VAV	Dual-Duct VAV
Throttling	Variable constant volume
Variable constant volume	Variable constant volume (ZEB)
Bypass	
Induction	
Fan-powered	

Throttling VAV terminal units (Figure 68) are sometimes pressure dependent; *i.e.*, the volume of air entering the conditioned space at any given space temperature varies as static pressure in the supply duct varies. The space thermostat controls the damper directly.

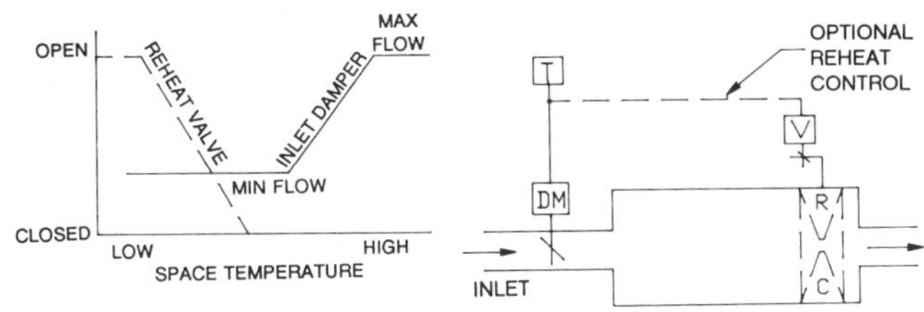

Fig. 68 Throttling VAV Terminal Unit

As an option, a reheat coil can be added. The space thermostat also controls the reheat coil valve in sequenced mode to open the valve after the damper has closed to its minimum flow position.

For perimeter areas, convectors or radiation with automatic control valves can meet reheat need. Functionally, these are controlled in a similar way to the reheat coil valves.

Variable constant volume terminal units (Figure 69) are sometimes pressure independent; *i.e.*, with varying supplying duct static pressure, the unit continues to deliver the same volume of air to the conditioned space at the same space temperature. Either a mechanical or receiver-controller-type airflow controller in the unit provides this function (Figure 69).

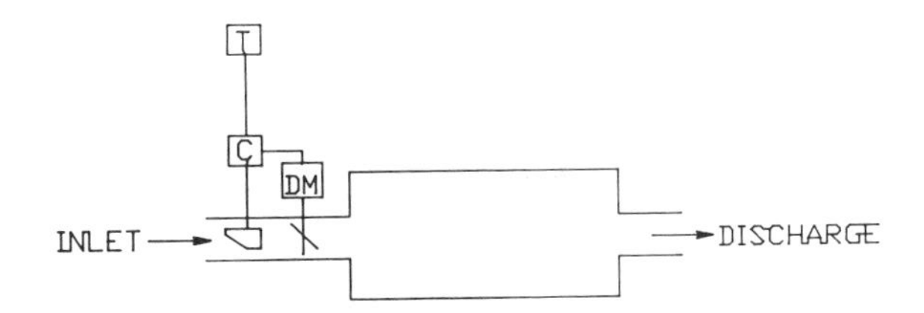

Fig. 69 Variable Constant Volume, VAV Terminal Unit

A flow sensor in the box airflow stream controls air volume. The room sensor resets airflow controller set point as thermal load on the area changes. The airflow controller can be set with a minimum flow condition to ensure comfortable distribution of supply air into the space under light loading. Maximum flow can be set to limit flow to that required for design conditions.

This unit can also have a reheat coil or separate convection, or radiation can provide the reheat function. Reheat control is similar to that for throttling VAV terminal units.

Bypass VAV terminal units (Figure 70) have a space thermostat-controlled diverting damper that proportions the amount of entering supply air between the discharge duct and the bypass opening into the return plenum. A manual balancing damper in the bypass is adjusted to match the resistance in the discharge duct. In this way, the supply air from the primary system remains at a constant volume.

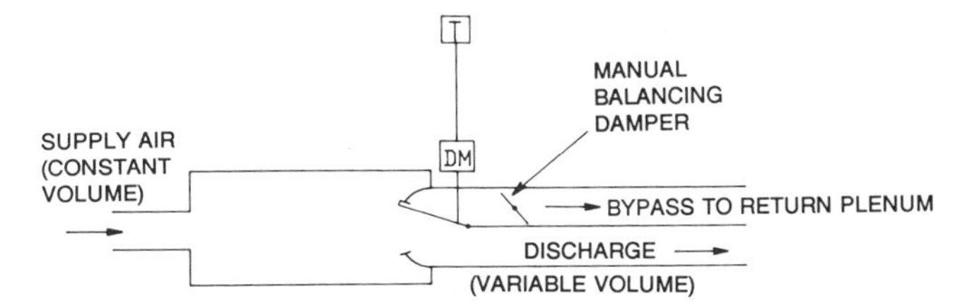

Fig. 70 Bypass VAV Terminal Unit

Induction VAV terminal units (Figure 71) provide return air reheat by routing air through the unit to induce air from the return air plenum into the stream of air being delivered to the conditioned space. In addition to the inlet damper, there is also a damper on the return air inlet. Both dampers are controlled simultaneously so that as the primary air opening decreases, the return air opening increases.

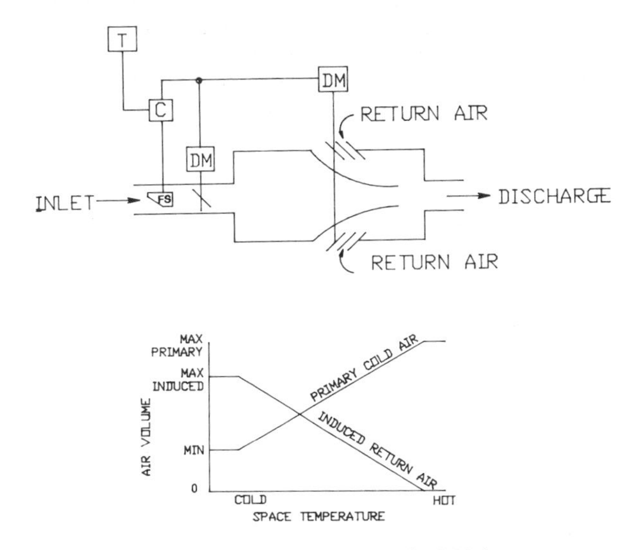

Fig. 71 Induction VAV Terminal Unit

An airflow controller in the unit controls the volume of air coming through the primary air damper. The space thermostat resets the set point of this controller, as required by thermal load on the conditioned space.

Fan-powered VAV terminal units (Figure 72) are similar to throttling VAV terminal units, except that they include an integral fan that recirculates space air at a constant volume. In addition to enhancing air distribution in the space, they also provide a reheat coil and a means of maintaining a lowered unoccupied temperature in the space when the primary system is off.

Figure 72 shows a typical control sequence. A space thermostat resets an integral airflow controller as it senses space load changes. As primary air decreases, the fan comes on to ensure adequate air circulation. The units serving the perimeter area of a building usually include a reheat coil, which is sequenced with the primary air damper to supply heat when required. When the primary air system is not operating (nighttime or unoccupied control mode), the "night" operating mode of the thermostat cycles the fan with the reheat coil valve open to maintain the lowered temperature in the space.

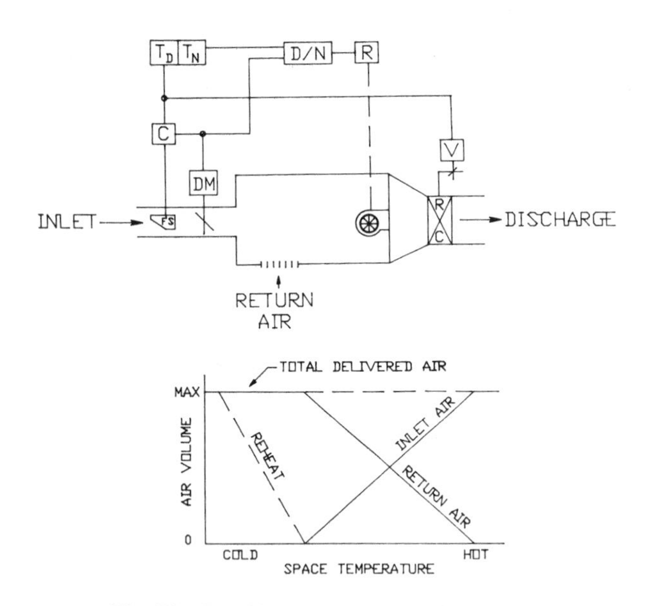

Fig. 72 Fan-Powered VAV Terminal Unit

Bypass fan induction terminal units (Figure 73) are similar to fan-powered VAV terminal units, except that the fan pulls air from the return plenum only. An alternate location for the reheat coil is in the return plenum opening. The control sequence is the same as the fan-powered VAV terminal unit.

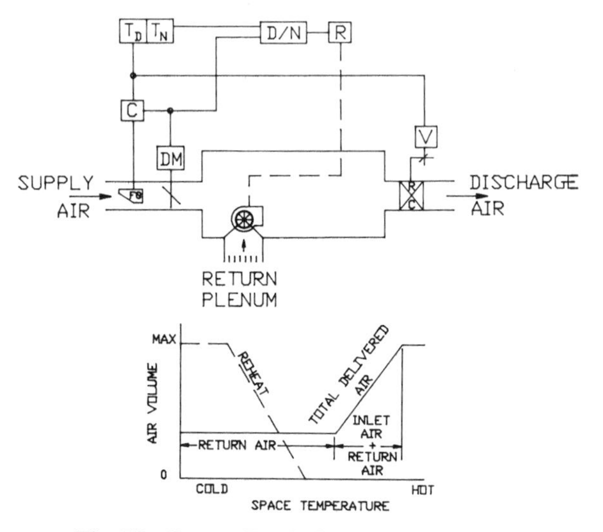

Fig. 73 Bypass Fan Induction Terminal Unit

Variable constant volume, dual-duct terminal units (Figure 74) have inlet dampers on the heating and cooling supply ducts. They are interlinked to operate in reverse of each other and require a single control actuator. There is also a total airflow volume damper with its own actuator.

The space thermostat controls inlet mixing dampers directly, and the airflow controller controls the volume damper. The space thermostat resets the airflow controller from maximum to minimum flow as thermal load on the conditioned area changes. Figure 74 shows the control schematic and damper operation. Note that in a portion of the control range there is a mixing of the heating and cooling supply air.

Variable constant volume (ZEB) dual-duct terminal units (Figure 75) have inlet dampers on the cooling and heating supply ducts with individual damper actuators and airflow controllers

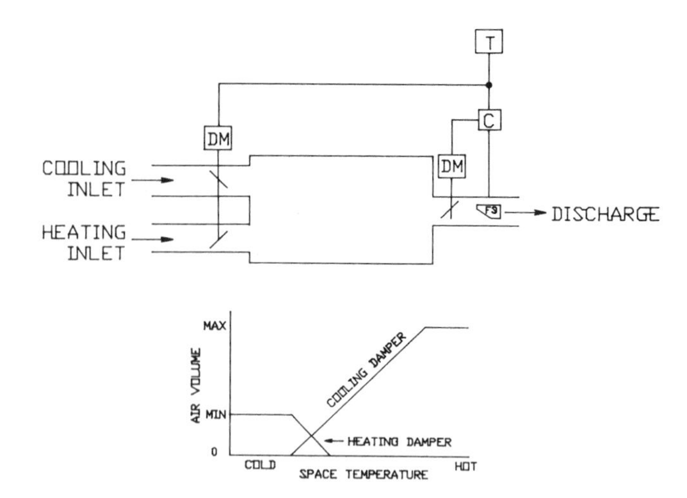

Fig. 74 Variable Constant Volume, Dual-Duct Terminal Unit

and no total airflow volume damper. The space thermostat (ZEB) resets airflow controller set points in sequence as space load changes. The airflow controllers maintain adjustable minimum flows for ventilation, with no overlap of damper operations, during the zero energy band when neither heating or cooling is required. Figure 75 shows the control schematic and damper operation. Energy consumption of variable constant volume dual-duct terminal units with and without ZEB is essentially the same.

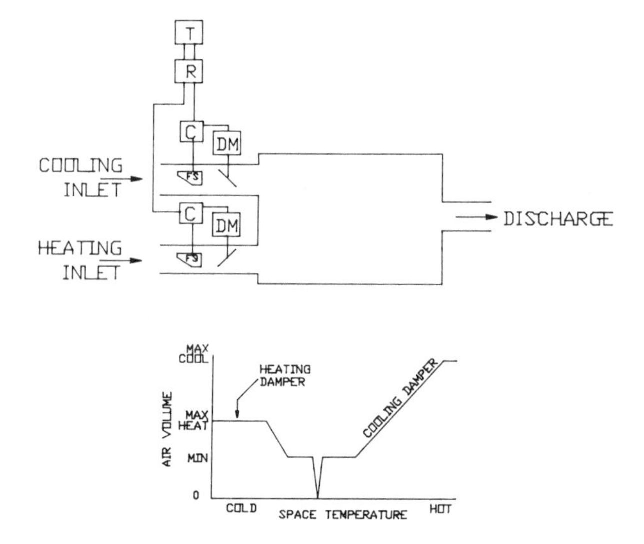

Fig. 75 Variable Constant Volume (ZEB), Dual-Duct Terminal Unit

Constant Volume Terminal Units

Multiple-zone systems using CV air distribution can be classified as:

Single-Duct	Dual-Duct
Zone reheat	Mixing box
Positive constant volume	Constant volume mixing box

Single-duct zone reheat systems have a heating coil (hot water, steam, or electric) in the branch supply duct to each zone. The central air-handling unit supplies constant-temperature air. The space thermostat positions the reheat coil valve (or electric heating elements) as required to maintain space condition (Figure 76).

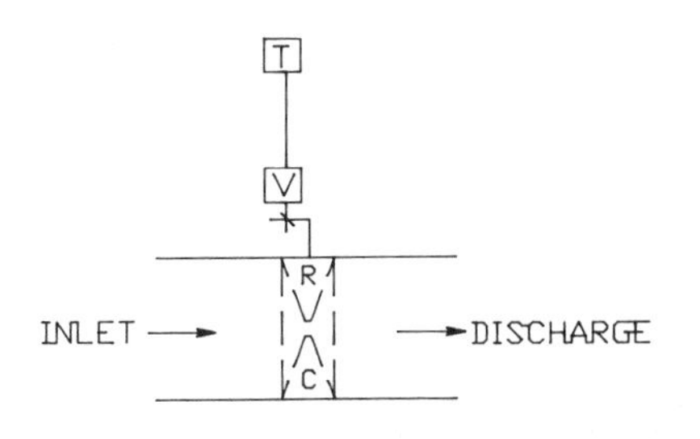

Fig. 76 Constant Volume, Single-Duct Zone Reheat

Single-duct positive constant volume terminal units supply a constant volume of distribution air to the space, even though static pressure varies in the supply duct system. This is accomplished by an integral mechanical constant volume regulator or an airflow constant volume control furnished by the terminal unit manufacturer. If a reheat coil comes with the unit, a space thermostat controls the reheat coil valve as required to maintain space condition (Figure 77).

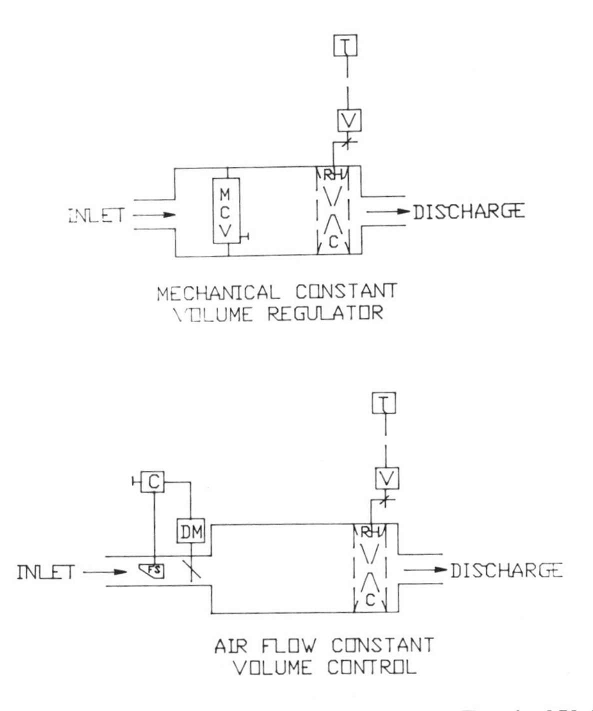

Fig. 77 Single-Duct, Positive Constant Volume Terminal Unit

Dual-duct mixing box terminal units generally apply to low-static-pressure systems that require large amounts of ventilation. The warm duct damper and the cool duct damper are linked to operate in reverse of each other. A space thermostat positions the mixing dampers through a damper actuator to mix warm and cool supply air to maintain space condition. Discharge air quantity depends on static pressure in each supply duct at that location. Static pressures in the supply ducts vary because of the varying airflow in each duct (Figure 78).

Dual-duct constant volume mixing box terminal units are typically used on high-static-pressure systems where the airflow quantity served into each space is critical. The units are the same as those described above, except that they include either an integral mechanical constant volume regulator or an airflow constant volume control furnished by the unit manufacturer (Figure 79).

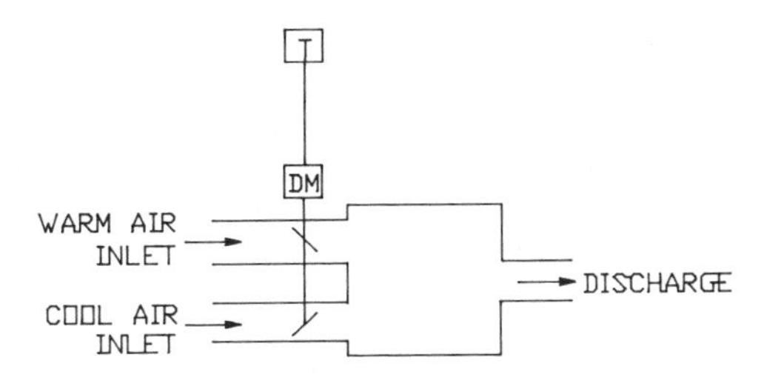

Fig. 78 Dual-Duct Mixing Box Terminal Unit

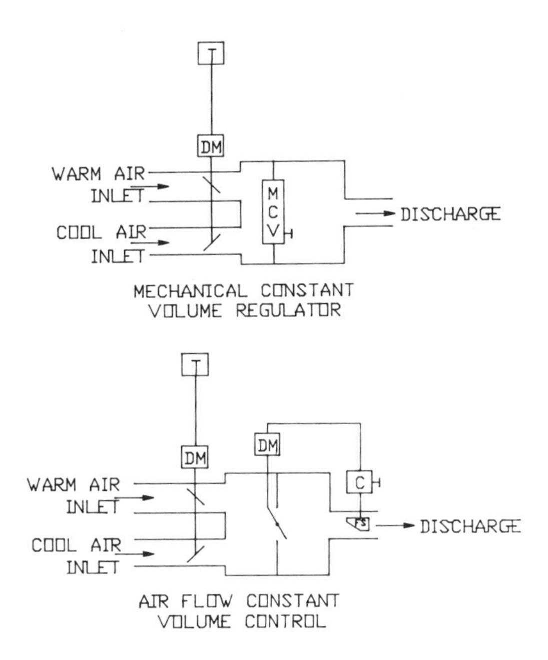

Fig. 79 Dual-Duct, Constant Volume Mixing Box Terminal Units

Perimeter Radiation and Convection

Radiators or convectors can provide either the total room heat or supplemental heat of the perimeter to offset building transmission losses. Control strategy depends on which function the radiation performs.

For a total room heating application, rooms are usually controlled individually. With individual room control, each radiator and convector is equipped with an automatic control valve. Depending on room size, one thermostat may control one or several valves in unison.

The thermostat can be located in the return air to the unit or on a wall at the occupant level. Return air control is generally the least accurate and results in the widest space temperature fluctuations. When space is controlled for comfort of seated occupants, wall-mounted thermostats give the best results.

For supplemental heating applications, where perimeter radiation is used to offset perimeter heat losses only (with a zone or space load handled separately by a zone air system), outdoor reset of the water temperature to the radiation should be considered. Radiation can be zoned by exposure and the compensating outdoor sensor can be located to sense compensated indoor (outdoor) temperature, solar load, or both.

Fan coil units can contain packaged controls for the fan and valves, or they can be field installed. These units can be categorized as follows:

Two-pipe heating
Two-pipe cooling
Two-pipe heating/cooling
Four-pipe heating/cooling, split coil
Four-pipe heating/cooling, single coil

Chapters 13, 14, and 15 of the 1987 ASHRAE *Handbook—HVAC Systems and Applications* describe design features and applications for the central plant and distribution systems associated with these fan coil units.

Typically, an integral return air thermostat or a wall-mounted one controls an automatic control valve on the heating (cooling) coil. Frequently, a local fan switch or a central time clock operates the fan.

Figure 80 shows typical control of a *two-pipe heating/cooling* fan coil unit. Since the single coil is used for heating or cooling, depending on the season, hot water temperatures can be much lower than with standard heating coils. The unit requires an integral return-air or a wall-mounted heating/cooling thermostat that reverses its action from a remote heating/cooling changeover signal. Alternately, a sensed change in supply water temperature from a pipe-mounted aquastat can cause the heating/cooling changeover. Frequently, a local fan switch or a central time clock operates the fan.

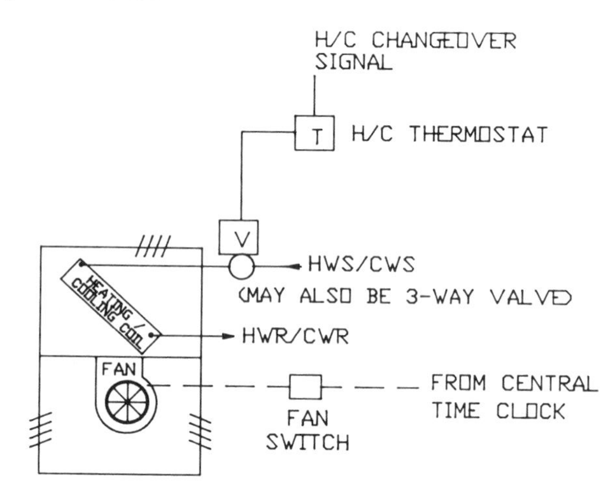

Fig. 80 Two-Pipe Heating/Cooling Fan Coil Unit

Four-pipe heating/cooling split coil systems hydraulically separate the heating and cooling systems, control precautions for two-pipe systems do not apply. The four-pipe fan coil unit typically has a split coil with one heating and two cooling rows. Hot water or steam can be used to heat, while direct expansion or chilled water can be used to cool. Figure 81 shows a room thermostat connected to the hot water valve and the chilled water valve. The hot water valve closes if room temperature increases. Further increases begin to open the chilled water valve. Valve ranges should be adjusted to provide a dead band within which both heating and cooling valves are closed. A local fan switch can control the fan, or a central time clock can operate it.

In *four-pipe heating/cooling single coil systems* (Figure 82), special control three-way valves are used on both the supply and the return water side of the coil. The space thermostat controls both valves in the following sequence: when the thermostat senses a need for heat, valves to the chilled water supply and return circuits remain closed while the supply water valve hot water port is throttled to maintain space temperature. When the thermostat senses a need for cooling, valves to the hot water supply and return circuits remain closed and the supply water valve cold water port is throttled to maintain space temperature. The supply water valve is adjusted so that both supply ports are closed in the dead band

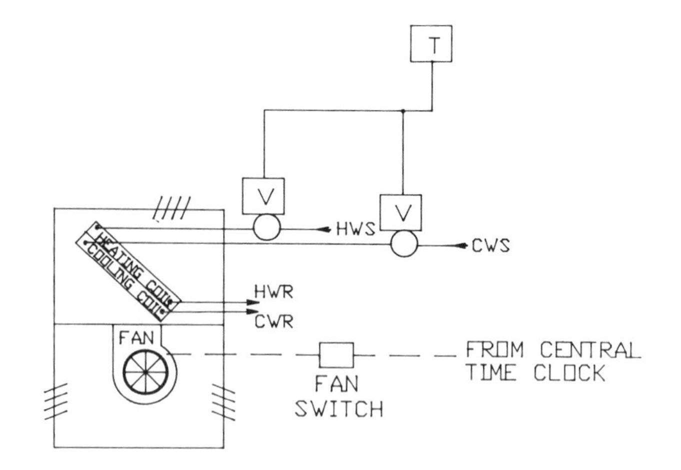

Fig. 81 Four-Pipe Heating/Cooling Split Fan Coil Unit

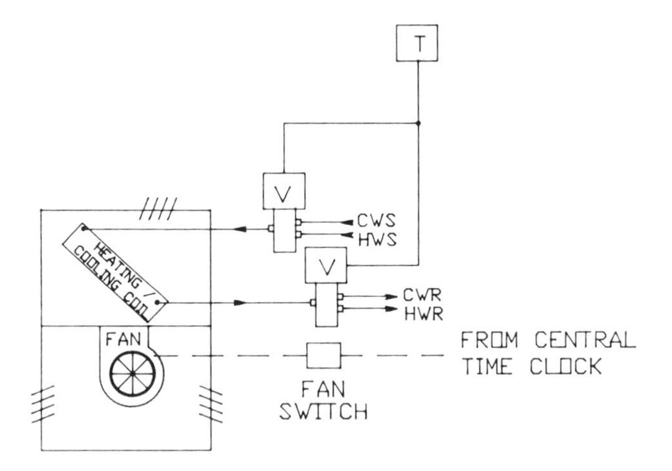

Fig. 82 Four-Pipe Heating/Cooling Single Fan Coil Unit

between heating and cooling operations. During this dead band, the return water valve is positively positioned to connect the correct return water circuit to the coil before either supply port opens. A local fan switch can control the fan, or a local time clock can operate it.

Unit ventilators are designed to heat, ventilate, and cool a space by introducing outdoor air in quantities up to 100%. Optionally, they can also cool and dehumidify with a cooling coil (either chilled water or DX). Heating can be by hot water, steam, or electric resistance. The control of these coils can be by valves or face and bypass dampers. Consequently, control systems applied to unit ventilators are many and varied. This section describes the three most commonly used control schemes: Cycle I, Cycle II, and Cycle III (Figures 83 and 84).

Cycle I. Except during the warm-up stage (Figure 83), Cycle I supplies 100% outdoor air at all times. During warm-up, the heating valve is open, the OA damper is closed, and the RA damper is open. As temperature rises into the operating range of the space thermostat, the OA damper fully opens, and the RA damper closes. To maintain space temperature, the heating valve is positioned, as required. The airstream thermostat can override space thermostat action on the heating valve to open the valve to maintain a minimum air discharge temperature into the space. Figure 85 shows the relative position of the heating valve and ventilation dampers in relation to space temperature.

Cycle II. During the heating stage, Cycle II supplies a set minimum quantity of outdoor air. Outdoor air is gradually increased,

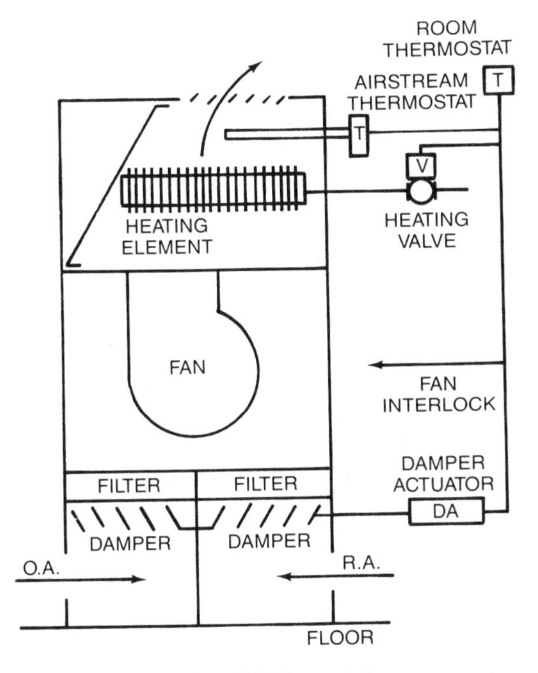

Fig. 83 Cycles I and II Control Arrangements

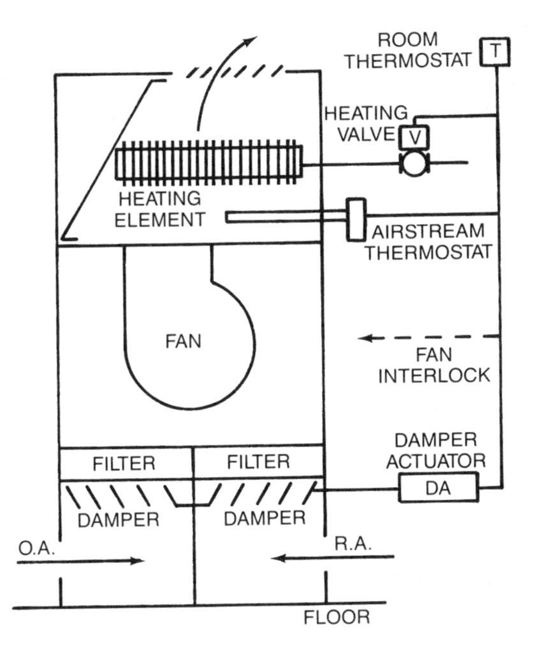

Fig. 84 Cycle III Control Arrangement

as required for cooling. During warm-up, the heating valve is open, the OA damper is closed, and the RA damper is open. As the space temperature rises into the operating range of the space thermostat, ventilation dampers move to their set minimum ventilation position. To maintain space temperature, the heating valve and ventilation dampers are operated in sequence, as required. The airstream thermostat can override space thermostat action of the heating valve and ventilation dampers to prevent discharge air from dropping below a minimum temperature. Figure 85 shows the relative position of the heating valve and ventilation dampers in relation to space temperature.

Cycle III. During the heating, ventilating, and cooling stages, Cycle III supplies a variable amount of outdoor air as required to

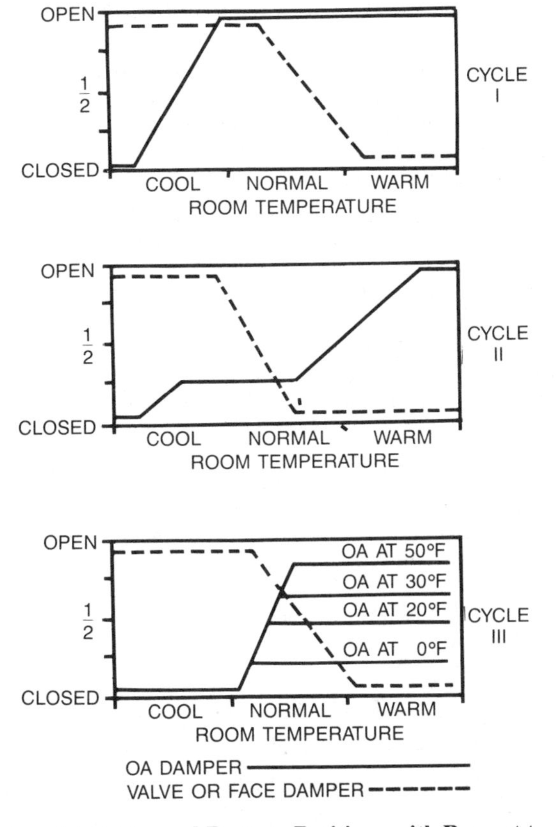

Fig. 85 Valve and Damper Positions with Respect to Room Temperature

maintain a fixed temperature (typically 55 °F) entering the heating coil (Figure 84). When heat is not required, this air is used for cooling. During warm-up, the heating valve is open, the OA air damper is closed, and the RA damper is open. As the space temperature rises into the operating range of the space thermostat, ventilation dampers control the air temperature entering the heat ing coil at the set temperature. Space temperature is controlled by positioning the heating valve, as required. Figure 85 shows the relative position of the heating valve and ventilation dampers in relation to space temperature.

Frequently, day/night thermostats are used with any of these control schemes to maintain a lower space temperature during unoccupied periods by cycling the fan with the outdoor air damper closed. Another common option is a freeze stat placed next to the heating coil that shuts the unit off when near-freezing temperatures are sensed.

Radiant panels combine controlled temperature room surfaces with central air conditioning and ventilation. The radiant panel can be in the floor, walls, or ceiling. Panel temperature is maintained by circulating water or air or by electric resistance. The central air system can be a basic one-zone, constant-temperature, CV system with a radiant panel operated by individual room control thermostats; or the central air system can include some or all the features of dual duct, reheat, multizone, or VAV systems, with the radiant panel operated as a one-zone, constant-temperature system. The one-zone radiant heating panel system is often operated from an outdoor temperature reset system to vary panel temperature as outdoor temperature varies.

Radiant panels for both heating and cooling applications require controls similar to those described for the four-pipe heating/cooling single-coil fan coil. During the cooling cycle, ventilation air supplied to the space should have a dew-point temperature below that of the radiant panel surface to prevent condensation.

CENTRAL PLANT HEATING AND COOLING SOURCES

The term central plant, as used here, refers not only to a plant supplying heating and/or cooling media to multiple buildings, but also to a single building having central chillers and boilers.

Control considerations for a central plant generally consist of the following broad categories:

- Control of individual units (boilers and chillers)
- Control of the plant (starting, stopping, and adjusting individual units making up the plant, including auxiliaries)
- Control of the hydronic distribution system

Control of Individual Units

Almost always, a manufacturer supplies boilers (both steam and hot water) and chillers (all types) with an automatic control package installed. Control functions fall in two categories—capacity and safety.

Capacity controls vary the thermal capacity of the unit as a function of the presented load. A designer needs to understand the operation of these controls to integrate them into control system design for a multiple-unit plant.

Safety controls generally shut down the unit and generate an alarm whenever an unsafe condition is detected. When a supervisory control system is used, the alarm should be retransmitted to the control center.

CONTROL OF THE PLANT

Load Control in Hydronic Heating Systems

Load affects the rate of heat input to a hydronic system. Rate control cycles and modulates the flame, and turns boilers on and off. The first two control items are in the boiler control package. The control system designer decides when to add or drop a boiler and at what temperature to control the boiler supply water.

Hot water distribution control must consider temperature control at the hot water boilers or converter, reset of heating water temperature, and the control method for multiple zones. Other factors requiring consideration include (1) minimum water flow through the boilers, (2) protection of boilers from temperature shock, and (3) coil freeze protection. If multiple or alternate heating sources (such as condenser heat recovery or solar storage) are used, the control strategy must also include the means for sequencing or selecting the most economical hot water source.

Figure 86 shows a system for load control of a gas- or oil-fired boiler. Boiler controls usually include combustion controls (flame failure, high temperature, and other safety cutouts) and capacity controls. Intermittent burner firing usually controls capacity, although fuel input modulation is common in larger systems. In most cases, the boiler is controlled to maintain a constant water temperature, although an outdoor air thermostat can reset temperatures when the boiler is not used for domestic water heating. A master-submaster arrangement resets supply water temperature, with the outdoor master thermostat resetting the submaster thermostat according to the reset schedule shown. Water temperature should not be reset below that recommended by the manufacturer, typically 140°F, to minimize condensation of flue gases and boiler damage.

In this example, three-way control valves at the heating coils will ensure minimum water flow through the boiler. Zone control is achieved by varying flow rate to the zone heating coil. A room thermostat controls a three-way valve that varies the coil output. Proper preheat coil control is achieved using a three-way control valve to vary the heating water flow from the boiler and maintaining constant flow through the preheat coil, by the recirculating pump, for freeze protection. Larger systems with sufficiently high pump operating costs can use variable-speed pump drives, pump discharge valves with minimum flow bypass valves, and two-speed drives to reduce secondary pumping capacity to match the load.

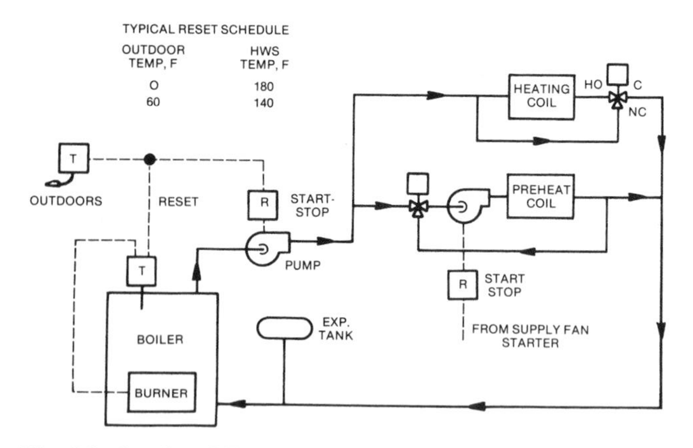

Fig. 86 Load and Zone Control in a Simple Hydronic System

Hot water heat exchangers or steam-to-water converters are sometimes used instead of boilers as hot water generators. Converters typically do not include a control package; therefore, the engineer must design the control scheme. The schematic in Figure 87 can be used with either low-pressure steam or boiler water ranging from 200 to 260°F as the heating source. The supply water thermostat controls a modulating two-way valve in the steam (or hot water) supply line. An outdoor thermostat usually resets the supply water temperature downward as load decreases. A flow switch interlock should close the two-way valve when the hot water pump is not operating.

Chapter 14 of the 1987 ASHRAE *Handbook—HVAC Systems and Applications* covers the central plant arrangement for a two-pipe changeover and three-pipe systems.

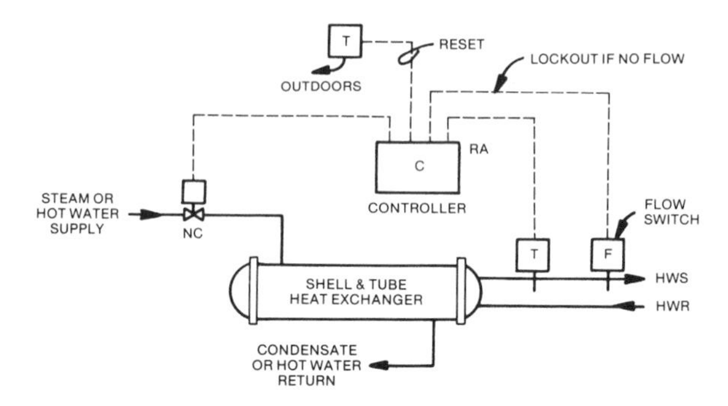

Fig. 87 Steam-to-Hot Water Heat Exchanger Control

Control of Chiller Plants

Because each central chiller plant is unique with a wide variety of chiller types, sizes, drives, manufacturers, piping configurations, pumps, cooling towers, distribution systems, and loads, it is almost inevitable that each installation, including its controls, is designed on a custom basis.

Section II (Chapters 12 through 21) of the 1988 ASHRAE *Handbook—Equipment* gives information on various types of chillers (*e.g.*, absorption, centrifugal, and reciprocating). Each type has specific characteristics that must match the requirements of the installation. Chapter 17 of the 1988 ASHRAE *Handbook—Equipment* covers variations in piping configurations (*e.g.*, series and parallel chilled water flow) and some associated control concepts.

Chiller plants are generally one of two types—*variable flow* (Figure 88) or *constant flow* (Figure 89). The examples show parallel flow configuration. The determining factor is generally the nature of control of the remote load. Throttling coil valves

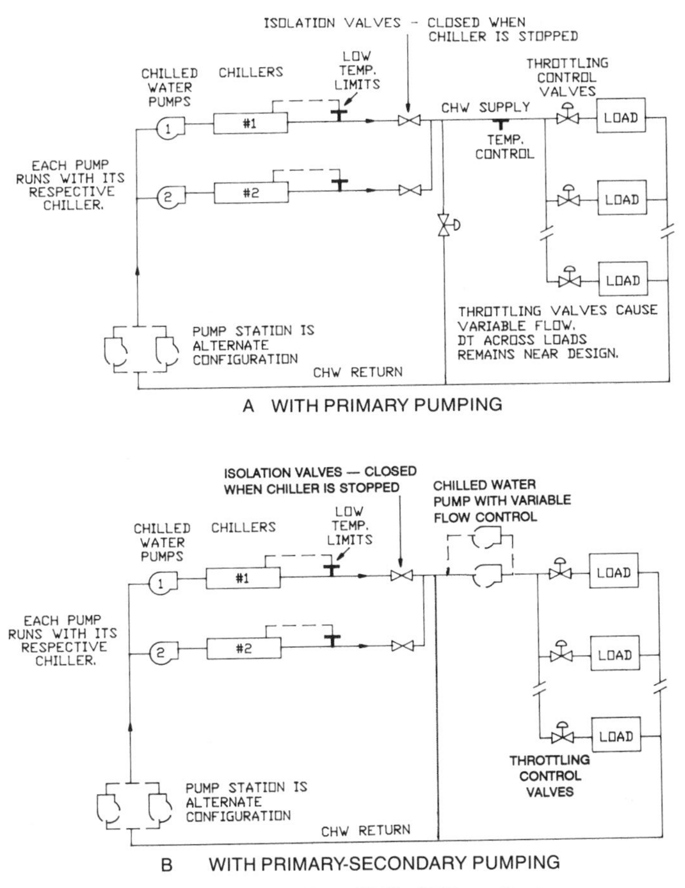

Fig. 88 Variable Flow Chilled Water Systems

produce a flow rate in the distribution system that varies with load and a temperature differential that tends to remain near the design temperature differential. Chilled water supply temperature typically controls such systems. To improve energy efficiency, the set point is reset based on the zone with the greatest load (load reset) or other variances. Note: There are different methods of piping and pumping to accomplish the desired task while maintaining the manufacturer's desired minimum flow through an operating chiller.

The constant flow system (Figure 89) is actually constant flow under each combination of chillers on line; a major upset always occurs whenever a chiller is added or dropped. The load reset function ensures that the zone with the largest load is satisfied, while supply or return water control treats average zone load.

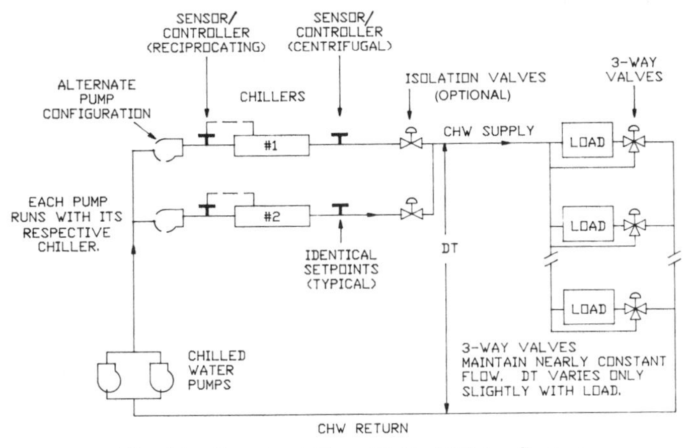

Fig. 89 Constant Flow Chilled Water System

Optimize Refrigerant Head. Chiller efficiency (kW/ton) is a function of the percent of full load on the chiller and the refrigerant head, which is the refrigerant pressure difference between the condenser and evaporator. In practice, the head is represented by condenser water leaving temperature minus chilled water supply temperature (Figure 90). To reduce the refrigerant head, either the chilled water supply temperature must be increased and/or the condenser water temperature decreased. The gain is 1 to 2% energy saving for each °F reduction in head.

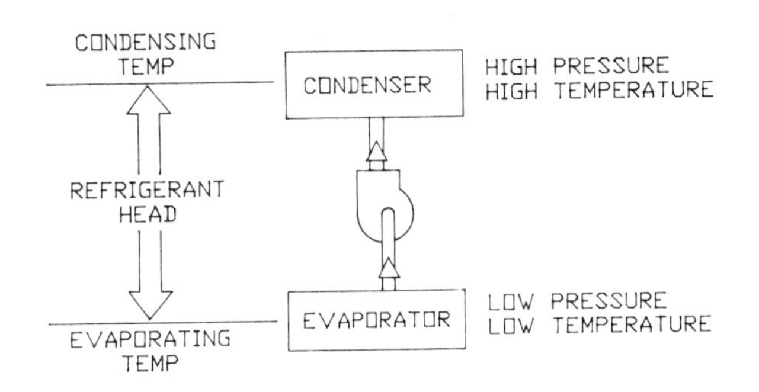

Fig. 90 Refrigerant Head Diagram

Two effective methods for reducing refrigerant head are as follows:

1. Use *chilled water load reset* to raise supply setpoint as load decreases. Figure 91 shows the basic principle of this function. Varying degrees of sophistication are available, especially with computer control.
2. *Lower condenser temperature* to the lowest safe temperature (use manufacturer's recommendations) by keeping the cooling tower bypass valve closed, operating at full condenser water pump capacity, and maintaining full airflow in all cells of the cooling tower until water temperature is within about 5 °F of outdoor air wet bulb. However, pumps and fans consume power. Consider all of them and the fan power of the VAV air handlers in calculating net energy savings.

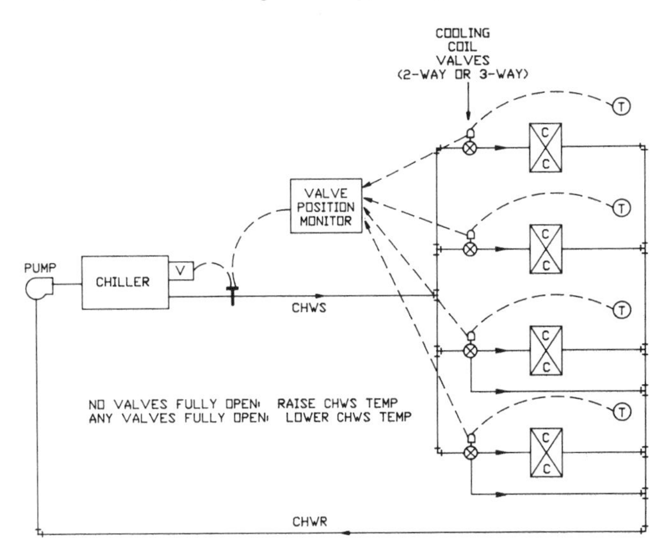

Fig. 91 Chilled Water Load Reset

Optimize Operation. Multiple chiller plants should be operated at the most efficient point on the part-load curve. Figure 92 shows a typical part-load curve for a centrifugal chiller operated at design conditions. Figure 93 shows similar curves at different head-limiting conditions. Figure 94 indicates the point at which a chiller

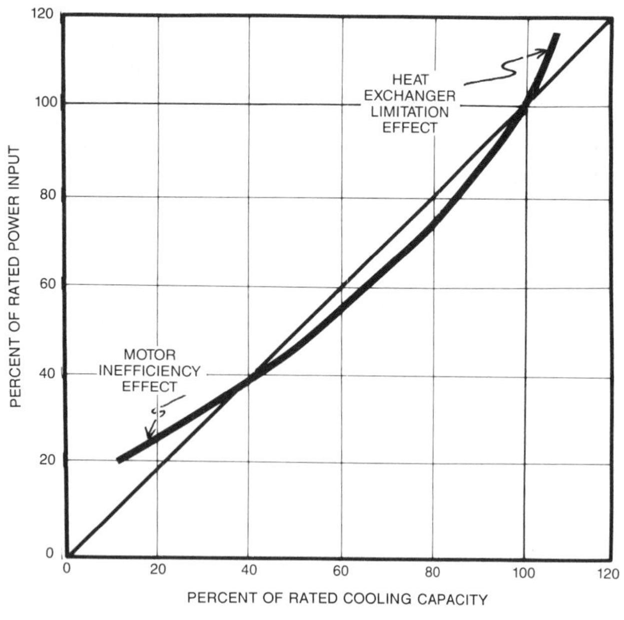

Fig. 92 Chiller Part-Load Characteristics at Design Refrigerant Head

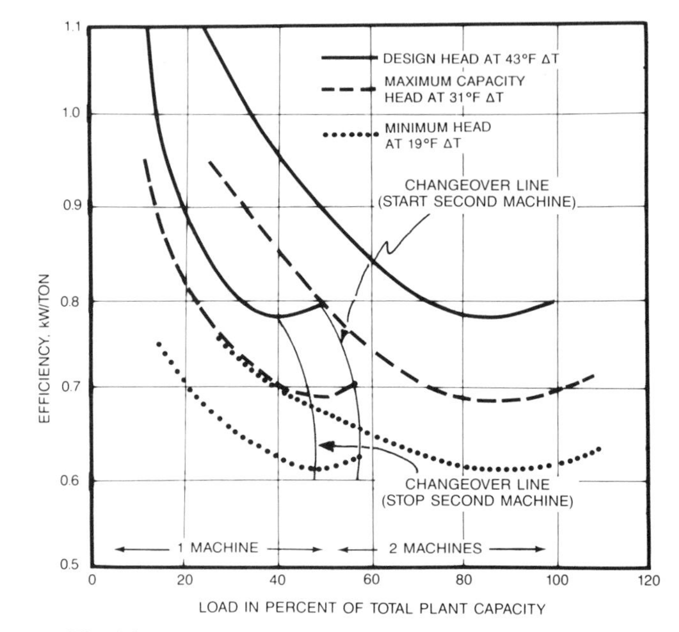

Fig. 94 Multiple Chiller Operation Changeover Point—Two Equal-Sized Chillers

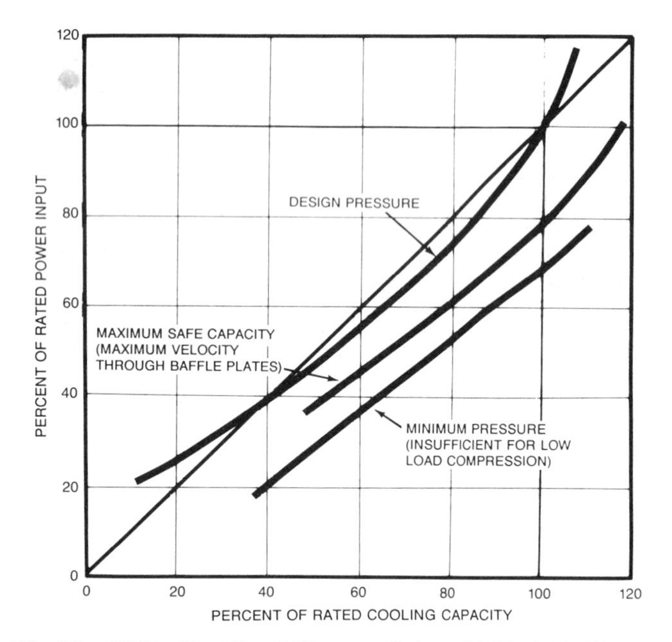

Fig. 93 Chiller Part-Load Characteristics with Variable Head

should be added or dropped in a two-unit plant. The part-load curves are plotted for all combinations of chillers; then the break-even point between n and $n + 1$ chillers can be determined.

Minimize Run Time. Daily start-up of the chiller plant based on start-up time of the air-handling units should be optimized. Generally, chillers may be started at the same time as the first fan system. Exceptions include chillers that are started early if the water distribution loop has great thermal mass; chillers may be started later when outside air can provide cooling to fan systems at start-up.

Control of Cooling Tower with Water-Cooled Condenser

Control system designers work with liquid chiller control when the equipment package is integrated into the central chiller plant. Typically, cooling tower and chilled and condenser pump's control must be considered if the overall plant is to be stable and energy-efficient. This section considers control of the condenser water circuit and the possible control arrangements for various central plants.

The most common packaged mechanical-draft cooling towers for comfort air-conditioning applications are counterflow induced-draft and forced-draft. Both are controlled similarly, depending on the manufacturer's recommendations. On larger towers, two-speed or variable-speed fans (and associated motor control circuitry) can reduce fan power consumption at part-load conditions and stabilize condenser water temperature.

Figure 95 shows bypass valve control of condenser water temperature. With centrifugal chillers, condenser supply water temperature is allowed to float as long as the temperature remains above a low limit. The manufacturer should specify the minimum entering condenser water temperature required for satisfactory performance of the particular chiller. (Minimum condenser water temperatures for centrifugal chillers usually range from 55 to 65°F.) The control schematic in Figure 95 works as follows: for a condenser supply temperature (*e.g.*, above a set point of 75°F), the valve is open to the tower, the bypass valve is closed, and the tower fan is operating. As water temperature decreases (*e.g.*, to 65°F), the tower fan speed can be reduced to low-speed operation if a two-speed motor is used. On a further decrease in condenser water supply temperature, the tower fan(s) stop and the bypass valve begins to modulate to maintain the acceptable minimum water temperature.

In colder areas that require year-round air-conditioning, cooling towers may require sump heating and continuous full flow over the tower to prevent ice formation. In that case, the cooling tower sump thermostat would control a hot water or steam valve to maintain water temperature above freezing.

Heat Pump, Heat Recovery, and Storage Systems

Heat pumps are refrigeration compressors in which the evaporator is used for cooling and the condenser is used for heating. Many conventional means can control heating and cooling cycles. The unique feature of a heat pump system is the heating/cooling changeover by which the desired evaporator/condenser circuit is selected. Any of three different media (refrigerant, air, or water) can be switched for changeover. Chapter 9 in the 1987 ASHRAE

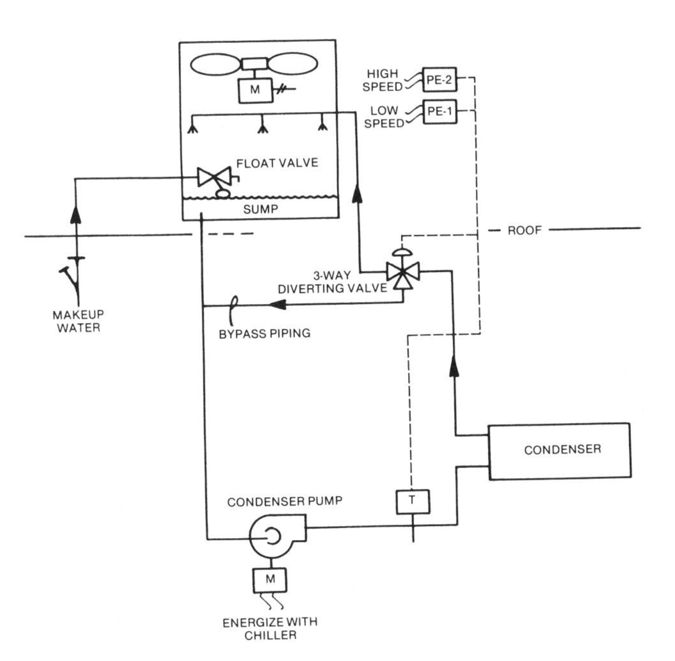

Fig. 95 Condenser Water Temperature Control

Handbook— HVAC Systems and Applications and Chapter 42 in the 1988 ASHRAE *Handbook—Equipment* have details.

Heat recovery and storage systems are almost always unique, so they require customized control systems to provide the desired sequence. Chapters 6 and 8 in the 1987 ASHRAE *Handbook—HVAC Systems and Applications* have further details.

DESIGN CONSIDERATIONS AND PRINCIPLES

Total building HVAC system selection and design considers the type and size of the structure and how it is used and operated. Subsystems, such as fan and water supply, are normally controlled by localized automatic control, or local loop control. A *local loop control system* includes the sensors, controllers, and controlled devices used with a single HVAC system and excludes any supervisory or remote functions such as reset and start-stop. However, extension of local control to a central control point is frequently included in the design when justified by operating economy through reduced labor and energy costs and need to diagnose system malfunction when it occurs, thus reducing any damage that might result from delay.

The growing popularity of distributed processing using microprocessors has extended computer use at many locations besides the central control point. Specifically, the local loop controller can be a microprocessor (DDC) instead of a pneumatic or electric thermostat, and some energy management functions may be performed by a local microprocessor.

Because heating or cooling systems are designed to meet maximum design conditions, they nearly always function at partial capacity. It is important to be able to control the system at all times; therefore, the HVAC system must facilitate control operation. Because the system must be adjusted and maintained in operation for many years, the simplest system that produces the necessary results is usually best.

Coordination Between Mechanical and Electrical Systems

Even when a system is basically pneumatic, the electrical engineer must design wiring, conduit, switchgear, and electrical distribution for many electrical devices involved.

The mechanical designer must tell the electrical designer the total electrical requirements of the control system if the controls are to be wired by the electrical contractor. These include the devices to be furnished, connected, or both; loads in watts or horsepower; location of electrical items; and a description of each control function. Proper coordination should develop a schematic control diagram that interfaces properly with other control elements to form a complete and usable system. As an option, the control engineer may develop a complete performance specification and require the control system contractor to perform all related wiring required to perform the specified sequence.

Coordination is essential. The control system designer must take the initiative and make the final checks of drawings and specifications. Both mechanical and electrical specifications must be checked to ensure compatability and uniformity of all documents.

Building and System Subdivision

Building and mechanical system subdivision considers:

- Heating and cooling loads as they vary—they may require the ability to heat or cool interior or exterior areas of a building at any time.
- Occupancy schedules and the flexibility to meet needs without undue initial and/or operating costs.
- Fire safety smoke control and possible compartmentation that matches the air-handling system layout and operation.

Control Principles for Energy Conservation

After the general needs of a building have been established and the building and system subdivision has been made based on similar needs, the mechanical system and its control approach can be considered. Designing systems that conserve energy requires (1) knowledge of the building, (2) its operating schedule, (3) the systems to be installed, and (4) knowledge of ASHRAE *Standard* 90.1-1989, Energy Conservation in New Building Design. The principles or approaches that conserve energy are as follows:

1. *Run equipment only when needed.* Schedule HVAC unit operation for occupied periods. Run heat at night only to maintain internal temperature between 50 and 55 °F to prevent freezing. Start morning warm-up as late as possible to achieve design internal temperature by occupancy time (optimum start control), considering residual temperature in space, outdoor temperature, and equipment capacity. Under most conditions, equipment can be shut down some time before the end of occupancy, depending on internal and external load and space temperatue (optimum stop control). Calculate shutdown time so that space temperature does not drift out of the selected comfort zone before the end of occupancy.
2. *Sequence heating and cooling.* Do not supply heating and cooling simultaneously. Central fan systems should use cool outdoor air in sequence between heating and cooling. The zoning and system selection should eliminate, or at least minimize, simultaneous heating and cooling. Also, humidification and dehumidification should not occur concurrently.
3. *Provide only the heating or cooling actually needed.* Generally, reset the supply temperature of hot and cold air (or water) according to actual need. This is especially important on systems or zones that allow simultaneous heating and cooling.
4. *Supply heating and cooling from the most efficient source.* Use free or low-cost energy sources first, then use higher-cost sources, as necessary.
5. *Apply outdoor air control.* Do not use outdoor air for ventilation until the building is occupied, and then use psychrometrically proper outdoor air quantities. When on minimum outdoor air, use no more than that recommended by ASHRAE *Standard* 62-1989, Standards for Natural and Mechanical Ventilation. In the cooling mode (in cost-effective areas), use enthalpy rather than dry bulb to determine whether outdoor or return air is the most energy efficient air source.

Automatic Control

System Selection. The mechanical system significantly affects how zones and subsystems can be controlled. The system selected and the number and location of zones further influence the amount of simultaneous heating and cooling that occurs. Systems for exterior building sections should control heating and cooling in sequence to minimize simultaneous heating and cooling. In general, the control system must be designed to accomplish this, since only a few mechanical systems have this inherent ability (*e.g.*, two-pipe systems and single-coil systems). Systems that require engineered control systems to minimize simultaneous heating and cooling include the following:

- *Cooling variable air volume with zone reheat.* Reduce cooling energy and/or air volume to a minimum before applying reheat.
- *Four-pipe heating and cooling for unitary equipment.* Sequence heating and cooling.
- *Double-duct systems.* Condition only one duct (either hot or cold) requiring one thermal energy at a time. The other duct should supply a mixture of outdoor and return air.
- *Single-zone heating/cooling systems.* Sequence heating and cooling.

Some exceptions will always exist, as in the case of dehumidification with reheat; therefore, the preceding principles are considered objectives.

Control zones are determined by location of the thermostat or temperature sensor that sets the requirements for heating and cooling supplied to the space. Typically, these control zones are for a room or an open area portion of a floor.

Many states no longer permit systems that reheat cold air or that mix heated and cooled air to heat and cool simultaneously. Such systems should be avoided. If selected, they should be designed for minimal use of the reheat function by zoning to match actual dynamic loads and reset cold and warm air temperatures based on the zone(s) with the greatest demand. Control details are shown later in this chapter. Heating and cooling supply zones should be structured to cover only areas of similar load. Different exterior exposures should have different supply zones.

Systems that provide changeover switching between heating and cooling prevent simultaneous heating and cooling. They include hot or cold secondary water for fan coils or single-zone fan systems. They usually require small operational zones, which have low load diversity, to permit changeover from warm to cold water without occupant dissatisfaction.

Systems for building interiors usually require year-round cooling and are somewhat simpler to control than exterior systems. These areas normally use all-air systems with constant supply air temperature, with or without variable air volume control. Proper control techniques and operational understanding can reduce the energy used to treat these areas. Reheat should be avoided.

General load characteristics of different parts of a building may lead to selecting different types of systems for each.

Load Matching. When individual room control is used, it is possible to control space more accurately and to conserve energy if the whole system can be controlled in response to the major factor influencing system load. Thus, water temperature in a hot water heating system, steam temperature or pressure in a steam heating system, or delivered air temperature in a central fan system can be varied as building load varies. This puts a reasonable control on the whole system, relieves individual space controls of part of their burden, and allows more accurate space control. Also, modifying the basic rate of heating or cooling input to the system in accordance with system load reduces losses in the distribution system.

The system must always satisfy the area or room with the greatest demand. Individual controls handle demand variations within the area the system serves. The more accurate the system zoning, the greater the control by the overall system, the smaller the system distribution losses, and the more effectively space conditions are maintained by individual controls.

Design Considerations

Size of Controlled Area. No individually controlled area should exceed about 5000 ft^2, because the difficulties of obtaining good distribution and of finding a representative location for the space controls increase with zone area. Each individually controlled area must have similar load characteristics throughout. For uniform conditions throughout an area, equitable distribution must be provided by competent engineering design, careful equipment sizing, and proper system balancing. The control can measure conditions only at its location; it cannot compensate for variable conditions throughout the area caused by improper distribution or inadequate design. Areas or rooms having dissimilar load characteristics, or different conditions to be maintained, should be individually controlled. The smaller the controller area, the better the control obtained and the more optimal the system performance and flexibility.

Location of Space Sensors. Space sensors and controllers must be located where they accurately sense the variables they control and where the condition is representative of the whole area (zone) they serve. In large open areas that have more than one zone, thermostats should be in the middle of their zone to prevent being affected by conditions in surrounding zones. There are three common locations for space temperature controllers or sensors.

Wall-mounted thermostats or *sensors* are usually placed on inside walls or columns in the occupied space they serve. Avoid outside wall locations. Mount thermostats where they will not be affected by heat from sources such as direct sun rays; wall pipes or ducts; convectors; and direct air currents from diffusers or equipment such as copy machines, coffee makers, or refrigeration cases. Air circulation should be ample and unimpeded by furniture or other obstructions, and there should be protection from mechanical injury. Thermostats located in spaces such as corridors, lobbies, or foyers should be used to control only those areas.

Return air thermostats can control floor-mounted unitary conditioners such as induction or fan-coil units and unit ventilators. On induction and fan-coil units, the sensing element is behind the return air grille. On classroom unit ventilators that use up to 100% outdoor air for natural cooling, however, a forced flow sampling chamber should be provided for the sensing element.

If return air sensing is used with central fan systems, locate the sensing element as near the space being controlled as possible to eliminate influence from other spaces and the effect of any heat gain or loss in the duct. Where combination supply/return light fixtures are used to return air to a ceiling plenum, the return air sensing element can be located in the return air opening of a light fixture. (Be sure to offset the set point to compensate for the heat from the light fixtures.)

The sensing element should be located carefully to avoid radiant effect and to ensure adequate air velocity across the element.

Diffuser-mounted thermostats usually have sensing elements mounted on circular or square ceiling supply diffusers and depend on aspiration of room air into the supply airstream. They should be used only on high-aspiration diffusers adjusted for a horizontal air pattern. The diffuser in which the element is mounted should be in the center of the occupied area of the controlled zone.

Cost Analysis

Due to the rising cost of energy, it has been popular to promote various types of control equipment by "months to pay back" based on projected energy savings. While this is one appropriate measure of the value of a purchase, many other factors can affect the true value. For example, if the estimated payback of an energy management system is 24 months, it does not necessarily mean an annual rate of return of 50%. For a detailed explanation of life cycle costing, see Chapter 33.

CONTROL APPLICATION LIMITATIONS AND PRECAUTIONS

Controls for Mobile Units

The operating point of any control that relies on pressure to operate a switch or valve varies as atmospheric pressure changes. Normal variations in atmospheric pressure do not noticeably change the operating point, but a change in altitude affects the control point to an extent governed by the change in absolute pressure. This is especially important when controls are selected for use in land and aerospace vehicles which, in normal use, are subject to wide variations in altitude. This effect can be substantial; for example, barometric pressure decreases by nearly one-third as the altitude increases from sea level to 10,000 ft.

In mobile applications, three detrimental factors are always present in varying degrees—vibration, shock, and G-forces. Controls selected for such service must qualify for the specific environment expected in the installation. In general, devices containing mercury switches, slow-moving or low-force contacts, or mechanically balanced components are unsuitable for mobile applications, while electronic solid-state devices are generally less susceptible to mobile forces.

Explosive Atmospheres

Sealed-in-glass contacts are not considered explosion-proof; therefore, other means must be provided to eliminate any possible spark within the atmosphere.

When using electric control systems, the designer can use an explosion-proof case to surround the control case and contacts, permitting only the capsule and the capillary tubing to extend into the conditioned space. It is often possible to use a long capillary tube and to mount the instrument case in a nonexplosive atmosphere. The latter method can be duplicated in an electronic control system by placing an electronic sensor in the conditioned space and feeding its signal to an electronic transducer placed in the nonexplosive atmosphere.

Because a pneumatic control system uses compressed air as its energy source, it is safe in otherwise hazardous locations. However, many pneumatic systems include E/P or P/E interfaces to electrical components. All electrical components require appropriate explosion-proof protection.

Sections 500 to 503 of the *National Electrical Code* include detailed information on electrical installation protection requirements for various types of hazardous atmospheres.

Limit Controls

When selecting automatic controls, it is important to consider the type of operation needed to control high or low limit (safety). This control may be inoperative for days, months, or years, and then operate immediately to prevent serious damage to equipment or property. Separate operating and limit controls are always recommended, even for the same functions.

Steam or hot water exchangers tend to be self-regulating and, in that respect, differ from electrical resistance heat transfer devices. For example, if airflow through a steam or hot water coil stops, coil surfaces approach the temperature of the entering steam or hot water, but cannot exceed it. Only convection or radiation losses from the steam or hot water to the surrounding area take place, and the coil is not usually damaged. Electric coils and heaters, on the other hand, can be damaged when no air flows around them. Therefore, control and power circuits must interlock with heat transfer devices (pumps and fans) to shut off the electrical energy when the device shuts down. Flow or differential pressure switches may be used for this purpose; however, they should be calibrated to energize only when airflow exists. This precaution shuts off power in case a fire damper closes or some duct lining blocks the air passage. Limit thermostats should also be installed to deenergize the heaters when temperatures exceed safe operating conditions.

Safety for Duct Heaters

The current in individual elements of electric duct heaters is normally limited to a maximum safe value established by the *National Electric Code* or local codes. In addition to the airflow interlock device, an automatic reset high limit thermostat and a manual reset backup high limit safety device are usually applied to duct heaters (Figure 96). The auto-reset high limit normally deenergizes the control circuit; however, if the control circuit has an inherent time delay or uses solid-state switching devices, a separate safety contactor may be desirable. The manual reset backup limit is generally arranged to interrupt all current to the heater independently, in case other control devices fail.

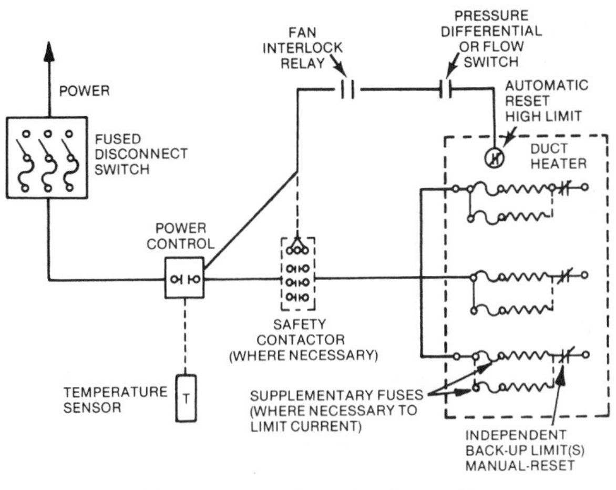

Fig. 96 Duct Heater Control

Cooling/Heating Changeover

With automatic changeover between heating and cooling, the control system should prevent operation of both modes for a time to prevent cycling between the two functions and/or unnecessary energy consumption. A controller with an adjustable dead band (differential) between heating and cooling easily accomplishes this function.

Lowered Night Temperature

When temperatures during unoccupied periods are lower than these normally maintained during occupied periods, an automatic timer often establishes the proper day and night temperature time cycle. Allow sufficient time in the morning to pick up the conditioning load well before there is any heavy load increase in conditioned spaces. Night setback temperatures are often monitored and controlled more closely with supervisory control systems (see Control of Central Subsystems). These supervisory systems take into account variables such as outdoor temperature, system capacity, and building mass to determine optimal start-up and shutdown times.

Multiple Thermostats per Zone

Buildings or zones with a modular arrangement can be designed for subdividing to meet occupant needs. Until subdividing is done, operating inefficiencies can occur if a zone has more than one thermostat. If a system allows one thermostat to turn on heating while another turns on cooling, control the two zones or terminals from a single thermostat until the area is subdivided properly.

COMMISSIONING

A successful control system receives a proper start-up and testing (commissioning), not merely the adjustment of a few parameters (set points and throttling ranges) and a few quick checks of the system. However, the controls used for many efficient HVAC systems present a formidable challenge for proper commissioning and require the services of experienced professionals. In general, the increased use of VAV systems and digital controls have increased the importance and complexity of commissioning.

Design and construction specifications should include specific commissioning procedures. In addition, commissioning should be coordinated with testing and balancing (Chapter 34), since each affects the other. The commissioning procedure begins by checking each control device to see that it is installed and connected according to approved shop drawings. Each electrical and pneumatic connection is verified, and all interlocks to fan and pump motors and primary heating and cooling equipment are checked.

Tuning—General Considerations

The systematic tuning of controllers improves the performance of all control systems and is particularly important for digital control. First, the controlled process should be controlled manually between various set points to evaluate the following questions:

- Is the process noisy (rapid fluctuations in controlled variable)?
- Is there appreciable hysteresis (backlash) in the actuator?
- How easy (or difficult) is it to maintain and change set point?
- In which operating region is the process most sensitive (highest gain)?

If the process cannot be controlled manually, the reason should be identified and corrected before tuning the controller.

Tuning selects control parameters that determine the steady-state and transient characteristics of the control system. HVAC processes are nonlinear, and characteristics change on a seasonal basis. Controllers tuned under one operating condition may become unstable as conditions change. A well-tuned controller will (1) minimize the steady-state error from set point, (2) respond quickly to disturbances, and (3) remain stable under all operating conditions. Tuning of proportional controllers is a compromise between minimizing steady-state error and maintaining margins of stability. Proportional plus Integral (PI) control minimizes this compromise because the integral action reduces steady-state error, while the proportional term determines the controller's response to disturbances.

Control loops should be tuned under conditions of highest process gain to ensure stability over a wide range of process conditions.

Tuning PI Controllers

Popular methods of determining PI controller tuning parameters include closed- and open-loop process identification methods and trial-and-error methods (Stoecker 1989). The closed-loop method increases the gain of the controller in a proportional-only mode until the system continuously cycles after a set point change (Figure 97). Proportional and integral terms are then computed from the loops period of oscillation and the proportional gain value that caused cycling. The open-loop method introduces a step change in input into the opened control loop. A graphical technique is used to estimate the process transfer function parameters. Proportional and integral terms are calculated from the estimated process parameters using a series of equations.

The trial-and-error method involves adjusting the gain of the proportional-only controller until the desired response to a set point change is observed. Conservative tuning dictates that this response should have a small initial overshoot and quickly damp to steady-state conditions. set point changes should be made in range where controller saturation, or output limits, is avoided. The integral term is then increased until changes in set point produce the same dynamic response as the controller under proportional-only control, but with the response now centered about the set point (Figure 98).

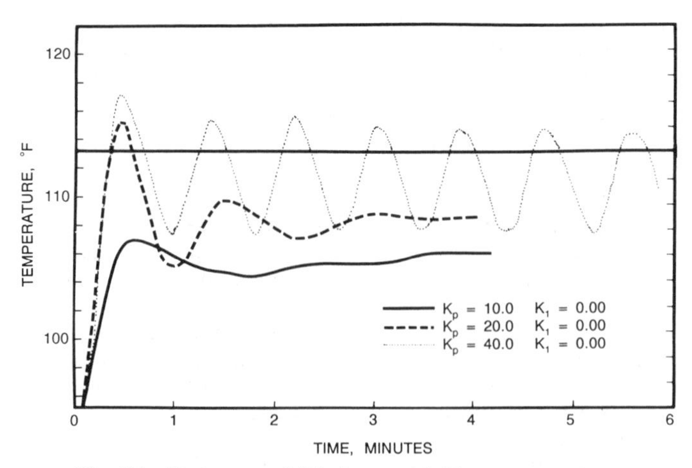

Fig. 97 Response of Discharge Air Temperature to a Step Change in Set Points at Various Proportional Constants and No Integral Action

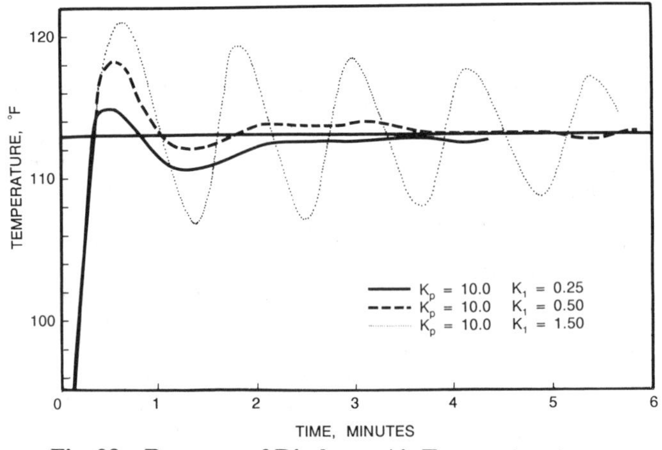

Fig. 98 Response of Discharge Air Temperature to a Step Change in Set Points at Various Integral Constants with a Fixed Proportional Constant

Tuning Digital Controls

In tuning digital controllers, additional parameters may need to be specified. The selection of a digital controller sampling interval is a compromise between computational resources and control requirements. A controller sampling interval about one-half of the time constant of the controlled process usually provides adequate control. Many digital control algorithms include an error dead band to eliminate unnecessary control actions when the process is near set point. Hysteresis compensation is possible with digital controllers, but it must be carefully applied as overcompensation can cause continuous cycling of the control loop.

COMPUTERS FOR AUTOMATIC CONTROL

Computers can perform the control schemes described in this chapter. Chapter 36 covers the computer components and some of the ways computers are being used in the HVAC industry. Other technical publications are available that describe the application of computers in the HVAC control industry.

OPERATION AND MAINTENANCE

Ultimately, the benefit from HVAC equipment depends on properly designed, installed, operated, and maintained automatic control systems. Control systems duplicate the action of the plant operator and, in certain situations, eliminate some of the functions that operators have historically performed. The trade-off is greater complexity and an increase in the number of items to be maintained.

Current construction practices complicate design because they shift responsibilities from one group of specialists to another. Often, each specialist has different objectives. Thus, direct communication to integrate all the related skills is essential to assure the HVAC system will provide the healthy and comfortable space conditions desired. The designer that accurately evaluates the design from the perspective of the installers, operators, and maintenance personnel has the best chance of assuring a successful control system.

OBJECTIVES

The following list of questions should be addressed by the system designer to clarify some operation and maintenance objectives.

Cost

1. Will trained personnel be employed to operate the equipment?
2. If not, will the automatic control system safely and cost-effectively operate all equipment as required to satisfy all comfort needs without trained operating personnel?
3. What is the relative cost of these alternative approaches? Are the owners aware of these costs in documented form, and are they in agreement with the analysis?
4. What is the estimated cost for maintaining the automatic control system? Has it been verified with the probable maintenance sources available to the owner? Have the owners been made aware of these costs in documented form, and are they in agreement with the estimate?
5. If the owners are going to hire their own operating and/or maintenance personnel, what has the designer provided (in the form of documentation and training recommendations) to ensure that the design intent will be available throughout the life of the facility? Has this been discussed with the owners, and do they concur with the plan?

General comment: Control system maintenance costs can be easily underestimated. They may range from 5 to 20% of their purchase or replacement cost annually. It is not unusual for the personnel requirement for the maintenance effort of automatic controls—furnished as stand-alone systems of integral components of HVAC equipment—to exceed that required to maintain the remainder of the HVAC components and their related electrical distribution system. The most common reason for inadequate control system maintenance is the lack of management understanding and emphasis relative to this cost.

Flexibility

1. How will standby, spare, or redundant system components (fans, pumps, switchgear) be interfaced with the control system to provide backup service when their paired components are taken out of service because of failure or scheduled maintenance requirements? If the system is not automatic, are operational instructions easy to understand and conveniently provided?
2. Has the designer or supplier looked at each component of the control system and evaluated its impact on system performance during normal failure modes? What is the appropriate operator function at each such occurrence? Is it documented?
3. Can the actuation devices (valves, dampers, relays) be manually manipulated by operators to facilitate control system component maintenance, replacement, or emergency operation? Are proper instructions for their emergency use provided?

General comment: Many designers provide two operator interface stations for each actuation device. These stations provide an automatic-manual selection and a positioning, or mode, selection capability for use during manual operation. This may be a desirable feature, even though there is added cost.

Maintainability

1. What provisions has the designer made to ensure that each control component is easily accessible for operation and maintenance? Are the number of components requiring ladders, supplemental lighting, special access panels, and so forth reduced to the lowest practical level? At the conclusion of the installation, will components be difficult to access because of conduit, piping, insulation, or proximity to other system components?
2. Are the system components that are most subject to failure of the quick-replacement, plug-in type?
3. Are piping and wiring practices specified? Are the materials subject to damage from normal maintenance functions or operation of other HVAC system components? Do they interfere with any other building maintenance function?
4. Are adequate gages, meters, information displays, and other diagnostic tools provided to facilitate convenient, detailed troubleshooting of system components?
5. Can components be replaced without draining hydronic systems, losing refrigerant charge or shutting down major system components?
6. Will repair parts or substitute components be readily available at a reasonable price during the economic life of the facility?
7. Is equipment maintenance and/or operation subject to the installation acquired knowledge of one or more key individuals, in lieu of broad-based documentation for new people?

Reliability

1. Have provisions been made to ensure a clean, dry, control air supply during all weather conditions and normal maintenance requirements?
2. Are power supplies protected from lightning and other electrical transients where applicable?
3. Are electronic components located in a clean, dry enclosure with adequate temperature regulation to avoid product damage?
4. Are the specified and supplied components adequately field-proven? If not, are both the manufacturer and the owner willing to assume the risk of added warranty and maintenance costs? If the components need replacement, what are the alternatives? Who pays? Who demonstrates that the components are not reliably functional, and who is responsible for those costs?
5. What are the quality control procedures used by the manufacturer? What type of functional tests are made at the factory? In the field? Is every component checked, or is random sampling the accepted practice?
6. Is the control system, as designed, subject to the accurate function and calibration of a large number of system components (*i.e.*, high- or low-temperature signal feedback, valve position feedback, occupancy feedback, etc.)? If so, what provisions have been made to detect any component malfunction, HVAC equipment application or installation problem, to ensure that the fault does not require the system to operate in the maximum energy consumption mode by default?

Documentation

1. Who is responsible for checking the completeness and accuracy of the project's documentated deliverables?
   a) "As-built" control diagrams?
   b) Laminated or framed sequences of operation and maintenance interface instructions?
   c) Component and subcomponent parts lists?
   d) Calibration and/or maintenance procedures for each component?
   e) Operating software, in field loadable form?
   f) Operating software, in engineering logic form for at least all operational sequences?
   g) Users manual, including software generation manual?
2. What is the schedule for documentation deliverables? Does it coincide with training schedules?
3. What provisions have been made to update documentation to accommodate future corrections or changes?
4. Who is responsible for furnishing the deliverables listed in item 1? Are systems and components that are an integral part of major items of equipment included?

Expandability or Modification

1. What are the alternatives for upgrading the control system to accommodate expansion or system modification?
2. Will the owner be required to accept the possibility of side-by-side systems to facilitate competitive bidding of system add-on components or services?
3. Have provisions been made to identify and control the future costs of software and hardware additions and modifications?

REFERENCES

Haines, R.W. 1982. Retrofit of existing control systems. *Heating/Piping/Air Conditioning*, Parts I and II (March and April).

Kinney, T.B. 1983. Tuning process controllers. *Chemical Engineering* (September).

Kuo, B.C. 1975. *Automatic control systems*, 3rd ed. Prentice Hall, Englewood Cliffs, NJ.

Stoecker, W.F. and D.A. Stoecker. 1989. Microcomputer control of thermal and mechanical systems. Van Nostrand Reinhold, New York.

Ziegler, J.G. and N.B. Nichols. 1942. Optimal settings for automatic controllers. ASME *Transactions* (64):759-65.

BIBLIOGRAHPY

Interfacing Sensors and Actuators

Baker, D.W. and C.W. Hurley. 1984. On-site calibration of flow metering systems installed in buildings. NBS Science Series Report 159, National Bureau of Standards, Washington, D.C.

Coad, W.J. 1985. Variable flow in hydronic systems for improved stability, simplicity and energy economics. ASHRAE *Transactions* 91(1B).

Hurley, C.W. and J.F. Schooley. 1984. Calibration of temperature measurement systems installed in buildings. NBS Science Series Report 153, National Bureau of Standards, Washington, D.C.

Hyland, R.W. and C.W. Hurley. 1983. General guidelines for the on-site calibration of humidity and moisture control systems in buildings. NBS Science Series Report 157, National Bureau of Standards, Washington, D.C.

Johnson, G.A. 1985. Retrofit of a constant volume air system for variable speed fan control. ASHRAE *Transactions* 91(1).

Kao, J.Y. and W.J. Snyder. 1982. Application information on typical hygrometers used in heating, ventilating, and air conditioning (HVAC) systems. National Bureau of Standards Report NBSIR 81-2460, Washington, D.C.

Treichler, W.W. 1985. Variable speed pumps for water chillers, water coils, and other heat transfer equipment. ASHRAE *Transactions* 91(1).

Zell, B.P. 1985. Design and evaluation of variable speed pumping systems. ASHRAE *Transactions* 91(1).

Automatic Computer Control Applications

Chapman, W.F. 1980. Microcomputers hail new era in controls. ASHRAE *Journal* (July):38.

Coggan, D.A. 1986. Control fundamentals apply more than ever to DDC. ASHRAE *Transactions* 92(1).

Doucet, P. 1982. Direct digital control: Next generation for building automation. *Specifying Engineer* (August).

Edwards, H.J. 1980. *Automatic controls for heating and air conditioning*. McGraw Hill, Inc., New York.

Haines, R.W. 1983. *Control systems for heating, ventilating and air conditioning*, 3rd ed. Van Nostrand Reinhold, New York.

Kirts, R.E. 1985. Users guide to direct digital control of heating, ventilating, and air conditioning equipment. Naval Civil Engineering Laboratory Report UG-0004, Port Hueneme, CA.

Lau, A.S., W.A. Beckman, and J.W. Mitchell. 1985. Development of computerized control strategies for large chilled water plants. ASHRAE *Transactions* 91(1B).

Levine, M. and L.W. Moll. 1981. Beyond setback: Energy efficiency through adaptive control. ASHRAE *Journal* (July):37.

May, W.B., B.A. Borresen, and C.W. Hurley. 1982. Direct digital control of a pneumatically actuated air-handling unit. ASHRAE *Transactions* 88(2).

Mills, S.J. 1983. The application flexibility of the EMCS-DDC combination. ASHRAE *Journal* (November):36.

Nesler, C.G. and W.F. Stoecker. 1984. Selecting the proportional and integral constants in the direct digital control of discharge air temperature. ASHRAE *Transactions* 90(2).

Walker, C.A. 1984. Application of direct digital control to a variable air volume system. ASHRAE *Transactions* 90(2).

Wichman, P.E. 1984. Improved local loop control systems. ASHRAE *Transactions* 90(2).

Williams, V.A. 1982. Better control through computers. ASHRAE *Transactions* 88(1).

Yaeger, G.A. 1986. Flow charting and custom programming. ASHRAE *Transactions* 92(1).

Communications

Davies, D.W., *et al.* 1979. *Computer networks and their protocols.* John Wiley and Sons, New York.

Digital Equipment Corporation 1981. *Terminals and communications handbook.* Maynard, MA.

Freeman, R.L. 1980. *Telecommunication system engineering.* John Wiley and Sons, New York.

Martin, J. 1976. *Telecommunications and the computer,* 2nd ed. Prentice-Hall, Englewood Cliffs, NJ.

Martin, J. 1981. *Computer networks and distributed processing.* Prentice-Hall, Englewood Cliffs, NJ.

McNamara, J.E. 1977. *Technical aspects of data communications.* Digital Equipment Corporation, Maynard, MA.

Newman, H.M. 1983. Data communications in energy management and control systems: Issues affecting standardization. ASHRAE *Transactions* 89(1).

Roden, M.S. 1982. *Digital and data communication systems.* Prentice-Hall, Englewood Cliffs, NJ.

Sapienza, G.R. 1986. The effect of EMCS architecture on direct digital controllers. ASHRAE *Transactions* 92(1).

Operator-Machine Interface

Dressel, L.J. 1982. Improved operator interface techniques. ASHRAE *Transactions* 88(1).

Schaefer, R.J. 1982. A technique for the use of color graphics and light pens in energy management applications. ASHRAE *Transactions* 88(1).

Westphal, 1982. Human engineering: The man/system interface. ASHRAE *Transactions* 88(1).

Review of Basic Control Theory

Deshpande, P.B. and R.H. Ash. 1981. Elements of computer process control. Instrument Society of America, Research Triangle Park, NC.

McMillan, G.K. 1983. Tuning and control loop performance. Instrument Society of America, Research Triangle Park, NC.

Murrill, P.W. 1981. Fundamentals of process control theory. Instrument Society of America, Research Triangle Park, NC.

Ogata, K. 1970. *Modern control engineering.* Prentice-Hall, Englewood Cliffs, NJ.

Ogata, K. 1978. *System dynamics.* Prentice-Hall, Englewood Cliffs, NJ.

Williams, T.J. 1984. The use of digital computers in process control. Instrument Society of America, Research Triangle Park, NC.

CHAPTER 42

SOUND AND VIBRATION CONTROL

A PROPER acoustical environment is as important for human comfort as other environmental factors controlled by air-conditioning systems. The objective of sound control is to achieve an appropriate sound level for all activities and people involved. Because of the wide range of activities and privacy requirements, appropriate indoor acoustical design sound levels may vary considerably from room to room. Appropriate outdoor sound levels depend on local ambient sound conditions or code requirements.

Sound and vibration are the result of a disturbance that propagates through an elastic medium. *Sound* is an oscillatory pressure pulsation in air; *vibration* is an oscillatory motion of a structure that can be felt or seen. Some of the energy required to operate HVAC systems is converted to acoustical as well as mechanical energy. HVAC noise and vibration control methods convert a portion of the energy to thermal energy (through duct system noise control) or mechanical energy (through vibration isolation).

This chapter is divided into six parts and is organized in much the same order in which mechanical systems are designed. Section topics include (1) Fundamental Concepts of Acoustics and Vibration, (2) Sound Control for Indoor Mechanical Systems, (3) Mechanical Equipment Room Noise Isolation, (4) Sound Control for Outdoor Equipment Installations and Outdoor Noise Control Barriers, (5) Vibration Isolation and Control, and (6) Troubleshooting.

FUNDAMENTALS

The human hearing response characterizes unlike sounds as being louder, of a different pitch, or varying with time. These subjective reactions are the result of changes in sound pressure level, spectral content, and temporal variation in the sound stimulus. Under experimental conditions, humans can detect small changes in sound level, but the human reaction describing halving or doubling of perceived loudness requires changes in sound pressure level of about 10 dB. In a typical environment for broadband sounds, 3 dB is a minimum perceptible change. This means that halving the power output of the source results in a barely noticeable change in sound pressure level, and the sound power output must be reduced by a factor of 10 before humans perceive that loudness has been halved. Subjective changes are shown in Table 1.

Table 1 Subjective Effect of Changes in Sound Pressure Level, Broadband Sounds

Change in Sound Pressure Level	Apparent Change in Loudness
3 dB	Just noticeable
5 dB	Clearly noticeable
10 dB	Twice (half) as loud

A more complete discussion of the fundamental concepts of sound and vibration is presented in Chapter 7 of the 1989 ASHRAE *Handbook—Fundamentals.*

ACOUSTICAL DESIGN CRITERIA AND RATING SYSTEMS FOR NOISE

Within the air-conditioning industry, the following methods of expressing acoustical design criteria are in common use: A-weighted sound level (dBA), noise criteria (NC) curves, and room criteria (RC) curves. Of the three, the use of RC curves to define system design goals is preferred.

A-Weighted Sound Level

The A-weighted sound level (dBA) has been widely used to express design goals because of its measurement simplicity and because the result can be expressed as a single number. However, the number has meaning only in terms of the loudness of the sound; it provides no clue to the frequency distribution of sound energy, which is an important element in judging the quality of the noise. In other words, although the loudness of the sound may be moderate, the quality may be objectionable and thus be judged unacceptable by the observer.

Criteria expressed in terms of A-weighted sound level should not be used to specify the noise control objectives of an HVAC sys-

The preparation of this chapter is assigned to TC 2.6, Sound and Vibration Control.

tem, because a sound level specified in dB(A) may not be acceptable to the occupant of a room supplied by the system. Also, noise should not be diagnosed with A-weighted sound level measurements alone because a spectrum shape of the noise is often required to discover and correct the reason for the complaint.

NC Curves

The family of NC curves shown in Figure 9 of Chapter 7 of the 1989 ASHRAE *Handbook—Fundamentals* is commonly used for specifying design goals in occupied buildings. The curve chosen for a given application depends on the type of space use and its sensitivity to the level of background noise (see Table 2). In principle, the chosen NC curve defines the recommended octave-band limits of an acceptable background noise spectrum for a particular type of space use. However, when interpreted strictly as a limit on the level not to be exceeded in any octave band, experience has shown that noise problems are not avoided unless the shape of the actual spectrum approximates that of the chosen NC curve over three to four contiguous octave bands; when the shape of the actual spectrum approximates that of the NC curve in only one or two octave bands, the quality of the background noise may be objectionably rumbly or hissy.

Unfortunately, in practice, both in an assessment of the noise-control design and in rating system noise in the field, the predicted (or measured) noise spectrum is not assigned an NC rating based on the highest NC curve tangent to the spectrum, regardless of where in the frequency range this occurs. This practice, while simple, has created both confusion and frustration in many field situations where a noise complaint has occurred. While an octave-band measurement of the noise may show that the specified NC limit has not been exceeded, the noise may sound unpleasant.

Figure 1 illustrates how using the tangent contour method to assign an NC rating to a noise can backfire in typical field applications. The three typical HVAC spectra shown have distinctly different sounding noises, yet each has the same NC tangent contour rating of NC 33. For example, spectrum H would generally be judged as hissy because of its excessive high-frequency imbalance. The rating of NC 33 was assigned because the highest numbered tangent contour is the NC 33 curve at 2000 Hz (the common convention permits extrapolation between curves in whole-number increments). A listener would probably rate spectrum R as rumbly because of its steeply rising low-frequency characteristics. The rating of NC 33 was assigned because the highest numbered tangent contour is the NC 33 curve at 250 Hz. Spectrum N would usually be classified as neutral; the NC 33 rating was assigned because the NC 33 curve is tangent to the noise spectrum at 500 Hz. However, spectrum N is the only one that would be judged acceptable by the occupant.

In view of the foregoing discussion, the NC tangent contour convention should not be used for field qualification of HVAC system background noise. A better method of specification requires that the spectrum shape of the background noise approximates that of an NC curve over at least three contiguous octave bands, without exceeding the limits defined by the specified NC curve.

RC Curves

RC curves are the preferred alternative to NC curves or A-weighted sound levels for defining system design goals and for qualifying or rating field installations. The advantages are that this rating system (1) accounts for the influence of both spectrum shape and level on the subjective assessment process, (2) includes data in the octave bands centered at 31.5 and 16 Hz, and (3) accounts for the ability of low-frequency acoustic energy to induce perceptible vibration in light building construction.

The family of RC curves illustrated in Figure 2 was derived from a study of acceptable background noises in a wide range of typi-

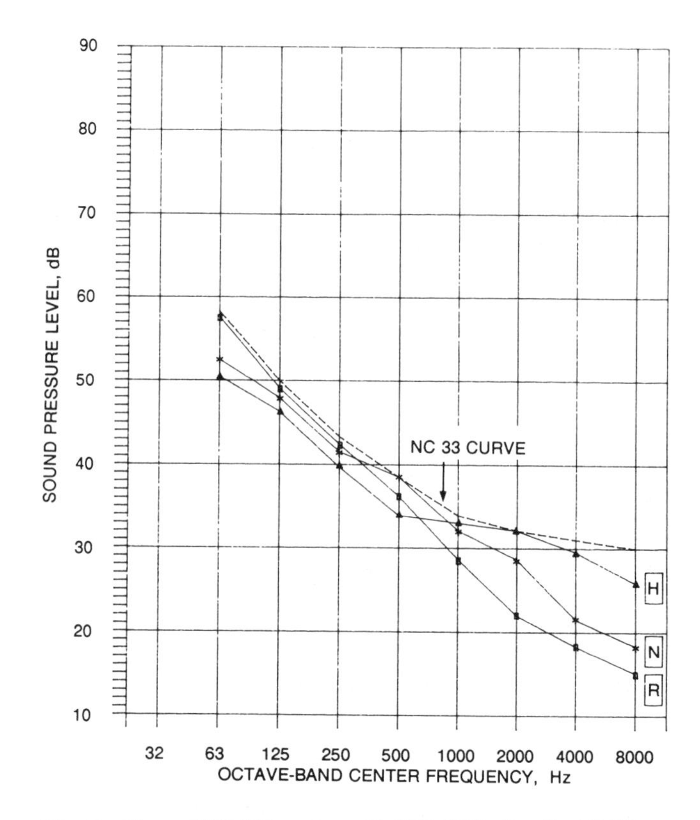

Fig. 1 Three HVAC Spectra with NC-33 Rating that Differ in the Way They Sound

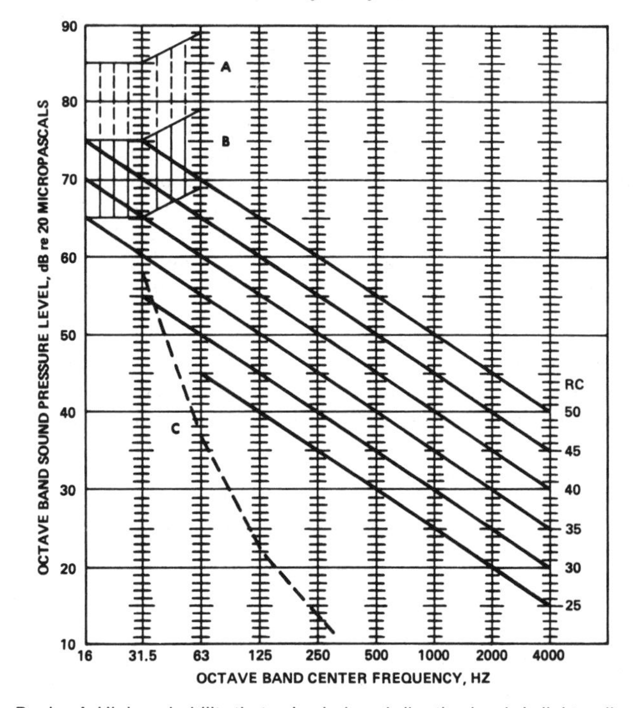

Region A: High probability that noise-induced vibration levels in light wall and ceiling constructions will be clearly feelable; anticipate audible rattles in low mass fixtures, doors, windows, etc.
Region B: Noise-induced vibration levels in light wall and ceiling constructions may be moderately feelable; slight possibility of rattles in low mass fixtures, doors, windows, etc.
Region C: Below threshold of hearing for continuous noise.

Fig. 2 RC (Room Criterion) Curves for Specifying Design Level for Balanced Spectrum

cal office environments. The −5 dB/octave slope of the RC curves corresponds to the average spectrum slope of the data measured in a large number of air-conditioned office environments rated as acceptable to the occupants. The slope of the RC curves do not change with level over the range shown. The numerical designation of each curve in the family corresponds to the level in the 1000 Hz octave band. Over the range of recommended applications, RC 25 to RC 50, these curves correspond to neutral-sounding spectra that are approximately balanced with respect to the relative loudness between the low- and high-frequency regions of the spectrum.

The low-frequency range of curves is influenced by the threshold of audibility at one extreme, and by the concern to avoid perceptible noise-induced vibration of room walls and ceilings found in typical light building construction, at the other. The cross-hatched regions shown in the upper left corner of the plot indicate the range of low-frequency sound pressure levels where the possibility of perceptible noise-induced vibration should be considered. In the upper cross-hatched region, a high probability that noise-induced vibration may be clearly perceptible and capable of causing acoustical ceiling systems and light fixtures to creak or rattle exists; in the lower cross-hatched region, the effects of noise-induced vibration may be marginal but still potentially capable of creating problems with secondary noise radiation.

RC NOISE RATING PROCEDURE

The RC rating of a noise requires the use of two descriptors—the first term is a number that represents the speech interference level of the spectrum (SIL), which is obtained by taking the arithmetic average of the noise levels in the 500, 1000, and 2000 Hz octave bands; the second term is a letter that denotes the quality of the sound, as it might subjectively be described by an observer. For example, a rating of RC 35(N) describes a noise having a speech interference level of 35 dB and a neutral balance of low- and high-frequency energy. A rating of RC 35(R) describes a noise with the same speech interference level, but which is rumbly sounding because of the spectral imbalance introduced by the presence of excess low-frequency energy. The letter H identifies a spectrum that sounds hissy (spectral imbalance at high frequencies), T identifies a spectrum that sounds tonal (audible pure-tone components present), and RV identifies a rumbly spectrum, which contains sufficient excess low-frequency energy to induce perceptible vibration in surfaces of light building construction.

The RC rating of a noise spectrum uses octave-band sound pressure level data at center frequencies from 31.5 to 4000 Hz. When acoustically induced vibration is of concern, an alternate form of the rating procedure requires the addition of the sound pressure level in the 16 Hz octave band. The RC rating is obtained by the following steps:

1. Plot the octave-band noise spectrum on a graph format similar to that illustrated in Figure 3.
2. Calculate the speech interference level of the spectrum. This is the arithmetic average, in dB, of the octave-band levels centered at 500, 1000, and 2000 Hz.
3. Draw a line with a slope of −5 dB per octave in the frequency range from 31.5 to 4000 Hz, passing through 1000 Hz at the speech interference level calculated in Step 2. This is the reference curve for use in evaluating the sound quality of the spectrum being rated.
4. Draw one line 5 dB above the reference curve extending from the 31.5 to the 500 Hz octave band. Draw a second line 3 dB above the reference curve, extending from the 1000 to the 4000 Hz octave band. The range between these two lines and the reference curve represents the maximum permitted deviation of the noise spectrum above the reference curve to receive a neutral (N) RC noise rating.
5. Determine the quality of the sound by observing how the shape of the spectrum deviates from the boundary limits of the reference curve established in Step 4. Use the criteria discussed below to choose the appropriate letter descriptor.
6. Assign the spectrum an RC rating. The numerical part of the rating corresponds to the level of the reference curve in the octave band centered at 1000 Hz; the appended letter descriptor is that determined in Step 5.

The following criteria are used to assign the appropriate letter descriptor of sound quality.

Neutral Spectrum (N). The levels in the octave bands centered at 500 Hz and below must not exceed the octave-band levels of the reference spectrum by more than 5 dB at any point in the range; the levels in the octave bands centered at 1000 Hz and above must not exceed the octave-band level of the reference spectrum by more than 3 dB at any point in the range.

Rumbly Spectrum (R). The level in the octave bands centered at 500 Hz and below exceeds the octave-band levels of the reference spectrum by more than 5 dB at one or more points in the range.

Hissy Spectrum (H). The level in the octave bands centered at 1000 Hz and above exceeds the octave-band level of the reference spectrum by more than 3 dB at one or more points in the range.

Tonal Spectrum (T). Noise with a tonal component usually has an abrupt change in sign of the spectrum slope that introduces a peak in a particular octave band, relative to the levels in the two adjacent bands. If the projection of this peak is more than 3 dB above a line constructed between the levels of the two adjacent bands, the tone may be audible to the ear. (Although a one-third octave-band analysis provides a more positive determination of the presence of a tonal component, a simple listening test may be sufficient to validate the presence of an audible pure tone.)

Acoustically Induced Perceptible Vibration (RV). The cross-hatched region of the RC curves in Figure 2 illustrates the frequency and amplitude range of octave-band sound pressure levels that may introduce perceptible vibration in the walls and ceilings of light building construction. The acoustically induced vibration may also cause secondary noise radiation, such as rattles in cabinet doors, pictures, ceiling fixtures, and other furnishings in contact with the walls. Sound pressure level data in the 16 Hz octave band are important in this analysis because HVAC noises capable of acoustically inducing wall and ceiling vibration usually contain significant energy in this frequency range.

Examples of RC Noise Rating

The figures in this section are developed from field data on HVAC system noise. With the exception of the figure illustrating a neutral spectrum, the other figures represent case histories in which the occupants registered strong complaints.

Neutral Spectrum. Figure 3 illustrates a noise spectrum that would receive an RC 35 (N) rating using the methodology described above. The computed speech interference level is 35 dB; the spectral deviations from the corresponding reference curve are within the specified boundaries of a neutral spectrum.

Rumbly Spectrum. Figure 4 illustrates a noise spectrum that would receive an RC 33 (R) rating. The computed speech interference level is 33 dB; the spectral deviations from the corresponding reference curve exceed the specified boundary conditions for a neutral spectrum in the octave bands below 125 Hz.

Hissy Spectrum. Figure 5 illustrates a spectrum that would receive a rating of RC 36 (H). The computed speech interference level is 36 dB; the spectral deviations from the corresponding reference curve exceed the specified boundary conditions in the octave bands above 1000 Hz.

Acoustically Induced Perceptible Vibration. Figure 6 illustrates a noise spectrum that would receive an RC 31 (RV) rating, according to the criteria listed above. Although the computed speech

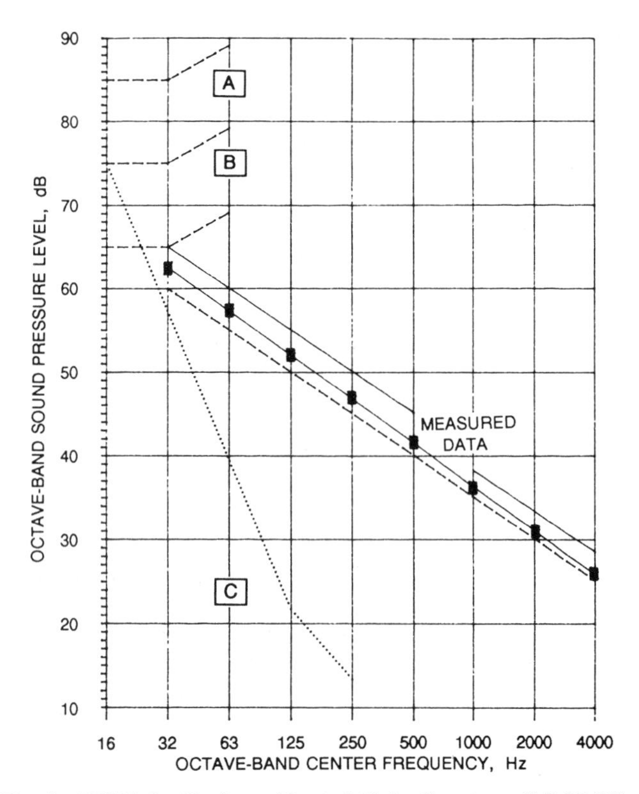

Fig. 3 RC Noise Rating—Neutral Noise Spectrum RC 35 (N)

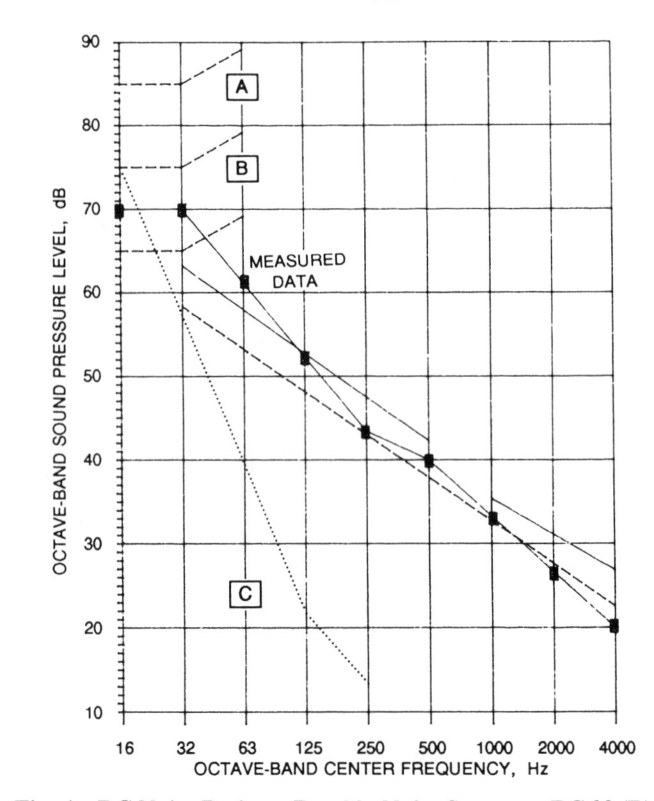

Fig. 4 RC Noise Rating—Rumbly Noise Spectrum RC 33 (R)

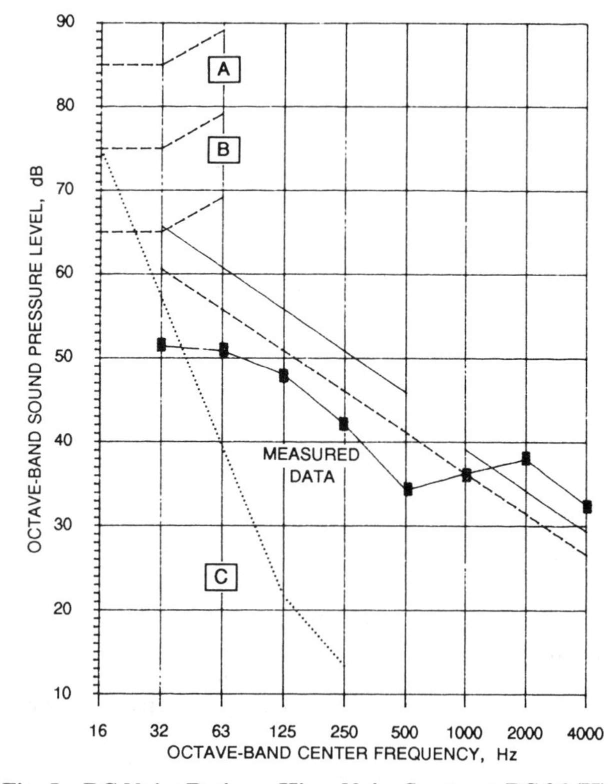

Fig. 5 RC Noise Rating—Hissy Noise Spectrum RC 36 (H)

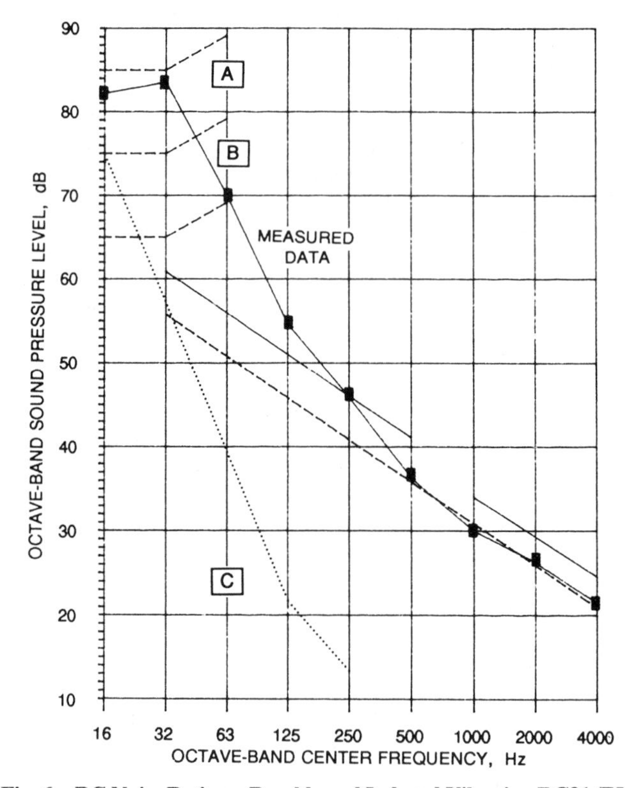

Fig. 6 RC Noise Rating—Rumbly and Induced Vibration RC31 (RV)

interference is only 31 dB, the spectrum of the noise greatly exceeds the assigned boundary limits in the low-frequency range; furthermore, the sound pressure levels in the octave bands centered at 16 and 31.5 Hz extend well into the upper cross-hatched region of the family of RC curves shown in Figure 2. For these reasons, the spectrum was assigned a rating of RC 31 (RV). In a typical office with a suspended acoustical ceiling, a spectrum of this character would not only sound unpleasantly rumbly, but would probably induce audible rattles in lighting fixtures, air diffusers, and return-air grilles.

CRITERIA FOR ACCEPTABLE HVAC NOISE LEVELS IN ROOMS

The noise due to HVAC systems is usually considered a part of the background sound of the environment. Therefore, to be judged acceptable it must neither noticeably mask sounds people want to hear, nor be otherwise intrusive or annoying in character. In an office environment, the acceptable level of background sound is generally established by speech requirements, which vary widely depending on the space use. In contrast, the acceptable level of background sound for unamplified performance spaces, such as recital and concert halls, is governed by the need to avoid masking the faintest of sounds that are likely to occur in a typical performance.

Table 2 lists recommended HVAC background noise criteria for a variety of space uses. The "Preferred" column is based on the RC curve noise-rating procedure. This criterion is recommended whenever the quality of space use dictates the need for a neutral, unobtrusive background sound. The "Alternate" column is based on the NC curve noise-rating procedure, which may be used when the quality of space use is not as demanding and can tolerate rumbly, hissy, or tonal characteristics in the background sound, as long as it is not too loud.

Table 2 Criteria for Acceptable HVAC Noise Levels in Unoccupied Rooms

Occupancy	Preferred	Alternate
Private residences	RC 25-30 (N)	NC 25-30
Apartments	RC 30-35 (N)	NC 30-35
Hotels/Motels		
Individual rooms or suites	RC 30-35 (N)	NC 30-35
Meeting/banquet rooms	RC 30-35 (N)	NC 30-35
Halls, corridors, lobbies	RC 35-40 (N)	NC 35-40
Service/support areas	RC 40-45 (N)	NC 40-45
Offices		
Executive	RC 25-30 (N)	NC 25-30
Conference rooms	RC 25-30 (N)	NC 25-30
Private	RC 30-35 (N)	NC 30-35
Open-plan areas	RC 35-40 (N)	NC 35-40
Business machines/computers	RC 40-45 (N)	NC 40-45
Public circulation	RC 40-45 (N)	NC 40-45
Hospitals and clinics		
Private rooms	RC 25-30 (N)	NC 25-30
Wards	RC 30-35 (N)	NC 30-35
Operating rooms	RC 25-30 (N)	NC 25-30
Laboratories	RC 35-40 (N)	NC 35-40
Corridors	RC 30-35 (N)	NC 30-35
Public Areas	RC 35-40 (N)	NC 35-40
Churches	RC 30-35 (N)	NC 30-35
Schools		
Lecture and classrooms	RC 25-30 (N)	NC 25-30
Open-plan classrooms	RC 35-40 (N)	NC 35-40
Libraries	RC 35-40 (N)	NC 35-40
Courtrooms	RC 35-40 (N)	NC 35-40
Legitimate theaters	RC 20-25 (N)	NC 20-35
Movie theaters	RC 30-35 (N)	NC 30-35
Restaurants	RC 40-45 (N)	NC 40-45
Concert and recital halls	RC 15-20 (N)	NC 15-20
Recording studios	RC 15-20 (N)	NC 15-20
TV studios	RC 20-25 (N)	NC 20-25

SOUND CONTROL FOR INDOOR MECHANICAL SYSTEMS

The air-conditioning system serving a room or space is often the major determinant of the background noise level. Noise reaches a listener by several paths, including airborne transmission of equipment noise to adjacent areas through the mechanical room construction; structure-borne transmission of equipment vibration through the building structure; and duct-borne noise created and transmitted by air-handling systems and their components.

The system noise level must be evaluated and controlled to achieve a satisfactory acoustical environment. Noise transmission paths should be evaluated during the design phase of a project, since remedial measures to reduce duct-borne noise are often expensive and only marginally effective.

SOURCE SOUND POWER LEVEL L_w AND ROOM SOUND PRESSURE LEVEL L_p

Single Sound Source

The sound pressure level L_p that occurs at a chosen point in a room when a given source sound power level L_w is introduced depends on (1) room volume, (2) furnishings, (3) source strength, and (4) distance of the sound source(s) from the point of observation.

Schultz (1985) updated prediction methods previously outlined in ASHRAE Handbooks of the relationship between sound power and sound pressure levels in typical furnished rooms of various sizes and shapes. Previous assumptions are only valid when dealing with empty rooms with hard acoustic surfaces, such as a laboratory reverberation chamber used for source sound power level determinations. Once any significant amount of acoustically absorptive material is introduced on any surface, the behavior of the sound field is altered significantly.

Most typical rooms contain distributed acoustical absorption and scattering elements, such as furniture. Although the amount of sound (acoustical) absorption and scattering varies from space to space (*e.g.*, hospital rooms versus executive offices), the effect does not drastically change the relationship between sound power and sound pressure levels.

The following empirical equation can be used to estimate the sound pressure level at a chosen distance from a sound source in normal rooms, as a function of the source sound power level, room size, and frequency. Predictions should be accurate to ±2 dB.

$$L_p = L_w - 5 \log V - 3 \log f - 10 \log r + 25 \text{ dB} \quad (1)$$

where

L_p = room sound pressure level at the chosen reference point, dB re 20 μPa
L_w = source sound power level, dB re 1 pW
V = room volume, ft^3
f = octave-band center frequency, Hz
r = distance from the source to the reference point, ft

Equation (1) applies directly to a single sound source in the room. When the room has more than one source, the total sound pressure level at the reference point is obtained by adding (on an energy basis) the contribution of each source, using the corresponding L_w and r for each source.

Distributed Ceiling Array

In most applications, air supply terminal diffusers are located in the ceiling of the conditioned space. The number of outlets within a given room are generally established by the floor area to be served by each terminal, which in turn varies as a function of the ceiling height and the load requirements. These considerations frequently lead to an approximately geometric distribution of air diffusers in the ceiling plane of the room, the individual sound power levels of which are nominally the same.

The total sound pressure level L_{pt} at a given reference point which results from the combination of individual sources in the room can be determined by using Equation (1) to compute the L_p due to each source, and then summing these on an energy basis. However, Equation (2) should be used for the frequently encountered situation in which the number and spacing of diffusers is influenced by the room geometric proportions and results in an approximately regular ceiling array.

Equation (2) may be used to estimate the resulting room sound pressure level L_{pt} in a plane 5 ft above the floor (standing head-height) due to a distributed ceiling array of nominally similar diffusers. For diffuser spacings on the order of the ceiling height, the variation in sound pressure level at any point within the reference plane above the floor should be about 1 dB, provided that there are at least four diffusers in the ceiling array. Equation (2) may also be used for an array of linear diffusers, by taking the source sound power level as that due to a single section and the number of sources as the number of sections in the array.

$$L_{pt}\,(5\text{ ft}) = L_{ws} - 5\log X - 28\log h + 1.3\log N - 3\log f + 31\text{ dB} \quad (2)$$

where

L_{pt} = average sound pressure level in a plane 5 ft above the floor, dB re 20 μPa
L_{ws} = sound power level of a single outlet in the array (*i.e.*, the combined sound power of that delivered by the distribution duct and that generated at the air terminal), dB re 1 pW
X = ratio of the floor area served by each outlet to the square of the ceiling height (X = 1 if the area served equals h^2)
N = number of ceiling outlets in the room (N should be at least 4)
f = octave-band center frequency, Hz
h = ceiling height, ft

Example 1. Calculate the average sound pressure level for the 500 Hz octave band in a plane 5 ft above the floor in an open-plan office served by a distributed array of ceiling outlets. The ceiling height is 9 ft, and the outlets are spaced on 9-ft centers. The total area served is 2000 ft^2, and each outlet serves approximately 133 ft^2. The sound power level of a single outlet at 500 Hz is 45 dB re 10 to 12 W.

Solution:

1. Each outlet serves 133 ft^2. Therefore,

$$N = 2000/133 = 15$$

2. The area served by each outlet is 133 ft^2; the ceiling height is 9 ft. Therefore,

$$133/9^2 = 1.64$$

3. From Equation (2):

$$L_{pt} = 45 - 5\log 1.64 - 28\log 9 + 1.3\log 15 - 3\log 500 + 31 = 41\text{ dB}$$

NOISE LIMITS FOR AIR DISTRIBUTION SYSTEMS

The distribution system, via supply and return openings, determines the airborne sound level in a room. In some cases, air valves and other devices radiate sound through the ceiling. Knowing this, the designer can specify components that do not exceed the design criterion. However, judgment of the acoustical acceptability of the overall final design is made by occupants of the conditioned spaces. For this reason, noise limits for components of the system should be established by working backwards from the selected room design criterion. Each component is chosen based on how much of the total sound power it is allowed to contribute to the room.

For example, the attenuation required between the fan discharge and room outlets can be determined by starting with the number of outlets in each conditioned space and computing what level of sound power must not be exceeded in the main supply ductwork ahead of the branch takeoffs. The difference between this sound power level and the level computed for the fan represents the amount of attenuation required on the inlet and discharge side of the fan, or in the main ductwork and branches. The same computations can be done for the fan inlet and room inlets. A limit specification can then be developed for the sound power output of the fan equipment by accounting for the amount of attenuation that can be incorporated external to the fan.

A similar procedure can be used to develop a specification for air terminal devices, such as VAV valves and fan-powered boxes. Adding up the sound power delivered at each outlet or inlet determines the total for such devices.

Sound power limits for air diffusers can be determined by starting with the room criterion and the number of diffusers serving the space to calculate the sound power per diffuser that should not be exceeded. Example 2 illustrates such a procedure.

Example 2. Using the data given in Example 1, calculate the sound power level spectrum of a single ceiling outlet that will allow the array not to exceed a room sound pressure level criterion of RC-40 at a level 5 ft above the floor.

Solution:

1. Determine [$L_{ws} - L_{pt}$ (5 ft)], the difference in dB between the sound power level of a single source in the array and the average sound pressure level at the 5-ft elevation. This is done by rearranging the terms in Equation (2):

$$L_{ws} - L_{pt}(5\text{ ft}) = 5\log 1.64 + 28\log 9 - 1.3\log 15 + 3\log f - 31\text{ dB} = 3\log f - 4.7$$

2. Determine [$L_{ws} - L_{pt}$(5 ft)] for each frequency band using the general result of Step 1.

RECOMMENDED NOISE CONTROL PROCEDURE

Figure 7 schematically illustrates a typical variable volume air-handling system. The numbers in the figure indicate the noise-generating and attenuating elements that may need consideration or evaluation. Not all paths shown in Figure 7 exist in every system; however, with the methodology presented in this section, the designer can apply the appropriate procedures to the specific system design.

In developing this procedure, the following assumptions have been made:

1. To achieve the design goals, adequate attenuation is obtained at the first critical outlet near the noise source. If this objective is achieved, all outlets further from the noise source are assumed to have adequate attenuation, unless the design goal for subsequent outlets is less than that for the first one.
2. The major noise source in an air-conditioning system is the fan. The procedure allows the designer to determine the attenuation required to eliminate fan noise to the desired degree. However, the designer must be aware of other noise sources in the system—regenerated noise created by fittings, terminal devices, and sound attenuators. Methods for evaluating these sources are also given in this chapter.

All prediction schemes and manufacturers' data assume uniform flow conditions. For example, for air distribution device sound power levels, manufacturers base their data on uniform velocity profile conditions and duct connection configurations. Corrections must be made and recognized where nonuniform flow conditions or turbulence exist. The system noise analysis procedure follows these basic steps:

1. Determine the noise generated by the source.
2. Determine the attenuation provided by all supply ductwork, fittings, and so forth. In doing this, calculate the resultant sound power level at each point of interest in the system. Determine if regeneration noise is a problem.
3. Calculate the resulting sound pressure level and compare it to the chosen design criteria.
4. Increase the attenuation in the system to eliminate deficiencies.
5. Determine if breakout noise is a problem and incorporate corrective measures.
6. Repeat the procedure for the return side of the system.
7. Determine if noise transmission through the mechanical equipment room enclosure is a problem.

All system components must be included in the procedure. Unlined ductwork, branch takeoffs, and elbows all provide atten-

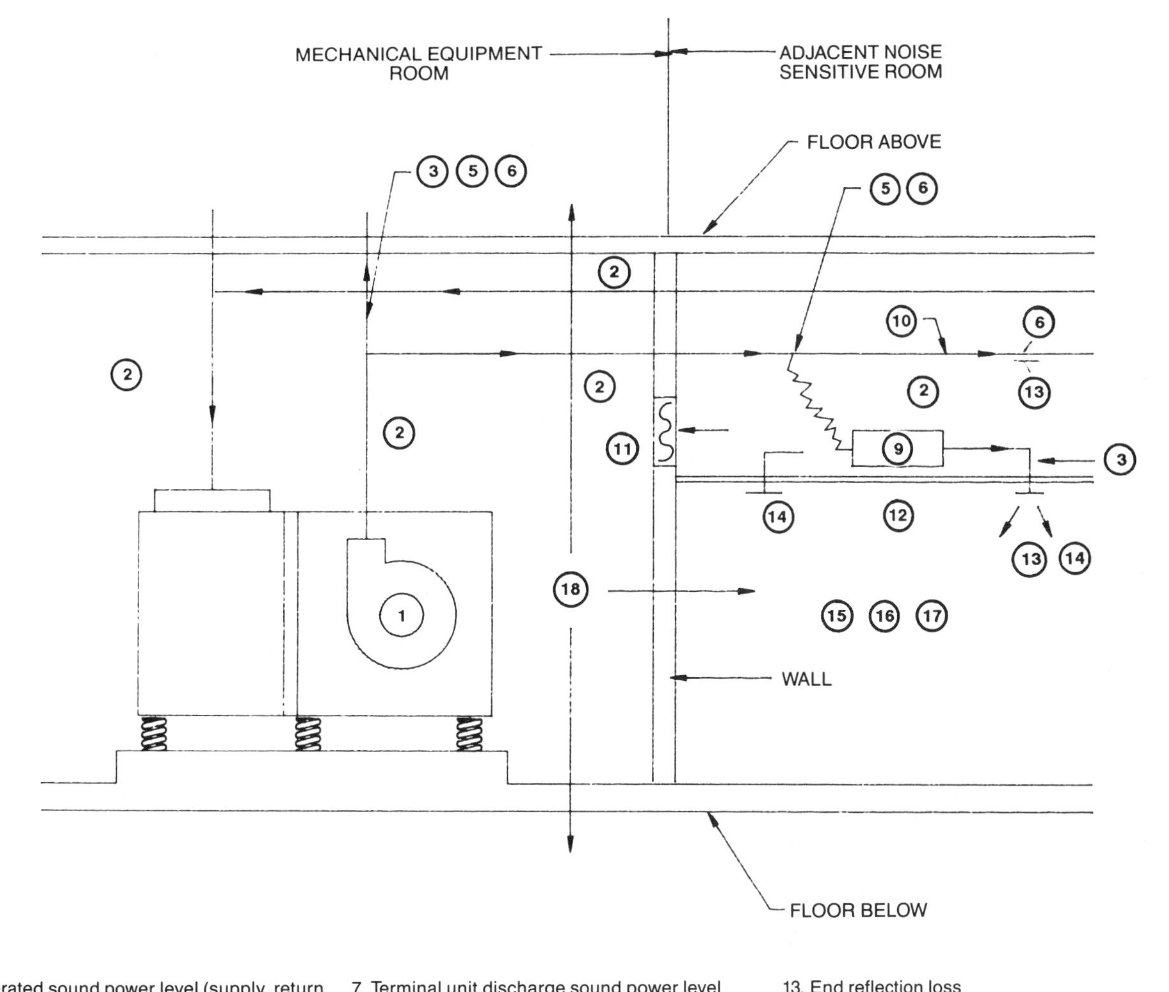

1. Fan-generated sound power level (supply, return, or casing radiated noise)
2. Attenuation in straight, lined, and unlined duct
3. Attenuation in lined and unlined fittings
4. Attenuation through plenum
5. Regenerated noise at fittings, etc.
6. Branch and outlet power division
7. Terminal unit discharge sound power level
8. Attenuation through terminal unit
9. Terminal unit casing radiated sound power level
10. Transmission loss though duct walls (breakout)
11. Transmission loss though return air opening in wall
12. Transmission loss through ceiling construction
13. End reflection loss
14. Air distribution device sound power level
15. Effect of multiple air distribution devices
16. Room effect
17. Room criterion
18. Transmission loss through mechanical equipment room walls, floor, or ceiling

Fig. 7 Typical Noise Paths in HVAC Systems

uation. Failure to consider the contribution of these components can result in too little background noise being generated by the system.

The system noise analysis demonstrates several useful facts. First, systems designed for low pressure and low velocities are less likely to have noise problems because of the lower sound power level of the fan, less regenerated noise at the fittings, and so forth. Second, location of the lining in a duct system is important. Lining in small ductwork provides more attenuation than lining in large ductwork. However, if only branch ductwork is lined for attenuation, spaces served by or located under major duct runs may be subjected to excessive noise because of unattenuated fan noise and any subsequent breakout.

Source Noise Levels

Source noise levels are often available from equipment manufacturers in the form of sound power level L_w or sound pressure level L_p at a specified distance from the noise source. In using these data, it is important to understand the test method used to gather them and to assess their relevance to the particular application. Where source noise levels are unavailable, Table 3 presents values that can be used to predict noise levels from selected types of equipment.

PREDICTION OF FAN SOUND POWER

The sound power generated by a fan performing at a given duty is best obtained from manufacturer's test data taken under approved test conditions (ASHRAE *Standard* 68-1986; also AMCA *Standard* 300-1986). However, if such data are not readily available, the octave-band sound power levels for various fans can be estimated by the procedure described here.

Fan noise can be rated in terms of the specific sound power level, defined as the sound power level generated by a fan operating at a capacity of 1 cfm and a pressure of 1 in. of water. By reducing all fan noise data to this common denominator, the specific sound power level serves as a basis for direct comparison of the octave-band levels of various fans and as the basis for a conventional method of calculating the noise levels of fans at actual operating conditions.

Table 3 Typical Noise Levels Produced by Mechanical Room Equipment

Noise Source	Typical Noise Level[a]	Comments	Standard Error of Estimate[b]	Estimated Uncertainty	Reference
Centrifugal chillers[c]	L_p (1 m) = 60 + 11 log (tons refrigeration) dB(A)	See table below for typical octave-band spectrum shape.	4 dB	—	Blazier (1972)
Reciprocating chillers	L_p (1 m) = 71 + 9 log (tons refrigeration) dB(A)		5 dB	—	Blazier (1972)
Absorption machines	L_p (1 m) < 85 dB(A)	Noise produced by solution pump and auxiliary equipment.	—	±5 dB	Blazier (1972)
Circulating pumps	L_p (1m) = 77 + 10 log (horsepower) dB(A)			±5 dB	Heinter (1968), Kugler *et al.* (1973)
Boilers	L_p (1 m) < 88 dB(A)	Forced draft type.	—	±5 dB	Blazier (1972)
Casing radiation, vaneaxial fans	L_w (case) = L_w (fan) − 10 dB	L_w (case) = A-weighted sound power of fan case.	—	±5 dB	
Casing radiation, central station centrifugal fans	L_w (case) = L_w (fan) − 15 dB	L_w (fan) = A-weighted sound power of fan inlet.	—	±5 dB	

Octave-band levels in dB referred to A-weighted levels	63	125	250	500	1000	2000	4000	Sessler (1973), Miller (1970)
Centrifugal chiller, internal geared, medium to full load	−8	−5	−6	−7	−8	−5	−8	
Centrifugal chiller, direct drive, medium to full load	−8	−6	−7	−3	−4	−7	−12	
Centrifugal chiller, >1000 ton, medium to full load	−11	−11	−8	−8	−4	−6	−13	
Reciprocating chiller, all loads	−19	−11	−7	−1	−4	−9	−14	

[a]Equations give the mean value to be anticipated for equipment of current manufacture at full load.
[b]The standard error of estimate is a measure of the variation to be anticipated from machine to machine due to differences in design, size, and point of operation. Statistically, two of the three machines sampled over a range of sizes among different manufacturers would be expected to have noise level differences within the range of ±1 standard error about the mean.
[c]During light load operation, centrifugal chiller noise levels can be expected to increase about 5 dB in all octave bands not containing either the compressor shaft frequency or the final stage blade passage frequency. These bands typically increase 10 to 13 dB (shaft frequency band) and 8 to 10 dB (blade pass frequency band).

On a specific sound power level basis, small fans are noisier than large fans. While size division is necessarily arbitrary, the size divisions indicated are practical for estimating fan noise. Fans generate a tone at the blade passage frequency, and the strength of this tone depends partly on the type of fan. To account for this blade passage frequency, an increase in sound pressure should be made in the octave band in which the blade frequency falls. The number of decibels added to the sound pressure level in this band is the blade frequency increment B_f. The blade frequency B_f is:

$$B_f = \text{rps} \times \text{No. of Blades or rpm} \times \text{No. of Blades}/60 \quad (3)$$

The number of blades and the fan rpm (rps) can be obtained from the fan selection catalog. Table 4 lists specific sound power levels and blade frequency increments. Table 5 lists the octave band in which the BFI occurs. At present, the sound power level data for forwardly curved fans varies widely; the specific sound power levels given in Table 4 are an average of that data. For critical applications, the sound power levels for a particular fan should be obtained from the manufacturer. For a more complete description of fan types, construction, and applications, see Chapter 3 of the 1988 ASHRAE *Handbook—Equipment*.

Sound power levels at actual operating conditions can be estimated by the actual fan volume flow rate and fan pressure, as:

$$L_w = K_w + 10 \log (Q/Q_1) + 20 \log (p/p_1) + C \quad (4)$$

where

L_w = estimated sound power level of fan, dB re 1 pW
K_w = specific sound power level (see Table 4)
Q = flow rate, cfm
Q_1 = 1 cfm
p = pressure drop, in. of water
p_1 = 1 in. of water
C = correction factor for point of fan operation, dB

Values of the estimated sound power level are calculated for all eight bands, and the BFI is added to the sound pressure level in the octave band in which the blade passage frequency falls.

Table 4 Specific Total Sound Power Levels of Typical Fans

Fan Type	63	125	250	500	1K	2K	4K	8K	BFI
Centrifugal									
AF, BC, or BI wheel diameter									
over 36 in.	40	40	39	34	30	23	19	17	3
under 36 in.	45	45	43	39	34	28	24	19	3
Forward Curved									
All wheel diameters	53	53	43	36	36	31	26	21	2
Radial Bladed									
Low pressure (4 to 10 in. of water)	56	47	43	39	37	32	29	26	7
Medium pressure (6 to 15 in. of water)	58	54	45	42	38	33	29	26	8
High pressure (15 to 60 in. of water)	61	58	53	48	46	44	41	38	8
Axial Fans									
Vaneaxial									
Hub ratio 0.3 to 0.4	49	43	43	48	47	45	38	34	6
Hub ratio 0.4 to 0.6	49	43	46	43	41	36	30	28	6
Hub ratio 0.6 to 0.8	53	52	51	51	49	47	43	40	6
Tubeaxial									
Over 40 in. wheel diameter	51	46	47	49	47	46	39	37	7
Under 40 in. wheel diameter	48	47	49	53	52	51	43	40	7
Propeller									
General ventilation	48	51	58	56	55	52	46	42	5

Note: Includes total sound power level in dB for both inlet and outlet. Values are for fans only—not packaged equipment.

Example 3. A forward-curved fan is selected to supply 8800 cfm at 1.5 in. of water. Sized for efficient operation, it has 24 blades and operates at 1170 rpm. The fan L_w at the fan discharge is determined as follows:

Step 1. Obtain specific sound power levels from the third line of Table 4, subtract 3 dB for outlet, and enter on Line 1 of Table 6.

Table 5 Octave Band in which Blade Frequency Increment (BFI) Occurs

Fan Type	Octave Band in which BFI Occurs, Hz
Centrifugal	
Airfoil, backward curved, backward inclined	125 - 250
Forward curved	500
Radial blade, pressure blower	125
Vaneaxial	125 - 250 - 500
Tubeaxial	63
Propeller	63

Note: Use for estimating purposes. For speeds of 1750 rpm (29 rps) or more, move the BFI to the next higher octave band. Where actual fan is known, use manufacturers' data.

Table 6 Sample Calculation for Example 3

Reference	Octave-Band Center Frequency, Hz							
	63	125	200	500	1000	2000	4000	8000
Table 4	50	50	40	33	33	28	23	18
Equation (4)	43	43	43	43	43	43	43	43
Equation (3) and Table 4	—	—		2		—		
Total (dB re:1 p W) (Line 4)	93	93	83	78	76	71	66	61

Add correction factor C. For off-peak operations, see Table 7.

Step 2. Use Equation (4) to find the additional sound power levels due to the volume flow rate and pressure, as:

$$L_w = K_w + 10 \log 8800 + 20 \log 3.5 + 0 = K_w + 43$$

Step 3. Since the fan has 24 blades and operates at 1170 rpm, B_f = 19.5 × 24 = 468 Hz, which falls in the 500 Hz octave band.

Step 4. Combining these steps results in Line 4 of Table 6.

Plug Fans

Unhoused plug and plenum fans such as illustrated in Figures 8 and 9 can literally be plugged into an air-handling system. This unhoused centrifugal fan may be installed horizontally or vertically in a plenum chamber. Air flows into the fan wheel through an inlet bell located in the chamber wall. The fan discharges directly into the chamber with no housing around the fan wheel. The discharge pressurizes the plenum chamber, and the chamber pressure then forces air through the attached ductwork. These fans can be provided with shaped inlet cones, nested inlet vanes, and extended shafts.

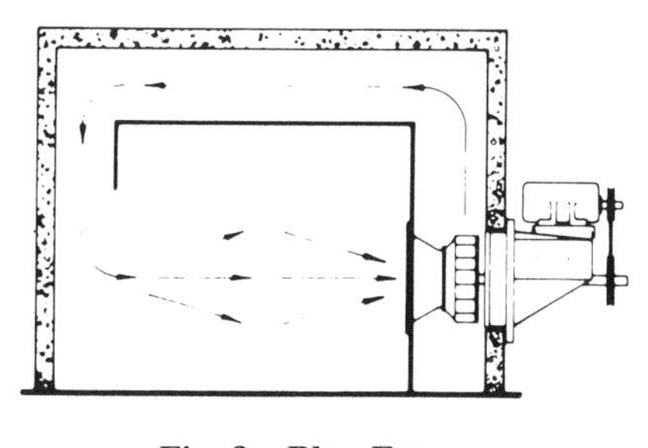

Fig. 8 Plug Fan

Plug fans require more power than housed centrifugal fans and generate a higher noise level, especially at low frequencies. However, the design occupies less space than conventional centrifugal fan designs. In present-day commercial buildings, space is at such a premium that this fan design can be justified. No sound power level data on the plug fan has been included in the table since insufficient data are available at this time.

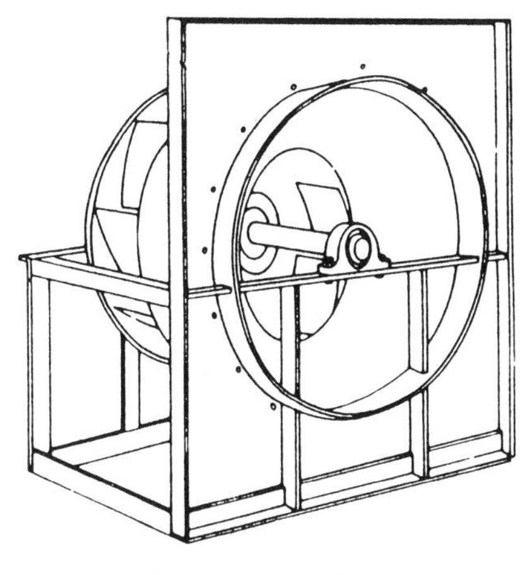

Fig. 9 Plenum Fan

Point of Operation

The specific sound power levels given in Table 4 are for fans operating at or near the peak efficiency point of the fan performance curve as shown in Figures 10 and 11. This conforms with the recommended practice of selecting fan size and speed so that operation falls at or near this point; it is advantageous for energy conservation and corresponds to the lowest noise levels for that fan. If, for any reason, a fan is not or cannot be selected optimally, the noise level produced will increase; correction factor C in Equation (4) accounts for this increase. This correction factor should be applied to all octave bands. For off-peak operation, see Table 7.

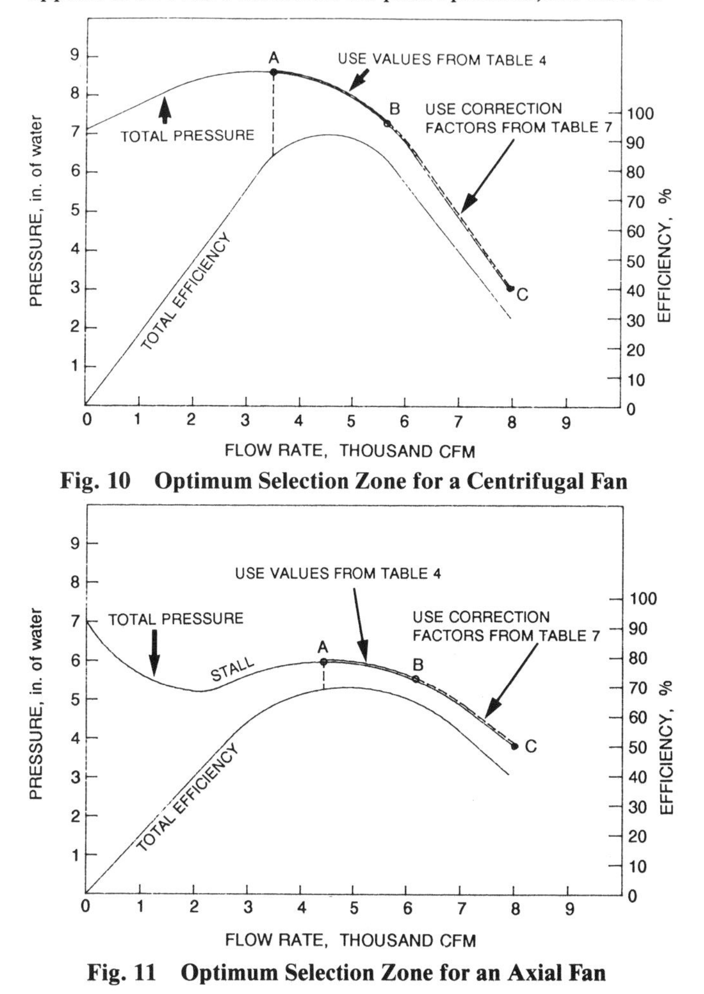

Fig. 10 Optimum Selection Zone for a Centrifugal Fan

Fig. 11 Optimum Selection Zone for an Axial Fan

Example 4. A fan will handle 27,000 cfm at 2.5 in. of water. For lower first-cost and space requirements, the engineer selects a 27-in. double-width air foil fan. Catalog performance data for this fan show a peak efficiency of 78%; however, for this selected duty, the fan curve shows an efficiency of only 49%. The percentage of peak static efficiency is (49/78) × 100 = 62%. According to Table 7, this will result in a 12 dB increase in the sound power level generated by the fan.

If operating cost optimization and mechanical efficiency were considered, the fan choice might be a 36-in. double-width unit. In this case, the efficiency is (71/78) × 100 = 91%, and the sound level does not increase because the fan is operating near its peak efficiency.

If a centrifugal fan is selected to operate between Point A and Point B in Figure 10, the specific sound power levels from Table 4 can be used; but if the operating point is between Point B and Point C, the correction factors from Table 7 should be used. Centrifugal fans designed for HVAC applications can operate at a point slightly to the left of peak efficiency and still be satisfactory. Industrial centrifugal fans should not be operated to the left of the peak efficiency point, however.

The same relationship applies to axial-flow fans as for centrifugal fans, except that an axial fan should not be operated to the left of peak pressure; it must be limited to operating points to the right of peak pressure from Point A to Point B (Figure 11). If the operating point is between Point B and Point C, the correction factors from Table 7 should be used. The noise levels in the stall zone can be 10 to 12 dB higher than the noise levels in the good selection zone.

The correction factor from Table 7 applies to all octave bands.

Table 7 Correction Factor C for Off-Peak Operation

% of Peak Static Efficiency	dB Correction Factor
90 to 100	0
85 to 89	3
75 to 84	6
65 to 74	9
55 to 64	12
50 to 54	15

VARIABLE-SPEED DRIVES

Three types of variable frequency drives are used with HVAC equipment—(1) current source invertor, (2) voltage source invertor, and (3) pulse width modulation (PWM). The current source invertor drive is usually the quietest of the variable frequency drives. The drive units of voltage source invertor drives are generally the noisiest component, and they impart noise to the cabinets in which they are contained, unless they are well isolated. Pulse width modulation units generally have noisy motors and quieter drives. The motors and drive units for pulse width modulation variable frequency drives should be compatibly matched for optimum performance. Certified noise data may be available from manufacturers, or data can be measured at actual installations. Although they are noisy while speeds are being changed, mechanical variable-speed drives do not usually represent a serious noise problem.

The type of vibration isolator and the isolator static deflection should be considered when variable-speed devices are selected for fans. Table 8 lists typical values of natural frequency f for steel spring isolators with different values of static deflection. When fans are provided with variable-speed control, it is not unusual to have speeds below 300 rpm; this may be close to the value where the transmissibility T of the isolated system reaches unity, thus providing no isolation. For low values of disturbing frequency, the transmissibility may exceed unity, resulting in amplification of vibration. When variable-speed devices are used, the designer should consider the static deflection requirements of the vibration isolator so that the system transmissibility is less than 1, and isolation is provided at the lower fan speeds (Table 8).

VAV and Fan Noise

Besides predicting fan sound power levels for fans serving constant volume systems, two additional areas should be considered when fans serve variable air volume (VAV) systems: (1) the efficiency and stability of the fan through the entire range of modulation, and (2) the acoustic impact of the means by which fan output is modulated.

Table 8 Steel Spring Static Deflection, Isolator Natural Frequency, and Equipment Operating Speed for Unity Transmissibility

Isolator Static Deflection, in.	Isolator Natural Frequency, rpm	Equipment Speed/ Disturbing Frequency, rpm
0.5	265	375
1.0	188	265
1.5	153	216
2.0	133	188
2.5	119	168

Fans must be selected for efficiency and stability through the entire range of modulation. A fan selected for peak efficiency at full output may surge at a normal operating point of 50% of maximum output. Similarly, a fan selected to operate at the 50% point may be so inefficient at full output that its noise levels are unacceptable. In general, fan selection for VAV systems is a compromise between fan surge and fan inefficiency, and the narrower the range of modulation, the more acceptable the compromise.

The way in which delivered air quantities are modulated is also important. Fan output is usually modulated in one of three ways: (1) through variable inlet vanes, which add resistance to the fan system, altering the operating point by doing so; (2) through variable-speed fan motors and mechanical drives, which slow the fan to match the operating requirements without altering the point of operation on the fan curve; and (3) by using variable-pitch fan blades, which change the fan geometry to increase or decrease fan output.

While fan modulation systems are often selected based on nonacoustic factors, such as initial cost of equipment, operating costs, or space requirements, acoustical factors should also be considered. Variable inlet vane systems may generate significant low-frequency noise as the vanes shut down. Additional attenuation with a corresponding additional pressure drop is required to attenuate the noise generated by the inlet vanes. Consequently, payback cost analysis must include the additional sound attenuation and energy costs to move the air through this sound attenuation.

Variable-speed motors and drives and variable-pitch fan blade systems are actually quieter at reduced air output than at full output. The designer has the option of designing for maximum output as if the system were constant volume, or selecting the sound attenuation for a more normal operating point and allowing fan noise to exceed the design criteria on the rare occasions when the fan operates at full output. In this case, the reduced cost of sound attenuation and corresponding pressure reduction is a trade-off against initial equipment costs.

General Discussion of Fan Sound

To minimize the required duct sound attenuation, the proper selection and installation of the fan (or fans) is vitally important. The following factors should be considered:

1. Design the air distribution system for minimum resistance, since the sound generated by a fan, regardless of type, increases by the square of the static pressure.
2. Examine the specific sound power levels of the fan designs for any given job. Different fans generate different levels of sound and produce different octave-band spectra. Select a fan that will generate the lowest possible sound level, commensurate with other fan selection parameters.
3. Fans with relatively few blades (less than 15) tend to generate tones, which may dominate the spectrum. These tones occur

at the blade passage frequency [Equation (3)] and its harmonics. The intensity of these tones depends on resonances within the duct system, fan design, and inlet flow distortions.

4. Design duct connections at both the fan inlet and outlet for uniform and straight airflow. Avoid unstable, gusting, and swirling inlet airflow. Deviation from accepted applications can severely degrade both the aerodynamic and acoustic performance of any fan and invalidate manufacturers' ratings or other performance predictions.
5. In variable air volume systems, consider the effect of changes in volume on the fan sound power. Reducing volume flow by changes in inlet vane settings may substantially increase the low frequency fan sound power levels of backwardly inclined and axial-flow fans.

NOISE CONTROL ALONG DUCT PATHS

ASHRAE offers several levels of acoustical design materials. The handbook, *A Practical Guide for HVAC System Noise and Vibration Control*, is for the designer who wants a basic knowledge of the dos and don'ts in HVAC acoustics. This chapter is for the more advanced designer who manually calculates acoustical predictions. *ASHRAE Algorithms for HVAC Acoustics* is a publication for advanced designers who prefer to let the computer handle the calculations involved with acoustical formulas.

Natural Attenuation in Ducts

Even if ductwork contains no sound attenuators (acoustical linings or sound traps), only a fraction of the acoustic energy generated by the fan, the duct fittings, and so forth reaches any one room because of (1) the combined effects of energy division at branch takeoffs and (2) energy losses because of duct wall vibration and sound reflections at elbows and duct outlets.

To avoid overdesigning the acoustic duct treatment of the duct system, natural attenuation should be considered. The natural attenuation for rectangular and circular ducts without internal insulation is given in Tables 9 and 10.

Selection of Absorptive Material

Sound-absorbing material can be arranged in a duct system in the following ways:

1. Line all fan suction and discharge plenums for an economical sound absorption system.
2. Line ducts with sound-absorbing material, which also serves as thermal insulation. Note that duct dimensions must be increased to compensate for area lost due to lining.
3. Line duct sections close to elbows to take advantage of the interaction of sound absorption and sound reflection.
4. Install prefabricated sound attenuators (sound traps, duct silencers), which contain specially shaped perforated baffles filled with sound-absorbing material.

Several arrangements are usually combined to achieve the required amount of sound attenuation.

Table 9 Natural Sound Attenuation in Unlined Rectangular Sheet Metal Ducts
(Ver 1978)

P/A Ratio,[a] in/in²	Octave-Band Center Frequency, Hz: 63	125	250 and Over
	Attenuation, dB/ft[b]		
Over 3.1	0	0.3	0.1
3.1 to 0.13	0.3	0.1	0.1
Under 0.13	0.1	0.1	0.1

[a]Perimeter divided by area.
[b]Double these values if the duct is externally insulated.

Table 10 Natural Sound Attenuation in Unlined Straight Round Ducts
(United Sheet Metal 1973)

Diameter, in.	Approximate Attenuation, dB/ft for 1 in. duct liner[a] Octave-Band Center Frequency, Hz: 63	125	250	500	1000	2000	4000
6	0.2	0.5	1.0	1.8	2.2	2.2	2.0
12	0.15	0.3	0.7	1.5	2.2	2.2	1.5
24	0.1	0.2	0.5	1.0	1.7	0.9	0.5
48	0.04	0.1	0.3	0.6	0.6	0.8	0.5

[a]Test data based on 26 to 22 gage spiral wound duct with perforated spiral wound steel liner 24 ft long. Diameter shown is free area. Data are for no airflow circumstance.

When a sound wave impinges on the surface of a porous material, the air within the small pores of the material is set into motion. The flow resistance within the pores converts a portion of the sound energy to heat. The decimal fraction representing the absorbed portion of the energy of the incident sound wave is the absorption coefficient.

Thin materials, particularly when mounted on hard, solid surfaces, do not absorb low-frequency sound. For significant sound absorption at frequencies below 500 Hz, the material must be 2 in. or thicker. Considerable absorption may also result from the flexural vibrations of duct walls, particularly in the low-frequency range.

Sound-absorbing materials suitable for use in air ducts are available in blankets and semirigid boards. Manufacturers' literature describe their selection and use. The following additional properties should be evaluated in the selection of acoustical materials: (1) adequate strength to avoid breakage and crumbling, (2) fire resistance and compliance with national and local code requirements, (3) resistance to erosion at high air velocities, and (4) freedom from odor when dry or wet.

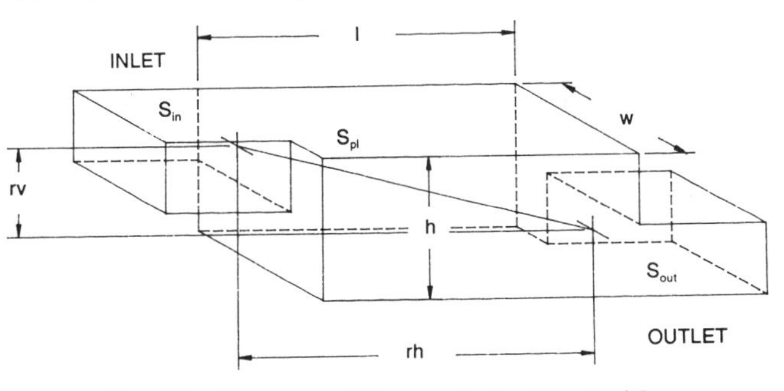

Fig. 12 Diagram of Plenum to be Lined with Sound Absorptive Material

Attenuation of Sound Using Lined Plenums

Sound absorption provided by a plenum on the fan discharge is often an economical arrangement (Figure 12). Both experiment and ray theory acoustics have led to the following expression for approximate acoustic plenum attenuation in dB (Wells 1958):

$$\text{Attenuation} = 10 \log \left[\frac{1}{S_e \left[\cos(\theta / 2\pi d^2) + (1-\alpha)/\alpha S_w \right]} \right] \quad (5)$$

where

α = absorption coefficient of the lining, dimensionless
S_e = plenum exit area, ft²
S_w = plenum wall area, ft²
d = distance between entrance and exit, ft (Figure 12)
θ = angle of incidence at the exit, *i.e.*, the angle that the direction d makes with the normal to the exit opening, degrees

For frequencies high enough to make the wavelength less than the plenum dimensions, Equation (5) is accurate to within a few decibels. At lower frequencies, Equation (5) is conservative, and the actual attenuation exceeds the calculated value by 5 to 10 dB because of sound reflection at the entrance and exit of the plenum.

For absorption coefficients of selected plenum materials, see Table 11.

Table 11 Absorption Coefficients for Selected Plenum Materials

	1/1 Octave-Band Center Frequency, Hz						
	63	125	250	500	1000	2000	4000
Non-Sound Absorbing Materials							
Concrete	0.01	0.01	0.01	0.02	0.02	0.02	0.03
Bare Sheet Metal	0.04	0.04	0.04	0.05	0.05	0.05	0.07
Sound Absorbing Materials							
1 in. 3.0 lb/ft² Fiberglass Insulation Board	0.02	0.03	0.22	0.69	0.91	0.96	099
2 in. 3.0 lb/ft² Fiberglass Insulation Board	0.18	0.22	0.82	1.00	1.00	1.00	1.00
3 in. 3.0 lb/ft² Fiberlass Insulation Board	0.48	0.53	1.00	1.00	1.00	1.00	1.00
4 in. 3.0 lb/ft² Fiberglass Insulation Board	0.76	0.84	1.00	1.00	1.00	1.00	0.97

Attenuation of Sound in Lined Ducts

Duct linings can be designed both to absorb sound and to insulate thermally. Thicknesses between 1 and 2 in. are usually adequate for thermal insulation. The sound absorption of such relatively thin linings is limited, especially at low frequencies below about 250 Hz.

Ver (1978) and Kuntz (1987) measured the attenuation of sound in lined rectangular ducts. From these studies, Equations (6) and (7) were developed to determine the attenuation provided by the acoustical lining in lined rectangular or square ducts. These equations do not apply to ducts with very low sound transmission walls, such as flexible ducts or ducts covered with thin plastic or metal foil. Most calculations of attenuation are for octave bands of noise rather than for specific tones, but attenuation is frequency-dependent. To compensate for this, Equation (6) can be evaluated one-third octave below, and Equation (7) one-third octave above, the center frequency of the octave.

At low frequencies (800 Hz and below), Equation (6) may be used to calculate the total attenuation or insertion loss for any duct length.

$$a = \frac{t^{1.08}\, h^{0.356}\, (P/A) L f^{(1.17\, +\, 0.19\, d)}}{1190\, d^{\,2.3}} \text{ dB} \qquad (6)$$

where

d = nominal density, lb/ft³
t = material thickness, in.
h = smallest inside cross dimension of the lined duct, in.
f = frequency, Hz
P = inside duct perimeter, in.
L = duct length, ft
A = inside duct area, in²

At high frequencies (800 Hz and above), Equation (7) may be used to calculate the total attenuation or insertion loss for duct lengths of 10 ft or less. For ducts longer than 10 ft, the 10 ft total attenuation values should be used.

$$a = \frac{2.11 \times 10^{9}\, (P/A)\, L f^{[-1.53\, -\, 1.61 \log (P/A)]}}{w^{2.5}\, h^{2.7}} \qquad (7)$$

where

a = total attenuation or insertion, loss dB
w = largest inside cross dimension of the duct, in.
L = duct length, ft (limited to 10 ft)

When lined ducts are used before and after turns, the high-frequency equation may be used for each section.

In one-third octave-band calculations, the smaller value calculated at 800 Hz should be taken. To prevent the prediction of excessive values, the insertion loss in any straight, lined section should be limited to 40 dB at any frequency. This limitation is due to structure-borne sound in the duct walls.

Tables 12 and 13 present the attenuation of random noise for frequently used, rectangular ducts with clear cross-sectional dimensions. The data are derived from Equations (6) and (7), with the frequency term taken at one-third octave below the center frequency of the octave for Equation (6) and at one-third octave above the center frequency of the octave for Equation (7). The high-frequency attenuation values are applicable to a maximum of 10 ft. Because of the differences in manufacturers' materials, the calculated values might be reduced by 10% for critical applications. The attenuation values listed apply when flow effects are negligible, as is usually the case of airflow below 2000 fpm.

The attenuation of sound in round ducts for four different duct diameters with no lining are presented in Tables 14 and 15.

Passive Duct Sound Attenuators

Prefabricated sound attenuators, which are available in various sizes, both rectangular and circular, constitute another method of obtaining noise attenuation in ducts. Rectangular units are generally available in lengths of 3, 5, 7, and 10 ft. The lengths of circular units are generally two to three times their diameter. They are also available with various percentages of open flow area to provide different degrees of attenuation and static pressure drop. When selecting sound attenuators, the following interrelating factors should be considered: (1) required insertion loss, (2) static pressure drop, and (3) self-noise, *i.e.*, the noise generated by air moving through the air passages of the attenuator. All three are related to the air velocity through the attenuator—the insertion loss in the low- and mid-frequency range is inversely related to the velocity, and the pressure drop and self-noise are directly related to the air velocity.

In most applications, prefabricated sound attenuators are of the dissipative type, containing sound-absorbing material, such as glass fiber, protected from the airflow by a perforated metal facing. As with most dissipative mufflers, the attenuation is highest in the mid-frequency range; it is limited in the low-frequency range (63 Hz octave), and limited to a lesser degree in the high-frequency range (8000 Hz octave). Attenuators are available with polymer sheeting encasing the dissipative material to prevent the glass fiber packing from being contaminated by chemical or particulate matter. They are also available without any dissipative packing where the polymer sheeting cannot be used.

General guidelines for locating duct silencers in ducted systems with fans and for minimum static pressure drop with maximum acoustical performance are as follows:

Centrifugal Fans

From fan discharge—One duct diameter for every 1000 fpm

From fan intake—0.75 duct diameter for every 1000 fpm

Axial Fans

From fan discharge—One duct diameter for every 1000 fpm

From fan intake—0.75 diameter for every 1000 fpm

Duct Elbows

Three duct diameters or equivalent both upstream and downstream

Mixing Boxes and VAV Terminals

One duct diameter upstream or downstream

Grilles, Registers, and Diffusers

One duct diameter upstream or downstream

Table 12 Insertion Loss for Rectangular Ducts with 1-in. Fiberglass Lining

Internal Cross-Sectional Dimensions, in.	Insertion Loss, dB/ft Octave-Band Center Frequency, Hz						
	63	125	250	500	1000	2000	4000
4 × 4	0.16	0.44	1.21	3.32	9.10	10.08	3.50
4 × 6	0.13	0.37	1.01	2.77	7.58	8.26	3.13
4 × 8	0.12	0.33	0.91	2.49	6.82	6.45	2.57
4 × 10	0.11	0.31	0.85	2.32	6.37	5.02	2.07
6 × 6	0.12	0.34	0.93	2.56	7.01	7.50	3.17
6 × 10	0.10	0.27	0.75	2.04	5.61	5.67	2.67
6 × 12	0.09	0.26	0.70	1.92	5.26	4.80	2.33
6 × 18	0.08	0.23	0.62	1.70	4.67	2.95	1.51
8 × 8	0.10	0.28	0.77	2.12	5.82	6.08	2.95
8 × 12	0.09	0.24	0.65	1.77	4.85	4.98	2.64
8 × 16	0.08	0.21	0.58	1.59	4.37	3.89	2.17
8 × 24	0.07	0.19	0.52	1.42	3.88	2.39	1.41
10 × 10	0.09	0.24	0.67	1.84	5.04	5.17	2.79
10 × 16	0.07	0.20	0.55	1.49	4.10	4.04	2.41
10 × 20	0.07	0.18	0.50	1.38	3.78	3.30	2.05
10 × 30	0.06	0.16	0.45	1.23	3.36	2.03	1.34
12 × 12	0.08	0.22	0.60	1.64	4.48	4.52	2.67
12 × 18	0.07	0.18	0.50	1.36	3.74	3.71	2.39
12 × 24	0.06	0.16	0.45	1.23	3.36	2.89	1.97
12 × 36	0.05	0.15	0.40	1.09	2.99	1.78	1.28
15 × 15	0.07	0.19	0.52	1.42	3.88	3.84	2.53
15 × 22	0.06	0.16	0.43	1.19	3.27	3.20	2.29
15 × 30	0.05	0.14	0.39	1.06	2.91	2.46	1.86
15 × 45	0.05	0.13	0.34	0.94	2.17	1.51	1.21
18 × 18	0.06	0.17	0.46	1.26	3.45	3.37	2.42
18 × 28	0.05	0.14	0.38	1.03	2.84	2.69	2.13
18 × 36	0.05	0.13	0.34	0.94	2.59	2.15	1.78
18 × 54	0.04	0.11	0.31	0.84	1.65	1.32	1.16
24 × 24	0.05	0.14	0.38	1.05	2.87	2.73	2.26
24 × 36	0.04	0.12	0.32	0.87	2.39	2.24	2.02
24 × 48	0.04	0.10	0.29	0.78	1.90	1.75	1.66
24 × 72	0.03	0.09	0.25	0.70	1.06	1.07	1.08
30 × 30	0.04	0.12	0.33	0.91	2.49	2.32	2.14
30 × 45	0.04	0.10	0.28	0.76	1.88	1.90	1.91
30 × 60	0.03	0.09	0.25	0.68	1.35	1.48	1.57
30 × 90	0.03	0.08	0.22	0.60	0.76	0.91	1.02
36 × 36	0.04	0.11	0.29	0.81	2.01	2.03	2.04
36 × 54	0.03	0.09	0.25	0.67	1.42	1.66	1.83
36 × 72	0.03	0.08	0.22	0.60	1.02	1.30	1.50
36 × 108	0.03	0.07	0.20	0.54	0.57	0.80	0.98
42 × 42	0.04	0.10	0.27	0.73	1.59	1.81	1.97
42 × 64	0.03	0.08	0.22	0.60	1.11	1.47	1.75
42 × 84	0.03	0.07	0.20	0.55	0.81	1.16	1.45
42 × 126	0.02	0.06	0.18	0.49	0.45	0.71	0.94
48 × 48	0.03	0.09	0.24	0.67	1.30	1.65	1.90
48 × 72	0.03	0.07	0.20	0.56	0.92	1.35	1.70
48 × 96	0.02	0.07	0.18	0.50	0.66	1.05	1.40
48 × 144	0.02	0.06	0.16	0.45	0.37	0.65	0.91

Notes:
1. Values based on measurements of surface-coated duct liners of 1.5 lb/ft^3 density.
2. For the specific materials tested, liner density had a minor effect over the nominal range of 1.5 to 3 lb/ft^3.
3. Add natural attenuation (Table 9) to obtain total attenuation.

Table 13 Insertion Loss for Rectangular Ducts with 2-in. Fiberglass Lining

Internal Cross-Sectional Dimensions, in.	Insertion Loss, dB/ft Octave-Band Center Frequency, Hz						
	63	125	250	500	1000	2000	4000
4 × 4	0.34	0.93	2.56	7.02	19.23	10.08	3.50
4 × 6	0.28	0.78	2.13	5.85	16.03	8.26	3.13
4 × 8	0.26	0.70	1.92	5.26	14.42	6.45	2.57
4 × 10	0.24	0.65	1.79	4.91	13.46	5.02	2.07
6 × 6	0.26	0.72	1.97	5.40	14.81	7.50	3.17
6 × 10	0.21	0.58	1.58	4.32	11.85	5.67	2.67
6 × 12	0.20	0.54	1.48	4.05	11.11	4.80	2.33
6 × 18	0.17	0.48	1.31	3.60	8.80	2.95	1.51
8 × 8	0.22	0.60	1.64	4.49	12.31	6.08	2.95
8 × 12	0.18	0.50	1.36	3.74	10.26	4.98	2.64
8 × 16	0.16	0.45	1.23	3.37	9.23	3.89	2.17
8 × 24	0.15	0.40	1.09	2.99	5.68	2.39	1.41
10 × 10	0.19	0.52	1.42	3.89	10.66	5.17	2.79
10 × 16	0.15	0.42	1.15	3.16	8.66	4.04	2.41
10 × 20	0.14	0.39	1.06	2.92	7.22	3.30	2.05
10 × 30	0.13	0.34	0.95	2.59	4.04	2.03	1.34
12 × 12	0.17	0.46	1.26	3.46	9.48	4.52	2.67
12 × 18	0.14	0.38	1.05	2.88	7.62	3.71	2.39
12 × 24	0.13	0.35	0.95	2.59	5.47	2.89	1.97
12 × 36	0.11	0.31	0.84	2.31	3.06	1.78	1.28
15 × 15	0.15	0.40	1.09	2.99	7.64	3.84	2.53
15 × 22	0.12	0.34	0.92	2.52	5.55	3.20	2.29
15 × 30	0.11	0.30	0.82	2.25	3.89	2.46	1.86
15 × 45	0.10	0.27	0.73	2.00	2.17	1.51	1.21
18 × 18	0.13	0.35	0.97	2.66	5.79	3.37	2.42
18 × 28	0.11	0.29	0.80	2.19	3.95	2.69	2.13
18 × 36	0.10	0.27	0.73	2.00	2.94	2.15	1.78
18 × 54	0.09	0.24	0.65	1.78	1.65	1.32	1.16
24 × 24	0.11	0.29	0.81	2.21	3.73	2.73	2.26
24 × 36	0.09	0.25	0.67	1.84	2.65	2.24	2.02
24 × 48	0.08	0.22	0.61	1.66	1.90	1.75	1.66
24 × 72	0.07	0.20	0.54	1.48	1.06	1.07	1.08
30 × 30	0.09	0.25	0.70	1.92	2.65	2.32	2.14
30 × 45	0.08	0.21	0.58	1.60	1.88	1.90	1.91
30 × 60	0.07	0.19	0.52	1.44	1.35	1.48	1.57
30 × 90	0.06	0.17	0.47	1.28	0.76	0.91	1.02
36 × 36	0.08	0.23	0.62	1.70	2.01	2.03	2.04
36 × 54	0.07	0.19	0.52	1.42	1.42	1.66	1.83
36 × 72	0.06	0.17	0.47	1.28	1.02	1.30	1.50
36 × 108	0.06	0.15	0.41	1.14	0.57	0.80	0.98
42 × 42	0.07	0.21	0.56	1.54	1.59	1.81	1.97
42 × 64	0.06	0.17	0.47	1.28	1.11	1.47	1.75
42 × 84	0.06	0.15	0.42	1.16	0.81	1.16	1.45
42 × 126	0.05	0.14	0.38	1.03	0.45	0.71	0.94
48 × 48	0.07	0.19	0.52	1.42	1.30	1.65	1.90
48 × 72	0.06	0.16	0.43	1.18	0.92	1.35	1.70
48 × 96	0.05	0.14	0.39	1.06	0.66	1.05	1.40
48 × 144	0.05	0.13	0.34	0.94	0.37	0.65	0.91

Notes:
1. Values based on measurements of surface-coated duct liners of 1.5 lb/ft^3 density.
2. For the specific materials tested, liner density had a minor effect over the nominal range of 1.5 to 3.0 lb/ft^3.
3. Add natural attenuation (Table 9) to obtain total attenuation.

Table 14 Insertion Loss for Circular Ducts with 1-in. Acoustical Lining

Diameter,	Insertion Loss, dB/ft 1/1 Octave-Band Center Frequency, Hz							
in.	63	125	250	500	1000	2000	4000	8000
6	0.38	0.59	0.93	1.53	2.17	2.31	2.04	1.26
8	0.32	0.54	0.89	1.50	2.19	2.17	1.83	1.18
10	0.27	0.50	0.85	1.48	2.20	2.04	1.64	1.12
12	0.23	0.46	0.81	1.45	2.18	1.91	1.48	1.05
14	0.19	0.42	0.77	1.43	2.14	1.79	1.34	1.00
16	0.16	0.38	0.73	1.40	2.08	1.67	1.21	0.95
18	0.13	0.35	0.69	1.37	2.01	1.56	1.10	0.90
20	0.11	0.31	0.65	1.34	1.92	1.45	1.00	0.87
22	0.08	0.28	0.61	1.31	1.82	1.34	0.92	0.83
24	0.07	0.25	0.57	1.28	1.71	1.24	0.85	0.80
26	0.05	0.22	0.53	1.24	1.59	1.14	0.79	0.77
28	0.03	0.19	0.49	1.20	1.46	1.04	0.74	0.74
30	0.02	0.16	0.45	1.16	1.33	0.95	0.69	0.71
32	0.01	0.14	0.42	1.12	1.20	0.87	0.66	0.69
34	0	0.11	0.38	1.07	1.07	0.79	0.63	0.66
36	0	0.09	0.35	1.02	0.93	0.71	0.60	0.64
38	0	0.06	0.31	0.96	0.80	0.64	0.58	0.61
40	0	0.03	0.28	0.91	0.68	0.57	0.55	0.58
42	0	0.01	0.25	0.84	0.56	0.50	0.53	0.55
44	0	0	0.23	0.78	0.45	0.44	0.51	0.52
46	0	0	0.20	0.71	0.35	0.39	0.48	0.48
48	0	0	0.18	0.63	0.26	0.34	0.45	0.44
50	0	0	0.15	0.55	0.19	0.29	0.41	0.40
52	0	0	0.15	0.46	0.13	0.25	0.37	0.34
54	0	0	0.12	0.27	0.09	0.22	0.31	0.29
56	0	0	0.10	0.28	0.08	0.18	0.25	0.22
58	0	0	0.09	0.17	0.08	0.16	0.18	0.15
60	0	0	0.08	0.06	0.10	0.14	0.09	0.07

Table 15 Insertion Loss for Circular Ducts with 2-in. Acoustical Lining

Diameter,	Insertion Loss, dB/ft 1/1 Octave-Band Center Frequency, Hz							
in.	63	125	250	500	1000	2000	4000	8000
6	0.56	0.80	1.37	2.25	2.17	2.31	2.04	1.26
8	0.51	0.75	1.33	2.23	2.19	2.17	1.83	1.18
10	0.46	0.71	1.29	2.20	2.20	2.04	1.64	1.12
12	0.42	0.67	1.25	2.18	2.18	1.91	1.48	1.05
14	0.38	0.63	1.21	2.15	2.14	1.79	1.34	1.00
16	0.35	0.59	1.17	2.12	2.08	1.67	1.21	0.95
18	0.32	0.56	1.13	2.10	2.01	1.56	1.10	0.90
20	0.29	0.52	1.09	2.07	1.92	1.45	1.00	0.87
22	0.27	0.49	1.05	2.03	1.82	1.34	0.92	0.83
24	0.25	0.46	1.01	2.00	1.71	1.24	0.85	0.80
26	0.24	0.43	0.97	1.96	1.59	1.14	0.79	0.77
28	0.22	0.40	0.93	1.93	1.46	1.04	0.74	0.74
30	0.21	0.37	0.90	1.88	1.33	0.95	0.69	0.71
32	0.20	0.34	0.86	1.84	1.20	0.87	0.66	0.69
34	0.19	0.32	0.82	1.79	1.07	0.79	0.63	0.66
36	0.18	0.29	0.79	1.74	0.93	0.71	0.60	0.64
38	0.17	0.27	0.76	1.69	0.80	0.64	0.58	0.61
40	0.16	0.24	0.73	1.63	0.68	0.57	0.55	0.58
42	0.15	0.22	0.70	1.57	0.56	0.50	0.53	0.55
44	0.13	0.20	0.67	1.50	0.45	0.44	0.51	0.52
46	0.12	0.17	0.64	1.43	0.35	0.39	0.48	0.48
48	0.11	0.15	0.62	1.36	0.26	0.34	0.45	0.44
50	0.09	0.12	0.60	1.28	0.19	0.29	0.41	0.40
52	0.07	0.10	0.58	1.19	0.13	0.25	0.37	0.34
54	0.05	0.08	0.56	1.10	0.09	0.22	0.31	0.29
56	0.02	0.05	0.55	1.00	0.08	0.18	0.25	0.22
58	0	0.03	0.53	0.90	0.08	0.16	0.18	0.15
60	0	0	0.53	0.79	0.10	0.14	0.09	0.07

Duct sound attenuators should be installed in or close to the mechanical equipment room wall opening. This helps keep the reverberant sound from the mechanical equipment room from bypassing the duct sound attenuator and entering into the ductwork exposed between the duct sound attenuator and the point where the duct leaves the mechanical equipment room.

Active Duct Sound Attenuators

Canceling the noise from one source by generating interfering sound waves from another source 180° out of phase is an old concept. Present-day systems use digital computers to reduce unwanted noise significantly at some critical frequencies.

Figure 13 illustrates how an active duct sound attenuator system works. An input microphone probe tube measures the noise in the duct. This signal is sent to a digital controller which inverts the sampled fan noise so that it is 180° out of phase. The output of the controller, which is out of phase from the noise coming down the duct, is boosted by an amplifier and put into the duct by a loudspeaker module. This secondary noise source cancels a portion of the fan noise, typically below 400 Hz. An error microphone is used to adjust the computer model so that the system performs at its optimum efficiency.

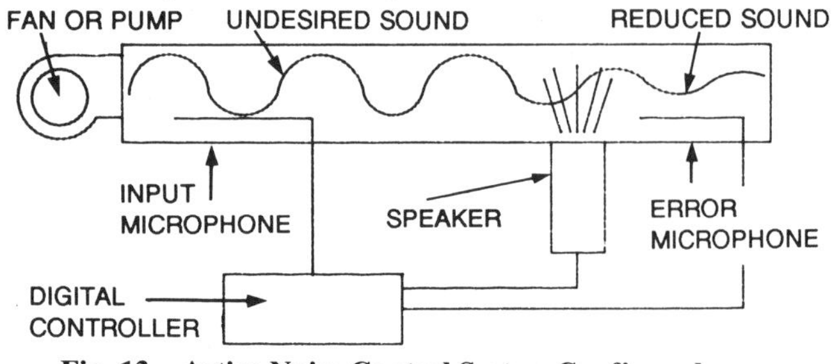

Fig. 13 Active Noise Control System Configured as Duct Sound Attenuators

Active systems can cancel low-frequency random broadband and/or tonal repetitive noise. Typical attenuations of 20 to 38 dB have been reported in the 40 through 400 Hz frequency range (Figure 14). When used as a duct silencer, the outside speaker mounting results in zero pressure drop and regenerated noise. Flexible membranes between the speakers and the duct allow the system to be used in harsh environments.

Figure 7a shows the dynamic insertion loss relationship between dissipative and actively enhanced dissipative sound attenuators.

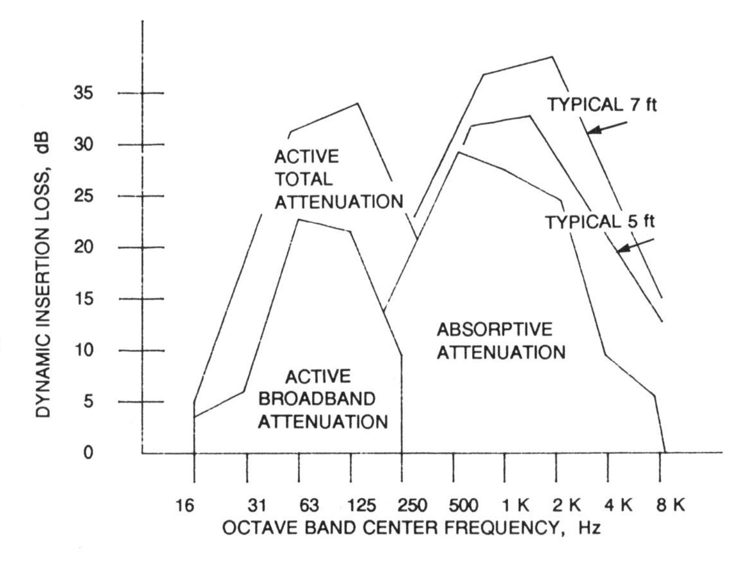

Fig. 14 Relative Performance of Active and Passive Duct Sound Attenuators

Attenuation of Duct Fittings

Sound is attenuated by lined or unlined elbows. The amount of the attenuation is a function of duct size, whether there is acoustical lining before and/or after the elbow, and whether or not turning vanes are provided in the elbow. Table 16 gives suggested insertion loss values for elbows without turning vanes. Table 17 gives values for unlined round duct elbows, and Table 18 gives values for square elbows in lined or unlined duct systems with turning vanes in the elbow.

In Tables 16, 17, and 18, $fw = f \times w$

where

f = frequency in kHz
w = rectangular duct dimension in plane of bend for rectangular duct
w = diameter for round duct

Table 16 Insertion Loss of Square Elbows without Turning Vanes

Duct Width, in.	63	125	250	500	1000	2000	4000
	Octave-Band Center Frequency, Hz — Insertion Loss, dB						
No Lining							
5 in.	—	—	—	1	5	7	5
10 in.	—	—	1	5	7	5	3
20 in.	—	1	5	7	5	3	3
40 in.	1	5	7	5	3	3	3
Lining[a] Ahead of Elbow							
5 in.	—	—	—	1	5	8	6
10 in.	—	—	1	5	8	6	8
20 in.	—	1	5	8	6	8	11
40 in.	1	5	8	6	8	11	11
Lining[a] After Elbow							
5 in.	—	—	—	1	6	11	10
10 in.	—	—	1	6	11	10	10
20 in.	—	1	6	11	10	10	10
40 in.	1	6	11	10	10	10	10
Lining[a] Ahead of and After Bend							
5 in.	—	—	—	1	6	12	14
10 in.	—	—	1	6	12	14	16
20 in.	—	1	6	12	14	16	18
40 in.	1	6	12	14	16	18	18

Note: For square elbows with short turning vanes, use average between Tables 17 and 16.
[a]Based on lining extending for a distance of at least two duct widths and lining thickness of 10% of duct width. For thinner lining, lined length must be proportionally longer.

Table 17 Insertion Loss of Unlined Round Elbows

Diameter, in.	63	125	250	500	1000	2000	4000
	Octave-Band Center Frequency, Hz — Insertion Loss, dB						
5 to 10	0	0	0	0	1	2	3
11 to 20	0	0	0	1	2	3	3
21 to 40	0	0	1	2	3	3	3
41 to 80	0	1	2	3	3	3	3

Branch Power Division

At branch takeoffs, acoustic energy is distributed between the branch and the main duct in accordance with the ratio of the cross-sectional area of the branch to the total cross-sectional area of all of the ducts following the takeoff. It can be assumed that the branch power division occurring at a branch takeoff is proportional to the decimal equivalent of the ratio of the area of the branch to the area of all ducts leaving the takeoff; it can be expressed by:

Table 18 Insertion Loss of Square Elbows with Turning Vanes

Duct Width, in.	63	125	250	500	1000	2000	4000	8000
	Octave-Band Center Frequency, Hz — Insertion Loss, dB							
Unlined Ducts								
6	0	0	0	1	8	12	17	17
12	0	0	1	8	12	17	17	17
24	0	1	8	12	17	17	17	17
48	1	8	12	17	17	17	17	17
1-in. Thick Lining								
6	0	0	0	3	14	22	33	33
12	0	0	1	10	18	30	30	30
24	0	1	8	14	24	24	24	24
48	1	8	12	18	18	18	18	18

Table 19 Power Level Division at Branch Takeoff

B/T	Division (dB)	B/T	Division (dB)
1.00	0	0.10	10
0.80	1	0.08	11
0.63	2	0.063	12
0.50	3	0.05	13
0.40	4	0.04	14
0.32	5	0.032	15
0.25	6	0.025	16
0.20	7	0.02	17
0.16	8	0.016	18
0.12	9	0.012	19

B = branch area; T = total cross-sectional area of all ducts following takeoffs.

Branch Power Division (dB) =
10 log (Branch area/Total area of all ducts after takeoff) (8)

Table 19 presents the branch power division for a number of branch area ratios.

Duct End Reflection Loss

When plane wave sound passes from a small space, such as a duct, into a large space, such as a room, a certain amount of sound is reflected back into the duct. The sound attenuation associated with duct end reflection losses for ducts terminated in free space can be approximated by:

$$\Delta L = 10 \log [1 + (c/\pi^{fD})\, 1.88] \qquad (9)$$

where

ΔL = attenuation, db
c = speed of sound in air, 1130 fps
D = diameter of a circular duct or effective diameter of a rectangular duct, ft
f = frequency, Hz

For ducts terminated flush with a wall or ceiling, sound attenuation associated with the duct end reflection loss can be approximated by:

$$\Delta L = 10 \log [1 + (0.8\, c/\pi^{fD}) 1.88] \qquad (10)$$

End reflection loss should not be included in the attenuation of any system where linear diffusers are tapped directly into plenums, where diffusers are connected to primary ductwork with curved elements (flexible ductwork), or where the distances between diffusers and the primary duct are less than 3 equivalent duct diameters.

The attenuation available through end reflection loss may be enough in critical situations to warrant designing the duct layout to maximize the loss. For instance, air can be dumped out of open ducts onto plaques suspended at least one duct diameter beneath the duct; a duct 3 to 5 diameters long can be connected to bar-type return air grilles (opposed blade dampers behind the bar grille can reduce end reflection loss); straight ductwork 3 to 5 diameters long can be connected to architectural openings.

Low-frequency attenuation from end reflection loss can easily equal 50 to 75 ft of lined ductwork or a 7-ft long standard pressure drop silencer. Expense and space is conserved when attenuation is part of the total building design. Table 20 presents optimal reflection losses for the configurations shown.

Table 20 End Reflection Loss

Mean Duct	Octave-Band Center Frequency, Hz				
Width, in.	63	125	250	500	1000
6	18dB	12dB	8dB	4dB	1dB
8	16	11	6	2	0
10	14	9	5	1	0
12	13	8	4	1	0
16	11	6	2	0	0
20	9	5	1	0	0
24	8	4	1	0	0
28	7	3	1	0	0
32	6	2	0	0	0
36	5	1	0	0	0
48	4	1	0	0	0
72	1	0	0	0	0

3-5 D
NOTE: NO DIFFUSER
TABLE ONLY APPLIES TO THIS CONFIGURATION

Note: Do not apply for linear diffusers or diffusers tapped directly into primary ductwork. If duct terminates in a diffuser, deduct at least 6 dB.

AIRFLOW NOISE

Aerodynamic Noise Generated at Duct Fittings

Although fans are a major source of the sound to be considered when designing quiet duct systems, they are not the only source. Aerodynamic noise is generated at elbows, dampers, branch takeoffs, air modulation units, sound attenuators, and other duct elements. The sound power levels in each octave band depend on the geometry of the device, the turbulence of the airflow, and the airflow velocity. The intensity of flow-generated noise is proportional to between the fifth and sixth power of the local flow velocity in the duct. Therefore, a small inaccuracy in predicting the local flow velocity can result in a serious noise problem.

Aerodynamic noise problems can be avoided by (1) sizing ductwork or duct configurations so that air velocities are low, (2) avoiding abrupt changes in area, and (3) attenuating noise generated at the fittings with sufficient sound attenuation between the fitting and the terminal device.

Sizing ductwork for low velocities is expensive and space-consuming. In practice, space is often insufficient to allow smooth airflow, area and direction change abruptly, and duct lengths between fittings and terminal devices are not ideal. Because duct elements can have an unlimited number of geometries, a universal law for aerodynamic noise generation has not been developed.

Sound Generated at Dampers, Elbows, and Junctions

Octave-band sound power levels generated at single- and multiblade dampers, at elbows, with and without turning vanes, and at junctions can be predicted by following this generalized equation:

$$PWL_{f_o} = K + 10 \log f_o + 50 \log U + 10 \log S + 10 \log D + \text{Special Parameters} \quad (11)$$

where

f_o = octave-band center frequency, Hz
K = characteristic spectrum of the fitting, based on Strouhal Number (Figures 15 and 16)
U = velocity factor; the velocity in the constricted part of the flow field or the velocity in the branch duct, fpm
S = cross-sectional area of the duct in which the dampers are installed, cross-sectional area of the elbow or cross-sectional area of the branch duct at a junction, ft²
D = duct height normal to the damper axis, the cord length of a typical turning vane, height of the elbow, or in the case of junctions, $D = \sqrt{4s/\pi}$, ft

Special parameters have the following values:

Dampers: -107 dB

Elbows with turning vanes: $10 \log n - 107$ dB

where n = number of turning vanes

For elbows without turning vanes and junctions: $-107 + \Delta r + \Delta T$

where

Δr = correction for rounding
ΔT = correction for upstream turbulence

Prior to solving Equation (11), perform the following preliminary calculations.

Preliminary Calculations for Dampers and Elbows with Turning Vanes

1. Determine pressure loss coefficient C

where $$C = 15.9 \times 10^6 \, \Delta p \, S^2/Q^2 \quad (12)$$

Δp = pressure drop across fitting, in. of water
Q = volume flow rate, cfm

2. Determine blockage factor BF

For multiblade dampers and elbows with turning vanes:

$$\text{BF} = 0.5 \text{ if } C = 1$$
$$\text{BF} = (C^{0.5} - 1)/(C - 1), \text{ for } C \neq 1 \quad (13)$$

For single-blade dampers:

$$\text{BF} = (C^{0.5} - 1)/(C - 1), \text{ if } C < 4$$
$$\text{BF} = 0.68\, C^{-0.15} - 0.22, \text{ if } C > 4 \quad (14)$$

3. Determine velocity term U

$$U = (Q/S)/\text{BF} \quad (15)$$

4. Determine Strouhal number St

$$\text{St} = 60 f_o D/U \quad (16)$$

5. Enter Figure 15 to determine the characteristic spectrum for dampers. Enter Figure 16 to determine the characteristic spectrum for elbows with turning vanes.

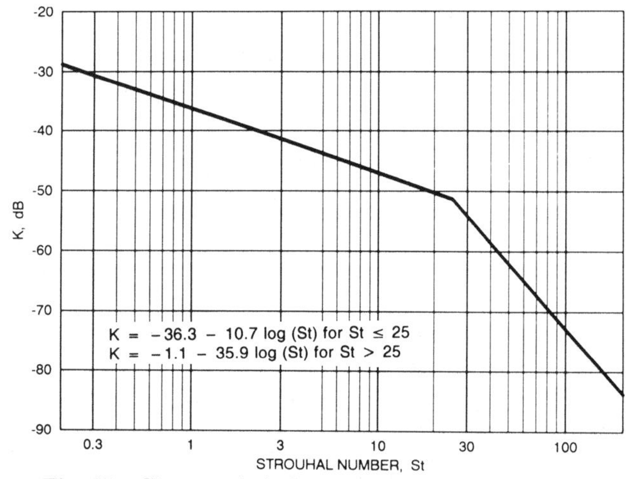

Fig. 15 Characteristic Spectrum K of Flow-Generated Noise of Dampers for Use with Equation (11)

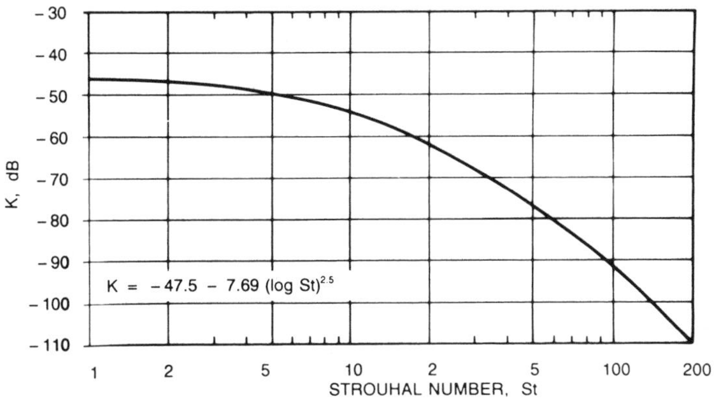

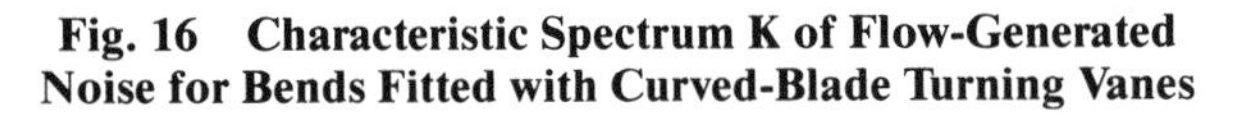

Fig. 16 Characteristic Spectrum K of Flow-Generated Noise for Bends Fitted with Curved-Blade Turning Vanes

Preliminary Calculations for Elbows and Junctions without Splitter Dampers

1. Determine velocity factor M

$$M = U_M/U_B \tag{17}$$

where

U_M = velocity in main duct, fpm
U_B = velocity in branch duct, fpm

2. Determine Strouhal number

$$\mathrm{St} = 60 f_o D/U_B \tag{18}$$

3. Enter Figure 17 to determine the characteristic spectrum K.
4. Determine the rounding correction from Figure 18. If the elbow is not rounded, then:

$$r/D_{BE} = 0$$

5. Determine the correction for upstream turbulence T from Figure 19. This correction should be applied only if duct upstream has dampers, turns, or takeoffs within a length of 5 main duct diameters.

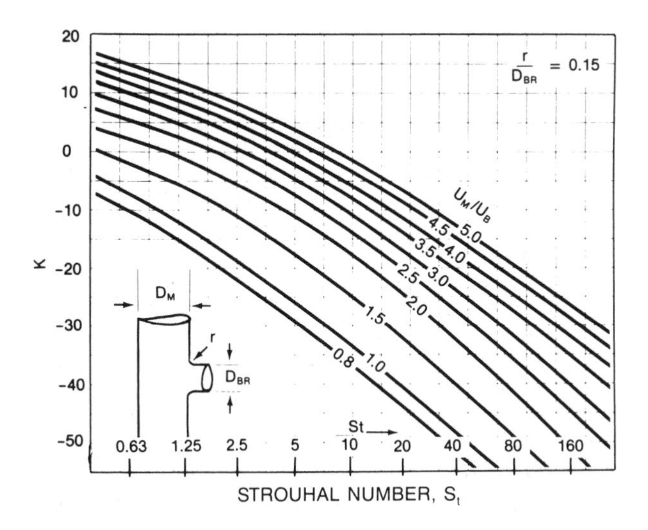

Fig. 17 Normalized Spectra for Flow-Generated Noise of Takeoffs and Junctions without Splitter Damper or Scoop
(Brockmeyer 1968)

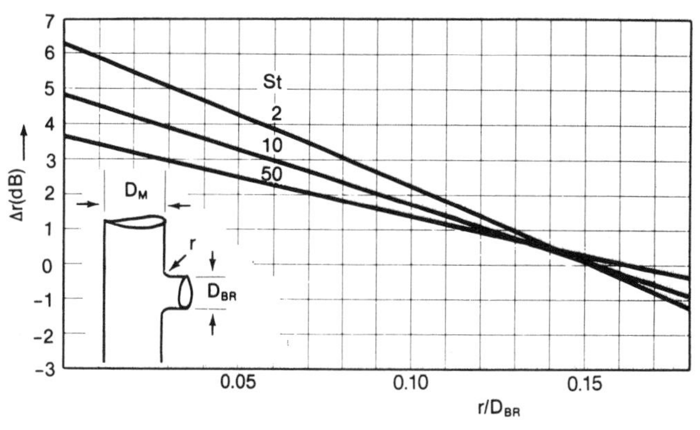

Fig. 18 Rounding Correction
(Brockmeyer 1968)

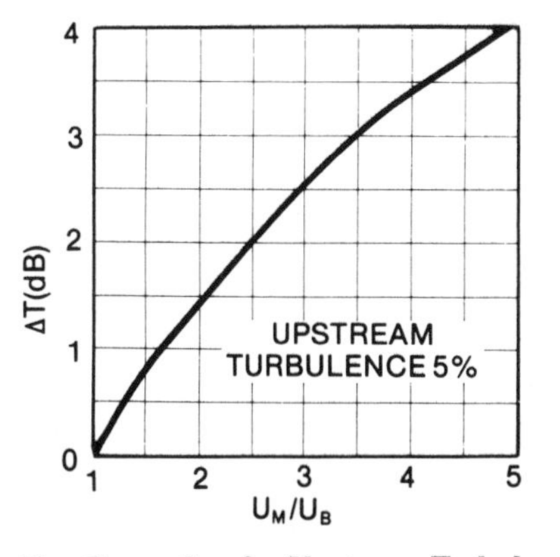

Fig. 19 Correction for Upstream Turbulence
(Brockmeyer 1968)

Prediction of Noise Along Other Branches at Junctions

Knowing the octave-band sound power level at a specific branch of a junction makes it possible to predict the flow-generated noise along other branches or in the main duct, using Figure 20.

Example 5. Determine the sound power level of a multiblade damper positioned in a 12-in. by 12-in. (1 ft by 1 ft) duct that drops the pressure by 0.5 in. of water at a volume flow rate of 4000 cfm.

1. Total pressure loss coefficient from Equation (12):

$$C = 15.9 \times 10^6 \times 0.5/[4000/(1 \times 1)]^2 = 0.5$$

2. Blockage factor from Equation (13):

$$\mathrm{BF} = (0.5^{0.5} - 1)/(0.5 - 1) = 0.586$$

3. Velocity factor from Equation (15):

$$U = 4000/(0.586 \times 1 \times 1) = 6830 \text{ fpm}$$

4. The remaining steps are tabulated in Table 21.

Example 6. Predict the octave-band sound power levels in the branch duct and main duct described below. The rectangular main duct is 12 in. by 36 in. (1 ft by 3 ft), and the volume flow rate in the main duct is 12,000 cfm. The branch duct is 10 in. by 10 in., and its flow rate is 1200 cfm. The 90° turn from the main duct to the branch duct has no rounding.

1. Area ratio: $S_M/S_B = 1 \times 3 \times 144/(10 \times 10) = 4.32$
2. Velocity ratio:
$$M = U_M/U_B = (12{,}000/3)/[1200 \times 144/(10 \times 10)] = 2.3$$
3. Velocity in branch: $U_B = 1200 \times 144/(10 \times 10) = 1730$ fpm
4. Strouhal number [Equation (18)] = St = $60 f_o\, 0.94/1730 = 0.0326 f_o$
5. The remaining steps are tabulated in Table 22.

Table 21 Sample Calculations for Example 5 (Multiblade Damper)

	Octave-Band Center Frequency, Hz								
f_o	31	63	125	250	500	1000	2000	4000	8000
St [from Equation (16)]	0.276	0.55	1.1	2.2	4.4	8.8	17.6	35.3	70.6
K	−30.3	−33.5	−36.7	−40	−43.2	−46.4	−49.6	−56.6	−67.5
10 log f_o	15	18	21	24	27	30	33	36	39
50 log U	191.8	191.8	191.8	191.8	191.8	191.8	191.8	191.8	191.8
10 log S	0	0	0	0	0	0	0	0	0
10 log D	0	0	0	0	0	0	0	0	0
Constant factor	−107	−107	−107	−107	−107	−107	−107	−107	−107
PWL (f_o)	69.5	69.3	69.1	68.8	68.6	68.4	68.2	64.1	56.3

Table 22 Sample Calculations for Example 6 (Junction)

	Octave-Band Center Frequency, Hz								
	31	63	125	250	500	1000	2000	4000	8000
St = $0.0326 f_o$	1	2	4	8	16	32	64	128	256
10 log f_o	15	18	21	24	27	30	33	36	39
K	+1.0	−3.0	−8.0	−14.5	−21	−28	−36.5	−44	−53.5
50 log U_B	162	162	162	162	162	162	162	162	162
10 log $S_B D_B$	−1.8	−1.8	−1.8	−1.8	−1.8	−1.8	−1.8	−1.8	−1.8
Δr	7	6.5	6	5	4.5	4	3.5	0	0
ΔT	—	—	—	—	—	—	—	—	—
Constant factor	−107	−107	−107	−107	−107	−107	−107	−107	−107
Branch duct PWL (f_o)	76.2	74.7	72.2	67.7	63.7	59.2	53.2	45.2	38.7
Correction to branch PWL to get to main duct PWL, 10 log S_M/S_B	+6.3	+6.3	+6.3	+6.3	+6.3	+6.3	+6.3	+6.3	+6.3
Main duct PWL	82.5	81.0	78.5	74.0	70.0	65.5	59.5	51.5	45.0

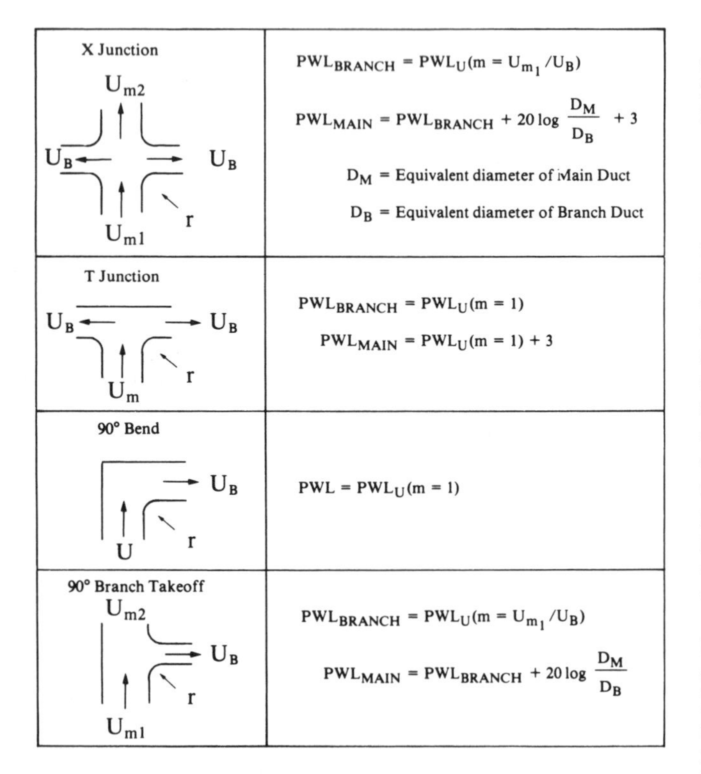

Fig. 20 Prediction of Flow-Generated Noise for Junctions and Turns

Aerodynamically Generated Duct Rumble

Any low-frequency noise that is radiated by ductwork can result in low-frequency noise, or rumble. The noise might be due to one or a combination of aerodynamic conditions or fan noise (Figure 21). The most difficult frequency region to attenuate is usually below 250 Hz.

NOISE GENERATED BY AIR TERMINALS

Air terminals—devices used to deliver air into a space and to control air volume and/or temperature by directional vanes, dampers or valves, heat exchangers, and fan control—fall into two categories. The first category, air terminal devices, comprises diffusers, grilles, and registers that radiate sound directly into an air-conditioned space. Allowing no opportunity for sound attenuation along the duct path, air terminal devices require selection for noise generation that meets the established room design criterion values as well as other factors. The second category, terminal valves, comprises terminal boxes and air valves that are separated from the space they serve by ductwork, ceiling plenums, or ceilings. These devices are selected so that their sound power ratings do not exceed the sum of the established room design criterion and the attenuation provided between the air-conditioned room and the device.

While fan-generated noise is most critical in the low-frequency bands, many air terminals such as diffusers, registers, grilles, and terminal boxes without fans do not contribute greatly to low-frequency noise levels. In these cases, an allowance for fan noise may be unnecessary in selecting the terminals. However, both fan and terminal unit noise must be considered to obtain the balanced sound spectrum implied by the RC curves.

Air Terminal Device Installation Factors

The sound level output of an air diffuser or grille depends not only on the air quantity and the size and design of the outlet, but also on the air approach configuration (Waeldner 1975). Manufacturers' ratings apply only to outlets installed as recommended, *i.e.*, with a uniform air velocity distribution throughout the neck

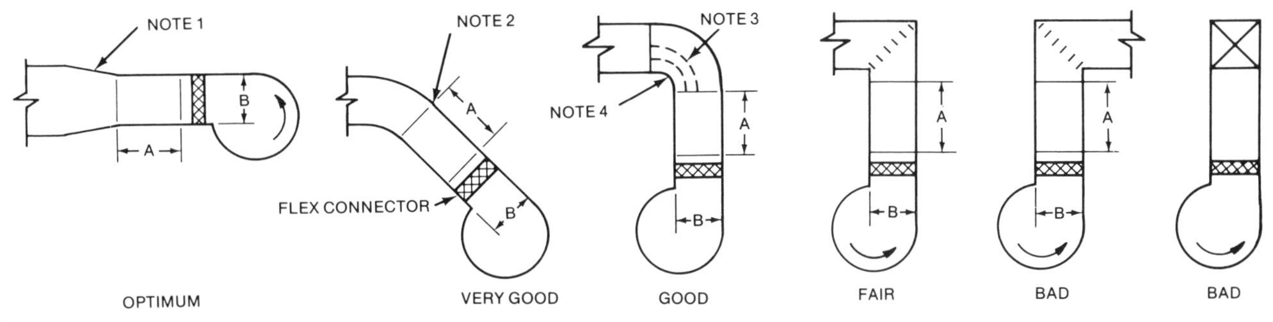

Notes:
1. Slopes of 1 in 7 preferred. Slopes of 1 in 4 permitted below 2000 ft/min.
2. Dimension A should be a least 1.5 times B, where B is the largest discharge duct dimension.
3. Rugged turning vanes should extend the full radius of the elbow.
4. Minimum 6-in. radius.

Fig. 21 Various Outlet Configurations and Their Possible Rumble Conditions

of the unit. Poor approach conditions can easily increase sound levels by 12 to 15 dB above the manufacturer's ratings (Figures 22 and 23). Poor approach conditions can sometimes be overcome with properly adjusted accessories such as turning vanes or equalizing grids. However, using these accessories in critical, low noise level projects should be avoided.

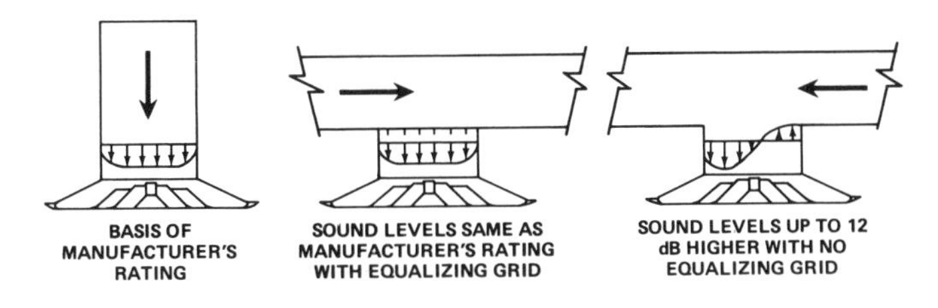

Fig. 22 Proper and Improper Airflow Conditions to an Outlet

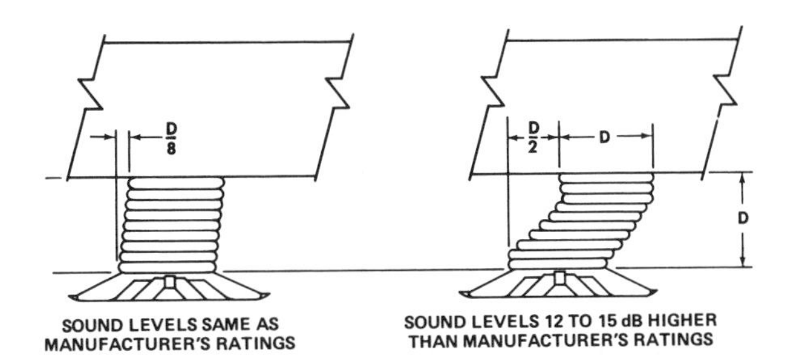

Fig. 23 Effect of Proper and Improper Alignment of Flexible Duct Connector

Flexible duct is often used to correct misalignment between the supply duct and the diffuser ceiling location. A misalignment or offset that exceeds one-fourth of a diffuser diameter in a diffuser collar length of 2 diameters significantly increases the diffuser sound level (Figure 23). There is no appreciable change in diffuser performance with an offset less than one-eighth the length of the collar.

When a volume control damper is installed close to an air outlet to achieve system balance, the acoustic performance of the air outlet must be based not only on the air volume handled, but also on the magnitude of the pressure drop across the damper. The sound level change is proportional to the pressure ratio (PR) of the throttled pressure drop to the catalog pressure drop of the outlet, as shown in Figure 24 and Table 23.

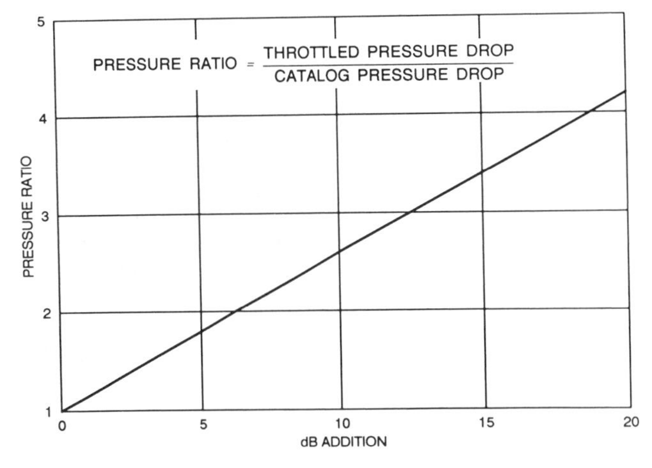

Fig. 24 Decibels to be Added to Outlet Sound for Throttled Damper Close to Outlet

Table 23 Decibels to be Added to Diffuser Sound Rating to Allow Throttling of Volume Damper

$$\text{Pressure Ratio (PR)} = \frac{\text{Throttled Pressure}}{\text{Minimum Pressure}}$$

Location of Volume Damper	1.5	2	2.5	3	4	6
(A) OB damper in neck of linear diffuser	5	9	12	15	18	24
(B) OB damper in plenum side inlet	2	3	4	5	6	9
(C) Damper in supply duct at least 5 ft from plenum	0	0	0	2	3	5

Balancing dampers, equalizers, and so forth should not be placed directly behind terminal devices or open-ended ducts in acoustically critical spaces, such as concert halls. They should be located 5 to 10 duct diameters from the opening, followed by lined duct to the terminal or open duct end.

Linear diffusers are often installed in distribution plenums so that the damper may be installed at the plenum entrance. The further a damper is installed from the outlet, the lower the resultant sound level will be (Table 23).

Terminal Valve Selection and Installation

Acoustic concerns for terminal valves include noise from the device, which radiates through the ceiling material, and sound, which discharges into the ductwork upstream and downstream from the device, arriving at the occupied space by way of the plenum and/or intervening ductwork. Most manufacturers provide radiated and discharge noise data, and use of ARI/ADC *Standard* 880 should improve the reliability of these data. However, this chapter can help to predict terminal valve noise levels in typical rooms (Blazier 1981).

Fan-Powered Terminal Valves

A fan in the terminal valve unit draws in heated air from the ceiling plenum which mixes with the primary air to control the delivered air temperature. In addition to the potential acoustic problems described in the preceding section, these devices can transmit fan noise into the occupied space by radiating noise into the ceiling plenum as well as along the duct path. The following general guidelines can be used to reduce the potential for noise problems.

1. NC or RC levels below 40 are difficult to achieve where fan-powered terminal valves are located immediately above or in a ceiling plenum shared by an air-conditioned space. Where possible, locate fan-powered terminal valves above nonsensitive spaces, such as corridors, utility rooms, toilets, or service closets. If this is not possible, consider the alternative in Item 8.
2. Install return air openings in acoustical ceilings away from fan-powered terminal valves. This precaution prevents the sound radiated from the valve from having a direct acoustical path to the listener.
3. Intermittent fan operation of parallel fan-powered terminal valves is often more annoying than a design with constant fan operation.
4. Fan speeds should be the minimum for required airflow. However, use judgment when considering catalog data as it pertains to fan speed.
5. Keep the static pressure at the valve as low as possible.
6. Installing lined duct elbows (boots) on the return air openings to fan-powered terminal valves can attenuate some fan noise, especially when the boots point up toward the structural deck.
7. Sound levels radiated by the bottoms of the boxes can sometimes be reduced by damping or stiffening the large radiating surfaces.
8. Noise radiated by fan-powered terminal valves located above noise-sensitive spaces, such as conference rooms, private offices, and classrooms, may need to be controlled with drywall ceilings or enclosures.

DUCTBORNE CROSSTALK

Ductborne crosstalk is sound transmitted between rooms by way of the duct system. It is controlled by using duct linings, splitters, prefabricated duct attenuators, and limiting straight duct routings. Generally, the requisite duct attenuation to prevent crosstalk is about 5 dB greater than the transmission loss (TL) of the intervening architectural construction, but if ceiling heights, room sound-absorbing treatments, or background noise levels are radically different between source and receiving rooms, the needed duct attenuation may be even greater. U.S. Gypsum (1971) and Hedeen *et al.* (1980) list transmission loss ratings for typical architectural constructions, and Table 24 presents a few ratings for more frequently encountered materials or construction.

TRANSMISSION LOSS THROUGH DUCT WALLS

Noise radiated through duct walls can be a problem if the noise traveling in the duct (usually fan or flow noise) is not adequately attenuated before the duct runs over an occupied space (Cummings 1985). This transmission path is known as breakout. Noise can also be transmitted into the duct in one space and radiated from the duct to another space. Transmission into a duct is known as breakin.

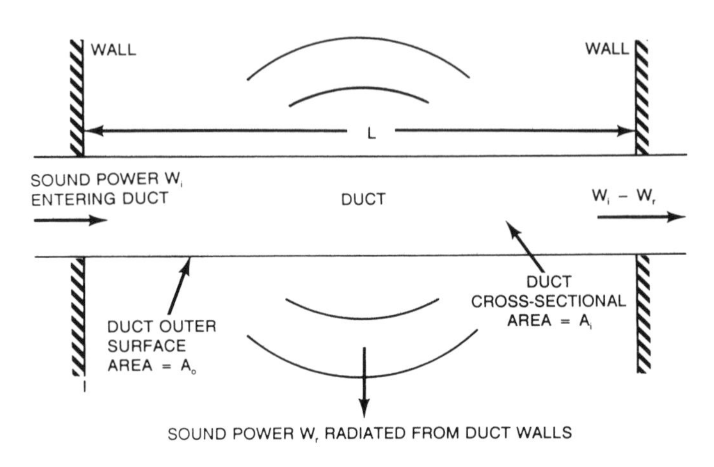

Fig. 25 Breakout Transmission

Transmission Loss of Rectangular Ducts

Breakout Transmission Loss. The breakout transmission loss TL_{out} of duct walls as shown in Figure 25, is defined as:

$$TL_{out} = 10 \log [(W_i A_o) / (A_i W_r)] \text{ dB} \quad (19)$$

where

W_i = sound power in the duct
W_r = sound power radiated from walls
A_i = ab
A_o = $2L(a + b)$
a = larger dimension of duct cross section
b = smaller dimension of duct cross section
L = duct length

To obtain W_r, Equation (19) may be rewritten as follows:

$$L_{w_r} = L_{w_i} + 10 \log (A_o/A_i) - TL_{out} \quad (20)$$

where

$L_{w_r} = 10 \log (W_r \times 10^{12})$
$L_{w_i} = 10 \log (W_i \times 10^{12})$

Table 25 lists values for TL_{out} in Equation (20) for selected duct sizes.

Since the sound power within the duct slowly decreases along the length of the duct, the actual TL_{out} as defined above increases slightly as the length increases for any type of duct. This effect does not apply to the prediction method and is usually minimal.

The transmission loss curve can be divided into two regions —one where the plane mode transmission within the duct predominates—and another where multimode transmission is dominant. The limiting frequency between these two regions is given by:

$$f_L = 24134/(ab)^{0.5} \quad (21)$$

where a and b are in inches.

For $f < f_L$, the plane mode breakout TL may be calculated from the formula

$$TL_{out} = 10 \log [fq^2/(a + b)] + 17 \text{ dB} \quad (22)$$

where

q = mass/unit area of the duct walls, lb/ft²
f = frequency, Hz
a,b = duct cross-section dimensions, in.

Table 24 Transmission Loss (Rows 1 through 17) and Insertion Loss (Rows 18 through 25) for Typical Construction Materials and Assemblies

	Assembly Octave-Band Center Frequency, Hz						
No.	63	125	250	500	1000	2000	4000
1	32	34	35	37	42	49	55
2	34	36	38	43	50	56	61
3	29	32	33	34	37	42	49
4	31	33	35	36	41	48	54
5	32	34	36	39	45	46	40
6	12	17	34	35	42	38	44
7	16	24	33	41	45	40	42
8	16	24	37	44	51	43	46
9	20	29	35	45	47	45	46
10	20	29	41	49	53	55	51
11	25	36	43	50	50	44	55
12	25	36	47	51	57	57	62
13	13	20	33	36	43	41	46
14	12	23	39	49	58	60	53
15	22	30	45	52	60	64	58
16	11	16	23	27	32	28	32
17	12	16	23	27	32	30	35
18	25	41	47	56	65	68	69
19	23	29	31	31	31	39	43
20	17	23	33	35	35	33	37
21	15	18	21	39	38	49	55
22	15	18	24	35	45	53	58
23	4	5	6	8	10	12	14
24	2	3	4	5	7	9	11
25	9	15	20	25	31	33	27

No.	Assembly
1	4 in. dense concrete or solid concrete block (48 lb/ft^2)
2	8 in. dense concrete or solid concrete block (96 lb/ft^2)
3	4 in. hollow core dense aggregate concrete block (28 lb/ft^2)
4	8 in. hollow core dense aggregate concrete block (44 lb/ft^2)
5	12 in. hollow core dense aggregate concrete block (60 lb/ft^2)
6	Standard drywall partition, 5/8 in. gypsum board on both sides of 2 × 4 in. wood studs
7	Standard drywall partition, 1 layer 5/8 in. gypsum board on each side of 3-5/8 in. metal studs
8	Standard drywall partition, 1 layer 5/8 in. gypsum board on each side of 3-5/8 in. metal studs, 2 in. fiberglass blanket in wall cavity
9	Standard drywall partition, 1 layer 1/2 in. gypsum board and 2 layers 1/2 in. gypsum board on either side of 3-5/8 in. metal studs
10	Standard drywall partition, 1 layer 1/2 in. gypsum board and 2 layers 1/2 in. gypsum board on either side of 3-5/8 in. metal studs, 2 in. fiberglass blanket in wall cavity
11	Standard drywall partition, two layers of 5/8 in. gypsum board on each side of 3-5/8 in. metal studs
12	Standard drywall partition, two layers of 5/8 in. gypsum board on each side of 3-5/8 in. metal studs, 1-1/2 in. fiberglass blanket in wall cavity
13	Cavity shaft wall, 1 layer 1 in. gypsum board and 2 layers 1/2 in. fire code gypsum board on either side of 2-1/2 in. metal C-H studs
14	Cavity shaft wall, 1 layer 1/2 in. fire code gypsum board on each side of 2-1/2 in. metal C-H studs, 1 in. gypsum board and 1 in. mineral fiber blanket in wall cavity
15	Cavity shaft wall, 1 layer 1/2 in. fire code gypsum board and 2 layers 1/2 in. fire code gypsum board on either side of 2-1/2 in. metal C-H studs, 1 in. gypsum board and 1 in. mineral fiber blanket in wall cavity
16	1/2 in. plate glass
17	Double glazing, two 1/2 in. panes, 1/2 in. airspace
18	Roof construction, 6 in. thick, 20 gage (0.0396 in.) steel deck with 4 in. lightweight concrete topping, 5/8 in. gypsum board ceiling on resilient hangers
19	1-3/4 in. thick solid core wood door
20	1-3/4 in. thick insulated hollow 16 gage (0.0635 in.) steel door with weather stripping
21	Acoustic equipment housing, 20 gage (0.0396 in.) steel outer shell, 2 in. thick acoustic insultion, 22 gage (0.0336 in.) perforated inner shell
22	Acoustic equipment housing, 20 gage (0.0396 in.) steel outer shell, 4 in. thick acoustic insulation, 22 gage (0.0336 in.) perforated inner shell
23	Typical mineral fiber lay-in acoustical ceiling—one pass
24	Typical fiberglass lay-in acoustical ceiling—one pass
25	Typical gypsum board ceiling

Notes:
1. The losses in Table 16 are the results of laboratory tests. Actual installations have losses up to 5 dB lower.
2. Data are compiled from several sources.
3. Some of the values in the 63 Hz octave band are extrapolated.

Table 25 TL_{out} versus Frequency for Various Rectangular Ducts

Duct Size		Octave-Band Center Frequency, Hz							
in.	Gage	63	125	250	500	1000	2000	4000	8000
12 × 12	24	21	24	27	30	33	36	41	45
12 × 24	24	19	22	25	28	31	35	41	45
12 × 48	22	19	22	25	28	31	37	43	45
24 × 24	22	20	23	26	29	32	37	43	45
24 × 48	20	20	23	26	29	31	39	45	45
48 × 48	18	21	24	27	30	35	41	45	45
48 × 96	18	19	22	25	29	35	41	45	45

Note: The data are from tests on 20-ft long ducts, but the TL values are for ducts of the cross section shown regardless of length.

The minimum possible value of TL_{out} occurs where $W_r = W_i$ and a lower limit is thus imposed on TL_{out}:

$$TL_{out}\ (min) = 10 \log [24L\ (1/a + 1/b)] \tag{23}$$

where L is in feet and a and b are in inches.

When $f \geqslant f_L$, the multimode TL may be found from the formula

$$TL_{out} = 20 \log [qf] - 31 \text{ dB} \tag{24}$$

where again q is in lb$_m$/ft^2 and f is in Hz.

To plot TL_{out} over the entire frequency range of interest on semilog paper, draw a baseline of $TL_{out}(min)$ from Equation (23) parallel to the (logarithmic) frequency axis. Next, draw a curve of Equation (22) down to the baseline [Equation (23)] and up to a frequency f_L. Erase the portion of the baseline to the right of the point where it meets this curve. Draw a curve of Equation (24), from f_L upward. The curves from Equation (22) and (24) generally do not meet at f_L and they should be joined by a vertical line. The wave coincidence effect and acoustic leaks at seams and joints combine to limit the high frequency TL to about 45 dB, so that finally, the upper portion of the curve should be leveled off at this value, in a horizontal line. Figure 26 shows the general form of the resultant curve.

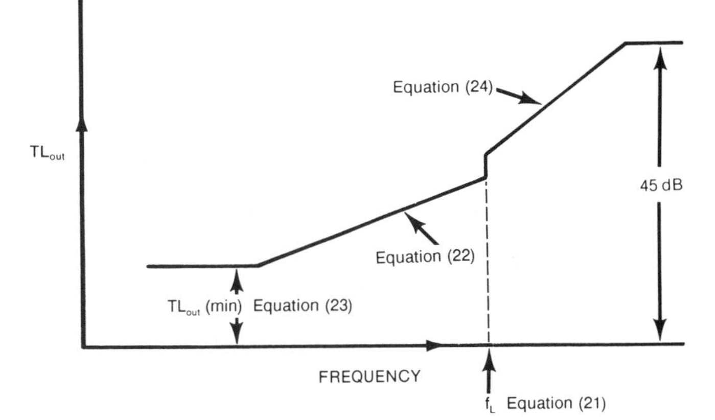

Fig. 26 Composite Curve of TL_{out} for a Rectangular Duct

This method is based on a mass law impedance for the duct walls and is only valid for:

$$f > 28{,}570\, t^{0.5}/a \tag{25}$$

where

t = duct wall thickness, in.
a = larger transverse wall dimension, in.

This frequency limit applies to galvanized steel ducts. Below this frequency, the *actual* TL will exhibit damped resonances, which produce progressively more pronounced maxima and minima in the TL curve as the frequency falls. The present method still predicts the overall behavior, however, and generally gives conservative predictions at low frequencies. The effects of wave coincidence are not explicitly included in the prediction scheme (aside from the 45 dB limit on TL_{out}), but these would occur, for ordinary ductwork, at frequencies above the range of interest.

Duct fittings, such as elbows, do not materially affect the TL but should be included in the effective radiating surface area when the actual sound power radiated from a length of ductwork is being calculated.

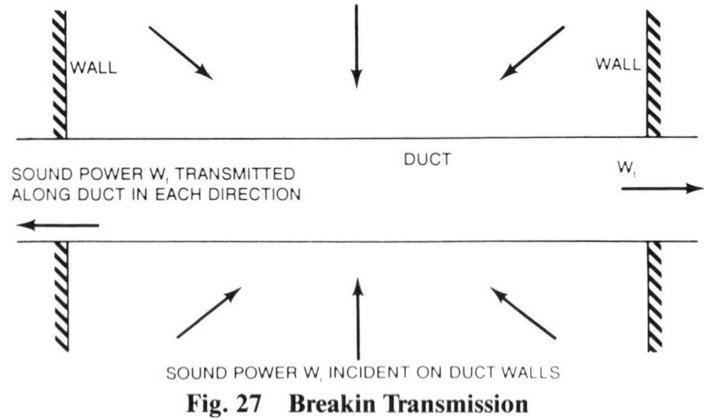

Fig. 27 Breakin Transmission

Breakin Transmission Loss. Figure 27 shows the breakin configuration. The incident sound power from the surrounding space is W_i, and sound power W_t travels out along the duct in *both* directions, as shown. The breakin TL can be defined as

$$TL_{in} = 10 \log (W_i/2W_t) \tag{26}$$

which is the logarithmic ratio of the total incident sound power W_i to the total transmitted power $2W_t$. To obtain sound power, W_t transmitted down a duct in a specified direction, Equation (26) must be rewritten as follows:

$$L_{W_t} = L_{W_i} - TL_{in} - 3\text{ dB} \tag{27}$$

This definition is more in line with the common definition of the TL of a partition than that of the breakout TL. An equivalent definition cannot be used in the latter case because there is no well-defined incident (as opposed to reflected) sound power. The incident sound power level L_{W_i} may be related to the sound pressure level in the reverberant field L_{prev} as follows:

$$L_{W_i} = 10 \log (W_i/10^{-12}) = L_{prev} + 10 \log A_o - 16\text{ dB} \tag{28}$$

where W_i is in watts and A_o is in square feet.

The breakin TL may be calculated in terms of TL_{out} (since it is difficult to calculate directly) by one of two formulas—the first applies at frequencies below the cut-off frequency f_L, for the lowest acoustic cross-mode in the duct, and the second, at frequencies above f_L. For rectangular ducts, this cut-off frequency is given by

$$f_L = 6764/a \tag{29}$$

where a (the larger duct dimension) is in inches.

For $f \leqslant f_1$, TL_{in} equals the larger of the following:

$$TL_{out} = 10 \log (a/b) + 20 \log (f/f_L) - 4\text{ dB} \tag{30}$$

or

$$10 \log [12L\,(1/a + 1/b)] \tag{31}$$

where L is in feet and a, b in inches.

For $f > f_L$, TL_{in} is given by

$$TL_{in} = TL_{out} - 3\text{ dB} \tag{32}$$

Table 26 shows some values of TL_{in}, corresponding to the ducts listed in Table 25.

Table 26 TL_{in} versus Frequency for Various Rectangular Ducts

Duct Size		Octave-Band Center Frequency, Hz							
in.	Gage	63	125	250	500	1000	2000	4000	8000
12×12	24	16	16	16	25	30	33	38	42
12×24	24	15	15	17	25	28	32	38	42
12×48	22	14	14	22	25	28	34	40	42
24×24	22	13	13	21	26	29	34	40	42
24×48	20	12	15	23	26	28	36	42	42
48×48	18	10	19	24	27	32	38	42	42
48×96	18	11	19	22	26	32	38	42	42

Note: The data are from tests on 20-ft long ducts, but the TL values are for ducts of the cross section shown regardless of length.

Transmission Loss of Circular Ducts

Breakout Transmission Loss. Simple prediction methods for the TL of circular ducts can disagree with actual measurements by 20 to 30 dB at low frequencies, so only general comments and data on selected ducts are given in this chapter.

An ideal circular duct has a high TL at low frequencies because uniform internal sound pressure fluctuations cause a uniform "breathing" in the pipe walls that presents a high impedence to the sound waves. Small departures from circularity in the ducts cause higher structural modes to be excited in the duct walls, even with plane wave transmission within the duct. Spiral-wound ducts are more nearly circular than long seam ducts and may have a significantly higher TL at low frequencies. However, both types of duct exhibit similar behavior at high frequencies. Figure 28 shows the general form of the TL curve for typical circular air-conditioning ducts. At very low frequencies, the TL continues to rise as the frequency falls.

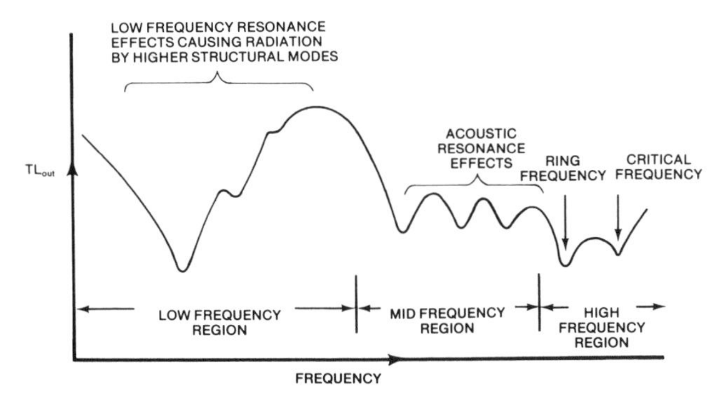

Fig. 28 General Form of TL_{out} Curve for Circular Duct

Low-frequency sound and vibration energy below 50 Hz can be important because it relates to motors running to shaft speeds of 29 Hz (1750 rpm) and fans running at shaft speeds of 15 to 25 Hz in typical HVAC equipment. A low-frequency resonance controlled region occurs at frequencies below 100 Hz, where the higher structural modes radiate sound. As the frequency rises, however, these modes radiate with decreasing efficiency as the mode order increases and the *TL* rises to a peak. The *TL* then begins to fall in a series of peaks and troughs corresponding to the acoustic modal resonances in the duct. This behavior continues until the *ring frequency* is reached and the entire duct cross section resonates at an almost uniform expansion and contraction. At an even higher frequency, the wave coincidence effect causes a further dip. The *TL* rises above this in a manner similar to that of a flat plate of the same material.

Although the *TL* curve is complex, it is highest at "low" frequencies. Its maximum value (discounting the very low frequency rise) is between 50 and 300 Hz, depending on the duct size. Here the *TL* can be 30 dB or more higher than that of a rectangular duct. The internal sound power levels in this region also tend to be highest. In this respect, circular ducts are far superior to either rectangular or flat oval ducts.

Duct fittings, such as elbows, appear to reduce the *TL* to some extent, although this effect appears to be most severe at low frequencies.

Table 27 gives measured *TL* data on a range of duct sizes and indicates the effect of elbows in reducing the *TL*. The *TL* definition is expressed in Equation (19), with appropriate expressions for A_o and A_i.

Table 27 Experimentally Measured TL_{out} versus Frequency for Circular Ducts

Duct Size and Type	63	125	250	500	1000	2000	4000	8000
	Octave-Band Center Frequency, Hz							
8 in. dia., 26 ga (0.022 in.) long seam, length = 15 ft	>45	(53)	55	52	44	35	34	26
14 in. dia., 24 ga (0.028 in.) long seam, length = 15 ft	>50	60	54	36	34	31	25	38
22 in. dia., 22 ga (0.034 in.) long seam, length =15 ft	>47	53	37	33	33	27	25	43
32 in. dia., 22 ga (0.034 in.) long seam, length =15 ft	(51)	46	26	26	24	22	38	43
8 in. dia., 26 ga (0.022 in.) spiral wound, length =10 ft	>48	>64	>75	>72	56	56	46	29
14 in. dia., 26 ga (0.022 in.) spiral wound, length = 10 ft	>43	>53	55	33	34	35	25	40
26 in. dia., 24 ga (0.028 in.) spiral wound, length = 10 ft	>45	50	26	26	25	22	36	43
26 in. dia., 16 ga (0.064 in.) spiral wound, length = 10 ft	>48	>53	36	32	32	28	41	36
32 in. dia., 22 ga (0.034 in.) spiral wound, length = 10 ft	>43	42	28	25	26	24	40	45
14 in. dia., 24 ga (0.028 in.) long seam with two 90° elbows, length = 15 ft plus elbows	>50	54	52	34	33	28	22	34

Note: In cases where background noise swamped the noise radiated from the duct walls, a lower limit on the *TL* is indicated by a > sign. Parentheses indicate measurements in which background noise has produced a greater uncertainty than usual in the data.

Breakin Transmission Loss. The definition of Equation (26) still applies for the breakin *TL* of circular ducts, and Equation (28) can still be used to relate W_i to the reverberant field sound pressure level, if the correct value of A_o is inserted.

In the case of circular ducts, the cutoff frequency for the lowest acoustic cross-mode is given by:

$$f_L = 7929/d \tag{33}$$

where *d* is the duct diameter in inches, and the breakin *TL* can be found, as before, in terms of TL_{out}.

Where $f \leqslant f_L$, TL_{in} is given by

$$TL_{in} = \text{the larger of} \begin{cases} TL_{out} + 20\log(f/f_1) - 4\text{ dB} \\ 10\log(2L/d) \end{cases} \tag{34}$$

where *L* and *d* are in inches; for $f > f_L$, TL_{in} is given by Equation (32).

Table 28 gives TL_{in} values for the ducts listed in Table 21.

Transmission Loss of Flat Oval Ducts

Breakout Transmission Loss. Vibration measurements indicate that at low to mid frequencies, most of the acoustic radiation emanates from the flat duct sides; at high frequencies, the duct radiates uniformly. The flat sides vibrate in much the same way as the walls of rectangular ducts at low frequencies, with the curved ends acting as springs. Resonant behavior, probably related to the ring frequency phenomenon of circular ducts, prevails at high frequencies. Available data show that the minimum in the *TL* curve is at the ring frequency of the equivalent circular duct, the diameter of which is equal to the smaller dimension of the flat oval duct.

Although flat oval ducts combine undesirable features of rectangular and circular ducts, they have one advantage—only the flat sides of the duct radiate acoustical energy, primarily at low to mid frequencies. This means that these ducts tend to have a higher *TL* than rectangular ducts in the low-frequency region, resulting in a difference of 8 to 10 dB in some cases. However, circular ducts are superior to flat oval and rectangular ducts at low frequencies (Figure 29).

Table 28 Experimentally Measured TL_{in} versus Frequency for Circular Ducts

Duct Size and Type	63	125	250	500	1000	2000	4000	8000
	Octave-Band Center Frequency, Hz							
8 in. dia., 26 ga (0.022 in.) long seam, length = 15 ft	>17	(31)	39	42	41	32	31	23
14 in. dia., 24 ga (0.028 in.) long seam, length = 15 ft	>27	43	43	31	31	28	22	35
22 in. dia., 22 ga (0.034 in.) long seam, length =15 ft	>28	40	30	30	30	24	22	40
32 in. dia., 22 ga (0.034 in.) long seam, length =15 ft	(35)	36	23	23	21	19	35	40
8 in. dia., 26 ga (0.022 in.) spiral wound, length =10 ft	>20	>42	>59	>62	53	53	43	26
14 in. dia., 26 ga (0.022 in.) spiral wound, length = 10 ft	>20	>36	44	28	31	32	22	37
26 in. dia., 24 ga (0.028 in.) spiral wound, length = 10 ft	>27	38	20	23	22	19	33	40
26 in. dia., 16 ga (0.064 in.) spiral wound, length = 10 ft	>30	>41	30	29	29	25	38	33
32 in. dia., 22 ga (0.034 in.) spiral wound, length = 10 ft	>27	32	25	22	23	21	37	42
14 in. dia., 24 ga (0.028 in.) long seam with two 90° elbows, length = 15 ft plus elbows	>27	37	41	29	30	25	19	31

Note: In cases where background noise swamped the noise radiated from the duct walls, a lower limit on the *TL* is indicated by a > sign. Parentheses indicate measurements in which background noise has produced a greater uncertainty than usual in the data.

The cross-sectional geometry of flat oval ducts, with *a* as the major axis and *b* as the minor axis, is illustrated in Figure 30. The breakout *TL* is defined by Equation (19) but A_i and A_o are now given by the expressions

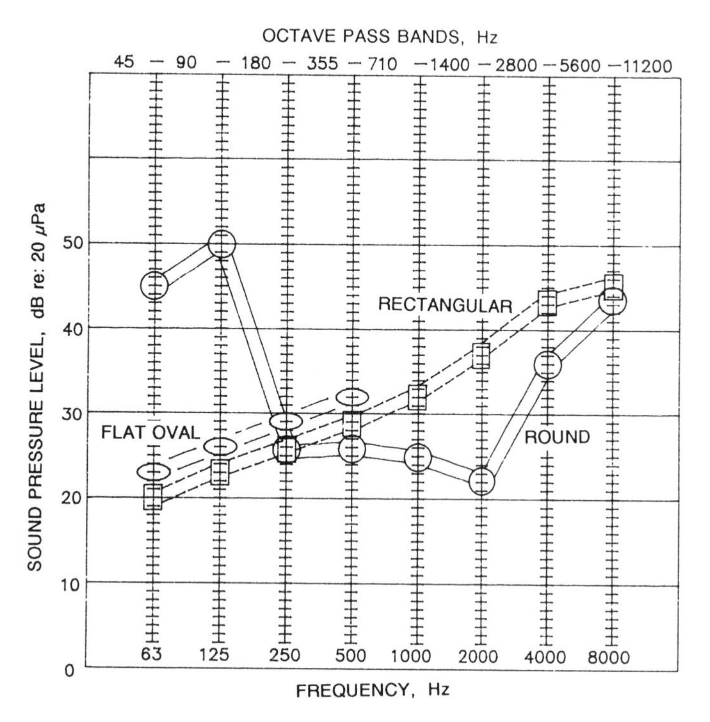

Fig. 29 Breakout Sound Transmission Loss (dB) Comparison

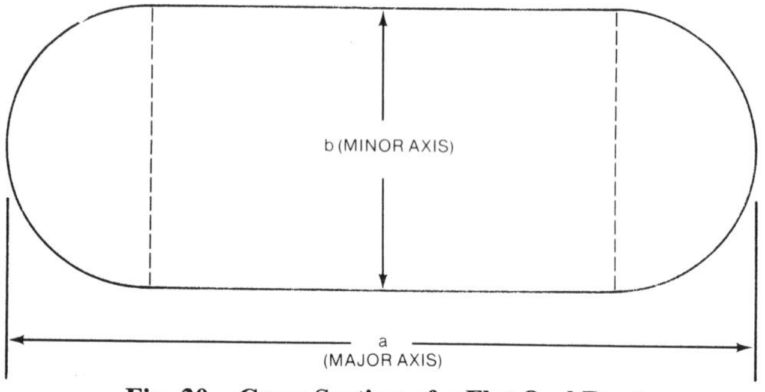

Fig. 30 Cross Section of a Flat Oval Duct

$$A_i = b(a - b) + \pi b^2/4 \quad (35)$$

$$A_o = L[2(a - b) + \pi b] \quad (36)$$

The duct perimeter P is given by

$$P = 2(a - b) + \pi b \quad (37)$$

and the fraction σ of P taken up by the flat duct sides is given by

$$\sigma = 1/[1 + \pi b/2(a - b)] \quad (38)$$

To estimate TL_{out} for flat oval ducts at low to mid frequencies, draw a baseline of $TL_{out}(min)$, where

$$TL_{out}(min) = 10 \log (A_o/A_i) \quad (39)$$

where A_o and A_i (in the same units) are given by Equations (35) and (36). Next, plot a curve of

$$TL_{out} = 10 \log (q^2 f/\sigma^2 P) + 20 \text{ dB} \quad (40)$$

where q is the mass/unit area of the duct walls in lb_m/ft^2 and f the frequency in Hz, from the baseline up to a frequency of

$$f_L = 8115/b \quad (41)$$

where b is the minor axis of the duct cross section in inches. The frequency f_L is the upper limit of applicability of Equation (40), and is equal to one-eighth of the ring frequency of the equivalent circular duct. Finally, erase that portion of the baseline to the right of the point where it meets this curve.

The comments on the low frequency TL_{out} of rectangular ducts should also apply to flat oval ducts; low-frequency resonance effects may be expected, but Equation (39) and (40) should still predict the overall behavior or give conservative predictions. The effects of fittings can be expected to be small at low frequencies; but, as in the case of rectangular ducts, they should be included in calculating the surface area of the duct.

Table 29 gives some values of TL_{out} for flat oval ducts of various sizes. The upper frequency limit on TL_{out} is imposed by Equation (41).

Breakin Transmission Loss. The breakin TL of flat oval ducts is defined by Equation (26); again, Equation (28) relates the reverberant field sound pressure level to W_i.

While there are no exact solutions for the cutoff frequency for the lowest acoustic cross-mode in flat oval ducts, an approximate solution exists as follows:

$$f_1 = \frac{6764}{(a - b)[1 + \pi b/2(a - b)]^{0.5}} \quad (42)$$

where a and b are in inches. Equation (42) is valid where $a/b \geq 2$, for $a/b < 2$, and deteriorates progressively as a/b approaches unity. Again, TL_{out} may be found in terms of TL_{in}, as follows:

Where $f \leq f_1$, TL_{in} is given by

$$TL_{in} = \text{the larger of} \begin{cases} TL_{out} + 10 \log (f^2 A_i) - 81 \text{ dB} \\ 10 \log (PL/A) \end{cases} \quad (43)$$

where A_i is in square inches and is given by Equation (35), P is in inches and is given by Equation (36), and L is in feet.

If $f > f_L$, TL_{in} is given by Equation (32). Table 29 gives TL_{out} and TL_{in} values for various flat oval ducts.

Acoustically Generated Duct Rumble

Low-frequency noise that breaks out of ductwork can give the perception of low-frequency noise, or rumble. The noise might be due to one or a combination of aerodynamic conditions or to fan noise.

Table 29 TL_{out} and TL_{in} versus Frequency for Various Flat Oval Ducts

Duct Size		TL_{out} Octave-Band Center Frequency, Hz								TL_{in} Octave-Band Center Frequency, Hz							
($a \times b$), in.	Gage	63	125	250	500	1000	2000	4000	8000	63	125	250	500	1000	2000	4000	8000
12 × 6	24	31	34	37	40	43	—	—	—	18	18	22	31	40	—	—	—
24 × 6	24	24	27	30	33	36	—	—	—	17	17	18	30	33	—	—	—
24 × 12	24	28	31	34	37	—	—	—	—	15	16	25	34	—	—	—	—
48 × 12	22	23	26	29	32	—	—	—	—	14	14	26	29	—	—	—	—
48 × 24	22	27	30	33	—	—	—	—	—	12	21	30	—	—	—	—	—
96 × 24	20	22	25	28	—	—	—	—	—	11	22	25	—	—	—	—	—
96 × 48	18	28	31	—	—	—	—	—	—	19	28	—	—	—	—	—	—

Note: The data are from tests on 20-ft long ducts, but the TL values are for ducts of the cross section shown regardless of length.

The reduction of energy passing through a typical rectangular duct wall is less than the *TL* of the duct up to the limit of $TL - 3$ dB for large *S/A* (long ducts). The measure of the breakout is the duct breakout noise reduction loss, and can be estimated from the tabulated *TL* and the duct dimensions as follows:

$$\text{Br Noise Reduction} = TL - 10 \log (S/A), \text{ dB} \quad (44)$$

where

TL = sound transmission loss in Table 25 for rectangular ducts
S = surface area of length of duct being considered
A = cross-sectional area of duct

If the sound power level entering the duct is L_{wd}, the sound power radiated through the duct surface is

$$L_{wr} = L_{wd} - \text{Br Noise Reduction (dB)} \quad (45)$$

For example the value for a 20,000 cfm fan might be 100 dB at 63 Hz for L_{wd}. Sound transmission loss might be 20 dB, with a breakout noise reduction of 12 dB for a 15-ft duct. The sound power radiated from the section is 88 dB (Harold 1986).

Insertion Loss of External Acoustic Wall Lagging on Rectangular Ducts

External acoustic lagging is sometimes applied to ductwork to reduce low-frequency noise radiation from the walls. This is common in rectangular ducts, since both flat oval and circular ducts have a higher *TL* at low frequencies than rectangular ducts.

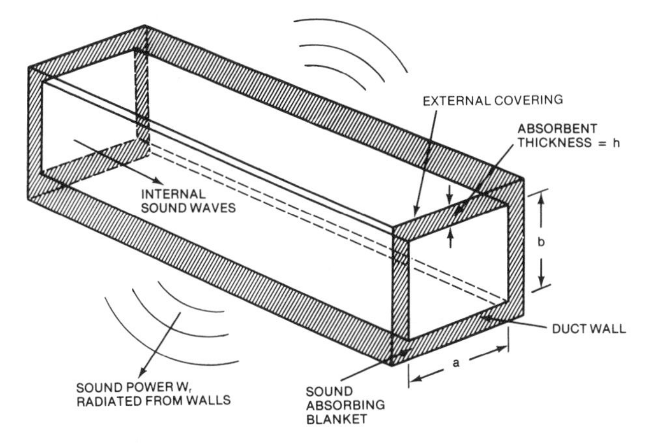

Fig. 31 Rectangular Duct with External Lagging

Prediction methods for noise reduction affected by external lagging have not been developed for flat oval and circular duct geometries. However, methods are given for estimating the insertion loss (*IL*) of lagging on rectangular ducts. Lagging consists of a layer of soft, flexible, porous material, such as glass fiber blanket covered with an outer impervious layer, *i.e.*, a relatively rigid material, such as gypsum board, or a limp material, such as sheet lead. The absorbent thickness is *h* and the duct cross section measures $a \times b$, as shown in Figure 31. The insertion loss is defined as

$$IL = 10 \log [W_r \text{ (without lagging)}/ W_r \text{ (with lagging)}] \quad (46)$$

Two prediction methods are required—the first is for rigid outer covering materials, and the second is for limp materials. In the case of rigid materials, a pronounced resonance effect occurs between the duct walls and the outer covering. With the limp materials however, the variation in the separation of the duct and its outer covering causes a "spreading" of this resonance so that it is no longer well-defined, and the two cases must be treated separately.

Rigid Covering Materials. This category includes any outer covering (sheet metal, gypsum board, and so forth) that maintains a relatively constant distance between the duct and covering.

Data Required:

Perimeter of duct = $P_1 = 2(a + b)$
Perimeter of outer covering = $P_2 = 2(a + b + 4h)$
Mass/unit area of duct = M_1
Mass/unit area of outer covering = M_2
Cross-sectional area of absorbent material = $S = 2h(a [b + 2h)$

1. Calculate the low-frequency insertion loss:

$$IL(lf) = 20 \log \left(1 + \frac{M_2}{M_1}\frac{P_1}{P_2}\right) \text{ dB} \quad (47)$$

On a semilog graph, draw a horizontal line parallel to the frequency axis (see Figure 32 for an example of a composite *IL* curve), representing a constant *IL* equal to IL_{lf}.

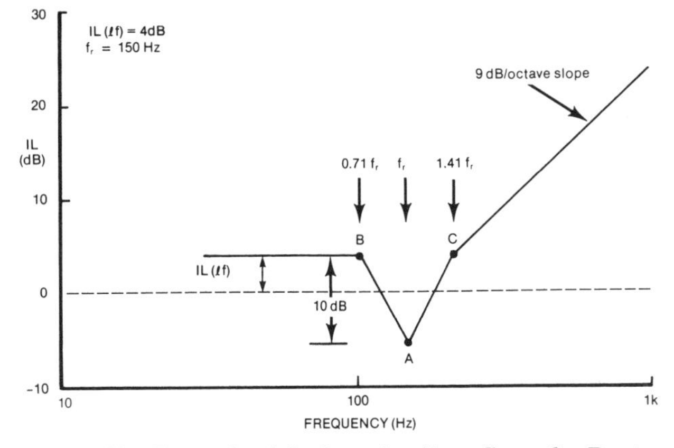

Fig. 32 Example of the Insertion Loss Curve for Duct Lagging with a Rigid Outer Covering

2. Calculate the resonance frequency as follows:

$$f_r = 156 [(P_2/P_2 + M_2/M_1) P_1/M_2S]^{0.5} \quad (48)$$

3. Mark point *A* at $IL_{lf} - 10$ dB and frequency f_r. Mark point *B* at 0.71 f_r and point *C*, at 1.41 f_r on the IL_{lf} line; join each of these to point *A* (see Figure 32). From point *C* draw a line to the right with a slope of +9 db/octave. The IL_{lf} line should be truncated, at its right end, in point *B*. The resulting curve, consisting of four straight-line segments, represents the *IL* of the duct lagging.

Limp Covering Materials. This category includes any material, such as sheet lead, sheet vinyl, and so forth, that does not retain its shape and sags, producing a variation in the separation of the duct and the outer covering. Data required are the same as those required for rigid covering materials.

a) Calculate IL_{lf} from Equation (47) and plot on a semilog graph as before.
b) Calculate f_r from Equation (48).
c) Mark a point on the IL_{lf} line at f_r, and from this draw a line to the right at +9 dB/octave. Truncate the right end of the IL_{lf} line at the point marked. This two-line segment curve represents the *IL* of the duct lagging.

Comments on the Method

The *IL* predictions are fairly reliable up to about 1 kHz for most ducts. Accurate predictions are not normally required above this frequency.

At low frequencies, some discrepancies can occur between the predicted and actual *IL* because of structural resonance effects in the duct and/or in the outer covering of the lagging, if the covering is rigid. The discrepancy should not exceed a few decibels. Duct lagging is not a particularly effective measure for reducing low-frequency noise (below about 100 Hz), unless the absorbent thickness is quite large and the outer covering is substantial. A more successful method of reducing duct-radiated noise is to use flat oval or, where space permits, circular ductwork, which has a higher transmission loss at low frequencies.

Rectangular versus Round or Flat Oval Ductwork

Compared with Table 27, Table 25 shows that round ductwork is much more effective in containing low-frequency noise than rectangular ductwork, regardless of whether the noise is flow-generated or due to acoustic excitation. In tight spaces, multiple round ducts in parallel can be used where space is not available for a single round duct of equivalent cross-sectional area. Flat oval ductwork falls between round and rectangular, and its breakout noise reduction depends on the area of flat surface.

Providing good airflow conditions at or near the fan is the most effective method of avoiding possible rumble conditions. Figure 21 shows various outlet conditions and their classifications with regard to possible rumble development (see Ebbing *et al.* 1978, Harold 1986).

Noise and turbulence inside a duct also cause the walls to vibrate and radiate noise to surrounding spaces. Rectangular ducts are more susceptible to this than are round ducts, and the problem is a function of duct size and air velocity. To prevent excessive noise radiation from duct walls, all fittings should be smooth and designed to avoid abrupt changes of direction or velocity. Whenever possible, medium and high-velocity ducts and terminal boxes should be located in noncritical areas (*e.g.*, above corridors).

RETURN AIR SYSTEMS

The fan return air system provides a sound path (through ductwork or through an unducted plenum) between the fan and the occupied spaces, typically with an opening in the mechanical equipment room. This condition can result in high sound levels in adjacent spaces due to the close proximity of the noise source and the low system attenuation between the mechanical equipment room and the adjacent spaces.

Sound propagates in the opposite direction to the airflow in a return air system, which results in a negligible effect on sound attenuation since the speed of sound (approximately 1100 ft/s) is much greater than typical return air velocities.

Noise in ducted return air systems is controlled by the fan intake sound power level. Unducted plenum return air systems include the sound power level of the intake and casing-radiated fan noise components. In certain installations, noise from other equipment located in the mechanical equipment room may also radiate through the wall opening and into adjacent spaces. Good design practice would keep the room return air system noise level approximately 5 dB below the room supply air system, so that it does not add significantly to the overall room sound pressure level.

Return Air System Calculations

Procedures for predicting the room sound pressure level in ducted return systems can make use of Equation (2). Unducted plenum return systems can use Equation (1), with adjustments made to the sound power level value to account for plenum losses. The composite transmission loss of the mechanical equipment room wall above the ceiling line and the loss associated with the ceiling plenum needs to be included. These values can be deducted from the sound power level to arrive at an adjusted sound power level at the ceiling return air opening. The composite transmission loss of the wall can be calculated by the following steps:

1. Determine the wall sound transmission loss from Table 24 or another source.
2. Define the opening as a percentage of the wall area above the ceiling line:

$$\frac{\text{Area of opening}}{\text{Wall area}} \times 100 \quad (49)$$

3. Calculate the composite transmission loss TL_c of the wall element with an opening using Equation (50):

$$TL_c = TL - 10 \log [1 - S_2/S_1 + (10^{TL/10}) S_2/S_1] \quad (50)$$

where

TL = transmission loss of the wall, by octave band
S_1 = area of the wall
S_2 = area of the opening

Alternately, the TL_c can be determined from Figure 33, based on the opening having a transmission loss of zero.

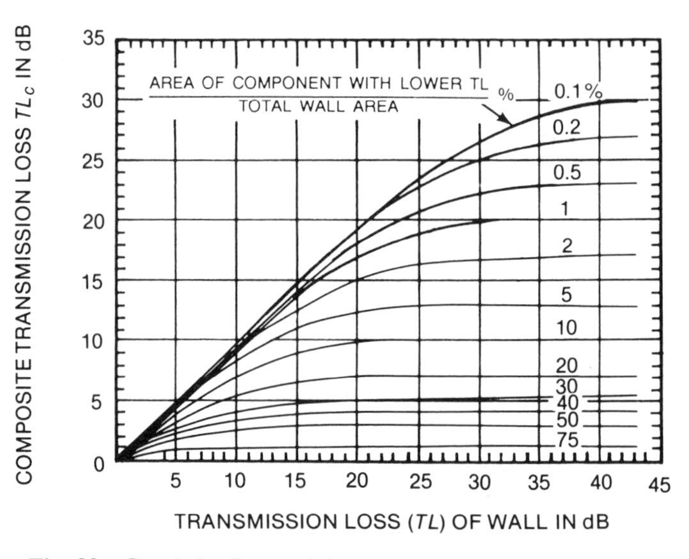

Fig. 33 Graph for Determining Effective Transmission Loss of a Composite Wall with Two Components Having Different Areas and Transmission Losses

Since an explicit equation for ceiling plenum losses does not exist, Equation (5) can be used as an approximation.

Noise Control Measures for Return Air Systems

When adjacent space noise levels are too high, noise control measures must be provided. Before specifying modifications, the controlling sound paths from adjacent spaces need to be identified. Ducted return air systems can be modified using methods applicable to ducted supply systems. Good planning is required with unducted plenum systems so that the mechanical equipment room is located away from noise-sensitive spaces.

Unducted plenum systems may still require several modifications to satisfy design goals. Prefabricated sound attenuators can be effective when installed at the mechanical equipment room wall opening or at the suction side of the fan. Improvements to the ceiling transmission loss are often limited by typical ceiling penetrations and lighting fixtures. Modifications to the mechanical equipment room wall can also be effective for some constructions. Adding acoustical absorption in the mechanical equipment room reduces the buildup of reverberant sound energy in this space; however, this typically reduces noise only slightly in the low-frequency range in nearby areas.

ROOFTOP CURB-MOUNTED AIR HANDLERS

Rooftop air handlers have unique noise control requirements because these units are often integrated into a lightweight roof deck construction. Large roof openings, within the periphery of the mounting curb, are required for supply and return duct connections. These ducts run directly from the noise-generating equipment to the building interior. Often there is insufficient space between the equipment location and the closest occupied spaces to apply adequate noise control treatment.

Rooftop units should be located above nonacoustically critical spaces to provide a buffer between the unit and the nearest occupied spaces. This measure can reduce the magnitude of noise control treatment necessary to achieve an acoustically acceptable installation. Figure 34 illustrates the four common airborne, structure-borne, and duct-borne paths by which noise can enter occupied spaces with a typical rooftop installation.

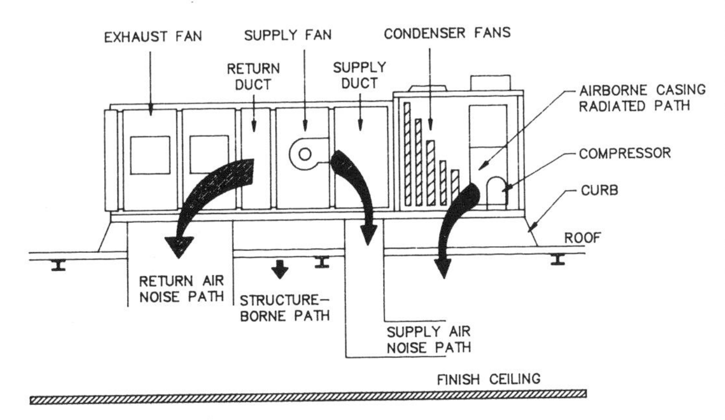

Fig. 34 Noise Paths in Typical Rooftop Installations

The airborne path is due to casing-radiated noise from the equipment components passing through the air handler enclosure and roof structure to the spaces below. Mounting the rooftop unit on a raised steel frame (dunnage) above the roof structure is recommended with this arrangement. The duct openings must no longer be located below the unit; they can be located as far away as needed for adequate noise control. The airborne path can also be controlled by increasing the transmission loss of the architectural construction either at the roof or the interior finish ceiling. In some installations, it may be possible to increase the transmission loss of the air handler enclosure or to install additional treatment in the steel framework or intervening airspace.

Structure-borne noise and vibration can be limited by using vibration isolators. Special curb mounting bases are available to support and provide vibration isolation for rooftop units. Typically, these bases are engineered to provide greater vibration isolation than would be achieved with standard vibration isolators.

The level of the return air noise path can be reduced through good duct design practices, lining ducts with acoustical insulation, and using sound attenuators. Particular attention should be paid to designing ducts to take advantage of end reflection loss factors, and limiting regenerated noise.

The supply air noise path consists of two major components—the noise delivered via the air distribution system, and the duct breakout noise. Supply duct airborne noise can be reduced in level by limiting regenerated noise through good duct design practices, lining ducts with acoustical insulation, and using sound attenuators. Experience has indicated that duct breakout noise below 250 Hz is often the major acoustical limitation for many rooftop installations. Excessive low-frequency noise results in duct rumble, which is difficult to attenuate. Often the regenerated noise, due to elbows and duct fittings close to the air handler, is as much as 10 dB lower than the fan sound power level.

Certain duct configurations can be used to contain the fan noise leaving the discharge side of the rooftop unit, thus reducing the level of duct breakout noise. Table 30 illustrates 12 possible discharge duct configurations with the associated low-frequency noise reduction potential. These values are adjusted relative to the noise reduction provided by a 90° unlined rectangular elbow and 12 ft of horizontal supply duct made of 22-gage steel, terminating in another room.

Table 30 shows a considerable spread in the potential low-frequency noise reduction for various discharge duct conditions. In general, a plenum drop and multiple supply ducts using round ductwork provides a substantial improvement in the noise reduction over conventional rectangular ductwork with external lagging. The greatest system attenuation at 63 Hz is provided by configuration 11, while the greatest attenuation at 125 Hz is provided by configuration 12. Large radius elbow round ductwork configurations, similar to those shown in configuration 12, may be difficult to install because of typical ceiling plenum space limitations. When this occurs, the arrangements shown in configurations 10 and 11 may be used. Turning vanes are recommended because they reduce the overall duct static pressure, thus lowering the sound power level produced by the fan. One disadvantage with round ductwork is that there is little inherent low-frequency sound absorption; thus, it may be necessary to provide sound attenuators or acoustical insulation to limit noise levels downstream of the initial duct run exiting the air handler unit.

MECHANICAL EQUIPMENT ROOM NOISE ISOLATION

Mechanical equipment is inherently noisy. Pumps, chillers, cooling towers, boilers, and fans create noise that is difficult to contain. Noise-sensitive spaces must be isolated from these sources. One of the best precautions is to locate mechanical spaces away from acoustically critical spaces. Buffer zones (storage rooms, corridors, or less noise-sensitive spaces) can be placed between mechanical equipment rooms and rooms requiring quiet. Noise transmission through roofs and exterior walls is usually less of a problem, making corner rooms and top-floor spaces reasonably good for housing mechanical equipment.

Construction enclosing a mechanical equipment room should be poured concrete or masonry units with enough surface weight to provide adequate sound transmission loss capability. Walls must be airtight and caulked at the edges to prevent sound leaks. Floors and ceilings should be concrete slabs, except for the ceiling or roof deck above a top-floor mechanical space.

Penetrations of the mechanical equipment room enclosure create potential paths for noise to escape into adjacent spaces. Therefore, wherever ducts, pipes, conduits, and the like penetrate the walls, floor, or ceiling of a mechanical equipment room, it is necessary to acoustically treat the opening for adequate noise control. A 1/2- to 5/8-in. clear space should be left all around the penetrating element and filled with fibrous material for the full depth of the penetration. Both sides of the penetration should be sealed airtight with a nonhardening resilient sealant (Figure 35).

Doors into mechanical equipment rooms are frequently the weak link in the enclosure. Where noise control is important, they should be as heavy as possible, gasketed around the perimeter, have no grilles or other openings, and be self-closing. If such doors lead to sensitive spaces, two doors separated by a 3- to 10-ft corridor may be necessary. Mechanical room doors should open out, not in, so that the negative pressure in the room will keep the door sealed against the jamb instead of holding it against the latch.

Penetration of Walls by Ducts

Ducts passing through the mechanical equipment room enclosure pose an additional problem. Noise can be transmitted to either side of the wall via the duct walls. Airborne noise in the mechanical room can be transmitted into the duct (breakin) and enter an adjacent space by reradiating (breakout) from the duct walls, even if the duct contains no grilles, registers, diffusers, or other openings.

Noise levels in ducts close to fans are usually high. Noise can come not only from the fan but also from pulsating duct walls,

Table 30 Duct Breakout Insertion Loss—Potential Low Frequency Improvement Over Bare Duct Elbow

Discharge Duct Configuration, 12 ft of Horizontal Supply Duct	Duct Breakout Transmission Loss in dB at Low Frequencies, Hz				
	63	125	250	Side View	End View
Rectangular duct: no turning vanes (reference).	0.0	0.0	0.0	AIR FLOW	22 GAGE
Rectangular duct: one dimensional turning vanes.	0.0	0.5	0.5	TURNING VANES	
Rectangular duct: two dimensional turning vanes.	0.0	0.5	1.0	TURNING VANES	
Rectangular duct: wrapped with foam and lead.	4.0	3.0	4.5	SEE END VIEW	FOAM INSUL. W/ 2 LAYERS LEAD
Rectangular duct: wrapped with glass fiber, one layer of 5/8 in. gypsum board.	4.0	7.0	5.5	SEE END VIEW	GLASS FIBER PRESSED FLAT AGAINST DUCT.
Rectangular duct: wrapped with glass fiber, two layers of 5/8 in. gypsum board.	7.5	8.5	9.0	SEE END VIEW	GYPSUM BOARD SCREWED TIGHT
Rectangular plenum drop: three parallel rectangular supply ducts.	1.0	2.0	3.5	12 GAGE	22 GAGE
Rectangular plenum drop: one round supply duct.	8.0	9.5	6.0	12 GAGE	18 GAGE
Rectangular plenum drop: three parallel round supply ducts.	10.5	13.5	8.0	12 GAGE	24 GAGE
Rectangular to multiple drop: round mitered with turning vanes, three parallel round supply ducts.	17.5	11.5	13.0	24 GAGE	14 GAGE
Rectangular to multiple drop: round mitered elbows with turning vanes, three parallel round lined double wall, 22 in. OD supply ducts.	18.0	13.0	16.0	24 GAGE	14 GAGE
Round drop: radius elbow, single, 37 in. diameter supply.	15.0	16.5	9.5	14 GAGE 18 GAGE	

Source: Robert Harold, The Trane Co.

excessive air turbulence, and air buffeting caused by tight or restricted airflow entrance of exit configurations. Controlling low-frequency noise propagation from these sources can be difficult. Thus, duct layout and airflow conditions should be carefully considered when there are nearby noise-sensitive areas.

Mechanical Chases

Mechanical chases and shafts should be treated the same way as mechanical equipment rooms, especially if they contain noise-producing equipment. The shaft should be closed at the mechanical equipment room, and shaft walls should have a surface weight sufficient to reduce sound transmission to noise-sensitive areas to acceptable levels. Chases should not be allowed to become speaking tubes between spaces requiring different acoustical environments. Any vibrating piping, duct conduits, or equipment should be isolated so that vibration is not transmitted to the shaft walls and the general building construction.

If mechanical equipment rooms are to be used as supply or return plenums, all openings into the equipment room plenum space may require noise control treatment. This is especially true if the ceiling space just outside the equipment room is used as a return air plenum and the ceiling is acoustical tile. Most acoustical tile is almost transparent acoustically, particularly at low frequencies, so that sound passes through it practically unimpeded.

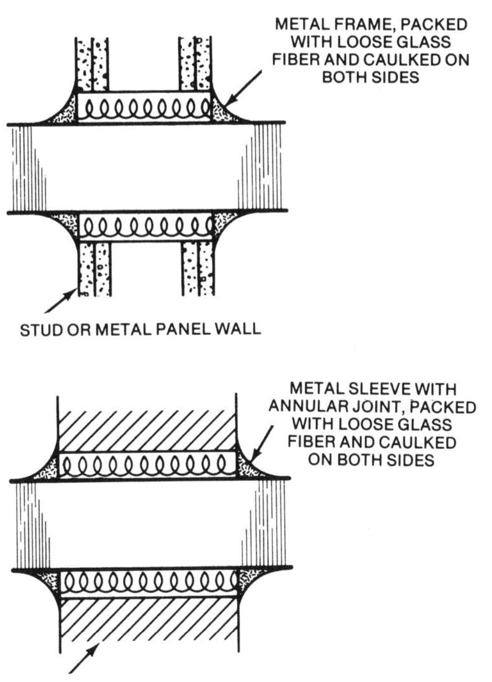

Fig. 35 Typical Duct, Conduit, and Pipe Wall Penetrations

Special Wall Construction

When mechanical equipment rooms must be placed adjacent to offices, conference rooms, or other noise-sensitive areas, the building construction around the equipment space should reduce the noise enough to satisfy the acoustical requirements of the nearby spaces. Depending on the degree of noise sensitivity, it may be necessary to consider one of the following treatments.

For walls, an additional skin that is totally separate and resiliently isolated from the heavy and dense inner wall may be provided outside the mechanical equipment room wall. It is necessary to provide an air cavity, at least 2 to 4 in. in depth, between the inner wall and the outer resilient skin. This outer skin may be similar to metal studs with two layers of 5/8 in. gypsum board or a surface with an equivalent mass attached and with all joints staggered and sealed airtight. Two to 4 in. of fiberglass insulation may be placed between studs. Optimum sound reduction can be achieved only if the outer skin resilient isolation is complete and cannot transmit structure-borne vibration or airborne sound from the mechanical equipment and the mechanical equipment room enclosure.

Other noisier situations may require the use of double-wall construction consisting of two heavy, dense masonry walls separated by an air cavity. Where floating floor construction is used, it is advantageous to support the inner wall on the floating slab.

Floating Floors

Typical floating floor construction can be used to further reduce noise between the mechanical room and noise-sensitive areas above or below it. It is composed of two reinforced concrete slabs—the floating or isolated wearing slab, and the structural floor slab. The floating floor slab is designed with a minimum thickness of 4 in. and with a 1- to 4-in. separation between the structural and the floating slabs. A 2-in. separation is most common. The connection between the slabs is made by supporting the floating slab on permanent load-bearing resilient material. The edges of the floating slab are isolated from the structure with resilient material, and the joint is resiliently caulked.

The isolation material supporting the floating slab must be resilient, have permanent dynamic and static properties, and safely support both the floating slab and the imposed live load for the life of the building. Isolation material for the floating slab should (1) have a natural frequency in the range of 7 to 15 Hz under the load conditions that exist in the building, (2) be tested for load versus deflection and for natural frequency with known aging properties and a history of applications in similar floating floor installations, (3) be tested in service for impact insulation class (IIC) and sound transmission class (STC) by an independent testing laboratory, and (4) be designed for the loads imposed on the isolator in service and during construction.

Following are two common methods for constructing floating floors:

1. Individual pads are spaced on 1- to 2-ft centers each way and covered with plywood or sheet metal. Low-density absorption material can be placed between the pads to reduce the tunneling effect. Reinforced concrete is placed directly on the waterproofed panels and cured; then the equipment is set (Figure 36A).
2. Cast-in-place canisters are placed on 2- to 4-ft centers each way on the structural floor. Reinforced concrete is poured over the canisters; after curing, the entire slab is raised into operating position, the canister access holes are grouted, and the equipment is set (Figure 36B).

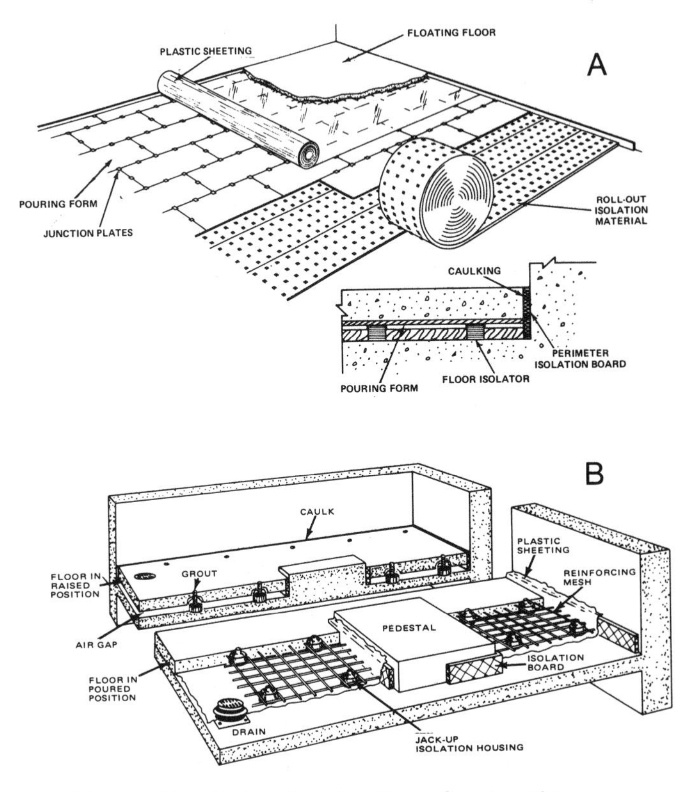

Fig. 36 Two Typical Floating Floor Constructions

With either system, mechanical equipment with vibration isolators can be placed on the floating slab unless the equipment operates at shaft rotation speeds of 0.7 to 1.4 times the resonance frequency of the floating slab system. In that case, the equipment should be supported with isolators on structural slab extensions that penetrate through the floating slab. This type of installation requires careful attention to detail to avoid acoustical flanking due to improper direct contact between the floating slab and structural penetrations. Floating floors primarily control airborne sound transmission; they are not intended to be used in place of vibration isolators.

The actual resonant frequency of the floated slab is determined by both the stiffness of the resilient elements used to support the floated slab and the stiffness of the airspace between the structural and floated slabs. With a 2-in. airspace and a 4-in. thick floating slab, the resonant frequency can be expected to be near 18 Hz. Thus, equipment operating between about 12.5 and 25 rps (750 and 1500 rpm) should be supported on structural slab extensions, unless the equipment is rated at about 5 hp or less.

Resilient material inserted between the floating and structural slabs results in an extremely resilient upper concrete slab. This slab must be designed to operate within the safe bending limits of the concrete. While floating floor tests have shown transmission loss ratings exceeding STC 75, these can only be realized in field installations where all flanking sound transmission paths or short circuits have been minimized. Table 31 shows improvement of sound transmission loss by floating floor slabs, with or without flanking noise control.

Table 31 Sound Transmission Loss through Typical Floor Construction, dB

	Frequency, Hz					
	125	250	500	1000	2000	4000
6 in. concrete structural floor	35	36	40	46	53	58
6 in. floor slab, 2 in. airspace, 4 in. floating floor, equipment noise impinging on walls and ceiling	44	47	58	68	75	87
6 in. floor slab, 2 in. airspace, 4 in. floating floor, equipment noise contained to avoid flanking	52	64	78	90	96	73

Note: A change in slab thickness, concrete mass, isolator stiffness, or flanking control will result in corresponding changes in the acoustical performance of the system.

Enclosed Air Cavity

If acoustically critical spaces are located over a mechanical equipment room, it may be necessary (1) to install a floating floor in the space above it, or (2) to form a totally enclosed air cavity between the spaces by installing a dense plaster or gypsum board ceiling resiliently suspended in the equipment room. Ideally, such a ceiling can be positioned between building beams, leaving the beam bottoms available for hanging equipment, piping, ducts, and so forth without penetrating the ceiling with hanger rods or other devices. In the case of bar joist-type construction, it may be easier to resiliently attach the ceiling to the underside of the joists and support all equipment, ducts, and pipes from the mechanical room floor or from wall supports, which are properly vibration isolated.

Transmission Loss through Ceiling Construction

When terminal units, fan-coil units, air-handling units, ductwork, or openings to mechanical equipment rooms are located in a ceiling plenum above an occupied room, noise transmission through the ceiling construction can be high enough to cause excessive noise levels in that room. Since there are no standard tests relating to the direct transmission of sound through ceiling construction, ceiling product manufacturers do not regularly publish data that can be used in calculations; data is usually for room-to-room sound transmission loss through a common ceiling plenum. Acoustical ceilings are rarely homogeneous surfaces; they usually have light fixtures, diffusers and grilles, speakers, and so forth, which reduce the transmission loss of the ceiling and must be considered.

To estimate the noise level in a room from sound transmission through the ceiling, the sound power level in the ceiling plenum must be reduced by the insertion loss of the ceiling material before applying Equation (1). In the absence of a recognized test standard, the values in Table 24 may be used as an estimate. This calculation should be repeated for each octave band of interest, and the results should be compared to the room design criterion.

If increased insertion loss is required, consider the following options:

1. Using a different type of ceiling with higher insertion loss, *i.e.*, gypsum board, high transmission loss acoustical tile, and so forth.
2. Increasing the insertion loss of the ceiling by applying gypsum board, sheet lead, or other secondary impervious barrier material on top of the ceiling.
3. Treating the source by lagging.

FUME HOOD DUCT SYSTEM DESIGN

Fume hood exhaust systems are often the major noise source in a laboratory and, as such, require noise control (Sessler and Hoover 1983). The exhaust system may consist of individual exhaust fans ducted to separate fume hoods, or a central exhaust fan connected through a distribution system to a large number of hoods. In either case, the noise levels produced in the laboratory space can be estimated using procedures described in this section.

To minimize static pressure loss and blower power consumption within a duct system, fume hood ducts should be of a sufficient size to permit the rated flow of air through the duct at a velocity no greater than 2000 ft/min.

Duct velocities in excess of 2000 ft/min should also be avoided for acoustical reasons and to conserve energy, unless the resulting increases in sound levels, static pressure, and blower power requirements are deemed acceptable for the spaces served.

Recommended noise level design criteria for laboratory spaces using fume hoods are as follows:

Laboratory Type	Noise Criteria (NC) Room Criteria (RC)
Testing or research with little requirement for speech or study	45 to 55
Research, with some communication, telephone, and study	40 to 50
Teaching	35 to 45

Noise control measures that have been successfully applied to a variety of fume hood systems consist of the following:

- Use backward inclined or forward curved, rather than radial blade fans where conditions permit.
- Select fan to operate at a low tip speed and maximum efficiency.
- Use prefabricated duct attenuators or sections of lined ductwork where conditions permit.

All potential noise control measures should be carefully evaluated for compliance with applicable codes, safety requirements, and corrosion resistance requirements of the specific system. In addition, vibration isolation for fume hood exhaust fans is generally required. (For centrifugal fans, see Table 35.) However, for some laboratory facilities, particularly those having electron microscopes, vibration control can be critical, and a vibration specialist should be consulted.

SOUND CONTROL FOR OUTDOOR EQUIPMENT AND OUTDOOR NOISE CONTROL BARRIERS

Outdoor mechanical equipment should be carefully selected, installed, and maintained so that the radiated sound is not annoying to people outdoors or to nearby building occupants and so that local noise codes are not violated. The difference in sound levels at the listening location with the equipment operating and the ambient conditions with the equipment not operating will determine if the noise from the equipment is annoying. Equipment with strong tonal components will more likely evoke complaints than equipment with a broadband noise spectrum.

Outdoor ambient noise is a complex phenomenon. It is primarily a function of traffic and fluctuates with time; it depends on such factors as the proximity of various types of roads, the traffic involved, and shielding by buildings. A complete description of the outdoor ambient noise in a given community requires statistical evaluation of noise levels occurring at different locations. While approximations for equipment selection purposes must still be made at specific locations, techniques are available to assess or predict such noise levels (Harris 1979).

DESIGN PROCEDURES FOR ACOUSTICAL BARRIERS

If the equipment sound power level spectrum and ambient sound pressure level spectrum are known, the contribution of the equipment to the sound level at any location can be estimated by analyzing the sound transmission paths involved. Outdoors, when there are no intervening barriers, the principal factors are reflections from buildings near the equipment and the distance to the specific location [see Figure 37 and Equation (51)].

The following equation may be used to estimate the decibel differences between the sound power level of the equipment and the sound pressure level at any distance from it, at any frequency:

$$L_p = L_w + 10 \log Q - 20 \log d - C \qquad (51)$$

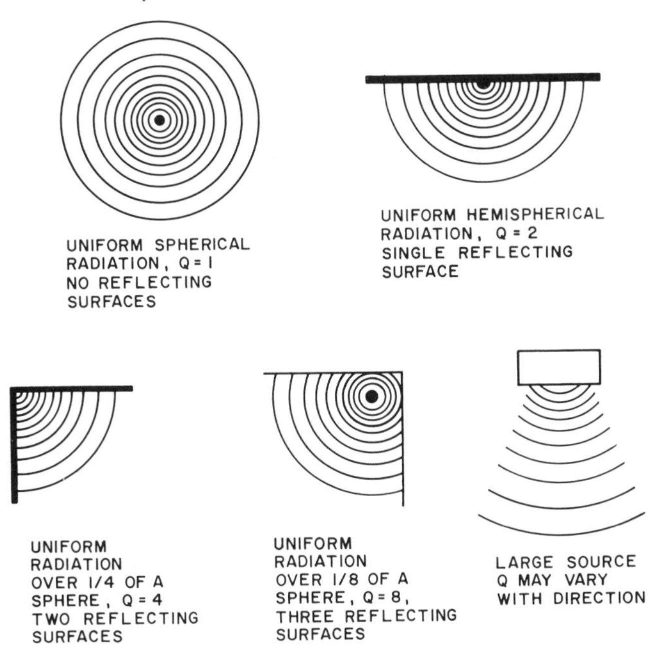

Fig. 37 Directivity Factors for Various Radiation Patterns

where

L_p = sound pressure level at a distance d from the source, ft
L_w = sound power level of the source, re 10-12 W
C = correction factor = 0 dB when d is in ft
Q = determined from Figure 37

Note: Equation (51) does not apply where d is less than twice the maximum dimension of the equipment; results may be low by up to 5 dB where d is between two and five times the maximum equipment dimension.

Acoustical Barriers

A barrier is a solid structure that intercepts the direct sound path from a source to a receiver. It reduces the sound pressure level within its shadow zone. Figure 38 illustrates the geometrical aspects of an outdoor barrier where no extraneous surfaces reflect sound into the protected area. Here the barrier is treated as an intentionally constructed noise control structure.

The insertion loss provided by an outdoor barrier is a function of the path length difference between the actual path traveled and the line-of-sight direct path distance. In Figure 38, the line-of-sight distance is D, while the defracted path traveled has a distance A + B. The insertion loss may be estimated as:

$$IL = 10 \log f(A + B - D) - 14 \qquad (52)$$

where

f = frequency, Hz
$A + B - D$ = defracted minus line-of-sight distance, ft

Equation (52) is valid when $f(A + B - D) \geqslant 250$ or dB > 10.

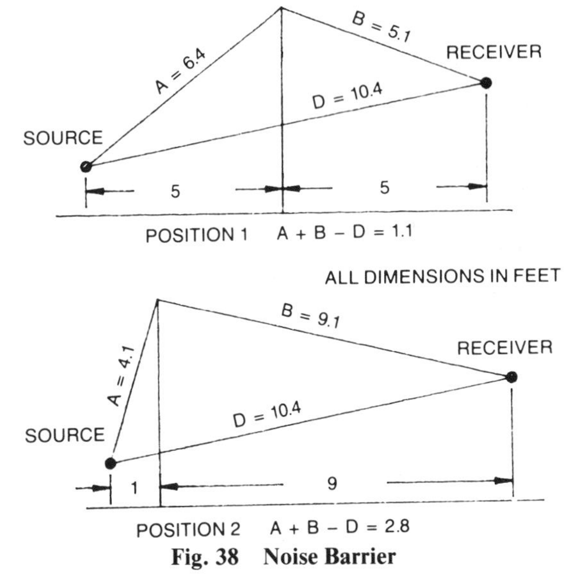

Fig. 38 Noise Barrier

Table 32 gives the insertion loss of an outdoor ideal solid barrier when (1) no surfaces reflect sound into the shadow zone, and (2) the sound transmission loss of the barrier wall or structure is at least 10 dB greater at all frequencies than the insertion loss expected of the barrier. The limiting value of about 24 dB is caused by scattering and refraction of sound into the shadow zone. For large distances outdoors, this scattering and bending of sound waves into the shadow zone reduces the effectiveness of the barrier. Quantitative data on the amount of reduction is unavailable, but the calculated insertion loss may be reduced by about 10% for each 1000-ft distance between the source and the receiver. For a conservative estimate, the source location should be taken as the topmost part of the radiating source, and the receiver location should be taken as the topmost location of a sound receiver (such as the top of the second-floor windows in a two-floor house or at 5 ft for standing human receptors).

Table 32 Insertion Loss of Ideal Solid Outdoor Barrier Wall
(See text for conditions)

Part Length Difference, ft	Insertion Loss, dB Octave Frequency Band, Hz								
	31	63	125	250	500	1000	2000	4000	8000
0.01	5	5	5	5	5	6	7	8	9
0.02	5	5	5	5	5	6	8	9	10
0.05	5	5	5	5	6	7	9	10	12
0.1	5	5	5	6	7	9	11	13	16
0.2	5	5	6	8	9	11	13	16	19
0.5	6	7	9	10	12	15	18	20	22
1	7	8	10	12	14	17	20	22	23
2	8	10	12	14	17	20	22	23	24
5	10	12	14	17	20	22	23	24	24
10	12	15	17	20	22	23	24	24	24
20	15	18	20	22	23	24	24	24	24
50	18	20	23	24	24	24	24	24	24

Other Reflecting Surfaces. There should be no other surfaces that can give specular reflection of sound around the ends or over the top of the barrier into the protected region (the shadow zone). Figure 39 shows examples of reflecting surfaces that can reduce the effectiveness of a barrier wall. These geometries should be avoided.

Width of Barrier. Each end of the barrier should extend horizontally beyond the line of sight from the outer edge of the source to the outer edge of the receiver position by a distance of at least three times the value of h used in the calculation. Near the end of the wall, the barrier effectiveness is reduced because some sound is diffracted over the top of the wall, some sound is diffracted around the end of the wall, and some sound is reflected or scattered from various nonflat surfaces along the ground near the end of the barrier. For critical problems, this degradation of the barrier near its end should be taken into account.

Figure 40 suggests a simplified procedure that gives the approximate insertion loss near the end of the barrier. Line A is drawn from the outer extremity of the sound source through the end of the wall. Line D is drawn from the source through a point at distance 3 h from the end of the wall, where h is the height used in the calculation of the barrier. The angle between lines A and D is divided into three equal parts, and lines B and C are drawn at the intermediate angles.

On line B at the receiver distance, the insertion loss will be approximately 50% of the full calculated value, and on line C at the receiver distance, the insertion loss will be approximately 80% of the full calculated value. Along line A there is no insertion loss, and on line D at the receiver distance, the full calculated insertion loss is achieved (within 1 or 2 dB). For other positions of interest near the receiver, interpolate between the values for these lines. This procedure applies to both ends of the wall.

Atmospheric Effects. For wind speeds in the range of 10 to 75 mph along the direction of the sound path from source to receiver and for distances over about 1000 ft between source and receiver, the wind bends the sound waves down over the top of the barrier. Under these conditions, the barrier is ineffective.

Reflection from a Barrier. A large flat reflecting surface, such as a barrier wall, may reflect more sound in the opposite direction than there would have been with no wall present. If there is no special focusing effect, the wall may produce only about 2 or 3 dB higher levels at most in the direction of the reflected sound.

VIBRATION ISOLATION AND CONTROL

Vibration and vibration-induced noise, major sources of occupant complaint, are increasing in modern buildings. Light

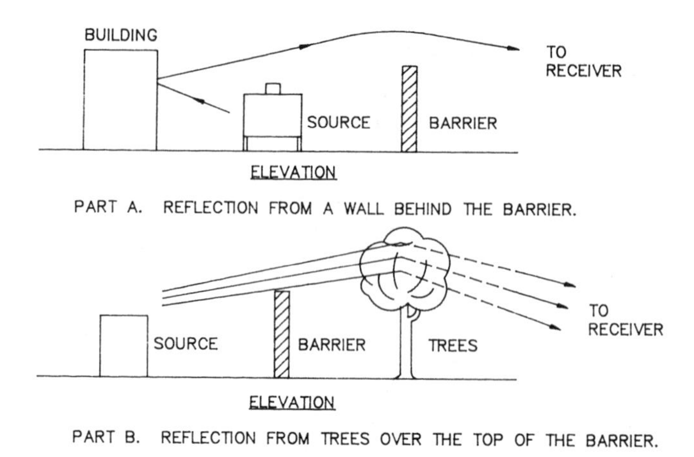

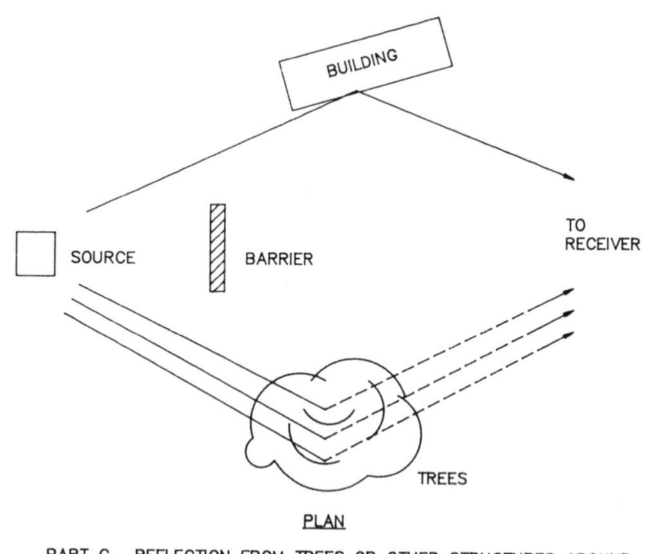

Fig. 39 Examples of Surfaces that Can Reflect Sound Around or Over a Barrier Wall

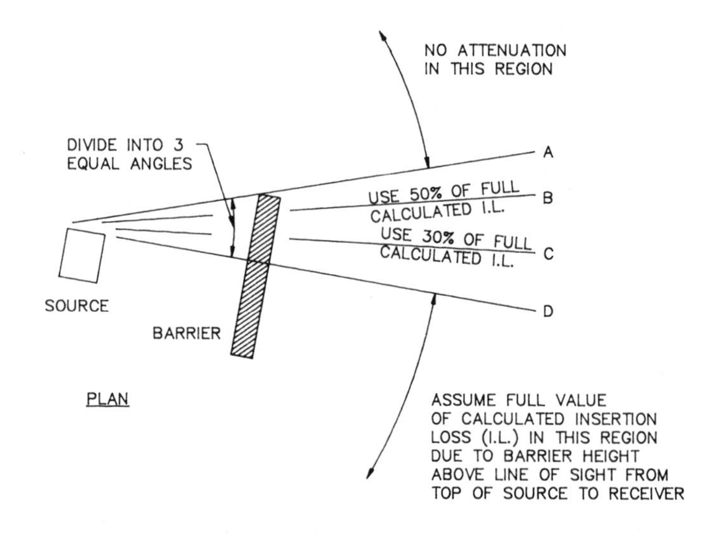

Fig. 40 Plan View Showing Procedure for Estimating Approximate Edge Effects at End of Outdoor Barrier Wall

construction and equipment located in penthouses or intermediate level mechanical rooms increase structure-borne vibration and noise transmission. Not only is the physical vibration disturbing, but the regenerated noise from structural movement can be heard in remote sections of the building.

Table 34 shows appropriate isolation systems for most mechanical equipment in actual buildings. The references cover special cases, describe the isolation system in detail, and identify possible problem areas. Piping and duct isolation and seismic protection are covered in separate sections.

Vibration can be isolated or reduced to a fraction of the original force with resilient mounts between the equipment and the supporting structure. The following sections present basic information to properly select and specify vibration isolators, and to analyze and correct field problems. Harris and Crede (1976), Den Hartog (1956), and Chapter 7 of the 1989 ASHRAE *Handbook—Fundamentals* provide more detailed information.

To determine the excessive forces that must be isolated or that adversely affect the performance or life of the equipment, criteria should be established for equipment vibration. Figures 41 and 42 show the relation between equipment vibration levels and vibration isolators that have a fixed efficiency. In this case, the magnitude of transmission to the building is a function of the magnitude of the vibration force.

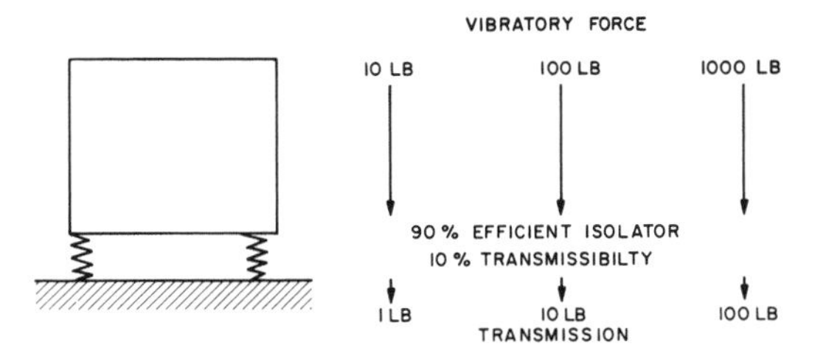

Fig. 41 Transmission to Structure Varies as Function of Magnitude of Vibratory Forces

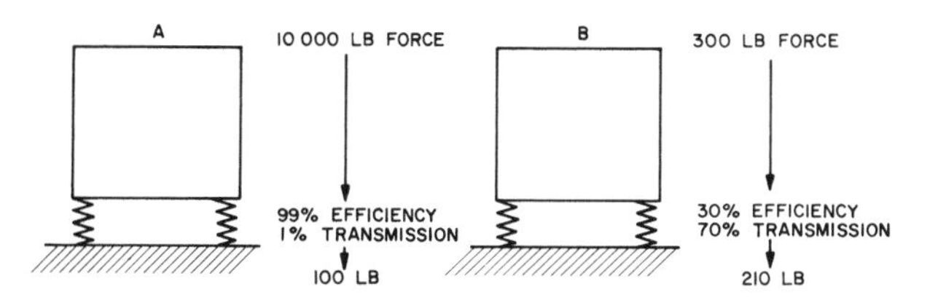

Fig. 42 Interrelationship of Equipment Vibration, Isolation Efficiency, and Transmission

Theoretically, an isolation system could be selected to isolate forces of extreme magnitude; however, isolators should not be used to mask a condition that should be corrected before it damages the equipment and its operation. If vibration is transmitted, isolators should indicate a faulty operating condition in need of correction.

Ideally, vibration criteria should (1) measure rotor unbalance as a function of type, size, mass, and stiffness of equipment; (2) consider the vibration generated by system components such as bearings and drives, as well as installation factors such as alignment, and (3) be verifiable by field measurement. Figure 43 presents some commonly used criteria; while they do not meet all the abovementioned requirements, they are generally satisfactory. A simpler approach is to use the criteria listed in Table 33, which have been developed by individuals and firms experienced in vibration testing of HVAC equipment. Table 33 shows the maximum allowable vibration levels for steady-state movement taken on the bearing or machine structure if it is sufficiently rigid. These criteria can be met by properly operating equipment; they will determine reasonable vibration levels to be isolated and make the equipment acceptable. These values represent levels that allow for misalignment, drive eccentricities, belt vibration, and similar factors affecting the overall vibration level and can be maintained throughout the life of the equipment.

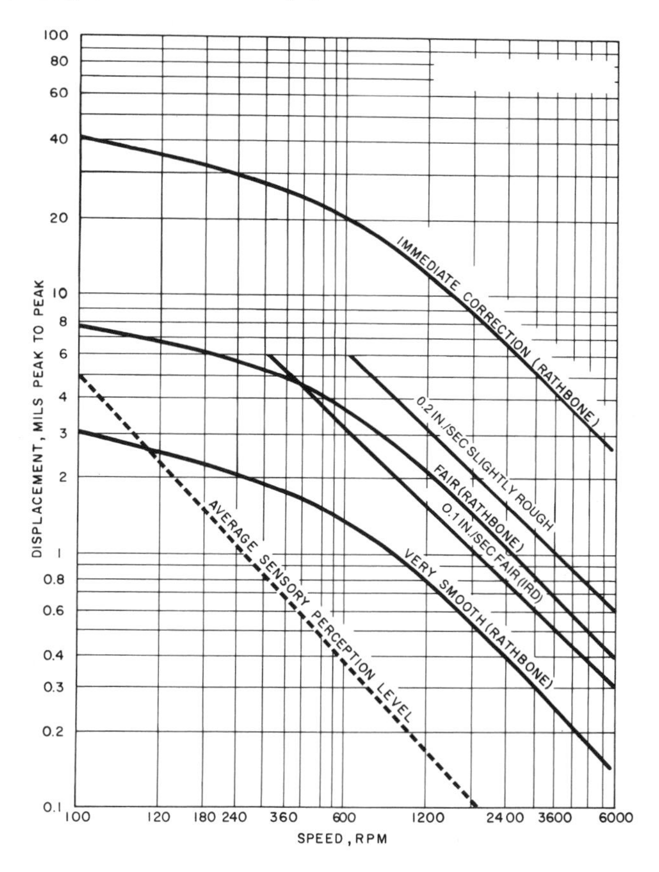

Fig. 43 Vibration Criteria

Table 33 Equipment Vibration Criteria

Equipment	Maximum Allowable Vibration Peak-to-Peak Displacement, mil (0.001 in.)
Pumps	
1800 rpm	2
3600 rpm	1
Centrifugal compressors	1
Fans (vent sets, centrifugal, axial)	
Under 600 rpm	4
600 to 1000 rpm	3
1000 to 2000 rpm	2
Over 2000 rpm	1

INSTRUCTIONS TO THE DESIGNER/SPECIFIER

Vibration isolators must be selected to compensate for floor deflection. Longer spans also allow the structure to be more flexible, permitting the building to be more easily set into motion. By using Table 34, building spans, equipment operating speeds, equipment horsepower, damping, and other factors are considered.

By specifying isolator deflection rather than isolation efficiency, transmissibility, or other theoretical parameters, a designer can compensate for floor deflection and building resonances by selecting isolators that provide minimum vibration transmission and that have more deflection than the supporting floor.

Table 34 Vibration Isolator Selection Guide

		Equipment Location (See Notes 3 and 11)														
		Grade Supported Slab			20 ft Floor Span			30 ft Floor Span			40 ft Floor Span			50 ft Floor Span		
Equipment Type	Notes	Base Type	Isolator Type	Min. Defl. in.	Base Type	Isolator Type	Min. Defl. in.	Base Type	Isolator Type	Min. Defl. in.	Base Type	Isolator Type	Min. Defl. in.	Base Type	Isolator Type	Min. Defl. in.
Refrigeration Machines	12,13															
Reciprocating compressors	2,3,13	C	3	0.75	C	3	0.75	C	3	1.50	C	3	1.50	C	3	2.50
Reciprocating condensing units and chilling units	2,3	A	2	0.25	A	4	0.75	A	4	1.50	A	4	2.50	A	4	2.50
Hermetic centrifugal chillers	2,3,4,12	A	1	0.25	A	4	0.75	A	4	1.50	A	4	1.50	A	4	1.50
Open centrifugal chillers	2,3,12	C	1	0.25	C	4	0.75	C	4	1.50	C	4	1.50	C	4	2.50
Absorption chillers	—	A	1	0.25	A	4	0.75	A	4	0.75	A	4	1.50	A	4	1.50
Air Compressors																
Tank mounted	3,15	A	3	0.75	A	3	0.75	A	3	1.50	A	3	2.50	A	3	2.50
Base mounted																
Up to 500 rpm	8,13,14,15	C	3	0.75	C	3	0.75	C	3	1.50	C	3	1.50	C	3	2.50
501 rpm and over	13,14,15	C	3	0.75	C	3	0.75	C	3	1.50	C	3	1.50	C	3	2.50
Pumps																
Close coupled, to 7½ hp	16	B/C	2	0.25	C	3	0.75	C	3	0.75	C	3	0.75	C	3	0.75
Close coupled, 10 hp and over																
Flexible coupled, to 40 hp	16	C	3	0.75	C	3	0.75	C	3	1.50	C	3	1.50	C	3	1.50
Flexible coupled, 50 to 125 hp	10,16	C	3	0.75	C	3	0.75	C	3	1.50	C	3	2.50	C	3	2.50
Flexible coupled, 150 hp and over	10,16															
Packaged Rooftop Air-Conditioning Units	5,6,8,17	(Not Applicable)			D	3	0.75	A/B	3	1.50	A/B	3	2.50	A/B	3	3.50
Ducts	7															
Piping	7															
Cooling Towers and Closed Circuit Coolers	5,18															
Up to 300 rpm	8															
301 to 500 rpm	—	A	1,2	0.25	A	4	2.50	A	4	2.50	A	4	2.50	A	4	3.50
501 rpm and over	—	A	1,2	0.25	A	4	0.75	A	4	1.50	A	4	1.50	A	4	2.50
Fans and Air-Handling Equipment	19															
Axial, Tubular, and Fan Heads	4,9															
Up to 22 in. wheel dia.	9	A/B	2	0.25	A/B	3	0.75	A/B	3	0.75	A/C	3	0.75	A/C	3	1.50
24 in. wheel dia. and over	9															
Up to 50 hp	9															
Up to 300 rpm	8															
301 to 500 rpm	8	B/C	3	0.75	C	3	1.50	C	3	2.50	C	3	2.50	C	3	2.50
501 rpm and over		B/C	3	0.75	C	3	1.50	C	3	1.50	C	3	1.50	C	3	2.50
Centrifugal Fans and Vent Sets	4,9,19															
Up to 22 in. wheel dia.	9	A/B	2	0.25	A/B	3	0.75	A/B	3	0.75	A/C	3	0.75	A/C	3	0.75
24 in. wheel dia. and over	9															
Up to 50 hp	9															
Up to 300 rpm	8															
301 to 500 rpm	8	B	3	1.50	B	3	1.50	B	3	1.50	B	3	2.50	B	3	2.50
501 rpm and over		B	3	0.75	B	3	0.75	B	3	0.75	B	3	1.50	B	3	2.50
50 hp and over	2,3,9															
Up to 300 rpm	8															
301 to 500 rpm	8	B/C	3	0.75	C	3	1.50	C	3	2.50	C	3	2.50	C	3	3.50
501 rpm and over		B/C	3	0.75	C	3	1.50	C	3	1.50	C	3	2.50	C	3	2.50
Packaged Air-Handling Equipment	4,19															
Up to 10 hp	4	A	2	0.25	A	3	0.75	A	3	0.75	A	3	0.75	A	3	1.50
15 hp and over	2,3,4,9															
Up to 500 rpm	8	A	2	0.25	A	3	0.75	A	3	1.50	A	3	1.50	A	3	2.50
501 rpm and over		A	2	0.25	A	3	0.75	A	3	1.50	A	3	1.50	A	3	2.50

Base Types:
A. No base, isolators attached directly to equipment (Note 27)
B. Structural steel rails or base (Notes 28 and 29)
C. Concrete inertia base (Note 30)
D. Curb-mounted base (Note 31)

Isolator Types:
1. Pad, rubber, or glass fiber (Notes 20 and 21)
2. Rubber floor isolator or hanger (Notes 20 and 25)
3. Spring floor isolator or hanger (Notes 22, 23, and 25)
4. Restrained spring isolator (Notes 22 and 24)
5. Thrust restraint (Note 26)

To apply the information from Table 34, base type, isolator type, and minimum deflection columns are added to the equipment schedule, then these isolator specifications are incorporated into mechanical specifications for the project.

The minimum deflections listed in Table 34 are based on the experience of acoustical and mechanical consultants and vibration control manufacturers. Recommended isolator type, base type, and minimum static deflection are reasonable and safe recommendations for 70 to 80% of HVAC installations. The selections are based on concrete equipment room floors 4 to 12 in. thick with typical floor stiffness. The type of equipment, proximity to noise-sensitive areas, and the type of building construction may alter these choices.

The following approach is suggested to develop isolator selections for specific applications:

1. Use Table 34 for floors specifically designed to accommodate mechanical equipment.
2. Use recommendations for the 20-ft span column for equipment on ground-supported slabs adjacent to noise-sensitive areas.
3. For roofs and floors constructed with open web joists, thin long span slabs, wooden construction, and any unusual light construction, evaluate all equipment weighing more than 300 lbs to determine the additional deflection of the structure caused by the equipment. Isolator deflection should be 15 times the additional deflection or the deflection shown in Table 34, whichever is greater. If the required spring isolator deflection exceeds commercially available products, consider air springs, stiffen the supporting structure, or change the equipment location.
4. When mechanical equipment is adjacent to noise-sensitive areas, isolate mechanical equipment room noise.

NOTES FOR VIBRATION ISOLATOR SELECTION GUIDE (TABLE 34)

The notes in this section are keyed to the numbers listed in the second column, Notes, in Table 34 and to other reference numbers throughout the table. While the guide is conservative, cases may arise where vibration transmission to the building is still excessive. If the problem persists after all short circuits have been eliminated, it can almost always be corrected by increasing isolator deflection, using low-frequency air springs, changing operating speed, reducing vibratory output by additional balancing or, as a last resort, changing floor frequency by stiffening or adding more mass.

Note 1. Isolator deflections shown are based on a floor stiffness that can be reasonably expected for each floor span and class of equipment.

Note 2. For large equipment capable of generating substantial vibratory forces and structure-borne noise, increase isolator deflection, if necessary, so isolator stiffness is at least 0.10 times the floor stiffness.

Note 3. For noisy equipment adjoining or near noise-sensitive areas, see Mechanical Equipment Room Noise Isolation.

Note 4. Certain designs cannot be installed directly on individual isolators (Type A), and the equipment manufacturer or a vibration specialist should be consulted on the need for supplemental support (Base Type).

Note 5. Wind load conditions must be considered. Restraint can be achieved with restrained spring isolators (Type 4), supplemental bracing, or limit stops.

Note 6. Certain types of equipment require a curb-mounted base (Type D). Airborne noise must be considered.

Note 7. See text for hanger locations adjoining equipment and in equipment rooms.

Note 8. To avoid isolator resonance problems, select isolator deflection so that natural frequency is 40% or less than the lowest operating speed of equipment.

Note 9. To limit undesirable movement, thrust restraints (Type 5) are required for all ceiling-suspended and floor-mounted units operating at 2 in. and more total static pressure.

Note 10. Pumps over 75 hp may require extra mass and restraining devices.

Note 11. See text for full discussion.

Isolation for Specific Equipment

Note 12. Refrigeration Machines. Large centrifugal, hermetic and reciprocating refrigeration machines generate very high noise levels and special attention is required when such equipment is installed in upper-story locations or near noise-sensitive areas. If such equipment is to be located near extremely noise-sensitive areas, confer with an acoustical consultant.

Note 13. Compressors. The two basic reciprocating compressors are (1) single- and double-cylinder vertical, horizontal or L-head, which are usually air compressors; and (2) Y, W, and multihead or multicylinder air and refrigeration compressors. Single- and double-cylinder compressors generate high vibratory forces requiring large inertia bases (Type C) and are generally not suitable for upper-story locations. If such equipment must be installed in an upper-story location or grade locations near noise-sensitive areas, unbalanced forces should always be obtained from the equipment manufacturer and a vibration specialist consulted for design of the isolation system.

Note 14. When using Y, W, and multihead and multicylinder compressors, obtain the magnitude of unbalanced forces from the equipment manufacturer so that the necessity for an inertia base can be evaluated.

Note 15. Base-mounted compressors through 5 hp and horizontal tank-type air compressors through 10 hp can be installed directly on spring isolators (Type 3) with structural bases (Type B) if required, and compressors 15 to 100 hp on spring isolators (Type 3) with inertia bases (Type C) weighing one to two times the compressor weight.

Note 16. Pumps. Concrete inertia bases (Type C) are preferred for all flexible-coupled pumps and are desirable for most close-coupled pumps, although steel bases (Type B) can be used. Close-coupled pumps should not be installed directly on individual isolators (Type A) because the impeller usually overhangs the motor support base, causing the rear mounting to be in tension. The primary requirement for Type C bases are strength and shape to accommodate base elbow supports. Mass is not usually a factor, except for pumps over 75 hp where extra mass helps limit excess movement due to starting torque and forces. Concrete bases (Type C) should be designed for a thickness of one-tenth the longest dimension with minimum thickness as follows: (1) for up to 30 hp, 6 in.; (2) for 40 to 75 hp, 8 in.; and (3) for 100 hp and higher, 12 in.

Pumps over 75 hp and multistage pumps may exhibit excessive motion at startup; supplemental restraining devices can be installed if necessary. Pumps over 125 hp may generate high starting forces, and it is recommended that a vibration specialist be consulted.

Note 17. Packaged Rooftop Air-Conditioning Equipment. This equipment is usually on lightweight structures that are susceptible to sound and vibration transmission. The noise problem is further compounded by curb-mounted equipment, which requires large roof openings for supply and return air.

The table shows Type D vibration isolator selections for all spans up to 20 ft, but extreme care must be taken for equipment located on spans of over 20 ft, especially if construction is open web joists or thin lightweight slabs. The recommended procedure is to determine the additional deflection caused by equipment in the roof. If additional roof deflection is 0.25 in. or under, the isolator can be selected for 15 times the additional roof deflection. If additional roof deflection is over 0.25 in., supplemental stiffening should be installed or the unit should be relocated.

For units, especially large units, capable of generating high noise levels, consider (1) mounting the unit on a platform above the roof deck to provide an air gap (buffer zone) and (2) locating the unit away from the roof penetration, thus permitting acoustical treatment of ducts before they enter the building.

Some rooftop equipment has compressors, fans, and other equipment isolated internally. This isolation is not always reliable because of internal short circuiting, inadequate static deflection, or panel resonances. It is recommended that rooftop equipment be isolated externally, as if internal isolation was not used.

Note 18. Cooling Towers. These are normally isolated with restrained spring isolators (Type 4) directly under the tower or tower dunnage. Occasionally, high deflection isolators are proposed for use directly under the motor-fan assembly, but this arrangement must be used with extreme caution.

Note 19. Fans and Air-Handling Equipment. The following should be considered in selecting isolation systems for fans and air-handling equipment:

a) Fans with wheel diameters of 22 in. and under and all fans operating at speeds to 300 rpm do not generate large vibratory forces. For fans operating under 300 rpm, select isolator deflection so that the isolator natural frequency is 40% or less than the fan speed. For example, for a fan operating at 275 rpm, 0.4 × 275 = 110 cpm. Therefore, an isolator natural frequency of 110 cpm or lower is required. This can be accomplished with a 3-in. deflection isolator (Type 3).

b) Flexible duct connectors should be installed at the intake and discharge of all fans and air-handling equipment to reduce vibration transmission to ductwork.

c) Inertia bases (Type C) are recommended for all Class 2 and 3 fans and air-handling equipment because extra mass permits the use of stiffer springs, which limit movement.

d) Thrust restraints (Type 5) which incorporate the same deflection as isolators should be used for all fan heads, all suspended fans, and all base-mounted and suspended air-handling equipment operating at 2 in. and over total static pressure.

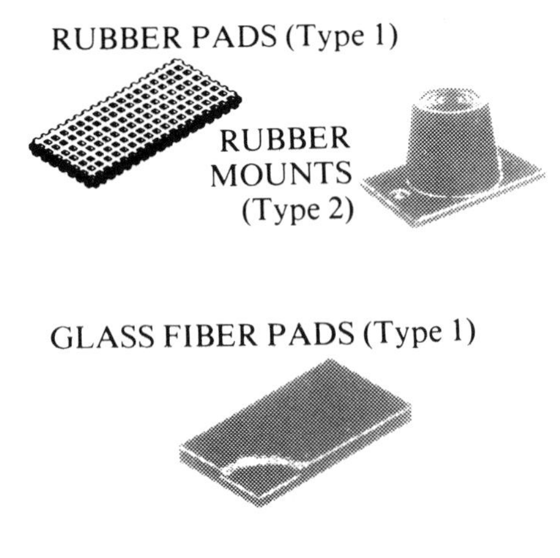

Note 20. Rubber isolators are available in pad (Type 1) and molded (Type 2) configurations. Pads are used in single or multiple layers. Molded isolators come in a range of 30 to 70 durometer (a measure of stiffness). Material in excess of 70 durometer is usually ineffective as an isolator. Isolators are designed for up to 0.5-in. deflection, but are used where 0.3 in. or less deflection is required. Solid rubber and composite fabric and rubber pads are also available. They provide high load capacities with small deflection and are used as noise barriers under columns and for pipe supports. These pad types work well only when they are properly loaded and the weight load is evenly distributed over the entire pad surface. Metal loading plates can be used for this purpose.

Note 21. Precompressed glass fiber isolation pads (Type 1) constitute inorganic inert material and are available in various sizes and thicknesses of 1 to 4 in. and in capacities of up to 500 psi. Their manufacturing process assures long life and a constant natural frequency of 7 to 15 Hz over the entire recommended load range. Pads are covered with an elastomeric coating to increase damping and to protect the glass fiber. Glass fiber pads are most often used for the isolation of concrete foundations and floating floor construction.

SPRING ISOLATOR (Type 3)

Note 22. Steel springs are the most popular and versatile isolators for HVAC applications because they are available for almost any deflection and have a virtually unlimited life. All spring isolators should have a rubber acoustical barrier to reduce transmission of high-frequency vibration and noise that can migrate down the steel spring coil. They should be corrosion-protected if installed outdoors or in a corrosive environment. The basic types include:

1. Note 23. Open spring isolators (Type 3) consist of a top and bottom load plate with an adjustment bolt for leveling. Springs should be designed with a horizontal stiffness at least 75% of the vertical stiffness to assure stability, 50% travel beyond rated load and safe solid stresses.

RESTRAINED SPRING ISOLATOR (Type 4)

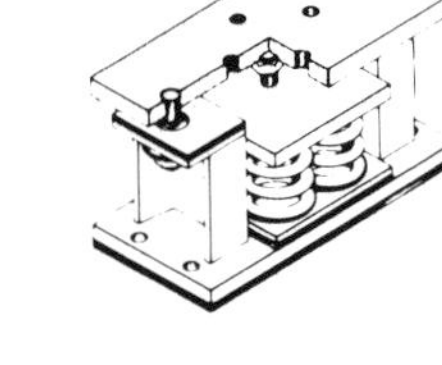

2. Note 24. Restrained spring isolators (Type 4) have hold-down bolts to limit vertical movement. They are used with (a) equipment with large variations in mass (boilers, refrigeration machines) to restrict movement and prevent strain on piping when water is removed, and (b) outdoor equipment, such as cooling towers, to prevent excessive movement because of wind load. Spring criteria should be the same as open spring isolators, and restraints should have adequate clearance so that they are activated only when a temporary restraint is needed.

3. Housed spring isolators consist of two telescoping housings separated by a resilient material. Depending on design and installation, housed spring isolators can bind and short circuit. Their use should be avoided.

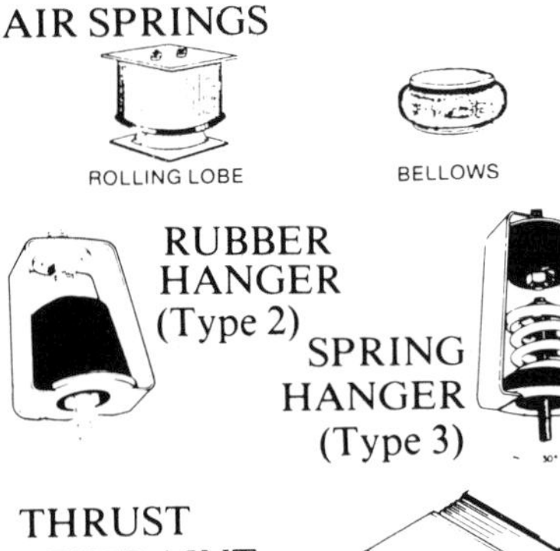

Air springs can be designed for any frequency, but are economical only in applications with natural frequencies of 1.33 Hz or less (6 in. or greater deflection). Their use is advantageous in that they do not transmit high-frequency noise and are often used to replace high deflection springs on problem jobs. Constant air supply is required, and there should be an air dryer in the air supply.

Note 25. Isolation hangers (Types 2 and 3) are used for suspended pipe and equipment and have rubber, springs, or a combination of spring and rubber elements. Criteria should be the same as for open spring isolators. To avoid short circuiting, hangers should be designed for 20 to 35° angular hanger rod misalignment. Swivel or traveler arrangements may be necessary for connections to piping systems subject to large thermal movements.

THRUST RESTRAINT (Type 5)

Note 26. Thrust restraints (Type 5) are similar to spring hangers or isolators and are installed in pairs to resist the thrust caused by air pressure.

Fig. 44 Vibration Isolators and Notes for Table 34

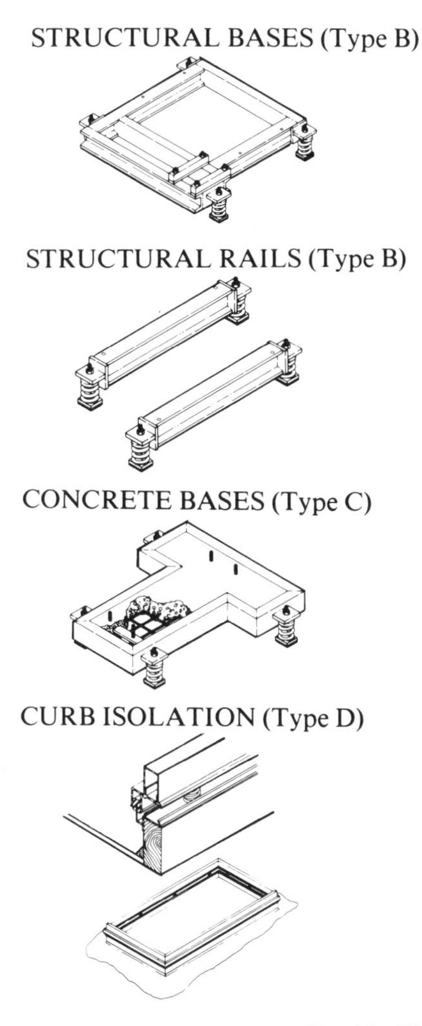

Note 27. Direct isolation (Type A) is used when equipment is unitary and rigid and does not require additional support. Direct isolation can be used with large chillers, packaged air-handling units, and air-cooled condensers. If there is any doubt that the equipment can be supported directly on isolators, use structural bases (Type B) or inertia bases (Type C), or consult the equipment manufacturer.

Note 28. Structural bases (Type B) are used where equipment cannot be supported at individual locations and/or where some means is necessary to maintain alignment of component parts such as fans. These bases can be used with spring or rubber isolators (Types 2 and 3) and should have enough rigidity to resist all starting and operating forces without supplemental hold-down devices. Bases are made in rectangular configurations using structural members with a depth equal to one-tenth the longest span between isolators, with a minimum depth of 4 in. Maximum depth is limited to 12 in., except where structural or alignment considerations dictate otherwise.

Note 29. Structural rails (Type B) are used to support equipment that does not require a unitary base or where the isolators are outside the equipment and the rails act as a cradle. Structural rails can be used with spring or rubber isolators and should be rigid enough to support the equipment without flexing. Usual industry practice is to use structural members with a depth one-tenth of the longest span between isolators with a minimum depth of 4 in. Maximum depth is limited to 12 in., except where structural considerations dictate otherwise.

Note 30. Concrete bases (Type C) consist of a steel pouring form usually with welded-in reinforcing bars, provision for equipment hold-down, and isolator brackets. Like structural bases, concrete bases should be rectangular or T-shaped and, for rigidity, have a depth equal to one-tenth the longest span between isolators with a minimum of 6 in. Base depth need not exceed 12 in., unless it is specifically required for mass, rigidity, or component alignment.

Note 31. Curb isolation systems (Type D) are specifically designed for curb-supported rooftop equipment and have spring isolation with a watertight and airtight curb assembly. The roof curbs are narrow to accommodate the small diameter of the springs within the rails, with static deflection in the 1- to 3-in. range to meet the design criteria described in Type 3.

Fig. 44 Vibration Isolators and Notes for Table 34 (*Concluded*)

Vibration Isolators: Materials, Types, and Configurations

Notes 20 through 31 are included in Figure 44 to assist in the evaluation of commercially available isolators for HVAC equipment. The isolator selected for a particular application depends on the required deflection, but life, cost, and suitability must also be considered.

ISOLATION OF VIBRATION AND NOISE IN PIPING SYSTEMS

All piping systems have mechanical vibration generated by the equipment and impeller-generated and flow-induced vibration and noise, which is transmitted by the pipe wall and the water column. In addition, equipment installed on vibration isolators exhibits some motion or movement from pressure thrusts during operation. Vibration isolators have even greater movement during start-up and shutdown, when the equipment goes through the isolators' resonant frequency. The piping system must be flexible enough to (1) reduce vibration transmission along the connected piping, (2) permit equipment movement without reducing the performance of vibration isolators, and (3) accommodate equipment movement or thermal movement of the piping at connections without imposing undue strain on the connections and equipment.

Flow noise in piping can be minimized by sizing pipe so that velocities are 4 fps maximum for pipe 2 in. and smaller and using a pressure drop limitation of 4 ft of water per 100 ft of pipe length with a maximum velocity of 10 ft/s for larger pipe sizes. Flow noise and vibration can be reintroduced by turbulence, sharp pressure drops, and entrained air. Care should be taken to avoid these conditions.

Resilient Pipe Hangers and Supports

Resilient pipe hangers and supports are necessary to prevent vibration and noise transmission from the piping system to the building structure and to provide flexibility in the piping.

Suspended Piping. Isolation hangers described in the vibration isolation section should be used for all piping in equipment rooms or for 50 ft from vibrating equipment, whichever is greater. To avoid reducing the effectiveness of equipment isolators, at least three of the first hangers from the equipment should provide the same deflection as the equipment isolators, with a maximum limitation of 2-in. deflection; the remaining hangers should be spring or combination spring and rubber with 0.75-in. deflection.

Good practice requires the first two hangers adjacent to the equipment to be the positioning or precompressed type, to prevent load transfer to the equipment flanges when the piping system is filled. The positioning hanger aids in installing large pipe, and many engineers specify this type for all isolated pipe hangers for piping 8 in. and over.

While isolation hangers are not often specified for branch piping or piping beyond the equipment room for economic reasons, they should be used for all piping over 2 in. in diameter and for any piping suspended below or near noise-sensitive areas. Hangers adjacent to noise-sensitive areas should be the spring and rubber combination Type 3.

Floor-Supported Piping. Floor supports for piping in equipment rooms and adjacent to isolated equipment should use vibration isolators as described in the vibration isolation section. They should be selected according to the guidelines for hangers. The first two adjacent floor supports should be the restrained spring type, with a blocking feature that prevents load transfer to equipment flanges as the piping is filled or drained. Where pipe is subjected to large thermal movement, a slide plate (Teflon, graphite, or steel) should be installed on top of the isolator, and a thermal barrier should be used when rubber products are installed directly beneath steam or hot water lines.

Riser Supports, Anchors, and Guides. Many piping systems have anchors and guides, especially in the risers, to permit expansion joints, bends, or pipe loops to function properly. Anchors and guides are designed to eliminate or limit (guide) pipe movement and must be rigidly attached to the structure; this is inconsistent with the the resiliency required for effective isolation. The engineer should try to locate the pipe shafts, anchors, and guides in noncritical areas, such as next to elevator shafts, stairwells, and toilets, rather than adjoining noise-sensitive areas. Where concern about vibration transmission exists, some type of vibration isolation support or acoustical support is required for the pipe support, anchors, and guides.

Since anchors or guides must be rigidly attached to the structure, the isolator cannot deflect in the sense previously discussed, and the primary interest is that of an acoustical barrier. Such acoustical barriers can be provided by heavy-duty rubber and duck and rubber pads that can accommodate large loads with minimal deflection. Figures 45, 46, and 47 show some arrangements for resilient anchors and guides. Similar resilient-type supports can be used for the pipe.

Resilient supports for pipe, anchors, and guides can attenuate noise transmission, but they do not provide the resiliency required to isolate vibration. Vibration must be controlled in an anchor

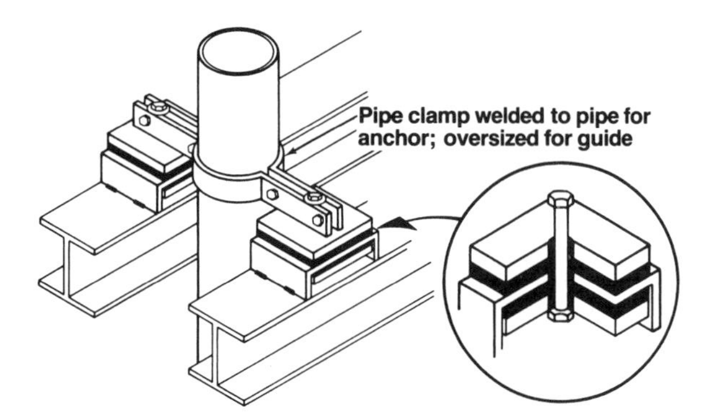

Fig. 45 Acoustical Barriers for Pipe Anchors and Guides

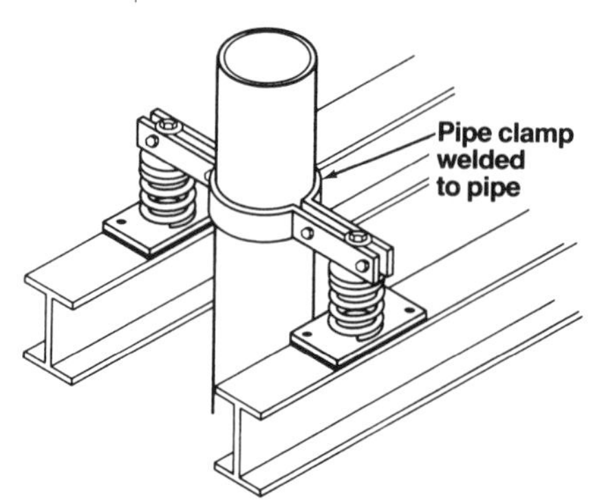

Fig. 46 Spring Isolated Riser System

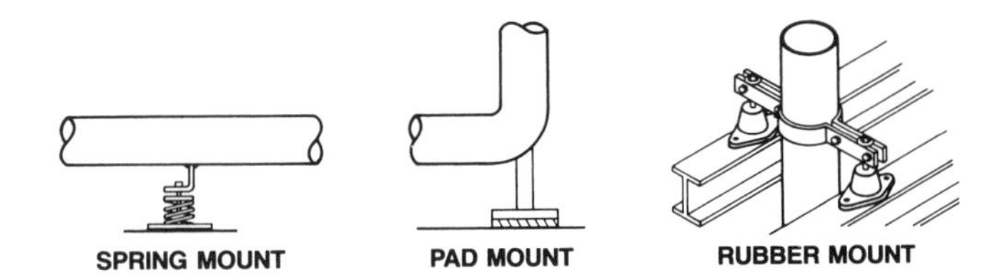

Fig. 47 Conventional Isolators as Pipe Supports for Lines with Expansion Joints

guide system by designing flexible pipe connectors and resilient isolation hangers or supports.

Completely spring-isolated riser systems that eliminate the anchors and guides have been used successfully in many instances and give effective vibration and acoustical isolation. In this type of isolation system, the springs are sized to accommodate thermal growth as well as to guide and support the pipe. Such systems require careful engineering to accommodate the movements encountered not only in the riser but also in the branch takeoff to avoid overstressing the piping.

Piping Penetrations. Most HVAC systems have many points at which piping must penetrate floors, walls, and ceilings. If such penetrations are not properly treated, they provide a path for airborne noise, which can destroy the acoustical integrity of the occupied space. Seal the openings in the pipe sleeves between noisy areas, such as equipment rooms, and occupied spaces with an acoustical barrier such as fibrous material and caulking or with engineered pipe penetration seals as shown in Figure 48.

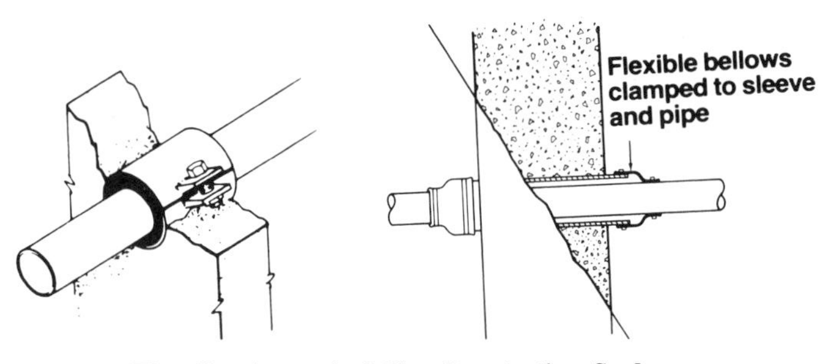

Fig. 48 Acoustical Pipe Penetration Seals

FLEXIBLE PIPE CONNECTORS

Flexible pipe connectors (1) provide piping flexibility to permit isolators to function properly, (2) protect equipment from strain from misalignment and expansion or contraction of piping, and (3) attenuate noise and vibration transmission along the piping (Figure 49). Connectors are available in two configurations: (1) hose type, a straight or slightly corrugated wall construction of either rubber or metal; and (2) the arched or expansion joint type, a short length connector with one or more large radius arches, of rubber, Teflon, or metal. Metal expansion joints are seldom used for vibration and sound isolation in HVAC systems, and their use is not recommended. All flexible connectors require end restraint to counteract the pressure thrust, which is (1) added to the connector, (2) incorporated by its design, (3) added to the piping system (anchoring), or (4) built in by the stiffness of the system. Connector extension caused by pressure thrust on isolated equipment should also be considered when flexible connectors are used. Overextension will cause failure. Manufacturers' recommendations on restraint, pressure, and temperature limitations should be strictly adhered to.

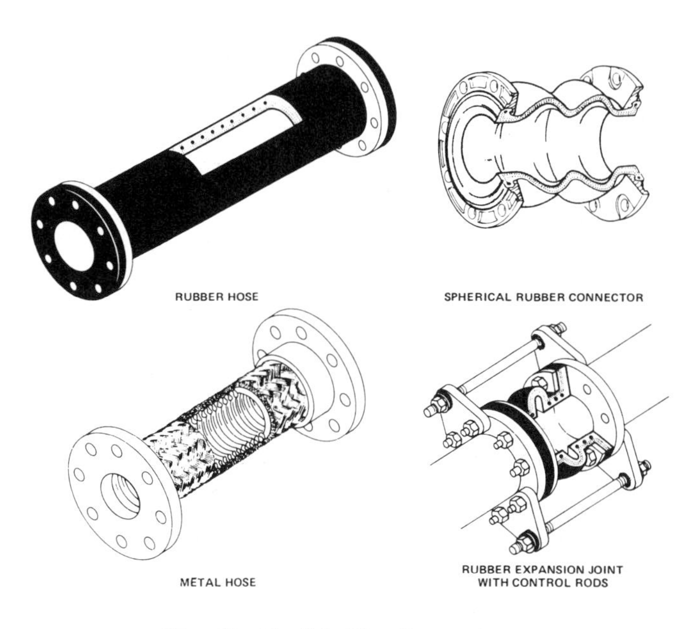

Fig. 49 Flexible Pipe Connectors

Hose Connectors

Hose connectors accommodate lateral movement perpendicular to the length and have very limited or no axial movement capability.

Rubber hose connectors can be of molded or handwrapped construction with wire reinforcing and are available with metal-threaded end fittings or integral rubber flanges. Application of threaded fittings should be limited to 3 in. and smaller pipe diameter. The fittings should be the mechanically expanded type to minimize the possibility of pressure thrust blowout. Flanged types are available in larger pipe sizes. Table 35 lists recommended lengths.

Table 35 Recommended Live Length[a] of Flexible Rubber and Metal Hose

Nominal Diameter, in.	Length,[b] in.	Nominal Diameter, in.	Length,[b] in.
0.75	12	4	18
1	12	5	24
1.5	12	6	24
2	12	8	24
2.5	12	10	24
3	18	12	36

[a]Live length is end-to-end length for integral flanged rubber hose and is end-to-end less total fitting length for all other types.
[b]Based on recommendations of Rubber Expansion Joint Division, Fluid Sealing Association.

Metal hose is constructed with a corrugated inner core and a braided cover, which helps attain a pressure rating and provides end restraints that eliminate the need for supplemental control assemblies. Short lengths of metal hose or corrugated metal bellows, or pump connectors, are available without braid and have built-in control assemblies. Metal hose is used to control misalignment and vibration rather than noise and is used primarily where temperature or the pressure of flow media preclude the use of other material. Table 35 provides recommended lengths.

Expansion Joint or Arched-Type Connectors

Expansion joint or arched-type connectors have one or more convolutions or arches and can accommodate all modes of axial, lateral, and angular movement and misalignment. These connectors are available in flanged rubber and Teflon construction. When made of rubber, they are commonly referred to as expansion joints, spool joints, or spherical connectors, and in Teflon, as couplings or expansion joints.

Rubber expansion joints or spool joints are available in two basic types: (1) handwrapped with wire and fabric reinforcing, and (2) molded with fabric and wire or with high-strength fabric only (instead of metal) for reinforcing. The handmade type is available in a variety of materials and lengths for special applications. Rubber spherical connectors are molded with high-strength fabric or tire cord reinforcing instead of metal. Their distinguishing characteristic is a large radius arch. The shape and construction of some designs permit use without control assemblies in systems operating to 150 psi. Where thrust restraints are not built in, they must be used as described for rubber hose joints.

Teflon expansion joints and couplings are similar in construction to rubber expansion joints with reinforcing metal rings.

In evaluating these devices, consider the temperature, pressure, and service conditions as well as the ability of each device to attenuate vibration and noise. Metal hose connections can accommodate misalignment and attenuate mechanical vibration transmitted through the pipe wall but do little to attenuate noise. This type of connector has superior resistance to long-term temperature effects. Rubber hose, expansion joints, and spherical connectors attenuate vibration and impeller-generated noise transmitted through the pipe wall. Because the rubber expansion joint and spherical connector walls are flexible, they have the ability to grow volumetrically and attenuate noise and vibration at blade passage frequencies. This is a particularly desirable feature in uninsulated piping systems, such as condenser water and domestic water, which may run adjacent to noise-sensitive areas. However, high pressure has a detrimental effect on the ability of the connector to attenuate vibration and noise.

Because none of the flexible pipe connectors control flow or velocity noise or completely isolate vibration and noise transmission to the piping system, resilient pipe hangers and supports should be used; these are shown in Figure 44 and are described in the Resilient Pipe Hangers and Supports section.

ISOLATING DUCT VIBRATION

Flexible canvas and rubber duct connections should be used at fan intake and discharge. However, they are not completely effective since they become rigid under pressure and allow the vibrating fan to pull on the duct wall. To maintain a slack position of the flexible duct connections, thrust restraints (see Figure 44) should be used on all equipment as indicated in Table 34 and in the Vibration Isolator Selection section.

While vibration transmission from ducts that are isolated by flexible connectors is not a common problem, flow pulsations within the duct can cause mechanical vibration in the duct walls, which can be transmitted through rigid hangers. Spring or combination spring and rubber hangers are recommended wherever ducts are suspended below or near a noise-sensitive area. These hangers are especially recommended for large ducts with velocities above 1500 fpm and for all size ducts when duct static pressures are 2 in. of water (gage) and over.

SEISMIC PROTECTION

Seismic restraint requirements are specified by applicable building codes that define the design forces to be resisted by the mechanical system, depending on the building location and occupancy, location of the system within the building, and whether it is used for life safety. Where required, seismic protection of resiliently mounted equipment poses a unique problem (Mason and Lama 1976), since resiliently mounted systems are much more susceptible to earthquake damage due to overturning forces and to resonances inherent in vibration isolators.

As a deficiency in seismic restraint design or anchorage would not become apparent until an earthquake occurs, with possible catastrophic consequences, the adequacy of the restraint system and anchorage to resist code design forces must be verified before the event. This verification should be either by equipment tests, calculations, or dynamic analysis, depending on the item, with calculations or dynamic analysis performed under the direction of a professional engineer. These items are often supplied as a package by the vibration isolation vendor.

The restraints for floor-mounted equipment should be designed with adequate clearances so that they are not engaged during normal operation of the equipment. Contact surfaces should be protected with resilient pad material to limit shock during an earthquake and the restraints should be sufficiently strong to resist the forces in any direction. Due to the difficulty of analyzing these devices, their integrity is normally verified by independent laboratory tests.

Chapter 49 in this volume covers seismic restraint design; but, calculations or dynamic analysis should have an engineer's seal to verify the input forces produced by the code or specified requirements. The anchorage calculations should also be made by a professional engineer in accordance with accepted standards.

TROUBLESHOOTING

In spite of all efforts taken by specifying engineers, consultants, and installing contractors, situations arise where there is disturbing noise and vibration. Fortunately, many problems can be readily identified and corrected by (1) determining which equipment or system is the source of the problem, (2) determining if the problem is one of airborne sound, vibration and structure-borne noise, or a combination of both, and (3) applying appropriate solutions.

DETERMINING PROBLEM SOURCE

The system or equipment that is the source of the problem can often be determined without instrumentation. Vibration and noise levels are usually well above the sensory level of perception and are readily felt or heard. A simple and accurate method of determining the problem source is to turn individual pieces of equipment on and off until the vibration or noise is eliminated. Since the source of the problem is more than one piece of equipment or the interaction of two or more systems, it is always good practice to double check by shutting off the system and operating the equipment individually.

DETERMINING PROBLEM TYPE

The next step is to determine if the problem is one of noise or vibration.

1. If vibration is perceptible, vibration transmission is usually the major cause of the problem. The possibility that lightweight wall or ceiling panels are excited by airborne noise should be considered. If vibration is not perceptible, the problem may still be one of vibration transmission causing structure-borne noise, which can be checked by following the procedure below.
2. If a sound level meter is available, check C-scale and overall scale readings. If the difference is greater than 6 dB, or if the slope of the curve is greater than 5 to 6 dB/octave in the low frequencies, vibration is probably the cause.
3. If the affected area is remote from source equipment, there is no problem in intermediary spaces, and noise does not appear to be coming from the duct system or diffusers, structure-borne noise is probably the cause.

Noise Problems

Noise problems are more complex than vibration problems and usually require the services of an acoustical engineer or consultant.

If the affected area adjoins the room where the source equipment is located, structure-borne noise must be considered as part of the problem, and the vibration isolation systems should be checked. A simple but reasonably effective test is to have one person listen in the affected area while another shouts loudly in the equipment room. If the voice cannot be heard, the problem is likely one of structure-borne noise. If the voice can be heard, check for openings in the wall or floor separating the areas. If no such openings exist, the structure separating the areas does not provide adequate transmission loss. In such situations, refer to the Mechanical Equipment Room Noise Isolation section of this chapter for possible solutions.

If ductborne sound, *i.e.*, noise from grilles or diffusers, and duct breakout noise is the problem, measure the sound pressure levels and compare them with the design goal RC curves. Where the measured curve differs from the design goal RC curve, the potential noise source can be narrowed down. Once the noise sources have been identified, the engineer can determine whether sufficient attenuation has been provided by analyzing each sound source using the procedures presented in this chapter.

If the sound source is a fan, pump, or similar rotating equipment, determine if it is operating at the most efficient part of its operating curve. This is the point at which most equipment operates best. Excessive vibration and noise can occur if a fan or pump is trying to move too little or too much air or water. In this respect, check that vanes, dampers, and valves are in the correct operating position and that the system has been properly balanced.

Vibration Problems

Vibration and structure-borne noise problems can occur from:

- Equipment operating with excessive levels of vibration, usually caused by unbalance
- Lack of vibration isolators
- Improperly selected or installed vibration isolators that do not provide the required isolator deflection
- Flanking transmission paths such as rigid pipe connections or obstructions under the base of isolated equipment
- Floor flexibility
- Resonances in equipment, the vibration isolation system, or the building structure

Most field-encountered problems are the result of improperly selected or installed isolators and flanking paths of transmission, which can be simply evaluated and corrected.

Floor flexibility and resonance problems are seldom encountered and usually require analysis by experts. However, the information provided below will identify such problems.

If the equipment lacks vibration isolators, isolators recommended in Table 27 can be added by using structural brackets without altering connected ducts or piping.

Testing Vibration Isolation Systems. Improperly functioning vibration isolation systems are the cause of most field-encountered problems and can be evaluated and corrected by the following procedures.

1. Ensure that the system is free-floating by bouncing the base, which should cause the equipment to move up and down freely and easily. On floor-mounted equipment, check that there are no obstructions between the base and the floor, which would short-circuit the isolation system. This is best accomplished by passing a rod under the equipment. A small obstruction might permit the base to "rock," giving the impression that it is free-floating when it is not. On suspended equipment, make sure that rods are not touching the hanger box. Rigid connections such as pipes and ducts can prevent equipment from floating freely, prohibit isolators from functioning properly, and provide flanking paths of transmission.
2. Determine if the isolator deflection is as specified or required, changing it if necessary, as recommended in Table 34. A common problem is inadequate deflection caused by underloaded isolators. Overloaded isolators are not generally a problem as long as the system is free-floating and there is space between the spring coils.

With the most commonly used spring isolators, determine the spring deflection by measuring the operating height and comparing it to the free height information available from the manufacturer. Once the actual isolator deflection is known, determine its adequacy by comparing it with the recommended deflection in Table 34.

If the natural frequency of the isolator is 25% or less than the disturbing frequency (usually considered the operating speed of the equipment), the isolators should be amply efficient except in situations where heavy equipment is installed on extremely long span floors or very flexible floors. If a transmission problem exists, it may be caused by (1) excessively rough equipment operation, (2) the system not being free-floating of flanking path transmission, or (3) a resonance or floor stiffness problem, as described below.

While it is easy to determine the natural frequency of spring isolators by height measurements, such measurements are difficult with pad and rubber isolators and are not accurate in determining their natural frequencies. Although such isolators can theoretically provide natural frequencies as low as 4 Hz, they actually provide higher natural frequencies and generally do not provide the desired isolation efficiencies for upper floor equipment locations.

Isolation efficiency cannot be checked by comparing the vibration amplitude level of the equipment to that of the structure as shown in Figure 50 (top). These levels seem to show 10% transmission (90% isolator efficiency) when, actually, the isolators might be in resonance, resulting in higher levels of vibration than if no isolators were installed. The only accurate way to measure isolator efficiency is to compare two sets of measurements on the structure—one having the equipment operating with isolators and the other without isolators or with isolators firmly blocked with wedges as shown in Figure 50 (bottom).

Floor Flexibility Problems. Floor flexibility is not a problem with most equipment and structures; however, such problems can occur with heavy equipment installed on long span floors or thin slabs and with rooftop equipment installed on light structures of open web joist construction. If floor flexibility is suspected, the isolators should be one-tenth or less as stiff as the floor to eliminate the problem. Floor stiffness can be determined by calculating the additional deflection in the floor caused by a specific piece of equipment.

For example, if a 10,000 lb piece of equipment causes floor deflection of an additional 0.1 in., floor stiffness is 100,000 lb/in., and an isolator of 10,000 lb/in. must be used. Note that the floor stiffness or spring rate, not the total floor deflection, is determined. In this example, the total floor deflection might be 1 in., but if the problem equipment causes 0.1 in. of that deflection, 0.1 in. is the important figure, and floor stiffness k is 100,000 lb/in.

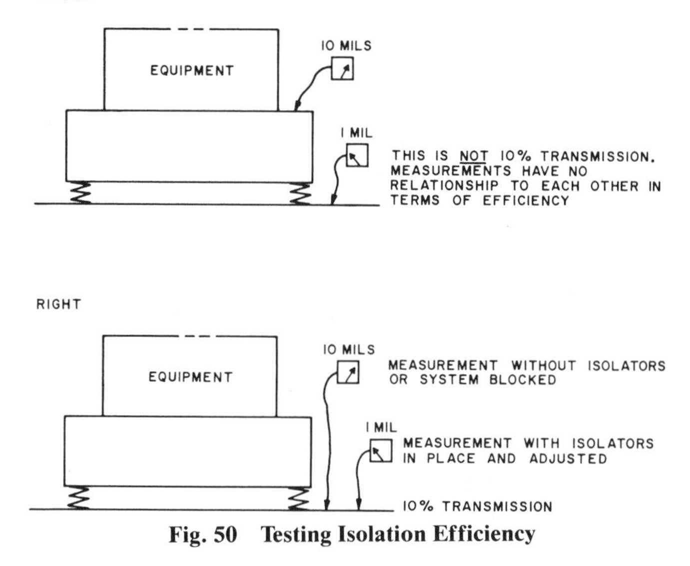

Fig. 50 Testing Isolation Efficiency

Resonance Problems

These problems occur when the operating speed of the equipment is the same or close to the natural frequency of (1) an equipment component such as a fan shaft or bearing support pedestal, (2) the vibration isolation system, or (3) the natural frequency of the floor or other building component such as a wall. Resonance can cause excessive equipment vibration levels as well as objectionable and possibly destructive vibration transmission. These conditions must always be determined and corrected.

Vibration Isolation System Resonance. Always characterized by excessive equipment vibration, vibration isolation system resonance usually results in objectionable transmission to the structure. However, transmission might not occur if the equipment is on-grade or on a stiff floor. Vibration isolation system resonance can be measured with instrumentation or, more simply, by determining the isolator natural frequency as described in the section Testing Vibration Isolation Systems and comparing this figure to the operating speed of the equipment.

When vibration isolation system resonance exists, the isolator natural frequency must be changed using the following guidelines:

1. If the equipment is installed on pad or rubber isolators, isolators with the deflection recommended in Table 34 should be installed.
2. If the equipment is installed on spring isolators and there is objectionable vibration or noise transmission to the structure, determine if the isolator is providing maximum deflection. For example, an improperly selected or installed nominal 2-in. deflection isolator could be providing only 1/8-in. deflection, which would be in resonance with equipment operating at 500 rpm. If this is the case, the isolators should be replaced with ones having enough capacity to provide 2-in. deflection. Since there was no transmission problem with the resonant isolators, it is not necessary to use greater deflection isolators than can be conveniently installed.
3. If the equipment is installed on spring isolators and there is objectional noise or vibration transmission, replace the isolators with spring isolators with the deflection recommended in Table 34.

Building Resonance. These problems occur when some part of the structure has a natural frequency the same as the disturbing frequency or the operating speed of some of the equipment. These problems can exist even if the isolator deflections recommended in Table 34 are used. The resulting objectionable noise or vibration should be evaluated and corrected. Often, the resonant problem is in the floor on which the equipment is installed, but it can also occur in a remotely located floor, wall, or other building component. If a noise or vibration problem has a remote source which cannot be associated with piping or ducts, resonance must be suspected.

Resonance problems can be resolved by the following:

1. Reducing the vibratory force F by balancing the equipment. This is not a practical solution for a true resonant problem; however, it is viable when the disturbing frequency equals the floor natural frequency, as evidenced by the equal displacement of the floor and the equipment, especially when the equipment is operating with excessive vibration.
2. Changing the isolator natural frequency f_n by increasing or decreasing the deflection. Only small changes are necessary to "detune" the system. Generally, increasing the deflection is preferred; if the initial deflection is 1 in., a 2- or 3-in. deflection isolator should be installed. However, if the initial isolator deflection is 4 in., it would be more practical and economical to replace it with 2- or 3-in. deflection isolators. Changing the natural frequency is only practical for true resonant conditions, not when the disturbing frequency equals the floor natural frequency ($f_d = f_f$). In such situations, the floor and equipment displacement are the same. Assuming that an effective isolation system is $f_d/f_n = 4$, it occurs in the asymptotic portion of the displacement curve and requires a tremendous increase in isolator deflection for even a small reduction in equipment displacement. In such situations, methods 1, 3, or 4 provide a more practical and economical solution.
3. Changing the structure stiffness k_f or the structure natural frequency f_f. A change in structure stiffness changes the structure natural frequency; the greater the stiffness, the higher the frequency. However, the structure natural frequency can also be changed by increasing or decreasing the floor deflection without changing the floor stiffness. While this approach is not recommended, it may be the only solution in certain cases.
4. Changing the disturbing frequency f_d by changing the equipment operating speed. This is practical only for belt-driven equipment, such as fans.

Troubleshooting is time-consuming, expensive, and often difficult. In addition, once a transmission problem exists, the occupants become more sensitive and require lower reduction of the sound and vibration levels than would initially have been satisfactory. Therefore, the need for troubleshooting should be avoided by carefully designing, installing, and testing the system as soon as it is operational and before the building is occupied.

BIBLIOGRAPHY

AMCA. 1985. Reverberation room method for sound testing of fans. AMCA *Standard* 300-85. Air Movement and Control Association, Inc., Arlington Heights, IL.

ASA. 1983. Specification for sound level meters. ASA *Standard* 47-83 (ANSI S1.4-83). Acoustical Society of America, New York.

ASHRAE. 1986. Laboratory method of testing in-duct sound power measurement procedure for fans. ASHRAE *Standard* 68-1986.

Beatty, J. 1987. Discharge duct configurations to control rooftop sound. *Heating/Piping/Air Conditioning* (July).

Beranek, L.L. 1957. Revised criteria for noise in buildings. *Noise Control* (January):19.

Beranek, L.L. 1988. Noise and vibration control. Institute of Noise Control Engineering, Poughkeepsie, NY.

Blazier, W.E. 1972. Chiller noise: Its impact on building design. ASHRAE *Symposium Bulletin No.* 3.

Blazier, W.E. 1981. Noise rating of variable air-volume terminal devices. ASHRAE *Transactions* 87(1).

Blazier, W.E. 1981. Revised noise criteria (RC curves) for application in the acoustical design of HVAC systems. *Noise Control Engineering* 16(2), March-April.

Brockmeyer, H. 1968. Stromungsakustische Untersuchungen an Kanalnetzelementen von Hochgeschwindigkeits-Klimaanlagen. Dissertation, Technische Universitat, Braunschweig, Germany.

Coad, W.J. 1988. Vibration isolation for fans. *Heating/Piping/Air Conditioning* (March).

Cummings, A. 1985. Acoustic noise transmission through duct walls. ASHRAE *Transactions* 91(2A):48-61.

Den Hartog, J.P. 1956. *Mechanical vibrations*. Dover Publications, New York.

Ebbing, C.E., D. Fragnito, and S. Inglis. 1978. Control of low frequency duct-generated noise in building air distribution systems. ASHRAE *Transactions* 84(2):191.

Egan, M.D. 1988. *Architectural acoustics*, Chapters 4 and 5. McGraw-Hill, New York.

Fry, A. and staff of Sound Research Laboratories. 1988. *Noise control in building services*. Pergamon Press, Oxford, U.K.

Harold, R.G. 1986. Round duct can stop rumble noise in air handling installations. ASHRAE *Transactions* 92(2).

Harris, C.M., ed. 1979. *Handbook of noise control*, Chapter 35. McGraw-Hill, New York.

Harris, C.M. and C.E. Crede, eds. 1976. *Shock and vibration handbook*. McGraw-Hill, New York.

Hedeen, R.A., *et al.* 1980. Compendium of materials for noise control. U.S. Department of Health, Education and Welfare, NIOSH, Cincinnati, OH.

Heinter, I. 1968. How to estimate plant noises. *Hydrocarbon Processing* 47(12):67.

Jones, R.S. 1984. *Noise and vibration control in buildings.* McGraw-Hill, New York.

Kugler, B.A., *et al.* 1973. Noise study of proposed SOCAL El Segundo refinery project. Bolt, Beranek & Newman, Inc. *Report No.* 2426.

McGavin, G.L. 1981. *Earthquake protection of essential building equipment*, Chapter 4. John Wiley & Sons, New York.

Macinante, J.A. 1984. *Seismic mountings for vibration isolation.* John Wiley & Sons, New York.

Mason, N.J. and P.J. Lama. 1976. Seismic control for floor mounted equipment. *Heating, Piping & Air Conditioning* (March).

Miller, L.N. 1970. Acquisition and study of the noise data of certain electrical and mechanical equipment used in buildings. Bolt, Beranek & Newman, Inc., *Report No.* 1778.

SAMA. 1980. Laboratory fume hoods. Scientific Apparatus Makers Association *Standard* LF-10-1980. Washington, D.C.

Schultz, T.J. 1983. Relationship between sound power level and sound pressure level in dwellings and offices. ASHRAE *Transactions* 91(1A):124-53.

Sessler, S.M. 1973. Acoustical and mechanical considerations for the evaluation of chiller noise. ASHRAE *Journal* 15 (October):39.

Sessler, S.M. and R.M. Hoover. 1983. Laboratory fume hood noise. *Heating/Piping/Air Conditioning* (September).

United Sheet Metal Co. 1973. *Bulletin* 40-1-672-B. Westerville, OH. (From KAL *Test Reports No.* KAL-985-1,2,3-70 and KAL 1507-7-73.)

U.S. Army. 1983. Noise and vibration control for mechanical equipment. U.S. Army TM5-805-4.

United States Gypsum. 1971. *Drywall construction handbook.* Chicago.

Ver, I.L. 1978. A review of the attenuation of sound in straight lined and unlined ductwork of rectangular cross section. ASHRAE *Transactions* 84(1):122.

Ver, I.L. 1984. Noise generation and noise attenuation of duct fittings—A review. ASHRAE *Transactions* 90(2A):354-90.

Waeldner, W.J. 1975. Acoustical considerations in air distribution. ASHRAE *Transactions* 81(2):504.

Wells, R.J. 1958. Acoustical plenum chambers. *Noise Control* (July):9.

CHAPTER 43

CORROSION CONTROL AND WATER TREATMENT

THIS chapter covers the fundamentals of corrosion, its prevention and control, and some of the common problems arising from corrosion in heating and air-conditioning equipment. Further information can be obtained from the references listed in the bibliography and in publications of the National Association of Corrosion Engineers and the American Water Works Association.

DEFINITIONS

Terms commonly used in the field of water treatment and corrosion control are defined as follows:

Alkalinity. The sum of the bicarbonate, carbonate, and hydroxide ions in water. Other ions, such as borate, phosphate, or silicate can also contribute to alkalinity.

Anion. A negatively charged ion of an electrolyte, which migrates toward the anode under the influence of a potential gradient.

Anode. The electrode of an electrolytic cell at which oxidation occurs.

Biological Deposits. Water-formed deposits of biological organisms or the products of their life processes, such as barnacles, algae, or slimes.

Cathode. The electrode of an electrolytic cell at which reduction occurs.

Cation. A positively charged ion of an electrolyte, which migrates toward the cathode under the influence of a potential gradient.

Corrosion. The deterioration of a material, usually a metal, by reaction with its environment.

Corrosivity. The capacity of an environment to bring about destruction of a specific metal by the process of corrosion.

Electrolyte. A solution through which an electric current can be made to flow.

Filtration. The process of passing a liquid through a suitable porous material in such a manner as to effectively remove the suspended matter from the liquid.

Galvanic Corrosion. Corrosion resulting from the contact of two dissimilar metals in an electrolyte or from the contact of two similar metals in an electrolyte of nonuniform concentration.

Hardness. The sum of the calcium and magnesium ions in water.

Inhibitor. A chemical substance that reduces the rate of corrosion, scale formation, or slime production.

Ion. An electrically charged atom or group of atoms.

Passivity. The tendency of a metal to become inactive in a given environment.

pH. The logarithm of the reciprocal of the hydrogen ion concentration of a solution. pH values below 7 are increasingly acidic, and those above 7 are increasingly alkaline.

Polarization. The deviation from the open circuit potential of an electrode resulting from the passage of current.

ppm. Parts per million by mass. In water, ppm are essentially the same as milligrams per litre (mg/L); 10 000 ppm (mg/L) = 1%.

Scale. The formation at high temperature of thick corrosion product layers on a metal surface. The deposition of water insoluble constituents on a metal surface.

Sludge. A sedimentary water-formed deposit, either of biological origin or suspended particles from the air.

Tuberculation. The formation of localized corrosion products scattered over the surface in the form of knob-like mounds.

Water-formed Deposit. Any accumulation of insoluble material derived from water or formed by the reaction of water on surfaces in contact with it.

CORROSION THEORY—MECHANISM

Corrosion is defined as destruction of a metal or alloy by chemical or electrochemical reaction with its environment. In most instances, this reaction is electrochemical in nature, much like that in an electric dry cell. The basic nature of corrosion is almost always the same: a flow of electricity between certain areas of a metal surface through a solution capable of conducting an electrical current. This electrochemical action causes destructive alteration (eating away) of a metal at areas, or anodes, where the electric current enters the solution. This is the critical step in a series of reactions associated with corrosion.

For corrosion to occur, there must be a release of electrons at the anode and a formation of metal ions through oxidation and disintegration of the metal. At the cathode, there must be a simultaneous acceptance of ions or formation of negative ions. Action at either electrode does not occur independently. According to Faraday's law, the two reactions must be at equivalent rates at the same time. Corrosion, or disintegration of the metal, occurs only at the anodes. Anode and cathode areas may shift from time to time as the corrosion process proceeds, resulting in uniform corrosion.

Reactions at the cathode surface most often control the rate of corrosion. Depending on the nature of the electrolyte, the hydrogen generated at the cathode surface may (1) accumulate to coat the surface and slow down the reaction (cathodic polarization); (2) form bubbles and be swept from the surface, thus allowing the reaction to proceed; and (3) react with oxygen in the electrolyte to form water or a hydroxyl ion. At the anode, the metal ion entering the solution may react with a constituent in the electrolyte to form a corrosion product. With iron or steel, the ferrous ion may react with the hydroxyl ion in water to form ferrous hydroxide and then with oxygen to produce ferric hydroxide (rust). These corrosion products may accumulate on the anode surface and slow down the reaction rate (anodic polarization).

The preparation of this chapter is assigned to TC 3.6, Water Treatment.

ACCELERATING OR INTENSIFYING FACTORS

Moisture

Corrosion does not occur in dry air. However, in most natural atmospheres, some moisture is present as water vapor. In pure air, almost no iron corrosion occurs at relative humidities up to 99%; however, when contaminants such as sulfur dioxide or solid particles of charcoal are present, corrosion could proceed at relative humidities of 50% or more. Pure air is seldom encountered in practice. During rains, exposed metal surfaces are completely wetted, allowing the corrosion reaction to proceed as long as the metal remains wet. This applies to iron and unalloyed steel. Many alloys develop thin corrosion product films or oxide coatings and are unaffected by moisture.

Oxygen

Since electrolytes consist of water solutions of salts or acids, the presence of oxygen in the media accelerates the corrosion rate of ferrous metals by depolarizing the cathodic areas through reaction with hydrogen generated at the cathode. This allows the anodic reaction to proceed. In many ferrous systems used for handling water, such as boiler systems or hot water heating systems, oxygen is removed to reduce the corrosion rate. This is done by adding oxygen scavenging chemicals, such as sulfites, to the system, or by using deaeration equipment to expel the dissolved oxygen.

The presence of oxygen in the media does not affect all alloys in the manner just described. In alloys that develop protective oxide films, such as stainless steels, oxygen can reduce corrosion by maintaining the oxide film. Free of oxygen, the media might cause some corrosion of the alloy; but with oxygen present, the oxide film becomes reinforced and prevents corrosion.

Solutes

In ferrous materials such as iron and steel, mineral acids accelerate the corrosion rate, whereas alkalis reduce it. Since the corrosion reaction at the cathode is related to the concentration of hydrogen ions present, the higher the concentration (the more acidic the media), the less likely the cathode area will become polarized. Alkaline solutions, which contain a much higher concentration of hydroxyl ions than hydrogen, promote polarization of the cathode areas, thus reducing the rate of dissolution at the anode. Relative acidity of alkalinity of a solution is defined as pH; a neutral solution has a pH of 7. Solutions increase in acidity as the pH decreases and increase in alkalinity as pH increases.

The corrosivity of most salt solutions depends on whether they are neutral, acidic, or alkaline when dissolved in water. For example, aluminum sulfate is the salt of a strong acid (sulfuric acid) and a weak alkali (aluminum hydroxide). When dissolved in water, a solution of aluminum sulfate is acidic in nature and reacts with iron as would a dilute acid. Since alkaline solutions are generally less corrosive to ferrous systems, it is practical in many closed water systems to minimize corrosion by adding alkali of alkaline salt to raise the pH to 9 or higher.

Differential Solute Concentration

A potential difference between anode and cathode areas is necessary for the corrosion reaction to proceed. Such potential difference can be established at different locations on a metal surface because of differences in concentration cell corrosion. Such cells can be metal ion of oxygen concentration cells.

In the metal ion cell, the metal surface in contact with the higher concentration of dissolved metal ion becomes the cathodic area, and the surface in contact with the lower concentration becomes the anode. The metal ions involved may be constituent of the media or may result from the corroding surface itself. The differences in concentration in the media may be caused by flow of the media sweeping away the dissolved metal ions at one location and not at another. Such differences could occur at crevices or be caused by deposits of one sort or another. The anodic area is outside the crevice or deposit where the metal ion concentration is least.

In the oxygen concentration cell, the surface area in contact with the media of higher oxygen concentration becomes the cathodic area, and the surface in contact with the media of lower oxygen concentration becomes the anode. Crevices or foreign deposits on the metal surface can produce conditions favorable to corrosion. The anodic area, where corrosion proceeds, will be in the crevice or under the deposit. Although they are actually manifestations of concentration cell corrosion, crevice corrosion and deposit attack are sometimes referred to as different types of corrosion.

Table 1 Galvanic Series of Metals and Alloys in Flowing Aerated Seawater at 40 to 80°F

Corroded End (Anodic or Least Noble)
Magnesium Alloys (1)
Zinc (1)
Beryllium
Aluminum Alloys (1)
Cadmium
Mild Steel, Wrought Iron
Cast Iron, Flake or Ductile
Low Alloy High Strength Steel
Ni-Resist, Types 1 & 2
Naval Brass (CA464), Yel. Brass (CA268), Al. Brass (CA687), Red Brass (CA230), Admlty Brass (CA443), Mn Bronze
Tin
Copper (CA102, 110), Si Bronze (CA655)
Lead-Tin Solder
Tin Bronze (G & M)
Stainless Steel, 12 to 14% Cr (AISI Types 410,416)
Nickel Silver (CA 732, 735, 745, 752, 764, 770, 794)
90/10 Copper-Nickel (CA 706)
80/20 Copper-Nickel (CA 710)
Stainless Steel, 16 to 18% Cr (AISI Type 430)
Lead
70/30 Copper-Nickel (CA 715)
Nickel Aluminum Bronze
INCONEL[a] Alloy 600
Silver Braze Alloys
Nickel 200
Silver
Stainless Steel, 18 Cr, 8 Ni (AISI Types 302, 304, 321, 347)
MONEL[a] Alloys 400, K-500
Stainless Steel, 18 Cr, 12 Ni-Mo (AISI Types 316, 317)
Carpenter 20[c] Stainless Steel, INCOLOY[a] Alloy 825
Titanium, HASTELLOY[b] Alloys C & C 276, INCONEL[a] Alloy 625
Graphite, Graphitized Cast Iron
Protected End (Cathodic or Most Noble)

[a] International Nickel Trademark
[b] Union Carbide Corp. Trademark
[c] The Carpenter Steel Co. Trademark

Galvanic or Dissimilar Metal Corrosion

Another factor that can accelerate the corrosion process is the difference in potential of dissimilar metals coupled together and immersed in an electrolyte. The following factors control the severity of corrosion resulting from such dissimilar metal coupling:

1. The relative differences in position (potential) in the galvanic series, with reference to a standard electrode. The greater the difference, the greater the driving force of the reaction. The galvanic series for metals in flowing aerated seawater is shown in Table 1.
2. The relative area relationship between anode and cathode areas. Since the amount of current flow and, therefore, total metal loss is determined by the potential difference and

resistance of the circuits, a small anodic area corrodes more rapidly; it is penetrated at a greater rate than a large anodic area.
3. Polarization of either the cathodic or anodic area can reduce the potential difference and thus reduce the rate of attack of the anode.
4. The mineral content of water will influence the rate of corrosion because of conductivity. If the water has a low electrical conductivity (low mineral content), the corrosion rate will be less severe than with water containing high mineral concentrations.

It is inadvisable to couple a small exposed area of a less noble metal with a large area of a more noble metal in media where the less noble material may tend to corrode by itself. If such couples cannot be avoided, but one of the dissimilar metals can be painted or coated with a nonmetallic coating, the cathodic material should be coated rather than the anodic one. If the two materials can be insulated from each other by using an insulated joint in piping systems, the galvanic couple can be avoided. Where this is not possible, a *waster* heavy wall nipple section of the less noble material can be used and readily replaced when it fails.

In piping systems handling natural waters, galvanic corrosion is not likely to extend more than three to five pipe diameters down the ID of the less noble pipe material. In metal components exposed to the atmosphere, galvanic effects are likely to be confined to the area immediately adjacent to the joint.

Stray Current Corrosion

Stray current corrosion is a form of galvanic corrosion. The electrical potential driving the corrosion reaction comes from stray electrical currents of an electric generator. This phenomenon can occur on buried or submerged metallic structures. The soil or submergence media provide the electrolyte. The anode, or structure suffering accelerated corrosion, may be located some distance from the cathode structure. The corrosion attack by stray currents is usually restricted to the structure surface in contact with the soil or other submergence media.

Effects of Stress

Stresses in metallic structures rarely have significant effects on the uniform corrosion resistance of metals and alloys. There have been a few instances of stress accelerating corrosion with some materials, where the more highly stressed areas, *i.e.*, pipe threads, are usually anodic to less severely stressed areas. On the other hand, stresses in specific metals and alloys can cause corrosion cracking when exposed to specific corrosive environments. The cracking can have catastrophic effects on the usefulness of the particular metal.

Almost all metals and alloys exhibit susceptibility to stress-corrosion cracking in one or more specific environments. Common examples are steels in hot caustic solutions, high zinc content brasses and ammonia, and stainless steels in hot chlorides. Metal producers' data has more details on specific materials. Stress-corrosion cracking can often be prevented by using the susceptible alloy in the annealed or stress-relieved condition, or by selecting a material resistant to attack by the specific media.

Temperature

According to studies of chemical reaction rates, it is believed that corrosion rates double for every 18°F rise in temperature. However, such a ratio cannot necessarily be applied to corrosion reactions, and the effect of temperature cannot be generalized.

In systems where the presence of oxygen in the media promotes corrosion, temperature increase may increase the corrosion rate to a point. Oxygen solubility will decrease as temperature increases and may approach zero at boiling in an open system. Therefore, the corrosion rate may decrease in oxygen solubility. However, in a closed system, from which oxygen cannot escape, the corrosion rate may continue to increase with temperature rise. For those alloys that depend on oxygen in the media for maintaining a protective oxide film, an increase in temperature and the corresponding reduction in oxygen content can accelerate the corrosion rate by preventing oxide film formation.

Temperature can also affect corrosion behavior by causing a dissolved salt in the media to precipitate on the surface as a scale, which can be protective. One example is calcium carbonate scale in hard waters. Temperature can affect the nature of the corrosion product, which may be relatively stable and protective in certain temperature ranges, and unstable and nonprotective in others. An example of this is zinc in distilled water; the corrosion product is nonprotective from 140 to 190°F but reasonably protective at other temperatures. Thus, the effect of temperature on a particular system is difficult to predict without specific knowledge of the characteristics of the metals involved and the constituents of the media.

Pressure

As with temperature, it is difficult to predict the effects of pressure on corrosion. Where dissolved gases such as oxygen and carbon dioxide affect the corrosion rate, pressure on the system may increase their solubility and thus increase corrosion. Similarly, a vacuum on the system reduces dissolved gas solubility, thus also reducing corrosion. In a heated system, pressure may rise with temperature. It is seldom possible or practical to control metallic system corrosion by pressure control alone.

Cavitation occurs in centrifugal pumps when the net positive suction head is not high enough. This causes either the release of gas (air) bubbles from the water or the generation of steam. The bubbles of gas or steam collapse rapidly as the pressure in the pump decreases. They cavitate (explode) on the metal surface and cause erosion. The attack is also speed related, and it is most severe at the outer edge of the pump impeller.

Velocity

The effects of flow velocity of the media in a system depend on the characteristics of the particular metal or alloy. In media where oxygen increases the corrosion rate, *e.g.*, iron or steel in water, flow velocity can increase the rate by making more oxygen available to the metal surface for reaction. Under the same circumstances, the films of alloys dependent on thin oxide films for corrosion resistance can be enhanced because of velocity, and maintain resistance up to high velocities. This is frequently true of stainless steels.

In metal systems where corrosion products retard corrosion by acting as a physical barrier, velocity of media flow may sweep away these products, permitting corrosion to proceed at its initial rate. In specific media and for specific metal, there may a critical velocity below the point that the corrosion product film is adherent and protective.

Turbulent media flow may cause uneven attack, involving localized erosion and corrosion. This attack is called erosion-corrosion and can occur in piping systems at sharp bends if the designed flow velocity is high. Copper and some of its alloys are subject to this type of attack.

Low velocities can allow suspended solids to precipitate on a metal surface, which initiates concentration cell corrosion.

PREVENTIVE AND PROTECTIVE MEASURES

Materials Selection

Almost any piece of heating or air-conditioning equipment may be constructed with materials that will not corrode significantly under service conditions. However, because of economic and physical limitations, this option is rarely possible.

When selecting construction materials for a piece of equipment, consider (1) the corrosion resistance of each proposed metal or other material to the service environment, (2) the nature of the corrosion products that may be formed and their effects on equipment operation, (3) the suitability of the materials to handling by standard fabrication methods, and (4) the effects of design and fabrication limitations on tendencies toward local corrosion (as discussed in the preceding section). The overall economic balance during the projected life of the equipment should also be considered. It may be less expensive in the long run to pay more for a corrosion-resistant material and avoid regular painting or other corrosion-control measures, than to use a less expensive material that requires regular corrosion control maintenance.

1. Use of dissimilar metals in any system should be avoided or minimized. Where dissimilar materials are used, insulating gaskets and/or organic coatings must be used to prevent galvanic couples.
2. Temperature and pressure-measuring equipment using mercury should be forbidden, since accidental breakage and mercury spillage into the system can cause severe damage.
3. Stagnation or sluggish fluid flow, unnecessarily high fluid velocities, impingement, and erosion should be eliminated through design of equipment and accessories.
4. Corrosion inhibitors and biological control additives must be compatible with the materials chosen for any given system.

Protective Coatings

The environment in which system components eventually operate largely determines the type of coating system needed. Even with a coating suited for the environment, the protection provided depends on the adhesion of the coating to the base material, which is decided by both surface preparation and application technique. Consult several reliable manufacturers of industrial coatings before choosing a coating system for any exposure and follow the manufacturer's recommendations for both surface preparation and application technique.

Shop-applied coatings offer the best results because better conditions for controlled surface preparation and coating application usually exist in the shop.

Field-applied coatings can be the same types as shop-applied coatings. The limiting factor is accessibility. Once equipment is in place, obstructions can make it difficult to use the type of surface preparation that will do the best job.

Maintenance of Protective Coatings. Defects in a coating are virtually inescapable. These defects can be caused either by coating flaws in the film during application or by mechanical damage after coating; the damage must be repaired to eliminate premature failure at these points. Depending on the severity of the service, an inspection should be scheduled after installation to detect mechanical damage. Damaged areas should be noted and repaired.

Cathodic Protection

Cathodic protection reduces or prevents corrosion by making the metal the cathode in a conducting medium by means of direct electric current that is impressed or galvanic. It is widely used to control corrosion in water storage tanks; heat exchanger water boxes; water and chemical processing equipment such as filters, reactors, and clarifiers; and the external surfaces of submerged and underground tanks, piping, and piling.

Although cathodic protection can be applied to bare structures underground, it is most widely used to provide corrosion control for exposed metal at the coating flaws that inevitably occur on the coating during and after installation. The rate of metal penetration at coating flaws is often greater than that on bare surfaces.

Cathodic protection may be used with iron, aluminum, lead, and stainless steel. On new steel structures, optimum corrosion control design is often achieved by applying cathodic protection in combination with coatings, environment conditioning, or both.

The cathodic protection principle is unique in that protective effects can be directed from distantly positioned anode current sources onto existing submerged or buried structures. Corrosion control can generally be accomplished by cathodic protection without taking the facilities out of service, exposing the surfaces for coating, or specially treating the surrounding environment.

Since this process superimposes an applied current onto an existing electrochemical corrosion system, its design must be adapted to meet the varying needs of the specific corrosion problem. The electrochemical corrosion mechanisms to which cathodic protection can be applied fall into the following two broad classifications:

1. Corrosion of a metal surface by stray direct currents, which flow in circuits grounded at more than one point or in subsurface structures purposely made part of a direct current circuit.
2. Corrosion of a metal surface in an electrolyte by galvanic currents originating between discreet areas of oxidation and reduction reactions. Galvanic currents are the effect rather than the cause of corrosion.

Complete corrosion control by cathodic protection is approached when the net current flow at any point on the metal surface either measures zero or is flowing from the corroding media into the metal. Generally, it is not feasible to measure current flow directly at all points on a metal surface. Full cathodic protection requires polarizing the cathode areas to the open circuit potential of the anodes. The criterion for cathodic protection of iron is met when all points on the metal surface are polarized to a potential of −0.85 V or more negative, measured against a copper sulfate reference electrode positioned on the metal surface.

The protective current requirements generally vary with factors influencing corrosion rates. Increasing oxygen concentration, temperature, and velocity increases protective current requirements. Resistive coatings, precipitated calcareous salts, adherent zinc or aluminum flocs, silt, and electrophoretically deposited particles all reduce the protective current requirements. Adequate protective current flow onto a surface from sacrificial or impressed current sources is equally effective.

Sacrificial Anodes. Coupling a more active metal to a structure results in galvanic current flow through the corroding electrolyte, providing a protection effect on the cathode surface. In providing the galvanic protective current flow, the more active metal is electrochemically consumed (sacrificed) and must be replaced. No outside power is required for protection. The properties of common sacrificial anode materials are listed in Table 2.

Table 2 Properties of Sacrificial Anode Materials

Material (and Specification)	Theoretical lb/A · yr	Actual lb/A · yr	Potential (Cu-$CuSO_4$) Volts
Magnesium (AS-63A)	8.5	17	−1.55
Zinc (MIL-A-18001)	23.0	24	−1.10[a]
Aluminum (B-605)	6.5	12	−1.05[a]
Aluminum (ERP-HP7)	6.5	8	−1.20[a]

[a] Seawater

The sacrificial anode consumption is greater than the theoretical electrochemical equivalent of its protective current output. This is attributable to the self-corrosion current flow superimposed on the protective current output. Sacrificial anode system design is limited by the available driving voltage between the sacrificial anode and the structure to be protected. The resistance of the circuit between the anodes and the structure mainly determines the protective current flow. A high resistivity electrolyte requires a larger number of anodes than a more conductive electrolyte to obtain the same amount of protective current flow.

Sacrificial anodes are cast in various forms and weights for optimum design and service life. Special backfills around the anodes are used in soils to maintain moisture at the anode surface and to increase efficiency. In condenser water boxes and similar exposures, sacrificial anodes are sometimes encased in perforated plastic containers for optimum service life and performance.

The zinc on galvanized iron protects the underlying metal until it is consumed in the protective process. Zinc and aluminum anodes can usually be used in combination with well-coated surfaces without accelerating coating damage.

Impressed Current. An external voltage source can be used to impress protective current flow from anodes through the conducting medium onto the corroding surface. Alternating current power, converted to direct current by an adjustable output selenium or silicon rectifier, is most commonly used.

Sacrificial anodes, such as scrap iron or aluminum, are sometimes used with impressed current. Nonsacrificial anodes (those not consumed by the electrochemical process) such as graphite, high silicon cast iron alloy, platinum or platinum plated or clad on titanium or tantalum, and silver-lead alloys in seawater are more widely used in impressed current systems. The distribution of protective current on the cathode, power costs, and stray current effects on neighboring structures should be considered when selecting number, form, and arrangement of anodes.

One method of impressed current cathodic protection is automatic potential control (Sudrabin 1963). In this system, the potential of the structure is continuously measured against a reference electrode by a monitoring system that regulates the protective current applied to maintain the structure at a preselected protective potential value. Since the protective current requirements vary with conditions, an automatic potential control system protects at all times and prevents excessive cathodic protection, which would accelerate (1) coating damage by electroendosmotic effect, (2) cathodic attack on amphoteric metals such as aluminum, and (3) power wastage.

Limitations. Full cathodic protection depends on adequate current flow reaching all surfaces to be protected. On bare surfaces, the relation between an anode and cathode configuration determines, in large part, the distribution of protective current. For example, relatively simple anode systems adequately distribute protective current on the outside of storage tanks, well-coated pipelines, tank interiors, and water boxes of heat exchangers. However, the protective current received on the interior of a condenser tube from anodes in the water box diminishes rapidly within a very short distance of the tube entrance.

Rapid attenuation of the protective effect occurs on the exterior of bare pipelines in conductive media. Therefore, protective current sources must be applied at more frequent intervals. Harmful effects of stray currents from a cathodic protection system on neighboring utilities or other isolated metallic systems must be considered. All underground cathodic protection installations should be reported to, and examined in cooperation with, local electrolysis committees.

Economics. Since the cathodic protection principle must be adapted to meet the specific needs of each corrosion problem, costs vary.

Environmental Change

Another approach to corrosion protection is to change the environment so that it is less aggressive to the equipment. A familiar example is the use of solid desiccants, such as silica gel, to maintain a low moisture level in refrigerant lines. Desiccants are applied to minimize atmospheric corrosion by maintaining a low relative humidity. The mothballing technique for preserving decommissioned ships and other heavy equipment is an example of this. These methods are employed for dry lay-up.

Vapor phase corrosion inhibitors can minimize atmospheric corrosion in certain cases. Adding alkali to water to raise the pH and using corrosion inhibitors, such as phosphates, are other examples of reducing corrosive attack by changing the environment. Corrosion inhibition in water systems is discussed more fully in the next section. Inhibitors for corrosion control are added to glycol antifreeze solutions, lubricants, lithium bromide absorption refrigeration brines, and other liquids.

The local environment at a metal surface can be made less corrosive by adding filming amines to steam to protect condensate lines and certain additives to fuel oil to reduce fireside corrosion in boilers.

Finally, equipment design modifications that reduce the likelihood of corrosion are also considered environmental changes. They include eliminating crevices and providing weep holes to prevent water accumulation.

UNDERGROUND CORROSION

Corrosion activity on subsurface structures must always be anticipated. In addition to the economic loss when the underground facilities are destroyed by corrosion, loss of valuable fluids and continuity of service, and creation of hazards to life and property must be considered. Corrosion control measures must be selected and adapted to meet the specific conditions in which the structure will exist. A corrosion survey is usually necessary before corrosion control measures can be designed for underground structures.

Types of Soils

The corrosivity of a soil is affected by its porosity (aeration), electrical resistivity, dissolved salts (depolarizers and inhibitors), moisture, and acidity or alkalinity. Although corrosion rates and characteristics cannot be related exactly to the individual factor, the attack on a metallic structure buried in nonuniform soil will be greatest on those surfaces in contact with the least porous (air-free), least resistive, most alkaline, or most acidic soil. Romanoff (1957) conducted a long-term study of underground corrosion from 1910 until 1955. These studies cover soil corrosivity, materials, coatings, cathodic protection, and stray currents.

Bacterial Activity

High rates of corrosion in some air-free environments are associated with the presence of sulfate-reducing bacteria (*Desulphovibrio spp*). Bacterial activity is not a special or new form of corrosion. The conventional use of well-applied coatings and cathodic protection will control corrosion on underground metallic structures in the presence of these bacteria. Coatings that disbond, allowing water and corrodent to reach the metal surface, will shield the exposed metal from the effects of applied cathodic protection.

Insulation Failures

Most thermal insulation does not protect against corrosion soils or water. Catastrophic corrosion occurs on hot pipe surfaces when intermittently contacted by water (Sudrabin 1956). This corrosion process cannot be controlled by cathodic protection. Insulation, such as glass fiber or magnesia fiber, must be kept dry in underground runs through conduits. Severe corrosion has been experienced within a few months after construction on magnesia-asbestos insulated, snow-melting pipe manifolds extending below the slab into the soil (Sudrabin and LeFebvre 1953). Natural asphalt materials used for underground pipe heat insulation exhibit some corrosion protective effect. However, they are relatively permeable to water, and repeated heating and cooling initiates attack under the insulation. Similarly, corrosion has occurred at locations where the water table fluctuates.

Radiant heat and snow-melting pipe embedded in slabs constructed with open expansion joints and vermiculite-filled concrete, brick, or concrete block supports have suffered severe corrosion (Sudrabin and LeFebvre 1953). Severe corrosion of radiant heating pipe embedded in sand or porous concrete under terrazzo flooring has also been experienced. In such heating systems, good construction practice requires a minimum thickness of 1.5 in. of cement-rich concrete to surround the piping, a waterproof membrane under the slab, and steel saddle pipe supports. The piping should be coated at expansion joints, and pipe should not contact the reinforcing steel. Lightweight or insulating concrete produces local cell activity at the nonuniformities in contact with the piping. The piping should be insulated from all metallic structures. Magnesium anodes buried in the adjoining soil may be attached to the piping for protection.

Cathodic Protection

Although cathodic protection has been applied to radiant heating piping, it is not always possible to direct adequate amounts of protective current to corroding surfaces. The best corrosion control measures are those integrated into the original design and construction of the heating system.

Galvanic anodes (magnesium or zinc) or impressed current systems can be used for cathodic protection of underground pipelines and tanks. Impressed current may be applied from anodes distributed alongside the structure, from remote anodes, or from deep well anodes.

The method used depends on the economics and the site conditions affecting the cathodic protection design requirements. Full cathodic protection is achieved when enough current is applied to the structure to prevent current flow into the soil from any and all points on the surface of the structure. Specifically, this protective current must be applied to the structure in an amount sufficient to maintain its external surface negative at every point by at least 0.85 V to a copper-saturated copper sulfate half-cell in its immediate proximity.

Electrically Insulating Protective Coatings

Protective coatings applied to underground structures isolate them from the soil environment and insulate them from electrical effects. Such coating materials must possess long-term qualities of high electrical resistance, inertness to the environment, low water absorption, high resistance to deformation by soil pressures and the temperatures at which they are operated, and good adhesion. Most coatings require reinforcing and/or shielding to resist soil stresses. The performance of all coating systems, even those of the best material specifications, depends on the metal surface preparation; application procedure and conditions; backfilling; and the physical, chemical, biological, and electrical stresses that occur in service.

The most common coatings used for underground structures include (1) hot-applied coal tar enamel; (2) hot-applied petroleum base coatings, such as asphalt enamels and waxes; (3) polyethylene or polyvinyl (tapes and extruded polyethylene); (4) coal tar epoxy resin; and (5) cold-applied bituminous and asphalt emulsions, cutback solvents, and greases. Cold-applied coatings are formulated by manufacturers for specific application and service conditions. The National Association of Corrosion Engineers (NACE) has prepared statements on minimum requirements for coal tar coatings, asphalt-type protective coatings, and prefabricated plastic films.

WATERSIDE CORROSION AND DEPOSITS

Since the occurrence and correction of corrosion in water systems are so closely related to the occurrence and correction of other water-caused pollution, it is impossible to consider them separately. The most common of these water problems in heating and cooling systems is one or more of the following: (1) corrosion, (2) scale formation, (3) biological growths, and (4) suspended solid matter. Some knowledge of these problems is important because each of them can reduce the cooling or heating efficiency of a system and can lead to premature equipment failure and widespread damage to people in the vicinity.

Controlling waterside problems, particularly in heating and cooling systems, is complex. Handling these problems involves water chemistry, engineering, economics, and personnel administration during each stage of the system development, design, construction, installation, and operation.

Water Characteristics

Between the time that water falls as rain, sleet, or snow and the time that it is pumped into a user's premises, it dissolves a small amount of almost every gas and solid substance with which it comes in contact. These dissolved impurities, rather than the water itself, are the primary cause of various water problems.

Table 3 indicates the complexity of water chemistry. Water received at a given location can vary widely from time to time, either because supplies from different sources are being used or because the composition of a single supply fluctuates, as in the case of river water.

Table 3 Analyses of Typical Public Water Supplies

		Location or Area[a]								
Substance	**Unit**	**(1)**	**(2)**	**(3)**	**(4)**	**(5)**	**(6)**	**(7)**	**(8)**	**(9)**
Silica	SiO_2	2	6	12	37	10	9	22	14	—
Iron	Fe_2	0	0	0	1	0	0	0	2	—
Calcium	Ca	6	5	36	62	92	96	3	155	400
Magnesium	Mg	1	2	8	18	34	27	2	46	1,300
Sodium	Na	2	6	7	44	8	183	215	78	11,000
Potassium	K	1	1	1	—	1	18	10	3	400
Bicarbonate	HCO_3	14	13	119	202	339	334	549	210	150
Sulfate	SO_4	10	2	22	135	84	121	11	389	2,700
Chloride	Cl	2	10	13	13	10	280	22	117	19,000
Nitrate	NO_3	1		0	2	13	0	1	3	—
Dissolved Solids		31	66	165	426	434	983	564	948	35,000
Carbonate Hardness	$CaCO_3$	12	11	98	165	287	274	8	172	125
Noncarbonate Hardness	$CaSO_4$	5	7	18	40	58	54	0	295	5,900

[a] All values are ppm of the unit cited to nearest whole number (Collins 1944). Numbers indicate location or area as follows:

(1) Catskill supply—New York City
(2) Swamp water (colored)—Black Creek, Middleburg, FL
(3) Niagara River (filtered)—Niagara Falls, NY
(4) Missouri River (untreated)—average
(5) Well waters—public supply—Dayton, OH—30 to 60 ft
(6) Well water—Maywood, IL—2090 ft
(7) Well water—Smithfield, VA—330 ft
(8) Well water—Roswell, NM
(9) Ocean water—average

Another often overlooked characteristic affecting problems caused by water is the change in composition of the water added to a system after the system starts to operate. Changes in chemical composition result from evaporation, aeration, corrosion, and scale formation. These chemical changes, temperature changes, formation of biological growths, and the accumulation of suspended matter all tend to produce operating results that can differ from those based on the chemical analysis of the makeup water.

Chemical Characteristics. The type and amount of dissolved inorganic materials, including gases, define the chemical characteristics of any water. Typical water analyses, such as those shown in Table 3, are not complete but give the major important constituents for municipal and average industrial water use. The many minor constituents present in most water supplies have little or no importance in most water uses.

In water analyses, values are usually given as parts per million. (Note: To obtain ppm, multiply gr/gal by 17.) When water analyses are reported in ppm, the chemical species must also be given. Thus, the same calcium concentration in a single water sample might be variously expressed as 100 ppm as CaCO, 56 ppm as CaO, or 40 ppm as Ca. Generally, lower pH water tends to be more corrosive and higher pH water tends to be more scale-forming, although many other factors affect both properties.

Most water analyses include only dissolved solids and omit the dissolved gases that are also present. Certain gases, such as nitrogen, have virtually no effect on any water use; others, such as oxygen, carbon dioxide, and hydrogen sulfide, produce important effects in water systems. Carbon dioxide either can be measured directly or estimated from the pH and total alkalinity.

To be meaningful, oxygen and hydrogen sulfide must be measured by special techniques at the time the sample is collected. Oxygen is important in the corrosion of metals and is readily dissolved on contact with air in many water systems. Thus, the water chemist often makes the conservative assumption that any water supply that contacts air will be saturated with oxygen according to its partial pressure in air and the water temperature.

Total hardness is the most commonly recognized chemical constituent in water analysis. In most cases, it corresponds to the calcium and magnesium content of the water. The hardness, particularly the calcium, is one of the factors influencing scale formation. The scaling potential increases with increasing hardness.

Alkalinity is a measure of the capacity of a water to neutralize strong acids. In natural waters, the alkalinity almost always consists largely of bicarbonate, although there may also be some carbonate present. In treated waters, borate, hydroxide, phosphate, and other constituents, if present, will be included in the alkalinity measurement. Alkalinity also contributes to scale formation. In water chemistry, factors that increase scale formation decrease corrosion, and vice versa.

Alkalinity is measured using two different end-point indicators. The phenolphthalein alkalinity (or P alkalinity) measures the strong alkali present; the methyl orange alkalinity (or M alkalinity), or total alkalinity, measures all of the alkalinity present in the water. Note that the total alkalinity includes the phenolphthalein alkalinity. For most natural waters, in which the concentration of phosphates, borates, and other noncarbonated alkaline materials is small, the actual chemical species present can be estimated from the two alkalinity measurements (Table 4).

Table 4 Alkalinity Interpretation for Natural Waters

P Alk[a]	Carbonate	Bicarbonate	Free Carbon Dioxide
0	0	= M Alk	Present
< 0.5M Alk[b]	= 2 P Alk	= M Alk − 2 P Alk	0
= 0.5M Alk	= 2 P Alk = M Alk	0	0
> 0.5M Alk[c]	= 2 (M Alk − P Alk)	0	0

[a] P Alk = Phenolphthalein Alkalinity.
[b] M Alk = Methyl Orange (Total) Alkalinity.
[c] Treated waters only. Hydroxide also present.

Alkalinity or acidity is often confused with pH. Such confusion may be avoided by keeping in mind that the pH is a measure of hydrogen ion concentration, in moles per litre, expressed as the logarithm of its reciprocal.

Dissolved solids also affect corrosion and scale formation. Low solids waters are generally corrosive because they have less tendency to deposit scale. If a high solids water is nonscaling, it will tend to produce more intensive corrosion because of its high conductivity. Dissolved solids are often referred to as total dissolved solids (TDS).

Specific conductance measures the ability of a water to conduct electricity. Conductivity increases with the TDS. Specific conductance can be used to estimate dissolved solids.

Sulfates also contribute to scale formation in high calcium waters. Calcium sulfate scale, however, forms only at much higher concentrations than the more common calcium carbonate scale. High sulfates also contribute to increased corrosion because of their high conductivity.

Chlorides have no effect on scale formation but do contribute to corrosion because of their conductivity and because the small size of the chloride ion permits the continuous flow of corrosion current when surface films are porous. Chlorides are a useful measuring tool in evaporative systems. Virtually all other constituents in the water increase or decrease as a result either of the addition of common treatment chemicals or of chemical changes that take place in the normal operation of the water system. With few exceptions, chlorides are changed only by evaporation, so that the ratio of chlorides in a water sample from an operating system to those of the makeup water provides a measure of the number of times that the water has been concentrated in the system. (Note: Chloride levels will change if the system is continuously chlorinated.)

Soluble iron can arise from metallurgy corrosion within the cooling water systems or as a contaminant in the makeup water supply. The iron can form heat-insulating deposits by precipitation as iron hydroxide or iron phosphate (if a phosphate-based water treatment product is used or if phosphate is present in the makeup water).

Silica can form particularly hard-to-remove scales if permitted to concentrate sufficiently. Fortunately, silicate scales are far less common than others. In addition to dissolved solids, waters (particularly unpurified waters from surface sources or those that have been circulating in open equipment) frequently contain suspended solids, both organic and inorganic. Organic matter in surface supplies may be present as colloidal solutions. Natural coloring matter usually occurs in this form. At high velocities, hard suspended particles can abrade equipment. Settled suspended matter of all types can contribute to concentration cell corrosion.

Biological Characteristics. Bacteria, algae, fungi, and protozoa are present in water systems, and their excessive growth can cause problems in the system's proper operation. Microorganisms are directly affected by the temperature of their habitat. Thus, for every 18 °F increase in temperature of the activity, the microorganism can increase 2 to 3 times, as long as that temperature does not exceed the optimum for the organism. Microbial examination of water and wastewater for sanitary and industrial purposes has established procedures for detecting the presence or absence of selected pathogenic microorganisms. In cooling systems, nonpathogenic microorganisms are the more important cause of operating difficulties. The vast majority of environmental microorganisms are rarely associated with operating problems in a cooling system, when temperatures exceed the optimum growth temperature—generally 100 °F—for these mesophiles, middle temperature-loving microorganisms. Useful microbiological tests have been developed for application in industrial water problems (APHA 1989).

Heating and Cooling Water Systems

To evaluate the probable effects of the water supply on the boiler or recirculating water, the treatment specialist needs information on the size, construction, operating pattern, and other characteristics of the system. All heating and cooling systems involve a temperature change, which influences the rates at which water-caused problems can develop.

In steam heating systems, the concentration of dissolved solids in the boiler water increases as boiler water is converted to steam. In the ordinary space heating system, this change is limited because all the steam is condensed and returned to the boiler. On the other hand, the nature of the water changes taking place within the boiler are drastically altered if the percentage of condensate returning to the boiler is reduced. The condensate has characteristics quite different from those of the boiler water, because it is substantially free of dissolved solids but can create different problems due to the presence of dissolved oxygen or carbon dioxide.

Hot water heating and other closed circulating systems generally have few water-caused problems because there should be virtually no makeup water and no opportunity for significant changes in the composition of the water within the system. Experience has shown, however, that most of these systems do not operate according to this theory (Sussman and Fullman 1953); thus engineers accept the need for water treatment in both ordinary hot water heating systems and high-temperature hot water systems.

Like hot water heating systems, closed cooling systems (chilled water, chilled-hot water, brine, or glycol) require more makeup in practice than in theory. As a result, treatment of the circulating water or solution in these systems is desirable. Monitoring these systems is advisable to ensure that treatment levels and corrosion rates meet specifications.

Condensers and other heat exchangers in refrigeration and air-conditioning systems can be cooled by water passing through the equipment and going to waste. In such cases, the water-caused problems are directly related to the chemical composition of the cooling water. As a result of increasing water shortages, such once-through cooling is becoming less common and is being replaced by open circulating systems that use cooling towers, evaporative condensers, or spray ponds. In these, the makeup water composition is drastically changed by evaporation, aeration, and other chemical and physical processes that depend on the contaminants present in the air to which the water is exposed in the open circuit.

Humidification and dehumidification equipment presents several types of water problems. During humidification, water circulating in the equipment is subjected to the same exposure conditions and composition changes as those that take place in open circulating systems. During dehumidification, however, water is removed from the surrounding air. Dissolved solids will be low except for the materials that may be scrubbed from the air. These dissolved solids may not contribute to conductivity, and the condensate may be relatively corrosive.

Materials of Construction. The seriousness of corrosion in heating and cooling systems depends, in part, on the nature of the materials of construction present, their degree of proximity, their relative areas, and other physical factors, as well as the chemical composition of the water.

Water-caused Problems

The operator of a heating or cooling system can recognize water-caused problems by the appearance of one or more of three symptoms, each of which can be produced by causes other than the water. Several different conditions can act together in the production of any given symptom, namely (1) reduction in heat transfer rate, in which case the formation of an insulating deposit on a heat transfer surface significantly reduces the cooling or heating efficiency of the equipment; (2) reduced water flow, which results from a partial or complete blockage of pipelines, condenser tubes, or other openings; and (3) damage to or destruction of the equipment, which can result from corrosion of metals or deterioration of wood or plastics. It can also be caused by excessively rapid wear rates of moving parts such as pumps, shafts, or seals.

Corrosion contributes to all three symptoms. Heat transfer rates can be reduced and water flow blockages can be created by deposits of corrosion products. The damage or destruction of equipment is most often the result of metallic corrosion and the somewhat parallel phenomenon of wood delignification.

Scale formation also contributes to all three symptoms. Even a small buildup of scale on a heat exchange surface reduces water flow. Finally, scale buildup in boilers may continue until it reaches the point at which heat transfer is so low that the metal overheats, permitting the tubes to rupture under the operating pressure. Scale particles can also accelerate the wear of moving parts.

Biological growths, such as slime, can act as insulators to reduce cooling efficiency. Bacterial and algae growths often accumulate to the point where they interfere with water flow in cooling towers. The poultice effect created by accumulations of organic matter can cause localized corrosive attack resulting in premature equipment failure. Destruction of cooling tower wood can also occur as a result of fungus action.

Suspended solid matter, such as dirt scrubbed from the air or finely divided mill scale, can also contribute to all three symptoms. These deposits can reduce heat transfer or water flow, depending on where they accumulate. Like local deposits of organic matter, dirt deposits tend to localize corrosion, and suspended solids cause rapid wear of moving parts.

Water Treatment

General Considerations. Since correcting water-caused problems in heating and cooling systems is complicated, it is important to consult a water treatment specialist early in the design stage of any system and regularly thereafter during design, construction, and operation. Such factors as the improper use of water treatment chemicals can cause more serious problems than those that would have occurred without treatment.

Frequently, more than one solution is possible for the same water treatment problem. The selection may be dictated by such considerations as economics, available space, or labor. A water treatment program selected for any given system should be based on the factors mentioned in this chapter. The selected treatment program must be followed diligently because it is only as effective as the consistency and control with which it is applied.

Using chemicals for water treatment requires certain safety precautions with which building, operating, and maintenance personnel are not ordinarily familiar. It is important that (1) the potential hazards associated with any particular water treatment chemical or program be well understood; (2) suitable safety rules be formulated; (3) appropriate safety equipment be supplied; and (4) the safety program be enforced at all times to avoid injury or equipment damage. Material Safety Data Sheets (MSDS) must be readily available to employees who handle chemicals. Also, contamination of drinking water by nonpotable treated or untreated water must be prevented by eliminating cross-connections between systems or by providing approved backflow preventers. Disposal of waters treated with some chemicals into municipal sewers or into streams or lakes may be restricted. Pollution control regulations should be consulted when selecting water treatment. Nonpolluting chemicals may not be as effective as those used in the past. In some cases, removal of treatment chemicals prior to discharging blowdown or drainage can be economically justified.

With very few exceptions, the proper control of water treatment programs that involve the addition of chemicals to the water depends on proportional feeding of the chemicals to maintain a desired concentration level at all times. Intermittent batch or slug feeding of water treatment chemicals cannot be relied on to produce satisfactory results, particularly in systems that have appreciable makeup rates.

Sound water treatment programs require care and consistent attention, yet devices appear on the market that allegedly prevent scale and corrosion without requiring the operator's attention. Various natural forces, such as electricity, magnetism, or catalysis, generally behaving in some new way, are said to be responsible for the effects claimed. However, independent investigations of these devices have found them to produce no significant effect in preventing or correcting corrosion and scale formation (Meckler 1974; Rosa 1988, 1989).

Corrosion Control. Corrosion damage to water systems can be minimized by using corrosion-resistant construction materials, providing protective coatings to separate the water from the metal surfaces of the equipment, removing oxygen from the water, or altering the water composition by adding corrosion inhibitors and pH control chemicals. Two or more of these methods are often used in the same system.

Corrosion control by the selection of corrosion-resistant materials of construction is the responsibility of the equipment or the system designer. Although it is technically possible to build equipment that shows no significant corrosion under almost any operating conditions, economic limitations usually make this impossible. On the other hand, investigation of corrosion failures in air-conditioning and heating equipment sometimes reveals design errors which prove that elementary principles of corrosion control have been ignored.

When controlling corrosion by water treatment, oxygen removal is effective for closed systems in which opportunities for the pick-up of additional oxygen are small. Thus, boiler feedwater can be deaerated mechanically in an open heater or deaerating heater. This process is based on the reduced solubility of oxygen in water at high temperatures (Table 5) and is made more effective by equipment with features that reduce the partial pressure of oxygen in the gas above the water. The last traces of oxygen are removed chemically by adding sodium sulfite or, at higher temperatures, hydrazine, or other types of organic oxygen scavengers. Hydrazine is a suspect carcinogen, however, and must be handled only with skin and respiratory protective equipment.

$$\text{Sulfite: } 2\,Na_2SO_3 + O_2 \rightarrow 2\,Na_2SO_4$$

$$\text{Hydrazine: } N_2H_4 + O_2 \rightarrow N_2 + 2H_2O$$

Oxygen is less often removed chemically from cold water circuits because of the slow rate of reaction of the sodium sulfite with the dissolved oxygen; however, a small amount of cobalt salt acts as a catalyst to speed the reaction (Pye 1947). In open-spray systems, chemical removal of oxygen would be too expensive because the circulating water is thoroughly oxygenated again with each passage through the spray equipment.

Table 5 Solubility of Oxygen from Air in Water at Different Temperatures (Nordell 1961)

Temperature, °F	Millilitres per Litre (mL/L)				
	Air	=	Oxygen	+	Nitrogen
32	28.64	=	10.19	+	18.45
41	25.21	=	8.91	+	16.30
50	22.37	=	7.87	+	14.50
59	20.11	=	7.04	+	13.07
68	18.26	=	6.35	+	11.91
77	16.71	=	5.75	+	10.96
86	15.39	=	5.24	+	10.15
104	13.15	=	4.48	+	8.67
122	11.40	=	3.85	+	7.55
140	9.78	=	3.28	+	6.50
176	6.00	=	1.97	+	4.03
212	0.00	=	0.00	+	0.00

Oxygen may be removed by vacuum deaeration to minimize corrosion on once-through cooling systems. It is particularly applicable for an appreciable carbon dioxide concentration, because the carbon dioxide as well as the oxygen is removed.

Corrosion control treatment of heating and cooling waters is most often a combination of controlling pH and maintaining a corrosion inhibitor in the water. Adjustment of the pH to 7.0 is not sufficient to stop corrosion, however. When the oxygen is removed, control of pH at certain levels is frequently adequate, as in the case of low-pressure heating boilers maintained at a pH above 10.5. In most other cases, both an inhibitor and pH control is required.

Historically, chromate was used as a highly effective and inexpensive corrosion inhibitor at concentrations from 200 to 2000 ppm as sodium chromate. Cooling system pH was controlled by sulfuric acid addition at pH 6.5 to 7.5 to minimize formation of calcium carbonate scale. Calcium carbonate and other scaling ions are more soluble at low pH than at high pH. Disposal of chromate-treated waters is subject to increasingly stringent pollution control regulations. State and federal regulations require chromate to be removed from the water before it is discharged to sewers or public waterways. The appropriate regional office of the Environmental Protection Agency provides these guidelines. Because chromate leaves the cooling tower as part of drift, and as chromate is a suspect carcinogen, the EPA has banned the use of chromate for comfort cooling systems in the United States.

Alternatives to chromate treatment for comfort cooling systems include blends of phosphate, phosphorate, molybdate, zinc, silicate, and various polymers for scale and corrosion control. Tolyltriazole or benzotriazole are added to these blends to protect copper and copper alloys from corrosion. When selecting a treatment product, consult local, state, and federal environmental guidelines.

In closed systems, chromates are becoming increasingly unacceptable because of their yellow color, disposal problems, and potential hazard if used carelessly. Sodium nitrite has been used as an inhibitor instead of chromate. With ferrous metals, it is nearly as effective as chromate but must be maintained at a pH above 7.0 to avoid breakdown and at a minimum concentration of about 500 ppm. It is also necessary to monitor nitrite, nitrate, and ammonia concentrations at frequent intervals, because certain bacteria may use the nitrate as an alternate electron acceptor, converting it to nitrogen gas. Alternatively, other bacteria may convert it to ammonia. As the nitrite disappears, corrosion protection decreases. Sodium nitrite has little or no protective effect on nonferrous metals, and other inhibitors must be included in the formulation to provide protection. A commonly used nitrite-base inhibitor includes borax as a pH buffer and sodium tolyltriazole as an inhibitor for nonferrous metals. Historically, hydrazine was used in closed loops, but it is a suspect carcinogen. Organic inhibitors are likely to see more use as chromates and hydrazine are used less.

Under the proper conditions, polyphosphates reduce tuberculation and pitting. For this purpose, polyphosphate concentrations higher than those used for scale control must be provided, as must close pH control, in the 6.0 to 7.0 range. In addition, polyphosphates must be combined with other inhibitors to control pitting corrosion of copper and steel.

Corrosion-control treatment of once-through cooling water is practical only with very inexpensive chemicals. Sodium silicate (water glass) or phosphate-silicate mixtures control corrosion in once-through systems, including potable water systems. Sufficient silicate is fed to increase the silica content of the water by about 8 ppm.

The delignification of wood in cooling towers refers to the loss of the primary structural component of wood, lignin. Over time, lignin can be dissolved by excessive doses of chlorine (greater than

1 ppm) or by high alkalinity. Fungal decomposition of wood has already been mentioned. Prevention of fungal deterioration is more effective than attempted cures. Manufacturers of cooling towers can provide wood that has been pressure treated with solutions of copper, chromates, and arsenic (CCA).

Scale Control. The methods used for scale control in heating and cooling systems include a variety of internal and external treatment procedures. Internal treatment, where chemicals are added directly to the water in the systems, is frequently used in smaller systems. External treatment means that the water received some modifications prior to the addition to the system; an example of this is softening the water prior to sending it to the boiler. The selection of an appropriate method for any one system requires the evaluation of many factors.

Scale control methods attempt to minimize the likelihood for precipitation of calcium carbonate, the least soluble common constituent of water. The solubility of calcium carbonate depends on the pH, temperature, and total solids content of water, in addition to the calcium and alkalinity (bicarbonate or carbonate). Using these items and one of several nomographs (such as that in Figure 1), the pH (*i.e.*, the pH at which any given water is in equilibrium with calcium carbonate), can be calculated. This pH can be used with the actual pH of the water in either of two calculations that indicate whether the water tends to precipitate calcium carbonate or to dissolve it. The older of these is the Langelier Saturation Index (1936).

$$\text{Saturation Index} = \text{pH} - \text{pH}_s$$

A positive Saturation Index shows a scale-forming tendency. The larger the index, the greater this tendency only. Other factors may inhibit scale formation under some circumstances. Usually, calcium carbonate precipitates as a scale when the Saturation Index exceeds +0.5 to +1.0. A negative Saturation Index indicates that calcium carbonate will dissolve and that bare metal will remain bare and thus accessible for corrosion.

Ryznar (1944) suggested a modified method for predicting scale formation based on operating performance, the Stability Index.

$$\text{Stability Index} = 2\,\text{pH}_s - \text{pH}$$

The Stability Index is always positive. When it falls below 6.0, scale formation is possible, and it becomes more probable the lower the numerical value of the index.

Similar but more difficult methods for estimating the scale-forming tendencies of calcium sulfate (Denman 1961) and calcium phosphate (Green and Holmes 1947) have been published but are not as commonly used. At cooling water temperatures and lower pressure boiler temperatures, calcium sulfate is far more soluble than calcium carbonate.

An effective method for preventing scale formation is to soften the water by passing it through an ion exchange water softener. This external treatment process removes all but 2 to 5 ppm of hardness. It is usually carried out in a closed vertical tank about two-thirds filled with small beads of a cation exchange resin. The resin preferentially removes calcium and magnesium ions from the water, while returning sodium ion to the water.

The calcium and magnesium content of the resin ultimately rises to the point where these elements are no longer completely absorbed from the water passing through the resin. The flow of water is then reversed, backwashing the resin to remove any dirt particles, and the resin is then regenerated by passing a much higher concentration of salt through it. The sodium ions of the strong salt solution displace the absorbed calcium and magnesium ions, restoring the resin to its initial sodium form. After rinsing out excess salt, the resin is ready for another softening cycle.

$$Na_2R + Ca^{++}\ (\text{or } Mg^{++}) \underset{\text{Regeneration}}{\overset{\text{Softening}}{\rightleftarrows}} CaR + 2Na^+$$

Regenerated Resin — Exhausted Resin

When it is possible to accept and control the presence of suspended matter in the water, hardness may be eliminated by precipitation within the operating equipment. This is commonly done in lower pressure process steam boilers. Alkalis are commonly used to precipitate calcium carbonate in accordance with Equation (1). Phosphates are added to remove the last of the hardness, by virtue of the lower solubility of calcium phosphate, as shown in Equation (2). Magnesium is most commonly precipitated as the hydroxide, as shown in Equation (3).

$$Ca(HCO_3)_2 + 2\,NaOH \rightarrow CaCO_3 + Na_2CO_3 + 2\,H_2O \quad (1)$$

$$3\,Ca(HCO_3)_2 + 66\,NaOH + 2\,Na_3PO_4 \rightarrow Ca_3(PO_4)_2 + 6\,Na_2CO_3 + 6\,H_2O \quad (2)$$

$$Mg(HCO_3)_2 + 4\,NaOH \rightarrow Mg(OH)_2 + 2\,Na_2CO_3 + H_2O \quad (3)$$

There are situations in which it is economically or otherwise impossible to remove the hardness from water. In these cases, other measures can be taken to control scale formation. One common method is to reduce the alkalinity of the water. This substantially reduces scale formation as the solubility of calcium carbonate is much less than that of other calcium salts.

Most of the alkalinity can be removed by an anion exchange resin in a process analogous to softening. In this case, the alkalinity (bicarbonate) is retained by the resin, which gives up to the water equivalent amounts of chloride. When the resin begins to pass larger amounts of alkalinity than desired, it is backwashed and regenerated by a salt solution, as described above. Usually, this salt solution contains a small percentage of alkali.

$$R'CL + HCO_3 \underset{\text{Regeneration}}{\overset{\text{Dealkalizing}}{\rightleftarrows}} R'HCO_3 + Cl$$

Regenerated Resin — Exhausted Resin

The process by which the hardness of the water remains unchanged but the alkalinity is reduced is called dealkalizing, rather than softening. Particularly in large cooling towers, sulfuric acid is commonly used to eliminate most of the alkalinity. The solubility of calcium sulfate is about 2000 ppm, in contrast to 35 ppm for calcium carbonate, thus permitting a much higher hardness to be present in the circulating water before scale can form. The acid feeding procedure involves a considerable risk without careful controls. The alkalinity of the circulating water is reduced to such a low figure that a slight overdose of acid produces a low pH, causing corrosive circulating water. Accordingly, it is good practice to restrict acid feed to systems that operate under constant conditions of makeup water composition and evaporation, or to use automatic pH control equipment.

For cooling waters that cannot use the above measures, two other measures are used in conjunction to minimize scale formation in cooling systems. First, the total dissolved solids in the circulating water are controlled by a continuous bleed or bleedoff to a maximum value proportional to the concentration of hardness, silica, or other limited-solubility constituent. Second, scale control adjuncts are added to the water to increase the apparent solubility of the calcium carbonate. Low concentrations (2 to 5 ppm) of sodium polyphosphates; organic dispersing agents such as various lignin derivatives; synthetic polymer polyelectrolytes; organic phosphates; or mixtures of these are used. These measures may also be used with acid feeding.

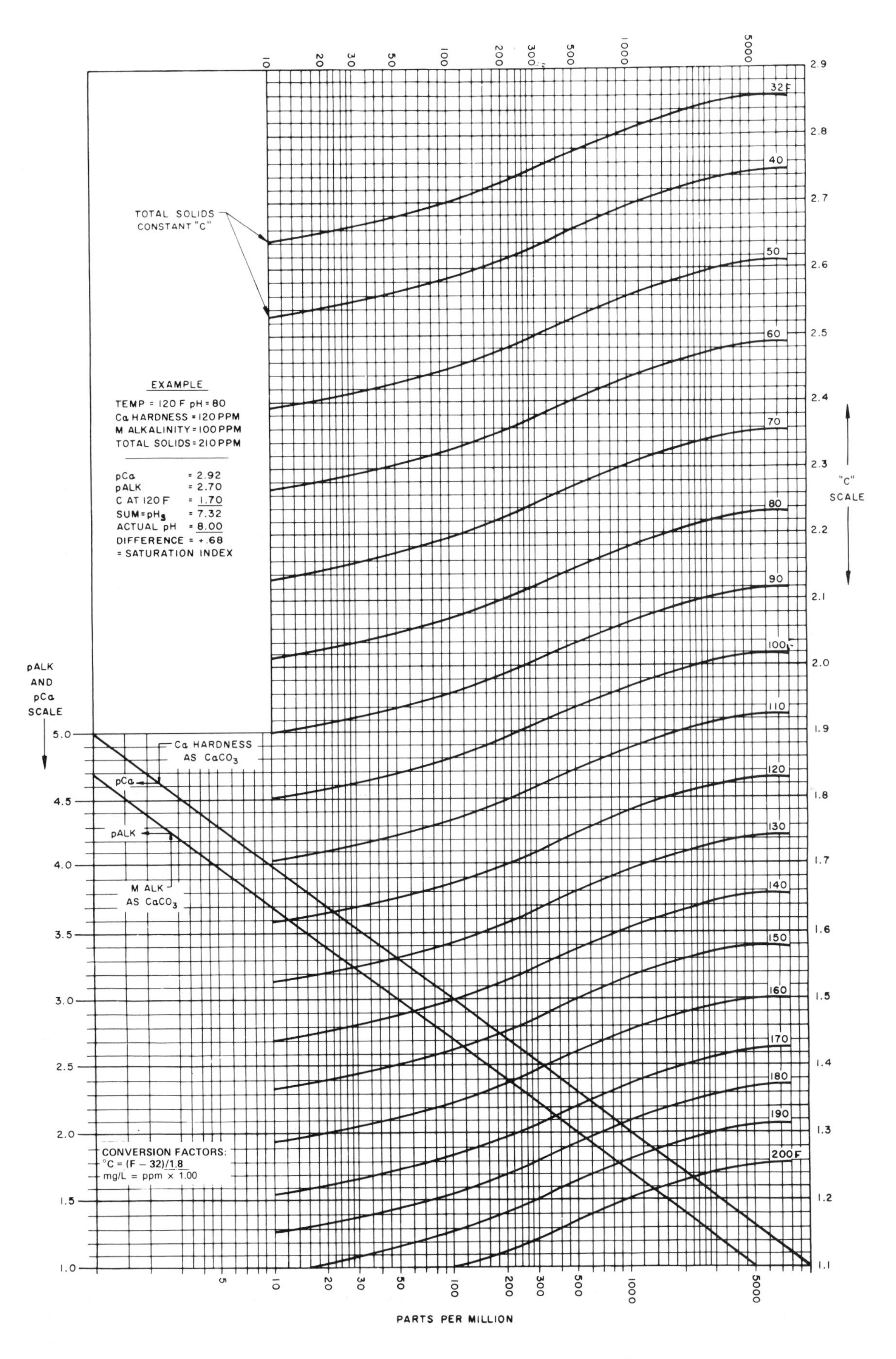

Fig. 1 Langelier Saturation Index Nomograph

Biological Growths. Algae, bacterial slimes, and fungi can interfere with the operation of cooling systems if they are able to clog distribution holes on the deck, spray nozzles, and water flow and heat transfer in heat exchangers, or weaken the structural integrity of cooling tower wood. Heating systems operate at temperatures outside biological limits and do not have microbial problems.

Algae use the energy from the sun to convert bicarbonate or carbon dioxide into biomass. Masses of algae can block piping, distribution holes, and nozzles when allowed to grow in sufficient profusion. A covered distribution deck is one of the most cost-effective additions that can be made to a cooling tower. Biocides are used to assist in the control of algae.

Most waters contain organisms capable of producing biological slime, but significant amounts of slime are produced only when conditions are favorable. Optimal conditions are poorly understood, but include sufficient nutrients and appropriate environmental conditions, such as temperature and pH. Equipment located near sources of nutrients is particularly susceptible to slime formation. Two common examples are air washers in printing plants where there is fine paper dust in the air, and in cooling towers located in food-packing plants which receive large amounts of contamination. Slimes can be formed by bacteria, algae, yeasts, or molds and frequently consist of a mixture of these organisms together with accumulations of organic and inorganic debris. Wherever slimes or other microbiological growths threaten to interfere with the efficient functioning of a cooling system, the use of biocidal materials is indicated.

In other systems, the effective control of slime and algae frequently requires a combination of mechanical and chemical treatments. For example, when a system already contains a considerable accumulation of slime, a preliminary mechanical or detergent cleaning will make the subsequent application of a microbicidal chemical more effective in killing the growths and more enduring in the prevention of further growths. Similarly, periodic mechanical removal of slimes from readily accessible areas reduces the chemical microbicide requirements and makes them more effective.

The organism associated with cooling systems that has received the most publicity in recent years is *Legionella pneumophila*, the bacterium that causes Legionnaire's disease. This organism is found in a variety of natural habitats, including surface waters and potable water distribution systems. The organism is frequently found in sites not associated with an outbreak of the disease; thus, a positive test for the organism does not mean that the disease is imminent. The question of what constitutes acceptable numbers of *Legionella* in a cooling system has not been determined, but the general consensus is that cooling systems should be maintained in a clean state to minimize the probability of high numbers of the organism occurring. The disease is contracted in susceptible individuals by breathing air containing aerosolized *Legionella*. Optimizing cooling tower design to minimize drift and appropriate architectural siting so that drift is not injected into the air handling system are other measures to decrease risk.

Chlorine and chlorine-yielding compounds, such as sodium hypochlorite and calcium hypochlorite, are among the most effective microbicidal chemicals. However, they do not always apply to the control of organic materials in cooling systems. In air washers, for instance, their odor may become offensive; in wood cooling towers, excessive concentrations of chlorine can cause rapid deterioration of wood construction, and in metal equipment, higher concentrations of chlorine can accelerate corrosion. In systems large enough to justify equipment for the controlled feeding of chlorine, its use may be both safe and economical. Some of its disadvantages can be avoided by combining the chlorine with isocyanuric acid, which minimizes the free chlorine concentration. Most chlorine programs can benefit from surfactant (chlorine helper) products or nonoxidizing, organic microbicides.

When selecting a microbicide, consider the pH of the circulating water and the chemical compatibility with the corrosion/scale inhibitor product. Many organic microbicides are available to allow for flexibility in these areas. Typical products include quaternary ammonium compounds, tributyl tin oxide, methylene bis (thiocyanate), isothiazolones, DBNPA, glutaraldehyde, bromochlorohydantoin, and many proprietary blends. All microbicides must be handled with care to ensure personal safety. Cooling water microbicides are approved and regulated through the Environmental Protection Agency and, by law, must be handled in accordance with labeled instructions. Maintenance staff handling the biocides should read the material safety data sheets and be provided with all the appropriate safety equipment to handle the substance. An automatic biocide feed system, such as a pump and timer, should be installed to minimize exposure to maintenance staff who would otherwise be dosing the cooling system with the aid of a bucket.

The manner of in which biocides are fed to a system is important. Often, the continuous feeding of low dosages is neither effective nor economical. Better results can be obtained by shock feeding larger concentrations to the system to achieve a toxic level of the chemical in the water for a sufficient time to kill the organisms present. Alternate shock feeding of two different types of microbicides usually gives the most effective results. Much larger biocide dosages are required when biomass has been permitted to accumulate than when they have been kept under control. At times, organisms appear to build up an immunity to the particular chemical being used, making it necessary to change the biocidal chemical occasionally to keep the organic growths under control.

In large systems, particularly once-through cooling systems using river, estuarine, or seawater, macroorganisms such as barnacles and mussels may accumulate. Although antifouling paints can prevent the growth of such macroorganisms on large diameter pipe surfaces, the paints must be renewed at frequent intervals and are not applicable to inaccessible areas such as the insides of smaller diameter piping. Generally, chlorine is used to destroy such organisms. In once-through systems, intermittent feeding of chlorine in the form of chlorine gas or hypochlorite solutions most effectively and economically controls microorganisms and slimes. Tin-containing paint and trihalomethane formation from the use of chlorine are both under increasing scrutiny from an environmental standpoint.

Suspended solids are undesirable in heating and cooling water systems. Settling out on heat exchange surfaces, they can be as effective as scale or slime in reducing heat transfer. They may accumulate enough to interfere with water flow, and localized corrosive attack can occur beneath such accumulations. Abrasive dirt particles can cause excessively rapid wear of such moving parts as pump shafts and mechanical seals. Treatment with polyelectrolytes, together with blowdown and filtration, will help control most harmful effects of suspended solids.

MECHANICAL FILTRATION

Strainers, filters, and separators may be used to reduce the suspended solids to an acceptably low level. Generally, if the screen is 200 mesh equivalent to 74 microns (.003 in.), it is called a strainer; if it is finer than 200 mesh, it is a filter.

Strainers. A strainer is a closed vessel with a cleanable screen element designed to remove and retain foreign particles down to 0.001 in. diameter from various flowing fluids. They extract material that is not wanted in the fluid; this can sometimes be a valuable product that may be saved. Strainers are available as single-basket or duplex units, manual or automatic cleaning units, and may be made of cast iron, bronze, stainless steel, monel, or plastic. Magnetic inserts are available where microscopic iron or steel particles are present in the fluid.

Cartridge Filters. These are typically used as final polishing filters to remove nearly all suspended particles from about 100 μm in size down to as small as 1 μm or less. Cartridge filters are, with a few exceptions, disposable (*i.e.*, once plugged, they must be replaced). The frequency of replacement, and thus the economic feasibility of their use, depends on the concentration of suspended solids in the fluid, the size of the smallest particle to be removed, and the removal efficiency level of the cartridge filter type selected.

In general, cartridge filtration is favored in systems where contamination levels are less than 0.01% by mass (<100 ppm). They are available in a number of different configurations and materials of construction. Filter media include yarns, felts, papers, nonwovens, resin-bonded fibers, woven wire cloths, sintered metal, and ceramic structures. The standard configuration is a cylinder with an outside diameter of about 2.5 to 2.75 in., an overall length of about 10 in. and an internal core, where the filtered fluid collects, with an inside diameter of about 1 in. Overall lengths from 3 to 40 in. are readily available.

Cartridges made of yarns or resin-bonded fibers normally have a structure with increasing density toward the center. These depth-type filters capture particles throughout the total media thickness. Thin media, such as pleated paper types, have a narrow pore size distribution designed to capture particulates at or near the surface of the filter. Surface-type filters can normally handle higher flow rates and provide higher removal efficiency than equivalent depth filters. Cartridge filters are rated in accordance with manufacturer's guidelines. Surface-type filters have an "absolute" rating, while depth-type filters have a "nominal" rating which reflects their general clarification function.

Sand Filters. A downflow filter is used to remove suspended solids from a water stream for various applications. The degree of suspended solids removal depends on the combinations and grades of the medium being used in the vessel.

During the filtration mode, service in-water enters at the top of the filter vessel. After passing through a flow impingement plate, it enters the quiescent freeboard area above the medium.

In multimedia downflow vessels, various grain sizes and types of media are employed to provide depth filtration primarily for increasing the suspended solids holding capacity of the system, thereby increasing the length of time between backwashing. Multimedia vessels might also be applied for low suspended solids applications, where chemical additives are required. In the multimedia vessel, the fluid enters the top layer of anthracite media which has an effective size of 1.0 mm. This relatively coarse layer will remove the larger suspended solids particles, a substantial portion of the smaller particles, and small quantities of free oil. Flow continues downward through the next layer of fine garnet material, which has an effective size of 0.3 mm. A more finely divided range of suspended solids is removed in this polishing layer. The fluid continues into the final layer, a coarse garnet material, which has an effective size of 2.0 mm. Contained within this layer is the header/lateral assembly which collects the filtered water.

When the vessel has retained enough suspended solids to achieve a substantial pressure drop, the unit must be backwashed. This is accomplished by reversing the direction of flow which carries the accumulated solids out through the top of the vessel.

Centrifugal-Gravity Type Separators. In this type of separator, liquids/solids enter the unit tangentially, which sets up a circular flow. Liquids/solids are drawn through tangential slots and accelerated into the separation chamber. Centrifugal action tosses the particles heavier than the liquid to the perimeter of the separation chamber. Solids gently drop along the perimeter and into the separator's quiescent (calm) collection chamber. Solids-free liquid is drawn to the separator's vortex (low-pressure area) and up through the separator's outlet. Solids are either purged periodically or continuously bled from the separator, as necessary with an appropriate valve system.

Bag-Type Filters. These filters comprise a bag of mesh or felt, supported by a removable perforated metal basket, placed in a closed housing with an inlet and outlet.

The housing is a welded, tubular pressure vessel with a hinged cover on top, for access to the bag and basket. Housings are made of carbon or stainless steel. The inlet can be in the cover, in the side (above the bag), or in the bottom (and internally piped to the bag). The side inlet is the simplest type. In any case, the liquid enters the top of the bag. The outlet is located at the bottom of the side (below the bag). Pipe connections can be threaded or flanged. Single-basket housings can handle up to 220 gpm, multibaskets up to 3500 gpm.

The support basket is usually of 304 stainless steel perforated with 1/8-in. holes. (Heavy wire mesh baskets also exist.) The baskets can be lined with fine wire mesh and used by themselves as strainers, without adding a filter bag. Some manufacturers offer a second, inner basket (and bag) that fits inside the primary basket. This provides for two-stage filtering: first, a coarse filtering stage, then a finer one. The benefits are longer service time and possibly eliminating a second housing to accomplish the same function.

The filter bags are made of many materials (cotton, nylon, polypropylene, and polyester) and micro ratings (from 1 to 840). Most common are felted materials, because of their depth-filtering quality which provides high dirt-loading capability, and their fine pores. Mesh bags are generally coarser, but are reusable and, therefore, less costly. The bags have a metal ring sewn into their opening; this holds the bag open and seats it on top of the basket rim.

In operation, the liquid enters the bag from above, flows out through the basket, and exits the housing cleaned of particulate down to the desired size. The contaminant is trapped inside the bag, making it easy to remove without spilling any downstream.

Localized areas frequently can be protected by special methods. Thus, pump-packing glands or mechanical shaft seals can be protected by fresh water makeup or by circulating water from the pump casing, through a cyclone separator or filter, and into the lubricating chamber.

In smaller equipment, a good dirt-control measure is the installation of backflush connections and shutoff valves on all condensers and heat exchangers so that they can be readily backflushed with makeup water or detergent solutions to remove accumulated settled dirt. These connections can also be used for acid cleaning, to remove calcium carbonate scale.

Selection of Water Treatment

Many methods are available for the correction of almost any water-caused difficulty. However, no one method of treatment applies to all cases. Selection of the proper water treatment method and details of the chemicals and equipment necessary to apply that method depend on many factors. The chemical characteristics of the water, which change with the operation of the equipment, are important. Other factors contributing to the selection of proper water treatment to a lesser degree are economics and other nonchemical influences, such as the design of individual major system components, *e.g.*, the cooling tower or boiler; equipment operation; and human factors, such as the quantity and quality of operation personnel available.

Once-Through Systems. Economics is the overriding consideration in the treatment of water in once-through systems. The quantities of water to be treated are usually so large that any treatment other than simple filtration or the addition of a few ppm of a polyphosphate, silicate, or other inexpensive chemical may not be feasible. Intermittent treatment with polyelectrolytes can help to maintain clean conditions when the cooling water is sediment-laden. In such systems, it is generally less expensive to invest more in corrosion-resistant construction materials than to attempt to treat the water.

Open Recirculating Systems. The selection of water treatment for open recirculating systems is affected considerably by the size of the system. At one extreme, large industrial cooling tower systems use such huge quantities of water that it is necessary to minimize the concentration of any treatment chemical. In these systems, sizable expenditures for chemical treatment and control equipment can be justified; so can the use of specially trained operating personnel. At the other extreme, in small air-conditioning and refrigeration cooling towers, the total amount of water used is so small that original and replacement costs for the system are lower and operating personnel need not be as highly trained.

Therefore, a typical water treatment scheme for a large industrial cooling tower system operating above a 10,000 gpm circulation rate might include scale control by a controlled bleed and alkalinity reduction via automatically pH-controlled sulfuric acid feed combined with corrosion control.

Cooling tower controllers can monitor these and other programs by controlling acid feed by pH, mineral content by conductivity, inhibitor concentration by direct or indirect measurement, and corrosion rate by test probes and corrosion coupons.

On the other hand, in a small air-conditioning or refrigeration cooling tower circulating at 100 gpm, treatment might consist of a controlled bleed to minimize scale formation, and the maintenance of 200 to 500 ppm of an inhibitor for corrosion control, plus the occasional application of a shock dose of microbicide. Maintaining a clean tower is important for efficiency, safety, and health reasons.

Air Washers and Sprayed Coil Units. A water treatment program for an air washer or a sprayed coil unit is usually complex and depends on the purpose and function of the system. Some systems, such as sprayed coils in office buildings, are used primarily for the control of temperature and humidity, while other systems are intended to remove dust, oil vapor, and other airborne contaminants from an airstream, in addition to supplying temperature and humidity control. Without proper chemical treatment, the fouling characteristics of the contaminants removed from the air will cause operational problems in the circulating water of the system. For proper system operation, it is imperative that a suitable water treatment program be initiated when the system is put into operation.

Scale control is important in air washers or sprayed coil systems providing humidification, since the minerals present in the water may become concentrated (through the process of water evaporation) to such a degree that they become a problem. Inhibitor/dispersant treatments commonly employed in cooling towers are often used in air washers to control scale formation and corrosion.

Using suitable dispersants and surfactants is often necessary to control oil and dust removed from the airstream. The type of dispersant depends on the nature of the contaminant and the degree of system contamination. For maximum operating efficiency, dispersants should produce a minimal amount of foam in the system.

Control of slime and bacterial growth is also important in the treatment of air washers and sprayed coil systems. Because these systems are constantly exposed to airborne microorganisms, the potential for biological growth is enhanced, especially if the water contains contaminants that are nutrients for the microorganisms. Because of the variations in conditions and applications of air-washing installations and the possibility of toxicity problems, individual cases should be studied by a chemical consultant before a water treatment program is begun. All microbicides applied in air washers must have specific Environmental Protection Agency approval.

Ice Machines. Lime scale formation, cloudy or "milky" ice, objectionable taste and odor, and sediment comprise the bulk of water problems encountered in ice machines. Lime scale formation is probably the most serious problem because it interferes with the harvest cycle by forming on the freezing surfaces where it prevents the smooth release of ice from the surface to the harvest bin.

Scale is caused by dissolved minerals in the water. Because water tends to freeze in a pure state, these dissolved minerals concentrate in the unfrozen water and some eventually deposit on the machine's freezing surfaces as a lime scale. Two major factors contributing to the problem are calcium hardness and total alkalinity (carbonate and bicarbonate).

The probability of scale formation in an ice machine varies directly with the concentrations of calcium hardness and total alkalinity in the water circulating in the machine. Table 6 shows how the problem can vary when using water with different hardness and alkalinity contents.

Table 6 Scale Formation in an Ice Machine

Total Alkalinity as Bicarbonate, ppm	Hardness as Calcium Carbonate			
	0 to 49	50 to 99	100 to 199	200 and up
0 to 49	No scale	Very light scale	Very light scale	Very light scale
50 to 99	Very light scale	Moderate scale	Moderate scale	Moderate scale
100 to 199	Very light scale	Troublesome scale	Troublesome scale	Heavy scale
200 and up	Very light scale	Troublesome scale	Heavy scale	Very heavy scale

To prevent dissolved minerals in the water from depositing on freezing surfaces during normal operation, the use of a slowly soluble food-grade polyphosphate is recommended. These products keep hardness minerals in solution, thereby inhibiting scale formation.

Dirty or scaled-up icemakers should be thoroughly cleaned before starting the water treatment program. Water distributor holes should be cleared and all loose sediment and other material flushed from the system. Existing scale formations can be removed by circulating an acceptable acid solution through the system. Several of these products are available from air-conditioning and refrigeration parts wholesalers.

Although polyphosphates can inhibit lime scale in the ice-making section and help prevent sludge deposits in the sump caused by loose particles of lime scale, they do not eliminate the soft, milky, or white ice caused by high concentrations of dissolved minerals in the water.

Even with proper chemical treatment, the maximum mineral content that can be carried in the recirculating water and still produce clear ice is about 500 to 1000 ppm. Increased bleedoff or reducing the thickness of the ice slab or the size of the cubes may help correct this condition. However, demineralizing or distillation equipment is needed to prevent this problem. This equipment is expensive and is usually not economical for use with small ice machines.

Another problem frequently encountered in servicing ice machines is objectionable taste or odor. When water containing an offensive taste or odor is used in an ice machine, the material causing the taste or odor is trapped in the ice. An activated carbon filter on the makeup water line can remove the objectionable taste or odor from the water. However, carbon filters need to be serviced or replaced regularly to avoid organic buildup in the carbon-bed itself.

Occasionally, slime growth is the cause of an odor problem in an ice machine. This problem can be controlled by regularly cleaning the machine with a food-grade acid. If the slime deposits persist, sterilization of the ice machine is helpful.

Feedwater often contains suspended solids such as mud, rust, silt, and dirt. To remove these contaminants, a sediment filter of appropriate size can be installed in the feed lines. In addition to improving the quality of the ice, installing a suspended solids filter protects the solenoid valves in the machine.

Closed recirculating systems are often defined as systems requiring less that 5% makeup per year. The need for water treatment in such systems (*i.e.*, hot water heat, chilled water, combined cooling and heating, and closed loop condenser water) is often ignored, because the total amount of scale that could be deposited from the water initially filling the system would be insufficient to interfere significantly with heat transfer. This rationalization leads to the erroneous conclusion that damage resulting from the corrosive factors of water required for the initial fill would not be serious. However, operating experience and tracer studies in many systems have shown that appreciable water losses are the rule, not the exception (Sussman and Fullman 1953). Losses of 25% of system volume per month are fairly common. Therefore, all systems should be adequately treated for corrosion control, and scale-forming waters should be softened before use as makeup. Many closed loop systems contain glycol or alcohol solutions, and the inhibitor products chosen for corrosion control must be chemically compatible. Monitoring water use, chemical consumption and concentration, and corrosion rates with corrosion coupons can be beneficial.

The selection of a treatment program for closed systems is often influenced by factors other than degree of corrosion protection offered. Chromate and hydrazine treatments used in the past are progressively less acceptable from a safety standpoint. Alternatives include buffered nitrite, molybdate, and organic blends. Nonconductive couplings of different metals should be employed throughout the construction of the system. The presence of aluminum or its alloys along with other metals makes corrosion prevention more difficult and usually requires supplementary additives such as nitrites or silicates. With buffered sodium nitrite inhibitors, a minimum of 500 ppm as sodium nitrite is required along with maintaining the pH in the range of 8.0 to 10.0.

Before treating new systems, they must be cleaned and flushed. Grease, oil, construction dust, dirt, and mill scale are always present in varying degrees and must be removed from the metallic surfaces to ensure adequate heat transfer and to reduce the opportunity for localized corrosion. Detergent cleaners with organic dispersants are available for proper cleaning and preparation of new closed systems.

Low-Temperature Secondary and Hot Water Heating Systems. Closed chilled water systems, which are usually converted to hot water heating during winter, and primary low-temperature hot water heating systems, all of which usually operate in the temperature range of 140 to 250 °F, require sufficient corrosion inhibitors to control corrosion to less than 0.005 in. per year. Organic proprietary inhibitors containing no chromates or nitrites are also available for such systems.

Medium and High-Temperature Hot Water Heating Systems. Medium temperature hot water heating systems (250 to 350 °F) and high-temperature, high-pressure hot water systems (above 350 °F) require careful consideration of treatment for corrosion and deposit control. Makeup water for such systems should be demineralized or softened to prevent scale deposits. For corrosion control, oxygen scavengers such as sodium sulfite or hydrazine should be introduced to remove dissolved oxygen, and neutralizing amines plus enough caustic soda should be used to maintain pH in the 8.0 to 10.0 range.

Electrode boilers are sometimes used to generate low- or high-temperature hot water. Such systems operate through heat generated because of the electrical resistance of the water between electrodes. The conductivity of the recirculating water is an important factor with specific requirements, depending on the voltage used. Treatment of this type of system for corrosion and deposit control varies. In some cases, sodium sulfite and caustic soda are used to maintain the desired specific conductance, sulfite residual, and pH value from 7.0 to 9.0. In other applications, hydrazine and neutralizing amines (which do not add to the conductivity) can remove dissolved oxygen and maintain the pH value at 7.0 to 9.0. The choice depends on the required conductivity of the electrode boiler and of the makeup water. Before selecting the water treatment for an electrode boiler, consult the boiler manufacturer or the operating manual for the boiler concerned.

Brines. Brine systems must be treated to control corrosion and deposits. The standard chromate treatment program is the most effective. Calcium chloride brines require a minimum of 1800 ppm of sodium chromate with pH 6.5 to 8.5. Sodium chloride brines require a minimum of 3600 ppm of sodium chromate and also pH to 8.5. Sodium nitrite at 3000 ppm in calcium brines or 4000 ppm in sodium brines while controlling pH between 7.0 and 8.5 should provide adequate protection. Organic inhibitors are available that may provide adequate protection where neither chromates nor nitrites can be used.

Ethylene glycol or propylene glycol are used instead of brine systems and as antifreeze in chilled or secondary hot water systems. Such glycols are available commercially, with inhibitors such as sodium nitrite, potassium phosphate, and organic inhibitors for nonferrous metals added by the manufacturer. These require no further treatment, but softened water should be used for all filling and makeup requirements. Samples from the systems should be checked periodically to ensure that the inhibitor has not been depleted. Analytical services are available from the glycol manufacturers and others for this purpose.

Boilers. The most important observations regarding boiler water treatment are that there is a wide range of treatment procedures and that the method selected must depend on the composition of the makeup water, the operating pressure of the boiler, and the makeup rate.

Minimum makeup water pretreatment for low-pressure boilers should consist of ion exchange softening with further consideration given to dealkalizers, desilicizers, or demineralizers.

In low-pressure systems where steam is used for cooking or heating, some makeup is required, and the use of inhibitors is paramount. In these cases, it is necessary to remove dissolved oxygen from the feedwater by using a deaerator or feedwater heater. This is then followed by treatment to raise the pH value above 10.5 and sodium sulfite or hydrazine to remove the last traces of dissolved oxygen (Blake 1968). Scale control may also be required. Steam used for cooking and humidification is under increasing scrutiny by the FDA, EPA, and USDA, and chemicals used in steam for these applications must follow guidelines.

As operating pressures and makeup rates go up, the corrosion problem does not decrease, but the scale problem becomes more important. Corrosion is controlled by using an open heater or, at higher operating pressures, a deaerating heater to remove oxygen from the boiler feedwater; the maintenance of a sulfite or hydrazine residual as an oxygen scavenger in the boiler water; and the maintenance of a sufficiently high pH in the boiler water. Scale is controlled by external softening and by internal treatment, usually with phosphates and organic dispersants, to precipitate residual traces of calcium, generally as a phosphate (hydroxy apatite), and magnesium (as magnesium hydroxide or a basic silicate). The ASME standards for boiler water quality are given in Table 7.

Other chemicals used for corrosion control in closed heating and cooling systems include volatile amines (such as morpholine and cyclohexylamine) and filming amines (such as octadecylamine) used for steam condensate line protection. Various proprietary chemical mixtures are used for glycol or alcohol antifreeze solutions in chilled water or snow-melting systems, where the inhibitor must be chemically compatible with the antifreeze agent. (Note: Volatile amines may present health hazards.)

Table 7 ASME Standards[a]

		Boiler Water Quality		
Drum Pressure, psig	Silica, ppm SiO_2	Total Alkalinity,[c] ppm $CaCO_3$	Neutralized Specific Conductance, micromhos/cm	Suspended Solids, ppm[e]
0-300	150	700[b]	7000	300
301-450	90	600[b]	6000	250
451-600	40	500[b]	5000	150
601-750	30	400[b]	4000	100
751-900	20	300[b]	3000	60
901-1000	8	200[b]	2000	40
1001-1500	2	0[d]	150	20
1501-2000	1	0[d]	100	10

[a] Source—ASME Research Committee on Water in Thermal Power Systems. (ASME Standards were proposed at the conclusion of 1978.)
[b] Alkalinity not to exceed 10% of specific conductance.
[c] Minimum level of OH alkalinity in boilers below 1000 psi must be individually specified with regard to silica solubility and other components of internal treatment.
[d] Zero in these cases refers to free sodium or potassium hydroxide alkalinity. Some small variable amount of total alkalinity will be present and measurable with the assumed coordinated control or volatile treatment employed at these high-pressure ranges.
[e] American Boiler and Affiliated Industries Manufacturers Association's maximum limits for boiler-water concentrations in units with a steam drum.

Treatment of the boiler water is often affected by the end use of the steam. This may range from reduction of alkalinity so that it can minimize condensate line corrosion by keeping carbon dioxide low in the steam, to silica removal for the protection of steam turbines against siliceous turbine blade deposits. Antifoam agents are used for improved operation and steam quality.

Return Condensate Systems. Three approaches are used to minimize corrosion in condensate systems: (1) eliminating alkalinity from all water entering the boiler to minimize the amount of carbon dioxide in the condensate lines; (2) the use of volatile amines, such as morpholine and cyclohexylamine, to neutralize carbon dioxide and thus raise the pH of the condensate; and (3) introducing filming amines, such as octadecylamine, into the steam to form a thin, hydrophobic film on the condensate line surfaces. The last approach is particularly effective in minimizing corrosion by oxygen. The need for chemical treatment can be minimized by designing return systems so that the condensate is still very hot when it reaches the boiler feed pump.

FIRESIDE CORROSION AND DEPOSITS

The surfaces of flues and boilers that are contacted by combustion products are seldom corroded while the equipment is operating, unless halogenated hydrocarbons, such as certain degreasing solvents, are present in the combustion air. Chimney connectors, smoke hoods, and canopies in contact with flue gas may, however, be subject to attack during warm-up periods or when the rate of operation is so low that the temperature of the flue gas is below its dew point. In stack sections where flue gas temperatures drop below the dew point during operations, corrosion is inevitable.

It is common to use cast iron or acid-resistant, vitreous enameled steel in flue gas connections to appliances to prolong the life of these parts. Metal surfaces with temperatures that do not exceed 400 °F can be protected by periodic applications of paints. Protective coatings with organic binders are destroyed rapidly above 400 °F because of the decomposition of the organic materials. Stacks and other surfaces that reach higher temperatures can be protected with special coatings, such as silicones, resistant to these higher surface temperatures.

Corrosion on the fireside of boilers is common, and certain precautions must be taken to prevent it. It has been reported that 15% of boiler tube failures have been caused by fireside attack (Hinst 1955). The most common cause of fireside corrosion is the condensation of sulfuric acid at cold ends of the furnace. All fossil fuels contain varying amounts of sulfur. Even low sulfur fuel oil containing less than 1.0% sulfur forms gaseous oxides of sulfur when burned. Sulfur dioxide has no particularly corrosive effect on the boiler when it remains in the gaseous state and is emitted from the stack, but it pollutes the atmosphere. However, in boilers that operate with fluctuating loads and varying amounts of excess air present in the combustion chamber, the sulfur dioxide can further oxidize to sulfur trioxide, which is corrosive to the metallic surfaces of the boiler. Metals such as vanadium and iron act as catalysts, encouraging oxidation of sulfur dioxide to sulfur trioxide.

Sulfur trioxide combines with moisture to form sulfuric acid. The condensation point of sulfuric acid is 330 °F. Where a boiler has a variable load, the metal surface temperature in the furnace may drop below 330 °F at low firing rates and condense sulfuric acid, causing serious corrosion. This may happen regularly throughout the normal operating period, but it is most likely to occur when the boiler is shut down for some period. Furthermore, moisture may condense on the metallic surfaces during shutdown because of the high relative humidity. In the presence of moisture, as well as acidic soot deposits, fireside corrosion will continue.

Fireside corrosion can be reduced by modifying the firing cycle to minimize the number of times the flue gas temperature drops below its dew point. Other methods include using low sulfur content fuel oils, regulating excess air, and using fuel oil additives that neutralize acidic condensate and deposits on fireside surfaces. Good maintenance practices are essential. A boiler should be fired to maintain a constant temperature. A boiler let down at the end of a day and left at low is not only conducive to condensation of sulfuric acid, causing corrosion, but also encourages soot and slag formation and stress corrosion cracking around the tube ends.

Additives, such as magnesium and calcium oxide slurries, have effectively neutralized sulfuric acid formed on firesides. Certain manganese-bearing compounds inhibit the formation of sulfur trioxide significantly and thereby are effective in reducing sulfuric acid corrosion. These compounds also reduce smoke, soot, and deposits (Belyea 1966). Some additives contain detergents, which are surface-active agents, and dispersants that prevent fuel from adhering to burner nozzles, a cause of sticking, fuel dribbling, improper spray pattern smoking, and soot deposits (Bausch and MacPherson 1965). The resulting deposits absorb acid gases and cause serious corrosion and combustion inefficiency.

Care after shutdown is equally important in preventing fireside corrosion. The fireside of a boiler should always be thoroughly brushed and cleaned to remove accumulations of soot and other deposits (Hinst 1955). This should be followed by air drying.

Moisture absorbents, such as silica gel and quicklime, can be spread on trays on top of the tubes or in the bottom of the boiler drum or shell. This will reduce the condensation of moisture and the resulting corrosion during idle periods.

REFERENCES

Belyea, A.R. 1966. Manganese additives reduces SO_3. *Power* (November).

Blake, R.T. 1968. Cure for pitting in low pressure boiler systems. *Air Conditioning, Heating and Ventilating* (November):45.

Hinst, H.F. 1955. Eleven ways to avoid boiler tube corrosion. *Heating, Piping and Air Conditioning* (January).

Meckler, M. 1974. H/P/A/C, Corrosion and scale formation (August).

Nordell, E. 1961. *Water treatment for industry and other uses.* Reinhold Publishing Co., New York.

Peabody, A.W. 1963. Pipeline corrosion survey techniques. *Materials Protection* 2(4):62.

Pye, D. Chemical fixation of oxygen. *Journal of the American Water Works Association* 39:121.

Romanoff, M. 1957. Underground corrosion. National Bureau of Standards Circular, No. 570.

Sudrabin, L.P. 1963. Designing automatic controls for cathodic protection. *Materials Protection* 2(2).

Sudrabin, L.P. 1956. An anomaly in pipe line corrosion diagnosis. *Corrosion* 12(3): 17.

Sudrabin, L.P. and F.J. LeFebvre. 1953. External corrosion of piping in radiant eating and snow melting systems. Heating, Piping and Air-Conditioning Contractors National Association Official *Bulletin* 60 (July).

Sudrabin, L.P. 1963. A review of cathodic protection theory and practice. *Materials Protection* 2(5).

Sussman, S. and J.B. Fullman. 1953. Corrosion in closed circulating water systems. *Heating and Ventilation* (October):77.

Weider, B.Q. and E.P. Partridge. 1954. Practical performance of water conditioning gadgets. *Industrial and Engineering Chemistry* 46:954.

BIBLIOGRAPHY

APHA. 1989. *Standard methods for the examination of water and wastewater*, 17th ed. American Public Health Association, Washington, D.C.

ARI. 1958. *Corrosion and its prevention.* Air-Conditioning and Refrigeration Institute, Arlington, VA.

ASTM. 1978. *Manual on water.* STP 442A. American Society for Testing Materials, Philadelphia.

Berk, A.A. 1962. *Handbook—Questions and answers on boiler feed water conditioning.* Bureau of Mines, U.S.Department of Interior, Washington, D.C.

Carrier Corp. 1963. *System design manual*, Part 5, Water conditioning. Syracuse, NY.

Evans, U.R. 1960. *The corrosion and oxidation of metals.* Edward Arnold, Ltd., London.

Evans, U.R. 1963. *An introduction to metallic corrosion*, 2nd ed. Edward Arnold, Ltd., London.

Hamer, P., J. Jackson, and E.F. Thurston. 1961. *Industrial water treatment practice.* Butterworth & Co., Ltd., London.

McCoy, J.W. 1974. *The chemical treatment of cooling water.* Chemical Publishing Company, New York.

McCoy, J.W. 1980. *Microbiology of cooling water.* Chemical Publishing Company, New York.

NACE. 1969. Control of external corrosion on underground or submerged metallic piping systems. NACE-RP-1-69. National Association of Corrosion Engineers, Houston, TX.

NACE. 1980. NACE *Corrosion engineer's reference book.* National Association of Corrosion Engineers, Houston, TX.

Nalco. 1979. *Nalco water handbook.* The Nalco Company, Oak Brook, IL.

Powell, S.T. 1954. *Water conditioning for industry.* McGraw-Hill, New York.

Rosa, F. 1985. *Water treatment specification manual.* McGraw-Hill, New York.

Rosa, F. 1988-1989. National engineer: A historical perspective of devices to control scale, corrosion and bio-growths in HVAC systems.

CHAPTER 44

SERVICE WATER HEATING

A SERVICE water-heating system has (1) a heat energy source, (2) heat transfer equipment, (3) a distribution system, and (4) terminal hot water usage devices.

Heat energy sources may be (1) fuel combustion, (2) solar energy collection, (3) electrical conversion, and/or (4) recovered waste heat from such sources as flue gases, ventilation and air-conditioning systems, refrigeration cycles, or process waste discharge.

Heat transfer equipment is either of the direct or indirect type. Direct heat is derived from the combustion of fuels, collection of solar energy, or direct conversion of electrical energy into heat and is applied directly in water-heating equipment.

Indirect heat energy is developed from remote heat sources, such as boilers, solar, cogeneration, refrigeration, or waste heat, and is then transferred to the water.

Distribution systems transport the hot water produced by the water-heating equipment and storage facilities to plumbing fixtures and other terminal points. The amount of water consumed must be replenished in the water-heating equipment and piping system under pressurized conditions from the building water service main. For locations where constant supply temperatures are desired, circulation piping or heat maintenance means must be provided from remote fixtures.

Terminal hot water usage devices are plumbing fixtures and equipment requiring hot water for specific uses. They typically exhibit intermittent periods of irregular, constant, or no-flow conditions. These patterns and their related water usage requirements vary for different buildings, process applications, and personal preference.

In this chapter, it is assumed that an adequate supply of service water is available. If this is not the case, alternate strategies, such as water accumulation, pressure control, and flow restoration, should be considered.

SYSTEM REASONING

Flow rate and temperature are the primary factors to be determined in the hydraulic and thermal design of water-heating and piping systems. Operating pressures and water quality are also mandatory considerations. Separate procedures are used to select water-heating equipment and to design the piping system.

Water-heating equipment, storage facilities, and piping should have enough capacity to (1) provide the required hot water without wasting energy or water, and (2) allow economical system installation, maintenance, and operation.

Hot water can be provided in many ways. The service water-heating system for a given application must be selected considering the overall design and energy demand for both the building's hot water and other mechanical systems.

Water-heating equipment types and designs are based on (1) the heat energy source, (2) application of the developed energy to heating the water, and (3) the control method used to deliver the necessary hot water at the required temperature under varying water demand conditions. Application of this equipment within the overall design of the hot water system is based on (1) location within the system, (2) related temperature requirements, and (3) the volume of water to be used.

HEAT ENERGY SOURCES

The choice among available energy sources, equipment types, and equipment locations should be made after evaluating initial purchase, operating, and maintenance costs. A life cycle analysis is highly recommended. In making energy conservation choices, current editions of the following energy conservation guides should be consulted: ANSI/ASHRAE/IES *Standard* 90 or the ANSI/ASHRAE/IES *Standard* 100 series sections on Service Water Heating (see also Design Considerations).

HEAT TRANSFER EQUIPMENT

Direct Gas-Fired or Oil-Fired

Residential water-heating equipment is usually the automatic storage type. For industrial and commercial use, common heaters are (1) automatic storage, (2) circulating tank, (3) instantaneous, (4) hot water supply boilers, and (5) immersion storage type.

Installation guidelines for gas-fired water heaters can be found in the National Fuel Gas Code, NFPA *Standard* 54-88 (ANSI Z223.1). This code also covers the installation of venting equipment and controls available for gas-fired water heaters. Installation guidelines for oil-fired water heaters can be found in NFPA *Standard* 31-87, Standard for the Installation of Oil Burning Equipment (ANSI Z95.1).

Automatic storage heaters incorporate the burner(s), storage tank, outer jacket, insulation, and controls in a single unit. They are normally installed without dependence on other hot water storage equipment.

Circulating tank heaters are classified in two types: (1) automatic, in which the thermostat is located in the water heater, and (2) nonautomatic, where the thermostat is located within the storage tank itself.

The preparation of this chapter is assigned to TC 6.6, Service Water Heating.

Instantaneous heaters have minimal water storage capacity, are self-contained, and are often similar to a circulating tank heater. They usually include a flow switch as part of the control system. Instantaneous heaters may have a modulating fuel valve that varies fuel flow as water flow changes.

Hot water supply boilers are capable of providing service hot water and are operated as a pressure vessel. They are typically applied as an alternative to circulating tank or instantaneous heaters.

Immersion storage-type heaters (Figure 1) have a power burner firing into a horizontal tube containing a finned-tube bundle. An intermediate heat transfer fluid, usually water, is pumped through the finned bundle and then to the water-heating bundle located below the fire tube in the shell or storage tank. The heat transfer fluid, continuously circulated at a controlled velocity, is the basic source of heat to the stored water, even though the firing tube is immersed in the water and serves as an additional heat surface. The heat transfer fluid is circulated at a maximum temperature of 250 °F, and the closed system is operated under pressure to prevent boiling.

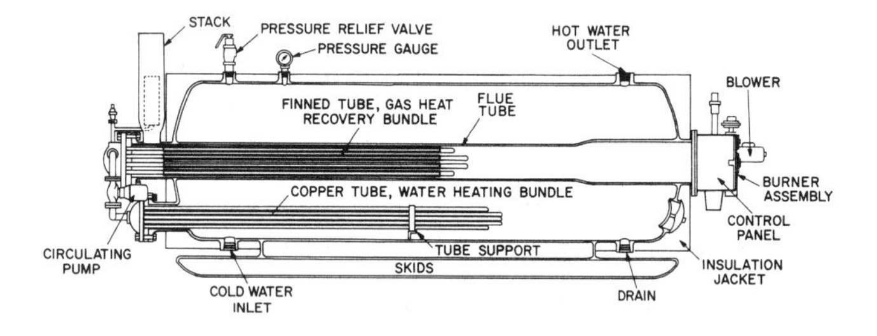

Fig. 1 Indirect Immersion-Fired Storage-Type Heater

Electric

Electric water heaters are generally the automatic storage type, consisting of a tank with one or more immersion heating elements attached to line voltages of 120, 208, 240, 277, 480, or 600 V. Element wattages are selected to meet recovery requirements or electrical demand considerations. Electric water-heating elements consist of resistance wire embedded in refractories with good heat conduction properties and electrical insulating values. Heating elements are generally sheathed in copper or alloy tubes, and are fitted into a threaded or flanged mounting for insertion into a tank. Thermostats controlling heating elements may be the immersion or surface-mounted type.

Residential storage water heaters range up to 120 gal with input up to 12 kW. They have a primary resistance heating element near the bottom, and often a secondary element located in the upper portion of the tank. In twin-element heaters, the thermostats are usually interlocked so that the lower heating element cannot operate if the top of the tank is cold.

Commercial electric water heaters are available in many combinations of power, voltage, and storage capacity. Tank types may be vertical or horizontal. Booster electric water heaters are commonly used for dishwasher rinse, swimming pool, or instantaneous applications.

Heat pump water heaters either include a storage tank or are meant to be connected to an existing storage water heater. They use a refrigeration cycle to extract heat from a source to heat water.

Demand controlled water heating can significantly reduce the cost of heating water electrically. Demand controllers operate on the principle that a building's peak electrical demand exists for a short period during which heated water can be supplied from storage rather than continuing electrical demand for hot water recovery. Avoidance of electrical use for service water heating during peak demand periods allows water heating at the lowest electric energy cost in many electric rate schedules. The essential functions of load detection and control involve sensing the building electrical load and comparing it with peak demand data. When the load is below peak demand, the device allows the water heater to operate. Some controllers can program deferred loads in steps as capacity is available. This priority sequence may involve each of several banks of elements in a water heater, multiple water heaters, or water-heating and other deferrable loads, such as pool heating and snow melting. When load controllers are used, hot water storage must be used.

Electric off-peak storage water heating is a water-heating equipment load management strategy whereby electrical demand to a water-heating system is controlled in relation to time, primarily the utility electrical load profile. This concept usually requires an increase in tank storage capacity to accommodate water use during peak periods.

Sizing recommendations in this chapter apply only to water-heating systems without demand or off-peak controlled systems. When demand control devices are used, the storage and recovery rate must be increased to supply all the hot water needed during the peak period and during the ensuing recovery period. Manian and Chakeris (1974) include a detailed discussion on load limited storage heating system design.

Indirect

Indirect water heating uses steam, hot water, or other fluids, heated in a separate generator or boiler, as the heating medium. The water heater extracts its heat through an external or internal heat exchanger.

When the heating medium is at a higher pressure than the service water, the service water may be contaminated by leakage of the heating medium through a damaged heat transfer surface. Some national, state, and local codes require double-wall, vented tubing in indirect water heaters to reduce the possibility of cross-contamination.

When steam is used as the heating medium, high rates of condensation occur, particularly when a sudden demand causes an inflow of cold water. The steam pipe and condensate return pipes should be of ample size. Condensate should drain by gravity without lifts to a vented condensate receiver located below the level of the heater. Otherwise, water hammer, reduced capacity, or heater damage may result. The condensate may be cooled by preheating the cold water supply to the heater.

Corrosion is minimized on the side with the heating medium because no makeup water, and hence no oxygen, is brought into the circulation fluid system. The metal temperature of the domestic water side of the heat exchanger is usually less than that of direct-fired water heaters. This minimizes scale formation from hard water.

Systems are classified as storage type, semi-instantaneous, or instantaneous.

Storage Type

Storage water heaters are designed for service conditions where hot water requirements are not constant, *i.e.*, when a large volume of heated water must be held in storage for periods of peak load. Residential equipment is usually a single tank with an immersed heat exchanger. For commercial applications, a number of units manifolded together or an individual large tank may be used. For the latter, the tank heating surface may consist of a tube bundle, usually copper tubing, attached to a tube sheet. The bundle is inserted into the tank through a flanged opening to which the bundle and bonnet are securely attached. To transfer heat to the water in the tank, the heating fluid is circulated through the tubes of the bundle. Adequate clearance is needed to remove the bundle for cleaning or replacement.

External Storage Type

Figure 2 shows an indirect, external water heater designed for connection to a separate tank. The boiler water circulates through the heater shell, while service water from the storage tank circulates through the tubes and back to the tank. Circulating pumps are usually installed in both the boiler water piping circuit and the circuits between the heat exchanger and the storage tank. Steam can also be used as the heating medium in a similar scheme.

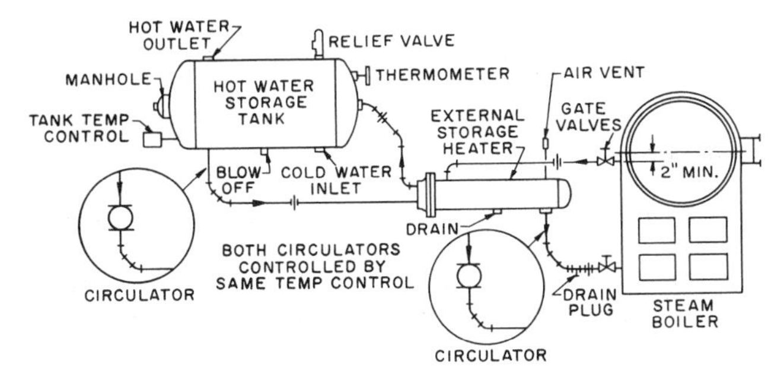

Fig. 2 Indirect, External Storage Water Heater

Instantaneous Type (Tankless Coil)

The instantaneous indirect water heater is best used for a steady, continuous supply of hot water. In this unit, the water is heated as it flows through the tubes. The heating medium flows through the shell, yielding a small ratio of hot water volume to heating medium volume. As a result, variable flow of the service water will cause uncertain temperature control. A thermostatic mixing valve could maintain a more uniform temperature of the hot water supply to the plumbing fixtures and reduce the danger of scalding.

Some indirect instantaneous water heaters are located inside a boiler. A special opening through which the coil can be inserted is provided in the boiler. While the coil can be placed in the steam space above the waterline of a steam boiler, it is usually placed below the water line. The water heater transfers heat from the boiler water to the service water. The gross output of the boiler must be sufficient to serve all the boiler loads.

Semi-Instantaneous Type

Semi-instantaneous water heaters have limited storage, determined by manufacturers to meet the average momentary surges of hot water. They usually consist of a heating element and control assembly devised for the close control of the leaving hot water temperature.

Circulating Tank Type

Instantaneous or semi-instantaneous types may be used with a separate storage tank and a circulating pump. The storage acts as a flywheel to surges in water flow.

Blending (Injected) Type

Steam or hot water can be injected directly into the process or volume of water to be heated. This is often associated with point-of-use applications, *e.g.*, certain types of commercial laundry, food, and process equipment. Caution: Cross-contamination could occur in potable uses.

Solar Water Heaters

Availability of solar energy at the building site, the efficiency and cost of solar collectors, and the availability and cost of other fuels determine whether solar energy collection units should be used as a primary heat energy source. Solar energy equipment can also be included to supplement other energy sources to conserve fuel or electrical energy.

The basic elements of a solar water heater include solar collectors, a storage tank, piping, controls, and a transfer medium. The system may use natural convection or forced circulation. Auxiliary heat energy sources may be added, if needed.

Collector design must allow operation in below-freezing conditions, where applicable. Antifreeze solutions in a separate collector piping circuit arrangement, as well as systems that allow water to drain back to heated areas when low temperatures occur, are often used.

Uniform flow distribution in a collector or in a bank of collectors is important for good performance. It is desirable to maintain stratification in the storage tank for better system performance.

The application of solar water heaters depends on: (1) auxiliary energy requirements; (2) collector orientation; (3) temperature of the cold water; (4) general site, climatic, and solar conditions; (5) installation requirements; (6) area of collectors; and (7) amount of storage.

Waste Heat Use

Waste heat can be recovered by using heat exchangers in gaseous or fluid streams. Heat recovered from this separate equipment is used frequently to preheat the water entering the service water heater. Heat can also be recovered from equipment, such as air-conditioning or refrigeration compressors. Waste heat recovery is an energy conservation method that can reduce (1) energy costs, (2) energy requirements of the building heating and service water-heating equipment, or (3) cooling requirements of applicable air-conditioning systems.

DISTRIBUTION SYSTEMS

Piping System Material

Traditional piping materials have included galvanized steel used with galvanized cast iron or galvanized malleable iron screwed fittings. Copper piping and copper tube types K, L, or M have been used with brass, bronze, or wrought copper solder fittings. Legislation or plumbing code changes have banned the use of any lead in solders or pipe-jointing compounds used in potable water piping because of possible lead contamination of the water supply.

Today, most potable water supplies require treatment before distribution; this may cause the water to become more corrosive. Therefore, depending on the water supply, traditional galvanized steel piping or copper tube may no longer be used satisfactorily, due to accelerated corrosion. Galvanized steel piping is particularly susceptible to corrosion when hot water is in the range of 140 to 180 °F, and also in locations where repairs have been made with copper tube replacing galvanized steel piping without a nonmetallic coupling.

Before selecting any water piping material or system, consult the local code authority. The local water supply authority should also be contacted for any history of water aggressiveness causing failures of any particular material.

Alternate piping materials that may be considered are (1) stainless steel tube, and (2) various plastic piping and tubes, which are extremely effective. Particular care must be taken to make sure that the application meets the design limitations set by the manufacturer and that the correct materials and methods of jointing are used.

These precautions are easily taken with new projects but become more difficult to control during repairs of existing work. The use of incompatible piping, fittings, and jointing methods or materials must be avoided, as they can cause severe problems.

Table 1 Hot Water Demand in Fixture Units (140 °F Water)

	Apartments	Club	Gymnasium	Hospital	Hotels and Dormitories	Industrial Plant	Office Building	School	YMCA
Basins, private lavatory	0.75	0.75	0.75	0.75	0.75	0.75	0.75	0.75	0.75
Basins, public lavatory	—	1	1	1	1	1	1	1	1
Bathtubs	1.5	1.5	—	1.5	1.5	—	—	—	—
Dishwashers	1.5	Five (5) fixture units per 250 seating capacity							
Therapeutic bath	—	—	—	5	—	—	—	—	—
Kitchen sink	0.75	1.5	—	3	1.5	3	—	0.75	3
Pantry sink	—	2.5	—	2.5	2.5	—	—	2.5	2.5
Service sink	1.5	2.5	—	2.5	2.5	2.5	2.5	2.5	2.5
Showers[a]	1.5	1.5	1.5	1.5	1.5	3.5	—	1.5	1.5
Circular wash fountain	—	2.5	2.5	2.5	—	4	—	2.5	2.5
Semicircular wash fountain	—	1.5	1.5	1.5	—	3	—	1.5	1.5

[a] In applications where the principal use is showers, as in gymnasiums or at end of shift in industrial plants, use conversion factor of 1.00 to obtain design water flow rate in gpm.

Pipe Sizing

Sizing of hot water supply pipes involves the same principles as for sizing cold water supply pipes. The water distribution system must be correctly sized for the total hot water system to function properly. Hot water needed in any building varies with the type of establishment, usage, occupancy, and time of day. The piping system should be capable of meeting peak demand within acceptable pressure losses.

Supply Piping System

Table 1, Figures 21 and 22, and manufacturer's specifications for fixtures and appliances allow hot water demands to be determined. These demands, together with procedures given in Chapter 33 of the 1989 ASHRAE *Handbook—Fundamentals*, are used to size the mains, branches, and risers.

Allowance for pressure drop through the heater should not be overlooked when sizing hot water distribution systems, particularly where instantaneous water heaters are used and where the available pressure is low.

Pressure Differentials

Sizing of both cold and hot water piping requires that pressure differentials at the point of use of blended hot and cold water be kept to a minimum. This required minimum differential is particularly important for tubs and showers, since sudden changes in fixtures cause discomfort and a possible scalding hazard. Pressure-compensating devices are available.

Return Piping System

Return piping systems are commonly provided for hot water supply systems in which it is desirable to have hot water available continuously at the fixtures. This includes cases where the hot water piping system exceeds 100 ft. The water circulation pump may be controlled by a thermostat (in the return line) set to start and stop the pump over an acceptable temperature range. Since hot water is corrosive, circulating pumps should be made of corrosion-resistant material.

For small installations, a simplified pump sizing method is to allow 1 gpm for every 20 fixture units in the system, or to allow 0.5 gpm for each 0.75 or 1 in. riser; 1 gpm for each 1.25 or 1.5 in. riser; and 2 gpm for each riser 2 in. or larger.

Werden and Spielvogel (1969) and Dunn *et al.* (1959) cover heat loss calculations for large systems. For larger installations, heat losses of lines become significant. For larger systems, a quick method to size the pump and return follows:

1. Determine the total length of all hot water supply and return piping.
2. Choose an appropriate value for piping heat losses from Table 2 or other engineering data (usually supplied by insulation companies, etc.). Multiply this value by the total length of piping involved.
   a) A rough estimation can be made by multiplying the total length of covered pipe by 30 Btu/h·ft or uninsulated pipe by 60 Btu/h·ft. Table 2 gives actual heat losses in pipes at a service water temperature of 140 °F and a 70 °F ambient temperature. The values of 30 or 60 Btu/h·ft are only recommended for ease in calculation.
3. Determine pump capacity as follows:

$$Q = q/(60 \rho c_p \Delta t) \quad (1)$$

where

Q = pump capacity, gpm
q = heat loss, Btu/h
ρ = density of water = 8.33 lb/gal
c_p = specific heat of water = 1 Btu/lb·°F
Δt = allowable temperature drop, °F

or for a 20 °F allowable temperature drop:

$$Q(\text{gpm}) = q/8.33 \times 1 \times 60 \text{ min/h} \times 20 = q/10{,}000 \quad (2)$$

Caution: This assumes a 20 °F temperature drop is acceptable at the last fixture.

4. Select a pump to provide the required flow rate and obtain from the pump curves the pressure created at this flow.
5. Multiply the head by 100 and divide by the total length of hot water return piping to determine the allowable friction loss per 100 ft of pipe.
6. Determine the required flow in each circulating loop and size the hot water return pipe based on this flow and the allowable friction loss from Step 5.

Where multiple risers or horizontal loops are used, balancing valves with test means are recommended in the return lines. A check valve should be placed in each return to prevent entry of cold water or reversal of flow, particularly during periods of high hot water demand.

Table 2 Heat Loss of Pipe (at 140 °F Inlet, 70 °F Ambient Temperatures)

Pipe Size, in.	Bare Copper Tubing, Btu/h·ft	0.5-in. Glass Fiber Insulated Copper Tubing, Btu/h·ft
0.75	30	17.7
1	38	20.3
1.25	45	23.4
1.5	53	25.4
2	66	29.6
2.5	80	33.8
3	94	39.5
4	120	48.4

Three common methods of arranging circulation lines are shown in Figure 3. Although the diagrams apply to multistory buildings, arrangements (A) and (B) are also used in residential designs. In circulation systems, air venting, pressure drops through the heaters and storage tanks, balancing, and line losses should be considered. In Figures 3A and 3B, air is vented by connecting the circulating line below the top fixture supply. With this arrangement, air is eliminated from the system each time the top fixture is opened. Generally, for small installations, a 0.5 or 0.75 in. hot water return is ample.

All storage tanks and piping on recirculating systems should be insulated as recommended by ASHRAE *Standard* 90A and the *Standard* 100 series.

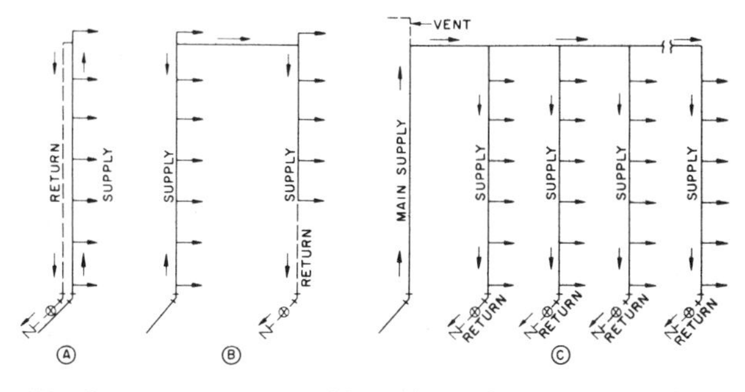

Fig. 3 Arrangements of Hot Water Circulation Lines

Heat Traced, Nonreturn Piping System

In this system, the fixtures can be as remote as in the foregoing return piping system. The hot water supply piping is heat traced with electric resistance heating cable preinstalled under the pipe insulation. Electrical energy input is self-regulated by the cable's construction to maintain the required water temperature at the fixtures. The return piping system and circulation pump are not required.

Special Piping—Commercial Kitchens

Adequate flow rates and pressures must be maintained for automatic dishwashers in commercial kitchens. To reduce operating difficulties, piping for automatic dishwashers should be installed according to the following recommendations:

1. The cold water feed line to the water heater should be no less than 1-in. ID pipe size.
2. The supply line for 180 °F water from the water heater to the dishwasher should not be less than 0.75-in. ID pipe size.
3. No auxiliary feed lines should connect to the 180 °F supply line to the dishwasher.
4. A return line should be installed if the source of 180 °F water is more than 5 ft from the dishwasher.
5. Forced circulation by a pump should be used if the water heater is installed on the same level as the dishwasher, if the length of return piping is more than 60 ft, or if the waterlines are trapped.
6. If a circulating pump is used, it is generally installed in the return line. It may be controlled by (a) the dishwasher wash switch, (b) a manual switch located near the dishwasher, or (c) an immersion or strap-on thermostat located in the return line.
7. A pressure-reducing valve should be installed in the low-temperature supply line to a booster water heater, but external to a recirculating loop. It should be adjusted, with the water flowing, to the value stated by the washer manufacturer (typically 20 psi).
8. A check valve should be installed in the return circulating line.
9. If a check valve type of water meter or a backflow prevention device is installed in the cold waterline ahead of the heater, it is necessary to install a properly sized diaphragm-type expansion tank between the water meter or prevention device and the heater.
10. National Sanitation Foundation (NSF) standards require the installation of a 0.25-in. IPS connection for a pressure gage mounted adjacent to the supply side of the control valve. They also require a waterline strainer ahead of any electrically operated control valve (Figure 4).
11. NSF standards do not accept copper waterlines that are not under constant pressure, except for the line downstream of the solenoid valve on the rinse line to the cabinet.

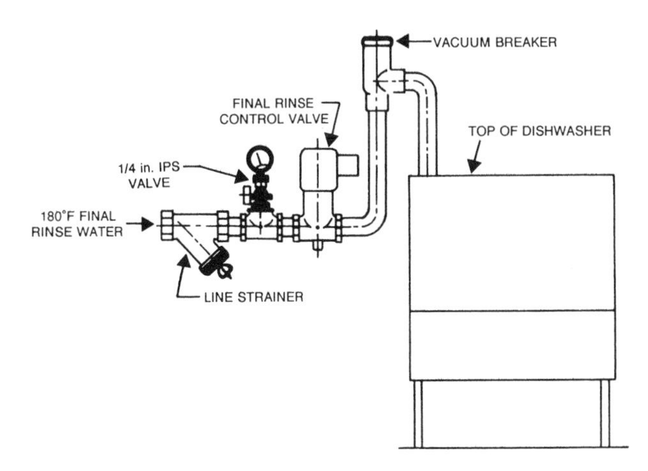

Fig. 4 National Sanitation Foundation (NSF) Plumbing Requirements for Commercial Dishwasher

Water Pressure—Commercial Kitchens

Proper flow pressure must be maintained to achieve efficient dishwashing. The standards of the National Sanitation Foundation for dishwasher water flow pressures are 15 psi gage minimum, 25 psi gage maximum, and 20 psi gage ideal. Flow pressure is the line pressure measured when water is flowing through the rinse arms of the dishwasher. Low flow pressure can be caused by undersized water piping, stoppage in piping, or excess pressure drop through heaters. Low water pressure causes an inadequate rinse, resulting in poor drying and sanitizing of the dishes. If flow pressure in the supply line to the dishwasher is below 15 psi gage, a booster pump or other means should be installed to provide supply water at 20 psi gage.

A flow pressure in excess of 25 psi gage will cause atomization of the 180 °F rinse water, resulting in an excessive temperature drop. Between the rinse nozzle and the dishes, the temperature drop can be as much as 15 °F. A pressure regulator should be installed in the supply waterline adjacent to the dishwasher and external to the return circulating loop if used. The regulator should be set to maintain a pressure of 20 psi gage.

Two-Temperature Service

Where multiple temperature requirements are met by a single system, the system temperature is determined by the maximum temperature needed. Lower temperatures can be obtained by mixing hot and cold water. Automatic blending to reduce hot water temperature available at certain outlets may be necessary to prevent injury or damage from excessive water temperatures (Figure 5). Check applicable codes for blending valve requirements.

Where predominate usage is at lower temperatures, the common approach is to heat all water to the lower temperature and then to use a separate booster heater to further heat the water for the higher temperature services (Figure 6). This method offers better protection against scalding.

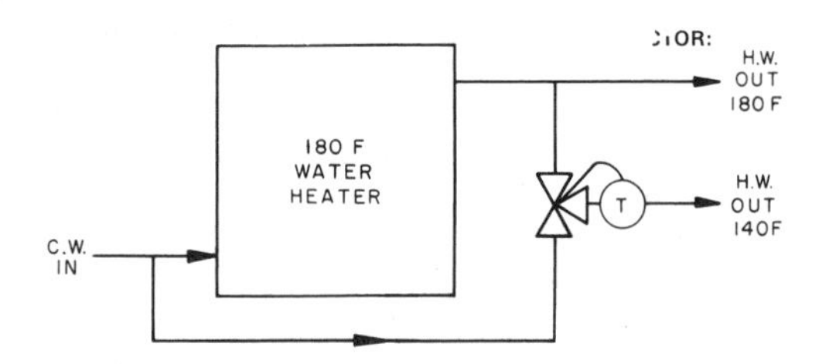

Fig. 5 Two-Temperature Service with a Three-Way Control Valve

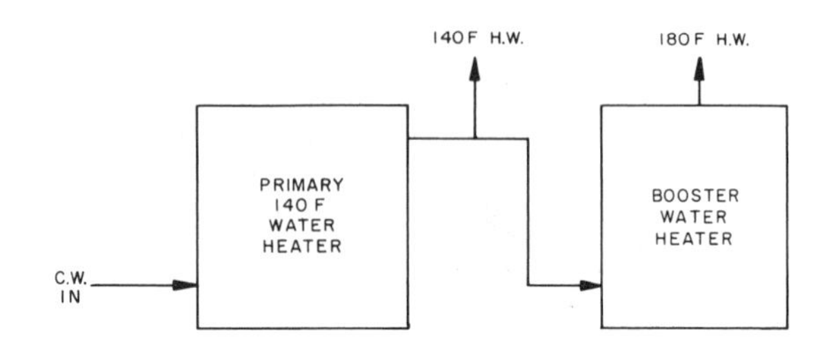

Fig. 6 Two-Temperature Service with a Primary Heater and Booster Heater in Series

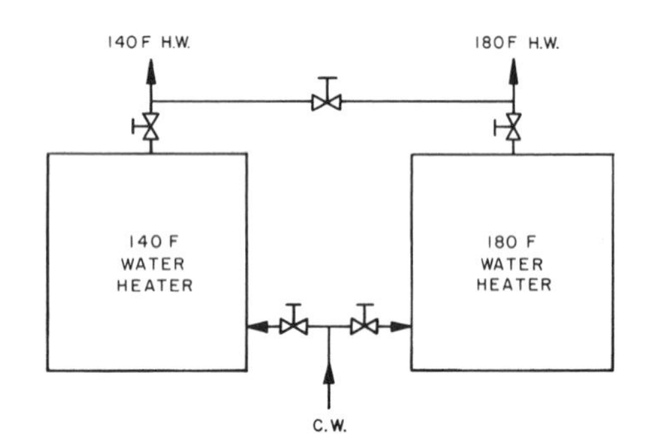

Fig. 7 Two-Temperature Service with Separate Heater for Each Service

A third method uses separate heaters for the higher temperature service (Figure 7). It is common practice to cross-connect the two heaters, so that one heater can serve the complete installation temporarily while the other is valved off for maintenance. Each heater should be sized for the total load unless hot water consumption can be reduced during maintenance periods.

Manifolding

Where one heater does not have the capacity to supply both needs, two or more water heaters may be installed in parallel. In such an installation, only one mixing valve of adequate capacity should be used. It is difficult to obtain even flow through parallel mixing valves.

Heaters installed in parallel should be identical, *i.e.*, have the same input and storage capacity, with inlet and outlet piping arranged so that an equal flow is received from each heater under all demand conditions.

Terminal Hot Water Usage Devices

Details on the vast number of devices using service hot water are beyond the scope of this chapter. Nonetheless, they are important to a successful overall design. Consult manufacturer's literature for information on required flow rates, temperature limits, and/or other operating factors for specific items.

DESIGN CONSIDERATIONS

WATER-HEATING EFFICIENCIES

The following terms apply to water heaters or water-heating systems.

Recovery efficiency. Heat absorbed by the water divided by heat input to the heating unit during the period that water temperature is raised from inlet temperature to final temperature under no water flow.

Thermal efficiency. Heat in the water flowing from the heater outlet divided by the heat input of the heating unit over a specific period of steady-state condition.

Energy factor. An indicator of the overall efficiency of a residential storage water heater representing heat in the estimated daily quantity of delivered hot water divided by the daily energy consumption of the water heater, as measured by U.S. Department of Energy test procedures.

Standby loss. The amount of heat lost from the water heater during periods of nonuse of service hot water.

Hot water distribution efficiency. Heat contained in the water at points of use divided by the heat delivered at the heater outlet at a given flow. Heater/system efficiency. Heat contained in the water at points of use divided by the heat input to the heating unit at a given flow rate (thermal efficiency times distribution efficiency).

Overall system efficiency. Heat in the water delivered at points of use divided by the heat supplied to the heater for any selected time period.

System standby loss. The amount of heat lost from the water-heating system during periods of nonuse of service hot water.

Hot water system design should consider the following:

- Water heaters of different sizes and insulation may have different standby losses, recovery efficiency, thermal efficiency, or energy factors.
- A properly designed, sized, and insulated distribution system is necessary to deliver adequate water quantities at temperatures satisfactory for the uses served. This reduces system standby loss and improves hot water distribution efficiency.
- Heat traps between recirculation mains and infrequently used branch lines reduce convection losses to these lines and improve heater system efficiency.
- Controlling circulating pumps to operate only as needed to maintain proper temperature at the end of the main reduces losses on return lines. This improves overall system efficiency.
- Provision for shutdown of circulators during building vacancy reduces system standby losses.

WATER QUALITY, CORROSION, AND SCALE

A complete water analysis and an understanding of system requirements are needed to protect water-heating systems from scale and corrosion. The analysis is needed because hard water may cause scale (fouling or liming of heat transfer and water storage surfaces). Soft water may aggravate corrosion problems.

Scale formation is also affected by system requirements and equipment. As shown in Figure 8, the rate of scaling increases with temperature and usage, because calcium carbonate and other scaling compounds lose solubility at higher temperatures. In water tube-type equipment, this can be offset by increasing the water flow over the heat transfer surfaces, as this reduces tube surface temperatures. Also, the turbulence of the flow, if high enough, works to keep any scale that does precipitate off the surface. When water hardness is over eight grains per gallon, water softening or other water treatments are often recommended.

Corrosion problems increase with temperature because corrosive oxygen and carbon dioxide gases are released from the water. Also, electrical conductivity increases with temperature, which

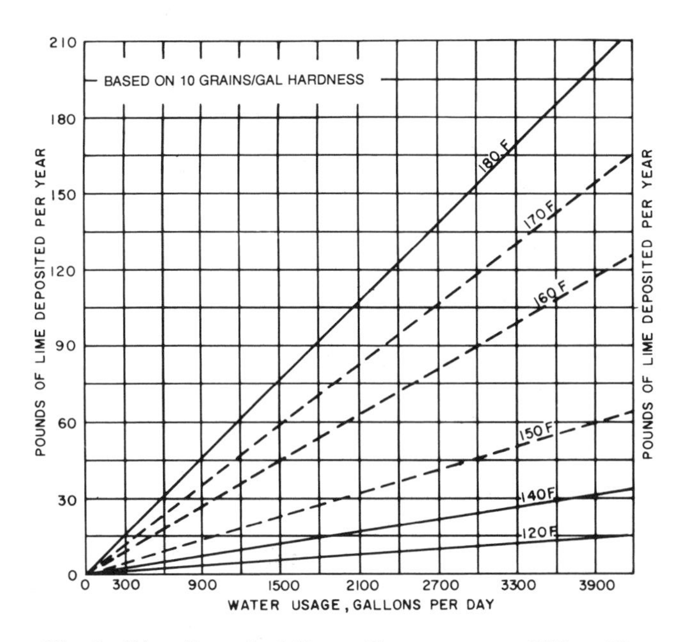

Fig. 8 Lime Deposited Versus Temperature and Water Use
(Purdue University Bulletin No. 74)

enhances electrochemical reactions such as rusting (Toaborek *et al.* 1972). A deposit of scale provides some protection from corrosion; however, this deposit also reduces the heat transfer rate, and it is not under the control of the system designer. Water heaters and hot water storage tanks constructed of stainless steel, copper, or other nonferrous alloys, provide corrosion protection. Some stainless steels, however, may be adversely affected by chlorides, while copper may be attacked by ammonia or carbon dioxide. Steel vessels can be protected to varying degrees by galvanizing or by lining them with copper, glass, cement, or other corrosion-resistant material. Glass-lined vessels are almost always supplied with electrochemical protection. Typically, a rod of magnesium alloy (the anode) is installed in the vessel by the manufacturer. This electrochemically active material sacrifices itself to reduce or prevent corrosion of the tank (the cathode). Higher temperatures, softened waters, and high water usage may lead to rapid anode consumption. Manufacturers recommend periodic replacement to prolong the life of the vessel. Conversely, some waters have very little electrochemical activity. In this instance, a standard anode will show little or no activity, and the vessel will not be adequately protected. If this condition is suspected, consult the equipment manufacturer on the possible need of a high potential anode.

SAFETY DEVICES FOR HOT WATER SUPPLY SYSTEMS

Various regulatory agencies differ as to the selection of protective devices and methods of installation. As a result, it is essential to check and comply with the manufacturer's instructions and the applicable local codes. In the absence of such instructions and codes, the following recommendations may be used as a guide.

- Thermal expansion control devices limit the pressure that results when the water in the tank is heated and expands in a closed system. While the water-heating system is under service pressure, the pressure will rise rapidly if backflow cannot occur during heating. Backflow can be prevented by devices such as a check valve, pressure-reducing valve, or backflow preventer in the cold waterline or by temporarily shutting off the cold water.

  In these cases, the pressure rise may rupture the tank or cause other damage. Such systems must be protected by a properly sized and located diaphragm-type expansion tank.
- Temperature limiting devices (energy cutoff/high limit) prevent water temperatures from exceeding 210 °F by stopping the flow of fuel or energy. These devices should be listed and labeled by Underwriter's Laboratories, the American Gas Association (AGA), or other recognized certifying agencies.
- Temperature and pressure relief valves open to prevent water temperature from exceeding 210 °F or when the pressure exceeds the valve setting. Combination temperature and pressure relief valves should be AGA or National Board listed and labeled and have a water discharge capacity equal to or exceeding the heat input rating of the water heater. Combination temperature and pressure relief valves should be installed so that the temperature-sensitive element is located in the top 6 in. of the tank, *i.e.*, where the water is hottest.
- Pressure relief valve opens when the pressure exceeds the valve setting. This valve should have a discharge capacity sufficient to relieve excess fluid pressure in the water heater. It should comply with current applicable American National Standards or the ASME Boiler and Pressure Vessel Code.

  A pressure relief valve should be installed in any part of the system containing any heat input device that can be isolated by valves. The heat input device may be solar water-heating panels, desuperheater water heaters, heat recovery devices, or similar equipment.

SPECIAL CONCERNS

Legionella pneumophila (Legionnaire's Disease)

The bacteria causing Legionnaire's disease when inhaled has been discovered in the service water systems of various buildings in the United States and abroad. Infection has often been traced to *Legionella pneumophila* colonies in shower heads. Ciesielki *et al.* (1984) determined that the *Legionella pneumophila* can colonize in hot water systems maintained at 115 °F or lower. Unrecirculated segments of the service water systems provide ideal breeding locations, *e.g.*, shower heads, faucet aerators, and uncirculated sections of storage-type water heaters.

To limit the potential of *Legionella pneumophila* growth, service water temperatures in the 140 °F range are recommended. This high temperature, however, increases the potential for scalding, so care must be taken. Supervised periodic flushing of fixture heads with 170 °F water is recommended in hospitals and health care facilities, since already weakened patients are generally more susceptible to infection.

Temperature Requirements

Typical temperature requirements for some services are shown in Table 3. In some cases, slightly lower temperatures may be satisfactory. Temperatures below 140 °F are usually obtained by blending hot and cold water at point of use.

Hot Water from Tanks and Storage Systems

With storage systems, 60 to 80% of the hot water in a tank is assumed to be usable before dilution by cold water lowers the temperature below an acceptable level. Thus, the hot water available from a self-contained storage heater is usually considered to be:

$$Q_t = R + MS_t/d \tag{3}$$

where

Q_t = available hot water, gph
R = recovery rate at the required temperature, gph
M = ratio of usable water to storage tank capacity
S_t = storage capacity of the heater tank, gal
d = duration of peak hot water demand, h

Usable hot water from an unfired tank in gallons is calculated from:

$$Q_a = MS_a \tag{4}$$

Table 3 Representative Hot Water Temperatures

Use	Temperature, °F
Lavatory	
Hand washing	105
Shaving	115
Showers and tubs	110
Therapeutic baths	95
Commercial or institutional laundry, dependent on fabric	up to 180
Residential dish washing and laundry	140
Surgical scrubbing	110
Commercial spray-type dish washing as required by NSF	
Single- or multiple-tank hood or rack type	
Wash	150 min
Final rinse	180 to 195
Single-tank conveyor type	
Wash	160 min
Final rinse	180 to 195
Single-tank rack or door type	
Single temperature wash and rinse	165 min
Chemical sanitizing types (see manufacturer for actual temperature required)	140
Multiple-tank conveyor type	
Wash	150 min
Pumped rinse	160 min
Final rinse	180 to 195
Chemical sanitizing glass washer	
Wash	140
Rinse	75 min

Table 4 Typical Residential Usage of Hot Water/Task

Use	High Flow, gal	Low Flow (Water Savers Used), gal
Food preparation	5	3
Hand dish washing	4	4
Automatic dishwasher	15	15
Clothes washer	32	21
Shower or bath	20	15
Face and hand washing	4	2

where

Q_a = usable water available from an unfired tank, gal
S_a = capacity of unfired tank, gal

Note: Assumes tank water at required temperature.

Hot water obtained from a water-heating system using a storage heater with an auxiliary storage tank can be determined by:

$$Q_z = dQ_t + Q_a = Rd + M(S_t + S_a) \tag{5}$$

where

Q_z = total hot water available from system during one peak, gal

Placement of Water Heaters

Water heaters not requiring combustion air may generally be placed in any location, as long as relief valve discharge pipes are open to a safe location.

Water heaters requiring ambient combustion air must be located in areas with air openings large enough to admit the required combustion/dilution air (see NFPA 54).

Gas or oil-fired water heaters should not be located in areas where flammable vapors are likely to be present without additional precautions to eliminate the probable ignition of flammable vapors (see sections 5.1.9 through 5.1.12 of NFPA 54).

Precaution should be taken against use, spillage, or release of combustible gases or liquids, such as LP gases, paint thinner, or gasoline where gas or oil-fired water heaters are located.

HOT WATER REQUIREMENTS AND STORAGE EQUIPMENT SIZING

Methods for sizing storage water heaters vary. Those using recovery versus storage curves are based on extensive research. All methods provide adequate hot water if the designer allows for unusual conditions.

RESIDENTIAL

Table 4 shows typical hot water usage in a residence. In its *Minimum Property Standards for One and Two Family Living Units*, No. 4900.1-1982, HUD-FHA has established minimum permissi-

Table 5 HUD—FHA Minimum Water Heater Capacities for One- and Two-Family Living Units

Number of Baths	1 to 1.5			2 to 2.5				3 to 3.5			
Number of Bedrooms	1	2	3	2	3	4	5	3	4	5	6
GAS[a]											
Storage, gal	20	30	30	30	40	40	50	40	50	50	50
1000 Btu/h input	27	36	36	36	36	38	47	38	38	47	50
1-h draw, gal	43	60	60	60	70	72	90	72	82	90	92
Recovery, gph	23	30	30	30	30	32	40	32	32	40	42
ELECTRIC[a]											
Storage, gal	20	30	40	40	50	50	66	50	66	66	80
kW input	2.5	3.5	4.5	4.5	5.5	5.5	5.5	5.5	5.5	5.5	5.5
1-h draw, gal	30	44	58	58	72	72	88	72	88	88	102
Recovery, gph	10	14	18	18	22	22	22	22	22	22	22
OIL[a]											
Storage, gal	30	30	30	30	30	30	30	30	30	30	30
1000 Btu/h input	70	70	70	70	70	70	70	70	70	70	70
1-h draw, gal	89	89	89	89	89	89	89	89	89	89	89
Recovery, gph	59	59	59	59	59	59	59	59	59	59	59
TANK-TYPE INDIRECT[b, c]											
I-W-H rated draw, gal in 3-h, 100°F rise		40	40		66	66[d]	66	66	66	66	66
Manufacturer-rated draw, gal in 3-h, 100°F rise		49	49		75	75[d]	75	75	75	75	75
Tank capacity, gal		66	66		66	66[d]	82	66	82	82	82
TANKLESS-TYPE INDIRECT[c, e]											
I-W-H-rated, gpm 100°F rise		2.75	2.75		3.25	3.25[d]	3.75	3.25	3.75	3.75	3.75
Manufacturer-rated draw, gal in 5 min, 100°F rise		15	15		25	25[d]	35	25	35	35	35

[a]Storage capacity, input, and recovery requirements indicated in the table are typical and may vary with each individual manufacturer. Any combination of these requirements to produce the stated 1-h draw will be satisfactory.
[b]Boiler-connected water heater capacities (180°F boiler water, internal, or external connection).
[c]Heater capacities and inputs are minimum allowable. Variations in tank size are permitted when recovery is based on 4 gph/kW at 100°F rise for electrical, AGA recovery ratings for gas heaters and IBR ratings for steam and hot water heaters.
[d]Also for 1 to 1.5 baths and 4 bedrooms for indirect water heaters.
[e]Boiler-connected heater capacities (200°F boiler water, internal, or external connection).

ble water heater sizes (Table 5). Storage water heaters may vary from the sizes shown in the table if combinations of recovery and storage that produce the 1-h draw required are used.

Over the last decade, the structure and life-style of a typical family has altered the household's hot water consumption. Due to variations in family size, age of family members, presence and age of children, hot water use volumes and temperatures, and other factors, demand patterns fluctuate widely in both magnitude and time distribution.

Perlman and Mills (1985) developed the overall and peak average of hot water use volumes shown in Table 6. Average hourly patterns and 95% confidence level profiles are illustrated in Figures 9 and 10. Samples of results from the analysis of similarities in hot water use are given in Figures 11 and 12.

COMMERCIAL AND INDUSTRIAL

Most industrial and commercial establishments use hot or warm water. The specific requirements vary in total volume, flow rate, duration of peak load period, and temperature needed. Water heaters and systems should be selected based on these requirements.

Table 6 Overall (OVL) and Peak Average Hot Water Use Volumes

	Average Hot Water Use, gal							
	Hourly		Daily		Weekly		Monthly	
Group	OVL	Peak	OVL	Peak	OVL	Peak	OVL	Peak
All families	2.6	4.6	62.4	67.1	436	495	1897	2034
"Typical" families	2.6	5.8	63.1	66.6	442	528	1921	2078

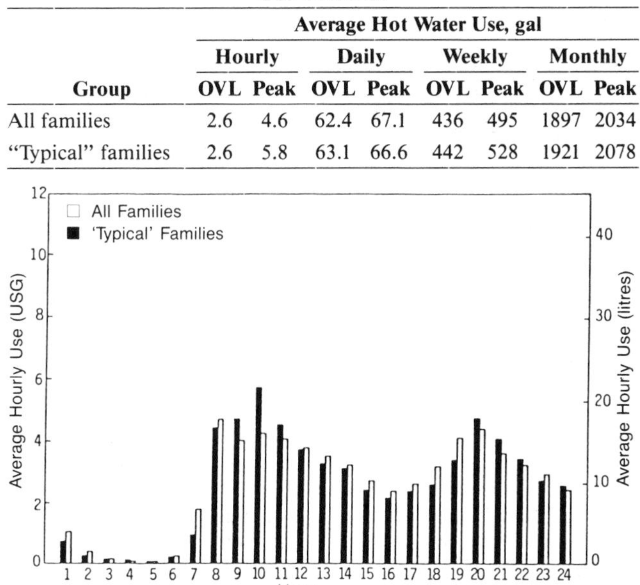

Fig. 9 Average Hourly Hot Water Use

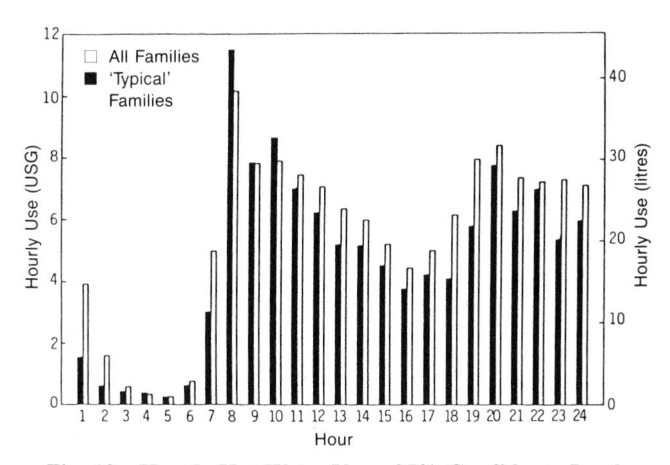

Fig. 10 Hourly Hot Water Use—95% Confidence Level

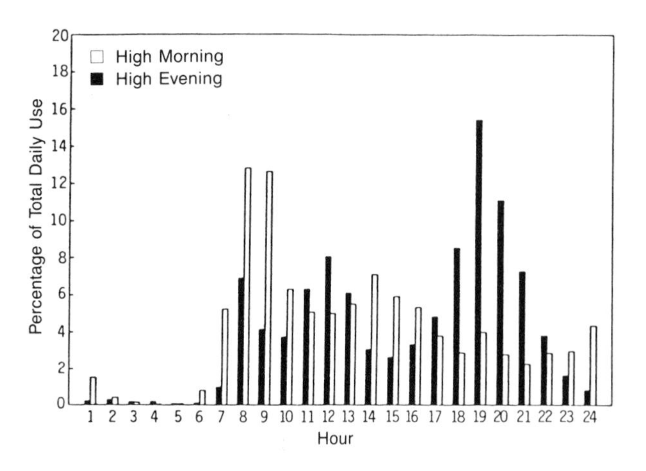

Fig. 11 Hourly Water Use Pattern for Selected High Morning and High Evening Users

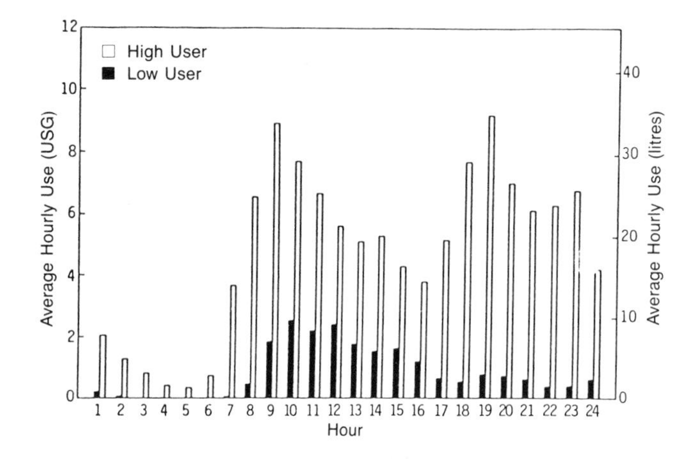

Fig. 12 Average Hourly Hot Water Use for Low and High Users

Commercial and Institutional Buildings

This section covers sizing recommendations for central storage water-heating systems. Hot water usage data and sizing curves for dormitories, motels, nursing homes, office buildings, food service establishments, apartments, and schools are based on EEI-sponsored research. Caution must be taken in applying these data to small buildings. Also, be aware that within any given category there may be significant variation. For example, the motel category encompasses standard, luxury, resort, and convention motels.

When additional hot water requirements exist, a designer should increase the recovery and/or storage capacity to account for this use. For example, if there is food service in an office building, recovery and storage capacities required for each use should be added when sizing a single central water-heating system.

Peak hourly and daily demands for various categories of commercial and institutional buildings are shown in Table 7. Demands for central storage-type hot water systems represent maximum flows metered in this 129-building study, excluding extremely high and very infrequent peaks. Table 7 also shows average hot water consumption figures for these types of buildings. Averages for schools and food service establishments are based on actual days of operation, while all others are based on total days. These averages can be used to estimate monthly consumption of hot water.

Dormitories

Hot water requirements for college dormitories generally include showers, lavatories, service sinks, and washing machines.

Table 7 Hot Water Demands and Use for Various Types of Buildings

Type of Building	Maximum Hour	Maximum Day	Average Day
Men's dormitories	3.8 gal/student	22.0 gal/student	13.1 gal/student
Women's dormitories	5.0 gal/student	26.5 gal/student	12.3 gal/student
Motels: Number of units[a]			
20 or less	6.0 gal/unit	35.0/unit	20.0 gal/unit
60	5.0 gal/unit	25.0 gal/unit	14.0 gal/unit
100 or more	4.0 gal/unit	15.0 gal/unit	10.0 gal/unit
Nursing homes	4.5 gal/bed	30.0/bed	18.4 gal/bed
Office buildings	0.4 gal/person	2.0 gal/person	1.0 gal/person
Food service establishments:			
Type A—full meal restaurants and cafeterias	1.5 gal/max meals/h	11.0 gal/max meals/h	2.4 gal/average meals/day[b]
Type B—drive-ins, grilles, luncheonettes, sandwich and snack shops	0.7 gal/max meals/h	6.0 gal/max meals/h	0.7 gal/average meals/day[b]
Apartment houses: Number of apartments			
20 or less	12.0 gal/apartment	80.0 gal/apartment	42.0 gal/apartment
50	10.0 gal/apartment	73.0 gal/apartment	40.0 gal/apartment
75	8.5 gal/apartment	66.0 gal/apartment	38.0 gal/apartment
100	7.0 gal/apartment	60.0 gal/apartment	37.0 gal/apartment
200 or more	5.0 gal	50.0 gal/apartment	35.0 gal/apartment
Elementary schools	0.6 gal/student	1.5 gal/student	0.6 gal/student[b]
Junior and senior high schools	1.0 gal/student	3.6 gal/student	1.8 gal/student[b]

[a]Interpolate for intermediate values.
[b]Per day of operation.

Peak demand usually results from the use of showers. Load profiles and hourly consumption data indicate that peaks may last 1 or 2 h and then taper off substantially. Peaks occur predominantly in the evening, mainly around midnight. The figures do not include hot water used for food service.

Military Barracks

Design criteria for military barracks are available from the engineering departments of the U.S. Department of Defense. Some measured data exists for hot water use in these facilities. For published data, contact the U.S. Army Corps of Engineers or Naval Facilities Engineering Command.

Motels

Domestic hot water requirements are for tubs and showers, lavatories, and general cleaning purposes. Recommendations are based on tests at low- and high-rise motels located in urban, suburban, rural, highway, and resort areas. Peak demand, usually from shower use, may last 1 or 2 h and then drop off sharply. Food service, laundry, and swimming pool requirements are not included.

Nursing Homes

Hot water is required for tubs and showers, wash basins, service sinks, kitchen equipment with food service for patients, and for general cleaning. These figures include hot water for kitchen use. When other equipment, such as that for heavy laundry and hydrotherapy purposes, is to be used, its additional hot water requirement should be added.

Office Buildings

Hot water requirements are primarily for cleaning and lavatory use by occupants and visitors. Hot water use for food service within office buildings is not included.

Food Service Establishments

Hot water requirements are primarily for dishwashing. Other uses include food preparation, cleaning pots and pans and floors, and hand washing for employees and customers. The recommendations of hot water requirements are for establishments serving food at table, counter, and booth seats, and to parked cars. Food service establishments that use disposable service exclusively are not included.

Dishwashing, as metered with other hot water requirements in these tests, is based on the normal practice of dishwashing after meals, but not on indiscriminate or continuous use of machines irrespective of the flow of soiled dishes. The recommendations include hot water supplied to dishwasher booster heaters.

Apartments

Hot water requirements for both garden-type and high-rise apartments are for one- and two-bath apartments, showers, lavatories, kitchen sinks, dishwashers, clothes washers, and general cleaning purposes. Clothes washers can be either in individual apartments or centrally located. These data apply to central water-heating systems only.

Elementary Schools

Hot water requirements are for lavatories, cafeteria and kitchen use, and general cleaning purposes. When showers are used, their additional hot water requirements should be added to those recommended. The recommendations include hot water for dishwashing machines but not for extended school operation, such as evening classes.

High Schools

Senior high schools, grades 9 or 10 through 12, require hot water for showers, lavatories, dishwashing machines, kitchens, and general cleaning. Junior high schools, grades 7 through 8 or 9, have requirements similar to those of the senior high schools. Where no showers are included, junior high schools follow the recommendations for elementary schools.

Requirements for high schools are based on daytime use. Recommendations do not include hot water usage for additional activities, such as night school. In such cases, the maximum hourly demand remains the same, but the maximum daily and the average daily usage is increased, usually by the number of additional people using showers and, to a lesser extent, eating and washing facilities.

SIZING EXAMPLES

Figures 13 through 20 show the relationships between recovery and storage capacity for the various building categories. Any combination of storage and recovery rates that falls on the proper curve will satisfy the building requirements. Using the minimum recovery rate and the maximum storage capacity on the curves yields the smallest hot water capacity capable of satisfying the building requirement. The higher the recovery rate, the greater the 24-h heating capacity and the smaller the storage capacity required.

These curves can be used to select water heaters that have fixed storage or recovery rates by adjusting recovery and storage requirements. Where hot water demands are not coincident with peak electric, steam, or gas demands, greater heater inputs can be selected if they do not create additional energy system demands, and the corresponding storage tank size can be selected from the curves.

Ratings of gas-fired water-heating equipment are based on sea level operation and need not be changed for operation at elevations up to 2000 ft. For operation at elevations above 2000 ft and, in the absence of specific recommendations from the local authority, equipment ratings shall be reduced at the rate of 4% for each 1000 ft above sea level before selecting appropriately sized equipment.

Recovery rates in Figures 13 through 20 represent the actual hot water required without considering system heat losses. Heat losses from storage tanks and recirculating hot water piping should be calculated and added to the recovery rates shown. With uninsulated storage tanks and uninsulated hot water piping, it is necessary to insulate the equipment.

The storage capacities shown are net usable requirements. On the assumption that 60 to 80% of the hot water in a storage tank is usable, the actual storage tank size should be increased by 25 to 66% to compensate for unusable hot water.

Examples

Example 1. Determine the required water heater size for a 300-student women's dormitory for the following criteria:

a. Storage system with minimum recovery rate.

b. Storage system with recovery rate of 2.5 gph per student.

c. With the additional requirement for a cafeteria to serve a maximum of 300 meals per hour for minimum recovery rate combined with item *a,* and for a recovery rate of 1.0 gph per maximum meal per hour, combined with item *b.*

Solution:

a. The minimum recovery rate from Figure 13 for women's dormitories is 1.1 gph per student, or a total of 330 gph recovery is required. Storage required is 12 gal per student or 3600 gal storage. On a 70% net usable basis, the necessary tank size is 1.43 × 3600 = 5150 gal.

b. The same curve also shows 5 gal storage per student at 2.5 gph recovery, or 300 × 5 = 1500 gal storage with recovery of 300 × 2.5 = 750 gph. The tank size will be 1.43 × 1500 = 2150 gal.

c. The additional requirement for a cafeteria can be determined from Figure 17, with the storage and recovery capacity added to that for the dormitory.

For the case of minimum recovery rate, the cafeteria requires 300 × 0.45 = 135 gph recovery rate and 300 × 7 × 1.43 = 3000 gal of additional storage. The entire building then requires 330 + 135 = 465 gph recovery and 5150 + 3000 = 8150 gal of storage.

With 1 gal recovery per maximum meal hour, the recovery required is 300 gph, with 300 × 2.0 × 1.43 = 860 gal of additional storage. Combining this with item *b,* the entire building requires 750 + 300 = 1050 gph recovery and 2150 + 860 = 3010 gal of storage.

Note: Recovery capacities shown are for heating water only. Additional capacity must be added to offset the system heat losses.

Example 2. Determine the water heater size and monthly hot water consumption for an office building to be occupied by 300 people:

a. Storage system with minimum recovery rate.

b. Storage system with 1.0 gal per person storage.

c. Additional minimum recovery rate requirement for a luncheonette, open 5 days a week, serving a maximum of 100 meals in 1 h, and an average of 200 meals per day.

d. Monthly hot water consumption.

Solution:

a. With minimum recovery rate of 0.10 gph per person from Figure 16 for office buildings, 30 gph recovery is required, while the storage is 1.6 gal per person, or 300 × 1.6 = 480 gal storage. The tank size will be 1.43 × 480 = 690 gal.

b. The curve also shows 1.0 gal storage per person at 0.175 gph per person recovery, or 300 × 0.175 = 52.5 gph. The tank size will be 1.43 × 300 = 430 gal.

c. The hot water requirements for a luncheonette are contained in Figure 17. The recovery versus storage curve shows that with a minimum recovery capacity of 0.25 gph per maximum meals per hour, 100 meals would require 25 gph recovery, while the storage would be 2.0 gal per meal, or 100 × 2.0 × 1.43 = 286 gal storage. The combined requirements with item *a* would then be 55 gph recovery and 976 gal storage.

Combined with item *b,* the requirement will be 77.5 gph recovery and 716 gal storage.

d. From Table 1, the office building will consume an average of 1.0 gal per person per day × 30 days × 300 people = 9000 gal per month, while the luncheonette will consume 0.7 gal per meal × 200 meals per day × 22 days per month = 3080 gal per month, for a total of 12,080 gal per month.

Note: Recovery capacities shown are for heating water only. Additional capacity must be added to offset the system heat losses.

Example 3. Determine the water heater size for a 200-unit apartment house:

a. Storage system with minimum recovery rate, with a single tank.

b. Storage system with 4 gph per apartment recovery rate, with a single tank.

c. Storage system for each of two 100-unit wings.

1. Minimum recovery rate.
2. Recovery rate of 4 gph per apartment.

Solution:

a. The minimum recovery rate, from Figure 18, for apartment buildings with 200 apartments is 2.1 gph per apartment, or a total of 420 gph recovery required. The storage required is 24 gal per apartment, or 4800 gal. Based on 70% of this hot water being usable, the necessary tank size is 1.43 × 4800 = 6860 gal.

b. The same curve also shows 5 gal storage per apartment at 4 gph recovery, or 200 × 4 = 800 gph. The tank size will be 1.43 × 1000 = 1430 gal.

c. Alternate solution for a 200-unit apartment house with two wings, each with its own hot water system. In this instance, the solution for each 100-unit wing would be:

1. With minimum recovery rate of 2.5 gph per apartment, from the curve, a 250 gph recovery is required, while the necessary storage is 28 gal per apartment, or 100 × 28 = 2800 gal. The required tank size is 1.43 × 2800 = 4000 gal for each wing.

2. The curve also shows that for a recovery rate of 4 gph per apartment, the storage would be 14 gal, or 100 × 14 = 1400 gal, with recovery of 100 × 4 = 400 gph. The necessary tank size is 1.43 × 1400 = 2000 gal in each wing.

Note: Recovery capacities shown are for heating water only. Additional capacity must be added to offset the system heat losses.

Example 4. Determine the water heater size and monthly hot water consumption for a 2000-student high school.

a. Storage system with minimum recovery rate.

b. Storage system with 4000-gal maximum storage capacity.

Solution:

a. With the minimum recovery rate of 0.15 gph per student from Figure 20 for high schools, 300 gph recovery is required. The storage required is 3.0 gal per student, or 2000 × 3.0 = 6000 gal storage. The tank size is 1.43 × 6000 = 8600 gal.

b. The net storage capacity will be 0.7 × 4000 = 2800 gal, or 1.4 gal per student. From the curve, 0.37 gph per student recovery capacity is required, or 0.37 × 2000 = 740 gph.

c. From Table 1, hot water consumed monthly is 2000 students × 1.8 gal per student per day × 22 days = 79,200 gal.

Note: Recovery capacities shown are for heating water only. Additional capacity must be added to offset the system heat losses.

Table 8 can be used to determine the size of water-heating equipment from the number of fixtures. To obtain the probable maximum demand, multiply the total quantity for the fixtures by the demand factor in line

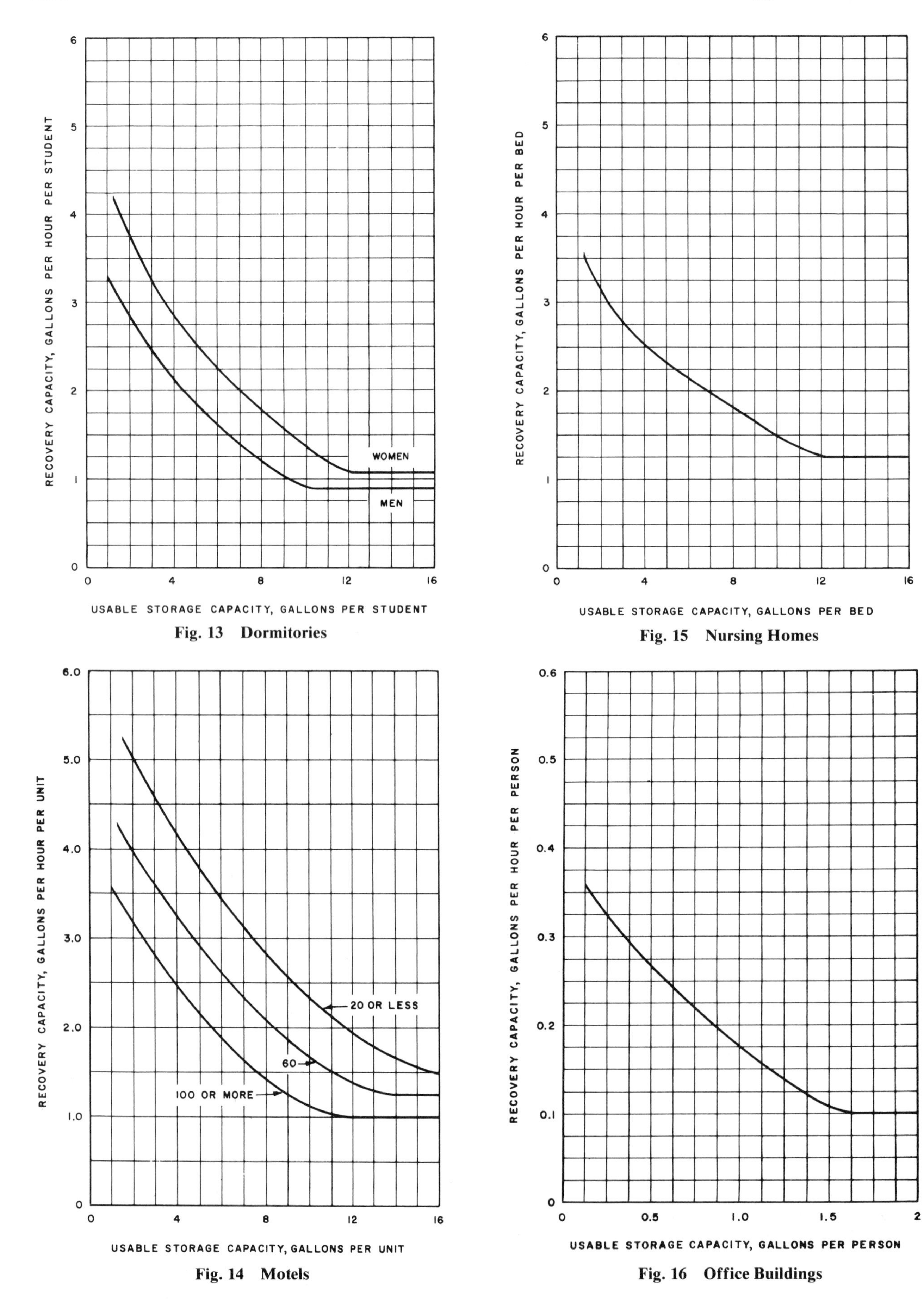

Fig. 13 Dormitories

Fig. 15 Nursing Homes

Fig. 14 Motels

Fig. 16 Office Buildings

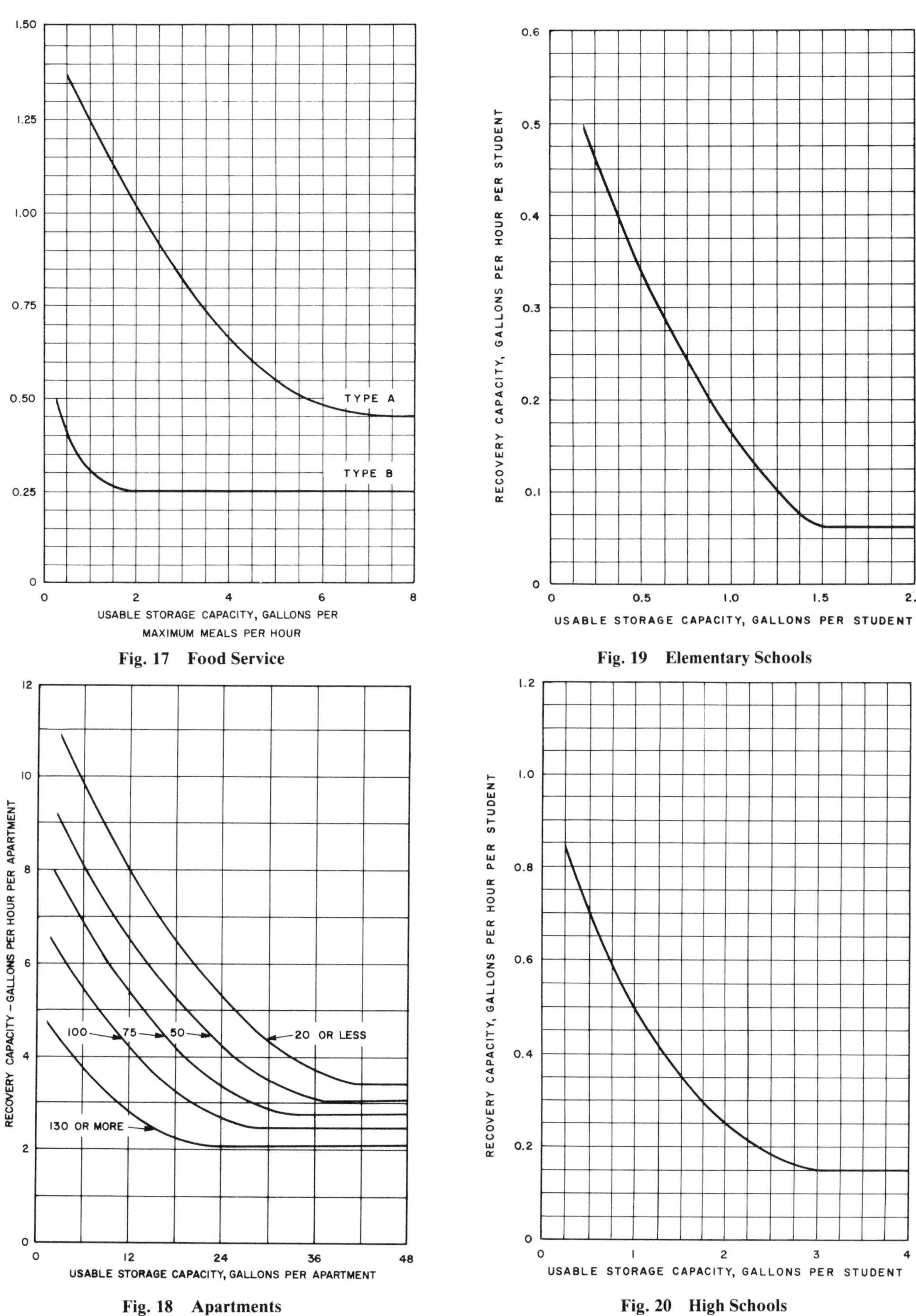

Fig. 17 Food Service

Fig. 19 Elementary Schools

Fig. 18 Apartments

Fig. 20 High Schools

Table 8 Hot Water Demand per Fixture for Various Types of Buildings
(Gallons of water per hour per fixture, calculated at a final temperature of 140 °F)

	Apartment House	Club	Gym-nasium	Hospital	Hotel	Industrial Plant	Office Building	Private Residence	School	YMCA
1. Basins, private lavatory	2	2	2	2	2	2	2	2	2	2
2. Basins, public lavatory	4	6	8	6	8	12	6	—	15	8
3. Bathtubs	20	20	30	20	20	—	—	20	—	30
4. Dishwashers[a]	15	50–150	—	50–150	50–200	20–100	—	15	20–100	20–100
5. Foot basins	3	3	12	3	3	12	—	3	3	12
6. Kitchen sink	10	20	—	20	30	20	20	10	20	20
7. Laundry, stationary tubs	20	28	—	28	28	—	—	20	—	28
8. Pantry sink	5	10	—	10	10	—	10	5	10	10
9. Showers	30	150	225	75	75	225	30	30	225	225
10. Service sink	20	20	—	20	30	20	20	15	20	20
11. Hydrotherapeutic showers				400						
12. Hubbard baths				600						
13. Leg baths				100						
14. Arm baths				35						
15. Sitz baths				30						
16. Continuous-flow baths				165						
17. Circular wash sinks				20	20	30	20		30	
18. Semicircular wash sinks				10	10	15	10	15		
19. DEMAND FACTOR	0.30	0.30	0.40	0.25	0.25	0.40	0.30	0.30	0.40	0.40
20. STORAGE CAPACITY FACTOR[b]	1.25	0.90	1.00	0.60	0.80	1.00	2.00	0.70	1.00	1.00

[a]Dishwasher requirements should be taken from this table or from manufacturers' data for the model to be used, if this is known.
[b]Ratio of storage tank capacity to probable maximum demand/h. Storage capacity may be reduced where an unlimited supply of steam is available from a central street steam system or large boiler plant.

19. The heater or coil should have a water-heating capacity equal to this probable maximum demand. The storage tank should have a capacity equal to the probable maximum demand multiplied by the storage capacity factor in line 20.

Example 5. Determine heater and storage tank size for an apartment building from a number of fixtures.

60 lavatories	× 2 gph	=	120 gph
30 bathtubs	× 20 gph	=	600 gph
30 showers	× 30 gph	=	900 gph
60 kitchen sinks	× 10 gph	=	600 gph
15 laundry tubs	× 20 gph	=	300 gph
Possible maximum demand		=	2520 gph

Solution:

Probable maximum demand = 2520 × 0.30 = 756 gph
Heater or coil capacity = 756 gph
Storage tank capacity = 756 × 1.25
= 945 gal

Showers

In many housing installations such as motels, hotels, and dormitories, the peak hot water load usually results from the use of showers. Tables 1 and 8 indicate the probable hourly hot water demand and the recommended demand and storage capacity factors applied to various types of buildings. Hotels could have a 3- to 4-h peak shower load. Motels require similar volumes of hot water, but the peak demand may last for only a 2-h period. In some types of housing, such as barracks, fraternities, or dormitories, all occupants may take showers within a very short period. In this case, it is best to find the peak load by determining the number of shower heads and the rate of flow per head; then estimate the length of time the shower will be on. It is estimated that the average time the shower will be on per individual is 1 to 3 min.

The flow rate from a shower head varies depending on the type, size, and water pressure. At 40 psi water pressure, available shower heads have nominal flow rates of blended hot and cold water from about 2.5 to 6 gpm. In multiple shower installations, flow control valves on shower heads are recommended, since they reduce the flow rate and maintain it, regardless of fluctuations in water pressure. The manufacturer's maximum flow rating can usually be reduced up to 50% on high flow shower heads without adversely affecting the spray pattern of the shower head. Flow control valves are commonly available in capacities from 1.5 to 4.0 gpm.

If the manufacturer's flow rate for a shower head is not available and a flow control valve is not used, the following may serve as a guide for sizing the water heater:

Small shower head	2.5 gpm
Medium shower head	4.5 gpm
Large shower head	6 gpm

Food Service

In a restaurant, bacteria are usually killed by rinsing the washed dishes with 180 to 195 °F water for several seconds. In addition, an ample supply of general-purpose hot water, usually 140 to

Table 9 NSF Final Rinse Water Requirements for Dishwashing Machines[a]

Type and Size of Dishwasher	Flow Rate, gpm	Hot Water Requirements at 180 to 195 °F Maximum: Heaters without Internal Storage,[b] gph	Heaters with Internal Storage to Meet Flow Demand[c], gph
Door type:			
16 × 16 in.	6.94	416	69
18 × 18 in.	8.67	520	87
20 × 20 in.	10.4	624	104
undercounter type	5	300	70
Conveyor type:			
single tank	6.94	416	416
multiple tank (dishes flat)	5.78	347	347
multiple tank (dishes inclined)	4.62	277	277
Silver washers	7	420	45
Utensil washers	8	480	75
Makeup water requirements	2.31	139	139

Note: Values are extracted from NSF *Standard* 5-83.
[a]Flow pressure at dishwashers is assumed to be 20 psi (gage).
[b]Based on the flow rate in gpm.
[c]Based on dishwasher operation at 100% of mechanical capacity.

150 °F, is required for the wash cycle of dishwashers. Although a water temperature of 140 °F in private dwellings is reasonable for dishwashing, in public places, sanitation regulations by the National Sanitation Foundation (NSF) or local health departments make 180 to 195 °F water mandatory in the rinsing cycle. However, the NSF allows lower temperatures when certain types of machines and chemicals are used. Because of the two-temperature need, the hot water requirements for food service establishments present special problems. The lower temperature water is distributed for general use, but the 180 °F water should be confined to the equipment requiring it and should be obtained by boosting the temperature. It would be dangerous to distribute 180 °F water for general use. NSF *Standard* 26-80 covers the design of dishwashing machines and water heaters used by restaurants. The following procedure for sizing water heaters for restaurants is recommended by the American Gas Association.

1. Types and sizes of dishwashers used
2. Required quantity of general-purpose hot water (manufacturers' data should be consulted to determine the initial fill requirements of the wash tanks)
3. Duration of peak hot water demand period
4. Inlet water temperature
5. Type and capacity of existing water-heating system
6. Type of water-heating system desired

Equation (6) may be used to size the required heater(s) after allowing for the quantity of hot water withdrawn from the storage tank each hour. The general-purpose and 180 to 195 °F water requirements are determined from Tables 9 and 10.

$$\text{Input (Btu/h)} = \frac{8.33\ (\text{gph})\,\Delta t}{\text{Thermal efficiency}} \tag{6}$$

To determine the quantity of usable hot water from storage, the duration (in hours) of consecutive peak demand must be estimated. This peak demand period usually coincides with the dishwashing period during and after the main meal and may last from 1 to 4 h.

Any hour in which the dishwasher is used at 70% or more of capacity should be considered as a peak hour. If the peak demand lasts for 4 h or more, the value of a storage tank is reduced, unless especially large tanks are used. Some storage capacity is desirable to meet momentary high draws.

The National Sanitation Foundation *Standard* No. 5-83 recommendations for hot water rinse demand are based on 100% operating capacity of the machines. The data provided in Table 8 are also based on 100% operating capacity. NSF 5-83 states that 70% of operating rinse capacity is all that is normally attained, except for rackless-type conveyor machines. Examples 6, 7, and 8 demonstrate the use of Equation (6) and Tables 9 and 10. The calculations assume a heater efficiency of 75%.

Table 10 General Purpose Hot Water (140 °F) Requirement for Various Kitchens Uses[a,b]

Equipment	gph
Vegetable sink	45
Single pot sink	30
Double pot sink	60
Triple pot sink	90
Prescrapper (open type)	180
Preflush (hand-operated)	45
Preflush (closed type)	240
Recirculating preflush	40
Bar sink	30
Lavatories (each)	5

Note: Values are extracted from Dunn *et al.* (1959).
[a]Supply water pressure at equipment is assumed to be 20 psi (gage).
[b]Dishwasher operation at 100% of mechanical capacity.

Example 6. Determine the hot water demand for a new water-heating system in a cafeteria kitchen with one vegetable sink, five lavatories, one prescrapper, one utensil washer, and one two-tank conveyor dishwasher (dishes inclined) with makeup device. The initial fill requirement for the tank of the utensil washer is 85 gph at 140 °F. The initial fill requirement for the dishwasher is 20 gph for each tank, or a total of 40 gph, at 140 °F. The maximum period of consecutive operation of the dishwasher at or above 70% capacity is assumed to be 2 h. The supply water temperature is 60 °F.

Solution: The required quantities of general purpose (140 °F) and rinse (180 °F) water for the equipment, from Tables 9 and 10, are shown in the following tabulation:

Item	Quantity Required[a] at 140 °F, gph	Quantity Required[b] at 180 °F, gph
Vegetable sink	45	—
Lavatories (5)	25	—
Prescrapper	180	—
Dishwasher	—	277
Initial tank fill	40	—
Makeup water	—	139
Utensil washer	—	75
Initial tank fill	85	—
Total requirements	375	491

[a]General-purpose hot water consumption, from Table 10.
[b]Water consumption when dishwasher is operated at 100% of mechanical capacity, from Table 9.

The total consumption of 140 °F water is 375 gph. The total consumption 180 °F depends on the type of heater to be used. For a heater that has enough internal storage capacity to meet the flow demand, the total consumption, based on the recommendation of the NSF, is 70% of the total calculated from Table 10, or approximately 350 gph (0.70×491 = 344 gph). For an instantaneous heater without internal storage capacity, the total quantity of 180 °F water consumed must be based on the flow demand. From Table 9, the quantity required for the dishwasher is 277 gph; for the makeup, 139 gph; and for the utensil washer, 480 gph. The total consumption of 180 °F water is 277 + 139 + 480 = 896 gph, or approximately 900 gph.

Example 7. Determine gas input requirements for heating water in the cafeteria kitchen described in Example 6, by the following systems, which are among many possible solutions:

a. Separate, self-contained, storage-type heaters.

b. Single instantaneous-type heater, having no internal storage to supply both 180 and 140 °F water through a mixing valve.

c. Separate instantaneous-type heaters, having no internal storage.

d. Combination of heater and external storage tank for 140 °F water, plus a booster heater for 180 °F water. The heater and external storage are to supply 140 °F water for both the general-purpose requirement and the booster heater. The booster heater is to have sufficient storage capacity to meet the flow demand of 180 °F rinse water.

Solution a: The temperature rise for 140 °F water is 140 − 60 = 80 °F. From Equation (6), the gas input required to produce 375 gph of 140 °F water with an 80 °F temperature rise at 75% efficiency is 334,000 Btu/h. One or more heaters with this total requirement may be selected.

From Equation (6), the gas input required to produce 350 gph of 180 °F water with a temperature rise of 120 °F (180 − 60) at 75% efficiency is 467,000 Btu/h. One or more heaters with this total requirement may be selected from manufacturers' catalogs.

Solution b: The correct sizing of instantaneous-type heaters depends on the flow rate of the 180 °F rinse water. From Example 6, the hourly consumption of 180 °F water based on the flow rate is 900 gph; hourly consumption of 140 °F water is 375 gph.

Gas input required to produce 900 gph of 180 °F water with a 120 °F temperature rise is 1,200,000 Btu/h. Gas input to produce 375 gph of 140 °F water with a temperature rise of 80 °F is 334,000 Btu/h. Total heater requirement is 1,200,000 + 334,000 = 1,534,000 Btu/h. One or more heaters meeting this total input requirement can be selected from manufacturers' catalogs.

Solution c: Gas input required to produce 140 °F water is the same as for Solution b, 334,000 Btu/h. One or more heaters meeting this total requirement can be selected.

Gas input required to produce 180 °F water is also the same as in Solution b, 1,200,000 Btu/h. One or more heaters meeting this total requirement can be selected.

Solution d: The *net* hourly hot water requirement must be determined to size the heater required to supply 140 °F water. From Equation (4), the quantity of usable hot water in storage Q_a is $0.7 \times 500 = 350$ gal. The total quantity of 140 °F water required for the 2 h is 2(375 + 350) = 1450 gal. From Equation (3), $Q_t = (1450 - 350)/2 = 550$ gph from the heater.

From Equation (6), the gas input required to produce 550 gph of 140 °F water with an 80 °F temperature rise is 489,000 Btu/h.

For systems involving storage tanks, it is assumed that water in the tanks has been brought up to temperature prior to the peak dishwashing period, and that enough time will elapse before the next peak period to permit recovery of the water temperature in the storage tank.

The booster heater is sized to heat 350 gph from 140 to 180 °F, a 40 °F rise. From Equation (6), the gas input required is 156,000 Btu/h.

Example 8. A luncheonette has purchased a door-type dishwasher that will handle 16 × 16-in. racks. The existing hot water system is capable of supplying the necessary 140 °F water to meet all requirements for general purpose use, plus supply to a booster heater that is to be installed. Determine the size of booster heater operating at 75% thermal efficiency required to heat 140 °F water to provide sufficient 180 °F rinse water for the dishwasher, using the following:

a. Booster heater with no storage capacity.
b. Booster heater with enough storage capacity to meet flow demand.

Solution a: Since the heater is the instantaneous type, it must be sized to meet the 180 °F water demand at a rated flow. From Table 9, this rated flow is 6.94 gpm, or 416 gph. From Equation (6), the required gas input, with a 40 °F temperature rise, is 185,000 Btu/h. A heater meeting this input requirement can be selected from manufacturers' catalogs.

Solution b: In designing a system with a booster heater having storage capacity, hourly flow demand of the dishwasher can be used instead of the flow demand used in Solution *a*. The flow demand from Table 9 is 69 gph when the dishwasher is operating at 100% mechanical capacity. However, the NSF states that 70% of operating rinse capacity is all that is normally attained for this type of dishwasher. Therefore, the hourly flow demand is 0.70 × 69 = 48 gph. From Equation (6), with a 40 °F temperature rise, the gas input required is 22,000 Btu/h. A booster heater with this input can be selected from manufacturers' catalogs.

Estimating Procedure. Hot water requirements for kitchens are sometimes estimated on the basis of the number of meals served (assuming eight dishes per meal). Dishwashing demand is either

$$R_1 = C_1 N/\theta \quad (7)$$

where

R_1 = 180 °F water for dishwater, gph
N = number of meals served
θ = hours of service
C_1 = 0.8 for single-tank dishwasher
C_1 = 0.5 for two-tank dishwasher

or

$$R_2 = C_2 V \quad (8)$$

where

R_2 = water for sink with gas burners, gph
C_2 = 3
V = sink capacity (15 in. depth), gal

General-purpose hot at 140 °F is

$$R_3 = C_3 N/(\theta + 2) \quad (9)$$

where

R_3 = general-purpose water, gph
C_3 = 1.2

Total demand is

$$R = R_1 + R_2 + R_3$$

For soda fountains and luncheonettes, use 75% of the total demand. For hotel meals or other elaborate meals, use 125%.

Schools

Service water heating in schools is needed for janitorial work, lavatories, cafeterias, shower rooms, and sometimes swimming pools.

Hot water used in cafeterias is about 70% of that usually required in a commercial restaurant serving adults and can be estimated by the method used for restaurants. Where NSF sizing is required, follow *Standard* No. 5-83.

Shower and food service loads will not ordinarily be concurrent. Each should be determined separately, and the larger load should determine the size of the water heater(s) and the tank. Provision must be made to supply 180 °F sanitizing rinse. The booster must be sized according to the temperature of the supply water. Where feasible, the same water-heating system can be used for both needs. Where the distance between the two points of need is great, different water-heating systems should be used.

A separate water-heating system for swimming pools can be sized as outlined in the section on swimming pools.

Domestic Coin-Operated Laundries

Small domestic machines in coin laundries or apartment house laundry rooms have a wide range of draw rates and cycle times. Domestic machines provide wash water temperatures (normal) as low as 120 °F. Some manufacturers recommend temperatures of 160 °F; however, the average appears to be 140 °F. The hot water sizing calculations must assure a supply to both the instantaneous draw requirements of a number of machines filling at one time and the average hourly requirements.

The number of machines that will be drawing at any one time varies widely; the percentage is usually higher in smaller installations. One or two customers starting several machines at about the same time has a much sharper effect in a laundry with 15 or 20 machines than in one with 40 machines. Simultaneous draw may be estimated as:

1 to 11 machines—100% of possible draw
12 to 24 machines— 80% of possible draw
25 to 35 machines— 60% of possible draw
36 to 45 machines— 50% of possible draw

Possible peak draw can be calculated from:

$$F = NPQ/T \quad (10)$$

where

F = peak draw, gpm
N = number of washers installed
P = number of machines drawing hot water divided by N
Q = quantity of hot water supplied to machine during hot wash fill, gal
T = wash fill period, min

Recovery rate can be calculated from:

$$R = 60NPQ/(\theta + 10) \quad (11)$$

where

R = total quantity of hot water (per machine) used for entire cycle (machine adjusted to hottest water setting), gph
θ = actual machine cycle time, min

Note: $(\theta + 10)$ is the cycle time plus 10 min for loading and unloading.

Commercial Laundries

Commercial laundries generally use a storage water-heating system. The water may be softened to reduce soap use and improve quality. The trend is toward installation of high capacity washer-extractor wash wheels, resulting in high peak demand.

Sizing Data. Laundries can normally be divided into five categories. The required hot water is determined by the weight of the material processed. Average hot water requirements at 180 °F are:

Institutional	2 gal/lb
Commercial	2 gal/lb
Linen supply	2.5 gal/lb
Industrial	2.5 gal/lb
Diaper	2.5 gal/lb

Total weight of the material times these values give the average hourly hot water requirements. The designer must consider peak requirements; for example, a 600-lb machine may have a 20 gpm average requirement, but the peak requirement could be 350 gpm.

In a multiple-machine operation, it is not reasonable to fill all machines at the momentary peak rate. Diversity factors can be estimated by using 1.0 of the largest machine plus the following balance:

	Total number of machines				
	2	3 to 5	6 to 8	9 to 11	12 and over
1.0 +	0.6	0.45	0.4	0.35	0.3

Types of Systems. Service water-heating systems for laundries are pressurized or vented. The pressurized system uses city water pressure, and the full peak flow rates are received by the softeners, reclaimer, condensate cooler, water heater, and the lines to the wash wheels. The flow surges and stops at each operation in the cycle. A pressurized system depends on an adequate water service.

The vented system uses pumps from a vented (open) hot water heater or tank to supply hot water. This water level fluctuates in the tank, from about 6 in. above the heating element to a point 12 in. from the top of the tank, the working volume. The level drops for each machine fill, while makeup water runs continuously at the average flow rate under water service pressure during the complete washing cycle. The tank is sized to have full working volume at the beginning of each cycle. Lines and softeners can be sized for this flow rate from the water service to the tank, not the peak machine fill rate as with a closed, pressurized system. The waste heat exchangers have a continuous flow across the heating surface at this low flow rate, with continuous heat reclamation from the wastewater and flash steam. Automatic flow-regulating valves on the inlet water manifold control this low flow rate. Rapid fill of machines will increase production, *i.e.*, more batches will be processed.

Heat Recovery. Commercial laundries are ideally suited for heat recovery because 135 °F waste temperature is discharged to the sewer. Fresh water can be conservatively preheated to within 15 °F of the wastewater temperature for the next operation in the wash cycle. Regions with an annual average temperature of 55 °F can increase to 120 °F the initial temperature of fresh water going into the hot water heater. For each 1000 gph or 8330 lb of water preheated 65 °F (55 to 120 °F), heat reclamation will be 540,000 Btu/h. This saves 655 ft^3 of natural gas per hour, or 3.92 gal of oil per hour.

Flash steam from a condensate receiving tank is often wasted to the atmosphere. The heat in this flash steam can be reclaimed with a suitable heat exchanger. Makeup water to the heater can be preheated 10 to 20 °F above existing makeup temperature with the flash steam. When condensate is otherwise wasted, this waste heat should also be considered.

Industrial Plants

Hot water is used in industrial plants for cafeteria, showers, lavatories, gravity sprinkler tanks, and industrial processes. If the same hot water system is used only for the cafeteria, employee cleanup, laundry, and small miscellaneous uses, the water heater can be sized to meet employee cleanup load (with additional provision for the sanitizing rinse needs of the cafeteria). Employee cleanup load is usually heaviest and not concurrent with other uses. The other loads should be checked, however, to be certain that this is true.

The employee cleanup load consists of one or more of the following: (1) wash troughs or standard lavatories, (2) multiple wash sinks, and (3) showers. Hot water requirements for employees using standard wash fixtures can be estimated at 1 gal of hot water for each clerical and light industrial employee per work shift and 2 gal for each heavy industrial worker.

The number of workers using multiple wash fountains is disregarded for sizing purposes. Hot water demand is based on full flow for the entire cleanup period. This usage over a 10-min period is indicated in Table 11. The shower load depends on the flow rate of the shower heads and their length of use. Table 11 may be used to estimate flow based on a 15 min period.

Table 11 Hot Water Usage for Industrial Wash Fountains and Showers

Multiple Wash Fountains		Showers	
Type, in.	Gal of 140 °F Water Required for 10-min Period[a]	Flow Rate, gpm	Gal of 140 °F Water Required for 15-min Period[b]
36 Circular	40	3	29.0
36 Semicircular	22	4	39.0
54 Circular	66	5	48.7
54 Semicircular	40	6	58.0

[a]Based on 110 °F wash water and 40 °F cold water at average flow rates.
[b]Based on 105 °F shower water and 40 °F cold water.

Water heaters used to prevent freezing in gravity sprinkler tanks or water storage tanks should be part of a separate system. The load depends on tank heat loss, tank capacity, and winter design temperature.

Process hot water load must be determined separately. Volume and temperature vary with the specific process. If the process load occurs at the same time as the shower or restaurant load, the system must be sized to reflect this total demand. Separate systems can also be used, depending on the size of the various loads and the distance between them.

Ready-Mix Concrete

In cold weather, ready-mix concrete plants need hot water to mix the concrete so that it will not be ruined by freezing before it sets. Operators prefer to place the mix at about 70 °F. With the cold aggregate, hot water must be used. Usually, about 150 °F water is considered proper for cold weather. When the water temperature is too high, some of the concrete will flash set.

Generally, 30 gal of hot water per cubic yard of concrete mix is used for sizing. To obtain the total hot water load, the number of trucks loaded each hour and the capacity of the trucks is calculated. The hot water is dumped into the mix as fast as possible at each loading, and ample hot water storage is required. If storage is not used, large heat exchangers must be used for the high draw rate. Table 12 shows a method of sizing for concrete plants.

Part of the heat may be obtained by heating the aggregate bin. This is done by circulating hot water through pipe coils in the walls or sides of the bin. If aggregate is warmed, the temperature of the mixing water may be lower, and the aggregate will flow easily from the bins. When aggregate is not heated, it often freezes into chunks, which must be thawed to go through the dump gates. If hot water is used for thawing, too much water would accumulate in the aggregate, and control of the final product might vary beyond allowable limits. Therefore, jets of steam supplied by a small boiler and directed on the large chunks are often used for thawing.

Table 12 Water Heater Sizing for Ready-Mix Concrete Plant (Input and Storage Tank Capacity to Supply 150 °F Water at 40 °F Inlet Temperature)

Time Interval between Trucks[a]	Capacity	Truck Capacity 6 yd	7.5 yd	9 yd	11 yd
50 min	1000 Btu/h	458	527	596	687
	gal	430	490	560	640
35 min	100 Btu/h	612	700	792	915
	gal	430	490	560	640
25 min	1000 Btu/h	785	900	1020	1175
	gal	430	490	560	640
10 min	1000 Btu/h	1375	1580	1790	2060
	gal	430	490	560	640
5 min	1000 Btu/h	1830	2100	2380	2740
	gal	430	490	560	640
0 min	1000 Btu/h	2760	3150	3580	4120
	gal	430	490	560	640

[a]This table assumes that there is 10-min loading time for each truck. Thus, for a 50-min interval between trucks, it is assumed that one truck/h is served. For 0-min between trucks, it is assumed that one truck loads immediately after the truck ahead has pulled away. Thus, 6 trucks/h are served.

It is also assumed that each truck carries a 120-gal storage tank of hot water for washing down at the end of dumping the load. This hot water is drawn from the storage tank and must be added to the total hot water demands. This has been included in the sizing table above.

Swimming Pools/Health Clubs

The desirable temperature for swimming pools is 80 °F. The U.S. Swimming Rules and Regulations (1987) state that for national championships and international competition, the pool temperature must be between 78 and 80 °F, and the air temperature within 8 ft above deck level for indoor facilities shall not be lower than 76 °F, with relative humidity maintained at about 60% and air velocity at about 25 fpm.

Most manufacturers of water heaters and boilers also offer specialized models for pool heating; these include a pool temperature controller and a water bypass to prevent condensation. The water-heating system is usually installed in the normal circulation of the pool, prior to the return of treated water to the pool. A circulation rate to generate a change of water every 8 h for residential pools and 6 h for commercial pools is acceptable. An indirect heating system, where piping is embedded in the walls or floor of the pool, has the advantage of reduced corrosion, scaling, and condensation, since pool water does not flow through these pipes. The disadvantage of this type of system is the high initial installation cost.

The installation should have a pool temperature control and a water pressure or flow safety switch. The temperature control should be installed at the inlet to the heater, while the pressure or flow switch can be installed at either the inlet or outlet, depending on the manufacturer's instructions. This will afford protection against inadequate water flow.

Pool use is of prime importance in sizing the water heater, and, for economical reasons, the heater should be on only during and just prior to the period of pool use. Sizing should be based on four considerations:

- Conduction through the pool walls
- Convection from the pool surface
- Radiation from the pool surface
- Evaporation from the pool surface

Except in aboveground pools and in rare cases where cold groundwater flows past the pool walls, conduction losses are small and can be ignored. Since convection losses depend on temperature differentials and wind speed, these losses can be greatly reduced by the installation of windbreaks, such as hedges, solid fences, and buildings.

Radiation losses occur when the pool surface is subjected to temperature differentials which frequently occur at night, when the sky temperature may be as much as 80 °F below ambient air temperature. This usually occurs on clear cool nights. During the daytime, however, an unshaded pool will receive a large amount of radiant energy, often as much as 100,000 Btu/h. These losses and gains may offset each other.

An easy solution to controlling nighttime temperature losses is to use a floating pool cover at night; this also substantially reduces evaporative losses.

Evaporative losses constitute the greatest heat loss of the pool—50 to 60% in most cases. If it is possible to cut evaporative losses drastically, the heating requirement of the pool may be cut as much as 50%. This reduction can be accomplished by using the floating pool cover.

Ideally, a pool heater with an input great enough to provide a heat-up time of 24 h would be the best solution. However, it may not be the most economical system for pools that are in continuous use during an extended swimming season. In this instance, an extended heat-up period of as much as 48 h can be used, requiring a less expensive unit. Pool water may be heated by several methods. Fuel-fired water heaters and boilers, electric boilers, tankless electric circulation water heaters, air source heat pumps, and solar heaters have all been used successfully. Air source heat pumps and solar systems would probably be used to extend a swimming season, rather than to allow intermittent use with rapid pickup.

At one time, pools were essentially rectangular in shape, but now the shape is limited only by the imagination of the pool designer. The following equations provides some assistance for determining the volume and area of irregularly shaped pools:

Elliptical

Area = $3.14\,AB$
A = short radius
B = long radius
Volume = 7.5 gal/ft^3 × Area × Average Depth

Odd Shapes

Area = $0.45\,BL + A$ (approximately)
L = length
Volume = 7.5 gal/ft^3 × Area × Average Depth

Oval

Area = $3.14\,R^2 + LW$
L = oval length
W = oval width
R_2 = radius
Volume = 7.5 gal/ft^3 × Area × Average Depth

Rectangular

Area = LW
L = length
W = width
Volume = 7.5 gal/ft^3 × Area × Average Depth

The following section presents an effective method of heating outdoor pools. Additional equations can be found in Chapter 4.

1. Obtain pool water capacity, in gallons, from the chart or by multiplying the length times the width times an average depth of 5.5 ft. Each cubic foot equals 7.5 gal.
2. Determine the desired heat pickup time, h.
3. Determine the desired pool temperature—if not known, use 80 °F.
4. Determine the average temperature of the coldest month of use.

The required heater output can now be determined by the following equations:

$$q_1 = 8.33V(t_f - t_i)/\theta \quad (12)$$

where

q_1 = pool heat-up rate, Btu/h
8.33 = density of water, lb/gal
V = pool volume, gal
t_f = desired temperature (usually 80°F)
t_i = initial temperature of pool, °F
θ = pool heat-up time, h

$$q_2 = 10.5A(t_p - t_a) \quad (13)$$

where

q_2 = heat loss from pool surface, Btu
10.5 = surface heat transfer coefficient, Btu/h·ft²·°F
A = pool surface area, ft²
t_p = pool temperature, °F
t_a = ambient temperature, °F

$$q_t = q_1 + q_2 \quad (14)$$

Notes: These heat losses assume a wind velocity of 3 to 5 mph. For pools sheltered by nearby fences, dense shrubbery, or buildings, an average wind velocity of less than 3.5 mph can be assumed. In this case, use 75% of the values calculated by Equation (13). For a velocity of 5 mph, multiply by 1.25; for 10 mph, multiply by 2.0.

Since Equation (13) applies to the coldest monthly temperatures, the calculated results may not be economical. Therefore, a value of one-half the surface loss plus the heat-up value yields a more viable heater output figure. The heater input then equals the output divided by the efficiency of the fuel source.

Whirlpools and Spas

Hot water requirements for whirlpool baths and spas depend on temperature, fill rate, and total volume. Water may be stored separately at the desired temperature or, more commonly, regulated at the point of entry by blending. If rapid filling is desired, provide storage at least equal to the volume needed; fill rate can then be varied at will. An alternative is to establish a maximum fill rate and provide an instantaneous water heater that will handle the flow.

SIZING INSTANTANEOUS AND SEMI-INSTANTANEOUS HEATERS

The methods for sizing storage water-heating equipment should not be used for instantaneous and semi-instantaneous heaters. The following is based on the Hunter method for sizing hot and cold water piping, with diversity applied for hot water and various building types.

Fixture units (Table 1) are selected for each fixture using hot water and are totalled. Maximum hot water demand is obtained from Figures 21 or 22 by matching total fixture units to the curve for the type of building. Hot water for fixtures and outlets that have constant flows should be added to demand.

The heater can then be selected with the total demand and temperature rise required. For critical applications such as hospitals, using multiple heaters with 100% standby is recommended. Consider multiple heaters for buildings in which continuity of service is important. The minimum recommended size for semi-instantaneous heaters is 10 gpm, except for restaurants, in which it is 15 gpm. When the flow for a system is not easily determined, the heater may be sized for the full flow of the piping system. Caution must be used when sizing heaters with low flows, and careful judgment should be applied to estimate diversities. Unusual hot water requirements in a building should be analyzed to determine if additional capacity is required. One example is a dormitory in a military school, where all showers and lavatories are used simultaneously when students return from a drill. In this case, the heater and piping should be sized for the full flow of the system.

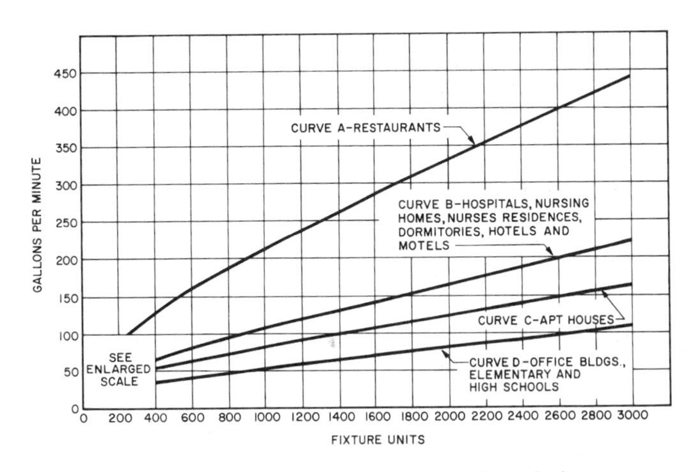

Fig. 21 Modified Hunter Curve for Calculating Hot Water Flow Rate

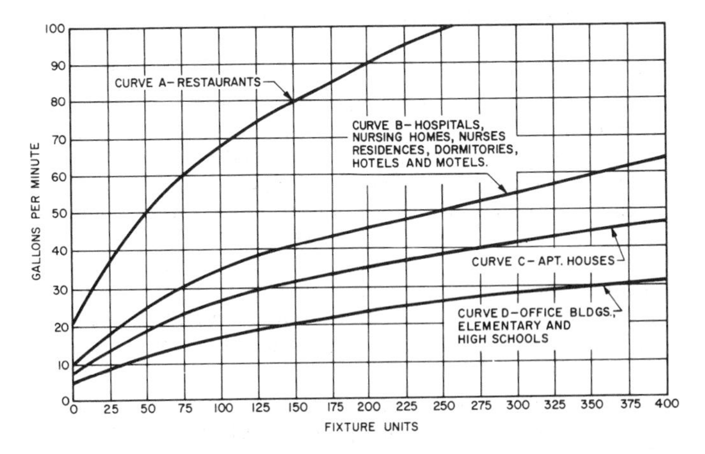

Fig. 22 Enlarged Section of Figure 21 (Modified Hunter Curve)

While the fixture count method bases heater size of the diversified system on hot water flow, hot water piping should be sized for the full flow to the fixtures. Recirculating hot water systems are adaptable to this type of heater. When these systems are installed, the heater capacity should be checked and increased if necessary to offset heat losses of the recirculating system.

To make preliminary estimates of hot water demand when the fixture count is not known, use Table 13 with Figures 21 or 22. The results will usually be higher than the demand determined from the actual fixture count. Actual heater size should be determined from Table 1. Hot water consumption over time can be assumed the same as that in the section on sizing storage heaters.

Example 9. Determine the hot water flow rate for sizing a semi-instantaneous heater for a 600-student elementary school with the following fixture count: 60 public lavatories, 6 service sinks, 4 kitchen sinks, 6 showers, and 1 dishwasher at 8 gpm.

Table 13 Preliminary Hot Water Demand Estimate

Type of Building	Unit	Fixture Units Per Unit
Hospital or nursing home	Bed	2.50
Hotel or motel	Room	2.50
Office building	Person	0.15
Elementary school	Student	0.30
Jr. and Sr. high school	Student	0.30[a]
Apartment house	Apartment	3.00

[a]Plus shower load.

Solution: For a preliminary estimate, use Table 13 to find estimated flow. The basic flow is determined from curve D of Figure 22, at 600 students × 0.3 fixture units per student = 180 fixture units, plus 6 showers × 1.5 fixtures units = 9, or 189 fixture units, for a total flow of 23 gpm.

To size the unit based on actual fixture count and Table 1, the calculation is as follows:

60 public lavatories	× 1 FU =	60 FU
6 service sinks	× 2.5 FU =	15 FU
4 kitchen sinks	× 0.75 FU =	3 FU
6 showers	× 1.5 FU =	9 FU
Subtotal		87 FU

At 87 fixture units, curve D of Figure 22 shows 16 gpm, to which must be added the dishwasher requirement of 8 gpm. Thus, the total flow is 24 gpm.

Comparing the flow based on actual fixture count to that obtained from the preliminary estimate shows the preliminary estimate to be slightly lower. It is possible that the preliminary estimate could have been as much as twice the final fixture count. To prevent oversizing the equipment, it is imperative to use the actual fixture count method to select the unit.

BOILERS FOR INDIRECT WATER HEATING

When service water heating is accomplished indirectly by using a space heating boiler, Figure 23 may be used to determine the additional boiler capacity required to meet the recovery demands of the domestic water-heating load. Such systems include arrangements using immersion coils in boilers as well as heat exchangers with space heating media used for service water heating.

Since the boiler capacity must meet not only the water supply requirement but also the space heating loads, the chart indicates the reduction of additional heat supply for water heating if the ratio of water-heating load to space heating load is low. This reduction is possible because:

1. Maximum space heating requirements do not occur at the time of day when the maximum peak domestic demands occur.
2. Space heating requirements are based on the lowest outdoor design temperatures, which may occur only for a few days of the total heating season.
3. An additional heat supply or boiler capacity is usual for pickup and radiation losses. The pickup load cannot occur at the same time as the peak hot water demand, since the building must be brought to a comfortable level before the occupants will be using hot water.

The factor obtained from Figure 23 is multiplied by the peak water-heating load to obtain the additional boiler output capacity required.

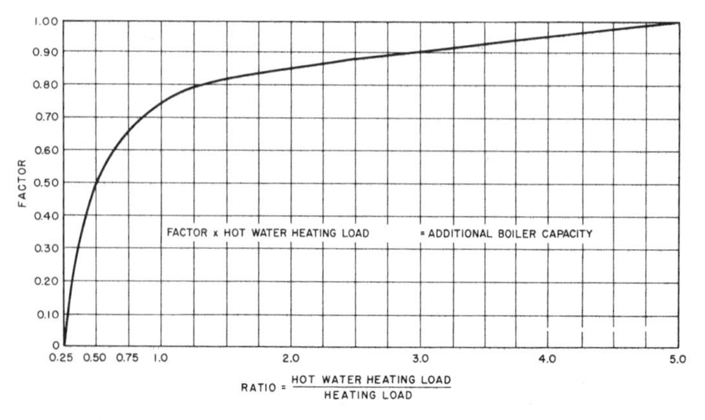

Fig. 23 Sizing Factor for Combination Heating and Water-Heating Boilers

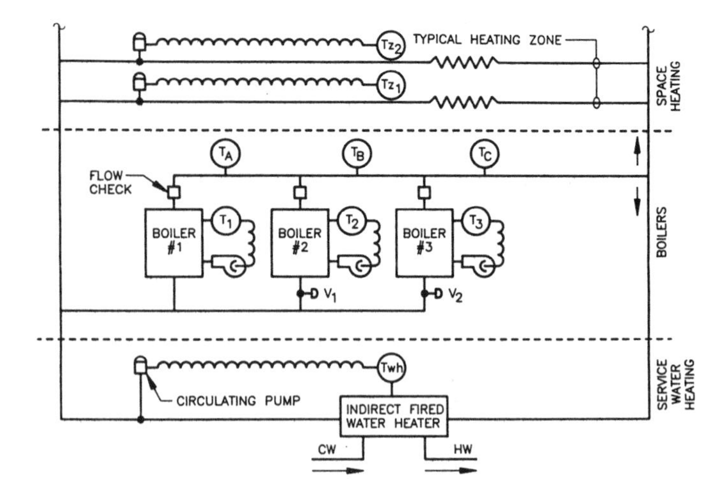

Fig. 24 Typical Modular Boiler for Combined Space Service Water Heating

For reduced standby losses in summer and improved efficiency in winter, step-fired modular boilers may be used. By arranging the piping and control package to fire only those boilers required to meet the load, the system's standby losses are reduced, and overall efficiency is increased. Units not in operation cool down and reduce or eliminate jacket losses. The installation should ensure that heated boiler water does not pass through an idle boiler. Figure 24 shows a typical modular boiler combination space and service water-heating arrangement.

Typical Control Sequence

1. Any control zone or indirectly fired water heater thermostat starts its circulating pump and supplies power to boiler No. 1 control circuit.
2. If T1 is not satisfied, burner is turned on, boiler cycles as long as any circulating pump is on.
3. After 5 min if TA is not satisfied, V1 opens and boiler No. 2 comes on line.
4. After 5 min if TB is not satisfied, V2 opens and boiler No. 3 comes on line.
5. If TC is satisfied and only two boilers or less are firing for a minimum of 10 min, V2 closes.
6. If TB is satisfied and only one boiler is firing for a minimum of 10 min, V1 closes.
7. If all circulating pumps are off, boiler No. 1 will shut down.

ANSI/ASHRAE/IES *Standard* 90 discusses combination service water-heating/space heating boilers and establishes restrictions on their use. The ANSI/ASHRAE/IES 100 series standards section, Service Water Heating, also has information on this subject.

CODES AND STANDARDS

Codes and standards that have special significance to water-heating applications follow.

ANSI Z-21.10.1-1990	American National Standard for Gas Water Heaters, Volume I: Automatic Storage Water Heaters with Inputs of 75,000 Btu per Hour or Less. (Also AGA 1631-0008.)
ANSI Z-21.10.3-1990	Gas Water Heaters, Volume III: Circulating Tank, Instantaneous and Large Automatic Storage Water Heaters. (Also AGA 1631-0110.)

ANSI Z-21.56-1986	Gas Fired Pool Heaters.
ANSI Z-21.22-1986	Relief Valve and Automatic Gas Shut-off Devices for Hot Water Supply Systems.
UL 174-1989	Household Electric Storage Tank Water Heaters.
UL 1453-1989	Electric Booster and Commercial Storage Tank Water Heaters.
UL 1261-81	Electric Water Heaters for Pools and Tubs.
UL 732-74	Oil-Fired Water Heaters.

ASME Boiler and Pressure Vessel Code Section IV (H and HLW) Code Construction. H label governs low-pressure heating boilers, and HLW label governs low-pressure, fired potable hot water heaters.

ASME Boiler and Pressure Vessel Code Section VIII. U label governs unfired water heaters and pressure vessels. (Note: Some states, cities, and counties require ASME-stamped equipment.)

NFPA *Standard* 54-84 (ANSI Z223.1-1988) National Fuel Gas Code. Governs installation of gas-fired water heaters.

NFPA 31-83 Installation of Oil Burning Equipment.

ANSI *Standard* 119 governs water-heating installation in mobile homes and recreational vehicles.

ASHRAE *Standard* 90 and 100 series provide recommended guidelines for energy conserving design of service water heating.

ASPE publishes a data book on plumbing and service water heating.

REFERENCES

AGA. Sizing and equipment data for specifying swimming pool heaters. American Gas Association, Catalog No. R-00995.

Ciesielki, C.A. *et al.* 1984. Role of stagnation and obstruction of water flow in isolation of *Legionella Pneumophila* from hospital plumbing. *Applied and Environmental Microbiology* (November):984-87.

Dunn, T.Z., R.N. Spear, B.E. Twigg and D. Williams. 1959. Water heating for commercial kitchens. *Air Conditioning, Heating and Ventilating* (May):70. Also published as a bulletin, *Enough hot water—hot enough.* American Gas Association (1959).

Manian, V.S. and W. Chackeris. 1974. Off peak domestic hot water systems for large apartment buildings. ASHRAE *Transactions* 80 (1):147.

NSF. 1983. Hot water generating and heat recovery equipment. National Sanitation Foundation *Standard* 5-83. Ann Arbor, MI.

Perlman, M. and B. Mills. 1985. Development of residential hot water use patterns. ASHRAE *Transactions* 91(2).

Toaborek, J. *et al.* 1972. Fouling—The major unresolved problem in heat transfer. *Chemical Engineering Progress* (February):59.

Werden, R.G. and L.G. Spielvogel. 1969. Sizing of service water heating equipment in commercial and institutional buildings, Part 1. ASHRAE *Transactions* 75:81.

BIBLIOGRAPHY

AGA. Comprehensive on commercial and industrial water heating. American Gas Association, Catalog No. R-00980.

AGA. 1962. Water heating application in coin operated laundries. American Gas Association, Catalog No. C-10540.

AGA. 1965. *Gas engineers handbook.* American Gas Association, Cleveland, OH.

Brooks, F.A. Use of solar energy for heating water. Smithsonian Institution, Washington, D.C.

Coleman, J.J. 1974. Waste water heat reclamation. ASHRAE *Transactions* 80(2):370.

Dawson, F.M. and A.A. Kalinski. Water-supply piping for plumbing pystems. National Association of Master Plumbers, *Technical Bulletin* No. 3.

Hebrank, E.F. 1956. Investigation of the performance of automatic storage-type gas and electric domestic water heaters. University of Illinois, *Engineering Experiment Bulletin* No. 436.

Jones, P.G. 1982. The consumption of hot water in commercial building. *Building Services Engineering, Research and Technology* 3:95-109.

Schultz, W.W., and V.W. Goldschmidt. 1978. Effect of distribution lines on stand-by loss of service water heater. ASHRAE *Transactions* 84(1):256-65.

Smith, F.T. 1965. Sizing guide for gas water heaters for in-ground swimming pools. American Gas Association, Catalog No. R-00999.

Talbert, S.G., G.H. Stickford, D.C. Newman, and W.N. Stiegelmeyer. 1986. Effect of hard water scale buildup and water treatment on residential water heater performance. ASHRAE *Transactions* 92(2).

Wetherington, T.I., Jr. 1975. Heat recovery water heating. *Building Systems Design* (December/January).

CHAPTER 45

SNOW MELTING

THE practicality of melting snow with heated coils has been demonstrated in a large number of installations including sidewalks, roadways, ramps, and runways. Melting eliminates the need for snow removal, provides greater safety for pedestrians and vehicles, and reduces the labor of slush removal.

There are three types of snow melting systems:

1. Hot fluid circulated in embedded pipes (hydronic)
2. Embedded electric heating resistance cable or wire (electric)
3. Overhead high intensity infrared radiant heating (infrared)

System design includes (1) heating requirement, (2) pavement design, (3) control, (4) hydronic system design, or (5) electrical system design.

HEATING REQUIREMENT (HYDRONIC AND ELECTRIC)

The heating requirement for snow melting is affected by four atmospheric factors: (1) rate of snowfall, (2) air temperature, (3) wind velocity, and (4) humidity. The effects of these factors can be evaluated by considering the action of snow falling on a warmed surface.

The first flakes fall on a dry, warm surface, where they are warmed to 32°F and melted. The water from the melted snow forms a film over the entire area and starts to evaporate. This evaporation is a mass heat transfer from the surface to the atmosphere. In addition, there is heat transfer from the film to the ambient air and surfaces.

Mass Transfer by Evaporation. The evaporation rate of the melted snow from the snow melting slab is affected by the wind speed and the vapor pressure difference between air and melted snow. The air vapor pressure, however, is fixed by the relative humidity and temperature of the air. If the slab surface temperature is fixed, the evaporation loss varies with changes in air temperature, relative humidity, and wind speed.

Heat Transfer by Convection and Radiation from Melted Snow to Ambient Air and Surfaces. A combined film coefficient is sufficiently accurate to determine the combined convection and radiation loss. This coefficient is based on heat transfer from a wetted surface, such as the film of melted snow, to the air. The coefficient is a function of wind speed alone. The heat transfer depends on the film coefficient and the temperature difference between the surface and the air. Since the surface temperature is fixed, convection and radiation losses vary with changes in air temperature and wind speed.

To determine evaporation and heat transfer from the melted snow to the air, three of these four climatic factors must be known: (1) wind speed, (2) air temperature, (3) relative humidity, and (4) rate of snowfall. Rate of snowfall determines the heat required to warm the snow to 32°F to melt it.

The preparation of this chapter is assigned to TC 6.1, Hydronic and Steam Equipment and Systems.

Free Area Ratio

Before deriving equations to give quantitative values for the effects of these four factors, consider the insulating effect of the unmelted snow. The first flakes fall on a dry, warm surface and are then warmed to 32°F and melted. While the flakes are being warmed and before they are completely melted, they act as tiny blankets or insulators. The effect of this insulation when measured can be large. Since the snowflakes cover a fraction of the surface area, it is convenient to think of the insulating effect as an area ratio. The area covered by snowflakes is the insulated area, and the uncovered area is the uninsulated area. The term *free area ratio* (A_r) represents the ratio of the uncovered, or free, area to the total area, and is expressed as:

$$A_r = A_f/A_t \tag{1}$$

where

A_r = free area ratio
A_f = free area, ft²
A_t = total area, ft²

therefore,

$$0 \leqslant A_r \leqslant 1$$

For $A_r = 1$, the system must melt the snow so rapidly that accumulation would be absolutely zero. This is impossible theoretically, but for practical purposes, it is permissible to have $A_r = 1$ as a maximum. For $A_r = 0$, the surface must be completely covered with snow to a depth sufficient to prevent evaporation and heat transfer losses. Research on the insulating effects of snow indicates that there are only three practical values for the free area ratio—0, 0.5, and 1.

Heating Equations

Chapman (1952) derives and explains equations for the heating requirements of a snow melting system. Chapman and Katunich (1956) derive the general equation for slab output q_o as:

$$q_o = q_s + q_m + A_r\,(q_e + q_h) \tag{2}$$

where

q_s = sensible heat transferred to snow, Btu/h·ft² (W/ft²)
q_m = heat of fusion, Btu/h·ft² (W/ft²)
A_r = ratio of snow-free area to total area, dimensionless
q_e = heat of evaporation, Btu/h·ft² (W/ft²)
q_h = heat transfer by convection and radiation, Btu/h·ft² (W/ft²)

The sensible heat q_s to bring the snow to 32°F is:

For hydronic (Btu/h·ft²) $\quad q_s = 2.6s\,(32 - t_a) \quad$ (3a)

For electric (W/ft²) $\quad q_s = 0.762s\,(32 - t_a) \quad$ (3b)

where

s = rate of snowfall, in. of water equivalent per hour
t_a = air temperature, °F

The heat of fusion q_m to melt the snow is:

For hydronic (Btu/h·ft²) $$q_m = 746s \quad (4a)$$

For electric (W/ft²) $$q_m = 218s \quad (4b)$$

The heat of evaporation q_e (mass transfer) is:

For hydronic (Btu/h·ft²)

$$q_e = h_{fg}\,(0.0201V + 0.055)(0.185 - p_{av}) \quad (5a)$$

For electric (W/ft²)

$$q_e = 0.293\,h_{fg}\,(0.0201V + 0.055)(0.185 - p_{av}) \quad (5b)$$

where

h_{fg} = heat of evaporation at the film temperature, Btu/lb
V = wind speed, mph
p_{av} = vapor pressure of moist air, in. of mercury

The heat transfer q_h (convection and radiation) is:

For hydronic (Btu/h·ft²)

$$q_h = 11.4\,(0.0201V + 0.055)(t_f - t_a) \quad (6a)$$

For electric (W/ft²)

$$q_h = 3.34\,(0.0201V + 0.055)(t_f - t_a) \quad (6b)$$

where

t_f = water film temperature, °F, usually taken as 33 °F

The designer can use Equations (2) through (6) to determine the heating requirement of a snow melting system. The solutions to these equations, however, require the simultaneous consideration of the four climatic factors: (1) wind speed, (2) air temperature, (3) relative humidity, and (4) rate of snowfall. Annual averages or maximums should not be used for the climatic factors, since there is no assurance that they will ever occur simultaneously. It is necessary, therefore, to make a frequency analysis of the solutions to Equation (2) for all the occurrences of snowfall for several years.

Weather Data

Table 1 shows analyses of 33 cities and operating information for applicable snow melting systems. In freezing temperatures (32 °F and below) without snowfall, the system may be idling, which means that some heat is supplied to the slab so that there will be immediate melting when snow starts to fall. Column 4 of Table 1 gives the mean temperature during freezing periods. This temperature together with wind speed is used to calculate the *idling load.* The column headed "Hours of Snowfall" indicates the number of hours that snow is falling at rates equal to or greater than 0.01 in. of water equivalent per hour. There are snowfalls of trace quantities about twice as often as there are for measurable quantities of 0.01 in. or more. Light falls can normally be handled by idling loads.

The remaining columns of Table 1 represent the frequency distribution of required heat output. This distribution is based on the solution to the basic equation for two values of the free area ratio A_r. This distribution represents the basis of the analyses and is also the basis for Tables 2 and 3.

Performance Classification

Snow melting installations are classified in Table 2 according to types as Class I, II, or III. Chapman (1957) discusses these classes, which are defined in the footnotes to Table 2. Snow melting systems are generally classified (as to the urgency for melting) as follows:

Class I (minimum): Residential walks or driveways and interplant areaways.

Class II (moderate): Commercial (stores and offices) sidewalks and driveways, and steps of hospitals.

Class III (maximum): Toll plazas of highways and bridges, and aprons and loading areas of airports and hospital emergency entrances.

These classifications depend on the allowable rate of snow melting. For example, a depth of snow of 1 in. for an hour during a heavy storm might not be objectionable for a residential system. On the other hand, a store manager might consider the system inadequate if 0.5 in. of snow accumulated on the sidewalk in front of the store. The difference between a Class I system and a Class II system is in the required ability of each system to melt snow. All classes must be adequate for some combination of weather factors. If equipment is selected with the capacity to melt snow whenever the conditions are milder than some of the critical values, it will be inadequate for a fraction of the time. In a residential system where initial cost must be kept at a minimum, the designer may have to accept more frequent snow accumulations.

Table 2 contains the design heat requirements for all classes of snow melting systems. Under Class I systems, the values after the slash are idling rates and, since they exceed the Class I design rates, should be taken as design output for this classification. The designer can alter design rates if a particular job has different design criteria from those given in the footnotes of Table 2. Any change in the design conditions used should be based on the frequency distribution given in Table 1.

Back and Edge Losses

In addition to determining the four heating requirements, back and edge losses must be included. Adlam (1950) demonstrates that these losses vary from 4 to 50%, depending on factors such as pavement construction, operating temperature, ground temperature, or back exposure (bridges). With construction shown in Figures 1 or 3, and ground temperature of 40 °F at a depth of 24 in., back losses are approximately 20%. Higher losses would be expected with (1) colder ground, (2) more cover over pipe or cable, or (3) exposed back, such as bridges or parking decks.

Heating Requirement Example

Example 1. An engineer has been retained to design snow melting systems for the service areas of a turnpike running from the eastern edge of the Wisconsin-Illinois border northwest to the Wisconsin-Minnesota border just east of St. Paul, Minnesota. It is decided that Chicago, Illinois, data will be adequate for the southern terminus and that Minneapolis-St. Paul data will be adequate for the northern terminus. Determine the heat and hydraulic requirements of the systems for service areas between Chicago and St. Paul.

Solution: Assume, for this example, that the city in question is Madison, Wisconsin. Weather bureau records indicate that the annual average number of days with snow cover of 1 in. or more would be 100, and that the engineer can assume an average snowfall of 40 in. In addition, about 11 days per year with a snowfall of 1 in. or more may be estimated (Chapman 1957).

For the walkways to the restaurant from the parking area, a Class I design rate could be used. This rate could be taken as 90 Btu/h·ft² (26 W/ft²). This agrees with data in Tables 2 and 3, which give the design rate at 89 (26) for Chicago and idling rate at 95 (28) for Minneapolis.

The lanes leading from the turnpike to the gasoline pumps and parking areas should be rated as Class II areas. A check of Tables 1 and 2 indicates that 160 Btu/h·ft² (47 W/ft²) would be adequate.

If an emergency area is included for a wrecking truck, ambulance, or police garage, it would be wise to consider a Class III rating for this area. An inspection of Table 1 for Chicago shows that a rate of 275 Btu/h·ft² (80 W/ft²) would be adequate for $A_r = 1$ for 99.4% of the time. Similarly, 275 (80) would be adequate 99.4% of the time in St. Paul. Therefore, 275 Btu/h·ft² (80 W/ft²) seems sufficient for the emergency areas. The Class III column in Table 2 lists 350 Btu/h·ft² (102 W/ft²) for Chicago and 254 Btu/h·ft² (74 W/ft²) for St. Paul, but for uses similar to the areas here, 275 Btu/h·ft² (80 W/ft²) should be adequate.

The above figures should be adjusted upward by 20 to 30% for back and edge losses.

Table 1 Data for Determining Operating Characteristics of Snow Melting Systems

City	Period of No Snowfall: Air Temperature[a]: Over 32°F, % winter hours with no snow	Period of No Snowfall: Air Temperature[a]: Below or Equal to 32°F, % winter hours with no snow	Mean during freezing period[g], °F	Wind speed freezing period[e], mph	Period of Snowfall[b]: Hours of Snowfall[a], %	Hours of Snowfall[a], Hour per Year	Free Area Ratio, A_r	Required Output,[c,d] Btu/h·ft²: 0 to 49	50 to 99	100 to 149	150 to 199	200 to 249	250 to 299	300 to 349	350 to 399	400 up	Max. Output[d] Btu/h·ft²
								Frequency distribution of snowfall hours at above outputs, %[f]									
Albuquerque, NM	74.7	24.7	26.2	8.5	0.6	22	1	62.0	25.4	7.6	4.2	0.0	0.8	—	—	—	259
							0	94.1	5.9	—	—	—	—	—	—	—	82
Amarillo, TX	73.1	26.0	24.6	13.3	0.9	33	1	33.7	35.4	15.4	10.7	3.0	1.8	—	—	—	260
							0	88.1	19.1	1.8	—	—	—	—	—	—	143
Boston, MA	64.6	31.4	24.7	14.2	4.0	145	1	51.5	30.0	12.3	4.3	1.2	0.6	0.1	—	—	320
							0	83.2	14.0	2.0	0.3	0.3	0.1	—	0.2	—	370[g]
Buffalo-Niagara Falls, NY	46.5	46.9	23.9	10.8	6.6	240	1	50.7	32.6	11.2	3.7	1.4	0.2	0.2	—	—	309
							0	95.9	3.4	0.2	0.5	—	—	—	—	—	192
Burlington, VT	39.0	54.5	19.6	10.8	6.5	236	1	53.7	29.9	13.2	2.5	0.6	0.1	—	—	—	280
							0	91.8	7.6	0.6	—	—	—	—	—	—	142
Caribou-Limestone, ME	21.4	70.6	16.5	10.0	8.0	290	1	35.0	39.7	16.0	5.7	2.0	1.0	0.5	0.1	—	378
							0	92.0	7.5	0.5	—	—	—	—	—	—	138
Cheyenne, WY	46.4	49.8	21.5	15.3	3.8	138	1	16.5	26.2	19.4	13.1	8.6	4.7	4.2	4.7	2.6	499
							0	94.3	5.4	0.3	—	—	—	—	—	—	129
Chicago, IL	45.4	50.9	21.4	11.5	3.7	134	1	45.8	37.4	11.4	3.1	1.4	0.6	0.2	0.1	—	368
							0	91.5	8.1	0.3	0.1	—	—	—	—	—	165
Colorado Springs, CO	54.3	43.6	22.1	11.5	2.1	76	1	26.8	36.3	19.0	7.5	4.4	5.5	0.5	—	—	311
							0	98.4	1.6	—	—	—	—	—	—	—	63
Columbus, OH	59.0	38.1	24.5	10.0	2.9	105	1	65.8	22.4	8.0	1.7	1.7	0.4	—	—	—	261
							0	97.7	2.3	—	—	—	—	—	—	—	72
Detroit, MI	47.0	49.3	24.1	10.6	3.7	134	1	60.4	27.7	9.3	1.5	0.8	0.3	—	—	—	278
							0	95.9	3.5	0.6	—	—	—	—	—	—	140
Duluth, MN	12.6	80.5	14.5	12.0	6.9	250	1	23.7	32.9	20.6	13.7	4.3	2.5	1.7	0.6	—	382
							0	94.8	4.7	0.0	0.3	0.2	—	—	—	—	206
Flamouth, MA	68.5	29.5	25.5	12.8	2.0	73	1	50.0	33.9	14.2	1.6	0.3	—	—	—	—	204
							0	91.5	7.4	1.1	—	—	—	—	—	—	144
Great Falls, MT	49.0	46.2	16.5	14.4	4.8	174	1	26.2	27.6	16.7	16.4	7.5	4.6	0.3	0.5	0.2	451
							0	94.6	4.8	0.6	—	—	—	—	—	—	138
Hartford, CT	56.4	38.9	24.4	8.2	4.7	171	1	48.4	34.6	11.2	4.3	0.8	0.7	—	0.1	—	396
							0	80.4	16.7	2.2	0.5	—	0.1	—	0.1	—	383
Lincoln, NB	45.0	52.5	20.8	10.1	2.5	91	1	32.7	26.2	20.0	13.9	5.7	1.5	—	—	—	293
							0	97.2	2.6	0.0	0.0	0.2	—	—	—	—	202
Memphis, TN	87.2	12.5	27.0	11.5	0.3	11	1	48.4	28.3	6.7	13.3	3.3	—	—	—	—	227
							0	85.0	8.3	6.7	—	—	—	—	—	—	144
Minneapolis-St. Paul, MN	23.6	70.8	16.9	11.1	5.6	203	1	28.4	31.4	21.7	14.1	3.5	0.6	0.3	—	—	313
							0	96.5	3.1	0.3	0.1	—	—	—	—	—	155
Mt. Home, ID	56.3	42.6	24.9	9.5	1.1	40	1	74.2	21.9	3.9	—	—	—	—	—	—	143
							0	98.1	1.9	—	—	—	—	—	—	—	90
New York, NY	55.7	42.2	24.2	11.8	2.1	76	1	53.1	31.8	9.4	2.2	1.5	1.7	—	0.3	—	385
							0	87.6	9.6	1.5	0.7	0.3	0.3	—	—	—	298
Ogden, UT	50.0	45.6	24.3	9.4	4.4	160	1	64.6	29.2	5.8	0.3	0.1	—	—	—	—	216
							0	88.8	9.4	1.4	0.3	0.1	—	—	—	—	216[g]
Oklahoma City, OK	79.0	19.8	24.6	15.8	1.2	44	1	27.8	18.7	17.0	12.6	14.3	5.9	2.7	1.0	—	394
							0	95.7	4.3	—	—	—	—	—	—	—	81
Philadelphia, PA	75.8	22.6	26.7	9.7	1.6	58	1	62.3	23.6	10.4	2.3	0.9	0.5	—	—	—	296
							0	84.3	14.0	1.1	0.2	0.4	—	—	—	—	229
Pittsburg, PA	55.2	39.8	24.3	11.6	5.0	182	1	53.6	30.8	8.4	4.6	1.9	0.7	—	—	—	282
							0	93.3	5.9	0.7	0.1	—	—	—	—	—	157
Portland, OR	92.9	6.1	28.9	8.4	1.0	36	1	78.0	16.9	5.1	—	—	—	—	—	—	125
							0	91.5	8.5	—	—	—	—	—	—	—	97
Rapid City, SD	45.2	51.6	19.3	12.9	3.2	116	1	29.7	29.0	16.0	8.4	6.3	3.6	1.9	2.0	3.1	581
							0	97.6	2.2	0.2	—	—	—	—	—	—	102
Reno, NV	56.0	41.6	24.3	5.6	2.4	87	1	82.6	15.4	1.8	0.2	—	—	—	—	—	152
							0	90.2	8.0	1.6	0.2	—	—	—	—	—	154[g]
St. Louis, MO	68.7	30.4	25.0	11.5	0.9	33	1	42.9	31.4	16.7	7.1	1.9	—	—	—	—	225
							0	85.2	11.6	2.6	0.6	—	—	—	—	—	152
Salina, KS	60.0	38.5	23.3	10.9	1.5	54	1	44.9	31.9	12.7	7.6	2.2	0.7	—	—	—	286
							0	93.5	6.2	0.3	—	—	—	—	—	—	120
Sault Ste. Marie, MI	21.3	69.2	18.6	9.4	9.5	345	1	45.7	32.8	14.3	5.7	1.4	0.1	—	—	—	262
							0	97.9	2.0	0.1	—	—	—	—	—	—	144
Seattle-Tacoma, WA	88.0	10.8	28.5	5.9	1.2	44	1	86.3	12.3	1.4	—	—	—	—	—	—	137
							0	91.0	8.1	0.9	—	—	—	—	—	—	128
Spokane, WA	48.5	46.1	25.7	10.7	5.4	196	1	62.6	28.7	7.4	1.1	2.0	—	—	—	—	205
							0	92.0	7.8	0.2	—	—	—	—	—	—	127
Washington, D.C.	77.9	21.2	26.8	9.6	0.9	33	1	59.0	29.8	10.6	0.6	—	—	—	—	—	154
							0	85.7	11.8	2.5	—	—	—	—	—	—	121

Source: *Air Conditioning, Heating and Ventilating,* August, 1957, p. 87.

Note: The period covered by this table is from November 1 to March 31, including February taken as a 28.25 day month. Total hours in period = 3630.

[a]The percentage in Columns 2 and 3 plus the percent under hours of snowfall total 100%. Note that "Hours of Snowfall" does not include idling time, and is not actual operating time. See text.

[b]Snowfalls of trace amounts are not included; hence, only those hours of 0.01 in. water equivalent per hour snowfall are listed.

[c]Output does not include allowance for back or edge losses since these depend on slab construction.

[d]Multiply output values by 0.293 to convert to W/ft².

[e]"Freezing Period" is that during "No Snowfall" when the air temperature is 32°F or below.

[f]Percentages total 100% of the number of hours of snowfall.

[g]*When heat output for* A_r = 0 equals or exceeds the heat output for A_r = 1, the heat transfer q_h is from the air *to* the slab. This occurs when snowfall is at temperatures above 32°F

Table 2 Design Data for Three Classes of Snow Melting System

City	Design Output, Btu/h · ft²[a] Class I System[a]	Class II System[b]	Class III System[c]
Albuquerque, NM	71	82	167
Amarillo, TX	98	143	241
Boston, MA	107	231	255
Buffalo-Niagara Falls, NY	80	192	307
Burlington, VT	90	142	244
Caribou-Limestone, ME	89/93	138	307
Cheyenne, WY	83	129	425
Chicago, IL	89	165	350
Colorado Springs, CO	49/63	63	293
Columbus, OH	52	72	253
Detroit, MI	69	140	255
Duluth, MN	83/114	206	374
Falmouth, MA	93	144	165
Great Falls, MT	84/112	138	372
Hartford, CT	115	254	260
Lincoln, NB	64/67	202	246
Memphis, TN	134	144	212
Minneapolis-St. Paul, MN	63/95	155	254
Mt. Home, ID	50	90	140
New York, NY	121	298	342
Ogden, UT	98	216	217
Oklahoma City, OK	66	81	350
Philadelphia, PA	97	229	263
Pittsburgh, PA	89	157	275
Portland, OR	86	97	111
Rapid City, SD	58/86	102	447
Reno, NV	98	154	155
St. Louis, MO	122	152	198
Salina, KS	85	120	228
Sault Ste. Marie, MI	52/78	144	213
Seattle-Tacoma, WA	92	128	133
Spokane, WA	87	127	189
Washington, DC	117	121	144

Source: *Air Conditioning, Heating and Ventilating,* August 1957, p. 92.
General Notes: Multiply design output values by 0.293 to convert to W/ft².
Output does not include allowance for back or edge losses since these depend on slab construction.
[a]For Class I (residential) systems, the design output is set at the required heat output (see Table 1) when $A_r = 0$ at the 98th percentile of the frequency distribution; that is, where 98% of the hours have this output or less. Where idling rate is greater than Class I design rate, idling rate value follows slash and should be used as Class I design output.
[b]For Class II (commercial) systems, the design output is the maximum output when $A_r = 0$ in Table 1 (last column).
[c]For Class III (industrial) systems, the design output is determined by the following four requirements: (1) output is never exceeded for two consecutive hours; (2) output for $A_r = 1$, Table 1, is at least 1 Btu/h · ft² greater than maximum output for $A_r = 0$; (3) A_r is greater than or equal to 0.5 maximum requirement shown in Table 1 for $A_r = 1$; that is, $q_o = q_s + q_m + 0.5\,(q_e + q_h)$ for the conditions where $q_t = q_s + q_m + q_h + q_e$ are a maximum, and (4) the free area ratio A_r is unity for at least 98% of the hours listed in Table 1.

Operating Cost Example

Example 2. A Class III snow melting system of 2000 ft² is installed in Chicago. Table 3 shows that annual output is 8700 Btu/ft². Assume a fossil fuel cost of $8.00 per million Btu and an electric cost of $0.07/kWh with back loss at 30%.

Solution: Operating cost O_c may be expressed as follows:

$$O_c = A\,A_o\,F_c/[1 - (B_l/100)]$$

where

O_c = operating cost, $/year
A = area, ft²
A_o = annual output, Btu/ft² · year or kWh/ft² · year
F_c = fuel cost, $/Btu or $/kWh
B_l = back loss, %

Operating cost for fossil fuel is:

$$O_c = (2000 \times 8700 \times 8) / [1 - (30/100)]\,(1{,}000{,}000)$$
$$= \$199/\text{year}$$

Operating cost for electric heat is:

$$O_c = [2000\,(8700 \times 0.293)(0.07)] / 1000\,[1 - (30/100)]$$
$$= \$510/\text{year}$$

PAVEMENT DESIGN (HYDRONIC AND ELECTRIC)

Concrete and asphalt may be used for snow melting systems. Thermal conductivity of asphalt is less than that of concrete; thus pipe or cable spacing and temperatures should be adjusted accordingly. Hot asphalt may be damaging to plastic or electric snow melting systems without adequate precautions. For specific recommendations, refer to the Hydronic and Electric System Design sections.

Concrete slabs containing hydronic or electric snow melting apparatus must be designed and constructed with subbase, expansion-contraction joints, reinforcement, and drainage to prevent slab cracking; otherwise, crack-induced shearing or tensile forces would break the pipe or cable. Pipe or cable must not run through expansion-contraction joints, keyed construction joints, or control joints (dummy grooves); however, pipe or cable may be run under 0.12-in. score marks (block and other patterns). Control joints must be placed wherever the slab changes size or direction. The maximum dimension between control joints for ground-supported slabs should be less than 15 ft, and the length should be no greater than twice the width, except for ribbon driveways or sidewalks. In ground-supported slabs, most cracking occurs during the early cure. Depending on the amount of water used in a concrete mix, shrinkage during cure will be up to 0.75 in. per 100 ft. During the early cure period, concrete does not have sufficient strength to overcome friction between the slab and under the bed while shrinking, if the slab is more than 15 ft long.

If the slabs containing snow melting apparatus are to be poured in two separate layers, the top layer containing the apparatus does not usually contribute toward total slab strength; therefore, the lower slab must be designed to provide total slab strength.

The concrete mix of the top layer of concrete should give maximum weatherability. Compressive strength should be 4000 to 5000 psi; recommended slump 3 in. maximum, 2 in. minimum. Air content and aggregate size should be as follows:

Maximum Size Crushed Rock Aggregate[a], in.	Air Content
2.5	5 ± 1%
1	6 ± 1%
0.5	7.5 ± 1%

[a]Do not use river gravel or slag.

Pipe or cable may be placed in contact with an existing sound pavement (either concrete or asphalt) and then covered as described in the Hydronic or Electric System Design sections. If there are signs of cracking or heaving, the pavement should be replaced. Pipe or cable should not be placed over existing expansion-contraction, control, or construction joints. The finest grade of asphalt is best for the top course; no stone should exceed a 0.38 in. diameter.

A moisture barrier should be placed between any insulation and the fill. The joints in the barrier should be mopped and the fill made smooth enough to eliminate holes or gaps for moisture transfer. Also, the edges of the barrier should be flashed to the surface of the pavement so that the ends are sealed.

Table 3 Yearly Operating Data

City	Idling				Melting						
						Annual Output,[f, g]					
	Time,[a] h/year	Rate,[b] Btu/h·ft²	Annual Output,[c,d] Btu/ft²	kWh/ft²	Time,[e] h/year	Class I Btu/ft²	Class I kWh/ft²	Class II Btu/ft²	Class II kWh/ft²	Class III Btu/ft²	Class III kWh/ft²
Albuquerque, NM	897	32.5	29,100	8.5	22	908	0.27	969	0.28	1,150	0.34
Amarillo, TX	944	51.1	48,000	14.1	33	2,150	0.63	2,520	0.74	2,770	0.81
Boston, MA	1,140	52.1	59,400	17.4	145	8,000	2.34	9,080	2.66	9,100	2.67
Buffalo-Niagara Falls, NY	1,702	50.2	85,500	25.0	240	11,800	3.46	14,600	4.28	14,900	4.37
Burlington, VT	1,978	76.9	152,000	44.6	236	11,800	3.46	13,500	3.96	13,800	4.04
Caribou-Limestone, ME	2,563	93.0	238,000	69.8	290	17,800[h]	5.22	20,600	6.04	22,500	6.59
Cheyenne, WY	1,808	77.7	140,000	41.2	138	9,730	2.85	13,200	3.87	20,200	5.92
Chicago, IL	1,848	67.8	125,000	36.7	134	7,200	2.11	8,390	2.46	8,700	2.55
Colorado Springs, CO	1,583	63.4	100,000	29.4	76	3,960[h]	1.16	3,960	1.16	7,390	2.17
Columbus, OH	1,383	45.0	62,200	18.2	105	3,590	1.05	4,180	1.22	5,350	1.57
Detroit, MI	1,790	49.0	87,600	25.7	134	5,540	1.62	6,850	2.01	7,070	2.07
Duluth, MN	2,922	113.8	332,000	97.4	250	33,200[h]	9.73	38,100	11.16	39,500	11.57
Falmouth, MA	1,071	44.2	47,400	13.9	73	3,830	1.12	4,250	1.25	4,290	1.26
Great Falls, MT	1,677	111.6	187,000	54.8	174	13,700[h]	4.01	15,400	4.51	19,100	5.60
Hartford, CT	1,412	41.9	59,200	17.3	171	9,830	2.88	10,800	3.16	10,810	3.17
Lincoln, NB	1,906	67.2	128,000	37.5	91	4,750[h]	1.39	8,350	2.45	8,520	2.50
Memphis, TN	454	32.0	14,500	4.3	11	702	0.21	721	0.21	792	0.23
Minneapolis-St. Paul, MN	2,570	95.1	244,000	71.6	203	14,200[h]	4.16	17,600	5.16	18,400	5.39
Mt. Home, ID	1,546	41.9	64,800	19.0	40	1,260	0.37	1,530	0.45	1,590	0.47
New York, NY	1,532	50.7	77,700	22.8	76	4,180	1.22	4,690	1.37	4,710	1.38
Ogden, UT	1,655	44.6	73,800	21.6	160	7,050	2.07	7,370	2.16	7,370	2.16
Oklahoma City, OK	719	56.2	40,400	11.8	44	2,380	0.70	2,800	0.82	5,400	1.58
Philadelphia, PA	820	31.4	25,700	7.5	58	2,710	0.79	3,100	0.91	3,110	0.91
Pittsburgh, PA	1,445	49.5	71,500	21.0	182	9,050	2.65	10,700	3.14	11,100	3.25
Portland, OR	221	17.4	3,840	1.1	36	1,300	0.38	1,330	0.39	1,360	0.40
Rapid City, SD	1,873	86.4	162,000	47.4	116	7,450[h]	2.18	8,250	2.42	13,400	3.93
Reno, NV	1,510	36.9	55,700	16.3	87	2,970	0.87	3,030	0.89	3,030	0.89
St. Louis, MO	1,104	44.8	4,950	14.5	33	2,190	0.64	2,290	0.67	2,380	0.70
Salina, KS	1,398	53.9	75,400	22.1	54	2,920	0.86	3,370	0.99	3,810	1.12
Sault Ste. Marie, MI	2,512	77.7	195,000	57.2	345	17,600[h]	5.16	22,200	6.50	23,200	6.80
Seattle-Tacoma, WA	392	17.2	6,750	2.0	44	1,410	0.41	1,430	0.42	1,430	0.42
Spokane, WA	1,673	39.1	65,500	19.2	196	8,650	2.53	9,350	2.74	9,560	2.80
Washington, DC	770	30.6	23,600	6.9	33	1,650	0.48	1,660	0.49	1,690	0.49

Source: *Air Conditioning, Heating and Ventilating*, August 1957, p. 94.
[a]From Table 1, Column 3 × 3630 (hour per year).
[b]Rate when idling, Btu/h·ft² = $(0.27\ V + 3.3)(32 - t)$, where V = wind speed from Column 5, Table 1, and t = air temperature from Column 4, Table 1.
[c]Multiply rate values listed by 0.293 to convert to W/ft².
[d]Product of the preceding columns.
[e]"Hours of Snowfall," from Column 7, Table 1.
[f]See footnotes for Table 2 for explanation of classes.
[g]Based on the condition that surface temperature is maintained at 33°F until required output exceeds designed output, at which time design output is used regardless of required output. Distribution of required output based on Table 1.
[h]Based on idling rate rather than design rate.

Snow melting systems should have good surface drainage. When the ambient air temperature is freezing, run-off from melting snow immediately freezes on leaving the heated area. Any water that is able to get under the pavement also freezes when the system is deenergized, causing extreme frost heaving. Runoff should be piped away in heated or below frost line drains.

The area to be protected by snow melting must first be measured and planned. For total snow removal, hydronic or electric heat must cover the entire area. In larger installations, it may be desirable to melt snow and ice from only the most frequently used areas, such as walkways and wheel tracks for trucks and autos. Planning for separate circuits should be considered so that individual areas within the system can be heated, as required.

When snow melting apparatus must be run around obstacles (*e.g.*, a storm sewer grate), the pipe or cable spacing should be reduced uniformly.

Because some drifting will occur on every system adjacent to a wall or vertical surface, extra heat should be added in these areas. If possible, heat should also be added in the vertical surface. Drainage carried to the area expected to be drifted tends to wash away some snow.

CONTROL (HYDRONIC AND ELECTRIC)

Snow melting systems can be controlled either manually or automatically.

Manual Control

Manual operation is strictly a two-position control—either on or off. An operator is required to activate and deactivate the system when snow falls. If the system is not turned off after snowfall, operating cost is increased.

Snow Detectors

Snow detectors monitor precipitation and temperature. They allow operation only when snow is present and may incorporate a delay off timer. Precipitation occurring below a preset temperature, usually 40°F, indicates presence of snow. Snow detectors located in the heated area activate the snow melting system when precipitation (snow) occurs below the preset pavement temperature. Another type of snow detector is mounted aboveground, adjacent to the heated area. This type may be installed without cutting into an existing system; however, it does not detect tracked or drifting snow. Locate both types of sensors so that they are not affected by overhangs, trees, blown snow, or other local conditions.

If the snow melting system is not turned on until the snow starts falling, it may not melt snow effectively for several hours, giving additional snowfall a chance to accumulate and increasing the time needed to melt the area. With adequate system heating capacity for the local conditions, automatic controls provide satisfactory operation because they turn on the system when light snow starts, allowing adequate warm-up before heavy snowfall develops. Operating costs are reduced with automatic turn-off.

Pavement Temperature Sensor

To limit excessive energy waste during normal and light snow conditions, it is common to include a remote temperature sensor installed midway between pipe or cable in the pavement; the setting is adjusted between 40 and 60 °F. Thus, during mild weather snow conditions, the system is automatically modulated or cycled on and off as the pavement temperature at the sensor reaches set point.

Outdoor Thermostat

The control system may have an outdoor thermostat that turns the system off when the outdoor ambient temperature rises above 35 to 40 °F as automatic protection against accidental operation in summer.

Control Selection

For optimum operating convenience and minimum operating cost, all the aforementioned controls should be incorporated in the snow melting system.

Operating Cost

To evaluate operating cost during idling or melting, use the "Annual Output" data from Table 3. Idling and melting data is based on pavement surface temperature control at 32 °F, which requires a pavement temperature sensor. During idling or melting, operating costs without the pavement temperature sensor will be substantially higher.

HYDRONIC SYSTEM DESIGN

Hydronic system design includes selection of the following components: (1) heat transfer fluid circulating through the system, (2) pipe system in the pavement, (3) fluid heater, (4) pump(s) to circulate the fluid, and (5) controls. The thermal stress in concrete pavement is also a design consideration.

Heat Transfer Fluid

A variety of fluids for transferring the heat from the fluid heater to the pavement, including brine, oils, and glycol-water, may be considered. Freeze protection is essential, since most systems will not be operated continuously in subfreezing weather. Power loss or pump failure could allow freeze damage to the pipe system and pavement.

Brine is the least costly heat transfer fluid, but it has lower specific heat than glycol. Brine may be discouraged from use because its corrosive potential requires more costly heating equipment.

While heat transfer oils are not corrosive, they are more expensive and have lower specific heat and higher viscosity than brine or glycol. Petroleum distillates suitable for fluids in snow melting systems are classified as nonflammable but have fire points between 300 and 350 °F. When using fluids of this type, any oil dripping from the seals on the pump should be collected. It is good practice to place a barrier between the oil lines and the boiler so that a flashback from the boiler will not ignite a possible oil leak. Other nonflammable fluids, such as those used in some transformers, can be used as antifreeze.

Glycols (ethylene and propylene) are most often used in snow melting systems because of moderate cost, high specific heat, low viscosity, and ease of corrosion control. Automotive glycols containing silicates are not recommended because the inhibitor can cause fouling, pump seal wear, fluid gelation, and reduced heat transfer. The glycol solution inhibitor should be designed for periodic maintenance by adding inhibitor. Glycols should be tested annually to determine any change in reserve alkalinity and freeze protection. Inhibitors obtained only from the manufacturer of the original glycol should be added. Heat exchanger surfaces should be kept below 285 °F, which corresponds to about 40 psig steam. Temperatures above 300 °F accelerate the deterioration of the inhibitors.

Since ethylene glycol and petroleum distillates are slightly toxic, the system should be installed and maintained independently. No permanent connection should be installed between the snow melting system and the drinking water supply.

Gordon (1950) discusses precautions to be taken during the installation of hydronic piping systems concerning internal corrosion, flammability, toxicity, cleaning, joints, and hook-up. Brine and glycol properties are discussed in Chapter 18 of the 1989 ASHRAE *Handbook—Fundamentals*. The effect of glycol on system performance is detailed in Chapter 13 of the 1987 ASHRAE *Handbook—HVAC Systems and Applications*.

Pipe System

Pipe systems may be metal or plastic. Metal pipes constructed of steel, iron, and copper have long been used. Corrosion may cause rapid failure of steel and iron if the pipe is not protected by coating and/or cathodic protection. The use of salts for deicing and the elevated temperature of the snow melting system accelerate corrosion of metallic components. NACE (1978) states that corrosion rate roughly doubles for each 18 °F rise in temperature.

Chapman (1952) derived the equation for the required fluid temperature to provide an output q_o. For construction similar to that shown in Figure 1, the equation is:

$$t_m = 0.5\, q_o + t_f \tag{7}$$

where

t_m = mean fluid temperature (antifreeze solution), °F

Equation (7) applies to 1 in. as well as 3/4-in. IPS pipe (Figure 1).

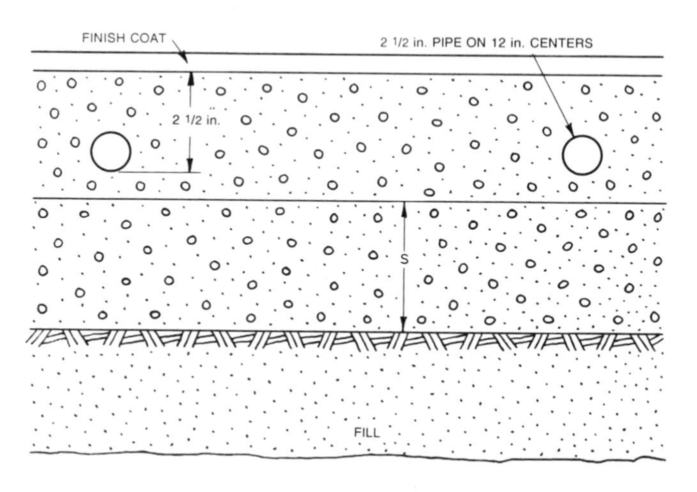

F = depth of finish coat—assumed as 0.5 in. of concrete. Finish coat may be asphalt, but then cover slab should be reduced from 3 in. Depth of slab should always keep thermal resistance equal to 3 in. of concrete.
S = depth required by structural design (should be a minimum of 2 in. of concrete).

Fig. 1 Detail of Hot Fluid Panel for Melting Snow

For specific conditions, or for cities other than those given in Table 1, Equations (2) and (7) are used. Table 4 gives solutions to these equations for relative humidities of 80%. For other values of relative humidity, Equations (8) and (9) can be used to determine the corrections.

Table 4 Snowfall Data for Various Cities

City	Number of Readings with Maximum Temperature in 6-h Period Below Freezing at Various Snowfall Rates — Snowfall Rate in Equivalent Inches of Water per 6 h: 0.00 to 0.24	0.25 to 0.49	0.50 to 0.75	0.75 to 0.99	Total Readings Taken	Assumed Design Rate of Snowfall[a] s in/h
Albany, NY	2052	29	5	1	3720	0.16
Asheville, NC	463	5	1	0	3536	0.08
Billings, MT	1640	4	0	0	3532	0.08
Bismarck, ND	2838	0	0	0	3720	0.08
Cincinnati, OH	1045	3	0	0	3720	0.08
Cleveland, OH	1569	2	0	0	3720	0.08
Evansville, IN	916	5	1	1	3720	0.08
Kansas City, MO	1189	12	2	1	3720	0.16
Madison, WI	2370	5	2	0	3720	0.08
Portland, ME	2054	33	4	1	3720	0.16

Note: Data from U.S. Weather Bureau. Based on readings taken 1:30 a.m., 7:30 a.m., 1:30 p.m., and 7:30 p.m. daily from November 15 to February 15 from 1940 to 1949. Where the total readings are less than 3720, the period of record is less than 10 years. The difference between Column 6 and the sum of readings in Columns 2, 3, 4, and 5 is the number of readings with a maximum temperature (in the 6-h period) above freezing.

[a]The design rate is found as follows: Proceed to left (on line from any city) from Column 5 until the column containing the tenth reading is found. Assume that the larger value in the heading of the selected column is an *average maximum* value, and should be multiplied by 2 to obtain the maximum rate for a 6-h period. This maximum rate divided by 6 is the design rate per hour. This is equivalent to dividing the larger value in the heading of the selected column by 3.

[b]For example: For Albany, New York, Columns 5 and 4 total six readings, and consequently the tenth reading is in Column 3, which has the larger value of 0.49 in the column heading. Dividing 0.49 by 3, the design water equivalent of 0.16 in. per hour is found, as listed in Column 7.

$$dq_o/dp_{av} = -A_r(0.0201V + 0.055)h_{fg} \quad (8)$$

$$dt_m/dp_{av} = -(A_r/2)(0.0201V + 0.055)h_{fg} \quad (9)$$

Appropriate climatic variables must be determined before solving Equation (2). The best procedure is to contact the local National Weather Service office and examine the Local Climatological Summary. An approximation can be made by using Table 4 for values of s, and solutions for Equation (2) taken from Table 5, with the following qualifications:

1. When designing a Class I system, use:

   $A_r = 1.0$, $t_a = 30$°F, and $V = 15$ mph

2. When designing a Class II system, use:

   $A_r = 1.0$, $t_a = 20$°F, and $V = 15$ mph

3. When designing a Class III system, use:

   $A_r = 1.0$, $t_a = 0$°F, and $V = 15$ mph

It is satisfactory to use 3/4-in. pipe or tube on 12-in. centers as a standard coil. If pumping loads require reduced friction, the pipe size can be increased to 1 in., but the pavement depth must be increased accordingly.

The piping should be supported by a minimum of 2 in. of concrete above and below the pipe. This requires a 5-in. pavement for 3/4-in. pipe and 5.4 in. for 1-in. pipe.

Plastic pipe (polyethylene, polybutylene) is popular due to lower material cost, lower installation cost, and corrosion resistance. Considerations for using plastic pipe should include stress crack resistance, temperature limitations, and thermal conductivity. Heat transfer coils should not be used with plastic pipe. Polyethylene and polybutylene pipe is furnished in coils. The smaller pipe sizes can be bent to form a variety of heating panel designs without elbows or joints. Mechanical compression connections can be used to connect the heating panel pipe to larger supply and return piping to the pump and fluid heater. Fusion joining of plastic pipe may be accomplished with appropriate fittings and fusion equipment. Fusion joining eliminates metallic components and the possibility of corrosion in the pipe system; it does, however, require considerable installation training.

When using plastic pipe, precaution needs to be taken so that the fluid temperature required will not damage the plastic pipe. Certain designs require a temperature above the capacity of plastic pipe, which will limit the heat output below design requirements. A logical solution would be to decrease the pipe spacing. Adlam (1950) addresses the parameter of pipe spacing, its effect on heat output, and the parameter of pipe size. A typical solution using plastic pipe in a snow melting system is summarized in Table 6, which shows a way of designing pipe spacing according to heating requirements. This table also shows adjustments for the effect of more than 2 in. of concrete or pavers over the pipe.

Good design practice avoids passing any embedded piping through a concrete expansion joint to avoid stressing and possibly rupturing the pipe. If the piping must pass through a concrete expansion joint, a method of protecting the pipe from stress under normal conditions is shown in Figure 2.

The pipe system should not be vented to the atmosphere. Introduction of air causes deterioration of the antifreeze. The pipe system should be divided into smaller zones to provide ease of filling and isolation if service is necessary.

Table 5 Heat Output and Fluid Temperature for Snow Melting System

s Rate of Snowfall in/h	A_r		$t_a = 0$°F, Speed V, mph: 5	10	15	$t_a = 10$°F, Speed V, mph: 5	10	15	$t_a = 20$°F, Speed V, mph: 5	10	15	$t_a = 30$°F, Speed V, mph: 5	10	15
0.08	1.0	q_o	151	205	260	127	168	209	102	128	154	75	84	94
		t_m	108	135	162	97	117	138	85	97	110	70	75	79
	0.0	q_o	66	66	66	64	64	64	62	62	62	60	60	60
		t_m	66	66	66	65	65	65	64	64	64	63	63	63
0.16	1.0	q_o	218	273	327	193	233	274	165	191	217	135	144	154
		t_m	142	169	198	129	149	170	117	129	142	100	105	109
	0.0	q_o	133	133	133	129	129	129	125	125	125	121	121	121
		t_m	99	99	99	97	97	97	95	95	95	93	93	93
0.25	1.0	q_o	292	347	401	265	305	346	235	261	287	203	212	221
		t_m	179	206	234	165	186	206	151	163	176	134	139	144
	0.0	q_o	208	208	208	202	202	202	195	195	195	188	188	188
		t_m	137	137	137	134	134	134	131	131	131	127	127	127

Note: Table developed from Equation (2). Relative humidity is assumed to be 80% at all air temperatures.

s = rate of snowfall, inches of water equivalent per hour
A_r = free area ratio
q_o = slab output
t_a = air temperature
t_m = fluid temperature. Based on construction as shown in Figure 1.
V = wind speed

Table 6 Plastic Pipe Specifications
(Mean fluid temperatures = 130°F)

Type	Heat Input[a] Btu/h·ft²	Snow Melting Rate[b] in/h	Minimum Air Temp.[c] °F	Pipe Circuit Spacing[d] in.
Moderate	200	1.0	10	12
Intermediate	250	1.5	5	9
Superior	300	2.5	0	6
Steps	400	1.0	10	4

[a]Heat input per unit area. Includes 30% back loss. Increase to 50% for bridges and structures with exposed back.
[b]Snow melting rate with 20°F air, 10 mph wind.
[c]Minimum air temperature at which surface can be maintained above 32°F with 10 mph wind, no snow.
[d]Plastic pipe circuit spacing. Space pipes 1 in. closer for each 1 in. of concrete cover over 2 in. Space pipes 2 in. closer for each 1 in. of brick paver and mortar.

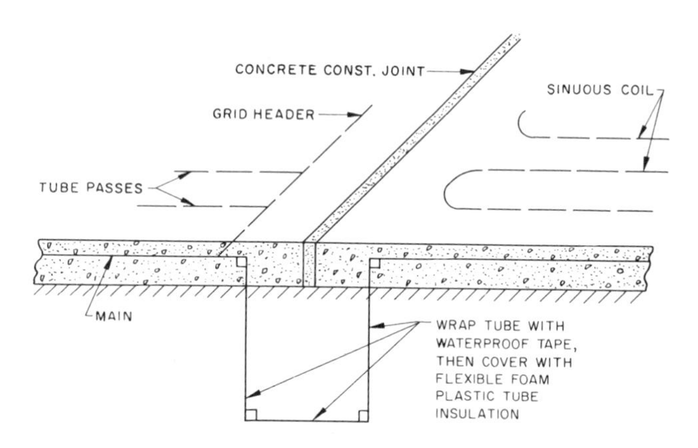

Fig. 2 Piping Detail for Concrete Construction Joints

After pipe installation, but before pavement installation, all piping should be air tested to about 100 psig. This pressure should be maintained until all welds and connections have been checked for leaks. Test only with air. Do not test with water for the following reasons: (1) small leaks may not be observed during pavement installation; (2) water leaks may damage concrete during installation; (3) the system may freeze before antifreeze is added; (4) it is difficult to add antifreeze when the system is filled with water.

Fluid Heater and Air Control

Heat transfer fluid can be heated using a variety of energy sources, which depend on availability at the location of the snow melting system. The fluid heater could use steam, hot water, gas, oil, or electricity. In some applications, heat may be available from other secondary sources, such as engine generators, computers, solar, and condensate.

Capacity of the fluid heater will be established by evaluating the data in the Heating Requirements section, usually 200 to 300 Btu/h·ft², allowing for back and edge losses.

Design of the fluid heater should follow standard practice, adjusting for film coefficient. Consideration should be given to flue gas condensation and thermal shock in boilers due to low fluid temperatures. Bypass flow controls may be necessary to maintain recommended boiler temperatures. Derate boilers for high altitude applications.

Air can be eliminated from the piping during initial filling by pumping the antifreeze from an open container into isolated zones of the pipe system. A properly sized pump and piping system that maintains an adequate fluid velocity, together with an air separator and expansion tank, will control air in the system during operation.

A strainer, sediment trap, or other means for cleaning the piping system may be provided. It should be placed in the return line ahead of the heat exchanger and must be cleaned frequently during the initial system operation to remove scale and sludge. The strainer should also be checked and cleaned, if necessary, at the start of and periodically during each new snow melting season.

An ASME safety relief valve of adequate capacity should be installed on a closed system.

Pump Selection

Proper pump selection depends on (1) fluid flow rate; (2) pressure requirements of the system; (3) specific heat of the fluid; and (4) viscosity of the fluid, particularly during a cold startup.

Pump Selection Example

Example 3. An area of 10,000 ft² is to be designed with a snow melting system. Using the heating requirement criteria of Equation (2), 250 Btu/h·ft² is selected. Thus the total fluid heater output must be 2,500,000 Btu/h. Size heater for fuel efficiency and heat transfer fluid used.

For a fluid with a specific heat of 0.85 Btu/lb·°F, the flow or temperature drop must be adjusted by 15%. For a 23°F temperature drop, flow would be 250 gpm. If glycol were used at 130°F, the effect on pipe pressure loss and pump performance would be negligible. If the system head were determined to be 40 ft and pump efficiency 60%, pump horsepower would be 4.2. Centrifugal pump design criteria may be found in Chapter 30 of the 1988 ASHRAE *Handbook—Equipment.*

Controls

Controls discussed earlier provide convenience and operating economy but are not required for operation. Hydronic systems require fluid temperature control for safety and component longevity. The temperature limits of the heat transfer fluid, pipe components, fluid heater, and pavement stress need to be considered. Certain plastic pipe materials should not operate above 140°F. A secondary fluid temperature sensor should deactivate the snow melting system if the primary control fails. The secondary control may also activate an alarm.

Thermal Stress

Chapman (1955) discusses the problems of thermal stress in concrete pavement. In general, thermal stresses will cause no problem if the following installation and operation rules are observed.

1. Minimize the temperature difference between the fluid and the slab surface by maintaining (a) close pipe spacing (see Figure 1), (b) low temperature drop in the fluid (less than 20°F), and (c) continuous operation (if economically feasible).
2. Keep pipe near surface to obtain about a 2-in. cover.
3. Use reinforcing steel designed for thermal stress if high structural loads are expected (such as on highways).

Thermal shock to the pavement may occur if heated fluid is introduced from a large source of residual heat such as storage tanks, large pipe systems, or other snow melting areas. The pavement should be brought up to temperature by maintaining the fluid temperature differential at less than 20°F.

ELECTRIC SYSTEM DESIGN

General

Snow melting systems using electricity as an energy source include the use of (1) mineral insulated (MI) cable, (2) resistance wire assembly to be embedded in the paving materials, or (3) high intensity infrared.

Heat Density

The basic load calculations for electric systems are the same as presented earlier in this chapter. However, electric system output is determined by installed resistance and the voltage impressed, and it cannot be overfired or altered by flow rates of fluid. Consequently, safety factors are not applied to requirements, nor are marginal capacity systems applied.

Heat intensity can be varied within a slab by altering the cable or wire spacing to compensate for anticipated drift areas or other high heat loss areas.

Switchgear and Conduit

Double-pole, single-throw switches or tandem circuit breakers to open both sides of the line should be used. The switchgear may be in any protected, convenient location. It is also advisable to include a pilot lamp on the load side of each switch so that there is a visual indication when the system is energized.

The power supply conduit is run underground, outside the slab, or in a prepared base. For concrete, this should be done before the reinforcing mesh is installed.

Mineral Insulated Cable

Mineral insulated (MI) heating cable is a seamless copper or stainless steel alloy sheath, magnesium oxide (MgO)-filled, die-drawn cable with one or two copper or copper alloy conductors. The metal outer sheath is protected from salts and other chemicals by a high density polyethylene jacket extruded over the cable. Whenever mineral insulated cable is embedded in a medium, a high density polyethylene jacket must be used. Although it is heavy-duty cable, its use is practical in any snow melting installation.

Cable Layout. To determine the MI heating cable needed for a specific area, the following must be known:

1. Heated area size
2. Watts per ft² area required
3. Voltage(s) available
4. Approximate cable length needed

To find approximate MI cable length, estimate 2 linear ft of cable per ft² of area for concrete. This corresponds to a 6-in. on-center spacing. Actual cable spacing will vary between 3 and 9 in. to allow for proper watt density.

Cable spacing is dictated primarily by the heat-conducting ability of the material in which it is embedded. Concrete has a higher heat transmission coefficent, which permits wider cable spacing. A procedure to determine the proper MI heating cable follows:

1. Determine total wattage

$$W_t = Aw$$

2. Determine total resistance

$$R = E^2/W_t$$

3. Determine calculated cable resistance per foot

$$r_1 = R/L_1$$

where

W_t = total wattage needed, W
A = heated area of each heated slab, ft²
w = desired watt density input, W/ft²
E = voltage available, V
r_1 = calculated cable resistance, ohms per foot of cable
L_1 = estimated cable length, ft
R = total resistance of heater cable, Ω
L = actual cable length needed, ft
r = actual cable resistance, ohms per ft
S = cable center to center spacing, in
I = total current per MI cable, A

Mineral insulated heating cables commercially available have actual total resistance values (total of two, if two conductor) ranging from 0.0016 to 0.6 ohm/ft. Manufacturing tolerances are ±10% on these values. Mineral insulated cables are die-drawn, with the internal conductor drawn to size indirectly via pressures transmitted through the mineral insulation. Special cables are not economical unless the quantity needed is 100,000 ft or more.

Table 7 Typical Heating Cable Resistances

ohm/ft Total at 77°F	Volt Rating	Conductor E, C[a]	Conductor American Wire Gage	Sheath E, C[a]	Sheath OD in.	lb/ft
Single Conductor						
0.610	300	E	26	E	0.120	15
0.391			25		0.128	20
0.300			23		0.145	34
0.200			21		0.150	38
0.100			18		0.165	45
0.050			15		0.175	56
0.030			13		0.200	72
0.020			11		0.215	85
0.010			8		0.246	120
0.200	600	E	21	E	0.183	53
0.105			18		0.205	69
0.060			16		0.210	74
0.035			16		0.215	85
0.025			12		0.240	100
0.020			11		0.253	112
0.013			9		0.277	143
0.006			6		0.240	202
0.00651	600	E	18	E	0.199	65
0.00409			16		0.215	73
0.00258			14		0.230	90
0.00162			12		0.246	105
0.00102			10		0.277	135
Two Conductor						
0.800	300	E	25	E	0.165	50
0.400			21		0.183	67
0.125			18		0.246	96
0.070	600	E	16	E	0.340	182
0.044		E	14	E	0.371	220
0.028		E	12	E	0.402	260
0.00818		C	16	C	0.340	182
0.00516		C	14	C	0.371	220

[a]E = Everdur; C = Copper

4. From manufacturers' literature, choose a cable closest in resistance to the calculated r_1. Table 7 illustrates typical cable resistances.
5. Determine actual cable length needed to give wattage desired.
6. Determine cable spacing within heated area.

$$S = 12A/L$$

Cable spacing for optimum performance should be within the following limits: in concrete, 3 in. minimum to 9 in. maximum; in asphalt, 3 in. minimum to 6 in. maximum.

Because the manufacturing tolerance on cable length is ±1%, and installation tolerances on cable spacing must be compatible with field conditions, it is usually necessary to adjust the installed cable as the end of the heating cable is rolled out. Cable spacing of the last several passes may have to be altered to give a uniform heat distribution.

The installed cable within the heated areas is a serpentine shape originating from a corner of the heated area (Figure 3). Since heat is evenly conducted from all sides of the heating cable, cables in a concrete slab can be run within half the spacing dimension of the heated area perimeter.

7. Determine current required for cable.

$$I = E/R, \text{ or } I = W/E$$

8. Choose cold lead cable as dictated by current requirements (see Table 8).

Cable Cold Lead. Every MI heating cable is factory fabricated with a nonheat-generating cold lead cable attached to the heating cable. The cold lead cable must be long enough to reach a dry

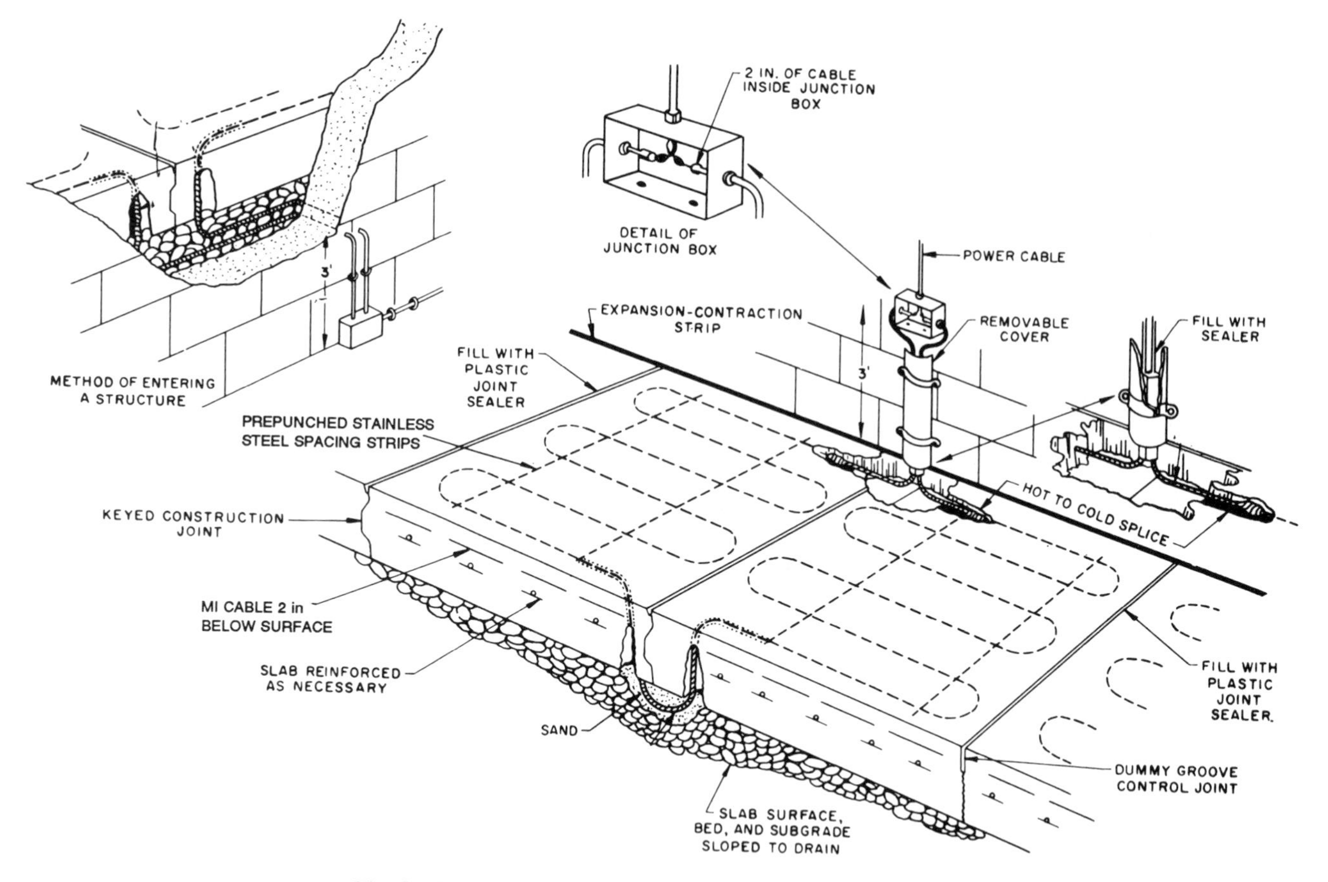

Fig. 3 Typical MI Heating Cable Installation in Concrete Slab

Table 8 Mineral Insulated Cold Lead Cables (Maximum Voltage—600 V)

Single-Conductor Cable		Two-Conductor Cable	
Current Capacity, A	American Wire Gage	Current Capacity, A	American Wire Gage
30	14	25	14/2
40	12	30	12/2
55	10	55	10/2
70	8	50	8/2
100	6	70	6/2
135	4	95	4/2
155	3		
180	2		
210	1		

location for termination, and of sufficient wire gage to comply with local and *National Electric Code* standards. Underwriters' Laboratories requires a minimum cold lead length of 7 in. Table 8 indicates the *National Electric Code* ratings of cold lead cables available.

Mineral insulated cable junction boxes must be in a location where the box will remain dry, and where at least 3 ft of cold lead cable is available at the end for any future service (Figure 3). Preferred junction box locations are indoors, on the side of a building, utility pole, or wall, or inside a manhole on the wall. Boxes should have a hole in the bottom to drain condensation. Outdoor boxes should be completely watertight, except for the condensation drain hole. Where junction boxes are mounted below grade, the cable end seals must be coated with an epoxy to prevent moisture entry. Cable end seals should extend into the junction box far enough to allow removal of the end seal, if necessary.

Magnesium oxide, used as insulation in MI cable, is hygroscopic. The only vulnerable part of the MI cable is the end seal. However, should moisture get into this end, it can be easily detected with a megohm meter, and driven out by applying a torch 2 to 3 ft from the end and working the flame toward the end.

Installation. Mineral insulated electric heating cable is installed in a concrete slab by pouring the slab in one or two pours. In single-pour application, the cable is hooked on top of the reinforcing mesh before the pour is started. In two-layer application, the cable is laid on top of the bottom structural slab and embedded in the finish layer. For a proper bond between the layers, the finish slab should be poured within 24 h of the first, and a bonding grout should be applied. The finished slab should be at least 2 in. thick. Cable should not run through expansion joints, control joints, or dummy joints (score or groove). If the cable must cross a joint of this type, the cable should exit the bottom of the slab at least 4 in. from one side of the joint and reenter the slab through the bottom at least 4 in. from the joint on the opposite side.

The cable is uncoiled from reels back and forth in a sinuous fashion with precalculated spacing between passes. Prepunched copper spacing strips, nailed to the lower slab, are often used for uniform spacing.

To protect the cold lead from chemical damage, a polyvinyl chloride jacket is extruded over the cable. The jacketing also provides excellent physical protection for the cable without adding excessive insulation.

Calcium chloride or other chloride additives should not be added to a concrete mix in winter, since chlorides are destructive to copper. Cinder or slag fill should also be avoided under snow melting panels. Where the cold lead cable is brought from the slab, it should be taped with polyvinyl chloride or polyethylene tape to protect it from corrosion from fertilizers and other ground attack. Within 2 ft of the heating section, only polyethylene should be used, since heat may break down the polyvinyl chloride tapes.

Underground, the leads should be installed in suitable conduits to protect them from physical damage.

To install MI cable in asphalt slabs, the cable is fixed in place on top of the base pour with prepunched stainless steel strips or 6 by 6 in. wire mesh. A coat of bituminous binder is applied over the base and the cable, to prevent them from floating when the top layer is applied. The top layer of asphalt over the cable should be 1.5 to 3 in. thick (Figure 4).

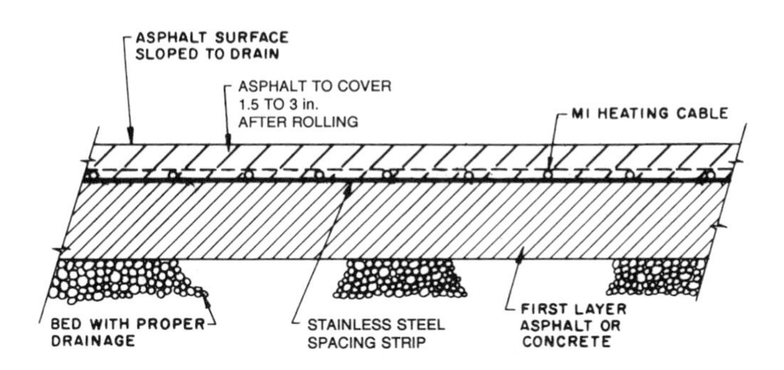

Fig. 4 Typical Section, Mineral Insulated (MI) Heating Cable in Asphalt

Testing. Mineral insulated heating cables should be thoroughly tested before, during, and after installation to ensure they have not been damaged, either in transit or installation.

Because of the hygroscopic nature of the magnesium oxide insulation, damage to the cable sheath is easily detectable with a 500-V field megohm meter. Cable insulation resistance should be noted on arrival. Any cable with insulation resistance less than 200 megohm (2×10^8 ohms) should not be used. Any cable with insulation resistance between conductor(s) and a metal sheath of less than 5 megohm (5×10 ohm) should not be used. Any cable that shows a marked loss of insulation resistance at the time of installation should be investigated for probable damage. Cable should also be checked for electrical continuity.

Embedded Wire Systems

In an embedded wire system, the resistance elements may consist of a length of copper wire or alloy with a given amount of resistance. When energized, this element will produce the required amount of heat. Witsken (1965) describes this system in further detail.

Elements are either solid-strand conductors or conductors spirally wrapped around a nonconducting fibrous material core. Both types are covered with an exterior layer of insulation, such as polyvinyl chloride or silicone rubber.

The heat-generating portion of an element is the conductive core. The resistance is specified in ohms per linear foot of core. Alternately, a manufacturer may specify the wire in terms of watts per foot of core, where the power is a function of ohms per foot of core, applied voltage, and total length of core.

The selection of insulating materials for heating elements is influenced by the watt density, application, and end use. Physical characteristics and chemical inertness are important qualities for an insulating shield.

Polyvinyl chloride is the lowest cost insulation and is widely used because of its inertness to oils, hydrocarbons, and alkalies. An outer covering of nylon is often added to increase physical strength and to withstand abrasion. Heat output of embedded polyvinyl chloride is limited to 5 watts per linear foot. Silicone rubber is not inert to oils or hydrocarbons. An additional covering—metal braid, conduit, or fiberglass braid—is needed for protection. This material can dissipate up to 10 watts per linear foot.

Lead can be used to encase the resistance element, insulated with glass fiber or asbestos. The lead sheath is then covered with a vinyl material. Output is limited to approximately 10 W/ft by the polyvinyl chloride jacket.

Teflon has good physical and electrical properties and can be operated at temperatures up to 500°F.

Low watt density (less than 10 W/ft) resistance wires may be attached to plastic or fiber mesh to produce a mat unit. Prefabricated factory-assembled mats are available to embed in specified paving materials in a variety of watt densities to match desired snow melting capacities. Mat lengths up to 60 ft are available for installation in asphalt sidewalks and driveways.

Preassembled mats are also available for stair steps, with appropriate tread melting section widths. Mats are seldom made larger than 60 ft^2, since larger areas are more difficult to install both mechanically and electrically. Mats can be tailored to follow contours of curves and around objects by making a series of cuts, as shown in Figure 5. Extreme care should be exercised to prevent damage to the heater wire (or lead) insulation during this operation.

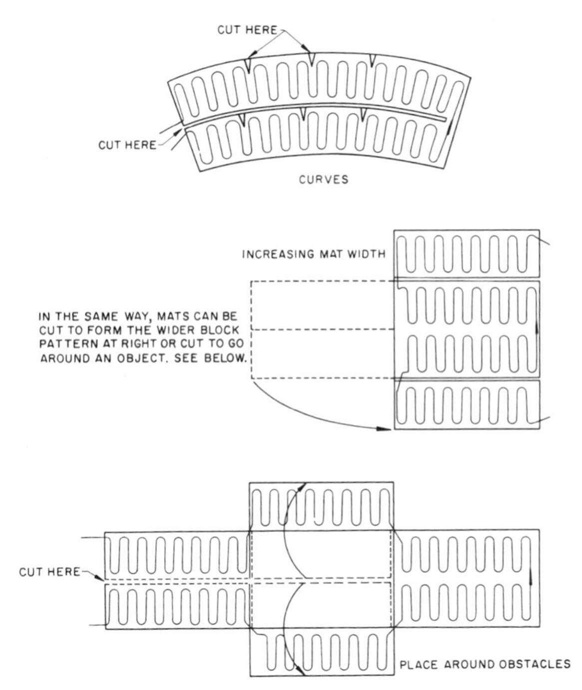

Fig. 5 Shaping Mats Around Curves and Obstacles

The mats should be installed from 1.5 to 3.0 in. below the finished surface of asphalt or concrete. Care should be taken in installing the mats deeper because the snow melting efficiency will be decreased. Mats that are not damaged by hot asphalt compaction should be used for asphalt paving.

Layout. Heating wires should be long enough to fit between the concrete slab dummy groove control or construction joints. Concrete forms may be inaccurate, thus, allow 2 to 4 in. between the edge of the concrete and the heating wire for clearance. Allow approximately 4 in. between the adjacent heating wires at the control or construction joints.

For asphalt, select the longest wire or largest heating mat that can be used on straight runs. Put the mats at least 12 in. in from the pavement edge. Adjacent mats must not overlap. Junction boxes should be located so that the maximum number of mats can be accommodated by each box. Wiring must conform to requirements of the *National Electric Code*. It is best to position junction boxes adjacent to or above the slab.

Installation

All Types

1. The wire or mats should be checked with an ohmmeter before, during, and after installation to prevent damage.
2. Temporarily lay the mats in position and install conduit feeders and junction boxes. Leave enough slack in the lead wires to permit temporary removal of the mats during the concrete or asphalt first pour. Ground all leads carefully using the grounding braids provided.
3. Secure all splices with approved crimped connectors or setscrew clamps. Tape all the power splices with plastic tape to make them waterproof. All junction boxes, fittings, and snug bushings must be approved for this class of application. The entire installation must be completely waterproof to ensure trouble-free operation.

Concrete

1. Each slab area between the expansion joints must be poured and finished individually. Pour the base slab and rough level to within 1.5 to 2 in. of the desired finish level. Place the mats in position and check for damage.
2. Pour the top slab over the mats while the rough slab is still wet and cover the mats to a depth of at least 1.5 in., but not more than 2 in.
3. Do not walk on or strike the mats with shovels or other tools.
4. Except for brief testing purposes, do not energize the mats until the concrete is completely cured.

Asphalt

1. Pour and level the base course. If units are to be installed on an existing asphalt surface, clean it thoroughly before placing the mats.
2. Apply a bituminous binder course to the lower base, install the mats, and apply a second binder coating over the mats. The finish topping over the mats should be applied in a continuous pour to a depth of 1.25 to 1.5 in. Note: Do not dump a large mass of hot asphalt on the mats because the heat could damage the insulation.
3. Check all circuits with an ohmmeter to be sure no damage occurred during the installation.
4. Do not energize the system until the asphalt has completely hardened.

Infrared Snow Melting Systems

While overhead infrared systems can be designed specifically for snow melting and pavement drying, they are usually installed for the additional features they offer. Infrared systems provide comfort heating for people, which can be particularly useful at plant, office building, and hospital entrances or on loading docks. Infrared lamps produce higher lighting levels, which can aid in the protection, safety, and attraction of a facility. These additional features sometimes justify the somewhat higher cost of infrared systems.

Infrared fixtures can be installed under entrance canopies, along building facades, and on poles away from other structures. Approved equipment is available for recess, surface, and pendant mounting.

Infrared Fixture Layout. The same infrared fixtures used for people comfort heating installations can be used for snow melting systems. The major differences in target area design and infrared fixture selection result from *horizontal* surfaces being emphasized in snow melting, whereas personal comfort emphasizes the need of the human body's *vertical* surfaces to be irradiated. When snow melting is the prime design concern, fixtures with narrow beam patterns confine the radiant energy within the target area for more efficient operation. Asymmetric reflector fixtures, which aim the radiation primarily to one side of the fixture centerline, are often used near the sidelines of the target area.

Infrared fixtures usually have a longer energy pattern parallel to the long dimension of the fixture than at right angles to it (Frier 1965). Therefore, fixtures should be mounted in a row parallel to the longest dimension of the area. Where the target area is 8 ft or more in width, it is best to locate the fixtures in two or more parallel rows. This arrangement also gives better personnel comfort heating, since radiation is directed from both sides across the target area, providing a more favorable incident angle.

Radiation Spill. In theory, the most desirable energy distribution would be uniform throughout the snow melting target area at a density equal to the design requirement. Heating fixture reflector design determines the percentage of the total fixture radiant output scattered outside of its target area design pattern.

Even the best controlled beam fixtures do not have a completely sharp cutoff at the beam edges. Therefore, if uniform distribution is maintained for the full width of the area, a considerable amount of radiant energy falls outside the target area. For this reason, infrared snow melting systems are designed so that the intensity on the pavement begins to decrease before reaching the edge of the area (Frier 1964). This design procedure minimizes stray radiant energy losses.

Figure 6 shows the watts per square foot values obtained in a sample snow melting problem (Frier 1965). The sample design average is 45 W/ft^2. It is apparent that the central part of the target area has an incident watt density above the design average value and that the peripheral area has radiation densities below the average. Figure 6 shows how the watt density and distribution in the snow melting area depends on the number, wattage, beam pattern, mounting height, and relative position of the heaters to the pavement (Frier 1964).

With distributions similar to the one in Figure 6, snow begins to collect at the edges of the area as the energy requirements for snow melting approach or exceed system capacity. As the snowfall lessens, the snow is melted again to the edges of the area and possibly beyond, if system operation is continued.

INFRARED INTENSITY ON THE PAVEMENT FROM FOUR FIXTURES — WATTS PER SQ FT —

14.7	16.0	19.75	23.7	25.5	27.5	28.2	28.0	28.2	27.5	25.5	23.7	19.75	16.0	14.7
23.7	24.8	28.7	31.7	35.7	38.2	38.5	39.4	38.5	38.2	35.7	31.7	28.7	24.8	23.7
25.7	31.7	37.5	42.7	46.4	49.4	52.2	53.0	52.2	49.4	46.4	42.7	37.5	31.7	25.7
28.2	34.3	42.8	46.7	51.2	55.7	58.5	63.0	58.5	55.7	51.2	46.7	42.8	34.3	28.2
28.2	34.3	42.8	46.7	51.2	55.7	58.5	63.0	58.5	55.7	51.2	46.7	42.8	34.3	28.2
25.7	31.7	37.5	42.7	46.4	49.4	52.2	53.0	52.2	49.4	46.4	42.7	37.5	31.7	25.7
23.7	24.8	28.7	31.7	35.7	38.2	38.5	39.4	38.5	38.2	35.7	31.7	28.7	24.8	23.7
14.7	16.0	19.75	23.7	25.5	27.5	28.2	28.0	28.2	27.5	25.5	23.7	19.75	16.0	14.7

Fig. 6 Typical Energy Levels per Unit Area for Infrared Snow Melting System

Target Area Watt Density. Theoretical target area wattage densities for snow melting with infrared systems are the same as those for commercial applications of embedded element systems; however, it should be emphasized that theoretical density values are for radiation incident to the pavement surface. Multiplying the recommended snow melting wattage density by the pavement area to obtain the total power input for the system does not result in good performance. An estimate, based on actual practice, to help estimate the total input wattage to be installed in the infrared fixtures above the target area is to multiply the commonly used snow melting watt densities shown in Table 6 by the target area; then, further increase the resulting product by a correction factor of 1.6. The resulting wattage compensates not only for the radiant efficiency involved, but also for the radiation falling outside the snow melting area. For small areas, or when the fixture mounting height exceeds 16 ft, the multiplying factor can be as large as 2.0; large areas with sides approaching equal lengths can have a factor of about 1.4.

The point-by-point method is the best way to calculate the heating fixture requirements for an installation. This method involves dividing the target area into a grid pattern of 1-ft squares, and cumulatively adding the watt per square foot radiant energy contributed by each infrared fixture on each square foot of grid (Figure 6). The recommended radiant energy distribution of a given infrared fixture can be obtained from the equipment manufacturer and should be followed for that fixture size and placement.

System Operation. With infrared energy, the target area can be preheated to snow melting temperatures in 20 to 30 min, unless the air temperature is well below 20 °F, or wind velocity is high (Frier 1965). This fast warm-up time makes it unnecessary to turn on the system until snow begins to fall. The equipment can be turned on either manually or automatically with a snow detector. A timer is sometimes used to turn the system off 4 to 6 h after the snow stops falling, allowing time to dry the pavement completely.

If the snow is allowed to accumulate before the infrared system is turned on, there will be a delay in clearing the pavement, as is the case with embedded systems. Since the infrared energy is absorbed in the top layer of snow rather than by the pavement surface, the length of time needed depends on the snow depth and atmospheric conditions. Generally, a system that maintains a clear pavement by melting 1 in. of snow an hour as it falls requires, under the same conditions, 1 h to clear 1 in. of accumulated snow.

To ensure maximum efficiency, fixtures should be cleaned at least once a year, preferably at the beginning of the winter season. Other maintenance is minimal.

Snow Melting in Gutters and Downspouts

Both MI cable and insulated wire are used to prevent heavy snow and ice accumulation on roof overhangs and to prevent ice dams from forming in gutters and downspouts (Lawrie 1966). Figure 7 shows a typical insulated wire layout to protect a roof edge and downspout. Wire or MI cable for this purpose is generally rated approximately 6 to 16 W/ft and about 2.5 ft of wire is installed per inear foot of roof edge. One foot heated wire length per linear foot of gutter or downspout is usually adequate.

If the roof edge or gutters (or both) are heated, downspouts must also be heated to carry away water from melted snow and ice. A heated length of wire is dropped inside the downspout to the bottom, even if it is underground, using weights, if necessary.

Lead wires should be spliced or plugged into the main power line in a waterproof junction box, and a ground wire should be installed from a downspout or gutter to a good electrical ground.

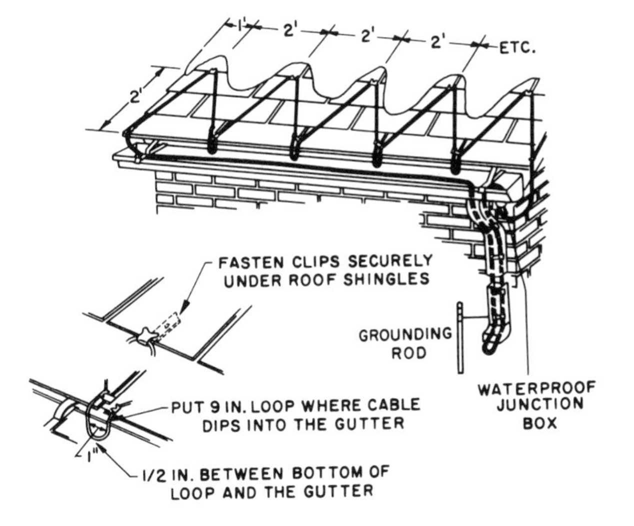

Fig. 7 Typical Insulated Wire Layout to Protect Roof Edge and Downspout

Manual switch control is generally used, although a protective thermostat, sensing outdoor temperature, should be used to prevent system operation above 40 °F.

REFERENCES

Adlam, T.N. 1950. *Snow melting*. The Industrial Press, New York; University Microfilms, Ann Arbor, MI.

Chapman, W.P. 1952. Design of snow melting systems. *Heating and Ventilating* (April):95; and (November):88.

Chapman, W.P. 1955. Are thermal stresses a problem in snow melting systems? *Heating, Piping and Air Conditioning* (June):104; and (August):92.

Chapman, W.P. 1957. Calculating the heat requirements of a snow melting system. *Air Conditioning, Heating and Ventilating* (September through August).

Chapman, W.P. and S. Katunich. 1956. Heat requirements of snow melting systems. ASHRAE *Transactions* 62:359.

Frier, J.P. 1964. Design requirements for infrared snow melting systems. *Illuminating Engineering* (October):686. Also discussion, December.

Frier, J.P. 1965. Snow melting with infrared lamps. *Plant Engineering* (October):150.

Gordon, P.B. 1950. Antifreeze protection for snow melting systems. *Heating, Piping and Air Conditioning Contractors National Association Official Bulletin* (February):21.

Hydronics Institute. 1968. Snow melting calculation and installation guide. Berkeley Heights, NJ.

Lawrie, R.J. 1966. Electric snow melting systems. *Electrical Construction and Maintenance* (March):110.

NACE. 1978. Basic corrosion course text (October). National Association of Corrosion Engineers, Houston, TX.

Witsken, C.H. 1965. Snow melting with electric wire. *Plant Engineering.* (September):129.

CHAPTER 46

EVAPORATIVE AIR COOLING

EVAPORATIVE cooling, using current technology and available equipment, can be an energy-efficient, environmentally benign, and cost-effective means of cooling. Applications are found for comfort cooling in commercial and institutional buildings, in addition to the traditional industrial applications for improvement of worker comfort in mills, foundries, power plants, and other hot operations. Several types of apparatus cool by evaporating water directly in the airstream. These include (1) evaporative coolers, (2) spray-filled and wetted-surface air washers, (3) sprayed coil units, and (4) humidifiers. Interest has increased in indirect evaporative cooling equipment that combines the evaporative cooling effect in a secondary airstream with heat exchange to produce cooling without adding moisture to the primary airstream.

In the past, indirect evaporative cooling equipment was considered too expensive when compared with assembly line refrigerated equipment, but energy conservation, increasing energy costs, concern for indoor air quality, and environmental concerns regarding chlorofluorocarbons have revived interest in both indirect and direct evaporative cooling systems.

Evaporative cooling reduces the dry-bulb temperature and provides a better environment for human occupancy and farm livestock. Evaporative cooling is also used to improve products grown or manufactured by controlling dry-bulb temperatures and/or relative humidity levels.

When temperature or humidity must be controlled within narrow limits, mechanical refrigeration can be combined with evaporative cooling in stages; it can also be used as a backup system. Evaporative cooling equipment, including unitary equipment and air washers, is covered in Chapter 4 of the 1988 ASHRAE *Handbook—Equipment.*

GENERAL APPLICATIONS

Cooling

Two common applications of evaporative cooling are (1) to improve the environment for people, animals, or processes, without attempting to control ambient temperature or humidity (spot cooling); and (2) to improve ambient conditions in a space (area cooling).

In a hot environment, where ambient heat control is difficult or impractical, cooling is accomplished by passing below skin-temperature air over the body. Evaporative coolers are suited to this purpose. The performance of evaporative cooling is directly related to climatic conditions. The entering wet-bulb temperature governs the final dry-bulb temperature of the air discharged from a direct evaporative cooler. The capability of the direct evaporative cooler is determined by how much the dry-bulb temperature exceeds the wet-bulb temperature. The performance of indirect evaporative coolers is also limited by the wet-bulb temperature of the secondary airstream.

Indirect applications using room exhaust as secondary air or those that incorporate precooled air in the secondary airstream may produce leaving dry-bulb temperatures approaching the wet-bulb temperature of the secondary airstream.

The direct evaporative cooling process is an adiabatic exchange of heat. For water to evaporate, heat must be added. The heat is supplied by the air into which water is evaporated. The dry-bulb temperature is lowered, and sensible cooling results. The amount of heat removed from the air equals the amount of heat absorbed by the water evaporated as heat of vaporization. If water is recirculated in the evaporative cooling apparatus, the water temperature in the reservoir will approach the wet-bulb temperature of the air entering the process. By definition, an adiabatic process is one during which no heat is added or extracted from the system. The initial and final conditions of an adiabatic process fall on a line of constant total heat (enthalpy), which nearly coincides with a line of constant wet-bulb temperatures.

The maximum reduction in dry-bulb temperature is the difference between the entering air dry- and wet-bulb temperatures. If the air is cooled to the wet-bulb temperature, it becomes saturated and the process is 100% effective. System effectiveness is the depression of the dry-bulb temperature of the air leaving the apparatus divided by the difference between the dry- and wet-bulb temperatures of the entering air. Evaporative cooling is less than 100% effective, although systems may be 85 to 90% or even more effective.

When a direct evaporative cooling unit cannot provide the desired conditions, several alternatives can satisfy application requirements and still be energy effective and economical to operate. The recirculating water supplying the evaporative cooling unit can be chilled by mechanical refrigeration to provide lower leaving wet-and dry-bulb temperatures and lower humidity. This arrangement reduces operating costs by as much as 25 to 40%, compared to the cost of using mechanical refrigeration only. Indirect evaporative precooling applied as a first stage, upstream from a second direct evaporative stage, makes it possible to reduce both the entering dry- and wet-bulb temperatures before the air enters the direct evaporative cooling unit. Indirect evaporative cooling systems may save as much as 60 to 75% or more of the total cost of operating a mechanical refrigeration system to produce the same cooling effect. Systems may combine indirect evaporative cooling, direct evaporative cooling, and mechanical refrigeration, or any two of these processes.

The psychrometric chart in Figure 1 illustrates what happens when air is passed through a direct evaporative cooling unit. In the example, assume an entering condition of 95 °F db and 75 °F wb. The initial difference is 20 °F ($95 - 75 = 20$). If the effectiveness

The preparation of this chapter is assigned to TC 5.7, Evaporative Cooling.

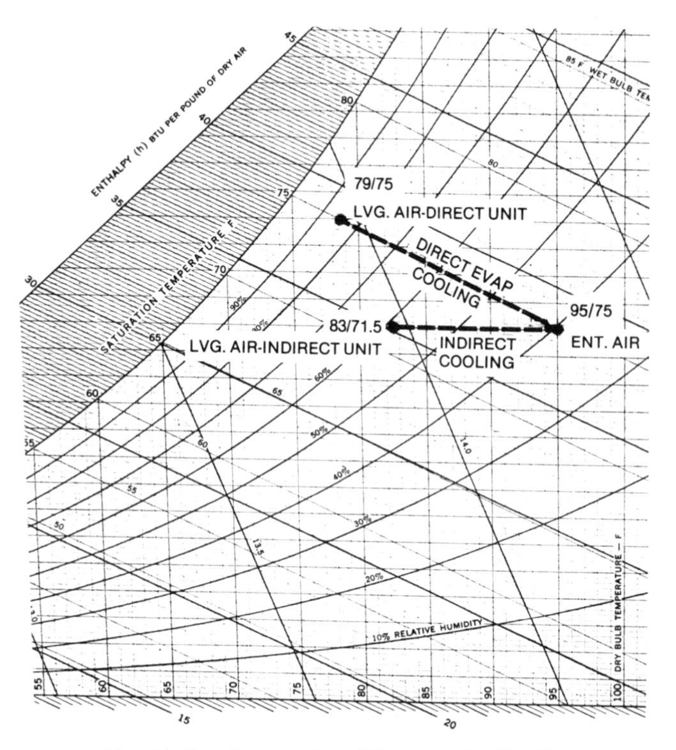

Fig. 1 Psychrometrics of Evaporative Cooling

is 80%, the depression is 16°F db or 0.80 × 20. The dry-bulb temperature leaving the evaporative cooler is 79°F (95 − 16 = 79). In the adiabatic evaporative cooler, only a portion of the water recirculated is assumed to evaporate and the water supply is recirculated. The recirculated water will reach an equilibrium temperature that is approximately the same as the wet-bulb temperature of the entering air.

The performance of an indirect evaporative cooling system can be shown on a psychrometric chart also. Many manufacturers of indirect evaporative cooling equipment use a similar definition of effectiveness as is used for a direct evaporative cooler. The term performance factor (PF) is also used. In indirect evaporative cooling, the cooling process in the primary airstream follows a line of constant moisture content (constant dew point). Performance factor (or effectiveness) is the dry-bulb depression in the primary airstream divided by the difference between the entering dry-bulb temperature of the primary airstream minus the entering wet-bulb temperature of the secondary air. Depending on the heat exchanger design and relative air quantities of primary and secondary air, effectiveness ratings may be as high as 85%.

Using the previous example, assuming an effectiveness of 60%, and assuming both primary air and secondary air enter the apparatus at the outdoor condition of 95°F db and 75°F wb, the dry-bulb depression is 0.60 (95 − 75) = 12°F. The dry-bulb temperature leaving the indirect evaporative cooling process is (95 − 12) = 83°F. Since the process cools without adding moisture, the wet-bulb temperature is also reduced. Plotting the psychrometric chart shows the final wet-bulb temperature to be 71.5°F. Since both the wet- and the dry-bulb temperatures in the indirect evaporative cooling process are reduced, indirect evaporative cooling can be used as a substitute for a portion of the refrigeration load in many applications.

Humidification

Air can be humidified with an evaporative cooler in three ways: (1) using recirculated water without prior treatment of the air, (2) preheating the air and treating it with recirculated water, and (3) using heated water. In any evaporative cooler installation, the air should not enter with a wet-bulb temperature of less than 39°F; otherwise, the water may freeze.

Recirculated Spray Water

Except for both the small amount of outside energy added by the recirculating pump in the form of shaft work and the small amount of heat leakage into the apparatus from outside (including the pump and its connecting piping), evaporative cooling is strictly adiabatic. Evaporation occurs from the recirculated liquid. Its temperature should adjust to the thermodynamic wet-bulb temperature of the entering air.

The whole airstream is not brought to complete saturation, but its state point should move along a line of constant thermodynamic wet-bulb temperature. The extent to which the leaving air temperature approaches the thermodynamic wet-bulb temperature of the entering air is expressed by a saturation effectiveness ratio. In humidifiers, this ratio is often called the humidifying effectiveness. Representative saturation, or humidifying effectiveness, of a spray-type air washer with various spray arrangements is listed in Table 1.

Table 1 Effectiveness of Spray Arrangements in a Spray-Type Air Washer

Bank	Arrangement	Length, ft	Effectiveness, %
1	downstream	4	50 to 60
1	downstream	6	60 to 75
1	upstream	6	65 to 80
2	downstream	8 to 10	80 to 90
2	opposing	8 to 10	85 to 95
2	upstream	8 to 10	90 to 98

The degree of saturation depends on the extent of the contact between air and water. Other conditions being equal, a low-velocity airflow is conducive to higher humidifying effectiveness.

Preheating Air

Preheating the air increases both the dry- and wet-bulb temperatures and lowers the relative humidity, but it does not alter the humidity ratio (mass ratio, water vapor to dry air). At a higher wet-bulb temperature, but with the same humidity ratio, more water can be absorbed per unit mass of dry air in passing through the evaporative cooler (if the humidifying effectiveness of the evaporative cooler is not adversely affected by operation at the higher wet-bulb temperature). The analysis of the process that occurs in the evaporative cooler is the same as that for recirculated water. The final preferred conditions are achieved by adjusting the amount of preheating to give the required wet-bulb temperature at the entrance to the evaporative cooler.

Heated Recirculated Water

Even if heat is added to the recirculated water, the mixing in the evaporative cooler may still be regarded as adiabatic. The state point of the mixture should move toward the specific enthalpy of the heated water. By elevating the water temperature, it is possible to raise the air temperature (both dry and wet bulb) above the dry-bulb temperature of the entering air.

The relative humidity of the leaving air may be controlled by (1) bypassing some of the air around the evaporative cooler and remixing the two airstreams downstream, or (2) automatically reducing the number of operating spray nozzles or sections of media that are being wetted by operating valves in the different recycle header branches.

Dehumidification and Cooling

Evaporative coolers are also used to cool and dehumidify air. Heat and moisture removed from the air raise the water temperature. If the entering water temperature is below the entering wet-bulb temperature, both the dry- and wet-bulb temperatures are lowered. Dehumidification results if the leaving water temperature is below the entering dew-point temperature. Moreover, the final water temperature is determined by the sensible and latent heat pickup and the amount of water circulated. However, this final temperature must not exceed the final required dew point, with one or two degrees below dew point being common.

The air leaving an evaporative cooler being used as a dehumidifier is substantially saturated. Usually, the spread between dry- and wet-bulb temperatures is less than 1 °F. The spread between leaving air and leaving water depends on the difference between entering dry- and wet-bulb temperatures and on certain design features, such as the length and height of a spray chamber, the cross-sectional area and depth of the media being used, air velocity, quantity of water, and the spray pattern. The rise in water temperature is usually between 6 and 12 °F, although higher increases have been used successfully. The lower temperature increases are usually selected when the water is chilled by mechanical refrigeration because of possible higher refrigerant temperatures. It is often desirable to make an economic analysis of the effect of higher refrigerant temperature compared to the benefits of a greater increase in water temperature. For systems receiving water from a well or other source at an acceptable temperature, it may be desirable to design on the basis of a high temperature increase and a minimum water flow.

Air Cleaning

Evaporative coolers of all types perform some air cleaning. Drip-type coolers are the least effective, removing particulates down to about 10 μm in size. Air washers can be highly effective air cleaners, provided they are equipped with high-pressure nozzles and are followed by well-designed, efficient entrainment separators.

The dust removal efficiency of evaporative coolers depends largely on the size, density, wettability, and solubility of the dust particles. Larger, more wettable particles are the easiest to remove. Separation is largely a result of the impingement of particles on the wetted surface of the eliminator plates or on the surface of the media. Since the force of impact increases with the size of the solid, the impact (together with the adhesive quality of the wetted surface) determines the cooler's usefulness as a dust remover. The standard low-pressure spray is relatively ineffective in removing most atmospheric dusts.

Evaporative coolers are of little use in removing soot particles because of the absence of an adhesive effect from a greasy surface. They are also ineffective in removing smoke, because the inertia of the small particles (less than 1 μm) does not allow them to impinge and be held on the media or the wet plates. Instead, the particles follow the air path between the media surfaces or the plates since they are unable to pierce the water film covering the plates.

Control of Gaseous Contaminants. When used in a makeup air system comprised of a mixture of outside air and recirculated air, evaporative coolers function as scrubbers and are effective in reducing the gaseous contaminants found in urban atmospheres.

Makeup Air

In most industrial plants and in all confined spaces (including office buildings), makeup air is required (for safe, effective operation) to replace the large volumes of air that must be exhausted to provide the required conditions for personnel comfort, safety, process operations, and to maintain high indoor air quality (IAQ). If makeup air is provided consistent with good air distribution practice, more effective cooling can be provided in the summer, and more efficient and effective heating will result in the winter. Windows or other inlets that cannot be used in stormy weather should be discouraged. Specifically, the most important needs for makeup air can be summarized as follows:

1. To secure the required exhaust for hoods, combustion, process, and building heat removal.
2. To eliminate cross drafts by proper arrangement of supply air and/or prevention of infiltration (through doors, windows, and similar openings), since these may make hoods unsafe or ineffective, defeat environmental control, bring in or stir up dust, or adversely affect processes by cooling or disturbance.
3. To obtain air from the cleanest source. Makeup (or supply) air can be filtered; infiltration air cannot.
4. To permit the control of building pressure and of airflow from space to space. Such control is necessary for the following reasons:
   a) To avoid positive or negative pressures which make it difficult or unsafe to open doors, and to avoid the conditions mentioned in Items 1, 2, and 3.
   b) To permit confinement of contaminants and positive control of temperature, humidity, and air movement. Heat is a contaminant, and unless it is properly confined and removed, it will spread to the work zones to be entrained with the supply air, making it impossible to provide a reasonable working environment. The buildup of a heat pocket results in a lowering of the building neutral zone, causing exfiltration under curtain walls or through openings to other spaces. For further information, refer to Chapter 23 of the 1989 ASHRAE *Handbook—Fundamentals.*
   c) To permit heat recovery and conservation.

Wherever possible, building pressure should be controlled by flow from one space to another rather than by an appreciable pressure differential; otherwise, large volumes of makeup air will be required. Note that even 0.01 in. of water creates a velocity of 400 fpm. Where air locks and adequate tight construction are provided to control toxic contaminants, higher pressure differentials of 0.02 to 0.06 in. of water may be required.

Ventilation rates for human occupancy should be determined by ASHRAE *Standard* 62 and applicable codes. Chapter 23 of the 1989 ASHRAE *Handbook—Fundamentals* also has further information.

INDIRECT EVAPORATIVE PRECOOLING

Outdoor Air Systems

Since there is no increase in absolute humidity in the primary airstream, indirect evaporative cooling is well suited to precooling the air entering a refrigerated coil. The cooling effect provided by the upstream indirect evaporative equipment is a sensible cooling load reduction to the downstream refrigerated coil and compression apparatus. This reduces the size of the required refrigeration system as well as energy and operating costs. By contrast, direct evaporative cooling equipment, when used with a refrigerated coil, will exchange latent heat for sensible heat, increasing the latent load on the coil in proportion to the sensible cooling achieved. The enthalpy of the air entering the coil is not changed.

The kilowatt per ton of cooling effect is substantially lower with indirect evaporative cooling than with conventional refrigerated equipment. The indirect equipment selected must result in only a minimal addition of static pressure loss in the primary air system. The added static pressure loss increases the primary air fan motor power, and the total effect on the system must be considered, even when continuous cooling is not required. Static pressure loss for a nominal selection may be as low as 0.2 in. of water, which represents a minimal addition to the supply fan power required. Furthermore, the equipment may also require additional

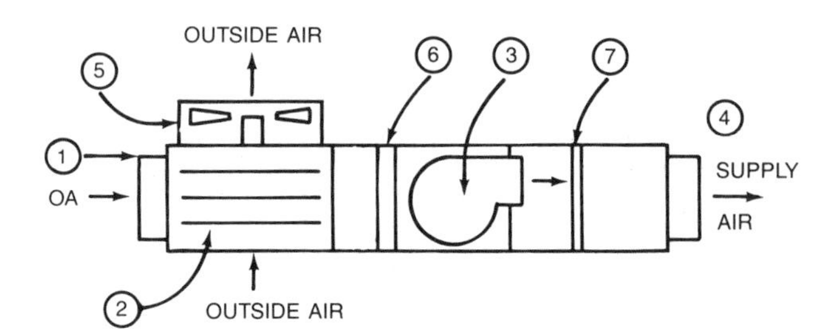

1 Screened air intake
2 Indirect evaporative cooler
3 Dry side, air moving device
4 Duct to convey air to conditioned space
5 Wet side, air moving device; may be part of building exhaust
6 Chilled water, direct expansion, or direct evaporative cooling coil section
7 Heating section, hot water coil, etc. (optional)

Fig. 2 Indirect Evaporative Cooling Configuration

energy for water pumping and for moving secondary air across the evaporative surfaces.

The cooling configuration is shown in Figure 2. The primary air side of the indirect unit is positioned at the intake to the refrigerated cooling coil. The secondary air to the unit can come from outdoor ambient air or from room exhaust air. Exhaust air from the space may have a lower wet-bulb temperature than outdoor ambient, depending on climate, time of year, and space latent load. Latent cooling may be possible in the primary airstream using room exhaust air as secondary air. This may occur if the dew-point temperature of the primary air is above the exhaust (secondary air) wet-bulb temperature. If this is possible, provision to drain the water condensed from the primary airstream may be necessary.

In many areas, an indirect precooler can satisfy more than one-half of the annual cooling load. For example, Supple (1982) showed that 30% of the annual cooling load for Chicago can be accomplished by indirect evaporative precooling. Indirect evaporative precooling systems using exhaust air as the secondary air may be equally effective in warm, humid climates as they are in drier areas.

Mixed Air Systems

Indirect evaporative cooling can also save energy in systems that use a mixture of return and outside air. The apparatus configuration is similar to that in Figure 2, except that a mixing section with outside- and return-air dampers is added upstream of the indirect precooling section. Outdoor air would be used as secondary air for the indirect evaporative cooler.

A typical indirect evaporative precooling stage can reduce the dry-bulb temperature by as much as 60 to 80% of the difference between the entering dry-bulb temperature and the wet-bulb temperature of the secondary air. When the dry-bulb temperature of the mixed air is more than a few degrees above the wet-bulb temperature of the secondary airstream, indirect evaporative precooling of the mixed airstream may reduce the amount of refrigerated cooling required.

The precooling contribution depends on the differential between mixed air dry-bulb temperature and secondary air wet-bulb temperature. As mixture temperature increases, or as secondary air wet-bulb decreases, the precooling contribution becomes more significant.

In variable air volume (VAV) systems, a decrease in supply air volume (during periods of reduced load) results in lower air velocity through the evaporative cooler; this increases equipment effectiveness. Lower static pressure loss reduces the energy consumed by the supply fan motor.

BOOSTER REFRIGERATION

Staged evaporative systems can totally cool office buildings, schools, gymnasiums, department stores, restaurants, factory space, and other buildings. These systems can control room dry-bulb temperature and relative humidity, even though one stage is a direct evaporative cooling stage. In many cases, booster refrigeration is not required. Supple (1982) showed that even in higher humidity areas with a 1% mean wet-bulb design temperature of 75 °F, 42% of the annual cooling load can be satisfied by two-stage evaporative cooling. Refrigerated cooling need supply only 58% of the load.

Figure 3 shows indirect/direct two-stage system performance for 16 cities in the United States. Performance is based on 60% effectiveness of the indirect stage and 90% for the direct stage. Supply air temperatures (leaving the direct stage) at the 1% design dry-bulb/mean wet-bulb condition range from 52.7 to 71 °F.

City	Outside Air Design db/wb, °F	Indirect/Direct Performance (Supply Air = 0.733 W/cfm) Indirect db/wb, °F	Supply Air db, °F	EER	EUC, %
Los Angeles, CA	93/70	79.2/65.8	67.1	26.4	30
San Francisco, CA	82/64	71.2/60.2	61.3	34.9	23
Seattle, WA	85/68	74.8/64.5	68.6	24.2	33
Albuquerque, NM	96/61	75.0/52.9	55.1	44.0	18
Denver, CO	93/59	72.6/50.5	52.7	47.6	17
Salt Lake City, UT	97/62	76.0/54.0	56.2	42.4	19
Phoenix,AZ	109/71	86.2/64.0	66.2	27.7	29
El Paso, TX	100/64	78.4/57.2	59.3	37.8	21
Santa Rosa, CA	99/68	80.4/61.6	63.5	31.7	25
Spokane, WA	93/64	75.6/57.8	59.6	37.4	21
Boise, ID	96/65	77.4/58.5	60.4	36.2	22
Billings, MT	94/64	76.0/57.5	59.4	37.7	21
Portland, OR	90/68	76.8/63.5	64.8	29.7	27
Sacramento, CA	101/70	82.4/64.0	64.8	28.3	28
Fresno, CA	102/70	82.8/63.8	65.7	28.4	28
Austin, TX	100/74	84.4/69.5	71.0	20.6	39

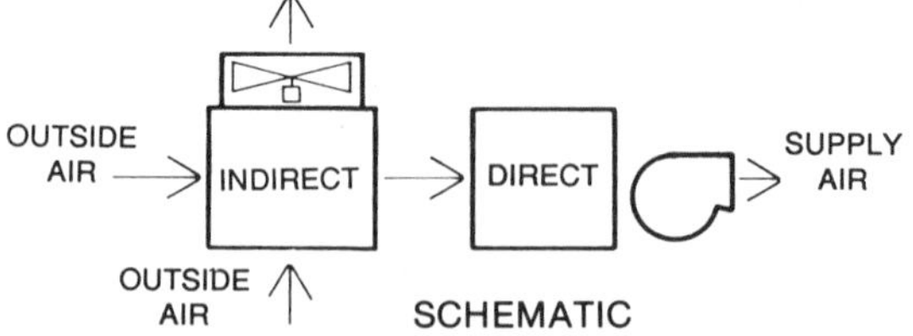

Outdoor air design condition: 1% dry bulb/mean coincident wet bulb.

EER = (Energy Efficiency Ratio) Btu/h cooling output per watt of electrical input. Comparison base to conventional system with 60 °F supply air and 25 °F temperature drop.

EUC = Energy Use Comparison to a conventional refrigeration system with EER = 8.

I/D efficiency: Indirect = 60% or 0.6 (db − wb); Direct = 90% or 0.9 (db − wb)

Note: Sea level psychrometric chart used. 5000-ft elevation will increase supply air temperature 3 to 4%.

Fig. 3 Indirect/Direct Two-Stage System Performance

Energy use ranges from 17 to 39%, compared to conventional refrigerated equipment.

Booster mechanical refrigeration allows the designer to provide indoor design comfort conditions regardless of the outdoor wet-bulb temperature without having to size the mechanical refrigeration equipment for the total cooling load. If the indoor design condition becomes questionable, the quantity of moisture introduced into the airstream must be limited in order to control room humidity. Where the upper relative humidity design level is critical, a life cycle cost analysis would favor a system design composed of an indirect cooling stage and a mechanical refrigeration stage.

RESIDENCE OR COMMERCIAL COOLING

In dry climates, evaporative cooling is effective, with lower air velocities than those required in humid climates. This makes it suitable for use in applications where low air velocity is desirable. Packaged evaporative coolers are commonly used for residential and commercial application. Cooler capacity requirements may be determined from standard heat gain calculation (see Chapter 26 of the 1989 ASHRAE *Handbook—Fundamentals*).

Detailed calculation of heat load, however, is usually not economically justified. Instead, one of several estimates, with proper consideration, will give satisfactory results.

One method of calculating heat load is to divide the difference between dry-bulb design temperature and coincident wet-bulb temperature by 10, and let this be the number of minutes needed for each air change. This or any other arbitrary rules for equating cooling capacity with airflow depends on evaporative cooler effectiveness of 70 to 80%. Obviously, it must be modified if unusual conditions, such as large unshaded glass areas, uninsulated roof exposure, or high internal heat gain, exist.

Such empirical methods make no attempt to predict air temperature at specific points in the system; they merely establish an air quantity for use in sizing equipment.

Example 1. An indirect evaporative cooling system is to be installed in a 50 by 80 ft one-story office building with a 10-ft ceiling and a flat roof. Outdoor design conditions are assumed to be 95 °F dry bulb and 65 °F wet bulb. Heat gains to be used in the cooling system design are:

	Gains, Btu/h
All walls, roof, and doors	78,500
Glass area	5,960
Occupants (sensible load)	17,000
Lighting	62,700
Total sensible heat load	164,160
Total latent load (occupants)	21,250
Total heat load	185,410

Find the required air quantity, the temperature and humidity ratio of the air leaving the cooler (entering the office), and the temperature and humidity ratio of the air leaving the office.

Solution: A temperature rise of 10 °F in the cooling air is assumed. The air volume required to be supplied by the evaporative cooler may be found in Equation (1):

$$Q_{ra} = q_s/1.08\,(t_1 - t_s) \tag{1}$$

$$Q_{ra} = 164{,}160/(1.08 \times 10) = 15{,}200 \text{ cfm}$$

where

Q_{ra} = required air quantity through equipment, cfm
q_s = instantaneous sensible heat load, Btu/h
t_1 = indoor air dry-bulb temperature, °F
t_s = room supply air dry-bulb temperature, °F

This air volume represents a 2.6-min air change for a building of this size. The evaporative air cooler is assumed to have a saturation effectiveness of 80%. This is the ratio of the reduction of the dry-bulb temperature to the wet-bulb depression of the entering air. The dry-bulb temperature of the air leaving the evaporative cooler is found from Equation (2):

$$t_2 = t_1 - e_h(t_1 - t')/100 \tag{2}$$
$$= 95 - 80\,(95 - 65)/100 = 71\,°\text{F}$$

where

t_2 = dry-bulb temperature of leaving air, °F
t_1 = dry-bulb temperature of entering air, °F
e_h = humidifying or saturating effectiveness, %
t' = thermodynamic wet-bulb temperature of entering air, °F

The humidity ratio of the cooler discharge air, W_2, is found (from the psychrometric chart) to be 0.01185 lb/lb dry air. The humidity ratio of the air leaving the space being cooled, W_3, is found from Equation (3):

$$W_3 = q_e/(4840\,Q_{ra}) + W_2 \tag{3}$$
$$= 21{,}250/(4840 \times 15{,}200) + 0.01185$$
$$= 0.01214 \text{ lb/lb dry air}$$

where

q_e = latent heat load, Btu/h

The remaining values of wet-bulb temperature and relative humidity for the problem may be found from the psychrometric chart. Figure 4 illustrates the various relationships of outdoor air, supply air to the space, and discharge air.

In the skeleton psychrometric chart in Figure 4, the problem is solved graphically.

The wet-bulb depression method to estimate airflow gives the following result:

$$\text{WBD}/10 = (95 - 65)/10 = 3.0 \text{ min per air change}$$

$$Q_{ra} = \text{Volume/Air change rate} = 80 \times 50 \times 10/3.0 = 13{,}300 \text{ cfm}$$

While not exactly alike, these two air volume calculations are close enough to select cooler equipment of the same size.

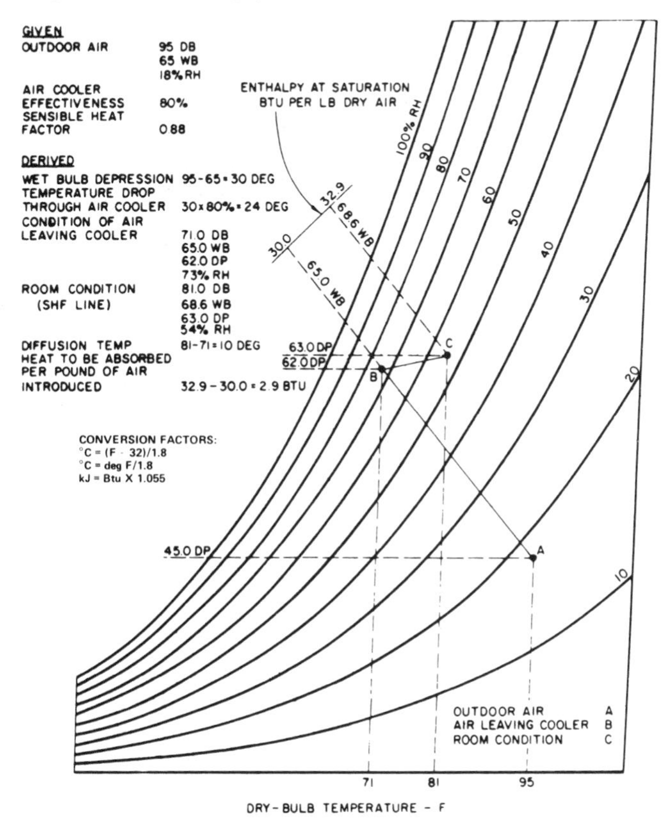

Fig. 4 Solution of Example 1

EXHAUST REQUIRED

If air is not exhausted freely, the static pressure buildup reduces airflow through the coolers. The result is a marked increase in the moisture and the heat absorbed per unit mass of air leaving the cooler, and a reduction in the room air velocity. These effects combine to produce almost certain discomfort. A good residential application includes attic exhausters and thermostatic operation.

For normal commercial systems with no internal heat emission, the exhaust should nearly equal the quantity of evaporatively cooled air introduced. Where heat emission occurs, additional exhaust capacity must be added. For most commercial and industrial installations, both power and gravity exhaust should be used. The power exhaust should be controlled so that it increases as the outdoor relative humidity increases. This reduces the buildup of excessive humidity in the space being cooled.

Direct System Energy Savings

Residences in arid or semiarid regions can use direct or indirect evaporative coolers. Direct, open systems are usually sized to meet the full sensible load using once-through airflows (no recirculation of return air). In such cases, the monthly or seasonal energy savings, compared to conventional vapor compression cooling, can be estimated using the bin method.

Indirect System Energy Savings

Commercial building sensible cooling loads can be met by direct or indirect evaporative cooling. Indirect cooling, or a combination of direct and indirect cooling, is customarily used with the central fan system because of humidity control requirements in commercial buildings. With indirect evaporative cooling, the wet-side (secondary) air can consist of full outside air, full return air, or any combination of the two. The monthly or seasonal energy savings resulting from, or the cooling produced by, an indirect evaporative system can be calculated using the bin method.

TWO-STAGE COOLING

Two-stage systems for commercial applications can extend the range of atmospheric conditions under which comfort requirements can be met as well as increase the energy savings. (For the same design conditions, two-stage cooling provides lower cool air temperatures, thereby reducing the required airflow rate.)

High internal load areas are similar to industrial environment conditions and are discussed in the following section.

INDUSTRIAL APPLICATIONS

In a factory with a large internal heat load, it is difficult to approach outdoor conditions during the summer without using an extremely large quantity of outdoor air. Using a more reasonable quantity of outdoor air, evaporative cooling can alleviate this heat problem and contribute to worker efficiency with improved employee morale. Without it, increased absenteeism, high labor turnover, and danger to health and safety can be expected during the summer months. Production declines in uncooled plants may range from 25 to 40% of normal on hot days.

An examination of the effective temperature chart reveals that a dry-bulb temperature reduction due to the evaporation of water always results in a lower effective temperature, regardless of the relative humidity level. Figure 5 shows an effective temperature chart for air velocities ranging from 20 to 700 fpm. Although the maximum velocity shown on the chart is 700 fpm, workers exposed to high heat-producing operations often request air

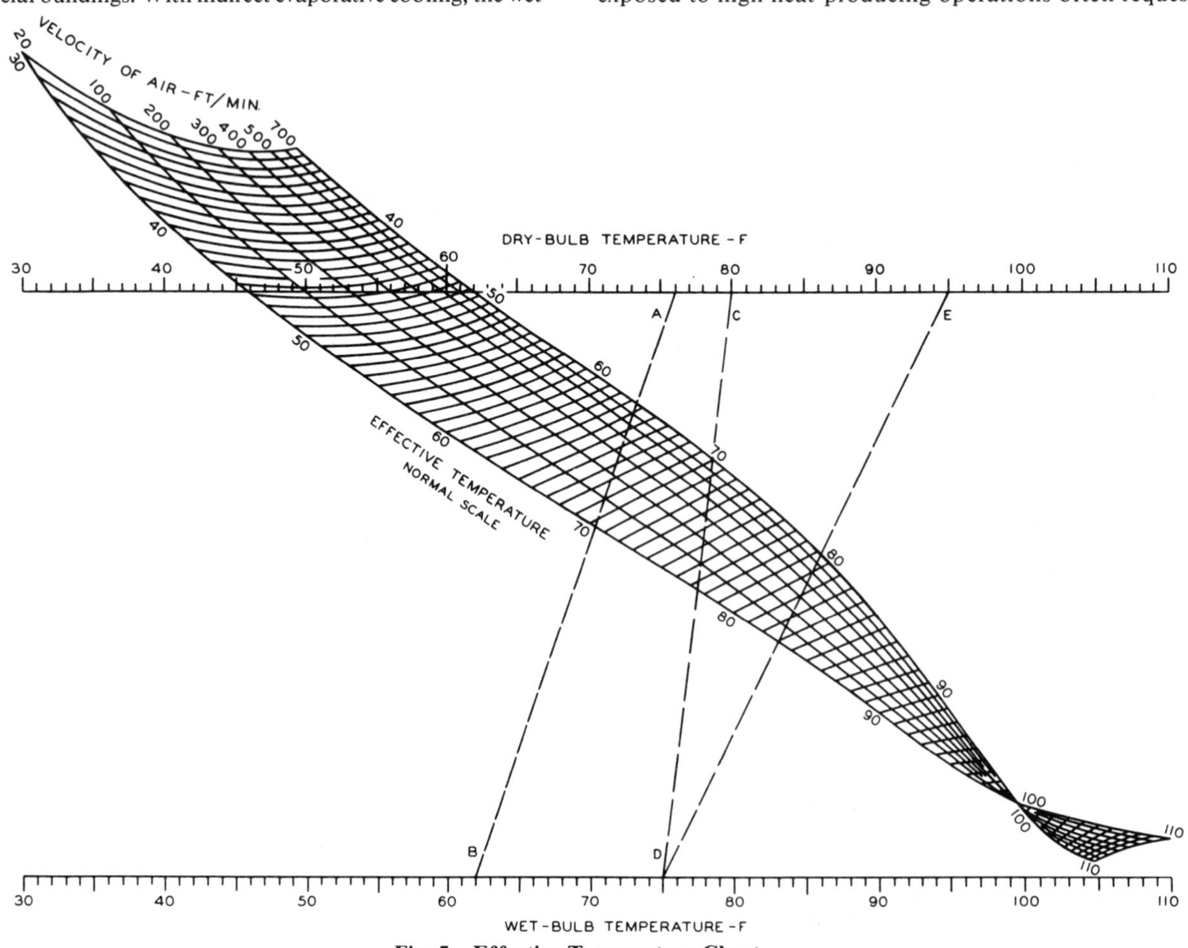

Fig. 5 Effective Temperature Chart

movement up to 4000 fpm. Since the working range of the chart is approximately midway between the vertical dry- and wet-bulb scales, each of these temperatures is equally responsible for the effective temperature. A reduction in either one will have the effect of decreasing the effective temperature by about one-half of the reduction. This is graphically illustrated by lines ED and CD on the chart.

Conditions of 95 °F dry-bulb and 75 °F wet-bulb were chosen as the original state, because this set of conditions is usually considered the summer design criteria in most areas. Reducing the temperature 15 °F by evaporating water adiabatically provides an effective temperature reduction of 5.5 °F for air moving at 20 fpm, and 9.5 °F for air moving at 700 fpm—an average of 4 °F.

Furthermore, the reduction in dry-bulb temperature through water evaporation increases the effectiveness of the cooling power of moving air in this example by over 100%. On line ED, the effective temperature varies from 83 °F at 20 fpm to 79.5 °F at 700 fpm, whereas line CD indicates an effective temperature of 77.5 °F at 20 fpm and 70 °F at 700 fpm. In the former case, increasing the air velocity from 20 to 700 fpm resulted in only a 3.5 °F decrease in effective temperature. This contrasts a 7.5 °F decrease in effective temperature for the same range of air movement when the dry-bulb temperature was lowered by water evaporation. Considering these relationships between water evaporative air cooling and effective temperatures can provide relief cooling of factories almost regardless of geographical location.

Several methods can be used to evaluate the environmental improvement that may be accomplished by evaporative coolers. One method is shown in Figure 6 in which temperature is plotted on the vertical axis and time of day on the horizontal. Curve A is plotted from maximum dry-bulb temperature recordings at the hours indicated. Curve B is a plot of the corresponding wet-bulb temperatures.

Curve C depicts the effective temperature when air is moved over a person at 300 fpm with all temperature conditions as stated previously. Note its relationship to the suggested maximum effective temperature of 80 °F. While the maximum suggested effective temperature is exceeded, both the differential and the total hours are substantially reduced from the conditions in still air.

If the air is passed through an 80% efficient evaporative cooler before being projected over the employee at 300 fpm, the effective temperature would, in all cases, be below the suggested maximum as shown by Curve D.

Curve E shows the additional decrease in effective temperature when air velocities of 700 fpm are provided. Higher velocities lower the effective temperature proportionally.

In spite of the high wet-bulb temperatures, the in-plant environment can be maintained below the suggested upper limit of 80 °F effective temperature; this should allow for full production without undue stress on individuals. This assumes that the combination of air velocity, duct length, and insulation between the cooler and the duct outlet is such that heat transfer between air in the ducts and warmer underroof air in the plant is only slight.

Another method of determining the effect of using evaporative coolers plots wet- and dry-bulb temperatures of the ambient temperature on an ASHRAE Effective Temperature Chart, as in Figure 7 (Crow 1972). The dashed lines show the improvement to expect when using an 80% efficient evaporative cooler. In these examples, a satisfactory effective temperature is provided when the air velocity around the occupant is 600 fpm.

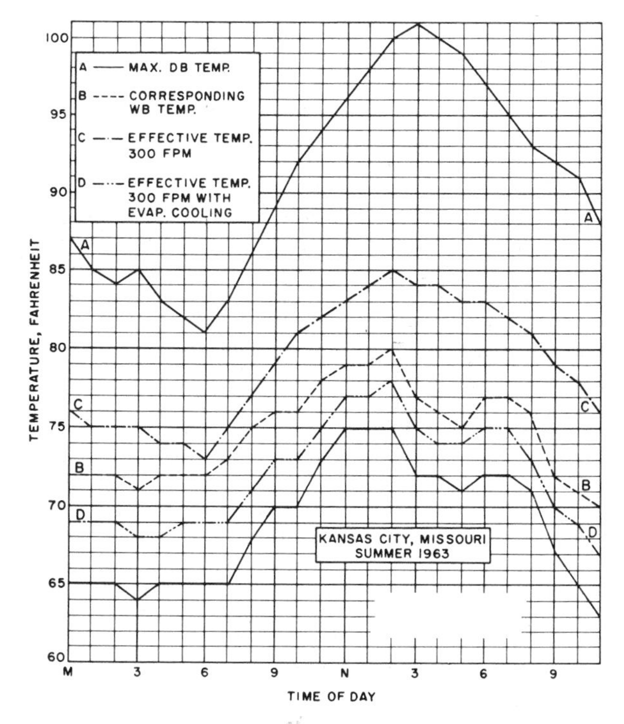

Fig. 6 Effective Temperature for Summer Day in Kansas City, Missouri
(Worst Case Basis)

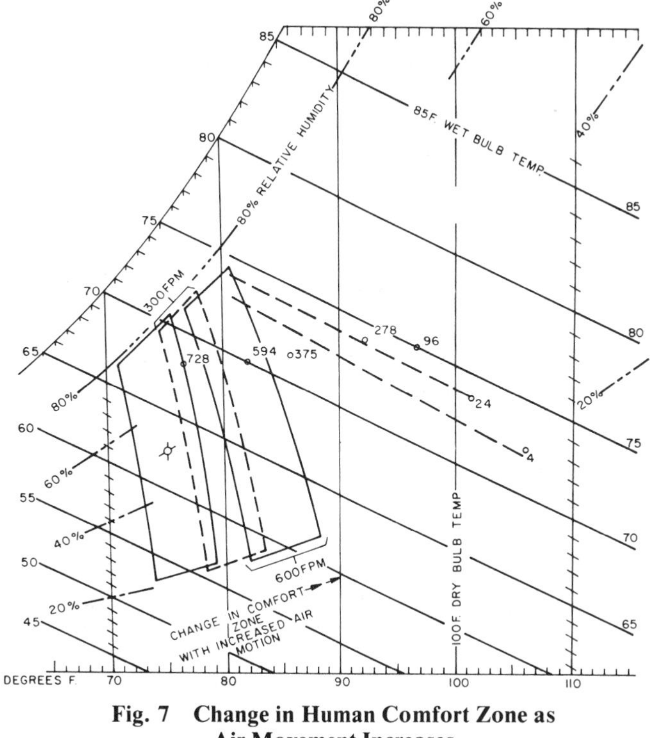

Fig. 7 Change in Human Comfort Zone as Air Movement Increases

Area Cooling

Evaporative cooling of industrial buildings may be accomplished on an area basis or by spot cooling. Evaporative coolers can be either automatically or manually controlled. The air-handling capacity of these systems can also be used to supply tempered fresh air during the fall, winter, and spring. Exhausting the spent air is accomplished by means of gravity and/or power roof ventilators. Ducts should be designed to distribute air in the lower 10 ft of the space to assure that air is supplied to where the workers are located.

Since cooling requirements vary from season to season (in fact, from day to day), adjustable discharge grilles should be provided.

If the incoming air temperature drops to a point where the cooling airstream becomes an annoying draft, the horizontal blades in the grille may be adjusted so that the air is discharged above the workers' heads and does not blow directly on them. In some cases, provision should be made for air volume control, either at the individual outlet or for the entire system. In some cases, it may be necessary to vary the exhaust volume accordingly. In a typical arrangement for area cooling, the evaporatively cooled air is brought in and down to the 10-ft level, where it is distributed from the duct in opposite directions.

Spot Cooling

While area cooling lends itself well to applications where personnel move about within a cooled space, spot cooling yields a more efficient use of the equipment when the personnel are relatively stationary. In these instances, however, cool air can be brought in at levels even lower than 10 ft; in some cases, it can even be introduced from ducts located under the floor. Selection of the height depends on the location of other equipment in the area. For best results, the throw should be kept to a minimum. Spot cooling is especially applicable in hot areas where cooling for individuals is needed, such as in chemical plants, pig and ingot casting, permanent mold, die casting shops, glass forming machines, and billet furnaces. Controls may be automatic or manual, with the fan often operating throughout the year.

The volume of cooling air supplied to each worker depends on the throw, amount of worker activity, and the degree of heat to be overcome. Usually, the air supply ranges from 2000 to 5000 cfm per worker with a target of 200 to 4000 fpm. If possible, air outlets should be no more than 4 to 10 ft from the workstations to avoid entrainment of warm air and to effectively blanket workers with cooler air. Provisions should be made so that workers can control the direction of discharge, because good hot weather air motion performance is too drastic for cool weather or even cool mornings. Volume controls should be considered for some installations to prevent overcooling of the building and to minimize time lost by excessive grille-blade adjustment.

These cooling systems are usually installed where there are elevated room temperatures; there are no climatic or geographical boundaries. When the dry-bulb temperature of air is below skin temperature, a body is cooled by convection rather than evaporation. An 80 to 85 °F airstream feels so good that its relative humidity is inconsequential.

Laundry

One of the most difficult or severest applications of evaporative air cooling is laundries, since heat is produced not only by the processing equipment, but by steam and water vapor as well. Figure 8 is a skeleton psychrometric chart which illustrates an evaporative cooling problem involving heat reduction brought about by airflow through a typical laundry. It indicates that a properly designed evaporative cooling system reduces the temperature in a laundry from 5 to 10 °F below outdoor temperature. With only fan ventilation, laundries usually exceed the outdoor temperature by at least 10 °F or more. Air distribution should be designed for a maximum throw of not more than 30 ft. A minimum circulated velocity of 100 to 200 fpm should prevail in the occupied space. Ducts can be located to discharge the air directly onto workers in the exceptionally hot areas, such as the pressing and ironing departments. For these outlets, there should be some means of manual control to direct the air where it is desired, with at least 500 to 1000 cfm at a target velocity of 600 to 900 fpm for each workstation.

Cooling of Large Motors

The rating of electrical generators and motors is generally based on a maximum ambient temperature of 104 °F. If this temperature is exceeded, excessive temperatures develop in the electrical windings unless the load on the motor or generator is reduced. If air is supplied to the windings from an evaporative cooler, this equipment may be safely operated without reducing the load, and in many cases an overload may be carried. Likewise, transformer capacity can be increased using evaporative cooling.

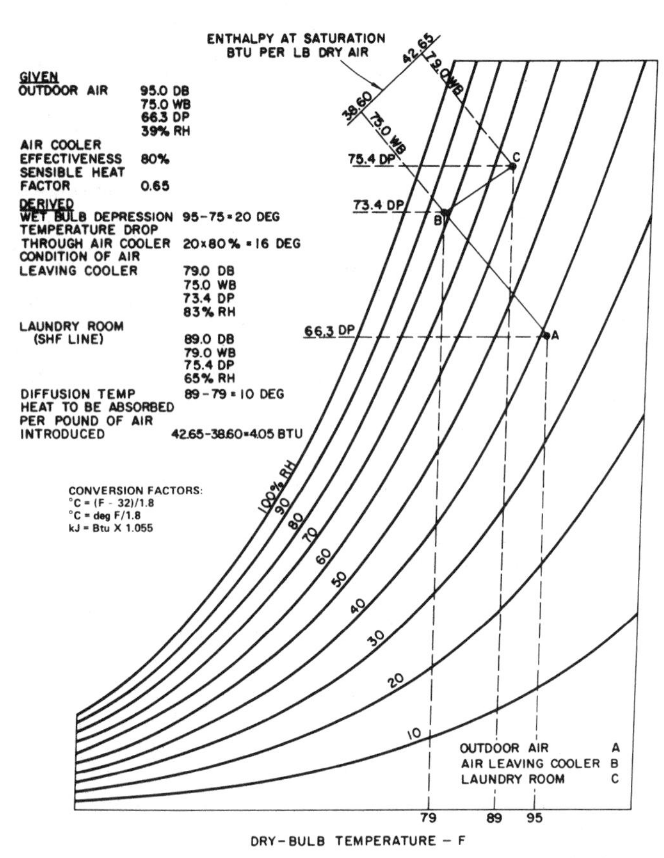

Fig. 8 Typical Evaporative Cooling Process for a Laundry

The heat emitted by high-capacity electrical equipment may also be sufficient to raise the ambient condition to an uncomfortable level. With mill drive motors, an additional problem is often encountered with the commutator. If the air used for motor ventilation is dry, the temperature rise through the motor results in a still lower relative humidity. Destruction of brush film with resultant brush and commutator wear as well as dusting occurs at low humidities. General design considerations require approximately 120 cfm ventilating air per kilowatt hour of losses with a temperature rise through the motor of 25 °F. Assuming 95 °F inlet air, the air leaving the motor would be 120 °F, which represents an average motor temperature of over 107 °F. This temperature is 3 °F higher than it should be for the normal 104 °F ambient. An evaporative cooling system with a 97% saturation effectiveness using the same quantity of air and in a 75 °F wet-bulb area would give an average motor temperature of 88 °F. This avoids the necessity of using special high-temperature insulation and improves the ability of the motor to absorb temporary overloads. By comparison, an air quantity of 185 cfm would be required if supplied by a cooler with 80% saturation efficiency. Figure 9 is a schematic of three basic arrangements for motor cooling systems.

The air from the cooler may be directed on the motor windings or into the room, which requires increased air volume to compensate for the building heat load. Operation of these evaporative cooling systems should be keyed to the motor operation to provide the following safeguards: (1) saturated or nearly saturated air should never be introduced into a motor until it has had time to

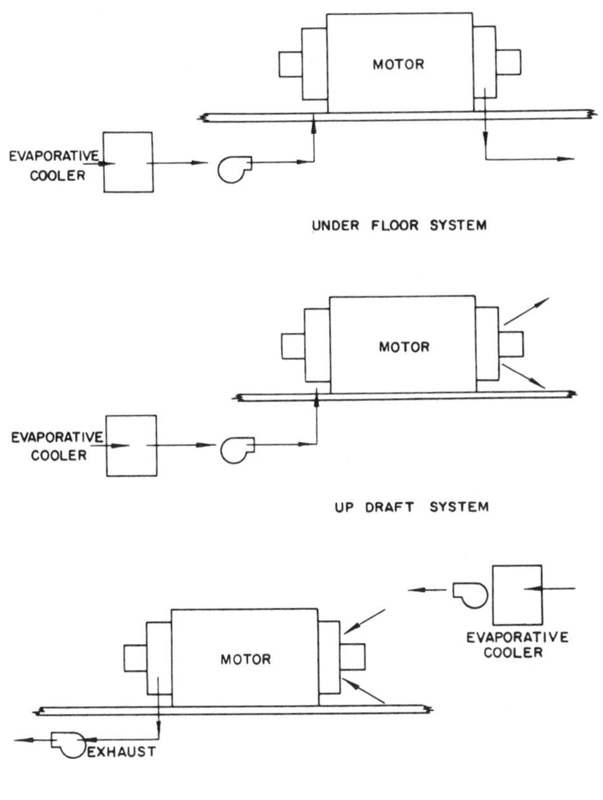

Fig. 9 Arrangements for Cooling Large Motors

warm up; and (2) if more than one motor is served by a single system, air circulation through idle motors should be prevented.

Gas Turbine Engines and Generators

Gas turbines that drive electric generators require large quantities—generally in the order of 80 to 90 lb of air per kilowatt hour—of clean cool air. With a 5000-kW generator, the use of about 95,000 cfm is required. Within limits, the temperature of the air supplied directly affects gas turbine output. If air is supplied at 100 °F db, output will be about 10% less than with air at 80 °F. Similarly, dropping the air temperature to 60 °F results in an increase of 10% over the performance at 80 °F. In an installation of this type, several precautions on the use of evaporative cooling must be taken: (1) the eliminators of the washer must be designed to stop entrainment of all free moisture; and (2) a temperature difference of at least 3 to 4 °F between dry- and wet-bulb temperatures must be maintained to prevent condensation resulting from the pressure drop at the compressor inlet.

Evaporative cooling to provide relief from the heat, increased generator capacity, and improved engine performance have been combined in primary aluminum reduction plants, where gas engine-generators are used to supply the heavy power requirements for this process. Evaporative coolers with 80% effectiveness and a capacity of 48,000 cfm have been provided for each 1600 hp engine-generator set; 9000 cfm is supplied to the scavenging air blower of the engine, 18000 cfm is drawn through the 1000 kW generator windings, and 22,000 cfm is supplied to the engine room to provide relief to the operators stationed at the control panels.

Process Cooling

In the tobacco, textile, spray coating, and other industries, where manufacturing requires accurate humidities, comfort cooling is also obtainable by evaporation. For moderate humidity conditions, the air-handling capacity of the evaporative cooling system is calculated on the basis of total heat load absorbed by the air. In the textile industry, where relatively high humidities are required and the machinery load is heavy, it is customary to use a split system whereby free moisture is introduced directly into the room; here, the air handled is reduced to approximately 60% of that normally required by an all-outdoor air evaporative cooling system. In cigar plants, humidities from 70 to 80% are required in some departments, depending on the type of tobacco used. Evaporative cooling can be used successfully where these high humidities are required.

Wood and Paper Products Facilities

Because of gases and particulates present in most paper plant atmospheres, most manufacturers prefer water-cooled over air-cooled systems. The most prevalent contaminants are chlorine gas, caustic soda, and sulfur compounds. With more efficient air-cleaning, the air quality in and about most mills is becoming adequate for properly placed air-cooled chillers or condensing units with well-analyzed and properly applied coil and housing coatings. Phosphor-free brazed coil joints are recommended in areas where sulfur compounds are present.

Heat is readily available from the processing operations and should be used whenever possible. Most plants have quality hot water and steam, which can be readily geared to unit heater, central station, or reheat use. Evaporative cooling should not be ignored. Newer plant air-conditioning methods, using energy conservation techniques (temperature stratification) apply themselves well to this type of large structure. As in most industrial applications, absorption systems can be considered for pulp and paper plants, since they afford some degree of energy recovery from high-temperature steam processes. Refer to Chapter 23 for further information on air conditioning in paper product facilities.

OTHER APPLICATIONS

Power Generation Facilities

Once the criteria are established and the preliminary heating and cooling loads are determined, an appropriate system can be selected. The same considerations for system selection apply for power generation facilities as for industrial facilities in terms of ventilation—natural or forced heating and adiabatic cooling or refrigeration cooling—as long as criteria for the individual facilities are met in terms of temperature, humidity, pressure, and airflow control.

Mine Cooling

Chapter 26 describes various evaporative cooling methods developed for cooling mines.

Survival Shelters

Cooling for survival shelters is covered in Chapter 11.

Animal Cooling

The design criteria for farm animals and the need for cooling of animal shelters are discussed in Chapter 37. Evaporative cooling is ideally suited to farm animal shelters since all outdoor air is used and, therefore, tight construction is not required. The fresh air also removes odors and reduces the harmful effects of ammonia fumes from urine and excrement. At night and in the spring and fall, the evaporative cooling system can also be used for ventilation.

These systems should be sized to change the air within the shelter in 1.5 to 2 min, assuming the ceiling height does not exceed 10 ft. This flow rate will keep the shelter at 85 °F or lower. In a typical milking parlor, packaged air coolers can improve working conditions for the operator, which makes handling the cows easier.

Evaporative cooling is also used to cool farrowing and gestation houses to improve production.

The production efficiency of egg layers and breeders can be enhanced in some areas by using evaporative coolers. Most applications require an air change every 0.75 to 1.50 min, with the majority at 1.0 min. Placement of the fans at the ends or the center of the house with the evaporative cooling system strategically placed, creates a tunnel ventilation system. The fans are generally selected for a total pressure drop of 0.125 in. of water, which means that the evaporative cooling media cannot have a pressure drop in excess of 0.075 in. of water. To guard against an inadequate volume of air being pulled through the poultry house, the designer must know the pressure drop through the evaporative media being selected.

Broiler operations are also being considered for evaporative cooling. Experiments have shown that while the feed conversion ratio is not necessarily improved, the average bird weighs more at the conclusion of the growth cycle. Most broiler houses today depend on lowering side curtains and natural draft to reduce the temperature within the house. However, when the ambient outside temperature exceeds 100°F, evaporative cooling is often the only way to keep the flock alive.

Produce Storage Cooling

Potatoes. Evaporative cooling systems for potato storage should pass air directly through and in contact with the potato pile. The capacity of the cooling system should be based on a 3-min air change to give quick and even cooling. The total volume of the storage should be considered in calculating the size of the system to provide sufficient cooling if the depth of storage is increased at a future time. Above-floor ducts can be used, but underfloor ducts are preferred as they can also be used for handling the potatoes with bin unloaders. Ducts are typically 20 in. wide and at least 14 in. deep to allow room for inserting the bin unloader conveyors. Duct tops must be removable and should be large enough to support a loaded truck. Boards should be spaced to provide 0.5-in. slots. Since distributing and delivery ducts act as extended plenums, it is unnecessary to reduce duct depth as the far end of the ducts is approached. Delivery ducts should extend to within 6 ft of the walls to provide uniform airflow through the storage. Since friction loss in the duct system is negligible, the resistance of the potato pile is the principal load on the fan in the evaporative cooler. The resistance is approximately 0.25-in. static pressure. The additional resistance of the ducts, inlets, dampers, and evaporative cooler will add another 0.25 in. of water for a total of 0.50 in. that the fan must overcome. In large storages, dividing the house into two sections keeps the cooling system within a reasonable size. A single system should not handle over 12000 cfm of air if main and distributing ducts are to be of reasonable size.

Apples. Evaporative cooling systems for apple storages without refrigeration should distribute cool air to all parts of the storage. The evaporative cooler may be floor-mounted or located near the ceiling in a fan room. The system should be designed to discharge air horizontally at the ceiling level. Since the degree of cooling is limited by the prevailing wet-bulb temperature, the maximum reasonable size system should be installed to bring down the storage temperature rapidly and as close to the wet-bulb temperature as possible. Generally, a system with a 3-min air change capacity is the largest system that can be installed. This capacity will result in a 1- to 1.5-min movement of air when the storage is loaded.

For further information on apple storage, see Chapter 16 of the 1990 ASHRAE *Handbook—Refrigeration.*

Citrus. The chief purpose of evaporative cooling as it is applied to fruits and vegetables is to provide an effective, yet inexpensive, means of improving common storages. However, it also serves a special function in the case of oranges, grapefruit, and lemons. Although mature and ready for harvest, citrus fruit have often not changed in color from green. Color change (degreening) is achieved through a sweating process in special rooms equipped with evaporative cooling. Air with a high relative humidity and a moderate temperature is circulated continuously during the operation. Ethylene gas, the concentration depending on the variety and intensity of green pigment in the rind, is discharged into the rooms. The ethylene mainly destroys the chlorophyll in the rind and allows the yellow or orange color to become evident. During the degreening operation, a temperature of 70°F and a relative humidity of 88 to 90% is maintained in the sweat room. (In the Gulf states, 82 to 85°F with 90 to 92% rh is recommended.) The evaporative cooling system is designed to deliver 4 cfm per field box of fruit (0.11 cfm/lb).

Evaporative cooling is also used as a supplement to refrigeration in the storage of citrus fruit. While citrus storage requires refrigeration in the summer, conditions can often be met with evaporative cooling during the fall, winter, and spring when the outdoor wet-bulb temperature is low. For further information, see Chapter 17 of the 1988 ASHRAE *Handbook—Refrigeration.*

Greenhouse Cooling

Proper regulation of greenhouse temperatures during the summer is essential for developing high quality crops. The principal load on a greenhouse is solar radiation, which at sea level at about noon in the temperate zone is approximately 200 Btu/h·ft². Smoke, dust, or heavy clouds reduce the radiation. Table 2 gives solar radiation loads for representative cities in the United States. Note that the values cited are average solar heat gains, not peak loads. Temporary rises in temperature inside a greenhouse can be tolerated; an occasional rise above design conditions is not likely to cause damage.

Table 2 Three-Year Average Hourly Solar Radiation for Horizontal Surface during Peak Summer Month

City	Btu/ft²	City	Btu/ft²
Albuquerque, NM	198	Lemont, IL	142
Apalachicola, FL	170	Lexington, KY	170
Astoria, OR	132	Lincoln, NE	150
Atlanta, GA	158	Little Rock, AR	148
Bismarck, ND	140	Los Angeles, CA	162
Blue Hill, MA	128	Madison, WI	138
Boise, ID	155	Medford, OR	170
Boston, MA	125	Miami, FL	153
Brownsville, TX	175	Midland, TX	177
Caribou, ME	115	Nashville, TN	154
Charleston, SC	152	Newport, RI	138
Cleveland, OH	152	New York, NY	140
Columbia, MO	153	Oak Ridge, TN	148
Columbus, OH	127	Oklahoma City, OK	165
Davis, CA	184	Phoenix, AZ	200
Dodge City, KS	184	Portland, ME	133
East Lansing, MI	132	Prosser, WA	176
East Wareham, MA	132	Rapid City, SD	152
El Paso, TX	195	Richland, WA	137
Ely, NV	175	Riverside, CA	176
Fort Worth, TX	176	St. Cloud, MN	132
Fresno, CA	188	San Antonio, TX	176
Gainesville, FL	156	Santa Maria, CA	188
Glasgow, MT	152	Sault Ste. Marie, MI	138
Grandby, CO	149	Sayville, NY	148
Grand Junction, CO	173	Schenectady, NY	117
Great Falls, MT	150	Seabrook, NJ	135
Greensboro, NC	155	Seattle, WA	117
Griffin, GA	164	Spokane, WA	139
Hatteras, NC	177	State College, PA	141
Indianapolis, IN	140	Stillwater, OK	167
Inyokern, CA	218	Tallahassee, FL	134
Ithaca, NY	145	Tampa, FL	167
Lake Charles, LA	160	Upton, NY	148
Lander, WY	177	Washington, D.C.	142
Las Vegas, NV	195		

Not all the solar radiation that reaches the inside of the greenhouse becomes a cooling load. Solar radiation is transformed into (1) heat of transpiration, (2) photosynthesis, and (3) sensible heat. Photosynthesis does not amount to more than about 2% of the total solar radiation. Transpiration of moisture varies from crop to crop, but is typically about 48% of the solar radiation. This leaves 50% to be removed by the cooling system. Example 2 shows how to calculate the size of greenhouse evaporative cooling systems.

Example 2. An evaporative cooling system is to be installed in a 50 by 100 ft greenhouse. Design conditions are assumed to be 92 °F db and 73 °F wb, and average hourly solar radiation is 138 Btu/ft². An indoor temperature of 90 °F db must not be exceeded at design conditions.

Solution: The evaporative air cooler is assumed to have a saturation effectiveness of 80%. For the conditions of this problem, Equation (2) becomes:

$$t_2 = 92 - 80\,(92 - 73\,°F)/100 = 77\,°F$$

Equation (4) may be used to calculate the air volume to be supplied by the evaporative cooler:

$$Q_{ra} = A\,I_t/[2(t_1 - t_2)] \qquad (4)$$

where

A = floor area of greenhouse, ft²

I_t = total incident solar radiation, Btu/h per ft² of receiving surface

For the conditions of this problem, Equation (4) becomes:

$$Q_{ra} = (50 \times 100 \times 138)/[2(90 - 77)] = 26{,}500 \text{ cfm}$$

Figure 10 graphically shows the psychrometry of this problem. Since 49% of the heat load is latent, the air absorbs heat along a 0.51 sensible heat factor line to the design indoor temperature of 90 °F. At this point the conditions in the greenhouse are 90 °F and 64% rh.

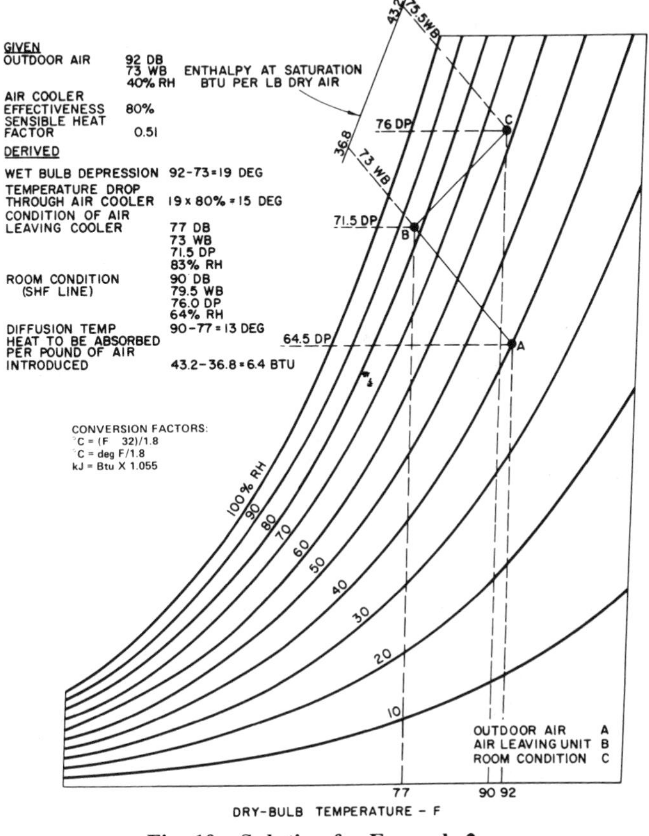

Fig. 10 Solution for Example 2

Horizontal illumination due to direct rays from a noonday summer sun with a clear sky is as much as 10,000 footcandles (ftc); under clear glass, this will be approximately 8,500 ftc. Crops such as chrysanthemums and carnations grow best in full sun, but many foliage plants, such as gloxinias and orchids, do not need more than 1500 to 2000 ftc. If shade is needed, it should be noted that solar radiation is approximately proportional to light intensity. Thus, the greater the shade, the smaller the capacity cooling system required. A value of 100 ftc per hour is approximately equivalent to 3 Btu/h·ft². Although atmospheric conditions, such as clouds and haze, affect the relationship, this is a safe conversion factor to use. For greenhouses where illumination can be determined by design or by actual observation, this relationship should be used instead of Table 2.

Evaporative cooling systems for greenhouses may be either the supply or extraction type. Regardless of the type of system used, the length of air travel should not exceed 125 ft. The rise in temperature of the cool air limits the throw to this value. In some cases, it is desirable to reduce the throw to 30 ft. Air movement must be kept at a low velocity to prevent mechanical damage to the plants; it should generally not exceed 100 fpm in the crop area.

WEATHER CONSIDERATIONS

The effectiveness of evaporative cooling depends on weather conditions. System design is affected by the prevailing outdoor dry-and wet-bulb temperatures as well as the application of the system. For example, a simple residential direct evaporative cooling system, with an effectiveness of 80%, will provide satisfactory room conditions (given an adequate quantity of outdoor supply air and a well-designed exhaust system) throughout the cooling season in areas such as Reno, Nevada. Here the 1% design dry-bulb and mean coincident wet-bulb temperatures are 96 and 61 °F, respectively. In the same location, additional cooling effect can be gained by the addition of an indirect evaporative precooling stage, which lowers the temperature (both dry- and wet-bulb) entering the direct evaporative cooling stage and, consequently, lowers the supply air temperature.

In a geographic location such as Atlanta, Georgia, with design temperatures of 94 and 74 °F, the same direct evaporative cooler could supply only 78 °F. This could be reduced to 72.4 °F by the addition of an indirect evaporative precooling stage that is 65% effective (Supple 1982).

A well-designed direct evaporative cooling system, including adequate powered exhaust and an air velocity of 100 to 300 fpm has provided comfort in a busy restaurant in Atlanta. This type of approach uses effective temperature and carefully designed exhaust and supply systems to provide a comfort level that is generally thought unattainable in a high wet-bulb design area. Long-term benefits to the owner(s) include 20 to 40% of the utility costs compared to mechanical refrigeration (Watt 1988).

ECONOMIC CONSIDERATIONS

Design of evaporative cooling systems and sizing of equipment should be based on the load requirements of the application and local dry- and wet-bulb design conditions, which may be found in Chapter 24 of the 1989 ASHRAE *Handbook—Fundamentals*.

Total energy use for a specific application during a set period may be forecast by using annual weather data. Dry-bulb and mean coincident wet-bulb temperatures, with the hours of occurrence, can be summarized and used in a modified bin procedure. The calculations must reflect the hours of occupancy. Results may vary for different types of buildings, even though the same weather station data may be used. When comparing systems, the cost analysis should include the annual energy reduction at the applicable electrical rate, plus the anticipated energy cost escalation over the expected life of the system.

Reducing air-conditioning kilowatt demand is even more important in areas with ratcheted demand rates (Scofield and Deschamps 1980). In these areas, the demand rate is not set monthly. A summer month with a heavy cooling peak energy demand can set the demand rate for the entire year.

Many areas have time-of-day electrical metering as an incentive to use energy during off-peak hours when rates are at their lowest. Thermal storage using ice banks or chilled water storage vessels may be used as part of a multistage evaporative-refrigerated cooling system to combine the energy-saving advantages of evaporative cooling; off-peak savings are accomplished using thermal storage. In addition to the cost savings because of off-peak energy rates, thermal storage systems may actually save energy because the refrigeration equipment operates at a time when the ambient temperature and resulting condensing temperatures are lower (Eskra 1980).

PSYCHROMETRICS

Figure 11 shows the two-stage process applied to nine western cities in the United States. The entering conditions to the first-stage indirect unit are at the recommended 1% design dry-/wet-bulb temperatures from Chapter 24 of the 1989 ASHRAE *Handbook—Fundamentals*. The effectiveness ratings are 60% for the first (indirect) stage and 90% for the second (direct) stage. The leaving air temperatures range from 52 to 70°F and are saturated (80% and higher). Figure 12 projects these second-stage supply temperatures (based on a 95% room sensible heat factor, *i.e.*, room sensible heat/room total heat) to a room condition at 78°F db. Given a 1% entering condition in the cities shown, room conditions are maintained within the comfort zone without a refrigerated third stage. However, Figures 11 and 12 point out the need to consider the following factors:

1. As the room sensible heat factor increases, the required supply air temperature decreases to maintain a given room condition.
2. As the supply air temperature increases, the supply air quantity increases, resulting in higher air-side system initial cost and increased supply air fan power.
3. A decrease in the required room dry-bulb temperature causes an increase in the supply air quantity. At a given room sensible heat factor, a decrease in room dry-bulb temperature may cause the relative humidity to exceed the comfort zone.

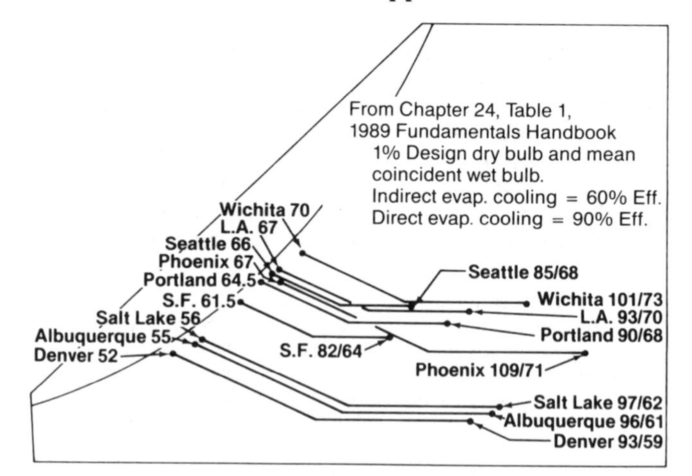

Fig. 11 Two-Stage Evaporative Cooling at 1% Design Condition in Various Cities

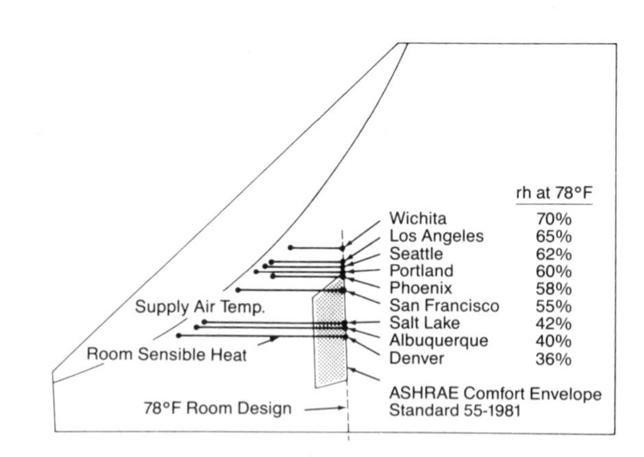

Fig. 12 Final Room Design Conditions after Two-Stage Evaporative Cooling

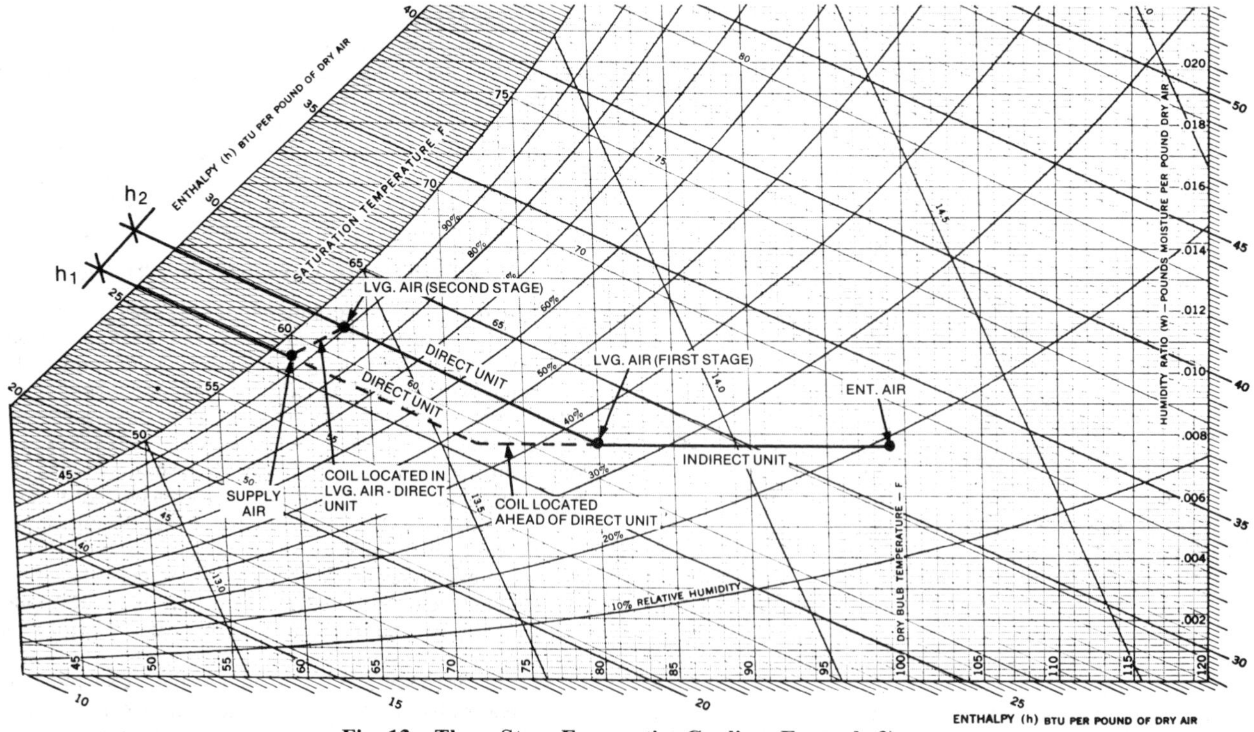

Fig. 13 Three-Stage Evaporative Cooling (Example 3)

4. The suggested 1% entering air (dry-bulb/mean wet-bulb) conditions to the system are only one concern. Partial load conditions must also be considered, along with the effect (extent and duration) of spike wet-bulb temperatures. Mean wet-bulb temperatures can be used to determine energy use of the indirect/direct system accurately. However, the higher wet-bulb temperature spikes should be considered to determine the operating results and affected room temperatures at these conditions.

An ideal condition for the maximum use with minimum energy consumption of a two- and three-stage indirect/direct system is a room sensible heat factor of 90% and higher, a supply air temperature of 60 °F, and 78 °F dry-bulb room temperature. In many cases, three-stage refrigeration is required to ensure satisfactory dry-bulb temperature and relative humidity. Refrigeration capacity for three-stage cooling is determined as in the following example. The calculations are reflected in Figure 13.

Example 3. Assume the following:

1. Supply air quantity = 24,000 cfm; supply air temperature = 60 °F
2. Design condition = 99 °F db and 68 °F wb
3. Effectiveness of indirect unit = 60%; effectiveness of direct unit = 90%
4. Indirect unit performance:

   $$99 - (99 - 68)(0.60) = 80.4\,°\text{F leaving dry-bulb } (61.8\,°\text{F wb})$$

   Direct unit performance:

   $$80.4 - (80.4 - 61.8)(0.90) = 63.7\,°\text{F db supply air temperature}$$

5. Calculate booster refrigeration capacity to drop the supply air temperature from 63.7 °F to the required 60 °F supply air temperature. Coil located ahead of direct unit:

$$\text{Btu/h cooling} = \frac{60(h_1 - h_2)(\text{Supply air, cfm})}{\text{Specific volume dry air at leaving air}}$$

With numeric values of enthalpies h_1 and h_2 (in Btu/lb) and the specific volume of air (in ft^3/lb dry air), taken from ASHRAE Psychrometric Chart No. 1, the cooling load is calculated as follows:

$$60(27.6 - 25.5)24{,}000/13.78 = 219{,}400 \text{ Btu/h (18.3 tons)}$$

Coil located in the leaving air of the direct unit:

$$60(27.8 - 25.7)24{,}000/13.43 = 225{,}000 \text{ Btu/h (18.8 tons)}$$

Depending on the location of the booster coil, the above calculations can be used to determine third-stage refrigeration capacity and to select a cooling coil.

By using this example, refrigeration sizing can be compared to a conventional, refrigerated system without staged evaporative cooling. Assuming mixed air conditions to the coil of 81 °F db and 66.5 °F wb, and the same 60 °F db supply air as shown in Figure 13, the refrigerated capacity is:

$$(60)(31.1 - 25.7)(24{,}000)/(13.31) = 584{,}200 \text{ Btu/h (48.7 tons)}$$

This represents an increase of 30.4 tons. The staged evaporative effect reduces the required refrigeration by 62.4%.

REFERENCES

Baschiere, R.J., C.E. Rathmann, and M. Lokmanhekim. 1968. Adequacy of evaporative cooling and shelter environmental prediction. General American Research Division, GATX, OCD, AD 697-874.

Crow, L.W. 1972. Weather data related to evaporative cooling. Research Report No. 2223. ASHRAE *Transactions* 78(1):153-64.

Eskra, N. 1980. Indirect/Direct evaporative cooling systems. ASHRAE *Journal* 22(5):22.

Scofield, M. and N. Deschamps. 1980. EBTR Compliance and comfort too. ASHRAE *Journal* 22(6):61.

Supple, R.G. 1982. Evaporative cooling for comfort. ASHRAE *Journal* 24(8):42.

Watt, J.R. 1988. Cost comparisons: Evaporative vs. refrigerative cooling. ASHRAE *Technical Data Bulletin* 4(4):41.

BIBLIOGRAPHY

Watt, J.R. 1986. *Evaporative air conditioning handbook*, 2nd ed. Chapman & Hall, New York.

CHAPTER 47

SMOKE CONTROL

IN building fires, smoke often flows to locations remote from the fire, threatening life and damaging property. Stairwells and elevators frequently become smoke-filled, thereby blocking or inhibiting evacuation. Smoke causes the most deaths in fires.

In the late 1960s, the idea of using pressurization to prevent smoke infiltration of stairwells began to attract attention. This concept was followed by the idea of the "pressure sandwich," *i.e.*, venting or exhausting the fire floor and pressurizing the surrounding floors. Frequently, a building's ventilation system is used for this purpose. Smoke control describes systems that use pressurization produced by mechanical fans to limit smoke movement in fire situations.

This chapter discusses fire protection and smoke control systems in buildings as they relate to the HVAC field. For a more complete discussion of this subject, refer to *Design of Smoke Control Systems for Buildings* (Klote and Fothergill 1983). The National Fire Protection Association (NFPA 1988) states that smoke consists of the airborne solid and liquid particulates and gases evolved when a material undergoes pyrolysis or combustion.

The objectives of fire safety are to provide some degree of protection for a building's occupants, the building and the property inside it, and neighboring buildings. Various forms of system analysis have been used to help quantify protection. Specific life safety objectives differ with occupancy; for example, nursing home requirements are different from those for office buildings.

Two basic approaches to fire protection are to prevent fire ignition and to manage fire impact. Figure 1 shows a decision tree for fire protection. The building occupants and managers have the primary role in preventing fire ignition. The building design team may incorporate features into the building to assist the occupants and managers in this effort. Because it is impossible to prevent fire ignition completely, managing fire impact has become significant in fire protection design. Compartmentation, suppression, control of construction materials, exit systems, and smoke management are examples. The NFPA *Fire Protection Handbook* (1986) contains detailed fire safety information.

Historically, fire safety professionals have considered the HVAC system as a potentially dangerous penetration of natural building membranes (wall, floors, and so forth) that can readily transport smoke and fire. For this reason, the systems have traditionally been shut down when fire is discovered. Although shutting down the system prevents fans from forcing smoke flow, it does not prevent smoke movement through ducts due to smoke buoyancy, stack effect, or the wind. To solve the smoke problem, the concept of smoke control has developed; it should be viewed as only one part of the overall building fire protection system.

The preparation of this chapter is assigned to TC 5.6, Control of Fire and Smoke.

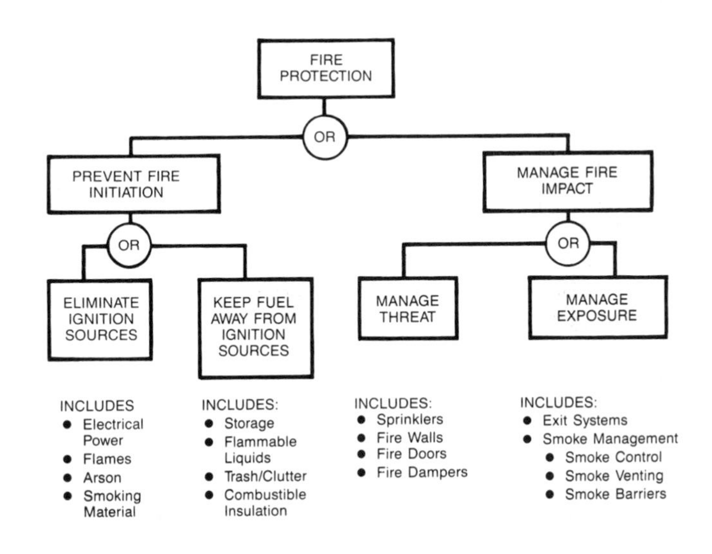

Fig. 1 Simplified Fire Protection Decision Tree

SMOKE MOVEMENT

A smoke control system must be designed so that it is not overpowered by the driving forces that cause smoke movement, including stack effect; bouyancy; expansion; wind; and the heating, ventilating, and air-conditioning system. In a fire, smoke is generally moved by a combination of these forces.

Stack Effect

When it is cold outside, air often moves upward within building shafts, such as stairwells, elevator shafts, dumbwaiter shafts, mechanical shafts, or mail chutes. Referred to as normal stack effect, this phenomenon occurs because the air in the building is warmer and less dense than the outside air. Normal stack effect is great when outside temperatures are low, especially in tall buildings. However, normal stack effect can exist even in a one-story building.

When the outside air is warmer than the building air, a downward airflow, or reverse stack effect, frequently exists in shafts. At standard atmospheric pressure, the pressure difference due to either normal or reverse stack effect is expressed as:

$$\Delta p = 7.64\,(1/T_o - 1/T_i)h \quad (1)$$

where

Δp = pressure difference, in. of water
T_o = absolute temperature of outside air, °R
T_i = absolute temperature of air inside shaft, °R
h = distance above neutral plane, ft

For a building 200-ft tall with a neutral plane at the midheight, an outside temperature of 0 °F, and an inside temperature of 70 °F, the maximum pressure difference due to stack effect would be 0.22 in. of water. This means that at the top of the building, a shaft would have a pressure of 0.22 in. of water greater than the outside pressure. At the bottom of the shaft, the shaft would have a pressure of 0.22 in. of water less than the outside pressure. Figure 2 diagrams the pressure difference between a building shaft and the outside. A positive pressure difference indicates that the shaft pressure is higher than the outside pressure, and a negative pressure difference indicates the opposite.

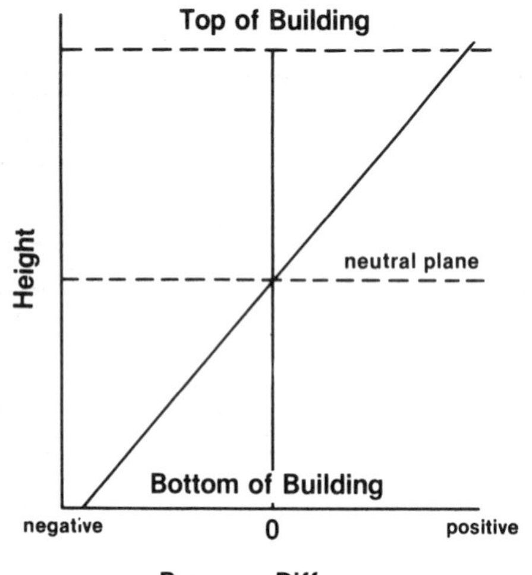

Fig. 2 Pressure Difference Between a Building Shaft and the Outside Due to Normal Stack Effect

Stack effect usually exists between a building and the outside. The air movement in buildings caused by both normal and reverse stack effect is illustrated in Figure 3. In this case, the pressure difference expressed in Equation (1) refers to the pressure difference between the shaft and the outside of the building.

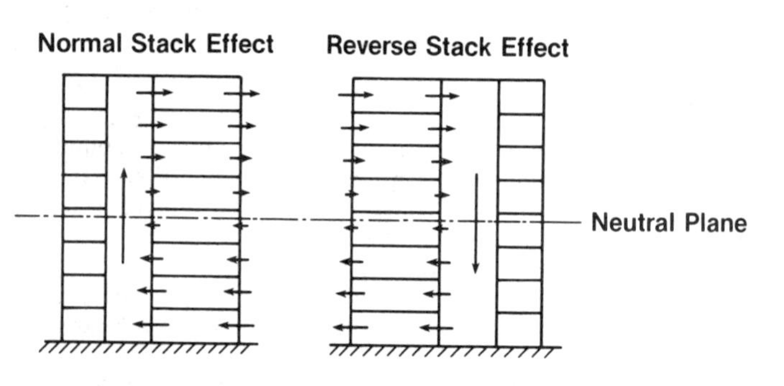

Fig. 3 Air Movement Due to Normal and Reverse Stack Effect

Figure 4 can be used to determine the pressure difference due to stack effect. For normal stack effect, $\Delta p/h$ is positive, and the pressure difference is positive above the neutral plane and negative below it. For reverse stack effect, $\Delta p/h$ is negative, and the pressure difference is negative above the neutral plane and positive below it.

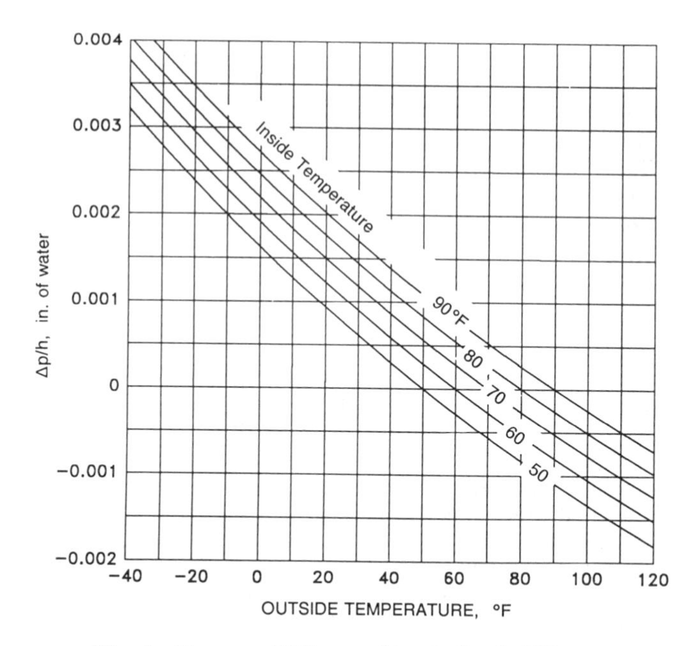

Fig. 4 Pressure Difference Due to Stack Effect

In unusually tight buildings with exterior stairwells, reverse stack effect has been observed even with low outside air temperatures (Klote 1980). In this situation, the exterior stairwell temperature is considerably lower than the building temperature. The stairwell represents the cold column of air, and other shafts within the building represent the warm columns of air.

If the leakage paths are uniform with height, the neutral plane is near the midheight of the building. However, when the leakage paths are not uniform, the location of the neutral plane can vary considerably, as in the case of vented shafts. McGuire and Tamura (1975) provide methods for calculating the location of the neutral plane for some vented conditions.

Smoke movement from a building fire can be dominated by stack effect. In a building with normal stack effect, the existing air currents (as shown in Figure 3) can move smoke considerable distances from the fire origin. If the fire is below the neutral plane, smoke moves with the building air into and up the shafts. This upward smoke flow is enhanced by buoyancy forces due to the temperature of the smoke. Once above the neutral plane, the smoke flows from the shafts into the upper floors of the building. If the leakage between floors is negligible, the floors below the neutral plane, except the fire floor, are relatively smoke-free until the quantity of smoke produced is greater than can be handled by stack effect flows.

Smoke from a fire located above the neutral plane is carried by the building airflow to the outside through exterior openings in the building. If the leakage between floors is negligible, all floors other than the fire floor remain relatively smoke-free until the quantity of smoke produced is greater than can be handled by stack effect flows. When the leakage between floors is considerable, the smoke flows to the floor above the fire floor.

The air currents caused by reverse stack effect (Figure 3) tend to move relatively cool smoke down. In the case of hot smoke, buoyancy forces can cause smoke to flow upward, even during reverse stack effect conditions.

Buoyancy

High-temperature smoke from a fire has a buoyancy force due to its reduced density. The pressure difference between a fire compartment and its surroundings can be expressed as follows:

$$\Delta p = 7.64\,(1/T_o - 1/T_f)h \qquad (2)$$

where

Δp = pressure difference, in. of water
T_o = absolute temperature of the surroundings, °R
T_f = absolute temperature of the fire compartment, °R
h = distance above the neutral plane, ft

The pressure difference due to buoyancy can be obtained from Figure 5 for the surroundings at 68 °F. The neutral plane is the plane of equal hydrostatic pressure between the fire compartment and its surroundings. For a fire with a fire compartment temperature at 1470 °F, the pressure difference 5 ft above the neutral plane is 0.052 in. of water. Fang (1980) studied pressures caused by room fires during a series of full-scale fire tests. During these tests, the maximum pressure difference reached was 0.064 in. of water across the burn room wall at the ceiling.

Much larger pressure differences are possible for tall fire compartments where the distance h from the neutral plane can be larger. If the fire compartment temperature is 1290 °F, the pressure difference 35 ft above the neutral plane is 0.35 in. of water. This causes a large fire, and the pressures it produces are beyond present smoke control methods. However, the example illustrates the extent to which Equation (2) can be applied.

In a building with leakage paths in the ceiling of the fire room, this buoyancy-induced pressure causes smoke to move to the floor above the fire floor. In addition, this pressure causes smoke to move through any leakage paths in the walls or around the doors of the fire compartment. As smoke travels away from the fire, its temperature drops due to heat transfer and dilution. Therefore, the effect of buoyancy generally decreases with distance from the fire.

Expansion

In addition to buoyancy, the energy released by a fire can move smoke by expansion. In a fire compartment with only one opening to the building, building air will flow in, and hot smoke will flow out. Neglecting the added mass of the fuel, which is small

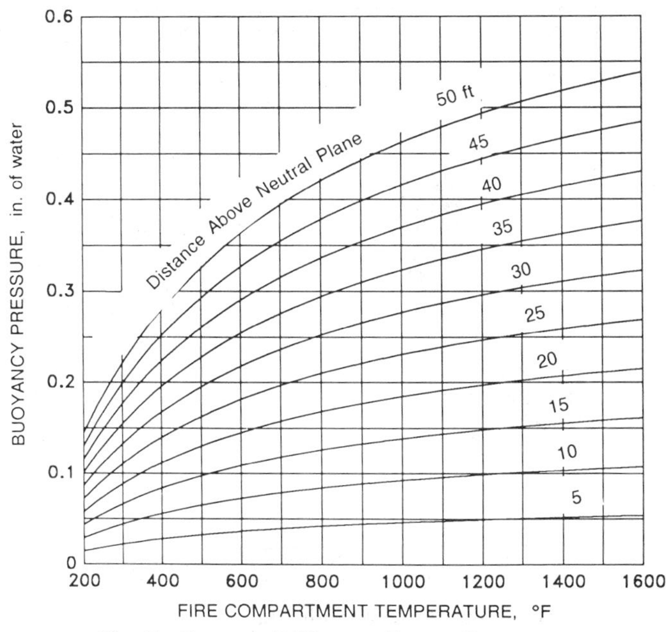

Fig. 5 Pressure Difference Due to Buoyancy

compared to the airflow, the ratio of volumetric flows can be expressed as a ratio of absolute temperatures.

$$Q_{out}/Q_{in} = T_{out}/T_{in}$$

where

Q_{out} = volumetric flow rate of smoke out of the fire compartment, cfm
Q_{in} = volumetric flow rate of air into the fire compartment, cfm
T_{out} = absolute temperature of smoke leaving fire compartment, °R
T_{in} = absolute temperature of air into fire compartment, °R

For a smoke temperature of 1290 °F, the ratio of volumetric flows would be 3.32. Note: Absolute temperature is used for calculation. In such a case, if the air flowing into the fire compartment is 3180 cfm, the smoke flowing out of the fire compartment would be 10,600 cfm. In this case, the gas has expanded to more than three times its original volume.

For a fire compartment with open doors or windows, the pressure difference across these openings due to expansion is negligible. However, for a tightly sealed fire compartment, the pressure differences due to expansion may be important.

Wind

In many instances, wind can have a pronounced effect on smoke movement within a building. The pressure p_w that the wind exerts on a surface can be expressed as:

$$p_w = 0.00643\, C_w \rho_o V^2 \qquad (3)$$

where

C_w = dimensionless pressure coefficient
ρ_o = outside air density, lb/ft^3
V = wind velocity, mph

The pressure coefficients C_w are in the range of -0.8 to 0.8, with positive values for windward walls and negative values for leeward walls. The pressure coefficient depends on building geometry and varies locally over the wall surface. In general, wind velocity increases with height from the surface of the earth. Sachs (1972), Houghton and Carruther (1976), Simiu and Scanlan (1978), and MacDonald (1975) give detailed information concerning wind velocity variations and pressure coefficients. Shaw and Tamura (1977) have developed specific information about wind data with respect to air infiltration in buildings.

A 35 mph wind produces a pressure on a structure of 0.47 in. of water with a pressure coefficient of 0.8. The effect of wind on air movement within tightly constructed buildings with all doors and windows closed is slight. However, the effects of wind can be important for loosely constructed buildings or for buildings with open doors or windows. Usually, the resulting airflows are complicated, and computer analysis is required.

Frequently in fire situations, a window breaks in the fire compartment. If the window is on the leeward side of the building, the negative pressure caused by the wind vents the smoke from the fire compartment. This reduces smoke movement throughout the building. However, if the broken window is on the windward side, the wind forces the smoke throughout the fire floor and to other floors, which endangers the lives of building occupants and hampers fire fighting. Pressures induced by the wind in this situation can be large and can dominate air movement throughout the building.

HVAC Systems

The system frequently transports smoke during building fires. For this reason, before the concept of smoke control developed, systems were shut down when fires were discovered.

In the early stages of a fire, the system can aid in fire detection. When a fire starts in an unoccupied portion of a building, the system can transport the smoke to a space where people can smell it and be alerted to the fire. However, as the fire progresses, the system transports smoke to every area it serves, thus endangering life in all those spaces. The system also supplies air to the fire space, which aids combustion. Although shutting the system down prevents it from supplying air to the fire, it does not prevent smoke movement through the supply and return air ducts, air shafts, and other building openings due to stack effect, buoyancy, or wind.

SMOKE MANAGEMENT

In this chapter, smoke management includes all methods that can modify smoke movement, either independently or in combination, for the benefit of occupants and fire fighters and for the reduction of property damage. Barriers, smoke vents, and smoke shafts have traditionally been used to manage smoke.

The effectiveness of a barrier in limiting smoke movement depends on the leakage paths in the barrier. Holes where pipes penetrate walls or floors, cracks where walls meet floors, and cracks around doors are examples of leakage paths. The pressure difference across these barriers depends on stack effect, buoyancy, wind, and the HVAC system.

The effectiveness of smoke vents and smoke shafts depends on their proximity to the fire, the buoyancy of the smoke, and the presence of other driving forces. When smoke is cooled due to sprinklers, the effectiveness of smoke vents and smoke shafts is reduced.

Elevator shafts have been used as smoke shafts. This prevents their use for fire evacuation, and these shafts frequently distribute smoke to floors far from the fire. Specially designed smoke shafts, which have essentially no leakage on floors other than the fire floor, can prevent the smoke shaft from distributing smoke to nonfire floors.

PRINCIPLES

Smoke control uses the barriers (walls, floors, doors) for traditional smoke management in conjunction with airflows and pressure differences generated by mechanical fans.

Figure 6 illustrates a pressure difference across a barrier acting to control smoke movement. Within the barrier is a door. The high-pressure side of the door can be either a refuge area or an escape route. The low-pressure side is exposed to smoke from a fire. Airflow through the cracks around the door and through other construction cracks prevents smoke infiltration to the high-pressure side. When the door is opened, air flows through the opening. When the air velocity is low, smoke can flow against the airflow into the refuge area or escape route, as shown in Figure 7. This smoke backflow can be prevented if the air velocity is sufficiently large (Figure 8). The magnitude of the velocity necessary to prevent backflow depends on the energy release rate of the fire.

The two basic principles of smoke control are:

1. Airflow by itself can control smoke movement if the average air velocity is of sufficient magnitude.
2. Air pressure differences across barriers can act to control smoke movement.

Pressurization results in airflow through the small gaps around closed doors and in construction cracks, thereby preventing smoke backflow through these openings. Therefore, in a physical sense, the second principle is a special case of the first principle. However, considering the two principles as separate is advantageous for smoke control design. For a barrier with one or more large openings, air velocity is the appropriate physical quantity for both design considerations and acceptance testing. However, when there are only small cracks, such as around closed doors, designing to and measuring air velocities is impractical. In this case, the appropriate physical quantity is pressure difference. Consideration of the two principles as separate also emphasizes the different considerations necessary for open and closed doors.

Because smoke control relies on air velocities and pressure differences produced by fans, it has the following advantages over the traditional methods of smoke management:

- Smoke control is less dependent on tight barriers. Allowance can be made in the design for reasonable leakage through barriers.

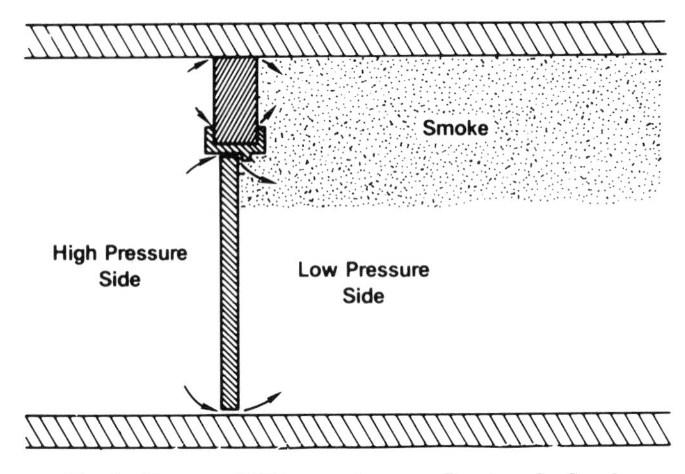

Fig. 6 Pressure Difference Across a Barrier of a Smoke Control System Preventing Smoke Infiltration to the High-Pressure Side of the Barrier

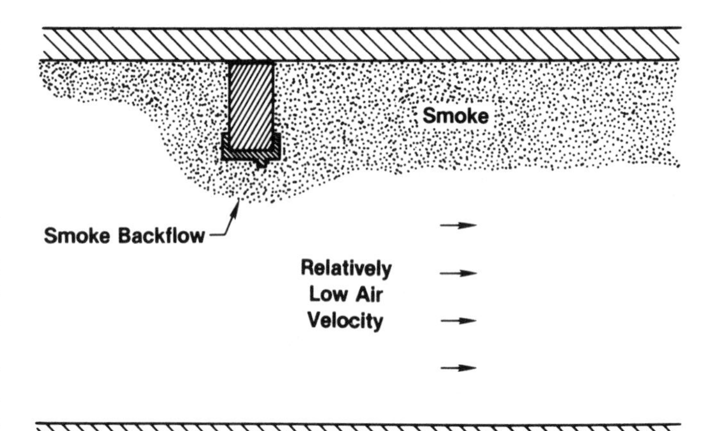

Fig. 7 Smoke Backflow Against Low Air Velocity Through an Open Doorway

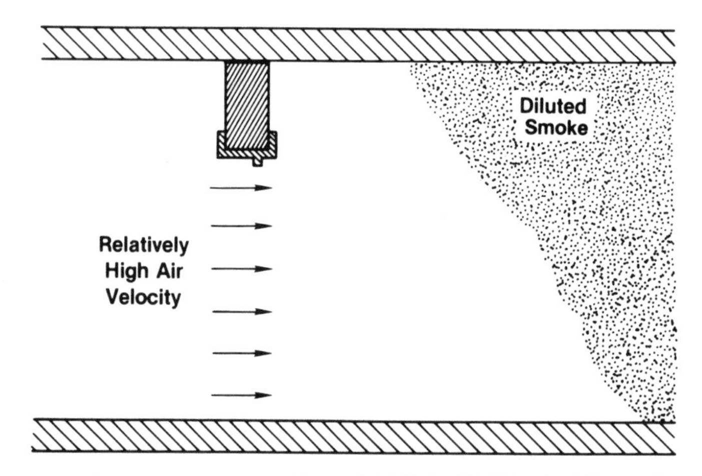

Fig. 8 No Smoke Backflow with High Air Velocity Through an Open Doorway

- Stack effect, buoyancy, and wind are less likely to overcome smoke control than with passive smoke management. In the absence of smoke control, these driving forces cause smoke movement to the extent that leakage paths allow. However, pressure differences and airflows of a smoke control system oppose these driving forces.
- Smoke control can use airflow to prevent smoke flow through an open doorway in a barrier. Doors in barriers are opened during evacuation and are sometimes accidentally left open or propped open throughout fires. In the absence of smoke control, smoke flow through these doors is common.

Smoke control systems should be designed so that a path exists for smoke movement to the outside; such a path relieves pressures of gas expansion due to the fire heat.

Dilution (or purging) of smoke in the fire space does not achieve smoke control, *i.e.*, smoke movement cannot be controlled by simply supplying and exhausting large quantities of air from the space or zone in which the fire is located. This supplying and exhausting of air is sometimes referred to as purging the smoke. Because a fire produces large quantities of smoke, purging cannot ensure breathable air in the fire space. In addition, purging cannot control smoke movement because it does not provide the needed airflow at open doors and the pressure differences across barriers. However, for spaces separated from the fire space by smoke barriers, purging can limit the level of smoke significantly.

Airflow

Theoretically, airflow can stop smoke movement through any space. However, the two places where air velocity is most commonly used to control smoke movement are open doorways and corridors. The following empirical relation for the critical velocity to prevent smoke from flowing upstream in a corridor is based on research by Thomas (1970):

$$V_k = 5.68\,(E/W)^{1/3} \tag{4}$$

where

V_k = critical air velocity to prevent smoke backflow, fpm
E = energy release rate into corridor, Btu/h
W = corridor width, ft

This relation can be used when the fire is located in the corridor or when the smoke enters the corridor through an open door, air transfer grille, or other opening. The critical velocities calculated

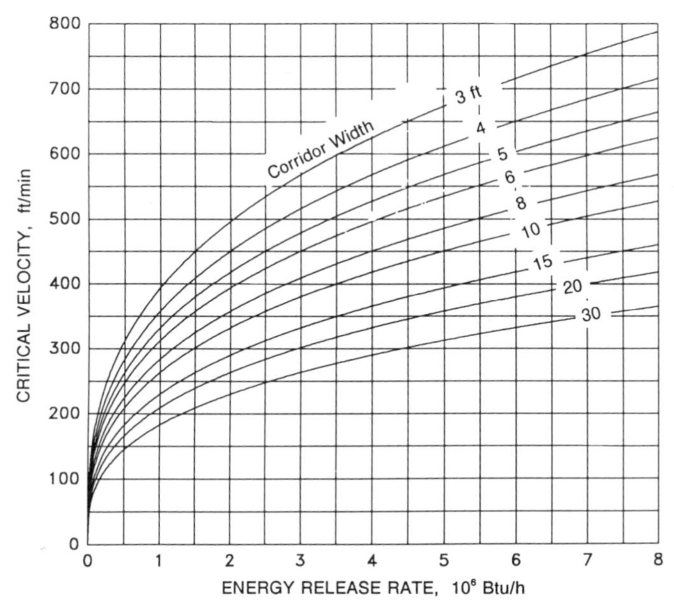

Fig. 9 Critical Velocity to Prevent Smoke Backflow

from Equation (4) are approximate because only an approximate value of K was used. However, critical velocities calculated from this relation indicate the air velocities required to prevent smoke backflow from fires of different sizes.

Equation (4) can be evaluated from Figure 9. For example, for an energy release rate of 0.512×10^6 Btu/h into a corridor 4.0 ft wide, the above relation yields a critical velocity of 286 fpm. However, for a larger energy release rate of 7.2×10^6 Btu/h, the relation yields a critical velocity of 690 fpm for a corridor of the same width.

In general, a high air velocity requires a smoke control system that is expensive and difficult to design. Airflow is most important in preventing smoke backflow through an open doorway that serves as a smoke control system boundary. Thomas (1970) indicates that Equation (4) gives an estimate of the airflow needed to prevent smoke backflow through a door. It is prohibitively expensive to design systems to maintain air velocities in doorways greater than 300 fpm.

Equation (4) is not appropriate for sprinklered fires with small temperature differences between the upstream air and downstream gases. Shaw and Whyte (1974) provide an analysis of a method to determine the velocity needed to prevent backflow of contaminated air through an open doorway. This analysis is specifically for small temperature differences and includes the effects of natural convection. For a sprinklered fire where the temperature difference is only 3.6 °F, an average velocity of 50 fpm would be the minimum velocity needed through a doorway to prevent smoke backflow. This temperature difference is small, and it is possible that larger values may be appropriate in many situations.

Even though airflow can control smoke movement, it is not the primary method of smoke control because of the large quantities of air required for such systems to be effective. The primary means is by air pressure differences across partitions, doors, and other building components.

Pressurization

The airflow rate through a construction crack, door gap, or other flow path is proportional to the pressure difference across that path raised to the power n. For a flow path of fixed geometry, n is theoretically in the range of 0.5 to 1. However, for all flow paths, except extremely narrow cracks, $n = 0.5$, and the flow can be expressed as:

$$Q = CA\sqrt{2\Delta p/\rho} \tag{5}$$

where

Q = volumetric airflow rate
C = flow coefficient
A = flow area (also called leakage area)
Δp = pressure difference across the flow path
ρ = density of air entering the flow path

The flow coefficient depends on the geometry of the flow path, as well as on turbulence and friction. In the present context, the flow coefficient is generally in the range of 0.6 to 0.7. For $\rho = 0.075$ lb/ft³ and $C = 0.65$, the flow equation above can be expressed as:

$$Q = 2610\,A\sqrt{\Delta p} \tag{6}$$

where

Q = volumetric flow rate, cfm
A = flow area, ft²
Δp = pressure difference across flow path, in. of water

The flow area is frequently the same as the cross-sectional area of the flow path. A closed door with a crack area of 0.11 ft² and a pressure difference of 0.01 in. of water would have an air leakage rate of approximately 29 cfm. If the pressure difference across the door is increased to 0.30 in. of water, the flow would be 157 cfm.

Frequently, in field tests of smoke control systems, pressure differences across partitions or closed doors have fluctuated by as much as 0.02 in. of water. These fluctuations have generally been attributed to wind, although they could have been due to the HVAC system or some other source. To control smoke movement, the pressure differences produced by a smoke control system must be sufficiently large so they are not overcome by pressure fluctuations, stack effect, smoke buoyancy, and the forces of wind. However, the pressure difference should not be so large that door opening problems result.

PURGING

As previously stated, purging cannot ensure breathable air in the fire space while the fire is burning and producing smoke. However, it can remove smoke from the fire space after fire extinction. After-fire purging is needed to allow the fire fighters to inspect and verify that the fire is totally extinguished. Traditionally, fire fighters purge by opening doors and breaking or opening windows. For spaces in which these techniques are not appropriate, an HVAC system with a purge mode of operation may be desirable.

While the systems discussed in this chapter are generally based on the two basic principles of smoke control, it is not always possible to maintain a sufficiently large airflow through open doors to prevent smoke from infiltrating a space that is intended to be protected. Ideally, open door occurrences will only happen for short periods during evacuation. Smoke that has entered such a space can be purged, or diluted, by supplying outside air to the space.

One example is a compartment isolated from a fire by smoke barriers and self-closing doors, so that no smoke enters the compartment when the doors are closed. However, when one or more of the doors is open, the airflow is insufficient to prevent smoke from flowing into the compartment from the fire space. For analysis, it is assumed that smoke is of uniform concentration throughout the compartment. When all the doors are closed, the concentration of contaminant in the compartment can be expressed as:

$$C/C_o = e^{-a\theta} \quad (7)$$

where

C_o = initial concentration of contaminant
C = concentration of contaminant at time θ
a = purging rate, number of air changes per minute
θ = time after doors closed, min
e = exponential base, approximately 2.718

The concentrations C_o and C must both be in the same units, and they can be any units appropriate for the particular contaminant being considered. McGuire *et al.* (1970) evaluated the maximum levels of smoke obscuration from a number of tests and a number of proposed criteria for tolerable levels of smoke obscuration. Based on this evaluation, the maximum levels of smoke obscuration are greater by a factor of 100 than those relating to the limit of tolerance. Thus, an area can be considered "reasonably safe" with respect to smoke obscuration if its atmosphere is not contaminated to an extent greater than 1% by the atmosphere prevailing in the immediate fire area. Such dilution would also reduce the concentrations of toxic smoke components. Toxicity is more complex, and no parallel statement has been made regarding the dilution needed to obtain a safe atmosphere with respect to toxic gases.

Equation (7) can be solved for the purging rate.

$$a = 1/\theta \ln (C_o/C) \quad (8)$$

For example, if the contaminant in a compartment is 20% of the burn room concentration when doors are open, and at 6 min after the door is closed, the contaminant concentration is 1% of the burn room, Equation (8) indicates that the compartment must be purged at a rate of one air change every 2 min.

In reality, the concentration of the contaminant is not uniform throughout the compartment. Because of buoyancy, higher concentrations of contaminant tend to be near the ceiling. Therefore, an exhaust inlet near the ceiling and a supply outlet near the floor would probably purge the smoke even faster than the above calculations indicate. Caution should be exercised in the location of the supply and exhaust points to prevent the supply air from blowing into the exhaust inlet, thus short-circuiting the purging operation.

DOOR-OPENING FORCES

The door-opening forces resulting from the pressure differences produced by a smoke control system must be considered. Unreasonably high door-opening forces can result in occupants having difficulty or being unable to open doors to refuge areas or escape routes.

The force required to open a door is the sum of the forces to overcome the pressure difference across the door and to overcome the door closer. This can be expressed as:

$$F = F_{dc} + 5.20\, W A\, \Delta p / 2(W-d) \quad (9)$$

where

F = the total door opening force, lb
F_{dc} = the force to overcome the door closer, lb
W = door width, ft
A = door area, ft^2
Δp = pressure difference across the door, in. of water
d = distance from the doorknob to the edge of the knob side of the door, ft

This relation assumes that the door-opening force is applied at the knob. Door-opening forces due to pressure difference can be determined from Figure 10 for a value of d = 3 in. The force to overcome the door closer is usually greater than 3 lb and, in some cases, can be as great as 20 lb. For a door that is 7 ft high and 36 in. wide, subject to a pressure difference of 0.30 in. of water, the total door-opening force is 30 lb, if the force to overcome the door closer is 12 lb.

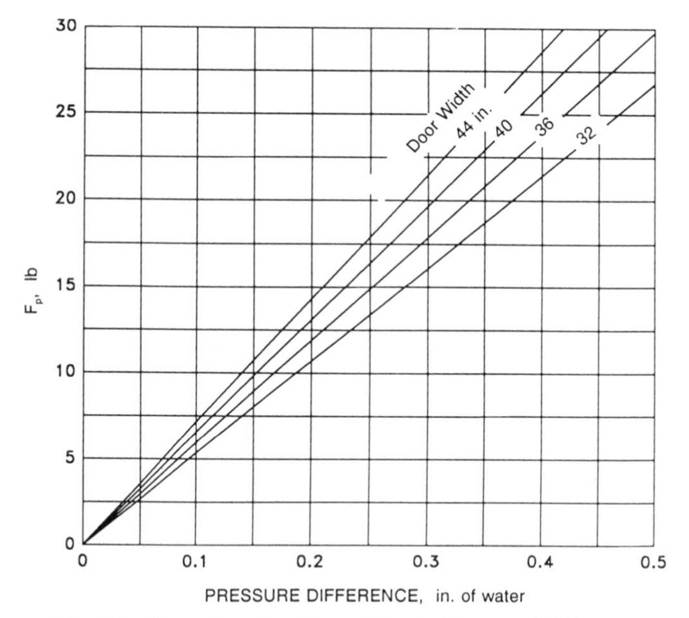

Fig. 10 Door-Opening Force Due to Pressure Differences

FLOW AREAS

In designing smoke control systems, airflow paths must be identified and evaluated. Some leakage paths, such as cracks around closed doors, open doors, elevator doors, windows, and air transfer grilles, are obvious. While construction cracks in building walls are less obvious, they are equally important.

The flow area of most large openings, such as open windows, can be calculated easily. However, flow areas of cracks are more difficult to evaluate. The area of these leakage paths depends on workmanship, *i.e.*, how well a door is fitted or how well weather stripping is installed. A door that is 36 in. by 7 ft with an average crack width of 1/8 in. has a leakage area of 0.21 ft^2. However, if this door is installed with a 3/4 in. undercut, the leakage area is 0.32 ft^2—a significant difference. The leakage area of elevator doors is in the range of 0.55 to 0.70 ft^2 per door.

For open stairwell doorways, Cresci (1973) found that complex flow patterns exist and that the resulting flow through open doorways was considerably below the flow calculated by using the geometric area of the doorway as the flow area in Equation (6). Based on this research, it is recommended that the design flow area of an open stairwell doorway be half that of the geometric area (door height times width) of the doorway. An alternate approach for open stairwell doorways is to use the geometric area as the flow area and use a reduced flow coefficient. Because it does not allow the direct use of Equation (6), this alternate approach is not used here.

Typical leakage areas for walls and floors of commercial buildings are tabulated as area ratios in Table 1. These data are based on a relatively small number of tests performed by the National Research Council of Canada (Tamura and Shaw 1976a, 1976b, 1978; Tamura and Wilson 1966). The area ratios are evaluated at typical airflows at 0.30 in. of water for walls and 0.10 in. of water for floors. Actual leakage areas depend primarily on workmanship rather than construction materials, and, in some cases, the flow areas in particular buildings may vary from the values listed. Data concerning air leakage through building components is also provided in Chapter 23 of the 1989 ASHRAE *Handbook—Fundamentals.*

Because the vent surface is usually covered by a louver and screen, the flow area of a vent is less than its area (vent height times width). Since the slats in louvers are frequently slanted, calculation of the flow area is further complicated. Manufacturers' data should be sought for specific information.

Table 1 Typical Leakage Areas for Walls and Floors of Commercial Buildings

Construction Element	Wall Tightness	Area Ratio A/A_w
Exterior building walls (includes construction cracks, cracks around windows and doors)	Tight	0.70×10^{-4}
	Average	0.21×10^{-3}
	Loose	0.42×10^{-3}
	Very Loose	0.13×10^{-2}
Stairwell walls (includes construction cracks, but not cracks around windows or doors)	Tight	0.14×10^{-4}
	Average	0.11×10^{-3}
	Loose	0.35×10^{-3}
Elevator shaft walls (includes construction cracks, but not cracks around doors)	Tight	0.18×10^{-3}
	Average	0.84×10^{-3}
	Loose	0.18×10^{-2}
		A/A_f
Floors (includes construction cracks and areas around penetrations)	Average	0.52×10^{-4}

A = leakage area; A_w = wall area; A_f = floor area

EFFECTIVE FLOW AREAS

The concept of effective flow areas is useful for analyzing smoke control systems. The paths in the system can be in parallel with one another, in series, or a combination of parallel and series paths. The effective area of a system of flow areas is the area that results in the same flow as the system when it is subjected to the same pressure difference over the total system of flow paths. This concept is similar to the effective resistance of a system of electrical resistances. The effective area for parallel leakage areas is the sum of the individual leakage paths:

$$A_e = \sum_{i=1}^{n} A_i \tag{10}$$

where n is the number of flow areas A_i in parallel.

For example, the effective area A_e for the three parallel leakage areas in Figure 11 is:

$$A_e = A_1 + A_2 + A_3 \tag{11}$$

If A_1 is 1.08 ft^2 and A_2 and A_3 are 0.54 ft^2 each, then the effective flow area, A_e, is 2.16 ft^2.

Three leakage areas in series from a pressurized space are illustrated in Figure 12. The effective flow area of these paths is:

$$A_e = (1/A_1^2 + 1/A_2^2 + 1/A_3^2)^{-1/2} \tag{12}$$

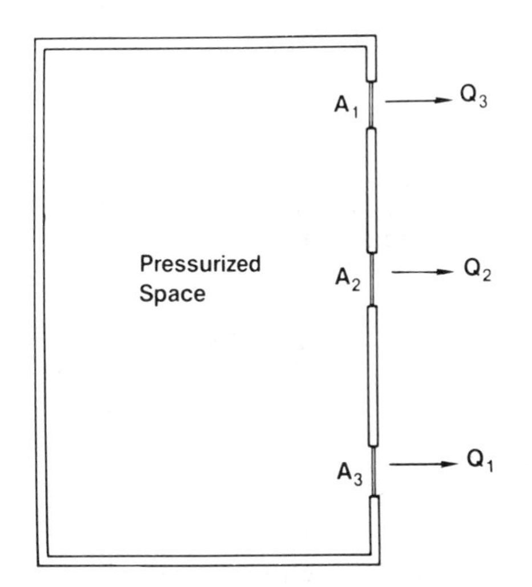

Fig. 11 Leakage Paths in Parallel

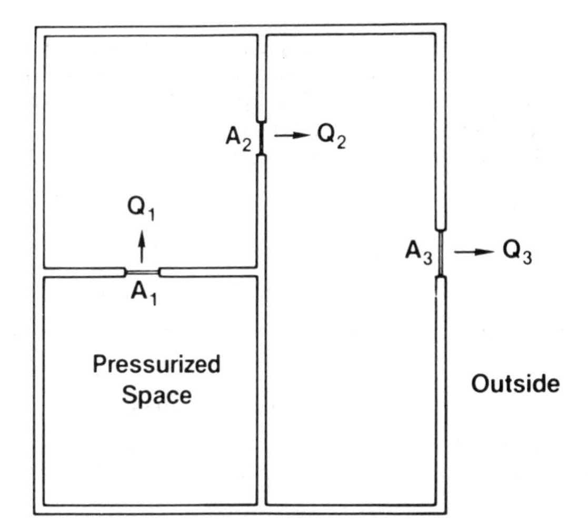

Fig. 12 Leakage Paths in Series

The general rule for any number of leakage areas in series is:

$$A_e = \left[\sum_{i=1}^{n} 1/A_i^2\right]^{-1/2} \tag{13}$$

where n is the number of leakage areas A_i in series. In smoke control analysis, there are frequently only two paths in series. Here the effective leakage area is:

$$A_e = A_1A_2/\sqrt{A_1^2 + A_2^2} \tag{14}$$

Example 1. Calculate the effective leakage area of two equal flow paths in series.

Let $A = A_1 = A_2 = 0.22$ ft²

$$A_e = A^2/(2A^2)^{0.5} = 2^{0.5} = 0.15 \text{ ft}^2$$

Example 2. Calculate the effective area of two flow paths in series, where $A_1 = 0.22$ ft² and $A_2 = 2.2$ ft².

$$A_e = A_1A_2\sqrt{A_1^2 + A_2^2} = 0.219 \text{ ft}^2$$

This example illustrates that when two areas are in series, and one is much larger than the other, the effective area is approximately equal to the smaller area.

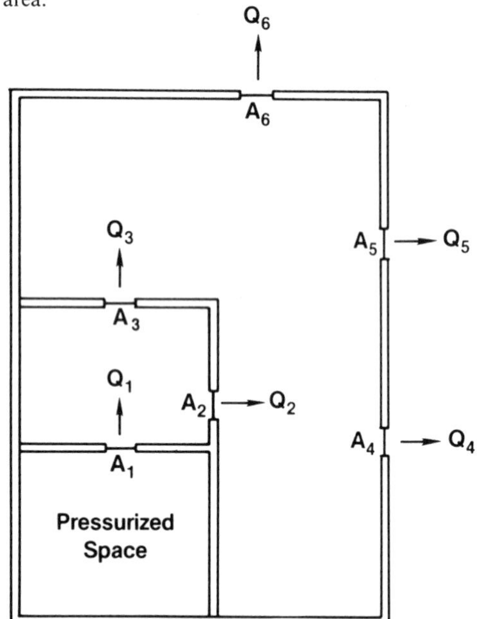

Fig. 13 Combination of Leakage Paths in Parallel and Series

The method of developing an effective area for a system of both parallel and series paths is to combine groups of parallel paths and series paths systematically. The system illustrated in Figure 13 is analyzed as an example. The figure shows that A_2 and A_3 are in parallel; therefore, their effective area is:

$$(A_{23})_{eff} = A_2 + A_3$$

Areas A_4, A_5, and A_6 are also in parallel, so their effective area is:

$$(A_{456})_{eff} = A_4 + A_5 + A_6$$

These two effective areas are in series with A_1. Therefore, the effective flow area of the system is given by:

$$A_e = [1/A_1^2 + 1/(A_{23})_{eff}^2 + (A_{456})_{eff}^2]^{-1/2}$$

Example 3. Calculate the effective area of the system in Figure 13, if the leakage areas are $A_1 = A_2 = A_3 = 0.22$ ft² and $A_4 = A_5 = A_6 = 0.11$ ft².

$(A_{23})_{eff} = 0.44$ ft²
$(A_{456})_{eff} = 0.33$ ft²
$A_e = 0.16$ ft²

SYMMETRY

The concept of symmetry is useful in simplifying problems. Figure 14 illustrates the floor plan of a multistory building that can be divided in half by a plane of symmetry. Flow areas on one side of the plane of symmetry are equal to corresponding flow areas on the other side. For a building to be treated in this manner, every floor of the building must be such that it can be divided in the same manner by the plane of symmetry. If wind effects are not considered in the analysis, or if the wind direction is parallel to the plane of symmetry, the airflow is only one-half of the total for the building analyzed. It is not necessary that the building be geometrically symmetric, as shown in Figure 14; it must be symmetric only with respect to flow.

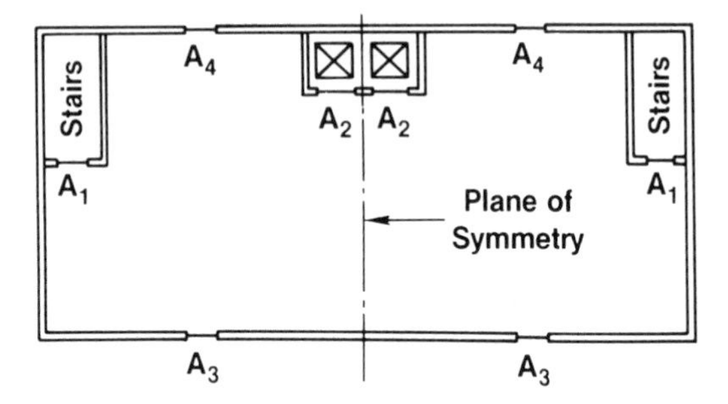

Fig. 14 Building Floor Plan Illustrating Symmetry Concept

DESIGN PARAMETERS

Ideally, codes should contain design parameters leading to the design of safe and economical smoke control systems. Unfortunately, because smoke control is a relatively new field, consensus has not been reached as to what constitutes reasonable design parameters. The designer must adhere to any smoke control criteria existing in appropriate codes or standards. If necessary, however, the designer should seek a waiver of local codes to ensure an effective smoke control system. In the absence of code requirements for specific parameters, the following discussion may be helpful.

Five areas for which design parameters must be established are (1) leakage areas, (2) weather data, (3) pressure differences, (4) airflow, and (5) number of open doors in the smoke control system.

Leakage areas have already been discussed. Windows in the fire compartment should be considered because they can affect pressure differences and airflow, depending on whether they are broken or not. Further, smoke control systems primarily aid in building evacuation and need to operate only during the time needed for evacuation.

Weather Data

Little weather data has been developed specifically for the design of smoke control systems. A designer may use the design temperatures for heating and cooling found in Chapter 24 of the 1989 ASHRAE *Handbook—Fundamentals.* In a normal winter, approximately 22 h are at or below the 99% design value, and approximately 54 h are at or below the 97.5% design value. Furthermore, extreme temperatures can be considerably lower than the winter design temperatures. For example, the 99% design temperature for Tallahassee, Florida, is 27 °F, but the lowest temperature observed there was −2 °F (NOAA 1979).

Temperatures are generally below the design values for short periods, and because of the thermal lag of building materials, these short intervals of low temperature usually do not cause problems with heating. However, there is no time lag for a smoke control system; thus it is subjected to all the extreme forces of stack effect that exist the moment it is operated. If the outside temperature is below the winter design temperature for which a smoke control system was designed, problems from stack effect may result. A similar situation can result with respect to summer design temperatures and reverse stack effect.

Wind data is needed for a wind analysis of a smoke control system. At present, no formal method of such an analysis exists, and the approach most generally taken is to design a smoke control system to minimize any effects of wind.

Pressure Differences

Both the maximum and minimum allowable pressure differences across the boundaries of smoke control should be considered. The maximum allowable pressure difference should not result in excessive door-opening forces.

A minimum allowable pressure difference across a boundary of a smoke control system might be that no smoke leakage occurs during building evacuation. In this case, the smoke control system must produce sufficient pressure differences to overcome forces of wind, stack effect, or buoyancy of hot smoke. The pressure differences due to wind and stack effect can be large in the event of a broken window in the fire compartment. Evaluation of these pressure differences depends on evacuation time, rate of fire growth, building configuration, and the presence of a fire-suppression system. The NFPA 92A (1988) suggests values of minimum and maximum design pressure difference.

Open Doors

Another design concern is the number of doors that could be opened simultaneously when the smoke control system is operating. A design that allows all doors to be open simultaneously may ensure that the system always works, but it probably adds to the cost of the system.

Deciding how many doors will be open simultaneously depends largely on building occupancy. For example, in a densely populated building, it is likely that all doors will be open during evacuation. However, if a staged evacuation plan or refuge area concept is incorporated in the building fire emergency plan, or if the building is sparsely occupied, only a few of the doors may be open during a fire.

FIRE AND SMOKE DAMPERS

Openings for ducts in walls and floors with fire resistance ratings should be protected by fire dampers and ceiling dampers, as required by local codes. Air transfer openings should also be protected. These dampers should be classified and labeled in accordance with UL 555.

A smoke damper can be used for either traditional smoke management (smoke containment) or for smoke control. In smoke management, a smoke damper inhibits the passage of smoke under the forces of buoyancy, stack effect, and wind. Generally, for smoke containment, smoke dampers should have low leakage characteristics at elevated temperatures. However, smoke dampers are only one of many elements (partitions, floors, doors) intended to inhibit smoke flow. In smoke management applications, the leakage characteristics of smoke dampers should be selected to be appropriate with the leakage of the other system elements.

In a smoke control system, a smoke damper inhibits the passage of air that may or may not contain smoke. Low leakage characteristics of a damper are not necessary when outside (fresh) air is on the high-pressure side of the damper, as is the case for dampers that shut off supply air from a smoke zone or that shut off exhaust air from a nonsmoke zone. In these cases, moderate leakage of smoke-free air through the damper does not adversely affect the control of smoke movement. It is best to design smoke control systems so that only smoke-free air is on the high-pressure side of a smoke damper.

Smoke dampers should be classified and listed in accordance with UL 555S. At locations requiring both smoke and fire dampers, combination dampers meeting the requirements of both can be used. Fire, ceiling, and smoke dampers should be installed in accordance with the manufacturer's instructions. NFPA *Standard* 90-A gives general guides regarding locations requiring these dampers.

PRESSURIZED STAIRWELLS

Many pressurized stairwells have been designed and built to provide a smoke-free escape route in the event of a building fire. They also provide a smoke-free staging area for fire fighters. On the fire floor, a pressurized stairwell must maintain a positive pressure difference across a closed stairwell door so that smoke infiltration is prevented.

During building fire situations, some stairwell doors are opened intermittently during evacuation and fire fighting, and some doors may even be blocked open. Ideally, when the stairwell door is opened on the fire floor, airflow through the door should be sufficient to prevent smoke backflow. Designing such a system is difficult because of the many combinations of open stairwell doors and weather conditions affecting airflow.

Stairwell pressurization systems may be single and multiple injection systems. A single injection system has pressurized air supplied to the stairwell at one location—usually at the top. Associated with this system is the potential of smoke entering the stairwell through the pressurization fan intake. Therefore, automatic shutdown during such an event should be considered.

For tall stairwells, single injection systems can fail when a few doors are open near the air supply injection point. Such a failure is especially likely when a ground-level stairwell door is open in bottom injection systems.

For tall stairwells, supply air can be supplied at a number of locations over the height of the stairwell. Figures 15 and 16 are two examples of multiple injection systems that can be used to overcome the limitations of single injection systems. In these figures, the supply duct is shown in a separate shaft. However, systems have been built that have eliminated the expense of a separate duct shaft by locating the supply duct in the stairwell itself. In such a case, care must be taken that the duct does not become an obstruction to orderly building evacuation.

STAIRWELL COMPARTMENTATION

Compartmentation of the stairwell into a number of sections is one alternative to multiple injection (Figure 17). When the doors between compartments are open, the effect of compartmentation

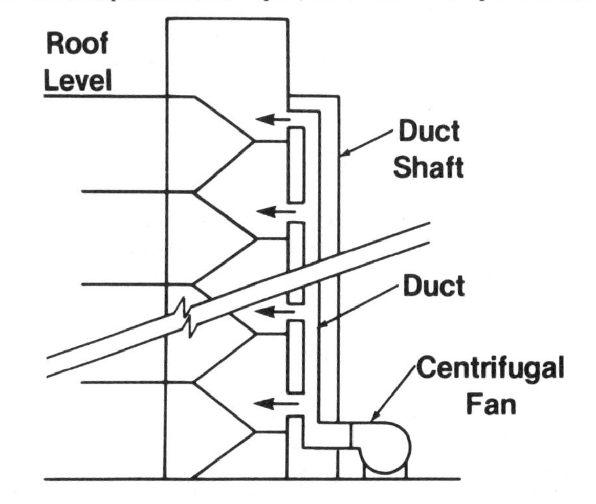

Fig. 15 Stairwell Pressurization by Multiple Injection with Fan Located at Ground Level

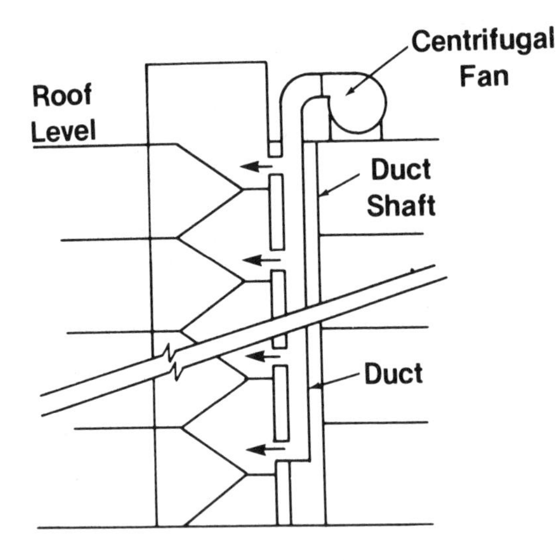

Fig. 16 Stairwell Pressurization by Multiple Injection with Roof-Mounted Fan

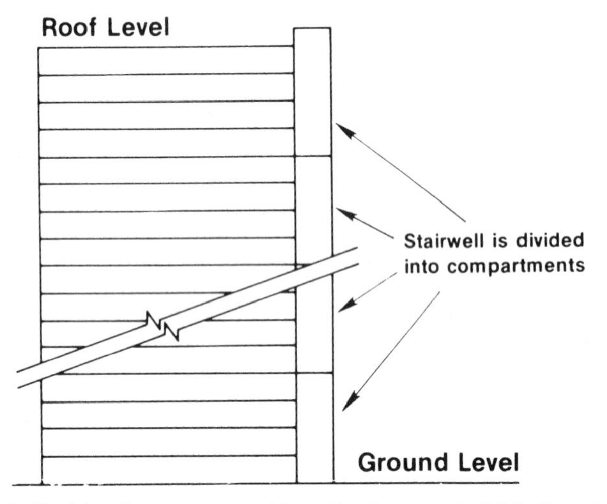

Note: Each four-floor compartment has at least one supply air injection point.

Fig. 17 Compartmentation of Pressurized Stairwell

is lost. For this reason, compartmentation is inappropriate for densely populated buildings where total building evacuation by the stairwell is planned in the event of fire. However, when a staged evacuation plan is used and when the system is designed to operate successfully when the maximum number of doors between compartments is open, compartmentation can be an effective means of providing stairwell pressurization for tall stairwells.

STAIRWELL ANALYSIS

This section presents an analysis for a pressurized stairwell in a building without vertical leakage. The performance of pressurized stairwells in buildings without elevators may be closely approximated by this method. It is also useful for buildings with vertical leakage in that it yields conservative results. Only one stairwell is considered in the building; however, the analysis can be extended to any number of stairwells by the concept of symmetry. For evaluation of vertical leakage through the building or with open stairwell doors, computer analysis is recommended. The analysis is for buildings where the leakage areas are the same for each floor of the building and where the only significant driving forces are the stairwell pressurization system and the temperature difference between the indoors and outdoors.

The pressure difference Δp_{sb} between the stairwell and the building can be expressed as:

$$\Delta p_{sb} = \Delta p_{sbb} + By/[1 + (A_{sb}/A_{bo})^2] \tag{15}$$

where

Δp_{sbb} = pressure difference from the stairwell to the building at the stairwell bottom, in. of water
y = distance above the stairwell bottom, ft
A_{sb} = flow area between the stairwell and the building (per floor), ft^2
A_{bo} = flow area between the building and the outside (per floor), ft^2
B = 7.64 $[1/(460 + t_o) - 1/(460 + t_s)]$
t_o = temperature of outside air, °F
t_s = temperature of stairwell air, °F

For a stairwell with no leakage directly to the outside, the flow rate of pressurization air is:

$$Q = 1740\, N A_{sb} \left(\frac{\Delta p_{sbt}^{3/2} - \Delta p_{sbb}^{3/2}}{\Delta p_{sbt} - \Delta p_{sbb}} \right) \tag{16}$$

where

Q = volumetric flow rate, cfm
N = number of floors
Δp_{sbt} = pressure difference from the stairwell to the building at the stairwell top, in. of water

Example 4. Each story of a 20-story stairwell is 10.8 ft high. The stairwell has a single-leaf door at each floor leading to the occupant space and one ground-level door to the outside. The exterior of the building has a wall area of 6030 ft^2 per floor. The exterior building walls and stairwell walls are of average leakiness. The stairwell wall area is 560 ft^2 per floor. The area of the gap around each stairwell door to the building is 0.26 ft^2. The exterior door is well gasketed, and its leakage can be neglected when it is closed.

For this example, the following design parameters are used: outside design temperatures, t_o = 14°F; stairwell temperature, t_s = 70°F; minimum design pressure differences when all stairwell doors are closed of 0.551 in. of water.

Using the leakage ratios for an exterior building wall of average tightness from Table 1, $A_{bo} = 6030(0.21 \times 10^{-3}) = 1.27$ ft^2. Using leakage ratios for a stairwell wall of average tightness from Table 1, the leakage area of the stairwell wall is $560(0.11 \times 10^{-3}) = 0.06$ ft^2. A^{sb} equals the leakage area of the stairwell wall plus the gaps around the closed doors. $A_{sb} = 0.06 + 0.26 = 0.32$ ft^2. The temperature factor B is calculated at 0.00170 in. of water/ft. The pressure difference at the stairwell bottom is selected as Δp_{sbb} = 0.080 in. of water to provide an extra degree of protection above the minimum allowable value of 0.052 in. of water. The pressure difference Δp_{sbt} is calculated from Equation (15) at 0.426 in. of water, using y = 217 ft. Thus, Δp_{sbt} does not exceed the maximum allowable pressure. The flow rate of pressurization air is calculated from Equation (16) at 8200 cfm.

The flow rate is highly dependent on the leakage area around the closed doors and on the leakage area that exists in the stairwell walls. In practice, these areas are difficult to evaluate and even more difficult to control. If the flow area A_{sb} in Example 4 were 0.54 ft^2 rather than 0.32 ft^2, a flow rate of pressurization air of 13,800 cfm would have been calculated from Equation (16). A fan with a sheave allows adjustment of supply air to offset for variations in actual leakage from the values used in design calculations.

STAIRWELL PRESSURIZATION AND OPEN DOORS

The simple pressurization system discussed previously has two limitations regarding open doors. First, when a stairwell door to the outside and doors to the building are open, the simple system cannot provide sufficient airflows through doorways to the building to prevent smoke backflow. Second, when stairwell doors are open, the pressure difference across the closed doors can drop to low levels. Two systems used to overcome these problems are overpressure relief and supply fan bypass. Research evaluating these systems is ongoing by Tamura (1990).

Overpressure Relief

The total airflow rate is selected to provide the minimum air velocity when a specific number of doors are open. When all the doors are closed, part of this air is relieved through a vent to prevent excessive pressure buildup, which could cause excessive door-opening forces. This excess air can be vented either to the building or to the outside. Since exterior vents can be subject to adverse effects of the wind, wind shields are recommended.

Barometric dampers that close when the pressure drops below a specified value can minimize the air losses through the vent when doors are open. Figure 18 illustrates a pressurized stairwell with overpressure relief vents to the building at each floor. In systems with vents between the stairwell and the building, the vents typically have one or more fire dampers in series with the barometric damper. As an energy conservation feature, these fire dampers are normally closed, but they open when the pressurization system is activated. This arrangement also reduces the possibility of the annoying damper chatter that frequently occurs with barometric dampers.

An exhaust duct can provide overpressure relief in a pressurized stairwell. This system is designed so that the normal resistance of a nonpowered exhaust duct maintains pressure differences within the design limits.

Exhaust fans can also relieve excessive pressures when all stairwell doors are closed. The fan should be controlled by a differential pressure sensor, so that it will not operate when the pressure difference between the stairwell and the building falls below a specific level. This control should prevent the fan from pulling smoke into the stairwell when a number of open doors have reduced stairwell pressurization. Such an exhaust fan should be specifically sized so that the pressurization system will perform within design limits. Because an exhaust fan can be adversely affected by the wind, a wind shield is recommended.

An alternate method of venting a stairwell is through an automatically opening stairwell door to the outside at ground level. Under normal conditions, this door would be closed and, in most cases, locked for security reasons. Provisions need to be made so that this lock does not conflict with the automatic operation of the system.

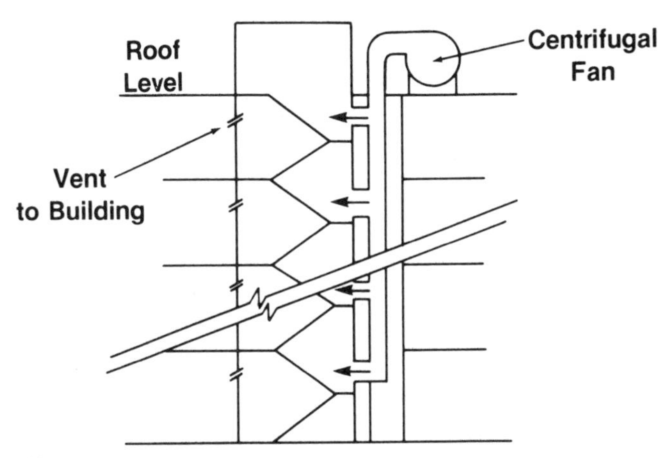

Notes:
1. Vents to the building have a barometric damper and one or two fire dampers in series.
2. A roof-mounted supply fan is shown; however, the fan may be located at any level.
3. A manually operated damper may be located at the stairwell top for smoke purging by the fire department.

Fig. 18 Stairwell Pressurization with Vents to the Building at Each Floor

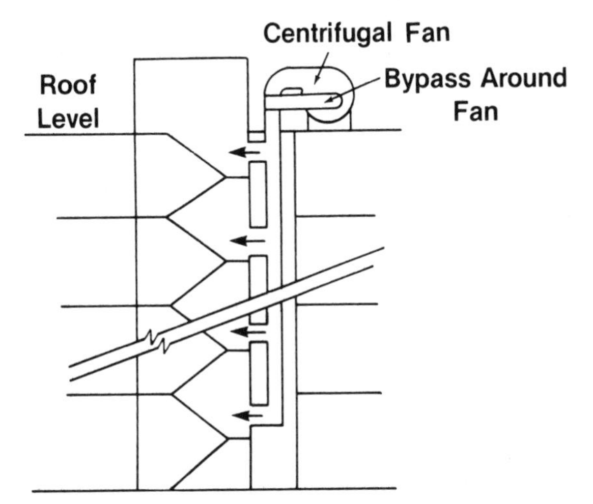

Notes:
1. Fan bypass controlled by one or more static pressure sensors located between the stairwell and the building.
2. A roof-mounted supply fan is shown; however, the fan may be located at any level.
3. A manually operated damper may be located at the stairwell top for smoke purging by the fire department.

Fig. 19 Stairwell Pressurization with Bypass Around Supply Fan

Possible adverse wind effects are also a concern with a system that uses an open outside door as a vent. Occasionally, high local wind velocities develop near the exterior stairwell door, and such winds are difficult to estimate without expensive modeling. Nearby obstructions can act as wind breaks or wind shields.

Supply Fan Bypass

In this system, the supply fan is sized to provide at least the minimum air velocity when the design number of doors are open. Figure 19 illustrates such a system. The flow rate of air into the stairwell is varied by modulating bypass dampers, which are controlled by one or more static pressure sensors that sense the pressure difference between the stairwell and the building. When all the stairwell doors are closed, the pressure difference increases and the bypass damper opens to increase the bypass air and decrease the flow of supply air to the stairwell. In this manner, excessive stairwell pressures and excessive pressure differences between the stairwell and the building are prevented.

ELEVATORS

Elevator smoke control systems intended for use by fire fighters should keep elevator cars, elevator shafts, and elevator machinery rooms smoke-free. Small amounts of smoke in these spaces are acceptable, provided that the environment is nontoxic and that the operation of the elevator equipment is not affected. Elevator smoke control systems intended for fire evacuation of the handicapped or other building occupants should also keep elevator lobbies smoke-free or nearly smoke-free. The long-standing obstacles to fire evacuation by elevators include (1) the logistics of evacuation, (2) the reliability of electrical power, (3) elevator door jamming, and (4) fire and smoke protection. All these obstacles, except smoke protection, can be addressed by existing technology (Klote 1984).

Klote and Tamura (1986) studied conceptual elevator smoke control systems for handicapped evacuation. The major problem was maintaining pressurization with open doors, especially doors on the ground floor. Of the systems evaluated, only one with a supply fan bypass with feedback control maintains adequate pressurization with any combination of open or closed doors. There are probably other systems capable of providing adequate smoke control, and the procedure used by Klote and Tamura can be viewed as an example of how to evaluate the performance of a system to meet the particular characteristics of a building under construction.

The transient pressures due to "piston effect" when an elevator car moves in a shaft have been a concern with regard to elevator smoke control. Piston effect is not a concern for slow-moving cars in multiple car shafts. However, for fast cars in single car shafts, the piston effect can be considerable.

ZONE SMOKE CONTROL

Klote (1990) conducted a series of tests on full-scale fires, which demonstrated that zoned smoke control can restrict smoke movement to the zone where a fire starts.

Pressurized stairwells are intended to prevent smoke infiltration into stairwells. However, in a building with only stairwell pressurization, smoke can flow through cracks in floors and partitions and through shafts to damage property and threaten life at locations remote from the fire. The concept of zone smoke control is intended to limit such smoke movement.

A building is divided into a number of smoke control zones, each zone separated from the others by partitions, floors, and doors that can be closed to inhibit smoke movement. In the event of a fire, pressure differences and airflows produced by mechanical fans limit the smoke spread from the zone in which the fire initiated. The concentration of smoke in this zone goes unchecked; thus, in zone smoke control systems, the occupants should evacuate the smoke zone as soon as possible after fire detection.

A smoke control zone can consist of one floor, more than one floor, or a floor can be divided into more than one smoke control zone. Some arrangements of smoke control zones are illustrated in Figure 20. All the nonsmoke zones in the building may be pressurized. The term *pressure sandwich* describes cases where only adjacent zones to the smoke zone are pressurized, as in Figures 20 (b) and (d).

Zone smoke control is intended to limit smoke movement to the smoke zone by the two principles of smoke control. Pressure differences in the desired direction across the barriers of a smoke zone can be achieved by supplying outside (fresh) air to nonsmoke zones, by venting the smoke zone, or by a combination of those methods.

Venting smoke from a smoke zone prevents significant overpressures due to thermal expansion of gases caused by the fire. However, venting results in only slight reduction of smoke concentration in the smoke zone. This venting can be accomplished by exterior wall vents, smoke shafts, and mechanical venting (exhausting).

COMPUTER ANALYSIS

Some design calculations associated with smoke control are appropriate for hand calculation. However, other calculations involve time-consuming, trial-and-error solutions that are more appropriately left to a computer. The National Institute of Standards and Technology has developed a computer program specifically for analysis of smoke control systems (Klote 1982). A number of other programs have also been developed. Some calculate steady-state airflow and pressures throughout a building (Sander 1974, Sander and Tamura 1973). Other programs go beyond this to calculate the smoke concentrations that would be produced throughout a building in the event of a fire (Yoshida *et al.* 1979, Evers and Waterhouse 1978, Wakamatsu 1977, Rilling 1978).

Each of these programs is different; however, the basic concepts are essentially the same. A building is represented by a network of spaces or nodes, each at a specific pressure and temperature.

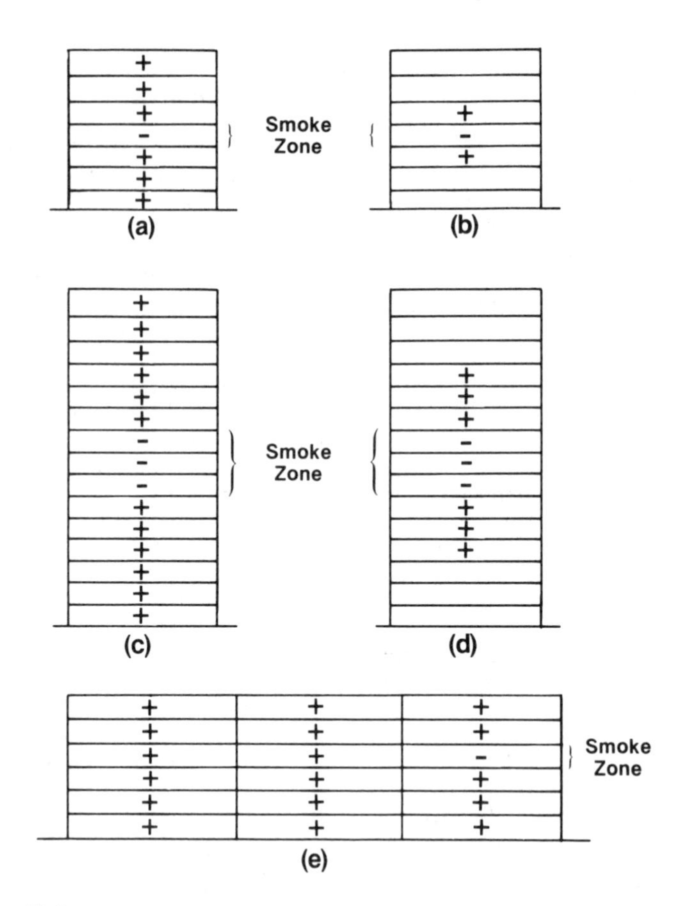

Note:
In the above figures, the smoke zone is indicated by a minus sign, and pressurized spaces are indicated by a plus sign. Each floor can be a smoke control zone as in (a) and (b), or a smoke zone can consist of more than one floor as in (c) and (d). All the non-smoke zones in a building may be pressurized as in (a) and (c) or only non-smoke zones adjacent to the smoke zone may be pressurized as in (b) and (d). A smoke zone can also be limited to a part of a floor as in (e).

Fig. 20 Some Arrangements of Smoke Control Zones

Stairwells and other shafts are modeled by a vertical series of spaces, one for each floor. Air flows through leakage paths from regions of high pressure to regions of low pressure. These leakage paths are doors and windows that may be opened or closed. Leakage can also occur through partitions, floors, and exterior walls and roofs. The airflow through a flow path is a function of the pressure difference across the path, as presented in Equation (5).

Air from outside the building can be introduced by a pressurization system into any level of a shaft or even into other building compartments. This allows simulation of stairwell pressurization. In addition, any building space can be exhausted, which allows simulation of zoned smoke control systems. The pressures throughout the building and flow rates through all the flow paths are obtained by solving the airflow network, including the driving forces such as wind, the pressurization system, or an inside-to-outside temperature difference.

ACCEPTANCE TESTING

Regardless of the care, skill, and attention to detail with which a smoke control system is designed, an acceptance test is needed as assurance that the system, as built, operates as intended.

An acceptance test should be composed of two levels of testing. The first is of a functional nature to determine if everything in the system works as it is supposed to, *i.e.*, an initial check of the system components. The importance of the initial check has become apparent because of the problems encountered during tests of smoke control systems. These problems include fans operating backward, fans to which no electrical power was supplied, and controls that did not work properly.

The second level of testing is of a performance nature to determine if the system performs adequately under all required modes of operation. This can consist of measuring pressure differences across barriers under various modes of smoke control system operation. In cases where airflows through open doors are important, these should be measured. Chemical smoke from smoke candles (sometimes called smoke bombs) is not recommended for performance testing because it normally lacks the buoyancy of hot smoke from a real building fire. Smoke near a flaming fire has a temperature in the range of 1000 to 2000°F. Heating chemical smoke to such temperatures to emulate smoke from a real fire is not recommended unless precautions are taken to protect life and property. The same comments about buoyancy apply to tracer gases. Thus, pressure difference testing is the most practical performance test. A guide specification for acceptance testing is available from the Smoke Control Association (1985).

REFERENCES

ASTM. 1985. Terminology relating to fire standards (rev. B). *Standard* E176-85. American Society for Testing and Materials, Philadelphia.

Cresci, R.J. 1973. Smoke and fire control in high-rise office buildings—Part II, Analysis of stair pressurization systems. Symposium on Experience and Applications on Smoke and Fire Control, ASHRAE Annual Meeting, June.

Evers, E. and A. Waterhouse. 1978. A computer model for analyzing smoke movement in buildings. Building Research Est., Fire Research Station, Borehamwood, Herts, England.

Fang, J.B. 1980. Static pressures produced by room fires. National Bureau of *Standards* (U.S.), NBSIR 80-1984.

Houghton, E.L. and N.B. Carruther. 1976. *Wind forces on buildings and structures.* John Wiley & Sons, New York.

Klote, J.H. 1980. Stairwell pressurization. ASHRAE *Transactions* 86(1):604-73.

Klote, J.H. 1982. A computer program for analysis of smoke control systems. National Bureau of *Standards* (U.S.), NBSIR 82-2512.

Klote, J.H. 1984. Smoke control for elevators. ASHRAE *Journal* 26(4):23-33.

Klote, J.H. and J.W. Fothergill. 1983. Design of smoke control systems for buildings. ASHRAE, Atlanta, GA.

Klote, J.H. and G.T. Tamura. 1986. Smoke control and fire evacuation by elevators. ASHRAE *Transactions* 92(1).

MacDonald, A.J. 1975. *Wind loading on buildings.* John Wiley & Sons, New York.

McGuire, J.H. and G.T. Tamura. 1975. Simple analysis of smoke flow problems in high buildings. *Fire Technology* 11(1):15-22 (February).

McGuire, J.H., G.T. Tamura, and A.G. Wilson. 1970. Factors in controlling smoke in high buildings. Symposium on Fire Hazards in Buildings. ASHRAE Semiannual Meeting, January.

NFPA. 1986. *Fire protection handbook*, 16th ed. National Fire Protection Association, Quincy, MA.

NFPA. 1985a. Installation of air-conditioning and ventilating systems. NFPA 90A-1985. National Fire Protection Association, Quincy, MA.

NFPA. 1985b. Code for safety to life from fire in buildings and structures. NFPA 101-1985, National Fire Protection Association, Quincy, MA.

NOAA. 1979. Temperature extremes in the United States. National Oceanic and Atmospheric Administration (U.S.), National Climatic Center, Asheville, NC.

Rilling, J. 1978. Smoke study, 3rd Phase, Method of calculating the smoke movement between building spaces. Centre Scientifique et Technique du Batiment (CSTB), Champs Sur Marne, France (September).

Sachs, P. 1972. Wind forces in engineering. Pergamon Press, New York.

Sander, D.M. and G.T. Tamura. 1973. FORTRAN IV Program to simulate air movement in multi-story buildings. DBR Computer Program No. 35. National Research Council, Canada.

Sander, D.M. 1974. FORTRAN IV Program to calculate air infiltration in buildings. DBR Computer Program No. 37. National Research Council, Canada (May).

Shaw, B.H. and W. Whyte. 1974. Air movement through doorways—The influence of temperature and its control by forced airflow. *Building Services Engineer* 42(December):210-18.

Shaw, C.Y. and G.T. Tamura. 1977. The calculation of air infiltration rates caused by wind and stack action for tall buildings. ASHRAE *Transactions* 83(2):145-58.

Simiu, E. and R.H. Scanlan. 1978. *Wind effects on structures: An introduction to wind engineering.* John Wiley & Sons, New York.

Smoke Control Association. 1985. *Smoke control system testing.* Buckingham, PA.

Tamura, G.T. and C.Y. Shaw. 1976a. Studies on exterior wall air tightness and air infiltration of tall buildings. ASHRAE *Transactions* 83(1):122-34.

Tamura, G.T. and C.Y. Shaw. 1976b. Air leakage data for the design of elevator and stair shaft pressurization systems. ASHRAE *Transactions* 83(2):179-90.

Tamura, G.T. and C.Y. Shaw. 1978. Experimental studies of mechanical venting for smoke control in tall office buildings. ASHRAE *Transactions* 86(1):54-71.

Tamura, G.T. and A.G. Wilson. 1966. Pressure differences for a 9-story building as a result of chimney effect and ventilation system operation. ASHRAE *Transactions* 72(1):180-89.

Thomas, P.H. 1970. Movement of smoke in horizontal corridors against an airflow. *Institution of Fire Engineers Quarterly* 30(77):45-53.

UL. 1981. Fire dampers and ceiling dampers, UL *Standard* 555. Underwriters Laboratories, Northbrook, IL.

UL. 1985. Leakage rated dampers for use in smoke control systems, UL *Standard* 555S. Underwriters Laboratories, Northbrook, IL.

Wakamatsu, T. 1977. Calculation methods for predicting smoke movement in building fires and designing smoke control systems. Fire *Standards* and Safety, ASTM STP 614. American Society for Testing and Materials, Philadelphia, 168-93.

Yoshida, H., C.Y. Shaw, and G.T. Tamura. 1979. A FORTRAN IV Program to calculate smoke concentrations in a multi-story building. DBR Computer Program No. 45. National Research Council, Canada.

CHAPTER 48

RADIANT HEATING AND COOLING

RADIANT heating and cooling applications are defined as low, medium, and high intensity if the surface or source temperature exceeds 300 °F; applications are considered low-temperature if they are below 300 °F.

LOW, MEDIUM, AND HIGH INTENSITY INFRARED

Low, medium, and high intensity infrared heaters are compact, self-contained direct heating devices used primarily in hangars, factories, greenhouses, and gymnasiums, as well as in areas such as loading docks, racetrack stands, under marquees, outdoor restaurants, and swimming pool lounge areas. Infrared heating is also used for snow control and in process heating, such as paint baking and drying. These units may be electric, gas-fired, or oil-fired. Classification of the equipment is identified by the source temperature as follows:

- Low intensity, with source temperatures to 1200 °F
- Medium intensity, with source temperatures to 1800 °F
- High intensity, with source temperatures to 5000 °F

The source temperature is determined by such factors as the source of energy, configuration, and size. Reflectors can be used to direct distribution of radiation in specific patterns.

Radiant energy is transmitted by electromagnetic waves that heat solid objects but not the air through which the energy is transmitted. Because of this characteristic, radiant energy is effective both for spot heating and for entire building space heating requirements.

As floors and solid objects are heated by infrared radiation, they in turn reradiate at lower source temperatures. Furthermore, the air in contact with the warm surfaces is heated, which raises the ambient temperature of the air by convection.

Additional information on radiant equipment is available in Chapter 16 of the 1987 ASHRAE *Handbook—HVAC Systems and Equipment* and in Chapter 29 of the 1988 ASHRAE *Handbook—Equipment*.

LOW-TEMPERATURE HEATING AND COOLING

Radiant panel heating systems include the following:

- Metal ceiling panels
- Embedded piping in ceilings, walls, or floors
- Air-heated floors
- Electric ceiling or wall panels
- Electric heating cable in ceilings or floors
- Deep heat, a modified storage system using electric heating cable in the ceiling or floor

With these systems, a controlled surface temperature radiates more than 50% of the heat transfer to other surfaces. They are used in residences, perimeter heating for office buildings, classrooms, hospital patient rooms, swimming pool lounge areas, repair garages, and industrial and warehouse applications. Additional information is available in Chapter 7 of the 1987 ASHRAE *Handbook—HVAC Systems and Applications*.

Some radiant panel systems combine controlled heating and cooling surfaces with central station air conditioning. The controlled temperature surfaces may be in the floor, walls, or ceiling, with the temperature maintained by circulating water, air, or electric resistance. The central station can be a basic, one-zone, constant temperature, or constant volume system, or it can incorporate some or all the features of dual-duct, reheat, multizone, or variable volume systems. When used in combination with other water/air systems, radiant panels provide zone control for temperature and humidity control.

Metal ceiling panels integrated into the central heating and cooling systems may be incorporated to provide individual room or zone heating and cooling. These panels can be designed as small units to fit the building module or they can be arranged as large continuous areas for economy. Room thermal conditions are maintained primarily by direct transfer of radiant energy, normally using four-pipe hot and chilled water. These metal ceiling panel systems have generally been used in hospital patient rooms.

The application of metal ceiling panel systems is discussed in Chapter 7 of the 1987 ASHRAE *Handbook—HVAC Systems and Applications*.

ELEMENTARY DESIGN RELATIONSHIPS

When using radiant heating for human comfort, five concepts describe the temperature and energy characteristics of the total radiant environment.

- *Mean radiant temperature* (MRT) $\bar{t}_r$ is the temperature of an imaginary isothermal black enclosure in which an occupant exchanges the same amount of heat by radiation as in an actual nonuniform environment.
- *Ambient air temperature* t_a is the temperature of the air surrounding the occupant.
- *Operative temperature* t_o is the temperature of a uniform isothermal black enclosure in which the occupant exchanges the same amount of heat by radiation and convection as in an actual nonuniform environment.

  At air velocities of 80 fpm or less and mean radiant temperatures less than 120 °F, operative temperature is approximately the average of the air and mean radiant temperatures, and equal to the adjusted dry-bulb temperature.
- *Adjusted dry-bulb temperature* is the average of the air temperature and the mean radiant temperature at a given location. The adjusted dry-bulb temperature is approximately equivalent to

The preparation of this chapter is assigned to TC 6.5, In-Space Heating and Cooling.

operative temperature at air motions less than 80 fpm, when mean radiant temperature is less than 120°F.

- *Effective radiant flux* (ERF) is defined as the *net radiant heat exchanged* at the ambient temperature t_a between an occupant, whose surface is hypothetical, and all enclosing surfaces and directional heat sources and sinks. Thus, ERF is the net radiant energy received by the occupant from all surfaces and sources whose temperatures *differ* from the ambient air t_a. This concept is particularly useful in high intensity radiant heating applications.

The relationship between the aforementioned factors is shown for an occupant at surface temperature t_{sf}, exchanging sensible heat H_m in a room with ambient air temperature t_a and mean radiant temperature $\bar{t}_r$. The linear radiative and convective heat transfer coefficients are h_r and h_c, respectively; the latter coefficient is a function of the air movement V present. The heat balance equation is:

$$q_m = h_r (t_{sf} - \bar{t}_r) + h_c (t_{sf} - t_a) \quad (1)$$

During thermal equilibrium, q_m is metabolic heat less work less evaporative cooling by sweating. By definition of operative temperature,

$$q_m = (h_r + h_c)(t_{sf} - t_o) \quad (2)$$

Thus

$$t_o = (h_r \bar{t}_r + h_c t_a)/(h_r + h_c) \quad (3)$$

and $h = h_r + h_c$. Thus, t_o is an average of $\bar{t}_r$ and t_a weighted by their respective heat transfer coefficients; it represents how people sense the thermal level of their total environment as a single temperature.

By rearranging Equation (1),

$$q_m + h_r(\bar{t}_r - t_a) = h (t_{sf} - t_a) \quad (4)$$

where $h_r (\bar{t}_r - t_a)$ is, by definition, the effective radiant field (ERF) and represents the radiant energy absorbed by the occupant from all temperature sources different from the ambient t_a.

The principal relationships between $\bar{t}_r$, t_a, t_o, and ERF are as follows:

$$\text{ERF} = h_r(\bar{t}_r - t_a) \quad (5)$$

$$\text{ERF} = h(t_o - t_a) \quad (6)$$

$$\bar{t}_r = t_a + \text{ERF}/h_r \quad (7)$$

$$t_o = t_a + \text{ERF}/h \quad (8)$$

$$\bar{t}_r = t_a + (h/h_r)(t_o - t_a) \quad (9)$$

$$t_o = t_a + (h_r/h)(\bar{t}_r - t_a) \quad (10)$$

In Equations (1) through (10), the radiant environment is treated as a blackbody with temperature $\bar{t}_r$. The effect of the emissivity of the source, radiating at absolute temperature in degrees Rankine, and the absorptance of the skin and clothed surfaces is reflected in the effective values of $\bar{t}_r$ or ERF and not in the linear radiation exchange coefficient h_r, which is given in general by:

$$h_r = 4\sigma f_{eff} [(\bar{t}_r + t_a)/2 + T]^3 \quad (11)$$

where

f_{eff} = ratio of radiating surface of the human body to its total DuBois surface area A_D = 0.71
σ = Stefan-Boltzmann constant = 0.1713×10^{-8} Btu/h·ft²·°R⁴
T = 460

The convection coefficient for an occupant is:

$$h_c = 0.107\ V^{0.5} \quad (12)$$

where

h_c = convection coefficient, Btu/h·°F
V = air movement, fpm

When $\bar{t}_r > t_a$, the radiant flux ERF adds heat to the body system; when $t_a > \bar{t}_r$, heat is lost from the body system by radiant cooling. ERF is independent of the surface temperature of the occupant and can be measured directly by a black globe thermometer or any blackbody radiometer or flux meter using the ambient air t_a as its heat sink.

In the above definitions, the body clothing and skin surface have been treated as a blackbody, exchanging radiation with an imaginary blackbody surface at temperature $\bar{t}_r$. However, the effectiveness of any individual radiating source on human occupants is governed by the absorptance α of the skin and clothing surface for the color temperature (in °R) of the radiating source. The relationship between α and temperature is illustrated in Figure 1. The values for α are those expected relative to a matte black surface normally used on the globe thermometer or any radiometer for measuring radiant energy. For radiators below 1700°F, skin and clothing surfaces may be considered essentially blackbodies for design purposes. A gas radiator usually operates at 1700°F; a quartz lamp, for example, radiates at 4000°F with 240 V; and the sun's radiating temperature is 10,000°F. The use of α in estimating the ERF and t_o caused by sources radiating at temperatures above 1700°F is discussed later.

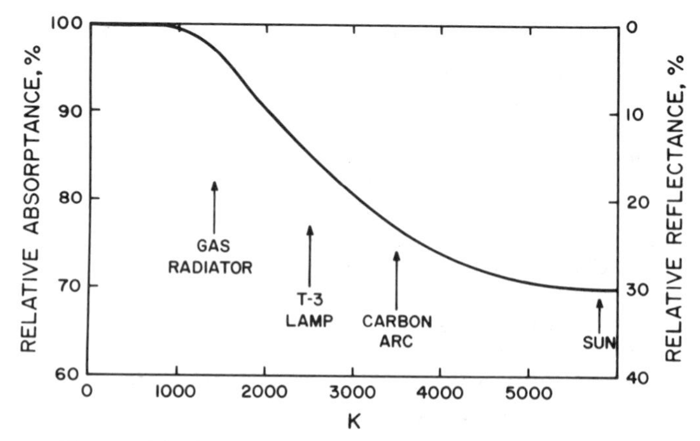

Fig. 1 Relative Absorptance and Reflectance of Skin and Typical Clothing Surfaces at Various Color Temperatures

DESIGN CRITERIA FOR ACCEPTABLE RADIANT HEATING

Perceptions of comfort, temperature, and thermal acceptability are related to activity, the transfer of body heat from the skin to the environment, and the resulting physiological adjustments and body temperatures. Heat transfer is affected by the ambient air temperature, thermal radiation, air movement, humidity, and clothing worn. Thermal sensation is described by feelings of hot, warm, slightly warm, neutral, slightly cool, cool, and cold. An acceptable environment is one in which at least 80% of the occupants would perceive a thermal sensation between "slightly cool" and "slightly warm." Comfort is associated with a neutral thermal sensation during which the human body regulates its internal temperature with a minimum of physiological effort in the activity concerned. Warm discomfort, in contrast, is related primarily to physiological strain necessary to maintain the body's thermal equilibrium rather than to the temperature sensation experienced. For a full discussion on the interrelation of the above physical, psychological, and physiological factors, refer to Chapter 8 of the 1989 ASHRAE *Handbook—Fundamentals.*

ASHRAE studies show a linear relationship between the clothing insulation worn and the t_o for comfort. This is illustrated for a sedentary subject in Figure 2, and Figure 3 shows how activity and clothing together affect the t_o for comfort. Both figures use a constant humidity at 40 to 60% rh. Figure 4 shows the effects of high and low humidity for a sedentary person wearing average clothing.

A comfortable t_o at 50% rh becomes slightly warmer as humidity rises or slightly cooler as humidity lowers below the 50% level. Changes in humidity have a much greater effect on "warm" and "hot" discomfort. In contrast, cold discomfort is only slightly affected by humidity and is very closely related to a "cold" thermal sensation. Figures 2, 3, and 4 are adapted from those in ANSI/ASHRAE *Standard* 55-1981, Thermal Environmental Conditions for Human Occupancy.

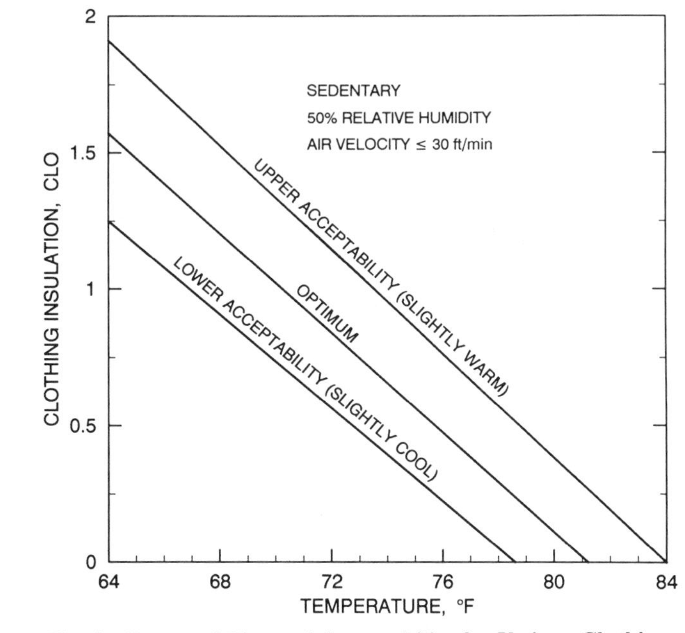

Fig. 2 Range of Thermal Acceptability for Various Clothing Insulations and Operative Temperatures

The desired specifications for any radiant heating installation designed for human occupancy and acceptability must involve the following steps:

1. Define the probable activity (metabolism) and clothing worn by the occupant and the air movement in the occupied space. The following are two examples:

Case 1: Sedentary (1.1 met)
Clothing insulation = 0.6 clo; air movement = 30 ft/min

Case 2: Light work (2 met)
Clothing insulation = 0.9 clo; air movement = 100 ft/min

2. Select from either Figure 2 or Figure 3 the optimum t_o for comfort and acceptability:

Case 1: $t_o = 74.5\,°F$; *Case* 2: $t_o = 62.5\,°F$

3. For the ambient air temperature t_a concerned, calculate the mean radiant temperature $\bar{t}_r$ and/or ERF necessary for comfort and thermal acceptability.

Case 1: For $t_a = 60\,°F$
Solve for h_r from Equation (11) by assuming $\bar{t}_r = 90\,°F$.

$$h_r = 4 \times 0.1713 \times 10^{-8} \times 0.71\,[(90 + 60)/2 + 460] = 0.745$$

Solve for h_c from Equation (12).

$$h_c = 0.107\,(30)^{0.5} = 0.586$$

and

$$h = h_r + h_c - 0.745 + 0.586 = 1.331\ \text{Btu/h}\cdot\text{ft}^2\cdot°\text{F}$$

By Equation (4), ERF for comfort equals:

$$1.331\,(74.5 - 60) = 19.3\ \text{Btu/h}\cdot\text{ft}^2$$

and by Equation (9):

$$\bar{t}_r = 60 + (1.331/0.745)(74.5 - 60) = 85.9\,°F$$

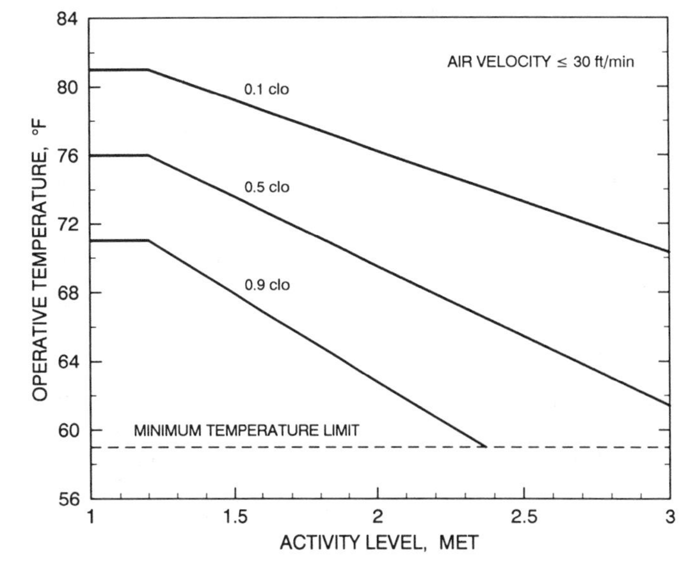

Fig. 3 Optimum Operative Temperatures for Active People in Low Air Movement Environments

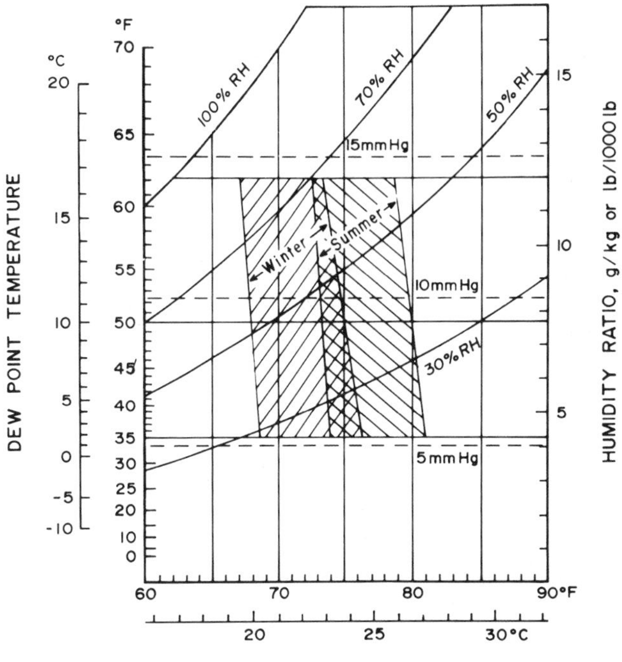

Fig. 4 ASHRAE Comfort Chart Modified for Radiant Heating

Case 2: For $t_a = 50\,°F$ at 50% rh and assuming $\bar{t}_r = 85\,°F$:

$$h_r = 4 \times 0.1713 \times 10^{-8} \times 0.71\,[(85 + 50)/2 + 460] = 0.714$$

$$h_c = 0.107\,(100)^{0.5} = 1.07$$

$$h = 0.724 + 1.07 = 1.784\ \text{Btu/h}\cdot\text{ft}^2\cdot°\text{F}$$

For comfort, $\text{ERF} = 1.784\,(62.5 - 50) = 22.3\ \text{Btu/h}\cdot\text{ft}^2$

$$\bar{t}_r = 50 + 22.3/0.714 = 81.2\,°F$$

The t_o for comfort, predicted by Figure 2, falls on the "slightly cool" side when humidity is low; for very high humidities, the predicted t_o for comfort appears to be "slightly warm." This small effect on comfort may be seen in the modified ASHRAE Comfort Chart (Figure 4).

For example, for high humidity at t_{dp} = 55.5°F, the t_o for comfort is

Case 1: t_o = 77°F, compared with 74.5°F at 50% rh.

Case 2: t_o = 59.9°F, compared with 62.5°F at 50% rh.

For installations where thermal acceptability is the primary consideration, humidity can be ignored in preliminary design specifications. However, for many conditions where radiant heating and the work level cause sweating and high heat stress, humidity is a major consideration.

DESIGN FOR BEAM RADIANT HEATERS

Spot beam radiant heat improves comfort at a specific location in a large, poorly heated work area. The design problem is specifying the type, capacity, and orientation of the beam heater.

Using the same reasoning for Equations (1) through (10), the radiant flux (ΔERF), which must be added to an unheated work space having operative temperature t_{uo} that results in a t_o for comfort (given by Figure 2 or 3), is:

$$\Delta\text{ERF} = h(t_o - t_{uo}) \tag{13}$$

or

$$t_o = t_{uo} + \Delta\text{ERF}/h \tag{14}$$

This equation is unaffected by air movement. The heat transfer coefficient h for the occupant in Equation (13) is always given by Equations (11) and (12).

The radiant field term ERF is, by definition, the energy absorbed per unit total body surface A_D (DuBois area) and *not* the total effective radiating area A_{eff} of the body.

Geometry of Beam Heating

Figure 5 illustrates the parameters that must be considered in specifying a beam radiant heater designed to produce the ERF, or $\bar{t}_r$, necessary for comfort at an occupant's workstation. They are as follows:

I_K = irradiance from beam heater, Btu/h·sr
K = absolute irradiating temperature of beam heater, °R
β = elevation angle of heater in degrees (at 0°, the beam is horizontal)
Φ = azimuth angle of heater in degrees (at 0°, the beam is facing the subject)
d = distance from beam heater to center of occupant, ft
A_p = projected area of occupant on a plane normal to direction of heater beam (Φ, β), ft²
α_K = absorptance of skin-clothing surface at emitter temperature, °F (see Figure 1)
Ω = solid angle of heater beam, steradians

For additional information on radiant flux distribution patterns and sample calculations of radiation intensity I_k and effective radiant fields (ERF), refer to Chapter 16 of the 1987 ASHRAE *Handbook—Systems and Applications.*

FLOOR RERADIATION

In most radiant heater installations, local floor areas are strongly irradiated. The floor absorbs most of this energy and warms to an equilibrium temperature t_f higher than that of the ambient air t_a and the unheated room enclosure surfaces. The temperature of the unheated areas (walls, ceiling, and floor distant from heaters) can be assumed to be the same as the air temperature. Part of the energy directly absorbed by the floor is transmitted by conduction to the cooler underside (or, for slabs-on-grade, to the ground), part is convected to room air, and the remainder is reradiated. The warmer floor will raise effective ERF or $\bar{t}_r$ over that caused by the heater alone.

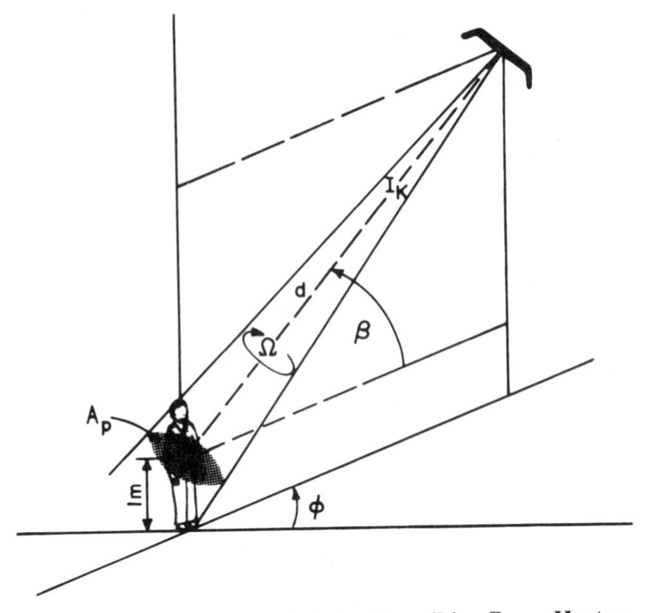

Fig. 5 Geometry and Symbols for Describing Beam Heaters

For a person standing on a large flat floor area raised to temperature t_f by direct radiation, the linearized $\bar{t}_r$, caused by the floor and unheated walls, on the occupant is:

$$\bar{t}_{rf} = F_{p-f}\,t_f + (1 - F_{p-f})t_a \tag{15}$$

where all unheated walls, ceiling, and ambient air are assumed to be t_a, and F_{p-f} is the angle factor governing the radiation exchange between the heated floor and the person.

The ERF_f from the floor affecting the occupant, caused by the $(t_f - t_a)$ difference, is:

$$\begin{aligned}\text{ERF}_f &= h_r(\bar{t}_{rf} - t_a)\\ &= h_r F_{p-f}(t_f - t_a)\end{aligned} \tag{16}$$

where h_r is again the linear radiation transfer coefficient for a person as given by Equation (11). Figures 50 and 53 in Fanger (1973) show F_{p-f} is 0.44 for a standing or sitting subject when walls are farther than 16 ft away. For an average size 16 ft by 16 ft room, a value of F_{p-f} = 0.35 is suggested for this preliminary estimation. For detailed information on floor reradiation, see Chapter 16 in the 1987 ASHRAE *Handbook—Systems and Applications.*

In summary, when radiant heaters warm occupants in a selected area of a poorly heated space, the radiation heat necessary for comfort consists of two additive components: (1) the ERF directly caused by the heater and (2) reradiation ERF_f from the floor. The effectiveness of floor reradiation can be improved by choosing flooring with a low specific conductivity. Flooring with high thermal inertia may be desirable during radiant transients, which may occur as the heaters are cycled by a thermostat set at the desired operative temperature t_o in the work area.

Asymmetric Radiant Fields

In the past, comfort heating has required flux distribution in occupied areas to be uniform, which is not possible with beam radiant heaters. Asymmetric radiation fields, such as lying in the sun on a cool day or standing in front of a warm fire, can be pleasant, however. While common sense suggests not to place a radiant beam heater too close to the subject, a limited amount of asymmetry is allowable for comfort heating. The phrase "reasonable uniform radiation distribution" is used as a design requirement. However, no fully satisfactory criteria are available for judging what degree of asymmetry is allowable.

For this purpose, Fanger (1980) proposed that radiant temperature asymmetry be defined as the difference between the plane radiant temperature of two opposing surfaces. Plane radiant temperature is the equivalent $\bar{t}_{r1}$, caused by radiation on one side of the subject, compared with the equivalent $\bar{t}_{r2}$, caused by radiation on the opposite side. For a subject sitting in a chair and heated by two lamps, a $\bar{t}_r - t_a$ asymmetry as high as 31 °F has been observed as comfortable for normally clothed subjects but only as high as 20 °F for unclothed subjects (Gagge *et al.* 1967). Both values applied to eight subjects.

For an unclothed subject lying on an insulated bed under a horizontal level bank of lamps, neutral temperature sensation occurred at a 72 °F t_o, which corresponded to a $t_o - t_a$ of 20 °F or $\bar{t}_r - t_a$ of 27 °F, both averaged for eight subjects (Stevens *et al.* 1969). In studies on heated ceilings, the asymmetry was 20 °F, when 80% of eight male and eight female clothed subjects voted conditions to be comfortable and acceptable. The latter case compared the floor and heated ceilings. For the combined direct versus floor reradiation, the resulting $\bar{t}_{rh} - \bar{t}_{rf}$ is about 1 °F, which indicates negligible asymmetry.

In general, the human body has a great ability to sum sensations spatially caused by radiant heat from many hot and cold sources on the skin surface. For example, Australian aborigines sleep unclothed next to open fires in the desert at night, where t_a is 43 °F. The $\bar{t}_r$ caused directly by the three fires alone is 171 °F, but the cold sky $\bar{t}_r$ is 30 °F; the resulting t_o during sleep is 82 °F, a value acceptable for human comfort when the subject is unclothed and acclimatized to cold (Scholander 1958).

From the limited field and laboratory data available, an allowable design radiant asymmetry of 22 ± 5 °F should cause little discomfort over the comfortable t_o range used by ANSI/ASHRAE *Standard* 55-1981 in Figures 2 and 3. Clothing insulation tends to raise the acceptable asymmetry, while increasing air movement reduces it. Increased activity also reduces human sensitivity to changing $\bar{t}_r$ or t_o and, consequently, increases the allowable asymmetry. However, the design engineer should use caution whenever asymmetry, either measured by a direct beam radiometer or estimated by calculation, causes a greater $\Delta\bar{t}_r$ than 27 °F from two opposing directions.

RADIATION PATTERNS

Figure 6 indicates the basic radiation patterns commonly used in design for radiation from point or line sources (Boyd 1962). The area radiated by a point source varies as the square of the distance from the source. The area from a (short) line source also varies substantially as the square of distance with about the same area as the circle actually radiated at that distance. For line sources, the width of the pattern is determined by the reflector shape and position of the element within the reflector. The rectangular area used for installation purposes as the pattern of radiation from a line source assumes a length equal to the width plus the fixture length. Although the length is often two or three times the pattern width, the assumed basis is satisfactory for design.

Electric infrared fixtures are often identified by their beam pattern (Rapp and Gagge 1967). The beam referred to is the radiation distribution normal to the line source element. The beam of a high intensity infrared fixture may be defined as that area in which the intensity is at least 80% of the maximum intensity encountered anywhere within the beam when measured in the plane in which maximum control of energy distribution is exercised.

The extent of the beam is usually designated in angular degrees and may be symmetrical or asymmetrical in shape. For adaptation to their design specifications, some manufacturers indicate beam characteristics based on 50% maximum intensity.

The control contemplated for an electric system affects the desirable maximum end-to-end fixture spacing. Since actual

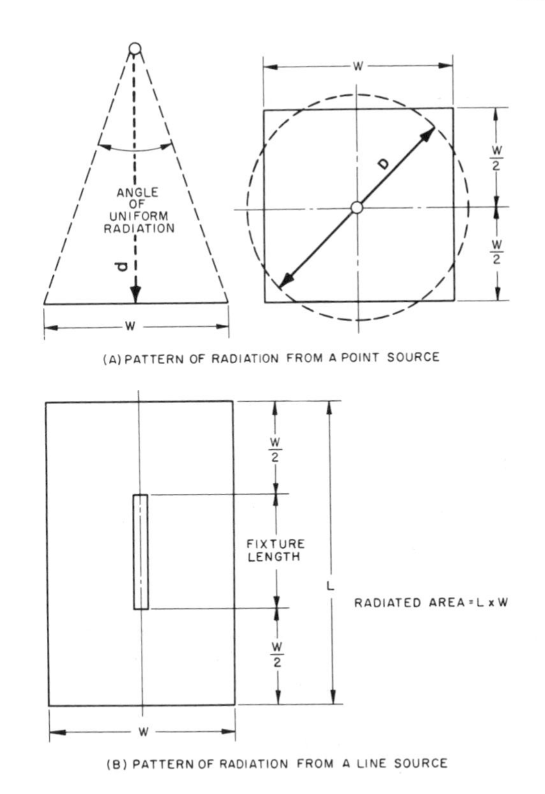

Note: The projected area W^2 normal to a beam, which is Ω steradians wide at distance d is Ωd^2. The floor area irradiated by a beam heater at an angle elevation β is $W^2/\sin\beta$. Fixture length L increases the area irradiated by the factor $(1 + L/W)$.

Fig. 6 Basic Radiation Patterns for System Design
(Boyd 1962)

pattern length is about three times the design pattern length, control in three equal stages is achieved by placing every third fixture on the same circuit. Where all fixtures are controlled by input controllers or variable voltage to electric units, end-to-end fixture spacing can be nearly three times the design pattern length. Side-to-side minimum spacing is determined by the distribution pattern of the fixture and is not influenced by the method of control.

Low intensity equipment typically consists of a steel tube hung near the ceiling and parallel to the outside wall. Combustion inside the tube and circulation of the products of combustion through the tube elevates the tube temperature and releases radiant energy. The tube is normally provided with a reflector to direct the radiant energy down into the space to be conditioned.

Radiant ceiling panels for heating only are installed in a narrow band around the perimeter of an occupied space and are usually the primary heat source for the space. For typical applications, the panel is installed in the ceiling plane. In these instances, the radiant source is identified as a (long) linear source. The flux density varies inversely with distance from the source.

The rate of radiation exchange between a panel and a particular object depends on the temperatures of the panel and the object, on their emissivity, and on their geometrical orientation (*i.e.*, shape factor). It also depends on the temperatures and configurations of all the other objects and walls within the space.

The energy flux for an ideal case of a line source is shown in Figure 7. All objects (except the radiant source) are at the same temperature, and the radiant source is suspended symmetrically in the room. For this case, as the distance from the source is doubled, the amount of flux is reduced to one-half.

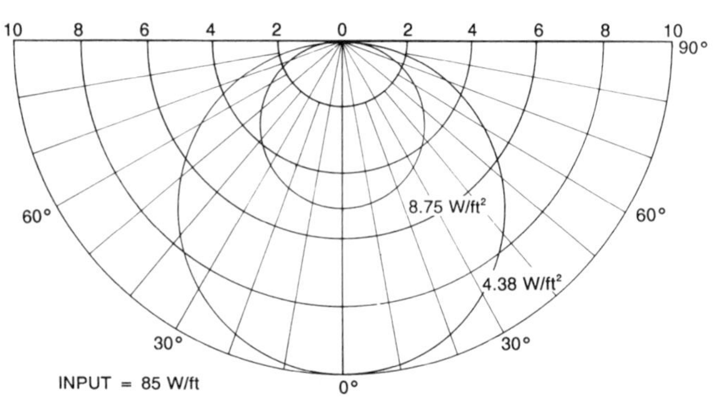

Fig. 7 Lines of Constant Radiant Flux for a Line Source

DESIGN FOR TOTAL SPACE HEATING

Radiant heating differs from conventional heating by a moderately elevated ERF, $\bar{t}_r$, or t_o over the ambient t_a. Standard methods of design are normally used, although informal studies indicate that radiant heating requires a lower heating capacity than convection heating. Buckley *et al.* (1987) demonstrated that a combination of elevated floor temperature, higher mean radiant temperature, and reduced ambient temperature results in lower thermostat settings and a reduced temperature differential across the building envelope. The result is a lower heat loss and lower heating load for the structure.

Most gas radiation systems for full building heating concentrate the bulk of capacity at mounting heights of 10 to 16 ft at the perimeter, directed at the floor near the walls. Mounting units considerably higher is not unusual. Successful application depends on supplying the proper amount of heat in the occupied area. Heaters should be located to take maximum advantage of the pattern of radiation produced. Exceptions to perimeter placement include walls with high transmission losses and extreme height, as well as large roof areas where roof heat losses exceed perimeter and other heat losses.

Electric infrared systems installed indoors for complete building heating have used layouts in which radiation is uniformly distributed throughout the area used by people, as well as layouts emphasizing perimeter placement, such as in ice hockey rinks. Some electric radiant heaters emit a significant amount of visible light, and in some cases, serve the combined purpose of heating and illumination.

For general area heating (large areas within larger areas), the orientation of equipment and people is less important than in spot heating. With reasonably uniform radiation distribution in work or living areas, exact orientation of the units can be ignored. To compensate for cold walls, higher intensities of radiation may be desirable near those walls with outside exposure. For frequently occupied work locations close to outside walls, radiation shields (preferably reflective to infrared) fastened a few inches from the wall and allowing for free air circulation between the wall and shield are effective.

In full building heating, units should be placed where radiant and convective output best overcome the structure's heat losses. The objective of a complete heating system should be to provide a warm floor with low conductance to the heat sink beneath the floor. This thermal storage may permit cycling of units with standard controls.

TEST INSTRUMENTATION FOR RADIANT HEATING

Accurate calculation of radiant heat exchange may involve some untested assumptions. Ultimately, the designer must test and readjust the installation in the field and evaluate it for human acceptability.

Black Globe Thermometer

The classic (Bedford) globe thermometer, a thin-walled, matte-black, hollow sphere with a thermocouple, thermister, or thermometer placed at the center, is the simplest physical instrument available that can directly measure $\bar{t}_r$, ERF, and t_o. When a black globe is in thermal equilibrium, the gain in radiant heat from various sources is balanced by the convective loss to ambient air. Thus, in terms of the globe's linear radiation h_{rg} and convection h_{cg} transfer coefficients, the heat balance after equilibrium is:

$$h_{rg}(\bar{t}_{rg} - t_g) = h_{cg}(t_g - t_a) \tag{17}$$

where $\bar{t}_{rg}$ is the mean radiant temperature measured by the globe.

In general, the $\bar{t}_{rg}$, as measured by Equation (17), also equals the $\bar{t}_r$ affecting a person when the globe is placed at the center of the occupied space and when the radiant sources are distant from the globe.

The ERF_g, as measured by a black globe, is

$$\text{ERF}_g = h_{rg} + (\bar{t}_{rg} - t_a) \tag{18}$$

which is analogous to Equation (5) for humans. From Equation (17) and (18), it follows that

$$\text{ERF}_g = (h_{rg} + h_{cg})(t_g - t_a) \tag{19}$$

The ERF_g, as measured by Equation (19), must be modified by two factors that describe the shape and skin-clothing absorptance of an occupant relative to the black sphere. The corresponding ERF affecting the occupant is

$$\text{ERF (for a person)} = f_{eff}\, \alpha_K\, \text{ERF}_g \tag{20}$$

where f_{eff} is the effective radiating area of the human body (approximately 0.71) and also equals the ratio (h_r/h_{rg}) [see Equation (11)]. The definition of t_o affecting a person, in terms of t_g and t_a, is given by:

$$t_o = Kt_g + (1 - K)t_a \tag{21}$$

where the coefficient K is:

$$K = \alpha_K f_{eff} (h_{rg} + h_{cg})/(h_r + h_c) \tag{22}$$

Ideally, when K is unity, the t_g of the globe would equal the t_o affecting a person.

For an average comfortable equilibrium temperature of 77 °F and noting that f_{eff} for the globe is unity, Equation (11) gives:

$$h_{rg} = 1.06\ \text{Btu/h} \cdot \text{ft}^2 \cdot {}^\circ\text{F} \tag{23}$$

and

$$h_{cg} = 0.345\, D^{-0.4} V^{0.5} \tag{24}$$

where

D = globe diameter, in.
V = air velocity, fpm

Equation (24) is Bedford's convective heat transfer coefficient for the 6-in. globe's convective loss, modified for D. For any radiating source below 1700 °F, the ideal diameter of a sphere to make $K = 1$ and to be independent of air movement is 8 in. Table 1 shows the value of K for various values of globe diameter D and ambient air movement V. The table shows that the uncorrected temperature of the traditional 6-in. globe would overestimate the true $t_o - t_a$ difference by 6% in V values up to 200 fpm, and the probable error of overestimating t_o by t_g uncorrected would be less than 0.9 °F. Globe diameters between 6 and 8 in. are optimum for using the uncorrected t_g measurement for t_o. The exact value for K may be used for the smaller sized globes when estimating for t_o from t_g and t_a measurement. The value of $\bar{t}_r$ may be found by substituting Equations (23) and (24) in Equation (17), since $\bar{t}_r$ (person) = $\bar{t}_{rg}$. The smaller the globe, the greater the variation in K caused by air movement. Globes with $D > 8$ in. will overestimate the importance of radiation gain versus convection loss.

Table 1 Value of K for Various Air Velocities and Globe Diameters ($\alpha_g = 1$)

Air Velocity, fpm	Globe Diameter, in.			
	≅ 2	≅ 4	≅ 6	≅ 8
50	1.35	1.15	1.05	0.99
100	1.43	1.18	1.06	0.99
200	1.49	1.21	1.07	1.00
400	1.54	1.23	1.08	1.00
800	1.59	1.26	1.09	1.00

For sources radiating at a high temperature (1340 to 10,000 °F), the ratio α_m/α_g may be set near unity by using a pink colored globe surface for α_g, whose absorptance for the sun is 0.7, a value similar to that of human skin and normal clothing (Madsen 1976). A globe with low mass and low thermal capacity is more useful, since its time to thermal equilibrium is shortened.

In summary, the black globe thermometer, which is simple and inexpensive, may be used to measure $\bar{t}_r$ [Equation (17)] and ERF [Figure 1 and Equations (19) and (20)]. The uncorrected t_g of a 6 to 8 in. black globe is an accurate measure of the t_o affecting occupants when the radiant heater temperature is less than 1700 °F. A "pink" globe extends its usefulness to sun temperature (10,000 °F). A globe with a low mass and low thermal capacity is more useful, since its time to thermal equilibrium is shortened.

Many instruments of various shapes, heated and unheated, have been designed using the previously described heat exchange principles. Madson (1976) developed an instrument that can measure the predicted mean vote (PMV) from the $t_g - t_a$ difference, as well as correct for clothing insulation, air movement, and activity (ISO 1984). All such instruments measure acceptability in terms of t_o, $\bar{t}_r$, and ERF, as sensed by their own geometric shape.

Directional Radiometer

In commercial radiometers, the angle of acceptance (in steradians) allows the engineer to point directly at a wall, floor, or high-temperature source and read the average temperature of the surface it senses. Directional radiometers are calibrated to measure either the radiant flux accepted by the radiometer or the equivalent blackbody radiation temperature of the emitting surface. Many are collimated to sense small areas of either body, clothing, wall, or floor surface. A directional radiometer allows rapid surveys and analyses of the important radiant heating factors in an installation such as temperature of skin, clothing surface, and walls and floors, as well as the radiation intensity I_K directed from heaters on the occupant. One radiometer for direct measurement of the equivalent radiant temperature has an angle of acceptance of 2.8° or 0.098 sr, *i.e.*, at 1 ft it would measure the average temperature over a projected circle about 3/8 in. in diameter.

APPLICATION CONSIDERATIONS

All bodies with a surface temperature above absolute zero emit rays with wavelengths that depend on the body surface temperature. Every facet of the surface emits rays in straight lines at right angles to the facet. When examined under a microscope, the surface of concrete or rough plaster is seen to be covered with numerous facets, each giving off radiant energy. Polished steel or similar polished surfaces show no such facets. Thus, a rough surface emits heat rays more efficiently than a polished surface.

The invigorating effect of radiant heat is experienced when the body is exposed to the sun's rays on a cool but sunny day. Some of these rays impinging on the body come directly from the sun and include the whole range of waves. Other rays coming from the sun and include the whole range of waves. Other rays coming from the sun impinge on surrounding objects, where they are increased in wavelength and reradiated to the body as low temperature radiation, producing a comfortable feeling of warmth. When a cloud passes over the sun, it instantly creates a sensation of cold; although in such a short interval, the air temperature does not vary at all.

No system can create the correct conditions compatible with the physiological demands of the human body unless it satisfies the three main factors controlling heat loss from the human body: radiation, convection, and evaporation. Sometimes, a radiant heating system is thought to be desirable only for certain buildings and only in some climates. However, wherever people live, these three factors of heat loss must be considered. Correct conditions are as important in very cold climates as in moderate climates. Maintaining the correct comfort conditions by low-temperature radiation is possible for even the most severe weather conditions.

Panel heating and cooling systems provide a comfortable environment by controlling surface temperatures and minimizing excessive air motion within the space. Thermal comfort, as defined by ASHRAE *Standard* 55-1981, is "that condition of mind which expresses satisfaction with the thermal environment." A person is not aware that the environment is being heated or cooled. The mean radiant temperature (MRT) strongly influences the feeling of comfort.

Radiant energy is also applied to control condensation on surfaces. One example is a large glass exposure in an airport terminal at Chicago.

When the surface temperature of the outside walls, particularly those with large areas of glass, deviates excessively from the room air temperature and from the temperature of other surfaces, simplified calculations for the load and the operative temperature may lead to errors in sizing and locating the panels. In such cases, more detailed radiant exchange calculations may be required, with separate estimation of heat exchange between the panels and each surface. A large window area may lead to significantly lower mean radiant temperatures than expected. For example, Athienitis and Dale (1987) reported a MRT 5.4 °F lower than room air temperature for a room with a glass area equivalent to 22% of its floor area.

Other factors to consider when placing radiant heaters in specific applications include the following:

- Both gas and electric high-temperature infrared heaters must not be placed where they might ignite flammable dust or vapors, or decompose vapors into toxic gases.
- Fixtures must be located with recommended clearances to ensure proper heat distribution. Stored materials must be kept far enough from the fixture to avoid hot spots. Manufacturers' recommendations must be followed.
- Unvented gas heaters inside tight, poorly insulated buildings may cause excessive humidity with condensation on cold surfaces. Proper insulation, vapor barriers, and ventilation prevent these problems.
- Combustion-type heaters in tight buildings may require makeup air to ensure proper venting of combustion gases. Some infrared heaters are equipped with induced draft fans to relieve this problem.
- Some partially transparent materials may be sensitive to uneven application of high intensity infrared. Infrared energy is transmitted without loss from the radiator to the absorbing surfaces. The system must produce the proper temperature distribution at the absorbing surfaces. Problems are rarely encountered with glass 0.25 in. or less in thickness.
- Comfort heating with infrared heaters requires a reasonably uniform flux distribution in the occupied area. While thermal

discomfort can be relieved in warm areas with high air velocity, such as on loading docks, the full effectiveness of a radiant heater installation is reduced by the presence of high air velocity. At specific locations in large working areas, direct radiant heating may be one practical means of gaining acceptability by the occupant.

APPLICATIONS

Low, Medium, and High Intensity Infrared Applications

Low, medium, and high intensity infrared is used extensively in industrial, commercial, and military applications. This equipment is particularly effective in large areas with high ceilings; *i.e.*, in buildings with large air volumes and in areas with high infiltration rates, large access doors, or large ventilation requirements.

Factories. Low intensity radiant equipment suspended near the ceiling around the perimeter of facilities with high ceilings enhances the comfort of employees because it warms floors and equipment in the work area. The cost for energy in older, uninsulated buildings is less than that required of other heating systems. High intensity infrared is particularly effective in providing spot heating in large facilities where the total facility is not heated.

Warehouses. Low and high intensity infrared are used for heating warehouses—facilities that usually have a large volume of air and are often poorly insulated and leaky. Low intensity infrared equipment is installed near the ceiling around the perimeter of the building. High intensity infrared equipment, also suspended near the ceiling, is arranged to control radiant intensity to provide uniform heating at the working level.

Highway Garages. Low intensity infrared provides greater comfort for mechanics working near or on the floor. With elevated MRT in the work area, comfort is provided at a lower ambient temperature.

The large overhead doors required to admit equipment for service, allows a substantial entry of cold outdoor air. On closing the doors, the combination of reradiation from the warm floor, as well as the radiant heat warming the occupants (not the air), provides rapid recovery of comfort. Radiant energy transmitted to the cold (perhaps snow-covered) vehicles results in rapid conditioning of the vehicles for service.

Low intensity equipment is suspended near the ceiling around the perimeter, often with special attention at the overhead doors to provide rapid recovery once the doors have been opened. High intensity equipment is also used to provide additional heat near the doors.

Aircraft Hangars. Equipment suspended near the roof of these high-ceilinged buildings with large access doors provides uniform radiant intensity throughout the working area. In these applications, the heated floor is particularly effective in restoring comfort when the large doors are closed after admitting an aircraft. The combination of reradiation from the warmed floor and radiation from the system heating the occupants and solid objects (not the air) provides rapid regain of comfort. Radiant energy also acts directly in heating the large mass of the aircraft moved into the work area.

Greenhouses. In greenhouse applications, a uniform flux density must be established throughout the facility to provide an acceptable growing condition. In a typical application, low intensity units are suspended near and running parallel to the peak of the greenhouse.

Outdoor Applications. Present-day applications include loading docks, racetrack stands, under marquees, and outdoor restaurants. Low, medium, and high intensity infrared are used in all these facilities, depending on their layout and use.

Other Applications. Radiant heat may be used in a variety of large, high-ceilinged facilities, such as churches, gymnasiums, swimming pools, enclosed stadiums, and facilities that are open to the outdoors.

Low, medium, and high intensity infrared are also used for a variety of other industrial applications—in process heating, such as component or paint drying ovens, humidity control in storage areas for corrosive metals, and snow control in parking or loading areas.

Panel Heating and Cooling

Residences. Embedded pipe coil systems, electric resistance panels, and forced warm air panel systems have all been used in residences. The embedded pipe coil system is most common, using grid coils in the floor slab or copper tubing systems in older plaster ceilings. These systems are well-suited to normally constructed residences with normal glass areas. Lightweight hydronic metal panel ceiling systems have been applied to residences. Prefabricated electric panels are also advantageous, particularly in rooms that have been added on.

Office Buildings. A panel system is usually applied as a perimeter heating system. Panels are typically piped to provide exposure control with one riser on each exposure and all horizontal piping incorporated in the panel piping. In these applications, the air system provides individual room control. Perimeter radiant panel systems have also been installed with individual zone controls. However, this type of installation is usually more expensive, provides minimal, if any, energy savings, and limited, if any, additional occupant comfort. Radiant panels can also be used for cooling as well as heating. Cooling installations are generally limited to retrofit or renovation jobs where ceiling space is insufficient for the required ductwork sizes. In these installations, the central air supply system provides ventilation air, dehumidification, and some sensible cooling. Water distribution systems using the 2- and 4-pipe concept may be used. Hot water supply temperatures should be reset by outside temperature, with additional reset or flow control to compensate for solar load. Panel systems are readily adaptable to accommodate most changes in partitioning. Electric panels in lay-in ceilings have been used for full perimeter heating.

Schools. Panels are usually selected for heating only, in all areas except gymnasiums and auditoriums. Heating-only panel systems may be used with any type of approved ventilation system. The panel system is usually sized to offset the transmission loads plus any reheating of the air. If the school is air conditioned by a central air system and has perimeter heating panels a single-zone piping system might be used to control the panel heating output, and the room thermostat would modulate the supply air temperature or supply volume of air delivered to the room. Heating and cooling panel applications are similar to those in office buildings. Another advantage of panel heating and cooling for classroom areas is that mechanical equipment noise does not interfere with instructional activities.

Hospitals. The principal application of heating and cooling radiant panel systems has been for hospital patient rooms. Perimeter radiant heating panel systems are typically applied in other areas in hospitals. The radiant heating and cooling system is well-suited to hospital patient rooms because it (1) provides a draft-free, thermally stable environment, (2) requires no mechanical equipment or bacteria and virus collectors in the space requiring maintenance, and (3) does not take up space within the room. Individual room control is usually achieved by throttling the water flow through the panel. The supply air system is often a 100% outdoor air system, and minimum air quantities delivered to the room are those required for ventilation and exhaust of the toilet room and soiled linen closet. The piping system is typically a 4-pipe design. Water control valves should be in the corridor outside the patient rooms so that they can be adjusted or serviced without entering the room. All piping connections above the ceiling should be soldered or welded and thoroughly tested. If cubicle tracks are applied to the ceiling surface, track installation should be coordinated with the radiant ceiling. Security panel ceilings are often used in areas of the hospital occupied by mentally disturbed patients, since there is no equipment accessible to the occupant for destruction or self-injury.

Swimming Pools. Panel heating systems are well-suited to swimming pools because a partially clothed body emerging from the water is very sensitive to the thermal environment. Floor panel temperatures are restricted so as not to cause foot discomfort. Ceiling panels are generally located around the perimeter of the pool, not directly over the water. Panel surface temperatures are higher to compensate for the increased ceiling height and to produce a greater radiant effect on the partially clothed body. Ceiling panels may also be placed over windows to reduce condensation.

Apartment Buildings. For heating, pipe coils are embedded in the masonry slab. The coils must be carefully positioned so as not to overheat one apartment when maintaining desired temperatures in another. The slow response of embedded pipe coils in buildings with large glass areas may prove unsatisfactory. Installations for heating and cooling have been made with pipes embedded in a hung plaster ceiling. A separate minimum volume dehumidified air system provides the necessary dehumidification and ventilation for each apartment. In recent years, there has been an increased application of electric resistance elements embedded in the floor or behind a skim coat of plaster at the ceiling. The electric panels are easy to install and have the advantage of simplified individual room control.

Industrial Applications. Panel systems are widely used for general space heating of industrial buildings in Europe. For example, one special application is an internal combustion engine test cell, where the walls and ceilings are cooled with chilled water. Although the ambient air temperature in the space ranges up to 95 °F, the occupants work in relative comfort when 55 °F water is circulated through the ceiling and wall panels.

Other Building Types. Metal panel ceiling systems can be operated as heating systems at elevated water temperatures and have been used in airport terminals, convention halls, lobbies, museums, and especially where large glass areas are involved. Cooling may also be applied. Because radiant energy travels through the air without warming it, ceilings can be installed at any height and remain effective. One particularly high ceiling installed for a comfort application is 50 ft above the floor, with a panel surface temperature of approximately 285 °F for heating. The ceiling panels offset the heat loss from a single-glazed, all-glass wall.

The high lighting levels in television studios make them well-suited to panel systems. The panels are installed for cooling only and are placed above the lighting system to absorb the radiation and convection heat from the lights and normal heat gains from the space. Besides absorbing heat from the space, the panel ceiling also improves the acoustical properties of the studio.

Metal panel ceiling systems are also installed in minimum and medium security jail cells and other areas where disturbed occupants are housed. The ceiling construction is made more rugged by increasing the gage of the ceiling panels and by using security clips so that the ceiling panels cannot be removed. Part of the perforated metal ceiling can be used for air distribution.

New Techniques

With the introduction of polybutylene tubing and new design techniques, interest has developed in radiant floor heating. The systems are energy efficient and use low water temperatures available from solar collector systems.

Metal radiant panels can be integrated into the ceiling design to provide a narrow band of radiant heating around the perimeter of the building. The radiant system offers advantages over baseboard or overhead air in appearance, comfort, operating efficiency and cost, and product life.

SYMBOLS

A_D area, total surface of person as measured by DuBois, ft^2
A_{eff} area, effective radiating of person, ft^2
A_p area, projected normal to the beam, ft^2
clo unit of clothing insulation and equal to 0.880 ft^2·°F·h/Btu
D diameter of globe thermometer, in.
ERF effective radiant flux (person), Btu/h·ft^2
ERF_f radiant flux caused by heated floor on occupant, Btu/h·ft^2
F_{p-f} angle factor between person and heater floor
F_{f-room} shape factor between floor and surrounding room
H_m net metabolic heat loss from body surface, Btu/h·ft^2
I_K irradiance from beam heater, Btu/h·sr
K coefficient that relates t_a and t_g to t_o [Equation (24)]
L fixture length, ft
met unit of metabolic energy equal to 18.4 Btu/h·ft^2
W width of a square equivalent to the projected area of a beam angle Ω steradians and d distance, ft
c specific heat, Btu/lb·°F
d distance of beam heater from occupant, ft
d_f average floor thickness, ft
F_{eff} ratio, radiating surface (person) to its total area (DuBois)
f_p fraction, total body surface irradiated by heater beam
h combined heat transfer coefficient (person), Btu/h·ft^2·°F
h_c convective transfer coefficient for person, Btu/h·ft^2·°F
h_{cf} convection coefficient between floor and air, Btu/h·ft^2·°F
h_{cg} convection coefficient for globe, Btu/h·ft^2·°F
h_r linear radiation transfer coefficient (person), Btu/h·ft^2·°F
h_{rf} linear radiation transfer coefficient from floor surface over 2π sr, Btu/h·ft^2·°F
h_{rg} linear radiation transfer coefficient for globe, Btu/h·ft^2·°F
k conductivity, Btu/h·ft·°F (floor)
k_f average conductivity of flooring material, Btu/h·ft·°F
t_a ambient air temperature (near occupant), °F
t_f temperature, floor, °F
t_{gr} temperature, ground or surface below floor, °F
t_o operative temperature, °F
$\bar{t}_r$ mean radiant temperature affecting occupant, °F
t_{sf} exposed surface temperature of occupant, °F
t_{uo} operative temperature of unheated workspace, °F
v air velocity, fpm
α relative absorptance of skin clothing surface to that of matte black surface
α_f absorptance of floor surface
β elevation angle of beam heater, degrees
Ω radiant beam width, sr
Φ azimuth angle of heater, degrees
ρ density (floor), lb/ft^3
σ Stefan-Boltzmann constant and equal to 0.1713×10^{-8} Btu/h·ft^2·°R

REFERENCES

Athienitis, A.K. and J.D. Dale. 1987. A study of the effects of window night insulation and low emissivity coating on heating load and comfort. ASHRAE *Transactions* 93(1A):279-94.

Boyd, R.L. 1962. Application and selection of electric infrared comfort heaters. ASHRAE *Journal* 4(10):57.

Buckley, N.A. and T.P. Seel. 1987. Engineering principles support an adjustment factor when sizing gas fired low-intensity infrared equipment. ASHRAE *Transactions* 93(1).

Fanger, P.O. 1973. *Thermal comfort.* McGraw-Hill Book Co., New York.

Fanger, P.O., L. Banhidi, B.W. Olesen, and G. Langkilde. 1980. Comfort limits for heated ceiling. ASHRAE *Transactions* 86(2):141-56.

Gagge, A.P., G.M. Rapp, and J.D. Hardy. 1967. The effective radiant field and operative temperature necessary for comfort with radiant heating. ASHRAE *Transactions* 73(1); and ASHRAE *Journal* 9:63-66.

Hardy, J.D., A.P. Gagge, and J.A.J. Stolwijk. 1970. *Physiological and behavioral temperature regulation.* C.C. Thomas, Springfield, IL.

ISO. 1984. Moderate thermal environments—Determination of the PMV and PPD indices and specifications of the conditions for thermal comfort. ISO *Standard* 7730-1984. International Standard Organization, Geneva.

Madsen, T.L. 1976. Thermal comfort measurements. ASHRAE *Transactions* 82(1):60-70.

Rapp, G.M. and A.P. Gagge. 1967. Configuration factors and comfort design in radiant beam heating of man by high temperature infrared sources. ASHRAE *Transactions* 73(3):1.1-1.8.

Scholander, P.E. 1958. Cold adaptation in the Australian Aborigines. *Journal of Applied Physiology* 13:211-18.

Stevens, J.C., L.E. Marks, and A.P. Gagge. 1969. The quantitative assessment of thermal comfort. *Environmental Research* 2:149-165.

BIBLIOGRAPHY

AGA. 1960. Literature review of infra-red energy produced with gas burners. Research *Bulletin* 83, Catalog No. 35/IR, American Gas Association.

Boyd, R.L. 1964. Beam characteristics of high intensity infrared heaters. IEEE Paper CP 64-123, Institute of Electrical and Electronics Engineers (February).

Boyd, R.L. 1963. Control of electric infrared energy distribution. *Electrical Engineering* (February):103.

Boyd, R.L. 1959. High intensity infrared radiant heating. *Heating, Piping and Air Conditioning* (November):140.

Boyd, R.L. 1961. How to apply high intensity infrared heaters for comfort heating. *Heating, Piping and Air Conditioning* (January): 230.

Frier, J.P. and W.R. Stephens. 1962. Design fundamentals for space heating with infrared lamps. *Illuminating Engineering* (December).

Griffiths, I.S. and D.A. McIntyre. 1974. Subjective response to overhead thermal radiation. *Human Factors* 16(3):415-22.

Hardy, J.S., A.P. Gagge, and J.A.J. Stolwijk. 1970. *Physiological and behavioral temperature regulation.* CC. Thomas, Springfield, IL.

McIntyre, D.A. 1974. The thermal radiating field. *Building Science* (9):247-62.

McNall, P.E., Jr. and R.E. Biddison. 1970. Thermal and comfort sensations of sedentary persons exposed to asymmetric radiant fields. ASHRAE *Transactions* 76(1).

Olesen, S., P.O. Fanger, P.B. Jensen, and O.J. Nielsen. 1972. Comfort limits for man exposed to asymmetric thermal radiation. Building Research Est. Report, *Thermal comfort and moderate heat stress.*

Walker, C.A. 1962. Control of high intensity infrared heating. ASHRAE *Journal* 4(10):66.

CHAPTER 49

SEISMIC RESTRAINT DESIGN

EARTHQUAKE damage to inadequately restrained heating, ventilating, air-conditioning, and refrigerating equipment can be extensive. High costs are incurred to replace or repair the damaged equipment, and the building cannot be occupied due to inadequate ventilation. The cost to restrain the equipment properly is relatively small compared to the possible damage that may occur.

This chapter covers restraint design to limit the movement of HVAC&R equipment during a seismic event (earthquake). Large or conservative safety factors are applied to reduce the complexity of earthquake response analysis and evaluation. However, the additional cost of higher capacity restraints is small.

Local building officials must be contacted for specific requirements that may be more stringent than presented in this chapter. Nearly all local codes in the United States are based on model building codes developed by three organizations: International Conference of Building Officials (ICBO), Building Officials and Code Administrators International (BOCAI), and Southern Building Code Conference, Inc. (SBCCI). Most seismic requirements adopted by local jurisdictions are based on the Uniform Building Code (UBC), which is developed by ICBO.

The Sheet Metal and Air Conditioning Contractors' National Association publishes HVAC *Systems—Duct Design* (SMACNA 1987), which includes seismic restraint guidelines for ductwork and piping. The National Fire Protection Association (NFPA) has developed standards on restraint design for fire protection systems. Restraint design for nuclear facilities, which is not discussed in this chapter, is covered in U.S. Department of Energy DOE 6430.1A and American Society of Mechanical Engineers publication ASME AG-1.

TERMINOLOGY

Base plate thickness. Thickness of the equipment bracket fastened to the floor.

Effective shear force (V_{eff}). Maximum shear force of one seismic restraint or tie-down bolt.

Effective tension force (T_{eff}). Maximum tension force or pullout force on one seismic restraint or tie-down bolt.

Equipment. Any component that must be restrained from movement during an earthquake.

Fragility level. Maximum lateral acceleration force that the equipment is able to withstand. This data may be available from the equipment manufacturer and is generally in the order of four times the force of gravity (4g) for mechanical equipment.

Liquefaction. Phenomenon of soil and sand behaving like a dense fluid during an earthquake.

Resilient support. Active seismic device (such as a spring with a bumper) to prevent equipment from moving beyond a specified amount.

Response spectra. Relationships between the acceleration response of the ground and the peak acceleration of the earthquake in a damped single degree of freedom at various frequencies. The ground motion response spectrum varies with different soil conditions.

Rigid support. Passive seismic device used to restrict any movement.

Shear force (V). Force generated at the plane of the seismic restraints; it acts to cut the restraint at the base.

Seismic restraint. Device designed to withstand an earthquake.

Snubber. Device made of steel-housed, resilient bushings arranged to prevent equipment from moving beyond an established gap.

Tension force (T). Force generated by overturning moments at the plane of the seismic restraints, acting to pull out the bolt.

UBC 88 seismic zones. Identify the seismic zone factor by the geographical location of a facility, as listed in the Uniform Building Code (1988) (see Figure 1).

CALCULATIONS

Both static and dynamic analyses reduce the force generated by the earthquake to an equivalent static force, which acts in a horizontal direction at the component's center of gravity. The resulting overturning moment is resisted by shear and tension (pullout) forces on the tie-down bolts. Static analysis is used for both rigid-mounted and resilient-mounted equipment.

Dynamic Analysis

A dynamic analysis is based on site-specific ground motions developed by a geotechnical, or soils engineer. A common approach assumes an elastic response spectrum. The results of the dynamic analysis are then scaled up or down as a percentage of the total lateral force obtained from the static analysis performed on the building. The scaling coefficient is established by the UBC or by the governing building official. The scaled acceleration calculated by the structural engineer at any level in the structure can be determined and compared to the force calculated by Equation (1). The greater of the two should be used in the anchor design. The horizontal force factor C_p should be multiplied by a factor of two in either case, as shown in Table 2.

Static Analysis

Equation (1) is a basic formula used to establish the lateral force developed by an earthquake:

$$F_p = ZIC_pW_p \tag{1}$$

The preparation of this chapter is assigned to Task Group for Seismic Restraint Design.

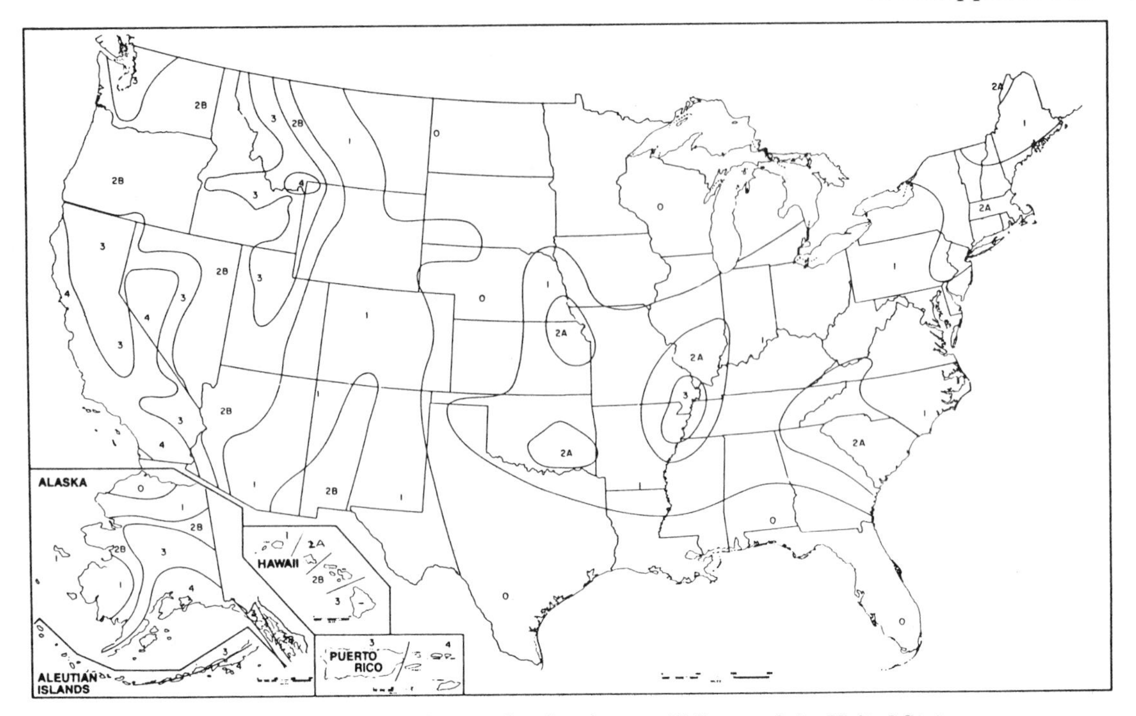

Fig. 1 Seismic Zone Factors for Contiguous 48 States of the United States

Figure 1 and Table 1 may be used to determine the seismic zone and seismic zone factor Z. The *Technical Manual—Seismic Design for Buildings* (Army, Navy, and Air Force 1982) can also be used to determine the seismic zone.

The importance factor I from the UBC (ICBO 1988) ranges from 1 to 1.25, depending on the building occupancy. For equipment, I should be conservatively set at 1.5.

The horizontal force factor C_p is determined from Table 2 based on the type of equipment, tie-down configuration, and type of base.

The mass W_p of the equipment should include all the items attached or contained in the equipment.

Table 1 Seismic Zone Factor Z

Zone	Z
1	0.075
2A	0.15
2B	0.20
3	0.30
4	0.40

Table 2 Horizontal Force Factor C_p

Equipment or nonstructural components	C_p
Mechanical equipment, plumbing, and electrical equipment and associated piping rigidly mounted	0.75
All equipment resiliently mounted (maximum 2.0)	$2C_p$

Note: Stacks and tanks should be evaluated to the applicable codes by a qualified engineer.

APPLYING STATIC ANALYSIS

The forces acting on the equipment are the lateral and vertical forces resulting from the earthquake, the mass of the equipment, and the forces of the restraint holding the equipment in place. Since the analysis assumes the equipment does not move during an earthquake, the sum of the forces and moments must be zero. When calculating the overturning moment, $F_p/3$ should be used as the vertical component at the center of gravity. This vertical component is generated by the seismic event simultaneous with the lateral force.

The forces of the restraint holding the equipment in position include shear and tension forces. It is important to determine the number of bolts that are affected by the earthquake forces. The direction of the lateral force should be evaluated in both horizontal directions as shown in Figure 2. All bolts or as few as a single bolt may be affected.

Figure 2 shows a typical rigid floor mount installation of a piece of equipment. To calculate the shear force, calculate F_p from Equation (1) and sum the forces in the horizontal plane as in Equation (2).

$$0 = F_p - V \tag{2}$$

The effective shear force (V_{eff}) is calculated by Equation (3).

$$V_{eff} = F_p/N_{bolt} \tag{3}$$

where

N_{bolt} = number of bolts in shear

The restraints shown in Figure 2 have two bolts on each side, so that four bolts are in shear. To calculate the tension force, sum the moments for overturning as follows:

$$F_p h_{cg} - (W_p - F_p/3)(D_o/2) - TD_o = 0$$

or,

$$T = [F_p h_{cg} - (W_p - F_p/3)(D_o/2)]/D_o \tag{4}$$

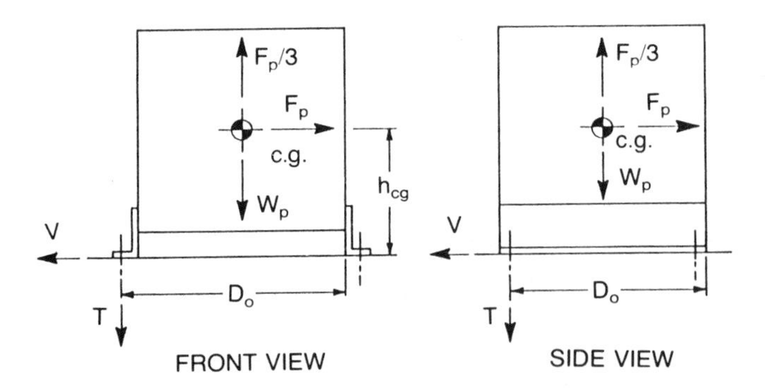

Fig. 2 Equipment with Rigidly Mounted Structural Bases

The effective tension force T_{eff}, where the overturning affects only one side, is calculated by Equation (5). In the example shown in Figure 2, two bolts are in tension.

$$T_{eff} = T/N_{bolt} \quad (5)$$

In the example in Figure 2, the evaluation of the shear and tension forces (T and V) should be independently calculated for both axes as shown in the front and side views. The worst case governs seismic restraint design; however, the direction of seismic loading that will govern the design is not always obvious. For example, if three bolts were installed on each side, the lateral force applied as shown in the side view of Figure 2 affects six bolts in shear and two in tension. The lateral force applied as shown on the front view results in six bolts affected in shear and three in tension. Also, D_o is different for each axis.

Equations (1), (2), and (3) may be applied to ceiling-mounted equipment. Equation (4) must be modified as in Equation (6), because the mass of the equipment adds to the overturning moment.

By summing the moments, Equation (6) may be used to determine the effective tension force.

$$T = [F_p h_{cg} + (W_p - F_p/3)(D_o/2)]/D_o \quad (6)$$

Interaction Formulas. To evaluate the combined effective forces (tension and shear) that act simultaneously on the bolt, the following equation applies:

$$(T_{eff}/T_{allow}) + (V_{eff}/V_{allow}) \leq 1.0 \quad (7)$$

For the allowable forces (T_{allow} and V_{allow}), refer to the generic allowable capacities given in Table 3 for wedge-type anchor bolts.

Table 3 Typical Allowable Loads for Wedge-Type Anchors

Diameter, in.	T_{allow}, lb	V_{allow}, lb
0.5	300	875
0.625	450	2200
0.75	675	3000

Notes:
1. The allowable tensile forces are for installations without special inspection (torque test) and may be doubled if the installation is inspected.
2. Additional tension and shear values may be obtained from published ICBO reports.

ANCHOR BOLTS

Several types of anchor bolts are manufactured. Wedge and sleeve anchors perform better than the self-drilling or drop-in types. Epoxy-type anchors are stronger than other anchors, but lose their strength at elevated temperatures, for example on rooftops and in areas damaged by fires.

Wedge-type anchors have a wedge on the end with a small clip around the wedge. After a hole is drilled, the bolt is inserted and the external nut tightened. The wedge expands the small clip which bites into the concrete.

A self-drilling anchor is basically a hollow drill bit. The anchor is used to drill the hole and is then removed. A wedge is then inserted on the end of the anchor, and the assembly is drilled back into place; the drill twists the assembly fully in place. The self-drilling anchor is weaker than other types, because it forms a rough hole.

Drop-in expansion anchors are hollow cylinders with a tapered end. After they are inserted in a hole, a small rod is driven through the hollow portion, expanding the tapered end. These anchors are recommended only for shallow installations since they have no reserve expansion capacity.

A sleeve anchor is a bolt covered by a threaded, thin-wall, split tube. As the bolt is tightened, the thin wall expands. Additional load tends to further expand the thin wall. The bolt must be properly preloaded or the friction force will not develop the required holding force.

Adhesive anchors may be in glass capsules or installed with various tools. Pure epoxy, polyester, or vinylester resin adhesives are used with a threaded rod supplied by the contractor or the adhesive manufacturer. Some adhesives have a problem with shrinkage; others are degraded with heat. However, some adhesives have been tested without protection to 1100°F before failure—all mechanical anchors will fail at this temperature. Where required, or if there is a concern, anchors should be protected with fire retardants similar to those applied to steel decks in high-rise buildings.

Follow the manufacturer's installation instructions. Performance test data published by manufacturers should include shock, fatigue, and seismic resistance. For allowable forces used in the design, refer to the International Conference of Building Officials (ICBO) reports. Add a safety factor of two (2) if the installation is not inspected.

WELD CAPACITIES

Weld capacities may be calculated to determine the size of welds needed to attach equipment to a steel plate or to evaluate raised support legs and attachments. A static analysis provides the effective tension and shear forces. The capacity of a weld is given per unit length of weld based on the shear strength of the weld material. For steel welds, the allowable shear strength capacity is 16,000 psi on the throat section of the weld. The section length is 0.707 times the specified weld size.

For a 1/16-in. weld, the length of shear in the weld is 0.707 (1/16) = 0.0442 in. The allowable weld force $(F_w)_{allow}$ for a 1/16-in. weld is:

$$(F_w)_{allow} = 0.0442 \times 16{,}000 = 700 \text{ lb per inch of weld}$$

For a 1/8-in. weld, the capacity is 1400 lb/in.

The effective weld force is the sum of the vectors calculated in Equations (3) and (5). Since the vectors are perpendicular, they are added by the method of square root of sum of the squares (SRSS).

$$(F_w)_{eff} = \sqrt{(T_{eff})^2 + (V_{eff})^2}$$

The length of weld required is given by Equation (8).

$$\text{Length} = (F_w)_{eff}/(F_w)_{allow} \quad (8)$$

SEISMIC SNUBBERS

Several types of snubbers are manufactured or field fabricated. All snubber assemblies should meet the following minimum requirements to avoid imparting excessive accelerations to the HVAC&R equipment. The impact surface should have a high quality elastomeric surface that is not cemented in place. The resilient material should be easy to inspect for damage and be replaceable if necessary. To ensure the stated load capacity and to avoid serious design flaws, snubbers should be tested by an independent test laboratory and should be analyzed by a registered engineer. The snubber types presently available are described as Types A through E.

Type A. All-directional with molded, replaceable neoprene element. Neoprene element of bridge-bearing quality is a minimum of 3/16 in. thick. Snubber must have a minimum of two anchor bolt holes (Figure 3).

Type B. All-directional with molded, replaceable neoprene element. Neoprene element of bridge-bearing quality is a minimum of 3/4 in. thick. Snubber must have a minimum of two anchor bolt holes (Figure 4).

Type C. Snubber built into a resilient mounting. All-directional, molded bridge-bearing quality neoprene element is a minimum of 1/8 in. thick. Mounting must have a minimum of two anchor bolt holes (Figure 5).

Type D. Neoprene mount capable of sustaining seismic loads in all directions. Mounting must have a minimum of two anchor bolt holes (Figure 6).

Type E. Aircraft wire rope with galvanized end connections that avoid bending the wire rope across sharp edges. This type of snubber must only be used with suspended pipe duct and equipment (Figure 7).

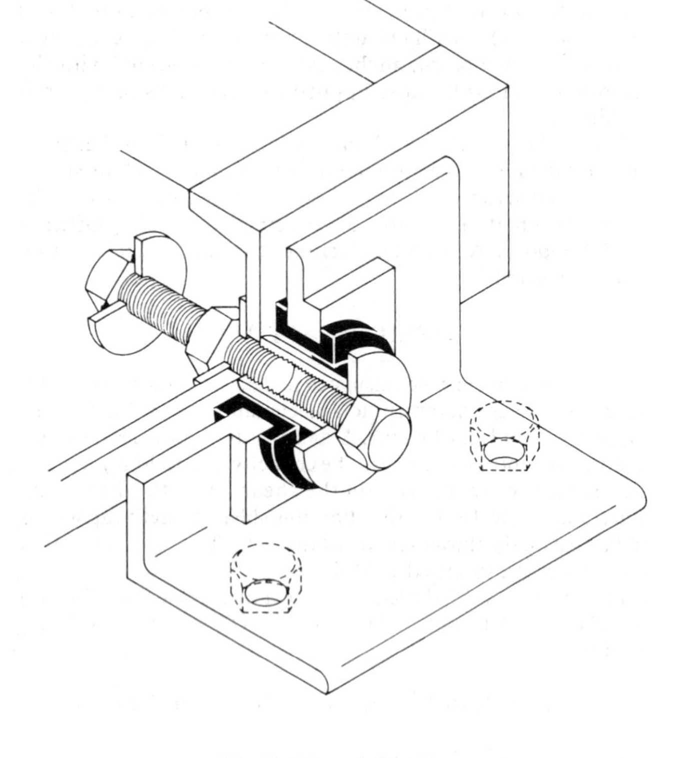

Fig. 3 Type A Snubber

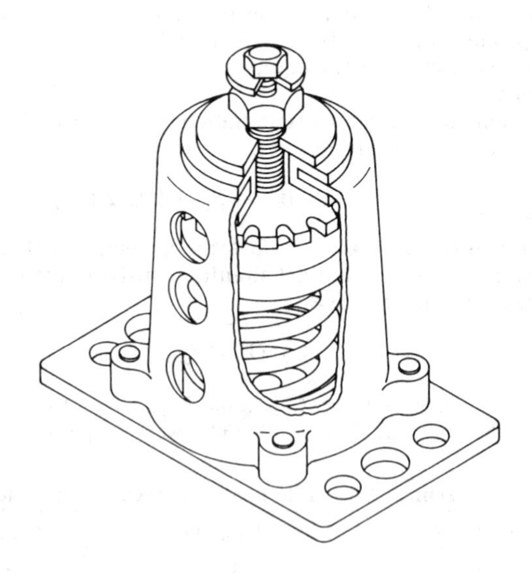

Fig. 5 Type C Snubber

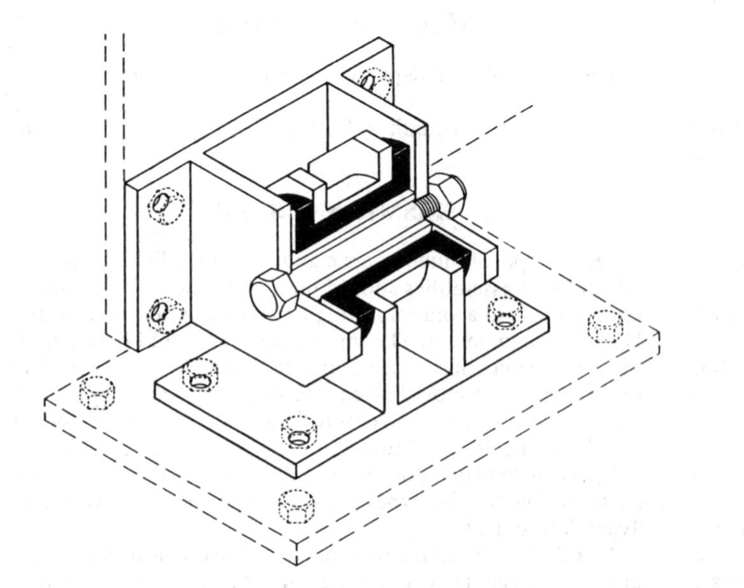

Fig. 4 Type B Snubber

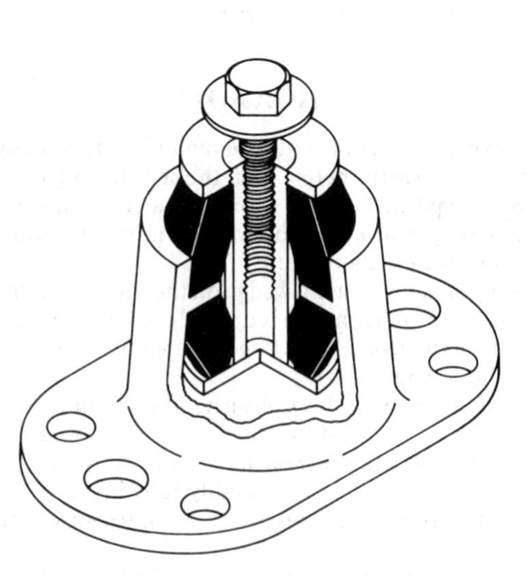

Fig. 6 Type D Snubber

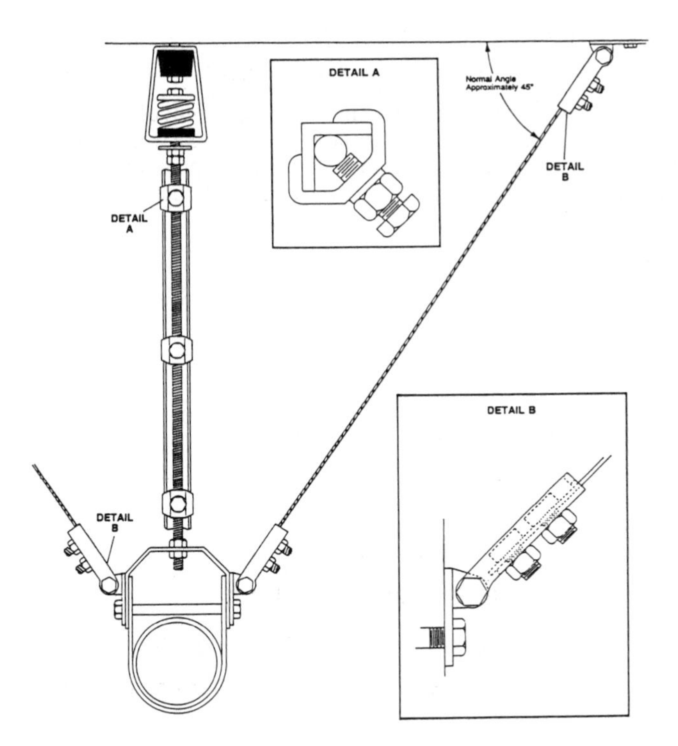

Fig. 7 Type E Snubber

Example 1

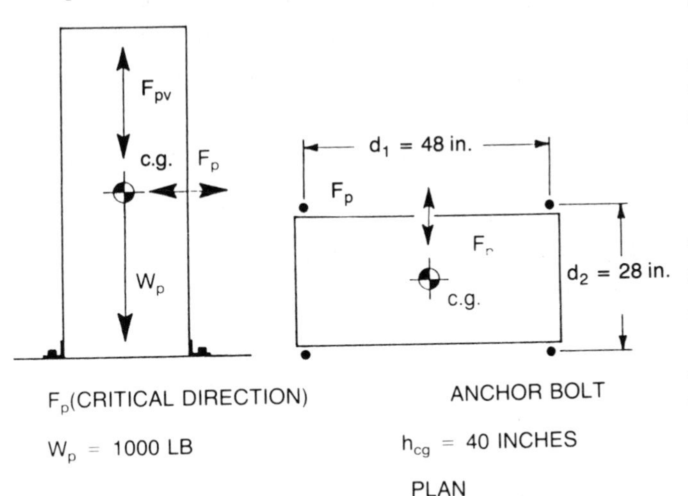

Fig. 8 Equipment Rigidly Mounted to Structure

EXAMPLES

The following examples are provided to assist in the design of equipment anchorage to resist seismic forces.

From Equation (1) (for Zone 4 areas):

$$F_p = 0.4 \times 1.5 \times 0.75 \times 1000 = 450 \text{ lb}$$

$$F_{pv} = F_p/3 = 150 \text{ lb}$$

Calculate the overturning moment (OTM):

$$\text{OTM} = F_p h_{cg} = (450 \times 40) = 18{,}000 \text{ in}\cdot\text{lb} \qquad (9)$$

Calculate the resisting moment (RM):

$$\text{RM} = (W_p \pm F_{pv})\, d_{min}/2 = (1000 \pm 150)28/2 \qquad (10)$$
$$= 16{,}100 \text{ or } 11{,}900 \text{ in}\cdot\text{lb}$$

Calculate the tension force T:

Use 11,900 in·lb to produce the maximum tension force for this case.

$$T = (\text{OTM} - \text{RM}_{min})/d_{min} = (18{,}000 - 11{,}900)/28 \qquad (11)$$
$$= 218 \text{ lb}$$

This force is the same as that obtained by Equation (4).

Calculate T_{eff} per bolt from Equation (5):

$$T_{eff} = 218/2 = 109 \text{ lb/bolt}$$

Calculate shear force per bolt from Equation (3):

$$V_{eff} = 450/4 = 112.5 \text{ lb/bolt}$$

Case 1. Equipment attached to a timber structure.

Reference: *National Design Specification for Wood Construction* (NDS)(NFPA 1986).

Note: Selected fasteners must be secured to solid lumber, not to plywood or other similar material. Check whether a 1/2-in. diameter, 4-in. long lag screw will hold the required load.

From Table 8.1A in the NDS, for Group IV (redwood) $G = 0.37$

From Table 8.6A of the NDS:

$$T_{allow} = (241 \text{ lb/in} \times 3.5\text{-in. penetration})2/3 = 562 \text{ lb}$$

From Table 8.6C, $V_{all} = 180$ lb

Note: The factor 2/3 accounts for the fact that about one-third of the length of a lag screw or bolt has no threads on the shank.

Other types of wood may be used with appropriate factors from Table 8 and/or other reductions as specified in Part II of NDS.

In timber construction, the interaction formula, Equation (7), does not apply per Section 8.6.8 of the NDS. The ratio of the calculated shear and tension values to the allowable values should each be less than 1.0.

$$T/T_{allow} = 218/562 = 0.39 < 1.0$$

$$V/V_{allow} = 112.5/180 = 0.63 < 1.0$$

Therefore, use a 1/2-in. diameter, 4-in. long lag screw at each corner of the equipment.

Case 2. Equipment attached to concrete with drill-in bolts.

Note: Good design practice is to specify a minimum of 1/2-in. diameter bolts to attach roof or floor-mounted equipment to the structure.

Try 1/2-in. wedge anchors with special inspection provisions:

From Table 3: $T_{allow} = 600$ lb $\qquad V_{allow} = 875$ lb

From Equation (7) $109/600 + 112.5/875 = 0.31 < 1.0$

Therefore, use 1/2-in. diameter, 4-in. long lag screws.

Case 3. Equipment attached to steel.

For the case where equipment is attached directly to a steel member, the analysis is the same as that shown in Case 1 above. The allowable values for the attaching bolts are given in the *Manual of Steel Construction* (AISC 1980).

The following values can be used for A307 bolts:

Table 4 Allowable Loads for A307 Bolts

Diameter, in.	T_{allow}, lb	V_{allow}, lb	A_b, in^2
1/2	3900	1950	0.196
5/8	6100	3100	0.307
3/4	8800	4400	0.442
1	15700	7900	0.785

The interaction formula, Equation (7), does not apply to steel-to-steel connections. The allowable tension load is modified by the following formula:

$$(T_{allow})_{mod} = F_t A_b$$

where

$$F_t = 26 - 1.8(V/N_{bolt}A_b) \leqslant 20(4/3) = 26.67$$

V/N_{bolt} is in kips (1 kip = 1000 lb) and F_t is in kip/in^2 (ksi). If F_t is less than 20 ksi, multiply the calculated $(T_{all})_{mod}$ by 1000 for equivalent values to those shown in Table 4. The 33% (4/3) stress increase is allowed for short-term loads such as wind or earthquakes.

Example 2

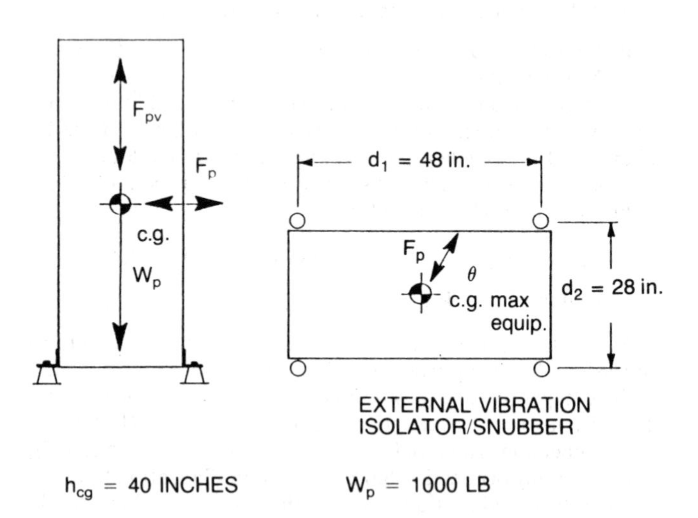

Fig. 9 Equipment Supported by External Spring Mounts

Note: A mechanical or acoustical consultant should choose the type of isolator/snubber or combination of the two. Then the product vendor should select the actual spring snubber.

Assume that the center of gravity (CG) of the equipment coincides with the CG of the isolator group.

T = maximum tension on isolator

C = maximum compression on isolator

$F_{pv} = F_p/3$

$$T = (-W_p + F_{pv})/4 + F_p h_{cg}\cos\theta/2b + F_p h_{cg}\sin\theta/2a$$

$$T = (-W_p + F_{pv})/4 + (F_p h_{cg}/2)(\cos\theta/b + \sin\theta/a)$$

To find maximum T or C, find $dt/d\theta = 0$

$$dt/d\theta = (F_p h_{cg}/2)(-\sin\theta/b + \cos\theta/a) = 0$$

Therefore, $\theta_{max} = \tan^{-1}(b/a)$ (12)

$$T = (-W_p + F_{pv})/4 + (F_p h_{cg}/2)(\cos\theta_{max}/b + \sin\theta_{max}/a) \quad (13)$$

$$C = (-W_p - F_{pv})/4 - (F_p h_{cg}/2)(\cos\theta_{max}/b + \sin\theta_{max}/a) \quad (14)$$

From Equation (1):

$$F_p = 0.4 \times 1.5 \times 2 \times 0.75 \times 1000 = 900 \text{ lb}$$

$$F_{pv} = 900/3 = 300 \text{ lb}$$

From Equation (12):

$$\theta = \tan^{-1}(28/48) = 30.26°$$

From Equations (13) and (14):

$$T = -175 + 744 = 569 \text{ lb}$$

$$C = -175 - 744 = -919 \text{ lb}$$

Calculate the shear force per isolator:

$$V = F_p/N_{iso} = 900/4 = 225 \text{ lb} \quad (15)$$

This shear force is applied at the operating height of the isolator. Uplift T on the vibration isolator provides the worst condition for the design of the anchor bolts. The compression force C must be evaluated to check the adequacy of the structure to resist the loads (Figure 10).

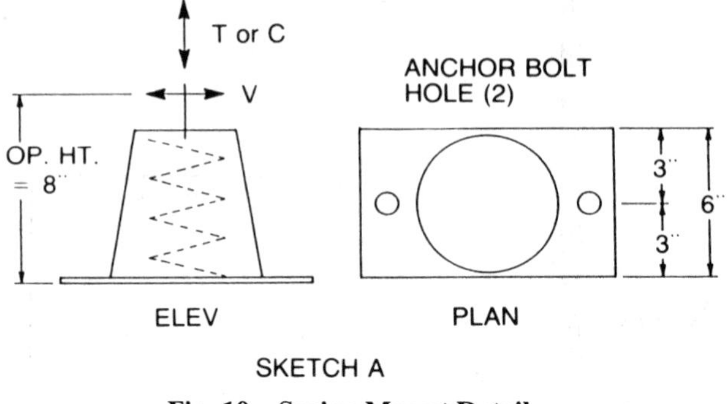

Fig. 10 Spring Mount Detail

$$(T_1)_{eff} \text{ per bolt} = T/2 = 569/2 = 284 \text{ lb}$$

The value of $(T_2)_{eff}$ per bolt due to overturning on the isolator is:

$$(T_2)_{eff} = V_{op\ ht}/0.85dN_{bolt}$$

where

d = distance from edge of isolator base plate to center of bolt hole

$$(T_2)_{eff} = 225 \times 8/(0.85 \times 3 \times 2) = 353 \text{ lb}$$

$$(T_{max})_{eff} = (T_1)_{eff} + (T_2)_{eff} = 284 + 353 = 637 \text{ lb}$$

$$V_{eff} = 225/2 = 113 \text{ lb}$$

Check if 5/8-in. drill-in bolts will handle this load. From Equation (7):

$$637/900 + 113/2200 = 0.76 < 1.0$$

Therefore, 5/8-in. bolts will carry the load.

Example 3

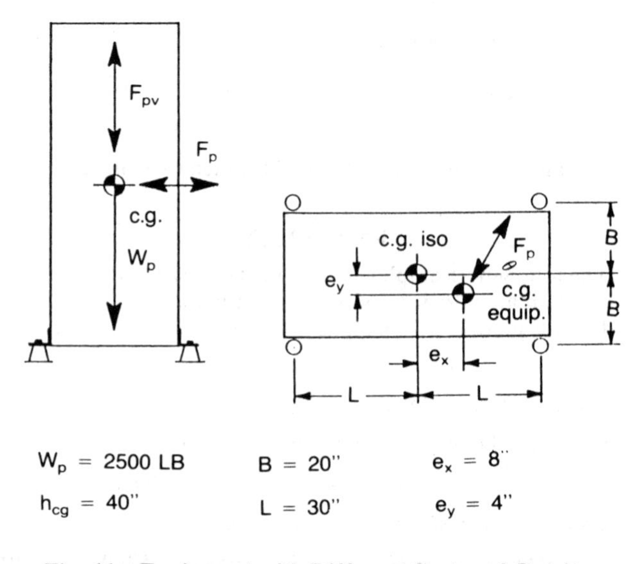

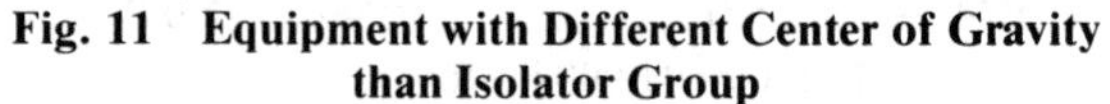

Fig. 11 Equipment with Different Center of Gravity than Isolator Group

Anchor properties: $I_x = 4B^2\ I_y = 4L^2$

Angles:

$$\theta = \tan^{-1}(B/L) \quad (16)$$

$$\alpha = \tan^{-1}(e_x/e_y) \quad (17)$$

$$\beta = 180 - |a - \theta| \quad (18)$$

$$\phi = \tan^{-1}(LI_x)/BI_y) \quad (19)$$

Vertical reactions:

$$(W_n)_{max/min} = W_p \pm F_{pv} \quad (20)$$

Vertical reaction due to overturning moment:

$$T_m = F_p h_{cg}[(B/I_x)\cos\phi + (L/I_y)\sin\phi] \quad (21)$$

Vertical reaction due to eccentricity:

$$(T_e)_{max/min} = (W_n)_{max/min}(Be_y/I_x + Le_x/I_y) \quad (22)$$

Vertical reaction due to W_p:

$$(T_w)_{max/min} = (W_n)_{max/min}/4 \quad (23)$$

$$T_{max} = T_m + (T_e)_{max} + (T_w)_{max} \quad (24)$$

$$T_{min} = T_m + (T_e)_{min} + (T_w)_{min} \text{ (tension if positive)} \quad (25)$$

Horizontal reactions:

Due to rotation: $V_{rot} = F_p[(e_x^2 + e_y^2)/16(B^2 + L^2)]^{0.5}$ (26)

$$V_{dir} = F_p/4 \quad (27)$$

$$V_{max} = (V_{rot}^2 + V_{dir}^2 - 2V_{rot}V_{dir}\cos\beta)^{0.5} \quad (28)$$

From Equation (1):

$F_p = 0.4 \times 1.5 \times 2 \times 0.75 \times 2500 = 2250$ lb

$F_{pv} = 2250/3 = 750$ lb

$I_x = 4(20)^2 = 1600 \quad I_y = 4(30)^2 = 3600$

From Equations (16) through (19):

$\theta = 33.69°$ $\quad \alpha = 63.43°$

$\beta = 150.26°$ $\quad \phi = 33.69°$

From Equation (20):

$(W_n)_{max/min} = 3000$ lb or 1500 lb

From Equation (21): $T_m = 90{,}000(0.01 + 0.005) = 1352$ lb

From Equation (22):

$$(T_e)_{max/min} = 0.1167(W_n)_{max/min} = 350 \text{ lb or } 175 \text{ lb}$$

From Equation (23): $(T_w)_{max/min} = 750$ lb or 375 lb

$$T_{max} = 1352 + 350 + 750 = 2452 \text{ lb}$$

$$T_{min} = 1352 + 175 - 375 = 1152 \text{ lb (tension)}$$

From Equation (26): $V_{rot} = 2250 \times 0.062 = 140$ lb

From Equation (27): $V_{dir} = 2250/4 = 563$ lb

From Equation (28): $V_{max} = 687$ lb

The values of T_{min} and V_{max} are used to design the anchorage of the isolators and/or snubbers. Use T_{max} to verify the adequacy of the structure to resist the vertical loads.

Example 4

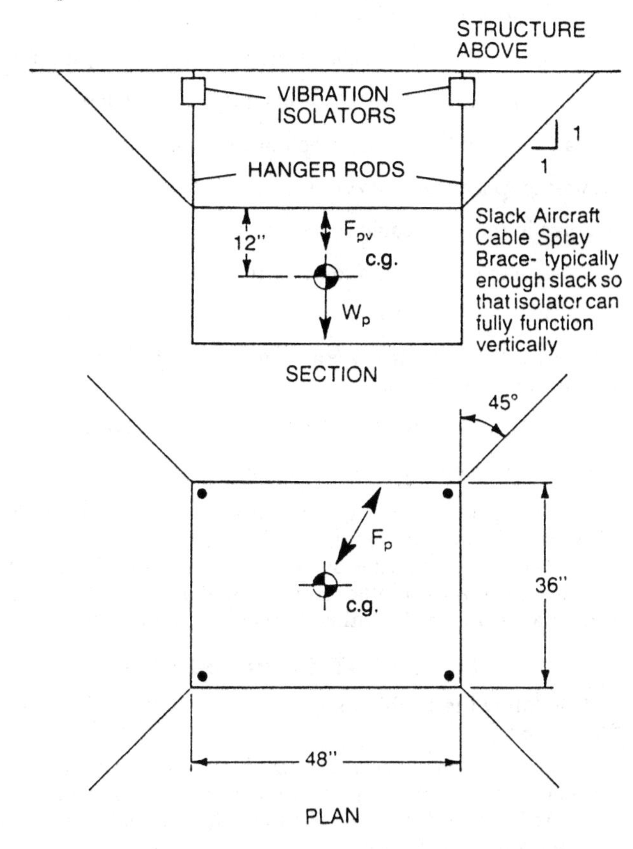

Fig. 12 Supports and Bracing for Suspended Equipment

Note: Vibration isolators should be used between the equipment and the structure to dampen out vibrations generated by the equipment, since drill-in bolts may not withstand published allowable static loads when subjected to vibratory loads.

$$W_p = 500 \text{ lb}$$

From Equation (1):

$F_p = 0.4 \times 1.5 \times 2 \times 0.75 \times 500 = 450$ lb

$F_{pv} = 450/3 = 150$ lb

From Equation (9):

$$\text{OTM} = 450 \times 12 = 5400 \text{ in}\cdot\text{lb}$$

From Equation (10):

$$\text{RM} = (500 \pm 150)36/2 = 11700 \text{ or } 6300 \text{ in}\cdot\text{lb}$$

Since RM > OTM, overturning is not critical.

Force to the hanger rods:

$$T_{eff} = (W_p + F_{pv})/4 = (500 + 150)/4 = 163 \text{ lb}$$

Force in the splay brace $= \sqrt{2}F_p = 636$ lb

Due to the force being applied at the critical angle, similar to Example 2, only 1 splay brace is effective in resisting the lateral load F_p. If eccentricities occur, as in Example 3, a similar method of analysis must be done to obtain the design forces.

Design of hanger rod/vibration isolator and connection to structure.

Note: When installing drill-in anchors into the underside of a concrete beam or slab, the allowable tension loads on the anchors must be reduced to account for the cracking of the concrete. A general rule is to use half the allowable load.

Using a 1/2-in. wedge anchor with special inspection provisions:

$$T_{allow} = 600 \times 0.5 = 300 \text{ lb} > T_{eff} = 163 \text{ lb}$$

Use a 1/2-in. rod and drill-in bolt at each corner of the unit.

Design of splay brace and connection to structure.

Force in the slack cable = 636 lb

Force in the connection to the structure:

$$V_{max} = 636/\sqrt{2} = 450 \text{ lb} \qquad T_{max} = F_p = 450 \text{ lb}$$

Check a 3/4-in. wedge-type anchor with special inspection provisions. From Table (3):

$$T_{allow} = 1350/2 = 675 \text{ lb} \qquad V_{allow} = 3000 \text{ lb}$$

From Equation (7): $450/675 + 450/3000 = 0.82 < 1.0$

A 3/4-in. anchor or multiple anchors of a smaller size bolted through a clip and to the structure is permissible.

Since the cable forces are relatively small, a 3/8-in. aircraft cable attached to clips with cable clamps should be used. The clips, in turn, may be attached to either the structure or to the equipment.

INSTALLATION PROBLEMS

The following problems with seismic restraint have been recognized.

- Anchor location affects the required strengths. Locate concrete anchors away from edges, stress joints, or existing fractures. Follow ASTM *Standard* E-488-90 as a guide for edge distances and center-to-center spacing.
- Concrete anchors too close together. Spacing of epoxy-type anchors can be closer together than expansion type anchors. Expansion-type anchors (self-drilling and drop-in anchors) can crush the concrete where they expand and impose internal stresses in the concrete. Carefully review spacing of all anchor bolts. (See manufacturer's recommendations.)
- Supplementary steel bases and frames, concrete bases, or equipment modifications may void some manufacturer's warranties. Snubbers, for example, should be properly attached to a subbase. Bumpers may be used with springs.
- Static analysis does not account for the effects of resonant conditions within a piece of equipment or its components. Because all equipment has different resonant frequencies during operation and nonoperation, the equipment itself might fail even if the restraints do not. Equipment mounted inside a housing should be seismically restrained to meet the same criteria as the exterior restraints.
- Snubbers used with spring mounts should withstand motion in all directions. Some snubbers are only designed for restraint in one direction; sets of snubbers or snubbers designed for multidirectional purposes should be used.
- Equipment must be strong enough to withstand the high deceleration forces developed by resilient restraints.
- Bumpers installed to limit horizontal motion should be outfitted with resilient neoprene pads to soften the potential impact loads of the equipment.
- Inspect the anchor installation; in many cases, damage occurs because bolts were not properly installed. To develop the rated restraint, bolts should be installed according to manufacturers' recommendations.
- Brackets in structural steel attachments should be matched to reduce bending and internal stresses at the joint. Rigid seismic restraints should not have slotted holes.

BIBLIOGRAPHY

AISC. 1980. *Manual of steel construction,* 9th ed. American Institute of Steel Construction, Chicago.

Army, Navy, and Air Force. 1982. Technical manual seismic design for buildings. TM5-809-10, NAVFAC P-355, AFN 88-3, Chapter 13 (February).

Associate Committee on the National Building Code. 1985. *National Building Code of Canada* 1985, 9th ed. National Research Council of Canada, Ottawa.

Associate Committees on the National Building Code. *Supplement to the National Building Code of Canada* 1985, 2nd ed. National Research Council of Canada, Ottawa. First errata, January 1986.

Astaneh, A., V.V. Bertero, B.A. Bolt, S.A. Mahin, J.P. Moehle, and R.B. Seed. 1989. Preliminary *report* on the seismological and engineering aspects of the October 17, 1989 Santa Cruz (Loma Prieta) Earthquake. Report No. UCB/EERC-89/14. Earthquake Engineering Research Center, College of Engineering, University of California at Berkeley (October).

BOCA. 1990. *The BOCA National Building Code.* Building Officials & Code Administrators International, Inc. Country Club Hills, Illinois.

Bolt, B.A. 1987, 1988. *Earthquakes.* W.H. Freeman & Co., New York. [Note: Earthquakes is a revision of *Earthquakes: A primer,* 1978.]

DOE. 1989. General design criteria. DOE Order 6430.1A. U.S. Department of Energy, Washington, D.C.

Guidelines for seismic restraints of mechanical systems and plumbing & piping systems, 1st ed. 1982. The Sheet Metal Industry Fund of Los Angeles and the Plumbing & Heating Industry Council, Inc.

ICBO. 1988. *Uniform Building Code.* International Conference of Building Officials, Whittier, California.

Jones, R.S. 1984. *Noise and vibration control in buildings.* McGraw-Hill, New York.

Kennedy, R.P., S.A. Short, J.R. McDonald, M.W. McCann, and R.C. Murray. 1989. Design and evaluation guidelines for the Department of Energy facilities subjected to natural phenomena hazards.

Lew, H.S., E.V. Leyendecker, and R.D. Dikkers. 1971. Engineering aspects of the 1971 San Fernando earthquake. *Building Science Series* 40. U.S. Department of Commerce National Bureau of Standards (December).

Maley, R., A. Acosta, F. Ellis, E. Etheredge, L. Foote, D. Johnson, R. Porcella, M. Salsman, and J. Switzer. 1989. Department of the Interior, U.S. geological survey. U.S. geological survey strong-motion records from the Northern California (Loma Prieta) earthquake of October 17, 1989. Open-file *report* 89-568.

Naeim, F. 1989. *The seismic design handbook.* Van Nostrand Reinhold International Company Ltd., London, England.

NFPA. 1986. National design specification, Wood construction. National Forest Products Association, Washington, D.C.

SBCCI. 1988. *Standard Building Code.* Southern Building Code Congress International, Birmingham, Alabama.

Weigels, R.L. 1970. *Earthquake engineering,* 10th ed. Prentice-Hall, Inc., Englewood Cliffs, New Jersey.

CHAPTER 50

CODES AND STANDARDS

THE Codes and Standards listed in Table 1 represent practices, methods, or standards published by the organizations indicated. They are valuable guides for the practicing engineer in determining test methods, ratings, performance requirements, and limits applying to the equipment used in heating, refrigerating, ventilating, and air conditioning. *Copies can usually be obtained from the organization listed in the Reference column.* These listings represent the most recent information available at the time of publication.

Table 1 Codes and Standards Published by Various Societies and Associations

Subject	Title	Publisher	Reference
Air Conditioners	Room Air Conditioners	CSA	C22.2 No. 117-1970
Room	Method of Testing for Rating Room Air Conditioners and Packaged Terminal Air Conditioners	ASHRAE	ANSI/ASHRAE 16-1983 (RA 88)
	Methods of Testing for Rating Room Fan-Coil Air Conditioners	ASHRAE	ASHRAE 79-1978 (RA 84)
	Room Air Conditioners (1982)	UL	ANSI/UL 484-1986
	Room Air Conditioners	AHAM	ANSI/AHAM
	Performance Standard for Room Air Conditioners	CSA	CAN/CSA C368.1-1990
	Method of Testing for Rating Room Air Conditioner and Packaged Terminal Air Conditioner Heating Capacity	ASHRAE	ANSI/ASHRAE 58-1986 (RA 90)
	Commercial and Residential Central Air Conditioners	CSA	C22.2 No. 119-M1985
Packaged Terminal	Packaged Terminal Air Conditioners	ARI	ARI 310-90
	Packaged Terminal Heat Pumps	ARI	ARI 380-90
Transport	Air Conditioning of Aircraft Cargo (1978)	SAE	SAE AIR806A
	Nomenclature, Aircraft Air-Conditioning Equipment (1978)	SAE	SAE ARP147C
Unitary	Air Conditioners, Central Cooling (1982)	UL	ANSI/UL 465-1984
	Load Calculation for Commercial Summer and Winter Air Conditioning, 4th ed. (1988)	ACCA	ACCA Manual N
	Methods of Testing for Rating Heat Operated Unitary Air-Conditioning Equipment for Cooling	ASHRAE	ANSI/ASHRAE 40-1980 (RA 86)
	Methods of Testing for Rating Unitary Air-Conditioning and Heat Pump Equipment	ASHRAE	ANSI/ASHRAE 37-1988
	Methods of Testing for Seasonal Efficiency of Unitary Air Conditioners and Heat Pumps	ASHRAE	ANSI/ASHRAE 116-1983
	Sound Rating of Outdoor Unitary Equipment	ARI	ARI 270-84
	Application of Sound Rated Outdoor Unitary Equipment	ARI	ARI 275-84
	Unitary Air-Conditioning and Air-Source Heat Pump Equipment	ARI	ARI 210/240-89
	Commercial and Industrial Unitary Air-Conditioning Equipment	ARI	ANSI/ARI 360-86
Air Conditioning	Automotive Air-Conditioning Hose (1989)	SAE	ANSI/SAE J51 MAY89
	Environmental System Technology (1984)	NEBB	NEBB
	Commercial Low Pressure, Low Velocity Duct System Design	ACCA	Manual Q
	Gas-Fired Absorption Summer Air Conditioning Appliances (with 1982 addenda)	AGA	ANSI Z21.40.1-1981
	Load Calculation for Residential Winter and Summer Air Conditioning, 7th ed. (1986)	ACCA	ACCA Manual J
	Duct Design for Residential Buildings	ACCA	ACCA Manual D
	Installation Standards for Residential Heating and Air Conditioning Systems (1988)	SMACNA	SMACNA
	HVAC Systems—Duct Design (1990)		
	HVAC Systems—Applications, 1st. ed. (1986)	SMACNA	SMACNA
Transport	Air Conditioning Equipment, General Requirements for Subsonic Airplanes (1961)	SAE	SAE ARP85D
	General Requirements for Helicopter Air Conditioning (1970)	SAE	SAE ARP292B
	Testing of Commercial Airplane Environmental Control Systems (1973)	SAE	SAE ARP217B
Unitary	Method of Rating Computer and Data Processing Room Unitary Air Conditioners	ASHRAE	ANSI/ASHRAE 127-1988
	Method of Rating Unitary Spot Air Conditioners	ASHRAE	ANSI/ASHRAE 128-1988

Table 1 Codes and Standards Published by Various Societies and Associations (*Continued*)

Subject	Title	Publisher	Reference
Air Curtains	Test Methods for Air Curtain Units	AMCA	AMCA 220-82
	Air Volume Terminals	ARI/ADC	ARI/ADC 880-89
	Air Distribution Basics for Residential and Small Commercial Buildings	ACCA	ACCA Manual T
	Residential Equipment Selection	ACCA	ACCA Manual S
	Method of Testing for Rating the Air Flow Performance of Outlets and Inlets	ASHRAE	ASHRAE 70-72
	High Temperature Pneumatic Duct Systems for Aircraft (1981)	SAE	ANSI/SAE ARP699D
	Metric Units and Conversion Factors	AMCA	AMCA 99-0100-76
	Laboratory Certification Manual	ADC	ADC 1062:LCM-83
	Residential Air Exhaust Equipment (1e)	CSA	C260.2-1976
	Test Code for Grilles, Registers and Diffusers	ADC	ADC 1062:GRD-84
	Standard Methods for Laboratory Air Flow Measurement	ASHRAE	ANSI/ASHRAE 41.2-1987
	Flexible Duct Performance and Installation Standards	ADC	ADC
Air Ducts and Fittings	Flexible Air Duct Test Code	ADC	ADC FD-72R1-1979
	Installation of Air Conditioning and Ventilating Systems (1989)	NFPA	ANSI/NFPA 90A-1989
	Installation of Warm Air Heating and Air-Conditioning Systems (1989)	NFPA	ANSI/NFPA 90B-1989
	HVAC Duct Construction Standards—Metal and Flexible, 1st ed. (1985)	SMACNA	SMACNA
	HVAC Air Duct Leakage Test Manual (1985)	SMACNA	9101 SMACNA
	Round Industrial Duct Construction (1977)	SMACNA	SMACNA
	Rectangular Industrial Duct Construction (1980)	SMACNA	SMACNA
	Ducted Electric Heat Guide for Air Handling Systems (1971)	SMACNA	SMACNA
	Thermoplastic Duct (PVC) Construction Manual (Rev. A, 1974)	SMACNA	SMACNA
	Marine Rigid and Flexible Air Ducting (1986)	UL	ANSI/UL 1136-1986
	Residential-Type Air-Conditioning Systems	CSA	B228.1-1968
	Factory-Made Air Ducts and Connectors (1990)	UL	UL 181
Air Filters	Test Performance of Air Filter Units (1987)	UL	ANSI/UL 900-1987
	Method of Testing Air-Cleaning Devices Used in General Ventilation for Removing Particulate Matter	ASHRAE	ASHRAE 52-1968 (RA 76)
	Commercial and Industrial Air Filter Equipment	ARI	ARI 850-84
	Air Filter Equipment	ARI	ARI 680-86
	High Efficiency, Particulate, Air Filter Units (1985)	UL	ANSI/UL 586-1990
	Methods of Test for Atmospheric Dust Spot Efficiency and Synthetic Dust Weight Arrestance	BSI	BS 6540 Part 1
	Method for Sodium Flame Test for Air Filters	BSI	BS 3928
	Electrostatic Air Cleaners (1981)	UL	ANSI/UL 867-1988
Air-Handling Units	Central Station Air-Handling Units	ARI	ARI 430-89
Air Leakage	Air Leakage Performance for Detached Single-Family Residential Buildings	ASHRAE	ANSI/ASHRAE 119-1988
Boilers	Recommended Design Guidelines for Stoker Firing of Bituminous Coals	ABMA	ABMA
	Boiler Water Limits and Steam Purity Recommendations for Watertube Boilers	ABMA	ABMA
	Boiler Water Requirements and Associated Steam Purity—Commercial Boilers	ABMA	ABMA
	A Guide to Clean and Efficient Operation of Coal Stoker-Fired Boilers	ABMA	ABMA
	Fluidized Bed Combustion Guidelines	ABMA	ABMA
	Guidelines for Industrial Boiler Performance Improvement	ABMA	ABMA
	Operation and Maintenance Safety Manual	ABMA	ABMA
	Matrix of Recommend Quality Control Requirements	ABMA	ABMA
	Thermal Shock Damage to Hot Water Boilers as a Result of Energy Conservation Measures	ABMA	ABMA
	Boiler and Pressure Vessel Code (11 sections) (1989)	ASME	ASME
	Boiler, Pressure Vessel, and Pressure Piping Code	CSA	B51-M1986
	Heating, Water Supply, and Power Boilers—Electric (1980)	UL	ANSI/UL834-1980
	Lexicon Boiler and Auxiliary Equipment	ABMA	ABMA
Cast-Iron	Testing and Rating Heating Boilers (1989)	HYD I	IBR/SBI
	Ratings for Cast-Iron and Steel Boilers (1989)	HYD I	IBR/SBI
Gas or Oil	Explosion Prevention of Fuel Oil and Natural Gas-Fired Single-Burner Boiler-Furnaces (1987)	NFPA	ANSI/NFPA 85A-1987
	Explosion Prevention of Natural Gas-Fired Multiple-Burner Boiler-Furnaces (1989)	NFPA	ANSI/NFPA 85B-1989
	Gas-Fired Low-Pressure Steam and Hot Water Boilers	AGA	ANSI Z21.13-1987; Z21.13a-1989
	Gas Utilization Equipment in Large Boilers (with 1972 and 1976 addenda; R-1983, 1989)	AGA	ANSI Z83.3-1971

Table 1 Codes and Standards Published by Various Societies and Associations (*Continued*)

Subject	Title	Publisher	Reference
Boilers (continued)	Oil Fired Boiler Assemblies (1990)	UL	ANSI/UL 726-1990
	Commercial-Industrial Gas Heating Equipment (1973)	UL	UL 795
	Prevention of Furnace Explosions in Fuel Oil-FiredMultiple-Burner Boiler-Furnaces (1989)	NFPA	ANSI/NFPA 85D-1989
	Control and Safety Devices for Automatically Fired Boilers	ASME	ANSI/ASME CSD.1-1988
	Oil Burning Appliance Standards (B140 Series)	CSA	B140.7.1-1976
	Oil-Fired Steam and Hot-Water Boilers for Commercial and Industrial Use (1a)	CSA	B140.7.2-1967
Watertube	Recommended Standard Instrument Connections Manual	ABMA	ABMA
Building Codes	BOCA National Building Code, 11th ed. (1990)	BOCA	BOCA
	CABO One- and Two-Family Dwelling Code (1989)	CABO	CABO
	Standard Building Code (1988) (with 1989/1990 revisions)	SBCCI	SBCCI
	Uniform Building Code (1988)	ICBO	ICBO
	Uniform Building Code Standards (1988)	ICBO	ICBO
	BOCA National Property Maintenance Code, 3rd ed. (1990)	BOCA	BOCA
	Model Energy Code (1989)	CABO	CABO
	Directory of Building Codes and Regulations, (1989 ed.)	NCSBCS	NCSBCS
Mechanical	BOCA National Mechanical Code, 7th ed. (1990)	BOCA	BOCA
	Safety Code for Elevators and Escalators (plus two yearly supplements)	ASME	ANSI/ASME A 17.1-1990
	Uniform Mechanical Code (1988) (with Uniform Mechanical Code Standards)	ICBO/ IAPMO	ICBO/IAPMO
	Standard Mechanical Code (1988) (with 1989/1990 revisions)	SBCCI	SBCCI
	Standard Gas Code (1988) (with 1989/1990 revisions)	SBCCI	SBCCI
Burners	Installation of Domestic Gas Conversion Burners	AGA	ANSI Z21.8-1984; Z21.8a-1990
	Domestic Gas Conversion Burners	AGA	ANSI Z21.17-1984; Z21.17a-1990
	Commercial-Industrial Gas Heating Equipment (1973)	UL	UL 795
	Oil Burners (1989)	UL	ANSI/UL 296-1989
	Installation Code for Oil Burning Equipment	CSA	B139-1976
	Supplement No. 1 to B139-1976, Installation Code for Oil Burning Equipment	CSA	B139S1-1982
	General Requirements for Oil Burning Equipment	CAN/CSA	B140.0-M1987
	Vaporizing-Type Oil Burners	CSA	B140.1-1966 (R 1980)
	Oil Burners, Atomizing Type	CSA	B140.2.1-1973
	Pressure Atomizing Oil Burner Nozzles	CSA	B140.2.2-1971 (R 1980)
	Replacement Burners and Replacement Combustion Heads for Residential Oil Burners	CSA	B140.2.3-M1981
	Guidelines for Burner Adjustments of Commercial Oil-Fired Boilers	ABMA	ABMA
Capillary Tubes	Method of Testing Flow Capacity of Refrigerant Capillary Tubes	ASHRAE	ANSI/ASHRAE 28-78
Chillers	Methods of Testing Liquid Chilling Packages	ASHRAE	ASHRAE 30-1988
	Absorption Water-Chilling Packages	ARI	ARI 560-82
	Centrifugal Water-Chilling Packages	ARI	ARI 550-90
	Reciprocating Water-Chilling Packages	ARI	ARI 590-86
Chimneys	Chimneys, Fireplaces, and Vents, and Solid Fuel Burning Appliances	NFPA	ANSI/NFPA 211-1988
	Chimneys, Factory-Built, Residential Type and Building Heating Appliance (1988)	UL	ANSI/UL 103-1988
	Chimneys, Factory-Built, Medium Heat Appliance (1986)	UL	ANSI/UL 959-1986
Cleanrooms	Procedural Standards for Certified Testing of Cleanrooms(1988)	NEBB	NEBB-1988
Coils	Forced-Circulation Air-Cooling and Air-Heating Coils	ARI	ARI 410-87
	Methods of Testing Forced Circulation Air Cooling and Air Heating Coils	ASHRAE	ASHRAE 33-78
Comfort Conditions	Thermal Environmental Conditions for Human Occupancy	ASHRAE	ANSI/ASHRAE 55-1981
Compressors	Compressors and Exhausters (reaffirmed 1986)	ASME	ANSI/ASME PTC 10-74-1965 (R 1986)
	Compressed Air and Gas Handbook, 5th ed. (1989)	CAGI	CAGI
	Safety Standard for Compressors for Process Industries	ASME	ASME/ANSI B19.3-1986
	Safety Standard for Air Compressor Systems	ASME	ANSI/ASME B19.1-1990
	Displacement Compressors, Vacuum Pumps and Blowers	ASME	ANSI/ASME PTC9-1974 (R 1985)

Table 1 Codes and Standards Published by Various Societies and Associations (*Continued*)

Subject	Title	Publisher	Reference
Refrigeration	Methods of Testing for Rating Positive Displacement Refrigerant Compressors	ASHRAE	ASHRAE 23-78
	Ammonia Compressor Units	ARI	ARI 510-87
	Hermetic Refrigerant Motor-Compressors (1990)	UL	ANSI/UL 984-1989
	Hermetic Refrigerant Motor-Compressors	CSA	CAN/CSA C22.2 No. 140.2-M89
	Positive Displacement Refrigerant Compressors and Condensing Units	ARI	ANSI/ARI 520-90
Computers	Protection of Electronic Computer/Data Processing Equipment	NFPA	ANSI/NFPA 75-1989
Condensers	Water-Cooled Refrigerant Condensers, Remote Type	ARI	ARI 450-87
	Methods of Testing for Rating Remote Mechanical-Draft Air-Cooled Refrigerant Condensers	ASHRAE	ASHRAE 20-70
	Methods of Testing for Rating Water-Cooled Refrigerant Condensers	ASHRAE	ASHRAE 22-1971 (RA 78)
	Methods of Testing Remote Mechanical-Draft Evaporative Refrigerant Condensers	ASHRAE	ASHRAE 64-1989
	Remote Mechanical Draft Air-Cooled Refrigerant Condensers	ARI	ARI 460-87
	Standards for Steam Surface Condensers, 8th ed. (1984)	HEI	HEI
	Addendum I, Standards for Steam Surface Condensers (1989)	HEI	HEI
Condensing Units	Methods of Testing for Rating Positive Displacement Condensing Units	ASHRAE	ASHRAE 14-80
	Refrigeration and Air-Conditioning Condensing and Compressor Units (1987)	UL	ANSI/UL303-1988
	Commercial and Industrial Unitary Air-Conditioning Condensing Units	ARI	ARI 365-87
	Heating and Cooling Equipment	CSA	CAN/CSA C22.2 No. 236-M90
Contactors	Definite Purpose Magnetic Contactors	ARI	ARI 780-86
	Definite Purpose Contactors for Limited Duty	ARI	ARI 790-86
Controls	Quick-Disconnect Devices for Use with Gas Fuel	AGA	ANSI Z21.41-1989; Z21.41a-1990
	Limit Controls (1989)	UL	ANSI/UL 353-1988
	Energy Management Control Systems Instrumentation	ASHRAE	ASHRAE 114-1986
	Primary Safety Controls for Gas- and Oil-Fired Appliances (1985)	UL	ANSI/UL 372-1985
	Temperature-Indicating and Regulating Equipment (1988)	UL	ANSI/UL 873-1987
	Industrial Control Equipment (1988)	UL	ANSI/UL 508-1988
	Temperature-Indicating and Regulating Equipment	CSA	C22.2 No. 24-1987
Residential	Automatic Gas Ignition Systems and Components	AGA	ANSI Z21.20-1989
	Temperature Limit Controls for Electric Baseboard Heaters	NEMA	NEMA DC 10-1983 (R 1989)
	Electrical Quick-Connect Terminals (1985)	UL	ANSI/UL 310-1986
	Quick Connect Terminals	NEMA	ANSI/NEMA DC 2-1982 (R 1988)
	Line-Voltage Integrally Mounted Thermostats for Electric Heaters	NEMA	NEMA DC 13-1979 (R 1985)
	Wall-Mounted Room Thermostats	NEMA	NEMA DC 3-1989
	Hot-Water Immersion Controls	NEMA	NEMA DC 12-1985
	Warm Air Limit and Fan Controls	NEMA	NEMA DC 4-1986
	Gas Appliance Thermostats	AGA	ANSI Z21.23-1989
	Gas Appliance Pressure Regulators (with 1989 addenda)	AGA	ANSI Z21.18-1987
	Manually-Operated Prego. Electric Spark Gas Ignition Systems and Components (and Addenda Z21.15a-1990)	AGA	ANSI Z21.77-1989
	Residential Controls—Surface Type Controls for Electric Storage Water Heaters	NEMA	NEMA DC 5-1989
	Residential Controls—Class 2 Transformers	NEMA	NEMA DC 20-1986
Coolers Air	Methods of Testing Forced Convection and Natural Convection Air Coolers for Refrigeration	ASHRAE	ASHRAE 25-1990
	Unit Coolers for Refrigeration	ARI	ARI 420-89
	Milk Coolers	CSA	C22.2 No. 132-1973
Bottled Beverage	Methods of Testing and Rating Bottled and Canned Beverage Venders and Coolers	ASHRAE	ANSI/ASHRAE 32-1986 (RA 90)
	Refrigerated Vending Machines (1989)	UL	ANSI/UL 541-1988
Drinking Water	Methods of Testing for Rating Drinking-Water Coolers with Self-Contained Mechanical Refrigeration Systems	ASHRAE	ASHRAE 18-1987
	Drinking Water Coolers (1987)	UL	ANSI/UL 399-1986
	Drinking Fountains and Self-Contained, Mechanically Refrigerated Drinking Water Coolers	ARI ANSI	ANSI/ARI 1010-84
	Application and Installation of Drinking Water Coolers	ARI	ANSI/ARI 1020-84
	Drinking Water Coolers and Beverage Dispensers	CSA	C22.2 No. 91-1971 (R 1981)
Liquid	Methods of Testing for Rating Liquid Coolers	ASHRAE	ASHRAE 24-1989
	Refrigerant-Cooled Liquid Coolers, Remote Type	ARI	ARI 480-87

Table 1 Codes and Standards Published by Various Societies and Associations (*Continued*)

Subject	Title	Publisher	Reference
Cooling Towers	Water-Cooling Towers	NFPA	ANSI/NFPA 214-1988
	Atmospheric Water Cooling Equipment	ASME	ANSI/ASME PTC 23-1986
	Acceptance Test Code for Water Cooling Towers: Mechanical Draft, Natural Draft Fan Assisted Types, Evaluation of Results, and Thermal Testing of Wet/Dry Cooling Towers (1986)	CTI	CTI ATC-105-1990
	Acceptance Test Code for Spray Cooling Systems (1985)	CTI	CTI ATC-133-1985
	Certification Standard for Commercial Water Cooling Towers (1986)	CTI	CTI STD-201-1986
	Code for Measurement of Sound from Water Cooling Towers	CTI	CTI ATC-128-1981
	Fiberglass-Reinforced Plastic Panels for Application on Industrial Water Cooling Towers	CTI	CTI STD-131-1986
	Nomenclature for Industrial Water-Cooling Towers	CTI	CTI NCL-109-1983
Dehumidifiers	Dehumidifiers	AHAM	ANSI/AHAM DH 1-1986
	Dehumidifiers (3a)	CSA	C22.2 No. 92-1971
	Dehumidifiers (1987)	UL	ANSI/UL 474-1987
Desiccants	Method of Testing Desiccants for Refrigerant Drying	ASHRAE	ASHRAE 35-1976 (RA 83)
Driers	Liquid Line Driers	ARI	ARI 710-86
	Method of Testing Liquid Line Refrigerant Driers	ASHRAE	ASHRAE 63.1-1988
Electrical	National Electric Code (1990)	NFPA	ANSI/NFPA 70-1990
	Canadian Electrical Code, Part 1 (16th ed.)	CSA	C22.1-1990
	Compatibility of Electrical Connectors and Wiring (1988)	SAE	SAE AIR1329 A
	Manufacturers' Identification of Electrical Connector Contacts, Terminals and Splices (1982)	SAE	SAE AIR1351 A
	Voltage Ratings for Electrical Power Systems and Equipment	ANSI	ANSI C84.1-1989
Energy	Air Conditioning and Refrigerating Equipment Nameplate Voltages	ARI	ARI 110-90
	Energy Conservation in New Building Design	ASHRAE	ANSI/ASHRAE/IES 90A-1980
	Energy Conservation in Existing Buildings—High Rise Residential	ASHRAE	ANSI/ASHRAE/IES 100.2-1981
	Energy Conservation in Existing Buildings—Commercial	ASHRAE	ANSI/ASHRAE/IES 100.3-1985
	Energy Conservation in Existing Facilities—Industrial	ASHRAE	ANSI/ASHRAE/IES 100.4-1984
	Energy Conservation in Existing Buildings—Institutional	ASHRAE	ANSI/ASHRAE/IES 100.5-1981
	Energy Conservation in Existing Buildings—Public Assembly	ASHRAE	ANSI/ASHRAE/IES 100.6-1981
	Energy Director	NCSBCS	NCSBCS
	Energy Efficient Design of New Buildings Except Low Rise Residential Buildings	ASHRAE	ASHRAE/IES 90.1-1989
	Energy Recovery Equipment and Systems, Air-to-Air (1978)	SMACNA	SMACNA
	Energy Conservation Guidelines (1984)	SMACNA	SMACNA
	Model Energy Code (MEC) (1989)	CABO	BOCA/ICBO/SBCCI
	Energy Management Equipment (1987)	UL	ANSI/UL 916-1987
	Retrofit of Building Energy Systems and Processes (1982)	SMACNA	SMACNA
Exhaust Systems	Method of Testing Performance of Laboratory Fume Hoods	ASHRAE	ANSI/ASHRAE 110-1985
	Installation of Blower and Exhaust Systems for Dust, Stock, Vapor Removal or Conveying (1990)	NFPA	ANSI/NFPA 91-1990
	Fundamentals Governing the Design and Operation of Local Exhaust Systems	ANSI	ANSI Z9.2-1979
	Safety Code for Design, Construction, and Ventilation of Spray Finishing Operations (reaffirmed 1971)	ANSI	ANSI Z9.3-1985
	Ventilation and Safe Practices of Abrasives Blasting Operations	ANSI	ANSI Z9.4-1985
	Compressors and Exhausters	ASME	ANSI/ASME PTC 10-1974 (R 1985)
	Mechanical Flue-Gas Exhausters	CSA/CAN	3-B255-M81
	Draft Equipment (1973)	UL	UL 378
Expansion Valves	Method of Testing for Capacity Rating of Thermostatic Refrigerant Expansion Valves	ASHRAE	ANSI/ASHRAE 17-1986 (RA 90)
	Thermostatic Refrigerant Expansion Valves	ARI	ARI 750-87
Fan Coil Units	Room Fan-Coil Air Conditioners	ARI	ARI 440-89
	Fan Coil Units and Room Fan Heater Units (1986)	UL	ANSI/UL 883-1986
	Methods of Testing for Rating Room Fan-Coil Air Conditioners	ASHRAE	ASHRAE 79-1978 (RA 84)

Table 1 Codes and Standards Published by Various Societies and Associations (*Continued*)

Subject	Title	Publisher	Reference
Fans	Standards Handbook	AMCA	AMCA 99-86
	Fans	ASME	ANSI/ASME PTC 11-1984
	Rating the Performance of Residential Mechanical Ventiliating Equipment	CSA	CAN/CSA C260-M90
	Electric Fans (1987)	UL	ANSI/UL 507-1987
	Laboratory Methods of Testing Fans for Rating	ASHRAE	ANSI/ASHRAE 51-1985 ANSI/AMCA 210-85
	Methods of Testing Dynamic Characteristics of Propeller Fans—Aerodynamically Excited Fan Vibrations and Critical Speeds	ASHRAE	ANSI/ASHRAE 87.1-1983
	Laboratory Methods of Testing Fans for Rating	AMCA	ANSI/AMCA 210-85
	Drive Arrangements for Centrifugal Fans	AMCA	AMCA 99-2404-78
	Designation for Rotation and Discharge of Centrifugal Fans	AMCA	AMCA 99-2406-83
	Motor Positions for Belt or Chain Drive Centrifugal Fans	AMCA	AMCA 99-2407-66
	Drive Arrangements for Tubular Centrifugal Fans	AMCA	AMCA 99-2410-82
	Site Performance Test Standard Power Plant and Industrial Fans	AMCA	AMCA 803-87
	Fans and Blowers	ARI	ARI 670-90
	Inlet Box Positions for Centrifugal Fans	AMCA	AMCA 99-2405-83
	Fans and Ventilators	CSA	C22.2 No. 113-M1984
	Performance of Ventilating Fans for Use in Livestock and Poultry Buildings	CSA	CAN/CSA C320-M86
Ceiling	AC Electric Fans and Regulators	AMCA	ANSI-IEC Pub. 385
Filters	Flow-Capacity Rating and Application of Suction-Line Filters and Filter Driers	ARI	ARI 730-86
	Grease Filters for Exhaust Ducts (1979)	UL	UL 1046
	Grease Extractors for Exhaust Ducts (1981)	UL	ANSI/UL 710-1981
Fire Dampers	Fire Dampers (1990)	UL	ANSI/UL 555-1989
Fireplaces	Factory-Built Fireplaces (1988)	UL	ANSI/UL 127-1985
Fire Protection	BOCA National Fire Prevention Code, 8th ed. (1990)	BOCA	BOCA
	National Fire Codes (issued annually)	NFPA	NFPA
	Fire Prevention Code	NFPA	ANSI/NFPA 1-1987
	Fire Protection Handbook, 16th ed.	NFPA	NFPA
	Flammable and Combustible Liquids Code	NFPA	ANSI/NFPA 30-1987
	Heat Responsive Links for Fire Protection Service (1987)	UL	ANSI/UL 33-1987
	Test Method for Surface Burning Characteristics of Building Materials	ASTM/ NFPA	ASTM E 84-89a NFPA 255-1984
	Fire Doors and Windows	NFPA	ANSI/NFPA 80-1990
	Fire Tests of Building Construction and Materials (1984)	UL	ANSI/UL 263-1986
	Life Safety	NFPA	ANSI/NFPA 101-1988
	Smoke Control Systems	NFPA	9ZA-1988
	Standard Fire Prevention Code (1988) (with 1989/1990 revisions)	SBCCI	SBCCI
	Standard Method of Fire Tests of Door Assemblies	NFPA	ANSI/NFPA 252-1990
	Uniform Fire Code (1988)	ICBO/WFCA	ICBO/WFCA
	Uniform Fire Code *Standards* (1988)	ICBO/WFCA	ICBO/WFCA
Fireplace Stoves	Fireplace Stoves (1988)	UL	ANSI/UL 737-1988
Flow Capacity	Method of Testing Flow Capacity of Suction Line Filters and Filter Driers	ASHRAE	ASHRAE 78-1985 (RA 90)
Freezers Household	Household Refrigerators and Freezers	CSA	C22.2 No. 63-M1987
	Household Refrigerators and Freezers (1983)	UL	ANSI/UL 250-1984
	Household Refrigerators, Combination Refrigerator-Freezers, and Household Freezers	AHAM	ANSI/AHAM HRF 1-1988
	Capacity Measurement and Energy Consumption Test Methods for Refrigerators, Combination Refrigerator-Freezers, and Household Freezers	CSA	CAN/CSA-C300-M89
Commercial	Ice Cream Makers (1986)	UL	ANSI/UL 621-1985
	Soda Fountain and Luncheonette Equipment	NSF	NSF-1
	Dispensing Freezers	NSF	ANSI/NSF-6-1989
	Commercial Refrigerators and Freezers (1985)	UL	ANSI/UL 471-1984
	Food Service Refrigerators and Storage Freezers	NSF	NSF-7
	Ice Makers (1958)	UL	ANSI/UL 563-1985
	Commercial Refrigerated Equipment	CSA	C22-2 No. 120-1970 (R 1981)
Furnaces	Gas-Fired Gravity and Fan Type Direct Vent Wall Furnaces (with 1989 and 1990 addenda)	AGA	ANSI Z21.44-1988
	Gas-Fired Central Furnaces (except Direct Vent)	AGA	ANSI Z21.47-1990
	Direct Vent Central Furnaces	AGA	ANSI Z21.64-1988; Z21.64a-1989; Z21.64b-1989

Table 1 Codes and Standards Published by Various Societies and Associations (*Continued*)

Subject	Title	Publisher	Reference
Furnaces (continued)	Gas-Fired Gravity and Fan Type Floor Furnaces	AGA	ANSI Z21.48-1989
	Commercial-Industrial Gas Heating Equipment (1973)	UL	UL 795
	Solid-Fuel and Combination-Fuel Central and Supplementary Furnaces (1981)	UL	ANSI/UL 391-1983
	Gas-Fired Gravity and Fan Type Vented Wall Furnaces	AGA	ANSI Z21.49-1989
	Methods of Testing for Rating Non-Residential Warm Air Heaters	ASHRAE	ANSI/ASHRAE 45-1978 (RA 86)
	Methods of Testing for Heating Seasonal Efficiency of Central Furnaces and Boilers	ASHRAE	ANSI/ASHRAE 103-1988
	Installation of Oil Burning Equipment	NFPA	NFPA 31-1987
	Oil-Fired Central Furnaces (1986)	UL	UL 727-1986
	Gas-Fired Duct Furnaces	AGA	ANSI Z83.9-1986; Z83.9a-1989; Z83.9b-1989
	Oil-Fired Floor Furnaces (1987)	UL	ANSI/UL 729-1987
	Oil-Fired Wall Furnaces (1987)	UL	ANSI/UL 730-1986
	Standard Gas Code (1988) (with 1989/1990 revisions)	SBCCI	SBCCI
	Oil Burning Stoves and Water Heaters (2a)	CSA	B140.3-1962 (R 1980)
	Oil-Fired Warm Air Furnaces (8a)	CSA	B140.4-1974
	Installation Code for Solid-Fuel-Burning Appliances and Equipment	CAN/CSA	B365-M87
	Solid Fuel-Fired Central Heating Appliances	CAN/CSA	B366.1-M87
	Residential Gas Detectors (1983)	UL	ANSI/UL 1484-1983
	Electric Central Warm-Air Furnaces	CSA	C22.2 No. 23-1980
	Heating and Cooling Equipment	CSA	CAN/CSA C22.2 No. 236-M90
Heat Exchangers	Standards of Tubular Exchanger Manufacturers Association, 7th ed. (1988)	TEMA	TEMA
	Liquid Suction Heat Exchangers	ARI	ARI 490-89
	Method of Testing Air-to-Air Heat Exchangers	ASHRAE	ASHRAE 84-78
	Standards for Power Plant Heat Exchangers, 2nd ed. (1990)	HEI	HEI
	Standard Methods of Test for Rating the Performance of Heat Recovery Ventilators	CSA	CAN/CSA C439-88
Heat Pumps	Heat Pumps (1985)	UL	ANSI/UL 559-1985
	Water-Source Heat Pumps	ARI	ANSI/ARI 320-86
	Ground Water-Source Heat Pumps	ARI	ARI 325-85
	Commercial and Industrial Heat Pump Equipment	ARI	ANSI/ARI 340-86
	Central Forced-Air Unitary Heat Pumps with or without Electric Resistance Heat	CSA	C22.2 No. 186.1 M1980
	Add-on Heat Pumps	CSA	C22.2 No. 186.2-M1980
	Performance Standard for Unitary Heat Pumps (1a)	CSA	C273.3-M1977
	Installation Requirements for Air-to-Air Heat Pumps (2a)	CSA	C273.5-1980
	Methods of Testing for Rating Unitary Air-Conditioning and Heat Pump Equipment	ASHRAE	ANSI/ASHRAE 37-1988
	Heating and Cooling Equipment	CSA	CAN/CSA C22.2 No. 236-M90
Heat Recovery	Energy Recovery Equipment and Systems, Air-to-Air (1978)	SMACNA	SMACNA
	Gas Turbine Heat Recovery Steam Generators	ASME	ANSI/ASME PTC 4.4-1981 (R 1987)
Heaters	Air Heaters	ASME	ANSI/ASME PTC 4.3-1968 (R 1985)
	Desuperheater/Water Heaters	ARI	ARI 470-87
	Electric Heaters for Use in Hazardous (Classified) Locations (1985)	UL	ANSI/UL 823-1990
	Standards for Closed Feedwater Heaters, 4th ed. (1984)	HEI	HEI
	Oil-Fired Air Heaters and Direct-Fired Heaters (1975)	UL	UL 733
	Oil-Fired Room Heaters (1973)	UL	UL 896
	Solid Fuel-Type Room Heaters (1988)	UL	ANSI/UL 1482-1988
	Gas-Fired Room Heaters, Vol. I, Vented Room Heaters (with 1989 and 1990 addenda)	AGA	ANSI Z21.11.1-1988
	Gas-Fired Room Heaters, Vol. II, Unvented Room Heaters (with 1990 addenda)	AGA	ANSIZ21.11.2-1989; Z83.6a-1989
	Gas-Fired Infrared Heaters	AGA	ANSI Z83.6-1990
	Gas-Fired Construction Heaters	AGA	ANSI Z83.7-1990
	Direct Gas-Fired Make-Up Air Heaters	AGA	ANSI Z83.4-1989
	Gas-Fired Unvented Commercial and Industrial Heaters (with 1984 and 1989 addenda)	AGA	ANSI Z83.16-1982
	Direct Gas-Fired Industrial Air Heaters	AGA	ANSI Z83.17-1988
	Gas-Fired Pool Heaters	AGA	ANSI Z21.56-1989
	Motor Vehicle Heater Test Procedure (1982)	SAE	SAE J638 JUN82
	Electric Heating Appliances (1987)	UL	ANSI/UL 499-1987
	Gas Heating Equipment, Commercial-Industrial (1973)	UL	UL 795
	Fuel-Fired Heaters—Air Heating—for Construction and Industrial Machinery (1989)	SAE	SAE J1024 MAY89

Table 1 Codes and Standards Published by Various Societies and Associations (*Continued*)

Subject	Title	Publisher	Reference
Heaters (continued)	Electric Air Heaters (1980)	UL	ANSI/UL 1025-1988
	Electric Central Air Heating Equipment (1986)	UL	ANSI/UL 1096-1985
	Direct Gas-Fired Door Heaters (with 1989 addenda)	AGA	ANSI Z83.17-1988
	Space Heaters for Use with Solid Fuels	CSA	B366.2 M1984
	Unvented Kerosene-Fired Room Heaters and Portable Heaters (1982)	UL	UL 647
	Electric Oil Heaters (1990)	UL	ANSI/UL 574-1990
Heating	Aircraft Electrical Heating Systems (1965) (reaffirmed 1983)	SAE	SAE AIR860
	Environmental System Technology (1984)	NEBB	NEBB
	Installation and Operation of Pulverized Fuel Systems	NFPA	ANSI/NFPA 85F-1988
	Manual for Calculating Heat Loss and Heat Gain for Electric Comfort Conditioning	NEMA	NEMA HE 1-1980 (R 1986)
	Heat Loss Calculation Guide (1984)	HYD I	IBR H-21
	Installation Guide for Residential Hydronic Heating Systems, 6th ed. (1988)	HYD I	IBR 200
	Advanced Installation Guide for Hydronic Heating Systems, 2nd ed.	HYD I	IBR 250
	Installation Standards for Residential Heating and Air Conditioning Systems (1988)	SMACNA	SMACNA
	HVAC Systems—Applications, 1st ed. (1986)	SMACNA	SMACNA
	Electric Baseboard Heating Equipment (1987)	UL	ANSI/UL 1042-1986
	Electric Central Air Heating Equipment (1986)	UL	ANSI/UL 1096-1985
	Portable Industrial Oil-Fired Heaters	CSA	B140.8-1967 (R 1980)
	Portable Kerosene-Fired Heaters	CSA	CAN 3-B140.9.3 M86
	Oil-Fired Service Water Heaters and Swimming Pool Heaters(7a)	CSA	B140.12-1976
	Automatic Flue-Pipe Dampers for Use with Oil-Fired Appliances	CSA	B140.14-M1979
	Determining the Required Capacity of Residential Space Heating and Cooling Appliances	CAN/CSA	CAN/CSA-F280-M86
	Performance Standard for Electrical Baseboard Heaters	CSA	C273.2-1971
	Performance Requirements for Electric Heating Line-Voltage Wall Thermostats	CSA	C273.4-M1978
	Electric Air Heaters	CSA	C22.2 No. 46-M1988
	Heater Elements	CSA	C22.2 No. 72-M1984
	Electric Duct Heaters	CSA	C22.2 No. 155-M1986
Humidifiers	Humidifiers (1987)	UL	ANSI/UL 998-1985
	Central System Humidifiers	ARI	ARI 610-89
	Self-Contained Humidifiers	ARI	ARI 620-89
	Appliance Humidifiers	AHAM	ANSI/AHAM HU 1-1987
	Humidifiers and Evaporative Coolers	CSA	C22.2 No. 104-M1983
Ice Makers	Ice Makers (1984)	UL	ANSI/UL 563-1985
	Methods of Testing Automatic Ice Makers	ASHRAE	ASHRAE 29-1988
	Automatic Commercial Ice Makers	ARI	ARI 810-87
	Split System Automatic Commercial Ice Makers	ARI	ARI 815-87
	Ice Storage Bins	ARI	ARI 820-88
	Automatic Ice-Making Equipment	NSF	NSF-12
	Ice-Making Machines	CSA	C22.2 No. 133-1964 (R 1981)
Incinerators	Residential Incinerators (1973)	UL	UL 791
	Incinerators, Waste and Linen Handling Systems and Equipment	NFPA	ANSI/NFPA 82-1990
	Incinerator Performance	CSA	Z103-1976
Induction Units	Room Air-Induction Units	ARI	ARI 445-87
Industrial Duct	Round Industrial Duct Construction (1977)	SMACNA	SMACNA
	Rectangular Industrial Duct Construction (1980)	SMACNA	SMACNA
Insulation	National Commercial and Industrial Insulation Standards	MICA	MICA 1988
	Test Method for Steady-State and Thermal Performance of Building Assemblies by Means of a Guarded Hot Box	ASTM	ASTM C236-89
	Test Method for Steady-State Heat Flux Measurements and Thermal Transmission Properties by Means of the Guarded Hot Plate Apparatus	ASTM	ASTM C177-85
	Test Method for Steady-State Heat Transfer Properties of Horizontal Pipe Insulations	ASTM	ASTM C335-89
	Test Method for Steady-State Heat Flux Measurements and Thermal Transmission Properties by Means of the Heat Flow Meter Apparatus	ASTM	ASTM C518-85
	Specification for Adhesives for Duct Thermal Insulation	ASTM	ASTM C916
	Specification for Thermal and Acoustical Insulation (Mineral Fiber, Duct Lining Material)	ASTM	ASTM C1071-86

Table 1 Codes and Standards Published by Various Societies and Associations (*Continued*)

Subject	Title	Publisher	Reference
	Thermal Insulation, Mineral Fiber, for Buildings	CSA	A101 M-1983
Louvers	Test Method for Louvers, Dampers, and Shutters	AMCA	AMCA 500-89
Lubricants	Test Method for Carbon-Type Composition of InsulatingOils of Petroleum Origin	ASTM	ASTM D2140-86
	Method for Conversion of Kinematic Viscosity to Saybolt Universal Viscosity or to Saybolt Furol Viscosity	ASTM	ASTM D2161-87
	Method for Calculating Viscosity Index from Kinematic Viscosity at 40 and 100°C	ASTM	ASTM D2270-86
	Method for Estimation of Molecular Weight of Petroleum Oils from Viscosity Measurements	ASTM	ASTM D2502-87
	Test Method for Molecular Weight of Hydrocarbons by Thermoelectric Measurement of Vapor Pressure	ASTM	ASTM D2503-82 (1987)
	Test Method for Mean Molecular Weight of Mineral Insulating Oils by the Cryoscopic Method	ASTM	ASTM D2224-78 (1983)
	Test Methods for Pour Point of Petroleum Oils	ASTM	ASTM D97-87
	Recommended Practice for Viscosity System for Industrial Fluid Lubricants	ASTM	ASTM D2422-86
	Test Method for Dielectric Breakdown Voltage of Insulating Liquids Using Disk Electrodes	ASTM	ASTMD877-87
	Test Method for Dielectric Breakdown Voltage of Insulating Oils of Petroleum Origin Using VDE Electrodes	ASTM	ASTM D1816-84a (90)
	Method for Separation of Representative Aromatics and Nonaromatics Fractions of High-Boiling Oils by Elution Chromatography	ASTM	ASTM D2549-85
	Method of Testing the Floc Point of Refrigeration Grade Oils	ASHRAE	ANSI/ASHRAE 86-1983
Measurements	Engineering Analysis of Experimental Data	ASHRAE	ASHRAE Guideline 2-1986 (RA 90)
	Standard Method for Pressure Measurement	ASHRAE	ANSI/ASHRAE 41.3-1989
	A Standard Calorimeter Test Method for Flow Measurement of a Volatile Refrigerant	ASHRAE	ANSI/ASHRAE 41.9-1988
	Standard Method for Temperature Measurement	ASHRAE	ASHRAE 41.1-1986
	Standard Method for Measurement of Proportion of Oil in Liquid Refrigerant	ASHRAE	ANSI/ASHRAE 41.4-1984
	Standard Method for Measurement of Moist Air Properties	ASHRAE	ANSI/ASHRAE 41.6-1982
	Standard Method for Measurement of Flow of Gas	ASHRAE	ASHRAE 41.7-1984
	Standard Methods of Measurement of Flow of Liquids in Pipes Using Orifice Flowmeters	ASHRAE	ASHRAE 41.8-78
	Standard Methods of Measuring and Expressing Building Energy Performance	ASHRAE	ASHRAE 105-1984
	Procedure for Bench Calibration of Tank Level Gaging Tapes and Sounding Rules	ASME	ANSI MC88.2-1974 (R 1987)
	Guide for Dynamic Calibration of Pressure Transducers	ASME	ANSI MC88-1-1972 (R 1987)
	Glossary of Terms Used in the Measurement of Fluid Flow in Pipes	ASME	ANSI/ASME MFC-1M-1979 (R 1986)
	Measurement Uncertainty for Fluid Flow in Closed Conduits	ASME	ANSI/ASMEMFC-2M-1983 (R 1988)
	Measurement of Fluid Flow in Pipes Using Orifice, Nozzle, and Venturi	ASME	ASME MFC-3M-1989
	Measurement of Gas Flow by Turbine Meters	ASME	ANSI/ASME MFC-4M-1986
	Measurement of Liquid Flow in Closed Conduits Using Transit-Time Ultrasonic Flowmeters	ASME	ANSI/ASME MFC-5M-1985 (R 1989)
	Measurement of Fluid Flow in Pipes Using Vortex Flow Meters	ASME	ASME/ANSI MFC-6M-1987
	Measurement of Gas Flow by Means of Critical Flow Venturi Nozzles	ASME	ASME/ANSI MFC-7M-1987
	Measurement Uncertainty	ASME	ANSI/ASME PTC 19.1-1985
	Pressure Measurement	ASME	ASME/ANSI PTC 19.2-1987
	Temperature Measurement	ANSI	ANSI/ASME PTC 19.3-1974 (R 1986)
	Measurement of Rotary Speed	ASME	ANSI/ASME PTC 19.13-1961 (R 1986)
	Measurement of Industrial Sound	ASME	ANSI/ASME PTC 36-1985
Mobile Homes and Recreational Vehicles	Plumbing System Components for Manufactured Homes and Recreational Vehicles	NSF	NSF-24
	Recreational Vehicle Parks	NFPA	NFPA 501C-1990
	Shear Resistance Tests for Ceiling Boards for Mobile Homes (1980)	UL	UL 1296
	Roof Trusses for Mobile Homes (1990)	UL	UL 1298

Table 1 Codes and Standards Published by Various Societies and Associations (*Continued*)

Subject	Title	Publisher	Reference
Mobile Homes and Recreational Vechicles (continued)	Gas Burning Heating Appliances for Mobile Homes and Recreational Vehicles (1965)	UL	UL 307B
	Gas Supply Connectors for Manufactured Homes	IAPMO	IAPMO TSC 9-1989
	Liquid-Fuel-Burning Heating Appliances for Mobile Homes and Recreational Vehicles (1990)	UL	ANSI/UL 307A-1989
	Recreational Vehicle Cooking Gas Appliances (with 1989 addenda)	AGA	ANSI Z21.57-1987
	Gas-Fired Cooking Appliances for Recreational Vehicles (1976)	UL	UL 1075
	Oil-Fired Warm-Air Heating Appliances for Mobile Housing and Recreational Vehicles (2a)	CSA	B140.10-1974 (R 1981)
	Mobile Homes	CSA	CAN/CSA-Z240 MH Series-M86
	Recreational Vehicles	CSA	CAN/CSA-Z240 RV Series-M86
	Manufactured Home Installations	NCSBCS	ANSI A225.1-87
	Mobile Home Parks	CSA	Z240.7.1-1972
	Recreational Vehicle Parks	CSA	Z240.7.2-1972
	Roof Jacks for Manufactured Homes and Recreational Vehicles (1990)	UL	ANSI/UL 311-1985
Motors and Generators	Motors and Generators	NEMA	NEMA MG 1-1987
	Impedance Protected Motors (1982)	UL	ANSI/UL 519-1989
	Electric Motors (1989)	UL	ANSI/UL 1004-1988
	Energy Efficiency Test Methods for Three-Phase Induction Motors/ Efficiency Quoting Method and Permissible Efficiency Tolerance	CSA	C390-1985
	Steam Generating Units	ASME	ANSI/ASME PTC 4.1-1964 (R 1985)
	Electric Motors and Generators for Use in Hazardous (Classified) Locations (1989)	UL	ANSI/UL 674-1984
	Nuclear Power Plant Air Cleaning Units and Components	ASME	ANSI/ASME N509-1989
	Testing and Nuclear Air-Cleaning Systems	ASME	ANSI/ASME N510-1989
	Motors and Generators	CSA	C22.2 No. 100-M1985
Outlets and Inlets	Method of Testing for Rating the Air Flow Performance of Outlets and Inlets	ASHRAE	ASHRAE 70-72
Pipe, Tubing, and Fittings	Power Piping	ASME	ANSI/ASME B31.1-1989
	Plastics Piping Components and Related Materials	NSF	NSF-14
	Scheme for the Identification of Piping Systems	ASME	ANSI/ASME A13.1-1981 (R 1985)
	National Fuel Gas Code	NFPA	ANSI/NFPA 54-1988
		AGA	ANSI Z223.1-1988
	Refrigeration Piping	ASME	ASME/ANSI B31.5-1987
	Refrigeration Tube Fittings (1977)	SAE	ANSI/SAE J513 OCT77
	Specification for Seamless Copper Pipe, Standard Sizes	ASTM	ASTM B42-89
	Specification for Acrylonitrile-Butadiene-Styrene (ABS) Plastic Pipe, Schedules 40 and 80	ASTM	ASTM D1527-89
	Specification for Poly (Vinyl Chloride) (PVC) Plastic Pipe, Schedules 40, 80, and 120	ASTM	ASTM D1785-89
	Specification for Polyethylene (PE) Plastic Pipe, Schedule 40	ASTM	ASTM D2104-90
	Standards of the Expansion Joint Manufacturers Association, Inc., 5th ed. (1980 with 1985 addenda)	EJMA	EJMA
	Rubber Gasketed Fittings for Fire Protection Service (1978)	UL	UL 213
	Tube Fittings for Flammable and Combustible Fluids, Refrigeration Service and Marine Use (1978)	UL	UL109
Plumbing	BOCA National Plumbing Code, 8th ed. (1990)	BOCA	BOCA
	Standard Plumbing Code (1988) (with 1989/1990 revisions)	SBCCI	SBCCI
	Uniform Plumbing Code (1988)	IAPMO	IAPMO
Pumps	Circulation System Components for Swimming Pools, Spas, or Hot Tubs	NSF	NSF-50
	Pumps for Oil-Burning Appliances (1986)	UL	ANSI/UL 343-1985
	Motor-Operated Water Pumps (1980)	UL	ANSI/UL 778-1983
	Electric Swimming Pool Pumps, Filters and Chlorinators (1986)	UL	ANSI/UL 1081-1985
	Hydraulic Institute Standards 14th ed. (1983)	HI	HI
	Hydraulic Institute Engineering Data Book, 1st ed. (1979)	HI	HI
	Performance Standard for Liquid Ring Vacuum Pumps, 1st ed. (1987)	HEI	HEI
	Centrifugal Pumps	ASME	ASME PTC 8.2-1965
	Displacement Compressors, Vacuum Pumps and Blowers	ASME	ANSI/ASME PTC 9-1974 (R 1985)
	Liquid Pumps	CSA	CAN/CSA C.22.2 No. 108-1989

Table 1 Codes and Standards Published by Various Societies and Associations (*Continued*)

Subject	Title	Publisher	Reference
Radiation	Testing and Rating Code for Baseboard Radiation, 6th ed. (1981)	HYDI	IBR
	Testing and Rating Code for Finned-Tube Commercial Radiation (1966)	HYDI	IBR
	Ratings for Baseboard and Fin-Tube Radiation (1989)	HYD I	IBR
Receivers	Refrigerant Liquid Receivers	ARI	ANSI/ARI 495-85
Refrigerant Containing Components	Refrigerant-Containing Components and Accessories, Non-electrical (1986)	UL	ANSI/UL 207-1986
	Refrigerant-Containing Components for Use in Electrical Equipment	CSA	C22.2 No. 140.3-M1987
Refrigerants	Number Designation and Safety Classification of Refrigerants	ASHRAE	ANSI/ASHRAE 34-1989
	Reducing Emission of Fully Halogenated Chlorofluorocarbon (CFC) Refrigerants in Refrigeration and Air-Conditioning Equipment and Applications	ASHRAE	ASHRAE Guideline 3-1990
	Refrigeration Oil Description	ASHRAE	ANSI/ASHRAE 99-1981 (RA 87)
	Methods of Testing Discharge Line Refrigerant-Oil Separators	ASHRAE	ASHRAE 69-71
	Sealed Glass Tube Method to Test the Chemical Stability of Material for Use Within Refrigerant Systems	ASHRAE	ANSI/ASHRAE 97-1983 (RA 89)
	Refrigerant Recovery/Recycling Equipment (1989)	UL	UL 1963
	Specifications for Fluorocarbon Refrigerants	ARI	ARI 700-88
Refrigeration	Safety Code for Mechanical Refrigeration	ASHRAE	ANSI/ASHRAE 15-1989
	Capacity Measurement of Field Erected Compression-Type Refrigeration and Air-Conditioning Systems	ASHRAE	ASHRAE 83-1985
	Refrigerated Medical Equipment (1978)	UL	UL 416
	Commercial Refrigerated Equipment	CSA	C22.2 No. 120-1974 (R 1981)
	Equipment, Design and Installation of Ammonia Mechanical Refrigeration Systems	IIAR	ANSI/IIAR 2-1984
Steam Jet	Standards for Steam Jet Vacuum Systems, 4th ed. (1988)	HEI	HEI
	Ejectors	ASME	ASME PTC 24-1976 (R 1982)
Transport	Safety Practices for Mechanical Vapor Compression Refrigeration Equipment or Systems Used to Cool Passenger Compartment of Motor Vehicles (1987)	SAE	SAE J639 JAN87
	Mechanical Refrigeration Installations on Shipboard	ASHRAE	ANSI/ASHRAE 26-1978 (RA 85)
	Mechanical Transport Refrigeration Units	ARI	ARI 1110-83
	General Requirements for Application of Vapor Cycle Refrigeration Systems for Aircraft (1973) (reaffirmed 1983)	SAE	SAE ARP731A
Refrigerators	Method of Testing Open Refrigerators for Food Stores	ASHRAE	ANSI/ASHRAE 72-1983
	Methods of Testing Self-Service Closed Refrigerators for Food Stores	ASHRAE	ASHRAE 117-1986
Commercial	Commercial Refrigerators and Freezers (1985)	UL	ANSI/UL 471-1984
	Food Service Refrigerators and Storage Freezers	NSF	NSF 7
	Food Carts	NSF	NSF 59
	Refrigerating Units (1989)	UL	ANSI/UL 427-1989
	Refrigeration Unit Coolers (1980)	UL	ANSI/UL 412-1984
	Soda Fountain and Luncheonette Equipment	NSF	NSF 1
	Food Service Equipment	NSF	NSF-2
Household	Refrigerators Using Gas Fuel (with 1984 and 1989 addenda)	AGA	ANSI Z21.19-1983
	Household Refrigerators and Household Freezers	AHAM	AHAM HRF 1-1988
	Household Refrigerators and Freezers (1983)	UL	ANSI/UL 250-1984
	Capacity Measurement and Test Methods for Household Refrigerators, Combination Refrigerator-Freezers and Household Freezers	CSA	CAN/CSA C300-M89
Roof Ventilators	Power Ventilators (1984)	UL	ANSI/UL 705-1984
Solar Equipment	Method of Testing to Determine the Thermal Performance of Solar Collectors	ASHRAE	ANSI/ASHRAE 93-1986
	Methods of Testing to Determine the Thermal Performance of Solar Domestic Water Heating Systems	ASHRAE	ASHRAE 95-1981 (RA 87)
	Methods of Testing to Determine the Thermal Performance of Unglazed Flat-Plate Liquid-Type Solar Collectors	ASHRAE	ANSI/ASHRAE 96-1980 (RA 90)
	Method of Testing to Determine the Thermal Performance of Flat-Plate Solar Collectors Containing a Boiling Liquid	ASHRAE	ANSI/ASHRAE 109-1986 (RA 90)
	Method of Measuring Solar-Optical Properties of Materials	ASHRAE	ASHRAE 74-73
Solenoid Valves	Solenoid Valves for Liquid Flow Use with Volatile Refrigerants and Water	ARI	ARI 760-87
Sound Measurement	Measurement of Sound from Boiler Units, Bottom-Supported Shop or Field Erected, 3rd ed.	ABMA	ABMA
	Sound Rating of Outdoor Unitary Equipment	ARI	ARI 270-84
	Sound Rating of Non-Ducted Indoor Air-Conditioning Equipment	ARI	ARI 350-86
	Sound Rating of Large Outdoor Refrigerating and Air-Conditioning Equipment	ARI	ARI 370-86
	Sound Level Prediction for Installed Rotating Electrical Machines	NEMA	NEMA MG 3-1974 (R 74, 84)

Table 1 Codes and Standards Published by Various Societies and Associations (*Continued*)

Subject	Title	Publisher	Reference
Sound Management (continued)	Rating the Sound Levels and Transmission Loss of Package Terminal Equipment	ARI	ARI 300-88
	Procedural Standards for Measuring Sound and Vibration	NEBB	NEBB-1977
	Sound and Vibration in Environmental Systems	NEBB	NEBB-1977
	Application of Sound Rated Outdoor Unitary Equipment	ARI	ARI 275-84
	Method of Measuring Machinery Sound within Equipment Rooms	ARI	ARI 575-87
	Laboratory Method of Testing In-Duct Sound Power Measurement Procedure for Fans	ASHRAE AMCA	ASHRAE 68-1986 AMCA 330-86
	Specification for Sound Level Meters (reaffirmed 1986)	ASA	ANSI S1.4-1983; ANSI S1.4A-1985
	Method for the Calibration of Microphones (reaffirmed 1986)	ASA	ANSI S1.10-1966 (R 1986)
	Reverberant Room Method for Sound Testing of Fans	AMCA	AMCA 300-85
	Guidelines for the Use of Sound Power Standards and for the Preparation of Noise Test Codes	ASA	ASA 10 ANSI S12.30-1990
	Measurement of Industrial Sound	ASME	ASME/ANSI PTC 36-1985
	Method of Measuring Sound and Vibration of Refrigerant Compressors	ARI	ARI 530-89
Space Heaters	Electric Air Heaters (1980)	UL	ANSI/UL 1025-1980
	Electric Air Heaters	CSA	C22.2 No. 46-M1988
Symbols	Graphic Electrical Symbols for Air-Conditioning and Refrigeration Equipment	ARI	ARI 130-88
	Graphic Symbols for Electrical and Electronic Diagrams	ASME	ANSI/IEEE 315-1975
	Graphic Symbols for Plumbing Fixtures for Diagrams used in Architecture Building Construction	ASME	ANSI Y32.4-1949 (R 1984)
	Symbols for Mechanical and Acoustical Elements as used in Schematic Diagrams	ASME	ANSI/ASME Y32.18-1972 (R 1985)
	Graphic Symbols for Pipe Fittings, Valves and Piping	ASME	ANSI/ASME Y32.2.3-1949 (R 1953)
	Graphic Symbols for Heating, Ventilating, and Air Conditioning	ASME	ANSI/ASME Y32.2.4-1949 (R 1984)
Testing and Balancing	Procedural Standards for Certified Testing of Cleanrooms (1988)	NEBB	NEBB-1988
	Procedural Standards for Testing, Adjusting, Balancing of Environmental Systems, 4th ed. (1983)	NEBB	NEBB-1983
	HVAC Systems—Testing, Adjusting and Balancing (1983)	SMACNA	SMACNA
	Site, Performance Test Standard-Power Plant and Industrial Fans	AMCA	AMCA 803-87
Terminals, Wiring	Equipment Wiring Terminals for Use with Aluminum and/or Copper Conductors (1988)	UL	ANSI/UL 486E-1987
Thermal Storage	Commissioning of HVAC Systems	ASHRAE	ASHRAE Guideline 1-1989
	Practices for Measurement, Testing and Balancing of Building Heating, Ventilation, Air-Conditioning, and Refrigeration Systems	ASHRAE	ANSI/ASHRAE 111-1988
	Method of Testing Active Latent Heat Storage Devices Based on Thermal Performance	ASHRAE	ANSI/ASHRAE 94.1-1985
	Methods of Testing Thermal Storage Devices with Electrical Input and Thermal Output Based on Thermal Performance	ASHRAE	ANSI/ASHRAE 94.2-1981 (RA 89)
	Metering and Testing Active Sensible Thermal Energy Storage Devices Based on Thermal Performance	ASHRAE	ANSI/ASHRAE 94.3-1986 (RA 90)
Turbines	Steam Turbines for Mechanical Drive Service	NEMA	NEMA SM23-1985
Unit Heaters	Oil-Fired Unit Heaters (1988)	UL	ANSI/UL 731-1987
	Gas Unit Heaters	AGA	ANSI Z83.8-1989
Valves	Methods of Testing Nonelectric, Nonpneumatic Thermostatic Radiator Valves	ASHRAE	ASHRAE 102-1983 (RA 89)
	Automatic Gas Valves for Gas Appliances (with 1989 addenda)	AGA	ANSI Z21.21-1987
	Manually Operated Gas Valves for Appliances, Appliance Connection Valves, and Hose End Valves	AGA	ANSI Z21.15-1989
	Relief Valves and Automatic Gas Shutoff Devices for Hot Water Supply Systems	AGA	ANSI Z21.22-1986
	Refrigerant Access Valves and Hose Connectors	ARI	ARI 720-88
	Refrigerant Pressure Regulating Valves	ARI	ARI 770-84
	Solenoid Valves for Use with Volatile Refrigerants	ARI	ARI 760-87
	Face-to-Face and End-to-End Dimensions of Ferrous Valves	ASME	ASME/ANSI B16.10-1986
	Manually Operated Metallic Gas Valves for Use in Gas Piping Systems up to 125 psig (Sizes 1/2 through 2)	ASME	ANSI B16.33-1981
	Valves—Flanged and Buttwelding End	ASME	ANSI/ASME B16.34-1989

Table 1 Codes and Standards Published by Various Societies and Associations (*Continued*)

Subject	Title	Publisher	Reference
Valves (continued)	Large Metallic Valves for Gas Distribution (Manually Operated, NPS-2 1/2 to 12, 125 psig Maximum)	ASME	ANSI/ASME B16.38-1985
	Manually Operated Thermoplastic Gas Shutoffs and Valves in Gas Distribution Systems	ASME	ANSI/ASME B16.40-1985
	Safety and Relief Valves	ASME	ANSI/ASME PTC25.3-1988
	Electrically Operated Valves (1982)	UL	ANSI-UL 429-1988
	Valves for Anhydrous Ammonia and LP-Gas (Other than Safety Relief) (1980)	UL	UL 125
	Safety Relief Valves for Anhydrous Ammonia and LP-Gas (1984)	UL	UL 132
	Pressure Regulating Valves for LP-Gas (1985)	UL	ANSI/UL 144-1985
	Valves for Flammable Fluids (1980)	UL	UL 842
Vending Machines	Refrigerated Vending Machines (1989)	UL	ANSI/UL 541-1988
	Coin-Operated Vending Machines	CSA	C22.2 No. 128-1963 (RA 89)
	Methods of Testing Pre-Mix and Post-Mix Soft Drink Vending and Dispensing Equipment	ASHRAE	ASHRAE 91-76
	Sanitation Ordinance and Code for Vending of Foods and Beverages (1965)	USDA	USDA 546
	Vending Machines for Food and Beverages	NSF	NSF-25
Vent Dampers	Automatic Vent Damper Devices for Use with Gas-Fired Appliances	AGA	ANSI Z21.66-1988
	Vent or Chimney Connector Dampers for Oil-Fired Appliances (1988)	UL	ANSI/UL 17-1988
Venting	Explosion Prevention Systems	NFPA	ANSI/NFPA 69-1986
	Chimneys, Fireplaces, Vents and Solid Fuel Burning Appliances	NFPA	ANSI/NFPA 211-1988
	Type L Low-Temperature Venting Systems (1986)	UL	ANSI/UL 641-1985
	Draft Hoods (with 1983 addenda)	AGA	ANSI Z21.12-1981
	Draft Equipment (1973)	UL	UL 378
	Gas Vents (1986)	UL	ANSI/UL 441-1985
	National Fuel Gas Code	AGA	ANSI Z223.1-1988
	Guide for Steel Stack Design and Construction (1983)	SMACNA	SMACNA
	Ventilation Directory	NCSBCS	NCSBCS
Ventilation	Removal of Smoke and Grease-Laden Vapors from Commercial Cooking Equipment	NFPA	ANSI/NFPA 96-1987
	Parking Structures; Repair Garages	NFPA	ANSI/NFPA 88A-1985; 88B-1985
	Ventilation for Acceptable Indoor Air Quality	ASHRAE	ASHRAE 62-1989
	Industrial Ventilation	ACGIH	ACGIH
	Food Service Equipment	NSF	NSF-2
	Method of Testing for Room Air Diffusion	ASHRAE	ANSI/ASHRAE 113-1990
	Ventilation Directory	NCSBCS	NCSBCS
	Residential Mechanical Ventilation Requirements	CSA	F326.1-M1989
	Residential Mechanical Ventilation System Requirements	CSA	F326.2-M1989
	Verification of the Performance of Residential Mechanical Ventilation Systems	CSA	F326.3-M1990
Water Heaters	Gas Water Heaters, Vol. I, Storage Water Heaters with Input Ratings of 75,000 Btu per Hour or Less	AGA	ANSI Z21.10.1-1990
	Gas Water Heaters, Vol. III, Storage, with Input Ratings Above 75,000 Btu per Hour, Circulating and Instantaneous Water Heaters	AGA	ANSI Z21.10.3-1990
	Household Electric Storage Tank Water Heaters (1989)	UL	ANSI/UL 174-1989
	Oil-Fired Storage Tank Water Heaters (1988)	UL	ANSI/UL 732-1987
	Commercial-Industrial Gas Heating Equipment (1973)	UL	UL 795
	Electric Booster and Commercial Storage Tank Water Heaters (1988)	UL	ANSI/UL-1987
	Hot Water Generating and Heat Recovery Equipment	NSF	NSF-5
	Construction and Test of Electric Storage Tank Water Heaters	CSA	C22.2 No. 110 M-1981
	Oil Burning Stoves and Water Heaters	CSA	B140.3-1962 (R 1980)
	Oil-Fired Service Water Heaters and Swimming Pool Heaters	CSA	B140.12-1976
	CSA Standards on Performance of Electric Storage Tank Water Heaters	CSA	C191-series-M90
	Methods of Testing to Determine the Thermal Performance of Solar Domestic Water Heating Systems	ASHRAE	ANSI/ASHRAE 95-1981 (RA 87)
	Construction and Test of Electric Storage-Tank Water Heaters	CSA	CAN/CSA C22.2 No. 110-M90
Woodburning Appliances	Installation Code for Solid Fuel Burning Appliances and Equipment	CSA	CAN/CSA-B365-M87
	Method of Testing for Performance Rating of Woodburning Appliances	ASHRAE	ANSI/ASHRAE 106-1984
	Solid-Fuel-Fired Central Heating Appliances	CSA	CAN/CSA-B366.1-M87
	Space Heaters for Use with Solid Fuels	CSA	B366.2-M1984
	Solid Fuel Type Room Heaters (1988)	UL	ANSI/UL 1482-1988
	Commercial Cooking and Hot Food Storage Equipment	NSF	NSF-4
	Chimneys, Fireplaces, Vents and Solid Fuel Burning Appliances	NFPA	ANSI/NFPA 211-1988

ABBREVIATIONS AND ADDRESSES

The Codes and Standards listed in Table 1 can be obtained from the organizations listed in the *Publisher* column.

ABMA	American Boiler Manufacturers Association, Suite 160, 950 N. Glebe Road, Arlington, VA 22203
ACCA	Air Conditioning Contractors of America, 1513 16th Street, NW, Washington, DC 20036
ACGIH	American Conference of Governmental Industrial Hygienists, 6500 Glenway Avenue, Building D-7, Cincinnati, OH 45211
ADC	Air Diffusion Council, 230 N. Michigan Avenue, Suite 1200, Chicago, IL 60601
AGA	American Gas Association, 1515 Wilson Boulevard, Arlington, VA 22209
AHAM	Association of Home Appliance Manufacturers, 20 N. Wacker Drive, Chicago, IL 60606
AIHA	American Industrial Hygiene Association, 345 White Pond Drive, Akron, OH 44320
AMCA	Air Movement and Control Association, Inc., 30 W. University Drive, Arlington Heights, IL 60004
ANSI	American National Standards Institute, 1430 Broadway, New York, NY 10018
ARI	Air-Conditioning and Refrigeration Institute, 1501 Wilson Boulevard, 6th Floor, Arlington, VA 22209
ASA	Acoustical Society of America, 335 E. 45 Street, New York, NY 10017-3483
ASHRAE	American Society of Heating, Refrigerating and Air-Conditioning Engineers, Inc. 1791 Tullie Circle, NE, Atlanta, GA 30329
ASME	The American Society of Mechanical Engineers, 345 E. 47 Street, New York, NY 10017 For ordering publications: ASME Marketing Department, Box 2350, Fairfield, NJ 07007-2350
ASTM	American Society for Testing and Materials, 1916 Race Street, Philadelphia, PA 19103
BOCA	Building Officials and Code Administrators International, Inc., 4051 W. Flossmoor Road, Country Club Hills, IL 60478-5795
BSI	British Standards Institution, 2 Park Street, London, W1A 2BS, England
CABO	Council of American Building Officials, 5203 Leesburg Pike, Suite 708, Falls Church, VA 22041
CAGI	Compressed Air and Gas Institute, Suite 1230, Keith Building, 1621 Euclid Avenue, Cleveland, OH 44115
CSA	Canadian Standards Association, 178 Rexdale Boulevard, Rexdale, Ontario M9W 1R3, Canada
CTI	Cooling Tower Institute, P.O. Box 73383, Houston, TX 77273
EJMA	Expansion Joint Manufacturers Association, Inc., 25 N. Broadway, Tarrytown, NY 10591
HEI	Heat Exchange Institute, Suite 1230, Keith Building, 1621 Euclid Avenue, Cleveland, OH 44115
HI	Hydraulic Institute, 30200 Detroit Rd., Cleveland, OH 44145-1967
HYD I	Hydronics Institute, 35 Russo Place, Berkeley Heights, NJ 07922
IAPMO	International Association of Plumbing and Mechanical Officials, 20001 Walnut Drive South, Walnut, CA 91789
ICBO	International Conference of Building Officials, 5360 S. Workman Mill Road, Whittier, CA 90601
IIAR	International Institute of Ammonia Refrigeration, 111 East Wacker Drive, Chicago, IL 60601
MICA	Midwest Insulation Contractors Association, 2017 South 139th Circle, Omaha, NE 68144
NCSBCS	National Conference of States on Building Codes and Standards, 505 Huntmar Park Dr., Suite 210, Herndon, VA 22070
NEBB	National Environmental Balancing Bureau, 1385 Piccard Drive, Rockville, MD 20850
NEMA	National Electrical Manufacturers Association, 2101 L Street, NW, Suite 300, Washington, D.C. 20037
NFPA	National Fire Protection Association, 1 Batterymarch Park, Quincy, MA 02269-9101
NSF	National Sanitation Foundation, Box 1468, Ann Arbor, MI 48106
SAE	Society of Automotive Engineers, 400 Commonwealth Drive, Warrendale, PA 15096
SBCCI	Southern Building Code Congress International, Inc., 900 Montclair Road, Birmingham, AL 35213
SMACNA	Sheet Metal and Air Conditioning Contractors' National Association, 4201 Lafayette Center Drive, Chantilly, VA 22021
TEMA	Tubular Exchanger Manufacturers Association, Inc., 25 N. Broadway, Tarrytown, NY 10591
UL	Underwriters Laboratories Inc., 333 Pfingsten Road, Northbrook, IL 60062-2096
WFCA	Western Fire Chiefs Association, Inc., 5360 S. Workman Mill Road, Whittier, CA 90601

ASHRAE HANDBOOK

ADDITIONS AND CORRECTIONS

This section supplements the current handbooks and notes technical errors found in the series. Occasional typographical errors and nonstandard symbol labels will be corrected in future volumes. The authors and editor encourage you to notify them if you find other technical errors. Please send corrections to: Handbook Editor, ASHRAE, 1791 Tullie Circle NE, Atlanta, GA 30329.

1988 Equipment

p. 6.7. Equation (2b) should read:

$w_a = 4.5 A_a V_a$ for I-P units and

$w_a = 1.2 A_a V_a$ for SI units.

p. 6.9, 2nd column. The values listed in the table at the bottom of the column are the true values of W_s multiplied by 10^3.

p. 6.10, 2nd column. The third inequality above Equation (15) should read:

(3) If $h_{ab} \leqslant h_{a2}$, all surface is completely dry.

p. 26.3. Equation (3) should read:

$$w = IM \tag{3}$$

where

w = mass flow, lb/h (g/s)
I = appliance heat input, Btu/h
M = mass flow input ratio, lb of chimney products per Btu of fuel burned (g/kJ)

1989 Fundamentals

p. 5.9, 2nd column, line 3 from bottom. The value of $(0.845)^{2/3}$ is 0.894.

p. 13.9, Equation (8). The velocity V is in ft/s, not fpm.

p. 20.18, 2nd column, sentence 2. The variable should read t_b, not D_b.

p. 22.8. Note under section on metals should read "(See Chapter 37, Table 3)."

p. 22.9. The conductance for Hardboard siding, 0.4375 in., should be 1.49 Btu/h·ft²·°F.

p. 24.13. Correct spelling for Greeneville, TN.

p. 27.31, 1st column, line 6. Change beginning of sentence to read Table 26, not Table 25.

p. 32.5, 1st column. The simplified formula for calculating friction factor should read:

$$f' = 0.11(12\epsilon/D_h + 68/\mathrm{Re})^{0.25}$$

p. 32.13, Figure 10. The arrow should not appear on the chart.

p. 32.19, Table 9. Duct section 9 should be 22 in. × 10 in., not 20 in. × 10 in.

p. 32.24, Example 4. The last line of the first paragraph should read:

through 90-in. diameters in 2-in. diameter increments.

p. 32.24, 2nd column. The last line of the paragraph should read "airflow rate using Equation (57)."

p. 32.25, 2nd paragraph The fan total pressure should be 8.86 in. water, and the equation for fan static pressure, using Equation (20), should read:

$$P_s = 8.86 - 0.86 = 8.0 \text{ in. water}$$

p. 32.28, Fitting 1-6. The value of C_o for $\theta = 0$ at $L/D = 0.4$ should read 1.39, not 0.39.

p. 32.32, Fitting 3-5. In the table on the top of the page, the first line should read:

0.5 1.5 1.4 1.3 1.2 1.1 1.0 1.0 1.1 1.1 1.2 1.2

p. 32.32, Fitting 3-6. The value of C_o' at $\theta = 30°$ and $H/W = 3.0$ should read 0.13, not 0.23.

p. 32.32, Fitting 3-7. The table presenting coefficients for elbows with one splitter vane (top of right column) should be replaced with the table shown in the Additions and Corrections chapter of the 1990 ASHRAE *Handbook—Refrigeration.*

p. 32.32, 2nd column. The figures for two splitter vanes and three splitter vanes should be reversed.

p. 32.33, Fitting 3-11. The second table should read:

Factors for Other Values of W/H

W/H	0.25	0.50	0.75	1.0	1.5	2.0	3.0	4.0	6.0	8.0
K	1.10	1.07	1.04	1.0	0.95	0.90	0.83	0.78	0.72	0.70

p. 32.33, Fittings 5-7 and 5-8. Footnotes for the main coefficients $C_{c,s}$ should refer to Fitting 5-6, not Fitting 5-3.

p. 32.34, Fittings 3-13 and 3-14. In the figures, all variables shown as D_o should read D.

p. 32.34, Fitting 3-13, 2nd column. In the first table, the value of K at $\theta = 45°$ and $L/D = 2$ should read 1.20, not 0.20.

p. 32.36, 2nd column. In the top table, the first value of B (column 3) should be 0.1, not 0.

p. 32.37, Fitting 4-7. In the table, the value of C_o at $A_o/A_1 = 0.25$, $L/D = 0.3$, and $\theta = 10°$ should read 0.27, not 0.17.

p. 32.37, Fitting 5-1, 2nd column. The heading of the first column on the 2nd table should read Q_b/Q_c, not Q_b/Q_s.

p. 32.37, Fitting 5-2, 2nd column. In the first table, the value of $C_{c,b}$ at $Q_b/Q_c = 0.2$ and $A_b/A_c = 0.6$ should read -0.21, not -1.21.

p. 32.38, Fitting 5-4, 2nd column. In the second table with the heading **Main,** C_s, the third value of A_b/A_c for $A_s/A_c = 0.5$ should read 0.4, not 0.3.

p. 32.38, Fitting 5-4, 2nd column. In the second table, the value for $C_{c,s}$ at $A_s/A_c = 0.6$ and $Q_b/Q_s = 0.8$ is -0.42, not -42.

p. 32.39, Fitting 5-6, 2nd column (top table). The first variable ratio in the heading should read A_b/A_s, not A_s/A_c.

p. 32.40, Fitting 5-9. The caption for this fitting should read "Tee, Converging, Rectangular Main and Tap...."

p. 32.40, Fitting 5-12. The value of $C_{c,b}$ at $V_b/V_c = 0.4$ should read 1.08, not 0.08.

p. 32.44, Fitting 5-28. The value of $C_{c,b}$ at $V_b/V_c = 0.6$ and $Q_b/Q_c = 0.3$ should read 1.05, not 0.05.

p. 32.44, Fitting 5-29. The value of $C_{c,b}$ at $V_b/V_c = 1.4$ and $Q_b/Q_c = 0.6$ should read 1.36, not 0.36.

p. 32.45, Fitting 5-34, 2nd table. The heading for the first column should read Q_{2b}/Q_{1b}, not A_{2b}/A_{1b}.

p. 32.51, Fitting 7-9. The value of C_o at $\theta = 15°$ and $A_1/A_o = 2.5$ should read 0.27.

p. 32.51, Fittings 7-9 and 7-10. Change footnotes to read Fitting 6-7, not 6-8.

p. 32.51, Fitting 7-11. The value of C_o at $\theta = 30°$ and $A_1/A_o = 2.5$ should read 0.29, not 0.39.

p. 33.5, Figure 3. Notes 1 through 4 in Figure 3 are notes for Figure 4.

p. 33.23, 1st column, line 6. Reference in last sentence should be to NFPA (1988), not Howell (1985).

1990 Refrigeration

p. 4.1, lines 3 and 4. Delete the words, "should be used to reduce hazards. Cold liquid refrigerant," so that the last two sentences of the paragraph read:

> Large quantities of ammonia should not be vented to enclosed areas near open flames or heavy sparks. A proportion of 16 to 25% by volume in air in the presence of an open flame burns rapidly and can explode.

p. 30.6, Figure 2. Note should read: "Return tanks may be omitted by venting through the gravity tank."

COMPOSITE INDEX

ASHRAE HANDBOOK SERIES

This index covers the current Handbook volumes published by ASHRAE. Listings from each volume are identified as follows:

A = 1991 HVAC Applications
R = 1990 Refrigeration
F = 1989 Fundamentals
E = 1988 Equipment
H = 1987 HVAC Systems and Applications (Chapters 1-16)

The index is alphabetized in a *word-by-word* format; for example, *air diffusers* is listed before *aircraft* and *heat flow* is listed before *heaters.*

Note that the code for a volume includes the chapter number followed by a decimal point and the page number(s) within the chapter. For example, F32.4 means the information may be found in the Fundamentals volume, chapter 32, page 4.